Encyclopedia of

Environmental Management

Volume I

Acaricides—Energy Conversion

Encyclopedia of

Environmental Management

Volume I

Acaricides—Energy Conversion

Edited by

Sven Erik Jørgensen

CRC Press
Taylor & Francis Group
Boca Raton London New York

CRC Press is an imprint of the
Taylor & Francis Group, an **informa** business

CRC Press
Taylor & Francis Group
6000 Broken Sound Parkway NW, Suite 300
Boca Raton, FL 33487-2742

Printed in the United States of America on acid-free paper
Version Date: 20121011

International Standard Book Number: 978-1-4398-2928-8 (Hardback)

Visit the Taylor & Francis Web site at
http://www.taylorandfrancis.com

and the CRC Press Web site at
http://www.crcpress.com

Brief Contents

Volume I

Acaricides ... 1
Acid Rain ... 5
Acid Rain: Nitrogen Deposition ... 20
Acid Sulfate Soils ... 26
Acid Sulfate Soils: Formation ... 31
Acid Sulfate Soils: Identification, Assessment, and Management ... 36
Acid Sulfate Soils: Management ... 55
Adsorption ... 61
Agricultural Runoff ... 81
Agricultural Soils: Ammonia Volatilization ... 85
Agricultural Soils: Carbon and Nitrogen Biological Cycling ... 88
Agricultural Soils: Nitrous Oxide Emissions ... 96
Agricultural Soils: Phosphorus ... 100
Agricultural Water Quantity Management ... 105
Agriculture: Energy Use and Conservation ... 118
Agriculture: Organic ... 125
Agroforestry: Water Use Efficiency ... 129
Air Pollution: Monitoring ... 132
Air Pollution: Technology ... 149
Alexandria Lake Maryut: Integrated Environmental Management ... 164
Allelochemics ... 175
Alternative Energy ... 179
Alternative Energy: Hydropower ... 201
Alternative Energy: Photovoltaic Modules and Systems ... 215
Alternative Energy: Photovoltaic Solar Cells ... 226
Alternative Energy: Solar Thermal Energy ... 241
Alternative Energy: Wind Power Technology and Economy ... 258
Aluminum ... 267
Animals: Sterility from Pesticides ... 277
Animals: Toxicological Evaluation ... 280
Antagonistic Plants ... 288
Aquatic Communities: Pesticide Impacts ... 291
Aral Sea Disaster ... 300
Arthropod Host-Plant Resistant Crops ... 304
Bacillus thuringiensis: Transgenic Crops ... 307
Bacterial Pest Control ... 321
Bioaccumulation ... 324
Biodegradation ... 328
Biodiversity and Sustainability ... 333
Bioenergy Crops: Carbon Balance Assessment ... 345
Biofertilizers ... 349
Bioindicators: Farming Practices and Sustainability ... 352
Biological Controls ... 366
Biological Controls: Conservation ... 370
Biomass ... 373
Biopesticides ... 381

Volume I (cont'd)

Bioremediation ... 387
Bioremediation: Contaminated Soil Restoration ... 401
Biotechnology: Pest Management ... 407
Birds: Chemical Control ... 413
Birds: Pesticide Use Impacts ... 416
Boron and Molybdenum: Deficiencies and Toxicities ... 424
Boron: Soil Contaminant ... 431
Buildings: Climate Change ... 435
Cabbage Diseases: Ecology and Control ... 442
Cadmium and Lead: Contamination ... 446
Cadmium: Toxicology ... 453
Carbon Sequestration ... 456
Carbon: Soil Inorganic ... 462
Chemigation ... 469
Chesapeake Bay ... 474
Chromium ... 477
Climate Policy: International ... 483
Coastal Water: Pollution ... 489
Cobalt and Iodine ... 502
Community-Based Monitoring: Ngarenanyuki, Tanzania ... 505
Composting ... 512
Composting: Organic Farming ... 528
Copper ... 535
Cyanobacteria: Eutrophic Freshwater Systems ... 538
Desertification ... 541
Desertification: Extent of ... 545
Desertification: Greenhouse Effect ... 549
Desertification: Impact ... 552
Desertification: Prevention and Restoration ... 557
Desertization ... 565
Developing Countries: Pesticide Health Impacts ... 573
Distributed Generation: Combined Heat and Power ... 578
Drainage: Hydrological Impacts Downstream ... 584
Drainage: Soil Salinity Management ... 588
Ecological Indicators: Eco-Exergy to Emergy Flow ... 592
Ecological Indicators: Ecosystem Health ... 599
Economic Growth: Slower by Design, Not Disaster ... 614
Ecosystems: Large-Scale Restoration Governance ... 624
Ecosystems: Planning ... 632
Endocrine Disruptors ... 643
Energy Commissioning: Existing Buildings ... 656
Energy Commissioning: New Buildings ... 665
Energy Conservation ... 677
Energy Conservation: Benefits ... 691
Energy Conservation: Industrial Processes ... 697
Energy Conservation: Lean Manufacturing ... 705
Energy Conversion: Coal, Animal Waste, and Biomass Fuel ... 714

Brief Contents (*cont'd*)

Volume II

Energy Efficiency: Low-Cost Improvements 735
Energy Efficiency: New and Emerging Technology 742
Energy Efficiency: Strategic Facility Guidelines 752
Energy Master Planning 763
Energy Sources: Natural versus Additional 770
Energy Use: Exergy and Eco-Exergy 778
Energy Use: U.S. 790
Energy: Environmental Security 798
Energy: Physics 808
Energy: Renewable 824
Energy: Solid Waste Advanced Thermal Technology 830
Energy: Storage 853
Energy: Walls and Windows 860
Energy: Waste Heat Recovery 869
Environmental Legislation: Asia 874
Environmental Legislation: EU Countries Solid Waste Management 892
Environmental Policy 914
Environmental Policy: Innovations 922
Erosion 930
Erosion and Carbon Dioxide 934
Erosion and Global Change 939
Erosion and Precipitation 943
Erosion and Sediment Control: Vegetative Techniques 947
Erosion by Water: Accelerated 951
Erosion by Water: Amendment Techniques 958
Erosion by Water: Assessment and Control 963
Erosion by Water: Empirical Methods 974
Erosion by Water: Erosivity and Erodibility 980
Erosion by Water: Process-Based Modeling 991
Erosion by Water: Vegetative Control 994
Erosion by Wind: Climate Change 1004
Erosion by Wind: Global Hot Spots 1010
Erosion by Wind: Principles 1013
Erosion by Wind: Source, Measurement, Prediction, and Control 1017
Erosion by Wind-Driven Rain 1031
Erosion Control: Soil Conservation 1034
Erosion Control: Tillage and Residue Methods 1040
Erosion: Accelerated 1044
Erosion: History 1050
Erosion: Irrigation-Induced 1053
Erosion: Snowmelt 1057
Erosion: Soil Quality 1060
Estuaries 1063
Eutrophication 1074
Everglades 1080
Exergy: Analysis 1083
Exergy: Environmental Impact Assessment 1092
Farming: Non-Chemical and Pesticide-Free (European Council Regulations [EED/209/91]) 1103
Farming: Organic 1109
Farming: Organic Pest Management 1115
Food Quality Protection Act 1118
Food: Cosmetic Standards 1120
Food: Pesticide Contamination 1124
Fossil Fuel Combustion: Air Pollution and Global Warming 1127
Fuel Cells: Intermediate and High Temperature 1138

Volume II (cont'd)

Fuel Cells: Low Temperature 1145
Genotoxicity and Air Pollution 1154
Geographic Information System (GIS): Land Use Planning 1163
Geothermal Energy Resources 1167
Giant Reed (*Arundo donax*): Streams and Water Resources 1182
Global Climate Change: Carbon Sequestration 1189
Global Climate Change: Earth System Response 1192
Global Climate Change: Gas Fluxes 1202
Global Climate Change: Gasoline, Hybrid-Electric and Hydrogen Fueled Vehicles 1206
Global Climate Change: World Soils 1213
Globalization 1218
Green Energy 1226
Green Processes and Projects: Systems Analysis 1242
Green Products: Production 1253
Groundwater: Arsenic Contamination 1262
Groundwater: Contamination 1281
Groundwater: Mining 1284
Groundwater: Mining Pollution 1289
Groundwater: Modeling 1295
Groundwater: Nitrogen Fertilizer Contamination 1302
Groundwater: Numerical Method Modeling 1312
Groundwater: Pesticide Contamination 1317
Groundwater: Saltwater Intrusion 1321
Groundwater: Treatment with Biobarrier Systems 1333
Heat Pumps 1346
Heat Pumps: Absorption 1356
Heat Pumps: Geothermal 1363
Heavy Metals 1370
Heavy Metals: Organic Fertilization Uptake 1374
Herbicides 1378
Herbicides: Non-Target Species Effects 1382
Human Health: Cancer from Pesticides 1394
Human Health: Chronic Pesticide Poisonings 1397
Human Health: Consumer Concerns of Pesticides 1401
Human Health: Hormonal Disruption 1405
Human Health: Pesticide Poisonings 1408
Human Health: Pesticide Sensitivities 1411
Human Health: Pesticides 1414
Hydroelectricity: Pumped Storage 1418
Industrial Waste: Soil Pollution 1430
Industries: Network of 1434
Inland Seas and Lakes: Central Asia Case Study 1436
Inorganic Carbon: Composition and Formation 1444
Inorganic Carbon: Global Carbon Cycle 1448
Inorganic Carbon: Modeling 1451
Inorganic Compounds: Eco-Toxicity 1455
Insect Growth Regulators 1459
Insecticides: Aerial Ultra-Low-Volume Application 1471
Insects and Mites: Biological Control 1474
Insulation: Facilities 1478
Integrated Energy Systems 1493
Integrated Farming Systems 1507
Integrated Nutrient Management 1510
Integrated Pest Management 1521
Integrated Weed Management 1524

Brief Contents (*cont'd*)

Volume III

Invasion Biology 1531
Irrigation Systems: Sub-Surface Drip Design 1535
Irrigation Systems: Sub-Surface History 1539
Irrigation Systems: Water Conservation 1542
Irrigation: Efficiency 1545
Irrigation: Erosion 1551
Irrigation: Return Flow and Quality 1557
Irrigation: River Flow Impact 1563
Irrigation: Saline Water 1568
Irrigation: Sewage Effluent Use 1570
Irrigation: Soil Salinity 1572
Lakes and Reservoirs: Pollution 1576
Lakes: Restoration 1588
Land Restoration 1600
Laws and Regulations: Food 1609
Laws and Regulations: Pesticides 1612
Laws and Regulations: Rotterdam Convention 1615
Laws and Regulations: Soil 1618
Leaching 1621
Leaching and Illuviation 1624
Lead: Ecotoxicology 1630
Lead: Regulations 1636
LEED-EB: Leadership in Energy and Environmental Design for Existing Buildings 1647
LEED-NC: Leadership in Energy and Environmental Design for New Construction 1654
Manure Management: Compost and Biosolids 1660
Manure Management: Dairy 1663
Manure Management: Phosphorus 1667
Manure Management: Poultry 1670
Mercury 1676
Methane Emissions: Rice 1679
Minerals Processing Residue (Tailings): Rehabilitation 1683
Mines: Acidic Drainage Water 1688
Mines: Rehabilitation of Open Cut 1692
Mycotoxins 1696
Nanomaterials: Regulation and Risk Assessment 1700
Nanoparticles 1711
Nanoparticles: Uncertainty Risk Analysis 1720
Nanotechnology: Environmental Abatement 1730
Natural Enemies and Biocontrol: Artificial Diets 1746
Natural Enemies: Conservation 1749
Nematodes: Biological Control 1752
Neurotoxicants: Developmental Experimental Testing 1755
Nitrate Leaching Index 1761
Nitrogen 1768
Nitrogen Trading Tool 1772
Nitrogen: Biological Fixation 1785
Nuclear Energy: Economics 1789
Nutrients: Best Management Practices 1805
Nutrients: Bioavailability and Plant Uptake 1817
Nutrient–Water Interactions 1823
Oil Pollution: Baltic Sea 1826
Organic Compounds: Halogenated 1841
Organic Matter: Global Distribution in World Ecosystems 1851

Volume III (cont'd)

Organic Matter: Management 1857
Organic Matter: Modeling 1863
Organic Matter: Turnover 1872
Organic Soil Amendments 1878
Ozone Layer 1882
Permafrost 1900
Persistent Organic Compounds: Wet Oxidation Removal 1904
Persistent Organic Pesticides 1913
Pest Management 1919
Pest Management: Crop Diversity 1925
Pest Management: Ecological Agriculture 1930
Pest Management: Ecological Aspects 1934
Pest Management: Intercropping 1937
Pest Management: Legal Aspects 1940
Pest Management: Modeling 1949
Pesticide Translocation Control: Soil Erosion 1955
Pesticides 1963
Pesticides: Banding 1983
Pesticides: Chemical and Biological 1986
Pesticides: Damage Avoidance 1997
Pesticides: Effects 2005
Pesticides: History 2008
Pesticides: Measurement and Mitigation 2013
Pesticides: Natural 2028
Pesticides: Reducing Use 2033
Pesticides: Regulating 2035
Pests: Landscape Patterns 2038
Petroleum: Hydrocarbon Contamination 2040
Pharmaceuticals: Treatment 2060
Phenols 2071
Phosphorus: Agricultural Nutrient 2091
Phosphorus: Riverine System Transport 2100
Plant Pathogens (Fungi): Biological Control 2107
Plant Pathogens (Viruses): Biological Control 2111
Pollutants: Organic and Inorganic 2114
Pollution: Genotoxicity of Agrotoxic Compounds 2123
Pollution: Non-Point Source 2136
Pollution: Pesticides in Agro-Horticultural Ecosystems 2139
Pollution: Pesticides in Natural Ecosystems 2145
Pollution: Point and Non-Point Source Low Cost Treatment 2150
Pollution: Point Sources 2166
Polychlorinated Biphenyls (PCBs) 2172
Polychlorinated Biphenyls (PCBs) and Polycyclic Aromatic Hydrocarbons (PAHs): Sediments and Water Analysis 2186
Potassium 2208
Precision Agriculture: Engineering Aspects 2213
Precision Agriculture: Water and Nutrient Management 2217
Radio Frequency Towers: Public School Placement 2224
Radioactivity 2234
Radionuclides 2243
Rain Water: Atmospheric Deposition 2249
Rain Water: Harvesting 2262
Rare Earth Elements 2266

Brief Contents (*cont'd*)

Volume IV

Remote Sensing and GIS Integration 2271
Remote Sensing: Pollution 2275
Rivers and Lakes: Acidification 2291
Rivers: Pollution 2303
Rivers: Restoration 2307
Runoff Water 2320
Salt-Affected Soils: Physical Properties and Behavior 2334
Salt-Affected Soils: Plant Response 2345
Salt-Affected Soils: Sustainable Agriculture 2349
Sea: Pollution 2357
Sodic Soils: Irrigation Farming 2364
Sodic Soils: Properties 2367
Sodic Soils: Reclamation 2370
Soil Degradation: Global Assessment 2375
Soil Quality: Carbon and Nitrogen Gases 2388
Soil Quality: Indicators 2393
Soil Rehabilitation 2396
Solid Waste Management: Life Cycle Assessment 2399
Solid Waste: Municipal 2415
Stored-Product Pests: Biological Control 2423
Strontium 2426
Sulfur 2431
Sulfur Dioxide 2437
Sustainability and Planning 2446
Sustainable Agriculture: Soil Quality 2457
Sustainable Development 2461
Sustainable Development: Ecological Footprint in Accounting 2467
Sustainable Development: Pyrolysis and Gasification of Biomass and Wastes 2482
Thermal Energy: Solar Technologies 2498
Thermal Energy: Storage 2508
Thermodynamics 2525
Tillage Erosion: Terrace Formation 2536
Toxic Substances 2542
Toxic Substances: Photochemistry 2547
Toxicity Prediction of Chemical Mixtures 2572
Vanadium and Chromium Groups 2582

Volume IV (cont'd)

Vertebrates: Biological Control 2595
Waste Gas Treatment: Bioreactors 2611
Waste: Stabilization Ponds 2632
Wastewater and Water Utilities 2638
Wastewater Treatment: Biological 2645
Wastewater Treatment: Conventional Methods 2657
Wastewater Treatment: Wetlands Use in Arctic Regions 2662
Wastewater Use in Agriculture 2675
Wastewater Use in Agriculture: Policy Issues 2681
Wastewater Use in Agriculture: Public Health Considerations 2694
Wastewater: Municipal 2709
Water and Wastewater: Filters 2719
Water and Wastewater: Ion Exchange Application 2734
Water Harvesting 2738
Water Quality and Quantity: Globalization 2740
Water Quality: Modeling 2749
Water Quality: Range and Pasture Land 2752
Water Quality: Soil Erosion 2755
Water Quality: Timber Harvesting 2770
Water Supplies: Pharmaceuticals 2776
Water: Cost 2779
Water: Drinking 2790
Water: Surface 2804
Water: Total Maximum Daily Load 2808
Watershed Management: Remote Sensing and GIS 2816
Weeds (Insects and Mites): Biological Control 2821
Wetlands 2824
Wetlands: Biodiversity 2829
Wetlands: Carbon Sequestration 2833
Wetlands: Conservation Policy 2837
Wetlands: Constructed Subsurface 2841
Wetlands: Methane Emission 2850
Wetlands: Petroleum 2854
Wetlands: Sedimentation and Ecological Engineering 2859
Wetlands: Treatment System Use 2862
Wind Farms: Noise 2867
Yellow River 2884
Appendixes 2887

Encyclopedia of Environmental Engineering

Editor-in-Chief

Sven Erik Jørgensen
Institute A, Section of Environmental Chemistry, Copenhagen University, Copenhagen, Denmark

Editorial Advisory Board

Reviewers

Reviewers (*cont'd*)

Richard Frankel
Alan J. Franzluebbers
Michael H. Glantz
Sabine Goldberg
Neva Goodwin
Tadeusz Górecki
Shela Gorinstein
Scott Grosse
Velma I. Grover
Silvio J. Gumiere
Heli Haapasaari
Douglas Hale
Ed Hanna
Chunyan Hao
P.S. Harikumar
Paul Hatcher
Michaela Hegglin
George Helz
Ivan Holoubek
A.R. Horowitz
Lucas Hyman
Harish Jeswani
Buzz Johnson
Jodi L. Johnson-Maynard
Bill Jokela
Jan Åke Jönsson
Sandeep Joshi
Puangrat Kajitvichyanukul
Ioannis K. Kalvrouziotis
Mehmet Kanoglu
Douglas L. Karlen
Gilbert Kelling
Ben Keraita
Thomas Kinraide
Holger Kirchmann
Andreas Klumpp
Gordana Kranjac-Berisavljevic
John M. Laflen
Paul J. Lamothe
Matthew Langholtz
E.F. Legner
Tarzan Legovic
Ronnie Levin
Guy Levy
Witold Lewandowski
Jianbing Li
Alessandro Ludovisi
Tapas Malick
Rob Malone
Andreas Mamolos
Stanley E. Manahan
Duncan Mara
Dan Marion
E.J.P. Marshall
David McBride
David McKenzie
Bernd Markert
Rick Miller
Lesley Mills
Roger Minear
Virginia Moser
Amitava Mukherjee
Deborah A. Neher
Kristian Fog Nielsen
Darrell Norton
Krystyna Olanczuk-Neyman
David Olszyk
A.D. Patwardhan
Janusz Pempkowiak
Robert Percival
Sandra Perez
Tanapon Phenrat
David Pimentel
Jon K. Piper
I. Popescu
Federico M. Pulselli
X.S. Qin
Philip S. Rainbow
Barnett Rattner
E. Remoundaki
Sergio Revah
Marc Ribaudo
H. Rodhe
Art Rose
Don Ross
Diederik Rousseau
John R. Ruberson
Mario Russo
Yvonne Rydin
Barbara J.S. Sanderson
Marvin Schaffer
James S. Schepers
Steven Sehr
Supapan Seraphin
Balwinder Singh Sra
Bogdan Skwarzec
Scott Slocombe
Shaul Sorek
Irena Staneczko-Baranowska
B.A. Stewart
Kristian Syberg
Piotr Szefer
Moses M. Tenywa
Khi V. Thai
Daniel L. Thomas
Abhishek Tiwary
José Torrent
Jeff N. Tullberg
Marco Vighi
Earl Vories
Maurizio Vurro
Maria Waclawek
J.Y. Wang
Waldemar Wardencki
Apichon Watcharenwong
Bernard Weiss
Fern Wickson
L.T. Wilson
Bofu Yu
Frank G. Zalom
Rick Zartman

Contributors

Diana Aga / *Department of Chemistry, State University of New York at Buffalo, Buffalo, New York, U.S.A.*

Matthew Agarwala / *Geography and Environment Department and Grantham Research Institute on Climate Change and the Environment, London School of Economics, London, U.K.*

Shaikh Ziauddin Ahammad / *School of Civil Engineering and Geosciences, Newcastle University, Newcastle, U.K.*

Imad A.M. Ahmed / *Lancaster Environment Center, Lancaster University, Lancaster, U.K.*

Erhan Akça / *Technical Programs, Adiyaman University, Adiyaman, Turkey*

Claude Amiard-Triquet / *French National Center for Scientific Research (CNRS) and University of Nantes, Nantes, France*

Ronald G. Amundson / *College of Natural Resources, University of California—Berkeley, Berkeley, California, U.S.A.*

Jirapat Ananpattarachai / *Center of Excellence for Environmental Research and Innovation, Faculty of Engineering, Naresuan University, Phitsanulok, Thailand*

Samson D. Angima / *Oregon State University, Oregon, U.S.A.*

Kalyan Annamalai / *Paul Pepper Professor of Mechanical Engineering, Texas A&M University, College Station, Texas, U.S.A.*

Massimo Antoninetti / *Institute for Electromagnetic Sensing of the Environment (IREA), National Research Council of Italy (CNR), Milan, Italy*

George F. Antonious / *Department of Plant and Soil Science, Water Quality/Environmental Toxicology Research, Kentucky State University, Frankfort, Kentucky, U.S.A.*

Ramón Aragues / *Agronomic Research Service, Government of Aragon, Zaragoza, Spain*

Senthil Arumugam / *Enerquip, Inc., Medford, Wisconsin, U.S.A.*

G.J. Ash / *E.H. Graham Center for Agricultural Innovation, Industry and Investment, NSW and Charles Sturt University, Wagga Wagga, New South Wales, Australia*

Muhammad Asif / *School of the Built and Natural Environment, Glasgow Caledonian University, Glasgow, U.K.*

William Au / *Department of Preventitive Medicine and Community Health, University of Texas Medical Branch, Galveston, Texas, U.S.A.*

Lu Aye / *Renewable Energy and Energy Efficiency Group, Department of Infrastructure Engineering, Melbourne School of Engineering, University of Melbourne, Melbourne, Victoria, Australia*

Thomas Backhaus / *Department of Plant and Environmental Sciences, University of Gothenburg, Gothenburg, Sweden*

Seungyun Baik / *Department of Chemistry, State University of New York at Buffalo, Buffalo, New York, U.S.A.*

Kenneth A. Barbarick / *Colorado State University, Fort Collins, Colorado, U.S.A.*

Javier Barragan / *Department of Agroforestry Engineering, University of Lleida, Lleida, Spain*

Simone Bastianoni / *Ecodynamics Group, Department of Chemistry, University of Siena, Siena, Italy*

Anders Baun / *Department of Environmental Engineering, Technical University of Denmark, Kongens Lyngby, Denmark*
Susana Bautista / *Department of Ecology, University of Alicante, Alicante, Spain*
Lindsay Beevers / *Lecturer in Water Management, School of the Built Environment, Heriot Watt University, Edinburgh, U.K.*
Richard W. Bell / *School of Environmental Science, Murdoch University, Perth, Western Australia, Australia*
Suha Berberoğlu / *Departments of Soil Science, Landscape Architecture, and Agricultural Engineering, University of Cukurova, Adana, Turkey*
Sanford V. Berg / *Director of Water Studies, Public Utility Research Center, University of Florida, Gainesville, Florida, U.S.A.*
Lars Bergström / *Department of Soil Science, Swedish University of Agricultural Sciences (SLU), Uppsala, Sweden*
Angelika Beyer / *Department of Analytical Chemistry, Chemical Faculty, Gdansk University of Technology, Gdansk, Poland*
Jerry M. Bigham / *Ohio State University, Columbus, Ohio, U.S.A.*
Marek Biziuk / *Department of Analytical Chemistry, Chemical Faculty, Gdansk University of Technology, Gdansk, Poland*
David L. Bjorneberg / *Northwest Irrigation and Soils Research Lab, Agricultural Research Service (USDA-ARS), U.S. Department of Agriculture, Kimberly, Idaho, U.S.A.*
John O. Blackburn / *Professor Emeritus of Economics, Duke University, Maitland, Florida, U.S.A.*
Frederick Paxton Cardell Blamey / *School of Land, Crop and Food Sciences, University of Queensland, St. Lucia, Queensland, Australia*
Elke Bloem / *Institute for Crop and Soil Science, Julius Kuhn Institute (JKI), Braunschweig, Germany*
W.E.H. Blum / *Institute of Soil Research, University of Natural Resources and Life Sciences, Vienna, Austria*
Pascal Boeckx / *Faculty of Agricultural and Applied Biological Sciences, University of Ghent, Ghent, Belgium*
Ben Boer / *School of Law, University of Sydney, Sydney, New South Wales, Australia*
Julia Boike / *Water and Environmental Research Center, University of Alaska—Fairbanks, Fairbanks, Alaska, U.S.A.*
Nadia Bernardi Bonumá / *Federal University of Santa Maria, Santa Maria, Brazil*
J.D. Booker / *Plant and Soil Science, Texas Tech University, Lubbock, Texas, U.S.A.*
John Borden / *Department of Biological Sciences, Simon Fraser University, Burnaby, British Columbia, Canada*
Virginie Bouchard / *School of Natural Resources, Ohio State University, Columbus, Ohio, U.S.A.*
Céline Boutin / *Science and Technology Branch, Environment Canada, Carleton University, Ottawa, Ontario, Canada*
John T. Brake / *Department of Soil Science, North Carolina State University, Raleigh, North Carolina, U.S.A.*
Vince Bralts / *Agricultural and Biological Engineering, Purdue University, West Lafayette, Indiana, U.S.A.*
James R. Brandle / *School of Natural Resource Sciences, University of Nebraska—Lincoln, Lincoln, Nebraska, U.S.A.*
D.R. Bray / *Department of Animal Sciences, University of Florida, Gainesville, Florida, U.S.A.*

Zachary T. Broome / *Bowen Rudson Schroth, P. A., Eustis, Florida, U.S.A.*
Dominic A. Brose / *University of Maryland, College Park, Maryland, U.S.A.*
Steven N. Burch / *University of Maryland, College Park, Maryland, U.S.A.*
Benjamin Burkhard / *Institute for the Conservation of Natural Resources, University of Kiel, Kiel, Germany*
David Burrow / *Agriculture Victoria (Tatura), Tartura, Victoria, Australia*
E.D. Burton / *Southern Cross GeoScience, Southern Cross University, Lismore, New South Wales, Australia*
Alan Busacca / *Department of Crop and Soil Sciences, Geology, Washington State University, Pullman, Washington, U.S.A.*
R.T. Bush / *Southern Cross GeoScience, Southern Cross University, Lismore, New South Wales, Australia*
Liana Buzdugan / *Center for Sustainable Exploitation of Ecosystems (CESEE), Alexandru Ioan Cuza University of Iasi, Iasi, Romania*
Nídia Sá Caetano / *Chemical Engineering Department, School of Engineering (ISEP), Polytechnic Institute of Porto (IPP), and Laboratory for Process, Environmental and Energy Engineering, Porto, Portugal*
James Call / *President, James Call Engineering, PLLC, Larchmont, New York, U.S.A.*
Carl R. Camp, Jr. / *Agricultural Research Service (USDA-ARS), U.S. Department of Agriculture, Florence, South Carolina, U.S.A.*
Kenneth L. Campbell / *Agricultural and Biological Engineering Department, University of Florida, Gainesville, Florida, U.S.A.*
Barney L. Capehart / *Department of Industrial and Systems Engineering, University of Florida College of Engineering, Gainesville, Florida, U.S.A.*
Kristi Denise Caravella / *Florida Atlantic University, Boca Raton, Florida, U.S.A.*
Jesus Carrera / *Technical University of Catalonia (UPC), Barcelona, Spain*
Vera Lucia S.S. de Castro / *Ecotoxicology and Biosafety Laboratory, Environment, Brazilian Agricultural Research Corporation (Embrapa), São Paulo, Brazil*
Nina Cedergreen / *Faculty of Life Sciences, University of Copenhagen, Frederiksberg, Denmark*
Carlos C. Cerri / *University of São Paulo, São Paulo, Brazil*
Dipankar Chakraborti / *Director (Research), School of Environmental Studies, Jadavpur University, Calcutta, India*
David Chandler / *Plant, Soils, and Biometeorology Department, Utah State University, Logan, Utah, U.S.A.*
Ni-Bin Chang / *Department of Civil, Environmental, and Construction Engineering, University of Central Florida, Orlando, Florida, U.S.A.*
Guangnan Chen / *Faculty of Engineering and Surveying, University of Southern Queensland, Toowoomba, Queensland, Australia*
Alexander H.-D. Cheng / *Department of Civil Engineering, University of Mississippi, Oxford, Mississippi, U.S.A.*
Angelique Chettiparamb / *School of Real Estate and Planning, University of Reading, Reading, U.K.*
Tarit Roy Chowdhury / *School of Environmental Studies, Jadavpur University, Calcutta, India*
Torben Røjle Christensen / *Climate Impacts Group, Department of Ecology, Lund University, Lund, Sweden*
Jock Churchman / *Land and Water, Commonwealth Scientific and Industrial Research Organization (CSIRO), Adelaide, South Australia, Australia*
Maria V. Cilveti / *Department of Entomology, Cornell University, Ithaca, New York, U.S.A.*

Ioan Manuel Ciumasu / *ECONOVING International Chair, University of Versailles Saint-Quentin-en-Yvelines, Guyancourt, France, and Center for Sustainable Exploitation of Ecosystems (CESEE), Alexandru Ioan Cuza University of Iasi, Iasi, Romania*
David E. Claridge / *Department of Mechanical Engineering, Energy Systems Laboratory, College Station, Texas, U.S.A.*
Sharon A. Clay / *Plant Science Department, South Dakota State University, Brookings, South Dakota, U.S.A.*
Matthew Cochran / *School of Natural Resources, Ohio State University, Columbus, Ohio, U.S.A.*
Gretchen Coffman / *Department of Environmental Health Sciences, University of California—Los Angeles, Ventura, California, U.S.A.*
Alexandra Robin Collins / *Department of Biology and Ecology, University of Fribourg/Perolles, Fribourg, Switzerland*
Ray Correll / *Commonwealth Scientific and Industrial Research Organization (CSIRO), Adelaide, South Australia, Australia*
Luca Coscieme / *Ecodynamics Group, Department of Chemistry, University of Siena, Siena, Italy*
Richard Cowell / *School of City and Regional Planning, Cardiff University, Cardiff, U.K.*
Robin Kundis Craig / *Attorneys' Title Professor of Law and Associate Dean for Environmental Programs, Florida State University College of Law, Tallahassee, Florida, U.S.A.*
Gemma Cranston / *Global Footprint Network, Geneva, Switzerland*
Eric T. Craswell / *Fenner School of Environment and Society, College of Medicine, Biology, and Environment, Australian National University, Canberra, Australian Capital Territory, Australia*
Richard Cruse / *Iowa State University, Ames, Iowa, U.S.A.*
Marc A. Cubeta / *Center for Integrated Fungal Research, Plant Pathology, North Carolina State University, Raleigh, North Carolina, U.S.A.*
Keith Culver / *Okanagan Sustainability Institute, University of British Columbia, Kelowna, British Columbia, Canada*
Marianna Czaplicka / *Institute of Non-Ferrous Metals, and Department of Analytical Chemistry , Silesian University of Technology, Gliwice, Poland*
Seth M. Dabney / *National Sedimentation Laboratory, U.S. Department of Agriculture, Agricultural Research Service (USDA-ARS), Oxford, Mississippi, U.S.A.*
Abhijit Das / *School of Environmental Studies, Jadavpur University, Calcutta, India*
Bhaskar Das / *School of Environmental Studies, Jadavpur University, Calcutta, India*
Franck Dayan / *National Center for Natural Products Research, Agricultural Research Service (USDA-ARS), U.S. Department of Agriculture, University, Missouri, U.S.A.*
Patrick De Clerq / *Department of Crop Protection, Ghent University, Ghent, Belgium*
Ana Maria Evangelista de Duffard / *Laboratorio de Toxicologia Expiremental, National University of Rosario, Rosario, Argentina*
J.R. de Freitas / *Department of Soil Science, University of Saskatchewan, Saskatoon, Saskatchewan, Canada*
Victor de Vlaming / *Aquatic Toxicology Laboratory, University of California—Davis, Davis, California, U.S.A.*
Bernd Delakowitz / *Faculty of Mathematics and Natural Sciences, University of Applied Sciences, Zittau, Germany*
Kathleen Delate / *Departments of Agronomy and Horticulture, Iowa State University, Ames, Iowa, U.S.A.*

Jorge A. Delgado / *Soil Plant Nutrient Research Unit, Agricultural Research Service (USDA-ARS), U.S. Department of Agriculture, Fort Collins, Colorado, U.S.A.*
Detlef Deumlich / *Institute of Soil Landscape Research, Leibniz Center for Agricultural Landscapre Research (ZALF), Muncheberg, Germany*
Malcolm Devine / *Aventis CropScience Canada Co., Saskatoon, Saskatchewan, Canada*
Harvey E. Diamond / *Energy Management International, Conroe, Texas, U.S.A.*
Jan Dich / *Unit of Cancer Epidemiology, Institution of Oncology-Pathology, Karolinska Institutet and Radiumhemmet, Karolinska University Hospital, Stockholm, Sweden*
Christina D. DiFonzo / *Department of Entomology, Michigan State University, East Lansing, Michigan, U.S.A.*
Peter Dillon / *Commonwealth Scientific and Industrial Research Organization (CSIRO), Adelaide, South Australia, Australia*
Ibrahim Dincer / *Faculty of Engineering and Applied Science, University of Ontario Institute of Technology (UOIT), Oshawa, Ontario, Canada*
Barbara Dinham / *Eurolink Center, Pesticide Action Network U.K., London, U.K.*
Craig Ditzler / *National Leader for Soil Classification and Standards, Lincoln, Nebraska, U.S.A.*
Jan Dolfing / *School of Civil Engineering and Geosciences, Newcastle University, Newcastle, U.K.*
Douglas J. Dollhopf / *Department of Land Resources and Environmental Sciences, Montana State University, Bozeman, Montana, U.S.A.*
Cenk Dönmez / *Departments of Soil Science, Landscape Architecture, and Agricultural Engineering, University of Cukurova, Adana, Turkey*
John W. Doran / *U.S. Department of Agriculture, Agricultural Research Service (USDA-ARS), Agronomy, University of Nebraska—Lincoln, Lincoln, Nebraska, U.S.A.*
Steve Doty / *Colorado Springs Utilities, Colorado Springs, Colorado, U.S.A.*
Pay Drechsel / *International Water Management Institute (IWMI), Colombo, Sri Lanka*
Svetlana Drozdova / *Institute of Chemical Technologies and Analytics, Vienna University of Technology, Vienna, Austria*
J.K. Dubey / *Regional Center, National Afforestation and Eco-Development Board, Dr. Y.S. Parmar University of Horticulture and Forestry, Solan, India*
Rathindra Nath Dutta / *Department of Dermatology, Institute of Post Graduate Medical Education and Research, SSKM Hospital, Calcutta, India*
Bahman Eghball / *Agricultural Research Service (USDA-ARS), U.S. Department of Agriculture, Lincoln, Nebraska, U.S.A.*
Reza Ehsani / *Ohio State University, Columbus, Ohio, U.S.A.*
Anna Ekberg / *Department of Ecology, Plant Ecology, Lund University, Climate Impacts Group, Lund, Sweden*
George Ekström / *Swedish National Chemicals Inspectorate (KEMI), Solna, Sweden*
Krisztina Eleki / *Department of Agronomy, Iowa State University, Ames, Iowa, U.S.A.*
Mehmet Akif Erdoğan / *Departments of Soil Science, Landscape Architecture, and Agricultural Engineering, University of Cukurova, Istanbul, Turkey*
Sarina J. Ergas / *Department of Civil and Environmental Engineering, University of South Florida, Tampa, Florida, U.S.A.*
Gunay Erpul / *Department of Soil Science, Ankara University, Ankara, Turkey*
Shannon Estenoz / *Office of Everglades Restoration Initiatives, U.S. Department of the Interior, Davie, Florida, U.S.A.*
Sara Evangelisti / *Interuniversity Research Center for Sustainable Develepment (CIRPS), Sapienza University of Rome, Rome, Italy*

Anna Eynard / *Ohio State University, Brookings, South Dakota, U.S.A.*
Delvin S. Fanning / *Department of Environmental Science and Technology, University of Maryland, College Park, Maryland, U.S.A.*
Norman R. Fausey / *Soil Drainage Research Unit, Agricultural Research Service (USDA-ARS), U.S. Department of Agriculture, Columbus, Ohio, U.S.A.*
David Favis-Mortlock / *Environmental Change Institute, University of Oxford, Oxford, U.K.*
Jolanta Fenik / *Department of Analytical Chemistry, Chemical Faculty, Gdansk University of Technology, Gdansk, Poland*
David N. Ferro / *Department of Entomology, University of Massachusetts, Amherst, Massachusetts, U.S.A.*
Charles W. Fetter / *C. W. Fetter, Jr. Associates, Oshkosh, Wisconsin, U.S.A.*
Maria Finckh / *Department of Ecological Plant Protection, University of Kassel, Witzenhausen, Germany*
Guy Fipps / *Agricultural Engineering Department, Texas A&M University, College Station, Texas, U.S.A.*
Dennis C. Flanagan / *National Soil Science Research Laboratory, Agricultural Research Service (USDA-ARS), U.S. Department of Agriculture, West Lafayette, Indiana, U.S.A.*
Stefan Fraenzle / *Department of Biological and Environmental Sciences; Research Group of Environmental Chemistry, International Graduate School, Zittau, Zittau, Germany*
Alan J. Franzluebbers / *Agricultural Research Service (USDA-ARS), U.S. Department of Agriculture, Watkinsville, Georgia, U.S.A.*
Gary W. Frasier / *U.S. Department of Agriculture (USDA), Fort Collins, Colorado, U.S.A.*
Dwight K. French / *Director, Energy Consumption Division, Energy Information Administration, U.S. Department of Energy, Washington, District of Columbia, U.S.A.*
John R. Freney / *Commonwealth Scientific and Industrial Research Organization (CSIRO), Campbell, Australian Capital Territory, Australia*
Martin V. Frey / *Department of Soil Science, University of Stellenbosch, Matieland, South Africa*
Brenda Frick / *Bluebur Fluent Organics, Saskatoon, Saskatchewan, Canada*
Monika Frielinghaus / *Institute of Soil Landscape Research, Leibniz Center for Agricultural Landscape Research (ZALF), Muncheberg, Germany*
W. Friesl-Hanl / *Health and Environment Department, Environmental Resources & Technologies, AIT Austrian Institute of Technology GmbH, Tulln, Austria*
Roger Funk / *Institute of Soil Landscape Research, Leibniz Center for Agricultural Landscape Research (ZALF), Muncheberg, Germany*
Donald Gabriels / *Department of Soil Management and Soil Care, Ghent University, Ghent, Belgium*
Wendy B. Gagliano / *Clark State Community College, Springfield, Ohio, U.S.A.*
Renata Gaj / *Institute of Soil Science, Agricultural University, Poznan, Poland*
Alessandro Galli / *Global Footprint Network, Geneva, Switzerland*
Ján Gallo / *Department of Plant Protection, Slovak University of Agriculture, Nitra, Slovak Republic*
Agniezka Gałuszka / *Division of Geochemistry and the Environment, Institute of Chemistry, Jan Kochanowski University, Kielce, Poland*
Anurag Garg / *Center for Environmental Science and Engineering, Indian Institute of Technology, Bombay, Mumbai, India*
Anja Gassner / *Institute of Science and Technology, University of Malaysia—Sabah, Kota Kinabalu, Malaysia*

David K. Gattie / *Biological and Agricultural Engineering Department, University of Georgia, Athens, Georgia, U.S.A.*

M.H. Gerzabek / *Institute of Soil Research, University of Natural Resources and Life Sciences, Vienna, Austria*

Stephen R. Gliessman / *Program in Community and Agroecology, Department of Environmental Studies, University of California—Santa Cruz, Santa Cruz, California, U.S.A.*

Fredric S. Goldner / *Energy Management & Research Associates, East Meadow, New York, U.S.A.*

Dan Golomb / *Department of Environmental, Earth and Atmospheric Sciences, University of Massachusetts—Lowell, Lowell, Massachusetts, U.S.A.*

Ragini Gothalwal / *Institute of Microbiology and Biotechnology, Barkatullah University, Bhopal, India*

Andrew S. Goudie / *St. Cross College, Oxford, U.K.*

Simon Gowen / *Department of Agriculture, University of Reading, Reading, U.K.*

David W. Graham / *School of Civil Engineering and Geosciences, Newcastle University, Newcastle, U.K.*

Timothy C. Granata / *Department of Civil and Environmental Engineering and Geodetic Science, Ohio State University, Columbus, Ohio, U.S.A.*

Alex E.S. Green / *Professor Emeritus, University of Florida, Gainesville, Florida, U.S.A.*

Ed G. Gregorich / *Easter Cereal and Oilseed Research Center, Agriculture and Agri-Food Canada, Ottawa, Ontario, Canada*

Simon Grenier / *Functional Biology, Insects and Interactions, National Institute for Agricultural Research (INRA), Villeurbanne, France*

Khara D. Grieger / *Department of Environmental Engineering, Technical University of Denmark, Kongens Lyngby, Denmark*

Lisa Guan / *School of Chemistry, Physics and Mechanical Engineering, Science and Engineering Faculty, Queensland University of Technology, Brisbane, Queensland, Australia*

Eliane Tigre Guimarães / *Experimental Air Pollution Laboratory, Department of Pathology, School of Medicine, University of São Paulo, São Paulo, Brazil*

Silvio J. Gumiere / *Department of Soil Science, Laval University, Quebec City, Quebec, Canada*

Umesh C. Gupta / *Crops and Livestock Research Center, Agriculture and Agri-Food Canada, Charlottetown, Prince Edward Island, Canada*

Andrew Paul Gutierrez / *Center for the Analysis of Sustainable Agricultural Systems, University of California—Berkeley, Berkeley, California, U.S.A.*

Ann E. Hajek / *Department of Entomology, Cornell University, Ithaca, New York, U.S.A.*

Ardell D. Halvorson / *U.S. Department of Agriculture (USDA), Fort Collins, Colorado, U.S.A.*

Denis Hamilton / *Department of Animal and Plant Health Service, Queensland Department of Primary Industries, Brisbane, Queensland, Australia*

Silvia Haneklaus / *Institute for Crop and Soil Science, Julius Kuhn Institute (JKI), Braunschweig, Germany*

Ian Hannam / *Center for Natural Resources, Department of Infrastructure, Planning and Natural Resources, Sydney, New South Wales, Australia*

Chris Hanning / *Sleep Medicine, University Hospitals of Leicester, Leicester, U.K.*

Lise Stengård Hansen / *Danish Pest Infestation Laboratory, Danish Institute of Agricultural Sciences, Konigs Lyngby, Denmark*

Steffen Foss Hansen / *Department of Environmental Engineering, Technical University of Denmark, Kongens Lyngby, Denmark*

Peter Harris / *Agriculture and Agri-Food Canada, Lethbridge, Alberta, Canada*

Kelsey Hart / *College of Veterinary Medicine, University of Georgia, Athens, Georgia, U.S.A.*

James D. Harwood / *Department of Entomology, University of Kentucky, Lexington, Kentucky, U.S.A.*

John V. Headley / *Water Science and Technology Directorate, Saskatoon, Saskatchewan, Canada*

Steven D. Heinz / *Good Steward Software, State College, Pennsylvania, U.S.A.*

Arif Hepbasli / *Department of Energy Systems Engineering, Faculty of Engineering, Yaşar University, Bornova, Izmir, Turkey*

Keith E. Herold / *Fischell Department of Bioengineering, University of Maryland, College Park, Maryland, U.S.A.*

James G. Hewlett / *Energy Information Administration, U.S. Department of Energy, Washington, District of Columbia, U.S.A.*

Robert W. Hill / *Biological and Irrigation Engineering Department, Utah State University, Logan, Utah, U.S.A.*

Philippe Hinsinger / *Sun and Environment Unit, National Institute for Agricultural Research (INRA), Montpellier, France*

Michael C. Hirschi / *University of Illinois, Urbana, Illinois, U.S.A.*

Rusty T. Hodapp / *Vice President and Sustainability Officer, Energy and Transportation Management, Dallas/Fort Worth International Airport Board, Dallas/Forth Worth Airport, Texas, U.S.A.*

Laurie Hodges / *Department of Agronomy and Horticulture, University of Nebraska—Lincoln, Lincoln, Nebraska, U.S.A.*

Glenn J. Hoffman / *Biological Systems Engineering, University of Nebraska—Lincoln, Lincoln, Nebraska, U.S.A.*

Heikki Hokkanen / *Department of Applied Biology, University of Helsinki, Helsinki, Finland*

John Holland / *Head of Entomology, Game Conservancy Trust, Hants, U.K.*

David J. Horn / *Department of Entomology, Ohio State University, Columbus, Ohio, U.S.A.*

Lloyd R. Hossner / *Soil and Crop Sciences Department, Texas A&M University, College Station, Texas, U.S.A.*

Terry A. Howell / *Conservation and Production Research Laboratory, Agricultural Research Service (USDA-ARS), U.S. Department of Agriculture, Bushland, Texas, U.S.A.*

Hei-Ti Hsu / *Floral and Nursery Plants Research, Agricultural Research Service (USDA-ARS), U.S. Department of Agriculture, Beltsville, Maryland, U.S.A.*

Zhengyi Hu / *Institute of Soil Science, Chinese Academy of Sciences, Nanjing, China*

Nathan E. Hultman / *University of Maryland, Maryland, U.S.A.*

Hayriye Ibrikci / *Soil Science and Plant Nutrition Department, Cukurova University, Adana, Turkey*

Craig Idso / *Center for the Study of Carbon Dioxide and Global Change, Tempe, Arizona, U.S.A.*

Keith E. Idso / *Center for the Study of Carbon Dioxide and Global Change, Tempe, Arizona, U.S.A.*

Sherwood Idso / *U.S. Department of Agriculture (USDA), Tempe, Arizona, U.S.A.*

R. Cesar Izaurralde / *Battelle Pacific Northwest National Laboratory, Washington, District of Columbia, U.S.A.*

Alexandra Izosimova / *St. Petersburg Agricultural Physical Research Institute, St. Petersburg, Russia*

Pierre A. Jacinthe / *Ohio State University, Columbus, Ohio, U.S.A.*
C. Rhett Jackson / *Daniel B. Warnell School of Forest Resources, University of Georgia, Athens, Georgia, U.S.A.*
Bruce R. James / *University of Maryland, College Park, Maryland, U.S.A.*
Philip M. Jardine / *Oak Ridge National Laboratory, Oak Ridge, Tennessee, U.S.A.*
David Jasper / *Center for Land Rehabilitation, University of Western Australia, Nedlands, Western Australia, Australia*
Julie D. Jastrow / *Environmental Research Division, Argonne National Laboratory, Argonne, Illinois, U.S.A.*
Ike Jeon / *Department of Animal Science and Industry, Kansas State University, Manhattan, Kansas, U.S.A.*
Kui Jiao / *Department of Mechanical Engineering, University of Waterloo, Waterloo, Ontario, Canada*
Blanca Jimenez / *Engineering Institute, National Autonomous University of Mexico (UNAM), Coyoacan, Mexico*
Sven Erik Jørgensen / *Institute A, Section of Environmental Chemistry, Copenhagen University, Copenhagen, Denmark*
Sandeep Joshi / *Shrishti Eco-Research Unit (SERI), Pune, India*
Sayali Joshi / *Shrishti Eco-Research Institute (SERI), Pune, India*
Puangrat Kajitvichyanukul / *Center of Excellence for Environmental Research and Innovation, Faculty of Engineering, Naresuan University, Phitsanulok, Thailand*
Gabriella Kakonyi / *Kroto Research Institute, Sheffield University, Sheffield, U.K.*
Inger Källander / *Swedish Ecological Farmers Association, Uppsala, Sweden*
Marion Kandziora / *Institute for the Conservation of Natural Resources, University of Kiel, Kiel, Germany*
Douglas Kane / *Water and Environmental Research Center, University of Alaska—Fairbanks, Fairbanks, Alaska, U.S.A.*
Chih-Ming Kao / *Institute of Environmental Engineering, National Sun-Yat Sen University, Kaohsiung, Taiwan*
Burçak Kapur / *Department of Biosystems Engineering, University of Yuzuncu Yil, Van, Turkey*
Selim Kapur / *Departments of Soil Science, Landscape Architecture, and Agricultural Engineeering, University of Çukurova, Adana, Turkey*
Gholamreza Karimi / *Department of Mechanical Engineering, University of Waterloo, Waterloo, Ontario, Canada*
Douglas L. Karlen / *U.S. Department of Agriculture (USDA), Ames, Iowa, U.S.A.*
Subhankar Karmakar / *Center for Environmental Science and Engineering (CESE), Indian Institute of Technology Bombay, Mumbai, India*
Marianne Karpenstein-Machan / *Institute of Crop Science, University of Kassel, Witzenhausen, Germany*
Janey Kaster / *Yamas Controls West, San Francisco, California, U.S.A.*
Anthony P. Keinath / *Coastal Research and Education Center, Clemson University, Charleston, South Carolina, U.S.A.*
Keith A. Kelling / *Professor Emeritus, Department of Soil Science, University of Wisconsin—Extension, Madison, Wisconsin, U.S.A.*
George Kennedy / *Department of Entomology, North Carolina State University, Raleigh, North Carolina, U.S.A.*
Rami Keren / *Agricultural Research Organization of Israel, Bet-Dagan, Israel*
Matthias Kern / *GTZ Pilot Project on Chemicals Management, Bonn, Germany*

Peter Kerr / *Commonwealth Scientific and Industrial Research Organization (CSIRO) Ecosystem Sciences, Canberra, Australian Capital Territory, Australia*
Gregory Kiker / *University of Florida, Gainesville, Florida, U.S.A.*
Peter I.A. Kinnell / *School of Resource, Environmental and Heritage Sciences, University of Canberra, Holt, Australian Capital Territory, Australia*
Ronald L. Klaus / *VAST Power Systems, Elkhart, Indiana, U.S.A.*
Peter Kleinman / *Pasture Systems and Watershed Management Research Unit, U.S. Department of Agriculture (USDA), University Park, Pennsylvania, U.S.A.*
Andreas Klik / *Department of Water, Atmosphere and Environment, University of Natural Resources and Life Sciences, Vienna, Austria*
Ewa Klugmann-Radziemska / *Chemical Faculty, Gdansk University of Technology, Gdansk, Poland*
Birgitta Kolmodin-Hedman / *Department of Public Health Sciences, Karolinska Institute, Stockholm, Sweden*
Rai Kookana / *Commonwealth Scientific and Industrial Research Organization (CSIRO), Adelaide, South Australia, Australia*
Monika Kosikowska / *Department of Analytical Chemistry, Chemical Faculty, Gdansk University of Technology, Gdansk, Poland*
John Kost / *Department of Agronomy, Iowa State University, Ames, Iowa, U.S.A.*
Andrey G. Kostianoy / *P.P. Shirshov Institute of Oceanology, Russian Academy of Sciences, Moscow, Russia*
Milivoje M. Kostic / *Department of Mechanical Engineering, Northern Illinois University, DeKalb, Illinois, U.S.A.*
William L. Kranz / *Northeast Research and Extension Center, University of Nebraska—Lincoln, Norfolk, Nebraska, U.S.A.*
David P. Kreutzweiser / *Canadian Forest Service, Natural Resources Canada, Sault Sainte Marie, Ontario, Canada*
Tore Krogstad / *Department of Plant and Environmental Sciences, Norwegian University of Life Science, Aas, Norway*
Atif Kubursi / *Department of Economics, McMaster University, Hamilton, Ontario, Canada*
Umesh Kulshrestha / *School of Environmental Sciences, Jawaharlal Nehru University, New Delhi, India*
Amit Kumar / *Research Group Environmental Organic Chemistry Group (EnVOC), Faculty of Bioscience Engineering, Ghent University, Ghent, Belgium, and Department of Environmental Engineering and Water Technology, Institute for Water Education (UNESCO-IHE), Delft, the Netherlands*
Klaus Kümmerer / *Institute of Environmental Medicine and Hospital Epidemiology, University Hospital Freiburg, Freiburg, Germany*
Yu-Chia Kuo / *Institute of Environmental Engineering, National Sun Yat-Sen University, Kaohsiung, Taiwan*
Witold Kurylak / *Institute of Non-Ferrous Metals, Gliwice, Poland*
John M. Laflen / *Agricultural Research Service (USDA-ARS), U.S. Department of Agriculture, Buffalo Center, and Iowa State University, Ames, Iowa, U.S.A.*
Rattan Lal / *School of Environment and Natural Resources, Ohio State University, Columbus, Ohio, U.S.A.*
Freddie L. Lamm / *Research Extension Center, Kansas State University, Colby, Kansas, U.S.A.*
Judith Lancaster / *Desert Research Institute, Reno, Nevada, U.S.A.*

Doug Landis / *Department of Entomology, Michigan State University, East Lansing, Michigan, U.S.A.*
Tomaz Langenbach / *Federal University of Rio de Janeiro, Rio de Janeiro, Brazil*
David B. Langston, Jr. / *Rural Development Center, University of Georgia, Tifton, Georgia, U.S.A.*
Robert J. Lascano / *Agricultural Research Service (USDA-ARS), U.S. Department of Agriculture, Lubbock, Texas, U.S.A.*
Jean-Claude Lefeuvre / *Laboratory of the Evolution of Natural and Modified Systems, University of Rennes, Rennes, France*
E.F. Legner / *Department of Entomology, University of California—Riverside, Riverside, California, U.S.A.*
Henry Noel LeHouerou / *Center for Functional and Evolutionary Ecology (CEFE), National Center for Scientific Research (CNRS), France*
Reynald Lemke / *Agriculture and Agri-Food Canada, Swift Current, Saskatchewan, Canada*
Rocky Lemus / *Mississippi State University, Starkville, Mississippi, U.S.A.*
Piet Lens / *Department of Environmental Engineering and Water Technology, Institute for Water Education (UNESCO-IHE), Delft, the Netherlands*
Xianguo Li / *Department of Mechanical Engineering, University of Waterloo, Waterloo, Ontario, Canada*
Shu-Hao Liang / *Institute of Environmental Engineering, National Sun Yat-Sen University, Kaohsiung, Taiwan*
Mingsheng Liu / *Architectural Engineering Program, Peter Kiewit Institute, University of Nebraska—Lincoln, Omaha, Nebraska, U.S.A.*
K.F. Andrew Lo / *Department of Natural Resources, Chinese Culture University, Taipei, Taiwan*
Hugo A. Loaiciga / *Department of Geography, University of California—Santa Barbara, Santa Barbara, California, U.S.A.*
Leslie London / *Occupational and Environmental Health Research Unit, University of Cape Town, Observatory, South Africa*
Richard Lowrance / *Agricultural Research Service (USDA-ARS), U.S. Department of Agriculture, Tifton, Georgia, U.S.A.*
John Ludwig / *Ecosystem Sciences, Commonwealth Scientific and Industrial Research Organization (CSIRO), Winnellie, Northern Territory, Australia*
Rory O. Maguire / *Virginia Tech, Blacksburg, Virginia, U.S.A.*
Barbara Manachini / *Departments of Environmental Biology and Biodiversity, University of Palermo, Palermo, Italy*
Kyle R. Mankin / *Department of Biological and Agricultural Engineering, Kansas State University, Manhattan, Kansas, U.S.A.*
Gamini Manuweera / *Secretariat of the Stockholm Convention, United Nations Environmental Program, Chatelaine, Switzerland*
Tek Narayan Maraseni / *Australian Center for Sustainable Catchments, School of Accounting, Economics and Finance, University of Southern Queensland, Toowoomba, Queensland, Australia*
Bernd Markert / *Environmental Institute of Scientific Networks (EISN), Haren-Erika, Germany*
Rudolf Marloth / *San Diego State University, San Diego, California, U.S.A.*
Jay F. Martin / *Department of Food, Agricultural, and Biological Engineering, Ohio State University, Columbus, Ohio, U.S.A.*

Jocilyn Danise Martinez / *University of South Florida, Tampa, Florida, U.S.A.*
Graça Martinho / *Department of Environmental Sciences and Engineering, Faculty of Sciences and Technology, New University of Lisbon, Caparica, Portugal*
Saburo Matsui / *Graduate School of Global Environmental Studies, Kyoto University, Kyoto, Japan*
Thomas J. Mbise / *Tanzania Association of Public Occupational and Environmental Health Experts, Dar-es-Salaam, Tanzania*
Ann McCampbell / *Chair, Multiple Chemical Sensitivities Task Force of New Mexico, Santa Fe, New Mexico, U.S.A.*
Donald K. McCool / *Agricultural Research Service (USDA-ARS), U.S. Department of Agriculture, Pullman, Washington, U.S.A.*
Richard McDowell / *AgResearch Ltd., Invermay Agricultural Center, Mosgiel, New Zealand*
Leslie D. McFadden / *Department of Earth and Planetary Sciences, University of New Mexico, Albuquerque, New Mexico, U.S.A.*
Sean McGinn / *Agriculture and Agri-Food Canada, Lethbridge, Alberta, Canada*
Ronald G. McLaren / *Soil, Plant, and Ecological Sciences Division, Lincoln University, Canterbury, New Zealand*
Mike J. McLaughlin / *Land and Water, Commonwealth Scientific and Industrial Research Organization (CSIRO), Glen Osmond, South Australia, Australia*
Agata Mechlińska / *Department of Analytical Chemistry, Chemical Faculty, Gdansk University of Technology, and Department of Environmental Toxicology, Interdepartmental Institute of Maritime and Tropical Medicine, Medical University of Gdansk, Gdansk, Poland*
Mallavarapu Megharaj / *Commonwealth Scientific and Industrial Research Organization (CSIRO), Adelaide, South Australia, Australia*
D. Paul Mehta / *Department of Mechanical Engineering, Bradley University, Peoria, Illinois, U.S.A.*
Michael D. Melville / *School of Biological, Earth, and Environmental Sciences, University of New South Wales, Sydney, New South Wales, Australia*
Neal William Menzies / *School of Land, Crop and Food Sciences, University of Queensland, St. Lucia, Queensland, Australia*
Gustavo Enrique Merten / *Federal University of Rio Grande do Sul, Porto Alegre, Brazil*
Andrea Micangeli / *Interuniversity Research Center for Sustainable Development (CIRPS), Sapienza University of Rome, Rome, Italy*
Małgorzata Michalska / *Institute of Maritime and Tropical Medicine, Gdynia, Poland*
Adnan Midili / *Department of Mechanical Engineering, Faculty of Engineering, Nigde University, Nigde, Turkey*
Zdzisław M. Migaszewski / *Division of Geochemistry and the Environment, Institute of Chemistry, Jan Kochanowski University, Kielce, Poland*
J. David Miller / *Department of Chemistry, Carleton University, Ottawa, Ontario, Canada*
Pierre Mineau / *Science and Technology Branch, Environment Canada, Ottawa, Ontario, Canada*
Jean Paulo Gomes Minella / *Federal University of Santa Maria, Santa Maria, Brazil*
I.M. Mishra / *Department of Chemical Engineering, Indian Institute of Technology, Roorkee, Mumbai, India*
J. Kent Mitchell / *Agricultural Engineering, University of Illinois, Urbana, Illinois, U.S.A.*
Luisa T. Molina / *Massachusetts Institute of Technology, Cambridge, Massachusetts, U.S.A.*

P. Mondal / *Department of Chemical Engineering, Indian Institute of Technology, Roorkee, Roorkee, India*
H. Curtis Monger / *Department of Agronomy and Horticulture, New Mexico State University, Las Cruces, New Mexico, U.S.A.*
Lynn E. Moody / *Earth and Soil Sciences Department, California Polytechnic State University, San Luis Obispo, California, U.S.A.*
David A. Mouat / *Division of Earth and Ecosystem Sciences, Desert Research Institute, Reno, Nevada, U.S.A.*
Martin A. Mozzo, Jr. / *M and A Associates Inc, Robbinsville, New Jersey, U.S.A.*
Subhas Chandra Mukherjee / *Department of Neurology, Medical College, Calcutta, India*
Felix Müller / *Ecology Center, University of Kiel, Kiel, Germany*
Heinz Müller-Schärer / *Department of Biology and Ecology, University of Fribourg/Perolles, Fribourg, Switzerland*
Tariq Muneer / *School of Engineering, Napier University, Edinburgh, U.K.*
Rafael Munoz-Carpena / *University of Florida, Florida, U.S.A.*
Joji Muramoto / *Program in Community and Agroecology, Department of Environmental Studies, University of California—Santa Cruz, Santa Cruz, California, U.S.A.*
Stephen D. Murphy / *Faculty of Environment, University of Waterloo, Waterloo, Ontario, Canada*
O.M. Musthafa / *Center for Pollution Control and Environmental Engineering, Pondicherry University, Pondicherry, India*
Ravendra Naidu / *Commonwealth Scientific and Industrial Research Organization (CSIRO), Adelaide, South Australia, Australia*
D.V. Naik / *Biofuels Division, Indian Institute of Petroleum, Dehradun, India*
Jacek Namieśnik / *Department of Analytical Chemistry, Chemical Faculty, Gdansk University of Technology, Gdansk, Poland*
Inés Navarro / *Engineering Institute, National Autonomous University of Mexico (UNAM), Coyoacan, Mexico*
Alexandra Navrotsky / *Department of Chemical Engineering and Materials Science, University of California—Davis, Davis, California, U.S.A.*
Mark A. Nearing / *Southwest Watershed Research Center, Agricultural Research Service (USDA-ARS), U.S. Department of Agriculture, Tucson, Arizona, U.S.A.*
Jerry Neppel / *Department of Agronomy, Iowa State University, Ames, Iowa, U.S.A.*
Elena Neri / *Ecodynamics Group, Department of Chemistry, University of Siena, Siena, Italy*
Aiwerasia V.F. Ngowi / *Tanzania Association of Public Occupational and Environmental Health Experts, and Department of Environmental and Occupational Health, Muhimbili University of Health and Allied Sciences (MUHAS), Dar-es-Salaam, Tanzania*
Valentina Niccolucci / *Ecodynamics Group, Department of Chemistry, University of Siena, Siena, Italy*
Niels Erik Nielsen / *Plant Nutrition and Soil Fertility Laboratory, Department of Agricultural Sciences, Royal Veterinary and Agricultural University, Frederiksberg, Denmark*
Egide Nizeyimana / *Department of Agronomy and Environmental Resources Research Institute, Pennsylvania State University, University Park, Pennsylvania, U.S.A.*
L. Darrell Norton / *National Soil Science Research Laboratory, Agricultural Research Service (USDA-ARS), U.S. Department of Agriculture, West Lafayette, Indiana, U.S.A.*
John J. Obrycki / *Department of Entomology, University of Kentucky, Lexington, Kentucky, U.S.A.*

Philip Oduor-Owino / *Department of Botany, University of Kenyatta, Nairobi, Kenya*
K.E. Ohrn / *Cypress Digital Ltd., Vancouver, British Columbia, Canada*
Michael T. Olexa / *Center for Agricultural and Natural Resource Law, Institute of Food and Agricultural Sciences, University of Florida, Gainesville, Florida, U.S.A.*
Jacob Opadeyi / *Department of Surveying and Land Information, Faculty of Engineering, University of the West Indies, St. Augustine, Trinidad and Tobago*
Bohdan W. Oppenheim / *U.S. Department of Energy Industrial Assessment Center, Loyola Marymount University, Los Angeles, California, U.S.A.*
Barron Orr / *Office of Arid Lands Studies, School of Natural Resources and the Environment, University of Arizona, Tucson, Arizona, U.S.A.*
Lloyd B. Owens / *U.S. Department of Agriculture (USDA), Coshocton, Ohio, U.S.A.*
Margareta Palmborg / *Swedish Poisons Information Center, Stockholm, Sweden*
Maurizio G. Paoletti / *Department of Biology, University of Padova, Padova, Italy*
J. Parikh / *Department of Chemical Engineering, S.V. National Institute of Technology, Surat, India*
David R. Parker / *Department of Soil and Environmental Sciences, University of California—Riverside, Riverside, California, U.S.A.*
Steven A. Parker / *Pacific Northwest National Laboratory, Richland, Washington, U.S.A.*
Shyamapada Pati / *Department of Obstetrics and Gynaecology, Calcutta National Medical College, Calcutta, India*
Timothy Paulitz / *Department of Plant Science, MacDonald Campus of McGill University, Ste-Anne-de-Bellevue, Quebec, Canada*
Judith F. Pedler / *Department of Environmental Sciences, University of California—Riverside, Riverside, California, U.S.A.*
Meir Paul Pener / *Department of Cell and Developmental Biology, Hebrew University of Jerusalem, Jerusalem, Israel*
Mark B. Peoples / *Plant Industry, Commonwealth Scientific and Industrial Research Organization (CSIRO), Canberra, Australian Capital Territory, Australia*
Kerry M. Peru / *Water and Technology Directorate, Saskatoon, Saskatchewan, Canada*
Julie A. Peterson / *Department of Entomology, University of Kentucky, Lexington, Kentucky, U.S.A.*
Mark A. Peterson / *Sustainable Success LLC, Clementon, New Jersey, U.S.A.*
Robert Pietrzak / *Department of Chemistry, Adam Mickiewicz University, Poznan, Poland*
David Pimentel / *Department of Entomology, Cornell University, Ithaca, New York, U.S.A.*
Ana Pires / *Department of Environmental Sciences and Engineering, Faculty of Sciences and Technology, New University of Lisbon, Caparica, Portugal*
Peter W. Plumstead / *North Carolina State University, Raleigh, North Carolina, U.S.A.*
Paola Poli / *Department of Genetics, Biology of Microorganisms, Anthropology, and Evolution, University of Parma, Parma, Italy*
Zaneta Polkowska / *Gdansk University of Technology, Gdansk, Poland*
Wendell A. Porter / *Department of Agricultural and Biological Engineering, University of Florida, Gainesville, Florida, U.S.A.*
Wilfred M. Post / *Oak Ridge National Laboratory, Oak Ridge, Tennessee, U.S.A.*
Franco Previtali / *Department of Environmental Science, University of Milan—Bicocca, Milan, Italy*
Odo Primavesi / *Brazilian Agricultural Research Corporation (Embrapa), São Paulo, Brazil*
Soyuz Priyadarsan / *Texas A&M University, College Station, Texas, U.S.A.*

Federico M. Pulselli / *Ecodynamics Group, Department of Chemistry, University of Siena, Siena, Italy*

Manzoor Qadir / *International Center for Agricultural Research in the Dry Areas (ICARDA), Aleppo, Syria, and International Water Management Institute (IWMI), Colombo, Sri Lanka*

X.S. Qin / *School of Civil and Environmental Engineering, Nanyang Technical University, Singapore*

Quazi Quamruzzaman / *Dhaka Community Hospital, Dhaka, Bangladesh*

Anna Rabajczyk / *Independent Department of Environment Protection and Modeling, Jan Kochanowski University of Humanities and Sciences in Kielce, Kielce, Poland*

Martin C. Rabenhorst / *University of Maryland, College Park, Maryland, U.S.A.*

Mohammad Mahmudur Rahman / *School of Environmental Studies, Jadavpur University, Calcutta, India*

Shafiqur Rahman / *Department of Agricultural and Biosystems Engineering, North Dakota State University, Fargo, North Dakota, U.S.A.*

Liqa Raschid-Sally / *International Water Management Institute (IWMI), Colombo, Sri Lanka*

Abdul Rashid / *Pakistan Atomic Energy Commission, Islamabad, Pakistan*

José Miguel Reichert / *Federal University of Santa Maria, Santa Maria, Brazil*

Pichu Rengasamy / *Department of Soil and Water Science, University of Adelaide, Waite, South Australia, Australia*

James D. Rhoades / *Agricultural Salinity Consulting, Riverside, California, U.S.A.*

Lisa A. Robinson / *Independent Consultant, Newton, Massachusetts, U.S.A.*

Mark Robinson / *Center for Ecology and Hydrology, Wallingford, U.K.*

Philippe Rochette / *Soils and Crops Research Center, Agriculture and Agri-Food Canada, Saint-Foy, Quebec, Canada*

Justyna Rogowska / *Department of Analytical Chemistry, Chemical Faculty, Gdansk University of Technology, and Department of Environmental Toxicology, Interdepartmental Institute of Maritime and Tropical Medicine, Medical University of Gdansk, Gdansk, Poland*

Alexandru V. Roman / *School of Public Administration, Florida Atlantic University, Boca Raton, Florida, U.S.A.*

Larama M.B. Rongo / *Muhimbili University of Health and Allied Sciences, Dar-es-Salaam, Tanzania*

Stephen A. Roosa / *Energy Systems Group, Inc., Louisville, Kentucky, U.S.A.*

Marc A. Rosen / *Faculty of Engineering and Applied Science, University of Ontario Institute of Technology (UOIT), Oshawa, Ontario, Canada*

Erwin Rosenberg / *Institute of Chemical Technologies and Analytics, Vienna University of Technology, Vienna, Austria*

Clayton Rubec / *Center for Environmental Stewardship and Conservation, Ottawa, Ontario, Canada*

Jennifer Ruesink / *Department of Zoology, University of Washington, Seattle, Washington, U.S.A.*

Gunnar Rundgren / *Grolink AB, Hoeje, Sweden*

John Ryan / *International Center for Agricultural Research in the Dry Areas (ICARDA), Aleppo, Syria*

D.W. Rycroft / *Formerly at Department of Civil and Environmental Engineering, Southampton University, Southampton, U.K.*

Paul G. Saffigna / *Tropical Forestry, University of Queensland, Gatton, Queensland, Australia*

Khitish Chandra Saha / *School of Environmental Studies, Jadavpur University, Calcutta, India*
Alka Sapat / *School of Public Administration, Florida Atlantic University, Boca Raton, Florida, U.S.A.*
Danilo Sbordone / *Interuniversity Research Center for Sustainable Develepment (CIRPS), Sapienza University of Rome, Rome, Italy*
Claus Schimming / *Institute for the Conservation of Natural Resources, University of Kiel, Kiel, Germany*
William H. Schlesinger / *Deptartment of Geology and Botany, Duke University, Durham, North Carolina, U.S.A.*
Ewald Schnug / *Institute for Crop and Soil Science, Julius Kuhn Institute (JKI), Braunschweig, Germany*
A. Paul Schwab / *Department of Agronomy, Purdue University, West Lafayette, Indiana, U.S.A.*
Cetin Sengonca / *Department of Entomology and Plant Protection, Institute of Plant Pathology, University of Bonn, Bonn, Germany*
Hamid Shahandeh / *Texas A&M University, College Station, Texas, U.S.A.*
A.V. Shanwal / *Department of Soil Science, Chaudhary Charan Singh Haryana Agricultural University, Hisar, India*
Andrew N. Sharpley / *University of Arkansas, Fayetteville, Arkansas, U.S.A.*
Brenton S. Sharratt / *Land Management and Water Conservation Research Unit, Agricultural Research Service (USDA-ARS), U.S. Department of Agriculture, Pullman, Washington, U.S.A.*
Daniel Shepherd / *Auckland University of Technology, Auckland, New Zealand*
Paul K. Sibley / *School of Environmental Science, University of Guelph, Guelph, Ontario, Canada*
Brian Silvetti / *CALMAC Manufacturing Corporation, Fair Lawn, New Jersey, U.S.A.*
Bal Ram Singh / *Department of Plant and Environmental Sciences, Norwegian University of Life Sciences, Aas, Norway*
S.P. Singh / *National Bureau of Soil Survey and Land Use Planning, Indian Agricultural Research Institute, New Delhi, India*
Johan Six / *Department of Agronomy and Range Science, University of California—Davis, Davis, California, U.S.A.*
Jeffrey G. Skousen / *Division of Plant and Soil Sciences, West Virginia University, Morgantown, West Virginia, U.S.A.*
Bogdan Skwarzec / *Faculty of Chemistry, University of Gdansk, Gdansk, Poland*
Edward H. Smith / *Cornell University, Ithaca, New York, U.S.A.*
Matt C. Smith / *Biological and Agricultural Engineering Department, University of Georgia, Athens, Georgia, U.S.A.*
Pete Smith / *Institute of Biological and Environmental Sciences, School of Biological Sciences, University of Aberdeen, Aberdeen, U.K.*
Sean M. Smith / *Ecosystem Restoration Center, Maryland Department of Natural Resources, Annapolis, Maryland, U.S.A.*
Hwat Bing So / *Center for Environmental Systems Research, Griffith University, Nathan, Queensland, Australia*
Robert E. Sojka / *Northwest Irrigation and Soils Research Lab, U.S. Department of Agriculture (USDA), Kimberly, Idaho, U.S.A.*
Leslie A. Solmes / *LAS and Associates, Mill Valley, California, U.S.A.*

Rolf Sommer / *International Center for Agricultural Research in the Dry Areas (ICARDA), Aleppo, Syria*

Roy F. Spalding / *Water Science Laboratory, University of Nebraska—Lincoln, Lincoln, Nebraska, U.S.A.*

Graham P. Sparling / *Landcare Research, Hamilton, New Zealand*

Gerd Sparovek / *College of Agriculture, Graduate School of Agriculture Luiz de Queiroz (ESALQ), University of São Paulo, São Paulo, Brazil*

Eric B. Spurr / *Department of Wildlife Ecology, Landcare Research New Zealand, Ltd., Lincoln, New Zealand*

Victor R. Squires / *Dryland Management Consultant, South Australia, Australia*

Amanda Staudt / *Climate Scientist, National Wildlife Federation, U.S.A.*

Nicolae Stefan / *Center for Sustainable Exploitation of Ecosystems (CESEE), Alexandru Ioan Cuza University of Iasi, Iasi, Romania*

Joshua Steinfeld / *Florida Atlantic University, Boca Raton, Florida, U.S.A.*

Larry D. Stetler / *Department of Geology and Geological Engineering, South Dakota School of Mines and Technology, Rapid City, South Dakota, U.S.A.*

F. Craig Stevenson / *University of Saskatchewan, Department of Soil Science, Saskatoon, Saskatchewan, Canada*

B.A. Stewart / *Dryland Agriculture Institute, West Texas A&M University, Canyon, Texas, U.S.A.*

Therese Stovall / *Oak Ridge National Laboratory, Oak Ridge, Tennessee, U.S.A.*

Evamarie Straube / *Institute of Occupational Medicine, University of Greifswald, Greifswald, Germany*

Sebastian Straube / *Department of Physiology, University of Oxford, Oxford, U.K.*

Wolfgang Straube / *Department of Gynecology and Obstetrics, University of Greifswald, Greifswald, Germany*

Tanja Strive / *Commonwealth Scientific and Industrial Research Organization (CSIRO) Ecosystem Sciences, Canberra, Australian Capital Territory, Australia*

Scott J. Sturgul / *Outreach Program Manager, Nutrient and Pest Management Program, University of Wisconsin, Madison, Wisconsin, U.S.A.*

Donald L. Suarez / *U.S. Salinity Laboratory, Agricultural Research Service (USDA-ARS), U.S. Department of Agriculture, Riverside, California, U.S.A.*

Praful Suchak / *Sneha Plastics Pvt. Ltd., Suchak's Consultancy Services, Mumbai, India*

L.A. Sullivan / *Southern Cross GeoScience, Southern Cross University, Lismore, New South Wales, Australia*

Matthew O. Sullivan / *Department of Food, Agricultural, and Biological Engineering, Ohio State University, Columbus, Ohio, U.S.A.*

Rao Y. Surampalli / *Department of Civil Engineering, University of Nebraska—Lincoln, Lincoln, Nebraska, U.S.A.*

Claus Svendsen / *Center for Ecology and Hydrology, Wallingford, U.K.*

John M. Sweeten / *Texas A&M University, Amarillo, Texas, U.S.A.*

Piotr Szefer / *Department of Food Sciences, Medical University of Gdansk, Gdansk, Poland*

Kenneth K. Tanji / *Department of Land, Air, and Water Resources, University of California—Davis, Davis, California, U.S.A.*

Maciej Tankiewicz / *Department of Analytical Chemistry, Chemical Faculty, Gdansk University of Technology, Gdansk, Poland*

Meena Thakur / *Department of Environment Sciences, Dr. Y.S. Parmar University of Horticulture and Forestry, Solan, India*

Daniel L. Thomas / *Louisiana State University, Baton Rouge, Louisiana, U.S.A.*
Bob Thorne / *Massey University, New Zealand*
Thomas L. Thurow / *Department of Renewable Resources, University of Wyoming, Laramie, Wyoming, U.S.A.*
Jill S. Tietjen / *Technically Speaking, Inc., Greenwood Village, Colorado, U.S.A.*
Ralph W. Tiner / *National Wetlands Inventory Program, U.S., Fish and Wildlife Service, Hadley, Massachusetts, U.S.A.*
Greg Tinkler / *RLB Consulting Engineers, Houston, Texas, U.S.A.*
Abhishek Tiwary / *Newcastle University, Newcastle, U.K.*
Marek Tobiszewski / *Department of Analytical Chemistry, Chemical Faculty, Gdansk University of Technology, Gdansk, Poland*
David Tongway / *Ecosystem Sciences, Commonwealth Scientific and Industrial Research Organization (CSIRO), Canberra, Australian Capital Territory, Australia*
Alberto Traverso / *Thermochemical Power Group, Department of Mechanical, Energy, Management and Transportation Engineering (DIME), University of Genoa, Genoa, Italy*
David Tucker / *National Energy Technology Laboratory, Department of Energy, Morgantown, West Virginia, U.S.A.*
W.D. Turner / *Department of Mechanical Engineering, Energy Systems Laboratory, College Station, Texas, U.S.A.*
Wayne C. Turner / *Industrial Engineering and Management, Oklahoma State University, Stillwater, Oklahoma, U.S.A.*
Robert E. Uhrig / *Department of Nuclear Engineering, University of Tennessee, Knoxville, Tennessee, U.S.A.*
Oswald Van Cleemput / *Faculty of Agricultural and Applied Biological Sciences, University of Ghent, Ghent, Belgium*
H.F. van Emden / *Department of Agriculture, University of Reading, Reading, U.K.*
H.H. Van Horn / *Department of Animal Science, University of Florida, Gainesville, Florida, U.S.A.*
Herman Van Langenhove / *Department of Environmental Engineering and Water Technology, Institute for Water Education (UNESCO-IHE), Delft, the Netherlands*
R. Scott Van Pelt / *Wind Erosion and Water Conservation Research Unit, Agricultural Research Service (USDA-ARS), U.S. Department of Agriculture, Big Spring, Texas, U.S.A.*
George F. Vance / *Department of Ecosystem Sciences and Management, University of Wyoming, Laramie, Wyoming, U.S.A.*
Andrea Nunes Vaz Pedroso / *Nucleus Research in Ecology, Institute of Botany, São Paulo, Brazil*
Peter Victor / *Faculty of Environmental Studies, York University, Toronto, Ontario, Canada*
Dalmo A.N. Vieira / *National Sedimentation Laboratory, Arkansas State University, U.S. Department of Agriculture, Agricultural Research Service (USDA-ARS), Jonesboro, Arkansas, U.S.A.*
Denise Vienne / *School of Public Administration, Florida Atlantic University, Boca Raton, Florida, U.S.A.*
Jan Vymazal / *Faculty of Environmental Sciences, Department of Landscape Ecology, Czech University of Life Sciences Prague, Prague, Czech Republic*
Leszek Wachowski / *Department of Chemistry, Adam Mickiewicz University, Posnan, Poland*
Mathis Wackernagel / *Global Footprint Network, Oakland, California, U.S.A.*
Joel T. Walker / *Department of Food, Agricultural, and Biological Engineering, Ohio State University, Columbus, Ohio, U.S.A.*

Doug Walsh / *Washington State University, Prosser, Washington, U.S.A.*
Ivan A. Walter / *Ivan's Engineering, Inc., Denver, Colorado, U.S.A.*
A. Wang / *E.H. Graham Center for Agricultural Innovation, Industry and Investment, NSW and Charles Sturt University, Wagga Wagga, New South Wales, Australia*
Xingxiang Wang / *Institute of Soil Science, Chinese Academy of Sciences, Nanjing, China*
Ynuzhang Wang / *Academy of Yellow River Conservancy Science, Zhengzhou, China*
Waldemar Wardencki / *Department of Chemistry, Gdansk University of Technology, Gdansk, Poland*
Andrew Warren / *Department of Geography, University College London, London, U.K.*
Apichon Watcharenwong / *School of Environmental Engineering, Suranaree University of Technology, Nakhon Ratchasima, Thailand*
Thomas R. Way / *National Soil Dynamics Laboratory, Agricultural Research Service (USDA-ARS), U.S. Department of Agriculture, Auburn, Alabama, U.S.A.*
Johannes Bernhard Wehr / *School of Land, Crop and Food Sciences, University of Queensland, St. Lucia, Queensland, Australia*
W.W. Wenzel / *Institute of Soil Research, University of Natural Resources and Life Sciences, Vienna, Austria*
Catharina Wesseling / *Central American Institute for Studies on Toxic Substances (IRET), National University, Heredia, Costa Rica*
Ian White / *College of Science, Fenner School of Environment and Society, College of Medicine, Biology and Environment, Australian National University, Canberra, Australian Capital Territory, Australia*
Dennis Wichelns / *Principal Economist, International Water Management Unit, Columbo, Sri Lanka*
Keith D. Wiebe / *Resource Economics Division, U.S. Department of Agriculture (USDA), Washington, District of Columbia, U.S.A.*
Gerald E. Wilde / *Department of Entomology, Kansas State University, Manhattan, Kansas, U.S.A.*
Larry P. Wilding / *Department of Soil and Crop Sciences, Texas A&M University, College Station, Texas, U.S.A.*
Wilhelm Windhorst / *Institute for the Conservation of Natural Resources, University of Kiel, Kiel, Germany*
Wanpen Wirojanagud / *Department of Environmental Engineering, Faculty of Engineering, Khon Kaen University, Khon Kaen, and Center of Excellence on Hazardous Substance Management, National Centers of Excellence (PERDO), Bangkok, Thailand*
Lidia Wolska / *Department of Analytical Chemistry, Chemical Faculty, Gdansk University of Technology, and Department of Environmental Toxicology, Interdepartmental Institute of Maritime and Tropical Medicine, Medical University of Gdansk, Gdansk, Poland*
Eric A. Woodroof / *Profitable Green Solutions, Plano, Texas, U.S.A.*
Brent Wootton / *Center for Alternative Wastewater Treatment, Fleming College, Lindsay, Ontario, Canada*
Steve D. Wratten / *Division of Soil, Plant and Ecological Sciences, Lincoln University, Canterbury, New Zealand*
I. Pai Wu / *College of Tropical Agriculture and Human Resources, University of Hawaii, Honolulu, Hawaii, U.S.A.*
Simone Wuenschmann / *Environmental Institute of Scientific Networks (EISN), Haren-Erika, Germany*

Y. Xu / *MOE Key Laboratory of Regional Energy Systems Optimization, S-C Energy and Environmental Research Academy, North China Electric Power University, Beijing, China*
Kazuyuki Yagi / *National Institute for Agro-Environmental Sciences, Japan*
Colin N. Yates / *Faculty of Environment, University of Waterloo, Waterloo, Ontario, Canada*
Bofu Yu / *Griffith School of Engineering, Griffith University, Nathan, Queensland, Australia*
Taolin Zhang / *Institute of Soil Science, Chinese Academy of Sciences, Nanjing, China*
X.-C. (John) Zhang / *Grazinglands Research Laboratory, Agricultural Research Service (USDA-ARS), U.S. Department of Agriculture, El Reno, Oklahoma, U.S.A.*
He Zhong / *Pesticide Environment Impact Section, Public Health Entomology Research and Education Center, Florida A&M University, Panama City, Florida, U.S.A.*
Xinhua Zhou / *School of Natural Resources, University of Nebraska—Lincoln, Lincoln, Nebraska, U.S.A.*
Yifei Zhu / *Gemune LLC, Fremont, California, U.S.A.*
Zixi Zhu / *Henan Institute of Meteorology, Zhengzhou, China*
Andrew R. Zimmerman / *Department of Geological Sciences, University of Florida, Gainesville, Florida, U.S.A.*
Claudio Zucca / *Department of Agriculture, University of Sassari, Sassari, Italy*

Contents

Editorial Advisory Board . . . . . . *ix*
Contributors . . . . . . *xi*
Topical Table of Contents . . . . . . *xli*
Foreword . . . . . . *lxvii*
Preface . . . . . . *lxix*
How to Use This Encyclopedia . . . . . . *lxxi*
About the Editor-in-Chief . . . . . . *lxxvii*

Volume I

Acaricides / *Doug Walsh* 1
Acid Rain / *Umesh Kulshrestha* 5
Acid Rain: Nitrogen Deposition / *George F. Vance* 20
Acid Sulfate Soils / *Delvin S. Fanning* 26
Acid Sulfate Soils: Formation / *Martin C. Rabenhorst, Delvin S. Fanning, and Steven N. Burch* 31
Acid Sulfate Soils: Identification, Assessment, and Management / *L.A. Sullivan, R.T. Bush, and E.D. Burton* 36
Acid Sulfate Soils: Management / *Michael D. Melville and Ian White* 55
Adsorption / *Puangrat Kajitvichyanukul and Jirapat Ananpattarachai* 61
Agricultural Runoff / *Matt C. Smith, David K. Gattie, and Daniel L. Thomas* 81
Agricultural Soils: Ammonia Volatilization / *Paul G. Saffigna and John R. Freney* 85
Agricultural Soils: Carbon and Nitrogen Biological Cycling / *Alan J. Franzluebbers* 88
Agricultural Soils: Nitrous Oxide Emissions / *John R. Freney* 96
Agricultural Soils: Phosphorus / *Anja Gassner and Ewald Schnug* 100
Agricultural Water Quantity Management / *X.S. Qin and Y. Xu* 105
Agriculture: Energy Use and Conservation / *Guangnan Chen and Tek Narayan Maraseni* 118
Agriculture: Organic / *Kathleen Delate* 125
Agroforestry: Water Use Efficiency / *James R. Brandle, Laurie Hodges, and Xinhua Zhou* 129
Air Pollution: Monitoring / *Waldemar Wardencki* 132
Air Pollution: Technology / *Sven Erik Jørgensen* 149
Alexandria Lake Maryut: Integrated Environmental Management / *Lindsay Beevers* 164
Allelochemics / *John Borden* 175
Alternative Energy / *Bernd Markert, Simone Wuenschmann, Stefan Fraenzle, and Bernd Delakowitz* 179
Alternative Energy: Hydropower / *Andrea Micangeli, Sara Evangelisti, and Danilo Sbordone* 201
Alternative Energy: Photovoltaic Modules and Systems / *Ewa Klugmann-Radziemska* 215
Alternative Energy: Photovoltaic Solar Cells / *Ewa Klugmann-Radziemska* 226
Alternative Energy: Solar Thermal Energy / *Andrea Micangeli, Sara Evangelisti, and Danilo Sbordone* 241
Alternative Energy: Wind Power Technology and Economy / *K.E. Ohrn* 258
Aluminum / *Johannes Bernhard Wehr, Frederick Paxton Cardell Blamey, and Neal William Menzies* 267
Animals: Sterility from Pesticides / *William Au* 277
Animals: Toxicological Evaluation / *Vera Lucia S.S. de Castro* 280
Antagonistic Plants / *Philip Oduor-Owino* 288

Volume I (*cont'd*)

Aquatic Communities: Pesticide Impacts / *David P. Kreutzweiser and Paul K. Sibley* 291
Aral Sea Disaster / *Guy Fipps* 300
Arthropod Host-Plant Resistant Crops / *Gerald E. Wilde* 304
***Bacillus thuringiensis*: Transgenic Crops** / *Julie A. Peterson, John J. Obrycki, and James D. Harwood* 307
Bacterial Pest Control / *David N. Ferro* 321
Bioaccumulation / *Tomaz Langenbach* 324
Biodegradation / *Sven Erik Jørgensen* 328
Biodiversity and Sustainability / *Odo Primavesi* 333
Bioenergy Crops: Carbon Balance Assessment / *Rocky Lemus and Rattan Lal* 345
Biofertilizers / *J.R. de Freitas* 349
Bioindicators: Farming Practices and Sustainability / *Joji Muramoto and Stephen R. Gliessman* 352
Biological Controls / *Heikki Hokkanen* 366
Biological Controls: Conservation / *Doug Landis and Steve D. Wratten* 370
Biomass / *Alberto Traverso and David Tucker* 373
Biopesticides / *G.J. Ash and A. Wang* 381
Bioremediation / *Ragini Gothalwal* 387
Bioremediation: Contaminated Soil Restoration / *Sven Erik Jørgensen* 401
Biotechnology: Pest Management / *Maurizio G. Paoletti* 407
Birds: Chemical Control / *Eric B. Spurr* 413
Birds: Pesticide Use Impacts / *Pierre Mineau* 416
Boron and Molybdenum: Deficiencies and Toxicities / *Umesh C. Gupta* 424
Boron: Soil Contaminant / *Rami Keren* 431
Buildings: Climate Change / *Lisa Guan and Guangnan Chen* 435
Cabbage Diseases: Ecology and Control / *Anthony P. Keinath, Marc A. Cubeta, and David B. Langston, Jr.* 442
Cadmium and Lead: Contamination / *Gabriella Kakonyi and Imad A.M. Ahmed* 446
Cadmium: Toxicology / *Sven Erik Jørgensen* 453
Carbon Sequestration / *Nathan E. Hultman* 456
Carbon: Soil Inorganic / *Donald L. Suarez* 462
Chemigation / *William L. Kranz* 469
Chesapeake Bay / *Sean M. Smith* 474
Chromium / *Bruce R. James and Dominic A. Brose* 477
Climate Policy: International / *Nathan E. Hultman* 483
Coastal Water: Pollution / *Piotr Szefer* 489
Cobalt and Iodine / *Ronald G. McLaren* 502
Community-Based Monitoring: Ngarenanyuki, Tanzania / *Aiwerasia V.F. Ngowi, Larama M.B. Rongo, and Thomas J. Mbise* 505
Composting / *Nídia Sá Caetano* 512
Composting: Organic Farming / *Saburo Matsui* 528
Copper / *David R. Parker and Judith F. Pedler* 535
Cyanobacteria: Eutrophic Freshwater Systems / *Anja Gassner and Martin V. Frey* 538
Desertification / *David Tongway and John Ludwig* 541
Desertification: Extent of / *Victor R. Squires* 545
Desertification: Greenhouse Effect / *Sherwood Idso and Craig Idso* 549
Desertification: Impact / *David A. Mouat and Judith Lancaster* 552
Desertification: Prevention and Restoration / *Claudio Zucca, Susana Bautista, Barron Orr, and Franco Previtali* 557

Desertization / *Henry Noel LeHouerou* 565
Developing Countries: Pesticide Health Impacts / *Aiwerasia V.F. Ngowi, Catharina Wesseling, and Leslie London* 573
Distributed Generation: Combined Heat and Power / *Barney L. Capehart, D. Paul Mehta, and Wayne C. Turner* 578
Drainage: Hydrological Impacts Downstream / *Mark Robinson and D.W. Rycroft* 584
Drainage: Soil Salinity Management / *Glenn J. Hoffman* 588
Ecological Indicators: Eco-Exergy to Emergy Flow / *Simone Bastianoni, Luca Coscieme, and Federico M. Pulselli* 592
Ecological Indicators: Ecosystem Health / *Felix Müller, Benjamin Burkhard, Marion Kandziora, Claus Schimming, and Wilhelm Windhorst* 599
Economic Growth: Slower by Design, Not Disaster / *Peter Victor* 614
Ecosystems: Large-Scale Restoration Governance / *Shannon Estenoz, Denise Vienne, and Alka Sapat* 624
Ecosystems: Planning / *Ioan Manuel Ciumasu, Liana Buzdugan, Ioan Manuel Ciumasu, Nicolae Stefan, and Keith Culver* 632
Endocrine Disruptors / *Vera Lucia S.S. de Castro* 643
Energy Commissioning: Existing Buildings / *David E. Claridge, Mingsheng Liu, and W.D. Turner* 656
Energy Commissioning: New Buildings / *Janey Kaster* 665
Energy Conservation / *Ibrahim Dincer and Adnan Midili* 677
Energy Conservation: Benefits / *Eric A. Woodroof, Wayne C. Turner, and Steven D. Heinz* 691
Energy Conservation: Industrial Processes / *Harvey E. Diamond* 697
Energy Conservation: Lean Manufacturing / *Bohdan W. Oppenheim* 705
Energy Conversion: Coal, Animal Waste, and Biomass Fuel / *Kalyan Annamalai, Soyuz Priyadarsan, Senthil Arumugam, and John M. Sweeten* 714

Volume II

Energy Efficiency: Low Cost Improvements / *James Call* 735
Energy Efficiency: New and Emerging Technology / *Steven A. Parker* 742
Energy Efficiency: Strategic Facility Guidelines / *Steve Doty* 752
Energy Master Planning / *Fredric S. Goldner* 763
Energy Sources: Natural versus Additional / *Marc A. Rosen* 770
Energy Use: Exergy and Eco-Exergy / *Sven Erik Jørgensen* 778
Energy Use: U.S. / *Dwight K. French* 790
Energy: Environmental Security / *Muhammad Asif* 798
Energy: Physics / *Milivoje M. Kostic* 808
Energy: Renewable / *John O. Blackburn* 824
Energy: Solid Waste Advanced Thermal Technology / *Alex E.S. Green and Andrew R. Zimmerman* 830
Energy: Storage / *Rudolf Marloth* 853
Energy: Walls and Windows / *Therese Stovall* 860
Energy: Waste Heat Recovery / *Martin A. Mozzo, Jr.* 869
Environmental Legislation: Asia / *Wanpen Wirojanagud* 874
Environmental Legislation: EU Countries Solid Waste Management / *Ni-Bin Chang, Ana Pires, and Graça Martinho* 892
Environmental Policy / *Sanford V. Berg* 914
Environmental Policy: Innovations / *Alka Sapat* 922
Erosion / *Dennis C. Flanagan* 930
Erosion and Carbon Dioxide / *Pierre A. Jacinthe and Rattan Lal* 934
Erosion and Global Change / *Taolin Zhang and Xingxiang Wang* 939

Volume II (*cont'd*)

Erosion and Precipitation / *Bofu Yu* 943
Erosion and Sediment Control: Vegetative Techniques / *Samson D. Angima* 947
Erosion by Water: Accelerated / *David Favis-Mortlock* 951
Erosion by Water: Amendment Techniques / *X.-C. (John) Zhang* 958
Erosion by Water: Assessment and Control / *José Miguel Reichert, Nadia Bernardi Bonumá, Gustavo Enrique Merten, and Jean Paolo Gomes Minella* 963
Erosion by Water: Empirical Methods / *John M. Laflen* 974
Erosion by Water: Erosivity and Erodibility / *Peter I.A. Kinnell* 980
Erosion by Water: Process-Based Modeling / *Mark A. Nearing* 991
Erosion by Water: Vegetative Control / *Seth M. Dabney and Silvio J. Gumiere* 994
Erosion by Wind: Climate Change / *Alan Busacca and David Chandler* 1004
Erosion by Wind: Global Hot Spots / *Andrew Warren* 1010
Erosion by Wind: Principles / *Larry D. Stetler* 1013
Erosion by Wind: Source, Measurement, Prediction, and Control / *Brenton S. Sharratt and R. Scott Van Pelt* 1017
Erosion by Wind-Driven Rain / *Gunay Erpul, L. Darrell Norton, and Donald Gabriels* 1031
Erosion Control: Soil Conservation / *Eric T. Craswell* 1034
Erosion Control: Tillage and Residue Methods / *Richard Cruse, Jerry Neppel, John Kost, and Krisztina Eleki* 1040
Erosion: Accelerated / *J. Kent Mitchell and Michael C. Hirschi* 1044
Erosion: History / *Andrew S. Goudie* 1050
Erosion: Irrigation-Induced / *Robert E. Sojka and David L. Bjorneberg* 1053
Erosion: Snowmelt / *Donald K. McCool* 1057
Erosion: Soil Quality / *Craig Ditzler* 1060
Estuaries / *Claude Amiard-Triquet* 1063
Eutrophication / *Sven Erik Jørgensen and Claude Amiard-Triquet* 1074
Everglades / *Kenneth L. Campbell, Rafael Munoz-Carpena, and Gregory Kiker* 1080
Exergy: Analysis / *Marc A. Rosen* 1083
Exergy: Environmental Impact Assessment / *Marc A. Rosen* 1092
Farming: Non-Chemical and Pesticide-Free (European Council Regulations [EED/209/91]) / *Inger Källander and Gunnar Rundgren* 1103
Farming: Organic / *Brenda Frick* 1109
Farming: Organic Pest Management / *Ján Gallo* 1115
Food Quality Protection Act / *Christina D. DiFonzo* 1118
Food: Cosmetic Standards / *David Pimentel and Kelsey Hart* 1120
Food: Pesticide Contamination / *Denis Hamilton* 1124
Fossil Fuel Combustion: Air Pollution and Global Warming / *Dan Golomb* 1127
Fuel Cells: Intermediate and High Temperature / *Xianguo Li, Gholamreza Karimi, and Kui Jiao* 1138
Fuel Cells: Low Temperature / *Xianguo Li and Kui Jiao* 1145
Genotoxicity and Air Pollution / *Eliane Tigre Guimarães and Andrea Nunes Vaz Pedroso* 1154
Geographic Information System (GIS): Land Use Planning / *Egide Nizeyimana and Jacob Opadeyi* 1163
Geothermal Energy Resources / *Ibrahim Dincer and Arif Hepbasli* 1167
Giant Reed (*Arundo donax*): Streams and Water Resources / *Gretchen Coffman* 1182
Global Climate Change: Carbon Sequestration / *Sherwood Idso and Keith E. Idso* 1189
Global Climate Change: Earth System Response / *Amanda Staudt and Nathan E. Hultman* 1192
Global Climate Change: Gas Fluxes / *Pascal Boeckx and Oswald Van Cleemput* 1202

Global Climate Change: Gasoline, Hybrid-Electric and Hydrogen Fueled Vehicles / *Robert E. Uhrig* 1206
Global Climate Change: World Soils / *Rattan Lal* 1213
Globalization / *Alexandru V. Roman* 1218
Green Energy / *Ibrahim Dincer and Adnan Midili* 1226
Green Processes and Projects: Systems Analysis / *Abhishek Tiwary* 1242
Green Products: Production / *Puangrat Kajitvichyanukul, Jirapat Ananpattarachai, and Apichon Watcharenwong* 1253
Groundwater: Arsenic Contamination / *Abhijit Das, Bhaskar Das, Subhas Chandra Mukherjee, Shyamapada Pati, Rathindra Nath Dutta, Khitish Chandra Saha, Quazi Quamruzzaman, Mohammad Mahmudur Rahman, Tarit Roy Chowdhury, and Dipankar Chakraborti* 1262
Groundwater: Contamination / *Charles W. Fetter* 1281
Groundwater: Mining / *Hugo A. Loaiciga* 1284
Groundwater: Mining Pollution / *Jeffrey G. Skousen and George F. Vance* 1289
Groundwater: Modeling / *Jesus Carrera* 1295
Groundwater: Nitrogen Fertilizer Contamination / *Lloyd B. Owens and Douglas L. Karlen* 1302
Groundwater: Numerical Method Modeling / *Jesus Carrera* 1312
Groundwater: Pesticide Contamination / *Roy F. Spalding* 1317
Groundwater: Saltwater Intrusion / *Alexander H.-D. Cheng* 1321
Groundwater: Treatment with Biobarrier Systems / *Chih-Ming Kao, Shu-Hao Liang, Yu-Chia Kuo, and Rao Y. Surampalli* 1333
Heat Pumps / *Lu Aye* 1346
Heat Pumps: Absorption / *Keith E. Herold* 1356
Heat Pumps: Geothermal / *Greg Tinkler* 1363
Heavy Metals / *Mike J. McLaughlin* 1370
Heavy Metals: Organic Fertilization Uptake / *Ewald Schnug, Alexandra Izosimova, and Renata Gaj* 1374
Herbicides / *Malcolm Devine* 1378
Herbicides: Non-Target Species Effects / *Céline Boutin* 1382
Human Health: Cancer from Pesticides / *Jan Dich* 1394
Human Health: Chronic Pesticide Poisonings / *Birgitta Kolmodin-Hedman and Margareta Palmborg* 1397
Human Health: Consumer Concerns of Pesticides / *George Ekström and Margareta Palmborg* 1401
Human Health: Hormonal Disruption / *Evamarie Straube, Sebastian Straube, and Wolfgang Straube* 1405
Human Health: Pesticide Poisonings / *Margareta Palmborg* 1408
Human Health: Pesticide Sensitivities / *Ann McCampbell* 1411
Human Health: Pesticides / *Kelsey Hart and David Pimentel* 1414
Hydroelectricity: Pumped Storage / *Jill S. Tietjen* 1418
Industrial Waste: Soil Pollution / *W. Friesl-Hanl, M.H. Gerzabek, W.W. Wenzel, and W.E.H. Blum* 1430
Industries: Network of / *Sven Erik Jørgensen* 1434
Inland Seas and Lakes: Central Asia Case Study / *Andrey G. Kostianoy* 1436
Inorganic Carbon: Composition and Formation / *Larry P. Wilding and H. Curtis Monger* 1444
Inorganic Carbon: Global Carbon Cycle / *William H. Schlesinger* 1448
Inorganic Carbon: Modeling / *Leslie D. McFadden and Ronald G. Amundson* 1451
Inorganic Compounds: Eco-Toxicity / *Sven Erik Jørgensen* 1455
Insect Growth Regulators / *Meir Paul Pener* 1459
Insecticides: Aerial Ultra-Low-Volume Application / *He Zhong* 1471
Insects and Mites: Biological Control / *Ann E. Hajek* 1474
Insulation: Facilities / *Wendell A. Porter* 1478
Integrated Energy Systems / *Leslie A. Solmes and Sven Erik Jørgensen* 1493

Volume II (*cont'd*)

Integrated Farming Systems / *John Holland* 1507
Integrated Nutrient Management / *Bal Ram Singh* 1510
Integrated Pest Management / *H.F. van Emden* 1521
Integrated Weed Management / *Heinz Müller-Schärer and Alexandra Robin Collins* 1524

Volume III

Invasion Biology / *Jennifer Ruesink* 1531
Irrigation Systems: Sub-Surface Drip Design / *Carl R. Camp, Jr. and Freddie L. Lamm* 1535
Irrigation Systems: Sub-Surface History / *Norman R. Fausey* 1539
Irrigation Systems: Water Conservation / *I. Pai Wu, Javier Barragan, and Vince Bralts* 1542
Irrigation: Efficiency / *Terry A. Howell* 1545
Irrigation: Erosion / *David L. Bjorneberg* 1551
Irrigation: Return Flow and Quality / *Ramón Aragues and Kenneth K. Tanji* 1557
Irrigation: River Flow Impact / *Robert W. Hill and Ivan A. Walter* 1563
Irrigation: Saline Water / *B.A. Stewart* 1568
Irrigation: Sewage Effluent Use / *B.A. Stewart* 1570
Irrigation: Soil Salinity / *James D. Rhoades* 1572
Lakes and Reservoirs: Pollution / *Subhankar Karmakar and O.M. Musthafa* 1576
Lakes: Restoration / *Anna Rabajczyk* 1588
Land Restoration / *Richard W. Bell* 1600
Laws and Regulations: Food / *Ike Jeon* 1609
Laws and Regulations: Pesticides / *Praful Suchak* 1612
Laws and Regulations: Rotterdam Convention / *Barbara Dinham* 1615
Laws and Regulations: Soil / *Ian Hannam and Ben Boer* 1618
Leaching / *Lars Bergström* 1621
Leaching and Illuviation / *Lynn E. Moody* 1624
Lead: Ecotoxicology / *Sven Erik Jørgensen* 1630
Lead: Regulations / *Lisa A. Robinson* 1636
LEED-EB: Leadership in Energy and Environmental Design for Existing Buildings / *Rusty T. Hodapp* 1647
LEED-NC: Leadership in Energy and Environmental Design for New Construction / *Stephen A. Roosa* 1654
Manure Management: Compost and Biosolids / *Bahman Eghball and Kenneth A. Barbarick* 1660
Manure Management: Dairy / *H.H. Van Horn and D.R. Bray* 1663
Manure Management: Phosphorus / *Rory O. Maguire, John T. Brake, and Peter W. Plumstead* 1667
Manure Management: Poultry / *Shafiqur Rahman and Thomas R. Way* 1670
Mercury / *Sven Erik Jørgensen* 1676
Methane Emissions: Rice / *Kazuyuki Yagi* 1679
Minerals Processing Residue (Tailings): Rehabilitation / *Lloyd R. Hossner and Hamid Shahandeh* 1683
Mines: Acidic Drainage Water / *Jerry M. Bigham and Wendy B. Gagliano* 1688
Mines: Rehabilitation of Open Cut / *Douglas J. Dollhopf* 1692
Mycotoxins / *J. David Miller* 1696
Nanomaterials: Regulation and Risk Assessment / *Steffen Foss Hansen, Khara D. Grieger, and Anders Baun* 1700
Nanoparticles / *Alexandra Navrotsky* 1711
Nanoparticles: Uncertainty Risk Analysis / *Khara D. Grieger, Steffen Foss Hansen, and Anders Baun* 1720
Nanotechnology: Environmental Abatement / *Puangrat Kajitvichyanukul and Jirapat Ananpattarachai* 1730

Natural Enemies and Biocontrol: Artificial Diets / *Simon Grenier and Patrick De Clerq* 1746
Natural Enemies: Conservation / *Cetin Sengonca* 1749
Nematodes: Biological Control / *Simon Gowen* 1752
Neurotoxicants: Developmental Experimental Testing / *Vera Lucia S.S. de Castro* 1755
Nitrate Leaching Index / *Jorge A. Delgado* 1761
Nitrogen / *Oswald Van Cleemput and Pascal Boeckx* 1768
Nitrogen Trading Tool / *Jorge A. Delgado* 1772
Nitrogen: Biological Fixation / *Mark B. Peoples* 1785
Nuclear Energy: Economics / *James G. Hewlett* 1789
Nutrients: Best Management Practices / *Scott J. Sturgul and Keith A. Kelling* 1805
Nutrients: Bioavailability and Plant Uptake / *Niels Erik Nielsen* 1817
Nutrient–Water Interactions / *Ardell D. Halvorson* 1823
Oil Pollution: Baltic Sea / *Andrey G. Kostianoy* 1826
Organic Compounds: Halogenated / *Marek Tobiszewski and Jacek Namieśnik* 1841
Organic Matter: Global Distribution in World Ecosystems / *Wilfred M. Post* 1851
Organic Matter: Management / *R. Cesar Izaurralde and Carlos C. Cerri* 1857
Organic Matter: Modeling / *Pete Smith* 1863
Organic Matter: Turnover / *Johan Six and Julie D. Jastrow* 1872
Organic Soil Amendments / *Philip Oduor-Owino* 1878
Ozone Layer / *Luisa T. Molina* 1882
Permafrost / *Douglas Kane and Julia Boike* 1900
Persistent Organic Compounds: Wet Oxidation Removal / *Anurag Garg and I.M. Mishra* 1904
Persistent Organic Pesticides / *Gamini Manuweera* 1913
Pest Management / *E.F. Legner* 1919
Pest Management: Crop Diversity / *Marianne Karpenstein-Machan and Maria Finckh* 1925
Pest Management: Ecological Agriculture / *Barbara Dinham* 1930
Pest Management: Ecological Aspects / *David J. Horn* 1934
Pest Management: Intercropping / *Maria Finckh and Marianne Karpenstein-Machan* 1937
Pest Management: Legal Aspects / *Michael T. Olexa and Zachary T. Broome* 1940
Pest Management: Modeling / *Andrew Paul Gutierrez* 1949
Pesticide Translocation Control: Soil Erosion / *Monika Frielinghaus, Detlef Deumlich, and Roger Funk* 1955
Pesticides / *Marek Biziuk, Jolanta Fenik, Monika Kosikowska, and Maciej Tankiewicz* 1963
Pesticides: Banding / *Sharon A. Clay* 1983
Pesticides: Chemical and Biological / *Barbara Manachini* 1986
Pesticides: Damage Avoidance / *Aiwerasia V.F. Ngowi and Larama M.B. Rongo* 1997
Pesticides: Effects / *Ana Maria Evangelista de Duffard* 2005
Pesticides: History / *Edward H. Smith and George Kennedy* 2008
Pesticides: Measurement and Mitigation / *George F. Antonious* 2013
Pesticides: Natural / *Franck Dayan* 2028
Pesticides: Reducing Use / *David Pimentel and Maria V. Cilveti* 2033
Pesticides: Regulating / *Matthias Kern* 2035
Pests: Landscape Patterns / *F. Craig Stevenson* 2038
Petroleum: Hydrocarbon Contamination / *Svetlana Drozdova and Erwin Rosenberg* 2040
Pharmaceuticals: Treatment / *Diana Aga and Seungyun Baik* 2060
Phenols / *Leszek Wachowski and Robert Pietrzak* 2071
Phosphorus: Agricultural Nutrient / *John Ryan, Hayriye Ibrikci, Rolf Sommer, and Abdul Rashid* 2091
Phosphorus: Riverine System Transport / *Andrew N. Sharpley, Peter Kleinman, Tore Krogstad, and Richard McDowell* 2100

Volume III (*cont'd*)

Plant Pathogens (Fungi): Biological Control / *Timothy Paulitz* 2107
Plant Pathogens (Viruses): Biological Control / *Hei-Ti Hsu* 2111
Pollutants: Organic and Inorganic / *A. Paul Schwab* 2114
Pollution: Genotoxicity of Agrotoxic Compounds / *Vera Lucia S.S. de Castro and Paola Poli* 2123
Pollution: Non-Point Source / *Ravendra Naidu, Mallavarapu Megharaj, Peter Dillon, Rai Kookana, Ray Correll, and W.W. Wenzel* 2136
Pollution: Pesticides in Agro-Horticultural Ecosystems / *J.K. Dubey and Meena Thakur* 2139
Pollution: Pesticides in Natural Ecosystems / *J.K. Dubey and Meena Thakur* 2145
Pollution: Point and Non-Point Source Low Cost Treatment / *Sandeep Joshi and Sayali Joshi* 2150
Pollution: Point Sources / *Ravendra Naidu, Mallavarapu Megharaj, Peter Dillon, Rai Kookana, Ray Correll, and W.W. Wenzel* 2166
Polychlorinated Biphenyls (PCBs) / *Marek Biziuk and Angelika Beyer* 2172
Polychlorinated Biphenyls (PCBs) and Polycyclic Aromatic Hydrocarbons (PAHs): Sediments and Water Analysis / *Justyna Rogowska, Agata Mechlińska, Lidia Wolska, and Jacek Namieśnik* 2186
Potassium / *Philippe Hinsinger* 2208
Precision Agriculture: Engineering Aspects / *Joel T. Walker, Reza Ehsani, and Matthew O. Sullivan* 2213
Precision Agriculture: Water and Nutrient Management / *Robert J. Lascano and J.D. Booker* 2217
Radio Frequency Towers: Public School Placement / *Joshua Steinfeld* 2224
Radioactivity / *Bogdan Skwarzec* 2234
Radionuclides / *Philip M. Jardine* 2243
Rain Water: Atmospheric Deposition / *Zaneta Polkowska* 2249
Rain Water: Harvesting / *K.F. Andrew Lo* 2262
Rare Earth Elements / *Zhengyi Hu, Gerd Sparovek, Silvia Haneklaus, and Ewald Schnug* 2266

Volume IV

Remote Sensing and GIS Integration / *Egide Nizeyimana* 2271
Remote Sensing: Pollution / *Massimo Antoninetti* 2275
Rivers and Lakes: Acidification / *Agniezka Gałuszka and Zdzisław M. Migaszewski* 2291
Rivers: Pollution / *Bogdan Skwarzec* 2303
Rivers: Restoration / *Anna Rabajczyk* 2307
Runoff Water / *Zaneta Polkowska* 2320
Salt-Affected Soils: Physical Properties and Behavior / *Hwat Bing So* 2334
Salt-Affected Soils: Plant Response / *Anna Eynard, Keith D. Wiebe, and Rattan Lal* 2345
Salt-Affected Soils: Sustainable Agriculture / *Pichu Rengasamy* 2349
Sea: Pollution / *Bogdan Skwarzec* 2357
Sodic Soils: Irrigation Farming / *David Burrow* 2364
Sodic Soils: Properties / *Pichu Rengasamy* 2367
Sodic Soils: Reclamation / *Jock Churchman* 2370
Soil Degradation: Global Assessment / *Selim Kapur, Suha Berberoğlu, Erhan Akça, Cenk Dönmez, Mehmet Akif Erdoğan, and Burçak Kapur* 2375
Soil Quality: Carbon and Nitrogen Gases / *Philippe Rochette, Sean McGinn, and Reynald Lemke* 2388
Soil Quality: Indicators / *Graham P. Sparling* 2393
Soil Rehabilitation / *David Jasper* 2396
Solid Waste Management: Life Cycle Assessment / *Ni-Bin Chang, Ana Pires, and Graça Martinho* 2399
Solid Waste: Municipal / *Angelique Chettiparamb* 2415

Stored-Product Pests: Biological Control / *Lise Stengård Hansen* 2423
Strontium / *Silvia Haneklaus and Ewald Schnug* 2426
Sulfur / *Ewald Schnug, Silvia Haneklaus, and Elke Bloem* 2431
Sulfur Dioxide / *Marianna Czaplicka and Witold Kurylak* 2437
Sustainability and Planning / *Richard Cowell* 2446
Sustainable Agriculture: Soil Quality / *John W. Doran and Ed G. Gregorich* 2457
Sustainable Development / *Mark A. Peterson* 2461
Sustainable Development: Ecological Footprint in Accounting / *Simone Bastianoni, Valentina Niccolucci, Elena Neri, Gemma Cranston, Alessandro Galli, and Mathis Wackernagel* 2467
Sustainable Development: Pyrolysis and Gasification of Biomass and Wastes / *P. Mondal, J. Parikh, and D.V. Naik* 2482
Thermal Energy: Solar Technologies / *Muhammad Asif and Tariq Muneer* 2498
Thermal Energy: Storage / *Brian Silvetti* 2508
Thermodynamics / *Ronald L. Klaus* 2525
Tillage Erosion: Terrace Formation / *Seth M. Dabney and Dalmo A.N. Vieira* 2536
Toxic Substances / *Sven Erik Jørgensen* 2542
Toxic Substances: Photochemistry / *Puangrat Kajitvichyanukul* 2547
Toxicity Prediction of Chemical Mixtures / *Nina Cedergreen, Claus Svendsen, and Thomas Backhaus* 2572
Vanadium and Chromium Groups / *Imad A.M. Ahmed* 2582
Vertebrates: Biological Control / *Peter Kerr and Tanja Strive* 2595
Waste Gas Treatment: Bioreactors / *Amit Kumar, Piet Lens, Sarina J. Ergas, and Herman Van Langenhove* 2611
Waste: Stabilization Ponds / *Sven Erik Jørgensen* 2632
Wastewater and Water Utilities / *Rudolf Marloth* 2638
Wastewater Treatment: Biological / *Shaikh Ziauddin Ahammad, David W. Graham, and Jan Dolfing* 2645
Wastewater Treatment: Conventional Methods / *Sven Erik Jørgensen* 2657
Wastewater Treatment: Wetlands Use in Arctic Regions / *Colin N. Yates, Brent Wootton, Sven Erik Jørgensen, and Stephen D. Murphy* 2662
Wastewater Use in Agriculture / *Manzoor Qadir, Pay Drechsel, and Liqa Raschid-Sally* 2675
Wastewater Use in Agriculture: Policy Issues / *Dennis Wichelns* 2681
Wastewater Use in Agriculture: Public Health Considerations / *Blanca Jimenez and Inés Navarro* 2694
Wastewater: Municipal / *Sven Erik Jørgensen* 2709
Water and Wastewater: Filters / *Sandeep Joshi* 2719
Water and Wastewater: Ion Exchange Application / *Sven Erik Jørgensen* 2734
Water Harvesting / *Gary W. Frasier* 2738
Water Quality and Quantity: Globalization / *Kristi Denise Caravella and Jocilyn Danise Martinez* 2740
Water Quality: Modeling / *Richard Lowrance* 2749
Water Quality: Range and Pasture Land / *Thomas L. Thurow* 2752
Water Quality: Soil Erosion / *Andreas Klik* 2755
Water Quality: Timber Harvesting / *C. Rhett Jackson* 2770
Water Supplies: Pharmaceuticals / *Klaus Kümmerer* 2776
Water: Cost / *Atif Kubursi and Matthew Agarwala* 2779
Water: Drinking / *Marek Biziuk and Małgorzata Michalska* 2790
Water: Surface / *Victor de Vlaming* 2804
Water: Total Maximum Daily Load / *Robin Kundis Craig* 2808
Watershed Management: Remote Sensing and GIS / *A.V. Shanwal and S.P. Singh* 2816
Weeds (Insects and Mites): Biological Control / *Peter Harris* 2821
Wetlands / *Ralph W. Tiner* 2824

Volume IV (*cont'd*)

Wetlands: Biodiversity / *Jean-Claude Lefeuvre and Virginie Bouchard* 2829
Wetlands: Carbon Sequestration / *Virginie Bouchard* 2833
Wetlands: Conservation Policy / *Clayton Rubec* 2837
Wetlands: Constructed Subsurface / *Jan Vymazal* 2841
Wetlands: Methane Emission / *Anna Ekberg and Torben Røjle Christensen* 2850
Wetlands: Petroleum / *John V. Headley and Kerry M. Peru* 2854
Wetlands: Sedimentation and Ecological Engineering / *Timothy C. Granata and Jay F. Martin* 2859
Wetlands: Treatment System Use / *Kyle R. Mankin* 2862
Wind Farms: Noise / *Daniel Shepherd, Chris Hanning, and Bob Thorne* 2867
Yellow River / *Zixi Zhu, Ynuzhang Wang, and Yifei Zhu* 2884
Appendixes 2887

Topical Table of Contents

The entries have been classified according to the presented procedure for integrated, holistic environmental and ecological management. The content uses the following topical classifications:

CLT: means that the solutions are based on cleaner technology
COV: indicates that the articles give comparative overviews of important topics for environmental management or background knowledge that is important for the evaluation of environmental problems
DIA: means that the articles are about diagnostic tools: monitoring, ecological modelling, ecological indicators and ecological services
ECT: covers solutions of the problems based on ecotechnology
ELE: focuses on the use of environmental legislation to solve environmental problems
ENT: refers to solutions of environmental problems by the use of environmental technology
IMS: are articles uncovering the possibilities to integrate the various tool boxes to make an integrated and holistic management
PSS: covers entries that focus on a pollution problem and its sources

The classical environmental classification indicates the sphere that is touched by the environmental problem. Is it a water problem? – The hydrosphere is involved. Is it a air pollution problem? – The atmosphere is involved. Or is a terrestrial problem? – The lithosphere is involved. This more traditional classification has been used as an additional topical classification, and the following abbreviations are used in this context:

AIR: air pollution problems
WAT: water pollution problems
TER: terrestrial pollution problems
GEN: general pollution problems that may involve more than one sphere
GLO: global pollution problems

All articles are marked with both topological classifications.

Agriculture (AGR)

Cleaner Technology (CLT)

Agriculture: Organic / *Kathleen Delate* 125
Arthropod Host-Plant Resistant Crops / *Gerald E. Wilde* 304
***Bacillus thuringiensis*: Transgenic Crops** / *Julie A. Peterson, John J. Obrycki, and James D. Harwood* 307
Cabbage Diseases: Ecology and Control / *Anthony P. Keinath, Marc A. Cubeta, and David B. Langston, Jr.* 442
Farming: Organic / *Brenda Frick* 1109
Land Restoration / *Richard W. Bell* 1600
Manure Management: Compost and Biosolids / *Bahman Eghball and Kenneth A. Barbarick* 1660
Manure Management: Dairy / *H.H. Van Horn and D.R. Bray* 1663
Manure Management: Poultry / *Shafiqur Rahman and Thomas R. Way* 1670
Nutrients: Best Management Practices / *Scott J. Sturgul and Keith A. Kelling* 1805
Nutrients: Bioavailability and Plant Uptake / *Niels Erik Nielsen* 1817

Agriculture (AGR) (*cont'd*)

Pesticide Translocation Control: Soil Erosion / *Monika Frielinghaus, Detlef Deumlich, and Roger Funk* 1955
Plant Pathogens (Fungi): Biological Control / *Timothy Paulitz* 2107
Plant Pathogens (Viruses): Biological Control / *Hei-Ti Hsu* 2111
Precision Agriculture: Engineering Aspects / *Joel T. Walker, Reza Ehsani, and Matthew O. Sullivan* 2213
Salt-Affected Soils: Sustainable Agriculture / *Pichu Rengasamy* 2349
Sustainable Agriculture: Soil Quality / *John W. Doran and Ed G. Gregorich* 2457

Comparative Overviews (COV)

Agricultural Runoff / *Matt C. Smith, David K. Gattie, and Daniel L. Thomas* 81
Agricultural Water Quantity Management / *X.S. Qin and Y. Xu* 105
Agriculture: Energy Use and Conservation / *Guangnan Chen and Tek Narayan Maraseni* 118
Erosion: History / *Andrew S. Goudie* 1050
Erosion: Soil Quality / *Craig Ditzler* 1060
Nitrate Leaching Index / *Jorge A. Delgado* 1761
Nitrogen / *Oswald Van Cleemput and Pascal Boeckx* 1768
Nitrogen: Biological Fixation / *Mark B. Peoples* 1785
Nutrients: Best Management Practices / *Scott J. Sturgul and Keith A. Kelling* 1805
Nutrients: Bioavailability and Plant Uptake / *Niels Erik Nielsen* 1817
Salt-Affected Soils: Physical Properties and Behavior / *Hwat Bing So* 2334
Salt-Affected Soils: Plant Response / *Anna Eynard, Keith D. Wiebe, and Rattan Lal* 2345
Sodic Soils: Properties / *Pichu Rengasamy* 2367
Soil Degradation: Global Assessment / *Selim Kapur, Suha Berberoğlu, Erhan Akça, Cenk Dönmez, Mehmet Akif Erdoğan, and Burçak Kapur* 2375

Diagnostic Tools (DIA)

Acid Sulfate Soils: Identification, Assessment, and Management / *L.A. Sullivan, R.T. Bush, and E.D. Burton* 36
Animals: Toxicological Evaluation / *Vera Lucia S.S. de Castro* 280
Bioindicators: Farming Practices and Sustainability / *Joji Muramoto and Stephen R. Gliessman* 352
Drainage: Soil Salinity Management / *Glenn J. Hoffman* 588
Erosion by Water: Assessment and Control / *José Miguel Reichert, Nadia Bernardi Bonumá, Gustavo Enrique Merten, and Jean Paolo Gomes Minella* 963
Erosion by Water: Empirical Methods / *John M. Laflen and John M. Laflen* 974
Erosion by Water: Erosivity and Erodibility / *Peter I.A. Kinnell* 980
Erosion by Water: Process-Based Modeling / *Mark A. Nearing* 991
Erosion by Wind: Source, Measurement, Prediction, and Control / *Brenton S. Sharratt and R. Scott Van Pelt* 1017
Nitrate Leaching Index / *Jorge A. Delgado* 1761
Soil Degradation: Global Assessment / *Selim Kapur, Suha Berberoğlu, Erhan Akça, Cenk Dönmez, Mehmet Akif Erdoğan, and Burçak Kapur* 2375
Soil Quality: Indicators / *Graham P. Sparling* 2393

Ecotechnology (ECT)

Acid Sulfate Soils: Management / *Michael D. Melville and Ian White* 55
Agricultural Water Quantity Management / *X.S. Qin and Y. Xu* 105
Agriculture: Organic / *Kathleen Delate* 125

Agroforestry: Water Use Efficiency / *James R. Brandle, Laurie Hodges, and Xinhua Zhou* 129
Arthropod Host-Plant Resistant Crops / *Gerald E. Wilde* 304
Biofertilizers / *J.R. de Freitas* 349
Biological Controls / *Heikki Hokkanen* 366
Biological Controls: Conservation / *Doug Landis and Steve D. Wratten* 370
Bioremediation / *Ragini Gothalwal* 387
Cabbage Diseases: Ecology and Control / *Anthony P. Keinath, Marc A. Cubeta, and David B. Langston, Jr.* 442
Composting / *Nídia Sá Caetano* 512
Composting: Organic Farming / *Saburo Matsui* 528
Desertification: Prevention and Restoration / *Claudio Zucca, Susana Bautista, Barron Orr, and Franco Previtali* 557
Erosion and Sediment Control: Vegetative Techniques / *Samson D. Angima* 947
Erosion by Water: Amendment Techniques / *X.-C. (John) Zhang* 958
Erosion by Water: Vegetative Control / *Seth M. Dabney and Silvio J. Gumiere* 994
Erosion Control: Soil Conservation / *Eric T. Craswell* 1034
Erosion Control: Tillage and Residue Methods / *Richard Cruse, Jerry Neppel, John Kost, and Krisztina Eleki* 1040
Farming: Organic / *Brenda Frick* 1109
Insects and Mites: Biological Control / *Ann E. Hajek* 1474
Land Restoration / *Richard W. Bell* 1600
Manure Management: Compost and Biosolids / *Bahman Eghball and Kenneth A. Barbarick* 1660
Manure Management: Dairy / *H.H. Van Horn and D.R. Bray* 1663
Manure Management: Phosphorus / *Rory O. Maguire, John T. Brake, and Peter W. Plumstead* 1667
Manure Management: Poultry / *Shafiqur Rahman and Thomas R. Way* 1670
Natural Enemies and Biocontrol: Artificial Diets / *Simon Grenier and Patrick De Clerq* 1746
Nutrients: Best Management Practices / *Scott J. Sturgul and Keith A. Kelling* 1805
Nutrients: Bioavailability and Plant Uptake / *Niels Erik Nielsen* 1817
Organic Soil Amendments / *Philip Oduor-Owino* 1878
Pesticide Translocation Control: Soil Erosion / *Monika Frielinghaus, Detlef Deumlich, and Roger Funk* 1955
Plant Pathogens (Fungi): Biological Control / *Timothy Paulitz* 2107
Plant Pathogens (Viruses): Biological Control / *Hei-Ti Hsu* 2111
Sodic Soils: Reclamation / *Jock Churchman* 2370
Soil Rehabilitation / *David Jasper* 2396
Sustainable Agriculture: Soil Quality / *John W. Doran and Ed G. Gregorich* 2457
Tillage Erosion: Terrace Formation / *Seth M. Dabney and Dalmo A.N. Vieira* 2536
Weeds (Insects and Mites): Biological Control / *Peter Harris* 2821

Environmental Legislation (ELE)

Laws and Regulations: Food / *Ike Jeon* 1609
Laws and Regulations: Rotterdam Convention / *Barbara Dinham* 1615
Laws and Regulations: Soil / *Ian Hannam and Ben Boer* 1618
Nitrogen Trading Tool / *Jorge A. Delgado* 1772

Environmental Technology (ENT)

Agricultural Water Quantity Management / *X.S. Qin and Y. Xu* 105
Agroforestry: Water Use Efficiency / *James R. Brandle, Laurie Hodges, and Xinhua Zhou* 129

Agriculture (AGR) (*cont'd*)

Precision Agriculture: Engineering Aspects / *Joel T. Walker, Reza Ehsani, and Matthew O. Sullivan* 2213

Integrated and Holistic Management (IMS)

Integrated Farming Systems / *John Holland* 1507
Integrated Nutrient Management / *Bal Ram Singh* 1510
Integrated Weed Management / *Heinz Müller-Schärer and Alexandra Robin Collins* 1524
Nutrients: Best Management Practices / *Scott J. Sturgul and Keith A. Kelling* 1805
Nutrients: Bioavailability and Plant Uptake / *Niels Erik Nielsen* 1817

Pollution Problems and Their Sources (PSS)

Acid Sulfate Soils / *Delvin S. Fanning* 26
Acid Sulfate Soils: Formation / *Martin C. Rabenhorst, Delvin S. Fanning, and Steven N. Burch* 31
Agricultural Soils: Ammonia Volatilization / *Paul G. Saffigna and John R. Freney* 85
Agricultural Soils: Nitrous Oxide Emissions / *John R. Freney* 96
Agricultural Soils: Phosphorus / *Anja Gassner and Ewald Schnug* 100
Animals: Toxicological Evaluation / *Vera Lucia S.S. de Castro* 280
Cabbage Diseases: Ecology and Control / *Anthony P. Keinath, Marc A. Cubeta, and David B. Langston, Jr.* 442
Desertification / *David Tongway and John Ludwig* 541
Desertification: Extent of / *Victor R. Squires* 545
Desertification: Greenhouse Effect / *Sherwood Idso and Craig Idso* 549
Desertification: Impact / *David A. Mouat and Judith Lancaster* 552
Erosion / *Dennis C. Flanagan* 930
Erosion and Carbon Dioxide / *Pierre A. Jacinthe and Rattan Lal* 934
Erosion and Global Change / *Taolin Zhang and Xingxiang Wang* 939
Erosion and Precipitation / *Bofu Yu* 943
Erosion by Water: Accelerated / *David Favis-Mortlock* 951
Erosion by Wind-Driven Rain / *Gunay Erpul, L. Darrell Norton, and Donald Gabriels* 1031
Erosion: Accelerated / *J. Kent Mitchell and Michael C. Hirschi* 1044
Erosion: Irrigation-Induced / *Robert E. Sojka and David L. Bjorneberg* 1053
Erosion: Snowmelt / *Donald K. McCool* 1057
Irrigation: Soil Salinity / *James D. Rhoades* 1572
Leaching / *Lars Bergström* 1621
Leaching and Illuviation / *Lynn E. Moody* 1624
Manure Management: Phosphorus / *Rory O. Maguire, John T. Brake, and Peter W. Plumstead* 1667
Manure Management: Poultry / *Shafiqur Rahman and Thomas R. Way* 1670
Nitrogen / *Oswald Van Cleemput and Pascal Boeckx* 1768
Phosphorus: Agricultural Nutrient / *John Ryan, Hayriye Ibrikci, Rolf Sommer, and Abdul Rashid* 2091
Phosphorus: Riverine System Transport / *Andrew N. Sharpley, Peter Kleinman, Tore Krogstad, and Richard McDowell* 2100
Sodic Soils: Irrigation Farming / *David Burrow* 2364
Soil Degradation: Global Assessment / *Selim Kapur, Suha Berberoğlu, Erhan Akça, Cenk Dönmez, Mehmet Akif Erdoğan, and Burçak Kapur* 2375
Soil Quality: Carbon and Nitrogen Gases / *Philippe Rochette, Sean McGinn, and Reynald Lemke* 2388

Air Pollution Problems (AIR)

Comparative Overviews (COV)

Soil Quality: Carbon and Nitrogen Gases / *Philippe Rochette, Sean McGinn, and Reynald Lemke*............2388

Diagnostic Tools (DIA)

Air Pollution: Monitoring / *Waldemar Wardencki*............132

Environmental Technology (ENT)

Adsorption / *Puangrat Kajitvichyanukul and Jirapat Ananpattarachai*............61
Air Pollution: Technology / *Sven Erik Jørgensen*............149
Waste Gas Treatment: Bioreactors / *Amit Kumar, Piet Lens, Sarina J. Ergas, and Herman Van Langenhove*............2611

Pollution Problems and Their Sources (PSS)

Acid Rain / *Umesh Kulshrestha*............5
Acid Rain: Nitrogen Deposition / *George F. Vance*............20
Agricultural Soils: Ammonia Volatilization / *Paul G. Saffigna and John R. Freney*............85
Agricultural Soils: Nitrous Oxide Emissions / *John R. Freney*............96
Fossil Fuel Combustion: Air Pollution and Global Warming / *Dan Golomb*............1127
Genotoxicity and Air Pollution / *Eliane Tigre Guimarães and Andrea Nunes Vaz Pedroso*............1154
Rain Water: Atmospheric Deposition / *Żaneta Polkowska*............2249
Rivers and Lakes: Acidification / *Agniezka Gałuszka and Zdzisław M. Migaszewski*............2291
Soil Quality: Carbon and Nitrogen Gases / *Philippe Rochette, Sean McGinn, and Reynald Lemke*............2388
Sulfur Dioxide / *Marianna Czaplicka, Marianna Czaplicka, and Witold Kurylak*............2437

Energy Issues (ENE)

Cleaner Technology (CLT): Alternative Energy

Agriculture: Energy Use and Conservation / *Guangnan Chen and Tek Narayan Maraseni*............118
Alternative Energy / *Bernd Markert, Simone Wuenschmann, Stefan Fraenzle, and Bernd Delakowitz*............179
Alternative Energy: Hydropower / *Andrea Micangeli, Sara Evangelisti, and Danilo Sbordone*............201
Alternative Energy: Photovoltaic Modules and Systems / *Ewa Klugmann-Radziemska*............215
Alternative Energy: Photovoltaic Solar Cells / *Ewa Klugmann-Radziemska*............226
Alternative Energy: Solar Thermal Energy / *Andrea Micangeli, Sara Evangelisti, and Danilo Sbordone*............241
Alternative Energy: Wind Power Technology and Economy / *K.E. Ohrn*............258
Carbon Sequestration / *Nathan E. Hultman*............456
Distributed Generation: Combined Heat and Power / *Barney L. Capehart, D. Paul Mehta, and Wayne C. Turner*............578
Energy Efficiency: New and Emerging Technology / *Steven A. Parker*............742
Energy Sources: Natural versus Additional / *Marc A. Rosen*............770
Energy: Renewable / *John O. Blackburn*............824
Energy: Solid Waste Advanced Thermal Technology / *Alex E.S. Green and Andrew R. Zimmerman*............830
Energy: Storage / *Rudolf Marloth*............853
Geothermal Energy Resources / *Ibrahim Dincer and Arif Hepbasli*............1167
Global Climate Change: Carbon Sequestration / *Sherwood Idso and Keith E. Idso*............1189

Energy Issues (ENE) (*cont'd*)

Global Climate Change: Gasoline, Hybrid-Electric and Hydrogen-Fueled Vehicles / *Robert E. Uhrig* ... 1206
Green Energy / *Ibrahim Dincer and Adnan Midili* ... 1226
Green Processes: Systems Analysis / *Abhishek Tiwary* ... 1242
Green Products: Production / *Puangrat Kajitvichyanukul, Jirapat Ananpattarachai, and Apichon Watcharenwong* ... 1253
Heat Pumps / *Lu Aye* ... 1346
Heat Pumps: Absorption / *Keith E. Herold* ... 1356
Heat Pumps: Geothermal / *Greg Tinkler* ... 1363
Hydroelectricity: Pumped Storage / *Jill S. Tietjen* ... 1418
Integrated Energy Systems / *Leslie A. Solmes and Sven Erik Jørgensen* ... 1493
Thermal Energy: Solar Technologies / *Muhammad Asif and Tariq Muneer* ... 2498
Thermal Energy: Storage / *Brian Silvetti* ... 2508

Cleaner Technology (CLT): Energy Conservation

Buildings: Climate Change / *Lisa Guan and Guangnan Chen* ... 435
Energy Commissioning: Existing Buildings / *David E. Claridge, Mingsheng Liu, and W.D. Turner* ... 656
Energy Commissioning: New Buildings / *Janey Kaster* ... 665
Energy Conservation / *Ibrahim Dincer and Adnan Midili* ... 677
Energy Conservation: Benefits / *Eric A. Woodroof, Wayne C. Turner, and Steven D. Heinz* ... 691
Energy Conservation: Industrial Processes / *Harvey E. Diamond* ... 697
Energy Conservation: Lean Manufacturing / *Bohdan W. Oppenheim* ... 705
Energy Conversion: Coal, Animal Waste, and Biomass Fuel / *Kalyan Annamalai, Soyuz Priyadarsan, Senthil Arumugam, and John M. Sweeten* ... 714
Energy Efficiency: Low-Cost Improvements / *James Call* ... 735
Energy Efficiency: Strategic Facility Guidelines / *Steve Doty* ... 752
Energy Master Planning / *Fredric S. Goldner* ... 763
Energy Use: U.S. / *Dwight K. French* ... 790
Energy: Environmental Security / *Muhammad Asif* ... 798
Energy: Walls and Windows / *Therese Stovall* ... 860
Energy: Waste Heat Recovery / *Martin A. Mozzo, Jr.* ... 869
Insulation: Facilities / *Wendell A. Porter* ... 1478
LEED-EB: Leadership in Energy and Environmental Design for Existing Buildings / *Rusty T. Hodapp* ... 1647
LEED-NC: Leadership in Energy and Environmental Design for New Construction / *Stephen A. Roosa* ... 1654
Sustainable Development: Pyrolysis and Gasification of Biomass and Wastes / *P. Mondal, J. Parikh, and D.V. Naik* ... 2482
Waste Gas Treatment: Bioreactors / *Amit Kumar, Piet Lens, Sarina J. Ergas, and Herman Van Langenhove* ... 2611
Wind Farms: Noise / *Daniel Shepherd, Chris Hanning, and Bob Thorne* ... 2867

Comparative Overviews (COV)

Agriculture: Energy Use and Conservation / *Guangnan Chen and Tek Narayan Maraseni* ... 118
Alternative Energy / *Bernd Markert, Simone Wuenschmann, Stefan Fraenzle, and Bernd Delakowitz* ... 179
Alternative Energy: Wind Power Technology and Economy / *K.E. Ohrn* ... 258
Bioenergy Crops: Carbon Balance Assessment / *Rocky Lemus and Rattan Lal* ... 345

Buildings: Climate Change / *Lisa Guan and Guangnan Chen* 435
Carbon Sequestration / *Nathan E. Hultman* 456
Cost Control: Consequences / *Peter Victor* 535
Energy: Physics / *Milivoje M. Kostic* 808
Energy: Renewable / *John O. Blackburn* 824
Fuel Cells: Intermediate and High Temperature / *Xianguo Li, Gholamreza Karimi, and Kui Jiao* 1138
Fuel Cells: Low Temperature / *Xianguo Li and Kui Jiao* 1145
Global Climate Change: Carbon Sequestration / *Sherwood Idso and Keith E. Idso* 1189
Global Climate Change: Earth System Response / *Amanda Staudt and Nathan E. Hultman* 1192
Global Climate Change: Gas Fluxes / *Pascal Boeckx and Oswald Van Cleemput* 1202
Global Climate Change: World Soils / *Rattan Lal* 1213
Globalization / *Alexandru V. Roman* 1218
Green Processes: Systems Analysis / *Abhishek Tiwary* 1242
Green Products: Production / *Puangrat Kajitvichyanukul, Jirapat Ananpattarachai, and Apichon Watcharenwong* 1253
Heat Pumps / *Lu Aye* 1346
Hydroelectricity: Pumped Storage / *Jill S. Tietjen* 1418
Inorganic Carbon: Composition and Formation / *Larry P. Wilding and H. Curtis Monger* 1444
Inorganic Carbon: Global Carbon Cycle / *William H. Schlesinger* 1448
Methane Emissions: Rice / *Kazuyuki Yagi* 1679
Permafrost / *Douglas Kane and Julia Boike* 1900
Thermal Energy: Solar Technologies / *Muhammad Asif and Tariq Muneer* 2498
Thermodynamics / *Ronald L. Klaus* 2525

Diagnostic Tools (DIA)

Bioenergy Crops: Carbon Balance Assessment / *Rocky Lemus and Rattan Lal* 345
Energy Use: Exergy and Eco-Exergy / *Sven Erik Jørgensen* 778
Inorganic Carbon: Composition and Formation / *Larry P. Wilding and H. Curtis Monger* 1444
Inorganic Carbon: Global Carbon Cycle / *William H. Schlesinger* 1448
Inorganic Carbon: Modeling / *Leslie D. McFadden and Ronald G. Amundson* 1451
Permafrost / *Douglas Kane and Julia Boike* 1900

Ecotechnology (ECT)

Bioenergy Crops: Carbon Balance Assessment / *Rocky Lemus and Rattan Lal* 345
Carbon Sequestration / *Nathan E. Hultman* 456
Desertification: Prevention and Restoration / *Claudio Zucca, Susana Bautista, Barron Orr, and Franco Previtali* 557
Global Climate Change: Carbon Sequestration / *Sherwood Idso and Keith E. Idso* 1189
Global Climate Change: Earth System Response / *Amanda Staudt and Nathan E. Hultman* 1192
Green Processes: Systems Analysis / *Abhishek Tiwary* 1242

Environmental Legislation and Policy (ELE)

Buildings: Climate Change / *Lisa Guan and Guangnan Chen* 435
Climate Policy: International / *Nathan E. Hultman* 483
Cost Control: Consequences / *Peter Victor* 535
Energy Commissioning: Existing Buildings / *David E. Claridge, Mingsheng Liu, and W.D. Turner* 656
Energy Commissioning: New Buildings / *Janey Kaster* 665
Nuclear Energy: Economics / *James G. Hewlett* 1789

Energy Issues (ENE) (*cont'd*)

Environmental Technology (ENT)

Alternative Energy / *Bernd Markert, Simone Wuenschmann, Stefan Fraenzle, and Bernd Delakowitz* ... 179
Alternative Energy: Hydropower / *Andrea Micangeli, Sara Evangelisti, and Danilo Sbordone* ... 201
Alternative Energy: Photovoltaic Modules and Systems / *Ewa Klugmann-Radziemska* ... 215
Alternative Energy: Photovoltaic Solar Cells / *Ewa Klugmann-Radziemska* ... 226
Alternative Energy: Solar Thermal Energy / *Andrea Micangeli, Sara Evangelisti, and Danilo Sbordone* ... 241
Alternative Energy: Wind Power Technology and Economy / *K.E. Ohrn* ... 258
Buildings: Climate Change / *Lisa Guan and Guangnan Chen* ... 435
Energy Conservation: Industrial Processes / *Harvey E. Diamond* ... 697
Energy Conservation: Lean Manufacturing / *Bohdan W. Oppenheim* ... 705
Energy Conversion: Coal, Animal Waste, and Biomass Fuel / *Kalyan Annamalai, Soyuz Priyadarsan, Senthil Arumugam, and John M. Sweeten* ... 714
Energy Efficiency: New and Emerging Technology / *Steven A. Parker* ... 742
Nuclear Energy: Economics / *James G. Hewlett* ... 1789

Integrated and Holistic Management (IMS)

Energy Efficiency: Low-Cost Improvements / *James Call* ... 735
Energy Efficiency: New and Emerging Technology / *Steven A. Parker* ... 742
Energy Efficiency: Strategic Facility Guidelines / *Steve Doty* ... 752
Energy Master Planning / *Fredric S. Goldner* ... 763
Energy Sources: Natural versus Additional / *Marc A. Rosen* ... 770
Energy: Solid Waste Advanced Thermal Technology / *Alex E.S. Green and Andrew R. Zimmerman* ... 830
Energy: Storage / *Rudolf Marloth* ... 853
Green Energy / *Ibrahim Dincer and Adnan Midili* ... 1226
Integrated Energy Systems / *Leslie A. Solmes and Sven Erik Jørgensen* ... 1493
LEED-EB: Leadership in Energy and Environmental Design for Existing Buildings / *Rusty T. Hodapp* ... 1647
LEED-NC: Leadership in Energy and Environmental Design for New Construction / *Stephen A. Roosa* ... 1654

Pollution Problems and Their Sources (PSS)

Desertification: Greenhouse Effect / *Sherwood Idso and Craig Idso* ... 549
Desertization / *Henry Noel LeHouerou* ... 565
Erosion and Global Change / *Taolin Zhang and Xingxiang Wang* ... 939
Erosion and Precipitation / *Bofu Yu* ... 943
Erosion by Wind: Climate Change / *Alan Busacca and David Chandler* ... 1004
Erosion by Wind: Global Hot Spots / *Andrew Warren* ... 1010
Methane Emissions: Rice / *Kazuyuki Yagi* ... 1679
Wind Farms: Noise / *Daniel Shepherd, Chris Hanning, and Bob Thorne* ... 2867

General Pollution Problems (GEN)

Cleaner Technology (CLT)

Industries: Network of / *Sven Erik Jørgensen* ... 1434
Nanotechnology: Environmental Abatement / *Puangrat Kajitvichyanukul and Jirapat Ananpattarachai* ... 1730

Comparative Overviews (COV): Pollution and Contamination

Rain Water: Atmospheric Deposition / *Zaneta Polkowska* 2249

Rivers and Lakes: Acidification / *Agniezka Gałuszka and Zdzisław M. Migaszewski* 2291

Rivers: Pollution / *Bogdan Skwarzec* 2303

Sea: Pollution / *Bogdan Skwarzec* 2357

Comparative Overviews (COV): Processes, Environmental Quality and Conditions

General Pollution Problems (GEN) (*cont'd*)

Vanadium and Chromium Groups / *Imad A.M. Ahmed* 2582
Water Quality: Range and Pasture Land / *Thomas L. Thurow* 2752
Water Quality: Soil Erosion / *Andreas Klik* 2755
Water: Cost / *Atif Kubursi and Matthew Agarwala* 2779
Water: Drinking / *Marek Biziuk and Małgorzata Michalska* 2790
Water: Surface / *Victor de Vlaming* 2804
Water: Total Maximum Daily Load / *Robin Kundis Craig* 2808

Diagnostic Tools (DIA)

Birds: Chemical Control / *Eric B. Spurr* 413
Community-Based Monitoring: Ngarenanyuki, Tanzania / *Aiwerasia V.F. Ngowi, Aiwerasia V.F. Ngowi, Larama MB Rongo, and Thomas J. Mbise* 505
Ecological Indicators: Eco-Exergy to Emergy Flow / *Simone Bastianoni, Luca Coscieme, and Federico M. Pulselli* 592
Ecological Indicators: Ecosystem Health / *Felix Müller, Benjamin Burkhard, Marion Kandziora, Claus Schimming, and Wilhelm Windhorst* 599
Exergy: Analysis / *Marc A. Rosen* 1083
Exergy: Environmental Impact Assessment / *Marc A. Rosen* 1092
Geographic Information System (GIS): Land Use Planning / *Egide Nizeyimana and Jacob Opadeyi* 1163
Inorganic Carbon: Composition and Formation / *Larry P. Wilding and H. Curtis Monger* 1444
Inorganic Carbon: Global Carbon Cycle / *William H. Schlesinger* 1448
Inorganic Carbon: Modeling / *Leslie D. McFadden and Ronald G. Amundson* 1451
Remote Sensing and GIS Integration / *Egide Nizeyimana* 2271
Remote Sensing: Pollution / *Massimo Antoninetti* 2275
Solid Waste Management: Life Cycle Assessment / *Ni-Bin Chang, Ana Pires, and Graça Martinho* 2399
Sustainable Development: Ecological Footprint in Accounting / *Simone Bastianoni, Valentina Niccolucci, Elena Neri, Gemma Cranston, Alessandro Galli, and Mathis Wackernagel* 2467

Ecotechnology (ECT)

Adsorption / *Puangrat Kajitvichyanukul and Jirapat Ananpattarachai* 61
Biofertilizers / *J.R. de Freitas* 349
Biological Controls / *Heikki Hokkanen* 366
Biological Controls: Conservation / *Doug Landis and Steve D. Wratten* 370
Bioremediation / *Ragini Gothalwal* 387
Ecosystems: Large-Scale Restoration Governance / *Shannon Estenoz, Denise Vienne, and Alka Sapat* 624
Ecosystems: Planning / *Ioan Manuel Ciumasu, Ioan Manuel Ciumasu, Liana Buzdugan, Nicolae Stefan, and Keith Culver* 632
Industries: Network of / *Sven Erik Jørgensen* 1434
Land Restoration / *Richard W. Bell* 1600
Manure Management: Phosphorus / *Rory O. Maguire, John T. Brake, and Peter W. Plumstead* 1667
Nutrients: Best Management Practices / *Scott J. Sturgul and Keith A. Kelling* 1805
Nutrients: Bioavailability and Plant Uptake / *Niels Erik Nielsen* 1817
Pollution: Point and Non-Point Source Low Cost Treatment / *Sandeep Joshi and Sayali Joshi* 2150

Environmental Legislation and Policy (ELE)

Cost Control: Consequences / *Peter Victor* 535
Environmental Legislation: Asia / *Wanpen Wirojanagud and Wanpen Wirojanagud* 874

Environmental Legislation: EU Countries Solid Waste Management / *Ni-Bin Chang, Ana Pires, and Graça Martinho* 892
Environmental Policy / *Sanford V. Berg* 914
Environmental Policy: Innovations / *Alka Sapat* 922
Food Quality Protection Act / *Christina D. DiFonzo* 1118
Food: Cosmetic Standards / *David Pimentel and Kelsey Hart* 1120
Laws and Regulations: Food / *Ike Jeon* 1609
Laws and Regulations: Rotterdam Convention / *Barbara Dinham* 1615
Water: Cost / *Atif Kubursi and Matthew Agarwala* 2779

Environmental Technology (ENT)

Adsorption / *Puangrat Kajitvichyanukul and Jirapat Ananpattarachai* 61

Integrated and Holistic Management (IMS)

Cost Control: Consequences / *Peter Victor* 535
Ecological Indicators: Ecosystem Health / *Felix Müller, Benjamin Burkhard, Marion Kandziora, Claus Schimming, and Wilhelm Windhorst* 599
Ecosystems: Large-Scale Restoration Governance / *Shannon Estenoz, Denise Vienne, and Alka Sapat* 624
Ecosystems: Planning / *Ioan Manuel Ciumasu, Ioan Manuel Ciumasu, Liana Buzdugan, Nicolae Stefan, and Keith Culver* 632
Geographic Information System (GIS): Land Use Planning / *Egide Nizeyimana and Jacob Opadeyi* 1163
Integrated Pest Management / *H.F. van Emden* 1521
Pest Management / *E.F. Legner* 1919
Solid Waste Management: Life Cycle Assessment / *Ni-Bin Chang, Ana Pires, and Graça Martinho* 2399
Sustainability and Planning / *Richard Cowell* 2446
Sustainable Development / *Mark A. Peterson* 2461
Sustainable Development: Ecological Footprint in Accounting / *Simone Bastianoni, Valentina Niccolucci, Elena Neri, Gemma Cranston, Alessandro Galli, and Mathis Wackernagel* 2467

Pollution Problems and Their Sources (PSS): Generally Point Sources

Aquatic Communities: Pesticide Impacts / *David P. Kreutzweiser and Paul K. Sibley* 291
Boron and Molybdenum: Deficiencies and Toxicities / *Umesh C. Gupta* 424
Cadmium and Lead: Contamination / *Gabriella Kakonyi and Imad A.M. Ahmed* 446
Cadmium: Toxicology / *Sven Erik Jørgensen* 453
Chromium / *Bruce R. James and Dominic A. Brose* 477
Coastal Water: Pollution / *Piotr Szefer* 489
Cobalt and Iodine / *Ronald G. McLaren* 502
Copper / *David R. Parker and Judith F. Pedler* 535
Endocrine Disruptors / *Vera Lucia S.S. de Castro* 643
Groundwater: Mining / *Hugo A. Loaiciga* 1284
Groundwater: Mining Pollution / *Jeffrey G. Skousen and George F. Vance* 1289
Heavy Metals: Organic Fertilization Uptake / *Ewald Schnug, Alexandra Izosimova, and Renata Gaj* 1374
Herbicides / *Malcolm Devine* 1378
Herbicides: Non-Target Species Effects / *Céline Boutin* 1382
Human Health: Hormonal Disruption / *Evamarie Straube, Sebastian Straube, and Wolfgang Straube* 1405

General Pollution Problems (GEN) (*cont'd*)

Irrigation: Erosion / *David L. Bjorneberg* ... 1551
Irrigation: Return Flow and Quality / *Ramón Aragues and Kenneth K. Tanji* ... 1557
Mercury / *Sven Erik Jørgensen* ... 1676
Petroleum: Hydrocarbon Contamination / *Svetlana Drozdova and Erwin Rosenberg* ... 2040
Phenols / *Leszek Wachowski and Robert Pietrzak* ... 2071
Pollution: Point Sources / *Ravendra Naidu, Mallavarapu Megharaj, Peter Dillon, Rai Kookana, Ray Correll, and W.W. Wenzel* ... 2166
Radio Frequency Towers: Public School Placement / *Joshua Steinfeld* ... 2224
Radioactivity / *Bogdan Skwarzec* ... 2234
Radionuclides / *Philip M. Jardine* ... 2243
Rare Earth Elements / *Zhengyi Hu, Gerd Sparovek, Silvia Haneklaus, and Ewald Schnug* ... 2266
Toxic Substances / *Sven Erik Jørgensen* ... 2542
Vanadium and Chromium Groups / *Imad A.M. Ahmed* ... 2582

Pollution Problems and Their Sources (PSS): Generally Non-Point and Diffuse Sources

Adsorption / *Puangrat Kajitvichyanukul and Jirapat Ananpattarachai* ... 61
Agricultural Soils: Carbon and Nitrogen Biological Cycling / *Alan J. Franzluebbers* ... 88
Allelochemics / *John Borden* ... 175
Aluminum / *Johannes Bernhard Wehr, Frederick Paxton Cardell Blamey, and Neal William Menzies* ... 267
Antagonistic Plants / *Philip Oduor-Owino* ... 288
Bioaccumulation / *Tomaz Langenbach* ... 324
Cyanobacteria: Eutrophic Freshwater Systems / *Anja Gassner and Martin V. Frey* ... 538
Erosion / *Dennis C. Flanagan* ... 930
Erosion: History / *Andrew S. Goudie* ... 1050
Estuaries / *Claude Amiard-Triquet* ... 1063
Fossil Fuel Combustion: Air Pollution and Global Warming / *Dan Golomb* ... 1127
Giant Reed (*Arundo donax*): Streams and Water Resources / *Gretchen Coffman* ... 1182
Groundwater: Contamination / *Charles W. Fetter* ... 1281
Inland Seas and Lakes: Central Asia Case Study / *Andrey G. Kostianoy* ... 1436
Methane Emissions: Rice / *Kazuyuki Yagi* ... 1679
Mycotoxins / *J. David Miller* ... 1696
Nanoparticles / *Alexandra Navrotsky* ... 1711
Phosphorus: Agricultural Nutrient / *John Ryan, Hayriye Ibrikci, Rolf Sommer, and Abdul Rashid* ... 2091
Pollutants: Organic and Inorganic / *A. Paul Schwab* ... 2114
Pollution: Genotoxicity of Agrotoxic Compounds / *Vera Lucia S.S. de Castro and Paola Poli* ... 2123
Pollution: Non-Point Source / *Ravendra Naidu, Mallavarapu Megharaj, Peter Dillon, Rai Kookana, Ray Correll, and W.W. Wenzel* ... 2136
Potassium / *Philippe Hinsinger* ... 2208
Rain Water: Atmospheric Deposition / *Zaneta Polkowska* ... 2249
Rivers and Lakes: Acidification / *Agniezka Gałuszka and Zdzisław M. Migaszewski* ... 2291
Rivers: Pollution / *Bogdan Skwarzec* ... 2303
Sea: Pollution / *Bogdan Skwarzec* ... 2357
Water Quality: Soil Erosion / *Andreas Klik* ... 2755

Global Pollution Problems (GLO)

Cleaner Technology (CLT)

Alternative Energy / *Bernd Markert, Simone Wuenschmann, Stefan Fraenzle, and Bernd Delakowitz* 179
Buildings: Climate Change / *Lisa Guan and Guangnan Chen* 435
Carbon Sequestration / *Nathan E. Hultman* 456
Energy Conservation / *Ibrahim Dincer and Adnan Midili* 677
Energy Conservation: Benefits / *Eric A. Woodroof, Wayne C. Turner, and Steven D. Heinz* 691
Energy Conservation: Industrial Processes / *Harvey E. Diamond* 697
Energy Efficiency: New and Emerging Technology / *Steven A. Parker* 742
Energy Master Planning / *Fredric S. Goldner* 763
Energy: Environmental Security / *Muhammad Asif* 798
Energy: Physics / *Milivoje M. Kostic* 808
Energy: Renewable / *John O. Blackburn* 824
Energy: Storage / *Rudolf Marloth* 853

Comparative Overviews (COV)

Agriculture: Energy Use and Conservation / *Guangnan Chen and Tek Narayan Maraseni* 118
Alternative Energy / *Bernd Markert, Simone Wuenschmann, Stefan Fraenzle, and Bernd Delakowitz* 179
Carbon Sequestration / *Nathan E. Hultman* 456
Energy: Environmental Security / *Muhammad Asif* 798
Energy: Physics / *Milivoje M. Kostic* 808
Energy: Renewable / *John O. Blackburn* 824
Energy: Storage / *Rudolf Marloth* 853
Energy: Waste Heat Recovery / *Martin A. Mozzo, Jr.* 869
Global Climate Change: Carbon Sequestration / *Sherwood Idso and Keith E. Idso* 1189
Global Climate Change: Earth System Response / *Amanda Staudt and Nathan E. Hultman* 1192
Global Climate Change: Gas Fluxes / *Pascal Boeckx and Oswald Van Cleemput* 1202
Global Climate Change: Gasoline, Hybrid-Electric and Hydrogen-Fueled Vehicles / *Robert E. Uhrig* 1206
Globalization / *Alexandru V. Roman* 1218
Green Processes: Systems Analysis / *Abhishek Tiwary* 1242
Oil Pollution: Baltic Sea / *Andrey G. Kostianoy* 1826
Organic Compounds: Halogenated / *Marek Tobiszewski and Jacek Namieśnik* 1841
Organic Matter: Global Distribution in World Ecosystems / *Wilfred M. Post* 1851
Organic Matter: Turnover / *Johan Six and Julie D. Jastrow* 1872
Ozone Layer / *Luisa T. Molina* 1882
Pesticides / *Marek Biziuk, Jolanta Fenik, Monika Kosikowska, and Maciej Tankiewicz* 1963
Pesticides: Effects / *Ana Maria Evangelista de Duffard* 2005
Rain Water: Atmospheric Deposition / *Zaneta Polkowska* 2249
Rivers and Lakes: Acidification / *Agniezka Gałuszka and Zdzisław M. Migaszewski* 2291
Rivers: Pollution / *Bogdan Skwarzec* 2303
Runoff Water / *Zaneta Polkowska* 2320
Sea: Pollution / *Bogdan Skwarzec* 2357

Global Pollution Problems (GLO) (*cont'd*)

Soil Degradation: Global Assessment / *Selim Kapur, Suha Berberoğlu, Erhan Akça, Cenk Dönmez, Mehmet Akif Erdoğan, and Burçak Kapur* 2375
Water Quality and Quantity: Globalization / *Kristi Denise Caravella and Jocilyn Danise Martinez* 2740
Water Quality: Range and Pasture Land / *Thomas L. Thurow* 2752
Water Quality: Soil Erosion / *Andreas Klik* 2755
Water: Drinking / *Marek Biziuk and Małgorzata Michalska* 2790
Water: Surface / *Victor de Vlaming* 2804
Water: Total Maximum Daily Load / *Robin Kundis Craig* 2808
Wetlands / *Ralph W. Tiner* 2824
Wetlands: Biodiversity / *Jean-Claude Lefeuvre and Virginie Bouchard* 2829
Wetlands: Carbon Sequestration / *Virginie Bouchard* 2833
Wetlands: Conservation Policy / *Clayton Rubec* 2837

Diagnostic Tools (DIA)

Community-Based Monitoring: Ngarenanyuki, Tanzania / *Aiwerasia V.F. Ngowi, Aiwerasia V.F. Ngowi, Larama MB Rongo, and Thomas J. Mbise* 505
Geographic Information System (GIS): Land Use Planning / *Egide Nizeyimana and Jacob Opadeyi* 1163
Inorganic Carbon: Composition and Formation / *Larry P. Wilding and H. Curtis Monger* 1444
Inorganic Carbon: Global Carbon Cycle / *William H. Schlesinger* 1448
Inorganic Carbon: Modeling / *Leslie D. McFadden and Ronald G. Amundson* 1451
Remote Sensing and GIS Integration / *Egide Nizeyimana* 2271
Remote Sensing: Pollution / *Massimo Antoninetti* 2275
Soil Degradation: Global Assessment / *Selim Kapur, Suha Berberoğlu, Erhan Akça, Cenk Dönmez, Mehmet Akif Erdoğan, and Burçak Kapur* 2375
Sustainable Development: Ecological Footprint in Accounting / *Simone Bastianoni, Valentina Niccolucci, Elena Neri, Gemma Cranston, Alessandro Galli, and Mathis Wackernagel* 2467
Water Quality and Quantity: Globalization / *Kristi Denise Caravella and Jocilyn Danise Martinez* 2740
Water Quality: Modeling / *Richard Lowrance* 2749

Ecotechnology (ECT)

Carbon Sequestration / *Nathan E. Hultman* 456
Nutrients: Best Management Practices / *Scott J. Sturgul and Keith A. Kelling* 1805
Nutrients: Bioavailability and Plant Uptake / *Niels Erik Nielsen* 1817
Wetlands / *Ralph W. Tiner* 2824
Wetlands: Biodiversity / *Jean-Claude Lefeuvre and Virginie Bouchard* 2829
Wetlands: Carbon Sequestration / *Virginie Bouchard* 2833
Wetlands: Conservation Policy / *Clayton Rubec* 2837
Wetlands: Constructed Subsurface / *Jan Vymazal* 2841
Wetlands: Sedimentation and Ecological Engineering / *Timothy C. Granata and Jay F. Martin* 2859
Wetlands: Treatment System Use / *Kyle R. Mankin* 2862

Environmental Legislation and Policy (ELE)

Climate Policy: International / *Nathan E. Hultman* 483
Environmental Legislation: Asia / *Wanpen Wirojanagud and Wanpen Wirojanagud* 874

Environmental Legislation: EU Countries Solid Waste Management / *Ni-Bin Chang, Ana Pires, and Graça Martinho* 892
Environmental Policy / *Sanford V. Berg* 914
Environmental Policy: Innovations / *Alka Sapat* 922

Integrated and Holistic Management (IMS)

Energy: Environmental Security / *Muhammad Asif* 798
Energy: Renewable / *John O. Blackburn* 824
Energy: Storage / *Rudolf Marloth* 853
Geographic Information System (GIS): Land Use Planning / *Egide Nizeyimana and Jacob Opadeyi* 1163
Global Climate Change: Carbon Sequestration / *Sherwood Idso and Keith E. Idso* 1189
Global Climate Change: Earth System Response / *Amanda Staudt and Nathan E. Hultman* 1192
Global Climate Change: Gas Fluxes / *Pascal Boeckx and Oswald Van Cleemput* 1202
Globalization / *Alexandru V. Roman* 1218
Green Processes: Systems Analysis / *Abhishek Tiwary* 1242
Oil Pollution: Baltic Sea / *Andrey G. Kostianoy* 1826
Organic Matter: Management / *R. Cesar Izaurralde and Carlos C. Cerri* 1857
Organic Matter: Turnover / *Johan Six and Julie D. Jastrow* 1872
Sustainability and Planning / *Richard Cowell* 2446
Sustainable Development / *Mark A. Peterson* 2461
Sustainable Development: Ecological Footprint in Accounting / *Simone Bastianoni, Valentina Niccolucci, Elena Neri, Gemma Cranston, Alessandro Galli, and Mathis Wackernagel* 2467

Pollution Problems and Their Sources (PSS)

Erosion by Wind: Climate Change / *Alan Busacca and David Chandler* 1004
Erosion by Wind: Global Hot Spots / *Andrew Warren* 1010
Fossil Fuel Combustion: Air Pollution and Global Warming / *Dan Golomb* 1127
Pesticides / *Marek Biziuk, Jolanta Fenik, Monika Kosikowska, and Maciej Tankiewicz* 1963
Pesticides: Effects / *Ana Maria Evangelista de Duffard* 2005
Rain Water: Atmospheric Deposition / *Zaneta Polkowska* 2249
Rivers and Lakes: Acidification / *Agniezka Gałuszka and Zdzisław M. Migaszewski* 2291
Rivers: Pollution / *Bogdan Skwarzec* 2303
Sea: Pollution / *Bogdan Skwarzec* 2357
Soil Degradation: Global Assessment / *Selim Kapur, Suha Berberoğlu, Erhan Akça, Cenk Dönmez, Mehmet Akif Erdoğan, and Burçak Kapur* 2375
Water Quality and Quantity: Globalization / *Kristi Denise Caravella and Jocilyn Danise Martinez* 2740
Water Quality: Range and Pasture Land / *Thomas L. Thurow* 2752
Water Quality: Soil Erosion / *Andreas Klik* 2755
Water: Drinking / *Marek Biziuk and Małgorzata Michalska* 2790
Water: Surface / *Victor de Vlaming* 2804

Terrestrial Pollution Problems (TER)

Cleaner Technology (CLT)

Acid Sulfate Soils: Management / *Michael D. Melville and Ian White* 55
Pesticide Translocation Control: Soil Erosion / *Monika Frielinghaus, Detlef Deumlich, and Roger Funk* 1955
Salt-Affected Soils: Sustainable Agriculture / *Pichu Rengasamy* 2349
Sustainable Agriculture: Soil Quality / *John W. Doran and Ed G. Gregorich* 2457

Terrestrial Pollution Problems (TER) (*cont'd*)

Comparative Overviews (COV)

Agricultural Soils: Carbon and Nitrogen Biological Cycling / *Alan J. Franzluebbers* 88
Erosion by Wind: Principles / *Larry D. Stetler* 1013
Erosion: History / *Andrew S. Goudie* 1050
Erosion: Soil Quality / *Craig Ditzler* 1060
Global Climate Change: World Soils / *Rattan Lal* 1213
Global Climate Change: World Soils / *Rattan Lal* 1213
Industrial Waste: Soil Pollution / *W. Friesl-Hanl, M.H. Gerzabek, W.W. Wenzel, and W.E.H. Blum* 1430
Salt-Affected Soils: Physical Properties and Behavior / *Hwat Bing So* 2334
Salt-Affected Soils: Plant Response / *Anna Eynard, Keith D. Wiebe, and Rattan Lal* 2345
Sodic Soils: Properties / *Pichu Rengasamy* 2367
Soil Degradation: Global Assessment / *Selim Kapur, Suha Berberoğlu, Erhan Akça, Cenk Dönmez, Mehmet Akif Erdoğan, and Burçak Kapur* 2375
Soil Quality: Carbon and Nitrogen Gases / *Philippe Rochette, Sean McGinn, and Reynald Lemke* 2388
Solid Waste: Municipal / *Angelique Chettiparamb* 2415

Diagnostic Tools (DIA)

Acid Sulfate Soils: Identification, Assessment, and Management / *L.A. Sullivan, R.T. Bush, and E.D. Burton* 36
Drainage: Soil Salinity Management / *Glenn J. Hoffman* 588
Erosion by Water: Assessment and Control / *José Miguel Reichert, Nadia Bernardi Bonumá, Gustavo Enrique Merten, and Jean Paolo Gomes Minella* 963
Erosion by Water: Empirical Methods / *John M. Laflen and John M. Laflen* 974
Erosion by Water: Erosivity and Erodibility / *Peter I.A. Kinnell* 980
Erosion by Water: Process-Based Modeling / *Mark A. Nearing* 991
Erosion by Wind: Source, Measurement, Prediction, and Control / *Brenton S. Sharratt and R. Scott Van Pelt* 1017
Soil Quality: Indicators / *Graham P. Sparling* 2393
Solid Waste Management: Life Cycle Assessment / *Ni-Bin Chang, Ana Pires, and Graça Martinho* 2399

Ecotechnology (ECT)

Acid Sulfate Soils: Management / *Michael D. Melville and Ian White* 55
Bioremediation: Contaminated Soil Restoration / *Sven Erik Jørgensen* 401
Composting / *Nídia Sá Caetano* 512
Composting: Organic Farming / *Saburo Matsui* 528
Desertification: Prevention and Restoration / *Claudio Zucca, Susana Bautista, Barron Orr, and Franco Previtali* 557
Erosion and Sediment Control: Vegetative Techniques / *Samson D. Angima* 947
Erosion by Water: Amendment Techniques / *X.-C. (John) Zhang* 958
Erosion by Water: Vegetative Control / *Seth M. Dabney and Silvio J. Gumiere* 994
Erosion Control: Soil Conservation / *Eric T. Craswell* 1034
Erosion Control: Tillage and Residue Methods / *Richard Cruse, Jerry Neppel, John Kost, and Krisztina Eleki* 1040
Organic Soil Amendments / *Philip Oduor-Owino* 1878

Toxic Substances in the Environment (TOX)

Cleaner Technology (CLT)

Bacterial Pest Control / *David N. Ferro* 321
Chemigation / *William L. Kranz* 469
Invasion Biology / *Jennifer Ruesink* 1531
Nanotechnology: Environmental Abatement / *Puangrat Kajitvichyanukul and Jirapat Ananpattarachai* 1730
Pest Management: Crop Diversity / *Marianne Karpenstein-Machan and Maria Finckh* 1925
Pest Management: Ecological Agriculture / *Barbara Dinham* 1930
Pest Management: Ecological Aspects / *David J. Horn* 1934
Pest Management: Intercropping / *Maria Finckh and Marianne Karpenstein-Machan* 1937
Pesticide Translocation Control: Soil Erosion / *Monika Frielinghaus, Detlef Deumlich, and Roger Funk* 1955
Pesticide Translocation Control: Soil Erosion / *Monika Frielinghaus, Detlef Deumlich, and Roger Funk* 1955
Pesticides: Damage Avoidance / *Aiwerasia V.F. Ngowi, Aiwerasia V.F. Ngowi, and Larama MB Rongo* 1997
Pesticides: Reducing Use / *David Pimentel and Maria V. Cilveti* 2033
Stored-Product Pests: Biological Control / *Lise Stengård Hansen* 2423

Comparative Overviews (COV)

Acaricides / *Doug Walsh* 1
Aluminum / *Johannes Bernhard Wehr, Frederick Paxton Cardell Blamey, and Neal William Menzies* 267
Antagonistic Plants / *Philip Oduor-Owino* 288
Bioaccumulation / *Tomaz Langenbach* 324
Biodegradation / *Sven Erik Jørgensen* 328
Biodiversity and Sustainability / *Odo Primavesi* 333
Biofertilizers / *J. R. de Freitas* 349
Biological Controls / *Heikki Hokkanen* 366
Biological Controls: Conservation / *Doug Landis and Steve D. Wratten* 370
Biomass / *Alberto Traverso and David Tucker* 373
Bioremediation / *Ragini Gothalwal* 387
Chemigation / *William L. Kranz* 469
Industrial Waste: Soil Pollution / *W. Friesl-Hanl, M.H. Gerzabek, W.W. Wenzel, and W.E.H. Blum* 1430
Inorganic Carbon: Composition and Formation / *Larry P. Wilding and H. Curtis Monger* 1444
Inorganic Carbon: Global Carbon Cycle / *William H. Schlesinger* 1448
Inorganic Compounds: Eco-Toxicity / *Sven Erik Jørgensen* 1455
Invasion Biology / *Jennifer Ruesink* 1531
Mercury / *Sven Erik Jørgensen* 1676
Nanoparticles / *Alexandra Navrotsky* 1711
Nanotechnology: Environmental Abatement / *Puangrat Kajitvichyanukul and Jirapat Ananpattarachai* 1730
Persistent Organic Pesticides / *Gamini Manuweera* 1913
Pest Management / *E.F. Legner* 1919
Pesticides / *Marek Biziuk, Jolanta Fenik, Monika Kosikowska, and Maciej Tankiewicz* 1963
Pesticides: Chemical and Biological / *Barbara Manachini* 1986

Pesticides: Damage Avoidance / *Aiwerasia V.F. Ngowi, Aiwerasia V.F. Ngowi, and Larama MB Rongo* ... 1997
Pesticides: Effects / *Ana Maria Evangelista de Duffard* ... 2005
Pesticides: History / *Edward H. Smith and George Kennedy* ... 2008
Toxic Substances / *Sven Erik Jørgensen* ... 2542
Toxic Substances: Photochemistry / *Puangrat Kajitvichyanukul* ... 2547
Toxicity Prediction of Chemical Mixtures / *Nina Cedergreen, Claus Svendsen, and Thomas Backhaus* ... 2572
Water Supplies: Pharmaceuticals / *Klaus Kümmerer* ... 2776

Diagnostic Tools (DIA)

Animals: Toxicological Evaluation / *Vera Lucia S.S. de Castro* ... 280
Birds: Chemical Control / *Eric B. Spurr* ... 413
Inorganic Carbon: Composition and Formation / *Larry P. Wilding and H. Curtis Monger* ... 1444
Inorganic Carbon: Global Carbon Cycle / *William H. Schlesinger* ... 1448
Inorganic Carbon: Modeling / *Leslie D. McFadden and Ronald G. Amundson* ... 1451
Nanomaterials: Regulation and Risk Assessment / *Steffen Foss Hansen, Khara D. Grieger, and Anders Baun* ... 1700
Nanoparticles: Uncertainty Risk Analysis / *Khara D. Grieger, Steffen Foss Hansen, and Anders Baun* ... 1720
Neurotoxicants: Developmental Experimental Testing / *Vera Lucia S.S. de Castro* ... 1755
Organic Matter: Management / *R. Cesar Izaurralde and Carlos C. Cerri* ... 1857
Organic Matter: Modeling / *Pete Smith* ... 1863
Organic Matter: Turnover / *Johan Six and Julie D. Jastrow* ... 1872
Pest Management: Modeling / *Andrew Paul Gutierrez* ... 1949
Pesticides: Measurement and Mitigation / *George F. Antonious* ... 2013
Polychlorinated Biphenyls (PCBs) and Polycyclic Aromatic Hydrocarbons (PAHs): Sediments and Water Analysis / *Justyna Rogowska, Agata Mechlińska, Lidia Wolska, Lidia Wolska, and Jacek Namieśnik* ... 2186

Ecotechnology (ECT)

Biological Controls / *Heikki Hokkanen* ... 366
Biological Controls: Conservation / *Doug Landis and Steve D. Wratten* ... 370
Bioremediation / *Ragini Gothalwal* ... 387
Bioremediation: Contaminated Soil Restoration / *Sven Erik Jørgensen* ... 401
Invasion Biology / *Jennifer Ruesink* ... 1531
Natural Enemies: Conservation / *Cetin Sengonca* ... 1749
Nematodes: Biological Control / *Simon Gowen* ... 1752
Pesticide Translocation Control: Soil Erosion / *Monika Frielinghaus, Detlef Deumlich, and Roger Funk* ... 1955
Stored-Product Pests: Biological Control / *Lise Stengård Hansen* ... 2423
Toxic Substances / *Sven Erik Jørgensen* ... 2542
Vertebrates: Biological Control / *Peter Kerr and Tanja Strive* ... 2595
Wetlands / *Ralph W. Tiner* ... 2824
Wetlands: Petroleum / *John V. Headley and Kerry M. Peru* ... 2854

Environmental Legislation and Policy (ELE)

Developing Countries: Pesticide Health Impacts / *Aiwerasia V.F. Ngowi, Aiwerasia V.F. Ngowi, Catharina Wesseling, and Leslie London* ... 573
Farming: Non-Chemical and Pesticide-Free (European Council Regulations [EED/209/91]) / *Inger Källander and Gunnar Rundgren* ... 1103

Toxic Substances in the Environment (TOX) (*cont'd*)

Food Quality Protection Act / *Christina D. DiFonzo* 1118
Food: Cosmetic Standards / *David Pimentel and Kelsey Hart* 1120
Food: Pesticide Contamination / *Denis Hamilton* 1124
Human Health: Cancer from Pesticides / *Jan Dich* 1394
Human Health: Chronic Pesticide Poisonings / *Birgitta Kolmodin-Hedman and Margareta Palmborg* 1397
Human Health: Consumer Concerns of Pesticides / *George Ekström and Margareta Palmborg* 1401
Laws and Regulations: Pesticides / *Praful Suchak* 1612
Lead: Regulations / *Lisa A. Robinson* 1636
Nanomaterials: Regulation and Risk Assessment / *Steffen Foss Hansen, Khara D. Grieger, and Anders Baun* 1700
Pest Management: Legal Aspects / *Michael T. Olexa and Zachary T. Broome* 1940
Pesticides: Banding / *Sharon A. Clay* 1983
Pesticides: Damage Avoidance / *Aiwerasia V.F. Ngowi, Aiwerasia V.F. Ngowi, and Larama MB Rongo* 1997
Pesticides: Regulating / *Matthias Kern* 2035

Environmental Technology (ENT)

Biopesticides / *G.J. Ash and A. Wang* 381
Biotechnology: Pest Management / *Maurizio G. Paoletti* 407
Farming: Organic Pest Management / *Ján Gallo* 1115
Lead: Ecotoxicology / *Sven Erik Jørgensen* 1630
Lead: Regulations / *Lisa A. Robinson* 1636
Minerals Processing Residue (Tailings): Rehabilitation / *Lloyd R. Hossner and Hamid Shahandeh* 1683
Mines: Rehabilitation of Open Cut / *Douglas J. Dollhopf* 1692
Nanotechnology: Environmental Abatement / *Puangrat Kajitvichyanukul and Jirapat Ananpattarachai* 1730
Persistent Organic Compounds: Wet Oxidation Removal / *Anurag Garg and I.M. Mishra* 1904
Pesticides: Natural / *Franck Dayan* 2028
Pests: Landscape Patterns / *F. Craig Stevenson* 2038
Pharmaceuticals: Treatment / *Diana Aga and Seungyun Baik* 2060

Integrated and Holistic Management (IMS)

Integrated Pest Management / *H. F. van Emden* 1521
Nanomaterials: Regulation and Risk Assessment / *Steffen Foss Hansen, Khara D. Grieger, and Anders Baun* 1700
Nanoparticles: Uncertainty Risk Analysis / *Khara D. Grieger, Steffen Foss Hansen, and Anders Baun* 1720
Organic Matter: Global Distribution in World Ecosystems / *Wilfred M. Post* 1851
Organic Matter: Management / *R. Cesar Izaurralde and Carlos C. Cerri* 1857
Organic Matter: Turnover / *Johan Six and Julie D. Jastrow* 1872
Pest Management / *E.F. Legner* 1919

Pollution Problems and Their Sources (PSS): Generally Point Sources

Aluminum / *Johannes Bernhard Wehr, Frederick Paxton Cardell Blamey, and Neal William Menzies* 267
Cadmium and Lead: Contamination / *Gabriella Kakonyi and Imad A.M. Ahmed* 446
Cadmium: Toxicology / *Sven Erik Jørgensen* 453

Chromium / *Bruce R. James and Dominic A. Brose* 477
Cobalt and Iodine / *Ronald G. McLaren* 502
Copper / *David R. Parker and Judith F. Pedler* 535
Endocrine Disruptors / *Vera Lucia S.S. de Castro* 643
Heavy Metals: Organic Fertilization Uptake / *Ewald Schnug, Alexandra Izosimova, and Renata Gaj* 1374
Human Health: Cancer from Pesticides / *Jan Dich* 1394
Human Health: Chronic Pesticide Poisonings / *Birgitta Kolmodin-Hedman and Margareta Palmborg* 1397
Human Health: Consumer Concerns of Pesticides / *George Ekström and Margareta Palmborg* 1401
Human Health: Hormonal Disruption / *Evamarie Straube, Sebastian Straube, and Wolfgang Straube* 1405
Human Health: Pesticide Poisonings / *Margareta Palmborg* 1408
Human Health: Pesticide Sensitivities / *Ann McCampbell* 1411
Human Health: Pesticides / *Kelsey Hart, David Pimentel, and David Pimentel* 1414
Lead: Regulations / *Lisa A. Robinson* 1636
Mercury / *Sven Erik Jørgensen* 1676
Methane Emissions: Rice / *Kazuyuki Yagi* 1679
Mycotoxins / *J. David Miller* 1696
Nanoparticles / *Alexandra Navrotsky* 1711
Neurotoxicants: Developmental Experimental Testing / *Vera Lucia S.S. de Castro* 1755
Organic Compounds: Halogenated / *Marek Tobiszewski and Jacek Namieśnik* 1841
Ozone Layer / *Luisa T. Molina* 1882
Persistent Organic Pesticides / *Gamini Manuweera* 1913
Petroleum: Hydrocarbon Contamination / *Svetlana Drozdova and Erwin Rosenberg* 2040
Phenols / *Leszek Wachowski and Robert Pietrzak* 2071
Polychlorinated Biphenyls (PCBs) / *Marek Biziuk and Angelika Beyer* 2172
Potassium / *Philippe Hinsinger* 2208
Radioactivity / *Bogdan Skwarzec* 2234
Radionuclides / *Philip M. Jardine* 2243
Rare Earth Elements / *Zhengyi Hu, Gerd Sparovek, Silvia Haneklaus, and Ewald Schnug* 2266
Strontium / *Silvia Haneklaus and Ewald Schnug* 2426
Sulfur / *Ewald Schnug, Silvia Haneklaus, and Elke Bloem* 2431
Sulfur Dioxide / Marianna Czaplicka, Marianna Czaplicka, and Witold Kurylak 2437
Toxic Substances / Sven Erik Jørgensen 2542
Toxic Substances: Photochemistry / *Puangrat Kajitvichyanukul* 2547
Vanadium and Chromium Groups / *Imad A.M. Ahmed* 2582
Water Supplies: Pharmaceuticals / Klaus Kümmerer 2776

Pollution Problems and Their Sources (PSS): Generally Non-Point and Diffuse Sources

Acaricides / *Doug Walsh* 1
Animals: Sterility from Pesticides / *William Au* 277
Animals: Toxicological Evaluation / *Vera Lucia S.S. de Castro* 280
Antagonistic Plants / *Philip Oduor-Owino* 288
Aquatic Communities: Pesticide Impacts / *David P. Kreutzweiser and Paul K. Sibley* 291
Bioaccumulation / *Tomaz Langenbach* 324
Birds: Pesticide Use Impacts / *Pierre Mineau* 416
Boron and Molybdenum: Deficiencies and Toxicities / *Umesh C. Gupta* 424

Toxic Substances in the Environment (TOX) (*cont'd*)

Boron: Soil Contaminant / *Rami Keren* 431
Carbon: Soil Inorganic / *Donald L. Suarez* 462
Developing Countries: Pesticide Health Impacts / *Aiwerasia V.F. Ngowi, Aiwerasia V.F. Ngowi, Catharina Wesseling, and Leslie London* 573
Food: Pesticide Contamination / *Denis Hamilton* 1124
Groundwater: Pesticide Contamination / *Roy F. Spalding* 1317
Herbicides / *Malcolm Devine* 1378
Herbicides: Non-Target Species Effects / *Céline Boutin* 1382
Inorganic Carbon: Composition and Formation / *Larry P. Wilding and H. Curtis Monger* 1444
Inorganic Compounds: Eco-Toxicity / *Sven Erik Jørgensen* 1455
Insect Growth Regulators / *Meir Paul Pener* 1459
Insecticides: Aerial Ultra-Low-Volume Application / *He Zhong* 1471
Lead: Ecotoxicology / *Sven Erik Jørgensen* 1630
Oil Pollution: Baltic Sea / *Andrey G. Kostianoy* 1826
Pesticides / *Marek Biziuk, Jolanta Fenik, Monika Kosikowska, and Maciej Tankiewicz* 1963
Pollutants: Organic and Inorganic / *A. Paul Schwab* 2114
Pollution: Genotoxicity of Agrotoxic Compounds / *Vera Lucia S.S. de Castro and Paola Poli* 2123
Pollution: Pesticides in Agro-Horticultural Ecosystems / *J.K. Dubey and Meena Thakur* 2139
Pollution: Pesticides in Natural Ecosystems / *J.K. Dubey and Meena Thakur* 2145
Toxicity Prediction of Chemical Mixtures / *Nina Cedergreen, Claus Svendsen, and Thomas Backhaus* 2572

Water Pollution Problems (WAT)

Cleaner Technology (CLT)

Agricultural Water Quantity Management / *X.S. Qin and Y. Xu* 105
Irrigation Systems: Water Conservation / *I. Pai Wu, Javier Barragan, and Vince Bralts* 1542
Irrigation: Sewage Effluent Use / *B.A. Stewart* 1570
Precision Agriculture: Water and Nutrient Management / *Robert J. Lascano and J.D. Booker* 2217
Wastewater and Water Utilities / *Rudolf Marloth* 2638
Wastewater Use in Agriculture / *Manzoor Qadir, Pay Drechsel, and Liqa Raschid-Sally* 2675
Wastewater Use in Agriculture: Policy Issues / *Dennis Wichelns* 2681
Water Harvesting / *Gary W. Frasier* 2738

Comparative Overviews (COV)

Adsorption / *Puangrat Kajitvichyanukul and Jirapat Ananpattarachai* 61
Agricultural Runoff / *Matt C. Smith, David K. Gattie, and Daniel L. Thomas* 81
Agricultural Water Quantity Management / *X.S. Qin and Y. Xu* 105
Agroforestry: Water Use Efficiency / *James R. Brandle, Laurie Hodges, and Xinhua Zhou* 129
Allelochemics / *John Borden* 175
Aquatic Communities: Pesticide Impacts / *David P. Kreutzweiser and Paul K. Sibley* 291
Coastal Water: Pollution / *Piotr Szefer* 489
Cyanobacteria: Eutrophic Freshwater Systems / *Anja Gassner and Martin V. Frey* 538
Estuaries / *Claude Amiard-Triquet* 1063
Giant Reed (*Arundo donax*): Streams and Water Resources / *Gretchen Coffman* 1182

Groundwater: Contamination / *Charles W. Fetter* 1281
Groundwater: Mining / *Hugo A. Loaiciga* 1284
Groundwater: Mining Pollution / *Jeffrey G. Skousen and George F. Vance* 1289
Inland Seas and Lakes: Central Asia Case Study / *Andrey G. Kostianoy* 1436
Irrigation Systems: Sub-Surface History / *Norman R. Fausey* 1539
Irrigation Systems: Water Conservation / *I. Pai Wu, Javier Barragan, and Vince Bralts* 1542
Irrigation: Efficiency / *Terry A. Howell* 1545
Irrigation: Erosion / *David L. Bjorneberg* 1551
Irrigation: Return Flow and Quality / *Ramón Aragues and Kenneth K. Tanji* 1557
Lakes and Reservoirs: Pollution / *Subhankar Karmakar and O.M. Musthafa* 1576
Rain Water: Atmospheric Deposition / *Zaneta Polkowska* 2249
Rivers and Lakes: Acidification / *Agniezka Gałuszka and Zdzisław M. Migaszewski* 2291
Rivers: Pollution / *Bogdan Skwarzec* 2303
Rivers: Restoration / *Anna Rabajczyk* 2307
Runoff Water / *Zaneta Polkowska* 2320
Sea: Pollution / *Bogdan Skwarzec* 2357
Wastewater and Water Utilities / *Rudolf Marloth* 2638
Wastewater Treatment: Conventional Methods / *Sven Erik Jørgensen* 2657
Wastewater Use in Agriculture / *Manzoor Qadir, Manzoor Qadir, Pay Drechsel, and Liqa Raschid-Sally* 2675
Wastewater Use in Agriculture: Public Health Considerations / *Blanca Jimenez and Inés Navarro* 2694
Wastewater: Municipal / *Sven Erik Jørgensen* 2709
Water Quality and Quantity: Globalization / *Kristi Denise Caravella and Jocilyn Danise Martinez* 2740
Water Quality: Range and Pasture Land / *Thomas L. Thurow* 2752
Water Quality: Soil Erosion / *Andreas Klik* 2755
Water: Cost / *Atif Kubursi and Matthew Agarwala* 2779
Water: Drinking / *Marek Biziuk and Małgorzata Michalska* 2790
Water: Surface / *Victor de Vlaming* 2804
Water: Total Maximum Daily Load / *Robin Kundis Craig* 2808

Diagnostic Tools (DIA)

Groundwater: Modeling / *Jesus Carrera* 1295
Groundwater: Numerical Method Modeling / *Jesus Carrera* 1312
Polychlorinated Biphenyls (PCBs) and Polycyclic Aromatic Hydrocarbons (PAHs): Sediments and Water Analysis / *Justyna Rogowska, Agata Mechlińska, Lidia Wolska, Lidia Wolska, and Jacek Namieśnik* 2186
Water Quality: Modeling / *Richard Lowrance* 2749

Ecotechnology (ECT)

Agricultural Water Quantity Management / *X.S.Q. and Y. Xu* 105
Agroforestry: Water Use Efficiency / *James R. Brandle, Laurie Hodges, and Xinhua Zhou* 129
Alexandria Lake Maryut: Integrated Environmental Management / *Lindsay Beevers* 164
Groundwater: Treatment with Biobarrier Systems / *Chih-Ming Kao, Shu-Hao Liang, Yu-Chia Kuo, and R.Y. Surampalli* 1333
Irrigation Systems: Sub-Surface Drip Design / *Carl R. Camp, Jr. and Freddie L. Lamm* 1535
Irrigation: Sewage Effluent Use / *B.A. Stewart* 1570
Lakes: Restoration / *Anna Rabajczyk* 1588
Rain Water: Harvesting / *K.F. Andrew Lo* 2262

Water Pollution Problems (WAT) (*cont'd*)

Rivers: Restoration / *Anna Rabajczyk* ... 2307
Waste: Stabilization Ponds / *Sven Erik Jørgensen* ... 2632
Wastewater Treatment: Wetlands Use in Arctic Regions / *Colin N. Yates, Brent Wootton, Sven Erik Jørgensen, and Stephen D. Murphy* ... 2662
Wastewater Use in Agriculture / *Manzoor Qadir, Manzoor Qadir, Pay Drechsel, and Liqa Raschid-Sally* ... 2675
Water and Wastewater: Filters / *Sandeep Joshi* ... 2719
Water and Wastewater: Ion Exchange Application / *Sven Erik Jørgensen* ... 2734
Water Harvesting / *Gary W. Frasier* ... 2738
Wetlands / *Ralph W. Tiner* ... 2824
Wetlands: Biodiversity / *Jean-Claude Lefeuvre and Virginie Bouchard* ... 2829
Wetlands: Carbon Sequestration / *Virginie Bouchard* ... 2833
Wetlands: Conservation Policy / *Clayton Rubec* ... 2837
Wetlands: Constructed Subsurface / *Jan Vymazal* ... 2841
Wetlands: Methane Emission / *Anna Ekberg and Torben Røjle Christensen* ... 2850
Wetlands: Petroleum / *John V. Headley and Kerry M. Peru* ... 2854
Wetlands: Sedimentation and Ecological Engineering / *Timothy C. Granata and Jay F. Martin* ... 2859
Wetlands: Treatment System Use / *Kyle R. Mankin* ... 2862

Environmental Legislation and Policy (ELE)

Wastewater Use in Agriculture: Policy Issues / *Dennis Wichelns* ... 2681
Wastewater Use in Agriculture: Public Health Considerations / *Blanca Jimenez and Inés Navarro* ... 2694
Water: Cost / *Atif Kubursi and Matthew Agarwala* ... 2779

Environmental Technology (ENT)

Adsorption / *Puangrat Kajitvichyanukul and Jirapat Ananpattarachai* ... 61
Wastewater Treatment: Biological / *Shaikh Ziauddin Ahammad, David W. Graham and Jan Dolfing* ... 2645
Wastewater Treatment: Conventional Methods / *Sven Erik Jørgensen* ... 2657
Wastewater: Municipal / *Sven Erik Jørgensen* ... 2709
Water and Wastewater: Filters / *Sandeep Joshi* ... 2719
Water and Wastewater: Ion Exchange Application / *Sven Erik Jørgensen* ... 2734
Water Harvesting / *Gary W. Frasier* ... 2738

Integrated and Holistic Management (IMS)

Alexandria Lake Maryut: Integrated Environmental Management / *Lindsay Beevers* ... 164
Chesapeake Bay / *Sean M. Smith* ... 474
Eutrophication / *Claude Amiard-Triquet and Sven Erik Jørgensen* ... 1074
Everglades / *Kenneth L. Campbell, Rafael Munoz-Carpena, and Gregory Kiker* ... 1080
Wastewater and Water Utilities / *Rudolf Marloth* ... 2638
Wastewater Use in Agriculture / *Manzoor Qadir, Manzoor Qadir, Pay Drechsel, and Liqa Raschid-Sally* ... 2675
Wastewater Use in Agriculture: Policy Issues / *Dennis Wichelns* ... 2681
Wastewater Use in Agriculture: Public Health Considerations / *Blanca Jimenez and Inés Navarro* ... 2694
Water: Cost / *Atif Kubursi and Matthew Agarwala* ... 2779
Watershed Management: Remote Sensing and GIS / *A.V. Shanwal and S.P. Singh* ... 2816
Wetlands: Biodiversity / *Jean-Claude Lefeuvre and Virginie Bouchard* ... 2829
Wetlands: Carbon Sequestration / *Virginie Bouchard* ... 2833
Yellow River / *Zixi Zhu, Ynuzhang Wang, and Yifei Zhu* ... 2884

Pollution Problems and Their Sources (PSS): Generally Point Sources

Aral Sea Disaster / *Guy Fipps* 300
Groundwater: Mining / *Hugo A. Loaiciga* 1284
Groundwater: Mining Pollution / *Jeffrey G. Skousen and George F. Vance* 1289
Irrigation: Erosion / *David L. Bjorneberg* 1551
Irrigation: Return Flow and Quality / *Ramón Aragues and Kenneth K. Tanji* 1557
Irrigation: River Flow Impact / *Robert W. Hill and Ivan A. Walter* 1563
Irrigation: Saline Water / *B.A. Stewart* 1568
Wastewater Use in Agriculture / *Manzoor Qadir, Manzoor Qadir, Pay Drechsel, and Liqa Raschid-Sally* 2675
Wastewater Use in Agriculture: Public Health Considerations / *Blanca Jimenez and Inés Navarro* 2694
Wastewater: Municipal / *Sven Erik Jørgensen* 2709
Water Quality: Timber Harvesting / *C. Rhett Jackson* 2770
Water Supplies: Pharmaceuticals / *Klaus Kümmerer* 2776

Pollution Problems and Their Sources (PSS): Generally Non-Point and Diffuse Sources

Acid Rain / *Umesh Kulshrestha* 5
Acid Rain: Nitrogen Deposition / *George F. Vance* 20
Alexandria Lake Maryut: Integrated Environmental Management / *Lindsay Beevers* 164
Aquatic Communities: Pesticide Impacts / *David P. Kreutzweiser and Paul K. Sibley* 291
Chesapeake Bay / *Sean M. Smith* 474
Coastal Water: Pollution / *Piotr Szefer* 489
Cyanobacteria: Eutrophic Freshwater Systems / *Anja Gassner and Martin V. Frey* 538
Drainage: Hydrological Impacts Downstream / *Mark Robinson and D.W. Rycroft* 584
Estuaries / *Claude Amiard-Triquet* 1063
Eutrophication / *Claude Amiard-Triquet and Sven Erik Jørgensen* 1074
Everglades / *Kenneth L. Campbell, Rafael Munoz-Carpena, and Gregory Kiker* 1080
Giant Reed (*Arundo donax*): Streams and Water Resources / *Gretchen Coffman* 1182
Groundwater: Arsenic Contamination / *Abhijit Das, Bhaskar Das, Subhas Chandra Mukherjee, Shyamapada Pati, Rathindra Nath Dutta, Khitish Chandra Saha, Quazi Quamruzzaman, Mohammad Mahmudur Rahman, Tarit Roy Chowdhury, and Dipankar Chakraborti* 1262
Groundwater: Contamination / *Charles W. Fetter* 1281
Groundwater: Nitrogen Fertilizer Contamination / *Lloyd B. Owens and Douglas L. Karlen* 1302
Groundwater: Pesticide Contamination / *Roy F. Spalding* 1317
Groundwater: Saltwater Intrusion / *Alexander H.-D. Cheng* 1321
Inland Seas and Lakes: Central Asia Case Study / *Andrey G. Kostianoy* 1436
Lakes and Reservoirs: Pollution / *Subhankar Karmakar and O.M. Musthafa* 1576
Lakes: Restoration / *Anna Rabajczyk* 1588
Mines: Acidic Drainage Water / *Jerry M. Bigham and Wendy B. Gagliano* 1688
Nutrient—Water Interactions / *Ardell D. Halvorson* 1823
Rain Water: Atmospheric Deposition / *Zaneta Polkowska* 2249
Rivers and Lakes: Acidification / *Agniezka Gałuszka and Zdzisław M. Migaszewski* 2291
Rivers: Pollution / *Bogdan Skwarzec* 2303
Sea: Pollution / *Bogdan Skwarzec* 2357
Water Quality and Quantity: Globalization / *Kristi Denise Caravella and Jocilyn Danise Martinez* 2740
Water Quality: Soil Erosion / *Andreas Klik* 2755
Yellow River / *Zixi Zhu, Ynuzhang Wang, and Yifei Zhu* 2884

Foreword

Environmental management started at the beginning of the Neolithic Era, when man began modifying his natural environment to deter predators and prepare land for agriculture. Nevertheless, it was not until the second half of the 20th century that environmental management was identified as a discipline in need of systematic study. During most of written history, management cases in agriculture have been researched and implemented, but these were isolated, local in character, and dependent upon the technology of the day. The international component was added at the 1972 Stockholm United Nations Conference on the Human Environment.

During the next four decades, university programs covering environmental management began to flourish, resulting in the full spectrum of university education available today—spanning undergraduate, graduate, and post-graduate programs. Parallel to increased academic activity, the International Organization for Standardization (ISO) formed Technical Committee No. 207 on Environmental Management, in response to objectives of sustainable development adopted at the United Nations Conference on Environment and Development (Rio de Janeiro, 1992). Gradually, over the next two decades, the committee issued the ISO 14000 family of international standards on environmental management. The work is not yet complete and more standards will be issued in the future.

Presently, environmental management is maturing into a truly comprehensive interdisciplinary activity including natural, medical, technical, economic, and social sciences. Although some cases have considered (and implemented) almost all of these aspects, the majority of previous studies propose a specific solution, which would have been different if the missing disciplines had been included.

This encyclopedia presents a unique collection of almost 400 issues, case studies, and practices. As such, it will be of critical importance for the sustainability of human development as examinations of current state-of-the-art techniques have at last been made available, thus making possible a much faster development towards more refined solutions in the future. Whether discussing the management of air, water, soil quality, or indeed any other issue, including those of global importance, this encyclopedia will be used frequently for years to come by all who hold a stake in the environment, including educational programs related to environmental management, industry, environmental regulatory bodies, and non-governmental organizations.

It is with greatest pleasure that I congratulate the editor S.E. Jorgensen, all contributors, and the editorial office for creating this monumental work.

Tarzan Legović
Secretary General, International Society for Ecological Modelling
Professor and Chairman, Division of Marine and Environmental Research, R. Bošković Institute
Zagreb, Croatia

Preface

The aims and scope of this encyclopedia are to present the basic knowledge for performance of an integrated environmental and ecologically sound management system. This encyclopedia cannot include all environmental problems, but it has been attempted to cover at least more than 90–95% of the most important problems and their sources. Environmental problems can be considered as health problems of ecosystems and solutions of the problems require therefore an as detailed as possible diagnosis, which we can develop by the use of the following toolboxes: ecological models, ecological indicators, and ecological services. The encyclopedia presents all three diagnostic tools, but a more comprehensive overview of the tool boxes can be found for ecological models in *Handbook of Ecological Models Used in Ecosystem and Environmental Management* (2011), edited by S.E. Jørgensen; if surface modelling would be beneficial to use, in *Surface Modeling, High Accuracy and High Speed Methods* (2011), edited by Tian-Xiang Yue; and for ecological indicators in *Handbook of Ecological Indicators for Assessment of Ecosystem Health* (2010), edited by S.E. Jørgensen, Fu-Liu Xu, and R. Costanza. It can therefore be recommended also to consult these handbooks to obtain more knowledge about the diagnostic tools. In addition to the problems and their sources, emphasis is also placed on solutions to the problems, where we distinguish four possibilities: environmental technology, ecological technology, cleaner technology, and environmental legislation. All four tools have been presented with reference to the environmental problems that they can solve, but the most comprehensive overview is given for the first three possibilities, while environmental legislation is covered with less detail, because it is more difficult due to significant variations from country to country.

Integrated and ecologically sound environmental management means that the definition of the problems, the sources and the ecosystems affected, the diagnosis, and the possible solutions are integrated and used as an entity to ensure that we do not solve one problem while creating two others and that we obtain a clear improvement of the health problems of the affected ecosystem. The encyclopedia also contains articles or entries that focus on the integration of the problem, the source, the ecosystem, the diagnosis, and the solution. Moreover, a few entries cover background knowledge needed to interpret the problem and the diagnosis. The details of the topical classification are given in the introduction to the Topical Table of Contents. The scope is that the user of the encyclopedia can find an overview of the above mentioned steps to integrated and ecologically sound environmental management. Each article has a comprehensive reference list, which the user of course should utilize to implement the proposed procedure.

To launch an encyclopedia requires a very wide knowledge of many details, which of course is impossible for one person to provide. This encyclopedia would therefore not have been possible without a very knowledgeable and skilled editorial board, covering different aspects of environmental management. Furthermore, without the contribution of all the authors, it would of course not have been possible to produce this encyclopedia. I would therefore underline my appreciation of the tremendous and important work by the editorial board and by the authors of the articles. In this context I also mention that the staff of Taylor & Francis has been very valuable in providing all the pieces and placing them correctly in the large mosaic of the encyclopedia.

Sven Erik Jørgensen
Editor
Copenhagen

How to Use This Encyclopedia

Integrated Ecological and Environmental Management

Integrated ecological and environmental management means that the environmental problems are viewed from a holistic angle considering the ecosystem as an entity and considering the entire spectrum of solutions, including all possible combinations of proposed solutions. Articles in the encylopedia focusing on *integrated* ecological and environmental management using *holistic* approaches are indciated with IMS. The experience gained from environmental management over the last forty years has clearly shown that it is important not to consider solutions of single problems but to consider *all* the problems associated with a considered ecosystem simultanously and evaluate *all* the solution possibilities proposed by the relevant disciplines at the same time, or, expressed differently: to observe the forest and not the single trees. The experience has clearly underlined that there is no alternative to *integrated* management, at least not on a long-term basis. Fortunately, new ecological sub-disciplines have emerged that offer tool boxes to perform integrated ecological and environmental management.

Integrated ecological and environmental mangement of today consists of a seven-step procedure that is proposed in Jørgensen and Nielsen:[1]

- Define the problem
- Determine the ecosystems involved
- Find and quantify all the sources to the problem
- Set up a diagnosis to understand the relation between the problem and the sources
- Determine all the tools that could be used to solve the problem(s)
- Select a solution or a combination of solutions and implement the selected solutions
- Follow the recovery process to ensure that the problems have been solved

When an environmental problem has been detected, it is necessary to determine and quantify the problem and all the sources to the problem. The entries dealing with this step of the procedure are marked with PPS. It requires the use of analytical methods or a monitoring program. To solve the problem a clear diagnosis has to be developed: what is the problem that the ecosystems are facing and what are the relationships between the sources and their quantities and the determined problem? Or expressed differently: to what extent do we solve the problems by reducing or eliminating the different sources to the problems? Entries dealing with these two questions are denoted DIA. A holistic integrated approach is needed in most cases because the problems and the corresponding ecological changes in the ecosystems are most often very complex, particularly when several environmental problems are interacting. When the first green wave started in the mid-1960s, the tools to answer these questions, which we today consider as very obvious questions in an environmental management context, were not yet developed. We were able to carry out the first three points on the above shown list but had to stop at point 4 and could at that time only recommend eliminating the source completely or almost completely by using the methods that were available at that time—that is, environmental technology (covered by entries marked with ENT) which at that time was at a slightly lower level than today.

Due to the development of several new ecological sub-disciplines, it is today possible to accomplish the fourth to sixth points, presented here tool boxes that we can apply today to carry out the the fourth to sixth points. They are the result of the emergence of six new ecological sub-disciplines: for a better diagnosis, ecological modelling, ecological indicators, ecological services (all three sub-disciplines are covered by the entries denoted DIA) and for more tools to solve the problems, ecological engineering (also denoted ecotechnology, covered by entries named ECT), cleaner production (covered by entries marked CLT), and environmental legislation (articles focusing on environmental legislation are named ELE).

Tool Boxes Available Today to Develop an Ecological-Environmental Diagnosis

A massive use of ecological models as an environmental management tool was initiated in the early 1970s. The idea was to answer the question: what is the relationship between a reduction of the impacts on ecosystems and the observable, ecological improvements? The answer could be used to select the pollution reduction that the society would require and could effort economically. Ecological models were developed as early as the 1920s by Steeter–Phelps and Lotka Volterra (see for instance Jørgensen and Fath),[2] but in the 1970s started a much more consequential use of ecological models and many more models of different ecosystems and different pollution problems were developed. Today we have at least a few models available for all combinations of ecosystems and environmental problems. The journal *Ecological Modelling* was launched in 1975 with an annual publication of 320 pages and about 20 papers. Today, the journal publishes 20 times as many papers. This means that ecological modelling has been adopted as a very powerful tool in ecological-environmental management to cover particularly the fourth point in the integrated ecological and environmental mangement procedure proposed previously. The encyclopedia contains several articles about the use of ecological and environmental models, but a more comprehensive overview of available models can be found in Jørgensen.[3] If surface modelling would be more beneficial to use, see Tian-Xiang Yue.[4]

Ecological models are powerful management tools but they are not easily developed. They require in most cases good data, which are resource- and time-consuming to provide. About 20 years ago, it was therefore proposed to use another tool box, one that required less resources to provide a diagnosis, namely, ecological indicators (see, for instance, Costanza, Norton and Haskell[5]). Ecological indicators can be classified as shown in Table 1 according to the spectrum from a more detailed or reductionistic view to a system or holistic view (see Jorgensen[16]). The reductionistic indicators can, for instance, be a chemical compound that causes pollution or specific species. A holistic indicator could, for instance, be a thermodynamic variable or biodiversity. Indicators can either be measured or they can be determined by the use of a model. In the latter case, time consumption is of course not reduced by the use of indicators instead of models, but the models get a more clear focus on one or more specific state variables, namely, the selected indicator, which best describes the problems. In addition, indicators are usually associated with very clear and specific health problems of the ecosystems, which of course is beneficial in environmental management. Several articles in the encyclopedia focus on the use of indicators for developing a diagnosis, but a more comprehensive overview of the application of indicators can be found in Jørgensen, Xu and Costanza.[7]

Over the last 10–15 years, the services offered by ecosystems to society have been discussed and attempts have been made to calculate the economic values of these services.[8] A diagnosis that would focus on the services actually reduced or eliminated due to environmental problems could be developed. Another possibility of using ecological services to assess the environmental problems and their consequences could be to determine the economic values of the overall ecological services offered by the ecosystems and then compare them with what is normal for the type of ecosystems considered. Jørgensen[9] has determined the values of all the services offered by various ecosystems by the use of the ecological holistic indicator eco-exergy expressing the total work capacity. It is a good measure of the total amount of ecological services as all services require a certain amount of free energy, i.e. energy that can do work. The values published in Jørgensen[9] are shown in Table 2 and can be used for the above indicated comparison. The eco-exergy is found as presented in the articles about this indicators, but see also Jørgensen et al.[10] and Jørgensen.[11] The use of eco-exergy as an indicator to find the value of the ecosystem services in this context is beneficial, because the development of sustainability can be described as maintenance of the total work capacity that is at our disposal.[12]

Assessments of ecosystem services frequently use ecological indicators. The indicators can be determined and followed by the use of models, and models can determine the reduced or lost ecological services of ecosystems. The three diagnostic tool boxes are closely related with other words and obviously the use of all three tool boxes will give the most complete diagnosis. They are, however, all based on observations, which means that they are dependent on a solid monitoring program. Articles in the encyclopedia about monitoring are denoted DIA, as monitoring works hand in hand with the diagnostic tools. On the other hand, the resources available for environmental management are always limited, which means that it is

Table 1 Classification of ecological indicators.

Level	Example
Reductionistic (single) indicators	PCB, Species present/absent
Semiholistic indicators	Odum's attributes
Holistic indicators	Biodiversity/ecological network
"Super-holistic"	Thermodynamic indicators as eco-exergy and energy

Table 2 Work capacity used to express the ecosystem services for various types of ecosystems.

Ecosystem	Biomass (MJ/m^2 y)	Information factor (β-value)	Work capacity (GJ/ha y)
Desert	0.9	230	2,070
Open sea	3.5	68	2,380
Coastal zones	7.0	69	4,830
Coral reefs, estuaries	80	120	960,000
Lakes, rivers	11	85	93,500
Coniferous forests	15.4	350	539,000
Deciduous forests	26.4	380	1,000,000
Temperate rainforests	39.6	380	1,500,000
Tropical rainforests	80	370	3,000,000
Tundra	2.6	280	7,280
Croplands	20.0	210	420,000
Grassland	7.2	250	18,000
Wetlands	18	250	45,000

It is calculated as biomass * the information factor.

hardly possible to apply all three tool boxes in all cases, and also because in most cases it would require a comprehensive monitoring program. It is therefore necessary in many cases to make a choice. If an ecological model is developed, anyhow, to be able to give more reliable prognoses, it is of course natural to apply the developed model and it may be beneficial in addition to select one or a few indicators to focus more specifically on a well-defined problem. If a model is not available but a monitoring program has to be developed, one would have to direct the observations to encompass the state variables that can be applied to assess the indicators that are closely related to the defined health problems. If society is dependent on specific ecological services of the ecosystem, it would be natural to assess to what extent these services are reduced or lost, possibly supplemented with health indicators that are particularly important for the maintenance of these services. The choice of tool boxes is therefore a question about the available resources and the specific case and problem.

Tool Boxes Available Today to Solve Environmental Problems

The tool box environmental technology (articles are denoted ENT) was the only methodological discipline available to solve environmental problems 45 years ago, when the first green waves started in the 1960s. This tool box was able to solve only point source problems, and sometimes at a very high cost. Today, fortunately, we have additional tool boxes that can solve diffuse pollution problems or find alternative solutions at lower costs when environmental technology would be too expensive to apply. As for diagnostic tool boxes, they are developed on basis of new ecological sub-disciplines.

To solve environmental problems today we have four tool boxes:

- Environmental technology (denoted ENT)
- Ecological engineering, also denoted ecotechnology (articles are marked with ECT)
- Cleaner production; under this heading we would also in this context include industrial ecology (see the articles denoted CLT)
- Environmental legislation (articles denoted ELE)

Environmental technology came about with the emergence of the first green waves about 45 years ago. Since then, several new environmental-technological methods (covered by articles named ENT) have been developed, and all the methods have been streamlined and are generally less expensive to apply today. There is and has been, however, an urgent need for other alternative methods to solve the entire spectrum of environmental problems at an acceptable cost. Environmental management today is more complicated than it was 45 years ago because of the many more tool boxes that should be applied to find the optimal solution and because global and regional environmental problems have emerged. The use of tool boxes and the more complex situation today is illustrated in Fig. 1.

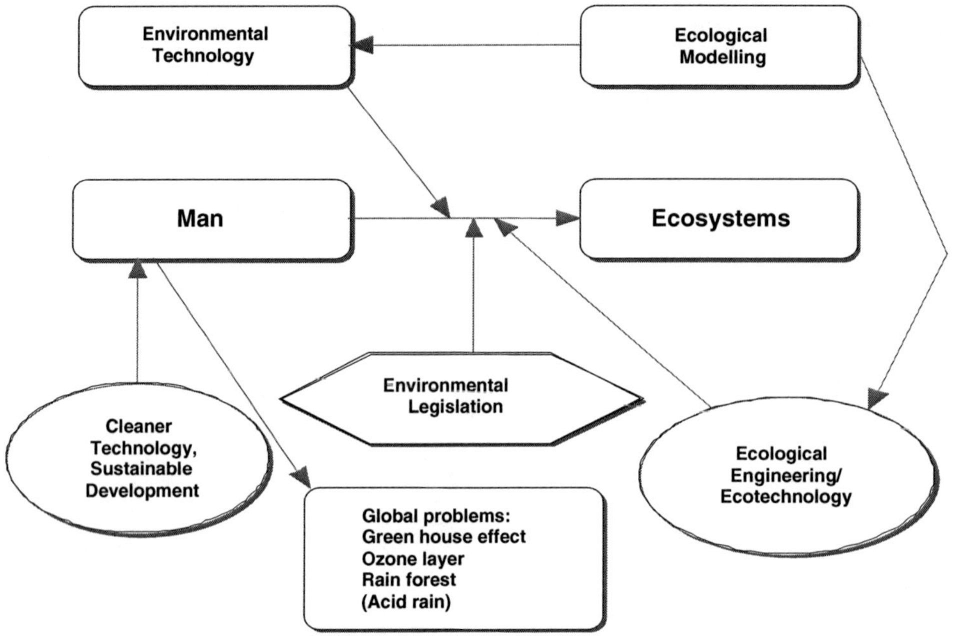
Environmental Technology
Ecological Modelling
Man
Ecosystems
Environmental Legislation
Cleaner Technology, Sustainable Development
Global problems: Green house effect Ozone layer Rain forest (Acid rain)
Ecological Engineering/ Ecotechnology

Fig. 1 Conceptual diagram of the complex ecological-environmental management of today, where there are various tool boxes available to solve the problems and where the problems are local, regional and global.

The tool box containing ecological engineering methods (the articles are indicated with ECT) was developed in the late 1970s. Ecological engineering is defined as the designing of sustainable ecosystems that integrate human society with its natural environment to the benefit of both.[13] It is an engineering discipline that operates in ecosystems, which implies that it is based on both design principles and ecology. The tool box contains four classes of tools:

- Tools that are based on the use of natural ecosystems to solve environmental problems (e.g., the use of wetlands to treat agricultural drainage water)
- Tool that are based on imitations of natural ecosystems (e.g., construction of wetlands to treat waste water)
- Tools that are applied to restore ecosystems (e.g., restoration of lakes by the use of biomanipulation)
- Ecological planning of the landscape (e.g., the use of agro-forestry)

The introduction of ecological engineering has made it possible to solve many problems that environmental technology could not solve, such as non-point pollution problems and a fast restoration of deteriorated ecosystems.

Some environmental problems, however, cannot be solved without more strict environmental legislation and for some problems a global agreement may be needed to achieve a proper solution, for instance by out-phasing the use of Freon to stop or reduce the destruction of the ozone layer. Notice also that environmental legislation (articles are de noted ELE) requires an ecological insight to assess the required reduction of the emission that is needed through the introduction of environmental legislation.

As environmental legislation has been tightened, it has been more and more expensive to treat industrial emissions, and the industry has of course considered whether it is possible to reduce the emissions by other methods at a lower cost. That has led to the development of what is called cleaner production (CLT is used to indicate the articles focusing on cleaner technology), which means the idea to produce the same product by a new method that would give a reduced emission and therefore less costs for the pollution treatment. New production methods have been developed by the use of innovative technology that has created a completely new method to produce the same product with less environmental problems. This has been the case particularly in energy technology, where a wide spectrum of methods are being developed as alternatives to the use of fossil fuel. Other emission reductions have been developed with the use of ecological principles on the industrial processes, for instance, recycling and reuse. In many cases it has also been possible to reduce environmental problems by identifying unnecessary waste. Industrial ecology could, in the author's opinion, be defined as the use of ecological principles in production, such as recycling, reusing and using holistic solutions to achieve a high efficiency in the general use of the resources. Industrial ecology today is, however, used to cover the use of waste from one production in another production.

Today, with the four tool boxes with environmental management solutions, it is possible to solve any environmental problem and often at a moderate cost and sometimes even at a cost which makes it beneficial to solve the problem properly. As is the case for diagnostic tool boxes, tool boxes with problem solution tools are, as indicated, rooted in recently developed ecological sub-disciplines that are named after the tools: ecological engineering, environmental legislation and cleaner technology.

Follow the Recovery Process

Environmental management is only complete if the environmental problem and the ecosystem are followed carefully after the tool boxes have been applied. It is usually not a problem because it is a question of providing the observations needed to follow the prognoses of the

- eventually developed ecological model
- the selected ecological indicators
- the recovery of the ecological services of the ecosystem (which can be done by focusing on a specific service or on the values of all the ecological services offered by the ecosystem)

Conclusion

From this review of the up-to-date integrated environmental management procedure, it is possible to conclude that the following steps are recommended:

- Define, preferably quantitatively, the problem(s), the ecosystem affected and the sources of the problem. Use articles here marked PSS. They correspond to the first three steps in the beginning of this introduction.
- Set up a diagnosis by using ecological models, ecological indicators and assessments of ecological services. Use the articles here marked by DIA.
- Go through all the possible tools that could be implemented to solve the problem. The encyclopedia presents here a particularly wide spectrum of articles, covering:
  - environmental technology, marked by ENT
  - ecotechnology, marked by ECT
  - cleaner technology, included industrial ecology, marked by CLT
  - environmental legislation, marked by ELE

Integrated, up-to-date environmental management requires the use of all seven of the presented tool boxes and would not be possible if these tool boxes were not developed as a result of recently emergent ecological sub-disciplines: ecological modelling, ecological engineering, application of ecological indicators, cleaner technology and industrial ecology. These ecological sub-disciplines are therefore crucial for environmental management today and they form an indispensable bridge between ecology and environmental management—between the basic science of ecology and its application in practical environmental management. IMS denotes articles that uncover the possibilities of integrating the various tool boxes to make an integrated and holistic management.

The encyclopedia gives significant background knowledge to be able to cover the seven presented steps to achieve an integrated, holistic environmental and ecological management of many of the actual environmental problems, from pollution problems rooted in our extensive use of fossil fuel and our water-, air- and solid-waste problems to the erosion and non-point agricultural pollution of pesticides and fertilizers. COV, which has not been mentioned yet, denotes articles that give comparative overviews of important topics for environmental management or background knowledge that are important for the evaluation of environmental problems.

References

1. Jørgensen, S.E.; Nielsen, S.N. Tool boxes for an integrated ecological and environmental management. Ecological Indicators, in press.
2. Jørgensen, S.E.; Fath, B. *Fundamentals of Ecological Modelling*, 4th ed., Elsevier: Amsterdam, 400 pp., 2011.
3. Jørgensen, S.E. *Handbook of Ecological Models Used in Ecosystem and Environmental Management*, 2011.
4. Yue, T.-X. Surface modeling. High Accuracy and High Speed Methods, 20.

5. Costanza, R.; Norton, B.G.; Haskell, B.D. *Ecosystem Health: New Goals for Environmental Management*, Island Press: Washington, 270 pp., 1992.
6. Jørgensen, S.E. *Integration of Ecosystem Theories: A Pattern*, Kluwer: Dordrecht, 386 pp.
7. Jørgensen, S.E.; Xu, F.-L.; Costanza, R. *Handbook of Ecological Indicators for assessment of Ecosystem Health*, 2nd ed., CRC: Boca Raton, FL, 484 pp., 2010.
8. Costanza, R.; d'Arge, R.; de Groot, R.; Farber, S.; Grasso, M.; Hannon, B.; Naeem, S.; Limburg, K.; Paruelo, J.; O'Neill, R.V.; Raskin, R.; Sutton, P.; van den Belt, M. The value of the world's ecosystem services and natural capital. Nature **1997**, *387*, 252–260.
9. Jørgensen, S.E. Ecosystem services, sustainability and thermodynamic indicators. Ecological Complexity **2010**, *7*, 311–313.
10. Jørgensen, S.E.; Ladegaard, N.; Debeljak, M.; Marques, J.C. Calculations of exergy for organisms. Ecological Modelling **2005**, *185*, 165–176.
11. Jørgensen, S.E. *Introduction of Systems Ecology*, CRC: Boca Raton, FL, 320 pp., 2012.
12. Jørgensen, S.E. *Eco-Exergy as Sustainability*, WIT: Southampton, 220 pp., 2006.
13. Mitsch, W.J.; Jørgensen, S.E. *Ecological Engineering and Ecosystem Restoration*, John Wiley: New York, 410 pp., 2004.

About the Editor-in-Chief

Sven Erik Jørgensen

Institute A, Section of Environmental Chemistry, Copenhagen University, Copenhagen, Denmark

Dr. Sven Erik Jørgensen is a professor of environmental chemistry at Copenhagen University. He received a doctorate of engineering in environmental technology and a doctorate of science in ecological modeling. He is an honorable doctor of science at Coimbra University, Portugal, and at Dar es Salaam University, Tanzania. He was editor-in-chief of *Ecological Modelling* from the journal's inception in 1975 to 2009. He has also been the editor-in-chief of the *Encyclopedia of Ecology*. In 2004 Dr. Jørgensen was awarded the prestigious Stockholm Water Prize and the Prigogine Prize. He was awarded the Einstein Professorship by the Chinese Academy of Science in 2005. In 2007 he received the Pascal medal and was elected a member of the European Academy of Science. He has written close to 350 papers, most of which have been published in international peer-reviewed journals. He has edited or written 64 books. Dr. Jørgensen has given lectures and courses in ecological modeling, ecosystem theory and ecological engineering worldwide.

Encyclopedia of Environmental Management

Volume I

Pages 1–734

Acaricides–Energy Conversion

Acaricides – Adsorption
Agricultural – Alexandria
Allelochemics – Alternative
Aluminum – Bioindicators
Biological – Cadmium
Carbon – Cyanobacteria
Desertification – Ecosystems
Endocrine – Energy Conversion

Acaricides

Doug Walsh
Washington State University, Prosser, Washington, U.S.A.

Abstract
Acaricides are pesticides applied to suppress populations of pest mites and ticks in the order Acari. Acaricide use has increased substantially in production agriculture, honeybee colonies, poultry, livestock, and domestic animals since the mid-20th century. Advances in production agriculture and consolidation of livestock and poultry operations and direct and indirect transport and introduction of exotic species have intensified agricultural crop and ornamental plant damage and honeybee colonies from mite infestation and propensity for mite and tick infestation to occur in poultry and livestock. The overwhelming majority of mites and ticks are economically benign. Unfortunately, some species of mites and ticks directly affect humans as parasites, vectors of disease, and producers of allergens. Mites and ticks are responsible for millions of dollars worth of economic losses each year worldwide. Acaricides come in many forms and can be applied as dusts, gases, or liquids. Some are insoluable while others are locally systemic or completely systemic in plants or animals. Some acaricides could be considered relatively harmless while others can be extremely toxic and long lasting. This entry details most of the major classes of acaricides used in plant agriculture, commercial bee keeping, and animal agriculture.

INTRODUCTION

Approximately 45,000 species of mites and hundreds of species of ticks are described worldwide. Many thousands of species still remain unidentified. About half are plant-feeding species, and among these, about half are in the superfamily Eriophyoidea (gall, bud, and rust mites). Most of the other plant-feeding mites are classified into the superfamilies Tetranychoidea and Tarsonemidae. The superfamily Tetranychoidea includes the economically important spider, flat, and fowl mites, and the superfamily Tarsonemidae includes the economically important broad, cyclamen mites and Varroa mites. Over another 3000 mite species are loosely classified in the order Astigmata. Economically important species include feather and scabies mites. Ticks are placed in the superfamily Ixodoidea and all are ectoparasites (blood feeders) of vertebrate animals.[1]

Most mites are small to minute and mites are universally cryptic, making them difficult to detect. Often infestations are overlooked. Mites are often colonizers of new or disturbed habitats, and once established on a new host, mites possess biological characteristics that permit rapid increases in population abundance. Factors in most mites' lifestyle that lead to rapid population buildup include high egg production, various modes of reproduction (parthenogenesis, pedogenesis, and sexual), short life cycles, a myriad of dispersal techniques, and adaptability to diverse ecological conditions.[1] These traits combined with an exponential increase in worldwide transport of humans and plant and animal products will likely contribute to increased concerns over mite pests in the future.

In plant-based agriculture, Van de Vrie et al.[2] observed that outbreaks of mite populations were uncommon historically in systems where productivity languished far below the levels achieved in modern production agriculture. Spider mite populations stayed below observable levels due to natural regulation by predators, disease, and poor nutrition from low-quality host plants. Van de Vrie et al.[2] went on to observe that mite populations often experienced outbreaks in agroecosytems where production levels were bolstered by the use of synthetic inputs including fertilizers and pesticides. When crop production is optimized (i.e., not limited by water, nutrients, competition from weeds, or predatory mites and insects), the plants in production become an excellent food source for mite pests. Under these conditions, the developmental rate, fecundity, and life span of mites are increased and contribute to population outbreaks.

SPIDER MITE PESTS

A number of mite species can achieve pest status at high population abundance. Spider mites develop through several stages: egg, six-legged larva, eight-legged protonymph, deutonymph, and adult. Males typically reach maturity before females and will position themselves near developing quiescent females. When an adult female emerges, copulation will often occur immediately. Under optimal conditions, most mite species can develop from egg to adult in 6 to 10 days. Egg laying can begin as soon as one or two days after maturing to adults. Most spider mite species

Encyclopedia of Environmental Management DOI: 10.1081/E-EEM-120046161

overwinter as mated adult females. A notable exception is the European red mite that overwinters as eggs.[3]

A BIG DRAIN FROM THE FEEDING OF SUCH SMALL PESTS

At the microscopic level, significant quantities (relative to mite size) of plant juices pass through the digestive tract of spider mites as they feed on leaf tissues. McEnroe[4] estimated this volume at 1.2×10^{-2} microliters per mite per hour. This quantity represents roughly 50% of the mass of an adult female spider mite. Leisering and Beitrag[5] calculated that the number of photosynthetically active leaf cells that are punctured and emptied per mite is 100 per minute. In gut content studies of two-spotted mites, Mothes and Seitz[6] observed only thylakoid granules inside their digestive tract following feeding. The thylakoid grana on which spider mites focus their feeding are key photosynthetic engines in plant cells. The grana consist of 45%–50% protein, 50%–55% lipid, and minute amounts of RNA and DNA.[7] Water and other low-density plant cell contents are directly excreted.[4]

At the macroscopic level, damage from mite feeding can cause leaf bronzing, stippling, or scorching. For most horticultural crops, economic loss is caused by a drop in yield or quality due to reduction in photosynthesis.

SPIDER MITE OUTBREAKS ARE PROMOTED BY HOT, DRY WEATHER

Water stress, wind, and dust all contribute to the potential for mite outbreaks. When mite outbreaks occur, acaricide treatments are often used for suppression.

Varroa mites *Varroa jacobsoni* provide an ideal example of how rapidly a mite species can spread and exploit a new habitat. First recorded in honeybee colonies in Southeast Asia in 1904, Varroa mites are now pandemic. Varroa mites feed parasitically on an individual bee's hemolymph fluid, weakening the bee and often causing premature death. Mites attach themselves to foraging workers in order to spread themselves from one hive to another. This mite can severely weaken bees, and an unchecked mite population will almost certainly lead to the premature death of a honeybee colony. Apiculturists speculate that Varroa mite has contributed substantially to the collapse of feral honeybee populations worldwide.[8]

The northern fowl mite *Ornithonyssus sylviarum* is a common pest of domestic fowl and other wild birds commonly associated with human settlements. The nymphs and adults have piercing mouthparts and seek blood meals. Mite populations build up rapidly and a generation can be completed in 5 to 12 days. Several generations occur each year. Northern fowl mite spends virtually its entire life on the host bird.[9]

Deer and dog ticks *Ixodes scapularis* and *Dermacentor variabilis* are two common ticks to which acaricides are applied for on a consistent basis, especially since both are parasitic feeders on mammals. Deer ticks are a significant concern since they are the primary vector for Lyme disease.[10]

Mange or scabies in livestock is a skin condition caused by microscopic mites in or on the skin. The mites cause intense itching and discomfort that is associated with decreased feed intake and production. Scratching and rubbing result in extensive damage to hides and fleece. Mange mites are able to cause mange on different species of livestock but are somewhat host specific, thus infecting some species more severely than others. The three most important types of mange are as follows: sarcoptic mange, caused by *Sarcoptes scabiei* feeding; psoroptic mange, caused by *Psoroptes ovis* feeding; and chorioptic mange, caused by *Chorioptes bovis* feeding.[11] Infestations of these mites on their respective livestock, domestic pet, or human host will cause skin irritation and itching and leave entry points for secondary infections. Weight gain can be reduced in livestock, pets can lose hair and itch persistently, and disfigurement can occur in humans. Acaracides are often applied to suppress mite populations parasitizing humans, pets, and livestock.

SMOTHERING AGENTS

Solutions containing petroleum-based horticultural oils, vegetable oils, or agricultural soaps are applied to many crops and, occasionally, livestock. Application of these types of products kills spider mites through suffocation. Unfortunately, oils and soaps can prove phytotoxic to crop plants and are typically not effective on mites or ticks infesting livestock, pets, or humans. Mites on animal hosts are typically cryptic or subcutaneous, so acaricide coverage is an impediment to effective control.

ORGANOCHLORINES

Endosulfan and dicofol are organochlorine miticides registered for use on many crops. Unlike many other organochlorine pesticides, endosulfan and dicofol are relatively non-persistent in the environment. Organochlorine acaricides interfere with the transmission of nerve impulses and disrupt the nervous system of pest mites. Organochlorine acaricides are more effective at killing mites at warmer temperatures. Overuse of organochlorine acaricides in commercial situations has resulted in the development of tolerance in many pest mite populations. Organochlorines were used substantially in the mid-20th century, but regulatory actions and public health and environmental concern have eliminated their use in most developed countries (though some continue to use them). Lindane was commonly used for mange mite in pets, livestock, and humans. Only in limited circumstances is lindane still permitted as a pharmaceutical second-line treatment. However, use of lindane continues in developing countries due to its low cost, effectiveness, and persistence.

Organophosphates and Carbamates

Many organophosphate and carbamate pesticides have acaricidal activity. Studies have demonstrated significant mite control with applications of parathion, TEPP (tetraethyl pyrophosphate), and aldicarb. Spider mites are listed as target pests on many organophosphate and carbamate products. However, many mite populations following long-term exposure have developed resistance to the toxic effects of organophosphates.[12] Carbaryl, a common carbamate, continues to be a mainstay for mite control on livestock and poultry, but its use on domestic pets and households is no longer permitted in most developed countries. The use of carbaryl continues extensively in many developing countries in domestic settings.

ORGANOTINS

Miticides in this category were synthesized in the 1960s and 1970s and registered for commercial use in the United States in the 1970s. They have been used extensively for their ability to quickly knock down spider mite populations through contact activity. Fenbutatin-oxide has been used extensively since the 1970s. Cyhexatin was used extensively in the 1970s and 1980s, but regulatory actions have now limited its use. Efficacy of the organotin acaricides is improved with warmer weather. Overuse of cyhexatin during the mid-1980s led to the development of resistance in several cropping and livestock production systems. However, populations of pest mites can regain susceptibility to organotins following a period of non exposure.[13]

PROPARGITE

This acaricide has been used since the 1960s. It provides effective suppression of pest mites on many crops. Regulatory constraints have resulted in the cancellation of a number of uses. Identification of propargite as a dermal irritant has led to substantial increases in time required following application before re-entry is permitted into the treated site.

AMIDINES

Amitraz is a miticide that once had significant use in plant and animal agriculture. At present, its use is restricted to only a small subset of the domestic pet care market.

OVICIDES

Clofentazine and hexythiazox are selective carboxamide ovicidal acaricides. Spider mite eggs exposed to either compound fail to hatch. These acaricides are selective and aid in the conservation of populations of beneficial arthropods. These acaricides are typically used relatively early in the production season before mite populations reach outbreak conditions.

ANTIMETABOLITES

A number of miticidal compounds have been developed within the past 30 years. These include avermectins, pyridazinones, carbazates, and pyrroles. Pest mortality results from disruption of metabolic pathways typically within the mitochondria of nerve cells of spider mites.[14] Avermectins, ivermectins, and related compounds are fermentation products derived from mycelial extracts of *Streptomyces* species (reviewed by Burg and Stapley).[17] Avermectins are locally systemic (translaminar) in plant tissues,[15] and ivermectins can be applied dermally, by injection, suppository, or in a bolus to livestock and domestic pets. The ivermectins are the predominant parasiticide used in livestock production today. A number of products have been commercialized in recent years. Pyridaben is a pyridazinone recently registered for use on ornamentals and some tree crops. Bifenazate is a carbazate acaricide. It has a new mode of action that is not clearly understood, but it has proven toxicologically safe in mammalian studies. Bifenazate is registered on ornamentals and food products. Chlorfenapyr is a synthetic pyrrole that has been commercially available on cotton. Other uses are pending.

Synthetic Pyrethroids

Fenpropathrin and bifenthrin are two synthetic pyrethroid insecticides registered for control of spider mites in plant agriculture. Permethrin is registered for mite control on livestock and poultry. Mites have a well-documented history of rapidly developing resistance to pyrethroid insecticides in both plant and animal production systems, and resurgence of spider mite populations following pyrethroid application is typical.[18]

TETRONIC ACIDS

Spiromesifen and spirotetramat are acaricides in a recently introduced class of selective chemistry tetronic acids that exhibit a broad-spectrum insecticidal acaricidal activity against mites. Their mode of action is by inhibition of lipid biosynthesis that affects the egg and immature stages of mites. Foliar sprays of spiromesifen are translaminar in plants and effective against mites in many cropping systems. Spirotetramat has a relatively unique property among currently registered acaricides in that it is phloem systemic within the plant it is applied to. These two acaricides have recently entered the acaricide market and are quickly gaining in use in production agriculture for mite control.

APPLICATION TECHNOLOGY

Mite pests can prove difficult to control with acaricides due to their potential for high population abundance, small size, and propensity to live on the bottom surfaces of leaves or within the folds of plant tissues. Good acaricide spray coverage is essential for mite control, particularly for acaricides that kill on contact with the pest mite.

COMBATING MITICIDE RESISTANCE

Following repeated exposure, spider mite populations have a history of rapidly developing resistance to acaricides.[16] Alternating acaricides that have different modes of action reduces the potential for development of resistance to acaricides within specific modes of activity. Other techniques to discourage resistance development include spraying only when necessary and treating only infested portions of the crop. Organophosphate, carbamate, and pyrethroid insecticide applications can induce spider mite outbreaks. If possible, avoid early-season insecticide application or apply insecticides that are less disruptive to beneficial arthropods. Careful selection and use of insecticides can potentially reduce the number of miticide applications required later in the season.

REFERENCES

1. Krantz, G.W.; Walter, D.E., Eds. *A Manual of Acarology*, 3rd Ed.; Texas Tech University Press: Lubbock, Texas, 2009; 807 pp.
2. Van de Vrie, M.; McMurtry, J.A.; Huffaker, C.B. Ecology of mites and their natural enemies. A review. III Biology, ecology, pest status, and host plant relations of tetranychids. Hilgardia **1972**, *41*, 345–432.
3. Bostanian, N.J. The relationship between winter egg counts of the European red mite *Panonychus ulmi* (Acari: Tetranychidae) and its summer abundance in a reduced spray orchard. Exp. Appl. Acarol. **2007**, *42*, 185–195.
4. McEnroe, W.D. The role of the digestive system in the water balance of the two-spotted spider mite. Adv. Acarol. **1963**, *1*, 225–231.
5. Leisering, R.; Beitrag, O. Beitrag zum phytopatologischen Wirkungsmeechanismus von Tetranychus urticae. Pflanzenschutz **1960**, *67*, 525–542.
6. Mothes, U.; Seitz, K.A. Functional microscopic anatomy of the digestive system of *Tetranychus urticae* (Acari: Tetranychidae). Acarologia **1981**, *22*, 257–270.
7. Noggle, G.R. The organization of plants. In *Introductory Plant Physiology*; Noggle, G.R., Fritz, G.J., Eds.; Prentice Hall: Englewood Cliffs, New Jersey, 1983; 9–38.
8. Mangum, W.A. Honey bee biology: The third annual report on the coexistence of my North Carolina bees with varroa mites. Am. Bee J. **2009**, *149*, 63–65.
9. Mullens, B.A. Temporal changes in distribution, prevalence and intensity of northern fowl mite (*Ornithonyssus sylviarum*) parasitism in commercial caged laying hens, with a comprehensive economic analysis of parasite impact. Vet. Parasitol. **2009**, *160*, 116–133.
10. Diuk-Wasser, M.A. Field and climate-based model for predicting the density of host-seeking nymphal *Ixodes scapularis*, an important vector of tick-borne disease agents in the eastern. U. S. Global Ecol. Biogeogr. **2010**, *19*, 504–514.
11. Vercruysse, J. World Association for the Advancement of Veterinary Parasitology (W.A.A.V.P.) guidelines for evaluating the efficacy of acaricides against (mange and itch) mites on ruminants [electronic resource]. Vet. Parasitol. **2006**, *136*, 55–66.
12. Smissaeret, H.R.; Voerman, S.; Oostenbrugge, L.; Reenooy, N. Acetylcholinesterases of organophosphate-susceptible and resistant spider mites *Tetranychus urticae*. J. Agric. Food Chem. **1970**, *18*, 66–75.
13. Hoy, M.A.; Conley, J.; Robinson, W. Cyhexatin and fenbutatin-oxide resistance in Pacific spider mite (Acari: Tetranychidae) stability and mode of inheritance. J. Econ. Entomol. **1988**, *81*, 57–64.
14. Hollingsworth, R.M.; Ahammadsahib, K.I.; Gadelhak, G.; McLaughlin, J.L. New inhibitors of Complex I of the mitochodrial electron transport chain with activity as pesticides. Biochem. Soc. Trans. **1994**, *22*, 230–233.
15. Walsh, D.B.; Zalom, F.G.; Shaw, D.V.; Welch, N. C. Effect of strawberry plant physiological status on the translaminar activity of avermectin B1 and its efficacy on the two-spotted spider mite *Tetranychus urticae* Koch (Acari: Tetranychidae). J. Econ. Entomol. **1996**, *89* (5), 1250–1253.
16. Leeuwen, T.V.; Dermauw, W.; Tirry, L.; Vontas, J.; Tsagkarakou, A. Acaricide resistance mechanisms in the two-spotted spider mite *Tetranychus urticae* and other important Acari. Insect Biochem. Mol. Biol. **2010**, *40*, 563–572.
17. Burg, R.W., and E.O. Stapley. Isolation and characterization of the producing organism. In W.C. Campbell (ed) Ivermectin and Abamectin. Springer-Verlag, New York, N.Y. 1989. pp. 24–32.
18. Leigh, T.F. Cotton. In W. Helle and M.W. Sabelis (eds.) World Crop Pests: Spider Mites. Elsevier Press, Amsterdam, the Netherlands. 1990. pp. 349–358.

Acid Rain

Umesh Kulshrestha
School of Environmental Sciences, Jawaharlal Nehru University, New Delhi, India

Abstract

After the industrial revolution, increased emissions of SO_2 and NO_x from fossil fuel combustion have resulted in an acid rain problem. Earlier, acid rain was observed as a common phenomenon in North America and Europe, but recent studies show its spread in East Asia too, covering China, Japan, and Thailand. European data show that most of the acidity was intensified during 1955–1970, with a sudden increase in the mid-1960s. Scandinavia and Central and Southern Germany were among the worst-hit areas, whereas northeastern United States and southeastern Canada were the most affected areas in North America. Acid rain resulted in loss of fish population in the lakes of Sweden, parts of southwest Norway, and eastern North America. In parts of Germany and other European countries, forest damage and loss of needles from pine and spruce trees was noticed. Beginning with the 1972 Conference on the Human Environment in Stockholm, successful efforts have been made by North America and Europe to control acid rain through SO_2 and NO_x emission control policies under various national and international cooperative programs. However, in the developing countries, SO_2 emissions are still on rise to achieve developmental targets. After the United States and Europe, China is the biggest consumer of fossil fuel where rapid increase in SO_2 and NO_x emissions is reported. South Asia is relatively safe from acid rain problems because of high buffering capacity of local dust in the atmosphere, which reacts with SO_2 and forms calcium sulfate. Ultimately, this results in higher pH of rain water. Similarly, acid rain is not an immediate problem in other parts of the world. However, consequences of increasing consumption of fossil fuel to meet energy demand in developing regions need to be monitored through national and international network programs. Apart from acidification of oceans by CO_2 rise, acid rain can also add to the process of acidification of coastal oceans, which might be damaging to the marine ecosystem in the future.

INTRODUCTION

Any form of precipitation (rain, snow, or hail) having high acidity is known as acid rain. The term "acid rain" was first used by Robert Angus Smith in his book *Air and Rain: The Beginnings of a Chemical Climatology*, published in England in 1872.[1] He had chemically analyzed the rainwater near Manchester and observed three types of rain composition—"that with carbonate of ammonia in the fields of distance, that with sulfate of ammonia in the suburbs and that with sulfuric acid or acid sulfate, in the town."

In broader perspectives, acid rain refers to wet deposition (rain, snow, hail, cloud water, fog, dew, or sleet) and dry deposition (absorption of SO_2, NO_x, other acidic gases and particles) of acidic compounds. High acidity is generally caused by higher levels of sulfuric and nitric acids. These acids are contributed by their precursor gases (SO_2 and NO_2), which are emitted by natural as well as anthropogenic sources. Natural sources include volcanoes, vegetation decay, various biological processes on the land, and oceans, while major anthropogenic sources of these gases are fossil fuel combustion and smelting of metal ores. In regions of North America, the rates of anthropogenic emissions of these two gases have gone up to 100 times more than the natural rates, adding to higher atmospheric acidity.[2] Acid rain has caused severe damage in Europe, North America, parts of China and Japan through acidification of lakes and other water bodies, decline of forests, acidification of soils, and corrosion of building materials.

HOW ACID RAIN HAPPENS

Pure water (H_2O) is neutral in nature, having a pH value of 7. Any aqueous solution having a pH higher than 7 is said to be alkaline, while one having a pH lower than 7 is known as an acidic solution. Rainwater in remote and unpolluted atmospheres has a slightly acidic pH of around 5.6 due to the presence of carbonic acid formed at equilibrium due to dissolution of atmospheric carbon dioxide in cloud water:[3]

$$CO_2 + H_2O \leftrightarrow H_2CO_3 \quad (1)$$

In water, carbonic acid is dissociated, forming bicarbonate ion:

$$H_2CO_3 \leftrightarrow HCO_3- + H^+ \quad (2)$$

Encyclopedia of Environmental Management DOI: 10.1081/E-EEM-120047106

Rainwater pH is further depressed to about 5.2 in unpolluted regions by organic acids. However, anthropogenic acid rain arises due to oxidation of SO_2 and NO_2 in the atmosphere to form sulfuric and nitric acids. There are a number of probable reactions for the oxidation of these gases involving both homogeneous and heterogeneous oxidation.[4] The gas-phase oxidation of these gases is initiated by reaction with hydroxyl radicals:

$$SO_2 + OH \rightarrow HOSO_2 \quad (3)$$

$$HOSO_2 + O_2 \rightarrow HO_2 + SO_3 \quad (4)$$

$$SO_3 + H_2O \rightarrow H_2SO_4 \quad (5)$$

Homogeneous aqueous-phase oxidation of SO_2 takes place by its dissolution and dissociation in water, forming equilibrium similar to CO_2:

$$SO_2 + H_2O \leftrightarrow SO_2H_2O \quad (6)$$

$$SO_2.H_2O \leftrightarrow HSO_3- + H^+ \quad (7)$$

$$HSO_3- \leftrightarrow SO_3^{2-} + H^+ \quad (8)$$

Gas-phase oxidation of NO_2 is faster than SO_2 by one order of magnitude:

$$NO_2 + OH \rightarrow HNO_3 \quad (9)$$

In addition, significant formation of nitric acid takes place through ozone and NO_3 radical reactions.

During daytime, NO_3 radical is formed as follows:

$$NO_2 + O_3 \rightarrow NO_3 + O_2 \quad (10)$$

NO_3 radical so formed reacts with NO_2 at nighttime, finally resulting in the formation of HNO_3:

$$NO_3 + NO_2 \leftrightarrow N_2O_5 \quad (11)$$

$$N_2O_5 + H_2O \rightarrow 2HNO_3 \quad (12)$$

At ambient levels of NO, its aqueous-phase oxidation is very slow due to its low solubility in water and also the dependence on NO_2 concentrations. It can be faster at higher NO_2 levels. However, the reaction follows the path

$$2NO_2 + H_2O \leftrightarrow 2H^+ + NO_3- + NO_2- \quad (13)$$

Heterogeneous oxidation of SO_2 and NO_2 involves gas–particle reactions. In the liquid phase, SO_2 is rapidly converted into sulfate by H_2O_2. SO_2 is also converted into sulfate on freshly emitted soot particles, but subsequently, the rate of oxidation is retarded due to saturation of soot particle surface. Preferable oxidation of SO_2 onto soil dust particles is reported in dusty regions where formation of calcium sulfate takes place instead of free sulfuric acid. NO_2 is also oxidized onto particles—for example, it reacts with NaCl of sea salt, forming $NaNO_3$ on the surface. However, over time, the surface is saturated, and the rate of oxidation becomes lower.

HISTORY OF ACID RAIN

The major cause of acid rain is the increased combustion of fossil fuels, which has been practiced at larger scale after the industrial revolution. The presence of sulfur compounds in the air of Sweden and England was realized in the 18th century. In fact, Robert Boyle, in 1692, mentioned in his book *A General History of the Air* the "nitrous and salino-sulphureous spirits" in the air.[4] The term "acid rain" was first used in 1872 by Robert Smith, who discovered acid rain in 1852 in the area surrounding Manchester. He referred to this term in a treatise on the chemistry of rain published in England in 1872. Robert Smith mentioned various factors such as coal combustion and the amount of rain affecting the precipitation. Unfortunately, this wonderful publication was overlooked until it was revisited and critiqued by Gorham in 1981.[1]

Acid rain attracted attention of the scientific community and society when Odén [6,7] reported that large-scale acidification of surface waters in Sweden could be attributed to pollution from the United Kingdom and central Europe. The worst-hit areas of acid rain were Scandinavia and Central and Southern Germany. In Europe, the rain pH was observed to be as low as 3.97 in Germany. Drastic loss of fish population was seen in lakes of Sweden and parts of southwest Norway. European data show that most of the acidity was intensified during 1955–1970, with a sudden increase in the mid-1960s. In parts of Germany and other European countries, forest damage and loss of needles from pine and spruce trees were noticed due to acid rain. Acid smog killed almost 4000 people in London in 1952.

SOURCES OF ACIDITY

As mentioned earlier, in natural conditions, atmospheric CO_2 when dissolved in water forms carbonic acid (H_2CO_3), which brings down the pH of water. Other gases such as SO_2 and NO_2 also form acids, viz., sulfuric (H_2SO_4) and nitric acid (HNO_3), respectively. The main cause of acid rain is excess contribution of H_2SO_4 and HNO_3 in precipitation due to anthropogenic sources, especially through combustion of coal and petroleum. Sulfur is present in significant amounts in fossil fuel (coal and petroleum), which is the major source of SO_2. Oxidation of SO_2 is accelerated by higher concentrations of H_2O_2 and O_3 found in polluted air. Martin and Barber[8] noticed that acidity of precipitation at several sites in England was the highest during spring, when O_3 concentrations were higher. During past century, huge consumption of fossil fuel in North America and Europe resulted in high SO_2 emissions.[9,10]

NATURAL ACIDITY CONTRIBUTED BY ORGANIC ACIDS

Apart from sulfuric and nitric acids, acidity in rainwater is also contributed by organic acids. Formic acid (HCOOH) and acetic acid (CH_3COOH) are the major species reported in rainwater, contributing around two-thirds of the total acidity at remote sites.[11] Generally, formic acid is found to dominate over acetic acid. A relatively higher contribution of organic acids is observed at tropical sites than at temperate ones. These organic acids are produced in gas phase by the oxidation of isoprene and terpenes emitted by the vegetation,[12,13] which are then scavenged by the rain. These are also formed through aqueous-phase oxidation of aldehydes. Sometimes, these acids are emitted by soils.[14] It is to be noted that organic acids may be important for pH in cloud and rainwater in some areas, but their contribution to acidification of soils and surface waters is small because these are quickly consumed by microorganisms.

SPREAD AND MONITORING OF ACID RAIN

Considering the degree of damage caused by acid rain, several efforts are made by European countries to monitor and control it. The European Air Chemistry Network was started in the early 1950s by Stockholm University in collaboration with the Swedish University of Agricultural Sciences. Both institutes served as centers for the network for the chemical analysis of samples. Originally, the purpose of this network was to study the depositions of plant nutrients to forest and agriculture systems. Under this network, continuous data related to chemical composition of precipitation have been available since 1955.[15] Later on, this network became part of the Swedish National Monitoring Programme. A Norwegian program called SNSF, "Acid Precipitation: Effects on Forests and Fish" was run up to 1980.[16] Immediately after the Stockholm conference in 1972, the European Organization for Economic Cooperation and Development (OECD), in 1978, established a network to monitor long-distance transport of pollutants and the impacts of European countries on their neighbors, known as the Cooperative Program for Monitoring and Evaluation of Long-Range Transmission of Air Pollutants in Europe (EMEP). Further, in 1983, the Convention on Long-Range Transboundary Air Pollution (CLRTAP) was signed by more than 30 countries, including the United States, Canada, and the European Union, to deal with transboundary air pollution.

The discovery of acid rain in Europe attracted attention of scientific community in North America too. Odén's study was followed up by the United Nations (UN).[17] This also led to the first international conference on acid rain in Columbus, Ohio, United States, in 1975. Later on, under the Acid Precipitation Act of 1980, U.S. Congress formed a national network called National Acid Precipitation Assessment Program (NAPAP), which supported the expansion of National Atmospheric Deposition Program (NADP) to monitor the trends in long-term precipitation chemistry and deposition. Further, the NADP was changed to the NADP National Trends Network, which has around 250 monitoring sites.

Similar to Europe and United States, Canada also experienced acid rains. Environment Canada has developed its program called the Canadian Air and Precipitation Monitoring Network (CAPMoN) to monitor the regional patterns and trends of atmospheric pollutants. including acid rain, smog, particulate matter, etc., Canada started the Canadian Network for Sampling Precipitation in 1978, which was renamed as the Air and Precipitation Network (APN). CAPMoN is the new name of APN (changed in 1983).

Later on, the spread of acid rain was also noticed in East Asia. After successful implementation of CLRTAP in Europe, the UN Conference on Environmental Development adopted to continue and share the experience gained from acid rain programs in Europe and North America and established the Acid Deposition Monitoring Network in East Asia (EANET) in 1993. This network includes Japan, Russia, China, the Republic of Korea, Mongolia, Thailand, Singapore, Cambodia, Lao People's Democratic Republic, Myanmar, Vietnam, the Philippines, Malaysia, and Indonesia.

Measurements through the long-term acid rain network have not been carried out extensively in other parts of the world such as India, Africa, and Latin America. However, programs such as the Composition of Atmospheric Aerosols and Precipitation in India and Nepal and the Composition of Asian Deposition (CAD), as part of the Regional Air Pollution in Developing Countries (RAPIDC) program funded by the Swedish International Development Cooperation Agency), were very effective in providing a summarized picture of the acid rain scenario in the Indian region.[18,19] The RAPIDC program was coordinated by the Stockholm Environment Institute, which facilitated international cooperation on air pollution issues to develop relevant knowledge to support decision making in Asia and Africa. The CAD program was a part of the International Global Atmospheric Chemistry/Deposition of Biogeochemically Important Trace Species (IGAC/DEBITS) activities of the International Geosphere–Biosphere Programme, which focused on good-quality measurements at rural sites in Asia to produce high-quality data so as to understand the Asian wet deposition scenario. In the African region, the IGAC/DEBITS–Africa program has its network of 10 stations for the measurement of wet and dry depositions at selected sites.

In Australia, acid rain studies have been carried out under the Commonwealth Scientific and Industrial Research Organisation (CSIRO) network of sites.[20] Globally, the 1989

Table 1 Change in pH of precipitation in western Europe during the 1950s and 1970s.

Country	pH in 1950s	pH in 1970s
Southern Norway	5.0–5.5	4.7
Northern Sweden	5.5–6.0	4.3
Southern Sweden	5.5–6.0	4.3
Southeast England	4.5–5.0	4.2

Source: Environmental Resources Ltd. Pearce.[24]

initiative of the World Meteorological Organization, under the Global Atmospheric Watch (GAW), is carrying out precipitation measurements at around 80 stations in different countries. Earlier, GAW used to be known as the Background Air Pollution Monitoring Network (BAPMoN).

REGIONAL ACIDITY OF PRECIPITATION

United States

As described by Gibson,[10] northeastern United States and southeastern Canada were the most affected areas by acid rain. Likens[21] was the first who evaluated the 1955–1956 and 1972–1973 data and found a significant increase in acidity in northeastern United States and southeastern Canada during the two decades. He also noticed a significant increase in the spread of acid rain in the areas of southeastern and Midwestern United States.[22]

Canada

Long-term measurements of acid deposition in Canada showed that other than local sources, acidity was also contributed by the long-range transport of oxides of sulfur and nitrogen from sources located southerly in the United States. According to Environment Canada, more than half of the acid deposition in eastern Canada is originated from the United States. Studies by the APN showed higher acidity in southern Ontario, having a pH of around 4.2.[23] Regionally representative sites Long Point and Chalk River experienced that, most of the time, wind parcels came from southerly source areas. These sites are the receptor sites to the major sources in Ontario and the lower Great Lakes region. Estimates of the year 1995 showed that 3.5–4.2 Tg per year of SO_2 was transported from the United States to Canada. Acid deposition in Canada can be reduced by the joint measures of the United States and Canada. Collaborative efforts in this direction are already in progress.

Europe

Areas affected by acid rain in Europe include northern and western Europe, southeast England, Germany, the Netherlands, and parts of Denmark. Table 1 shows a drastic decrease of precipitation pH in western Europe during the 1950s and 1970s.[24]

OECD 1977 estimates indicate that anthropogenic emissions of sulfur in Europe increased by 50% during 1955–1970. However, later on (1972–1982), many European countries, viz., the United Kingdom, West Germany, the Netherlands, Sweden, Norway, and Denmark reduced their sulfur emissions. These reduction measures improved the situation in Europe.[15]

Asia

In the Asian region, much of the acid rain problem prevails in East Asia, covering China, Japan, North Korea, and Thailand. Among these, China is the biggest polluter. According to estimates, China's sulfur emissions will triple between

Table 2 Average pH of rainwater at various sites in India.

Site	Nature of site	pH	Reference
Calcutta	Urban	6.8	Das[31]
Nainital	High altitude	6.2	Hegde et al.[32]
Iqbalpur	Rural	7.1	Jain et al.[33]
Mumbai (Colaba)	Urban	5.9	Khemani et al.[34]
Darjiling	High altitude semiurban	6.4	Kulshrestha[35]
Haflong	High altitude rural	7.3	Kulshrestha[35]
Delhi	Urban	5.7	Kulshrestha et al.[36]
Hyderabad	Urban	6.4	Kulshrestha et al.[37]
Jorhat	Rural	5.8	Kulshrestha et al.[38]
Hudegadde	Reserve forest	6.0	Kulshrestha et al.[38]
Agra (Dayalbagh)	Semiurban	7.1	Kumar et al.[39]
Malikadevi	Remote	6.4	Mahadevan et al.[40]
Allahabad[a]	Urban	7.1	Mukhopadhyay et al.[41]
Jodhpur[a]	Rural	8.3	Mukhopadhyay et al.[41]
Kodaikanal[a]	Rural	6.1	Mukhopadhyay et al.[41]
Mohanbari[a]	Rural	6.4	Mukhopadhyay et al.[41]
Nagpur[a]	Rural	6.3	Mukhopadhyay et al.[41]
Srinagar[a]	Rural	7.0	Mukhopadhyay et al.[41]
Pune	Urban	6.3	Pillai et al.[42]
Sinhagad	High altitude rural	6.2	Pillai et al.[42]
Silent Valley	Reserve forest	5.3	Rao et al.[43]
Agra (Tajganj)	Semiurban	7.0	Saxena et al.[44]
Indian Ocean (during Jan–Mar)	Northern and central	Below 5.6	Kulshrestha et al.[45]

[a]BAPMoN sites.

Table 3 Average pH and major ions (μeq/L) in precipitation at Banizaumbou during 1994–2005.

Parameter	Value
pH	6.05
SO_4^{2-}	9.4
NO_3^-	11.6
Ca^{2+}	27.3
NH_4^+	18.1

Source: Galy-Lacaux et al.[47]

1990 and 2020.[25] After the United States and Europe, China is the biggest consumer of fossil fuel. Rapid increase in SO_2 and NO_x emissions from 2000 onward is a major reason for the spread of acid rain in China.[26] The area most affected by acid rain in China is south of the Yangtze River, where average pH is recorded to be less than 4.5.

Acidity levels in precipitation in Japan show seasonality. During the summer season, most of the acidity is observed to be due to local sources of sulfur oxides, whereas during the winter season, increased level of acidity is due to long-range transport from the Asian continent, which results in higher acidity of precipitation at the sites in western Japan.[27] Similarly, Thailand also experiences acid rain. Around 50% of rain events are reported acidic due to high concentration of sulfate and nitrate. Of these oxides, 70%–80% are contributed by Thai sources.[28]

In the Indian subcontinent, precipitation is reported to have relatively higher pH (>5.6).[19,29] The pH of rainwater at some of the continental sites in India is as high as 8.3 (Table 2) due to interference of soil dust (rich in calcium carbonate) suspended in the atmosphere. Abundance of soil dust in air is a common feature of the Indian atmospheric environment. The pH of most soils of India is very high as compared with the pH of soils in acidified regions of the world. Generally, in India, the pH of rainwater is the mirror image of the pH of soil in the region. The acidity generated by the oxidation of gases like SO_2 and NO_x is buffered by soil-derived particles. Acidity of SO_2 is buffered by $CaCO_3$ of soil dust forming calcium sulfate, which is removed by below-cloud scavenging (Fig. 1). Due to this, the spread of acid rain at continental sites in India is controlled by the continuous suspension of loose soil during prevailing dry weather conditions. However, a bigger number of hot spots of higher wet deposition of non–sea salt sulfate (nss SO_4) are reported in urban and industrial areas than in rural areas. Several of these larger hot spots lie in the Indo-Gangetic region (Fig. 2).

Although rainwater pH higher than 5.6 is more frequently recorded in India, sometimes, occurrence of acid rain (pH <5.6) is also reported. Fig. 3 shows the frequency of acid rain reported from various sites in India.[46] Rainwater is noticed to be acidic in India if any of following applies:

1. Rain continues for a long time, washing off soil dust from the atmosphere. A similar situation prevails over the Indian Ocean, where soil dust interference is at a minimum.
2. In the areas where a large part of ground is covered with vegetation.
3. In the areas where soil itself is acidic (northeast, east, and southwest India).
4. Near heavy sources of SO_2 (e.g., thermal power plants).

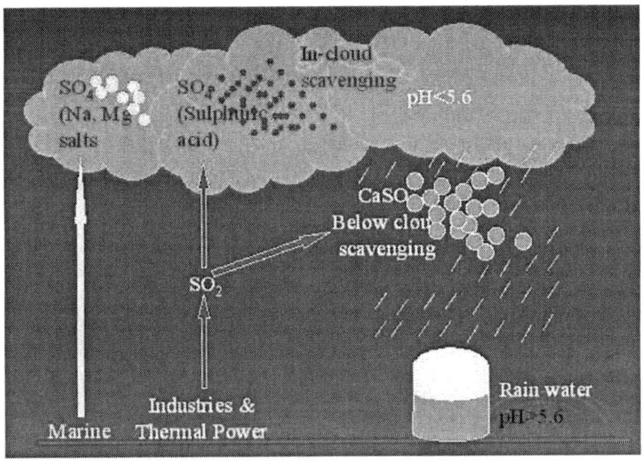
In-cloud
scavenging
pH<5.6
salts
acid)
SO2
scavenging
Industries &
Thermal Power
Marine
Rain water
pH>5.6

Fig. 1 Schematic diagram showing alkaline rains by removal of soil dust during below-cloud scavenging process in India.
Source: Kulshrestha.[30]

In the African region, the pH of precipitation is reported to be nearly similar to that in the Indian region. Long-term data (1994–2005) showed that the acidity of precipitation at Banizaumbou, a regional representative site in the semiarid savanna region, has high interference of soil dust, resulting in higher pH.[47] In addition, neutralization by ammonium ion is also partly responsible for elevated pH in Africa.

At a glance, the model based global distribution of pH of rain water is shown in Fig. 4.[48] The distribution of pH shows high acidity in Europe, eastern North America, and East Asia. To some extent, acidity is seen in the west coast of South America and Africa also. High acidity in these areas is due to free acidity contributed by sulfuric acid. However, in areas such as north of South America, northern Africa, South Asia, and part of China, the acidity is lower which is due to neutralization by ammonia and soil dust.

TRENDS IN ACIDITY

United States and Europe

After 1980, significant reduction in SO_2 emissions has improved the situation of acidity in Europe.[49] Similarly, in Canada and the United States, effective steps of reduction in SO_2 emissions have contributed to improved acidity

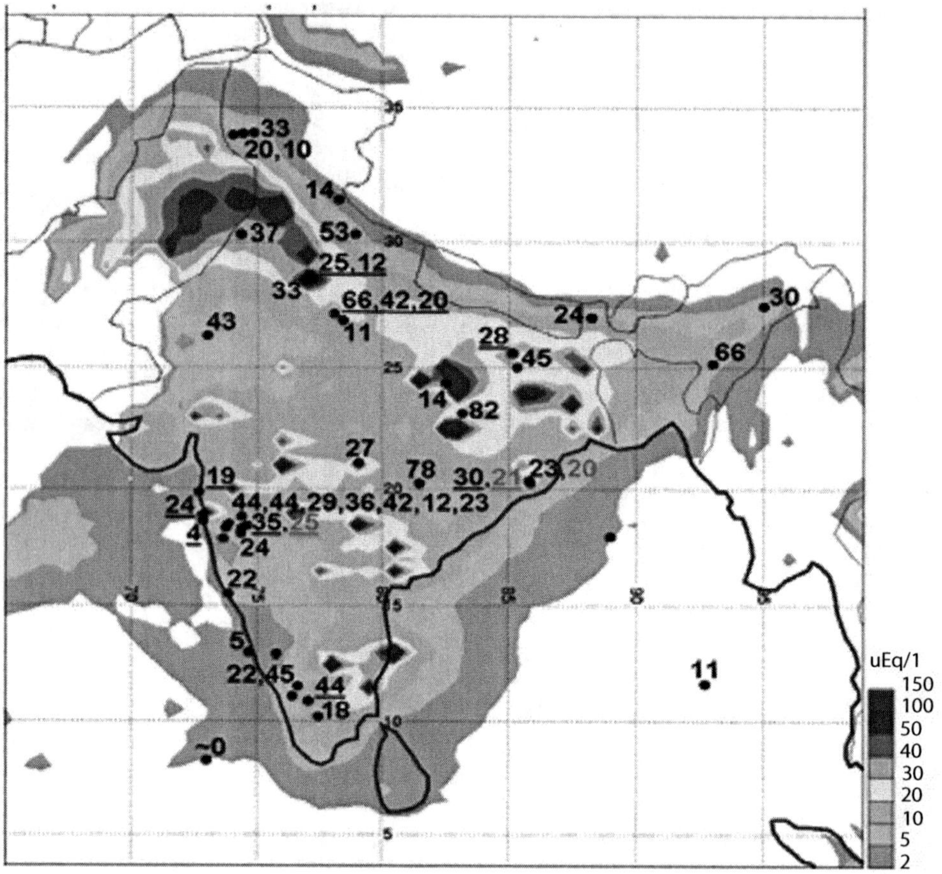
uEq/1
150
100
50
40
30
20
10
5
2

Fig. 2 Concentration of nss SO_4 in rainwater. Data from measurements at rural and suburban (underlined) sites obtained with bulk (black) and wet (red) collectors only (scaled to year 2000) compared with the concentration field obtained with the Multiscale Atmospheric Transport and Chemistry Model (MATCH) for the year 2000.
Source: Kulshrestha et al.[19]

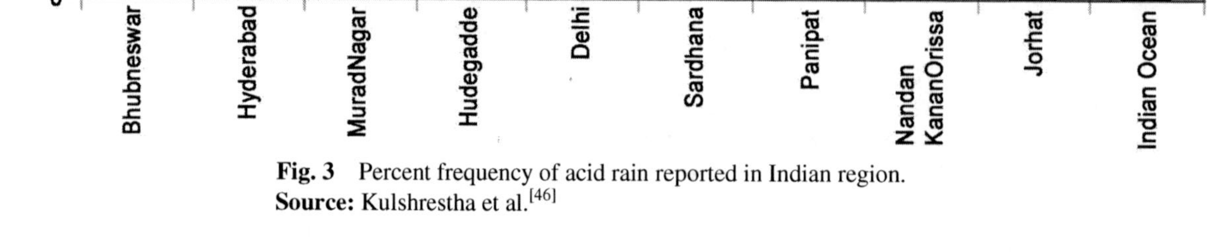
Acidity
% Acid Rain
100
80
60
40
20
0
Bhubneswar
Hyderabad
MuradNagar
Hudegadde
Delhi
Sardhana
Panipat
Nandan KananOrissa
Jorhat
Indian Ocean

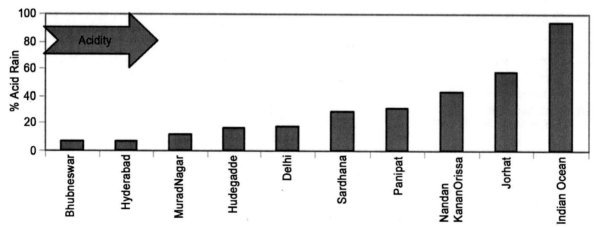

Fig. 3 Percent frequency of acid rain reported in Indian region.
Source: Kulshrestha et al.[46]

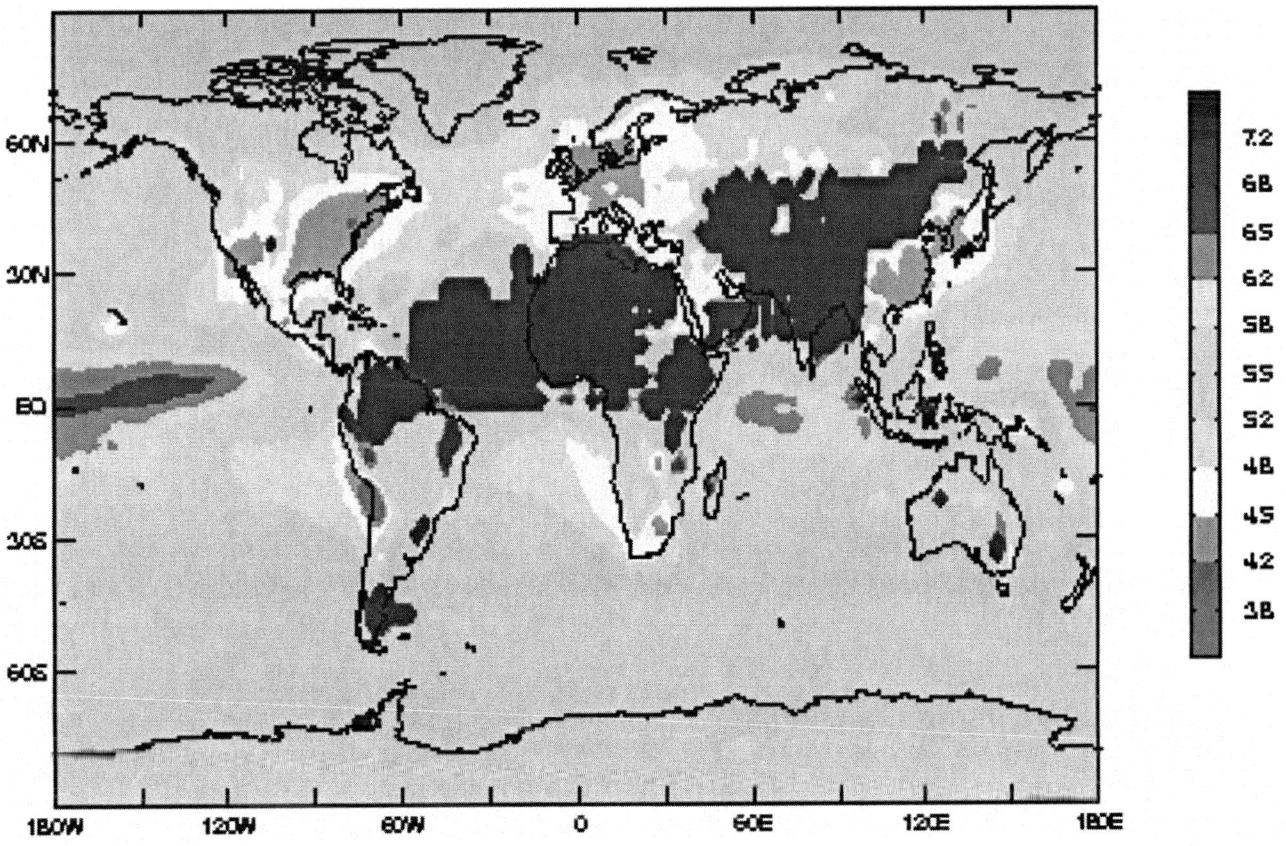
60N
30N
EQ
30S
60S
180W
120W
60W
0
60E
120E
180E
7.2
6.8
6.5
6.2
5.8
5.5
5.2
4.8
4.5
4.2
3.8

Fig. 4 Estimated global distribution of pH of precipitation.
Source: Rodhe et al.[48]

levels. In the United States, the average reduction in SO_2 emissions from 1980 to 2008 was around 54%, as shown in Fig. 5. Overall, the reduction measures have resulted in a decrease in H^+ in precipitation over these regions. Trends of acidity in the United States show a significant improvement after implementation of the 1995 Clean Air Act Amendment, which forced them to reduce SO_2 emissions. Fig. 6 is an example of trends of acidity of precipitation in North Carolina during 1985–2005,[52] which shows around 50% reduction in H^+ during two decades. In Europe, sulfate concentrations increased by approximately 50% between the 1950s and the late 1960s but have been

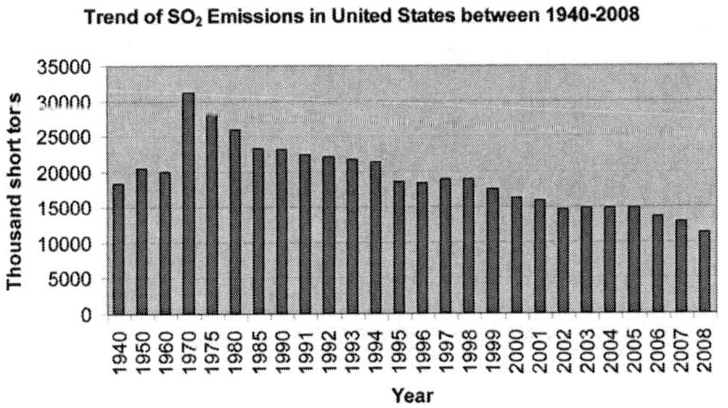
Trend of SO2 Emissions in United States between 1940-2008
Thousand short tons
35000
30000
25000
20000
15000
10000
5000
0
1940
1950
1960
1970
1975
1980
1985
1990
1991
1992
1993
1994
1995
1996
1997
1998
1999
2000
2001
2002
2003
2004
2005
2006
2007
2008
Year

Fig. 5 Trend of SO_2 Emissions in the United States from 1940 to 2008.
Source: Adapted from the USEPA Web site[50] and Stensland.[51]

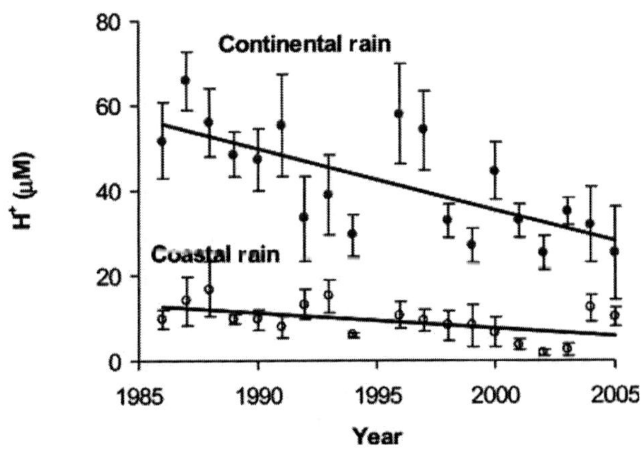
Continental rain
Coastal rain
H+ (μM)
80
60
40
20
0
1985
1990
1995
2000
2005
Year

Fig. 6 Annual volume-weighted average H^+ concentration in Wilmington, North Carolina, continental (filled circles) and coastal (open circles) precipitation from 1985 to 2005.
Source: Willey et al.[52]

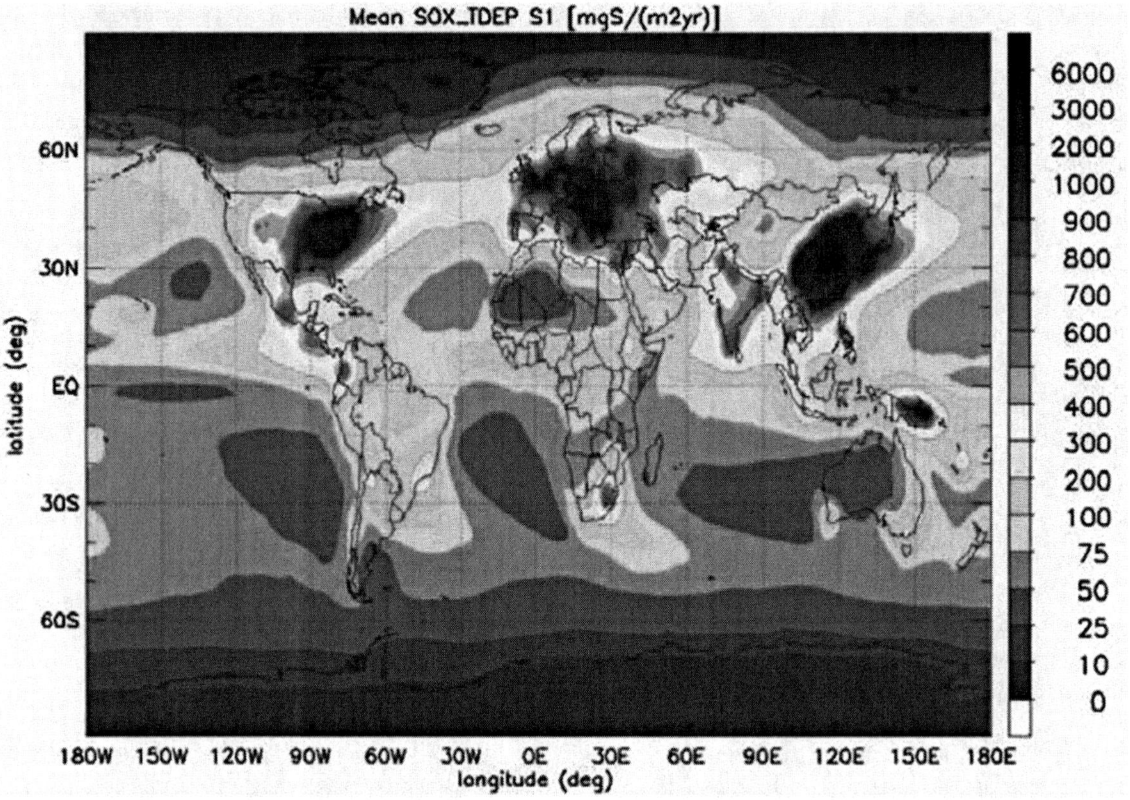

Mean SOX_TDEP S1 [mgS/(m2yr)]
latitude (deg)
60N
30N
EQ
30S
60S
180W 150W 120W 90W 60W 30W 0E 30E 60E 90E 120E 150E 180E
longitude (deg)
6000
3000
2000
1000
900
800
700
600
500
400
300
200
100
75
50
25
10
0

(a)

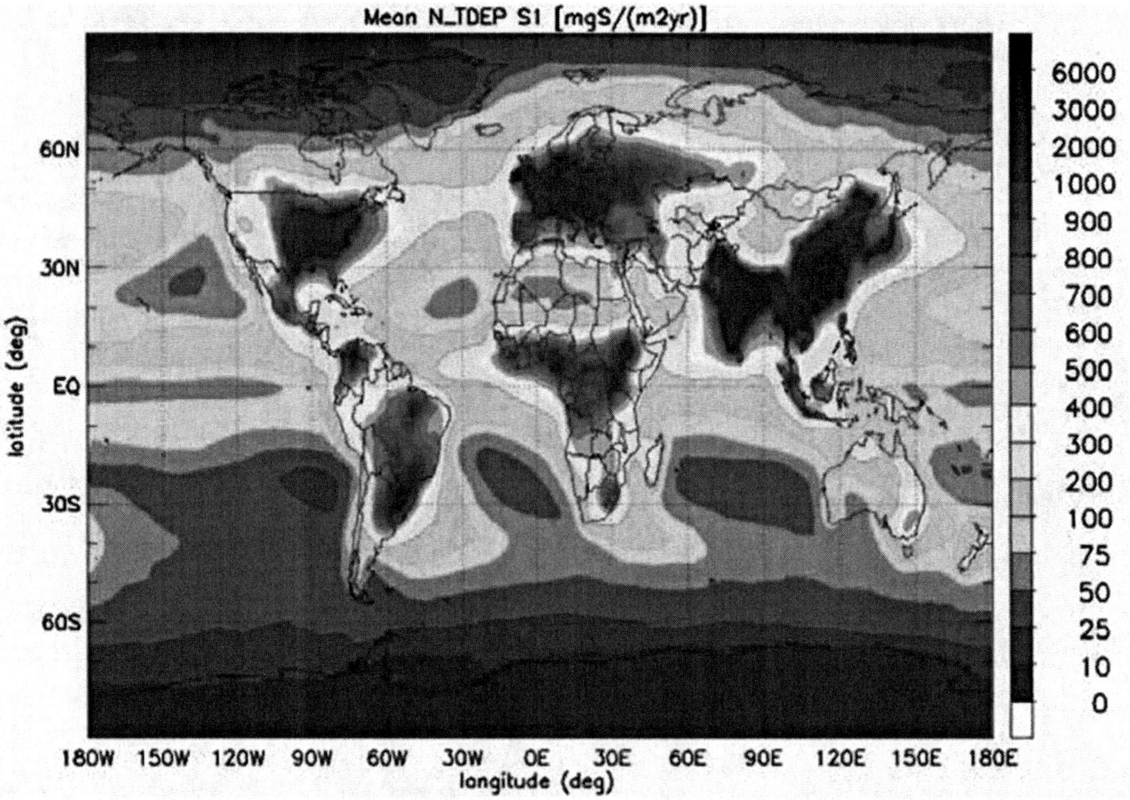

Mean N_TDEP S1 [mgS/(m2yr)]
latitude (deg)
60N
30N
EQ
30S
60S
180W 150W 120W 90W 60W 30W 0E 30E 60E 90E 120E 150E 180E
longitude (deg)
6000
3000
2000
1000
900
800
700
600
500
400
300
200
100
75
50
25
10
0

(b)

Fig. 7 Estimated total deposition of (a) S (mg m^{-2} yr^{-1}) of SO_x and (b) N (mg m^{-2} yr^{-1}) of reactive nitrogen for year 2000.
Source: Dentener et al.[55]

declining since the mid-1970s. In Sweden and Norway, an average of 20% reduction in SO_4 levels has been achieved since the 1970s,[15] followed by higher reductions in more recent decades.

Asia

China

In the developed countries, efforts have been made to control SO_2 emissions, but in the developing countries, SO_2 emissions continue to be high. In Asia, China is the biggest SO_2 emitter. Total Chinese SO_2 emission increased from 21.7 Tg to 33.2 Tg (53% increase) from 2000 to 2006, showing an annual growth rate of 7.3% per year.[53]

Long-term acid precipitation observations show that the temporal and spatial distribution of rain acidity in China has changed remarkably since 2000.[26] Future estimates of acidification potential, using a dynamic soil acidification model, indicate that sensitive soils in south China and Southeast Asia may reach a critical threshold within a few decades.[54] Model-based estimates (Fig. 7a and b) indicate higher levels of total depositions of $S\text{-}SO_x$ and $N\text{-}N_r$ in East Asia.

India

SO_2 emissions in India are relatively less, but the rate of increase is almost doubled from 1985 to 2005. In 2000, Indian SO_2 emission was estimated to be 4.26 Mt.[53] Precipitation studies in India lack long-term measurements. Most of the studies were carried out by individual scientists/groups. A few sites under the GAW network are in operation. In addition, one study from Pune[56] reports that during 1984–2002, there was significant increase in SO_4 and NO_3 concentrations (Fig. 8), which resulted in decrease in pH of rainwater from 6.9 in 1982 to 6.5 in 2002. These changes are due to increase in industrial and vehicular activities in the region. Another long-term network study in a rural area of Nandankanan (Orissa state in east India) reported 57% frequency of acid rain events during 1997–1998,[57] which was reduced to 40% during 2005–2007.[58] SO_4/Na ratios were also reduced drastically from 1.58 in 1997–1998 to 0.519 during 2005–2007. In a review compiled by Kulshrestha and coworkers,[19] it is reported that most of the Indian precipitation measurements lack quality assurance (QA) and quality control (QC) in sampling, storage, and analysis of samples. Hence, this region really needs quality-controlled measurements of wet and dry depositions at a few selected sites in order to get an idea about the trends of acidity with the growing emissions of oxides of S and N.

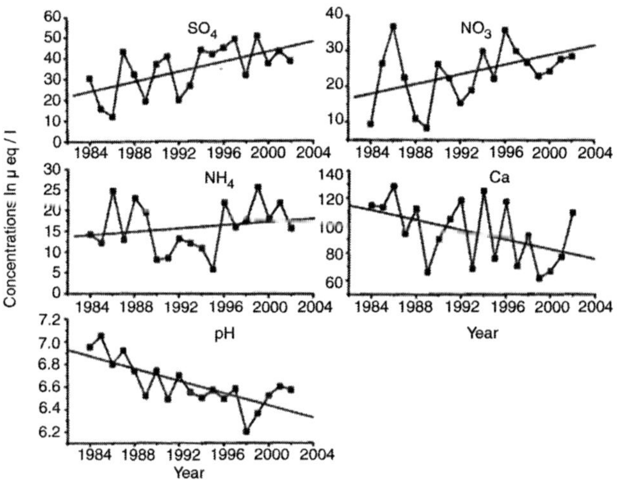
Concentrations in μ eq / l
SO4
NO3
NH4
Ca
pH
1984 1988 1992 1996 2000 2004
Year

Fig. 8 Trends of pH, SO_4, NO_3, Ca, and NH_4 in rainwater at Pune during 1984–2000.
Source: Rao et al.[56]

ACIDIFICATION OF OCEANS

Oceans are the biggest sinks for atmospheric CO_2. Increasing emissions of CO_2 due to anthropogenic sources will lead to ocean acidification through excess dissolution of CO_2 in seawater. Since the industrial revolution, the acidity of the ocean has increased by 30% (from a pH of 8.2 to 8.1). Future projections show that under a business-as-usual scenario, surface ocean pH will be lowered by 0.4 pH units by the end of the century. Acidification of the ocean affects the nitrification process, which further affects marine biota. Apart from CO_2 rise, acid deposition can add to the acidification of oceans. Precipitation having very low pH contributes a significant amount of hydrogen ions in seawater, which in the long term may alter the pH of seawater. Results from Indian Ocean Experiment showed the pH of rainwater to be between 3.8 and 5.6 over the Indian Ocean.[45] The acidic nature of rainwater over the Indian ocean is due to insignificant influence of soil dust and the dominance of anthropogenic sulfate contributed by long-range transport.[59,60] This aspect needs to be investigated in the future in order to protect the marine ecosystem.

GLOBAL SENSITIVITY TOWARD ACIDIFICATION

Global precipitation acidity and mapping of soil sensitivity to acid deposition suggests three main problematic areas. These are North America, Europe, and southern China (Fig. 9), where acid rain control is necessary. Already, in North America and Europe, steps have been taken to reduce SO_2 emissions. Other parts of the world also need to take appropriate steps to reduce sulfur emissions.[62]

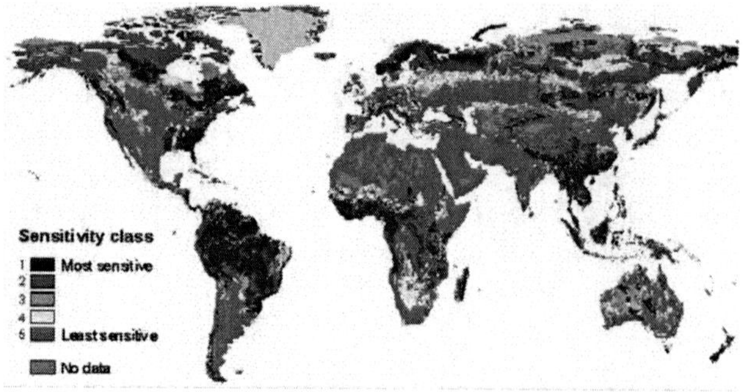
Sensitivity class
1 Most sensitive
2
3
4
5 Least sensitive
No data

Fig. 9 Global sensitivity toward acidification.
Source: Kuylenstierna et al.[63]

GLOBAL SCENARIO: FUTURE PROJECTIONS THROUGH MODELING

Recently, the Regional Air Pollution Information and Simulation model has been developed by International Institute for Applied Systems Analysis (IIASA) as a tool for the integrated assessment of alternative strategies to reduce acid deposition in Europe and Asia.[63] Dentener and coworkers[55] have attempted simulation of the global future scenario (up to 2030) of deposition of oxides of nitrogen and sulfur by using 23 atmospheric chemistry transport models. The stud focused mainly upon three emission scenarios: 1) current legislation (CLE); 2) case of the maximum emission reductions (MFR); and 3) pessimistic IPCC SRES A2 scenario. The model output showed a good agreement with observations in Europe and North America primarily because of quality-controlled measurements reported from these regions. The study suggested that in the future, deposition fluxes are going to be controlled mainly by the changes in emissions, with atmospheric chemistry and climate having a very limited role.

REGIONAL COMPARISON OF PRECIPITATION SCENARIO

Sulfate in the Atmosphere

Normally, free acidity is contributed by the acids of SO_2 and NO_2. In case of free acidity of H_2SO_4, pH value decreases with increasing concentration of SO_4 ions. A comparison of pH and SO_4 in rainwater at different sites in the United States, Sweden, and India is shown in Figs. 10–12. The United States and Sweden are examples of developed and acidified countries, whereas India is a developing country having high pH of rainwater. From Figs. 10 and 11, pH decreases with increase of SO_4 concentration, while at Indian sites (Fig. 12), even at higher SO_4 levels, higher pH of rainwater is observed. This indicates that in the United States and Sweden, the SO_4 is present as H_2SO_4, which gives free

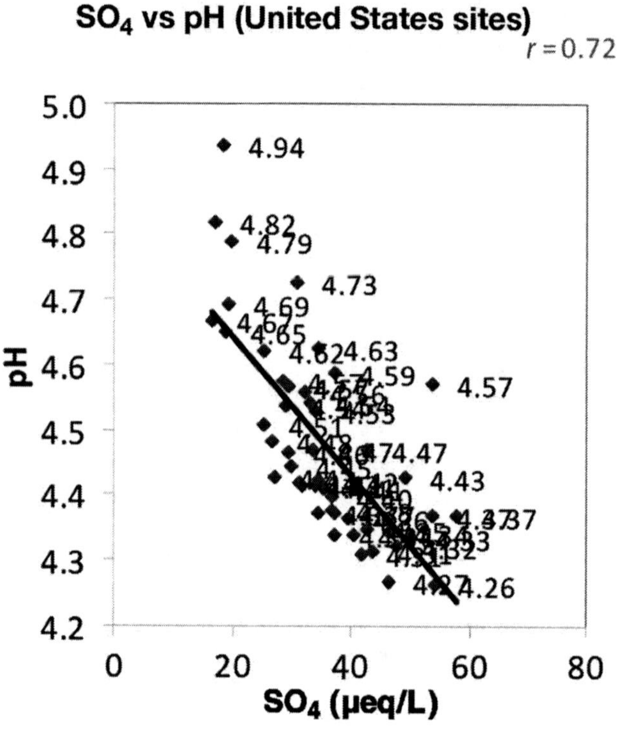
SO4 vs pH (United States sites)
r = 0.72
5.0
4.9
4.8
4.7
4.6
4.5
4.4
4.3
4.2
pH
0
20
40
60
80
SO4 (μeq/L)

Fig. 10 Variation of SO_4 and pH in rainwater at the sites in the United States.
Source: *Trends in Precipitation Chemistry in the United States, 1983–94: An Analysis of the Effects in 1995 of Phase I of the Clean Air Act Amendments of 1990, Title IV*.[61]

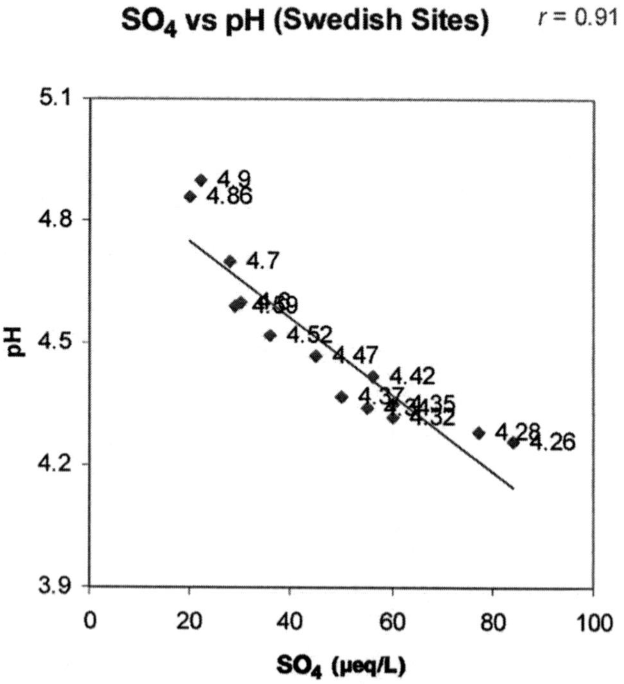
SO4 vs pH (Swedish Sites)
r = 0.91
5.1
4.8
4.5
4.2
3.9
pH
0
20
40
60
80
100
SO4 (μeq/L)

Fig. 11 Variation of SO_4 and pH in rainwater at Swedish sites.
Source: Granat.[65]

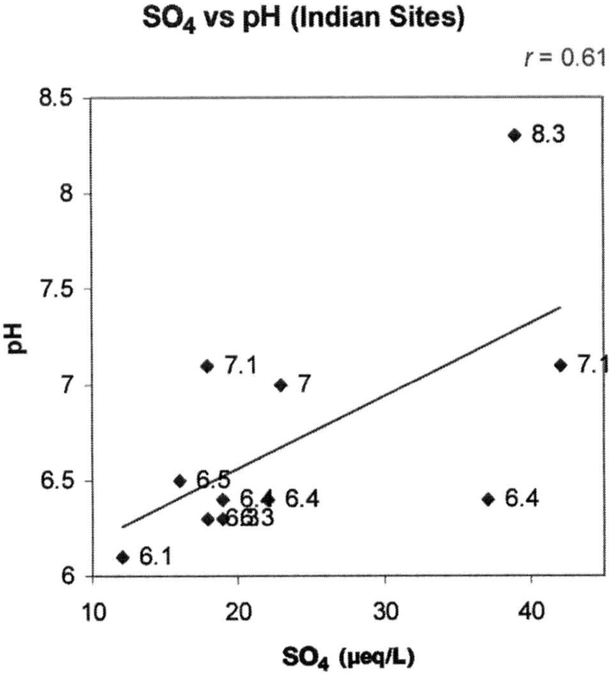
SO4 vs pH (Indian Sites)
r = 0.61
pH
SO4 (µeq/L)
8.5
8
7.5
7
6.5
6
10
20
30
40
8.3
7.1
7
6.5
6.4
6.1
6.4

Fig. 12 Variation of SO_4 with pH at Indian sites.
Source: Granat.[65]

H^+ in rain, but in India, it is present in a different form that does not contribute free H^+ in rainwater. The Indian dusty atmosphere is rich in $CaCO_3$, which allows SO_2 to form calcium sulfate, due to which pH of rainwater in India and other dusty regions is observed to be higher. In the Indian region, the possible mechanism of SO_2 removal is the adsorption of SO_2 onto the $CaCO_3$-dominated dust particles forming calcium sulfate.[62]

$$CaCO_3 + SO_2 + 1/2\ O_2 + 2H_2O \rightarrow CaSO_4.2H_2O + CO_2 \quad (14)$$

A comparison of typical composition of precipitation in an acidified region and a dusty region has been reported by Rodhe and coworkers,[48] establishing such differences very clearly.

CONTROL OF ACID RAIN

Acid Rain Control Policy of the United States

Due to pressure from the public, the federal government of the United States adopted the Clean Air Act in 1970. Under this act, emission standards were set for SO_2 and NO_x, and states were directed to compliance with the National Ambient Air Quality Standards. Congress formed a 10 years program, the NAPAP, and mandated it to conduct scientific, technological, and economic study of the acid rain.

In 1990, under Clean Air Act Amendments, the National Deposition Control Program was implemented. This legislation was to control adverse effects of acidic deposition through reductions in emissions of SO_2 and NO_x. It was targeted to achieve a 50% reduction in annual SO_2 emissions by the year 2000. In 2001, SO_2 emissions from utilities subject to the provisions of the acid rain program were 39% below their 1980 levels, and total emissions from all sources were 50% less than their 1980 levels.[67] These implementation steps will help in environment protection, in particular, to check deterioration of historic buildings, to reduce fine particulate matter (sulfates, nitrates) and ground-level ozone (smog), and to improve public health.[68]

Since Canada is affected by transboundary pollution, an air quality accord was signed between the United States and Canada in 1991 under the framework of the UN Economic Commission for Europe. This bilateral accord facilitates the United States' and Canada's meeting their emissions targets for SO_2 and NO_x and putting coordinated efforts into atmospheric modeling and monitoring the effects of transboundary air pollution.[69]

European Policy to Control Acid Rain

Among European countries, most of the scientific research on acid rain effects was conducted in Norway and Sweden. In the beginning, acidification of lakes through transboundary pollution was the primary issue in Scandinavian countries. In 1972, at the UN Conference on the Human Environment in Stockholm, Sweden's case study on the effects of long-range transport of sulfur compounds was presented, which emphasized the need for international agreement to reduce damage from acid deposition. Soon after the Stockholm conference, in 1978, OECD initiated EMEP to monitor long-range pollution.

The 1985 Helsinki Protocol was the first binding commitment on the reduction of sulfur emissions or their transboundary fluxes by at least 30%. Later, a group of 12 countries decided to sign a declaration to reduce by 30% NO_x emissions by 1998 as compared with 1986 (base year). The latest agreement is the 1999 Gothenburg Protocol, which deals with SO_2, NO_x, NH_3, and non-methane volatile organic compounds (VOCs), aiming to mitigate the problem of acidification, eutrophication, and ground-level ozone. The Gothenburg Protocol targets the reduction of Europe's sulfur emissions by at least 63%, NO_x emissions by 41%, VOC emissions by 40%, and NH_3 emissions by 17% by the year 2010 from their 1990 levels.[70]

EFFECTS OF ACID RAIN

Acid rain is very harmful to the environment as it damages many living and non-living things over a period of time. Acid rain affects both terrestrial as well as aquatic life. There are many inevitable impacts of acid rain, which affect natural as well as anthropogenic environments.

Effects on Aquatic Ecosystem

The aquatic ecosystem is visibly affected by acid rain as it directly falls into water bodies like lakes, streams, and rivers. Also, the extra acidic rainwater from other terrestrial places like forests and roads flows into nearby water bodies. Although the acidic effect of the rain may get diluted after it is mixed into the water bodies, it may lower the average pH of the aquatic system over the period, and in case the water body has low base cation supply or buffering capacity, it can become acidic faster. Charles and Norton[71] have reviewed the situation in lakes in United States and Canada and found that around the 1920s–1950s onward, weakly buffered lakes in some regions became more acidic. There are estimated to be around 50,000 lakes in the United States and Canada with a pH below 5.3. Out of these, hundreds of lakes have very low pH and are unable to support aquatic life, eliminating many existing insect and fish species. At pH lower than 5, the life of aquatic animals is threatened as they are unable to absorb oxygen from water. At pH lower than 4.8, fishes, frogs, and aquatic insects experience increased mortality. However, constructive steps, such as SO_2 emission reduction and liming of lakes, have significant potential for reversibility.[72,73]

Effects on Vegetation and Soil

In addition to the aquatic ecosystem, acid rain significantly affects trees, plants, forests, and other vegetation. Acid rain can damage the leaves and stems of the trees and affect their growth by getting absorbed through roots via soil. Acid rain reacts with leaves and stem wax coatings and allows acidic water to enter the leaves, thereby damaging the trees and plants. Experimental studies have established that acidic deposition causes some physiological effects in plants.[74] Especially at high altitude, the forests are surrounded by clouds carrying acidic water. The moisture of the clouds passing through the forests leads to severe damaging effects on the forests. In addition to individual effects on trees and forests, there are also effects on the soil, which contains the necessary nutrients and microorganisms for the healthy growth of the trees, plants, forests, and other vegetation. Virtual effects of acids rain have been observed all over the world, especially in Europe and eastern North America.[75,76] Hedin and coworkers[77] have reported evidence of steep decline of base cations in precipitation, which is based on long-term quality-controlled measurements, in Europe and North America. These measurements support that decline in base cations in precipitation might have resulted in increased sensitivity of a weakly buffered ecosystem, affecting forests and vegetation. Nitrification of ammonia (NH_3) and ammonium (NH_4^+) leads to the acidification of soil, which also adds to nutrient leaching:

$$NH_3 + O_2 \rightarrow NO_2- + 3H^+ + 2e- \qquad (15)$$

$$NO_2- + H_2O \rightarrow NO_3- + 2H^+ + 2e- \qquad (16)$$

Effects on Buildings and Monuments

Acid rain can damage buildings and historic monuments as well by reacting with their paints and construction material. The damaged walls of the buildings and monuments leave a rough surface along with the moisture, which is a favorable place for the growth of microorganisms. Acid rain can affect sculpture and architecture adversely by corrosion. Acid rain can even corrode railway tracks, paints of cars, and joints of bridges and flyovers. A BERG report[78] gives more details of evidence of damage to stone and other materials by acid deposition. Under CLRTAP, the International Cooperative Programme on Effects on Materials, Including Historic and Cultural Monuments, has been set up, with 39 test sites, three of which are in the United States and Canada. According to a report, the corrosion rates of carbon steel, paint on steel, limestone, and bronze have decreased to about 60% from 1987 values in Europe due to decrease in SO_2 levels.[79]

CONCLUSION

Acid rain, which was a problem of North America and Europe, is now spreading in East Asia due to increased emissions of SO_2 from fossil fuel combustion. Experiences of the global community show that acid rain is spread through transboundary pollution. Scandinavia, Canada, and Japan are such examples. Acid rain damages water bodies, forests, vegetation, buildings, human health, etc., costing a huge loss to the economy. Although appreciable steps are taken by the United States, Canada, and Europe to control SO_2 and NO_x emissions, the considerable increase in SO_2 emission rates in China is of concern in the Asian region. During the past two decades, Chinese SO_2 emissions increased tremendously. Data show that spread of acid rain is very much controlled by suspended atmospheric soil dust in Indian and African regions. At a glance, the following are concluded:

1. Decreasing trends of SO_2 emissions in Europe and North America will be helpful in improving the pH of precipitation in coming decades.
2. Increasing trends of acidity in China in Asia may result in more acid rains in the region.
3. Systematic monitoring networks are established in North America, Europe, and East Asia to monitor acid deposition.

4. There is a strong need for long-term studies on acid deposition, including wet and dry depositions at selected sites in South Asia, Africa, and South America, through extensive networking.

REFERENCES

1. Cowling, E.B. Acid precipitation in historical perspectives. Environ. Sci. Technol. **1982**, *16*, 110A.
2. Galloway, J.N. Acidification of the world: Natural and anthropogenic. Water Air Soil Pollut. **2001**, *130*, 17–24.
3. Charlson, R.J.; Rodhe, H. Factors controlling the acidity of natural rainwater. Nature **1982**, *295*, 683–685.
4. Seinfeld, J.; Pandis, S. *Atmospheric Chemistry and Physics*, 2nd Ed.; Wiley: New Jersey, 2006.
5. Brimblecombe, P. Interest in air pollution among early fellows of the Royal Society. Notes Rec. R. Soc. **1978**, *32*, 123.
6. Odén, S. Dagens Nyheter (Sweden), October 24, 1967.
7. Odén, S. The acidification of air precipitation and its consequences in natural environment. Ecology Committee Bulletin No. 1; Swedish Sciences Research Council, Stockholm. Translation Consultants Ltd.: Arlington, VA, 1968.
8. Martin, A.; Barber, F.R. Acid gases and acid in rain monitored for over 5 years in rural east-central England. Atmos. Environ. **1984**, *18*, 1715–1724.
9. Vermeulen, A.J. The acidic precipitation phenomenon. A study of this phenomenon and of a relationship between the acid content of precipitation and the emission of sulphur dioxide and nitrogen oxides in the Netherlands. In *Polluted Rain*; Toribara, T.Y., Miller, M.W., Morrow, P.E., Eds.; Plenum Press: New York, 1979; 7–60.
10. Gibson, J.H. Evaluation of wet chemical deposition in north America. In *Deposition Both Wet and Dry*, Acid Precipitation Series; Hicks, B.B., Ed.; Butterworth Publishers: London, 1984; Vol. 4, 1–13.
11. Keene, W.C.; Galloway, J.N.; Holden, J.D. Measurement of weak organic acidity in precipitation from remote areas of the world. J. Geophys. Res. **1983**, *88*, 5122.
12. Jacob, D.J.; Wofsy, S.C. Photochemistry of biogenic emissions over the Amazon forest. J. Geophys. Res. **1988**, *93*, 1477–1486.
13. Chameides, W.L.; Davis, D.D. Aqueous phase source for formic acid in clouds. Nature **1983**, *304*, 427–429.
14. Sanhueza, E.; Andreae, M.O. Emission of formic and acetic acids from tropical Savanna soils. Geophys. Res. Lett. **1991**, *18*, 1707–1710.
15. Rodhe, H.; Granat, L. An evaluation of sulfate in European precipitation 1955–1982. Atmos. Environ. **1984**, *18*, 2627–263.
16. Overrein, L.; Seip, H.M.; Tollan, A. *Acid Precipitation—Effects on Forest and Fish*, Final report of the SNSF project 1972–1980. FR 19/80; Norwegian Institute for Water Research: Oslo, Norway, 1980.
17. Bolin, B., Ed. The impact on the environment of sulphur in air and precipitation. Sweden's National Report to the United Nations Conference on the Human Environment; Air Pollution Across Boundaries; Norstadt, Stockholm, 1971.
18. Parashar, D.C.; Granat, L.; Kulshrestha, U.C.; Pillai, A.G.; Naik, M.S.; Momim, G.A.; Prakasa Rao, P.S.; Safai, P.D.; Khemani, L.T.; Naqvi, S.W.A.; Narverkar, P.V.; Thapa, K.B.; Rodhe, H. Chemical composition of precipitation in India and Nepal. A preliminary report on an Indo-Swedish project on atmospheric chemistry. Report CM 90; IMI, Stockholm University: Sweden, 1996.
19. Kulshrestha, U.C.; Granat, L.; Engardt, M.; Rodhe, H. Review of precipitation chemistry studies in India—A search for regional patterns. Atmos. Environ. **2005**, *39*, 7403–7419.
20. Ayers, G.P.; Gillet, R.W. Acidification in Australia. In *Acidification in Tropical Countries*, SCOPE 36; Rodhe, H., Herrera, R., Eds.; John Wiley and Sons: Chichester, England, 1988; 347–400.
21. Likens, G.E.; Acid precipitation. Chem. Eng. News, November 22, 1976; 29–43.
22. Likens, G.E.; Butler, T.J. Recent acidification of precipitation in North America. Atmos. Environ. **1981**, *15*, 1103–1109.
23. Barrie, L.A.; Anlauf, K.; Wiebe, H.A.; Fellin, P. Acidic pollutants in air and precipitation at selected rural locations in Canada. In *Deposition of Both Wet and Dry*, Acid Precipitation Series; Hicks, B.B., Ed.; Butterworth Publishers: Boston, 1984; Vol. 4, 15–36.
24. Environmental Resources Ltd Pearce, *Acid Rain—A Review of the Phenomenon in the EEC and Europe*; Graham and Trotman: London, 1983.
25. Streets, D.G.; Carmichael, G.R.; Amann, M.; Arndt, R.L. Energy consumption and acid deposition in northeast Asia. Ambio **1999**, *28*, 135–143.
26. Jie, T.; Xiao, X.; Bin, B.; Jin, A.; Feng, W.S. Trends of the precipitation acidity over China during 1992–2006. Chin. Sci. Bull. **2010**, *55*, 1800–1807. doi: 10.1007/s11434-009-3618-1.
27. Hara, H.; Akimoto, H. National level variations in precipitation chemistry in Japan. In Proceedings of the International Conference on Regional Environment and Climate Changes in East Asia, Taipei. November 30–December 3, 1993.
28. Garivaita, H.; Yoshizumib, K.; Morknoya, D.; Chanatorna, D.; Meepola, J.; Mark-Maia, A. Characterization of wet deposition in suburban area of Bangkok, Thailand. In Proceedings of CAD Workshop, Hyderabad, India, Nov–Dec 2006.
29. Khemani, L.T.; Momin, G.A.; Prakash Rao, P.S.; Safai, P.D.; Singh, G.; Kapoor R.K. Spread of acid rain over India. Atmos. Environ. **1989**, *23*, 757–762.
30. Kulshrestha, U.C. Air quality assessment through atmospheric depositions: A comparison of measurements and model calculations for India. Indian J. Environ. Manage. **2007**, *34*, 51–55.
31. Das, D.K. Chemistry of monsoon rains over Calcutta, West Bengal. Mausam **1988**, *39*, 75–82.
32. Hegde, P.; Kulshrestha, U.C.; Dumka, U.C.; Naza, M.; Pant, P. Chemical characteristics of rainwater and atmospheric aerosols at an elevated site in Himalayan Ranges in India. In Proceedings of CAD Workshop, Hyderabad, India, Nov–Dec 2006.
33. Jain, M.; Kulshrestha, U.C.; Sarkar, A.K.; Parashar, D.C. Influence of crustal aerosols on wet deposition at urban and rural sites in India. Atmos. Environ. **2000**, *34*, 5129–5137.

34. Khemani, L.T.; Momin, G.A.; Rao P.S.P.; Pillai, A.G.; Safai, P.D.; Mohan, K.; Rao, M.G. Atmospheric pollutants and their influence on acidification of rain water at an industrial location on the west coast of India. Atmos. Environ. **1994**, *28*, 3145–3154.
35. Kulshrestha, U.C. Chemistry of atmospheric depositions in India. In Proceedings of Symposium Science at High Altitudes, Darjiling, May 7–11, 1997; 1998. 187–194.
36. Kulshrestha, U.C.; Sarkar, A.K.; Srivastava, S.S.; Parashar, D.C. Investigation into atmospheric deposition through precipitation studies at New Delhi (India), Atmos. Environ. **1996**, *30*, 4149–4154.
37. Kulshrestha, U.C.; Kulshrestha, M.J.; Sekar, R.; Sastry, G.S.R.; Vairamani, M. Chemical characteristics of rain water at an urban site of south-central India. Atmos. Environ. **2003**, *37*, 3019–3026.
38. Kulshrestha, M.J.; Reddy, L.A.K.; Satyanarayana, J.; Duarah, R.; Rao, P.G.; Kulshrestha, U.C. Chemical characteristics of rain water at two rural sites of NE and SW India. In Proceedings of CAD Workshop, Hyderabad, India, Nov–Dec 2006.
39. Kumar, N.; Kulshrestha, U.C.; Saxena, A.; Kumari, K.M.; Srivastava, S.S. Formate and acetate in monsoon rain water of Agra. J. Geophys. Res. **1993**, *98*, D3, 5135.
40. Mahadevan, T.N.; Negi, B.S.; Meenakshi, V. Measurements of elemental composition of aerosol matter and precipitation from a remote background site in India. Atmos. Environ. **1989**, *23*, 869–874.
41. Mukhopadhyay, B.; Datar, S.V.; Srivastava, H.N. Precipitation chemistry over the Indian region. Mausam **1992**, *43*, 249–258.
42. Pillai, A.G.; Naik, M.S.; Momin, G.A.; Rao, P.D.; Safai, P.D.; Ali, K.; Rodhe, H.; Granat, L. Studies of wet deposition and dustfall at Pune, India. Water Air Soil Pollut. **2001**, *130*, 475–480.
43. Rao, P.S.P.; Momin, G.A.; Safai, P.D.; Pillai, A.G.; Khemani, L.T. Rain water and throughfall chemistry in the Silent Valley forest in South India. Atmos. Environ. **1995**, *29*, 2025–2029.
44. Saxena, A.; Sharma, S.; Kulshrestha, U.C.; Srivastava, S.S. Factors affecting alkaline nature of rain water in Agra (India). Environ. Pollut. **1991**, *74*, 129–138.
45. Kulshrestha, U.C.; Jain, M.; Mandal, T.R.; Gupta, P.K.; Sarkar, A.K.; Parashar, D.C. Measurements of acid rain over Indian Ocean and surface measurements of atmospheric aerosols at New Delhi during INDOEX pre-campaigns. Curr. Sci. **1999**, *76*, 968–972.
46. Kulshrestha, U.C.; Rodhe, H. Precipitation chemistry metadata from Asia. Presented in the 2nd Steering Committee Meeting of CAD programme, Bangkok, Thailand, Nov 26–27, 2007.
47. Galy-Lacaux, C.; Laouali, D.; Descroix, L.; Gobron, N.; Liousse, C. Long term precipitation chemistry and wet deposition in a remote dry savanna site in Africa (Niger) Atmos. Chem. Phys. **2009**, *9*, 1579–1595.
48. Rodhe, H.; Dentener, F.; Schulz, M. The global distribution of acidifying wet deposition. Environ. Sci. Technol. **2002**, *36*, 4382–4388.
49. Zhu, Q. Trends in SO_2 emissions. Report of IEA Clean Coal Center, October 2006. PF 06-09, London.
50. USEPA Web site, available at http://www.epa.gov/air/emissions/so2.htm. (accessed March 2, 2011).
51. Stensland, G.J. Precipitation chemistry trends in the northern United States. In *Polluted Rain*; Toribara, T.Y., Miller, M.W., Morrow, P.E., Eds.; Plenum Press: New York, 1979; 87–108.
52. Willey, J.; Kiber, R.; Avery, G.B., Jr. Changing chemical composition of precipitation in Wilmington, North Carolina, U.S.A.: Implications for the continental U.S.A. Environ. Sci. Technol. **2006**, *40*, 5675–5680.
53. Lu, Z.; Streets, D.G.; Zhang, Q.; Wang, S.; Carmichael, G.R.; Cheng, Y.F.; Wei, C.; Chin, M.; Diehl, T.; Tan, Q. Sulfur dioxide emissions in China and sulfur trends in East Asia since 2000. Atmos. Chem. Phys. Discuss. **2010**, *10*, 8657–8715.
54. Hicks, K.; Kuylenstierna, J.; Owen, A.; Rodhe, H.; Seip, H.; Dentener, F. Assessing the time development of acidification damage in Asian soils at regional scale. In Proceedings of CAD Workshop, Hyderabad, India, Nov–Dec 2006.
55. Dentener, F.; Drevet, J.; Lamarque, J.F.; Bey, I.; Eickhout, B.; Fiore, A.M.; Hauglustaine, D.; Horowitz, L.W.; Krol, M.; Kulshrestha, U.C.; Lawrence, M.; Galy-Lacaux, C.; Rast, S.; Shindell, D.; Stevenson, D.; Van Noije, T.; Atherton, C.; Bell, N.; Bergman, D.; Butler, T.; Cofala, J.; Collins, B.; Doherty, R.; Ellingsen, K.; Galloway, J.; Gauss, M.; Montanaro, V.; Müller, J.F.; Pitari, G.; Rodriguez, J.; Sanderson, M.; Solmon, F.; Strahan, S.; Schultz, M.; Sudo, K.; Szopa, S.; Wild, O. Nitrogen and sulfur deposition on regional and global scales: A multimodel evaluation. Global Biogeochem. Cycles **2006**; *20* (4), GB4003, doi:10.1029/2005GB002672.
56. Rao, P.S.P.; Safai, P.D.; Momin, G.A.; Ali, K.; Chate, D.M.; Praveen, P.S.; Tiwari, S. Precipitation chemistry at different locations in India. In Proceedings of CAD Workshop, Hyderabad, India, Nov–Dec 2006.
57. Das, R.; Das, S.N.; Misra, V.N.; Chemical composition of rainwater and dust fall at Bhubaneswar in the east coast of India. Atmos. Environ. **2005**, *34*, 5908–5916.
58. Das, N.; Das, R.; Chaudhury, G.R.; Das, S.N. Chemical composition of precipitation at background level. Atmos. Res. **2010**, *95*, 108–113.
59. Kulshrestha, U.C.; Jain, M.; Sekar, R.; Vairamani, M.; Sarkar, A.K.; Parashar, D.C. Chemical characteristics and source apportionment of aerosols over Indian Ocean during INDOEX-1999. Curr. Sci. **2001**, *80*, 180–185.
60. Granat, L.; Norman, M.; Leck, C.; Kulshrestha, U.C.; Rodhe, H. Wet scavenging of sulfur and other compounds during INDOEX. J. Geophys. Res. **2002**, *107*, D19, 8025, doi:1011029/ 2001JD000499.
61. Trends in Precipitation Chemistry in the United States, 1983–94: An Analysis of the Effects in 1995 of Phase I of the Clean Air Act Ammendments of 1990, Title IV, http://pvbs.usgs.gov/acidrain/index.html#tables.
62. Kulshrestha, M.J.; Kulshrestha, U.C.; Parashar, D.C.; Vairamani, M. Estimation of SO_4 contribution by dry deposition of SO_2 onto the dust particles in India. Atmos. Environ. **2003**, *37*, 3057–3063.
63. Kuylenstierna, J.C.I.; Rodhe, H.; Cinderby, S.; Hicks, K. Acidification in developing countries: Ecosystem sensitivity and the critical load approach on a global scale. Ambio **2001**, *30*, 20–28.

64. Alcamo, J.; Shaw, R.; Hordijk, L., Eds. *The RAINS Model of Acidification. Science and Strategies in Europe*. Kluwer Academic Publishers: Dordrecht, Netherlands, 1990.
65. Granat, L. Luft-och nederbordskemiska stationsnatet inom PMK. Report 3942, Meteorological Institute at Stockholm University: Stockholm, 1990.
66. Hara, H. Temporal variation of wet deposition in the EANET region during 2000–2004. In Proceedings of CAD Workshop, Hyderabad, India, Nov–Dec 2006.
67. U.S. Environmental Protection Agency (USEPA). *Acid Rain Program, 2001 Progress report*; U.S. Environmental Protection Agency: Washington, DC, available at http://www.epa.gov/airmarkets/cmprpt/arp01/2001report.pdf.
68. U.S. Environmental Protection Agency (USEPA). *Acid Rain Program: Overview, Environmental Benefits*, available at http://www.epa.gov/airmarkets/arp/overview.html. Retrieved on March 2, 2011.
69. U.S. Environmental Protection Agency (USEPA). Canada–United States agreement, 1994 Progress report, EPA/430/R-94/013; Washington, DC. Retrieved on March 2, 2011.
70. Menz, F.C.; Seip, F.M. Acid rain in Europe and the United States: An update. Environ. Sci. Policy **2004**, *7*, 253–265.
71. Charles, D.F.; Norton, S.A. Paleolimnological evidence for trends in the atmospheric deposition of acids and heavy metals. In *Atmospheric Deposition: Historic Trends and Spatial Patterns*; Norton, S.A., Ed., National Academy Press: Washington, DC, USA, 1985; 86–105.
72. Battarbee, R.W.; Flower, R.J.; Stevenson, A.C.; Harriman, R.; Appleby, P.G. Diatom and chemical evidence for reversibility of acidification of Scottish Lochs. Nature (London) **1988**, *332*, 530–532.
73. Jenkins, A.; Whitehead, P.G.; Cosby, B.J.; Birks, H.J.B. Modelling long term acidification: A comparison with diatom reconstructions and the implications for reversibility. Philos. Trans. R. Soc. London **1990**, *B327*, 435–440.
74. Evans, L.S.; Lewin, K.F.; Parri, M.J.; Cunningham, E.A Productivity of field-grown soybeans exposed to simulated acid rain. New Phytol. **1983**, *93*, 377–388.
75. Rehfuess, K.E. On the impact of acid precipitation in forest ecosystems. Forstwiss. Centralbl. **1981**, *100*, 363–370.
76. Cowling, E.B. Regional declines on forests in Europe and North America: The possible role of airborne chemicals. In *Aerosols, Research Risk Assessment and Control Strategies*; Lewis: Chelsea, MI, USA, 1986; 855–864.
77. Hedin, L.O.; Granata, L.; Likens, G.E.; Buishand, T.A.; Galloway, J.N.; Butler, T.J.; Rodhe, H. Steep decline in atmospheric base cations in regions of Europe and north America. Nature **1994**, *367*, 351–367.
78. BERG, Building Effects Review Group. *The Effects of Acid Deposition on Building Materials in U.K.*; Department of Environment, HMSO: London, 1989.
79. Kucera, V. *Atmospheric Corrosion in Urban Air Pollution in Asia and Africa: The Approach of the RAPIDC Programme*; Kuylenstierna, J., Hicks, K., Eds.; Stockholm Environment Institute (SEI): Stockholm, Sweden, 2002.

Acid Rain: Nitrogen Deposition

George F. Vance
Department of Ecosystem Sciences and Management, University of Wyoming, Laramie, Wyoming, U.S.A.

Abstract
Air pollution has occurred naturally since the formation of the Earth's atmosphere; however, the industrial era has resulted in human activities greatly contributing to global atmospheric pollution. One of the more highly publicized and controversial aspects of atmospheric pollution is that of acidic deposition. Acidic materials can be transported long distances, some as much as hundreds of kilometers. Acidic deposition can impact buildings, sculptures, and monuments that are constructed using weatherable materials like limestone, marble, bronze, and galvanized steel. While acid soil conditions are known to influence the growth of plants, agricultural impacts related to acidic deposition are of less concern due to the buffering capacity of these types of ecosystems. When acidic substances are deposited in natural ecosystems, a number of adverse environmental effects are believed to occur, including damage to vegetation, particularly forests, and changes in soil and surface water chemistry.

INTRODUCTION

Air pollution has occurred naturally since the formation of the Earth's atmosphere; however, the industrial era has resulted in human activities greatly contributing to global atmospheric pollution.[1,2] One of the more highly publicized and controversial aspects of atmospheric pollution is that of acidic deposition. Acidic deposition includes rainfall, acidic fogs, mists, snowmelt, gases, and dry particulate matter.[3] The primary origin of acidic deposition is the emission of sulfur dioxide (SO_2) and nitrogen oxides (NO_x) from fossil fuel combustion; electric power generating plants contribute approximately two-thirds of the SO_2 emissions and one-third of the NO_x emissions.[4]

Acidic materials can be transported long distances, some as much as hundreds of kilometers. For example, 30%–40% of the S deposition in the northeastern U.S. originates in industrial midwestern U.S. states.[5] After years of debate, U.S. and Canada have agreed to develop strategies that reduce acidic compounds originating from their countries.[5,6] In Europe, the small size of many countries means that emissions in one industrialized area can readily affect forests, lakes, and cities in another country. For example, approximately 17% of the acidic deposition falling on Norway originated in Britain and 20% in Sweden came from eastern Europe.[5]

The U.S. EPA National Acid Precipitation Assessment Program (NAPAP) conducted intensive research during the 1980s and 1990s that resulted in the "Acidic Deposition: State of the Science and Technology" that was mandated by the Acid Precipitation Act of 1980.[6] NAPAP Reports to Congress have been developed in accordance with the 1990 amendment to the 1970 Clean Air Act and present the expected benefits of the Acid Deposition Control Program,[6,7] http://www.nnic.noaa.gov/CENR/NAPAP/. Mandates include an annual 10 million ton or approximately 40% reduction in point-source SO_2 emissions below 1980 levels, with national emissions limit caps of 8.95 million tons from electric utility and 5.6 million tons from point-source industrial emissions. A reduction in NO_x of about 2 million tons from 1980 levels has also been set as a goal; however, while NO_x has been on the decline since 1980, projections estimate a rise in NO_x emissions after the year 2000. In 1980, the U.S. levels of SO_2 and NO_x emissions were 25.7 and 23.0 million tons, respectively.

Acidic deposition can impact buildings, sculptures, and monuments that are constructed using weatherable materials like limestone, marble, bronze, and galvanized steel,[7,8] http://www.nnic.noaa.gov/CENR/NAPAP/. While acid soil conditions are known to influence the growth of plants, agricultural impacts related to acidic deposition are of less concern due to the buffering capacity of these types of ecosystems.[2,5] When acidic substances are deposited in natural ecosystems, a number of adverse environmental effects are believed to occur, including damage to vegetation, particularly forests, and changes in soil and surface water chemistry.[9,10]

SOURCES AND DISTRIBUTION

Typical sources of acidic deposition include coal- and oil-burning electric power plants, automobiles, and large industrial operations (e.g., smelters). Once S and N gases enter the earth's atmosphere they react very rapidly with moisture in the air to form sulfuric (H_2SO_4) and nitric (HNO_3) acids.[2,3] The pH of natural rainfall in equilibrium with atmospheric CO_2 is about 5.6; however, the pH of rainfall is less than 4.5 in many industrialized areas. The nature of acidic deposition is controlled largely by the

Encyclopedia of Environmental Management DOI: 10.1081/E-EEM-120001785

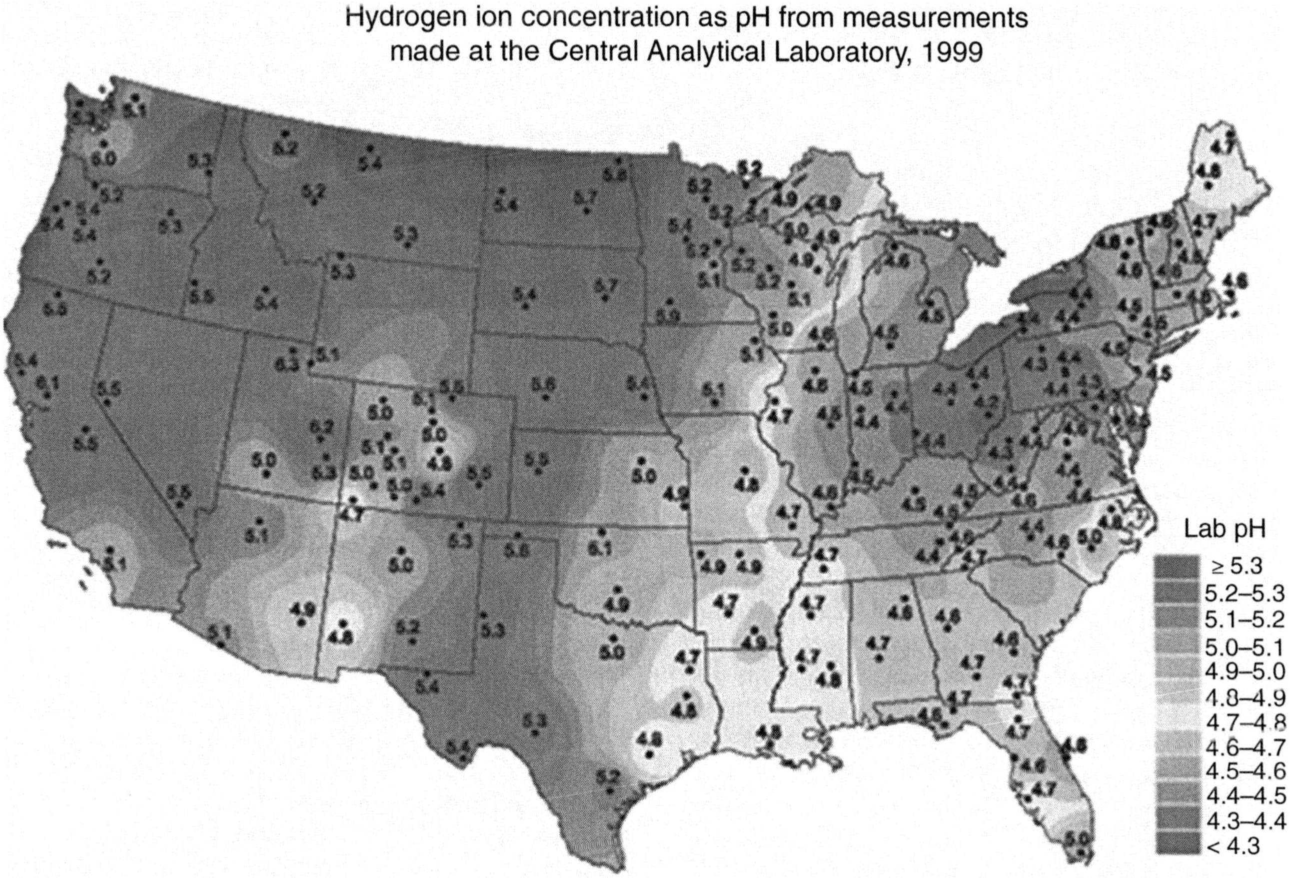
Hydrogen ion concentration as pH from measurements made at the Central Analytical Laboratory, 1999
Lab pH
≥ 5.3
5.2–5.3
5.1–5.2
5.0–5.1
4.9–5.0
4.8–4.9
4.7–4.8
4.6–4.7
4.5–4.6
4.4–4.5
4.3–4.4
< 4.3

Fig. 1 Acidic deposition across the U.S. during 1999.
Source: National Atmospheric Deposition Program/National Trends Network http://nadp.sws.uiuc.edu.[11]

geographic distribution of the sources of SO_2 and NO_x (Fig. 1). In the midwestern and northeastern U.S., H_2SO_4 is the main source of acidity in precipitation because of the coal-burning electric utilities.[2] In the western U.S., HNO_3 is of more concern because utilities and industry burn coal with low S contents and populated areas are high sources of NO_x.[2]

Emissions of SO_2 and NO_x increased in the 20th century due to the accelerated industrialization in developed countries and antiquated processing practices in some undeveloped countries. However, there is some uncertainty as to the actual means by which acidic deposition affects our environment,[11,12] http://nadp.sws.uiuc.edu/isopleths/maps1999/. Chemical and biological evidence, however, indicates that atmospheric deposition of H_2SO_4 caused some New England lakes to decrease in alkalinity.[13,14] Many scientists are reluctant to over-generalize cause and effect relationships in an extremely complex environmental problem. Although, the National Acid Deposition Assessment Program has concluded there were definite consequences due to acidic deposition that warrant remediation[6,7] http://www.nnic.noaa.gov/CENR/NAPAP/. Since 1995, when the 1990 Clean Air Act Amendment's Title IV reduction in acidic deposition was implemented, SO_2 and NO_x emissions have, respectively, decreased and remained constant during the late 1990s.[4]

Both H_2SO_4 and HNO_3 are important components of acidic deposition, with volatile organic compounds and inorganic carbon also components of acidic deposition-related emissions. Pure water has a pH of 7.0, natural rainfall about 5.6, and severely acidic deposition less than 4.0. Uncontaminated rainwater should be pH 5.6 due to CO_2 chemistry and the formation of carbonic acid. The pH of most soils ranges from 3.0 to 8.0.[2] When acids areadded to soils or waters, the decrease in pH that occurs depends greatly on the system's buffering capacity, the ability of a system to maintain its present pH by neutralizing added acidity. Clays, organic matter, oxides of Al and Fe, and Ca and Mg carbonates (limestones) are the components responsible for pH buffering in most soils. Acidic deposition, therefore, will have a greater impact on sandy, low organic matter soils than those higher in clay, organic matter, and carbonates. In fresh waters, theprimary buffering mechanism is the reaction of dissolved bicarbonate ions with H^+ according to the following equation:

$$H^+ + HCO_3^- = H_2O + CO_2 \quad (1)$$

HUMAN HEALTH EFFECTS

Few direct human health problems have been attributed to acidic deposition. Long-term exposure to acidic deposition precursor pollutants such as ozone (O_3) and NO_x, which are respiratory irritants, can cause pulmonary edema.[5,6] Sulfur dioxide (SO_2) is also a known respiratory irritant, but is generally absorbed high in the respiratory tract.

Indirect human health effects due to acidic deposition are more important. Concerns center around contaminated drinking water supplies and consumption of fish that contain potential toxic metal levels. With increasing acidity (e.g., lower pH levels), metals such as mercury, aluminum, cadmium, lead, zinc, and copper become more bioavailable.[2] The greatest human health impact is due to the consumption of fish that bioaccumulate mercury; freshwater pike and trout have been shown to contain the highest average concentrations of mercury.[5,15] Therefore, the most susceptible individuals are those who live in an industrial area, have respiratory problems, drink water from a cistern, and consume a significant amount of freshwater fish.

A long-term urban concern is the possible impact of acidic deposition on surface-derived drinking water. Many municipalities make extensive use of lead and copper piping, which raises the question concerning human health effects related to the slow dissolution of some metals (lead, copper, zinc) from older plumbing materials when exposed to more acidic waters. Although metal toxicities due to acidic deposition impacts on drinking waters are rare, reductions in S and N fine particles expected by 2010 based on Clean Air Act Amendments will result in annual public health benefits valued at $50 billion with reduced mortality, hospital admissions and emergency room visits.[16]

STRUCTURAL IMPACTS

Different types of materials and cultural resources can be impacted by air pollutants. Although the actual corrosion rates for most metals have decreased since the 1930s, data from three U.S. sites indicate that acidic deposition may account for 31%–78% of the dissolution of galvanized steeland copper,[7,8] http://www.nnic.noaa.gov/CENR/NAPAP/. In urban or industrial settings, increases in atmospheric acidity can dissolve carbonates (e.g., limestone, marble) in buildings and other structures. Deterioration of stone products by acidic deposition is caused by: 1) erosion and dissolution of materials and surface details; 2) alterations (blackening of stone surfaces); and 3) spalling (cracking and spalling of stone surfaces due to accumulations of alternation crusts.[8] Painted surfaces can be discolored or etched, and there may also be degradation of organic binders in paints.[8]

ECOSYSTEM IMPACTS

It is important to examine the nature of acidity in soil, vegetation, and aquatic environments. Damage from acidification is often not directly due to the presence of excessive H^+, but is caused by changes in other elements. Examples include increased solubilization of metal ions such as Al^{3+} and some trace elements (e.g., Mn^{2+}, Pb^{2+}) that can be toxic to plants and animals, more rapid losses of basic cations (e.g., Ca^{2+}, Mg^{2+}), and the creation of unfavorable soil and aquatic environments for different fauna and flora.

Soils

Soil acidification is a natural process that occurs when precipitation exceeds evapotranspiration.[2] "Natural" rainfall is acidic (pH of ~ 5.6) and continuously adds a weak acid (H_2CO_3) to soils. This acidification results in a gradual leaching of basic cations (Ca^{2+} and Mg^{2+}) from the uppermost soil horizons, leaving Al^{3+} as the dominant cation that can react with water to produce H^+. Most of the acidity in soils between pH 4.0 and 7.5 is due to the hydrolysis of Al^{3+},[17,18] http://www.epa.gov/airmarkets/acidrain/effects/index.html. Other acidifying processes include plant and microbial respiration that produces CO_2, mineralization and nitrification of organic N, and the oxidation of FeS_2 in soils disturbed by mining or drainage.[2] In extremely acidic soils (pH < 4.0), strong acids such as H_2SO_4 are a major component.

The degree of accelerated acidification depends both upon the buffering capacity of the soil and the use of the soil. Many of the areas subjected to the greatest amount of acidic deposition are also areas where considerable natural acidification occurs.[19] Forested soils in the northeastern U.S. are developed on highly acidic, sandy parent materials that have undergone tremendous changes in land use in the past 200 years. However, clear-cutting and burning by the first European settlers have been almost completely reversed and many areas are now totally reforested.[5] Soil organic matter that accumulated over time represents a natural source of acidity and buffering. Similarly, greater leaching or depletion of basic cations by plant uptake in increasingly reforested areas balances the significant inputs of these same cations in precipitation.[20,21] Acidic deposition affects forest soils more than agricultural or urban soils because the latter are routinely limed to neutralize acidity. Although it is possible to lime forest soils, which is done frequently in some European countries, the logistics and cost often preclude this except in areas severely impacted by acidic deposition.[5]

Excessively acidic soils are undesirable for several reasons. Direct phytotoxicity from soluble Al^{3+} or Mn^{2+} can occur and seriously injure plant roots, reduce plant growth, and increase plant susceptibility to pathogens.[21] The relationship between Al^{3+} toxicity and soil pH is complicated by the fact that in certain situations organic matter can form complexes with Al^{3+} that reduce its harmful effects on plants.[18] Acid soils are usually less fertile because of a lack of important basic cations such as K^+, Ca^{2+}, and Mg^{2+}. Leguminous plants may fix less N_2 under very acidic conditions due to reduced rhizobial activity and greater soil adsorption of Mo by clays and Al and Fe oxides.[2] Mineralization of N, P, and S can also be reduced because of the lower metabolic activity of bacteria. Many plants and microorganisms have adapted to very acidic conditions (e.g., pH<5.0). Examples include ornamentals such as azaleas and rhododendrons and food crops such as cassava, tea, blueberries, and potatoes.[5,22] In fact, considerable efforts in plant breeding and biotechnology are directed towards developing Al- and Mn-tolerant plants that can survive in highly acidic soils.

Agricultural Ecosystems

Acidic deposition contains N and S that are important plant nutrients. Therefore, foliar applications of acidic deposition at critical growth stages can be beneficial to plant development and reproduction. Generally, controlled experiments require the simulated acid rain to be pH 3.5 or less in order to produce injury to certain plants.[22] The amount of acidity needed to damage some plants is 100 times greater than natural rainfall. Crops that respond negatively in simulated acid rain studies include garden beets, broccoli, carrots, mustard greens, radishes, and pinto beans, with different effects for some cultivars. Positive responses to acid rain have been identified with alfalfa, tomato, green pepper, strawberry, corn, lettuce, and some pasture grass crops.

Agricultural lands are maintained at pH levels that are optimal for crop production. In most cases the ideal pH is around pH 6.0–7.0; however, pH levels of organic soils are usually maintained at closer to pH 5.0. Because agricultural soils are generally well buffered, the amount of acidity derived from atmospheric inputs is not sufficient to significantly alter the overall soil pH.[2] Nitrogen and S soil inputs from acidic deposition are beneficial, and with the reduction in S atmospheric levels mandated by 1990 amendments to the Clean Air Act, the S fertilizer market has grown. The amount of N added to agricultural ecosystems as acidic deposition is rather insignificant in relation to the 100–300 kg N/ha/yr required of most agricultural crops.

Forest Ecosystems

Perhaps the most publicized issue related to acidic deposition has been widespread forest decline. For example, in Europe estimates suggest that as much as 35% of all forests have been affected.[23] Similarly, in the U.S. many important forest ranges such as the Adirondacks of New York, the Green Mountains of Vermont, and the Great Smoky Mountains in North Carolina have experienced sustained decreases in tree growth for several decades.[6] Conclusive evidence that forest decline or dieback is caused solely be acidic deposition is lacking and complicated by interactions with other environmental or biotic factors. However, NAPAP research[6] has confirmed that acidic deposition has contributed to a decline inhigh-elevation red spruce in the northeastern U.S. In addition, nitrogen saturation of forest ecosystems from atmospheric N deposition is believed to result in increased plant growth, which in turn increases water and nutrient use followed by deficiencies that can cause chlorosis and premature needle-drop as well as increased leaching of base cations from the soil.[24]

Acidic deposition on leaves may enter directly through plant stomates.[1,22] If the deposition is sufficiently acidic (pH ~ 3.0), damage can also occur to the waxy cuticle, increasing the potential for direct injury of exposed leaf mesophyll cells. Foliar lesions are one of the most common symptoms. Gaseous compounds such as SO_2 and SO_3 present in acidic mists or fogs can also enter leaves through the stomates, form H_2SO_4 upon reaction with H_2O in the cytoplasm, and disrupt many metabolic processes. Leaf and needle necrosis occurs when plants are exposed to high levels of SO_2 gas, possibly due to collapsed epidermal cells, eroded cuticles, loss of chloroplast integrity and decreased chlorophyll content, loosening of fibers in cell walls and reduced cell membrane integrity, and changes in osmotic potential that cause a decrease in cell turgor.

Root diseases may also increase in excessively acidic soils. In addition to the damages caused by exposure to H_2SO_4 and HNO_3, roots can be directly injured or their growth rates impaired by increased concentrations of soluble Al^{3+} and Mn^{2+} in the rhizosphere,[2,25] http://nadp.sws.uiuc.edu. Changes in the amount and composition of these exudates can then alter the activity and population diversity of soil-borne pathogens. The general tendency associated with increased root exudation is an enhancement in microbial populations due to an additional supply of carbon (energy). Chronic acidification can also alter nutrient availability and uptake patterns.[8,22]

Long-term studies in New England suggest acidic deposition has caused significant plant and soil leaching of base cations,[1,21] resulting in decreased growth of red spruce trees in the White Mountains.[6] With reduction in about 80% of the airborne base cations, mainly Ca^{2+} but also Mg^{2+}, from 1950 levels, researchers suggest forest growth has slowed because soils are not capable of weathering at a rate that can replenish essential nutrients. In Germany, acidic deposition was implicated in the loss of soil Mg^{2+} as an accompanying cation associated with the downward leaching of SO_4^{2-}, which ultimately resulted in forest decline.[2] Several European countries have used helicopters to fertilize and lime forests.

Aquatic Ecosystems

Ecological damage to aquatic systems has occurred from acidic deposition. As with forests, a number of interrelated factors associated with acidic deposition are responsible for undesirable changes. Acidification of aquatic ecosystems is not new. Studies of lake sediments suggest that increased acidification began in the mid-1800s, although the process has clearly accelerated since the 1940s.[15] Current studies indicate there is significant S mineralization in forest soils impacted by acidic deposition and that the SO_4^{2-} levels in adjacent streams remain high, even though there has been a decrease in the amount of atmospheric-S deposition.[24]

Geology, soil properties, and land use are the main determinants of the effect of acidic deposition on aquatic chemistry and biota. Lakes and streams located in areas with calcareous geology resist acidification more than those in granitic and gneiss materials.[16] Soils developed from calcareous parent materials are generally deeper and more buffered than thin, acidic soils common to granitic areas.[2] Land management decisions also affect freshwater acidity. Forested watersheds tend to contribute more acidity than those dominated by meadows, pastures, and agronomic ecosystems.[8,14,20] Trees and other vegetation in forests are known to "scavenge" acidic compounds in fogs, mists, and atmospheric particulates. These acidic compounds are later deposited in forest soils when rainfall leaches forest vegetation surfaces. Rainfall below forest canopies (e.g., throughfall) is usually more acidic than ambient precipitation. Silvicultural operations that disturb soils in forests can increase acidity by stimulating the oxidization of organic N and S, and reduced S compounds such as FeS_2.[2]

A number of ecological problems arise when aquatic ecosystems are acidified below pH 5.0, and particularly below pH 4.0. Decreases in biodiversity and primary productivity of phytoplankton, zooplankton, and benthic invertebrates commonly occur.[15,16] Decreased rates of biological decomposition of organic matter have occasionally been reported, which can then lead to a reduced supply of nutrients.[20] Microbial communities may also change, with fungi predominating over bacteria. Proposed mechanisms to explain these ecological changes center around physiological stresses caused by exposure of biota to higher concentrations of Al^{3+}, Mn^{2+}, and H^+ and lower amounts of available Ca^{2+}.[15] One specific mechanism suggested involves the disruption of ion uptake and the ability of aquatic plants to regulate Na^+, K^+, and Ca^{2+} export and import from cells.

Acidic deposition is associated with declining aquatic vertebrate populations in acidified lakes and, under conditions of extreme acidity, of fish kills. In general, if the water pH remains above 5.0, few problems are observed; from pH 4.0 to 5.0 many fish are affected, and below pH 3.5 few fish can survive.[23] The major cause of fish kill is due to the direct toxic effect of Al^{3+}, which interferes with the role Ca^{2+} plays in maintaining gill permeability and respiration. Calcium has been shown to mitigate the effects of Al^{3+}, but in many acidic lakes the Ca^{2+} levels are inadequate to overcome Al^{3+} toxicity. Low pH values also disrupt the Na^+ status of blood plasma in fish. Under very acidic conditions, H^+ influx into gill membrane cells both stimulates excessive efflux of Na^+ and reduces influx of Na^+ into the cells. Excessive loss of Na^+ can cause mortality. Other indirect effects include reduced rates of reproduction, high rates of mortality early in life or in reproductive phases of adults, and migration of adults away from acidic areas.[16] Amphibians are affected in much the same manner as fish, although they are somewhat less sensitive to Al^{3+} toxicity. Birds and small mammals often have lower populations and lower reproductive rates in areas adjacent to acidified aquatic ecosystems. This may be due to a shortage of food due to smaller fish and insect populations or to physiological stresses caused by consuming organisms with high Al^{3+} concentrations.

REDUCING ACIDIC DEPOSITION EFFECTS

Damage caused by acidic deposition will be difficult and extremely expensive to correct, which will depend on our ability to reduce S and N emissions. For example, society may have to burn less fossil fuel, use cleaner energy sources and/or design more efficient "scrubbers" to reduce S and N gas entering our atmosphere. Despite the firm conviction of most nations to reduce acidic deposition, it appears that the staggering costs of such actions will delay implementation of this approach for many years. The 1990 amendments to the Clean Air Act are expected to reduce acid-producing air pollutants from electric power plants. The 1990 amendments established emission allowances based on a utilities' historical fuel use and SO_2 emissions, with each allowance representing 1 ton of SO_2 that canbought, sold or banked for future use,[4,6,7] http://www.nnic.noaa.gov/CENR/NAPAP/. Short-term remedial actions for acidic deposition are available and have been successful in some ecosystems. Liming of lakes and some forests (also fertilization with trace elements and Mg^{2+}) has been practiced in European counties for over 50 years.[16,23] Hundreds of Swedish and Norwegian lakes have been successfully limed in the past 25 years. Lakes with short mean residence times for water retention may need annual or biannual liming; others may need to be limed every 5–10 years. Because vegetation in some forested ecosystems has adapted to acidic soils, liming (or over-liming) may result in an unpredictable and undesirable redistribution of plant species.

REFERENCES

1. Smith, W.H. Acid rain. In *The Wiley Encyclopedia of Environmental Pollution and Cleanup*; Meyers, R.A., Dittrick, D.K., Eds.; Wiley: New York, 1999; 9–15.
2. Pierzynski, G.M.; Sims, J.T.; Vance, G.F. *Soils and Environmental Quality*; CRC Press: Boca Raton, FL, 2000; 459 pp.
3. Wolff, G.T. Air pollution. In *The Wiley Encyclopedia of Environmental Pollution and Cleanup*; Meyers, R.A., Dittrick, D.K., Eds.; Wiley: New York, 1999; 48–65.
4. U.S. Environmental protection agency. In *Progress Report on the EPA Acid Rain Program*; EPA-430-R-99-011; U.S. Government Printing Office: Washington, DC, 1999; 20 pp.
5. Forster, B.A. *The Acid Rain Debate: Science and Special Interests in Policy Formation*; Iowa State University Press: Ames, IA, 1993.
6. *National Acid Precipitation Assessment Program Task Force Report*, National Acid Precipitation Assessment Program 1992 Report to Congress; U.S. Government Printing Office: Pittsburgh, PA, 1992; 130 pp.
7. *National Science and Technology Council*, National Acid Precipitation Assessment Program Biennial Report to Congress: An Integrated Assessment; 1998 (accessed July 2001).
8. Charles, D.F., Ed. The acidic deposition phenomenon and its effects: critical assessment review papers. In *Effects Sciences*; EPA-600/8-83-016B; U.S. Environmental Protection Agency: Washington, DC, 1984; Vol. 2.
9. McKinney, M.L.; Schoch, R.M. *Environmental Science: Systems and Solutions*; Jones and Bartlett Publishers: Sudbury, MA, 1998.
10. United Nations, World band and World resources institute. In *World Resources: People and Ecosystems—The Fraying Web of Life*; Elsevier: New York, 2000.
11. *National Atmospheric Deposition Program (NRSP-3)/National Trends Network*. Isopleth Maps. NADP Program Office, Illinois State Water Survey, 2204 Griffith Dr., Champaign IL, 61820, 2000 (accessed July 2001).
12. Council on environmental quality. In *Environmental Quality, 18th and 19th Annual Reports*; U.S. Government Printing Office: Washington, DC, 1989.
13. Charles, D.F., Ed.; *Acid Rain Research: Do We Have Enough Answers*? Proceedings of a Speciality Conference. Studies in Environmental Science #64, Elsevier: New York, 1995.
14. Kamari, J. *Impact Models to Assess Regional Acidification*; Kluwer Academic Publishers: London, 1990.
15. Charles, D.F., Ed. *Acidic Deposition and Aquatic Ecosystems*; Springer-Verlag: New York, 1991.
16. Mason, B.J. *Acid Rain: Its Causes and Effects on Inland Waters*; Oxford University Press: New York, 1992.
17. U.S. environmental protection agency. Effects of acid rain: human health. In *EPA Environmental Issues Website*. Update June 26, 2001 (accessed July 2001).
18. Marion, G.M.; Hendricks, D.M.; Dutt, G.R.; Fuller, W.H. Aluminum and silica solubility in soils. Soil Science 1976, 121, 76–82.
19. Kennedy, I.R. *Acid Soil and Acid Rain*; Wiley: New York, 1992.
20. Reuss, J.O.; Johnson, D.W. *Acid Deposition and the Acidification of Soils and Waters*; Springer-Verlag: New York, 1986.
21. Likens, G.E.; Driscoll, C.T.; Buso, D.C. Long-term effects of acid rain: response and recovery of a forest ecosystem. Science 1996, 272, 244–246.
22. Linthurst, R.A. *Direct and Indirect Effects of Acidic Deposition on Vegetation*; Butterworth Publishers: Stoneham, MA, 1984.
23. Bush, M.B. *Ecology of a Changing Planet*; Prentice- Hall: Upper Saddle River, NJ, 1997.
24. Alawell, C.; Mitchell, M.J.; Likens, G.E.; Krouse, H.R. Sources of stream sulfate at the hubbard brook experimental forest: long-term analyses using stable isotopes. Biogeochemistry 1999, 44, 281–299.
25. National Atmospheric Deposition Program. *Nitrogen in the Nation's Rain.* NADP Brochure 2000–01a (accessed July 2001).

Acid Sulfate Soils

Delvin S. Fanning
Department of Environmental Science and Technology, University of Maryland, College Park, Maryland, U.S.A.

Abstract

The term *acid sulfate soils* was introduced by L.J. Pons and others prior to the first international acid sulfate soils symposium at Wageningen, the Netherlands, in 1972. According to Pons, these soils were known for centuries and were recognized with terms such as *argilla vitriolacea*, meaning "clay with sulfuric acid" by Linnaeus in the 18th century. They were recognized prior to the acid sulfate soils terminology by colloquial terms such as the Dutch *Kattekleigronden* or *Katte Klei*, or the English *cat clay soils*, or the German *Maibolt* to imply hayfields affected by an evil spirit, or *Gifterde* for poison earth, to connote mysterious and evil circumstances surrounding the difficulty in producing crops on them. Five subsequent international symposia/conferences, held in Thailand/Malaysia in 1981, West Africa (Senegal, Gambia, Guiné Bissau) in 1986, Vietnam in 1992, Australia in 2002, and China in 2008, with a seventh planned for Finland in 2012, show the wide geographic distribution of these environmentally sensitive soils in coastal regions of the world; however, *sulfidic materials* lurk in the unoxidized zone of the soil-geologic columns, ready to give rise to active acid sulfate soils upon exposure by land disturbance, in many inland/upland regions of the earth over sediments and sedimentary rocks influenced by sulfidization during their deposition under marine or estuarine environments in the geologic past. This entry references nomenclature schemes for these soils based on current *soil taxonomy* definitions (provided) of *sulfidic materials* (potential acid sulfate soil materials) and the *sulfuric horizon* (active acid sulfate soil materials) and discusses potential, active, and postactive acid sulfate soils utilizing *soil taxonomy* terms and concepts and references some taxonomic great group and subgroup classes that represent the genetic stages of development. The Australian soil classification system and the WRB (World Reference Base) system of IUSS (International Union of Soil Science) define *sulfidic materials* (Australia) or *sulphidic materials* (WRB) and *sulfuric materials* (Australia) or the *thionic horizon* (WRB) in ways similar to the equivalent *soil taxonomy* terms. Also, the geographic distinction between Coastal as opposed to Inland/Upland acid sulfate soils is briefly explained.

INTRODUCTION

Pons[1] reported that acid sulfate soils have been known for ages and began to receive scientific attention in the 18th century when Linnaeus recognized them in the Netherlands with terms such as *argilla vitriolacea*, meaning "clay with sulfuric acid." Terms such as the Dutch *Kattekleigronden* or *Katte Klei*, or the English *cat clay soils*, or the German *Maibolt* to imply hayfields affected by an evil spirit, or *Gifterde* for poison earth, were applied to these soils to connote mysterious and evil circumstances surrounding the difficulty in producing crops on them. The *acid sulfate soils* term is of more recent usage, introduced a little before the first international symposium on acid sulfate soils in 1972. It has gained popularity with the six international acid sulfate soils symposia/conferences[2–7] that have been held under the sponsorship of the International Land Reclamation Institute (ILRI), Wageningen, the Netherlands, and other organizations, particularly for the fifth (Australia) and sixth (China) conferences, and usage of the term is apparent in many other publications (e.g., Dent,[8] Kittrick et al.,[9] and Sammut[10]) on these soils.

DEFINITIONS AND MAIN KINDS

In the broad sense, acid sulfate soils include all soils in which sulfuric acid may be produced, is being produced, or has been produced in amounts that have a lasting effect on main soil characteristics.[1,11] This definition includes potential, active, and postactive acid sulfate soils, which are broad genetic kinds as described in succeeding paragraphs.

Potential acid sulfate soils are anaerobic soils, commonly occurring in, or at one time formed in, coastal (tidal) sedimentary environments affected by sulfidization (see Fanning and Fanning[12] and Fanning[13] for explanations of this process). Potential acid sulfate soils contain sulfide minerals at such levels in near-surface horizons/layers that they are expected to generate, upon exposure to oxidizing conditions, sufficient sulfuric acid to drive the pH of these

Encyclopedia of Environmental Management DOI: 10.1081/E-EEM-120046326

horizons/layers to ultralow levels. Under these conditions, most plants would be unable to grow on the soils and an active acid sulfate soil would then be recognized.

Active acid sulfate soils form where sulfide minerals (most typically iron sulfides and the mineral pyrite) have oxidized in near-surface horizons and formed enough sulfuric acid, with insufficient neutralization, to have made the pH drop to ultralow levels. Most commonly, the pH, as measured in water, is 3.5 or less, or 4.0 or less with sulfide minerals still present, such that a *sulfuric horizon*, as defined later, is recognized. In postactive acid sulfate soils, weathering and pedogenesis have proceeded beyond the active stage to where sulfide minerals are no longer present in surface and near-surface soil horizons, and the pH in these horizons typically has risen to levels above that which would cause them to be recognized as a *sulfuric horizon*.

A simpler definition of acid sulfate soils[10] is simply that they are soils that contain iron sulfides. This definition calls attention to the minerals that cause acid sulfate soils to become acid; however, it neglects to consider that how acidic the soils become depends on the balance between the acid-forming substances (mainly iron sulfides) and the substances (minerals) that neutralize acidity, most commonly calcium carbonate minerals. This simpler definition also does not recognize the genetic distinctions among the potential, active, and postactive stages of development.

CLASSIFICATION IN SOIL TAXONOMY

Potential, active, and early postactive acid sulfate soils receive special recognition in *soil taxonomy*.[14,15] The taxonomic definition of *sulfidic materials* also permits the recognition of these acid-forming, potential acid sulfate soil materials, regardless of their depth in the soil-geologic column. This is useful for their recognition in construction activities such as mining, highway construction, and dredging operations, which can bring active acid sulfate soils into existence by exposing (to oxidation) *sulfidic materials* that previously occurred deep in the column or under deep water.

DIAGNOSTIC CHARACTERISTICS

Sulfidic Materials

Current *soil taxonomy*[15] gives the following definition (required characteristics): *Sulfidic materials* have *one or both* of the following:

1. A pH value (1:1 in water) of more than 3.5, and, when the materials are incubated at room temperature as a layer 1 cm thick under moist aerobic conditions (repeatedly moistened and dried on a weekly basis), the pH decreases by 0.5 or more units to a value of 4.0 or less (1:1 by weight in water or in a minimum of water to permit measurement) within 16 weeks or longer until the pH reaches a nearly constant value if the pH is still dropping after 16 weeks.
2. A pH value (1:1 in water) of more than 3.5 and 0.75% or more S (dry mass) mostly in the form of sulfides, and less than 3 times as much calcium carbonate equivalent as S.

Sulfuric Horizon

The following is the current definition (required characteristics)[15] of the *sulfuric horizon* as given in *soil taxonomy*:

The *sulfuric horizon* is 15 cm (or more) thick and is composed of either mineral or organic soil material that has a pH value (1:1 by weight in water or in a minimum of water to permit measurement) of 3.5 or less or less than 4.0 if sulfide or other S-bearing minerals that produce sulfuric acid upon their oxidation are present. The horizon shows evidence that the low pH value is caused by sulfuric acid.

The evidence is *one or both* of the following:

1. The horizon has
   a. Concentrations of jarosite, schwertmannite, or other iron and/or aluminum sulfates or hydroxysulfate minerals, or
   b. 0.05 or more water-soluble sulfate.
2. The layer directly underlying the horizon consists of *sulfidic materials* (defined above).

The definition of the *sulfuric horizon* permits recognition of active acid sulfate soils and soil materials currently affected by active sulfuricization (see Fanning and Fanning[12] and Fanning[13] for explanations of this process).

CLASSES

Potential and active acid sulfate soils are classified in *soil taxonomy* in the orders of *Entisols*, *Inceptisols*, and *Histosols*. Within these orders, acid sulfate soils are recognized at the great group and subgroup levels, which are intermediate categories—third and fourth in level from the top (order) level of the six categories of *soil taxonomy*. Mineral potential acid sulfate soils that have *sulfidic materials* within their profiles with no overlying *sulfuric horizon* are classified in various great groups and subgroups of *Entisols* (e.g., in *Sulfaquents* or *Sulfowassents* at the great group level or in subgroups such as *Sulfic Psammowassents* at the subgroup level). The *Wassents* suborder, which has only recently (2010) been added to *soil taxonomy* for subaqueous soils, has several great groups and subgroups that contain *sulfidic materials* where *Wassents* occur in coastal regions. The sediments that occur in these places commonly offer good conditions for the formation of sulfide minerals from the sulfate of seawater by sulfidization as the sediments are deposited. Thus, dredged materials from dredging in coastal regions are likely to give rise to active acid sulfate soils if they are deposited in situations (e.g., upland disposal sites) under which the *sulfidic materials* can undergo oxidation. Consult *Keys to Soil Taxonomy*[15] for details of the various classes of *Entisols* that contain *sulfidic materials*. Organic potential acid sulfate soils that contain a zone of *sulfidic materials* 15 cm (or more) thick within 1 m of their surface are classified in various great groups or subgroups of the suborders of *Hemists*, *Saprists*, or *Wassists*. See *Keys to Soil Taxonomy*[15] for details.

Postactive acid sulfate soils are recognized in *soil taxonomy* only if they are early postactive. *Sulfic Endoaquepts* is an example subgroup of early postactive acid sulfate soils. These soils have a *sulfuric horizon* or *sulfidic materials* or both above a depth of 2 m, but too deep to qualify as a potential or active acid sulfate soil as explained above. Many other soils beyond those recognized as special classes in *soil taxonomy* are considered to be postactive by acid sulfate soil experts familiar with such soils, such as certain *Alfisols* and *Ultisols* in Texas[16] and certain *Ultisols* in Maryland.[17] It is important to recognize such soils that can have *sulfidic materials* or a *sulfuric horizon* at depths of greater than 2 m, because disturbance of these soils to these greater depths may expose *sulfidic materials* to oxidation and set off a new cycle of active sulfuricization and new active acid sulfate soils on new human-constructed land surfaces with the many environmental problems associated with these soils as described in other entries in this encyclopedia or elsewhere.[18,19] If avoidance of exposure of *sulfidic materials* to oxidizing conditions is to be practiced as a management strategy, *sulfidic materials* need to be recognized at whatever depth they may occur that may be affected by land disturbance activities. Places that have postactive acid sulfate soils on natural land surfaces are places where such situations are likely to pertain.

CLASSIFICATION IN OTHER SYSTEMS

Dent[8] presented the ILRI (which unfortunately no longer exists, at least by the ILRI title) system for the classification of acid sulfate soils, which considered acidity and potential acidity, salinity, soil composition and texture, degree of physical ripening, and profile form (the depth zone at which various properties occur).

Other modern systems that consider acid sulfate soils[20,21] use modifications of the *soil taxonomy* approach for the classification of acid sulfate soils. For diagnostic characteristics (terms for properties roughly equivalent to *sulfidic materials* and the *sulfuric horizon* of *soil taxonomy*), WRB (World Reference Base) defines *sulphidic materials* and the *thionic horizon* (differs from *sulfuric horizon* in that the pH is only required to be 4.0 or less, regardless of whether or not sulfide minerals are still present in the soil material), much like the *soil taxonomy* equivalent terms, whereas the Australian[20] system uses *sulfidic materials* and *sulfuric materials* (instead of *sulfuric horizon*), but with no depth restriction on the thickness of the *sulfuric materials*, compared to the 15 cm or more thickness required by *soil taxonomy* and the WRB systems, for the *sulfuric horizon* and the *thionic horizon*, respectively. Classes in the various systems are named differently than those of *soil taxonomy*.

The Food and Agriculture Organization (FAO) of the United Nations grouped potential and active acid sulfate soils together in *Thionic* classes for purposes of showing these soils on the FAO/UNESCO soil map of the world,[22] as it is not practical to separate potential and active acid sulfate soils on small-scale maps. *Thionic Fluvisols*, *Thionic Gleysols*, and *Thionic Histosols* were recognized.

INLAND/UPLAND AS OPPOSED TO COASTAL ACID SULFATE SOILS

Although official soil classification schemes do not separate coastal as opposed to inland/upland acid sulfate soils, most scientific research on soils traditionally called acid sulfate soils, such as much of those reported at the six international conferences, has been conducted on what are generally reported to be (geographically) coastal acid sulfate soils—where most of the sediments, in which the soils exist and develop, were deposited geologically in recent (e.g., Holocene) times. Soil scientists have less commonly recognized acid sulfate soils and utilized acid sulfate soil concepts and terminology in inland/upland situations. Part of the reason for this is that the extensive deep land disturbance that has brought *sulfidic materials* to land surfaces and set off new cycles of sulfuricization is a relatively recent phenomenon in human history. Since about the time of World War II (1940s), huge earth-moving equipment (drag lines, bulldozers, etc.) have been increasingly developed and employed in surface mining, in highway and airport construction, in dredging of shipping channels and other transportation-connected activities, and in making flatter land in hilly places for housing developments, shopping centers, etc. This deep land disturbance in uplands often occurs far from sea coasts, thus inland in peoples' minds. *Sulfidic materials* have been exposed or deposited in inlands/uplands, often from 2 m to many meters below original land surfaces and below depths normally considered in soil classification schemes. It has become increasingly recognized that new soils that develop, where *sulfidic materials* that prior to exposure were geologic formations at depth, give rise following exposure to active acid sulfate soils with *sulfuric horizons* at their surfaces that may be classified as *Sulfudepts* or *Sulfaquepts* by *soil taxonomy*. To draw attention to the geographic and/or geologic distinction of these soils compared to coastal acid sulfate soils, terms such as *inland acid sulfate soils* and *upland acid sulfate soils* are sometimes employed by authors of entries and books about these soils.[18,23]

Acid mine drainage (AMD) and acid rock drainage (ARD) have been recognized in areas of highly human-disturbed lands for many years. Now, it is increasingly recognized that the soils brought into existence by land disturbance that give rise to AMD and ARD are acid sulfate soils and that the biogeochemical processes that take place that give rise to the acid and metal charged AMD or ARD waters are close to, if not identical with, those that take place in acid sulfate soils. Humankind should not think of acid sulfate soils as a phenomenon of only coastal regions of the earth, although many of the geologic formations that give rise to acid sulfate soils in inland/upland regions may have been deposited as sediments and have been influenced by seawater and sulfidization during their deposition in coastal regions. For elaboration on land/sea and coastal/inland/upland models, pertaining to acid sulfate soils in the Chesapeake Bay region (in the United States), see Fanning et al.[24] and Fanning.[25] Coastal acid sulfate soils are commonly much softer with much lower bulk densities (with high *n* values indicating a low bearing capacity for machines or animals that may traffic over them) than dense/firm soil materials, including underlying *sulfidic materials* that are commonly found in inland/upland situations.

REFERENCES

1. Pons, L.J. Outline of the genesis, characteristics, classification and improvement of acid sulphate soils. In *Proceedings of the 1972 (Wageningen, Netherlands) International Acid Sulphate Soils Symposium, Volume 1*; Dost, H., Ed.; International Land Reclamation Institute Publication 18: Wageningen, the Netherlands, 1973; 3–27.
2. Dost, H., Ed. *Proceeding of the International Symposium on Acid Sulphate Soils*, Wageningen, the Netherlands, Aug. 13–29, 1972, Volumes I and II; International Land Reclamation Institute Pub. 18: Wageningen, the Netherlands. 1973.
3. Dost, H.; van Breemen, N., Eds. *Proceedings of the Bangkok Symposium on Acid Sulphate Soils, Jan. 18–24, 1981*. International Land Reclamation Institute Pub. 31: Wageningen, the Netherlands, 1982.
4. Dost, H., Ed. *Selected Papers of the Dakar Symposium on Acid Sulphate Soils. Dakar, Senegal, Jan. 1986*. International Land Reclamation Institute Pub. 44: Wageningen, the Netherlands, 1988.
5. Dent, D.; van Mensvoort, M.E.F., Eds. *Selected Papers of the Ho Chi Minh City Symposium on Acid Sulphate Soils, Ho Chi Minh City, Vietnam, Mar. 1992*. International Land Reclamation Institute Pub. 53: Wageningen, the Netherlands, 1993.
6. *Australian Journal of Soil Research, Special Issue: Sustainable Management of Acid Sulfate Soils*. Selected Papers of 5th International Acid Sulfate Soils Conference, Tweed Heads, Australia, Aug. 25–30, 2002. Aust. J. Soil Res. **2004**, *42* (5 and 6).
7. Lin, C.; Huang, S.; Li, Y., Eds. *Proceedings of the Joint Conference of the 6th International Acid Sulfate Soil Conference and the Acid Rock Drainage Symposium*; Guangdong Science and Technology Press: Guangzhou, China, 2008.
8. Dent, D. *Acid Sulfate Soils: A Baseline for Research and Development*. International Land Reclamation Institute Pub. 39: Wageningen, the Netherlands, 1993.
9. Kittrick, J.A.; Fanning, D.S.; Hossner L.R., Eds. *Acid Sulfate Weathering*; Soil Science Society of America Special Publication No. 10; Soil Science Society of America: Madison, WI, 1982.
10. Sammut, J. *An Introduction to Acid Sulfate Soils*; New South Wales Department of Agriculture: Wollongbar, New South Wales, Australia, 1997; 23 pp.
11. Fanning, D.S.; Burch, S.N. Coastal acid sulfate soils. In *Reclamation of Drastically Disturbed Lands*; Barnhisel, R.I., Daniels, W.L., Darmody, R.G., Eds.; Agronomy Monograph 41, American Society of Agronomy: Madison, WI, 2000; 921–937.
12. Fanning, D.S.; Fanning, M.C.B. *Soil: Morphology, Genesis, and Classification*; John Wiley and Sons: New York, 1989.
13. Fanning, D.S.; Rabenhorst, M.C.; Burch, S.N.; Islam K.R.; Tangren, S.A. Sulfides and sulfates. In *Soil Mineralogy with Environmental Applications*; Dixon, J.B.; Schulze, D.G., Eds.; Soil Science Society of America: Madison, WI, 2002; 229–260.
14. Soil Survey Staff. *Soil Taxonomy*, 2nd Ed.; U.S. Department of Agriculture Handbook No. 436; U.S. Government Printing Office: Washington, DC, 1999.
15. Soil Survey Staff. *Keys to Soil Taxonomy*, 11th Ed.; USDA Natural Resources Conservation Service: Washington, DC, 2010, available at ftp://ftp-fc.sc.egov.usda.gov/NSSC/Soil_Taxonomy/keys/2010_Keys_to_Soil_Taxonomy.pdf (accessed March 7, 2011).
16. Carson, C.D.; Fanning, D.S.; Dixon, J.B. Alfisols and ultisols with acid sulfate weathering features in Texas. In *Acid Sulfate Weathering*; Kittrick, J.A., Fanning, D.S., Hossner, L.R., Eds.; Soil Science Society of America Special Publication No. 10; Soil Science Society of America: Madison, WI, 1982; 127–146.
17. Wagner, D.P.; Fanning, D.S.; Foss, J.E.; Patterson, M.S.; Snow, P.A. Morphological and mineralogical features related to sulfide oxidation under natural and disturbed land surfaces in Maryland. In *Acid Sulfate Weathering*; Kittrick, J.A., Fanning, D.S., Hossner, L.R., Eds.; Soil Science Society of America Special Publication No. 10; Soil Science Society of America: Madison, WI, 1982; 109–125.
18. Fanning, D.S.; Coppock, C.; Orndorff, Z.W.; Daniels, W.L.; Rabenhorst, M.C.; Upland active acid sulfate soils from construction of new Stafford County, Virginia, USA, Airport. Austr. J. Soil Res. **2004**, *42*, 527–536.
19. Fanning, D.S.; Burch, S.N. Acid sulfate soils and some associated environmental problems. In *Soils and Environment*; Auerswald, K., Stanjek, H., Bigham, J.M., Eds.; Advances

in Geoecology 30, Catena Verlag: Reiskirchen, Germany, 1997; 145–158.
20. Isbell, R.F. *The Australian Soil Classification,* Revised Ed.; CSIRO Publishing: Melbourne, 2002; Available at http://www.clw.csiro.au/aclep/asc_re_on_line/soilhome.htm (accessed March 7, 2011).
21. IUSS Working Group WRB. *World Reference Base for Soil Resources 2006, first update 2007.* World Soil Resources Reports No. 103.; FAO: Rome, 2007; Available at http://www.fao.org/ag/agl/agll/wrb/doc/wrb2007_corr.pdf (accessed March 7, 2011).
22. FAO/UNESCO/ISRIC (1988, reprinted 1990) Revised Legend, Soil Map of the World. FAO World Soil Resources Reports # 60. FAO Rome. Available at http://www.iiasa.ac.at/Research/LUC/External-World-soil-database/HTML/ (accessed 2008)
23. Fitzpatrick, R.; Shand, P., Eds. *Inland acid Sulfate Soils Systems across Australia. Thematic Volume. Covering: Distribution, Properties, Significance and Biogeochemical Processes of Inland Acid Sulfate Soils (ASS) across Australia and Overseas*; CRC LEME (Cooperative Research Centre for Landscape Environments and Mineral Exploration) Open File Report 249, CRC LEME: Perth, Australia, 2008; 303 pp. Individual chapters and case studies are downloadable as pdf files from http://crcleme.org.au/Pubs/monographs.html (accessed March 7, 2011) or from http://www.clw.csiro.au/acidsulfatesoils/index.html (accessed March 7, 2011).
24. Fanning, D.S.; Rabenhorst, M.C.; Balduff, D.M.; Wagner, D.P.; Orr, R.S.; Zurheide, P.K. An acid sulfate perspective on landscape/seascape soil mineralogy in the U.S. Mid-Atlantic region. Geoderma **2009**, *154*, 457–464.
25. Fanning, D.S. (Tour Leader, Guidebook Editor). Acid sulfate soils of the U.S. Mid-Atlantic/Chesapeake Bay Region. Guidebook for tour, July 6–8, 2006 for 18th World Congress of Soil Science. University of Maryland: College Park, MD, 2006; Available at http://www.sawgal.umd.edu/mapss/Documents_MAPSS/WCSS_Guidebook.pdf (accessed March 7, 2011).

Acid Sulfate Soils: Formation

Martin C. Rabenhorst
University of Maryland, College Park, Maryland, U.S.A.
Delvin S. Fanning
Department of Environmental Science and Technology,
University of Maryland, College Park, Maryland, U.S.A.
Steven N. Burch
University of Maryland, College Park, Maryland, U.S.A.

Abstract
Soils containing sulfide minerals that have not yet been oxidized through acid sulfate weathering are referred to as potential acid sulfate soils (potential acid SS). This entry first discusses the processes involved in the formation and accumulation of sulfide minerals in soils leading to the formation of potential acid SS. Processes related to the oxidation of sulfides in the formation of active acid SS are then examined.

INTRODUCTION

Soils containing sulfide minerals that have not yet been oxidized through acid sulfate weathering are referred to as potential acid sulfate soils (potential acid SS). The processes involved in the formation and accumulation of sulfide minerals in soils leading to the formation of potential acid SS will be discussed first. Subsequently, processes related to the oxidation of sulfides in the formation of active acid SS will be examined. The extent of acid SS worldwide has been estimated to be approximately 12–15 MHa.[1]

POTENTIAL ACID SULFATE SOILS AND SULFIDE MINERAL FORMATION—SULFIDIZATION

Biogeochemistry of Sulfide Mineral Formation

Several factors are required for sulfate reduction. These include a source of sulfate, a source of oxidizable carbon, reducing conditions and the presence of sulfate reducing bacteria.[2] Any of these components could theoretically limit sulfate reduction. In saturated soil or sedimentary environments where the required factors are present, heterotrophic microbes utilize sulfate as an electron acceptor that becomes reduced to sulfide according to Eq. (1).

$$SO_4^{2-} + 10H^+ + 8e^- \rightarrow H_2S + 4H_2O \qquad (1)$$

Sulfate

Provided that the other required factors are met, the quantity of sulfate may limit the rate of sulfate reduction. Goldhaber and Kaplan[3] reported sulfate reduction to be independent of concentration when sulfate levels are above 10 mM (320 mg/L). Work by Haering,[4] in Chesapeake Bay, indicated that sulfate levels may begin to limit sulfur accumulation in marsh soils when levels drop below 1 mM (32 mg/L). Some degree of sulfate reduction will continue as long as sulfate is present at minimal levels (>5–20 μM, 0.16–0.6 mg/L).[5] Because sulfate-reducing bacteria are better able to complete for electron-donating substrates than are methane-generating bacteria, methanogenesis is of minimal significance so long as sulfate levels are above 0.03–0.4 mM.[6] Therefore, sulfate reduction dominates in brackish systems. In freshwater, sulfate reduction may become overshadowed by methanogenesis as sulfate is depleted.

Oxidizable Organic Carbon

The oxidation of organic matter provides the energy microorganisms need to facilitate sulfate reduction. Plant materials rich in labile components are more easily decomposed than humified soil organic matter or peat. In sediments low in organic matter, sulfate reduction may be limited by the paucity of oxidizable carbon. This can be demonstrated in thin sections from mineral horizons in tidal marsh soils where iron sulfide minerals have accumulated in pores occupied by decaying plant roots (Fig. 1). The intimate association of pyrite with the decomposing organic minerals, and its near absence from the surrounding soil matrix suggests that organic matter is limiting the formation of sulfide.[7]

Reducing/Saturated Conditions

Because diffusion of gases through saturated soils and sediments is very slow, oxygen becomes depleted under saturated conditions and microbes which utilize other electron acceptors become active. Nitrate, Mn(IV) and Fe(III) are so utilized as the environment becomes progressively reduced (Fig. 2). If the conditions permit the entry of oxygen, then redox potentials may never become sufficiently

Encyclopedia of Environmental Management DOI: 10.1081/E-EEM-120001777

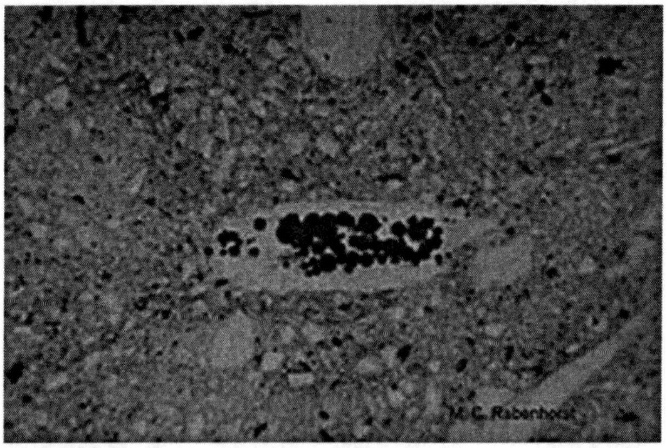
M.C. Rabenhorst

Fig. 1 Micrograph of a thin section from the mineral (Cg) horizon of a tidal marsh soil illustrating accumulation of pyrite framboids in the channel occupied by decaying plant roots. Plane polarized light; frame length 1.2 mm.

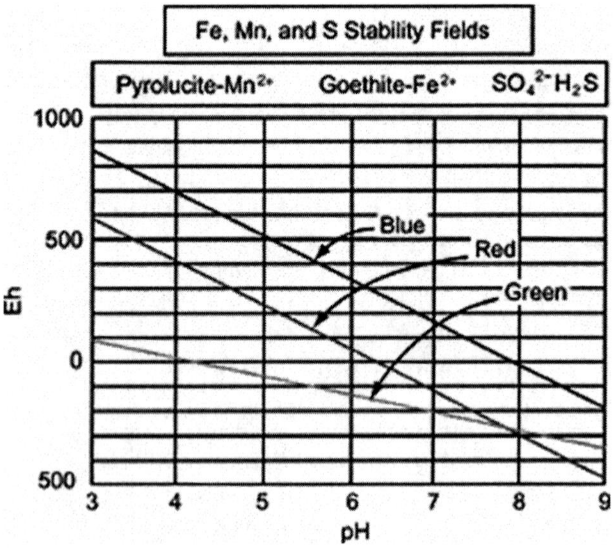
Fe, Mn, and S Stability Fields
Pyrolucite-Mn2+
Goethite-Fe2+
SO4 2- -H2S
1000
500
0
500
Eh
Blue
Red
Green
3
4
5
6
7
8
9
pH

Fig. 2 pe-pH diagram illustrating location of stability fields for redox sensitive components. The sulfate-sulfide lines (Green) is based on a $(SO)_4{}^{2-}$ concentration of 10 mM and a pH_2S of 0.001 atm. (Blue line separates the pyrolucite-Mn^{2+} stability fields; red line separates the goethite-Fe^{2+} stability fields.)

low to foster sulfate reduction. More typically, diffusion of oxygen into a saturated soil or sediment is sufficiently slow, and if other necessary factors are present, sulfate reduction will occur. Fig. 2 illustrates that pH, as well as E_h, must be specified in assessing sulfur phase equilibria. For example, as the pH increases from 5 to 7, the minimum E_h at which sulfate reduction is expected decreases from approximately −50 to −200 (based on a $SO_4{}^{2-}$ concentration of 10 mM and a pH_2S of 0.0001 atm).

Sulfate-Reducing Bacteria

Some 15 genera of bacteria have been recognized as sulfate reducers including *Desulfovibrio*, *Desulfotomaculum*, and *Desulfobacter*.[8,9] These organisms thrive under strongly reducing conditions, but many are able to persist in aerobic conditions for significant periods of time. Thus, if the other factors necessary for sulfate reduction are present, sulfate reducing bacteria will also become active.

As with most heterotrophic bacteria, rates of sulfate reduction are temperature dependent. Optimum temperature for most sulfate reducers is 30–40°C,[8] and the rate of sulfate reduction generally increase with temperature across this range. Some groups of sulfate reducers are thermophyllic and can function at temperature up to 85°C. Thus, in tropical coastal wetlands, sulfate reduction occurs all year round. In higher latitudes, where soil and sediment temperatures may approach biological zero, rates may become very slow during winter.

Reactive Iron

Once formed, sulfide is available to form a variety of minerals provided there is adequate reactive iron present. Most of the iron enters coastal environments as iron oxides sorbed to the surface of clay and silt particles. When iron oxides in the sediments and marsh soils become reduced to Fe(II), they can form iron sulfide minerals. While monosulfide species may form first [Eq. (2)], and minerals such as greigite (Fe_3S_4) may persist in recent sediments, disulfide forms such as pyrite are energetically more stable and will form at the expense of the monosulfides.

$$Fe^{2+} + S^{2-} \rightarrow FeS \qquad (2)$$

$$\begin{aligned} &FeS \text{ and } (+S_x^{y-}, \text{ loss of } e^-, \text{ or } + H_2S) \\ &\rightarrow FeS_2 + \text{various} \end{aligned} \qquad (3)$$

Mechanisms for pyrite formation may follow several possible pathways including 1) reaction of monosulfide with polysulfide; 2) partial oxidation of monosulfide; and 3) reaction of monosulfides with H_2S[10] (Eq. 3). Sulfide itself has the ability to reduce Fe(III) to Fe(II) on the surface of iron oxides.[11] Pyrite can occur either as small (<2 μm) individual crystals or as spherical clusters of crystals called framboids. In low organic mineral sediments, reactive iron is usually present in excess, resulting in a low degree of pyritization.[12] However, in organic-rich soils iron may limit the accumulation of sulfide minerals, and the degree of pyritization is generally high. This has been demonstrated experimentally in salt marsh Histosols.[13]

Environments of Sulfide Formation and Accumulation

It is clear that in environments which provide a source of oxidizable carbon and sulfate and which are sufficiently saturated to enhance reducing conditions, sulfate reducing bacteria will generate sulfide. If reactive iron is present, then solid phase ferrous minerals will accumulate. This process of *sulfidization*[14] is shown schematically in Fig. 3. The obvious settings for these processes are coastal marine and brackish environments, where sulfate is abundant. Under

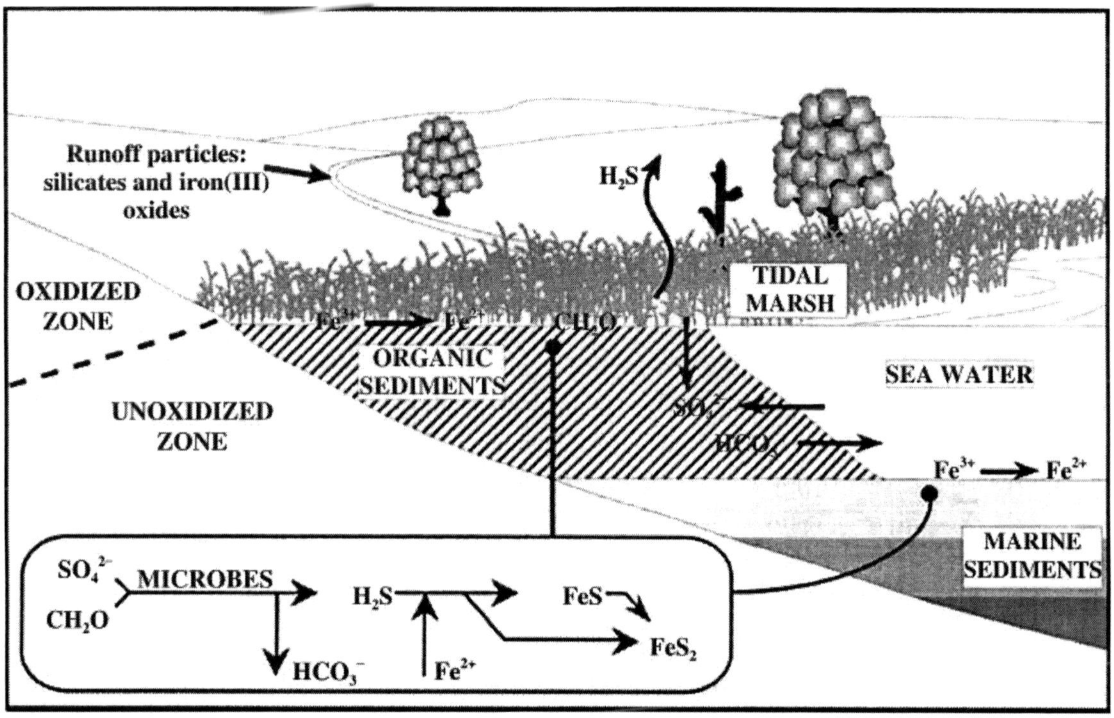
Runoff particles: silicates and iron(III) oxides
OXIDIZED ZONE
UNOXIDIZED ZONE
ORGANIC SEDIMENTS
TIDAL MARSH
SEA WATER
MARINE SEDIMENTS
H2S
Fe3+ → Fe2+
SO4 2-
CH2O
MICROBES
H2S
FeS
FeS2
HCO3-
Fe2+

Fig. 3 Schematic diagram illustrating the generalized process of sulfidization which leads to the formation of iron sulfide minerals and potential acid SS.
Source: Fanning et al.[26]

permanently submersed conditions, detrital carbon is added by flora and fauna to the sediment. In shallow water settings (<3 m) where various pedogenic processes are at work, these accumulated sediments have been recognized as subaqueous soils[15] and are classified to reflect the sulfide components.

The soils of coastal marshes (in temperate environments) and mangroves (in tropical settings) also are ideal for sulfide formation and accumulation. The high primary productivity of these ecosystems (up to $3\,kg\,m^{-2}\,yr^{-1}$ in marshes and up to $5\,kg\,m^{-2}\,yr^{-1}$ in mangroves)[16] makes these an exceptionally good environment for sulfate reduction. Such soils may contain up to 20–30 g/kg of pyrite sulfur, and estimates of pyrite S accumulation rates in estuarine marshes are as high as $7\,g\,m^{-2}\,yr^{-1}$.[17]

Sulfate reduction can occur in other settings, so long as a source of sulfate is available. While generally small, atmospheric deposition of sulfate may be enough to induce sulfate reduction in the sediments of some interior freshwater lakes. Sulfate reduction has also been documented in prairie potholes where sulfate has apparently been contributed by the weathering of sulfur-bearing shales.[18]

FORMATION OF ACTIVE ACID SULFATE SOILS—SULFURICIZATION

Chemistry of Sulfide Oxidation

Sulfides begin to oxidize once they are exposed to more oxidizing conditions. This occurs most often as a result of such human activities as drainage or dredging of sulfide-bearing soils or sediments, or the mining of sulfide bearing coal, but may also occur due to tectonic uplift or oceanic regression. Under humid or moist aerobic conditions, sedimentary sulfide minerals can oxidize chemically[19] but this is a slow process, probably due to particular rate-limiting reactions. Various microorganisms are adapted to oxidize sulfides either directly through sulfur transformations or by facilitating (catalyzing) such rate-limiting reactions as the oxidation of Fe(II) to Fe(III).[20] While there are many possible intermediate reactions in the oxidation of pyrite, the overall reaction is summarized in Eq. (4). One mole of pyrite eventually yields two moles of sulfuric acid and a mole of iron hydroxide.

$$FeS_2 + 3\tfrac{3}{4}O_2 + 3\tfrac{1}{2}H_2O \rightarrow 2H_2SO_4 + Fe(OH)_3 \qquad (4)$$

The oxidation of pyrite proceeds along two fronts (Fig. 4). First the S is oxidized (through intermediates) to sulfate yielding sulfuric acid and the remaining Fe(II). The generated Fe(II) sulfate salts are very soluble and potentially mobile. Secondly Fe(II) is oxidized to Fe(III), which when hydrolyzed produces additional acid. At high pH, the oxidation of sulfide is accomplished with oxygen, but under low pH conditions sulfide is oxidized by Fe(III). Microorganisms such as *Thiobacillus ferrooxidans* facilitate this reaction by oxidizing Fe(II) to Fe(III). For more details refer to Gagliano and Bigham.[21]

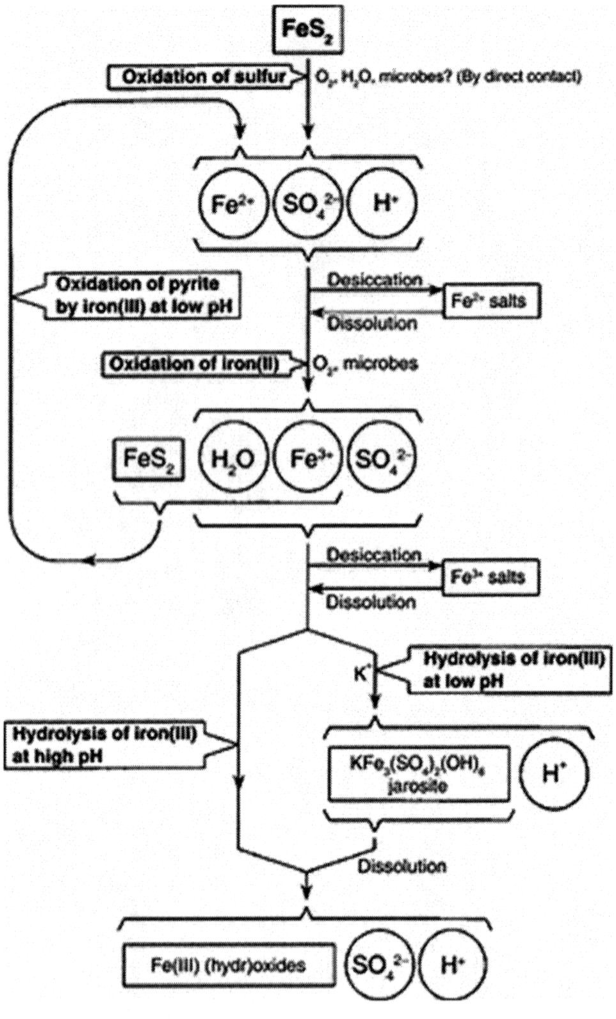
FeS2
Oxidation of sulfur
O2, H2O, microbes? (By direct contact)
Fe2+
SO42-
H+
Oxidation of pyrite by iron(III) at low pH
Desiccation
Fe2+ salts
Dissolution
Oxidation of iron(II)
O2, microbes
FeS2
H2O
Fe3+
SO42-
Desiccation
Fe3+ salts
Dissolution
Hydrolysis of iron(III) at low pH
K+
Hydrolysis of iron(III) at high pH
KFe3(SO4)2(OH)6 jarosite
H+
Dissolution
Fe(III) (hydr)oxides
SO42-
H+

Fig. 4 Schematic diagram illustrating the generalized process of sulfuricization which involves the oxidation of iron sulfide minerals and the production of acidity and the formation of new sulfate and other minerals.
Source: Fanning et al.[26]

Other Aspects of Sulfuricization and Properties of Acid Sulfate Soils

Sulfuricization is the overall process by which sulfide-bearing minerals are oxidized, minerals are weathered by the sulfuric acid produced and new mineral phases are formed from the dissolution products.[14,21] When $CaCO_3$ minerals are present, the sulfuric acid reacts with them to form the mineral gypsum, according to Eq. (5).

$$CaCO_3 + H_2SO_4 + H_2O \rightarrow CaSO_4 \bullet 2H_2O + CO_2 \quad (5)$$

As long as sufficient $CaCO_3$ is present, the pH is prevented from becoming very low and the soil does not become acid. When insufficient acid neutralizing minerals are present, the oxidation of pyrite in soils will lower the pH. The pH of active SS commonly drops to below four and in extreme can go below two. As iron is oxidized and hydrolyzed, various iron minerals form in the soil including ferrihydrite, schwestmannite and goethite. If the soil pH falls below four while maintaining an oxidizing environment (E_h > 400 mV) then the mineral jarosite ($KFe_3(SO_4)_2(OH)_6$) can form.[23] Because jarosite forms under conditions of high E_h and very low pH, which can only develop from the generation of sulfuric acid, it is considered a diagnostic mineral for acid SS.[24]

Jarosite has been reported in soils, which are not extremely acid and which may even contain carbonates.[22] These are interpreted to be "postactive" acid SS, meaning that earlier in their pedogenic history, they had undergone acid sulfate weathering. Subsequently, the soil pH has risen due to weathering of silicate minerals or addition of eolian carbonates. Because the redox potential has remained strongly oxidized, the jarosite has persisted as a metastable species. A recent review of acid sulfate soils including a discussion of the modeling of associated processes was recently completed.[25]

REFERENCES

1. Andriesse, W. Acid sulfate soils: Global distribution. In *Encyclopedia of Soil Science*; Lal, R., Ed.; Marcel Dekker: New York, 2001.
2. Rabenhorst, M.C.; James, B.R.; Magness, M.C.; Shaw, J.N. Iron removal from acid mine drainage in wetlands by optimizing sulfate reduction. In *The Challenge of Integrating Diverse Perspectives in Reclamation*. Proc. Am. Soc. Surf. Mining Reclam; Spokane: Washington, 1993; 678–684.
3. Goldhaber, M.B.; Kaplan, I.R. Controls and consequences of sulfate reduction rates in recent marine sediments. Soil Sci. **1975**, *119*, 42–55.
4. Haering, K.C. *Sulfur Distribution and Partitionment in Chesapeake Bay Tidal Marsh Soils*; M.S. thesis University of Maryland: College Park, MD, 1986; 172 pp. 5.
5. Ingvorsen, K.; Zehnder, A.J.B.; Jorgensen, B.B. Kinetic of sulfate and acetate uptake by delulfobacter postgatei. Appl. Environ. Microbiol. **1984**, *47*, 403–408.
6. Smith, D.W. Ecological actions of sulfate-reducing bacteria. In The *Sulfate-Reducing Bacteria: Contemporary Perspectives*; Odom, J.M., Singleton, R., Jr., Eds.; Springer: New York, 1993; 161–188.
7. Rabenhorst, M.C.; Haering, K.C. Soil micromorphology of a Chesapeake Bay Tidal Marsh: implications for sulfur accumulation. Soil Sci. **1989**, *147*, 339–347.
8. Fauque, G.D. Ecology of sulfate-reducing bacteria. In *Sulfate-Reducing Bacteria*; Barton, L.L., Ed.; Plenum Press: New York, 1995; 217–241.
9. Postgate, J.R. The sulphate-reducing bacteria (Ch. 7). In *Ecology and Distribution*, 2nd Ed.; Cambridge University Press: London, 1984; 107–122.
10. Rickard, D.; Schoonen, M.A.A.; Luther, G.W., III Chemistry of iron sulfides in sedimentary environments. In *Geochemical Transformations of Sedimentary Sulfur*; Vairavamurthy, M.A., Schoonen, M.A.A., Eds.; American Chemical Society: Washington, DC, 1995; 168–193.

11. Ghiorse, W.C. Microbial reduction of manganese and iron. In *Biology and Anaerobic Microorganisms*; Zehnder, A.J.B., Ed.; John Wiley and Sons: New York, 1988; 305–331.
12. Griffin, T.M.; Rabenhorst, M.C. Processes and rates of pedogenesis in some Maryland Tidal Marsh soils. Soil Sci. Soc. Am. J. **1989**, *53*, 862–870.
13. Rabenhorst, M.C. Micromorphology of induced iron sulfide formation in a Chesapeake Bay (USA) Tidal Marsh. In *Micromorphology: A Basic and Applied Science*; Douglas, L.A., Ed.; Elsevier: Amsterdam, 1990; 303–310.
14. Fanning, D.S.; Fanning, M.C.B. *Soil Morphology, Genesis, and Classification*; John Wiley and Sons: New York, 1989; 395 pp.
15. Demas, G.P.; Rabenhorst, M.C. Subaqueous soils: pedogenesis in a submersed environment. Soil Sci. Am. J. **1999**, *63*, 1250–1257.
16. Mitch, W.J.; Gosselink, J.G. *Wetlands*, 2nd Ed.; Van Nostrand Reinhold: New York, 1993; 722 pp.
17. Hussein, A.H.; Rabenhorst, M.C. Modeling of sulfur sequestration in coastal marsh soils. Soil. Sci. Soc. Am. J. **1999**, *63*, 1954–1963.
18. Arndt, J.L.; Richardson, J.L. Geochemistry of hydric soil salinity in a recharge-throughflow-discharge Prairie-Pothole wetland system. Soil. Sci. Soc. Am. J. **1989**, *53*, 848–855.
19. Borek, S.L. Effect of humidity of pyrite oxidation. In *Environmental Geochemistry of Sulfide Oxidation*; Alpers, C.N., Blowes, D.W., Eds.; American Chemical Society: Washington, DC, 1994; 31–44.
20. Nordstrom, D.K. Aqueous pyrite oxidation and the consequent formation of secondary iron minerals. In *Acid Sulfate Weathering*; Kittrick, J.A., Fanning, D.S., Hossner, L.R., Eds.; Soil Sci. Soc. Am. Spec. Pub. No. 10: Madison, WI, 1982; 37–56.
21. Gagliano, W.B.; Bigham, J.M. Acid mine drainage. In *Encylopedia of Soil Science*; Lal, R., Ed.; Marcel Dekker: New York, 2001.
22. Carson, C.D.; Fanning, D.S.; Dixon, J.B. Alfisols and ultisols with acid sulfate weathering features in Texas. In *Acid Sulfate Weathering*; Kittrick, J.A., Fanning, D.S., Hossner, L.R., Eds.; Soil Sci. Soc. Am. Spec. Pub. No. 10: Madison, WI, 1982; 127–146.
23. van Breeman, N. Genesis, Morphology, and classification of acid sulfate soils in coastal plains. In *Acid Sulfate Weathering*; Kittrick, J.A., Fanning, D.S., Hossner, L.R., Eds.; Soil Sci. Soc. Am. Spec. Pub. No. 10: Madison, WI, 1982; 95–108.
24. Fanning, D.S. Sulfate and Sulfide Minerals. In *Encyclopedia of Soil Science*; Lal, R., Ed.; Marcel Dekker: New York, 2001.
25. Ritsema, C.J.; van Mensvoort, M.E.F.; Dent, D.L.; Tan, Y.; van den Bosch, H.; van Wijk, A.L.M. Acid sulfate soils. In *Handbook of Soil Science*; Sumner, M.E., Ed.; CRC Press: Boca Raton, 1999; G121–G154.
26. Fanning, D.S.; Rabenhorst, M.C.; Burch, S.N.; Islam, K.R.; Tangren, S.A. Sulfides and sulfates. In *Soil Mineralogy with Environmental Applications*; Dixon, J.B., Schulze, D.G., Eds.; Soil Science Society of America Book Series #7; Madison, WI, 2002; 229–260.

Acaricides – Adsorption

Acid Sulfate Soils: Identification, Assessment, and Management

L.A. Sullivan
R.T. Bush
E.D. Burton
Southern Cross GeoScience, Southern Cross University, Lismore, New South Wales, Australia

Abstract

The presence and behavior of acid sulfate soil materials need to be appreciated by environmental managers if the severe environmental hazards that these soil materials present when mismanaged are to be avoided or minimized. Such environmental hazards include severe acidification of waterways and can impact on the build environment—bridges, drains, pipes, and roadways—as well as severely impact on the ecology of landscapes containing these soil materials, endangering aquatic life and public health. Many land uses—but especially agriculture, forestry, and aquaculture—may be severely affected by these acid sulfate soil–related hazards. Management of acid sulfate soil landscapes needs to take into account the hazards that these materials pose and include appropriate precautions to avoid if possible or, at the very least, to minimize the severe environmental degradation that mismanagement of these soil materials can cause. This entry provides information to enable the recognition of acid sulfate soil materials and the environmental hazards these materials pose as well as summarizes their properties and the processes that operate within them. The broad range of strategies available for the environmentally sustainable management of these materials, whether for intensive developments such as building sites or for rural areas used for agriculture, etc., are also addressed.

INTRODUCTION

Acid sulfate soil materials are distinguished from other soil materials by the dominating effect of redox-driven sulfur and iron geochemistry on their behavior and properties. Acid sulfate soil materials have either properties and behavior that *have been* affected considerably by the oxidation of reduced inorganic sulfur (RIS) or the capacity *to be* affected considerably by the oxidation of their RIS constituents. Similarly, reductive processes leading to the formation of RIS can also substantially affect the properties and behavior of soils and sediments.

A suite of environmental hazards can arise from the oxidation/reduction of acid sulfate soil materials. These include the following: 1) severe acidification of soil and drainage waters (pH below 4 and often less than 3); 2) mobilization of metals and metalloids, nutrients, and rare earth elements; 3) deoxygenation of water bodies; 4) production of noxious gases; and 5) scalding of landscapes (Fig. 1a). Some of these hazards are caused directly or indirectly by the severe acidification that can occur as a result of the oxidation of RIS contained in these soil materials, whereas some may arise from the reductive processes that these materials can experience after, for example, prolonged wetting.

Waters draining from acid sulfate soil materials may adversely impact on aquatic life, agriculture, aquaculture, infrastructure (especially concrete and steel structures such as bridges), and public health. For example, drainage waters from acid sulfate soil landscapes can often have a pH lower than 3.5 (Fig. 1b) and can be the cause of massive fish kills, the death of invertebrates and benthic organisms, the development of chronic fish diseases, and impaired fish recruitment.[1] Furthermore, acid sulfate soil materials can present health hazards to people living in landscapes containing these soil materials. Ljung et al.[2] found that acid sulfate soil materials could impact detrimentally on human health. Human health issues were related mainly to the increased mobility of acid and metals from these soils, affecting drinking water quality and food production and quality, but also to other issues, such as increased dust generation causing respiratory health issues and acidic pools of surface water in acid sulfate soil landscapes providing suitable environments for mosquito breeding. Consequently, management of developments located on, or impacting on, these soil materials needs to be aimed at avoiding if possible or, at the least, minimizing any such untoward effects on the immediate and surrounding landscapes and the environmental, social, and economic values contained therein.

Encyclopedia of Environmental Management DOI: 10.1081/E-EEM-120046327

CHARACTERISTICS, OCCURRENCE, AND GENERAL IMPACTS

Characteristics

The required conditions for the formation and accumulation of the RIS that characterize acid sulfate soil materials are as described simply in Eq. 1 below: 1) a supply of organic matter (exemplified in Eq. 1 as CH_2O); 2) reducing (usually waterlogged) conditions sufficient for sulfate reduction; 3) a supply of sulfate (SO_4^{2-}) from tidewater or other saline waters; and 4) a supply of iron from the soil (e.g., Fe_2O_3) sufficient for the accumulation of RIS, a process often called "sulfidization."

$$Fe_2O_3 + 4SO_4^{2-} + 8CH_2O + \frac{1}{2}O_2 \rightarrow FeS_2 + 8HCO_3^- + 4H_2O \quad (1)$$

Environments that commonly provide conditions conducive for sulfidization include marine and estuarine environments, brackish waterways, soils inundated by brackish water, or water bodies such as drains or wetlands affected by salinity (when sulfate is an appreciable component of that salinity). Tidal swamps, salt marshes, mangroves, and salt-affected seepage areas, to name a few examples, commonly meet these preconditions.[3,4] The RIS usually comprises either masses of framboids or individual crystals of FeS_2 (e.g., Fig. 2a) and tend to accumulate preferentially in voids.

Disturbance of sulfidic soils by, for example, drainage or excavation, often causes oxidation of RIS and the production of acidity (Eq. 2).

$$FeS_2 + 3.75O_2 + 3.5H_2O \rightarrow Fe(OH)_3 + 4H^+ + 2SO_4^{2-} \quad (2)$$

If there are insufficient effective neutralizing materials (e.g., fine-grained calcium carbonate particles) in the sediment to neutralize this acidity, low pHs can develop within weeks, resulting in sulfuric soil material (Fig. 3). Sulfuric soil material is characterized by acidic pHs (e.g., pHs < 4) and accumulations of products of RIS oxidation. Fig. 1c shows an acid sulfate profile from the scald in Fig. 1a, highlighting yellow segregations of jarosite [$KFe_3(SO_4)_2(OH)_6$] characteristic of acid sulfate soil materials in tropical and temperate climates. Red-brown segregations of schwertmannite [$Fe_8O_8(OH)_6(SO_4)$] and goethite ($\alpha FeOOH$) in the

(a)
(b)
(c)
(d)

Fig. 1 (a) An acid sulfate soil scald in eastern Australia. (b) Acidic (pH 3.4) iron-stained (by schwertmannite and goethite) drain in an acid sulfate soil landscape. (c) Profile of acid sulfate soil located in scald in (a). (d) Profile of a sandy acid sulfate soil from Lake Alexandrina in South Australia showing a surficial oxidized horizon with iron segregations, overlying a gleyed sulfidic horizon. The width of the soil exposures in (c) and (d) are 35 and 32 cm, respectively.

(a)
(b)
(c)
(d)

Fig. 2 (a) Scanning electron microscope (SEM) image of pyrite framboids in a root remnant of a sulfidic soil material. (b) SEM image of segregations of schwertmannite with characteristic "ball-and-whisker" morphology. (c) Monosulfidic black ooze accumulating beneath a red iron mineral-rich surficial layer in an acid sulfate soil landscape drain. (d) A trial showing successful early remediation of an acid sulfate scald wherever plots have been treated by ridging and the application of liming and mulch materials. The bar in (a) indicates 5 μm, and in (b), 1 μm.

uppermost sulfuric layers of acid sulfate soil materials are also common. Sulfidic layers vary greatly in appearance but often have the gleyed colors (i.e., blue-gray) typical of reduced soil materials that have experienced prolonged waterlogging (e.g., Fig. 1d).

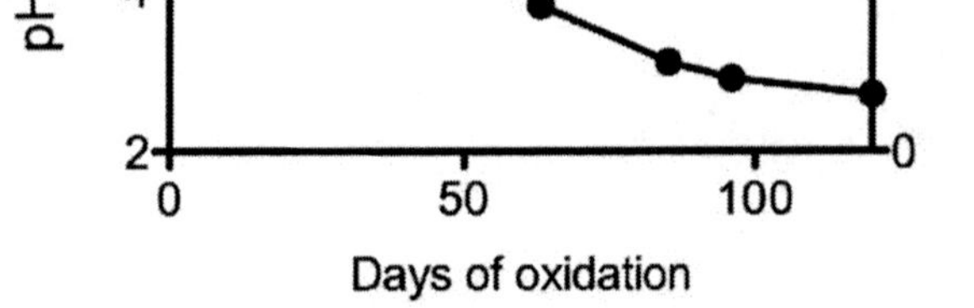
pH (1:1 soil:water)
8
6
4
2
0
50
100
Days of oxidation
S_{Cr}
pH
S_{Cr} (%)
3
2
1
0

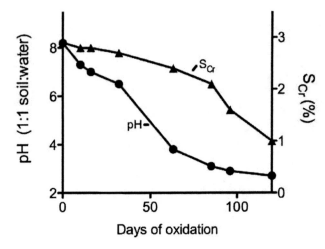

Fig. 3 Typical change in S_{Cr} content (mainly pyritic sulfur) and pH (1:1 soil:water) during moist oxidation of 10 mm thick slabs of a clayey gel-like sulfidic soil material from McLeods Creek in eastern Australia.
Source: Adapted from Word et al.[5]

Waters draining from acid sulfate soil landscapes exhibit a range of colors and clarity. For example, they can be red/brown from precipitation of iron flocs (e.g., Fig. 1b); white from the formation of aluminum flocs (especially if the pH is suddenly increased, for example, by the introduction of drainage waters into tidal waters); blue-green from dissolved ferrous iron compounds; or crystal clear (low-pH conditions and the presence of appreciable Al^{3+} can effectively flocculate suspended clays).

Sediments that form acid sulfate soil materials are usually deposited under saturated conditions such as those found in estuaries. Consequently, many of these sediments contain up to 70% water and have a gel-like consistency. These extremely soft sediments pose significant challenges due to their capacity to shrink on drying, deform under load, as well as acidify when disturbed sufficient to cause appreciable oxidation.

Occurrence

The global extent of land containing acid sulfate soil materials is approximately 50 million ha.[6] As well as being commonly found in tropical regions (e.g., Vietnam, Thailand, Indonesia, Malaysia, and northern Australia), acid sulfate soil materials are also commonly found in temperate regions (e.g., United States, Scandinavia, and southern

Australia). As acid sulfate soil materials are often located in densely settled coastal areas and floodplains where little suitable alternative land exists for the expansion of farming or urban and industrial development, they are soil materials that often are subject to intense development pressures.

Although acid sulfate soil materials have often been considered as almost exclusively a coastal issue, acid sulfate soil materials are also widely distributed in areas far from coastal influence. In Australia, for example, large areas affected by human-induced salinity (such as saline water disposal basins and saline seepage areas) have also been found to be areas affected by the contemporary formation of acid sulfate soil materials.[7–9] Additionally, acid sulfate soil materials also commonly form from pyritic soil parent materials in upland locations in the eastern United States.[10]

IDENTIFICATION AND ASSESSMENT

Accurate identification and assessment of acid sulfate soil materials is critical if the environmental hazards that these soil present are to be avoided or controlled. The basic accepted concept of a sulfuric soil material is one that is strongly acidic as a result of the oxidation of RIS. The concept of what constitutes a sulfidic soil material is generally that the soil material has the potential to become a sulfuric material following oxidation of the RIS it contains. However, uniform widely accepted definitions of "acid sulfate soil material," and of related terms such as "sulfidic" and "sulfuric" soil materials, do not exist, and in the case of the term "sulfidic," two distinctly different technical uses of this term exist, both within science, as will be discussed later.

Additionally, as will be clear from the following text, the type of methods used for identification and assessment of acid sulfate soil materials usually varies according to whether the main purpose of any particular exercise is for the classification of those materials per se or for the management of those materials.

Identification and Assessment of Sulfuric Material

The central concept of sulfuric soil (material or horizon) varies among soil taxonomies, and this is embodied in the definition from Isbell[11] below:

> Soil material that has a pH less than 4 (1:1 by weight in water, or in a minimum of water to permit measurement) when measured in dry season conditions as a result of the oxidation of sulphidic materials (defined above). Evidence that low pH is caused by oxidation of sulphides is one of the following:
>
> - Yellow mottles and coatings of jarosite (hue of 2.5Y or yellower and chroma of about 6 or more)
> - Underlying sulphidic material

Minor variations in definitions of sulfuric soil materials exist, but these are mainly in the choice of the critical pH. For example, the pH range chosen in the *Australian Soil Classification*[11] definition of sulfuric material is less than 4, whereas in the U.S. Department of Agriculture's *Soil Taxonomy*[12] the critical pH range for a sulfuric horizon is 3.5 or less, but it can also be "less than 4.0 if sulfide or other S-bearing minerals that produce sulfuric acid upon their oxidation are present." In the *World Reference Base*,[13] the critical pH range for thionic horizons (essentially synonymous with sulfuric horizons) is less than 3.5.

For assessment and management purposes, the following two acidity-related measures of sulfuric soil materials usually need to be assessed

1. *pH*. Determined either in water or dilute solutions, pH measures the "intensity" of the H^+ and is the most relevant measure of the geochemical environment.
2. *Titratable actual acidity* (*TAA*). Determined usually by titration of a soil suspension up to a defined pH (usually 6.5), the TAA is a measure of the capacity of the soil material to supply acidity. In acid sulfate soil materials, the main components of the TAA are Fe^{3+}, Al^{3+}, as well as H^+.[14]

Identification and Assessment of Sulfidic Material

There are several approaches that have been used for the identification and assessment of sulfidic soil materials for management purposes. For soil classification purposes, the most widely used is the incubation method,[15] as discussed later, which allows oxidation of field-moist soil to occur over prolonged durations (i.e., weeks to months) with periodic measurement of pH. Field methods for assessing the likely presence of sulfidic materials, such as reaction with concentrated hydrogen peroxide,[14] can be very useful to indicate the likely presence of sulfidic materials and to inform field sampling strategies, but such field methods are not considered to yield data sufficiently rigorous or quantitative to underpin satisfactory management strategies: laboratory methods have this role.[14] The most used and widely accepted approach for management purposes is the acid-base accounting (ABA) approach, based on detailed laboratory assessment of the relevant components of a soil's alkalinity and acidity. Acid–base accounting methods have benefited from recent technical improvements in methods determining the RIS content of acid sulfate soil materials. These RIS quantification methods—such as chromium-reducible sulfur methods[16,17]—are simple, rapid, accurate, and inexpensive.

Importantly, the concept of "sulfidic" (as currently widely used in soil science) implies not simply the presence of appreciable sulfides, but rather that there is a surplus of sulfides in that material [cf. acid neutralizing capacity (ANC)], such that when oxidized, the soil material is capable of becoming extremely acidic. Thus, soil materials containing even high sulfide mineral contents cannot be classified as sulfidic in many soil taxonomies if there are sufficient neutralizing materials also present in the soil materials to prevent acidification. Although identifying soil materials that can severely acidify as a result of sulfide oxidation is clearly a very useful concept, the term sulfidic for this concept as currently used in several soil taxonomies[11–13] has been misleading to the broader scientific community, who almost uniformly use sulfidic to denote "containing sulfides."[18] Recent developments in classification of acid sulfate soil materials that realign the use of the term "sulfidic" with its usage in the broader scientific community (i.e., that the material contains appreciable sulfide[18,19]) have been adopted for environmental management purposes in Australia.[20–22]

Acid–Base Accounting

The long period of time usually required to gain results from the incubation method (i.e., at least several weeks) along with the need for quantitative analyses for environmental assessment and management purposes has resulted in the widespread adoption of ABA approaches for management.[14] One of the benefits of ABA is that it is quantitative and provides data on the acidification hazard that are suitable for purposes such as acidity hazard prioritization, determination of liming requirements prior to oxidation, verification of liming quantities post treatment, etc.

While many ABA approaches have been used for acid sulfate soil material assessment, they all share a common underlying principle whereby the acidity hazard is the difference between alkalinity and acidity sources, as shown below.

$$\text{Acidity hazard} = \text{acidity} - \text{alkalinity} \qquad (3)$$

There are several sources of acidity and alkalinity in soil materials, and in practice, the determination of several of these acidity sources is determined and expressed separately in the ABA. A commonly used ABA for acid sulfate soil materials is the one of Ahern et al.,[14] as shown below:

$$\text{Net acidity} = \text{potential sulfidic acidity} + \text{existing acidity} - \text{acid neutralizing capacity} \qquad (4)$$

Net acidity in this ABA represents the acidity hazard of the soil material. The potential sulfidic acidity refers to the potential for acidity to develop from oxidation of pyrite. The potential sulfidic acidity is estimated from the RIS determination and assumes both that the RIS is pyritic sulfur and that the following overall oxidation reaction occurs to completion as for Eq. 2 [i.e., 1 mole of pyrite produces 4 moles of (H^+) acidity].

In this ABA, the existing acidity is defined as follows:

$$\text{Existing acidity} = \text{actual acidity} + \text{retained acidity} \qquad (5)$$

The existing acidity comprises both actual acidity and retained acidity. Actual acidity is a measure of the readily available acidity in the soluble and exchangeable fractions. Retained acidity is a measure of the more slowly available acidity contained within minerals such as jarosite and schwertmannite. Retained acidity can be realized and released when these minerals decompose. Both Jones et al.[23] and Johnston et al.[24] demonstrate that jarosite can rapidly dissolve, releasing acidity when jarositic sediments are submitted to reducing conditions. For example, 1 mole of jarosite releases 3 moles of acidity as described by the following reaction:

$$KFe_3(SO_4)_2(OH)_6 + 3H_2O \rightarrow 3Fe(OH)_3 + 2SO_4^{2-} + 3H^+ + K^+ \qquad (6)$$

The ANC refers to the quantity of effective neutralizing sources. In acid sulfate soil materials, acceptable sources of ANC include calcium and magnesium carbonates, exchangeable alkalinity, and organic matter; however, sources of buffering that do not act above a pH of 6.5 (e.g., the bulk of buffering provided by aluminosilicate mineral dissolution) are considered ineffective.[14]

In Australia, the minimum critical net acidity levels that initiate the development of detailed management plans vary according to soil texture and the amount of soil materials disturbed. For sandy soil materials, and/or where large amounts of soil materials are to be disturbed, the critical net acidity level is more than 0.03% S [or alternatively >18 moles (H^+) Mg^{-1} when expressed as acidity].[14]

Limitations of the current ABA methods stem from both our incomplete understanding of the acidifying and neutralizing processes that take place in these soil materials and the lack of reliable methods to quantify retained acidity and ANC. These limitations include the following:

1. It is not clear what proportion of the potential acidity capable of being produced by pyrite oxidation eventually becomes expressed.
2. Our understanding of the kinetics of acidification and neutralization processes is limited. The current ABAs consider only the size of the acidity and alkalinity pools. However, if during oxidation of a sulfidic soil material, the acidity pool is realized much more quickly than alkalinity is made available, then an environmental acidification hazard may eventuate from this soil material during oxidation, although the size of its acidity pool is less than that of its alkalinity pool.
3. The currently available methods for quantifying ANC in acid sulfate soil materials require improvement to provide accurate determinations.[14] Current methods may either overestimate or underestimate

the "real" ANC due to a number of reasons, including the following:

a. Overestimation may be due to the inclusion of finely ground shell materials in the test sample deriving from large shell materials in the field samples. Large shell materials are generally ineffective as a neutralizing agent, but if finely ground for analysis, such components will be included in the ANC determination, inflating the true capacity of the original field samples.[14]
b. Overestimation of the ANC may also result from the imposition of extremely low pHs (e.g., pHs < 2) when using acid back-titration methods,[14] hence the inclusion of acid neutralizing mechanisms such as clay mineral dissolution, which may not occur in natural acid sulfate soil landscapes with less extreme pH.
c. The current lack of reliable quantification procedures for retained acidity can lead to inaccurate estimation of the acidity hazard posed by acid sulfate soil materials.

Despite these limitations, the quantitative capability of the ABA method provides distinct advantages for the purposes of managing acid sulfate soil materials over the other acid sulfate soil material identification methods used mainly for soil taxonomic purposes.

Incubation Methods

This approach simulates natural acidification behavior and has long been used for the assessment of these materials. This method is considered to be direct, allowing the soil to "speak for itself"[25] with respect to whether or not the soil material will acidify upon oxidation. However, incubation can also be a protracted method requiring 2 to 3 months to give a determination.[26]

In the current soil classifications using the incubation method for the recognition of sulfidic materials, both the critical pH target and the duration of incubation have been standardized. For example, the definition used in the Australian Soil Classification[11] for sulfidic materials is as below:

> A subsoil, waterlogged, mineral or organic material that contains oxidisable sulphur compounds, usually iron disulphide (e.g., pyrite, FeS_2), that has a field pH of 4 or more but which will become extremely acid when drained. Sulphidic material is identified by a drop in pH by at least 0.5 unit to 4 or less (1:1 by weight in water, or in a minimum of water to permit measurement) when a 10mm thick layer is incubated at field capacity for 8 weeks.

Some problems have emerged regarding the duration of incubation being limited to 8 weeks in these definitions. For example, the results of Ward et al.[5] indicated that 8 weeks of incubation allowed only minor oxidation of the RIS and expression of acidification in some gel-like clayey pyritic soil materials. To overcome these issues and to allow "the soil to speak without being interrupted in mid-sentence," Sullivan et al.[15] used 2 mm thick incubation slabs to hasten acidification and proposed changing the recommended duration of incubation to the following: a) until the soil pH changes by at least 0.5 pH unit to below 4; or b) until a stable pH is reached after at least 8 weeks of incubation. These changes have been incorporated into recent protocols for recognition of sulfidic soil materials in Australia.[21,22]

Sequential Extraction Methods

The oxidative dissolution of pyrite that occurs when acid sulfate soil materials are disturbed results in the release of metals,[27] the rate of which is dependent upon the geochemical regime experienced. When used within their well documented limitations,[28] sequential extraction procedures can provide useful assessments of changes in metal mobility, and in the case of acid sulfate soil material, assessing the potential metal hazard that the unoxidized acid sulfate soil materials pose. Sequential extraction methods have been utilized in a comparative fashion to examine metal mobilization in acid sulfate soil materials. Both Åström[29] and Sohlenius and Öborn[30] compared metal mobility in different fractionations in the oxidized and unoxidized zones of in situ profiles containing acid sulfate soil materials. These studies showed that a range of metals were mobilized from the oxidized zones, especially from the pyritic and residual soil metal fractions. More recently, a sequential extraction method featuring the separate partitioning of metals in the pyritic fraction has been specifically designed for acid sulfate soil materials.[31]

CHARACTERISTIC MINERALS AND GEOCHEMICAL PROCESSES

Reduction Processes

Reduced inorganic sulfur in sulfidic acid sulfate soil materials includes iron disulfides [most commonly pyrite (FeS_2)][27,32,33] and smaller amounts of other minerals such as greigite (Fe_3S_4),[34] mackinawite (FeS),[35] and elemental sulfur (S_8).[36,37]

The vast majority of RIS in sulfidic acid sulfate soil materials has formed at earth surface temperatures and pressures under waterlogged, anoxic conditions. Under such conditions, accumulation of RIS species depends on microbially mediated sulfate reduction, which is itself dependent on organic carbon availability, supply of sulfate, and on the amount of competing electron acceptors including reactive Fe^{III} minerals.[38] These variables influence the activity of dissimilatory sulfate-reducing microorganisms that oxidize simple organic compounds or hydrogen using sulfate as the electron acceptor. The overall process of dissimilatory sulfate reduction can be shown, for example, by the following:

$$CH_3COO^- + SO_4^{2-} + H^+ \rightarrow H_2S + 2HCO_3^- \quad (7)$$

During this process, the sulfur in sulfate is reduced from the S^{6+} oxidation state to S^{2-}. Conditions that are conducive to microbially mediated sulfate reduction occur in organic-rich coastal and estuarine sediments, such as in tidal marshes and swamps. In such systems, tidal exchange of pore water supplies sulfate and removes the resultant HCO_3^- produced via the reaction in Eq. 7.

In contrast, in soils containing Fe^{2+}, often produced by the activity of ferric iron–reducing microorganisms, H_2S may react rapidly to form monosulfide (FeS) precipitates as below:

$$H_2S + Fe^{2+} \rightarrow FeS + 2H^+ \quad (8)$$

Monosulfides have been found in a wide variety of acid sulfate soils and often in substantial concentrations[39,40] Mackinawite and greigite are often described as "iron-monosulfide" minerals because they have an Fe–S ratio that is close to 1:1.[41] These mineral species are defined analytically by their dissolution in HCl to yield H_2S gas and described as acid-volatile sulfur. There are only few studies that conclusively document the occurrence of the monosulfide mineral greigite[34] and only one that has so far demonstrated the presence of mackinawite in acid sulfate soil materials.[35]

Pyrite is by far the most commonly observed RIS species in sulfidic acid sulfate soil materials. In these materials, pyrite presents a range of distinct crystal morphologies (e.g., Fig. 1a). The most remarkable of these morphologies are framboids (from the French term for raspberry, *frambois*). Pyrite framboids consist of spheroidal aggregates of densely packed, individual microcrystals. Accumulation of pyrite in soil can occur rapidly under suitable field conditions[42,43] via a range of processes. One example is the hydrogen sulfide pathway of Rickard and Luther[44] involving the oxidation of FeS by H_2S to form pyrite (FeS_2) as described below:

$$FeS + H_2S \rightarrow FeS_2 + H_2 \quad (9)$$

Oxidation Processes

Pyrite and the iron-sulfide minerals persist in soils only under anoxic, waterlogged conditions. If conditions become oxic by, for example excavation of the soils, the iron-sulfide components can undergo a series of oxidation reactions. For example, in the presence of oxygen (and water), pyrite oxidizes to ultimately yield sulfuric acid, and poorly soluble Fe^{III} precipitates.

The oxidation of FeS_2 depends on factors including the supply of O_2, the availability of water, and the physical properties of FeS_2. Pyrite oxidation generates acid and releases heat; consequently, the acidity released into the surrounding solution will affect the overall reaction rates. The oxidation of FeS_2 in the environment is usually ultimately determined by the supply of O_2. Temperature, which influences both chemical and microbial oxidation, is also an important factor in determining the oxidation rate of pyritic materials.

While exposure to oxygen under moist conditions is the driving force for pyrite oxidation, there are a number of reaction steps in the overall oxidation process. This includes a number of possible final iron phases as well as the formation of intermediate sulfoxyanions and elemental S. Fe- and S-oxidizing bacteria play an important role in mediating various steps in the overall oxidation process and in determining the formation and persistence of intermediate S species.

Under continuation of oxidizing conditions, the Fe^{2+} released by pyrite oxidation is also subject to oxidation to Fe^{3+}. While this simple oxidation process consumes some acidity, the subsequent hydrolysis of the resulting Fe^{3+} leads to the liberation of acidity. At low pH (e.g., <4), Fe^{3+} is sufficiently soluble that it may serve as a very effective electron acceptor driving further pyrite oxidation.[45] For this reason, it has been often suggested that rate of Fe^{2+} oxidation to Fe^{3+} may be the rate-determining step in pyrite oxidation.

The Fe^{3+} produced via pyrite oxidation also commonly precipitates as a range of Fe^{III}-bearing minerals. In acid sulfate soil conditions at $pH < 3$, and/or in the presence of abundant K^+, jarosite appears to be the predominant Fe^{III} phase, whereas in the pH range of 3–4, schwertmannite is an important Fe^{III} phase in acid-sulfate soil landscapes.[46,47] The widespread occurrence of schwertmannite in acid-sulfate soils has been confirmed only relatively recently.[47] Schwertmannite has poor crystallinity and often presents a "ball-and-whisker" micromorphology (Fig. 2b).

Schwertmannite is metastable and over time transforms, via dissolution–reprecipitation, to form a range of Fe^{III} oxyhydroxides,[48] including ferrihydrite, lepidocrocite, and goethite, with the latter being most stable. The transformation of schwertmannite (an Fe^{III} oxyhydroxysulfate) to these Fe^{III} oxyhydroxides involves the hydrolysis of Fe^{3+} and the liberation of acidity. As a consequence, schwertmannite transformation can suppress pH long after the initial source of acidification (i.e., pyrite) has been consumed.

The type of secondary minerals formed from the Fe released during pyrite oxidation determines to a large extent the amount of acidity expressed.[49,50] For example, if the released Fe produced by sulfide oxidation precipitates as goethite or ferrihydrite, then 3.0 moles of H^+ are formed for every mole of Fe^{3+} hydrolyzed from pyrite. However, if hydrolysis is incomplete and jarosite is formed, only around 2 moles of H^+ is released for every mole of Fe^{3+} hydrolyzed from pyrite.[51] If schwertmannite is formed, then approximately 2.575 moles of H^+ is released for every mole of Fe^{3+} hydrolyzed from pyrite.[52] The "stored" acidity in these two minerals is important as the Fe in both jarosite and schwertmannite can undergo further hydrolysis and result in the release of acidity into the surrounding environment.[47,49,50]

HAZARDS ARISING FROM DISTURBANCE OF ACID SULFATE SOIL MATERIALS

Acidification

Oxidation of RIS is the primary cause of the extreme acidification that characterizes sulfuric acid sulfate soil materials. By definition, the pH of sulfuric acid sulfate soil is lower than 3.5 to 4 (depending on the particular soil taxonomy being employed), but values of pH lower than 3 in actively oxidizing soils are frequently observed.[53] Such extreme acidification significantly alters the soil chemistry, can render it hostile to plants, and can create a source of contamination to groundwater and surface water runoff. The acid produced can react with clay minerals and oxides to release silica and metal ions, principally aluminum, iron, potassium, sodium, and magnesium.[54] Other ions such as other metals and metalloids can also be released.[27,55,56]

The impacts of severe acid sulfate soil acidification on agricultural crops have been well documented.[53] Many crop plants are highly sensitive to low-pH soil conditions, and acidification can greatly reduce yields and, in extreme cases, cause complete crop failure. In addition, the formation of acidic secondary iron minerals such as jarosite and schwertmannite can significantly reduce the availability of nutrients such as phosphorus, potassium, and nitrogen. Farmers have tried many different approaches to ameliorate acidity by techniques such as the addition of neutralizing agents, soil amendments, organic mulch, and reconfiguring plant beds to enhance the leaching of acidic products from the soil.[53]

Aluminum toxicity is a significant issue linked to acidification of acid sulfate soil materials used by terrestrial plants[53] and downstream aquatic flora and fauna.[55,57] The solubility of Al is critically dependent on pH, becoming soluble at environmentally significant levels only at a pH of approximately lower than 5. Severe environmental impacts can occur when acidic Al-rich leachate from acid sulfate soil materials enters water bodies. The more acute ecological impacts of acidification of waters draining from acid sulfate soil landscapes into waterways include fish kills,[55,57,58] loss of native aquatic macrophytes and fauna followed by invasion by acid-tolerant species,[57] mass mortality of crustaceans and shell fish,[59] and loss of benthic communities.[60] Sublethal exposure of fish to acidity has also been linked to an increased susceptibility to fish skin diseases,[58] whereas depletion of alkalinity has been linked to poor shell development in crustaceans.[61]

A range of potentially longer-term impacts on aquatic ecosystems arising from acid sulfate soil leachate include the following: disturbance to fish reproduction and recruitment, acidity barriers to fish migration, decline of primary food web, reduction of species diversity, and long-term habitat degradation.[55,57] In assessing the likely impacts of acidification of waters draining from acid sulfate soil landscapes into downstream aquatic environments, it is necessary to consider the vulnerability of the aquatic ecosystems, the duration and frequency of acidification episodes, and the potential intensity of acidification based on the properties and quantities of the acidic leachate.

Iron Mobilization

Fe^{2+} is a primary product of pyrite oxidation. At high pH values (pH > 7), Fe^{2+} is chemically rapidly oxidized to Fe^{3+}.[62] At lower pHs (i.e., pH < 4.5), the oxidation of Fe^{2+} to Fe^{3+} is catalyzed by acidophilic bacteria such as *Acidithiobacillus ferrooxidans*,[63] *Thiobacillus ferrooxidans*, and *Leptospirillum ferrooxidans*.[64] The hydrolysis of Fe^{2+} has direct environmental consequences arising from the liberation of acidity and the formation of secondary iron minerals that can control soil and water geochemistry.

Accumulations of iron minerals are ubiquitous in acid sulfate soil landscapes. The precipitation and mineralogy of secondary iron minerals have been reviewed in detail by Alpers and Nordstrom[65] and Cornell and Schwertmann.[62] Understanding the types of iron precipitates that form in acid sulfate soil landscapes during oxidation is important as particular iron mineral phases can exercise a major influence on the environment.[47,50] In a study of surface iron precipitate accumulations associated with waterways in acid sulfate soil landscapes, Sullivan and Bush[47] found schwertmannite was the dominant secondary iron mineral. The schwertmannite occurred as coatings on vegetation, accumulations in low depressions, and iron flocs adhering to surfaces in acidified waterways. The potential acidity within the schwertmannite was high, up to 2580 moles (H^+) Mg^{-1}, indicating that the schwertmannite was a substantial intermediate store of acidity within these acid sulfate soil landscapes. The acidity retained within both schwertannite and jarosite has recently been included into the quantitative assessment of the net acidity of sulfate soil materials.[14]

Iron precipitates in the form of iron flocs within the water column also are known to directly affect gilled organisms, smother benthic communities and aquatic flora,[55,57] diminish the aesthetic values of recreational waterways, and threaten estuarine and marine environments.[66]

Heavy Metal and Metalloid Mobilization

Mobilization of metals and metalloids (e.g., arsenic) to soil pore waters from acid sulfate soil materials can constitute a major environmental hazard.[67–69] Metals and metalloids that have been reported at levels exceeding accepted environmental protection thresholds in acid sulfate soil materials include Al, As, Ba, Cd, Co, Cr, Cu, Fe, Mn, Ni, Pb, Sb, V, and Zn.[67,68,70] Acidification can greatly enhance the solubility of metals, promoting their subsequent release from mineral phases by dissolution or cation exchange. The pH dependence of metal release has received considerable attention,[55,67,70–73] and there are strong similarities in metal release between acid sulfate soil landscapes and acid mine drainage systems.[74]

Numerous studies have documented the impacts of soluble metals on crop production[53] and terrestrial habitats,[27] and more recently, attention has turned to their impact on aquatic environments.[55,57,58,71,75] Gilled organisms are particularly vulnerable to soluble metals, and metal mobilization can lead to rapid mortality rates in these species.[1,55,57,59] Studies of the effects of metals on shellfish (oysters) revealed longer-term, more chronic impacts on their growth and survival.[61] However, the longer-term impacts of metal release from acid sulfate soils to surrounding aquatic environments are poorly understood.

Most reports on the impacts arising from metal release from acid sulfate soil materials focus on the consequences of metal mobilization under oxic acidified conditions. However, metals can also be mobilized when sulfuric acid sulfate soils are subject to prolonged inundation, resulting in the development of anoxic reducing conditions. Acid sulfate soil materials occur in low-lying floodplain environments and therefore are subject to periodic waterlogging and oscillating redox conditions. The processes of metal mobilization and behavior of metals are very different under anoxic conditions. The release of iron and arsenic is a good example of metal mobilization from acid sulfate soil materials following inundation. Accumulations of iron minerals in acid sulfate soils are often concentrated at the ground surface and include goethite, ferrihydrite, jarosite, and schwertmannite. These iron minerals often have a large surface area and are a significant sink for the sorption of metals. Under reducing conditions, these iron oxides are prone to microbial reductive dissolution.[27,76]

Microbial iron reduction triggers three major changes that affect metal mobilization. Firstly, it results in the dissolution of Fe^{3+} and transformation to Fe^{2+}, causing the co-release of other metals sorbed to the Fe mineral surfaces. Secondly, the microbial reduction process is proton consuming and, when accompanied by the formation of bicarbonate as a by-product of microbial respiration, can result in in situ neutralization.[77] The increase in pore water pH generally reduces the solubility of divalent metals and aluminum. It also facilitates the recently identified Fe^{2+}-catalyzed transformation of poorly crystalline iron oxide minerals to more crystalline phases (e.g., rapid transformation of schwertmannite to goethite[76,78]). Although the overall consequences of these rapid mineral transformations on metal mobility are yet to be quantified,[79] the mobility of some metals and metalloids can increase under these conditions. For example, when associated with iron oxides in acid sulfate soil materials, arsenic is readily mobilized at the onset of microbially mediated iron reduction.[69]

It is important to recognize that metals and metalloids can have a significant impact in acid sulfate soil landscapes both when acid sulfate soil materials are allowed to oxidize and acidify, and following the prolonged inundation of previously oxidized, iron-enriched acid sulfate soil material.

Deoxygenation of Water Bodies

Acute deoxygenation of estuaries, lakes, rivers, and drainage channels is a major contributor to catastrophic fish kills.[80–82] Many potential factors contribute to deoxygenation events, and they are known to impact a very wide range of environments. Severe deoxygenation of waterways within acid sulfate soil landscapes has been linked directly to the behavior of acid sulfate soil materials.[83]

Deoxygenation results when solids and aqueous compounds with a capacity to react with dissolved oxygen enter water bodies and consume oxygen more rapidly than it can be replenished. The magnitude of deoxygenation depends on the spatial scale of the event, its persistence, and its intensity. Aquatic ecosystems require dissolved oxygen concentrations generally greater than 85% saturation for lowland rivers.[84] Native fish and other large aquatic organisms are known to survive on dissolved oxygen concentration of as little as 2 mg L^{-1} but may become stressed below 4–5 mg L^{-1}.[85] In recent studies of a major estuarine river system in eastern Australia affected by deoxygenation, Wong et al.[86] found that deoxygenation was confined to downstream confluences of subcatchments with appreciable acid sulfate soil materials and occurred during the latter phases of the flood recession. During hypoxic events, the river water had elevated concentrations of redox-sensitive species associated with acid sulfate soil material (e.g., Fe^{2+}, dissolved Mn, and elemental sulfur[86]), further implicating a role of acid sulfate soil and monosulfidic black ooze (MBO) materials in deoxygenation events.

Anaerobic decomposition of floodplain vegetation in backswamps can be a primary process leading to the deoxygenation of large volumes of waters in acid sulfate soil landscapes.[86,87] Decomposition of flood-intolerant vegetation in drained acid sulfate floodplains can lead to the formation of "blackwater"—a colloquial term used to describe anoxic stagnant floodplain water that develops a distinctive dark color as a result of the accumulation of dissolved organic carbon compounds. Blackwater is typically anoxic, has a high chemical oxygen demand and high dissolved Fe concentrations, and rapidly consumes dissolved oxygen when it discharges to main water bodies.[87]

The propensity for MBO (Fig. 2c) to accumulate and be mobilized by floodwaters in drainage channels has also been identified as a contributing factor to deoxygenation in acid sulfate soil areas.[8,37,88,89] The role of MBO in deoxygenation and latter acidification in acid sulfate landscapes has only recently been discovered.[8,16] Burton et al.[68] have described the oxidation dynamics of MBO when mobilized into oxygenated water. Elevated elemental sulfur concentrations in deoxygenated waterways in acid sulfate soil landscapes may be a useful indicator of MBOs as a contributing cause to deoxygenation, although elemental sulfur can also form as a primary product of H_2S oxidation and may be present within MBOs prior to flood events.[36,37,86]

Production of Noxious Gases

Anthropogenic and biogenic sulfur-containing gases have important impacts on global climate change[90,91] and atmospheric acid–base chemistry.[92] Coastal estuarine and marine environments are major emitters of biogenic H_2S.[93,94] Emissions of H_2S, and more recently, sulfur dioxide (SO_2), from floodplains have been linked to the management of acid sulfate soil materials.[95]

H_2S is a highly noxious gas that causes distress to humans[96,97] and threatens aquatic organisms.[98,99] As described by Eq. 7, H_2S is produced by sulfur-reducing bacteria under anoxic conditions. Even at very low concentrations, H_2S can be detected by its characteristic rotten-egg odor. In acid sulfate soil landscapes, periodically inundated soil surfaces, shallow waterways, depressions, and field drains where stratified anoxic conditions can develop are all situations conducive to sulfate reduction and the formation of H_2S.[53] However, H_2S is an unstable phase, and its persistence in water and soil and ultimate gaseous emission is highly constrained by a wide range of oxidants in natural sediments and water bodies.[100] These oxidants include O_2, NO_3, Mn and Fe oxyhydroxides.[101,102] Due to their abundance in acid sulfate soil, iron oxides[103] are a particularly effective oxidant of H_2S, a process leading to the formation of iron sulfides as described previously. H_2S becomes a problem when the rate of its formation exceeds the catalytic oxidative capacity of the sediments and water bodies to eliminate its gaseous emission. An excess of labile carbon and stagnant water bodies creates conditions that favor H_2S emissions in acid sulfate soil landscapes.[104]

Partially oxidized RIS-containing acid sulfate soil materials are a known source of SO_2. Macdonald et al.[95] showed that the rates of SO_2 emission from agricultural acid sulfate soils were closely linked to soil moisture and evaporative flux, leading the authors to conclude that acidic dissociation of sulfite (SO_3^{2-}) occurring within the near-surface soil pore water was probably the major source of SO_2. From relatively few measurements, Macdonald et al.[95] estimated global SO_2 emissions from acid sulfate soil materials to be 3.0 Tg S yr^{-1}, approximately 3% of global anthropogenic emissions.

PHYSICAL BEHAVIOR

Consistence and Strength

Estuarine and marine sediments are usually deposited under saturated conditions and consequently may contain up to 70% water and often have a gel-like consistence. Such extremely "soft" sediments pose significant challenges, due to their capacity to shrink on drying, deform under load, as well as oxidize and acidify when disturbed. The unique physical properties of sulfidic gel-like soil materials have been well described.[53,105] These soft sulfidic soils are fundamentally different from rigid soils, with both hydraulic conductivities and consolidation coefficients for the gel material being generally very low (e.g., hydraulic conductivities <1 m day^{-1}).

Drainage and oxidation of sulfidic soils usually results in irreversible shrinkage, development of fissures, and structure. "Soil ripening" is a term used to embrace the physical, chemical, and biological processes that transform unconsolidated sediment to a structured, dry soil.[53] As summarized by Dent,[53] profiles containing acid sulfate soil materials often consist of oxidized, acidic, sulfuric horizons overlying unoxidized, sulfidic horizons, with a transition zone of variable thickness in between. The overlying sulfuric horizon generally has far greater structural development than the underlying sulfidic horizon due to physical ripening processes.

At the time of deposition, clayey marine sediments can have an open "house-of-cards" microporous architecture with water-filled pore space. This type of clay microstructure confers minimal frictional and cohesive strength to the sediment. Even though on a microscale the materials are highly porous, hydraulic conductivities tend to be low due to the lack of connected macropore networks.[106] The removal of water by drainage, evaporation, and evapotranspiration causes the microstructure to collapse, increasing cohesive strength and initiating shrinkage-induced fissures. The development of such structure ultimately enhances the potential oxidation of iron sulfides and further microstructural collapse as a result of acidic weathering. Unripe soils are soft, whereas ripe soils are firm and not overly "sticky."[53] The process of ripening is irreversible and, as a consequence, can cause land subsidence of up to 2 m depending on the depth of unripe soil affected by oxidation (e.g., Fig. 4a).

(a)
(b)
(c)
(d)

Fig. 4 (a) Deeply embedded pylons of a building indicate subsidence of the ground surface in an acid sulfate soil landscape. The ground surface when the building was constructed is indicated by the concrete collars around each pylon: The tops of these collars were originally level with the ground surface. (b) Sheet pilings are used in this construction site to control water tables and water flows during dewatering. (c) Aquaculture ponds in an acid sulfate soil landscape in Asia. (d) A small weir installed across a drain to retain water within a low-lying acid sulfate soil landscape.

Permeability

The hydraulic gradient and hydraulic conductivity (permeability) govern the overall groundwater seepage rates in acid sulfate soil landscapes. Factors that affect hydraulic gradient include rainfall, topography, and the water level at the discharge site. The hydraulic conductivity of acid sulfate soil materials varies greatly, both vertically within the soil profile and laterally across the landscape.[105–107] Ripening processes and spatially variable properties can induce considerable variability in soil structure and consequently, large differences in hydraulic conductivity in sulfuric horizons. Lateral seepage of acidic ground waters is often the major route for the discharge of contaminants from acid sulfate soil materials to nearby waterways.[67,71,87]

The scale of the potential range in the hydraulic conductivity of acid sulfate soil landscapes was demonstrated in a recent study by Johnston et al.,[106] who examined seven coastal floodplains in eastern Australia. The measured saturated hydraulic conductivity (K_{sat}) ranged from less than 1 m day^{-1} to more than 500 m day^{-1}. These large differences in K_{sat} reflect difference in the development, orientation, and density of macropore networks, with the highest K_{sat} occurring in soils with large interconnected tubular macropore networks.[106]

MANAGEMENT OF ACID SULFATE SOILS

Management for Intensive Developments

Where acid sulfate soils are likely to be directly or indirectly disturbed by a development, risk assessment should be conducted to determine the nature and magnitude of any risks associated with the development. Once the risk has been assessed, then appropriate management of acid sulfate soil material disturbance can be developed.

General Principles

Dear et al.[108] have outlined the following eight management principles—here slightly modified—that should be followed when intensive developments are being considered in areas that are likely to contain acid sulfate soils:

1. The disturbance of acid sulfate soils should be avoided wherever possible.
2. Where disturbances of acid sulfate soil materials are unavoidable, preferred management strategies are the following:
   - Minimization of disturbance
   - Neutralization

- Hydraulic separation of sulfides either on its own or in conjunction with dredging
- Strategic reburial (reinterment)

3. Works should be performed in accordance with best practice environmental management when it has been demonstrated that the potential impacts of works involving acid sulfate soils are manageable to ensure that the potential short- and long-term environmental impacts are minimized.
4. The material being disturbed (including the in situ acid sulfate soils) and any potentially contaminated waters associated with acid sulfate soil material disturbance must be considered in developing a management plan for acid sulfate soil.
5. Receiving marine waters, estuarine waters, brackish waters, or freshwaters are not to be used as a primary means of diluting and/or neutralizing acid sulfate soils or associated contaminated waters.
6. Management of disturbed acid sulfate soils is to occur if the acid sulfate soil material action criteria listed in Dear et al.[108]—and briefly discussed under Oxidation Processes—are reached or exceeded.
7. Stockpiling of untreated acid sulfate soil material above the permanent groundwater table with (or without) containment is not an acceptable long-term management strategy. For example, soils that are to be stockpiled, disposed of, used as fill, placed as temporary or permanent cover on land or in waterways, sold or exported off the treatment site, or used in earth bunds and that exceed the acid sulfate soil material action criteria listed in Dear et al.[108] (and briefly discussed under Oxidation Processes) should be treated/managed as acid sulfate soil materials.
8. The following issues should be considered when formulating acid sulfate soil material environmental management strategies:
    - The sensitivity and environmental values of the receiving environment. This includes the conservation, protected, or other relevant status of the receiving environment (e.g., fish habitat area, marine park, coastal management district, and protected wildlife).
    - Whether groundwaters and/or surface waters are likely to be directly or indirectly affected.
    - The heterogeneity, geochemical, and textural properties of soils on site.
    - The management and planning strategies of relevant regulating agencies.

Avoidance of disturbance of acid sulfate soil materials is the preferred method for management, and this depends on a sufficiently detailed acid sulfate soil material survey of the proposed site being undertaken and inclusion of the resulting information into the design of those developments. Where disturbance cannot be avoided, there are a suite of techniques, in addition to those listed above, that can be utilized to manage impacts of disturbed acid sulfate soils. Of course, the appropriateness of any of these techniques for any particular situation involving acid sulfate soil materials depends on the precise nature of that situation.

Covering with Fill

Where sulfidic materials are not disturbed by excavation or dewatering, covering soils with non-sulfidic fill materials can provide a base for developments such as building foundations.[109] It should be realized that such filling activities may still disturb acid sulfate soil materials, raising water tables and bringing groundwater into contact with previously unsaturated sulfuric soil materials, causing potential mobilization of acidity. The low bearing capacity of many gel-like acid sulfate soil materials may also cause problems such as subsidence and failure when loaded by fill materials. This may cause problems in infrastructure built on these areas unless appropriate geotechnical precautions (such as preloading and dewatering) have occurred.

Hydraulic Separation

Hydraulic separation utilizes differences in the particle size (and hence hydraulic behavior) between sulfidic and non-sulfidic fractions that exist in most soil materials. To be successful, these two fractions must be easily partitioned from each other in flowing water, and the absence of this often precludes the use of this technique for the treatment of acid sulfate soils with appreciable clay contents (i.e., >15% clay by mass). A major aim of hydraulic separation is to isolate the sulfidic fractions (generally in the finer fraction) such that the volumes of sulfidic soil materials required to be treated can be greatly minimized and, concurrently, to create a large volume of sulfide-free soil materials that can be considered as non-acid sulfate soil material.

The most commonly used technique for the hydraulic separation of acid sulfate materials is sluicing, but hydrocycloning has also been used successfully.

Sluicing refers to the process whereby sulfidic fines are hydraulically separated from sands at the discharge point of a dredge. If the right combination of discharge rate from the dredge outlet and slope down which the dredged materials flow from the dredge outlet is achieved, then the larger grained sandy materials settle first, creating a relatively clean fill material, whereas the smaller sulfidic "fines" continue to flow with the water into a still water location where they can be isolated, concentrated, and treated separately as an acid sulfate soil material. The correct combination of discharge rate and slope varies from soil material to soil material and requires vigilance and appropriate testing to ensure that combination ensures that the "clean" fill material remains unpolluted by sulfidic fines.

Containment of Contaminants

Sites on which acid sulfate materials are likely to be processed or stored for appreciable times should have in place strategies that maximize the containment and treatment of acidic waters on site. Such strategies usually include the isolation of these acid sulfate soil materials from waterways by the use of constructed bunds, drains, and collection basins that capture runoff and allow regular assessment and, if required, treatment (e.g., by neutralizing materials) of any surface waters exiting the site. The use of impermeable materials to construct and line storage or treatment areas to minimize the possibility of loss of acidity in groundwaters is also required.

Adding Neutralizing Materials

Perhaps the most commonly used technique for management of acid sulfate soil materials if disturbance of these soil materials cannot be avoided is the use of alkaline materials to neutralize any acidity that exists or that may develop upon oxidation (i.e., this includes actual and retained acidities as well as the acidity that could develop upon complete oxidation of RIS). Net acidity from the ABA is generally used to calculate the appropriate liming rate required to maintain soil pH above pH 5.5. A safety factor of 1.5 is applied to this calculated liming rate to compensate for factors such as ineffective lime mixing, the possible development of passivating coatings of gypsum, etc., on lime particles.

A wide variety of liming materials have been used including $CaCO_3$ (either geological or biological [e.g., shell materials]), dolomite ($[Ca,Mg]CO_3$), magnesite ($MgCO_3$) and burnt magnesite (MgO), and hydrated lime ($Ca[OH]_2$). Particular attention must be paid, especially when using the more soluble and caustic liming materials, to ensure the following: 1) these materials do not cause pH to become excessively alkaline; as well as 2) the health and safety of workers applying these hazardous liming materials are guarded. All liming materials can be environmentally hazardous and must be used judiciously.

The particle size of the liming materials must be sufficiently small[110] to allow timely neutralization of the acidity contained within sulfuric materials or that develops upon oxidation of sulfidic materials. These liming materials generally require thorough mixing (often on specially prepared treatment pads[108]) with the acid sulfate soil materials to be effective—this poses practical issues when working with wet "sticky" clay materials whose consistence often precludes uniform mixing with liming materials.

Neutralizing materials can also be applied to acidic waters draining from development sites (or affected acid sulfate soil landscapes) to neutralize these waters prior to their flow off-site. A range of technologies is available for dosing waters with appropriate quantities of lime for this purpose. Consideration should also be given to the metal flocs that will generally precipitate as often metal-laden acidic drainage waters become alkaline.

Strategic Reburial

Strategic reburial involves placing acid sulfate soil materials into created voids in water bodies or beneath water tables. Sulfidic materials should be placed deeply in locations where anaerobic conditions sufficient to preclude oxidation and acidification of RIS can be reasonably guaranteed. While sulfidic materials do not require the addition of neutralizing materials for this technique, sulfuric materials require such additions prior to burial. For this reason, stockpiling of unoxidised sulfidic soil materials needs to be optimized to ensure that minimal oxidation and acidification occur prior to burial.

Stockpiling

Stockpiling untreated acid sulfate soil materials creates conditions conducive to oxidation of RIS and to the mobilization of acidity. If the acid sulfate soil materials drain easily (e.g., sandy, peaty, or well-structured materials) these processes can be rapid. Therefore, the duration of stockpiling needs to be minimized, and water containment practices need to be in place wherever stockpiling of these materials is undertaken for extended periods. Dear et al.[108] give indicative durations of stockpiling considered appropriate for untreated acid sulfate soil materials in relation to factors such as soil texture and the level of containment precautions implemented.

Dewatering

Dewatering is a practice that is often required in areas that contain acid sulfate soil materials (i.e., they are often located in low-lying waterlogged wetland areas). Infrastructure such as drains, service pipes, footings, etc., in these areas often needs to be installed in soil that has been dewatered to obtain sufficient strength to allow any excavations to retain their integrity and that will not collapse. Consequently, as the pumping lowers the water table, the opportunity for oxidation of any RIS formerly under the water table can occur, leading to acidification of both sulfidic soil material now located above the lowered water table and the water exiting the pump as a result of soil acidification. It is important to note that unless suitable dewatering management practices are employed, the effects of dewatering can be located far away (up to several kilometers) from the site being dewatered.

The guiding principles for dewatering in acid sulfate soil landscapes have been described by the WA DoE[111] as below

- Wherever possible, RIS below the water table should not be exposed to air and allowed to oxidize as a result of changing the elevation of the water table.

- Where disturbance is unavoidable, the disturbance should be minimized or managed to prevent long-term environmental problems caused by oxidation of RIS. Management will need to consider the potential impact in the entire area to be affected by the groundwater cone of depression.
- Where oxidation of RIS has resulted from dewatering, these problems should be remediated wherever possible, or otherwise, risk-based strategies be implemented to prevent potential impacts on human health and the environment.

The effects of dewatering can be minimized by practices such as the following:

- The use of physical barriers—such as sheet piling (e.g., Fig. 4b) or slurry walls—to isolate the site to be dewatered from the surrounding aquifer.
- Developing the site in stages so as to minimize the extent of the groundwater depression.
- Use of groundwater recharge trenches around the perimeter of the site to be dewatered to minimize the extent of the groundwater depression.

The water pumped from the site to be dewatered and not used in perimeter recharge trenches should be monitored to ensure that the quality of the water (in terms of properties such as pH, toxicants, nutrients, odor, etc.,) is suitable for the intended use/disposal of this water.

Other Methods

There are other management methods that have been used or suggested to avoid any untoward effects of disturbance of acid sulfate soil materials. These include the following: 1) vertical mixing whereby sulfidic or sulfuric soil layers are mixed with soil materials lower down in the profile that contain appreciable neutralizing materials (e.g., fine-sized carbonates); and 2) hastened oxidation whereby acid sulfate soils are placed on a containment pad and allowed (or encouraged) to oxidize—after oxidation of all RIS, these soil materials are treated with neutralizing materials and considered to then be non-acid sulfate soil materials.

Management for Agriculture, Forestry, and Aquaculture

The successful use of acid sulfate soil landscapes for agriculture, forestry, or aquaculture involves overcoming a number of unique challenges presented by these soils. These challenges have already been described and include the following: severe acidification of the soil and waters, nutrient deficiencies, element toxicities, waterlogged conditions, and salinity (in tidally affected areas). These challenges notwithstanding, a wide range of land uses occupy acid sulfate soil landscapes, including the cultivation of rice, sugar cane, horticultural crops including vegetable and fruit, prawning and other aquaculture activities (e.g., Fig. 4c), grazing, and forestry.

Our knowledge of suitable management approaches for these soil materials has developed apace but still requires much effort. As an example in forestry, *Melaleuca* spp. have been considered as useful and versatile trees to grow on these soil areas.[112] However, later research has shown that although these tree species can grow well on these soil types, that they also cause greatly enhanced mobilization of acidity in these soils.[80]

There are a number of general strategies that can be used to make acid sulfate soils productive, including the following:

- *Neutralizing acidity*. The low pHs of sulfuric soil materials can be neutralized by the application of liming materials. The array of liming materials available is wide and includes carbonate materials such as $CaCO_3$; alkaline waste materials such as fly ash and red mud; and organic materials (usually waste matter such as rice husks or sawdust). These materials can be costly to purchase and apply especially given the large amounts of acidity that can exist in areas containing acid sulfate soil materials. For example, Tan[113] found that 90 Mg ha^{-1} of $CaCO_3$ was required to treat severely acidified soils for aquaculture. As mentioned previously, all liming materials can be environmentally hazardous and must be used judiciously.
- *Leaching acidity and salt* from soil by growing crops on raised mounds. In low-lying waterlogged areas where acid sulfate soils are often found, raised beds or mounds can be employed[53,114] to enhance soil conditions by "lifting" the crop up from the water table: this also enhances the leaching of acidity and related toxins from the raised soil.
- *Water table management*. Drainage has often been used in acid sulfate soil landscapes to create suitable conditions for the growth of agricultural crops and pastures that are not adapted to prolonged waterlogged conditions. As described previously, such "land improvement" has led to severe degradation of many acid sulfate soil landscapes. While logic indicates that over time, the acidity created from the oxidation of RIS in these soils may be leached from the drained acid sulfate soil materials, resulting in better quality soil materials, the use of acid budgets has shown that the characteristic rates of acid exports from these landscapes, although sufficient to cause considerable environmental damage, are relatively low compared with the acidity stored in the landscapes, and it will require many hundreds of years before this acidity stored is depleted.[55,57]

If intensive drainage programs have often caused these problems, then it might be supposed that the lessening of drainage intensity may reduce the magnitude of the problem.

Indeed, studies have been undertaken that show that reversal of the intensive drainage regimes to wetter regimes via tidal exchange or weirs in drains (Fig. 4d) can result in rapid reversal of the acidic geochemical regimes.[107] Of course, the implementation of wetter conditions in these landscapes will usually also dictate changes to land use in these areas.

Revegetation of Scalded Acid Sulfate Soil Landscapes

Scalded (i.e., non-vegetated) land surfaces are an extreme symptom of land degradation and, in low-lying acid sulfate soil landscapes, scalds can extend for hundreds of hectares, impacting the environment and those who live and rely on these areas. Land scalded by the disturbance of acid sulfate soil materials is environmentally damaging, agriculturally unproductive, and difficult to rehabilitate (Fig. 1a). There are a multitude of causes for the complete and prolonged failure of vegetation to establish. In acid sulfate soil landscapes, extreme acidification and/or salinization are often involved with the initiation of scalds.[43,115] Rosicky et al.[43,115] found that the surface soil layers of scalds experienced extreme acidification ($pH < 3$), evaporative accumulation of acidic salts and metals (Al, Fe), high salinities caused by the accumulation of evaporative salts (e.g., gypsum), and accumulations of iron minerals (e.g., schwertmannite, ferrihydrite, goethite, and jarosite). Combined with other stresses such as grazing pressure and frosts, such soil conditions generally prevent the long-term establishment of vegetation.

The primary management goal for scalds is usually to establish persistent vegetation. Strategies for revegetating scalds generally revolve around improving the surface soil layers by practical agricultural intervention. Techniques that have been demonstrated to work include the following: the exclusion of stock, the use of ridges and furrows, mulching, liming, addition of fertilizer, pretreating seed with nutrients and neutralizing agents, and more recently, water management practices that create and maintain wetter conditions. Of particular interest are the simpler interventions such as ridging and furrowing. This remediation involves the forming of ridges and furrows using cultivation and, especially when combined with a mulch layer (e.g., straw), has proven very effective in facilitating the establishment of vegetation (e.g., Fig. 2d). Ridges and furrows establish different microhabitats, with the water-tolerant species occupying the wetter furrows.[116]

More recently, landholders have begun experimenting with water table manipulation to provide more persistently wet conditions to enable plant establishment on scalds. Excessive drainage is generally the most important primary driver of acid sulfate soil scald formation, and strategies that reduce evaporation from bare areas and maintain or raise water tables in the near vicinity of scalds can contribute to their restoration and revegetation. The shallow ponding of freshwater can trigger rapid and complete revegetation of scalds.[115]

CONCLUSION

Acid sulfate soils can greatly affect the environment within landscapes that contain these materials. A wide range of environmental hazards are presented by these soil materials, including the following: 1) severe acidification of soil and drainage waters; 2) mobilization of many metals, metalloids, nutrients, and rare earth elements; 3) deoxygenation of water bodies; 4) production of noxious gases; and 5) scalding of landscapes.

These environmental hazards can impact on infrastructure such as bridges, drains, pipes, and roadways. Waters draining from acid sulfate soil materials may be enriched in a wide range of potential toxicants, including metals and metalloids, endangering aquatic life and public health. Crops, trees, and pastures may all be severely affected by acid sulfate soil materials, as can aquaculture. The extent, location, properties, and often-intense demands made on acid sulfate soils for water and food supply dictate that the development of our understanding of these soils will need to continue apace.

The management of acid sulfate soil landscapes needs to take into account the hazards that these materials pose. Precautions should be made to avoid if possible or, at the very least, to minimize the environmental degradation that can be caused by the mismanagement of these soil materials to acceptable levels.

REFERENCES

1. Sammut, J.; Callinan, R.B.; Fraser, G.C. The impact of acidified water on freshwater and estuarine fish populations in acid sulphate soil environments. In Proceedings of the National Conference on Acid Sulphate Soils, Coolangatta, Qld, Jun 24–25, 1993; Bush, R.T., Ed.; CSIRO, NSW Agriculture, Tweed Shire Council, 1993; 26–40.
2. Ljung, K.; Maley, F.; Cook, A.; Weinstein, P. Acid sulfate soils and human health—A millennium ecosystem assessment. Environ. Int. **2009**, *35*, 1234–1242.
3. Pons, L.J.; van Breemen, N. Factors influencing the formation of potential acidity in tidal swamps. In Proceedings of the Bangkok Symposium on Acid Sulfate Soils; Dost, H., van Breemen, N., Eds.; ILRI Pub. 31; International Institute for Land Reclamation and Improvement: Wageningen, Netherlands, 1982; 37–51.
4. Dent, D.L.; Pons, L.J. A world perspective on acid sulphate soils. Geoderma **1995**, *67*, 263–276.
5. Ward, N.J.; Sullivan, L.A.; Bush, R.T. Soil pH, oxygen availability and the rate of sulfide oxidation in acid sulfate soil materials: implications for environmental hazard assessment. Aust. J. Soil Res. **2004**, *42*, 509–514.
6. Ritsema, C.J.; Van Mensvoort, M.E.F.; Dent, D.L.; Tan, Y.; Van den Bosch, H.; Van Wijk, A.L.M. Acid sulfate soils. In *Handbook of Soil Science*; Sumner, M.E., Ed.; CRC Press: Boca Raton, FL, 2000; G121–G154.
7. Fitzpatrick, R.W.; Fritsch, E.; Self, P.G. Interpretation of soil features produced by ancient and modern processes

in degraded landscapes. V. Development of saline sulfidic features in non-tidal seepage areas. Geoderma **1996**, *69*, 1–29.

8. Sullivan, L.A.; Bush, R.T.; Fyfe, D. Acid sulfate soil drain ooze: Distribution, behaviour and implications for acidification and deoxygenation of waterways. In *Acid Sulfate Soils in Australia and China*; Lin, C., Melville, M., Sullivan, L.A., Eds.; Science Press: Beijing, China, 2002; 91–99.
9. Fitzpatrick, R.W.; Shand, P.; Merry, R.H. Acid sulfate soils. In *Natural History of the Riverlands and Murraylands*; Jennings, J.T., Ed.; Royal Society of South Australia Inc.: Adelaide, South Australia, 2009; 65–111.
10. Fanning D.S.; Rabenhorst, M.C.; Sullivan, L.A. Some history of the recognition of and the development of knowledge about acid sulfate soils—Internationally and in the U.S. (especially in Maryland and in nearby States). In *Acid Sulfate Soils of the U.S. Mid-Atlantic/Chesapeake Bay Region*, Field Tour Book of the 198th World Soils Congress, Philadelphia; Fanning, D.S., Ed.; University of Maryland: MD, 2006.
11. Isbell, R.F. *The Australian Soil Classification*; CSIRO Publishing: Melbourne, Australia, 1996.
12. Soil Survey Staff. *Keys to Soil Taxonomy*, 11th Ed.; U.S. Department of Agriculture Natural Resources Conservation Service: Washington DC, 2010.
13. IUSS Working Group WRB. *World Reference Base for Soil Resources 2006*, World Soil Resources Reports No. 103; FAO: Rome, 2006.
14. Ahern, C.R.; McElnea, A.E.; Sullivan, L.A. *Acid Sulfate Soils Laboratory Methods Guidelines*; Queensland Department of Natural Resources, Mines and Energy: Indooroopilly, Queensland, Australia, 2004.
15. Sullivan, L.A.; Ward, N.J.; Bush, R.T.; Burton, E.D. Improved identification of sulfidic soil materials by a modified incubation method. Geoderma, **2009**, *149*, 33–38.
16. Sullivan, L.A.; Bush, R.T.; McConchie, D.M. A modified chromium-reducible sulfur method for reduced inorganic sulfur: Optimum reaction time for acid sulfate soil. Aust. J. Soil Res. **2000**, *38*, 729–734.
17. Burton, E.D.; Sullivan, L.A.; Bush, R.T.; Johnston, S.G.; Keene, A.F. A simple and inexpensive chromium-reducible sulfur method for acid-sulfate soils. Appl. Geochem. **2008** *23*, 2759–2766.
18. Sullivan, L.A.; Fitzpatrick, R.W.; Bush, R.T.; Burton, E.D.; Shand, P.; Ward. N.J. *Modifications to the Classification of Acid Sulfate Soil Materials*, Southern Cross GeoScience Technical Report No. 309; Southern Cross University: Lismore, NSW, Australia, 2009.
19. Sullivan, L.A.; Fitzpatrick, R.W.; Bush, R.T.; Burton, E.D.; Shand, P.; Ward, N.J. *The Classification of Acid Sulfate Soil Materials: Further Modifications*, Southern Cross GeoScience Technical Report No. 310; Southern Cross University: Lismore, NSW, Australia, 2010.
20. Fitzpatrick, R.W.; Powell, B.; Marvanek, S. Atlas of Australian acid sulfate soils. In *Inland Acid Sulfate Soil Systems Across Australia*; Fitzpatrick, R.W., Shand, P., Eds.; CRC LEME Open File Report No. 249. CRC LEME: Perth, Australia, 2008; 90–97.
21. MDBA. *Detailed Assessment of Acid Sulfate Soils in the Murray–Darling Basin: Protocols for Sampling, Field Characterisation, Laboratory Analysis and Data Presentation*; Murray Darling Basin Authority, Publication 57/10; Canberra, Australia, 2010.
22. EPHC and NRMMC. *National Guidance for the Management of Acid Sulfate Soils in Inland Aquatic Ecosystems*; Environment Protection and Heritage Council and the Natural Resource Management Ministerial Council: Canberra, Australia, 2010.
23. Jones, E.P.J.; Nadeau, T.L.; Voytek, M.A.; Landa, E.R. Role of microbial iron reduction in the dissolution of iron hydroxysulfate minerals. J. Geophys. Res. **2006** *111*, G01012.
24. Johnston S.G.; Keene A.F.; Burton E.D.; Bush R.T.; Sullivan L.A.; Isaacson L.S.; McElnea, A.E.; Ahern, C.R.; Smith, C.D.; Powell, B. Iron geochemical zonation in a tidally inundated acid sulfate soil wetland. Chem. Geol. **2011**, *280*, 257–270.
25. van Breemen, N. Genesis, morphology, and classification of acid sulfate soils in coastal plains. In *Acid Sulfate Weathering*; Kitrrick, J.A., Fanning, D.S., Hossner, L.R., Eds.; Special Publication 10; Soil Science Society of America: Madison, WI, USA, 1982; 95–108.
26. Andriesse, W. Acid sulphate soils: Diagnosing the ills. In *Selected Papers of the Ho Chi Minh City Symposium on Acid Sulphate Soils*; Dent, D.L., van Mensvoort, M.E.F., Eds.; International Institute for Land Reclamation and Improvement: Wageningen, Netherlands, 1993; 11–29.
27. van Breemen, N. Soil forming processes in acid sulphate soils. In Proceedings of the International Symposium on Acid Sulphate Soils, Wageningen, Netherlands, Aug 13–20, 1972; Dost, H., Ed.; International Institute for Land Reclamation and Improvement: Wageningen, Netherlands, 1973; 66–129.
28. Rao, C.R.M.; Sahuquillo, A.; Sanchez, J.E. A review of different methods applied in environmental geochemistry for single and sequential extraction of trace elements in soils and related materials. Water Air Soil Pollut. **2008**, *189*, 291–33.
29. Åström, M. Partitioning of transition metals in oxidised and reduced zones of sulphide-bearing fine-grained sediments. Appl. Geochem. **1998**, *13*, 607–617.
30. Sohlenius, G.; Öborn, I. Geochemistry and partitioning of trace metals in acid sulphate soils in Sweden and Finland before and after sulphide oxidation. Geoderma **2004**, *122*, 167–175.
31. Claff, S.R.; Sullivan, L.A.; Burton, E.D.; Bush, R.T. A sequential extraction for acid sulfate soils: Partitioning of iron. Geoderma **2010**, *155*, 244–230.
32. Pons, L.J. Outline of the genesis, characteristics, classification and improvement of acid sulphate soils. In *Acid Sulphate Soils*, Proceedings of the International Symposium on Acid Sulphate Soils, Wageningen, Netherlands, Aug 13–20, 1972; Dost, H., Ed.; Publication No. 18, International Institute for Land Reclamation and Improvement: Wageningen, Netherlands, 1973; 3–27.
33. Bloomfield, C.; Coulter, J.K. Genesis and management of acid sulfate soils. Adv. Agron. **1973**, *25*, 265–326.
34. Bush, R.T.; Sullivan, L.A. Morphology and behaviour of greigite from a Holocene sediment in eastern Australia. Aust. J. Soil Res. **1997**, *35*, 853–861.
35. Burton E.D.; Bush, R.T.; Sullivan, L.A.; Hocking, R.K.; Mitchell, D.R.G.; Johnston, S.G.; Fitzpatrick, R.W.;

Raven, M.; McClure, S.; Jang, L.Y. Iron-monosulfide oxidation in natural sediments: resolving microbially mediated S transformations using XANES, electron microscopy, and selective extractions. Environ. Sci. Technol. **2009**, *43*, 3128–3134.

36. Burton, E.D.; Bush, R.T.; Sullivan, L.A. Elemental sulfur in drain sediments associated with acid sulfate soils. Appl. Geochem. **2006**, *21*, 1240–1247.
37. Burton, E.D.; Bush, R.T.; Sullivan, L.A. Reduced inorganic sulfur speciation in drain sediments from acid-sulfate soil landscapes. Environ. Sci. Technol. **2006**, *40*, 888–893.
38. Fanning D.S.; Rabenhorst, M.C.; Burch, S.N.; Islam, K.R.; Tangren, S.A. Sulfides and sulfates. In *Soil Mineralogy with Environmental Applications*; Dixon, J.B., Schulze, D.G., Daniels, W.L., Eds.; Soil Science Society of America: Madison, WI, 2002.
39. Georgala, D. Paleoenvironmental studies of post-glacial black clays in Northern Sweden. PhD Thesis; University of Stockholm: Sweden, 1980.
40. Bush, R.T.; Sullivan, L.A.; Lin, C. Iron monosulfide distribution in three coastal floodplain acid sulfate soils, eastern Australia. Pedosphere **2000**, *10*, 237–246.
41. Rickard D.; Morse, J.W. Acid volatile sulfide (AVS). Mar. Chem. **2005**, *97*, 141–197.
42. Howarth, R.W. Pyrite: Its rapid formation in a salt marsh and its importance in ecosystem metabolism. Science **1979**, *203*, 49–51.
43. Rosicky, M.A.; Sullivan, L.A.; Slavich, P.G.; Hughes, M. Soil properties in and around acid sulfate soil scalds in the coastal floodplains of New South Wales, Australia. Aust. J. Soil Res. **2004**, *42*, 587–594.
44. Rickard, D.; Luther, G.W. Kinetics of pyrite formation by the H2S oxidation of iron (II) monosulfide in aqueous solutions between 25 and 125°C: The mechanism. Geochim. Cosmochim. Acta **1997**, *61*, 135–147.
45. Moses, C.O.; Nordstrom, D.K.; Hermann, J.S.; Mills, A.L. Aqueous pyrite oxidation by dissolved oxygen and by ferric iron. Geochim. Cosmochim. Acta **1987**, *51*, 1561–1571.
46. Bigham J.M.; Schwertmann, U.; Carlson, L. Mineralogy of precipitates formed by the biogeochemical oxidation of Fe(II) in mine drainage. In *Biomineralization Processes of Iron and Manganese—Modern and Ancient Environments*; Skinner, H.C.W., Fitzpatrick, R.W., Eds.; Catena, Supplement 21, 1992; 219–232.
47. Sullivan, L.A.; Bush, R.T. Iron precipitate accumulations associated with waterways in drained coastal acid sulfate landscapes of eastern Australia. Mar. Freshwater Res. **2004**, *55*, 727–736.
48. Bigham J.M.; Schwertmann, U.; Traina, S.J.; Winland, R.L.; Wolf, M. Schwertmannite and the chemical modeling of iron in acid sulfate waters. Geochim. Cosmochim. Acta **1996**, *60*, 2111–2121.
49. Dold, B. Dissolution kinetics of schwertmannite and ferrihydrite in oxidized mine samples and their detection by differential x-ray diffraction (DXRD). Appl. Geochem. **2003**, *18*, 1531–40.
50. Dold, B.; Fontbote, L. Element cycling and secondary mineralogy in porphyry copper tailings as a function of climate, primary mineralogy, and mineral processing. J. Geochem. Explor. **2001**, *74*, 3–55.
51. van Breemen, N. *Genesis and solution chemistry of acid sulphate soils in Thailand*. Agric. Res. Rep. 848. PUDOC; Wageningen, Netherlands, 1976.
52. Peine, A.; Tritschler, A.; Kusel, K.; Peiffer, S. Electron flow in an iron-rich acidic sediment—Evidence for an acidity-driven iron cycle. Limnol. Oceanogr. **2000**, *45*, 1077–87.
53. Dent, D.L. *Acid Sulfate Soils: A Baseline for Research and Development*, ILRI Publ. 39; International Institute for Land Reclamation and Improvement: Wageningen, Netherlands, 1986.
54. Nriagu, J.O. *Biogeochemistry of Lead in the Environment. Ecological Cycles*; Elsevier: Amsterdam, 1978.
55. Sammut J.; White, I.; Melville, M.D. Acidification of an estuarine tributary in eastern Australia due to drainage of acid sulphate soils. Mar. Freshwater Res. **1996**, *47*, 669–684.
56. Åström, M.; Corin, N. Abundance, source and speciation of trace elements in humus-rich streams affected by acid sulfate soils. Aquat. Geochem. **2000**, *6*, 367–383.
57. Sammut, J.; Callinan, R.B.; Fraser, G.C. An overview of the ecological impacts of acid sulfate soils in Australia. In *Proceedings of the 2nd National Conference on Acid Sulfate Soils*; Smith, R.J., Ed.; R.J. Smith and Associates and ASSMAC: Alstonville, Australia, 1996; 140–143.
58. Callinan, R.B.; Sammut, J.; Fraser, G.C. Dermatitus, bronchitis and mortality in empire gudgeon *Hypseleotris compressa* exposed naturally to run-off from acid sulfate soil. Dis. Aquat. Org. **2005**, *63*, 247–253.
59. Simpson, H.J.; Pedini, M. *Brackishwater Aquaculture in the Tropics: the Problem of Acid Sulfate Soils*; Food and Agriculture Organisation of the United Nations: Rome, Italy, 1985.
60. Corfield, J. The effects of acid sulfate soil run-off on a subtidal estuarine macrobenthic community in the Richmond River, NSW, Australia. ICES J. Mar. Sci. **2000**, *57*, 1517–1523.
61. Dove, M.C.; Sammut, J. Impacts of estuarine acidification on survival and growth of Sydney rock oysters *Saccostreaglomerata* (Gould, 1850). J. Shellfish Res., **2007**, *26*, 519–527.
62. Cornell, R.M.; Schwertmann, U. *The Iron Oxides*; Wiley-VCH: Weinheim, 2003.
63. Pronk, J.; Johnson, D.B. Oxidation and reduction of iron by acidophilic bacteria. Geomicrobiol. J. **1992**, *10*, 153–171.
64. Johnson D.B. Biogeochemical cycling of iron and sulfur in leaching environments. FEMS Microbiol. Rev. **1993**, *11*, 63–70.
65. Alpers C.N.; Nordstrom, D.K. Geochemical modelling of water-rock interactions in mining environments. In *The Environmental Geochemistry of Mineral Deposits*; Plumee, G.S., Logsdon, M.J., Eds.; Rev. Econ. Geol. **1999**; Vol. 6, 289–323.
66. Powell, B.; Martens, M. A review of acid sulfate soil impacts, actions and policies that impact on water quality in the Great Barrier Reef catchments, including a case study on remediation at East Trinity. Mar. Poll. Bull. **2005**, *51*, 149–164.
67. Åström, M. Effect of widespread severely acidic soils on spatial features and abundance of trace elements in steams. J. Geochem. Explor. **2001**, *73*, 181–191.
68. Burton, E.D.; Bush, R.T.; Sullivan, L.A. Acid-volatile sulfide oxidation in coastal floodplain drains: Iron-sulfur cy-

cling and effects on water quality. Environ. Sci. Technol. **2006**, *40*, 1217–1222.

69. Burton, E.D.; Bush, R.T.; Sullivan, L.A.; Johnston, S.G.; Mitchell, D.R.G. Mobility of arsenic and selected metals during re-flooding of iron- and organic-rich acid-sulfate soil. Chem. Geol. **2008**, *253*, 64–73.
70. Macdonald, B.C.T.; Smith, J.; Keene, A.K.; Tunks, A.K.; White, I. Impacts from runoff from sulphuric soils on sediment chemistry in an estuary lake. Sci. Total. Environ. **2004**, *329*, 115–130.
71. Wilson, B.P.; White, I.; Melville, M.D. Floodplain hydrology, acid discharge and changing water quality associated with a drained acid sulfate soil. Mar. Freshwater Res. **1999**, *50*, 149–157.
72. Preda, M.; Cox, M.E. Trace metals in acid sediments and waters, Pimpama catchment, southeast Queensland, Australia. Environ. Geol. **2001**, *40*, 755–768.
73. Simpson, S.L.; Fitzpatrick, R.W.; Shand, P.; Angel, B.M.; Spadaro, D.A.; Mosley, L. Climate-driven mobilisation of acid and metals from acid sulfate soils. Mar. Freshwater Res. **2010**, *61*, 129–138.
74. Evangelou, V.P.; Zhang, Y.L. A review: Pyrite oxidation mechanism and acid mine drainage prevention. Crit. Rev. Environ. Sci. Technol. **1995**, *25*, 141–199.
75. Johnston, S.G.; Slavich, P.G.; Hirst, P. The acid flux dynamics of two artificial drains in acid sulfate soil backswamps on the Clarence River floodplain, Australia. Aust. J. Soil Res. **2004**, *42*, 623–637.
76. Burton, E.D.; Bush, R.T.; Sullivan, L.A.; Mitchell, D.R.G. Reductive transformation of iron and sulfur in schwertmannite-rich accumulations associated with acidified coastal lowlands. Geochim. Cosmochim. Acta **2007**, *71*, 4456–4473.
77. Blodau, C. A review of acidity generation and consumption in acidic coal mine lakes and their watersheds. Sci. Total Environ. **2006**, *369*, 307–332.
78. Burton, E.D.; Bush, R.T.; Sullivan, L.A.; Mitchell, D.R.G. Schwertmannite transformation to goethite via the Fe(II) pathway: Reaction rates and implications for iron-sulfide formation. Geochim. Cosmochim. Acta **2008**, *72*, 4551–4564.
79. Burton, E.D.; Johnston, S.G.; Watling, K.; Bush, R.T.; Keene, A.F.; Sullivan, L.A. Arsenic effects and behavior in association with the Fe(II)-catalysed transformation of schwertmannite. Environ. Sci. Technol. **2010**, *44*, 2016–2021.
80. Johnston, S.G.; Slavich, P.G.; Hirst, P. Alteration of groundwater and sediment geochemistry in a sulfidic backswamp due to *Melaleuca quinquenervia* encroachment. Aust. J. Soil Res. **2003**, *41*, 1343–1367.
81. Howitt, J.A.; Baldwin, D.S.; Rees, G.N.; Williams, J.L. Modelling blackwater: Predicting water quality during flooding of lowland river forests. Ecol. Modell. **2007**, *203*, 229–242.
82. Hamilton, S.K.; Sippel, S.J.; Calheiros, D.F.; Melack, J.M. An anoxic event and other biogeochemical effects of the Pantanal wetland on the Paraguay River. Limnol. Oceanogr. **1997**, *42*, 257–272.
83. Sullivan, L.A.; Bush, R.T. The behaviour of drain sludge in acid sulfate soil areas: Some implications for acidification of waterways and drain maintenance. In Proceedings of the Workshop on Remediation and Assessment of Broadacre Acid Sulfate Soils; Slavich, P., Ed.; Acid Sulfate Soil Management Advisory Committee (ASSMAC): Southern Cross University, Lismore, 2000; 43–48.
84. ANZECC/ARMCANZ. *Australian and New Zealand Guidelines for Fresh and Marine Water Quality*; Australian and New Zealand Environment and Conservation Council, Agricultural and Resource Management Council of Australia and New Zealand, 2000.
85. Hladyz, S.; Watkins, S. *Understanding Blackwater Events and Managed Flows in the Edward-Wakool River System*; Fact Sheet; Murray-Darling Freshwater Research Centre: Wodonga, Vic., 2009; 2.
86. Wong, V.N.L.; Johnston, S.G.; Bush, R.T.; Sullivan, L.A.; Clay, C.; Burton, E.D.; Slavich, P.G. Spatial and temporal changes in estuarine water quality during a post-flood hypoxic event. Estuarine, Coastal Shelf Sci. **2010**, *87*, 73–82.
87. Johnston, S.G.; Slavich, P.G.; Sullivan, L.A.; Hirst, P. Artificial drainage of floodwaters from sulfidic backswamps: Effects on deoxygenation in an Australian estuary. Mar. Freshwater Res. **2003**, *54*, 781–795.
88. Bush, R.T.; Fyfe, D.; Sullivan, L.A. Occurrence and abundance of monosulfidic black ooze in coastal acid sulfate soil landscapes. Aust. J. Soil Res. **2004**, *42*, 609–616.
89. Burton, E.D.; Bush, R.T.; Sullivan, L.A. Sedimentary iron geochemistry in acidic waterways associated with coastal lowland acid sulfate soils. Geochim. Cosmochim. Acta **2006**, *70*, 5455–5468.
90. Charlson, R.J.; Lovelock, J.E.; Andreae, M.O.; Warren, S.G. Oceanic phytoplankton, atmospheric sulfur, cloud albedo and climate. Nature **1987**, *326*, 655–61.
91. Lohmann, U.; Feichter, J. Global indirect aerosol effects: A review. Atmos. Chem. Phys. **2005**, *5*, 715–37.
92. Berresheim, H.; Wine, P.H.; Davis, D.D. Sulfur in the atmosphere. In *Composition, Chemistry, and Climate of the Atmosphere*; Singh, H.B., Ed.; Van Nostrand Reinhold: New York, 1995; 251–307.
93. Aneja, V.P. Natural sulfur emissions into the atmosphere. JAPCA J. Air Waste Manage. **1990**, *40*, 469–76.
94. Bates, T.S.; Lamb, B.K.; Guether, A.; Dignon, J.; Stoiber, R.E. Sulfur emissions to the atmosphere from natural sources. J. Atmos. Chem, **1992**, *14*, 315–37.
95. Macdonald, B.C.T.; Denmead, O.T.; White, I.; Melville, M.D. Natural sulfur dioxide emissions from sulfuric soils. Atmos. Environ. **2004**, *38*, 1473–1480.
96. Luther, G.W., III; Glazer, B.; Ma, S.; Trouwborst, R.; Shultz, B.R.; Drushcel, G.; Kraiya, C. Iron and sulfur chemistry in a stratified lake: Evidence for iron-rich sulfide complexes. Aquat. Geochem. **2003**, *9*, 87 110.
97. EPA. Toxicological Review of Hydrogen Sulfide, 2003, available at http://www.epa.gov/iris/toxreviews/0061-tr.pdf (accessed August 16, 2010).
98. Diaz, R.J.; Rosenberg, R. Marine benthic hypoxia: A review of its ecological effects and the behavioural responses of benthic macrofauna. Oceanogr. Mar. Biol.: Annu. Rev. **1995**, *33*, 245–303.
99. Rabalais, N.N. Nitrogen in aquatic ecosystems. Royal Swedish Academy of Sciences. Ambio **2002**, *31*, 102–112.
100. Jorgensen, B.B.; Fossing, H.; Wirsen, C.O.; Jannasch, H.W. Sulfide oxidation in the anoxic Black Sea chemocline. Deep-Sea Res. **1991**, *38*, 1083–1103.

101. Froelich, P.N.; Klinkhammer, G.P.; Bender, M.L.; Luedtke, N.A.; Heath, G.R.; Cullen, D.; Dauphin, P.; Hammond, D.; Hartman, B.; Maynard, V. Early oxidation of organic matter in pelagic sediments of the eastern equatorial Atlantic: Suboxic diagenesis. Geochim. Cosmochim. Acta **1979**, *43*, 1075–1090.
102. Luther, G.W., III; Sundby, B.; Lewis, B.L.; Brendel, P.J.; Silverberg, N. Interactions of manganese with the nitrogen cycle: alternative pathways to dinitrogen. Geochim. Cosmochim. Acta, **1997**, *61*, 4043–4052.
103. Millero, F.J.; Hubinger, S.; Fernandez, M.; Garnett, S. Oxidation of H_2S in seawater as a function of temperature, pH and ionic strength. Environ. Sci. Technol. **1987**, *21*, 439–443.
104. Rozan, T.F.; Taillefert, M.; Trouwborst, R.E.; Glazer, B.T.; Ma, S.; Herszage, J.; Valdes, L.M.; Price, K.S.; Luther, G.W., III. Iron, sulfur and phosphorus cycling in the sediments of a shallow coastal bay: Implications for sediment nutrient release and benthic macroalgal blooms. Limnol. Oceanogr. **2002**, *47*, 1346–1354.
105. White, I. Swelling and hydraulic properties and management of acid sulfate soils. In *Acid Sulfate Soils in Australia and China*; Lin, C., Melville, M.D., Sullivan, L.A., Eds.; Science Press: Beijing, 2000; 24–48.
106. Johnston, S.G.; Slavich, P.G.; Hirst, P.; Bush, R.T.; Aaso, T. Saturated hydraulic conductivity of sulfuric horizons in coastal floodplain acid sulfate soils: Variability and implications. Geoderma **2009**, *151*, 387–394.
107. Johnston, S.J.; Keene, A.F.; Bush, R.T.; Burton, E.D.; Sullivan, L.A.; Smith, D.; McElnea, A.E.; Martens, M.A.; Wilbraham, S. Contemporary pedogenesis of severely degraded tropical acid sulfate soils after introduction of regular tidal inundation. Geoderma **2009**, *149*, 335–346.
108. Dear, S.E.; Moore, N.G.; Dobos, S.K.; Watling, K.M.; Ahern, C.R. *Soil Management Guidelines. Queensland Acid Sulfate Soil Technical Manual*, Version 3.8 (2002). Department of Natural Resources and Mines: Indooroopilly, Queensland, Australia, 2002.
109. Ahern, C.R.; Stone, Y.; Blunden, B. Acid sulfate Soils management guidelines. In *Acid Sulfate Soil Manual*; Acid Sulfate Soil Management Advisory Committee: Wollongbar, NSW, Australia, 1998.
110. Watling, K.M.; Sullivan, L.A.; McElnea, A.; Ahern, C.; Burton, E.D.; Johnston, S.G.; Keene, A.F.; Bush, R.T. Effectiveness of lime particle size in the neutralisation of sulfidic acid sulfate soil materials. In Proceedings of the 19th World Congress of Soil Science, Brisbane, Australia, Aug 1–6, 2010; Gilkes, B., Ed.; 2010.
111. WA DoE. *Acid Sulfate Soils Guideline Series: Guidance for Groundwater Management in Urban Areas on Acid Sulfate Soils*; 2004, available at http://www.dec.wa.gov.au/component/option,com_docman/Itemid,849/gid,1172/task,doc_details/ (accessed May 1, 2009).
112. Brinkman, W.J.; Xuan, V. *Melaleuca leucodendron* S.L., a useful and versatile tree for acid sulfate soils and some other poor environments. Int. Tree Crop J. **1991**, *6*, 61–274.
113. Tan, E.O. Coastal aquaculture in the Philippines. In *Coastal Aquaculture in Asia, Part A. Taipei City*; Republic of China, Food and Fertilizer Technology Center: Taiwan, 1983.
114. Tri, L.Q. Developing management packages for acid sulfate soils based on farmer and expert knowledge. Field study in the Mekong delta, Vietnam. PhD Thesis; Department of Soil Science and Geology, Wageningen Agricultural University: Wageningen, Netherlands, 1996.
115. Rosicky, M.A.; Sullivan, L.A.; Slavich, P.G.; Hughes, M. Factors contributing to the acid scalding process in the coastal floodplains of New South Wales. Aust. J. Soil Res. **2004**, *42*, 587–594.
116. Rosicky, M.A.; Sullivan, L.A.; Slavich, P.G.; Hughes, M. Techniques for the revegetation of acid sulfate soil scalds in the coastal floodplains of New South Wales, Australia: Ridging, mulching, and liming in the absence of stock grazing. Aust. J. Exp. Agr. **2006**, *46*, 1589–1600.

Acid Sulfate Soils: Management

Michael D. Melville
School of Biological, Earth, and Environmental Sciences, University of New South Wales, Sydney, New South Wales, Australia

Ian White
College of Science, Fenner School of Environment and Society, College of Medicine, Biology and Environment, Australian National University, Canberra, Australian Capital Territory, Australia

Abstract

The issue of acid sulfate soils (ASSs) and their management must now encompass those landscapes containing sulfidic material, both soft sediments and hard rocks, located on the coast and inland of land masses. In the past 20 years around the world, there has been a marked shift in understanding the necessity of this broader view. This entry covers a large topic and for this reason touches only lightly in some detailed aspects of ASS characteristics and processes. The references will, however, lead readers to more details on many aspects of the subject. In this entry, several concepts and case studies involving policy, regulation, and best practice from Australia are discussed because Australia now probably provides much of the impetus and methods for global adoption of best management practices, which can be applied worldwide. There are three fundamental questions underlying the use of ASS landscapes. Firstly, how best to use the existing environment productively while minimizing any sulfide mineral oxidation. Secondly, how to effectively neutralize any existing or newly created acidity formed during a particular land use. Thirdly, how to eliminate or minimize downstream environmental impacts from ASS drainage waters, which may include deoxygenation of soil or surface waters and the release of metals and nutrients. Therefore, best management of ASS must address these questions.

INTRODUCTION

Acid sulfate soils (ASSs) is the name given to all those soils or unconsolidated sediments that contain reduced sulfur minerals and compounds that have been, or have the potential to be, oxidized and thereby produced acidity.[1–5] They occur globally and are an important problem requiring careful management. In our earlier publication on these soils,[6] we stressed the general coastal location of these soils, which contain a range of materials due to the sulfate in seawater that provides a large reservoir of dissolved sulfur that can be reduced biochemically in the presence of organics to form various monosulfides, and eventually the disulfide, pyrite (FeS_2). However, it is now clear that we should also recognize materials and problems from sulfidic hard rocks, generally freshwater bodies, and landscapes in inland areas. This need has been admirably illustrated by Fitzpatrick and Shand,[7,8] working on a wide range of sites and issues in Australia, but the principles and examples that they illustrate have global application.

The biochemical processes in ASSs are analogous to those of acid mine drainage (AMD)/acid rock drainage (ARD), and the rain-induced drainage from some mine sites with these materials also have analogues in the hydrology of acid sulfate soil (ASSs). Nevertheless, there are important differences between problems and management with AMD/ARD and other ASS landscapes, mostly because of their respective scales, material, and mineral characteristics, but particularly in their respective economics that affect the viable management options.

There are three fundamental questions underlying the use of ASS landscapes. Firstly, how best to use the existing environment productively while minimizing any sulfide mineral oxidation. Secondly, how to effectively neutralize any existing or newly created acidity formed during a particular land use. Thirdly, how to eliminate or minimize downstream environmental impacts from ASS drainage waters, which may include deoxygenation of soil or surface waters and the release of metals and nutrients. Therefore, best management of ASS must address these questions.

EXISTING AND EMERGING PROBLEMS

Much of the early research on ASS, as illustrated in the contents of the early series (from 1973 until 1992) of international symposia organized by ILRI (International Institute for Land Reclamation and Improvement) from Wageningen, the Netherlands, was directed towards improving their agronomic usefulness. However, the 2002 international ASS conference in Australia emphasized the range of environmental problems with ASS and their best management, and the most recent (2008) conference

Encyclopedia of Environmental Management DOI: 10.1081/E-EEM-120046329

in Guangzhou, China, continued that theme and was combined with research and management for ARD.

While ASSs were recognized in Europe more than 250 years ago,[1] appreciation of their existence and importance in terms of environmental impacts in Australia was limited.[9] The most important ecological impact that changed this appreciation was the 1987 acidity discharge event on the Tweed River in northeastern New South Wales. Twenty-three kilometers of the Tweed River were completely clarified and sterilized of fish, crustaceans, and most benthic organisms after the flooding caused acidic drainage discharges from ASSs of the river floodplain. It was the astute observations of a local entomologist, published in an amateur fisherman's magazine,[10] which identified the true cause of this devastating event that took up to 18 months to heal.[6]

While Australia was slow to become involved in ASS research and management, the Tweed River fish kill initiated broad interest and involvement of stakeholders, researchers, industry groups, regulators, and politicians. The pathway of change with these groups on the issue of ASS in Australia has not always been straightforward without conflict; neither has it been simple.[9] However, a generally very positive outcome has been achieved and Australia now probably provides much of the impetus and methods for global adoption of best management practices with respect to ASS. Each State and the National Commonwealth of Australia have adopted policies and practices for ASS best management. Much of the materials necessary for the dissemination of information on ASS are freely available on the web.[7,23–25] The Queensland Government publications are updated from time to time. A quarterly newsletter, *ASSAY*, which has now been in publication for about 13 years, is available for free electronic subscription.[11]

As we pointed out in White,[9] many coastal floodplains around the world are composed of Holocene-age (<10,000 years BP) sediments that contain iron sulfide minerals,[2] mostly as pyrite, but in some regions, monosulfides are important.[12] It is estimated that the global extent of such ASS is between 10^7 and 10^8 ha, with large deposits in Southeast Asia, the Far East, Africa, and North and South America.[13] The total extent of ASS now mapped in Australia[14] is about 2.2×10^6 ha, of which about 0.6×10^6 ha is coastal and 1.6×10^6 ha is inland.[12] When exposed to air by natural or human-induced drainage, sulfides in these materials oxidize to sulfuric acid that may leach iron, aluminium, and other metals into sediment pore waters. Drainage of these pore waters has significant detrimental ecological impacts,[2,10,15,16] corrosion of engineering infrastructure,[17,18] blooms of cyanobacteria,[19–21] and emission of toxic and greenhouse gases.[22,23] The existence and problems of non-coastal ASS, and the need for considering hard rock sulfidic materials, increase the scope of management strategies.[8,25]

There are many global issues that relate to ASS and that are likely to emerge in the future, but there are two that seem to us are particularly problematic. Firstly, with global warming is the possible oxidation of sulfidic material in organic-rich landscapes presently preserved or protected by permafrost in high latitudes of the Northern Hemisphere (e.g., around James and Hudson Bays in Canada). Drainage of oxidation products and emission of greenhouse and sulfurous gases are likely. Secondly, perhaps also to some extent due to global warming but from population growth and political decisions for development, the increasing upstream abstraction of water resources from major rivers so that delivery of freshwater to coastal deltas or estuaries decreases markedly and wetlands underlain by sulfidic materials or sediments drain and oxidize and/or experience subsidence and increased saline intrusion (e.g., the Mekong Delta of Vietnam, and with the Brahmaputra and Ganges Rivers into Bangladesh). Any sea level rise associated with global warming increases this tidal inundation risk.

In southeastern Australia, the severe drought from 2005 to about 2009 markedly reduced the delivery of discharge to the outflow lakes area of the Murray–Darling Basin (MDB). This caused oxidation of the sulfidic materials and sediments that had accumulated during the various water management regimes that were initiated in the lakes and supply rivers following European settlement (Fitzpatrick and Shand,[8] pp. 38–43). Many different management regimes have been proposed and begun to be implemented to overcome these problems. However, it seems clear that licensing of water abstraction, predominantly for agricultural irrigation in the midcatchment during the multidecadal "flood-dominated rainfall regime" after 1946,[27] exceeded the capacity of the river system to supply water to the outflow lakes and maintain their health when a "drought-dominated rainfall regime"[27] possibly began.

In western Finland, the rapid isostatic uplift (<12 mm/yr) from retreat of the last northern European glacial sheet is exposing sulfidic sediments to natural oxidation, but this is being exacerbated by artificial drainage systems to enable agricultural land uses.[28] Drainage from these ASS landscapes is markedly polluting streams and waters of the Gulf of Bothnia, and it is estimated that the heavy metal pollution from this source is equal to that from all Finnish secondary industry.

It is clear that management of landscapes in which ASSs occur must make due cognizance of the landscape's ability to accommodate human impacts and probable variability in climatic or geomorphic conditions.

In recent years, the distinction between the characteristics, processes, problems, and management options of ASS and ARD have almost disappeared. Soil scientists such as Fitzpatrick and his colleagues are researching on what might previously have been considered ARD issues [e.g., see Fitzpatrick and Shand[8] (pp. 31–37) and Skwarnecki and Fitzpatrick[29]], and management techniques used commonly in AMD sites, such as capping to prevent sulfide oxidation and constructed wetlands for acidity discharge neutralization,[30] have been used successfully for ASS management.

MANAGEMENT APPROACHES

We have previously described[9] how early major problems from ASS on East Coast Australia created significant conflict between fishermen, sugarcane farmers, and environmentalists, and how this was resolved through a cooperative learning and coastal stewardship approach. This began the process that has underlain Australia's becoming a global leader in the understanding and management of ASS.

In complex environmental situations and problems, such as are likely to occur with ASS, significant knowledge gaps are likely to exist as to what is the problem and how best to manage it. Nevertheless, action is thought necessary. Adaptive management is sometimes proposed[9] on the basis that any mistakes due to incorrect information can be corrected provided there is rigorous monitoring. However, adaptive management assumes linear processes where any mistakes can easily be reversed. Unfortunately, with many situations of environmental management, including with ASS, processes are not linear (and are probably generally hysteretic) and environments are susceptible to dramatic collapse. Therefore, environmental degradation is not easily reversed.[31,32]

Many of the problems arising with ASS are the outcome of environmental change from natural processes, anthropogenic activities, or some combination of these. Within the population affected by such environmental changes, management of problems arising is best achieved if the reasons for the change are understood. Appropriate, reliable information, communicated in a relevant way, can provide an important catalyst for changed attitudes. The challenge is to collect and communicate such complex information in a manner that is trusted and accepted by all. Therefore, the first step is to describe the nature and cause of the problem and indicate some likely consequences, but without causing further and new concerns. The next step needs to identify and bring together all stakeholders in a participatory way so that consensus and compromise agreement can be reached.[33] This may require the input of some external, trusted facilitators. Strangely, because many see these personnel as rather "ivory towerish," universities can fulfill this role. Unfortunately, government employees are often unable to take this role because they have a regulatory role in environmental management.[25] In Australia, the problems of managing coastal resources have come from the many top–down, conflicting visions and disparate goals of successive governments and individuals. These goals, past policies, and legislation have concerned issues such as environmental protection and/or rehabilitation, economic development, and regional employment and growth. Politics is an ever-present factor.

Environmental stewardship has been proposed as one way by which conflicts might be reduced so that ownership and pride in the environment and its heritage are encouraged. In Australia, the indigenous original owners of the land have a culture of custodial stewardship and knowledge that can be used for best management practices, but the early European settlers were unable or unwilling to accept this. Now, theirs is a voice that must be heard in environmental management, including that with ASS. Stewardship involves voluntary compliance, strong commitment, and a willing participation in the sustainable use of resources through wise practices. There are probably well-founded concerns that without strong underpinning regulations, voluntary compliance agreements do not have effective mechanisms to address persistent breaches of agreements.

The best management of ASS involves a range of activities that address the issues of minimizing the export of acidity, metals, and nutrients into downstream environments. This involves two issues: 1) the creation of new acidity by sulfide mineral oxidation and 2) the management of existing acidity in the landscape. The latter has been frequently ignored. Across the range of land uses of northern NSW, the existing acidity in the sulfuric layer averages approximately 50 tons of sulfuric acid per hectare.[34] Potential acidity represented in the sulfidic minerals of the deeper subsoil is many times this amount. Nevertheless, the annual discharge of acidity is <0.5 tons/ha.[35] The degree to which soil acidification has been caused by artificial drainage is uncertain but natural processes are also involved.[36] It is clear, however, that artificial drainage networks, which decrease by orders of magnitude the time of inundation of floodplain backswamps from their natural conditions, provide the conduit by which acidity is transferred rapidly to the estuary and potentially causes downstream impacts.[37] Decreasing the density of drains has a major impact on the export of acidity. Laser grading provides a technique for removing surface waters and decreasing drainage density and allows more land to be planted to crops.[37]

The actual impact caused by rain-induced acidic discharges depends upon the magnitude of the discharge relative to the dilution by upland flows and neutralization capacity of receiving waters. Land users on coastal ASS floodplains must appreciate that other parts of the estuarine ecosystem depend on the dissolved alkalinity of the receiving water, consumed during acidity neutralization. Floodplain land uses should be undertaken so as to avoid creating any new acidification and minimizing the export of any acidity.

Overall, best management of ASS involves four possible approaches: 1) education and assessment; 2) avoidance; 3) oxidation prevention; and 4) acidity containment and neutralization. These approaches for ASS management have been included to a varying extent by cases described in Jiggins[31] and Harris[32] and detailed in Queensland's

ASS Management Guidelines.[24] The latter work also provides insight to the risks associated with the various management approaches. ASS management is not necessarily straightforward and amenable to a prescriptive set of firm rules. Each project must align with the particular existing and proposed environmental situations.

Education and Assessment

Best management of ASSs requires knowledge of their distribution; the depth of the sulfidic material from the soil surface; the acidity stored in sulfuric material; the hydrological behavior of the soil profiles, landscape, and drainage system; the climatic regime; and the magnitude, tidal characteristics, and water quality of the receiving waters.[37] All of these are infrequently considered.

Australian soil scientists were generally slow in appreciating both the presence and the environmental problems from ASSs, and in raising the awareness of the public, land managers, and policy makers. The first national symposium on ASSs was only in 1993.[8] Nevertheless, in the past 20 years, a major shift has occurred, and now all States and Territories, and the Commonwealth of Australia, have included ASS management in their environmental policies. Australia is one of the few countries with a national strategy on ASSs. The shift in New South Wales (NSW) has been greatly helped by publication of "Acid Sulphate Soil Risk Maps"[38] for the entire NSW coastline (scale 1:25,000). These maps are being used as the basis for land-use planning instruments ("Local Environment Plans," or LEPs) of local government authorities. These LEPs require submission of a Development Application for any activity that disturbs more than 1 ton of ASS.

The sugar industry, a major user of ASSs in NSW, has been granted State-wide exemption from these requirements for normal farming and drain cleaning activities because each of the 700 cane growers has signed a contract to comply with an approved code of best management practice for ASSs. Less than 20 years ago, the sugar industry was in a state of denial about the existence and problems of ASSs in their cane lands. The industry's new sense of land stewardship with ASSs is exemplary and indications are that economic rewards from up to 30% increase in yields are occurring. Education and capacity building of the cane growers has been helped through an ASS survey demonstration and assessment on each grower's land. Compliance to the agreement is independently audited each year, and to date, there have been very few inadvertent non-compliances found and these were quickly rectified.

The existence and characteristics of ASS across Australia are now provided in the Atlas of Australian Acid Sulfate Soils[14] that is accessible online through the Australian Soil Resource Information System (ASRIS: http://www.aris.gov.au). The management of complex data sets, such as arise in studies of ASS landscapes, can be managed using a system such as that described in Baker.[39] The methods for sampling, laboratory analysis, and assessment of ASS are provided by the ever-evolving Queensland Acid Sulfate Soils Manual.[24] Many organizations across Australia now provide training courses specifically for various aspects about ASS, and these are used for training groups such as stakeholders and regulators (e.g., see various issues of *ASSAY*).[11]

Avoidance

A primary preventative consideration with ASSs should be an avoidance strategy. Such strategies include the decision not to drain ASS wetlands with sulfidic material and to divert or relocate a proposed land use to an alternative site. If use of the site is unavoidable, then treatment so as to prevent sulfide mineral oxidation and export of any existing acidity is necessary.

This seems a relatively easy and sensible first step and one that we advocated as the first principle in an acid sulfate best management consultancy report to the NSW Roads Traffic Authority (RTA), prior to their undertaking major construction of the 1000 km divided highway between Sydney and Brisbane.[40] However, the RTA ASS Guidelines document did not include "avoidance" but most other parts of our report were included verbatim. Nevertheless, we believe that RTA has considered avoidance in assessing the final route of this construction but otherwise has ensured that appropriate treatment has been included in their ASS management plans for the various highway sections that encounter ASS. In assessing the need for avoidance or otherwise, land-use planners will consider relative land values and political advice as well as any engineering issues. This is a complex assessment process.

Oxidation Prevention

This might involve separation of the sulfidic material and its capping with material that is impermeable to oxygen and rainfall infiltration. The use of other sulfidic clays as capping materials must be avoided.[4] Decisions to not drain coastal wetlands are intended not only to avoid the creation of new acidity from sulfide mineral oxidation but also to address the issue of retaining existing acidity.[36] Management techniques such as raising water table elevations are seen as a means of protecting sulfide minerals from oxidation, but this should also consider the hydrological effects with possible increase of existing acidity and the social and economic impact of any subsequently necessary changed land use.

Acidity Containment and Neutralization

Neutralization of ASS acidity by the alkalinity of seawater is the ultimate natural process in the geochemical sulfur cycle. The use of this approach with human-induced acidification and increased acidity discharge needs to be approached with great caution. In eastern Australia, where existing acidity concentrations can exceed 50 tons/ha,[34] application of sufficient crushed limestone or other neutralizing material is impracticable for agriculture. Far better options are land-use practices that minimize acidity export and the strategic application of crushed limestone to drains through which the acidity is exported. These practices can be incorporated into normal activities at the individual paddock scale of intensive agriculture such as with sugarcane. In the NSW sugarcane industry, the adoption of laser grading on estuary floodplains has allowed the marked reduction in the number of field drains that are the main source of acidity export.[36] This has also improved the productivity of the cane land. Low-value land uses, such as with grazing, will be precluded from these new developments, and a more communal approach with some external funding may be necessary. The use of emerging technologies from AMD experience may be useful.[30,41]

CONCLUSION

The understanding of the need and techniques for best management of ASSs has shifted markedly in the past 20 years. This has been particularly noticeable in Australia and the Baltic coasts and now in inland problems arising in Australia as well. The boundaries between what were previously considered "acid sulfate soil" and "acid rock drainage" or "acid mine drainage" issues have almost disappeared as researchers from both discipline areas have worked together in understanding the chemical and hydrological processes that are associated with their common sulfidic mineral oxidation, and acidity and dissolved metal drainage processes. The issue of ASSs and their problems has involved not only scientific and engineering researchers but also the stakeholders managing land uses and the general public. Across Australia, at least, ASSs and their best management have become incorporated into government policy, regulation, and practice with all three levels of government. It is hoped that adoption by all stakeholders and land managers of a culture of environmental stewardship will greatly reduce the problems arising from ASSs.

REFERENCES

1. Pons, L.J. Outline of the genesis, characteristics, classification and improvement of acid sulphate soils. In Proceedings of the 1972 International Acid Sulphate Soils Symposium, Wageningen, the Netherlands, Aug. 13–29, 1972; Dost, H., Ed.; International Institute for Land Reclamation and Improvement, ILRI Pub. No. 18, 1973; Vol. 1, 3–27.
2. Dent, D.L. *Acid Sulphate Soils: A Baseline for Research and Development*; International Institute for Land Reclamation and Improvement: Wageningen, the Netherlands, 1986; ILRI Pub. No. 39, 1–204.
3. Dent, D.L.; Pons, L.J. A world perspective on acid sulfate soils. Geoderma **1995**, *67*, 263–276.
4. Fanning, D.S.; Burch, S.N. Coastal acid sulfate soils. In *Reclamation of Drastically Disturbed Lands*; Barnhisel, R.I.; Darmody, R.G.; Daniels, W.L., Eds.; Agronomy 41, American Society of Agronomy: Madison, WI, 2000; 921–937.
5. Fanning, D.S. Acid sulfate soils, definitions, and classification. In *The Encyclopedia of Soil Science*; Lal, R., Ed.; Marcel Dekker: New York, 2002; 11–13.
6. Melville, M.D.; White, I. Acid sulfate soils: Problems and management. In *The Encyclopedia of Soil Science*; Lal, R., Ed.; Marcel Dekker: New York, 2002; 19–22.
7. Fitzpatrick, R.; Shand, P., Eds. *Inland acid sulfate soil systems across Australia*, CRC LEME Open File Report 249, CRC LEME: Perth, Australia, 2008; 1–303. Available online at http://crcleme.org.au/Pubs/OFRSSindex.html (accessed June 2010).
8. Fitzpatrick, R.; Shand, P. Inland acid sulfate soils in Australia; overview and conceptual models (Ch. 1). In *Inland Acid Sulfate Soil Systems across Australia*; Fitzpatrick, R., Shand, P., Eds.; CRC LEME Open File Report 249, CRC LEME: Perth, Australia, 2008; 6–74.
9. White, I.; Melville, M.; Macdonald, B.; Quirk, R.; Hawken, R.; Tunks, M.; Buckley, D.; Beattie, R.; Williams, J.; Heath, L. From conflicts to wise practice agreement and national strategy: Cooperative learning and coastal stewardship in estuarine floodplain management, Tweed River, eastern Australia. J. Cleaner Prod. **2007**, *15*, 1545–1558.
10. Easton, C. The trouble with the Tweed. Fishing World, **1989**, *3*, 58–59.
11. Walsh, Ed. ASSAY; Quarterly publication. Issues 1–52 available at http://www.dpi.nsw.gov.au/aboutus/resources/periodicals/newsletters/assay/. Subscription free from simon.walsh@industry.nsw.gov.au (accessed June 2010) with subject: "subscribe ASSAY".
12. Bush, R.T.; McGrath, R.; Sullivan, L.A. Occurrence of marcasite in an organic-rich Holocene estuarine mud. Aust. J. Soil Res. **2005**, *42*, 617–621.
13. Brinkman, R. Social and economic aspects of reclamation of acid sulphate soil areas. In Proceedings of the Bangkok Symposium on Acid Sulphate Soils, Bangkok, Thailand, Jan. 8–24, 1981; Dost, H., van Breemen, N., Eds.; Institute for Land Reclamation and Improvement: Wageningen, the Netherlands, 1982; ILRI Pub. No. 31, 21–36.
14. Fitzpatrick, R.; Marvanek, S.; Powell, B. Atlas of Australian acid sulfate soils (Ch. 2). In *Inland Acid Sulfate Soil Systems across Australia*; Fitzpatrick, R., Shand, P., Eds.; CRC LEME Open File Report 249, CRC LEME: Perth, Australia, 2008; 75–89.
15. Sammut, J.; Melville, M.D.; Callinan, R.D.; Fraser, G.C. Estuarine acidification: Impacts on aquatic biota of draining acid sulphate soils. Aust. Geogr. Stud. **1995**, *33*, 89–100.
16. Degens, B.; Shand, P.; Fitzpatrick, R.; George, R.; Rogers, S.; Smith, M. Avon Basin, WA wheatbelt: Acidification and formation of inland ASS materials in salt lakes by acid drainage and regional groundwater discharge (Ch. 11). In *Inland Acid Sulfate Soil Systems across Australia*; Fitzpatrick, R., Shand, P., Eds.; CRC LEME Open File Report 249, CRC LEME: Perth, Australia, 2008; 176–188.
17. White, I.; Melville, M.D.; Sammut, J.; Wilson, B.P.; Bowman, G.M. Downstream impacts from acid sulfate soils. In *Downstream Impacts of Landuse*; Hunter, H., Eyles, A.,

Raymont, G., Eds.; Department of Natural Resources: Brisbane, 1996, 165–172.

18. Shand, P.; James-Smith, J.; Hodgkin, T.; Fitzpatrick, R.; McLure, S.; Raven, M.; Stadter, M.; Hill, T. Ancient acid sulfate soils in Murray Basin sediments: Impacts on borehole clogging by $Al(OH)_3$ and salt interception scheme efficiency (Ch. 5). In *Inland Acid Sulfate Soil Systems across Australia*; Fitzpatrick, R., Shand, P., Eds.; CRC LEME Open File Report 249, CRC LEME: Perth, Australia, 2008; 129–136.
19. Dennison, W.C.; O'Neil, J.M.; Duffy, E.; Oliver, P.; Shaw, G. Blooms of the alga cyanobacterium *Lyngbya majuscula* in coastal waters of Queensland. In Proceedings of the International Symposium on Marine Cyanobacteria, Paris, Nov. 24–28, 1997; Charpy, L., Larkum, A.W.D., Eds.; Bulletin de L'Institut Oceanographique, Monaco, 1999; 632.
20. Albert, S.; O'Neil, J.M.; Udy, J.W.; Ahern, K.S.; O'Sullivan, C.M.; Dennison, W.C. Blooms of the cyanobacterium *Lyngbya majuscula* in coastal Queensland, Australia: Disparate site, common factors. Mar. Pollut. Bull. **2005**, *51* (1–4), 428–437.
21. Abram, N.J.; Gagan, M.K.; McCulloch, M.T.; Chappell, J.; Hantoro, W.S. Coral reef death during the 1997 Indian Ocean dipole linked to Indonesian wildfires. Science **2003**, *301*, 952–955.
22. Macdonald, B.C.T.; Denmead, O.T.; White, I.; Melville, M.D. Natural sulfur dioxide emissions from sulfuric soils. Atmos. Environ. **2004**, *38*, 1473–1480.
23. Hicks, W.; Fitzpatrick, R. Greenhouse gas emissions and toxic gas missions from soil organic matter and carbonates associated with acid sulfate soils (Ch. 6). In *Inland Acid Sulfate Soil Systems across Australia*; Fitzpatrick, R., Shand, P., Eds.; CRC LEME Open File Report 249, CRC LEME: Perth, Australia, 2008; 137–148.
24. Ahern, C.R.; McElnea, A.E.; Sullivan, L.A. Acid sulfate soils laboratory guidelines (Version 2.1). In *Queensland Acid Sulfate Soil Technical Manual*; Queensland Department of Natural Resources, Mines and Energy: Indooroopilly, Queensland, Australia, 2004; i-I2-4. (June 2010, current Queensland publication on ASS available at http://www.derm.qld.gov.au/land/ass/).
25. Dear, S.E.; Moore, N.G.; Dobos, S.K.; Wattling, K.M.; Ahern, C.R. Soil management guideline (Version 3.8). In *Queensland Acid Sulfate Soil Technical Manual*; Queensland Department of Natural Resources and Mines: Indooroopilly, Queensland, Australia, 2002; 1–63.
26. Dear, S.E.; Moore, N.G.; Wattling, K.M.; Fahl, D.; Dobos, S.K. Legislation and policy guidelines (Version 2.2). In *Queensland Acid Sulfate Soil Technical Manual*; Queensland Department of Natural Resources and Mines: Indooroopilly, Queensland, Australia, 2004; 1–64.
27. Erskine, W.D.; Warner, R.F. Geomorphic effects of alternating flood and drought dominated regimes on NSW coastal rivers. In *Fluvial Geomorphology of Australia*; Warner, R.F., Ed.; Academic Press: Sydney, 1988; 223–244.
28. Astrom, M.E.; Nystrand, M.; Gustafsson, J.P.; Osterholm, P.; Nordmyr, L.; Reynolds, J.K.; Peltola, P. Lanthanoid behavior in an acidic landscape. Geochim. Cosmochim. Acta **2010**, *74* (3), 829–845.
29. Skwarnecki, M.; Fitzpatrick, R. Geochemical dispersion in acid sulfate soils: Implications for mineral exploration in the Mount Torrens-Strathalbyn area, South Australia (Ch. 4). In *Inland Acid Sulfate Soil Systems across Australia*; Fitzpatrick, R., Shand, P., Eds.; CRC LEME Open File Report 249, CRC LEME: Perth, Australia, 2008; 98–128.
30. Quirk, R.; Melville, M.; Kinsela, A.; Zwemer, T.; Hancock, M.; Macdonald, B.; White, I. Treatment of drainage from acidic canelands using a constructed wetland. Sugar Technol. **2009**, *11* (1), 73–76.
31. Jiggins, J. Interagency learning process. In The OECD Workshop on: An Interdisciplinary Dialogue: Agricultural Production and Integrated Ecosystem Management of Soil and Water, Ballina, NSW, Nov. 10–16, 2002. OECD, University of Western Sydney, NSW Agriculture, 2002, 86–94.
32. Harris, G.P. Comparison of the biogeochemistry of lakes and estuaries: Ecosystem processes, functional groups, hysteresis effects and interactions between macro- and microbiology. Mar. Freshwater Res. **1999**, *50*, 791–811.
33. Thom, B.G.; Harvey, N. Triggers for late 20th century reform of Australian coastal management. Aust. Geogr. Stud. **2000**, *38*, 275–290.
34. Smith, J.; Marston, H.; Melville, M.D.; Macdonald, B.C.T. Spatial distribution and management of actual acidity in an acid sulfate soil environment, McLeods Creek, northeastern NSW, Australia. Catena **2003**, *51*, 61–79.
35. Wilson, B.P.; White, I.; Melville, M.D. Floodplain hydrology, acid discharge and change in water quality associated with a drained acid sulfate soil. Mar. Freshwater Res. **1999**, *50*, 149–157.
36. Kinsela, A.; Melville, M.D. Mechanisms of acid sulfate soil oxidation and leaching under sugarcane cropping. Aust. J. Soil Res. **2004**, *42*, 569–578.
37. White, I.; Melville, M.D.; Wilson, B.P.; Sammut, J. Reducing acidic discharges from coastal wetlands in eastern Australia. Wetlands Ecol. Manage. **1997**, *5*, 55–72.
38. Naylor, S.D.; Chapman, G.A.; Atkinson, G.; Murphy, C.L.; Tulau, M.J.; Flewin, T.C.; Milford, H.B.; Morand, D.T., Eds. *Guidelines for the Use of Acid Sulphate Soil Risk Maps.* NSW Soil Conservation Service, Department of Land and Water Conservation: Sydney, 1995.
39. Baker, A.; Fitzpatrick, R. A new web-based approach for the acquisition, collation and communication of complex inland acid sulfate soil data (Ch. 9). In *Inland Acid Sulfate Soil Systems across Australia*; Fitzpatrick, R., Shand, P., Eds.; CRC LEME Open File Report 249, CRC LEME: Perth, Australia, 2008; 162–168.
40. White, I.; Melville, M.D. Treatment and containment of potential acid sulphate soils: Formation, properties and management of potential acid sulphate soils. In *Report for Roads and Traffic Authority, NSW*; Technical Report T53; CSIRO Centre for Environmental Mechanics: Canberra, 1993.
41. Waite, T.D.; Desmier, R.; Melville, M.D.; Macdonald, B.C.T. Preliminary investigations into the suitability of permeable reactive barriers for the treatment of acid sulfate soil discharge. In *Handbook of Groundwater Remediation of Trace Metals, Radionuclides and Nutrients with Permeable Reactive Barriers*; Academic Press imprint of Elsevier Science: San Diego, 2002; 68–106.

Adsorption

Puangrat Kajitvichyanukul
Jirapat Ananpattarachai
Center of Excellence for Environmental Research and Innovation, Faculty of Engineering, Naresuan University, Phitsanulok, Thailand

Abstract
Adsorption is a well-known equilibrium separation process and an effective method for water and wastewater treatment applications. Several types of adsorbents, such as activated carbon, zeolite, chitin, or chitosan, and various agricultural wastes such as wood, peat, saw dust, etc., are applied for contaminant removal from the waste stream. This entry provides an overview of adsorption theory, kinetics and mechanism, and recent applications of various adsorbents for contaminant removal from water and wastewater. A summary of recent information concerning adsorption process with an extensive list of adsorbents from many research papers has been provided. The future progress in this technology is also described.

INTRODUCTION

Water is essential for life. Currently, quality of water becomes a major problem in many areas owing to pollution emission from industrial, agricultural, and domestic activities to the water bodies. These activities generate wastewater, which contains both inorganic and organic pollutants. Some of the common pollutants are phenols, dyes, detergents, insecticides, pesticides, and heavy metals. These pollutants are often toxic and cause adverse effects on human life. To avoid pollution of natural water bodies, treating wastewater from the originated source by removing pollutants before being discharging is necessary. Various treatment techniques and processes such as coagulation, membrane process, adsorption, dialysis, foam flotation, osmosis, and biological methods have been used to remove the pollutants from contaminated water. Among all the approaches proposed, adsorption is one of the most popular methods and is currently considered as an effective, efficient, and economic method for water purification.

Adsorption is a well-known equilibrium separation process and an effective method for water and wastewater treatment applications.[1–4] It is the process in which molecules accumulate in the interfacial layer, but desorption denotes the converse process. The fundamental concept in adsorption science is the equilibrium relation between the quantity of the adsorbed material and the pressure or concentration in the bulk fluid phase at constant temperature. The material adsorbed on the surface of "adsorbent" is defined as the "adsorbate." The penetration by the adsorbate molecules into the bulk solid phase is determined as "absorption." The terms "sorption," "sorbent," "sorbate," and "sorptive" are also used to denote both adsorption and absorption, when both occur simultaneously or cannot be distinguished.[1]

Recently, many works related to contaminant removal by adsorption process have been reviewed.[5–7] Several types of adsorbents applied in wastewater treatment are activated carbon, zeolite, chitin or chitosan, and various agricultural wastes such as wood, peat, saw dust, etc. These adsorbents can take a broad range of chemical forms and different geometrical surface structures and properties. This is reflected in the range of their applications in industry for both water and wastewater treatment. Compared with alternative technologies, adsorption is attractive for its relative simplicity of design, operation, and scale-up; high capacity and favorable rate; insensitivity to toxic substances; ease of regeneration; and low cost. Additionally, it avoids using toxic solvents and minimizes degradation.[6]

This review highlights and provides an overview of adsorption theory, kinetics and mechanisms, and recent applications of various adsorbents for contaminant removal from water and wastewater. The main aim of this review is to provide a summary of recent information concerning adsorption process. In this entry, an extensive list of adsorbents from many research papers has been provided. It is strongly encouraged to refer to these original papers for more information on experimental conditions.

ADSORPTION THEORY

In solid–liquid interface, adsorption of a species on a solid surface follows three steps:[7]

1. Transport of the adsorbate from the bulk to the external surface of the adsorbent.

Encyclopedia of Environmental Management DOI: 10.1081/E-EEM-120047363

2. Passage through the liquid film attached to the solid surface.
3. Interactions with the surface atoms of the solid leading to chemisorption. (strong adsorbate–adsorbent interactions equivalent to covalent bond formation) or weak adsorption (weak adsorbate–adsorbent interactions, very similar to van der Waals forces).

If step 1 is the slowest, the adsorption will be a transport-limited process. This step is usually the rate-limiting process in systems that are characterized by poor mixing, dilute concentration of adsorbate, small particle size of the adsorbent, etc.

When step 2 is the rate-determining slowest step, the physical process of diffusion through the liquid film influences the outcome of the process, and the efficiency of the solid as an adsorbent can hardly be improved.

When step 3 is the slowest, the adsorption is controlled by a chemical process, and the efficiency of the adsorbent can be influenced by suitably controlling the interactions.[7]

For porous solids, when the adsorbate passes through the liquid film attached to the external surface, it will slowly diffuse into the pores and get trapped or adsorbed. The pore diffusion plays an important role when the adsorbate is present in higher concentration, the adsorbent is made of large particles, and good mixing is ensured.[7–9]

Adsorption capacity is highly dependent on several factors such as properties of the adsorbent (porosity, surface area, particle size) and adsorbate (structure, water solubility, ionic charge, functional groups, pKa, polarity, functionality, molecular weight, and size); solution conditions (solvent, pH, temperature, ionic strength, solute concentration, and competition between solutes); interactions at the solid–liquid interface; and type of experimental setup. Generally, the adsorption capacity of a sorbent for a solute increases with increase in liquid-phase concentration. Physicochemical properties of the adsorbent also play a major role in adsorption. It must have high surface area, particularly internal surface area, high internal volume, and a good pore size distribution. The chemical properties of the adsorbent include degree of ionization at the surface and types of functional groups present. The adsorbent should also have good mechanical properties such as strength and resistance to destruction.

Temperature and pressure have an effect on increasing or decreasing of adsorption capacity. Under low-temperature conditions, the adsorption increases in the forward condition and liberates heat as it is exothermic in nature. With the increasing of pressure, adsorption increases up to a certain extent until saturation level is achieved. After it reaches the equilibrium level, no more adsorption takes place no matter how high the pressure applied.

The coverage of the adsorbent surface by the adsorbate leads to the adsorption process. It assumes that the surface consists of "sites" onto which the adsorbate can adsorb. Accordingly, the adsorption between the adsorbate molecule and the adsorbent surface can be depicted with the chemical reaction below:

$$\text{Molecule} + \text{surface site} \leftrightarrow \text{adsorbed molecule} + \text{heat } (\Delta H) \qquad (1)$$

According to the above reaction, adsorption characteristic can be expressed in thermodynamic parameters such as ΔG, ΔH, and ΔS. A negative ΔG value stands for the adsorption to take place. Change in enthalpy (ΔH) gives an indication of the bonding strength. The higher the value of heat of adsorption, the weaker the bond between adsorbate and adsorbent. The sign of ΔS indicates the direction: for adsorption, ($+\Delta S$), and for desorption, ($-\Delta S$).

The change in Gibbs free energy is given by the following expression:

$$\Delta G^\circ = -RT \ln K_L \qquad (2)$$

where R is the universal gas constant, T is the absolute temperature, and K_L (L/mg) is the affinity constant of Langmuir model. Negative ΔG values confirm the feasibility of the adsorption, and its absolute values measure the adsorption driving force. Negative or positive values of ΔG indicate the exothermic or endothermic nature of adsorption, respectively. Positive ΔS values reveal a random organization of the adsorbate at the solid/solution interface. The sorption entropy can be calculated from the Gibbs–Helmholtz equation:

$$\Delta G = \Delta H - T\Delta S \qquad (3)$$

For a reaction or process to be spontaneous, there must be decreases in free energy of the system. ΔG of the system must have a negative value. During the adsorption, randomness of the molecule decreases, so that ΔS is negative. The above equation can be rewritten as follows:

$$\Delta G = \Delta H + T\Delta S \qquad (4)$$

Therefore, for a reaction to be spontaneous, ΔH has to be negative, and $|\Delta H| > |T\Delta S|$.

The adsorption process occurs when adsorbate is adsorbed on adsorbent. Heat energy developing during the adsorption process is released as it is an exothermic process.

During adsorption, forces of attraction play an important role between adsorbate and adsorbent. Van der Waals forces of attraction are weak forces, while forces from chemical bonding are strong forces. Accordingly, the adsorption can be classified into two types: physical adsorption or chemical adsorption.

Physisorption or physical adsorption is a type of adsorption in which the adsorbate adheres to the surface through van der Waals (weak intermolecular) interactions. Physical adsorption takes place with formation of a multilayer

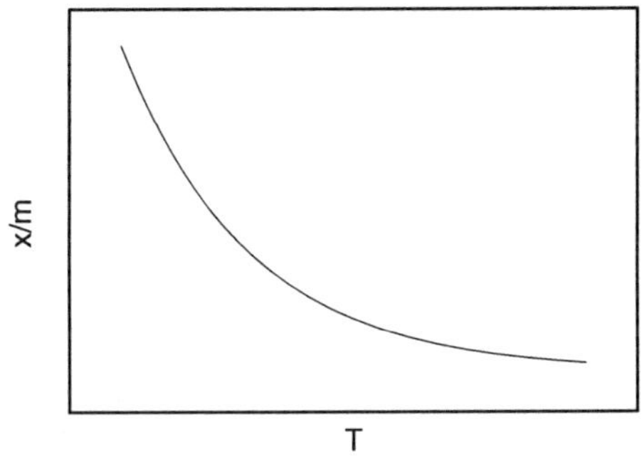
x/m
T

Fig. 1 Relation of physical adsorption with temperature.

of adsorbate on adsorbent. The chemical identity of the adsorbate remains intact without breaking of the covalent structure of the adsorbate. To be a spontaneous thermodynamic process, ΔG has to be a negative value. As the translational degrees of freedom of the adsorbate are lost upon deposition onto the adsorbent, ΔS is negative for the process, and ΔH for physical adsorption must be exothermic. The energy released upon accommodation to the surface is of the same order of magnitude as an enthalpy of condensation (in the order of 20 kJ/mol). This process occurs at low temperature, below the boiling point of adsorbate. As the temperature increases, the process of physical adsorption decreases, as shown in Fig. 1.

Chemisorption is a type of adsorption whereby a molecule adheres to a surface through the formation of a chemical bond, as opposed to the van der Waals forces. Chemical adsorption takes place when adsorbate attaches on the adsorbent surface with formation of a unilayer by chemical bonding. This interaction is much stronger and has higher enthalpy of adsorption, ΔH (200–400 kJ/mol), than physical adsorption. As the adsorbates can interact with each other when they lie upon the surface, the energy of adsorption relies on the extent to which the available surface is covered with adsorbate molecules. Thus, the chemical bonds in chemical adsorption may be stronger than the bonds internal to the free adsorbate, resulting in the dissociation of the adsorbate upon adsorption. This type of adsorption can take place at all temperatures. Chemical adsorption first increases and then decreases with the increases in temperature, as shown in Fig. 2.

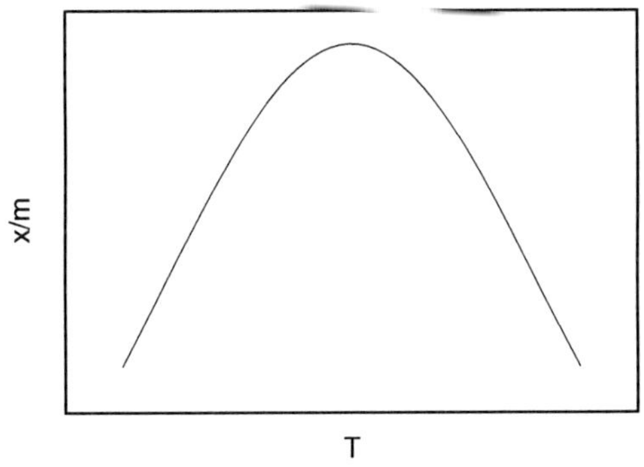
x/m
T

Fig. 2 Relation of chemical adsorption with temperature.

Adsorption Isotherm

Adsorption isotherms are defined by the adsorbate–adsorbent interactions. Generally, it is constructed by measuring the concentration of the adsorbate in the medium before and after adsorption, at a fixed temperature. This is the practical way to investigate the interaction between the adsorbate and the surface of the adsorbent and to obtain information about the structure of the adsorbed layer. Four classes of adsorption isotherms [S type with an initial convexing to the concentration axis, Langmuir (L) type with an initial concavity to the concentration axis, H type with an intercept on the ordinate, and C type with an initial linear portion] and subgroups have been defined according to their configuration,[10–13] as shown in Fig. 3. The shape of isotherm provides qualitative information on the nature of solute–surface interaction. The Langmuir class (L) is widespread in the case of adsorption of many organic contaminants from water. The organic contaminant adsorbs parallel to the surface, and no strong competition exists between the adsorbate and the solvent to occupy the adsorption sites. However, the H class (high affinity) results from

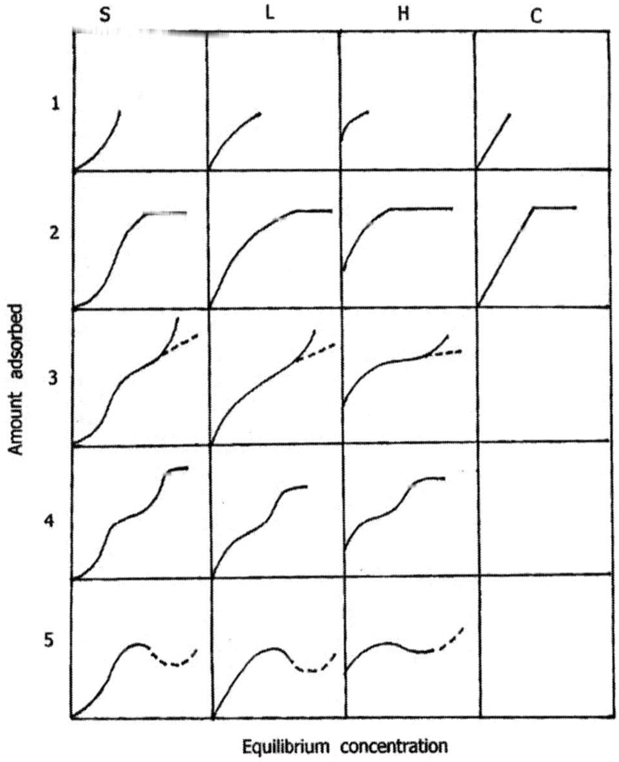
S
L
H
C
1
2
3
4
5
Amount adsorbed
Equilibrium concentration

Fig. 3 Four classes of adsorption isotherms (S, L, H, C).
Source: Giles et al.[11]

extremely strong adsorption at very low concentrations, giving rise to an apparent intercept on the ordinate.[12]

In the adsorption process, when a mass of adsorbent and a waste stream of adsorbate are in contact for a sufficiently long time, equilibrium between the amount of pollutant adsorbed and the amount remaining in solution will develop. Under equilibrium conditions, the amount of material adsorbed onto the media can be calculated using the mass balance of Eq. 5:

$$q_e = \frac{(C_0 - C_e)V}{W} \tag{5}$$

where q_e (mg pollutant/g adsorbent) is the mass of pollutant per mass of adsorbent used, C_0 is the liquid-phase concentration of pollutant in solution at the beginning, C_e is the concentration of the pollutant in solution after equilibrium has been reached, V is the volume of the solution, and W is the mass of dry adsorbent used.

Adsorption isotherms are related to the amount of adsorbate on the adsorbent, with its concentration. Several models describing process of adsorption are Freundlich isotherm, Langmuir isotherm, Brunauer–Emmett–Teller (BET) isotherm, etc., with details described below.

Langmuir Isotherm

The Langmuir equation initially derived from kinetic studies was based on the assumption that on the adsorbent surface, there is a definite and energetically equivalent number of adsorption sites. Consequently, the hypotheses for this isotherm are uniform adsorption energies along the homogeneous adsorbent surface, equal solute affinity in all the adsorption sites, no interaction among adsorbed molecules, a single adsorption mechanism, and formation of a monolayer on the free surface.[14] This isotherm is concerned with the monolayer (single-layer) coverage of the solid surface by the adsorbate. The bonding to the adsorption sites can be either chemical or physical, but it must be sufficiently strong to prevent displacement of adsorbed molecules along the surface.

The Langmuir isotherm is useful in determining the interactions between the adsorbate and the adsorbent. Adsorption data for a wide range of adsorbate concentrations are most conveniently described by the Langmuir isotherm:

$$q_e = \frac{QK_LC_e}{1 + K_LC_e} \tag{6}$$

where Q (mg/g) is the maximum amount of the adsorbate per unit weight of the adsorbent to form a complete monolayer on the surface, whereas K_L (L/mg) is the Langmuir constant related to the affinity of the binding sites.

The essential characteristics of Langmuir isotherm can be expressed by a dimensionless constant called separation factor or equilibrium parameter, R_L, defined by Weber and Chakkravorti[15] as follows:

$$R_L = \frac{1}{1 + K_LC_0} \tag{7}$$

The parameter R_L indicates the shape of isotherm as shown in Table 1.

The limitations of this isotherm come from the fact that several types of adsorption sites exist, the adsorption mechanism is not the same for the first and for the last molecules adsorbed, and models based on the monomolecular surface layer without interactions could be not realistic. Consequently, many other isotherms such as Freundlich, BET, Temkin, and Dubinin–Radushkevich isotherms have been proposed to explain adsorption behavior.

The Freundlich Isotherm

The Freundlich isotherm is an empirical equation that is more suitable than the Langmuir isotherm when the amount of adsorbent in contact with the solid surface is relatively low. In addition, if the binding energy changes continuously from site to site on solid surfaces, the expression used for the Freundlich isotherm is much more applicable to explaining the adsorption behavior. It is also generally used in nonideal systems with highly heterogeneous surfaces or surfaces supporting sites of varied affinities. The assumption for this isotherm is that the stronger binding sites are occupied first and that the binding strength decreases with an increasing degree of site occupation.[16] It does not imply the formation of a monolayer and frequently gives good interpretation of data over a restricted concentration range.[12] The Freundlich isotherm is expressed as follows:

$$q_e = K_FC_e^{1/n} \tag{8}$$

where q_e (mg/g) is the equilibrium value for removal of adsorbate per unit weight of adsorbent, C_e (mg/L) is the equilibrium concentration of metal ion in solution, and K_F and n are Freundlich isotherm constants that are related to the adsorption capacity (or the bonding energy) and intensity of the sorbent, respectively. K_F can be defined as the adsorption or distribution coefficient and represents the ad-

Table 1 Relation between the value of R_L and the type of isotherm.

Value of R_L	Type of isotherm
$R_L > 1$	Unfavorable
$R_L = 1$	Linear
$0 < R_L < 1$	Favorable
$R_L = 0$	Irreversible

sorbent onto adsorbate for a unit equilibrium concentration. A value for $1/n < 1$ indicates a normal Langmuir isotherm, while $1/n > 1$ is indicative of cooperative adsorption:[17]

$$\log q_e = \left(\frac{1}{n}\right)\log C_e + \log K_F \tag{9}$$

Experimentally it was determined that extent of adsorption varies directly with pressure until saturation pressure Ps is reached. Beyond that point, the rate of adsorption saturates even after applying higher pressure. Thus, Freundlich adsorption isotherm fails at higher pressure.

BET Isotherm

This isotherm is widely applied in the gas–solid equilibrium systems. It has a theoretical background based on multilayer adsorption, which is the true picture of physical adsorption. The BET isotherm was derived from computer-based calculations as the nonlinear isotherm modeling, usually based on algorithms dealing with error distribution.[18]

Under the high pressure, the BET theory is more feasible than the Langmuir adsorption, which is applicable only under the conditions of low pressure. With high pressure and low temperature, thermal energy of gaseous molecules decreases, and more and more gaseous molecules would be available per unit surface area. Consequently, the multilayer adsorption would occur on the adsorbent surface. This multilayer formation is explained by BET theory. The BET equation is given as

$$V_{total} = \frac{V_{mono}C\left(\frac{P}{P_0}\right)}{\left(1-\frac{P}{P_0}\right)\left(1+C\left(\frac{P}{P_0}\right)-\frac{P}{P_0}\right)} \tag{10}$$

Another form of the BET equation is

$$\frac{P}{V_{total}(P-P_0)} = \frac{1}{V_{mono}C} + \frac{c-1}{V_{mono}C}\left(\frac{P}{P_0}\right) \tag{11}$$

where V_{mono} is the adsorbed volume of gas at high-pressure conditions so as to cover the surface with a unilayer of gaseous molecule.

Temkin Isotherm

The Temkin isotherm is applicable when the behavior of an adsorption system occurs on heterogeneous surfaces. It is expressed by the following equation.[19,20]

$$q_e = \frac{RT}{b_T}\ln(K_T C_e) \tag{12}$$

The linear form of the Temkin isotherm is represented by the following equations:

$$q_e = A + B\ln C_e \tag{13}$$

$$A = \frac{RT}{b_T}\ln K_T \tag{14}$$

$$B = \frac{RT}{b_T} \tag{15}$$

where R is the gas constant (8.341 J mol^{-1} K^{-1}), T is the absolute temperature (K), and A and B represent isotherm constants. K_T is the equilibrium binding constant (L/g); b_T is related to the heat of adsorption (J/mol).

Dubinin–Radushkevich Isotherm

The Dubinin–Radushkevich isotherm is applicable to the adsorption mechanism based on the potential theory assuming a heterogeneous surface, and its linearized form is given in the following equations:[20,21]

$$\ln q_e = \ln q_m - K\varepsilon^2 \tag{16}$$

$$\varepsilon = RT\ln\left(1+\frac{1}{Ce}\right) \tag{17}$$

where K, a constant, is related to the adsorption energy and ε (kJ/mol) is used to estimate the type of adsorption process (chemical or physical adsorption) and calculated from the following equation:

$$\varepsilon = (-2K)^{-1/2} \tag{18}$$

If the ε values of the isotherm fall between 8 and 16 kJ/mol, the adsorption reaction can be explained by an ion-exchange mechanism, whereas if its value is less than 8 kJ/mol, it indicates that the adsorption process has a physical nature.[22]

Redlich–Peterson Isotherm

This isotherm is widely used as a compromise between the Langmuir and Freundlich systems.[23] In this equation, three parameters have been used to incorporate the advantageous significance of both models. The model can be represented as follows:[24]

$$q_e = \frac{K_{RP}C_e}{1+(\alpha C_e)^{\beta}} \tag{19}$$

where K_{RP} (L/g), α (L/mg), and β are Redlich–Peterson (R–P) isotherm constants, whereas β is the exponent that lies between 0 and 1. The R–P model has two limiting cases. When $\beta = 1$, the Langmuir equation results, whereas when $\beta = 0$, the R–P equation transforms to Henry's law equation.

Adsorption Kinetics

Kinetics is the study of the rates of chemical processes by monitoring the experimental conditions that influence the speed of a chemical reaction and help attain equilibrium in a reasonable length of time. The possible mechanism of adsorption and the different transition states of the final adsorbate–adsorbent complex are normally studied, and the appropriate mathematical models to describe the interactions are applied to the reaction.

The adsorption kinetics is useful in understanding the complex dynamics of the adsorption process.

Generally, four consecutive steps describing the occurrence of adsorbate transport are as follows[6]

1. Transport of the adsorbate from the bulk solution to the boundary layer surrounding the adsorbent particles
2. Transport of solute across the boundary layer
3. Intraparticle solute diffusion into the pores
4. Adsorption and desorption of adsorbate

One of the above steps, or a combination of them, controls the adsorption mechanism. Factors influencing the rate-limiting step include characteristics of the adsorbent, adsorbate, and solution, for instance, adsorbent particle size, adsorbate concentration, degree of mixing, affinity between adsorbate and adsorbent, and diffusion coefficients of the adsorbate.

Adsorption kinetics can be determined by the following stages: diffusion of molecules from the bulk phase toward the interface space or external diffusion, diffusion of molecules inside the pores or internal diffusion, diffusion of molecules in the surface phase or surface diffusion, and adsorption–desorption elementary processes.

The most popular models used to establish the controlling adsorption mechanism can be grouped as follows:[25]

1. Those assuming that the controlling step is mass transfer (homogeneous surface diffusion, pore diffusion, and heterogeneous diffusion models).
2. Those assuming that adsorption is governed by surface phenomena.

The models involved in an adsorption kinetics study are the Lagergren pseudo–first-order and pseudo–second-order model, the Elovich equation, intraparticle diffusion, and liquid film diffusion.

Lagergren Pseudo–First-Order Model

This equation describing the rate of adsorption in the liquid-phase systems is the most used equation, particularly for pseudo–first-order kinetics.[26] The linearized Lagergren equation considers a reversible equilibrium of organic molecules between a liquid and a solid phase. Many different adsorption situations can be described by pseudo–first-order kinetics, including the following: 1) systems close to equilibrium; 2) systems with time-independent solute concentration or linear equilibrium adsorption isotherm; and 3) special cases of more complex systems.[6]

The Lagergren equation as pseudo–first-order kinetics is described as follows:[7]

$$\frac{dq_t}{dt} = k_1(q_e - q_t) \qquad (20)$$

where k_1 (min^{-1}) is the pseudo–first-order adsorption rate coefficient. The integrated form of this equation for the boundary conditions of $t = 0$, $q_t = 0$ and $t = t$, $q_t = q_t$ is

$$\ln(q_e - q_t) = \ln q_e - k_1 t \qquad (21)$$

where q_e and q_t are the values of amount adsorbed per unit mass at equilibrium and at any time t. The values of k_1 can be obtained from the slope of the linear plot of $\ln(q_e - q_t)$ vs. t.

McKay et al.[27] reported that the Lagergren pseudo–first-order model is found suitable only for the initial 20 to 30 min of interaction and not fit for the whole range of contact time. As the value of k_1 depends on the initial concentration of the adsorbate, it usually decreases with the increasing initial adsorbate concentration in the bulk phase.[28–31]

Pseudo–Second-Order Model

The pseudo–second-order equation has been explained as a special kind of Langmuir kinetics with the following assumptions: 1) the adsorbate concentration is constant in time; and 2) the total number of binding sites depends on the amount of adsorbate adsorbed at equilibrium.[32]

The pseudo–second-order equation based on equilibrium adsorption is expressed as

$$\frac{dq_t}{dt} = k_2(q_e - q_t)^2 \qquad (22)$$

The initial adsorption rate of a second-order process as $t \to 0$ is defined as

$$\frac{dq}{(q_e - q_t)^2} = k_2 dt \qquad (23)$$

The integrating equation with respect to boundary conditions $q_t = 0$ at $t = 0$ and $q = q_e$ at $t = t$ is

$$\frac{t}{q} = \frac{1}{k_2 q_e^2} + \frac{1}{q_e} t \tag{24}$$

where k_2 (g/mg h) is the rate constant of second-order adsorption. It depends on the applied operating conditions, namely, initial metal concentration, pH of solution, temperature and agitation rate, etc.[33–34] The integral form of the model predicts that the ratio of the time/adsorbed amount should be a linear function of time.[35] The linear plot of t/q_t vs. t gives $1/q_e$ as the slope and $1/k_2 q_e^2$ as the intercept. This procedure tends to predict the behavior over the whole range of adsorption.

Several research works[30–31,34,36] reported that the rate coefficient, k_2, decreases with the increasing initial adsorbate concentration in the bulk phase. The higher the initial concentration of adsorbate, the longer the time required to reach an equilibrium and, consequently, the k_2 value decreases.

The second-order rate constants were used to calculate the initial sorption rate given by

$$h = k_2 q_e^2 \tag{25}$$

where h is the initial adsorption rate, q_e is the adsorption capacity, and k_2 is the pseudo–second-order rate coefficient. The value of k_2 can be determined experimentally from the slope and intercept of a plot of t/q_t against t.

It has been reported[24] that the initial adsorption rate, h, was found to increase with initial concentration; however, the value started to decrease when the high initial concentration was applied to the system. The possible reason might be that too-high solute concentrations would slow down the adsorption process. The h value could also be influenced by the characteristics of the adsorbent and adsorbate.

One of the advantages of the pseudo–second-order equation for estimating the q_e values is its small sensitivity for the influence of the random experimental errors.[7]

Elovich Equation

The Elovich equation is one of the most widely used models for describing chemical adsorption. It assumes that the actual solid surfaces are energetically heterogeneous and that neither desorption nor interactions between the adsorbed species could substantially affect the kinetics of adsorption at low surface coverage.[7] The Elovich equation is also restricted to the initial sorption stages, when the system is relatively far from equilibrium.

The Elovich equation can be expressed as follows:[37]

$$\frac{dq_t}{dt} = \alpha \exp(-\beta q_t) \tag{26}$$

where α and β represent the initial adsorption rate (g/mg min^2) and the desorption coefficient (mg/g min), respectively. Assuming $\alpha\beta q_t \gg 1$, and $q_t = 0$ at $t = 0$ and $q_t = q_t$ at $t = t$, the linear form of the above equation is given by[38–40]

$$q_t = \beta \ln(\alpha\beta) + \beta \ln t \tag{27}$$

The Elovich coefficients could be computed from the plots of q_t vs. ln t. It is suggested that Elovich equation is restricted to the initial part of the adsorbate–adsorbent interaction process, when the system is relatively far from equilibrium.[7] The pseudo–second-order and the Elovich equations exhibit closely related behaviors for values of fractional surface coverage up to 0.7.[35]

Intraparticle Diffusion Model

This model is mostly used with the porous adsorbent. As the adsorbate molecules or ions diffuse into the pores, the adsorption kinetics relating to this intraparticle diffusion has to be studied with the proper model. The intraparticle diffusion model can be derived from Fick's second law. The simplified form of this equation as a dimensionless equation was proposed to assess the initial adsorption stages.[41–43] The intraparticle diffusion equation can be expressed as follows:

$$\frac{q_t}{q_e} = 1 - \left(\frac{6}{\pi^2}\right) \Sigma \left(\frac{1}{n^2}\right) \exp\left(\frac{-n^2 \pi^2 D_C t}{r^2}\right) \tag{28}$$

where D_C is the intracrystalline diffusivity, r is the particle radius, t is the reaction time, and the summation is carried out from $n = 1$ to $n = \alpha$. The ratio, q_t/q_e, is the fractional approach to equilibrium.

The simplified form is rewritten below:

$$1 - \frac{q_t}{q_e} = \left(\frac{6}{\pi^2}\right) \exp\left(\frac{-\pi^2 D_C}{r^2}\right) t \tag{29}$$

or

$$\ln\left(1 - \frac{q_t}{q_e}\right) = \left(\frac{-\pi^2 D_C}{r^2}\right) t + \ln\left(\frac{6}{\pi^2}\right) \tag{30}$$

This equation can be expressed in the linear line with the plot of $\ln\left(1 - \frac{q_t}{q_e}\right)$ vs. t. A slope of $\left(\frac{-\pi^2 D_C}{r^2}\right)$ is the value of the diffusion time constant, and it is expressed as follows:

$$k' = \frac{\pi^2 D_C}{r^2} \tag{31}$$

where k' is the overall rate constant, inversely proportional to the square of the particle radius. A simpler expression to obtain the diffusion rate coefficient, k_i, is written as follows:

$$q_t = k_i t^{0.5} \quad (32)$$

The linear plots of q_t vs. $t^{0.5}$ should pass through the origin (zero intercept) indicating a controlling influence for the diffusion process on the kinetics. The slope of the plot is the rate coefficient, k_i (mg/g-min$^{0.5}$). This equation represents a simplistic approximation of the pore diffusion kinetics without considering the possible impacts of the pore dimensions.

Liquid Film Diffusion Model

This equation is applied when the flow of the reactant through the liquid film surrounding the adsorbent particles is the slowest process. This equation determines kinetics of the rate process, and it is written as follows:[44]

$$\ln(1 - F) = -k_{fd} t \quad (33)$$

where F is the fractional attainment of equilibrium ($=q_t/q_e$) and k_{fd} (min^{-1}) is the film diffusion rate coefficient. A linear plot of $-\ln(1-F)$ vs. t with zero intercept suggests that the kinetics of the adsorption process is controlled by diffusion through the liquid film.

ADSORBENTS AND THEIR APPLICATIONS IN WATER AND WASTEWATER TREATMENT

Carbon Adsorbents

There are several types of carbon adsorbents used in the adsorption process, for example, activated carbon, activated carbon fibers, fullerene, etc. Activated carbon is the most widely used carbon adsorbent. It is a crude form of graphite with a random or amorphous highly porous structure with a broad range of pore sizes, from visible cracks and crevices to crevices of molecular dimensions.[45] Generally, it is the carbon material derived mostly from charcoal. Besides coal, agricultural by-products are conventional sources of commercial activated carbon. Many types of agricultural wastes were proposed as raw materials for producing carbon adsorbents, including cork;[46–47] sucrose chars;[48] corncob;[49–50] jackfruit peels;[51–52] wood;[53–54] oil palm;[18] stones; fruit shells, coats, and husks;[48, 55–64] and wastes from cherries,[65–66] plums,[67] coconut, apricot, almond, and nuts.[66–68] Conditions in synthesizing activated carbon from these materials are provided in Table 2. The comparison of adsorption capacity of each type of activated carbon is shown in Table 3. Sources of the raw materials used and the preparation and treatment conditions such as pyrolysis temperature and activation time are

Table 2 Conditions in synthesizing activated carbon from natural materials.

Natural material	Conditions	Refs.
Algerian coal	930°C with KOH/NaOH	Alvim Ferraz[69]
Almond and pecan shells	Chemical activation with H_3PO_4/physical CO_2	Tancredi et al.[70]
Almond shell, olive stones, and peach stones	Heating in CO_2 at 606°C	Ferro-García et al.[71]
Bituminous coal	N_2/400–700°C with $ZnCl_2$	Hall and Holmes[72]
Coal or coconut shell	Phosgene or chlorine gas at 180°C	Otowa et al.[73]
Coconut shell	Parts by weight H_2SO_4 for 24 h at 150°C	Manju et al.[74]
Coconut shell	450°C with H_3PO_4	Laine et al.[75]
Coconut shells and coconut shell fibers	Carbonized with H_2SO_4 and activated at 600°C for 1 hr	Mohan et al.,[76] Mohan et al.,[77] Mohan et al.,[78] and Mohan et al.[79]
Eucalyptus wood chars	CO_2 activation, 400–800°C	Kumar and Sivanesan[80]
Fertilizer slurry	450°C, 1 hr with H_2O_2/H_2O, N_2	Marungrueng and Pavasant[81]
Fly ash	Froth flotation, hydrophobic char was separated from hydrophilic ash with the help of methyl isobutyl ketone	Basava Rao and Mohan Rao[82]
Lignite	Inert atmosphere/600–800°C with $Na_2MoO_4/NaWO_4/NH_4VO_3/(NH_4)_2MoO_4/FeCl_3/Fe(NO_3)_3$	El Qada et al.[83]
Oat hulls	Fast pyrolysis at 500°C with inert nitrogen	Tamai et al.[84]
Palm tree cobs	730°C, 6 hr with H_3PO_4/H_2SO_4	Banat et al.[85]
Petroleum coke	700–850°C, 4 hr with KOH/H_2O	McKay et al.[86]
	KOH dehydration at 400°C followed by activation at 500–900°C	McKay et al.[87]
Pine sawdust	850°C, 1 hr; 825°C, 6 hr with $Fe(NO_3)_3/CO_2$	Kannan and Sundaram[88]
Raffination earth	10% (v/v), 350°C with H_2SO_4	Bestani et al.[89]
Solvent-extracted olive pulp and olive stones	Under vacuum and atmospheric pressure; 60°C/min; 800°C; activation under N_2 at 10°C/min with K_2CO_3	Stavropoulos and Zabaniotou[90]

major factors that have an effect on the adsorption capacity of activated carbon. Many other factors can also affect the adsorption capacity in the same sorption conditions such as surface chemistry (heteroatom content), surface charge, and pore structure.

The adsorbent properties of activated carbon depend on their composition, physicochemical properties, and mechanical strength. Activation by physical means, by chemical means, or by a combination of both has been employed to control the pore size and distribution of activated carbon and/or to increase porosity, surface modification, and improvement of carbonization.[92–98] Normally, activated carbons are made up of small hydrophobic graphite layers with disordered, irregular, and heterogeneous surfaces bearing hydrophilic functional groups. The surface chemistry of activated carbon depends mainly upon the activation conditions and temperatures employed. The activated carbon has strong heterogeneous surfaces. Its geometrical heterogeneity is the result of differences in size and shape of pores, and cracks, pits, and steps. The chemical heterogeneity is involved with different functional groups, mainly oxygen groups that are located most frequently at the edges of the crystallites among with various surface impurities. These heterogeneity surfaces contribute to the unique sorption properties of activated carbon.[12,99–100] The presence of oxygen and hydrogen in surface groups affects strongly the adsorptive properties of the activated carbon. The apparent chemical character of an activated carbon surface is determined by functional groups and delocalized electrons of the graphitic structure.[101] Oxygen on an activated carbon surface may be present in various forms, such as carboxyls, aldehydes, ketones, phenols, lactones, quinines, hydroquinones, anhydrides, or ethereal structures. Some of the groups, e.g., carbonyl, carboxyl, phenolic hydroxyl, and lactonic ones, are acidic. Consequently, the pH value of the liquid bulk phase can have an effect on the acidic and/or basic functional groups of the carbon surface. Thus, the surface charge of carbon is a function of pH of the solution. Considering the point of zero charge (PZC) and the isoelectric point (IEP), the surface is positively charged at $pH < pH_{PZC}$ and negatively charged at $pH > pH_{PZC}$. In practice, pH_{IEP} is usually close to pH_{PZC}, but it is lower than pH_{PZC} for activated carbon.[102] For $pH < pK_a$ adsorption of non-ionized organics does not depend on the surface charge of activated carbon. However, for $pH > pK_a$, the adsorption of its ionic form depends on the surface charge. As a result, the activated carbon possesses perfect adsorption ability for relatively low-molecular-weight organic compounds from drinking water and wastewater streams.

Table 3 Adsorption capacity of each type of activated carbon.

Adsorbent	Adsorption capacity (mg/g)	Refs.
Activated carbon	400	Kumar and Sivanesan[80]
	238	Marungrueng and Pavasant[81]
	9.81	Basava Rao et al.[82]
Activated carbon produced from New Zealand coal	588	El Qada et al.[83]
Activated carbon produced from Venezuelan bituminous coal	380	El Qada et al.[83]
Bituminous coal	176	Tamai et al.[84]
Charcoal	62.7	Banat et al.[85]
Coal	323.68	McKay et al.[86]
	230	McKay et al.[87]
Commercial activated carbon	980.3	Kannan and Sundaram[88]
	200	Bestani et al.[89]
Peat	324	Fernandes et al.[91]

Several methods have been used to removal organic pollutants from water. However, the use of activated carbons is perhaps the best broad-spectrum technology available at present.[1] Accordingly, the use of activated carbons in water treatment has increased tremendously. Generally, the three main physical carbon types are granular, powder, and extruded (pellet). The granular activated carbon (GAC) adsorption, the most widely used type, is an effective treatment technology for organic contaminant removal from drinking water to improve taste and odor. The use of GAC for treatment of municipal and industrial wastewaters has developed rapidly in the last three decades from small size for household units to large scale for industrial wastewater application. Moving beds, downflow fixed beds, and upflow expanded beds have been widely used for water purification for industry.

It is well known that activated carbon can remove several types of pollutants including metal ions,[102–104] phenols,[46,105] pesticides,[106] chlorinated hydrocarbons,[107] detergents,[108] and many other chemicals and organisms. Application of activated carbon in removal of various heavy metals and organic contaminants with the Langmuir and Freundlich capacities is shown in Table 4.

Clay Minerals

Clay minerals are hydrous aluminosilicates composed of minerals that make up the colloid fraction (<2 μm) of soils, sediments, rocks, and water[117] and may be composed of mixtures of fine-grained clay minerals and clay-sized crystals of other minerals such as quartz, carbonate, and metal oxides. Their structures are similar to micas with the formation of flat hexagonal sheets. Clay minerals and oxides are widespread and abundant in aquatic and terrestrial environments.

Table 4 Adsorption capacities of activated carbon for heavy metal and organic contaminant removal from water and wastewater.

Pollutant	Activated carbon	Adsorption capacity (mg/g)	Isotherm	Refs.
Cr(VI)	Commercial activated carbon	4.7	Langmuir	Babel and Kurniawan[109]
	Commercial activated carbon oxidized with H_2SO_4	8.9	Langmuir	Babel and Kurniawan[109]
	Commercial activated carbon oxidized with HNO_3	10.4	Langmuir	Babel and Kurniawan[109]
Fe(III)	Granular activated carbon	0.1	Freundlich	Kim[110]
Ni(II)	Granular activated carbon	6.5	Langmuir	Satapathy and Natarajan[111]
	Modified activated carbon	7.0	Langmuir	Satapathy and Natarajan[111]
Catechol	Activated charcoal	320	Langmuir	Richard et al.[112]
Gallic acid	Activated charcoal	408-488	Langmuir	Figaro et al.[113]
Tannin	Activated charcoal	0.39	Langmuir	Mohan and Karthikeyan[114]
Vanillin	Activated charcoal	93.18–121.72	Langmuir	Michailof et al.[63]
Phenol	Rice husk activated carbon	27.58	Langmuir	Kalderis et al.[115]
Nonylphenol	Activated charcoal	83.1	Langmuir	Lang et al.[116]

Clay contains various types of exchangeable ions on its surface. The prominent ions found on the clay surface are Ca^{2+}, Mg^{2+}, H^+, K^+, NH^{4+}, Na^+, SO_4^{2-}, Cl^-, PO_4^{3-}, and NO_3^-. These ions can be exchanged with other ions easily without affecting the structure of the clay mineral.[118] Clay can adsorb the cationic, anionic, and neutral metal species. They act as a natural scavenger of pollutants by taking up cations and/or anions through either ion exchange or adsorption, or both.

Currently, several types of clay minerals such as montmorillonite and kaolinite are widely used in the water purification process. Because of their low cost, abundance in most continents of the world, high sorption properties, and potential for ion exchange, clay materials are strong adsorbents. Montmorillonite is a clay mineral with substantial isomorphic substitution. It is composed of units made up of two silica tetrahedral sheets with a central alumina octahedral sheet. The theoretical composition without the interlayer material is SiO_2, 66.7%; Al_2O_3, 28.3%; and H_2O, 5%. There is substitution of Si^{4+} by Al^{3+} in the tetrahedral layer and of Al^{3+} by Mg^{2+} in the octahedral layer. Exchangeable cations in the 2:1 layers balance the negative charges generated by isomorphic substitution. The uptake kinetics of cation exchange is fast, and the cations such as Na^+ and Ca^{2+} form outer-sphere surface complexes, which are easily exchanged with solute ions by varying the cationic composition of the solution.

Kaolinite is the least reactive clay. It has the theoretical composition of SiO_2, 46.54%; Al_2O_3, 39.50%; and H_2O, 13.96%, expressed in terms of the oxides. It has a small net negative charge, which is responsible for the surface not being completely inert. Its high pH dependency enhances or inhibits the adsorption of metals according to the pH of the environment.[119] The metal adsorption is usually accompanied by the release of hydrogen (H^+) ions from the edge sites of the mineral. The substitution of H^+ ions for metal ions could influence the van der Waals force within the kaolinite structure.

Their applications are mainly found in dye and heavy metal removal. From previous research, it was reported that the sorption capacity of clay minerals can vary strongly with pH. Gupta and Bhattacharyya[120–121] used kaolinite and montmorillonite along with their poly(oxo zirconium) and tetrabutylammonium derivatives for Cd(II) removal from water. The adsorption of Cd(II) was influenced by pH of the aqueous medium, and the amount adsorbed increased with gradually decreasing acidity. By increasing the solution pH from 1.0 to 10.0, the extent of adsorption increased from 4.3% to 29.5% for kaolinite and 74.7% to 94.5% for montmorillonite. In dye removal, Bagane and Guiza[122] reported an adsorption capacity of 300 mg/g and suggested that clay is a good adsorbent for methylene blue removal due to its high surface area. Almeida et al.[123] studied the removal of methylene blue from synthetic wastewater by using montmorillonite and described it as an efficient adsorbent where the equilibrium was attained in less than 30 min. The adsorption of dyes on kaolinite was also studied by Ghosh and Bhattacharyya,[124] who reported that its adsorption capacity can be improved by purification and by treatment with NaOH solution.

The adsorption capacities vary from metal to metal and also depend on the type of clay used.[118] When a comparison is made with other low-cost adsorbents, the clays have been found to be either better or equivalent in adsorption capacity. Type of pollutant and adsorption capacity of each clay mineral are summarized in Table 5.

Natural Zeolites

Zeolites are highly porous aluminosilicates with different cavity structures. They consist of a three-dimensional framework, having a negatively charged lattice. A well-

Table 5 Adsorption capacities of clay minerals for heavy metal removal from water and wastewater.

Pollutant	Clay mineral	Langmuir capacity	Freundlich capacity	Refs.
Cd(II)	Kaolinite	9.9	0.5	Gupta and Bhattacharyya[125]
	Montmorillonite	32.7	8.6	Gupta and Bhattacharyya[125]
Ni(II)	Acid-activated montmorillonite	29.5	6.0	Bhattacharyya and Gupta[126]
	Kaolinite	10.4	1.1	Gupta and Bhattacharyya[127]
	Montmorillonite	28.4	4.5	Gupta and Bhattacharyya[127]
Cr(VI)	Kaolinite	11.6	–	Bhattacharyya and Gupta[128]
	Acid-activated kaolinite	13.9	–	Bhattacharyya and Gupta[128]
Co(II)	Raw kaolinite	11.5	–	Yavuz et al.[129]
	Kaolinite	11.2	1.1	Bhattacharyya and Gupta[130]
	Acid-activated kaolinite	12.1	1.5	Bhattacharyya and Gupta[130]
	Montmorillonite	28.6	4.6	Bhattacharyya and Gupta[130]
	Acid-activated montmorillonite	29.7	6.0	Bhattacharyya and Gupta[130]
Pb(II)	Kaolinite	11.2	0.7	Gupta and Bhattacharyya[131] and Bhattacharyya and Gupta[132]
	Acid-activated kaolinite	12.1	1.0	Gupta and Bhattacharyya[131] and Bhattacharyya and Gupta[132]
	Montmorillonite	33.0	8.9	Gupta and Bhattacharyya[131] and Bhattacharyya and Gupta[132]
	Acid-activated montmorillonite	34.0	11.3	Gupta and Bhattacharyya[131] and Bhattacharyya and Gupta[132]
Fe(III)	Kaolinite	11.2	1.3	Bhattacharyya and Gupta[133]
	Acid-activated kaolinite	12.1	1.7	Bhattacharyya and Gupta[133]
	Montmorillonite	28.9	5.2	Bhattacharyya and Gupta[133]
	Acid-activated montmorillonite	30.0	6.4	Bhattacharyya and Gupta[133]
Cu(II)	Kaolinite	4.4	1.1	Bhattacharyya and Gupta[134]
	Acid-activated kaolinite	5.6	1.3	Bhattacharyya and Gupta[126]
	Montmorillonite	25.5	9.2	Bhattacharyya and Gupta[134]
	Acid-activated montmorillonite	28.0	12.4	Bhattacharyya and Gupta[126]

Note: Units of Langmuir capacity and Freundlich capacity are mg/g and $mg^{1-1/n}\ L^{1/n}/g$, respectively.

defined pore structure in the microporous range of zeolite can accommodate a wide variety of cations such as Na^+, K^+, Ca^{2+}, Mg^{2+}, and others. These charge-compensating cations are free to migrate in and out of zeolite structures, and they are rather loosely held so that they can readily be exchanged for others in a contact solution. Accordingly, zeolites are not only good adsorbates but also good ion exchangers. This property can be used to introduce different cations into the structure, creating selective sites for adsorption purposes or catalysis. Their narrow pore size and tuneable affinity for certain molecules make them ideal adsorbents for selective purification to encapsulate hazardous compounds. Zeolites are characterized not only by a high selectivity separation mechanism but also by the ability to separate substances based on differences in sizes and shapes of molecules' steric separation mechanism.

Zeolites have been widely used for pollution control due to their ion exchange and adsorption properties. They have been used for the selective separation of cations from aqueous solution. The diffusion, adsorption, and ion exchange in zeolites have been extensively reviewed in many previous works.[135–137] Kesraoui-Ouki, Cheeseman, and Perry[138] reviewed natural zeolite utilization in metal effluent treatment applications. Dewatered zeolites produce channels that can adsorb molecules small enough to access the internal cavities while excluding larger species. Zeolites, modified by ion exchange, can be used for adsorption of different metal ions according to requirements and costs. The characteristics and applications of zeolites have been extensively reviewed by Ghobarkar, Schaf, and Guth.[139] High ion-exchange capacity and relatively high specific surface areas, and more importantly, their relatively cheap prices, make zeolites more attractive adsorbents.

Besides zeolite, other siliceous materials such as perlite and glass have been proposed for contaminant removal. The use of natural siliceous adsorbents such as silica, glass fibers, and perlite for wastewater is increasing because of their high abundance, easy availability, and low cost. The other commonly applied inorganic sorbents are silica gels, activated alumina, and oxide and hydroxide metals. Perlite is another siliceous material that exhibits a good adsorbent for decontamination purposes. It has been used as a low-cost adsorbent for the removal of methylene blue.[140,141] Methylene blue is physically adsorbed onto the perlite.

However, perlites of different types (expanded and unexpanded) and of different origins have different properties because of the differences in composition. Chakrabarti and Dutta[142] also investigated glass fiber for the adsorption of methylene blue. They stated that a considerable amount of the dye is adsorbed on soft glass even at ambient temperature. Accordingly, several siliceous materials become widely used as adsorbate materials in the adsorption process.

Currently, a new family of mesopore materials, so-called MCM materials or Mobil Composition of Matter (MCM), was developed by Mobil Oil Corporation, which proposed a revolutionary synthesis method to obtain such materials that comprise strictly uniform pores. An organic surfactant like an alkyltrimethylammonium bromide in an aqueous medium forms rod-like micelles, which are used as templates to form two or three monolayers of silica or alumina particles encapsulating the micelles' external surface. By removing the organic species from a well-ordered organic–inorganic condensed phase, a porous silicate or alumina material with uniformly porous structure remains. The mesopore size can be controlled by the molecular size template of the surfactant. Nowadays, MCM materials have been widely used in heavy metal removal, and they are currently the adsorption material that plays an important role in water and wastewater treatment.

Chitin and Chitosan

Chitin is a nontoxic, biodegradable polymer of high molecular weight. It contains 2-acetamido-2-deoxy-β-D-glucose through a β (1→4) linkage. Chitin is the most abundant natural fiber next to the cellulose and is similar to cellulose in many respects. The most abundant source of chitin is the shell of crab and shrimp. Chitin has presented exceptional chemical and biological qualities that can be used in water and wastewater purification through the adsorption process.

Chitin and chitosan have their chemical structures in common. Chitin is made up of a linear chain of acetylglucosamine groups. Chitosan is obtained by removing enough acetyl groups (CH_3–CO) for the molecule to be soluble in most diluted acids. This process, called deacetylation, releases amine groups (NH) and gives the chitosan a cationic characteristic. Chitosan contains 2-acetamido-2-deoxy-β-D-glucopyranose and 2-amino-2-deoxy-β-D-glucopyranose residues. Chitosan is known as an ideal natural support for enzyme immobilization because of its special characteristics such as hydrophilicity, biocompatibility, biodegradability, non-toxicity, adsorption properties, etc.[143]

Chitosan has drawn particular attention as an effective biosorbent due to its high content of amino and hydroxyl functional groups, giving it high adsorption potential for various aquatic pollutants.[143–147] This biopolymer represents an attractive alternative to other biomaterials because of its physicochemical characteristics, chemical stability, high reactivity, excellent chelation behavior. and high selectivity toward pollutants. Chitin and chitosan derivatives have been extensively investigated as adsorbents for the removal of organic molecules and metal ions from water and wastewater. The high adsorption potential of chitosan can be attributed to the following: 1) high hydrophilicity due to a large number of hydroxyl groups of glucose units; 2) presence of a large number of functional groups; 3) high chemical reactivity of these groups; and 4) flexible structure of the polymer chain.[148,149]

To enhance the adsorption capacity for pollutant removal, chitosan has been modified by several methods, either physical or chemical processes. Different shapes of chitosan, e.g., membranes, microspheres, gel beads, and films, have been synthesized and tested for their performance in pollutant removal from water and wastewater.[143–147] A cross-linked chitosan bead is one type of chemical modification for chitosan to increase the uptake capacity in the adsorption process.[150] This method using the chemical reaction of ethylenediamine and carbodiimide in modifying chitosan provided a high uptake capacity for mercury (Hg^{2+}) ions, which is considered to be one of the highest uptake capacities among various biosorbents.

Beads of 1 and 3 mm diameter were prepared as one type of modified chitosan.[151] The gelled chitosan beads were cross-linked with glutaraldehyde and then freeze-dried. Beads of 1 mm diameter possessed surface areas exceeding 150 m^2/g and mean pore sizes of 560 Å and were insoluble in acid media at pH 2. A new composite chitosan biosorbent was also prepared by coating chitosan onto perlite ore. It was used in the removal of Cu(II) and Ni(II) from aqueous solution.[152] The magnetic chitosan nanocomposites were synthesized on the basis of amine-functionalized magnetite nanoparticles.[153] These nanocomposites provide a very efficient, fast, and convenient tool for removing Pb^{2+}, Cu^{2+}, and Cd^{2+} from water. It was suggested that synthesized magnetic chitosan nanocomposites can be used as a recyclable tool for heavy metal ion removal. Several types of heavy metals and organic contaminants removed by chitosan are shown in Table 6.

Agricultural-Based Waste Materials

Agricultural by-products usually are composed of lignin and cellulose as major constituents that have the ability to some extent to bind some type of pollutants, for example, heavy metals, by donation of an electron pair from these groups to form complexes with the metal ions.[166] Currently, many types of agricultural-based waste materials play a significant role in the adsorption process. They are normally organic materials from plants, trees, crops, and algae. Two larger carbohydrate categories that play a significant role in the adsorption process are cellulose and hemicelluloses (holocellulose). Cellulose is a remarkable

Table 6 Adsorption capacities of chitosan and its composite for removal of heavy metals and some organic contaminants from water and wastewater.

Pollutant	Chitosan	Adsorption capacity (mg/g)	Isotherm	Refs.
Hg(II)	Chitosan/cotton fibers	104.31	Langmuir	Qu et al.[154]
Cd(II)	Chitosan/cotton fibers	15.74	Langmuir	Zhang et al.[155]
Cr(VI)	Magnetic chitosan	69.40	Langmuir	Huang et al.[156]
	Chitosan/cellulose	13.05	Langmuir	Sun et al.[157]
	Chitosan/perlite	153.8	Langmuir	Shameem et al.[158]
	Chitosan/ceramic alumina	153.8	Freundlich	Veera et al.[159]
Pb(II)	Chitosan/cotton fibers	101.53	Freundlich	Zhang et al.[155]
	Chitosan/magnetite	63.33	Langmuir	Tran et al.[160]
	Chitosan/cellulose	26.31	Langmuir	Sun et al.[157]
	Chitosan/sand	12.32	Langmuir	Rorrer et al.[151]
Cu(III)	Chitosan/cellulose	26.50	Langmuir	Sun et al.[157]
	Chitosan/perlite	196.07	Langmuir	Kalyani et al.[152]
	Chitosan/polyvinylchloride	87.9	Langmuir	Srinivasa et al.[162]
Ni(II)	Chitosan/magnetite	52.55	Langmuir	Tran et al.[160]
	Chitosan/cellulose	13.21	Langmuir	Sun et al.[157]
	Chitosan/perlite	114.94	Langmuir	Kalyani et al.[152]
	Chitosan/silica	254.3	Langmuir	Vijaya et al.[163]
Phenol	Chemically modified chitosan	2.22–151.50	Langmuir	Li et al.[164]
	Chitosan/calcium alginate beads	108.69	Langmuir	Nadavala et al.[165]
4-Chlorophenol	Chemically modified chitosan	2.58–179.73	Langmuir	Li et al.[164]
Nonylphenol	Chitosan	56.3	Langmuir	Lang et al.[116]

pure organic polymer, consisting solely of units of anhydroglucose held together in a giant straight-chain molecule.[168] These anhydroglucose units are bound together by β-(1,4)-glycosidic linkages. Hemicelluloses consist of different monosaccharide units. The polymer chains of hemicelluloses have short branches and are amorphous. Hemicelluloses are derived mainly from chains of pentose sugars and act as the cement material holding together the cellulose micelles and fiber.[168] Hemicelluloses are partially soluble in water. Currently, chemical modification is widely used to alter the biochemical component of the biomaterials to obtain higher efficiency in pollutant removal by biosorption process.[169] Biomass chemical modifications include delignification, esterification of carboxyl and phosphate groups, methylation of amino groups, and hydrolysis of carboxylate groups. Sawamiappan and Krishnamoorthy[170] replaced phenol–formaldehyde cationic matrices with sulfonated bagasse. Odozi et al.[171] polymerized corncob, sawdust, and onion. However, the disadvantages of chemical modification are a high expense to pay and unwanted problems, such as bleeding of excessive quantities of colored organic compounds, odor, and further pollution through the use of toxic chemicals. Several types of agricultural wastes have been used in the adsorption process, with the differences in adsorption capacity as shown in Table 7.

Mechanisms involved in the biosorption process include chemical adsorption, complexation, adsorption–complexation on surfaces and in pores, ion exchange, microprecipitation, heavy metal hydroxide condensation onto the biosurface, and surface adsorption.[172–174] In the adsorption process, functional groups are responsible for pollutant binding on the surface of biomaterial. Most of the functional groups involved in the binding process are found in cell walls. Plant cell walls are generally considered as structures built by cellulose molecules, organized in microfibrils and surrounded by hemicellulosic materials (xylans, mannans, glucomannans, galactans, arabogalactans), lignin, and pectin along with small amounts of protein.[175] During biosorption, water is able to permeate the noncrystalline portion of cellulose and all of the hemicellulose and lignin. The aqueous solution comes into contact with a very large surface area of different cell wall components. The disordered structure of amorphous cellulose allows easier access to reagents than highly structured crystalline cellulose. While water penetrates through the cell wall components, water adsorption of fibers causes swelling. The bigger the amount of water adsorption, the bigger the swelling. Swelling also depends on the fiber's structure, on the degree of crystallinity, and on the amorphous and void regions.[176] Swelling occurs when polar solvents such as water and alcohols come into contact with

Table 7 Adsorption capacities of agricultural waste for heavy metal and organic contaminant removal from water and wastewater.

Pollutant	Agriculture waste	Adsorption capacity (mg/g)	Isotherm	Refs.
Cd(II)	Juniper fiber	9.2	Langmuir	Min et al.[181]
	Base-treated juniper fiber	29.5	Langmuir	Min et al.[181]
Cr(VI)	Cactus	7.1	Langmuir	Dakiky et al.[182]
	Coconut shell carbon	2.2	Langmuir	Babel and Kurniawan[109]
	Coconut shell carbon oxidized with H_2SO_4	4.1	Langmuir	Babel and Kurniawan[109]
	Coconut shell carbon oxidized with HNO_3	10.9	Langmuir	Babel and Kurniawan[109]
	Sawdust	15.8	Langmuir	Dakiky et al.[182]
Pb(II)	Carbonaceous adsorbent	25.0	Langmuir	Bhatnagar et al.[183]
	Sawdust	22.2	Langmuir, Freundlich	Taty-Costodes et al.[184]
Fe(III)	Maize cobs	2.5	Langmuir	Nassar et al.[185]
Cu(II)	Tree fern	7.6	Langmuir	Ho et al.[186]
Ni(II)	Peat	28.3	Langmuir, Freundlich	Chen et al.[187]
Phenol	Banana pith	49.9–129.4	Langmuir	Sathishkumar et al.[188]
	Banana peel	688.9	Langmuir	Achak et al.[189]
	Corn grain	256	Langmuir	Park et al.[190]
2-Nitrophenol	*Lessonia nigrescens*	71.28	Langmuir	Navarro et al.[191]
	Macrocystis integrifolia	97.37	Langmuir	Navarro et al.[191]
2,4-Dicholorophenol	Pomegranate peel	65.7	Langmuir	Bhatnagar and Minocha[192]
Nonylphenol	*Rhizopus arrhizus*	4.5–43.7	Langmuir	Lang et al.[116]

wood.[177] These polar solvent molecules are attracted to the dry solid matrix and held by hydrogen bonding forces between the –OH or –COOH groups in the wood structure and cause the biosorption of pollutants in aqueous solution. Many research works[178–180] have reported the wide use of biosorption process in heavy metal removal. Thus, the agricultural-based waste materials become the adsorption material that plays an important role in water and wastewater treatment nowadays (Table 7).

PROGRESS IN RESEARCH ON ADSORPTION PROCESS IN WATER PURIFICATION

Adsorbents and adsorption processes have been widely studied and applied in different aspects for a long time. Owing to its effective, efficient, and economic approach to water purification, this process has been applied in removal of several contaminants, such as pesticides, halogenated carbon, dyes, phenol and its derivatives, and heavy metals. The most widely used adsorbent in the adsorption process is activated carbon. This adsorbent is highly inert and thermally stable, and it can be used over a broad pH range. Although it has a great capacity for adsorbing various organic compounds and can be easily modified by chemical treatment to increase its adsorption capacity, activated carbon has several disadvantages.[193] Owing to the process mechanism, adsorption transfers pollutants from one phase to another rather than eliminating them from the environment. Thus, after adsorption, the contaminants in liquid phase absorb on the surface of adsorbent, which has to be separated from aquatic system when it becomes exhausted or the effluent reaches the maximum allowable discharge level. Furthermore, the regeneration of exhausted activated carbon by a chemical and thermal procedure is also expensive and results in loss of the sorbent.

Recently, a lot of novel adsorption processes have been developed for enhancing the efficiency of removing organic and inorganic contaminants from water. The development of cheaper and more effective novel composite adsorbents[194–197] in comparison with the classical adsorbents has been investigated by researchers from many countries all over the world. These adsorbents are metal oxide–based composite adsorbents such as TiO_2 and MnO_2, surface-modified Fe_3O_4 adsorbent, magnetic particle–modified carbon adsorbent, magnetic particle–modified clay mineral adsorbent, and magnetic particle–modified biopolymer adsorbent. These composite materials deserve particular attention because they combine the properties and advantages of each of their components. They represent an interesting and attractive alternative as adsorbents and/or catalysts due to their high reactivity and excellent selectivity toward specific pollutant compounds. To obtain the anticipated function and enhance the efficiency of water purification, these adsorbents should be designed and modified in their compositions, structures, surfaces, and preparation methods to obtain the requirement of physicochemical properties for the purpose of adsorption. Extensive research in synthesizing of new adsorbents and investigating of adsorption mechanism is needed. Advances in development of new adsorbents for the adsorption process will be the progress of future technology in water purification.

REFERENCES

1. Dabrowski, A. Adsorption, from theory to practice. Adv. Colloid Interface Sci. **2001**, *93*, 135–224.
2. Ahmad, A.; Rafatullah, M.; Danish, M. Removal of Zn(II) and Cd(II) ions from aqueous solutions using treated sawdust of sissoo wood. Holz Roh Werkst **2007**, *65*, 429–436.
3. Ahmad, A.; Rafatullah, M.; Sulaiman, O.; Ibrahim, M.H.; Chii, Y.Y.; Siddique, B.M. Removal of Cu(II) and Pb(II) ions from aqueous solutions by adsorption on sawdust of Meranti wood. Desalination **2009**, *247*, 636–646.
4. Rafatullah, M.; Sulaiman, O.; Hashim, R.; Ahmad, A. Adsorption of copper (II), chromium (III), nickel (II) and lead (II) ions from aqueous solutions by meranti sawdust. J. Hazard. Mater. **2009**, *170*, 969–977.
5. Gupta, V.K.; Carrott, P.J.M; Ribeiro Carrott M.M.L; Suhas. Low-cost adsorbents: Growing approach to wastewater treatment—A review. Crit. Rev. Environ. Sci. Technol. **2009**, *39*, 783–842.
6. Soto, M.L.; Moure, A.; Domínguez, H.; Parajó, J.C. Recovery, concentration and purification of phenolic compounds by adsorption: A review. J. Food Eng. **2011**, *105*, 1–27.
7. Gupta, S.S.; Bhattacharyya, K.G. Kinetics of adsorption of metal ions on inorganic materials: A review. Adv. Colloid Interface Sci. **2011**, *162*, 39–58.
8. Gupta, V.K.; Mohan, D.; Sharma, S. Removal of lead from wastewater using bagasse fly ash—A sugar industry waste material. Sep. Sci. Technol. **1998**, *33*, 1331–1343.
9. Gupta, V.K.; Jain, C.K.; Ali, I.; Sharma, M.; Saini, V.K. Removal of cadmium and nickel from wastewater using bagasse fly ash—A sugar industry waste. Water Res. **2003**, *37*, 4038–4044.
10. Parfitt, D.; Rochestor, H. Adsorption of small molecules. In *Adsorption from Solution at the Solid/Liquid Interface*; Parfitt, G.D., Rochester, C.H., Eds.; Academic Press: New York, 1983; 3.
11. Giles, C.H.; Smith, D.; Huitson, A. A general treatment and classification of the solute adsorption isotherm. I. Theoretical. J. Colloid Interface Sci. **1974**, *47*, 755–765.
12. Dabrowski, A.; Podkościelny, P.; Hubicki, Z.; Barczak, M. Adsorption of phenolic compounds by activated carbon—A critical review. Chemosphere **2005**, *58*, 1049–1070.
13. Bansal, R.C.; Goyal, M. *Activated Carbon Adsorption*; CRC Press: Boca Raton, FL, 2005.
14. Bretag, J.; Kammerer, D.R.; Jensen, U.; Carle, R. Evaluation of the adsorption behavior of flavonoids and phenolic acids onto a food-grade resin using a D-optimal design. Eur. Food Res. Technol. **2009**, *228*, 985–999.
15. Weber, T.W.; Chakkravorti, R.K. Pore and solid diffusion models for fixed bed adsorbers. AIChE J. **1974**, *20*, 228–238.
16. Freundlich, H.M.F. Over the adsorption in solution. J. Phys. Chem. **1906**, *57*, 385–470.
17. Fytianos, K.; Voudrias, E.; Kokkalis, E. Sorption–desorption behavior of 2,4-dichlorophenol by marine sediments. Chemosphere **2000**, *40*, 3–6.
18. Foo, K.Y.; Hameed, B.H. Insights into the modeling of adsorption isotherm systems. Chem. Eng. J. **2010**, *156*, 2–10.
19. Temkin, M.I.; Pyzhev, V. Kinetics of ammonia synthesis on promoted iron catalysts. Acta Phys. Chem. U.R.S.S. **1940**, *12*, 327–356.
20. Rashidi, F.; Sarabi, R.S.; Ghasemi, Z.; Seif, A. Kinetic, equilibrium and thermodynamic studies for the removal of lead (II) and copper (II) ions from aqueous solutions by nanocrystalline TiO_2. Superlattices Microstruct. **2010**, *48*, 577–591.
21. Dubinin, M.M.; Zaverina, E.D.; Radushkevich, L.V. Sorption and structure of active carbons. I. Adsorption of organic vapors. J. Phys. Chem. **1947**, *21*, 1351–1362.
22. Unlu, N.; Ersoz, M. Adsorption characteristics of heavy metal ions onto a low cost biopolymeric sorbent from aqueous solutions. J. Hazard. Mater. **2006**, *136*, 272–280.
23. Redlich, O.; Peterson, D.L. A useful adsorption isotherm. J. Phys. Chem. **1959**, *63*, 1024–1029.
24. Hameed, B.H.; Tan, I.A.W.; Ahmad, A.L. Adsorption isotherm, kinetic modeling and mechanism of 2,4,6 trichlorophenol on coconut husk–based activated carbon. Chem. Eng. J. **2008**, *144*, 235–244.
25. Ho, Y.S.; Ng, J.C.Y.; McKay, G. Kinetics of pollutant sorption by biosorbents: Review. Sep. Purif. Methods **2000**, *29*, 189–232.
26. Lagergren, S. Zur theorie der sogenannten adsorption gelöster stoffe. Kungliga Svenska Vetenskapsakademiens. Handlingar, Vetensk. Handl. **1898**, *24*, 1–39.
27. McKay, G.; Ho, Y.S.; Ng, J.C.Y. Biosorption of copper from wastewaters: A Review. Sep. Purif. Methods **1999**, *28*, 87–125.
28. Allen, S.J.; Gan, Q.; Matthews, R.; Johnson, P.A. Kinetic modeling of the adsorption of basic dyes by kudzu. J. Colloid Interface Sci. **2005**, *286*, 101–109.
29. Febrianto, J.; Kosasih, A.N.; Sunarso, J.; Ju, Y.-H.; Indraswati, N.; Ismadji, S. Equilibrium and kinetic studies in adsorption of heavy metals using biosorbent: A summary of recent studies. J. Hazard. Mater. **2009**, *162*, 616–645.
30. Al-Ghouti, M.A.; Khraisheh, M.A.M.; Ahmad, M.N.M.; Allen, S. Adsorption behaviour of methylene blue onto Jordanian diatomite: A kinetic study. J. Hazard. Mater. **2009**, *165*, 589–598.
31. Nandi, B.K.; Goswami, A.; Purkait, M.K. Adsorption characteristics of brilliant green dye on kaolin. J. Hazard. Mater. **2009**, *161*, 387–395.
32. Lin. C. I.; Wang, L. H. Rate equations and isotherms for two adsorption models. J. Chin. Inst. Chem. Eng. **2008**, *39*, 579–585.
33. Plazinski, W.; Rudzinski, W.; Plazinska, A. Theoretical models of sorption kinetics including a surface reaction mechanism: A review. Adv. Colloid Interface Sci. **2009**, *152*, 2–13.

34. Ho, Y.S.; McKay, G. The kinetics of sorption of divalent metal ions onto sphagnum moss peat. Water Res. **2000**, *34*, 735–742.
35. Rudzinski, W.; Plazinski, W. On the applicability of the pseudo–second order equation to represent the kinetics of adsorption at solid/solution interfaces: A theoretical analysis based on the statistical rate theory. Adsorption **2009**, *15*, 181–192.
36. Ho, Y.S.; McKay, G. Kinetic model for lead(II) sorption on to peat. Adsorpt. Sci. Technol. **1998**, *16*, 243–255.
37. Aharoni, C.; Tompkins, F.C. Kinetics of adsorption and desorption and the Elovich equation. In *Advances in Catalysis and Related Subjects*; Eley, D.D., Pines, H., Weisz, P.B., Eds.; Academic Press: New York, 1970; Vol. 21, 2–50.
38. Ho, Y.S.; McKay, G. Application of kinetic models to the sorption of copper(II) onto peat. Adsorpt. Sci. Technol. **2002**, *20*, 797–813.
39. Rudzinski, W.; Panczyk, T. The Langmuirian adsorption kinetics revised: A farewell to the 20th century theories? Adsorption **2002**, *8*, 23–34.
40. Chien, S.; Clayton, W.R. Application of Elovich equation to the kinetics of phosphate release and sorption in soils. Soil Sci. Soc. Am. J. **1980**, *44*, 265–268.
41. Ruthven, D.M. *Principles of Adsorption and Adsorption Processes*; Wiley-Interscience: New York, 1984.
42. Banerjee, K.; Cheremisinoff, P.N.; Cheng, S.L. Adsorption kinetics of *o*-xylene by fly ash. Water Res. **1997**, *31*, 249–261.
43. Manju, G.N.; Anoop Krishnan, K.; Vinod, V.P.; Anirudhan, T.S. An investigation into the sorption of heavy metals from wastewaters by polyacrylamide-grafted iron(III) oxide. J. Hazard. Mater. **2002**, *91*, 221–238.
44. Boyd, G.E.; Adamson, A.W.; Myers Jr., L.S. The Exchange adsorption of ions from aqueous solutions by organic zeolites. II. Kinetics. J. Am. Chem. Soc. **1947**, *69*, 2836–2848.
45. Hamerlinck, Y.; Mertens, D.H. In *Activated Carbon Principles in Separation Technology*; Vansant, E.F., Ed.; Elsevier: New York, 1994.
46. Mourão, P.A.M.; Carrott, P.J.M.; Ribeiro Carrott, M.M.L. Application of different equations to adsorption isotherms of phenolic compounds on activated carbons prepared from cork. Carbon **2006**, *44*, 2422–2429.
47. Mestre, A.S.; Pires, J.; Nogueira, J.M.F.; Parra, J.B.; Carvalho, A.P.; Ania, C.O. Waste-derived activated carbons for removal of ibuprofen from solution: Role of surface chemistry and pore structure. Bioresour. Technol. **2009**, *100*, 1720–1726.
48. Evans, M.J.B.; Halliop, E.; MacDonald, J.A.F. The production of chemically activated carbon. Carbon **1999**, *37*, 269–274.
49. Wu, F.C.; Tseng, R.L.; Juang, R.S. Adsorption of dyes and phenols from water on the activated carbons prepared from corncob wastes. Environ. Technol. **2001**, *22*, 205–213.
50. Tseng, R.L.; Tseng, S.K. Characterization and use of high surface area activated carbons prepared from cane pith for liquid-phase adsorption. J. Hazard. Mater. **2006**, *136*, 671–680.
51. Prahas, Y.D.; Kartika, N.; Indraswati, S.; Ismadji, S. Activated carbon from jackfruit peel waste by H3PO4 chemical activation: Pore structure and surface chemistry characterization. Chem. Eng. J. **2007**, *140*, 32–42.
52. Jain, S.; Jayaram, R.V. Adsorption of phenol and substituted chlorophenols from aqueous solution by activated carbon prepared from jackfruit (*Artocarpus heterophyllus*) peel-kinetics and equilibrium studies. Sep. Sci. Technol. **2007**, *42*, 2019–2032.
53. Tancredi, N.; Medero, N.; Möller, F.; Píriz, J.; Plada, C.; Cordero, T. Phenol adsorption onto powdered and granular activated carbon, prepared from Eucalyptus wood. J. Colloid Interface Sci. **2004**, *279*, 357–363.
54. Mudoga, H.L.; Yucel, H.; Kincal, N.S. Decolorization of sugar syrups using commercial and sugar beet pulp based activated carbons. Bioresour. Technol. **2008**, *99*, 3528–3533.
55. Philip, C.A.; Girgis, B.S. Adsorption characteristics of microporous carbons from apricot stones activated by phosphoric acid. J. Chem. Technol. Biotechnol. **1996**, *67*, 248–254.
56. Ahmadpour, A.; Do, D.D. The preparation of activated carbon from Macadamia nutshell by chemical activation. Carbon **1997**, *35*, 1723–1732.
57. Toles, C.A.; Marshall, W.E.; Johns, M.M.; Wartelle, L.H.; McAloon, A. Acid activated carbons from almond shells: Physical, chemical and adsorptive properties and estimated cost of production. Bioresour. Technol. **2000**, *71*, 87–92.
58. Toles, C.A.; Marshall, W.E.; Wartelle, L.H.; McAloon, A. Steam- or carbon dioxide–activated carbons from almond shells: Physical, chemical and adsorptive properties and estimated cost of production. Bioresour. Technol. **2000**, *75*, 197–203.
59. Galiatsatou, P.; Metaxas, M.; Arapoglou, D.; Kasselouri-Rigopoulo, V. Treatment of olive mill waste water with activated carbons from agricultural by-products. Waste Manage. **2002**, *22*, 803–812.
60. Rengaraj, S.; Moon, S.H.; Sivabalan, R.; Arabindoo, B.; Murugesan, V. Agricultural solid waste for the removal of organics: Adsorption of phenol from water and wastewater by palm seed coat activated carbon. Waste Manage. **2002**, *22*, 543–548.
61. Qi, J.; Li, Z.; Guo, Y.; Xu, H. Adsorption of phenolic compounds on micro- and mesoporous rice husk–based active carbons. Mater. Chem. Phys. **2004**, *87*, 96–101.
62. Tan, I.A.W.; Ahmad, A.L.; Hameed, B.H. Preparation of activated carbon from coconut husk: Optimization study on removal of 2,4,6-trichlorophenol using response surface methodology. J. Hazard. Mater. **2008**, *153*, 709–717.
63. Michailof, C.; Stavropoulos, G.G.; Panayiotou, C. Enhanced adsorption of phenolic compounds, commonly encountered in olive mill wastewaters, on olive husk derived activated carbons. Bioresour. Technol. **2008**, *99*, 6400–6408.
64. Muñoz-González, Y.; Arriagada-Acuña, R.; Soto-Garrido, G.; García-Lovera, R. Activated carbons from peach stones and pine sawdust by phosphoric acid activation used in clarification and decolorization processes. J. Chem. Technol. Biotechnol. **2009**, *84*, 39–47.
65. Lussier, M.G.; Shull, J.C.; Miller, D.J. Activated carbons from cherry stones. Carbon **1994**, *32*, 1493–1498.
66. Shopova, N.; Minkova, V.; Markova, K. Evaluation of the thermochemical changes in agricultural by-products and in the carbon adsorbents obtained from them. J. Therm. Anal. **1997**, *48*, 309–320.
67. Marsh, H.; Iley, M.; Berger, J.; Siemieniewska, T. The adsorptive properties of activated plum stone chars. Carbon **1975**, *13*, 103–109.

68. Mohanty, K.; Jha, M.; Meikap, B.C.; Biswas, M.N. Preparation and characterization of activated carbons from *Terminalia arjuna* nut with zinc chloride activation for the removal of phenol from wastewater. Ind. Eng. Chem. Res. **2005**, *44*, 4128–4138.
69. Alvim Ferraz, M.C. Preparation of activated carbon for air pollution control. Fuel **1988**, *67*, 1237–1241.
70. Tancredi, N.; Cordero, T.; Rodríguez-Mirasol, J.; Rodríguez, J.J. Activated carbons from Uruguayan eucalyptus wood. Fuel **1996**, *75*, 1701–1706.
71. Ferro-García, M.A.; Rivera-Utrilla, J.; Rodríguez-Gordillo, J.; Bautista-Toledo, I. Adsorption of zinc, cadmium, and copper on activated carbons obtained from agricultural by-products. Carbon **1988**, *26*, 363–373.
72. Hall, C.R.; Holmes, R.J. The preparation and properties of some activated carbons modified by treatment with phosgene or chlorine. Carbon **1992**, *30*, 173–176.
73. Otowa, T.; Nojima, Y.; Miyazaki, T. Development of KOH activated high surface area carbon and its application to drinking water purification. Carbon **1997**, *35*, 1315–1319.
74. Manju, G.N.; Raji, C.; Anirudhan, T.S. Evaluation of coconut husk carbon for the removal of arsenic from water. Water Res. **1998**, *32*, 3062–3070.
75. Laine, J.; Calafat, A.; Labady, M. Preparation and characterization of activated carbons from coconut shell impregnated with phosphoric acid. Carbon **1989**, *27*, 191–195.
76. Mohan, D.; Singh, K.P.; Ghosh D. Removal of α picoline, β-picoline, and γ-picoline from synthetic wastewater using low cost activated carbons derived from coconut shell fibers. Environ. Sci. Technol. **2005**, *39*, 5076–5086.
77. Mohan, D.; Singh, K.P.; Sinha, S.; Ghosh, D. Removal of pyridine derivatives from aqueous solution by activated carbons developed from agricultural waste materials. Carbon **2005**, *43*, 1680–1693.
78. Mohan, D.; Singh, K.P.; Singh, V.K. Trivalent chromium removal from wastewater using low cost activated carbon derived from agricultural waste material and activated carbon fabric cloth. J. Hazard. Mater. **2006**, *135*, 280–295.
79. Mohan, D.; Singh, K.P.; Sinha, S.; Ghosh, D. Removal of pyridine from aqueous solution using low cost activated carbons developed from agricultural waste materials. Carbon **2004**, *43*, 2409–2421.
80. Kumar, K.V.; Sivanesan, S. Equilibrium data, isotherm parameters and process design for partial and complete isotherm of methylene blue onto activated carbon. J. Hazard. Mater. **2006**, *134*, 237–244.
81. Marungrueng, K.; Pavasant, P. High performance biosorbent (*Caulerpa lentillifera*) for basic dye removal. Bioresour. Technol. **2007**, *98*, 1567–1572.
82. Basava Rao, V.V.; Mohan Rao, S.R. Adsorption studies on treatment of textile dyeing industrial effluent by fly ash. Chem. Eng. J. **2006**, *116*, 77–84.
83. El Qada, E.N.; Allen, S.J.; Walker, G.M. Adsorption of basic dyes from aqueous solution onto activated carbons. Chem. Eng. J. **2008**, *135*, 174–184.
84. Tamai, H.; Kakii, T.; Hirota, Y.; Kumamoto, T.; Yasuda, H. Synthesis of extremely large mesoporous activated carbon and its unique adsorption for giant molecules. Chem. Mater. **1996**, *8*, 454–462.
85. Banat, F.; Al-Asheh, S.; Al-Ahmad, R.; Bni-Khalid, F. Bench-scale and packed bed sorption of methylene blue using treated olive pomace and charcoal. Bioresour. Technol. **2007**, *98*, 3017–3025.
86. McKay, G.; Porter, J.F.; Prasad, G.R. The removal of dye colours from aqueous solutions by adsorption on low-cost materials. Water Air Soil Pollut. **1999**, *114*, 423–438.
87. McKay, G.; Ramprasad, G.; Pratapamowli, P. Equilibrium studies for the adsorption of dyestuffs from aqueous solution by low-cost materials. Water Air Soil Pollut. **1986**, *29*, 273–283.
88. Kannan, N.; Sundaram, M.M. Kinetics and mechanism of removal of methylene blue by adsorption on various carbons—A comparative study. Dyes Pigments **2001**, *51*, 25–40.
89. Bestani, B.; Benderdouche, N.; Benstaali, B.; Belhakem, M.; Addou, A. Methylene blue and iodine adsorption onto an activated desert plant. Bioresour. Technol. **2008**, *99*, 8441–8444.
90. Stavropoulos, G.G.; Zabaniotou, A.A. Production and characterization of activated carbons from olive-seed waste residue. Microporous Mesoporous Mater. **2005**, *82*, 79–85.
91. Fernandes, A.N.; Almeida, C.A.P.; Menezes, C.T.B.; Debacher, N.A.; Sierra, M.M.D. Removal of methylene blue from aqueous solution by peat. J. Hazard. Mater. **2007**, *144*, 412–419.
92. Jones, D.A.; Lelyveld, T.P.; Mavrofidis, S.D.; Kingman, S.W.; Miles, N.J. Microwave heating applications in environmental engineering—A review. Resour. Conserv. Recycl. **2002**, *34*, 75–90.
93. Marsh, H.; Rodríguez-Reinoso, F. *Activated Carbon*; Elsevier: Oxford, 2006.
94. Ioannidou, O.; Zabaniotou, A. Agricultural residues as precursors for activated carbon production—A review. Renew. Sustainable Energy Rev. **2007**, *11*, 1966–2005.
95. Yin, C.Y.; Aroua, M.K.; Daud, W.M.A.W. Review of modifications of activated carbon for enhancing contaminant uptakes from aqueous solutions. Sep. Purif. Technol. **2007**, *52*, 403–415.
96. Paraskeva, P.; Kalderis, D.; Diamadopoulos, E. Production of activated carbon from agricultural by-products. J. Chem. Technol. Biotechnol. **2008**, *83*, 581–592.
97. Yuen, F.K.; Hameed, B.H. Recent developments in the preparation and regeneration of activated carbons by microwaves. Adv. Colloid Interface Sci. **2009**, *149*, 19–27.
98. Menéndez, J.A.; Arenillas, A.; Fidalgo, B.; Fernández, Y.; Zubizarreta, L.; Calvo, E.G.; Bermúdez, J.M. Microwave heating processes involving carbon materials. Fuel Process. Technol. **2010**, *91*, 1–8.
99. Laszlo, K.; Podkoscielny, P.; Dabrowski, A. Heterogeneity of polymer-based active carbons in adsorption of aqueous solutions of phenol and 2,3,4-trichlorophenol. Langmuir **2003**, *19*, 5287–5294.
100. Podkoscielny, P.; Dabrowski, A.; Marijuk, O.V. Heterogeneity of active carbons in adsorption of phenol aqueous solutions. Appl. Surf. Sci. **2003**, *205*, 297–303.
101. León Y León, C.A.; Radovic, L.R. Interfacial chemistry and electrochemistry of carbon surfaces. In *Chemistry and Physics of Carbon*; Thrower, P.A., Ed.; Marcel Dekker: New York, 1994; Vol. 24, 214–310.
102. Boehm, H.P. Surface oxides on carbon and their analysis: A critical assessment. Carbon **2002**, *40*, 145–149.
103. Gabaldón, C.; Marzal, P.; Seco, A.; Gonzalez, J.A. Cadmium and copper removal by a granular activated carbon

in laboratory column systems. Sep. Sci. Technol. **2000**, *35*, 1039–1053.

104. Carrott, P.J.M.; Ribeiro Carrott, M.M.L.; Nabais, J.M.V.; Ramalho, J.P.P. Influence of surface ionization on the adsorption of aqueous zinc species by activated carbons. Carbon **1997**, *35*, 403–410.
105. Carrott, P.J.M.; Mourao, P.A.M.; Ribeiro Carrott, M.M.L.; Goncalves, E.M. Separating surface and solvent effects and the notion of critical adsorption energy in the adsorption of phenolic compounds by activated carbons. Langmuir **2005**, *21*, 11863–11869.
106. Hu, J.; Aizawa, T.; Ookubo, Y.; Morita, T.; Magara, Y. Adsorptive characteristics of ionogenic aromatic pesticides in water on powdered activated carbon. Water Res. **1998**, *32*, 2593–2600.
107. Urano, K.; Yamamoto, E.; Tonegawa, M.; Fujie, K. Adsorption of chlorinated organic compounds on activated carbon from water. Water Res. **1991**, *25*, 1459–1464.
108. Malhas, A.N.; Abuknesha, R.A.; Price, R.G. Removal of detergents from protein extracts using activated charcoal prior to immunological analysis. J. Immunol. Methods **2002**, *264*, 37–43.
109. Babel, S.; Kurniawan, T.A. Cr(VI) removal from synthetic wastewater using coconut shell charcoal and commercial activated carbon modified with oxidizing agents and/or chitosan. Chemosphere **2004**, *54*, 951–967.
110. Kim, D.S. Adsorption characteristics of Fe(III) and Fe(III)–NTA complex on granular activated carbon. J. Hazard. Mater. **2004**, *106*, 67–84.
111. Satapathy, D.; Natarajan, G.S. Potassium bromate modification of the granular activated carbon and its effect on nickel adsorption. Adsorption **2006**, *12*, 147–154.
112. Richard, D.; Delgado Nunez, M.L.; Schweich, D. Adsorption of complex phenolic compounds on active charcoal: adsorption capacity and isotherms. Chem. Eng. J. **2010**, *158*, 213–219.
113. Figaro, S.; Louisy-Louis, S.; Lambert, J.; Ehrhardt, J.J.; Ouensanga, A.; Gaspard, S. Adsorption studies of recalcitrant compounds of molasses spentwash on activated carbons. Water Res. **2006**, *40*, 3456–3466.
114. Mohan, S.V.; Karthikeyan, J. Removal of lignin and tannin colour from aqueous solution by adsorption onto activated charcoal. Environ. Pollut. **1997**, *97*, 183–187.
115. Kalderis, D.; Koutoulakis, D.; Paraskeva, P.; Diamadopoulos, E.; Otal, E.; Valle, J.O.; Fernandez-Pereira, C. Adsorption of polluting substances on activated carbons prepared from rice husk and sugarcane bagasse. Chem. Eng. J. **2008**, *144*, 42–50.
116. Lang, W.; Dejma, C.; Sirisansaneeyakul, S.; Sakairi, N. Biosorption of nonylphenol on dead biomass of *Rhizopus arrhizus* encapsulated in chitosan beads. Bioresour. Technol. **2009**, *100*, 5616–5623.
117. Pinnavaia, T. Intercalated clay catalysts. J. Sci. **1983**, *220*, 365–371.
118. Bhattacharyya, K.G.; Gupta, S.S. Adsorption of a few heavy metals on natural and modified kaolinite and montmorillonite: A review. Adv. Colloid Interface Sci. **2008**, *140*, 114–131.
119. Mitchell, J.K. *Fundamentals of Soil Behavior*, 2nd Ed; Wiley: New York, 1993.
120. Gupta, S.S.; Bhattacharyya, K.G. Removal of Cd(II) from aqueous solution by kaolinite, montmorillonite and their poly(oxo zirconium) and tetrabutylammonium derivatives. J. Hazard. Mater. **2006**, *128*, 247–257.
121. Bhattacharyya, K.G.; Gupta S.S. Influence of acid activation of kaolinite and montmorillonite on their adsorptive removal of Cd(II) from water. Ind. Eng. Chem. Res. **2007**, *46*, 3734 –3742.
122. Bagane, M.; Guiza, S. Removal of a dye from textile effluents by adsorption. Ann. Chim. **2000**, *25*, 615–626.
123. Almeida, C.A.P.; Debacher, N.A.; Downs, A.J.; Cottet, L.; Mello, C.A.D. Removal of methylene blue from colored effluents by adsorption on montmorillonite clay. J. Colloid Interface Sci. **2009**, *332*, 46–53.
124. Ghosh, D.; Bhattacharyya, K.G. Adsorption of methylene blue on kaolinite. Appl. Clay Sci. **2002**, *20*, 295–300.
125. Gupta, S.S.; Bhattacharyya, K.G. Removal of Cd(II) from aqueous solution by kaolinite, montmorillonite and their poly(oxo zirconium) and tetrabutylammonium derivatives. J. Hazard. Mater. **2006**, *128*, 247–257.
126. Bhattacharyya, K.G.; Gupta, S.S. Influence of acid activation on adsorption of Ni(II) and Cu(II) on kaolinite and montmorillonite: Kinetic and thermodynamic study. Chem. Eng. J. **2008**, *136*, 1–13.
127. Gupta, S.S.; Bhattacharyya, K.G. Adsorption of Ni(II) on clays. J. Colloid Interface Sci. **2006**, *295*, 21–32.
128. Bhattacharyya, K.G.; Gupta, S.S. Adsorption of chromium(VI) from water by CLAYS. Ind. Eng. Chem. Res. **2006**, *45*, 7232–7240.
129. Yavuz, O.; Altunkaynak, Y.; Guzel, F. Removal of copper, nickel, cobalt and manganese from aqueous solution by kaolinite. Water Res. **2003**, *37*, 948–952.
130. Bhattacharyya, K.G.; Gupta, S.S. Adsorption of Co(II) from aqueous medium on natural and acid activated kaolinite and montmorillonite. Sep. Sci. Technol. **2007**, *42*, 3391–3418.
131. Gupta, S.S.; Bhattacharyya, K.G. Interaction of metal ions with clays: I. A case study with Pb(II). Appl. Clay Sci. **2005**, *30*, 199–208.
132. Bhattacharyya, K.G.; Gupta, S.S. Pb(II) uptake by kaolinite and montmorillonite in aqueous medium: Influence of acid activation of the clays. Colloids Surf., A **2006**, *277*, 191–200.
133. Bhattacharyya, K.G.; Gupta, S.S. Adsorption of Fe(III) from water by natural and acid activated clays: Studies on equilibrium isotherm, kinetics and thermodynamics of interactions. Adsorption **2006**, *12*, 185–204.
134. Bhattacharyya, K.G.; Gupta, S.S. Kaolinite, montmorillonite, and their modified derivatives as adsorbents for removal of Cu(II) from aqueous solution. Sep. Purif. Technol. **2006**, *50*, 388–397.
135. Townsend, R.P. Ion exchange in zeolites. In *Introduction to Zeolite Science and Practice*; Bekkum, H.V., Flannigen, E.M., Janmsen, J.C., Eds.; Elsevier: Amsterdam, 1991; 359–390.
136. Dyer, A. *An Introduction to Zeolite Molecular Sieves*; John Wiley and Sons: Chichester, 1988; Vol. 5–8.
137. Barer, R.M. *Zeolites and Clay Minerals as Sorbent and Molecular Sieves*; Academic Press: New York, 1978.
138. Kesraoui-Ouki, S.; Cheeseman, C.R.; Perry, R. Natural zeolite utilization in pollution control: A review of application to metal's effluents. J. Chem. Technol. Biotechnol. **1994**, *59*, 121–126.
139. Ghobarkar, H.; Schaf, O.; Guth, U. Zeolites from kitchen to space. Prog. Solid State Chem. **1999**, *27*, 29–73.

140. Dogan, M.; Alkan, M.; Onager, Y. Adsorption of methylene blue from aqueous solution onto perlite. Water Air Soil Pollut. **2000**, *120*, 229–248.
141. Dogan, M.; Alkan, M.; Turkyilmaz, A.; Ozdemir, Y. Kinetics and mechanism of removal of methylene blue by adsorption onto perlite. J. Hazard. Mater. **2004**, *109*, 141–148.
142. Chakrabarti, S.; Dutta, B.K. Note on the adsorption and diffusion of methylene blue in glass fibers. J. Colloid Interface Sci. **2005**, *286*, 807–811.
143. Kumar, M.N.V.R. A review of chitin and chitosan applications. React. Funct. Polym. **2000**, *46*, 1–27.
144. Guibal E. Interactions of metal ions with chitosan-based sorbents: A review. Sep. Purif. Technol. **2004**, *38*, 43–74.
145. Varma, A.J.; Deshpande, S.V.; Kennedy, J.F. Metal complexation by chitosan and its derivatives: A review. Carbohydr. Polym. **2004**, *55*, 77–93.
146. Gerente, C.; Lee, V.K.C.; Cloirec, P.L.; McKay, G. Application of chitosan for the removal of metals from wastewaters by adsorption—mechanisms and models review. Crit. Rev. Environ. Sci. Technol. **2007**, *37*, 41–127.
147. Crini, G.; Badot, P.-M. Application of chitosan, a natural aminopolysaccharide, for dye removal from aqueous solutions by adsorption processes using batch studies: A review of recent literature. Prog. Polym. Sci. **2008**, *33*, 399–447.
148. Crini, G. Recent developments in polysaccharides-based materials used as adsorbents in wastewater treatment. Prog. Polym. Sci. **2005**, *30*, 38–70.
149. Bhatnagar, A.; Sillanpää, M. Applications of chitin- and chitosan-derivatives for the detoxification of water and wastewater—A short review. Adv. Colloid Interface Sci. **2009**, *152*, 26–38.
150. Jeon, C., Höll, W.H. Chemical modification of chitosan and equilibrium study for mercury ion removal. Water Res. **2003**, *37*, 4770–4780.
151. Rorrer, G.L.; Hsien, T.-Y.; Way, J.D. Synthesis of porous-magnetic chitosan beads for removal of cadmium ions from wastewater. Ind. Eng. Chem. Res. **1993**, *32*, 2170–2178.
152. Kalyani, S.; Priya, J.A.; Rao, P.S.; Krishnaiah, A. Removal of copper and nickel from aqueous solutions using chitosan coated on perlite as biosorbent. Sep. Sci. Technol. **2005**, *40*, 1483–1495.
153. Liu, X.; Hu, Q.; Fang, Z.; Zhang, X.; Zhang, B. Magnetic chitosan nanocomposite: A useful tool for heavy metal ion removal. Langmuir **2009**, *25*, 3–8.
154. Qu, R.; Sun, C.; Ma, F.; Zhang, Y.; Ji, C.; Xu, Q.; Wang, C.; Chen, H. Removal of recovery of Hg(II) from aqueous solution using chitosan-coated cotton fibers. J. Hazard. Mater. **2009**, *167*, 717–727.
155. Zhang, G.; Qu, R.; Sun, C.; Ji, C.; Chen, H.; Wang, C.; Niu, Y. Adsorption for metal ions of chitosan coated cotton fiber. J. Appl. Polym. Sci. **2008**, *110*, 2321–2327.
156. Huang, G.L.; Zhang, H.Y.; Jeffrey, X.S.; Tim, A.G.L. Adsorption of chromium(VI) from aqueous solutions using cross-linked magnetic chitosan beads. Ind. Eng. Chem. Res., **2009**, *48*, 2646–2651.
157. Sun, X.Q.; Peng, B.; Jing, Y.; Chen, J.; Li, D.Q. Chitosan(chitin)/cellulose composite biosorbents prepared using ionic liquid for heavy metal ions adsorption. Separations **2009**, *55*, 2062–2069.
158. Shameem, H.; Abburi, K.; Tushar, K.G.; Dabir, S.V.; Veera, M.B.; Edgar, D.S. Adsorption of chromium(VI) on chitosan-coated perlite. Sep. Sci. Technol. **2003**, *38*, 3775–3793.
159. Veera, M.B.; Krishnaiah, A.; Jonathan, L.T.; Edgar, D.S. Removal of hexavalent chromium from wastewater using a new composite chitosan biosorbent. Environ. Sci. Technol. **2003**, *37*, 4449–4456.
160. Tran, H.V.; Tran, L.D.; Nguyen, T.N. Preparation of chitosan/magnetite composite beads and their application for removal of Pb(II) and Ni(II) from aqueous solution. Mater. Sci. Eng., C **2010**, *30*, 304–310.
161. Wan, M.W.; Kan, C.C.; Buenda, D.R.; Maria, L.P.D. Adsorption of copper(II) and lead(II) ions from aqueous solution on chitosan-coated sand. Carbohydr. Polym. **2010**, *80*, 891–899.
162. Srinivasa, R.P.; Vijaya, Y.; Veera, M.B.; Krishnaiah, A. Adsorptive removal of copper and nickel ions from water using chitosan coated PVC beads. Bioresour. Technol. **2009**, *100*, 194–199.
163. Vijaya, Y.; Srinivasa, R.P.; Veera, M.B.; Krishnaiah, A. Modified chitosan and calcium alginate biopolymer sorbents for removal of nickel (II) through adsorption. Carbohydr. Polym. **2008**, *72*, 261–271.
164. Li, J.M.; Meng, X.G.; Hu, C.W.; Du, J. Adsorption of phenol, *p*-chlorophenol and *p*-nitrophenol onto functional chitosan. Bioresour. Technol. **2009**, *100*, 1168–1173.
165. Nadavala, S.K.; Swayampakula, K.; Boddu, V.M.; Abburi, K. Biosorption of phenol and *o*-chlorophenol from aqueous solutions on to chitosan–calcium alginate blended beads. J. Hazard. Mater. **2009**, *162*, 482–489.
166. Pagnanelli, F.; Mainelli, S.; Veglio, F.; Toro, L. Heavy metal removal by olive pomace: biosorbent characterization and equilibrium modeling. Chem. Eng. Sci. **2003**, *58*, 4709–4717.
167. Demirbas, A. Mechanisms of liquefaction and pyrolysis reactions of biomass. Energy Convers. Manage. **2000**, *41*, 633–646.
168. Theander, O. Cellulose, Hemicellulose, and Extractives. In *Fundamentals of Thermochemical Biomass Conversion*; Overand, R.P., Mile, T.A., Mudge, L.K., Eds.; Elsevier Applied Science Publisher: New York, 1985; 35–60.
169. Morita, M.; Higuchi, M.; Sakata, I. Binding of heavy metal ions by chemically modified Woods. J. Appl. Polym. Sci. **1987**, *34*, 1013–1023.
170. Sawamiappan, N.; Krishnamoorthy, S. Phenol formaldehyde cationic matrices substitutes by bagasse-charcoal. Res. Ind. **1984**, *29*, 293–297.
171. Odozi, T.O.; Okeke, S.; Lartey, L.B. Studies on binding metal ions with polymerized corncob and a composite resin with sawdust and onion. Agric. Waste **1985**, *12*, 13–21.
172. Gardea-Torresdey, J.L.; de la Rosa, G.; Peralta-Videa, J.R. Use of phytofiltration technologies in the removal of heavy metals: a review. Pure Appl. Chem. **2004**, *76*, 801–813.
173. Volesky, B. Detoxification of metal-bearing effluents: Biosorption for the next century. Hydrometallurgy **2001**, *59*, 203–216.
174. Brown, P.A.; Gill, S.A.; Allen, S.J. Metal removal from wastewater using peat. Water Res. **2000**, *34*, 3907–3916.
175. Nobel, P. *Physicochemical and Environmental Plant Physiology*; Academic Press: New York, 1991.
176. Rowell, R.M. In *Removal of Metal Ions from Contaminated Water Using Agricultural Residues*, Proceedings of

ECOWOOD 2006 2nd International Conference on Environmentally Compatible Forest Products, Fernando Pessoa University, Oporto, Portugal, Sept. 20–22, 2006.
177. Mantanis, G.I.; Young, R.A.; Rowell, R.M. Swelling of wood. Part III. Effect of temperature and extractives on rate and maximum swelling. Holzforschung **1995**, *49*, 239–248.
178. Dahiya, S.; Tripathi, R.M.; Hegde, A.G. Biosorption of lead and copper from aqueous solutions by pre-treated crab and arca shell biomass. Biores. Technol. **2008**, *99*, 179–187.
179. Hashem, A.; Abdel-Halim, E.; Maauof, H.A.; Ramadan, M.A.; Abo-Okeil, A. Treatment of sawdust with polyamine for wastewater treatment. Energy Edu. Sci. Technol. **2007**, *19*, 45–58.
180. Demirbas, A. Heavy metal adsorption onto agro-based waste materials: A review. J. Hazard. Mater. **2008**, *157*, 220–229.
181. Min, S.H.; Han, J.S.; Shin, E.W.; Park, J.K. Improvement of cadmium ion removal by base treatment of juniper fiber. Water Res. **2004**, *38*, 1289–1295.
182. Dakiky, M.; Khamis, M.; Manassra, A.; Mer'eb, M. Selective adsorption of chromium(VI) in industrial wastewater using low-cost abundantly available adsorbents. Adv. Environ. Res. **2002**, *6*, 533–540.
183. Bhatnagar, A.; Jain, A.K.; Minocha, A.K.; Singh, S. Removal of lead ions from aqueous solutions by different types of industrial waste materials: equilibrium and kinetic studies. Sep. Sci. Technol. **2006**, *41*, 1881–1892.
184. Taty-Costodes, V.C.; Fauduet, H.; Porte, C.; Delacroix, A. Removal of Cd(II) and Pb(II) ions, from aqueous solutions, by adsorption onto sawdust of *Pinus sylvestris*. J. Hazard. Mater. **2003**, *105*, 121–142.
185. Nassar, M.M. Adsorption of Fe^{+3} and Mn^{+2} from ground water onto maize cobs using batch adsorber and fixed bed column. Sep. Sci. Technol. **2006**, *41*, 943–959.
186. Ho, Y.S.; Huang, C.T.; Huang, H.W. Equilibrium sorption isotherm for metal ions on tree fern. Process Biochem. **2002**, *37*, 1421–1430.
187. Chen, B.; Hui, C.W.; McKay, G. Film-pore diffusion modeling for the sorption of metal ions from aqueous effluents onto peat. Water Res. **2001**, *35*, 3345–3356.
188. Sathishkumar, M.; Vijayaraghavan, K.; Binupriya, A.R.; Stephan, A.M.; Choi, J.G.; Yun, S.E. Porogen effect on characteristics of banana pith carbon and the sorption of dichlorophenols. J. Colloid Interface Sci. **2008**, *320*, 22–29.
189. Achak, M.; Hafidi, A.; Ouazzani, N.; Sayadi, S.; Mandi, L. Low cost biosorbent "banana peel" for the removal of phenolic compounds from olive mill wastewater: Kinetic and equilibrium studies. J. Hazard. Mater. **2009**, *166*, 117–125.
190. Park, K.H.; Balathanigaimani, M.S.; Shim, W.G.; Lee, J.W.; Moon, H. Adsorption characteristics of phenol on novel corn grain-based activated carbons. Microporous Mesoporous Mater. **2010**, *127*, 1–8.
191. Navarro, A.E.; Cuizano, N.A.; Lazo, J.C.; Sun-Kou, M.R.; Llanos, B.P. Comparative study of the removal of phenolic compounds by biological and non-biological adsorbents. J. Hazard. Mater. **2009**, *164*, 1439–1446.
192. Bhatnagar, A.; Minocha, A.K. Adsorptive removal of 2,4-dichlorophenol from water utilizing *Punica granatum* peel waste and stabilization with cement. J. Hazard. Mater. **2009**, *168*, 1111–1117.
193. Jiuhui, Q.U. Research progress of novel adsorption processes in water purification: A review. J. Environ. Sci. **2008**, *20*, 1–13.
194. Oliveira, L.C.A.; Petkowicz, D.I.; Smaniotto, A.; Pergher, S.B.C. Magnetic zeolites: A new adsorbent for removal of metallic contaminants from water. Water Res. **2004**, *38*, 3699–3704.
195. Gu, Z.; Fang, J.; Deng, B.L. Preparation and evaluation of GAC-based iron-containing adsorbents for arsenic removal. Environ. Sci. Technol. **2005**, *39*, 3833–3843.
196. Machado, L.C.R.; Lima, F.W.J.; Paniago, R. Polymer coated vermiculite–iron composites: Novel floatable magnetic adsorbents for water spilled contaminants. Appl. Clay Sci. **2006**, *31*, 207–215.
197. Zhang, G.S.; Qu, J.H.; Liu, H.J.; Liu, R.P.; Wu, R.C. Preparation and evaluation of a novel Fe–Mn binary oxide adsorbent for effective arsenite removal. Water Res. **2007**, *41*, 1921–1928.

Agricultural Runoff

Matt C. Smith
David K. Gattie
Biological and Agricultural Engineering Department, University of Georgia, Athens, Georgia, U.S.A.
Daniel L. Thomas
Louisiana State University, Baton Rouge, Louisiana, U.S.A.

Abstract
Agricultural runoff is surface water leaving farm fields as a result of receiving water in excess of the infiltration rate of the soil. Agricultural runoff is grouped into the category of non-point source pollution because the potential pollutants originate over large, diffuse areas and the exact point of entry into water bodies cannot be precisely identified. Non-point sources of pollution are particularly problematic in that it is difficult to capture and treat the polluted water before it enters a stream. Because of the non-point source nature of agricultural runoff, efforts to minimize or eliminate pollutants are, by necessity, focused on practices to be applied on or near farm fields themselves.

INTRODUCTION

Agricultural runoff is surface water leaving farm fields as a result of receiving water in excess of the infiltration rate of the soil. Excess water is primarily due to precipitation, but it can also be due to irrigation and snowmelt on frozen soils. In the early 20th century, there was considerable concern about erosion of farm fields due to rainfall. The concern was primarily related to the loss of valuable topsoil from the fields and the resulting loss in productivity. With the passage of the Federal Water Pollution Control Act Amendments of 1972, the potential for pollution of surface water features such as rivers and lakes due to agricultural runoff was officially recognized and an assessment of the nature and extent of such pollution was mandated.[1,2]

Agricultural runoff is grouped into the category of non-point source pollution because the potential pollutants originate over large, diffuse areas and the exact point of entry into water bodies cannot be precisely identified (see *Pollution: Point Sources*, p. 2190). Non-point sources of pollution are particularly problematic in that it is difficult to capture and treat the polluted water before it enters a stream. Point sources of pollution such as municipal sewer systems usually enter the water body via pipes and it is comparatively easy to collect that water and run it through a treatment system prior to releasing it into the environment. Because of the non-point source nature of agricultural runoff, efforts to minimize or eliminate pollutants are, by necessity, focused on practices to be applied on or near farm fields themselves. In other words, we usually seek to prevent the pollution rather than treating the polluted water.

Due to the great successes made in treating polluted water from point sources such as municipal and industrial wastewater treatment plants, the relative significance of pollution from agricultural runoff has increased. Agricultural runoff is now considered to be the primary source of pollutants to the streams and lakes in the United States. It is also the third leading source of pollution in U.S. estuaries.[3] The water pollutants that occur in agricultural runoff include eroded soil particles (sediments), nutrients, pesticides, salts, viruses, bacteria, and organic matter.

AGRICULTURAL RUNOFF QUANTITY

Agricultural runoff occurs when the precipitation rate exceeds the infiltration rate of the soil. Small soil particles that have been dislodged by the impact of raindrops can fill and block soil pores with a resulting decrease in infiltration rate throughout the duration of the storm. As the excess precipitation builds up on the soil surface it flows in thin layers from higher areas of the field towards lower areas. This diffuse surface runoff quickly starts to concentrate in small channels called rills. The concentrated flow will generally have a higher velocity than the flow in thin films over the surface. The concentrated flow velocity may become rapid enough to cause scouring of the soil that makes up the channel sides and bottom. The dislodged soil particles can then be carried by the flowing water to distant locations in the same field or be carried all the way to a receiving water body. If the quantity of flow and the velocity of flow are large enough, the rills

Encyclopedia of Environmental Management DOI: 10.1081/E-EEM-120010101

can grow so large that they cannot be easily repaired by typical earth moving machinery. When this happens, the rill has become a gulley.

The quantity of runoff from agricultural fields is not usually listed explicitly as a concern separate from the quality of the runoff. However, it should be considered because it transports the pollutants and can cause erosion of receiving streams due to excessive flows. If less runoff is allowed to leave a field, there is less flow available to transport pollutants to the stream. Also, if more water is retained on the field, there is likely to be a corresponding reduction in the amount of supplemental water that will need to be added through irrigation. Runoff quantity varies significantly due to factors such as soil type, presence of vegetation and plant residue, physical soil structures such as contoured rows and terraces, field topography, and the timing and intensity of the rainfall event.

Some agricultural practices increase the infiltration capacity of the soil while other practices can result in decreases. The presence of vegetation and plant residues on a field reduce runoff due to improving and maintaining soil infiltration capacity. Actively growing plants also reduce the amount of water in the soil due to evapotranspiration, thus making more room for infiltrating water to be stored in the soil profile. Bare soils increase runoff because there is nothing except the soil surface to absorb the energy of the falling raindrops. The rain, therefore, dislodges soil particles that will tend to seal the surface and reduce infiltration.

SOIL EROSION AND ASSOCIATED POLLUTANTS

One of the primary pollutants in agricultural runoff is eroded soil. In 1975, 223 million acres of cropland produced 3700 million tons of eroded sediments or an average of 17 tons of soil lost per acre of cropland per year (see various *Erosion* entries, pp. 967–1103). It is estimated that cropland, pasture, and rangeland contributed over 50% of the sediments discharged to surface waters in 1977.[4] As noted above, the energy of raindrops can dislodge and transport soil particles. In the aquatic environment the eroded soil is called sediment. There are several concerns related to excessive sediments in aquatic systems. These include loss of field productivity, habitat destruction, reduced capacity in reservoirs, and increased dredging requirements in shipping channels.

Eroded sediments represent a loss of fertile topsoil from the field, which can reduce the productivity of the field itself. Soil formation is an extremely slow process occurring over periods ranging from decades to centuries.[5] Possible results to a grower from excessive erosion of their fields include increasing fertilizer and water requirements, planting more tolerant crops, and possibly abandoning the field for agricultural production.

A second concern is that many of these sediments are heavy and will settle out in slow moving portions of streams or in reservoirs. The settled sediments can dramatically alter the ecology of the streambed. Aquatic plants, insects, and fish all have specific requirements related to composition of the streambed for them to live and reproduce.[6] Sediments in reservoirs reduce the volume of the reservoir available to store water. This may result in reduced production of hydroelectric power, reduced water availability for municipal supply, interference with navigation and recreation, and increased dredging requirements to maintain harbor navigability.

Another concern with eroded sediments is that they can transport other pollutants into receiving waters. The plant nutrient phosphorus, for example, is most often transported from the fields where it was applied as fertilizer by chemically bonding to clay minerals. Many agricultural pesticides also bond to eroded clays and organic matter. Once these chemicals have entered the aquatic ecosystem, many processes occur that can result in the release of the pollutants from their sediment carriers. Phosphorus, when released, can contribute to the eutrophication of lakes and reservoirs (see the entry *Eutrophication*, p. 1115). Pesticides and their degradation products can be toxic to aquatic life and must be removed from municipal water supplies.

Erosion from animal agriculture such as feedlots and pastures can also result in the transport of sediments composed of animal manures (see the various *Manure Management* entries, pp. 1680–1695). These sediments can transport significant quantities of potential pathogens (viruses and bacteria). The animal manures are primarily organic in nature and can serve as a food source for natural bacteria in the receiving water. When these naturally occurring bacteria begin to utilize the organic matter in this way they may lower or deplete the water of dissolved oxygen as they respire and multiply. This use of oxygen by aquatic bacteria is known as biochemical oxygen demand (BOD). High levels of BOD can reduce stream oxygen level to the point that fish and other organisms that require dissolved oxygen suffer, die, or relocate, when possible, to more suitable habitats.[6]

DISSOLVED POLLUTANTS

Agricultural runoff can carry with it many pollutants that are dissolved in the runoff water itself. These may include plant nutrients, pesticides, and salts. Since these pollutants are dissolved in the runoff, control measures are most often aimed at reducing the volume of runoff leaving an agricultural field, or making the pollutants less available to be dissolved into the runoff water.

One of the major pollutants of concern in agricultural runoff is the plant nutrient nitrogen. Nitrogen is a relatively cheap component of most fertilizers and is necessary for plant growth. Unfortunately, nitrogen in the form of nitrate is highly soluble in water. Thus nitrate can be easily dissolved in runoff water. Just as it does in an agricultural field, nitrogen can promote growth of aquatic vegetation. Excess nitrogen and phosphorus in runoff can lead to the eutrophication of lakes, reservoirs, and estuaries (see the entry *Eutrophication*, p. 1115). Nitrogen in the form of ammonia can be dissolved into runoff from pastures and feedlots. Ammonia is toxic to many aquatic organisms, thus it is important to minimize ammonia in runoff.[7]

Many agriculturally applied pesticides are also soluble in water. They can be dissolved in runoff and transported into aquatic ecosystems where there is a potential for toxic effects. These pesticides must also be removed from drinking water supplies and, if concentrations are high or persistent, such treatment can be difficult and expensive. Stable, persistent pesticides can bioaccumulate in the food chain with the result that consumers of fish from contaminated waters might be exposed to higher concentrations than exist in the water itself.[8]

Runoff from agricultural fields can contain significant concentrations of dissolved salts. These salts originate in precipitation, irrigation water, fertilizers and other agricultural chemicals, and from the soil minerals. Plants generally exclude ions of chemicals that they do not need. In this way, dissolved salts in irrigation water, for example, can be concentrated in the root zone of the growing crop. Runoff can redissolve these salts and transport them into aquatic ecosystems where some, naturally occurring selenium for example, can be toxic to fish and other wildlife.[9]

Transport of fertilizers and pesticides from their point of application can result in significant environmental costs. This transport, or loss from the field, can also have significant negative economic impacts on the grower. Fertilizers lost from the field are not available to promote crop growth. Agricultural chemicals lost from the field, likewise, are not available to protect the plants from pests and diseases. In both cases the grower is paying for expensive inputs and paying to apply them. It is always in the growers' and the environment's best interests, therefore, to keep agricultural chemicals in the field where they are needed and where they were applied.

CONTROL OF AGRICULTURAL RUNOFF

One of the most direct methods of controlling pollution by agricultural runoff is to minimize the potential for runoff to occur. Other methods can be employed to reduce the amounts of sediments and dissolved chemicals in runoff. As a whole, management practices designed to minimize the potential for environmental damage from agricultural runoff are called best management practices (BMPs), (see the entry *Nutrients: Best Management Practices*, p. 1829). Many times, practices aimed at controlling one aspect of agricultural runoff are also effective at reducing other components. This is due to the interrelationships between runoff volume, erosion, transport, dissolution, and delivery.

Maintaining good soil tilth and healthy vegetation can minimize runoff. This will promote increased infiltration and a resultant decrease in runoff. Other management practices such as terracing, contour plowing, and using vegetated waterways to convey runoff can result in decreased quantities of runoff by slowing the water leaving the field and allowing more time for infiltration to occur. Construction of farm ponds to receive runoff can result in less total runoff from the farm, lowered peak rates of runoff, and storage of runoff for use in irrigation or livestock watering.[2]

Control of water pollution by the mineral and organic sediments and associated chemicals in agricultural runoff is most effectively achieved by reducing erosion from the field. The primary method of reducing erosion is by maintaining a vegetative or plant residue cover on the field at all times or minimizing areas of the field that are bare. Techniques utilized to accomplish these tasks include conservation tillage, strip tillage, and the use of cover crops (see the entry *Erosion Control: Tillage and Residue Methods*, p. 1081). Additional measures that can be employed at the edge of the field, or off-site include vegetative filter strips and farm ponds.

Methods to control the loss of nitrogen and other plant nutrients from cropland include applying nitrogen in the quantity required by the crop and at the time the crop needs it (see the entry *Nutrients: Best Management Practices*, p. 1829). This requires multiple applications and can be difficult for tall crops. For this reason, most, or all, of the nitrogen required by the crop is often applied at planting. Nitrogen fertilizers have often been applied based on general recommendations for the type of crop to be grown. Since nitrogen fertilizers are relatively inexpensive, growers have tended to over apply rather than under apply. Soil tests can tell a grower how much nitrogen is already in the soil and how much needs to be applied for a specific crop. Efforts have been made to make the nitrogen less soluble by changing the form of nitrogen applied to the field so that it becomes available to the plants (and, thus available for loss in runoff) more slowly.[10]

One method of controlling the loss of agricultural chemicals is to minimize their solubility in water. Another is to minimize their use through programs such as integrated pest management (IPM) where some crop damage is allowed until it reaches a point that it becomes economically justified to apply pesticides.[11] And a third approach is to make the chemicals more easily degraded so that they do their job and then degrade into other, less harmful, chemicals so that they do not stay around long enough to be influenced by runoff-producing rainfall events.

CONCLUSION

Agricultural runoff is one of the leading causes of water quality impairment in streams, lakes, and estuaries in the United

States. It can transport large quantities of sediments, plant nutrients, agricultural chemicals, and natural occurring minerals from farm fields into receiving water bodies. In many cases the loss of these substances from the field represent an economic loss to the grower as well as a potential environmental contaminants. There are many methods by which the quantity of agricultural runoff can be reduced. Many of these methods are referred to generically as BMPs. Adoption of BMPs can also improve the quality (reduce contaminant concentrations) of the runoff that does leave the farm. By reducing the quantity and improving the quality of agricultural runoff, it will be possible to improve the water quality in our streams, river, lakes, and estuaries.

REFERENCES

1. U.S. Environmental Protection Agency. *EPA Releases Guidelines for New Water Quality Standards*; 2002; http://www.epa.gov/history/topics/fwpca/02.htm (accessed July 2002).
2. Stewart, B.A.; Woolhiser, D.A.; Wischmeier, W.H.; Caro, J.H.; Frere, M.H. *Control of Water Pollution from Cropland, Volume II—An Overview*, EPA-600/ 2-75-026b; U.S. Environmental Protection Agency: Washington, DC, 1976.
3. U.S. Environmental Protection Agency. *Nonpoint Source Pollution: The Nation's Largest Water Quality Problem*; 2002; http://www.epa.gov/OWOW/NPS/ facts/point1.htm (accessed July 2002).
4. Leeden, Van der. *The Water Encyclopedia*; Lewis Publishers: Chelsea, MI, 1990.
5. Foth, H.D. *Fundamentals of Soil Science*, 8th Ed.; John Wiley and Sons, Inc.: New York, 1990.
6. Gordon, N.D.; McMahon, T.A.; Finlayson, B.L. *Stream Ecology: An Introduction for Ecologists*; John Wiley and Sons Inc.: New York, 1992.
7. Abel, P.D. *Water Pollution Biology*, 2nd Ed.; Taylor and Francis, Inc.: Bristol, 1996.
8. U.S. Environmental Protection Agency. *The Persistent Bioaccumulators Project*; 2002; http://www.epa.gov/ chemrtk/persbioa.htm (accessed July 15, 2002).
9. U.S. Geological Survey. *Public Health and Safety: Element Maps of Soils*; http://minerals.cr.usgs.gov/ gips/na/0elemap.htm#elemap (accessed July 15, 2002).
10. Owens, L.B. Impacts of soil N management on the quality of surface and subsurface water. In *Soil Process and Water Quality*; Lal, R., Stewart, B.A., Eds.; Lewis Publishers, Inc.: Boca Raton, FL, 1994.
11. U.S. Department of Agriculture. *National Integrated Pest Management Network*; 2002; http://www.reeusda.gov/agsys/nipmn/ (accessed July 15, 2002).

Agricultural Soils: Ammonia Volatilization

Paul G. Saffigna
Tropical Forestry, University of Queensland, Gatton, Queensland, Australia

John R. Freney
Commonwealth Scientific and Industrial Research Organization (CSIRO), Campbell Australian Capital Territory, Australia

Abstract
The exchange of ammonia between soils, plants, waters, and the atmosphere is an important part of the terrestrial nitrogen cycle. The process by which ammonia is lost from the Earth's surface to the atmosphere is termed volatilization. The primary source of ammonia for loss is the natural microbial decomposition of amino acids and proteins in dead plants, animals and microorganisms in soils and waters, but substantial amounts come from the excreta of animals and the use of nitrogen fertilizers. Volatilization is a complex process affected by a combination of biological, chemical, and physical factors and the loss process may be hindered.

INTRODUCTION

The exchange of ammonia between soils, plants, waters, and the atmosphere is an important part of the terrestrial nitrogen cycle. The process by which ammonia is lost from the Earth's surface to the atmosphere is termed volatilization. The primary source of ammonia for loss is the natural microbial decomposition of amino acids and proteins in dead plants, animals and microorganisms in soils and waters, but substantial amounts come from the excreta of animals and the use of nitrogen fertilizers.[1,2] Ammonia (NH_3) has a strong affinity for water and it readily dissolves in it to form ammonium (NH_4^+) hydroxide, viz.

ammonia (gas) ⇔ ammonia (dissolved) + water
⇔ ammonium hydroxide
⇔ ammonium ions + hydroxyl ions

The ammonia and ammonium ions are in equilibrium and the reaction may be displaced to the left or right depending on the conditions. For example, if the pH of the system is increased by addition of alkali (hydroxyl ions), the reaction is displaced to the left and ammonia gas is formed and lost to the atmosphere. As ammonia is a gas at normal temperatures and pressures, and as the concentration in the atmosphere is usually low, it can be readily lost to the atmosphere. However, volatilization is a complex process affected by a combination of biological, chemical, and physical factors and the loss process may be hindered.[1,2]

THE MECHANISM OF AMMONIA VOLATILIZATION

Before volatilization can occur there must be a source of ammonia. This can be in the form of native organic matter, which decomposes to release ammonia, or fertilizer such as anhydrous ammonia, ammonium salts, or urea. Urea, either from animal urine or fertilizer, is rapidly hydrolyzed to ammonium carbonate in soil. This urease catalyzed reaction results in localized areas of high pH. Apart from the application of anhydrous ammonia, these sources tend to add ammonium ions rather than ammonia to the soil. Therefore, the conversion of ammonium ions to ammonia controls the loss of ammonia.[1,2]

The relative concentrations of ammonium and ammonia in solution are strongly affected by pH (acidity or alkalinity) and temperature. For example, the percentage of ammonia present at pH 6, 7, 8, and 9 is approximately 0.1, 1, 10, and 50.[7] Thus the higher the pH, the greater is the potential for ammonia loss from soil. The loss of ammonia from an application of ammonium sulfate increased from nil at pH 7, to 87% at pH 10.5.

The main driving force for ammonia volatilization is the difference in concentration between ammonia gas in the soil and ammonia in the atmosphere. Increasing windspeed increases the rate of volatilization by promoting more rapid transport of ammonia away from the soil surface. A four-fold increase in wind speed resulted in a ten-fold increase in ammonia loss. Ammonia volatilization and water loss from soils are directly related; no ammonia is emitted until evaporation commences.

Any factor which affects the ammonium ion concentration in soil will also affect the ammonia gas concentration and the loss process. Consequently, plant uptake, immobilization by microorganisms, nitrification, and leaching will reduce the amount of nitrogen available for volatilization, whereas increasing the rate of application of ammonium or ammonium producing fertilizers or organic residue will increase the potential for volatilization.[1,2]

Other factors which control ammonia volatilization are cation exchange capacity, buffer capacity, presence of calcium carbonate, water content of soil, soil texture, plant

Encyclopedia of Environmental Management DOI: 10.1081/E-EEM-120001781

residues, fertilizer form, radiation, and atmospheric ammonia concentration.[7] A number of models incorporating these factors have been developed to describe the volatilization of ammonia.[3]

MEASUREMENT

A number of workers have used canopies over field crops and pastures, combined with acid traps, to measure ammonia loss, but the canopies affect the temperature, moisture and wind speed in the immediate environment of the plant, and thus the result obtained may not reflect ammonia loss from the natural environment. Simplified micrometeorological techniques have been developed to measure ammonia volatilization in the unconfined field situation, thus allowing the assessment of the importance of ammonia loss, and the factors controlling loss in different agricultural systems with minimum labor, equipment and skills.[1,3]

AMMONIA EMISSIONS

Animals and Their Wastes

Waste from farm animals is the principal source of atmospheric ammonia. Ammonia concentrations in the air range from as low as $1\,\mu g\,N\,m^{-3}$ over oceans to 5 in rural areas, 15 in urban areas, 50 in areas of intense animal husbandry, and 1000 over a field shortly after spreading animal waste as a slurry. The importance of animals as a source of ammonia is well illustrated by the situation in Europe, where nearly three quarters of the total emissions are from animals and their wastes (stables 34%, surface spreading 32%, grazing 8%), and only one quarter comes from a combination of fertilizer application (12%), industry (0.5%), crops (5%) and miscellaneous(e.g., treatment of waste water and sludge, pets, humans, refrigeration, 8%).[1]

The large loss of ammonia from livestock systems is due to the low conversion of dietary nitrogen into animal protein. More than 75% of the nitrogen intake is excreted in forms that give rise to ammonia emissions. Nitrogen is excreted mostly in urine with some present in droppings as microbial cell constituents and undigested food. Urine contains 70%–90% urea which is rapidly hydrolyzed to ammonia by naturally occurring urease. The amount excreted depends on the feed composition—the better the quality of the food the less nitrogen excreted. Factors influencing ammonia emissions include manure properties, weather, soil attributes, and application measures (incorporation, dilution, soil preparation). Where animals are grazing pastures, about 10% of the excreted nitrogen is lost as ammonia—mostly from urine.[1]

Cropping Systems

There has been a widespread move to urea as the major form of nitrogen fertilizer used in cropping systems because of its relatively low manufacturing cost and low transportation cost per unit of nitrogen. However, large losses of ammonia have been detected from rainfed and irrigated crops, and flooded rice in many countries following applications of urea (Table 1). Ammonia loss from cropping systems is affected by many of the factors discussed above, fertilizer composition, rate, time and method of application, and factors unique to the crop.

In flooded rice, up to 56% of the applied nitrogen is lost from the system by ammonia volatilization as a result of the growth of photosynthetic algae, which markedly increases the pH of the floodwater during daytime. The new practice of retaining tops and leaves of cut sugar cane plants on the soil surface following green cane harvesting has created problems for farmers when they apply urea. So as not to disturb the residue cover, which has many advantages, including weed control, many farmers apply urea by broadcasting onto the surface of the residue. The sugar cane residue has high urease activity and low ammonia retention capacity, so when the residue layer is moistened by dewfall, rainfall or condensation of evaporated soil water, some of the urea dissolves, is hydrolyzed, and when the water evaporates, between 30% and 40% of the applied nitrogen is lost as ammonia. Bananas have a very high requirement for nitrogen, frequent applications are made and more than $500\,kg\,N\,ha^{-1}\,yr^{-1}$ is applied. Direct measurements of ammonia volatilization from a banana crop in tropical Australia showed that, when urea was applied onto wet soil, 20% of the applied N was lost even though 90 mm of rain fell during the study. The extensive canopy of banana plants restricted rainwater from falling on the fertilized area and washing the urea into the soil.

Plants can absorb ammonia from the air or release it to the atmosphere. It has been established that plants have an ammonia compensation point, which is a finite ammonia concentration in the intercellular air spaces of plant leaves. Plants absorb or lose ammonia depending on whether the ambient ammonia concentration is above or below the compensation

Table 1 Ammonia volatilized (% of N applied) from different cropping systems fertilized with urea.

Plant	Treatment	Loss
Bananas	Surface applied	20
Corn	Surface applied	22
Rice	Broadcast into floodwater	10–56
	Broadcast into floodwater and incorporated	10–43
	Incorporated before flooding	5–16
	Broadcast 12 d after transplanting	21
	Broadcast at panicle initiation	3
Sugar Cane	Broadcast on trash	30–40
Wheat	Surface applied	36
	Buried	7

Source: Adapted from Peoples et al.[7]

point. Ammonia losses of about 1 kg N ha^{-1} yr^{-1} have been determined from fields of barley, maize, and wheat.[1]

Biomass Burning

During combustion, considerable plant nitrogen is converted into gaseous forms including ammonia. Most biomass burning (~90%) occurs in the tropics as a result of forest clearing, savanna and sugar cane fires, and burning of agricultural wastes and firewood. According to IPCC,[4] 8700 Mt of biomass is burned every year, and agriculture accounts for half of this. About 4% of the biomass nitrogen is released as ammonia during combustion, with the result that biomass burning contributes between 4.5 and 5.9 Mt N per yr^{-1} globally to the atmosphere.[5,6]

GLOBAL SIGNIFICANCE

The global emission of ammonia was estimated to be 54 Mt N yr^{-1} in 1990.[6] The contributions from the major sources were given as: 1) excreta from animals, 21.7 Mt; 2) synthetic fertilizers, 9.0 Mt; 3) oceans, 8.2 Mt; 4) biomass burning, 5.9 Mt; 5) crops, 3.6 Mt; 6) human population and pets, 2.6 Mt; 7) soils under natural vegetation, 2.4 Mt; 8) industrial processes, 0.2 Mt; and 9) fossil fuels, 0.1 Mt. About half of the global emission originates in Asia, and approximately 70% is associated with food production.

When ammonia is emitted into the atmosphere, some is absorbed by vegetation, some is dissolved in atmospheric water, converted to aerosols and transported long distances (>1000 km), and some is deposited nearby.[3,8] Model estimates indicate that about 50% of the emitted ammonia is deposited within 50 km of the source.[8] High ammonia concentrations close to point sources such as cattle feedlots can damage vegetation, and deposition of ammonia and ammonium can result in acidification of soils and lakes, increased carbon storage in pristine areas, and increased emission of the greenhouse gas nitrous oxide;[9] measures need to be instituted to reduce losses.

MITIGATION

Techniques proposed for reducing loss of ammonia from animal slurries applied to soils include incorporation or injection of the slurry into the soil, application with trail hoses, acidification before application, applying during rainfall, at night, or in winter, and matching nitrogen supply to the demand of the crop. Decreasing the water content of the slurry, and delaying application until a substantial canopy has developed (to reduce wind speeds) would also appear to have a large impact on ammonia loss.[10] A logical option for limiting ammonia volatilization from cropping systems is to drill the fertilizer into the soil. Other recommendations include spreading the fertilizer just prior to rain, application in irrigation water, optimizing split application schemes, changing the fertilizer type to suit the conditions, and better matching of nitrogen supply with crop demand. Controlled release fertilizers provide an opportunity to match supply and demand while protecting the remainder of the fertilizer from release. Adoption of some or all of these practices will allow the farmer to reduce inputs and reduce the impact on the environment.[4]

REFERENCES

1. ECETOC (European Centre for Ecotoxicology and Toxicology of Chemicals). *Ammonia Emissions to Air in Western Europe*, Technical Report No. 62; ECETOC: Brussels, Belgium, 1994.
2. Nelson, D.W. Gaseous losses of nitrogen other than through denitrification. In *Nitrogen in Agricultural Soils*; Stevenson, F.J., Ed.; American Society of Agronomy: Madison, WI, 1982; 327–363.
3. Denmead, O.T. Progress and challenges in measuring and modelling gaseous nitrogen emissions from grasslands: an overview. In *Gaseous Nitrogen Emissions from Grasslands*; Jarvis, S.C., Pain, B.F., Eds.; CABI: Wallingford, U.K., 1997; 423–438.
4. Watson, R.T., Zinyowera, M.C., Moss, R.H., Eds.; IPCC (Intergovernmental Panel on Climate Change). In *Climate Change 1995. Impacts, Adaptations and Mitigation of Climate Change: Scientific–Technical Analyses*; Cambridge University Press: Cambridge, 1996; 1–878.
5. Schlesinger, W.H.; Hartley, A.E. A global budget for atmospheric NH_3. Biogeochemistry **1992**, *15*, 191–211.
6. Bouwman, A.F.; Lee, D.S.; Asman, W.A.H.; Dentener, F.J.; Van der Hoek, K.W.; Olivier, J.G.J. A global high-resolution emission inventory for ammonia. Global Biogeochem. Cycles **1997**, *11*, 561–587.
7. Peoples, M.B.; Freney, J.R.; Mosier, A.R. Minimizing gaseous losses of nitrogen. In *Nitrogen Fertilization in the Environment*; Bacon, P.E., Ed.; Marcel Dekker: New York, 1995; 565–602.
8. Ferm, M. Atmospheric ammonia and ammonium transport in europe and critical loads—a review. Nutr. Cycl. Agroecosyst. **1998**, *51*, 5–17.
9. Smil, V. Nitrogen in crop production: an account of global flows. Global Biogeochem. Cycles **1999**, *13*, 647–662.
10. Jarvis, S.C.; Pain, B.F. *Gaseous Nitrogen Emissions from Grasslands*; CABI: Wallingford, U.K., 1997.

Agricultural Soils: Carbon and Nitrogen Biological Cycling

Alan J. Franzluebbers
Agricultural Research Service (USDA-ARS), U.S. Department of Agriculture, Watkinsville, Georgia, U.S.A.

Abstract
Carbon and nitrogen are two key elements of global significance, playing large roles in the production of food, feed, fiber, and fuel for human existence, as well as providing numerous other ecosystem services. Although nitrogen is often a limiting element in natural systems, it can become a polluting element as a by-product of agricultural management due to the many pathways for it to be lost from the point of application (high input of nitrogen fertilizers has occurred during the last half century to achieve high production goals). Nitrogen loss from leaching, runoff, volatilization, and denitrification can pollute water and air resources. The conundrum of agriculture is to get enough of nitrogen without releasing it to the environment. Continuous biological cycling of carbon and nitrogen with conservation or ecologically based agricultural approaches can significantly reduce the environmental pollution with nitrogen. Limiting the time that inorganic nitrogen is present in soil will limit its loss, and this can be achieved by limiting soil disturbance, maintaining continuous plant and residue cover on soil, and creating a diversified cropping system to balance ecological stability.

INTRODUCTION

Carbon (C) and nitrogen (N) are two of the most important elements that affect soil productivity and environmental quality.[1] Carbon is found throughout nature in a wide variety of forms and particularly in soil as 1) complex organic compounds (e.g., carbohydrates—$C_xH_{2x}O_x$, lignin, etc.,) derived from living organisms; 2) carbonate minerals such as calcite ($CaCO_3$) and dolomite [$CaMg(CO_3)_2$]; and 3) carbon dioxide (CO_2) and methane (CH_4) as decomposition end products. Nitrogen is an essential element of plants, animals, and microorganisms—a part of chlorophyll, enzymes, amino acids, and proteins, which are necessary for growth and development of organisms. In typical unpolluted soil, quantity of N in organic matter and fixed as ammonium (NH_4^+) in clay minerals far exceeds quantities in plant-available forms of soluble nitrate (NO_3^-) and NH_4^+. Among several soils in North America, total N in 1 m depth of soil was 16.0 ± 6.9 Mg N ha^{-1} with 13% ± 15% fixed as NH_4^+ in clay minerals and <1% as soluble NO_3^-.[2]

Agriculture, i.e., the growing of plants and animals for human and livestock consumption, is a widespread land usage throughout the world. Globally, agriculture occupies approximately 38% of the total land area with 1.5 billion ha in cropland and 3.4 billion ha in perennial grassland.[3] Addition of N, phosphorus (P), potassium (K), and other nutrients to soil is often needed to satisfy the demands by high-production crops and forages. A portion of these nutrients is naturally supplied through plant residue and soil organic matter decomposition, but amendment with inorganic or organic fertilizers is often needed to achieve high production. Unfortunately, there are many pathways for nutrients to escape from the agricultural landscape into nearby streams, lakes, groundwater, and the atmosphere. Preventing these losses is one of the goals of sustainable, ecologically based approaches to agricultural production.

CARBON AND NITROGEN CYCLES

Both C and N are biologically fixed from inorganic atmospheric forms to organically bound plant and microbial forms. Photosynthesis converts inorganic CO_2 from the atmosphere into organic carbohydrates in plants, algae, and cyanobacteria. Biological N fixation is a unique transformation carried out by a number of bacteria, which convert N_2 gas into ammonia (NH_3) for biological utilization. N-fixing bacteria are most prevalent in symbiotic relationships with plants, such as *Rhizobium* that forms nodules on the roots of clovers where the nitrogenase enzyme catalyzes the reaction. Fertilizer manufacturing converts N_2 gas into NH_3 in a similar manner without an enzyme, but rather large quantities of energy necessary to create the pressure required for the transformation.

Under certain conditions, both inorganic C and N can be chemically fixed in the subsoil. Carbon dioxide forms carbonic acid in water, which can precipitate with the basic cations, Ca^{2+}, Mg^{2+}, and Na^+, to form pedogenic carbonates. Inorganic C is most abundant in soils of the semiarid

Encyclopedia of Environmental Management DOI: 10.1081/E-EEM-120046347

and arid regions. Ammonium can be fixed as nonexchangeable components of the lattice structure of 2:1-type clay minerals, which are especially prevalent in the subsoil of many younger soils.

Carbon and N occur in various forms and undergo transformations from one form to another, primarily through biochemical manipulations involving enzymes.[4,5] Enzymes are proteins, functioning to catalyze very specific reactions either 1) intracellularly within plants, microorganisms, or soil animals; or 2) extracellularly in soil solution or attached to soil colloids. Some major enzyme categories and their reactions with C and N substrates in soil are 1) hydrolases, such as amylase and cellulase, which hydrolyze various carbohydrate and macromolecular compounds; 2) oxyreductases, which catalyze various electron transfer reactions; 3) proteinases, which convert proteins to amino acids; 4) lignocellulases, which catalyze the ecologically resistant step of lignin breakdown; and 5) lyases, which form double bonds through reactions other than hydrolysis or oxidation. Two key enzymes involved in the fixation of C and N into organic forms are 1) ribulose bisphosphate carboxylase (rubisco), which is the photosynthetic enzyme catalyzing the transformation of CO_2 from the atmosphere into carbohydrates; and 2) nitrogenase, which catalyzes the biological N fixation reaction in symbiotic bacteria associated with leguminous plant roots:

$$N_2 + 6\,H + \text{energy} \rightarrow 2\,NH_3$$

Forms and fluxes of an element are commonly illustrated in a cycle following the principles of conservation of mass (i.e., elements are transferred from one molecule to another) (Fig. 1). Carbon and N cycles have global dimensions with terrestrial, aquatic, and atmospheric components of major significance.[6,7] The sun initiates a chain of energy reactions, which drive elemental cycles. The elemental cycles of C and N interact closely with the water cycle, as water is a fundamental internal component of life and a major transport mechanism of nutrients.

In natural systems without significant import of N from fertilizers, the cycling of N is largely dependent upon the cycling of C. Since growth of plants is often limited in N supply due to the strong competition for N by soil microorganisms, which have a steady supply of C-rich substrates at the surface of undisturbed soil, N losses from natural systems are typically low.[8] The need for additional N in agricultural systems can be historically derived from two major pathways: (1) high protein harvest of grain, forage, and animal products that requires supplemental N to replace the already limited N supply in natural systems (and eventual lack of recycling waste and manure by-products from harvested food products back to the land); and (2) loss of soil-surface residue cover and soil organic matter with intensive tillage that initially stimulates N release to crops, but that eventually exhausts the soil resource in its ability to supply N to crops. Loss of C-rich surface residue and soil organic matter essentially removes the C stimulus needed to conserve N in soil, thereby resulting in major losses of N from agricultural systems with time and creating a system that relies heavily on external N inputs to supply crops with not only the N removed from harvested crops but also the N lost via leaching, runoff, volatiliza-

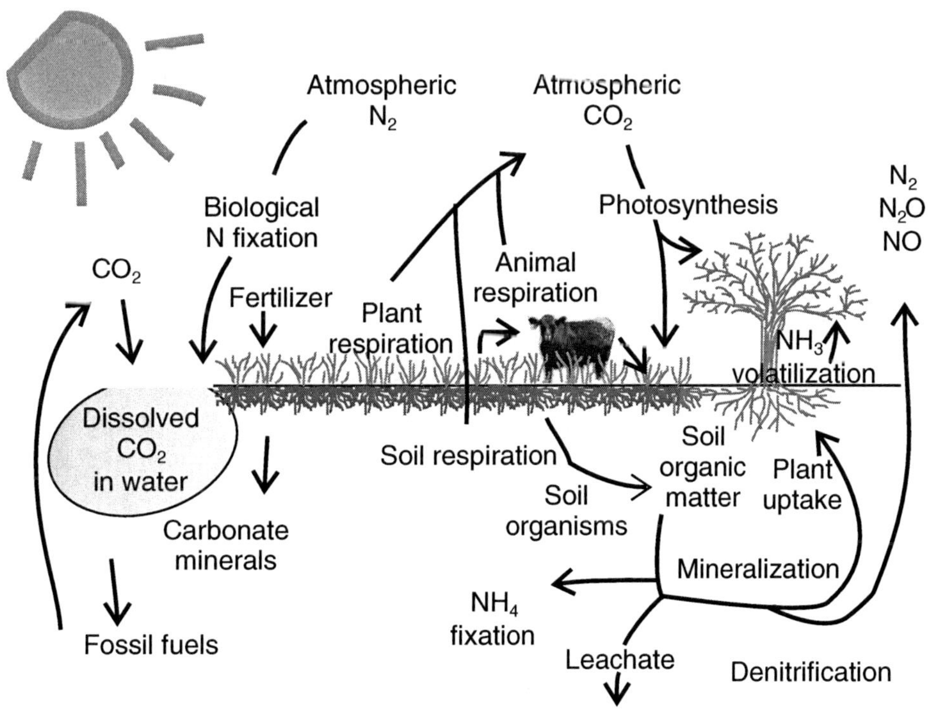

Atmospheric N2
Atmospheric CO2
Biological N fixation
Photosynthesis
N2 N2O NO
CO2
Fertilizer
Plant respiration
Animal respiration
NH3 volatilization
Dissolved CO2 in water
Soil respiration
Soil organic matter
Plant uptake
Soil organisms
Carbonate minerals
Mineralization
NH4 fixation
Fossil fuels
Leachate
Denitrification

Fig. 1 Generalized diagram of the C and N cycles in soil.

tion, and denitrification. Estimated global production of N and P fertilizers was 100 and 41 Tg (10^{12} g), respectively, in 2007,[3] which compared to about 10 Tg of production for each nutrient in 1960.[9]

Autotrophic fixation of atmospheric CO_2 by plants captures the energy of the sun within organic compounds via the process of photosynthesis (Fig. 1):

$$6\ CO_2 + 6\ H_2O + \text{energy} \rightarrow C_6H_{12}O_6 + 6\ O_2$$

Inorganic N is taken up by plant roots and synthesized into amino acids and proteins during plant development. Plants are eventually consumed by animals or microorganisms, transferring portions of this stored energy through biochemical processes into various cellular components. Once in soil, the C cycle is dominated by the heterotrophic process of decomposition, i.e., the breakdown of complex organic compounds into simple organic constituents. Mineralization is the complete decomposition of organic compounds into mineral constituents:

$$C_6H_{12}O_6 + 6O_2 \rightarrow 6\ CO_2 + 6H_2O + \text{energy}$$

$$R\text{-}NH_2 + H_2O \rightarrow R\text{-}OH + NH_3$$

Immobilization of N occurs simultaneously with N mineralization when soil organisms require additional inorganic N to meet the high demand for new body tissue while decomposing C-rich substrates low in available N. Net N mineralization occurs when gross N mineralization exceeds that of N immobilization.

ENVIRONMENTAL INFLUENCES ON SOIL MICROBIAL ACTIVITY

Organisms predominantly responsible for decomposition of organic matter and associated mineralization of C and N are soil microorganisms, composed of bacteria, actinomycetes, fungi, and protozoa.[10,11] Soil fauna are larger soil organisms, such as beetles and earthworms (macrofauna, >2 mm width × >10 mm length), collembolan and mites (mesofauna, 0.1–2 mm width × 0.2–10 mm length), and protozoa and nematodes (microfauna, <0.1 mm width × <0.2 mm length), that also indirectly affect C and N cycling by 1) comminuting plant residues and exposing a greater surface area to soil microorganisms; 2) transporting plant and animal residues to new locations in the soil to facilitate decomposition, interaction with soil nutrients, or isolation from environmental conditions; 3) inoculating partially digested organic substrates with specific bacteria and enzymes; and 4) altering physical characteristics of soil by creating burrows, fecal pellets, and distribution of soil particles that influence water, air, nutrient, and energy retention and transport. With suitable environmental conditions, soil microorganisms grow rapidly in response to the availability of organic substrates rich in C and N.

Soil Temperature

Temperature controls both plant and soil microbial activity, although not at the same level (Fig. 2). Plant and soil microbial activity are limited by low temperature resulting in low photosynthetic potential, as well as low decomposition potential. For many plants, net photosynthetic activity is optimized between 20 and 30°C, because at higher temperatures, plant respiration consumes energy for maintenance. In many temperate soils, microbial activity is maximized between 30 and 35°C and decreases at higher temperatures. An intermediate temperature is often ideal for maximizing C retention in soil, because optimum plant activity competes well against soil microbial activity.

Soil Water Content

Diversity of soil microorganisms is greatest under aerobic conditions, where maximum energy is obtained. However,

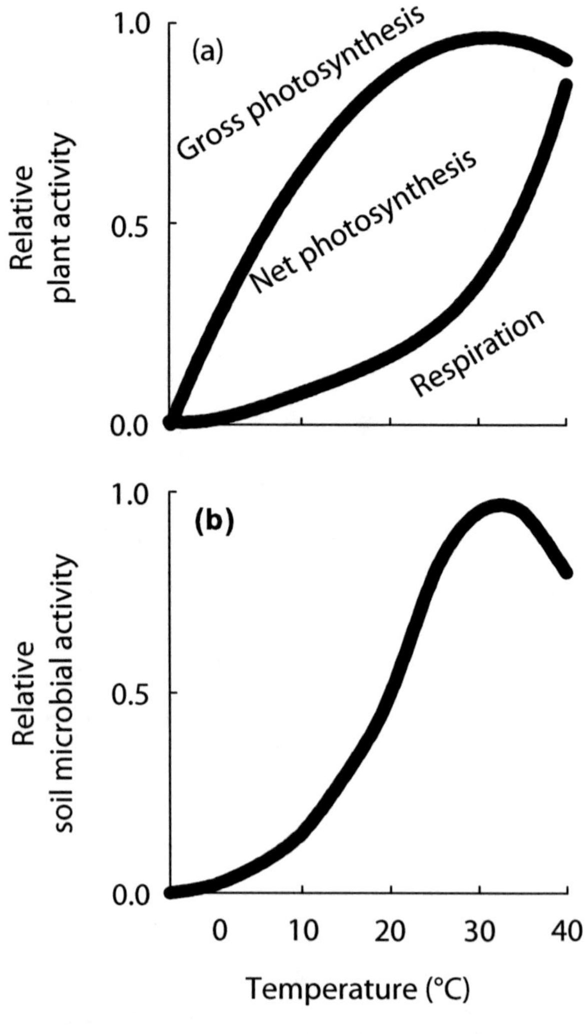
1.0
(a)
Gross photosynthesis
Net photosynthesis
Respiration
Relative plant activity
0.5
0.0
1.0
(b)
Relative soil microbial activity
0.5
0.0
0
10
20
30
40
Temperature (°C)

Fig. 2 Typical responses of plant and soil microbial activities to temperature.

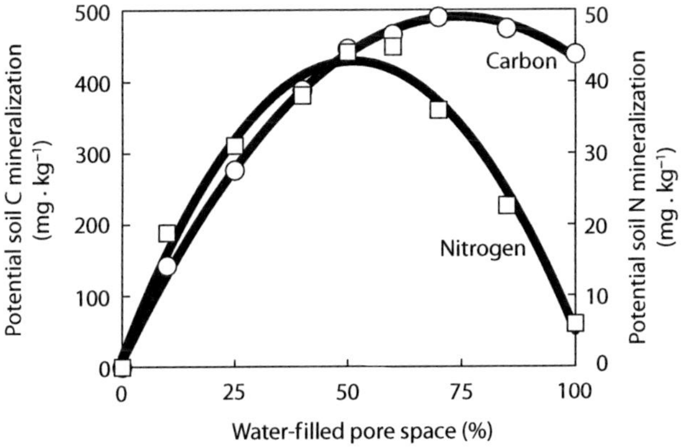
Potential soil C mineralization (mg · kg⁻¹)
Potential soil N mineralization (mg · kg⁻¹)
Carbon
Nitrogen
Water-filled pore space (%)
0
25
50
75
100
100
200
300
400
500
10
20
30
40
50

Fig. 3 Responses of potential soil C and N mineralization to water-filled pore space in Typic Kanhapludults in Georgia, USA [air-filled pore space would be 100 – (water-filled pore space)]. **Source:** Franzluebbers.[12]

there are a number of soil bacteria that thrive under anaerobic conditions, in which alcohols, acetic acid, lactic acid, and CH_4 become C end products via fermentation and nitrate is converted to N gases (e.g., N_2, N_2O, NO) via the process of denitrification. Soil C and net N mineralization are maximized at an optimum balance between soil moisture and oxygen availability (Fig. 3). Significant denitrification occurs at water-filled pore space >70%, resulting in low availability of inorganic N to plants.

Soil Texture

Soil texture can influence both the quantity of C and N accumulation in soil and their potential mineralization. Potential C mineralization is often greater in coarse-textured soils than in fine-textured soils, due to both increased microbial predation by soil fauna and greater accessibility of organic substrates in coarse-textured soils. Organic C and N can also be protected from decomposition when bound within soil aggregates. Water-stable aggregates are a coherent assemblage of primary soil particles (i.e., sand, silt, clay) cemented through natural forces and substances derived from root exudates and soil microbial activity.

Spatial Distribution of Organic Substrates

Distribution of organic substrates in soil has a major impact on potential C and N mineralization. Potential C mineralization is often several-fold greater in the rhizosphere (i.e., 0–5 mm zone surrounding roots) than in bulk soil. However, because of the high demand for N by plant roots and the stimulated soil microflora, net N mineralization is often initially lower in the rhizosphere because of immobilization of N. Keeping soil active with roots whenever conditions are conducive for plant growth will 1) keep inorganic N at low levels (as well as keep soil covered with protective plant cover to guard against soil erosion); 2) stimulate soil biological activity; and 3) create a richly diverse soil microbial community, all of which prevent nutrients from being lost from the soil.

Surface soil often contains greater quantities of organic matter than at lower depths due to surface deposition of plant residues, as well as greatest plant root activity. Surface soil usually undergoes the most extreme drying/wetting cycles and has the greatest exchange of gases, both of which contribute to enhanced soil microbial biomass and activity. Tillage of soil with traditional agriculture redistributes organic substrates uniformly within the plow layer, often resulting in immediately stimulated soil microbial activity from disruption of organic substrates protected within stable soil aggregates. Minimum soil disturbance with conservation tillage practices can reduce oxidation of soil organic matter and preserve more C within soil, which can have implications for potentially mitigating the greenhouse effect.[13]

Stratification of soil organic matter with depth is common in natural ecosystems and in conservation agricultural systems (Fig. 4). Conservation agricultural systems are defined as those that 1) minimize soil disturbance with tillage; 2) maximize soil-surface cover with continuous plant and/or residue cover; and 3) stimulate biological activity through diverse crop rotations and integrated nutrient and pest management.

Depth stratification of soil organic matter with time occurs when soils remain undisturbed from tillage (e.g., with conservation tillage and pastures) and sufficient organic materials are supplied to the soil surface (e.g., with cover crops, sod rotations, and diversified cropping systems). Depth stratification with time can be viewed as an improvement in soil quality, because several key soil functions are enhanced, including water infiltration, conservation and cycling of nutrients, and sequestration of C from

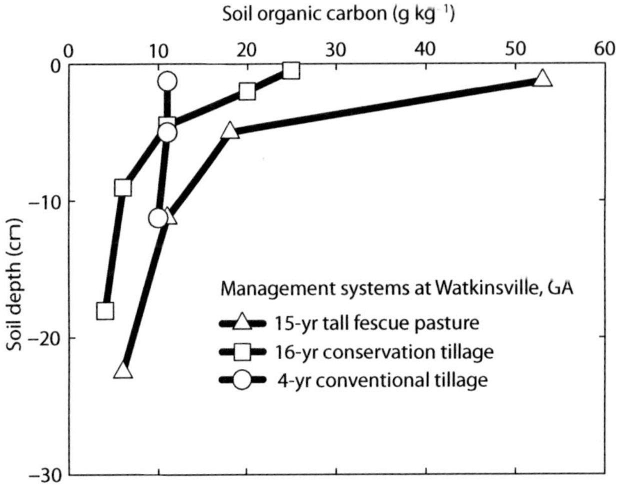
Soil organic carbon (g kg⁻¹)
0
10
20
30
40
50
60
Soil depth (cm)
0
−10
−20
−30
Management systems at Watkinsville, GA
15-yr tall fescue pasture
16-yr conservation tillage
4-yr conventional tillage

Fig. 4 Depth distribution of soil organic C under pastured grassland, conservation-tillage cropland, and conventional-tillage cropland on a Typic Kanhapludult in Georgia, USA. **Source:** Schnabel et al.[14]

the atmosphere.[15] Depth stratification of soil organic C generally reduces water runoff volume and soil loss from agricultural fields. Grasslands often reduce water runoff volume and soil loss even further than with conservation-tilled cropland due to even greater accumulation of surface soil organic matter. Total runoff loss of nutrients is often lower with conservation tillage than with conventional tillage, because of a reduction in sediment-borne nutrients (Fig. 5). Soluble (or dissolved) N and P in water runoff can be a threat to water quality with excessive nutrient applications from fertilizers and manures (even under conservation management), and therefore, further research is being conducted to identify ways of reducing nutrient loss.[31]

Stratification ratio of soil organic C has been proposed as an index of soil quality, because soil-surface enrichment of organic matter is important for improving water-stable aggregation, water infiltration and storage, nutrient cycling, and soil microbial biomass, activity, and diversity.[32] In a land-use survey in the southeastern United States, stratification ratio of soil organic C was related to the total stock of soil organic C in the surface 20 cm depth (Fig. 6). This relationship indicates that the majority of C stored with conservation management in these Ultisols and Alfisols of the region occurred within the surface 5 cm.

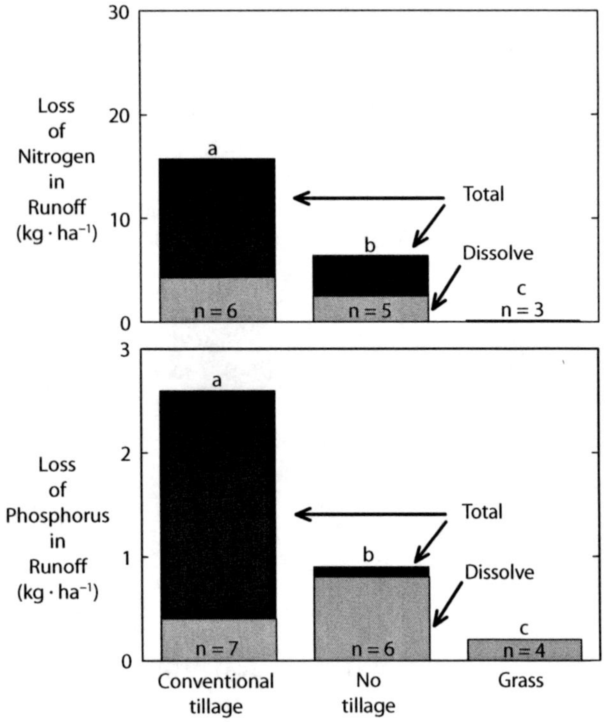
Loss of Nitrogen in Runoff (kg · ha⁻¹)
30
20
10
0
a
n = 6
b
n = 5
c
n = 3
Total
Dissolve
Loss of Phosphorus in Runoff (kg · ha⁻¹)
3
2
1
0
a
n = 7
b
n = 6
c
n = 4
Total
Dissolve
Conventional tillage
No tillage
Grass

Fig. 5 Mean loss of N and P in water runoff across several water catchment studies in the USA.
Source: Data from Van Doren et al.,[16] Langdale et al.,[17] Blevins et al.,[18] Seta et al.,[19] Sharpley and Smith,[20] Shipitalo and Edwards,[21] Endale et al.,[22] Endale et al.,[23] Endale et al.,[24] Ross et al.,[25] Rhoton et al.,[25] Rhoton et al.,[26] Sharpley and Kleinman,[27] Truman et al.,[28] Harmel et al.,[29] and Bosch.[30]

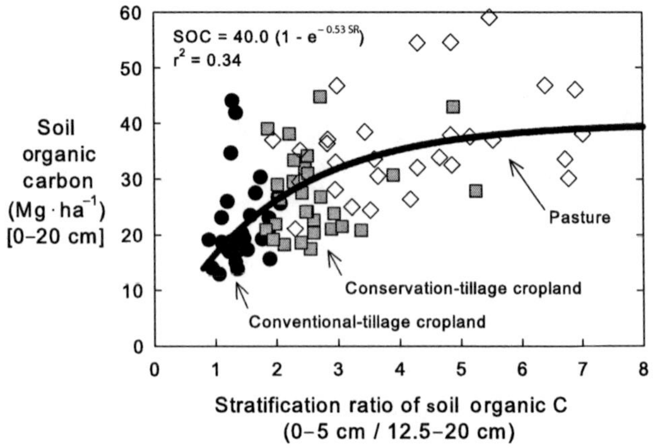
SOC = 40.0 (1 - e^−0.53 SR)
r² = 0.34
Soil organic carbon (Mg · ha⁻¹) [0–20 cm]
60
50
40
30
20
10
0
Pasture
Conservation-tillage cropland
Conventional-tillage cropland
0 1 2 3 4 5 6 7 8
Stratification ratio of soil organic C (0–5 cm / 12.5–20 cm)

Fig. 6 Relationship of soil organic C storage at a depth of 0–20 cm to the stratification ratio of soil organic C among conventional-tillage, conservation-tillage, and pasture land uses on different soils throughout the southeastern USA.
Source: Causarano et al.[33]

ORGANIC SUBSTRATE QUALITY

Quality of organic substrates has a major influence on the rate of decomposition and the transformations that occur in soil. Plant residues do not vary greatly in total C concentration on a dry-weight basis (e.g., 37–47 mg g^{-1}), but do vary enormously in the type of C compounds, which determine its quality or conversely its resistance to degradation. The diversity of organic compounds attacked by soil microorganisms is extensive (e.g., organic acids, polysaccharides, lignins, aromatic and aliphatic hydrocarbons, sugars, alcohols, amino acids, purines, pyrimidines, proteins, lipids, and nucleic acids). Almost all naturally occurring organic compounds, and even most synthetic organic compounds, are susceptible to decomposition given the appropriate environment, microbial community, and time.[34,35] Generally, the primary components of plants can be categorized according to relative rate of decomposition: rapid (sugars, starches, fats, and proteins), intermediate (cellulose and hemicellulose), and slow (lignin and lignocellulose). Young plants are of high quality and low resistance to decomposition, whereas with aging, lignin and polyphenolic concentrations increase, resulting in greater resistance to decomposition. Low N concentration of organic amendments usually results in temporary net N immobilization into microbial biomass, which grows rapidly in response to the availability of organic C. Soil microbial biomass typically maintains a C-to-N ratio of 10 ± 5. Following a proliferation of microbial biomass that depletes the source of readily decomposable organic C, N in excess of microbial demands becomes mineralized and available for plant uptake (Fig. 7). In general, plant

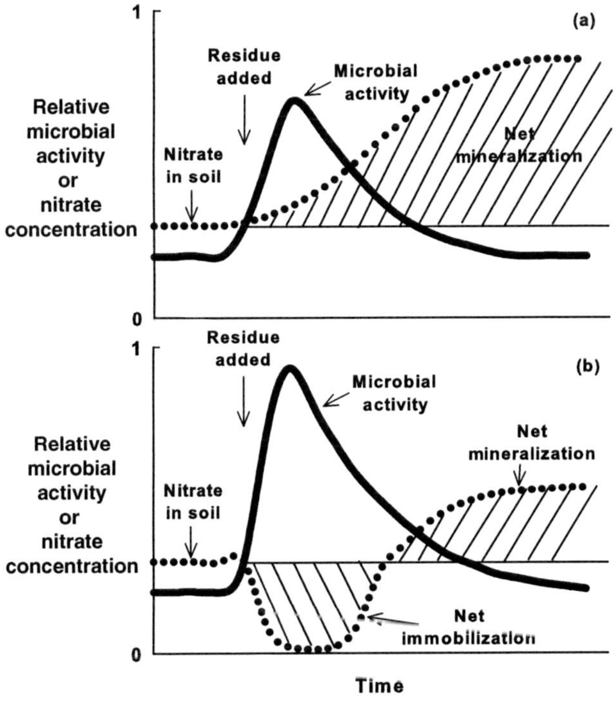

1
(a)
Residue added
Microbial activity
Relative microbial activity or nitrate concentration
Nitrate in soil
Net mineralization
0
1
Residue added
(b)
Microbial activity
Net mineralization
Relative microbial activity or nitrate concentration
Nitrate in soil
Net immobilization
0
Time

Fig. 7 Typical responses in soil microbial activity and soil nitrate concentration with the addition of plant residue of (a) high N concentration and (b) low N concentration.

residues with C-to-N ratio >40 will result in longer periods of net N immobilization.

Soil Organic Matter

Soil organic matter is composed of a large diversity of organic compounds that can be characterized in many ways. A useful separation of soil organic matter for modeling is based on turnover times, whereby at least three pools can be defined: 1) active (composed of microbial biomass and light fraction material with a turnover time of <1 year); 2) passive (composed of macroorganic matter and protected organic matter with a turnover time of 3–10 years); and 3) slow (stable humus fraction with a turnover time of >100 years). Fractions of soil organic matter have also been methodologically characterized and can be tied to above-mentioned kinetic pools.[36] Active fractions have been characterized using soil microbial biomass techniques (chloroform fumigation-incubation, chloroform extraction, substrate-induced respiration, microwave irradiation, phospholipids, and flush of CO_2) and determination of labile substrates (mineralizable C and N during incubation, hot-water extractable C and N, dilute permanganate oxidizable C, carbohydrates, light-fraction C and N, amino sugars). Passive fractions have been estimated using particulate organic matter, glycoproteins, and humic materials. Slow fractions of soil organic matter have been characterized with various resistant components, such as charcoal, lignins, aliphatic macromolecules, humin, nonhydrolyzable organic matter, and silt- and clay-associated organic matter.

LOSSES OF NITROGEN FROM SOIL

Nitrogen cycling in soil is different from that of C because of the more numerous transformations that can occur upon mineralization to an inorganic form.[2] Mineralization of N from organic matter results in NH_4^+ released into soil solution. In the presence of nitrifying bacteria, NH_4^+ is converted to NO_3^-, a process called nitrification. The fate of NO_3^- in soil depends upon environmental conditions. Active plant growth in natural and agricultural systems would provide enough demand to remobilize N into organic forms. However, NO_3^- can be used as an electron acceptor in place of O_2 under anaerobic conditions, resulting in gaseous loss of N to the atmosphere via denitrification. In temperate soils with a net negative charge on colloidal surfaces, NO_3^- can readily leach into the vadose zone and contaminate groundwater if not taken up by plants or denitrified. In tropical soils with a net positive charge, NO_3^- can be retained on anion exchange sites. Opposite behavior of the cations, NH_4^+ and NO_3^-, occurs with respect to clay mineralogy.

Humans, as well as roving animals, impose great demands on the C and N cycles. Management of agricultural and forest land for food and fiber often removes nutrients from soil for consumption and utilization elsewhere. Return of these nutrients to soil is possible when municipal and agricultural solid wastes and wastewater are applied to land. Losses of C and N from managed lands also occurs through soil erosion, which transports nutrients via 1) water from overland flow into streams, lakes, and oceans and 2) air as dust from bare land surfaces.

Volatilization of NH_3 to the atmosphere is possible when NH_4^+ is exposed to alkaline soil conditions. Significant ammonia volatilization can occur with surface application of urea fertilizer to non-acidic soils, from animal manures, and from green plant materials not incorporated into the soil.

Strategies to Mitigate Loss of Nitrogen from Agricultural Soils

Loss of nitrogen from agricultural soils occurs through harvest, runoff, leaching, volatilization, and denitrification. These losses are potential threats to the environmental quality of water and air resources, as well as to the quality of soil itself. Harvest losses of N can be accepted as practically necessary, but such losses can be partially mitigated by returning waste by-products from food processing and from animal and human consumption back to the soil as organic amendments. This recycling is an important step towards global sustainability of nutrient use and protection of the environment from nutrient loading, particularly relevant in industrialized countries, which seem to have abandoned this age-old practice.

Runoff losses of N can be controlled by protecting the soil surface from soil loss via erosion and by creating a porous surface to allow rainfall to infiltrate rather than runoff. Various conservation agricultural approaches are available to mitigate runoff losses, including reduced or no tillage (i.e., conservation tillage), intensified crop rotations to avoid bare fallow periods, winter cover cropping, crop–pasture rotation, moderate grazing of perennial pastures, and timely and deep placement of fertilizers. Loss of N in runoff can occur as part of N-rich sediment (both organic and inorganic N) and as soluble inorganic N in overland flow of water. Stopping sediment loss is most effective by protecting the soil surface with plants and surface residue, both of which mitigate the energy of rainfall impact and, therefore, avoid soil detachment. Surface application of fertilizers, whether inorganic or organic, is susceptible to runoff loss as water flows over the landscape even if soil is not detached. Therefore, timing fertilizer application to when the plant needs it the most will be an effective N-loss mitigation strategy, as well as placing fertilizer into the soil in proximity with roots with deep banding will limit access to overland flow potential.

Leaching of N through the soil profile is a concern in well-drained soils. Conditions for significant leaching often occur due to overapplication of N fertilizer and long bare-fallow periods that limit plant uptake of residual fertilizer in the soil profile. Mitigation of N leaching can be through continuous plant growth with diverse crop rotations and winter cover cropping. Perennial pastures are also often effective in mitigating N leaching, because of the extended growing season of diverse assemblages of forages. Limiting the quantity of N fertilizer applied is still a basic principle to avoid potential N leaching.

Volatilization of NH_3 can occur when sufficient NH_4^+ accumulates at the soil surface from partial decomposition of animal manures and N-rich plant residues and from ammonium-based fertilizer sources. Volatilization is most prevalent in soils with high pH and when soil is rather dry. Mitigation of ammonia volatilization is possible by getting animal manure and N-rich plant residues in close proximity with moist soil and avoiding surface application of N-rich organic residues and NH_4-based fertilizers on soils with high pH.

Denitrification of NO_3^- to N_2 (and to N_2O or NO) is a concern in soils under the following conditions: 1) low oxygen composition of the soil atmosphere, due to either high water content or rapid consumption of oxygen by vigorous microbial activity and/or poor air exchange; 2) readily decomposable source of organic C compounds for energy; 3) abundant supply of NO_3^-; and 4) suitable temperature for microbial activity. Mitigation of N loss through denitrification is possible by keeping soil well aerated by avoiding compaction and reducing the accumulation of NO_3^-.

Chemical nitrification inhibitors have been developed to inhibit or slow down the nitrification process that converts NH_4^+ to NO_3^-. Biological approaches to control susceptibility of NO_3^- to loss via leaching and denitrification focus more on maintaining a continuously growing and diverse plant root system penetrating a soil that develops excellent tilth from a healthy and diverse soil biological community, resulting in stable aggregate structure and vigorous cycling of C and N.

CONCLUSION

Nitrogen is a key element for the production of high-value crops and animal products around the world. High input of N in the current industrialized model of agricultural production often leaves behind a significant amount of inorganic N, such that loss from leaching, runoff, volatilization, and denitrification can occur. Ecologically based strategies of agricultural production recognize the importance of having sufficient N for production, but rely on biological cycling among plants, animals, and soil microorganisms and fauna to synchronize the release of organically bound N into inorganic forms and avoid environmental pollution. Ecological principles of high biological diversity, continuous plant growth and soil cover, and limited soil disturbance can be used in many different climatic and ecological conditions to avoid environmental pollution by excessive N entering water bodies and air systems.

REFERENCES

1. Follett, R.F.; Stewart, J.W.B.; Cole, C.V., Eds. *Soil Fertility and Organic Matter as Critical Components of Production Systems*; Spec. Publ. No. 19, Soil Science Society of America: Madison WI, 1987; 166 pp.
2. Stevenson, F.J., Ed. *Nitrogen in Agricultural Soils*; Agronomy Monograph 22; American Society of Agronomy: Madison WI, 1982; 940 pp.
3. Food and Agriculture Organization of the United Nations, available at http://faostat.fao.org (accessed March 2010).
4. Burns, R.G., Ed. *Soil Enzymes*; Academic Press: London, 1978; 380 pp.
5. Tabatabai, M.A. Soil enzymes. In *Methods of Soil Analysis, Part 2. Microbiological and Biochemical Properties*; Weaver, R.W., Angle, J.S., Bottomley, P.S., Eds.; Book Series No. 5; Soil Science Society of America: Madison WI, 1994; 775–833.
6. Stevenson, F.J.; Cole, M.A. *Cycles of Soil: Carbon, Nitrogen, Phosphorus, Sulfur, Micronutrients*, 2nd Ed.; John Wiley and Sons, Inc.: New York, 1999; 427 pp.
7. Schlesinger, W.H. *Biogeochemistry: An Analysis of Global Change*, 2nd Ed.; Academic Press: San Diego CA, 1997; 588 pp.
8. Woodmansee, R.G. Comparative nutrient cycles of natural and agricultural ecosystems: A step toward principles. In

Agricultural Ecosystems: Unifying Concepts; Lowrance, R., Stinner, B.R., House, G.J., Eds.; John Wiley and Sons: New York, 1984; 145–156.
9. Tilman, D.; Cassman, K.G.; Matson, P.A.; Naylor, R.; Polasky, S. Agricultural sustainability and intensive production practices. Nature **2002**, *418*, 671–677.
10. Alexander, M.A. *Introduction to Soil Microbiology*, 2nd Ed.; Krieger Publ. Co.: Malabar, FL, 1991; 467 pp.
11. Sylvia, D.M.; Fuhrmann, J.J.; Hartel, P.G.; Zuberer, D.A., Eds. *Principles and Applications of Soil Microbiology*; Prentice-Hall: Upper Saddle River, NJ, 1998; 550 pp.
12. Franzluebbers, A.J. Microbial activity in response to water-filled pore space of variably eroded southern Piedmont soils. Appl. Soil Ecol. **1999**, *11*, 91–101.
13. Lal, R.; Kimble, J.M.; Follett, R.F.; Cole, C.V., Eds. *The Potential of U.S. Cropland to Sequester Carbon and Mitigate the Greenhouse Effect*; Ann Arbor Press: Chelsea, MI, 1998; 128 pp.
14. Schnabel, R.R.; Franzluebbers, A.J.; Stout, W.L.; Sanderson, M.A.; Stuedemann, J.A. The effects of pasture management practices. In *The Potential of U.S. Grazing Lands to Sequester Carbon and Mitigate the Greenhouse Effect*; Follett, R.F., Kimble, J.M., Lal, R., Eds.; Lewis Publishers: Boca Raton, FL, 2001, 291–322.
15. Franzluebbers, A.J. Linking soil and water quality in conservation agricultural systems. J. Integr. Biosci. **2008**, *6* (1), 15–29.
16. Van Doren, D.M., Jr.; Moldenhauer, W.C.; Triplett, G.B., Jr. Influence of long-term tillage and crop rotation on water erosion. Soil Sci. Soc. Am. J. **1984**, *48*, 636–640.
17. Langdale, G.W.; Leonard, R.A.; Thomas, A.W. Conservation practice effects on phosphorus losses from Southern Piedmont watersheds. J. Soil Water Conserv. **1985**, *40*, 157–161.
18. Blevins, R.L.; Frye, W.W.; Baldwin, P.L.; Robertson, S.D. Tillage effects on sediment and soluble nutrient losses from a Maury silt loam soil. J. Environ. Qual. **1990**, *19*, 683–686.
19. Seta, A.K.; Blevins, R.L.; Frye, W.W.; Barfield, B.J. Reducing soil erosion and agricultural chemical losses with conservation tillage. J. Environ. Qual. **1992**, *22*, 661–665.
20. Sharpley, A.N.; Smith, S.J. Wheat tillage and water quality in the Southern Plains. Soil Tillage Res. **1994**, *30*, 33–48.
21. Shipitalo, M.J.; Edwards, W.M. Runoff and erosion control with conservation tillage and reduced-input practices on cropped watersheds. Soil Tillage Res. **1998**, *46*, 1–12.
22. Endale, D.M.; Cabrera, M.L.; Radcliffe, D.E.; Steiner, J.L. Nitrogen and phosphorus losses from no-till cotton fertilized with poultry litter in the Southern Piedmont. In *Proceedings of the Georgia Water Resources Conference*, Athens GA, 26–27 March 2001; 408–411.
23. Endale, D.M.; Schomberg, H.H.; Jenkins, M.B.; Cabrera, M.L.; Radcliffe, D.R.; Hartel, P.G.; Shappell, N.W. Tillage and N-fertilizer source effects on yield and water quality in a corn-rye cropping system. In *Proceedings of the Southern Conservation Tillage Conference*, Raleigh, NC, June 8–9, 2004; 37–48.
24. Endale, D.M.; Schomberg, H.H.; Steiner, J.L. Long term sediment yield and mitigation in a small Southern Piedmont watershed. Int. J. Sediment Res. **2000**, *14*, 60–68.
25. Ross, P.H.; Davis, P.H.; Heath, V.L. *Water Quality Improvement Resulting from Continuous No-Tillage Practices: Final Report*; Virginia Polytechnic Institute and State University: Blacksburg, VA, 2001.
26. Rhoton, F.E.; Shipitalo, M.J.; Lindbo, D.L. Runoff and soil loss from midwestern and southeastern US silt loam soils as affected by tillage practice and soil organic matter content. Soil Tillage Res. **2002**, *66*, 1–11.
27. Sharpley, A.; Kleinman, P. Effect of rainfall simulator and plot scale on overland flow and phosphorus transport. J. Environ. Qual. **2003**, *32*, 2172–2179.
28. Truman, C.C.; Reeves, D.W.; Shaw, J.N.; Motta, A.C.; Burmester, C.H.; Raper, R.L.; Schwab, E.B. Tillage impacts on soil property, runoff, and soil loss variations from a Rhodic Paleudult under simulated rainfall. J. Soil Water Conserv. **2003**, *58*, 258–267.
29. Harmel, R.D.; Torbert, H.H.; Haggard, B.E.; Haney, R.; Dozier, M. Water quality impacts of converting to a poultry litter fertilization strategy. J. Environ. Qual. **2004**, *33*, 2229–2242.
30. Bosch, D.D.; Potter, T.L.; Truman, C.C.; Bednarz, C.W.; Strickland, T.C. Surface runoff and lateral subsurface flow as a response to conservation tillage and soil-water conditions. Trans. Am. Soc. Agric. Eng. **2005**, *48*, 2137–2144.
31. Sharpley, A.N.; Herron, S.; Daniel, T. Overcoming the challenge of phosphorus-based management in poultry farming. J. Soil Water Conserv. **2007**, *62* (6), 375–389.
32. Franzluebbers, A.J. Soil organic matter stratification as an indicator of soil quality. Soil Tillage Res. **2002**, *66*, 95–106.
33. Causarano, H.J.; Franzluebbers, A.J.; Shaw, J.N.; Reeves, D.W.; Raper, R.L.; Wood, C.W. Soil organic carbon fractions and aggregation in the Southern Piedmont and Coastal Plain. Soil Sci. Soc. Am. J. **2008**, *72*, 221–230.
34. Tate, R.L., III. *Soil Organic Matter: Biological and Ecological Effects*; John Wiley and Sons, Inc.: New York, 1987; 291 pp.
35. Alexander, M. *Biodegradation and Bioremediation*; Academic Press: San Diego, CA, 1994; 302 pp.
36. Wander, M.M. Soil organic matter fractions and their relevance to soil function. In *Soil Organic Matter in Sustainable Agriculture*; Magdoff, F., Weil, R.R., Eds.; CRC Press: Boca Raton, FL, 2004; 67–102.

Agricultural Soils: Nitrous Oxide Emissions

John R. Freney
Commonwealth Scientific and Industrial Research Organization (CSIRO), Campbell, Australian Capital Territory, Australia

Abstract
Nitrous oxide is a gas that is produced naturally by many different micro-organisms in soils and waters, and as a result of human activity. It is apparent that most nitrous oxide is derived from soils. Because of the intimate connection between the Earth and the atmosphere, much of the nitrous oxide produced enters the atmosphere and affects its chemical and physical properties.

INTRODUCTION

Nitrous oxide is a gas that is produced naturally by many different micro-organisms in soils and waters, and as a result of human activity associated with agriculture, biomass burning, stationary combustion, automobiles, and the production of nitric and adipic acids for industrial purposes. According to the Intergovernmental Panel on Climate Change (IPCC),[1] ~23.1 million metric tons (Mt) of nitrous oxide is emitted each year, 14.1 Mt as a result of natural processes (~4.7Mt from the oceans, ~6.3 Mt from tropical soils, and ~3.1 Mt from temperate soils); and ~9 Mt as a result of human activities (5.5 Mt from agricultural soils, 0.6 Mt from cattle and feedlots, 0.8 Mt from biomass burning, and 2.1 Mt from mobile sources and industry). While there is considerable uncertainty associated with each of these estimates, it is apparent that most nitrous oxide is derived from soils.

Because of the intimate connection between the Earth and the atmosphere, much of the nitrous oxide produced enters the atmosphere and affects its chemical and physical properties. Nitrous oxide contributes to the destruction of the stratospheric ozone layer that protects the Earth from harmful ultraviolet radiation, and is one of the more potent greenhouse gases that trap part of the thermal radiation from the Earth's surface. The atmospheric concentration of nitrous oxide is ~313 parts per billion. It is increasing at the rate of ~0.7 parts per billion each year, and its lifetime is ~166 years.[2] It seems that the increased atmospheric concentration results from the increased use of synthetic fertilizer nitrogen, biologically fixed nitrogen, animal manure, crop residues, and human sewage sludge in agriculture to produce food and fiber for the rapidly increasing world population.[3]

NITROUS OXIDE EMISSION FROM AGRICULTURE

All soils are deficient in nitrogen for the growth of plants, but the deficiency can be overcome by adding fertilizer nitrogen. When the fertilizer (e.g., urea or ammonia-based compounds) is applied to soil, it is transformed by micro-organisms as follows:

$$\begin{aligned}\text{Fertilizer nitrogen} &\xrightarrow{1} \text{Ammonium} \xrightarrow{2} \text{Nitrite}\\ &\xrightarrow{3} \text{Nitrate} \xrightarrow{4} \text{Nitrite}\\ &\xrightarrow{5} \text{Nitric oxide} \xrightarrow{6} \text{Nitrous oxide}\\ &\xrightarrow{7} \text{Dinitrogen}\end{aligned} \qquad (1)$$

When the soil is aerobic (i.e., when oxygen is present) ammonium is oxidized to nitrite and nitrate (Steps 2 and 3). This process is called nitrification. After addition of irrigation water or rain, the soil may become anaerobic (devoid of oxygen). The nitrate is then reduced by soil organisms to nitrite and the gases nitric oxide, nitrous oxide, and dinitrogen (Steps 4–7) in a process termed denitrification.[4]

When atmospheric scientists first expressed concern that nitrous oxide emission into the atmosphere, as a result of fertilizer use, would lead to destruction of the ozone layer, it was thought that nitrous oxide was produced mainly from the microbiological reduction of nitrate in poorly aerated soils. However, research in the latter part of the 1970s showed that significant nitrous oxide was emitted from aerobic soils during nitrification of ammonium, and subsequent work has shown that nitrification is a major source of nitrous oxide.[4]

Nitrous Oxide from Denitrification

Certain micro-organisms in the absence of oxygen have the capacity to reduce nitrate (or other nitrogen oxides). Most denitrifying bacteria are heterotrophs—that is, they require a source of organic matter for energy—but denitrifying organisms that obtain their energy from light or inorganic compounds also occur in soils. The capacity to denitrify has been reported in more than 20 genera of bacteria, and almost all are aerobic organisms that can only grow anaerobically in the presence of nitrogen oxides. The dominant denitrifying organisms in soil are *Pseudomonas*

Encyclopedia of Environmental Management DOI: 10.1081/E-EEM-120001564

and *Alcaligenes*. In addition to the free-living denitrifiers, Rhizobia living symbiotically in root nodules of legumes are able to denitrify nitrate and produce nitrous oxide.[4]

The general requirements for biological denitrification include the presence of micro-organisms with denitrifying capacity, nitrate (or other nitrogen oxides) and available organic matter, the absence of oxygen, and a suitable pH and temperature environment. In aerobic soils, denitrification can occur in anaerobic microsites in soil aggregates or in areas of high carbon content, where active microbial activity rapidly consumes all of the available oxygen.[4]

Nitrous Oxide from Nitrification

The process of nitrification is normally defined as the biological oxidation of ammonium to nitrate with nitrite as an intermediate.[4] The first step in the reaction, the oxidation of ammonium to nitrite, is carried out mainly by the micro-organism *Nitrosomonas*. The second step, oxidation of nitrite to nitrate, is carried out by *Nitrobacter*. It has been shown in a number of publications that *Nitrosomonas europaea* produces nitrous oxide during the oxidation of ammonium.[4]

The possibility that significant nitrous oxide can be produced in soils by nitrifying organisms was indicated by studies that showed that soils incubated under aerobic conditions with ammonium produced more nitrous oxide than soils amended with nitrate.[4] In addition, treatment of aerobic soils with nitrapyrin, which inhibits nitrification of ammonium but has little effect on denitrification, markedly reduced the emission of nitrous oxide.[4] Production of nitrous oxide by nitrification in soils is increased by increasing temperature, pH, water content, available carbon, and the addition of ammonium-based fertilizers, plant residues, and animal manure.

Flooded Soils

In the past few years, increased attention has been given to nitrous oxide emission from paddy soils. The concern is that the introduction of management practices to reduce methane emissions from flooded soils may result in increased emissions of nitrous oxide.

Flooded soils are characterized by an oxygenated water column overlying an oxidized layer at the soil–water interface, an aerobic zone around each root, and anaerobic conditions in the remainder of the soil. This differentiation of the flooded soil into oxidized and reduced zones has a marked effect on the transformation of nitrogen.[5] The resulting reactions are as follows:

1. Ammonium in the reduced zone diffuses to the oxidized zone;
2. Ammonium is oxidized to nitrate by nitrifying organisms;
3. The nitrate formed diffuses to the anaerobic zone;
4. Denitrification occurs with the production of nitrous oxide and dinitrogen;
5. The gaseous products diffuse through the soil and water layers to the atmosphere.[6]

It is apparent that the rate of diffusion of ammonium to an oxidized layer and the rate of nitrification in the oxidized layer are factors controlling the production of nitrous oxide in flooded soils. The rate of diffusion of nitrous oxide through the soil and water layers will control its rate of emission to the atmosphere, or its further reduction to dinitrogen.[5]

A number of mechanisms have been identified for the transfer of nitrous oxide from the soil to the atmosphere.[3] Nitrous oxide may diffuse from the zone of production through the saturated soil and water layer to the atmosphere. It may also enter the roots of the rice plant and move by diffusion through the plant to the atmosphere in the same way as methane. Bhadrachalam et al.[6] studied the importance of the two pathways in intermittently flooded rice in the field in India and found that nitrogen gas fluxes were ~30% greater when transfer through the plants was included.

In the tropics, rice is usually transplanted and fertilized some time after flooding. Because of the anaerobic conditions that develop before fertilization and the slow rate of diffusion of nitrous oxide in flooded soils, most of the nitrous oxide is reduced to dinitrogen and very little escapes to the atmosphere. Nitrous oxide emission from intermittently flooded rice was relatively large compared with that from permanently flooded rice, reflecting the different oxidation states of intermittently and continuously flooded soils.[6] Studies of nitrous oxide emission from rice fields from the time the soils were drained for harvest, through to flooding the soil in preparation for planting the next crop, showed that nitrous oxide was emitted continuously while the soil was not flooded. Overall, the rate of emission of nitrous oxide from floodedsoils was less than that from upland soils after application of nitrogen fertilizer.[3]

Biomass Burning

During combustion the nitrogen in the fuel can be converted into gaseous forms such as ammonia, nitric oxide, nitrous oxide, dinitrogen, and hydrogen cyanide. It is estimated that biomass burning contributes between 0.3 and 1.6 Mt nitrous oxide per year globally to the atmosphere.[3] Most of the biomass burning (~90%) takes place in the tropics as a result of forest clearing, savanna and sugar cane fires, and burning of agricultural wastes and firewood.[7]

Biomass burning is not only an instantaneous source of nitrous oxide, but it results in a longer-term enhancement of the production of this gas. Measurements of nitrous oxide emissions from soils, before and after burning showed that significantly more nitrous oxide was exhaled after the burn through alteration of the chemical, biological, and physical processes in soil.[7]

Fertilizer Consumption and Nitrous Oxide Production

Nitrous oxide emissions from agricultural soils are generally greater and more variable than those from uncultivated land. Application of fertilizer nitrogen, animal manure, and sewage sludge usually results in enhanced emissions of nitrous oxide.[7] Generally, there is a large emission of nitrous oxide immediately after the application of fertilizer. After about 6 weeks, the emission rate falls and fluctuates around a low value. Mosier[8] concluded that interactions between the physical, chemical, and biological variables are complex, that nitrous oxide fluxes are variable in time and space, and that soil management, cropping systems, and variable rainfall appear to have a greater effect on nitrous oxide emission than the type of nitrogen fertilizer. Consequently, Mosier et al.[9] recommend the use of one factor only for calculating the emission of nitrous oxide from different fertilizer types:

$$N_2O \text{ emitted} = 125\% \text{ of N applied (kgN/ha)} \quad (2)$$

This equation is based on data from long-term experiments with a variety of mineral and organic fertilizers, and encompasses 90% of the direct contributions of nitrogen fertilizers to nitrous oxide emissions.

Mosier et al.[3] developed a methodology to estimate agricultural emissions of nitrous oxide, taking into account all of the nitrogen inputs into crop production. They included direct emissions from agricultural soils as a result of synthetic fertilizer addition, animal wastes, increased biological nitrogen fixation, cultivation of mineral and organic soils through enhanced organic matter mineralization, and mineralization of crop residues returned to the field. Indirect nitrous oxide emissions resulting from deposition of ammonia and oxides of nitrogen, leaching of nitrate, and introduction of nitrogen into sewage systems were also included. They concluded that in 1989, 9.9 Mt of nitrous oxide was emitted into the atmosphere directly or indirectly, as a result of agriculture (Table 1).

Table 1 Calculated emission of nitrous oxide from agricultural activities.

	Mt nitrous oxide per year
Direct soil emissions	
Synthetic fertilizer	1.4 (0.28–2.5)
Animal waste	0.9 (0.19–1.7)
Biological nitrogen fixation	0.16 (0.03–0.3)
Crop residue	0.6 (0.11–1.1)
Cultivation of Histosols	0.16 (0.03–0.3)
Total	3.3 (0.6–5.9)
Animal production	
Waste management systems	3.3 (0.9–4.9)
Indirect emissions	
Atmospheric deposition	0.47 (0.09–0.9)
Nitrogen leaching and runoff	2.5 (0.2–12.1)
Human sewage	0.3 (0.06–4.1)
Total	3.3 (0.35–17.1)
Total	9.9 (1.9–27.9)

Source: Modified from Mosier et al.[3]

MANAGEMENT PRACTICES TO DECREASE NITROUS OXIDE EMISSION

The low efficiency of fertilizer nitrogen in agricultural systems is primarily caused by the large losses of mineral nitrogen from those systems by gaseous loss: nitrous oxide emission is directly linked to the loss processes. It is axiomatic that any strategy that increases the efficiency of nitrogen fertilizer use will reduce emissions of nitrous oxide, and this has been directly demonstrated for a number of strategies.[3]

The IPCC[1] reported that some combination of the following management practices, if adopted worldwide, would improve the efficiency of the use of synthetic fertilizer and manure nitrogen, and significantly reduce nitrous oxide emission into the atmosphere:

1. Match nitrogen supply with crop demand.
2. Tighten nitrogen flow cycles by returning animal wastes to the field and conserving residues instead of burning them.
3. Use controlled-release fertilizers, incorporate fertilizer to reduce volatilization, use urease and nitrification inhibitors, and match fertilizer type to precipitation.
4. Optimize tillage, irrigation, and drainage.

The potential decrease in nitrous oxide emissions from synthetic fertilizer, as a result of the mitigation techniques, could amount to 20%.[1]

REFERENCES

1. IPCC (Intergovernmental Panel on Climate Change). *Climate Change 1995. Impacts, Adaptations and Mitigation of Climate Change*: Scientific–Technical Analyses; Watson, R.T., Zinyowera, M.C., Moss, R.H., Eds.; Cambridge University Press: Cambridge, England, 1996; 1–878.
2. Hengeveld, H.; Edwards, P. 1998. An assessment of new research developments relevant to the science of climate change. Climate Change Newsletter **2000**, *12*, 1–52.
3. Mosier, A.; Kroeze, C.; Nevison, C.; Oenema, O.; Seitzinger, S.; van Cleemput, O. Closing the global N2O budget: nitrous oxide emissions through the agricultural nitrogen cycle. Nutri. Cycling Agroecosys. **1998**, *52*, 225–248.
4. Bremner, J.M. Sources of nitrous oxide in soils. Nutri. Cycling Agroecosys. **1997**, *49*, 7–16.
5. Patrick, W.H., Jr. Nitrogen transformation in submerged soils. In *Nitrogen in Agricultural Soils*; Stevenson, F.J.,

Ed.; American Society of Agronomy: Madison, WI, 1982; 449–465.
6. Bhadrachalam, A.; Chakravorti, S.P.; Banerjee, N.K.; Mohanty, S.K.; Mosier, A.R. Denitrification in intermittently flooded rice fields and N-gas transport through rice plants. Ecol. Bull. **1992**, *42*, 183–187.
7. Granli, T.; Bøckman, O.C. Nitrous oxide from agriculture. Norwegian J. Agric. Sci. **1994**, Supplement No. 12, 1–128.
8. Mosier, A.R. Chamber and isotope techniques. In *Exchange of Trace Gas between Terrestrial Ecosystems and the Atmosphere*; Andreae, M.O., Schimel, D.S., Eds.; John Wiley and Sons: Chichester, England, 1989; 175–187.
9. Mosier, A.R.; Duxbury, J.M.; Freney, J.R.; Heinemeyer, O.; Minami, K. Nitrous oxide emissions from agricultural fields: assessment, measurement and mitigation. Plant Soil **1995**, *181*, 95–108.

Agricultural Soils: Phosphorus

Anja Gassner
Institute of Science and Technology, University of Malaysia—Sabah, Kota Kinabalu, Malaysia

Ewald Schnug
Institute for Crop and Soil Science, Julius Kuhn Institute (JKI), Braunschweig, Germany

Abstract

The management of phosphate, a finite nonrenewable resource, is far from being in accordance with the principles of sustainability. This is partly because of the still prevailing misconception that soil is a homogenous, static entity. Phosphorus is regarded traditionally as immobile, but its chemical reactivity with environmental factors, acting at different spatial scales and over different time periods, results in the formation of phosphorus species, which not only differ in their plant availability but also in their spatial distribution throughout the field. This spatial speciation is the key for a new approach to assess soil analysis methods.

INTRODUCTION

Phosphorus mirabilis, the light-bearing nutrient, was probably discovered by an Arabian alchemist named Alchid Bechil, although its discovery is usually attributed somewhat later to Henning Brandt.[1] Since the early work of Justus Liebig (1803–1873), agricultural researchers have tried to tackle the mystery of phosphorus (P) availability to plants. Phosphorus research intuitively reminds one of Heisenberg's theory: "The more you see, the less you know." Despite the substantial amount of information available about the transformation products of P fertilizer within temperate[2] as well as tropical soils[3] and numerous attempts to conduct elaborate soil test methods to predict P availability to crops and algae,[2] today the management of phosphate, a finite nonrenewable resource, is far from being in accordance with the principles of sustainability.[4]

This is partly because of the still prevailing misconception that soil is a homogenous, static entity. Phosphorus is regarded traditionally as immobile, but its chemical reactivity with environmental factors, acting at different spatial scales and over different time periods, results in the formation of P species, which not only differ in their plant availability but also in their spatial distribution throughout the field. This spatial speciation[5] is the key for a new approach to assess soil analysis methods.

CHEMICAL SPECIATION OF PHOSPHORUS

With the exception of small molecular organic P fractions, plants can only utilize P in its soluble form as orthophosphate.[6] Unlike carbon (C) and nitrogen (N), which can be added to the soil system from the atmosphere, the P status of natural systems is essentially controlled by the occurrence of primary apatite minerals.[7,8] Consequently, the P enrichment of soils depends directly on P inputs by mineral fertilizers and manure. Although the fate of P when applied to soil remains something of an enigma, it is widely accepted that more than 80% is immobilized by the soil because of precipitation and sorption processes,[3] whereby the limiting step to furnish crop requirements is the dissolution of initial reaction products during the cropping season.[9]

For a more practical approach, the soil P continuum is thought of as three functional pools: a readily available pool, a reversibly available pool, and a sparingly available P pool,[10] whereby the readily available P pool is related to the so-called "intensity factor," which is a measure of the gradient in the electrochemical potential of the phosphate ions across the adsorbing surfaces of plant roots and, in its simplest form, can be regarded as the P concentration in the soil solution.[11] This pool represents P that is readily accessed by plant roots. The reversibly available P denotes the soil P reserve that can be converted into soluble (readily available) P, by either living organisms or by weathering during the growth season. This pool relates to the so-called "quantity factor" or "richness factor."[12] Whereas the sparingly available P is not available on a short time scale such as one or more crop cycles, a small fraction of this pool may become available during long-term soil transformation.

MOBILITY OF PHOSPHORUS IN SOILS

As a tetrahedral oxyanion, phosphate has a very low solubility in soils and in general does not move with solvent fluxes, apart from small distance diffusion.[6] In this sense, P is regarded as an "immobile" nutrient. Thus the physical movement of P is restricted to the movement of P associated with soil particles and large-molecular-weight organic matter (particulated P) by either bioturbation, soil tillage activities, or soil erosion during flow

Encyclopedia of Environmental Management DOI: 10.1081/E-EEM-120018611

events.[13,14] This behavior of soil P was recognized long ago by European geographers and since then has been used in archaeology to trace back ancient settlements.[15] Conway[16] introduced the use of total P distribution patterns for the analysis of small-scale occupation deposits. For example, one building showed evidence of having been demolished and partially reincorporated into the courtyard of a subsequent structure. The floor area of the remnant original structure was protected by a layer of small stones and contained high levels of P. That portion of the floor, which was subsequently converted into a courtyard, unprotected by stones, had less total P, having lost it by exposure and erosion. In an agricultural context, this translates to the following conclusions. First, as far as its total amounts are concerned, P applied with fertilizers may not move from the place it is applied to, and, second, keeping in mind that most soils are naturally poor in P, nearly all the spatial distribution of total P in agricultural soils should be more or less random, reflecting only the spatial sum of distribution faults of past anthropogenic activities (e.g., fertilization, animal husbandry). Spatial relationships may only have developed under the influence of erosion processes.[16]

SPATIAL SPECIATION OF PHOSPHORUS

During the last 10 years, the study of spatial variation of soil fertility parameters has expanded considerably, but studies that investigated the spatial distribution of soil P generally focused solely on the distribution of so-called plant-available P. The results of these studies showed that plant-available P does not fluctuate randomly, but shows distribution patterns with well-defined lag ranges, where the ranges differ, depending on the sampling procedure and scale of investigation (Table 1). For an introduction into geostatistical terminology one may consult Gassner and Schnug[17] in this issue.

Additionally, it is generally accepted that the distribution of available P does not necessarily resemble the distribution of total P.[20] It appears that the speciation of soil P is dependent on site-specific factors and, as such, is a spatial process.[21–23] In this sense, "spatial speciation" is defined as the chemical reactivity of a nutrient with site-specific environmental factors, and the subsequent formation of geochemical species that display different spatial dependencies (Fig. 1).[5]

Insights into the environmental processes that result in the spatial speciation of soil P and, as such, govern the behavior of applied fertilizer are necessary to predict the interconversion and equilibrium distribution of different soil P pools under specific conditions such as geomorphology, field management, and soil types.

The ratio between the readily available P pool (intensity factor) and the reversible available P pool (quantity factor), reflecting the ease of P withdrawal by the plant, is expressed as the "capacity factor."[12] The capacity factor, the P adsorption capacity of the soil, is predominantly dependent on the negative surface charges as well as the specific surface area of soil particles, and is mainly a func-

Table 1 Parameters of autocorrelation for soil P extracted by different P methods in selected investigations.

Reference	Soil texture	Sampling design (m)	Extraction method	Variogram model	Range of autocorrelation (m)
Trangmar[a]	—	1.5×1.5 m	Truog	Spherical	5.6
Boyer et al.[a]	μL	2 m transects	Bray I	Spherical	37
Doberman et al.[a]	T	5-m triangular grid	Olsen	Nested	48
Simard et al.[a]	μT—T	12×15 triangular	Mehlich III	Exponential	139
Karlen et al.[a]	—	15×15 grid	Bray I	Spherical	70
Webster and McBratney[a]	—	16×16 random	Morgan	Spherical	241
Gupta et al.[a]	sT	20×20 grid	Mehlich I	Exponential	29
Romanokov[a]	μL	20×20 grid	0.2 N HCl	Spherical	50–60
Nolin et al.[a]	μT—T	30×30 grid	Mehlich III	Exponential	39
Haneklaus et al.[a]	lS—sL	30×30 grid	CAL	Spherical	153
Gassner et al.[19]	μL	30×30 grid	AAC–EDTA	Spherical	110
Haneklaus et al.[a]	sL—μL	50×50 grid	DL	Spherical	115
Haneklaus et al.[a]	lS—sL	50×50 grid	DL	Spherical	131
Gassner et al.[5]	lS—sL	50×50 grid	CAL	Spherical	253
Chien et al.[a]	sL—μL	250×250 grid	Mehlich III	Spherical	580
Yost et al.[a]	—	1–2 km transect	Olsen	Exponential	1000

[a]Cited in Grassner[18]

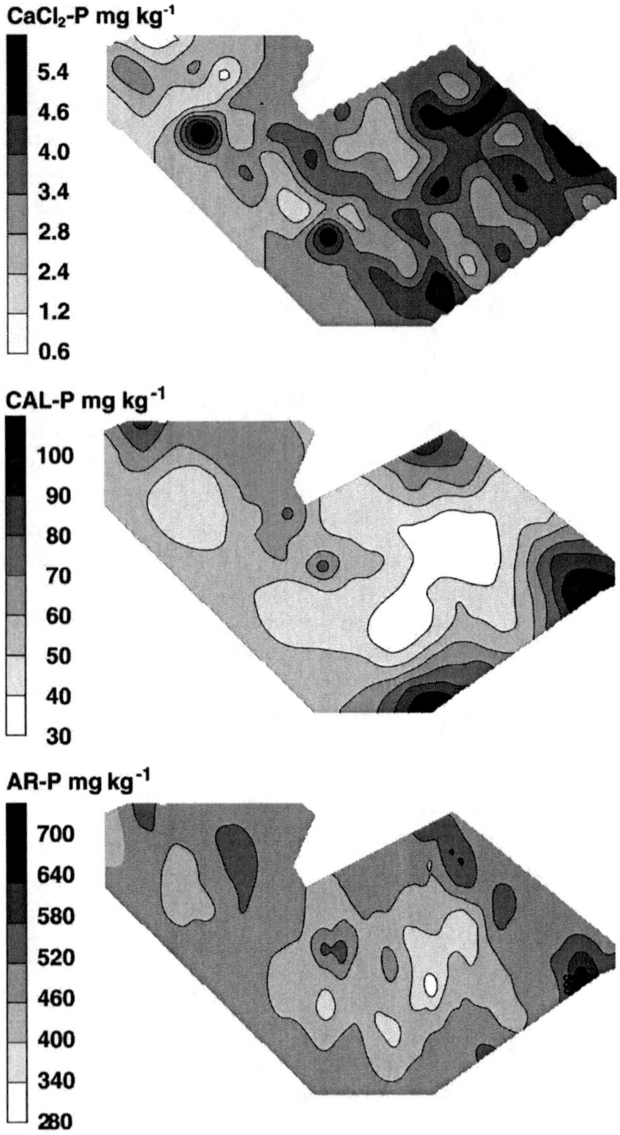
CaCl2-P mg kg-1
5.4
4.6
4.0
3.4
2.8
2.4
1.2
0.6
CAL-P mg kg-1
100
90
80
70
60
50
40
30
AR-P mg kg-1
700
640
580
520
460
400
340
280

Fig. 1 Spatial distribution of three different P fractions: $CaCl_2$ extractable P (upper map), Ca-lactate extractable P (middle map), and Aqua Regia extractable P (bottom map), at Kassow (E12°06 , N53°10), Northern Germany.

tion of the amount and nature of available adsorption sites in the soil.[24]

For a homogeneous soil, most of the potential adsorption sites will have a similar bonding energy for P. Thus the speciation of P will mainly be a function of the soil pH and the aging of initial reaction products of freshly applied fertilizer P. In this case, the spatial speciation is assumed to be low (Fig. 2). For a heterogeneous soil, the nature of the chemical reactions between P and particle surface is more diverse and the chemical speciation will result in a differentiated distribution of P among soil components. In this case, the spatial speciation is assumed to be more pronounced as the distribution of different soil components with a particular bonding energy for P will result in a spatial differentiation of P species.

Apart from the site-specific adsorption capacity of a field, site-specific anthropogenic and environmental factors such as management, biological, and physical factors can result in a spatial speciation of P. Whereas the total P—and, therefore, the sparingly available P pool—is relatively inert to short-term environmental impacts, such as recent fertilizer placement or changes in tillage practices,[5,25] the reversibly available P pool is most affected by management.[26,27] Although the readily available P pool is directly influenced by crop selection and management, McDowell and Sharpley[28] showed that under similar management conditions, the specific adsorption capacity of the soil (capacity factor) controlled the rate of P release into solution.

In a study investigating the spatial speciation of P at three different study sites, which differed in climate, parent material, topography, P fertilizer regime, and land management, the main environmental factors that controlled the spatial distribution of individual P pools were: soil texture for the readily available P; degree of aging of fresh soil P fractions, precipitated from application of soluble fertilizer P, for the reversibly available P; and primary P minerals and geomorphology for the sparingly available P.[18]

CONCLUSIONS

Phosphorus is present in soils not as definite and easily separated species, but as a continuum of compounds of different compositions and plant availability, in equilibrium with each other. This heterogeneous equilibrium is constantly disturbed by the uptake of the growing plant and by physical, chemical, or biochemical changes in the soil as well as fertilizer input. Within fields and across short distances, these factors can vary significantly in well-defined patterns, resulting in the spatial speciation of P. Consequently, soil P tests, evaluating the P status of a field, have to be adjusted to the site-specific soil characteristics. Furthermore, the interpretation of these results for subsequent management practices have to consider site-specific factors such as topography and past management history. Finally, the concept of spatial speciation should be applied to other nutrients.

REFERENCES

1. Shapiro, J. Introductory lecture at the international symposium 'phosphorus in freshwater ecosystems,' Uppsala, Sweden, October 1985. Hydrobiologia **1985**, *170*, 9–17.
2. Frossard, E.; Condron, L.M.; Oberson, A.; Sinaj, S.; Fardeau, J.C. Processes governing phosphorus availability in temperate soils. J. Environ. Qual. **2000**, *29*, 15–23.
3. Buehler, S.; Oberson, A.; Rao, I.M.; Friesen, D.K.; Frossard, E. Sequential phosphorus extraction of a P-33-labeled

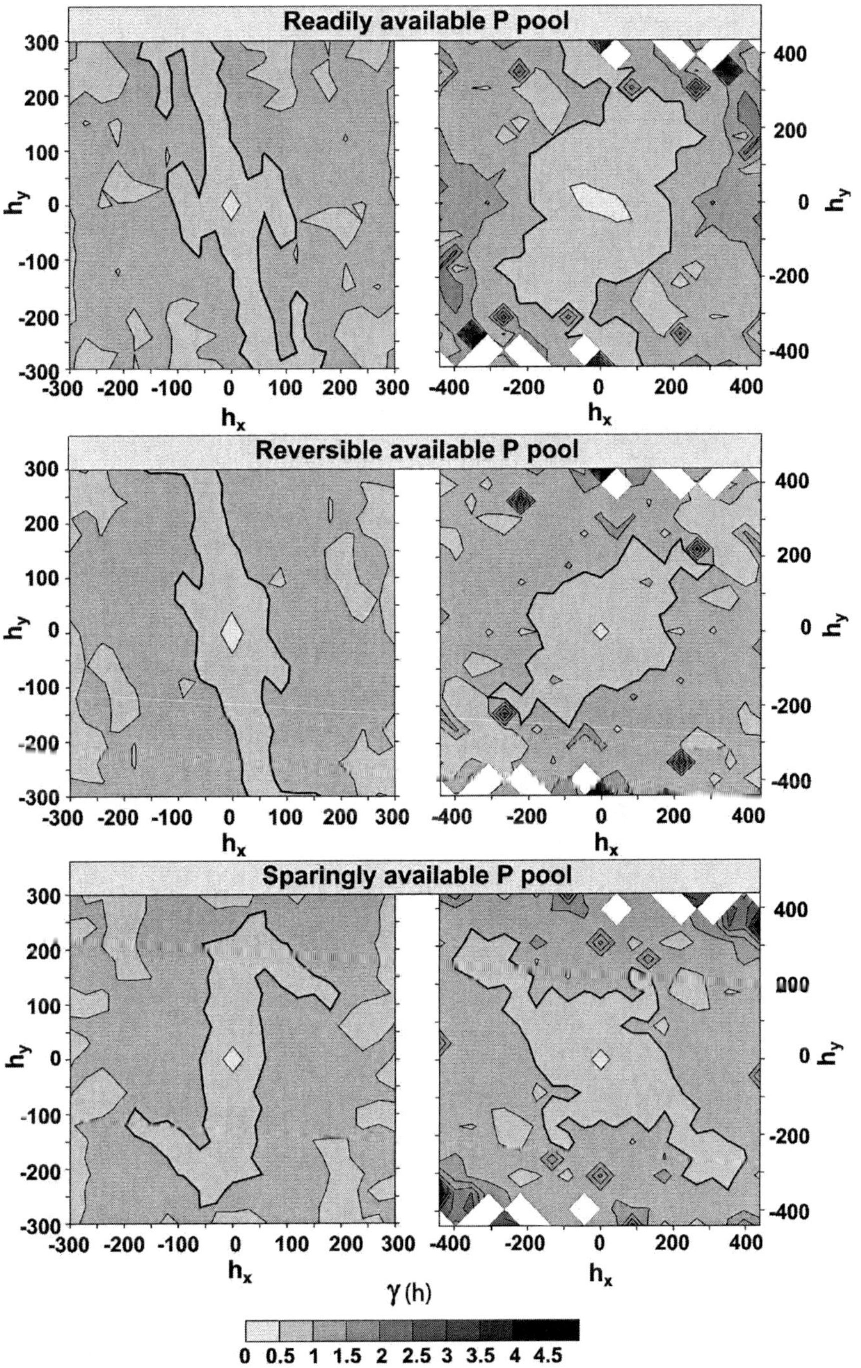
Readily available P pool
Reversible available P pool
Sparingly available P pool
h_x
h_y
300
200
100
0
-100
-200
-300
-400
400
γ(h)
0 0.5 1 1.5 2 2.5 3 3.5 4 4.5

Fig. 2 Variogram maps showing the spatial distribution of the semivariance of different P pools in a loamy sand (right) and a loamy clay (left). The semivariance is a measure of the average degree of dissimilarity between two data points.
Source: Gassner and Schnug.[17]

oxisol under contrasting agricultural systems. Soil Sci. Soc. Am. J. **2002**, *66* (3), 868–877.

4. Steen, I. Phosphorus availability in the 21st century—management of a non-renewable resource. Phosphorus Potassium 1998, 217, 25–31. British Sulphur Publishing.
5. Gassner, A.; Fleckenstein, J.; Haneklaus, S.; Schnug, E. Spatial speciation—A new approach to assess soil analysis methods. Commun. Soil Sci. Plant Anal. **2002**, *33* (15–18), 3347–3357.
6. Barber, S.A. Soil–plant interactions in the phosphorus nutrition of plants. In *The Role of Phosphorus in Agriculture*; Khasawneh, F.E., Sample, E.C., Kamprath, E.J., Eds.; ASA CSSA and SSSA: Madison, WI, 1980; 591–613.
7. Walker, T.W.; Syers, J.K. The fate of phosphorus during pedogenesis. Geoderma **1976**, *15*, 1–19.
8. Bowman, R.A.; Rodriguez, J.B.; Self, J.R. Comparison of methods to estimate occluded and resistant soil phosphorus. Soil Sci. Soc. Am. J. **1998**, *62*, 338–342.

9. Schnug, E.; Rogasik, J.; Haneklaus, S. Die ausnutzung von phosphor aus düngemitteln unter besonderer berücksichtigung des ökologischen landbaus—the utilization of fertiliser with special view of organic farming. Landbauforsch. Voelkenrode (FAL Agric. Res.) **2003**, *53* (1), 1–11.
10. Guo, F.; Yost, R.S. Partitioning soil phosphorus into three discrete pools of differing availability. Soil Sci. **1998**, *163* (10), 822–833.
11. Olsen, S.R.; Khasawneh, F.E. Use and limitation of physical–chemical criteria for assessing the status of phosphorus in soils. In *The Role of Phosphorus in Agriculture*; Khasawneh, F.E., Sample, E.C., Kamprath, E.J., Eds.; ASA CSSA and SSSA: Madison, WI, 1980; 361–410.
12. Williams, D.E.G. *The Intensity and Quality Aspects of Soil Phosphate Status and Laboratory Extraction Values*; Anales de Edafología y Agrobiología, 1966.
13. Catt, J.A.; Johnston, A.E.; Quinton, J.N. Phosphate losses in the woburn erosion reference experiment. In *Phosphorus Loss from Soil to Water*; Tunney, H., Carton, O.T., Brookes, P.C., Johnson, A.E., Eds.; CAB International, 1997; 374–377.
14. Sharpley, A.; Foy, B.;Withers, P. Practical and innovative measures for the control of agricultural phosphorus losses to water: an overview. J. Environ. Qual. **2000**, *29*, 1–9.
15. Arrhenius, O. Die phosphatfrage. Z. Pflanzenernähr. Düng. Bodenkd. **1929**, *14* (A), 185–194.
16. Conway, J.S. An investigation of soil phosphorus distribution within occupation deposits from a Romano-British hut group. J. Archaeol. Sci. **1983**, *10*, 117–128.
17. Gassner, A.; Schnug, E. Geostatistics for soil science. In *Encyclopedia of Soil Science*; Lal, R., Ed.; Marcel Dekker: New York, 2003, *in press*.
18. Gassner, A. Factors controlling the spatial speciation of phosphorous in agricultural soils. Landbauforsch. Völkenrode (Braunschweig) **2003**, *244*.
19. Gassner, A.; Habib, L.; Haneklaus, S.; Schnug, E. Significance of the spatial speciation of phosphorous in agricultural soils for the interpretation of variability. Landbauforsch. Völkenrode (FAL Agric. Res.) **2003**, *53* (1), 19–25.
20. Kamprath, E.J.; Watson, M.E. Conventional soil and tissue tests for assessing the phosphorus status of soils. In *The Role of Phosphorus in Agriculture*; Khasawneh, F.E., Sample, E.C., Kamprath, E.J., Eds.; ASA CSSA and SSSA: Madison, WI, 1980; 433–469.
21. Nowack, K.-H. *Phosphorversorgung Biologisch Bewirtschafteter Äcker und Möglichkeiten der Bioindikation*; University of Göttingen: Göttingen, 1990.
22. Strohbach, B. Gesetzmäßigkeiten der arealen verteilung des gesamtphosphorgehaltes auf landwirtschaftlich genutzten flächen. Arch. Acker-Pflanzenbau Bodenkd. (Berlin) **1986**, *30* (9), 573–580.
23. Jentsch, U. *Kapazitäts-, Quantitäts-, Intensitäts- und Kinetikparameter des Phosphats in verschiedenen Böden der DDR*; Humbold-Univ. Berlin: Berlin, 1986.
24. Schwertmann, U.; Taylor, R.M. Iron oxides. In *Minerals in Soil Environments*; Dixon, J.B., Weed, S.B., Eds.; SSSA: Madison, WI, 1989; 379–427.
25. Beckett, P.H.T.; Webster, R. Soil variability: a review. Soils Fert. **1971**, *34*, 1–15.
26. Schepers, J.S.; Schlemmer, M.R.; Ferguson, R.B. Site-specific considerations for managing phosphorus. J. Environ. Qual. **2000**, *29*, 125–130.
27. Mallarino, A.P. Spatial variability patterns of phosphorus and potassium in no-tilled soils for two sampling scales. Soil Sci. Soc. Am. J. **1996**, *60*, 1473–1481.
28. McDowell, R.; Sharpley, A. Availability of residual phosphorus in high phosphorus soils. Commun. Soil Sci. Plant Anal. **2002**, *33* (7–8), 1235–1246.

Agricultural Water Quantity Management

X.S. Qin
School of Civil and Environmental Engineering, Nanyang Technical University, Singapore

Y. Xu
MOE Key Laboratory of Regional Energy Systems Optimization, S-C Energy and Environmental Research Academy, North China Electric Power University, Beijing, China

Abstract

Over the past decades, increasing awareness has been raised regarding the development of sophisticated mathematical models for supporting agricultural water management. These models aim to cope with the problems of aggravating water shortage and serious water quality degradation and at the same time satisfy the requirement of socioeconomic development. However, the intrinsic uncertainties associated with the agricultural water management systems would bring significant difficulties in formulating and solving the related models and lead to dilemma of decision making. The objective of this study is to give a general introduction of how to conceptualize and formulate inexact optimization models for supporting integrated agricultural water quantity and quality management under uncertainty. A number of inexact optimization approaches based on stochastic, fuzzy, and interval programming models are introduced. Each of them is capable of handling a certain type of uncertain information, which depends on the uncertain features of the system components. Finally, the applicability of various optimization methods is demonstrated through an agricultural water management example.

INTRODUCTION

During the past decades, integrated water resource management has been widely recognized as a major challenge to many countries around the world.[1,2] With the rapid economic development and population growth, the water consumption rates have kept rising, leading to serious water shortage problems; meanwhile, the problems are further compounded by a drastic decrease in water quality, resulting from contamination by industrial wastes, domestic sewage water, and agricultural fertilizers and pesticides.[3] Hence, the degradation of water quality is closely linked to the reduction of available water resources. Among various economic activities, the agricultural system has received intensive concerns from government and public, due to the fact that it not only provides necessities for human living but also poses a threat to ambient environmental quality.[4] In a typical agricultural system, the crops' growth requires the nutrients provided by fertilizer and manure. As a consequence, a significant amount of nitrogen and phosphorus may be released from the applied fertilizer and manure, leading to the increase of nutrient levels and deterioration of water quality in the receiving water bodies. In addition, the irrigation activities would consume a huge amount of water, which could exacerbate water shortage problems under disadvantageous conditions.[5] Therefore, an integrated water management strategy becomes important for agricultural development.

Previously, many measures have been proposed and implemented to save water resources or protect water quality for agricultural activities, including promulgation of environmental regulations, reduced application of fertilizers or manures, adoption of soil/water conservation practices, cultivation of low-water-consumption crops, and improvement of farming practices. However, the water shortage and contamination problems have not been significantly ameliorated.[6,7,8,4] One of the major reasons is that management of agricultural water systems is a rather complex issue, involving many interacting components (e.g., type of crops, tillable land areas, and livestock numbers) or influencing factors (e.g., available water resources, environmental loading capacities, and cost implications); single control measures can hardly solve the problem as they are trying to tackle the problem from an unilateral viewpoint instead of a holistic one. On the one hand, the interactive relationships among various system components would lead to difficulties in decision making. For example, the application of the manures and fertilizers provide necessary nutrients to the crops, which are the main source of system benefits; meanwhile, it is related to the applied cost and release of pollutants. On the other hand, an agricultural system involves many social, economic, and environmental factors. For example, a variety of agricultural activities (e.g., crop cultivation and livestock breeding) require enormous infrastructural investment and resources (i.e., land and water); meanwhile, they also pose influences on

Encyclopedia of Environmental Management DOI: 10.1081/E-EEM-120047678

the socioeconomic activities, which are the main receivers of agricultural products; the socioeconomic activities could also affect agricultural activities through adopting various policies and strategies, such as prescribed pollution-control standards and fertilizer-application practices. Another challenging issue is the existence of uncertainties, which are associated with many system components in an agricultural water management system, such as the economic benefit of crop cultivation, the applied cost of fertilizer and manure, allowable pollutant discharge amounts, and the environmental capacities of receiving water bodies. These uncertainties could be derived from the random nature of the system, errors in measurement, or subjective human judgment, and would directly affect the generation of rational decision schemes.

Over the past decades, many optimization approaches were advanced for handling the above complexities and helping managers for better decision making. Traditionally, to quantify available water resources and identify the relationships among crop yield, crop quality, and irrigation water requirement, the local water authorities should conduct a thorough survey to obtain full meteorological and hydrological information, such as rainfall, surface water inflow and outflow, land areas, crop type and yield, and irrigation frequencies and quotas. Afterward, simulation models could be used to describe the interrelationships among various components within the agricultural system.[9–13] To generate cost-efficient water management strategies, optimization models could be applied.[14–17] Generally, the main objective is to maximize the economic benefit through crop cultivation and livestock breeding; meanwhile, the constraints, such as available water resources and allowable pollutant discharges, should be satisfied.

However, many of the earlier versions of optimization models did not take uncertainty into consideration, leading to oversimplified or inaccurate solutions. In recent years, a large number of inexact optimization techniques were proposed for dealing with uncertainties in agricultural water management fields.[8,4,18,5] A majority of these methods focused on stochastic mathematical programming (SMP), fuzzy mathematical programming (FMP), and interval linear programming (ILP). The selection of uncertainty-analysis techniques depends on the data availability and uncertainty characteristics associated with the system components. Stochastic mathematical programming is used mainly to tackle uncertainties expressed as random variables with probabilistic distribution functions (PDFs). Fuzzy mathematical programming uses fuzzy sets to handle uncertain data showing features of vagueness. The fuzzy membership functions are estimated empirically and can be obtained through human subjective judgments. Interval linear programming can express uncertainties as interval numbers and is effective in situations when little parameter information is available. Conclusively, a variety of optimization models were proposed in the agricultural water management field. In this entry, we aim to systematically introduce a number of the related methodologies, with a focus on demonstration of their applicability through a simplified agricultural water management case. The obtained results will effectively reflect the advantages of various uncertain optimization algorithms. Some implications of the findings and the potential research directions in this field will also be discussed.

GENERAL INEXACT OPTIMIZATION MODELS

In order to realize an integrated agricultural water quantity and quality management, the simulation models should be used to identify the interactive relationships among various system components. The model should be calibrated and verified based on the collection and analysis of long-term historical data. Then, depending on data availability and uncertain features of system components, specific optimization models can be formulated for generating cost-effective management schemes for decision makers. The alternatives of different simulation models or equations can be seen in the work of Zhang et al.,[19] Doorenbos and Kassam[9], Wang[20] Sang,[21] and Lin and Liang.[22] In this entry, the general forms of a number of inexact optimization models will be introduced.

Stochastic Mathematical Programming

Stochastic mathematical programming is used mainly to tackle uncertainties expressed as random variables with PDFs. A series of solutions under different probability levels can be obtained. However, the applicability of SMP is restricted by its rigorous data requirement to specify parameter probability distributions and intensive computational burden.[23] A general SMP model can be written as follows:

$$\text{Minimize } f = \sum_{j=1}^{J} c_j(s) x_j \tag{1a}$$

subject to

$$\sum_{j=1}^{J} a_{ij}(s) x_j \le b_i(s) \quad \forall i, s \tag{1b}$$

$$\sum_{j=1}^{J} d_{rj} x_j \le e_r, \quad \forall r \tag{1c}$$

$$x_j \ge 0, \quad \forall j \tag{1d}$$

$$c_j(s), a_{ij}(s), d_{rj} \ne 0, \quad \forall i, r, j, s \tag{1e}$$

where x_j are decision variables; $c_j(s)$, $a_{ij}(s)$, and $b_i(s)$ are sets with random elements defined on a probability space s, $s \in S$;[24] and d_{rj} and e_r are fixed coefficients for deterministic constraints. The main step of solving Eq. 1 is to transform the random variables into deterministic ones based on different algorithms. Currently, SMP handles random variables in different ways, such as stochastic chance-constrained programming (SCCP),[8,24–26] two-stage stochastic programming (TSP),[27,28] multistage stochastic programming,[29,30] and stochastic robust optimization.[31,26,28,32] The detailed solution procedures can be found in the related references.

Fuzzy Mathematical Programming

Fuzzy mathematical programming is derived through incorporating the fuzzy set theory into the conventional mathematical programming framework. Fuzzy mathematical programming uses fuzzy sets to handle uncertain data, which are characterized by vagueness or imprecision. The fuzzy membership functions are estimated empirically and can be obtained based on subjective judgments. A general FMP can be written as follows:

$$\text{Minimize } \tilde{f} = \sum_{j=1}^{J} \tilde{c}_j x_j \tag{2a}$$

subject to

$$\sum_{j=1}^{J} \tilde{a}_{ij} x_j \leq \tilde{b}_i \quad \forall i \tag{2b}$$

$$\sum_{j=1}^{J} d_{rj} x_j \leq e_r, \quad \forall r \tag{2c}$$

$$x_j \geq 0, \quad \forall j \tag{2d}$$

$$\tilde{c}_j, \tilde{a}_{ij}, d_{rj} \neq 0, \quad \forall i, r, j \tag{2e}$$

where $\tilde{f}$ is the objective function presented in a fuzzy form; x_j are decision variables; $\tilde{c}_j$, $\tilde{a}_{ij}$, and $\tilde{b}_i$ are assumed to be different types of fuzzy sets (e.g., triangular, trapezoidal, and exponential forms) with fuzzy membership functions being denoted as $\mu(\tilde{c}_j)$, $\mu(\tilde{a}_{ij})$, and $\mu(\tilde{b}_i)$, respectively; and d_{rj} and e_i are fixed coefficients. To solve Eq. 2, the fuzzy sets are also required to be transformed into deterministic variables. Currently, FMP could be sorted into the following categories based on different modeling structures and solution algorithms: fuzzy flexible programming,[33,34] fuzzy possibilistic programming,[34] fuzzy robust programming (FRP),[23,4] and fuzzy chance-constrained programming (FCCP).[35,3] The detailed solution procedures can be referred to in the related references.

Interval Linear Programming

Interval linear programming is capable of handling the interval-type uncertainties in optimization problems, with

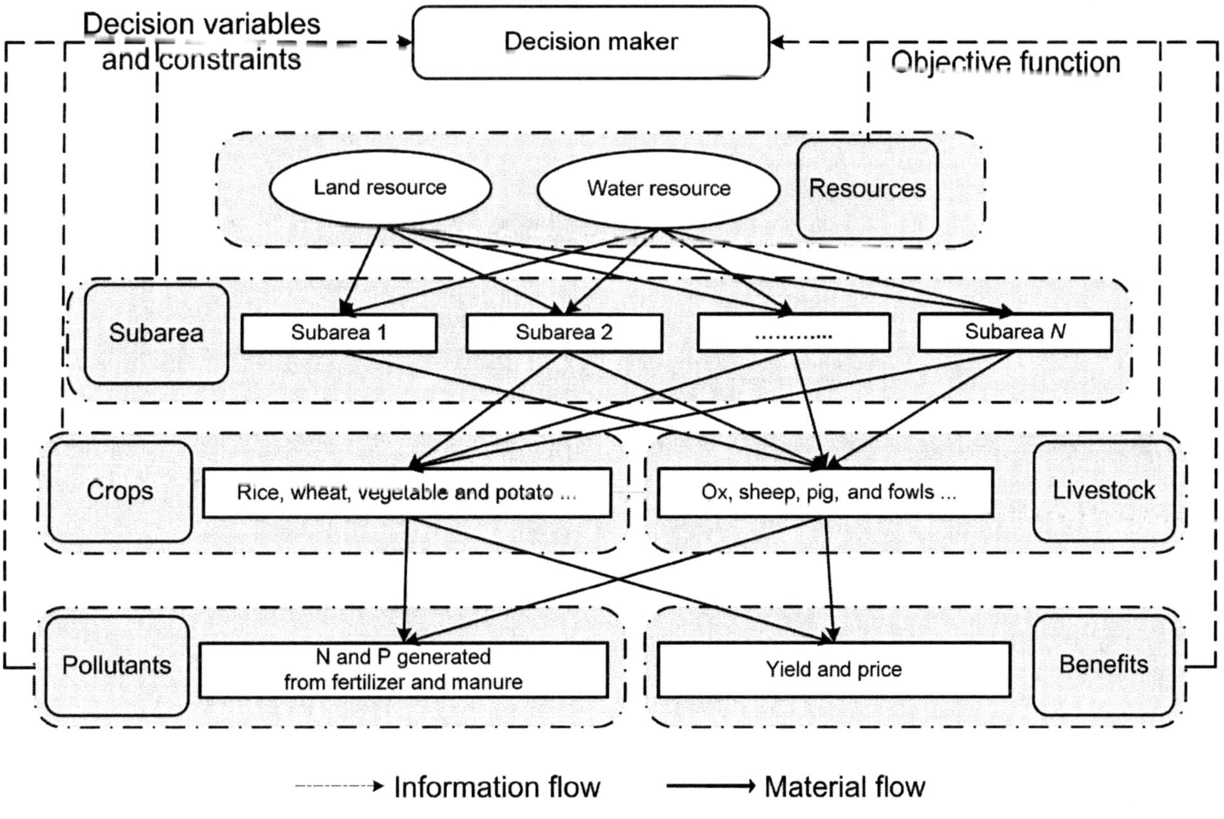

Decision variables and constraints
Decision maker
Objective function
Land resource
Water resource
Resources
Subarea
Subarea 1
Subarea 2
....
Subarea N
Crops
Rice, wheat, vegetable and potato ...
Ox, sheep, pig, and fowls ...
Livestock
Pollutants
N and P generated from fertilizer and manure
Yield and price
Benefits
Information flow
Material flow

Fig. 1 Overall structure of the studied system.

all or part of the model parameters being expressed as interval numbers. The model could be solved by the interactive two-step algorithm proposed by Huang et al.[36] A general ILP model can be formulated as follows:

$$\text{Minimize } f^{\pm} = \sum_{j=1}^{n} c_j^{\pm} x_j^{\pm} \tag{3a}$$

subject to

$$\sum_{j=1}^{n} a_{ij}^{\pm} x_j^{\pm} \leq b_i^{\pm} \quad \forall i \tag{3b}$$

$$x_j^{\pm} \geq 0, \quad \forall j \tag{3c}$$

$$c_j^{\pm}, a_{ij}^{\pm} \neq 0, \quad \forall i, j \tag{3d}$$

where $x_j^{\pm}$ are the decision variables presented in interval forms; ± is a boundary symbol for an interval number, representing a spectrum of possible values between the lower and upper bounds; and $a_{ij}^{\pm}$, $b_i^{\pm}$, and $c_j^{\pm}$ are coefficients represented as interval numbers. The term "interval number" is expressed as $a_{ij}^{\pm} = [a_{ij}^{-}, a_{ij}^{+}]$ where the items a_{ij}^{-} and a_{ij}^{+} are the lower and upper bounds of $a_{ij}^{\pm}$, respectively. The objective value and decision variables are obtained as $f_{\text{opt}}^{\pm} = \left[f_{\text{opt}}^{-}, f_{\text{opt}}^{+}\right]$ and $x_{j,\text{opt}}^{\pm} = \left[x_{j,\text{opt}}^{-}, x_{j,\text{opt}}^{+}\right]$, respectively. Previously, many applications of ILP in dealing with environmental management problems were reported.[33,36–39, 26,28,32]

CASE STUDY

Overview of the Study Case

To demonstrate the applicability of the introduced inexact optimization methods, a simplified agricultural water quantity and quality management case is adopted.[7,8] The study system is conceptualized as a typical rural area (as shown in Fig. 1), consisting of two subareas, where crop cultivation (including wheat, vegetable, and potato) and livestock breeding (including swine and poultry) are the two major economic activities. The water consumption is mainly required by crop irrigation; the pollutants generated from manure/fertilizer applications are discharged into the receiving water bodies. Each subarea has a canal to supply water. The decision makers are responsible for designing optimal farming patterns within a multiperiod horizon, in light of the economic return from the agricultural products, the local land/water resource conditions, pollutant discharge standards, and environmental loading capacities of the receiving water bodies. Fig. 2 shows the management framework of the system. For uncertain or deterministic parameters that will be used in different methodologies, they are drawn from literature.[4,5,7,8] In order to better demonstrate methodologies, some parameters are directly assumed based on literature, and the scale of the problem (e.g., the type of agricultural activities and the number of subareas) is considerably simplified.

General Model Formulation

A general model for agricultural water quantity–quality management can be formulated as follows:[4,5,7,8]

$$\text{Maximize } f = \sum_{t=1}^{T}\sum_{i=1}^{I} \delta_i Y_i S_{it} + \sum_{j=1}^{J} \eta_j Z_j - \sum_{t=1}^{T}\sum_{i=1}^{I} G_i S_{it} - G_M \sum_{i=1}^{I} M_i - G_F \sum_{i=1}^{I} F_i - \sum_{t=1}^{T}\sum_{i=1}^{I} v_{it} W_{it} S_{it} \tag{4a}$$

subject to

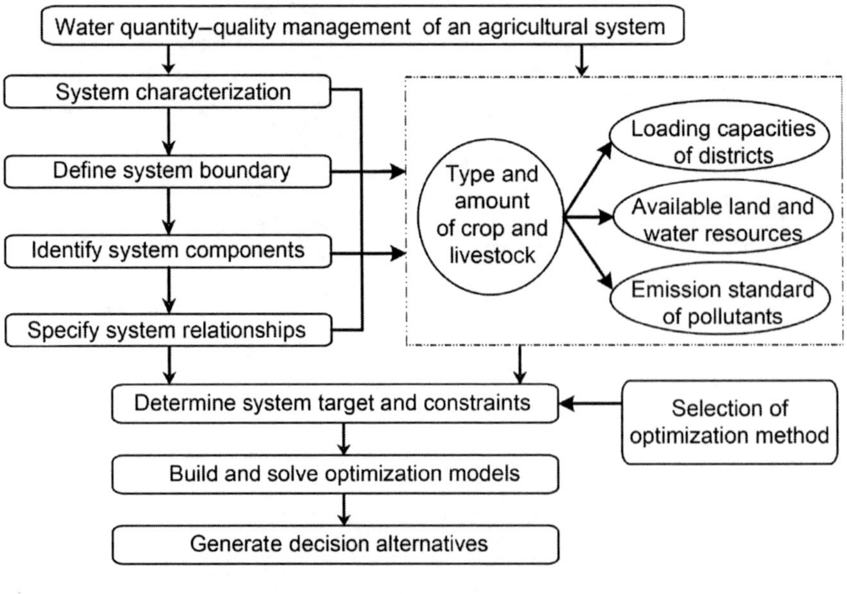

Water quantity–quality management of an agricultural system
System characterization
Define system boundary
Identify system components
Specify system relationships
Type and amount of crop and livestock
Loading capacities of districts
Available land and water resources
Emission standard of pollutants
Determine system target and constraints
Selection of optimization method
Build and solve optimization models
Generate decision alternatives

Fig. 2 Procedures of agricultural water quantity and quality management.

1. Manure mass balance:

$$\sum_{i=1}^{I} M_i - \sum_{j=1}^{J} B_j Z_j = 0 \tag{4b}$$

2. Crop nutrient balances:

$$(1-p_1)g_i M_i + (1-p_2)F_i - \sum_{t=1}^{T} q_i S_{it} \geq 0, \quad \forall i \tag{4c}$$

3. Energy and digestive protein requirements:

$$\sum_{t=1}^{T}\sum_{i=1}^{I} Y_i \beta_i S_{it} - \sum_{j=1}^{J} E_j Z_j \geq 0 \tag{4d}$$

$$\sum_{t=1}^{T}\sum_{i=1}^{I} Y_i \gamma_i S_{it} - \sum_{j=1}^{J} D_j Z_j \geq 0 \tag{4e}$$

4. Pollutant losses:

$$\sum_{i=1}^{I}\left(M_i g + F_i - q_i \sum_{t=1}^{T} S_{it}\right) \leq a \sum_{t=1}^{T} K_t \tag{4f}$$

$$\sum_{i=1}^{I} S_{it} \leq K_t \quad \forall t \tag{4g}$$

$$\sum_{t=1}^{T}\sum_{i=1}^{I} L_i S_{it} \leq b \sum_{t=1}^{T} K_t \tag{4h}$$

$$\sum_{t=1}^{T}\sum_{i=1}^{I} h_1 L_i S_{it} \leq c_1 \sum_{t=1}^{T} K_t \tag{4i}$$

$$\sum_{t=1}^{T}\sum_{i=1}^{I} h_2 L_i S_{it} \leq c_2 \sum_{t=1}^{T} K_t \tag{4j}$$

$$\sum_{t=1}^{T}\sum_{i=1}^{I} (R_{1i}N_{1i} + R_{2i}N_{2i}) S_{it} \leq u_1 \sum_{t=1}^{T} K_t \tag{4k}$$

$$\sum_{t=1}^{T}\sum_{i=1}^{I} (R_{1i}P_{1i} + R_{2i}P_{2i}) S_{it} \leq u_2 \sum_{t=1}^{T} K_t \tag{4l}$$

5. Water quantity constraints:

$$\sum_{i=1}^{I} W_{it} S_{it} < Q_t, \quad \forall t \tag{4m}$$

6. Nonnegativity constraints:

$$S_{it}, M_i, F_i, T_j \geq 0, \quad \forall i, j, t \tag{4n}$$

where f = net system income (\$); t, i, and j (t = 1, 2, ..., T; i = 1, 2, ..., I; j = 1, 2, ..., J) are indexes of subarea, crops, and livestock, respectively; T, I, and J are numbers of subarea, crops, and livestock, respectively; Y_i = unit yield of crop i (kg/ha); S_{it} = area of crop i in subarea t (ha); Z_j = number of livestock j in the study area; M_i = amount of manure applied to crop i (t); F_i = amount of fertilizer nitrogen applied to crop i (kg); W_{it} = flow rate of irrigation water required by crop i in subarea t [(m^3/s)/ha]; B_j = unit amount of manure generated by livestock j that needs to be disposed (t/unit); p_1 = nitrogen volatilization/denitrification rate of manure (%); g = nitrogen concentration of manure (kg/t); p_2 = nitrogen volatilization/denitrification rate of fertilizer (%); q_i = unit nitrogen requirement of crop i (kg/ha); β_i = net energy potential of crop i (Mcal/kg); E_j = unit net energy requirement of livestock j (Mcal/unit); γ_i = digestible protein content of crop i (%); D_j = unit digestible protein requirement of livestock j (kg/unit); K_t = tillable area in subarea t (ha); a = maximum allowable total nitrogen loss rate (kg/ha); h_1 = nitrogen content of soil (%); h_2 = phosphorus content of soil (%); N_{1i} = dissolved nitrogen concentration in wet season runoff from land planted to crop i (mg/l); N_{2i} = dissolved nitrogen concentration in dry season runoff from land planted to crop i (mg/l); R_{1i} = wet season runoff from land planted to crop i (mm); R_{2i} = dry season runoff from land planted to crop i (mm); u_1 = maximum allowable loss rate of dissolved nitrogen by runoff (kg/ha); u_2 = maximum allowable loss rate of dissolved phosphorus by runoff (kg/ha); p_{1i} = dissolved phosphorus concentration in wet season runoff from land planted to crop i (mg/l); p_{2i} = dissolved phosphorus concentration in dry season runoff from land planted to crop i (mg/l); Q_t = maximum canal flow within subarea t (m^3/s); δ_i = unit price of crop i (\$/kg); η_j = unit average return from livestock j (\$/unit); G_i = unit farming cost for crop i (\$/ha); G_M = unit cost of manure collection and disposal (dollars/t); G_H = unit cost of fertilizer application (\$/kg); v_{it} = unit cost to deliver water to S_{it} {\$/(m^3/s)}; L_i = soil loss rate from land planted to crop i (kg/ha); b = maximum allowable soil loss rate (kg/ha); c_1 = maximum allowable solid-phase nitrogen loss rate (kg/ha); and c_2 = maximum allowable solid-phase phosphorus loss rate (kg/ha).

As discussed in the "Introduction" section, a variety of uncertainties are associated with the agricultural water management system. The optimization model can effectively reflect the relationships between the agricultural and social–economic system and among various system components. The uncertainties can be expressed as random variables, fuzzy sets, or discrete intervals based on the uncertain features of parameters and data availability. In the next section, three inexact optimization models will be used to tackle problem.

Interval Parameter Water Quality Management Model

Huang[7] presented an interval parameter water quality management model (IPWM) for agricultural water quantity and quality planning. The model allowed uncertain information of parameters, expressed as interval numbers, to be effectively communicated into the optimization process and the obtained solutions. Compared with SMP and FMP, the generation of interval numbers has the least data requirement. In a typical agricultural system, many hard-to-get data could be assumed as interval numbers. Table 1 shows part of the interval parameters for the study case. The interval constraints of an IPWM model for the pollutant losses part could be written as follows:

$$\text{Maximize } f^{\pm} = \sum_{t=1}^{r}\sum_{i=1}^{m}\delta_i^{\pm}Y_i^{\pm}S_{it}^{\pm} + \sum_{j=1}^{n}\eta_j^{\pm}Z_j^{\pm} - \sum_{t=1}^{r}\sum_{i=1}^{m}G_i^{\pm}S_{it}^{\pm} - G_M^{\pm}\sum_{i=1}^{m}M_i^{\pm} - G_F^{\pm}\sum_{i=1}^{m}F_i^{\pm} - \sum_{t=1}^{r}\sum_{i=1}^{m}v_{it}^{\pm}W_{it}^{\pm}S_{it}^{\pm} \quad (5a)$$

subject to

$$\sum_{i=1}^{m}M_i^{\pm} - \sum_{j=1}^{n}B_j^{\pm}Z_j^{\pm} = 0 \quad (5b)$$

$$(1-p_1^{\pm})g_i^{\pm}M_i^{\pm} + (1-p_2^{\pm})F_i^{\pm} - \sum_{t=1}^{T}q_i^{\pm}S_{it}^{\pm} \geq 0, \quad \forall i \quad (5c)$$

$$\sum_{t=1}^{T}\sum_{i=1}^{I}L_i^{\pm}S_{it}^{\pm} \leq b^{\pm}\sum_{t=1}^{T}K_t \quad (5d)$$

$$\sum_{t=1}^{T}\sum_{i=1}^{I}h_1^{\pm}L_i^{\pm}S_{it}^{\pm} \leq c_1^{\pm}\sum_{t=1}^{T}K_t \quad (5e)$$

$$\sum_{t=1}^{T}\sum_{i=1}^{I}h_2^{\pm}L_i^{\pm}S_{it}^{\pm} \leq c_2^{\pm}\sum_{t=1}^{T}K_t \quad (5f)$$

$$\sum_{t=1}^{T}\sum_{i=1}^{I}(R_{1i}^{\pm}N_{1i}^{\pm} + R_{2i}^{\pm}N_{2i}^{\pm})S_{it}^{\pm} \leq u_1^{\pm}\sum_{t=1}^{T}K_t \quad (5g)$$

$$\sum_{t=1}^{T}\sum_{i=1}^{I}(R_{1i}^{\pm}P_{1i}^{\pm} + R_{2i}^{\pm}P_{2i}^{\pm})S_{it}^{\pm} \leq u_2^{\pm}\sum_{t=1}^{T}K_t \quad (5h)$$

$$\sum_{i=1}^{I}W_{it}^{\pm}S_{it}^{\pm} \leq Q_t^{\pm}, \quad \forall t \quad (5i)$$

As proposed by Huang et al.[36,38], the solution for Eq. 5 can be obtained through an interactive two-step method. A submodel corresponding to f^+ (when the objective function is to be maximized) is first formulated and solved, and then the relevant submodel corresponding to f can be formulated based on the solution of the first submodel. Finally, the objective value and decision variables as discrete intervals can be obtained as $f_{\text{opt}}^{\pm} = \left[f_{\text{opt}}^{-}, f_{\text{opt}}^{+}\right]$ and $x_{j,\text{opt}}^{\pm} = \left[x_{j,\text{opt}}^{-}, x_{j,\text{opt}}^{+}\right]$. The interactive method does not lead to complicated intermediate models and thus has low computational requirements.

Table 2 shows the solutions from IPWM for the study case. Some results are represented as interval numbers. For example, the objective function value (i.e., net income) would range from \$121.38 × 10^3 to \$310.99 × 10^3. The upper bound of the objective function represents an optimal decision alternative with the highest net income. Correspondingly, the obtained decision variables of cropping area and livestock amount would reach their upper bounds, and the manure and fertilizer would achieve their lower-bound values. For example, the cropping area of the vegetables in subarea 1 is 39.60 km^2; the number of swine is

Table 1 Part of the model parameters presented as discrete intervals.

Model parameters	Values
Nitrogen content of manure (kg/t)	[10, 13]
Unit cost of manure collection/disposal (\$/t)	[4, 5]
Unit cost of fertilizer application (\$/kg)	[0.7, 0.9]
Nitrogen content of soil (%)	[0.0020, 0.0025]
Phosphorus content of soil (%)	[0.0009, 0.0011]
Nitrogen volatilization/denitrification rate of manure (%)	[0.3, 0.35]
Phosphorus volatilization/denitrification rate of manure (%)	[0.10, 0.12]
Maximum allowable dissolved nitrogen loss rate by runoff (kg/ha)	[2.2, 2.3]
Maximum allowable dissolved phosphorus loss rate by runoff (kg/ha)	[0.20, 0.22]
Maximum allowable nitrogen losses (kg/ha)	[38, 40]
Maximum allowable soil loss (kg/ha)	[3650, 5600]
Maximum allowable soil-phase nitrogen loss rate (kg/ha)	[10, 11]
Maximum allowable soil-phase phosphorus loss rate (kg/ha)	[4.3, 4.7]

Source: Adapted from Huang.[7]

Table 2 Interval solutions of IPWM model.

Model solutions	Values
Cropping area of wheat in subarea 1 (km^2)	0
Cropping area of wheat in subarea 2 (km^2)	0
Cropping area of vegetable in subarea 1 (km^2)	[22.38, 39.60]
Cropping area of vegetable in subarea 2 (km^2)	[10.96, 19.23]
Cropping area of potato in subarea 1 (km^2)	0
Cropping area of potato in subarea 2 (km^2)	0
Applied manure amounts of wheat (10^6 t)	[0, 8.19]
Applied manure amounts of vegetable (10^6 t)	606.12
Applied manure amounts of potato (10^6 t)	0
Applied fertilizer amounts of wheat (kg)	0
Applied fertilizer amounts of vegetable (kg)	4004.15
Applied fertilizer amounts of potato (kg)	0
Swine amounts (unit)	[311, 324]
Poultry amounts (unit)	[379, 557]
Net income (10^3 \$)	[121.38, 310.99]

324 units; the applied manure and fertilizer amounts of the vegetables are 606.12×10^6 t and 4004.15 kg, respectively. These upper-bound solutions are obtained under an advantageous condition. For example, the discharge standards of pollutants are relatively loose, and the loading capacities of the receiving water bodies and the available water amounts are relatively larger. This also means that higher system benefits correspond with higher levels of pollutant discharge and/or more serious water shortage problems. Conversely, the lower-bound solution of net income is under conservative consideration, where the conditions are stricter (e.g., the allowable pollutant amounts and provided water amounts are lower). In such a case, the water quality could be maintained at a higher level, and the water shortage crisis could be alleviated somewhat. Based on the obtained interval solutions, a variety of decision alternatives can be generated through adjusting continuously within their solution intervals. In fact, decision variables presented as discrete intervals are rather flexible for decision makers, since they are capable of assisting decision makers in making the decision based on their preferences (e.g., economic development or environmental protection).

Although the ILP method has many advantages, it also has limits. Firstly, the representation of uncertainties by discrete intervals is relatively simple, where much useful information is neglected. Secondly, although ILP has been proven as an effective approach in dealing with uncertainties, it may lead to the difficulties when the model's right-hand-side coefficients are highly uncertain. Therefore, further investigations are needed to improve the applicability of ILP. In fact, many other types of uncertainty methods, such as SCCP, TSP, and FRP, have potential to be further integrated into an ILP framework for mitigating the drawbacks of ILP and reflecting more complex conditions.

Inexact Stochastic Water Management Model

Stochastic chance-constrained programming is one of the SMP methods that can tackle uncertain parameters presented as PDFs. Its main advantage is that it does not require that all of the constraints be strictly satisfied. Instead, they can be satisfied in a proportion of cases with given probabilities.[40] However, SCCP is restricted by data availability to generate stochastic distributions.[37,38] Interval linear programming is widely applicable for addressing all uncertain parameters; however, it does not allow any violation of system constraints and might be infeasible when the right-hand-side parameters in constraints are highly uncertain. The two methods have varied strengths and weaknesses, with a potential for compensating each other when they are combined.

Huang[8] proposed an interval-stochastic chance-constrained programming (ISCCP) model for agricultural water quantity and quality management. The ISCCP model is an integration of ILP and SCCP but improves upon both. It allows the right-hand-side parameters in the constraints to be expressed as PDFs and the coefficients in the objective function and the left-hand-side parameters in the constraints to be described as interval numbers. For the introduced study case, the available water resource amount, the maximum allowable soil loss, the maximum allowable soil-phase nitrogen loss, and the maximum allowable soil-phase phosphorus loss are assumed as random variables (expressed as PDFs). Table 3 lists the random parameters used in the ISCCP model. Other uncertain parameters are tackled as discrete intervals, similar to those in the IPWM model.

Thus, the interval-stochastic part of an ISCCP model can be written as follows:

$$\text{Maximize } f^{\pm} = \sum_{t=1}^{r}\sum_{i=1}^{m}\delta_i^{\pm}Y_i^{\pm}S_{it}^{\pm} + \sum_{j=1}^{n}\eta_j^{\pm}Z_j^{\pm} - \sum_{t=1}^{r}\sum_{i=1}^{m}G_i^{\pm}S_{it}^{\pm} - G_M^{\pm}\sum_{i=1}^{m}M_i^{\pm} - G_F^{\pm}\sum_{i=1}^{m}F_i^{\pm} - \sum_{t=1}^{r}\sum_{i=1}^{m}v_{it}^{\pm}W_{it}^{\pm}S_{it}^{\pm} \tag{6a}$$

subject to

$$\sum_{i=1}^{m}M_i^{\pm} - \sum_{j=1}^{n}B_j^{\pm}Z_j^{\pm} = 0 \tag{6b}$$

$$\sum_{t=1}^{T}\sum_{i=1}^{I}L_i^{\pm}S_{it}^{\pm} \le b(s)\sum_{t=1}^{T}K_t \tag{6c}$$

$$\sum_{t=1}^{T}\sum_{i=1}^{I}h_1^{\pm}L_i^{\pm}S_{it}^{\pm} \le c_1(s)\sum_{t=1}^{T}K_t \tag{6d}$$

$$\sum_{t=1}^{T}\sum_{i=1}^{I}h_2^{\pm}L_i^{\pm}S_{it}^{\pm} \le c_2(s)\sum_{t=1}^{T}K_t \tag{6e}$$

$$\sum_{i=1}^{I}W_{it}^{\pm}S_{it}^{\pm} \le Q_t(s), \quad \forall t \tag{6f}$$

In the solution process of Eq. 6, the constraints in Eq. 6c–f are required to be satisfied with at least a probability of $1-p_h$. According to Charnes et al.,[24] these constraints can be written as[24]

Table 3 Random parameters in ISCCP model.

Model parameters	Values
Supplied water amounts of canal 1 (m^3/ha-s)	(3.6, 0.12)
Supplied water amounts of canal 2 (m^3/ha-s)	(0.9, 0.04)
Maximum allowable soil loss (kg/ha)	(4600, 150)
Maximum allowable soil-phase nitrogen loss (kg/ha)	(10.5, 0.5)
Maximum allowable soil-phase phosphorus (kg/ha)	(4.5, 0.2)

Note: (a_1, a_2) represents a random variable where a_1 and a_2 are mean and standard deviation, respectively.

Agricultural – Alexandria

$$P_r\left[\sum_{t=1}^{T}\sum_{i=1}^{I} L_i^{\pm} S_{it}^{\pm} \le b(s)\sum_{t=1}^{T} K_t\right] \ge 1-p_h, \quad \forall t,h \qquad (7a)$$

$$P_r\left[\sum_{t=1}^{T}\sum_{i=1}^{I} h_1^{\pm} L_i^{\pm} S_{it}^{\pm} \le c_1(s)\sum_{t=1}^{T} K_t\right] \ge 1-p_h, \quad \forall t,h \qquad (7b)$$

$$P_r\left[\sum_{t=1}^{T}\sum_{i=1}^{I} h_2^{\pm} L_i^{\pm} S_{it}^{\pm} \le c_2(s)\sum_{t=1}^{T} K_t\right] \ge 1-p_h, \quad \forall t,h \qquad (7c)$$

$$P_r\left[\sum_{i=1}^{I} W_{it}^{\pm} S_{it}^{\pm} \le Q_t(s)\right] \ge 1-p_h, \quad \forall t,h \qquad (7d)$$

where p_h means the allowable probability of violation given by the decision makers. In this entry, the values of p_h are considered to be 0.01 and 0.05, respectively. Referring to Charnes et al.,[24] the constraints in Eq. 7a–7d can be transformed to their equivalent forms as follows:

$$\sum_{t=1}^{T}\sum_{i=1}^{I} L_i^{\pm} S_{it}^{\pm} \le b(s)^{p_h}\sum_{t=1}^{T} K_t \qquad (8a)$$

$$\sum_{t=1}^{T}\sum_{i=1}^{I} h_1^{\pm} L_i^{\pm} S_{it}^{\pm} \le c_1(s)^{p_h}\sum_{t=1}^{T} K_t \qquad (8b)$$

$$\sum_{t=1}^{T}\sum_{i=1}^{I} h_2^{\pm} L_i^{\pm} S_{it}^{\pm} \le c_2(s)^{p_h}\sum_{t=1}^{T} K_t \qquad (8c)$$

$$\sum_{i=1}^{I} W_{it}^{\pm} S_{it}^{\pm} \le Q_t(s)^{p_h}, \quad \forall t,h \qquad (8d)$$

where $a(s)^{p_h} = F_h^{-1}[a(s)]$, given the cumulative distribution function of $a(s)$ and the probability of violating constraint h (i.e., p_h). Finally, Eq. 8 can be transformed into pure ILP models under various constraint-violation levels and then solved by an interactive two-step method.[36] Fig. 3 shows the solution procedures of an ISCCP model.

Table 4 lists the solutions from the ISCCP model under probability levels of 0.01 and 0.05. Since ISCCP is an integration of ILP and SCCP, the obtained solutions reflect characteristics of both methods. Firstly, due to the existence of ILP, a majority of solutions present as interval numbers, which are feasible and stable in the given decision space; thus, the decision schemes can be generated in the ranges of their solution intervals. For example, at a significance level of $p_h = 0.01$, the total system benefits are [124.38, 256.03] $\times 10^3$. The upper bounds of objective function are preferable under advantageous conditions; conversely, the lower bounds of objective function value are more desirable under conservative conditions. The local managers can make decisions based on their preferences.

The solutions of ISCCP also possess features of SCCP where the trade-off between cost and risk can be evaluated (reflected by adjusting p_h). For example, when p_h increases from 0.01 to 0.05, the total system benefit would increase from [124.38, 256.03] to [127.40, 262.25] $\times 10^3$. This is due to the fact that, as p_h increases, the allowable degree of violating constraints would increase (e.g., looser pollutant discharge standards and environmental loading capacities). This indicates that a higher system benefit could lead to a higher system risk. Conversely, if the planner aims toward a lower system benefit, a more reliable management alternative would be generated. The p_h values can be used to help analyze the trade-off between system economy and failure risk and offer local water managers a spectrum of alternatives for decision making.

ISCCP also shows some limitations. Firstly, the conversion from a stochastic constraint into an "equivalent" deterministic one applies only in conditions when the right-hand-side items are stochastic. The interval treatment for the left-hand-side uncertainties is a compromised approach, which may lead to the loss of valuable information if additional distribution information of the interval parameters is available.

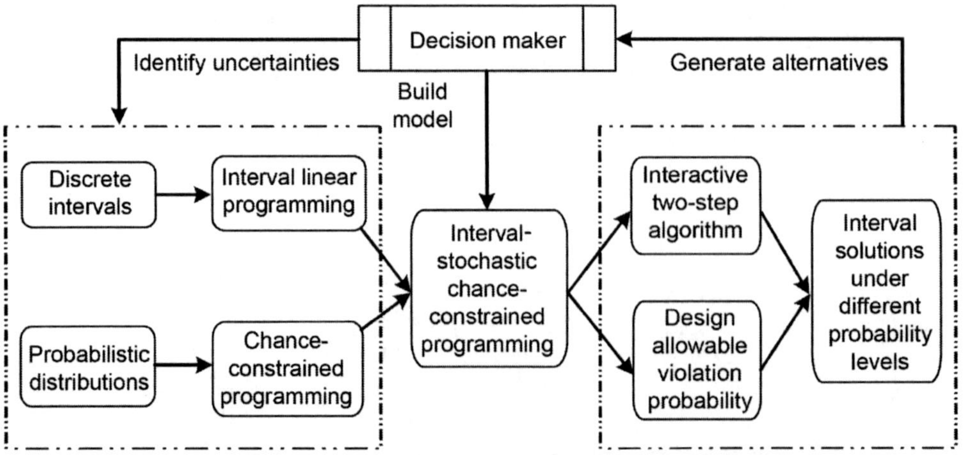

Decision maker
Identify uncertainties
Generate alternatives
Build model
Discrete intervals
Interval linear programming
Probabilistic distributions
Chance-constrained programming
Interval-stochastic chance-constrained programming
Interactive two-step algorithm
Design allowable violation probability
Interval solutions under different probability levels

Fig. 3 The framework for applying the ISCCP model.

Table 4 Solutions of ISCCP model at probability levels of 0.01 and 0.05.

Model solutions	Probability (p_i)	Values
Cropping area of wheat in subarea 1 (km^2)	0.01	0
	0.05	0
Cropping area of wheat in subarea 2 (km^2)	0.01	0
	0.05	0
Cropping area of vegetable in subarea 1 (km^2)	0.01	[23.22, 32.87]
	0.05	[23.78, 33.66]
Cropping area of vegetable in subarea 2 (km^2)	0.01	[11.10, 15.58]
	0.05	[11.37, 15.96]
Cropping area of potato in subarea 1 (km^2)	0.01	0
	0.05	0
Cropping area of potato in subarea 2 (km^2)	0.01	0
	0.05	0
Applied manure amounts of wheat (10^6 t)	0.01	0
	0.05	0
Applied manure amounts of vegetable (10^6 t)	0.01	499.11
	0.05	511.24
Applied manure amounts of potato (10^6 t)	0.01	0
	0.05	0
Applied fertilizer amounts of wheat (kg)	0.01	0
	0.05	0
Applied fertilizer amounts of vegetable (kg)	0.01	3297.27
	0.05	3377.35
Applied fertilizer amounts of potato (kg)	0.01	0
	0.05	0
Swine amounts (unit)	0.01	267
	0.05	273
Poultry amounts (unit)	0.01	458
	0.05	470
Net income (10^3 \$)	0.01	[124.38, 256.03]
	0.05	[127.40, 262.25]

Notes: [a,b] represents an interval number, where a and b are the lower and upper bounds, respectively.

Inexact Double-Sided Fuzzy Chance-Constrained Programming

The fuzzy approach is another alternative to deal with uncertainties. Referring to Xu and Qin,[5] an inexact double-sided fuzzy chance-constrained programming (IDFCCP) model was used to solve the agricultural water quantity and quality management problem. The IDFCCP model is formulated through incorporating ILP into a double-sided fuzzy chance-constrained programming (DFCCP) framework and could be used to deal with uncertainties expressed as fuzzy sets and interval parameters. Recently, FCCP has been presented as a novel FMP method through incorporating predefined confidence levels of fuzzy constraint satisfaction into optimization models. Similar to SCCP models, FCCP requires that the fuzzy chance constraints be transformed to deterministic equivalents at predetermined confidence levels. Fuzzy chance-constrained programming is sorted into single-sided FCCP and DFCCP. In many real-world problems, it is more common that both sides of model constraints be associated with uncertainties. Interval linear programming is capable of handling the uncertainties associated with other parameters in the optimization model if they do not have either fuzzy or stochastic distribution information.

For the introduced study case, the environmental loading capacities (such as allowable soil loss and pollutant discharge amounts) are subject to human judgment and described by fuzzy sets. Table 5 lists the related fuzzy parameters. Other uncertain parameters are treated as discrete intervals. The specific constraints related to the IDFCCP model can be formulated as follows:[5]

$$\text{Maximize } f^{\pm} = \sum_{t=1}^{r}\sum_{i=1}^{m}\delta_i^{\pm}Y_i^{\pm}S_{it}^{\pm} + \sum_{j=1}^{n}\eta_j^{\pm}Z_j^{\pm} - \sum_{t=1}^{r}\sum_{i=1}^{m}G_i^{\pm}S_{it}^{\pm} - G_M^{\pm}\sum_{i=1}^{m}M_i^{\pm} - G_F^{\pm}\sum_{i=1}^{m}F_i^{\pm} - \sum_{t=1}^{r}\sum_{i=1}^{m}v_{it}^{\pm}W_{it}^{\pm}S_{it}^{\pm} \tag{9a}$$

subject to

$$Pos\left\{\tilde{L}_i,\tilde{b}\,\middle|\,\sum_{t=1}^{r}\sum_{i=1}^{m}\tilde{L}_i S_{it}^{\pm} \le \tilde{b}\sum_{t=1}^{r}K_t\right\} \ge \alpha \tag{9b}$$

$$Pos\left\{\tilde{l}_i,\tilde{c}_1\,\middle|\,\sum_{t=1}^{r}\sum_{i=1}^{m}h_1^{\pm}\tilde{L}_i S_{it}^{\pm} \le \tilde{c}_1\sum_{t=1}^{r}K_t\right\} \ge \alpha \tag{9c}$$

$$Pos\left\{\tilde{L}_i,\tilde{c}_2\,\middle|\,\sum_{t=1}^{r}\sum_{i=1}^{m}h_2^{\pm}\tilde{L}_i S_{it}^{\pm} \le \tilde{c}_2\sum_{t=1}^{r}K_t\right\} \ge \alpha \tag{9d}$$

$$Pos\left\{\tilde{Q}_t\,\middle|\,\sum_{i=1}^{m}W_{it}^{\pm}S_{it}^{\pm} \le \tilde{Q}_t\right\} \ge \alpha, \quad \forall t \tag{9e}$$

where $Pos\{\cdot\}$ denotes possibility of events in $\{\cdot\}$; and α is a predetermined confidence level, where the values of α are 0.3, 0.6, and 0.9, respectively. According to predetermined confidence levels, a fuzzy chance constraint can be converted to two crisp equivalents as follows:[41]

Table 5 Fuzzy parameters in IDFCCP model.

Model parameters	Values
Supplied water amounts of canal 1 (m^3/ha-s)	(0.8, 0.88, 1)
Supplied water amounts of canal 2 (m^3/ha-s)	(3.2, 3.5, 4.0)
Maximum allowable soil loss (kg/ha)	(3650, 4600, 5600)
Maximum allowable soil-phase nitrogen loss (kg/ha)	(10, 10.4, 11)
Maximum allowable soil-phase phosphorus (kg/ha)	(4.3, 4.5, 4.7)

Notes: (a, b, c) represents a triangular fuzzy set, where a and c are the minimum and maximum possible values, and b is the most likely value.

1. Confidence levels under the minimum reliability.

$$\sum_{t=1}^{r}\sum_{i=1}^{m} L_i^L(\alpha) S_{it} \le b^R(\alpha)\sum_{t=1}^{r} K_t \quad (10a)$$

$$\sum_{t=1}^{r}\sum_{i=1}^{m} h_1^{\pm} L_i^L(\alpha) S_{it}^{\pm} \le c_1^R(\alpha)\sum_{t=1}^{r} K_t \quad (10b)$$

$$\sum_{t=1}^{r}\sum_{i=1}^{m} h_2^{\pm} L_i^L(\alpha) S_{it}^{\pm} \le c_2^R(\alpha)\sum_{t=1}^{r} K_t \quad (10c)$$

$$\sum_{i=1}^{I} W_{it}^{\pm} S_{it}^{\pm} \le \underset{\sim}{Q}_t^R(\alpha), \quad \forall t \quad (10d)$$

2. Confidence levels under the maximum reliability.

$$\sum_{t=1}^{r}\sum_{i=1}^{m} L_i^R(1-\alpha) S_{it}^{\pm} \le b^L(1-\alpha)\sum_{t=1}^{r} K_t \quad (11a)$$

$$\sum_{t=1}^{r}\sum_{i=1}^{m} h_1^{\pm} L_i^R(1-\alpha) S_{it}^{\pm} \le c_1^L(1-\alpha)\sum_{t=1}^{r} K_t \quad (11b)$$

$$\sum_{t=1}^{r}\sum_{i=1}^{m} h_2^{\pm} L_i^R(1-\alpha) S_{it}^{\pm} \le c_2^L(1-\alpha)\sum_{t=1}^{r} K_t \quad (11c)$$

$$\sum_{i=1}^{I} W_{it}^{\pm} S_{it}^{\pm} \le \underset{\sim}{Q}_t^L(1-\alpha), \quad \forall t \quad (11d)$$

The solutions at different confidence levels for IDFCCP can be then obtained through the two-step method.[36] Finally, two groups of objective value and decision variables (presented as interval numbers) corresponding to two scenarios of reliabilities can be obtained. Tables 6–8 list the solutions from IDFCCP at confidence levels of 0.3, 0.6, and 0.9 with two reliabilities. It can be found that the objective function value and part of the decision variables would present as discrete intervals. The solutions also have considerable variations under different combinations of α-cut levels and reliability scenarios. Generally, the system income would decrease as α-cut level increases; meanwhile, the system income under the minimum reliability would be lower than that under the maximum one. For example, at confidence levels of 0.3, 0.6, and 0.9 under the minimum reliability, the system incomes are [145.53, 299.48], [139.94, 287.98], and [134.36, 276.47] ($\times 10^3$), respectively. Under the

Table 6 Solutions of ISCCP model under $\alpha = 0.3$.

Model solutions	Reliability	Values
Cropping area of wheat in subarea 1 (km^2)	Minimum	0
	Maximum	0
Cropping area of wheat in subarea 2 (km^2)	Minimum	0
	Maximum	0
Cropping area of vegetable in subarea 1 (km^2)	Minimum	[26.92, 38.12]
	Maximum	[23.85, 33.76]
Cropping area of vegetable in subarea 2 (km^2)	Minimum	[13.21, 18.54]
	Maximum	[11.73, 16.46]
Cropping area of potato in subarea 1 (km^2)	Minimum	0
	Maximum	0
Cropping area of potato in subarea 2 (km^2)	Minimum	0
	Maximum	0
Applied manure amounts of wheat (10^6 t)	Minimum	0
	Maximum	0
Applied manure amounts of vegetable (10^6 t)	Minimum	583.68
	Maximum	517.41
Applied manure amounts of potato (10^6 t)	Minimum	0
	Maximum	0
Applied fertilizer amounts of wheat (kg)	Minimum	0
	Maximum	0
Applied fertilizer amounts of vegetable (kg)	Minimum	3855.96
	Maximum	3418.12
Applied fertilizer amounts of potato (kg)	Minimum	0
	Maximum	0
Swine amounts (unit)	Minimum	312
	Maximum	276
Poultry amounts (unit)	Minimum	536
	Maximum	475
Net income (10^3 \$)	Minimum	[145.53, 299.48]
	Maximum	[129.01, 265.48]

Table 7 Solutions of ISCCP model under $\alpha = 0.6$.

Model solutions	Reliability	Values
Cropping area of wheat in subarea 1 (km^2)	Minimum	0
	Maximum	0
Cropping area of wheat in subarea 2 (km^2)	Minimum	0
	Maximum	0
Cropping area of vegetable in subarea 1 (km^2)	Minimum	[25.87, 36.63]
	Maximum	[23.22, 32.87]
Cropping area of vegetable in subarea 2 (km^2)	Minimum	[12.71, 17.85]
	Maximum	[11.40, 16.00]
Cropping area of potato in subarea 1 (km^2)	Minimum	0
	Maximum	0
Cropping area of potato in subarea 2 (km^2)	Minimum	0
	Maximum	0
Applied manure amounts of wheat (10^6 t)	Minimum	0
	Maximum	0
Applied manure amounts of vegetable (10^6 t)	Minimum	561.25
	Maximum	503.47
Applied manure amounts of potato (10^6 t)	Minimum	0
	Maximum	0
Applied fertilizer amounts of wheat (kg)	Minimum	0
	Maximum	0
Applied fertilizer amounts of vegetable (kg)	Minimum	3707.76
	Maximum	3326.06
Applied fertilizer amounts of potato (kg)	Minimum	0
	Maximum	0
Swine amounts (unit)	Minimum	300
	Maximum	269
Poultry amounts (unit)	Minimum	516
	Maximum	462
Net income (10^3 \$)	Minimum	[139.94, 287.98]
	Maximum	[125.53, 258.33]

maximum reliability, the system incomes are [129.01, 265.48], [125.53, 258.33], and [122.06, 251.17] ($\times 10^3$), respectively. This is due to the fact that, when α-cut level increases, the confidence levels of constraint satisfaction would increase, and the constraints of pollutant discharge standards, loading capacities, and available water resources would become stricter. This indicates that a high economic benefit could lead to a high system failure risk. Conversely, a conservative decision alternative is more desirable for mitigating water quality and quantity crisis. A trade-off between system benefit and reliability of satisfying model constraints needs to be analyzed in order

Table 8 Solutions of ISCCP model under $\alpha = 0.9$.

Model solutions	Reliability	Values
Cropping area of wheat in subarea 1 (km^2)	Minimum	0
	Maximum	0
Cropping area of wheat in subarea 2 (km^2)	Minimum	0
	Maximum	0
Cropping area of vegetable in subarea 1 (km^2)	Minimum	[24.83, 35.15]
	Maximum	[22.59, 31.98]
Cropping area of vegetable in subarea 2 (km^2)	Minimum	[12.22, 17.15]
	Maximum	[11.07, 15.54]
Cropping area of potato in subarea 1 (km^2)	Minimum	0
	Maximum	0
Cropping area of potato in subarea 2 (km^2)	Minimum	0
	Maximum	0
Applied manure amounts of wheat (10^6 t)	Minimum	0
	Maximum	0
Applied manure amounts of vegetable (10^6 t)	Minimum	538.82
	Maximum	489.54
Applied manure amounts of potato (10^6 t)	Minimum	0
	Maximum	0
Applied fertilizer amounts of wheat (kg)	Minimum	0
	Maximum	0
Applied fertilizer amounts of vegetable (kg)	Minimum	3559.57
	Maximum	3234
Applied fertilizer amounts of potato (kg)	Minimum	0
	Maximum	0
Swine amounts (unit)	Minimum	288
	Maximum	261
Poultry amounts (unit)	Minimum	495
	Maximum	450
Net income (10^3 \$)	Minimum	[134.36, 276.47]
	Maximum	[122.06, 251.17]

to gain an in-depth insight into the characteristics of the management system.

One of the limitations associated with IDFCCP is that the determination of the confidence levels is rather arbitrary. There are currently no effective tools or guidelines to help choose proper confidence levels for satisfying constraints. Secondly, IDFCCP generates many sets of solutions under different confidence levels and reliability scenarios, which may complicate the decision-making process. Multicriteria decision analysis tools may be a potential tool to mitigate such a problem.

Further Discussions

The methodologies introduced in this study were based on integration of SMP, FMP, and ILP techniques. The major reason is that different types of uncertainty-analysis tools have varied strengths and weaknesses, in terms of the sensitivity to data quality, accuracy of information interpretation, and requirement of computational efforts. Integration of different algorithms could better deal with dual or multiple uncertainties. In agricultural water management, many types of uncertainties may exist. For example, the available water amount is affected by hydrological and meteorological conditions and may show random features; it can normally be expressed as a stochastic variable; the parameters related to allowable pollutant discharge amounts and environmental loading capacities are subject to human judgments and could be expressed by fuzzy membership functions; parameters related to the benefit and the crop yield may fluctuate in a range where their distribution information may be difficult to obtain due to a lack of sufficient historical data, and therefore, they are more suitable to be described by discrete intervals. Depending on the quality of data, how to choose proper uncertainty-analysis methods becomes a critical challenge for decision makers.

The study case used in this entry is only for demonstration purpose, and it is therefore highly simplified. The water quantity and quality levels are conceptualized into simple mass-balance expressions (i.e., the left-hand sides of model constraints). In real-world cases, the related processes are not necessity linear, and rigid simulation models are required to quantify the related relationships. For example, the prediction of nonpoint sources of pollution on a watershed scale requires a sophisticated hydrological model. Thus, a coupling of simulation and optimization models is desired to tackle more complicated problems. However, how to effectively realize such integration is a challenging topic and deserves further investigations.[42] Another issue is about the acquisition of data. Many of the parameters used in this study were obtained through subjective assumption or referred to in the literature. In actual applications, the data quality significantly determines the results of models and thus influences the decision-making processes. Although uncertainty-analysis tools are available, they cannot be considered as alternatives to good-quality data. Efforts are still needed to acquire as much accurate information as possible in order to ensure more reliable modeling results and robust management strategies.

CONCLUSION

This entry gave a general introduction of how to conceptualize and formulate inexact optimization models for supporting integrated agricultural water quantity and quality management under uncertainty. A number of inexact optimization approaches based on stochastic, fuzzy, and interval programming models were introduced. Each of them is capable of handling a certain type of uncertain information, which is dependent on the uncertain features of the system components. A typical agricultural water management problem was used to demonstrate the applicability of different methods. The study results effectively showed the characteristics of various optimization approaches. Limitations associated with different modeling algorithms were also discussed.

REFERENCES

1. Johnson, C.; Handmer, J. Water supply in England and Wales: Whose responsibility is it when things go wrong? Water Policy **2002**, *4* (4), 345–366.
2. Al-Salihi, A.H.; Himmo, S.K. Control and management study of Jordan's water resource. Water Int. **2003**, *28*(1), 1–10.
3. Xu, Z.X.; Li, J.Y. Sustainable water resource management in China: A great challenge. Alliance for Global Sustainability Bookseries **2010**, *18* (1), 3–10.
4. Nie, X.H.; Huang, G.H.; Wang, D.; Li, H.L. Robust optimisation for inexact water quality management under uncertainty. Civ. Eng. Environ. Syst. **2008**, *25*(2), 167–184.
5. Xu, Y.; Qin, X.S. Rural effluent control under uncertainty: An inexact double-sided fuzzy chance-constrained model. Adv. Water Resour. **2010**, *33* (9), 997–1014.
6. Haith, D.A. *Environmental Systems Optimization*; John Wiley and Sons, Inc.: New York, 1982.
7. Huang, G.H. IPWM: An interval parameter water quality management model. Eng. Optim. **1996**, *26*, 79–103.
8. Huang, G.H. A hybrid inexact-stochastic water management model. Eur. J. Oper. Res. **1998**, *107*, 137–158.
9. Doorenbos, J.; Kassam, A.H. *Yield Response to Water*, FAO Irrigation and Drainage Paper 33; FAO: Rome, 1979.
10. Rao, N.H. Field test of a simple soil-water balance model for irrigated areas. J. Hydrol. **1987**, *91*, 179–186.
11. Howitt, R.E. Positive mathematical-programming. Am. J. Agric. Econ. **1995**, *77* (2), 329–342.
12. Doll, P.; Siebert, S. Global modeling of irrigation water requirements. Water Resour. Res. **2002**, *38* (4), 1037.
13. Yang, Y.M.; Yang, Y.H.; Moiwo, J.P.; Hu, Y.K. Estimation of irrigation requirement for sustainable water resources reallocation in North China. Agric. Water Manage. **2010**, *97*, 1711–1721.

14. Amir, I.; Fisher, F.M. Analyzing agricultural demand for water with an optimizing model. Agric. Syst. **1999**, *61*, 45–56.
15. Salman, A.Z.; Al-Kaeablieh, E.K.; Fisher, F.M. An inter-seasonal agricultural water allocation system (SAWAS). Agric. Syst. **2001**, *68*, 233–252.
16. Montesinos, P.; Camacho, E.; Alvarez, S. Seasonal furrow irrigation model with genetic algorithms (optimec). Agric. Water Manage. **2001**, *52*(1), 1–16.
17. Pais, M.S.; Ferreira, J.C.; Teixeira, M.B.; Yamanaka, K.; Carrijo, G.A. Cost optimization of a localized irrigation system using genetic algorithms. Intell. Data Eng. Autom. Learn. **2010**, *6283*, 29–36.
18. Zhang, X.D.; Huang, G.H.; Nie, X.H. Optimal decision schemes for rural water quality management planning with imprecise objective. Agric. Water Manage. **2009**, *96*, 1723–1731.
19. Zhang, W.S.; Wang, Y.; Peng, H.; Li, Y.T.; Tang, J.S.; Wu, K.B. A coupled water quantity–quality model for water allocation analysis. Water Resour. Manage. **2010**, *24*, 485–511.
20. Wang, J. The estimation of evapotranspiration with Penman–Monteith and evaporator methods. Agric. Res. Arid Areas **2002**, *20*(4), 67–71.
21. Sang, S. Simulation–optimization method for crop irrigation scheduling with limited water supply. J. Tsinghua Univ. **2005**, *45* (9), 1179–1183.
22. Lin, R.; Liang, Y. Characteristics of dry matter accumulation and partitioning in the process of yield formation in different rice cultivate. China Agric. Sci. Bull. **2006**, *22* (2), 185–190.
23. Liu, L.; Huang, G.H.; Liu, Y.; Fuller, G.A. A fuzzy-stochastic robust optimization model for regional air quality management under uncertainty. Eng. Optim. **2003**, *35* (2), 177–199.
24. Charnes, A.; Cooper, W.W.; Kirby, P. Chance constrained programming: An extension of statistical method. In *Optimizing Methods in Statistics*; Academic Press: New York, 1972; 391–402.
25. Qin, X.S.; Huang, G.H. An inexact chance-constrained quadratic programming model for stream water quality management. Water Resour. Manage. **2009**, *23*, 661–695.
26. Xu, Y.; Huang, G.H.; Qin, X.S.; Cao, M.F. SRCCP: A stochastic robust chance-constrained programming model for municipal solid waste management under uncertainty. Resour., Conserv. Recycl. **2009**, *53*, 352–363.
27. Huang, G.H.; Loucks, D.P. An inexact two-stage stochastic programming model for water resources management under uncertainty. Civ. Eng. Environ. Syst. **2000**, *17*, 95–118.
28. Xu, Y.; Huang, G.H.; Qin, X.S. An inexact two-stage stochastic robust optimization model for water resources management under uncertainty. Environ. Eng. Sci. **2009**, *26*, 1765–1776.
29. Li, Y.P.; Huang, G.H.; Nie, S.L. An interval-parameter multi-stage stochastic programming model for water resources management under uncertainty. Adv. Water Resour. **2006**, *29*, 776–789.
30. Li, Y.P.; Huang, G.H.; Yang, Z.F.; Nie, S.L. IFMP: Interval-fuzzy multistage programming for water resources management under uncertainty. Resour., Conserv. Recycl. **2008**, *52*, 800–812.
31. Watkins, D.W., Jr.; McKinney, D.C. Finding robust solutions to water resource problems. J. Water Resour. Plann. Manage. **1997**, *123*, 49–58.
32. Xu, Y.; Huang, G.H.; Qin, X.S.; Cao, M.F.; Sun, Y. An interval-parameter stochastic robust optimization model for supporting municipal solid waste management under uncertainty. Waste Manage. **2010**, *30* (2), 316–327.
33. Huang, G.H.; Baetz, B.W.; Patry, G.G. A grey fuzzy linear programming approach for municipal solid waste management planning under uncertainty. Civ. Eng. Syst. **1993**, *10*, 123–146.
34. Zimmermann, H.J. *Fuzzy Set Theory and Its Applications*, 3rd Ed; Kluwer Academic Publishers, 1995.
35. Liu, B.D.; Iwamura, K. Chance-constrained programming with fuzzy parameters. Fuzzy Sets Syst. **1998**, *94*, 227–237.
36. Huang, G.H.; Baetz, B.W., Patry, G.G. A grey linear programming approach for municipal solid waste management planning under uncertainty. Civ. Eng. Syst. **1992**, *9*, 319–335.
37. Huang, G.H.; Baetz, B.W.; Patry, G.G. Grey integer programming: An application to waste management planning under uncertainty. Eur. J. Oper. Res. **1995**, *83*, 594–620.
38. Huang, G.H.; Baetz, B.W.; Patry, G.G. Grey fuzzy integer programming: An application to regional waste management planning under uncertainty. Socio-Economic Plann. Sci. **1995**, *29*, 17–38.
39. Qin, X.S.; Huang, G.H.; Zeng, G.M.; Chakma, A.; Huang, Y.F. An interval-parameter fuzzy nonlinear optimization model for stream water quality management under uncertainty. Eur. J. Oper. Res. **2007**, *180* (3), 1331–1357.
40. Loucks, D.P.; Stedinger, J.R.; Haith, D.A. *Water Resource Systems Planning and Analysis*; Prentice-Hall: Englewood Cliffs, NJ, 1981.
41. Fiedler, M.; Nedoma, J.; Ramík, J.; Rohn, J.; Zimmermann, K. Linear Optimization Problems with Inexact Data; Springer Science Business Media, Inc., New York, 2006.
42. Wardlaw, R.; Bhaktikul, K. Application of a genetic algorithm for water allocation in an irrigation system. Irrig. Drain. **2001**, *50*, 159–170.

Agriculture: Energy Use and Conservation

Guangnan Chen
Faculty of Engineering and Surveying, University of Southern Queensland, Toowoomba, Queensland, Australia

Tek Narayan Maraseni
Australian Center for Sustainable Catchments, School of Accounting, Economics and Finance, University of Southern Queensland, Toowoomba, Queensland, Australia

Abstract

Effective energy use and conservation are becoming increasingly important in the context of rising energy costs and concerns over greenhouse gas (GHG) emissions. This entry reviews the methods of energy audit and compares the energy use and conservation for different farming systems and practices, including machinery operation, tillage practice, and irrigation methods. Opportunities for adopting new technologies such as renewable energy and biotechnology to reduce farming energy inputs and impacts of GHG emissions are also discussed.

INTRODUCTION

Farming is often an energy-intensive operation. With the rising energy costs and increasing concern over greenhouse gas (GHG) emissions, effective energy use and conservation are becoming increasingly important.[1,2] In many parts of the world, energy inputs represent a major and rapidly growing cost input for growers, with on-farm fuel requirement in cereal cropping systems often exceeding 100 L/ha for tillage, seeding, fertilizing, spraying, and harvesting.[3–5] In the United States, it has been found that the operations of food systems, including agricultural production, food processing, packaging, and distribution, accounted for approximately 19% of America's national fossil fuel energy use.[6]

METHODS OF ENERGY AUDIT

Energy audits and assessments are a first step and crucial part of the energy and environmental management process. Energy audits refer to the systematic examination of an entity, such as a firm, organization, facility, or site, to determine whether, and to what extent, it has used energy efficiently.[7] They may also assess opportunities of potential energy savings through fuel switching, tariff negotiation, and demand-side management.

Energy audits may be broadly classified into the following three levels:[7,8]

- Energy audit level 1, or preliminary audit. The main purpose of this level of energy audits is to overview the total energy consumption of a farm. This is the simplest and cheapest form of energy audits. It usually involves collating all the energy use data from the farm, including the total fuel usage from diesel, petrol, and other fuels and the total electricity energy consumed, in order to derive basic energy performance indices such as GJ/ha and GJ/t.
- Energy audit level 2, or standard/general audit. Level 2 energy audits generally involve breaking down the total energy usage into each individual major processes. A level 2 audit is usually process-based and involves some specific measurements for the key processes. It may also involve considerable farmers' interviews to identify the major energy usage.
- Energy audit level 3, or detailed specific operation investigation. This is the highest and the most expensive level of energy audits. It may involve investigating ways to improve the efficiency of a specific operation so that the investment return can be accurately predicted. This will usually involve the uses of a range of different sensors to measure the performance of a variety of machines. Data loggers may be used to record the data for a considerable period of time.

Underpinning the energy audit assessment process is the identification and development of a set of toolkits (software, hardware, and other supporting information and resources) that can be applied to different farm production systems. A number of farm energy calculators[4,7–9] have already been developed to evaluate the energy uses from agricultural systems. Various hardware and technologies such as electricity power meters, fuel flow meters, data logging, and monitoring equipment may also be used for undertaking field measurements.[9–11] By itemizing farm energy usage from each operation, farmers will be able to identify where energy is mostly consumed and therefore explore ways to reduce energy use.

Encyclopedia of Environmental Management DOI: 10.1081/E-EEM-120046114

Brown and Elliot[12] found that although the currently available data of agricultural energy use may be sufficient for general policy development, the quality of existing energy end-use data is often unsatisfactory. They suggested that further research be conducted to achieve a clear and consistent definition of farm types and energy end uses. Energy uses of individual operations will also need to be more accurately measured.

EFFECT OF FARMING SYSTEMS ON ENERGY USE

Considerable research has been conducted on energy use and conservation both in agriculture[13–16] and in other industries.[17] In the cropping sector, it has been identified that a number of practice changes and technology developments may be adopted to reduce fuel/energy use or energy use intensity. Examples of these may include more effective machinery operation, conservation farming practices, improved irrigation methods and water use efficiency measures, precision agriculture, and, where appropriate, the use of renewable energy and planting of alternative crops.

Machinery Operation

Farm machinery is integral to many aspects of modern farming systems. Without it, we may not be able to effectively implement the farming systems. It was estimated that since the Second World War, more than 25% of the increase in grain productivity may be attributed to machinery innovations. When the ownership cost and timeline loss are taken into consideration, it is estimated that machinery may consist of up to 40%–50% of total farm input costs.[18]

Farm machinery often consumes a large amount of energy and therefore produces considerable emissions, especially carbon dioxide. Fig. 1 shows the average percentage contribution of the direct energy input for different farming processes in European tillage-based systems. Chen et al.[19] and Smith[20] showed that by changing gear selection and engine speed, up to 30% of energy can often be saved for the same tractor power output. Careful selection of energy-efficient farming equipment and proper matching

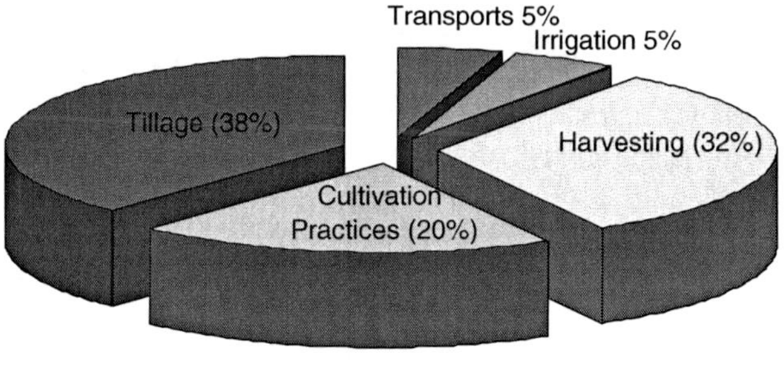
Transports 5%
Irrigation 5%
Tillage (38%)
Harvesting (32%)
Cultivation Practices (20%)

Fig. 1 Direct on-farm energy inputs in Europe.

Table 1 Average fuel use for different tillage methods.

Soil tillage methods	Average fuel (diesel) use (L/ha)
Subsoiling	18
Discing	12
Chisel ploughing	7
Power harrowing	8
Light harrowing/rolling	4

Source: Adapted from Chen and Baillie[4] and Downs and Hansen.[22]

of tractor and implement are also important.[21] At the systems level, it is possible that much larger improvements can be achieved by changing the cropping practices and systems. Table 1 shows the estimated average fuel use for different tillage operations,[22] where a ratio of up to 3:1 or 4:1 from the highest to the lowest energy use can be found.

Conservation Agriculture

Conservation agriculture is the generally accepted label for farming systems developed with the specific objective of improving productivity and sustainability, by reducing input waste and resource degradation. Conservation tillage involves reduced tillage, in which a crop is planted in the residue from a previous crop with either minimal (minimum tillage) or zero (zero tillage) soil disturbance. The weed control function may be achieved by using a combination of herbicides and agronomic measures (cover crops, rotation, seed bank depletion) and soil damage minimized by better management (soil health, traffic intensity, controlled traffic).

Based on 2 years experiments in Croatia, Kosutic et al.[23] demonstrated that compared with the conventional "intensive" tillage, up to 82.6% of energy could be saved by the adoption of no-till conservation farming system, and without the significant yield reduction. Gulden and Entz[24] found that this was 36% in Canada. Bailey et al.[25] showed that in terms of total energy used, when compared with conventional farming systems on a per-hectare basis, the integrated arable systems had the potential to reduce overall energy consumption by about 8% in the United Kingdom. However, in terms of energy use per kilogram of output, their results were less conclusive. A comparative study of conventional tillage and conservation farming was also carried out by Smith[26] to compare the impact of these practices on soil characteristics, crop performance, and economic outcomes.

While many no-till seeders have been marketed in recent years, there is still no one universally applicable machine that guarantees success under all conditions. In fact, better seed and fertilizer placement and good depth control and stubble handling have been identified by farmers as the highest priority needs in the design of planting equipment, particularly on "unprepared" soil with heavy stubble in direct drilling/conservation farming.[27] Difficulties of no-till seeding are therefore the major stumbling block to the adoption of full conservation agriculture. Despite these issues, effective machine/system combinations for no-till seeding have been evolved for many cropping environments, particularly in the subtropical grain production, where conventional tillage-based systems were causing major erosion damage and loss of organic matter. Where soil degradation problems are less obvious, for instance, in the long-established plough-based systems of Europe, adoption of conservation agriculture has been much slower,[28] and the same is true in most horticultural and intensive cropping areas, worldwide. This situation is expected to change slowly as cropping inputs, particularly fuel, fertilizer, and water resources, become more expensive, improving the attractiveness of system change.

Controlled traffic is a farming system in which the wheel tracks of all operations are confined to fixed paths so that re-compaction of soil by traffic (traction or transport) does not occur outside the selected paths. It has been shown that in addition to benefits of improved productivity and soil and water conservation, this method could also provide up to 50% reduction in power and fuel requirements of field operations.[29,30] Tullberg[31] demonstrated the effects of system change from the conventional tillage-based to no-till and to controlled traffic no-till in Australian grain production, clearly illustrating the significance of the reduction of energy and GHG emissions. It was shown that the total GHG emissions per ton of grain production can be reduced from 314 to 306 and to 175 kg, respectively. One survey indicated that in 2003, some form of controlled traffic was in use over 1 million ha in Australia.[32] To achieve further widespread adaptation, it is identified that the lack of compatibility between different machinery would have to be overcome.[31]

Irrigation Methods

Irrigation can be broadly defined as the practice of applying additional water (beyond what is available from rainfall) to soil to enable or enhance plant growth and yield. The water source could be groundwater pumped to the surface or surface water diverted from one location to another. On a global scale, it was estimated that 17% of irrigated cropland produces 40% of the total production.[33]

Farm irrigation systems may be broadly classified as surface irrigation, pressured sprinkler systems, trickle, and subsurface systems. In Australia, surface irrigation systems, such as border check and furrow irrigation, are the most widely used irrigation systems, covering some 70% of the total irrigated area.[34] This is then followed by the pressurized spray irrigation methods (22%) and trickle and subsurface systems (8%).

Irrigation is however often a very energy-intensive operation. It was estimated that 23% of the on-farm energy use for crop production in the United States was for on-farm pumping.[35] Fig. 2 shows the estimated total life-cycle energy consumption for different irrigation methods.[34] It can be seen that the pressurized irrigation system can use significantly more energy over its life cycle than a gravity-fed system, with a ratio of about 1.5 or 2. The increased water-use efficiency of pressured irrigation systems will therefore need to be balanced against the higher cost of the energy needed.[36] At the current market condition, 1 GJ of energy would typically cost around $10–$40. The corresponding GHG emissions range from 0 (using renewable energy) to around 78 kg CO_2/GJ (using diesel fuel).

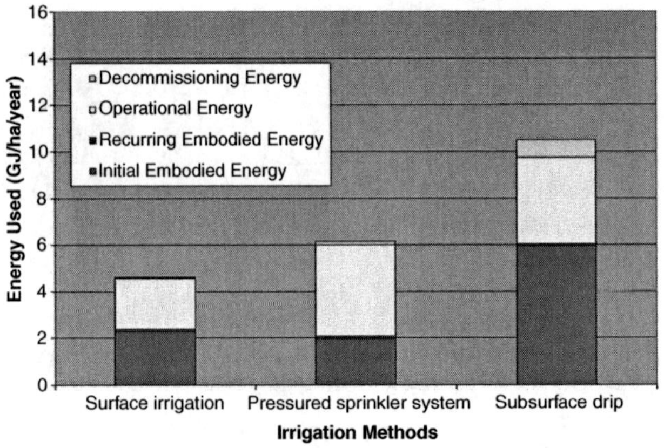
Energy Used (GJ/ha/year)
16
14
12
10
8
6
4
2
0
Decommissioning Energy
Operational Energy
Recurring Embodied Energy
Initial Embodied Energy
Surface irrigation
Pressured sprinkler system
Subsurface drip
Irrigation Methods

Fig. 2 Life-cycle energy consumption for different irrigation systems.

Brown and Elliot[12] found that the largest on-farm energy savings might be available in motorized systems, especially in irrigation pumping. Barron[37] also identified that poor system design contributed significantly to low pumping efficiency. Pathak and Bining[38] showed that for irrigation, fuel saving of more than 50% was feasible through improvement in the irrigation equipment and improved water management practices. Chen et al.[19] demonstrated that with suitable operation adjustments, up to 55% of pumping energy can be saved. It has also been reported in the United States that energy efficiency audits on irrigation systems have, on average, identified savings of at least 10% of the energy bill—and in many instances up to 40%.[38] Very often, the irrigators who owned these inefficient systems were unaware of any problems.[39]

Overall Systems Comparison

Table 2 summarizes some of the published energy use data for different crops in different countries.[8,39–44] It can be seen that significant variation in energy input costs occurs between different crops and between different growers due to the adopted production system and a grower's performance within that system. Depending on the farming systems adopted, the direct on-farm energy input may range from 2.5 GJ/ha for wheat production in Europe to 53.4 GJ/ha for greenhouse tomato production in Turkey. This is a ratio of approximately 20.

Table 2 Energy input for different crops in different countries.

Crops	Direct energy input (GJ/ha)	Indirect energy input (GJ/ha)	Total energy input (GJ/ha)	Country/ Region
Wheat	2.5–4.3			Europe
Wheat			16–32	Greece
Cotton	21.14	28.59	49.73	Turkey
Cotton			82.6	Greece
Cotton	3.7–15.2			Australia
Maize	4.7–5.0			Europe
Rice			64.89	United States
Pea			2.5–5.4	Canada
Dairy pasture	14.6	3.6	18.2	New Zealand
Greenhouse tomato	53.4	53.3	106.7	Turkey

Source: Chen and Baillie.[8]

It is noted that on-farm energy uses of crop production is only a proportion of total energy usage. The others include the post-farm processing, transport, distribution, and energy used to produce agricultural inputs such as fertilizers and pesticides.[44–46] Table 3 shows the estimated embodied energy content for manufacturing various chemical fertilizers and herbicides. It can be seen that with the current manufacturing technology, the production of 1 kg of nitrogen fertilizer would typically require the energy input equivalent to 1.5–2 kg of fuel, while 1 kg of pesticides would require the energy input equivalent to up to 5 kg of fuel.[44] It is therefore important to reduce not only the on-farm energy uses but also the embodied energy and post-harvest energy uses. This is especially important, as the embodied energy of farm chemicals may account for up to 50%–70% of the total energy input in agricultural production. In particular, the cost of nitrogen fertilizer is mostly a reflection of energy costs, so maximizing the efficiency of fertilizers or using legumes to produce nitrogen can help reduce energy costs on farms. It was demonstrated that energy input in the integrated forage–grain rotation was approximately 40% lower than that in the grain-based rotation.[47]

Yaldiz et al.[48] reported that fertilizers and irrigation energy dominated the total energy consumption in Turkish cotton production. Yilmaz et al.[41] showed that the energy intensity in agricultural production was closely related with production techniques. They estimated that cotton production in Turkey consumed a total of 49.73 GJ/ha energy, consisting of 21.14 GJ/ha (42.5%) direct energy input and 28.59 GJ/ha (57.5%) indirect energy input. Modern rice production in the United States also requires a total energy input of 64.89 GJ/ha for a yield of 5.8 tons/ha or 11.19 GJ energy input per ton of grain produced.[43]

To save energy and cost, the precision-agriculture technologies such as controlled traffic farming, high spatial and temporal precision in input placement, sensor-controlled spraying, and variable-rate fertilizer applications may offer significant benefits. Pimentel et al.[3] also showed that by appropriate technology changes in food production, processing, packaging, transportation, and consumption, fossil energy use in the U.S. food system could be reduced by about 50%. Using corn production as a model crop, it was estimated that total energy in corn production could be reduced by more than 50% by combining the following changes: 1) using smaller machinery and less fuel; 2) replacing commercial nitrogen applications with legume cover crops and livestock manure; and 3) reducing soil erosion through alternative tillage and conservation techniques. Hoeppner et al.[47] and Ziesemer[49] also showed that energy input in the organic management system was approximately 50% lower than that in the conventional management system.

Table 3 Embodied energy content for various agricultural fertilizers and herbicides.

Chemicals	Energy content (MJ/kg element or active ingredient)
N	65
P	15
K	10
S	5
Generic herbicide	270

DEVELOPMENT AND APPLICATIONS OF NEW TECHNOLOGY

With the advance of biotechnology, new crop varieties with higher yield and improved performance such as reduced herbicide and pesticide usage are being continuously introduced.[50] This will reduce not only the amount of her-

Table 4 Estimated costs of various electricity generation methods.

Energy type	Cost (¢/kWh)
Hydropower	4–10
Coal-fired power	2–4
Gas-fired power	5
Wind energy	5–8
Solar PV electricity	15–25

bicide and pesticide used but also the fuel usage associated with the crop spraying operations. For example, it was reported that with the adoption of genetically modified cotton, pesticide use has been reduced by up to 70%–90% in the Australian cotton industry.[51] To address the public's concern, the introduction of biotechnology to food crops may require a more careful approach.

Where the opportunities are appropriate, renewable energy such as solar, wind, and biofuel may also be integrated into the farming operations to save energy costs and reduce GHG emissions.[52–55] Examples of direct applications include solar crop drying, solar space and water heating, and using biomass for heating purpose. Other applications include generation of off-grid electricity for electric fences, lighting, irrigation, livestock water supply, wastewater treatment, pond aeration, communication, and remote equipment operation. Table 4 shows the estimated costs for various electricity generation technologies. Overall, renewable energy at present may not be viable to serve in large scale as a primary energy source, and in many cases, additional government support and financial incentives would still be required to compete with more polluting fossil fuels. However, because of the increasing concern over the GHG emissions and declining reserves of fossil fuel, use of renewable energy is on the rapid rise globally.[56] Various user-friendly design and cost estimation tools have also been developed to promote specific applications.[57]

CONCLUSION

Energy use efficiency and conservation are of significant interest to the agricultural industry. Within highly mechanized agricultural productions systems, operational energy inputs (i.e., diesel, petrol, gas, and electricity) often represent a major cost to the growers. In the United States, it has been found that the operations of food systems, including agricultural production, food processing, packaging, and distribution, accounted for approximately 19% of America's national fossil fuel energy use. A total of up to 64.89 GJ/ha of energy is spent in the rice production in the United States.

It has been shown that significant variation in energy input costs can occur between growers due to the adopted production system and a grower's performance within that system. In the cropping sector, it has been identified that a number of practice changes and technology developments may be adopted to reduce fuel/energy use in agriculture. These may include more effective machinery operation, conservation farming practices, improved irrigation methods and water use efficiency measures, precision agriculture, and, where appropriate, the use of renewable energy and planting of alternative crops or crop rotation. It has been demonstrated that by changing the farming systems (e.g., changing to conservation tillage system, controlled traffic, or organic management system) and undertaking the same operation more efficiently (e.g., more precise targeting and use of fertilizers and irrigation), up to 50% of overall energy use may be realistically saved. It has also been found that from the highest to the lowest energy use, the estimated average fuel use for different tillage operations may be up to a ratio of 3:1 or 4:1. In many cases, the largest on-farm energy savings might be available in motorized systems, especially in irrigation pumping. The energy efficiency audits on irrigation systems in the United States have, on average, identified savings of at least 10% of the energy bill—and in many instances up to 40%. Because the embodied energy of various chemicals and machinery may make up 50%–70% of the total energy input and GHG emissions in agricultural production, the applications of variable-rate fertilizer technology and replacement of commercial nitrogen applications with legume cover crops and livestock manure would bring significant benefits.

With the development and applications of new technology and the increasing cost of fuel, fertilizer, and water resources, it is expected that various energy-saving measures will be increasingly adopted by the farmers. In particular, with the increasing concern over the GHG emissions and the depleting of fossil fuel, renewable energy such as solar, wind, and biofuel may be integrated into the different farming operations, including solar crop drying and heating, off-grid electricity supply for fences, lighting, irrigation, livestock water supply, wastewater treatment, pond aeration, communication, and remote equipment operation. With the adoption of genetically modified cotton, pesticide use has been reduced by up to 70%–90% in the Australian cotton industry.

ACKNOWLEDGMENTS

The authors wish to extend special thanks to Dr. Jeff N. Tullberg for his helpful advice and comments to this entry.

REFERENCES

1. Fischer, J.R.; Johnson, S.R.; Finnell, J.A. Energy, agriculture and the food—Cutting costs, improving efficiency, and

stewarding natural resources. Resour. Mag. **2009**, *16* (2), 16–19.
2. Zegada-Lizarazu, W.; Matteucci, D.; Monti, A. Critical review on energy balance of agricultural systems. Biofuels, Bioprod. Biorefin. **2010**, *4* (4), 423–446.
3. Pimentel, D.; Williamson, S.; Alexander, C.; Gonzalez-Pagan, O.; Kontak, C.; Mulkey, S. Reducing energy inputs in the US food system. Hum. Ecol. **2008**, *36*, 459–471.
4. Chen, G.; Baillie, C. Development of EnergyCalc—A tool to assess cotton on-farm energy uses. Energy Convers. Manage. **2009**, *50* (5), 1256–1263.
5. Safa, M.; Tabatabaeefar, A. Fuel consumption in wheat production in irrigated and dry land farming. World J. Agric. Sci. **2008**, *4* (1), 86–90.
6. Pimentel, D. *Impacts of Organic Farming on the Efficiency of Energy Use in Agriculture*; The Organic Center, Cornell University, 2006.
7. Standards Australia. *Australian/New Zealand Standards. AS/NZS 3598:2000. Energy Audits*.
8. Chen, G.; Baillie, C. Agricultural applications: Energy uses and audits. In *Encyclopedia of Energy Engineering and Technology*; Capehart, B., Ed.; Taylor & Francis Books: London, 2009; 1: 1, 1–5.
9. Hill, H. *Farm Energy Calculators: Tools for Saving Money on the Farm*; National Center for Appropriate Technology: USA. Available at http://attra.ncat.org/attra-pub/PDF/farmenergycalc.pdf.
10. Svejkovsky, C. *Conserving Fuel on the Farm*; National Center for Appropriate Technology: USA. Available at http://attra.ncat.org/attra-pub/PDF/consfuelfarm.pdf.
11. *A Guide to Energy Efficiency Innovation in Australian Wineries*. Available at http://www.ret.gov.au/energy/Documents/best%20practice%20guides/energy_bpg_wineries.pdf.
12. Brown, E.; Elliot, R.N. *Potential Energy Efficiency Savings in the Agriculture Sector*; The American Council for an Energy-Efficient Economy: Washington, D.C., 2005. Available at http://www.aceee.org/pubs/ie053.htm.
13. Pellizzi, G.; Cavalchini, A.G.; Lazzari, M. *Energy Savings in Agricultural Machinery and Mechanization*; Elsevier Science Publishing Co.: New York, USA, 1988.
14. Stout, B.A. *Handbook of Energy for World Agriculture*; Elsevier Science Publications Ltd: London, 1990.
15. Adams, S. *Direct Energy Use in Agriculture—Opportunities for Reducing Fossil Fuel Inputs*, 2007. Available at http://randd.defra.gov.uk/Document.aspx?Document=AC0401_6343_FRP.pdf.
16. Wilhem, L.R.; Suter, D.A.; Brusewitz, G.H. Energy use in food processing (Ch. 11). In *Food and Process Engineering Technology*; ASAE: St. Joseph, Michigan, 2004; 285–291.
17. Eastop, T.D.; Croft, D.R. *Energy Efficiency for Engineers and Technologists*; Longman Publishing Group, 1990.
18. Kastens, T. *Farm Machinery Cost Calculations*; Kansas State University. Available at http://fieldcrop.msu.edu/documents/mf2244%20Farm%20Machinery%20Operation%20Costs.pdf.
19. Chen, G.; Kupke, P.; Baillie, C. Evaluating on-farm energy performance in agriculture. Aust. J. Multi-Discip. Eng. **2009**, *7* (1), 55–61.
20. Smith, L.A. Energy requirements for selected crop production implements. Soil Tillage Res. **1993**, *25*, 281–299.
21. McLaughlin, N.B.; Drury, C.F.; Reynolds, W.D.; Yang, X.M.; Li, Y.X.; Welacky, T.W.; Stewart, G. Energy inputs for conservation and conventional primary tillage implements in a clay loam soil. Trans. ASABE. **2008**, *51* (4), 1153–1163.
22. Downs, H.W.; Hansen, R.W. *Estimating Farm Fuel Requirements*, Colorado State University, 2007. Available at http://www.ext.colostate.edu/PUBS/FARMMGT/05006.html#top.
23. Kosutic, S.; Filipovic, D.; Gospodaric, Z. Maize and winter wheat production with different soil tillage systems on silty loam. Agric. Food Sci. Finl. **2001**, *10* (2), 81–90.
24. Gulden, R.H.; Entz, H.E. *A Comparison of Two Manitoba Farms with Contrasting Tillage Systems*; University of Manitoba: Canada, 2005. Available at http://umanitoba.ca/outreach/naturalagriculture/articles/energy.html.
25. Bailey, A.P.; Basford, W.D.; Penlington, N.; Park, J.; Keatinge, J.D.H.; Rehman, T.; Tranter, R.B.; Yates, C. A comparison of energy use in conventional and integrated arable farming systems in the U.K. Agric. Ecosyst. Environ. **2003**, *97*, 241–253.
26. Smith, I. Conservation or conventional tillage? Nat. Resour. Manage. **2001**, *4* (2), 9–15.
27. White, B.; Ryan, W. *A Scoping Study—Potential Investment in Agricultural Engineering Research and Development for the Australian Grains Industry*; Kondinin Group: Australia, 2004.
28. Holland, J.M. The environmental consequences of adopting conservation tillage in Europe: Reviewing the evidence. Agric. Ecosyst. Environ. **2004**, *103*, 1–125.
29. Tullberg, J.N.; Yule, D.F.; McGarry, D. Controlled traffic farming—From research to adoption in Australia. Soil Tillage Res. **2007**, *97*, 272–281.
30. Tullberg, J.N. Traffic effects on tillage energy. J. Agric. Eng. Res. **2000**, *75* (4), 375–382.
31. Tullberg, J.N. Tillage, traffic and sustainability—A challenge for ISTRO. Soil Tillage Res. *in press*.
32. Tullberg, J.N.; Yule, D.F.; McGarry, D. *'On Track' for sustainable cropping in Australia*, International Soil Tillage Research Organisation Conference (Invited plenary paper). ISTRO Proceedings CD, University of Queensland, 2003.
33. Available at http://www.pollutionissues.com/A-Bo/Agriculture.html.
34. Jacobs, S. *Comparison of Life Cycle Energy Consumption of Alternative Irrigation Systems*; University of Southern Queensland, 2006. Available at http://eprints.usq.edu.au/2471/.
35. Lal, R. Carbon emission from farm operations. Environ. Int. **2004**, *30*, 981–990.
36. Jackson, T.M.; Khan, S.; Hafeez, M. A comparative analysis of water application and energy consumption at the irrigated field level. Agric. Water Manage. **2010**, *97*, 1477–1485.
37. Barron, G. *Centre Pivot Irrigation in the Riverina*; Primefact 98, NSW DPI: Orange, NSW, Australia. Available at http://www.dpi.nsw.gov.au/agriculture/resources/water/irrigation/systems/pressurised/centre-pivot-irrigation.
38. Pathak, B.S.; Bining, A.S. Energy use pattern and potential for energy saving in rice-wheat cultivation. Energy Agric. **1985**, *4*, 271–278.

39. Morris, M.; Lynne, V. *Energy Saving Tips for Irrigators*; ATTRA Publication #IP278, 2006. Available at http://attra.ncat.org/attra-pub/energytips_irrig.html.
40. Tsatsarelis, C.A. Energy inputs and outputs for soft winter wheat production in Greece. Agric. Ecosyst. Environ. **1993**, *43* (2), 109–118.
41. Yilmaz, I; Akcaoz, H.; Ozkan, B. An analysis of energy use and input costs for cotton production in Turkey. Renewable Energy **2005**, *30*, 145–155.
42. Tsatsarelis, C.A. Energy requirements for cotton production in central Greece. J. Agric. Eng. Res. **1991**, *50*, 239–246.
43. Pretty, J.N. *Regenerating Agriculture: Policies and Practice for Sustainability and Self-reliance*; Earthscan Publications Ltd.: London, 1995.
44. Wells, C. *Total Energy Indicators of Agricultural Sustainability: Dairy Farming Case Study*, Final Report to MAF Policy; University of Otago: New Zealand, 2001.
45. Hatirli, S.A.; Ozkan, B.; Fert, C. Energy inputs and crop yield relationship in greenhouse tomato production. Renewable Energy **2006**, *31* (4), 427–438.
46. Chen, G.; Maraseni, T.; Yang, Z. Life-cycle energy and carbon footprint assessments: Agricultural and food products. In *Encyclopedia of Energy Engineering and Technology*; Capehart, B., Ed.; Taylor & Francis Books: London, 2010; 1: 1, 1–5.
47. Hoeppner, J.W.; Entz, M.H.; McConkey, B.G.; Zentner, R.P.; Nagy, C.N. Energy use and efficiency in two Canadian organic and conventional crop production systems. Renewable Agric. Food Syst. **2006**, *21* (1), 60–67.
48. Yaldiz, O.; Ozturk, H.H.; Zeren, Y.; Bascetincelik, A. *Energy usage in production of field crops in Turkey*, 5th Int. Congress on Mechanization and Energy Use in Agriculture, Kusadasi, Turkey, 1993.
49. Ziesemer, J. *Energy use in organic food systems*; Food and Agriculture Organization of the United Nations. available at http://www.fao.org/docs/eims/upload/233069/energy-use-oa.pdf.
50. Tribe, D. *Sustaining the food supply of a developing world: Genetically modified crops enter their second decade*; University of Melbourne.
51. Available at http://www.hexima.com.au/pdf_files/articles/Support_grows_for_GM-Australian.pdf.
52. Available at http://www.usaid.gov/our_work/economic_growth_and_trade/energy/publications/empowering_agriculture.pdf.
53. Svejkovsky, C. *Renewable Energy Opportunities on the Farm, National Center for Appropriate Technology*; ATTRA Publication #IP304, USA. Available at http://attra.ncat.org/attra-pub/energyopp.html.
54. Fischer, J.R.; Johnson, S.R.; Finnell, J.A.; Price, R.P. Renewable energy technologies in agriculture: Solar, wind, geothermal, and anaerobic digestion. Resour. Mag. **2009**, *16* (3), 4–9.
55. Hansen, A.C.; He, B. Biodiesel fuels for off-road vehicles. In *Encyclopedia of Agricultural, Food, and Biological Engineering*; Heldman, D.R., Ed.; 2009; 1: 1, 1–3.
56. Fischer, J.R.; Buchanan, G.A.; Orbach, R.; Harnish III, R.L.; Jena, P. Renewable energy gains global momentum. Resour. Mag. **2008**, *15* (7), 9–11.
57. Available http://www.solar-estimate.org/index.php?verifycookie=1&page=estimatoroverview&subpage=.

Agriculture: Organic

Kathleen Delate
Departments of Agronomy and Horticulture, Iowa State University, Ames, Iowa, U.S.A.

Abstract
Organic agriculture is based on minimal use of off-farm inputs and on management practices that restore, maintain, or enhance ecological harmony. The term "organic" is defined by law, but the labels "natural," "eco-friendly," and similar statements do not guarantee complete adherence to organic practices as defined by law.

INTRODUCTION

The USDA National Organic Standards Board (NOSB) defines organic agriculture as "an ecological production management system that promotes and enhances biodiversity, biological cycles, and soil biological activity.[1] It is based on minimal use of off-farm inputs and on management practices that restore, maintain, or enhance ecological harmony. The primary goal of organic agriculture is to optimize the health and productivity of interdependent communities of soil life, plants, animals and people."[1] The term "organic" is defined by law. The labels "natural," "eco-friendly," and similar statements do not guarantee complete adherence to organic practices as defined by law.

HISTORY

In 1990, the U.S. Congress passed the Organic Food Production Act (OFPA). This law was heralded as the first U.S. law established to regulate a system of farming. The OFPA requires that anyone selling products as "organic" must follow a set of prescribed practices that include avoidance of synthetic chemicals in crop and livestock production, and in the manufacturing of processed products. "Certified organic" crops must be raised on land to which no synthetic chemical (any fertilizers, herbicides, insecticides, or fungicides) inputs were applied for three years prior to the crops' sale. Organic certification agencies became established in the United States to deal with a required "third-party certification." There are at least 20 private certification agencies and 15 state agencies certifying organic production and processing in the United States. Proposed rules implementing the federal OFPA law were promulgated in 1997, after seven years of revisions. Unfortunately, these rules did not meet private certification agencies' standards, and a record number of complaints (275,000) were issued in the public comment period. Now that the federal rules are established (released in 2001), all certifiers must utilize the federal standards as the minimum standard for the "certified organic" label in the United States. European regulation is under the auspices of the International Federation of Organic Agriculture Movements (IFOAM) with national certification agencies in each country.[2] Japan currently certifies under the Ministry of Agriculture and Forestry. Certification for the European Union and Japan is extended to several U.S. certifiers that meet international standards.

Organic agriculture is the oldest form of agriculture on earth. Farming without the use of petroleum-based chemicals (fertilizers and pesticides) was the norm for farmers in the developed world until post–World War II. The war era led to technologies that were adapted for agricultural production. Ammonium nitrate used for munitions during WWII evolved into ammonium nitrate fertilizer; organophosphate nerve gas production led to the development of powerful insecticides. These technical advances since WWII have resulted in significant economic benefits, as well as unwanted environmental and social effects. Organic farmers seek to utilize those advances that yield benefits (e.g., new varieties of crops, more efficient machinery) while discarding those methods that have led to negative impacts on society and the environment, such as pesticide pollution and insect pest resistance.[3] Instead of using synthetic fertilizers and pesticides, organic farmers utilize crop rotations, cover crops, and naturally based products to maintain or enhance soil fertility.[4,5] These farmers also rely on biological, cultural, and physical methods to limit pest expansion and increase populations of beneficial insects on their farms. By managing their ecological capital through efficient use of on-farm natural resources, organic farmers produce for diverse and specialized markets that provide premium prices.

Because genetically modified organisms (GMOs) constitute synthetic inputs and pose unknown risks, GMOs, such as herbicide-resistant seeds, plants, and product ingredients are disallowed in organic agriculture. Organic livestock, like organic crops, must be fed 100% organic food or feed in their production. Synthetic hormones and antibiotics are dis-

Encyclopedia of Environmental Management DOI: 10.1081/E-EEM-100000879

allowed in organic livestock production. Traditional farmers throughout the world have relied on natural production methods for centuries, maintaining consistent yields within their local environment. While "green revolution" technologies have led to increased yields in many less developed countries, many farmers have seen an increase in pest problems with new varieties and high input–based systems.

Motivations for organic production include economic, food safety, and environmental concerns. All organic farmers avoid the use of synthetic chemicals in their farming systems, but philosophies differ among organic farmers regarding methods to achieve the ideal system. Organic farmers span the spectrum from those who completely eschew external inputs, create on-farm sources of compost for fertilization, and encourage the activity of beneficial insects through conservation of food and natural habitats, to those farmers who import their fertility and pest management inputs. A truly sustainable method of organic farming would seek to eliminate, as much as possible, reliance on external inputs.

WORLDWIDE STATISTICS

USDA does not publish systematic reports on organic production in the United States. The most recent census in 1994 identified 1.5 million acres of organic production in the United States with 4050 farmers reporting organic acreage.[6] This figure underrepresents current production because many organic farmers opt to sell their products as organic without undergoing certification. The U.S. organic industry continues to grow at a rate of 20% annually. The industry was listed as a $4.5 billion industry in 1998, with predicted future growth to $10 billion by 2003. The organic industry is a consumer–driven market. According to industry surveys, the largest purchasers of organic products are young people and college–educated consumers. Worldwide consumption of organic products has experienced tremendous growth, often surpassing U.S. figures of 20% annual gain. Much of the increase in consumption worldwide has been fueled by consumers' demand for GMO–free products. Because GMOs are disallowed in organic production and processing, organic products are automatically segregated as GMO–free at the marketplace. European consumers have led the demand for organic products, particularly in countries such as the Netherlands and Scandinavia. Two percent of all German farmland, 4% of Italian farmland, and 10% of Austrian farmland is managed organically.[2] Prince Charles of England has developed a model organic farm and established a system of government support for transitioning organic farmers. Major supermarket chains and restaurants in Europe offer a wide variety of organic products in their aisles and on their menus. Industry experts predict that the establishment of federal rules will advance organic sales in the United States. Although the organic industry began as a niche market, steady growth has led to its place in a "segment" market since 1997. The organic dairy industry, for example, expanded by 73% from 1996 to 1997, and continues to grow today. Organic markets can be divided into indirect and direct markets. Indirect or wholesale markets include cooperatives, wholesale produce operations, brokers, and local milling operations. Many supermarket chains buy directly from farmers or from wholesalers of organic products. Because meat can now be labeled as "organic," as of 1999, the marketing of organic beef, pork, chicken, and lamb has been significantly simplified. Roadside stands, farmers' markets and community supported agriculture (CSA) farms constitute the direct marketing end of the organic industry. Most consumers relate their willingness to pay premium prices for food that has been raised without synthetic chemicals because of their concern for food safety and the environment. Supporting local family farmers also enters into their purchasing decisions.

CROP AND PEST PERFORMANCE IN ORGANIC SYSTEMS

The basis for all organic farming systems is the health of the soil.[4,7] In addition to maintaining adequate fertility, organic farmers strive for biologically active soil, containing microbial populations required for nutrient cycling.[8,9] Crop rotations (required for all organic operations) provide nutrients such as nitrogen in the case of legume crops (alfalfa and clover) and carbonaceous biomass upon which beneficial soil microorganisms depend for survival.[10,11] A crop rotation plan is required as protection against pest problems and soil deterioration.[12–15] Ideally, no more than four out of six years should be in agronomic crops, and the same row crop cannot be grown in consecutive years on the same land. Legumes (alfalfa, clovers, and vetches) alone, or in combination with small grains (wheat, oats, and barley), must be rotated with row crops (corn, soybeans, amaranth, vegetables, and herbs) to ensure a healthy system. A typical six–year rotation in the Midwestern United States would be corn (with a cover of winter rye)–soybeans–oats (with an underseeding of alfalfa)–alfalfa–corn–soybeans.[16,17] Horticultural crops must be rotated with a leguminous cover crop at least once every five years.

Pest management in organic farming systems is based on a healthy plant able to withstand some pest injury and on the inherent equilibrium in nature, as most insect pests have natural enemies that regulate their populations in unperturbed environments.[18,19] Because only naturally occurring materials are allowed in organic production, insect predators, parasites, and pathogens exist without intervention from highly toxic insecticides.[20] Most organic farmers rely on naturally occurring beneficial insects on their farms, but some farmers purchase and release lacewings and other natural enemies every season, for example. There are also commercial preparations of natural insect pathogens, such as *Bacillus thuringiensis* (Bt), which are used to manage pestiferous larvae, such as corn borers. Botanical insecticides, such as neem and ryania, are also allowable in organic production, but as with all insecticides, sprays should be used only as a last resort. Although these materials are naturally based, some materials may affect natural enemies. Prevention is a cornerstone of organic farming.[21] Pest–free seeds and transplants, along with physical and cultural methods, are used to prevent pest infestations. Physical methods include the use of row covers for protection against insects such as cabbage butterflies and aphids. Cultural methods include sanitation and resistant varieties. Plant varieties are used that have been bred traditionally (i.e., no manipulated gene insertion or engineering involved) for insect, disease, and nematode resistance or tolerance.

Most organic farmers rely on multiple tactics for their weed management.[22,23] Allelopathic crops, cultivation, mulching, and flame burning are all methods available for organic farmers. Allelopathic crops, such as rye and oats, produce an exudate that mitigates against small weed seed germination. Depending on the crop, cultivation offers the least labor–intensive method of organic weed management. Timely cultivation is key; without specific schedules, weeds proliferate. Propane flame burning is generally used in conjunction with cultivation, particularly during times of high field moisture. Mulching with straw or wood chips is commonly used in many organic horticultural operations.

Yields comparable to conventional crops have been shown for organic crops in three university long–term experiments in the United States (South Dakota State University,[24] Iowa State University,[16] and the University of California–Davis.[25] and in many European studies.[26] Factoring in an organic premium (ranging from 50 to 400%, depending upon crop and season), organic systems consistently out–performed conventional systems in terms of economics.[17,27,28] Pest problems were not a critical factor in these organic systems. Other studies have shown the benefits of organic practices, such as composting, in mitigating root–borne diseases.[8]

KEY ISSUES REQUIRING ADDITIONAL RESEARCH

Continued verification of the long–term benefits of organic versus conventional farming in terms of soil quality,[29] pest management, and nutritional benefits.[30] is needed. Key issues include the development of management practices to increase nutrient cycling for maintenance of crop yields and optimize biological control of plant pests and diseases[31] Economic analysis, including risks of the three–year transition required for organic certification, will provide useful information for growers interested in alternative systems.[26,32] Appropriate tillage systems, which protect soil quality and provide adequate soil preparation, remain as important issues for organic producers. The improvement of natural parasiticide formulations, such as diatomaceous earth, is required for optimum organic livestock production. Marketing and support needs include the availability of reliable statistics for organic operations and prices. Although many European countries support their farmers in their organic production practices through environmental subsidies,[33] the United States has made small gains in this area. Some state agencies (Minnesota Department of Agriculture) and the USDA Natural Resources Conservation Services (NRCS) through the Environmental Quality Indicators Program (EQIP) offer financial incentives to organic farmers during their transitioning years. More of these support services are needed to encourage farmers interested in the conversion to alternative production.[34]

REFERENCES

1. National Organic Program USDA Agricultural Marketing Service Washington, DC, 1. http://www.ams.usda.gov/nop/ (accessed June 5, 2000).
2. Lampkin N.H.; *The Policy and Regulatory Environment for Organic Farming in Europe*; Universität Hohenheim, Institut für Landwirtschaftliche Betriebslehre: Stuttgart, Germany, 1999; 1–379.
3. Altieri, M. *Agroecology*; Second Ed., Westview Press: Boulder, CO, 1995; 1–433.
4. Lampkin, N.H.; Measures, M. *1999 Organic Farm Management Handbook*, 3rd Ed. Welsh Institute of Rural Studies, University of Wales, Hamstead Marshall, Berkshire, 1999; 1–163.
5. Lockeretz, W.; Shearer, G.; Kohl, D. Organic farming in the corn belt. *Science,* **1981**, *211*, 540–547.
6. Lipson, M. *Searching for the "O–Word": analyzing the USDA current research information system for pertinence to organic farming*; Organic Farming Research Foundation: Santa Cruz, CA, 1997; 1–85.
7. Macey, A.; Kramer, D. *Organic Field Crop Handbook*; Canadian Organic Growers, Inc.: Ottawa, Canada, 1992; 1–256.
8. Drinkwater, L.E.; Letourneau, D.K.; Shennan, C. Fundamental differences between conventional and organic tomato agroecosystems in California. Ecological Appl. **1995**, *5* (4), 1098–1112.
9. Wander, M.M.; Traina, S.J.; Stinner, B.R.; Peters, S.E. Organic and conventional management effects on biologically active soil organic matter pools. Soil Sci. Soc. Am. J. **1994**, *58*, 1130–1139.
10. Drinkwater, L.E.; Wagoner, P.; Sarrantonio M. Legume–based cropping systems have reduced carbon and nitrogen losses. Nature **1998**, *396*, 262–265.
11. Yeates, G.W.; Bardgett, R.D.; Cook, R.; Hobbs, P.J.; Bowling, P.J.; Potter, J.F. Faunal and microbial diversity in three welsh grassland soils under conventional and organic management regimes. J. Appl. Ecol. **1997**, *34* (3) 453–470.
12. Adee, E.A.; Oplinger, E.S.; Grau, C.R. Tillage, rotation sequence, and cultivar influences on brown stem rot and soybean yield. J. Prod. Agric. **1994**, *7* (3), 341–347.

13. Karlen, D.L.; Varvel, G.E.; Bullock, D.G.; Cruse, R.M. Crop rotations for the 21st century. Adv. in Agron. **1994**, *53*, 1–45.
14. Lipps, P.E.; Deep, I.W. Influence of tillage and crop rotation on yield, stalk rot, and recovery of Fusarium and Trichoderma spp. from corn. Plant Dis. **1991**, *75* (8), 828–833.
15. Stinner, D.H.; Stinner B.R.; Zaborski, E.R.; Favretto M.R.; McCartney, D.A. *Ecological Analyses of Ohio Farms under Long–term Sustainable Management*; Ecological Society of America, August: Knoxville, TN, 1994;7–11.
16. Delate, K. *Comparison of Organic and Conventional Rotations at the Neely–Kinyon Long–Term Agroecological Research (LTAR) Site*; Leopold Center for Sustainable Agriculture Annual Report, Iowa State University: Ames, IA, 1999; 1–12.
17. Welsh, R. *The Economics of Organic Grain and Soybean Production on the Midwestern United States*; Henry, A., Ed.; Wallace Institute for Alternative Agriculture: Greenbelt, MD, 1999.
18. Neher, D.A. Nematode communities in organically and conventionally managed agricultural soils. J. Nematol. **1999**, *31*, 142–154.
19. Pfiffner, L.; Niggli, U. Effects of bio–dynamic, organic and conventional farming on ground beetles and other epigaeic arthropods in winter wheat. Biol. Agri. Hort. **1996**, *12*, 353–364.
20. Kromp, B.; Meindel, P.; Harris, P.J.C. *Entomological Research in Organic Agriculture*; Academic Publishers: Bicester, England, 1999, 1–386.
21. Lockeretz, W. *Environmentally Sound Agriculture: Selected Papers from the Fourth International Conference of the International Federation of Organic Agriculture Movements (IFOAM)*; Praeger: New York, 1983; 1–426, Cambridge, MA, Aug 18–20, 1982.
22. Lanini, W.T.; Zalom, F.; Marois, J.; Ferris, H. Researchers find short–term insect problems, long–term weed problems. Ca. Agric. **1994**, *48*, 27–33.
23. Liebman, M.; Ohno, T. Crop Rotation and Legume Residue Effects on Weed Emergence and Growth: Applications for Weed Management. In *Integrated Weed and Soil Management*; Hatfield, J.L.; Buhler, D.D.; Stewart, B.A., Eds.; Ann Arbor Press: Chelsea, MI, 1997; 181–221.
24. Dobbs, T. Report on Organic and Conventional Grain Trials at South Dakota State University. *USDA–ERS Conference on The Economics of Organic Farming Systems,* USDA: Washington, DC, 1999; 1–4.
25. Klonsky, K.; Livingston, P. Alternative systems aim to reduce inputs, maintain profits. Ca. Agric. **1994**, *48* (5), 34–42.
26. Lampkin, N.H.; Padel, S. *The Economics of Organic Farming: An International Perspective*; CAB International: Wallingford, Oxon, U.K., 1994; 1–468.
27. Hanson, J.C.; Lichtenberg, E.; Peters, S.E. Organic versus conventional grain production in the Mid–Atlantic: an economic and farming system overview. Am. J. Alt. Agric. **1997**, *12*, 2–9.
28. Stanhill, G. The comparative productivity of organic agriculture. Agric. Ecosys. Env. **1990**, *30*, 1–26.
29. Lockeretz, W.; Shearer, G.; Klepper, R.; Sweeney, S. Field crop production on organic farms in the midwest. J. Soil Water Conserv. **1978**, *33*, 130–134.
30. Woese, K.; Lange, D. Boess, C.; Bögl, K.W. A comparison of organically and conventionally grown foods—results of a review of the relevant literature. J. Sci. Food Agric. **1997**, *74*, 281–293.
31. Walz, E. *Final Results of the Third Biennial National Organic Farmers' Survey*; Organic Farming Research Foundation: Santa Cruz, CA, 1999; 1–75.
32. Chase, C.; Duffy, M. An economic comparison of conventional and reduced–chemical farming systems in Iowa. Amer. J. Alt. Agric. **1991**, *6* (4), 168–73.
33. Zygmont J. International Issues Pertaining to Organic Agriculture, Proceedings of the Workshop: Organic Farming and Marketing Research—New Partnerships and Priorities, Lipson, M., Hammer, T., Eds., Organic Farming Research Foundation: Santa Cruz, CA, 1999; 317–379.
34. D'Souza G.; Cyphers, D.; Phipps, T. Factors affecting the adoption of sustainable agricultural practices. Agric. and Resource Econ. Rev. **1993**, *22* (2), 159–205.

Agroforestry: Water Use Efficiency

James R. Brandle
School of Natural Resource Sciences, University of Nebraska—Lincoln, Lincoln, Nebraska, U.S.A.

Laurie Hodges
Department of Agronomy and Horticulture, University of Nebraska—Lincoln, Lincoln, Nebraska, U.S.A.

Xinhua Zhou
School of Natural Resources, University of Nebraska—Lincoln, Lincoln, Nebraska, U.S.A.

Abstract

Agroforestry is the intentional integration of trees and shrubs into agricultural systems. In agricultural systems, water is often the major factor limiting growth. When water availability is limited as a result of limited supply or high cost, its efficient use becomes critical to successful production systems. This entry focuses on water use efficiency (WUE), defined as the amount of biomass (or grain) produced per unit of land area for each unit of water consumed. Because agroforestry practices alter the microclimate of adjacent fields, they affect WUE of plants growing in those fields.

INTRODUCTION

Agroforestry is the intentional integration of trees and shrubs into agricultural systems. Windbreaks, riparian forest buffers, alley-cropping, silvopastoral grazing systems, and forest farming are the primary agroforestry practices found in temperate regions of North America.[1] Placing trees and shrubs on the landscape changes the surface energy balance, influences the surrounding microclimate, and has the potential to alter water use and productivity of adjacent crops.[2,3]

In agricultural systems, water is often the major factor limiting growth. When water availability is limited as a result of limited supply or high cost, its efficient use becomes critical to successful production systems. For example, proper irrigation at the appropriate stage of crop development minimizes pumping costs and increases yield; reducing soil tillage conserves soil water and may enhance yield, and reducing surface runoff or trapping snow improves soil water storage for future crop use. These water conservation efforts contribute to the efficient use of available water and are determined primarily by management practices. In contrast, Tanner and Sinclair[4] distinguish between the efficient use of water and water use efficiency (WUE). WUE is primarily a function of physiological responses of plants to environmental conditions. This review focuses on WUE defined as the amount of biomass (or grain) produced per unit of land area for each unit of water consumed.[4]

Soil water may be consumed by evaporation from the soil surface or by the transport of water through the plant and subsequent evaporation from the leaf surface. The rate of water consumption is determined by the microclimate of the crop. Because agroforestry practices alter the microclimate of adjacent fields, they affect WUE of plants growing in those fields.

DISCUSSION

Windbreaks, riparian forest buffers or alley-cropping systems are the practices most likely to be integrated into crop production systems. In all three practices, trees and shrubs tend to be arranged in narrow barriers adjacent to the crop field. Microclimate responses downwind of any of these types of barriers are similar and the following discussion applies to all three types of barriers. As wind approaches these barriers, it is diverted up and over the barrier creating two zones of protection, a larger zone to the lee of the barrier (the side away from the wind) and a smaller zone on the windward side of the barrier. In these zones, wind speed is reduced and turbulence and eddy structure in the vicinity of the barrier are altered. As a result of these changes, the transfer coefficients for heat and mass between the crop and the atmosphere are altered; the gradients of temperature, humidity, and carbon dioxide concentration above the soil and canopy are changed;[5] and the plant processes of transpiration and photosynthesis are altered.[6]

McNaughton[5] defined two regions within the leeward zone of protection: the *quiet zone*, extending from the top of the barrier down to a point in the field located approximately $8H$ leeward (H is the height of the barrier) and a *wake zone*, lying beyond the quiet zone and extending from approximately $8H$ to a distance of $20H$ to $25H$ from the barrier. Within the quiet zone where turbulence is reduced, we expect conditions to be such that the canopy is "uncoupled" from the atmospheric conditions above the sheltered zone, while in the wake zone where turbulence is increased, we expect the canopy to become more strongly "coupled" to the atmosphere above. In both locations we would expect the rates of photosynthesis and transpiration to be altered and WUE to change.

Encyclopedia of Environmental Management DOI: 10.1081/E-EEM-120010098

The magnitude of change in wind speed, as well as the extent of microclimate modifications within the quiet and wake zones, are largely determined by the structure of the windbreak or barrier and the underlying meteorological conditions. Structure refers to the amounts of solid material and open space and their arrangement within the barrier. Dense barriers, for example, multiple rows of conifers, generally result in greater wind speed reduction but more turbulence. More porous barriers, for example, single rows of deciduous species, result in less wind speed reduction but also less turbulence. The downwind extent of the protected area is generally greater for more porous barriers. As a result, narrow, less dense barriers (40%–60% density) are typically used to protect crop fields.

The overall influence of wind protection on plant water relations is complex and linked to temperature, humidity, wind speed, and other meteorological conditions found in the protected zone, the amount of available soil water, crop size, and stage of development.[2,3,7] Until recently, the major effect of wind protection and its influence on crop growth and yield were assumed to be due primarily to soil water conservation and reduced water stress of sheltered plants.[8,9] There is little question that the evaporation rate from bare soil is reduced in the protected zone.[3] However, the effect of reduced wind speed on transpiration, evaporation from the plant canopy, and overall plant water status is less clear.[2,3,7]

According to Grace,[9] transpiration rates may increase, decrease, or remain unaffected by wind protection depending on wind speed, atmospheric resistance, and saturation vapor pressure deficit. Cleugh[3] suggests that as stomatal resistance increases, evaporation from the canopy may actually be increased with a reduction in wind speed. When stomatal resistance is high and water is limited, stomatal resistance controls the rate of evaporation from the leaf surface, not the amount of turbulence. Under these conditions a decrease in wind speed and turbulent mixing may increase the potential for evaporation from the leaf surface.[3]

Evaporation from the leaf surface consists of two phases, an energy driven phase and a diffusion driven phase. Movement of water through the plant and out the stomata is driven by the water potential gradient within the plant. This gradient is influenced by the plant's energy balance. On the lee side of the buffer, reduced wind speed and turbulent mixing lead to increases in leaf temperature and transpiration to meet the increased energy load on the plant. If adequate water is available, it is moved through the plant to the leaf surface and the potential for evaporation from the leaf surface is increased. If water is limited, the stomata partially or completely close, transpiration is reduced, and evaporation from the leaf surface declines.

In contrast, movement of water vapor across the leaf boundary layer is controlled by the vapor pressure gradient and the thickness of the leaf boundary layer. As wind speed decreases, the thickness of this boundary layer increases, the vapor pressure gradient decreases, and the rate of evaporation from the leaf surface decreases. The relative magnitude of the two processes determines whether or not transpiration and subsequent evaporation from the canopy are increased, decreased, or remain unchanged.[7,9,10]

While these theoretical considerations are important in understanding the process, several studies[11–13] have demonstrated a good correlation between wind protection, conservation of soil water, and enhanced crop yield. Even so, the effect of wind protection on WUE is neither constant throughout the growing period[7] nor is it consistent over varying meteorological conditions.

Agroforestry practices impact the water relations of the crop by affecting the loss of water through damaged leaves. On soils subject to wind erosion, windbreaks or other agroforestry buffers provide significant reductions in the amount of wind blown soil and subsequent abrasion of plant parts and cuticular damage.[9,14] Loss of cuticular integrity or direct tearing of the leaves[15] reduces the ability of the plant to control water loss.

Agroforestry buffers have a direct effect on the distribution of precipitation, both rain and snow. In the case of snow, a porous barrier will result in a more uniform distribution of snow across the field, providing additional soil water for the crop.[16] In the case of rain, the barrier has minimal influence on the distribution of precipitation across the field; however, in the area immediately adjacent to the barrier a rain shadow may occur on the leeward side. On the windward side, the barrier may lead to slightly higher levels of measured precipitation at or near the base of the trees due to increased stem flow or dripping from the canopy.

Trees and shrubs used in agroforestry practices also consume a portion of the available water. In the area immediately adjacent to the barrier, competition for water between the crop and the barrier has a negative impact on yield. These same areas are also subject to some degree of shading depending on the orientation of the barrier. These changes in radiation load influence the energy balance and thus the growth and development of the crop and the utilization of water.[2]

SUMMARY

In summary, agroforestry practices such as windbreaks, riparian forest buffers and alley-cropping systems generally improve both the efficient use of water by the agricultural system and the WUE of the individual crop. In the case of efficient water use, the evidence is clear. In the case of crop WUE, the evidence leaves some unanswered questions. How do we account for the varied crop yield responses reported in the literature? In many cases yields are increased but no clear relationship to crop water budget is shown. In other cases crop yield response is minimal. Under what meteorological conditions are the effects of agroforestry practices most valuable to water balance questions? Final crop yield is a integration of the environmental conditions over the entire growing season. Many different combinations of environmental conditions may result in similar plant responses. How do we address the numerous combinations of plant stress and plant growth to determine "a response" to wind protection? To answer many of these questions it will be necessary to intensify the numerical modeling methods developed by Wilson[17] and Wang and Takle.[18] With a better model to describe the turbulence fields and the transport of water, heat, and carbon dioxide as influenced by agroforestry practices, it should be possible to assess the numerous combinations of environmental factors influencing crop growth in these systems.

REFERENCES

1. Lassoie, J.P.; Buck, L.E. Development of agroforestry as an integrated land use management strategy. In *North American Agroforestry: An Integrated Science and Practice*; Garrett, H.E., Rietveld, W.J., Fisher, R.F., Eds.; American Society of Agronomy, Inc.: Madison, WI, 2000; 1–29.
2. Brandle, J.R.; Hodges, L.; Wight, B. Windbreak practices. In *North American Agroforestry: An Integrated Science and Practice*; Garrett, H.E., Rietveld, W.J., Fisher, R.F., Eds.; American Society of Agronomy, Inc.: Madison, WI, 2000; 79–118.
3. Cleugh, H.A. Effects of windbreaks on airflow, microclimates and crop yields. Agrofor. Syst. **1998**, *41*, 55–84.
4. Tanner, C.B.; Sinclair, T.R. Efficient water use in crop production: research or re-search? In *Limitations to Efficient Water Use in Crop Production*; Taylor, H.M., Jordan, W.R., Sinclair, T.R., Eds.; American Society of Agronomy, Inc.: Madison, WI, 1983; 1–27.
5. McNaughton, K.G. Effects of windbreaks on turbulent transport and microclimate. Agric. Ecosyst. Environ. **1988**, *22/23*, 17–39.
6. Grace, J. Some effects of wind on plants. In *Plants and Their Atmospheric Environment*; Grace, J., Ford, E.D., Jarvis, P.G., Eds.; Blackwell Scientific Publications: Oxford, 1981; 31–56.
7. Nuberg, I.K. Effect of shelter on temperate crops: a review to define research for Australian conditions. Agrofor. Syst. **1998**, *41*, 3–34.
8. Caborn, J.M. *Shelterbelts and microclimate*, Forestry Commission Bulletin No. 29; Her Majesty's Stationery Office: Edinburgh, 1957; 135 pp.
9. Grace, J. Plant response to wind. Agric. Ecosyst. Environ. **1988**, *22/23*, 71–88.
10. Thornley, J.H.M.; Johnson, I.R. *Plant and Crop Modeling: A Mathematical Approach to Plant and Crop Physiology*; Clarendon Press: New York, 1990; 669 pp.
11. Song, Z.M.; Wei, L. The correlation between windbreak influenced climate and crop yield. In *Agroforestry Systems in China*; Zhu, Z.H., Cai, M.T., Wang, S.J., Jiang, Y.X., Eds.; International Development Research Centre (IDRC, Canada), Regional Office for Southeast and East Asia, published jointly with the Chinese Academy of Forestry: Singapore, 1991; 21–115.
12. Wu, Y.Y.; Dalmacio, R.V. Energy balance, water use and wheat yield in a Paulownia-wheat intercropped field. In *Agroforestry Systems in China*; Zhu, Z.H., Cai, M.T., Wang, S.J., Jiang, Y.X., Eds.; International Development Research Centre (IDRC, Canada), Regional Office for Southeast and East Asia, published jointly with the Chinese Academy of Forestry: Singapore, 1991; 54–65.
13. Huxley, P.A.; Pinney, A.; Akunda, E.; Muraya, P. A tree/crop interface orientation experiment with a *Grevillea robusta* hedgerow and maize. Agrofor. Syst. **1994**, 26, 23–45.
14. Kort, J. Benefits of windbreaks to field and forage crops. Agric. Ecosyst. Environ. **1988**, *22/23*, 165–190.
15. Miller, J.M.; Böhm, M.; Cleugh, H.A. *Direct Mechanical Effects of Wind on Selected Crops: A Review*, Technical Report Number 67; CSIRO Center for Environmental Mechanics: Canberra, Australia, 1995; 68 pp.
16. Scholten, H. Snow distribution on crop fields. Agric. Ecosyst. Environ. **1988**, *22/23*, 363–380.
17. Wilson, J.D. Numerical studies of flow through a windbreak. J. Wind Eng. Ind. Aerodyn. **1985**, *21*, 119–154.
18. Wang, H.; Takle, E.S. A numerical simulation of boundary-layer flows near shelterbelts. Boundary-Layer Meteorol. **1995**, *75*, 141–173.

Agricultural – Alexandria

Air Pollution: Monitoring

Waldemar Wardencki
Department of Chemistry, Gdansk University of Technology, Gdansk, Poland

Abstract
This entry presents an overview of the issues in the field of air pollution monitoring. At the beginning, the general objectives of air monitoring, ambient air quality standards for so-called criteria pollutants, and their sources are discussed. In the next part, both analytical methods and instruments for monitoring of ambient air and stack gases are briefly presented. Additionally, other approaches used in air pollution monitoring, such as biomonitoring, geographical information system (GIS), or remote monitoring, are also briefly characterized.

INTRODUCTION

Concern about air quality is not new. The first reports of air pollution problems appear to have been made by writers in ancient Rome who were aware of its adverse effects on human health. Air pollution and its consequences had originally been considered to be relatively local phenomena associated with urban and industrial centers. Complaints were recorded in the 13th century when coal was first used in London. Now, it has become apparent that pollutants may be transported long distances in the air, causing adverse effects in environments far removed from the source emission. Scientific research, conducted over 200 years, has evidently shown that polluted air has a negative influence on health and, in some cases, may lead to death. The World Health Organization (WHO) appraises that air pollution causes approximately 2 million premature deaths worldwide per year. The levels of pollutants, which have a negative influence on life, are nowadays well defined. Because current thresholds set by national or global air quality guideline values are frequently exceeded, further reductions of emissions are necessary.

The first essential step in controlling and mitigating air pollution is to quantify the emissions of air pollutants. Most countries entail controlling a range of key pollutants at their point of discharge. The most important tool in environmental protection is monitoring. Environmental monitoring is the general term for systematic observations of what is going on in the environment. In the broadest context, environmental monitoring is defined as a system of measurements, evaluations, and forecasts of environmental states, and the collecting, processing, and spreading of information on the environment.

Air pollution and its control are a global issue demanding international cooperation. Monitoring of air pollution is a very important source of data. However, measurement of air pollutant concentrations, in comparison to monitoring of other elements of the environment, is the most difficult. This is related to the dynamics of the atmosphere, making it the main route of pollutant transport between the remaining environmental compartments. Unlike the case of water and soil pollution, environmental pollution is not geographically restricted, as a result of which large human populations can be exposed to it. Another problem is low concentration of air pollutants and their interaction with other gases.

This entry reviews the issues in the field of air pollution monitoring. At the beginning, the general objectives of air monitoring, ambient air quality standards for so-called criteria pollutants, and their sources are presented. In the next part, both analytical methods and instruments for monitoring of ambient air and stack gases are briefly presented. Additionally, other approaches applied in air pollution monitoring, such as biomonitoring, geographical information system (GIS), or remote monitoring, are also briefly characterized.

OBJECTIVES OF AIR MONITORING

Collecting information on the presence and concentration of pollutants in the environment, both naturally occurring or from anthropogenic sources, may be achieved by measurements of such substances or phenomenon of interest. For realistic assessment, temporal and spatial variations of concentrations in the particular environmental compartment, repeated measurements rather than single ones, are made.

The general aim of monitoring is to provide information about the actual levels of harmful or potentially harmful pollutants to indicate areas in which the quality of air does not fulfill proper standards. The main objectives of air monitoring are as follows:

- To measure pollutant mixing ratios and their interactions, patterns, and fate in the environment.

Encyclopedia of Environmental Management DOI: 10.1081/E-EEM-120046061

- To carry out ecotoxicological studies and assessment of the effects of pollution on man and the environment, to identify possible cause-and-effect relationship between pollutant concentration and health effects.
- To assess emission sources and the need for legislative controls on emissions of pollutants and to ensure compliance with emission standards.
- To activate emergency procedures in areas prone to acute pollution episodes.
- To obtain a historical record of air quality to provide a database for future use.

The area of applications of air monitoring data is presented in Fig. 1.

When the objectives of monitoring are clearly defined, several decisions should be made to generate suitable data for the intended use. Decisions on what to monitor, when and where to monitor, and how to monitor are usually undertaken at the beginning. More difficult are next decisions, e.g., establishing the number and location of sampling sites, the duration of the survey, and the time resolution of sampling. All the steps in the design of a monitoring program are presented in Fig. 2.

HISTORY OF AIR POLLUTION LEGISLATION

A growing concern over the influence of different air pollutants on human health was the main driving force to develop and implement air quality criteria and standards. Impetus was given to the development of air quality standards in 1958 when it was realized that photochemical problems could not be resolved without control of motor vehicle emission.

Air Quality Standards

Efforts to regulate air quality by law were discussed and undertaken in the 1960s. One of the first proposition was presented by Atkisson and Gaines[1] in 1970. The initial regulations were set in California in response to concerns over human health.[2] After that, similar air quality programs were soon adopted nationally. In 1967, the U.S. Congress enacted the Air Quality Acts, the first modern environmental law.

The Clean Air Act,[3] which was last amended in 1990, requires the United States Environmental Protection Agency (US EPA) to set National Ambient Air Quality Standards (NAAQS) for pollutants considered harmful to public health and the environment. The Clean Air Act established two types of national air quality standards. *Primary standards* set limits to protect public health, including the health of "sensitive" populations such as asthmatics, children, and the elderly. *Secondary standards* set limits to protect public welfare, including protection against decreased visibility and damage to animals, crops, vegetation, and buildings. EPA has established NAAQS for six principal pollutants, which are called criteria pollutants: sulfur dioxide, particulate matter (PM), nitrogen oxide, carbon monoxide, ozone, and lead. These standards are threshold concentrations based on a detailed review of scientific information related to effects.

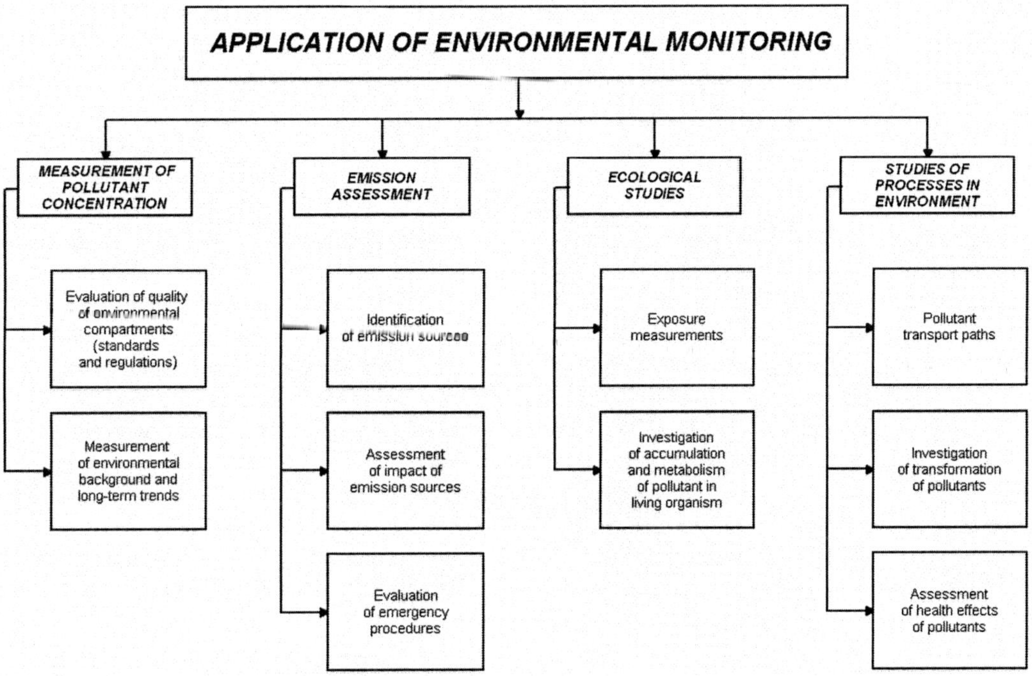
APPLICATION OF ENVIRONMENTAL MONITORING
MEASUREMENT OF POLLUTANT CONCENTRATION
EMISSION ASSESSMENT
ECOLOGICAL STUDIES
STUDIES OF PROCESSES IN ENVIRONMENT
Evaluation of quality of environmental compartments (standards and regulations)
Measurement of environmental background and long-term trends
Identification of emission sources
Assessment of impact of emission sources
Evaluation of emergency procedures
Exposure measurements
Investigation of accumulation and metabolism of pollutant in living organism
Pollutant transport paths
Investigation of transformation of pollutants
Assessment of health effects of pollutants

Fig. 1 The detailed goals of activities in air monitoring.

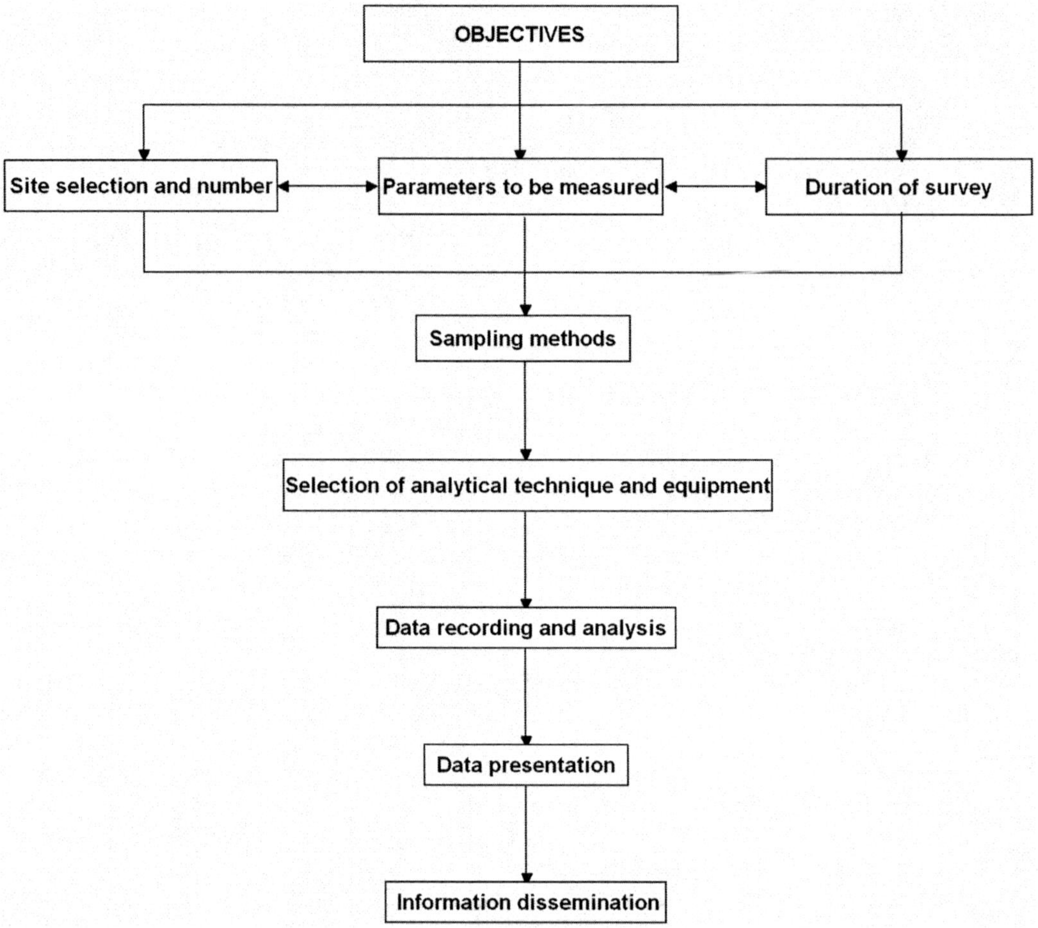
OBJECTIVES
Site selection and number
Parameters to be measured
Duration of survey
Sampling methods
Selection of analytical technique and equipment
Data recording and analysis
Data presentation
Information dissemination

Fig. 2 Steps in the design of a monitoring program.

In Europe, the first international air quality standards were introduced by the European Commission in 1980 for SO_2 and suspended particulates, mainly aimed at protecting human health. A few years later, the WHO, recognizing ecological damage as being relevant to human health, introduced air quality guidelines for Europe, which include the former as well as the latter revision in 2000.[4] The newest directive on ambient air quality and cleaner air of the European Union (EU) entered into force in June 2008.[5] It merges four earlier directives and one Council decision into a single directive on air quality. The new directive of the EU on air quality takes into account concerns from latest WHO air quality guidelines[6] on fine particles. Reflecting the latest WHO air quality guidelines that identify fine particles (PM2.5) as one of the most dangerous pollutants for human health, the new EU directive sets objectives and target dates for reducing population exposure to PM2.5. It also maintains limits for concentration of coarser particles known as PM10 and other main pollutants already subject to legislation.

Table 1 presents examples of air quality standards issued by the EPA, the WHO, and some states.

Air Quality Index

The Air Quality Index (AQI), also known as the Air Pollution Index (API) or Pollutant Standard Index (PSI), is a number used by different government agencies to characterize the quality of the air at a given location. The index aims to help the public easily understand air quality level and protect the health of people from air pollution. As the AQI increases, an increasingly large percentage of the population is likely to experience increasingly severe adverse health effects. Computing the AQI requires an air pollutant concentration from a monitor or model. The function used to convert air pollutant concentration to AQI varies by pollutant and is different among countries. AQI values expressed in different values (most frequently from 0 or 1 to 10, 100, or to 500) are divided into ranges (from 4 to 10), and each range is assigned a descriptor and a color code. Standardized public health advisories are associated with each AQI range. An agency might also encourage members of the public to take public transportation or work from home when AQI levels are high.

Table 1 Comparison of limit values (μm/m^3) for a given averaging time and the number of exceedances per year issued by different countries and organizations.

Pollutant	Averaging time	WHO	EPA	EU	UK	France	Germany	Poland
SO_2	10–15 min	500	–	–	266 (not more than 35 times)	–	–	–
	30 min	–	–	–	–		–	–
	1 hr	–	–	350 (not more than 24 times)	350 (not more than 24 times)	350 (not more than 24 times)	350	350 (not more than 24 times)
	3 hr	–	1,300	–	–	–	–	–
	24 hr	125	365	125 (not more than 3 times)	125 (not more than 3 times)	125 (not more than 3 times)	125	125 (not more than 3 times)
	Year	50	80	20	20	20	20	20
NO_2	30 min	–	–	–	–	–	200	–
	1 hr	200	–	200 (not more than 18 times)	200 (not more than 18 times)	230 (not longer than 0.2% of time)	–	200 (not more than 18 times)
	24 hr	–	–	–	–	–	100	–
	Year	40	100	40	40	46	–	40
PM10	30 min	–	–	–	–	–	–	–
	24 hr	20	150	50 (not more than 35 times)	50 (not more than 35 times)	50 (not more than 35 times)	50	50 (not more than 35 times)
	Year	30		40	40	40	40	40
CO	10–15 min	100,000	–	–	–		–	–
	30 min	60,000	–	–	–	–	–	–
	1 hr	30,000	4,000	–	–	–	–	–
	8 hr	10,000	1,000	1,000	1,000	1,000	1,000	1,000
	24 hr	–		–	–	–	–	–
	Year	–	–	–	–		–	–
O_3	30 min	–	–	–	–	–	–	–
	1 hr	–	235	–	–	–	–	–
	8 hr	100	157	120	100 (not more than 10 times)	120	120	120 (not more than 25 days)
Pb	24 hr	–	–	–	–		5	–
	3 months	–	1.5	–	0.5	–	–	–
	Year	0.5	–	0.5	–	0.5	0.5	0.5
Benzene	Year	–	–	5	16.25	8	–	5

Source: WHO.[6]

Not all air pollutants are characterized by AQI. Many countries monitor only some pollutants, e.g., ground-level ozone, sulfur dioxide, carbon monoxide, and nitrogen dioxide, and calculate AQIs for these pollutants.[7]

The EPA in the United States measures air quality in all parts of the country and publishes a daily AQI based on the data obtained. The following six priority pollutants are measured regularly in order to generate the AQI: carbon monoxide, nitrogen dioxide, particulates, sulfur dioxide, ozone, and lead. EPA has assigned a specific color to each AQI category to make it easier for people to understand quickly whether air pollution is reaching unhealthy levels in their communities. For example, the color orange means that conditions are "unhealthy for sensitive groups," while red means that conditions may be "unhealthy for everyone," and so on.

In Canada, API has values of 0 up to 100+ and is divided into four categories (from good to very poor).

CITEAIR (Common Information to European Air) has developed the first AQIs in Europe.[8] An important feature of the indices is that they differentiate between traffic and city background conditions. The Common Air Quality Index (CAQI) is designed to present and compare air quality in near real time on an hourly or daily basis. The CAQI has five levels, using a scale from 0 (very low) to >100 (very high) and the matching colors range from light green to dark red. The Year Average Common Air Quality Index (YACAQI) uses a different approach adopting the difference to target's principle. If the index is higher than 1.0, it means that for one or more pollutants, the limit values are not met. If the index is below 1, it means that on average the limit values are met. Both indices are practically implemented on a Common Operational Webpage (COW).[9] The project CITEAIR II will further develop the AQIs.[10]

The EU's Sixth Environment Action Programme (EAP), "Environment 2010: Our Future, Our Choice," includes Environment and Health as one of the four main target areas requiring greater effort—air pollution is one of the issues highlighted in this area. The Sixth EAP aims to achieve levels of air quality that do not result in unacceptable impacts on, and risks to, human health and the environment.

The EU is acting at many levels to reduce exposure to air pollution: through EC legislation, through work at the international level to reduce cross-border pollution, through cooperation with sectors responsible for air pollution, through national and regional authorities and NGOs, and through research. The Clean Air for Europe (CAFE) initiative has led to a thematic strategy setting out the objectives and measures for the next phase of European air quality policy.[11]

REGULATED AIR POLLUTANTS

The contaminants in ambient air that are of concern are basically categorized as criteria and non-criteria pollutants.[12]

Criteria air pollutants are those air contaminants for which numerical concentration limits have been set as the dividing line between acceptable air quality and poor or unhealthy air quality. Criteria pollutants include five gases/vapors and two solids: nitrogen oxides (NO_x), sulfur dioxide (SO_2), carbon monoxide (CO), ozone (O_3), benzene, PM10, and lead (Pb). Non-criteria pollutants are those contaminants designated as toxic or hazardous by legislation or regulation. They fall in two further subcategories, depending on the legislation that defines them. In general, hazardous air pollutants may pose a variety of health effects, whereas toxic ones focus on one physiological response.

Main Sources of Air Pollutants

Air pollution may be defined as a situation in which substances change the qualitative composition of air in relation to average composition of troposphere sufficiently high above their normal ambient air levels to produce a measurable effect on humans, animals, vegetation, or material.[13–17]

Pollutants (both organic and inorganic) may be present in different forms such as gases, aerosols (liquid, solid), and sorbates and have a very broad range of concentration.

The concentrations of ambient air pollutants are expressed in terms of either a mass per unit volume ratio, such as $\mu g/m^3$, or a volumetric ratio (i.e., volumes of contaminant per million or billion volumes of air). The conversion between mass units and volumetric ratios at standard temperature or pressure is:[12]

$$\mu g/m^3 = \text{ppm x MW}/0.02445 = \text{ppb x MW}/24.45,$$

where:

$\mu g/m^3$—micrograms per cubic meter

ppm—parts per million by volume ($1{:}10^6$)

MW—molecular weight of the contaminant

ppb—parts per billion by volume ($1{:}10^9$).

Some of the most important atmospheric pollutants, their sources, and impacts on the environment and human health are presented in Table 2.

Urban traffic has become the most important cause of air pollution in the cities.[15] Road traffic is responsible for emission of several air pollutants; the most important of which are nitrogen oxides (NO and NO_2), sulfur dioxide (SO_2), PM, carbon monoxide (CO), and volatile organic compounds (VOCs), all of which can pose a health hazard.

Characteristics of Criteria Air Pollutants

Air pollutants arise from a wide variety of sources although they are mainly a result of the combustion process.[15] The largest sources include power generation, motor vehicles, and industries. The emissions of pollutants to the atmosphere badly influence vegetation, human and animal life, agriculture, and climate. Emissions of carbon monoxide (CO), nitrogen oxides (NO_x), and hydrocarbons are controlled by catalytic converters on new gasoline-driven cars. Emissions of sulfur oxides are being reduced through a lower sulfur content in gasoline. However, emissions of PM are not decreasing. Any successful strategies for controlling or countering these problems must be based on reliable air quality monitoring data for management, to make informed decisions on air pollution control.

Volatile organic compounds is a collective name for a very large number of different chemical species, including hydrocarbons, halocarbons, and oxygenates that have different physicochemical properties and are directly emitted from both anthropogenic and natural sources, and which can contribute to the formation of secondary pollutants with different efficiencies. For vehicular emissions, the list of compounds is long and variable depending on fuel, type of engine, and operating conditions. Hydrocarbons such as ethane, ethyne, higher aliphatic hydrocarbons, benzene, toluene, and xylenes are typical emissions in most cases. Each of these compounds can be released unreacted or can undergo oxidation reactions. One of them, benzene, is found in highest concentrations. Ambient concentrations are typically between 1 and 50 ppb, but close to major

Table 2 Atmospheric pollutants and their sources and effects.

Pollutant	Sources	Impact
Sulfur dioxide (SO_2)	Power generation, industry	Acid deposition, smog formation, threat to human health, smog formation
Nitrogen oxides (NO, N_2O, NO_2)	Transport, power generation, industry	Acid deposition, smog formation, O_3 precursor, threat to human health
Carbon dioxide (CO_2)	Combustion processes, power generation, transport, landfills	Global warming
Carbon monoxide (CO)	Combustion processes, power generation	Toxic to humans
Particulate matter (PM)	Power generation, industry, transport	Threat to human health, reduced visibility
Volatile organic compounds (VOCs)	Transport, industry	Photochemical smog, O_3 precursor, global warming
Ozone (O_3)	Photochemical reactions between VOCs and NO_x	Photochemical smogs, respiratory irritant, crop damage
Methane (CH_4)	Landfills, agriculture, gas industry	Global warming
Benzene, 1,3-butadiene	Transport industry	Carcinogenic
Ammonia (NH_3)	Industry, farming, refrigeration, power plant	Toxic to humans and wildlife
Heavy metals	Industry, transport	Toxic to humans and wildlife
Dioxins and furans	Incineration, electrical equipment	Toxic to humans

Source: Bogue.[18]

emissions can be as high as several hundred parts per billion. In the unreacted state, it has undesirable ecotoxicological properties. Besides causing annoying physiological reactions such as dizziness and membrane irritation, it is known to be a human carcinogen.

The two nitrogen oxides, NO and NO_2 (together called NO_x), from anthropogenic sources are present as a consequence of various combustion processes from both stationary sources, i.e., power generation (21%), and mobile sources, i.e., transport (44%). These species have very short atmospheric lifetimes, around 5 days, and have been ultimately converted to nitric acid and removed in rainfall. However, nitrogen oxide is important because it is a precursor to tropospheric ozone. Whereas NO does not affect climate, ozone does. A typical sea-level mixing ratio of NO is 5 ppt (parts per trillion, $1{:}10^{12}$), but in urban regions, NO mixing ratios reach 0.1 ppm in the early morning, but it decreases to zero by midmorning due to reaction with ozone. A major source of NO_2 is oxidation of NO, with NO_2 being intermediary between NO emission and O_3 formation. Nitrogen dioxide is one of the six criteria air pollutants for which ambient standards are set by the US EPA under CAAA70 (Clean Air Act Amendments of 1970). In the urban regions, the mixing ratio of NO_2 ranges from 0.1 to 0.25 ppm. It is more prevalent during midmorning than during midday or afternoon because sunlight breaks down most NO_2 past midmorning. Exposure to high concentrations of NO_2 harms the lungs and increases respiratory infections. It may trigger asthma by damaging or irritating and sensitizing the lungs, making people more susceptible to allergens. At higher concentrations, it can result in acute bronchitis or death.

Sulfur dioxide (SO_2) is a strong-smelling, colorless gas that is formed by the combustion of fossil fuels, smelting, manufacture of sulfuric acid, conversion of wood pulp to paper, incineration of refuse, and production of elemental sulfur. Power plants, which may use coal or oil high in sulfur content, can be major sources of SO_2, accounting for about 50% of annual global emissions. SO_2 and other sulfur oxides contribute to the problem of acid deposition and can be major contributors to smog. Natural background levels of SO_2 are about 2 ppb. Hourly peak values can reach 750 ppb on infrequent occasions. Sulfur dioxide can lead to lung diseases. SO_2 is a criteria air pollutant.

Ozone (O_3) is not directly emitted from both anthropogenic and natural sources. Its only source into air is chemical reaction. In the urban air, ozone mixing ratios range from less than 0.01 ppm at night to 0.5 ppm (during afternoons in the most polluted cities worldwide), with typical values of 0.15 ppm during moderately polluted afternoons. Ozone causes headaches at concentrations greater than 0.15 ppmv, chest pains at mixing ratios greater than 0.25 ppm, and sore throat and cough at mixing ratios greater than 0.30 ppm. Exceeding the level 0.30 ppm, it decreases lung functions. Symptoms of a respiratory condition include coughing and breathing discomfort. Ozone can also accelerate the aging of lung tissue. It also interferes with the growth of plants and deteriorates organic materials, such as rubber, textiles, and some paints and coatings. Furthermore, ozone increases plant and tree stress and their susceptibility to disease, infestation, and death.

PM, frequently described simply as particle pollution, in the atmosphere arise from natural sources, such as windborne dust, sea spray, and volcanoes, and from anthropogenic activities, such as combustion of fuels. Particle pollution in the air includes a mixture of solids and liquid droplets and come in a wide range of sizes. Those less than 10 micrometers (μm) in diameter (PM10) are so small that they can get into the lungs, potentially causing serious health problems. Particles less than 2.5 μm in diameter are called fine particles. These particles are so small that they can be detected only with an electron microscope. Sources

24/05/2010

Fig. 3 Stationary monitoring station of the ARMAAG network.

of fine particles include all types of combustion, including motor vehicles, power plants, residential wood burning, forest fires, agricultural burning, and some industrial processes. Particles between 2.5 and 10 μm in diameter are referred to as coarse. Sources of coarse particles include crushing or grinding operations and dust stirred up by vehicles traveling on roads. After releasing into air, particles can change their size and composition by condensation of vapor species or by evaporation, by coagulating with other particles, or by chemical reaction. Particles with a diameter smaller than 1 μm generally have atmospheric concentrations in the range from around 10 to several thousands per cubic meter; those exceeding 1 μm diameter are usually found at concentration less than 1 cm^{-3}.

Carbon monoxide (CO) is a colorless, odorless gas that is produced by the incomplete burning of carbon-based fuels including petrol, diesel, and wood. It is also produced from the combustion of natural and synthetic products such as cigarettes. Natural background levels of CO fall in the range of 10–200 ppb. Levels in urban areas are highly variable, depending upon weather conditions and traffic density. Eight-hour mean values are generally less than 10 ppm, but sometimes, they can be as high as 500 ppm. Carbon monoxide lowers the amount of oxygen that enters the blood. It can slow human reflexes and make people confused and sleepy.

AIR QUALITY MONITORING

Design of Monitoring Networks for Air Pollution

For the purpose of monitoring and reporting air pollution, most industrialized countries have been divided into regions or zones and urban areas or agglomeration, e.g., in Europe, in accordance with EC Directive 96/62/EC.[19] This Directive sets a framework for ways how to monitor and report ambient levels of air pollutants. Other directives set ambient air limit values for particular pollutants:

- Directive 99/30/EC for nitrogen dioxide and oxides of nitrogen, sulfur dioxide, lead, and PM.
- Directive 2000/69/EC for benzene and carbon monoxide.
- Directive 2002/3/EC for ozone.

The monitoring sites are organized into automatic and non-automatic networks (regional and national) that gather a particular kind of information using a particular method. For example, across the U.K., there are more than 1500 monitoring sites that monitor air quality, and these are organized into networks (automatic and non-automatic) that gather a particular kind of information, using a particular method. The pollutants measured and method used by each network depend on the reason for setting up the network, and what the data are to be used for. In Poland, monitoring of air quality has been performed systematically since 1992, mainly by using automatic air monitoring stations. Air quality monitoring data are used at national, voivodship (provincial), and local scale. An exemplary air automated monitoring station, together with a general view of analyzers situated in them, is shown in Figs. 3 and 4. This station belongs to the Agency of Regional Air Moni-

24/05/2010

Fig. 4 General view of an interior of a monitoring station.

toring of the Gdańsk Agglomeration (ARMAAG), which is Poland's first local government-owned air monitoring network. Foundation ARMAAG provides information of air condition in Gdańsk agglomeration in real time from the automatic measurement network.

The obtained data from air monitoring are used in air quality inventories and bulletins.

Types of Information Obtained from Air Monitoring

The obtained information concerns different types of concentration of investigated pollutants depending on applied sampling techniques and measuring period. The results of measurements may be referred to real time (instantaneous concentrations) or to a selected period of time (e.g., 1 hr, 8 hr, 24 hr, 1 month, 1 year). Final measurements represent averaged concentrations.

Fig. 5 presents schematically different forms of concentration obtained as a function of sampling time.[20]

Considering the frequency of sampling, discrete, periodic, and instantaneous measurements are distinguished. Taking into account space, parameter measurements are divided to a point, averaged along a defined part of space, and averaged on the selected area. Point monitoring is inadequate to measure poorly mixed gases such as fugitive emissions over large areas. If the point instrument is wrongly placed, measurement results are not representative. Final measurements enable determination of weighted average concentrations over the sampling period.

General Requirements of the Instruments Used in Monitoring

Monitoring of air pollution is a prerequisite of air quality control and is carried out by a wide variety of analytical methods employing different measuring instruments (analyzers) that have different sensitivities and specificities.[21]

The basic requirement of the analyzers for air monitoring is high measurement sensitivity, i.e., the low limit of detection (LOD) and the low limit of quantitation (LOQ). It gives a chance to detect the pollutants at required levels. The instruments that should acquire analytical data in real

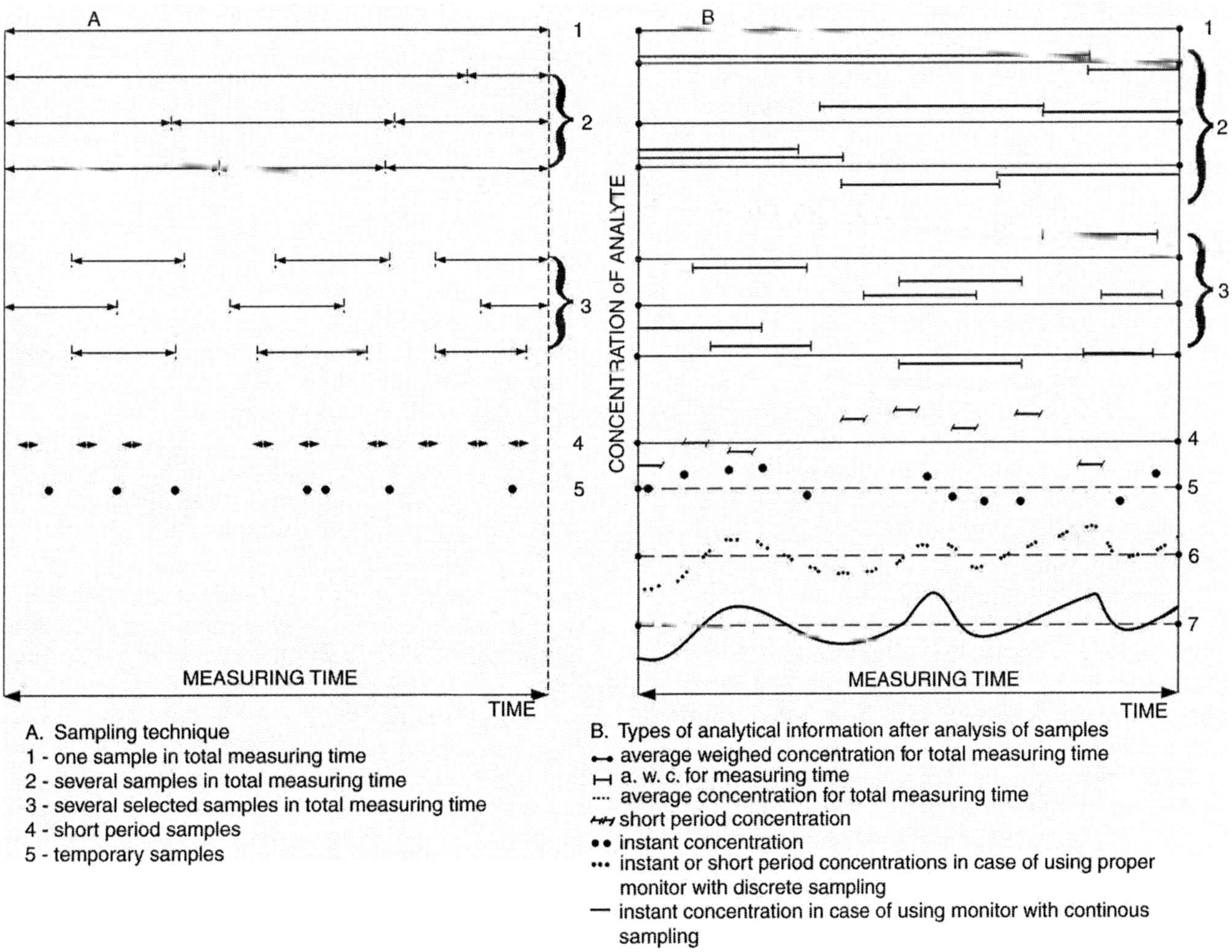
A
B
1
2
3
4
5
6
7
CONCENTRATION oF ANALYTE
MEASURING TIME
TIME
A. Sampling technique
1 - one sample in total measuring time
2 - several samples in total measuring time
3 - several selected samples in total measuring time
4 - short period samples
5 - temporary samples
B. Types of analytical information after analysis of samples
average weighed concentration for total measuring time
a. w. c. for measuring time
average concentration for total measuring time
short period concentration
instant concentration
instant or short period concentrations in case of using proper monitor with discrete sampling
instant concentration in case of using monitor with continous sampling

Fig. 5 Schematic diagram of different sampling techniques used for getting information on concentration of pollutants in determined measuring time.

time or only with a small time delay have to possess the following additional capabilities:

- Providing high data resolution (characterized by low response time).
- Providing automatic calibration and zeroing.
- Long functioning without service.

The last demand means that they should be equipped with an independent power supply and be able to automatically regenerate or exchange worn-out filters and, depending on the type of detector used (sensor), fulfill special demands, e.g., for electrochemical sensors, exchange or supplement the working solution and reagents, and in devices with Flame Ionisation Detector (FID) or Flame Photometric Detector (FPD) detection, protect against flame extinguishing.

Depending on the number of analytes that an instrument can determine in a single sample, they can be single-parameter (single gas) or multiparameter (multigas) instruments.

Based on sampling frequency, analyzers can be discrete (for single measurement), periodic (for measurements at preset intervals), or continuous (for permanent monitoring).

Classification of the Instruments Used for the Detection and Monitoring of Air Pollutants

The analytical instruments currently used for the detection and determination of atmospheric pollutants can be classified according to various criteria.

Recently, measuring techniques based on a physical (or physicochemical) principle are more frequently used in the assessment of air quality. Such methods involve direct determination of a physical property of a pollutant, sometimes after its interaction with another compound. In this approach, better stability, sensitivity, and reliability may be easily achieved. Furthermore, the practical application requires less maintenance. Instruments based on this principle can be easily automated, which enables their use in providing continuous measurements needed for up-to-date assessments of air quality. It is especially relevant to environmental monitoring because many existing standards refer to specified period of time, i.e., 1 hr, 24 hr, or 1 year.

According to the location where measurements are taken, instruments can be stationary or *on-site*. In the first case, analysis is performed in the laboratory and sophisticated instruments are applied, such as mass, electron mobility, or x-ray fluorescence spectrometers. *On-site* systems enable measuring of pollution levels in the field. Since access to a sophisticated laboratory is not required, the devices (usually uncomplicated, relatively cheap, and portable) hold great promise for use in remote locations. The main advantage of on-site analysis is the potential for rapid assessment and response to a particular problem.

All monitoring systems can be classified as mobile or stationary. Most existing systems monitoring gaseous pollutants of atmospheric air and ambient aerosols, both automatic and manual, usually perform stationary measurements; i.e., they are directly linked to a specific point or space in the vicinity of that point. Basing on the data obtained from single monitoring sites, it is not possible to assess spatial and temporal variations of air pollutants.

Mobile refers to a continuous-monitoring instrument that is portable or transportable. They are usually designed to perform analytical measurements without preliminary operations. Portable refers to self-contained, battery-operated, or worn or carried by the person using it, or may require the use of special vehicles for placement in a specific area to be monitored. Transportable gas monitors can be mounted on a vehicle such as a car, plane, balloon, ship, or space shuttle, but not to a mining machine or industrial truck.

For mobile systems, the registered values of pollutants have to be correlated with information about the geographical site and actual meteorological conditions (temperature and humidity).

Portable systems for field measurements should meet the following requirements:

- Compactness and robustness
- Ease of handling
- Adaptability to on-site measurements
- Automated operation with a long-lasting power supply
- Stability under aggressive environmental conditions

Several contributions published during the last decade have proven the advantages of mobile systems in getting information concerning the spatial and temporal distribution of atmospheric trace gases, without the need of a dense network of stationary stations. Most of the proposed systems are based on application of mobile laboratories,[21–23] equipped with appropriate monitors. There are also systems that allow to measure pollutants in a stream of vehicles but the measuring unit is installed on any vehicle[24–27] rather than attached to a dedicated van.[28,29]

The general trend in the field of creating instruments for air quality assessment is combining several instruments into one system and forming so-called hybrid multisensor systems, controlled by a microprocessor capable of transferring the obtained data to a central station, frequently using a wireless system. In the central station, the data are collected both from single objects (houses, plants) and from large areas. Many systems are equipped with devices for testing the sensors and for providing diagnosis of the whole instrument. Frequently, they have alarms that warn the user of any dangerous situation due to the breaching of some value limit. Such systems are battery powered and able to work continuously for several days or months.

The environments in which analyzers are used differ from the relative calm of the laboratory. Analyzers have to withstand wide ambient temperatures, fluctuations, and vibrations. Due to this, many systems are completely sealed so as to operate independent of outside conditions and be able to withstand the onslaught of monkey-wrench mechanics.

GENERAL CHARACTERISTICS OF THE METHODS AND ANALYTICAL INSTRUMENTS FOR AIR MONITORING

Due to the complexity of environmental problems and the variety of pollutants and their different concentrations (typically parts per million or percentage levels in stack gases and parts per billion in air), there is a wide range of methods and instruments used in measuring ambient air quality.

Based on physicochemical principles, the monitoring instruments involve direct determination of the different physical properties of the pollutant or following its interaction with another compound. These methods allow determination of air pollutants in a continuous and automatic way. Such approach requires extremely sensitive instrumentation. Therefore, the most advanced techniques, comprising chemistry, physics, and microelectronics, should be used. As a result, the instruments are combinations of many different devices giving one measuring system. In developing such a system, it should be remembered that it will be exposed to environmental impact, such as changes in temperature, dustiness, humidity of air, aggressive components of air, vibrations, and transportation stress.

The typical instruments used for atmospheric ambient monitoring are based on optical, electrochemical, and semiconductor principles. Among spectroscopic techniques, chemiluminescence, infrared (IR), and fluorescence are the most frequently applied.

The range of typical measuring methods and techniques used in air monitoring is shown in Table 3.

Among the many different optical spectroscopic methods, differential optical absorption spectroscopy (DOAS) has found wide use in atmospheric research and air quality monitoring. The technique is based on the measurements of absorption features of gas molecules along a path of known length in the open atmosphere. The DOAS systems, due to the calibration-free absolute measurements and the unequivocal identification of many trace pollutants, such as CO, SO, NO, and VOCs, and highly reactive radicals, e.g., OH, NO, and halogen oxide radicals, are exploited by air quality monitoring agencies around Europe and in the United States. The physical and chemical principles, the current state of this measurement method, and details for users are broadly presented in a recently published book.[30]

In Europe, standard/reference methods (EN) are provided by the European standardization body (CEN). The standard methods use the following principles:

Chemiluminescence for NO EN 14211[31]

Nitrogen oxide reacts with ozone, generated within the instrument, produces an excited molecule of nitrogen dioxide, which emits light returning to its original state. A photomultiplier tube measures the emitted light that, if the volumes of sample gas and excess ozone are carefully controlled, is proportional to the concentration of NO in the gas sample.

The chemiluminescent method used for nitrogen oxides is based on the following reactions:

$$NO + O_3 \rightarrow NO_2^* + O_3$$

$$NO_2^* \rightarrow NO_2 + h\nu$$

Table 3 Air quality monitoring techniques used in air monitoring.

Pollutant	Emissions	Ambient air
CO_2	FTIR, NDIR, TDLAS	NDIR, DOAS
CO	FTIR, NDIR, TDLAS, DOAS	NDIR (gas filter correlation variant)
SO_2		UV fluorescence
NO_x		Chemiluminescence
PM	Triboelectric, opacity, beta ray attenuation	Beta ray attenuation, oscillating microbalance, gravimetric
VOCs	FID, GC	FID, GC
CH_4	NDIR, FTIR,TDLAS, FID	FID, GC
O_3		UV absorption spectroscopy, DOAS, electrochemical sensors
NH_3	FTIR,TDLAS, chemiluminescence GC	Chemiluminescence, DOAS GC
Benzene, 1,3-butadiene	In situ GC–MS with continuous sampling	MS, GC-MS
Dioxins and furans	XRF, LIBS, cold vapor AFS, atomic	Sampling plus MS or GC–MS
Metals	emission spectrometry	Sampling plus ICP-MS, DOAS (for Hg)

Note: FTIR, Fourier transform infrared absorption spectroscopy; NDIR, non-dispersive infrared absorption; TDLAS, tunable laser diode absorption spectroscopy; DOAS, differential optical absorption spectroscopy; FID, flame ionization detector; GC, gas chromatography; MS, mass spectrometry; XRF, x-ray fluorescence; LIBS, laser-induced breakdown spectroscopy; ICP, inductively coupled plasma; AES, atomic emission spectrometry; AFS, atomic fluorescence spectrometry.

Source: Bogo.[29]

The chemiluminescence technique may be used to measure total oxides of nitrogen (NO_x) by passing the sample over a heated catalyst to reduce all oxides of nitrogen to NO. This is done within the instrument, just prior to the reaction chamber. Some instruments can perform the automatic switching of the catalyst in and out of the sample path so that the resulting signals may be compared to indirectly measure NO_2.

Ultraviolet Fluorescence for SO_2 EN 14212[32]

Sulfur dioxide is measured without chemical pretreatment by gas-phase fluorescence spectrometry in the UV region. Molecules of SO_2 are excited by UV radiation (200–220 nm) into unstable forms, which return to a basic state, emitting radiation in the range of 240–420 nm according to following reactions:

$$SO_2 + h\nu \rightarrow SO_2^*$$

$$SO_2^* \rightarrow SO_2 + (UV)$$

The intensity of fluorescence radiation is proportional to sulfur dioxide in the sample.

Non-dispersive Infrared for CO EN 14626[33]

The analytical principle is based on absorption of IR light by the CO molecule. NDIR-GFC (non-dispersive infrared–gas filter correlation) analyzers operate on the principle that CO has a sufficiently characteristic IR absorption spectrum such that the absorption of IR by the CO molecule can be used as a measure of CO concentration in the presence of other gases. CO absorbs IR maximally at 2.3 and 4.6 μm. Because many other molecules also absorb IR radiation in practice, different technical designs are proposed. The following approaches are typically applied:

- Measurement of IR absorption at specific wave for CO (2.3 or 4.6 μm)
- Analyzers with two cells, one of which is filled with pure air (compensation of drift)
- Analyzers with turning circle (GFC)

Ultraviolet Photometry for O_3 EN 14625[34]

Upon exposure to UV light, ozone absorbs some of the light and the intensity difference is directly proportional to the concentration of ozone. Frequently, the UV light source is a 254 nm emission line from a mercury discharge lamp.

Online Gas Chromatography for Benzene EN 14662—Part 3[35]

A measured volume of sample air is drawn or forced through a sorbent tube. Provided suitable sorbents are chosen, benzene is retained by the sorbent tube and thus is removed from the flowing air stream. The collected benzene (on each tube) is desorbed by heat and is transferred by inert carrier gas into a gas chromatograph equipped with a capillary column and a flame ionization detector or another suitable detector, where it is analyzed. Prior to entering the column, the sample is concentrated either on a cryo trap, which is heated to release the sample into the column, or on a pre-column, where higher boiling hydrocarbons are removed from the pre-column by back flush. Two general types of instruments are used. One is equipped with a single sampling trap and the other is equipped with two or more traps. The single-trap instrument samples for only part of the time in each cycle, whereas the multitrap instrument samples continuously. Typical cycling times are between 15 min and 1 hr.

PM is usually determined using active filter method by gravimetry as the reference method.[36] In this method, the air is passed through a filter that stops particles above 10 μm (PM10) or 2.5 μm (PM2.5). Measurements are made over a period of 24 hr or longer. The filters are collected and the adsorbed particles are measured in the laboratory. Other methods use beta ray absorption or tapered element oscillating microbalance (TEOM) of PM. In beta gauge instruments, which are used for real-time measurements of particulate emissions from stationary sources, the mass of the sample deposited on the filter tape is automatically measured by beta ray attenuation. The measurement is made first on a blank, then on the particulate-laden filter. The range is 2–4000 mg/m^3 without interference or effect from color, size, or atomic mass of the dust.

An interesting review on *online* analyzers for monitoring of VOCs was recently published.[37]

OTHER MONITORING APPROACHES

Biomonitoring Using Plants

Modern air instruments cannot measure the effects air pollution has on living cells and are limited to measuring the present conditions. Biological materials can be an excellent basis for establishing a biomonitoring network on large areas for a long time. Biomonitoring, as a continuous observation of an area with the help of bioindicators, can allow a qualitative survey and quantitative estimation of the pollutants in the environment. Since the 19th century, biological monitoring as a rather simple observation has turned into a serious alternative if not a useful complement to the traditional methods of assessing contamination, from both natural and anthropogenic sources. In the case of airborne pollution, its heyday really began after World War II. The expensive growth of biomonitoring research works has gained momentum mainly from lower organisms such as lichens, bryophytes, and, to a lesser degree, fungi. The use of cosmopolite organisms for assessing pollution has developed notably during the last two decades.

Bioindicators are organisms or organs of such organisms that respond to a certain level of pollution by the change in their life cycle or accumulation of the particular pollutant. Bioindicators, in contrast to direct analysis, reflect complex effects of harmful substances, as such organisms show not only the synergistic effects of a sum of parameters but also a time-integrated picture of the history of their life span. Another advantage is the selective uptake of such substances, as an organism exposed to an environmental pollutant, either through air or via direct uptake, absorbs the bioavailability fraction only. They readily reflect the proportion of the pollutants, which may be dangerous to human beings as well.

In air monitoring, two organisms, i.e., lichens and mosses, have become the most popular bioindicators.

Lichens are unusual organisms because they consist of fungal threads and microscopic green alga living together and functioning as a single organism. Lichens grow on rocks, soil, trees, or artificial structures in unpolluted habitats. Lichens act like sponges, taking in everything that is dissolved in the rainwater and retaining it. Different species of lichens vary in sensitivity to air pollution. Lichens are commonly used as air quality indicators since some species are more pollution tolerant than others. The most sensitive lichens are shrubby and leafy, whereas the most tolerant lichens are crustose. In city centers, lichens may be entirely absent. If the air is clean, shrubby, hairy, and leafy lichens colonize every available surface.

Lichens may be used as bioindicators in two different ways:

- By mapping all species present in a specific area.
- Through the individual sampling of lichen species and measurement of the accumulated pollutants or by transplanting lichens from a clean environment to a contaminated one.

Lichens are used as biomonitors in the examination of the level of pollutants such as sulfur, nitrogen, and phosphorus compounds, as well as ozone, fluorides, chlorides, and heavy metals. Several biomonitoring methods using lichens have been described since the 1970s when Hawksworth and Rose[38,39] proposed a method based on the determination of zones (from 0—strong pollution to 10—clear air) with selected epiphytic lichens (on two different kinds of tree bark) that relate to levels of sulfur dioxide pollution. This method was widely adopted (both in the original scale and in relation to the real concentrations of sulfur dioxide) in many countries mainly due to its simplicity. Over the last decade, new techniques like the European method for mapping lichen diversity (LDV), as an indicator of environmental stress/quality, have been proposed. The procedure is based on the fact that epiphytic lichen diversity is impaired by air pollution and environmental stress. It provides a rapid, low-cost method to define zones of different environmental quality. In addition to information on the long-term effects of air pollutants, data on eutrophication, anthropization, and climatic change on sensitive organisms may likewise be obtained. Data quality depends on the uniformity of growth conditions, and usually standardization in sampling procedure is necessary.

The relative ease of sampling, the absence of any need for complicated and expensive equipment, and the accumulative and time-integrative behavior of the monitor organisms that make biomonitoring of atmospheric pollutants possible could be continued for the foreseeable future, especially in large-scale surveys.[40–45]

GIS in Air Quality Monitoring

Reliable information on air quality is needed not only on temporal trends in air pollution (as, for example, provided by data from fixed-site monitoring stations) but also on geographical variations. Maps are needed, for example, to identify so-called hot spots, to define at-risk groups, to show changes in spatial patterns of pollution, and to provide improved estimates of exposure for epidemiological studies. The development of GIS techniques offers considerable potential to mapping air pollution. GIS technology enables obtaining statistical and spatial data on air quality by estimation of environmental levels of regulated contaminants.[46]

Remote Monitoring Techniques

Remote monitoring techniques enable the measurement of atmospheric pollutants in remote, poorly accessible, and dangerous regions.[47–49] Remote sensing is especially recommended for the detection of diffuse emissions that are hard to quantify with typical ground-point measurements, but its use is restricted to specialized monitoring demands due to very considerable cost of the equipment.

The cheapest and most widely used methods are those of aerial photography, including IR sensing and optical spectroscopy. Different spectrometers, whether they are ground based in a mobile laboratory or airborne, are the most common instruments used in air pollution for determination of SO_2 and NO_2 concentrations in plumes from tall stacks. The results provide reliable data for studying the transport and dispersion of a plume.

Typical gaseous pollutants such as NO_x, SO_2, CO, and O_3 may be monitored using different types of lasers, which, due to long-path absorption measurements, enable the determination of pollutants at very low concentrations. In laser absorption methods, a detector is used to monitor absorption of specific wavelengths in light paths. Lidar (light detection and ranging) transmits light out to a target and part of this light is reflected/scattered back to the instrument where it is analyzed. The time for the light to travel out to the target and back to the lidar is used to determine the range to the target, allowing spatial resolution of pollutant concentration data within the light path; by monitoring back-scatter intensity at two close wavelengths, the species concentration as well as its spatial distribution may be inferred. This method has been successfully used for measurements of SO_2 up to a range of 2 km.

There are three basic generic types of lidars: range finder lidars, differential absorption lidars (dial), and Doppler lidars.

Range finder lidars are used to measure the distance from the lidar instrument to a solid or hard target. Dial is used to measure pollutant concentration in the atmosphere on the basis of the difference of the two return signals having two different wavelengths (one is absorbed by the molecule of interest while the other is not absorbed). The Doppler lidar is used to measure the velocity of a target. The target can be either a hard target or an atmospheric target.

Another type of device used for remote monitoring is sodar (sound detection and ranging). The Doppler sodar sends out sound pulses of several frequencies in slightly different directions. The acoustic signals are back-scattered by inhomogeneities in the atmosphere.

Further significant developments of laser techniques use the Raman back-scatter, which is highly characteristic of the scattering molecule.

MONITORING OF FLUE AND EXHAUST GAS EMISSIONS

Different industrial branches, e.g., coal-fired power plants, chemical plants, petrochemical plants, oil refineries, PVC factories, heavy industries, and incinerators, are principal stationary pollution sources emitting, usually by chimneys, stack (flue) gases, containing different pollutants. Monitoring of such emission sources needs continuous, automatically acting systems that are usually a multielement, integrated, and cooperative set of measuring devices, auxiliary equipment, and calibration appliance.[50,51]

Systems for Continuous Monitoring of Stack Gases

The monitoring systems for continuous monitoring can be classified on the basis of different criteria. Depending on the way in which measurement is made, and especially on the applied sampling mode, extractive and in situ systems are distinguished.[52]

In extractive systems, as the name implies, the sample is extracted continuously from a duct or stack from a representative volume of stack gases and transported by transfer line to analyzers (one or more single-component analyzers or one multicomponent analyzer). However, in most cases, some conditioning of the sample is required to remove water vapor and PM. The two main types of extractive systems are fully extractive (sometimes called simply "extractive") and dilution extractive (also known as "dilution").

A typical extractive system (Fig. 6) has a stainless steel probe, with a filter to remove coarse particulates. After filtration, a heated, unchanged sample is transferred to a sample conditioner located in the system enclosure. Calibration gas is delivered from the enclosure to the probe and back through the sample tubing to calibrate the system. The simplest sampling conditioning method is cooling the sample and allowing the moisture to condense and drain out of the system.

Instruments based on spectroscopic techniques (mainly UV and IR), paramagnetic properties, and solid electrolytes (zirconium dioxide) for oxygen determination are frequently used to monitor CO, CO_2, NO_x, and SO_2. For low concentrations of NO_x and SO_2, a chemiluminescent method can be applied. Fully extractive systems can be sometimes used without moisture removal, especially when the sample contains components that are easily soluble in water.

Fully extractive systems are recommended for monitoring of pollutants in stack gases with different physicochemical parameters of compounds. Another advantage of such systems is the possibility of monitoring several locations using one analyzer (time-share systems).

Dilution of the sample gas (Fig. 7) with clean, dry air to the sample (usually from 50:1 to 250:1) considerably facilitates sample handling and reduces the dew point of the sample gas so that the sampling line can be unheated. Furthermore, the diluted sample is similar in respect of pollutant concentration to ambient air, enabling the use of ambient analyzers. Relatively small amounts of sampled gases increase the time between cleaning the filters. Because most dilution-extractive systems are affected by changes in temperature and barometric pressure, it is recommended that temperature and pressure sensors be installed at the sampling location to compensate these effects. The dilution

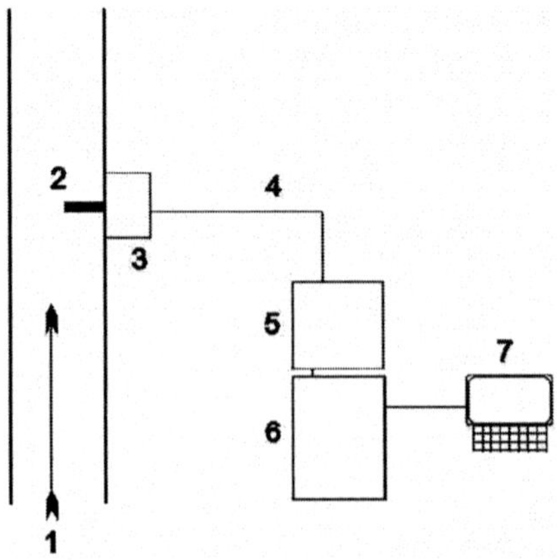
2
4
3
5
7
6
1

Fig. 6 Schematic diagram of a fully extractive system for continuous emission monitoring of stack gases: 1, stack gases; 2, probe; 3, filter; 4, heated sampling line; 5, moisture removal; 6, analyzers; 7, storage unit.

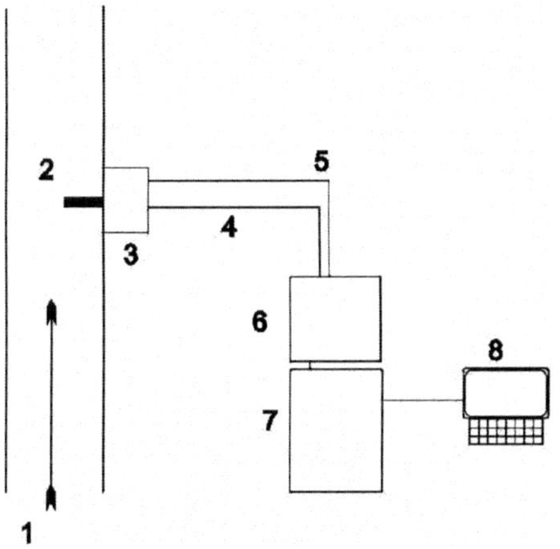
1
2
3
4
5
6
7
8

Fig. 7 Schematic diagram of a dilution-extractive system for continuous emission monitoring of stack gases: 1, stack gases; 2, probe; 3, filter; 4, clean and dry dilution air; 5, diluted sample to analyzers; 6, dilution control unit; 7, analyzers; 8, storage unit.

systems are recommended for plants fueled with carbons when high levels of particulates are present in stack gases (0.1 g/m^3) and corrosive substances (e.g., HCl or SO_3).

In in situ systems (Fig. 8), a gas probe is inserted into the wall. It allows monitoring the sample without removing it from the source and does not require sample conditioning or transport of the sample gas, thus minimizing the measurement errors during sampling, transferring, and conditioning the sample. An optical beam is contained within the probe. This optical beam represents the absorption path that enables the analysis. The sample is drawn into the probe but remains under the conditions found in the stack. The sample is never removed or extracted from the stack.

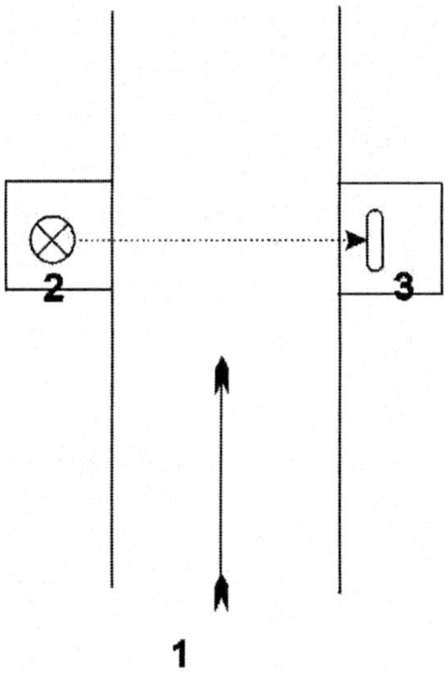
1
2
3

Fig. 9 Schematic diagram of a single-path-type in situ system for continuous emission monitoring of stack gases: 1, stack gases; 2, light source; 3, receiver.

In practice, two types of in situ systems are used: point and path monitoring systems (Figs. 9 and 10). In point monitoring systems, a sample probe and analyzer are installed inside the stack. They are also called in-stack monitors and measure gas at a single point. Therefore, it is important to choose a location that is representative in terms of the components of interest. As analyzers in such systems, spectroscopic instruments are used (based on absorption of UV and IR radiation) as well as electrochemical devices. In situ monitoring systems are recommended for locations with easy stack access and for measuring SO_2 and O_2 in

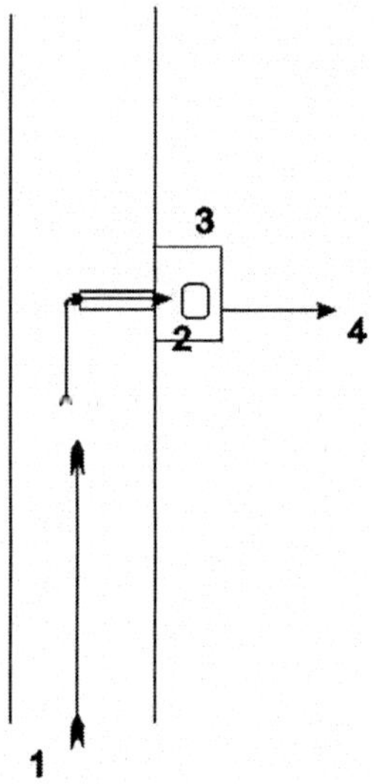
1
2
3
4

Fig. 8 Schematic diagram for a point-type in situ system for continuous emission monitoring of stack gases (sensor mounted in the box with the sensor electronics): 1, stack gases; 2, sensor; 3, electronics in enclosure; 4, signal.

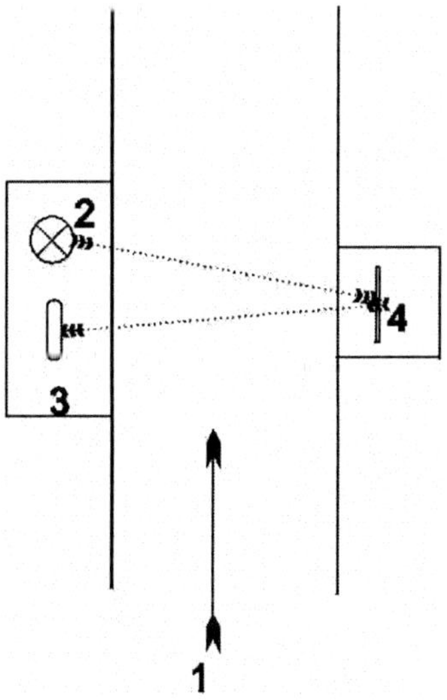
1
2
3
4

Fig. 10 Schematic diagram of a dual-path-type in situ system for continuous emission monitoring of stack gases: 1, stack gases; 2, light source; 3, receiver; 4, reflector.

combustion sources because point monitors are very cost-effective for measuring only one or two components.

Path monitoring systems minimize errors that can arise when the location of a measuring point is not representative and due to the disturbances in the flow of stack gases. They measure gas concentration along a path, usually across the diameter of the stack or duct. A light source is mounted on one side of the stack and a beam is passed through to the other side. A single-pass system measures the light that reaches the other side of the stack, whereas a double-pass system uses a reflector and passes the light back across the stack before performing the measurements. Two parameters are limited in these systems: the length of the measuring path (no less than 0.5 m, no more than 8–10 m) and the temperature of stack gases (no more than 300°C).

In situ systems are usually mounted in the ducts after electrostatic precipitators or in chimney ducts.

The systems for stack sampling can be easily adapted to process monitoring or even ambient (closed or open path) monitoring.

CONCLUSIONS

Air quality has unquestionably adverse effects on human health. For example, air pollution is increasingly being cited as the main cause of lung conditions such as asthma—twice as many people suffer from asthma today compared to 20 years ago.

This is the main reason that the issue of air quality is now a major concern for many countries that have been working to improve air quality by controlling emissions of harmful substances into the atmosphere, improving fuel quality, and integrating environmental protection requirements into the transport and energy sectors. Despite these improvements in air quality over the last few years, the problem of air pollution still remains. Therefore, more needs to be done at the local, national, and international level. For example, a wide interest is observed to establish common criteria (e.g. AQIs– Air Quality Criteria (AQC)) to compare the state of the air for different countries that follow different directives.

During the last decade, different types of AQC were proposed in literature[53,54] and/or adopted by governments. An interesting review on AQI published recently online by the Italian Group of Environmental Statistics (GRASPA) (http://www.graspa.org) shows the lack of a common strategy to compare the state of the air for cities that follow different directives.[55] The major differences between the indices in the literature are found in the number of index classes (and their associated color) and relative descriptive terms (e.g., considered pollutants, averaging time, frequency). Also, the guidelines themselves are sometimes consistently different from state to state, not only in indicating the pollutants to be monitored but also in setting the threshold values and the number of exceedances per year. Furthermore, the way air quality is interpreted on the basis of a country- or city-specific AQI differs considerably.

Monitoring of air pollution is a prerequisite of air quality control and is carried out by a wide variety of analytical methods employing different measuring instruments that have different sensitivities and specificities. Monitoring plays a critical role in protecting the environment and is a key element of all actions related with management and protection of ambient air.

Every developed country has legislation to control or limit emissions of atmospheric pollution. The air quality standards that could be regulated by law and achieved were established. Concentrations of selected pollutants would have to be determined, and reliable analytical methods would be required to measure the levels of the pollutants. Monitoring actions are based on using stationary networks of measuring stations and/or mobile laboratories equipped with proper instruments.

REFERENCES

1. Atkisson, A.; Gaines R.S., Eds.; *Development of Air Quality Standards*; Charles E. Merril Publ. Co.: Columbus, Ohio, 1970.
2. California State Department of Public Health, Health and Safety Code, Section 426.1,1959.
3. Available at http://www.epa.gov/air/caa/ (accessed May 15, 2012).
4. World Health Organisation. *Air Quality Guidelines for Europe*, 2nd Ed.; WHO Regional Publications European Series No.91: Copenhagen, 2000.
5. Directive 2008/50/EC of the European Parliament and of the Council of 21 May 2008 on ambient air quality and cleaner air for Europe. Official Journal of the European Union 11.6.2008, L. 152/1–152/44.
6. WHO air quality guidelines for particulate matter, ozone, nitrogen dioxide and sulfur dioxide—Global update 2005—Summary of risk assessment.
7. Available at http://armaag.gda.pl/normy.html (accessed May 15, 2012).
8. Available at http://citeair.rec.org/products.html (accessed May 15, 2012).
9. Available at http://www.airqualitynow.eu (accessed May 15, 2012).
10. Available at http://www.citeair.eu (accessed May 15, 2012).
11. Available at http://europa.eu/legislation_summaries/environment/air_pollution/l28026_en.htm (accessed May 15, 2012).
12. Griffin, R.D. *Principles of Air Quality Management*; Taylor & Francis Group: Boca Raton, 2007.
13. Jacobsen, M.Z. *Atmospheric Pollution. History, Science and Regulation*; Cambridge University Press: Cambridge, U.K., 2002.
14. Hobbs, P.V. *Introduction to Atmospheric Chemistry*; Cambridge University Press: Cambridge, U.K., 2000.
15. Friedrich, R.; Reis, S., Eds. *Emission of Air Pollutants*; Springer-Verlag: Berlin Heidelberg, 2004.
16. Seinfeld, J.H.; Pandis, S.N., Eds. *Atmospheric Chemistry and Physics. From Air Pollution to Climate Change*, 2nd Ed.; John Wiley and Sons, Inc.: New York, 2006.

17. Bell, J.N.B.; Treshow, M., Eds. *Air Pollution and Plant Life*; John Wiley & Sons, Ltd.: New York, 2002.
18. Bogue, R. Environmental sensing: Strategies, technologies and applications. Sens. Rev. **2008**, *28* (4), 275–282.
19. Council Directive 96/62/EC of 27 September 1996 on ambient air quality assessment and management. Official Journal L *296*, 21/11/1996 P, 0055–0063.
20. Michulec, M.; Wardencki, W.; Partyka, M.; Namieśnik, J. Analytical techniques used in monitoring of atmospheric air pollutants. Crit. Rev. Anal. Chem. **2005**, *35*, 1–17.
21. Wardencki, W.; Katulski, R.; Stefański, J.; Namieśnik, J. The state of the art in the field of non-stationary instruments for the determination and monitoring of atmospheric pollutants. Crit. Rev. Anal. Chem. 2008, *38*, 1–10.
22. Bukowiecki, N.; Dommen, J.; Prevot, A.S.H.; Richter, R.; Weingartner, E.; Baltensperger, U. A mobile pollutant measurement laboratory—Measuring gas phase and aerosol ambient concentrations with high spatial and temporal resolution. Atmos. Environ. **2002**, *36*, 5569–5579.
23. Gouriou, F.; Morin, J.-P.; Weill, M.-E. On-road measurements of particle number concentrations and size distributions in urban and tunnel environments. Atmos. Environ. **2004**, *38* (18), 2831–2840.
24. Pirjola, L.; Parviainen, H.; Hussein, T.; Valli, A.; Hameri, K.; Aaalto, P.; Virtanen, A.; Keskinen, J.; Pakkanen, T.A.; Makela, T., Hillamo, R.E. "Sniffer"—A novel tool for chasing vehicles and measuring traffic pollutants. Atmos. Environ. **2004**, *38*, 3625–3635.
25. Katulski, R.; Stefański, J.; Wardencki, W.; Żurek, J. Concept of the mobile monitoring system for chemical agents control in the air, Proceedings of the IEEE Conference on Technologies for Homeland Security—Enhancing Transportation Security and Efficiency, Boston, USA, June 7–8, 2006; 181–184.
26. Katulski, R.; Stefański, J.; Wardencki, W.; Żurek J.The Mobile Monitoring System (MMS)—A useful tool for assessing air pollution in cities, Proceedings of the Pittsburgh Conference on Analytical Chemistry and Applied Spectroscopy PITTCON'2007, February 25–March 2, 2007, Chicago, USA, 2007.
27. Katulski, R.; Namieśnik, J.; Sadowski, J.; Stefański, J.; Szymańska, K.; Wardencki, W. Mobile system for on-road measurements of air pollutants. Rev. Sci. Instrum. **2010**, *81*, 045104.
28. Seakins, P.W.; Lansley, D.L.; Hodgson, A.; Huntley, A.; Pope, F. New directions: Mobile laboratory reveals new issues in urban air quality. Atmos. Environ. **2002**, *36*, 1247–1248.
29. Bogo, H.; Negri, R.M.; San Roman, E. Continuous measurement of gaseous pollutants in Buenos Aires city. Atmos. Environ. **1999**, *33*, 2587–2598.
30. Platt, U.; Stutz, J. *Differential Optical Absorption Spectroscopy. Principles and Applications*; Springer-Verlag: Berlin, Heidelberg, 2008.
31. EN 14211. *Ambient air quality*. Standard method for the measurement of the concentration of nitrogen oxide by chemilumiscence, 2005.
32. EN 14212. *Ambient air quality*. Standard method for the measurement of the concentration of sulphur dioxide by ultraviolet fluorescence, 2005.
33. EN 14626. *Ambient air quality*. Standard method for the measurement of the concentration of carbon monoxide by nondispersive infrared spectroscopy, 2005.
34. EN 14625. *Ambient air quality*. Standard method for the measurement of the concentration of ozone by ultraviolet photometry, 2005.
35. EN 14662–3. *Ambient air quality*. Standard method for measurement of benzene concentrations—Part 3: Automated pumped sampling with in situ gas chromatography, 2005.
36. Larssen, S.; de Leew, F. *PM10 and PM2.5 concentration in Europe as assessed from monitoring data reported to AirBase*, ETC/ACC; 2007.
37. Król, S.; Zabiegała, B.; Namieśnik, J. Monitoring of volatile organic compounds (VOCs) in atmospheric air. Part I. On-line gas analyzers. Trends Anal. Chem. **2010**, *29* (9), 1092–1100.
38. Hawksworth, D.L.; Rose F. Quantitative scale for estimating sulphur dioxide air pollution in England and Wales using epiphytic lichens. Nature, **1970**, *227*, 145–148.
39. Nash, T.H.; Gries C. Lichens as bioindicators of sulphur dioxide. Symbiosis **2002**, *33* (1), 1–21.
40. Svoboda, D. Evaluation of the European method for mapping lichen diversity (LDV) as an indicator of environmental stress in the Czech Republic. Biologia **2007**, *62* (4), 424–431.
41. Agraval, M.; Sing, B.; Rajput, M.; Marshall, F.; Bell, J.N.B. Effect of air pollution on peri-urban agriculture: A case study. Environ. Pollut. **2003**, *126*, 323–329.
42. Orendovici, T.; Skelly, J.M.; Ferdinad, J.A.; Savage, J.E.; Sanz, M.-J.; Smith, G.C. Response of native plants of northeastern United States and southern Spain to ozone exposures; determining exposure/response relationships. Environ. Pollut. **2003**, *125* (1), 31–40.
43. Wolseley, P.A. Using lichens on twigs to assess changes in ambient atmospheric conditions. In *Monitoring with lichens—Monitoring Lichens*; Nimis, P.L., Scheidegger, C., Wolseley, P.A., Eds.; NATO Science Series IV; Kluwer: Dordrecht, 1999; Vol. 7, 291–294.
44. Van Haluwyn, C.; van Hert, C.M. Bioindication: the community approach. In *Monitoring with lichens—Monitoring Lichens*; Nimis, P.L., Scheidegger, C., Wolseley, P.A., Eds.; NATO Science Series IV; Kluwer: Dordrecht, 2002; Vol. 7, 39–64.
45. Asta, J.; Erhard, W.; Ferretti, M.; Fornasier, F.; Kirschbaum, U.; Nimis, P.L.; Purvis, O.W.; Pirintsos, S.; Scheidegger, C.; van Haluwyn, C.; Wirth, V. Mapping lichens diversity as an indicator of environmental quality. In *Monitoring with Lichens—Monitoring Lichens*; Nimis, P.L., Scheidegger, C., Wolseley, P.A., Eds; NATO Science Series IV; Kluwer: Dordrecht, 2002; Vol. 7, 273–279.
46. Korte, G.B. *The GIS Book. How to Implement, Manage and Asses the Value of Geographic Information System*; Onward Press: Albany, NY, 2000.
47. Schroter, M.; Obermeier, A.; Bruggemann, D.; Plechschmidt, M.; Klemm, O. Remote monitoring of air pollutant emissions from point sources by a mobile Lidar/Sodar System. J. Air Waste Manage. Assoc. **2003**, *53*, 716–723.
48. Available at http://www.ghcc.msfc.nasa.gov/sparcle_tutorial.html (accessed May 15, 2012.)
49. Weibring, P.; Andersson, M.; Edner, H.; Svanberg, S. Remote monitoring of industrial emissions by combination of lidar and plume velocity measurements. Appl. Phys. B, **1998**, *66*, 383–388.

50. Jahnke, J.A. *Continuous Emissions Monitoring*; Van Nostrand Reinhold: New York, 1993.
51. White, J.R. Technologies for enhanced monitoring. Pollut. Eng. **1995**, *27* (6), 46–50.
52. Walker, K. Select a continuous emissions monitoring system. Chem. Eng. Progress. **1996**, *92* (2), 28–34.
53. Mayer, H.; Holst, J.; Schindler, D.; Ahrens, D. Evolution of the air pollution in SW Germany evaluated by the long-term air quality index LAQx. Atmos. Environ. **2008**, *42*, 5071–5078.
54. Shooter, D.; Brimblecombe, P. Air quality indexing. Int. J. Environ. Pollut. **2009**, *36* (1–3), 19–29.
55. Plaia, A.; Rugierri, M. *Air quality indices: A review*; GRASPA Working paper no. 39; June 2010.

Air Pollution: Technology

Sven Erik Jørgensen
Institute A, Section of Environmental Chemistry, Copenhagen University, Copenhagen, Denmark

Agricultural – Alexandria

Abstract
The sources of pollution, pollution problems, and environmental–technological methods to reduce and control air pollution are presented for particulate pollution, pollution by carbon monoxide, carbon hydrides, sulfur dioxide, nitrogenous gases, and heavy metals. A section is devoted to the wide spectrum of industrial air pollutants, their problems, and how this air pollution can be reduced and controlled.

SOURCES OF PARTICULATE POLLUTION

When considering particulate pollution, the source should be categorized with regard to the contaminant type. Inert particulates are distinctly different from active solids in the nature and type of their potentially harmful human health effects. Inert particulates comprise solid airborne material, which does not react readily with the environment and does not exhibit any morphological changes as a result of combustion or any other process. Active solid matter is defined as particulate material that can be further oxidized or react chemically with the environment or the receptor. Any solid material in this category can, depending on its composition and size, be considerably more harmful than inert matter of similar size.

A closely related group of emissions come from aerosols of liquid droplets, generally below 5 μm. They can be oil or other liquid pollutants (e.g., freon) or may be formed by condensation in the atmosphere. Fumes are condensed metals, metal oxides, or metal halides, formed by industrial activities, predominantly as a result of pyro-metallurgical processes: melting, casting, or extruding operations. Products of incomplete combustion are often emitted in the form of particulate matter. The most harmful components in this group are particulate polycyclic organic matter (PPOM), which are mainly derivatives of benz[a]pyrene.

Natural sources of particulate pollution are sandstorms, forest fires, and volcanic activity. The major sources in towns are vehicles, combustion of fossil fuel for heating and electricity production, and industrial activity.

The total global emission of particulate matter is in the order of 10^7 t per year. Deposition of particles may occur by three processes:

1. Sedimentation (Stokes law may be applied, particles >20 μm)
2. Impaction (determined by differences in concentrations by use of Fick's law, particles between 5 and 20 μm)
3. Diffusion (particles <5 μm)

Particles <20 μm are identified as suspended particulate matter. Particles >20 μm may be denoted dust, which will be deposited close to the source due to the high sedimentation rate. Dry deposition consists of gases or dry particles. Wet deposition is raindrops containing gases and particles. Particles may consist of minor concentrations of dissolved salts in water drops, crystals, or a combination of the two.

PARTICULATE POLLUTION PROBLEM

Particulate pollution is an important health factor, most crucially the toxicity and size distribution. Many particles are highly toxic, such as asbestos and those of metals such as beryllium, lead, chromium, mercury, nickel, and manganese. In addition, particulate matter is able to absorb gases, which enhances the effects of these components. In this context, the particle size distribution is of particular importance, as particles greater than 10 μm are trapped in the human upper respiratory passage and the specific surface (expressed as square meter per gram of particulate matter) increases with *l*/*d*, where *d* is the particle size. The adsorption capacity of particulate matter, expressed as grams adsorbed per gram of particulate matter, will generally be proportional to the surface area. Table 1 lists some typical particle size ranges. However, size as well as shape and density must be considered. Furthermore, particle size is determined by two parameters: the mass median diameter, which is the size that divides the particulate sample into two groups of equal mass, i.e., the 50% point on a cumulative frequency versus particle size plot, and the geometric standard deviation.

CONTROL METHODS APPLIED TO PARTICULATE POLLUTION

Particulate pollutants have the ability to adsorb gases including sulfur dioxide, nitrogen oxides, carbon monoxide, and so on. The inhalation of these toxic gases is frequently

Encyclopedia of Environmental Management DOI: 10.1081/E-EEM-120048396

Table 1 Typical particle size ranges.

	μm
Tobacco smoke	0.01–1
Oil smoke	0.05–1
Ash	1–500
Ammonium chloride smoke	0.1–4
Powdered activated carbon	3–700
Sulfuric acid aerosols	0.5–5

Source: Jørgensen.[10]

associated with this adsorption, as the gases otherwise would be dissolved in mouthwash and spittle before entering the lungs. Particulate pollution may be controlled by modifying the distribution pattern. This method is described in detail below. In principle, it represents an obsolete philosophy of pollution abatement, dilution, but it is still widely used to reduce the concentration of pollutants at ground level and thereby minimizes the effect of air pollution. Particulate control technology can offer a wide range of methods aimed at the removal of particulate matter from gas. These methods are settling chambers, cyclones, filters, electrostatic precipitators, wet scrubbers, and the modification of particulate characteristics.

Modifying the Distribution Patterns

Although emissions, gaseous or particulate, may be controlled by various sorption processes or mechanical collection, the effluent from the control device must still be dispersed into the atmosphere. Atmospheric dispersion depends primarily on horizontal and vertical transport. The horizontal transport depends on the turbulent structure of the wind field. As the wind velocity increases, so does the degree of dispersion and there is a corresponding decrease in the ground-level concentration of the contaminant at the receptor site.

The emissions are mixed into larger volume of air, and the diluted emission is carried out into essentially unoccupied terrain away from any receptors. Depending on the wind direction, the diluted effluent may be funneled down a river valley or between mountain ranges. Horizontal transport is sometimes prevented by surrounding hills that form a natural pocket for locally generated pollutants. This particular topographical situation occurs for instance in the Los Angeles area, which suffers heavily from air pollution.

The vertical transport depends on the rate of change of ambient temperature with altitude. The dry adiabatic lapse rate is defined as a decrease in air temperature of 1°C per 100 m. This is the rate at which, under natural conditions, a rising parcel of unpolluted air will decrease in temperature with elevation into the troposphere up to approximately 10,000 m. Under so-called isothermal conditions, the temperature does not change with elevation. Vertical transport can be hindered under stable atmospheric conditions, which occur when the actual environmental lapse rate is less than the dry adiabatic lapse rate. A negative lapse rate is an increase in air temperature with latitude. This effectively prevents vertical mixing and is known as inversion.

The dispersion from a point source (a chimney for instance) may be calculated from the Gaussian plume model (see, for instance, Reible.[1])

These different atmospheric conditions are illustrated in Fig. 1 where stack gas behavior under the various conditions is shown. Further explanations are given in Table 2. The distribution of particulate material is more effective the higher the stack. The maximum concentration, C_{max}, at ground level can be shown to be approximately proportional to the emission and to follow approximately this expression:

$$C_{max} = kQ/H^2 \quad (1)$$

where Q is the emission (expressed as grams per particulate matter per unit of time), H is the effective stack height, and k is a constant. The effective height is slightly higher than the physical height and can be calculated from information about the temperature, the stack exit velocity, and the stack inside diameter.

These equations explain why a lower ground-level concentration is obtained when many small stacks are replaced by one very high stack. In addition to this effect, it is always easier to reduce and control one large emission than

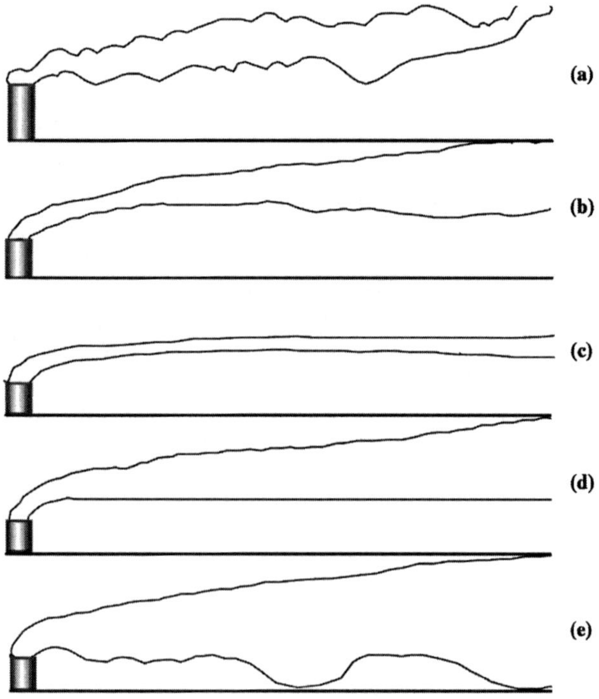
(a)
(b)
(c)
(d)
(e)

Fig. 1 Stack gas behavior under various conditions. (a) Strong lapse (looping); (b) weak lapse (coning); (c) inversion (fanning); (d) inversion below, lapse aloft (lofting); (e) lapse below, inversion aloft (fumigation).

Table 2 Various atmospheric conditions.

Strong lapse (looping)	Environmental lapse rate > adiabatic lapse rate
Weak lapse (coning)	Environmental lapse rate < adiabatic lapse rate
Inversion (fanning)	Increasing temperature with height
Inversion below, lapse aloft (lofting)	Increasing temperature below, App. adiabatic lapse rate aloft
Lapse below, inversion aloft (fumigation)	App. adiabatic lapse rate below, increasing temperature aloft

Source: Jørgensen.[10]

many small emissions, and it is more feasible to install and apply the necessary environmental technology in one big installation.

Particulate Pollution Control Equipment

Environmental technology offers several solutions to the problem of particulate matter removal. The methods and their optimum particle size and efficiency are compared in Table 3. The cost of the various installations varies of course from country to country and is dependent on several factors (material applied, standard size or not standard size, automatized, and so on). Generally, electrostatic precipitators are the most expensive solution and are mainly applied for large quantities of air. Wet scrubbers also belong among the more expensive installations, while settling chamber and centrifuges are the most cost-effective solutions.

Settling Chambers

Simple gravity settling chambers depend on gravity or inertia for the collection of particles. Both forces increase in direct proportion to the square of the particle diameter, and the performance limit of these devices is strictly governed by the particle settling velocity. The pressure drop in mechanical collectors is low to moderate, 1–25 cm water in most cases. Most of these systems operate dry, but if water is added, it performs the secondary function of keeping the surface of the collector clean and washed free of particles.

The settling or terminal velocity can be described by the following expression, which has general applicability:

$$V_t = (\partial_P - \partial) g \frac{d_P^2}{17\mu} \quad (2)$$

where V_t is terminal velocity, ∂_p is particle density, ∂ is gas density, d_p is particle diameter, and μ is gas viscosity.

This is the Stokes' law and is applicable to $N_{Re} < 1.9$, where

$$N_{Re} = d_p^* V_t^* \frac{\partial}{\mu} \quad (3)$$

The intermediate equation for settling can be expressed as:

$$V_t = \frac{0.153 * g^{0.71} * d_p^{1.14} (\partial p - \partial)^{0.71}}{\partial^{0.29} * \mu^{0.43}} \quad (4)$$

This equation is valid for Reynolds numbers between 1.9 and 500, while the following equation c an be applied for $N_{Re} \geq 500$ and up to 200,000:

$$V_t = 1.74 \left(d_p^* g \frac{(\partial p - \partial)}{\partial} \right)^{1/2} \quad (5)$$

The settling velocity in these chambers is often in the range 0.3–3 m/sec. This implies that for large volumes of emission, the settling velocity chamber must be very large in order to provide an adequate residence time for the particles to settle. Therefore, the gravity settling chambers are not generally used to remove particles smaller than 100 μm (= 0.1 mm). For particles measuring 2–5 μm, the collection efficiency will most probably be as low as 1%–2%. A variation of the simple gravity chamber is the baffled separation chamber. The baffles produce a shorter settling distance, which means a shorter retention time. The shown equations can be used to design a settling chamber.

Table 3 Characteristics of particulate pollution control equipment.

Device	Optimum Particle size (μm)	Optimum concentration (g^{-3})	Temperature limitations (°C)	Air resistance (mm H_2O)	Efficiency (% by weight)
Settling chambers	>50	>100	−30 to 350	<25	<50
Centrifuges	>10	>30	−30 to 350	<50–100	<70
Multiple centrifuges	>5	>30	−30 to 350	<50–100	<90
Filters	>0.3	>3	−30 to 250	>15–100	>99
Electrostatic precipitators	>0.3	>3	−30 to 500	<20	<99
Wet scrubbers	>2–10	>3–30	0 to 350	>5–25	<95–99

Source: Jørgensen.[10]

Cyclones

Cyclones separate particulate matter from a gas stream by transforming the inlet gas stream into a confined vortex. The mechanism involved in cyclones is the continuous use of inertia to produce a tangential motion of the particles towards the collector walls. The particles enter the boundary layer close to the cyclone wall and lose kinetic energy by mechanical friction. The forces involved are the centrifugal force imparted by the rotation of the gas stream and a drag force, which is dependent on the particle density, diameter, shape, etc. A hopper is built at the bottom. If the cyclone is too short, the maximum force will not be exerted on some of the particles, depending on their size and the corresponding drag forces.[2] If, however, the cyclone is too long, the gas stream might reverse its direction and spiral up the center.

It is therefore important to design the cyclone properly. The hopper must be deep enough to keep the dust level low. The efficiency of a cyclone is described by a graph similar to Fig. 2, which shows the efficiency versus the relative particle diameter, i.e., the actual particle diameter divided by D_{50}, which is defined as the diameter corresponding to 50% efficiency. D_{50} can be found from the following equation:

$$D_{50} = K * \left(\frac{\mu Dc}{V_c * \partial_p} \right)^{1/2} \qquad (6)$$

where D_c is the diameter of cyclone, V_c is inlet velocity, ∂_p is the density of particles, μ is gas viscosity, and K is a constant dependent on cyclone performance.

If the distribution of the particle diameter is known, it is possible to calculate the total efficiency from a graph such as Fig. 2:

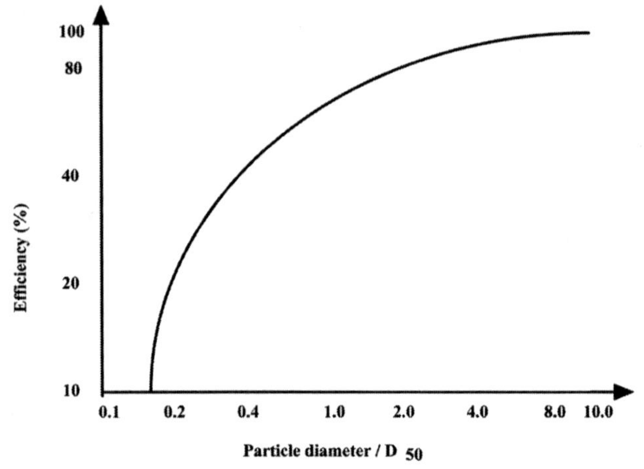
100
80
40
20
10
Efficiency (%)
0.1
0.2
0.4
1.0
2.0
4.0
8.0
10.0
Particle diameter / D 50

Fig. 2 Efficiency plotted against relative particle diameter. Notice that it is a log–log plot.

$$\text{eff} = \sum \text{m}_i^* \text{eff}_i \qquad (7)$$

where m_i is the weight fraction in the ith particle size range and eff_i is the corresponding efficiency.

The pressure drop for cyclones can be found from:

$$\Delta p = N * \frac{V_c^2}{2_g}$$

From these equations, it can be concluded that higher efficiency is obtained without increased pressure drop if D_c can be decreased while velocity V_c is maintained. This implies that a battery of parallel coupled small cyclones will work more effectively than one big cyclone. Such cyclone batteries are available as blocks and are known as multiple cyclones. Compared with settling chambers, cyclones offer higher removal efficiency for particles below 50 μm and above 2–10 μm, but involve a greater pressure drop.

Filters

Particulate materials are collected by filters by the following three mechanisms:

Impaction, where the particles have so much inertia that they cannot follow the streamline around the fiber and thus impact on its surface.

Direct interception, where the particles have less inertia and can barely follow the streamlines around the obstruction.

Diffusion, where the particles are so small (below 1 μm) that their individual motion is affected by collisions on a molecular or atomic level. This implies that the collection of these fine particles is a result of random motion.

Different flow patterns can be used, as demonstrated in Fig. 3. The types of fibers used in fabric filters range from natural fibers, such as cotton and wool, to synthetics (mainly polyesters and nylon), glass, and stainless steel.

Some properties of common fibers are summarized in Table 4. As can be seen, cotton and wool have a low temperature limit and poor alkali and acid resistance, but they are relatively inexpensive. The selection of filter medium must be based on the answer to several questions:[3,4]

What is the expected operating temperature?

Is there a humidity problem that necessitates the use of a hydrophobic material, such as, e.g., nylon?

How much tensile strength and fabric permeability are required?

How much abrasion resistance is required?

Permeability is defined as the volume of air that can pass through 1 m^2 of the filter medium with a pressure drop of no more than 1 cm of water.

The filter capacity is usually expressed as cubic meter of air per square meter of filter per minute. A typical capacity ranges between 1 and 5 $m^3/m^2/min$.

The pressure drop is generally larger than for cyclones and will in most cases be 10–30 cm of water, depending

Table 4 Properties of fibers.

Fabric	Acid resistance	Alkali	Fluoride strength	Tensile	Abrasion resistance
Cotton	Poor	Good	Poor	Medium	Very good
Wool	Good	Poor	Poor	Poor	Fair
Nylon	Poor	Good	Poor	Good	Excellent
Acrylic	Good	Fair	Poor	Medium	Good
Polypropylene	Good	Fair	Poor	Very good	Good
Orlon	Good	Good	Fair	Medium	Good
Dacron	Good	Good	Fair	Good	Very good
Teflon	Excellent	Excellent	Good	Good	Fair

on the nature of the dust, the cleaning frequency, and the type of cloth.

There are several specific methods of filter cleaning. The simplest is backwash, where dust is removed from the bags merely by allowing them to collapse. This is done by reverting the airflow through the entire compartment. The method is remarkable for its low consumption of energy. Shaking is another low-energy filter-cleaning process, but it cannot be used for sticky dust. The top of the bag is held still and the entire tube sheath at the bottom is shaken. The application of blow rings involves reversing the airflow without bag collapse. A ring surrounds the bag; it is hollow and supplied with compressed air to direct a constant stream of air into the bag from the outside.

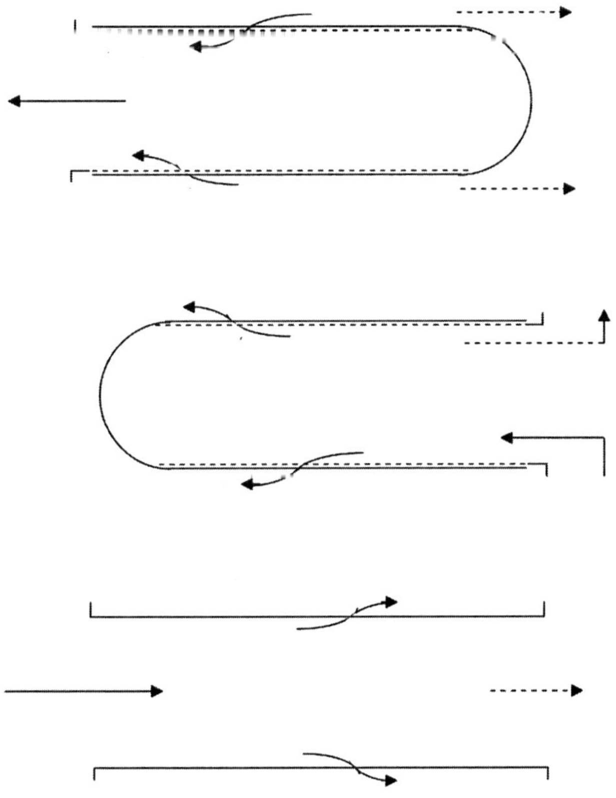

Fig. 3 Flow pattern of filters.

The pulse and improved jet cleaning mechanism involves the use of a high-velocity, high-pressure air jet to create a low pressure inside the bag and induce an outward airflow, cleaning the bag by sudden expansion and reversal of flow.

In some cases, as a result of electrostatic forces, moisture on the surface of the bag, and a slight degree of hygroscopicity of the dust itself, the material forms cakes that adhere tightly to the bag. In this case, the material must be kept drier, and a higher temperature on the incoming dirty airstream is required. Filters are highly efficient even for smaller particles (0.1–2 m), which explains their wide use as particle collection devices.

Electrostatic Precipitators

The electrostatic precipitator consists of four major components:

1. *A gas-tight shell with hoppers* to receive the collected dust, inlet and outlet, and an inlet gas distributor
2. *Discharge electrodes*
3. *Collecting electrodes*
4. *Insulators*

The principles of electrostatic precipitators are illustrated in Fig. 4. The dirty airstream enters a filter, where a high, 20–70 kV, usually negative voltage exists between discharge electrodes. The particles accept a negative charge and migrate towards the collecting electrode. The efficiency is usually expressed by use of Deutsch's equation (see the discussion including correction of this equation in the work of Gooch and Francis[5]). This equation entails that the relationship between migration velocity and particle diameter has a minimum between 0.1 and 1.0 μm.

The operation of an electrostatic precipitator can be divided into three steps:

1. The particles *accept a negative charge.*
2. The charged particles *migrate towards the collecting electrode* due to the electrostatic field.
3. The collected dust *is removed from the collecting electrode* by shaking or vibration, and is collected in the hopper.

r, the specific electrical resistance, measured in ohm meter, determines the ability of a particle to accept a charge. The practical specific resistance can cover a wide range of about four orders of magnitude, in which varying degrees of collection efficiencies exist for different types of particles. The specific resistance depends on the chemical nature of the dust, the temperature, and the humidity.

Electrostatic precipitators have found wide application in industry. As the cost is relatively high, the airflow should be at least 20,000 m^3/hr; volumes as large as 1,500,000 m^3/hr have been treated in one electrostatic precipitator.

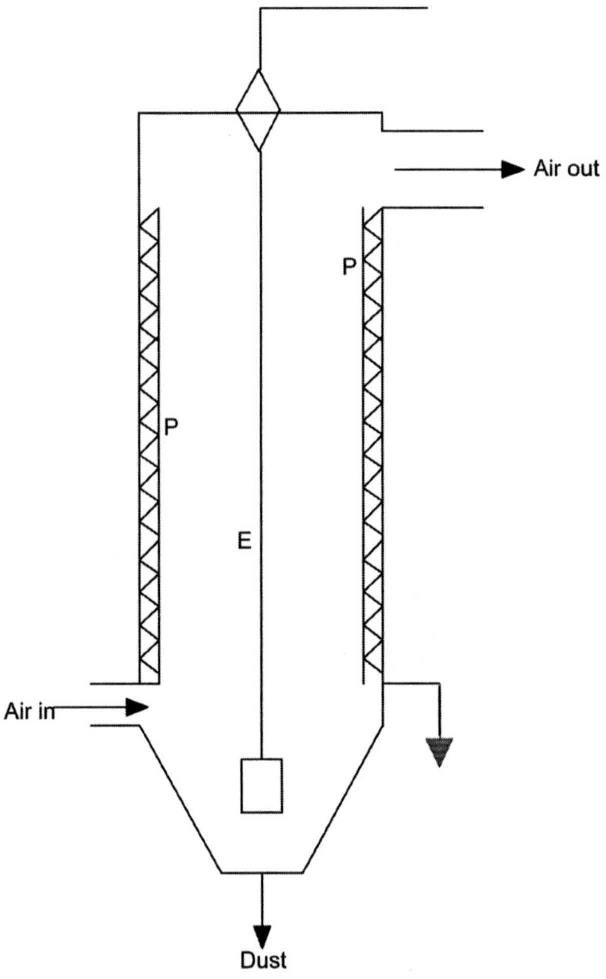
Air out
P
P
E
Air in
Dust

Fig. 4 The dust is precipitated on the electrode P. E has a high, usually negative voltage and emits a great number of electrons that give the dust particles a negative charge. The dust particles will therefore be attracted to P.

Very high efficiencies are generally achieved in electrostatic precipitators and emissions as low as 25 mg/m^3 are quite common. The pressure drop is usually low compared with other devices—25 mm water at the most. The energy consumption is generally 0.15–0.45 Wh/m^3/hr.

Wet Scrubbers

A scrubbing liquid, usually water, is used to assist separation of particles, or a liquid aerosol from the gas phase. The operational range for particle removal includes material less than 0.2 μm in diameter up to the largest particles that can be suspended in air. Gases soluble in water are also removed by this process. Four major steps are involved in the collection of particles by wet scrubbing. First, the particles are moved to the vicinity of the water droplets, which are 10–1000 times larger. Then, the particles must collide with the droplets. In this step, the relative velocity of the gas and the liquid phases is very important: If the particles have an excessively high velocity in relation to liquid, they cannot be retained by the droplets unless they can be wetted and thus incorporated into the droplets. The last step is the removal of the droplets containing dust particles from the bulk gas phase. Scrubbers are generally very flexible. They are able to operate under peak loads or reduced volumes and within a wide temperature range.[6] They are smaller and less expensive than dry particulate removal devices, but the operating costs are higher. Another disadvantage is that the pollutants are not collected but transferred into water, which means that the related water pollution problem must also be solved.[7]

Several types of wet scrubbers are available and their principles are outlined below:

1. *Chamber scrubbers* are spray towers and spray chambers that can be either round or rectangular. Water is injected under pressure through nozzles into the gas phase.
2. *Baffle scrubbers* are similar to a spray chamber but have internal baffles that provide additional impingement surfaces. The dirty gas is forced to make many turns to prevent the particles from following the airstream.
3. *Cyclonic scrubbers* are a cross between a spray chamber and a cyclone. The dirty gas enters tangentially to wet the particles by forcing its way through a swirling water film onto the walls. There, the particles are captured by impaction and are washed down the walls to the sump. The saturated gas rises through directional vanes, which are used solely to impart rotational motion to the gas phase. As a result of this motion, the gas goes out through a demister for the removal of any included droplets.
4. *Submerged orifice scrubbers* are also called gas-induced scrubbers. The dirty gas is accelerated over an aerodynamic foil to a high velocity and directed into

a pool of liquid. The high velocity impact causes the large particles to be removed into the pool and creates a tremendous number of spray droplets with a high amount of turbulence. These effects provide intensive mixing of gas and liquid and thereby a very high interfacial area. As a result, reactive gas absorption can be combined with particle removal.

5. The *ejector scrubber* is a water jet pump (see Fig. 5). The water is pumped through a uniform nozzle and the dirty gas is accelerated by the action of the jet gas. The result is aspiration of the gas into the water by the Bernoulli principle and, accordingly, a lowered pressure. The ejector scrubber can be used to collect soluble gases as well as particulates.

6. The *venturi scrubber* involves the acceleration of the dirty gas to 75–300 m/min through a mechanical constriction. This high velocity causes any water injected just upstream of or in the venturi throat to be sheared off the walls or nozzles and atomized. The droplets are usually 5–20 μm in size and form into clouds from 150 to 300 μm in diameter, depending on the gas velocity. The scrubber construction is similar to that of the

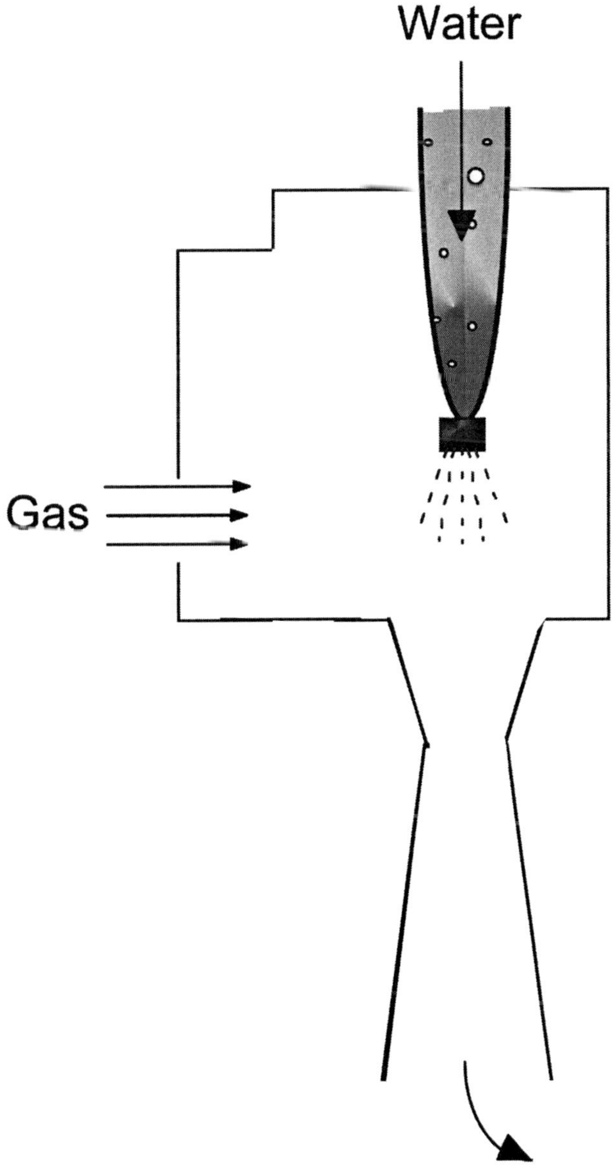
Water
Gas

Fig. 5 Principle of ejector scrubber.

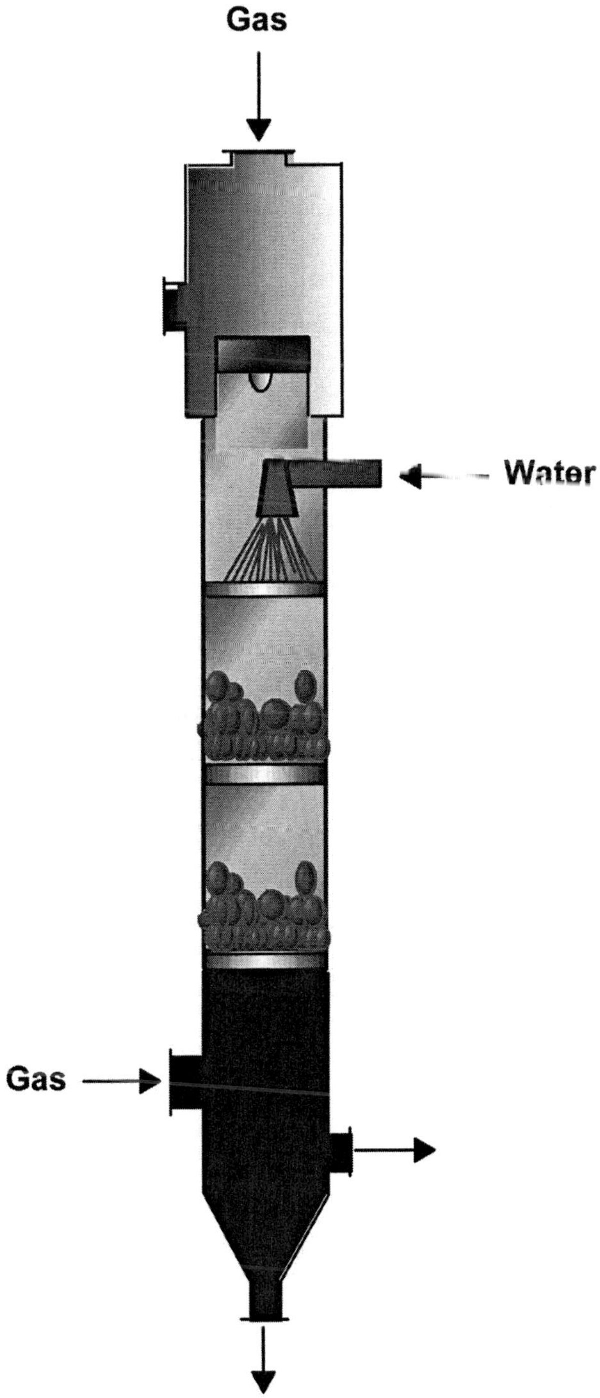
Gas
Water
Gas

Fig. 6 Packed-bed scrubber.

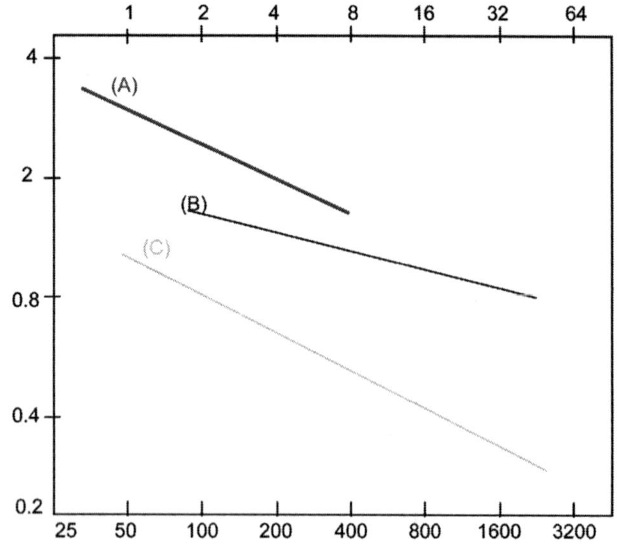
1
2
4
8
16
32
64
4
2
0.8
0.4
0.2
25
50
100
200
400
800
1600
3200
(A)
(B)
(C)

Fig. 7 Relationship between D_{50} (μm), pressure drop (mm H_2O, lower axis), and energy consumption (kW/m^3/sec; upper axis). A: Packed-bed scrubber. B: Baffled scrubber. C: Venturi scrubber.

ejector scrubber, but the jet pump is replaced by a venturi constriction.

7. *Mechanical scrubbers* have internal rotating that which break up the scrubbing liquid into small droplets and simultaneously create turbulence.
8. *Charged-droplet scrubbers* have a high voltage ionization section where the corona discharge produces air ions (as in electrostatic precipitators). Water droplets are introduced into the chamber by the use of spray nozzles or similar devices. The additional collection mechanism provided by the induction of water droplets increases the collection efficiency.
9. *Packed-bed scrubbers* have a bottom support grid, and a top retaining grid (see Fig. 6). The fluid (often water or a solution of alkali or acid) is distributed as shown in the figure over the top of the packed section, while the gas enters below the packing.

The flow is normally counter current. Packed-bed scrubbers offer the possibility of combining gas absorption with the removal of particulate material. The pressure drop is often in the order of 3 cm water per meter of packing. If the packing consists of expanded fiber, the bed scrubber is known as a fiber-bed scrubber. The packed-bed scrubber has a tendency to clog under high particulate loading, which is its major disadvantage.

Common packing includes saddles, rings, etc., like those used in absorption towers. Some important parameters for various scrubbers are plotted in Fig. 7, which demonstrates the relationship between pressure drop, energy consumption, and D_{50} (the diameter of the particles removed at 50% efficiency).

AIR POLLUTION PROBLEMS OF CARBON HYDRIDES AND CARBON MONOXIDE: SOURCES OF POLLUTANTS

All types of fossil fuel will produce carbon dioxide on combustion, which is used in the photosynthetic production of carbohydrates. As such, carbon dioxide is harmless and has no toxic effect, whatever the concentration levels, but see the entries about the greenhouse effect of carbon dioxide. An increased carbon dioxide concentration in the atmosphere will increase absorption of infrared radiation and the heat balance of the earth will be changed.

Carbon hydrides are the major components of oil and gas, and incomplete combustion will always involve their emission. Partly oxidized carbon hydrides, such as aldehydes and organic acids, might also be present.

The major source of carbon hydrides pollution is motor vehicles.

In reaction with nitrogen oxides and ozone, they form so-called photochemical smog, which consists of several rather oxidative compounds, such as peroxyacyl nitrates and aldehydes. In areas where solar radiation is strong and the atmospheric circulation is weak, the possibility of smog formation increases as the processes are initiated by ultraviolet radiation.

Incomplete combustion produces carbon monoxide. By regulation of the ratio of oxygen to fuel, more complete combustion can be obtained, but the emission of carbon monoxide cannot be totally avoided.

Motor vehicles are also a major source of carbon monoxide pollution. On average, 1 L of gasoline (petrol) will produce 200 L of carbon monoxide, while it is possible to minimize the production of this pollutant by using diesel instead of gasoline.

The annual production of carbon monoxide is more than 200 million tons, of which 50% is produced by the United States alone.

In most industrial countries, more than 75% of this pollutant originates from motor vehicles.

POLLUTION PROBLEM OF CARBON HYDRIDES AND CARBON MONOXIDE

Carbon hydrides, partly oxidized carbon hydrides, and the compounds of photochemical smog are all more or less toxic to man, animals, and plants. Photochemical smog reduces visibility, irritates the eyes, and causes damage to plants, with immense economic consequences, for example, for fruit and tobacco plantations. It is also able to decompose rubber and textiles.

Carbon monoxide is strongly toxic as it reacts with hemoglobin and thereby reduces the blood's capacity to take up and transport oxygen. Ten percent of the hemoglobin occupied by carbon monoxide will produce symptoms such as headache and vomiting. It should be mentioned here that smoking also causes a higher carboxyhemoglobin concentration. An examination of policemen in Stockholm has shown that non-smokers had 1.2% carboxyhemoglobin, while smokers had 3.5%.

CONTROL METHODS APPLIED TO CARBON DIOXIDE, CARBON HYDRIDES, AND CARBON MONOXIDE POLLUTION

Carbon dioxide pollution is inevitable with the use of fossil fuels. Therefore, it can only be solved by the use of other sources of energy.

Legislation plays a major role in controlling the emission of carbon hydrides and carbon monoxide. As motor vehicles are the major source of these pollutants, control methods should obviously focus on the possibilities of reducing vehicle emission. The methods available today are as follows:

1. Motor technical methods
2. Afterburners
3. Alternative energy sources

The first method is based upon a motor adjustment according to the relationship between the composition of the exhaust gas and the air/fuel ratio. A higher air/fuel ratio results in a decrease in the carbon hydrides and carbon monoxide concentrations, but to achieve this, a better distribution of the fuel in the cylinder is required, which is only possible through the construction of another gasification system. This method may be considered cleaner technology.

At present, two types of afterburners are in use—*thermal* and *catalytic afterburners*. In the former type, the combustible material is raised above its autoignition temperature and held there long enough for complete oxidation of carbon hydrides and carbon monoxide to occur. This method is used on an industrial scale[8,9] when low-cost purchased or diverted fuel is available; in vehicles, a manifold air injection system is used.

Catalytic oxidation occurs when the contaminant-laden gas stream is passed through a catalyst bed, which initiates and promotes oxidation of the combustible matter at a lower temperature than would be possible in thermal oxidation. The method is used on an industrial scale for the destruction of trace solvents in the chemical coating industry. Vegetable and animal oils can be oxidized at 250–370°C by catalytic oxidation. The exhaust fumes from chemical processes, such as ethylene oxide, methyl methacrylate, propylene, formaldehyde, and carbon monoxide can easily be catalytically incinerated at even lower temperatures. The application of catalytic afterburners in motor vehicles presents some difficulties due to poisoning of the catalyst with lead. With the decreasing lead concentration in gasoline, it is becoming easier to solve that problem, and the so-called double catalyst system is now finding wide application. This system is able to reduce nitrogen oxides and oxidize carbon monoxide and carbon hydrides simultaneously. New catalysts are currently coming on the market and offer a higher efficiency.

Lead in gasoline has been replaced by various organic compounds to increase the octane number. Benzene has been applied, but it is toxic and causes air pollution problems because of its high vapor pressure. MTBE (methyl tertiary butyl ether) is another possible compound for increasing the octane number. It is, however, very soluble and has been found as a groundwater contaminant close to gasoline stations.

Application of alternative energy sources is still at a preliminary stage. The so-called Sterling motor is one alternative, as it gives more complete combustion of the fuel, but there have been several improvements of the efficiency of the motor during the last years due to the increasing cost of fossil fuel. Much interest has, however, been devoted to electric and hybrid vehicles.

AIR POLLUTION OF SULFUR DIOXIDE: SOURCES, PROBLEMS, AND SOLUTIONS

Fossil fuel contains approximately 2%–5% sulfur, which is oxidized by combustion to sulfur dioxide. Although fossil fuel is the major source, several industrial processes produce emissions containing sulfur dioxide, for example, mining, the treatment of sulfur containing ores, and the production of paper from pulp. The total global emission of sulfur dioxide has been decreasing during the last 25 years due to the installation of pollution abatement equipment, particularly in North America, the European Union, and Japan. The concentration of sulfur dioxide in the air is relatively easy to measure, and sulfur dioxide has been used as an indicator component. High values recorded by inversion are typical.

Sulfur dioxide is oxidized in the atmosphere to sulfur trioxide, which forms sulfuric acid in water. Since sulfuric acid is a strong acid, it is easy to understand that sulfur dioxide pollution indirectly causes the corrosion of iron and other metals and is able to acidify aquatic ecosystems.[10]

The health aspects of sulfur dioxide pollution are closely related to those of particulate pollution. The gas is strongly adsorbed onto particulate matter, which transports the pollutant to the bronchi and lungs. There is a clear relationship between concentration, effect, and exposure time, which is reflected in the emission standards for sulfur dioxide (see Table 5).

Clean Air Acts were introduced in all industrialized countries during the 1970s and 1980s. Table 5 illustrates some typical sulfur dioxide emission standards, although these may vary slightly from country to country.

Table 5 SO_2 emission standards.

Duration	Concentration (ppm)	Comments
Month	0.05	
24 hr	0.10	Might be exceeded once a month
30 min	0.25	Might be exceeded 15 times/month

Source: Jørgensen.[10]

The approaches used to meet the requirements of the acts as embodied in the standards can be summarized as follows:

1. Fuel switching from high to low sulfur fuels.
2. Modification of the distribution pattern—use of tall stacks.
3. Abandonment of very old power plants that have a particular high emission.
4. Flue gas cleaning.

Desulfurization of liquid and gaseous fuel is a well-known chemical engineering operation. In gaseous and liquid fuels, sulfur either occurs as hydrogen sulfide or reacts with hydrogen to form hydrogen sulfide. The hydrogen sulfide is usually removed by absorption in a solution of alkanolamine and then converted to elemental sulfur. The process in general use for this conversion is the so-called Claus process. The hydrogen sulfide gas is fired in a combustion chamber in such a manner that one-third of the volume of hydrogen sulfide is converted to sulfur dioxide. The products of combustion are cooled and then passed through a catalyst-packed converter, in which the following reaction occurs:

$$2H_2S + SO_2 = 3S + 2H_2O \tag{8}$$

The elemental sulfur has commercial value and is mainly used for the production of sulfuric acid.

Sulfur occurs in coal both as pyritic sulfur and organic sulfur. Pyritic sulfur is found in small discrete particles within the coal and can be removed by mechanical means, e.g., gravity separation methods. However, 20%–70% of the sulfur content of coal is present as organic sulfur, which can hardly be removed today on an economical basis. Since sulfur recovery from gaseous and liquid fuels is much easier than that from solid fuel (which has other disadvantages as well), much research has been and is being devoted to *the gasification or liquefaction of coal.* It is expected that this research will lead to an alternative technology that will solve most of the problems related to the application of coal, including sulfur dioxide emission. Approach (2) listed above has been mentioned earlier in this entry, while approach (3) needs no further discussion. The next subsection is devoted to (4) flue gas cleaning.

Flue Gas Cleaning of Sulfur Dioxide

When sulfur is not or cannot be economically removed from fuel oil or coal prior to combustion, removal of sulfur oxides from combustion gases will become necessary for compliance with the stricter air pollution control laws.

The chemistry of sulfur dioxide recovery presents a variety of choices and five methods should be considered:

1. Adsorption of sulfur dioxide on active metal oxides with regeneration to produce sulfur.
2. Catalytic oxidation of sulfur dioxide to produce sulfuric acid.
3. Adsorption of sulfur dioxide on charcoal with regeneration to produce concentrated sulfur dioxide.
4. Reaction of dolomite or limestone with sulfur dioxide by direct injection into the combustion chamber. A lime slurry is injected into the flue gas beyond the boilers.
5. Fluidized bed combustion of granular coal in a bed of finely divided limestone or dolomite maintained in a fluid-like condition by air injection. Calcium sulfite is formed as a result of these processes.

In particular, the two latter methods have found wide application, particularly to large industrial installations. It is possible to recover the sulfur dioxide or elemental sulfur from these processes, making it possible to recycle the spent sorbing material.

AIR POLLUTION PROBLEMS OF NITROGENOUS GASES: SOURCES, PROBLEMS, AND CONTROL

Seven different compounds of oxygen and nitrogen are known: N_2O, NO, NO_2, NO_3, N_2O_3, N_2O_4, and N_2O_5—often summarized as NO_x. From the point of view of air pollution, it is mainly NO (nitrogen oxide) and NO_2 (nitrogen dioxide) that are of interest. Nitrogen oxide is colorless and is formed from the elements at high temperatures. It can react further with oxygen to form nitrogen dioxide, which is a brown gas. The major sources of the two gases are combustion of gasoline and oil (nitrogen oxide) and combustion of oil, including diesel oil (nitrogen dioxide). The production of NO is favored by high temperature. In addition, a relatively small emission of nitrogenous gases originates from the chemical industry. The total global emission is approximately 10 million tons per year. This pollution has only local or regional interest, as the natural global formation of nitrogenous gases in the upper atmosphere by the influence of solar radiation is far more significant than the man-controlled emission.

As mentioned above, nitrogen oxide is oxidized to nitrogen dioxide, although the reaction rate is slow—in the order of 0.007/hr. However, it can be accelerated by solar radiation. Nitrogenous gases take part in the formation of smog, as the nitrogen in peroxyacyl nitrate originates from nitrogen oxides. They are highly toxic but as their contribution to global pollution is insignificant, local and regional problems can partially be solved by changing the distribution pattern (see the section on *Control Methods Applied to Particulate Pollution*).

The emission from motor vehicles can be reduced by the same methods as mentioned for carbon hydrides and carbon monoxide. The air/fuel ratio determines the concentration of pollutants in exhaust gas. An increase in the ratio will reduce the emission of carbon hydrides and carbon monoxide, but unfortunately will increase the concentration of nitrogenous gases. Consequently, the selected air/fuel ratio will be a compromise. A double catalytic afterburner is applied today, and it is able to reduce nitrogenous gases and simultaneously oxidize carbon hydrides and carbon monoxide. The application of alternative energy sources will, as for carbon hydrides and carbon monoxide, be a very useful control method for nitrogenous gases at a later stage.

Between 0.1 and 1.5 ppm of nitrogenous gases, of which 10%–15% consists of nitrogen dioxide, are measured in urban areas with heavy traffic. On average, the emission of nitrogenous gases is approximately 15 g per liter of gasoline and 25 g per liter of diesel oil.

Nitrogenous gases in reaction with water form nitrates that are washed away by rainwater. In some cases, this can be a significant source of eutrophication. For a shallow lake, for example, the increase in nitrogen concentration due to the nitrogen input from rainwater will be rather significant. In a lake with a depth of 1.7 m and an annual precipitation of 600 mm, which is normal in many temperate regions, the annual input will be as much as 0.3 mg/L.

The methods used for control of industrial emission of nitrogenous gases, including ammonia, will be discussed in the next section that discusses industrial air pollution, but as pointed out above, industrial emission is of less importance, even though it might play a significant role locally. The emission of nitrogenous gases by combustion of oil for heating and the production of electricity can hardly be reduced.

INDUSTRIAL AIR POLLUTION, OVERVIEW, AND CONTROL METHODS

The rapid growth in industrial production during recent decades has exacerbated the industrial air pollution problem, but due to increased application of continuous processes, recovery methods, air pollution control, use of closed systems, and other technological developments, industrial air pollution has, in general, not increased in proportion to production.

Industry displays a wide range of air pollution problems related to a large number of chemical compounds in a wide range of concentrations.

It is not possible in this context to discuss all industrial air pollution problems; instead, we shall touch on the most important problems and give an overview of the control methods applied today. Only the problems related to the environment will be dealt with in this context.

A distinction should be made between air quality standards, which indicate that the concentration of a pollutant in the atmosphere at the point of measurement should not be greater than a given amount, and emission standards, which require that the amount of pollutant emitted from a specific source should not be greater than a specifically indicated amount.

The standards reflect, to a certain extent, not only the toxicity of the particular component but also the possibility for its uptake.

Here, the distribution coefficient for air/water (blood) plays a role. The more soluble the component is in water, the greater the possibility for uptake. For example, the air quality standard for acetic acid, which is very soluble in water, is relatively lower than the toxicity of aniline, which is almost insoluble in water.

Since industrial air pollution covers a wide range of problems, it is not surprising that all three classes of pollution control methods mentioned previously have found application: modification of the distribution pattern, alternative (cleaner) production methods, and particulate and gas/vapor control technology.

All the methods mentioned in the section on *Control Methods Applied to Particulate Pollution* also apply for industrial air pollution control.

In gas and vapor technology, a distinction has to be made between condensable and non-condensable gaseous pollutants. The latter must usually be destroyed by incineration, while the condensable gases can be removed from industrial effluents by absorption, adsorption, condensation, or combustion.

Recovery is feasible by the first three methods.

Gas Absorption

Absorption is a diffusion process that involves the mass transfer of molecules from the gas state to the liquid state along a concentration gradient between the two phases. Absorption is a unit operation that is enhanced by all the factors generally affecting mass transfer, i.e., high interfacial area, high solubility, high diffusion coefficient, low liquid viscosity, increased residence time, turbulent contact between the two phases, and possibilities for reaction of the gas in the liquid phase. This last factor is often very significant and almost 100% removal of the contaminant is the result of such a reaction. Acidic components can easily be removed from gaseous effluents by absorption in alkaline solutions, and, correspondingly, alkaline gases can easily be removed from effluent by absorption in acidic solutions (Table 6).

Table 6 Absorber reagents.

Reagents	Applications
$KMnO_4$	Rendering, polycyclic organic matter
NaOCl	Protein adhesives
C12	Phenolics, rendering
Na_2SO_3	Aldehydes
NaOH	CO_2, H_2S, phenol, Cl_2, pesticides
$Ca(OH)_2$	Paper sizing and finishing
H_2SO_4	NH_3, nitrogen bases

Carbon dioxide, phenol, and hydrogen sulfide are readily absorbed in alkaline solutions in accordance with the following processes:

$$CO_2 + 2NaOH \rightarrow 2Na^+ + CO_3^{2-} \tag{9}$$

$$H_2S + 2NaOH \rightarrow 2Na^+ + S_{2-} + 2H_2O \tag{10}$$

$$C_6H_5OH + NaOH \rightarrow C_6H_5O^- + Na^+ + H_2O \tag{11}$$

Ammonia is readily absorbed in acidic solutions:

$$2NH_3 + H_2SO_4 \rightarrow 2NH_4 + SO_4^{2-} \tag{12}$$

Gas Adsorption

Adsorption is the capture and retention of a component (adsorbate) from the gas phase by the total surface of the adsorbing solid (adsorbent). In principle, the process is the same as when dealing with wastewater treatment; the theory is equally valid for gas adsorption.

Adsorption is used to concentrate (often 20 to 100 times) or store contaminants until they can be recovered or destroyed in the most economical way. Fig. 8 illustrates some adsorption isotherms applicable to practical gas adsorption problems. These are often described as either Langmuir's or Freundlich's adsorption isotherms. Adsorption is dependent on temperature: increased temperature means that the molecules move faster and therefore it is more difficult to adsorb them. There are four major types of gas adsorbents, the most important of which is activated carbon, but aluminum oxide (activated aluminum), silica gel, and zeolites are used as well.

The selection of adsorbent is made according to the following criteria:

1. High selectivity for the component of interest.
2. Easy and economical to regenerate.
3. Availability of the necessary quantity at a reasonable price.
4. High capacity for the particular application so that the unit size will be economical. Factors affecting capacity include total surface area involved, molecular weight, polarity activity, size, shape, and concentration.
5. Pressure drop, which is dependent on the superficial velocity.
6. Mechanical stability in the resistance of the adsorbent particles to attrition. Any wear and abrasion during use or regeneration will lead to an increase in bed pressure drop.
7. Microstructure of the adsorbent should, if at all possible, be matched to the pollutant that has to be collected.
8. The temperature, which has a profound influence on the adsorption process, as already mentioned.

Regeneration of the adsorbents is an important part of the total process. A few procedures are available for regeneration:

1. *Stripping* by use of steam or hot air.
2. *Thermal desorption* by raising the temperature high enough to boil off all the adsorbed material.
3. *Vacuum desorption* by reducing the pressure enough to boil off all the adsorbed material.
4. *Purge gas stripping* by using a non-adsorbed gas to reverse the concentration gradient. The purge gas may be condensable or non-condensable. In the latter case, it might be recycled, while the use of a condensable gas has the advantage that it can be removed in a liquid state.
5. *In situ oxidation* based on the oxidation of the adsorbate on the surface of the adsorbent.

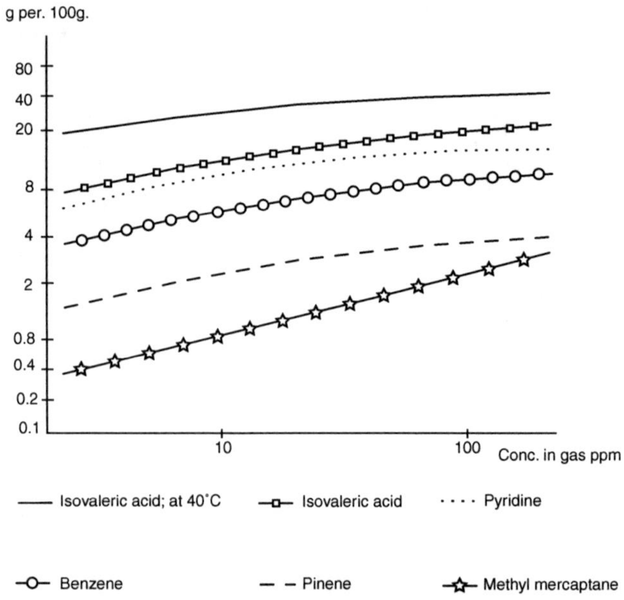
g per. 100g.
80
40
20
8
4
2
0.8
0.4
0.2
0.1
10
100
Conc. in gas ppm
Isovaleric acid; at 40°C
Isovaleric acid
Pyridine
Benzene
Pinene
Methyl mercaptane

Fig. 8 Adsorption isotherms, at 20°C.

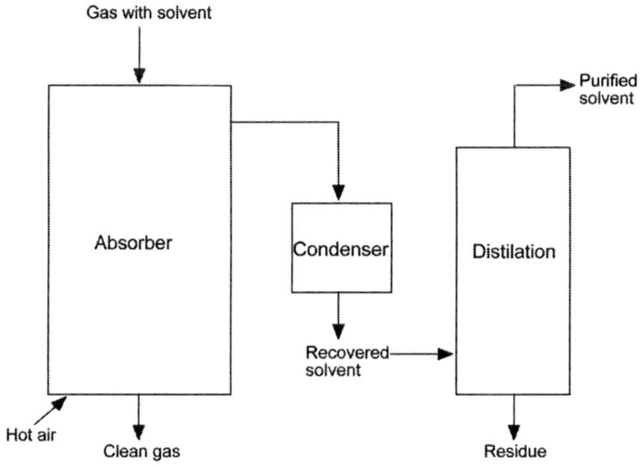
Gas with solvent
Purified solvent
Absorber
Condenser
Distilation
Recovered solvent
Hot air
Clean gas
Residue

Fig. 9 Flow chart of solvent recovery by the use of activated carbon.

6. *Displacement* by use of a preferentially adsorbed gas for the desorption of the adsorbate. The component now adsorbed must, of course, also be removed from the adsorbent, but its removal might be easier than that of the originally adsorbed gas, for instance, because it has a lower boiling point.

Although the regeneration is 100%, the capacity of the adsorbent may be reduced 10%–25% after several regeneration cycles, due to the presence of fine particulates and/or high molecular weight substances that cannot be removed in the regeneration step. A flowchart of solvent recovery using activated carbon as an adsorbent is shown in Fig. 9 as an illustration of a plant design.

Combustion

Combustion is defined as rapid, high-temperature gas-phase oxidation. The goal is the complete oxidation of the contaminants to carbon dioxide and water, sulfur dioxide, and nitrogen dioxide.

The process is often applied to control odors in rendering plants, paint and varnish factories, rubber tire curing, and petrochemical factories. It is also used to reduce or prevent an explosion hazard by burning any highly flammable gases for which no ultimate use is feasible. The efficiency of the process is highly dependent not only on temperature and reaction time but also on turbulence or the mechanically induced mixing of oxygen and combustible material. The relationship between the reaction rate, r, and the temperature can be expressed by Arrhenius' equation:

$$r = A * e - E/RT \tag{13}$$

where A is a constant, E is the activation energy, R is the gas constant, and T is the absolute temperature. A distinction is made between combustion, thermal oxidation, and catalytic oxidation, the latter two being the same in principle as the vehicle afterburners.

HEAVY METALS AS AIR POLLUTANTS

Heavy metals, which may be defined as the metals with a specific gravity >5.00 kg/L, comprise 70 elements. Most of them are, however, only rarely found as pollutants. The heavy metals of environmental interest form very heavy soluble compounds with sulfide and phosphate and form very stable complexes with many ligands present in the environment. It means, fortunately, that most of the heavy metals are not very bioavailable in most environments (see also *Bioremediation*, p. 408).

A number of enzymes activated by metal ions and metalloenzymes are known. Members of the first mentioned group, comprising iron, cobalt, chromium, vanadium, selenium, copper, zinc, iron, cobalt, and molybdenum, are able with a stronger bond to form metalloenzymes: metalloproteins, metalloporphyrins, and metalloflavins.

Heavy metals are emitted to the atmosphere by energy production and a number of technological processes (see Table 7). It makes the atmospheric deposition of heavy metals originating from human activities the dominant pollution source for the vegetation of natural ecosystems—forests, wetlands, peat lands, and so on. The heavy metal content in sludge and fertilizers plays a more important role for agricultural land where also the inputs of heavy metals by irrigation, natural fertilizers, and application of chemicals including pesticides may add to the overall pollution level. The atmosphere and hydrosphere both have a well-developed ability for "self-purification"—for heavy metals by removal processes, for instance, sedimentation. The lithosphere has a high buffer capacity toward the effects of most pollutants and also has an ability to self-purify, for instance, by runoff and uptake by plants, although the rates usually are much lower than in the two other spheres. Table 8 illustrates the removal rates.

Heavy metals are bound to clay particles due to their ion-exchange capacity and to hydratized metal oxides, such

Table 7 Important atmospheric pollution sources of heavy metals.

Source	Heavy metals
Incineration of oil	V, Ni
Incineration of coal	Hg, V, Cr, Zn, As
Gasoline	Pb (leaded gasoline)
Metal industry	Fe, Cu, Mn, Zn, Cr, Pb, Ni, Cd, and others
Application of pesticides	Hg, Cr, Cu, As
Incineration of solid waste	Hg, Zn, Cd, and others

Source: Jørgensen.[10]

Table 8 Removal of heavy metals by runoff and drainage from a typical cultivate clay soil.

Metal	Removal ($mg/m^2/yr$)	Removal % of pool
Pb	0.5	1–3
Cu	1.2	2–3
Zn	15.9	30–50
Cd	0.07	15–30

Source: Waid.[8]

as iron sesquioxide (As, Cr, Mo, P, Se, and V) and manganese sequioxides (Co, Ba, Ni, and lanthanides). Calcium phosphate is further better able to bind As, Ba, Cd, and Pb in alkaline soil. Fulvic acids (molecular weight about 1000) and humic acid (molecular weight about 150,000) are able to form complexes with a number of heavy metals, Hg(II), Cu(II), Pb(II), and Sn(II). The mobility of heavy metals is dependent on a number of factors. The soil pore water contains soluble organic compounds (acetic acid, citric acid, oxalic acid, and other organic acids), partly excreted by the roots. These small organic molecules form chelated, soluble compounds with metal ions such as Al, Fe, and Cu. Activity of living organisms in soil may also enhance the mobility of heavy metal ions. Fungi and bacteria may utilize phosphate and thereby release cations. Formation of insoluble metal sulfide under anaerobic conditions from sulfate implies a reduced mobility. The lower oxidation stages of heavy metals are generally more soluble than the higher oxidation stages, implying increased mobility.

The many possibilities of binding heavy metals in soil explain the long residence time. Cadmium, calcium, magnesium, and sodium have the most mobile metal ions with a residence time of about 100 years. Mercury has a residence time of about 750 years, while copper, lead, nickel, arsenic, selenium, and zinc have residence times of more than 2000 years under temperate conditions. Tropic residence times are typically lower (for all heavy metals, about 40 years).

The biological effect of heavy metal pollution occurs in accordance with Sections 4.4 and 4.5 on two levels: on an organism level and on the higher level—the ecosystem level.

Plant toxicity is very dependent on the presence of other metal ions. For instance, Rb and Sr are very toxic to many plants, but the presence of the biochemically more useful K and Ca is able to reduce or eliminate toxicity. The toxicity of arsenate and selenate can be reduced in the same manner by sulfate and phosphate.

Formation of complexes by reaction with organic ligands also reduces toxicity due to reduced bioavailability. The plant toxicity of heavy metals in soil is consequently also correlated with the concentration of heavy metal ions in the soil solution.

The heavy metals that are most toxic to plants are silver, beryllium, copper, mercury, tin, cobalt, nickel, lead, and chromium. With the exception of silver and chromium, the divalent form is most toxic. For silver, it is Ag^+, and for chromium, it is chromate and dichromate that are most toxic. Silver and mercury ions are very toxic to fungus spores, and copper and tin ions are very toxic to green algae; lethal concentrations may be as low as 0.002–0.01 mg/L.

One of the key processes on ecosystem level is the mineralization process, because it determines the cycling of nutrients. Heavy metals can inhibit the mineralization due to the blocking of enzymes. The effect is known not only for the enzymes produced in the organisms but also for extracellular enzymes—exo-enzymes—originated from dead cells or excreted from roots and living microorganisms. As the various processes forming the cycling of nutrients are coupled, the entire mineralization cycle is disturbed if only one process is reduced. It is therefore possible to determine the change of the mineralization cycle by measuring the respiration, the transformation of nitrogen, and the release of phosphorus. As low a concentration of copper as 3–4 times the background concentration may imply a reduced soil respiration. A few hundred milligrams of copper per kilogram of soil is furthermore able to diminish the nitrogen release rate by one half.

The most sensitive mineralization process is phosphorus cycling. Biological material binds phosphorus as esters of phosphoric acid. The phosphate is released by the hydrolysis of the ester bond, a process catalyzed by phosphatase. This process is inhibited by the presence of heavy metals. The inhibition is decreasing in the following sequence: molybdenum (VI) > wolframate (VI) > vanadate (V) > nickel (II) > cadmium > mercury (II) > copper (II) > chromate (VI) > arsenate (V) > lead (II) > chromium (III).

The inhibition of exo-enzymes by heavy metals does not form a clear pattern. It is therefore difficult to generalize. Most experiments, however, give a clear picture of the influence of heavy metals on mineralization: the rate of mineralization may be reduced significantly with a consequent reduction of the productivity of the entire ecosystem.

In Denmark (a country with relatively little heavy industry and good pollution control), atmospheric deposition causes an average annual increase of the total content of heavy metal in soil between 0.4% and 0.6%, but it varies very much from location to location. In accordance with the many possibilities for side reactions of heavy metals in soil, including adsorption to the soil particles, the amount of heavy metal ions that are available to plants is only a fraction of the total content. If only the bioavailable heavy metals are used as the basis, the annual percentage increase in the soil concentration due to atmospheric deposition is probably higher.

Most lead in soil is not mobile and cannot be transported via the root system to the leaves and stems. This is in contrast to cadmium, which is very mobile. About 50% of the cadmium in soil will be found in the plants after the growth season, although the concentration may be very different in different parts of the plants. The cadmium in grains for instance has not increased parallel to the increased atmospheric deposition of cadmium.

The heavy metal pollution of soil is one of the major challenges in environmental management in industrialized countries. Due to the many diffuse sources of heavy metal pollution, the solution of the problem requires a wide spectrum of methods, the first of which is application of cleaner technology (see the section on *Industrial Air Pollution, Overview, and Control Methods*). It is in other words necessary to reduce the total emission of heavy metals. Dilution (for instance, higher chimneys) is not an applicable solution. Moreover, as pollution, particularly air pollution, has no borders, it is necessary to take international initiatives and agree on international standards, particularly for the most problematic heavy metals, i.e., cadmium, mercury, nickel, chromium, and vanadium. A three-point program must be adopted:

- A nationally and internationally accepted environmental strategy.
- Agreed international standards and long-term goals.
- A monitoring program to assess the pollution level and compare the measured concentrations with standards.

REFERENCES

1. Reible, D.D. *Fundamentals of Environmental Engineering*; Lewis Publ.: Boca Raton, New York, Washington, DC, 1998; 526 pp.
2. Leith, D.; Licht, W. The collection of cyclone type particles—A new theoretical approach. AICHE Symp. Ser. **1975**, *68*, 196–206.
3. Pring, R.T. Speciation considerations for fabric collectors. Pollut. Eng. **1972**, *4*, 22–24.
4. Rullman, D.H. Backhouse technology: A perspective. J. Air Pollut. Control Assoc. **1976**, *26*, 16–18.
5. Gooch, N.P.; Francis, N.L. A theoretically based mathematical model for calculations of electrostatic precipitators performance. J. Air Pollut. Control Assoc. **1975**, *25*, 106–113.
6. Onnen, J.H. Wet scrubbers tackle pollution. Environ. Sci. Technol. **1972**, *6*, 994–998.
7. Hanf, E.B. A guide to a scrubber selection. Environ. Sci. Technol. **1970**, *4*, 110–115.
8. Waid, D.E. Controlling pollutants via thermal incineration. Chem. Eng. Prog. **1972**, *68*, 57–58.
9. Waid, D.E. *Thermal Oxidation or Incineration*; Pollut. Control Assoc.: Pittsburgh, PA, 1974; 62–79.
10. Jørgensen, S.E. *Principles of Pollution Abatement*; Elsevier: Amsterdam, 2000; 520 pp.

Alexandria Lake Maryut: Integrated Environmental Management

Lindsay Beevers
Lecturer in Water Management, School of the Built Environment, Heriot Watt University, Edinburgh, U.K.

Abstract
Lake Maryut is a shallow, closed lake, located on the Mediterranean coast of Egypt. The lake is fed by the Rosetta branch of the River Nile and serves several important functions for the adjacent city of Alexandria, namely, fisheries production, water supply for irrigation, navigation routes to the River Nile, receiving body for industrial and domestic waste, and other human activities such as farming. Currently, the lake receives pollution from a number of different sources and is highly eutrophic. The lake area is dominated by vegetation, principally *Phragmites australis* and *Eichornia crassipes,* which, if left unmanaged, impacts upon the fisheries production capability of the lake. A significant population in the surrounding area rely directly on the lake as a source of income and food. Currently, the management of the lake is controlled by several authorities. If the lake is to continue to provide the services it currently supports, while allowing the desired level of economic and urban development, environmental intervention is required to improve the water quality. This entry details the analysis of the current ecological state of Lake Maryut and proposes actions for the integrated environmental management of the lake.

INTRODUCTION

Lake Maryut is located on the Mediterranean coast of Egypt, in the delta of the River Nile and defines the southern boundary of the city of Alexandria.[1] The lake extends for about 80 km along the northwest coast of Alexandria and 30 km south and has a wetted surface area of 65 km^2 (Fig. 1). It is a shallow water lake[2] with a water depth of approximately 1.5 m across the different basins and, unlike any of the other Nile deltaic lakes, is not directly connected to the Mediterranean Sea. The water level of the lake is kept below mean sea level, and freshwater is supplied, through irrigation canals, from the Rosetta branch of the Nile.[3] Throughout literature, the lake is referred to variously as Mariout, Mariut, or Maryut Lake. For this entry, the lake will be identified as Lake Maryut.

The lake has been in existence for more than 6000 years, as part of the Nile deltaic formation. In an extensive study of the evolution of the lake,[4] several drivers for the current form of the lake were established, namely, eustatic sea level, climate oscillations, compaction, sediment transport, and more recent anthropogenic influences such as land reclamation, irrigation, and agricultural practices. It is a mixture of these anthropogenic influences that has shaped the development of the lake in more recent years, along with its connection to the city of Alexandria.

During the Greco-Roman period, Alexandria was an active port, serving as a key navigational route from the Mediterranean, through Lake Maryut, up the Nile to Cairo. Canals linked the lake with the sea to the north and the Nile to the east. However, the siltation of the canopic mouth of the Nile in the 12th century[5,6] severed the freshwater influx to the lake and with it the navigational links. During this period, the city of Alexandria went into a phase of decline, and the lake went through several phases, which included coastal connection and influx and coastal disconnection and drying out.

In 1892, the irrigation system of the Beheirah district was established, which restored a flow of freshwater to the lake. The lake remained disconnected from the coast, with excess water pumped to the bay through El Mex station (Fig. 1), and navigation permitted through sea locks.[7] This is similar to the form the lake takes today, where the principal drainage inflows arise from agricultural irrigation channels fed by the main River Nile.

Over the years, Lake Maryut has provided Alexandria and Egypt with a source of fish, and currently, part of the lake is dedicated to aquaculture (Fisheries Basin, Fig. 1). Over 7000 fishermen have rights to fish in the lake, and recent pollution levels and increased vegetation growth affect their catch.[8] *Tilapia* species are the predominant catch, representing approximately 90% of the total.[9] In addition to fisheries, the lake acts as a source of water for irrigation and for raising animals and provides areas for dwelling beside its shores.

In its current form, canals divide the lake water body into several basins. The earth embankments along the canals have several breaches that allow water to flow from the canals into the basins and vice versa, creating interaction between water bodies. There are three main inflows to the Lake Maryut: the Qala drain located to the northeast part of the lake, the Omum drain located at the east of

Encyclopedia of Environmental Management DOI: 10.1081/E-EEM-120046808

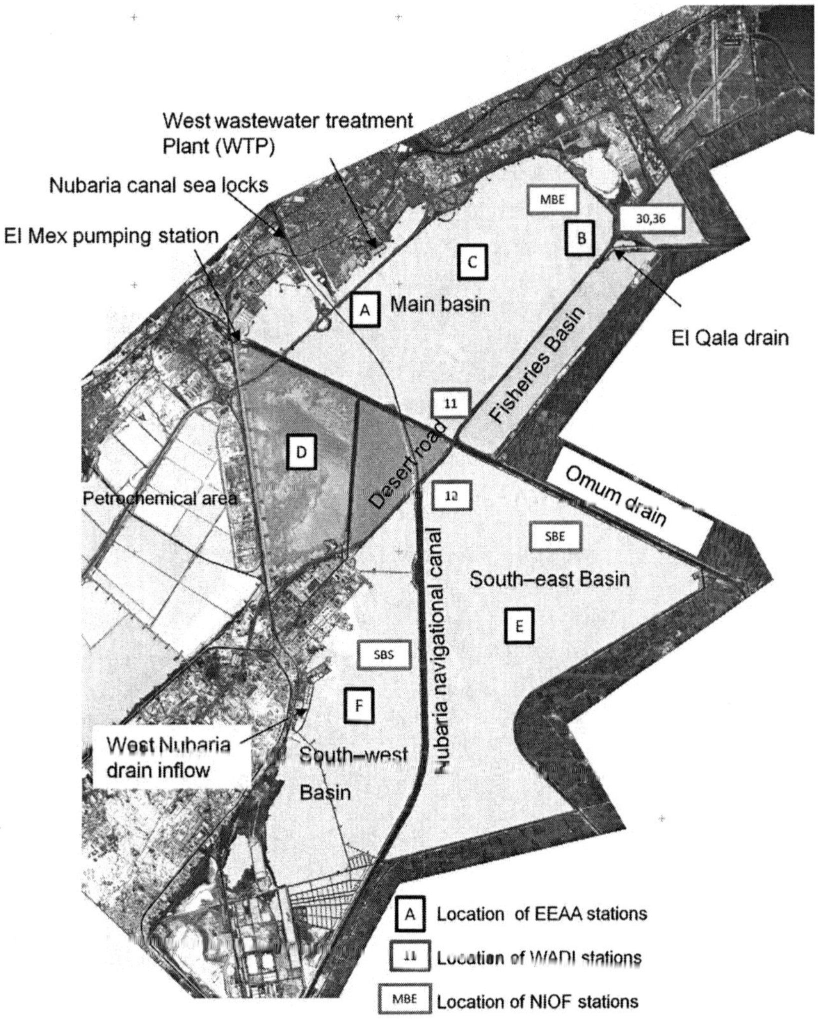
West wastewater treatment Plant (WTP)
Nubaria canal sea locks
El Mex pumping station
MBE
30,36
B
C
A
Main basin
Fisheries Basin
El Qala drain
11
D
Desert road
Petrochemical area
12
Omum drain
SBE
Nubaria navigational canal
South–east Basin
E
SBS
F
West Nubaria drain inflow
South–west Basin
Location of EEAA stations
Location of WADI stations
Location of NIOF stations

Fig. 1 Lake Maryut: location map.

the lake, and the Nubaria navigational canal located at the south of the lake (Fig. 1). In addition to this, there are three minor inflows: one from the West Nubaria drain, one from the petrochemical industrial area, and the third from the West Wastewater Treatment Plant (WWWTP) (see Fig. 1). The main outflow from the Lake Maryut is El Mex pumping station, which consists of two buildings, each housing six pumps with nominal capacities of 12.5 m^3/sec.[7] Small flows are lost to the system through the navigational locks as well as through evaporation.

As a consequence of high nutrient loadings due to the agricultural and industrial activities upstream and in its surrounds, Lake Maryut has become a highly polluted and eutrophic lake. Eutrophication of lakes is a natural process that can be accelerated by human activities (Fig. 2) that introduce an excess of nutrients together with other pollutants.[10] The main sources of nutrients and pollutants to Lake Maryut are human sewage, industrial waste, farm, and urban runoff.[11,12]

Eutrophic lakes often experience an excessive growth of algae and larger aquatic plants. Such growth consumes dissolved oxygen (DO), vital for the fish and other animal life.[10] This growth and subsequent proliferation of vegetation occurs in Lake Maryut due to the high nutrient loadings that has entered the lake for a number of years. Currently, 60% of the surface area of the lake is covered by vegetation (*Phragmites australis* and *Eichornia crassipes*), which reduces the DO concentrations in the lake, especially in the main basin. Significant discharge of domestic sewage (with basic primary treatment) from the Qala drain and the East Wastewater Treatment Plant (EWWTP) enter the lake. The Omum and Nubaria Canals are less polluted than the Qala drain; however, they also contribute nutrient loadings to the lake, albeit to a lesser extent. Finally, nonpoint sources such as agricultural runoff containing pesticides and fertilizers contribute to the deterioration of the environmental quality of the lake.

Fig. 2 Fishing activities on the lake.

The lake also suffers from other anthropogenic pressures including urbanization, unplanned settlements, and land reclamation (which has reduced the surface area of the lake gradually). An analysis of the stakeholders directly responsible for managing the lake was undertaken in 2007.[13] This showed that Alex Company for Sanitary and Drainage, the Fishing Authority, and The Ministries of Industry, Water Resources, and Environment are the main authorities responsible for the lake. The coordination and communication regarding the management of the lake between these authorities are known to be poor, with each developing management plans in isolation.

Between 2007 and 2009, a European Union (EU)-funded SMAP III project was set up to analyze the current functioning of the lake in more detail and to develop an integrated action plan for the environmental management of the lake. This entry will cover the development of this plan, specifically focusing on the scientific models set up to support the plan development and to investigate the results of different interventions.

CURRENT STATE OF LAKE MARYUT

The current functioning of the lake has been investigated through a mixture of studies commissioned to gather specific information and ongoing monitoring surveys. Hydraulic information was gathered through a United States Agency for International Development (USAID) project in 1996,[7] hydrologic information was sourced from the local airport weather station (Nouzha airport), and a longer data series was obtained from the weather station at Port Alexandria.[14] Bathymetric information of the lake came from a survey completed in 2008 by the National Institute of Oceanography and Fisheries (NIOF).[15] The ecosystem functioning and health of the basin were constructed using a number of surveys, principally the long-term monitoring strategy undertaken annually by the Egyptian Environmental Affairs Authority (EEAA),[16] a comprehensive sampling campaign by NIOF[15] in 2008 and the EU-funded Water Demand Integration (WADI) project.[8]

Using these data, a comprehensive picture of the lake can be assembled.

Hydraulic and Hydrological Functioning

Table 1 shows the monthly average meteorological data for the lake. It is clear that, generally, precipitation in the region is very low, with the lowest values recorded between April and October. During the same period, along with significant sunlight hours recorded, evapotranspiration is highest.

Table 2 summarizes the main inflows to the lake (Fig. 1) over the year. These flows are known to remain reasonably

Table 1 Average monthly meteorological data.

Month	Precipitation (mm/day)[14]	Evapotranspiration	Wind speed (m^3/sec)[15]	Solar radiation (J/m^2/day)[20]
January	1.73	1.4	3.23	10,000,000
February	0.92	2.2	3.97	12,000,000
March	0.44	1.7	3.57	17,000,000
April	0.12	4.0	4.17	22,000,000
May	0.04	4.0	3.60	24,000,000
June	0.00	5.6	3.54	26,000,000
July	0.00	6.4	3.32	25,000,000
August	0.01	7.7	3.04	23,000,000
September	0.04	9.3	2.15	20,000,000
October	0.25	5.5	2.58	16,000,000
November	1.15	3.4	2.92	12,000,000
December	1.82	1.3	3.77	10,000,000
Average		1.4	3.32	18,083,333

Table 2 Monthly flow discharges.

	Discharges (m³/day)		
Month	**Qala**	**Omum**	**Nubaria**
January	529,920	3,368,640	959,040
February	466,560	2,478,640	1,270,080
March	475,200	3,248,640	462,240
April	532,224	3,318,640	666,144
May	671,328	3,680,640	691,200
June	414,720	3,453,038	243,792
July	673,920	3,136,320	285,120
August	734,400	3,412,800	838,030
September	676,500	4,275,690	492,480
October	873,500	4,813,000	559,800
November	853,500	5,353,000	559,800
December	693,500	4,103,000	559,800

Source: USAID.[7]

steady throughout the year, though rising slightly in the months of August to February due to the connection with the main river (Nile) and agricultural requirements. The Omum drain carries the greatest flow to the basin. Direct inputs to the lake from the WWWTP, West Nubaria drain, and the petrochemical area (Fig. 1) are detailed in Table 3. These flows are considered to be constant inflows to the lake. El Mex pumping station manages the outflow of the lake. Twelve pumps with nominal capacities of 12.5 m³/sec[7] maintains the water level in the lake to –2.8 m below sea level.[7,17] In addition, small flows are lost from the lake through the navigational locks and of course through evaporation. Water depths vary throughout the lake, with the deepest occurring in the canals due to the navigational requirements. Depths in the basins vary between 0.5 and 1.2 m,[15] and in places, this is maintained by dredging and vegetation removal.

The flow direction in the lake is generally concentrated down the canals. Interaction of the canal water with the basin water occurs through the breached bunds; however, velocities and, hence, mixing in these sections are low. The point where the greatest interaction of canal and basin flow can be observed is at the junction between the Omum and Nubaria Canals in the main basin.

Water Quality and Ecosystems

Table 4 shows the time period available for each of the data sources, and Fig. 1 shows the location of the sampling points. As a whole, the data of these sources were consistent and a similar magnitude was observed between sources. Table 5 presents the seasonal values available from the EEAA[16] data. Typical effluent concentrations from treatment plants with primary treatment can be found in literature.[18,19]

Table 3 Constant flow discharges.

Discharges (m³/day)		
WWWTP[17]	**West Nubaria Drain**	**Petrochemical area[16]**
410,325	259,200	47,398

The data clearly show that the most polluted basin is the main basin, and the source of that pollution comes from the Qala drain, which carries with it effluent from the EWWTP. The highest. Biological oxygen demand (BOD) values and the lowest DO% values are found at station B (Fig. 1) at the entrance of the Qala drain. Similarly, the highest nitrogen loading enters the main basin through the Qala drain, along with the highest phosphorus loading. Station C shows generally lower values of the N and P as the water from the drain enters the lake and becomes mixed.

In general, the proportion of total N arising from ammonia was higher than that from nitrates, particularly in the main basin. It is thought that this is due in part to the source of pollution arising from the Qala drain and the low DO values recorded. In the West Basin, the proportion of ammonia is lower than that in the main basin, which is corroborated by the findings of El Rayis.[21] Since the levels of DO are also higher in this basin, it is believed that the nitrification process is more evident here.

This poor water quality has led to a drastic reduction in biodiversity.[8] The eutrophic nature of the lake results in a significant dominance of *P. australis,* while at the outfalls from the WWTPs, significant growths of the lead tolerant species *E. crassipes* is found. Periodic cutting of the *Phragmites* is practiced to increase the nitrogen removal from the lake. This intervention is necessary to prevent the lake area being dominated by vegetation, which in turn reduces the area available for fish habitat.

It is clear that in its present state, Lake Maryut is severely polluted with reduced biodiversity. A continual deterioration in the status of the basin will lead to a significant reduction in economic and industrial development potential.

Social Considerations and Governance

There are no specific studies that survey those living by the lake. However, the Integrated Action Plan research[22] identifies three main groups, namely, fishermen, poorer communities, and a scattered population, totaling approxi-

Table 4 Water quality data sources.

Source	Sampling period	Number of measurements	Number of stations
EEAA[16]	2004–2008	2–3 per year	6
NIOF[15]	March–May 2008	5	10
WADI[8]	March 2007	1	31

Table 5 Water quality parameters from selected basins.

Water quality parameters (EEAA 2004–2008)																
Yearly averages																
EEAA stations	Year	DO (mg/L)	DO %	BOD	COD	NH_3-N (mg/L)	NO_2-N (mg/L)	NO_3-N (mg/L)	Total N (mgN/L)	PO_4-P (mg/L)	Temperature (°C)	Conductivity (mS/cm)	pH	TDS	TS	TSS
Main basin in front of Qala drain (B)	2004	3.8	43.7	38.0	110.0			3.6	3.6		21.5		8.0	2454.0	2479.0	25.0
	2005	2.0	22.5	30.0	98.0	12.6	0.8	1.2	14.2	5.1	19.4	3.0	7.5	1828.0		
	2006	2.6	29.2	40.0	137.0	21.0		1.8	22.8	4.5	21.0	3.6	7.2	1825.0	1852.0	27.0
	2007					20.0	0.1	1.3	21.4	4.8						
	2008	1.0			133.0	27.8		1.5	29.3		24.0		7.8	1923.0	1958.0	35.0
	Ave	2.4	31.8	36.0	119.5	20.3	0.4	1.9	18.2	4.8	21.5	3.3	7.6	2007.5	2096.3	29.0
Middle of south basin (E)	2004	7.2	83.5	2.0	11.8	0.3	0.2	4.0	4.3	0.5	23.0	6.5	7.7	5134.0	5139.3	5.3
	2005	6.7	76.9	1.0	29.3	0.3	0.2	2.2	2.7	0.8	21.2	7.5	8.1	5266.3	5349.0	6.5
	2006	7.4	82.5	4.3	34.8	0.6	0.1	1.9	2.1	0.4	22.0	8.0	7.8	4289.8	4307.5	17.8
	2007					0.1	0.3	1.0	1.2	0.7						
	2008	9.2		2.0	55.0	0.4	0.0	2.9	3.3		27.5		8.1	5141.0	5153.5	12.5
	Ave	7.6	81.0	2.3	32.7	0.3	0.2	2.4	2.7	0.6	23.4	7.3	7.9	4957.8	4987.3	10.5
Middle of west basin (F)	2004	5.4	63.8	6.0	99.0	0.8	0.2	5.5	5.9	3.8	23.5	8.1	7.5	6953.3	6965.3	12.0
	2005	7.3	84.9	2.0	49.7	0.4	0.3	2.6	2.4	0.3	21.0	8.5	8.0	5085.0	5339.0	20.0
	2006	6.7	73.9	4.3	84.5	0.3	0.1	2.4	2.1	0.8	22.2	9.4	7.8	5885.8	5907.0	21.3
	2007					0.1	0.1	0.9	1.1	1.2						
	2008	7.9		5.0	81.5	0.4	0.1	1.9	2.4		27.0		8.2	7254.0	7274.0	20.0
	Ave	6.8	74.2	4.3	78.7	0.4	0.2	2.7	2.8	1.6	23.4	8.7	7.9	6294.5	6371.3	18.3
Middle of main basin (C)	2004	4.5	52.3	21.0	97.0			1.2	1.2		24.2	3.6	8.0	2415.0	2415.0	29.0
	2005	3.3	36.3	44.0	99.0	6.8	0.1	1.0	7.9	4.2	19.0	3.1	7.9	1961.0		17.0
	2006	3.4	38.3	29.7	136.7	15.6	0.1	4.9	18.8	3.0	23.7	4.6	7.8	1935.7	1965.3	29.7
	2007					9.0	0.1	3.9	13.0	4.7						
	2008	7.5			67.0	16.5		0.2	16.7		25.0		8.0	1941.0	1973.0	32.0
	Ave	4.7	42.3	31.6	99.9	12.0	0.1	2.2	11.5	4.0	23.0	3.8	7.9	2063.2	2117.8	26.9

mately 30,000. It is estimated that 62% of the population is aged between 15 and 55, which is the average employment age in Egypt. There are high unemployment rates (up to 15%–20%) within this group, leading to low annual incomes. High illiteracy, poor health services, high mortality rates, high crime levels, and a tendency to marry young are also characteristics of the demographic.

A significant proportion of the population are living in settlements and accommodation that are temporary or informal. In these scattered settlements, some common issues arise: 1) lack of, or poor infrastructure coverage, particularly water supply and sanitation/wastewater networks and paved roads; 2) inadequate local services, especially health care, education, and youth facilities; 3) poor housing conditions; 4) lack of secure land tenure in the squatter settlements; and 5) high unemployment.

An analysis of the stakeholders directly responsible for managing the lake was undertaken in 2007.[13] This showed that the Alex Company for Sanitary and Drainage, the Fishing Authority, and The Ministries of Industry, Water Resources, and Environment are the main authorities responsible for the lake. The coordination and communication regarding the management of the lake between these authorities are known to be poor. Each authority makes strategic plans for the lake in isolation, with little discussion; hence, an integrated vision of the lake did not exist.

INTEGRATED ENVIRONMENTAL MANAGEMENT

Integrated Plan Concept

The integrated action plan for Lake Maryut draws on the principles of Integrated Coastal Zone Management (ICZM) to develop integrated actions for the sustainable development of Lake Maryut. Central to the ICZM approach is participation from stakeholders in the development of a plan that integrates strategic actions to address the environmental, economic, administrative, social, and urban issues of the lake.[22] The outcomes of this process defined four specific objectives of the action plan:

- Improve the ecological and chemical status of Lake Maryut.
- Enhance levels of economic, urban, and social development in a sustainable manner.
- Increase the potential for industrial activities around the lake and their environmental management.
- Adapt the governance system of the lake to execute the plan.

Strategic actions were proposed for each of the objectives, following analysis of the current state of the lake, and where feasible, these actions were tested. This entry will focus on the process completed for the first objective in detail and will present a short summary of the actions identified for the other three objectives.

Scenario Identification

The geographic scope of the plan was taken as the general boundary of the lake (Fig. 1) and the natural extension to these boundaries according to the interaction of the lake. To the north, the plan includes the navigational locks and the El Mex pumping stations (necessary connections to the sea). To the east, the limit is bounded by the main road to Alexandria from Cairo, plus the position of the EWWTP and the Qala drain. To the west, the limit was the industrial area and a 500 m perimeter to the lake, and finally, the 500 m perimeter was extended to the south of the lake as the limit to the plan.

Through a participatory process, which included four training events and six participation workshops, a series of scenarios and potential actions for the improvement of the water ecosystem were defined by key stakeholders. These can be summarized as follows:

- Redevelopment of the waterfront and potential development of islands in the lake for urban regeneration projects.
- Improve lake mixing.
- Removal or improvement of the WWTP discharges into the lake.
- Vegetation management and sediment management.
- Industrial effluent improvement.
- Installation of wetland areas for nutrient removal.

Model Development

To assess the impact of certain actions, tools were developed to predict the hydrodynamic functioning and ecosystem functioning of the lake. Given the level of information available to create these tools and the complex nature of the processes in the lake, a mixture of approaches was proposed. To investigate the hydrodynamics and mixing of Lake Maryut, a two-dimensional (2D) hydrodynamic model was developed. To investigate the ecological improvements as a result of intervention, a point model of ecosystem functioning of different basins was set up.[23]

Hydraulic Model

Using the data set out in Tables 1–3, a 2D hydraulic model of the lake was built using MIKE21 software.[24] The computational mesh is shown in Fig. 3, which defines the principal flow routes through the canal. A roughness map was developed to represent vegetation growth in the basins.[25] The model was calibrated for water levels but not velocities due to lack of data. A comprehensive sensitivity study was undertaken, which showed that the main control on

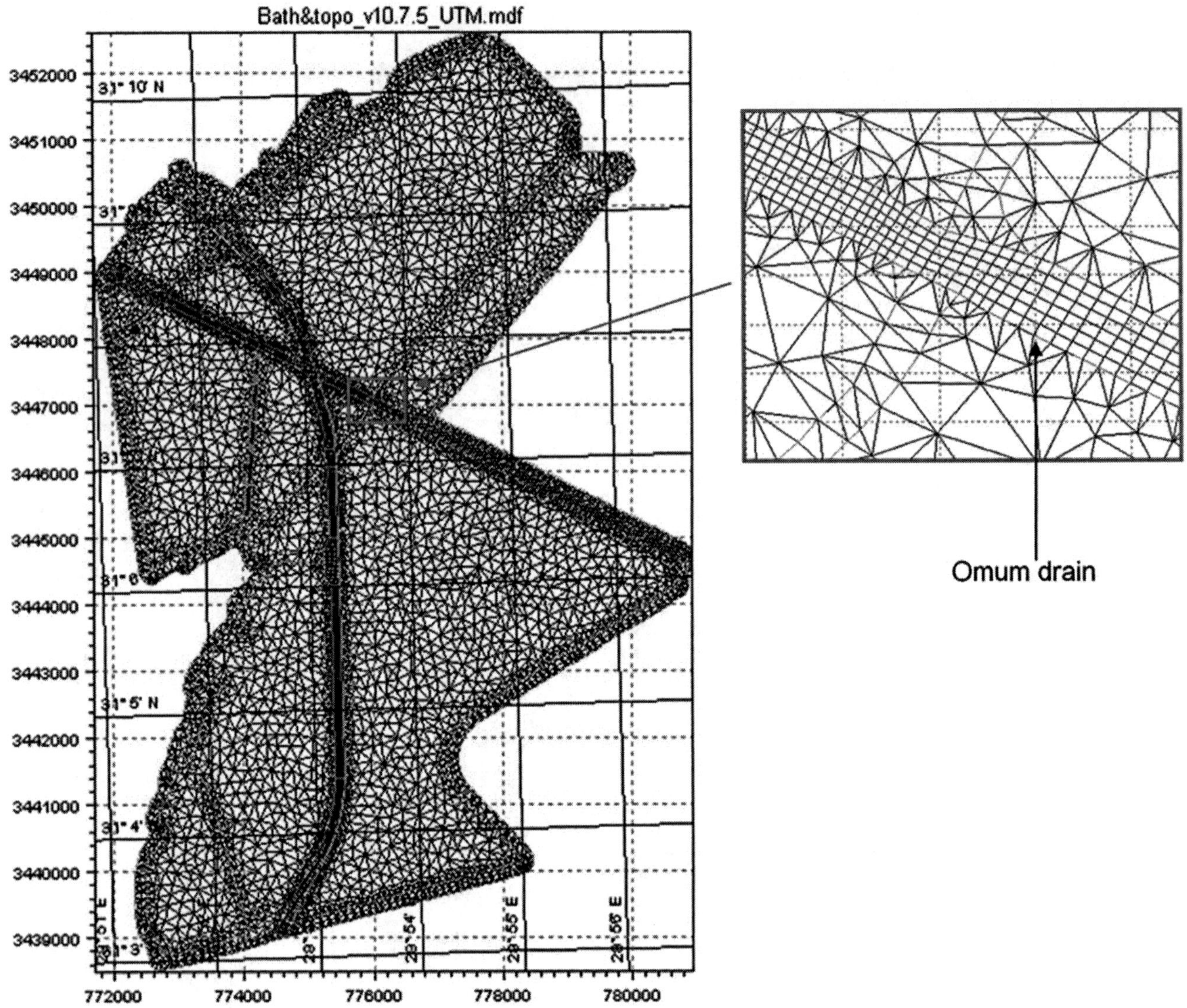
Bath&topo_v10.7.5_UTM.mdf
3452000
3451000
3450000
3449000
3448000
3447000
3446000
3445000
3444000
3443000
3442000
3441000
3440000
3439000
772000
774000
776000
778000
780000
31° 10' N
31° 5' N
Omum drain

Fig. 3 Computational mesh: Hydraulic model.
Source: Duran and Beevers.[26]

the lake's hydraulic system is the pumping station at El Mex.[26] Fig. 4 shows the general circulation patterns predicted by the model. The interactions are mainly driven by the drain inflows and clear areas of the lake where no vegetation is found.

To investigate the impact of particular actions on the hydraulics, a number of simulations were undertaken (Fig. 5). These included the land reclamation from the lake and addition of islands to promote urban regeneration of the lakeshore, as well as the design of different entry configurations for the Qala drain to promote mixing. Fig. 6 shows the magnitude of velocities predicted in the basin following these interventions. The change in velocities is highest in the main basin where most interventions are proposed; however, the magnitude of change is only in the order of a 9% increase.[26] While areas in the main basin become better mixed, areas of stagnation occur, most notably around the islands. From these results, it becomes clear that increasing mixing significantly would be difficult unless an increase to the inflow to the lake could be facilitated.

Ecological Model

Using the data set out in Tables 1–5, a dynamic 0-dimension ecological model was built for the different basins of Lake Maryut. The software PCLAKE describes the dominant ecological interaction in a shallow lake ecosystem.[23] This model was chosen for its ability to model the complex processes and allow a number of different proposed actions to be modeled, including vegetation management on a seasonal basis, dredging, wetland installation, and improvements to discharges. It should be noted that the model assumes mixing across a basin, which, for Lake Maryut, is a simplification; however, the model is detailed enough to give an indication of the potential change caused by an action.

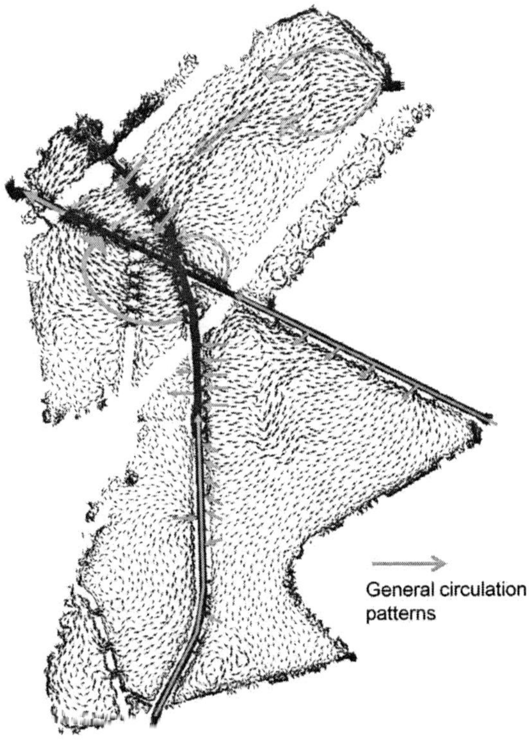
General circulation patterns

Fig. 4 General circulation patterns: Hydraulic model.
Source: Duran and Beevers.[26]

The model was built using the seasonal data available from the EEAA monitoring for the inflows, and this was complemented by the WADI and NIOF data. Calibration was possible, comparing monitoring data available in the lake itself.

A significant number of actions were tested in the ecological model. Available literature was used to represent the actions in the lake:

- Secondary treatment of effluent from the WWTPs.[27]
- Addition of aluminum to the WWTP process.[18]
- Installation and removal efficiencies of constructed wetlands (for the Qala drain).[28–31]
- Vegetation and sediment management.

Table 6 shows the predicted improvement to different parameters if the following actions are taken:

- Removal of the WWWTP from the lake (outfall to the sea) after upgrade to secondary treatment.
- Upgrade of EWWTP to secondary treatment plus additional treatment using aluminum.
- Upgrade of industrial effluent through improved environmental management.
- Dredge sediments in the Qala drain and installation of a constructed wetland.
- Aquatic vegetation management.

The oxygen consumption in Lake Maryut is mainly attributed to the respiration of aquatic organisms including

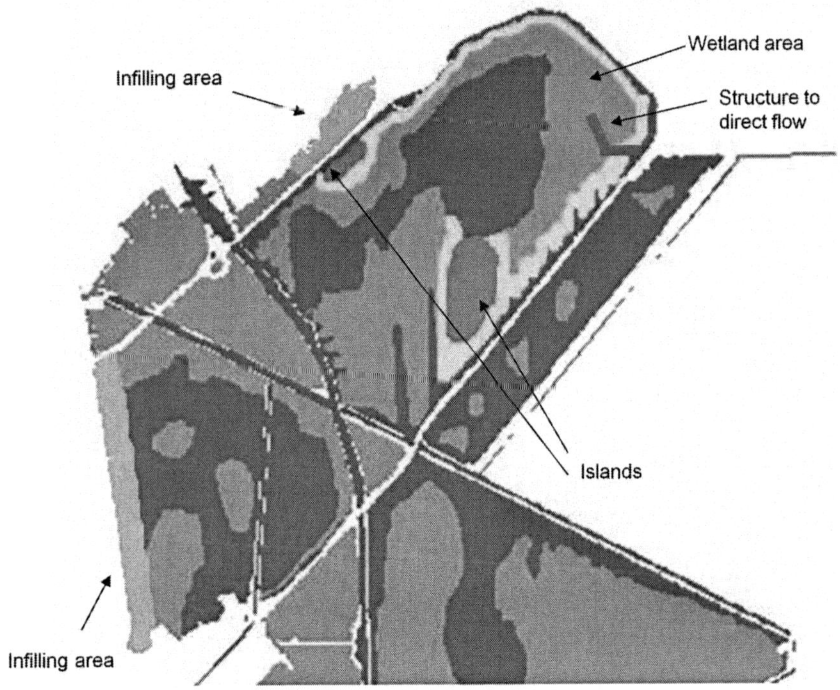
Wetland area
Infilling area
Structure to direct flow
Islands
Infilling area

Fig. 5 Scenario modifications: Hydraulic model.
Source: Duran and Beevers.[26]

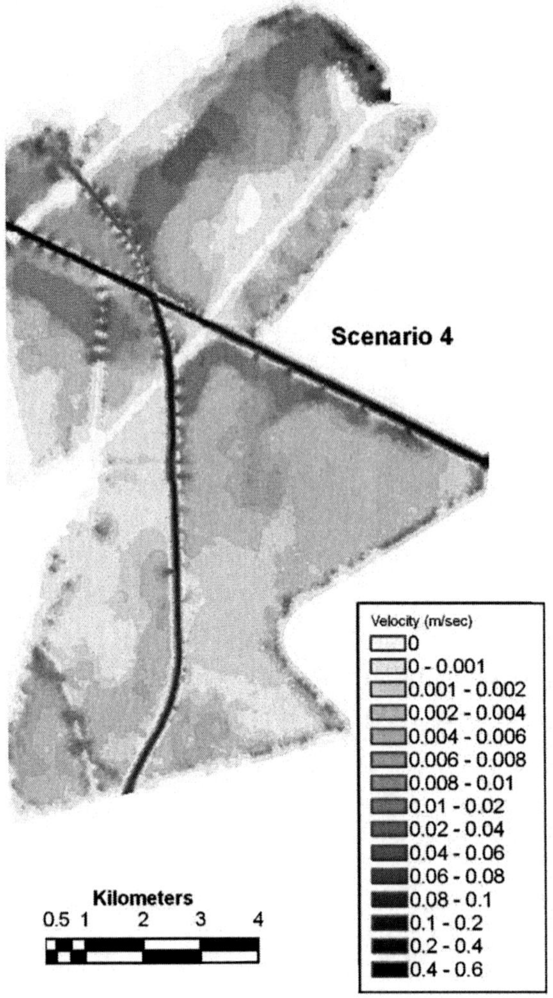
Scenario 4
Velocity (m/sec)
0
0 - 0.001
0.001 - 0.002
0.002 - 0.004
0.004 - 0.006
0.006 - 0.008
0.008 - 0.01
0.01 - 0.02
0.02 - 0.04
0.04 - 0.06
0.06 - 0.08
0.08 - 0.1
0.1 - 0.2
0.2 - 0.4
0.4 - 0.6
Kilometers
0.5 1 2 3 4

Fig. 6 Velocity predictions: Hydraulic model.
Source: Duran and Beevers.[26]

plankton and aerobic bacteria and the oxidation of organic matter.[32] High values of organic carbon have been reported in sediments close to the treatment plants.[21] The oxidation of these sediments, rich in organic matter, is one of the factors that increases the DO consumption and therefore decreases the DO concentrations in the water column. Improving the treatment processes for the treatment plants has a significant impact on the model results, indicating up to a 300% improvement to DO levels in the main basin.

Ammonium is generated by heterotrophic bacteria as the primary nitrogenous end product of decomposition of organic matter. NH_4-N concentrations are usually low in oxygenated waters of oligotrophic lakes since it is utilized by plants in the nitrification process. However, at low DO concentrations, nitrification ceases, the absorptive capacity of the sediments is reduced, and an increase of the release of NH_4-N from the sediments occurs. As a result, the NH_4-N concentration of shallow lakes would increase.[33] Increasing the levels of DO through improved treatment of wastewater and a reduction in pollutant loads arriving to the lake as a result of constructed wetlands has the potential to reduce overall ammonia levels in the main basin by up to 47%.

Chlorophyll-a is a primary productivity indicator and a very good estimate for monitoring and assessing the eutrophication status of lakes.[34] The supply of P and N is considered to be one of the main factors that determine the magnitude of the primary production.[35] In cases where pollution is caused by domestic wastewater with a high nutrient content, the algae production in the recipient watercourse increases considerably.[34] With the improvement to N and P and the increase in the DO predicted, a decrease in chlorophyll-a of up to 85% could be achieved for the lake.

Table 6 Ecological model results.

			Improvement with stated actions	
Parameter	**Unit**	**Baseline value (average)**	**Increase %**	**Decrease %**
DO	mg/L	1.3	303	
NH_4-N	mg N/L	13.4		47
NO_3-N	mg N/L	0.8	120	
Organic N	mg N/L	6.4		83
N_{total}	mg N/L	20.5		52
P_{total}	mg P/L	4.9		72
Chlorophyll-a	mg/m^3	93.4		85
Phyto-biomass	mg DW/L	5.9		85
Zoo biomass	g DW/m^2	0.53	39	
Detritus	mg/L	83.5		82
Secchi depth	m	0.1	377	

Source: Alvarez-Mieles and Beevers.[36]

Finally, the Secchi depth is a measure of the transparency of a water column. Secchi readings between 0.1 and 0.4 m were reported in the main basin,[32] which are directly related to the high production rate of phytoplankton. Eutrophic lakes present low values of Secchi depth due to the high concentrations of algae and detritus that increase turbidity and therefore decrease transparency. With the proposed measures and reduction in detritus, an increase in the transparency of the lake is predicted to be greater than 370%.[36]

Proposed Actions

The modeling component tested a number of actions for the improvement of the water ecosystem and showed that a significant improvement could be achieved. In addition to these, actions were proposed to address the other strategic objectives of the plan.[22] The following is a summary of such objectives:

- Prepare a Lake Maryut land and water plan.
- Promote urban regeneration around the lake shoreline and create ports and markets in the main basin.
- Promote ecotourism and encourage new ecological economic activities.
- Upgrade informal settlement areas and integrate social housing into urban development projects.
- Plan and develop fisheries activities.
- Prepare online monitoring systems along with water quantity and quality maps.
- Regulate specific limits for industrial water discharge into Lake Maryut.
- Establish a Lake Maryut Authority and set up a management unit with a clear structure and purpose.
- Establish a robust monitoring system for water quality.

The plan aims to address all four strategic objectives defined at the outset. To monitor the effect of any potential implemented action as part of the plan, a monitoring program is suggested. The implementation of such a monitoring plan would allow the incorporation of adaptive management techniques should unexpected impacts become evident.

CONCLUSIONS

An integrated action plan for Lake Maryut is proposed, using a participatory process and supported by a technical modeling study that was used to predict the impact of potential actions. It is clear from the study that significant improvement to the lake can be achieved, thus encouraging economic development and urban regeneration. However, this improvement requires substantial investment. The plan suggests that to achieve this, a financial strategy that links the urban and industrial future development potential with the environmental interventions is required to improve the lake, thereby linking potential future economic gains to the required lake renovation.

Since the completion of the proposed plan in 2009, there have been follow-up activities that move towards the adoption of a sustainable development strategy for the lake. A lake management unit is established and receives technical input from Centre for Environment and Development for the Arab Region and Europe (CEDARE) for Geographical Information System (GIS) support and NARSS (National Authority for Remote Sensing and Space Sciences of Egypt) for model development. In addition, the Ministry for Housing has proposed to use the outcomes of the study for the development of a new urban territorial plan for Alexandria that includes an integrated sustainable development vision for the Lake Maryut zone.

The study has shown that, with a coordinated and integrated approach to environmental management, the current deteriorating trend of Lake Maryut could be reversed. However, to achieve this goal, actions addressing environmental, social, and governance issues must be implemented.

ACKNOWLEDGMENTS

The author would like to acknowledge the official partners of the ALAMIM project: MEDCITIES, Barcelona Metropolitan Area, CEDARE, Alexandria Governate, the EEAA, NARSS, City of Marseilles, Environment and Housing Ministry of the Catalan Government, UNESCO-IHE, and the EUCC—The Coastal Union. In particular, the author is grateful to Mr. Joan Parpal, the project leader, and Ms Roxan Duran and Ms. Gabriela Alvarez-Mieles.

REFERENCES

1. Saad, M. Distribution of phosphates in Lake Maryut, a heavily polluted lake in Egypt. Water, Air, Soil Pollut. **1973**, *2*, 515–522.
2. Mesnage, V.; Picot, B. The distribution of phosphate in sediments and its relation with eutrophication of a Mediterranean coastal lagoon. Hydrobiologia **1995**, *297* (1), 29–41, DOI: 10.1007/BF00033499.
3. Hughes, R.; Hughes, S. *A Directory of African Wetlands*; IUCN: Gland, Switzerland and Cambridge, UK/UNEP: Nairobi, Kenya/WCMC: Cambridge, U.K., 1992; 820 pp., ISBN 2-88032-949-3.
4. Warne, A.; Stanley, D. Late quaternary evolutions of the Northwestern Nile Delta and adjacent coast in the Alexandria region, Egypt. J. Coastal Res. **1993**, *9* (1), 26–64.
5. Stanley, J.; Warne, A.; Schnepp, G. Geoarchaeological interpretation of the Canopic, largest of the relict Nile Delta distributaries, Egypt. J. Coastal Res. **2004**, *20* (3), 920–930, doi: 10.2112/1551-5036(2004)20[920:GIOTCL]2.0.CO;2.
6. Stanley, D.; Warne, A. Nile Delta: Recent geological evolution and human impact. Science **1993**, *260* (5108), 628–634 DOI: 10.1126/science.260.5108.628.

7. USAID. Hydrologic Studies of Lake Mariout Technical Memorandum N° 8. 1996 USAID Project N° 263-0100.
8. Mateo, M. Lake Mariut: An Ecological Assessment. Final report of the Water Demand Integration (WADI) project: INCO-C-2005—15226, 2009. Available for download at http://www.medcities.org/ocuments section (accessed April 29, 2011).
9. Bakhoum Shnoudy, A. Comparative study on length weight relationship and condition factor of the genus *Oreochromis* in polluted and non-polluted parts of Lake Mariut Egypt. Bull. Natl. Inst. Oceanogr. Fish. **1994**, *20* (1), 201–282.
10. Khan F.A.; Ansari, A.A. Eutrophication: An ecological vision. The Botanical Review. **2005**, *71* (4), 449–482. doi: 10.1663/00068101(2005)071[0449:EAEV]2.0.CO;2
11. Wahby, S.; Kinawy, S.; El-Tabbach, T.; Abdel Moneim, M. Chemical characteristics of Lake Maryût, a polluted lake south of Alexandria, Egypt. Estuarine Coastal Mar. Sci. **1978**, *7* (1), 17–28.
12. Amr, H.; El-Tawila, M.; Ramadam, M. Assessment of pollution in fish and water of main basin, Lake Maryut. J. Egypt. Public-Health Assoc. **2005**, *80* (1–2).
13. Kafafi, A. *Stakeholder analysis report,* 2007. Final report for the ALAMIM project available for download at http://www.medcities.org/ documents section (accessed April 29, 2011).
14. World Climate. Port Alexandria, Egypt Weather History and Climate Data, http://www.worldclimate.com/cgi-bin/data.pl?ref=N31E029+2100+62315W (accessed April 2009).
15. National Institute of Oceanography and Fisheries (NIOF). *Lake Maryut Data Acquisition.* Final report for the ALAMIM project available for download at http://www.medcities.org/ documents section (accessed April 29, 2011).
16. Egyptian Environmental Assessment Authority (EEAA). *Monitoring Data Lake Maryut*, 2004–2008 reports.
17. CEDARE. *Alexandria Lake Mariout Integrated Management—Stocktaking Analysis report*, 2007. ALAMIM report available for download at http://www.medcities.org/ documents section (accessed April 29, 2011).
18. El-Bestawy, E.; Hussein, H. Baghdadi, H.; El-Saka, M. Comparison between biological and chemical treatment of wastewater containing nitrogen and phosphorous. J. Industrial Microbiol. Biotechnol. **2005**, *32*, 195–203.
19. Metcalf, A.; Eddy, G.; Tchobanoglous, F.; Burton, F.; Stensel, H. *Wastewater Engineering, Treatment and Reuse*, 4th Ed.; McGraw Hill Education: New York, 2003; 1329, ISBN: 0070418780.
20. Trabea, A.; Salem, I. Empirical relationship for ultraviolet solar radiation over Egypt. Egypt J. Sol. Energy **2001**, *24* (1), 123–132.
21. El Rayis, O. Impact of man's activities on a closed fishing lake, Lake Maryout in Egypt, as a case study. Mitigation Adapt. Strategies Global Change **2005**, *10,* 145–157.
22. El-Refaie, M.; Rague, X. *ALAMIM Integrated Action Plan*, 2009. ALAMIM report available for download at http://www.medcities.org/ documents section (accessed April 29, 2011).
23. Janse, J. Model studies on the eutrophication of shallow lakes and ditches, Ph.D. Thesis, Wageningen University: Wageningen, the Netherlands, 2005.
24. Warren, I.; Back, H. MIKE21: A modelling system for estuarine, coastal waters and seas. Environ. Software **1992**, *7* (4), 229–240.
25. Wang, C.; Zhu, P.; Wang, P.; Zhang, W. Effects of aquatic vegetation on flow in the Nansi Lake and its flow velocity modeling. J. Hydrodyn. **2006**, *18* (6), 640–648.
26. Duran, R.; Beevers, L. *Alexandria Lake Mariout Integrated Management: Hydrodynamic Model Report*, 2009. ALAMIM report available for download at http://www.medcities.org/ documents section.
27. Knight, R.; Rubles, R.; Kadlec, R.; Reed, S. Wetlands for wastewater treatment performance data base. In *Constructed Wetlands for Water Quality Improvement*; Moshiri, G., Ed.; CRC Press; Boca Raton, FL, 1993; 35–49.
28. Vymazal, J. Removal of nutrients in various types of constructed wetlands. Sci. Total Environ. **2007**, *380*, 48–65.
29. Moortel van de, A.; Rousseau, D.; Tack, F.; De Pauw, N. A comparative study of surface and subsurface flow constructed wetlands for treatment of combined sewer overflows: A greenhouse experiment. Ecol. Eng. **2009**, *35*, 175–183.
30. Verhoeven, J.; Meuleman, A. Wetlands for wastewater treatment: Opportunities and limitations. Ecol. Eng. **1999**, *12*, 5–12.
31. Lu, S.; Wu, F.; Lu, Y.; Xiang, C.; Zhang, P.; Jin, C. Phosphorous removal from agricultural run off by constructed wetland. Ecol. Eng. **2009**, *35*, 402–409.
32. Anwar, A.; Samaan, A. Productivity of Lake Mariut, Egypt. Part I. Physical and chemical aspects. Int. Rev. Hydrobiol. **1969**, *54* (3), 313–355.
33. Quiros, R. The relationship between nitrate and ammonia concentrations in the pelagic zone of lakes. Limnetica **2003**, *22* (1–2), 37–50.
34. Heinonen, P.; Ziglio, G.; Beken van der, A. *Hydrological and Limnological Aspects of Lake Monitoring*; Wiley: Chichester, England, 2000, ISBN 0-471-89988-7.
35. Fathi, A.; Abdelzaher, H.; Flower, R.; Ramdani, M.; Kraiem, M. Phytoplankton communities of North African wetland lakes: The Cassarina Project. Aquat. Ecol. **2001**, *35*, 303–318.
36. Alvarez-Mieles, G.; Beevers, L. *Alexandria Lake Mariout Integrated Management: Ecological Modelling for Lake Maryut*, 2009. ALAMIM report available for download at http://www.medcities.org/ documents section (accessed May 29, 2011).

Allelochemics

John Borden
Department of Biological Sciences, Simon Fraser University, Burnaby, British Columbia, Canada

Abstract
Allelochemics comprise a vast number of known compounds, and hundreds more yet to be discovered, that mediate behavioral or physiological interactions between organisms of different species. They may benefit the emitter, the receiver, or both. Most are still mainly of scientific interest, but a few (principally attractants or repellents for insect pests) have been incorporated into operational pest management protocols.

TERMINOLOGY

To clarify in part the emerging maze of newly discovered message-bearing chemicals, or *semiochemicals* (Gk. *semeion*, sign or signal), the term *allelochemic* (Gk. *allelon*, of each other) was coined in 1970 to embrace any semiochemical with interspecific activity.[1] Thus allelochemics are distinguished from *pheromones* (Gk. *pherum*, to carry; *horman*, to excite) that convey a message between organisms of the same species. Three categories of allelochemics are commonly recognized.

Kairomones (Gk. *kairos*, opportunistic) are allelochemics that provide an adaptive advantage to the perceiver. In most cases there is no benefit, or even harm, to the emitter, for example, the attraction of predators to the odor of their prey. The evolution of such chemicals as true biological signals would be disadaptive, and therefore unlikely. Therefore some semantic purists remove all evolutionary implications pertaining to kairomones by referring to them as infochemicals.

Allomones (Gk. *allos*, other) are allelochemics that convey an adaptive advantage to the emitter. The repellent odor of an alarmed skunk is often used as an example. However, in an evolutionary sense it may also be adaptive for the receiver to be able to detect and avoid the skunk's odor, for example, for a predator not to be "tagged" with an aroma that warns potential prey of its presence. Therefore, a skunk's odor is more aptly termed a *synomone* (Gk. *syn*, with), an allelochemic that conveys a mutual advantage to both the emitter and the receiver.

Most allelochemics have a *releaser* effect, in which behavioral responses are evoked. However, they may also have a *primer* effect, in which a physiological or biochemical function is stimulated or inhibited.

NATURAL OCCURRENCE

Table 1 provides a small window on the thousands of allelochemic interactions that occur in nature. The compounds that mediate these interactions are equally diverse (Fig. 1). Very few interactions are mediated by a single compound; most involve relatively simple blends; and some, for example, floral fragrances, comprise dozens of compounds in a single blend. Although Table 1 provides examples of allelochemic interactions among terrestrial plants, arthropods, and vertebrates, many are also found among aquatic organisms, and examples occur in all Kingdoms and Phyla.

Kairomonal interactions include the attraction of many species of phytophagous insects to their host plants, entomophagous insects to their prey or to insects that they parasitize, and blood-feeding diptera to their vertebrate hosts.[1–3] They also include the avoidance by prey species of odors associated with predators, a phenomenon found in five animal phyla, but curiously not yet among terrestrial insects.[4] Sometimes more than one type of semiochemical may be involved; for example, the aggregation of bark beetles necessary to mass attack and kill a tree is mediated by a blend of aggregation pheromones synergized by host tree kairomones.

Allomonal interactions may employ "trickery",[5] for example, bolas spiders that emit moth sex pheromones that lure mate-seeking male moths to their death, and many species of myrmecophiles (ant lovers) that gain access to ant nests by chemically mimicking the cuticular recognition compounds of ants on which they prey. The most well-known allomones are released by or are contained in plants, and have a primer effect. Allelopathic allomones are often leached from the leaves (or other parts) of plants of one species, and inhibit the germination of seeds or growth of plants in other species that could be potential competitors.[6] Some plants may also produce defensive allomones against insect herbivores, for example, hormones or analogues of hormones that disrupt the growth and metamorphosis of their insect enemies.

Among the many examples of synomones are repellents that ensure reproductive isolation between closely related species of insects, or mitigate against the occurrence of interspecific exploitative competition for a limited host

Encyclopedia of Environmental Management DOI: 10.1081/E-EEM-100000865

Table 1 Examples of natural occurrence and function of allelochemics.[a]

Type of allelochemic and source	Example
Kairomone:	
Plants	Attraction of Colorado potato beetles, *Leptinotarsa decemlineata*, to 6-carbon leaf volatiles, e.g., (*E*)-2-hexen-1-ol-(**2**), from solanaceous plants
	Attraction of ambrosia beetles, *Trypodendron lineatum*, *Gnathotrichus sulcatus* and *G. retusus*, to ethanol (**1**) from moribund coniferous trees, logs, and stumps
	Attraction and stimulation of oviposition by onion maggots, *Delia antigua*, in response to mono- and disulfides, e.g., dipropyl disulfide (**3**) from onions, *Allium cepa*
	Employment of host tree kairomones, e.g., α-pinene (**8**) from conifers and α-cubene (**10**) from elms as synergists of aggregation pheromones that mediate mass attack of trees by bark beetles e.g., the Douglas-fir beetle, *Dendroctonus pseudotsugae*, and the smaller European elm bark beetle, *Scolytus multistriatus*, respectively
Insects	Stimulation in a parasitic chalcidoid wasp, *Trichogramma evanescens*, of searching for and oviposition in corn earworm eggs, *Helicoverpa zea*, by tricosane (**15**) in moth scales adhering to newly laid eggs
	Attraction of predaceous clerid beetles, *Thanasimus* and *Enoclerus* spp. to aggregation pheromones, e.g., ipsenol (**19**), ipsdienol (**20**), and frontalin (**22**) of their bark beetle hosts
Vertebrates	Attraction of blood-feeding mosquitoes to CO_2 (**23**) exhaled by mammals
	Attraction of tsetse flies, *Glossina* spp., to volatiles, e.g., acetone (**24**) and 1-octen-3-ol (**25**) from bovine animals on which they feed
	Avoidance by voles, *Microtus* spp., of volatile chemicals in the urine of mustellid predators, e.g., 2-propylthiotane (**26**) and 3-propyl-1,2-dithiolane (**27**) from short-tailed weasels, *Mustella erminea*
Allomone:	
Plants	Inhibition of germination of growth of one species of plant by allelopathic chemicals produced in the leaves, roots, or other tissues of another plant, e.g., by juglone (**7**) leaching from the leaves of black walnut trees, *Juglans nigra*
	Disruption of growth, metamorphosis, or reproduction of insect herbivores by producing insect hormones or hormone analogues, e.g., juvabione (**11**), a juvenile hormone mimic in true firs, *Abies* spp.
Spiders	Emission of moth sex pheromones, e.g., (*Z*)-9-tetradecenyl acetate (**18**) by bolas spiders, *Mastophora* spp., to attract male moths as prey
Insects	Mimicking the cuticular recognition compounds, e.g., 11-methyl penta-cosane (**16**) of larval ants by caterpillars of lycenid butterflies, syrphid fly larvae, and scarab beetle larvae, thereby gaining entry into and acceptance within ant nests, where they prey on ant brood
Synomone:	
Plants	Avoidance of nonhost plants by insects in response to volatiles emitted by the plants, e.g., repellency of coniferophagous bark beetles, e.g., the mountain pine beetle, *Dendroctonus ponderosae*, to conophthorin (**13**) (also a repellent pheromone of cone and twig beetles) in the bark of birches, *Betula* spp.
	Repellency of black bean aphids, *Aphis fabae*, to methyl salicylate (**4**) in the volatiles of nonhost plants
	Tritophic interaction in which corn plants, *Zea mays*, respond to volicitin (**14**) (a kairomone) in the saliva of beet armyworm caterpillars, *Spodoptera exigua*, feeding on them by producing specific blends of volatiles (synomones), e.g., (*E*)-4, 8-dimethyl-1,3,7-nonatriene (**9**) that attract females of a parasitic wasp, *Cotesia marginiventris*, which in turn oviposit on the feeding caterpillars
	Attraction of honey bees, *Apis mellifera*, to multicomponent blends of floral volatiles, e.g., geraniol (12), of many species of angiosperm plants, with mutual benefit of pollination to the plant and a pollen and nectar source to the bees
Insects	Antagonists in the blends of moth sex pheromones that repel males of related species, ensuring reproductive isolation even though the major pheromone components are attractive to males of both species, e.g., (*Z*)-9-tetradecanal (**17**), a pheromone component of threelined leafrollers, *Pandemis limitata*, that inhibits response of obliquebanded leafroller males, *Choristoneura rosaceana*, to threelined leafroller females
	Mutual repellency between aggregation pheromones of two species of bark beetles, e.g., *(R)-(* –)-ipsdienol (**20**) produced by pine engravers, *Ips pini,* and (S)-(–)-ipsenol (**19**) produced by California fivespined ips, *I. paraconfusus*, reserving the host phloem resource for the first-arriving species, and avoiding interspecific exploitative competition

[a]Numbers in parentheses correspond with numbered compounds in Fig. 1.

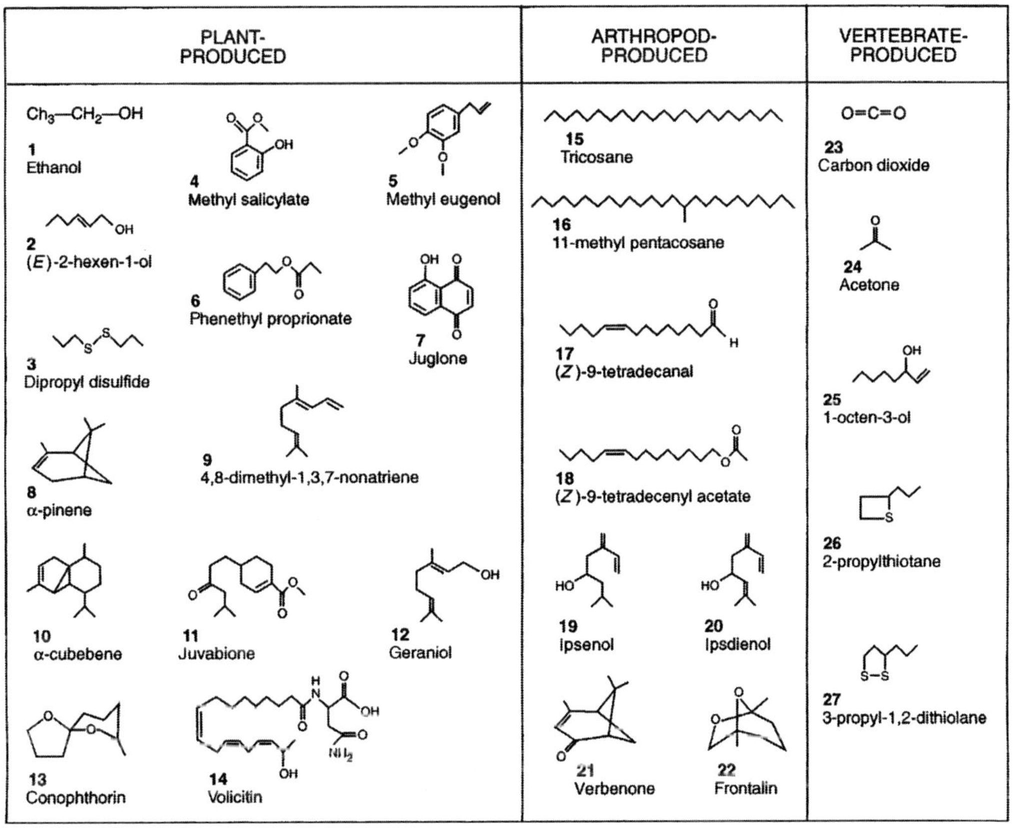
PLANT-PRODUCED
ARTHROPOD-PRODUCED
VERTEBRATE-PRODUCED
Ch3—CH2—OH
1
Ethanol
4
Methyl salicylate
5
Methyl eugenol
2
(E)-2-hexen-1-ol
6
Phenethyl proprionate
7
Juglone
3
Dipropyl disulfide
8
α-pinene
9
4,8-dimethyl-1,3,7-nonatriene
10
α-cubebene
11
Juvabione
12
Geraniol
13
Conophthorin
14
Volicitin
15
Tricosane
16
11-methyl pentacosane
17
(Z)-9-tetradecanal
18
(Z)-9-tetradecenyl acetate
19
Ipsenol
20
Ipsdienol
21
Verbenone
22
Frontalin
O=C=O
23
Carbon dioxide
24
Acetone
25
1-octen-3-ol
26
2-propylthiotane
27
3-propyl-1,2-dithiolane

Fig. 1 Structural formulae of compounds given in Table 1 exemplifying some of the chemical diversity among allelochemics as follows: primary alcohol (**1**, **2**), secondary alcohol (**25**), disulfide (**3**), aromatic ester (**6**), unsaturated ester (**18**), monoterpene (**8**), sesquiterpene (**9**, **10**), sesquiterpenoid (**11**), terpene alcohol (**12**, **19**, **20**), terpene ketone (**21**), spiroacetal (**13**), fatty acid derivative conjugated to an amino acid (**14**), straight chain hydrocarbon (**15**), branched hydrocarbon (**16**), unsaturated aldehyde (**17**), bicyclic ketal (**22**), atmospheric gas (**23**), ketone (**24**), thiotane (**26**), and thiolane (**27**).

resource. Often synomones are the same compounds as one or more of the components that convey a pheromonal message, for example, to attract mates or to aggregate on or near a food source. There is increasing evidence that host-seeking phytophagous insects not only use kairomones to find their host plants, but also use synomones to avoid nonhost plants on which their fitness would be greatly reduced. Synomonal floral scents provide a mutual benefit to flowering plants that gain from pollination by insects that in turn are attracted to a nutritious nectar or pollen source.

Another type of allelochemic interaction involves three trophic levels and the action of both kairomonal and synomonal stimuli.[3] In one remarkable example of this type of tritrophic interaction, corn plants being fed on by beet armyworm caterpillars are exposed to minute amounts of a kairomone called volicitin in the insect's saliva. Volicitin has a primer effect, eliciting the plant to synthesize a specific blend of volatile synomonal compounds that attract females of a parasitic wasp. The wasp oviposits in the beet armyworm larvae, benefiting by finding its host, and in turn providing an advantage to the plant by parasitizing and killing the caterpillar.

PRACTICAL APPLICATIONS

Knowledge about the natural occurrence and role of allelochemics opens up a huge, but relatively untapped, potential for exploiting them as pest management tools.[2] In some cases the knowledge itself is important. Plant breeders may seek varieties of plants that contain or release chemicals that deter feeding or development by phytophagous insects. Plants with allelopathic characteristics, for example, *Eucalyptus* spp., may be useful in landscaping to reduce weed problems. Species or varieties rich in attractive kairomones may be used as trap crops for various insect pests. In the production of transgenic agricultural crops it is critical not to loose the capacity for tritrophic interaction that will ensure parasitism of herbivorous insects, lest the genetically modified plants be more vulnerable to insect pests than unmodified plants.

In other cases, the capacity to use allelochemics as pest management tools may be demonstrated, but technological, economic, or social limitations may prevent their use. Allelopathic allomones from plants are under consideration for development as a new class of biodegradable herbicides.[7] But to date none can compete with conventional chemical herbicides with regard to ease of synthesis, capacity for formulation, efficacy, and/or safety. If used widely, some allelopathogens may pose an unacceptable threat to environmental or human health. Recent investigations show considerable promise for using nonhost volatiles to "disguise" herbivorous host plants or trees as nonhosts, but commercial formulations have not yet appeared on the market, in part because of the challenge and expense of registering those allelochemic products as pesticides. Similar problems beset the development and use of predator volatiles to protect plants from damage by herbivorous vertebrates such as deer and voles.

Despite the above limitations, a few kairomones have found widespread commercial use.[2,8] Among them is methyl engenol, which is used worldwide for capturing tephritid fruit flies, both for detection of unwanted introductions and for direct suppression of populations in a lure and kill tactic employing an insecticide-laced substrate baited with methyl engenol. A lure and kill tactic is also used effectively for control of tsetse flies that are drawn by acetone and 1-octen-3-ol baits to insecticide-treated "target" traps that simulate the silhouette of a large vertebrate. Other applications combine kairomones with pheromones. A combination of phenethyl proprionate and methyl engenol with the sex pheromone of the Japanese beetle is used in many thousands of traps in the United States each year. Similarly the kairomones ethanol and α-pinene have been used since 1981 in combination with aggregation pheromones in commercial mass trapping programs for three species of ambrosia beetles in British Columbia.

Allelochemics may also find use in the future in the application of "push–pull" tactics, in which one repellent volatile treatment is used to protect a plant or group of plants from attack by insects (push), and another attractive treatment is used in baited traps or trap plants to pull the insects away. One outstanding example of a successful push–pull application saved a rare stand of endangered Torrey pines in California from being killed by the California five spined ips. Two repellent synomones, verbenone, produced by western pine beetles, and (−)-ipsdienol, produced by pine engravers, were deployed inside the uninfested portion of the stand, and traps baited with attractive aggregation pheromone were arrayed in an adjacent area of beetle-killed pines. Over 86 weeks beginning in May 1999, 330,717 beetles were captured.

CONCLUSION

Despite many studies, most natural allelochemic interactions are yet to be discovered. The adoption of allelochemics as pest management tools has been limited. However, there is great potential for judicious selection and commercial development of allelochemics, particularly in integrated pest management programs that will combine a number of alternative ecologically based tactics with the reduced use of conventional chemical pesticides.

REFERENCES

1. Whittaker, R.; Feeney, P. Allelochemics: chemical interactions between species. Science **1971**, *171*, 757–770.
2. Metcalf, R.; Metcalf, E. *Plant Kairomones in Insect Ecology and Control*; Chapman and Hall: New York, 1992; 168.
3. Vet, L.; Dicke, M. Ecology of infochemical use by natural enemies in a tritrophic context. Annu. Rev. Entomol. **1992**, *37*, 141–172.
4. Kats, L.; Dill, L. The scent of death: chemosensory assessment of predation risk by prey animals. Ecoscience **1998**, *5*, 361–394.
5. Stowe, M.; Turlings, T.; Loughrin, J.; Lewis, W.; Tumlinson, J. The Chemistry of Evesdropping, Alarm, and Deceit. In *Chemical Ecology: The Chemistry of Biotic Interaction*; Eisner, T., Meinwald, J., Eds.; National Academy Press: Washington, DC, 1995; 51–65.
6. Zindahl, R. *Fundamentals of Weed Science*; Academic Press: New York, 1993; 135–146.
7. Cutler, G. Allelopathy for Weed Suppression. In *Pest Management: Biologically Based Technologies*; Lumsden, R., Vaughn, J., Eds.; American Chemical Society: Washington, DC, 1993; 290–302.
8. Borden, J. Disruption of Semiochemical-Mediated Aggregation in Bark Beetles. In *Insect Pheromone Research: New Directions*; Cardé, R., Minks, A., Eds.; Chapman and Hall: New York, 1997; 421–438.

Alternative Energy

Bernd Markert
Simone Wuenschmann
Environmental Institute of Scientific Networks (EISN), Haren-Erika, Germany

Stefan Fraenzle
Department of Biological and Environmental Sciences; Research Group of Environmental Chemistry, International Graduate School, Zittau, Zittau, Germany

Bernd Delakowitz
Faculty of Mathematics and Natural Sciences, University of Applied Sciences, Zittau, Germany

Abstract

From the industrial revolution onwards, certain kinds of energy sources are being used while new ones and novel methods of energy conversion are developed [electric motors, generators, and AC (alternating current) grids in the 19th century; fuel cell, nuclear fission energy, and semiconductor devices for converting various kinds of energy into electricity in the 20th century]. Now, fossil fuels run scarce—as does uranium-235—and meet increasing criticism due to the climatic (greenhouse) effects of both methane and the principal combustion product CO_2. Hence, regenerative resources like wind and solar energies, running and falling water (hydropower), biomass processed in some way, and geothermic energy are in a position to replace crude oil, natural gas, coal, and uranium step by step, depending on both technological innovations and political decisions. Thus, we will contrast the problems associated with conventional sources of energy to the challenges and chances linked to renewable ones.

INTRODUCTION

Regenerative resources like wind and solar energies, running and falling water (hydropower), biomass processed in some way and geothermic energy are in a position to replace crude oil, natural gas, coal, and uranium step by step depending on both technological innovations and political decisions. Here, technological innovations are not meant to require fundamental inventions to render some renewable resource useful for energy (electric current) "harvesting." Rather, this is about making existing technologies cheaper and overcoming specific material problems, like device corrosion and fast "blocking" of underground heat exchange pathways with geothermic energy, storage in an easy-to-handle form (methanol?) for both solar energy and biomass, or using less toxic and less brittle semiconductors in much thinner layers in photovoltaics [thin-film cells based on either copper indium dichalcogenides (chalkopyrites) or organic semiconductors].

Given the relativeness of time and the notorious "difficulty to make predictions which refer to the future," what does it say about classical fossil resources running out in the foreseeable future? Putting this into proper context means to distinguish between "reserve" and "resource": "reserve" just encompasses those deposits of energy carriers that are actually known and can be really accessed following both technical and economic criteria. In the pre-1973 world of %2.70 per barrel of crude oil (some $19 per ton), it would have been considered a fancy to try to extract oil from shales or sand and drill deep below the seafloor (now always performed in the Atlantic Ocean and the Mexican Gulf, off Brazil and Angola) or in remote arctic regions. Currently, the meaning of underground hard coal production or making access to very deep natural gas deposits is doubtful. While oil production from oil sands (Alberta Province, Canada) is now economically viable—but still an ecological disaster in a sensitive surrounding—other methods of accessing certain fossil resources will probably never be viable: it simply takes more energy to extract and process traces of ^{235}U dissolved in seawater than can be obtained from its fission afterwards. As the term "reserves" by definition ("share of total potential which can be mined and exploited economically reasonable by currently available technical means") includes both the present level of technology and current pricing, reserves are subject to changes other than due to ongoing prospection and exploitation. For example, when disregarding the environmental and safety issues associated with either kind of fossil energy carrier for this moment, both oil sands and shale gas are close to the lower limit of economic feasibility given the current crude and natural gas prices. However, available amounts are not settled with shale gas, not permitting to include it into the reserves, while oil sands—although a blueprint

Encyclopedia of Environmental Management DOI: 10.1081/E-EEM-120048462

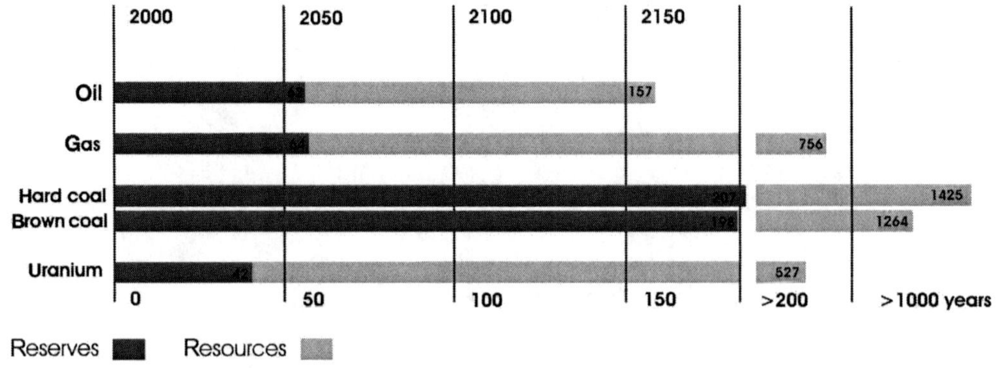
2000
2050
2100
2150
Oil
157
Gas
756
Hard coal
1425
Brown coal
1264
Uranium
42
527
0
50
100
150
>200
>1000 years
Reserves
Resources

Fig. 1 Predicted reserves/resources of fossil energy carriers from year 2000 onwards
Source: German Federal Institute for Geosciences and Natural Resources, 2002.

for ecological disaster—are economically and technically feasible to produce and hence are part of (Canada's and global) crude reserves. Matters are different once again if the technology to actually obtain energy from some fossil or far-spread reservoir around is not yet at hand or it is doubtful whether energy required to mine and concentrate that particular source might even exceed the energetic payback obtained thereafter: consider amounts/concentrations of deuterium, ^{6}Li, and ^{235}U in seawater. The former two (^{6}Li being a precursor to fusion "fuel" tritium) would combine to an inexhaustible source of energy if "only" net-energy-yield nuclear fusion would be at hand already, while there are just some 15 ng (!) of ^{235}U that can be actually extracted from 1 L of ocean water, providing a few hundred joules ($\approx 10^{-4}$ kWh) of electric energy in nuclear (now, fission) power plants thereafter. Even though the corresponding extraction was already demonstrated in Japan, using ion-exchange resins, the energy for producing the resins and running the device is so large that ocean-derived uranium is not a viable resource either and thus cannot be counted among the reserves.

"Resources," on the other hand, are those energy carriers either already discovered (not a single "elephant field" of crude oil was spotted after the 1960s anywhere in the world!) or *reasonably believed* to exist from geological arguments in Earth's crust, but cannot be exploited right now for either technological or economic reasons. Due to the difficulties and risks associated with drilling either below some 9 km underground or into active magma regions, most of the huge amounts of geothermic energy can never be actually used. Hence, for fossil resources as well, we are left with what is in the crust or the ocean (floor). Given this distinction, the residual economic lifetimes for traditional forms of energy carriers such as oil, natural gas, brown and hard coal, or uranium given in Fig. 1 are obtained.

Energy Depletion of Fossil Fuels

Counting from year 2000 onwards, the German Federal Institute for Geosciences and Natural Resources estimated in 2002 that reserves of oil, natural, gas and uranium will last for just another 40 to 65 years. Reserves of both hard and brown coal will still last up to 200 years, whereas the resources of coal and gas and uranium are to last more than 200 years. Things are more critical with crude oil, reserves of which will be gone within 60 years while resources are estimated to last for 160 years at best. Data and predictions in Fig. 1 do not cover and include the energy consumption of current growth regions [BRICS states like Brazil, Russia, India, China (PR), and other Latin American countries], leaving us with the conclusion that we are left with much less time to change our bases of energy supply altogether.[1–4] The resources of crude oil are much disputed. Apparently, most of the additional stockpiles—beyond established reserves still considered a few years ago—simply do not exist. Speculation on oil term in markets is considerably influenced by this insecurity of affairs while the big oil companies obviously produce an exaggerated picture of resources rather than make speculation go its way.

Another matter is the regional distribution of these reserves/resources all around the globe (Fig. 2; Federal Institute for Geosciences and Raw Material Research.[5]): the problem is obvious with crude oil, and everybody is aware of it, but there are also biases/imbalances with hard coal. Considering the total energy stored in it, some 60% of it rests with brown or hard coal; the smaller part is included in liquid and gaseous hydrocarbons. Note that efficiencies of power plants differ considerably among these energy carriers, with hard coal and gas steam plants being superior to others ($\eta_{el} > 55\%$). Crude oil, which may be produced by current technologies, reasonably amounts to 6682 EJ, a little less than with natural gas (7136 EJ). 1000 EJ (exajoule) = 10^{21} J. Standard heats of formation of the compounds/mixture/combustion products involved are as follows:

CH_4	–50 kJ/mol
C (graphite)	zero [by definition (standard state of an element)]
"CH_2" (fraction of crude oil)	≈ –20 kJ/mol
CO_2	–394 kJ/mol
H_2O	–237 kJ/mol

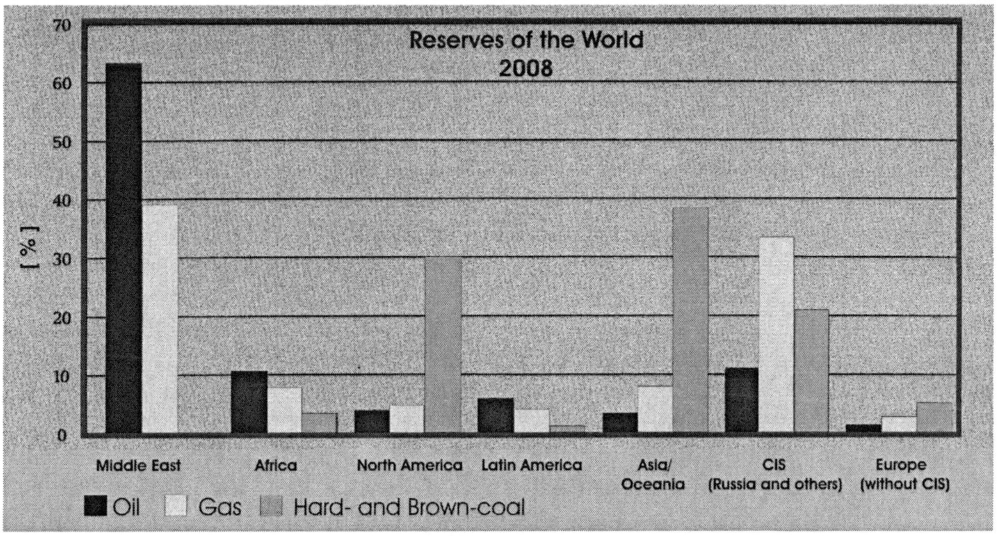
Reserves of the World
2008
70
60
50
40
30
20
10
0
[%]
Middle East
Africa
North America
Latin America
Asia/
Oceania
CIS
(Russia and others)
Europe
(without CIS)
Oil
Gas
Hard- and Brown-coal

Fig. 2 Estimate of the regional/continental distribution of oil/gas/coal reserves.[5] As for the Commonwealth of Independent States (former USSR except of Baltic states and Georgia), the biggest share is with just three of them: Russia (the current biggest oil producer in the world), Azerbaijan (both oil and gas), and Turkmenistan (almost gas only).

Thus, 1 mol (44 g) of CO_2 produced from combustion of natural gas (CH_4, essentially), crude oil, and coal yields 818 kJ, 611 kJ, and 394 kJ, respectively. The annual global anthropogenic CO_2 output is of the order of gigatons (1015 g, Pg), with the atmosphere containing some 770 Pg of it now, considerably more than which is tied up in living biota (about 610 Pg). One gigaton of CO_2 from hard coal, crude oil, and natural gas natural translates into somewhat less than 9, 13.9, and 18.6 EJ of thermal energy, respectively. Thus, actually combusting the above-estimated resources would leave us with some 50,000 gigatons, that is, 65 times (!) the present CO_2 inventory of the atmosphere.

The total energy from conventional fossil energy carriers (resources combined) would be some 448,289 EJ, 97% of these resources being hard or brown coal. Besides the above-mentioned conventional energy carriers, there are substantial amounts of other energy carriers such as oil sands, oil shales, tars, and other kinds of the heaviest, most condensed oils, natural gas from dense storage sites, carbon deposits (adsorbed), or aquifer waters, plus gas (CH_4) hydrates on deep shelf (below some 350 m of seawater or in the uppermost sediments down there) reserves and resources that correspond to an energy equivalent of 2368 EJ and 116,270 EJ, respectively.[6]

Most industrial activities now depend on oil, as do almost all traffic systems, be it airplanes, ships, or cars. Oil getting scarce thus does not only cause prices to rise; there will also be ramifications on workforce in petrochemical branches and political implications. The largest share of oil (2005), some 742 billion barrels, is located in Central and Southern Asia (Bangladesh, Bhutan, India, Maldives, Nepal, Pakistan, Afghanistan, Sri Lanka, and parts of Iran). Considering the global oil reserves according to British Petrol in 2005, this translates into 62% of global stockpiles, whereas North America commands just 5%, being one of the metaregions scarcest in oil besides Asia/Pacific and Europe.

With oil getting scarce, a one-sided economic dependence on the Middle East poses increasing political risks, causing everybody to consider oil and gas resources located elsewhere and how they and additional geological goods might be obtained. Thus, there is a recent growing geopolitical interest in the Arctic region, which will become void of drifting ice during the next 20 or 30 years due to climate change. According to the U.S. Geological Survey Institute,[7] some 30% of natural gas and 13% of crude oil are located there. However, most of these are located far offshore [the remainder already exploited for decades in Russia (Taimyr Peninsula) and Alaska], with most of the gas belonging to the Russian Federation while oil is scattered among Canada, Alaska, and Greenland (still partly governed by Denmark). Though very large, experts say[7] that these deposits are not enough to consider relocation of most production activities from the Middle East. In addition, long transport distances (thus, costs) and still adverse weather and climate conditions pose grave problems. Everybody is still aware of what can happen with deep sea-based oil production considering the Deepwater Horizon catastrophe of April 2010, environmental concerns translating into larger political obstacles, higher insurance fees, and eventually less consumer acceptance.

That the so-called peak-oil level where global oil production has reached its maximum ever was already achieved in the beginning of the 21st century is evident from a depiction of development of oil production from 1930 until (pre-

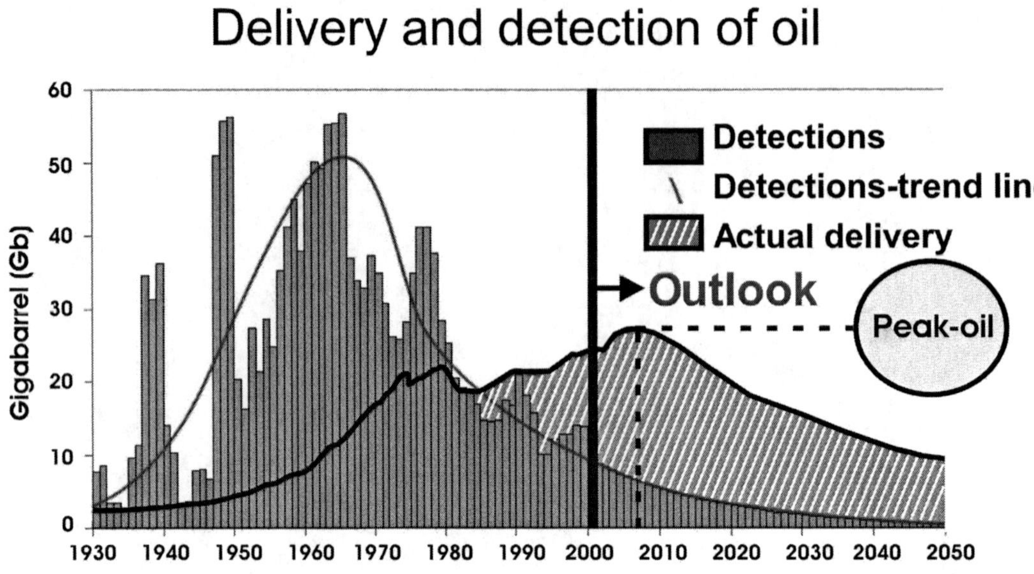
Delivery and detection of oil
Gigabarrel (Gb)
60
50
40
30
20
10
0
1930 1940 1950 1960 1970 1980 1990 2000 2010 2020 2030 2040 2050
Detections
Detections-trend line
Actual delivery
Outlook
Peak-oil

Fig. 3 Oil findings and delivery rates from 1930 to 2050 with the outlook after Campbell (2006).[8] The delivery rate per year is calculated by the Association for Peak-Oil and Gas Studies, Germany. Data from ExxonMobil (2002). Peak oil: point of time when the maximum rate of global petroleum extraction is reached.
Source: Figure modified from Blum (2005).[9]

dicted) 2050 (Fig. 3): scarce oil means there will be no more cheap oil. Until the beginning of the 21st century, one barrel of oil commanded between \$20 and a maximum of \$40. The financial crisis made it (Brent) rise up to \$147 after 2007. Declining somewhat, it now (2011) stabilizes around \$100. In the long term, it is more likely to increase again, having severe drawbacks on economic conditions in industrialized countries especially.

Fig. 1 tells us that hard and brown coal reserves are to last another about 200 years, rendering coal-fired power plants most attractive if it were not for coal being one of the most polluting sources of energy. Combustion of coal produces plenty of CO_2, and therefore, it contributes to the anthropogenic part of greenhouse effect warming of our atmosphere.

Climate Protection

CO_2 is one of the most prominent greenhouse gases in the atmosphere, contributing to heating the atmosphere and thus the Earth's surface.[10,11] There is both a natural and an anthropogenic (share of) greenhouse effect. CO_2, produced by animal respiration, wildfires, and volcanoes, is a natural (<300 ppm) component of the atmosphere, heating our planet from a radiation equilibrium ("blackbody") value of some −18°C to a global average of +15°C, that is, by a considerable 33°C.

A global increase of average atmospheric temperatures is now seen for decades, with Fig. 4 displaying the distribution of this effect for 2000–2009 as compared to the reference time frame of 1951–1980. Satellite measurements by the NASA Earth Observatory showed the largest increases of the average temperature in the Arctic parts of the Northern Hemisphere, besides some parts of Antarctica. Between the beginning and the end of the 20th century, the average increase was 0.74 ± 0.18°C.

Time is imminent for rethinking energy production with a focus on sustainability in all ecological, economic terms and social acceptance. Man-made radiative greenhouse forcing is established for more than 20 years now. It causes enhanced glacier meltdown rates in the Arctic (Greenland actually turns green again), the Alps, and other mountainous regions, as well as sea level rising (both due to meltdown waters and to thermal expansion of warmed surface water), with more frequent droughts causing limnetic waters to evaporate often completely. We also would be urged and obliged to seriously think about what we are doing if, as some scientists still maintain to argue, this would be a normal climate excursion as occurred in the later Middle Ages. It does not take much proficiency in the natural sciences to acknowledge the problems associated with the industry emitting greenhouse gases[12] while striving to maintain and enhance our common well-being. The destruction of the environment concomitant with the exploitation of the developing Third World countries has already been an issue some 40 years ago, prompting the first Conference on the Human Environment at Stockholm (Sweden) on June 16, 1972, initiated and organized by the United Nations. Twenty years after, in 1992, the most comprehensive global conference on such issues ever held, the United Nations Conference on Environment and Development (UNCED), took place in Rio de Janeiro. One hundred years ago, Wilhelm Ostwald mentioned in his approach "Die Mühle des Lebens,"[13] "Do not waste energy, rather use it wisely," This is in opposite to the Kantian

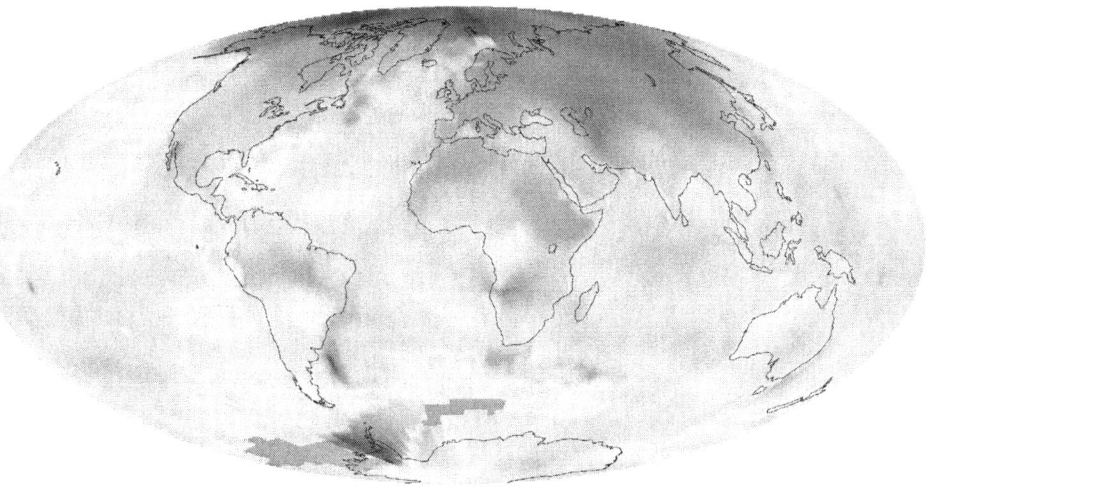

Fig. 4 Regional distribution of relative greenhouse effect warming. The map shows the 10 years average (2000–2009) global mean temperature anomaly relative to the 1951–1980 mean. The largest temperature increases are in the Arctic and the Antarctic Peninsula. The darker the gray shades in the picture of NASA,[14] the larger the increase in local average temperatures.
Source: NASA Earth observatory.

imperative at this time, and the first energetic step to a so-called sustainable chemistry of today.[15]

A total of 178 countries dedicated themselves to Agenda 21 to oblige the 118 more developed countries to embrace environmental restoration, preservation, and social development. Their aims are to meet the challenge of global warming, pollution, and biodiversity and to solve the interrelated social problems of poverty, health, and population. The Agenda 21 program furthermore encompasses constructing a network of "new and equitable global partnership through the creation of new levels of cooperation among States, key sectors of societies and people, while working towards international agreements which respect the interests of all and protect the integrity of the global environmental and developmental system."

In 1997, the World Climate Summit was held at Kyoto, Japan, focusing on the future climate protection policy and measures. The 158 states that had so far signed and ratified the Framework Convention on Climate Protection and 6 "observer states" had sent a total of almost 2300 delegates, plus another 3900 observers from non-governmental and other international organizations and 3700 journalists from global media, making up a total of close to 10,000.[16] The Kyoto process unleashed by this huge meeting now describes the attempts to maintain climate protection agreements beyond the Kyoto protocol running out in 2012 and to do so in a way that still is safeguarded by agreements of international law.

The industrialized states listed in Appendix 1 of the Kyoto Protocol agreed to reduce their collective greenhouse gas emissions by 5.2% from the 1990 levels by the year 2012. The common (purported) aim of the international community is to limit the increase of global average temperature within 2°C. By now, it has become most doubtful whether this "two-degree aim" can be kept at all. In addition, it was proven that even an aggressive forestation policy could not cope by photosynthetic absorption (assimilation) with the present upsurge of CO_2 produced. The progress made so far is modest at best, with real successes still to be waited for. The subsequent UN Climate Conference at Copenhagen (Denmark) in late 2009 saw just a communiqué of minimal consent without any mandatory aims in CO_2 reduction although the "two-degree aim" was once again acknowledged to be worthwhile.

Role of Nuclear Power

The Fukushima (Japan) sequence of catastrophic events began on March 11, 2011 [the strongest earthquake ever recorded in Japan (Richter magnitude 9.1)], causing coolant pipes to break and starting a sequence of reactor core meltdowns in at least three adjacent nuclear power plants (NPPs). The area was then hit by a tsunami; any semblance of control of the preceding events was lost. Finally, the uncooled reactor systems exploded one after the other, releasing large amounts of radioactivity and visibly destroying reactor block 3. People reconsidered the risks of nuclear energy all over the world, although with grossly differing political consequences.

Now introduced into IAEA (International Atomic Energy Agency) accident level VII, the events at Fukushima were set equal to the Chernobyl disaster [April 1986, Ukraine (officially: Ukrainian Soviet Socialist Republic)] and considered worse than the explosion of a nuclear fission waste storage tank at Kyshtym in 1957 [near Chelyabinsk, Urals, then RSFSR (Russian Soviet Federative Socialist Republic), USSR (Union of Soviet Socialist Republic); level VI accident], which delivered very large amounts of ^{90}Sr [several EBq (exabecquerel: 10^{18} Bq)] that an area of some 15,000 km^2 had to be abandoned for any human use up to now. Japan is known to be tectonically most active, prompting the authorities and engineers to consider earthquakes up to Richter magnitude 7.9 in design. Two matters demand further consideration here:

a. What causes, e.g., main coolant pipes to break is acceleration and the mere amplitude of dislocation combined with inertia, that is, parameters rather covered by the "old-fashioned" Mercalli scale of earthquake intensity, with Mercalli 13 denoting accelerations larger than that of the Earth's gravity. Such quakes actually happened, e.g., the Easter quake of 1964 in Central Alaska (there is a very impressive movie showing cars and humans and even some homes losing contact with the ground during the quake). In contrast, Richter scaling gives the energy unleashed by a quake [or a landslide or an underground explosion (be it chemical or nuclear in origin)] in a logarithmic way: Depending on the depth of the epicenter, i.e., the amount of matter located above the vibrating or breaking sample of crustal matter, accelerations due to a given amount of tectonic energy can be quite different: the "fairly moderate" Haiti quake of February 2010 (Richter 7.1) became a real killer not because of its energy but since its epicenter was located just a few kilometers below the surface, producing a massive acceleration, thus aggravating the effects of non-adapted architecture. Inertia also is a matter of dislocation of the ground, which provides a kind of reference frame, physically speaking, to which any device that can vibrate, etc., will respond: this dislocation was about 2.5 m in Japan on March 11, 2011, that is, much larger than in the even stronger (more energetic) Chile 1960 (Richter 9.5) and Sumatra Christmas 2004 (Richter 9.3) earthquakes.
b. At least two NPPs, one of them in Japan, had already experienced earthquakes exceeding these limits of design: at Niigata (NW Japan) and at then Leninakan in Armenian Soviet Socialist Republic (SSR) in 1988. Hence, the design was obviously short of reasonable expectations [Central Europe, German, Swiss, and French NPPs next to the Upper Rhine Rift Valley do not take account of the fact that Basle City was almost completely destroyed by a quake in historic times (in 1356)].

The Fukushima earthquake was far stronger than this. The tragedy was worsened by a tsunami caused by this quake; it took thousands of lives in cities and villages near the coast, but most likely, the crucial damage to the Fukushima NPPs was caused by the earthquake, though the NPP area was inundated soon after. Certain radionuclides, mainly highly volatile ones were released; that core meltdowns had occurred was first contested and then confirmed by Japanese authorities only 2 1/2 mo later. These highly volatile radionuclides include ^{131}I, $^{134;137}$Cs and some noble gases, and traces of ^{132}Te but not Sr, Ba, and rare earth elements. Apparently, the temperatures during core meltdown did not yet suffice to vaporize alkaline earths or rare earth element compounds (cesium becomes highly volatile as a hydroxide, CsOH molecules sublimating as easily as NaOH and KOH do), neither did volatile chlorides form after NaCl from seawater pumped inside to cool the reactor cores solidified. Including those still missing, their whereabouts unaccounted for but unlikely still to be alive, there were some 23,000 fatalities due to the quake and tsunami, and several workers received massive radiation damages during cleanup, with a nuclear catastrophe still looming ahead. The number of people forced to abandon their homes and jobs (permanently) at Fukushima and Iitate provinces [40 km NW Fukushima (the NPP site, not the town of Fukushima)] is more than 80,000 now, that is, close to that of people who became homeless due to the Chernobyl accident. Responses from different countries were different: while politicians and citizens in certain countries started to abandon energetic uses of nuclear fission (Germany and Switzerland) or reaffirmed prior decisions of this kind [New Zealand, referendum in Italy with a 94% (2011) turnout, Belgium], others chose to keep a pro-nuclear course or even kept up decisions to build national first-ever NPPs (Poland).

On a global scale, the issue of nuclear energy is controversial in all developing countries, classical industrialized states, and "official" and "unofficial" nuclear-arms-possessing states. While the United States plans to commence building the first new NPPs after a moratorium of some third of a century,[17] Italy officially revoked the schedule to abandon nuclear energy adopted by a 1994 referendum (but only so in the old Berlusconi regime, the Monti administration being undecided on this issue), and the spring 2012 presidential election campaign in France triggered a broad discussion on the future of NPPs there for the first time. Several developing states extend their present NPP program apparently as a pretext to obtain more fissionable material for nuclear arms, like India and Iran, regardless of whether they signed the Non-Proliferation treaty (Iran), did not sign (India and Israel), or left the protocol some time later (North Korea). In "established" nuclear powers, on the other hand, NPP extension also is a means to get rid of excess amounts of weapons-grade or "special" nuclear fuels (United States, Russia, U.K.). The most interesting case is some 2 tons of ^{233}U left over from the thorium/uranium breeding cycle, which is/was used only in India for energy production and formerly in the United States for making arms. Now, the United States is left with a stockpile of this nuclide, a nuclide that actually would be usable by very-low-technology groups or states to produce "efficient" and reliable-yield warheads (different from that with any plutonium isotope [-mixture]!) and correspondingly must safeguard it to a level that would not be necessary with any other isotope—unless they "burn" it in reactors.

Hence, the ambiguity of nuclear technologies continues to influence decisions on introducing or maintaining it as an energy source, far beyond, and competing with, considerations of supply reliability (the number of states that provide/export significant amounts of uranium is considerably smaller than those that provide all natural gas, crude oil, and hard coal). Political decisions on increased or renewed use of nuclear energy in certain countries hence cannot be considered a signal that is globally significant, except for their drawbacks to the issue of proliferation.

The classical argument in favor of nuclear energy production is their releasing much less CO_2 as compared to oil, coal, or natural gas power plants, enabling economies to meet Kyoto ends more easily. For certain countries that produce nuclear fuels but command few other natural resources, there is the bonus of apparent independence; this includes Japan. In the EU, the Czech Republic presently is the only state to mine domestic uranium ores, globally large producers being differently organized and reliable in political terms [Australia, Canada, the United States, Kazakhstan, Niger, Gabon (Central Africa)].

Nuclear energy is rather "compact," a GW-class power plant taking a few hectares at most, which is important in countries as densely populated as Japan (or India, or Belgium), with Japan covering a total area hardly larger than that of Germany (372,000 km^2). Next to the United States (104) and France (59), Japan has the largest number of NPPs [53, from which 6 at Fukushima and 3 at another highly earthquake-prone site (scheduled for shutdown) must be subtracted; thus, effectively it is only 44] due to an experience from the 1970s: the OPEC oil crisis in 1973, rapidly increasing the prices of crude, of course hit all the already then industrialized countries except for those producing sufficient fossil fuels of their own (Canada, Australia) or relying on other sources mainly for different reasons (Norway, New Zealand, South Africa) but worst so in Japan: in fact, there were large-scale electricity shutdowns badly hurting Japan's economy. Hence, Japan—once again in its history—became eager for independent supply, planning to build yet more NPPs before Fukushima happened. By now, unlike France, Belgium, the United States, or formerly Lithuania, which obtains 60% or more of its electricity from NPPs, Japan's rate is 30%. During the period of Soviet dominance, Lithuania even had the largest NPP in the world, a 2400 MW_{el} plant at Ignalina (now Visaginas) of the notorious (Chernobyl) RbmK (*reaktor bolshoy moshchnosti kanalniy*: high-power channel-type reactor) construction type. Relying on nuclear energy by >70% shortly after regaining independence, Lithuania had to close down this plant as a precondition for joining the EU, as Bulgaria had with Kozloduj. Now, Lithuanians plan to erect a new, similarly huge NPP at the same site next to the border triangle with Latvia and Belarus to supply all of the three Baltic states even though Latvia is now the EU's champion in regenerative energy supply, producing almost 40% of its (admittedly rather limited) demand mainly from hydropower, wind, and some biomass use, and suffers from financial crisis (much like Ireland and Greece but less perceived to do so since they still maintain their own currency, the Lat). Fig. 5 shows the global pattern of distribution of NPPs that is almost complete [note that there are 1) considerable differences in electric outputs of individual plants and 2) some plants were primarily meant to produce radionuclides rather than electricity, although they are connected to the public current grids (e.g., Dimona in Southern Israel)]. The total number is about 450 NPPs.

As shown above (Fig. 1), uranium reserves are to last for just another about 30 years from now (2011), adding

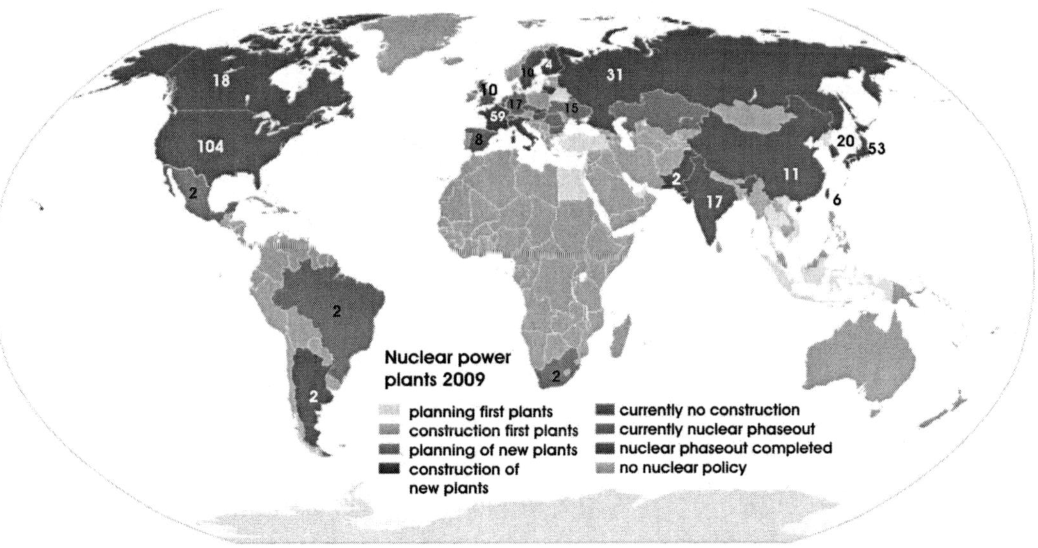
Nuclear power plants 2009
planning first plants
construction first plants
planning of new plants
construction of new plants
currently no construction
currently nuclear phaseout
nuclear phaseout completed
no nuclear policy
18
104
2
10
59
17
8
15
31
20
53
11
2
17
6
2
2
2

Fig. 5 Nuclear power worldwide in 2009 (the figure was transferred from Wikipedia; the number of nuclear power plants has been added by collecting information separately by Wikipedia). The number of nuclear power plants is not complete.

another problem to safety concerns and disposal of radioactive wastes (see below) associated with nuclear (fission) energy. The increase in prices during the last 5 years even relatively surpassed that of oil, the process probably going on as demands for uranium did not yet reduce considerably on a global scale after Fukushima. Fig. 6 shows a map indicating the 10 most prolific uranium-mining countries (by 2008).

Let us turn to the third issue associated with nuclear energy: radioactive wastes, which are produced by all nuclear fission, then located in "used" (irradiated) fuel rods, neutron impact on non-fissionable ^{238}U, and on construction materials from Zircalloy to concrete, and their first decay products, sometimes dissolved in activated wastewaters including dissolution residues (nitric acid) in nuclear fuel reprocessing (now banned in both the United States and Germany), must be disposed of and stored until radioactive towards stable products is essentially complete. A fuel rod in an NPP is a metal (Zr mainly) tube filled with cylindrical sintered pellets of UO_2 or some mixture of UO_2 and PuO_2; metal alloys or other compounds (carbide UC_2, hydride UH_3) are used in minireactors only. After a while, a fuel rod is "spent," reducing the ^{235}U content from the original 3.5% to some 1.3% while 1–1.5% of plutonium—most of it fissionable also—were produced from ^{238}U in situ. Although use could be continued by further pulling out the control rods that absorb excess neutrons, it would be no longer safe to work with such rods. Thus, they are replaced, usually a third of them every year or so. The "spent" rods are stored for several decades, usually next to "their" reactor to get rid of the highly active short-lived isotopes of high yields, e.g., $^{141;144}Ce$ and $^{103;106}Ru$ and ^{91}Y, then either processed or put to a final depository (if there is one, by now only in Finland).

The case is not at all settled; there are no operating final deposit sites but only such ones meant to contain (withhold) the waste for a few decades. Obviously, radioactive waste solutions were discarded into the open sea or rivers both after accidents and routinely during nuclear reprocessing [Windscale/Sellafield (U.K.) and LaHague (France), Mayak near Chelyabinsk (Russian Federation)], not to mention simply dumping entire discarded reactors from nuclear-powered submarines into the sea, with or without the rest of the vessels . . . What can be done responsibly with nuclear waste instead?

The periods of time over which radionuclides from fission reactors must be stored and safeguarded are outright unimaginable and far beyond any other time frame of political or economic planning: there are nuclides with half-lives of several million years (^{94}Nb, ^{129}I, ^{237}Np); thus, they will create a danger for 10^7 year or the like. With ^{237}Np (tons of which currently exist) and ^{243}Am (americium, tens of kilograms of which exist), there is an additional problem: during very long storage, enriched samples of either nuclide will spontaneously turn into fissionable materials (^{233}U and ^{239}Pu, respectively), causing heat and neutron release to increase after millennia.

Here are some examples:

Uranium	^{238}U	4.468 billion years
Uranium	^{235}U	704 million years
Iodine	^{129}I	15 million years
Neptunium	^{237}Np	2.144 million years
Plutonium	^{239}Pu	24,110 years

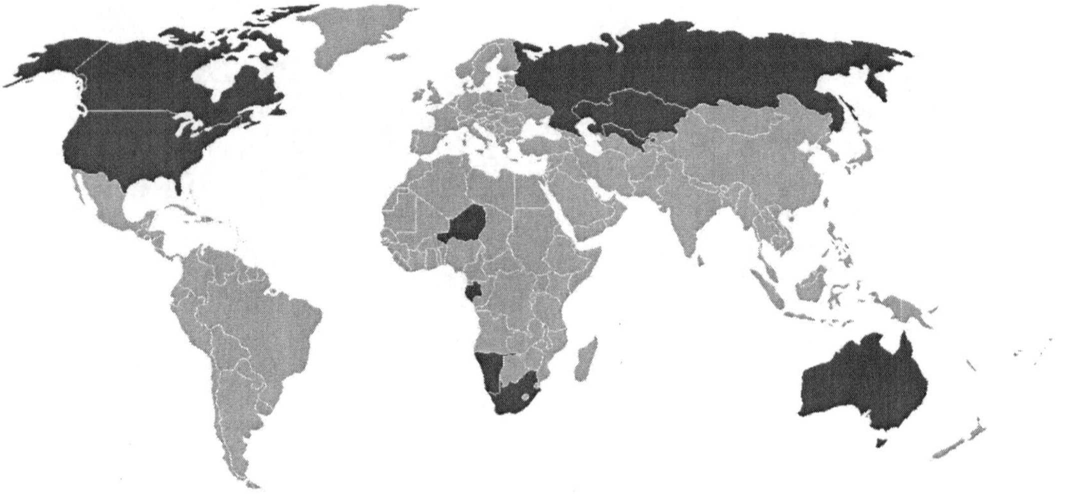

Fig. 6 The 10 states (dark gray) that produced the largest amounts of uranium ores (in 2008). Besides the spatially largest states outside of South America and except for China and India, most are located in Western Africa, including Gabon (Central Africa) whose ultrahigh-grade (>60% U) deposits at Oklo and neighboring sites gave rise to natural reactors some 2 billion years ago (map from Wikipedia.de).

While actinoides [from the third, protactinium (Pa, Z = 91) onwards] generally are extremely chemotoxic, this does only matter for long-lived nuclides like the natural uranium isotopes, ^{237}Np, $^{242;244}Pu$, or $^{247;248}Cm$. In the other cases, say at $T_{1/2} \leq 10^5$ years for α emitters with negligible spontaneous fission shares, radiotoxicity, i.e., the effects caused by particles of ionizing radiation emitted during decay, prevails even against this chemotoxicity. Among these nuclides, $^{239;240}Pu$ (which are produced together in a reactor given there is substantial irradiation of a uranium sample) are peculiar in their radiotoxicity but are rather long lived as they do reside in the body at very sensitive points: the marrow and mucosa around or in bones and liver, while radioactive ("hot") particles may be inhaled and reside in the lung, exposing it to radiation. Note that some 6 tons (!) of ^{239}Pu that escaped fission during nuclear bomb tests were spread up into the stratosphere as an aerosol that still keeps on being deposited ("fallout") apart from some other transuranic nuclides: from analyses of fallout composition, 1100 kg of ^{240}Pu and substantial amounts of $^{241;243}Am$ (produced in nuclear weapons containing ^{241}Pu or ^{242m}Am as fissionable nuclides, which have much smaller critical masses with fast neutrons than all $^{233;235}U$ and ^{239}Pu) are determined. Estimated amounts of the latter are 25 kg ^{241}Pu, 70 kg of its decay daughter ^{241}Am, and 2.5 kg ^{238}Pu [most of it from breakup of a RIG (radioisotope thermoelectric generator) during re-entry in a 1960s space probe mission; data calculated from Breban et al.[18] Besides this, there are natural contributions of plutonium in the biosphere: some 5 kg of ^{244}Pu is still left over from the origins of the solar system owing to its 83-milion-year half-life and trace amounts (<1 g) of ^{238}Pu from the double-β-decay branch of ^{238}U. Most fallout radioactivity is still due to ^{241}Pu ($T_{1/2}$ = 14 years), but this is less destructive in radiotoxicological terms because it does undergo β-decay, rather than α decay, followed by each comparable activities of $^{239;240}Pu$ and ^{241}Am ($T_{1/2}$ = 433 years). This compromises sensitive tissues.

Cancer rates, e.g., bone skin sarcomas and their metastases, will thus increase after a Super-GAU (German: *größter anzunehmender Unfall* means worst case, meltdown) if plutonium is released, like in the 1951 Mayak accident (Russian Federation) or after a Pu-based nuclear fission bomb was destroyed 1) in an airplane crash at Thule (Greenland) in 1968 or 2) by mis-ignition ("fizzle") in the *Hardtack Quince* test (explosion yield but 0.02 kt TNT) at Runit Island, Enewetak Atoll (now some part of Republic of Marshall Islands) in 1958. Both latter events spread several kilograms each of ^{239}Pu over a very restricted area, which made effectively cleaning it impossible. The doubtful results of (2 out of 5) nuclear bomb tests in Pakistan (1998) and the first one in North Korea (2006)—neither yet precluding burst of Pu and fission products to the surface though these were underground test "fizzles"—must be added to this list. Runit Island (Pacific Ocean) is now used as a dumpsite for highly contaminated materials from nearby test craters, the hole capped by a concrete dome of some 100 m diameter. This is the gravest possible kind of catastrophe from an ecological and ecotoxicological point of view, going beyond even what happened at Fukushima or Chernobyl. The environment, including groundwater, is so polluted that access is strongly impeded for centuries or even longer.

As with Chernobyl and the Bikini Atoll test site, animals and plants apparently adapted to the harsh radiological conditions there. Starting soon after the Chernobyl accident, researchers noted that diverse animals (including wolves, foxes, lynx, moose, hares, and many kinds of birds)—and some humans—returned to the off-limits area around Chernobyl, including the so-called Red Forest and the decontamination lake. Among these, there are species that try to avoid man, prompting them to invade an area where there are (almost) no humans left behind. In addition, biodiversity is lower than in comparable areas, brain sizes tend to be reduced in both mammals and birds, and there are other malformations, tumors, and evidence of genetic alterations.[19] Likewise, there are lots of aquatic life in Bikini Atoll [except for the crater basin of the largest ever U.S. test, *Castle Bravo* (1954, some 15 MT)], but once again, there is reduced biodiversity.

At least since 9/11 some 10 years ago, but actually ever since the first threat to attack a NPP by a flying passenger airplane abducted before (in 1972 over Oak Ridge, Tennessee) citizens and public authorities of countries which make use of nuclear fission power: while older NPPs—all over the world—must be considered to be improperly protected against even the impact of a smaller airliner there is virtually none anywhere to withstand the perpendicular impact of a fighter plane. The jet engine compressor axle in a military plane is about 3–3.5 m long, made of dense metals such as niobium alloys, then hitting a concrete ($\rho \approx 2.8$ g/cm^3) shell at about the speed of sound. This will suffice to penetrate some 10 m (!) of concrete, as compared to an actual containment thickness of less than 2 m. In addition, global warming and eutrophication combine to increase the likelihood that river or ocean water cooling can no longer be taken for granted in summer at least: in 2003, rivers became so warm all over Central Europe (probably related to global warming) that NPPs from France to Lithuania had to be shut down for many weeks. In the same summer, cooling water entries at the Russian Sosnovy Bor plant were blocked by algae excessively growing in the adjacent Bight of Helsinki. Either problem is likely to occur more often in the future.

Transmutation[20,21] is discussed as a means of faster disposal of fissiogenic radionuclides, exposing the nuclide mixture obtained by fuel reprocessing to a high-energy proton beam. Fission products that are distinguished by a considerable excess of neutrons in their nuclei hence are brought closer to stability while actinoides will either undergo fission directly or produce at least nuclides of much shorter lifetimes. It is completely unsettled and doubtful whether this can be done on a scale of tons of material.

RETHINKING IN THE WAY FOR ECOLOGICAL ECONOMICS

The previous entry was mainly concerned with the problems of conventional and nuclear energy sources from an economic, ecological, and political point of view, which will challenge us more and more in the near future. There are several quite different—and mutually independent—reasons to switch to alternative,[4,22–25] that is, regenerative, resources of energy, including resources of fossil fuels becoming scarce(r), constraining greenhouse gas emissions, and striving for geopolitical risk management (spreading suppliers among most diverse regions and political systems, if you are not in a position to produce the materials related to your energy demands domestically). Although there are numerous incentives, this transformation will take several decades not only on a global scale but also in national dimensions, and it depends on both economic interest positions and technical innovations. While wind energy is, at present, technically "ripe," the size of rotors being no longer limited by mechanical problems with generators, etc., exposed to vibrations and bending along a horizontal axis, but by the size of available cranes required to erect them (maximum: some 200 m rotor diameter, 7–9 MW peak power), and photovoltaic devices that reasonably benefit from technologies of semiconductor processing—which is now advanced much beyond anything that would ever be required in solar energy conversion—becoming a large-scale-business as well [the combined area of highly integrated chips that make it to the market in computers and numerous other devices (some 70 of them alone in a common car) annually is in the square kilometer region also], extracting energy from the oceans actually still takes an engineer's ingenuity to make it reasonably work. For solar energy, the "surviving" technical innovation challenges are restricted to thin-layer and organic semiconductor systems and to conversion/storage in a convenient chemical storage form, like methanol.

The following part of this entry will deal with both those energy forms already present on the global market for energy conversion devices, plus a discussion of how energy can be harvested from the open sea, keeping in mind which will be the challenges for Germany or Italy who both decided either to abandon (D) the use of nuclear energy until 2022 or not to restart it (I). This prompts the question whether redesigning the energy supply of a medium-sized highly industrialized country, including consumers which are really demanding, within some 10 years, merits further consideration which is added, too.

Globally View of Renewable Energy

Available energy resources are "renewable" when they are sustainable and environmentally benign (which need not coincide) if

- They renew themselves in the short term by itself (e.g., biomass).
- Their use does not contribute to the depletion of the source (e.g., wind, sun, water).
- They do not have an impact on the environment.

Sustainability in that sense means a triad related to ecology, economy, and social affairs. Fig. 7 gives an overview of different energy sources like sun, wind, water, geothermal energy, and biomass for the establishment of technical supported renewable energies.[4,22,24,26,27]

In 2010, the Global Status Report on Renewable Energies[28] was published, parts of which are now to be quoted. In its preface, El Ashry (UN Foundation) already noted more than a hundred states to have political agendas or strategies in 2010 in favor of spreading and widening the use of renewable energy sources, about a doubling from just 55 five years before. In 2009, the largest increases were observed with wind power and photovoltaics, investing some $150 billion into extending capacities and producing and implementing renewable energies vs. only $30 billion in 2004. Except sometimes for advancement of nuclear energy, more money is allocated into capacities and growth of renewable energy sources than into fossil types. By reaching substantial shares of energy input (particularly in some less-populated countries like Denmark, Latvia, or Mongolia), renewable energies did have a turning point, which renders them significant, as they also address the problems of climate change. As implied before, acceptance and broad application of renewables are not (no longer) restricted to highly developed industrialized countries but extend to developing nations now displaying more than half of implemented and operating renewable energy supply systems. However, it remains to be seen as to how far this is actually in favor of domestic development, better situations especially in rural areas, or larger autonomy towards globalized oil and coal markets or is another blueprint for export of goods to the North: growing sugar cane and oil palms for biofuel exports in Brazil or Indonesia threatens local primary forests if not even the food supply there. The Desertec blueprint (http://www.desertec.org) for producing lots of hydrogen from solar sources in Northern Africa revives existing supplier roles with countries like Algeria once again.

Fig. 8 gives a more precise idea of the present relevance of renewables, their share being some 19% of total in 2008. Share of final energy means: at the point of end use, as electricity, heat, and directly used fuels. This method counts all forms of electricity equally, regardless of origin. The European Commission adopted this method in 2007 when setting the EU target of a 20% renewables share of energy by 2020. Thus, it could be called the "EC method."[29] This includes all traditional sources of biomass energy carriers (biogas, ethanol, wood, etc.), large hydropower plants, and the "novel" kinds of renewables such as small-scale hydropower

(a)
(b)
(c)
(d)
(e)
(f)

Fig. 7 Renewable energy sources: onshore and offshore wind power, biomass, solar power, geothermal, hydropower. (a) Onshore wind turbines located outside of Palm Springs, California (photograph: Wikipedia, Tim1337); (b) sugar cane is a major supplier of biomass, which is used either as food or as energy supplier (photograph: Wikipedia, Culture sugar cane, Avaré, São Paulo. José Reynaldo da Fonseca, 2006); (c) newly constructed offshore windmills on the Thornton Bank, 28 km offshore, on the Belgian part of the North Sea (photograph: Wikipedia, Hans Hillewaert, 2008); (d) photovoltaic array near Freiberg (Germany) (photograph: Wikipedia, I, Eclipse.sx); (e) Krafla Geothermal Station (2006), North Iceland (photograph: Wikipedia, Mike Schiraldi); (f) Grand Coulee Dam is a hydroelectric gravity dam on the Columbia River in the U.S. state of Washington. It is the largest electric-power-producing facility in the United States (photograph: Wikipedia, (http://users.owt.com/chubbard/gcdam/html/photos/exteriors.html) U.S. Bureau of Reclamation). Larger dams are located in Russia, Brazil, and China.

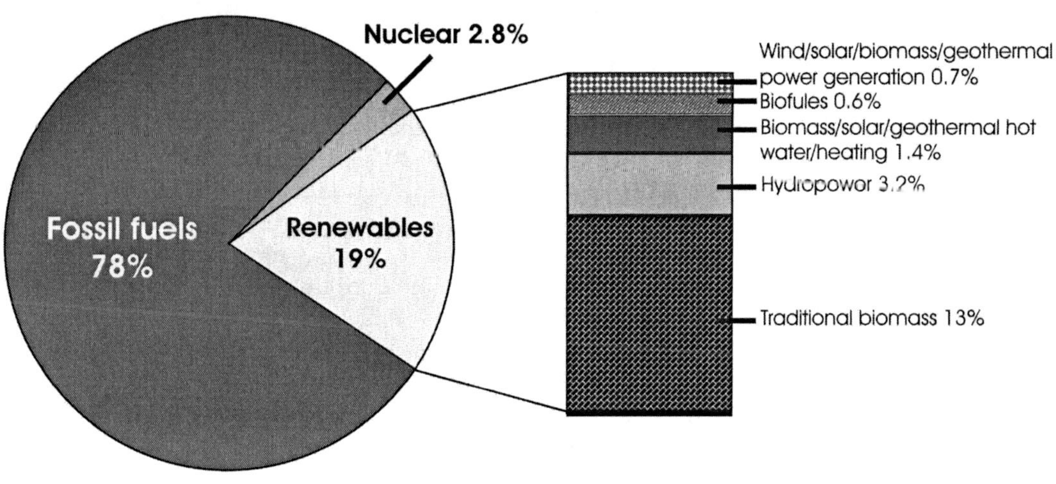
Nuclear 2.8%
Fossil fuels 78%
Renewables 19%
Wind/solar/biomass/geothermal power generation 0.7%
Biofules 0.6%
Biomass/solar/geothermal hot water/heating 1.4%
Hydropower 3.2%
Traditional biomass 13%

Fig. 8 Global final energy consumption of conventional and renewable energy in 2008 after REN 21 (2010). Also consider footnote 327 of REN 21 (2010).[28]

plants, modern biomass sources, wind and solar energy, geothermics, and biofuels. To be exact, it should be noted that some of the "modern" ones actually are rather old, from small-scale water mills also delivering electricity up to biofuels [Rudolf Diesel demonstrated his engine to be run with peanut oil fuel already at the 1900 World Fair (EXPO) at Paris].

Most of these 19% (13%, i.e., relatively 70% among the renewables) rests with traditional biomass used for cooking and heating. This, however, is subject to some change and even cutbacks, especially because there is virtually no possible firing wood (not even small branches) left over around (and this is to say, up to 50–70 km from the outer edge of "informal suburban settlements"!) megacities[30] all over developing and threshold countries, causing to replace with more advanced (e.g., solar cooking devices) or at least more efficiently use these resources. Hydropower represents 3.2% (17% out of renewables) and is growing modestly but from an already advanced level. The other renewable energies contribute 2.7% (relative share: some 14%) and undergo very fast growth in both industrialized and certain developing countries.

In developing countries, renewable energies have a particular role in national support and corresponding policies as well, contributing to their present share of renewable energy capacities of more than 50% [but compare this to both their share of global population, which, including the BRICS (Brazil, Russia, India, China) states except Russia, is >80%, on one hand, and their share of global energy consumption, on the other hand]. China now leads in several indicators of market growth. Among suppliers of wind energy, India now ranks No. 5 globally, owing to its substantial expansion of biogas and photovoltaic capacities especially in the countryside (India also plans to install another 15 NPPs owing to its being populated as densely as, e.g., Belgium or the Netherlands in total and citing growing industrial energy demands). Brazil is now a key supplier of ethanol from sugar cane, adding to other kinds of biomass and wind power in its renewable energy portfolio. Concerning the complete array of renewable energy sources, Argentina, Costa Rica, Egypt, Indonesia, Kenya, Tanzania, Thailand, Tunisia, Uruguay, and others boast high growth rates, often even surpassing their sometimes impressive gross economic growth rates.

In a nutshell, the geographical distribution and political significance of renewable energies have considerably been altered in favor of near-global distribution and application. As compared to quite a few countries in the 1990s, there are at least 82 countries presently operating wind power plants. As for manufacturing these devices, production was relocated from Europe to Asia, with the largest shares in China, India, and South Korea, which additionally become more devoted to renewable energy applications. China not only is the "production yard" of the world for conventional items including consumer electronics but also made, in 2009, 40% of PV devices (from solar cells to complete panels with control and DC/AC converter units), 30% of the world's wind power plants, and even 77% of solar-driven water-heating collectors. All over Latin America, countries like Argentina, Brazil, Colombia, Ecuador, and Peru increase their production, including exports of biofuels, additionally investing in other kinds and technologies of renewable energies. More than 20 countries in the "Solar Belt" of the Middle, Northern, and sub-Saharan Africa are involved in renewable energy markets.

Large economic gains and considerable further technological changes are to be achieved from this state of affairs and developments beyond the principal players of the highly developed world such as the European Union (EU), the United States, Australia, Canada, or Japan. With renewable energies gaining a truly global footage and application area, there is internationally growing confidence into renewable energies being less susceptible towards perturbations by either political turnovers or market changes such as financial crises than "classical" energy carriers (which readily, like other raw materials, become an item of speculation undergoing tremendous price level oscillations in such conditions).

Another impetus to renewable energy development—as to every other kind of large-scale technical innovation—is its inherent potential to create entire new industries (or at least create vast uncharted fields of opportunities for existing technical branches) and thus millions of new workplaces. In photovoltaics, much larger amounts of semiconductor-grade to medium-purity silicon, SiH_x, and GaAs are used than in integrated (chip) solid-state microelectronics (e.g., GaAs chips are applied in cellphones), whereas technologies required to obtain control of a large rotating propeller in changing conditions, originally developed for helicopters, now increase reliability and output of wind power systems. Creating new workplaces is, by the way, the positive, friendly side of the "dual-use" problem: both engineers and blue-collar technicians who were mainly concerned with development and production of arms systems had to strive to save and "humanize" their workplaces after the Cold War came to an unanticipatedly happy end around 1990; these highly skilled metal workers and engineers at Kiel harbor (Germany) then started to focus on wind power plants (with rotor techniques borrowed from military helicopters) and integrated current-heat support systems (using diesel engines originally designed for tank propulsion to consume and convert plant and waste oils to produce some 1 MW each of electricity and heat). Hence, although not creating environmentally benign branches from scratch, this "Arbeitskreis für Alternative Produktion" (workgroup planning alternative production) and similar endeavors in U.S., British, Australian, and French arms enterprises effectively increased the array of possible customers much beyond the state (i.e., department of defense and sometimes police forces) and in the same turn reduced political–military dependences even though

Table 1 Jobs worldwide from Renewable Energy (REN 21[28] with added information from the United Nations Environment Programme report 2010).

Industry	Estimated jobs worldwide	Selected national estimates
Biofuels	>1,500,000	Brazil, 730,000 for sugar cane and ethanol production
Wind power	>500,000	Germany, 100,000; United States, 85,000; Spain, 42,000; Denmark, 22,000; India, 10,000
Solar hot water	~300,000	China, 250,000
Solar PV	~300,000	Germany, 70,000; Spain, 26,000; United States, 7,000
Biomass power	–	Germany, 110,000; Unites States, 66,000; Spain, 5,000
Hydropower	–	Europe, 20,000; United States, 8,000; Spain, 7,000
Geothermal	–	Germany, 9,000; United States, 9,000
Solar thermal power	~2,000	Spain, 1,000; United States, 1,000
Total	**>3,000,000**	

Note: Further information about the evaluation of the data are reported in REN 21,[28] p. 75, note 226. The table is incomplete.

most of the respective employers did not really like the idea of employees considering what should (better) be produced on their own.

This is part of the ethical issues and bonuses associated with alternative energy production: there is a considerable bonus in terms of both workplace safety and numbers of workers required to install and run 1 GW_{el} of alternative energies as compared to fossil and nuclear types; besides, there are fewer risks associated with making PV devices than with coal or uranium mining for the miners themselves, counting and comparing, for example, mine accidents and cancer fatalities per MWy. In 2009, there were an estimated three million workforce directly related and devoted to renewable energies, about half of them concerned with biofuels, and many more than this in branches indirectly connected with renewables (Table 1).

What are the recent performances of renewable energies as of 2009 (Fig. 9, source: REN 21[28])? For the second consecutive year, in 2009, in both the United States and the EU, the newly installed renewable energy capacities exceeded those of combined conventional fossil energies and nuclear power. Renewables accounted for 60% of newly installed power capacity in Europe in 2009, and nearly 20% of annual power production.

Christopher Flavin (Worldwatch Institute) pointed out in his entry "Renewable Energy at the Tipping Point" within the REN 21[28] report that China's recent leader role in producing wind rotors and photovoltaic devices just gives proof of the political prerogatives in favor of renewable energy exploitation, including both laws and funding, to be successful. Although there were initial problems, the important reforms in China starting with the national legislation on renewable energies of 2005 caused fast and efficient development there. China since then increased its efforts to become a leading innovative power as well as key producer of renewable energy technologies.

Table 2 shows the five countries that are the most important players concerning renewable energies, as of 2009.

Owing to the well-known variabilities of renewable energy output, which are due to weather, time of day, and longer-period (tidal power plants) periodic or aperiodic changes, there are much larger theoretical (peak power output) renewable energy technical potentials than average yields. The present (as of end of 2009) capacity of a global 1.23 TW (1230 GW) that now constitutes just more than 25% of total electric generating capacity worldwide thus is considerably larger than the actual share/contribution of produced electricity.

What about Ocean-Related Energy [Waves, Ocean Currents, Tidal Power Plants, Osmotic Energy Conversion, Ocean Thermal Energy Conversion (OTEC)]?

The power associated with flowing water is impressive and has motivated people to use it many centuries ago in mills located at running creeks and rivers. It was an obvious idea to extend this technique to tapping ocean currents, like

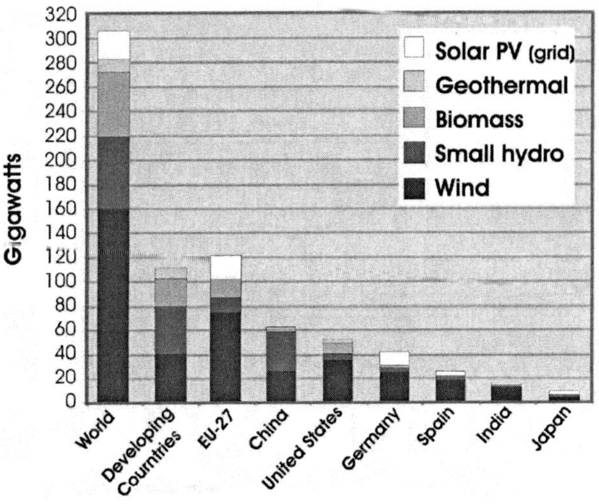
Gigawatts
320
300
280
260
240
220
200
180
160
140
120
100
80
60
40
20
0
Solar PV (grid)
Geothermal
Biomass
Small hydro
Wind
World
Developing Countries
EU-27
China
United States
Germany
Spain
India
Japan

Fig. 9 Renewable power capacities in 2009 without inclusion of large-scale hydropower: Developing World, European Union, and Top Six Countries (REN 21 2010).[28]
Note: Only includes small hydropower <10MW.

Table 2 The five countries that are the most important players concerning renewable energies [as of 2009[28]].

	#1	#2	#3	#4	#5
Annual amounts in 2009					
New capacity investment	Germany	China	United States	Italy	Spain
Wind power added	China	United States	Spain	Germany	India
Solar PV added (grid connected)	Germany	Italy	Japan	United States	Czech Republic
Ethanol production	United States	Brazil	China	Canada	France
Biodiesel production	France/Germany		United States	Brazil	Argentina
Existing capacity as of end of 2009					
Renewables' power capacity (including only small hydro)	China	United States	Germany	Spain	India
Renewables' power capacity (including all hydro)	China	United States	Canada	Brazil	Japan
Wind power	United States	China	Germany	Spain	India
Biomass power	United States	Brazil	Germany	China	Sweden
Geothermal power	United States	Philippines	Indonesia	Mexico	Italy
Solar PV (grid connected)	Germany	Spain	Japan	United States	Italy

the Gulf current, as well as tidal water flows[31] that can reach speeds much above those seen in most rivers [e.g., some 6 m/sec (11 knots/hr) along the Welsh coast of the Atlantic ocean (rather than perpendicular to it)], providing concentrated energy as flow speeds are close to those in air (wind), with water being 800 times as dense; thus, a rotor of equal diameter exposed to a water flow of equal speed delivers 800 times as much power [or the same power at $^3\sqrt{(1/800)}$ this speed, i.e., some 10.8%]. Basins that are filled with water at maximum level differences of a few meters are commonplace in electric storage (in German: Pumpspeicherwerk), with the same being offered by tidal changes of ocean and estuary water levels [the largest tides are seen in river mounds, e.g., River Severn (Wales) or River Rance (France, Normandy) at some 10 m], and not just after pumping water into them by electrical power obtained otherwise, but for free twice a day (actually even 4 times daily using differences of either levels).

Concerning periodical filling and uncharging of storage basins connected to flowing water turbines, there is but one large power plant (240 MW_{el}) in the world still connected to the French grid back in 1967. It is located at La Rance in the mounding of River Rance, next to the famous island Mont St. Michel (Fig. 10).

Fig. 10 Aerial photograph of the world's biggest tidal power station. The Rance Tidal Power Station is located at La Rance (NW France) in the mounding of River Rance, next to the famous island Mont St. Michel.
Source: Photograph: Wikipedia.

Here, the tidal water level differences are far larger than the global average of about 3.5 m (isolated ocean basins, such as the Black Sea and the Baltic Sea, and even the Mediterranean Sea, tend to lack any significant tides, e.g., average tides in the Baltic Sea are some 20 cm); the minimum tidal heights required for a meaningful operation of a tidal power plant are estimated to be some 5 m.[32] Obviously, the gain of energy from a basin of given size interacting with the tide flows increases by the square of tidal water level changes: the amount of water flowing in and out is proportional to tidal height and so is the energy gain from a given mass of water flowing through the turbines. Now being operated for more than 40 years (it was connected to the French grid in 1967), effects from this plant and its overall performance can be well evaluated:

- Local tidal height decreased from some 14 m (!) to <8 m due to deposition and relocation of sediments.
- Corrosion is an issue, requiring both to avoid combination of different alloys in the construction to exclude galvanic effects in seawater, and active electrochemical protection measures.
- There were some ecological side effects concerning distributions of limnetic and marine fishes in and on either side of the basin, with the power plant shifting the limits of marine and limnetic populations of fishes somewhat downstream into the estuary. The turbines rotate so slowly that both fishes and squids can pass through them without being hurt, much unlike classical running-water power plants.
- Actual average output to the grid is some 540 GWh/yr, which is an average of 62 MW, about 25% of nominal power production.

Nevertheless, the La Rance plant remained unique, being the only large-scale plant existing, although sites having even higher tidal heights such as Bay of Fundy (Canada, up to 19 m) were used for yet smaller installations.

Waves[33] may be destructive, but their energy can also be technically exploited, e.g., with transducers based on either bending of some part of the device [hydraulically or by piezoelectric (piezoelectricity means electricity resulting from applying pressure, squeezing) devices]. Other devices employ either systems operating on periodically compressing and expanding air by by-passing volumes in a volume (some hollow metal or concrete bunker) under which the waves pass through and pass this air over a rotating turbine or using the alternating flow directions of water or directly converting the water level oscillation into electricity by electromagnetic induction.

It is estimated that one can gain some 15 kW_{el} (130,000 kWh/yr) from a single meter of coastline around the North Sea, which is comparable to the gain earned by covering about 1 km of land behind this shoreline by photovoltaics in the same climate (this sounds stunning if not paradoxical but consider the width of the seas where the waves could gain energy before from wind rather efficiently converting solar energy). So far, wave energy converters were constructed at Western ocean coasts in the Northern Hemisphere, including SW Norway (Toftestallen, near Bergen), the northernmost Scotch coast, open to the Atlantic Ocean and Oregon. A most straightforward way to convert mechanical movements into electricity, besides piezoelectric techniques, is by magnetic induction. A buoy is towed to the seafloor, with a magnetic shaft tethered to it, while the floating part of the buoy contains some induction coil that hence moves up and down around the shaft within the magnetic field produced by the latter, directly producing an alternating current even though its voltage and frequency still have to be adjusted.

Of course, such interferences with natural water flows may alter and affect the ecological situation in estuaries especially; in the best cases, using both tidal and wave energy reduces coastal erosion.

Less obvious sources of energy associated with ocean and shoreline,[34] respectively, include the temperature difference between surface waters and the deep sea (particularly in the tropics, except for the Red Sea and its annexes like the Gulf of Aqaba/Eilat), some 25–30°C there and the gradient in salt content and, thus, the osmotic pressure difference of some 26 bar between the ocean and freshwater from rivers and creeks running into it (the latter being identical to have this water falling through a Pelton turbine from about 260 m of height).

At a surface temperature of some 300–305°K in tropical waters, the Carnot (most efficient heat engine) maximum efficiency would be about 8% 10%. Usually, in OTEC plants, ammonia is used as a heat-transferring working medium; actual efficiencies are about 3% due to long hoses required to exchange hot and cold waters. Rather than steep cliffs selected earlier, OTEC plants are now located on either ships or floating constructions much like those employed in crude oil production. Given the problems associated with coral bleaching due to surface waters exceeding 30°C, some larger use of the OTEC technology should even provide a real ecological bonus while mixing nutrients between deep and surface waters might cause problems of eutrophication.

The osmotic pressure of some solution is about the pressure that would be exerted if the same concentration of dissolved entities (i.e., ions from, say, $MgCl_2$ solutions, to be counted individually, hence producing 3 times the osmotic pressure in water or acetonitrile than if dissolved in non-ionizing solvent like a long-chain alcohol); 1 M/kg of solvent (here water) hence is equivalent to a gas compressed to 1 M/L, which is tantamount to a pressure of some 24 bar at 20°C. A unimolar $MgCl_2$ solution in water would thus produce an osmotic pressure of more than 70 bar, while the corresponding value for seawater—which is 0.55 M NaCl solution to a first approximation—is about 26 bar [the actual value for Norway is a little smaller because of 1) the water being colder and 2) some dilution by the very freshwater discharges into the Fjord site used, which is rather remote from the open ocean].

Before dreaming of what the latter source could deliver when passing but substantial parts of the freshwater flows of giant rivers like River Amazon, Mississippi, Yangtsekiang (China's largest river), Kongo, or the Siberian rivers like Yenisei through such devices, be reminded of some problems that are not yet solved. The only demonstration plant existing so far is located at a creek mounding into a fjord in Norway, producing some 3–4 kW of electric power since November 2009. There are three quite different kinds of converting osmotic potentials into electricity:

- Using a water-permeable yet pressure-proof membrane (this is done in Norway). The saltwater content of this closed membrane volume is going to absorb pure water from the freshwater passed along outside (i.e., in the mounding of the river). Due to the osmotic pressure, the water level inside will increase considerably, allowing to pass it over a falling-water turbine and produce electricity eventually. This process is a somewhat periodical one: as the saltwater gets diluted during the process, you need to discard it into the ocean sooner or later and refill the chamber with "pure" (3.5% salt) ocean water. By now, electric power output is about 1 W/m^2 of membrane interface area; the aim is to achieve 5 W/m^2 soon.
- Electrochemical settings using the diffusion potential: concentration differences leveled off by diffusion create electric potentials even though no redox reactions are involved in charge transfer. Likewise, second-type electrochemical cells draw upon concentration potentials: an electrode made of combined silver and AgCl (both solid and mixed among each other) will adjust its potential according to the concentration level of chloride ions. This can be used both in analytical chemistry and for producing electric currents from chloride solutions that differ in concentration (freshwater typical values being 1–2 mM/L as opposed to 0.55 M/L in seawater, giving an open-cell voltage of 150–170 mV).
- Theoretically speaking, the zeta (electrokinetic) potential could also be used in osmotic energy conversion: by osmotic pressure differences forcing water through some membrane, or a column filled with a packed solid, it will produce an electrical potential difference between either side of the interface. This potential is due to selective adsorption of cations or anions onto a typically charged particle, charging being caused by oxide/hydroxide particles (say, wet alumina) behaving as an acid (adsorbing hydroxide) or base (adsorbing protons) depending on local pH, co-sorbents, and material (point of zero zeta potential). Then, the other, non-adsorbed ions will be passed through along the solvent, and a potential of typically half a V forms. The effect is reversible (electrokinetic water pumping).

The problems are with membrane stability and, more generally, also affecting electrochemical systems, clogging of the interfaces by biomass (mainly phytoplankton) or even mineral concretions. The Norwegian success terminated a history of decades of failed experiments on osmotic power production.

Things became a little different—and better, rather more advanced—because of OTEC. The first plant ever of this kind was constructed by French Georges Claude (1870–1960) at the coast of Cuba back in the late 19th century. OTEC devices are simple thermal power plants using rather small T differences, more like a steam engine compared to turbines, coal-fired plants, or internal combustion engines.

The global distribution of chances for this way of harvesting is solar energy indirectly (having hot water from insolation in the tropics while cooling is provided by Arctic or Antarctic undercurrents at some 1000–2000 m of depth).

Eventually, there can be integrated offshore energy parks making use of almost all the energy sources discussed above combined on, or beneath, a tethered floating island.

Geothermal Energy

For many decades, *geothermal energy*[35,36] has been well established for heating purposes in countries like Iceland, New Zealand, and several developing countries in Central America such as El Salvador. Although electrical uses, with geothermal (fumarole) vapors directly run through a steam turbine, were first tried in Italy more than a century ago (at Larderello in 1904, delivering about 200 W), corrosion and clogging problems remain severe until this day. The obvious reason is the "contamination" of fumaroles with both clogging agents like boric acid and hydrolyzable volatile metal chlorides, besides the large shares of corrosive gaseous acid precursors like SO_2, HCl, and HF. Thus, one has to create a primary heat exchange cycle directly exposed to these corrosive items, as well as a secondary one linked to the heat/mechanical/electrical conversion systems, much like in NPPs and mainly for the same reasons (if not even worse here), and worse, due to the rather small heat difference, this decreases total efficiency of conversion. When obtaining the vapor from underground wells drilled several kilometers into the Earth's crust rather than operating close to active volcanoes, clogging of drilling holes or rock fractures required to circulate some operating medium also remains critical. Hence, it is safe to predict that geothermics will remain more concerned with heating (houses, swimming pools, etc.,) than with electricity production in the near future as well.

Wind Power

Wind power now is an established source of energy, with average production costs per energy unit (e.g., cent per kilowatt-hour) coming close to those by conventional (fossil) energy sources.[37–40] In certain countries, the share of wind power in total electricity production exceeds 20% (Denmark, Mongolia), and there is a broad international consensus that state subsidies are no longer required nor given to enhance the rate of implementation. Rather, as with all kinds of renewable energies that are subject to considerable periodic (sunlight) or non-periodic changes of supply, the optimum strategy of storage becomes imminent. Hence, electrolysis of water (and possibly secondary production of methane or methanol) by "excess" wind power (excesses being produced by mismatches with the grid also), storage of H_2 or CH_3OH, and use of the latter energy carriers in either vehicles or stationary fuel cells connected to the electric grid are gaining importance.

The size of individual plants is limited by the necessity to erect them and thereby place 100-ton items more than 100 m above the ground within millimeter precision, that is, by size of the available cranes. The largest wind power systems thus now have 200 m rotor diameter and deliver some 8 MW while arrays of them ("wind parks") can produce outputs in the size of classical power plants and NPPs both on- and offshore. However, in either case, interactions with local fauna may become significant.

Solar

Solar energy is the key source of almost all the biological and meteorological processes operating on Earth.[41,42] Semiconductor solid-state devices allow for a remarkably efficient exploitation of this source, with an additional role for thermal processes that latter rely on focusing and, thus, on non-scattered sunlight, that is, on clear skies. These thermal processes include production of fuels by cycles involving zinc or cerium oxides as well as metallurgical transformations.

Solar thermal power plants, augmented by natural gas or biogas combustion during night and other dark times, are now realized in a scale of hundreds of megawatts, while photovoltaics in 2011 first yielded an all-year average of more than 2000 MW (2 GW) electrical output in Germany (some 3% of total current). The price breakdown in production and processing of semiconductors (which need not be that pure or advanced than with electronic microdevices) supports the "boom" furthermore, regardless of fast cuts in state subsidies paid for supplying PV current to the public grid in all the EU member countries now.

When considering very large plants such as in the Desertec initiative, the increase of radiation absorption however becomes likely to influence the performance of solar parks by itself: large volumes of heated air will rise right above the plants, causing an increased dust advection to the panels as well as clouds to form on their top (both sailplane pilots and birds of prey look for typical kinds of clouds to spot regions of upwind over hotter surface areas!). High-yield photovoltaics by thin-layer solar cells [a few micrometers' total thickness, unlike 0.3–0.5 mm with polycrystalline SiH_x ("blue silicon")] demands rare (e.g., In) and/or highly toxic (Cd, As, Se, Te) elements, causing problems in all mining, processing them and eventually abandoning old PV devices when their performance sharply decreases after several decades.

Biomass

Biomass can be used in a variety of ways as a source of energy, with combustion of wood and vegetable or animal oils for both heating and illumination purposes dating back as far as the Stone Age.[43,44] More recently, plant oils were introduced into internal combustion engines (first with peanut oil; diesel, 1900), while other engines were fed with either wood distillates (containing mainly CH_3OH and acetone besides H_2 and CO) or ethanol produced by microbial activity, much after steam engines had been powered by either wood, wood-processing residues (e.g., sawdust), or peat to replace hard coal (which is biogenic in itself, of course).

Conversion methods of brown coal—lignite—by hydrogenation, gasification (steam gas process), and liquefaction can also be readily applied to biomass, including less obvious representatives of biomass such as sewage sludge (chiefly containing heterotrophic bacteria), and then mostly even take less vigorous conditions in terms of all temperature, H_2 pressure (3–10 bar rather than hundreds of bars), and needed catalysts. Finally, motivated by the fact that biomasses, especially scrap biomasses, became an item of fuel production (and waste treatment/compaction) once again, the very former coal liquefaction plants in, e.g., South Africa are now used for this purpose. Regionally, in Germany, success and economic performance were poorer, however.

Using scrap or digestible waste fractions relieves an ethical problem from the competition among food/fodder and "energy plants" for the same agrarian areas, but even after avoiding this, one must bear in mind the poor area productivity of photosynthesis—the only economically viable source of biomass energy carriers in a large scale—which typically is 0.5% or a few kilograms of reduced $C/m^2{*}a$, i.e., far short of photovoltaics. While there will be a role in waste processing, a large-scale use of biomass for energy purposes such as in Brazil poses a lot of difficult problems, including ecological ones associated with monocultures, possible fertilization/eutrophication, and high water requirements.

Renewable Energy in Germany and the Planned Nuclear Exit

Concerning its gross domestic product (GDP), Germany is the largest national economy in Europe and No. 4 in the world. In 2009, it was second in export and third in import values. Like with the GDP, Germany ranks No. 4 in energy consumption [measured in fossil fuel (hard

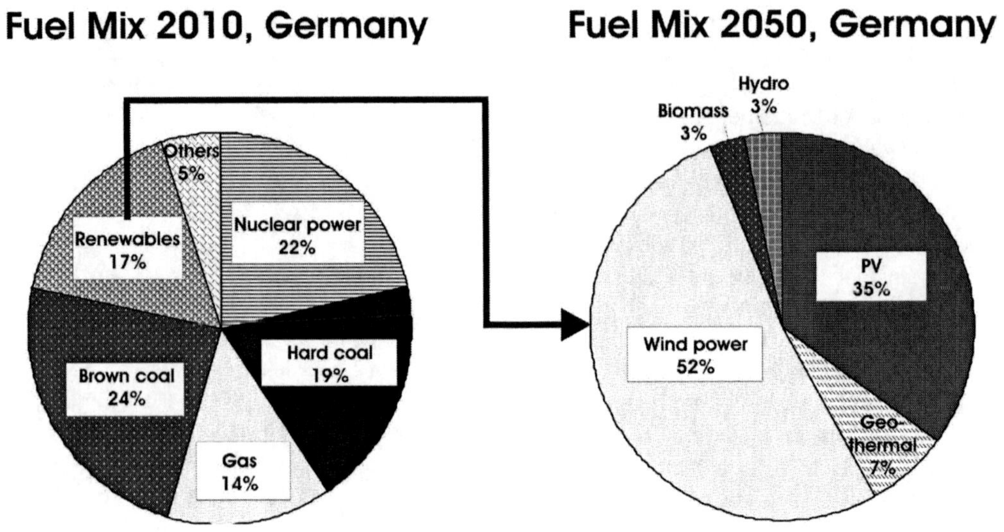
Fuel Mix 2010, Germany
Others 5%
Renewables 17%
Nuclear power 22%
Hard coal 19%
Brown coal 24%
Gas 14%
Fuel Mix 2050, Germany
Hydro 3%
Biomass 3%
PV 35%
Wind power 52%
Geo-thermal 7%

Fig. 11 Energy mix contributions in Germany in 2010 and the probable future in 2050.[45]

coal) mass equivalents (BTU: British thermal unit)] but just No. 21 among the energy producers in the world).[46] The intention of the Federal Government of Germany in fulfilment of Kyoto Protocol obligations is a reduction of greenhouse gas production of about 40% by 2020 and up to about 80%–95% reduction by 2050. Quite recently, the greenhouse gas issue and the risk of NPPs (disaster of Fukushima in 2011) were aggravated by the decision to abandon nuclear energy use in Germany in the early 2020s. Fig. 11 shows the fuel mix in 2010 and the aimed fuel mix in 2050 in Germany, which mainly will be supplied by wind and solar power.

By steadily replacing fossil energy sources with renewables, the share of the latter will increase, even allowing for a slight increase in energy consumption (which, in Germany, like most other highly developed countries, is rather constant for decades now, notwithstanding a slight decrease in population happening soon). Hydropower is fully established now except for reactivation of very small local plants, many of which had been in operation since the early 20th century. Hence, the present 3% share will remain almost the same; 35% from photovoltaics corresponds to an average output of some 23 GW, which, in our climates, is tantamount to an introduced peak power of 130–150 GW, more than twice that what now is funneled into the entire grid by all kinds of power plants. Dealing with this excess energy on sunny summer afternoons, possibly by chemical storage (water electrolysis, then linked to fuel cells), remains to be figured out. The total area required to produce this amount of PV electricity is about 1000 km^2 [<0.3% of Germany's total area (356,000 km^2) and <10% of the fields on which "energy plants" are now grown (>12,000 km^2 are covered with rape alone)], even assuming no further improvement in today's Si hydride polycrystallinic or $CuIn(S; Se)_2$ thin-layer solar cells (some 13% efficiency).

Growth and Booming Region Ems-Axis, Lower Saxony (NW Germany)

The previous entries dealt with the global-scale relevance of renewable energies. Besides the environmental issue, there are both economic and social surpluses produced by creating novel workplaces.

This can turn a formerly "just" agrarian region into some diversified boom area as will be shown by the example of the so-called River Ems-Axis (Lower Saxony, Northwestern Germany). For more than a decade now, the region keeps increasing its workforce by 3% per year—no "Mc jobs" but fully qualified jobs that produce social security and modest earnings, with >10,000 enterprises that keep expanding and creating new jobs one year after another. The regional motto reads: "Powerful, innovative and ready to achieve by unconventional solutions—these are our region's benchmarks."

As the Ems-Axis is located next to the North Sea, maritime-related activities are prominent by locating shipyards, shipping companies, and wind power plant producers, among suppliers of other renewable energies. This model region was created through a combination of prudent political support, improvement of infrastructures, synergy among regionally active enterprises, and finally the support of the public. It is located near the Dutch border in Central Europe, making use of already existing East–West connections, and, in addition, links the North Sea shores to the German megalopolis Ruhr district, which is the most populated part of the most populated and economically prolific *Bundesland* of Germany. The Ems-Axis includes the counties Wittmund, Aurich, Leer, Emsland, Grafschaft Bentheim, and Emden City with its large harbor (see Fig. 12).

There are six permanent workgroups concerned with energy, integrated maritime economy, tourism, produc-

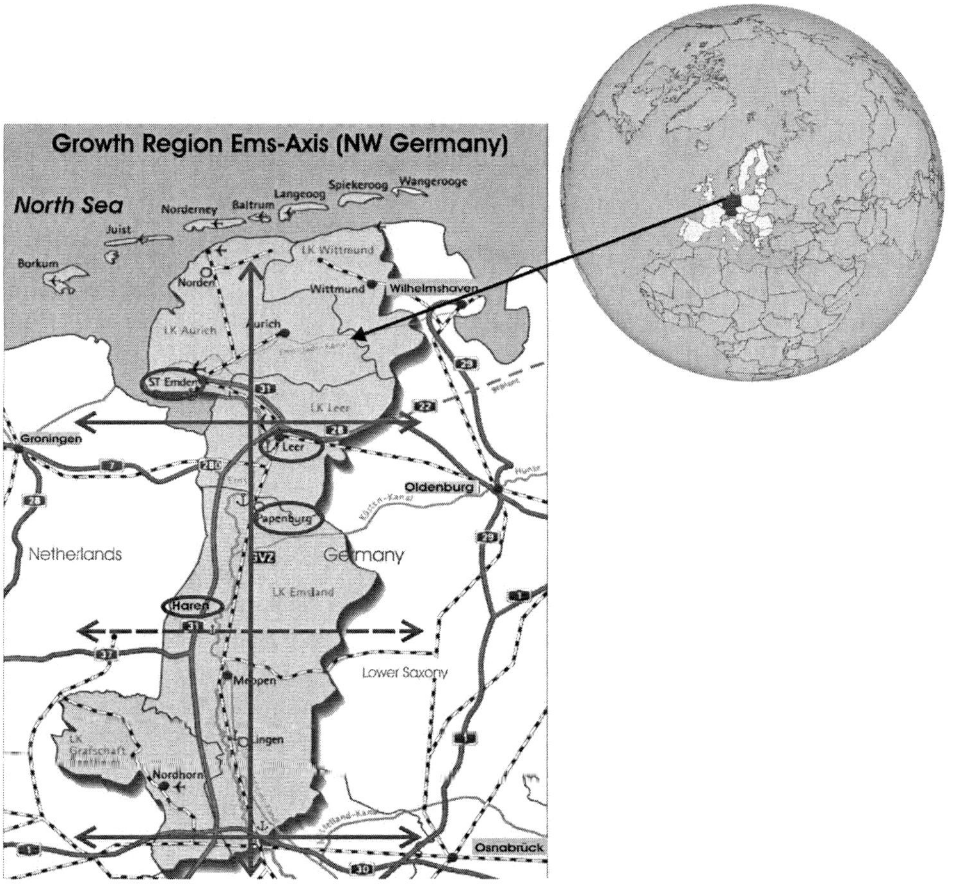
Growth Region Ems-Axis (NW Germany)
North Sea
Norderney
Baltrum
Langeoog
Spiekeroog
Wangerooge
Juist
Borkum
LK Wittmund
Wittmund
LK Aurich
ST Emden
LK Leer
Groningen
Leer
Oldenburg
Netherlands
Germany
LK Emsland
Haren
Lower Saxony
Nordhorn
Osnabrück

Fig. 12 The growth region Ems-Axis in Northwestern Germany. This figure shows the excellent infrastructure that will imminently cause new enterprises to settle and expand here. The East–West and South–North highways (motorways) are marked with Nos. 7, 28, 30, 31, and 37, while rivers and channels for ship travel are marked medium gray. These are the River Ems, which is deep and wide enough to permit economically meaningful transportation by ship, and the Dortmund–Ems channel, which extends almost parallel to it. Railway tracks are outlined in black and white. Framed: the cities of Leer and Haren/Ems are among the most important locations for shipowners all over Germany. In Papenburg, there is the Meyer shipyards, among the largest in Europe and moreover the one producing the biggest ships (passenger and cruise ships). Other notable shipyards located next to the shore at Emden recently rather switched to producing windpower plants. The Ems-Axis is distinguished by intense economic activities covering all energy supply, integrated maritime activities, agriculture, and tourism; processing plastics and metals; building vehicles and machines; and providing logistical infrastructure.
Source: The main figure of the Ems-Axis is modified after http://www.emsachse.de. The figure on the right is from Wikipedia, TUBS.

tion of plastics items, vehicles and machines, and finally logistics to initiate and run projects. It is the aim of these workgroups and the economic region to make Ems-Axis an independent axis along which economic, travel, and transport activities will organize. This implies strengthening economy-related infrastructure and creating networks for regional economy.

The cities of Haren and Leer combine to be the second-largest shipowner's site in Germany. A total of 750 ships are run from here, making these two special and significant players in running ship travel and dockyards and providing additional maritime goods and items, together with Papenburg. The existing travel infrastructure allows to process materials inshore and, using local logistics, build huge ocean liners such as that for Disney Cruise Line (340 m long and 37 m wide, 128,000 tons, can accommodate 2500 passengers) at Papenburg's Meyer dockyards (Fig. 13). The latter commands the world's most advanced instrumentation and facilities for building ships, its workforce being about 2500.

One should mention that there is minimal required bureaucracy used to acquire these infrastructures. This enabled Motorway 31—a crucial NorthSouth connection—to be completed years ahead of planning, with the region providing the required funds itself by joint and coordinated action. Another ambitious project was Euro-harbour Emsland at Haren (operated jointly with nearby Meppen city), the construction of which began in 2007. In August 2011, construction of the plant of ENERCON wind power devices began here. ENERCON is the manufacturer of the most advanced wind rotors (the actual propellers), producing blades that are aimed to deliver 3 MW per unit at the Emsland Euroharbour site. The principal administrative person (*Landrat*) of the largest of the involved counties and cities of Ems-Axis, Emsland itself, uses to call

Fig. 13 Meyer shipyards at Papenburg (NW Germany). The largest dock is incredibly 504 m long. Right figure: Norwegian Jewel in front of the 70 m tall Meyerwerft Hall.
Source: Pictures courtesy of Wikipedia: left, C. Walther; right, satermedia.de, C. Brinkmann.

this a "pro-climate climate," stressing that currently, an impressive 82% of the energy consumed in the county are derived from renewable energy. The location of Euroharbour, the 24,000-population town of Haren, even boasts a 100% renewable electric current production. Among the renewables, wind is most important for the Ems-Axis region. With the shore nearby and little terrain roughness, it is most suited to create onshore wind plants; thus, NW Germany outcompetes the southern parts of the country in this respect.

Yet, there are also offshore wind power parks in the region now. In 2010, the first one in German domestic waters, "Alpha Ventus," was erected and connected to the grid. As for crucial parts of wind power technology, BARD Energy at Emden both produces rotor blades specifically designed for offshore application (there are special criteria to withstand salt corrosion, impact of water drops on the fastly moving blades, etc.,) and likewise constructs entire power plants at offshore sites (Fig. 14).

Suffice this to show features of the booming Ems-Axis economic region, which additionally sports, for example, the Transrapid (maglev) testbed at Lathen, and notably a big plant at Werlte, which will be the first in the world to convert excess wind power energy via hydrogen and hydrogenation of CO_2 into methane for energy storage purposes (to be combined with natural gas CH_4 and biogas). With regard to issues of energy use efficiency and extending the amount of renewable energy supply, the Ems-Axis consortium stated in May 2011:

"Partners in growing region Ems-Axis consider big chances for local and regional economy to be obtained from making energy supply a cornerstone of economical politics. Simultaneously they respond to their environmental responsibility by making energy use more efficient and increasing the share of renewable energy sources.

Growing region Ems-Axis is capable of becoming a model (blueprint) energy supply region for the future. Concerning Germany, this region both has the largest concentration of wind power plants and is the site of globally active producers of wind power devices. In addition, renewable energy is earned here from all biomass, sun and geothermal resources. So there is a bandwidth of competence in energy supply which yields new impetus to the region by enhanced cooperation and thus advantages in

WIND LIFT I

Fig. 14 BARD Emden Energy GmbH & Co. KG produces rotor parts, etc., for offshore windpower plants, then mounting them at sea also. Located at Emden, it belongs to the economic region Ems-Axis. Photos: (Left) German special crane ship for the setups of offshore wind farms called Wind Lift I (BARD) in the harbor of Emden.
Source: Wikipedia, photographer Carschten. (Right) BARD offshore 1 (Mai, 2011). Courtesy of the BARD Group.

competition which in turn once more improves the economic performance of the local enterprises".

More pieces of information on the Ems-Axis region, including pertinent enterprises, can be obtained via http://www.emsachse.de.

CONCLUSION

The present mix of renewable resources used in both thermal and electrical energy delivery represents a superposition of both technical problems still to be overcome (the less so) and political decisions, many of which are made in favor of protecting the respective domestic industries for both producing energies and the very power plants required to obtain and convert them: this partly is a quite reasonable and, to some extent, even responsible industrial policy. Now, there are "old" energy sources, exploitation of which has become so costly that it is worthwhile only in certain most simple conditions, including hard coal and, in another way, oil sands. This statement refers to all economic costs of exploitation, ecological side effects (as well as cultural ones such as destruction of villages and first-nation settlements in favor of open pits), and risk production causes to the workers. The renewables make it to the market step by step with their increasing ability to compete economically and the perspective to relieve old dependences, in addition to avoiding the above risks by offering genuine technical alternatives.

Of course, this might produce problems for countries that have virtually nothing else to offer to today's global markets than their fossil energy carriers, including uranium, but not to some of the "big shots" in fossil fuel mining—highly industrialized countries such as the Unites States, Canada, Australia, and Russia. Apparently, however, there is no convincing perspective of sustainable development by which the common population might benefit from exploitation of fossil energy carriers alone for countries such as Niger in Western Africa (uranium) or Yemen in the Middle East (oil). Other large uranium suppliers like Gabon (West Central Africa) or Kazakhstan (Central Asia/Eastern Europe) have a more diversified supply portfolio. Several of the Arab oil-producing countries are very aware of what might happen to them, their regimes, their population, and their common welfare (which is often truly restricted to some indigenous minorities) when oil continues to get scarce, and there are cautionary economic examples of countries, societies, and national economies running out of the single, principal minable resource the entire economy was based on, such as the tiny South Pacific Republic of Nauru (phosphate) and Bolivia in Central South America (tin, silver).

Nevertheless, the exchange of our joint economic basis for energy production appears feasible globally within some 50 years from now. It remains to be seen whether this is fast enough both to control climate effects from fossil combustion within acceptable limits and to reorganize completely our strategies of personal transportation while avoiding yet more catastrophes like those in Chernobyl or Fukushima [as well as the failure of a hydropower plant in Longarone (Friaul, NE Italy) which took some 2000 lives in 1963]. Besides this, nuclear power plants—like other technical systems—can run into operation states where they almost or entirely escape control. If a catastrophic accident then can be avoided due to self-regulation or simply luck, it is by no means satisfying or consoling that, e.g., nuclear reactors arrived at states that were not even known to their own operators for extended periods of time (like in Forsmark, Sweden, in 2006), let alone these people would be able to influence it anymore.

The future awaits us but is notoriously hard to predict, but we should take chances, even severe ones, if we decide either way, and we should be aware that doing nothing is tantamount not only to taking chances but also pursuing ways that we know for sure to be not sustainable, not even in the shorter term.

ACKNOWLEDGMENTS

We are deeply thankful to the thousands of comments given by colleagues, students, friends, and especially opposite-thinking people during attractive and highly motivating discussions during the past decades. For intensive support during the preparation of this entry, we would like to thank Prof. Michael Tomaschek, University of Applied Sciences Emden/Leer.

This entry corresponds in parts with chapter 4.4 (Energy—One of the Biggest Challenges of the 21st Century) of the textbook by Fraenzle, S., Markert, B., and Wuenschmann, S. (2012) on an "Introduction to Environmental Engineering" published by Wiley/VCH, Weinheim.

REFERENCES

1. Shafiee, S.; Topal, E. When will fossil fuel reserves be diminished? Energy Policy **2009**, *37* (1), 181–189.
2. Kutz, M. *Environmentally Conscious Fossil Energy Production*; John Wiley and Sons, Hoboken, NJ, 2010.
3. Kreith, F.; Goswami, D. *Principles of Sustainable Energy*; CRC Press, Boca Raton, FL, 2011.
4. Fraenzle, S.; Markert, B.; Wuenschmann, S. *Introduction to Environmental Engineering—Innovative Technologies for Soil, Air and (Ground)water Remediation and Pollution Control*; Wiley-VCH: Weinheim, 2012.
5. Federal Institute for Geosciences and Raw Material Research. *Reserven und Verfügbarkeit von Energierohstoffen (Kurzstudie)*; Bundesanstalt für Geowissenschaften und Rohstoffe: Hannover, 2009.
6. Federal Ministry of Economics and Technology. *Energie in Deutschland (Energy in Germany)*; BMWI: Berlin, 2010.
7. U.S. Geological Survey. *Circum-Arctic Resource Appraisal: Estimates of Undiscovered Oil and Gas North of the Arctic Circle*; Washington, DC, 2008.

8. Campbell, C. The Rimini Protocol: An oil depletion protocol. Heading off economic chaos and political conflict during the second half of the age of oil. Energy Policy **2006**, *34* (12), 1319–1325.
9. Blum, A. Die finale Ölkrise—fossile Brennstoffe, vor allem Erdöl, sind endlich und eine Verknappung ist absehbar, 2005, available at http://www.raize.ch/Geologie/erdoel/oil.html (accessed on April 11, 2005).
10. Kondratyev, K.; Krapivin, V.; Varostos, C. *Global Carbon Cycle and Climate Change*; Springer: Berlin, 2003.
11. Leroux, M.; Comby, J. *Global Warming. Myth or Reality*; Springer: Berlin, 2005.
12. Hoel, M.; Kverndokk, S. Depletion of fossil fuels and the impacts of global warming. Resour. Energy Econ. **1996**, *18* (2), 115–136.
13. Ostwald, W. *Die Mühle des Lebens*; Theod. Thomas: Leipzig, 1911.
14. Voiland, A. *2009: Second Warmest Year on Record; End of Warmest Decade*; NASA Goddard Institute for Space Studies (accessed January 22, 2010).
15. Reschetilowski, W. Vom energetischen Imperativ zur nachhaltigen Chemie. Nachr. Chem. **2012**, 134–136.
16. United Nations. United Nations Framework Convention on Climate Change: FCCC/CP/1997/INF. 5, List of participants (COP 3), available at http://en.wikipedia.org/wiki/United_Nations_Framework_Convention_on_Climate_Change (accessed January 21, 2011).
17. Romberg, B. Reaktorneubau: Amerika wagt Renaissance der Kernenergie. Süddeutsche Zeitung, Febr. 13th, 2012.
18. Breban, D.C.; Moreno, J.; Mocanu, N. Activities of Pu radionuclides and ^{241}Am in soil samples from an alpine pasture in Romania. J. Radioanal. Nucl. Chem. **2003**, *258*, 613–617.
19. Gill, V. Chernobyl zone shows decline in biodiversity. BBC News online, Science and Environment, 2010, available at http://www.bbc.co.uk/news/science-environment-10819027 (accessed January 20, 2010).
20. Etspüler, M. *Transmutation: Die zauberhafte Entschärfung des Atommülls*; Frankfurter Allgemeine Zeitung: Frankfurt, 2011.
21. De Bruyn, D. European fast neutron transmutation reactor projects (MYRRHA/XT-ADS). IAEA Review Paper, 2009, available at http://www-pub.iaea.org/MTCD/publications/PDF/P1433_CD/datasets/summaries/Sum_SM-ADS.pdf (accessed November 2011).
22. Hoffert, M.; Caldeira, K.; Benford, G.; Criswell, D.; Green, C.; Herzog, H.; Jain, A.; Kheshgi, H.; Lackner, K.; Lewis, J.; Lightfoot, D.; Manheimer, W.; Mankins, J.; Mauel, M.; Perkins, J.; Schlesinger, M.; Volk, T.; Wigley, T. Advanced technology paths to global climate stability: Energy for a greenhouse planet. Science **2002**, *298* (5595), 981–987.
23. Turner, J. A realizable renewable energy future. Science **1999**, *285* (5428), 687–689.
24. Kaltschmitt, M.; Streicher, W.; Wiese, A. *Renewable Energy: Technology, Economics and Environment*; Springer: Berlin Heidelberg, 2010.
25. Pimentel, D., Ed. *Biofuels, Solar and Wind as Renewable Energy System. Benefits and Risks*; Springer, Science+Business Media B.V., Dordrecht, 2008.
26. Olah, G.; Goeppert, A.; Prakash, C.K. *Beyond Oil and Gas: The Methanol Economy*; Wiley-VCH: Weinheim, 2006.
27. REN 21. Renewables 2010 Global Status Report (Paris: REN 21 Secretariat). Deutsche Gesellschaft für Technische Zusammenarbeit (GTZ) GmbH, 2010, http://www.ren21.net (accessed January 22, 2010).
28. REN 21. Renewables 2007 Global Status Report (Paris: REN 21 Secretariat and Washington, DC: Worldwatch Institute). Deutsche Gesellschaft für Technische Zusammenarbeit (GTZ) GmbH, 2008; 21, available at http://www.ren21.net (accessed January 22, 2010).
29. Markert, B.; Wuenschmann, S.; Fraenzle, S.; Figueiredo, A.; Ribeiro, A.P.; Wang, M. Bioindication of trace metals—With special reference to megacities. Environ. Pollut. **2011**, *159*, 1991–1995.
30. Charlier, R.; Finkl, C. *Ocean Energy: Tide and Tidal Power*; Springer: Berlin, Heidelberg, 2009.
31. Hoffmann, V. Energie aus Sonne, Wind und Meer. Harri Deutsch Thun (SUI): Frankfurt/Main, 1990.
32. Cruz, J. *Ocean Wave Energy. Current Status and Future Perspectives*; Springer: Heidelberg, 2008.
33. Multon, B. *Marine Renewable Handbook*; Wiley and Sons, Weinheim, 2011.
34. Glassley, W. *Geothermal Energy. Renewable Energy and the Environment*; CRC Press, Taylor and Francis Group, Boca Raton, FL, 2010.
35. Ghosh, T.; Prelas, M. Geothermal energy. In *Energy Resources and Systems*; Gosh, T., Prelas, M., Eds.; Springer, Weinheim, 2011; 217–266.
36. European Wind Energy Association. *Wind Energy—The Facts: A Guide to the Technology, Economics and Future of Wind Power*; Taylor and Francis Ltd., Oxford, 2009.
37. Nelson, V. *Wind Energy: Renewable Energy and the Environment*; CRC Press, Boca Raton, Florida, 2009.
38. Hau, E. *Wind Turbines: Fundamentals, Technologies, Application, Economics*, 3rd Ed.; Springer: Berlin, 2012.
39. Maki, K.; Sbragio, R.; Vlahopoulos, N. System design of a wind turbine using a multi-level optimization approach. Renewable Energy, **2012**, *43*, 101–110.
40. Goetzberger, A.; Hoffmann, V. *Photovoltaic Solar Energy Generation*; Springer: Berlin, Heidelberg, 2005.
41. Foster, R.; Ghassemi, M.; Cota, A. *Solar Energy: Renewable Energy and the Environment*; CRC Press, Taylor and Francis Group, Boca Raton, FL, 2009.
42. Goldemberg, J.; Coelho, S. Renewable energy—Traditional biomass vs. modern biomass. Energy Policy, **2004**, *32* (6), 711–714.
43. Fraenzle, S.; Markert, B. Metals in biomass: From the biological system of elements to reasons of fractionation and element use. Environ. Sci. Pollut. Res. **2007**, (6), 404–413.
44. Burkhardt, M.; Weigand, T. Der Strom-Mix in Deutschland (the fuel mix in Germany). German television ZDF, 2011, available at http://www.heute.de/ZDFheute/inhalt/24/0,3672,8233016,00.html (accessed January 21, 2011).
45. U.S. Energy Information Administration. Total Primary Energy. Statistics on Germany, available at http://www.eia.gov/countries/country-data.cfm?fips=GM (accessed January 22, 2010).
46. Flavin, C. Last word: Renewable energy at the tipping point. In *Renewable Energy Policy Network for the 21st Century (REN 21) (2010) Renewables 2010 Global Status Report (Paris: REN 21 Secretariat)*; Deutsche Gesellschaft für Technische Zusammenarbeit (GTZ) GmbH, 2010; 52–53.

Alternative Energy: Hydropower

Andrea Micangeli
Sara Evangelisti
Danilo Sbordone
Interuniversity Research Center for Sustainable Development (CIRPS), Sapienza University of Rome, Rome, Italy

Abstract
Hydropower generates electricity by using water, and it is one of the cheapest and eco-friendly ways, especially the use of small hydropower plants. It exploits the vast global water cycle: the water constantly evaporates from lakes and oceans, forming clouds, precipitating as rain or snow, and then flowing back down to the oceans. In this way, it is possible to use water's potential energy in its natural flow to produce power. The water cycle is an endless, constantly recharging system, and therefore, hydropower is considered a renewable energy. As stated by International Energy Agency (IEA) in 2008, the total installed capacity of hydropower is about 850 GW, and hydro sources produce about 3000 TWh of electricity annually, supplying about 15% of total world's electricity. The IEA projects that hydro will grow up to 63% for the period 2002–2030. The agency predicts that new hydro plants will continue to be built, not at a rate high enough to maintain hydro's current percentage of total electricity generation. As a result, hydropower is projected to fall to 13% by 2030, from the present 15%. It is estimated that two-thirds of the world's economically feasible potential is still to be exploited and it is mainly concentrated in developing countries such as Africa, Asia, and South America. China is using only about one-quarter of its huge hydro potential of 450 GW, and it is the main developer of hydro technology today. Figures from the Chinese government suggest that it will add more than 12 GW of new capacity each year until 2020 to reach 300 GW. This entry highlights the basics of different hydro technologies, with a special focus on small hydro run-of-river plants.

INTRODUCTION

Approximately 70% of the earth's surface is covered with water, a resource that has been exploited for many centuries. Hydropower is currently the most common and the most important renewable energy source: throughout the world, it produces 3288 TWh, just over 17% of global production and the 84% of energy produced by renewable energy sources[1] from an installed capacity of about 850 GW.[2]

The International Energy Agency (IEA) has developed a number of scenarios that describe the efforts needed to reduce carbon dioxide emissions. The "business-as-usual" baseline scenario foreshadows the situation in the absence of policy change and major supply constraints leading to increases in oil demand and CO_2 emissions. The "**BLUE**" scenario is the most ambitious, bringing emissions at 50% of the 2005 level by 2050. This implies of course higher investment costs, as well as greater needs in technological and policy developments. In *Energy Technology Perspectives 2010*, it states that hydro could produce up to 6000 TWh in 2050.

The main characteristics that make hydropower a successful energy source are its plant storage capacity and fast responses to meet sudden fluctuations in electricity demand. Global hydropower generation has increased by 50% since 1990, with the highest absolute growth in China, as shown in Fig. 1.

IEA estimates the *global technically exploitable hydropower potential* (the *technically exploitable potential* is the annual energy potential of all natural water flows that can be exploited within the limits of current technologies[3]) at more than 16,400 TWh per year.[1] However, hydroelectric plants of big dimensions, with million cubic meter water basins, have negatively affected the natural and social environment of the territories. Small plants are characterized by a different management, distributed on the territory, managed in small communities, integrated in the multiple and balanced use of the water resources.

The contribution of *small hydropower* (*SHP*) *plants*, defined as those with installed capacity of up to 10 MW, to the worldwide electrical supply is about 1%–2% of the total one, amounting to about 61 GW.[2] Europe with about 13 GW installed capacity has the second biggest contribution to the world's installed capacity, just behind Asia. Moreover, the SHP potential is estimated in 180,000 MW.

SHP has a key role to play in the development of renewable energy resources and an even greater role in developing countries. In the face of increasing electricity demand, international agreements to reduce greenhouse gases,

Encyclopedia of Environmental Management DOI: 10.1081/E-EEM-120047189

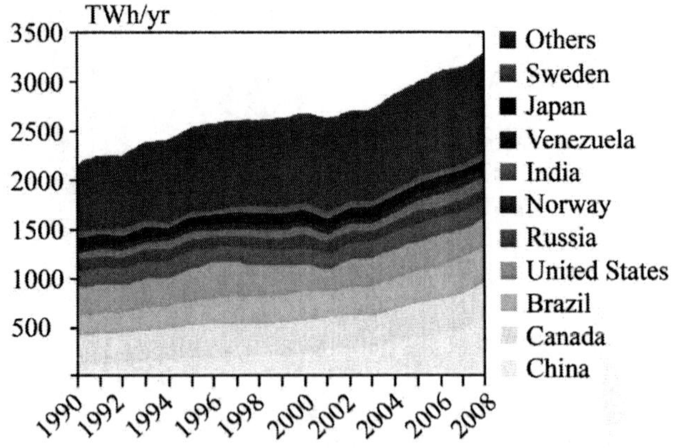
TWh/yr
3500
3000
2500
2000
1500
1000
500
1990
1992
1994
1996
1998
2000
2002
2004
2006
2008
Others
Sweden
Japan
Venezuela
India
Norway
Russia
United States
Brazil
Canada
China

Fig. 1 Evolution of global hydropower generation, 1990–2008. **Source:** International Energy Agency.[1]

environmental degradation from fossil fuel extraction and use, and the fact that, in many countries, large hydropower sites have been mostly exploited, there is an increasing interest in developing SHP. Indeed, SHP has a huge, as yet largely untapped potential, which will enable it to make a significant contribution to future energy needs, offering a very good alternative to conventional sources of electricity, not only in the developed world but also in developing countries.

A hydropower sector technological maturity has already been reached during the last century, but only big plants have received all the benefits from technological development, while those of smaller dimension have been neglected. Nowadays the economy of scale, social and environmental implications, suggest this solution due to their economical feasibility and environmental respectful, allowing sustainable distributed production with an easy installation and great applicability in developing countries.

This entry is organized as follows: first, a classification of the hydropower plants is given. Then, the basics of the technology of mini and micro-hydro plants are illustrated, together with a description of the main civil works that occurred in a hydropower scheme. Finally, a conclusion on the potential and shortcomings of the hydropower technology is drawn.

CLASSIFICATION OF THE HYDROELECTRICAL PLANTS

Hydropower plants can be generally classified in terms of power outputs:

- *Micro-hydro plant*, with a nominal power lower than 100 kW, subdivided into *low-head plants*, when the vertical drop is lower than 50 m, and *low-flow rate plants*, when the water flow is lower than 10 m^3/sec.
- *Mini-hydro plant*, with a nominal power between 100 kW and 1000 kW, subdivided into *mini-head plants*, when the vertical drop is between 50 and 250 m, and *mini-flow rate plants*, if the water flow is between 10 and 100 m^3/sec.
- *Small hydro plants*, with a nominal power between 100 kW and 10 MW, subdivided into *medium-head plants*, when the drop is between 250 and 1000 m, and *medium flow-rate plants*, when the water flow is between 100 and 1000 m^3/sec.
- *Big hydro plants*, as shown in Fig. 2, with a nominal power of more than 10 MW, defined as *high-head plants* if the drop is higher than 1000 m and as *high-flow rate plants* with a water flow of more than 1000 m^3/sec.

Another important way to classify hydroelectric plants is on the basis of their typology. In particular, they can be classified as follows:

- *Run-of-river scheme*: or fluent water plants, they take a portion of a river through a canal or penstock. They do not require the use of a dam or catch basin. Because of that, they aim to affect upstream water levels and downstream stream flow less than any other power plants. Electricity generation from these plants could change in the amount of water flowing in the river.
- *Storage scheme*: an impound water behind a dam, as a reservoir. Water is released through turbine generators to produce electricity. The water storage and release cycles can be relatively short, for instance, storing water at night for daytime power generation, or the cycles can be long, storing spring runoff for generation in the summer, when air conditioner use

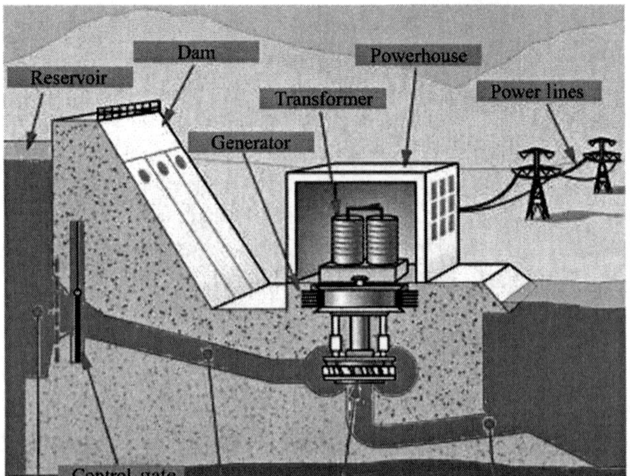
Dam
Powerhouse
Reservoir
Transformer
Power lines
Generator
Control gate
Intake
Penstock
Turbine
Outflow

Fig. 2 Big hydro storage scheme.

increases power demand. Some projects operate on multiyear cycles carrying over water in a wet year to offset the effects of dry years.

– *Pumped-storage scheme*: these plants use off-peak electricity to pump water from a lower reservoir to an upper reservoir. During periods of high electrical demand, the water is released back to the lower reservoir to generate electricity.

The following sections focus on mini and micro-hydropower plants and run-of river scheme.

MINI-HYDRO PLANTS

Mini-hydropower (MHP) plants have an installed power lower than 1 MW actually in Europe is 3 MW but in many country (in particular USA 5 MW) it may be more. Generally, these plants need less civil works, consistently reducing the costs connected to the realization of the plant and justifying their realization also under an economic point of view. If the plants are well planned and placed, their environmental impact is reduced for their limited dimensions. The simplicity of construction allows them to be introduced in contexts where the technology of the sector is not yet developed and there is a strong need for mechanical or electrical power. Also in those cases, MHP plants can be operated and maintained locally, even with less-specialized technicians. Changing hydropower plant size or typology, many things are the same, such as a turbine installation. This entry focuses on mini-hydro due to its low environmental impact and its opportunities of developing in the future.

WATER RESOURCE

Hydraulic energy, as almost every forms of energy on the earth, comes from the sun, which is the "engine" of the hydrological cycle. The sun, irradiating and warming up the atmosphere, makes seas and lakes evaporate; the water vapor rises up and thickens the clouds that move because of the wind, also generated by the sun; the clouds then produce precipitations in the form of snow, hail, and rain. When the rainfall ends up in the natural basin situated at a higher level, energy is transformed to *potential energy*. This energy is naturally stocked in rivers and in creeks that flow into the sea, closing in this way the hydrological cycle (see Fig. 3).

The amount of available energy, which the water basin can produce at a given height, comes from the water level reached at the end of the cycle. In other words, to know the potential energy of a basin, it is necessary to evaluate the available rise, depending on the orography of the territory and on specific water works such as dams or small barriers.

The amount of water available is defined as the mass of water flowing per time unit (*flow rate*). In general, the

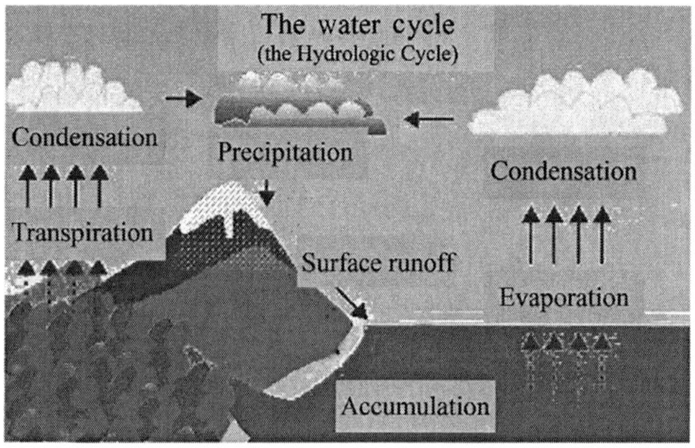

The water cycle
(the Hydrologic Cycle)
Condensation
Precipitation
Condensation
Transpiration
Surface runoff
Evaporation
Accumulation

Fig. 3 Hydrological cycle.
Source: Harvey et al.[4]

potential power from a reservoir can be calculated by the following equation:

$$P_O = \rho \cdot g \cdot Q \cdot H_O \quad (1)$$

where
P_0 = theoretical power (W)
ρ = water density (1000 kg/m^3)
g = gravitational acceleration (9.8 m/sec^2)
Q = flow rate (m^3/sec)
H_0 = net head (m)

The output power of the plant is a percentage of it, due to mechanical, electrical, and friction losses.

Hydrology and Rain Measurements

In order to exploit the energy of water for power purposes, a *hydro geological analysis* of the territory is needed, particularly for catch basins. The analysis is based on the evaluation of the supply of the basin and its outflow. To obtain a balanced catch basin, it has to take into consideration the *meteoric flow rate*, *evaporation*, and both *superficial* and *underground circulation*. The quantity of water from the rainfall to the basin, the *meteoric supply*, must be evaluated not only on the water surface of the basin itself but also on the whole area in which the rainfall is collected as well as the streams towards the basin (see Eq. 2). The flows, which depend on soil permeability, are essentially of two types: *superficial* and *underground*, i.e., when the water filters through the soil and supplies underground basins and water-bearing stratum.

It is possible to evaluate the *meteoric supply*—superficial flow, as:

$$P = E + D + (I - C) \quad (2)$$

where P is the *meteoric supply* to the basin, E is the contribution given to the evaporation, D is the outflows,

and I and C are the increase and decrease of the basins, respectively.
The term E is the amount of the following different contributions:

- Evaporation of water from the soil
- Transpiration of plants
- Evaporation of water intercepted by vegetation
- Evaporation of internal basins

Similarly, the term in relation to draining—underground flow—can be subdivided into the following:

- Natural water draining underground toward the external (*groundwater*)
- Artificial draining water toward the external (*inversion*)
- Natural superficial water inflow from the external (*water flows*)
- Underground natural water inflow (*water-bearing stratum*)
- Artificial inflow from the external (*adduction*)

Joining of different rivers has to be considered as well. The evaluation of the meteoric intake is usually performed through specialized devices, such as *rain gauges*, very common all over the world (see Fig. 4).

Fig. 4 Example of a rain gauge.

The intensity of the rainfall flow rate is not constant through time and it has to be referred to different periods of the year (usually a multiyear). Not all the water from the rainfall ends up in the catch basin as shown before. Generally, the phenomenon is estimated by introducing a *coefficient of draining*, depending on the waterfall and the water collected into the catch basin, as shown in Eq. 3:

$$C = \frac{V}{V_0} \quad (3)$$

where V is the real caught volume and V_0 is the waterfall.

Once V is determined, it is possible to calculate directly the energy exploitable from the plant in a given site:

$$E = 0.00273 \cdot \eta \cdot H_0 \cdot V \quad (4)$$

where H_0 indicates the *net drop*, in other words, the available drop minus the losses in the work of adduction with η output efficiency of the turbine.

Within a natural *hydrological basin*, it is necessary to analyze the head and the flow rate available along the whole river bed, through the *hydrodynamic curve* (see Fig. 5). It shows the surfaces of the catch basin on the horizontal axes and the height of the water flow on the vertical one. Through the hydrodynamic curve, it is possible to optimize the entire use of a catch basin, while for the realization of a single plant, without having the intention to optimize the use of the resource along the whole river bed, it is enough to measure the flow rate of the river and the consequent evaluation of the quantity of water that can be taken from the basin or from the available drop.

MEASURE OF THE WATER FLOW

The determination of the water flow of a catch basin can be done by using specific devices, but it has to be undertaken only in absence of historical data of the course along the years. Different methods exist.[4] However, for each of them, it would be necessary to repeat the measures along a period of time to obtain the variation that occur throughout the year.

One of the simplest methods that can be used is to force the flow to get into a container of known dimensions, measuring the necessary time to fill it up. This method is known for its simplicity; however, it is limited in that it can be applied only in rivers with small water flows.

A second method (Fig. 6) consists of the realization of a weir of known dimensions, in which the river is forced to get into—the *weir method*. This method can be used to bring up to around 1 m^3/sec. To measure how much is carried, the second level reached by the water is taken into consideration as illustrated in Fig. 7.

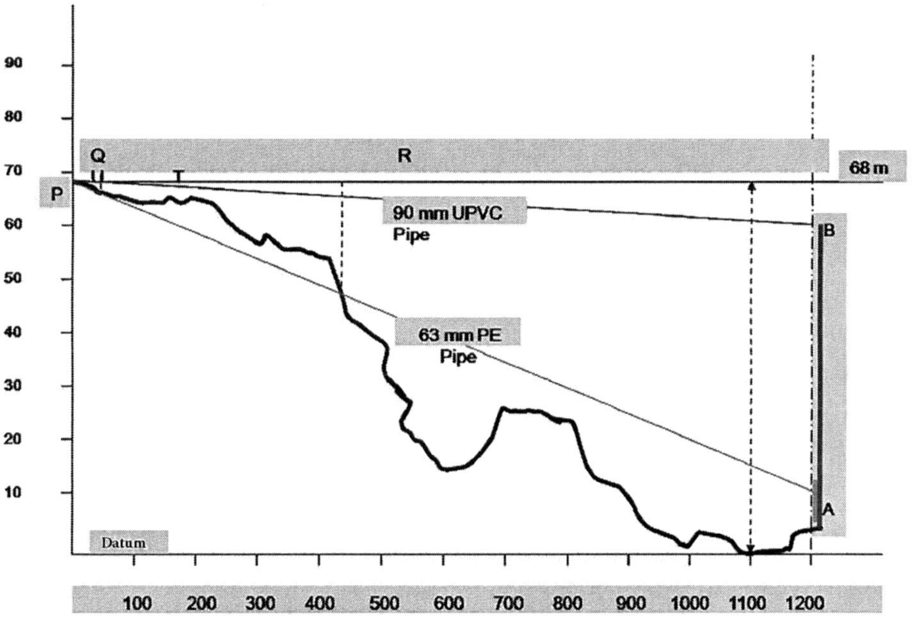
90
80
70
60
50
40
30
20
10
Q
R
T
P
68 m
90 mm UPVC Pipe
63 mm PE Pipe
B
A
Datum
100 200 300 400 500 600 700 800 900 1000 1100 1200

Fig. 5 Example of a hydrodynamic curve.

The water flow Q can be calculated using:

$$Q = 0.41 \cdot B \cdot H \cdot \sqrt{2 \cdot g \cdot H} \quad (5)$$

In this formula, H can be calculated as the difference between H_2 and H_1 (see Fig. 7), B is the width of the weir, and g is the gravitational acceleration.

Another method of measuring water flow involves the evaluation of water velocity on the cross-sectional area. Velocity measurements can be made through a very simple method: with a floating, not too light in order to avoid the friction with the air, in a place where the river is pretty regular and flat, measuring the time that the floating takes to cover a specific distance. The measurement must be repeated more than once and the sought value taken into consideration must be the average between the distance covered and the time spent to cover it. The final velocity must be corrected with a factor between 0.75 and 0.85, given by the losses due to the friction with the sides of the canal.

The velocity can be obtained also through different methods, such as the use of "titled solutions," taking into account the variations on the electricity conductivity of the river when it flows with a known quantity of salt inside.

Once the cross section is evaluated, the water flow can be then calculated through the *formula of Manning*:[5]

$$Q = \frac{A \cdot R^{\frac{2}{3}} \cdot S^{\frac{1}{2}}}{n} \quad (6)$$

where A represents the cross section, R is the hydraulic radius, and S is the slope of the water surface. The value of n can be obtained from Table 1.

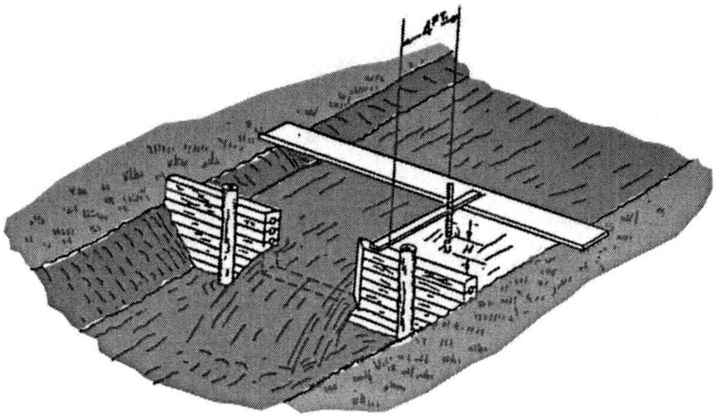

Fig. 6 Weir method.

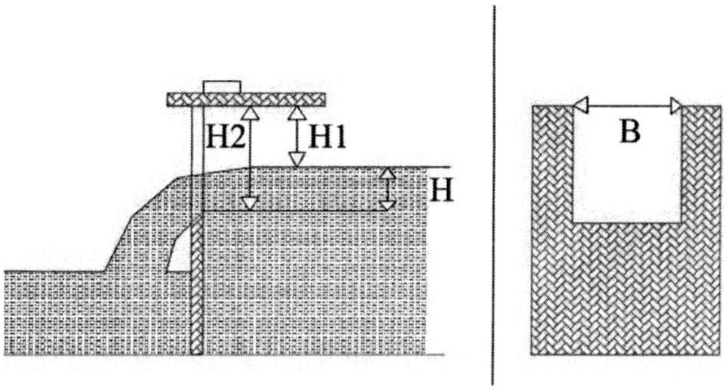
H2
H1
H
B

Fig. 7 Weir realization.

Table 1 Value of the *factor of Manning* (*n*) for different river bed typologies.

River bed typology	*n*
Regular river bed with a minimum annual flow	0.030
Stable flow condition	0.035
River with stagnant water, aquatic vegetation, and meanders	0.045
River with stones and shrubs with shallow pools and lush vegetation	0.060

Flow Duration Curve

Water flow measurements are always referred to a specific period, as the water of the river changes during the year, passing several times from huge quantities of water to smaller ones. In general, the curve is uneven as it reports the water flow rate throughout the year as shown in Fig. 8.

To organize the collected measurement data, it is possible to use another graphic that puts them all together. Indeed, in Fig. 8, it is possible to note that there are two evident points of absolute maximum and absolute minimum, corresponding to the maximum quantity of water that occurs for a very short period and the minimum quantity of water that is the quantity available all year. Moreover, Fig. 9 shows a *flow duration curve*, which presents the duration of each amount of the flow rate.

MEASUREMENTS OF THE GROSS HEAD

To measure the height difference between two points, it is necessary to utilize a level and to follow the scheme in Fig. 10. The operator must simply read the values of each ruler to come out with the height by computing the difference between the two values. This procedure can be repeated until the final point is reached. If a level is not available, it is possible to use a table with a carpenter level—even if it requires a lot of patience—or to proceed with a plastic transparent pipe filled up with water, which fulfills the same characteristics of a level. While measuring from the available head, it is wise to also calculate the length of the *forced penstock* as distance from the hold point to the arrival of the penstock itself. This is essential both for the choice of the material of the penstock and for the evaluation of the pressure drop.

INSTREAM FLOW AND ENVIRONMENTAL IMPACT

The balance of any catch basin is connected not only to the water balance but also to the real possibility of exploitation, characterized by other aspects:

- Rivers can be used not only for power purposes. Before proceeding to the derivation of the outflow, it is necessary to be sure that it will not have a negative impact on further communities. It is necessary to verify all the aspects connected with the multiple uses of the water resource.
- The subsistence of the natural balances involved in the river and in the catch basin itself. An example, to adduce water to a riverbed of a catch basin, can be useful for the fish fauna, although the flow rate is high and it comes from nearby.

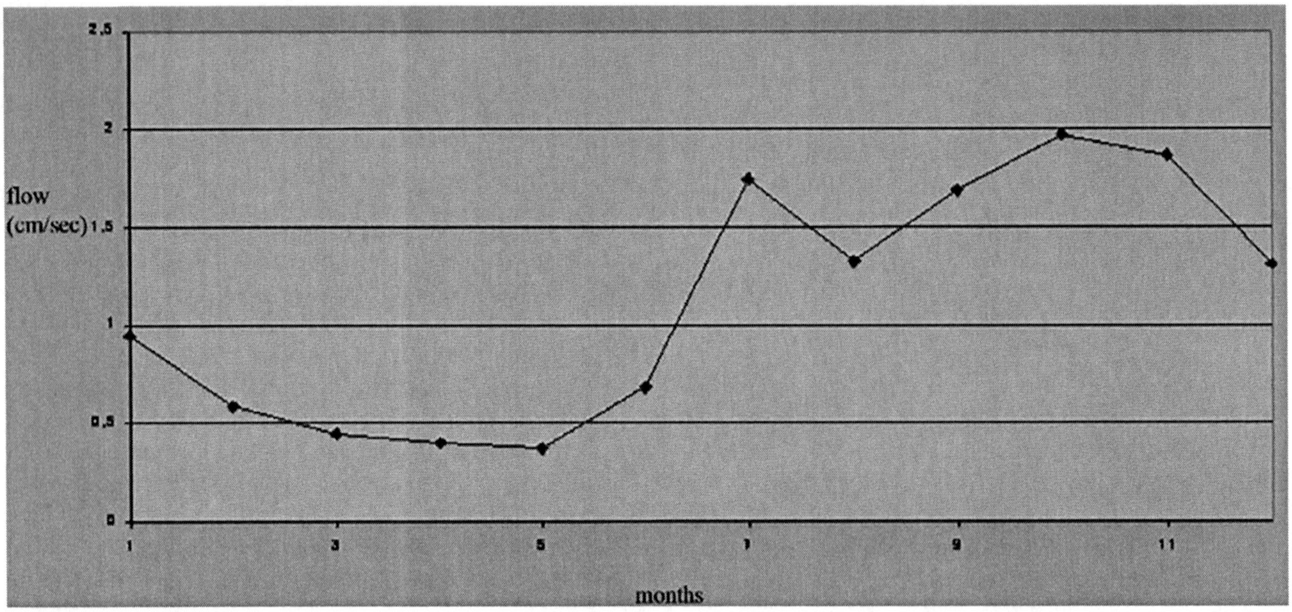
2.5
2
flow
(cm/sec) 1.5
1
0.5
0
1
3
5
7
9
11
months

Fig. 8 Example of a daily flow curve.

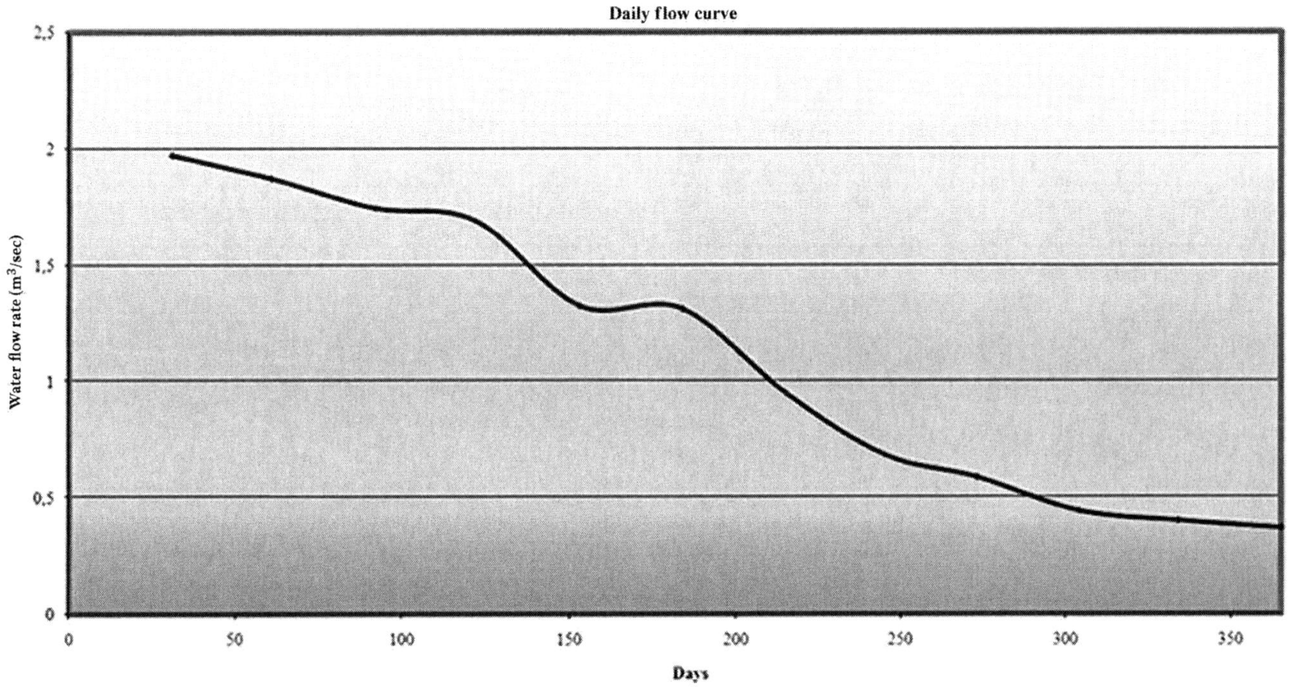
Daily flow curve
Water flow rate (m³/sec)
2,5
2
1,5
1
0,5
0
0
50
100
150
200
250
300
350
Days

Fig. 9 Example of a flow duration curve.

The use of the water resource of a catch basin refers to a wider issue that generally is approached letting a minimum natural course of the river, generally indicated as *minimun instream flow* (*IF*), defined as "the minimum height of water needed to maintain the values of the basin at an acceptable level." The calculation of the IF is essential: in fact, if the minimum flow rate is lower than the IF during the planning stage, the no-working periods of the plant can be estimated.

To guarantee a minimum flow means to preserve the biological balance and the need of the use of a civil work as a caption downriver. The derivation established on the basis of this context must be lower than the limit beyond which it may influence the river ecosystem and may cause the entering in crisis however the natural water regime must be guaranteed.

The river regime model must therefore take into account the following aspects:

H1
H2
ΔH = H2 − H1

Fig. 10 Example of measurement of the gross head.

- Biological species that would suffer from the uncontrolled derivation of water.
- Hydrologic characterization for the protection of water balance (equilibrium) and the defense of soil.
- The use of the water resource represented by the social and economic wardship of needs.

Two different relations are commonly used for the calculation of the minimum IF:

$$Q_{\mathrm{IF}} = \frac{15 \cdot \alpha \cdot Q_{\mathrm{media}}}{\left[\ln\left(\alpha \cdot Q_{\mathrm{media}}\right)\right]^2} \tag{7}$$

where α is the *coefficient of perpetuity* given by the relation between low intake and the medium IF values, expressed in liters per second:

$$\mathrm{IF}_{\mathrm{hydrol.}} = 6 \cdot \mathrm{IF}_{\mathrm{microhabitat}} \tag{8}$$

The equation shows that conserving the IF from a hydrological point of view also means, with a big margin, that the conservation of IF from a biological point of view is connected with the microhabitat. If the value of the low intake is unknown and only the average is known, it is possible to assume that the coefficient of perpetuity α is equal to 0.24, showing the outflow of hydropower plant that guarantees the preservation of the fish fauna and hydrology. Fig. 11 shows a typical example of the realization of a ca-

Fig. 11 Example of a system to conserve fish fauna.

nal to guarantee the conservation of hydrology and of the fish fauna in a hydropower plant.

CIVIL WORKS IN MHP PLANTS

The *civil works* for *hydropower facilities* have the functions to capture, to exploit, and to return the water downriver. In small rivers, the realization of those works is done by deflecting the water flow for a short period of time, to operate in dry conditions. To deflect the river, a provisional river bed has to be built or the plant must be realized during the dry season, if feasible.

Dams and Weirs

The choice of the typology of the dam must be made taking into consideration the orography of the territory and the water resource to exploit.[6] Generally, hydraulic works must create a good drop between mountain and valley, to have the possibility to exploit the potential energy from the mass of water. The structure can be *dams* (Fig. 12) or *catch basins* (Fig. 13).

A *dam* is a "lung" of water available that is able to compensate, partially and totally depending on the volume created, the variation of the flow rate of the river during the year. Although the creation of a dam needs a big investment, this makes it practically not usable in MHP.[7]

A type of barrier—often installed in small plants—is the *derivation weir*. The derivation weir is smaller and more cost-effective than a traditional dam and generally the water overflows the traverse, reversing in the natural path of the river. If the plant is big enough, the derivation weir turns

Fig. 12 Example of a dam.

into a small dam that basically does not function like a water accumulator, but it aims at raising the level of the water flow. The derivation weir can be realized with several different materials and, when possible, it is suggested that locally available material, such as rocky materials, be used. Otherwise, it is possible to use blocks of flat rocksand soil kept altogether by metal nets to constitute the *barrage*.

It is important that the barrage is realized with a central waterproof nucleus, made with clay, and supported with soil or minerals. If there sand or gravel is available, a barrier made of concrete can be also taken into consideration.

If the regimes of full flow are huge, it is necessary that the intake must be drained through ad hoc dischargers not easy to realize when the barrage has been made in the river. Instead, in places with much seismic activities or in very cold climates, it is better to avoid rigid structures, so floor barriers are preferred.

Stability of the Dam

In small plants, the stability of the dam generally depends on the weight of the dam itself. The strains that the dam applies to the ground through its own weight can be higher than those that the ground can take. It is necessary to minimize the infiltration of water underneath the dam with the use of spillway drain and erosion or diaphragms.

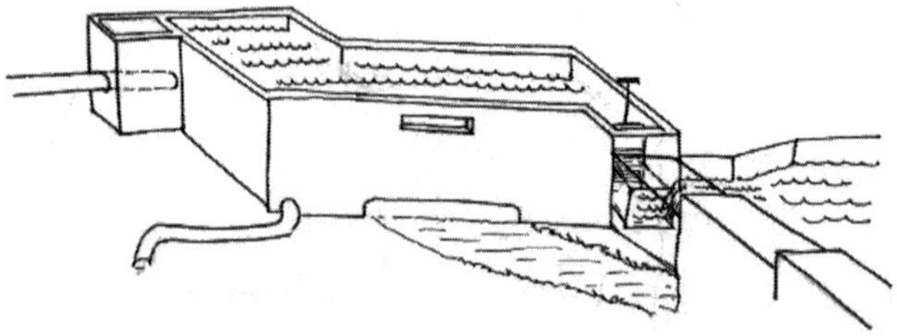

Fig. 13 Example of a catch basin.

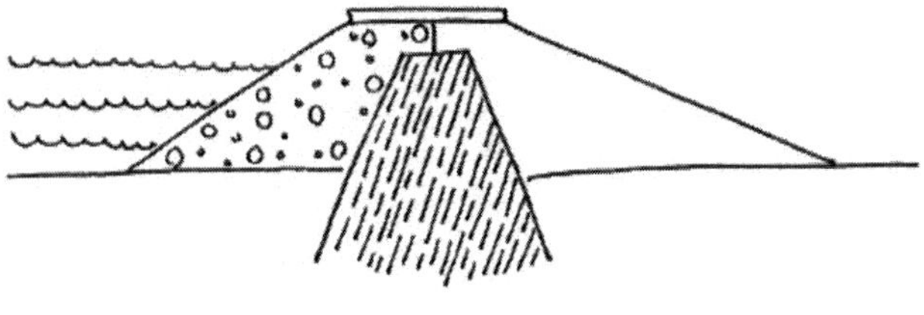

Fig. 14 Stability of the dam.

Also, the stability to the overturning of the dam must be guaranteed in all the conditions of charge concerning the hydrostatic push related to the dragging of solid material or to seismic waves. In order to guarantee that, it is enough to ensure that all the strains against the ground are not negative, in order to avoid dangerous situations that can bring the structure to overturn. This entails designing a dam where all the horizontal and vertical strains fall internally in the central part of its own basement (see Fig. 14).

As the hydrostatic strains generally go from upstream to downstream, the side of the dam facing downstream is more sloping as compared to the perpendicular of the basement; thus, the profile of the dam upstream has to be more vertical than the downstream.

System of Elevating the Free Water Surface

To control the water flow, *floodgates* are generally used. Given that most of these devices are used to control the free water surface elevation being stored or routed, they are also known as *crest gates.*

A removable type of *floodgates* is *flashbooks gates*—wooden panels usually installed on the cap of the barrier that allow the increase of the water surface and that can be removed during floods, avoiding the inundation of the upstream fields.[8]

Fig. 15 Example of fusegates: "Sant'Antonio 1" IdroPower Station (Italy).

Generally, to avoid the manual intervention and to check upstream flooding, it is possible to install a gate that can be opened progressively during full intakes. Another solution is to install *fusegates*, concrete crates that flip over when the water level is reached (Fig. 15). Finally, another type is the *dinghy*, anchored to the cap of the barrier that blows up during full intakes of water, only to deflate again during the rest of the year.

Spillway

If the flow rate varies during the year, it could be that the floodgates systems are not enough to guarantee the integrity of the plant. In this case, it is necessary to foresee a system that takes the surplus flow rate to downstream—the *spillway*. Except during flood periods, water does not normally flow over a spillway. The surplus water usually flows at a high velocity and often it is necessary to insert a system to reduce its kinetic energy. In small plants, the introduction of a drainage, which allows emptying the loading tank to help in the maintenance of the plant, is always considered.

The spillway is also used in times of emergency, i.e., shutting down the plant.

Intake

The *water intake* (see Fig. 16) is a structure in which water is adduced in order to bring it to the *forebay tank*. It must

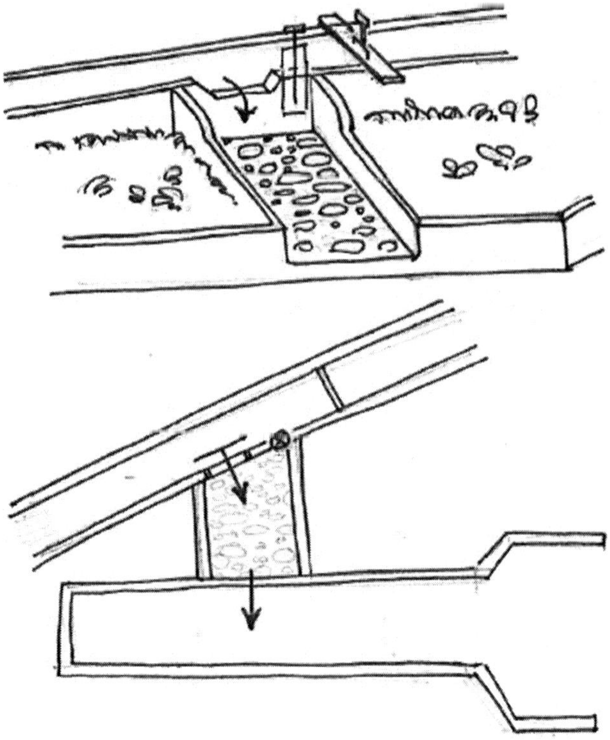

Fig. 16 Example of intake works.

be able to address in the penstock or in the drainage canal the amount of water estimated in the project. It is pleonastic to underline how the civil work has to be studied and realized in a way to minimize friction losses and the impact on the environment. It must also be designed to minimize maintenance and to reduce the costs. Practical aspects of the project related to civil work therefore must follow the following criterion: Hydraulics and structural works have to guarantee the resistance of the pipes to minimize the waste of energy and to be cost-effective, to avoid transportation of solid material inside the pipes to the powerhouse and for easier maintenance, and to reduce the passage of fishes and to not compromise the ecosystem of the area in which it has been realized.

The water intake works can be realized in different ways, depending on the peculiarity of the orography of the territory. Often it is a channel that brings the water to the point where the forced pipe is. In this case, the channel is realized in a way that the water flows slowly with a contained slope of the pipe in order to contain the losses and the erosion of the walls and to preserve the jump.

The choice of the dimension of the intake channel and the water velocity are results of a compromise to avoid frequent maintenance due to the deposits of sand and slime, to avoid losses and erosion against the wall. Generally, the water flows along the channel with a velocity between 0.1 and 0.4 m/sec. If the quantity of solid materials to be transported is huge, then a bigger tank, where the deposits end up due to the reduction of the speed of the water as a result of a bigger section, is needed, based on the fundamental hypothesis of the continuity of the intake.

The *sedimentation tank* is necessary if the channel has not an open surface, due to the costs connected to the maintenance of the closed channel. In small plants, the drainage tank usually works as a sedimentation tank as well; however it must be cleaned more frequently and it will have a bigger dimension due to the sedimentation process. The transportation of solid material is very deleterious not only for turbine performance but also for the life of the device itself.[5]

The *orientation* of the *intake* is crucial when choosing a project, as it can reduce the accumulation of material over the grid itself and the frequency of the intervention of maintenance. The best position is parallel to the flow letting the full flows the task of removing the material stuck in front of the grid. Anyway, it does not have to be located in areas of stagnant water as the whirlpool and the parasite flows tend to accumulate solid material in front of the grill. If there is a discharger, it is good to place the grill next to it to simplify maintenance as the deposits can be pushed to the discharger as well.

Forebay Tank

At the end of the channel or coincident with the intake, replacing the channel of charge, a little tank, known as *forebay tank* or *basin of charge*, has to be realized. Its function is to guarantee the presence of upstream water in the penstock, in order to avoid the entrance of air along the pipe and the formation of whirlpools.

Channel

If the plant needs it, the sampling of the flow to adduce in the central, is done by using a channel. The channel can be realized both as open channel and as under pressure pipe. In the small plant, the technical solution is oriented towards an open surface channel, the sizing of which is done considering first the intake to derive. The intake is the function of the section of the channel as well as of the slope and the roughness that depends on the material used and on the degree of finishing of the wall. The channel can be made from different materials such as soil, wood, and concrete.

Generally, for small plants where the banks have an inclination of 45° with the base, if the width is L, the width of water surface is $2L$ and the height is $L/2$. Concerning concrete structures, using one that is rectangular shaped, which helps in the cleaning of the channel, is usually preferred.

Penstock and Pressure Drop

The *penstock* (see Fig. 17) takes the water from the load tank and pushes it to the turbine. It can be realized with the use of different materials. One should take into account the cost, weight, type of joints, and the conditions of the ground when choosing the type of material to be used. The *penstock* is also characterized by the diameter of the conduit itself that must contain the loss of load.

The choice of the pipe diameter has to be made as a compromise between three needs:

1. Keeping the costs down and therefore realizing a small-diameter pipe.
2. Containing load losses.
3. Realizing a bigger pipe to increase the energy.

Fig. 17 Example of a penstock.

The first head measure can be seen as a gross head, keeping into consideration the losses inside the conduits and all the other works of adduction.

The real *head* exploited by the turbine is lower than the first value above. Indeed, the definition of drop goes together with the *net head* that identifies the usable jump by the turbine; thus, the gross head minus the losses of the adduction works. Such a definition allows dividing the losses of the hydraulic parts from those related to the turbine. The value of the net head obviously depends on the pressure drops occurring inside the penstock. Through the definition of *net head*, it is also possible to define the output of the section of the hydraulics work simply as:

$$\eta_{\text{idr}} = \frac{H_{\text{n}}}{H_0} \quad (9)$$

where n and 0 refer to the net head and the gross head, respectively. The bigger is the hydraulic output of the plant, the better will be the exploitation of the water resource as higher power can be obtained with the same load or the same power can be obtained with less load. Moreover, if the entity of the load losses compared to the available drop is modest in high- and very-high-fall plants, in the low-fall plant with 6 m of available drop, a load loss of 1 m is almost 20% of the produced power. Thus, the hydraulic works of an adduction channel have to be realized with focus on MHP plants.

The penstock is the part of the plant in which the water flows faster. Considering that the losses are proportional to the square of the speed, the realization of the penstock is very important in terms of hydraulic performances of work of adduction.

The amount of water that flows inside the penstock is functional to the section of the pipe, its diameter and the water velocity. Once the flow is designed, a relation between speed flooding and penstock diameter is needed.

The problem can be solved with a dimensional analysis that puts into evidence how the *Fanning factor* is a function of the *Reynolds number* and *relative roughness*, known from fluid dynamic theory. The Fanning factor is connected to load losses and it represents their adimensionalization; the Reynolds number comes from the relation between inertia forces, viscosity, and velocity. Relative roughness is connected to the choice of material and the level of superficial finishing. Generally speaking, it is verified that, depending on the fluid regime, the Fanning factor tends to depend only on one of the variables.[8]

Walls

The thickness of the walls and veins are subject to the pressure of the impulse load, which also includes a water hammer. Nevertheless, in the case of a water hammer, plastic pipelines react better than iron ones, because the elasticity of the plastic tends to absorb overpressure better than other materials.

Once the ideal material is selected, the formula of the thickness can be found using the *Mariotte's formula*:[9]

$$t = \frac{P \cdot D}{2\sigma_{\text{f}}} \quad (10)$$

where t is the thickness of the pipe, P stands for the hydrostatic pressure, D is the diameter, and σ is the allowable stress.

Eq. 10 is only valid for stationary systems, where both capacity reductions and closure operations are not verified. Moreover, it does not take into account the problems that occur in iron pipes. Therefore, Eq. 11 should be amended, and, taking into account the types of joints, it becomes:

$$t = \frac{P \cdot D}{2\sigma_{\text{f}} \cdot k_{\text{f}}} + t \cdot s \quad (11)$$

where k_{f} is the efficiency of the welding and $t \times s$ represents the overpressure due to corrosion. k_{f} can be derived using Table 2.

In general, the value obtained as the thickness is always corrected when it is too low to take into account other factors: the tube must have achieved a sufficient rigidity to be moved without deformation. If the plant has a high fall, then a conduct with variable thicknesses (based on the pressure) can be used in order to reduce the cost of the materials. In addition to resistance to pressure increases, a conduct has also to withstand internal depressions to avoid collapsing:

$$P_{\text{c}} = 882.500 \cdot \left(\frac{t}{D}\right)^3 \quad (12)$$

where P_{c} is the pressure of collapse.

The depressions can be avoided through an *aerophore* with a minimum diamater:

$$d = 7.47\sqrt{\frac{Q}{\sqrt{P_{\text{c}}}}} \quad (13)$$

where d is the diameter of the aerophore.

Finally, to conclude the calculation of the wall thickness, a water hammer has to be considered. The *Allievi–Michaud*

Table 2 Value of k_{f} for different types of joints.

Type of joint	k_{f}
Without welding	1
Welding checked with x-ray	0.9
Welding checked with x-ray and subjected to a relaxation	1

Allelochemics – Alternative

formula can be modified if the pressure is expressed in water column, as:

$$\Delta P = c\frac{\Delta V}{g} \tag{14}$$

where c is the propagation speed in the middle of the pressure wave that depends on the water density and the elasticity of the material:

$$c = \sqrt{\frac{k}{\left(1+\frac{kD}{Et}\right)}} \tag{15}$$

where k is the water cubic compression module (2.1 × 10^9 MPa); E is Young's modulus of the conducting material; t and D are the thickness and diameter of the tube, respectively; and ρ is the water density. By applying the relationships (Eqs. 14 and 15) to PVC and iron pipes, it is possible to calculate for an instant closure (d of 400 mm, PVC thickness of 14 mm, and iron thickness of 4 mm):

$$c_{\text{pvc}} = 305 \text{ m/sec}$$

$$c_{\text{acciaio}} = 1024 \text{ m/sec}$$

$$\Delta P_{\text{pvc}} = 123 \text{ m}$$

$$\Delta P_{\text{acciaio}} = 417 \text{ m}$$

This provides a quantitative demonstration of the previously described nature of the two materials. If the operating time increases, the pressure is drastically reduced. Indeed, the maneuvering speed plays a crucial role in the generation of the overpressure. In large systems, it is common not to install pipes that can withstand overpressure, or water hammer, but rather to install mechanisms for the exclusion of the load to prevent the turbine from going into "overspeed."

Another device that serves to absorb the pressure waves that can occur within the pipes is the *piezometric borehole*. To evaluate if its installation is needed, the following formula can be considered:

$$I = \frac{V \cdot L}{gH} \tag{17}$$

where I is the constant acceleration. If I is less than 3, it can be assumed that the piezometric borehole is not necessary.[5]

The penstocks are also anchored to the ground, or supported on special works such as anchor blocks or saddles. The distance between two saddles or between two anchor blocks has to be as much as to make the pipe's arrow acceptable when it is full.

WATER TURBINES

A turbine converts energy in the form of falling water into rotary shaft power. The selection of the best turbine depends on several factors: the net head of the plant, the nominal flow, the power rating, and the shape of the turbine. The most installed turbine models, mainly used in big hydro plants, are three different typologies: Pelton, Francis, and Kaplan. In MHP, the turbine choice is made on different considerations, not only economical. Indeed, its construction and operational simplicity become essential, especially in developing contexts. Often, it is possible to install simpler versions of big hydro plant turbines—not for the Pelton model, which is the simplest one yet, for example, changing the blade edges.

As stated before, the choice of the turbine depends on the net head and the flow rate, as well as on the available water resources and the plant typology. A good criterion to select the turbine is resumed in the following well-known diagram, where they have a range in head and flow (Fig. 18).

It should be stressed that the fields of employment are not very narrow and are only suggestions for the best choice. In fact, a Pelton turbine could be installed in a low head–high flow rate plant, even if the turbine efficiency will be strongly penalized. This results in areas of the diagram where different typologies of turbines can be used at the same time. The final choice has to be taken considering also other factors, such as the operation of the plant.

CONCLUSION

Hydropower has been used as far back as the Roman empire and through history has been used to power water mills, textile machines, sawmills, and irrigation systems. In the early 1800s, however, people started to see that the use of water to power small factories and machines is but a minor application of its potential. As early as the 19th century, waterpower was being used as a source of electricity. Though primitive hydropower technology only consisted of wheels, buckets, and river flow, it was from this point on that waterpower's potential as one of the most efficient and abundant sources of renewable energy became apparent.

Mini-hydropower is probably the least common of the three readily used renewable energy sources (i.e., water, sun, and wind), but it has the potential to produce the most power, more reliably than solar or wind power if you have the right site. Small-scale hydro is in most cases run-of-river, without dam or water storage, and is one of the most cost-effective and environmentally benign energy technologies for developing countries and further hydro development in Europe.[17]

In this entry, a summary of the main advantages and shortcomings of small-scale hydropower has been presented. The hydro resource is a much more concentrated

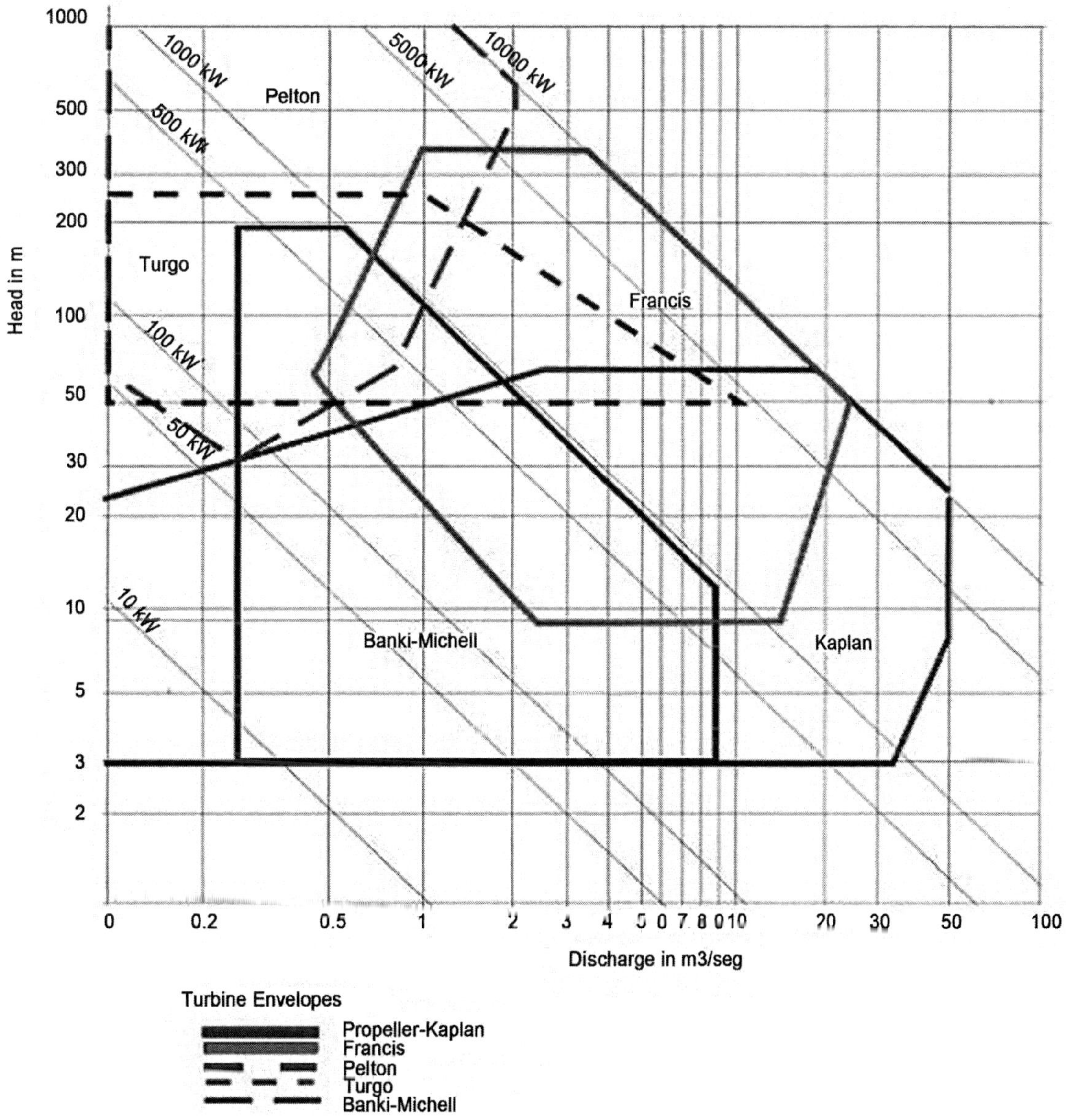

1000
500
300
200
100
50
30
20
10
5
3
2
Head in m
1000 kW
5000 kW
10000 kW
500 kW
100 kW
50 kW
10 kW
Pelton
Turgo
Francis
Banki-Michell
Kaplan
0
0.2
0.5
1
2
3
4
5 0 7 8 0 10
20
30
50
100
Discharge in m3/seg
Turbine Envelopes
Propeller-Kaplan
Francis
Pelton
Turgo
Banki-Michell

Fig. 18 Values of flow and height for each field of application of different turbines.

energy resource than either wind or solar power and the energy available is readily predictable. Moreover, no fuel and only limited maintenance are required and, if well designed, it has almost no environmental impact. On the other hand, it has to be considered that it is a site-specific technology and no general consideration has to be taken in the design of the plant; otherwise, environmental, social, or economical problems could occur, such as conflicts with fisheries interests on low-head plants and with irrigation needs on high-head plants. Furthermore, river flows often vary considerably with the seasons, especially where there are monsoon-type climates, and this can limit the firm power output to quite a small fraction of the possible peak output.

However, where a hydropower resource exists, experience has shown that there is no more cost-effective, reliable, and environmentally sound means of providing power than a hydropower system. Even with the various advantages of hydropower, it is still an underused alternative energy source. As of 2008, only 6% of the United States' electricity production came from hydropower, while nearly 50% came from the non-renewable source that is coal.[12] Due to a lack of economic speculation, a vast amount of potential for renewable hydropower remains untapped. Third world countries and underdeveloped areas have many areas that would be highly conducive to hydropower. The construction and use of hydropower facilities in these countries/areas, along with an increase in hydropower in the United States, could result in a great increase in renewable, affordable, and non-polluting energy.

However, if the prospected potential has to be realized, significant challenges have to be faced, in terms of *decision-making process*, establishing an equitable, credible, and effective environmental assessment procedure that takes into account both environmental and social concern and that takes into consideration the share of the benefits with local communities, both in the short term and in the long term. Finally, increasing efficiency, developing high-tech turbines, and reducing the costs of very low-head schemes, along with proper technology transfer of appropriate turbines to local manufacturers and technical support to the developers, will help realize our long-term objectives.

REFERENCES

1. International Energy Agency. *Renewable Energy Essentials: Hydropower*; OECD/IEA, 2010.
2. *State of the art of small hydro power in the EU 25*; Thematic Network of Small Hydropower Project, European Small Hydro Power Association, 2005; 1–20.
3. *2007 Survey of Energy Resources*; World Energy Council, ISBN: 0 946121 26 5, 2007; 1–600.
4. Harvey, A., et al. *Micro-Hydro Design Manual*; IT Publications Ltd.: London, 1993.
5. Caputo, *C. Gli impianti convertitori di energia;* Casa editrice Ambrosiana, ristampa, 2008.
6. *International Journal of Hydropower and Dams, World Atlas*; Aquamedia Publications: Sutton, 2000.
7. Khennas, S.; Barnett, A. Best practices for sustainable development of micro-hydro in developing countries.
8. Available at http://www.friulanacostruzioni.it/pages/ita/prodotti/paratoie/paratoia-a-tenuta-su-tre-lati-manuale.php.
9. Churchill, S.W. Friction factor equations spans all fluid-flow ranges. Chem. Eng. **1977**, 91.
10. Arrighetti, C. *Dispense del corso di Macchine II*; 2007.
11. Paish, O. Micro-Hydro Power: Status and Prospects. Journal of Power and Energy, Professional Engineering Publishing, 2002.
12. Energy Technology Perspective 2010, IEA, Paris.

Alternative Energy: Photovoltaic Modules and Systems

Ewa Klugmann-Radziemska
Chemical Faculty, Gdansk University of Technology, Gdansk, Poland

Abstract
Use of solar energy does not contribute to global warming. The light-to-current conversion (photovoltaic conversion) takes place within solar cells, which in most cases are made of silicon. Solar module consists of many solar cells, which are electrically connected and placed between glass or Tedlar® and framed by an aluminium frame. A number of solar modules and other components form photovoltaic systems. In this entry, a brief overview of the construction of solar modules and systems is presented.

INTRODUCTION

A whole series of determinants are favoring the development of the energy sector based on renewable resources, including solar energy, which photovoltaic modules convert into electricity: increasing social awareness of the need to limit emissions of harmful substances, legislation, pro-environmental policies of governments, bylaws, and support in the form of programs and financial mechanisms, not to mention the rising costs of energy from conventional sources and the need to ensure energy security.

Photovoltaics is an attractive technology for dependable, non-polluting power generation. Growth in the demand for solar cell modules has been especially strong in the past 10 years. Photovoltaic technology is used worldwide to provide reliable and cost-effective electricity for industrial, commercial, residential, and community applications.

The European Commission, the International Energy Agency, the U.S. Department of Energy, and institutions and policy makers worldwide recognize photovoltaic solar energy as a key technology to address environmental and climate change challenges, particularly energy safety, security, sustainability, access, and affordability for all. Under favorable conditions, photovoltaics can make a substantial contribution to the European Union (EU) electricity supply already by 2020—it may be more than 10%—as an important first step towards an even much larger share. Already in 2009, photovoltaics represented the third largest net new energy-generating capacity installed in the EU27, after wind energy and gas.[1]

According to the EPIA SET for 2020 Study, a global annual market of 163 GW_p could be reached with adequate support by 2020. PV system prices will decrease by at least 5% annually over the next two decades. This will further foster the attractiveness of solar power and support market growth.[2]

To create truly sustainable buildings, the long-term goal must be to design and construct buildings that do not need more energy over their entire life than they can produce. Rooftop-mounted PV systems are expected to produce during their whole lifetime between 8 and 17.9 times the amount of energy that was needed for their manufacture, installation, and dismantling, with the best case in Perth, Australia, and the worst case in Edinburgh, U.K. PV facades are expected to produce during their whole lifetime between 5.4 and 10.1 times the amount of energy that was needed for their manufacture, installation, and dismantling, with the best case in Perth, Australia, and the worst case in Brussels, Belgium.[3]

The energy payback time of a complete PV system is, depending on the solar irradiation of its location, in the range of 19–40 mo for a roof-mounted system and from 32 to 56 mo for a PV facade. Based on a commonly admitted 30 years long commercial life cycle, the energy return factor (ERF) is between 8 and 18 for roof-mounted systems and between 5.4 and 10 for facades. Varying widely from one country to another, using a single kilowatt of PV panel (roughly 10 m^2) can avoid up to 40 tons of CO_2 during its whole commercial life cycle (23.5 for a facade).[4]

The basic building block of the photovoltaic module (array) is the photovoltaic cell (PV). When exposed to solar radiation, it becomes a source of direct current (DC). A single cell generates a voltage of ca. 0.6 V and supplies from 1 to 2 W of electrical power. In order to achieve higher voltages, cells are connected in series or in parallel in photovoltaic modules. Modules are hermetically sealed to protect them from corrosion, moisture, contaminants, and the elements in general. In addition, housings are usually rigid, since PV modules are expected to have a life span of 20–30 years. The voltage obtainable from such a module is dependent to only a small degree on the level of insolation. With series-connected cells, a PV module can be designed to work at almost any voltage, even up to several hundred volts. For certain applications, PV modules can function at a constant voltage of 12 or 14 V, whereas for energy applications, solar panels can operate at the grid voltage. PV modules are a basic element of PV systems, also known as PV generators.

Encyclopedia of Environmental Management DOI: 10.1081/E-EEM-120046054

STRUCTURE OF PHOTOVOLTAIC MODULES

The most important part of a photovoltaic system is the solar module and its parameters (current–voltage characteristic, spectral characteristic, energy conversion efficiency), which determine the final amount of energy obtainable.

In a photovoltaic module, several solar cells are interconnected to achieve greater power. Here, two types are possible: series and parallel cell interconnection. In PV modules, the solar cells are mostly connected in series to create a higher voltage, since the cell voltages add up while the current remains the same. The front contacts of each cell are soldered to the rear contacts of the next cell in order to connect the negative pole (front) of each cell with the positive pole (rear) of the following cell. The start and end of each string are extended outwards for later electrical connection (Fig. 1).

To protect the cells against mechanical stress, weathering and humidity, the cell strings are embedded in a transparent bonding material that also isolates the cells electrically. For structural stabilisation, the bonding system is applied to a substrate, which is usually glass, but it is also possible to use acrylic plastic, metal or plastic sheeting (Fig. 2). To enable as much incident solar energy as possible to hit the solar cell, low-iron solar glass is generally used as the front substrate, which allows up to 91% of the light to penetrate. Recently developed solar antireflective glass, which has an additional antireflective coating applied with a caustic process or by dip coating, attains light transmissions up to 96%.

A PV module has different electrical parameters, depending on the number of cells connected within it and the type of connection (series, parallel, series–parallel). In practical applications, the dimensions of the module and its weight are also important (Fig. 2).

Cell connections in the module can be rigid or flexible; they are usually the latter, however, in order to avoid movements and tensions within the module as a result of thermal expansion and other factors. All electrical connections and contacts must offer the smallest possible series resistance. Wires must therefore be short and have an appropriate cross section. Cells are connected by wire: the front contacts of one cell are soldered to the rear contacts of the next one. The wires at the beginning and end of such a series array are made longer so that other electrical connections can be made. Industrial manufacturers machine-link the cells into strings. Only in custom-built modules is the soldering done by hand. The electrical connection of thin-film cells is made by the deposition of layers, while the conduction bands in the individual layers are cut with a laser or milled mechanically.

Thin layers of cells of $CuInSe_2$ (CIS) and of amorphous silicon are deposited on an elastic substrate, starting with the rear contact layer of the cell. In contrast, thin-layer cells of cadmium telluride (CdTe) are deposited in the reverse order, starting with the transparent, conducting oxide layer (TCO).

The dimensions of cells are a crucial aspect of the design of a module because they affect its electrical characteristic. The current intensity is directly proportional to cell area, but the voltage is independent of this factor. The voltage is a function of the height of the potential barrier in the p–n junction region and of the number of cells connected in series. For building a module that supplies a higher output voltage, smaller cells are usually connected in series. A module is regarded as optimally filled if the cells cover 90% of its surface.

Standard modules contain from 36 to 216 series-connected cells, which are arranged in two or three strings connected in parallel. In this way, a suitable voltage and current intensity can be obtained in the junction box. By connecting cells in different ways, modules with a nominal power from 100 to 300 W_p can be produced.

The current flowing through each cell in the string is the same ($I_1 = I_2 = ... = I_{36}$), but the voltages are summed: $U = n \cdot U_i$. When generators with the same voltage are connected in parallel, the intensities of the currents flowing into the node are summed: $I = n \cdot I_i$.

To ensure the module's stability, it is fixed to a substrate, which is usually glass, plastic, or a metal plate. The standard thickness of the low-iron solar glass used in module production is 4 mm, but for the production of large

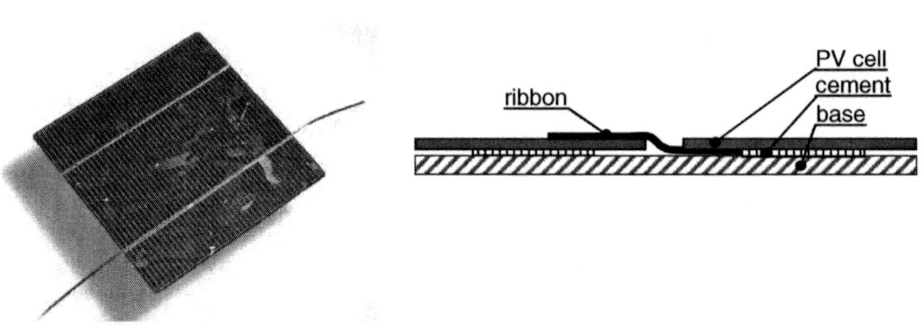
ribbon
PV cell
cement
base

Fig. 1 Connecting cells in series with the aid of conducting tape (not to scale).

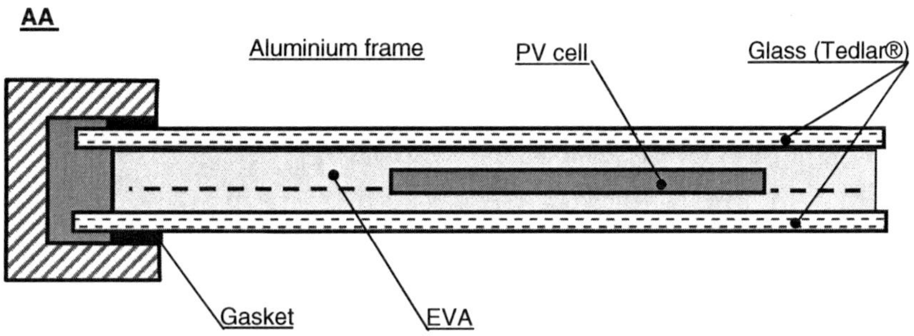
AA
Aluminium frame
PV cell
Glass (Tedlar®)
Gasket
EVA

Fig. 2 Structure of a photovoltaic module (not to scale).

modules, 10 mm thick glass is used. This has to be toughened to make it more resistant to thermal stress.

To protect the cells and the electrical connections from mechanical stress, moisture, and atmospheric precipitation, the module is inserted in a transparent envelope that doubles as an electrical insulator.

Since about 8% of energy losses are due to reflection, the next step in module manufacture involves applying an antireflective coating to the front side of the module; this reduces reflection from this side by 3%–5%.

In systems where modules are connected in series, bypass diodes are used to prevent hot spots from occurring, usually 1 for every 18 series-connected cells (a better effect is obtained if each cell in the module is protected by a bypass diode, but such modules are rare). When the insolation incident on the module is wholly uniform, the bypass diode is polarized in the reverse direction, but if the module is partially shaded, the bypass diode becomes polarized in the forward (conducting) direction and the current flowing through it avoids the shaded cells. Modules with cells connected in parallel are less susceptible to shading and bypasses are not used.

The most important module parameters include a short-circuit current, an open-circuit voltage, and a nominal voltage at 1000 W/m^2 solar radiation, current and rated power at 1000 W/m^2 solar radiation value. Module parameters are measured at standard test conditions (STC)—solar radiation, 1000 W/m^2; air mass, 1.5; and temperature, 25°C.

The following parameters can usually be found in module datasheets:

- Maximum power (maximum power point)—P_{MPP} [W_p].
- Open-circuit voltage—V_{oc}.
- Short-circuit current—I_{sc}.
- Voltage at maximum power—V_{MPP}.
- Current at maximum power—I_{MPP}.
- Nominal operating cell temperature (NOCT) [°C].
- Wind loading of surface pressure [N/m^2] (km/h).
- Maximum system voltage—V_{max}.
- Storage and operating temperature [°C].

PHOTOVOLTAIC SYSTEMS

Photovoltaic systems are capable of supplying electrical energy practically anywhere, from the poles to the tropics. Such systems have the following advantages:

- No fuel is required to power PV cells, so questions regarding the transport and storage of fuel and the disposal of waste products are irrelevant.
- Installation is simple and quick.
- PV arrays are modular, so systems can be of any size—sufficient to power a pocket calculator and watches or to provide floodlighting for sports stadia.
- PV cells can also utilize scattered solar radiation, incident on the Earth on misty or cloudy days throughout the year.
- During energy production, no chemical wastes that could pollute the environment are formed.
- Stationary systems do not need staffing, and breakdowns are extremely rare occurrences.
- The efficiency of PV systems does not decrease with time, because most commercial PV cells do not age during their operation.
- The life span of PV modules is usually 20–30 years (irreversibly destructive factors include high temperatures of the order of 500 K and elementary particle radiation).
- Stationary arrays have no moving parts that could wear out: they therefore need no spare parts or regular maintenance.
- They produce no noise.
- They generate more energy than is used to produce them (the latter amount of energy is "repaid" within 2–5 years, depending on the location and type of system).
- All the parts of a PV system are recyclable.
- They raise the awareness of energy consumers, who become interested in using energy-saving devices.
- They are a good advertisement for all renewable sources of energy.
- They improve the visual aspects of the buildings into which they are integrated.

- Photovoltaics is a rapidly developing market, in which all kinds of investors can find a place.

Photovoltaic systems are used to supply power to various devices in the following fields:

- Navigation: navigational signs at sea, on land, and for aircraft; for charging batteries on sea-going yachts
- Agriculture and forestry: electrical devices for protecting pastures and woodland, as well as irrigation, drainage, and fire prevention equipment
- Telecommunications: relay stations, remote radio stations, and mobile phone installations
- Transport: the illumination of airport signs, and signs on roads and railway lines, especially where sections on these routes are dangerous or under repair, etc
- The military: electrical devices used in the field (radio stations, range stations, measurement devices, lights, etc.)
- Meteorology: remote weather stations
- The household: household equipment ranging from calculators and watches, through radio and television sets, all the way to an entire house
- Medicine: field ambulances in Third World Countries (especially important for refrigerators where vaccines and medicines are stored)
- Tourism: autonomous power systems in caravans and in mountain huts
- Automation: autonomous data acquisition systems

There are two basic types of photovoltaic systems—grid connected and stand-alone—whereby the modules in both can be fixed (stationary system) or track the sun through the day (solar tracking system). PV arrays with fixed mounts (immobile with respect to the sun) should be tilted at an optimal angle and face due south (in the northern hemisphere). The tracking system uses a mechanical arrangement that responds to two variables (movement in two planes) characterizing the position of the Sun in such a way that its rays always impinge perpendicularly on the array.

Grid-connected systems are usually used in the urban and industrialized areas of developed countries; stand-alone systems come into their own in sparsely populated regions, or in remote areas, where there is no grid electricity supply.

Depending on the practicalities of using a building or land, PV modules can be mounted on the roofs of buildings, either as an independent system or integrated with the roof cover, on the facades of buildings, or on the ground.

Stand-Alone Photovoltaic Systems

Stand-alone systems were the first ones commercially available, used wherever the supply of grid-electricity was neither possible nor economically desirable. Examples of these situations are remote houses, mobile devices, horticulture and water supply systems. These installations require energy storage to compensate for the time lapse between the production of energy and the energy requirement (Fig. 3).

Today, the most common method for electricity storage is the lead-acid battery, and the main reason for this is cost. The car industry in particular prefers lead batteries. So-called solar batteries have a slightly modified structure, compared with car batteries, in order to achieve longer lifetimes. The simplest battery system consists only of a PV generator; however, these systems are not protected against deep discharge or overcharging. As a consequence, most battery systems use a charge controller; today, most of these are parallel charge controllers. This charge controller measures the battery voltage and disconnects the load if the battery is nearly empty; if the battery is full, the PV generator is short-circuited, and in this case, a blocking diode avoids battery discharge. A good charge controller has very low internal power consumption and includes a load cutoff switch that protects the battery against discharge. Power conditioning may be needed to adapt the voltage level of the photovoltaic system to that of the load. For PV-powered appliances, this is usually a DC/DC converter, which transforms one DC voltage to another. If alternating current (AC) voltage is needed by a consumer, an inverter

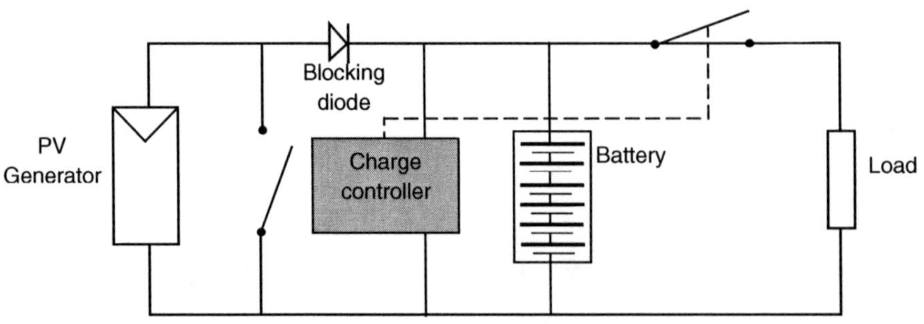
Blocking
diode
PV
Generator
Charge
controller
Battery
Load

Fig. 3 Stand-alone installation scheme.

must be used. This converts the DC voltage delivered by the solar generator or the battery to an AC voltage.

Grid-Connected PV Systems

Grid-connected systems instead rely on the versatility of the electricity grid, which acts as the "storage" of the system, providing the excess of energy to the grid and taking energy from it when the production is lower than the demand.

Distributed grid-connected PV systems offer many benefits to both the owner of the system and the utility network. For many owners, the main attractions of such a system are self-sufficiency and the environmental benefits of using renewable energy. Being a modular system, it can also expand easily as requirements or available capital grow.

In recent years, small rooftop-mounted systems have become increasingly popular, as improved technology has enabled the advantages of such systems to be exploited. It is now becoming increasingly common for homeowners to install a small PV system on their roof to supply some or all of their electricity needs.

The modularity of PV systems offers further benefits. The production costs for some PV system components are related to volume of production, meaning that a large number of small identical components can be cheaper to make than one big component. This means that a small PV system can be as cheap as or in some cases cheaper than a large system. Furthermore, many small systems can be distributed throughout an electricity network rather than centralized in one location. This allows the electricity utility to take advantage of locations where the value of electricity is greater, such as at the end of a long and inefficient transmission line.

On many electricity networks, peak loads coincide with peak solar power production. Peak loads are more expensive to satisfy than other loads, and thus the electricity generated by a PV system during a peak load is of greater value than that generated at times of low demand.

A grid-connected PV system offers other potential cost advantages when placed at the end of a transmission line, since it reduces transmission and distribution losses and helps stabilize line voltage. PV systems can also be used to improve the quality of supply by reducing "noise" or providing reactive power conditioning on a transmission line. When all these advantages are considered, well-positioned grid-connected PV systems are already economically viable, even though further cost reductions are required to make PV systems economic over the entire electricity network.

The main technical advance that has made grid connection of small PV systems feasible is the availability of low-cost, high-quality inverters. These inverters convert the DC electricity generated by the PV system into AC grid electricity. Recent developments have been towards even smaller low-cost units that can be individually incorporated into PV modules. Built-in electronics would then allow such "AC modules" to be interconnected and grid connected with a minimum of costly external circuitry or protection equipment.

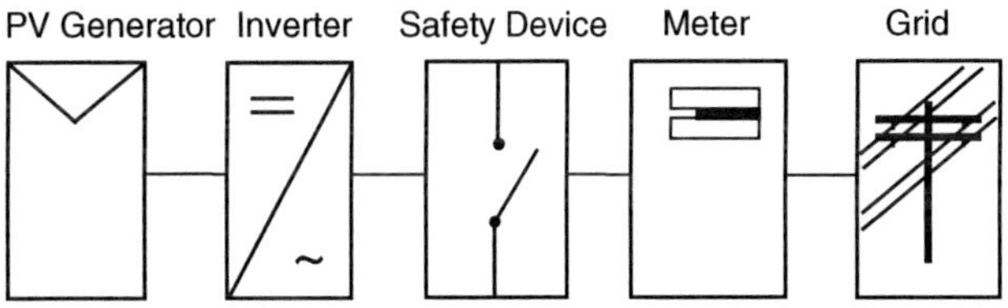
PV Generator
Inverter
Safety Device
Meter
Grid

Fig. 4 Grid-connected installation scheme.

This configuration consists mainly of the following components: the PV generator, the inverter, the safety devices, and, in many cases, the electric meter as shown in Fig. 4.

COMPONENTS OF PHOTOVOLTAIC SYSTEMS

Photovoltaic Modules

The output voltage of the photovoltaic module is only slightly dependent on light radiation, while the current intensity increases with higher luminosity. A very important point in the *I–V* characteristic of solar modules is the maximum power point (MPP).

MPP occurs at a certain voltage that depends on the temperature and irradiance. The maximum power and the required voltage decrease with the temperature and increase with the irradiance, while the required intensity increases with temperature and decreases with irradiance. The way they are modified by these parameters is described by the temperature coefficients given by the manufacturer for P_{MPP}, V_{MPP}, I_{MPP}, V_{oc}, and I_{sc}.

Generator Junction Box, String Diodes, and Fuses

The individual strings are connected together in the generator junction box. The string cables, the DC main cable, and, if required, the equipotential bonding conductor are connected.

The generator junction box contains terminals, breaks, and, if required, string fuses and string diodes. Surge arresters are often installed in generator junction boxes to divert overvoltage to earth. That is why the equipotential bonding or earth conductor is run into the generator junction box. The DC main switch is also sometimes housed in the generator junction box. The generator junction box should be executed to Protection Class II and demonstrate a clear separation of the positive and negative sides within the box. If mounted externally, it should be protected to at least IP 54.

String fuses must be installed for all arrays formed of four or more strings connected in parallel. The string fuses protect the wiring against overloading. They should be designed for DC operation.

To decouple the individual module strings, string diodes can be connected in series with every string. Should a short circuit or shading occur in a string, the other strings can continue to work undisturbed.

Without the string diodes, a current would flow through the failed string in the consumer direction (reverse current). If string diodes are used, their blocking voltage must be designed to be twice the open-circuit voltage of the PV string at STC.

When operating the PV system, string diodes are connected in the forward-biased direction. This enables the complete string current to flow through the string diodes (heat sinks are usually necessary). This current flow results in power losses (approximately 0.5%–2.0%), caused by the drop in forward voltage, at the string diodes of about 0.5–1.0 V. Thus, even with shaded systems and using string diodes, the annual energy yield is not substantially greater than with systems without string diodes. The losses from the reverse currents are compensated by the losses from the voltage drop. The failure of string diodes has proved to be problematic.

Inverters

The solar inverter is the link between the PV generator and the AC grid or AC load. Its basic task is to convert the solar DC electricity generated by the PV generator into AC electricity and to adjust this to the frequency and level of the building grid if it is grid connected. This type of solar inverter is also known as a DC–AC converter.

Depending on their use, a distinction between inverters that are used in grid-connected systems (grid inverters) or in stand-alone systems (stand-alone inverters) is made. In stand-alone systems, inverters enable the operation of conventional AC loads. In grid-connected PV systems, the inverter is linked to the mains electricity grid directly or via the building's grid. With a direct connection, the generated electricity is fed only into the mains grid. With a coupling to the building's grid, the generated solar power is first consumed in the building, and then any surplus is fed to the mains electricity grid.

In order to feed the maximum power into the electricity grid, the inverter must work at the MPP of the PV generator. In the inverter, an MPP tracker ensures that the inverter is adjusted to the MPP. The MPP tracker essentially consists of a DC converter. This is connected in series with the actual inverter unit and adjusts the inverter's input voltage to the voltage of the MPP.

The inclusion of a transformer in an inverter allows the electrical isolation of the input and output voltages. This reduces electromagnetic interferences and makes potential equalization unnecessary, since the operation point does not fluctuate but increases the magnetic and ohmic losses and the size and weight. Transformers are generally used in centralized inverters and module inverters and string inverters are generally transformerless. However, transformerless inverters are prohibited in some countries like the United States, where the approach to grounding differs from the European one, and a grounded negative lead is required. This prevents the use of transformerless inverters since the grounded negative lead prevents it from operating properly.

Modern solar inverters are able to perform the following functions:

- Converting DC generated by the PV generator into AC to comply with the grid requirements. Rotating generators such as conventional turbines automatically generate AC power with a sine wave voltage fluctuation due to the rotation involved (think of the sine wave as a partially unrolled set of circles). In PV, an inverter has to manufacture a waveform; for some simple off-grid applications, a square wave will do, but for grid-connected applications, the wave form needs to be close to a sine wave, i.e., with a low total harmonic distortion.
- Adjusting the operating point of the inverter to the MPP of the PV generator (MPP tracking). This system uses a control algorithm to keep the PV modules operating close to their peak power point while the incoming solar radiation level varies. It was originally a separate component but is now integrated as standard within inverters.
- Recording operational data and signaling (e.g., display, data storage, data transfer). At its simplest, this can be a set of lights indicating correct operation, warning light, or system disconnected light; at the other end of the scale, inverters can include a small display of various parameters, can be interrogated, and can send monitoring data to a data-logger or remote computer via a data link, possibly even via satellite.
- DC and AC protective devices (e.g., automatic disconnection in the event of a loss of mains or the mains supply falling outside the permitted voltage/frequency window). Some disconnection and isolation functions are integrated as standard within inverters; however, in some countries, regulations may require additional isolation and disconnection components to be installed. The inverter needs to trip out if the voltage, current, or frequency goes outside acceptable ranges on the DC or AC side; it also detects grounding faults and provides protective functions such as detecting any superposition of DC on the AC side in systems employing transformerless inverters, overheating, and protection against transient overvoltage from lightning and surge voltage. The inverter electronics also include components that are responsible for the daily operation mode.

Many inverters adjust the corresponding operation voltage continuously to the actual MPP. This operation is called maximum power point tracking (MPPT). The method most commonly used to perform MPPT is to change the actual input voltage in such a way that maximum power is obtained. The effect of continuous MPPT is often overestimated. Simulation has shown that for grid-connected PV systems, continuous voltage operation leads to losses of only 1%–2% when properly adjusted.

Since the power factor in modern grid-connected inverters can be adjusted by internal control, this kind of inverter can be used to compensate reactive power flow in the grid, which otherwise must be performed by extra compensation units such as inductors or capacitors. This ability can either be fixed to a constant value or, in the case of an appropriate communication system, be controlled by the grid operator according to actual needs.

As a further means of power quality improvement, high-quality inverters are able to compensate deviations in the sinusoidal voltage of the grid. In a later stage of PV use, inverters will have to prevent grid overloading. Grid-connected inverters can easily handle this kind of power control by changing the DC input voltage from the MPP in such a way that the PV generator reduces power production to the desired level. This request may come in a situation in which several hundreds of megawatts of PV power are fed into a local system. The grid operator must be able to communicate with these inverters. A few inverters on the market already possess these features.

Islanding

A particular safety issue raised by the connection of local generation to the utility is what happens when there is a loss of supply from the utility. If the loss of connection is not detected, and the inverter shut down, the local network may remain energized. If the inverter is able to supply sufficient power to match the load, this situation may continue for some time. This is referred to as islanding.

Islanding presents a potential safety hazard to utility and other personnel and also potential problems if the connection is remade while the inverter is supplying power. Basic protection measures such as shutting down operation if the voltage or frequency lies outside normal values will usually be sufficient to ensure that islanding is detected and the inverter is shut down. However, there are circumstances where such measures are inadequate, notably with multiple inverters, a matched resonant load, or a rotating machine. Other measures that can be taken to improve the prevention of islanding include monitoring the frequency for rapid changes and measuring the impedance of the network by injecting a suitable current impulse.

Passive Methods

Passive islanding detection methods detect the characteristic features of the islanding mode, and the operation of the inverter is stopped. This method is suitable for integration into the inverter. Passive techniques do not, by themselves, alter the operation of the power system in any way: they detect loss of grid by deducing it from measurements of system parameters. Passive techniques are suitable for all types of generators. Typical methods are as follows: overvoltage, undervoltage, overfrequency, underfrequency, and rate of change of frequency. Other methods are detection of phase difference jumps between inverter output voltage and current or detection of harmonics, which will increase if the impedance of the grid increases heavily.

Active Methods

Active methods for detecting the island introduce deliberate changes or disturbances to the connected circuit, following which the response is monitored to determine whether the utility grid with its stable frequency, voltage, and impedance is still connected. If a small perturbation is able to affect the parameters of the load connected within prescribed limits, the active circuit causes the inverter to shut down.

Types of Inverters

For a long time, it was usual to have a central inverter for the entire PV system. It is now increasingly common to have a number of small inverters (decentralized concept). Inverters have been developed for strings, called string inverters, and for individual modules, so-called module inverters (Fig. 5). Each of the three concepts mentioned has advantages and disadvantages. Which concept is chosen depends on the type of application.

Decentralized inverter concepts should be considered for systems consisting of subgenerator areas with different orientations and tilts or for systems that are partially shaded.

Central Inverter Concept

In the low-voltage range ($V < 120$ V), only a few modules are connected in series in a string. The disadvantage of this concept is the resulting high currents. Relatively high cable cross sections have to be deployed to reduce the ohmic losses. For concepts with longer strings and the associated higher voltages ($V > 120$ V), Protection Class II (protective insulation) is required. The advantage of this concept is that smaller cable cross sections can be used because of the lower currents. Larger PV systems often use a central inverter concept based on the master–slave principle. This concept uses several central inverters (mostly two to three). For the sizing, the total power is divided by the number of inverters. One inverter is the master device and operates in lower irradiance ranges. With increasing irradiance, the power limit of the master device is reached and the next inverter (slave) is connected. In order to load the inverters equally, the master and slave are swapped over (rotating master) in a specific cycle.

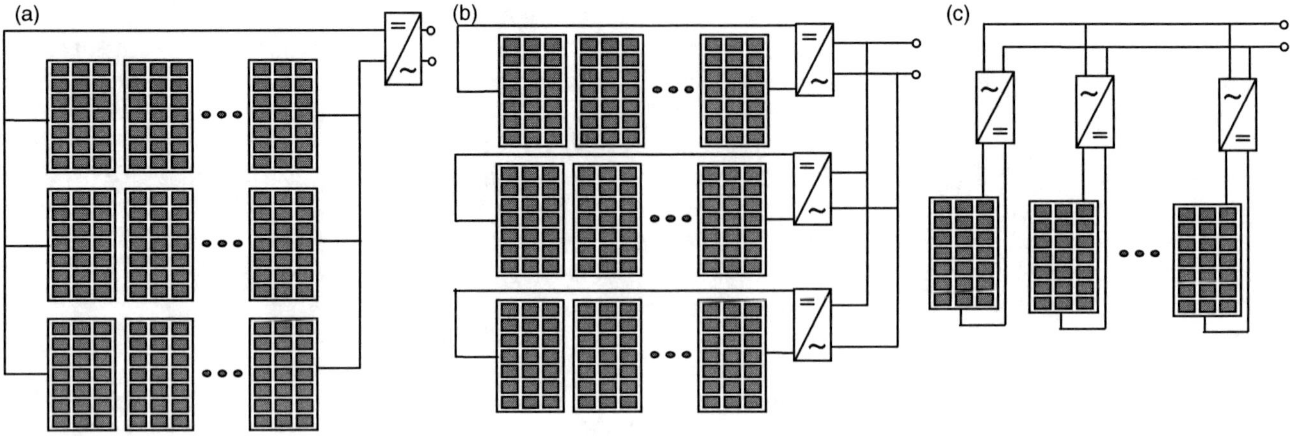
(a)
(b)
(c)

Fig. 5 Central (a), string (b), and module (c) inverter concept.

The advantage of this concept is that with lower irradiance, only one inverter operates (master), and thus, the efficiency—particularly in lower power ranges—is greater than if only one central inverter is used. However, the investment costs increase in comparison with one central inverter.

Array and String Inverter Concept

With a system with differently oriented arrays or with shading, an array or string inverter concept enables better correlation with the different irradiance conditions. One inverter is used per array or per string. Care should be taken to ensure that only modules with similar ambient conditions (orientation, shading) are connected together in a string.

Module Inverter Concept

A prerequisite for high system efficiency is that the inverters are optimally adjusted to the PV modules. Ideally, every module would be operated permanently at its MPP. The MPP matching is more successful if PV modules and inverters are conceived as a unit. These module-inverter units are also called AC modules. The inverters deployed are called module inverters. Some devices are so small that they can be stored in the module junction boxes.

Module inverters enable PV systems to be extended as desired, even with only one inverter module. It is often claimed that the disadvantage of module inverters is their lower efficiency. In addition, this lower efficiency is compensated for by the greater yield that results from improved matching with the respective module's MPP.

Cables

The electrical wiring between the individual modules of a solar generator and the generator junction box is termed *module cables* or *string cables.* These cables are generally used outdoors. In order to ensure earthing-fault-proof and short-circuit-proof cable laying, the positive and the negative conductors may not be laid together in the same cable. Single-wire cables with double insulation have proven to be a practicable solution and offer high reliability. The cables must be UV and weather resistant and suitable for a large temperature range (cables routed behind a PV array must be rated for a minimum of 80°C).

The DC main cable connects the generator terminal box with the inverter and should be selected so as to minimize the risk of short circuits and earthing faults. This is achieved through the use of single-core cable that is both insulated and sheathed or by using suitable conduit/trunk. Steel-wire armored cable may also be used where additional mechanical protection is also appropriate. The type and size of cable specified in national electrical codes should be used.

For PV installations exposed to a lightning risk, screened cables should be used. It should be possible to free all conductors from the voltage of the DC main line. The DC main switch and the isolation points in the generator terminal box are used to do this.

The AC connecting lead line links the inverter via the protective equipment to the electricity grid. In the case of three-phase inverters, the connection to the low-voltage grid is made using a five-core cable. For single-phase inverters, a three-core cable is used.

In the event of faults, or to carry out maintenance and repair work, it must be possible to isolate the inverter from the PV generator. The DC load switch is used for this. According to the IEC 60364-7-712 standard, *Electrical Installations of Buildings—Requirements for Special Installations or Locations—Solar Photovoltaic (PV) Power Supply Systems*, an accessible load switch is required between the PV generator and the inverter.[5]

The DC main switch should be double pole and have load-switching capability. It must be rated for the maximum open-circuit voltage of the solar generator (at -10°C) and for the maximum generator current (short-circuit current under STC). When selecting the switch, one should ensure that it can switch the relevant DC.

The DC main switch is often housed in the generator junction box. For safety reasons, it is better to install it directly before the inverter. Touch-proof plug connectors should not be used as the means of electrical isolation. As long as the irradiance is sufficient, the PV system supplies energy and is therefore under load. When separating a plug connector under load, the DC can result in a long burning arc, which is a serious safety and fire risk.

The AC switch disconnector must be double pole, clearly labeled, and lockable in the "off" position only. A second isolator is required adjacent to the inverter if the inverter is mounted in a separate room.[5]

A circuit breaker is an automatically operated electrical switch designed to protect an electrical circuit from damage caused by overload or short circuit. Its basic function is to detect a fault condition and, by interrupting continuity, to immediately discontinue electrical flow. They automatically isolate the PV system from the electricity grid if an overload or short circuit occurs. Automatic circuit breakers are often used as AC isolators.

Earth leakage or residual current devices (RCDs) monitor the current flowing in the forward and return conductors in the electrical circuit. If the difference between the two currents exceeds 30 mA, the RCD isolates the circuit within 0.2 sec. The RCD will trip if there is an insulation fault or if there is earth or body contact with one of the conductors.[5] This is not always mandatory, however, as most inverters switch off when there is an earth fault.

Grid-connection Configurations

In the single-phase configuration, the grid connection is made to the incoming cable through two meters: one records the electricity taken from the grid and the other registers the energy put back into the grid. Alternatively, one import–export meter can be used. In the three-phase system, three single-phase PV generators with three individual meters are connected to a three-phase import–export meter or directly to one three-phase meter, which records the electrical energy transmitted to the grid, while another three-phase meter, connected between the building loads and grid, registers the electricity taken from the grid.[6]

LARGE-SCALE PHOTOVOLTAIC POWER PLANTS

In 2008, more than 1000 large-scale PV plants were constructed and put into service worldwide and large-scale PV plants with cumulative power more that 2 GW_p were connected to the grid, most of them in Spain. Some other EU countries (Italy, Czech Republic) also significantly increased cumulative power installed (Fig. 6). Among Asian countries, it is worth to mention Korea with almost

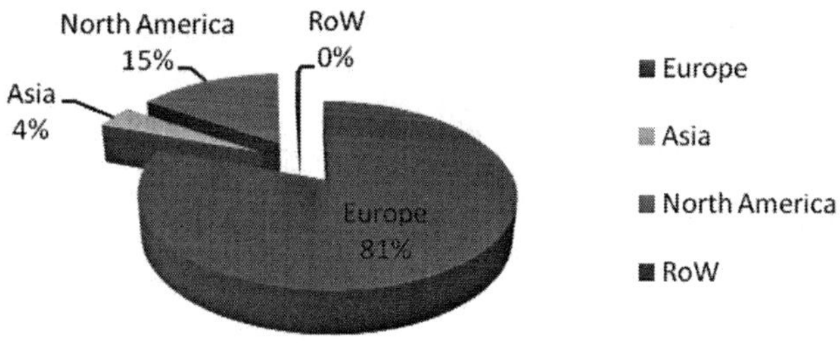
North America
15%
RoW
0%
Asia
4%
Europe
81%
Europe
Asia
North America
RoW

Fig. 6 Large-scale photovoltaic power plants—power capacity by region.
Source: Based on Lenardic.[8]

100 MW power capacity installed. Many of these plants consist of several stages where each stage can be considered as a unique power plant, so the actual number is even higher.[7]

At the end of 2007, almost 70% of all large-scale photovoltaic power plants (power related) were ground mounted and 29% were roof mounted; other plants (about 1%) include photovoltaic power plants integrated into building envelopes, noise barriers, and similar applications. Twenty-seven percent of all power plants (power related) have tracking arrays (single- or double-axis tracking), and 73% have fixed arrays (Fig. 7).

BIPV

BIPV (building-integrated photovoltaics) refers to photovoltaic systems integrated with an object's building phase. This means that they are built/constructed along with an object. They are also planned together with the object. Yet, they could be installed later on. In view of the specifics of this task, close cooperation between many different experts, such as architects, civil engineers, and PV system designers, is called for. According to how and where such systems are built, whether into the facade or in the roof, the following BIPV systems are recognized:

- Facade or roof systems added after the completion of the building.
- Facade-integrated photovoltaic systems built along with an object.
- Roof-integrated photovoltaic systems built along with an object.
- "Shadow-Voltaic"–PV systems, also used as shadowing systems, built along with an object or added later.

In the case of facade or roof systems, the photovoltaic system is added to the building after it has been built. These low-powered systems of up to some 10 kW are usually integrated into the south facade. Facade-integrated photovoltaic systems could consist of different transparent module types, such as crystalline and microperforated amorphous transparent modules. In such a case, part of the natural light is transferred to the building via the modules.

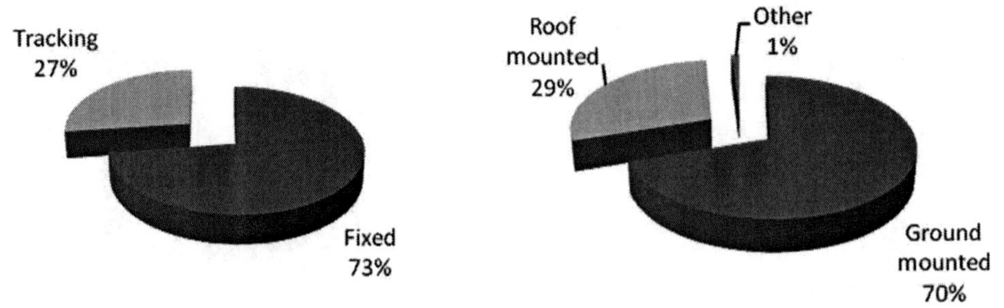
Tracking
27%
Fixed
73%
Roof
mounted
29%
Other
1%
Ground
mounted
70%

Fig. 7 Large-scale photovoltaic power plants—market shares of fixed and tracking arrays and market shares of ground- and roof-mounted arrays.
Source: Based on Lenardic.[8]

Solar cells are available in different colors; therefore, there is no limitation to the imagination of the architect or designer. Buildings constructed in this way give a whole new meaning to the term *architecture*. Roof-integrated photovoltaic systems are integrated into the roof; the roof is covered with transparent photovoltaic modules, or they are added to the roof later. Such systems are added to a flat roof or to a tilted roof usually only if the building is small. It is possible to use tiles, which integrate solar cells.

Photovoltaic systems can be used for shading, where photovoltaic modules serve as venetian blinds. In some of these cases, the tilt angle of the photovoltaic modules can be adjusted manually or automatically, optimizing the shading of the building and/or photovoltaic module efficiency. Such systems are also known as Shadow-Voltaic systems.

The best results and efficiency can be achieved with systems that are closely integrated into passive solar buildings; however, the use of active solar systems is an additional possibility. This is where the modules are partially transmissive, allowing natural light to penetrate the building. Undoubtedly, such systems challenge even the best of architects. A high level of expertise is required for successful BIPV systems planning, not only with regard to architecture but also with regard to civil and photovoltaic engineering.

Past projects have shown that successful BIPV systems design is based heavily on technical experience and knowledge. Poorly designed systems usually have to be redesigned or repaired later, consequently swelling maintenance costs and lowering the system efficiency rate.

COSTS OF INVESTING IN PHOTOVOLTAIC SYSTEMS

Photovoltaics is one of the most rapidly developing industries in the world. For example, in 2007, more than $100 billion were invested in the renewable energy sector, with 30% of this in photovoltaics. In 2007, photovoltaic production rose by more than 60% (to 4 GW) in comparison with the previous year. Similarly, the installation market recorded a growth of more than 60%, reaching 2825 MW of grid-connected installations and 1200 MW of stand-alone systems, consumer products, and warehouse reserves. Constant growth in the power of PV systems is predicted—up to 9.7 GW in 2013, an average of 22% per annum. In monetary terms, this means an increase in investment from $30 billion in 2008 to an expected $60 billion in 2013.[9]

In 2007, 4 TWh of electrical energy was produced by photovoltaic systems in the 27 EU member states. The world's 10 leading manufacturers of PV cells and modules are as follows: Q-Cells (Germany), Sharp (Japan), Suntech (China), Kyocera (Japan), First Solar (United States), Motech (Taiwan), Sanyo (Japan), SunPower (United States), Yingli Solar (China), and Solarworld (Germany/United States).

In 2008, 99% of PV modules were mounted in installations connected to the European electricity grid—globally, 70% of all systems supply excess electricity to the grid. This is the result of the deliberate governmental policies of a number of countries, which have encouraged this development with feed-in tariffs.

It is becoming more common for smaller systems to be designed, as well as for private houses, a development that is generally regarded as the future for photovoltaics. Especially on the outskirts of large cities, where there is no

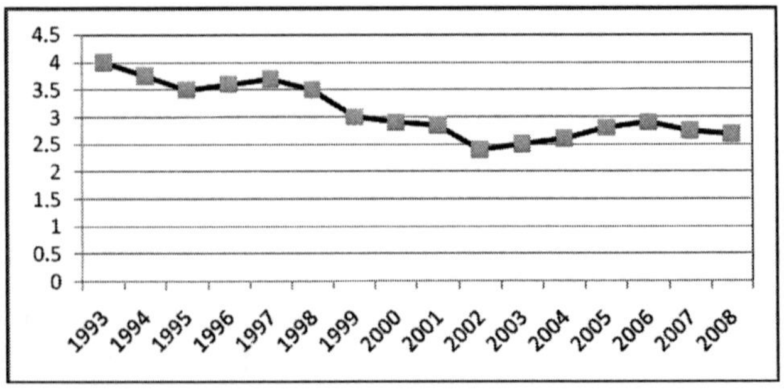
4.5
4
3.5
3
2.5
2
1.5
1
0.5
0
1993
1994
1995
1996
1997
1998
1999
2000
2001
2002
2003
2004
2005
2006
2007
2008

Fig. 8 Prices of PV modules in 1993–2008 in dollars per watt.
Source: Based on PV Technology, Production and Cost.[10]

Table 1 Photovoltaic system installation costs.

System type	System power	Investment costs
Off grid	100–500 W_p	10–20 €/W_p
	1–4 kW_p	8–15 €/W_p—developed countries 20–40 €/W_p—worldwide
On grid	1–4 kW_p	4–6 €/W_p
	10–50 kW_p	4–6 €/W_p
	>50 kW_p	4–6 €/W_p

Source: Photovoltaic Economics.[11]

room for installations of other types, this may be the only way of utilizing renewable energy.

A photovoltaic system is expensive to install (see Table 1), but its operating costs are minimal. The cost of purchasing a photovoltaic system includes the following:

- The photovoltaic modules
- System elements (accumulator/battery, inverter, controller, cable connections, etc.)
- Transport and installation
- Design

The prices of PV modules are falling steadily, possibly to 2.60 dollars/W by the end of 2009 (Fig. 8). For installations in areas with good insolation, the cost of energy generation in 2010 will therefore fall to $0.17/kWh for crystalline silicon modules and to %0.13/kWh for thin-film modules.

The overall cost of a PV system can vary widely depending on its size and location, whether it can be connected to the grid, and the cost of its component parts and installation.

Typical investment costs are from 4000 €/kW_p for thin-film PV power plants up to 8000 €/kW_p for c-Si tracking power plants.

CONCLUSION

The increasing cost of conventional fuels and environmental concerns have contributed to the intensification of research and development in the field of renewable energy sources.

The increasing interest in PV cells and modules worldwide is due mainly to the fact that they convert solar radiation directly into electricity without generating any pollution, noise, or other environmentally harmful side effects. At the same time, production costs of PV system components, especially the PV cells, are falling steadily. The governments of many countries are actively encouraging the development of renewable energy sources, and a whole range of governmental programs are dedicated to photovoltaics. This has considerably enhanced the profitability of investments in this field.

Over the past 30 years, the photovoltaics industry has grown at an annual rate of 34%. Photovoltaics is therefore the most rapidly developing technology, besides wind turbines, for the conversion of energy from renewable sources.

REFERENCES

1. Sinke, W.C. European Photovoltaic Technology Platform comment on "Towards a new Energy Strategy for Europe 2011–2020", Belgium, July 2010.
2. The Status and Future of the Photovoltaic Market, European Photovoltaic Technology Platform, fact sheet LB-32-10-253-En-D, available at http://www.eupvplatform.org/publications/misperceptions-fact-sheets.html#c934 (accessed May 5, 2012).
3. Gaiddon, B.; Jedliczka, M. Environmental benefits of PV systems in OECD cities. Proceedings of 21st European Photovoltaic Solar Energy Conference and Exhibition, 4–8 September 2006, Dresden, Germany.
4. Compared Assessment of Selected Environmental Indicators of Photovoltaic Electricity in OECD cities, EU PV platform brochure, available at http://www.eupvplatform.org/publications/brochures.html (accessed May 5, 2012).
5. *Planning and Installing Photovoltaic Systems, A Guide for Installers, Architects and Engineers*, 2nd Ed., Deutsche Gesellshaft Fur Sonnenenergie, Earthscan, 2008.
6. Antony, F.; Dürschner, C.; Remmers, K.H. *Photovoltaic for Professionals*; Solarpraxis AG, Berlin, 2007.
7. Lenardic, D. Large-Scale Photovoltaic Power Plants, Annual and Cumulative Installed Power Output Capacity; Annual Report, 2008, available at http://www.sunenergysite.eu/download/AnnualReview_FreeEdition.pdf (accessed May 5, 2012).
8. Lenardic, D. Large-Scale Photovoltaic Power Plants, Annual and Cumulative Installed Power Output Capacity; Annual Report, 2007, available at http://ss1.spletnik.si/000/000/128/96a/AnnualReport2007.pdf (accessed May 5, 2012).
9. Trends in Photovoltaic Applications; IEA PVPS, 2009, available at http://www.iea-pvps.org/index.php?id=37 (accessed May 5, 2012).
10. PV Technology, Production and Cost, 2009. Forecast PV THROUGH 2012: The Anatomy of a Shakeout, GTM Research, available at http://www.gtmresearch.com/report/pv-technology-production-and-cost-2009-forecast (accessed May 5, 2012).
11. Photovoltaic Economics, available at http://www.pvresources.com/Economics.aspx. (accessed May 5, 2012).

Alternative Energy: Photovoltaic Solar Cells

Ewa Klugmann-Radziemska
Chemical Faculty, Gdansk University of Technology, Gdansk, Poland

Abstract
Solar radiation is a fuel for photovoltaic cells and modules, which are a source of electric energy. Viable solar technologies that have been developed in the past are starting to be adopted on a broader and more meaningful global scale across different sectors with ever-increasing market penetration. In this entry, a brief overview of the material properties relevant to solar cell operation, the structure of the solar cell, and electrical characteristics is presented.

INTRODUCTION

Nowadays, fossil fuels are the main sources of energy from which electricity is obtained, but these sources will not last forever, so in due course renewable energies will have to replace them in this role. One of these new sources is solar energy. Each year, the Earth receives around 1×10^{18} kWh of solar energy, which is more than 1000 times the current global energy demand. This is therefore a vast source of energy that can be tapped to satisfy human energy requirements. To generate electricity from sunlight, solar cells (photovoltaic cells) are used. These devices are based on the photovoltaic effect, in which a p–n semiconductor is exposed to light, and photons are absorbed by electrons, providing an electric current. The electrons that are set free are pulled through the electric field and into the n-area. The holes produced move in the other direction, into the p-area.

The use of solar energy releases no CO_2, SO_2, or NO_2 gases and does not contribute to global warming. Photovoltaics is now a proven technology that is inherently safe, as opposed to some dangerous electricity-generating technologies. Over its estimated life, a photovoltaic module will produce much more electricity than was used in its production. A 100 W module will prevent the emission of more than 2 tons of CO_2. Photovoltaic systems make no noise and cause no pollution while in operation.

PHOTOVOLTAIC EFFECT

The solar cells now in use are the practical application of fundamental physical phenomena observed already in the 19th century (see Table 1).

The absorption of light in semiconductors takes place when electrons are released from interatomic chemical bonds. In order to produce a free electron in a given semiconductor material, a certain quantity of energy must be supplied, equal at least to that of the energy band gap, which in the case of silicon at a temperature of 300 K is $E_g = 1.12$ eV. The liberated electron leaves behind it a hole that can move about by diffusion or drift under the influence of an electrical field owing to its being positively charged (Fig. 1).

Table 1 The beginnings of photovoltaics worldwide.

Year	Achievement
1839	Alexander Edmund Becquerel observed the photovoltaic effect in a circuit of two illuminated electrodes immersed in an electrolyte.
1843	Fritts produced the first tin–selenium solar cell.
1879	Adams and Day observed the photovoltaic effect at the interface between two solid bodies (selenium–platinum).
1930	The first copper/copper oxide solar cell.
1941	Ohl patented the silicon cell (monocrystalline silicon doped during its growth).
1954	At the Bell laboratory (United States), Chapin, Fuller, and Pearson developed a cell on monocrystalline silicon with a diffusion p–n junction of 6% efficiency, which was subsequently manufactured by two companies.
1954	Lindmayer and Allison obtained a cell with an efficiency of 16% (radiation intensity, 1000 w/m^2).
1954	Reynolds produced the first multiple-junction Cu_2S/CdS cell.
1958	Monocrystalline solar cells were used for the first time in the Vanguard I satellite, where modules of six cells supplying 5 W of electrical power were installed; the traditional batteries ran out after a few months, but the photovoltaic panels powered the satellite's transmitter for another 6 years.
1962	The first thin-layer Cu_2S/CdS photocell was obtained.

Encyclopedia of Environmental Management DOI: 10.1081/E-EEM-120046053

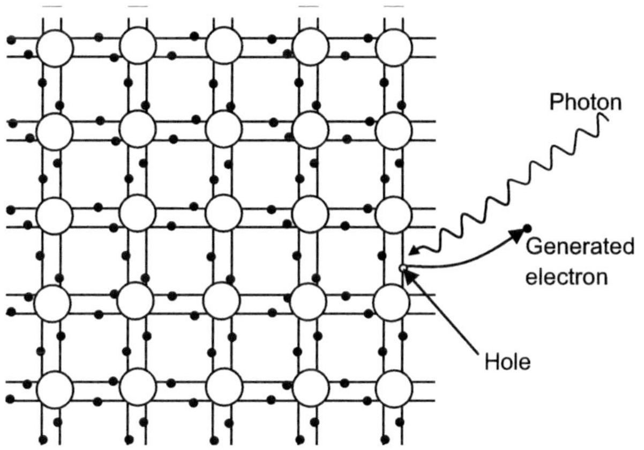
Photon
Generated electron
Hole

Fig. 1 The generation of an electron–hole pair by a photon of energy $h\nu > E_g$.

The introduction of other atoms in place of the parent atoms (or at interstitial positions) of a pure intrinsic semiconductor considerably improves its electrical conductivity. Energy levels of elements with one valence electron more than the semiconductor atoms form donor levels in the neighborhood of the conduction band (n-type). Energy levels of elements with one valence electron less than the semiconductor atoms form acceptor levels in the vicinity of the valence band (p-type).

If a p–n junction is formed from the p- and n-type areas of the semiconductor, then the charge carriers move around in such a way that the Fermi level will be identical throughout the crystal (Fig. 2b). At room temperature (300 K), practically all donor and acceptor dopants are ionized; hence, the concentrations of majority carriers (electrons in the n type area and holes in the p-type area) are approximately equal to the concentrations of the relevant dopants (Fig. 2a).

At the instant these two areas are brought into intimate contact, a very large concentration gradient of electrons and holes across the boundary between them comes into existence. This gradient causes electrons to diffuse from the n-type area to the p-type area and the holes to move in the opposite direction. As a result of this diffusion, a space charge region comes into being near the junction: on the n-type side, this is positive, since electrons have left this area, while the uncompensated positive charges of immobile donor ions remain along with the holes newly arrived from the p-type area; on the p-type side, it is negative, because in the same way carrier diffusion has given rise to an area of negative charge consisting of immobile acceptor ions and electrons newly arrived from the n-type area. In this way, a dipole space charge layer is formed in the area around the p–n junction (Fig. 3). A potential barrier and electric field are formed within this layer that counteract further diffusion and restrict the diffusion current. Apart from majority carriers, there are minority carriers in the two areas on either side of the junction, which come about as a result of the thermal generation of electron–hole pairs. The potential barrier formed as a result of majority carrier diffusion favors the outflow of minority carriers from both areas. The movement of these carriers creates a dark current, which flows in the opposite direction to that of the diffusion current.

If the p–n junction is illuminated by photons with an energy equal to or greater than the band gap width E_g ($h\nu \geq E_g$), then electron–hole pairs form on either side of the junction, as in the case of thermal generation (Fig. 4).

Carriers forming no farther from the potential barrier than the diffusion length of minority carriers will diffuse towards the potential barrier and will be distributed there by the electric field due to the presence of the junction (the diffusion length is the mean distance that minority carriers have to move before they recombine with majority carriers). This field causes the carriers to move in opposite directions—electrons to the n-type area and holes to the p-type area. If an electron–hole pair forms on the p-type side of the junction, the electron reaches the junction before it has any chance of recombining with the hole (if

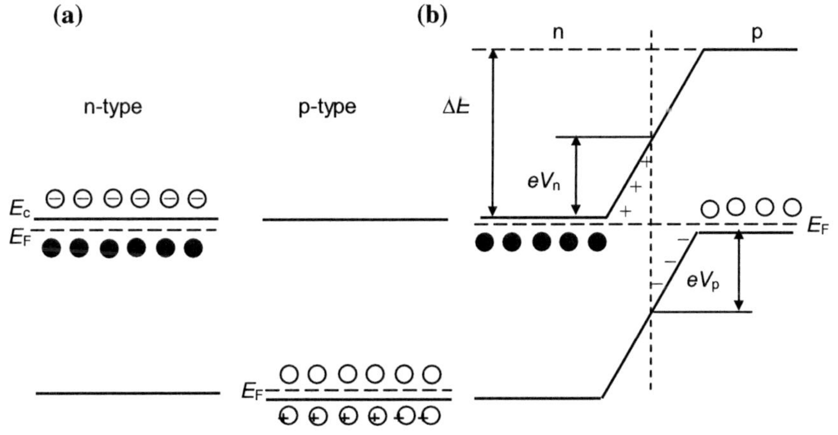
(a)
n-type
p-type
E_c
E_F
E_F
(b)
n
p
ΔE
eV_n
E_F
eV_p

Fig. 2 Formation of an abrupt p–n junction (b) as a result of the juxtaposition of n- and p-type areas (a), $\Delta E = e(V_n + V_p)$.

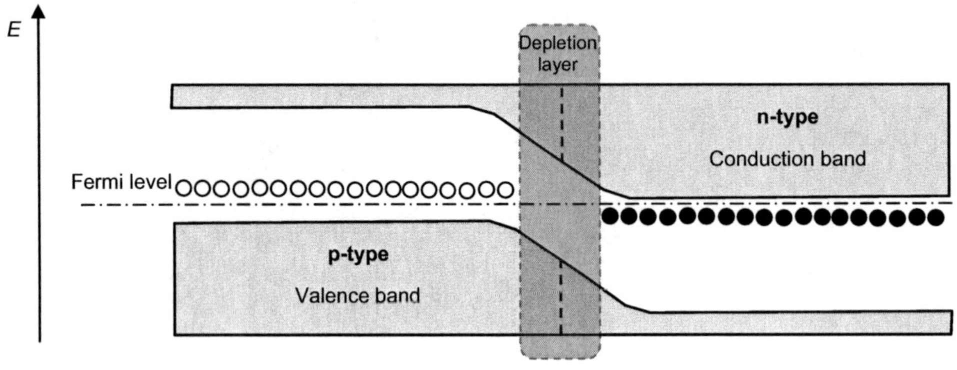
E
Depletion layer
n-type
Conduction band
Fermi level
p-type
Valence band

Fig. 3 Equilibrium in the p–n junction region.

recombination does occur, the resultant energy is emitted in the form of heat, and the effect is entirely useless as far as the photovoltaic effect is concerned), and the hole in this pair stays on the p-type side since it is repelled by the barrier in the junction. There is no danger of recombination here as there is an excess of holes in this region. The same thing happens when an electron–hole pair is generated by light on the n-type side of the junction. Then, the liberated electron remains on the n-type side, as it is repelled by the barrier. On the p-type side, however, we now have very few free electrons capable of recombining. This causes an increase in negative charge on the n-type side and of positive charge on the p-type side, which leads to a charge imbalance in the cell. This charge separation gives rise to a potential difference across the junction. As a result, a photoelectric current I_f comes into being in a closed circuit, regardless of the height of the potential barrier.

The generation of a photoelectric current I_f by a stream of photons in a solar cell can be demonstrated using the model of a current generator connected in parallel with a diode representing the p–n junction of the cell. As Fig. 5 shows, the output current I flowing through the series resistance r_s of the cell and the load resistance is equal to the difference between the generated photoelectric current I_f and the diode current I_d.

$$I = I_f - I_d = I_f - I_s\left(\exp\frac{eU}{mkT} - 1\right),$$

where I_s is the saturation current and m is the diode ideality factor.

It emerges from the above equation that for cell operated at open circuit ($I = 0$):

$$I_d = I_f \quad \text{and} \quad U = U_{oc} = \frac{mkT}{e}\ln\left(\frac{I_f}{I_s} + 1\right).$$

From this last relationship, we obtain:

$$I_f = I_s\left(\exp\frac{eU_{oc}}{mkT} - 1\right) \text{ and}$$

$$\text{finally: } I = I_s\left(\exp\frac{eU_{oc}}{mkT} - \exp\frac{eU}{mkT}\right).$$

For an exact description, however, we replace the single-diode electrical model with a two-diode equivalent circuit, which has two resistors: r_s is the series resistance of the cell, which consists of a number of components, and r_p is the effect of all defects in the crystal in the p–n junction area and is a shunt resistor (Fig. 6).

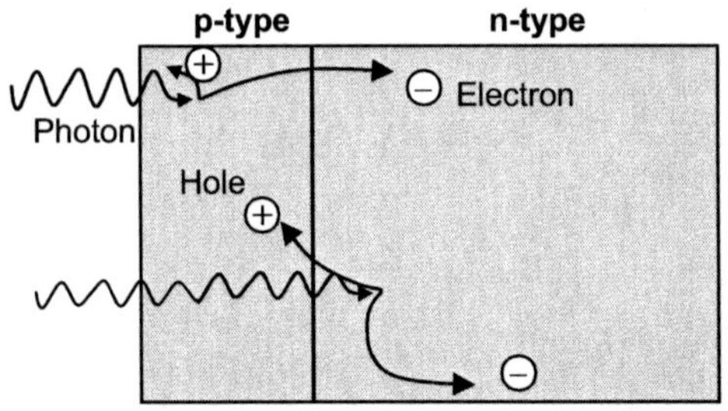
p-type
n-type
Electron
Photon
Hole

Fig. 4 The potential barrier in a solar cell distributes the charge carriers generated by light.

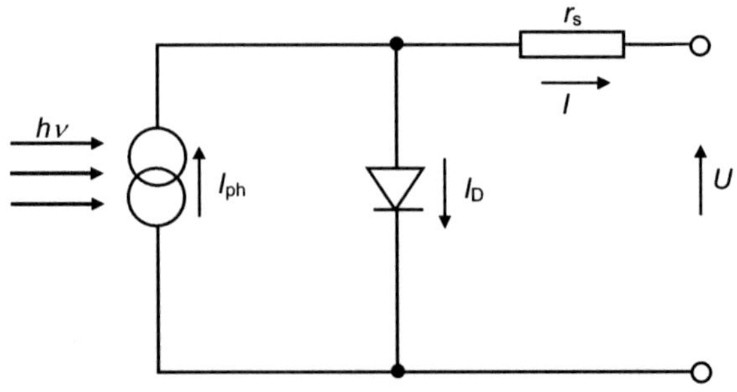
r_s
I
$h\nu$
I_{ph}
I_D
U

Fig. 5 Electrical model of a solar cell.

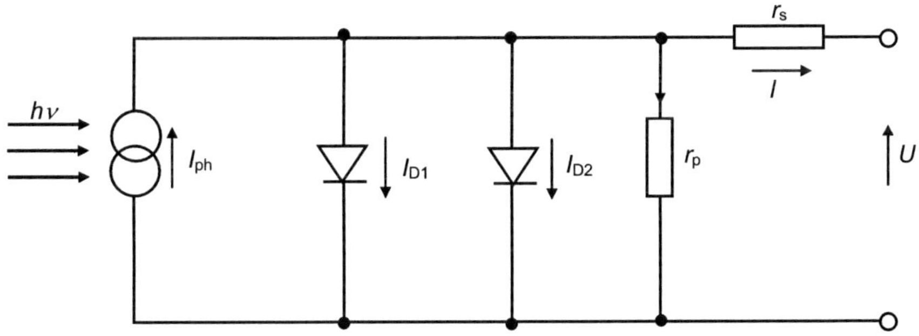

r_s
$h\nu$
I_{ph}
I_{D1}
I_{D2}
r_p
I
U

Fig. 6 Equivalent circuit of a two-diode model of a solar cell.
Source: Stutenbaeumer and Masfin.[1]

In this model, the current generated is described as a function of the cell voltage as follows:[2]

$$I = I_{s1}\left(\exp\frac{e\left(U - Ir_s\right)}{m_1 kT} - 1\right) + I_{s2}\left(\exp\frac{e\left(U - Ir_s\right)}{m_2 kT} - 1\right) + \frac{U - Ir_s}{r_p} - I_{ph}$$

where I_{ph} is the photoelectric current, I_{s1} and I_{s2} are saturation currents, and m_1 and m_2 are non-ideality factors of the characteristics of the two diodes.

The parameters of the model are defined in such a way as to ensure that the above equation gives a good description of the real characteristic of a photovoltaic cell. The first exponential term in the characteristic equation represents the diffusion current, whereas the second one represents the recombination currents in the entire cell, particularly in the space charge region. The characteristics, which also enable the parameters of the two-diode model to be determined, are measured when the cell is polarized in the forward direction and in the complete absence of any illumination, the dark current being measured as a function of the external voltage.

SOLAR CELL CHARACTERISTICS

The usable voltage from solar cells depends on the semiconductor material. In silicon cells, it amounts to approximately 0.6 V.

Under illumination, the fourth quadrant of the light *I–U* is the region of interest (Fig. 7), and the figures of merit for the device are the following:

1. The open-circuit voltage (U_{oc}) is the maximum voltage obtainable under open-circuit conditions (Fig. 8).
2. The short-circuit current (I_{sc}) is the maximum current through the load under short-circuit conditions (Fig. 8).
3. Fill factor (*FF*).

The output voltage of the photovoltaic cell is only slightly dependent on irradiance, while the current intensity increases with intensity of insolation. The working point of the solar cell therefore depends on load and insolation. In addition, the output voltage of a solar cell is temperature dependent. A higher cell working temperature leads to lower output and, hence, to lower efficiency (Fig. 9).

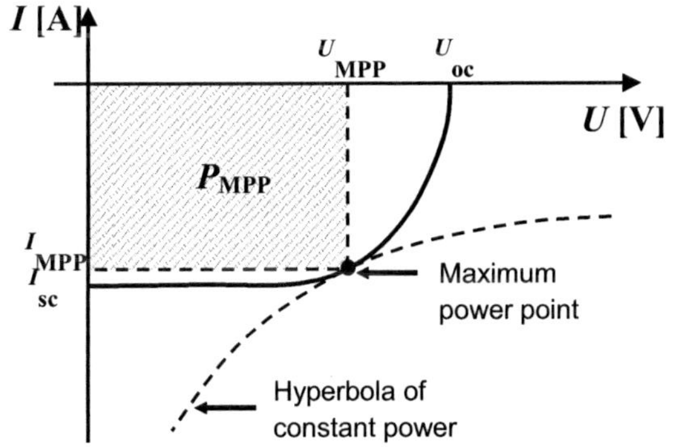

I [A]
U MPP
U oc
U [V]
P MPP
I MPP
I sc
Maximum power point
Hyperbola of constant power

Fig. 7 Current–voltage characteristic of illuminated photovoltaic solar cell.

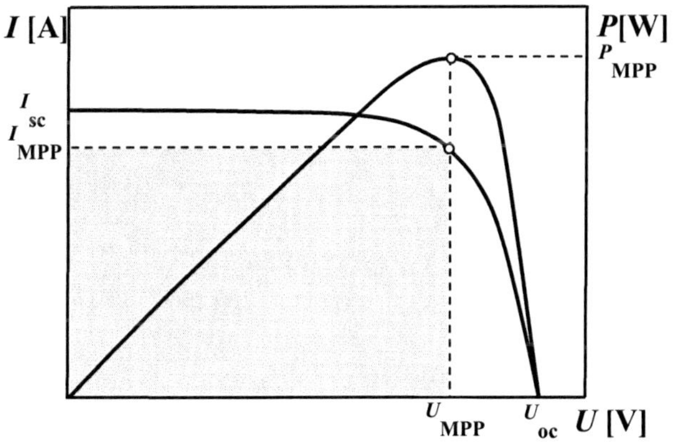

I [A]
P[W]
P MPP
I sc
I MPP
U MPP
U oc
U [V]

Fig. 8 Current–voltage and power–voltage characteristics of the solar cell.

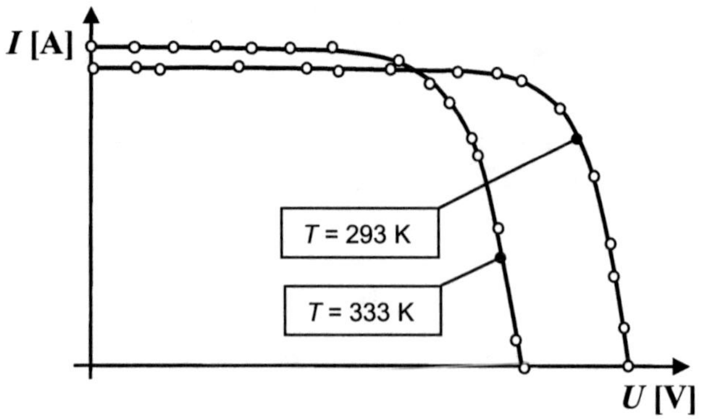
I [A]
T = 293 K
T = 333 K
U [V]

Fig. 9 Temperature dependence of the solar cell characteristic.
Source: Radziemska and Klugmann.[3]

An important parameter as regards the application of a PV module in photovoltaics is the peak power obtainable from the module at the load resistance R_{opt}, where the rectangle under the characteristic $I(U)$ has the maximum area equal to the maximum power $P_{MPP} = I_{MPP} \cdot U_{MPP}$, and the point of intersection with the curve of $I(U)$ is in this case the maximum power point (MPP). The load resistance R in the cell circuit or PV module should be chosen such that the power it generates takes the maximum value, i.e., $P = P_{MPP}$.

The MPP is the point at which the coordinates I_{MPP} and U_{MPP} form a rectangle with the largest possible area under the $I(U)$ curve.

The level of efficiency indicates how much of the radiated quantity of light is converted into usable electrical energy. The photovoltaic conversion efficiency of the cell η_{PV} is calculated from the maximum output power point (MPP) in the $I(U)$ curve:

$$\eta_{PV} = \frac{I_{MPP} U_{MPP}}{E \cdot S_C} \cdot 100\%,$$

where S_C is the total surface of solar cell and E is the irradiance (W/m^2).

To describe solar cell quality, a special parameter—the fill factor (FF)—is used. It can be calculated from the following equation:

$$FF = \frac{I_{MPP} U_{MPP}}{I_{sc} \cdot U_{oc}},$$

where I_{MPP} is the MPP current, U_{MPP} is the MPP voltage, I_{sc} is the short-circuit current, and U_{oc} is the open-circuit voltage.

For the ideal solar cell, the fill factor is a function of open-circuit parameters and can be calculated as follows:

$$FF \approx \frac{v_{oc} - \ln(v_{oc} + 0.72)}{v_{oc} + 1},$$

where v_{oc} is the voltage, calculated from the equation:

$$v_{oc} = U_{oc} \frac{e}{mkT},$$

where k is the Boltzmann constant = 1.38×10^{-23} J K^{-1}, T is the temperature (K), e is the charge on an electron = 1.6×10^{-19} C, and m is the diode ideality factor (–).

SOLAR CELL MATERIALS: PRODUCTION AND FEATURES

Silicon

The most important material for solar cell production is silicon. At the present time, it is almost the only material used for the mass production of solar cells. Being the most often used semiconductor material, it has some important advantages.

In nature, it is readily found in large quantities. Silicon dioxide forms one-third of the Earth's crust. It is environmentally friendly and not poisonous, and its waste does not cause any problems. It is easily melted and handled and it is fairly easy cast into its monocrystalline form. Its electrical properties, which remain unchanged up to temperatures of 125°C, allow the use of silicon semiconductor devices even in the harshest environments and applications.

In technology, pure silicon is the only widely used chemical element produced at such a high level of purity. The percentage of pure silicon in "pure silicon" is at least 99.9999999%. The concentration of silicon is 5×10^{22} atoms/cm^3, which means 5×10^{13} impure atoms/cm^3. Quantities of impure atoms are measured using sophisticated physical methods like mass spectrometry.

Pure silicon is produced from sand (silicon dioxide—SiO_2) by reduction at carbon electrodes at 1800°C in specially designed furnaces. The final material contains 98%–99% pure silicon. The complete reaction is:

$$SiO_2 + C \rightarrow Si + CO_2.$$

Such silicon is the raw material for the production of pure silicon. It is also used in steel and aluminium production as a supplementary material. The most important producers of raw silicon are Canada, Norway, and Brazil. Fifteen to twenty-five kilowatt-hours of electrical energy is needed to produce 1 kg of silicon. Silicon tetrachloride (tetrachlorosilane) gas is obtained by the chlorination of finely ground metallurgical-grade silicon in a special reactor. Additions or impurities are eliminated in the form of chlorine salt.

$$Si + 2Cl \rightarrow SiCl_4.$$

The following reaction produces trichlorosilane gas:

$$SiCl_2 + HCl \rightarrow SiHCl_3.$$

This gas is then further purified with the removal of any remaining tetrachlorosilane and other silanes. The purification is followed by reduction in a hydrogen atmosphere at 950°C:

$$4SiHCl_3 + H_2 \rightarrow 2Si + SiCl_4 + SiCl_2 + 6HCl.$$

Besides pure silicon, the procedure yields a number of gaseous by-products, which condense outside the reactor. Tetrachlorosilane is one of these by-products. At 1200°C, it can be converted into trichlorosilane using the following reaction:

$$SiCl_4 + H_2 \rightarrow SiHCl_3 + HCl.$$

This example illustrates one possible way of producing pure silicon. There are other procedures using different chemical reactions, but the end product is the same—pure silicon.

Crystalline Silicon Solar Cells

Polycrystalline as well as monocrystalline solar cells belong to this group. The basic form for crystalline solar cell production is the silicon ingot (see the description of the production procedure above). The ingot (block of silicon), cut with a diamond saw into thin wafers, is the basis of solar cell production. One-millimeter-thick wafers sawn accurate to 1/10 mm are placed between two plane-parallel metal plates rotating in opposite directions. This procedure enables the wafer thickness to be adjusted to within 1/1000 mm. The subsequent procedure for solar cell production consists of the following steps:

- Doped wafers are first etched some micrometers deep. The procedure removes crystal structure irregularities caused by sawing and cleans the wafer. During the extraction of pure silicon, the material is doped either as powdered polycrystalline silicon or by the addition of a suitable gas. This is then followed by diffusion. Phosphorus, supplied inside the material in gaseous form, diffuses at 800°C. The n-doped layer and the p-rich oxide layer form on top of the wafers as a result of reaction with oxygen.
- The wafers are then folded to form a cube and etched in oxygen plasma, which removes the n-doped layer from the edges. Wet chemical etching then removes the oxide layers from the top of the wafer.
- At the rear, the contact surface is produced from silver containing 1% aluminium. Special procedures enable silver to be printed over mask on cell surface. The pressed cells are then sintered at high temperatures. A similar procedure is used to print the contacts on the front cell surface, and the anti-reflex layer is applied likewise. In this case, titanium paste is used, which, on sintering, forms titanium dioxide (TiO_2) or silicon nitride (Si_3N_4).

Polycrystalline Silicon Production

The extraction of pure polycrystalline silicon from trichlorosilane can be carried out in special furnaces, such as those developed by Siemens. The furnaces are heated by electric current, which, in most cases, flows through silicon electrodes. These 2 m long electrodes are 8 mm in diameter. The current flowing through the electrodes can be as much as 6000 A. The furnace walls are cooled to prevent the formation of unwanted reactions producing gaseous by-products. The procedure yields pure polycrystalline silicon, which is used as a raw material for solar cell production. Polycrystalline silicon can be extracted from silicon by heating it up to 1500°C and then cooling it down to 1412°C, which is just above the melting point of the material. As material cools, a 40 × 40 × 30 cm ingot of fibrous polycrystalline silicon forms.

Monocrystalline Silicon Production

Two different technological procedures are used to produce monocrystalline silicon from pure silicon.

Czochralski's Method

In 1918, the Polish scientist Jan Czochralski discovered a method for producing monocrystalline silicon, from which monocrystalline solar cells could be manufactured. The first monocrystalline silicon solar cell was constructed in 1941. In Czochralski's method, silicon is extracted from the melt in a graphite-lined induction oven at a temperature of 1415°C. A silicon crystal with a set orientation is placed on a rod. Spinning the rod in the melt makes the crystal grow. The rod spins at 10 to 40 revolutions per minute and grows in length at a rate of between 1 μm and 1 mm per second. This allows the production of rods measuring 30 cm in diameter and several meters in length. The whole process takes place in an inert atmosphere. Possible impurities are burnt or eliminated in the melt.

Float Zone Method

With this method, monocrystalline silicon is produced from polycrystalline silicon. The main advantage of this procedure over the previous one is the better yield of pure silicon.

The silicon rods produced measure 1 m in length and 10 cm in diameter. This procedure, in which an induction heater travels along the rod melting the silicon, also takes place in an inert atmosphere. Monocrystalline silicon is produced during the cooling stage. Monocrystalline or polycrystalline silicon ingots are then sawn and the wafers are worked upon until they can serve as a foundation for solar cell production. Sawing causes approximately 50% of the material to be wasted.

Amorphous Solar Cells

Amorphous silicon is produced in high-frequency furnaces under partial vacuum. In the presence of a high-frequency electrical field, gases like silane, B_2H_6, or PH_3 are blown through the furnaces, supplying silicon with boron and phosphorus.

Amorphous solar cells are produced with technologies similar to those used in the manufacture of integrated circuits. Due to this procedure, these modules are also known as thin-film solar cells (thin-film modules). Here is a brief summary of amorphous solar cell production:

- The glass substrate is cleaned thoroughly.
- The lower contact layer is applied.
- The surface is then structured—it is divided into bands.
- The amorphous silicon layer is applied under vacuum and in the presence of a high-frequency electric field.
- The surface is rebanded.
- The upper metal electrodes are fixed.

Other Solar Cells

The less frequently used solar cell types include solar cells produced by the EFG (edge-defined film-fed growth) method, as well as Apex solar cells made from silicon, cadmium telluride (CdTe) solar cells, and copper-indium selenide (CIS) solar cells. EFG monocrystalline solar cells are produced directly from the silicon melt, which eliminates wafer sawing; production costs are thus lower and material is saved since there is no "sawdust." In the EFG procedure, an octagonal tube of silicon, several meters long, is extruded from the silicon melt. The flat sides of this tube are then laser sawn into separate solar cells. Most solar cells are square in shape and 100 × 100 mm in size. Consequently, the module power is greater with a smaller surface compared to crystal modules of square cells with truncated sides. Contacts take the form of copper bands. The separate cells are then combined in the same way as other cell types.

EFG cells are produced by Schott Solar. In contrast to EFG cells, Apex cells are polycrystalline. Their production procedure is protected by patent. The production procedure was developed by Astropower Inc.

Cadmium Telluride

Thin-film material produced by deposition or by sputtering is a promising low-cost foundation for photovoltaic applications in the future. The disadvantage of this procedure, however, is that the materials used in production are toxic. The efficiency of solar cells in the laboratory is as high as 16%, but that of commercial types is only 8%.

Copper-Indium-Diselenide ($CuInSe_2$, or CIS)

$CuInSe_2$ is a thin-film material with an efficiency of up to 17%. The material is promising, but not yet widely used owing to production-specific problems. CdTe and CIS cells have so far been used mostly in laboratory research. Commercial modules made from these materials are still hard to find.

Gallium Arsenide (GaAs)

GaAs is used in the production of high-efficiency solar cells. It is often utilized in concentrated PV systems and space applications. Their efficiency is as good as 25%, and even 28% at concentrated solar radiation. Special types have an efficiency of more than 30%.

Structure and Manufacture of Photovoltaic Cells from Crystalline Silicon

In principle, a photovoltaic cell consists of the following elements (see Fig. 10):

- A mono- or polycrystalline silicon wafer in which a p–n junction has been formed.
- Contacts, i.e., the front and rear electrodes; the front one should be shaped in such a way that the maximum amount of incident radiation can reach the junction region, the depth of which is limited by the permeability of silicon to radiation.
- An antireflective coating (ARC) on the front side of the cell.

The manufacture of a crystalline cell takes place in the following stages.

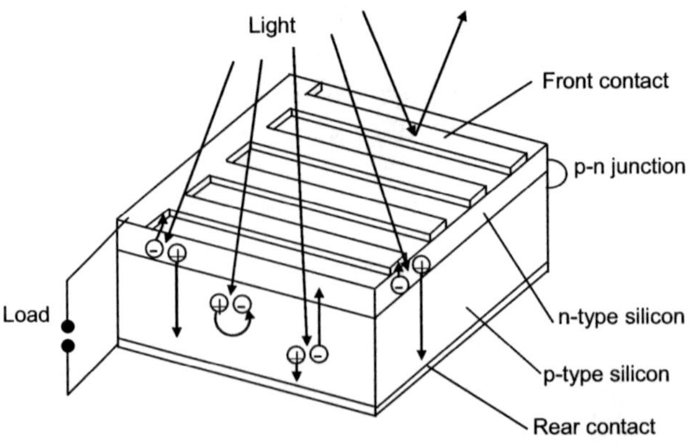
Light
Front contact
p-n junction
Load
n-type silicon
p-type silicon
Rear contact

Fig. 10 The structure and functioning of a photovoltaic cell (not to scale).

Surface Preparation

The surfaces of silicon wafers, cut from monocrystalline ingots, are degreased, cleaned, polished (mechanically or chemically), and etched in an aqueous solution of sodium hydroxide (40% NaOH) at a temperature of 383 K. A pyramidal surface structure is thereby obtained, which is then rinsed in hydrochloric or nitric acid. Dry (plasma) etching is also possible.

Diffusion Formation of a p–n Junction

A dopant, usually an acceptor, is added to the silicon base during crystal growth, whereas the n–p junction is produced by the diffusion of a dopant (usually a donor) to the p-type base wafer across one of its surfaces. If the donor concentration in the subsurface layer of the silicon (initially p-type) is greater than the acceptor concentration, this layer then becomes an n-type semiconductor.

The source of the dopant may be a solid or a gas. There are a number of techniques involving diffusion from the solid phase:

- Vacuum deposition of a thin layer of dopant
- Doping a vacuum-deposited layer of SiO_2
- Doping a mechanically deposited or screen-printed layer of SiO_2
- Coating the silicon wafer with a material containing phosphorus and silicon dioxide

If the diffusion of phosphorus to silicon is carried out at a temperature of 1220 K, 10 min are sufficient to obtain an n-type layer 0.25 μm in thickness.

Formation of an p–n Junction by Ion Implantation

Ion implantation is a technique for obtaining shallow, abrupt n–p junctions; it allows precise control of junction depth and dopant concentration. It is based on the implantation of dopant ions into silicon when its surface is bombarded with a beam of ions (of the order of 10^{16} ions/cm^2) of energy 5–300 keV.

The depth of the junction depends on the energy of the ions. A drawback of this method it that its gives rise to a large number of structural defects in the silicon; these can be removed after implantation by either heating the silicon wafer or irradiating it with an electron or laser beam. Ion implantation is quite a costly method—the necessary equipment (ion source, ion separator and accelerator, vacuum) is expensive.

Passivation of the Silicon Surface

Both the quantum efficiency for blue light and the open circuit voltage can be considerably enhanced if the front surface of the silicon is passivated. This is easily done with a layer of silicon dioxide, which forms on the surface when this oxidizes. Such a thin layer of SiO_2 of controlled thickness can be obtained by heating the silicon wafer in a stream of neutral gas like nitrogen containing a small quantity of dry oxygen. In the cells designed by Green, there is also a very thin layer of SiO_2 (ca. 2 nm thick) underneath the metal front electrode.

Metal Contacts

Since cells are an integral part of an electrical installation, metal contacts are made on either side of the cell. The front electrode is a fine grid, so as to reduce shading of the light-sensitive surface to a minimum. The "fingers" are ca. 0.1 mm wide, and the bus bar is from 1.5 to 2 mm wide. During the module's construction, the bus bar is connected to the rear electrode of the adjacent cell by means of copper strips soldered to it.

The metallic layers in a cell should be in ohmic contact with the silicon and have a low contact resistance, good adhesion to the silicon surface, and good soldering properties. Various techniques are used worldwide for producing metal contacts fulfilling these requirements.

Unlike the front electrode, the rear one can cover the entire area of the wafer. To improve efficiency, the rear electrode is made from a layer of aluminium vacuum deposited on the silicon surface between silver contacts in the form of strips or squares 2.5–6 mm wide. When this electrode is heated at temperatures from 770 to 1070 K, aluminium diffuses to the silicon forming a thin p^+ layer. To obtain a p–p^+ junction, the p^+ layer must be far thicker (0.2 μm); this is achieved by heating the cell to a temperature of 970–1070 K.

A very much cheaper method of making the two electrodes is the chemical deposition of nickel, or screen printing using a paste containing silver, aluminium, copper, or some other metal. Screen printing is a method that was used to produce lettering with a stencil over a thousand years ago. In 1975, it was first used in silicon cell technology to deposit the front and rear electrodes, thus replacing the costly vacuum deposition technique.

Silver paste is used for producing the front electrode, while aluminium paste combined with a small quantity of silver is usually used for the rear one. The silver pastes used for screen printing metal contacts on a silicon surface consist of a conducting phase (powdered silver of grain size 1–3 μm), an auxiliary phase (assisting the sintering of solid-phase grains—an enamel formed from the melting of a mixture of inorganic oxides), and an organic carrier facilitating the screen-printing process, which is burnt off when the layer is fired. The silver layers are usually sintered at 850°C.

About 86% of the silicon cells produced in 2006 had screen-printed metal contacts. At present, standard silicon cells with screen-printed electrodes achieve an efficiency of around 15% if they are polycrystalline and ca. 18% if they are monocrystalline (produced by Czochralski's method). During the screen printing of metal contacts, the mesh must be placed at a constant distance from the front side of the wafer. Silver paste is applied to the mesh and then imprinted using two squeegees. To ensure that the paste properly fills all the openings in the mesh, one squeegee moves along it, spreading the paste down its whole length. The other squeegee then applies just enough pressure to force the paste out of the mesh openings onto the wafer surface. After drying at 120°C for ca. 60 min, the printed layer consists of an aggregation of loose grains 1–2 μm in size; this must now be fired in order to impart stability to the layer.[4]

Cells with Rear Contacts

In cells of this type, both the positive and negative contacts are made on the rear surface of the cell. In this way, the whole of the front light-sensitive surface can be used to harvest light and the space between the cells can be minimized. The SunPower company produces commercial cells and modules of this type from n-type monocrystalline silicon—the efficiency of the cells is 21.5%, and that of the modules is 18.6%. The p–n junction is produced in the lower layer in the form of bands. This means that the photogenerated charges have to cover quite a long distance to reach the junction region, so only high-quality silicon is suitable for this type of cell.

Deposition of the ARC

One of the most significant parameters affecting the efficiency of a cell is the coefficient of light reflection from its surface—in the case of silicon, this is 33%–54%. This can be minimized by applying a transparent ARC to the cell's active surface.

ARCs can be applied in various ways: chemical vapor deposition (CVD), spraying, spin-on, and screen printing. The spin-on technique is the simplest one, as it does not require expensive equipment and is very efficient, but it can only be usefully applied to silicon wafers with a smooth, polished surface. If the wafer surface is textured, the ARC obtained in this way will not be of uniform thickness. Plasma-enhanced chemical vapor deposition (PECVD) produces ARCs with very good refractive index, photonic band gap, homogeneity, chemical composition, and controlled thickness. The surfaces of silicon wafers are usually coated with one or two antireflective layers.

The presence of an ARC is responsible for the color of the cells: polycrystalline cells are blue and monocrystalline ones are dark blue to black. By optimizing the thickness of the ARC, it is now possible to produce cells that are green, gold, brown, and violet in color, but this is only at the expense of their efficiency. One can, of course, do without the ARC and apply the cells in their original silver (polycrystalline) or dark gray (monocrystalline) colors; depending on architectural requirements, solar panels without an ARC can be integrated, for example, into the façades of buildings.

Dye-Sensitized Solar Cells

Dye-sensitized cells (DSCs) imitate the way that plants and certain algae convert sunlight into energy. The cells are inexpensive, easy to produce, and can withstand long exposure to light and heat compared with traditional silicon-based solar cells. This is a relatively new class of low-cost solar cell. It is based on a semiconductor formed between a photosensitized anode and an electrolyte, a photoelectrochemical system.

The fruits, flowers, and leaves of plants are tiny factories in which sunlight converts carbon dioxide gas and water into carbohydrates and oxygen. Although not very efficient, they are very effective over a wide range of sunlight conditions. Despite the low efficiency and the fact that the leaves must be replaced, the process has worked for hundreds of millions of years and forms the primary energy source for all life on earth. On the basis of this principle, there were early attempts to cover crystals of semiconductor titanium dioxide with a layer of chlorophyll. Unfortunately, the efficiency of the first solar cells sensitized in this way was about 0.01%. In 1991, Michael Grätzel and Brian O'Regan at the École Polytechnique Fédérale de Lausanne (Switzerland) used a sponge of small particles, each about 20 nm in diameter, coated with an extremely thin layer of pigment to obtain the effective surface area available for absorbing light.[5] These dye-sensitized solar cells (DSSCs or DSCs) are also known as Grätzel cells. Following much academic research, the energy conversion efficiency of laboratory cells made on glass substrates with liquid electrolytes has steadily increased to around 10% at air mass 1.5, 1 Sun conditions (for testing, the cells, regardless of design and active material, are typically insolated at a constant density of roughly 1000 W/m^2, which is defined as the standard 1 Sun value).

Dye-sensitized photoelectrochemical solar cells differ from conventional photovoltaic solar cells in that they separate the function of light absorption from charge carrier transport. The cells are made up of a porous film of tiny (nanometer sized) white pigment particles of titanium dioxide. These are covered with a layer of dye that is in contact with an electrolyte solution. Photoexcitation of the dye results in the injection of an electron into the conduction band of the oxide. The original state of the dye is subsequently restored by electron donation from a redox system, such as the iodide/tri-iodide couple. This process results in the conversion of sunlight into electrical energy.

In the case of the original Grätzel design, the cell has three primary parts. On the top is a transparent anode made of fluorine-doped tin dioxide (SnO_2:F) deposited on the back of a (typically glass) plate. On the back of the conductive plate, a thin layer of titanium dioxide (TiO_2) is deposited, which forms into a highly porous structure with an extremely high surface area. The plate is then immersed in a mixture of a photosensitive ruthenium-polypyridine dye and a solvent. After the film has been soaked in the dye solution, a thin layer of the dye is left covalently bonded to the surface of the TiO_2. A separate backing is made with a thin layer of the iodide electrolyte spread over a conductive sheet, typically platinum metal. The front and back parts are then joined and sealed together to prevent the electrolyte from leaking. The construction is simple enough that there are hobby kits available for hand-constructing them. Although they use a number of advanced materials, these are inexpensive compared to the silicon needed for normal cells because they require no expensive manufacturing steps. TiO_2, for instance, is already widely used as a paint base.

Principles

DSSCs separate the two functions typical of a traditional silicon cell design. In the crystalline silicon solar cells, the silicon acts as the source of photoelectrons and provides the electric field to separate the charges and produce the current.

In the DSSC, the bulk of the semiconductor is used solely for charge transport, while the photoelectrons are provided from a separate photosensitive dye. Charge separation occurs at the semiconductor/dye/electrolyte interface.

Although photoelectrochemical cells can operate without an organic dye, the efficiency of such cells is very low due to the low light-harvesting ability of inorganic n-conductors, which normally absorb light only from the high-energy ultraviolet region of the solar spectrum. The introduction of an organic dye makes for a significant increase in the absorption ability of the cells that extends across almost the entire solar spectrum.

DSC Structure

The structure of currently produced DSSCs is similar to those produced by Grätzel and O'Regan: on the top, there is a transparent anode made of indium tin oxide (ITO), deposited on the back of a glass plate. ITO (or tin-doped indium oxide) is a solid solution of indium (III) oxide (In_2O_3) and tin (IV) oxide (SnO_2), typically 90% In_2O_3 + 10% SnO_2 by weight. In thin layers, it is transparent and colorless, but in bulk form, it is yellowish to gray. In the infrared region of the spectrum, it is a metal-like mirror. ITO's main feature is the combination of electrical conductivity and optical transparency. A compromise has to be reached during film deposition, as a high concentration of charge carriers will increase the material's conductivity but decrease its transparency. Thin films of ITO are most commonly deposited on surfaces by electron beam evaporation or a range of sputter deposition techniques.

The thin oxide coating on one side of the glass makes the glass surface electrically conducting. On the back of the conductive plate, a thin layer of titanium dioxide (TiO_2) is deposited, which forms into a highly porous structure with an extremely high surface area—a 10 μm layer of randomly stacked nanoparticles (ca. 20 nm in diameter) (Fig. 11).

The plate is then immersed in a mixture of a photosensitive ruthenium–polypyridine dye and a solvent. After the film has been soaked in the dye solution, a thin layer of the dye is left covalently bonded to the surface of the TiO_2. A separate backing is made with a thin layer of the iodide electrolyte spread over a conductive sheet, typically platinum metal. The front and back parts are then joined and sealed together to prevent the electrolyte from leaking. Sunlight enters the cell through the transparent top contact, striking the dye on the surface of the TiO_2.

Titanium Dioxide

Titanium dioxide, also known as titanium(IV) oxide or titania, is the naturally occurring oxide of titanium TiO_2. When used as a pigment, it is called titanium white.

Titanium dioxide occurs in nature as the well-known, naturally occurring minerals rutile (the most common and stable form), anatase, and brookite. Crude titanium dioxide is purified via titanium tetrachloride in the chloride process. In this process, the crude ore (containing at least 90% TiO_2) is reduced with carbon, oxidized with chlorine to give titanium tetrachloride.

This titanium tetrachloride is distilled and then reoxidized with oxygen to give pure titanium dioxide. Another widely used process utilizes ilmenite as the titanium dioxide source, which is digested in sulfuric acid (as in Millenium Inorganic Chemicals). The by product, iron(II)

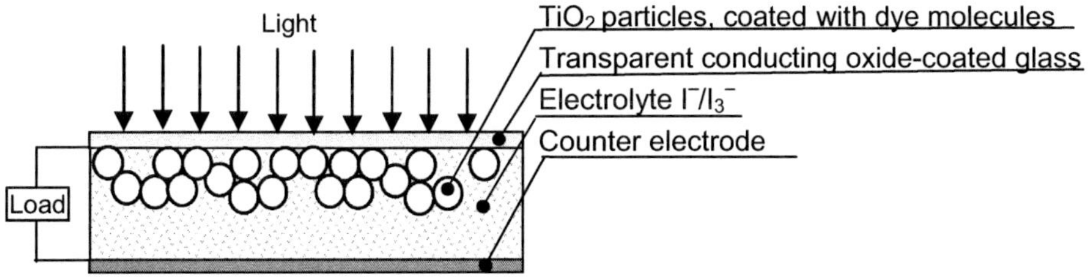
Light
TiO_2 particles, coated with dye molecules
Transparent conducting oxide-coated glass
Electrolyte I^-/I_3^-
Counter electrode
Load

Fig. 11 DSC structure.

sulfate, is crystallized and filtered off to yield only the titanium salt in the digestion solution, which is processed further to yield pure titanium dioxide.

The TiO_2 semiconductor has three functions in the DSSC:

1. It provides the surface for the dye adsorption.
2. It functions as electron acceptor for the excited dye.
3. It serves as electron conductor.

Colloid preparation and layer deposition have been developed to optimize the TiO_2 for these functions. Most important for the performance of desensitized solar cells was the development of a mesoporous semiconductor structure. This becomes evident considering the limited light capture of a dye monolayer on a flat surface.

The conductivity of nanophase TiO_2 films in a vacuum has been found to be very low, ~10^{-9} $(\Omega\ cm)^{-1}$ at room temperature.[6] However, on exposure to UV light, the conductivity is much increased, indicating that the low conductivity in the dark is due to the low electron concentration in the conduction band rather than to poor electrical contacts between the particles.

Untreated TiO_2 is an insulator that becomes "photodoped" and therefore conductive following electron injection of the adsorbed dye. Electronic contact between the nanoparticles is established by sintering the nanoparticles together, which enables the entire surface-adsorbed molecular layer to be accessed electronically. The interconnection of the nanoparticles by the sintering process allows the deposition of a mechanically stable, transparent film, typically a few microns thick. It is not necessary to increase the free electron concentration in the dark; indeed, this may even be detrimental to the photoelectrochemical behavior of the TiO_2. Among several semiconductors studied for photoelectrochemical applications, TiO_2 is by far the most commonly used, because of its energetic properties, its stability, and the ability to attach dyes. It is, furthermore, a low-cost material that is widely available. TiO_2 is used in its low-temperature stable form anatase (pyramid-like crystals), as rutile shows non-negligible absorption in the near-UV region (350–400 nm). This excitation within the band gap leads to the generation of holes, which are strong oxidants and cause long-term instability issues in the solar cell.

Mechanism of Operation

Photons striking the dye with enough energy to be absorbed will create an excited state in the dye, from which an electron can be injected directly into the conduction band of the TiO_2, and from where it moves by diffusion (as a result of an electron concentration gradient) to the clear anode on top (Fig. 12).

Meanwhile, the dye molecule has lost an electron and the molecule will decompose if another electron is not provided. The dye strips one from the iodide in the electrolyte below the TiO_2, oxidizing it to tri-iodide. This reaction occurs quite quickly compared to the time that it takes for the injected electron to recombine with the oxidized dye molecule, thus preventing the recombination reaction that would effectively short-circuit the solar cell. The tri-iodide then recovers its missing electron by mechanically diffusing to the bottom of the cell, where the counter electrode reintroduces the electrons flowing through the external circuit.

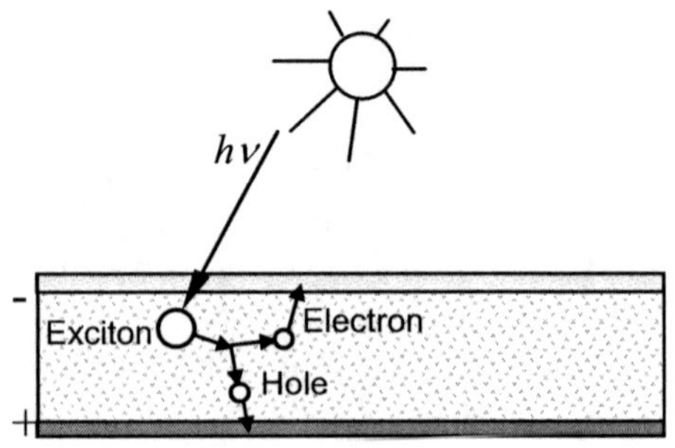
hv
-
Exciton
Electron
Hole
+

Fig. 12 Photogeneration of charge carriers in DSC.

The dye-sensitized oxide is usually deposited on a highly doped transparent conducting oxide (TCO), which allows light transmission while providing sufficient conductivity for current collection. Recently, high-conductivity organic polymers deposited onto plastic foil have found increasing application as substrates for flexible devices. The conductivities of metal foils are superior to those of TCOs and polymers. Because of their opacity, the illumination of the cell has to be established through the counter electrode. The surface of TCO should make good mechanical and electrical contact with the porous TiO_2 film.

To reduce dark current losses due to the short-circuiting of electrons in the substrate with holes in the hole conductor, a thin underlayer of TiO_2 is introduced between the SnO_2 layer and the nanocrystalline TiO_2 layer. This thin compact layer improves the mechanical adhesion of the porous TiO_2 film to the substrates, especially to SnO_2 layers of low haze, i.e., layers with less surface roughness and thus less contact area. Fig. 13 illustrates a model of charge carrier separation and charge transport in a nanocrystalline film.[7] The electrolyte is in contact with the individual nanocrystallites. Illumination produces an electron–hole pair in one crystallite. The hole transfers to the electrolyte and the electron traverses several crystallites before reaching the substrate.

The photogenerated hole always has a short distance (roughly equal to the particle radius) to cover before reaching the semiconductor–electrolyte interface whenever an electron–hole pair is created in the nanoporous film. However, the probability that the electron will recombine depends on the distance between the photoexcited particle and the back contact.

Allelochemics – Alternative

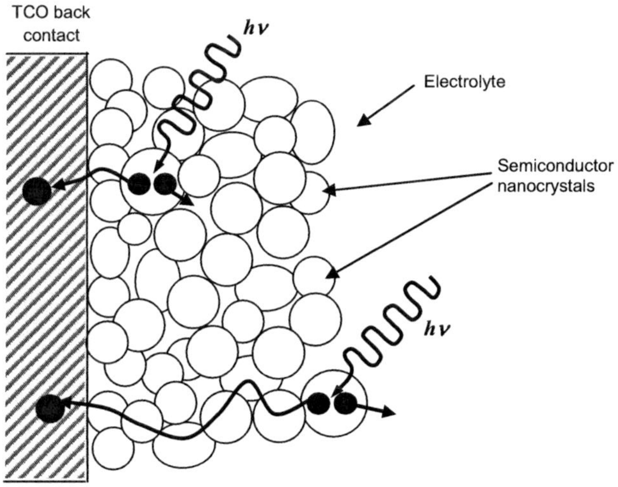
TCO back contact
hv
Electrolyte
Semiconductor nanocrystals
hv

Fig. 13 Qualitative model of photocurrent generation in nanocrystalline films.

The mechanism for converting solar energy into electrical energy in a DSSC is a five-step process:

1. Solar energy (photons of light $h\nu$) causes electrons in the molecular orbitals within the adsorbed dye sensitizer (S) molecules to become photoexcited (S*):

$$\text{S (adsorbed on TiO}_2) + h\nu \rightarrow \text{S* (adsorbed on TiO}_2)$$

The trapping of solar energy by a sensitizer molecule is analogous to the light-absorbing chlorophyll molecule found in nature, which converts carbon dioxide and water to glucose and oxygen.

2. The excited electrons escape from the dye molecules:

$$\text{S* (adsorbed on TiO}_2) \rightarrow \text{S}^+ \text{(adsorbed on TiO}_2) + e^-$$

3. The free electrons then move through the conduction band of TiO_2, gather at the anode (the dyed TiO_2 plate), and then start to flow as an electric current through the external load to the counter electrode.
4. The oxidized dye (S^+) is reduced to the original form (S) by regaining electrons from the organic electrolyte solution that contains the iodide/tri-iodide redox system, with the iodide ions being oxidized (loss of electrons) to tri-iodide ions:

$$\text{S}^+ \text{(adsorbed on TiO}_2) + 3/2\ I^- \rightarrow$$
$$\text{S (adsorbed on TiO}_2) + 1/2\ I_3^-$$

5. To restore the iodide ions, free electrons at the counter (graphite) electrode (which have travelled around the circuit) reduce the tri-iodide molecules back to their iodide state. The dye molecules are then ready for the next excitation/oxidation/reduction cycle.

Initially, the solutions of iodine–iodide mixtures in volatile solvents, usually acetonitrile, were used as redox electrolytes:

Electrolyte:	$I_2 + I^- \leftrightarrow I_3^-$
Anode (Dye):	$2\ Dye^+ + 3\ I^- \rightarrow 2\ Dye^0 + I_3^-$
Cathode:	$I_3^- + 2e^- \rightarrow 3\ I^-$

On the basis of different measurements, it is possible to indicate the orders of magnitude for the rate of the reaction steps. Upon illumination, the sensitizer is photoexcited in a few femtoseconds and electron injection is ultrafast from S* to TiO_2 on the subpicosecond time scale, where they are rapidly (less than 10 fs) thermalized by lattice collisions and phonon emissions. The nanosecond-ranged relaxation of S* is rather slow compared to injection, which ensures that the injection efficiency is unity. The ground state of the sensitizer is then recuperated by I^- in the microsecond domain, effectively annihilating S^+ and intercepting the recombination of electrons in TiO_2 with S^+, which happens in the millisecond time range. This is followed by the two most important processes—electron percolation across the nanocrystalline film and redox capture of the electron by the oxidized relay, I_3^-, within milliseconds or even seconds.[8]

Recently, solvent-free redox electrolytes, prepared from ionic liquids (liquid ionic organic compounds) or from ionically conducting polymer–nanocrystal blends were found to be very efficient. Fig. 14 shows the processes

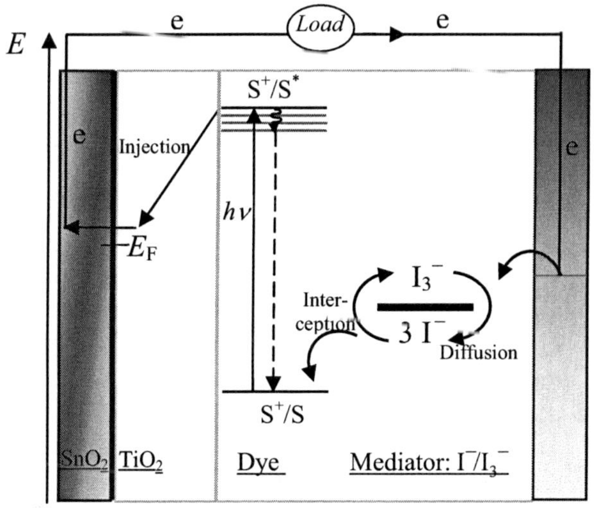
E
e
Load
e
e
Injection
e
S+/S*
hv
EF
I3−
Inter-
ception
3 I−
Diffusion
S+/S
SnO2
TiO2
Dye
Mediator: I−/I3−

Fig. 14 The electron processes taking place in an illuminated DSSC under open-circuit conditions.
Source: Anandan.[9]

Allelochemics – Alternative

taking place during the conversion of light into electrons in a DSSC.

Upon excitation, the dye injects electrons into the conduction band of the titanium oxide. The photo-induced electrons diffuse through the porous TiO_2 network and are extracted at the SnO_2 substrate. The dye itself is regenerated by the electrolyte containing the I^-/I_3^- redox pair. In the most efficient DSCs, the dye is regenerated by I^-, present in an external electrolyte at high (~0.45 M) concentrations.[10] The electronic circuit is closed by the reduction of the iodide couple at the platinized SnO_2 counter electrode.

Titania Solar Cells: Manufacturing Process

Manufacturing processes utilize the production methodologies and equipment already in use in the manufacture of electronic components. Production of cells needs a relatively low capital investment, the equipment does not rely on highly skilled operators, and—most importantly—the technology is environmentally friendly. A number of these manufacturing processes are protected by patents.

During production, a piece of glass is coated with fluorine-doped tin oxide and then sputtered onto a 500 nm thick layer of titanium. The layer is then anodized by placing it in an acidic bath with a mild electric current, and the titanium dioxide nanotube arrays grow to about 360 nm. The tubes are then heated in oxygen so that they crystallize. The process turns the opaque coating of titanium into a transparent coating of nanotubes.

In the next step, the nanotube array is coated with a commercially available dye—this becomes the negative electrode. The cell is sealed with a positive electrode, which contains an iodized electrolyte.

The most important advantages of the DSC manufacturing processes are as follows:

- The materials can be produced cheaply in large quantities.
- They use standard processing and assembly equipment.
- They are not very energy intensive (about 32 kWh/m^2).
- They can be automated.
- The equipment consists of a number of processing stations, each easily reprogrammable to adopt a wide range of DSC designs.

DSSCs can be connected in series or parallel and integrated into solar panels (modules). The tiles are normally ochre, but other colors such as gray, green, and blue are to be introduced shortly.

Advantages of DSC in Comparison with Other Solar Cells

The important features that distinguish a DSC from a conventional silicon crystalline solar cell are as follows:

- It is a photoelectrochemical cell: charge separation occurs at the interface between a wide band gap semiconductor and an electrolyte.
- It is a nanoparticulate titania cell: it is not a dense film like amorphous silicon, but a "light sponge."
- In a DSC, the dye monolayer chemically absorbed on the semiconductor is a primary absorber of sunlight; free charge carriers are generated by electron injections from a dye molecule, excited by visible radiation.
- In a DSC, light absorption and charge carrier transport are separate, whereas in a conventional cell, both processes are performed by the semiconductor.
- An electric field is necessary for charge separation in the p–n junction cell. Nanoparticles in the DSC are too small to sustain a built-in field; accordingly, charge transport occurs mainly via diffusion.
- Inside a p–n junction, minority and majority charge carriers coexist in the same bulk volume. This makes conventional solar cells sensitive to bulk recombination and demands the absence of any recombination centres such as trace impurities. DSCs are majority charge carrier devices in which the electron transport occurs in the TiO_2 and the hole transport in the electrolyte. Recombination processes can therefore only occur in the form of surface recombination at the interface.
- The maximum voltage generated by DSC in theory is the difference between the Fermi level of the TiO_2 and the redox potential of the electrolyte, equal to about 0.7 V (V_{oc})—it is slightly higher than V_{oc} for silicon, which has an open cell voltage equal to 0.6 V. However, the most important differences are dominated by current production.

Compared to other solar cells, the titania solar cell has the following advantages:

- It is much less sensitive to the angle of incidence of radiation—it can therefore utilize refracted and reflected light.
- It performs over a much wider range of light conditions owing to the high internal surface of the titania (light sponge)—it can thus be designed for operation under very poor light conditions.
- It can be designed to operate optimally over a wide range of temperatures; unlike silicon solar cells, whose performance declines with increasing temperature, DSSC devices are only negligibly influenced when the operating temperature increases from ambient to 60°C.
- There is an option for transparent modules—they can be used for daylighting, roof lighting, and displays.
- DSC production needs only commonly available (non-vacuum) processing equipment, making it appreciably cheaper to set up facilities.
- DSC has a significantly lower embodied energy than all other forms of solar cell.

- Crystalline silicon PV modules are suited to full sun applications, particularly sun tracking systems and roofs, and are best suited to cold climates and clear sky conditions.
- In contrast, the DSC is particularly suited to target markets in temperate and tropical climates, because of its good temperature stability and excellent performance under indirect radiation, during cloudy conditions, and when temporarily or permanently partially shaded.
- The DSC is also particularly suited to indoor applications that require stability of voltage and power output over a wide range of low-light conditions.

COMPARISON OF DIFFERENT TYPES OF SOLAR CELL

Table 2 compares the different solar cell types.

CONCLUSION

In many countries, the decentralization of power production, the increasing proportion of energy generated from renewable sources, and the development of cogenerative systems for producing heat and electricity are viewed as

Table 2 Comparison of different types of solar cell (*copyright: ©pvresources.com*).

Material	Thickness	Efficiency (%)	Color	Features
Monocrystalline Si solar cells	0.3 mm	15–18	Dark blue, black with AR coating; gray, without AR coating	Lengthy production procedure, wafer sawing necessary. Best researched solar cell material—highest power/area ratio
Polycrystalline Si solar cells	0.3 mm	13–15	Blue, with AR coating; silver-gray, without AR coating	Wafer sawing necessary. Most important production procedure, at least for the next 10 years
Polycrystalline transparent Si solar cells	0.3 mm	10	Blue, with AR coating, silver-gray, without AR coating	Lower efficiency than monocrystalline solar cells. Attractive solar cells for different BIPV (Building Integrated Photovoltaics) applications
EFG	0.28 mm	14	Blue, with AR coating	Limited use of this production procedure. Very fast crystal growth; no wafer sawing necessary
Polycrystalline ribbon Si solar cells	0.3 mm	12	Blue, with AR coating; silver-gray, without AR coating	Limited use of this production procedure; no wafer sawing necessary. Decrease in production costs expected in the future
Apex (polycrystalline Si) solar cells	0.03 to 0.1 mm + ceramic substrate	9.5	Blue, with AR coating; silver-gray, without AR coating	Production procedure used only by one producer; no wafer sawing necessary; production in band form possible. Significant decrease in production costs expected in the future
Monocrystalline dendritic web Si solar cells	0.13 mm including contacts	13	Blue, with AR coating	Limited use of this production procedure; no wafer sawing necessary; production in band form possible
Amorphous silicon	0.0001 mm + 1 to 3 mm substrate	5–8	Red blue, black	Lower efficiency; shorter life span. No sawing necessary; possible production in band form
CdTe	0.008 mm + 3 mm glass substrate	6–9 (module)	Dark green, black	Toxic raw materials; significant decrease in production costs expected in the future
CIS	0.003 mm + 3 mm glass substrate	7.5–9.5 (module)	Black	Limited supply of indium in nature. Significant decrease in production costs possible in the future
Hybrid silicon (HIT) solar cell	0.02 mm	18	Dark blue, black	Limited use of this production procedure; higher efficiency; better temperature coefficient and lower thickness

the paths to be taken by energy production in the future. A secure, long-term plan for energy development that takes environmental conservation requirements into account should therefore fulfill two conditions:

- The supply of power from conventional sources should decrease.
- The proportion of energy generated from renewable sources should increase.

Electricity consumption is increasing by 1% per annum in developed countries and by ca. 5% per annum in developing countries.[11] This implies the necessity to look for sources of electrical energy other than the traditional ones. Photovoltaics is one such non-traditional source, which satisfies all the criteria now required of energy sources: solar energy is universally available, and photovoltaic cells and modules are some of the environmentally safest devices for energy conversion.

REFERENCES

1. Stutenbaeumer, U.; Masfin, B. Equivalent model of monocrystalline, polycrystalline and amorphous silicon solar cells. Renewable Energy **1999**, *18*, 501.
2. Abdel Rassoul, R.A. Analysis of anomalous current–voltage characteristics of silicon solar cells. Renewable Energy **2001**, *23*, 409.
3. Radziemska, E.; Klugmann, E. Photovoltaic maximum power point varying with illumination and temperature. J. Sol. Energy Eng. ASME **2006**, *128/1*, 34–39.
4. Panek, P.; Lipiński, M.; Beltowska-Lehman, E.; Drabczyk, K.; Ciach, R. Industrial technology of multicrystalline silicon solar cells. Opto-Electron Rev. **2003**, *11* (4), 269–275.
5. Lund, J.W.; Freeston, D. World-wide direct uses of geothermal energy 2000. Geothermics **2001**, *30*, 29–68.
6. Hagfeldt, A.; Grätzel, M. Light redox reactions in nanocrystalline systems. Chem. Rev. **1995**, *95*, 49–68.
7. O'Regan, B.; Grätzel, M. A low-cost, high-efficiency solar cell based on dye-sensitized colloidal TiOz films. Nature **1991**, *353*, 737–740.
8. Zhang, Z. PhD Thesis N°4066, École Polytechnique Fédérale de Lausanne: Suisse, 2008.
9. Anandan, S. Recent improvements and arising challenges in dye-sensitized solar cells. Sol. Energy Mater. Sol. Cells **2007**, *91*, 843–846.
10. Marton, C.; Clark, C.C.; Srinivasan, R.; Freundlich, R.E.; Narducci Sarjeant, A.A.; Meyer, G.J. Static and dynamic quenching of Ru(II) polypyridyl excited states by iodide. Inorg. Chem. **2006**, *45*, 362–369.
11. Muneer, T.; Asif, M.; Munawwar, S. Sustainable production of solar electricity with particular reference to the Indian economy. Renewable Sustainable Energy Rev. **2005**, *9*, 444.

Alternative Energy: Solar Thermal Energy

Andrea Micangeli
Sara Evangelisti
Danilo Sbordone
Interuniversity Research Center for Sustainable Development (CIRPS), Sapienza University of Rome, Rome, Italy

Abstract

Exploitation of non-renewable resources can create non-sustainable conditions and environmental degradation resulting in a significant loss of final product quality. Nowadays, in many industrialized countries, including the United States, the heating, cooling, ventilation, and lighting of buildings represent approximately 40% of the annual nation's energy consumption. Therefore, it is first important to seek solutions that improve our quality of life, while reducing energy and environment consumption, then a fast transition to a structure based on renewable energy is of utmost importance. Solar thermal energy is an important alternative to fossil fuels with a huge potential. At the end of 2006, the solar thermal collector capacity in operation worldwide equalled 127.8 GW_{th}, corresponding to 182.5 million m^2. Of this, 102.1 GW_{th} were accounted for by flat-plate and evacuated tube collectors and 24.5 GW_{th} for unglazed plastic collectors. The installed air collector capacity was 1.2 GW_{th}. In terms of the total capacity in operation of flat-plate and evacuated tube collectors, installed at the end of the year 2006, China (65.1 GW_{th}), Turkey (6.6 GW_{th}), Germany (5.6 GW_{th}), Japan (4.7 GW_{th}), and Israel (3.4 GW_{th}) are the leading countries. The typical share of solar thermal energy necessary to meet the heating and cooling demands of a single building will be increased dramatically by more than 50%, and up to 100%. Without any doubt, solar thermal technology is already a mature technology, and its 30 years development has led to efficient and long life systems. In this entry, some different types of solar thermal collectors are presented, along with the design phases of a complete flat-plate collector plant. Although mature solar thermal technologies are already available, further developments are needed to provide adjusted products and applications, reduce the systems costs, and increase market deployment. Thus, innovative and cost-effective solar thermal systems are here presented and a description of a large solar thermal plant is given.

INTRODUCTION

The sun produces energy through nuclear fusion at its core, where tremendous amounts of energy are released by the fusion of nuclei into more massive nuclei under extreme conditions such as extreme pressures and temperatures. In this way, the conversion of hydrogen nuclei into the much heavier helium causes neutrinos and photons to be discharged. This energy travels through space by radiation that is a form of energy transmitted in electromagnetic waves. The energy is transferred from the core to the photosphere through the convection zone. The photosphere, or the radiating surface of the sun, is where the energy will be radiated into space. Electromagnetic radiation is mostly emitted here, though sometimes small amounts of microwave, radio, and x-ray emissions can also be emitted. A typical photon journey lasts about 100,000 years from the core of the sun to its surface, while a photon will take only 8 min from the sun's surface to reach the Earth.

Solar intensity can be measured by the Inverse Square Law, where the intensity of radiation hitting or striking objects in space, such as planets, asteroids, or dust, can be quantitatively assessed. The total energy given off by the sun is very high, owing through the tremendous energy released by nuclear fusion. Scientists estimate the energy output at 63,000,000 W/m^2 (watts per square meter).

Given the distance of the Earth from the sun, and due to the intercepting atmosphere, this figure is definitely much lower. Radiation in the outer atmosphere amounts to approximately 1367 W/m^2. Of these, only about 40% will reach the surface of the Earth as shown in Fig. 1.

The solar energy that reaches the surface of the Earth can be used by a photovoltaic system to produce electricity or by a thermal system to produce heat. When the solar energy is used to produce heat, it is called *solar thermal energy* technology. In this case, the solar radiation is used to warm up a fluid (water, air, solutions appropriated to each system) that can circulate, mainly:

- In heat exchangers at the beginning of the circuit that will use the returned available heat.
- In pipes and radiant objects put in place to warm up.
- In the refrigeration cycles to evaporate the volatile substances that are used in the condensation phase.

Through the surface, rarely adjustable, the solar energy must be accumulated and transformed into thermal energy, so it is possible:

Encyclopedia of Environmental Management DOI: 10.1081/E-EEM-120047274

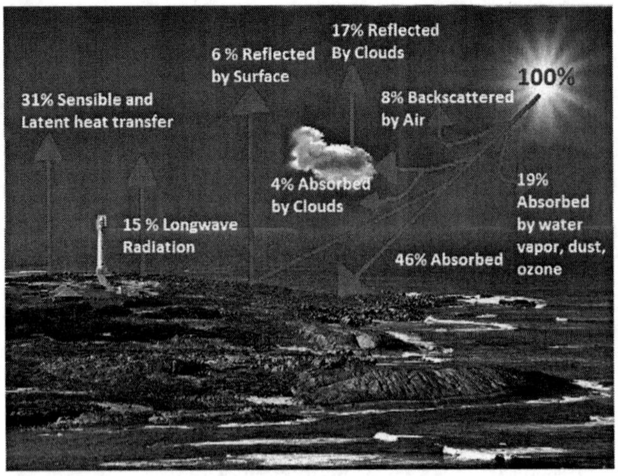
17% Reflected
By Clouds
6 % Reflected
by Surface
100%
31% Sensible and
Latent heat transfer
8% Backscattered
by Air
4% Absorbed
by Clouds
19%
Absorbed
by water
vapor, dust,
ozone
15 % Longwave
Radiation
46% Absorbed

Fig. 1 Earth's energy budget.

- To concentrate it through mirrors or lenses that reflect the radiation towards panels and a boiler to be able to use directly to warm water or to the production of vapor in the pipe and into a turbine.
- To collect it from applied panels or integrated panels in closure of buildings (walls, roofs, parapet, etc.)

The most important application that can be used by everyone is a *low-temperature solar thermal technology*. It consists of a system that uses a solar collector to warm up a liquid. The purpose is to capture and to transfer solar energy to produce *domestic hot water* (DHW) or to control the temperature inside a building. The term "low temperature" refers to the temperature that a working fluid can reach, generally up to 100–120°C.

This technology is ideal for application on a small scale, because it is cost-effective and it is simple to install and to manage. Low-temperature solar collectors can be classified as follows:

- *Flat plate collectors*, the most common type that consists of a dark flat-plate absorber of solar energy, a transparent cover that allows solar energy to pass through but reduces heat losses, a heat-transport fluid (air, antifreeze, or water) to remove heat from the absorber, and a heat-insulating backing.
- *Unglazed collectors* that are realized with tubes in plastic materials such as propylene, neoprene, synthetic rubber, and PVC. These are cheap because there is no insulation or transparent coverage. They have good performance only during summertime. They are recommended only if thermal energy for open swimming pools (and the like) is requested for.
- *Integrated collector storage*, recommended only for temperate climates, where the collector itself is the storage.
- *Evacuated tube panels*, where it is possible to eliminate the air between the capturing plate (reduced to a strip) and the evolved transparent sheet in a glass cylinder to resist the pressure difference.

The principal elements that are used by solar collectors are as follows:

- The *capturing plate*
- The *insulated material*
- The *transparent coverage*
- The *external casing*
- The *working fluid*

SOLAR COLLECTOR TECHNOLOGY

Flat-Plate Solar Collector

The *capturing plate* is realized with copper or steel and it is treated with satin and dark paint to reduce the reflection losses to bost the absorption capability to the wavelength of solar radiations with a low emissivity in the infrared radiations.

Normally, the canalizations on the plate are built to resist a pressure of about 6–7 bar; some collectors guarantee the resistance to a pressure of up to 10 bar.

The *insulated material* is a barrier against the conduction losses of the plate toward the external part of the collector. The materials used are always characterized by a porous or alveolar conformation in order to create microscopic motionless air spaces (that constitute a perfect barrier to the heat transmission). Polyurethane, polyester wool, glass wool, and rock wool are the most used materials. An enemy of the insulating material is humidity, which can appear for many reasons inside the collector (moisture, rain due to the gasket caused by little leakage in the pipes); often, the above-mentioned materials are covered with a thin aluminum layer that acts as a barrier to humidity and, at the same time, reflects towards the absorption plate.

The *transparent coverage* has the dual function of limiting the loss of energy towards the outside of the collector and facilitating the penetration of the radiation inside the collector. To satisfy this request, the coverage should be the most transparent as possible to the wavelength typical of the solar radiation (approximately between 0.2 and 2.5 mm) and at the same time it should be matt to the infrared radiation coming from the pipe-table while their temperature rises. The material that meets these qualifications is glass, above all if treated to acquire more transparency. Sometimes, due to some adverse elements, such as fragility and weight, plates made of plastic materials (polycarbonate) are preferred to glass plates.

The *external casing* has the dual function of contributing consistency and mechanical solidity to the collector and protecting the internal elements from dirt and atmospheric agents. It can be made of stainless steel (zinced) or aluminium.

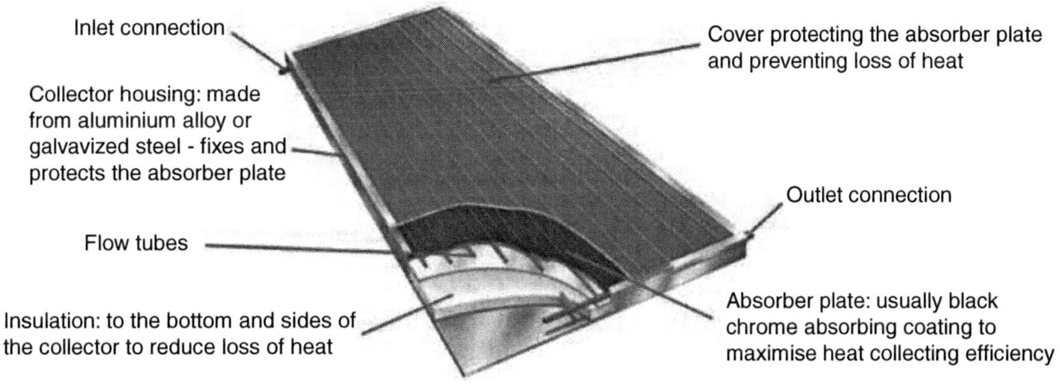
Inlet connection
Collector housing: made from aluminium alloy or galvavized steel - fixes and protects the absorber plate
Flow tubes
Insulation: to the bottom and sides of the collector to reduce loss of heat
Cover protecting the absorber plate and preventing loss of heat
Outlet connection
Absorber plate: usually black chrome absorbing coating to maximise heat collecting efficiency

Fig. 2 Flat plate solar collector scheme.

The *working fluid* that flows along the pipe system must take the largest quantity as much as possible. The fluid should also have a high density even at high temperature. It is important that it does not have a corrosive effect along the wall of the circuit; it must be inert and stable at temperatures below 100°C, and it should also have limited hardness to avoid limestone deposits. The hardness of the water refers to the quantity of magnesium and calcium salts in the water. The fluid should have a low freezing point. The option used among the producers of solar panels is a water solution of propylene glycol (not toxic and has a good anti-freezing action). In Fig. 2, a section of a flat plate collector with all components is shown.

Unglazed or Open Collectors

An *unglazed collector* is a simple form of flat-plate collector without a transparent cover. Typically, polypropylene, Ethylene-Propylene Diene Monomer (EPDM) rubber, or silicone rubber is used as an absorber as shown in the left panel of Fig. 3. Used for pool heating, they can work quite well when the desired output temperature is near the ambient temperature (i.e., when it is warm outside); moreover, they are cost-effective. As the ambient temperature gets cooler, these collectors become less effective. They can be used as preheat make-up ventilation air in commercial, industrial, and institutional buildings with a high ventilation load. They are called "transpired solar panels," and they employ a painted perforated metal solar heat absorber that also serves as the exterior wall surface of the building. Heat conducts from the absorber surface to the 1 mm thick thermal boundary layer of air on the outside of the absorber and to air that passes behind the absorber. The boundary layer of air is drawn into a nearby perforation before the heat can escape by convection to the outside air. The heated air is then drawn from behind the absorber plate into the building's ventilation system (Fig. 3, right panel).

Integrated Collector Storage

An *integrated collector* storage system is constituted by a unique element that assumes the role of capturing plate, absorber, and external accumulation. In this type of solar collector, the storage is located in the collector itself, and it is exposed to a slow heating process.

The water is placed inside the insulated tank and in this case the heat losses due to the exposure surface cannot be

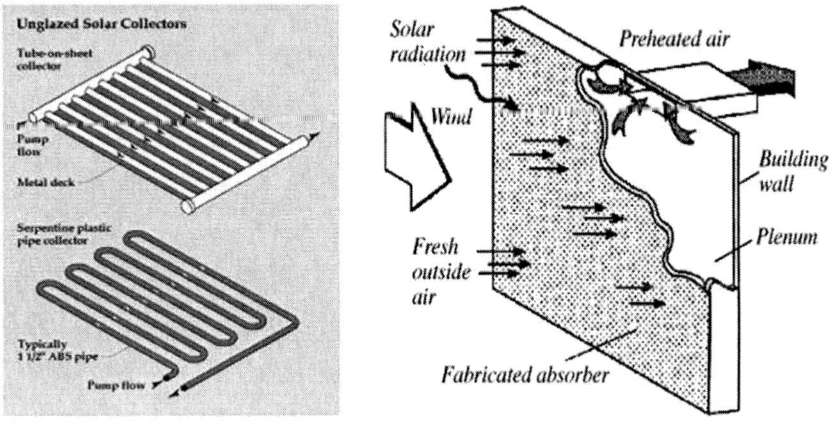
Unglazed Solar Collectors
Tube-on-sheet collector
Pump flow
Metal deck
Serpentine plastic pipe collector
Typically 1 1/2" ABS pipe
Pump flow
Solar radiation
Preheated air
Wind
Building wall
Plenum
Fresh outside air
Fabricated absorber

Fig. 3 Unglazed solar collector scheme.

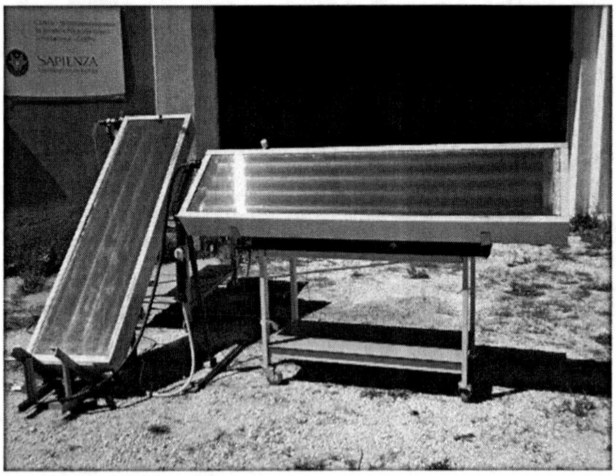

Fig. 4 Example of an integrated collector storage system.

efficient although more complex. Taking advantage of this type of heat exchange, the pipe system where the thermal vector liquid flows should be depressurized to decrease the evaporation temperature. A collector is built by the plate in long metal cylinders (copper) superficially covered with black and selected paint; this tube is in a second glass tube, in a way that surface of the first one is perfectly tallying with the internal of the second one; it is concentrically inserted in a second glass pipe. Fig. 5 shows a section of the collector.

The air between the two tubes is vacuumed out, until a pressure of $P = 5 \times 10^{-3}\ Pa$ is reached. A small tube goes lengthwise through the copper cylinder following a U path, inside the thermal vector fluid flows reaching a temperature close to 100°C. In order to maximize the use of the heat pipe, a bigger pipe can be used in a concentric position; the heating of the fluid in the pipe will increase because the captured radiation increases. Heater exchangers are placed at the ends of the pipes, in order to transfer heat to the users. The evacuated tube collectors are divided into two main types:

ignored. The integrated collector storage works without pumps or electrical devices: the panel absorbs the solar energy and the water inside the collector rises up for conduction, moving towards the outlet pipe and reaching the domestic network when hot water is required (see Fig. 4). The performance of such equipment is not fully satisfactory, because during the discharge phase, the water temperature decreases rapidly, reducing the overall usability of the collector.

1. *Evacuated pack collectors* with direct circulation of thermal vector fluid.
2. *Heat pipe evacuated pack collectors*: the fluid inside the pipe system evaporates along the way and give its heat, due to the condensation process.

Evacuated Tube Collectors

The *evacuated tube collectors* are obtained, reducing the presence of air in the space between the plate and the transparent cover, thus avoiding losses caused by convective movements. In spite of their higher cost, these collectors are able to perform well even when the environment temperature is low. Among the possible technological solutions that can be used to build these collectors, the *heat pipe* technology does not limit the exchange of heat between liquids but in the case the fluid flows in thin tube pipes, also between vapor and liquid, taking advantage of the heat for the condensation along its way, resulting more

In the first type, the plate is divided in long metal cylinders superficially treated with selected black paint, with each of these tubes inserted in a glass tube, which is also inserted in a larger glass tube, and then vacuum packed. A small tube goes through the copper cylinder following a U path.

In the heat pipe evacuated pack collectors, the little tube under pressure (heat pipe) receives heat from the capturing plate. The tube contains water or alcohol that evaporates at around 25°C under pressure. The vapor goes up until the head, where it exchanges heat through the condensation phase giving heat to another external fluid.

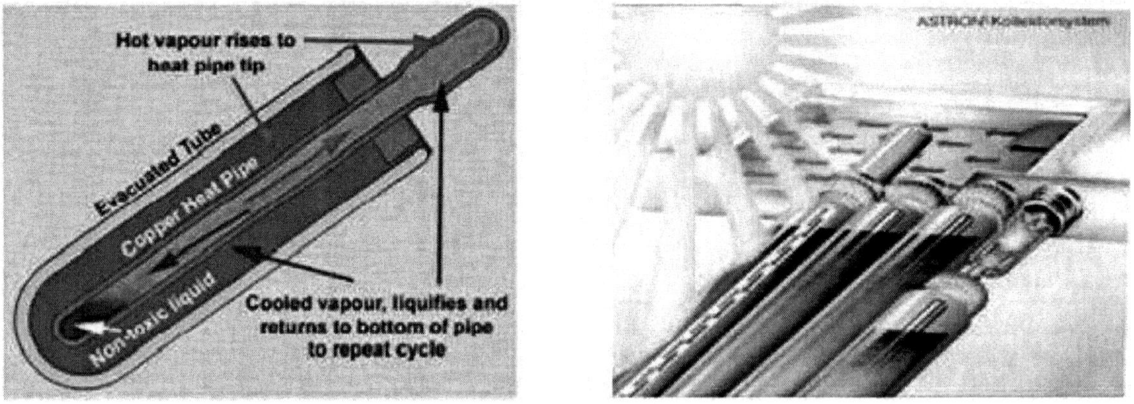
Hot vapour rises to heat pipe tip
Evacuated Tube
Copper Heat Pipe
Non-toxic liquid
Cooled vapour, liquifies and returns to bottom of pipe to repeat cycle

Fig. 5 Section of an evacuated tube collector.

ENERGY BALANCE OF A COLLECTOR

The phenomena that interact within a manifold are multiple and interconnected: the main energy exchanges between solar radiation and the various elements of the system are here described.

As shown in Fig. 6, the solar radiation (E_0) hits the glass cover. A small amount of radiation (E_1) is reflected and absorbed by the transparent cover. The copper absorber does not absorb all the remaining radiation into useful heat and partly reflects and dissipates heat (E_2) by convection, conduction, and radiation to the outside. Treatment with selective coatings, as mentioned above, reduces leakages. On one side, the transparent cover prevents the reflection of the solar radiation from the plate to disperse outwards favoring the greenhouse effect inside the collector, and on the other side, it limits the heat convection dispersion (Q_1). If a good thermal insulation in the back and sides of the collector is designed, with standard insulating materials such as rock wool or polyurethane foam, the energy losses by thermal conduction are reduced to a minimum (Q_2).

Only a portion of the incident solar energy (E_0) is transferred to the fluid as useful heat (Q_3) due to the different energy losses (E_1, E_2, Q_1, and Q_2).

Collector Efficiency Curves

The thermal energy transferred to the fluid per time unit is calculated as the difference between solar radiation captured by the plate and converted into heat, and the heat losses by convection, conduction, and radiation.

$$Q_n = E_c - Q_p \qquad (1)$$

E_c takes into account the absorbance and transmittance of the glass plate, and it is calculated as the product of the irradiance E, the transmittance τ, and absorbance of the plate α, as follows:

$$E_c = E * \tau * \alpha \qquad (2)$$

The heat loss depends on the temperature difference between the plate and the environment. As a first approximation (for low temperatures of the plate), this relationship is linear and it can be described through the coefficient of total losses:

$$Q_p = k * \Delta T \qquad (3)$$

where $\Delta T = T_p - T_a$, T_p is the average temperature of the plate, and T_a is the ambient temperature.

By substituting these relations in the efficiency of the collector, the following are obtained:

$$\eta = (E * \tau * \alpha - k * \Delta T)/E, \qquad (4)$$

$$\eta = (E * \tau * \alpha/E) - (k * \Delta T/E), \qquad (5)$$

and

$$\eta = \tau * \alpha - (k * \Delta T/E), \qquad (6)$$

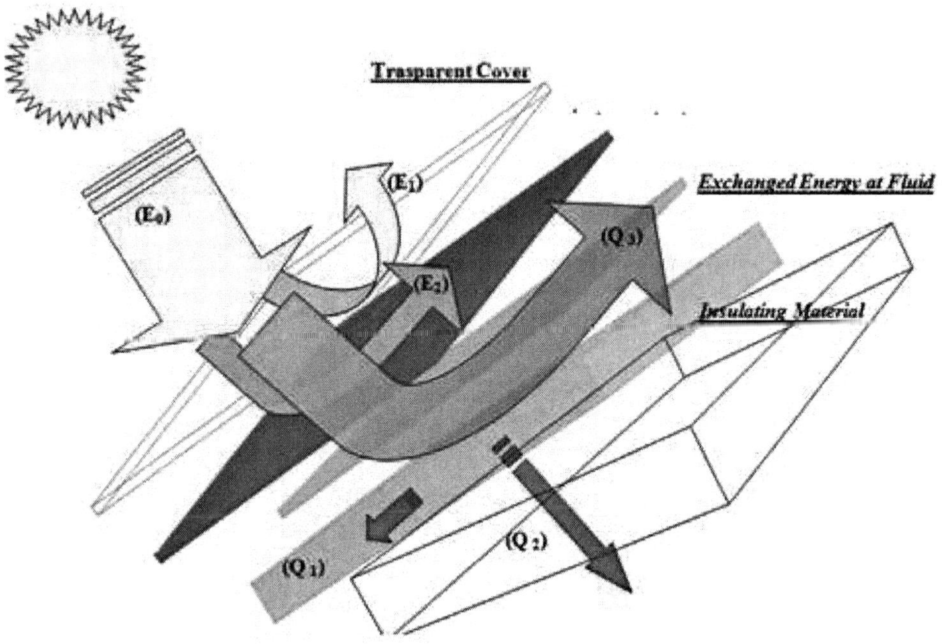
Trasparent Cover
Exchanged Energy at Fluid
Insulating Material
(E0)
(E1)
(E2)
(Q3)
(Q1)
(Q2)

Fig. 6 Energy balance of a solar thermal collector.

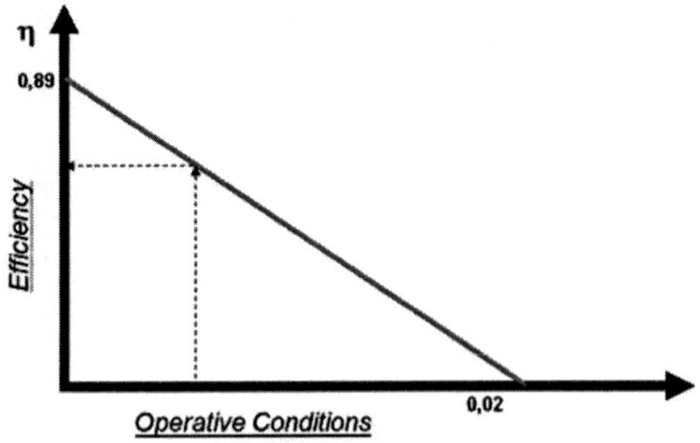
η
0,89
Efficiency
Operative Conditions
0,02

Fig. 7 Efficiency of a solar thermal collector at different operative conditions.

whereas

$$\eta_0 = \tau * \alpha. \quad (7)$$

It follows that the efficiency of the collector is equal to:

$$\eta = \eta_0 - (k * \Delta T/E). \quad (8)$$

Conventionally, it is defined:

$$\Delta T^* = (T_{mf} - T_a)/E. \quad (9)$$

Finally, the efficiency of a collector can be expressed as:

$$\eta = F' * \eta_0 - K\Delta T^*. \quad (10)$$

In Fig. 7, the efficiency curve of a collector is shown.

Instantaneous Efficiency of a Collector

The instantaneous efficiency depends on the optical losses (E_1 and E_2) and on the temperature (Q_1 and Q_2).

The overall losses of the total heat occurring in the collector by conduction, convection, and radiation can be expressed as the coefficient of total loss in K (W/m^2 * °C).

The graph shows that, with constant irradiance, the higher is the difference between the average temperature of the fluid and the ambient temperature, the heat losses also increase and, consequently, the efficiency of the collector decreases.

COMPARISON BETWEEN DIFFERENT TYPES OF COLLECTORS

By comparing the two efficiency curves, the principal characteristics of the collectors can be classified:

- *No glass collector* is the one with the best possibility to absorb the incidental radiations; its efficiency though decreases fast until zero in situations where the other collectors still have valid performances.
- *Flat-plate collector* with a selective plate has a better performance than the open one, practically in every working condition.
- *The evacuated collector* has a more stable efficiency curve, and it guarantees good performance even during bad weather conditions.

From Fig. 8, which shows the efficiency curve of different collector types, it is possible to observe that the unglazed collectors have better optics than others. In fact, the absence of covering helps eliminate the untranspar-

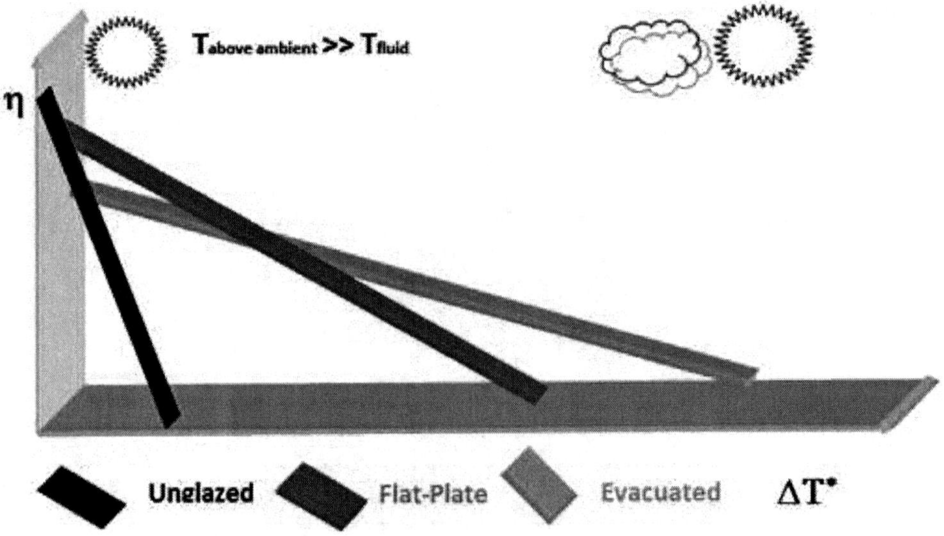
T above ambient >> T fluid
η
Unglazed
Flat-Plate
Evacuated
ΔT*

Fig. 8 Efficiency curves of different collector systems.

ency losses and reflection that often occur with the glass covering.

NATURAL CIRCULATION SYSTEMS

Once the differences between an open and a closed circuit have been examined, it is possible to analyze the circulation of the thermal vector fluid of the plant.

Natural circulation and closed circulation systems are based on the convective movements flown from the thermal vector fluid caused by a difference of temperature in the fluid itself. Indeed, the fluid warms up inside the serpentine of the capturing plate and it naturally goes up to the top of the collector. It will need an inclination compared to the floor to maximize the quantity of transferred energy.

Over the collector, a "storage tank" is placed with the heat exchanger inside. In the tank for the accumulation, there are two separate flows: the closed circuit of the collector with the thermal vector fluid and the water net of the running system designated to the final users.

The water, in contact with the heat flow, has a lower temperature, with higher heat absorption capacity, thanks to its stratification. While liberating heat to the running water, the fluid cools off moving to the lower part of the capturing plate; meanwhile, the part of the fluid that was at a colder temperature, being at this point at a higher temperature, tends to go up to the accumulation tank, cooling off while transferring heat to the running water. The quantity of fluid inside the solar collector remains constant and does not need any regulation pump for the circulation of the water, because a self-regulation natural mechanism works thanks to the trigger of the convective movement (see Fig. 9).

FORCED CIRCULATION SYSTEMS

In the system where the circulation is forced, the presence of the circulation pump, driven by a differential thermostat, allows the fluid to circulate inside the pipes. After a selected difference between the water temperature and the fluid, it activates automatically. The tube in which the thermal vector fluid circulates represents a primary circuit. In the highest part of the accumulation tank, there is an exchange of heat coming from an auxiliary traditional circuit that starts functioning when the water temperature designated to the users does not arrive to the requested one. The fluid that liberates its heat inside the tank of accumulation cools off through a pump of circulation that is sent to the lower part of the collector usually sited on the roof. See Fig. 10 for the scheme.

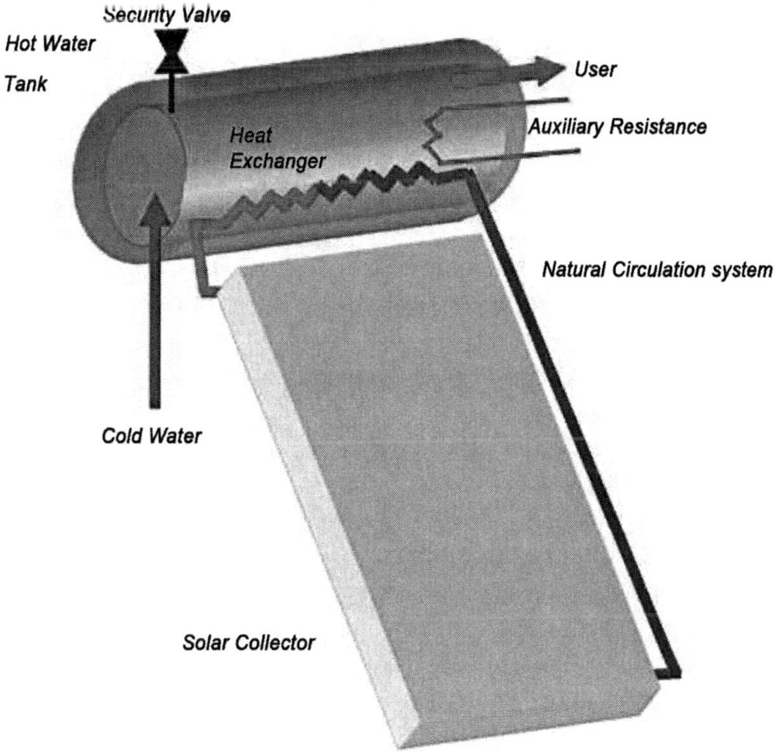
Security Valve
Hot Water
Tank
User
Heat
Exchanger
Auxiliary Resistance
Natural Circulation system
Cold Water
Solar Collector

Fig. 9 Natural circulation system.

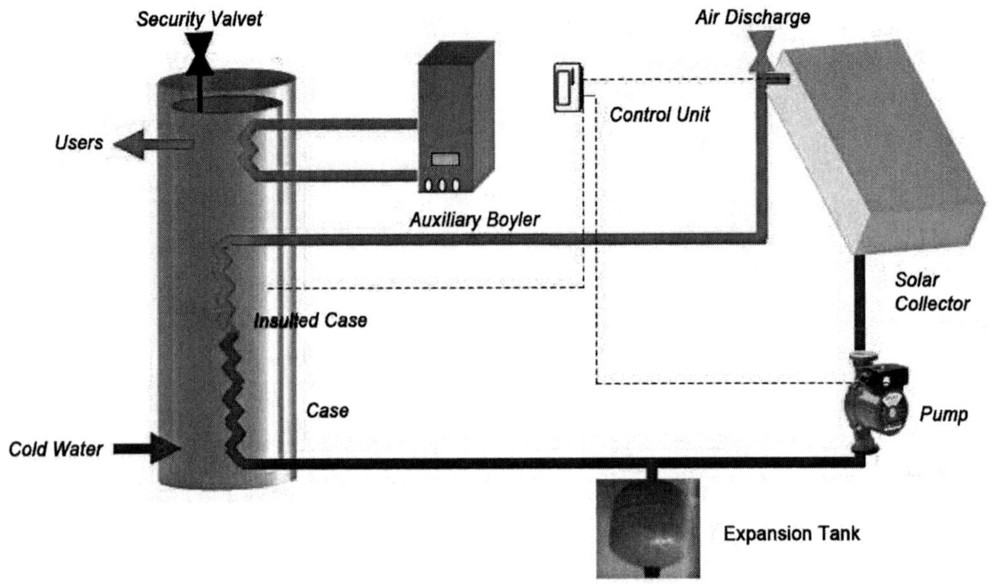
Security Valvet
Air Discharge
Control Unit
Users
Auxiliary Boyler
Insulted Case
Solar
Collector
Case
Pump
Cold Water
Expansion Tank

Fig. 10 Forced circulation system.

OPEN CIRCUIT SYSTEMS

Open circuit systems have some advantages, such as the simplicity of the hydraulic circuit realization and the lowest thermal losses that always occur when heat goes from one fluid to another. There are two problems that put limitations to this type of plant: the possible freezing of the water in case the temperature reaches values below 0°C and the calcium deposits along the tube system of the collector. In both cases, the collectors could go out of service.

CLOSED CIRCUIT SYSTEMS

The closed circuit is the most common solution. In this case, there are two different hydraulic systems: the main one, in which the thermal vector fluid circulates, and the secondary one, where the water coming from the hydro net is used (a third circuit is foreseen in the large-scale plans for the heat storage).

The thermal vector fluid must respond to specific functions:

1. To increase density and specific heat capacity to be able to use the smaller pipe systems.
2. To avoid limestone deposits due to hardness of the water.
3. To reduce freezing points and viscosity.
4. To be not toxic (in the case of plant for sanitary warm water).
5. To be chemically inert, stable, and not corrosive.

The option adopted by most of the producers is a solution of water and polyethylene glycol (usually 25%–45% of glycol).

The purpose of the thermal vector fluid is to take thermal energy captured by plane and transfer it to the water to heat up. In this energy transfer, the fluid can give some heat to the cold water through the exchanger in a proportional measure depending on the difference of temperature between the two fluids. In Fig. 11, the plate exchanger of heat is shown together with the tank with a double exchanger and a serpentine plunged. The larger the interface, the greater the amount of energy exchanged. The differences in temperature are of the utmost importance. To unify the need of big exchange surfaces with compact exchangers, the immersion serpentines are used.

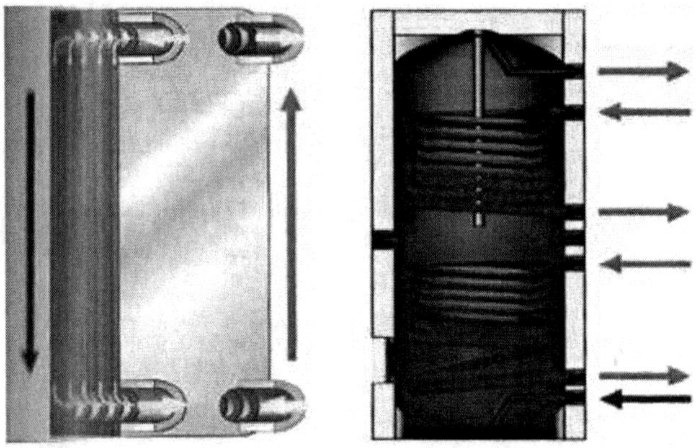

Fig. 11 The plate exchanger of heat is shown together with a tank with a double exchanger and a serpentine plunged.

SOLAR COOLING

The physical principle of generating solar cooling power is almost similar to the operating principle of conventional air-conditioning systems (condenser-compressor type) air-conditioning. Both systems rely on systems full vacuum single glass heat pipe collector picture liquid-to-vapor phase change energy of the refrigerant to attract heat (i.e., to produce cooling effect). The way the two systems achieve this is quite different because the condenser machine achieves a cooling effect by expanding compressed refrigerant into a low-pressure chamber, while the solar cooling machine relies on the absorbing action of the absorbent to create near-vacuum inside its chamber. In near-vacuum, the refrigerant will evaporate at a very low temperature, removing latent heat from the refrigerant (i.e., producing a cooling effect). This happens at a temperature significantly lower than the refrigerant's evaporation temperature at atmospheric pressure. A cooling effect is thus achieved at usefully low refrigerant temperatures, making this principle practical for commercial use.

Heat supply from a field solar collector to the solar air-conditioning system is required not to directly provide the cooling action directly but to maintain the absorbent concentration. This ensures that low chamber pressure and low evaporation temperature of the refrigerant are maintained.

PRELIMINARY ANALYSIS AND SOLAR THERMAL PLANT DESIGN

Due to the abundance and benefits of solar energy, solar thermal plants can be very useful especially if there is a willingness to invest some capital into them, but the choice should be made taking into account the quality of the project and the devices.

The covered need can be high (60%–70% or more), for various reasons such as the randomness of the sunshine and the urban or market situations; it will not be able to meet the demand of the users. Most of the time, the plant will limit the use of fuel for heat production.

In the supply of DHW (domestic home water) through this type of plant, it must take into consideration that the weather conditions will not always be able to provide a sufficient quantity of energy to satisfy the demand and the complete comfort of the users. Demand and supply of heat will not be equal at the same time, and so to obtain independent plants in an annual scale, the capturing surface should be overestimated for summer (to supply the energy needed during winter).

Nowadays, investing in large seasonal storage is still in an experimental phase and involves very interesting cases. Fig. 12 shows the offset of the thermal charge compared to the availability of the energetic source as the solar radiation incident. The latter is present in a solar thermal plant.

Matching Energy Availability and Thermal Energy Need

The way to estimate the energy needed to consume hot water is simple, but it is important to pay attention to the details involved.

Previously collected data can be used, but it is important to verify who collected them and what the original purpose was. If purposes are not reliable, it is necessary to go through monitoring to carry out data collection using a manual or digital device (calorie counter or a simple flow meter), gathering the data daily for at least 1 mo. In the sizing of the solar plants, the most important element is the energy need that will be accumulated and not the power need, which will be obtained at the right moment, from the mix of the solar heat storage and the auxiliary heater. Moreover, it is worth to invest on a thermal solar plant, as in all the renewable energies, only after a strong initiative to reduce the energy consumption, for example the aerator to be put at the final tap, can reduce up to 50% of the consume of hot water.

The Design Phase

The main design activities for a thermal solar system will be described here. The parameters and the different components that must be chosen for the plant and its dimensioning

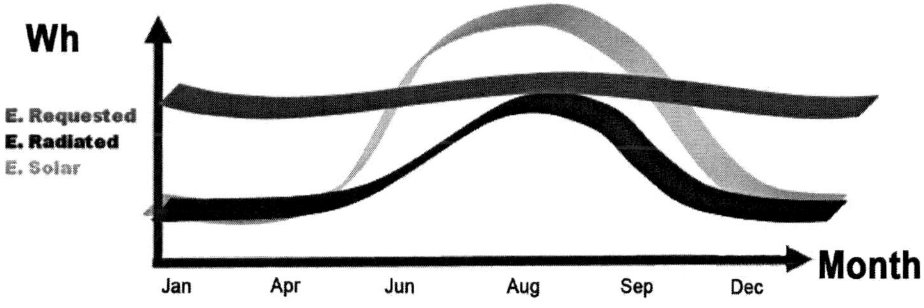
Wh
E. Requested
E. Radiated
E. Solar
Jan
Apr
Jun
Aug
Sep
Dec
Month

Fig. 12 Offset of the thermal charge compared to the availability of the energetic source as the solar radiation incident.

will be analyzed by the on-site procedure. It is important to note that in order to obtain the estimation of the costs of a solar plant, there is no need to proceed with detailed dimensioning, but it is sufficient to make an estimation of the following data:

- The surface of the collectors
- The volume of the tanks
- The thermal energy needs of the user

The phases of the project can be organized in the following way:

1. On-site investigation
2. Selection of the typology of plant and collectors
3. Dimensioning of the capturing surface
4. Dimensioning of the different components

On-site Investigation

This is the first step in the design of thermal solar plants that includes three main targets:

- A detailed analysis of user energy consumption to estimate energy need (in joules, kilowatt-hours, or calories).
- A set of possible solutions to reduce the consumptions through energy saving.
- A check on possible solutions considering the logistic realization of the plant, with particular reference to the collectors' colocation.

Analysis of the Users' Consumptions

The first step toward a correct project is to estimate the exact heat consumptions; this allows the dimensioning of the solar system to satisfy the user energy needs through solar energy. There are two rapid assessment methods through which the evaluation of the users on thermal energy consumption is made:

1. The study of energy bills of the previous years
2. The study of the consumption habit of the users

Saving Energy Interventions

The dimensioning of a solar system cannot be based exclusively on the previous analysis; it needs to examine the opportunity to minimize each and every cost, to utilize clean energy after dispersion and waste reduction. Saving energy represents an objective of main importance, which can be realized in two ways:

1. Working on the demand level, boosting the users to modify their habits (this is called energy sacrifice), not always accepted but free of cost.
2. Working on the level of the offer, promoting the substitution of old and common devices with high efficiency products having the same performance but lower consumption.

The realization of a recovery intervention implies almost all the time an economic investment, the convenience of which must be analyzed before comparing the saved energy that would be obtained. It has to be found out *how much is saved.*

In the first case, the person who is taking care of the project must underline the waste, applying devices of energy saving. For example, in the case of DHW, both flow reducers and an air–water mixer limit the flow of hot water. In this way, users have the impression that the jet of water is the same although the quantity is much lower.

Logistic Aspects

The purpose of the on-site visit is to analyze some logistic aspects because it is very important to evaluate the feasibility of the solar plant. This phase is often ignored but very important. The first examination should verify if any historical constraint, related to the buildings or to the landscape in the area, exists. There could be situations where it is not possible at all to install a solar plant because it should be a waste of time to proceed. Under a technical point of view, it is necessary to find out a free area in which the solar panel could be installed, on the surface of which it is necessary to measure:

- Typology and material of the surface (characteristic of the roof, for instance)
- Gross area available
- Obstacles (antenna, chimney, others)
- Shade elements nearby or far away (buildings, trees, etc.)
- Accessibility for the installation and for the following maintenance operations
- Azimuth (orientation compared to the south) and inclination compared to the horizon

With reference to the specific problem of the shade, it can be useful in areas that are growing, to evaluate the local urbanization policies on the short and medium terms and to avoid that some years after the installation, the panels might fall in the shade of a new building.

Choice of Solar Plant Type

From an initial analysis, it is possible to distinguish the solar thermal plants on the basis of the hydraulic circuit. In an open circuit, the thermal vector fluid coincides with the thermal vector fluid used by the final users. A closed circuit consists of three hydraulic circuits divided into

1. Thermal vector fluid
2. Accumulation fluid
3. Water from the user

Analyzing the system circulation two categories of plants can be underlined: systems of natural circulation and systems of forced circulation. In most cases, the plant is designed with a closed circuit because there is the possibility of using a thermal vector fluid different from simple water.

The use of water as a fluid has some disadvantages like the presence of calcium and a low freezing point on one hand, and the open circuit is more economic and rapid to be installed on the other hand.

Estimation of the DHW Need

The production of DHW represents one optimum solution for the use of solar energy because it allows using solar radiation also when it is at its maximum power. A reason that has increased their new diffusion is the economic return in terms of capital investment due to the costs relatively contained for the installation and the good reliability of the plant. The evaluation of the theoretical thermal need in the systems of DHW can be quantified through the following expression:

$$Q_{ac} = V^*(t_{ut} - t_{al})^*C_s \quad (11)$$

where Q_{ac} is the needed daily thermal energy (kcal/g), V is the requested water volume (L/g), t_{al} is the temperature of water source (°C), t_{ut} is the temperature of water output (°C), and C_s is the specific heat (kcal/kg * °C).

The supplied temperature depends on the place of the plant installation and on the water net from which it is taken, as it is possible to have differences of temperature depending on the period of the year. The temperature of warm water supply depends on the user considering that the maximum temperature of the water to the thermal generation should be t_{ut} = 48 + 5°C, where the second term indicates the maximum tolerance. To be able to guarantee this temperature until the water reaches the user, it is necessary that the distribution system must be suitably insulated. For the buildings that have no isolated distribution system, the temperature of supply would be higher than 50–55°C with peaks of 60°C.

The effective calculation takes into consideration the thermal losses that occur in the distribution network that brings the water from the production point to where it should be used. These losses make one think of the necessity of an effective thermal load higher than the theoretical one, to be able to satisfy the users' needs.

The total output of the distribution network depends on the grade of the insulation of the network itself. The average of the final output can have values in the following ranges:

- 0.85 to 0.90 in the case of recent constructions with insulated pipe system and with the recirculation.
- 0.75 to 0.85 for the constructions with plants without recirculation or with a cycle that works only during the day.
- 0.65 to 0.705 for the construction of plants with a working recycle during full time.

Sizing the Collector Field Surface

The first step to designing a solar plant system is to evaluate the surface needed for the plant as a compromise between the technical need and the economic need: unless a solar roof is used, it is necessary to think about the tilt

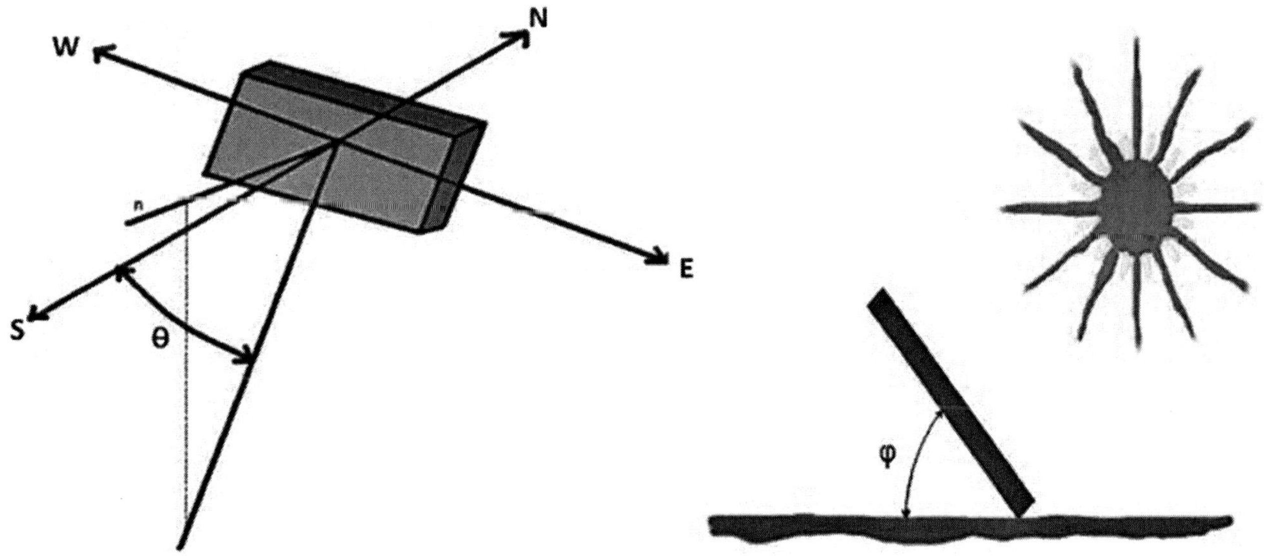
W
N
E
S
n
θ
φ

Fig. 13 Azimuth and tilt angles.

Table 1 Increment of the surface of the collectors under suboptimal conditions.

	Tilt angle						
	0°	**15°**	**30°**	**45°**	**60°**	**75°**	**90°**
South	12%	3%	0%	1%	8%	20%	45%
South/East or South/West	12%	6%	3%	5%	11%	23%	43%
East and west	12%	14%	15%	20%	28%	41%	61%

angle and its right positioning. Basically, the exposure should not face north and the surface should be enough to have all the collectors needed. That surface of collectors must avoid getting over economic convenience, even only on log term. The dimension of the capturing also depends on the type of chosen collectors, particularly on their performance and on the orientation of the tilt in the roof.

The quantity of energy on the surface of the collectors varies in function of two angles: the azimuth angle θ and the tilt angle φ. The azimuth angle θ is the projection on the horizon plane surface and the tilt angle φ defines the inclination of the collector to the surface as shown in Fig. 13.

The tilt angle depends on the building on which the plant is installed and on the orientation of the building itself as well. Since it is often not economically and aesthetically convenient, they are fixed directly on the top of the roof to create an ad hoc structure to hold the collectors.

About the typology of collectors, the choice depends on the conditions of the performance that will influence the output, and so from

- The internal temperature: the temperature to which the water is heated up.
- The external temperature: it depends on the period of the year in which it will be used.

The most important data (for the output) are the difference between the temperatures of the collector and the external environment because it characterizes the thermal losses. If data on consumptions are not yet available (for example, the building has just been built), an estimation on the energy needed (referring to average values) must be done. The corrections match the increases of the surface in bad orientation conditions, as seen in Table 1.

For combined plants, used for house heating and for the production of DHW would be appropriate to increase the value of the inclination to reduce the difference between the summer and winter production as underlined in Fig. 14.

HEAT STORAGE SYSTEM

The energy needed for a large number of applications depends on the time, but often in a different way from solar energy. As a consequence, it is necessary that the accumulation of thermal energy produced with solar panels firstly in those cases in which the solar energy must cover an important fraction of consumes. The optimal skill of the heat storage system depends on

1. The availability of solar energy.
2. The nature of the loads.

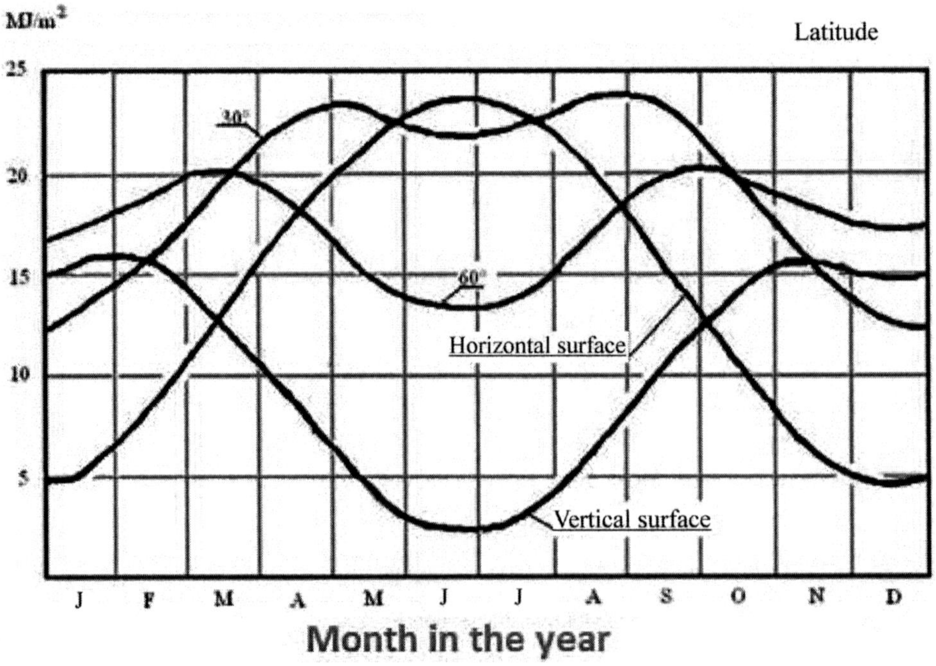
MJ/m²
Latitude
25
20
15
10
5
30°
60°
Horizontal surface
Vertical surface
J F M A M J J A S O N D
Month in the year

Fig. 14 Monthly average solar radiation during a year at 40° latitude.

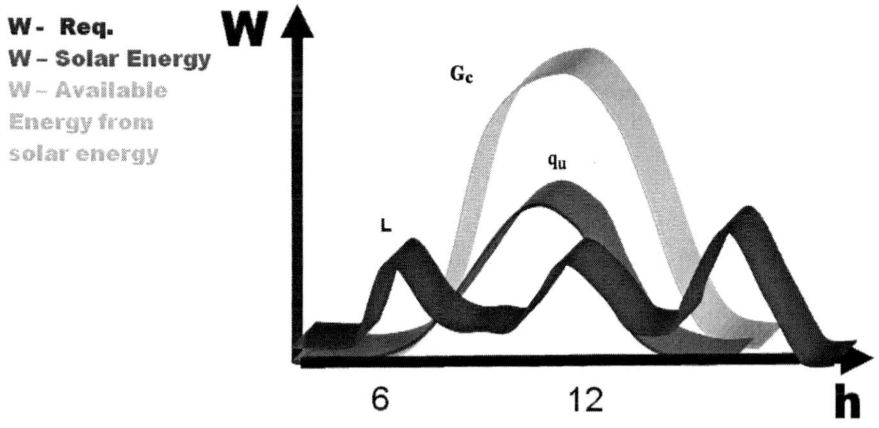
W - Req.
W - Solar Energy
W - Available Energy from solar energy
W
Gc
qu
L
6
12
h

Fig. 15 Energies during the day for a typical plant with storage.

3. The level of reliability needed in supply processes.
4. The economic analysis that sets the optimal percentage of traditional means.

Normally, a thermally isolated tank and a boiler are used. The boiler should store the solar energy when available and give it back when requested.

The most common energy discrepancy include alternation of day and night, sunny days as against cloudy days, and the summer season compared to the winter season. The latter is very important under a technical point of view because it reduces the technical problems connected to the summer overproduction as well.

A typical situation is a plant with accumulation. Paste Figs. 9 and 10:

Heat Storage Systems Dimensioning

Considering the time course of thermal loads and solar energy available:

- G_c represents the incident solar power.
- q_u represents the power delivered to the fluid.
- L is the power required by the loads that vary over time.

The areas above the red line and below the green one represent the time intervals that exceed the energy needs and must be accumulated. In contrast, the areas above the line in red and green show the time intervals in which the heat must be supplied to the loads from the storage system as in Fig. 15.

The storage systems can be divided into two broad categories:

- Sensible heat storage systems
- Latent heat storage systems

In the sensitive system, energy is stored by raising the temperature of a suitable material. Latent heat storage systems take advantage of the latent energy at the phase change (usually the liquefaction) of a substance, and in this case, the process takes place at constant temperature.

In the first type, energy is taken by decreasing the temperature of the substance in storage while in the second type, stored energy is made available, causing a phase change.

In the build-sensitive systems, adopted in most cases, the energy is proportional to the temperature change of the substance contained in the batteries themselves:

$$Q_s = M\, c_p\, \Delta T_s \quad (12)$$

where Q_s represents the accumulated energy, M is the mass storage, c_p is the specific heat of substance accumulation, and ΔT_s is the temperature variation.

Often a single temperature does not characterize the accumulation, because it cannot be considered perfectly mixed; it is actually layered. The top tank is the hottest part and the bottom is colder. The stratification is beneficial because of the collectors; reducing the average temperature of the absorber plate improves the efficiency to capture solar energy.

The best storage material in liquid systems is certainly water, since it has a low cost, has high specific heat, and is not toxic, and its boiling temperature at atmospheric pressure is high enough. The size of the storage system depends on the absorbing surface. Recommended values for solar systems for hot water are 100 L/m^2. The most important aspect to consider is, as already mentioned, the stratification of the water tank. As an index to assess the extent of the stratification of the water in a reservoir, the extraction efficiency is defined as:

$$\eta = (Q\, t^*)/V \quad (13)$$

where Q is the volumetric tank drain, V is the volume of accumulation, and t^* is the time required for starting from a completely mixed storage; the temperature difference of input–output has fallen to 90% of initial value.

Since the tank has a considerable cost, mainly due to heat insulation, sizing is necessary to make technical evaluations of economics. The storage tanks can be recharged in a sunny day or in an entire season. The first are those most commonly used and consist of an insulated tank to maintain hot water temperature. The degree of isolation of the accumulation should be as high as possible if the tank is installed outside, as often happens in systems with forced circulation. The storage tanks must also possess other important characteristics:

- They must be suitable for containment of potable water and must have had internal anticorrosion treatment.
- They must be resistant to high pressure (6 bar).
- They must be equipped with safety devices such as air vent, expansion vessel, and safety valve.
- They must be equipped with the following measuring devices: temperature gauge and pressure gauge for measuring pressure.

When the heat demand is roughly constant during the day, usually about 20 L/m^2 of collector is sufficient, while this figure varies between 50 and 100 L/m^2 of collector daily batch loads such as residential use, where consumption of water for showers are concentrated at certain times of day (morning or evening).

An important phenomenon is the *stagnation* of the system. In summer, the energy produced is often greater than the amount of thermal energy users and, if the accumulation is too small or absent, the excess energy increases the temperature inside the collector to allow evaporation of the fluid heat transfer. This leads to the thermal gradients and overpressure under serious problem the hydraulic components of the system, and it can compromise the integrity of the plant. Also, bear in mind that the plant also has air release valves to eliminate the air that enters the piping system itself, inevitably compromising the thermal exchanges and thus the operation. If it is used, an automatic valve (Jolly) is strongly discouraged, due to the stagnation of the primary because the emptying of the liquid transfer medium such as air comes out unwanted in automatic air vents. The sizing of the accumulation is performed for the reasons outlined above, depending on the surface of the manifold: in practice, it takes an average of 50 to 90 L of storage per square meter of collector area (see Fig. 16).

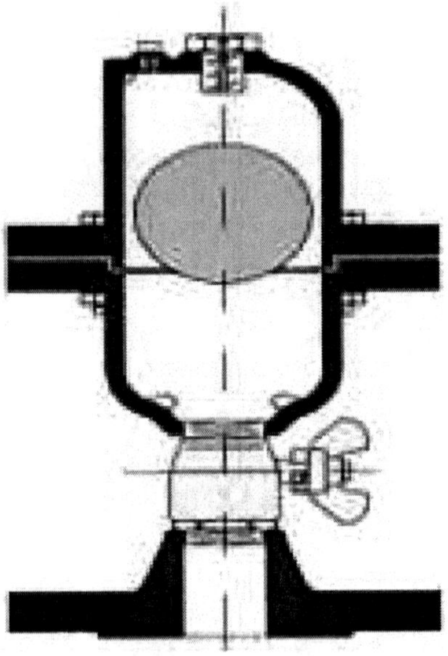

Fig. 16 Example of an expansion tank.

SIZE OF THE AUXILIARY

The sizes of the auxiliary are dependent on the diameter pipes that are obviously related to flow values (Table 2).

Another very important element, which has a major impact on system performance, is the pressure drop. The volume flow can be calculated by the following equation:

$$P_v = Q/(c_g\ \Delta T\ m_v) \qquad (13)$$

where Q is the thermal power made available by the solar collector (W/m^2); c_g = 1.3 Wh/(kg °C), the specific heat of the fluid; m_v is the density of the fluid (1 kg/L,); and ΔT is the temperature difference between inlet and outlet collectors (10°C).

The section of pipe can be calculated from the volumetric flow rate and velocity of the fluid P_v.

PREVENTION AND CONTROL OF LEGIONELLA EXPOSURE RISK

A problem that needs to be focused on in water systems for large facilities (hotels, prisons, hospitals, etc.,) is the presence of *Legionella*.

During the design phase of some facilities, a plant configuration to eventually prevent *Legionella bacteria* has already been furnished. Legionnaires' disease

Table 2 The size of the pipes according to the scale.

Flow (L/hr)	External diameter per thickness (mm)
<240	15 × 1
240–410	18 × 1
410–570	22 × 1
570–880	28 × 1.5
880–1450	35 × 1.5

is contracted by breathing, by inhalation, or by aerosol microaspiration that contains the bacterium. The aerosol is formed by the droplets generated by water spray or the impact of water on solid surfaces. Most droplets are small; the more dangerous water droplets are those with a diameter less than 5 microns as they can more easily reach the lower respiratory tract. To ensure a reduction in the risk of legionellosi, the following preventive measures should be followed:

- Always keep hot water at a temperature above 55°C.
- Slide the water (either hot or cold) taps and showers that are not used for a few minutes at least once a week.
- Keep showers, jet showers, and speaker of the taps clean and free of fouling; replace as needed.
- Clean and disinfect all water filters regularly every 1–3 mo.
- Ensure that any changes made to the system, including new installations, do not create dead arms or pipes with no water or stream flows intermittently.
- Adhering to a point, the aim is to ensure a continuous thermal disinfection of hot water.
- Heating the water to 55°C results to, in fact, short-term elimination of *Legionella bacteria*, as evidenced by the Hodgson Casey diagram taken as a point of reference for setting temperature and the time needed to implement thermal disinfection against *Legionella*.

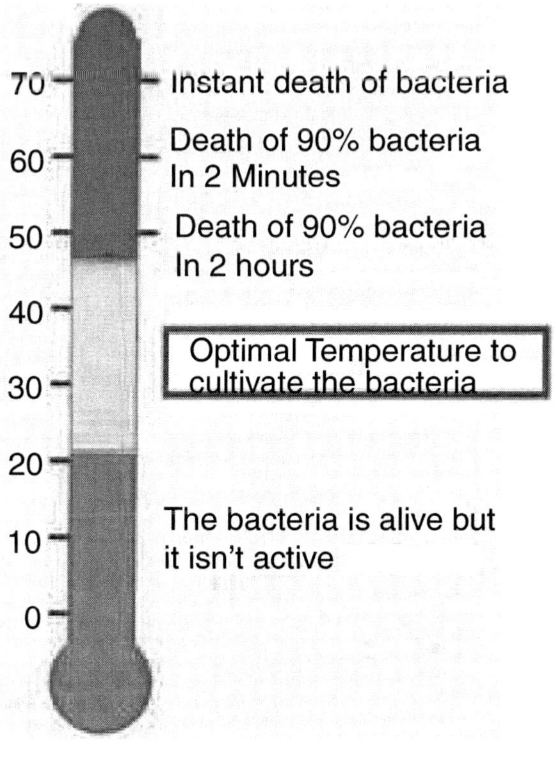
70
60
50
40
30
20
10
0
Instant death of bacteria
Death of 90% bacteria
In 2 Minutes
Death of 90% bacteria
In 2 hours
Optimal Temperature to cultivate the bacteria
The bacteria is alive but it isn't active

Fig. 17 Behavior of the *Legionella* bacteria, depending on different temperatures.

However, in order to properly set up a continuous thermal disinfection, it is necessary that the water plant is at least maintained at 55°C.

It should be considered, in fact, that if the thermal head high facilities are not well insulated and well balanced between supply points and certain areas of the circulation thermal head high can be determined, as shown in the diagram below (Fig. 17). So, even if hot water is supplied at a temperature below 50°C, there may be circulating in the networks of temperatures and can encourage the development of Legionella.

LARGE SYSTEMS

Solar systems are designed to be large to produce hot water for space heating. Possible applications include the following (see Fig. 18): hospitals, home care for seniors, military barracks, hotels, gyms, prisons, and residential complexes—situations where the demand for heat is constant both during the day and at different seasons of the year.

They are called large systems when the surface of the solar field is greater than 100 m^2. The investments made to realize these plants are vital because of the current costs of the collectors; thus, before spending a lot of money, an in-depth energy audit to evaluate the requests of the users and assess the economic breakeven point should be undertaken. These systems, of course, are not able to meet the total energy demand, so they must be integrated into a conventional source.

In support of these investments often involved in campaign financing from the state, such as loans to grants and tax deductions on the cost of the material. In recent years, the problem of pollution and continuous increases in the cost of traditional energy sources have caused people to engage in new ways, i.e., try to change their old habits, often met by indiscriminate consumption of energy.

Fig. 18 Abruzzo region post-earthquake emergency camps (2009).

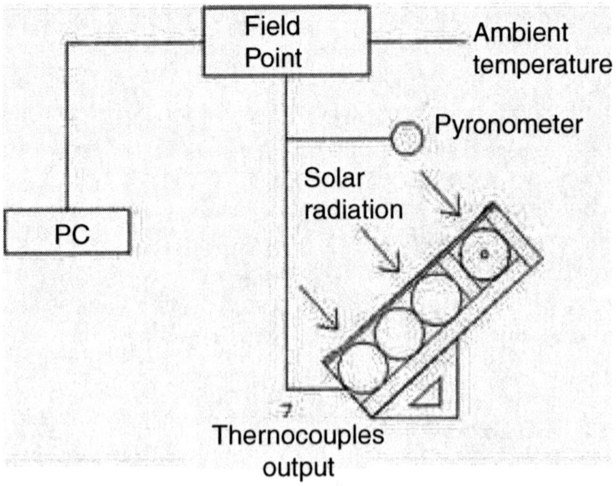
Field
Point
Ambient
temperature
Pyronometer
Solar
radiation
PC
Thernocouples
output

Fig. 19 Experimental device used for the ICS prototype.

SOLAR INTEGRATED COLLECTOR STORAGE SYSTEM INNOVATIONS

The use of a solar integrated collector storage (ICS) system (see Fig. 19) represents a well-established technology for heat storage in civil and industrial applications.

An innovative solar thermal device has been used as an integrated collector storage providing DHW (up to 50–80°C). Here, the collector also acts as a storage unit, without requiring an external vessel. This device was successfully used in several circumstances, especially in extreme situations such as in the post-earthquake tent cities or to feed remote users in developing countries.

The efficiency of this device is strongly related to the draw-off curve. In fact, it is strongly desirable that during the draw-off, the water temperature remains as high as possible. Presently, this aspect is not optimal, leading to a strong reduction of the water temperature (more than 50% when 50% of the hot water filling the collector was discharged).

The collector was designed for use in emergency situations or for feeding remote users. Then, a very simple configuration was adopted.

A series of eight J-type thermocouples have been placed to measure the temperature during the thermal energy storage phase and during discharge. The accuracy of the thermocouples was ±1.5°C and the acquisition frequency was 0.88 Hz. Solar radiation fluctuations have been followed by a pyranometer system with an accuracy of ±10 μV/(W m^2). The eight thermocouples (seven have been used for monitoring the pipe and one for the ambient temperature) and the pyranometer have been connected through a thermocouple and RTD modules, with a National Instrument field point with Ethernet connection, and connected to a data processing system through LabVIEW, as shown in Fig. 19.

During the charge phase, the pipe has been placed in vertical position in order to be completely filled with water and to eliminate the air. Then, it has been placed on its supports, with a tilt of 42°. The experimental analysis demonstrates that in the present configuration, the ISC has a reasonable performance in supplying hot water during the discharge phase.

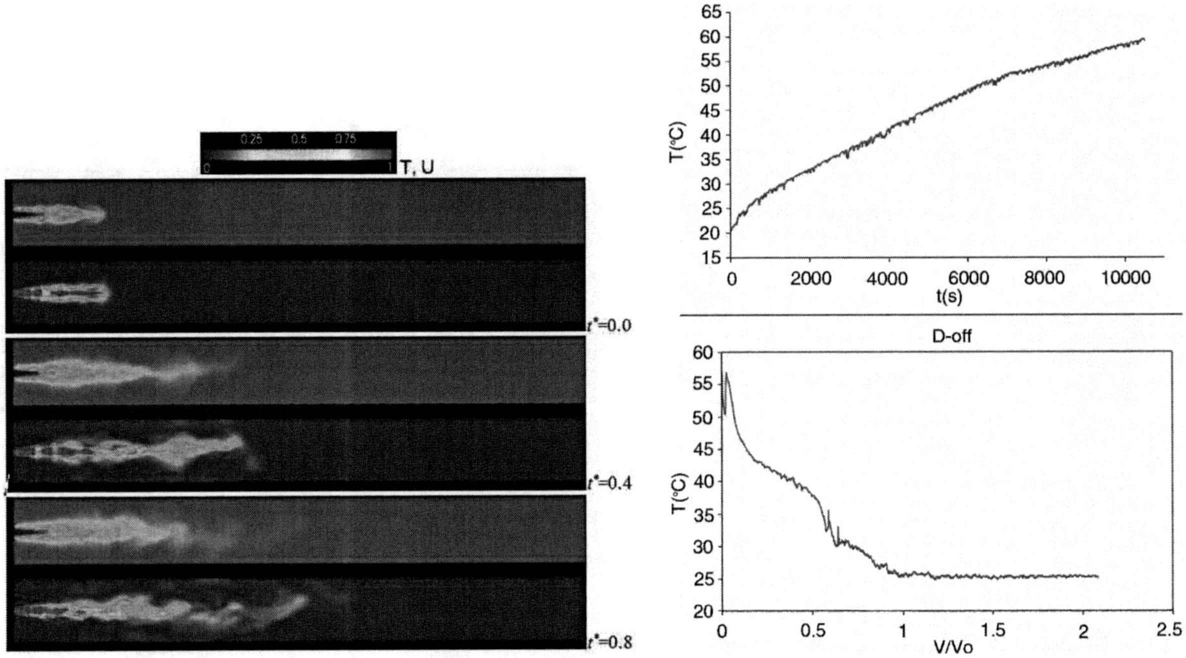
T, U
0.25
0.5
0.75
t*=0.0
t*=0.4
t*=0.8
T(°C)
t(s)
D-off
T(°C)
V/Vo

Fig. 20 Longitudinal view of the temperature (top) and velocity magnitude (bottom) fields in three successive time steps.

CONCLUSIONS

The investments made to realize the plants are important because of the current costs of the collectors; therefore, before spending such quantity of money, it is better to undertake an in-depth energy audit to evaluate not only the requests of the users but also the breakeven point.

These plants are not able to supply the total energy demand; therefore, they should be integrated with a traditional source.

To support such investments, often there are financial aids from the state, such a free grant loan or allowances on the V.A.T. on the costs of the material.

In recent years, the pollution problem and the constant rise in the cost of traditional energy sources have resulted to users trying to find new solutions, changing old habits, often entailing excessive energy consumption.

REFERENCES

1. Weiss, W.; Biermayr, P. Potential of solar thermal in Europe. European Solar Thermal Industry Federation (ESTIF), RESTMAC 6th Framework EU founded project, 2009, available at http://www.solarthermalworld.org/node/878.
2. Kalogirou, S.A. Solar thermal collectors and applications. Progress Energy Combustion Sci. **2004**, *30*, 231–295.
3. Hazami, M.; Kooli, S.; Lazaar, M.; Farhat, A.; Belghith, A. Performance of a solar storage collector. Desalination **2005**, *183*, 167–172.
4. Battisti, R.; Corrado, A. Environmental assessment of a solar thermal collectors with integrated water storage. J. Cleaner Prod. **2005**, *13*, 1295–1300.
5. Micangeli, A. *Design and implementation of innovative systems for the exploitation of renewable sources in social conflict areas: Desalination with heat recovery and hydroelectric energy production.* Sapienza University Doctoral Thesis, 2000.
6. Lesieur, M.; Metais, O.; Comte, P. *Large Eddy Simulations of Turbulence*; Cambridge University Press: New York, 2005; ISBN 0-521-78124-8.
7. Borello; Corsini; Delibra; Evangelisti; Micangeli. *Experimental and computational investigation of a new solar integrated collector storage system*, Third International Conference on Applied Energy, May 16–18, 2011, Perugia, Italy.

Alternative Energy: Wind Power Technology and Economy

K.E. Ohrn
Cypress Digital Ltd., Vancouver, British Columbia, Canada

Abstract
This entry discusses a broad range of topics as an overview of the field of wind power. The topics include current penetration, market shares, technology, costs, governments, and regulation.

INTRODUCTION (WEB PREVIEW)

This entry is a broad overview of wind power. It covers a range of topics, each of which could be expanded considerably. It is intended as an introductory reference for engineers, students, policy-makers, and the lay public.

Wind power is a small but growing source of electrical energy. Its economics are well known; there are several large and competent manufacturers, and technical problems are steadily being addressed. Wind power can now be considered as a financially and operationally viable alternative when planning additional electrical capacity. However, as shown below, wind power is only a minor component of present energy sources.

World final energy consumption (2002)[1]: 100%
World final electrical consumption (2002)[2]: 16.1%
World final wind power consumption (2004)[3]: 0.15%

Wind power's main deficiency as a power source is variability. Since wind velocity cannot be controlled or predicted with pinpoint accuracy, alternatives must be available to meet demand fluctuations.

Wind power carries few environmental penalties and makes use of a renewable resource. It has the potential to become a major but not dominant part of the future energy equation.

History

People have used wind to move boats, grind grains, and pump water for thousands of years. Wind-powered flour mills were common in Europe in the 12th century. In the 1700s, the Dutch added technical sophistication to their windmills with improved blades and a method to follow the prevailing wind. Isolated farms in the last century used windmills to generate electricity until the availability of the electrical power grid became widespread.

Past interest in wind power has tended to rise and fall with fuel prices for the predominant method of electrical production—thermal plants burning oil, natural gas, and coal.

CURRENT

Electrical Production

Although small, wind power is a fast-growing part of the energy picture. Since 1990, worldwide installed capacity has grown about 27% (Table 1).[3]

Business is good for the leading manufacturers of wind power devices. Sales have increased; the technology is stable and predictable, with low maintenance and high availability.

Geographical Distribution (Countries)

The European Union had around 72% of installed capacity in 2004, and Germany, Spain, and the United States accounted for 66.1%. Denmark, Spain, and Germany had by far the largest 2004 capacity in terms of MW per million populations, and 10.6% of the world's population had 81.9% of its wind power capacity. In 2004, Denmark produced about 20% of its electrical power from wind power in 2004 (Table 2).[3]

Manufacturing capacity in 2002 was largely confined to this group of countries, with five big vendors accounting for 76% of sales. European Union vendors accounted for 85% of manufacturing market share.[3]

ECONOMICS

Cost per kWh

Wind power is a viable method of producing electricity that is capital intensive with low operating costs. The cost of production compares favorably with traditional fossil fuel or nuclear plant costs.

Encyclopedia of Environmental Management DOI: 10.1081/E-EEM-120042087

Table 1 Capacity growth.

Year	Capacity (MW)	Growth rate year-over-year (%)
1990	1,743	13.8
1991	1,983	17.0
1992	2,321	20.7
1993	2,801	26.1
1994	3,531	36.5
1995	4,821	26.6
1996	6,104	25.1
1997	7,636	33.0
1998	10,153	33.9
1999	13,594	27.7
2000	17,357	66.3
2001	28,857	7.9
2002	31,128	26.9
2003	39,500	20.3
2004	47,500	13.8
	Average Growth Rate	26.7

Source: Reprinted with permission from European Wind Energy Association.[3]

The major cost elements of a modern wind power installation are as follows[3,5,9].

Capital
- Onshore: 1200–1500 USD/kW
- Offshore: 1700–2200 USD/kW

Operating: usually about 1.5%–2.0% of capital cost per year.[12]

Capital costs include wind capacity survey and analysis, land surveying, permits, roads, foundations, towers and turbines, sensors and communications systems, cabling to transformers and substations, maintenance facilities, testing, and commissioning. By far the largest individual capital cost is the turbine (up to 75%).

Operating and maintenance costs include management fees, insurance, property taxes, rent, and both scheduled and unscheduled maintenance.

Financing costs are a major portion of energy production costs, making them very sensitive to interest rates, incentives and subsidization.

Energy production cost estimates vary considerably. Optimists in the industry, such as the British Wind Energy Association, quote a low of 4.8 USD cents per kWh for an onshore plant in an optimal location. Pessimists, like the Royal Academy of Engineering in the U.K., quote up to 13.2 USD cents per kWh for an offshore plant, partly by including a controversial 3.1 USD cents per kWh cost for "standby capacity" required to supply demand when wind power is not available (Table 3).

Capacity Factor

The power generated by a wind turbine depends on the speed of the wind, and on how often it is available. At any given site, this is measured by the capacity factor, or the ratio of actual generated energy to the theoretical maximum. Wind power turbine electrical output rises as the cube of the wind speed. When wind speed doubles, energy output increases eightfold. A typical turbine begins to turn when wind speed is at 9 MPH and will cut out at 56 MPH for safety reasons.

Capacity factors vary by site but are typically in the range of 20%–30% with occasional very good offshore

Table 2 Capacity distribution.

Country	Wind power installed capacity (MW) 2001	2002	2003	2004	One-year (%)	Three-year (%)	Population (millions)	Capacity (MW/Million)	Percent of world capacity
Denmark	2,456	2,880	3,076	3,083	0.2	7.9	5.4	570.9	6.4
Spain	3,550	5,043	6,420	8,263	28.7	32.5	40.3	205.0	17.2
Germany	8,734	11,968	14,612	16,649	13.9	24.0	82.4	202.1	34.7
Netherlands	523	727	938	1,081	15.2	27.4	16.4	65.9	2.3
USA	4,245	4,674	6,361	6,750	6.1	16.7	295.8	22.8	14.1
Italy	700	806	922	1,261	36.8	21.7	58.1	21.7	2.6
UK	525	570	759	889	17.1	19.2	60.4	14.7	1.9
Japan	357	486	761	991	30.2	40.5	127.4	7.8	2.1
India	1,456	1,702	2,125	3,000	41.2	27.2	1080.4	2.8	6.3
China	406	473	571	769	34.7	23.7	1306.4	0.6	1.6
Total	22,952	29,329	36,545	42,736	16.9	23.0	3073		89.2

Source: Global Wind Energy Council—"Wind Force 12".[17]

Table 3 Wind power costs.

	Cents (US) per kWh			
Wind power costs	**Wind onshore**	**Wind offshore**	**Coal CFB**	**Gas CCGT**
RAE [15]	6.8	9.9	4.8	4.0
RAE 2[a]	10.1	13.2	9.2	6.0
EWEA [3]	5.5	8.5		
AWEA (5)	6.6			
AWEA 2[b]	5.4			
BWEA (minimum) [4]	4.8	6.9	4.8	4.8
BWEA (maximum)[c]	6.8	9.1	6.8	5.9
Euro to USD conversion (2004)				1.22
GBP to USD conversion (2004)				1.83

[a]Adds 3.1 cents per kWh for wind power backup capacity and 1.9–4.6 cents/kWh for coal and gas carbon capture.
[b]AWEA figures adjusted to delete 1.8 cents US/kWh production tax credit and are for onshore sites with different average wind speeds.
[c]BWEA figures for a range of site types in November 2004 and 1–2 cents/kWh for carbon capture.
Source: RAE, Royal Academy of Engineering[15]; BWEA, British Wind Energy Association[4]; EWEA, European Wind Energy Association[3]; AWEA, American Wind Energy Association.[5]

sites reaching 40%. The yearly energy output from a wind farm is given by the following formula:

$$\text{Output/year (kWh)} = [\text{Capacity(kW)}] \times [8760 \text{ hours/years}] \times [\text{Capacity factor}]$$

Site

Power production costs, site size, site design, and energy output and variability will depend mainly on details about the wind. These details include wind speed, wind direction, and the geographical distribution of favorable wind profiles. During analysis of potential sites, most planners use high (60 m plus) anemometer towers—often several of them—to gather at least one year's data per site. These data are usually correlated with national meteorological observations, if these are available and suitable. If not available, it would be prudent to gather site data for a longer period of up to three years.

Investors and regulators are increasingly aware of the crucial nature of wind data in estimating the quantity and timing of potential power production at a specific site. This research is crucial to the financial analysis of a potential wind power venture.

Other site analysis factors are accessibility via road for heavy equipment, electrical grid proximity and capacity, land ownership, and environmental impact.

LOCATION

Favored Geography

Many countries have developed wind charts of broad areas based on meteorological data gathered for weather and aviation purposes. These charts show potential areas for investigation, where wind strength is high and constant over long periods of the year. Once potential sites, and their extent, have been identified, on-site data measurements provide the basis for analysis and modeling of potential energy production for a specific site.

After wind modeling, the site's geographical, environmental, financing, and ownership issues can be explored in detail.

Generally, sites are either onshore or offshore. Onshore sites are cheaper to construct, but have lower capacity factors due to wind turbulence from nearby hills, trees, and buildings. Offshore sites can have more potential energy available due to higher wind speeds and lower turbulence, which also reduces turbine component wear. Good offshore sites can be near high-demand load areas such as coastal cities, which also increases transmission options. Aesthetic and noise concerns are often fewer offshore, and sea-bed environmental concerns can be lower than land-use concerns for an onshore site.

Sizing a Location

Wind farm towers are usually spread over a large area in order to minimize wake losses. A spacing of five rotor diameters is often recommended. In a typical wind farm, the land physically occupied by tower foundations, buildings, and roads is often less than 2% of overall land area.[6] The remaining land is quite suitable for agriculture and other uses.

Limits to Maximum Production

How much capacity exists to generate electricity from wind? Is it possible that we will require more energy than wind can provide? After surveying wind patterns in the United States and applying energy density and extraction calculations, Elliott and Schwartz[14] concluded in 1993 that 6% of the available U.S. land mass could provide 150% of then-current U.S. electrical consumption. Furthermore, the needed land would be sparsely affected by the wind farm installation, with the vast majority of it (95%–98%) unoccupied by tower foundations, roads, or ancillary equipment and suitable for farming, ranching, and other uses. This study excluded land that is environmentally or otherwise unsuitable, such as cities, forests, parks, wildlife refuges, and environmental exclusion areas.

In the European Union, potential wind power capacity is also larger than current electrical consumption.

FUTURE

Projected Growth

Thanks to increasing concern over the environmental effects of greenhouse gas emissions, the rising cost of fossil fuels, and the impending decrease in availability of oil and natural gas, wind power has a bright future.

Current 25%–30% growth rates are likely not sustainable, due to equipment production volume constraints and limits to perceived need for further capacity. Given the Eurocentric, highly clustered nature of current installed capacity, there is significant potential for high-rate growth elsewhere. However, even in European countries like Germany and Denmark, steady growth will be driven by predicted cost reductions in the 10%–20% range and by regulatory and governmental initiatives aimed at reducing emissions from electrical energy production and transitioning to renewable resources.

Projected Cost

Wind power technology is well down the cost improvement curve, with costs having fallen to present levels, below ten cents USD per kWh, from over $1.00 U.S. in 1978. Costs for a medium-sized turbine have dropped 50% since the mid-1980s, reflecting increasing maturity in the market. Cost projections range from a further 9%–17% drop as installed capacity doubles in the near-term future.

Projected Production

With increasing governmental policy support and commitment, growth rates of 15%–20% appear achievable in five to ten years. But there is likely an upper limit to the amount of electrical energy that can be produced from the wind.

Reaching Maximum Production

Production limits for wind power are based on its variable nature. Other types of electrical production capacity will be needed to provide base-load electrical capacity in the event that there is little wind available. Wind power will then become one player in a mix of generating technologies.

Hydrogen Economy

As wind power becomes a larger portion of electrical supply, occasionally its supply will exceed demand. Rather than simply curtailing wind plant production, it is attractive to think of using this excess electrical power capacity to generate hydrogen via electrolysis. This has the effect of storing wind energy that would otherwise not be harvested. This energy, in the form of hydrogen, can be used directly as a non-polluting fuel or as an input source to fuel cells to produce electricity at a later time.

When there is a significant hydrogen economy, with transmission lines, storage, and fuel cell capacity, this use for wind power will become a very attractive scenario.

Other Issues for the Future

Learning more about wind and forecasting—predicting the best locations, wind farm output, gusts, and directional shear.[10] This will help reduce financing costs when wind power plant output and impact on the grid are better understood and more predictable.

- Improving the control of demand through incentives around end-user load shedding, rescheduling and simple conservation methods. This could be used to offset wind power production shortfalls as an alternative to other forms of generation.
- Advancing aerodynamics specific to wind turbine blades and control systems.
- Designing extremely large wind tunnels to study wake effects minimization, structural load prediction, and energy output maximization at lower wind speeds.
- Enhancing power system capacity planning models to include wind farm components.
- Re-planting, or upgrading older mechanical and electrical components at existing wind farms.
- Wind farm siting further offshore and on floating platforms.
- Combining wind power and hydroelectric capacity by using surplus wind power to pump water behind dams and so store power that might otherwise be wasted.
- Determining how and whether to allocate full costs of environmental impact to fossil and nuclear plants.

STRENGTHS AND WEAKNESSES

Strengths

Environment

Wind power installations do not emit air pollution in the form of carbon dioxide, sulfur dioxide, nitrogen oxides, or other particulate matter such as heavy metal air toxins. Wind power installations do not use water or discharge any hazardous waste or heat into water. Conventional coal, oil, and gas electric power plants produce significant emissions of all kinds. Nuclear power plants produce dangerous and long-lasting radioactive waste. Greater use of wind power means less impact on health and the environment, particularly regarding climate change due to greenhouse gas emissions.

Renewable

Wind power produces energy from a resource that is constantly renewed. The energy in wind is derived from the sun, which heats different parts of the earth at different rates during the day and over the seasons. Unlike fossil and nuclear plants, the source of energy is essentially inexhaustible.

Costs

Wind power's costs are well known and are dropping to the point at which this technology is very competitive with other means of production. Fuel costs are nil, meaning that fuel costs have no uncertainty. Wind power costs should be more stable and predictable over the lifetime of the plant than power costs for fossil fuel plants.

Local and Diverse

Wind power plants provide energy source diversity and reduce the need to find, develop, and secure sources of fossil or nuclear fuel. This reduces foreign dependencies in energy supply, and reduces the chances of a political problem or natural disaster interfering with and diminishing the supply of electricity.

Quick to Build, Easy to Expand

Wind power plants of significant capacity can be constructed and installed within a year, a much shorter time than conventional plants. The planning time horizon is similar to conventional plants, given the need to accurately survey site wind characteristics and deal with normal environmental and related site issues. This means that capacity can be increased in closer step with demand than with conventional plants. With the right site and design, a wind power plant can be incrementally expanded very quickly.

Weaknesses

Natural Variability

A single wind farm produces variable amounts of energy, and its output is not yet as predictable as a traditional plant. As the geographical distribution and number of wind plants increases, and as research into predicting wind continues, these problems should be minimized, allowing cost-effective and orderly scheduling and dispatch of total grid capacity sources—but it is difficult to see traditional power sources disappearing entirely.

Connection to Grid

As the amount of electrical power supplied by wind power plants increases, concern increases over its effects on the electrical grid.

In order to maintain a reliable supply of electricity that matches demand, utility operators maintain emergency reserve capacity in order to deal with plant outages (failures) and unexpected demand across their entire system. This reserve is in the form of purchased power, unused capacity at conventional plants running below their maximum, or quick-start plants such as gas-fired turbines. Often, conventional plants on the grid are allocated a cost to cover this reserve based on their capacity (large plant, large reserve) and reliability (more outages, more reserve).

The industry is working on ways to determine and allocate this reserve cost for wind power plants. Yet to be agreed upon is the statistical basis for calculating such wind power plant reserve costs. Improvements in day-ahead wind forecasting will greatly reduce the uncertainty around wind plant output, and so decrease the cost burden to provide this reserve.

Several current estimates prepared for U.S. utilities show this reserve cost burden (or ancillary services cost) to increase with the amount of capacity provided by wind power, and to be in the range of 0.1–0.5 cents USD per kWh for penetrations between 3.5 and 29%. In no case was it thought necessary to allocate a reserve equal to 100% of the wind power capacity.[11] German experience is similar,[8] with no additional reserve capacity required for the 14% wind energy share of the national electrical consumption forecast for 2015.

When wind power supplies less than 20% of electrical consumption, these problems are not severe. At larger penetrations, reserves become a major issue. Interestingly, wind power plants may be subject to shutdown or voluntary power reductions in the event of coincident high wind, low demand situations. This is occasionally the case today in Denmark and Spain.

In some cases, wind power sites are situated far from the location of high electrical power demand, placing strain or potential overload on existing transmission facilities. In these cases, there are often cost, ownership and responsibility issues yet to be resolved.

Power quality problems around power factor, harmonic distortions, and frequency and voltage fluctuations are being successfully addressed in modern large production wind farms.[8]

This is one of the most difficult sets of issues facing the future of wind power as it matures from small-scale and local to large-scale penetration.

Local Resource Shortage

In a few places, high-quality wind power sites are not available or are already in production, leaving these places to import electrical power or use traditional sources.

Noise

Noise levels have decreased and are now confined to blade noise in modern units. Generator and related mechanical noise has been effectively eliminated. Noise, however, will always be a significant factor. Blade noise is described as a "whoosh, whoosh" sound, and is in the 45–50 decibel range at a distance of 200–300 m. This noise level is consistent with many national noise level regulations. However, this noise buffer zone adds to the overall land requirement for a wind power plant and so increases costs.

Visual Impact

Onshore wind farms are highly visible due to the height of towers and the size of the blades and generator. The impact of this varies with each person. Each wind plant operator needs to determine the levels of support and opposition from those who live and work within sight of the plant. Offshore plants attract fewer detractors than onshore plants—one of the reasons for their increasing popularity.

Offshore wind plants are less likely to cause unwanted noise since they are far from human habitation. This reduces turbine and blade design constraints and can lead to higher capacity factors.

Bird Impact

Bird deaths are a regrettable reality. The bird death rate at a specific wind farm project is quite variable. Several early wind farms (Altamont Pass, California, and La Tarifa, Spain) caused concern over death rates. The California Energy Commission estimates the death rate at Altamont (5400 turbines) to be 0.33–0.87 bird deaths per turbine per year.[16] The overall recent U.S. national average[13] is 2.3 bird deaths per turbine per year. Prudently located sites are off migration routes and not in nesting, over-wintering, or feeding areas. Their tower designs do not offer nesting or even roosting places. In such locations, death rates are lower, and overall impact is much lower than that caused by other types of human activity.

Since climate change is a very serious environmental problem faced by bird populations, wind power and other renewables are an important part of the solution.

TECHNOLOGY

Overview

Wind turbine design has three major components, and there are large economies of scale in design.

- *Tower height:* Wind turbine energy output is proportional to the cube of wind speed. Since moving air (wind) is subject to drag and turbulence from its contact with the earth and the objects on the earth, wind speed increases with height (vertical shear). The higher the tower, the more advantage there is for power generation. The tradeoff is between tower costs and increase in power generation. Typically, tower heights are rising, and are currently in the 100 m range. Off shore, vertical wind shear is generally less than onshore, so towers can be shorter, with wave height clearance being the factor that determines tower height (Fig. 1).
- *Blade diameter:* The power capacity (watts) of a wind turbine varies with the square of its blade diameter, because a blade with a larger diameter has a larger area available for harvesting the wind energy passing through it. The coefficient of performance defines the actual power capacity compared to the maximum—how much energy can be extracted from the wind compared to the available energy. Modern wind turbines can achieve a coefficient of performance approaching 0.5, very close to the theoretical maximum of 0.59 derived by Betz.[3,18] This maximum is derived from the concept that if 100% of wind energy were extracted, the wind exiting the turbine would be at zero speed, so no new air could enter the turbine. Larger capacity turbines benefit significantly

Fig. 1 Typical large wind turbine. Note entrance steps and utility pole at base for scale.
Source: Photo courtesy of Suncor Energy, Inc.

Fig. 2 Site assembly of large wind turbine nacelles and blades.
Source: Photo courtesy of Suncor Energy, Inc.

from economies of scale in foundation and support costs as well as swept area (Fig. 2).

- *Controls and generating equipment:* The turbine's hub (or nacelle) is the most costly component and contains the generator, gear boxes, yaw controls, brakes, cooling mechanism, computer controls, anemometer, and wind directional vane.

Generators

As the blades turn, they drive a generator to produce electricity. Generating capacity ranges from a few hundred kW to over 3 MW. In older designs, there is usually a 40:1 gearbox to match low, fixed rotor speeds (~30 RPM) to required generator speed (1200 RPM for a 60 Hz output, 6-pole generator). The gearbox often incorporates brakes as a part of the overall wind turbine control system. Generators that operate at low RPM are available and are called direct drive generators. These would eliminate the gearbox.

In more modern designs, rotor speed is variable and controlled to optimize power extraction from the available wind. Generator output is converted to d.c. and then back to a.c. at the required grid frequency and voltage. The conversion equipment is sometimes located at a central part of the wind power plant.

This is an active area for ongoing technical innovation.

Blades

In order to maximize power capacity through size, blades are very long, up to 50 m. To minimize noise, they must turn slowly so as to reduce tip speed, the primary blade noise source. Typical rotation speeds are in the 10–30 RPM range. Blades are increasingly made from composites (carbon fibre reinforced epoxy resins).

Rotor blade aerodynamics[19] have much in common with the aerodynamics of a propeller or a helicopter blade, but they are sufficiently different that the aerodynamics of wind turbine blades is an evolving field. The difference is that wind turbine airflows are unsteady due to gusting, turbulence, vertical shear, turbine tower upstream shadow, yaw correction lag, and the effects of rotation on flow development. For example, at present it can be difficult to predict rotor torque (and therefore power output) accurately for normal turbine operating conditions. Further development of theory and modeling tools should allow the industry to improve rotor strength, weight, power predictability, power output, and plant longevity while controlling cost and structural life.

For a given site wind velocity, the rotor blade's tip has very different air flow than its root, requiring the blade to be designed in a careful twist. The outer third of the blade generally produces two-thirds of the rotor's power. The third nearest the hub provides mechanical strength to support the tip, and also provides starting torque in startup situations.

Each blade generally has lightning protection in the form of a metallic piece on the tip and a conductor running to the hub.

Some manufacturers place Whitcomb winglets at the blade tips to reduce induced drag and rotor noise, in common with aircraft wing design.

In order to control blades during high wind speeds, some are designed with a fixed pitch that will progressively stall in high wind speeds. Others incorporate active pitch control mechanisms at the hub. Such control systems use hydraulic actuators or electric stepper motors and must act very quickly to be effective.

Wind Sensors

Wind turbines incorporate an anemometer to measure wind speed and one or more vanes to measure direction. These are primary inputs to the control mechanism and data gathering systems usually incorporated into a wind turbine.

Control Mechanisms (Computer Systems)

Control systems are used to yaw the wind turbine to face into the wind, and in some designs to control blade pitch angle or activate brakes when wind gets too strong. In sophisticated cases, the controllers are redundant closed-loop systems that operate pumps, valves and motors to achieve optimum wind turbine performance. They also monitor and collect data about wind strength and direction; electrical voltage, frequency and current; nacelle and bearing temperatures; hydraulic pressure levels; and rotor speeds, vibration, yaw, fluid levels, and blade pitch angle. Some designs provide warnings and alarms to central site operators via landline or radio. Manufacturers do not release much detail about these systems, since they are a critical contributor to a wind turbine's overall effectiveness, safety, and mechanical longevity.

ROLE OF GOVERNMENTS AND REGULATORS

Governments play a large part in determining the role and scale of wind energy in our future mix of energy production capabilities.

Subsidies, Tax Incentives

As part of programs to encourage wind power production, the following are used in varying ways[7]:

- Outright subsidies, grants and no-interest loans.
- Tax incentives such as accelerated depreciation.
- Fixed prices paid for produced electrical power.
- Renewable energy quantity targets imposed on power utility operators.

Grid Interconnection and Regulatory Issues

Since many power utility operators are owned by governments, and most are regulated heavily, governments have a role to play in encouraging solutions to grid interconnection issues. There must be a political will to address issues, find solutions, and develop practices and different management strategies that will allow greater penetration of wind power into the electrical supply.

Improving Wind Information

Climate and environmental information is most often collected and supplied by national governments in support of weather and aviation services. Wind atlases are an invaluable resource to the wind plant planning process. National efforts to improve long- and short-term wind forecasting, atmospheric modeling tools, and techniques will benefit wind power projects' ability to forecast power output for long-term and short term planning purposes.[10]

Environmental Regulation

In this controversial area, government can tighten its regulation of air quality, carbon emissions and other environmental areas. This would have the effect of increasing the apparent cost of conventional thermal electrical power, which is responsible for significant emissions. It is often argued that wind power would already be cost competitive if environmental and health costs were to be fully allocated to conventional oil, gas, and particularly coal-powered plants, or if such plants were required to make investments to significantly reduce emissions.

CONCLUSION

Western societies depend on a steady supply of energy, much of it in the form of electricity. Most of that supply comes from thermal plants that burn oil, natural gas, and coal, or from nuclear plants. Where will our electrical energy come from in the future? How will we keep our environment livable and healthy?

One part of this answer lies in wind power. Its costs are within reason; the technology has matured with some gains yet to be realized; it carries little environmental penalty; and the source of its energy is renewable. As long as the sun heats the earth, wind power will be available.

Wind power will not likely be the complete answer; it is an intermittent source because the wind doesn't always blow. But there is a very large amount of it available for us to harvest. As wind power moves quickly from small-scale to large-scale, its future path depends on governments and regulators as much as it does on technical innovators and manufacturers.

REFERENCES

1. Annual Energy Review 2003, DOE/EIA-0384(2003), Energy Information Administration, U.S. Department of Energy, Washington, DC, September 2004.
2. Key World Energy Statistics - 2004 Edition, International Energy Agency, OECD/IEA – 2, rue André-Pascal, 75775 Paris Cedex 16, France or 9, rue de la Fédération, 75739 Paris Cedex 15, France, 2004.
3. Wind Energy, The Facts, Analysis of Wind Energy In the EU-25, European Wind Energy Association, Brussels, Belgium, 2004.
4. BWEA Briefing Sheet, Wind and the U.K.'s 10% Target, The British Wind Energy Association, http://www.bwea.com/energy/10percent.pdf (accessed July 2005).
5. The Economics of Wind Energy, American Wind Energy Association, http://www.awea.org/pubs/factsheets/EconomicsOfWind-Feb2005.pdf (accessed July 2005).
6. Saddler, H., Dr.; Diesendorf, M., Dr.; Denniss, R. A Clean Energy Future For Australia, A Study By Energy Strategies for the Clean Energy Future Group, Australia, March 2004.
7. Policies to Promote Non-hydro Renewable Energy in the United States and Selected Countries, Energy Information Administration, Office of Coal, Nuclear, Electric and Alternate Fuels, United States Department of Energy, Washington, DC, February 2005.
8. BRIEFING May 10th, 2005. German Energy Agency DENA study demonstrates that large scale integration of wind energy in the electricity system is technically and economically feasible. http://www.ewea.org/documents/0510_EWEA_BWE_VDMA_dena_briefing.pdf (accessed July 2005).
9. Offshore Wind Experiences, International Energy Agency, 9, rue de la Fédération, Paris Cedex 15, 2004.
10. Milborrow, D. Forecasting for Scheduled Delivery, Windpower Monthly, December 2003.
11. Utility Wind Interest Group—Wind Power Impacts On Electric-Power-System Operating Costs, November 2003.
12. Danish Wind Industry Association, http://www.windpower.org/en/tour.htm (accessed July 2005).
13. Wind Turbine Interactions With Birds and Bats: A Summary of Research Results and Remaining Questions, Fact

Sheet, Second Edition, National Wind Coordinating Committee, Washington, DC, November 2004. http://www.nationalwind.org/publications/avian/wildlife_factsheet.pdf (accessed July 2005).
14. Elliott, D.L.; Schwartz, M.N., Wind Energy Potential In the United States, Pacific Northwest Laboratory, PNL-SA-23109, Richland, Washington, USA, September 1993.
15. The Cost of Generating Electricity, Royal Academy of Engineering, http://www.raeng.org.uk/news/publications/list/reports/Cost_of_Generating_Electricity.pdf (accessed July 2005).
16. Developing Methods to Reduce Bird Mortality in the Altamont Pass Wind Resource Area, August 2004, http://www.energy.ca.gov/reports/500-04-052/500-04-052_00_EXEC_SUM.PDF (accessed July 2005).
17. Wind Force 12, Global Wind Energy Council, June 2005, http://www.ewea.org/03publications/WindForce12.htm (accessed July 2005).
18. Proof of Betz Theorem, Danish Wind Industry Association, http://www.windpower.org/en/stat/betzpro.htm (accessed July 2005).
19. National Wind Technology Center, http://www.nrel.gov/wind/about_aerodynamics.html (accessed July 2005).

Aluminum

Johannes Bernhard Wehr
Frederick Paxton Cardell Blamey
Neal William Menzies
School of Land, Crop and Food Sciences, University of Queensland, St. Lucia, Queensland, Australia

Abstract
Aluminum (Al) is a ubiquitous light metal found in the earth's crust in aluminosilicate minerals. There is no known biological role for Al in plants, animals, or microorganisms. Aluminosilicate minerals can dissolve in acidic soils, releasing soluble trivalent Al, which is toxic in the micromolar concentration range. Soluble Al in the soil environment is strongly toxic to plants and limits plant production on acidic soils. In the aqueous environment, soluble Al is strongly toxic to fish and algae. Uptake of Al by humans from food, water, and pharmaceuticals has been implicated in some neurological diseases in humans.

INTRODUCTION

Aluminum is a metallic element characterized by its low density (2.7 g/cm^3) and resistance to corrosion due to the formation of a protective oxide layer on the surface. It is the most common (8%) metallic element of the earth's crust, and the third most common element (after oxygen and silicon) on earth.[1] The element reacts strongly with oxygen-containing ligands and never occurs naturally as pure metal. It forms both octahedral and tetrahedral coordination compounds and occurs only in the trivalent oxidation state. In the soil, Al-containing minerals can dissolve at low pH and release Al into the soil solution and aquatic environment. Al can also occur in the atmosphere as aeolian dust or ash.

Soluble forms of Al can have toxic effects on plants, as well as soil and aquatic organisms, but alleviation of Al toxicity can be achieved by increasing the pH to above 4.5. Aluminum present in water and food is not readily taken up by humans. Nano-sized Al particles in the atmosphere may result in some human health effects but more research is needed.

The chemistry and uses of Al are discussed, as well as the sources and alleviation of Al toxicity in soils, water, and atmosphere. Finally, the environmental toxicology of Al to plants, microorganisms, and animals is briefly discussed.

CHEMISTRY OF Al

Dissolution of Al minerals at low pH releases octahedral Al^{3+}, which is hexa-coordinated with water molecules.[2] Soluble Al^{3+} is classified as a hard acid due to its small ionic radius of 0.53 Å and reacts strongly with "hard" ligands such as oxygen and fluoride.[2,3] It also forms stable complexes with di- and multidentate ligands (chelate effect).[3,4] As the pH of an acidic Al solution is increased to >4, hydrolysis of Al occurs, giving rise to a series of Al-hydroxy species [$AlOH^{2+}$, $AlOH_2^{+}$, and $Al(OH)_3$],[2] which may undergo aggregation (polymerization) via OH bridges if the Al concentration is sufficiently high (>10 μM[5]). As the pH is raised above pH 6, Al changes its coordination number to 4 and yields tetrahedral aluminate ($AlOH_4^{-}$).[2] Depending on the pH and Al concentration in solution, hydrolyzed Al species may aggregate and form polymeric (polycationic) Al species such $Al_2(OH)_2^{4+}$, $Al_3(OH)_4^{5+}$, $Al_8(OH)_{20}(H_2O)_x^{4+}$, the "gibbsite fragment" model forms, $Al_6(OH)_{12}(H_2O)_{12}^{6+}$ through $Al_{54}(OH)_{144}(H_2O)_{36}^{18+}$, $Al_{13}O_4(OH)_{24}(H_2O)_{12}^{7+}$ (Al_{13}), and $Al_2O_8Al_{28}(OH)_{56}(H_2O)_{26}^{18+}$ species.[6–10] While many more Al polymers have been proposed, the experimental evidence in support of these species is limited.

Al USES AND PRODUCTION

Due to its corrosion resistance, light weight, and excellent thermal and electrical conductivity, the metal is extensively used in the building and construction industries (window frames, doors, external cladding, A/C ducts, thermal insulation), automotive (engine blocks, car bodies), shipping (hulls) and aerospace industries (aircraft bodies), manufacture of power lines, and food packaging (cans and other containers, foil). Aluminum salts and compounds are used for water purification (Alum), as catalysts in the chemical industry, and as ingredients in cosmetics (antiperspirants), pharmaceuticals (antacids, vaccine adjuvant), and foods (baking powder, spreading agent) (Table 1).

The main producer of bauxite ore is Australia (40% of world production, 67 Mt in 2007), with China, Brazil, Guinea, Jamaica and India producing lesser quantities.[11] The main

Encyclopedia of Environmental Management DOI: 10.1081/E-EEM-120046331

Table 1 Use of Al in Europe in 2010.

Use	
Building and Construction	25%
Transport industry (car, ship, plane)	37%
Engineering	14%
Packaging	17%
Other	7%

Source: European Aluminium Association. Activity Report 2011, page 3. http://www.alueurope.eu/wpcontent/uploads/2012/03/EAA-Activity-Report-2011-HI-RES_V1_FINAL.pdf (last accessed May 14, 2012).

producers of alumina in 2006 were Australia and China,[11] whereas smelting (refining of bauxite) is occurring in regions with cheap electrical energy.[12] Table 2 lists the worldwide production of alumina and primary Al in 2009.

SOURCES OF Al IN SOIL

Al Minerals

The main Al-containing primary minerals are feldspars and micas. Several gemstones (ruby, sapphire, tourmaline) also contain Al. The primary Al-containing minerals weather initially to 2:1 layer aluminosilicate clay minerals (e.g., vermiculite, montmorillonite, smectite), which upon further weathering form 1:1 layer aluminosilicates such as kaolinite. Further weathering of clay minerals leads to leaching of silica and base cations (calcium and magnesium), leaving behind hydrous aluminum oxide (e.g., gibbsite, boehmite, and diaspore) in the form of bauxite, which is an important aluminum ore.[1]

Forms of Soluble Al in Soil

In acidic soils (pH <4.5), aluminosilicate clay minerals are unstable and dissolve, releasing the trivalent Al^{3+} cation. The trivalent Al^{3+} ion is toxic to plant roots at concentrations of 5–50 μM.[12–14] The Al^{3+} ion is readily complexed by soil organic matter and the soil organic matter controls the Al availability in most soils containing sufficient organic matter.[14–20] In mineral soils low in organic matter, the availability of Al is determined by the cation exchange capacity, ionic strength, and pH.[20] The polymeric Al species (e.g., Al_{13}) are metastable with a half-life for Al_{13} of several hundred hours,[21,22] which implies that the species undergo depolymerization or crystallization with time. The Al_{13} species is very rhizotoxic to plant roots, with concentrations between 0.1 and 2 μM inhibiting root growth. It is generally accepted that sulfate, phosphate, fluoride, silicate, and organic acids prevent or reverse Al_{13} formation,[8,23–25] and crystalline gibbsite may induce crystallization and depolymerization of Al_{13}.[26] Since sulfate and silicate ions are prevalent even in acid soils, the natural occurrence of Al_{13} in soil solution is considered unlikely.[8,27,28] However, there is evidence that Al_{13} may form in the cell wall of plant roots[29–31] even if the conditions in bulk solution are not favorable for Al_{13} formation. A report by Hunter and Ross[32] claiming that up to 30% of acid forest soils contained Al_{13} could not be replicated by other research groups. In alkaline soils, the aluminate anion $[Al(OH)_4^-]$ can form but this species is not very toxic to plants[33] but may be toxic to aquatic organisms.[34,35]

Table 2 Worldwide production of alumina and primary Al metal (in kilotons) in 2009.

Geographic region	Alumina	Primary Al metal
Africa	530	1,681
North America	4,279	4,759
Latin America	13,275	2,508
Asia	6,005	4,400
Western Europe	4,665	3,722
Eastern/Central Europe	4,558	4,117
Oceania	20,262	2,211
Total	53,570	23,398

Source: International Aluminium Institute:
https://stats.world-aluminium.org/iai/stats_new/formServer.asp?form=1 (last accessed May 14, 2012),
https://stats.world-aluminium.org/iai/stats_new/formServer.asp?form=5 (last accessed May 14, 2012).

Mobilization of Al in Soil

Aluminosilicate clay minerals slowly dissolve at pH <4.5 and release Al and Si into the soil solution. Currently, more than 30% of potentially arable soils are acidic and affected by Al toxicity.[36] Soil acidification and consequential Al toxicity can be exacerbated by anthropogenic acid deposition (acid rain), use of ammonium-containing fertilizers, use of legumes in crop rotations, and the removal of crops from agricultural land.[36,37] In unfarmed soils, atmospheric acid inputs are the main causes of acidification,[38,39] whereas acidification in fertilized soils is mainly caused by nitrification of ammonium N-fertilizer.[39,40] The effect of Al on soil organic matter is still debated with researchers claiming either that Al can increase humification of organic matter[41] or that Al prevents microbial degradation of Al–soil organic matter,[42–46] leading to an increase in soil organic matter.

Aluminum can also be released from soil minerals in strongly acidic conditions found in acid sulfate soils,[47,48] monosulfidic black ooze,[49] and acid mine drainage.[50,51] Acid sulfate soils are limited in occurrence to permanently or temporarily waterlogged low-lying areas (up to 5 m above sea level),[47] whereas organic matter in irrigation channels can lead to the formation of monosulfidic black ooze.[49] Exposure of these sulfide-containing materials to air leads to the formation of sulfuric acid and release of Al.[50–54]

Fluoride ions (F^-), present as impurity in phosphate fertilizers, can form strong complexes with Al and the resultant lesser charged complexes, AlF^{2+}, AlF_2^+, and AlF_3, are mobile in soil solution[2,55] and can be either removed from soil by drainage or seepage water, or taken up by plants.[56]

Organic acids form complexes with Al and can mobilize Al in the soil.[57,58] In the field, movement of organic acid–Al complexes gives rise to podzolization of soils and downward movement of Al until conditions are favorable for dissociation of Al complexes deeper in the profile or export to waterways.[59]

ALLEVIATION OF Al IN SOILS

Since solubility of Al in soil solution is lowest around pH 6–7,[60] controlling the pH is the most effective way in reducing potential Al toxicity in ecosystems. Application of calcitic ($CaCO_3$) or dolomitic ($CaMgCO_3$) limestone,[61] slaked lime [$Ca(OH)_2$],[62] wood ash,[63] alkaline poultry manure,[64,65] charcoal,[66] fly ash,[67,68] or waste cement[68] has been used to increase the soil pH of acidic soils. Any waste product with high ash alkalinity can be used to ameliorate acid soil by increasing the pH.[69] Surface application of liming agents with subsequent incorporation into the soil (ploughing, harrowing, etc.,) is commonly used. The mobility of limestone in the soil profile is very low.[70–72] Therefore, rapid alleviation of subsoil acidity can only be achieved by deep placement of limestone (deep ripping). This, however, is currently not economically possible for most low-value broadacre crops. Depending on the severity of soil acidity, and the buffer capacity of the soil as determined by the cation exchange capacity of the soil, between 0.1 and 1 ton limestone/ha/yr are needed to maintain soil pH in agricultural production systems.[61,70] Numerous methods have been developed to determine the required liming rate based on laboratory soil analyses. The liming of soil needs to be repeated every few years depending on the land management system. To overcome the low mobility of limestone in soils, higher rates of gypsum can be used instead. The Ca from gypsum moves more readily down the soil profile[73] and the increase in Ca and ionic strength can lower the Al toxicity, despite increasing the overall concentration of Al (as $AlSO_4^+$, which is less toxic than Al^{3+}). Liming has greater beneficial effect on soil bacterial communities than on soil fungal communities.[74,75]

Application of organic matter (green manure, farm yard manure) can lower Al toxicity by complexing Al.[76,77] However, application of organic matter requires higher application rates (>10 tons/ha) than limestone (1 ton/ha) and is, therefore, not as effective as limestone. Humic acid has been shown to either precipitate or bind Al_{13}[78] and Al^{3+},[79] thereby reducing root growth inhibition. Sewage sludge can also be alkaline and the presence of organic acids in sludge can bind and immobilize Al.[80]

Alkaline bauxite refinery waste (red mud) poses a challenge for plant growth due to the high Fe oxide concentration leading to phosphate immobilization, absence of diverse microbial communities, and lack of organic matter.[81] The presence of aluminate in these waste materials is not limiting to plant root growth.[31] The wastes can be ameliorated with gypsum, sewage sludge, and Ca-phosphate to overcome nutrient deficiencies induced by the high pH.[82]

SOURCES OF Al IN WATER

Acid Deposition

Atmospheric gases such as NO_2, SO_2, and CO_2 dissolve in rainwater and form acid rain.[38,83] The atmospheric acid inputs in Europe are considered to be around 0.2–4 kmol/ha/yr.[84] The acids can either dissolve Al minerals in soil (see above) or dissolve aquatic sediments in water bodies receiving acidic water. Furthermore, acid deposition can release Al complexed to organic matter. Introduction of gaseous emission standards in the 1980s in Europe, and later in the United States, has lowered acid rain, with a consequential decrease in Al in surface water.[85,86] Podzolization of lateritic soils was found to export vast quantities of Al to waterways.[59] The concentration of Al in drainage water from forest watersheds is highly variable. In coastal areas subjected to atmospheric salt deposition, the salt can displace Al from cation exchange sites, leading to an increase in Al in drainage water.[58,87] Concentrations of Al in groundwater are governed by pH since Al solubility is pH dependent: less than 0.4 μM Al is observed at pH 7, which increases to 1940 μM at pH <4.[88]

Acid Mine Drainage and Acid Sulfate Soils

Sulfidic mine wastes and acid sulfate soils release acid when exposed to air, resulting in leaching of acid drainage water and soluble Al into water bodies. While there are indications that the high Fe concentration in acid mine drainage can protect aluminosilicate clay mineral from dissolution and minimize Al release,[54] no such mechanism has been reported for acid sulfate soils. At Trinity inlet in North Queensland, Australia, Al concentrations in drainage water from acid sulfate soils were measured in the range 4,200–10,000 μg/L.[89]

Aluminum from Water Purification

Water clarification relies on settling of fines (suspended colloids and microorganisms) with alum (potassium aluminum sulfate). Since drinking water will be near neutral (pH 6.5–8.5[90]), solubility of Al is low.[88,91] The European, Australian, New Zealand, and WHO regulations for drinking water propose a concentration of Al of less than

0.1 mg L (3.7 μM). Traces of Al are found in raw and alum-treated potable water,[92] and alum treatment of water at pH 7 does not increase Al concentration in drinking water.[92] Furthermore, the alum binds mercury[93] and phosphate,[94] thereby improving the quality of treated water. The amount of Al taken up from drinking water has been estimated at 1%–2% of total Al uptake per day,[95] and less than 2% of Al in drinking water is bioavailable.[96]

Water treatment residues (alum sludges) have been used to immobilize phosphate and heavy metals in soils and wastes.[97–99] Land application of these sludges has not resulted in elevated Al concentrations in plants grown in these soils,[98] especially when the sludges were aged to reduce Al availability.[100]

ALLEVIATION OF Al IN WATER

Raising the pH of water to pH 6–7 will precipitate Al and establish an equilibrium concentration of Al determined by the solubility product of Al-hydroxide in water. At pH <4, up to 1940 μM Al was found in water, but at pH >7, the concentration dropped to less than 0.4 μM.[88] Al remaining in potable water after purification can only be removed by ion exchange resins.[101]

SOURCES OF Al IN THE ATMOSPHERE

Atmospheric dust derived from raised soil particles is the main source of Al in the atmosphere, but volcanic activity can also release vast quantities of Al-containing particles into the atmosphere. In the atmosphere, Al occurs generally as silicate, sulfate, or oxide compounds. Flue ash and flue gases can also contribute Al to the atmosphere.[102] Dust contains around 2.6% Al.[103] Concentrations of Al in the atmosphere vary between geographic locations (Table 3). Inhalation of Al from the atmosphere has been estimated as 4–20 μg Al/day.[95]

Table 3 Concentrations of Al (nanograms per cubic meter) in the atmosphere.

Region	Al concentration
South Pole	0.3–0.8
Greenland	240–380
Shetland Islands	60
Norway	32
Germany	160–2,900
Japan	40–10,600
China	2,000–6,000
North America	600–2,330
Central America	760–880
South America	460–15,000

Source: Adapted from Kabata-Pendias and Kabata[102] and Wang et al.[156]

SOURCES OF Al IN FOODSTUFF, COSMETICS, PHARMACEUTICALS, AND WORKPLACES

The concentration of Al in foodstuff is generally very low since little Al is taken up by plants and translocated to the aboveground parts (Table 4),[102,104–106] with the exception of tea. Most of the Al in infused tea is complexed with phenolics and organic acids, but addition of lemon or lime juice may increase Al availability.[106,107]

Preparation of acidic foodstuff in Al containers or covering with Al foil may increase Al in foodstuff.[108,109] The addition of sodium aluminum phosphate as a spreading agent to processed foodstuff (e.g., cheese spread) can also contribute to the dietary Al intake.[95] Salts of Al are also used as a food additive for clarification of beverages, as anticaking agent, as baking powder, and to enhance color stability.[110] The Al in food contributes around 95% of daily oral Al uptake and the daily dietary Al intake has been estimated as 4–16 mg, of which 0.1%–0.3% is bioavailable[95,107] (Table 5). The tolerable weekly intake of Al has been set by the European Food Safety Authority to 1 mg Al per kilogram of bodyweight per week.[111]

Cosmetics, especially antiperspirants, contain 25% Al as Al-chlorohydrate.[112] Daily antiperspirant use can contribute around 70 mg Al per day, yet dermal absorption of Al is only 0.02–0.2%.[95,113] Buffered aspirin and antacids often contain Al-hydroxide suspensions and can contribute up to 5 g Al/day.[95]

Al-phosphate and Al-hydroxide are used as an adjuvant to enhance vaccine effectiveness. It has been estimated that the amount of Al taken up from Al-containing adjuvant in vaccine is in the range 1–8 μg Al/day,[95] with a lifetime intake of 15 mg, of which most is likely bioavailable.[96]

ENVIRONMENTAL TOXICOLOGY OF Al

Al Toxicity in Plants

The leading cause of poor fertility of acidic soils is Al toxicity.[114] Toxicity of Al in plants is mainly manifested by an inhibition of root growth, and while it is considered

Table 4 Concentration of Al in foodstuff (expressed as milligram per kilogram of dry weight).

Foodstuff	Al concentration
Grains	3–135
Leafy vegetables	9–104
Root crops	8–76
Fruits	7–20
Tea (*Camellia sinensis*) leaves	1000

Source: Adapted from Kabata-Pendias and Kabata,[102] Yi and Cao,[104] and Fung et al.[106]

Table 5 Estimated sources and daily exposure to Al, its bioavailability, and estimated daily intake.

Environmental exposure	Exposure (μg/day)	Availability (%)	Intake (μg/day)
Air	2–200	2	0.04–4
Industrial air	25,000	2	
Water	200–1,000	<2	2–10
Food	8,000–16,000	<0.3	8–48
Cosmetics	<70,000	0.2	140
Vaccines	1–8	95	1–8
Antacids	<5,000,000	<2	100,000

Source: Adapted from Yokel et al.,[95] Krewski et al.,[107] and Gourier-Frery and Frery.[157]

that Al^{3+} and Al_{13} are the most toxic species,[12,33,115] studies correlating Al species and their activity with degree of root growth inhibition have often given poor results. This is due to the several Al-hydroxy species that coexist within a narrow pH band and cannot be investigated in isolation. Furthermore, the activities of individual species must be calculated from equilibrium data that may be uncertain.[116] The critical Al concentration at which root elongation is inhibited is as low as 5–20 μM in solutions of low ionic strength, representative of acid soils. Exact critical values depend on the plant species and the conditions in which the plants are grown. Generally, root hairs are more sensitive to Al toxicity than roots.[117] Root growth inhibition results in short stubby roots and absence of root hairs, leading to poor water and nutrient utilization by plants. The main site affected by Al is the elongation zone of roots (1–4 mm behind the root tip).[118,119] The reaction of aluminum with phosphate anions in the soil may result in the precipitation of Al-phosphate minerals (e.g., variscite), lowering phosphorus availability to plants. This effect is accentuated by the lower phosphorus uptake capacity of Al-damaged roots.

In plants, detoxification of Al is generally achieved by organic acid exudation, but coprecipitation of Al with Si in the cell wall to form insoluble pytholith has also been reported.[120,121]

Many *Rhizobium* strains necessary for nodulation in legumes are inhibited by soil Al, with 1 μM Al at pH 5 being toxic.[122] Therefore, nodulation and nitrogen fixation are decreased in acidic soils. Soil bacteria are considered to be more sensitive to low pH and Al than soil fungi.[123] Swarming of some bacteria can also be inhibited by Al.[124]

Nano-sized aluminum had little effect on root growth and plant Al uptake. Also, it had no effect on soil respiration,[125] but this is contradicted by Mishra and Kumar[126] and Jiang et al.[127] who found that nanoparticles were detrimental to bacteria in soil and these discrepancies may be attributed to differences in particle sizes and experimental approaches.

Al Toxicity in Aquatic Organisms

In solution, Al is highly toxic to aquatic organisms in fresh waters.[128] The freshwater organism *Daphne* is sensitive to nano-sized Al.[129] In freshwater crayfish, Al impairs gill function by inducing secretion of mucus and causes oxygen stress.[130] No biomagnification of Al was observed in aquatic food webs.[131]

The main environmental risk of Al in water bodies is to aquatic organisms such as fish and molluscs. Maximal Al accumulation occurs on gills at pH 6–8,[130,132] but complexation with organic matter can prevent Al binding to gills.[132]

Al Toxicity in Humans

Aluminum has been implicated in Alzheimer's disease and dialysis dementia in humans.[133,134] Alzheimer's disease is a multifactorial disease, with environmental and genetic factors playing a role in the pathogenesis.[135–137] Age, but not Al exposure, is the most important determinant of developing Alzheimer's disease.[138] It has been proposed that Al acts by interfering with the inositol phosphate system and Ca signaling pathways in cells, leading to the formation of reactive oxygen species,[133] thereby affecting mitochondria[139,140] and triggering neurodegenerative disorders.[141] It was also shown that Al can cause lipid peroxidation in rat brains.[142] It has been suggested that Al may affect the metal homeostasis in the brain,[143] leading to oxidative damage.[144]

Parenterally and intramuscularly applied Al is 100% bioavailable.[145] By contrast, availability of Al via oral, inhalation, or dermal routes is less than 2%.[145] In plasma, Al is bound to transferrin[145] and citric acid.[96] Most of the Al taken up by humans is excreted in the urine[96,146] and feces.[96,145]

Bauxite mining in open-cut mines does not appear to affect the respiratory health of mine workers.[147] Bauxite mine and Al refinery workers have no increased risk of cancer,[148,149] but some bauxite may contain radioactive elements (radium and thorium), which may pose a long-term radiation exposure risk.[150] In Al smelter workers, exposure to fluoride may affect lung function more than Al itself[151] and no congenital abnormalities have been observed in Al smelter workers.[149,152] Likewise, Al welders are not showing increased incidence of neurobehavioral problems.[152]

Nanoparticles of Al oxide have been shown to cause neurotoxicity, and they can cross the blood–brain barrier.[127,153] Nano-sized Al_2O_3 is more biotoxic than micro-sized Al_2O_3.[127,154] Therefore, more research is needed into the environmental threats posed by nano-sized Al.

CONCLUSION

Aluminum has no known biological benefit to living organisms and is toxic in micromolar concentrations. Since solubility of Al is minimal at circumneutral pH, alleviation of Al toxicity can be achieved by pH adjustment. Complexation of Al by organic acids and humic substances also lowers biotoxicity of Al. While Al toxicity in soil can be easily overcome by incorporating lime into the soil by tilling, new management strategies need to be developed to counteract soil acidification and consequent Al toxicity in zero-till farming systems. The decrease in atmospheric acid inputs (acid rain) over the last two decades has resulted in lower Al solubility and less Al damage to aquatic ecosystems. Uptake of Al in humans through foodstuff and water poses an uncertain risk, but air pollution with nano-particulate Al may result in human health problems. The effect of nano-sized Al particles on the functioning of soil and water ecosystems, as well as human health, needs more research.

ACKNOWLEDGMENTS

This research was supported under the Australian Research Council's Discovery Projects Funding scheme (project number DP0665467).

REFERENCES

1. Wehr, J.B.; Blamey, F.P.C.; Menzies, N.W. Aluminum. In *Encyclopedia of Soil Science*; Lal, R., Ed.; Taylor and Francis: New York, 2007; Vol. 1, 1–6.
2. Martin, R.B. Ternary complexes of Al^{3+} and F^- with a third ligand. Coord. Chem. Rev. **1996**, *141*, 23–32.
3. Martell, A.E.; Hancock, R.D.; Smith, R.M.; Motekaitis, R.J. Coordination of Al(III) in the environment and in biological systems. Coord. Chem. Rev. **1996**, *149*, 311–328.
4. Salifoglou, A. Synthetic and structural carboxylate chemistry of neurotoxic aluminum in relevance to human diseases. Coord. Chem. Rev. **2002**, *228*, 297–317.
5. Furrer, G.; Trusch, B.; Muller, C. The formation of polynuclear aluminum under simulated natural conditions. Geochim. Cosmochim. Acta **1992**, *56*, 3831–3838.
6. Brown, P.L.; Sylva, R.N.; Batley, G.E.; Ellis, J. The hydrolysis of metal ions. Part 8. Aluminium (III). J. Chem. Soc., Dalton Trans. **1985**, *1985*, 1967–1970.
7. Orvig, C. The aqueous coordination chemistry of aluminum. In *Coordination Chemistry of Aluminum*; Robinson, G.H., Ed.; VCH: New York, 1993, 85–121.
8. Bertsch, P.M.; Parker, D.R. Aqueous polynuclear aluminum species. In *The Environmental Chemistry of Aluminum*, 2nd Ed.; Sposito, G., Ed.; CRC Lewis Publishers: Boca Raton, 1996, 117–168.
9. Wang, M.; Muhammed, M. Novel synthesis of Al_{13}-cluster based alumina materials. Nanostruct. Mater. **1999**, *11*, 1219–1229.
10. Sarpola, A.T.; Hietapelto, V.K.; Jalonen, J.E.; Jokela, J.; Ramo, J.H. Comparison of hydrolysis products of $AlCl_3.6H_2O$ in different concentrations by electrospray ionization time of flight mass spectrometer (ESI TOF MS). Int. J. Environ. Anal. Chem. **2006**, *86*, 1007–1018.
11. OECD Global Forum on Environment. 25–27 October 20120, Mechelen Belgium. Materials Case Study 2: Aluminium. Working Document, OECD Directorate 2010. Available at http://www.oecd.org/dataoecd/52/42/46194971.pdf (last accessed May 14, 2012)
12. Available at http://www.aac.aluminium.qc.ca/documents/newsletter11_F.pdf (accessed June 20, 2010).
13. Poschenrieder, C.; Gunse, B.; Corrales, I.; Barcelo, J. A glance into aluminum toxicity and resistance in plants. Sci. Total Environ. **2008**, *400*, 356–368.
14. Horst, W.J.; Wang, Y.; Eticha, D. The role of the root apoplast in aluminium-induced inhibition of root elongation and in aluminium resistance of plants: A review. Ann. Bot. **2010**, *106*, 185–197.
15. Brown, T.T.; Koenig, R.T.; Huggins, D.R.; Harsh, J.B.; Rossi, R.E. Lime effects on soil acidity, crop yield, and aluminum chemistry in direct-seeded cropping systems. Soil Sci. Soc. Am. J. **2008**, *72*, 634–640.
16. Simonsson, M. Interactions of aluminium and fulvic acid in moderately acid solutions: Stoichiometry of the H^+/Al^{3+} exchange. Eur. J. Soil Sci. **2000**, *51*, 655–666.
17. Skyllberg, U. pH and solubility of aluminium in acidic forest soils: A consequence of reactions between organic acidity and aluminium alkalinity. Eur. J. Soil Sci. **1999**, *50*, 95–106.
18. Adams, M.L.; Hawke, D.J.; Nilsson, N.H.S.; Powell, K.J. The relationship between soil solution pH and Al^{3+} concentrations in a range of South Island (New Zealand) soils. Aust. J. Soil Res. **2000**, *38*, 141–153.
19. Lofts, S.; Woof, C.; Tipping, E.; Clarke, N.; Mulder, J. Modelling pH buffering and aluminium solubility in European forest soils. Eur. J. Soil Sci. **2001**, *52*, 189–204.
20. van Hees, P.; Lundstrom, U.; Danielsson, R.; Nyberg, L. Controlling mechanisms of aluminium in soil solution—An evaluation of 180 podzolic forest soils. Chemosphere **2001**, *45*, 1091–1101.
21. Guo, J.H.; Zhang, X.S.; Vogt, R.D.; Xiao, J.S.; Zhao, D.W.; Xiang, R.J.; Luo, J.H. Evaluating main factors controlling aluminum solubility in acid forest soils, southern and southwestern China. Appl. Geochem. **2007**, *22*, 388–396.
22. Furrer, G.; Gfeller, M.; Wehrli, B. On the chemistry of the Keggin Al_{13} polymer: Kinetics of proton-promoted decomposition. Geochim. Cosmochim. Acta **1999**, *63*, 3069–3076.
23. Etou, A.; Bai, S.Q.; Saito, T.; Noma, H.; Okaue, Y.; Yokoyama, T. Formation conditions and stability of a toxic tridecameric Al polymer under a soil environment. J. Colloid Interface Sci. **2009**, *337*, 606–609.
24. Casey, W.H. Large aqueous aluminum hydroxide molecules. Chem. Rev. **2005**, *106*, 1–16.
25. Yamaguchi, N.U.; Hiradate, S.; Mizoguchi, M.; Miyazaki, T. Formation and disappearance of Al tridecamer in the

presence of low molecular weight organic ligands. Soil Sci. Plant Nutr. **2003**, *49*, 551–556.

26. Masion, A.; Thomas, F.; Tchoubar, D.; Bottero, J.Y.; Tekely, P. Chemistry and structure of Al(OH)/organic precipitates. A small-angle x-ray scattering study. 3. Depolymerization of the Al_{13} polycation by organic ligands. Langmuir **1994**, *10*, 4353–4356.
27. Sanjuan, B.; Michard, G. Aluminum hydroxide solubility in aqueous solutions containing fluoride ions at 50°C. Geochim. Cosmochim. Acta **1987**, *51*, 1823–1831.
28. Hiradate, S.; Taniguchi, S.; Sakurai, K. Aluminum speciation in aluminum–silica solutions and potassium chloride extracts of acidic soils. Soil Sci. Soc. Am. J. **1998**, *62*, 630–636.
29. Gerard, F.; Boudot, J.P.; Ranger, J. Consideration on the occurrence of the Al-13 polycation in natural soil solutions and surface waters. Appl. Geochem. **2001**, *16*, 513–529.
30. Xia, H.; Rayson, G.D. Investigation of aluminum binding to a *Datura innoxia* material using ^{27}Al NMR. Environ. Sci. Technol. **1998**, *32*, 2688–2692.
31. Masion, A.; Bertsch, P.M. Aluminium speciation in the presence of wheat root cell walls: A wet chemical study. Plant, Cell Environ. **1997**, *20*, 504–512.
32. Kopittke, P.M.; Menzies, N.W.; Blamey, F.P.C. Rhizotoxicity of aluminate and polycationic aluminium at high pH. Plant Soil **2004**, *266*, 177–186.
33. Hunter, D.; Ross, D.S. Evidence for a phytotoxic hydroxy-aluminum polymer in organic soil horizons. Science **1991**, *251*, 1056–1058.
34. Kinraide, T.B. Identity of the rhizotoxic aluminium species. In *Plant–Soil Interactions at Low pH*; Wright, R.J., Baligar, V.C., Murrmann, R.P., Eds.; Kluwer Academic Publishers: Dordrecht, 1991, 717–728.
35. Griffitt, R.J.; Luo, J.; Gao, J.; Bonzongo, J.C.; Barber, D.S. Effects of particle composition and species on toxicity of metallic nanomaterials in aquatic organisms. Environ. Toxicol. Chem. **2008**, *27*, 1972–1978.
36. Sjostedt, C.; Wallstedt, T.; Gustafsson, J.P.; Borg, H. Speciation of aluminium, arsenic and molybdenum in excessively limed lakes. Sci. Total Environ. **2009**, *407*, 5119–5127.
37. von Uexküll, H.R.; Mutert, E. Global extent, development and economic impact of acid soils. In *Plant–Soil Interactions at Low pH*; Date, R.A., Grundon, N.J., Rayment, G.E., Probert, M.E., Eds.; Kluwer Academic: Dordrecht, 1995, 5–19.
38. Lesturgez, G.; Poss, R.; Noble, A.; Grunberger, O.; Chintachao, W.; Tessier, D. Soil acidification without pH drop under intensive cropping systems in Northeast Thailand. Agric., Ecosyst. Environ. **2006**, *114*, 239–248.
39. Lapenis, A.G.; Lawrence, G.B.; Andreev, A.A.; Bobrov, A.A.; Torn, M.S.; Harden, J.W. Acidification of forest soil in Russia: From 1893 to present. Global Biogeochem. Cycles **2004**, *18*.
40. Bergholm, J.; Berggren, D.; Alavi, G. Soil acidification induced by ammonium sulphate addition in a Norway spruce forest in Southwest Sweden. Water, Air, Soil Pollut. **2003**, *148*, 87–109.
41. Guo, J.H.; Liu, X.J.; Zhang, Y.; Shen, J.L.; Han, W.X.; Zhang, W.F.; Christie, P.; Goulding, K.W.T.; Vitousek, P.M.; Zhang, F.S. Significant acidification in major Chinese croplands. Science **2010**, *327*, 1008–1010.
42. Liu, C.; Huang, P.M. Role of hydroxy-aluminosilicate ions (proto-imogolite sol) in the formation of humic substances. Org. Geochem. **2002**, *33*, 295–305.
43. Mulder, J.; De Wit, H.A.; Boonen, H.W.J.; Bakken, L.R. Increased levels of aluminium in forest soils: Effects on the stores of soil organic carbon. Water, Air, Soil Pollut. **2001**, *130*, 989–994.
44. Scheel, T.; Jansen, B.; van Wijk, A.J.; Verstraten, J.M.; Kalbitz, K. Stabilization of dissolved organic matter by aluminium: a toxic effect or stabilization through precipitation? Eur. J. Soil Sci. **2008**, *59*, 1122–1132.
45. Scheel, T.; Pritsch, K.; Schlater, M.; Kalbitz, K. Precipitation of enzymes and organic matter by aluminum—Impacts on carbon mineralization. J. Plant Nutr. Soil Sci. **2008**, *171*, 900–907.
46. Curtin, D. Possible role of aluminum in stabilizing organic matter in particle size fractions of Chernozemic and Solonetizic soils. Can. J. Soil Sci. **2002**, *82*, 265–268.
47. Scheel, T.; Dorfler, C.; Kalbitz, K. Precipitation of dissolved organic matter by aluminum stabilizes carbon in acidic forest soils. Soil Sci. Soc. Am. J. **2007**, *71*, 64–74.
48. Dent, D.L.; Pons, L.J. A world perspective on acid sulfate soils. Geoderma **1995**, *67*, 263–276.
49. Faltmarsch, R.M.; Astrom, M.E.; Vuori, K.M. Environmental risks of metals mobilised from acid sulphate soils in Finland: A literature review. Boreal Environ. Res. **2008**, *13*, 444–456.
50. Bush, R.T.; Fyfe, D.; Sullivan, L.A. Occurrence and abundance of monosulfidic black ooze in coastal acid sulfate soil landscapes. Aust. J. Soil Res. **2004**, *42*, 609–616.
51. Pu, X.X.; Vazquez, O.; Monnell, J.D.; Neufeld, R.D. Speciation of aluminum precipitates from acid rock discharges in Central Pennsylvania. Environ. Eng. Sci. **2010**, *27*, 169–180.
52. Liang, H.C.; Thomson, B.M. Minerals and mine drainage. Water Environ. Res. **2009**, *81*, 1615–1663.
53. Dsa, J.V.; Johnson, K.S.; Lopez, D.; Kanuckel, C.; Tumlinson, J. Residual toxicity of acid mine drainage-contaminated sediment to stream macroinvertebrates: Relative contribution of acidity vs. metals. Water, Air, Soil Pollut. **2008**, *194*, 185–197.
54. Soucek, D.J.; Cherry, D.S.; Zipper, C.E. Impacts of mine drainage and other nonpoint source pollutants on aquatic biota in the upper Powell River system, Virginia. Hum. Ecol. Risk Assess. **2003**, *9*, 1059–1073.
55. Dubikova, M.; Cambier, P.; Sucha, V.; Caplovicova, M. Experimental soil acidification. Appl. Geochem. **2002**, *17*, 245–257.
56. Martinent-Catalot, V.; Lamerant, J.-M.; Tilmant, G.; Bacou, M.-S.; Ambrosi, J.P. Bauxaline: A new product for various applications of Bayer process red mud. Light Met. **2002**, *2002*, 125–131.
57. Manoharan, V.; Loganathan, P.; Tillman, R.W.; Parfitt, R.L. Interactive effects of soil acidity and fluoride on soil solution aluminium chemistry and barley (*Hordeum vulgare* L.) root growth. Environ. Pollut. **2007**, *145*, 778–786.
58. Takahashi, T.; Mitamura, A.; Ito, T.; Ito, K.; Nanzyo, M.; Saigusa, M. Aluminum solubility of strongly acidified allophanic Andosols from Kagoshima prefecture, southern Japan. Soil Sci. Plant Nutr. **2008**, *54*, 362–368.

59. Lange, H.; Solberg, S.; Clarke, N. Aluminum dynamics in forest soil waters in Norway. Sci. Total Environ. **2006**, *367*, 942–957.
60. Bardy, M.; Bonhomme, C.; Fritsch, E.; Maquet, J.; Hajjar, R.; Allard, T.; Derenne, S.; Calas, G. Al speciation in tropical podzols of the upper Amazon basin: A solid-state Al-27 MAS and MQMAS NMR study. Geochim. Cosmochim. Acta **2007**, *71*, 3211–3222.
61. McBride, M.B. *Environmental Chemistry of Soils*; Oxford University Press: New York, 1994.
62. Machacha, S. Comparison of laboratory pH buffer methods for predicting lime requirement for acidic soils of eastern Botswana. Commun. Soil Sci. Plant Anal. **2004**, *35*, 2675–2687.
63. Sun, B.; Poss, R.; Moreau, R.; Aventurier, A.; Fallavier, P. Effect of slaked lime and gypsum on acidity alleviation and nutrient leaching in an acid soil from Southern China. Nutr. Cycling Agroecosyst. **2000**, *57*, 215–223.
64. Materechera, S.A.; Mkhabela, T.S. The effectiveness of lime, chicken manure and leaf litter ash in ameliorating acidity in a soil previously under black wattle (*Acacia mearnsii*) plantation. Bioresour. Technol. **2002**, *85*, 9–16.
65. Tang, Y.; Zhang, H.; Schroder, J.L.; Payton, M.E.; Zhou, D. Animal manure reduces aluminum toxicity in an acid soil. Soil Sci. Soc. Am. J. **2007**, *71*, 1699–1707.
66. Mokolobate, M.S.; Haynes, R.J. Comparative liming effect of four organic residues applied to an acid soil. Biol. Fertil. Soils **2002**, *35*, 79–85.
67. Steiner, C.; Teixeira, W.G.; Lehmann, J.; Nehls, T.; de Macedo, J.L.V.; Blum, W.E.H.; Zech, W. Long term effects of manure, charcoal and mineral fertilization on crop production and fertility on a highly weathered Central Amazonian upland soil. Plant Soil **2007**, *291*, 275–290.
68. Tarkalson, D.D.; Hergert, G.W.; Stevens, W.B.; McCallister, D.L.; Kackman, S.D. Fly, ash as a liming material for corn production. Soil Sci. **2005**, *170*, 386–398.
69. Morikawa, C.K.; Saigusa, M. Si amelioration of Al toxicity in barley (*Hordeum vulgare* L.) growing in two Andosols. Plant Soil **2002**, *240*, 161–168.
70. Naramabuye, F.X.; Haynes, R.J. Effect of organic amendments on soil pH and Al solubility and use of laboratory indices to predict their liming effect. Soil Sci. **2006**, *171*, 754–763.
71. Scott, B.J.; Fenton, I.G.; Fanning, A.G.; Schumann, W.G.; Castleman, L.J.C. Surface soil acidity and fertility in the eastern Riverina and western slopes of southern New South Wales. Aust. J. Exp. Agric. **2007**, *47*, 949–964.
72. Conyers, M.K.; Heenan, D.P.; McGhie, W.J.; Poile, G.P. Amelioration of acidity with time by limestone under contrasting tillage. Soil Tillage Res. **2003**, *72*, 85–94.
73. Godsey, C.B.; Pierzynski, G.M.; Mengel, D.B.; Lamond, R.E. Management of soil acidity in no-till production systems through surface application of lime. Agron. J. **2007**, *99*, 764–772.
74. Liu, J.; Hue, N.V. Amending subsoil acidity by surface applications of gypsum, lime, and composts. Commun. Soil Sci. Plant Anal. **2001**, *32*, 2117–2132.
75. Mota, D.; Faria, F.; Gomes, E.A.; Marriel, I.E.; Paiva, E.; Seldin, L. Bacterial and fungal communities in bulk soil and rhizospheres of aluminum-tolerant and aluminum-sensitive maize (*Zea mays* L.) lines cultivated in unlimed and limed Cerrado soil. J. Microbiol. Biotechnol. **2008**, *18*, 805–814.
76. Nelson, D.R.; Mele, P.M. The impact of crop residue amendments and lime on microbial community structure and nitrogen-fixing bacteria in the wheat rhizosphere. Aust. J. Soil Res. **2006**, *44*, 319–329.
77. Vieira, F.C.B.; He, Z.L.; Bayer, C.; Stoffella, P.J.; Baligar, V.C. Organic amendment effects on the transformation and fractionation of aluminum in acidic sandy soil. Commun. Soil Sci. Plant Anal. **2008**, *39*, 2678–2694.
78. Qin, R.J.; Chen, F.X. Amelioration of aluminum toxicity in red soil through use of barnyard and green manure. Commun. Soil Sci. Plant Anal. **2005**, *36*, 1875–1889.
79. Yamaguchi, N.; Hiradate, S.; Mizoguchi, M.; Miyazaki, T. Disappearance of aluminum tridecamer from hydroxyaluminum solution in the presence of humic acid. Soil Sci. Soc. Am. J. **2004**, *68*, 1838–1843.
80. Matthias, A.; Maurer, M.; Parlar, H. Comparative aluminium speciation and quantification in soil solutions of two different forest ecosystems by Al-27-NMR. Fresenius Environ. Bull. **2003**, *12*, 1263–1275.
81. Lopez-Diaz, M.L.; Mosquera-Losada, M.R.; Rigueiro-Rodriguez, A. Lime, sewage sludge and mineral fertilization in a silvopastoral system developed in very acid soils. Agroforestry Syst. **2007**, *70*, 91–101.
82. Wehr, J.B.; Menzies, N.W.; Fulton, I. Revegetation strategies for bauxite refinery residue: A case study of Alcan Gove in Northern Territory, Australia. Environ. Manage. **2006**, *37*, 297–306.
83. Xenidis, A.; Harokopou, A.D.; Mylona, E.; Brofas, G. Modifying alumina red mud to support a revegetation cover. JOM **2005**, *57*, 42–46.
84. Lawrence, G.B. Persistent episodic acidification of streams linked to acid rain effects on soil. Atmos. Environ. **2002**, *36*, 1589–1598.
85. de Vries, W.; Reinds, G.J.; Vel, E. Intensive monitoring of forest ecosystems in Europe 2: Atmospheric deposition and its impacts on soil solution chemistry. For. Ecol. Manage. **2003**, *174*, 97–115.
86. Hrkal, Z.; Prchalova, H.; Fottova, D. Trends in impact of acidification on groundwater bodies in the Czech Republic; an estimation of atmospheric deposition at the horizon 2015. J. Atmos. Chem. **2006**, *53*, 1–12.
87. Skjelkvale, B.L.; Torseth, K.; Aas, W.; Andersen, T. Decrease in acid deposition—Recovery in Norwegian waters. Water, Air, Soil Pollut. **2001**, *130*, 1433–1438.
88. Andersen, D.O. Labile aluminium chemistry downstream a limestone treated lake and an acid tributary: Effects of warm winters and extreme rainstorms. Sci. Total Environ. **2006**, *366*, 739–748.
89. Fest, E.P.M.J.; Temminghoff, E.J.M.; Griffioen, J.; Van Der Grift, B.; Van Riemsdijk, W.H. Groundwater chemistry of Al under Dutch sandy soils: Effects of land use and depth. Appl. Geochem. **2007**, *22*, 1427–1438.
89. Hicks, W.S.; Bowman, G.M.; Fitzpatrick, R.W. *Environmental Impact of Acid Sulfate Soils near Cairns, QLD*; CSIRO Land and Water Technical Report 15/99; Commonwealth Scientific and Industrial Research Organisation: Canberra, 1999; 1–8.
90. *Australian Drinking Water Guidelines*; National Water Quality Management Strategy; National Health and Medical Research Council: Canberra, 1996.

91. McCrohan, C.R.; Campbell, M.M.; Jugdaohsingh, R.; Ballance, S.; Powell, J.J.; White, K.N. Bioaccumulation and toxicity of aluminium in the pond snail at neutral pH(+). Acta Biol. Hung. **2000**, *51*, 309–316.
92. Srinivasan, P.T.; Viraraghavan, T. Characterisation and concentration profile of aluminium during drinking-water treatment. Water SA **2002**, *28*, 99–106.
93. Hovsepyan, A.; Bonzongo, J.C.J. Aluminum drinking water treatment residuals (Al-WTRs) as sorbent for mercury: implications for soil remediation. J. Hazard. Mater. **2009**, *164*, 73–80.
94. Malecki-Brown, L.M.; White, J.R.; Sees, M. Alum application to improve water quality in a municipal wastewater treatment wetland. J. Environ. Qual. **2009**, *38*, 814–821.
95. Yokel, R.A.; Hicks, C.L.; Florence, R.L. Aluminum bioavailability from basic sodium aluminum phosphate, an approved food additive emulsifying agent, incorporated in cheese. Food Chem. Toxicol. **2008**, *46*, 2261–2266.
96. Yokel, R.A.; McNamara, P. Aluminium toxicokinetics: An updated mini review. Pharmacol. Toxicol. **2001**, *88*, 159–167.
97. Warren, J.G.; Penn, C.J.; McGrath, J.M.; Sistani, K. The impact of alum addition on organic P transformations in poultry litter and litter-amended soil. J. Environ. Qual. **2008**, *37*, 469–476.
98. Oladeji, O.O.; Sartain, J.B.; O'Connor, G.A. Land application of aluminum water treatment residual: Aluminum phytoavailability and forage yield. Commun. Soil Sci. Plant Anal. **2009**, *40*, 1483–1498.
99. Mahdy, A.M.; Elkhatib, E.A.; Fathi, N.O. Drinking water treatment residuals as an amendment to alkaline soils: Effects on bioaccumulation of heavy metals and aluminum in corn plants. Plant Soil Environ. **2008**, *54*, 234–246.
100. Agyin-Birikorang, S.; O'Connor, G.A. Aging effects on reactivity of an aluminum-based drinking-water treatment residual as a soil amendment. Sci. Total Environ. **2009**, *407*, 826–834.
101. Othman, M.; Abdullah, M.; Abd Aziz, Y. Removal of aluminium from drinking water. Sains Malays. **2010**, *39*, 51–55.
102. Kabata-Pendias, A.; Kabata, H. *Trace Elements in Soils and Plants*, 3rd Ed.; CRC Press: Boca Raton, 2001.
103. Schussler, U.; Balzer, W.; Deeken, A. Dissolved Al distribution, particulate Al fluxes and coupling to atmospheric Al and dust deposition in the Arabian Sea. Deep-Sea Res., Part II **2005**, *52*, 1862–1878.
104. Yi, J.; Cao, J. Tea and fluorosis. J. Fluorine Chem. **2008**, *129*, 76–81.
105. Chen, R.F.; Shen, R.F.; Gu, P.; Wang, H.Y.; Xu, X.H. Investigation of aluminum-tolerant species in acid soils of South China. Commun. Soil Sci. Plant Anal. **2008**, *39*, 1493–1506.
106. Fung, K.F.; Carr, H.P.; Poon, B.H.T.; Wong, M.H. A comparison of aluminum levels in tea products from Hong Kong markets and in varieties of tea plants from Hong Kong and India. Chemosphere **2009**, *75*, 955–962.
107. Krewski, D.; Yokel, R.A.; Nieboer, E.; Borchelt, D.; Cohen, J.; Harry, J.; Kacew, S.; Lindsay, J.; Mahfouz, A.M.; Rondeau, V. Human health risk assessment for aluminium, aluminium oxide, and aluminium hydroxide. J. Toxicol. Environ. Health, Part B **2007**, *10*, 1–269.
108. Turhan, S. Aluminium contents in baked meats wrapped in aluminium foil. Meat Sci. **2006**, *74*, 644–647.
109. Scancar, J.; Stibilj, V.; Milacic, R. Determination of aluminium in Slovenian foodstuffs and its leachability from aluminium-cookware. Food Chem. **2004**, *85*, 151–157.
110. Walton, J.R. A longitudinal study of rats chronically exposed to aluminum at human dietary levels. Neurosci. Lett. **2007**, *412*, 29–33.
111. Schafer, U.; Jahreis, G. Update on regulations of aluminium intake—Biochemical and toxicological assessment. Trace Elem. Electrolytes **2009**, *26*, 95–99.
112. Shen, J.S.; Nardello-Rataj, V. Deodorants and antiperspirants: Chemistry under arms. Actual. Chim. **2009**, 8–18.
113. Flarend, R.; Bin, T.; Elmore, S.; Hem, S. A preliminary study of the dermal absorption of aluminium from antiperspirants using aluminium-26. Food Chem. Toxicol. **2001**, *39*, 163–168.
114. Asher, C.J.; Grundon, N.J.; Menzies, N.W. *How to Unravel and Solve Soil Fertility Problems*; ACIAR Monograph No. 83; Australian Centre for International Agricultural Research: Canberra, 2002.
115. Kinraide, T.B. Reconsidering the rhizotoxicity of hydroxyl, sulphate, and fluoride complexes of aluminium. J. Exp. Bot. **1997**, *48*, 1115–1124.
116. Boudot, J.P.; Maitat, O.; Merlet, D.; Rouiller, J. Occurrence of non-monomeric species of aluminium in undersaturated soil and surface waters: Consequences for the determination of mineral saturation indices. J. Hydrol. **1996**, *177*, 47–63.
117. Brady, D.J.; Edwards, D.G.; Asher, C.J.; Blamey, F.P.C. Calcium amelioration of aluminium toxicity effects on root hair development in soybean *Glycine max*. New Phytol. **1993**, *123*, 531–538.
118. Sivaguru, M.; Horst, W.J. The distal part of the transition zone is the most aluminum-sensitive apical root zone of maize. Plant Physiol. **1998**, *116*, 155–163.
119. Blamey, F.P.C.; Nishizawa, N.K.; Yoshimura, E. Timing, magnitude, and location of initial soluble aluminum injuries to mungbean roots. Soil Sci. Plant Nutr. **2004**, *50*, 67–76.
120. Sangster, A.G.; Hodson, M.J. Silicon and aluminium codeposition in the cell wall phytoliths of gymnosperm leaves. In *Phytoliths: Applications in Earth Sciences and Human History*; Meunier, J.D., Colin, F., Eds.; AA Balkema Publ.: Leiden, 2001, 343–355.
121. Exley, C.; Schneider, C.; Doucet, F.J. The reaction of aluminium with silicic acid in acidic solution: An important mechanism in controlling the biological availability of aluminium? Coord. Chem. Rev. **2002**, *228*, 127–135.
122. Kinraide, T.B.; Sweeney, B.K. Proton alleviation of growth inhibition by toxic metals (Al, La, Cu) in rhizobia. Soil Biol. Biochem. **2003**, *35*, 199–205.
123. Shirokikh, I.G.; Shirokikh, A.A.; Rodina, N.A.; Polyanskaya, L.M.; Burkanova, O.A. Effects of soil acidity and aluminum on the structure of microbial biomass in the rhizosphere of barley. Eurasian Soil Sci. **2004**, *37*, 839–843.
124. Illmer, P.; Schinner, F. Influence of aluminum on motility and swarming of *Pseudomonas* sp. and *Arthrobacter* sp. FEMS Microbiol. Lett. **1997**, *155*, 121–124.
125. Doshi, R.; Braida, W.; Christodoulatos, C.; Wazne, M.; O'Connor, G. Nano-aluminum: Transport through sand columns and environmental effects on plants and soil communities. Environ. Res. **2008**, *106*, 296–303.

126. Mishra, V.K.; Kumar, A. Impact of metal nanoparticles on the plant growth promoting *Rhizobacteria*. Dig. J. Nanomater. Biostruct. **2009**, *4*, 587–592.
127. Jiang, W.; Mashayekhi, H.; Xing, B.S. Bacterial toxicity comparison between nano- and micro-scaled oxide particles. Environ. Pollut. **2009**, *157*, 1619–1625.
128. Gensemer, R.W.; Playle, R.C. The bioavailability and toxicity of aluminum in aquatic environments. Crit. Rev. Environ. Sci. Technol. **1999**, *29*, 315–450.
129. Strigul, N.; Vaccari, L.; Galdun, C.; Wazne, M.; Liu, X.; Christodoulatos, C.; Jasinkiewicz, K. Acute toxicity of boron, titanium dioxide, and aluminum nanoparticles to *Daphnia magna* and *Vibrio fischeri*. Desalination **2009**, *248*, 771–782.
130. Ward, R.J.S.; McCrohan, C.R.; White, K.N. Influence of aqueous aluminium on the immune system of the freshwater crayfish *Pacifastacus leniusculus*. Aquat. Toxicol. **2006**, *77*, 222–228.
131. Winterbourn, M.J.; McDiffett, W.F.; Eppley, S.J. Aluminium and iron burdens of aquatic biota in New Zealand streams contaminated by acid mine drainage: Effects of trophic level. Sci. Total Environ. **2000**, *254*, 45–54.
132. Winter, A.R.; Nichols, J.W.; Playle, R.C. Influence of acidic to basic water pH and natural organic matter on aluminum accumulation by gills of rainbow trout (*Oncorhynchus mykiss*). Can. J. Fish. Aquat. Sci. **2005**, *62*, 2303–2311.
133. Yokel, R.A. The toxicology of aluminum in the brain: A review. Neurotoxicology **2000**, *21*, 813–828.
134. Nayak, P. Aluminum: Impacts and disease. Environ. Res. **2002**, *89*, 101–115.
135. Oyanagi, K. The nature of the parkinsonism–dementia complex and amyotrophic lateral sclerosis of Guam and magnesium deficiency. Parkinsonism Relat. Disord. **2005**, *11*, S17–S23.
136. Frisardi, V.; Solfrizzi, V.; Capurso, C.; Kehoe, P.; Imbimbo, B.; Santamato, A.; Dellegrazie, F.; Seripa, D.; Pilotto, A.; Capurso, A.; Panza, F. Aluminum in the diet and Alzheimer's disease: From current epidemiology to possible disease-modifying treatment. J. Alzheimer's Dis. **2010**, *20*, 17–30.
137. Mutter, J.; Naumann, J.; Schneider, R.; Walach, H. Mercury and Alzheimer's disease. Fortschr. Neurol. Psychiatr. **2007**, *75*, 528–540.
138. Dartigues, J.F.; Berr, C.; Helmer, C.; Letenneur, L. Epidemiology of Alzheimer's disease. Med. Sci. **2002**, *18*, 737–743.
139. Kumar, V.; Gill, K.D. Aluminium neurotoxicity: Neurobehavioural and oxidative aspects. Arch. Toxicol. **2009**, *83*, 965–978.
140. El-Demerdash, F.M. Antioxidant effect of vitamin E and selenium on lipid peroxidation, enzyme activities and biochemical parameters in rats exposed to aluminium. Journal of Trace Elements in Medicine and Biology **2004**, *18*, 113–121.
141. Campbell, A.; Bondy, S. Aluminum induced oxidative events and its relation to inflammation: a role for the metal in Alzheimer's disease. Cellular and Molecular Biology **2000**, *46*, 721–730.
142. Nehru, B.; Anand, P. Oxidative damage following chronic aluminium exposure in adult and pup rat brains. J. Trace Elem. Med. Biol. **2005**, *19*, 203–208.
143. Michalke, B.; Halbach, S.; Nieschwitz, V. JEM Spotlight: Metal speciation related to neurotoxicity in humans. J. Environ. Monit. **2009**, *11*, 939–954.
144. Liu, G.; Garrett, M.R.; Men, P.; Zhu, X.W.; Perry, G.; Smith, M.A. Nanoparticle and other metal chelation therapeutics in Alzheimer disease. Biochim. Biophys. Acta **2005**, *1741*, 246–252.
145. Schafer, U.; Jahreis, G. Exposure, bioavailability, distribution and excretion of aluminum and its toxicological relevance to humans. Trace Elem. Electrolytes **2006**, *23*, 162–172.
146. Barabasz, W.; Albinska, D.; Jaskowska, M.; Lipiec, J. Ecotoxicology of aluminium. Pol. J. Environ. Stud. **2002**, *11*, 199–203.
147. Beach, J.R.; de Klerk, N.H.; Fritschi, L.; Sim, M.R.; Musk, A.W.; Benke, G.; Abramson, M.J.; McNeil, J.J. Respiratory symptoms and lung function in bauxite miners. Int. Arch. Occup. Environ. Health **2001**, *74*, 489–494.
148. Fritschi, L.; Hoving, J.L.; Sim, M.R.; Del Monac, A.; MacFarlane, E.; McKenzie, D.; Benke, G.; de Klerk, N. All cause mortality and incidence of cancer in workers in bauxite mines and alumina refineries. Int. J. Cancer **2008**, *123*, 882–887.
149. Sakr, C.J.; Taiwo, O.A.; Galusha, D.H.; Slade, M.D.; Fiellin, M.G.; Bayer, F.; Savitz, D.A.; Cullen, M.R. Reproductive outcomes among male and female workers at an aluminum smelter. J. Occup. Environ. Med. **2010**, *52*, 137–143.
150. Abbady, A.G.E.; El-Arabi, A.M. Naturally occurring radioactive material from the aluminium industry—A case study: The Egyptian Aluminium Company, Nag Hammady, Egypt. J. Radiol. Prot. **2006**, *26*, 415–422.
151. Fritschi, L.; Sim, M.R.; Forbes, A.; Abramson, M.J.; Benke, G.; Musk, A.W.; de Klerk, N.H. Respiratory symptoms and lung-function changes with exposure to five substances in aluminium smelters. Int. Arch. Occup. Environ. Health **2003**, *76*, 103–110.
152. Kiesswetter, E.; Schaper, M.; Buchta, M.; Schaller, K.H.; Rossbach, B.; Kraus, T.; Letzel, S. Longitudinal study on potential neurotoxic effects of aluminium: II. Assessment of exposure and neurobehavioral performance of Al welders in the automobile industry over 4 years. Int. Arch. Occup. Environ. Health **2009**, *82*, 1191–1210.
153. Sharma, H.S.; Ali, S.F.; Hussain, S.M.; Schlager, J.J.; Sharma, A. Influence of engineered nanoparticles from metals on the blood-brain barrier permeability, cerebral blood flow, brain edema and neurotoxicity. An experimental study in the rat and mice using biochemical and morphological approaches. J. Nanosci. Nanotechnol. **2009**, *9*, 5055–5072.
154. Stanley, J.K.; Coleman, J.G.; Weiss, C.A.; Steevens, J.A. Sediment toxicity and bioaccumulation of nano and micron-sized aluminum oxide. Environ. Toxicol. Chem. **2010**, *29*, 422–429.
155. Available at https://stats.world-aluminium.org/iai/stats_new/formServer.asp?form=1 (accessed July 1, 2010).
156. Wang, P.; Bi, S.P.; Zhou, Y.P.; Tao, Q.S.; Gan, W.X.; Xu, Y.; Hong, Z.; Cai, W.S. Study of aluminium distribution and speciation in atmospheric particles of different diameters in Nanjing, China. Atmos. Environ. **2007**, *41*, 5788–5796.
157. Gourier-Frery, C.; Frery, N. Aluminium. EMC-Toxicol. Pathol. **2004**, *1*, 79–95.

Animals: Sterility from Pesticides

William Au
Department of Preventitive Medicine and Community Health, University of Texas Medical Branch, Galveston, Texas, U.S.A.

Abstract
Pesticides belong to a unique group of synthetic chemicals that have high biological activities and that are released legally and extensively into our environment. Therefore, there has been continued concern about their potential hazard to living organisms. A major concern is whether prolonged exposure to pesticides can cause reproductive problems leading to the extinction of species. A variety of studies have been conducted to elucidate the reproductive hazards of pesticides in native species, experimental animals, and human beings.

MECHANISMS OF ACTION OF PESTICIDES

Pesticides can be subdivided into several major categories: organophosphates, carbamates, organochlorines, synthetic pyrethroids, and others. Their principal mechanisms of action include inhibition of cholinesterase, perturbation of microsomal enzyme production, and damage to nervous systems.[1] Therefore, it is possible that excessive exposure to pesticides can interfere with gametogenesis, sexual activities, and reproduction leading to the expression of sterility.

OBSERVED EFFECTS IN NATIVE ANIMALS AND IN EXPERIMENTAL SYSTEMS

Among the pesticides, organochlorines are characterized by their persistence in the environment, and the potential for both bioaccumulation and transfer of the pesticides up the food chain. Therefore, the widespread use of organochlorines in the past has been documented to cause contamination of wildlife and reduction of their populations. For example, the exposure is associated with a significant reduction of fish populations such as trout and salmon.[2] Subsequently, the populations of predatory birds were significantly reduced.[2] The devastating effects in migratory birds and in other wildlife were also demonstrated.[3,4] In these cases, failure to reproduce appropriately has been shown to be a major cause for the decline of the populations.

In studies using experimental animals under controlled exposure conditions to pesticides, organochlorine pesticides such as methoxychlor have been reported to cause reduction of fertility and litter size,[5,6] and kepone to cause anovulation.[7,8] Organophosphate pesticides have been shown to induce premature ovulation and to perturb oocyte development.[9]

OBSERVED EFFECTS IN HUMANS

Very few pesticides have been shown systematically and consistently to cause sterility in humans. An exception is the exposure to a nematocide, 1,2-dibromo-3-chloropropane (DBCP). In the 1970s, workers in several pesticide manufacturing plants were reported to have fertility problems. From a systematic investigation, the infertility based on reduced sperm counts was shown to be associated with testicular function alteration and with exposure to DBCP rather than to other pesticides.[10] Subsequently, the same group of scientists found that the reduction of sperm count was associated with an occupational exposure as short as 3 months[11] and with an employment duration-dependent effect (Table 1A). As shown in the table, as many as 76.5% of the workers who had been exposed to DBCP for more than 42 months were oligospermic. Among all the affected workers, many were azospermic or sterile.

Long-term follow-up studies of DBCP production workers showed that some of the affected workers did regain fertility and testicular function, and many of them were able to have children. Their offspring appeared normal and healthy. However, these workers predominantly had female offspring (Table 1B), ranging from 58.6 to 84.6% for the recovery duration from 5 to 17 years.[12] As shown in the table, the highest female to male offspring ratio was found among workers within 5 years of recovery.[13,14] It appears that the recovery is a slow process and complete recovery with respect to the sex ratio was achieved only after 17 years.[15] The observation confirms the previous recommendation of using altered sex ratios as an indication of reproductive hazards associated with pesticides.[16] In another study, males infertile due to poor sperm quality were more likely than expected to be in the agricultural occupations with exposure to pesticides.[17] Papaya fumigant workers with exposure to ethylene dibromide were reported to have significantly reduced sperm count per ejaculate (the percentage of viable and motile sperm) and increases in

Encyclopedia of Environmental Management DOI: 10.1081/E-EEM-100001025

Table 1 Reproductive problems from exposure to dibromochloropropane (DBCP).

A. Oligospermia	
Months of exposure to DBCP[a]	**% Workers with oligospermia**
0	2.9
1–6	8.3
7–24	28.6
25–42	66.7
>42	76.5

B. Offspring	
Years after recovery from oligospermia[b]	**% Females in offspring**
5	84.6
8	78.9
17	58.6

[a]Data derived from 10.
[b]Data derived from 13–15.

the proportion of sperm with abnormalities.[18] Abnormal pregnancy outcomes (miscarriages and preterm deliveries) were associated with exposure to a variety of chemicals in combination with pesticides (atrazine, glyphosate, organophaosphates, 4-[2,4-dichlorophenoxy]butyric acid) in males.[19] On the other hand, fertility in traditional male farmers, compared with organic farmers (who do not use pesticides), was not influenced by exposure to pesticides, based on the time taken to have the youngest child.[20]

Among females, exposure to DBCP in pesticide manufacturing plants appears to have no effects on their fertility based on a limited survey.[21] A study was conducted to investigate the relationship between the plasma level of organochlorine pesticides and the diagnosis of endometriosis, and no association was found.[22] On the other hand, among women with medically confirmed infertility, exposure to pesticides was shown to be a significant contributing factor.[23] Furthermore, the mechanism appears to be due to abnormal ovulation.

FUTURE CONSIDERATIONS

Based on the mechanisms of action of pesticides[1,24] and on observations in animals, it is highly likely that overexposure and/or prolonged exposure to pesticides can cause reproductive problems in human. However, adverse reproductive effects have not been demonstrated unequivocally with modern pesticides. One reason is that the human population is usually exposed to much lower doses of pesticides, except in accidental exposure conditions, than those used in animals that have been shown to cause sterility. Under this condition, any adverse effects in human would be very small. Therefore, investigations using inappropriate study protocols may have generated inconsistent observations. Future studies should be conducted by using large enough populations and by minimizing multiple confounding factors. At this stage of our knowledge, it is fair to state that the potential impact of modern pesticides on sterility in humans has not been clearly demonstrated yet. However, based on the known biological activities of pesticides, they should be considered hazardous chemicals and should be handled with extreme caution.

REFERENCES

1. Kaloyanova, F.P.; el Batawi, M.A. *Human Toxicity to Pesticides*; CRC Press: Boca Raton, FL, 1991.
2. Pimentel, D. *Ecological Effects of Pesticides on Non-target Species*; U.S. Government Printing Office: Washington, DC, 1971.
3. Gard, N.; Hooper, M. An Assessment of Potential Hazards of Pesticides and Environmental Contaminants. In *Ecology and Management of Neotropical Migratory Birds*; Martin, T., Finch, D., Eds.; Oxford University Press: Oxford, 1995; 294–310.
4. Stinson, E.; Bromely, P. *Pesticides and Wildlife: A Guide to Reducing Impacts on Animals and Their Habitat*; Publication No. 420-004, Virginia Department of Game and Inland Fisheries: Virginia, 1991.
5. Gray, L.E.; Ostby, J.S.; Ferrell, J.M.; Sigman, E.R.; Goldman, J.M. Methoxychlor induces estrogen-like alterations of behavior and the reproductive tract in the female rat and hamster: effects on sex behavior, running wheel activity, and uterine morphology. Toxicol. Appl. Pharmacol. **1988**, *96*, 525–540.
6. Gray, L.E.; Ostby, J.S.; Ferrel, J.M.; Rehnberg, G.; Lindler, R.; Cooper, R.; et al. A dose-response analysis of methoxychlor-induced alterations of reproductive development and function in the rat. Fundam. Appl. Toxicol. **1989**, *12*, 92–108.
7. Eroschenko, V.P. Estrogenic activity of the insecticide chlordecone in the reproductive tract of birds and mammals. J. Toxicol. Environ. Health **1981**, *8*, 731–742.
8. Guzelian, P.S. Comparative toxicology of chlordecone (kepone) in humans and experimental animals. Annu. Rev. Pharmacol. Toxicol. **1982**, *22*, 89–113.
9. Rattner, B.A.; Michael, S.D. Organophosphorous insecticide induced decrease in plasma luteinizing hormone concentration in white-footed mice. Toxicol. Lett. **1985**, *24*, 65–69.
10. Whorton, D.; Milby, T.H.; Krauss, R.M. Testicular function in DBCP exposed pesticide workers. J. Occup. Med. **1979**, *21*, 161–66.
11. Whorton, D.; Krauss, R.M.; Marshall, S. Infertility in male pesticide workers. Lancet **1977**, *ii*, 1259–1261.
12. Goldsmith, J.R. Dibromocholorpropane: epidemiological findings and current questions. Ann. of the New York Academy of Sci. **1997**, *831*, 300–306.
13. Potashnik, G.; Goldsmith, J.; Insler, V. Dibromochloropropane-induced reduction of the sex-ratio in man. Andrologia **1984**, *16*, 213–218.
14. Potashnik, G.; Yanai-Inbar, H. Dibromochloropropane: an eight-year re-evaluation of testicular function and reproductive performance. Fertil. Steril. **1987**, *47*, 317–322.

15. Potashnik, G.; Porath, A. Dibromochloropropane: a 17-year reassessment of testicular function and reproductive performance. J. Occup. Environ. Med. **1995**, *37*, 1287–1292.
16. James, W.H. Offspring sex ratio as an indicator of reproductive hazards associated with pesticides. Occup. Environ. Med. **1996**, *52*, 429–430.
17. Strohmer, H.; Boldizsar, A.; Plockinger, B.; Feldner-Busztin, M.; Feichtinger, W. Agricultural work and male infertility. Am. J. Ind. Med. **1993**, *24*, 587–592.
18. Ratcliff, J.M.; Schrader, S.M.; Steenland, K. Semen quality in papaya workers with long term exposure to ethylene dibromide. Br. J. Ind. Med. **1987**, *44*, 317–326.
19. Savitz, D.A.; Arbuckle, T.; Kaczor, D.; Curtis, K.M. Male pesticide exposure and pregnancy outcome. Am. J. Epidemiol. **1997**, *146*, 1025–1036.
20. Larsen, S.B.; Joffe, M.; Bonde, J.P. The asclepiod study group. Time to pregnancy and exposure to pesticides in Danish farmers. Occup. Environ. Med. **1998**, *55*, 278–283.
21. Marshall, S.; Whorton, D.; Krauss, R.M.; Palmer, W.S. Effect of pesticides on testicular function. Urology **1978**, *11*, 257–259.
22. Lebel, G.; Dodin, S.; Ayotte, P.; Marcoux, S.; Ferron, L.A.; Dewailly, E. Organochlorine exposure and the risk of endometriosis. Fertil. Steril. **1998**, *69*, 221–228.
23. Smith, E.M.; Hammonds-Ehlers, M.; Clark, M.K.; Kirchner, H.L.; Fuortes, L. Occupational exposures and risk of female infertility. J. Occup. Environ. Med. **1997**, *39*, 138–147.
24. Sharara, F.I.; Seifer, D.B.; Flaws, J.A. Environmental toxicants and female reproduction. Fertil. Steril. **1998**, *70*, 613–622.

Animals: Toxicological Evaluation

Vera Lucia S.S. de Castro
Ecotoxicology and Biosafety Laboratory, Environment, Brazilian Agricultural Research Corporation (Embrapa), São Paulo, Brazil

Abstract
Maternal exposure to toxic chemicals during pregnancy can disrupt the development, or even cause death, of the embryo or fetus. The developing organism is generally more vulnerable to injury caused by different classes of chemicals than the adult. Developmental toxicants are agents that cause adverse effects on developing organisms. This notion refers to any effect interfering with normal development, both before and after birth. A series of experimental methods are useful in studying developmental toxicity. Different international organizations have developed protocols for testing the reproductive and developmental toxicity of chemicals. Their results related to toxic effects produced by environmental pollutants are used worldwide. This entry shows some evidence on the impact of exposure to chemical pollutants during the perinatal period and presents some data on the adverse impacts of this exposure.

INTRODUCTION

Developmental toxicants may be a chemical, microorganism, physical agent, or deficiency state that alters the morphology or a physiological process of a developing organism, both before and after birth. Exposure of the developing embryo or fetus to some environmental agents like gamma irradiation and thalidomide is known to produce anatomical anomalies leading to in utero death or structural birth defects. Developmental toxicology studies the causes, mechanisms, manifestation, and prevention of developmental deviations produced by developmental toxicants. Several environmental agents are established as causing developmental toxicity in humans, while many others are suspected of causing developmental toxicity in humans on the basis of data from experimental animal studies.

Nowadays, all the possible manifestations of developmental toxicity (death, structural abnormalities, growth alterations, and behavioral and functional deficits) are of concern, while in the past there has been a tendency to consider only malformations and death as end points of concern. Developmental toxicity usually results from prenatal exposures to toxicants experienced by the mother, but it can also result from paternal exposures, e.g., in rural workers exposed to pesticides. Effects can include birth defects, reduced body weight at birth, growth and developmental retardation, organ toxicity, death, abortion, and functional dysfunctions. It can also lead to behavioral deficits that become manifest as the organism develops since the chemicals can impair postnatal development up to pubertal development. Further considerations related to the efforts to assemble an internationally harmonized source of common nomenclature for use in describing observations of fetal and neonatal external, visceral, and skeletal abnormalities can be found in Makris.[1]

In May 2001, more than 90 nations adopted the Stockholm Convention on Persistent Organic Pollutants (POPs), with significant contributions from non-governmental organizations, trade unions, and private companies. The Stockholm Convention is a global treaty created to protect human health and the environment from chemicals that remain intact in the environment for long periods, become widely distributed geographically, accumulate in the fatty tissue of humans and wildlife, and have adverse effects to human health or to the environment. Exposure to POPs can lead to serious health effects, including certain cancers, birth defects, dysfunctional immune and reproductive systems, greater susceptibility to disease, and even diminished intelligence. Given their long-range transport, no one government acting alone can protect is citizens or its environment from POPs. In response to this global problem, the Stockholm Convention entered into force in 2004, requiring parties to take measures to eliminate or reduce the release of POPs into the environment. The convention is administered by the United Nations Environment Programme and is based in Geneva, Switzerland. The substances covered initially are eight pesticides (aldrin, chlordane, DDT, dieldrin, endrin, heptachlor, mirex, and toxaphene), two industrial chemicals (hexachlorobenzene and polychlorinated biphenyls), and two POP byproducts (dioxins and furans). In 2009, the Conference of the Parties decided to undertake a work program to provide guidance to parties on how best to restrict and eliminate nine newly listed POPs and invited parties to support work on the evaluation of alternatives and other work related

Encyclopedia of Environmental Management DOI: 10.1081/E-EEM-120047300

to the restriction and elimination of these new POPs [α-hexachlorocyclohexane, β-hexachlorocyclohexane, lindane, pentachlorobenzene, perfluorooctane sulfonic acid (PFOS), PFOS salts, perfluorooctane sulfonyl fluoride, tetrabromodiphenyl ether, and pentabromodiphenyl ether (commercial pentabromodiphenyl ether)] (http://chm.pops.int/default.aspx).[2]

Exposure of the developing embryo or fetus to some environmental agents like gamma irradiation and thalidomide is known to produce anatomical anomalies leading to in utero death or structural birth defects, commonly termed teratogenesis. Perhaps less well appreciated is that such environmental exposures also can cause functional disorders that persist postnatally and into adult life. The spectrum of such postnatal consequences is growing, and more recently is thought to include disorders of the immune system, brain function, obesity, and diseases such as diabetes and cancer, to name a few.[3]

Importance of Oxidative Stress

Xenobiotics such as phenytoin and benzo[*a*]pyrene can be bioactivated by enzymes like the cytochromes P450 (CYPs); however, the developing embryo and fetus have relatively low levels of most CYP isozymes. The xenobiotic bioactivation to a free radical intermediate by enzymes associated with peroxidase activities within the embryo or fetus can be a critical determinant of teratogenesis. In the developing embryo and fetus, enhanced formation of reactive oxygen species by xenobiotics may adversely alter development by oxidatively damaging cellular lipids, proteins, and DNA, and/or by altering signal transduction. The postnatal consequences may include an array of birth defects, postnatal functional deficits, and diseases.[3]

Also, estrogen-like substances, such as several organochlorine pesticides, have been demonstrated to induce defeminization, miscarriages, malformations, and transplacental carcinogenesis. This variety of effects results from the interference of these substances with the metabolism of steroid and protein hormones, therefore altering a whole spectrum of complex developmental functions.[4]

Importance of Toxic Effects on Nervous System Development

Only about 200 chemicals out of more than 80,000 registered with the U.S. Environmental Protection Agency (EPA) have undergone extensive neurotoxicity testing, and many chemicals found in consumer goods are not required to undergo any neurodevelopmental testing. The cumulative effects of co-contaminants and the difficulties in analyzing biomarkers of exposure in human tissues have complicated comprehensive risk assessment. Furthermore, population-based studies that measure subtle effects on neurobehavioral outcomes are challenging to interpret and costly to conduct.[5]

The effects observed following developmental exposure may be different both quantitatively and qualitatively from adult exposure because of the potential to affect processes in the developing child that have no parallel process in adults. Central nervous system development consists of a series of processes that occur in sequence and are dependent on each other, such that interference with one stage may also affect later stages of development. This makes the timing of a potential environmental neurotoxicant a critical parameter in the risk for subsequent neurologic effects. The sequence includes proliferation, migration, differentiation, synaptogenesis, apoptosis, and myelination.

There are multiple windows of vulnerability during which environmental exposures can interfere with normal development. For many developmental toxicants, there is a spectrum of adverse outcome depending on dose timing of exposure, maternal and fetal susceptibility, and interactions with other environmental factors.[6]

Therefore, environmental exposure to a toxic agent that affects synaptogenesis, such as lead, may affect brain areas differentially depending on timing of exposure. Since different brain areas develop on different time lines during prenatal and postnatal life, an environmental neurotoxic agent may produce impairment in different functional domains depending on the time of exposure. For example, the same exposure at different points in development could result in an adverse effect on motor systems versus memory or executive functions. Similarly, exposures at different concentrations or for different lengths of time could potentially produce differential effects. Therefore, the constellation of observed effects should not be expected to be the same in different children exposed to the same neurotoxic agent.[7]

CONCERN ON DEVELOPMENTAL IMPAIRMENT IN CHILDREN

The findings of the some studies imply that children's exposure to pesticides may bring about impairments in their neurobehavioral development. The long-term neurotoxicity risks caused by prenatal exposures to pesticides are sometimes unclear. Effective control of exposure is complicated by variable exposure sources and variable contaminant levels in food and environment. This awareness has also been extended to effect(s) of toxic contaminants on breastfeeding women and their children.[8]

The information deriving from epidemiological studies indicate a need to increase awareness among people and children exposed to pesticides about the association between the use of pesticides and neurodevelopmental impairments. There are modest epidemiological evidence on occupational exposures of female workers to industrial chemicals and the consequences in regard to the child's neurodevelopment. The majority of the occupational studies identified aimed to assess organic solvents and organophosphate pesticide effects in the offspring, and neurobehavioral impairments were reported. In some reports, however, the evidence suffers from a variety of shortcomings and sources of imprecision. These problems would tend to cause an underestimation of the true extent of the risks. Due to the vulnerability of the brain during early development, a precautionary approach to neurodevelopmental toxicity needs to be applied in occupational health.[9,10]

With increasing evidence of the high prevalence of pesticide use and the considerable risk it poses to children, it is of concern that there has been little research into the health implications of household pesticide use. Children are exposed to pesticides in various ways, not only environmentally, but also through food and through use of pesticides at home and in public areas. The exposure depends on a large variety of factors related to chemical characteristics and use, and children's activities. In spite of its potential health and environmental risks and contribution to agribusiness, the use of agricultural chemicals for yard care has not been well studied. The probability that a household chooses a mix of do-it-yourself and hired applications of synthetic chemicals increases with income, age, and the presence of preschoolers.[11]

Also, children who live in farming communities are furthermore exposed to both agricultural and household pesticides. Farmworkers bring home pesticide residues on their clothing, boots, and skin, placing other household members at risk, particularly children.[12] Recent studies on in utero exposure to the organochlorine pesticide dichlorodiphenyltrichloroethane and its breakdown product, dichlorodiphenyldichloroethene, indicate that exposure is associated with poorer infant (6 mo and older) and child neurodevelopment depending on the end point evaluated. Research on organophosphate pesticide exposure and neurodevelopment also suggests some negative association of exposure and neurodevelopment at certain ages. About abnormal reflexes in neonates and in young children (2–3 years), two separate studies observed an increase in maternally reported pervasive developmental disorder with increased levels of organophosphate exposure.[13]

For example, children whose mothers worked in the flower industry during pregnancy scored lower on communication and fine motor skills, and had higher odds of having poor visual acuity, compared with children whose mothers did not work in the flower industry during pregnancy, after adjusting for potential confounders. These facts showed that maternal occupation in the cut-flower industry during pregnancy may be associated with delayed neurobehavioral development of children aged 3–23 mo. However, possible hazards associated with working in the flower industry during pregnancy include pesticide exposure, exhaustion, and job stress.[14]

The increased emphasis on children's exposures to pesticides and other organic pollutants has led to a surge in recent years in the number of research studies aimed at this specific susceptible population. Continuous strong investment in research plus strong preventive action by the government is required for further progress in environmental pediatrics and for better control of the diseases caused in children by environmental toxic exposures. This research will have high costs and demanding long-term multiyear studies.[15]

DEVELOPMENTAL ANIMAL EXPERIMENTAL STUDIES

Before 1960, governmental recommendations for the assessment of chemical effects on the reproductive cycle involved limited animal testing. During the early 1960s, the thalidomide disaster evidenced, on the one hand, the greater vulnerability of the embryo and fetus, and on the other, that the complexity of the mother–child unity warranted special consideration. Thalidomide was used as a sedative drug in the 1950s, also in pregnant women. Later, it was found that it could cause malformations in newborns of mothers who ingested the drug during the sensitive period.[16] This disaster fostered the establishment of formal laboratory animal-testing procedures for assessing fetal development.

In contrast to most other toxicological tests, developmental studies are usually required in rodent and nonrodent subjects. One of the reasons for this requirement is the thalidomide disaster. When the developmental toxicity of thalidomide was studied in experimental animals, large interspecies differences were found in effective doses and in the types of effects. However, the discrimination between direct and indirect (i.e., as a consequence of maternal toxicity) developmental effects was often doubtful, and is one of the factors that could explain the apparent differences between species.[16]

Despite the limitations, animals can be useful predictors of chemical hazards to humans. A specific animal model might be chosen for any conjunction of widely varying reasons. Accessibility of embryos, cost of acquiring or maintaining animals, availability of genomic analyses or probes, and/or close similarity to human physiology might factor in the design of a laboratory experiment. Growth and development are compressed into a shorter period in animals, which makes interpretation of animal testing inherently more difficult. Each experimental species has its own advantages and the use of laboratory animal models is based on diverse practical grounds arising from an assumption of generalizability across species. The conjunction of evolutionary and developmental biology shows that the timing and sequence of early events in brain development are remarkably conserved across mammals.[17] During mammalian development, the fetal organism is exposed to its own gonadal hormones, placental steroids, and maternal hormones that may cross the placental barrier.[18]

Manifestations of developmental toxicity observed in humans are not always reproduced in experimental animals, and in general there is at least one experimental species that mimics the types of effects seen in humans. The fact that every species may not react in the same way could be due to species-specific differences in critical periods, differences in timing of exposure, metabolism, developmental patterns, placentation, or mechanisms of action.[6]

The vast majority of the laboratory studies on the developmental effects of environmental contaminants use the pregnant or lactating animal as the conduit to deliver the contaminant to the developing offspring. The traditional approach in developmental toxicology adopts a linear perspective for the interpretation of the effects of contaminants on the offspring as a particular one is given to the mother in order to expose the fetus in utero or postnatally via lactation. Variations of this approach may manipulate time of exposure or use cross-fostering strategies to separate the effects of in utero from those of lactational exposure.

Maternal behavior is not always monitored during the treatment period, even though there is evidence of the effects of these compounds on adult behaviors, particularly on behaviors that are sensitive to hormonal manipulations. As is the case with all mammals, maternal care in female rodents comprises very specific behaviors that help ensure the survival of the offspring by providing nourishment, warmth, sensory stimulation, cleaning, and protection. Maternal behavior begins even before parturition as the dam builds a nest in order to provide warmth and protection for the coming offspring. The fact that many developmental outcomes are determined or modulated by the amount and quality of maternal care raises the question of the importance of possible changes in maternal behavior in determining the consequences of their exposure during early development.[19] There are some guidelines proposed in the literature in order to evaluate the maternal behavior and its effects on litter development.[20,21]

Reproductive and Developmental Protocols

Developmental toxicology bioassays are designed to identify agents with the potential to induce adverse effects and include dose levels that induce maternal toxicity. In reproduction toxicity studies, the determination of the high dose is important since changes in body weight are often used as an index of toxicity. The highest dose level should be chosen with the aim to induce some parental toxicity (e.g., clinical signs, decreased body weight gain, not more than 10%) and/or evidence of toxicity in a target organ. Consequently, the knowledge on maternal (and to some degree, paternal) toxicity is important as a natural limiter to prevent underdosing. A comparison between doses causing effects in adults and offspring can also be used, although a direct comparison is difficult since the level of observation applied in offspring is often much higher than in adults. It is also useful in order to obtain information on the influence of pregnancy and/or lactation on the susceptibility to a test compound. However, if dosing was high enough to cause maternal toxicity, these doses often also cause some effects in the offspring.[22]

Prenatal developmental toxicity studies are designed to provide general information concerning the effects of exposure to the pregnant test animal on the developing organism. Although exposures during a typical guideline prenatal developmental toxicity study are designed to include either the entire period of gestation or limited species-specific gestation periods, developmental end points are considered to be an integral concern in the assessment of potential health effects from continuous lifetime exposures to a toxicant. Pregnancy and fetal development are thus considered to represent a potentially susceptible life stage that should be considered in lifetime or chronic assessments. If studies and information on reproductive or developmental toxicity are absent in a health-effect assessment, specific uncertainty factors (e.g., database) may be applied to the final point of departure used in risk calculations. It is also well established, however, that developmental toxicity may occur in response to single exposure, such as during specific developmental windows of susceptibility. Whereas this circumstance does not influence the relevance of typical guideline developmental studies to the evaluation of chronic or lifetime assessments of health effects, it does signify that developmental end points observed in these repeated dose studies are relevant in health-effect assessments of shorter-term exposures, including acute exposures.[23] Results from developmental toxicology bioassays have significant predictive value in identifying potential health risks to the human embryo/fetus.[24]

Data Quality Control of Experimental Studies

To accomplish good experimental planning, some points before the beginning of the study should be observed for adequate data interpretation in view of the experimental delineation.[25,26] Animals to be used in laboratory research should experience an acceptable welfare (for ethical reasons) and show normal behavioral and physiological reaction patterns to guarantee the quality of research.[27] In this sense, the performing laboratory should maintain a historical control database to track any changes in the data over time in the animals and/or in the equipment. The value of historical data depends on its quality and its reliability at the side of contemporary controls.

INTERNATIONAL PROTOCOLS

Different international organizations have developed protocols for testing the reproductive and developmental toxicity of chemicals. The information produced by them, related to toxic effects produced by environmental pollutants, is very useful worldwide. Besides, new protocols are being evaluated viewing to reduce the number of animals used and to improve the predictability for human health hazards identification.

Developmental study methodology has been extensively reviewed and evaluated over the last 25 years. This has included the conduct of a number of meetings and collaborative studies involving experts from academic, industry, regulatory, and public interest groups. For example, in recent years, the International Life Sciences Institute (ILSI), under a cooperative agreement with the EPA, established a working group of scientists from government, industry, and academia, to discuss developmental neurotoxicity test protocol ending with a public workshop in which occasion the conclusions of the working group were presented.[28]

Food Quality Protection Act

Developmental and reproductive toxicity testing protocols such as those recommended by the EPA, Food and Drug Administration, and Organization for Economic Cooperation and Development are useful for characterizing toxicity in developing animals and for assessing risks to children that might arise from in utero and postnatal exposures.[7] However, there is a global interest in reducing, refining, and replacing (3Rs) the use of animals in research.[29]

The Food Quality Protection Act (FQPA) of 1996 enactment by U.S. Congress amended the Federal Insecticide, Fungicide, and Rodenticide Act and the Federal Food, Drug, and Cosmetic Act by fundamentally changing the way the EPA regulates pesticides. The major requirements of the FQPA include stricter safety standards, especially for infants and children, and a complete reassessment of all existing pesticide tolerances. They include an additional safety factor to account for developmental risks and incomplete data when considering a pesticide's effect on infants and children, and any special sensitivity and exposure to pesticide chemicals that infants and children may have (http://www.epa.gov/opp00001/regulating/laws/fqpa/).

The FQPA mandated that all pesticides in the United States undergo re-registration with a focus on reducing cumulative risk of exposure to pesticides sharing a common mode of action. Enforcement of FQPA has resulted in the modification of use patterns and removal (or pending removal) of many organophosphate insecticides that had previously seen wide use.[30]

The FQPA also requires the EPA to consider the cumulative effects of exposure to pesticides having a common mechanism of toxicity, considering for a cumulative risk assessment the exposure to all chemicals that act by a common mechanism of toxicity, as well as the exposure to each chemical via various routes and sources in an aggregate risk assessment. To support the grouping of different chemicals together for purposes of cumulative risk assessment, there must be sufficient evidence to support a common adverse effect that is associated with a common mechanism of action in specific target tissues. However, the criteria that are required to establish a common mechanism of toxicity with a specific toxic effect have not always been achieved for various pesticides as the common mechanism of toxicity of organophosphorus and carbamate insecticides (inhibition of acetylcholinesterase activity) that can be associated with adverse effects (cholinergic signs of intoxication). For example, a determination of common mechanism of toxicity in mammals is complicated by the number of potential biological target sites and effects expressed by various pyrethroid insecticides on these targets. Probably, the differences of action on neuronal ion channels among the pyrethroid insecticides contribute to the diversity of neurologic and behavioral manifestations of acute toxicity that are evident in the whole animal.[31]

EUROPEAN UNION'S REACH LEGISLATION

REACH is the European Community Regulation on chemicals and their safe use. It deals with the Registration, Evaluation, Authorization, and Restriction of Chemicals. The law entered into force on June 1, 2007. The REACH regulation places greater responsibility on industry to manage the risks from chemicals and to provide safety information on the substances. Manufacturers and importers are required to gather information on the properties of their chemical substances, which will allow their safe handling, and to register the information in a central database in Helsinki (http://ec.europa.eu/environment/chemicals/reach/reach_intro.htm). It transfers responsibility for risk assessment from government to the manufacturers and importers, and includes downstream uses in the registration and management process. It introduces authorization and restriction procedures for the most hazardous chemicals and creates a new European Chemicals Agency. The legal permission to market products is conditional on the firms testing them for toxicity. If firms do not provide data required by the program, their products will not be permitted to enter (or remain in) the market. This program holds some promise for detecting developmental toxicants before they enter commerce and cause adverse effects. Whether this will work for subclinical neurotoxic and other developmental effects depends on tests the European Union requires. The REACH testing strategy is to require fewer tests for products produced in lesser amounts and to require more tests and more detailed tests as the production volume increases.[32]

In particular, large numbers of industrial chemicals are unlikely to be tested under this paradigm, and there is no specific requirement for developmental neurotoxicological testing under the new European Union law governing chemical regulation (REACH), which was passed in 2006 and went into effect in 2007. In response to the need for broader screening for developmental neurotoxicity, efforts are under way to develop additional developmental neurotoxicity screening paradigms. Additional efforts will be needed to focus on identifying possible chemical class-specific targets and biomarkers of effect, and on ways to differentiate normal variability in response from changes that are adverse.[33]

PERSPECTIVES

There is a need to expand risk assessment paradigms to evaluate exposures relevant to children from preconception to adolescence, taking into account the specific susceptibilities at each developmental stage. Risk assessment approaches for exposures in children must be linked to life stages. Establishing causal links between specific environmental exposures and complex, multifactorial health outcomes is difficult and challenging, particularly in children. For children, the stage in their development when the exposure occurs may be just as important as the magnitude of exposure. Very few studies have characterized exposures during different developmental stages. Some examples of health effects resulting from developmental exposures include those observed prenatally and at birth (e.g., miscarriage, stillbirth, low birth weight, birth defects), in young children (e.g., infant mortality, asthma, neurobehavioral and immune impairment), and in adolescents (e.g., precocious or delayed puberty). Emerging evidence suggests that an increased risk of certain diseases in adults (e.g., cancer, heart disease) can result in part from exposures to certain environmental chemicals during childhood. Advancing technology and new methodologies now offer promise for capturing exposures during these critical windows. This will enable investigators to detect conceptions early and estimate the potential competing risk of early embryonic mortality when considering children's health outcomes that are conditional on survival during the embryonic and fetal periods.[34]

Furthermore, emerging technologies such as gene expression, electrical activity measurements, and metabonomics have been identified as promising tools for evaluating neurotoxicity. In a combination with other assays, the in vitro approach could be included into a developmental neurotoxicological intelligent testing strategy to speed up the process of developmental neurotoxicity evaluation mainly by initial prioritization of chemicals with developmental neurotoxicity potential for further testing. Also, emerging nano/microtechnologies (piezoelectric spotting and microcontact printing of different biomolecules to create protein microarrays) can be used to promote cell differentiation and make the model suitable for developmental neurotoxicity screening.[35]

DEVELOPMENTAL TOXIC EFFECTS ON NON-TARGET ORGANISMS IN ENVIRONMENT

In addition to their potential role as human reproductive toxicants, pesticides are also implicated in reproductive failure of wildlife species exposed to pesticide sprays and residues. One main example of the ecological consequences of teratogenic pesticides is related to the organochlorines, which, besides inducing malformations in embryos, cause calcification problems in eggshells and impair reproduction in several wild bird species. This problem still remains in numerous areas, due to organochlorine residue accumulation through food chains and to wild populations exposed to organochlorine-contaminated sites.[36] Also, organochlorine pesticides can be maternally transferred to the developing eggs of alligators. This maternal exposure is associated with reduced clutch success and increased embryonic mortality.[37]

Toxicity assays are available for the evaluation of pesticide impact on wildlife or their surrogates, and many compounds have been shown to cause teratogenesis in fish, amphibian, avian, and mammalian species. Some chemicals that have been detected in the environment may be disrupting of both target and non-target systems in exposed populations of wildlife and fish.[38] Environmental compounds can also interfere with the endocrine systems of wildlife. Surface waters are the main sinks of endocrine disrupters, which are mainly of anthropogenic origin. Thus, aquatic organisms, especially lower vertebrates such as fish and amphibians, are the main potential targets for endocrine disrupters at direct or indirect risk via ingestion and accumulation of endocrine disrupters, direct exposure, or via the food chain. The impact of these compounds on reproductive biology can be mediated through four principal mechanisms of action: estrogenic, anti-estrogenic, androgenic, and anti-androgenic.[39]

Furthermore, population studies have revealed disruptions in crustacean growth, molting, sexual development, and recruitment that are indicative of environmental endocrine disruption. However, environmental factors other than pollution (i.e., temperature, parasitism) also can elicit these effects, and definitive causal relationships between endocrine disruption in field populations of crustaceans and chemical pollution is generally lacking.[40]

Amphibians are considered reliable indicators of environmental quality, in particular due to their biphasic life (aquatic and terrestrial) and semipermeable skin. These vertebrates are sensitive to a great number of pollutants dispersed in the environment, such as pesticides, heavy metals, and polychlorinated biphenyls. Field studies on frogs from polluted and reference sites have provided information on the effects of chronic exposure to contaminants.[41]

Despite the fact that almost all environmental chemical exposure is to mixtures, the current understanding of environmental health risks is based almost entirely on the evaluation of chemicals studied in isolation. Consequently, it is essential to develop and validate methods to accurately predict effects of endocrine-disrupting mixtures beyond the individual exposure to a single chemical, in order to protect humans and wildlife from the risk associated with potentially cumulative effects of these mixtures.[36,42] Its focus should be on the biological system or the target tissue rather than on the mechanism of toxicity or even a single signaling pathway.[43]

Although the importance of multiple stressors is widely recognized in aquatic ecotoxicology, pesticide mixture studies pose some major challenges such as experimental design difficulties (e.g., near-insurmountable factorial complexity for large numbers of chemicals), poorly understood pathways for chemical interaction, potential differences in response among species, and the need for more sophisticated statistical tools for analyzing complex data.[44]

CONCLUSION

The ability of a species to reproduce successfully requires the careful orchestration of developmental processes during critical time points, particularly the late embryonic and early postnatal periods. Standard developmental toxicology bioassays are designed to identify agents with the potential to induce adverse effects. Government agencies that regulate the use of pesticides and various industrial chemicals evaluate the toxicity of these agents to the developing embryo/fetus as an integral part of the testing protocol used to assess potential dangers to the public. However, an adequate experimental delineation is important to the data interpretation.

Humans and other non-target organisms are exposed to a mixture of chemicals. Various chemicals may target the organism's development during the same critical developmental period. Although developmental studies have been historically conducted on a chemical-by-chemical basis, the interest on considering cumulative risks of chemicals is growing. Chemicals as pesticides represent a risk for the reproduction and development of children and non-target organisms of terrestrial and aquatic ecosystems.

REFERENCES

1. Makris, S.L.; Solomon, H.M.; Clark, R.; Shiota, K.; Barbellion, S.; Buschmannf, J.; Emag, M.; Fujiwarah, M.; Grotei, K.; Hazeldenj, K.P.; Hewk, K.; Horimotol, M.; Ooshimam, Y.; Parkinsonn, M.; Wiseo, L.D. Terminology of developmental abnormalities in common laboratory mammals (version 2). Reprod. Toxicol. **2009,** *28*, 371–434.
2. Rodan, B.D.; Pennington, D.W.; Eckley, N.; Boethlin, R.S. Screening for persistent organic pollutants: Techniques to provide a scientific basis for POPs criteria in international negotiations. Environ. Sci. Technol. **1999**, *33*, 3482–3488.
3. Wells, P.G.; McCallum, G.P.; Chen, C.S.; Henderson, J.T.; Lee, C.J.J.; Perstin, J.; Preston, T.J.; Wiley, M.J.; Wong, A.W. Oxidative stress in developmental origins of disease: Teratogenesis, neurodevelopmental deficits, and cancer. Toxicol. Sci. **2009**, *108* (1), 4–18.
4. Chernoff, N. The reproductive toxicology of pesticides. In *Toxicology of Pesticides: Experimental, Clinical and Regulatory Perspectives*; Costa, L.G., Galli, C.L., Murphy, S.D., Eds.; NATO ASI-Cell Biology; Springer-Verlag: Berlin, Germany, 1987; H13, 109–123.
5. Miodovnik, A. Environmental neurotoxicants and developing brain. Mt. Sinai J. Med. **2011**, *78*, 58–77.
6. U.S. EPA. *Guidelines for Developmental Toxicity Risk Assessment*; U.S. Environmental Protection Agency, Risk Assessment Forum: Washington, D.C.; EPA/600/FR-91/001, 1991.
7. Mendola, P.; Selevan, S.G.; Gutter, S.; Rice, D. Environmental factors associated with a spectrum of neurodevelopmental deficits. Mental retardation and developmental disabilities. Res. Rev. **2002**, *8,* 188–197.
8. Lakind, J.S.; Berlin, C.M.; Naiman, D.Q. Infant exposure to chemicals in breast milk in the United States: What we need to learn from a breast milk monitoring program. Environ. Health Perspect. **2001**, *109*, 75–88.
9. Jurewicz, J.; Hanke, W. Prenatal and childhood exposure to pesticides and neurobehavioral development: Review of epidemiological studies. Int. J. Occup. Med. Environ. Health **2008**, *21* (2), 121-132.
10. Julvez, J.; Grandjean, P. Neurodevelopmental toxicity risks due to occupational exposure to industrial chemicals during pregnancy. Ind. Health **2009**, *47*, 459–468.
11. Templeton, S.R.; Zilberman, D.; Yoo, S.J.; Dabalen, A.L. Household use of agricultural chemicals for soil-pest management and own labor for yard work. Environ. Resour. Econ. **2008**, *40*, 91–108.
12. Strong, L.L.; Starks, H.E.; Meischke, H.; Thompson, B. Perspectives of mothers in farmworker households on reducing the take-home pathway of pesticide exposure. Health Educ. Behav. **2009**, *36* (5), 915–929.
13. Rosas, L.G.; Eskenazi, B. Pesticides and child neurodevelopment. Curr. Opin. Pediatr. **2008**, *20* (2), 191–197.
14. Handal, A.J.; Harlow, S.D.; Breilh, J.; Lozoff, B. Occupational exposure to pesticides during pregnancy and neurobehavioral development of infants and toddlers. Epidemiology **2008**, *19* (6), 851–859.
15. Landrigan, P.J.; Miodovnik, A. Children's health and the environment: An overview. Mt. Sinai J. Med. **2011**, *78*, 1–10.
16. Janer, G.; Slob, W.; Hakkert, B.C.; Vermeire, T.; Piersma, A.H. A retrospective analysis of developmental toxicity studies in rat and rabbit: What is the added value of the rabbit as an additional test species? Regul. Toxicol. Pharmacol. **2008**, *50*, 206–217.
17. Clancy, B.; Finlay, B.L.; Darlington, R.B.; Anand, K.J.S. Extrapolating brain development from experimental species to humans. NeuroToxicology **2007**, *28*, 931–937.
18. Gore, A.C. Developmental programming and endocrine disruptor effects on reproductive neuroendocrine systems. Front. Neuroendocrinol. **2008**, *29*, 358–374.

19. Cummings, J.A.; Clemens, L.G.; Nunez, A.A. Mother counts: How effects of environmental contaminants on maternal care could affect the offspring and future generations. Front. Neuroendocrinol. **2010**, *31*, 440–451.
20. Champagne, F.; Francis, D.; Mar, A.; Meaney, M. Variations in maternal care in the rat as a mediating influence for the effects of environment on development. Physiol. Behav. **2003**, *79*, 359-371.
21. Slamberová, R.; Charousová, P.; Pometlová, M. Maternal behavior is impaired by methamphetamine administered during pre-mating, gestation and lactation. Reprod. Toxicol. **2005**, *20*, 103–110.
22. Buschmann, J. Critical aspects in reproductive and developmental toxicity testing of environmental chemicals. Reprod. Toxicol. **2006**, *22*, 157–163.
23. Davis, A.; Gift, J.S.; Woodall, G.M.; Narotsky, M.G.; Foureman, G.L. The role of developmental toxicity studies in acute exposure assessments: Analysis of single-day vs. multiple-day exposure regimens. Regul. Toxicol. Pharmacol. **2009**, *54*, 134–142.
24. Chernoff, N.; Rogers, E.H.; Gage, M.I.; Francis, B.M. The relationship of maternal and fetal toxicity in developmental toxicology bioassays with notes on the biological significance of the "no observed adverse effect level". Reprod. Toxicol. **2008**, *25* (2), 192–202.
25. Slikker, Jr., W.; Acuff, K.; Boyes, W.; Chelonis, J.; Crofton, K.; Dearlove, G.; Li, A.; Moser, V.; Newland, C.; Rossi, J.; Schantz, S.; Sette, W.; Sheets, L.; Stanton, M.; Tyl, S.; Sobotka, T. Behavioral test methods workshop. Neurotoxicol. Teratol. **2005**, *27*, 417–427.
26. Festing, M.; Altman, D. Guidelines for the design and statistical analysis of experiments using laboratory animals. ILAR J. **2002**, *43* (4), 244–258.
27. Olsson, I.A.S.; Westlund, K. More than numbers matter: The effect of social factors on behaviour and welfare of laboratory rodents and non-human primates. Appl. Anim. Behav. Sci. **2007**, *103*, 229–254.
28. Tyl, R.W.; Crofton, K.; Moretto, A.; Moser, V.; Sheets L.P.; Sobotka, T.J. Identification and interpretation of developmental neurotoxicity effects. A report from the ILSI Research Foundation/Risk Science Institute expert working group on neurodevelopmental endpoints. Neurobehav. Toxicol. **2008**, *30*, 349–381.
29. Matthews, E.J.; Kruhlak, N.L.; Benz, R.D.; Contrera, J.F. A comprehensive model for reproductive and developmental toxicity hazard identification: I. Development of a weight of evidence QSAR database. Regul. Toxicol. Pharmacol. **2007**, *47*, 115–135.
30. Jones, V.P.; Steffan, S.A.; Hull, L.A.; Brunner, J.F.; Biddinger, D.J. Effects of the loss of organophosphate pesticides in the US: Opportunities and needs to improve IPM Programs. Outlooks Pest Manage. **2010**, *21* (4), 161–166.
31. Weiner, M.L.; Nemec, M.; Sheets, L.; Sargent , D.; Breckenridge, C. Comparative functional observational battery study of twelve commercial pyrethroid insecticides in male rats following acute oral exposure. NeuroToxicology **2009**, *30S*, S1–S16.
32. Cranor, C. The legal failure to prevent subclinical developmental toxicity. Basic Clin. Pharmacol. Toxicol. **2008**, *102*, 267–273.
33. Raffaele, K.C.; Rowland, J.; May, B.; Makris, S.L.; Schumacher, K.; Scarano, L.J. The use of developmental neurotoxicity data in pesticide risk assessments. Neurotoxicol. Teratol. **2010**, *32*, 563–572.
34. Louis, G.B.; Damstra, T.; Díaz-Barriga, F.; Faustman, E.; Hass, U.; Kavlock, R.; Kimmel, C.; Kimmel, G.; Krishnan, K.; Luderer, U.; Sheldon, L. Principles for evaluating health risks in children associated with exposure to chemicals. Environ. Health Criteria **2006**, *237*, 1–327.
35. Bal-Price, A.K.; Hogberg, H.T.; Buzanska, L.; Lenas, P.; van Vliet, E.; Hartung T. In vitro developmental neurotoxicity (DNT) testing: Relevant models and endpoints NeuroToxicology **2010**, *31*, 545–554.
36. Hotchkiss, A.K.; Rider, C.V.; Blystone, C.R.; Wilson, V.S.; Hartig, P.C.; Ankley, G.T.; Foster, P.M.; Gray, C.L.; Gray, L.E. Fifteen years after "Wingspread"—Environmental endocrine disrupters and human and wildlife health: Where we are today and where we need to go. Toxicol. Sci. **2008**, *105* (2), 235–259.
37. Rauschenberger, R.H.; Wiebe, J.J.; Buckland, J.E.; Smith, J.T.; Sepulveda, M.S.; Gross, T.S. Achieving environmentally relevant organochlorine pesticide concentrations in eggs through maternal exposure in *Alligator mississippiensis*. Mar. Environ. Res. **2004**, *58*, 851–856.
38. Ankley, G.T.; Brooks, B.W.; Huggett, D.B.; Sumpter, J P. Repeating history: Pharmaceuticals in the environment. Environ. Sci. Technol. **2007**, *41*, 8211–8217.
39. Kloas, W.; Lutz, I. Amphibians as model to study endocrine disrupters. J. Chromatogr. A **2006**, *1130*, 16–27.
40. LeBlanc, G.A. Crustacean endocrine toxicology: A review. Ecotoxicology **2007**, *16*, 61–81.
41. Falfushinska, H.I.; Romanchuk, L.D.; Stolyar O.B. Different responses of biochemical markers in frogs (*Rana ridibunda*) from urban and rural wetlands to the effect of carbamate fungicide. Comp. Biochem. Physiol. C **2008**, *148*, 223–229.
42. Jia, Z.; Misra H.P. Developmental exposure to pesticides zineb and/or endosulfan renders the nigrostriatal dopamine system more susceptible to these environmental chemicals later in life. NeuroToxicology **2007**, *28*, 727–735.
43. Rider, C.V.; Furr, J.R.; Wilson, V.S.; Gray, Jr., L.E. Cumulative effects of in utero administration of mixtures of reproductive toxicants that disrupt common target tissues via diverse mechanisms of toxicity. Int. J. Androl. **2010**, *33*, 443–462.
44. Laetz, C.A.; Baldwin, D.H.; Collier, T.K.; Hebert, V.; Stark, J.D.; Scholz, N.L. The synergistic toxicity of pesticide mixtures: Implications for risk assessment and the conservation of endangered pacific salmon. Environ. Health Perspect. **2009**, *117* (3), 348–353.

Antagonistic Plants

Philip Oduor-Owino
Department of Botany, University of Kenyatta, Nairobi, Kenya

Abstract
Control of nematodes has been mainly through the use of chemicals and host resistance. However, the existence of physiological races in the pathogen's population has complicated efforts to breed for resistant cultivars. Interest in developing alternative control measures that are safe and economically attractive has now intensified worldwide. The use of antagonistic plants is viewed as a viable nematode management option.

INTRODUCTION

Plant parasitic nematodes cause significant crop losses in Africa and other parts of the world. Infected plants suffer from water deficiency and low yields, and have necrotic and/or galled roots. Control of nematodes has been mainly through the use of chemicals and host resistance. However, the existence of physiological races in the pathogen's population has complicated efforts to breed for resistant cultivars. Chemical control is effective but difficult to sustain for long-term benefits. The high cost of nematicides and their toxic effects also make them less attractive. Some nematicides such as Nemagon and Fumazon have now been banned from the world market and this has placed severe constraints on strategies for nematode control. Interest in developing alternative control measures that are safe and economically attractive has now intensified worldwide. The use of antagonistic plants is viewed as a viable nematode management option.[1–3]

NATURE OF ANTAGONISTIC PLANTS

Antagonistic plants are defined as plants that produce chemicals in their roots that are toxic and/or repellant to phytonematodes in the soil ecosytem.[4] These plants include *Tagetes erecta* L; *Tagetes patula* L; *Datura stramonium* L; *Ricinus communis* L; and *Asparagus officinalis*. Fresh roots of asparagus produce asparaguric acid glycoside that is toxic, even when diluted, to most plant parasitic nematodes. Root exudates from *Tagetes*, *Datura*, and *Ricinus* spp. induces premature hatching of nematode eggs, blocks the processes of mitosis and meoisis, and reduces galling intensity on roots of susceptible plants. This has been attributed, in part, to the toxic effects of the alkaloids terthienyl, hyosine, and ricinine present in *Tagetes*, *Datura*, and *Ricinis* spp., respectively. These compounds may also disrupt female taxis to roots or male taxis to female.[4] Other plants with antagonistic properties include some crucifers and citrus. Root diffusates from crucifers reduce the pathogenicity of nematodes on potato, while a compound in citrus roots is toxic to *Tylechulus semipenetrans*.[1]

Antagonistic Plants in Cultural Pest Control

Antagonistic plants may have a great nematode-control potential in agriculture if properly utilized in crop rotation and intercropping systems.[2] For example, intercropping food crops with nematicidal plants is now a nematode management strategy in Tanzania, India, and Zimbabwe and has also been recommended for Pakistan.[5] Field trials with *T. minuta*, *D. stramonium*, and *R. communis* are promising.[2,5,6] These plants reduce galling intensity and enhance tomato performance significantly. In India, a rotation of *D. stramonium*, maize, tomato, and pepper reduced the population of root-knot nematodes by 30% but the level of nutrient depletion by the antagonistic plant was 15%.[4] Integration of these plants with the biological agent, *Paecilomyces lilacinus* Thom (Sam), gave better results in Kenya.[3,7,8] Tomato plants grown in soils planted with the various antagonistic plants in combination with *P. lilacinus* develop significantly heavy shoots and roots and relatively fever root galls compared to controls (Table 1). Cases where antagonistic plants are used in crop rotation or intercropping systems are now increasing.[4] For instance, in Indonesia, *Tagetes* sp. *Crotalaria usaramoensis*, corn, and sweet potato (*Ipomea batatas*) are used to reduce *Meloidogyne* spp. density in the soil. For cereal-based cropping systems, the following crop sequences for root-knot nematode control are recommended in the Philippines: rice–mung bean (*Phaseolus aureus*)–corn–cabbage–rice, rice–tobacco (*Nicotiana tobacum*)–rice and rice–tobacco and *Tagetes* spp. There is also considerably less galling by *Meloidogyne* spp. on potato (*Solanum tuberosum*) roots intercropped with onion (*Allium* sp.), corn, and marigold compared with galling found on potatoes alone. Although antagonistic plants are gaining popularity as pest management tools, their benefits and risks must be understood

Encyclopedia of Environmental Management DOI: 10.1081/E-EEM-100000639

Table 1 Effect of tomato intercropping with *Datura stramonium*, *Ricinus communis*, and *Tagetes minuta*, and soil treatment with Captafol and Aldicarb, on gall index, number of juveniles, tomato growth and fungal parasitism of *Meloidogyne javanica* eggs by the fungus *Paecilomyces lilacinus*, 50 days after inoculation.

Soil treatment[a]	Egg parasitism (%)	No. of juveniles/ 300 cm^3 soil	Shoot dry weight (g)	Shoot height (cm)	Gall[b] index (0–4)
Ne "only" untreated	1.0de	670a	1.5f	26.6f	4.0a
Soil "only" untreated	0.0e	0d	3.5b	43.lab	0.0e
F + Ne	23.2c	660a	1.6f	30.le	3.0ab
F + Cap + Ne	1.3de	635a	1.9f	31.2e	3.9a
F + Ald + Ne	27.6b	12de	4.6a	45.3a	1.4d
F + Tag + Ne	29.8ab	161c	2.4e	38.4b	2.1c
F + Dat + Ne	28.3ab	173c	2.5de	36.4cd	2.2c
F + Ric + Ne	30.9a	210c	2.9cd	35.4cde	2.4c
Tag + Ne	2.1de	209c	2.8d	33.4cde	2.9b
Dat + Ne	3.0d	183c	3.0cd	36.6c	2.8b
Ric + Ne	2.8de	204c	3.2bc	37.7bc	3.0ab
Ald + Ne	1.2de	14d	4.5a	46.1a	3.8a
Cap + Ne	0.0e	460b	1.9f	36.1d	4.0a

Note: Numbers are means of 10 replications. Means followed by different letters within a column are significantly different ($P = 0.05$) according to Duncan's Multiple Range Test.
[a]Ne = nematode; F = P. *lilacinus*; Cap = Captafol: Ald = Aldicarb: Tag = *T. minuta;* Ric = *R. communis*.
[b]Gall index was based on a 0–4 rating scale, where 0 = no galls and 4 = 76–100% of the root system galled.
Source: Oduor-Owino.[7]

thoroughly before one can exploit their potential in pest control.

Benefits and Risks of Antagonistic Plants

Benefits and risks associated with the utilization of antagonistic plants in agriculture are varied. Phytonematoxic plants such as *R. communis*, *D. stramonium*, *Tagetes* spp. *Crotalaria* spp. *A. najus*, and *Datura metel* L. are traditionally gaining popularity due to their medicinal significance.[9] The flowers of *D. metel* are used against asthma, while *Crotalaria* spp. is a nitrogen-fixing legume. Castor oil from *R. communis* is used for making soaps and waxes: rinolecic acid from castor seeds is a valuable laxative.[9] Despite these attributes, antagonistic plants may pose a serious threat to food production if not well utilized. They may compete with economically important crops for space and nutrients. In addition they are slow in action, an attribute that makes them less attractive for use in a commercial setting. Because of this scenario, it is important that scientific disciplines work together in order to develop a viable pest control system. What is good for the nematologist may not be good for either the agronomist or economist.

Table 2 Effect of soil treatment with Aldicarb, *Tagetes minuta*, *Datura metel*, and *Datura stramonium* on root-knot nematodes in tomatoes (greenhouse test).

Soil treatment[a]	Shoot height (cm)	Shoot dry weight (g)	Gall index (0–5)	Galls ($no.g^{-1}$) root weight	Nematodes, no. $(300\ mL)^{-1}$ soil
Soil + Ne[b], untreated (control)	12.4c[c]	1.3e	4.4a	510.0a	564.1a
Soil only, untreated	97.8a	4.9b	0.0c	0.0d	0.0e
Soil + Ne + Aldicarb	116.3a	6.1a	1.0c	23.3d	18.4d
Soil + Ne + *D. metel*	73.6d	3.7c	2.1b	77.4c	176.3c
Soil + Ne + *T. minuta*	65.4b	2.6d	2.0b	134.9b	310.4b
Soil + Ne + *D. stramonium*	73.4b	3.1c	2.2b	88.4c	170.4c
Soil + Ne + *R. communis*	69.0b	3.3c	2.4b	90.0c	173.0c

[a]Autoclaved soil used
[b]Ne, nematode eggs added to soil
[c]Means followed by the same letter within each column are not significantly different at the 5% level (Duncan's Multiple Range Test).
Source: Oduor-Owino, P. Effects of Aldicarb and Selected Medicinal Plants of Kenya on Tomato Growth and Root-Knot Severity, unpublished data, 1992.

Table 3 Effect of soil treatment with Aldicarb, *Tagetes minuta*, *Datura metel*, and *Datura stramonium* on infection of tomato by root-knot nematodes (field test).

Treatment	Shoot height (cm)	Shoot dry weight (g)	Fruit yield (g)	Galls (no. g^{-1} root weight)	Nematodes, no. $(300\ mL)^{-1}$ soil
Untreated (control)	80.3d[a]	40.5d	380.3e	69.1a	150.4a
Aldicarb	187.3a	135.1a	3800.4a	4.50d	6.4d
D. metel	157.1b	89.3b	2590.1b	6.4c	17.3bc
T. minuta	107.1c	45.1d	761.1c	11.4b	21.1b
D. stramonium	150.1b	69.4c	2030.4b	9.6b	18.4c

[a]Means followed by the same letter within each column are not significantly different at the 5% level (Duncan's Multiple Range Test).
Source: Oduor-Owino.[2]

FUTURE CONCERNS

There is increasing internal awareness of the value of natural plants and their products in the development of new drugs and formulation of materials that can be used for pest control. Since some of the antagonistic plants can also be used to treat human ailments,[9] they may attract intensive scientific evaluation, recognition, and funding. However, more work should be done to reexamine the future of antagonistic plants in nematode control and in the drug industry. Efficacy of these plants against nematodes and their utilization in the pharmaceutical industry will depend highly on the concentrations of the active ingredients in their tissues.[10] It will also depend on whether they can stimulate activity of most biocontrol fungi and plant growth consistently. They have so far enhanced tomato growth in the greenhouse and in the field significantly[2,7] (Tables 2 and 3), but more trials are needed in order to understand the relationship between antagonistic plants, natural enemies, and crop performance.

See also *Biological Control of Nematodes*, pages 61–63; *Risks of Biological Control*, pages 720–722; *Toxins in Plants*, pages 840–842; *Pest–Host Plant Relationships*, pages 593–594; *Intercropping for Pest Management*, pages 423–425.

REFERENCES

1. Caswell, E.P.; Tan, C.S.; De Frank, J.; Apt, W.J. The influence of root exudates of *Chloris gayana* and *Tagetes patula on Rotylenchulus reniformis*. Revue de Nematologie **1991**, *14* (2), 581–587.
2. Oduor-Owino, P. Effects of Aldicarb, *Datura stramonium, Datura metel* and *Tagetes minuta* on the pathogenicity of root-knot nematodes in Kenya crop protection. Organic Soil Amendment **1993**, *12* (4), 315–317.
3. Oduor-Owino, P.; Sikora, R.A.; Waudo, S.W.; Schuster, R.P. Effects of aldicarb and mixed cropping with *Datura stramonium, Ricinus communis* and *Tagetes minuta* on the biological control and integrated management of *Meloidogyne javanica*. Nematologica **1996**, *42*, 127–130.
4. Yeates, G.W. How plants affect nematodes. Advances in Ecological Research **1987**, *17* (2), 61–137.
5. Oduor-Owino, P.; Waudo, S.W. Effects of antagonistic plants and chicken manure on the biological control and fungal parasitism of root-knot nematode eggs in naturally infested field soil. Pakistan Journal of Nematology **1995**, *13* (2), 109–117.
6. Oduor-Owino, P.; Waudo, S.W. Comparative efficacy of nematicides and nematicidal plants on root-knot nematodes. Trop. Agric. **1994**, *71* (4), 272–274.
7. Oduor-Owino, P. *Fungal Parasitism of Root-knot Nematode Eggs and Effects of Organic Matter, Selected Agrochemicals and Intercropping on the Biological Control of Meloidogyne javanica on Tomato*; Ph.D. Thesis, Kenyatta University: Nairobi, Kenya, 1996; 132.
8. Oduor-Owino, P.; Sikora, R.A.; Waudo, S.W.; Schuster, R.P. Tomato growth and fungal parasitism of *Meloidogyne javanica* eggs as affected by nematicides, time of harvest and intercropping. East African Agricultural and Forestry Journal **1995**, *61* (1), 23–30.
9. Oduor-Owino, P.; Waudo, S.W. Medicinal plants of Kenya: effects on *Meloigogyne incognita* and the growth of okra. Afro-Asian Journal of Nematology **1992**, *2* (1), 64–66.
10. Oduor-Owino, P. Effects of marigold leaf extract and captafol on fungal parasitism of root-knot nematode eggskenyan isolates. Nematologia Mediteranea **1992**, *20*, 211–213.
11. Oduor-Owino, P.; Waudo, S.W.; Sikora, R.A. Biological control of *Meloidogyne javanica* in Kenya: effect of plant residues, benomyl, and decomposition products of mustard (*Brassica campestris*). Nematologica **1993**, *39*, 127–134.

Aquatic Communities: Pesticide Impacts

David P. Kreutzweiser
Canadian Forest Service, Natural Resources Canada, Sault Sainte Marie, Ontario, Canada
Paul K. Sibley
School of Environmental Sciences, University of Guelph, Guelph, Ontario, Canada

Abstract
Freshwater aquatic communities are usually contained within distinct boundaries or systems, and this generates a high degree of connectivity among species, thereby increasing their susceptibility to pesticide-induced disturbances at the community level. Through direct and (or) indirect effects, pesticides can disrupt interactions and linkages among species and impair their ecological function, causing large changes in community stability or balance. The risk of harm posed by pesticides to aquatic communities will depend on the exposure concentration, bioavailability, exposure duration, rate of uptake, inherent species sensitivities, community composition, and other community attributes. Recent advances in pesticide risk assessment for aquatic communities have improved the ecological relevance and predictive capabilities for determining, and thus mitigating, potential harmful impacts. These are explored with some examples of pesticide impacts on aquatic communities, suggestions are provided to minimize the potential for adverse effects of pesticides in aquatic ecosystems, and outstanding issues are identified as a basis for future research.

INTRODUCTION

A biotic community can be defined as an assemblage of plant or animal species utilizing common resources and cohabiting a specific area. Examples could include a fish community of a stream, an insect community of a forest pond, or a phytoplankton community of a lake. Interactions among species provide ecological linkages that connect food webs and energy pathways, and these interconnections provide a degree of stability, or balance, to the community. Community balance can be described as a state of dynamic equilibrium in which species and their population dynamics within a community remain relatively stable, subject to changes through natural adjustment processes. Toxic effects of pesticides can disrupt these processes and linkages and thereby cause community balance upsets. For example, this can occur when a pesticide has a direct impact on a certain species in a community and reduces its abundance while other unaffected species increase in abundance in response to the reduced competition for food resources or increased habitat availability. Some of the best examples of pesticide impacts on biological communities are found in freshwater studies. Freshwater aquatic communities are usually contained within distinct boundaries or systems, and this generates a high degree of connectivity among species, thereby increasing their susceptibility to pesticide-induced disturbances at the community level.

We examine traditional and developing methods for measuring pesticide impacts on freshwater communities, with emphasis on recent improvements in risk assessment approaches and analyses, and provide some examples for illustration. We then describe some advances in impact mitigation strategies and discuss some ongoing issues pertaining to understanding, assessing, and preventing pesticide impacts including probabilistic risk assessment (PRA), population and ecological modeling, and pesticide interactions with multiple stressors. The integration of improved risk assessment and mitigation approaches and technologies together with information generated from the numerous impact studies available will provide a sound scientific basis for decisions around the use and regulation of pesticides in and near water bodies.

MEASURING IMPACTS ON AQUATIC COMMUNITIES

Changes in aquatic communities can be measured directly in water bodies by a number of quantitative and qualitative sampling methods. Descriptions of those methods can be found in any up-to-date text or handbook (e.g., Hauer and Lamberti[1]). Measurements can be in terms of community structure (species composition) or community function (a measurable ecosystem process attributable to a biotic community that causes a change in condition) and can include both direct and indirect effects.[2,3] Community structure is a measure of biodiversity in its most general sense, that is, the number of species or other taxonomic units and their relative abundances. Some community functions are referred to as environmental or ecosystem services. Examples include organic matter breakdown and nutrient cycling that is largely mediated by microbial communities, or water

Encyclopedia of Environmental Management DOI: 10.1081/E-EEM-120046189

uptake, filtration, and flood control mediated by shoreline plant communities.[4] Both community structure (biodiversity) and function (ecosystem services) are being increasingly valued by society and global economies,[5,6] and therefore sustaining healthy aquatic communities will be an important driver of pesticide impact mitigation efforts.

Detecting impacts of pesticides typically involves repeated sampling and a comparison of community attributes among contaminated and uncontaminated test units over time, or across a gradient of pesticide concentrations. The test units can range from petri dishes to natural ecosystems, with a trade-off between experimental control in small test units and environmental realism in field-level testing and whole ecosystems.[7] In an effort to incorporate both experimental control and environmental realism in pesticide impact testing, the use of microcosms or model ecosystems for measuring impacts on aquatic communities has increased over the past couple of decades.[8,9] Model ecosystems for community-level pesticide testing can be quite simple at lower-trophic levels such as with microbial communities (e.g., Widenfalk et al.[10]) but will necessarily be more complex for testing higher-order biological communities (e.g., Wojtaszek[11]). Regardless of the test units, an important consideration for measuring pesticide impacts will be an assessment of the duration of impact or rate of recovery. A rapid return to pre-pesticide or reference (no-pesticide) community condition will reduce the long-term ecological consequences of the pesticide disturbance.[12]

Traditional measures of community-level impacts have focused on structure and have usually been expressed in terms of single-variable indices such as species richness, diversity, or abundance. These indices are useful descriptors of community structure but suffer from the fact that they reduce complex community data to a single summary metric and may miss subtle or ecologically important changes in species composition across sites or times. Over the last couple of decades, ecotoxicologists have increasingly turned to multivariate statistical techniques for analyzing community response data.[13] A variety of multivariate statistical techniques and software are available and are usually considered superior for the analysis of community data because they retain and incorporate the spatial and temporal multidimensional nature of biological communities.[14] This includes various ordination techniques that can provide graphical representation of spatiotemporal patterns in community structure in which points that lie close together in the ordination plot represent communities of similar composition (richness, abundance), while communities with dissimilar species composition are plotted further apart.

Fig. 1 illustrates the use of an ordination plot generated by nonmetric multidimensional scaling for detecting differences among aquatic insect communities in four control and eight insecticide-treated streams. These data have been adjusted for illustrative purposes but are based on real invertebrate community responses to an insecticide in outdoor stream channels.[15] At both concentrations of the insecticide, the community structure of stream insects clearly shifted away from the natural community composition in control streams as depicted by the separation of treated streams (T1 and T2) from controls (C) in the ordination bi-plot. The plot also illustrates that the variability among treated streams (relative distance between points) was greater than that among control streams, that the low-concentration streams (T1) and high-concentration streams (T2) tended to separate along axis 1, and that the T2 streams were further removed from controls than the T1 streams, indicating a differential response by the insect communities to the two test concentrations. Canonical correspondence analysis and redundancy analysis have also been commonly used to assess aquatic community responses to pesticide contamination.[16,17] A useful refinement of an ordination technique for detecting and interpreting pesticide impacts on aquatic communities is principal response curves (PRCs).[18] PRC is derived from redundancy analysis, and time-dependent responses in the treatments are expressed as deviations from the control or reference system allowing for clear visualization of pesticide effects.

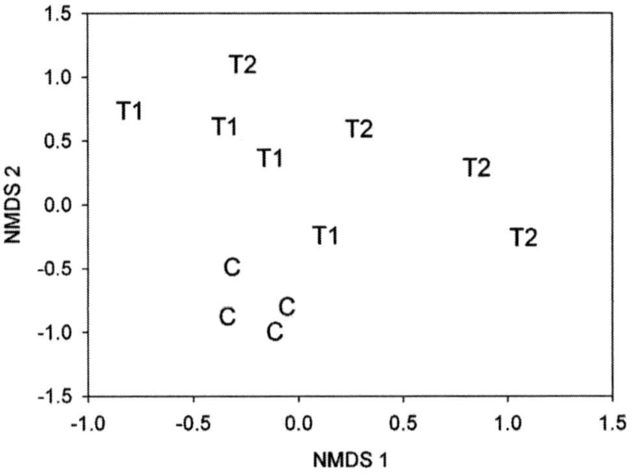
1.5
1.0
0.5
0.0
-0.5
-1.0
-1.5
NMDS 2
T2
T1
T1
T2
T1
T2
T1
T2
C
C
C
C
-1.0
-0.5
0.0
0.5
1.0
1.5
NMDS 1

Fig. 1 Ordination by nonmetric multidimensional scaling of aquatic insect communities in stream channels. Each point represents the community structure of control channels (C) and channels treated with a neem-based insecticide at a low (T1) and high (T2) concentration.
Source: Adapted from Kreutzweiser et al.[15]

ASSESSING RISK OF PESTICIDE IMPACTS ON AQUATIC COMMUNITIES

The likelihood or risk of harmful effects on aquatic communities from exposure to pesticides will depend on the exposure concentration, bioavailability, exposure duration, rate of uptake, inherent species sensitivities, community composition, and other community attributes. All of these must be measured, estimated, modeled, or predicted to derive an assessment of risk to aquatic communities for any given pesticide. Formalized risk assessment frameworks and guidelines for pesticides have been developed in the United States,[19] the European Union,[20] Canada,[21] and elsewhere and can be consulted for detailed information on the various components of a risk assessment. In brief, pesticide risk assessments typically include the following phases: 1) defining the problem by determining the pesticide use patterns and developing conceptual models and hypotheses around how it is expected to behave, the anticipated exposure regimes, the kinds of organisms that are likely to be at risk, the community or entity that is to be protected, and the level of protection that will be acceptable; 2) developing the measurement endpoints for assessing risk of harm by establishing which response measurements are relevant and applicable, and how the measurements will be made; 3) outlining the risk assessment process by specifying the kinds of data to be used and how they will be derived including simulation modeling, empirical laboratory, microcosm or field testing, their appropriate spatial and temporal scales, and their statistical analyses; 4) applying the risk assessment by running models or collecting data, completing analyses, summarizing outputs, and providing risk estimates; 5) conducting risk communication and management by answering questions posed in the problem formulation, suggesting risk mitigation strategies if necessary, and communicating those to appropriate users; and 6) conducting follow-up monitoring to evaluate the success of mitigation strategies and to implement adaptive management to address deficiencies if or when necessary.[22,23]

Traditionally, pesticide risk assessments have relied on standardized, single-species toxicity tests to predict effects on communities, the underlying assumption being that protecting the most sensitive species will protect whole communities. In this case, the selection and relevance of test species are critically important to a successful and meaningful risk assessment.[24] However, the accuracy and relevance of estimating the potential risk to aquatic communities can be greatly improved by consideration of specific species or community attributes. In particular, attribute information can improve the ecological relevance and predictive capabilities of conceptual models and the generation of hypotheses in the risk assessment process. Insofar as these attributes affect exposure, sensitivity, or both, they can increase or decrease risk beyond what could be determined from toxicity estimates or species sensitivity distributions alone.

Behavioral attributes can elevate the risk of pesticide effects on species by increasing the likelihood of intercepting the stressor. For example, young-of-the-year bluefish (*Pomatomus saltatrix*) typically feed in estuaries during their early life stages where agricultural runoff can elevate concentrations of pesticides in food items. This feeding behavior can result in bioaccumulation and in adverse effects such as reduced migration, overwinter survival, and recruitment success in fish communities.[25] Incorporating this kind of information into conceptual models and risk hypotheses will generate more realistic risk assessments. In addition, behavioral attributes themselves can be relevant measurement endpoints if the pesticide mode of action indicates risk of sublethal behavioral effects at expected concentrations. For example, some pesticides have been shown to impair the ability to capture prey in fish[26] and the ability to avoid predators in zooplankton.[27] These types of adverse effects can disrupt trophic linkages and reduce survival or reproduction, thus impacting community balance.

Inclusion of life history information into conceptual models and risk hypotheses can also refine and improve the risk assessment process. Life history strategies can influence a species susceptibility to a stressor through effects on a population's resilience or ability to recover from disturbance.[28] Different species exposed to the same pesticide and experiencing similar levels of effect in terms of population declines do not necessarily recover at the same rates when recovery is dependent on reproduction or dispersion. Populations of organisms with short regeneration times (e.g., several generations per year) and/or high dispersal capacity have higher likelihood of recovery from pesticide-induced population declines than those with longer regeneration periods and limited dispersal capacity. These differential life history strategies and their influences on community response and recovery from pesticide effects have been demonstrated empirically (e.g., van den Brink et al.[29] and Kreutzweiser et al.[30]) and through population modeling.[31] These community balance upsets could not have been predicted from screening-level toxicity data or from species sensitivity data; thus, inclusion of life history information in conceptual models can improve risk hypotheses and direct the assessment to focus on species at higher risk owing to specific life history strategies.

Life history attributes can also influence the risk of pesticide effects through differential life-stage sensitivity or susceptibility. Early life stages are often (but not exclusively) more sensitive to pesticides than later stages. An organism's life stage can also influence its susceptibility to a pesticide by increasing or decreasing the likelihood of intercepting the stressor. If a contaminant is present in the environment at effective concentrations during a period in which the particular life stage of a species is present, then the risk to that species is increased. For some amphibians, aquatic (larval) stages could be at higher risk of direct and indirect effects of pesticides than their terrestrial (adult) life stages when their larval stage coincides with pesticide contamination of water bodies.[32] Thus, while a species sensitivity and geographical distribution may indicate potential risk, the life-stage information coupled with pesticide use pattern, timing, or fate information may indicate little likelihood of exposure to the pesticide and the risk assessment can be adjusted accordingly.

Functional attributes may also be important for refining or improving pesticide risk assessments. Protection goals for populations and communities often include the safeguarding of critical biological processes or ecosystem function. Measuring ecosystem function integrates responses of component populations and can be a relevant measurement endpoint when species loss affects ecosystem function such as energy transfer and organic matter cycling.[33] However, most ecosystems are complex and it may not be clear which functional attributes are critical for sustaining ecological processes or the extent to which they can sustain changes in structural properties (e.g., population levels, diversity) without adversely affecting ecosystem function. Neither is it clear if functional endpoints are more or less sensitive than structural endpoints for detecting ecosystem disturbance. Some studies investigating the relationship between species diversity and ecosystem function have indicated that ecosystems can tolerate some species loss because of functional redundancy.[34] Functional redundancy is thought to occur when several species perform similar functions in ecosystems such that some may be eliminated with little or no effect on ecosystem processes. Others have suggested that redundant species are required to ensure ecosystem resilience to disturbance as a form of biological insurance, especially at large spatial scales.[35]

Given these discrepancies, measurement endpoints based on functional attributes are not typically used in pesticide risk assessments because it is generally accepted that protection of community structure will protect ecosystem function. However, when specific functional attributes can be identified and are known or suspected to be at risk from a pesticide, they can be included in the data requirements for a risk assessment. An example would be the risk of adverse effects on leaf litter decomposition (a critical ecosystem function in forest soils and water bodies) posed by a systemic insecticide for control of wood-boring insects in trees.[36] In that case, the protection goal was maintaining leaf litter decomposition, the community at risk was decomposer invertebrates feeding on leaves from insecticide-treated trees, and the selection of test species was directed to a specific functional group because of the unique route of exposure to decomposer organisms identified in the risk hypotheses.

SOME EXAMPLES OF PESTICIDE IMPACTS ON AQUATIC COMMUNITIES

A few examples will serve to illustrate how pesticides can cause disruptions to aquatic communities. DeNoyelles et al.[37] reviewed studies into pesticide impacts on aquatic communities and reported that herbicides like atrazine, hexazinone, and copper sulfate were directly toxic to most species of phytoplankton (waterborne algae). After herbicide applications, reductions in phytoplankton caused secondary reductions in herbivorous zooplankton, resulting from a depleted food source for the zooplankton. They further showed that direct adverse effects on phytoplankton can also cause disruptions to the bacterial-based energy pathways by reducing carbon flow from phytoplankton to bacteria, and ultimately to grazing protozoans and zooplankton. Boyle et al.[38] found that applications of the insecticide diflubenzuron to small ponds reduced populations of several aquatic invertebrate species. This in turn resulted in indirect effects on algae (increased productivity because of release from grazing pressure by the invertebrates) and on juvenile fish populations (reduced production because of limited invertebrate prey availability). George et al.[39] used a novel approach to predict effects of pesticide mixtures on zooplankton communities and then tested the predictions in outdoor microcosms. Responses among zooplankton populations within the community differed, depending on the pesticide mixture, and those differences appeared to reflect the relative susceptibilities among specific taxa within groups. Cladocerans declined but were less sensitive than copepods to a chlorpyrifos-dominated mixture, while rotifers actually increased after application in response to release from competition or predation pressures.

Kreutzweiser et al.[40] applied a neem-based insecticide to forest pond enclosures and measured effects on zooplankton community structure, respiration, and food web stability. Significant concentration-dependent reductions in numbers of adult copepods were observed, but immature copepods and cladocerans were unaffected (Fig. 2). There was no evidence of recovery of adult copepods within the sampling season. During the period of maximal impact (about 4 to 9 weeks after the applications), total plankton community respiration was significantly reduced, and this contributed to significant concentration-dependent increases in dissolved oxygen and decreases in specific conductance. The reductions in adult copepods resulted in negative effects on zooplankton food web stability through elimination of a trophic link and reduced interactions and connectance.

Van Wijngaarden et al.[41] evaluated the responses of aquatic communities in indoor microcosms to a suite of pesticides used for bulb crop protection. At pesticide concentrations equivalent to 5% spray drift deposition, zooplankton taxa within communities showed significant changes relative to non-treated controls, reflecting taxon-specific sensitivities. Some copepods and rotifers in particular showed significant declines for at least 13 weeks, while many other rotifers and cladocerans were unaffected or increased. Several macroinvertebrate taxa were negatively affected, and this contributed to significant declines in leaf litter decomposition among treated microcosms. The herbicide asulam was among the suite of pesticides, and it induced significant reduction of the macrophyte *Elodea nuttallii*. This in turn caused significant changes in water chemistry (decreases in dissolved oxygen and pH, increases in alkalinity and specific conductance) and increases in phytoplankton biomass from decreased competition for nutrients. Increased phytoplankton and reduced zooplankton predators combined to support higher abun-

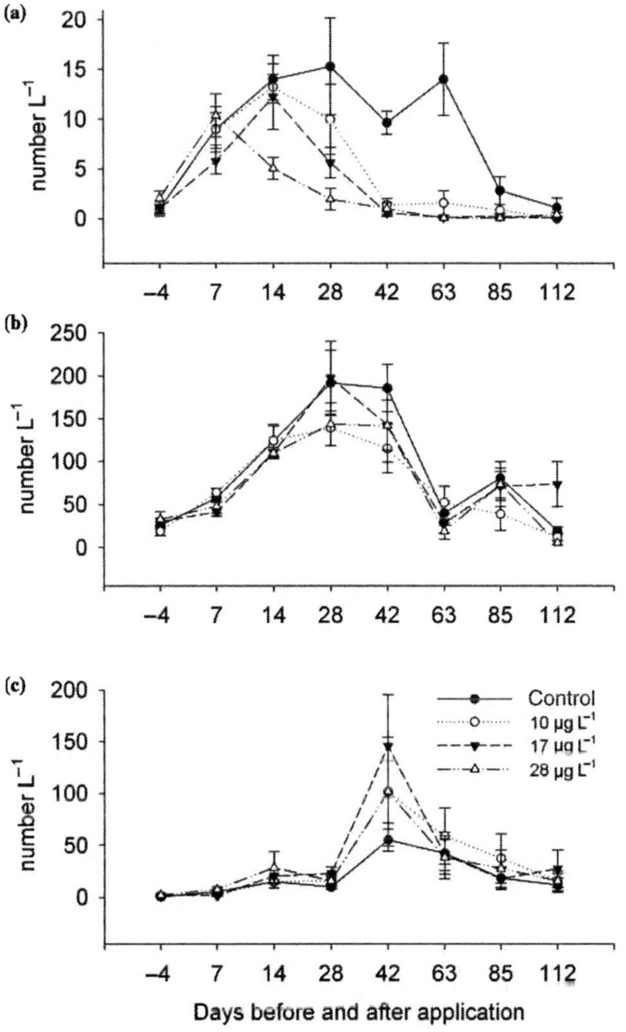
(a)
number L⁻¹
(b)
number L⁻¹
(c)
number L⁻¹
Control
10 µg L⁻¹
17 µg L⁻¹
28 µg L⁻¹
Days before and after application

Fig. 2 Mean abundance (±1 SE, $n = 5$) of (a) adult copepods, (b) immature copepods, and (c) cladocerans in natural pond microcosms (controls) and microcosms treated at three different rates of a neem-based insecticide.
Source: Taken from Kreutzweiser et al.[40]

dance of less sensitive zooplankton taxa. The authors point out that most of these effects were not measurable at more realistic rates of spray drift deposition.

Relyea and Hoverman[42] investigated impacts of the insecticide malathion on aquatic communities in microcosms designed to mimic a simple aquatic food web that can be found in ponds and wetlands. The insecticide generally reduced zooplankton abundance, and these reductions stimulated increases in phytoplankton, decreases in periphyton (attached algae), and decreases in growth of frog tadpoles. While invertebrate predator survival was not affected, amphibian prey survival increased with insecticide concentration, apparently the result of insecticide-induced impairment of predation success by the invertebrates. Overall, the study demonstrated that realistic concentrations of an insecticide can interact with natural predators to induce large changes in aquatic community balance.

REDUCING RISK OF PESTICIDE IMPACTS ON AQUATIC COMMUNITIES

For pesticides applied to crops and forests, exposure to aquatic communities can be minimized by the implementation of vegetated spray buffers or setbacks to intercept off-target spray drift and runoff.[43] Pesticide runoff can be further reduced by using formulations that are less prone to wash-off, leaching, and mobilization. Recent advances in spray drift reduction and improved spray guidance systems can also significantly reduce the off-target movement of pesticides to water bodies.[44] Examples include new technologies in map-based automated boom systems for row crops[45] and Geographical Information System (GIS)-based landscape analysis for predicting off-target pesticide movement.[46]

The risk of adverse effects on aquatic communities may also be decreased by intentional selection and use of pesticides that are inherently safer to the environment. This would include so-called reduced-risk pesticides that are bioactive compounds usually with unique modes of action and derived from microbial, plant, or other natural sources. These are generally thought to be less persistent and toxic to non-target organisms than conventional synthetic pesticides.[47] Examples include the bacteria-derived insecticide *Bt* (*Bacillus thuringiensis*), the plant-derived insecticide neem, and the microbe-derived herbicide phosphinothricin. However, Thompson and Kreutzweiser[48] caution that it cannot be assumed that this group of pesticides is inherently safer or more environmentally acceptable than synthetic counterparts and that full environmental risk evaluations must be conducted to ensure their environmental safety.

These types of technologies combined with the use of non-pesticide approaches to pest management form the basis of integrated pest management (IPM) strategies. IPM strategies are those in which the judicious use of pesticides is only one of several concurrent methods to control or manage losses from pest damage. This can include the use of natural enemies and parasites, biological control agents, insect growth regulators, confusion pheromones, sterile male releases, synchronizing with weather patterns known to diminish pest populations, and cultivation methods and crop varieties to improve conditions for natural enemies or degrade conditions for pest survival.[49] Increasing the use of IPM approaches can reduce reliance on pesticides and thus reduce the risk of pesticide impacts overall.

RECENT ADVANCES AND OUTSTANDING ISSUES

Pesticide risk assessments and risk reductions have recently been advanced in terms of ecological realism and effectiveness through some developing methods and techniques. Traditional risk assessments have estimated hazards from pesticides by comparing the expected environmental concentration (often predicted from worst-case scenarios) to the toxic threshold for the most sensitive test species. When the expected concentration is higher than the toxicity threshold, the pesticide is considered to have potential for environmental effects. These so-called hazard or risk quotient approaches are still widely used in pesticide risk assessment and regulation, but more recently, PRA and probabilistic hazard assessment (PHA) approaches are being adopted. In these approaches, pesticide exposure levels and the likelihood of toxic effects are estimated from probability distributions based on all reliable data available.[50] In PRA, exposure and effects distributions are developed from modeling or measurements in laboratory, microcosm, or field studies and used to improve the accuracy and relevance of the estimated likelihood of environmental effects compared to the traditional worst-case (hazard/risk quotient) approach (e.g., Solomon[51]). In PHA, a distribution approach is also used, except that the probability of hazard is estimated from distributions built on the relative sensitivity of interspecies endpoints rather than species sensitivity itself.[52] Fig. 3 illustrates the principles of PRA (Fig. 3a) and PHA (Fig. 3b). Regardless of the approach, one important aspect of PRA that is ongoing is the development and use of uncertainty analysis to quantify variability and uncertainty in exposure and effects estimates. Characterizing and quantifying uncertainty will provide more meaningful risk assessments and improved decision making for minimizing potential risk of pesticide impacts in or near water.[53]

Efforts at incorporating population or ecological modeling into pesticide risk assessments have also improved their accuracy and relevance for predicting, and therefore mitigating, risk of harm to aquatic communities.[54] The use of ecological models to incorporate a suite of factors including lethal and sublethal effects and their influences on the risks to organisms, populations, or communities can provide useful insights into receptor/pesticide interactions and can thereby improve risk assessments and direct mitigation measures. Population models that account for differential demographics and population growth rates within communities have been shown to provide more accurate assessments of potential pesticide impacts on populations and communities than what conventional lethal concentration estimates can provide.[55] Ecological and population modeling combined with pesticide exposure modeling and case-based reasoning (drawing on past experience or information from similar chemical exposures) can provide further refinements and improve risk assessment for aquatic communities.[56] Another recent advancement in ecological modeling to predict pesticide effects is the use of trait-based information such as organism morphology, life history, physiology, and feeding ecology in risk assessments.[57] This approach includes some of the functional attributes and concepts described above in the section on “Assessing Risk of Pesticide Impacts on Aquatic Communities” and has the advantage of formally expressing com-

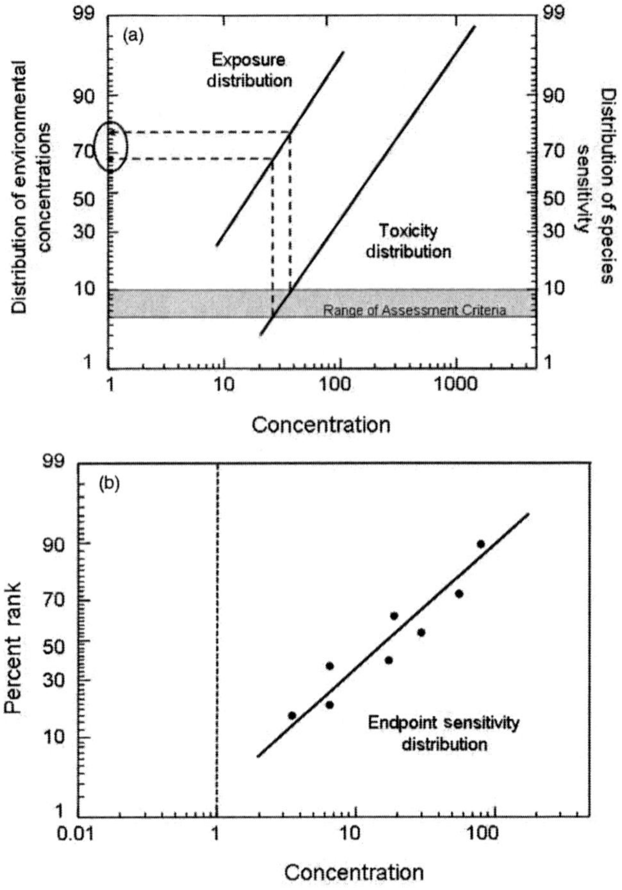
(a)
Exposure distribution
Toxicity distribution
Range of Assessment Criteria
Distribution of environmental concentrations
Distribution of species sensitivity
Concentration
(b)
Percent rank
Endpoint sensitivity distribution
Concentration

Fig. 3 Schematic illustrating the principle of PRA (a) and PHA (b). PRA is based on a comparison of exposure and effects distributions using a predetermined criterion typically in the range of 5%–10% (shaded area and dashed lines in panel A) to determine the probability of exceeding the criterion (ellipse on *y*-axis); PHA is based on a comparison of an endpoint-derived sensitivity distribution within a test species to a threshold value such as a hazard quotient (dashed line in panel B).

munities as combinations of functional traits rather than as groups of species, thereby yielding a more meaningful description of community structure and function. Taken together, these modeling approaches that incorporate probability distributions, toxicological sensitivities, population dynamics, ecological information, and functional trait attributes can be integrated into improved risk assessments that will inform mitigation and prevention strategies for pesticide use.[58]

Two additional issues that present challenges to pesticide risk assessment and mitigation are pesticide mixtures and the combined or cumulative effects of multiple stressors on pesticide impacts. Pesticides frequently occur as mixtures in aquatic systems, particularly in agricultural regions, and methods to assess and/or predict pesticide mixture toxicity under laboratory conditions have been relatively well developed. However, there are still large uncertainties associated with the prediction of pesticide mixture toxicity, and additional studies are needed to evaluate the performance of mixture models when evaluating community-level endpoints and toxicity thresholds over long-term exposures.[59] Secondly, whereas most pesticide assessment data are derived from tests or experiments under controlled or semicontrolled environmental conditions, pesticides in natural environments may interact with a number of other natural or human-caused stressors that can substantially alter the likelihood and magnitude of pesticide impacts.[60] Other stressors could include overarching effects of climate change that can influence water temperature and quality; land use activities that result in chemical, sediment, and nutrient pollution of waterways; and biotic interactions with invasive species in aquatic communities. A number of studies have examined the combined effects of a pesticide with other stressors, but they have usually been single stressor effects tested at the single-species level. Examples of studies that examined combined effects include pesticide interactions with water temperature,[61] pH,[62] dissolved organic matter,[63] UV radiation,[64] predators,[65] competitors,[66] food availability,[67] elevated sediments,[68] and other chemical stressors.[69] However, potential multiple stressors and their interactions with pesticides can be myriad and testing or extrapolating to community-level impacts is onerous at best. Sorting out and mitigating pesticide impacts from among these multiple stressors continues to be a challenge, and the suggestion by Laskowski et al.[70] to include studies of toxicant interactions with a range of environmental conditions in risk assessments seems warranted.

CONCLUSIONS

Because of the high degree of connectivity among species in an aquatic community, pesticides pose a risk of harm to the community stability or balance. The community structure can be altered by direct effects, indirect effects, or both, and this can cause disruptions to the interactions and linkages among species and to their ecological function. This risk of harm will depend on exposure concentration, bioavailability, exposure duration, rate of uptake, species sensitivities, community composition, and other community attributes. Recent advances in pesticide risk assessment for aquatic communities have improved the ecological relevance and predictive capabilities for determining, and thus mitigating, potential harmful impacts. Pesticide impacts on aquatic communities can be minimized by the use of improved application technologies to reduce application rates and to decrease off-target movement to water bodies. Potential impacts can be further minimized through the selection and use of pesticides that are demonstrated to be inherently safer to the environment and through the application of IPM strategies. Given the preponderance of pesticide impact studies in freshwater aquatic ecosystems, the improved risk assessment frameworks and regulatory requirements for pesticide evaluations, and the recent advances in mitigation technologies, many decisions around the use of pesticides can be made on a sound scientific basis rather than on misinformed perceptions or politically driven agendas. Integrated, science-based pest management strategies including the prudent use of appropriate pesticides will contribute to ensuring the sustainability of aquatic communities in areas subjected to pest management programs.

REFERENCES

1. Hauer, F.R.; Lamberti, G.A. *Methods in Stream Ecology*, 2nd Ed.; Elsevier: Amsterdam, 2006.
2. Fleeger, J.W.; Carman, K.R.; Nisbet, R.M. Indirect effects of contaminants in aquatic ecosystems. Sci. Tot. Environ. **2003**, *317*, 207–233.
3. Rohr, J.R.; Crumrine, P.W. Effects of an herbicide and an insecticide on pond community structure and processes. Ecol. Appl. **2005**, *15*, 1135–1147.
4. Daily, G.C. Introduction: What are ecosystem services? In *Nature's Services: Societal Dependence on Natural Ecosystems*; Daily, G.D., Ed.; Island Press: Washington, DC, 1997; 1–10.
5. Anderson, L.; Gomez, W.C.; McCarney, G.; Adamowicz, W.; Chalifour, N.; Weber, M.; Elgie, S.; Howlett, M. *Ecosystem Service Valuation, Market-based Instruments and Sustainable Forest Management: A Primer*; Sustainable Forest Management Network: Edmonton, Alberta, 2010.
6. Bayon, R.; Jenkins, M. 2010. The business of biodiversity. Nature **2010**, *466*, 184–185.
7. Sibley, P.K.; Chappel, M.J.; George, T.K.; Solomon, K.R., Liber, K. Integrating effects of stressors across levels of biological organization: Examples using organophosphorus insecticide mixtures in field-level exposures. J. Aquat. Ecosyst. Stress Recovery **2000**, *7*, 117–130.
8. Campbell, P.J.; Arnold, D.J.S.; Brock, T.C.M.; Grandy, N.J.; Heger, W.; Heimbach, F.; Maund, S.J.; Streloke, M. *Guidance Document on Higher-Tier Aquatic Risk Assessment for Pesticides*; SETAC Europe: Brussels, Belgium, 1999.
9. Kennedy, J.H.; LaPoint, T.W.; Balci, P.; Stanley, J.K.; Johnson, Z.B. Model aquatic ecosystems in ecotoxicological research: Considerations of design, implementation, and analysis. In *Handbook of Ecotoxicology*, 2nd Ed.; Hoffman, D.J.; Rattner, B.A.; Burton, G.A., Jr.; Cairns, J., Jr., Eds.; Lewis Publishers: Boca Raton, Florida, 2003; 45–74.
10. Widenfalk, A.; Svensson, J.M.; Goedkoop, W. Effects of the pesticides captan, deltamethrin, isoproturon, and pirimicarb on the microbial community of a freshwater sediment. Environ. Toxicol. Chem. **2004**, *23*, 1920–1927.
11. Wojtaszek, B.F.; Buscarini, T.M.; Chartrand, D.T.; Stephenson, G.R.; Thompson, D.G. Effect of Release® herbicide on morality, avoidance response, and growth of amphibian larvae in two forest wetlands. Environ. Toxicol. Chem. **2005**, *24*, 2533–2544.
12. Barnthouse, L.W. Quantifying population recovery rates for ecological risk assessment. Environ. Toxicol. Chem. **2004**, *23*, 500–508.
13. Maund, S.; Chapman, P.; Kedwards, T.; Tattersfield, L.; Matthiessen, P.; Warwick, R.; Smith, E. Application of

multivariate statistics to ecotoxicological field studies. Environ. Toxicol. Chem. **1999**, *18*, 111–112.
14. Clarke, K. R.; Warwick, R.M. *Change in Marine Communities: An Approach to Statistical Analysis and Interpretation*, 2nd Ed.; PRIMER-E: Plymouth, U.K., 2001.
15. Kreutzweiser, D.P.; Capell, S.S.; Scarr, T.A. Community-level responses by stream insects to neem products containing azadirachtin. Environ. Toxicol. Chem. **2000**, *19*, 855–861.
16. Frieberg, N.; Lindstrom, M.; Kronvang, B.; Larsen, S.E. Macroinvertebrate/sediment relationships along a pesticide gradient in Danish streams. Hydrobiologia **2003**, *494*, 103–110.
17. Berenzen, N.; Kimke, T.; Schulz, H.K.; Schulz, R. Macroinvertebrate community structure in agricultural streams: Impact of run-off-related pesticide contamination. Ecotoxicol. Environ. Saf. **2005**, *60*, 37–46.
18. van den Brink, P.J.; ter Braak, C.J.F. Principal response curves: Analysis of time-dependent multivariate responses of biological community to stress. Environ. Toxicol. Chem. **1999**, *18*, 138–148.
19. USEPA. *Guidelines for Ecological Risk Assessment*; United States Environmental Protection Agency, Risk Assessment Forum: Washington, DC, 1998.
20. EUFRAM. *Introducing Probabilistic Methods into the Ecological Risk Assessment of Pesticides, Version 6*; European Framework for Risk Assessment of Pesticides (EUFRAM): York, U.K., 2005.
21. Delorme, P.; François, D.; Hart, C.; Hodge, V.; Kaminski, G.; Kriz, C.; Mulye, H.; Sebastien, R.; Takacs, P.; Wandelmaier, F. *Final Report for the PMRA Workshop: Assessment Endpoints for Environmental Protection*; Environmental Assessment Division, Pest Management Regulatory Agency, Health Canada: Ottawa, Ontario, 2005.
22. Suter, G.W.; Barnthouse, L.W.; Bartell, S.M.; Mill, T.; Mackay, D.; Patterson, S. *Ecological Risk Assessment*; Lewis Publishers: Boca Raton, Florida, 1993.
23. Reinert, K.H.; Bartell, S.M.; Biddinger, G.R., Eds. *Ecological Risk Assessment Decision-Support System: A Conceptual Design*; SETAC Press: Pensacola, Florida, 1998.
24. Maltby, L.; Blake, N.; Brock, T.C.M.; van den Brink, P.J. Insecticide species sensitivity distributions: Importance of test species selection and relevance to aquatic ecosystems. Environ. Toxicol. Chem. **2005**, *24*, 379–388.
25. Candelmo, A.C.; Deshpande, A.; Dockum, B.; Weis, P.; Weis, J.S. The effect of contaminated prey on feeding, activity, and growth of young-of-the-year bluefish, *Pomatomus saltatrix*, in the laboratory. Estuaries Coasts **2010**, *33*, 1025–1038.
26. Baldwin, D.H.; Spromberg, J.A.; Collier, T.K.; Scholz, N.L. A fish of many scales: Extrapolating sublethal pesticide exposures to the productivity of wild salmon populations. Ecol. Appl. **2009**, *19*, 2004–2015.
27. Pestana, J.L.T.; Loureiro, S.; Baird, D.J.; Soares, A.M.V.M. Pesticide exposure and inducible antipredator responses in the zooplankton grazer, *Daphnia magna* Straus. Chemosphere **2010**, *78*, 241–248.
28. Stark, J.D.; Banks, J.E.; Vargas, R.I. How risky is risk assessment: The role that life history strategies play in susceptibility of species to stress. Proc. Natl. Acad. Sci. U. S. A. **2004**, *101*, 732–736.
29. van den Brink, P.J.; Hattink, J.; Bransen, F.; van Donk, E.; Brock, T.C.M. Impact of the fungicide carbendazim in freshwater microcosms. II. Zooplankton, primary producers and final conclusions. Aquat. Toxicol. **2000**, *48*, 251–264.
30. Kreutzweiser, D.P.; Back, R.C.; Sutton, T.M.; Pangle, K.L.; Thompson, D.G. Aquatic mesocosm assessments of a neem (azadirachtin) insecticide at environmentally realistic concentrations—2: Zooplankton community responses and recovery. Ecotoxicol. Environ. Saf. **2004**, *59*, 194–204.
31. Wang, M.; Grimm, V. Population models in pesticide risk assessment: Lessons for assessing population-level effects, recovery, and alternative exposure scenarios from modeling a small mammal. Environ. Toxicol. Chem. **2010**, *29*, 1292–1300.
32. Brodman, R.; Newman, W.D.; Laurie, K.; Osterfeld, S.; Lenzo, N. Interaction of an aquatic herbicide and predatory salamander density on wetland communities. J. Herpetol. **2010**, *44*, 69–82.
33. Rosenfeld, J.S. Functional redundancy in ecology and conservation. Oikos **2002**, *98*, 156–162.
34. Lawton, J.H.; Brown, V.K. Redundancy in ecosystems. In *Biodiversity and Ecosystem Function*; Schulze, E.D., Mooney, H.A., Eds.; Springer: New York, 1993; 255–268.
35. Naeem, S.; Li, S. Biodiversity enhances ecosystem stability. Nature **1997**, *390*, 507–509.
36. Kreutzweiser, D.P.; Good, K.P.; Chartrand, D.T.; Scarr, T.A.; Thompson, D.G. Are leaves that fall from imidacloprid-treated maple trees to control Asian longhorned beetles toxic to non-target decomposer organisms? Journal of Environmental Quality **2008**, *37*, 639–646.
37. deNoyelles, F., Jr.; Dewey, S.L.; Huggins, D.G.; Kettle, W.D. Aquatic mesocosms in ecological effects testing: Detecting direct and indirect effects of pesticides. In *Aquatic Mesocosm Studies in Ecological Risk Assessment*; Graney, R.L., Kennedy, J.H., Rodgers, J.H., Jr., Eds.; Lewis Publishers: Boca Raton, 1994; 577–603.
38. Boyle, T.P.; Fairchild, J.F.; Robinson-Wilson, E.F.; Haverland, P.S.; Lebo, J.A. Ecological restructuring in experimental aquatic mesocosms due to the application of diflubenzuron. Environ. Toxicol. Chem. **1996**, *15*, 1806–1814.
39. George, T.K.; Liber, K.; Solomon, K.R.; Sibley, P.K. Assessment of the probabilistic ecological risk assessment-toxic equivalent combination approach for evaluating pesticide mixture toxicity to zooplankton in outdoor microcosms. Archives of Environmental Contamination and Toxicology **2003**, *45*, 453–461.
40. Kreutzweiser, D.P.; Sutton, T.M.; Back, R.C.; Pangle, K.L.; Thompson, D.G. Some ecological implications of a neem (azadirachtin) insecticide disturbance to zooplankton communities in forest pond enclosures. Aquat. Toxicol. **2004**, *67*, 239–254.
41. van Wijngaarden, R.P.A.; Cuppen, J.G.M.; Arts, G.H.P.; Crum, S.J.H.; van den Hoorn, M.W.; van den Brink, P.J.; Brock, T.C.M. Aquatic risk assessment of a realistic exposure to pesticides used in bulb crops: a microcosm study. Environ. Toxicol. Chem. **2004**, *23*, 1479–1498.
42. Relyea, R.A.; Hoverman, J.T. Interactive effects of predators and a pesticide on aquatic communities. Oikos **2008**, *117*, 1647–1658.
43. Zhang, X.; Liu, X.; Zhang, M.; Dahlgren, R.A.; Eitzel, M. A review of vegetated buffers and a meta-analysis of their

mitigation efficacy in reducing nonpoint source pollution. J. Environ. Qual. **2010**, *39*, 76–84.
44. van de Zande, J.C.; Porskamp, H.A.; Michielsen, J.M.; Holterman, H.J.; Juijsmans, J.M. Classification of spray applications for driftability to protect surface water. Aspects Appl. Biol. **2000**, *57*, 57–64.
45. Luck, J.D.; Zandonadi, R.S.; Luck, B.D.; Shearer, S.A. Reducing pesticide over-application with map-based automatic boom section control on agricultural sprayers. Trans. Am. Soc. Agric. Biol. Eng. **2010**, *53*, 685–690.
46. Pfleeger, T.G.; Olszyk, D.; Burdick, C.A.; King, G.; Kern, J.; Fletcher, J. Using a geographical information system to identify areas with potential of off-target pesticide exposure. Environ. Toxicol. Chem. **2006**, *25*, 2250–2259.
47. PMRA. *Regulatory Directive DIR2002-02: The PMRA Initiative for Reduced Risk Pesticides*; Pest Management Regulatory Agency, Health Canada Information Services: Ottawa, Ontario; 2002, availabe at http://www.hc-sc.gc.ca/pmra-arla/english/pdf/dir/dir2002-02-e.pdf. (accessed September 2010).
48. Thompson, D.G.; Kreutzweiser, D.P. A review of the environmental fate and effects of natural "reduced-risk" pesticides in Canada. In *Crop Protection Products for Organic Agriculture: Environmental, Health, and Efficacy Assessment, ACS Symposium Series 947*; Felsot, A.S., Racke, K.D., Eds.; American Chemical Society: Washington, DC, 2007; 245–274.
49. van Emden, H. Integrated pest management. In *Encyclopedia of Pest Management*; Pimentel, D., Ed.; Marcel Dekker Inc.: New York, 2002; 413–415.
50. Solomon, K.R.; Takacs, P. Probabilistic ecological risk assessment using species sensitivity distributions. In *Species Sensitivity Distributions in Ecotoxicology*; Posthuma, L., Suter, G.W., Traas, T.P., Eds.; Lewis Publishers: Boca Raton, Florida, 2002; 285–314.
51. Solomon, K.R.; Baker, D.B.; Richards, R.P.; Dixon, K.R.; Klaine, S.J.; La Point, T.W.; Kendall, R.J.; Weisskopf, C.P.; Giddings, J.M.; Giesy, J.P.; Hall, L.W.; Williams, W.M. Ecological risk assessment of atrazine in North American surface waters. Environ. Toxicol. Chem. **1996**, *15*, 31–76.
52. Hanson, M.L.; Solomon, K.R. New technique for estimating thresholds of toxicity in ecological risk assessment. Environ. Sci. Technol. **2002**, *36*, 3257–3264.
53. Warren-Hicks, W.J.; Hart, A., Eds. *Application of Uncertainty Analysis to Ecological Risks of Pesticides*; CRC Press: Boca Raton, Florida, 2010.
54. Thorbek, P.; Forbes, V.E.; Heimbach, F.; Hommen, U.; Thulke, H.; van den Brink, P.; Wogram, J.; Grimm, V. *Ecological Models for Regulatory Risk Assessments of Pesticides*; SETAC Press: Pensacola, Florida, 2010.
55. Stark, J.D.; Banks, J.E. Population-level effects of pesticides and other toxicants on arthropods. Annu. Rev. Entomol. **2003**, *48*, 505–519.
56. van den Brink, P.J.; Roelsma, J.; van Nes, E.H.; Scheffer, M.; Brock, T.C.M. PERPEST model: A case-based reasoning approach to predict ecological risks of pesticides. Environ. Toxicol. Chem. **2002**, *21*, 2500–2506.
57. Baird, D.J.; Baker, C.J.O.; Brua, R.B.; Hajibabaei, M.; McNicol, K.; Pascoe, T.J.; de Zwart, D. Towards a knowledge infrastructure for traits-based ecological risk assessment. Integr. Environ. Assess. Manage. **2010**, online DOI 10.1002/ieam.129 (accessed September 2010).
58. van den Brink, P.J. Ecological risk assessment: From book-keeping to chemical stress ecology. Environ. Sci. Technol. **2008**, *42*, 8999–9004.
59. Beldon, J.B.; Gilliom, R.J.; Lydy, M.J. How well can we predict the toxicity of pesticide mixtures to aquatic life? Integr. Environ. Assess. Manage. **2007**, *3*, 364–372.
60. Heugens, E.H.W.; Hendricks, A.J.; Dekker, T.; van Straalen, N.M.; Admiraal, W. A review of the effects of multiple stressors on aquatic organisms and analysis of uncertainty factors for use in risk assessment. Crit. Rev. Toxicol. **2001**, *31*, 247–285.
61. Lydy, M.J.; Lohner, T.W.; Fisher, S.W. Influence of pH, temperature and sediment type on the toxicity, accumulation and degradation of parathion in aquatic systems. Aquat. Toxicol. **1990**, *17*, 27–44.
62. Howe, G.E.; Marking, L.L.; Bills, T.D.; Rach, J.J.; Mayer, F.L. Jr. Effects of water temperature and pH on toxicity of terbufos, trichlorfon, 4-nitrophenol and 2,4-dinitrophenol to the amphipod *Gammarus pseudolimnaeus* and rainbow trout (*Oncorhynchus mykiss*). Environ. Toxicol. Chem. **1994**, *13*, 51–66.
63. Yang, W.; Spurlock, F.; Liu, W.; Gan, J. Effects of dissolved organic matter on permethrin bioavailability to *Daphnia* species. J. Agric. Food Chem. **2006**, *54*, 3967–3972.
64. Puglis, H.J.; Boone, M.D. Effects of technical-grade active ingredient vs. commercial formulation of seven pesticides in the presence or absence of UV radiation on survival of green frog tadpoles. Arch. Environ. Contam. Toxicol. **2010**, online DOI 10.1007/s00244-010-9528-z (accessed September 2010).
65. Sandland, G.J.; Carmosini, N. Combined effects of a herbicide (atrazine) and predation on the life history of a pond snail, *Physa gyrina*. Environ. Toxicol. Chem. **2006**, *25*, 2216–2220.
66. Davidson, C.; Knapp, R.A. Multiple stressors and amphibian declines: Dual impacts of pesticides and fish on yellow-legged frogs. Ecol. Appl. **2007**, *17*, 587–597.
67. Barry, M.J.; Logan, D.C.; Ahokas, J.T.; Holdway, D.A. Effect of algal food concentration on toxicity of tow agricultural pesticides to *Daphnia carinata*. Ecotoxicol. Environ. Saf. **1995**, *32*, 273–279.
68. Wu, Q.; Riise, G.; Pflugmacher, S.; Greulich, K.; Steinberg, C.E.W. Combined effects of the fungicide propiconazole and agricultural runoff sediments on the aquatic bryophyte *Vesicularia dubyana*. Environ. Toxicol. Chem. **2005**, *24*, 2285–2290.
69. Boone, M.D.; Bridges, C.M.; Fairchild, J.F.; Little, E.E. Multiple sublethal chemicals negatively affect tadpoles of the green frog, *Rana clamitans*. Environ. Toxicol. Chem. **2005**, *24*, 1267–1280.
70. Laskowski, R.; Bednarska, A.J.; Kramarz, P.E.; Loureiro, S.; Schell, V.; Kudlek, J.; Holmstrup, M. Interactions between toxic chemicals and natural environmental factors—A meta-analysis and case studies. Sci. Tot. Environ. **2010**, *408*, 3763–3774.

Aral Sea Disaster

Guy Fipps
Agricultural Engineering Department, Texas A&M University, College Station, Texas, U.S.A.

Abstract
The Aral Sea is one of the worst ecological disasters on our planet. What was once the world's fourth largest inlet sea, the Aral Sea has lost over 60% of its surface area, two-third of its volume, declined 40 m in depth, and has fallen to the eighth largest inland body of water in the world.

INTRODUCTION

The cause is attributed to a vast expansion of irrigation in the Central Asian Republics beginning in the 1950s, which greatly reduced inflows to the Sea. The diversion of water for massive irrigation development was done deliberately by Soviet Union officials, unconcerned about the consequences of their actions.

The environmental, social, and economic damage has been immense. Winds pick up dust from the dry seabed and deposit it over a large populated area. The dust likely contains pesticide and chemical residues that are blamed for the serious rise in mortality and health problems in the region. The Sea, and the now exposed dry seabed, may also be contaminated by runoff from a former Soviet military base and a biological weapons lab. The ecosystem of the Aral Sea has collapsed, and climate changes in the Aral Sea Basin have been documented. Hundreds of agreements have been signed since 1980s on programs designed to address the "Aral Sea Problem" which, to date, have not been effective at preventing the continuing shrinking of the sea.

ARAL SEA BASIN

The Aral Sea is located in Central Asia and lies between Uzbekistan and Kazakhstan in a vast geological depression, the Turan lowlands, in the Kyzylkum and Karakum Deserts. In the 1950s, the sea covered 66,000 km^2, contained about 1090 km^3 of water, and had a maximum depth of about 70 m. The Aral Sea supported vast fisheries and shipping industries. At that time the sea was fed by two rivers, the Amu Darya (2540 km) and the Syr Darya (2200 km), which originate in the mountain ranges of central Asia and flow through the five republics of Uzbekistan, Kazakhstan, Kyrgyzstan, Tajikistan, and Turkmenistan.

The two rivers provide most of the fresh water used in Central Asia. In the last 50 years, about 20 dams and reservoirs and 60 major irrigation schemes have been constructed. About 82% of river diversions are for agricultural use and 14% is for municipal and industrial use (Table 1).

Water demand due to population growth and industrial expansion continues to increase (Table 2). Since 1960, the population of the Central Asian republics has increased 140% and totals over 50 million. Likewise, industrial production using large amounts of water has also increased. Examples include steel production which rose 200%, cement production by 170%, and electricity generation by a factor of 12.

The total inflows to the Aral Sea began decreasing rapidly in the 1960s, and by 1990 the storage volume of the sea has decreased by 600 km^3 (Table 3). As the water level fell, salinity levels have tripled, rising from about 1000 ppm to just under 3000 ppm today. By the 1980s, as the Aral Sea problem became well known in the Soviet Union, government officials proposed ambitious projects to divert water from other rivers, including ones in South Russia and Siberia, to be transported to the Aral Sea in massive

Table 1 Average water supply and demand in the Aral Sea Basin.

Total water available	km^3	%
Amu Darya Basin	84.3	64
Syr Darya Basin	47.8	36
Total	132.1	100
Water demand		
Agriculture		
Amu Darya Basin	44.8	81.6
Syr Darya Basin	34.6	
Municipal Water		
Amu Darya	3	6.5
Syr Darya	3.3	
Industry		
Amu Darya	3	8.2
Syr Darya	5	
Livestock		
Amu Darya	0.2	0.2
Syr Darya	0	
Fishery		
Amu Darya	2.6	3.5
Syr Darya	0.8	
Total	97.3	100

Encyclopedia of Environmental Management DOI: 10.1081/E-EEM-120010131

Table 2 General statistics of the Aral Sea Basin countries in 1995.

	Kazakhstan	Uzbekistan	Turkmenistan	Kyrgyzstan	Tajikistan
Area, km^2	2,717,300	447,400	488,100	198,500	143,100
Irrigated land, km^2	23,080	41,500	12,450	10,320	6.940
Population	17,376,615	23,089,261	4,075,316	4,769,877	6,155,474
Population growth rate, %	0.62	2.08	2.5	1.5	2.6

canals. However, these plans died with the breakup of the Soviet Union.

The decrease in sea level has now split the Aral Sea into two separate water bodies: the Small and Large Aral Seas (Maloe More and Bol'shiye More) each separately fed by the Syr Darya and the Amu Darya, respectively. The once vast Amu Darya delta which once covered 550,000 ha has now shrunk to less than 20,000 ha.

IRRIGATION AND COTTON

For thousands of years, Central Asian farmers diverted water from the Amu Darya and Syr Darya Rivers, transforming desert into green oases and supporting great civilizations. Historically, irrigation water use was conducted at a sustainable level. The creation of the Soviet Union and the collectivization of farmlands resulted in the end for traditional agricultural practices. Beginning as early as 1918, Soviet leaders began expanding irrigated land in Central Asia for export and hard currency. Cotton was known as "white gold." The USSR became a net exporter of cotton by the 1930s, and by the 1980s, was ranked fourth in the world in cotton production.

The policy of emphasizing cotton production was accelerated in the 1950s as Central Asia's irrigated agriculture was expanded and mechanized. In 1956, the Kara Kum Canal was opened, diverting one-third of the flow in the Amu Darya to new cultivated areas in the deserts of Uzbekistan and Turkmenistan. The year 1960 represents the critical junction when the Aral Sea began to drop. Irrigated cotton production and water diversions continued to be expanded until the break-up of the Soviet Union (Table 4).

Estimates are that upwards of 80% of the workforce is employed in agriculture. The main agricultural crops in the basin are cotton (6.4 million ha), forage (1.7 million ha), rice (0.4 million ha), and tree crops (0.4 million ha).

Some Central Asian irrigation experts estimate that only 20%–25% of the water diverted from the rivers is actually used by the crops, the rest being lost in the canals that transport the water to the fields and due to inefficient irrigation practices used on-farm. It is believed that over the past decade, adequate maintenance, repair, and renovation of the irrigation infrastructure were not performed at a meaningful level, and water losses from deteriorating canals, gates, and other facilities have increased.

Most land is under furrow irrigation, with drip irrigation accounting for about 5% of the irrigated cropland (used primarily on orchard crops), and sprinkler irrigation accounts for about 3%. Even though the water saving benefits of gated pipe are well known in the region, less than one-sixth of the farms use this technology. Reasons may include costs and product availability. Most farms follow the centuries' old practice of cutting earthen canals with shovels in order to divert water into the field. The volume of water available at these farm ditches is not sufficient to provide an even distribution of water over the field. As a result, water logging and soil salinity now affects about 40% of all the cultivated land in the region.

Table 3 Decline of the Aral Sea during the 1980s and total estimated inflows from the Amu Darya and Syr Darya rivers.

		Aral Sea	
Year	Inflows (km^3)	Volume (km^3)	Surface area (km^2)
1911–1960	56.0	1064	66,100
1981	6.0	618	50,500
1982	0.04	583	49,300
1983	2.3	539	47,700
1984	7.9	501	46,100
1986	0.0	424	41,100
1987	9.0		
1988	23.0		41,000
1989		300	30,000

MUYNAK AND ARALSK

Of all the villages affected by the drying of the Aral Sea, Muynak is the best known. Historically, Muynak was located on an island of the vast Aral Sea delta at the convergence of the Amu Darya River in Karakalpakstan (a semi-autonomist republic in Uzbekistan). In 1962, the island became a peninsula. By 1970 the former seaport was 10 km from the sea. The retreat of the sea accelerated and the town was 40 km from the sea by 1980, 70 km in 1995, and close to 100 km today.

Table 4 Cultivated land along the Amu Darya and Syr Darya rivers.

	Before 1917	1960	1980	1992
Millions of hectares	5.2	10	15	18.3

Over 3000 fishermen once worked the abundant waters around Muynak which supported 22 different commercial species of fish. In 1957, Muynak fishermen harvested 26,000 tons of fish, about half of the total catch that year taken from the Aral Sea. Muynak also produced 1.1 million farmed muskrat skins which were used to produce coats and hats.

The Kazakhstan city of Aralsk, was once located on the northern edge of the Aral Sea, and like Muynak, had major fisheries and commerce industries. A major shipping and transport industry existed between these two cities. As the Aral Sea skunk, Aralsk found itself farther and farther from the shore which had retreated nearly 129 km by the 1980s. In the early 1990s, a dam was built just to the south of the mouth of the Syr Darya, to protect the northern part of the Aral Sea, letting the southern portion of the Aral Sea evaporate. Although only 10% of the water in the Syr Darya River reaches the northern part of the Aral Sea, the Little Aral has risen 3 m since the construction of the dam, and the shoreline has crept to within 16 km of the town.

ENVIRONMENTAL PROBLEMS

The Aral Sea is an unfortunate example of an old Uzbek proverb: "at the beginning you drink water, at the end you drink poison." As the rivers flow through cultivated areas, they pick up fertilizers, pesticides, and salts from runoff, drainage water and groundwater flow. In the 1960s, it was common for about 550 kg ha^{-1} of chemicals to be applied to cotton fields in Central Asia, compared to an average of 25 kg used for other crops in the Soviet Union. Residues of these chemicals are now found on the dry seabed. Estimates are that millions of tons of dust are picked up from the seabed and distributed over the Aral Sea region.

The Sea may have been contaminated from runoff from by two former USSR military installations in the area. A chemical weapons testing facility was located on the Ust-Jurt Plateau (north shore), and was closed in the mid-1980s. Renaissance Island (Vorzrozhdeniya Island), located in the central Aral Sea, was the site of the former USSR Government's Microbiological Warfare Group which produced the deadly Anthrax virus. Some scientists believe that some containers holding the virus were not properly stored or destroyed. As the Aral Sea continues to dry and water levels recede, the ever-expanding island will soon connect to the surrounding land. Scientists fear that reptiles, including snakes that have been exposed to the various viruses, will move onto the surrounding land and possibly infect the humans living around the shores of the Aral Sea.

The Aral Sea once supported a complex ecosystem, an oasis in the vast desert. Over 20 species of fish are now extinct. Karakalpakstan scientists believe that a total of about 100 species of fish and animals that once flourished in the region are now extinct, as are many unique plants.

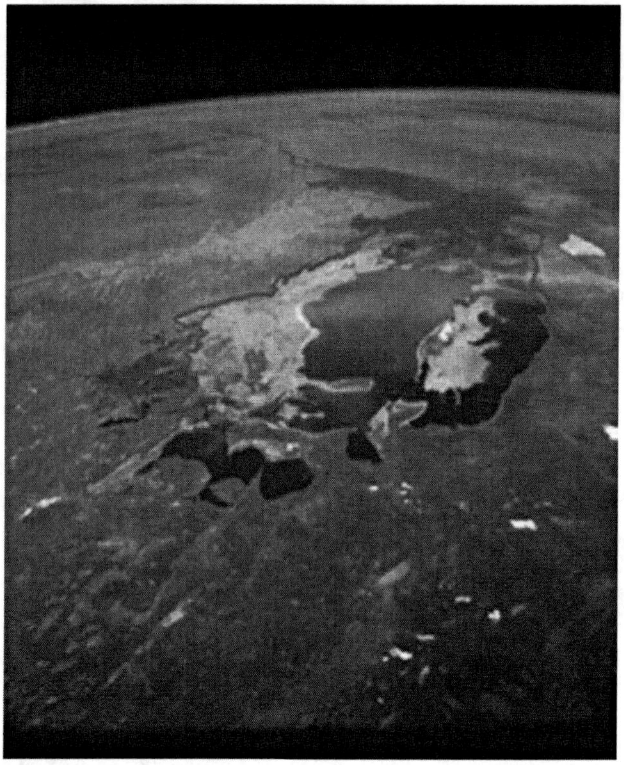

Fig. 1 This NASA photograph (STS085-503-119) was taken in August 1997 and looks toward the southeast. The Amu Darya River is visible to the right and the Syr Darya on the left. The Aral Sea is now separated into the Small Aral to the north and the Large Aral to the south. Shown are the approximate extent of the Aral Sea in 1957 before a massive expansion of irrigation diversions from the rivers.

Residents believe that there is a direct correlation between the drying of the sea and changes in climate of the Aral Sea Basin. The moderating effect of the sea has diminished and temperatures are now about 2.5°C higher in the summer and lower in the winter. Rainfall in the already arid basin has decreased by about 20 mm.

HUMAN TRAGEDY

Over the last 50 years, there has been a large increase in mortality, illnesses, and poor health in the region. Some estimate that 70%–90% of the population of Karakalpakstan suffer some an environmentally induced malady. Tuberculosis is rampant. Hardest hit are women and children. Common health problems include kidney diseases, thyroid dysfunctions, anemia, bronchitis, and cancers.

CONCLUSION

Some accounts are that since 1984, hundreds of international agreements have been signed to address Aral Sea problems.

The early agreements had the goals of first stabilizing the Sea, then slowly increasing flows to restore its ecosystem. In 1992, the Interstate Commission for Water Coordination was formed by the five central Asian republics, which also accepted, in principle, to adhere to the limits on water diversions as set during the Soviet era in 1984 and 1987. To date, however, no progress has been made on stabilizing or reversing the declining inflows. With no water reaching the Aral Sea from the Amu Darya, scientists predict that this portion of the sea (the Large Aral Sea) will disappear by 2020 (Fig. 1).

BIBLIOGRAPHY

1. The Aral Sea Homepage: Available at http://www.dfd.dlr.de/app/land/aralsee/
2. Requiem for a dying sea: Available at http://www.oneworld.org/patp/pap_aral.html/
3. Disappearance of the Aral Sea: Available at http://www.grida.no/aral/maps/aral.htm
4. Earth from Space: Available at http://earth.jsc.nasa.gov/categories.html

Arthropod Host-Plant Resistant Crops

Gerald E. Wilde
Department of Entomology, Kansas State University, Manhattan, Kansas, U.S.A.

Abstract
Resistance of plants to pest attack is defined as the relative amount of heritable qualities possessed by the plant that influence the ultimate amount of damage caused by the pest. The use of plant resistance to manage arthropod pest populations provides an ideal approach to integrated pest management because it is biologically and economically sound, environmentally friendly, and generally compatible with other management tactics or strategies. The cultivar forms the foundation on which all pest management programs and tactics are applied, and its effects are specific, cumulative, and persistent.

INTRODUCTION

Because of the many advantages plant resistance offers, virtually every cultivated crop has been evaluated for this trait and one or more resistant sources have been identified. The challenge has been to incorporate these resistant sources into agronomically adapted and consumer acceptable, high-yielding cultivars. In addition to traditional breeding methods, the use of modern breeding techniques and genetic transformation of crops has opened the door to other ways of identifying, incorporating, and employing pest-resistance genes to effectively and economically manage arthropod pest populations. The use of resistant cultivars contributes significant economic and social benefits and sustainable agricultural systems to the world's farmers. The positive effects of resistant cultivars have been demonstrated repeatedly in crops as diverse as wheat, alfalfa, grape, sorghum, maize, rice, apple, and cotton.

PERCENTAGE OF CROPS THAT HAVE SOME DEGREE OF PEST RESISTANCE

Plant resistance has been employed to a greater or lesser degree in practically all of the major food, feed, and fiber crops. Table 1 lists a number of major crops grown in the world and the number of pests for which resistance has been employed to at least some extent in the field. Hectarage planted to resistant cultivars varies for each pest and crop and over time as new varieties and hybrids (both susceptible and resistant) are grown and, in some instances, as new pest biotypes (pest populations that are capable of damaging previously resistant sources) develop. For example, most of the modern rice varieties and hybrids grown in China, India, and other countries are resistant to one or more major pests. Resistant American grape rootstocks have been used extensively over the world to control *Phylloxera vittifolae* (Fitch). A large percentage of the alfalfa planted in the United States is comprised of varieties resistant to aphid species. Sorghum hybrids with resistance to the greenbug have occupied up to 80% of the hectarage in the United States. Significant hectarages of wheat and barley in the United States, Canada, and North and South Africa have resistance to at least one pest. Most commercial soybean varieties are resistant to the potato leafhopper. Several cotton varieties carrying genes for resistance to jassids (*Empoasca* sp) are grown widely in Africa, India, and the Philippines. In the United States, more than 65% of commercial maize hybrids have some resistance to corn leaf aphid, >90% have some resistance to first generation European corn borer, and >75% have some resistance to second generation corn borer.

However, many more resistance genes have been identified in all crops than have been used in modern commercial varieties and hybrids, because incorporating them into high yielding cultivars acceptable to growers has been difficult. Recently, transgenic crops have been utilized to combat major insect pests. Hybrids or varieties with insect-resistance genes have been developed in cotton, maize, and potato. An estimated 6.7 million hectares of transgenic corn resistant to the European corn borer, 2.5 million hectares of transgenic cotton resistant to several pests, and 20,000 hectares of transgenic potato resistant to Colorado potato beetle were grown in the world in 1998. The hectares planted to transgenic crops are likely to increase as additional countries register these products and this technology is used on additional crops. For example, specific biotechnology applications are being field tested for rice and wheat, which together occupy 400 million hectares globally.

EFFECT OF PLANT RESISTANCE ON PEST POPULATIONS

The growing of pest-resistant cultivars can be used as a major control tactic or adjunct to other measures. Historically, the use of resistant cultivars combined with other

Encyclopedia of Environmental Management DOI: 10.1081/E-EEM-100001006

Table 1 List of some major crops grown in the world and number of arthropod pests for which resistant cultivars have been used in the field by growers for pest management.

Crop	No. of pests
Alfalfa	6
Apple	1
Asparagus	1
Barley	3
Bean	1
Cassava	2
Chickpea	0
Cotton	6
Grape	1
Lettuce	1
Maize	10
Millet	1
Oat	1
Pea	1
Peanut	4
Potato	1
Raspberry	1
Rice	14
Rye	1
Sorghum	6
Soybean	1
St. Augustine grass	1
Sugar beet	1
Sugarcane	3
Sunflower	1
Sweet clover	1
Sweet potato	1
Wheat	7

tactics has resulted in a reduction of many pest species to subeconomic levels. Even small increases in resistance enhance the effectiveness of cultural, biological, and insecticidal controls. The extent to which growing resistant plants affects pest populations is dependent upon the level of resistance expressed, the mechanisms of resistance involved, and the number of hectares grown. The growing of resistant wheat on 50% of the hectarage in Kansas has been shown to reduce Hessian fly populations to extremely low levels. Resistance in wheat to wheat curl mite (ca. 25% of the hectarage) was effective in limiting the spread of wheat streak mosiac virus, which the mite transmits. The incorporation of leaf and stem pubescence into most commercial soybean varieties has resulted in population suppression of the potato leafhopper over the past 60 years. As the hectarage of sorghum resistant to the greenbug increased to >50%, the area of sorghum treated with insecticide was reduced by 50%. Tenfold reductions in pest populations have been observed where insect-resistant rice cultivars have been grown widely.

ECONOMIC AND SOCIAL BENEFITS

Assessing the economic benefits of plant resistance is difficult in the context of integrated pest management programs and is likely to be underestimated frequently and substantially. Even determining the obvious advantages (yield benefits and reduced production costs) may be difficult over a large area where pest populations vary from locality to locality and year to year. Other environmental benefits, such as cleaner water and food, reduced risks to farmers, more flexibility in planting and cropping systems, reduced disease transmission, and reduced secondary pest outbreaks, also are difficult to quantify. Nevertheless, some specific estimates are available. In the United States alone the estimated valued of using arthropod-resistant alfalfa, barley, corn, sorghum, and wheat cultivars is more than $1.4 billion each year. The net economic benefit of greenbug resistance in U.S. sorghum production is estimated at close to $400 million annually. The global economic value of arthropod-resistant wheat has been estimated at $250 million annually. The value of resistance to aphids in alfalfa in the major alfalfa-producing states of the United States is estimated at more than $100 million annually. Breeding for pest resistance in rice has been estimated to be responsible for one-third of recent yield increases and $1 billion of additional annual income to rice producers. The net return of insect-resistant Bt maize in the United States and Canada has been estimated in some studies at $42.00–$67.30 per hectare, but other studies have indicated less of an economic return. The average net economic return of insect-resistant Bt cotton in 1997 was $133 per hectare.

BIBLIOGRAPHY

1. Antle, J.M.; Pingali, P.L. Pesticides, productivity and farmer health: A philippine case study. Am. J. Agric. **1994**, *76*, 418–430.
2. *Global Plant Genetic Resources for Insect-Resistant Crops*; Clement, S.L., Quisinberry, S.S., Eds.; CRC Press: New York, 295.
3. Harvey, T.L.; Martin, T.J.; Seifers, D.L. Importance of plant resistance to insect and mite vectors in controlling virus diseases of plants: resistance to the wheat curl mite (Acari: Eriophyidae). J. Agric. Entomol. **1994**, *11*, 271–277.
4. Hyde, J.; Martin, M.A.; Preckel, P.V.; Edwards, L.R. The economics of Bt corn: valuing protection from the European corn borer. Rev. Agric. Econ. **1999**, *21*, 442–454.
5. James, C. *Global Review of Commercialized Transgenic Crops*; ISAAA Briefs No. 8, ISAA: Ithaca, NY, 1998, 43.
6. In *Insect Resistant Maize: Recent Advances and Utilization*, Proceedings of an International Symposium, International.

7. Maize and Wheat Improvement Center (CIMMYT), Nov 27–Dec 3, 1994; Mihm, J.A., Ed.; CIMMYT: Mexico, D.F., 1997; 302.
8. Painter, R.H. *Insect Resistance in Crop Plants*; University of Kansas Press: Lawrence, KS, 1968, 520.
9. Smith, C.M. *Plant Resistance to Insects. A Fundamental Approach*; John Wiley and Sons: New York, 1989, 286.
10. Smith, C.M.; Quisinberry, S.S. Value and use of plant resistance to insects in integrated pest management. J. Agric. Entomol. **1994**, *11*, 189–190.
11. van Emden, H.F. Host-Plant Resistance to Insect Pests. In *Techniques for Reducing Pesticides: Environmental and Economic Benefits*; Pimentel, D., Ed.; John Wiley: Chichester, England, 1997, 124–132.
12. In *Economic, Environmental, and Social Benefits of Resistance in Field Crops*, Proceedings, Thomas Say Publications in Entomology, Wiseman, B.R., Webster, J.A., Eds.; Entomological Society of America: Lanham, MD, 1999, 189.
13. Contribution No. 00-252-B of the Kansas Agricultural Experiment Station.

Bacillus thuringiensis: Transgenic Crops

Julie A. Peterson
John J. Obrycki
James D. Harwood
Department of Entomology, University of Kentucky, Lexington, Kentucky, U.S.A.

Abstract

Bacillus thuringiensis (Bt) crops, genetically modified to express insecticidal toxins that target key pests of corn, cotton, rice, potato, and other crops, have been rapidly adopted and have become dominant fixtures in agroecosystems throughout the world. Due to the constitutive nature of Bt toxin expression, insecticidal proteins may be found in nearly all plant tissues, presenting multiple sources for Bt toxins to enter the environment, thus creating complex direct and indirect pathways for non-target organisms to be exposed to insecticidal proteins. The environmental impacts of Bt crops have been widely debated, although both benefits and risks do exist. Benefits of Bt crop adoption include reduced risks to non-target organisms when compared with conventional spray applications of insecticides, as well as economic savings to growers and increased global food security. Conversely, impacts on non-target organisms, presence in the human food supply, pleiotropic effects of genetic transformation, and gene escape to wild plant populations are all considered as viable risks of Bt technology. To address the potential risks of Bt crop technology, proposed approaches to the environmental management of Bt crops are discussed, including within-plant modifications, reduction in Bt toxin and transgene escape, and large-scale integration into integrated pest and resistance management programs. Additionally, continued study of the effects of Bt toxins on non-target organisms at multiple tiers is necessary for intelligent use of this valuable pest management tool. The global area planted to Bt crops is expanding, and new Bt products and combinations are in various stages of development. Although Bt technology may offer an environmentally superior alternative to many insecticide applications, further risk assessment research addressing the impacts of Bt crops on agroecosystem function are needed to promote environmental safety.

INTRODUCTION

Genetically modified organisms have been widely adopted in many parts of the world, prompting debate about the implications that this technology may have for environmental health. Transgenic crops have been genetically engineered to incorporate genes derived from another species that confer nutritional and agronomic benefits, such as resistance to insect pests, viruses, herbicides, or environmental conditions, such as low water availability. Among insect-resistant transgenic crops, the most widespread are those that express Bt toxins, coded for by genes from the naturally occurring soil bacterium *Bacillus thuringiensis*. Commercialized Bt crops include corn, cotton, and rice that are protected against Coleoptera and Lepidoptera pests. Bt toxins are recognized as having a narrower range of toxicity than many insecticides, including pyrethroids and neonicotinoids, and may therefore pose less risk to non-target organisms; however, potential environmental impacts of Bt toxins need to be examined and documented. This entry will therefore examine the environmental risk assessment of Bt crops, focusing on sources and fate of Bt toxins in exposure pathways for non-target organisms, impact of Bt crops on the environment, and approaches to environmental management of Bt crops.

WHAT ARE Bt CROPS?

Transgenic Bt crops are genetically engineered to express insecticidal proteins that cause mortality of several common agricultural pests. The genes that code for these proteins, from a naturally occurring bacterium, *Bacillus thuringiensis* (Berliner) (Bacillaceae: Bacillales), are inserted into the genome of the desired crop plant. Genetic transformation is achieved by insertion of the target gene, its promoter and termination sequences, and a marker gene into the crop genome using the microprojectile bombardment method ("gene gun") or the *Agrobacterium tumefaciens* (Smith and Townsend) (Rhizobiales: Rhizobiaceae) bacterium (vector-mediated transformation).

Encyclopedia of Environmental Management DOI: 10.1081/E-EEM-120046904

Bt Toxins

Bacillus thuringiensis bacterial strains can produce a series of different toxins; however, only a few have been bioengineered into agricultural crops, including crystalline (Cry) and vegetative insecticidal (VIP) proteins.[1,2] These Bt toxins vary in their range of toxicity to invertebrates, with targeted pests dominated by larval insects in the orders Lepidoptera (moths) and Coleoptera (beetles). The insecticidal mode of action occurs when the Bt toxins bind to receptors on the midgut lining of susceptible insects, causing lysis of epithelial cells on the gut wall and perforations in the midgut lining. This damage to the insect's digestive tract induces cessation of feeding and death by septicemia. An important component of the insecticidal mechanism is its specificity, which is greater than that of many currently used insecticides. Additionally, Bt toxins degrade rapidly in the digestive tract of vertebrates,[3] contributing to their selective nature.

Bt Crops and their Targeted Pests

Many crop plants have been genetically engineered to express Bt toxins, including field and sweet corn, cotton, potato, rice, eggplant, oilseed rape (canola), tomato, broccoli, collards, chickpea, spinach, soybean, tobacco, and cauliflower. However, only corn and cotton have seen widespread commercialization. Bt potatoes were grown commercially in the United States starting in 1995, but were withdrawn from the market in 2001 following pressure from anti-biotechnology groups and the decision of the global fast-food chain McDonalds to ban the use of genetically modified potatoes in their products.[4] This crop may see a resurgence in planting in Russia and eastern Europe in the near future,[5] as small-scale and subsistence farmers in these regions seek alternatives to expensive insecticide applications.[4] Bt rice has also been approved in certain regions of China,[5] thereby facilitating increased production worldwide.

Global Prevalence

The planting of Bt crops has increased dramatically since the mid-1990s, becoming a prevalent component of agroecosystems worldwide[5–10] (Table 1). For example, Bt cotton and corn in the United States comprised just 1% of total area planted in 1996, their first year of commercial release; however, planting rates have increased rapidly, with areas of Bt cotton and corn in 2010 comprising 73% and 63% of total U.S. production, respectively.[11] Genetically modified crops are grown on 134 million hectares of land in 25 countries by 14.0 million farmers[5]; approximately 40% of that area is planted to corn and cotton expressing Bt insecticidal toxins.[12]

SOURCES AND FATE OF Bt TOXINS IN THE ENVIRONMENT

Toxin distribution and expression levels within a transgenic plant vary depending on the type of Bt protein, transformation event, gene promoter used, crop phenology, and environmental and geographical effects.[13–17] Most Bt crops employ a constitutive promoter, such as the cauliflower mosaic virus (CaMV 35S), that expresses insecticidal proteins throughout the life of the plant in nearly all tissues, which may include foliage, roots and root exudates, phloem, nectar, and pollen, creating the potential for a multitude of sources for environmental exposure. These pathways to exposure of non-target organisms include, but are not limited to, direct consumption of Bt toxins via ingestion of live or detrital plant material, as well as indirect consumption of Bt toxins via soil contamination from root exudates and persistence in the soil, or consumption of Bt-containing prey in tritrophic interactions (Fig. 1). These pathways allow for multiple routes to exposure, even potentially within a given taxonomic group, such as ground beetles (Coleoptera: Carabidae), which have been documented to take up Bt toxins in the field.[18] Certain agronomic practices may also create unexpected routes to exposure. For example, following harvest in China, cottonseed hulls may be used as substrate for growing edible oyster mushrooms before being incorporated into cattle feed.[19] Other cotton gin by-products from transgenic plants are used in a variety of ways, including as catfish feed,[20] mulch, and fuel for wood-burning stoves.[21] Although transfer of Bt toxins from cottonseed hulls into mushrooms or cattle feed was not detected,[22] investigation of these complex and non-conventional pathways for Bt toxin movement is critically important.

Direct Consumption of Bt Toxins

Consumption of Live Plant Tissue

Ingestion of plant material, including foliage, roots, phloem, nectar, or pollen may be the most obvious pathway to Bt toxin exposure for targeted pests species, as well as non-target herbivores and natural enemies. Uptake of Bt toxins by herbivores feeding on transgenic plants is well documented (e.g., Dutton et al.,[23] Harwood et al.[24] Meissle et al.,[25] Obrist et al.,[26] and Obrist[27]). However, ingestion of Bt crop tissue may not always result in exposure to toxins. For example, phloem-feeding insects and their honeydew have tested positive for Bt toxins in some transgenic agroecosystems, including certain rice, oilseed rape, and corn events,[28–30] while failing to take up toxins from selected Bt corn events.[31] Exposure pathways of Bt toxins to herbivorous arthropods in transgenic agroecosystems are variable and may therefore be difficult to predict.

Table 1 Commercialized Bt crops, years marketed, Bt toxins most commonly expressed in commercial lines, their targeted pests, and countries that have adopted this technology.

Crop	Marketed	Bt toxins expressed	Targeted pest/s	Countries
Corn	1996–present	Cry1Ab, Cry1A.105, Cry1F, Cry2Ab2, Cry9C (withdrawn in 2000), VIP3A	European corn borer *Ostrinia nubilalis* Hübner, southwestern corn borer *Diatraea grandiosella* Dyar (Lepidoptera: Pyralidae), corn earworm *Helicoverpa zea* (Boddie), fall armyworm *Spodoptera frugiperda* Smith (Lepidoptera: Noctuidae)	United States, Brazil, Argentina, Canada, South Africa, Uruguay, Philippines, Spain, Chile, Honduras, Czech Republic, Portugal, Romania, Poland, Egypt, Slovakia
	2003–present	Cry3Bb1, Cry34Ab1, Cry35Ab1, Cry3Aa	Corn rootworm *Diabrotica* spp. (Coleoptera: Chrysomelidae)	
Cotton	1996–present	Cry1Ac, Cry1F, Cry2Ab, VIP3A	Bollworm complex: *Heliothis*, *Helicoverpa* (Lepidoptera: Noctuidae), and *Pectinophora* (Lepidoptera: Gelechiidae)	United States, Brazil, Argentina, India, China, South Africa, Australia, Burkina Faso, Mexico, Colombia, Costa Rica
Potato	1995–2000	Cry3Aa	Colorado potato beetle *Leptinotarsa decemlineata* Say (Coleoptera: Chrysomelidae)	United States, Canada, Romania

Source: Data from James[5] and Duan et al.[146]

Many natural enemies are facultatively phytophagous during some or all of their life stages, consuming plant material or feeding on plant liquids to meet their nutritional and moisture requirements (reviewed in Lundgren).[32] Despite an abundant supply of moisture and prey items, many predatory insects, including ground beetles (Coleoptera: Carabidae), damsel bugs (Hemiptera: Nabidae), stink bugs (Hemiptera: Pentatomidae), and ladybird beetles (Coleoptera: Coccinellidae) will also ingest plant leaf tissue, nectar, or phloem to supplement a prey-based diet.[33]

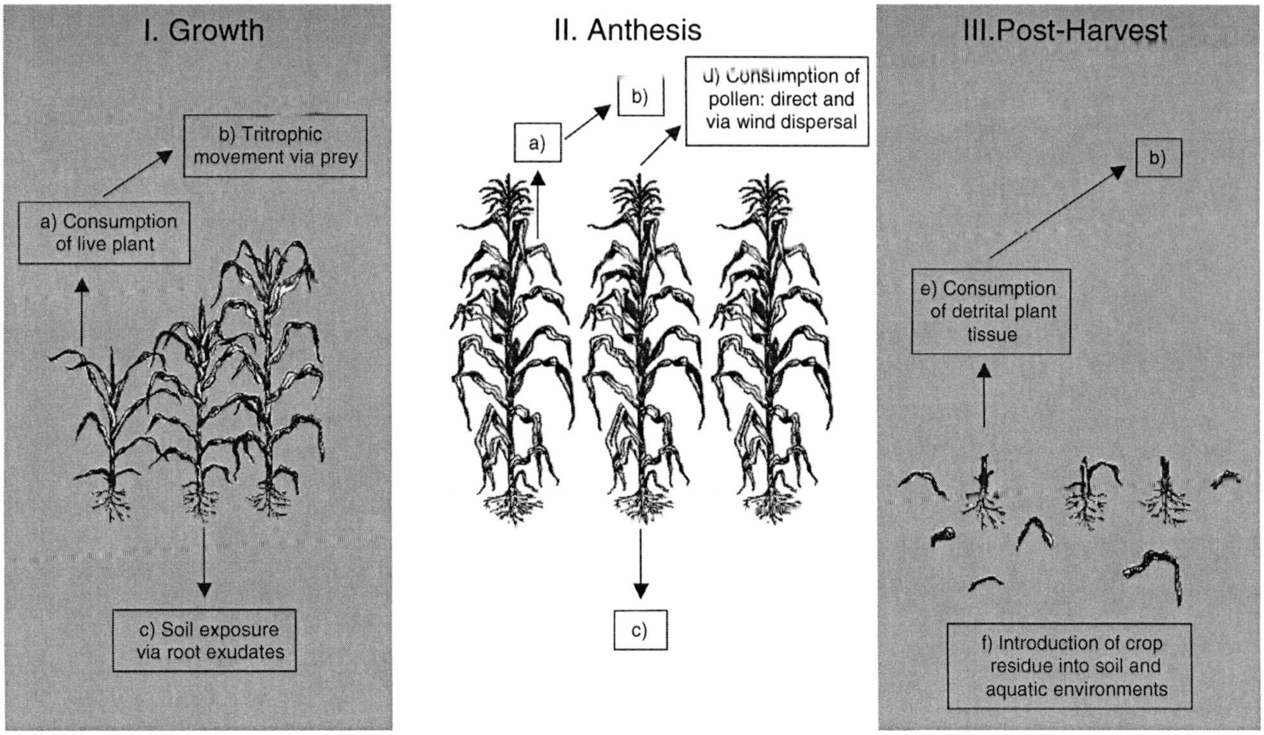
I. Growth
b) Tritrophic movement via prey
a) Consumption of live plant
c) Soil exposure via root exudates
II. Anthesis
b)
a)
d) Consumption of pollen: direct and via wind dispersal
c)
III.Post-Harvest
b)
e) Consumption of detrital plant tissue
f) Introduction of crop residue into soil and aquatic environments

Fig. 1 Sources for Bt toxin movement in a transgenic corn agroecosystem over the course of a growing season, including (I) growth, (II) anthesis, and (III) post-harvest time periods.

Pollen Feeding

Another potential route of Bt toxin flow in the environment is through direct pollen feeding or consumption of pollen-contaminated material. Pollen is a component of the diets of many organisms, including springtails (Collembola)[34,35] and Western corn rootworms *Diabrotica virgifera virgifera* LeConte (Coleoptera: Chrysomelidae),[36] as well as natural enemies, including ladybird beetles (Coleoptera: Coccinellidae),[37] ground beetles (Coleoptera: Carabidae),[38] green and brown lacewings (Neuroptera: Chrysopidae, Hemerobiidae),[39] hoverflies (Diptera: Syrphidae),[40] and spiders (Araneae).[41,42] In wind-pollinated Bt crops, such as corn, pollen is an abundant resource during anthesis and is deposited in large quantities (up to 1400 grains/cm^2 on plant surfaces[43] and more than 250 grains/cm^2 in ground-based spider webs[44]). Some pollen-feeding omnivores, such as *Orius insidiosus* (Say) (Hemiptera: Anthocoridae), may also maximize their exposure to Bt toxins by aggregating at corn silks and leaf axils, where pollen grains accumulate during anthesis.[45,46] Pollen consumption can therefore represent a significant direct and indirect (through consumption of pollen-feeding prey) route of exposure for both predators and prey in transgenic agroecosystems, particularly during periods of crop anthesis.

Consumption of Detritus

Bt toxins can persist in plant detritus beyond a single growing season[47,48] thereby exposing detritivores, such as earthworms, slugs, nematodes, protozoa, bacteria, and fungi, to Bt toxins through the consumption of such litter.[49–51] Crop detritus may also enter aquatic environments; for example, in agricultural systems where crop detritus is left in the field to prevent erosion, plant residues may account for up to 40% of non-woody vegetation entering streams.[52] Bt-containing crop tissue may then be consumed by aquatic detritivores, such as larval caddisflies (Trichoptera), crane flies (Diptera: Tipulidae), midges (Diptera: Chironomidae), and isopods. However, the bioactivity of Bt toxins in senesced plant material may be relatively short; lepidopteran-specific toxins were absent after 2 weeks in aquatic systems, while coleopteran-specific toxins decayed in as few as 6 days.[53] The harsh environmental conditions and constant physical abrasion experienced by plant tissue in flowing water were suggested as mechanisms stimulating such rapid breakdown.[54] Thus, while detritus provides a potential route of exposure, the functional consequence of Bt toxins in detritivore food webs remains unclear. However, what is evident is the persistence of Bt toxins in the environment following harvest and the possibility for long-term exposure of non-target organisms to this material.

Indirect Consumption of Bt Toxins

Soil Contamination via Root Exudates

One potential pathway of indirect exposure to Bt toxins is through contamination of the soil and therefore to soil-dwelling arthropods via root exudates. Bt corn, potato, and rice all release transgenic proteins from their roots during plant growth.[55,56] The soil-dwelling fauna, including beneficial non-target organisms, may therefore be exposed to Bt toxins via their secretion in plant root exudates. Bt toxin exposurc to epigeal predators, ground beetle larvae and adults, and certain spiders [e.g., wolf spiders (Araneae: Lycosidae)] may also occur because of their feeding habits. Several studies have quantified the persistence of Bt toxins in the soil,[47,57–59] with results indicating persistence of these insecticidal proteins ranging from 2 to 32 weeks after introduction into the soil. This wide discrepancy may partially reflect differences in microbial activity of soils,[57,60,61] which is in turn affected by pH and mineral content.[59] Bt toxins may bind to humic acids, organic supplements, or soil particles, protecting the toxins from degradation by microbes and extending the persistence of insecticidal activity in the soil.[2] Thus, the persistence of Bt toxins may vary significantly due to their differential rate of decay based on microbial activity, soils, and environmental factors.

Consumption of Prey Containing Bt Toxins

The movement of Bt toxins from plant tissue into herbivores and subsequently into their natural enemies has been well documented. Concentrations of Bt toxins typically decrease as they move through a food chain, indicating little evidence for bioaccumulation effects as seen in other insecticidal compounds[62]; however, two-spotted spider mites, *Tetranychus urticae* Koch (Acari: Tetranychidae), show evidence for the bioaccumulation of Bt toxins.[63] Although in a more typical example, Cry1Ac proteins expressed in transgenic cotton are ingested by beet armyworm caterpillars *Spodoptera exigua* (Hübner) (Lepidoptera: Noctuidae) and are also detectible, but at lower concentrations, in predatory stink bugs *Podisus maculiventris* (Say) (Hemiptera: Pentatomidae) when these prey are consumed.[63] However, not all tritrophic pathways facilitate the uptake of Bt toxins; Cry1Ab toxins are present in the marsh slug *Deroceras laeve* (Müller) (Pulmonata: Agriolimacidae) following consumption of Bt corn tissue, but are not taken up by the predatory ground beetle *Scarites subterraneus* (F.) (Coleoptera: Carabidae) in laboratory studies[64]; accordingly, field-collected specimens of this species did not test positive for Bt toxins.[18] Additionally, the concentration of Bt toxins transferred via trophic connections may vary based on the identity of the prey. In a laboratory experiment, two prey species of the wolf spider *Pirata subpiraticus* (Bösenberg and Strand) (Araneae: Lycosidae), the striped stem borer *Chilo suppressalis* (Walker) (Lepidoptera: Crambidae), and the Chinese brushbrown caterpillar *Mycalesis gotama* Moore (Lepidoptera: Nymphalidae) were allowed to feed on transgenic rice expressing Cry1Ab Bt toxins. These prey were subsequently fed to the wolf spider, and assays of each trophic level indicated that Bt toxins were transferred up the food chain; Cry1Ab concentration diminished with each additional trophic step, and the two prey species transferred Cry1Ab with significantly different efficiencies, having approximately 60 times the Cry1Ab concentration in brushbrown caterpillar-fed spiders compared with striped stem borer-fed spiders.[65] Adult ladybird beetles (Coleoptera: Coccinellidae) showed greatest uptake of Bt toxins in a corn agroecosystem post-anthesis, indicating that tritrophic movement of toxins was a greater pathway for toxin uptake than direct pollen consumption.[66] It is therefore clear that consumption of Bt-containing prey could be a major source of Bt toxin flow in non-target food webs, although the extent of toxin uptake and its concentration will depend on the strength of specific trophic pathways that occur within a given food web in the field.

IMPACTS OF BT CROPS ON THE ENVIRONMENT

Bt crops have become a dominant fixture in selected agroecosystems worldwide. Their planting on cultivated lands globally allows for large potential impacts of this technology on the environment. These impacts include both benefits and potential risks, the consequences to the environment of using Bt technology are intensely debated.

Benefits of Bt Technology

Reduced Risk Compared with Conventional Insecticides

The insecticidal toxins produced by transgenic Bt crops are considered to have fewer non-target effects than many insecticides due to their narrow range of toxicity and, therefore, to be advantageous to traditional methods of control. For example, populations of many natural enemies responded negatively to foliar applications of broad-spectrum pyrethroids compared with more selective insecticides, such as Bt toxins, indoxacarb, and spinosad, used to combat lepidopteran pests in sweet corn agroecosystems.[67] Field studies comparing Bt crops with their non-transgenic isolines that have been treated with broad-spectrum insecticides almost always reveal higher populations of beneficial arthropods in the Bt crops. A meta-analysis of these studies found that total non-target invertebrate abundance was higher in lepidopteran-targeting corn and cotton compared with non-transgenic crops managed with insecticides; however, no differences for coleopteran-targeting corn were reported.[68] Non-transgenic control plots treated with insecticides had lower predator and herbivore abundance compared with unsprayed Bt fields; this result was particularly strong for predator populations in non-transgenic plots treated with pyrethroids, such as lambda-cyhalothrin, cyfluthrin, and bifenthrin.[69] Similarly, spiders were more abundant in Bt corn, cotton, and potato when compared with conventionally managed crops employing a range of insecticides, including foliar pyrethroid sprays, systemic neonicotinoid seed treatments, and organophosphate soil applications at planting.[70] Due to their selectivity, Bt crops are therefore safer for non-target organisms when compared with many insecticides, particularly those with broad-spectrum action.

Economic Savings

A reduction in the quantity and frequency of insecticide applications are economically beneficial, in addition to reduced exposure to chemical insecticides for farm workers and the environment. Bt cotton has significantly reduced insecticide inputs in numerous regions of the world, including the United States,[71,72] China,[73] and South Africa.[74] The adoption of Bt corn in the midwestern United States has provided an estimated $6.9 billion in benefits to growers of both Bt and non-Bt corn in the past 14 years, due to area-wide suppression of European corn borer *Ostrinia nubilalis* (Hübner) (Lepidoptera: Pyralidae), a key pest of this crop.[75] With more than 53 million hectares of Bt crops now planted worldwide, there are significant economic considerations, and it is evident that Bt-based production systems are not only more sustainable in the context of pest management but also have the capacity to enhance agricultural diversity through reduced chemical inputs.

Global Food Security

The human population is projected to reach 10 billion by 2050, and concomitant to this is the need for augmented global food security and production.[76] The employment of Bt crop technology may aid in this goal by increasing quantity and consistency of crop yields; for example, corn yields are increased or protected because of season-long control of European corn borer.[71] Additionally, stored corn grain is protected against lepidopteran pests[77] and mycotoxin levels, which pose a threat to the health of humans and livestock if introduced into the food supply,[78] are lower because of reduced feeding activity of European corn borer, which are associated with the fungal causal agents.[71,79] Bt crops may therefore confer significant beneficial effects for the global drive to increase agricultural productivity and safety.

Potential Risks of Bt Crops

Impacts on Non-Target Organisms

Despite the specificity of Bt toxins toward target pests, questions have been raised concerning their effects on abundance, diversity, or fecundity of some components of the non-target food web, including beneficial species such as pollinators, natural enemies, and/or detritivores. Given the important ecosystem services provided by the abovementioned non-target organisms, the risk assessment of these groups is essential in the context of understanding environmental health. Lundgren et al.[17] identified four main approaches that risk assessment researchers have used to study the impact of Bt crops on non-target invertebrates: direct toxicity, tritrophic interactions, community level studies, and meta-analyses of data.

Direct toxicity. Feeding non-target organisms a diet that contains Bt toxins and measuring resulting parameters of development, fitness, and fecundity are done to examine the potential for directly toxic effects of Bt crops. The literature (reviewed in Lundgren et al.[17] and Lövei and Arpai[80]) provides contrasting evidence of non-target effects, ranging from no discernable effects of consumption of transgenic crops (e.g., Harwood et al.,[64] Pilcher et al.,[81] Al-Deeb et al.,[82] Lundgren and Wiedenmann,[83] and Anderson et al.[84]) to reports of a variety of negative effects (e.g., increased mortality, delay in development, reduction in weight gain, or changes in behavior) on beneficial organisms, such as pollinators,[85] predators,[86] parasitoids,[87] and other non-target arthropods.[50,88–91] Differing results of studies of direct toxicity of Bt toxins to non-target organisms exist for many groups; for example, in caddisflies (closely related to the target order Lepidoptera), studies have been published that report both sublethal negative effects[91] and the absence of negative impacts of Bt toxins.[54] Such laboratory toxicity studies may be extrapolated to the field, although toxicity studies should address all ecologically relevant routes to exposure for non-target organisms.[92]

Tritrophic interactions. These studies test for effects of Bt crops on natural enemies via consumption of Bt-containing prey; any observed effects may be due to ingestion of toxins or through prey-quality-mediated effects. Several studies have reported no tritrophic effects of Bt crops on natural enemies[63,93–95]; however, negative effects have been observed in other cases,[96,97] although these results are often attributed to prey-mediated effects whereas prey quality is lower when fed Bt crop tissue. Meta-analyses of tritrophic studies revealed that using prey items that were totally or partially susceptible to Bt toxins (and therefore were likely to be of lesser quality) had a negative effect on the performance of natural enemies, while using non-susceptible prey (whose quality should be unaffected by consuming Bt toxins) had no effect on the performance of the natural enemies that consumed it.[98,99]

Community level. To study the effect of Bt crops on non-target organisms at the community level, arthropods are sampled from Bt and conventional crops to observe any differences in abundance, diversity, or community structure. Such studies have examined a variety of non-target organism populations, including soil microarthropods, nematodes, decomposers, pollinators, and natural enemies.[81,100–110] Results of such studies often report no significant differences between populations of non-target organisms in Bt and non-Bt crops; however, a lack of taxonomic resolution in some studies can weaken these results.[70]

Meta-analysis data. This quantitative method addresses effects of Bt crops across multiple published studies and has been widely used to infer the consequence of Bt crops on a series of different parameters. For example, a meta-analysis of 42 field experiments revealed that the overall mean abundance of non-target invertebrates was significantly lower in lepidopteran-targeting Bt corn fields compared with non-transgenic fields when neither is treated with insecticides; no differences were found between coleopteran-targeting Bt and non-transgenic corn.[68] Unsurprisingly, the abundance of non-target arthropods was significantly higher in transgenic corn versus non-transgenic corn that had been treated with insecticides.[68] Additional meta-analyses have reported the effects of Bt crops on functional guilds of non-target organisms,[69] honey bees,[111] and spiders,[70] generally finding no differences in non-target arthropod populations between Bt and non-Bt crops. When examined at further taxonomic resolution, such analyses may reveal differential responses of functionally distinct taxa to Bt crops, as is the case with spider families. Meta-analysis revealed positive effects of Bt crops on the abundance of certain groups (Clubionidae, Linyphiidae, Thomisidae), no effect on others (Lycosidae, Oxyopidae, Araneidae), and negative effects on several families (Anyphaenidae, Philodromidae) relative to non-transgenic crops untreated with insecticides.[70]

Presence in Human Food Supply

Concerns about the presence of Bt toxins in the human food supply do not stem from any direct toxic effects, as vertebrates lack the midgut receptors for binding of Bt toxins, but from the possibility that a portion of the population will exhibit an allergic reaction to ingestion of Bt proteins.[112] Most Bt toxins will readily break down in the acidic environment of a vertebrate digestive tract.[3] Bt corn expressing Cry9C proteins, marketed under the commercial name StarLink™, was planted in the United States from 1998 to 2000, but approved only for animal feed and ethanol production due to the persistence of Cry9C in the vertebrate gut.[113] When traces of Cry9C proteins were found in cornmeal destined for human consumption, several food items were recalled, including Taco Bell® taco shells, and StarLink was voluntarily removed from the market.[114] However, no confirmed allergenic reactions due to Cry9C contamination were reported. Despite the lack of evidence for any true risk to humans based on consumption of Bt food products, sentiment against transgenic agricultural products destined for human consumption exists, especially in Europe, and has influenced the commercial acceptance of some products such as Bt potatoes.[4] Therefore, despite these limited effects on the human (and vertebrate) population, safeguards need to be in place to prevent the presence of unapproved genetically modified products entering the human food chain.

Pleiotropic Effects of Genetic Transformation

Insertion of a Bt gene complex into a crop plant may result in unpredicted and unintended pleiotropic effects that change the plant from its non-transgenic counterpart in ways beyond just the expression of Bt toxins.[49,115–117] For example, a reported pleiotropic effect in Bt corn is an increase in the lignin content in transgenic plant tissue,[49] a trait that could lead to reduced decomposition rates in the soil.[118] However, other studies have contested this conclusion and shown no differences in rate of decomposition.[119] An additional pleiotropic effect of transformation in Cry1F corn may be an increase in attractiveness as an oviposition site for corn leafhoppers *Dalbulus maidis* (DeLong and Wolcott) (Hemiptera: Cicadellidae), a pest that is not targeted by Bt toxins, possibly due to altered plant traits that influence oviposition, such as leaf vein characteristics, foliar pubescence, or plant chemistry.[120] There is a lack of understanding of how these pleiotropic effects will affect ecosystem processes, although the potential consequences merit further examination in the context of their environmental impacts.

Gene Escape

The transfer of genes from populations of domesticated crops into wild plants has been documented for many years.[121] The "escape" of Bt transgenes into wild plants could have undesirable effects by reducing genetic diversity and fecundity in wild populations or increasing fecundity and creating an invasive weed through reduction or elimination of herbivory. The presence of transgenic material from the CaMV 35S promoter used in Bt crops was reported in native maize landraces grown in remote areas of Oaxaca, Mexico, in 2001.[122] However, these results have been highly debated[123–125] and additional studies are conflicting, reporting both the presence[126] and absence[127] of transgenic DNA in traditional maize lines in Mexico. Additionally, transgene escape into weedy rice may increase the fecundity of this plant, as well as its ecological interactions with surrounding organisms.[128] The implications of transgene escape are yet to be fully understood, particularly in the context of ecological risk assessment.

APPROACHES TO ENVIRONMENTAL MANAGEMENT OF BT CROPS

To safely incorporate Bt crop technology into agroecosystems, approaches to environmental management should address the issue at multiple scales. These include engineering at the level of the individual plant genome, field- and farm-level modifications to reduce exposure of Bt toxins and escape of transgenes, and large-scale incorporation of Bt technology into integrated pest and resistance management programs. Finally, continued research concentrating on the non-target impacts of Bt crops should be conducted at multiple tiers across crop and toxin types, geographic regions, non-target organism taxa, and temporal and spatial scales, studying non-target organisms at the greatest taxonomic resolution possible. Regulation of transgenic crops in the effort to mitigate risk is complex; further recommendations and discussion of this topic can be found elsewhere.[129]

Within-Plant Modifications

Selection of Low-Risk Promoters

As the gene promoter used in a transgenic event can have a strong impact on the eventual concentration and distribution of Bt toxins within the plant, the choice of promoter should be made within the context of environmental safety. Certain promoters have been identified as having greater non-target risks than others; for example, harmful effects of Bt corn event 176 on non-target Lepidoptera larvae [monarchs *Danaus plexippus* (L.) (Nymphalidae) and black swallowtails *Papilio polyxenes* F. (Papilionidae)] have been reported, while other events expressing the same Cry toxin (e.g., Bt11 and MON810) had no effect.[90] Event 176 has increased expression of Bt proteins in the pollen compared with the other events[130] and therefore poses a greater risk to non-target organisms.

Tissue- and Time-Specific Expression

The use of gene promoters that are tissue- or time-specific to express toxins only in plant tissues when they are susceptible to feeding has been introduced.[131] This technique has been employed in the transgenic expression of snowdrop lectin, a plant-derived protein with insecticidal properties, in rice. To target phloem-feeding pests such as brown planthoppers, lectins are selectively expressed in the vascular tissue.[132,133] Such selective expression of Bt toxins in tissue and time to target susceptible pests and reduce exposure to non-target beneficial arthropods could potentially increase environmental safety, thereby reducing the pathways for Bt toxin movement through non-target food webs.

Reduction in Bt Toxin and Transgene Escape

At the field or farm level, management practices may be implemented that reduce the movement of Bt toxins or transgenes from their source (Bt plants) into surrounding habitats. Current practices may depend on the crop and agronomic aims of the grower; for example, large quantities of crop residue may be incorporated into the soil during the harvesting process, although this is not the practice when crop material is removed for ethanol production or under reduced-tillage practices.[134] Although Bt toxins may degrade quickly following the incorporation of Bt crop plant detritus into aquatic systems, this potential pathway for transgenic protein movement may be avoided through the employment of practices that prevent movement of transgenic crop tissue beyond field borders. The establishment of riparian buffer zones and filter strips may reduce the quantity of crop detritus and other compounds originating in cropland (e.g., fertilizers, insecticides) that enter nearby streams and waterways.[135] Similarly, reducing exposure pathways for gene flow into wild plant populations via physical methods, such as isolation of crops or plant destruction, may delay transgene escape. However, controlling gene flow via pollen and seeds in the environment can be very difficult; a physical separation of 200 m between transgenic corn still yields contamination levels of 0.1% between plant populations due to cross-pollination.[136] Seeds are additionally difficult to control owing to their persistence in the soil seed bank, as well as ability to sometimes germinate and persist outside of cultivated fields.[137] Management of the movement of Bt toxins and genetic material from cultivated fields into the surrounding environments warrants additional research. Interestingly, technology that could have reduced the spread of transgenes, called the "terminator gene," was abandoned in 1999 because of the criticism that the gene prevents farmers from harvesting viable seed and thereby exclusively benefits the seed companies.[138]

Large-Scale Integrated Pest and Resistance Management

Although Bt crops allow reductions in the application of certain insecticides compared with conventionally managed crops (while other insecticidal practices persist, such as neonicotinoid seed treatments on corn[139]), it should not be assumed that this technology will readily fit into integrated pest management practices.[17] Considerations of compatibility with biological control and delaying resistance in pest populations are also necessary.

Compatibility with biological control

Integrated pest management practices attempt to incorporate mechanical, physical, chemical, and cultural controls; host resistance (including transgenic crops); and autocidal, biochemical, and biological controls in a synergistic manner. Increased attention has focused on conservation biological control: the modification of the environment or existing practices to protect and enhance specific natural enemies or other organisms to reduce effects of pests (e.g., Landis et al.[140] and Eilenberg et al.[141]). Natural enemies can be abundant in agricultural systems and often play an essential role in pest suppression. Maintenance of relevant natural enemy populations via conservation biological control is a practical and sustainable option for high-acreage field crops, such as corn and cotton,[142] which are dominated by Bt varieties. Any negative effects of Bt toxins on natural enemies could reduce their effectiveness as biological control agents and therefore limit natural pest suppression in agroecosystems. Understanding the potential impacts of transgenic crops on non-target arthropods is essential in order to provide a framework for integrating natural enemies into sustainable methods of pest control in the agricultural environment.

Resistance Management Techniques

The development of resistance to Bt toxins by pest populations is a major concern. Integrated resistance management programs must continue to be developed and followed to promote the sustainable use of Bt crops. This is of critical importance given that resistance to Bt sprays has occurred in multiple populations of the pestiferous diamond back moth *Plutella xylostella* (L.) (Lepidoptera: Plutellidae)[143] and three instances of field-evolved resistance to transgenic Bt crops have been reported in moth larvae: *Spodoptera frugiperda* (Smith) (Lepidoptera: Noctuidae) to Cry1F corn in Puerto Rico, *Busseola fusca* (Fuller) (Lepidoptera: Noctuidae) to Cry1Ab corn in South Africa, and *Pectinophora gossypiella* (Saunders) (Lepidoptera: Gelechiidae) to Cry1Ac in the southwestern United States.[144] Current resistance management employs structured refuges and high-dose toxin crops, as well as monitoring for resistance development in the field and monitoring for compliance of growers to refuge protocol. Additional attempts to delay resistance include creating transgenic plants that express more than one type of Bt toxin that targets the same pest, called gene pyramiding.[131] Improved resistance management would include increased education for growers and the public about the importance of resistance management and refuge compliance, as well as continued monitoring of field populations for the development of resistance. Future strategies to passively achieve resistance compliance include mixed seed refuges, in which transgenic and non-transgenic seeds are sold in combination within the bag.[145]

CONCLUSIONS

The sources and fate of Bt toxins in the environment can be complex and variable depending on crop, transgenic event, geography, and other environmental variables. The effects of Bt crops and their toxins on the environment have been widely debated, particularly the potential implications associated with ecological impacts such as gene escape and non-target risks. Approaches to the environmental management of Bt crops and their integration into integrated pest and resistance management systems warrant further study. Despite the concerns associated with Bt crops, significant reductions in chemical input are evident and this technology is environmentally safer when compared with many approaches to pest suppression, particularly those using broad-spectrum insecticides.

Future of Bt Technology

The focus of current transgenic technology has been on stacking and pyramiding of events. Stacking incorporates multiple transgenic traits into the crop genome in order to express more than one type of insecticidal toxin, therefore targeting multiple pest species. Pyramiding of transgenes allows for the crop to express multiple types of Bt toxins that target the same pest. Additionally, several other Bt crops are expected to be approved for commercial availability by 2015, including potatoes for planting in eastern Europe and eggplant in India.[5] The global adoption of biotechnology in agriculture is projected to continue with estimates that genetically modified crops will reach 200 million hectares, grown by 20 million farmers in 40 countries by 2015.[5]

ACKNOWLEDGMENTS

Funding for this project was provided by the U.S. Department of Agriculture Cooperative State Research Education and Extension Service Biotechnology Risk Assessment Grant 2006-39454-17446. JDH is supported by the University of Kentucky Agricultural Experiment Station State Project KY008043. This is publication number 10-08-132 of the University of Kentucky Agricultural Experiment Station.

REFERENCES

1. Yu, C.G.; Mullins, M.A.; Warren, G.W., Koziel, M.G.; Estruch, J.J. The *Bacillus thuringiensis* vegetative insecticidal protein Vip3 lyses midgut epithelium cells of susceptible insects. Appl. Environ. Microbiol. **1997**, *63*, 532–536.
2. Glare, T.R.; O'Callaghan, M. *Bacillus thuringiensis: Biology, Ecology, and Safety*; John Wiley & Sons, Ltd.: West Sussex, U.K., 2000.
3. Mendehlson, M.; Kough, J.; Vaituzis, Z.; Matthews, K. Are *Bt* crops safe? Nat. Biotechnol. **2003**, *21* (9), 1003–1009.
4. Kaniewski, W.K.; Thomas, P.E. The potato story. AgBioForum **2004**, *7*, 41–46.
5. James, C. *Global Status of Commercialized Biotech/GM Crops: 2009*; ISAAA Brief No. 41. ISAAA: Ithaca, NY, 2009.
6. Cannon, R.J.C. Bt transgenic crops: Risks and benefits. Integr. Pest Manag. Rev. **2000**, *5*, 151–173.
7. Pray, C.E.; Huang, J.; Hu, R.; Rozelle, S. Five years of Bt cotton in China—The benefits continue. Plant J. **2002**, *31*, 423–430.
8. Shelton, A.M.; Zhao, J.Z.; Roush, R.T. Economic, ecological, food safety, and social consequences of the deployment of Bt transgenic plants. Annu. Rev. Entomol. **2002**, *47*, 845–881.
9. Lawrence, S. AgBio keeps on growing. Nat. Biotechnol. **2005**, *63*, 3561–3568.
10. James, C. *Executive Summary of Global Status of Commercialized Biotech/GM Crops: 2006*; ISAAA Brief No. 36. ISAAA: Ithaca, NY, 2006.
11. United States Department of Agriculture National Agricultural Statistics Service. Acreage report, 2010, available at http://usda.mannlib.cornell.edu/usda/current/Acre/Acre-06-30-2010.pdf. (Accessed, November 2010).
12. GMO Compass. *Rising Trend: Genetically Modified Crops Worldwide on 125 Million Hectares*, 2009, available at http://www.gmocompass.org/eng/agri_biotechnology/gmo_planting/257.global_gm_planting_2008.html.
13. Fearing, P.L.; Brown, D.; Vlachos, D.; Meghji, M.; Privalle, L. Quantitative analysis of CryIA(b) expression in *Bt* maize plants, tissues, and silage and stability of expression over successive generations. Mol. Breed. **1997**, *3* (3), 169–176.
14. Duan, J.J.; Head, G.; McKee, M.J.; Nickson, T.E.; Martin, J.H.; Sayegh, F.S. Evaluation of dietary effects of transgenic corn pollen expressing Cry3Bb1 protein on a non-target ladybird beetle, *Coleomegilla maculata*. Entomol. Exp. Appl. **2002**, *104*, 271–280.
15. Grossi-de-Sa, M.F., Lucena, W.; Souza, M.L.; Nepomuceno, A.L.; Osir, E.O.; Amugune, N.; Hoa, T.T.C.; Hai, T.N.H.; Somers, D.A.; Romano, E. Transgene expression and locus structure of Bt cotton. In *Environmental Risk Assessment of Genetically Modified Organisms. Methodologies for Assessing Bt Cotton in Brazil*; Hilbeck, A., Andow, D.A., Fontes, E.M.G., Eds., CAB International: Wallingford, U.K., 2006; 93–107.
16. Obrist, L.B.; Dutton, A.; Albajes, R.; Bigler, F. Exposure of arthropod predators to Cry1Ab toxin in Bt maize fields. Ecol. Entomol. **2006**, *31* (2), 143–154.
17. Lundgren, J.G.; Gassman, A.J.; Bernal, J.; Duan, J.J.; Ruberson, J. Ecological compatibility of GM crops and biological control. Crop Prot. **2009**, *28*, 1017–1030.
18. Peterson, J.A.; Obrycki, J.J.; Harwood, J.D. Quantification of Bt-endotoxin exposure pathways in carabid food webs across multiple transgenic events. Biocontrol Sci. Technol. **2009**, *19* (6), 613–625.
19. Li, X.; Pang, Y.; Zhang, R. Compositional changes of cotton-seed hull substrate during *P. ostreatus* growth and the effects on the feeding value of spent substrate. Bioresour. Technol. **2001**, *80*, 157–161.
20. Li, M.H.; Hartnell, G.F.; Robinson, E.H.; Kronenberg, J.M.; Healy, C.E.; Oberle, D.F.; Hoberg, J.R. Evaluation of cottonseed meal derived from genetically modified cotton as feed ingredients for channel catfish, *Ictalurus punctatus*. Aquac. Nutr. **2008**, *14* (6).

21. Robertson, R. *Cotton Gin Trash Now Valuable By-Product*, Southeast Farm Press, 2009, available at http://southeast farmpress.com/grains/cotton-gin-trash-now-valuable-product.
22. Jiang, L.; Tian, X.; Duan, L.; Li, Z. The fate of Cry1Ac Bt toxin during oyster mushroom (*Pleurotus ostreatus*) cultivation on transgenic Bt cottonseed hulls. J. Sci. Food Agr. **2008**, *88*, 214–217. Accessed November 1, 2010.
23. Dutton, A.; Klein, H.; Romeis, J.; Bigler, F. Uptake of Bt-toxin by herbivores feeding on transgenic maize and consequences for the predator *Chrysoperla carnea*. Ecol. Entomol. **2002**, *27* (4), 441–447.
24. Harwood, J.D.; Wallin, W.G.; Obrycki, J.J. Uptake of Bt endotoxins by nontarget herbivores and higher order arthropod predators: Molecular evidence from a transgenic corn agroecosystem. Mol. Ecol. **2005**, *14* (9), 2815–2823.
25. Meissle, M.; Vojtech, E.; Poppy, G.M. Effects of Bt maize-fed prey on the generalist predator *Poecilus cupreus* L. (Coleoptera: Carabidae). Transgenic Res. **2005**, *14* (2), 123–132.
26. Obrist, L.B.; Klein, H.; Dutton, A.; Bigler, F. Effects of Bt maize on *Frankliniella tenuicornis* and exposure of thrips predators to prey-mediated Bt toxin. Entomol. Exp. Appl. **2005**, *115* (3), 409–416.
27. Obrist, L.B.; Dutton, A.; Romeis, J.; Bigler, F. Biological activity of Cry1Ab toxin expressed by Bt maize following ingestion by herbivorous arthropods and exposure of the predator *Chrysoperla carnea*. BioControl **2006**, *51* (1), 31–48.
28. Raps, A.; Kehr, J.; Gugerli, P.; Moar, W.J.; Bigler, F.; Hilbeck, A. Immunological analysis of phloem sap of *Bacillus thuringiensis* corn and of the nontarget herbivore *Rhopalosiphum padi* (Homoptera: Aphididae) for the presence of Cry1Ab. Mol. Ecol. **2001**, *10* (2), 525–533.
29. Bernal, C.C.; Aguda, R.M.; Cohen, M.B. Effect of rice lines transformed with *Bacillus thuringiensis* toxin genes on the brown planthopper and its predator *Cyrtorhinus lividipennis*. Entomol. Exp. Appl. **2002**, *102* (1), 21–28.
30. Burgio, G.; Lanzoni, A.; Accinelli, G.; Dinelli, G.; Bonetti, A.; Marotti, I.; Ramilli, F. Evaluation of *Bt*-toxin uptake by the non-target herbivore, *Myzus persicae* (Hemiptera: Aphididae), feeding on transgenic oilseed rape. Bull. Entomol. Res. **2007**, *97* (2), 211–215.
31. Head, G.; Brown, C.R.; Groth, M.E.; Duan, J.J. Cry1Ab protein levels in phytophagous insects feeding on transgenic corn: Implications for secondary exposure risk assessment. Entomol. Exp. Appl. **2001**, *99* (1), 37–45.
32. Lundgren, J.G. *Relationships of Natural Enemies and Non-prey Foods*; Springer International: Dordrecht, the Netherlands, 2009.
33. Hagen, K.S.; Mills, N.J.; Gordh, G.; McMurtry, J.A. Terrestrial arthropod predators of insect and mite pests. In *Handbook of Biological Control*; Bellows, T.S., Fisher, T.W., Eds.; Academic Press: San Diego, CA, 1999; 383–504.
34. Kevan, P.G.; Kevan, D.K.M. Collembola as pollen feeders and flower visitors with observations from the high Arctic. Quaest. Entomol. **1970**, *6*, 311–326.
35. Chen, B.; Snider, R.J.; Snider, R.M. Food consumption by Collembola from northern Michigan deciduous forest. Pedobiologia **1996**, *40*, 149–161.
36. Kim, J.H.; Mullin, C.A. Impact of cysteine proteinase inhibition in midgut fluid and oral secretion on fecundity and pollen consumption of western corn rootworm (*Diabrotica virgifera virgifera*). Arch. Insect Biochem. **2003**, *52*, 139–154.
37. Lundgren, J.G.; Wiedenmann, R.N. Nutritional suitability of corn pollen for the predator *Coleomegilla maculata* (Coleoptera: Coccinellidae). J. Insect Physiol. **2004**, *50* (6), 567–575.
38. Larochelle, A.; Larivière, M.C. *A Natural History of the Ground-Beetles (Coleoptera: Carabidae) of America North of Mexico*; Pensoft: Sofia, Bulgaria, 2003.
39. Canard, M. Natural food and feeding habits of lacewings. In *Lacewings in the Crop Environment*; McEwen, P.K., New, T.R., Whittington, A.E., Eds.; Cambridge University Press: Cambridge, U.K. 2001; 116–129.
40. Olesen, J.M.; Warncke, E. Predation and potential transfer of pollen in a population of *Saxifraga hirculus*. Holarctic Ecol. **1989**, *12*, 87–95.
41. Smith, R.B.; Mommsen, T.P. Pollen feeding in an orb-weaving spider. Science **1984**, *226* (4680), 1330–1332.
42. Ludy, C. Intentional pollen feeding in the garden spider *Araneus diadematus*. Newsl. Br. Arachnol. Soc. **2004**, *101*, 4–5.
43. Pleasants, J.M.; Hellmich, R.L.; Dively, G.P.; Sears, M.K.; Stanley-Horn, D.E.; Mattila, H.R.; Foster, J.E.; Clark, P.; Jones, G.D. Corn pollen deposition on milkweeds in and near cornfields. Proc. Natl. Acad. Sci. U. S. A. **2001**, *98*, 11919–11924.
44. Peterson, J.A.; Romero, S.A.; Harwood, J.D. Pollen interception by linyphiid spiders in a corn agroecosystem: Implications for dietary diversification and risk-assessment. Arthropod Plant Interact. **2010**, *4* (4), 207–217.
45. Isenhour, D.J.; Marston, N.L. Seasonal cycles of *Orius insidiosus* (Hemiptera: Anthocoridae) in Missouri soybeans. J. Kans. Entomol. Soc. **1981**, *54*, 129–142.
46. Coll, M.; Bottrell, D.G. Microhabitat and resource selection of the European corn-borer (Lepidoptera, Pyralidae) and its natural enemies in Maryland field corn. Environ. Entomol. **1991**, *20*, 526–533.
47. Zwahlen, C.; Hilbeck, A.; Gugerli, P.; Nentwig, W. Degradation of the Cry1Ab protein within transgenic *Bacillus thuringiensis* corn tissue in the field. Mol. Ecol. **2003**, *12* (3), 765–775.
48. Zwahlen, C.; Andow, D.A. Field evidence for the exposure of ground beetles to Cry1Ab from transgenic corn. Environ. Biosafety Res. **2005**, *4* (2), 113–117.
49. Saxena, D.; Stotzky, G. *Bacillus thuringiensis* (Bt) toxin released from the root exudates and biomass of Bt corn has no apparent effect on earthworms, nematodes, protozoa, bacteria, and fungi in soil. Soil Biol. Biochem. **2001**, *33* (9), 1225–1230.
50. Zwahlen, C.; Hilbeck, A.; Howald, R.; Nentwig, W. Effects of transgenic Bt corn litter on the earthworm *Lumbricus terrestris*. Mol. Ecol. **2003**, *12* (4), 1077–1086.
51. Harwood, J.D.; Obrycki, J.J. The detection and decay of Cry1Ab Bt-endotoxins within non-target slugs, *Deroceras reticulatum* (Mollusca: Pulmonata), following consumption of transgenic corn. Biocontrol Sci. Technol. **2006**, *16* (1), 77–88.
52. Stone, M.L.; Whiles, M.R.; Webber, J.A.; Williard, K.W.J.; Reeve, J.D. Macroinvertebrate communities in agriculturally impacted southern Illinois streams: Patterns with ripar-

ian vegetation, water quality, and in-stream habitat quality. J. Environ. Qual. **2005**, *34*, 907–917.

53. Prihoda, K.R.; Coats, J.R. Aquatic fate and effects of *Bacillus thuringiensis* Cry3Bb1 protein: Toward risk-assessment. Environ. Toxicol. Chem. **2008**, *27* (4), 793–798.
54. Jensen, P.D.; Dively, G.P.; Swan, C.M.; Lamp, W.O. Exposure and nontarget effects of transgenic Bt corn debris in streams. Environ. Entomol. **2010**, *39* (2), 707–714.
55. Saxena, D.; Stotzky, G. Insecticidal toxin from *Bacillus thuringiensis* is released from roots of transgenic Bt corn in vitro and in situ. FEMS Microbiol. Ecol. **2000**, *33* (1), 35–39.
56. Saxena, D.; Stewart, C.N.; Altosaar, I.; Shu, Q.Y.; Stotzky, G. Larvicidal Cry proteins from *Bacillus thuringiensis* are released in root exudates of transgenic *B-thuringiensis* corn, potato, and rice but not of *B-thuringiensis* canola, cotton, and tobacco. Plant Physiol. Biochem. **2004**, *42* (5), 383–387.
57. Koskella, J.; Stotzky, G. Microbial utilization of free and clay-bound insecticidal toxins from *Bacillus thuringiensis* and their retention of insecticidal activity after incubation with microbes. Appl. Environ. Microb. **1997**, *63* (9), 3561–3568.
58. Stotzky, G. Persistence and biological activity in soil of the insecticidal proteins from *Bacillus thuringiensis*, especially from transgenic plants. Plant Soil **2004**, *266* (1–2), 77–89.
59. Icoz, I.; Stotzky, G. Fate and effects of insect-resistant *Bt* crops in soil ecosystems. Soil Biol. Biochem. **2008**, *40*, 559–586.
60. Palm, C.J.; Schaller, D.L.; Donegan, K.K.; Seidler, R.J. Persistence in soil of transgenic plant-produced *Bacillus thuringiensis* var. *kurstaki* Δ-endotoxin. Can. J. Microbiol. **1996**, *42*, 1258–1262.
61. Crecchio, C.; Stotzky, G. Insecticidal activity and biodegradation of the toxin from *Bacillus thuringiensis* subsp. *kurstaki* bound to humic acids from soil. Soil Biol. Biochem. **1998**, *30*, 463–470.
62. Skarphedinsdottir, H.; Gunnarsson, K.; Gudmundsson, G.A.; Nfon, E. Bioaccumulation and biomagnification of organochlorines in a marine food web at a pristine site in Iceland. Arch. Environ. Contam. Toxicol. **2009**, *58* (3), 800–809.
63. Torres, J.B.; Ruberson, J.R. Interactions of *Bacillus thuringiensis* Cry1Ac toxin in genetically engineered cotton with predatory heteropterans. Transgenic Res. **2008**, *17* (3), 345–354.
64. Harwood, J.D.; Samson, R.A.; Obrycki, J.J. No evidence for the uptake of Cry1Ab Bt-endotoxins by the generalist predator *Scarites subterraneus* (Coleoptera: Carabidae) in laboratory and field experiments. Biocontrol Sci. Technol. **2006**, *16* (4), 377–388.
65. Jiang, Y.-H.; Fu, Q.; Cheng, J.-A.; Zhu, Z.-R.; Jiang, M.-X., Ye, G.-Y.; Zhang, Z.-T. Dynamics of Cry1Ab protein from transgenic Bt rice in herbivores and their predators. Acta Entomol. Sin. **2004**, *47*, 454–460 [in Chinese with English abstract].
66. Harwood, J.D.; Samson, R.A.; Obrycki, J.J. Temporal detection of Cry1Ab-endotoxins in coccinellid predators from fields of *Bacillus thuringiensis* corn. Bull. Entomol. Res. **2007**, *97*, 643–648.
67. Musser, F.R.; Shelton, A.M. Bt sweet corn and selective insecticides: Impacts on pests and predators. J. Econ. Entomol. **2003**, *96* (1), 71–80.
68. Marvier, M.; McCreedy, C.; Regetz, J.; Kareiva, P. A meta-analysis of effects of Bt cotton and maize on nontarget invertebrates. Science **2007**, *316* (5830), 1475–1477.
69. Wolfenbarger, L.L.; Naranjo, S.E.; Lundgren, J.G.; Bitzer, R.J.; Watrud, L.S. Bt crop effects on functional guilds of non-target arthropods: A meta-analysis. PLoS One **2008**, *3* (5), e2118.
70. Peterson, J.A.; Lundgren, J.G.; Harwood, J.D. Interactions of transgenic *Bacillus thuringiensis* insecticidal crops with spiders (Araneae). J Arachnol., **2011**, 39(1), 1–21.
71. Betz, F.S.; Hammond, B.G.; Fuchs, R.L. Safety and advantages of *Bacillus thuringiensis*-protected plants to control insect pests. Regul. Toxicol. Pharmacol. **2000**, *32* (2), 156–173.
72. Gianessi, L.P.; Carpenter, J.E. Agricultural biotechnology: Insect control benefits; National Center for Food and Agricultural Policy, 1999, available at https://research.cip.cgiar.org/confluence/download/ attachments/3443/AG7.pdf.
73. Pray, C.; Ma, D.; Huang, J.; Qiao, F. Impact of Bt cotton in China. World Dev. **2001**, *29*, 813–825.
74. Thirtle, C.; Beyers, L.; Ismael, Y.; Piesse, J. Can GM-technologies help the poor? The impact of Bt cotton in Makhathini Flats, KwaZulu-Natal. World Dev. **2003,** *31*, 717–732.
75. Hutchinson, W.D.; Burkness, E.C.; Mitchel, P.D.; Moon, R.D.; Leslie, T.W.; Fleischer, S.J.; Abrahamson, M.; Hamilton, K.L.; Steffey, K.L.; Gray, M.E.; Hellmich, R.L.; Kaster, L.V.; Hunt, T.E.; Wright, R.J.; Pecinovsky, K.; Rabaey, T.L.; Flood, B.R.; Raun, E.S. Areawide suppression of European corn borer with Bt maize reaps savings to non-Bt maize growers. Science **2010**, *330*, 222–225.
76. Lutz, W.; Samir, K.C. Dimensions of global population projections: What do we know about future population trends and structures? Philos. Trans. R. Soc. Lond. B **2010**, *365* (1554), 2779–2791.
77. Giles, K.L.; Hellmich, R.L.; Iverson, C.T.; Lewis, L.C. Effects of transgenic *Bacillus thuringiensis* maize grain on *B-thuringiensis*-susceptible *Plodia interpunctella* (Lepidoptera: Pyralidae). J. Econ. Entomol. **2000**, *93* (3), 1011–1016.
78. Hussein, H.S.; Brasel, J.M. Toxicity, metabolism, and impact of mycotoxins on humans and animals. Toxicology **2001**, *167* (2), 101–134.
79. Munkvold, G.P.; Hellmich, R.L.; Rice, L.G. Comparison of fumonisin concentrations in kernels of transgenic Bt maize hybrids and nontransgenic hybrids. Plant Dis. **1999**, *83* (2), 130–138.
80. Lövei, G.L.; Arpaia, S. The impact of transgenic plants on natural enemies: A critical review of laboratory studies. Entomol. Exp. Appl. **2005**, *114*, 1–14.
81. Pilcher, C.D.; Obrycki, J.J.; Rice, M.E.; Lewis, L.C. Preimaginal development, survival, and field abundance of insect predators on transgenic *Bacillus thuringiensis* corn. Environ. Entomol. **1997**, *26*, 446–454.
82. Al-Deeb, M.A.; Wilde, G.E.; Higgins, R.A. No effect of *Bacillus thuringiensis* corn and *Bacillus thuringiensis* on the predator *Orius insidiosus* (Hemiptera: Anthocoridae). Environ. Entomol. **2001**, *30* (3), 625–629.
83. Lundgren, J.G.; Wiedenmann, R.N. Coleopteran-specific Cry3Bb1 toxin from transgenic corn pollen does not affect the fitness of a nontarget species, *Coleomegilla maculata*

DeGeer (Coleoptera: Coccinellidae). Environ. Entomol. **2002**, *31*, 1213–1218.

84. Anderson, P.L.; Hellmich, R.L.; Sears, M.K.; Sumerford, D.V.; Lewis, L.C. Effects of Cry1Ab-expressing corn anthers on monarch butterfly larvae. Environ. Entomol. **2004**, *33*, 1109–1115.

85. Ramirez-Romero, R.; Desneux, N.; Decourtye, A.; Chaffiol, A.; Pham-Delègue, M.H. Does Cry1Ab protein affect learning performances of the honey bee *Apis mellifera* L. (Hymenoptera, Apidae)? Ecotoxicol. Environ. Saf. **2008**, *70* (2), 327–333.

86. Hilbeck, A.; Moar, W.J.; Pusztai-Carey, M.; Filippini, A.; Bigler, F. Toxicity of *Bacillus thuringiensis* Cry1Ab toxin to the predator *Chrysoperla carnea* (Neuroptera: Chrysopidae). Environ. Entomol. **1998**, *27*, 1255–1263.

87. Ramirez-Romero, R.; Bernal, J.S.; Chaufaux, J.; Kaiser, L. Impact assessment of Bt-maize on a moth parasitoid, *Cotesia marginiventris* (Hymenoptera: Bracondiae), via host exposure to purified Cry1Ab protein or Bt-plants. Crop Prot. **2007**, *26*, 953–962.

88. Losey, J.E.; Rayor, L.S.; Carter, M.E. Transgenic pollen harms monarch larvae. Nature **1999**, *399*, 214.

89. Jesse, L.C.H.; Obrycki, J.J. Field deposition of *Bt* transgenic corn pollen: Lethal effects on the monarch butterfly. Oecologia **2000**, *1125*, 241–248.

90. Zangerl, A.R.; McKenna, D.; Wraight, C.L.; Carroll, M.; Ficarello, P.; Warner, R.; Berenbaum, M.R. Effects of exposure to event 176 *Bacillus thuringiensis* corn pollen on monarch and black swallowtail caterpillars under field conditions. Proc. Natl. Acad. Sci. U. S. A. **2001**, *98* (21), 11908–11912.

91. Rosi-Marshall, E.J.; Tank, J.L.; Royer, T.V.; Whiles, M.R.; Evans-White, M.; Chambers, C.; Griffiths, N.A.; Pokelsek, J.; Stephen, M.L. Toxins in transgenic crop byproducts may affect headwater stream ecosystems. Proc. Natl. Acad. Sci. U. S. A. **2007**, *104* (41), 16204–16208.

92. Duan, J.J.; Lundgren, J.G.; Naranjo, S.E.; Marvier, M. Extrapolating non-target risk of *Bt* crops from laboratory to field. Biol. Lett. **2010**, *6*, 74–77.

93. Zwahlen, C.; Nentwig, W.; Bigler, F.; Hilbeck, A. Tritrophic interactions of transgenic *Bacillus thuringiensis* corn, *Anaphothrips obscurus* (Thysanoptera: Thripidae), and the predator *Orius majusculus* (Heteroptera: Anthocoridae). Environ. Entomol. **2000**, *29* (4), 846–850.

94. Lundgren, J.G.; Wiedenmann, R.N. Tritrophic interactions among Bt (CryMb1) corn, aphid prey, and the predator *Coleomegilla maculata* (Coleoptera: Coccinellidae). Environ. Entomol. **2005**, *34* (6), 1621–1625.

95. Ferry, N.; Mulligan, E.A.; Majerus, M.E.N.; Gatehouse, A.M.R. Bitrophic and tritrophic effects of Bt Cry3A transgenic potato on beneficial, non-target, beetles. Transgenic Res. **2007**, *16*, 795–812.

96. Hilbeck, A.; Baumgartner, M.; Fried, P.M.; Bigler, F. Effects of transgenic *Bacillus thuringiensis* corn-fed prey on mortality and development time of immature *Chrysoperla carnea* (Neuroptera: Chrysopidae). Environ. Entomol. **1998**, *27*, 480–487.

97. Hilbeck, A.; Moar, W.J.; Pusztai-Carey, M.; Filippini, A.; Bigler, F. Prey-mediated effects of Cry1Ab toxin and protoxin and Cry2A protoxin on the predator *Chrysoperla carnea*. Entomol. Exp. Appl. **1999**, *91*, 305–316.

98. Naranjo, S.E. *Risk Assessment: Bt Crops and Invertebrate Non-Target Effects—Revisited*; ISB News Report: Agricultural and Environmental Biotechnology, December 2009, 1–4.

99. Naranjo, S.E. Impacts of *Bt* crops on non-target invertebrates and insecticide use patterns. CAB Rev. Perspect. Agric. Vet. Sci. Nutr. Nat. Resour. **2009**, *4* (11), 1–23.

100. Orr, D.B.; Landis, D.A. Oviposition of European corn borer (Lepidoptera: Pyralidae) and impact of natural enemy populations in transgenic versus isogenic corn. J. Econ. Entomol. **1997**, *90*, 905–909.

101. Reed, G.L.; Jensen, A.S.; Riebe, J.; Head, G.; Duan, J.J. Transgenic Bt potato and conventional insecticides for Colorado potato beetle management: Comparative efficacy and non-target impacts. Entomol. Exp. Appl. **2001**, *100*, 89–100.

102. Al-Deeb, M.A.; Wilde, G.E.; Blair, J.M.; Todd, T.C. Effect of *Bt* corn for corn rootworm control on nontarget soil microarthropods and nematodes. Environ. Entomol. **2003**, *32*, 859–865.

103. Sisterson, M.S.; Biggs, R.W.; Olson, C.; Carriére, Y.; Dennehy, T.J.; Tabashnik, B.E. Arthropod abundance and diversity in Bt and non-Bt cotton fields. Environ. Entomol. **2004**, *33*, 921–929.

104. Bhatti, M.A.; Duan, J.; Head, G.P.; Jiang, C.; McKee, M.J.; Nickson, T.E.; Pilcher, C.L.; Pilcher, C.D. Field evaluation of the impact of corn rootworm (Coleoptera: Chrysomelidae)-protected *Bt* corn on foliage-dwelling arthropods. Environ. Entomol. **2005**, *34*, 1336–1345.

105. de la Poza, M.; Pons, X.; Farinós, G.P.; López, C.; Ortego, F.; Eizaguirre, M.; Castañera, P.; Albajes, R. Impact of farm-scale Bt maize on abundance of predatory arthropods in Spain. Crop Prot. **2005**, *24*, 677–684.

106. Meissle, M.; Lang, A. Comparing methods to evaluate the effects of *Bt* maize and insecticide on spider assemblages. Agric. Ecosyst. Environ. **2005**, *107*, 359–370.

107. Naranjo, S.E. Long-term assessment of the effects of transgenic *Bt* cotton on the abundance of nontarget arthropod natural enemies. Environ. Entomol. **2005**, *34*, 1193–1210.

108. Naranjo, S.E. Long-term assessment of the effects of transgenic Bt cotton on the function of the natural enemy community. Environ. Entomol. **2005**, *34*, 1211–1223.

109. Ludy, C.; Lang, A. A 3-year field-scale monitoring of foliage-dwelling spiders (Araneae) in transgenic *Bt* maize fields and adjacent field margins. Biol. Control **2006**, *38*, 314–324.

110. Torres, J.B.; Ruberson, J.R. Abundance and diversity of ground-dwelling arthropods of pest management importance in commercial Bt and on-Bt cotton fields. Ann. Appl. Biol. **2007**, *150* (1), 27–39.

111. Duan, J.J.; Marvier, M.; Huesing, J.; Dively, G.; Huang, Z.Y. A meta-analysis of effects of Bt crops on honey bees (Hymenoptera: Apidae). PLoS One **2008**, *3* (1), e1415.

112. Bernstein, J.A.; Bernstein, I.L.; Bucchini, L.; Goldman, L.R.; Hamilton, R.G.; Lehrer, S.; Rubin, C.; Sampson, H.A. Clinical and laboratory investigation of allergy to genetically modified foods. Environ. Health Perspect. **2003**, *111* (8), 1114–1121.

113. Environmental Protection Agency. *Cry9C Food Allergenicity Assessment Background Document*, 1999, available at http://

www.epa.gov/pesticides/biopesticides/cry9c/cry9c-peer review.htm.
114. Carter, C.A.; Smith, A. Estimating the market effect of a food scare: The case of genetically modified StarLink corn. Rev. Econ. Stat. **2007**, *89* (3), 522–533.
115. Picard-Nizou, A.L.; Pham-Delegue, M.-H.; Kerguelen, V.; Douault, P.; Marillaeu, R.; Olsen, L.; Grison, R.; Toppan, A.; Masson, C. Foraging behaviour of honey bees (*Apis mellifera* L.) on transgenic oilseed rape (*Brassica napus* L. var. *oleifera*). Transgenic Res. **1995**, *4*, 270–276.
116. Birch, A.N.E.; Geoghegan, I.E.; Griffiths, D.W.; McNicol, J.W. The effect of genetic transformations for pest resistance on foliar solanidine-based glycoalkaloids of potato (*Solanum tuberosum*). Ann. Appl. Biol. **2002**, *140*, 143–149.
117. Faria, C.A.; Wäckers, F.L.; Pritchard, J.; Barrett, D.A.; Turlings, T.J.C. High susceptibility of Bt maize to aphids enhances the performance of parasitoids of lepidopteran pests. PLoS One **2007**, *2*, e600.
118. Flores, S.; Saxena, D.; Stotzky, G. Transgenic *Bt* plants decompose less in soil than non-*Bt* plants. Soil Biol. Biochem. **2005**, *37*, 1073–1082.
119. Zurbrugg, C.; Honemann, L.; Meissle, M.; Romeis, J.; Nentwig, W. Decomposition dynamics and structural plant components of genetically modified Bt maize leaves do not differ from leaves of conventional hybrids. Transgenic Res. **2010**, *19*, 257–267.
120. Virla, E.G.; Casuso, M.; Frias, E.A. A preliminary study on the effects of a transgenic corn event on the non target pest *Dalbulus maidis* (Hemiptera: Cicadellidae). Crop Prot. **2010**, *29* (6), 635–638.
121. Ellstrand, N.C.; Prentice, H.C.; Hancock, J.F. Gene flow and introgression from domesticated plants into their wild relatives. Annu. Rev. Ecol. Syst. **1999**, *30*, 539–563.
122. Quist, D.; Chapela, I.H. Transgenic DNA introgressed into traditional maize landraces in Oaxaca, Mexico. Nature **2001**, *414* (6863), 541–543.
123. Christou, P. No credible scientific evidence is presented to support claims that transgenic DNA was introgressed into traditional landraces in Oaxaca, Mexico. Transgenic Res. **2002**, *11* (1), III–V.
124. Kaplinsky, N.; Braun, D.; Lisch, D.; Hay, A.; Hake, S.; Freeling, M. Biodiversity (communications arising): Maize transgene results in Mexico are artefacts. Nature **2002**, *416* (6881), 601.
125. Quist, D.; Chapela, I.H. Maize transgene results in Mexico are artefacts—Reply. Nature **2002**, *416* (6881), 602.
126. Serratos-Hernández, J.-A.; Gómez-Olivares, J.-L.; Salinas-Arreortua, N.; Buendía-Rodríguez, E.; Islas-Gutiérrez, F.; de-Ita, A. Transgenic proteins in maize in the soil conservation area of Federal District, Mexico. Front. Ecol. Environ. **2007**, *5* (5), 247–252.
127. Ortiz-García, S.; Ezcurra, E.; Schoel, B.; Acevedo, F.; Soberón, J.; Snow, A.A. Absence of detectable transgenes in local landraces of maize in Oaxaca, Mexico (2003–2004). Proc. Natl. Acad. Sci. U. S. A. **2005**, *102* (35), 12338–12343.
128. Xia, H.; Lu, B.R.; Su, J.; Chen, R.; Rong, J.; Song, Z.P.; Wang, F. Normal expression of insect-resistant transgene in progeny of common wild rice crossed with genetically modified rice: Its implication in ecological biosafety assessment. Theor. Appl. Genet. **2009**, *119* (4), 635–644.
129. Committee on Genetically Modified Pest-Protected Plants, Board on Agriculture and Natural Resources, National Research Council. *Genetically Modified Pest-Protected Plants: Science and Regulation*; National Academy Press: Washington, D.C., 2000.
130. Wraight, C.L.; Zangerl, A.R.; Carroll, M.J.; Berenbaum, M.R. Absence of toxicity of *Bacillus thuringiensis* pollen to black swallowtails under field conditions. Proc. Natl. Acad. Sci. U. S. A. **2000**, *97* (14), 7700–7703.
131. Gould, F. Integrating pesticidal engineered crops into Mesoamerican agriculture. In *Transgenic Plants in Mesoamerican Agriculture: Bacillus thuringiensis*; Hruska, A.J., Pavón, M.L., Eds.; Zamorano Academic Press: Zamorano, Honduras, 1997; 6–36.
132. Shi, Y.; Wang, W.B.; Powell, K.S.; Van Damme, E.; Hilder, V.A.; Gatehouse, A.M.R.; Boulter, D.; Gatehouse, J.A. Use of the rice sucrose synthase-1 promoter to direct phloem-specific expression of β-glucoronidase and snowdrop lectin genes in transgenic tobacco plants. J. Exp. Bot. **1994**, *45*, 623–631.
133. Rao, K.V.; Rathore, K.S.; Hodges, T.K.; Fu, X.; Stoger, E.; Sudhaker, D.; Williams, D.; Christou, P.; Bharathi, M.; Brown, D.P.; Powell, K.S.; Spence, J.; Gatehouse, A.M.R.; Gatehouse, J.A. Expression of snowdrop lectin (GNA) in transgenic rice plants confers resistance to rice brown planthopper. Plant J. **1998**, *15*, 469–477.
134. Giampietro, M.; Ulgiati, S.; Pimentel, D. Feasibility of large scale biofuel production—Does an enlargement of scale change the picture? Bioscience **1997**, *47*, 587–600.
135. Mayer, P.M.; Reynolds, S.K.; McCutchen, M.D.; Canfield, T.J. *Riparian Buffer Width, Vegetative Cover, and Nitrogen Removal Effectiveness: A Review of Current Science and Regulations*; EPA/600/R-05/118; U.S. Environmental Protection Agency: Cincinnati, OH, 2006.
136. National Academy of Sciences. *Genetically Modified Pest-Protected Plants: Science and Regulation*; National Academy Press: Washington D.C., 2000.
137. Pessel, F.D.; Lecomte, J.; Emeriau, V.; Krouti, M.; Messean, A.; Gouyon, P.H. Persistence of oilseed rape (*Brassica napus* L.) outside of cultivated fields. Theor. Appl. Genet. **2001**, *102*, 841–846.
138. Terminator gene halt a 'major U-turn.' *BBC News* 1999, available at http://news.bbc.co.uk/2/hi/science/nature/465222.stm.
139. Leslie, T.W.; Biddinger, D.J.; Mullin, C.A.; Fleischer, S.J. Carabidae population dynamics and temporal partitioning: Response to couples neonicotinoid-transgenic technologies in maize. Environ. Entomol. **2009**, *38* (3), 935–943.
140. Landis, D.A.; Wratten, S.D.; Gurr, G.M. Habitat management to conserve natural enemies of arthropod pests in agriculture. Annu. Rev. Entomol. **2000**, *45*, 175–201.
141. Eilenberg, J.; Hajek, A.; Lomer, C. Suggestions for unifying the terminology in biological control. Biocontrol **2001**, *46*, 387–400.
142. Thorbek, P.; Sunderland, K.D.; Topping, C.J. Reproductive biology of agrobiont linyphiid spiders in relation to habitat, season and biocontrol potential. Biol. Control **2004**, *30*, 193–202.
143. Tabashnik, B.E. Genetics of resistance to *Bacillus thuringiensis*. Annu. Rev. Entomol. **1994**, *34*, 47–79.
144. Tabashnik, B.E.; Patin, A.L.; Dennehy, T.J.; Liu, Y.-B.; Carrière, Y.; Sims, M.A.; Antilla, L. Frequency of

resistance to *Bacillus thuringiensis* in field populations of pink bollworm. P. Natl. Acad. Sci. U. S. A. **2000**, *97* (24), 12980–12984.

145. Environmental Protection Agency, Office of Pesticide Programs, Biopesticides and Pollution Prevention Division. *Optimum AcreMax Bt Corn Seed Blends: Biopesticides Registration Action Document*, 2010, available at http://www.epa.gov/oppbppd1/biopesticides/ingredients/tech_docs/brad_006490_oam.pdf.

146. Duan, J.J.; Head, G.; Jensen, A.; Reed, G. Effects of *Bacillus thuringiensis* potato and conventional insecticides for Colorado potato beetle (Coleoptera: Chrysomelidae) management on the abundance of ground-dwelling arthropods in Oregon potato ecosystems. Environ. Entomol. **2004**, *33*, 275–281.

Bacterial Pest Control

David N. Ferro
Department of Entomology, University of Massachusetts, Amherst, Massachusetts, U.S.A.

Abstract
Bacteria are the most widely used microbial agents for controlling insect pests. Bacteria that replicate within their hosts and that persist in the environment by maintaining an infection cycle are biological control agents in the traditional sense. However, some products produce toxins that kill insect pests and are applied the way an insecticide would be applied are often not considered to be biological control agents. Nematodes serve as vectors that mechanically penetrate into the insect hemocoel and deposit the bacteria; the bacteria then replicate and kill the host quickly by causing septicemia.

INTRODUCTION

Although many genera of bacteria are found to be associated with insects—such as *Clostridium*, *Strategus*, *Pseudomonas*, *Proteus*, *Diplococcus*, *Serratia*, *Bacillus*, and *Enterobacter*—only *Bacillus* and *Serratia* represent agents that cause suppression of insect populations, i.e., that perform as biological control agents. Bacteria are the most widely used microbial agents for controlling insect pests. Some species of *Bacillus* and *Serratia* kill by replicating within the host, while strains of *Bacillus thuringiensis* produce protein toxins that kill soon after being ingested. Bacteria that replicate within their hosts and that persist in the environment by maintaining an infection cycle are biological control agents in the traditional sense. However, products of *B. thuringiensis* that produce toxins that kill insect pests and are applied the way an insecticide would be applied are often not considered to be biological control agents. Bacteria in the genera *Photorhabdus* and *Xenorhabdus* are symbiotic with nematodes in the families Heterorhabditidae and Steinernematidiae, respectively. The nematodes serve as vectors that mechanically penetrate into the insect hemocoel and deposit the bacteria. The bacteria then replicate and kill the host quickly by causing septicemia. The only commercially available bacterial products are from strains of *B. popillae*, *B. thuringiensis*, and *Serratia entomophila*.

PAENIBACILLUS (FORMERLY *BACILLUS*) *POPILLIAE* (DUTKY)

Milky disease was first observed in Japanese beetle larvae (grubs) in New Jersey in 1933. *P. popilliae* is an obligate pathogen of larvae in the family Scarabaeidae, as it is only found associated with its host or in the soil surrounding its host. *P. popilliae* and *Paenibacillus lentimorbus* (Dutky) both cause milky disease of scarab beetles; however, most discussions of milky disease refer to strains of *P. popilliae*. *P. popilliae* produces a crystal or parasporal body, which allows it to survive for many years in the soil in the absence of its host. Although there are dozens of strains of *P. popilliae* that infect scarab hosts, only *P. popilliae* has been used commercially as a biological control agent of the Japanese beetle, *Popillia japonica* (Newman), a major pest of turf.

The term "milky disease" describes the advanced stages of infection in scarab larvae where the host is turned a milky white by the build-up of *Bacillus* spores in the hemolymph. The infection process begins with the scarab larvae ingesting spores while feeding on roots and organic matter in the soil. The spores then undergo germination and outgrowth in the cells of the lumen of the alimentary canal. The vegetative rods penetrate the epithelial cells of the midgut, and then move into the hemolymph where they multiply and sporulate. Death often occurs a month or more after ingestion. It is unclear what the role of the proteinaceous parasporal body is in the infection process.

Culture and Control

Many attempts have been made to rear *P. popillae* on an artificial diet. Even though spores and vegetative rods from field-collected larvae can be plated on agar media, the inability of the milky disease bacteria to grow and sporulate on standard microbiological media has made it extremely costly to produce for commercial purposes. Products, to date, are made from milky larvae, primarily from naturally infected larvae collected from the field.

The spores are formulated on talc and contain 10^8 spores/g of powder. The powder is applied at about 20 kg/ha using a fertilizer spreader or by punching holes in the soil and adding bacteria. Infection can occur in all three larval stages. For optimal replication to occur, soil temperatures need to exceed 20°C. Large overwintered larvae usually pupate before soil temperatures are high enough in late spring. For this reason, applications are targeted against small larvae

Encyclopedia of Environmental Management DOI: 10.1081/E-EEM-120003820

late in the summer when the small larvae are actively feeding near the soil surface. Control seems to be greatest when larval densities exceed 300/m^2; however, economic losses in turf occur at densities above 100/m^2. Unless a more virulent strain is found or a more cost-effective way to produce spores is developed, the use of this bacterium is likely to be restricted to lawns and playing fields that can tolerate higher densities of larvae.

SERRATIA ENTOMOPHILA (GRIMONT ET AL.)

Amber disease of the New Zealand grass grub *Costelytra zealandica* (White) is a chronic infection of the larval gut caused by *S. entomophila*. This disease was first observed in New Zealand in 1981. Following ingestion of bacterial cells while feeding on grass roots, the bacteria adhere to the foregut and multiply in the region of the cardiac valve; the larvae cease feeding after 2–5 days and become amber colored due to clearance of the gut. Death does not occur until 1–3 mo after ingestion. As the disease progresses, the larvae become shrunken due to a general degradation of the fat cells. Invasion of the hemocoel does not occur until late stages of the disease, when general septicemia is accompanied by death of the insect.

Culture and Control

S. entomophila is produced in large fermentors as nonspore-forming bacteria to be applied as a live microbial pesticide. Recently, the Industrial Processing Division of DSIR, New Zealand produced 4×10^{10} bacteria/mL, and field trials have shown that $>4 \times 10^{13}$/ha are needed for control. The problem with using live bacteria (vs. spores) is the difficulty of maintaining viability on the shelf and in the field prior to ingestion. Currently, refrigerated product can be kept for only 3 mo.

Grass grub larvae live in the soil as pests of low-value grasslands. Because *S. entomophila* is applied as live bacteria rather than as spores, it is more vulnerable to UV light and desiccation. For this reason, it is important to place the formulated material 2–5 cm below the soil surface using a subsurface applicator, such as a modified seed drill. This approach allows for 90% survival of the bacteria. Bacteria applied in this way quickly start an epizootic, which then spreads through the grass grub population.

BACILLUS THURINGIENSIS (BERLINER)

B. thuringiensis is a spore-forming bacterium that produces a parasporal crystal (protein delta-endotoxin). After the susceptible insect larva ingests the endotoxin, in the absence or presence of the spore, the crystal is solubilized and activated by alkaline (pH 10.5) gut proteases. The toxic subunits bind to receptor sites on the midgut epithelium within minutes of ingestion. This is quickly followed by lysis of these cells, causing a cessation of feeding within 10–15 min of ingestion. Although the spores pass into the hemocoel through pores in the epithelium of the midgut, it is the starvation in conjunction with infection that kills the insect. The toxins from these bacteria are formulated in much the same way as a synthetic toxin, and do not cause an epizootic.

There are several subspecies (= strains) of *B. thuringiensis* based on the serotype of flagellar antigens, and these subspecies produce different endotoxins, or at least different amounts of endotoxins that are relatively host specific. For example, *B.t. israelensis* is effective against Nematocera (Diptera) larvae such as mosquito larvae, *B.t. kurstaki* against Lepidoptera, *B.t. aizawai* against Lepidoptera, and *B.t. tenebrionis* against Chrysomelidae (Coleoptera). Notation for the gene that encodes for the toxin is in lowercase; for example, Cry3A gene regulates the production of the Cry3A toxin. Table 1 includes a list of some of the subspecies and toxins they produce. Because these bacteria are so

Table 1 *B. thuringiensis* subspecies and crystal protein toxins.

	***B.t.* subspecies**			
Crystal protein	***B.t aizawai***	***B.t. kurstaki***	***B.t. tenebrionis***	***B.t. israelensis***
CrylAa	*	*		
CrylAb	*	*		
Cry1Ac		*		
CrylC	*			
Cry ID	*			
Cry2A		*		
Cry2B		*		
Cry3A			*	*
Cry4A				*
Cry4B				*
Cry4C				*
Cry4D				*
CytA				*

Table 2 Stage-specific larval mortality for the Colorado potato beetle fed foliage treated with *B. thuringiensis san diego* (= *tenebrionis*).

			95% CI	
Larval stage	**LC_{50} (mg/l)**	**Larval weight (mg)**	**Lower**	**Upper**
Early 1st instar	2.03	1.0	1.46	2.60
Late 1st instar	3.92	2.3	2.02	6.27
Early 2nd instar	4.35	4.0	3.30	5.56
Late 2nd instar	14.45	7.8	10.75	19.50
Early 3rd instar	14.86	15.6	9.95	20.48

host-specific, they can be quickly incorporated into a pest management program in which biological control agents are an integral component.

Culture and Control

B. thuringiensis can be produced in large quantities using commercial fermentors. Formulations can be applied to foliage or other larval substrates in the same manner as most insecticides. However, several operative factors affect the effectiveness of these bacterial agents.

*B.t.*s are most effective against early instars (Table 2). Their effectiveness is very dependent upon ambient temperatures; the protein endotoxin is not very persistent; thorough coverage of foliage is necessary; and they are host-specific. This host specificity allows for control of the target pest without killing other insect biological control agents; however, in many cropping systems, there is a complex of insect pests and often these need to be controlled at the same time, which may require using the *B.t.* product with a synthetic insecticide, if natural controls fail. Novel ways have been developed to deliver the toxin for ingestion by the pest.

One of the genes that control the production of the toxin has been inserted into *Pseudomonas fluorescens*. After the fermentation has been completed, the broth is chemically treated and heated to kill the bacteria. During this process, the protein toxin becomes encapsulated by the bacterial cell wall. The encapsulation process appears to protect the toxin from degradation in the field, making it more persistent. Several genes have also been inserted into plants that express the toxin in its tissues. In the case of potatoes, the transgenic plants are highly resistant to the Colorado potato beetle, which has considerably reduced the insecticide load on potatoes.

POTENTIAL BIOLOGICAL CONTROL AGENTS

Bacillus sphaericus (Neide) has been shown to be toxic only to larvae of culicid Diptera mosquitoes. This bacterium can be easily produced via fermentation. Insecticidal activity is due to crystalline toxins associated with the cell wall. The toxin is released by digestion after the host insect has consumed the bacteria. *B. alvei* and *B. brevis* are infectious for larvae of several mosquito species. There is no evidence that these species are significant biocontrol agents. The success of these bacteria in the field is likely to be dependent on selection of strains that are more virulent and that can persist in a range of aquatic environments.

BIBLIOGRAPHY

1. Crickmore, N.; Zeigler, D.R.; Feitelson, J.; Schnepf, E.; Van Rie, J.; Lereclus, D.; Baum, J.; Dean, D.H. Revision of the nomenclature for the *Bacillus thuringiensis* pesticidal crystal proteins. Am. Soc. Micro. **1998**, *62*, 807–813.
2. Glare, T.R.; Jackson, R.A. *Use of Pathogens in Scarab Pest Management*; Intercept: Andover, England, 1992; 43–61, 179–198.
3. Jackson, T.A.; Huger, A.M.; Glare, T.R. Pathology of amber disease in the New Zealand grass grub *Costelytra zealandica* (Coleoptera: Scarabaeidae). J. Invert. Pathol. **1993**, *61*, 123–130.
4. Tanada, Y.; Kaya, H.K. *Insect Pathology*; Academic Press: New York, 1993; 83–146.
5. Van Driesche, R.G.; Bellows, T.S., Jr. *Biological Control*; Chapman and Hall: New York, 1996.

Bioaccumulation

Tomaz Langenbach
Federal University of Rio de Janeiro, Rio de Janeiro, Brazil

Abstract
Bioaccumulation is the equilibrium process in which the uptake of substances in biota reaches much higher concentrations than those occurring in the environments. In literature, a clear definition of the characteristics that identify the substances that are able or unable to bioaccumulate is not found. The most common use of this term is related to hazardous substances such as heavy metals or predominantly nonpolar xenobiotics. Many of these substances are in low concentrations in the environment and the severity of the poisoning effect is mainly due to bioaccumulation up to high concentrations in the organism. The poisoning effects in animals are quite diverse but the most common ones observed are related to the nervous system, which can lead to death.

MOLECULAR PROPERTIES FOR BIOACCUMULATION

Bioaccumulation occurs only with molecules with low degradability and correlates with their grade of lipophilicity.[1] Organic substances with main bonds of aliphatic and aromatic C—C, C—H, and C—Cl (or other halogens) are predominantly nonpolar molecules (Lipophilic) with low water solubility and high stability. They are less susceptible to chemical reactions of hydrolysis, oxidation, and enzymatic attack.[2] On the other hand, bonds with different functional groups with O, P, N, and other elements turn molecules more polar, soluble, and degraded more easily. Bioaccumulation can occur with molecules between 100 and 600 units of molecular weight with the maximum of 350.[3] Probably this is related to membrane permeability capacity.

A common feature of bioaccumulation is the molecular stability of lipophilic organic substances and the nondegradability of heavy metals. The severity of heavy metals is due to many factors.

1. Metals with Hg, Cd, Zn, Cu, and Pb are the most toxic and most studied types followed by metals containing Ni, Al, As, Cr, and other elements.[4] Bioaccumulation can also occur also with essential metals such as Fe, Zn, Cu, Mo, Na, and Ca.
2. Speciation is the anions or other components that constitute the heavy metal molecules. This is important in defining solubility that, for example, is high for sulfate and low for sulphide.[5] Heavy metals bound to organic molecules such as methyl, ethyl or other aliphatic or aryl groups increase penetration capacity through membranes and consequently, have a poisoning effect.
3. The sensivity to the toxic effects of heavy metals and other xenobionts is dependent on the biological material being a microorganism, plant, animal, or type of tissue.

BIOACCUMULATION AND THE ENVIRONMENT

The pollution sources can be released by discharge of substances with uneven distribution in air, water, and soil. The movement of these substances up to bioaccumulation can occur by different routes mainly mediated by the food chain. This process can involve water, suspended particles, sediments, food, soil, and air particles (Fig. 1). An important part of these substances can be concentrated in nonliving components. From these sources, persistent organic pollutants (POP) or heavy metals can be released to biota.[6] The final distribution presented by the mass balance in the environment with a group of organochlorines and polyaromatics, shows that most are found in soil or sediment, whereas for highly volatile substances, most remain in the atmosphere. Less than 0.7% of the total remains in vegetation and no more than $2 \times 10^{-3}\%$ can be found in the aquatic biota.[7] The relationship between biota and environment shows that concentrations of bioaccumulated heavy metals in organisms are always higher than in water, but are usually lower than in sediments.[4]

Water Environment

After pollution reaches water bodies, different processes can occur to incorporate it into nonliving components as sediments and biota represented by microorganisms, plants, crustaceans, fishes, etc. The route of pollutant uptake in the biota if from waterborne, adsorption, filtration, or by food chain is an important factor in bioaccumulation.[8] Along the food chain the step-wise increase of concentration from lower to higher trophic levels, called biomagnification, can reach the bioconcentration factor (BCF) up to 100,000. In this process terrestrial animals as well as birds can be heavily contaminated by eating polluted fish.

In the global marine environment the apparent final fate of persistent organic pollutants (POP) is in the flora, fauna,

Encyclopedia of Environmental Management DOI: 10.1081/E-EEM-100200060

Fig. 1 Bioaccumulation in the environment. The black points represent the pollutant molecules.

and sediment of the abyss.[9] The main transport of POP follows the downstream movement of the organic flow in the water and in the long term these chemicals are incorporated in the sediment that function as final sink. It was observed that the bioaccumulation in the deep water fishes are up to 10 times higher than in surface water fishes.[9]

In the flora and fauna some heavy metals can bioaccumulate up to threshold values and others maintain a correlation with the concentration in the environment.[4] In aquatic plants, fish, and other metazoarians the distribution of the substances are quite distinct between tissues.[10] Lipophilic substances are preferentially found in adipose tissues with high lipid content.

In the Soil

Soil is polluted in large areas by pesticides application or by discharge as final disposal in landfills of industrial products. These lipophilic substances move in soil rather slowly by leaching, runoff, and volatilization. The main factor that influences bioaccumulation process in the soil is the biodisponibility. This property is conditioned by the adsorption/desorption capacity of the different soils and by the chemical nature of the pollutant. This process is driven by the stronger or weaker binding forces involved, which influence the amount that is bioavailable for plant uptake of these lipophilic pollutants. The main flow of POP generally occurs toward organic matter from soil particles and not to biota, a process called preferential partition. A negative correlation was observed between the adsorbtion coefficient related to soil organic carbon (*Koc*) and the bioaccumulation factor by plants. This means that in organic rich soils, bioaccumulation in plants is rather small.[2] A similar situation occurs with microorganisms in which previous bioaccumulated organochorines can move out from the cell to the soil.[11] The preferential partition toward soil can be the reason why a lack of toxicity on soil microorganisms by pesticide applications was frequently observed, even in high concentrations of pesticides. Little information about bioaccumulation in soil could be observed, but nevertheless cotransport of some organochlorine accumulated in microorganisms in sand aquifers with low organic content was reported.[12]

Soil invertebrates such as earthworms, beetles, slugs, and others can bioaccumulate lipophilic pesticides. The bioaccumulation process could be seen as a soil to soil–water equilibrium followed by a soil–water to worm equilibrium.[2] Consumers of this biota in animals of higher trophic levels such as birds can biomagnificate these chemicals. Plants can adsorb and bioaccumulate products from the soil with incorporation of residues mainly in the root. The translocation from root to foliage depends on plant species and on the chemical properties of the pollutant. Several evidences show that lipophilic compounds are sorbed onto the outer surface of roots of several plants, and in this case translocation is very low.

Bioaccumulation in plants can also occur with heavy metals. As safety rules, domestic waste and sludge from wastewater treatment stations with heavy metal contamination can be applied on soil for agriculture up to limited amounts to avoid pollution with hazardous toxicological effects. Contaminated grass, grain, and fruits can be accumulated by biomagnification when consumed by mammals, insects, and birds. Terrestrial animals have a plant mediated relationship with soil contamination.

In the Air

The main sources of atmospheric pollution are pesticide spraying with the reverse process of evaporation from soil to air, poliaromatics produced by burning of fuels, and plastic incineration. The rate of entry to the atmosphere and the distance of movement are principally dependent on the vapor pressure of the pollutants and metereological conditions. In some cases movement occurs on a global scale. The dynamic nature of the atmosphere can dilute pollutants in the air to exceptionally low concentrations and in these cases no significant bioaccumultion occurs.[13] Nevertheless urban and industrial areas, as well as the margins of roads with intense traffic, can have high concentrations of pollutants. Plants exposed to xenobiotics in the form of vapor, particules, aerosol or larger droplets, can undergo a passive process of foliage adsorption with an uptake mainly in the wax cuticle.[2] Bioaccumulation in plants can result in damage and can also affect higher trophic levels that consume these vegetables.

Direct contact of animals with these chemicals can enter by the respiratory organs, in mammals, or the outer body surface, mainly in insects.

BIOACCUMULATION MECHANISM IN BIOTA

In terrestrial animals, pollutants can enter by dermal contact, respiration, and food consumption. Atmospheric pollutants move to the lungs, where an equilibrium is difficult to be established, while generally atmosphere dilutes pollutant concentrations, unless there is an exposure to constant pollution sources. Chemicals move from lungs to circulatory fluid (plasma) and can be metabolized with further excretion. Another route is the storage mainly in rich lipid bodies such as brain and eggs in bird's.[2] If the entrance is by food consumption, the gastrointestinal tract can eliminate[10] these substances or degrade than to more polar compounds with further excretion, or can promote adsorption by plasma following the same route described earlier.

The uptake of heavy metals in microorganisms can occur by bioadsorbtion in capsular polysaccharides and cell-wall polymers, or cross these layers and cell membranes by an active enzymatic process involving phosphatases, reaching to the cell interior.[14] Some authors define bioaccumulation as only the process in which molecules reach cell interior. Many cells from animals, plants, fungi, yeast, and bacteria have metal-binding proteins with low molecular weight called metallothioneins. These proteins bind mainly to Cd and Zn and constitute a protection mechanisms to the toxic effects of these substances. Some other cell protective mechanisms exist such as enhancement of efflux from cell to the outside. Metals bind on different macromolecules and change enzymatic activities with inhibition or stimulation effects.

Lipophilic hydrocarbons in microorganisms cross polysaccharides from capsule and cell wall polymers with adsorption mainly by the lipids of the membranes.[15] Compounds that are inserted in cell lipids are more difficult to be degraded by chemical or enzymatic processes getting higher persistence.[16] The probable mechanism of action seems to be nonspecific, this means not related to a specific target.

APPLICATIONS AND FUTURE PERSPECTIVES

From the scientific point of view a better understanding of the integration between the different environment compartments and biota including modeling systems is an important approach that needs more development. Another possibility is to use bioaccumulation for environmental monitoring, based on the accumulation capacity of many pollutants in specific plants or animals, allowing chemical measurements that otherwise in water or air are below the analytic detection capacity.[17] This method has the possibility to integrate all pollutant exposure of plants or animals in a specific environment and can be in the future an important parameter for ecotoxicological evaluations.

After the disaster of the mercury pollution in the Minamata Bay in Japan in which more than 630 people died and many became physically and mentally disabled, the magnitude of poisoning effects due to bioaccumulation was recognized for the first time.[18] This was the beginning of a scientific research that produced a large amount of information. With this knowledge it became clear that bioaccumulation is a natural process that cannot be stopped by man but can be avoided with a more efficient control of pollutant release. To overcome the economic, social, and political difficulties for better pollution control together with the development of more ecological technologies are our challenge for today and for the future.

REFERENCES

1. Connell, D.W. General Characteristics of Organic Compounds which Exhibit Bioaccumulation. In *Bioaccumulation of Xenobiotic Compounds*; Connell, D.W., Ed.; CRC Press: Boca Raton, Florida, 1990; 47–57.
2. Connell, D.W. Bioamagnification of Lipophilic Compounds in Terrestrial and Aquatic Systems. In *Bioaccumulation of Xenobiotics Compounds*; Connell, D.W., Ed.; CRC Press: Boca Raton, Florida, 1990; 145–185.
3. Brooke, D.N.; Dobbs, A.J.; Williams, N. Octanol: Water partition coefficients (*P*): measurement estimation and interpretation, particularly for chemicals with *P* > 105. Ecotoxicol. Eviron. Saf. **1986**, *11*, 251.
4. Goodyear, K.L.; McNeill, S. Bioaccumulation of heavy metals by aquatic macro-invertebrates of different feeding guilds: a review. Sci. Total Environ. **1999**, *229*, 1–19.
5. Bourg, A.C.M. Speciation of Heavy Metals in Soils and Groundwater and Implications for Their Natural and Provoked Mobility. In *Heavy Metals: Problems and Solutions*; Salomons, N., Fo¨rstner, U., Mader, P., Eds.; Springer Verlag: Berlin, 1995; 17–31.
6. Tsezos, M.; Bell, J.P. Comparison of the biosorption and desorption of hazardous organic pollutants by life and dead bioamass. Wat. Res. **1989**, *23* (5), 561–568.
7. Connell, D.W.; Hawker, D.W. Predicting the distribution of persistent organic chemicals in the environment. Chem. Aust. **1986**, *53*, 428.
8. Carbonell, G.; Ramos, C.; Pablos, M.V.; Ortiz, J.A.; Tarazona, J.V. A system dynamic model for the assessment of different exposure routes in aquatic ecosystems. Sci. Total Environ. **2000**, *247*, 107–118.
9. Froescheis, O.; Looser, R.; Cailliet, G.M.; Jarman, W.M.; Ballschmiter, K. The deep-sea as a final global sink of

semivolatile persistent organic pollutants part I: PCDs in surface and deep-sea dwelling fish of the north and south atlantic and the Monterey bay canion (California). Chemosphere **2000**, *40*, 651–660.
10. Lin, K.H.; Yen, J.H.; Wang, Y.S. Accumulation and elimination kinetics of herbicides butachlor, thiobencarb and Chlomethoxyfen by *Aristichthys nobilis*. Pestic. Sci. **1997**, *49*, 178–184.
11. Brunninger, B.M.; Mano, D.M.S.; Scheunert, I.; Langenbach, T. Mobility of the organochlorine compound dicofol in soil promoted by *Pseudomonas fluorescens*. Ecotoxic. Environ. Safety **1999**, *44*, 154–159.
12. Jenkins, M.B.; Lion, L.W. Mobile bacteria and transport of polynuclear aromatic hydrocarbons in porous media. Appl. Environ. Microbiol. **1993**, *59* (10), 3306–3313.
13. Connel, D.W. Environmental Routes Leading to the Bioaccumulation of Lipophilic Chemicals. In *Bioaccumulation of Xenobiotic Compounds*; Connell, D.W., Ed.; CRC Press: Boca Raton, FL, 1990; 59–73.
14. Gomes, N.C.M.; Mendonca-Hagler, L.C.S.; Savvaidis, I. Metal bioremediation by microorganisms. Rev. Microbiol. **1998**, *29*, 85–92.
15. Mano, D.M.S.; Langenbach, T. [14C]Dicofol association to cellular components of azospirillum. Pestic. Sci. **1998**, *53*, 91–95.
16. Mano, D.M.S.; Buff, K.; Clausen, E.; Langenbach, T. Bioaccumulation and enhanced persistence of the acaricide dicofol by *Azospirillum lipoferum*. Chemosphere **1996**, *33* (8), 1609–1619.
17. Maagd, P.G.J. Bioaccumulation test applied in whole effluent assessment: a review. Env. Toxic. Chem. **2000**, *19* (1), 25–35.
18. Takashi, H.; Tsubaki, T. Epidemiology: Mortality in Minamata Disease. In *Recent Advance in Minamata Disease Studies; Methylmercury Poisoning in Minamata and Niigata Japan*; Kodansha Ltd.: Tokyo, 1986; 1–23.

Biodegradation

Sven Erik Jørgensen
Institute A, Section of Environmental Chemistry, Copenhagen University, Copenhagen, Denmark

Abstract
Section 1 presents an overview of the most important properties for toxic chemical compounds, and it is concluded that biodegradability is a very important property to use for the estimation of the environmental effects of toxic substances. Definitions of units of biodegradability are given in the section "Biodegradation." A simple estimation method for biodegradability is presented in the section "Estimation of Biodegradation," with reference to available software for a first, fast but coarse estimation of the properties of environmentally threatening chemicals.

INTRODUCTION: OVERVIEW OF THE PROPERTIES OF TOXIC CHEMICAL COMPOUNDS OF PARTICULAR IMPORTANCE FOR ENVIRONMENTAL MANAGEMENT

Slightly more than 100,000 chemicals are produced in such an amount that they threaten or may threaten the environment. They cover a wide range of applications: household chemicals, detergents, cosmetics, medicines, dye stuffs, pesticides, intermediate chemicals, auxiliary chemicals in other industries, additives to a wide range of products, chemicals for water treatment, and so on. They are (almost) indispensable in modern society and cover many more or less essential needs in the industrialized world, which has increased the production of chemicals about 40-fold during the last five decades. A minor or major proportion of these chemicals are inevitably reaching the environment through their production, during their transportation from the industries to the end user, or by their application. In addition, the production or use of chemicals may cause more or less unforeseen waste or by-products, for instance, chloro-compounds from the use of chlorine for disinfection. As we would like to have the benefits of using the chemicals but cannot accept the harm they may cause, this conflict raises several urgent questions, which we already have discussed in other entries.

We cannot answer these crucial questions without knowing the properties of the chemicals. Organization for Economic Cooperation and Development (OECD) has made a review of the properties that we should know for all chemicals. We need to know the boiling point and melting point to know in which form (solid, liquid, or gas) the chemical will be found in the environment. We must know the dispersion of the chemicals in the five spheres: the hydrosphere, the atmosphere, the lithosphere, the biosphere, and the technosphere (the part of the earth that is controlled and under the influence of human technology). This will require knowledge about their solubility in water; the water/lipid partition coefficient; Henry's constant (the constant in Henry's Law, which indicates the distribution of the chemical between air and water); the vapor pressure; the rate of degradation by hydrolysis, photolysis, chemical oxidation, and microbiological processes; and the adsorption equilibrium between water and soil—all as functions of the temperature. We need to discover the interactions between living organisms and the chemicals, which implies that we should know the biological concentration factor (BCF), the magnification through the food chain, the uptake rate and the excretion rate by the organisms, and where in the organisms the chemicals will be concentrated, not only for one organism but for a wide range of organisms. We must also know the effects on a wide range of different organisms. It means that we should be able to find the Lethal Concentration causing 50% mortality oif the test animals (LC50) and Lethal Dose causing 50% mortality of the test animals (LD50) values; the Maximum Allowable Concentration (MAC) and Non-effect Concentration (NEC) values (for the abbreviations and the definitions used, see Appendix 5); the relationship between the various possible sublethal effects and concentrations; the influence of the chemical on fecundity; and the carcinogenic and teratogenic properties. We should also know the effect on the ecosystem level. How do the chemicals affect populations and their development and interactions, i.e., the entire network of the ecosystem? A reduction of one population may for instance influence the entire ecosystem because all populations are bound together in an ecological network.

Table 1 gives an overview of the most relevant physical–chemical properties of organic compounds and their interpretation with respect to the behavior of the environment. It is clear from the table that a high water solubility is not desirable as it implies that the chemical compound is very mobile. On the other hand, a low water solubility means a high solubility in fat tissue and,therefore, a high bioaccumulation and a high biomagnification. The

Encyclopedia of Environmental Management DOI: 10.1081/E-EEM-120048392

Table 1 Overview of the most relevant environmental properties of organic compoundsand their interpretation.

Property	Interpretation
Water solubility	High water solubility corresponds to high mobility.
K_{ow}	High K_{ow} means that the compound is lipophilic. It implies that it has a high tendency to bioaccumulate and be sorbed to soil sludge and sediment. BCF and Koc are correlated with K_{ow}.
Biodegradability	This is a measure of how fast the compound is decomposed to simpler molecules. A high biodegradation rate implies that the compound will not accumulate in the environment, while a lowbiodegradation rate may create environmental problems related to the increasing concentration in the environment and the possibilities of a synergistic effect with other compounds.
Volatilization, vapor	A high rate of volatilization (high vapor pressure) implies that thepressure compound will cause an air pollution problem.
Henry's constant, H	H determines the distribution between the atmosphere and the hydrosphere.
pK	If the compound is an acid or a base, pH determines whether the acid or the corresponding base is present. As the two forms havedifferent properties, pH becomes important for the properties of the compounds.

Note: K_{ow} = Ratio solubility in octanol (represent fat tissue) divided by the solubility in water; Koc = express the adsorption ability to soil consisting of 100% organic carbon, can also with good approximation be considered as the concentration in soil with 1100% orgniac carbon divided with the concentration in water at equilibrium; H = Henry's constant in Henry's Law; pK = – log (equilibrium constant for the dissociation process of acids: $HA = A^{-} + H^{+}$

biodegradability may, however, frequently be considered an even more important property than water solubility or K_{ow}. If a compound is biodegraded fast, it will be decomposed before it harms the environment. It means that a fast-biodegradation will, so to speak, neutralize the (harmful) effect of a high water solubility and a high solubility in fat tissue. On the other hand, if a compound is biodegraded slowly, it will stay in the environment for a very long time, which implies that a high mobility and a high risk of bioaccumulation and biomagnifications will be harmful. Therefore, it is almost possible to conclude that a compound with high biodegradability will clearly be much less harmful than a compound with a low biodegradability. Biodegradability is therefore a very crucial property for the estimation of a chemical compound's environmental effects.

The list of properties needed to give an adequate answer to the six questions mentioned above could easily be made longer (see, for instance, the list recommended by OECD). To provide all the properties corresponding to the list given here is already a huge task. More than 10 basic properties should be known for all 100,000 chemicals and organisms, which would require 1,000,000 pieces of information. In addition, we need to know at least 10 properties to describe the interactions between 100,000 chemicals and organisms. Let us say, modestly, that we use 10,000 organisms to represent the approximately 10 million species on earth. This gives a total of 1,000,000 + 100,000*10,000*10 = in the order of 10^{10} properties to be quantified! Today, we have determined less than 1% of these properties by measurements, and with the present rate of generating new data, we can be certain that during the 21st century, we shall not be able to reach 10% even with an accelerated rate of ecotoxicological measurements.

Environmental risk assessments require, among other inputs, information about the properties of the chemicals and their interactions with living organisms. It is maybe not necessary to know the properties with the very high accuracy that can be provided by measurements in a laboratory, but it would be beneficial to know the properties with sufficient accuracy to make it possible to utilize the models for management and risk assessment. Therefore, estimation methods have been developed as an urgently needed alternative to measurements. They are to a great extent based on the structure of the chemical compounds, the so-called QSAR and SAR methods (Quantitative Structure-Activity Relationship, it means estimation methods of chemical properties based on the chemical structure), but it may also be possible to use allometric principles to transfer rates of interaction processes and concentration factors between a chemical and one or a few organisms to other organisms.[1]

It may be interesting in this context to discuss the obvious question: why is it sufficient to estimate a property of a chemical in an ecotoxicological context with 20%, or sometimes 50% or higher, uncertainty? Ecotoxicological assessment usually gives an uncertainty of the same order of magnitude, which means that the indicated uncertainty may be sufficient from, for instance, the viewpoint of ecological modeling or ecological indicators, but can results with such an uncertainty be used at all? The answer in most (many) cases is "yes," because we want in most cases to assure that we are (very) far from a harmful or very harmful level. We use a safety factor of 100–1000 in many cases. When we are concerned with very harmful effects, such as, for instance, complete collapse of an ecosystem or a health risk for a large human population, we will inevitably select a safety factor that is very high. In addition, our lack of knowledge about synergistic effects and the presence of many compounds in the environment at the same time force us to apply a very high safety factor. In such a context, we will usually go for a concentration in the environment that is magnitudes lower than that corresponding to a slightly harmful effect or considerably lower than the NEC. It is analogous to civil engineers constructing bridges. They make very sophisticated calculations (develop models) that account for wind, snow, temperature changes, and so on, and afterward, they multiply the results by a safety factor of 2–3 to ensure that the bridge will not collapse. They use safety factors because the consequences of a bridge collapse are unacceptable.

The collapse of an ecosystem or a health risk to a large human population is also completely unacceptable. Thus, we should use safety factors in ecotoxicological modeling to account for the uncertainty. Duc to thc complcxity of the system, the simultaneous presence of many compounds, and our present knowledge, or rather, lack of knowledge, we should, as indicated above, use 10–100 or sometimes even 1000 as safety factor. If we use safety factors that are too high, the risk is only that the environment will be less contaminated at maybe a higher cost. Besides, there are no alternatives to the use of safety factors. We can, step by step, increase our ecotoxicological knowledge, but it will take decades before it may be reflected in considerably lower safety factors. A measuring program of all processes and components is impossible due to the high complexity of the ecosystems. This does not, of course, imply that we should not use the information of measured properties available today. Measured data will almost always be more accurate than the estimated data. Furthermore, the use of measured data within the network of estimation methods will improve the accuracy of estimation methods. Several handbooks on ecotoxicological parameters are, fortunately, available. References to the most important are given below. Estimation methods for the physical–chemical properties of chemical compounds were already applied 40–60 years ago, as they were urgently needed in chemical engineering. They are, to a great extent, based on contributions to a focal property by molecular groups and the molecular weight: the boiling point, the melting point, and the vapor pressure as a function of the temperature. In addition, a number of auxiliary properties result from these estimation methods, such as the critical data and the molecular volume. These properties may not have a direct application as ecotoxicological parameters in environmental risk assessment but are used as intermediate parameters, which may be used as a basis for estimation of other parameters.

The water solubility, the octanol/water partition coefficient, K_{ow}, and Henry's constant are crucial parameters in our network of estimation methods, because many other parameters are well correlated with these three parameters. The three properties can fortunately be found for a number of compounds or be estimated with reasonably high accuracy by use of knowledge of the chemical structure, i.e., the number of various elements, rings, and functional groups. In addition, there is a good relationship between water solubility and K_{ow}.[2] Particularly in the last decade, many good estimation methods for these three core properties have been developed.

During the last couple of decades, several correlation equations have been developed based upon a relationship between the water solubility, K_{ow}, or Henry's constant on the one hand and physical, chemical, biological, and ecotoxicological parameters for chemical compounds on the other. The most important of these parameters are the following: the soil/water adsorption isotherms; the rate of the chemical degradation processes (hydrolysis, photolysis, and chemical oxidation); the BCF; the ecological magnification factor (EMF); the uptake rate; the excretion rate; and a number of ecotoxicological parameters. The ratio of concentrations both in the sorbed phase and in water at equilibrium Ka and BCF may often be estimated with a relatively good accuracy from expressions like Ka or BCF = $a \log K_{ow} + b$. Numerous expressions with different a and b values have been published.[3–5]

BIODEGRADATION

It was concluded in the previous section that biodegradation is probably the most important property for the estimation of a chemical compound's environmental effect. This section is therefore devoted to the presentation of this important property.

Biodegradation rates may be expressed in several ways. Microbiological biodegradation may, with good approximation, be described as a Monod equation:[5]

$$dc/dt = - dB/Y dt = - \mu max^* \, Bc/Y\,(Km + c) \qquad (1)$$

where c is the concentration of the compound considered, Y is the yield of biomass B per unit of c, B is the biomass concentration, μmax is the maximum specific growth rate, andKm is the half saturation constant. If $c \ll$ Km, the expression is reduced to a first-order reaction scheme:

$$dc/dt = - K'\, B\, c \qquad (2)$$

where $K' = \mu max/(Y\,Km)$. B is in nature determined by the environmental conditions. In aquatic ecosystems, B is for instance highly dependent on the presence of suspended matter. B may therefore under certain conditions[5] be considered a constant, which reduces the rate expression to

$$dc/dt = -k\, c \qquad (3)$$

An indication of k in the unit 1/hr, 1/24hr, 1/week, 1/mo, or 1/yr can therefore be used to describe the rate of biodegradation. If the biological half-life time is denoted t, we get the following relation:

$$\ln 2 = 0.7 = k\, t \qquad (4)$$

This implies that the biological half-life time also can be used to indicate the biodegradation rate.

The biodegradation in waste treatment plants is often of particular interest, in which case the % of the Theoretical Oxygen Demand (ThOD) or the theoretical biological oxygen demand (BOD) may be used as a suitable reference. Most often, however, the 5-day BOD as percentage of the theoretical BOD is used. It may also be indicated as the BOD5fraction. For instance, a BOD5fraction of 0.7 will mean that BOD5 corresponds to 70% of the theoretical BOD. It is, however, also possible to find an indication of percentage removal in an activated sludge plant. The biodegradation is, however, in some cases very dependent on the concentration of microorganisms as expressed in the above-shown equations. Therefore K' indicated in the unit mg/(g dry wt 24 hr) will in many cases be more informative and correct.

In the microbiological decomposition of xenobiotic compounds, an acclimatization period from a few days to 1–2 mo should be foreseen before the optimum biodegradation rate can be achieved. We distinguish between primary and ultimate biodegradation. Primary biodegradation is any biologically induced transformation that changes the molecular integrity. Ultimate biodegradation is the biologically mediated conversion of organic compounds to inorganic compounds and products associated with complete and normal metabolic decomposition.

To conclude, the biodegradation rate is expressed by application of a wide range of units:

1. As a first-order rate constant (1/24hr).
2. As half-life time (days or hours).
3. mg perg sludge per 24hr[mg/(g 24hr)].
4. mg per g bacteria per 24 hr[mg/(g 24hr)].
5. mL of substrate per bacterial cell per 24hr[mL/(24hr cells)].
6. mg COD per g biomass per 24 hr[mg/(g 24hr)].
7. mL of substrate per gram of volatile solids inclusive microorganisms [mL/(g 24hr)].
8. BODx/BOD∞, i.e., the biological oxygen demand in x days compared with complete degradation (-), named the BODxcoefficient.
9. BODx/COD, i.e., the biological oxygen demand in x days compared with complete degradation, expressed by means of COD (-).

ESTIMATION OF BIODEGRADATION

The biodegradation rate in water or soil is difficult to estimate because the number of microorganisms varies several orders of magnitudes from one type of aquatic ecosystem to the next and from one type of soil to the next. Artificial intelligence has been used as a promising tool to estimate this important parameter. However, a (very) rough, first estimation can be made on the basis of the molecular structure and the biodegradability. The following rules can be used to set up these estimations:

1. Polymer compounds are generally less biodegradable than monomer compounds; 1 point for a molecular weight >500 and ≤1000, 2 points for a molecular weight >1000.
2. Aliphatic compounds are more biodegradable than aromatic compounds; 1 point for each aromatic ring.
3. Substitutions, especially with halogens and nitro groups, will decrease the biodegradability; 0.5 point for each substitution, although 1 point if it is a halogen or a nitro group.
4. The introduction of a double or triple bond will generally mean an increase in the biodegradability (double bonds in aromatic rings are of course not included in this rule); -1 point for each double or triple bond.
5. Oxygen and nitrogen bridges [– O – and – N – (or=)] in a molecule will decrease the biodegradability; 1 point for each oxygen or nitrogen bridge.
6. Branches (secondary or tertiary compounds) are generally less biodegradable than the corresponding primary compounds; 0.5 point for each branch.

Find the number of points and use the following classification:

<1.5 points: the compound is readily biodegraded. More than 90% will be biodegraded in a biological treatment plant.

2.0–3.0 points: the compound is biodegradable. Probably about 10–90% will be removed in a biological treatment plant. BOD5 is 0.1–0.9 of the theoretical oxygen demand.

3.5–4.5 points: the compound is slowly biodegradable. Less than 10% will be removed in a biological treatment plant. BOD10 ≤ 0.1 of the theoretical oxygen demand.

5.0–5.5 points: the compound is very slowly biodegradable. It will hardly be removed in a biological treatment plant, and a 90% biodegradation in water or soil will take ≥6 mo.

≥6.0 points: the compound is refractory. The half-life time in soil or water is counted in years. The structure of dichlorodiphenyltrichloroethane (DDT) corresponds, for instance, to about 7 points, and the biological half-life of DDT in soil is about 14 years.

Several useful methods for estimation of biological properties are based upon the similarity of chemical structures. The idea is that if we know the properties of one compound, it may be used to find the properties of similar compounds. If for instance we know the properties of phenol, which is named the parent compound, it may be used to give more accurate estimation of the properties of monochloro-phenol, dichloro-phenol, trichloro-phenol, and so on and for the corresponding cresol compounds. Estimation approaches based on chemical similarity give generally more accurate estimation but of course are also more cumbersome to apply, as they cannot be used generally in the sense that each estimation has a different starting point, namely, the compound, named the parent compound, with known properties.

Allometric estimation methods presume[6] that there is a relationship between the value of a biological parameter and the size of a considered organism.

The various estimation methods, including estimation methods applicable for biodegradation, may be classified into two groups:

1. General estimation methods based on an equation of general validity for all types of compounds, although some of the constants may be dependent on the type of chemical compound, or they may be calculated by adding contributions (increments) based on chemical groups and bonds.
2. Estimation methods valid for a specific type of chemical compound, for instance, aromatic amines, phenols, aliphatic hydrocarbons, and so on. The property of at least one key compound is known. Based upon the structural differences between the key compound and all other compounds of the considered type(for instance, two chlorine atoms have substituted hydrogen in phenol to get 2,3-dichloro-phenol) and the correlation between the structural differences and the differences in the considered property, the properties for all compounds of the considered type can be found. These methods are based on chemical similarity.

Methods of class 2 are generally more accurate than methods of class 1, but they are more cumbersome to use as it is necessary for each type of chemical to find for each property the right correlation. Furthermore, the requested properties should be known for at least one key component, which sometimes may be difficult when a series of properties are needed. If estimation of the properties for a series of compounds belonging to the same chemical class is required, it is tempting to use a suitable collection of class 2 methods.

Methods of class 1 form a network that facilitates possibilities of linking the estimation methods together in a computer software system, for instance,WINTOX. [1] An updated version named Estimation of Ecotoxicological Properties (EEP) is now available. EEP can estimate the biodegradability, in contrast to WINTOX. The software is easy to use and can rapidly provide estimations. Each relationship between two properties is based on the average result obtained from a number of different equations found in the literature. There is, however, a price for using such "easy-to-go" software. The accuracy of the estimations is not as good as with more sophisticated methods based upon similarity in chemical structure, but in many, particularly modeling, contexts, the results found by WINTOX and EEP can offer sufficient accuracy. In addition, it is always useful to come up with a first intermediate guess. It could, for instance, be used to estimate whether a chemical compound would be decomposed by biological treatment.

The software also makes it possible to start the estimations from the properties of the chemical compound already known. The accuracy of the estimation from use of the software can be improved considerably by having knowledge about a few key parameters, for instance, the boiling point and Henry's constant. WINTOX and EEP are based on average values of results obtained by simultaneous use of several estimation methods for most of the parameters. It implies increased accuracy of the estimation, mainly because it gives a reasonable accuracy for a wider range of compounds. If several methods are used in parallel, a simple average of the parallel results has been used in some cases, while a weighted average is used in other cases where it has been found beneficial for the overall accuracy of the program. When parallel estimation methods are giving the highest accuracy for different classes of compounds, use of weighting factors seems to offer a clear advantage. It is generally recommended to apply as many estimation methods as possible for a given case study to increase the overall accuracy. If the estimation by WINTOX and EEP can be supported by other recommended estimation methods, it is strongly recommended to do so.

REFERENCES

1. Jørgensen, S.E.; Halling-Sørensen, B.; Mahler, H. *Handbook of Estimation Methods in Ecotoxicology and Environmental Chemistry*; Taylor and Francis Publ.: Boca Raton, FL, 1998; 230 pp.
2. Jørgensen, S.E. *Principles of Pollution Abatement*; Elsevier: Amsterdam, 2000; 520 pp.
3. Jørgensen, S.E.; Jørgensen, L.A.; Nors Nielsen, S. *Handbook of Ecological and Ecotoxicological Parameters*; Elsevier: Amsterdam, 1991; 1380 pp.
4. Jørgensen, L.A.; Jørgensen, S.E.; Nors Nielsen, S. *Ecotox*, CD; Elsevier: Amsterdam, 2000; corresponding to 4000 pp.
5. Jørgensen, S.E.; Bendoricchio, G. *Fundamentals of Ecological Modelling*, 3rd Ed.; Elsevier: Amsterdam, 2001; 530 pp.
6. Peters, R.H. *The Ecological Implications of Body Size*; Cambridge University Press: Cambridge, 1983; 329 pp.

Biodiversity and Sustainability

Odo Primavesi
Brazilian Agricultural Research Corporation (Embrapa), São Paulo, Brazil

Abstract
This entry deals with the practical meaning of objectives like biodiversity, environmental health, and sustainability. It provides examples of how nature works to reach these objectives for the benefit of the whole food web, where human participates as the main predator. It also points out some ideas on how they can be improved.

INTRODUCTION

One of the greatest concerns related to sustainability today is the degradation and switching off of essential ecosystem services caused by anthropogenic activities like land degradation and pollution. These human activities, by producing solid, liquid, gaseous, and radiating (heat, light, radioactive emissions, and others) wastes and contaminants, lead to physical, chemical, and biological pollution, affecting the whole planet. For example, global warming and climate change result from a combination of air pollutants, such as the excess of greenhouse gases (carbon dioxide, methane, nitrous oxide, and others) and the excess of long-wave infrared or heat radiation (>300 W/m^2) from degraded and dry landscapes, and other factors.

Another example is the pollution of water due to poor management of soils (erosion, siltation of water bodies, and contamination with nitrates, phosphates, and other pollutants) or direct discharge of household, laboratory, or industrial waste. Water pollution and air pollution to a great extent contribute to levels of different human sicknesses and death rates in the very contaminated areas. Lack of good management practices regarding the conservation of natural resources (soil, water, air, and biodiversity) and ecosystem services and the lack of adequate management of waste are also very important issues.

In this entry, the goals are to improve the awareness and understanding of the importance of concepts or objectives like biodiversity, environmental health, and sustainability, and to give some suggestions on how the reader may contribute to individual and communal well-being.

BIODIVERSITY

What Is Biodiversity?

Biodiversity or biological diversity is defined by the United Nations Convention on Biological Diversity as "the variability among living organisms from all sources, including terrestrial, marine, and other aquatic ecosystems and the ecological complexes of which they are part; this includes diversity within species, among species, and of natural and altered ecosystems,"[1,2] including species, ecosystem, morphology, gene, and molecular diversity.

Ecosystem means a dynamic complex of plant, animal, and microorganism communities and the inorganic environment, interacting as a functional unity. Different levels of diversity are considered, among individuals, subspecies, species (most useful level), biological communities, and ecosystems. Species richness increases from colder to warmer latitudes. This is also true for the deep-sea species diversity.[1]

The number of species on Earth is estimated to be 5 to 30 million, from which 1.4 to 2 million were identified in a formal system. The majority are invertebrates (1 million; mainly insects and myriapods), followed by microorganisms (5760; e.g., fungi), chelicerates, protists, nematodes, plants (250,000; around 50,000 as trees), molluscs, crustaceans, and vertebrates (19,100 fishes, 9000 birds, 6300 reptiles, 4200 amphibians, and 4000 mammals).

Occurrence of species diversity follows in this decreasing order: tropical and subtropical moist broadleaf forests >>> tropical and subtropical grasslands, savannas, and shrub lands > deserts and xeric shrub lands = tropical and subtropical dry broadleaf forests > mountain grassland and shrub lands > temperate broadleaf and mixed forests > flooded grasslands and savannas = tropical and subtropical coniferous forests > temperate grasslands, savannas, and shrub lands = mangroves = temperate coniferous forests > Mediterranean forests, woodland, and scrub >> boreal forests or taiga > tundra.[1]

The tropics are the home of most of the species. An example is the Amazonian rainforest, where 60% of all lifeforms (e.g., 60,000 plant species) reside.[3] In one hectare of Atlantic rainforest in the southern Bahia state in Brazil, there are up to 454 tree species recorded.[4]

The reason for the species richness of the tropics is not well known. Some ideas proposed are the longer time

Encyclopedia of Environmental Management DOI: 10.1081/E-EEM-120046332

available to develop new species and potentially greater supply of solar energy, allowing more biomass production or more organisms per unit of area.[1]

Furthermore, soils in temperate climates show greater chemical fertility, water-holding capacity, and clay activity, and the cold switch off of biological activity controls these processes. Under tropical conditions, deep soils with mostly low chemical fertility, low water-holding capacity, and low clay activity, and the higher temperatures throughout the year allow for a greater number of interactions of water/drought × water table depth × nutrient availability × salinity × temperature/altitude/shade × strong rains × wind × fire × plant residues × organic matter content × photoperiod × oxygen (because of faster respiration rates and heat).

Therefore, habitat variability occurs, with specificities settled in by the different plant species, the first component of the food web and net.

Biological diversity is organized in a food web, with plants as base, harvesting sun energy freely available to them, and humans as top of the pyramidal net, where the individuals act as producer or consumer, or as recycler or decomposer.[5] The diversity of litter, defense substances, and root exudates produced by these different plant species, and correlated fauna, need to be chopped up by, e.g., invertebrates and decomposed by a greater number of microorganisms in soil, because of their specificity in producing degradation enzymes. The great recycling activities in soil need to be considered, because of the big importance of organic material as nutrient source for higher plants, as energy source for the microbial activity, and as a factor in improving soil structure and its water-holding capacity. In the tropics, with the great variability of habitats, the diversity of species is the keystone for high biomass yield per unit of area, making the food net of an ecosystem very complex.

Biodiversity reaches the maximum level when the environment offers enough water and energy and low to medium level of nutrients, such as nitrogen and phosphorus. With high levels of nitrogen and phosphorus occurs a very stiff inter- and intraspecific competition, mainly for light, from some more responsive or demanding species, similar to that occurring in high-fertility and very high-fertility soils or in eutrophic water bodies. There are also growth-restrictive conditions like in very low-fertility soils, with very low phosphorus and/or high aluminum content, or saline soils.

What Is the Importance of Biodiversity?

The greatest importance of biodiversity is still the optimized ecosystem services it provides.[1,6–9] Ecosystem services are flows of material, energy, and information on environmental structures (natural capital), which, combined with services, products, and human capital, generate human well-being. Usually, most of these processes could not be substituted in needed scale by any human technology, and if the ecosystem is extinguished, reversion is in general very difficult to realize in an economic way. The best way is conservation, including the costs in the price of products and services. Usually, ecosystem services occur in an imperceptible way, similarly to the involuntary and vital processes in our organism such as pumping of oxygenated blood or breathing.

Biodiversity provides three functional ecosystem services (production, regulation, and cultural) and a support service. Production and supply include mainly food, freshwater, timber, fiber, fuel, energy, genetic resources, medicines, wildlife, and others. Regulation includes maintenance of climate, carbon sequestration, soil conservation, wind and sea wave power, biodegradation and recycling of wastes, biological remediation of soils, and decontamination and cleaning up of water, cleaning up of air, maintaining soil permeability, and others. Cultural services could include cultural and heritage diversity; aesthetic, ethical, medicinal, and health knowledge; inspiration; educational, spiritual, and religious values; leisure; ecotourism; and so on. Support of life is carried out by maintaining a stratospheric oxygen-ozone layer to filter ultraviolet radiation and a greenhouse gas layer to filter infrared sun radiation and to retain partial infrared or heat radiation from surface, as well as by clouds, which reduce sunshine incidence on Earth surface, avoiding the burning of life. It is also carried out by maintaining a long water cycle (rain–interception–infiltration–storage–internal flow–evaporation and transpiration–air humidity–clouds) with distributed soft rains, by stabilizing air temperature and air humidity, and others.[1,2,10]

Biological diversity also provides resilience or a stabilizing effect on the food web, which sustains the human species, when an environmental disturbance occurs.[11]

Nature uses biodiversity to produce the maximum of life and biomass per square meter and year by optimal use of the available sun energy. Biodiversity is also the result of this settlement process of nature, with a great variety of abiotic conditions, mainly in the tropics.

At the same time, nature, by developing a food chain into a complex food net, allows for greater food availability and diversity for the individuals on the top of the food web or food pyramid, such as the human species, and also ensures their sustainability. However, when biodiversity of the food web is disrupted by the establishment of a monoculture (industrial cropping system) and/or when the environment is under subjection of a degradation process, a population outbreak of the more resistant or adapted members of the food web may occur such as the so-called parasites and pathogens.[12]

The soil is one of the most diverse habitats on Earth and contains one of the most diverse assemblages of living organisms, mainly in the humid tropics,[13] due to its plant diversity. In a broader view, it is advisable to consider soil as the undisassociable soil–plant interaction, mainly in the tropics and subtropics. Soil without a permanent living plant cover will lose its main function of harvesting rain and storing resident available freshwater, essential for life and biological production (food, fiber, wood, biofuel, etc.). This interaction will improve the degree of soil biodiversity, including also the rooting system architecture of the so-called weeds. Their rooting structure, as well as that of crops, may be used as a visible indicator of the degree of soil health.[14,15]

In both natural and agricultural ecosystems, the different groups of soil biota interacting with plants and their debris are responsible for, or strongly influence, the soil properties and optimize processes or ecosystem services such as soil genesis, soil structure, carbon, nutrient and water cycles, agrochemical movement or breakdown, plant protection, growth, and production.[5,16]

Soil organisms act in processes of synthesis or production, transformation and decomposition, or consumption of organic material, affecting abiotic and biotic components, transportation, and soil engineering. Therefore, soil biodiversity is a keystone for sustainable agriculture and it could be used as a good indicator of agro-ecosystem or soil health. Soil biodiversity does not necessarily refer to the number of individuals or species, but to the ratio of functional groups,[17,18] and the result or the tool of their activities, such as the presence and intensity of enzymatic activity.[12] It is necessary to remember the importance of soils.

The settlement of terrestrial environments by life was only possible by storing rainwater in rocks that nature developed to permeable soils, and these soils were maintained permeable by a triple-protection layer: plant canopy, litter, and surface rooting system. To succeed in our activities and also to meet our quality of life, it is necessary to maintain, restore, or mimic this structure and the processes involved using artificial technology.

It can be said that the disappearance of several ancient human civilizations with populations concentrated in cities was partially caused by food insecurity due to soil degradation, combined with freshwater shortage due to destruction of forests and soil permeability, therefore reducing the long water cycle, and by lack of sanitation and waste disposal affecting public health.[19] These problems of slowing down or disrupting ecosystem services are at present global in scale, with the emergence of a new problem—global warming and climate change. Besides, there occurs programmed or accidental inclusion of poisons, toxic substances, heavy metals, nitrates, phosphates, hormones, etc.

To have an idea of the current importance of the whole biodiversity, the monetary value of 17 ecosystem services required to sustain life and the biological production capacity of landscapes was estimated to be about $33 trillion/yr, against the $18 trillion of the global gross domestic product (GDP).[9]

What Destroys Biodiversity and Ecosystem Services?

This can be answered by the simple elimination of plants covering the soil and the prevention of their complete regrowth and occupation of soil surface, as well as by turning the soil impervious, by crusting, compaction, pavement, etc.

How Do We Take Care of Biodiversity and Essential Ecosystem Services?

An interesting point is that, instead of considering the natural climax ecosystem as reference for good environmental practices, we use the primary natural ecosystem (rocky landscape) as reference for characteristics we do not want in our agricultural or urban ecosystems. The following are environmental characteristics we have to avoid in our management program: the primary environment has no capacity to store water; it has no biological carrying capacity for higher species; it sustains no food chain nor web; it presents a very short water cycle (rain, evaporation and runoff); and it shows high temperature and air humidity amplitude during the day.

Hence, primary environments are unsuitable for life and production. For example, land and soil degradation will turn life- and production-friendly environmental characteristics into life- and production-unsuitable conditions (with impervious, compacted, dry, hard soil, like a rock), similar to that occurring in primary natural ecosystems.

Considering the growing soil and landscape degradation, for example, due to further erosion and salinity, we need to be aware that the process of desertification[23–25] is a great challenge in dry lands. These dry lands support 44% of all cultivated systems and are the origin of 30% of world's cultivated plants.[26,27]

Thus, first of all, we need to stop the landscape and ecosystem services degradation, by conservation practices. The second step is to recover soils and landscapes by simple harvesting and storing rainwater, and reducing water losses, e.g., by runoff or evaporation in excess, by protecting the soil surface against erosion, and by allowing the growth of a diverse plant cover, to turn the soil permeable and to stabilize air temperature and humidity.

It is necessary to prevent the destruction of the whole plant cover and their debris. The biologically diverse green areas in rural and urban environments are mandatory. Under tropical and subtropical conditions, this means that we have to avoid large-scale areas of pasture or cropland or buildings (cities) without trees. A permanent tree cover is important to maintain the evapotranspiration and windbreak service and to stabilize air temperature and humidity.[8]

Considering that the greatest land-use change, with massive deforestation, is mainly for agricultural purposes (as well as for wood harvest, coal production, and mining activity), it will significantly alter the micro- and mesoclimate in a region, increasing infrared radiation and heating up of lower atmosphere, and because it will promote the significant degradation of other essential ecosystem services due to its scale, agricultural practices based on ecological principles are advisable.

For large areas, the first step should be the practice of conservation agriculture[20] complemented by windbreaks and vaporizing tree cluster. For small areas, agroforestry[21,22] production systems are useful, mainly under tropical and subtropical conditions, considering that the tropics are the engine of global climate dynamics. Without trees, the heat production over terrestrial areas will be greater, the cloud production will be smaller, and the climate dynamics will be faster or stronger, more dangerous.

Examples of successful ecosystem services restoration are those in conservation agriculture,[20,28–30] agro-ecological production systems,[31–33] and agroforestry,[34–39] aside from forest management[40] and those biodiversity conservation practices[41,42] that substitute paid environmental services for free ecosystem functions or services, by using ecological principles.

Also, the need to reduce the use of toxic substances, as well as the production of wastes and their random release to the environment, is urgent. Gaseous wastes affect human and biodiversity health, especially when considering the troposphere ozone and acid rain production, the increasing carbon footprint, and the large amount of smog production, brought about by landscape fire, fire from furnaces (used for coal production), or fire from household wood-burning stoves. Liquid waste degrades freshwater, turns it dangerous for human health, and increases freshwater shortages. Solid waste released in large amounts on the landscape will reduce land and water quality and will increase the ecological footprint.

The best indicator of biodiversity that provides adequate ecosystem services is when we have a greater number of plants per unit of area. Plants are associated with fauna and microorganism.[6]

Which Tools Should Be Used to Improve Biodiversity?

Considering the process nature uses to develop a certain site, or to recover a degraded soil or land under fallow, it could be seen that plant diversity is the key tool used. This is because it allows the complementary activity of individuals with different structures, functions, wastes (debris, root exudations, and others), and needs to flourish in one of the different habitats created by the diverse interaction of the abiotic and emergent biotic factors occurring.

Therefore, vegetative techniques to recover the permanent diverse plant cover together with the improvement of environmental legislation and education are in place.

ENVIRONMENTAL HEALTH

What Is Environmental Health?

According to the World Health Organization,[43,44] environmental health addresses all the physical, chemical, and biological factors external to a person, and all the related factors impacting behavior. It encompasses the assessment and control of those environmental factors that can potentially affect health. It is targeted towards preventing disease and creating health-supportive environments.

Which Environmental Conditions Will Affect Health?

Ecosystem degradation or disruption can impact on health in a great variety of ways.[43–46] The most important ones are freshwater pollution (lack of sanitation), contamination (heavy metals, hormones, poisons, and excess of medicines), and degradation (siltation and increase of nitrates and phosphates).

Perhaps we need to have a broader view of what is environmental health, considering that persons are also participants of our environment. In a health environment, for example, aside from waterborne diseases due to pollution and acute and chronic respiratory diseases due to air pollution, we need to also consider malnutrition due to food lacking micronutrients, extreme heat and very low air humidity, lack of education and training, low (or no) income, and so forth.

What Is the Importance of Environmental Health?

A healthy environment, with its natural resources and main structures conserved or improved, will allow the ecosystem services to run in order to benefit our well-being. A healthy environment will provide enough clean, fresh water (150 to 200 L/person/day); it will secure food (1500 to 2000 kcal/person/day, without toxic or dangerous substances and contaminants); it will provide clean air with enough humidity (around 10 g water as vapor/m^3 air, or 40%–60% relative humidity in a range of 20–24°C, the comfort range, without solid microparticles, smog, dangerous gases and substances, and inconvenient odor); and it will lead to stabilized temperature.

An excess or low level of air humidity will bring about and increase the occurrence of several diseases.[47] Unprotected surface/soil will produce higher temperature (Fig. 1) and therefore greater amount of long-wave infrared radiation, heating up the lower atmosphere. In addition, increas-

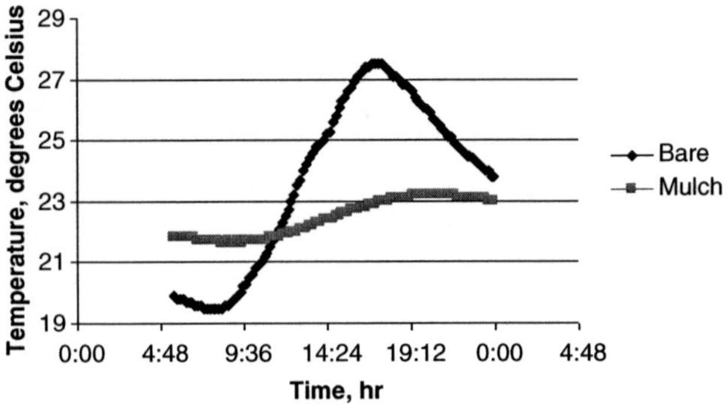
Temperature, degrees Celsius
29
27
25
23
21
19
0:00
4:48
9:36
14:24
19:12
0:00
4:48
Time, hr
Bare
Mulch

Fig. 1 Temperature variation (amplitude), at 15 cm depth, in a bare and mulch-covered soil, under tropical conditions (November). The graph shows that it is possible to manage temperature extremes using mulching technique.
Source: Adapted from Torres (1997) in Primavesi.[8]

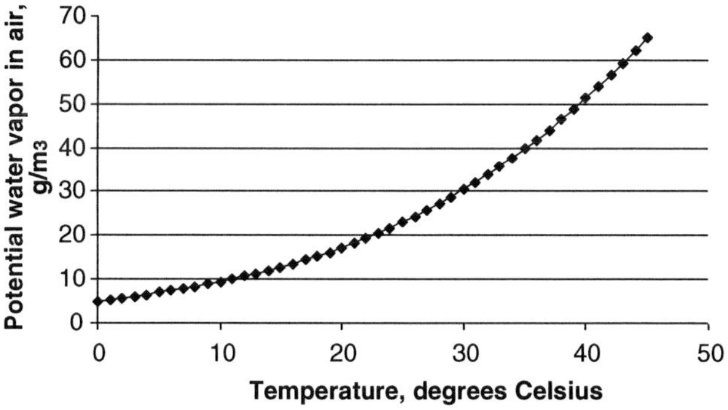
Potential water vapor in air, g/m3
70
60
50
40
30
20
10
0
0
10
20
30
40
50
Temperature, degrees Celsius

Fig. 2 Relation between air temperature and potential water demand for air saturation.
Source: Adapted from Addams et al.[52]

ing air temperature will mainly result in decreasing air humidity, because of the increase of atmospheric demand for water, when no water is available to be evaporated or transpired by plants (Fig. 2).

Heat in excess, above 30°C, will sharply increase the productivity loss of labor and increase the mistakes or errors in the production line,[48] while polluted and dry air will increase respiratory diseases.[47] Polluted water will cause several gastroenteritis processes.[49–51] Human health degradation will cause problems for the production system, the health care service, and to our well being, with expectation of an earlier death.

What to Do?

The reactivation of the ecosystem services is needed, with the restoration of biodiversity, on a permeable soil, increasing rainwater yield and storage. It is also necessary to reduce waste production, by reduction, reuse, and recycling, as well as by neutralizing and adequate disposal of all harmful materials. We need to use water more efficiently.[52]

Another very important point is to stabilize air temperature (avoiding extremes) and air humidity amplitude, by restoring green evapotranspiring areas, by establishing shade (Table 1) or soil cover (Fig. 1), or by managing the surface albedo.[54–62] Nature manages albedo or sunshine reflectivity in function of available liquid water. Water bodies, humid soils, or live plants are darker (with lower albedo) and may absorb more sun energy than dry surfaces, dry leaves (straw), or solid water (snow and ice) fields (with greater albedo).

Table 1 Temperature of shaded and unshaded surfaces in the tropics.

	Temperature		
	Shaded	Unshaded	Increase
Surface	°C		%
Green lawn	32	35	9
Dry lawn	35	52	48
Concrete	37	52	40
Asphalt	37	57	54

Source: Addams et al.[52]

A notable complementary problem is when fire use is routine in landscape management,[63] especially in dry periods, turning the albedo very low (black surfaces), increasing the heat, and therefore knocking down air humidity and quality, affecting health. Since we have global areas experiencing water shortage[64] [being dry for moderate (3 to 5 months) to long (>6 months) periods], this may result in heating up of soils and surfaces, with temperatures above 52°C,[65] and this will produce atmospheric heat in excess (>300 W/m^2),[66] surprisingly where there were no forests.[67]

Trees, due to their darker color, will absorb more sun energy and produce more heat, warming up a cold environment, but when air temperature rises above a certain level, it starts the evapotranspiration process, keeping away the heat in excess. Trees may warm up and cool down an environment, depending on the need. Therefore, it is advisable to make a global effort to plant trees (forests) and manage agroforestry systems to restore ecosystem services such as temperature and air humidity stabilization, as well as to maintain a longer water cycle with more and better-distributed rain.[68–76]

It is necessary to take into account that there are biophysical limits for economic growth (adjusted to nature's biological carrying capacity), which are mainly responsible for pollution as well as for biodiversity and land degradation. From the nine main limitations (biodiversity loss, nitrogen and phosphorus cycle, climate change, acidification of seawater, ozone reduction in stratosphere, freshwater availability, change of soil use, chemical pollution, and aerosol pollution in atmosphere), we did trespass the first three, endangering our health and livelihood.[77]

Indicators of Improvement

The presence of biodiversity developing on permeable soils and the absence of pollution, waste disposal in landscapes, and contaminants result in clean freshwater, soil, air, and food.

Tools to Be Used

Processes and tools based on ecological principles and processes to recover soil permeability (Figs. 3 and 4), biodiversity, and ecosystem services need to be used. The keystone of the food web we depend on is plants, especially those with healthy roots. Roots need to be in aerated and humid soil, protected (against temperatures above 33°C, which harms root health), and rich in organic material.

The concept of integrated natural resource management considering the watershed scale, aside from integrated and efficient water, fertilizer, and pest management, is advisable.

Fig. 3 Crusting of an unprotected prepared seedbed after tropical rainshower, turning soil impervious to rainwater and aeration.
Source: Author's personal archive.

In relation to waste disposal, pollution, and contamination, environmental technology, eco-technology, and cleaner technology are all in place. In the case of fossil fuel use, the reduction of wasteful use and the substitution by alternative renewable energy sources are advisable. In all cases, improvement in environmental legislation and education is necessary.

Fig. 4 Left: permeable healthy soil from a natural ecosystem, plenty of visible roots. Right: the same soil type, compacted, impervious, and dry from the inter row of a sugarcane field after 5 years of continuous intensive cultivation.
Source: Author's personal archive.

SUSTAINABILITY

What Is Sustainability?

In biology, sustainability means the processes running to maintain biological systems diverse and productive over time.[5] In 1987, the World Commission on Environment and Development established a definition of sustainability, known as the Brundtland Report. It stated that sustainable development should reach the needs of the present without compromising the ability of future generations to meet their own needs. Although this definition has become widely publicized, the term *sustainability* is not limited to one precise definition.[78]

Several authors have discussed the real meaning of sustainability and sustainable development.[16,28,79] Munoz[80] has presented a theoretical model of true sustainability or sustainable development ideas with global to local implications based on the need to balance social, economic, and environmental goals; stakeholders' interests; and issues to induce or determine fairer or more appropriate development solutions, options, and actions.

However, the social component can be seen as part of the environmental component, and the interaction of a health and developed environment (essential ecosystem services running) with educated, trained, organized, and health persons will generate a long-lasting economic component. Hence, the environmental component is the keystone, to sustain the virtual world as well, and only its improvement and quality will allow us to reach a stable social welfare and a sustainable economic profit.

What Is the Importance of Sustainability?

Historically, the first known examples of worry on sustainability, and the establishment of a definition, were from those growing up in the forestry sector, and this is because of wood shortage for a salt mining activity in Germany or for house building and coal production in Japan. Both cases resulted from poor forest management and overexploitation. The question was, "What should we do to maintain a constant wood yield and cash flow for current and future forest owners', managers', and local workers' livelihood?" This question was discussed since 1442 in various regions of Germany, where wood-consuming salt-mining activities did occur, and the concept was formulated firstly in 1650 in Saxonia (Germany) and later, independently, in 1666 in some deforested regions of Japan.[81]

The idea was that regenerative living resources, like wood/trees, could only be used/harvested in the same amount of the natural regrowth/recover/refill (with no use of external inputs, like water, energy, or fertilizer), and this by maintaining productivity, vitality, rejuvenation capacity, and biological diversity, in a time span of around 120 years (the time needed by trees to grow for cutting), to avoid natural resource shortage, labor and cash shortage, and livelihood and health problems (mainly related to erosion and flooding).[81]

In 1732, the idea on sustainable use of forest was published for the first time by von Carlowitz, and in 1795, Georg Ludwig Hartig described how to manage a sustainable forest.[81] History shows us also that a sustainable activity may last for 5000 years and more.[82]

Related to sustainable forest management, the current definition is as follows: "The stewardship and use of forests and forest lands in a way, and at a rate, that maintains their biodiversity, productivity, regeneration capacity, vitality and their potential to fulfill, now and in the future, relevant ecological, economic and social functions, at local, national, and global levels, and that does not cause damage to other ecosystems."[40]

Nature teaches us that the real development process of a natural primary environment occurs with the development or restoration of a permeable soil, protected by a diversified vegetation cover with an active rooting system, and the return to soil of diversified organic material, the energy source for the diversified and active soil life.

The soil–vegetation and associated biodiversity interaction could improve the available resident water of a site and also a longer water cycle. The more resident water, the more vegetation and the more permeable the soil, in a growing feedback loop. With greater amounts of available resident water-permeable soil-diversified vegetation and soil life, there is also an improvement in micro- and mesoclimate, with an increase in relative air humidity and a decrease in the maximum temperature and thermal amplitude, characteristic for desert environments.

This friendlier mesoclimate helps more sensible plant and animal species to establish and improves biodiversity, with their additive and emergent characteristics, mainly observed in the humid tropics. Ecology considers desert ecosystems sometimes as sustainable as drier tropical forest ecosystems, due to their richness in biodiversity. Thus, what level of environmental sustainability is desirable? It depends on the biological carrying capacity we want.

The biological carrying capacity represents the concept of primary productivity of an ecosystem, or the rate and amount in which energy is stored by photosynthetic or chemosynthetic activity of the producer organisms as organic substances, food for the food web.[5] The biological carrying capacity also considers the feeding capacity of grazing cattle, or grain equivalent available for humans (4 or 16 persons/ha/yr, with a minimum need of 1000 kcal/day), calculated as available digestive energy or calories per surface unit and year. The biological carrying capacity depends on the recovering capacity of a site (resilience) to produce biomass, after yield, extraction, degradation, or pollution activities.

Hence, considering the exuberant flora and biomass production by the Amazonian forest, the question arises: What is the biological carrying capacity of the around 40% sandy soils (<15% clay content) in the Amazonian basin without that great vegetation and the aggregated or dependent mesoclimate? Something similar to the Sahara? This example shows that the biological carrying capacity may be managed in a certain range by improving the natural resource structural tripod of resident water (harvested and stored rainwater) in permeable soil (organic matter rich) under biodiverse, permanent, alive plant cover and the ecosystem services.

Some agricultural production systems—although their processes result in improvement in environmental characteristics, income, and social inclusion—are not sustainable, because of their great dependence on external inputs [fertilizer, energy, water (including fossil underground water), and technical support]. We could observe, however, that, with time, it is possible to reduce part of this dependence, by switching to more organic and biological processes, after building up a minimum of fertility and organic matter level in soil, and introducing nitrogen-fixing leguminous trees. The goal of such process is to be more efficient and productive while maintaining or improving environment (natural resources) quality.[29]

What Needs to Be Focused on Sustainability?

Considering that the main objective of all human activity, from a global to a local scale, is to promote life and its quality, environmental sustainability will be reached when the biological carrying capacity is adequate to supply the minimal health life requirements of a given human population.

An increase in the biological carrying capacity level will allow a rise in human population density. Instead of this, what occurs now is the destruction of the main natural resource structures and functions (or ecosystem services), a decrease in the biological carrying capacity with an increase in human population density, and an increase in the production of solid, liquid, gaseous, and radiative wastes, thrown randomly in landscapes and marine ecosystems. We are currently watching a global ecological regression process of terrestrial and marine ecosystems, back to conditions unsuitable for human health, livelihood, and life.[8]

We need to reduce losses and wasting of materials, as well as to avoid pollution and contamination of natural resources and products.

Which Indicators Are Usable?

First, we need to reduce our ecological footprint (optimal land use),[83–88] our water footprint (optimal and efficient water use),[89–93] and our carbon footprint (carbon equivalent production in processes and carbon cycle)[94–96] by turning our production systems or life system more efficient and adequate to the global ecosystem carrying capacity. The measure of total sun energy use[97–100] is perhaps the best way to measure and turn the processes more sustainable.

Which Tools Should Be Used?

The following are the main tools we need to use: primarily, knowledge on ecological principles and processes;[5,97] a better understanding of the processes using energy, land, water, and primary products; a retooling of the processes to turn them more efficient, using materials more harmless to the environment and health; and a reduction of waste disposal in ecosystems. All of these are geared towards improving environmental legislation and education.

CONCLUSION

It is necessary to stop the degradation or regression and pollution processes of terrestrial and marine ecosystems and restore their health. The processes nature uses to develop complex resilient ecosystems with great biological (including human) carrying capacity are known. To conserve, restore, or improve a sustainable and healthy planet with the immense carrying capacity of the human species, biodiversity and ecosystem services should be considered, based on the ecological knowledge of essential natural structures and processes.

Moreover, as in nature, all kinds of wastes need to be reduced and recycled. Our technologies and processes need to maintain, restore, or mimic natural structures and processes to succeed and improve sustainability. A global awareness on the ecological principles and processes for a regional planning of the integrated local participatory activities to restore, maintain, or even improve the environmental health for sustainability of human societies is necessary.

REFERENCES

1. Millennium Ecosystem Assessment. *Ecosystems and Human Well-Being. Biodiversity Synthesis*; World Resources Institute: Washington, 2005; 1–100. Available at http://www.millenniumassessment.org/documents/document.354.aspx.pdf (accessed March 2010).
2. Millennium Ecosystem Assessment. *Ecosystem and Human Well-Being: Synthesis*; Island Press: Washington, DC, 2005; 1–155. Available at http://www.millenniumassessment.org/en/Synthesis.aspx and http://www.millenniumassessment.org/documents/document.356.aspx.pdf (accessed March 2010).
3. Peneireiro, F.M.; Rodrigues, F.Q.; Brilhante, M.O.; Ludewigs, T. *Apostila do Educador Agroflorestal: Introdução aos Sistemas Agroflorestais—Um Guia Técnico*; Universidade Federal do Acre, Parque Zoobotânico: Rio Branco-AC, Brazil, 1999; 1–76. Available at http://www.agrofloresta.net/artigos/apostila_do_educador_agroflorestal-arboreto.pdf and http://www.slideshare.net/FlaviaCremonesi/apostila-do-educador-agroflorestal-arboreto-1353733 (accessed March 2010) (in Portuguese).
4. Girardi, E.P. *Atlas da Questão Agrária Brasileira*; Unesp: Presidente Prudente-SP, Brazil, 2008; 1–259. Available at http://www4.fct.unesp.br/nera/atlas/configuracao_territorial.htm (accessed March 2010) (in Portuguese).
5. Odum, E.P.; Barrett, G.W. *Fundamentals of Ecology*, 5th. Ed.; Brooks/Cole: New York, 2005; 1–612.
6. Convention on Biological Diversity. COP 6 Decision VI/7. Identification, monitoring, indicators and assessments; UNEP: Montreal, Canada, 2002. Available at http://www.cbd.int/decision/cop/?id=7181 (accessed March 2010).
7. Chivian, E., Ed. Biodiversity: Its importance to human health. *Interim Executive Summary*; Center for Health and Global Environment/Harvard Medical School/WHO/UNDP/UNEP: Harvard, USA, 2003; 1–59. Available at http://chge.med.harvard.edu/publications/documents/Biodiversity_v2_screen.pdf (accessed March 2010).
8. Primavesi, O.; Arzabe, C.; Pedreira, M.S. *Mudanças Climáticas: Visão Tropical Integrada das Causas, Dos Impactos e de Possíveis Soluções para Ambientes Rurais ou Urbanos*; Embrapa Pecuaria Sudeste: Sao Carlos-SP, Brazil, 2007; 1–200 (Embrapa Pecuaria Sudeste. Documentos, 70). Available at http://www.cppse.embrapa.br/080servicos/070publicacaogratuita/documentos/Documentos70.pdf (accessed March 2010) (in Portuguese, with a 3-page Executive Summary in English).
9. Costanza, R.; D'Arge; Groot, R.; Farber, S.; Grasso, M.; Hannon, B.; Limburg, K.; Naeem, S.; O'Neill, R.V.; Paruelo, J.; Raskin, R.G.; Sutton, P.; Belt, M. The value of the world's ecosystem services and natural capital. Nature **1997**, *15* (387, May), 253–260. Partially available at http://myweb.facstaff.wwu.edu/~medlerm/classes/08_09/502/nature-paper.pdf and http://earthmind.net/marine/docs/session2c-on-costanza-global-valuation.ppt (accessed March 2010).
10. Daily, G.C.; Alexander, S.; Ehrlich, P.R.; Goulder, L.; Lubchenco, J.; Matson, P.; Mooney, H.A.; Postel, S.; Schneider, S.H.; Tilman, D.; Woodwell, G.M. Ecosystem services: Benefits supplied to human societies by natural ecosystems. Issues Ecol. **1997**, *1* (2), 1–18. Available at http://www.esa.org/science_resources/issues/FileEnglish/issue2.pdf (accessed March 2010).
11. Fischer, J.; Lindenmayer, D.B.; Manning, A.D. Biodiversity, ecosystem function, and resilience: Ten guiding principles for commodity production landscapes. Front. Ecol. Environ. **2006**, *4* (2), 80–86. Available at http://people.anu.edu.au/adrian.manning/ten_guiding_principles.pdf (accessed March 2010).
12. Primavesi, A.M. *Manejo Ecologico do Solo: A Agricultura em Regioes Tropicais*; Nobel: São Paulo-SP, Brazil, 1980; 1–541 (in Portuguese).
13. Bunning, S.; Jiménez, J.J. *Soil Biodiversity Portal: Conservation and Management of Soil Biodiversity and Its Role in Sustainable Agriculture*; FAO: Rome, Italy, 2004. Available at http://www.fao.org/ag/agl/agll/soilbiod/default.stm (accessed March 2010).
14. Food and Agriculture Organization. *Biological Management of Soil Ecosystem for Sustainable Agriculture*; Report of the International Technical Workshop: Londrina-PR, Brazil, 2002. FAO: Rome, Italy, 2003; 1–37 (World Soil Resources Report, 101). Available at ftp://ftp.fao.org/docrep/fao/006/Y4810E/Y4810E00.pdf (accessed March 2010).

15. Food and Agriculture Organization. *Integrated Crop Management: An International Technical Workshop Investing in Sustainable Crop Intensification—The Case for Improving Soil Health*; FAO: Rome, Italy, 2008; Vol. 6, 1–139. Available at http://www.fao.org/docrep/012/i0951e/i0951e.pdf (accessed March 2010).
16. Dumanski, J.; Gameda, S.; Pieri, C. *Indicators of Land Quality and Sustainable Land Management: An Annotated Bibliography*; The World Bank: Washington, DC, 1998; 1–126. Available at http://books.google.com.br/books?id=vY3HjKEJitkC&printsec=frontcover&dq=Indicators+of+Land+Quality+and+Sustainable+Land+Management:+An+Annotated+Bibliography+dumanski&source=bl&ots=y67bbN2_L-&sig=hveLwdTFCNXfoPEpXFRTYdNRIMo&hl=pt-BR&ei=iIvxS9HQLsX_lgfW-cGzCA&sa=X&oi=book_result&ct=result&resnum=1&ved=0CBoQ6AEwAA (accessed March.2010).
17. Tilman, D.; Cassman, K.G.; Matson, P.A.; Naylor, R.; Polansky, S. Agricultural sustainability and intensive production practices. Nature **2002**, *418*, 671–677. Available at http://pangea.stanford.edu/research/matsonlab/members/PDF/TilmanNaylorMatson2002.pdf (accessed March 2010).
18. Scherer-Lorenzen, M.; Palmborg, C.; Prinz, A.; Schulze, E.-D. The role of plant diversity and composition for nitrate leaching in grasslands. Ecology **2003**, *84* (6), 1539–1552. Abstract available at http://www.jstor.org/pss/3107974 (accessed March 2010).
19. Liebmann, H. *Ein Planet wird unbewohnbar: ein Suenderegister der Menschheit von der Antike bis zur Gegenwart*; R. Piper & Co: Munchen, Germany, 1973; 1–181 (in German).
20. United Nations Environment Programme. *The United Nations Convention to Combat Desertification: A New Response to an Age-Old Problem*; UNDPI: New York, 1997. Available at http://www.un.org/ecosocdev/geninfo/sustdev/desert.htm (accessed March 2010).
21. United Nation Convention to Combat Desertification. GBO-3's Significant Findings Also in What's Unsaid about Landbased Biological Diversity. Press Release, 12/10, 2010. Available at http://www.unccd.int/publicinfo/pressrel/showpressrel.php?pr=press11_05_10 (accessed March 2010).
22. Menne, B.; Bertollini, R. The health impacts of desertification and drought. Down to Earth (Newsletter of the Convention to Combat Desertification) **2000** (14), 4–7. Available at http://www.google.com.br/url?sa=t&source=web&ct=res&cd=1&ved=0CB4QFjAA&url=http%3A%2F%2Fwww.unccd.int%2Fpublicinfo%2Fnewsletter%2Fno14%2Fnews14eng.pdf&rct=j&q=The+health+impacts+of+desertification+and+drought&ei=EpnxS5y2KoL6lwf-9Ki0CA&usg=AFQjCNHNGLIR1oJFAQZQzDiEUeqHzC9k-Q (accessed March 2010).
23. Food and Agriculture Organization. *Carbon Sequestration in Dryland Soils. Chapter 2. The World's Dry Land* (World Resources Report, 102). FAO/Natural Resources Management and Environment Department: Rome, Italy, 2004; 1–129. Available at http://www.fao.org/docrep/007/y5738e/y5738e06.htm and http://www.fao.org/ag/agl/agll/wrb/soilres.stm (accessed March 2010).
24. United Nation Convention to Combat Desertification. *Only One Earth—Drylands Are Vital*; UNCCD: Bonn, Germany, 2010. Available at http://www.unccd.int/publicinfo/announce/earth_day.php (accessed March 2010).
25. Food and Agriculture Organization. *Conservation Agriculture*. Available at http://www.fao.org/ag/ca/ (accessed March 2010).
26. Food and Agriculture Organization. Realizing the economic benefits of agroforestry: Experience, lessons and challenges. In *State of the World's Forests*; FAO: Rome, Italy, 2005; 88–97. Available at ftp://ftp.fao.org/docrep/fao/007/y5574e/y5574e09.pdf (accessed March 2010).
27. Hailu, M.; Landford, K.; Selvarajah-Jaffery, R.; Vanhoutte, K. (Coord.) *Agroforestry—A Global Land Use*. World Agroforestry Centre: Nairobi, Quenia, 2009; 1–53 (Annual Report 2008–2009). Available at http://www.worldagroforestry.org/downloads/publications/PDFs/B16416.PDF (accessed March 2010).
28. Dumanski, J.; Peiretti, R.; Benites, J.R.; McGarry, D.; Pieri, C. The paradigm of conservation agriculture. Proceedings of the World Association on Soil and Water Conservation, 2006, 58–64. Available at http://www.unapcaem.org/publication/ConservationAgri/ParaOfCA.pdf (accessed March 2010).
29. García-Torres, L.; Martínez-Vilela, A.; Holgado-Cabrera, A.; Gónzalez-Sánchez, E. Conservation agriculture, environment and economic benefits. International Congress of the European Society for Soil Conservation, 2000, 4, Valencia-Spain. Proceedings, 2000; 1–10. Available at http://www.unapcaem.org/publication/ConservationAgri/CA1.pdf (accessed March 2010).
30. Ngandwe, T. Conservation agriculture boosts yields and incomes. SciDev.Net **2006** (January 26). Available at http://www.scidev.net/en/news/conservation-agriculture-boosts-yields-and-incom.html (accessed March 2010).
31. Altieri, M.A. The ecological role of biodiversity in agroecosystems. Agric., Ecosyst. Environ. **1999**, *74*, 19–31. Available at http://comunidades.mda.gov.br/o/1540391 (accessed March 2010).
32. Uphoff, N. *Agroecologically Sound Agricultural Systems: Can They Provide for the World's Growing Populations?* International Research of Food Security, Natural Resource Management and Rural Development: Tropentag, 2005. University of Hohenheim: Stuttgart, Germany, 2005. Available at http://www.tropentag.de/2005/proceedings/node181.html and http://www.tropentag.de/2005/proceedings/node3.html (accessed March 2010).
33. Uphoff, N. Agricultural futures: What lies beyond modern agriculture? Trop. Agric. Assoc., Newsl., London, U.K., **2007**, *27* (3; September), 13–19. Available at http://www.slideshare.net/SRI.CORNELL/0906-agricultural-development-what-comes-after-modern-agriculture (accessed March 2010).
34. Wilkinson, K.M.; Elevitch, C.R. *Integrating Understory Crops with Tree Crops: An Introductory Guide for Pacific Islands*; Permanent Agriculture Resources: Holualoa, Hawaii, 2000; 1–50. Available at http://www.agroforestry.net/pubs/Understory.pdf (accessed March 2010).
35. Elevitch, C.R.; Wilkinson, K.M. *Agroforestry Guides for Pacific Islands*; Permanent Agriculture Resources: Holualoa, Hawaii, 2000; 1–239. Available at http://www.agroforestry.net/afg/ (accessed March 2010).
36. Beetz, A. *Agroforestry Overview*; ATTRA: Fayetteville, Arkansas, 2002; 1–16. Available at http://attra.ncat.org/attra-pub/PDF/agrofor.pdf and http://attra.ncat.org/attra-pub/agroforestry.html (accessed March 2010).

37. Davies, K. *Indian Agroforestry: Some Ecological Aspects of Northeastern India Agroforestry Practices*, 1984. Available at http://www.daviesand.com/Papers/Tree_Crops/Indian_Agroforestry/ (accessed March 2010).
38. Dixon, R.K. Agroforestry systems: Sources or sinks of greenhouse gases? Agroforestry Syst. **1995**, *31* (2), 99–116. Abstract available at http://www.springerlink.com/content/g106873871666629/ (accessed March 2010).
39. Leakey, R.R.B.; Tchoundjeu, Z. Diversification of tree crops: Domestication of companion crops for poverty reduction and environmental services. Exp. Agric. **2001**, *37* (3), 279–296. Available at http://www.wanatca.org.au/acotanc/Papers/Leakey-2/index.htm (accessed March 2010).
40. Convention on Biological Diversity. *A Good Practice Guide: Sustainable Forest Management, Biodiversity and Livelihoods*; CBD: Montreal, Canada, 2009; 1–47. Available at http://www.cbd.int/development/doc/cbd-good-practice-guide-forestry-booklet-web-en.pdf (accessed March 2010).
41. Hopper, K.; Summers, D., Eds. *Protecting the Source: Land Conservation and the Future of America's Drinking Water*; The Trust for Public Land and American Water Works Association: San Francisco, 2004; 1–56 (Water Protection Series). Available at http://earthtrends.wri.org/pdf_library/feature/eco_fea_value.pdf (accessed March 2010).
42. Rand Corporation. *New York City Depends on Natural Water Filtration*; Rand Co.: Santa Monica, California, 2007. Available at http://www.rand.org/scitech/stpi/ourfuture/NaturesServices/sec1_watershed.html (accessed March 2010).
43. World Health Organization. *Environmental Health*; WHO: Geneva, Switzerland, 2010. Available at http://www.who.int/topics/environmental_health/en/ (accessed March 2010).
44. World Health Organization. *Ecosystems and Health*; WHO: Geneva, Switzerland, 2010. Available at http://www.who.int/globalchange/ecosystems/en/ (accessed March.2010).
45. World Health Organization. *Water, Sanitation and Hygiene*; WHO: Geneva, Switzerland, 2010. Available at http://www.who.int/water_sanitation_health/en/ (accessed March 2010).
46. World Health Organization. *Water for Health: Taking Charge*; WHO: Geneva, Switzerland, 2001; 1–40. Available at http://www.who.int/water_sanitation_health/takingcharge/en/ (accessed March 2010).
47. Skuttle. *Impact of Relative Humidity on Air Quality*; Skuttle: Marietta, OH, USA, 2010. Available at http://www.skuttle.com/pdfs/skuttlehumidity.pdf (accessed March 2010).
48. Ciocci, M.V. *Reflexos do excesso de calor na saúde, e na redução da produtividade*; Cabano Engenharia: Belém-PA, Brazil, 2003. Available at http://www.cabano.com.br/excesso_de_calor.htm (accessed March 2010) (in Portuguese).
49. Corocoran, E.; Nellermann, C.; Baker, E.; Bos, R.; Osborn, D.; Savelli, H., Eds. *Sick Water? The Central Role of Wastewater Management in Sustainable Development. A Rapid Response Assessment*; United Nations Environment Programme: Nairobi, Quenia, 2010; 1–88. Available at http://www.cbsnews.com/htdocs/pdf/SickWater_screen.pdf (accessed March 2010).
50. Palaniappan, M.; Gleick, P.H.; Aallen, L.; Cohen, M.J.; Christian-Smith, J.; Smith, C.; Ross, N. *Clearing the Waters: A Focus on Water Quality Solutions*; United Nations Environment Programme: Nairobi, Quenia, 2010; 1–91. Available at http://www.unep.org/PDF/Clearing_the_Waters.pdf (accessed March 2010).
51. United Nations Environment Programme. *World Water Day 2010 Highlights Solutions and Calls for Action to Improve Water Quality Worldwide*; UNEP: Nairobi, Quenia, 2010 (March). Available at http://www.unep.org/Documents.Multilingual/Default.asp?DocumentID=617&ArticleID=6505&l=en&t=long (accessed March 2010).
52. Addams, L.; Boccaletti, G.; Kerlin, M.; Stuchtey, M. *Charting our Water Future: Economic Frameworks to Inform, Decision-Making*; WRG—2030. Water Resources Group/McKinsey and Company: Chicago, USA, 2009; 1–198. Available at http://www.mckinsey.com/App_Media/Reports/Water/Charting_Our_Water_Future_Full_Report_001.pdf and http://www.nestle.com/InvestorRelations/Events/AllEvents/2030+Water+Resources.htm (accessed March 2010).
53. Gonçalves, C.E.C. *Ruas confortáveis, ruas com vida.: proposição de diretrizes de desenho urbano bioclimático para vias públicas. Av. Juscelino Kubitscheck, Palmas, TO*; UNB: Brasília, Brazil, 2009; 1–149. (MSc Dissertation). Available at http://repositorio.bce.unb.br/bitstream/10482/3901/2/2009_CarlosEduardoCavalheiroGoncalves_pag_82_ate_final.pdf and http://repositorio.bce.unb.br/bitstream/10482/3901/1/2009_CarlosEduardoCavalheiroGoncalves_ate_pag_81.pdf (accessed March 2010).
54. Gash, J.H.C.; Shuttleworth, W.J. Tropical deforestation: Albedo and the surface-energy balance. Clim. Change **1991**, *19* (1–2), 123–133. Partially available at http://books.google.com.br/books?id=XB16EtxyZPUC&pg=PA123&lpg=PA123&dq=Tropical+deforestation:+albedo+and+the+surface-energy+balance&source=bl&ots=2PFRYbSs0l&sig=LOW2DnQ-d4ZNgthrnFY4BxzYhu0&hl=pt-BR&ei=ccX1S9S3JoL-8Aavm9njCg&sa=X&oi=book_result&ct=result&resnum=5&ved=0CDwQ6AEwBA#v=onepage&q=Tropical%20deforestation%3A%20albedo%20and%20the%20surface-energy%20balance&f=false (accessed March 2010).
55. Gordeau, J. *Albedo*; Environmental Science Published for Everybody Round the Earth—ESPERE; Educational Network on Climate: Cracow, Poland, 2004. Available at http://www.atmosphere.mpg.de/enid/25w.html (accessed March 2010).
56. Russell, R. *Global Warming, Clouds, and Albedo: Feedback Loops*; National Earth Science Teachers Association: Boulder, CO, USA, 2007. Available at http://www.windows.ucar.edu/tour/link=/earth/climate/warming_clouds_albedo_feedback.html (accessed March 2010).
57. Hartmann, D.L. *Global Physical Climatology*; Academic Press: California, USA, 1994; 1–386 (International Geophysics, v. 56). Partially available at http://books.google.com.br/books?id=Zi1coMyhlHoC&pg=PA90&lpg=PA90&dq=albedo+straw&source=bl&ots=_SiR7JL67l&sig=8XHCaS4O43ICJR0XkagXhXN_5rs&hl=pt-BR&ei=urRgS9vqMs-WtgeipuzYDQ&sa=X&oi=book_result&ct=result&resnum=10&ved=0CDgQ6AEwCQ#v=onepage&q=albedo%20straw&f=false (accessed March 2010).
58. Yang, Z.L. *Physical Climatology: The Global Energy Balance*, 2006; 1–22. Available at http://www.geo.utexas.edu/courses/387H/Lectures/chap2.pdf and http://www.geo.utexas.edu/courses/387H/default.htm (accessed March 2010).
59. Sharratt, B.S.; Campbell, G.S. Radiation balance of a soil-straw surface modified by straw color. Agron. J. **1994**, *86*, 200–203. Abstract available at http://agron.scijournals.org/cgi/content/abstract/86/1/200 (accessed March 2010).

60. Triparthi, R.P.; Katiyara, T.P.S. Effect of mulches on the thermal regime of soil. Soil Tillage Res. **1984**, *4* (4), 381–390. Abstract available at http://www.sciencedirect.com/science?_ob=ArticleURL&_udi=B6TC6-48XDCKB-1B&_user=10&_coverDate=07%2F31%2F1984&_rdoc=1&_fmt=high&_orig=search&_sort=d&_docanchor=&view=c&_searchStrId=1183152286&_rerunOrigin=google&_acct=C000050221&_version=1&_urlVersion=0&_userid=10&md5=19d963847840c192c30e24092fba1ad2 (accessed March 2010).
61. Heat Island Group. *Cool Roofs*; University of California: Berkeley, USA, 2000. Available at http://eetd.lbl.gov/HeatIsland/ (at cool roofs) (accessed March 2010).
62. Levinson, R.; Berdahl, P.; Akbari, H. Solar spectral optical properties of pigments—Part II: Survey of common colorants. Sol. Energy Mater. Sol. Cells **2005**, *89* (4), 351–389. Available at http://www.odulo.com/eric/Enviroglobal/email%20documents/Pigments2.pdf (accessed March 2010).
63. European Space Agency. World fire maps now available online in near-real time. ESA Observing the Earth, News, 2006 (May 26). Available at http://www.esa.int/esaEO/SEMRBH9ATME_index_0.html (accessed March 2010).
64. United States Department of Agriculture. *Soil Moisture Regimes Map*. USDA/Natural Resources Conservation Service/Soil Survey Division/World Soil Resources: Washington, DC, 1999. Available at http://soils.usda.gov/use/worldsoils/mapindex/smr.html (accessed March 2010).
65. Prata, F. *Global Distribution of Maximum Land Surface Temperature Inferred from Satellites*; CSIRO Atmospheric Research, Aspendale: Victoria, 2000; 1–13 (see page 8). Available at http://www.eoc.csiro.au/associates/aatsr/lst_atlas.pdf (accessed March 2010).
66. The Centre for Australian Weather and Climate Research. *Current Climate Charts: Outgoing Longwave Radiation (OLR) Products—Latest Day and Night-Pass Data*; Bureau of Meteorology Research Centre: Australia, 2009. Available at http://cawcr.gov.au/bmrc/clfor/cfstaff/matw/maproom/index.htm (accessed March 2010).
67. Greenpeace. The World's Last Intact Forest Landscapes (poster A1), 2006. Available at http://www.intactforests.org/pdf.publications/World.IFL.2006.poster_low.pdf and http://www.greenpeace.org/international/campaigns/forests/our-disappearing-forests (accessed March 2010).
68. Natural Environment Research Council. *Satellites Reveal that Green Means Rain in Africa*; ScienceDaily, Sept. 27, 2006. Available at http://www.sciencedaily.com/releases/2006/09/060925064922.htm (accessed March 2010).
69. Coghla, A. More crops for Africa as trees reclaim the desert. New Sci. **2006** (October 14) (Article preview). Available at http://www.newscientist.com/article/dn10293-more-crops-for-africa-as-trees-reclaim-the-desert.html (accessed March 2010).
70. Makarieva, A.M.; Gorshkov, V.G.; Li, B.-L. Conservation of water cycle on land via restoration of natural closed-canopy forests: Implications for regional landscape planning. Ecol. Res. **2006**, *21*, 897–906. Available at http://www.biotic-regulation.pl.ru/offprint/wat_pr1.pdf (accessed March 2010).
71. Makarieva, A.M.; Gorshkov, V.G. Biotic pump of atmospheric moisture as driver of the hydrological cycle on land. Hydrol. Earth Syst. Sci. **2007**, *11*, 1013–1033. Available at http://www.biotic-regulation.pl.ru/offprint/hess07.pdf (accessed March 2010).
72. Rowntree, P.R. Review of general circulation models as a basis for predicting the effects of vegetation change on climate (Ch. 8). In *Forests, Climate, and Hydrology: Regional Impacts*; Reynolds, E.R.C., Thompson, F.B., Eds.; The United Nations University: Tokyo, Japan, 1988. Available at http://www.unu.edu/unupress/unupbooks/80635e/80635E00.htm#Contents (accessed March 2010).
73. Sheil, D.; Murdiyarso, D. How forests attract rain: An examination of a new hypothesis. BioScience **2009**, *59* (4), 341–347. Abstract available at http://www.bioone.org/doi/abs/10.1525/bio.2009.59.4.12?journalCode=bisi and http://www.ncriverwatch.org/wordpress/2009/12/14/forests-attract-rain-an-examination-of-a-new-hypothesis/ (accessed March 2010).
74. Bonan, G.B. Forests and climate change: Forcing feedbacks and the climate benefits of forests. Science **2008**, *320* (5882), 1444–1449. Abstract available at http://www.sciencemag.org/cgi/content/abstract/sci;320/5882/1444 (accessed March 2010).
75. Takata, K.; Saito, K.; Yasunari, T. Changes in the Asian monsoon climate during 1700–1850 induced by pre-industrial cultivation. PNAS Online **2009**, *106* (24; June). Available at http://www.nagoya-u.ac.jp/en/pdf/research/activities/090529_hyarc_yasunari.pdf (accessed March 2010).
76. Pearce, F. Rainforests may pump wind worldwide. New Sci. **2009** (2702; April 1), 1–3. Available at http://www.ideastransformlandscapes.org/media/uploads/File/Rainforests%20may%20pump%20winds%20worldwide.pdf (accessed March 2010).
77. Rockström, J.; Steffen, W.; Noone, K.; Persson, A.; Chapin, F.S.; Lambin, E.F.; Lenton, T.M.; Scheffer, M.; Folke, C.; Schellnhuber, H.J.; Nykvist, B.; de Wit, C.A.; Hughes, T.; van der Leeuw, S.; Rodhe, H.; Sörlin, S.; Snyder, P.K.; Costanza, R.; Svedin, U.; Falkenmark, M.; Karlberg, L.; Corell, R.W.; Fabry, V.J.; Hansen, J.; Walker, B.; Liverman, D.; Richardson, K.; Crutzen, P.; Foley, J.A. A safe operating space for humanity. Nature **2009**, *461* (September), 472–475. Abstract available at http://www.nature.com/news/specials/planetaryboundaries/index.html (accessed March 2010).
78. Towers, P.T.; Lumper, K. *Definitions of Sustainability*; The University of Reading: Reading, U.K., 2010. Available at http://www.ecifm.rdg.ac.uk/definitions.htm (accessed March 2010).
79. Food and Agriculture Organization. *Land Quality Indicators and Their Use in Sustainable Agriculture and Rural Development*. Land and Water Bulletin 5, FAO, UNDP, UNEP, World Bank: Rome, 1996; 1–217. Available at http://www.mpl.ird.fr/crea/taller-colombia/FAO/AGLL/pdfdocs/landqual.pdf (accessed March 2010).
80. Munoz, L. Understanding sustainability versus sustained development by means of a WIN Development Model. In *Sustainability Review*; Flint, W., Ed.; Five E's Unlimited: Pungoteague, VA, Canada, 1999 (1; September 6); 1–18. Available at http://theomai.unq.edu.ar/artmunoz001.htm (accessed March 2010).
81. Wikipedia. *Nachhaltigkeit* (Sustainability). Available in German at: http://de.wikipedia.org/wiki/Nachhaltigkeit and http://de.wikipedia.org/wiki/Nachhaltigkeit_%28Forstwirtschaft%29 (accessed March 2010).
82. Kanshie, T.K. *Five Thousand Years of Sustainability? A Case Study on Gedeo Land Use (Southern Ethiopia)*; Kanshie:

Wageningen, Netherlands, 2002; 1–295 (Treebook, 5). Available at http://www.treemail.nl/books (accessed March 2010).
83. Chambers, N.; Simmons, C.; Wackernagel, M. *Sharing Nature's Interest: Ecological Footprints as an Indicator of Sustainability*; James and James, Earthscan: London, 2001; 1–200. Partially available at http://en.book2down.com/Sharing-Natures-Interest-Using-Ecological-Footprints-As-An-Indicator-Of-Sustainability/186751 (accessed March.2010).
84. Wikipedia. *List of Countries by Ecological Footprint*. 2008. Available at http://en.wikipedia.org/wiki/List_of_countries_by_ecological_footprint (accessed March 2010).
85. Schaefer, F.; Luksch, U.; Steinbach, N.; Cabeça, J.; Hanauer, J. *Ecological Footprint and Biocapacity: The World's Ability to Generate Resources and Absorb Wastes in a Limited Time Period* (Working paper and studies). Office for Official Publications of the European Communities; Luxembourg, 2006; 1–11. Available at http://epp.eurostat.ec.europa.eu/cache/ITY_OFFPUB/KS-AU-06-001/EN/KS-AU-06-001-EN.PDF (accessed March 2010).
86. Global Footprint Network. *Footprint for Nations. 2009*; Global Footprint Network: Oakland, CA, USA, 2010. Available at http://www.footprintnetwork.org/en/index.php/GFN/page/footprint_for_nations/ (accessed March 2010).
87. World Wide Fund for Nature. *Living Planet Report 2008*; WWF International: Gland, Suisse, 2008; 1–48. Available at http://assets.panda.org/downloads/living_planet_report_2008.pdf (accessed March 2010).
88. Best, A.; Giljun, S.; Simmons, C.; Blobel, D.; Lewis, K.; Hammer, M.; Cavalieri, S.; Lutter, S.; Magguirre, C. *Potential of the Ecological Footprint for Monitoring Environmental Impacts from Natural Resources Use: Analysis of the Potential of the Ecological Footprint and Related Assessment Tools for use in the EU's Thematic Strategy on the Sustainable Use of Natural Resources*; Report to the European Commission, DG Environment, 2008; 1–312. Available at http://ec.europa.eu/environment/natres/pdf/footprint.pdf (accessed March 2010).
89. Waterfootprint Network. *Water Footprint versus Water Scarcity, Self-Sufficiency and Water Import Dependency per Country*; Water Footprint Network/University of Twente: Enschede, Netherlands, 2001. Available at http://www.waterfootprint.org/?page=files/NationalStatistics (accessed March 2010).
90. Waterfootprint Network. *Water Footprints of Crop and Livestock Products (m3/t) for Some Selected Countries (1997–2001)*; Water Footprint Network/University of Twente: Enschede, Netherlands, 2001. Available at http://www.waterfootprint.org/?page=files/Productwaterfootprint-statistics (accessed March 2010).
91. Hoekstra, A.Y.; Chapagain, A.K. Water footprint of nations: Water use by people as a function of their consumption pattern. Water Resour. Manage. **2007**, *21*, 35–48. Available at http://www.waterfootprint.org/Reports/Report18.pdf and http://www.waterfootprint.org/Reports/Hoekstra_and_Chapagain_2007.pdf (accessed March 2010).
92. Hoekstra, A.Y. *Water Neutral: Reducing and Offsetting the Impacts of Water Footprints*. Research report series nr. 28—Value of Water, UNESCO-IHE, 2008; 1–42. Available at http://www.waterfootprint.org/Reports/Report28-WaterNeutral.pdf (accessed March 2010).
93. Hoekstra, A.Y.; Chapagain, A.K.; Aldaya, M.M.; Mekonnen, M.M. *Waterfootprint Manual: State of the Art 2009*; Netherlands: Water Footprint Network: Enschede, Netherlands, 2009; 1–131. Available at http://www.waterfootprint.org/downloads/WaterFootprintManual2009.pdf (accessed March 2010).
94. Wikipedia. *Carbon Footprint*. Available at http://en.wikipedia.org/wiki/Carbon_footprint (accessed March 2010).
95. Wikipedia. *List of Countries by Carbon Dioxide Emission per Capita*. Available at http://en.wikipedia.org/wiki/List_of_countries_by_carbon_dioxide_emissions_per_capita (accessed March 2010).
96. Global Footprint Network. *Carbon Footprint*; Global Footprint Network: Oakland, CA, USA, 2009. Available at http://www.footprintnetwork.org/en/index.php/GFN/page/carbon_footprint/ (accessed March 2010).
97. Odum, H.T. *Environmental Accounting, Emergy and Decision Making*; John Wiley: New York, USA, 1996; 1–370. Partially available at http://dieoff.org/page170.htm (accessed March 2010).
98. Hau, J.L.; Bakshi, B.R. *Promise and Problems of Emergy Analysis*; Ohio State University: Ohio, USA, 2003; 1–13. Available at http://www.che.eng.ohio-state.edu/~bakshi/EcolModel3.pdf (accessed March 2010).
99. Ferreyra, C. Emergy analysis of one century of agricultural production in the rolling pampas of Argentina. Int. J. Agric. Resour. Governance Ecol. **2006**, *5* (2–3), 185–205. Available at http://snre.ufl.edu/graduate/files/publicationsbyalumni/Ferreyra%202006.pdf (accessed March 2010).
100. Martin, J.F.; Tilley, D.R. Accounting for environmental sustainability with emergy analysis. In: *Encyclopedia of Soil Science*; Lal, R., Ed.; Taylor and Francis Group: New York, 2007. Abstract available at http://www.informaworld.com/smpp/content~db=all~content=a788663399 (accessed March 2010).

Bioenergy Crops: Carbon Balance Assessment

Rocky Lemus
Mississippi State University, Starkville, Mississippi, U.S.A.
Rattan Lal
School of Environment and Natural Resources, Ohio State University, Columbus, Ohio, U.S.A.

Abstract
Bioenergy crops are perennial crops with potential to sequester carbon (C) into the biomass and soil while providing an alternative source of energy. The decrease in crop productivity due to soil degradation and the increased risks of global warming warrant assessing alternative sources of energy produced on degraded soils. New strategies to mitigate atmospheric carbon dioxide (CO_2) by increasing soil organic matter (SOM) and biomass productivity are some of the alternatives through bioenergy crops.

OVERVIEW

Since the 1990s, several species have been assessed as potential energy sources and for C sequestration. Switchgrass (*Panicum virgatum* L.), a herbaceous species, and two short-rotation woody crops (SRWCs), hybrid poplar (*Populs* spp.) and willow (*Salix* spp.), are promising for C sequestration and biofuel production. Using these species as energy crops can partly offset CO_2 emitted by fossil fuel combustion.

There are approximately 60 million ha (Mha) of degraded soil (severely eroded and mined land) in the United States that can be sown to bioenergy crops to decrease soil erosion and increase C sequestration.[1] The Midwest and the Southeast regions of the United States are the potential areas where bioenergy crops are most likely to compete with other traditional crops for land resources.[2] The use of bioenergy crops in these areas offers an opportunity to replenish the soil organic C (SOC) pool depleted by tillage and soil degradation.

Biomass fuels used in a sustainable manner can result in no net increase in atmospheric CO_2. Indeed, sustainable use of biomass can result in a net decrease in the rate of enrichment of atmospheric CO_2. This is based on the assumption that all the CO_2 emitted by the use of biomass fuels is absorbed from the atmosphere by photo-synthesis. Increased substitution of fossil fuels with biomass-based fuels reduces the risk of global warming by enrichment of atmospheric CO_2. This entry discusses the importance of energy crops in offsetting fossil fuel combustion through C sequestration in biomass and soil.

INFLUENCE OF BIOENERGY CROPS AT THE TEMPORAL SCALE

Long-term tillage and continuous cropping can reduce SOC pool[3] and increase atmospheric concentration of CO_2 (Fig. 1).[4] Conversion of natural to agricultural ecosystems results in the net release of C into the atmosphere.[6] Conservation practices, such as the introduction of perennial crops on degraded agricultural soils and growing bioenergy crops, constitute a direct link between sink (SOC) and the source of CO_2 (fossil fuel). Bioenergy crops are the sink/source transition because the C incorporated into their biomass and root system has a high potential of being incorporated into the SOM pool. Most of the C is incorporated into the biomass and soil by the deep and extensive root systems. The SOC pool attains a new equilibrium after grassland restoration, but the time required for SOC to reach equilibrium is variable, especially under diverse climatic conditions.[7] Most models used to estimate C sequestration in bioenergy crops suggest that a period longer than 10 years may be necessary for pronounced soil-quality improvements.

BENEFITS OF BIOENERGY CROPS

The SOC pool and its dynamics are the major indicators of soil quality and crop productivity.[8] Bioenergy crops can improve soil quality through increase in SOM, nutrient dynamics, erosion control, and improvement in soil structure and porosity. Most perennial crops and SRWCs have extensive root systems that reduce soil erosion and non-point-source pollution.

A shift from traditional agricultural crops into perennial bioenergy crops stabilizes agricultural soils, reduces erosion, and improves water quality.[9] Most of these benefits are due to the elimination of tillage leading to a significant decrease in both erosion and chemical runoff especially nitrate (NO_3–N). Soil erosion and nitrate runoff are reduced in 2-year-old stands of switchgrass when compared to no-till corn (Table 1). Conservation effectiveness of switchgrass is attributed to high root biomass, which stabilizes the soil,[11] depletes excess N and P,[12] and increases microbial activity.[13]

Encyclopedia of Environmental Management DOI: 10.1081/E-EEM-120038525

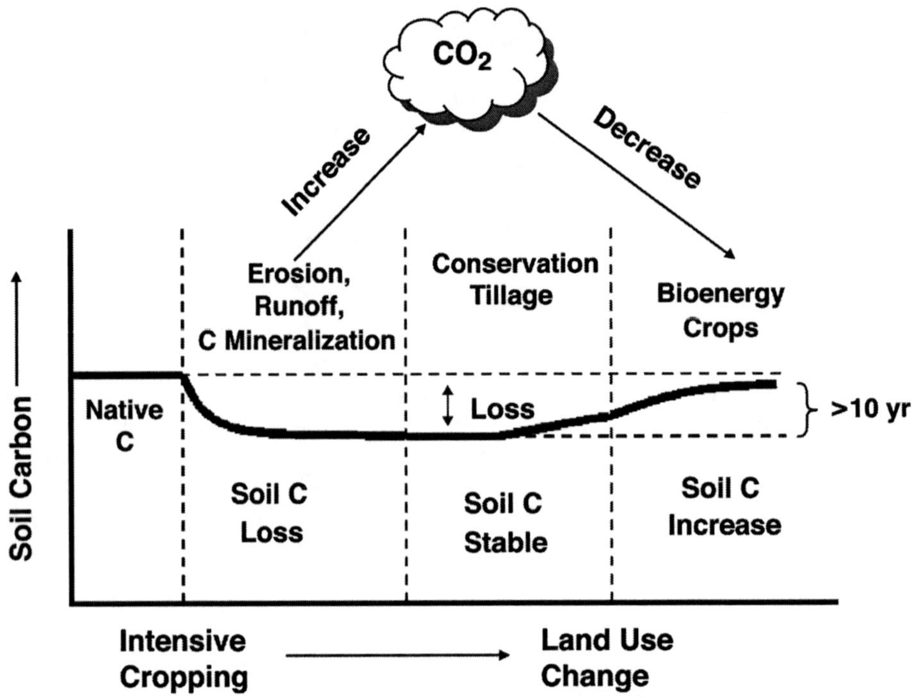
CO2
Increase
Decrease
Erosion,
Runoff,
C Mineralization
Conservation
Tillage
Bioenergy
Crops
Native
C
Loss
>10 yr
Soil Carbon
Soil C
Loss
Soil C
Stable
Soil C
Increase
Intensive
Cropping
Land Use
Change

Fig. 1 Influence of bioenergy crops on C sequestration and CO_2 mitigation with changes in agricultural practices.
Source: Adapted from Janzen et al.[5]

POTENTIAL OF C SEQUESTRATION

Bioenergy crops have the potential to store large quantities of C. Exploring their biomass potential through N fertilization and soil management are some of the proposed strategies to offset CO_2 emissions by fossil fuel combustion. However, a bioenergy crop is not a closed system and only some portion of the C sequestered might be conserved through the production–utilization cycle. Most of the C sequestered in the biomass is utilized for energy production allowing some C release back into the atmosphere. This system does not cause negative impacts as most of the C returned to the atmosphere is recaptured by the plants during the subsequent season.

Assessing the net C sequestration of bioenergy compared to agricultural crops requires analyses of C sequestration across the soil profile. Carbon sequestration is influenced by biological processes such as root biomass and crop species, and by soil physical and chemical aspects such as texture, bulk density, and pH. The amount of C in these types of management options is affected by soil perturbation which could increase SOM oxidation, exacerbating losses, especially at the soil surface. Estimations of C sequestration onland under the Conservation Reserve Program (CRP) have ranged from 0.6 to 1.0 Mg C $ha^{-1} yr^{-1}$ in SOM,[14] compared to 2.0 Mg C $ha^{-1} yr^{-1}$ in SRWC.[15] Root biomass and its relationship to SOC are always positive but may be influenced by soil depth and plant species.[16] Switchgrass improves soil quality via reduced nutrient loss (especially NO_3–N) and increases C sequestration through its deep rooting system, high biomass production, and perenniality.[17] Switchgrass root biomass can be as high as 16.8 Mg ha^{-1} up to 3.3 m deep with some root mass variability among soil type.[13]

The sustainability of soil and crop systems is typically affected by changes in the SOC pool. The largest gain in SOC pool occurs in the upper 30-cm layer. There are differences in the amount of C being sequestered by herbaceous crops and SRWCs, but they havethe potential to

Table 1 Environmental benefits of switchgrass (SWG) compared to no-till corn (NTC) overtime.

	Soil erosion			Water runoff			Nitrate concentration		
Year	SWG ($Mg\ ha^{-1}$)	NTC ($Mg\ ha^{-1}$)	Ratio SWG/NTC[a]	SWG ($Mg\ ha^{-1}$)	NTC ($Mg\ ha^{-1}$)	Ratio SWG/NTC	SWG (ppm)	NTC (ppm)	Ratio SWG/NTC
1	2.80	0.70	4.00	10.70	2.60	4.11	3.41	0.57	5.98
2	0.14	0.19	0.74	0.70	1.40	0.50	0.72	2.18	0.33
3	0.06	0.08	0.75	0.30	0.90	0.33	0.77	0.90	0.86

[a]The ratio SWG/NTC of <1 indicates conservation effectiveness of a SWG system during the second and third year of its establishment.
Source: Adapted from McLaughlin et al.[10]

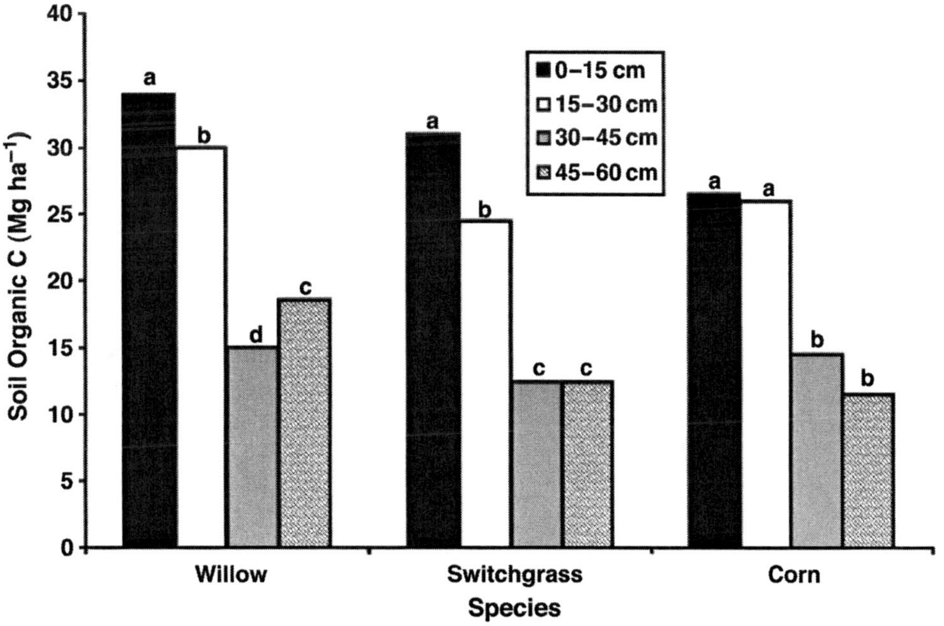
Soil Organic C (Mg ha⁻¹)
40
35
30
25
20
15
10
5
0
a
b
d
c
a
b
c
c
a
a
b
b
0–15 cm
15–30 cm
30–45 cm
45–60 cm
Willow
Switchgrass
Corn
Species

Fig. 2 Comparison of C sequestration between bioenergy crops and corn at different soil depths 5 years after establishment.
Source: Mehdi et al.[18]

increase C both in the soil and aboveground biomass. The data in Fig. 2 show that after 5 years SOC pool at 60-cm depth was 97.5 $Mg\,ha^{-1}$ under willow, 80.5 $Mg\,ha^{-1}$ under switchgrass, and 78.5 $Mg\,ha^{-1}$ under corn. The mean rate of SOC sequestration, vis-à-vis corn, over 5 years was 3.8 $Mg\,C\,ha^{-1}\,yr^{-1}$ under willow and 0.4 $Mg\,C\,ha^{-1}\,yr^{-1}$ under switchgrass (Fig. 2). The SOC fluctuations are more drastic in the top soil layers, probably because of the greater effect of precipitation, soil temperature, larger root biomass, and greater microbial activity.

CONCLUSIONS

Bioenergy crops can influence the global C cycle. An important challenge in using bioenergy crops is assessing how much C is being sequestered and how much is being put back into the atmosphere by the cofiring process to produce electricity or other sources of energy. Available data indicate that perennial crops are moving in the right direction in improving soil quality while providing an alternative energy source. A combination of soil C sequestration and bioenergy crops with high biomass capability and deep perennial root systems is a useful strategy of reducing the rate of enrichment of atmospheric CO_2. The use of bioenergy crops increases soil C and improves soil quality by eliminating C losses associated with annual cultivation. However, these improvements depend on the rate of soil C additions, the long-term capacity of the soil for C storage, and the stability or permanence of the sequestered C overtime.

REFERENCES

1. Kort, J.; Collins, M.; Ditsch, D. A review of soil erosion potential associated with biomass crops. Biomass Bioenergy **1998**, *14* (4), 351–359.
2. Walsh, M.E.; Ince, P.J.; De la Torre Ugarte, D.G.; Alig, R., Mills, J.; Spelter, H.; Skog, K.; Slinsky, S.P.; Ray, D.E.; Graham, R.L. Potential of short-rotation wood crops as a fiber and energy source in the U.S. In *Proceedings of the Fourth Biomass Conference of the Americas*; Elsevier Science Ltd.: Oxford, 1999, 195–206.
3. Kern, J.S.; Johnson, M.G. Conservation tillage impacts on national soil and atmospheric carbon levels. Soil Sci. Soc. Am. J. **1993**, *57*, 200–210.
4. Schlamadinger, B.; Marland, G. The role of forest and bioenergy strategies in the global carbon cycle. Biomass Bioenergy **1996**, *10*, 275–300.
5. Janzen, H.H.; Campell, C.A.; Izaurralde, R.C.; Ellert, B.H.; Juma, N.; McGill, W.B.; Zenter, R.P. Management effects on soil C storage on the Canadian prairies. Soil Tillage Res. **1998**, *47*, 181–195.
6. Schulze, E.D.; Wirth, C.; Heimann, M. Managing forests after Kyoto. Science **2000**, *289*, 2058–2059.
7. Potter, K.N.; Tobert, H.A.; Johnson, H.B.; Tischler, C.R. Carbon storage after long-term grass establishment on degraded soils. Soil Sci. **1999**, *164*, 718–725.
8. Mann, L.; Tolbert, V. Soil sustainability in renewable biomass plantings. Ambio **2000**, *29* (8), 492–498.
9. McLaughlin, S.B.; Walsh, M.E. Evaluating environmental consequences of producing herbaceous crops for bioenergy. Biomass Bioenergy **1998**, *14* (4), 317–324.
10. McLaughlin, S.B.; De La Torre Ugarte, D.G.; Garten, C.T.; Lynd, L.R.; Sanderson, M.A.; Tolbert, V.R.; Wolf, D.D.

High-value renewable energy from prairie grasses. Environ. Sci. Technol. **2002**, *36*, 2122–2129.

11. Gilley, J.E.; Eghball, B.; Kramer, L.A.; Moorman, T.B. Narrow grass hedge effects on runoff and soil loss. J. Soil Water Conserv. **2000**, *55*, 190–196.
12. Eghball, B.; Gilley, J.E.; Kramer, L.A.; Moorman, T.B. Narrow grass hedge effects on phosphorus and nitrogen in runoff following manure and fertilizer application. J. Soil Water Conserv. **2000**, *55*, 172–176.
13. Ma, Z.; Wood, C.W.; Bransby, D.I. Soil management on soil C sequestration by switchgrass. Biomass Bioenergy **2000**, *18*, 469–477.
14. Follet, R.F. Soil management concepts and carbon sequestration in zin cropland soils. Soil Tillage Res. **2001**, *61*, 77–92.
15. Tuskan, G.A. Short-rotation woody crop supply systems in the United States: What do we know and what do we need to know?. Biomass Bioenergy **1998**, *14* (4), 307–315.
16. Slobodian, N.; Van Rees, K.; Pennock, D. Cultivationinduced effects on belowground biomass and organic carbon. Soil Sci. Soc. Am. J. **2002**, *66*, 924–930.
17. Sladden, S.E.; Bransby, D.I.; Aiken, G.E. Biomass yield, composition, and production costs for eight switchgrass varieties in Alabama. Biomass Bioenergy **1991**, *1* (2), 119–122.
18. Mehdi, B.; Zan, C.; Girouard, P.; Samson, R.; Soil Carbon Sequestration Under Two Dedicated Perennial Bioenergy Crops. Research Reports. Resource Efficient Agricultural Production (REAP)-Canada, 1998. Available at http://www.reap-canada.com/Reports/reportsindex.htm (accessed April 6, 2004).

Biofertilizers

J.R. de Freitas
Department of Soil Science, University of Saskatchewan, Saskatoon, Saskatchewan, Canada

Abstract
Biofertilizers include microorganisms and their metabolites that are capable of enhancing soil fertility, crop growth, or yield. Biofertilizers will become an increasingly important area of research and development. Biofertilizers provide an alternative to agricultural chemicals as more sustainable and ecologically sound practices to increase crop productivity.

INTRODUCTION

Biofertilizers include microorganisms and their metabolites that are capable of enhancing soil fertility, crop growth, and/or yield. These include both indigenous microbes and microbial inoculants, that is, microorganisms that replace fertilizers or increase a crop's fertilizer use efficiency. Soil microorganisms such as bacteria, ectomycorhiza, arbuscular mycorrhizal fungi, and soil algae, especially the N_2-fixing cyanobacteria have potential as biofertilizers. Nitrogen-fixing inoculants based on *Rhizobium* species were among the first biofertilizers introduced into agroecosystems back in the 19th century. In the 21th century, biofertilizers will become an increasingly important area of research and development.[1] The use of fertilizers and pesticides has increased steadily since the 1970s; consequently, concerns about the impacts of these chemicals on land, air, and water have become significant environmental issues. Biofertilizers provide an alternative to agricultural chemicals as more sustainable and ecologically sound practices to increase crop productivity. Biofertilizer sales forecasts in the United States for the years 2001 and 2006 represent $690 million and $1.6 billion, respectively. Examples of some biofertilizers currently in use worldwide are shown in Table 1.[2]

MARKET FOR BIOFERTILIZERS

The market potential for biofertilizers includes the high value vegetable industry. A comparison of the base value of various crops and the increased value that can be obtained as the crop yield rises is illustrated in Table 2. Due to high nutrient requirements and high susceptibility to diseases, vegetable growers spend substantial amounts to protect this valuable produce. For example, average broccoli and tomato crops grown in California require ca. $62 and $170 worth of fertilizer and/or fungicide per acre, respectively. When the U.S. government prohibits the use of methyl bromide as a soil fumigant, as anticipated in 2005, development of biological products will be stimulated as an alternative to the use of chemicals.[3]

MECHANISMS OF GROWTH PROMOTION

The mechanisms covered in this entry are those that show commercial market potential; thus, it does not include all modes of action by which biofertilizers promote crop growth. Biofertilizers promote crop growth using several mechanisms with the primary one varying as a function of environmental conditions. Although the mechanisms of commercially available biofertilizers are not always entirely understood, growth promotion has been classified as the result of indirect or direct mechanisms. Indirect plant growth promotion may be associated with the repression of negative effects caused by phytopathogenic organisms, that is, biological control. Conversely, direct growth promotion mechanism may either provide some compound essential to crop development and/or stimulate nutrient uptake. Biofertilizers based on biological control agents *Mycorrhizae* and *Rhizobium* will be discussed in more detail elsewhere in this encyclopedia (e.g., Biological Pest Controls; Mycorrhiza; Rhizobia).

Phytohormones

Production of phytohormones is a commonly noted direct mechanism of plant growth promotion.[4] The nature of growth response may be the result of phytohormone production in the rhizosphere. Phytohormones are produced by many biofertilizers and include a list of plant growth regulators that are important in the plant's metabolism. For example, auxins such as indole-3-acetic acid are known for their ability to stimulate root cell division, differentiation, and promote cell elongation. Other phytohormones such as cytokinin, gibberellin,

Encyclopedia of Environmental Management DOI: 10.1081/E-EEM-100000878

Table 1 Organisms, mode of action, crops, and producers of biofertilizers currently in use for agriculture.

Type	Mode of action	Crop	Used in
Rhizobium spp.	N_2 fixation	Legumes	Russia; several countries
Cyanobacteria	N_2 fixation	Rice	Japan; several countries
Azospirillum spp.	N_2 fixation	Cereals	Several countries
Mycorrhizae	Nutrient acquisition	Conifers	Several countries
Penicillium bilaii	P solubilization	Cereals, legumes	Western Canada
Directed compost	Soil fertility	All plants	Several countries
Earthworm	Humus formation	Vegetables, flowers	Cottage industry

Source: Adapted from Tengerdy and Szakacs.[2] Copyright 1998 Elsevier Science.

and ethylene also play key roles in plant development and have been reported to increase the growth of various commercial crops. The horticulture market for biofertilizer products based on gibberellin and other auxins is currently estimated at $600 million per year.

Plant Nutrient Acquisition

Several direct mechanisms are responsible for increased nutrient acquisition.

Biological Nitrogen Fixation (BNF)

Nitrogen (N) is an essential macronutrient, that is, it is the key building block of proteins, thus an indispensable component of the protoplasm of microorganisms, animals, and plants. The supply of biologically available N to agriculture through BNF represents ca. 140×10^6 ton/year, globally.[1] Therefore, BNF represents an economy of millions of dollars. N_2-fixation by free-living bacteria such as *Azospirillum, Azotobacter, Bacillus,* and *Derxia* species have been exploited in agricultural systems for many decades and constitute an important source of N input into agroecosystems.[5] Other BNF associations include the water fern *Azolla* that forms a symbiosis with the heterocystous cyanobacterium *Anabaena azollae.* The *Azolla–Anabaena* system has been used as a biofertilizer in Vietnam and China for rice production and has the potential to supply the entire N requirement (30–50 kg N/ha) for a rice crop during the growing season.[6] Another diazotroph, the N_2-fixing actinomycete *Frankia,* forms nodules (actinorrhizae) in ca. 17 genera of nonlegume wood species with *Alnus* (alder) and the genus *Casuarina* being the most important for forestry and agriculture. Estimates of total N_2 fixed range between 50–250 kg/ha/year, depending on the plant species and region. However, in some cases, inoculation with *Frankia* is necessary for nodulation to occur. Actinorrhizal plant species have been successfully inoculated with *Frankia* on a large scale. For example, millions of actinorrhizal trees, especially *Alnus* spp., inoculated with *Frankia* were used in land reclamation programs established in Canada.[7]

Phosphorus Solubilization

Certain microorganisms are very effective in solubilizing phosphorus (P) from insoluble phosphate compounds such as hydroxyapatite through the action of organic acids. Numerous claims have been made about biofertilizers that can enhance plant growth by solubilizing P. A classical example is the bacterium *Bacillus megaterium,* which was formulated into an inoculant under the name of Phosphobacterin in the former Soviet Union. A similar biofertilizer based on P-solubilizing fungi is currently marketed in Canada as JumpStart™ for use on wheat, canola, mustard, and N_2-fixing legumes.

Table 2 Market price and potential price increments with yield increases of 5% and 25% in selected vegetable crops.

		Price increments ($ per acre)	
Vegetable crop	Market price ($ per acre)	5% yield increase	25% yield increase
Carrot	4,520	226	1130
Cauliflower	4,179	209	1045
Celery	10,132	507	2533
Cucumber	3,296	165	824
Lettuce	5,882	294	1471
Tomato	9,966	498	2492

Source: Adapted from USDA.[3]

Microbial Siderophore Uptake

Iron (Fe) is an important plant micronutrient. Plants assimilate iron by acidifying the rhizosphere and/or secreting phyto-siderophores with subsequent reassimilation of the iron–siderophore complex. However, plants also may benefit from the direct uptake of microbial siderophore–iron complexes. For example, some biofertilizers synthesize siderophores that can solubilize and sequester Fe from soil and provide it to plant cells, thus contributing to the nutrition and development of crops. In fact, studies demonstrate that ferric pseudobactin 358 may stimulate chlorophyll synthesis in carnation and barley.[8]

Other Nutrients

Studies with *Azospirillum* spp. and plant growth-promoting rhizobacteria (PGPR) have demonstrated the ability of these biofertilizers to promote enhancement of nutrient and water uptake into the plant. For example, inoculation of winter wheat seeds with pseudomonad PGPR stimulated the uptake of soil-Fe and fertilizer-^{15}N by winter wheat cultivated in two Canadian soils.[9] Similarly, inoculation of canola seeds with a *Pseudomonas putida* increased phosphate uptake from nutrient solution.[10] In these cases, the authors speculated that plant growth regulators produced by the biofertilizers in the plant's rhizosphere stimulated root development which, in turn, enhanced nutrient acquisition.

FUTURE RESEARCH DIRECTIONS

It is clear that commercial crops can benefit directly from biofertilizers. Certainly, with the development of molecular biology and genetic manipulation of biofertilizers to improve N_2-fixation, rhizosphere competence and ability to be used together with specific chemicals, will contribute to an integrated strategy to reduce the total amount of chemicals used in agriculture. Although biofertilizer products are currently available on the market, consistency is still the major factor that limits their use. Elucidation of mechanisms, development of stable formulations, effective delivery systems, and field demonstration of effective biofertilizers will definitely improve reliability and enhance their use as commercial biofertilizers.

REFERENCES

1. Killham, K. *Soil Ecology*; Cambridge University Press: Cambridge, U.K., 1994; 242.
2. Tengerdy, R.P.; Szakács, G. Perspectives in agrobiotechnology. J. Biotechnol. **1998**, *66*, 91–99.
3. USDA. USDA Economics and Statistics System. National Agricultural Statistics Service; Cornell University. http://mann77.mannlib.cornell.edu/reports/nassr/fruit/pvg-bban/vegetables_annual_summary-01.16.98 (accessed June 1999).
4. Glick, B.R. The enhancement of plant growth by freeliving bacteria. Can. J. Microbiol. **1995**, *41*, 109–117.
5. Pankhurst, C.E.; Lynch, J.M. The role of soil microbiology in sustainable intensive agriculture. Adv. Plant Pathol. **1995**, *11*, 230–247.
6. Zuberer, D.A. Biological Dinitrogen Fixation: Introduction and Nonsymbiotic. In *Principles and Applications of Soil Microbiology*; Sylvia, D.M., Fuhrmann, J.J., Hartel, P.G., Zuberer, D.A., Eds.; Prentice Hall: Upper Saddle River, NJ, 1998; 295–321.
7. Périnet, P.; Brouillette, J.G.; Fortin, J.A.; Lalonde, M. Large scale inoculation of actinorrhizal plants with frankia. Plant Soil **1985**, *87*, 175–183.
8. Duiff, B.J.; de Kogel, W.J.; Bakker, P.A.H.M.; Schipper, B. Significance of Pseudobactin 358 for the Iron Nutrition of Plants. In *Improving Plant Productivity with Rhizosphere Bacteria*; Ryder, M.H., Stephens, P.M., Bowen, G.D., Eds.; Third International Workshop on Plant-Growth Promoting Rhizobacteria, Adelaide, Australia, March 7–11, CSIRO Division of Soils: South Australia, 1994; 142–144.
9. De Freitas, J.R.; Germida, J.J. Growth promotion of winter wheat by fluorescent pseudomonads under growth chamber conditions. Soil Biol. Bioch. **1992**, *24*, 1127–1135.
10. Lifshitz, R.; Kloepper, J.W.; Kozlowiski, M.; Simonson, C.; Carlson, J.; Tipping, E.M.; Zaleska, I. Growth promotion of canola (rapeseed) seedlings by a strain of *pseudomonas putida* under gnotobiotic conditions. Can. J. Microbiol. **1987**, *33*, 390–395.

Aluminum – Bioindicators

Bioindicators: Farming Practices and Sustainability

Joji Muramoto
Stephen R. Gliessman
Program in Community and Agroecology, Department of Environmental Studies, University of California—Santa Cruz, Santa Cruz, California, U.S.A.

Abstract

This entry summarizes the background, definitions, procedures, and current status of developing bioindicators for assessing sustainability of farming practices. Biodiversity in agroecosystems, or "agrobiodiversity," can serve as a basis for sustainable food production. Biotic parameters have begun to be used as bioindicators during the last decade to evaluate the status of biodiversity, ecosystem functions, and sustainability of agroecosystems. International organizations, such as the Organisation for Economic Co-operation and Development (OECD) and the European Union (EU), encourage the use of bioindicators as evaluative tools to promote agro-environmental policies. Bioindicators have been used commercially to certify "environmentally sound" products. Compared to abiotic indicators, however, the development of bioindicators has multiple challenges. Bioindicators developed in European agroecosystems and Latin American shaded coffee systems are reviewed. To implement biodiversity-based sustainable agriculture policy, participation of both the farmer and the consumer is crucial. An example of such activity from Japan is presented. In conclusion, the development of bioindicators for estimating the sustainability of farming practices is in its early stages but making good progress. Directions for future study include the following: 1) standardization of sampling methods; 2) expansion of databases and improvement of the statistical techniques to minimize potential bias of indicators; 3) improved understanding of the mechanisms linking the status of biodiversity, Earth system processes, human decisions, and ecosystem services impacting human welfare; 4) use of participatory approach in the processes of developing bioindicators; 5) development of multiple sets of bioindicators tailored to different end users (general public, farmers, policy makers, and scientists); and 6) constructing a hierarchical system that integrates different types of bioindicators and other ecological, social, cultural, and economic indicators to evaluate sustainability of farming practices.

INTRODUCTION

The growth of larger-scale monocultures that heavily depend on use of pesticides, chemical fertilizers, and fossil-fuel-based agricultural machinery has significantly increased crop yield. However, this has also brought about both direct and indirect negative consequences on ecological (e.g., soil erosion, nitrate and pesticide contamination of groundwater, loss of agrobiodiversity), economic (e.g., increase of production and marketing costs and decrease of the net income of farmers), and social (e.g., loss of agricultural communities due to farm consolidations) sustainability of agriculture.[1,2] Especially, agricultural intensification is one of the major drivers of global biodiversity loss as a result of associated habitat fragmentation, land conversion, and agrochemical applications.[3,4] Since the 1970s, and more so the 1980s, therefore, various alternative farming practices have been developed in pursuit of sustainable agriculture.[5–12] Farmers around the world have adopted these practices to varying degrees, but evaluation of their success needs to be conducted locally. To implement sustainable agricultural practices and policies, managers and policy makers need tools for assessing changes in agroecosystems in various time and spatial scales. Numerous indicators for agricultural sustainability including physicochemical, socioeconomic, and biological indicators, or bioindicators, have been developed.[13–17] Bioindicators are important because they are direct measures of the desired outcome, i.e., sustained or increased biodiversity. They also are living, dynamic, and active indicators, often responding quickly to the way farming is carried out. As evidence regarding the role of biodiversity in maintaining agroecosystem structure and function accumulates, its role as a bioindicator increases.

This entry will first briefly review the background, definitions, concepts, and history of bioindicators. Then, we discuss the current status of developing bioindicators with two case studies: Europe and Latin America (shaded coffee systems). The former case represents temperate agroecosystems with 4000 years of farming history. The second case characterizes upland tropical agroecosystems known to occur in regions of the world with some of the highest, yet most threatened, biodiversity. Furthermore, an example of the participatory approach in implementing agrobiodiversity policy in Japan is presented.

Encyclopedia of Environmental Management DOI: 10.1081/E-EEM-120046173

AGRICULTURAL SUSTAINABILITY AND AGROBIODIVERSITY

Although many definitions of sustainable agriculture exist, most of them address ecological, economic, and social goals.[18,19] From an ecological perspective, sustainable farming practices 1) maintain their natural resource base; 2) rely on minimum artificial inputs from outside the farm system; 3) manage pests and diseases through internal regulating mechanisms; and 4) allow the recovery from the disturbances caused by cultivation and harvest.[20] Ecological characteristics of an agroecosystem, such as diversity, trophic structure, energy flow, nutrient cycling, population-regulating mechanisms, stability, and resilience, are metrics to help determine if a particular farming practice, input, or management decision is sustainable.[2]

Functions of biodiversity in agroecosystems, or agrobiodiversity, are foundations of sustainable farming practices. Ever since agriculture began some 10,000 to 12,000 years ago, biodiversity has allowed farming systems to evolve by providing genetic resources including edible plants, crop species, and livestock species.[21] However, diversities of crops and livestock have been decreasing rapidly since the introduction of monocultures of high-yielding cultivars and industrialization.[22] Insect and plant biodiversity in an agroecosystem provide multiple ecological services for agriculture. Examples include pollination,[23,24] insect pest management by intercropping,[25] and disease control by mixed planting of multiple cultivars.[26] Soil biodiversity, even though perhaps fewer than 10% of the species have been identified, is vital to soil fertility, decomposition of organic matters, soil structure, and soil health.[27,28] With higher diversity, there is greater microhabitat differentiation; there are more opportunities for coexistence and beneficial interference between species; there are more possible kinds of beneficial interactions between herbivores and their predators; there is more efficient use of resources of soil, water, and light; and there is reduced risk for the farmer. Examples of farming practices that can enhance agrobiodiversity and agroecosystem sustainability are listed in Table 1. Studies show that species and taxonomic groups respond to varying degree to environmentally friendly management.[29,30]

Table 1 Farming practices that can enhance agrobiodiversity and agroecosystem sustainability.

Habitat diversification	
Spatial	Temporal
Intercropping	Rotations
Trap crops	Fallow
Hedgerows	Cover crops
Shelterbelts	
Windbreaks	
Agroforestry	
Mosaic landscape	
Organic amendment applications	
Compost, organic mulch	
Green manure	
Conservation or minimum tillage	
Biological pest management	
No or reduced use of pesticides, fungicides, herbicides, and fumigants	
Use of beneficial insects	
Biofumigation	
Physical pest management	
Solarization	
Flooding	
Plant resistance	

BIOINDICATORS

Heink and Kowarik[31] proposed an all-encompassing definition of indicators: "An indicator in ecology and environmental planning is a component or a measure of environmentally relevant phenomena used to depict or evaluate environmental conditions or changes or to set environmental goals. Environmentally relevant phenomena are pressures, states, and responses as defined by OECD (2003)"[32] (see Fig. 1).

Bioindicators are based on biota that serve as indicators of the quality of the environment, the biotic component, or human impacts within an ecosystem.[33] The concept of bioindicators has been around for a long time—for example, the use of canaries to detect deadly carbon monoxide and methane gas buildup in mines has a long history. In the U.K., it had been used since 1911 until it was completely replaced with gas detectors in 1986.[34] Other early uses of bioindicators include aquatic organisms to evaluate water quality in the United States[35] and the melanic form of moths to detect air pollution in the U.K.[36,37] The number of publications about environmental indicators including bioindicators has been increasing since the 1980s.[38] The inaugural issues of the journals Ecological Indicators and Environmental Bioindicators were published in 2001 and 2006, respectively. At the 2002 Johannesburg World Summit on Sustainable Development, representatives from 190 countries committed to achieving a significant reduction of the current rate of biodiversity loss at the global, regional, and national levels by 2010.[39] Since then, the use of bioindicators for evaluating the status and functions of ecosystems and sustainability of agroecosystems has risen considerably,[40,41] yet the majority of recent studies of bioindicators has focused on the detection of pollution.[38]

Successful bioindicators should have not only biological and methodological relevance but also societal relevance because indicators require long-term monitoring

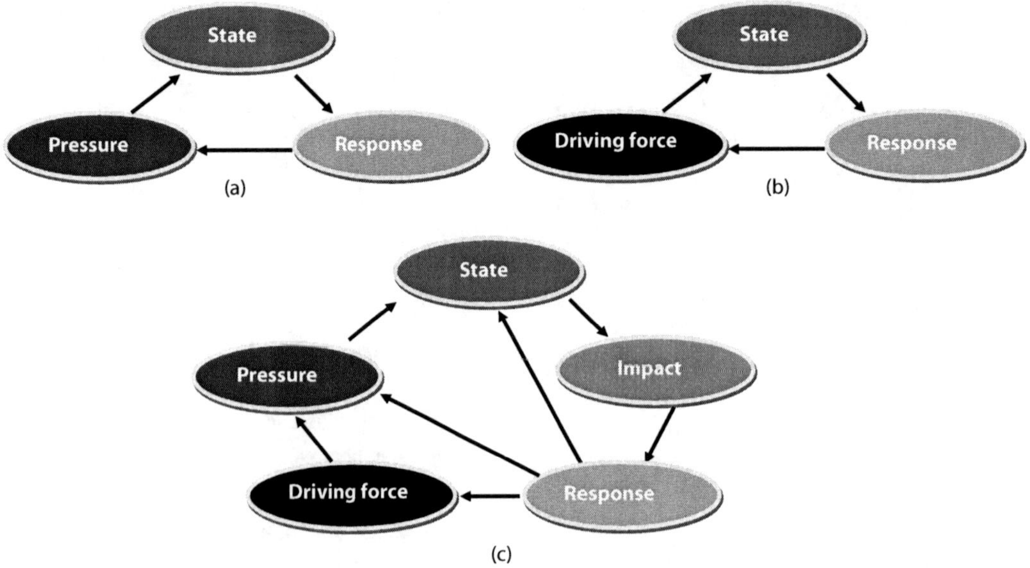
State
Pressure
Response
(a)
State
Driving force
Response
(b)
State
Pressure
Impact
Driving force
Response
(c)

Fig. 1 Varying causal chain frameworks. (a) PSR (pressure–state–response), (b) DSR (driver–state–response), and (c) DPSIR (driving force–pressure–state–impact–response).
Source: Niemeijer and de Groot.[52]

to separate real change from natural fluctuations. Features of bioindicators are listed in Table 2. Inventory and classification are the foundations for developing indicators.[42] Developing an indicator then involves several steps: define objectives; determine end uses; construct indicators; determine norms and thresholds; and test sensitivity, probability, and usefulness.[43] Compared to abiotic indicators, however, developing a bioindicator has greater challenges for several reasons. First, biotic parameters are highly variable both temporally and spatially. There is a paucity of

Table 2 Features of bioindicators for environmental and ecological health assessment.

Biological relevance	• Provides early warning • Exhibits changes in response to stress • Changes are measured on appropriate time scale • Intensity of changes related to intensity of stressors • Change occurs when effect is real • Changes are biologically important and occur early enough to prevent catastrophic effects • Changes can be attributed to a cause • Changes indicate effects on ecosystem services • Can be used as sentinels for humans
Methodological relevance	• Easy to use in the field • Can be used by nonspecialists • Easy to analyze and interpret data • Measures what it is supposed to measure • Useful to test management questions • Can be used for hypothesis testing • Can be conducted in a reasonable time • Does not require expensive or complicated equipment • Easily repeatable with little training
Societal relevance	• Of interest to the public • Of interest to regulators and public policy makers • Easily understood by the public • Methods transparent to the public • Measure related to environment, ecological integrity, and human health • Cost-effective • Adds measurably to other indicators • Complements other indicators

Source: Modified from Burger.[33]

information on temporal and spatial variability of most natural species populations. Secondly, due to high variability, it costs more to collect additional replications to achieve the same statistical power in data. Standardized sampling methods for bioindicators are often lacking. Databases for bioindicators are critically limited especially in developing countries.[44] Lastly, it is more difficult with bioindicators to define background levels, norms, and thresholds.[41]

Nevertheless, demand for bioindicators as evaluative tools at diverse levels is increasing. Some of these are policy driven and others are market driven. For example, international organizations, such as the Organisation for Economic Co-operation and Development (OECD)[45–47] and the European Union (EU),[48] promote development of standardized bioindicators as a policy to compare biodiversity worldwide. Among commercial sectors, bioindicators have been used to certify "environmentally sound" products, such as migratory birds for shade-grown coffee.[49] Ecologists are using bioindicators and various other indicators at different levels of organization to evaluate the health of ecosystems (Table 3) and make recommendations for future management.[50,51] Bioindicators are often integrated into the frameworks of sustainability indicators of agroecosystems.[52,53]

Table 3 Usefulness of indicators at different biological levels of organization to ecological health.

Ecological level	Type of indicator	Ecological health
Individual	Contaminant levels Lesions Disease Tumors Infertility Growth Longevity Reproduction Age of reproduction Hormonal balance Proper development and maturation	Used to evaluate health of individuals For evaluation of risk to higher-level consumers As an indicator of health of its foods, including prey
Population	Reproductive rates Growth rates Survival rates Movements Population genetics related to the breadth of the gene bank	Used to evaluate health of populations of species, particularly endangered or threatened species For comparison among populations For temporal comparisons Sources of resistance and pressure of natural selection
Community	Foraging guilds Breeding guilds (groups of related species) Predator–prey interactions Competitive interactions Pathogen–host, pest–host relationships Species richness	Measures health of species using the same niche, such as colonial birds nesting in a colony, or foraging animals such as dolphins and tuna Indicates relationship among different species within guilds or assemblages For spatial and temporal comparisons For evaluating efficacy of management options
Ecosystem	Decomposition rates Erosion rates Primary productivity Energy transfer Nutrient flow Relationship among different trophic levels Biomass Energy flow	Measure changes in relative presence of species, how fast nutrients and energy will become available, how fast nutrients in soil will no longer be available, how much photosynthesis is occurring Examines overall structure of the ecosystem in terms of the relationships among trophic levels For evaluating efficacy of management options
Landscape	Relative amounts of different habitats Patch size Corridors between habitat types or different ecosystems The extent of uniform genetics	Measure dispersion of different habitat types, indicates relative species diversity values Measures the difference among habitats Measures distribution of corridors and refugia within the landscape Also can measure the relationship between developed and natural areas For evaluating the importance of specific ecosystems within the landscape

Source: Modified from Burger.[33]

BIOINDICATOR DEVELOPMENT IN EU AGROECOSYSTEMS

The importance of agro-environmental indicators has been highlighted by the EU[48] where current agricultural policies aim to increase multifunctionality of agricultural production. Intensive studies on bioindicators have been conducted in European countries. Bioindicators demonstrated to be sensitive to farm management intensities in European and some other agroecosystems are listed in Table 4. Generally, it is observed among invertebrate species that with less intensive management, there are more specialists and less generalists (as a result of ecological succession), greater biodiversity, and greater resilience.[41] An example of a bioindicator based on these correlations is European spiders; habitat preferences of spiders, particularly the ratio of "pioneer species (mostly Linyphiidae)" versus "wolf spiders (Araneae: Lycosidae)," can be a sensitive indicator for the assessment of farming intensity.[54] Many bioindicators listed in Table 4, however, have critical use limitations due to technically complex sampling methods and greater temporal and spatial variability. Although special instruments are required, recent advancements in molecular techniques may make some highly sensitive bioindicators such as nematodes much more accessible.[55,56]

To practically implement agro-environmental policy in the EU, efforts have been made to develop relatively easy to measure surrogate indicators (e.g., length of borders, farm size, area managed with organic farming).[41] Another practical bioindicator is a list of indicator plant species to evaluate species richness of a farm. A total of 28 indicator flower species for meadows and pastures that can be easily identified by local farmers were selected in Baden-Württemberg, Germany. Agro-environmental payments are granted to farms that have at least 4 of these 28 indicator species in all of the meadows and pastures on the farm.[57] In designing more efficient agro-environmental schemes, advantages of result-oriented remuneration (e.g., payment towards species-rich meadow) over action-oriented remuneration (e.g., payment towards manuring and mowing once per year) have been discussed.[58] It has been further recognized that the preservation of biodiversity is only possible through the (re)establishment of a mosaic of habitat patches at the landscape level. To meet this need, Geographical Information System (GIS)-based landscape-oriented indicators have been examined.[59] GIS approaches are powerful tools for special analysis of bioindicators such as studies on correlations of bioindicator distributions (e.g., birds) and environmental and social factors at a landscape level.[60] Landscape context can have significant effect on biodiversity.[61,62]

To select objective, broad-scale, and unbiased indicators, conceptual frameworks for agro-environmental indicators were proposed. Varying causal chain frameworks such as pressure–state–response (PSR), driving force–state–response (DSR),[63] and driving force–pressure–state–impact–response (DPSIR)[64] have been developed (Fig. 1). Enhanced DPSIR (eDPSIR) was proposed as a way to provide improved conceptual guidance in indicator selection based upon the DPSIR approach, systems analysis, and causal networks.[52] Sustainability Assessment of Farming and the Environment (SAFE), a hierarchical framework for assessing the sustainability of agricultural systems in Belgium, was created as a consistent and comprehensive framework of principles, criteria, and indicators.[53] These frameworks integrate bioindicators as a component of sustainability indicators of agroecosystems.

Biodiversity has multiple dimensions, and it is difficult to measure by a single indicator. Feest et al. (2010) proposed the "biodiversity quality" approach using multiple indicators (e.g., species richness, evenness/dominance, density/population, relative biomass, and species conservation value index) in combination to assess the balance between a range of indices and their relative magnitude.[65] Four different types of indicators of biodiversity change were suggested as a tool to facilitate indicator development (Fig. 2).[66] With a goal of stable or increasing populations in all species associated with agricultural landscapes, Butler et al. developed a cross-taxonomic index for quantifying the health of farmland biodiversity by which the detrimental impacts of agricultural change to a broad range of taxonomic groupings can be assessed.[67]

For nationwide soil monitoring in the U.K., a framework in selecting soil bioindicators for balancing scientific and technical opinion to assist policy development was established.[56] This semiobjective approach using "logical sieve" yielded 17 bioindicators that cover a range of genotypic-, phenotypic-, and functional-based indicators for different trophic groups out of 183 candidate bioindicators. This framework allows transparency in the decision-making process in selecting soil bioindicators as well as flexibility in including other indicators depending on priorities of the monitoring. The need for unambiguous and broad definitions of terms such as *indicators* and *biodiversity* was addressed for improving communication among interdisciplinary researchers, policy makers, and stakeholders.[68] To make indicators relevant to potential users and to have long-term public support, the necessity of stakeholder participation in developing ecological indicators was also emphasized.[69]

BIOINDICATOR DEVELOPMENT IN SHADED COFFEE IN LATIN AMERICA

Coffee is the second most traded commodity in the world after petroleum and forms the principal economic activity

Table 4 Examples of potential bioindicators for sustainability of farming practices in European and other agroecosystems.

Bioindicator	Parameter	Comments	References
Arthropods			
Ground beetles (Carabidae)	Abundance	• Sensitive to management intensity but needs intensive data collection	[81–83]
Spiders (Araneae)	Habitat preferences Percent pioneer species	• Highly sensitive to management intensity and database is available on ecological characteristics of central European spiders	[84–91]
Hoverflies (Syrphidae)	Percent stenotopic species	• Diversity of landscape structure adjacent to the field enhances species numbers	[92–96]
Pollinators	Individuals Populations Ecological guilds	• Environmental stress brought about by pesticides and habitat modification reduce pollinators	[97,98]
Arthropod community	Abundance of key species	• Diverse arthropod communities are affected by landscape diversity and environmental stress	[85,99]
Soil fauna			
Ants	Diversity Community composition	• Good indicator for rangeland monitoring (first developed in Australia)	[100,101]
Earthworm	Biomass Species number Ecological guilds	• Suitable indicator for soil structure or compaction, tillage practice, heavy metals, and pesticides	[102–106]
Collembola Protozoa Nematode Micro arthropods Mites Soil animal	Physiotype Biodiversity Trophic index Maturity index Food web	• Highly sensitive to management intensity but time consuming and special skills are required for identification. Recent advancement in molecular approach makes these more accessible	[56,107–126]
Soil microbiota			
Soil enzymes (e.g., glucosinases for cellulose decomposition, phenol oxidases for lignin decomposition)	Activities	• Moderately sensitive to management intensity and relatively easy to measure	[56,127–129]
Soil algae	Species diversity Photosynthetic activity	• Less studied but sensitive to pesticides	[130]

(*Continued*)

Table 4 Examples of potential bioindicators for sustainability of farming practices in European and other agroecosystems. (Continued)

Bioindicator	Parameter	Comments	References
Soil microbiota			
Microbial communities	Composition Functional diversity PLFA[a] profiles qPCR[b]	• Moderately sensitive to management intensity but special skills and equipment are required and difficult to interpret	[56,128,129,131,132]
Functional groups	Mycorrhizae Nitrification Root pathogens	• Highly sensitive to management intensity but special skills and instruments are required. DNA-based approaches (e.g., TRFLP[c], qPCR) becoming popular	[56,133–135]
Microbial activity	Soil respiration Mineralization Multiple substrate-induced respiration (MSIR)	• Relatively easy to measure but highly variable both temporally and spatially	[56,128,129,136]
Microbial biomass	C, N, and P biomass	• Relatively easy to measure but highly variable both temporally and spatially	[128,129]
Birds			
Terrestrial birds Farmland birds	Species abundance of focal species and taxonomic composition at landscape scales	• Bird communities can be used to evaluate the effect of agriculture on surrounding ecosystems and hydrology and to explore sustainable land-use scenarios at regional scales	[60,137–139]
Plants			
Higher plants	Numbers of "characteristic" species, weeds, functional groups, and endangered species Cover of litter in vegetation Diversity Evenness indices Habitat age	• Capable of being integrated into sophisticated floristic diversity at the habitat scale but requires intensive data collection	[140–143]

[a] PLFA, phospholipid fatty acid.
[b] qPCR, quantitative polymerase chain reaction.
[c] TRFLP, terminal restriction fragment length polymorphism.

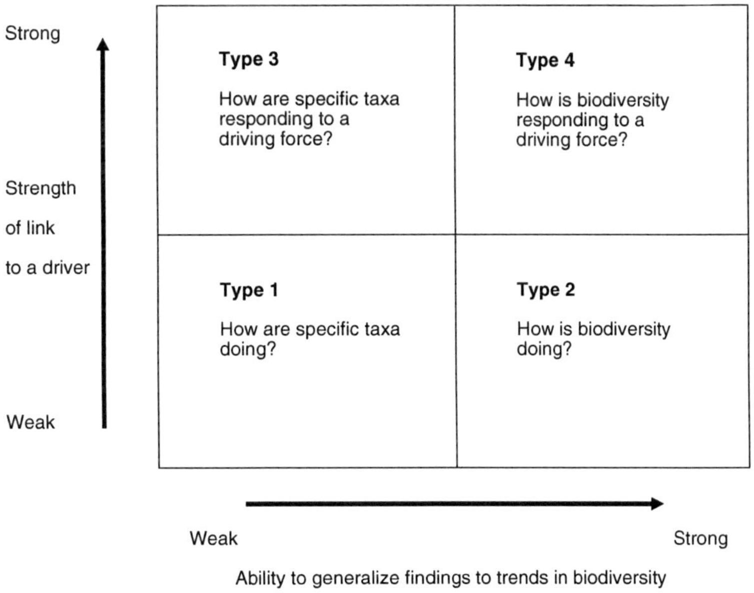
Strong
Strength
of link
to a driver
Weak
Type 3
How are specific taxa responding to a driving force?
Type 4
How is biodiversity responding to a driving force?
Type 1
How are specific taxa doing?
Type 2
How is biodiversity doing?
Weak
Strong
Ability to generalize findings to trends in biodiversity

Fig. 2 Typology of biodiversity change indicators.
Source: van Strien et al[66]

of more that 20 million people in farming communities throughout much of the developing world. Traditionally grown in the understory of forest cover or planted shade trees, throughout the 1970s and 1980s many coffee farmers adopted more modern production practices, planting higher-yielding varieties in full sun, eliminating shade trees, and increasing pesticide and fertilizer applications.[71] The loss of shade cover, and associated biodiversity, has led to many environmental problems such as soil erosion, loss of water capture and recharge ability, and contamination from the excessive fertilizer and pesticide use associated with sun-grown coffee.

Recent research has documented the high levels of agrobiodiversity and ecosystem services (such as pollination, soil conservation, and natural pest control) associated with diverse shade coffee production.[72] From this research, several very important bioindicators are being developed. The number of tree species in the shade canopy is an important indicator of both conservation potential of coffee plantations and increased options for farmer livelihood alternatives.[65] The species richness of ants and insect-feeding birds are greater in shade coffee than in sun coffee, especially ground-foraging ants that act as important predators and birds that are bark gleaners and leaf surface foragers. It also appears that the higher diversity of predaceous ants restricts the development of pest ants such as the fire ant *Solenopsis* sp. (Hymenoptera), a very common pest in open landscapes and sun-grown coffee. Associated species such as orchids in the shade tree canopy can also be important bioindicators. Local farmers can be trained to recognize orchids, demonstrating how the development of bioindicators of sustainability can be an accessible and local methodology that adds value to more sustainable farming practices. Orchids have an intrinsic value from a conservation perspective and, at the same time in this region of northern Nicaragua, have added an attractive value for an emerging agro-ecotourism industry associated with shade-grown coffee landscapes.[73]

IMPROVING AWARENESS OF FARMERS AND CONSUMERS

Usually, farmers themselves are not fully aware of the biodiversity in their farms. Moreover, to economically sustain environmentally friendly farming practices, recognition and support from consumers are necessary. To implement sustainable agricultural policy, therefore, improved awareness of both farmers and consumers on agrobiodiversity is required.

An example of such activity is the participatory biodiversity inventory in paddy rice ecosystems in Japan, where paddy fields occupy ~50% of the cultivated lands of the country. Japanese paddy agroecosystems contain a diverse group of organisms: Recent inventory found 5668 species including birds (189 spp.), fish (143 spp.), reptiles and amphibian (61 spp.), arthropods (1726 spp.), and plants (2075 spp.).[74] Since the 1960s, however, many native species inhabiting paddy fields have decreased in abundance mainly due to non-target effects of pesticide applications and the construction of concrete ditches. Consequently, 3 mammal species, 15 insect species, 5 bird species, 27

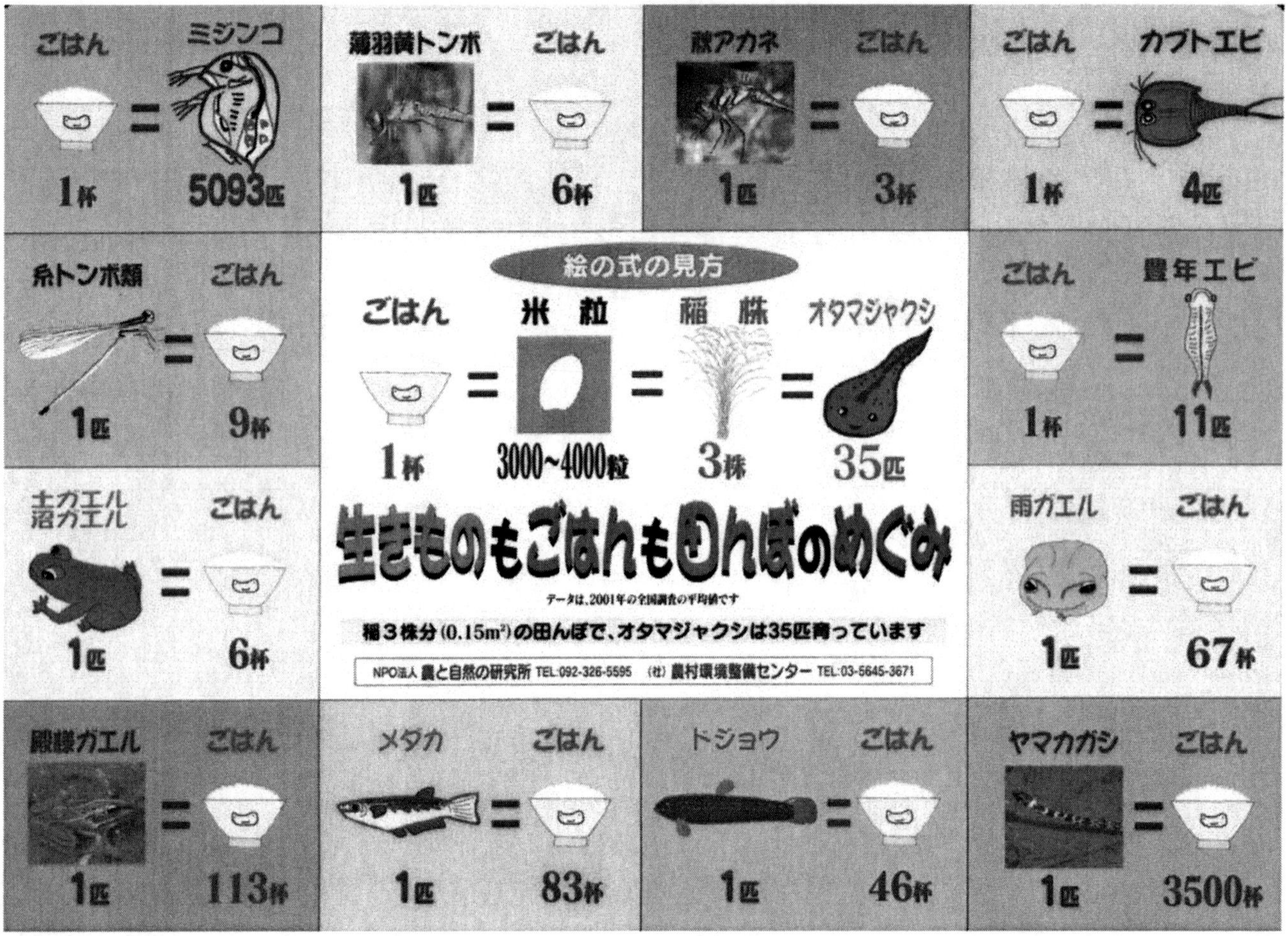
ごはん
ミジンコ
1杯
5093匹
薄羽黄トンボ
ごはん
1匹
6杯
秋アカネ
ごはん
1匹
3杯
ごはん
カブトエビ
1杯
4匹
糸トンボ類
ごはん
1匹
9杯
絵の式の見方
ごはん
米粒
稲株
オタマジャクシ
1杯
3000~4000粒
3株
35匹
ごはん
豊年エビ
1杯
11匹
土ガエル
沼ガエル
ごはん
1匹
6杯
生きものもごはんも田んぼのめぐみ
データは、2001年の全国調査の平均値です
稲3株分(0.15m²)の田んぼで、オタマジャクシは35匹育っています
NPO法人 農と自然の研究所 TEL:092-326-5595 (社) 農村環境整備センター TEL:03-5645-3671
雨ガエル
ごはん
1匹
67杯
殿様ガエル
ごはん
1匹
113杯
メダカ
ごはん
1匹
83杯
ドジョウ
ごはん
1匹
46杯
ヤマカガシ
ごはん
1匹
3500杯

Fig. 3 A portion of Japanese poster for improving farmer and consumer awareness of agrobiodiversity in paddy fields. Central figure indicates a bowl of rice being equivalent to 3000–4000 grains of rice, 3 stubs of rice plants, and 35 tadpoles in a paddy. This can be interpreted as eating a bowl of rice supports 35 tadpoles in a paddy. Numbers of species are averages from the national survey of agrobiodiversity in paddy fields in Japan. Clockwise from the upper left, the species listed are *Daphnia pulex* (water flea), *Pantala flavescens* (Globe Skimmer), *Sympetrum frequens* (Autumn Darter), *Triops longicaudatus* (longtail tadpole shrimp), *Branchinella kugenumaensis* (fairy shrimp), *Hyla japonica* (Japanese tree frog), *Rhabdophis tigrinus* (tiger keelback), *Misgurnus anguillicaudatus* (Dojo Loach), *Oryzias latipes* (Medaka), *Rana nigromaculata* (Dark-spotted frog), *Rana rugosa* (Japanese wrinkled frog), and *Ceriagrion melanurum* (damselfly).
Source: Une.[144]

species of fish, 3 amphibian species, 3 mollusks, and 26 plant species are listed as endangered species inhabiting paddy fields.[74] Nongovernmental organizations (NGOs), scientists, farmers, the general public, and administrators collaborated to create nationwide inventories of biodiversity in paddy agroecosystems and proposed a broad range of bioindicators (237 animal species[75] and 223 plant species[76]) to enhance the awareness of both farmers and consumers on paddy field biodiversity (Fig. 3).

CONCLUSIONS

Indicators represent a compromise between scientific knowledge of the moment and simplicity of use.[43] Compared to assessment systems for natural ecosystems,[42,77] the current status of developing bioindicators for sustainability of farming practices appears to still be in its early stages but is making good progress. For using bioindicators to implement sustainable agricultural policy, however, we need not only more research on the science of bioindicators but also better awareness on agrobiodiversity among farmers and consumers especially in places outside of Europe.

Future studies on bioindicators for sustainable farming practices should address the following: 1) the importance of stability and reproductive potential of not only pest and beneficial species but also other species typical of agroecosystems;[41,67,78] 2) standardization of sampling methods;[41] 3) expansion of databases and improvement of the statistical techniques to minimize potential bias of indicators;[66,79] 4) improved understanding of the mechanisms linking the status of biodiversity, Earth system processes, human decisions, and ecosystem services impacting human welfare through interdisciplinary studies;[52,79] 5) use of a participatory approach in the processes of developing bioindicators;[69,80] 6) development of multiple sets of bioindicators tailored to different end users (general public, farmers, policy makers, and scientists);[41] and 7) constructing a hierarchical system that integrates different types of bioindicators and ecological, social, cultural, and economic indicators to evaluate sustainability of farming practices.[53]

ACKNOWLEDGMENTS

We thank the anonymous reviewers for their valuable comments to a previous version of the manuscript.

REFERENCES

1. Pretty, J. Agricultural sustainability: Concepts, principles and evidence. Philos. Trans. R. Soc., B **2008**, *363* (1491), 447–465.
2. Gliessman, S.R. *Agroecology: The Ecology of Sustainable Food Systems*, 2nd Ed.; CRC Press/Taylor and Francis: Boca Raton, FL, 2007.
3. Green, R.E.; Cornell, S.J.; Scharlemann, J.P.W.; Balmford, A. Farming and the fate of wild nature. Science **2005**, *307* (5709), 550–555.
4. Tscharntke, T.; Klein, A.M.; Kruess, A.; Steffan-Dewenter, I.; Thies, C. Landscape perspectives on agricultural intensification and biodiversity—Ecosystem service management. Ecol. Lett. **2005**, *8* (8), 857–874.
5. International Federation of Organic Agriculture Movements (IFOAM). *Towards a Sustainable Agriculture: Papers*; Wirz: Aarau, Switzerland, 1978.
6. Douglass, G.K. *Agricultural Sustainability in a Changing World Order*; Westview Press: Boulder, Colorado, 1984.
7. Jackson, W.; Berry, W.; Colman, B. *Meeting the Expectations of the Land: Essays in Sustainable Agriculture and Stewardship*; North Point Press: San Francisco, 1984.
8. Dover, M.J.; Talbot, L.M. *To Feed the Earth: Agro-ecology for Sustainable Development*; World Resources Institute: Washington, DC, 1987.
9. Altieri, M.A. Beyond agroecology: Making sustainable agriculture part of a political agenda. Am. J. Altern. Agric. **1988**, *3* (4), 142–143.
10. Andow, D.A.; Hidaka, K. Experimental natural history of sustainable agriculture: Syndromes of production. Agric. Ecosyst. Environ. **1989**, *27*, 447–462.
11. MacRae, R.J.; Hill, S.B.; Henning, J.; Mehuys, G.R. Agricultural science and sustainable agriculture—A review of the existing scientific barriers to sustainable food-production and potential solutions. Biol. Agric. Hort. **1989**, *6* (3), 173–219.
12. Soule, J.D.; Piper, J.K. *Farming in Nature's Image: An Ecological Approach to Agriculture*; Island Press: Washington, DC, 1992.
13. Liverman, D.M.; Hanson, M.E.; Brown, B.J.; Merideth, R.W.J. Global sustainability toward measurement. Environ. Manage. **1988**, *12* (2), 133–144.
14. Jansen, D.M.; Stoorvogel, J.J.; Schipper, R.A. Using sustainability indicators in agricultural land use analysis: An example from Costa Rica. Neth. J. Agric. Sci. **1995**, *43* (1), 61–82.
15. Food Agriculture Organization of the United Nations (FAO). *Land Quality Indicators and Their Use in Sustainable Agriculture and Rural Development: Proceedings of the Workshop*; FAO: Rome, Italy, 1997.
16. MAFF U.K. Toward Sustainable Agriculture: A Pilot Set of Indicators; 2000, available at http://lis4.zalf.de/home_zalf/institute/lsa/lsa/ergebnisse/agstruk/indikatoren/pdfs%5Cind_sustainable_LU.pdf (accessed December 2010).
17. Bell, S.; Morse, S. *Sustainability Indicator: Measuring the Immeasurable*, 2nd Ed.; Earthscan: London, 2008.
18. Spedding, C.R.W. *Agriculture and the Citizen*, 1st Ed.; Chapman and Hall: London, New York, 1996.
19. Ikerd, J.E. Rethinking the first principles of agroecology: Ecological, social, and economic. In *Sustainable Agroecosystem Management: Integrating Ecology, Economics, and Society*; Bohlen, P.J., House, G.J., Eds.; CRC Press/Taylor and Francis: Boca Raton, FL, 2009; 41–52.
20. Edwards, C.A. Sustainable agricultural practices. In *Encyclopedia of Pest Management*; Pimentel, D., Ed.; Marcel Dekker/Taylor and Francis: Boca Raton, FL, 2002; 812–814.
21. Hillel, D.; Rosenzweig, C. The role of biodiversity in agronomy and agroecosystem management in the coming century. In *Sustainable Agroecosystem Management: Integrating Ecology, Economics, and Society*; Bohlen, P.J., House, G.J., Eds.; CRC Press/Taylor and Francis: Boca Raton, FL, 2009; 167–193.
22. Thrupp, L.A. The importance of biodiversity in agroecosystems. J. Crop Improv. **2004**, *11* (1/2), 315–338.
23. Kearns, C.A.; Inouye, D.W.; Waser, N.M. Endangered mutualisms: The conservation of plant–pollinator interactions. Annu. Rev. Ecol. Syst. **1998**, *29*, 83–112.
24. Buchmann, S.L.; Nabhan, G.P. *The Forgotten Pollinators*; Island Press/Shearwater Books: Washington, DC, 1996.
25. Ogol, C.; Spence, J.R.; Keddie, A. Maize stem borer colonization, establishment and crop damage levels in a maize-leucaena agroforestry system in Kenya. Agric. Ecosyst. Environ. **1999**, *76* (1), 1–15.
26. Zhu, Y.Y.; Chen, H.R.; Fan, J.H.; Wang, Y.Y.; Li, Y.; Chen, J.B.; Fan, J.X.; Yang, S.S.; Hu, L.P.; Leung, H.; Mew, T.W.; Teng, P.S.; Wang, Z.H.; Mundt, C.C. Genetic diversity and disease control in rice. Nature **2000**, *406* (6797), 718–722.
27. Pankhurst, C.E.; Doube, B.M.; Gupta, V.V.S.R., Eds. *Biological Indicators of Soil Health*; CAB International: Wallingford, U.K., 1997.
28. Wall, D.H. *Sustaining Biodiversity and Ecosystem Services in Soils And Sediments*; Island Press: Washington, DC, 2004.
29. Critchley, C.N.R.; Fowbert, J.A.; Sherwood, A.J.; Pywell, R.F. Vegetation development of sown grass margins in ar-

able fields under a countrywide agri-environment scheme. Biol. Conserv. **2006**, *132* (1), 1–11.
30. Marshall, E.J.P.; West, T.M.; Kleijn, D. Impacts of an agri-environment field margin prescription on the flora and fauna of arable farmland in different landscapes. Agric. Ecosyst. Environ. **2006**, *113* (1–4), 36–44.
31. Heink, U.; Kowarik, I. What are indicators? On the definition of indicators in ecology and environmental planning. Ecol. Indic. **2010**, *10* (3), 584–593.
32. Organisation for Economic Co-operation and Development (OECD). *Core Environmental Indicators: Development, Measurement, and Use*; OECD: Paris, France, 2003.
33. Burger, J. Bioindicators: Types, development, and use in ecological assessment and research. Environ. Bioindic. **2006**, *1* (1), 22–39.
34. BBC. 1986: Coal Mine Canaries Made Redundant. On This Day: 30th of December, available at http://news.bbc.co.uk/onthisday/hi/dates/stories/december/30/newsid_2547000/2547587.stm (accessed December 2010).
35. Richardson, R.E. The bottom fauna of the middle Illinois river, 1913–1925. Ill. Nat. Hist. Surv. Bull. **1928**, *17*, 387–475.
36. Kettlewell, H.B.D. How industrialisation can alter species. Discovery **1955**, *16* (12), 507–511.
37. Heck, W.W. The use of plants as indicators of air pollution. Air Water Pollut. **1966**, *10* (2), 99–111.
38. Burger, J. Bioindicators: A review of their use in the environmental literature 1970–2005. Environ. Bioindic. **2006**, *1* (2), 136–144.
39. United Nations Environment Programme (UNEP). Decisions Adopted by the Conference of the Parties to the Convention on Biological Diversity at Its Sixth Meeting, The Hague, 7–19 April 2002. UNEP/CBD/COP/6/20, available at http://www.biodiv.org/doc/meetings/cop/cop-06/official/cop-06-20-part2-en.pdf (accessed December 2010).
40. Paoletti, M.G. Using bioindicators based on biodiversity to assess landscape sustainability. Agric. Ecosyst. Environ. **1999**, *74* (1–3), 1–18.
41. Buchs, W. Biodiversity and agri-environmental indicators—General scopes and skills with special reference to the habitat level. Agric. Ecosyst. Environ. **2003**, *98* (1–3), 35–78.
42. Innis, S.A.; Naiman, R.J.; Elliott, S.R. Indicators and assessment methods for measuring the ecological integrity of semi-aquatic terrestrial environments. Hydrobiologia **2000**, *422–423*, 111–131.
43. Girardin, P.; Bockstaller, C.; Van der Werf, H. Indicators: Tools to evaluate the environmental impacts of farming systems. J. Sustainable Agric. **1999**, *13* (4), 5–21.
44. Dudley, N.; Baldock, D.; Nasi, R.; Stolton, S. Measuring biodiversity and sustainable management in forests and agricultural landscapes. Philos. Trans. R. Soc., B **2005**, *360* (1454), 457–470.
45. Neher, D.; Meyer, J.R.; Campbell, C.L.; Heck, W. Monitoring environmental sustainability in agricultural systems. In *Towards Sustainable Agricultural Production: Cleaner Technologies*; Organisation for Economic Co-operation and Development (OECD): Paris, France, 1994; 53–55.
46. Organisation for Economic Co-operation and Development (OECD). *Environmental Indicators for Agriculture: Methods and Results.* OECD Publishing: Paris, France, 2001; Vol. 3.
47. Organisation for Economic Co-operation Development (OECD). *Agriculture and Biodiversity : Developing Indicators for Policy Analysis : Proceedings from an OECD Expert Meeting, Zurich, Switzerland.* OECD Publishing: Paris, France, 2001.
48. The Commission of the European Communities. Evaluation of Agri-environment Programmes. State of Application of Regulation. (EEC) no. 2078/92. DGVI Commission Working Document VI/7655/98, 1998, available at http://ec.europa.eu/agriculture/envir/programs/evalrep/text_en.pdf (accessed December 2010).
49. Mas, A.H.; Dietsch, T.V. Linking shade coffee certification to biodiversity conservation: Butterflies and birds in Chiapas, Mexico. Ecol. Appl. **2004**, *14* (3), 642–654.
50. The Millennium Assessment Board. The Millennium Ecosystem Assessment. 2005, available at http://www.maweb.org/en/index.aspx (accessed December 2010).
51. The H. John Heinz III Center for Science, Economics and the Environment. *The State of the Nation's Ecosystems 2008: Measuring the Lands, Waters, and Living Resources of the United States*; Island Press: Washington, DC, Sept. 2008. Partially available online at: http://www.heinzcenter.org/ecosystems/ (accessed December 2010).
52. Niemeijer, D.; de Groot, R.S. A conceptual framework for selecting environmental indicator sets. Ecol. Indic. **2008**, *8* (1), 14–25.
53. Van Cauwenbergh, N.; Biala, K.; Bielders, C.; Brouckaert, V.; Franchois, L.; Cidad, V.G.; Hermy, M.; Mathijs, E.; Muys, B.; Reijnders, J.; Sauvenier, X.; Valckx, J.; Vanclooster, M.; Van der Veken, B.; Wauters, E.; Peeters, A. SAFE—A hierarchical framework for assessing the sustainability of agricultural systems. Agric. Ecosyst. Environ. **2007**, *120* (2–4), 229–242.
54. Buchs, W.; Harenberg, A.; Zimmermann, J.; Wei, B. Biodiversity, the ultimate agri-environmental indicator?: Potential and limits for the application of faunistic elements as gradual indicators in agroecosystems. Agric. Ecosyst. Environ. **2003**, *98* (1–3), 99–123.
55. Donn, S.; Griffiths, B.S.; Neilson, R.; Daniell, T.J. DNA extraction from soil nematodes for multi-sample community studies. Appl. Soil Ecol. **2008**, *38* (1), 20–26.
56. Ritz, K.; Black, H.I.J.; Campbell, C.D.; Harris, J.A.; Wood, C. Selecting biological indicators for monitoring soils: A framework for balancing scientific and technical opinion to assist policy development. Ecol. Indic. **2009**, *9* (6), 1212–1221.
57. Oppermann, R. Nature balance scheme for farms-evaluation of the ecological situation. Agric. Ecosyst. Environ. **2003**, *98* (1–3), 463.
58. Matzdorf, B.; Kaiser, T.; Rohner, M.-S. Developing biodiversity indicator to design efficient agri-environmental schemes for extensively used grassland. Ecol. Indic. **2008**, *8* (3), 256–269.
59. Osinski, E. Operationalisation of a landscape-oriented indicator. Agric. Ecosyst. Environ. **2003**, *98* (1–3), 371.
60. Gottschalk, T.K.; Dittrich, R.; Diekötter, T.; Sheridan, P.; Wolters, V.; Ekschmitt, K. Modelling land-use sustainability using farmland birds as indicators. Ecol. Indic. **2010**, *10* (1), 15–23.
61. Merckx, T.; Feber, R.E.; Riordan, P.; Townsend, M.C.; Bourn, N.A.D.; Parsons, M.S.; Macdonald, D.W. Optimiz-

ing the biodiversity gain from agri-environment schemes. Agric. Ecosyst. Environ. **2009**, *130* (3–4), 177–182.
62. Rundlof, M.; Bengtsson, J.; Smith, H.G. Local and landscape effects of organic farming on butterfly species richness and abundance. J. Appl. Ecol. **2008**, *45* (3), 813–820.
63. Organisation for Economic Co-operation and Development (OECD). *Environmental Indicators for Agriculture: Volume 1, Concepts and Frameworks*; OECD Publisher: Paris, France, 1999. Available at http://www.oecd.org/dataoecd/24/37/40680795.pdf (accessed December 2010).
64. Smeets, E.; Weterings, R. *Environmental Indicators: Typology and Overview*; European Environmental Agency: Copenhagen, 1999. Available at http://www.brahmatwinn.uni-jena.de/fileadmin/Geoinformatik/projekte/brahmatwinn/Workshops/FEEM/Indicators/EEA_tech_rep_25_Env_Ind.pdf (accessed December 2010).
65. Feest, A.; Aldred, T.D.; and Jedamzik, K. Biodiversity quality: A paradigm for biodiversity. Ecol. Indic. **2010**, *10* (6), 1077–1082.
66. van Strien, A.J.; van Duuren, L.; Foppen, R.P.B.; Soldaat, L.L. A typology of indicators of biodiversity change as a tool to make better indicators. Ecol. Indic. **2009**, *9* (6), 1041–1048.
67. Butler, S.J.; Brooks, D.; Feber, R.E.; Storkey, J.; Vickery, J.A.; Norris, K. A cross-taxonomic index for quantifying the health of farmland biodiversity. J. Appl. Ecol. **2009**, *46* (6), 1154–1162.
68. Sieg, C.H.; Flather, C.H. Applicability of Montreal Process Criterion 1—Conservation of biological diversity—to rangeland sustainability. Int. J. Sustainable Dev. World Ecol. **2000**, *7* (2), 81.
69. Turnhout, E.; Hisschemöller, M.; Eijsackers, H. Ecological indicators: Between the two fires of science and policy. Ecol. Indic. **2007**, *7* (2), 215–228.
70. Moguel, P.; Toledo, V.M. Biodiversity conservation in traditional coffee systems of Mexico. Conserv. Biol. **1999**, *13*, 11–21.
71. Perfecto, I.; Mas, A.; Dietsch, T.; Vandermeer, J. Conservation of biodiversity in coffee agroecosystems: A tri-taxa comparison in southern Mexico. Biodiversity Conserv. **2003**, *12*, 1239–1252.
72. Mendez, V.E.; Gliessman, S.R.; Gilbert, G.S. Tree biodiversity in farmer cooperatives of a shade coffee landscape in western El Salvador. Agric. Ecosyst. Environ. **2007**, *119* (1–2), 145–159.
73. Bacon, C.M. Confronting the coffee crisis: Nicaraguan farmers' use of cooperative, fair trade, and agroecological networks to negotiate livelihoods and sustainability. PhD Dissertation, Department of Environmental Studies. University of California Santa Cruz, California, 2005.
74. Kiritani, K.; Ed. *A Comprehensive List of Organisms Associated with Paddy Ecosystems in Japan*, Revised Ed.; Nou-to shizen-no kenkyujo, Seibutsu-tayousei-nougyou shien senta: Fukuoka, Tokyo, Japan, 2010 (in Japanese).
75. Kiritani, K., Ed. *Bioindicators for Paddy Ecosystems*; Nou-to shizen-no kenkyujo, Seibutsu-tayousei-nougyou shien senta: Fukuoka, Tokyo, Japan, 2009 (in Japanese).
76. Ito, K.; Mineta, T., Eds. *Plant Bioindicators for Paddy Ecosystems*; Nou-to shizen-no kenkyujo, Seibutsu-tayousei-nougyou shien senta: Fukuoka, Tokyo, Japan, 2009 (in Japanese).
77. Karr, J.R.; Fausch, K.D.; Angermeier, P.L.; Yant, P.R.; Schlosser, I.J. *Assessing Biological Integrity in Running Waters: A Method and Its Rationale*. Special Publication 5; Illinois Natural History Survey: Urbana, Illinois, 1986.
78. Kiritani, K. Integrated biodiversity management in paddy fields: Shift of paradigm from IPM toward IBM. Integr. Pest Manage. Rev. **2000**, *5* (3), 175–183.
79. Balmford, A.; Crane, P.; Dobson, A.; Green, R.E.; Mace, G.M. The 2010 challenge: Data availability, information needs and extraterrestrial insights. Philos. Trans. R. Soc., B **2005**, *360* (1454), 221–228.
80. Burger, J. Stakeholder involvement in indicator selection: Case studies and levels of participation. Environ. Bioindic. **2009**, *4* (2), 170–190.
81. Irmler, U. The spatial and temporal pattern of carabid beetles on arable fields in northern Germany (Schleswig-Holstein) and their value as ecological indicators. Agric. Ecosyst. Environ. **2003**, *98* (1–3), 141–151.
82. Heyer, W.; Huelsbergen, K.J.; Wittmann, C.; Papaja, S.; Christen, O. Field related organisms as possible indicators for evaluation of land use intensity. Agric. Ecosyst. Environ. **2003**, *98* (1–3), 453–461.
83. Doring, T.F.; Hiller, A.; Wehke, S.; Schulte, G.; Broll, G. Biotic indicators of carabid species richness on organically and conventionally managed arable fields. Agric. Ecosyst. Environ. **2003**, *98* (1–3), 133.
84. Oberg, S. Influence of landscape structure and farming practice on body condition and fecundity of wolf spiders. Basic Appl. Ecol. **2009**, *10* (7), 614–621.
85. Fernandez, D.E.; Cichon, L.I.; Sanchez, E.E.; Garrido, S.A.; Gittins, C. Effect of different cover crops on the presence of arthropods in an organic apple (*Malus domestica* Borkh) orchard. J. Sustainable Agric. **2008**, *32* (2), 197–211.
86. Birkhofer, K.; Fliessbach, A.; Wise, D.H.; Scheu, S. Generalist predators in organically and conventionally managed grass-clover fields: Implications for conservation biological control. Ann. Appl. Biol. **2008**, *153* (2), 271–280.
87. Clough, Y.; Holzschuh, A.; Gabriel, D.; Purtauf, T.; Kleijn, D.; Kruess, A.; Steffan-Dewenter, I.; Tscharntke, T. Alpha and beta diversity of arthropods and plants in organically and conventionally managed wheat fields. J. Appl. Ecol. **2007**, *44* (4), 804–812.
88. Isaia, M.; Bona, F.; Badino, G. Influence of landscape diversity and agricultural practices on spider assemblage in Italian vineyards of Langa Astigiana (northwest Italy). Environ. Entomol. **2006**, *35* (2), 297–307.
89. Schmidt, M.H.; Roschewitz, I.; Thies, C.; Tscharntke, T. Differential effects of landscape and management on diversity and density of ground-dwelling farmland spiders. J. Appl. Ecol. **2005**, *42* (2), 281–287.
90. Hough-Goldstein, J.A.; Vangessel, M.J.; Wilson, A.P. Manipulation of weed communities to enhance ground-dwelling arthropod populations in herbicide-resistant field corn. Environ. Entomol. **2004**, *33* (3), 577–586.
91. Marc, P.; Canard, A.; Ysnel, F. Spiders (Araneae) useful for pest limitation and bioindication. Agric. Ecosyst. Environ. **1999**, *74* (1–3), 229–273.
92. Kleijn, D.; van Langevelde, F. Interacting effects of landscape context and habitat quality on flower visiting insects in agricultural landscapes. Basic Appl. Ecol. **2006**, *7* (3), 201–214.
93. Schweiger, O.; Maelfait, J.P.; Van Wingerden, W.; Hendrickx, F.; Billeter, R.; Speelmans, M.; Augenstein, I.;

Aukema, B.; Aviron, S.; Bailey, D.; Bukacek, R.; Burel, F.; Diekotter, T.; Dirksen, J.; Frenzel, M.; Herzog, F.; Liira, J.; Roubalova, M.; Bugter, R. Quantifying the impact of environmental factors on arthropod communities in agricultural landscapes across organizational levels and spatial scales. J. Appl. Ecol. **2005**, *42* (6), 1129–1139.

94. Thomas, J.A. Monitoring change in the abundance and distribution of insects using butterflies and other indicator groups. Philos. Trans. R. Soc., B **2005**, *360* (1454), 339–357.
95. Henle, K.; Dziock, F.; Foeckler, F.; Follner, K.; Husing, V.; Hettrich, A.; Rink, M.; Stab, S.; Scholz, M. Study design for assessing species environment relationships and developing indicator systems for ecological changes in floodplains—The approach of the RIVA project. Int. Rev. Hydrobiol. **2006**, *91* (4), 292–313.
96. Dormann, C.F.; Schweiger, O.; Augenstein, I.; Bailey, D.; Billeter, R.; de Blust, G.; DeFilippi, R.; Frenzel, M.; Hendrickx, F.; Herzog, F.; Klotz, S.; Liira, J.; Maelfait, J.P.; Schmidt, T.; Speelmans, M.; van Wingerden, W.; Zobel, M. Effects of landscape structure and land-use intensity on similarity of plant and animal communities. Global Ecol. Biogeogr. **2007**, *16* (6), 774–787.
97. Kevan, P.G. Pollinators as bioindicators of the state of the environment: Species, activity and diversity. Agric. Ecosyst. Environ. **1999**, *74* (1–3), 373–393.
98. Pe'er, G.; Settele, J. The rare butterfly *Tomares nesimachus* (Lycaenidae) as a bioindicator for pollination services and ecosystem functioning in northern Israel. Isr. J. Ecol. Evol. **2008**, *54* (1), 111–136.
99. Paoletti, M.G.; Hu, D.X.; Marc, P.; Huang, N.X.; Wu, W.L.; Han, C.R.; He, J.H.; Cai, L.W. Arthropods as bioindicators in agroecosystems of Jiang Han Plain, Qianjiang City, Hubei China. Crit. Rev. Plant Sci. **1999**, *18* (3), 457–465.
100. de Bruyn, L.A.L. Ants as bioindicators of soil function in rural environments. Agric. Ecosyst. Environ. **1999**, *74* (1–3), 425–441.
101. Hoffmann, B.D. Using ants for rangeland monitoring: Global patterns in the responses of ant communities to grazing. Ecol. Indic. **2010**, *10* (2), 105–111.
102. Springett, J.A.; Gray, R.A.J.; Bakker, L. Influence of agriculture on Enchytraeidae fauna of soils in the south-west of the north island of New Zealand. Pedobiologia **1996**, *40* (5), 461–466.
103. Buckerfield, J.C.; Lee, K.E.; Davoren, C.W.; Hannay, J.N. Earthworms as indicators of sustainable production in dryland cropping in southern Australia. Soil Biol. Biochem. **1997**, *29* (3–4), 547–554.
104. Paoletti, M.G. The role of earthworms for assessment of sustainability and as bioindicators. Agric. Ecosyst. Environ. **1999**, *74* (1–3), 137–155.
105. Suthar, S. Earthworm communities a bioindicator of arable land management practices: A case study in semiarid region of India. Ecol. Indic. **2009**, *9* (3), 588–594.
106. Bartlett, M.D.; Briones, M.J.I.; Neilson, R.; Schmidt, O.; Spurgeon, D.; Creamer, R.E. A critical review of current methods in earthworm ecology: From individuals to populations. Eur. J. Soil Biol. **2010**, *46* (2), 67–73.
107. Filser, J.; Fromm, H.; Nagel, R.F.; Winter, K. Effects of previous intensive agricultural management on microorganisms and the biodiversity of soil fauna. Plant Soil **1995**, *170* (1), 123–129.
108. Heisler, C. Collembola and Gamasina: Bioindicators for soil compaction. Acta Zool. Fenn. **1995**, (196), 229–231.
109. Wardle, D.A.; Yeates, G.W.; Watson, R.N.; Nicholson, K.S. The detritus food-web and the diversity of soil fauna as indicators of disturbance regimes in agro-ecosystems. Plant Soil **1995**, *170* (1), 35–43.
110. Porazinska, D.L.; McSorley, R.; Duncan, L.W.; Graham, J.H.; Wheaton, T.A.; Parsons, L.R. Nematode community composition under various irrigation schemes in a citrus soil ecosystem. J. Nematol. **1998**, *30* (2), 170–178.
111. Porazinska, L.; McSorley, R.; Duncan, L.W.; Gallaher, R.N.; Wheaton, T.A.; Parsons, L.R. Relationships between soil chemical status, soil nematode community, and sustainability indices. Nematropica **1998**, *28* (2), 249–262.
112. van Straalen, N.M. Evaluation of bioindicator systems derived from soil arthropod communities. Appl. Soil Ecol. **1998**, *9* (1–3), 429–437.
113. Enami, Y.; Shiraishi, H.; Nakamura, Y. Use of soil animals as bioindicators of various kinds of soil management in northern Japan. JARQ **1999**, *33* (2), 85–89.
114. Foissner, W. Soil protozoa as bioindicators: Pros and cons, methods, diversity, representative examples. Agric. Ecosyst. Environ. **1999**, *74* (1–3), 95–112.
115. Koehler, H.H. Predatory mites (Gamasina, Mesostigmata). Agric. Ecosyst. Environ. **1999**, *74* (1–3), 395–410.
116. Paoletti, M.G.; Hassall, M. Woodlice (Isopoda: Oniscidea): Their potential for assessing sustainability and use as bioindicators. Agric. Ecosyst. Environ. **1999**, *74* (1–3), 157–165.
117. Yeates, G.W.; Bongers, T. Nematode diversity in agroecosystems. Agric. Ecosyst. Environ. **1999**, *74* (1–3), 113–135.
118. Lenz, R.; Eisenbeis, G. Short-term effects of different tillage in a sustainable farming system on nematode community structure. Biol. Fertil. Soils **2000**, *31* (3–4), 237–244.
119. Neher, D.A. Role of nematodes in soil health and their use as indicators. J. Nematol. **2001**, *33* (4), 161–168.
120. Li, Y.; Wu, J.; Chen, H.; Chen, J. Nematodes as bioindicator of soil health: Methods and applications. Yingyong Shengtai Xuebao **2005**, *16* (8), 1541–1546.
121. Briar, S.S.; Jagdale, G.B.; Cheng, Z.; Hoy, C.W.; Miller, S.A.; Grewal, P.S. Indicative value of soil nematode food web indices and trophic group abundance in differentiating habitats with a gradient of anthropogenic impact. Environ. Bioindic. **2007**, *2* (3), 146–160.
122. Gulvik, M.E. Mites (Acari) as indicators of soil biodiversity and land use monitoring: A review. Pol. J. Ecol. **2007**, *55*, 415–440.
123. Menta, C.; Leoni, A.; Bardini, M.; Gardi, C.; Gatti, F. Nematode and microarthropod communities: Comparative use of soil quality bioindicators in covered dump and natural soils. Environ. Bioindic. **2008**, *3* (1), 35–46.
124. Gergocs, V.; Hufnagel, L. Application of oribatid mites as indicators (review). Appl. Ecol. Environ. Res. **2009**, *7* (1), 79–98.
125. Aspetti, G.P.; Boccelli, R.; Ampollini, D.; Del Re, A.A.M.; Capri, E. Assessment of soil-quality index based on microarthropods in corn cultivation in northern Italy. Ecol. Indic. **2010**, *10* (2), 129–135.
126. Sánchez-Moreno, S.; Jiménez, L.; Alonso-Prados, J.L.; García-Baudín, J.M. Nematodes as indicators of fumigant effects on soil food webs in strawberry crops in southern Spain. Ecol. Indic. **2010**, *10* (2), 148–156.

127. Antonious, G.F. Impact of soil management and two botanical insecticides on urease and invertase activity. J. Environ. Sci. Health, Part B **2003**, *B38* (4), 479–488.
128. Benintende, S.M.; Benintende, M.C.; Sterren, M.A.; De Battista, J.J. Soil microbiological indicators of soil quality in four rice rotations systems. Ecol. Indic. **2008**, *8* (5), 704–708.
129. Lagomarsino, A.; Moscatelli, M.C.; Di Tizio, A.; Mancinelli, R.; Grego, S.; Marinari, S. Soil biochemical indicators as a tool to assess the short-term impact of agricultural management on changes in organic C in a Mediterranean environment. Ecol. Indic. **2009**, *9* (3), 518–527.
130. Berard, A.; Rimet, F.; Capowiez, Y.; Leboulanger, C. Procedures for determining the pesticide sensitivity of indigenous soil algae: A possible bioindicator of soil contamination? Arch. Environ. Contam. Toxicol. **2004**, *46* (1), 24–31.
131. Stenberg, B. Monitoring soil quality of arable land: Microbiological indicators. Acta Agric. Scand., Sect. B **1999**, *49* (1), 1–24.
132. Kaur, A.; Chaudhary, A.; Choudhary, R.; Kaushik, R. Phospholipid fatty acid—A bioindicator of environment monitoring and assessment in soil ecosystem. Curr. Sci. **2005**, *89* (7), 1103–1112.
133. Jordan, D.; Miles, R.J.; Hubbard, V.C.; Lorenz, T. Effect of management practices and cropping systems on earthworm abundance and microbial activity in Sanborn field: A 115-year-old agricultural field. Pedobiologia **2004**, *48* (2), 99.
134. Subbarao, K.V.; Kabir, Z.; Martin, F.N.; Koike, S.T. Management of soilborne diseases in strawberry using vegetable rotations. Plant Dis. **2007**, *91* (8), 964–972.
135. Smukler, S.M.; Jackson, L.E.; Murphree, L.; Yokota, R.; Koike, S.T.; Smith, R.F. Transition to large-scale organic vegetable production in the Salinas Valley, California. Agric. Ecosyst. Environ. **2008**, *126* (3–4), 168–188.
136. Elmholt, S. Microbial activity, fungal abundance, and distribution of *Penicillium* and *Fusarium* as bioindicators of a temporal development of organically cultivated soils. Biol. Agric. Hort. **1996**, *13* (2), 123–140.
137. Padoa-Schioppa, E.; Baietto, M.; Massa, R.; Bottoni, L. Bird communities as bioindicators: The focal species concept in agricultural landscapes. Ecol. Indic. **2006**, *6* (1), 83–93.
138. Bar, A.; Loffler, J. Ecological process indicators used for nature protection scenarios in agricultural landscapes of SW Norway. Ecol. Indic. **2007**, *7* (2), 396–411.
139. Robledano, F.; Esteve, M.A.; Farinós, P.; Carreño, M.F.; Martínez-Fernández, J. Terrestrial birds as indicators of agricultural-induced changes and associated loss in conservation value of Mediterranean wetlands. Ecol. Indic. **2010**, *10* (2), 274–286.
140. Albrecht, H. Suitability of arable weeds as indicator organisms to evaluate species conservation effects of management in agricultural ecosystems. Agric. Ecosyst. Environ. **2003**, *98* (1–3), 201–211.
141. Sturz, A.; Matheson, B.; Arsenault, W.; Kimpinski, J.; Christie, B. Weeds as a source of plant growth promoting rhizobacteria in agricultural soils. Can. J. Microbiol. **2001**, *47*, 1013–1024.
142. Aavik, T.; Liira, J. Agrotolerant and high nature-value species—Plant biodiversity indicator groups in agroecosystems. Ecol. Indic. **2009**, *9* (5), 892–901.
143. Höft, A.; Müller, J.; Gerowitt, B. Vegetation indicators for grazing activities on grassland to be implemented in outcome-oriented agri-environmental payment schemes in north-east Germany. Ecol. Indic. **2010**, *10* (3), 719–726.
144. Une, Y. *Illustrated Poster of Rice and Living Organisms*; Nou-to-shizen-no-kenkyujo: Fukuoka, Japan, 2004 (in Japanese).

Biological Controls

Heikki Hokkanen
Department of Applied Biology, University of Helsinki, Helsinki, Finland

Abstract
Biological control of pests has been actively practiced for more than 100 years, and it has had some 150 spectacular successes, which in economic terms have been just as impressive as in ecological terms. However, the obtained successes are only the tip of the iceberg of all the work carried out in the field. A major ecological and economic challenge is to improve the ratio of successes in biological control, while retaining the excellent safety record of this approach to pest control.

INTRODUCTION

Biological control of pests has been actively practiced for the control of pests, weeds, and plant diseases for more than 100 years, and it has had some 150 spectacular successes,[1] which in economic terms have been just as impressive as in ecological terms: the calculated return for investment is 32:1, while for other control methods the ratio is around 2.5:1.[2,3] However, the obtained successes are only the tip of the iceberg of all the work carried out in the field. To date, more than 6000 introductions of alien natural enemies have been carried out, worldwide.[4] It is estimated that only about 35% of all introduced biocontrol agents have become ecologically established in the target ecosystem, and only 60% of these have provided any economic or biocontrol success.[3,5] Of all the individual biocontrol projects, only 16% have resulted in complete control of the target pest.[6] A major ecological and economic challenge is to improve the ratio of successes in biological control, while retaining the excellent safety record of this approach to pest control.

GENERAL PRINCIPLES

While it has been shown that biological control can be effective in any climate, ecosystem, and crop, the factors determining success or failure remain largely unknown, and often are economic rather than ecological in nature.[1] Very few general principles to improve the efficacy and predictability of biological control have emerged; these include better ecological background knowledge, genetic improvement (in particular, genetic engineering) of biocontrol agents, and the utilization of new ecological associations in selecting the biocontrol agents.[7] The genetic engineering of biocontrol agents, especially insects, is still in its infancy and cannot be expected to improve the success ratios in the foreseeable future. In contrast, the new association principle has—usually unknowingly—been used for a long time, and is increasingly employed to find more effective natural enemies for current biological control programs.

NEW ASSOCIATIONS

The standard biological control principle is to reestablish the ecological balance between an exotic pest and its natural enemies occurring in their country of origin (the "old association approach").[2,3] It has been argued, however, that this is an inefficient way of practicing biological control, because due to an evolved long-term equilibrium between the pest and the natural enemy, the control agent only seldom is very efficient.[6] To find more effective enemies one should search among agents that do not share an evolutionary history with the target pest (the "new association" approach). Such natural enemies can be found, for example, for the target pest in areas where the pest has been introduced only recently, or among enemies attacking related species in other geographical areas.[6]

EVIDENCE FOR IMPROVED EFFICACY

Analysis of past biocontrol successes and failures have indicated that when employing the new association principle it is possible to increase the success ratio by at least 75%.[6] More detailed studies showed that some natural enemy groups may be particularly attractive as new association agents (Table 1). Such analyses are, however, often confounded by the fact that new association agents seldom have been considered as the primary choice in biological control, and consequently, usually five to seven old association agents are introduced before a new association agent is tried. In addition, on average much greater numbers (two- to fourfold) of old association agents are normally introduced (Table 1), which further increases the probability of biocontrol

Encyclopedia of Environmental Management DOI: 10.1081/E-EEM-100000602

Table 1 Comparisons of biological control introductions with old and new association control agents utilizing Tachinidae, Braconidae and Eulopidae.

	Proportion (%) of successes of all cases (introductions)		Total number of cases		Bias in the release numbers[a]
	Old	New	Old	New	
Tachinidae	10.9	17.1	92	41	3.8-fold
Braconidae	17.2	14.4	169	97	1.6-fold
Eulopidae	28.6	35.7	56	28	1.7-fold
Overall	17.4	18.7	317	166	

[a]Indicates how many more individuals on average of old association agents were released in the introduction projects, compared with new association agents. In the case of Tachinidae and Braconidae the mean number of released new association agents was below 5000 individuals, which is considered to be the necessary number to ensure a fair chance for the natural enemies to establish themselves.
Source: Hokkanen, H. M. T., unpublished data.

success, and biases the analyses against new association agents. Therefore, the estimate for improving the success ratio appears to be conservative.

Several spectacular, well-documented biocontrol successes that have employed new association control agents are known, and these include the complete control of serious pests such as the sugarcane borer *Diatraea saccharalis* in the Caribbean, coconut spike moth *Levuana iridescens* in Fiji, southern green stink bug *Nezara viridula* in Hawaii, the moth *Oxydia trychiata* in Colombia, and several scale insect species in California, Greece, and Australia.[7] Further, more detailed examples will be given below on new research with good prospects of success utilizing this approach.

RECENT CASES EMPLOYING NEW ASSOCIATIONS

Eurasian Watermilfoil

The Eurasian watermilfoil (*Myriophyllum spicatum*) was introduced into North America several decades, possibly 100 years, ago. It grows rapidly, forms a dense canopy on the water surface, and often interferes with recreation, inhibits water flow, and impedes navigation. Herbicides and mechanical harvesting have been used to control infestations, costing \$150–\$2000 per acres annually in Minnesota.[8]

Sometimes naturally occurring declines of the watermilfoil have been observed. The main causal agent proved to be a native beetle *Euhrychiopsis lecontei*, the milfoil weevil, which subsequently has shown control potential in controlled field experiments. The weevil is a specialist herbivore of watermilfoils, but prefers the Eurasian to its native host, the northern watermilfoil (*M. sibiricum*). Research is in progress to use the milfoil weevil effectively as a biocontrol agent against the Eurasian watermilfoil in North America.[8]

Lantana

Lantana camara is a serious weed of Mexican or Caribbean origin, affecting cropping lands and forest areas in 47 countries. Lantana was the focus of the first weed biocontrol effort in history (1902), and there is an enormous literature on Lantana biocontrol. Several complexes of herbivores have been credited for exerting some degree of biocontrol of the weed (e.g., in Hawaii), many employing new association agents jointly with old association agents. Latest research gives data on the good efficacy and release in Australia of the moth *Ectaga garcia* originating from South America, where it feeds on the related weed *Lantana montevidensis*.[9]

Triffid (Siam) Weed

The triffid weed (*Chromolaena odorata*) is a perennial shrub native of tropical America. In recent decades it has become a serious pest of humid tropics around the world.[10] It spreads rapidly in lands used for forestry, pasture, and plantation crops and can reach a height of three meters in open situations and up to eight meters in forests. For more than two decades the triffid weed has been the subject of intensive research as a target for biological control. However, so far all attempts at biocontrol of *C. odorata* have failed. Recently the new association biological control agent, arctiid moth *Pareuchaetes aurata aurata* collected from *C. jujuensis* in South America, was considered as more promising than the related moth *P. pseudoinsulata*, an old association control agent previously thought of as one of the best biocontrol candidates.[10]

Southern Green Stink Bug

The biological control of the southern green stink bug (the green vegetable bug) (*Nezara viridula*) in Australia, New Zealand, and Hawaii has been heralded as a landmark example of classical biological control.[11] An egg parasitoid—old association agent—*Trissolcus basalis* and a tachinid fly—new association agent—*Trichopoda pennipes* have jointly provided these successes. Control by the fly has been considered as relatively more important, and indeed, in Australia where the fly has failed to establish, the control is poor and the bug remains a serious pest. Currently in Australia another new association tachinid fly, *Trichopoda giacomellii*, is being released after research showed it has excellent potential for control.[12]

Citrus Leafminer

Citrus agroecosystems have numerous potentially damaging pests often maintained under substantial to complete biological control by both old and new association agents. The citrus leafminer *Phyllocnistis citrella*, native to Asia, has spread rapidly throughout the citrus growing areas of the world in recent years.[13] It arrived in Florida in 1993 and in less than one year invaded and colonized the entire state. An old association parasitic wasp *Ageniaspis citricola* was introduced in 1994, and after establishment it has held the pest under control with significant help from native parasitoids such as *Pnigalio minio* (new association agent).[13] In some other areas native parasitoids similarly have shown significant control effect on the citrus leafminer (e.g., in Italy). This example illustrates well the fact that invading species often do not become pests, because effective local natural enemies keep them in check.

Tarnished Plant Bug

An ongoing study in the United States has identified as the most important parasitoid of the native pest *Lygus lineolaris*, the tarnished plant bug, the exotic species *Peristenus digoneutis*, originally introduced for the control of related introduced mirid plant bugs.[14] This example serves well to point out the importance of native pests, which in most if not all areas form the majority of all pest species. As old association biological control agents seldom can be utilized for the control of native pests, their biocontrol by introduced natural enemies has attracted relatively little attention and, indeed, only three decades ago was considered an impossible task. Several recent examples, usually utilizing new association control agents, show that biological control can work against native pests just as well as against exotic ones.

FUTURE PROSPECTS

Compared with chemical control, the success rates of biological control are outstanding. While only about one out of 15,000 tested chemicals ends up as a chemical pesticide meeting the requirements of efficacy and safety, approximately one out of seven introductions of natural enemies has been successful using old associations.[15] Using new association control agents this rate could still be increased to about one out of four, while the array of potential natural enemies is also substantially larger providing a wider choice. In addition, the potential uses for natural enemy introductions are broadened to include the control of native pests.

A major concern with respect to all biological control introductions is the question of nontarget safety. Biological control has an excellent record of safety[3,16] and it covers the new association agents as well: there have been some 1500–2000 introductions already (out of 6000) that have involved new association agents.[7] Those extremely few cases where a negative nontarget effect has been suspected as a result of biological control, all involve old association agents; therefore it is clear that new associations can safely be used to help obtain biological control successes at an increasing rate.

REFERENCES

1. Hokkanen, H.M.T. Success in classical biological control. CRC Crit. Rev. Plant Sci. **1985**, *3*, 35–72.
2. Cullen, J.M.; Whitten, M.J. Economics of Classical Biological Control: A Research Perspective. In *Biological Control: Benefits and Risks*; Hokkanen, H.M.T., Lynch, J.M., Eds.; Cambridge University Press: Cambridge, U.K., 1995; 270–276.
3. *Biological Control: Benefits and Risks*; Lynch, J.M., Hokkanen, H.M.T., Eds.; Cambridge University Press: Cambridge, U.K., 1995.
4. Waage, J. In *Agendas, Aliens and Agriculture*; Global Biocontrol in the Post UNCED Era, Cornell Community Conference on Biological Control, http://www.nysaes.cornell.edu/ent/bcconf/talks/waage.html (accessed Jan 5, 1999).
5. Hokkanen, H.M.T. Pest Management, Biological Control. In *Encyclopedia of Agricultural Science*; Academic Press, Inc.: San Diego, 1994; 3, 155–167.
6. Hokkanen, H.M.T.; Pimentel, D. New associations in biological control: theory and practice. Can. Entomol. **1989**, *121*, 829–840.
7. Hokkanen, H.M.T. New Approaches in Biological Control. In *CRC Handbook of Pest Management in Agriculture*, 2nd Ed., Pimentel, D., Ed.; CRC Press: Boca Raton, FL, 1991; II, 185–198.
8. Newman, R.M. Biological Control of Eurasian Watermilfoil. http://www.fw.umn.edu/research/milfoil/milfoilbc.html (accessed Feb 5, 1999).
9. Day, M.D.; Wilson, B.W.; Latimer, K.J. The life history and host range of *Ectaga garcia*, a biological control agent for *Lantana camara* and *L. montevidensis* in Australia. BioControl **1998**, *43*, 325–338.
10. Kluge, R.L.; Caldwell, P.M. The biology and host specificity of *Pareuchaetes aurata aurata* (Lepidoptera: Arctiidae), a "New Association" biological control agent for *Chromolaena odorata* (Compositae). Bull. Entomol. Res. **1993**, *83*, 87–94.
11. Caltagirone, L.E. Landmark examples in classical biological control. Annu. Rev. Entomol. **1981**, *26*, 213–232.
12. Coombs, M. *Biological Control of Green Vegetable Bug in Australia and PNG*; Pest Management Current Programs and Projects. http://www.ento.csiro.au/research/pestmgmt/pmp16.htm (accessed Oct 5, 1999).

13. Timmer, L.W. *Citrus Leafminer Proves to be an IPM Success*; IPM Florida, 1996 *Winter*. http://www.ias.ufl.edu/~FAIRSWEB/IPM/IPMFL/v2n4/leafminer.htm (accessed Feb 5, 1999).
14. Day, W.H. Host preferences of introduced and native parasites (Hymenoptera: Braconidae) of phytophagous plant bugs (Hemiptera: Miridae) in Alfalfa-Grass Fields in the Northeastern USA. BioControl **1999**, *44*, 249–261.
15. Hokkanen, H.M.T. Role of Biological Control and Transgenic Crops in Reducing Use of Chemical Pesticides for Crop Protection. In *Techniques for Reducing Pesticide Use*; Pimentel, D., Ed.; John Wiley and Sons: New York, 1997; 103–127.
16. *Evaluating Indirect Ecological Effects of Biological Control*; Scott, J.K., Quimby, P.C., Wajnberg, E., Eds.; CABI Publishing: Wallingford, U.K., 2001; 261.

Biological Controls: Conservation

Doug Landis
Department of Entomology, Michigan State University, East Lansing, Michigan, U.S.A.

Steve D. Wratten
Division of Soil, Plant and Ecological Sciences, Lincoln University, Canterbury, New Zealand

Abstract
Biological control is the use of living natural enemies to control pest species and is generally accomplished via importation, augmentation or conservation of natural enemies. Conservation biological control seeks to manipulate the environment to enhance the survival, fecundity, longevity, and behavior of natural enemies to increase their effectiveness in controlling pests. This differentiates it from the numerical augmentation of natural enemies through inundative or inoculative releases, or the importation of exotic natural enemies sometimes referred to as classical biological control. All forms of biological control ultimately depend on natural enemies finding a favorable environment in which to impact the pest, thus, conservation of natural enemies should be a critical consideration in all biological control efforts.

REDUCING HARMFUL CONDITIONS

Conservation efforts may be focused on either reducing factors that are harmful to natural enemies, or on enhancing those environmental attributes that are favorable to natural enemies.[2] Reducing harmful conditions is the first step in an attempt at conservation biological control. This must be considered at the time natural enemies first enter the environment as well as throughout the remainder of their life cycle.

Pesticides

Pesticides are perhaps the most well-known factor inhibiting natural enemy effectiveness. Insecticides, herbicides, fungicides, and antibiotic products may all interfere with the success of natural enemies. Insecticide applications can directly kill insect predators or parasitoids and may have prolonged toxicity due to their residual activity. For example, in greenhouse production systems where insect natural enemies are routinely augmented, producers can consult information on the relative toxicity of many pesticides to common natural enemies and estimates of the length of time before natural enemies can be successfully re-established. However, this type of information is less available for other types of agroecosystems, particularly those that rely on naturally occurring enemies.[4]

Insecticides may also have indirect effects on the success of arthropod natural enemies. Even if natural enemies survive initial pesticide application(s), they may find unfavorable conditions due to a reduction in the number or availability of their hosts or prey. Some insecticides may have sub-lethal impacts that reduce fecundity or induce sterility in natural enemies. Others may alter the behavior of arthropod natural enemies so that they are less effective at searching for hosts or are repelled from treated surfaces.

Fungicides and other chemical agents may have direct negative effects on beneficial fungi or bacteria that help to control pests. Poorly timed fungicide applications can exacerbate certain pest populations via suppression of insect pathogenic fungi that would naturally suppress pest numbers. Herbicides can also have negative impacts on certain arthropod natural enemies. While only a few are directly toxic, others may act as repellents or irritants increasing natural enemy emigration from treated areas. Perhaps more important, by reducing weed density, herbicides alter the crop environment in other ways that may be detrimental to natural enemies (see *Alternative Food Sources* and *Shelter and Microclimate* below).

Cultural Practices

Various cultural practices such as tillage or burning of crop debris can kill natural enemies or make the crop habitat unfavorable.[8] Ceasing to burn crop residues increased parasitism of the sugarcane leafhopper and reduced pest density in India. In annual crops, the repeated disturbances of plowing, planting, and cultivation may make the habitat unsuitable for ground-dwelling predators such as spiders and carabid beetles. Providing small habitats as refuges from these disturbances has been suggested as one effective conservation measure. In orchards, repeated tillage may create dust deposits on leaves, killing small predators and parasites and causing increases in certain insect and mite pests.[9]

Encyclopedia of Environmental Management DOI: 10.1081/E-EEM-100000400

Host Plant Effects

Host plant effects, such as chemical defenses that are harmful to natural enemies but to which pests are adapted, can reduce the effectiveness of biological control. Some pests are able to sequester toxic components of their host plant and use them as defense against their own enemies. In other cases, physical characteristics of the host plant, such as leaf hairiness, may reduce the ability of the natural enemy to find and attack its prey or hosts. Selection of appropriate cultivars can enhance natural enemy survival and reduce pest damage.[4]

Secondary Enemies

All natural enemies are themselves attacked by other predators, pathogens, and parasitoids. These organisms can in some cases suppress the natural enemy population sufficiently to inhibit effective biological control. Controlling secondary enemies such as hyperparasitoids is yet another conservation tool. Perhaps the most effective means of doing so is to prevent secondary enemies from becoming established in the first place. This is a regular part of importation biological control programs, where quarantine of imported material is required to assure it is free of unwanted secondary enemies.

ENHANCING FAVORABLE CONDITIONS FOR NATURAL ENEMIES

Ensuring that the ecological requirements of the natural enemy are met in the agricultural ecosystem is the other major means of conserving natural enemies. While this may seem self-evident, it is not a trivial matter. To be effective, natural enemies may need access to alternative prey or hosts, adult food resources, overwintering habitats, constant/alternative food supply, and appropriate microclimates. Identifying the ecological factor(s) necessary to favor even a few key natural enemies can be a time-consuming research endeavor. Ensuring that these resources are available in the agroecosystem at the correct time(s) and spatial scale(s) is then required to complete the task.[6,7]

Alternative Prey or Hosts

Generalist natural enemies may require a succession of different prey/hosts in order to remain in a given environment, and manipulating the presence of alternate prey/hosts within a crop before the arrival or seasonal increase of targeted pests may be necessary. This may be done by managing alternative prey in surrounding vegetation, on weeds within the crop, or in plant residue/organic matter maintained on the soil surface. Providing alternative hosts for more host- or habitat-specific natural enemies may be more challenging. The classic example is *Anagrus* (Hymenoptera: Mymaridae) parasitoids of the grape leafhopper that must overwinter on plant hosts outside the grape vineyard. Provision of wild or cultivated plants, which support overwintering eggs of alternative leafhopper hosts, increases *Anagrus* parasitism and contributes to control of grape leafhopper. Manipulating the dispersal of natural enemies from habitats containing alternative prey via carefully timed mowing or herbicide treatment of these habitats may also be necessary to encourage natural enemy movement into the desired crop.[6]

Alternative Food Sources

Many natural enemies benefit from the availability of food sources other than their prey/hosts. Plant nectar is consumed by many parasitoid species and can enhance activity, longevity, and rates of parasitism. The presence of honeydew-producing insects is another means by which some parasitoids obtain access to plant sugars. Many natural enemies including parasitoids, certain lady beetles, syrphid fly adults, and predaceous mites consume pollen. For some species pollen may simply provide nutrition during periods of low prey availability, while in other species access to pollen may be necessary to complete egg development.

Provision of flowering plants that provide pollen and nectar to enhance natural enemies has received considerable attention. In Europe, the North American annual plant *Phacelia tanacetifolia* Bentham has been extensively used since it produces large quantities of pollen and nectar, and can enhance the activity and fecundity of syrphid flies that are important aphid predators. Lists of plants that are known to be attractive to a variety of predators and parasitoids are available. Often these are perennial species that flower over extended periods of time, produce large quantities of pollen, or have accessible nectaries. While the presence of flowering plants almost always augments natural enemy populations, there is less evidence that they increase biological control in adjacent crop areas, due in part to the difficulties in conducting appropriate field experiments. An alternative to using plants to attract and enhance natural enemies is that of artificial food sprays. While this approach has proven effective in some cases, it may be more costly and most appropriate for high-value crops.[6]

Shelter and Microclimate

Most natural enemies require sheltered sites in which to overwinter, or to escape disruptive cultural practices and pesticide applications. Effective conservation may require that such habitats are present within the agroecosystem. An early example of habitat management was the provision of a refuge for natural enemies of alfalfa pests displaced by cutting. Strip-harvesting of alfalfa was used to provide temporary refuges from the disruption of harvesting and encourage recolonization of the regrowing crop. Overwinter survival of many natural enemies can be encouraged by providing undisturbed habitats in or adjacent to crop fields. The example of grassy "beetle banks" for the conservation of ground-dwelling arthropods has been adopted in several European countries. Overwintering predator populations exceeding 1500 individuals/m^2 have been reported after two years of beetle bank establishment.[10] Intercrops during the growing season and cover crops during the noncrop period can also be used to enhance natural enemy habitat.[7]

ECOLOGICAL INFRASTRUCTURE IN AGROECOSYSTEMS

Many of the difficulties that natural enemies experience in annual crops can be traced to the frequent and intense disturbances that characterize these agricultural ecosystems.[9–11] In these simplified ecosystems, many ecological services associated with the maintenance or enhancement of biodiversity, such as soil and water erosion prevention, nutrient cycling, biological control, pollination etc., are also compromised. The concept of restoring these functions by managing the ecological infrastructure of landscapes shows promise in alleviating some of these problems. For example, landscape features such as windbreaks or riparian buffers that are primarily used to reduce soil erosion, prevent pesticide or nutrient runoff, and maintain water quality may also be managed to provide needed resources to support natural enemies for biological control. The cost of this ecological infrastructure is thus spread across several segments of society. Taking a broader approach and seeking to manage agricultural landscapes to provide multiple ecosystem benefits may be one key to the further implementation of conservation biological control.

REFERENCES

1. Van Driesche, R.G.; Bellows, T.S., Jr. *Biological Control*; Chapman Hall: New York, 1996; 539.
2. Rabb, R.L.; Stinner, R.E.; van den Bosch, R. Conservation and Augmentation of Natural Enemies. In *Theory and Practice of Biological Control*; Huffaker, C.B., Messenger, P.S., Eds.; Academic Press: New York, 1976; 233–254.
3. Van den Bosch, R.; Telford, A.D. Environmental Modification and Biological Control. In *Biological Control of Pests and Weeds*; DeBach, P., Ed.; Reinhold: New York, 1964; 459–488.
4. Barbosa, P. *Conservation Biological Control*; Academic Press: San Diego, 1998; 396.
5. Gurr, G.M.; Wratten, S.D. *Biological Control: Measures of Success*; Kluwer Academic Publishers: Dordrecht, the Netherlands, 2000; 429.
6. Landis, D.; Wratten, S.D.; Gurr, G. Habitat manipulation to conserve natural enemies of arthropod pests in agriculture. Annu. Rev. Entomol. **2000**, *45*, 173–199.
7. Pickett, C.H.; Bugg, R.L. *Enhancing Biological Control: Habitat Management to Promote Natural Enemies of Agricultural Pests*; University of California Press: Berkley, 1998; 422.
8. Altieri, M.A.; Letourneau, D.K. Vegetation management and biological control in agroecosystems. Crop Protection **1982**, *1*, 405–430.
9. Landis, D.A.; Marino, P.C. Landscape Structure and Extrafield Processes: Impact on Management of Pests and Beneficials. In *Handbook of Pest Management*; Ruberson, J., Ed.; Marcel Dekker, Inc.: New York, 1999; 74–104.
10. Thomas, M.B.; Wratten, S.D.; Sotherton, N.W. Creation of island habitats in farmland to manipulate populations of beneficial arthropods: predator densities and species composition. J. Appl. Ecol. **1992**, *29*, 524–531.
11. Menalled, F.D.; Marino, P.C.; Gage, S.H.; Landis, D.A. Does agricultural landscape structure affect parasitism and parasitoid diversity? Ecol. Applic. **1999**, *9*, 634–641.

Biomass

Alberto Traverso
Thermochemical Power Group, Department of Mechanical, Energy, Management and Transportation Engineering (DIME), University of Genoa, Genoa, Italy

David Tucker
National Energy Technology Laboratory, Department of Energy, Morgantown, West Virginia, U.S.A.

Abstract

This entry deals with biomass as energy source. Different types of biomass are described from the energy perspective, focusing on those more interesting for energy application. The main energy conversion technologies available are outlined, as well as the properties of their main products. Finally, an overview over the benefits that come from biomass exploitation for energy purposes is provided. Global potential for CO_2 emission reduction is described.

INTRODUCTION

This work is organized into seven main sections. The first section provides the reader with a general overview on biomass, including definition, environmental benefits, energetic properties, and a short list of biomass types that can be used as energy sources. The second section illustrates the mechanical processes to produce standardized solid biomass fuels. The third section describes one of the major technologies for converting biomass into energy, combustion. The fourth section analyzes pyrolysis and gasification as promising techniques for efficient exploitation of biomass, still in a precommercial phase. The fifth section is concerned with biochemical processes for producing biogas and biofuels for transportation. The sixth section outlines the major benefits from biomass exploitation for energy purposes. The seventh section reports about the potential carbon dioxide (CO_2) emission reduction due to extensive use of biomass as a renewable energy resource. The eighth section concludes this entry.

GENERALITIES ABOUT BIOMASS

In general, biomass is whatever substance produced or by-produced by biological processes. Commonly, biomass refers to the organic matter derived from plants and generated through photosynthesis. Biomass provides not only food but also construction materials, textiles, paper, medicines, and energy. In particular, biomass can be regarded as solar energy stored in the chemical bonds of the organic material or as a reduced state of carbon. CO_2 from the atmosphere and water absorbed by the plant roots are combined in the photosynthetic process to produce carbohydrates (or sugars) that form the biomass. The solar energy that drives photosynthesis is stored in the chemical bonds of the biomass structural components. During biomass combustion, oxygen from the atmosphere combines with the carbon and hydrogen in biomass to produce CO_2 and water. The process is therefore cyclic because the carbon dioxide is then available to produce new biomass. This is also the reason why bioenergy is potentially considered as carbon-neutral, although some non-recoverable CO_2 emissions occur due to the use of fossil fuels during the production and transport of biofuels.

Biomass resources can be classified according to the supply sector, as shown in Table 1.

The chemical composition of plant biomass varies among species. Yet, in general terms, plants are made of approximately 25% lignin and 75% carbohydrates or sugars. The carbohydrate fraction consists of many sugar molecules linked together in long chains or polymers. Two categories are distinguished: cellulose and hemicellulose. The lignin fraction consists of non-sugar-type molecules that act as a glue holding together the cellulose fibers.

Energy Content of Biomass

Bioenergy is energy of biological and renewable origin, normally derived from purpose-grown energy crops or by-products of agriculture and forestry. Examples of bioenergy resources are wood, straw, bagasse, and organic waste. The term *bioenergy* encompasses the overall technical means through which biomass is produced, converted, and used. Fig. 1 summarizes the variety of processes for energy production from biomass.

Encyclopedia of Environmental Management DOI: 10.1081/E-EEM-120046117

Table 1 Typical types of biomass for energy use.

Supply sector	Type	Example
Forestry	Dedicated forestry	Short rotation plantations (e.g., willow, poplar, eucalyptus)
	Forestry by-products	Wood blocks, wood chips from thinnings
Agriculture	Dry lignocellulosic energy crops	Herbaceous crops (e.g., miscanthus, reed canary-grass, giant reed)
	Oil, sugar and starch energy crops	Oil seeds for methylesters (e.g., rape seed, sunflower)
		Sugar crops for ethanol (e.g., sugar cane, sweet sorghum)
		Starch crops for ethanol (e.g., maize, wheat)
	Agricultural residues	Straw, prunings from vineyards and fruit trees
	Livestock waste	Wet and dry manure
Industry	Industrial residues	Industrial waste wood, sawdust from sawmills
		Fibrous vegetable waste from paper industries
Waste	Dry lignocellulosic	Residues from parks and gardens (e.g., prunings, grass)
	Contaminated waste	Demolition wood
		Organic fraction of municipal solid waste
		Biodegradable landfilled waste, landfill gas
		Sewage sludge

Source: Adapted from European Biomass Industry Association[1] and DOE Biomass Research and Development Initiative.[2]

The calorific value of a fuel is usually expressed as higher heating value (HHV) and/or lower heating value (LHV). The difference results from the vaporization of water formed from the combustion of hydrogen in the material and the original moisture.

The most important property of biomass feedstocks with regard to combustion—and to the other thermochemical processes—is the moisture content, which influences the energy content of the fuel. Wood, just after falling, has a typical 55% water content and an LHV of approximately 7.1 MJ/kg; logwood after 2–3 years of air-drying may present a 20% water content and an LHV of 14.4 MJ/kg; pellets show a quite constant humidity content of about 8% with an LHV equal to 17 MJ/kg.

MECHANICAL PROCESSES FOR ENERGY DENSIFICATION

Some practical problems are associated with the use of biomass material (sawdust, wood chips, or agricultural residues) as fuel. Those problems are mainly related to the high bulk volume, which results in high transportation costs and requires large storage capacities, and to the high moisture content, which can result in biological degradation as well as in freezing and blocking the in-plant transportation systems. In addition, variations in moisture content make difficult the optimal plant operation and process control. All those problems may be overcome by standardization and densification. The former consists in processing the

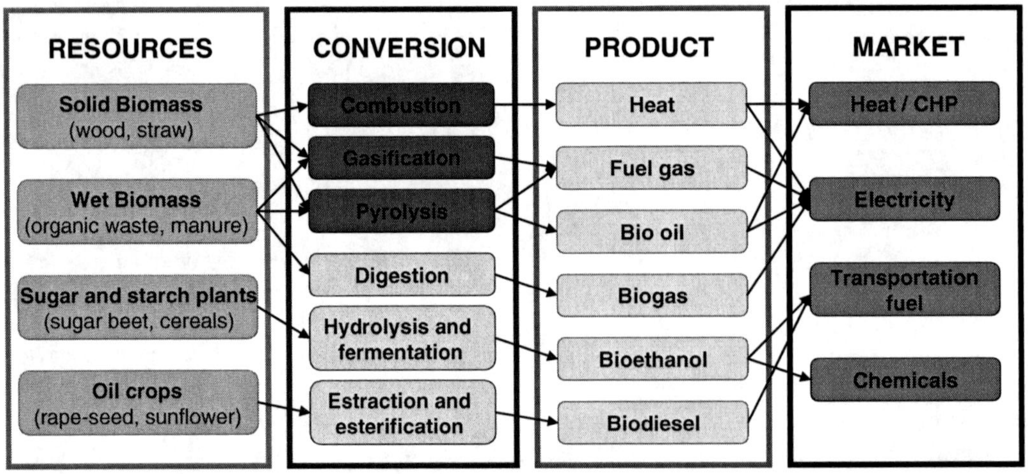
RESOURCES
CONVERSION
PRODUCT
MARKET
Solid Biomass
(wood, straw)
Wet Biomass
(organic waste, manure)
Sugar and starch plants
(sugar beet, cereals)
Oil crops
(rape-seed, sunflower)
Combustion
Gasification
Pyrolysis
Digestion
Hydrolysis and fermentation
Estraction and esterification
Heat
Fuel gas
Bio oil
Biogas
Bioethanol
Biodiesel
Heat / CHP
Electricity
Transportation fuel
Chemicals

Fig. 1 Processes to convert biomass into useful energy, i.e., bioenergy.
Source: European Biomass Industry Association,[1] Overend et al.[3] and Risoe National Laboratory.[6]

Table 2 Comparison of different solid wood fuels.

	Pellets	Briquettes	Chips
Appearance			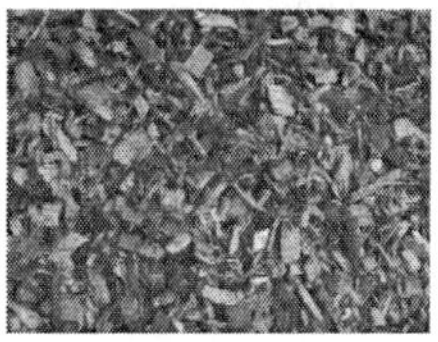
Raw material	Dry and ground wood or agricultural residues	Dry and ground wood or agricultural residues. Raw material can be more coarse than for pelleting, due to the larger dimensions of final product	Dry wood logs
Shape	Cylindrical (generally Ø 6 to 12 mm, with a length 4 to 5 times the Ø).	Cylindrical (generally Ø 80 to 90 mm) or parallelepiped (150 × 70 × 60 mm)	Irregularly parallelepiped (70 × 30 × 3 mm)

original biomass in order to obtain fuels with standard size and heating properties, while the latter consists in compressing the material, which needs to be available in the sawdust size, to give it more uniform properties.

Table 2 reports the main features of pellets, briquettes, and chips.

BIOMASS COMBUSTION

The burning of wood and other solid biomass is the oldest energy technology used by man. Combustion is a well-established commercial technology with applications in most industrialized and developing countries, and development is concentrated on resolving environmental problems, improving the overall performance with multifuel operation, and increasing the efficiency of the power and heat cycles (combined heat and power, CHP).

The devices used for direct combustion of solid biomass fuels range from small domestic stoves (1 to 10 kW) to the large boilers used in power and CHP plants (>5 MW). Intermediate devices cover small boilers (10 to 50 kW) used in single family houses heating, medium-sized boilers (50 to 150 kW) used for multifamily house or building heating, and large boilers (150 kW to more than 1 MW) used for district heating. Cofiring in fossil-fired power stations enables the advantages of large-sized plants (>100 MWe) that are not applicable for dedicated biomass combustion due to limited local biomass availability.

To achieve complete burnout and high efficiencies in small-scale combustion, downdraft boilers with inverse flow have been introduced, applying the two-stage combustion principle. An operation at very low load should be avoided as it can lead to high emissions. Hence, it is recommended to couple log wood boilers to a heat storage tank. Since wood pellets are well suited for automatic heating at small heat outputs, as needed for buildings nowadays, pellet furnaces are an interesting application with increasing propagation. Thanks to the well-defined fuel at low water content, pellet furnaces can easily achieve high combustion quality. They are applied both as stoves and as boilers and find increased acceptance in urban areas, due to the high efficiency of modern pellet stoves now used for home heating. While a conventional fireplace is less than 10% efficient at delivering heat to a house, an average modern pellet stove achieves 80%–90% efficiency. Technology development has led to the application of strongly improved heating systems, which are automated and have catalytic gas cleaning equipment. Such systems significantly reduce the emissions from fireplaces and older systems while at the same time significantly improving the efficiency.

Understoker furnaces are mostly used for wood chips and similar fuel with relatively low ash content, while grate furnaces can also be applied for high ash and water content. Special types of furnaces have been developed for straw, a very low density material that is usually stored in bales. Beside conventional grate furnaces operated with whole bales, cigar burners and other specific furnaces are in operation. Stationary or bubbling fluidized bed (SFB) as well as circulating fluidized bed (CFB) boilers are applied for large-scale applications and often used for waste wood or mixtures of wood and industrial wastes, e.g., from the pulp and paper industry.

Co-combustion

Bioenergy production might be hampered by limitations in the supply and/or fuel quality. In those cases, cofiring of several types of biomass or of biomass with coal ensures flexibility in operation, both technically and economically. Several concepts have been developed:

- Co-combustion or direct cofiring. The biomass is directly fed to the boiler furnace, if needed after physical preprocessing of the biomass such as drying, grinding, torrefaction, or metal removal is applied. This typically takes place in SFB or CFB combustors. Such technologies can be applied to a wide range of fuels, even for very wet fuels like bark or sludge. Multifuel fluidized bed boilers achieve efficiencies of more than 90%, while flue gas emissions are lower than for conventional grate combustion due to lower combustion temperatures.

- Indirect cofiring. Biomass is first gasified and the fuel gas is then cofired in the main boiler. Sometimes, the gas has to be cooled and cleaned, which is more challenging and implies higher operation costs.
- Parallel combustion. The biomass is burnt in a separate boiler for steam generation. The steam is used in a power plant together with the main fuel.

Problems in Biomass Combustion

Biomass has a number of characteristics that makes it more difficult to handle and combust than fossil fuels. The low energy density, the high water content, and the tendency to "bridge" in tanks or pipes are the main problems in handling and transport of the biomass, while the difficulty in using biomass as fuel relates to its content of inorganic constituents. Some types of biomass used contain significant amounts of chlorine, sulfur, and potassium. The salts, KCl and K_2SO_4, are quite volatile, and the release of these components may lead to heavy deposition on heat transfer surfaces, resulting in reduced heat transfer and enhanced corrosion rates. Severe deposits may interfere with operation and cause unscheduled shutdowns.

In order to minimize these problems, various fuel pretreatment processes have been considered, including washing the biomass with hot water or using a combination of pyrolysis and char treatment.

THERMOCHEMICAL CONVERSION OF BIOMASS

Pyrolysis and gasification are the two most typical thermochemical processes because they convert the original bioenergy feedstock into more convenient energy carriers such as producer gas, oil, methanol, and char,[3] instead of producing useful energy directly.

Pyrolysis

Pyrolysis is a process for thermal conversion of solid fuels, like biomass or wastes, in the complete absence of oxidizing agent (air/oxygen) or with such limited supply that gasification does not occur to any appreciable extent. Commercial applications are focused on either the production of charcoal or the production of a liquid product, the bio-oil, and pyro-gas. Charcoal is a very ancient product. Traditional processes (partial combustion of wood covered by a layer of earth) were very inefficient and polluting, but modern processes presently used in industry such as rotary kiln carbonization have been optimized to maximize efficiency and minimize environmental impact. Bio-oil production (or wood liquefaction) is potentially very interesting as a substitute for fuel oil and as a feedstock for production of synthetic gasoline or diesel fuel. Pyro-gas has higher energy density than gasification gas (syngas) because it has been created without oxygen (and nitrogen, if air is employed); hence, it does not contain the gaseous products of partial combustion.

Pyrolysis takes place at temperatures in the range 400–800°C, and during this process, most of the cellulose and hemicellulose and part of the lignin will disintegrate to form smaller and lighter molecules, which are gases at the pyrolysis temperature. As these gases cool, some of the vapors condense to form a liquid, which is the bio-oil and tar. The remaining part of the biomass, mainly parts of the lignin, is left as a solid, i.e., the charcoal. It is possible to influence the product mix through a control of heating rate, residence time, pressure, and maximum reaction temperature, so that either gases, condensable vapors, or the solid charcoal is promoted.

Gasification

Modern gasification technology has been developed since the 18th century (first historical hints about gasification date from the Chinese Han Dynasty, between 206 BC and AD 220), but it is still at a development phase.[4,5] Gasification is a conversion process that involves partial oxidation at elevated temperature. It is intermediate between combustion and pyrolysis: in fact, oxygen (or air) is present but it is not enough for complete combustion. This process can start from carbonaceous feedstock such as biomass or coal and convert them into a gaseous energy carrier. The overall gasification process may be split into two main stages: the first is the pyrolysis stage, i.e., where oxygen is not present but temperature is high; typical pyrolysis reactions take place here; the second stage is the partial combustion, where oxygen is present and reacts with the pyrolysed biomass to release heat necessary for the process. In the latter stage, the actual gasification reactions take place, which consist of almost complete charcoal conversion into lighter gaseous products (e.g., carbon monoxide and hydrogen), through the chemical oxidizing action of oxygen, steam, and carbon dioxide: such gases are injected into the reactor near the partial combustion zone (normally, steam and carbon dioxide are mutually exclusive). Gasification reactions require temperature in excess of 800°C to minimize tar and maximize gas production. The gasification output gas, called "producer gas," is composed of hydrogen (18%–20%), carbon monoxide (18%–20%), carbon dioxide (8%–10%), methane (2%–3%), trace amounts of higher hydrocarbons like ethane and ethene, water, nitrogen (if air is used as oxidant agent), and various contaminants such as small char particles, ash, tars, and oils. The incondensable part of producer gas is called "syngas," and it represents the useful product of gasification. If air is used, syngas has a high heating value in the order of 4–7 MJ/m^3, which is exploitable for boiler, engine, and turbine operation, but, due to its low energy density, it is not suitable for pipeline transportation. If pure oxygen is used, the syngas high heating value almost doubles (approximately 10–18 MJ/m^3 high heating value). Such a syngas is suitable for limited pipeline distribu-

Table 3 Qualitative comparison of technologies for energy conversion of biomass.

Process	Technology	Economics	Environment	Market potential	Present deployment
Combustion—heat	+++	€	+++	+++	+++
Combustion—electricity	++(+)	€€	++(+)	+++	++
Gasification	+(+)	€€€	+(++)	+++	(+)
Pyrolysis	(+)	€€€	(+++)	+++	(+)

Note:+, low; +++, high; €, cheap; €€€, expensive.
Source: Adapted from European Biomass Industry Association[1] and Risoe National Laboratory.[6]

tion as well as for conversion to liquid fuels (e.g., methanol and gasoline). However, the most common technology is air gasification because it avoids the costs and the hazards of oxygen production and usage. With air gasification, the syngas efficiency, describing the energy content of the cold gas stream in relation to that of the input biomass stream, is on the order of 55%–85%, and typically 70%.

Comparison of Thermal Conversion Methods of Biomass

Table 3 reports a general overview on specific features of the conversion technologies analyzed here, showing the related advantages and drawbacks.

BIOCHEMICAL CONVERSION OF BIOMASS

Biochemical conversion of biomass refers to processes that decompose the original biomass into useful products. Commonly, the energy product is either in the liquid or in the gaseous form; hence, it is called "biofuel" or "biogas," respectively. Biofuels are very promising for the transportation sector, while biogas is used for electricity and heat production. Normally, biofuels are obtained from dedicated crops (e.g., biodiesel from seed oil), while biogas production results from concerns over environmental issues such as elimination of pollution, treatment of waste, and control of landfill greenhouse gas (GHG) emissions.

Biogas from Anaerobic Digestion

Biogas is produced most commonly by anaerobic digestion of biomass. Anaerobic digestion refers to the bacterial breakdown of organic materials in the absence of oxygen. This biochemical process produces a gas called biogas, principally composed of methane (30%–60% in volume) and carbon dioxide. Such a biogas can be converted to energy in the following ways:

- Biogas converted by conventional boilers for heating purposes at the production plant (house heating, district heating, industrial purposes).
- Biogas for CHP generation.
- Biogas and natural gas combinations and integration in the natural gas grid.
- Biogas upgraded and used as vehicle fuel in the transportation sector.
- Biogas utilization for hydrogen production and fuel cells.

An important production of biogas comes from landfills. Anaerobic digestion in landfills is brought about by the microbial decomposition of the organic matter in refuse. Landfill gas is, on average, 55% methane and 45% carbon dioxide. With waste generation increasing at a faster rate than economic growth, it makes sense to recover the energy from that stream, through thermal or fermentation processes.

Biofuels for Transport

A wide range of chemical processes may be employed to produce liquid fuels from biomass. Such fuels can find a very high level of acceptance by the market, thanks to the relatively easy adaptation to existing technologies (i.e., gasoline and diesel engines). The main potential biofuels are outlined below.

- Biodiesel is a methyl-ester produced from vegetable or animal oil to be used as alternative to conventional petroleum-derived diesel fuel. Compared to pure vegetable or animal oil, which can be used in adapted diesel engines as well, biodiesel presents lower viscosity and slightly HHV.
- Pure vegetable oil is produced from oil plants through pressing, extraction, or comparable procedures, crude or refined but chemically unmodified. Usually, it is compatible with existing diesel engines only if blended with conventional diesel fuel, at rates not higher than 5%–10% in volume. Higher rates may lead to emission and engine durability problems.
- Bioethanol is ethanol produced from biomass and/or the biodegradable fraction of waste. Bioethanol can be produced from any biological feedstock that contains appreciable amounts of sugar or other matter that can be converted into sugar, such as starch or cellulose. Also, lignocellulosic materials (wood and straw) can be used,

but their processing into bioethanol is more expensive. Application to modified spark ignition engines is possible.

- Bio-ETBE (ethyl-tertio-butyl-ether) is ETBE produced on the basis of bioethanol. Bio-ETBE may be effectively used for enhancing the octane number of gasoline (blends with petrol gasoline).
- Biomethanol is methanol produced from biomass. Methanol can be produced from gasification syngas (a mixture of carbon monoxide and hydrogen) or wood dry distillation (old method with low methanol yields). Virtually all syngas for conventional methanol production is produced by steam reforming of natural gas into syngas. In the case of biomethanol, a biomass is gasified first to produce a syngas from which the biomethanol is produced. Application to spark ignition engines and fuel cells is possible. Compared to ethanol, methanol presents more serious handling issues, because it is corrosive and poisonous for human beings.
- Bio-MTBE (methyl-tertio-butyl-ether) is a fuel produced on the basis of biomethanol. It is suitable for blends with petrol gasoline.
- Biodimethylether (DME) is dimethylether produced from biomass. Bio-DME can be formed from syngas by means of oxygenate synthesis. It has emerged only recently as an automotive fuel option. Storage capabilities are similar to those of LPG. Application to spark ignition engines is possible.

BENEFITS FROM BIOMASS ENERGY

There is quite a wide consensus that, over the coming decades, modern biofuels will provide a substantial source of alternative energy. Nowadays, biomass already provides approximately 11%–14% of the world's primary energy consumption (data vary according to sources).

There are significant differences between industrialized and developing countries; in particular, in many developing countries, bioenergy is the main energy source, even if it is used in very low efficient applications (e.g., cooking stoves have an efficiency of about 5%–15%). Furthermore, inefficient biomass utilization is often associated with the increasing scarcity of hand-gathered wood, nutrient depletion, and the problems of deforestation and desertification.

One of the key drivers to bioenergy deployment is its positive environmental benefit regarding the global balance of GHG emissions. This is not a trivial matter, because biomass production and use are not entirely GHG neutral. In general terms, the GHG emission reduction as a result of employing biomass for energy is as reported in Table 4.

Since the energy cost associated with collection and transport of biomass is a significant portion, bioenergy is a decentralized energy option whose implementation presents positive impacts on rural development by creating business and employment opportunities. Jobs are created all along the bioenergy chain, from biomass production or procurement, to its transport, conversion, distribution, and marketing.

Bioenergy is a key factor for the transition to a more sustainable development.

Table 4 Benefits in reduction of GHG emissions.

+	Avoided mining of fossil resources
–	Emission from biomass production
+	Avoided fossil fuel transport (from producer to user)
–	Emission from biomass fuel transport (from producer to user)
+	Avoided fossil fuel utilization

Note: +, positive; –, neutral.
Source: Risoe National Laboratory.[6]

POTENTIAL FOR CO_2 EMISSION REDUCTION

When biomass is used for energy production, the carbon contained in it is ultimately transformed into CO_2. In fact, such a biomass-derived CO_2 does not contribute to global warming, as it equals the CO_2 absorbed by the biomass during its growth; the relatively short time of such carbon cycle makes the biomass a carbon-neutral energy resource.

Abundant resources and favorable policies[7] enable bio-power to expand in Northern Europe (mostly cogeneration from wood residues), in the United States, and in countries producing sugar cane bagasse (e.g., Brazil).

In the short term, cofiring remains the most cost-effective use of biomass for power generation, along with small-scale, off-grid use. In the mid-long term, gasification plants and biorefineries for biofuel production could expand significantly (mainly ethanol, lignocellulosic ethanol, biodiesel). International Energy Agency projections suggest that the biomass share in electricity production may increase from the current 1.3% to some 3%–5% by 2050, depending on assumptions.[8] This is a small contribution compared to the estimated total biomass potential, but biomass are also used for heat generation and to produce fuels for transport.

Today, biomass supplies some 50 EJ/yr (1 EJ = 10^{18} joules [J] = 10^{15} kilojoules [kJ] = 24 million tons of oil equivalent [Mtoe]) globally, which represents 10% of global annual primary energy consumption (Fig. 2). This is mostly traditional biomass used for cooking and heating.

Based on this diverse range of feedstocks, the technical potential for biomass is estimated in the literature to be possibly as high as 1500 EJ/yr by 2050, although most biomass supply scenarios that take into account sustainability constraints indicate an annual potential of between 200 and

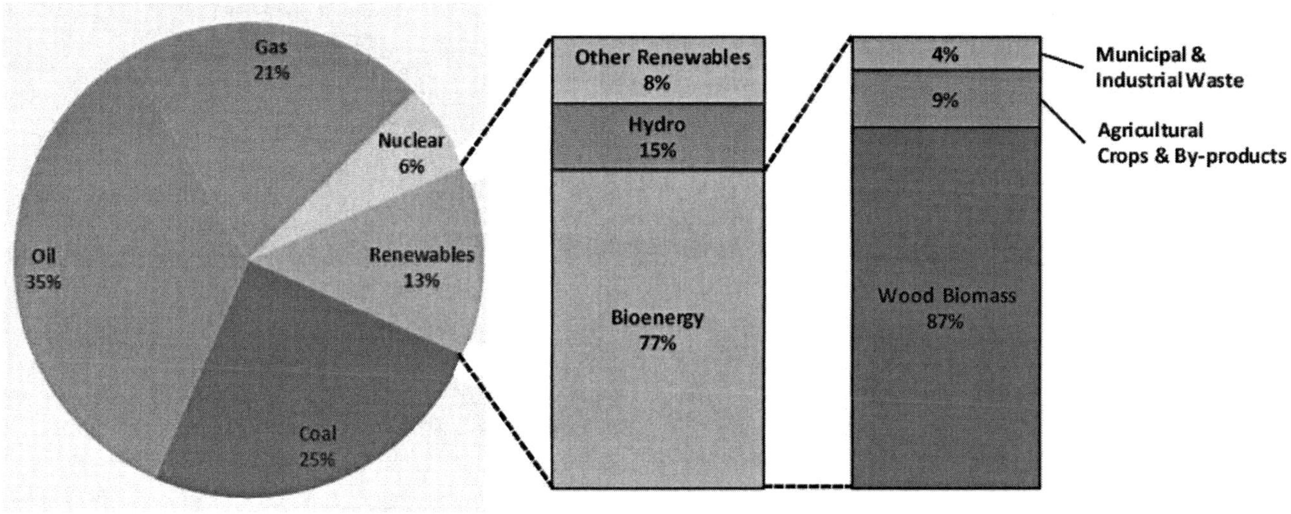
Gas
21%
Nuclear
6%
Oil
35%
Renewables
13%
Coal
25%
Other Renewables
8%
Hydro
15%
Bioenergy
77%
4%
9%
Wood Biomass
87%
Municipal & Industrial Waste
Agricultural Crops & By-products

Fig. 2 Share of bioenergy in the world primary energy mix.
Source: International Energy Agency.[9]

500 EJ/yr (excluding aquatic biomass). Forestry and agricultural residues and other organic wastes (including municipal solid waste) would provide between 50 and 150 EJ/yr, while the remainder would come from energy crops, surplus forest growth, and increased agricultural productivity.

Projected world primary energy demand by 2050 is expected to be in the range of 600 to 1000 EJ, compared to about 500 EJ in 2008.[9] Scenarios looking at the penetration of different low-carbon energy sources indicate that future demand for bioenergy could be up to 250 EJ/yr. This projected demand falls well within the sustainable supply potential estimate, so it is reasonable to assume that biomass could sustainably contribute between a quarter and a third of the future global energy mix. Whatever is actually realized will depend on the cost competitiveness of bioenergy and on future policy frameworks, such as GHG emission reduction targets.

Given the CO_2-neutral nature of biomass, and assuming that biomass will primarily substitute fossil fuels, the potential for reduction in CO_2 emission in 2050 can then be estimated as the same figure (i.e., around 20%–30% of anthropogenic CO_2), compared to a business-as-usual scenario. Definitely, biomass will play a determinant role towards a CO_2-free development.

CONCLUSIONS

Biomass refers to a very wide range of substances produced by biological processes. In the energy field, special focus has been and will be placed on vegetable biomass, such as wood and agricultural by-products, because of the energy potential as well as economic and environmental benefits. Size and humidity standardization of biomass is a necessary step to make it suitable for effective domestic and industrial exploitation: chips, briquettes, and pellets are modern examples of standard solid fuels.

Biomass can be converted into energy in three pathways: combustion, thermochemical processing, and biochemical processing. The combustion of solid biomass for the production of heat or electricity and heat is the most viable technology, while pyrolysis and gasification still face economic and reliability issues. Among biochemical processes, anaerobic digestion is often used to reduce the environmental impact of hazardous waste and landfills. Biochemical processes are also concerned with the conversion of biomass into useful fuels for transportation, such as biodiesel, bioethanol, and biomethanol. All of them can effectively contribute to the transition to a more sustainable transportation system at zero GHG emissions.

Biomass represents a viable option for green energy resources of the 21st century.

REFERENCES

1. European Biomass Industry Association, available at http://www.eubia.org/.
2. DOE Biomass Research and Development Initiative, available at http://www.bioproducts-bioenergy.gov/.
3. Overend, R.P.; Milne, T.A.; Mudge, L.K. *Fundamentals of Thermochemical Biomass Conversion*; Elsevier Applied Science Publishers Ltd.: New York, 1985.
4. Bridgewater, A.V. The technical and economic feasibility of biomass gasification for power generation. Fuel J. **1995**, *74* (6–8), 557–564.
5. Franco, A.; Giannini, N. Perspectives for the use of biomass as fuel in combined cycle power plants. Int. J. Thermal Sci. **2005**, *44*, (2), 163–177.

6. Risoe National Laboratory, Denmark, available at http://www.risoe.dk/.
7. Biomass for Power Generation and CHP, International Energy Agency (IEA), Energy Technology Essentials, ETE03, Jan 2007.
8. World Energy Outlook; International Energy Agency (IEA), 2002.
9. Bioenergy—A Sustainable and Reliable Energy Source, International Energy Agency (IEA), available at http://www.ieabioenergy.com, 2009.

Biopesticides

G.J. Ash
A. Wang
E.H. Graham Center for Agricultural Innovation, Industry and Investment, NSW and Charles Sturt University, Wagga Wagga, New South Wales, Australia

Abstract

Biopesticides are environmentally sensitive alternatives to the use of synthetic pesticides for the management of pests. The broadest definition of biopesticides includes the use of naturally occurring chemicals, pesticidal compounds produced from genetically engineered plants, and microbially based products. This latter group has shown considerable promise for the management of a wide range of target pests. Although they are often thought of as being analogous to synthetic pesticides, biopesticides (including their production, handling, application, and commercialization) need to be treated in an alternative paradigm.

INTRODUCTION

Inappropriate use of synthetic pesticides in agriculture can lead to environmental degradation, health risks, and loss of biodiversity. Additionally, overuse of particular pesticides may also lead to a loss in their effectiveness through pesticide resistance. Several studies have highlighted the accumulation of pesticides in soils in agricultural systems and the resultant environmental damage.[1] The short-term effects of pesticides on mammalian health are relatively well known and documented through pesticide approval processes. However, the longer-term effects are less well understood but may include cancer, neurotoxic effects, and reproductive disorders.[2] In developing countries, it is estimated that there may be 25 million cases of occupational pesticide poisoning annually.[3] Biopesticides are environmentally sensitive alternatives to the use of synthetic pesticides for the management of a wide range of agricultural pests including weeds, insects, pathogens, and other invertebrate pests. Biopesticides are a class of pesticides that contain plant, animal, or microbial products or organisms.[4] The International Union of Pure and Applied Chemistry[5] further classifies biopesticides as plant-incorporated protectants (protectants produced from genetically modified plants), biochemical pesticides (natural products that affect pests), and microbial biopesticides. Microbial biopesticides are a type of biological control in which a living organism is included as the active ingredient in an inundative application of a formulated product. A range of organisms have been used in biopesticide applications including fungi, nematodes, bacteria, and viruses for the control of weeds, insects, acarines, plant diseases, and molluscs.[6–16]

Formulations of biopesticides contain the organism, a carrier, and adjuvants, which may contain compounds such as nutrients and/or chemicals that aid in the survival of the pathogen or help in protecting the active ingredient from adverse environmental conditions.[17] The formulation of the biocontrol agents is affected by the type of biocontrol organism but must ensure that the agent is delivered in a form that is viable, virulent, and with sufficient inoculum potential to be effective in the field. Furthermore, the formulation may also dictate the means of delivery of the final product, for example, as a seed dressing or a foliar-applied formulation. To be successful, a formulation must be effective, economical, and practical to use.[18] Formulation of biopesticides can be in the form of dry products (dusts, granules, pellets, wettable powders, and encapsulated products) and liquid formulations (suspensions, emulsions, and encapsulated products).[19] The biopesticide is usually packaged, handled, stored, and applied in a similar way to traditional, synthetic pesticides.[20]

Due to the enormous costs involved in the production of synthetic pesticides, global companies have tended to focus on the registration of pesticides in the major crop/cropping systems. This has led to a number of attempts to develop biopesticides in these non-core or niche markets.[21] These markets could be of considerable size and include those that have been created by synthetic pesticide withdrawal and the organic food movement. Philosophically, the use of biopesticides is compatible with organic food production, provided the agent had not been genetically modified, the carriers and adjuvants are natural products, and the host range of the biopesticide is not considered to be too wide.[22]

The use of biopesticides as a strategy in pest management can be applied to both native and introduced pests. However, the success of this type of biocontrol revolves around the costs of production, the quality of the inoculum, and, most importantly, the field efficacy of the product.[18] Biopesticides are usually developed by collaboration with commercial companies in an expectation that they will recoup their costs through the sale of the product.

Encyclopedia of Environmental Management DOI: 10.1081/E-EEM-120046186

The broad range of biopesticides may be further subdivided based on the type of target pest and/or the active ingredient. For example, a bioherbicide is a biopesticide developed for weed management whereas a mycoherbicide is a bioherbicide that contains a fungus as an active ingredient. The goal of this entry is to introduce the broad range of biopesticides available and their targets.

BIOINSECTICIDES AND BIOACARICIDES

More than 1500 species of pathogens have been shown to attack arthropods and include representatives from bacteria, viruses, fungi, protozoa, and nematodes.[23] Diseases caused by insects have been known since the early 1800s with the first attempts at inundative applications of fungi to control insects being developed in 1884, when the Russian entomologist Elie Metchnikoff mass produced the spores of the fungus *Metarhizium anisopliae*. A mycoacaricide for the management of citrus rust mite was first registered in the United States in 1981.[24] This was based on the fungus *Hirsutella thompsonii.* Currently, the main fungal species formulated as mycoinsecticides and mycoacaricides include *M. anisopliae* (33.9%), *Beauveria bassiana* (33.9%), *Isaria fumosorosea* (5.8%), and *Beauveria brongniartii* (4.1%).[15] De Faria and Wraight[15] reported a total of 171 fungi-based products with the majority being for the control of insects (160) and mites.[28] Their review indicated that the main formulation types are fungus-colonized substrates, wettable powders, and oil dispersions containing conidia (asexual spores). In the United States, there are seven bioinsecticide products based on various strains of *B. bassiana* registered. These include products such as BotaniGard® and Mycotrol® (available as emulsifiable suspensions and wettable powders).

Bacteria that attack insects can be divided into non-spore-forming and spore-forming bacteria. The non-spore-forming bacteria include species in the Pseudomonadaceae and the Enterobacteriaceae. The spore-forming bacteria belong to the Bacillaceae and include species such as *Bacillus popilliae* and *Bacillus thuringiensis. B. thuringiensis* (Bt) have primarily been developed as biopesticides to control Lepidopteran larva. However, other serotypes of Bt produce toxins that kill insects in the Coleoptera and Diptera as well as nematodes. The bacterium produces δ-endotoxin, an insecticidal crystal protein, which is converted into proteolytic toxins on ingestion. Up to nine different toxins, which have different host ranges, have been described.[25] Commercial formulations of the bacteria contain living spores of the bacteria. Biopesticides based on Bt are the most widely available of the bacterially based products.[25] Up to 90% of the Microbial Pest Control Agent market is Bt or Bt-derived products. The most well known is Dipel.

Entomopathogenic nematodes of the families Steinernematidae and Heterorhabditidae, in conjunction with bacteria of the genus *Xenorhabdus*, have been successfully deployed as biopesticides, for example, BioVECTOR.[23] They are usually applied to control insects in cryptic and soil environments. The nematodes harbor the bacteria in their intestines. The infective third-stage larvae enter the host through natural openings and penetrate into the hemocoel. The bacteria are voided in the insect and cause septicemia, killing the insect in approximately 48 hr.

Entomopathogenic viruses have also shown promise as bioinsecticides. They were first used to control populations of *Lymantria monacha* in pine forests in Germany in 1892 (Huber, 1986, from Moscardi[26]), but the first commercial viral insecticide registered was called Viron/H for the control of *Helicoverpa* (*Heliothis*) *zea* in 1971.[27] Viruses from the family Baculoviridae have been isolated from more than 700 invertebrates, with the virus group not common outside of the Lepidoptera and Hymenoptera.[26,28] The nucleopolyhedroviruses (NPVs) are rod-shaped, double-stranded DNA viruses that are produced in polyhedral proteinaceous occlusion bodies[28] that are ingested by the insect. Granulosis viruses (GVs) are also members of the Baculoviridae but are restricted to the Lepidoptera and have capsular proteinaceous occlusion bodies.[26] These authors[26] provide a table of products that have been developed for the control of insects using these two viral groups.

BIOFUNGICIDES

Biological control of fungi that cause plant disease can be accomplished by a number of mechanisms including antibiosis, hyperparasitism, or competition. Additionally, weak pathogens may induce systemic acquired resistance in the host, giving a form of cross-protection. Biofungicides have been used in both the phylloplane and rhizosphere to suppress disease. A biological control agent for the control of foliar pathogens in the phylloplane must have a high reproductive capacity, the ability to survive unfavorable conditions, and the ability to be a strong antagonist or be very aggressive. A wide range of bacteria and fungi are known to produce antibiotics that affect other microorganisms in the infection court. Most often, these organisms are sought from a soil environment, as this environment is seen as the richest source of antibiotic-producing species. Species of *Bacillus* and *Pseudomonas* have been successfully used as seed dressings to control soil-borne plant diseases.[29] Serenade®, marketed by BASF, is a formulation of *Bacillus subtilis* (strain QST713), which has claimed activity against a wide range of plant diseases.[30] It is applied as a foliar spray to crops such as cherries, cucurbits, grapes, leafy vegetables, peppers, potatoes, tomatoes, and walnuts.[31] Fluorescent pseudomonads are also often seen as a component of suppressive soils. These bacteria may prevent the germination of fungi by the induction of iron competition through the production of siderophores (iron-chelating compounds). These are effective only in those soils where the availability of iron is low. Control of foliar and fruit pathogens such as *Botrytis cinerea*, a pathogen of strawberries, has been accomplished by the foliar application of the soil-inhabiting fungus *Trichoderma viride*.[32] This fungus inhibits *Botrytis* using a combination of antibiosis and competition. On grapevines, *Trichoderma harzianum* competes with *B. cinerea* on senescent floral parts, thus preventing the infection of the ovary. It has also been shown to coil around the hyphae of the pathogen during hyperparasitism.[33] *T. harzianum* has also been reported to induce systemic resistance in plants.[34] One of the earliest commercial successes using *T. harzianum* is the product Rootshield®. Rootshield contains the T-22 strain of *T. harzianum* and is produced and marketed by Bioworks Inc. This strain of the fungus was first registered by the U.S. Environmental Protection Agency in 1993. The product is available as a granular formulation and is usually applied to soil mixtures in glasshouse situations.[35–37]

BIOHERBICIDES

Fungi are the most important group of pathogens causing plant disease. Therefore, fungi (or oomycetes) are most commonly used as the active ingredient in bioherbicides and as such the formulated organism is referred to as a mycoherbicide.[38] However, there are examples of bacteria[7–11] and viruses being used or proposed to be used as bioherbicides.[39,40] The aim of bioherbicide development is to overcome the natural constraints of a weed–pathogen interaction, thereby creating a disease epidemic on a target host.[41] For example, the application of fungal propagules to the entire weed population overcomes the constraint of poor dissemination. After removal of the host weed, the pathogen generally returns to background levels because of natural constraints on survival and spread.

The first commercially available biopesticide for the control of weeds was DeVine®, a bioherbicide for the control of strangler vine in citrus groves in the United States. It was released in 1981.[42] In 1982, a formulation of *Colletotrichum gloeosporioides* f. sp. *aeschynomene* was released to control northern jointvetch in soybean crops in the United States. Since then, there have been a number of products commercialized[12,21] as well as numerous examples of pathogen–weed combinations that had been reported as having potential as bioherbicides in countries including in Canada, United States, Europe, Japan, Australia, and South Africa.[18,21] Necrotrophic or hemibiotrophic fungi are usually used as the basis of mycoherbicides, as they can be readily cultured on artificial media and so lend themselves to mass production. Other desirable characteristics of fungi under consideration as mycoherbicides include the ability to sporulate freely in artificial culture, limited ability to spread from the site of application, and genetic stability. In most cases, these biopesticides are applied in a similar fashion to chemical herbicides using existing equipment, although the development of specialized application equipment and formulation may improve their efficacy and reliability. Since 2000, there have been two successful registrations for bioherbicides in Canada. In 2002, a product called Chontrol®, based on the fungus *Chondrostereum purpureum* for the control of trees and shrubs, was registered.[12] This was based on the research of Hintz and colleagues.[43–45] A more recent success in the area of bioherbicides includes the registration of Sarritor® for dandelion control by the company of the same name in Canada. This product is based on the phytopathogenic fungus *Sclerotinia minor*, which has been extensively researched by Professor Alan Watson at McGill University.[46–50]

BIOMOLLUSCICIDES

Biomolluscicides are a type of molluscicide derived from natural materials such as animals, plants, and microorganisms (e.g., bacterium, fungus, virus, protozoan, or nematode). They are usually used in the fields of agriculture and gardening to control pest slugs and snails. In some circumstances, biomolluscicides are also used in the health area to control molluscs acting as vectors of harmful parasites to human beings.

Currently, the most widely used biomolluscicide is Nemaslug®, a successful biomolluscicide developed by Becker Underwood (U.K.). The active ingredient of Nemaslug is *Phasmarhabditis hermaphrodita*, a nematode species from the family of Rhabditidae. The pathogenicity of *P. hermaphrodita* against slugs had not been recognized until 1994 when Wilson et al.[51] discovered that *P. hermaphrodita* could infect and kill a wide variety of pest slugs under laboratory conditions. Like entomopathogenic nematodes, *P. hermaphrodita* kills slugs by penetrating into the hemocoel of hosts through natural openings and releasing its associated bacteria, which kill the host eventually.

P. hermaphrodita was found to be associated with several different bacteria rather than one particular species, but its association with *Moraxella osloensis* proved to be highly pathogenic to gray garden slug (*Deroceras reticulatum*).[51] This bacterium was used in the mass production of *P. hermaphrodita* via monoxenic culture.[51]

The host of *P. hermaphrodita* is not restricted to only one slug species (*D. reticulatum*). It can attack and kill several species of slugs including *Arion ater*, *Arion intermedius*, *Arion distinctus*, *Arion silvaticus*, *Deroceras caruanae*, *Tandonia budapestensis*, and *Tandonia sowerbyi*.[51] Moreover, *P. hermaphrodita* can also parasitize several species of snails including *Cernuella virgata*, *Cochlicella acuta*, *Helis aspersa*, *Monacha cantiana*, *Lymnaea stagnalis*, and *Theba pisana*.[52]

Nemaslug is now sold in many European countries, including U.K., Ireland, France, the Netherlands, Belgium,

Germany, Denmark, Norway, Finland, Poland, Spain, the Czech Republic, Italy, and Switzerland. In 2005, the retail sale of this biomolluscicide was up to £1 million in Europe and approximately 500 ha horticultural crops (e.g., lettuce and strawberries) and field crops (e.g., wheat, potatoes, and oilseed) were treated with this biomolluscicide. At the dose rate of 3×10^9/ha, *P. hermaphrodita* provides protection against slug damage similar to, if not better than, methiocarb pellets.[16]

Bacteria-based biomolluscicides are now in the process of development. *Streptomyces violaceoruber* and *Xanthobacter autotrophicus* have been examined for their molluscicidal activity against *Oncomelania hupensis* (a unique host of schistosomiasis blood fluke parasite) under laboratory conditions.[53] The results revealed that both bacteria were effective in killing *O. hupensis*, with *S. violaceoruber* causing more snail mortality than *X. autotrophicus* (90% vs. 85%).

Biomolluscicides of plant origin have also been studied extensively in recent years when the environmental pollution caused by chemical molluscicides was realized increasingly. More than 1400 plant species have been screened for their molluscicidal properties against pest snail species.[54] Several groups of compounds present in various plants have been found to be poisonous to snails at acceptable doses, ranging from <1 to 100 ppm, including saponins, tannins, alkaloids, alkenyl phenols, glycoalkaloids, flavonoids, sesquiterpene lactones, and terpenoids.[54] The molluscicidal activity of the dried root latex powder of *Ferula asafoetida*, the flower-bud powder of *Syzygium aromaticum*, and the seed powder of *Carum carvi* against the snail *Lymnaea acuminata* was proved.[55] Similarly, acetogenin (extracted from the seed powder of custard apple) presented promising and stable molluscicidal activity against *L. acuminata*.[54] When sodium alginates was used as a binding matrix for the formulation of acetogenin, the release of this biomolluscicide extended over 25 days, which set up a good example for the development of biomolluscicide delivery system.[54]

The combination of bacteria-based biomolluscicides and plant-based biomolluscicides may lead to a synergistic effect between plant and microbe extracts as molluscicides. Zhang and coworkers[56] reported that higher snail mortality was produced when a mixture of *Arisaema erubescens* tuber extracts and *S. violaceoruber* dilution was applied against the snail *O. hupensis*. The mechanisms of snail toxicosis might be that the combination of *A. erubescens* tuber extract and *S. violaceoruber* dilution reduced the detoxification ability of liver and increased the oxidative damage in liver cells of snails.

CONCLUSION

Biopesticides are a viable alternative to synthetic pesticides in a number of crops. The development of microbial biopesticides relies on agent discovery and selection, development of methods to culture the pathogen, creation of formulations that protect the organism in storage as well as aid in its delivery, studies of field efficacy, and methods of storage. Each microbial biopesticide is unique, in that not only will the organism vary but so too will the host, the environment in which it is being applied, and economics of production and control.

There are a number of advantages of the use of biopesticides over the use of conventional pesticides, including the minimal residue levels, control of pests already showing resistance to conventional pesticides, host specificity, and the reduced chance of resistance to biopesticides. This indicates an emerging, strong role for biopesticides in any integrated pest management strategy and an important involvement in sustainable farming production systems in the future.

There have been some spectacular successes in the use of microbial biopesticides, despite the perceived constraints to their deployment.[57] In the past, biopesticides have been expected to behave in the same way as synthetic pesticides. For the ultimate success of biopesticides, microorganisms developed for biological control must be viewed by researchers, manufacturers, and end users in a biological paradigm rather than a chemical one. The business model for the commercialization of the products may also vary significantly from that used for traditional synthetic pesticides.[18]

The efficacy and reliability of many microbial biopesticides may be affected by environmental parameters as well as the aggressiveness of the pathogen. Furthermore, the narrow host range of many pathogens may restrict their commercial attractiveness. Both of these issues can be addressed by research into the use of genetic engineering and formulation.[18,58–60] As research into the molecular basis of host specificity and pathogenesis continues, it will become possible to produce more aggressive pathogens with the desired host range for biological control. The survival and efficacy of these pathogens will be enhanced through the use of novel formulations.

REFERENCES

1. Alletto, L.; Coquet, Y.; Benoit, P.; Heddadj, D.; Barriuso, E. Tillage management effects on pesticide fate in soils. A review. Agron. Sustainable Dev. **2010**, *30*, 367–400.
2. Komarek, M.; Cadkova, E.; Chrastny, V.; Bordas, F.; Bollinger, J.C. Contamination of vineyard soils with fungicides: A review of environmental and toxicological aspects. Environ. Int. **2010**, *36*, 138–151.
3. Jeyaratnam, J. Acute pesticide poisoning: A major global health problem. World Health Stat. Q. **1990**, *43*, 139–144.
4. Available at http://www.epa.gov/opp00001/biopesticides/whatarebiopesticides.htm (accessed July 30, 2010).
5. Available at http://agrochemicals.iupac.org/index.php?p=biopesticides (accessed July 30, 2010).
6. Jaronski, S.T. Ecological factors in the inundative use of fungal entomopathogens. Biocontrol **2010**, *55*, 159–185.

7. Imaizumi, S.; Nishino, T.; Miyabe, K.; Fujimori, T.; Yamada, M. Biological control of annual bluegrass (*Poa annua* L.) with a Japanese isolate of *Xanthomonas campestris* pv *poae* (JT-P482). Biol. Control **1997**, *8*, 7–14.
8. Daigle, D.J.; Connick, W.J.; Boyetchko, S.M. Formulating a weed-suppressive bacterium in "Pesta". Weed Technol. **2002**, *16*, 407–413.
9. Weissmann, R.; Uggla, C.; Gerhardson, B. Field performance of a weed-suppressing *Serratia plymuthica* strain applied with conventional spraying equipment. Biocontrol **2003**, *48*, 725–742.
10. Anderson, R.C.; Gardner, D.E. An evaluation of the wilt-causing bacterium *Ralstonia solanacearum* as a potential biological control agent for the alien kahili ginger (*Hedychium gardnerianum*) in Hawaiian forests. Biol. Control **1999**, *15*, 89–96.
11. DeValerio, J.T.; Charudattan, R. Field testing of *Ralstonia solanacearum* [Smith] Yabuuchi et al. as a biocontrol agent for tropical soda apple (*Solanum viarum* Dunal). Weed Sci. Soc. Am. Abstr. **1999**, *39*, 70.
12. Bailey, K.L.; Boyetchko, S.M.; Langle, T. Social and economic drivers shaping the future of biological control: A Canadian perspective on the factors affecting the development and use of microbial biopesticides. Biol. Control **2010**, *52*, 221–229.
13. Bailey, K.L. Canadian innovations in microbial biopesticides. Can. J. Plant Pathol. **2010**, *32*, 113–121.
14. Hartman, C.L.; Markle, G.M. IR-4 biopesticide program for minor crops. In *Biopesticides*; Humana Press Inc.: Totowa, 1999; Vol. 5, 443–452.
15. de Faria, M.R.; Wraight, S.P. Mycoinsecticides and mycoacaricides: A comprehensive list with worldwide coverage and international classification of formulation types. Biol. Control **2007**, *43*, 237–256.
16. Rae, R.; Verdun, C.; Grewal, P.; Robertson, J.F.; Wilson, M.J. Biological control of terrestrial molluscs using *Phasmarhabditis hermaphrodita*—Progress and prospects. Pest Manage. Sci. **2007**, *63*, 1153–1164.
17. Hynes, R.K.; Boyetchko, S.M. Research initiatives in the art and science of biopesticide formulations. Soil Biol. Biochem. **2006**, *38*, 845–849.
18. Ash, G.J. The science, art and business of successful bioherbicides. Biol. Control **2010**, *52*, 230–240.
19. Auld, B.A.; Hertherington, S.D.; Smith, H.E. Advances in bioherbicide formulation. Weed Biol. Manage. **2003**, *3*, 61–67.
20. Van Driesche, R.G.; Bellows, T.S. *Biological Control*; Chapman and Hall: New York, 1996.
21. Charudattan, R. Biological control of weeds by means of plant pathogens: Significance for integrated weed management in modern agro-ecology. Biocontrol **2001**, *46*, 229–260.
22. Rosskopf, E.; Koenig, R. Are bioherbicides compatible with organic farming systems and will businesses invest in the further development of this technology? VI International Bioherbicide Group Workshop, Canberra, Australia, 2003; 2003.
23. Kaya, H.K.; Gaugler, R. Entomopathogenic nematodes. Annu. Rev. Entomol. **1993**, *38*, 181–206.
24. McCoy, C.W. Factors governing the efficacy of *Hirsutella thompsonii* in the field. In *Fundemental and Applied Aspects of Invertebrate Pathology*; Samson, R.A., Vlak, J.M., Peters, D., Eds.; Foundation of the Fourth International Colloquim of Inverstbrate Pathology: Wageningen, the Netherlands, 1986; 171–174.
25. Rosell, G.; Quero, C.; Coll, J.; Guerrero, A. Biorational insecticides in pest management. J. Pestic. Sci. **2008**, *33*, 103–121.
26. Moscardi, F. Assessment of the application of baculoviruses for control of lepidoptera. Annu. Rev. Entomol. **1999**, *44*, 257–289.
27. Ignoffo, C.M.; Rice, W.C.; Mcintosh, A.H. Inactivation of occluded baculoviruses and baculovirus-DNA exposed to simulated sunlight. Environ. Entomol. **1989**, *18*, 177–183.
28. Fuxa, J.R. Ecology of insect nucleopolyhedroviruses. Agric. Ecosyst. Environ. **2004**, *103*, 27–43.
29. Johnsson, L.; Hokeberg, M.; Gerhardson, B. Performance of the *Pseudomonas chlororaphis* biocontrol agent MA 342 against cereal seed-borne diseases in field experiments. Eur. J. Plant Pathol. **1998**, *104*, 701–711.
30. Available at http://www.agro.basf.com/agr/AP-Internet/en/content/solutions/solution_highlights/serenade/bacillus-subtilis (accessed December 1, 2010).
31. Available at http://www.epa.gov/pesticides/biopesticides/ingredients/factsheets/factsheet_006479.htm (accessed December 1, 2010).
32. Sutton, J.C.; Peng, G. Biocontrol of *Botrytis cinerea* in strawberry leaves. Phytopathology **1993**, *83*, 615–621.
33. Oneill, T.M.; Elad, Y.; Shtienberg, D.; Cohen, A. Control of grapevine grey mould with *Trichoderma harzianum* T39. Biocontrol Sci. Technol. **1996**, *6*, 139–146.
34. Shoresh, M.; Harman, G.E. The relationship between increased growth and resistance induced in plants by root colonizing microbes. Plant Signaling Behav. **2008**, *3*, 737–9.
35. Larkin, R.P.; Fravel, D.R. Efficacy of various fungal and bacterial biocontrol organisms for control of fusarium wilt of tomato. Plant Dis. **1998**, *82*, 1022–1028.
36. Brewer, M.T.; Larkin, R.P. Efficacy of several potential biocontrol organisms against *Rhizoctonia solani* on potato. Crop Prot. **2005**, *24*, 939–950.
37. Gravel, V.; Menard, C.; Dorais, M. Pythium root rot and growth responses of organically grown geranium plants to beneficial microorganisms. HortScience **2009**, *44*, 1622–1627.
38. Crump, N.S.; Cother, E.J.; Ash, G.J. Clarifying the nomenclature in microbial weed control. Biocontrol Sci. Technol. **1999**, *9*, 89–97.
39. Ferrell, J.; Charudattan, R.; Elliott, M.; Hiebert, E. Effects of selected herbicides on the efficacy of tobacco mild green mosaic virus to control tropical soda apple (*Solanum viarum*). Weed Sci. **2008**, *56*, 128–132.
40. Charudattan, R.; Hiebert, E. A plant virus as a bioherbicide for tropical soda apple, *Solanum viarum*. Outlooks Pest Manage. **2007**, *18*, 167–171.
41. TeBeest, D.O., Ed. Biological control of weeds with plant pathogens and microbial pesticides. In *Advances in Agronomy*; Academic Press Inc.: San Diego, 1996; Vol. 56, 115–137.
42. Tebeest, D.O.; Yang, X.B.; Cisar, C.R. The status of biological control of weeds with fungal pathogens. Annu. Rev. Phytopathol. **1992**, *30*, 637–657.
43. Harper, G.J.; Comeau, P.G.; Hintz, W.; Wall, R.E.; Prasad, R.; Hocker, E.M. *Chondrostereum purpureum* as a biological control agent in forest vegetation management. II. Efficacy on Sitka alder and aspen in western Canada. Can. J. Forest Res. **1999**, *29*, 852–858.

44. de la Bastide, P.Y.; Zhu, H.; Shrimpton, G.; Shamoun, S.F.; Hintz, W.E. *Chondrostereum purpureum:* An alternative to chemical herbicide brush control. Seventh International Symposium on Environmental Concerns in Rights-of-Way-Management **2002**, 665–672.
45. Becker, E.M.; Ball, L.A.; Hintz, W.E. PCR-based genetic markers for detection and infection frequency analysis of the biocontrol fungus *Chondrostereum purpureum* on Sitka alder and Trembling aspen. Biol. Control **1999**, *15*, 71–80.
46. Shaheen, I.Y.; Abu-Dieyeh, M.H.; Ash, G.J.; Watson, A.K. Physiological characterization of the dandelion bioherbicide, *Sclerotinia minor* IMI 344141. Biocontrol Sci. Technol. **2010**, *20*, 57–76.
47. Li, P.; Ash, G.J.; Ahn, B.; Watson, A.K. Development of strain specific molecular markers for the *Sclerotinia minor* bioherbicide strain IMI 344141. Biocontrol Sci. Technol. **2010**, *20*, 939–959.
48. Abu-Dieyeh, M.H.; Watson, A.K. Efficacy of *Sclerotinia minor* for dandelion control: Effect of dandelion accession, age and grass competition. Weed Res. **2007**, *47*, 63–72.
49. Abu-Dieyeh, M.H.; Watson, A.K. Effect of turfgrass mowing height on biocontrol of dandelion with *Sclerotinia minor*. Biocontrol Sci. Technol. **2006**, *16*, 509–524.
50. Abu-Dieyeh, M.H.; Watson, A.K. The significance of competition: Suppression of *Taraxacum officinale* populations by *Sclerotinia minor* and grass overseeding. Can. J. Plant Sci. **2006**, *86*, 1416–1416.
51. Wilson, M.J.; Glen, D.M.; Hughes, L.A.; Pearce, J.D.; Rodgers, P.B. Laboratory tests of the potential of entomopathogenic nematodes for the control of field slugs (*Deroceras reticulatum*). J. Invertebr. Pathol. **1994**, *64*, 182–187.
52. Coupland, J.B. Susceptibility of helicid snails to isolates of the nematode *Phasmarhabditis hermaphrodita* from southern France. J. Invertebr. Pathol. **1995**, *66*, 207–208.
53. Li, Y.D.; Yang, J.M. The study on effect of microbe and microbial pesticides killing *Oncomelania hupensis*. Acta Hydro. Sin. **2005**, *29*, 203–205.
54. Singh, A.; Singh, D.K.; Kushwaha, V.B. Alginates as binding matrix for bio-molluscicides against harmful snails *Lymnaea acuminata*. J. Appl. Polym. Sci. **2007**, *105*, 1275–1279.
55. Kumar, P.; Singh, D.K. Molluscicidal activity of *Ferula asafoetida*, *Syzygium aromaticum* and *Carum carvi* and their active components against the snail *Lymnaea acuminata*. Chemosphere **2006**, *63*, 1568–1574.
56. Zhang, Y.; Ke, W.S.; Yang, J.L.; Ma, A.N.; Yu, Z.S. The toxic activities of *Arisaema erubescens* and *Nerium indicum* mixed with *Streptomycete* against snails. Environ. Toxicol. Pharmacol. **2009**, *27*, 283–286.
57. Auld, B.A.; Morin, L. Constraints in the development of bioherbicides. Weed Technol. **1995**, *9*, 638–652.
58. Wang, C.S.; St Leger, R.J. A scorpion neurotoxin increases the potency of a fungal insecticide. Nat. Biotechnol. **2007**, *25*, 1455–1456.
59. Ash, G.J. Biological control of weeds with mycoherbicides in the age of genomics. Pest Technol. Rev. **2010**, *in press*.
60. St Leger, R.J.; Joshi, L.; Bidochka, M.J.; Roberts, D.W. Construction of an improved mycoinsecticide overexpressing a toxic protease. Proc. Natl. Acad. Sci. U. S. A. **1996**, *93*, 6349–6354.

Bioremediation

Ragini Gothalwal
Institute of Microbiology and Biotechnology, Barkatullah University, Bhopal, India

Abstract
Bioremediation is a key area of white biotechnology because the elimination of a wide range of pollutants from water and soil is an absolute requirement for sustainable development. Bioremediation makes a significant contribution in remediation efforts and in decontamination of industrial effluents where conventional physico-chemical treatment methods may be neither adequate nor feasible. The different types of bioremediation include bacterial remediation, mycoremediation, rhizoremediation, phycoremediation, and phytoremediation. Phytoremediation techniques use "hyperaccumulator plants," which store 10–500 pollutants in their leaves and stem which are thereafter harvested and incinerated, and metals recovered from ashes can be reused in metallurgy. Biodegradation of chemical warfare agents is carried out using the organophosphorus-insecticide-hydrolyzing enzyme, organophosphate hydrolase. Heap soil washing technology using leaching microbes is receiving much attention these days for the remediation of large volumes of heavy metals and radioactive elements in contaminated soil. Successful bioremediation requires not only the knowledge of which microorganism degrades a particular compound but also an understanding of the pathways involved in degradation both at the physiological and molecular levels. The exploration and exploitation of the untrapped genetic diversity in the field of bioremediation is in its infancy and needs a lot more investigation. Furthermore, systematic efforts are required to determine the effects of bioremediation and phytoremediation on the food chain and natural recycling, as well as for making the process safer, economical, and ecofriendly.

INTRODUCTION

After the industrial revolution, a significant number of industries were set up, which released organic compounds in the environment. These chemical compounds (pesticides, fertilizers, plastic, dyes, and heavy metals) are present in the form of metal, non-metal, metalloid, inorganic compounds, and organic compounds. Organic compounds are divided into aliphatic, alicyclic, aromatic, polyaromatic, and mixed types. These compounds have a tendency to transform into another compound and reach the ecosystem. Owing to the inadequacy and increased cost of physical and chemical methods available, the need for ecofriendly and cheaper methods has been strongly realized. Bioremediation has the potential to lower the cost of remediation by orders of magnitude over alternative technologies, and governmental agencies and the public as taxpayers are interested in cost-effective cleanup technologies.

Biodegradation and bioremediation are matching processes, to the extent that they are both based on the conversion or metabolism of pollutants by microorganisms. The difference between these two is that biodegradation is a natural process whereas bioremediation is a technology. Bioremediation requires an efficient bacterial strain that can degrade the largest pollutant to a minimum level. Microbial diversity offers an immense field of environmental friendly options for mineralization of contaminants or their transformations into less harmful or non-hazardous compounds. Molecular biology methods are now being employed to study bioremediation. Conjugative gene transfer occurs, through plasmid-borne catabolic genes, between bacteria in oil and the competitive, indigenous bacterial population. Therefore, it is important to understand the role of catabolic genes, by molecular cloning and characterization, in the degradation of a particular organic compound, so that it can be applied for bioremediation. This review emphasizes the distribution and extent of environmental contaminants from the ecosystem. It also accentuates the current status of research in the area of biodegradation and bioremediation and presents the new approaches available for microbial and phyto tracking in the environment.

PRINCIPLE OF BIOREMEDIATION

The understanding of the bioremediation process is derived from the combination of biochemical and microbiological processes,[1,2] summarized as follows:

1. The relationship of comparative biochemistry applies equally to the axenic culture of microorganisms and those in field soil.
2. Microbial growth requirements are the same whether in laboratory culture or in the field.
3. Limitations resulting from ecological interactions include the need to accommodate the presence of other microbes as well as adaptation to the physical and chemical properties of the microsite wherein the microbes live.

Encyclopedia of Environmental Management DOI: 10.1081/E-EEM-120046187

4. A microbe amended in a soil ecosystem not only should pass the requisite genetic information and be capable of expressing that capability *in situ*, but it must also have the capacity to become a part of the overall soil microbial community.[1] Such microbes are referred to as competent rhizospheric bacteria with intrinsic bioremediation potential.

Targets of bioremediation must be based not only in terms of structure but also in terms of the matrix containing the target. Thus, the applicability of bioremediation can be considered for each of the environmental status of matter: 1) solid (soil, sediment, sludge); 2) liquid (ground water, industrial wastewater); 3) gas (industrial air emission); and 4) subsurface environment (saturated and vadose zones). The general approaches to bioremediation (Fig. 1) are as follows:

1. Bioaugmentation;
2. Biostimulation (environmental modification, through the nutrient application process);
3. Phytoremediation (addition of microbes).

The biological community for bioremediation generally consists of the natural soil microflora. However, higher plants can also be manipulated to enhance toxicant removal (phytoremediation), especially for remediation of metal-contaminated soils.[3,4]

Microbial Degradation of Organic Pollutants

The soil microbial diversity is regarded as a major factor responsible for many biochemical transformations, including degradation of diverse organic compounds.[5–7] The chemical structures of organic pollutants exert profound influence on the metabolic abilities of microbes. Thus, harvesting the microbiological activities for biodegradation of recalcitrant hazardous chemicals and its implementation forms the basis for the increased use of biotreatment systems (Table 1). Predominantly, noxious chemicals, such as polychlorinated biphenyls (PCBs), trichloroethylene (TCE), polycyclic aromatic hydrocarbons (PAHs), and pesticide residue in soils and water are the targets. Removal of unwanted residues, as well as pesticide efficacy, is ultimately dependent on the presence, number, and enzymatic capability of soil microbes. Mineralization of these compounds takes place only when environmental condition, water activity, presence of O_2, temperature, and pH are suitable for the growth and survival of the organism (Table 2). Degradation fails when the target compound is either very concentrated or much diluted. Successful biological cleanup of soil and water contaminants takes place in the presence of aerobic microorganisms. A dichlorinating anaerobic microbial population can also be helpful in cleanup processes. Genes for complete mineralization of some of the haloaromatic compounds are also reported and effectively utilized for recruitment of microbes and environmental applications. Degradation of naphthalene is more difficult than that of bionuclear compounds such as biphenyl, dibenzofuran, and dibenzo-*p*-dioxin, and of mononuclear aromatic compounds such as aniline, benzene, salicylate, phenoxyacetate, and toluene. Peripheral or funneling, central degradative and oxoadipate pathway sequences are necessary for the complete degradation of the compounds.[8] Pentachlorophenol (PCP) and PCB are

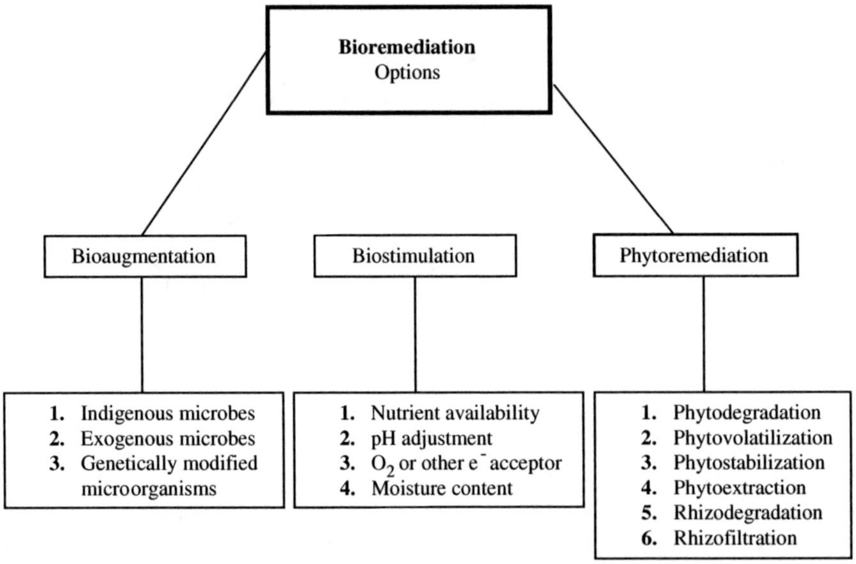
Bioremediation Options
Bioaugmentation
Biostimulation
Phytoremediation
1. Indigenous microbes
2. Exogenous microbes
3. Genetically modified microorganisms
1. Nutrient availability
2. pH adjustment
3. O2 or other e⁻ acceptor
4. Moisture content
1. Phytodegradation
2. Phytovolatilization
3. Phytostabilization
4. Phytoextraction
5. Rhizodegradation
6. Rhizofiltration

Fig. 1 Bioremediation approaches for environmental cleanup.
Source: Musarrat and Zaidi.[1]

Table 1 Examples of bacteria, actinomycetes, fungi, and cyanobacteria engaged in pesticide degradation.

Organisms	Organic compounds or pesticides
Bacteria and actinomycetes	
Alcaligenes denitrificans	Fluoranthene (PAH)
Alcaligenes faecalis	Arylacetonitriles
Archromobacter	Carbofuran
Arthrobacter	EPTC, glyphosate, pentachlorophenol (PCP)
Bacillus sphaericus	Urea herbicides
Brevibacterium oxydans IH35A	Cyclohexylamine
Burkholderia sp. P514	1,2,4,5-TeCB
Clostridium	Quinoline, glyphosate
Comamonas testosteroni	Arylacetonitriles
Corynebacterium nitrophilous	Acetonitrile, carboxylic acid, ketones
Dehalococcoides ethenogenes 195	Trichloroethylene (TCE)
Desulfitobacterium dehalogenans	Hydroxylated PCBs
Desulfovibrio sp.	Nitroaromatic compound
Flavobacterium	PCP
Geobacter sp.	Aromatic compound
Klebsiella pneumoniae	3 and 4 Hydroxybenzoate
Methylococcus capsulatus (Bath)	Trichloroethylene
Methylosinus trichosporium OB 3b	1,1,1-Trichloroethane (TCA)
Moraxella	Quinoline, glyphosate
Nitrosomonas europaea	TCA
Nocardia	Quinoline, glyphosate
Pseudomonas aeruginosa	Nitriles, biphenyl, parathion
Pseudomonas sp.	Quinoline, glyphosate
Pseudomonas stutzeri	Parathion
Pseudomonas cepacia	2,4,5-T
Pseudomonas paucimobilis	PCP
Pseudomonas putida 6786	Propane
Pseudomonas striata	Propham, chloropham
Rhodococcus chlorophenolicus	PCP
Rhodococcus corallinus	*S*-triazines
Rhodococcus rhodochrous	Propane
Rhodococcus sp.	Propane, TCA
Rhodococcus UM1	Pyrene
Fungi	
Aspergillus flavus	DDT
Aspergillus parasiticus	DDT
Aspergillus niger	2,4-D
Candida tropicalis	Phenol
Chrysosporium lignorum	3,4-Dichloroaniline
Fusarium solani	Acylamilide
Fusarium oxysporum	DDT

Table 1 (*Continued*)

Organisms	Organic compounds or pesticides
Hendersonula toruloidea	2,4-D
Hydrogenomonas + *Fusarium* sp.	DDM, nitrile
Mucor alternans	DDT
Penicillium	Acylamilide
Penicillium megasporum	2,4-D
Phallinus weirii	DDT
Phanerochaete chrysosporium	PAH, 2,4,6-trinitrotoluene, PCP, DDT, 2,4,5-T, lindane
Pleurotus ostreatus	DDT
Polyporus versicolor	DDT
Pullularia	Acylamilide
Rhodotorula	Benzaldehyde
Stereum hirsutum	Phenanthrene
Trametes versicolor	Dieldrin
Trichoderma sp.	Nitrile
Trichoderma viride	DDT
Trichosporon cutaneum	Phenol
Yeast	Paraquat
Ectomycorrhizal fungi	
Tylospora fibrillosa	Mefluidide
Thelephora terrestris	
Suillus variegatus	
Suillus granulatus	
Suillus luteus	
Hymenoscyphus ericae	
Paxillus involutus	
Cyanobacteria	
Cylindrospermum sp. *Aulosira fertilissima* *Plectonema boryanum*	BHC lindane diazinon, endrin
Nostoc muscorum *Wollea bhardlvajae*	Carbofuran hexachlorocyclohexane (HCH)
Nostoc musorum	HCH
Mastigocladus laminosus *Tolypothrix tenuis*	Tolkan fluchloralin
Anabaena doliolum, *Nostoc muscorum*	Butachlor
Anabaena ARM 286 *Anabaena* ARM 310	BHC Ekalux
Anabaena variabilis *Aulosira fertilissima* *Scytonema chiastum* *Scytonema stuposum*	Bavistin eenlate captan dithane cyathion

(*continued*)

Table 1 Examples of bacteria, actinomycetes, fungi, and cyanobacteria engaged in pesticide degradation (*Continued*).

Organisms	Organic compounds or pesticides
Anabaena khannae	Butachlor
Calothrix marchica	benthiocarb
Nostoc calcicola	pandimethalin
Tolypothrix limbata	oxadiazon
Aulosira fertilissima ARM 68 *Nostoc muscorum* ARM 221	Monocrotophos malathion, dichlorovos, phosphomidon, quinolphos

Source: Data from Adhikary[9] and Gothalwal and Bisen.[10]

major recalcitrant compounds. The major problems of PCP degradation are the formation of toxic end products during the metabolism and substitutions of chlorine.

Organochlorine pesticides (DDT, endosulfan, hexachlorobenzene, and hexachlorocyclohexane) are degraded by microbes by means of reductive dechlorination, dehydrochlorination, oxidation, and isomerization of the parent molecule. The principal reactions involved in the breakdown of phosphotriesters (organophosphate insecticides) are hydrolysis, oxidation, alkylation, and dealkylation.[11] Microbial degradation through hydrolysis of p-O-alkyl and p-O-aryl bonds is considered to be the most significant step in the degradation of parathions, methyl parathion, and p-nitrophenol. Reductive dechlorination of organochlorine is an important microbial reaction. A classic example of this is the conversion of DDT to DDD; this reaction occurs in several species of bacteria such as *Pseudomonas*, *Bacillus*, *Arthrobacter*, *Clostridium*; in soil actinomycetes; in yeasts; and in fungi such as *Trichoderma viridae*, *Mucor alterans*, white rot fungi, *Pleurotus australis*, *Phellinus weirii*, and *Polyporus versicolor* (Table 3). The formation of DDE from DDT through dehydrochlorination is commonly observed in algae, diatoms, and phytoplankton (*Cylindrotrea dentorium*, *Cyclotella nana*, *Isochryeii gabana*, *Nitzschia* spp.). Microbial isomerization reactions involve the conversion of γ-BHC to α-BHC, diedrin to photodieldrin, and D-keto andrin to endrin by *Pseudomonas putida*.[12]

Lindane is very persistent in the environment and resistant to microbial degradation.[13] It is degraded aerobically as well as anaerobically. *Pseudomonas paucimobilis* UT26 is a unique microbe that utilizes hexachlorocyclohexane (HCH) as its sole source of carbon and energy under aerobic conditions.[14] Five structural genes (*lin A*, *lin B*, *lin C*, *lin D*, and *lin E*) and one regulatory gene (*lin R*) are involved in degradation of α-HCH in UT 26. *lin A*, *lin B*, *lin C*, and *lin D* codes for dehydrochlorinase, halidohydralone, dehydrogenase, and reductive dehalogenase, respectively, and *lin E* encodes ring cleavage oxygenase. Microbial degradation of γ-HCH has also been reported in *Anabaena* sp. PCC 7120 and *Nostoc ellipsosorum*.[16] The enzyme responsible for catalyzing the hydrolysis step in parathion is organophosphate hydrolase (OPH), which is encoded by the *opd* gene. This gene has been isolated from several bacteria. A naturally occurring variant of OPH enzyme, designated as opdA, capable of degrading a broad range of organophosphates was isolated from an *Agrobacterium radiobacter* strain.[17] The slow growth and low culture yields of native OPH-producing strains make them uneconomical for practical use. Therefore, efforts have been made by several researches to improve the applicability of OPH for pesticide bioremediation. A consortium comprising two bacteria that were genetically engineered (*E. coli* and *P. putida* KT 2440) efficiently worked together to break down parathion and prevent accumulation of p-nitrophenol [PNP].[18]

Table 2 Major factors affecting bioremediation.

Microbial
- Growth until critical biomass is reached
- Mutation and horizontal gene transfer
- Enrichment of the capable microbial populations
- Production of toxic metabolites

Environmental
- Depletion of preferential substrates
- Lack of nutrients
- Inhibitory environmental conditions

Substrate
- Too low concentration of contaminants
- Chemical structure of contaminants
- Toxicity of contaminants
- Solubility of contaminants

Biological aerobic vs. anaerobic process
- Oxidation/reduction potential
- Availability of electron acceptors
- Microbial population present in the site

Growth substrate vs. co-metabolism
- Type of contaminants
- Concentration
- Alternate carbon source present
- Microbial interaction (competition, succession, and predation)

Physico-chemical bioavailability of pollutants
- Equilibrium sorption
- Irreversible sorption
- Incorporation into humic matter

Mass transfer limitations
- Oxygen diffusion and solubility
- Diffusion of nutrients
- Solubility/miscibility in/with water

Source: Data from Boopathy.[15]

Table 3 Mechanisms of radionuclide bioremediation.

S.N.	Mechanism	Microorganisms	Remediated radio
1.	Biosorption	*Rhizopus arrhizus*, *S. cerevisae*, *Penicillium americium*, *Aspergillus*, *Aeromonas hydrophila*, *Candida utilis*	Uptake
2.	Engineering biosorption	Eukaryotic metallothien, *E. coli* Lan B	
3.	Bioaccumulation	*Citrobacter* sp.	$UO_2{}^{2+}$
		Rhodococcus erythropali CS98	Cesium
		Rhodococcus sp. strain CS 402	Cesium
		Micrococcus luteus	Neptunium
		Radioresistant bacteria	
		Deinococcus radiodurans	U (Vi), Tc (Vii)
		D. geothermalis	Radioactive waste

Catechol is a terminal metabolite formed during the degradation pathways of various compounds, and a variety of potential degraders of catechol have been reported by Kim et al.[19] Biodegradation of phenol and toluene by *Pseudomonas* sp., *Bacillus* sp., and *Staphylococcus* sp. was studied by Prasanna et al.[20] These strains were isolated from pharmaceutical industrial effluents. Mixed cultures showed more efficient degradation than pure strains within 5–7 hr at lower concentrations. Polycyclic aromatic hydrocarbon-degrading bacteria and ligninolytic and non-ligninolytic fungi are ubiquitously distributed in the natural environment such as soils and woody materials. The principal mechanism for aerobic bacterial metabolism of PAH is the initial oxidation of the benzene ring by the action of dioxygenase enzyme to form *cis*-dihydrodiol intermediates, which can then be further metabolized via catechols to CO_2 and H_2O.[21] Many bacterial, fungal, and algal strains have been shown to degrade a wide variety of PAHs (Table 1). There are limited reports on degradation of high molecular weight PAHs with more than four benzene rings. In general, high molecular weight PAHs are degraded slowly by indigenous or augmented microorganisms, as the persistence of PAHs increases with their molecular size.

The recalcitrant nature of chloroaromatic haloalkanes is due to the low electron density at the aromatic ring which makes the enzyme oxygenase unable to attack this compound. Many soil microorganisms (*Pseudomonas* and *Alcaligenes*) which synthesize the halogenase can utilize Halogenated alkonic acids [HAA].[22] Anaerobic methane-oxidizing bacteria can degrade TCE in pure *Pseudomonas* culture through a co-metabolic process.[23] The PCP-degrading *Pseudomonas* sp. strain IST 103 has been isolated, which was found to be capable of utilizing PCP as a carbon source. The enzyme PCP-4 mono-oxygenase was found to be responsible for the dechlorination of PCP.[24] The gene for the degradation may be plasmid encoded or present on the chromosome.[25,26] Attempts have been made to enhance PCB biodegradation by modifying oxygenase.[27] One of the most efficient methods of biodegradation consists of sequential anaerobic and aerobic treatments for highly chlorinated compounds. Biochemical and genetic engineering approaches for dehalogenase and oxygenase could lead to "super bugs" that could be used for the bioremediation of chlorinated compounds.[28] Modified degradative genes could be introduced into the original strain and/or major indigenous strains isolated from contaminated sites, and it is hoped that these super bugs could have application in bioremediation in the near future, confirming their usefulness and safety. Raji et al.[29] have isolated a bacterial culture able to grow on benzoate and useful for remediation of PCB-contaminated sites. *Arthrobacter* sp. IFL YN 10 demonstrated mineralization of C^{14} ring-labeled atrazine. This isolates can be used to develop a consortium for bioremediation of pesticides.[30]

To improve the biodegradation efficiency and implementation, integrating various components such a microbial strain in consortium, solid O_2 source, and appropriate role of nutrients with controlled release pattern into a granule formulation with an oleophilic matrix, may provide an ideal approach to improve remediation of crude oil pollutants.[31] Abed et al.[32] reported that salinity and temperature are important environmental parameters that influence the degradation process of petroleum compounds. The inhibitory effect of salinity was shown to be more pronounced for aromatic than for aliphatic compounds.[33] Higher temperatures also reduce the viscosity of crude oil, which increases its diffusion through sediments, a process that render oil components accessible to bacteria. The possibility of the involvement of catabolic plasmid in the degradation of anthracene by *Pseudomonas* sp. isolated from an oil filling station was investigated by Kumar et al.[34] Many γ and β proteobacterial groups (*Halophaga*, *Geothrix*, *Acidobacterium*) and green non-sulfur bacteria with a strong potential to degrade hydrocarbons were present in benthic cyanobacterial mats.[35] The aliphatic fraction of petroleum hydrocarbon[36] is degraded by *Arthrobacter*,

Alcaligenes, *Flavobacterium*, and *Bacillus*. Kniemeyer et al.[37] came across a green *Methanospirillium* that is able to degrade aromatic hydrocarbons. Singh and Lin[38] isolated 10 indigenous microorganisms from oil-containing soil; five isolates achieved 86.94% diesel degradation in 2 weeks. The results strongly indicate that the environmental condition of the contaminated site plays a crucial role in the degradation. Cohen[39] reported the development of cyanobacterial mats in oil-contaminated courts. Cyanobacterial polysaccharides play a major role in the emulsification of oil, actually breaking the oil into small droplets, which are subsequently attacked by the heterotrophs. Bioremediations of high fat and oil wastewater by selected lipase-producing bacteria such as *Bacillus subtilis*, *B. lichenformis*, *B. amyloliquifaciens*, *Serratia marcescens*, *P. aeruginosa*, and *Staphylococcus aureus* were carried out in wastewater from palm oil mill, dairy, slaughterhouse, soap industry, and domestic wastewater. After 12 days of bioremediation, the least biological oxygen demand and lipid content was observed in consortia, and the lipid degradation capacity of *P. aeruginosa* was higher than that of other bacteria.[40] Verma et al.[41] proved the biotechnological importance and advantage of using *P. aeruginosa* SL72 and *Acinetobacter* sp. SL-3 individually or as a consortium for waste treatment, resulting in substantial removal of the crude oil within a week, using a low-cost, efficient, and environment-friendly technique.

Spent wash is dark brown due to the recalcitrant melanoidin pigment. *Pseudomonas* sp. was selected for degradation of the pigment by Chavan et al.[42] Chuphal and Thakur[43] have characterized an alkalophilic bacterial consortium (*Micrococcus luteus*, *Deinococcus radiothilus*, *Micrococcus diversue*, *P. syringae P. myricure*) for *ex situ* bioremediation of color and adsorbable organic halogens in pulp and paper mill effluent. Nanda et al.[44] employed *Nostoc* sp. for bioremediation of tannery effluents; the main economic advantage of this system is the lack of a serious sludge disposal problem, consequently resulting in a much cheaper operating cost. No microorganism has been found to degrade polythene without an additive such as starch. The discovery of new enzymes and the cloning of genes for synthetic polymer–degrading enzyme from *Pseudomonas* sp. were reviewed by Premraj and Doble.[45]

Chemotaxis has been postulated to play an important role in enhancing biodegradation as it increases the bioavailability of pollutants to bacteria. Some toxic organic compounds are chemoattractants for different bacterial species, which can lead to improved biodegradation of these compounds. A *Ralstonia* sp. was chemotactic toward different Nitro aeromatic compounds (NACs), i.e., p-nitrophenol (PNP), p-nitrobenzoate (PNB), and o-nitrobenzene (ONB).[46]

Mycoremediation

The key to mycoremediation is determining the right fungal species to target specific pollutants. Certain strains have been reported to successfully degrade the nerve gases VX and sarin. Battelle in a plot of soil contaminated with diesel oil was inoculated with mycelia of oyster mushrooms; within weeks, more than 95% of PAH had been reduced to non-toxic components. Mycofiltration is a similar process using fungal mycelia to filter toxic waste and microorganisms from waste in soil. Breakdown (70%–100%) of anthracene oil found in PAH was reported in 27 days in an N_2-limited culture of *Phaenerochaete chrysosporium*. Pentachlorophenol is an important constituent of paper mill effluents, and *Phaenochaste chrysorperiucm* immobilized on rotating biological contactor disk efficiently degrades 2,4-dichlorophenol, 2,4,6-trichlorophenol (TCP), polychlorinated quiacole, and several chlorinated vanillins.[47]

Bioremediation of Inorganic Contaminants

Microbes encounter metals such as Cr, Mn, Fe, Co, Ni, Cu, Zn, Ni, Ag, Cd, Pb, and Au, and metalloids such as As, Se, and Sb, having a diverse nature, in the environment. Microbes can detoxify metals by valence transformation, extracellular chemical precipitation, or volatilization. Such microbes combat high concentrations of heavy metals by the following processes:

1. Inactivation of metals;
2. Alteration of the site of inhibition;
3. Enhancement in impermeability of metals; and
4. Other by-pass mechanisms.[48]

Bacterial biomass can also bioaccumulate heavy metals both in live and dead states through intracellular accumulation and extracellular absorption, respectively, giving an effective alternative for small-scale remediation purposes.[49] Bacteria can remediate heavy metals by a variety of mechanisms, including bioaccumulation, biosorption, and bioremediation (Fig. 2). Bioaccumulation is the retention and concentration of a substance by an organism through the cell membrane into the cytoplasm, where the metal is separated and immobilized. However, in biosorption, the negatively charged metal ions are separated through adsorption to the negative ionic groups on the cell surface (carboxyl residue, phosphate residue, SH groups, or hydroxyl group) such as capsule or slim layers. The charged functional groups[50] serve as nucleation sites for the deposition of various metal-bearing precipitates.[51] *Bacillus* SJ-101 exhibits a much higher capacity of intracellular Ni accumulation, which is attributed to the anionic nature of its cell surface. In bioremediation, biologically catalyzed redox reactions lead to immobilization of metals. Microorganisms are known to reduce a wide variety of multivalent metals that pose environmental problems.[52,53] The reduced species are highly insoluble and precipitate out from solution.

Bioremediation is one such promising option that harnesses the impressive capabilities of microbes associated with roots to degrade organic pollutants and transform toxic metals. Since it is a plant-based *in situ* phytorestoration technique, it is proven to be economicaly efficient and easy to implement under field conditions.[54] All plant growth-promoting rhizobacterial strains (*Azotobacter chroococcum, Bacillus megatorium, B. mucilaginosus, B. subtilis* SJ-101, *Pseudomonas* sp., *P. fluorescens, Rhizobium leguminosarum, Kluyvera ascorbata* SUD165)[55] can be used for bioremediation of metals. Rhizobacteria associated with hyperaccumulators (*B. subtilis, B. pumilus, P. pseudoalcaligens, and Brevibacterium halotolerans*) are also widely used in bio- and rhizoremediation of multimetal-contaminated sites.[56] The rhizobacteria are used or manipulated with three main objectives for bioremediation of metal-contaminated soils: 1) hyperaccumulation of metals in plants; 2) reduction of the uptake of metals; and 3) *in situ* stabilization of the metals as organocomplexes (Fig. 2). The chemical conditions of the rhizosphere differ from bulk soil as a consequence of various processes induced by plant roots as well as by rhizobacteria,[57] such as secretion of organic acids followed by reduction in pH and production of siderophores, phytochelains, amino acid, and ACC deaminase. *Pseudomonas maltophilia* was shown to reduce the mobile and toxic Cr (VI) to non-toxic and immobile Cr (III) and also minimize the environmental mobility of other toxic ions (Hg, Pb, Cd).[58] Al Ayely et al[59] studied the effect of increasing levels of As and P on fern injected with mycorrhiza. The greater diversity of plant species may be responsible in part for the greater bacterial diversity in the bulk soils.

Redox Reaction Leading to Immobilization

The direct enzymatic reduction of soluble uranium (Vi), Cr (Vi), and Tc (Vii) to insoluble species is well documented.[60,61] A number of Cr (Vi)-reducing microbial strains, including *Oscillatoria* sp., *Arthrobacter* sp., *Agrobacter* sp., *Pseudomonas ambigua, Chlamydomonas* sp., *Chlorella vulgaris, Zoogloea ramigera, P. aeruginosa,*[62] and anaerobic sulfate-reducing bacteria have been isolated from chromate contaminated soil water and sediment.[63]

Redox Reaction Leading to Solubilization

Solubilization of adsorbed and co-precipitated metals may occur by direct or indirect microbial processes. The solubilization of toxic heavy metals and radionuclide from coprecipitates requires at least partial solubilization of oxide minerals.[64] Presumably, the organic acids formed by the metabolic activity of microbes lowers the pH of the system so that it interferes with the electrostatic forces that hold heavy metals and radionuclides on the surface of iron or Mn oxide minerals. Extracellular polymeric substances serve as biosorbing agents by accumulating nutrients from the surrounding environment, and also play a crucial role in biosorption of heavy metals. Being polyanionic in nature, exopolysaccharide forms complexes with metal cations, resulting in metal immobilization with the exopolymeric matrix.[65] Biofilm formation is a strategy that microorganisms might use to survive a toxic flux in these inorganic compounds; biofilm populations are protected from toxic metals by the combined action of chemical, physical, and physiological phenomena, which are in some instances linked to phenotypic variations among the constituent biofilm cells. Harrison et al.[67] have prepared a multifunctional model by which a biofilm population can withstand metal toxicity by the process of cellular diversification.

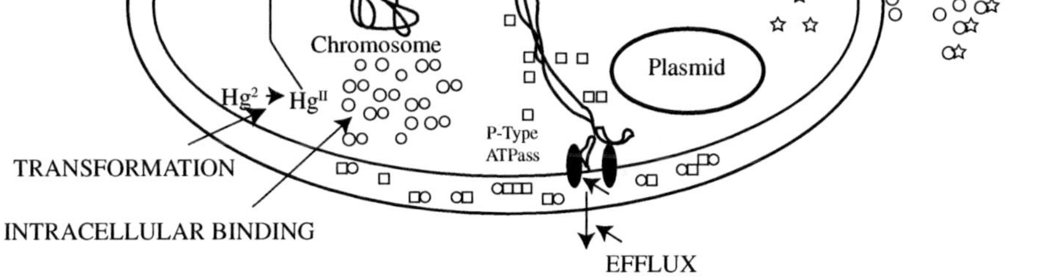
REDUCED UPTAKE
EXTRACELLULAR BINDING
Periplasmic Space
Chromosome
Plasmid
Hg2 → HgII
P-Type ATPass
TRANSFORMATION
INTRACELLULAR BINDING
EFFLUX

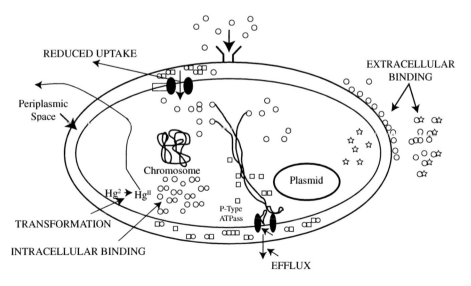

Fig. 2 Different metal resistance mechanisms in bacteria.
Source: Singh and Srivastava.[66]

An effective ecofriendly approach for the removal of Mn from e-waste by the fungus *Helminthosporium solani* was shown by Savitha et al.[68] through the process of non-metabolism-dependent biosorption under different environmental conditions of metal concentration, pH, and dry biomass concentration. "Metallothionein" has become a generic term applied to low molecular weight proteins or polypeptides that bind metal ions in metal thiolate clusters, and whose synthesis increases in response to elevated concentrations of certain metals.[69] The maximum cobalt removal efficacy (1 μg of ^{60}Co/g, dry wt) of bacterial mass (*B. megaterium*, *P. putida*, *Flavobacteriuam devorans*, *Salmonella typhimurium*, *Streptomyces gresius*, *Rhizopus* sp., *Rhodococcus* sp., *E. coli*) could be achieved within 6 hr[70] compared with the 8–500 ng/g attained after 24-48 hr. The gum kondagogu, a natural carbohydrate polymer, was investigated[71] for its adsorptive removal of the toxic metal ions Cd^{2+}, Cu^{2+}, Fe^{2+}, Pb^{2+}, Ni^{2+}, Zn^{2+}, and Hg^{2+} present in industrial effluents. Kondogogu has a potential to be used as an effective, non-toxic, economical, and efficient biosorbent cleanup matrix for the removal of toxic ions with re-adsorption capacity at 90% level even after three cycles of desorption. *Klebsiella oxytoca* was able to biodegrade cyanide to a non-toxic end product (ammonia) using cyanide as the sole N_2 source, which might preceed using ammonia as an assimilatory substrate.[72]

Microorganisms interact with radionuclides via several mechanisms (Table 3), some of which are used as the basis of potential bioremediation strategies. Based on the different types of effluents generated, a single technology cannot be suitable to address the problems. Cyanobacteria have the capacity to utilize nitrogenous compounds as well as phosphates; in addition, they pick up metal ions such as Cr, Co, Cu, and Zn very effectively. It has been observed that immobilized cyanobacteria have greater potential than their free cells. Various natural polymers such as alginate agar and carrageenan, and synthetic polymers such polyacrylamide and polyurethane have already been tried.[73] Novel methods of immobilization, including co-immobilizations of various species, are required for the symbiotic interaction among themselves, which will result in synergistic enhancement or removal capabilities. A gene cluster composed of nine open reading frames involved in Ni^{2+}, Co^{2+}, and Zn^{2+} sensing and tolerance in *Synechocystis* sp. PCC 6803 has been identified by Mario Garcia et al.[74] The biosorption of Hg^{2+} by *Spirulina platensis* and *Aphanothece flocculosa* was studied under a batch stirred reaction system,[75] and more than 90% from *A. flocculosa* and 100% mercury recovery from *S. platensis* can be achieved for 4 and 1 cycle, respectively. Protein and total non-protein thiols were measured as stress-responsive metabolites in response to Ni in *Anabaena doliolum*.[76] As *A. doliolum* is a high-biomass-producing strain, it can be conveniently separated from the solution by filtration and can be used at pilot-scale removal of Ni from wastewater.

Biological – Cadmium

Phytoremediation: A Beneficial Alternative

Phytoremediation is a promising green technology for accelerated decontamination of soil and water. It is a natural process in which plants are used to remove pollutants (pesticides, solvents, explosives, crude oil, PAHs) from soil and water. Plants can degrade or transform both the organic and metal (landfill leachates and radionuclide) contaminants by acting as filters or traps. Plants and associated microbes can degrade the pollutants or at least limit their spread in the environment. There are several ways in which plants can be used for the phytoremediation of organic contaminants viz. phytoextraction, phytodegradation, rhizodegradation, and phytovolatization (Fig. 3). They can be compared to solar driven pumps for extraction and concentration of elements from the environment.[4] The plants can also be used for "phytomining." Rugh et al.[77] inserted an altered mercuric ion reductive gene (*mer A*) into *Arabidopsis thaliana*, creating a transgenic plant that volatilizes mercury to the atmosphere. Phytoremediation depends on a variety of factors, including physical conditions, chemical properties of the contaminant, and relative tolerance of the plant to the contaminant. Morphological characteristics of plant uptake are controlled by soil factors such as clay content, organic matter, soil moisture, and pH. The plant root can alter the pH in the rhizosphere by secreting organic acids, thus effecting the bioavailability of certain compounds, i.e., ionic pollutants whose solubility and desorption from soil colloids are pH dependent. Certain plants also produce biosurfactants that may increase the solubility of more lipophilic compounds.

Rugh[78] reported laboratory model plants such as *Thale cress* and tobacco to enhance phytoremediation of organomercurials, TCE, and nitroaromatic explosives that have been engineered with a non-plant transgene. Hyperaccumulator plants are also grown along with non-hyperaccumulators to enhance the heavy metal uptake by intermingling the roots and induce the colonization of efficient rhizobacteria. Recombinant *Mesorhizobium huakii*, by incorporating the phytochelatinase gene from *Arabidopsis thaliana* into *M. huakii* subsp. *rengei* 133, increased Cd accumulation by 1.5-fold in *Astragalus sinicus*.[79] The phytoremediation potential of water hyacinth (*Eichhornia crassipes*) in the treatment of tannery effluents was evaluated by Athaullah et al.[80] who found a promising risk reduction, cost-effective technology for water sanitation and conservation.

BIOREMEDIATION TECHNOLOGY

Bioremediation technology using microorganisms was invented by George M. Robinson in 1960. Bioremediation technologies can be generally classified as *in situ* or (microbial ecological approach) or *ex situ* (microbial approach) (Fig. 3); some examples of these technologies

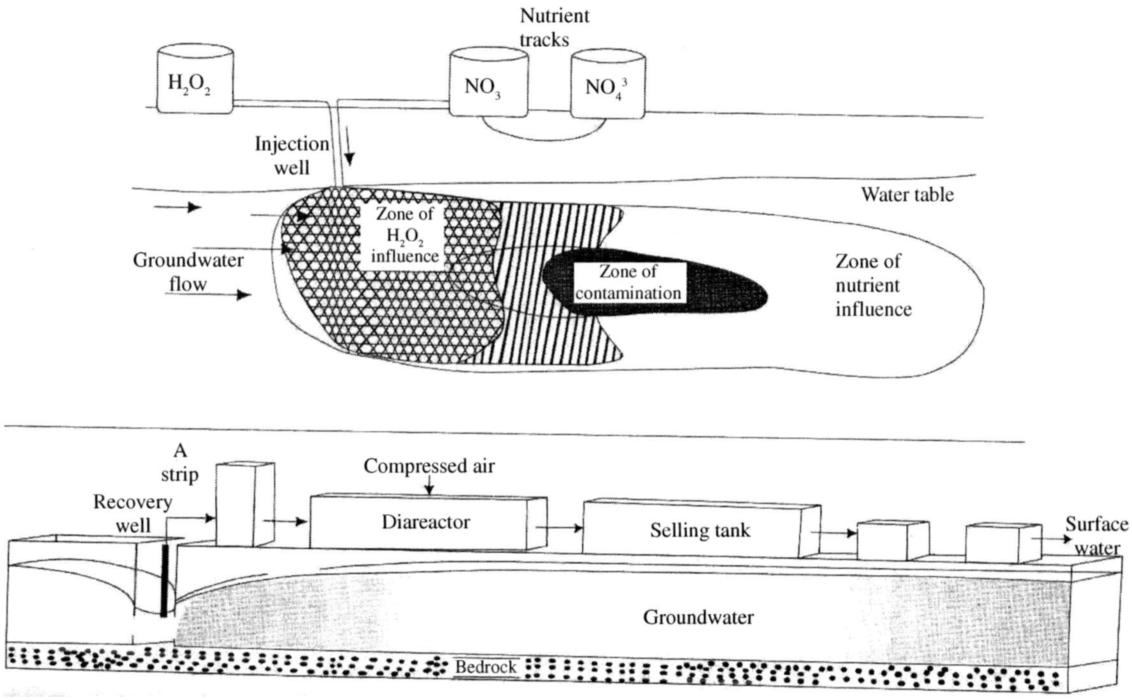
Nutrient
tracks
H_2O_2
NO_3
NO_4^3
Injection
well
Water table
Zone of
H_2O_2
influence
Groundwater
flow
Zone of
contamination
Zone of
nutrient
influence
A
strip
Compressed air
Recovery
well
Diareactor
Selling tank
Surface
water
Groundwater
Bedrock

Fig. 3 *In situ* bioremediation processes of bioaugmentation and biostimulation.
Source: Thakur.[81]

include bioventing, land farming, bioreactor, composting, bioaugmentation, rhizofiltration, biostimulation, and solid phase bioremediation. In the development of the technology, the following points should be considered:

1. Heterogeneity of the contaminant
2. Concentration of the contaminant and its effect on the biodegradative microbe
3. Persistence and toxicity of the contaminant
4. Behaviour of the contaminant in soil
5. Conditions favorable for the biodegradative microbe or microbial population

Bioremediation is one of the most promising technologies for treating military sites, industrial wastes, municipal/urban wastes, mining waste, chemical spills and hazardous wastes, etc. *In situ* bioremediation can be implemented in many treatment modes, including aerobic, anoxic, anaerobic, and co-metabolic. The aerobic mode has proved to be most effective in reducing contaminant levels of aliphatic and aromatic compounds. Biofiltration is best suited for airstream containing volatile organic compounds.

Pollution remediation of tannery effluents is very complex. A multiprong treatment is thus required for a combination of nanotechnology and microbial technology, with a prior proceeding such as cycloning, flotation, microflotation, or electroflotation.

Bioremediation can be done on site, is often cheap, with minimal site disruption, eliminates waste permanently, eliminates long-term liability, has greater public acceptance, and can be coupled with other physical or chemical methods.[15] Table 2 shows the major factors affecting bioremediation. Another technique employed for bioremediation involves bioaugmentation, i.e., pumping genetically engineered microorganisms or microorganisms with enhanced degradation ability into the subsurface. Timely bioremediation of petroleum-contaminated soils is possible with innovative engineering and environmental manipulation to enhance microbial activity beyond the natural effective season.[82]

The first field slurry process used to remediate the nitroaromatic herbicide dinoterb used SABRE (sequential anaerobic bioremediation process), which was renamed as the FAST (fermentative anaerobic soil treatment) process.[83] Hence, the optimized remediation strategies exploring microbial diversity are being executed *in situ* successfully.

The aim is to translate research findings from the laboratory into viable technologies for remediation in the field. In the injection method, bacteria and nutrients are injected directly into the contaminated aquifer, or nutrient or enzymes are often referred fertilizer, that stimulate the activity of the bacteria are added. Bioreactors using immobilized cells have several advantages over conventional effluent-treatment technologies. Degradation of 4-chlorophenol by anaerobes attached to granular activated carbon in a biofilm reactor was evaluated during both open and closed modes of operation. Continuous flow fluidized bed reactor and bench scale continuous flow activated sludge reactor were used to study the removal of TCP, TeCP and PCP[84] which are used as sole source of carbon and energy. The ability of *Arthrobacter* cells to degrade PCP was evaluated for immobilized, non-immobilized, and co-immobilized cells. Fixed film bioreactor has been used with mixed bacterial communities for the treatment of pulp and paper mill effluent. The effluent after treatment showed a removal of color (80%), Chemical oxygen demand (71%), and chlorinated organic compound (68%).[85] The fungal slurry was successfully applied for compost preparation and biomediation of the Cr-contaminated tannery soil. The treated effluent was used for seed germination of crops without any phytotropic influence.[86]

Biocolloid formation methods have been adapted for bioremediation of metals by bacteria and fungi without the need

Table 4 *In situ* ground water biodegradation examples of process and companies.

S.N.	Name of company	Technology used	Types of environmental pollutant remediated
1.	Bio-genesis Technology	Custom-blended microbial culture (GT-1000 series)	Oil, BTEX, diesel fuel
2.	Biopim	Biological sand filter	BTEX, TPH, phenol monochlorobenzene, metals
3.	ENSR Consulting and Engineering Technology (remediation beneath building foundation)	Steam injection, soil vapor extraction, ground water extraction, and air stripping	TCA, TCE, DWAPLS
4.	Petro Clean Bioremediation System	Indigenous microorganisms	Gasoline, diesel fuel, aviation fuel, solvents, PNAs, VOCs, and other organic compounds
5.	Kemron Environmental Services	Engineered, site-specific, groundwater recirculation system	Petroleum products solvents, halogenated volatiles and semivolatiles, BTEX, polynuclear aromatics, and organic acids
6.	Remediation Technologies Inc.	Treating groundwater in saturated zone injection, recovery well, monitoring wells	Dissolved contaminants
7.	SBP Technologies	Encapsulated cell inserted into a well	PAH (high mol. wt.), chlorinated aromatics (PCP), and pesticides
8.	OHM Remediation Services Corporation	Aquifer	Petroleum hydrocarbons, BTEX, chlorinated and non-chlorinated solvents
9.	Electrokinetics Inc. (no need to add microorganism)	Electro osmosis or electro chemical migration	TCE, BTEX, PAH
10.	Geo Microbial Technologies Inc.	Removal of H_2S anaerobic process	—
11.	EODT Services Inc.	Use of biodispersant (also same as high-energy nutrients for microorganisms)	—
12.	Ecology Technologies International Inc.	Use of FyreZyme (multifactorial liquid agents)	Organic contaminants
13.	Gaia Resource Inc.	—	Hazardous and radioactive waste
14.	IT Corporation	Pump and treat system	Industrial effluents (hazardous organic compounds
15.	Ground Water Tech Inc.	Destructive technology	Hazardous compounds
16.	Yellow Stone Environmental Science Inc.	Pump and treat, aerobic processes, denitrification, sulfate reduction	Aromatic hydrocarbon, halogenated hydrocarbon, VOC, BTEX, phenol, cresol, CCl_4, PCE, vinyl chloride
17.	Waste Stream Tech. Inc.	Bioaugmentation	Organic compounds
18.	Micro Bac International Inc.	Batch and continuous feed treatment using M-1000 microbial consortium	Host specific
19.	Kuzanci Environmental Techniques (Fig. 4)[87]	Microlife DCB series	Petroleum derivatives, stops foul odor

Source: Data from Bioremediation 1999.[88]

MICROLIFE BIOREMEDIATION MICROBIAL PRODUCTS:
Before
After
Before
After

Fig. 4 Treatment of soil hydrocarbon-polluted soils by Microlife DCB series bioremediation products. The progress is clearly seen in the photos.
Source: http://www.microbial-products.com/microbial-bacterial products bioremedation.asp/geli (accessed December 2011).[87]

ex situ treatment. Electrokinetically enhanced *in situ* soil decontamination and dispersing by chemical reaction together with microbes. Chlorophenol-contaminated saw mill soil used composting without bioaugmentation in a cheap and feasible method.[89] The use of biphenyl as an *in situ* co-substrate is expensive and environmentally problematic,[90] thus there was a need to investigate the ability of alternative co-substrates to support the co-metabolic degradation to PCBs.

Biocapsules have been tested for various applications and can be produced for site-specific applications (Bioremediation Applied Bioscience http://www.bioprocess.com). *In situ* groundwater biodegradation in the United States has been carried out through various processes by numerous companies, as shown in Table 4. The removal effectiveness can reach 100% (Fig. 4).

MOLECULAR PROBES IN BIOREMEDIATION

The traditional method of bacterial enumeration is often insufficient for monitoring the specific microbe's biochemical reaction in mixed microbial communities. It has became apparent that a significant number of microbes in these systems are viable but non-culturable. The catabolic enzymes, genes, and proteins expressed in the microbes can be exploited for the detection of the fate and effect of microbes in the bioremediation process.[91] Antibody- and fatty acid-based probes, nucleic acid sequences, and DNA probes can be used to detect genes in the bacterial genome or on plasmid, or to detect mRNA or tRNA. Other relevant techniques employed are PCR, repetitive sequence-based PCR, 16S rDNA, random amplified polymorphic DNA, and fluorescence *in situ* hybridization.[81] In a previous study, a specific synthetic PCR-amplifiable DNA fragment was introduced into a *Pseudomonas* chromosome to allow genetically engineered microbes to be identified easily.

The development of a new field of metabolic engineering involves the improvement of cellular activities by manipulation of enzymatic, transport, and regulatory function of the cell by using recombinant DNA technology. Advances in the field of genetic engineering, sequencing of the whole genome of several organisms, and developments in bioinformatics have speed up the process of gene cloning and transformation. Furthermore, many powerful analytical techniques have been developed for metabolic pathway analysis and analysis of cellular functions, such as gas chromatography (GC), gas chromatography-mass spectrometry (GC-MS), nuclear magnetic resonance, 2D gel electrophoresis, matrix-assisted laser desorption/ionization time of flight (MALDI-TOF), liquid chromatography-mass spectrometry (LC-MS), and DNA chips. Metabolic engineering is, therefore, an effort to improve the ability of microorganisms. Bioremediation require the integration of huge amount of data from different sources. Pazos et al.[92] developed "Meta Router," a system for maintaining heterogeneous information related to bioremediation in a framework that allows its query, administration, and mining. The system can be accessed and administered through a web interface for studying and representing the global properties of the bioremediation network. Bioinformatics require the study of microbial genomics, proteomics, systems biology, computational biology, phylogenetic trees, and data mining, and the application of major bioinformatics tools for determining the structure and biodegradation pathway of xenobiotic compounds. Bioinformatics has taken on a new glittering by entering the field of bioremediation.[93] The limitations of bioremediation has paved the way for the development of Genetically engineered microorganisms (GEMs), or designer biocatalysts harboring artificially designed catabolic pathways.[94] Database such as the University of Minnesota Biocatalysts/Biodegradation database provide a scope for *in silico* designing of biocatalysts for in vivo construction followed by *in situ* application. In the era of functional genomics, it is easy to construct GEMs by reshuffling the gene(s), promoter, etc., to enhance their performance in situ.

CONCLUSION

The popularity of bioremediation is further enhanced because it is perceived as being more "green" than other remediation technologies. As a result, bioremediation companies have a viable future regardless of the long-term effectiveness of the process. Special emphasis is required on the exploitation of biotechnological innovations to improve presently available biocatalysts, and for the evaluation of future effects of microorganisms and their proper application in the optimization of *in situ* bioremediation. The use of enzymes for degradation of pesticides can be developed as a technology for bioremediation. A super strain can be created to achieve the required result in a short time frame. One important characteristic of this technology is that it is carried out in a non-sterile open environment, which contains a host of microbes. Therefore, a strategy should be tailored in such a manner that due consideration be given to the various environmental constraints (type and amount of pollutant, climatic condition, hydrogeodynamics) that affect a particular location. Feasibility studies are essential and can have an enormous impact on the cost of full-scale remediation. Rhizoremediation can contribute to the restoration of polluted sites. Phytoremediation will require an integration of activities by plant scientists, microbiologists, chemists, and engineers, so that these systems that can be used to prevent and remediate pollution can become a reality. Environmental friendly processed need to be developed to clean up the environment without creating harmful waste products. For the development of economically usable technologies, scientists and technologists would have to offer creative solutions for either introducing new capabilities or enhancing current efficiencies.

REFERENCES

1. Musarrat, J.; Zaidi, S. Bioremediation of agrochemicals and heavy metals in soil. In *Biotechnological Applications of Microorganisms. A Techno-Commercial Approach*; Maheshwari, D.K., Dubey, R.C., Eds.; I.K. International Publishing House, Pvt. Ltd.: New Delhi, India, 2006; 311–331.
2. Tate, R.L. *Soil Microbiology*; John Wiley & Sons: New York, 2000; 464–494.
3. Salt, D.E.; Smith, R.D.; Raskin, I. Phytoremediation. Annu. Rev. Plant Physiol. Plant Mol. Biol. **1998**, *49*, 643–668.
4. Alkorta, I.; Garbisu, C. Biores. Technol. **2001**, *77*, 229–236.
5. Spain, J.C. *Biodegradation of Nitro Aromatics Compounds*; Plenum Press: New York, 1995.
6. Stoner, D.L. *Biotechnology for the Treatment of Hazardous Waste*; Lewis Publisher: Boca Raton, FL, 1994.
7. Unterman, R. A history of PCB biodegradation. In *Bioremediation Principle and Application*; Crawford, R.L., Crawford, D.L., Eds.; University Press, Cambridge, U.K., 1996; 209–253.
8. Timmis, K.N.; Steffen, R.J.; Unterman, R. Degrading microorganisms for the treatment of toxic waste. Annu. Rev. Microbiol. **1994**, *48*, 527–557.
9. Adhikary, S.P. General introduction. In *Blue Green Algae Survival Strategies in Extreme Environment*; Printer Publication: Jaipur, India, **2006**; 7–17.
10. Gothalwal, R.; Bisen, P.S. Bioremediation. In *Encyclopedia of Pest Management*; Pimentel, D., Ed.; Marcel Dekker Pub. Ltd.: New York, 2002; 89–93.
11. Singh, N.; Asthana, R.K.; Kayastha, A.M.; Pandey, S.; Chaudhary, A.K.; Singh, S.P. Thiol: An exopolysaccharide production in a cyano bacterium under heavy metal stress. Process. Biochem. **1999**, *35*, 63–68.
12. Benezet, H.J.; Matsumura, F. Isomerization of γ-BHC to α-BHC in the environment. Nature, **1973**, *243*, 480–481.
13. Agnihotri, N.P.; Gajbhiye, V.T.; Kumar. M.; Mahapatra, S.P. Environ. Manage. Assess. **1994**, *30*, 105–112.
14. Imai, R.; Nagata, Y.; Fukuda, M.; Takagi, M.; Yano, K. Molecular cloning of *Pseudomonas paucimobilis* gene encoding a 17-kilodalton polypeptide that eliminates HCL molecules from γ-hexachlorodihexane. J. Bacteriol. **1989**, *173*, 6811–6819.
15. Boopathy, R. Factor limiting bioremediation technologies. Bioresour. Technol. **2000**, *74*, 63–67.
16. Kuritz, T.; Wolk, P. Use of filamentous cyanobacteria for biodegradation of organic pollutants. Appl. Environ. Microbiol. **1995**, *61*, 234–238.
17. Horne, I.; Sutherland, T.D.; Harcourt, R.L.; Rusell, R.J.; Oakes Hott, J.G. Identification of an *opd* (organophosphate degradation) gene in an *Agrobacterium* isolate. Appl. Environ. Microbiol. **2002**, *68*, 3371–3376.
18. Gilbert, E.S.; Walker, A.W.; Keasling, J.D. A constructed microbial consortium for biodegradation of the organophosphorus insecticide parathion. Appl. Microbiol. Biotechnol. **2003**, *61*, 77–81.
19. Kim, K.P.; Lee, J.S.; Park, S.I.; Rhee, M.S.; Kim, C.K. Isolation and identification of *Klebsiella oxytoca* C302 and its degradation of aromatic hydrocarbons. Korean J Microbiol. **2000**, *36*, 58–63.
20. Prasanna, N.; Sarvanan, N.; Geetha, P.; Shanmugaprakash, M.; Rajasekardan, P. Biodegradation of phenol and toluene by *Bacillus* sp., *Pseudomonas* sp. and *Staphylococcus* sp. isolated from pharmaceutical industrial effluent. Adv. Biotech. **2008**, *7*, 20–24.
21. Mueller, J.G.; Cerniglia, C.E.; Pritchard, P.H. Bioremediation of environments contaminated by polycyclic aromatic hydrocarbons. In *Bioremediation: Principle and Applications*; Crawford, R.L., Crawford, D.L., Eds.; Cambridge Univ. Press: Idaho, 1996; 125–194.

22. Hardman, D.J.; Gowland, P.C.; Slater, J.H. Large plasmids from soil bacteria enriched on halogenated alkonic acids. Appl. Environ. Microbiol. **1986**, *51*, 44–51.
23. Little, C.D.; Palumbo, A.U.; Herbes, S.E.; Lidstrom, M.E.; Tyndall, R.L.; Gilmer, P.J. Trichloroethylene biodegradation by a methane-oxidizing bacterium. Appl. Environ. Microbiol. **1988**, *54*, 951–956.
24. Thakur, I.S.; Verma, P.K.; Upadhyaya. K.C. Molecular cloning and characterization of pentachlorophenol-degrading monooxygenase genes of *Pseudomonas* sp. from the chemostat. Biochem. Biophys. Res. Commun. **2002**, *290*, 770–774.
25. Khan, A.; Tewari, R.; Walia, A. Molecular cloning of 3-phenyl catechol dioxygenase involved in the catabolic pathway of chlorinated biphenyl from *Pseudomonas putida* and its expression in *E. coli.* Appl. Environ. Microbiol. **1988**, *54*, 2664–2671.
26. Pritchard, P.H. Fate of pollutants. J. Water Pollut. Control Fed. **1986**, *58*, 635–645.
27. Furukawa, K. Engineering dioxygenase for efficient degradation of environmental pollutants. Curr. Opin. Biotechnol. **2000**, *11*, 244–249.
28. Furakawa, K. 'Superbugs' for bioremediation. Trends Biotechnol. **2003**, *21*, 187–190.
29. Raji, S.; Mitra, S.; Sumathi, S. Dechlorination of chlorobenzoates by an isolated bacterial culture. Curr. Sci. **2007**, *93*, 1126–1129.
30. Sagarkar, S.; Nouriainen, A.; Bijorklot, K.; Purohit, H.J.; Jargensen, R.S.; Kapley, A. Bioremediation of atrazine in agricultural soil. In *52nd Annual Conference of AMI.* International Conference of Microbial Biotechnology for Sustainable Development, Nov 3–6, 2011, Chandigarh, 140.
31. Wang, Q.; Zhang, S.; Li, Y.; Klassen, W. Potential approaches to improving biodegradation of hydrocarbon for bioremediation of crude oil pollution. J. Environ. Protect. **2011**, 2, 47–55.
32. Abed, R.M.M.; Thukain, A.A.; de Beer, D. Bacterial diversity of a cyanobacterial mat degrading petroleum compounds at elevated salinities and temperatures. FEMS Microbiol. Ecol. **2006**, *57*, 290–301.
33. Milli, G.; Almallah, M.; Bianchi, M.; Wambeke, F.V.; Bertrand, J.C. Effect of salinity on petroleum biodegradation. Fresenius J. Anal. Chem. **1991**, *339*, 788–791.
34. Kumar, G.; Singla, R.; Kumar, R. Plasmid associated anthracene degradation by *Pseudomonas* sp. isolated from filling station site. Nat. Sci. **2010**, *8*, 89–94.
35. Margesin, R.; Labbe, D.; Schinner, F.; Green, C.W.; Whyte, L.G. Characterization of hydrocarbon degrading microbial population in contaminated and pristine alpine soils. Appl. Environ. Microbiol. **2003**, *69*, 3985–3092.
36. Mishra, S.; Jyot, J.; Kuhad, R.C.; Lal, B. Evaluation of inoculum addition to stimulate in situ bioremediation of oily sludge contaminated soil. Appl. Environ. Microbiol. **2001**, *67*, 1675–1681.
37. Kniemeyer, O.; Fischer, T.; Wilkes, H.; Glockner, F.O.; Widdel, F. Anaerobic degradation of ethylbenzene by a new type of marine sulfate—Reducing bacterium. Appl. Environ. Microbiol. **2003**, *69*, 760–768.
38. Singh, C.; Lin, J. Isolation and characterization of diesel oil degrading indigenous microorganisms in Kwazulu-Natal, South Africa. Afr. J. Biotechnol. **2008**, *7*, 1927–1932.
39. Cohen, Y. Bioremediation of oil by marine microbial mats. Int. Microbiol. **2002**, *5*, 189–193.
40. Prasad, M.P.; Manjunath, K. Comparative study on biodegradation of lipid rich waste water using lipase producing bacterial species. Indian J. Biotechnol. **2011**, *10*, 121–124.
41. Verma, S.; Lata; Saxena, J.; Sharma, V. Bioremediation of crude oil contaminated waste using mixed consortium of biosurfactant and lipase producing strains. In *52 Annual Conference of AMI*; Int. Conf. on Microbial Biotechnology for Sustainable Development, Nov 3–6, 2011, Chandigarh, 48.
42. Chavan, M.N.; Kulkarni, M.V.; Zope, V.P.; Mahulikar, P.P. Microbial degradation of melanoidine in distillery spent wash by an indigenous isolate. Indian J. Biotechnol. **2006**, *5*, 416–421.
43. Chuphal, Y.; Thakur, I.S. Characterization of alkalophilic bacterial consortium for *ex-situ* bioremediation of color and adsorbable organic halogen in pulp and paper mill effluent. In *Microbial Diversity Current Perspective and Potential Application*; Satyanarayan, T., Johri, B.N., Eds.; I.K. International Pvt. Ltd.: New Delhi, India, 2005; 573–584.
44. Nanda, S.; Sarangi, P.K.; Abraham, J. Cyanobacterial remediation of industrial effluents. N. Y. Sci. J. **2010**, *3*, 32–36.
45. Premraj, R.; Doble, M. Biodegradation of polymers. Indian J. Biotechnol. **2005**, *4*, 186–193.
46. Pandey, G.; Chauhan, A.; Samanta, S.K.; Jain; R.K. Chemotoxin of a *Ralstonia* sp. SJ98 toward co-metabolizable nitroaromatic compounds. Biochem. Biophys. Res. Commun. **2002**, *299*, 404–409.
47. Arora, D.S.; Chander, M. Biotechnological application of white rot fungi in biodegradation of various pollutants. In *The Environment in Biotechnological Applications of Microbes*; Verma, A., Podila, K.I.K., Eds.; International Pvt. Ltd.: New Delhi, India, 2007; 262–280.
48. Belliveau, B.H.; Trevors, J.T. Mercury resistance and detoxification in bacteria. Appl. Organometal. Chem. **1989**, *3*, 283–294.
49. Chang, J.S.; Hang, J. Biotechnol. Bioeng. **1944**, *44*, 999–1006.
50. Beveridge, T.J. Biotechnol. Bioeng. Sym. N. Y. **1986**, *52*, 127–140.
51. Volseky, B.; Chang, K.H. Biotechnol. Bioeng. **1995**, *42* , 451–460.
52. Wildung, R.E.; Gorby, Y.A.; Krupka, K.M.; Hess, N.J.; Li, S.W.; Plymale, A.E.; McKinloj, J.P.; Fredrickson, J.K. Effect of electron donor and solution chemistry on products of dissimilarity reduction of technetium by *Shewanella putrefaciens*. Appl. Environ. Microbiol. **2000**, *66*, 2451–2460.
53. Weilingo, B.; Mizuba, M.M.; Hansel, C.M.; Fendrof, S. Environ Sci. Technol. **2006**, *34*, 522–527.
54. Kamaludeen, S.P.B.; Ramasamy, K. Rhizoremediation of metals: Harnessing microbial communities. Indian J. Microbiol. **2008**, *40*, 80–88.
55. Zhuang, X.; Chen, J.; Shim, H.; Bai, Z. New advances in plant growth promoting rhizobacteria for bioremediation. Environ. Int. **2007**, *33*, 406–413.
56. Abou-Shanab, R.A.; Ghanem, K.; Ghanem, N.; Al-Kalaibe, A. The role of bacteria on heavy metal extraction and uptake by plants growing on multi-metal-contaminated soil. World J. Microbiol. Biotechnol. **2008**, *24*, 253–262.
57. Marschner, H. *Mineral Nutrition of Higher Plants*; Academic Press: London, 1995; 889.

58. Blake, R.C.; Choate, D.M.; Bardhan, S.; Revis. N.; Barton, L.L; Zocco, T.G. Chemical transformation of toxic metals by a *Pseudomonas* strain from a toxic waste site. Environ. Toxicol. Chem. **1993**, *12*, 1365–1376.
59. Al Ayely, A.; Sylvia, D.M.; Ma, L. Mycorrhiza increase arsenic uptake by the hyper accumulator Chinese brake fern (*Pleris vittata* L.). J. Environ. Qual. **2005**, *34*, 2181–2186.
60. Henrot, J. Health Phys. **1989**, *57*, 239–245.
61. Sabaty, M.; Avazeri, C.; Pignol, D.; Vermeiglio, A. Characterisation of the reduction of selenate and tellurite by nitrate reductases. Appl. Eviron. Microbiol. **2001**, *67*, 5122–5126.
62. Chatterjee, S.; Ghosh, I.; Mukherjea, K.K. Uptake and removal of toxic Cr (V1) by *Pseudomonas aeruginosa*: Physico-chemical and biological evaluation. Curr. Sci. **2011**, *101*, 645–652.
63. Mabbett, A.N.; Lloyd, J.R.; Macaskie, L.E. Effect of complexing agents on reduction of Cr(VI) by *Desulfovibrio vulgaris* ATCC 29579. Biotechnol. Bioengg. **2002**, 79, 389–397.
64. Lovely, D.R.; Phillips, E.J.P. Bioremediation of uranium contamination with enzymatic uranium reduction. Environ. Sci. Technol. **1992**, *26*, 2228–2234.
65. Pal, A.; Paul, A.K. Microbial extracellular polymeric substances; central elements in heavy metal bioremediation. Indian J. Microbiol. **2008**, *48*, 49–64.
66. Singh, S; Srivastava, S. Rhizobia and its role in rhizoremediation. In *Microbial Diversity: Current Perspective and Potential Applications.* Satynarayana, T., Johri, B.N., Eds.; I.K. International Pvt. Ltd., New Delhi, India, 2005; 655–676.
67. Harrison, J.J.; Ceri, H.; Turner, R.J. Multimetal resistance and tolerance in microbial biolfilms. Nature **2007**, *5*, 928–938.
68. Savitha, J.; Sahana, N.; Prakash, V.K. Metal biosorption by *Helminthosporium solani*—A simple microbiological technique to remove metal from e-waste. Curr. Sci. **2010**, *98*, 903–904.
69. Kojima, Y. Definitions and nomenclature of metallothioneins. Methods Enzymol. **1991**, *205*, 8–10.
70. Rashmi, K.; Haritha, A.; Balaji, V.; Tripathi, V.S.; Venkateswaran, G.; Maruthi Mohan, P. Bioremediation of ^{60}Co from simulated spent decontamination solution of nuclear power reactors by bacteria. Curr. Sci. **2007**, *92*, 1407–1409.
71. Vinod, V.T.P.; Sashidhar, R.B. Bioremediation of industrial toxic metals with gum kondagogu (*Cochlospermum gassypium*): A natural carbohydrate biopolymer. Indian J. Biotechnol. **2011**, *10*, 113–120.
72. Kao, C.M.; Liu, J.K.; Lou, H.R.; Lin, C.S.; Chen, S.C. Biotransformation of cyanide to methane and ammonia by *Klebsiella oxytoca*. Chemosphere **2003**, *50*, 1055–1061.
73. Prakasham, R.S.; Ramakrishna, S.V. The role of cyanobacterium in effluent treatment. J. Sci. Ind. Res. **1998**, *57*, 258–265.
74. Mario Garcia, D.; Luis, L.M.; Francisco, J.F.; Jose, C.R. A gene cluster involved in metal homeostasis in the cyanobacterium *Synechocystis* sp. strain PCC 6803. J. Bacteriol. **2000**, *182*, 1507–1514.
75. Cain, A.; Vannela, R.; Keith Woo, L. Cyanobacteria as a biosorbent for mercuric ion. Bioresour. Technol. **2008**, *99*, 6578–6586.
76. Shukla, M.K.; Tripathi, R.D.; Sharma, N.; Dwivedi, S.; Mishra, S.; Singh, R.; Shukla, O.P.; Rai, U.N. Responses of cyanobacteria *Anabaena doliolum* during nickel stress. J. Environ. Biol. **2009**, *30*, 871–876.
77. Rugh, C.L.; Senecoff, J.F.; Meagher, R.B.; Merkle, S.A. Development of transgenetic yellow poplar for mercury phytoremediation. Nat. Biotechnol. **1998**, *16*, 925–928.
78. Rugh, C.L. Genetically engineered phytoremediation: One man's trash in another man's transgene. Trends Biotechnol. **2004**, *22*, 496–498.
79. Sriprang, R.; Hayashi, M.; Ono, H.; Takayi, M.; Hirata, K.; Murooka, Y. Enhanced accumulation of Cd (2) by a *Mesorhizobium* sp. transformed with a gene from *Arabidopsis thaliana* coding for phytochelation synthase. Appl. Environ. Microbiol. **2003**, *69*, 1791–1796.
80. Athaullah, A.; Asrarsheriff, M.; Sultan Mohideen, A.K. Phytoremediation of tannery effluent using *Eichhornia crassipes* (Mart.) Solms. Adv. Bio Tech. **2011**, *3*, 10–12.
81. Thakur, I.S. Microbial bioremediation of pollutant chlorinated in the environment. In *Biotechnological Application of Microbes*; Verma, A., Podila, K.I.K., Eds.; International Pvt. Ltd.: New Delhi, India, 2007; 239–261.
82. Filler, D.M.; Lindstrom, J.E.; Braddock, J.F.; Johnson, R.A.; Nickalashi, R. Integral biopile components for successful bioremediation in the Arctic. Cold Reg. Sci. Technol. **2001**, *32*, 143–156.
83. Crawford, R.L. The microbiology and treatment of Nitro aromatic compounds. Curr. Opin. Biotechnol. **1995**, *6*, 329–336.
84. Thakur, I.S.; Verma, P.; Upadhyaya, K. Involvement of plasmid in degradation of pentachlorophenol by *Pseudomonas* sp. from a chemostat. Biochem. Biophys. Res. Commun **2001**, *286*, 109–113.
85. Thakur, I.S. Screening and identification of microbial strains for removal of colour and adsorbable organic halogen in pulp and paper mill effluent. Process Biochem. **2004**, *39*, 1693–1699.
86. Shah, S.; Thakur, I.S. Enrichment and characterization of a microbial community from tannery effluent for degradation of pentachlorophenol. World J. Microbiol. Biotechnol. **2002**, *18*, 693–698.
87. Available at http://www.microbial-products.com/microbial-bacterial-products-bioremedation.asp (accessed December 2011).
88. Bioremediation 1999, available at http://www.inweh.unu.edu/447/lectures/bioremediation.htm (accessed December 2011).
89. Hamada, M.F; Haddad, A.I.; Abd-E-L-Bury, M. Treatment of phenolic wastes in an aerated submerged fixed-film (ASFF) bioreactor. J. Biotechnol. **1987**, *5*, 279–292.
90. Lajoie, C.A.; Zylstra, G.J.; Deflaun, M.F.; Strom, P.F. Development of field application vector for bioremediation of soil contaminated with polychlorinated biphenyls. Appl. Environ. Microbiol. **1993**, *59*, 1735–1741.
91. Thakur, I.S. Structural and functional characterization of a stable, 4-chlorosalicyclic acid degrading bacterial community in a chemostat. World J. Microbiol. Biotechnol. **1995**, *119*, 643–645.
92. Pazos, F.; Gurjas, D.; Valencia, A.; Lorenzo, V.D. Meta Router: Bioinformatics for bioremediation. Nucleic Acid Res. **2005**, *33*, D.588–D592.
93. Fluekar, M.H.; Sharma, J. Bioinformatics applied in bioremediation. Innov. Rom. Food Biotechnol. **2008**, 2, 28–36.
94. Urbance, J.W.; Coli, J.; Saxena, P.; Tiedje, J.M. Nucleic Acid Res. **2003**, *31*, 152–155.

Bioremediation: Contaminated Soil Restoration

Sven Erik Jørgensen
Institute A, Section of Environmental Chemistry, Copenhagen University, Copenhagen, Denmark

Abstract
Restoration of contaminated soil is made possible by ecotechnological methods. Organic toxic substances can be removed by plants or by microbiological decomposition. It is possible in many cases to obtain a higher level of removal efficiency by the use of adapted microorganisms. The removal processes are dependent on the bioavailability of the toxic organic compound. The bioavailability is dependent on a number of factors such as water solubility, sorption to the soil of the compound, physical structure of the soil (pore size), and, as already mentioned, adaptation of the microorganisms to the toxic compounds. Heavy metals can be removed by plants that are better able to take up heavy metal the higher their protein content is. Models for predicting the uptake of heavy metals by plants have been developed.

INTRODUCTION

Contaminated soil is an increasing problem in industrialized countries, and the high cost of remediation has driven the interest in the direction of ecological engineering applications of bioremediation technologies. They apply biological processes, mainly microorganisms or plants. It is often possible to solve the pollution problem satisfactorily by this ecotechnologically based methodology without the hazard and expense involved in removing polluted materials for treatment elsewhere for the use of traditional environmental technological methods.

Ecotechnological bioremediation may be applied for both organic waste and heavy metals, although the methods applied in practice may differ and they are therefore treated below in two different sections, on organic compounds and heavy metals. The success of any bioremediation technology depends on a number of factors including site characteristics, environmental factors such as temperature, pH, redox potential, concentrations of nutrients, the contaminant, the presence of microorganisms, and bioavailability. It is therefore necessary to look into these factors to comprehend the applicability of these methods.

BIOAVAILABILITY OF TOXIC ORGANIC COMPOUNDS

Bioavailability is a crucial factor for the application of bioremediation. It is defined as the amount of contaminant present that can be readily taken up by organisms. The bioavailability controls the biodegradation rates for organic contaminants because microbial cells must expend energy to induce the catabolic processes used in biodegradation. If the contaminant concentration is too low, induction will not occur. Soil microbial populations are typically slow-growing organisms and often exposed to nutrient-poor environments.[1] Bioavailability also determines the toxicity of both organic and inorganic contaminants to organisms other than those applied for bioremediation. There is therefore an increased need for bioremediation when bioavailability is high, which fortunately makes bioremediation more attractive. Three cases can be envisioned that would result from different bioavailabilities of contaminants:[2]

1. Biodegradation will not occur because the concentration of the bioavailable contaminant is insufficient and/or the biodegradation rate for the contaminant is too low to justify the energy expenditure to induce biodegradation.
2. Microbial cells may degrade the contaminant at low bioavailable concentrations and/or low biodegradation rates, but in a resting or maintenance stage rather than in a growing stage.
3. At a sufficient bioavailability and biodegradation rate, there is enough bioavailable contaminant to induce biodegradation in a growing stage. That will allow for optimal rates of remediation.

The biodegradability of organic contaminants is highly dependent on the physical and chemical structure[3] of the contaminant and the soil. The section on "Biodegradation" will discuss this topic. A coarse but still applicable rule for a very first estimation of the biodegradability of organic compounds is given in the entry entitled "Biodegradation." Moreover, the software EEP (Estimation of Ecotoxicological Parameters) is able to give some first estimation of biodegradability.

The bioavailability of heavy metals is also a significant factor for the applicability of bioremediation. Heavy metals are of course not degraded but removed, mainly by plant uptake. The uptake by organisms of heavy

Encyclopedia of Environmental Management DOI: 10.1081/E-EEM-120048393

metals, which determines the overall removal efficiency, is entirely controlled by the bioavailable amount of heavy metals. An ecological model presented in the section on *Uptake of Heavy Metals by Plants* will illustrate the strong dependence of the bioavailability of heavy metals.[4]

Bioavailability is influenced by a number of factors:

1. Low water solubility
2. Sorption on the solid phase of soil
3. Physical makeup of the soil (pore size distribution)
4. Microbial adaptations

Low water solubility can limit availability of the substrate to bacterial cells and hence constrain biodegradation.[5,6] Microbial cells are 70%–90% water, and the food they utilize comes from the water surrounding the cells. Plants take up water to cover the evapotranspiration needed for the maintenance of their life functions. Therefore, uptake and transport are only feasible for water-soluble material. If first-order biodegradation kinetics is presumed, the biodegradation rate becomes proportional to the concentration in the water phase. It means for components with low water solubility that they are biodegraded very slowly. There are clear relationships between the water solubility of an organic compound and the chemical structure that can be utilized to estimate the water solubility.[3] EEP (a software containing many equations to estimate ecotoxicological parameters) and other estimation equations utilize these relationships to make estimation of the water solubility and of K_{ow}. Side reactions may change the water solubility. This is of particular interest for heavy metal ions, which can increase the solubility by the formation of complexes either with organic or with inorganic compounds. The formation of complexes with humic acid and fluvic acid plays a major role for the solubility of metal ions in soil water. Hydrocarbons that are frequently found as soil contaminant have a low water solubility: 2–6 μg/L for pentacyclic aromatic hydrocarbons and *n*-alkanes of chain length 18–30. The solubility decreases with increasing molecular weight.[3]

The state of the contaminant in combination with the water solubility is also of importance. There is evidence that liquid-phase hydrocarbons are more bioavailable than solid-phase hydrocarbons.[7] In practical terms, this means that the maximum growth rate occurs in different solubility ranges for liquid-phase (0.01–1 mg/L) and solid-phase (1–10 mg/L) components. Degradation can be described by a Michaelis–Menten expression. Water solubility increases with increasing temperature and usually an Arrhenius expression can be applied with the temperature coefficient 1.06 or 1.07.

Many authors have found that surfactants increase mineralization rate due to increased dissolution. Also, surfactants may provide an additional carbon source, which is preferentially utilized by the bacteria. There may, however, also be a negative effect by surfactants due to their toxicity to the bacterial population.

Sorption on the solid phase of soil may be a limiting factor for biodegradation of microorganisms and uptake by plants. There are several reports that suggest that organic chemicals are not mineralized while associated with solid phases.[8,9] Experiments by Robinson et al.[10] show that sorbed-phase substrate was not degraded and that long-term biodegradation was limited by the slowly desorbing fraction of substrate. These results suggest that rate-limited mass transfer processes (primary desorption) may significantly affect the rate at which a compound is degraded in the presence of a solid phase.

The model presented in the section on "Uptake of Heavy Metals by Plants" uses the fraction soluble in the soil water of heavy metal ions to determine the uptake. The sorption is dependent of pH, redox potential, and humus, clay, and sand fractions in the soil. The relationship between these factors and the sorption is included in the model. If the sorption of organic compounds to soil is not known, the soil–water partition coefficient, K_{oc}, can be estimated from the octanol–water partition coefficient by the following equations:

$$\log K_{oc} = -0.006 + 0.937 \log K_{ow} \quad (1)$$

$$\log K_{oc} = -0.35 + 0.99 \log K_{ow} \quad (2)$$

In the case that the carbon fraction of organic carbon in soil is *f*, the distribution coefficient, K_D, for the ratio of the concentration in soil and in water can be found as $K_D = K_{oc}f$.

It has been suggested that there are different stages of sorption processes and that newly sorbed material is more labile and therefore more bioavailable than aged sorbed material. Numerous experiments have demonstrated that aging affects bioavailability in soil due to changes in the soil structure, resulting in slower desorption processes. The sorption can frequently be described by either Freundlich or Langmuir adsorption isotherm, expressed respectively by the following equations:

$$a = kc^b \quad (3)$$

$$a = k'c/(c + b') \quad (4)$$

where *a* is the concentration in soil, *c* is the concentration in water, and *k*, *k*′, *b*, and *b*′ are constants. Eq. 3, corresponding to Freundlich adsorption isotherm, is a straight line with slope *b* in a log–log diagram, since $\log a = \log k + b \log c$.[11] The Langmuir adsorption isotherm is an expression similar to the Michaelis–Menten equation. If $1/a$ is plotted versus $1/c$,[11] we obtain a straight line, Lineweaver–Burk's plot, as $1/a = 1/k' + b'/k'c$. When $1/a = 0$, $1/c = -1/b'$ and when $1/c = 0$, $1/a = 1/k'$. This plot can be applied to assess the expression of the type used in Michaelis–Menten's equation and in Langmuir's adsorption isotherm; it is observed that b is often close to 1 and c is for most environmental problems small. This implies that the two adsorption isotherms get close to $a/c = k$, and k becomes a distribution coefficient. k for 100% organic carbon is usually denoted K_{oc} (see above).

The sorption determines the uptake of organic contaminants by plants as it is expressed in the following equation:[11]

$$\mathrm{BCF} = f_{\mathrm{lipid}} K_{\mathrm{ow}}{}^{b} / h f K_{\mathrm{ow}}{}^{a}, \tag{5}$$

where BCF, the biological concentration factor, expresses the ratio between the concentration in soil and in the plant (or the microorganisms); f_{lipid} is the lipid fraction in the plant; f is, as shown above, the fraction of organic carbon in the soil; and a, b, and h are constants. The denominator expresses the fraction of the organic matter that is dissolved in the soil water. h is therefore the constant determined by Eqs. 1 and 2. If we use Eq. 1, Eq. 5 may be reformulated to the following equation:

$$\mathrm{BCF} = 1.01\, f_{\mathrm{lipid}} K_{\mathrm{ow}}{}^{0.063} / f \tag{6}$$

As it is seen, BCF becomes almost independent of K_{ow} and mainly dependent on the ratio between f_{lipid} and f. A high BCF means that the concentration in the plants (eventual microorganisms) is high and a significant amount of the toxic compound is removed; the lipid fraction in the plants has to be high and the carbon content of the soil has to be low.

Physical makeup of the soil (pore size distribution) is of importance for bioavailability. Bacteria may be excluded from the microporous domain since most bacteria range from 0.5 to 2 μm. If such an exclusion occurs, biodegradation cannot take place in the microporous domain. The degradation rate is therefore limited by the diffusion of solute from the microporous to the macroporous domain. This is obviously of particular importance for organic contaminants with a high molecular weight. In a field situation, it is difficult to separate the effects of sorption and micropore exclusion, as some residues are protected from biodegradation by both mechanisms.

Microbial adaptations: Microorganisms have developed several strategies to increase the bioavailability of organic contaminant. One strategy is the development of increased cell affinity for hydrophobic surfaces. It allows the microorganisms to attach to the hydrophobic substrate and directly adsorb it. A second strategy is the production and release of surface active agents or biosurfactants.[12,13]

The biological adaptation is a current change of the properties of the microbial population by a selection of the microorganisms that are best fitted to survive and grow under the prevailing conditions. They are determined by the properties of the environment including the concentrations and characteristics of the contaminants. A biological adaptation is widely used to prepare a microorganism population for bioremediation. It is often possible, although not general, that a 10 times faster decomposition can be achieved by the use of adapted microorganisms.[2]

BIODEGRADATION

See the entry entitled "Biodegradation," where a general presentation of this process included methods for estimation of the biodegradation rate from the chemical structure of the toxic organic compounds.

The usual applied procedure to follow for the utilization of microbiological biodegradation to reduce the concentration of a toxic organic matter in contaminated soil has six steps:

1. Spatial mapping of quantitative distribution: the contaminant is developed by analytical chemistry intensively. Analyses of pore water are often applied to evaluate the extent of environmental risks.
2. Laboratory test/treatability studies to verify the applicability of bioremediation.
3. Calculation (often by development of a model) to assess the feasibility of the method in situ.
4. Production of an adapted strain of the microorganisms in sufficient amount.
5. Implementation in situ. If the groundwater table is high, it is usually lowered. Injection pits are introduced into the soil, and air is blown into the soil to reinforce the decomposition of organic matter. In case of chlorinated compounds, a mixture of methane and air may be applied.
6. The results are followed by use of a wide spectrum of analytical methods including radioactive tracers, detection of intermediary metabolites, and respiration rate.

UPTAKE OF HEAVY METALS BY PLANTS

Plants are contaminated by heavy metals originating from deposition of heavy metals (waste sites), air pollution, the application of sludge from municipal wastewater plant as a soil conditioner, and the use of fertilizers. The uptake of heavy metals from municipal sludge by plants has previously been modeled.[4] The model is based on a mass balance for cadmium in a typical Danish soil (see Fig. 1).

The model can briefly be described as follows: Depending on the soil composition, it is possible to find for various heavy metal ions a distribution coefficient, i.e., the fraction of the heavy metal that is dissolved in the soil water relative to the total amount. The distribution coefficient was found by examination of the dissolved heavy metals

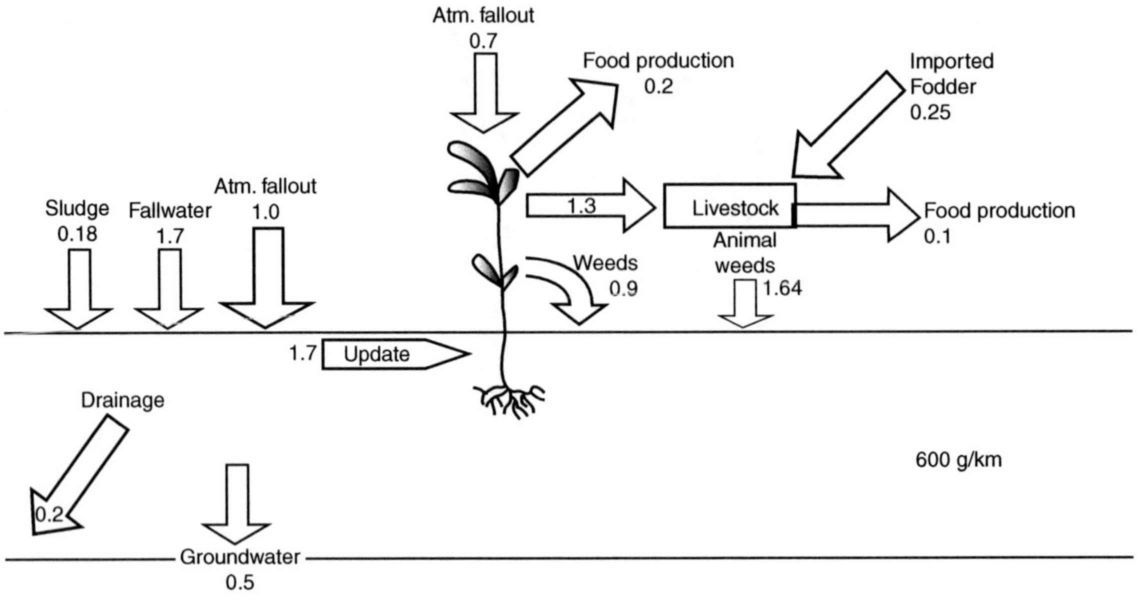
Atm. fallout
0.7
Food production
0.2
Imported
Fodder
0.25
Atm. fallout
1.0
Sludge
0.18
Fallwater
1.7
1.3
Livestock
Food production
0.1
Animal
weeds
1.64
Weeds
0.9
1.7
Update
Drainage
600 g/km
0.2
Groundwater
0.5

Fig. 1 Cadmium balance of 1 ha of an average Danish agriculture land. All rates are expressed as grams of Cd per hectare per day.

relative to the total amount for several different types of soil. Correlation between pH, the concentration of humic substances, clay, and sand in the soil on the one hand, and the distribution coefficient on the other, was also determined. The uptake of heavy metals was considered a first-order reaction of the dissolved heavy metal. It is, however, also possible to test acid volatile sulfide and organic carbon to describe the metal binding capacity of sediment in constructed wetlands. This will give approximately the same ratio "bound" to "bioavailable" heavy metals as the above-mentioned correlation. The basic idea is the same, namely, to find easily measurable soil properties that determine the metal binding capacity, which is crucial for the uptake of heavy metals by plants.

In addition to the uptake from soil water, the model presented below considers the following:

1. Direct uptake from atmospheric fallout onto the plants.
2. Other sources of contamination such as fertilizers and the long-term release of heavy metal bound to the soil and the non-harvested parts of the plants.

Published data on lead and cadmium contamination in agriculture are used to calibrate and validate the model that is intended to be used for the following:

1. Risk assessment for the use of fertilizers and sludge that contain heavy metals as contaminants.
2. A risk involved in the use of plants harvested from a waste site.
3. Determining the possibilities of removal of heavy metals by plants that have a particular ability to take up heavy metals. This last intended application of the model makes it useful for determination of the result of application of bioremediation.

Fig. 2 shows a conceptual diagram of the Cd version of the model. The STELLA software was applied. As can be seen, it has four state variables: Cd-bound, Cd-soil, Cd-detritus, and Cd-plant. An attempt was made to use one or two state variables for cadmium in the soil, but to get acceptable accordance between data and model output, three state variables were needed. This can be explained by the presence of several soil components that bind the heavy metal differently.

The loss covers transfer to the soil and groundwater below the root zone. It is expressed as a first-order reaction with a rate coefficient dependent on the distribution coefficient that is found from the soil composition and pH, according to the correlation found by Jørgensen.[14] The transfer from Cd-bound to Cd-soil indicates the slow release of cadmium due to a slow decomposition of the more or less refractory material to which cadmium is bound. The cadmium uptake by plants is expressed as a first-order reaction, where the rate is dependent on the distribution coefficient, as only dissolved cadmium can be taken up. It is furthermore dependent on the plant species. As will be seen, the uptake is a step function that, here (grass), is 0.0005 during the growing season, and, of course, zero after the harvest and until the next growing season starts. Cd-waste covers the transfer of plant residues to detritus after harvest. It is therefore a pulse function, which here is 60% of the plant biomass, as the remaining 40% has been harvested. Cd-detritus covers a wide range of biodegradable matter and the mineralization and is therefore accounted for in the model by use

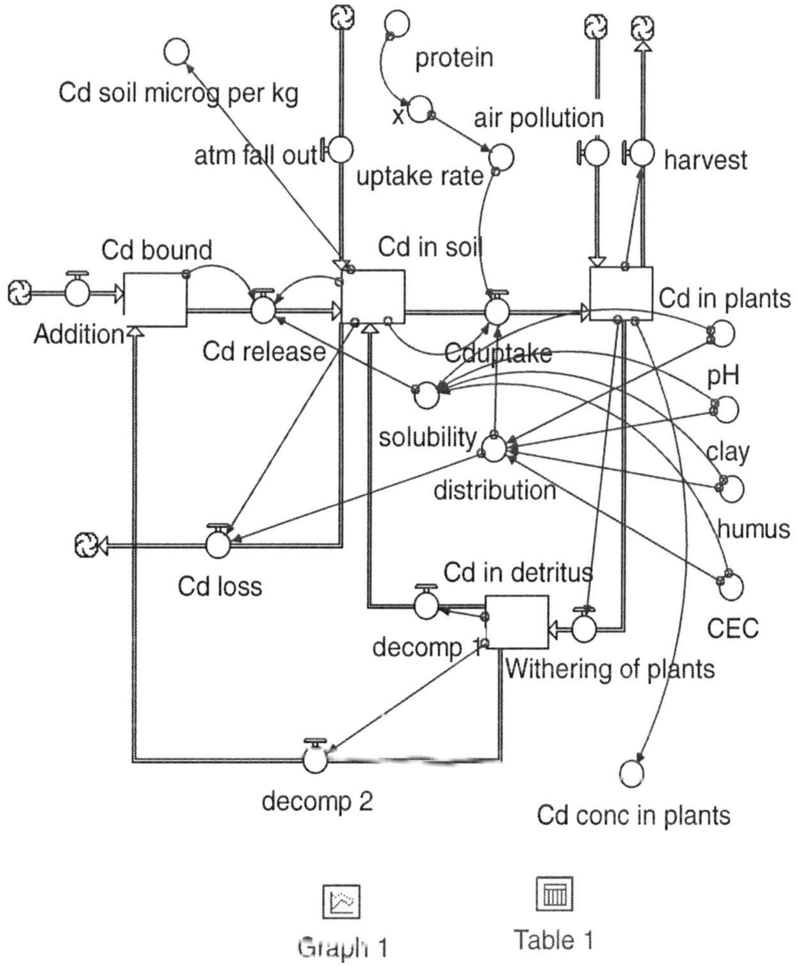
protein
Cd soil microg per kg
x
air pollution
atm fall out
uptake rate
harvest
Cd bound
Cd in soil
Cd in plants
Addition
Cd release
Cd uptake
pH
solubility
clay
distribution
humus
Cd loss
Cd in detritus
CEC
decomp 1
Withering of plants
decomp 2
Cd conc in plants
Graph 1
Table 1

Fig. 2 Conceptual diagram of the model. Boxes show state variables, double-line arrows denote flows, circles show functions, and single-line arrows show feedback mechanisms.

of two mineralization processes: one to Cd-soil and one to Cd-bound. The first one is rapid and is given a higher rate for the first 180 days as the addition of municipal sludge in this case is at day 0. The second one is at about the same rate, but as the cadmium is transferred to the Cd-bound, the slow release rate is considered by the very slow transfer from Cd-bound to Cd-soil. Similar models can be erected for other heavy metals, but the distribution coefficient is of course different for the other heavy metals.[14]

Plants with a high protein content will generally take up heavy metals with a higher efficiency. Several applications of plants with a high uptake efficiency have been reported in the literature; see Tongbin et al.,[15] where a very effective uptake of arsenic is reported, and Feng et al.,[16] where simultaneous removal of arsenic and antimony is reported. The use of algae for removal of heavy metals has also been successfully tested.[17]

Models are used increasingly to solve the problems of contaminated soil (see UNEP-EITC and Copenhagen University[18]). Spatial models have been developed to consider the distribution of the contaminants in eco-spatial time scale. These models are very useful in setting up and optimizing time-bound action plans of bioremediating the contaminated soil.

REFERENCES

1. Roszak, D.B.; Colwell, R. Survival strategies of bacteria in natural environment. Microbiol. Rev. **1987**, *51*, 365–379.
2. Maier, R.M. Biovailability and its importance for bioremediation. In *Bioremediation*; Valdes, J.J., Ed.; Kluwer: Dordrecht, the Netherlands, 2000; 58–79.
3. Jørgensen, S.E.; Halling-Sørensen, B.; Mahler, H. *Handbook of Estimation Methods in Ecotoxicology and Environmental Chemistry*; Taylor and Francis Publ.: Boca Raton, FL, 1997; 230 pp.
4. Jørgensen, S.E.; Fath, B. *Fundamentals of Ecological Modelling*, 4th Ed.; Elsevier: Amsterdam, 2011; 398 pp.

5. Fogel et al. EPA Report 560/5-82-015 Washington, DC, 1981.
6. Zhang, Y.; Miller, R.M. Enhanced octadecane dispersion and biodegration by *Pseudomonas rhamnolipid* surfactant. Appl. Environ. Microbiol. **1992**, *58*, 3276–3282.
7. Miller, R.M. Surfactant-enhanced bioavailability of slightly soluble organic compounds. In *Bioremediation: Science and Applications*; Skipper, H., Turco, R., Eds.; Specila Publications. Soil Science Society of America: Madison, WI, 1995; 33–54.
8. Greer, L.E.; Shelton, D.R. Surfactant-enhanced bioavailability of slightly soluble organic compounds. In *Bioremediation*; Skipper, H., Turco, R., Eds.; Soil Science Society of America: Madison, WI, 1992; 33–54.
9. Miller, M.E.; Alexander, M. Kinetics of bacterial degradation of benzylamine in montmorillonic suspension. Environ. Sci. Technol. **1991**, *25*, 240–245.
10. Robinson K.G. et al. Availability of sorbed toluene in solids for biodegradation by acclimatized bacteria. Water Res. **1990**, *24*, 345–350.
11. Jørgensen, S.E. *Principles of Pollution Abatement*; Elsevier: Amsterdam, 2000; 520 pp.
12. Rosenberg, E. Microbial surfactants. Crit. Rev. Biotechnol. **1986**, *3*, 109–132.
13. Fiechter, A. Biosurfactant: moving towards industrial application. Trends Biotechnol. **1992**, *10*, 208–217.
14. Jørgensen, S.E. Do heavy metals prevent the agricultural use of municipal sludge? Water Res. **1975**, *9*, 163–170.
15. Tongbin, C. et al. Arsenic hyperaccumulator *Pteris vittata* L. and its arsenic accumulation. Chin. Sci. Bull. **2002**, *47*, 902–905.
16. Feng R. et al. Simultaneous hyper-accumulation of arsenic and antimony in Cretan brake fern: Evidence of plant uptake and sub-cellular distributions. Microchem. J. **2011**, *97*, 38–43.
17. Mitsch, W.J.; Jørgensen, S.E. *Ecological Engineering and Ecosystem Restoration*; John Wiley: New York, 2004; 410 pp.
18. UNEP-IETC and Copenhagen University. *Handbook of Phytotechnology for Water Quality Improvement and Wetland Management through Modelling Applications*, 2005.

Biotechnology: Pest Management

Maurizio G. Paoletti
Department of Biology, University of Padova, Padova, Italy

Abstract
Disease and insect pest resistance to various pests has been slowly bred into crops for the past 13,000 years; current techniques in biotechnology now offer opportunities to further and more rapidly improve the non-chemical control of disease and insect pests of crops. However, relying on a single factor, like the *Bacillus thuringiensis* (BT) toxin that has been inserted into corn, cotton, and a few other crops for insect control, leads to various environmental problems, including insect resistance, and a serious threat to beneficial biological control insects and endangered species. A major environmental and economic cost associated with genetic engineering applications in agriculture relates to the use of herbicide-resistant crops (HRCs). In general, HRC technology results in increased herbicide use, pollution of the environment, and weed control costs for farmers that may be twofold greater than standard weed control costs. Therefore, pest control with both pesticides and genetic engineering methods can be improved for effective, safe, and economical pest control.

BENEFITS OF GENETIC ENGINEERING IN PEST CONTROL

Since 1987, many crops have been genetically modified for features such as resistance to insects, resistance to pathogens (including viruses) and herbicides, and for improved features such as longer-lasting ripening, higher nutritional status, protein content, seedless fruit, and sweetness. Up to 34 new genetically engineered crops have been approved to enter into the market.

In 1998, 27.8 million ha of engineered crops were planted in countries such as the United States, Argentina, Canada, and Australia. The United States alone contains 74% of the modified crop land-planted. Globally, 19.8% of this area has been planted with herbicide-tolerant crops, 7.7% with insect-resistant crops, and 0.3% with insect and HRCs. Five crops—soybean, corn, cotton, canola, and potato—cover the largest acreage of engineered crops.[1,2]

DISEASE RESISTANCE IN CROPS

The crops currently on the market that have been engineered for resistance to plant pathogens are listed in Table 1. Disease-resistant engineered crops have some potential advantages because few current pesticides can control bacterial and viral diseases of crops. In addition, these engineered plants help reduce problems from pesticides.

The large-scale cultivation of plants expressing viral and bacterial genes might lead to adverse ecological consequences. The most significant risk is the potential for gene transfer of disease resistance from cultivated crops to weed relatives. For example, it has been postulated that a virus-resistant squash could transfer its newly acquired virus-resistant genes to wild squash (*Cucurbita pepo*), which is native to the southern United States. If the virus-resistant genes spread, newly disease-resistant weed squash could become a hardier, more abundant weed. Moreover, because the United States is the origin for squash, changes in the genetic make-up of wild squash could conceivably lessen its value to squash breeders.

Some plant pathologists have also suggested that development of virus-resistant crops could allow viruses to infect new hosts through transencapsidation. This may be especially important for certain viruses, e.g., luteoviruses, where possible heterologus encapsidation of other viral RNAs with the expressed coat protein is known to occur naturally. With other viruses, such as the PRV that infects papaya, the risk of heteroencapsidation is thought to be minimal because the papaya crop itself is infected by very few viruses.

Virus-resistant crops may also lead to the creation of new viruses through an exchange of genetic material or recombination between RNA virus genomes. Recombination between RNA virus genomes requires infection of the same host cell with two or more viruses. Several authors have pointed out that recombination could also occur in genetically engineered plants expressing viral sequences of infection with a single virus, and that large-scale cultivation of such crops could lead to increased possibilities of combinations. It has recently been shown that RNA transcribed from a transgene can recombine with an infecting virus to produce highly virulent new viruses.

A strategy for reduced risk would include: 1) identification of potential hazards; 2) determination of frequency of recombination between homologous, but nonidentical sequences in crops and weeds; and 3) determination of whether or not such recombinants can have selective advantage.

Encyclopedia of Environmental Management DOI: 10.1081/E-EEM-120009903

Table 1 Plants genetically-engineered for virus resistance that have been approved for field tests in the United States from 1987 to July 1995.

Crop	Disease(s)	Research organization
Alfalfa	Alfalfa mosaic virus, Tobacco mosaic virus (TMV), Cucumber mosaic virus (CMV)	Pioneer Hi-Bred
Barley	Barley yellow dwarf virus (BYDV)	USDA
Beets	Beet necrotic yellow vein virus	Betaseed
Cantelope and/or squash	CMV, papaya ringspot virus (PRV) Zucchini yellow mosaic virus (ZYMV), Watermelon mosaic virus II (WMVII)	Upjohn
	CMV	Harris Moran Seed
	ZYMV	Michigan State University
	ZYMV	Rogers NK Seed
	Soybean mosaic virus (SMV)	Cornell University
	SMV, CMV	New York State Experiment Station
Corn	Maize dwarf mosaic virus (MDMV) Maize chlorotic mottle virus (MCMV), Maize chlorotic dwarf virus (MCDV)	Pioneer Hi-Bred
	MDMV	Northup King
	MDMV	DeKalb
	MDMV	Rogers NK Seed
Cucumbers	CMV	New York State Experiment Station
Lettuce	Tomato spotted wilt virus (TSWV)	Upjohn
Papayas	PRV	University of Hawaii
Peanuts	TSWV	Agracetus
Plum Trees	PRV, plum pox virus	USDA
Potatoes	Potato leaf roll virus (PLRV), Potato virus X (PVX), Potato virus Y (PVY)	Monsanto
	PLRV, PVY, late blight of potatoes	Frito-Lay
	PLRV	Calgene
	PLRV, PRY	University of Idaho
Potatoes	PLRV, PVY	USDA
	PYV	Oregon State University
Soybeans	SMV	Pioneer Hi-Bred
Tobacco	ALMV, tobacco etch virus (TEV), Tobacco vein mottling virus	
	TEV, PVY	University of Florida
	TEV, PVY	North Carolina State University
	TMV	Oklahoma State University
	TEV	USDA
Tomatoes	TMV, tomato mosaic virus (TMV) CMV, tomato yellow leafcurl virus	Monsanto
	TMV, ToMV	Upjohn
	ToMV	Rogers NK Seed
	CMV	PetoSeed
	CMV	Asgrow
	CMV	Harris Moran Seeds
	CMV	New York State Experiment Station
	CMV	USDA

Source: Krimsky and Wrubel[4] and McCullum et al.[5]

Biological – Cadmium

ASSESSMENT OF TRANSGENIC VIRUS-RESISTANT POTATOES IN MEXICO

An in-depth assessment of potential socioeconomic implications related to the introduction of some genetically modified varieties of virus-resistant potatoes (PVY, PVX, PlRV) in Mexico underscores the importance of this technology. This type of genetic modification could prove especially beneficial to large-scale farmers, but only marginally beneficial to small-scale farmers, because most small farmers use red potato varieties that are not considered suitable for transformation. In addition, 77% of the seeds that small farmers use come from informal sources, not from the seed providers that could sell the new resistant varieties.

The mycoplasma and virus diseases in Mexico are not currently controlled with pesticides, and rank second and third in economic damages. The major pest, the fungus *Phytophtora infestans*, ranks first in economic damages and requires, in some cases, up to 30 fungicide applications. Thus, the interesting new genetically altered varieties of potatoes are of little benefit to crop production for small farmers.

HRCS

Several engineered crops that include herbicide resistance are commercially available; 13 other key crops in the world are ready for field trials (Table 2). In addition, some crops (e.g., corn) are being engineered to contain both herbicide (glyphosate) and biotic insecticide resistance (BT α-endotoxin).

Herbicides adopted for herbicide-resistant crops employ lower doses when compared with atrazine, 2,4-D, and alachlor. However, the resistance of the crop to the target herbicide would, in practice, suggest to the farmer to apply dosages higher than recommended. In addition, costs for this new technology of HRCs are about 2-times higher in corn than the recommended herbicide use and cultivation weed control program.

Integrated pest management (IPM) could benefit from some HRCs, if alternative non-chemical methods can be applied first to control weeds and the target herbicide could be used later, only when and where the economic threshold of weeds is surpassed. Generally, though, the use of herbicide resistant crops will lead to increased use of herbicides and environmental and economic problems. Most HRCs were developed for Western agriculture. For example, in Northern African countries, most crops, such as sorghum, wheat, and canola (oilseed rape), have wild weed relatives, thereby increasing the risk that genes from the herbicide-resistant crop varieties could be transferred to wild weed relatives.

The risk of herbicide-resistant genes from a transgenic crop variety being transferred to weed relatives has been demonstrated for canola (oilseed rape) and sugar beet.

Repeated use of herbicides in the same area creates problems of weed herbicide resistance. For instance, if glyphosate is used with HRCs crops on about 70 million ha, this might accelerate pressure on weeds to evolve herbicide resistant biotypes. Sulfonylureas and imidazolinones in HRCs are particularly prone to rapid evolution of resistant weeds. Extensive adoption of HRCs will increase the hectarage and surface treated, thereby exacerbating the resistance problems and environmental pollution problems.

Bromoxynil has been targeted in herbicide resistant cotton by Calgene and Monsanto (Table 2). This herbicide has been used on winter cereals, cotton, corn, sugarbeets, and onions to control broad leaf weeds. Drift of bromoxynil has been observed to damage nearby grapes, cherries, alfalfa, and roses. In addition, legumious plants can be sensitive to this herbicide, and potatoes can be damaged by it. Herbicide residues above the accepted standards have been detected in soil and groundwater, and as drift fallout. Rodents demonstrate some mutagenic responses to bromoxynil. Beneficial *Stafilinid* beetles show reduced survival and egg production, even at recommended dosages of bromoxynil. Crustaceans (*Daphnia magna*) have also been severely affected by this herbicide.

Toxicity of Herbicides and HRCs

Toxic effects of herbicides to humans and animals also have been reported. For example, the Basta surfactant (sodium polyoxyethylene alklether sulfate) has been shown to have strong vasodialatative effects in humans and cardiostimulative effects in rats. Treated mice embryos exhibited specific morphological defects.

Most HRCs have been engineered for glyphosate resistance. Although adverse effects of herbicide-resistant soybeans have not been observed when fed to animals such as cows, chickens, and catfish, genotoxic effects have been demonstrated on other non-target organisms. Earthworms have been shown to be severely injured by the glyphosate herbicide at 2.5–10.1/ha. For example, *Allolobophora caliginosa*, the most common earthworm in European, North American, and New Zealand fields, is killed by this herbicide. In addition, aquatic organisms, including fish, can be severely injured or killed when exposed to glyphosate. The beneficial nematode, *Steinerema feltiae*, a useful biological control organism, is reduced by 19%–30% by the use of glyphosate.

There are also unknown health risks associated with the use of low doses of herbicides. Due to the common research focus on cancer risk, little research has been focused on neurological, immunological, developmental, and reproductive effects of herbicide exposures. Much of this problem is due to the fact that scientists may lack the methodologies and/or the diagnostic tests necessary to properly evaluate the risks caused by exposure to many toxic chemicals, including herbicides.

Table 2 Herbicide-resistant crops (HRCs) approved for field tests in the United States from 1987 to July 1995.

Crop	Herbicide	Research organization
Alfalfa	Glyphosate	Northrup King
Barley	Glufosinate/Bialaphos	USDA
Canola (oilseed rape)	Glufosinate/Bialaphos	University of Idaho
	Glyphosate	Hoechst-Roussel/AgrEvo
		InterMountain Canola
		Monsanto
Corn	Glufosinate/Bialaphos	Hoechst-Roussel/AgrEvo
		ICI
		UpJohn
		Cargill
		DeKalb
		Holdens
		Pioneer Hi-Bred
		Asgrow
		Great Lakes Hybrids
		Ciba-Geigy
		Genetic Enterprises
	Glyphosate	Monsanto
		DeKalb
	Sulfonylurea	Pioneer Hi-Bred
		Du Pont
	Imidazolinone	American Cyanamid
Cotton	Glyphosate	Monsanto
		Dairyland Seeds
		Northrup King
	Bromoxynil	Calgene
		Monsanto
		Rhone Poulenc
	Sulfonylurea	Du Pont
		Delta and Pine Land
	Imidazolinone	Phytogen
Peanuts	Glufosinate/Bialaphos	University of Florida
Potatoes	Bromoxynil	University of Idaho
		USDA
	2,4-D	USDA
	Glyphosate	Monsanto
	Imidazolinone	American Cyanamid
Rice	Glufosinate/Bialaphos	Louisiana State University
Soybeans	Glyphosate	Monsanto
	Glyphosate	UpJohn
		Pioneer Hi-Bred
		Northrup King
		Agri-Pro
	Glufosinate/Bialaphos	UpJohn
	Sulfonylurea	Hoechst/AgrEvo
		Du Pont
Sugar Beets	Glufosinate/Bialaphos	Hoechst-Roussel
	Glyphosate	American Crystal Sugar
Tobacco	Sulfonylurea	American Cyanmid
Tomatoes	Glyphosate	Monsanto
	Glufosinate/Bialaphos	Canners Seed
Wheat	Glufosinate/Bialaphos	AgrEvo

Source: Krimsky and Wrubel,[4] McCullum et al.,[5] and Agribusiness.[6]

While industry often stresses the desirable characteristics of their HRCS, environmental and agricultural groups, and other scientists, have indicated the risks. For example, research has shown that the application of glyphosate can increase the level of plant estrogens in the bean, *Vicia faba*. Feeding experiments have shown that cows fed transgenic glyphosate-resistant soybeans had a statistically significant difference in daily milk-fat production as compared to control groups. Some scientists are concerned that the increased milk-fat production by cows fed these transgenic soybeans may be a direct consequence of higher estrogen levels in these transgenic soybeans.

Economic Impacts of HRCs

Some analysts project that switching to bromoxynil for broadleaf weed control in cotton could result in savings of 37 million dollars each year. Furthermore, recent problems with use of glyphosate-resistant cotton in the Mississippi Delta region—crop losses resulting in up to $500,000 of this year's cotton crop—suggest that this technology needs to be further developed before some farmers will reap economic benefits. In addition, a recent study of herbicide-resistant corn suggests that the costs of weed control might be about two times more expensive than normal herbicide and cultivation weed control in corn.

While some scientists suggest that use of HRCs will cause a shift to fewer broad spectrum herbicides, most scientists conclude that the use of HRCs will actually increase herbicide use.

BT for Insect Control

More than 40 BT crystal protein genes have been sequenced, and 14 distinct genes have been identified and classified into six major groups based on amino acids and insecticidal activity. Many crop plants have been engineered with the BT α-endotoxin, including alfalfa, corn, cotton, potatoes, rice, tomatoes, and tobacco (Table 3). The amount of toxic protein expressed in the modified plant is 0.01%–0.02% of the total soluble proteins.

Some trials with corn demonstrate a high level of efficacy in controlling corn borers. Corn engineered with BT endotoxin has the potential to reduce corn borer damage by 5%–15% over 28 million ha in the US, with a potential economic benefit of $50 million annually. Some suggest that corn engineered with BT toxin will increase yields by 7% over similar varieties. However, it is too early to tell if all these benefits will be realized consistently. Potential negative environmental effects also exist because the pollen of engineered plants contains BT, which is toxic to bees, beneficial predators, and endangered butterflies like the Karka Blue and Monarch Butterflies.

Cotton was the first crop plant engineered with the BT α-endotoxin. Caterpillar pests, including the cotton bollworm and budworm, cost U.S. farmers about $171 million/

Table 3 Transgenic insect resistant crops containing BT δ-endotoxins. Approved field tests in United States from 1987 to July 1995.

Crop	Research organization
Alfalfa	Mycogen
Apples	Dry Creek
	University of California
Corn	Asgrow
	Cargill
	Ciba-Geigy
	Dow
	Genetic Enterprises
	Holdens
	Hunt-Wesson
	Monsanto
	Mycogen
	NC + Hybrids
	Nortrup King
	Pioneer Hi-Bred
	Rogers NK Seed
Cotton	Calgene
	Delta and Pineland
	Jacob Hartz
	Monsanto
	Mycogen
	Northrup King
Cranberry	University of Wisconsin
Eggplant	Rutgers University
Poplar	University of Wisconsin
Potatoes	USDA
	Calgene
	Frito-Lay
	Michigan State University
	Monsanto
	Montana State University
	New Mexico State University
	University of Idaho
Rice	Louisiana State University
Spruce	University of Wisconsin
Tobacco	Auburn University
	Calgene
	Ciba-Geigy
	EPA
	Mycogen
	North Carolina State University
	Roham and Haas
Tomatoes	Campbell
	EPA
	Monsanto
	Ohio State University
	PetoSeeds
	Rogers NK Seeds
Walnuts	University of California, Davis
	USDA

Source: Krimsky and Wrubel[4] and Agribusiness.[8]

yr as measured in yield losses and insecticide costs. Benedict et al.[3] predict that the widespread use of BT cotton could reduce insecticide use and thereby reduce costs by as much as 50% to 90%, saving farmers $86 to $186 million/yr.

The development of insect resistance to transgenic crop varieties is one highly possible risk associated with the use of BT D-endotoxin in genetically engineered crop varieties. Resistance to BT has already been demonstrated in the cotton budworm and bollworm. If BT-engineered plants become resistant, a key insecticide that has been utilized successfully in IPM programs could be lost. Therefore, proper resistance management strategies with use of this new technology are imperative. Another potential risk is that the BT α-endotoxin could be harmful to non-target organisms. For example, it is not clear what potential effect the BT D-endotoxin residues that are incorporated into soils will have against an array of non-target useful invertebrates living in the rural landscape. It has also been demonstrated that predators, such as the lacewing larvae (*Crysoperla carnea*) that feed on corn borers (*Ostrinia nubilalis*), grown on engineered BT corn have consistently higher mortality rates when compared to specimens fed with non-engineered corn borers. In addition, the treated larvae need three more days to reach adulthood than lacewings fed on prey from non-BT corn.

DISCUSSION

Both pesticides and biotechnology have definite advantages in reducing crop losses to pests. At present, pesticides are used more widely than biotechnology, and thus are playing a greater role in protecting world food supplies. In terms of environmental and public health impacts, pesticides probably have a greater negative impact at present because of this more widespread use.

Genetically engineered crops for resistance to insect pests and plant pathogens could, in most cases, be environmentally beneficial, because these more resistant crops could allow a reduction in the use of hazardous insecticides and fungicides in crop production. In time, there may also be economic benefits to farmers who use genetically engineered crops; this will depend, however, on the prices charged by the biotechnology firms for these modified, transgenic crops.

There are, however, some environmental problems associated with the use of genetically engineered crops in agriculture. For example, adding BT to crops like corn for insect control can result in any of the following negative environmental consequences: 1) development of resistance to BT by pests species in corn and other crops; 2) health risks from exposure to the BT toxin to humans in their food and to livestock in feed; 3) the toxicity of the pollen from the BT-treated corn to honey bees, beneficial natural enemies, and endangered species of insects that feed on the modified corn plants or come into contact with the drifting pollen; engineered plant residues incorporated into soil can produce undesirable effects on soil micros and mesofauna.

A major environmental and economic concern associated with genetically engineered crops is the development of HRCs. Although in rare instances HRCs may result in a beneficial reduction of toxic herbicide use, it is more likely that the use of HRCs will increase herbicide use and environmental pollution. In addition, farmers will suffer because of the high costs of employing HRCs—in some instances, weed control with HRCs may increase weed control costs for the farmer threefold.

More than 40% of the research by biotechnology firms is focused on the development of HRCs. This is not surprising, because most of the biotechnology firms are also chemical companies who stand to profit if herbicide resistance in crops result in greater pesticide sales. Theoretically, the acceptance and use of engineered plants in sustainable and integrated agriculture should consistently reduce current use of pesticides, but this is not the current trend. In addition, most products and new technologies are designed for Western agriculture systems, not for poor or developing countries. For instance, if terminator genes enter into the seed market, there will be no possibility of traditional and small farmers using their plants to produce their seeds. Thus, genetic engineering could promote improvements for the environment; however, the current products—especially the herbicide-resistant plants and the BT-resistant crops—do have serious environmental impacts, similar to the consequences of pesticide use.

REFERENCES

1. James, C. Global Review of Commercialized Transgenic Crops: 1998. In *ISAAA Briefs*; Cornell University: Ithaca, New York, 1998; 8-1998.
2. Moff, A.S. Toting up the early harvest of transgenic plants. Science **1998**, *282*, 2176–2178.
3. Benedict, J.H.; Ring, D.R.; Sachs, E.S.; Altman, D.W.; DeSpain, R.R.; Stone, T.B.; Sims, J.R. Influence of Transgenic BT Cottons on Tobacco Budworm and Bollworm Behavior Survival, and Plant Injury. In *Proceedings Beltwide Cotton Council*; Herber, J., Richter, D.A., Eds.; National Cotton Council: Memphis, Tennessee, 1992; 891–895.
4. Krimsky, S.; Wrubel, R.P. *Agricultural Biotechnology and the Environment*; University of Illinois Press: Urbana, Illinois, 1996.
5. McCullum, C.; Pimentel, D.; Paoletti, M.G. Genetic Engineering in Agriculture and the Environment: Risks and Benefits. In *Biotechnology and Safety Assessment*; Thomas, J.A., Ed.; Taylor and Francis: Washington, D.C., 1998; 177–217.
6. Agribusiness. The Gene Exchange, Fall 1997; http://www.ucsusa.org/Gene/F97.agribusiness.html (accessed July 5, 2001).
7. Agribusiness. The Gene Exchange, Winter 1996; http://www.ucsusa.org/Gene/W96.agribusiness.html (accessed July 5, 2001).

Birds: Chemical Control

Eric B. Spurr
Department of Wildlife Ecology, Landcare Research New Zealand, Ltd., Lincoln, New Zealand

Abstract

Chemicals are used to control bird populations when they cause damage to crops, stock-food, buildings and other structures, and when they are hazards at places such as airports, public parks, golf courses, and rubbish dumps. They include lethal toxicants (avicides) and stressing agents, and nonlethal immobilizing agents, repellents, and reproductive inhibitors. Lethal methods of control attempt to reduce bird numbers, whereas nonlethal methods generally attempt to modify bird behavior without causing mortality, as a means of reducing damage.

TOXICANTS

The earliest toxicants were formulations containing arsenic, antimony, phosphorus, and various botanical extracts.[1] Other toxicants used previously include chlorinated hydrocarbons such as endrin, metallic salts such as thallium, organometallic salts such as sodium monofluoroacetate (1080), alkaloids such as nicotine, and anticoagulants such as coumatetralyl and brodifacoum. Most are highly toxic to both birds and mammals. More than 2000 chemicals were evaluated as avicides between the 1940s and the 1980s, and some were found that were selectively toxic to birds. Some were even selectively toxic to certain species of birds. Since the 1980s, however, little effort has been put into finding new toxicants. Instead, most effort has been spent gathering toxicological and environmental data to ensure continued registration of existing products. Recently, international attention has focussed on the animal welfare aspects of toxicants.[2]

Strychnine was once used widely as an oral toxicant for control of birds such as rock pigeons (*Columba livia*) and house sparrows (*Passer domesticus*), and is still used by certified operators in some countries today (Table 1). It is mainly applied in grain baits (e.g., Sanex Poison Corn in Canada). Strychnine is highly toxic to both birds and mammals, and poses a high risk of both primary and secondary poisoning to nontarget species. Time to death varies from 5 to 50 minutes. It causes extreme pain in poisoned animals and is considered inhumane.

Fenthion was previously used as an oral and dermal toxicant, but is currently used only as a dermal toxicant. Its use is restricted to certified operators. It is applied to wicks in artificial perches or other surfaces, for control of birds such as rock pigeons, house sparrows, and starlings (*Sturnus vulgaris*). It was available previously as Rid-A-Bird® in the United States, and currently as Control-A-Bird® and Avigrease® in Australia. Fenthion (Queletox®) is also aerially sprayed onto birds, especially red-billed quelea (*Quelea quelea*), in their nighttime roosts, to protect ripening grain crops in some African countries. It is highly toxic to birds and moderately toxic to mammals. Death occurs in 3 to 12 hours. The risk of nontarget bird mortality (from both primary and secondary poisoning) and environmental contamination is high, especially following aerial application. The symptoms of poisoning (e.g., convulsions) indicate that fenthion is likely to be inhumane.

4-Aminopyridine (Avitrol® in the U.S. and Canada, Avis Scare® and Scatterbird® in Australia) is often described as a frightening agent, but it is also an oral toxicant. Birds that ingest it die, but before dying they exhibit erratic behavior and alarm calling (often termed distress behavior) that supposedly frightens away other birds in the flock before they are able to ingest it. Time to death ranges from 15 min to 3 days. It is used to control birds such as rock pigeons, house sparrows, starlings, and in the U.S., red-winged blackbirds (*Agelaius phoeniceus*). It is available as a concentrate or as ready-to-use treated grain to certified operators. It is highly toxic to both birds and mammals, and may cause both primary and secondary poisoning of nontarget species. Despite appearances to the contrary, it has been claimed that death from the compound is relatively painless. However, this needs to be verified because severe symptoms of intoxication may last for up to 3 days.

DRC-1339 (3-chloro-*p*-toluidine hydrochloride) (Starlicide®) is an oral toxicant used for the control of birds such as rock pigeons, starlings, and in the U.S., red-winged blackbirds. It is available as a concentrate or as a ready-to-use cereal-based bait to certified operators. Time to death varies from 3 to 50 hr, depending upon the amount of toxicant ingested. DRC-1339 is not suitable as a toxicant for all pest bird species because it is not highly toxic to all species. For example, it has only low toxicity to sparrows (Ploceidae) and finches (Fringillidae). It also has low toxicity to most mammals. This selective toxicity is unique. DRC-1339 is rapidly metabolized, so there is little risk of secondary poisoning. The death of birds from DRC-1339 has been described as painless, but symptoms such as difficult breathing indicate that this might not be so.

Alpha-chloralose is used in some countries (e.g., Australia and New Zealand) as an oral toxicant, but in other

Encyclopedia of Environmental Management DOI: 10.1081/E-EEM-120009966

Table 1 Chemicals currently used for bird control in United States of America (U.S.A.), Canada, United Kingdom (U.K.), France, Israel, Australia, and New Zealand (N.Z.).

Compound	Activity	Countries
Strychnine	Oral toxicant	Canada, Australia
Fenthion	Oral and dermal toxicant	Some African countries, Australia
4-Aminopyridine	Oral toxicant, frightening agent	U.S.A., Canada, Australia
DRC-1339	Oral toxicant	U.S.A., N.Z.
Alpha-chloralose	Oral toxicant, immobilizing agent	U.S.A., France, U.K., Israel, Australia, N.Z.
Seconal (+ alpha-chloralose)	Immobilizing agent	U.K.
Polybutene	Tactile repellent	U.S.A., Canada, U.K., Israel, Australia, N.Z.
Denatonium saccharide	Taste repellent	U.S.A., Canada
Aluminium ammonium sulfate	Taste repellent	U.K., Australia
Thiram	Taste repellent	France, Israel
Endosulfan	Taste repellent	France
Triacetate guazatine	Taste repellent	France
Methyl anthranilate	Irritant	U.S.A., Canada
Capsaicin	Irritant	U.S.A.
Naphthalene	Irritant	U.S.A.
Methiocarb	Secondary repellent	U.S.A., Canada, Israel, Australia, N.Z.
Ziram	Secondary repellent	U.K., France
Anthraquinone	Secondary repellent	U.S.A., France, N.Z.
Azacosterol	Reproductive inhibitor	Canada
Corn oil	Reproductive inhibitor	U.S.A.
Paraffin oil	Reproductive inhibitor	U.K.

Source: Adapted from Schafer,[1] Ministry of Agriculture[2] and Clark.[3]

countries only as an immobilizing agent (see below). It is available to certified operators as a concentrate or as ready-to-use treated grain, for the control of birds such as rock pigeons and house sparrows. It is generally more toxic to birds than to mammals, and is relatively fast-acting. The first signs of narcosis may occur 10 min after ingestion, and immobilization may last for up to 27 hr, though it generally lasts less than 1 hr, after which birds may recover. However, death may result from hypothermia if sufficient active ingredient is ingested, and/or the weather is inclement. Alpha-chloralose is only slowly metabolized, and so may cause secondary poisoning of nontarget species. It is considered to be relatively humane on the basis of the generally short time to insensitivity.[2]

LETHAL STRESSING AGENTS

PA-14 (Tergitol®) is a surfactant that was used as a lethal stressing agent in the U.S., but is no longer available for this purpose. It was sprayed onto birds, such as starlings and red-winged blackbirds, in their nighttime roosts, resulting in a break-down of the oil in the birds' feathers, destroying their natural waterproofing, and causing death from hypothermia.

IMMOBILIZING AGENTS

Immobilizing agents, administered in baits, are used to make birds easier to capture for removal from areas where they cause problems, or for killing humanely by other methods (e.g., by breaking their necks, or gassing them with carbon dioxide). Nontarget birds that become immobilized can be revived and released. However, the effectiveness of immobilizing agents depends upon the amount ingested and environmental conditions. All known immobilizing agents are lethal to birds if they ingest a sufficient quantity. The most commonly used immobilizing agent worldwide is alpha-chloralose, which is also used in some countries as a lethal toxicant (see above). In the U.S., it is available as an immobilizing agent only to approved operators, mainly to capture rock pigeons and waterfowl in nuisance situations. In the U.K., seconal is also used as an immobilizing agent, in combination with alpha-chloralose, to enhance its speed of action.

REPELLENTS

Chemical repellents can be primary or secondary in effect. Primary repellents are avoided reflexively because of an unpleasant sensation (e.g., touch, taste, smell, irritation). Tactile repellents include polybutene-based products (e.g., 4 The Birds®, Hot Foot®, and Tanglefoot® in the U.S., Bird-X, Buzz-Off®, Shoo, Super Hunter, and Waco in Canada). They are applied to buildings and other structures, modifying the surface so that it becomes sticky or slippery and discouraging birds from landing or roosting. They are all available to the general public.

Taste repellents, which discourage birds from eating potential food sources to which they are applied, include denatonium saccharide (Ro-Pel® in the U.S. and Canada) and aluminium ammonium sulfate (Curb, Guardsman, and Rezist in the U.K., D-ter, Gaard, and Scat in Australia). Ro-Pel® also contains thymol, a fungicide that imparts a secondary repellent effect. Irritants include methyl anthranilate (ReJeX-iT® and Bird Shield® in the U.S., Avigon in Canada), capsaicin (Sevana), and naphthalene (Dr. T's), although there is no evidence that the latter two, by themselves, are effective.[3] Methyl anthranilate may be applied to grassy areas such as parks and golf courses to deter feeding by birds such as Canada geese, and also to ripening fruit to deter birds such as house sparrows and starlings.

Secondary repellents cause post-ingestional illness, resulting in conditioned aversion to the treated food source. Examples include methiocarb (Mesurol®), ziram (AAprotect), and anthraquinone (Flight Control™ in the U.S., Avex™ in New Zealand). Methiocarb and ziram are moderately toxic to birds and mammals. In some countries, methiocarb may be applied to seeds and seedlings, but in the U.S. it may be used only in dummy egg baits to condition crows (*Corvus* spp.) not to prey on the eggs of endangered birds. Ziram and anthraquinone may be sprayed onto grass, field crops, ornamentals, conifers, and dormant fruit trees, but not onto products for immediate human consumption.

4-Aminopyridine is sometimes described as a frightening agent, and classified as a repellent, because it induces behavioral changes in birds. However, it is highly toxic to birds, and should be considered as a toxicant (see above).

REPRODUCTIVE INHIBITORS

Reproductive inhibitors have the potential to reduce bird populations by preventing or reducing the production of young. Azocosterol (Ornitrol®) is one of a number of chemicals that have been investigated for this purpose. It is applied to baits and fed to females daily for 10 to 15 days before egg-laying. It is no longer available in the U.S., but is still available for the control of rock pigeons in Canada.

Corn oil (in the U.S.) and paraffin oil (in the U.K.) are two chemicals used to destroy the eggs of birds, such as gulls (*Larus* spp.) and Canada geese (*Branta canadensis*), after they have been laid. The oil may be sprayed onto the eggs in the nest, or the eggs may be temporarily removed, immersed in oil, and then returned to the nest. The oil occludes the pores in the shell, asphyxiating the developing embryo. The technique is considered humane.[2]

FUTURE DEVELOPMENTS

No existing products are ideal for the control of pest birds. Toxicants are becoming increasingly publicly unacceptable worldwide from environmental and animal welfare perspectives. Currently, research is being done on the effectiveness of an oral toxicant/anaesthetic combination that reduces the time to unconsciousness, as a means of improving the animal welfare aspects of lethal bird control. Research is also being done on potential new repellents, including other derivatives of anthranilate, acetophenone, benzoate, cinnamamide, and d-pulegone. The use of nonlethal methods of bird control, especially repellents, may be a better option for the future than the use of toxicants.

REFERENCES

1. Schafer, E.W., Jr. Bird Control Chemicals—Nature, Modes of Action, and Toxicity. In *CRC Handbook of Pest Management in Agriculture*, 2nd Ed.; Pimentel, D., Ed.; CRC Press. Boca Raton, Florida, USA, 1991; Vol. 2, 599–610.
2. Ministry of Agriculture, Fisheries and Food. *Assessment of Humaneness of Vertebrate Control Agents*; Ministry of Agriculture, Fisheries and Food, Pesticides Safety Directorate: York, United Kingdom, 1997.
3. Clark, L. Review of Bird Repellents. In *Proceedings of the 18th Vertebrate Pest Conference*; Baker, R.O., Crabb, A.C., Eds.; University of California: Davis, California, USA, 1998; 330–337.

Birds: Pesticide Use Impacts

Pierre Mineau
Science and Technology Branch, Environment Canada, Ottawa, Ontario, Canada

Abstract
Pesticides are known to affect birds lethally and non-lethally, directly through intoxication and indirectly through their food supply. This entry reviews the types of impacts that have been recorded as well as the ways in which birds are exposed, how these impacts can be measured and how they can be reduced.

Birds inhabiting our farmland are in decline. We know this to be the result of agricultural intensification in which pesticide use plays a large direct and indirect role. Most pesticides used in developed countries no longer accumulate in birds, but they can poison birds and make them more susceptible to other causes of mortality.

More than 30 pesticides registered in North America or Europe have been known to result in kills of wild terrestrial vertebrates even when used according to the relatively stringent regulations in force in those countries. Among the species affected, birds figure prominently in the kill record and this, for several reasons. Birds are ubiquitous and visible. In North America, as in several other countries, most species are federally protected from unlicensed taking or kill. Birds are extremely mobile and cannot be excluded from areas that have been treated with pesticides. Some bird species are attracted to agricultural fields, and many are economically important to the control of agricultural pests, notably insects. Finally, birds, as a group, are particularly sensitive to some of the more toxic classes of pesticides such as the organophosphorus and carbamate insecticides (fortunately, the use of these compounds is in decline), and their reproduction has been found to be potentially affected by a wide range of pesticides. New pesticides developed in part for their relative safety to humans have been found to be especially toxic to birds.

In addition, pesticides are known to alter birds' basic requirements of food and shelter. Loss of food, especially, has been linked to population declines of several farmland bird species in Europe where this has been studied extensively.

Several different strategies are employed to study bird impacts, ranging from monitoring of pesticide applications to surveys of birds in farms subject to different pesticide regimes. Modeling has now given us an estimate of the yearly losses to acutely toxic pesticides; the full impact of pesticide use on birds is more difficult to assess and remains controversial.

HOW SERIOUS IS THE IMPACT?

Most of our farmland bird species appear to be declining globally and even common species are experiencing long-term declines, both in North America and in Europe.[1,2] For example, 76% of common grassland species in Canada are declining. The proportion of species declining or showing range contractions in the United Kingdom is higher still. Less is known about common farmland bird species outside of Europe or North America, but farming worldwide has been implicated in declines of specific groups of birds such as raptors. It is difficult to isolate the specific factors responsible for these declines: it is likely that a combination of factors is to blame and each species must be considered on a case-by-case basis. Agricultural landscapes have changed dramatically in the 20th century. There has been a shift from mixed agriculture including row crops, field crops, and livestock to more specialized farming where monocultures are regionally dominant. Field size has increased to accommodate larger machinery, and this increase has been often at the expense of marginal non-crop habitats such as fencelines, ditches, hedgerows, windbreaks, and remnant woodlots. In Europe especially, a shift to autumn sowing of grain crops has meant that much of the waste grain traditionally available to birds postharvest and throughout the lean winter months is no longer available.

Agricultural inputs in the form of synthetic fertilizers and pesticides have increased dramatically also and, increasingly, have been found to be contributing to bird declines. Two decades ago, a long-term study of declining grey partridge (*Perdix perdix*) populations in the United Kingdom identified insect prey reductions resulting from both insecticide and herbicide use as the main contributing factor. More recently, researchers in several regions of North America and Europe have shown that organic farms tend to support a higher diversity and abundance of birds even when matched for habitat characteristics. The reproductive success of some farmland species such as the Eurasian skylark (*Alauda arvensis*) is higher on organic or

Encyclopedia of Environmental Management DOI: 10.1081/E-EEM-120046191

reduced input farms than it is on more "conventional" ones. The use of toxic granular insecticides for oilseed production has contributed to grassland bird population declines in the Canadian prairies.[3] It has been recently suggested[4] that broad aquatic contamination by the new neonicotinoid class of seed-treatment insecticides could be the main reason behind the decline of insectivorous bird species; this is expected to be a hot research area and point of debate in coming years. Taken as a whole, these results implicate current agricultural practices, and pesticide use in particular, in the decline of several farmland species. This is doubly unfortunate because, with a few exceptions, birds can play a useful role in integrated pest management systems.[5]

TYPES OF BIRD IMPACTS

There are several mechanisms through which pesticides can affect birds. The case of the grey partridge, Eurasian skylark, and other European species has shown that the effect can be an indirect one, mediated through "weed" removal and loss of insect biomass at critical times of the breeding season.[6] Herbicide use has increased dramatically in the past decades, and herbicide sales far surpass insecticide sales in North America and Europe at least. However, several direct mechanisms through which birds are impacted are also recognized.

Persistent Organochlorine Pesticides

Historically, several species of raptors such as the Eurasian sparrowhawk (*Accipiter nisus*) and peregrine falcon (*Falco peregrinus*) as well as fish-eating species such as the brown pelican (*Pelecanus occidentalis*) faced serious difficulties and regional extinction as a result of persistent organochlorine pesticides such as DDT (dichlorodiphenyltrichloro ethane), aldrin/dieldrin, chlordane, and heptachlor. These were poorly metabolized and poorly excreted by birds and accumulated in fatty tissue. The impact of such substances was twofold. Some, such as aldrin and dieldrin, caused frequent poisonings, especially during lean times when birds metabolized their fat reserves and the pesticides reached extreme concentrations in the brain. Others, such as DDE (dichlorodiphenyldichloroethylene), a breakdown product of DDT, interfered with the bird's ability to lay eggs with normal shells. These substances were banned or severely restricted in most of the developed world in the early 70s. Yet, lower reproduction of birds breeding in areas with high historical usage is still being documented because of long persistence in soils. Several of these pesticides continue to be used massively in parts of the world such as the Indian subcontinent although current impacts on bird life are poorly documented. By and large, modern pesticides do not show such extreme persistence, at least in warm-blooded organisms.

Lethal Effects

The acute oral toxicity (LD_{50}) of a pesticide and the extent of its use are good predictors of wildlife kills.[7] The dietary toxicity test currently carried out on young birds (dietary LC_{50}) can seriously mislead however. As a rule, insecticides and vertebrate control agents are much more likely than herbicides or fungicides to give rise to wildlife kills. Two groups of pesticides, the organophosphorus and carbamate insecticides, were initially introduced to replace persistent organochlorines. Unfortunately, they proved particularly toxic to birds. Their mode of action (inhibition of the enzyme acetylcholinesterase in the nervous system and at neuromuscular junctions) is not specific to the pests and affects a broad range of vertebrates and invertebrates alike. Birds are especially vulnerable because their ability to detoxify these pesticides is generally much lower than that of mammals. The more toxic products such as the carbamate insecticide carbofuran killed thousands of individuals in a single application. Reports of large numbers of North American birds being poisoned on their wintering grounds in Latin America by the insecticide monocrotophos have emphasized the need to consider bird impacts in a hemispheric, if not global, context.[8] The poisoning of birds is largely inevitable where acutely toxic pesticides are registered at high rates of application and used broadly. Of particular concern have been granular insecticides and seed treatments because birds are often attracted to them.[9] Pesticide poisonings can be a significant source of mortality relative to other factors, especially in the case of long-lived species such as birds of prey.[10]

There have been very few attempts to estimate the total incidental take resulting from direct intoxications following the use of toxic pesticides anywhere. Pimentel,[11] in an oft-cited study, estimated that pesticide-induced direct mortality totaled approximately 67 million per year in the United States. He based this estimate on the fact that 160 million ha of cropland received a very heavy dose of pesticides per year (3 kg a.i./ha on average—including a number of very toxic pesticides), a breeding density of 4.2 birds/ha (from census plot data), and a conservative kill estimate of 10% of exposed birds. This estimate ignores kills of wintering birds, which could be substantial. Also, some of the largest kills recorded in North America have been of migrants (e.g., Lapland longspurs (*Calcarius lapponicus*)), which would not be captured in estimates based on breeding densities in farmland.

The carbamate insecticide carbofuran (Furadan™) was very broadly used in North America and has been studied more than any other insecticide.[12] The manufacturer's own studies on a granular formulation of carbofuran as well as search efficiency and scavenging studies were used to provide an estimate of bird mortality per treated surface.[13] Two major field studies, both from the United States, were retained for purposes of extrapolation. Estimated kill rates were 3.05 birds/ha for an Iowa site (once raw carcass counts were corrected for scavenging and for unsearched areas of the field) and 15.9 birds/ha for an Illinois site with better off-field habitat nearby. A third study gave estimates that were simply too high to lead to a kill rate that could safely be extrapolated; fully 799 carcasses of a single species (horned lark—*Eremophila alpestris*) were recovered from slightly more than 100 ha of crop. Based on the two lesser kill rates, it was estimated that, at the height of its popularity, in the late 70s to mid-80s, this single pesticide was killing approximately 17 to 91 million songbirds annually in the 32 million ha of U.S. corn (maize) fields alone.[13]

Fortunately, several of the more toxic organophosphorous and carbamate insecticides are being phased out and replaced in North America and Europe—although their use may still be increasing in the developing world. Their cancellation was not out of a concern for birds[14] but rather an attempt to reduce risks to consumers and applicators under new legislation that demanded the assessment of cumulative impacts from pesticides with the same mode of action. The result of these product cancellations, however, has been a definite reduction in the proportion of our crop area where birds are at risk of lethal poisoning.[15] The authors estimated that the cumulative number of cropped hectares over which avian mortality was probable decreased from about 17 million ha in 1997 to about 6 million ha 5 years later.

The measurement of brain cholinesterase levels was an extremely useful (although certainly not foolproof) diagnostic tool for bird mortality from these classes of pesticides. The test has the advantage of being economical and relatively easy to carry out. Wildlife kills resulting from newly developed insecticides will be harder to elucidate in the absence of such a convenient biomarker. Diagnosis will hinge on sophisticated and costly residue analyses without the benefit of the "smoking gun," which cholinesterase titers represented.

Secondary Poisoning

Secondary poisoning occurs when predators, such as hawks or owls, consume prey contaminated by pesticides. Such predators are few because of their position at the top of the food chain. Therefore, the death of one predator may constitute a significant reduction in the local population of that species. Historically, researchers have associated secondary poisoning with persistent organochlorine insecticides and other substances that are not readily metabolized and that accumulate in tissues. However, other currently registered pesticides can cause secondary poisoning when the predator encounters the pesticide in a high concentration on the surface or in the gastrointestinal tract of its prey. Also, predators capture birds debilitated by insecticides much more easily.

Sublethal Effects and Delayed Mortality

Many pesticides can affect the normal functioning of exposed individuals at doses insufficient to kill them directly. At high doses, the organophosphorus and carbamate pesticides previously described cause respiratory failure and death. However, wild birds exposed to these agents in lesser amounts have experienced impaired coordination, weight loss, an inability to maintain body temperature, and loss of appetite. Also, exposed birds may spend less time at the nest, provide less food for their young, be less able to escape predation, and be more aggressive with their mates. Finally, exposure to some pesticides may reduce resistance to disease.

Effects on Reproduction

A high proportion of pesticides currently registered have the potential to affect reproduction by reducing egg production, hatching, or fledging success, although the extent to which this actually happens in the wild is not known.[16,17] A few products cause embryonic mortality when sprayed directly onto eggs.

ROUTES OF PESTICIDE EXPOSURE IN BIRDS

Birds ingest pesticides through their food or through preening or grooming. Despite being feathered, they absorb pesticides through their skin, encountering droplets directly or by rubbing against foliage and other contaminated surfaces. Birds are also exposed through their feet. Finally, they have a very high ventilation rate and inhale vapor and fine droplets. The degree to which each of these routes of exposure contributes to the total dose depends on the crop being sprayed, the chemical, the species exposed, and environmental factors. Evidence to date suggests that the dietary route is not necessarily the dominant route of exposure in birds under most situations. Yet, this is the only route currently assessed by regulators worldwide—a mind-set that clearly needs to change if we want to be proactive in protecting birds from pesticides.

Although the relative importance of different exposure routes is difficult to ascertain on a case-by-case basis, it is possible to recognize different situations that arise where birds are massively exposed to pesticides and are often poisoned as a result.

Abuse and Misuse

Pesticide abuse is the deliberate use of a pesticide in a non-authorized fashion, usually to poison wildlife species considered to be pests. In the United Kingdom, as well as in several European countries, officials estimate that deliberate bird kills due to pesticide abuse outnumber cases where label instructions were strictly followed. Between 1978 and 1986, officials in the United Kingdom estimate that, on average, 71% of incidents were the result of abuse. For birds of prey alone, more than 90% of cases recorded between 1985 and 1994 in the United Kingdom were abuse cases. On the other hand, for raptors in the United States during the same period, kills involving labeled uses of pesticides were almost as frequent as abuse cases. This difference appears to be wholly attributable to the high toxicity of insecticides used in the United States. Abuse generally involves baits of some kind, the only limit being the imagination of the perpetrator. Typically, liquid insecticides are poured or injected and applied to seed, bread, meat, etc., and granules are sprinkled or mixed into a paste. Because of the high concentration of pesticide involved in abuse cases, carcasses are usually found in close proximity to the site of baiting, thus biasing the kill record through a higher recovery of carcasses. The choice of chemicals used in abuse cases reflects availability and toxicity. Pesticides typically used in deliberate poisoning attempts include carbofuran, aldicarb, monocrotophos, parathion, mevinphos, diazinon, and fenthion, chemicals that are all recognized as being inherently very toxic to vertebrates in general and birds in particular. The main problem of course is that the baits are often indiscriminate in the species that they kill. Secondary poisoning is also frequent when predators or scavengers take dead or debilitated prey with highly concentrated bait in their gut.

Pesticide misuse refers to a pesticide application that is not exactly as specified by the label. This may be an application at a rate that is higher than specified or an application to a crop or pest other than those listed. Alternatively, the user may not have the legal permission to use a certain product even if he followed label directions to the letter. Pesticide misuse is difficult to establish, especially after the fact. In many cases, it becomes very difficult to distinguish a misuse from a normal agronomic use when the label contains instructions that are vague, difficult, or impossible to follow. What constitutes a misuse in one jurisdiction may indeed be an approved use elsewhere.

Granular Formulations and Treated Seed

Granular insecticides and treated seeds are frequent routes of exposure and intoxication in birds. Granular insecticides were designed for convenience, safety to applicators, and time release of the chemical, yet for birds, granular formulations of the more toxic insecticides such as aldicarb, parathion, carbofuran, fensulfothion, phorate, terbufos, fonofos, disulfoton, diazinon, and bendiocarb are repeatedly associated with bird mortality. Several bird and small mammal species have a fatal attraction to granular formulations, mistaking them either as dietary grit or as a food source. The most attractive granules are those made of sand (silica) or an organic base such as dried corn (maize) cobs. Somewhat less attractive are clay, gypsum, and coal granules. Exposure can also occur via invertebrates, especially earthworms to which granules easily adhere. Secondary toxicity is likely in predators and scavengers that eat their prey whole or ingest their gastrointestinal tract contents. In Canada and the United States, there have been cases of poisoning of waterfowl foraging in puddles in fields more than 6 months after applications of granular insecticides because of specific soil conditions that may retard breakdown.[18] Granules that are friable and disintegrate quickly when exposed to moisture are best for birds, but they are the products least convenient to farmers. Regardless of the type of carrier, a pesticide granule is likely to be a problem if a lethal dose can be obtained in a few granules only.

To date, *no* agricultural machinery or application technique can achieve complete incorporation of granular insecticides below the soil horizon. Birds have also been known to probe the soil for granules or to pull up germinating seeds with granules attached. The worst applications are those made above the soil surface and "banded" or "side dressed" over or to the side of the seed furrow. In carefully controlled engineering trials, between 6% and 40% of applied granules were left on the soil surface. The same equipment can achieve radically different soil incorporation when used by different individuals under different conditions.

Treated seeds present a similar engineering problem. As with granules, more seeds are left on the surface wherever the seeders have to turn or negotiate obstacles. Small spills are part of normal farming practice and can occur anywhere depending on topography and soil conditions but more often at field edge. Historically, seed dressings were one of the main sources of bird exposure to organochlorine and mercurial compounds. Poisoning incidents with seed dressings are still relatively frequent because several bird species make heavy use of waste (or even planted) grain in fields. The size and type of seed dictate which bird species are at risk. Since use of organochlorines and organomercurials has declined, kills have been recorded with cholinesterase-inhibiting insecticides such as carbophenothion, chlorfenvinphos, isofenphos, bendiocarb, disulfoton, furathiocarb, and fonofos. Some kills have been recorded with newer insecticides as well, e.g., the neonicotinoid imidacloprid, although it is not yet known how serious or frequent a problem this will become.

Liquid Formulations on Vegetation: The Grazing Problem

Grazing birds are particularly vulnerable to foliar applications of pesticides. Kills have been recorded with several cholinesterase-inhibiting pesticides such as parathion, diazinon, carbofuran, isofenphos, dimethoate, and triazophos. Grazers typically include geese, ducks, and coots (families Anatidae and Rallidae). These birds eat large quantities of foliage because they do not digest cellulose. Fertilized areas are particularly attractive to grazing species that can detect the high nitrogen levels. Golf courses attract grazers because the turf is cut frequently, watered, and fertilized, and courses often have other attractions such as ponds and drainage streams. More than 100 cases of waterfowl mortality were recorded due to the use of diazinon on turf[19] before the pesticide was withdrawn from golf courses and sod farms in the United States. Other well-documented problems are kills of ducks and geese in alfalfa fields treated with carbofuran and of sage grouse (*Centrocercus urophasianus*) feeding on alfalfa crops treated with dimethoate or on potato foliage and weeds in potato fields sprayed with methamidophos.

Liquid Formulations on Insect Prey: The Gorging Problem

Bird species that feed on agricultural pests such as grasshoppers, leatherjackets (larvae of the crane fly), grubs, and cutworms are at high risk of poisoning. Kills of these species are all the more tragic because they are beneficial to agriculture. Some species are particularly vulnerable because they specialize in insect outbreaks. These birds take advantage of pest control operations that result in insects becoming either debilitated or more visible following treatment. In a well-studied case in Argentina, approximately 20,000 Swainson's hawks were poisoned within the span of a few weeks after feeding on grasshoppers sprayed with monocrotophos. As with carbofuran, the extreme toxicity of this product means that it is difficult to find use patterns that do not result in bird kills.

Vertebrate Control Agents: Unintended Victims

Rodenticides as a rule are not specific to their intended targets and cause direct impacts to non-target species. Only a detailed knowledge of the habits of the target species and use of specific baiting locations or specialized bait holders can reduce kills of non-target species. More problematic is secondary poisoning. Historically, the use of thallium and endrin to control rodents has had disastrous consequences on raptors. Recently, the trend has been to use more efficacious "single feed" anticoagulants; these present a greater hazard to predators than the older products (e.g., warfarin, diphacinone, chlorophacinone). The new "super coumarin"-type products include compounds such as difenacoum, brodifacoum, bromadiolone, difethialone, and flocoumafen—all extremely toxic and very long lived in liver tissue, thereby increasing the likelihood of secondary poisoning. A recent analysis of Canadian data suggests that approximately 11% of the great horned owl (*Bubo virginianus*) population of southern inhabited Canada is at risk of fatal poisoning from anticoagulants.[20]

Fenthion, an organophosphorus "insecticide" used to control pest birds in Africa (e.g., *Quelea quelea*) and in North America (e.g., by means of the Rid-a-Bird™ perch system), has given rise to frequent secondary poisoning.[21] Secondary poisoning is also very likely following the use of toxic organophosphorous or carbamate products for the control of parakeets, doves, and other seed eaters. The use of organophosphorus pesticides such as famphur and fenthion for the treatment of parasites in livestock frequently leads to wildlife kills. Famphur, which was one of the leading causes of eagle poisonings in the American Southwest, persists on the hair of cattle up to 100 days after treatment. Magpies are poisoned when they eat the hair, and eagles are poisoned when they scavenge the magpies. Medicated feed at livestock feed yards is another high-exposure situation. Sparrows, starlings, and other birds pick up the feed and subsequently are scavenged by hawks and eagles.

Forestry Insecticides

Forestry uses of toxic insecticides deserve special consideration because the terrain and method of application result in kills being difficult or impossible to detect. In a forestry situation, critical wildlife habitat is sprayed directly, and a large number of individuals of many species are exposed to the chemical. In Canada, the forestry insecticides phosphamidon and fenitrothion were canceled after impacts on birds were judged unacceptable. Although fenitrothion is not as acutely toxic as a number of other anti-cholinesterase insecticides used in agriculture, its use in forestry led to severe and widespread inhibition of brain acetylcholinesterase in a number of songbird species as well as some reports of kills.

MEASURING BIRD IMPACTS

Incident Monitoring

Incident monitoring refers to the capacity of competent authorities to investigate reported kills or conduct spot checks of use conditions. Even if a pesticide has been studied extensively under controlled conditions, unforeseen problems and situations often arise following commercialization of the product. An absence of incident reports does not necessarily mean there are no problems but, conversely, well-investigated incidents and kills can reveal unforeseen aspects of a pesticide or reinforce a suspicion that arose in the course of laboratory or field testing. An incident monitoring scheme will require a network of individuals trained in carrying out pesticide investigations and in proper handling of carcasses and tissue samples, as well as access to a laboratory equipped to perform the required chemical and biochemical analyses.[22]

Even where relatively efficient incident monitoring systems are in place, only a very small proportion of kills are ever uncovered. There are several reasons for this: affected wildlife are often dispersed and at relatively low density in farm fields, they often leave the treated area to die, they are likely to seek cover and hide when overcome by the pesticide, they are often cryptic and hard to see, and their carcasses are scavenged rapidly after death. Typical rates of carcass removal by scavengers are 40%–90% in the first 24 hr. Farm fields are large; the mechanization and sheer size of modern agricultural machinery often remove the farmer from any "close contact" with the land. The increasing size of farms also means that, when kills occur, often in the few days that follow a pesticide application, the farmer is busy elsewhere, treating another part of the farm. Pesticide intoxication may be a causal factor in a kill visibly caused by something else—e.g., intoxicated birds hitting fences, utility wires, cars, or buildings—and not be recognized as a pesticide kill. Also, there is a large difference between casual searching of fields and a well-organized intensive search effort. An intensive search effort consisting of several trained individuals, transects, and repeated, well-timed searches have produced between 10- and 500-fold improvements in carcass detection rates over field inspections carried out once or a few times only by single individuals. Equally important is the motivation and training of the search teams. Finally, a proper investigation of kills can be expensive and out of the reach of many jurisdictions despite the availability of inexpensive biomarkers—currently at least.[23]

Even when incidents are uncovered, they are often not reported. If the kills involve only one or a few individuals, not much importance is attached to the incident even though, for reasons just outlined, these few carcasses likely represent the "tip of the iceberg." Even if the kills are reported, the information is often not centralized and made available to national pesticide regulatory bodies. It is important to understand and recognize biases inherent in any incident reporting system. Some of those biases will depend on how the incident monitoring system is set up and which persons/organizations are responsible. Some biases can be reduced over time, but others are unavoidable. Common biases relate to body size and color of the casualties, numbers and density of the species in any given area, "status" of the species, and individual and institutional interests and sensitivities. We expect most kills to be of small-bodied birds widely dispersed in field margins. Yet, such kills are seldom reported.

Despite these limitations, it is important for countries to investigate wildlife kills and make the information available.[24] Registration decisions are made on the basis of very limited information. There are large differences in toxicological and ecological vulnerability among species. The ways in which wildlife species are exposed to pesticides are varied and sometimes difficult to predict or study. The behavior of pesticides depends on local conditions although pesticides are often tested under standardized conditions only. The outcome of exposure is also much more variable in the wild. Pesticide exposure can interact with weather, the condition or health of the animal, etc. Therefore, whether or not pesticides are routinely field tested to look for environmental impacts, it is essential to have a good incident monitoring system in place. An incident monitoring system can also be useful to warn manufacturers if their products are abused or used incorrectly.

Field Testing: Active Monitoring

Carrying out a field study to measure the impact (or lack of impact) of a specific pesticide usually consists of the surveillance of a group of birds prior to, during, and after the application of the pesticide according to label instructions. Researchers observe or count individuals of one or more species within and outside the treated area and record their behavior. Frequently, they search for carcasses in order to determine the extent of pesticide-induced mortality.[25] for an example of how these data can yield useful predictive models.) They may capture birds to ascertain the health of individuals or to collect samples, for example, blood or brain tissue for biochemical assays or feathers and foot rinses for residue determinations. Agricultural engineering studies (e.g., measurements of granular insecticides or treated seed remaining on the surface) or monitoring of pesticide residues remaining on avian food items over time provides valuable information on expected exposure levels. The most sophisticated field studies will involve monitoring nests, as well as banding, marking, or radio-tagging individuals in order to assess turnover rates and help locate sick and dead birds. Rare, vulnerable, or ecological keystone species can be used as indicator species where relevant and feasible.

It is not always feasible to investigate the effects of a single pesticide. In a number of cropping situations, several pesticides are used as a mix or in quick succession, making the identification of compound-specific impacts difficult. In agricultural systems, the mosaic of treated fields can be so complex as to make it difficult to assess exposure to any one pesticide. Two approaches then suggest themselves: 1) treated sites or landscapes are compared to non-treated areas, provided those can be found, and 2) the "severity" of treatment (the *a priori* expectation of toxicity) for any given site is used as a variable against which a number of different parameters (such as reproductive success) are regressed. Great care must be taken in comparing treated to non-treated areas because they are likely to differ in other ways as well.

Surveys

Data from regional or national surveys of bird population levels are rarely adequate to demonstrate specific pesticide impacts, although surveys can point to a general situation of bird declines in farmland. In order to carry out wildlife monitoring in treated areas, it is necessary to have a good knowledge of the normal complement of species for the area of concern and to be able to assess the vulnerability of each of these species during and after pesticide treatments. The diversity or abundance of species may already have been affected by previous pesticide use so that only a complement of the more insensitive species remains available for testing.

Regardless of the strategy employed, more attention needs to be paid to the impact of pest control practices on bird species if we are to reverse the current trends of population declines.

Modeling

The probability of finding a bird kill (of any size) following a pesticide application was derived from models based on a large sample of empirical field studies where known insecticides were applied, and searching was carried out to detect casualties.[25] Models were developed for field and orchard crops separately. Because few of the studies were quantitative in nature, logistic modeling was used and the output of the models is the likelihood that a kill of undefined size would occur and be found assuming an adequate search effort.

Species most frequently implicated in kills are those that are cosmopolitan, closely associated with agriculture, and reasonably common, e.g., mourning doves (*Zenaida macroura*); several sparrows, horned larks, and meadowlarks (*Sturnella* spp.), American robins (*Turdus migratorius*); house sparrows (*Passer domesticus*); and several blackbird species. However, the sheer diversity of birds potentially killed by pesticides is impressive and suggests that toxicological or ecological susceptibility is less important than being simply in the wrong place at the wrong time.

CONCLUSION: THE WAY FORWARD

The pesticide industry has clearly shown that it is incapable of policing itself when it comes to reducing or eliminating impacts on birds. Elsewhere,[14,26] I have reviewed some of the more egregious cases, showing how it often takes decades of legal wrangling to remove clear problem pesticides from the market in developed countries while the use of those same compounds continues to increase in the developing world, sometimes affecting the same bird populations (e.g., neotropical migrants). The extent to which birds are protected from pesticide impacts in various countries depends very much on public opinion and on the effectiveness of bird conservation groups. The pesticides responsible for most of the impacts on birds around the world (at least the direct impacts) tend to be the same group of depressingly familiar products. Fortunately, they usually can be replaced by better alternatives without risk to the livelihood of farmers or food security. Adopting better laws and regulations to protect the environment against pesticide use is part of the answer; enforcing those regulations in the face of a very strong pesticide lobby is undoubtedly the biggest hurdle.

REFERENCES

1. Askins, R.A. Population trends in grassland, shrubland, and forest birds in eastern North America. Curr. Ornithol. **1993**, *11*, 1–34.
2. Sirawardena, G.; Baillie, S.R.; Buckland, S.T.; Fewster, R.M.; Marchant, J.H.; Wilson, J.D. Trends in the abundance of farmland birds: A quantitative comparison of smoothed Common Birds Census indices. J. Appl. Ecol. **1998**, *35*, 24–43.
3. Mineau, P.; Downes, C.M.; Kirk, D.A.; Bayne, E.; Csizy, M. Patterns of bird species abundance in relation to granular insecticide use in the Canadian prairies. Ecoscience **2005**, *12* (2), 267–278.
4. Tennekes, H. The systemic insecticides: a disaster in the making. Weevers Walburg Communicatie, Zutphen, Netherlands, 2010, 72 pp.
5. Kirk, D.A.; Evenden, M.D.; Mineau, P. Past and current attempts to evaluate the role of birds as predators of insect pests in temperate agriculture. Curr. Ornithol. **1997**, *13*, 175–269.
6. Campbell, L.H.; Avery, M.I.; Donald, P.; Evans, A.D.; Green, R.E.; Wilson, J.D. A review of the indirect effect of pesticides on birds. Joint Nature Conservation Committee Report No. 227. JNCC: Peterborough, England, U.K., 1997.
7. Mineau, P.; Baril, A.; Collins, B.T.; Duffe, J.; Joerman, G.; Luttik, R. Reference values for comparing the acute toxicity of pesticides to birds. Rev. Environ. Contam. Toxicol. **2001**, *170*, 13–74.
8. Hooper, M.J.; Mineau, P.; Zaccagnini, M.E.; Winegrad, G.W.; Woodbridge, B. Monocrotophos and the Swainson's hawk. Pestic. Outlook **1999**, *10* (3), 97–102.
9. Stafford, T.R.; Best, L.B. Bird response to grit and pesticide granule characteristics: Implications for risk assessment and risk reduction. Environ. Toxicol. Chem. **1999**, *18* (4), 722–733.
10. Mineau, P.; Fletcher, M.R.; Glazer, L.C.; Thomas, N.J.; Brassard, C.; Wilson, L.K.; Elliott, J.E.; Lyon, L.A.; Henny, C.J.; Bollinger, T.; Porter, S.L. Poisoning of raptors with organophosphorous pesticides with emphasis on Canada, U.S. and U.K. J. Raptor Res. **1999**, *33*, (1), 1–37.
11. Pimentel, D.; Acquay, H.; Biltonen, M.; Rice, P.; Silva, M.; Nelson, J.; Lipner, V.; Giordano, S.; Horowitz, A.; D'Amore, M. Environmental and economic costs of pesticide use. BioScience **1992**, *42*, 750–760.
12. Richards, N., Ed. *Carbofuran and Wildlife Poisoning: Global perspectives and forensic approaches*; Wiley-Blackwell, Chichester, U.K. 2012, 277 pp.
13. Mineau, P. Direct losses of birds to pesticides—Beginnings of a quantification. In *Bird Conservation Implementation and Integration in the Americas*, Proceedings of the Third International Partners in Flight Conference 2002; Ralph, C.J., Rich, T.D., Eds.; U.S.D.A. Forest Service, GTR-PSW-191, Albany, CA, 2005; Vol. 2, 1065–1070.
14. Mineau, P. Birds and pesticides: Are pesticide regulatory decisions consistent with the protection afforded migratory bird species under the Migratory Bird Treaty Act? William Mary Environ. Law Policy Rev. **2004**, *28* (2), 313–338.
15. Mineau, P.; Whiteside, M. The lethal risk to birds from insecticide use in the U.S.—A spatial and temporal analysis. Environ. Toxicol. Chem. **2006**, *25* (5), 1214–1222.
16. Mineau, P.; Boersma, D.C.; Collins, B. An analysis of avian reproduction studies submitted for pesticide registration. Ecotoxicol. Environ. Saf. **1994**, *29*, 304–329.
17. Mineau, P. A review and analysis of study endpoints relevant to the assessment of "long term" pesticide toxicity in avian and mammalian wildlife. Ecotoxicology **2005**, *14* (8), 775–799.

18. Elliott, J.E.; Wilson, L.K.; Langelier, K.M.; Mineau, P.; Sinclair, P. Secondary poisoning of birds of prey by the organophosphorus insecticide, phorate. Ecotoxicology **1996**, *5*, 1–13.
19. Frank, R.; Mineau, P.; Braun, H.E.; Barker, I.K.; Kennedy, S.W.; Trudeau, S. Deaths of Canada geese following spraying of turf with diazinon. Bull. Environ. Contam. Toxicol. **1991**, *46*, 852–858.
20. Thomas, P.J.; Mineau, P.; Shore, R.F.; Champoux, L.; Martin, P.; Wilson, L.; Fitzgerald, G.; Elliott, J.E. Second generation anticoagulant rodenticides in predatory birds: Probabilistic characterisation of toxic liver concentrations and implications for predatory bird populations in Canada. Environ. Int. **2011**, *37* (5), 914–920.
21. Hunt, K.A.; Bird, D.M.; Mineau, P.; Shutt, L. Secondary poisoning hazard of fenthion to American Kestrels. Arch. Environ. Contam. Toxicol. **1991**, *21*, 84–90.
22. ASTM. *Standard Guide for Fish and Wildlife Incident Monitoring and Reporting*; American Society for Testing and Materials: Philadelphia, PA, Standard E 1997; 1849–1896.
23. Mineau, P.; Tucker, K.R. Improving detection of pesticide poisoning in birds. J. Wildl. Rehab. **2002**, Part 1: *25* (2), 4–13; Part 2: *25* (3), 4–12.
24. Greig-Smith, P.W. Understanding the impact of pesticides on wild birds by monitoring incidents of poisoning. In *Wildlife Toxicology and Population Modeling*, SETAC Special Publication Series; Kendall, R.J., Lacher, T.E., Eds.; CRC Press, Inc.: Boca Raton, FL, 1994; 301–319.
25. Mineau, P. Estimating the probability of bird mortality from pesticide sprays on the basis of the field study record. Environ. Toxicol. Chem. **2002**, *24* (7), 1497–1506.
26. Mineau, P. Birds and pesticides: Is the threat of a silent spring really behind us? 2009 Rachel Carson Memorial Lecture, Pesticide News **2009**, *86*, 12–18, available at http://www.pan-uk.org/pestnews/Free%20Articles/PN86/Birds%20and%20pesticides.pdf.

Boron and Molybdenum: Deficiencies and Toxicities

Umesh C. Gupta
Crops and Livestock Research Center, Agriculture and Agri-Food Canada, Charlottetown, Prince Edward Island, Canada

Abstract

There is an extensive body of information available, which is related to boron (B) and molybdenum (Mo) in crops and soils. This entry places major emphasis on the toxicity aspect of these two micronutrients. In nature, mineral toxicities are mainly due to overfertilization and due to the addition of waste materials, such as sewage sludge, mine tailings, and fly ash, containing high amounts of these elements. Toxicity of these elements is reported in regions derived from rocks containing high amounts of these two elements. Certain environmental factors and soil physical properties can cause accumulation of these elements in quantities toxic to crops, livestock, and humans. Data reported in this investigation include information on sources and management of toxicities by methods such as phytoremediation, judicious use of waste materials, use of crop cultivars less sensitive to toxicity, and development of varieties tolerant to excesses of these elements.

INTRODUCTION

Boron and Mo are among the chief micronutrients essential for plant growth, which exist in soil solutions as anions. Because of their anionic nature, B and Mo tend to leach rapidly from soils because they are not retained by the cation exchange complex. Boron deficiencies occur in many countries of the world on coarse textured soils in humid regions where leaching and heavy cropping have diminished the soil B reserves. Boron toxicity is not as widely distributed, as is B deficiency. Boron toxicity occurs in soils, which are inherently high in B and as a result of high B fertilization, for example, if a B-sensitive crop were to follow a crop, which received higher rates of B, or through the use of irrigation water high in B. Crops sensitive to B, e.g., peanuts (*Arachis hypogaea* L.), could exhibit B toxicity[1] at levels as low as 2 kg B ha^{-1}. Details on B deficiency and toxicity have been described elsewhere.[2]

Molybdenum deficiencies are common in many countries where soils are acidic in nature and coarse in texture. Molybdenum toxicities in crops tend to be rare in distribution and occur only when unusually high quantities of Mo are present. High Mo concentrations have been reported in some poorly drained soils, e.g., in Australia, Scotland, and the United States. Readers are referred to a book on Mo[3] for details of Mo deficiencies and toxicities.

This entry contains a condensed version of factors such as soil, water, environment, and genetics that affect B and Mo toxicities in crops and livestock. Some suggestions will be made on the management and reduction of toxicities due to B and Mo.

DISCUSSION

Boron

Boron is biologically an essential element but is toxic to plants at high concentrations (approximately above >1 mg B L^{-1} in irrigation water) and probably to humans.[4] The European Union (EU) Drinking Water Directive fixes a threshold of 1 mg L^{-1} and the World Health Organization (WHO) earlier set a recommended limit at 0.3 mg L^{-1}, which now has been increased to 0.5 mg L^{-1}.[4]

The major sources of B for humans include fruits, vegetables, nuts, and legumes, which are naturally rich in B. Human dietary consumption of B is below the estimated tolerable intake levels established in these recent risk assessments.[5]

Soil pH is an important factor affecting the availability of B in soil and crops. Generally, B becomes less available to plants with increasing soil pH or when Ca is added in excess.[6]

Higher quantities of available B are generally found in fine-textured rather than in coarse-textured soils.[7] The lower amounts of B in sandy soils are associated with higher leaching of B. Increasing soil salinity levels, particularly at high B levels, have been found to decrease B concentration in plants.[8] Irrigation waters with high B can result in B toxicity in crops such as chickpeas (*Cicer arietinum* L.).[8] Generally, groundwater emanating from light textured soils is higher in B than that from heavy textured soils. On clay loam soils, rates of up to 8.8 kg B ha^{-1} are not toxic, while on sandy loam soils, this rate

Encyclopedia of Environmental Management DOI: 10.1081/E-EEM-120046335

was toxic to cauliflower (*Brassica oleracea* var. Botrytis L.).[9] Other factors include genotypes and plant species; e.g., legume species can accumulate and tolerate more B than grasses.[10] Details on soil and environmental factors, such as light intensity, temperature, and moisture supply, affecting plant B availability can be found elsewhere.[7] However, on crops such as peanuts, sensitive to excess B, an application of only 2 kg B ha^{-1} was found to be toxic.[1] Grasses are highly tolerant of B; e.g., 45 kg B ha^{-1} caused only 12% yield reduction in tall fescue (*Festuca arundinacea* cv. Kentucky 31).[11]

Boron Toxicity and Management

In nature, B toxicity is not as widespread as B deficiency. Toxicity can occur under three main conditions: 1) in soils inherently high in B or in which B has accumulated naturally; 2) as a result of overfertilization with waste materials high in B; and 3) the use of irrigation waters high in B leading to B accumulation and concentration in soil. Although of considerable agronomic importance, our understanding of B toxicity mechanism in plants is still incomplete. The objective of this entry is to report the best available methods to detoxify soils having excess B.

Sources and Causes of Toxicity

Boron toxicity is a serious concern in irrigated agriculture throughout the world for sustainable crop production.[12] It is an important constraint to the use of saline irrigation waters in agriculture.[13] Excess B in irrigation waters has negative impacts, on the growth of processing tomato (*Lycopersicon esculentum*).[14] Boron is often found in higher amounts in polluted and desalinated waters. Boron toxicity is an important disorder that can limit plant growth of tomatoes and peppers (*Capsicum annuum* L.) on soils of arid and semiarid environments throughout the world.[15] Boron toxicity is increasingly being recognized as a problem in arid areas of West Asia, where lentil (*Lens culinaris* Medic) is widely grown.[16] Depending upon the 77 lines, significant differences have been reported in B toxicity tolerance in lentil. On average, accessions from Afghanistan were the most tolerant, followed by those from India, Iraq, Syria, Europe, Ethiopia, and Nepal.[16] Waste materials such as fly ash could be a source of B toxicity, unless managed properly, e.g., one concern for rice (*Oryza sativa*) plants grown on soils amended with fly ash is the toxicity of B because most of the fresh fly ash contains considerable B.[17]

Boron Toxicity Symptoms and Levels in Crops

In a study conducted on barley (*Hordeum vulgare* L.),[18] plant samples were free from B toxicity symptoms in spite of the fact that their B concentrations were up to 44.4 μg g^{-1}.[18]

Boron toxicity symptoms generally are infrequent at 10 to 20 mg B kg^{-1} levels. Under excessive B conditions, B toxicity symptoms in tomatoes included markedly reduced growth of plants, development of marginal necrosis in old leaves, and reduction in number and size of lamina.[19] The concentrations of reducing, non-reducing, and total sugars and phenols were high in fruits. The concentration of ascorbic acid accumulated and lycopene content of tomatoes decreased due to excess B.[19]

Boron levels of deficiency and toxicity were found to be 13.5 and 310 μg B g^{-1}, respectively, in the leaves of gram.[20] Toxicity symptoms in cotton (*Gossypium hirsutum* L.) appeared as yellowing and necrosis in patches between veins and tips and margins of leaves. They first appeared on older leaves and increased in severity with increasing B levels.[12] The deficiency and toxicity levels were 5 and 250 μg B g^{-1} dry matter, respectively, in the young leaves of tomatoes[19] where high B reduced dry matter, fruit yield, and chlorophyll content.[19,21]

Toxicity symptoms in gram (*Cicer arietinum* L.) appeared as marginal necrosis of old leaflets along with growth depression.[20] Subsequently, the necrotic leaflets became completely dry and shed. Excess B (>0.33 mg L^{-1}) reduced the quality of gram seeds by lowering the seed yield, starch, and protein concentrations and by increasing the accumulation of phenols and sugars.

Boron toxicity symptoms in a variety of crops such as soybeans (*Glycine max* L. Merr.), potato tubers (*Solanum tuberosum* L.), and tomatoes appeared on old leaves as chlorotic spots along with the margins.[22] These spots enlarged, coalesced, and turned necrotic. High B levels reduced the economic yield and deteriorated the quality of the produce by lowering the starch and protein concentration in seeds and storage organs. Also, excess B reduced the lycopene and ascorbic acid in tomato fruits and reduced oil in oil seeds.[22]

Measures to Mitigate Boron Toxicity

Boron-Tolerant Cultivars

There are genetic variations in genotype tolerance to B toxicity in soils.[23] It has been hypothesized that the internal mechanisms (e.g., adsorption to cell walls and compartmentation of B in vacuoles) could be an explanation for B tolerance in crops such as durum wheat (*Triticum aestivum* L.).[23]

The low frequency of B toxicity problems is generally attributed to common cultivation of B-tolerant barley cultivars, which have been improved under Central Anatolia and Transitional Zones of Turkey conditions over the last decade.[18] As it is neither practical nor easy to detoxify high-B soils by agronomic means in most circumstances, selecting or breeding crop cultivars with high B toxicity tolerance is probably the best approach to control toxicity and increase yields on high-B soils.[24]

Phytoremediation

Studies have been conducted at Bangladesh Agricultural University, Mymensingh, to identify naturally grown B hyperaccumulating weed species in B-contaminated soils.[25] Considering the absorption pattern, biomass, and toxicity tolerance, Joina (*Fimbristylis miliacea*) and Barnyard grass (*Echinochloa crus-galli*) were found to be the best performers and could be considered to mitigate the B-contaminated soil problem due to irrigation water. It was further suggested that Water cress (*Enhydra fluctuans*) and Malancha (*Alternanthera philoxeroides*) can be used for remediation of stagnant B-contaminated water as a phytoremediation technology.[25]

The duckweed (*Lemna gibba*) has been shown to remove toxic elements from water; therefore, it has been examined for its tolerance to B in water and its B removal efficiency.[26] It was found that B content in plants at the end of the experiment ranged between 930 and 1900 mg kg^{-1} dry weight and was comparable to that of wetland plants, which are reported to be good B accumulators. Boron removal by duckweed may therefore be a suitable option for the treatment of water containing B concentrations below 2 mg L^{-1}.

Wetland microcosms have been used to evaluate the ability of constructed wetlands to remove extremely high concentrations of B and other minerals from wastewater generated by a coal gasification plant in Indiana.[27] The wetland microcosms were found to significantly reduce B concentrations by 31%. Of the number of plant species tested, cattail (*Typha latifolia*), genus *Thalia* (Bog, swamp plants), and rabbit foot grass (*Polypogon monspeliensis*) were highly tolerant of the contaminants and exhibited no growth retardation. The data from the wetland microcosms support the view that constructed wetlands can be used to successfully reduce the toxicity of aqueous effluent contaminated with extremely high B concentrations.

Use of Minerals such as Phosphorus, Zinc and Silica

It has been reported that supplementary P can mitigate the adverse effects of high B on fruit yield and growth in tomato fruits.[21] Likewise, studies on maize (*Zea mays* L.) genotypes showed that B is more toxic in the absence of rather than in the presence of P, and that this toxicity could be alleviated with applications of P in calcareous soils of semiarid areas.[28]

Boron toxicity symptoms in tomato plants were found to be somewhat lower when grown with applied P.[21] Zinc treatments partially depressed the inhibitory effect of B on the growth of beans (*Phaseolus* spp.).[29] Based on experiments conducted on bean plants, it was concluded that Zn supply alleviates B toxicity by preventing oxidative membrane damage due to increased activities of antioxidant enzymes catalase and ascorbate peroxidase.[29] Results from another study indicated that supplemental Zn has the potential and is of practical importance in controlling B absorption by plants and B toxicity in soil, e.g., where lemon (*Citrus limon*) plants were grown under Zn deficiency and B toxicity.[30]

Based on past research, it has been concluded that silica is also effective in alleviating B toxicity of wheat by preventing oxidative membrane damage by reducing the activity of, e.g., proline, hydrogen peroxide, and malondialdehyde, and also the translocation of B from root to shoot and/or soil to plants.[31]

Recirculating Leachate with High Boron

In several countries, leachates containing toxic amounts of B are successfully treated by recirculating it to the vegetated landfill cover, as it contains several micro- and macronutrients for plant growth.[32] Results of these studies have shown that the leachate application increased plant growth during the observation period and there were no toxic effects on plant leaves, although B concentrations were elevated compared to the usual concentrations found in the natural environment.[32]

Judicial Use of Waste Materials

In all fly ash treatments, B content in rice leaves and available B in soil at all growth stages were found to be higher than those of controls but did not exceed the toxicity levels.[17] Boron occluded in amorphous Fe and Al oxides comprised about 20%–39% of total B and was not affected by fly ash application. Most of the B, which was accumulated from fly ash application, as a residual B, which is plant-unavailable form, comprised >60% of the total B in soils. Thus, fly ash can be a good soil amendment for rice production without B toxicity.[17]

Molybdenum

The chief sources of Mo for human consumption are cereal products, sugar- and starch-rich foods, bread, pulses, cake, spices, and most fruits, which provide humans 10–400 μg Mo kg^{-1} dry matter.[33] Herbs and vegetables are generally rich in Mo, e.g., cucumber (*Cucumis sativus* L.) can store up to 4000 μg Mo kg^{-1} dry matter. Animal products, with the exception of liver and kidney, are poor in Mo concentration.[33]

In 1930, Bortels showed that Mo is required for nitrogen fixation in *Azotobacter*, and in 1939, Arnon and Stout reported that Mo is essential for life in higher plants as reviewed by other researchers.[34] Nitrogenase is the nitrogen-fixing enzyme complex, while nitrate reductase requires Mo for its activity.

The weathering soils of granite, porphyry, gneiss, and Rotliegendes origin produce Mo-rich vegetation.[34] However, Mo toxicity in plants is rare. Soils formed on sandstones that have experienced heavy leaching losses are most likely to be low in total Mo. Well-drained soils, e.g., podzols, as well as severely eroded and/or heavily weathered soils, are low in Mo. Soybeans grown on coarse-textured and silty loam soils, relatively low in organic matter, have been found to respond to Mo applications.[35] The amount of Mo adsorbed was found closely related to the soil's organic matter content.[36] One possible explanation for the relationship of organic matter to Mo is that soils high in organic matter contain the readily exchangeable mobile MoO_4^{2-}.[37]

Alfisols derived from mixed shale have been found to be higher in Mo than spodosols formed from mixed schist. Lateritic soils derived from granite and basalt rocks are very high in total Mo.

Soil pH is one of the most important factors affecting the Mo availability in plants. Soil solution MoO_4^{-2} is the most available form to plants. Its availability increases a hundredfold for each unit increase in pH. Soil pH is of utmost importance in affecting the Mo availability in alkaline soils, because Mo is only weakly adsorbed on soils and hydrous oxides of Fe at high pH.

Alkaline soils have a relatively large proportion of Mo in the soil solution phase. Recent studies[38] indicated that the severity of Mo deficiency in lentil, for example, could be reduced by liming of Mo-deficient acid alluvial soils. For details, the reader is referred to a Mo textbook.[3]

Soil moisture is one of the chief factors affecting the availability of Mo. Peat and mucks are products of a wet environment and have been associated with high Mo in certain regions of the world. Availability of Mo to plants is also reduced by S fertilization due to anion competition. Detailed information on factors affecting Mo in soils and crops can be found elsewhere.[3]

Methods of controlling Mo deficiency in crops include application in bands or broadcast (generally Mo contained in P or NPK fertilizers) to the soil, as foliar sprays, or as seed treatment. Amounts of Mo applied to soil range from 50 g ha^{-1} to 1 kg ha^{-1} depending upon the crop, soil pH, and the application method. However, seed treatment is also a common method because the recommended rates are low, ranging from 50 to 400 g ha^{-1}. Molybdenum sources can be applied to seeds as liquid or slurry, and some type of sticking or conditioning agents may be included.

Because legume seeds, such as soybeans, peas (*Pisum sativum* L.), alfalfa (*Medicago sativa* L.), and clovers (*Trifolium* spp.) are treated with a bacterial inoculant, the Mo source for seed treatment must be compatible with the inoculants. For details on correcting Mo deficiency, readers are referred to two entries.[39,40] It should also be pointed out that on soils containing adequate total Mo, liming soils to pH 6–6.5 can overcome a Mo deficiency in crops.[41]

Molybdenum Toxicity and Management

Sources and Causes of Molybdenum Toxicity

The chief cause of Mo toxicity in livestock animals is an imbalance between the minerals Cu, Mo, and S in pastures and or livestock feeds. Mine tailings and waste materials such as sewage sludge and coal fly ash can also be a source of Mo toxicity. In some regions, the Mo concentration in soils is inherently high due to high Mo in the parent material or because of excessive reducing conditions.

Thus, in general, the Mo concentration in wetland sediments is highest under more reducing conditions and lowest under more oxidizing conditions.[42] Most of the accumulated Mo (73%) becomes water soluble on drying of samples. This has important implications for systems undergoing changes in redox status; for instance, if these wetland sediments are dried, potentially large amounts of Mo may be solubilized.[42]

Molybdenum Toxicity Symptoms and Levels

Molybdenum in excess of >2 mg L^{-1} not only resulted in reduced growth of chickpea (*Cicer arietinum* L.) but also caused chlorosis (iron deficiency type) of young leaves, which intensified with increasing age.[43] Leaves, flowers, and pods were reduced in number and size. Levels of deficiency and toxicity were 0.38 and 15 μg Mo kg^{-1}, respectively, in chickpea leaves.[43]

Likewise, excess Mo deteriorated the quality of wheat grain by reducing the content of starch, sugars, protein, non-protein, and total nitrogen (N), as well as prolamin, glutelin, and globulin fractions of seed proteins and increasing the content of albumin and electrical conductivity (EC) of the seed leachate.[44] Excess Mo resulted in the production of lightweight immature seeds, poor in vigor and germination potential. The Mo levels of sufficiency and toxicity in leaves were 0.13 and 1.15 μg Mo kg^{-1} dry matter of wheat (*Triticum aestivum* L.).[44]

Increased Mo uptake was demonstrated by elevated Mo levels in the rumen contents and feces.[45] Clinical signs of Mo toxicity were observed in less than half of the cows and in only a few calves. These included a stiff shuffling gait, watery diarrhea, and a rough hair coat.[45] Lameness, the primary sign of Mo toxicity, was resolved in all animals by the end of each trial. Diarrhea was also controlled by the end of the trial, and hair coats returned to normal by the following spring. The onset and severity of the affliction appeared to be related to the prevailing moisture conditions, which may have affected Mo availability in forages.[45]

Liver biopsies and serum samples showed marginal to adequate Cu levels but potentially toxic levels of Mo.[45] In the third-year trial, Cu-containing boluses were employed, but they did not prevent the onset of clinical signs of Mo toxicity. The clinical signs of a disorder known locally as "shakeback disease" in yaks in the North of the Qinghai-Tibetan Plateau of China included emaciation, unsteady gait, a "shivering" back, and deprived appetites.[46] Copper deficiency in yaks was most severe during pregnancy and lactation, but oral administration of copper sulfate prevented and cured the disease. It was concluded that shakeback disease of yaks in this region was probably caused by a secondary Cu deficiency, mainly caused by the high Mo content in soils and forages.[46]

Copper: Molybdenum: Sulfur Ratios in Feeds and Rations

Waste materials applied to agricultural land can contain significant concentrations of bioavailable Mo. It has been concluded that Mo, Cu, and sulfur (S) bioavailability remains elevated in the soil several decades after sewage sludge application.[47]

For ruminants, there is a narrow span between nutritional deficiency of Mo and its potential toxicity.[48] Molybdenum toxicity also known as molybdenosis occurs among cattle feeding on forages with Mo concentrations above 10 $\mu g\ g^{-1}$ or a Cu:Mo ratio of <2. In the area under investigation, forage Mo contents in the valley were as high as 180 $\mu g\ g^{-1}$ due to industrial pollution, while the alpine pastures where cattle graze during summer are deficient in Cu.[48] When driven to the valley pastures in the fall, the animals often fell ill with molybdenosis, and several died. Phosphorus fertilization and vermiculite were thus recommended for the severely contaminated sites to enhance phytoremediation through Mo export, and Mn-humate and sewage sludge application appeared suited to remediate the less severely contaminated sites.[48]

Forages that accumulate excessive Mo from excess Mo application to the land can cause molybdenosis, a Cu-induced deficiency in ruminants.[49] Limited information is available on the effect of land-applied biosolids, Mo content, and quality of winter wheat forages. These results suggest that winter wheat forage produced using biosolids presents minimal risk of molybdenosis to livestock. Ingestion of soil on unwashed forage increases forage Cu:Mo ratio and offers more protection from molybdenosis.[49]

A case of Cu deficiency or Mo toxicosis in cattle, sheep, and horses after heavy pollution of a pasture with fly ash has been described.[50] It is argued that Mo intoxication, although seldom seen in non-ruminants, was the cause of death of horses.[50] It was suggested that the bioavailability of Mo in fly ash is high, and therefore, it can cause equine intoxication. It may be of interest to note that the best indicators of Mo intoxication in cattle are liver, kidney, blood, and milk.[51] The intrauterine storage of Mo in mammals is low.[51]

Monitoring Molybdenum in Soils and Waste Materials

Molybdenum at elevated concentrations in non-acid soils is readily taken up into forages, particularly legumes, and can result in secondary Cu deficiency or molybdenosis in ruminants.[52] Because sewage sludge products are commonly higher in total and available Mo than soils, amendment of soils with sludges could cause health problems in livestock. The results from a study revealed that residual plant-available Mo in sludge-amended soils can persist for decades with some types of sludge materials, but leaching losses of Mo may reduce the impact of residual Mo in soils.[52] The need for stronger regulation and monitoring of Mo in waste materials intended for forage, pasture, and range land application is indicated.

Sustainable Grazing and Reducing Iron Intake

Revegetation and sustainable cattle grazing are major objectives in the reclamation of mine tailings at the Highland Valley Copper mine in British Columbia, Canada.[45] Prevention of Mo toxicity can also be achieved by reducing the Fe intake of the animal and by supplementation with Cu that is sacrificial in the rumen and not primarily supplied for absorption.[53]

Irrigation

Irrigation has been proposed as a possible disposal method for large quantities of water having high concentrations of Mo (5–100 mg L^{-1}) resulting from mining and reclamation activities.[54] There appears to be a strong relationship between the laboratory sorption data and the equilibrium solution Mo in the soil and that grass grown on soils irrigated with waters high in Mo may reach levels of Mo that are expected to be toxic to animals.[54]

CONCLUSIONS

This entry includes several topics related to B and Mo. There is a great deal of information now available on crop responses to these nutrients and symptoms of deficiency and toxicity in various crops. Nutrient sampling of the appropriate part of the plant and stage of growth for crops should be standardized, which could be used internationally. While reports on the level of Mo in feed crops, which cause toxicity to livestock, are available, little data are found on the levels of B in feeds that could be toxic to livestock or human consuming food crops. Information is necessary to develop methods of detoxifying soils high in B and Mo, which could be adopted universally.

Some of the established methods to control B toxicity include using B-tolerant cultivars; phytoremediation; use of minerals such as phosphorus, zinc, and silica; leaching; and judicial use of waste materials. Molybdenum toxicity remedial measures available consist of maintaining a proper ratio between Mo, Cu, and S in livestock feeds. Use of waste materials high in Mo, such as coal fly ash and sewage sludge, as a source of nutrients should be used with care, keeping the amount of Mo applied in the safe range. Livestock grazing on pastures containing high Mo should be limited to avoid excess intake of Mo. Leaching with large quantities of water has also been used in controlling Mo toxicities.

Future studies on B and Mo should be considered in cooperation with animal clinicians and nutritionists.

REFERENCES

1. Rashid, A.; Rafique, E.; Ali, N. Micronutrient deficiencies in rainfed calcareous soils of Pakistan II. Boron nutrition of the peanut plant. Commun. Soil Sci. Plant Anal. **1997**, *28* (1/2), 149–159.
2. Gupta, U.C.; Jame, Y.W.; Campbell, C.A.; Leyshon, A.J.; Nicholaichuk, N. Boron toxicity and deficiency in crops: A review. Can. J. Soil Sci. **1985**, *65* (3), 381–409.
3. Gupta, U.C. *Molybdenum in Agriculture*; Gupta U.C., Ed.; Cambridge University Press: U.K., 1997; 276.
4. Dotsika, E.; Poutoukis, D.; Michelot, J.L.; Kloppmann, W. Stable isotope and chloride, boron study for tracing sources of boron contamination in groundwater: Boron contents in fresh and thermal water in different areas in Greece. Water, Air, Soil Pollut. **2006**, *174* (1/4), 19–32.
5. Murray, F.J.; Schlekat, C.E. Comparison of risk assessments of boron: Alternate approaches to chemical-specific adjustment factors. Hum. Ecol. Risk Assess. **2004**, *10* (1 special issue), 57–68.
6. Chatterjee, C.; Sinha, P.; Nautiyal, N.; Agarwala, S.C.; Sharma, C.P. Metabolic changes associated with boron-calcium interaction in maize. Soil Sci. Plant Nutr. **1987**, *33* (4), 607–617.
7. Gupta, U.C. *Boron and Its Role in Crop Production*; Gupta, U.C., Ed., CRC Press: Boca Raton, Florida, 1993; 237.
8. Yadav, H.D.; Yadav, O.P.; Dhankar, O.P.; Oswal, M.C. Effect of chloride salinity and boron on germination, growth and mineral composition of chickpea (*Cicer arietinum* L.). Ann. Arid Zone **1989**, *28* (1/2), 63–67.
9. Batal, K.M.; Granberry, D.M.; Mullinix, Jr., B.G. Nitrogen, magnesium, and boron applications affect cauliflower yield, curd mass, and hollow stem disorder. Hort. Sci. **1997**, *32* (1), 75–78.
10. Adarve, M.J.; Hernandez, A.J.; Gil, A.; Pastor, J. Boron, zinc, iron, and manganese content in four grassland species. J. Environ. Qual. **1998**, *27* (6), 1286–1293.
11. Wilkinson, S.R. Response of Kentucky-31 tall fescue to broiler litter and composts made from broiler litter. Commun. Soil Sci. Plant Anal. **1997**, *28* (3/5), 281–299.
12. Niaz, A.; Muhammad, A.; Fiaz, A. Boron toxicity in irrigated cotton (*Gossypium hirsutum* L.). Pak. J. Bot. **2008**, *40* (6), 2443–2452.
13. Brown, P.; Hu, H. Mechanisms of boron toxicity in crop plants. Rep.—Univ. Calif., Water Resour. Cent. **2008**, *11*, 65–66.
14. Brown, P.H.; Wu, Y.; Hu, H. Mechanism of boron toxicity in crop plants. Rep.—Univ. Calif., Water Resour. Cent. **2009**, *112*, 33–34.
15. Eraslan, F.; Inal, A.; Gunes, A.; Alpaslan, M. Boron toxicity alters nitrate reductase activity, proline accumulation, membrane permeability, and mineral constituents of tomato and pepper plants. J. Plant Nutr. **2007**, *30* (4/6), 981–994.
16. Yau, S.; Erskine, W. Diversity of boron-toxicity tolerance in lentil growth and yield. Genet. Resour. Crop Evol. **2000**, *47* (1), 55–61.
17. Lee, S.; Lee, Y.; Lee, C.; Hong, C.; Kim, P.; Yu, C. Characteristics of boron accumulation by fly ash application in paddy soil. Bioresour. Technol. **2008**, *99* (13), 5928–5932.
18. Avci, M.; Akar, T. Severity and spatial distribution of boron toxicity in barley cultivated areas of Central Anatolia and Transitional Zones. Turkish J. Agric. For. **2005**, *29* (5), 377–382.
19. Pratima, S.; Dube, B.K.; Singh, M.V.; Chatterjee, C. Effect of boron stress on yield, biochemical parameters and quality of tomato. Indian J. Hortic. **2006**, *63* (1), 39–43.
20. Chatterjee, C.; Sinha, P.; Dube, B.K. Biochemical changes, yield, and quality of gram under boron stress. Commun. Soil Sci. Plant Anal. **2005**, *36* (13/14), 1763–1771.
21. Maya, C.; Tuna, A.L.; Debilities, M.; Muhammad, A.; Koskeroglu, S.; Guneri, M. Supplementary phosphorus can alleviate boron toxicity in tomato. Sci. Hortic. **2009**, *121* (3), 284–288.
22. Pratima, S.; Neena, K.; Nirmala, N. Boron stress influences economic yield and quality in crop species. Indian J. Plant Physiol. **2009**, *14* (2), 200–204.
23. Torun, A.A.; Yazici, A.; Erdem, H.; Ãakmak, I. Genotypic variation in tolerance to boron toxicity in 70 durum wheat genotypes. Turkish J. Agric. For. **2006**, *30* (1), 49–58.
24. Yau, S.; Ryan, J. Boron toxicity tolerance in crops: A viable alternative to soil amelioration. Crop Sci. **2008**, *48* (3), 854–865.
25. Riffat, S.; Arefin, M.T.; Mahmud, R. Phytoremediation of boron contaminated soils by naturally grown weeds. J. Soil Nature **2007**, *1* (1), 1–6.
26. Marin, C.M.d.C.; Oron, G. Boron removed by the duckweed (*Lemna gibba*): A potential method for the remediation of boron-polluted waters. Water Res. (Oxford). **2007**, *41* (20), 4579–4584.
27. Ye, Z.H.; Lin, Z.Q.; Whiting, S.N.; De Souza, M.P.; Terry, N. Possible use of constructed wetland to remove selenocyanate, arsenic, and boron from electric utility wastewater. Chemosphere **2003**, *52* (9), 1571–1579.
28. Gunes, A.; Alpaslan, M. Boron uptake and toxicity in maize genotypes in relation to boron and phosphorus supply. J. Plant Nutr. **2000**, *23* (4), 541–550.
29. Gunes, A.; Inal, A.; Bagci, E.G. Recovery of bean plants from boron-induced oxidative damage by zinc supply. Russ. J. Plant Physiol. **2009**, *56* (4), 503–509.
30. Rajaie, M.; Ejraie, A.K.; Owliaie, H.R.; Tavakoli, A.R. Effect of zinc and boron interaction on growth and mineral composition of lemon seedlings in a calcareous soil. Int. J. Plant Prod. **2009**, *3* (1), 39–50.
31. Gunes, A.; Inal, A.; Bagci, E.G.; Coban, S.; Sahin, O. Silicon increases boron tolerance and reduces oxidative damage of wheat grown in soil with excess boron. Biol. Plant **2007**, *51* (3), 571–574.
32. Justin, M.Z.; Zupancic, M. Boron in irrigation water and its interactions with soil and plants: An example of municipal landfill leachate reuse. Acta Agric. Slov. **2007**, *89* (1), 289–300.
33. Seifert, M.; Dorn, W.; Muller, R.; Holzinger, S.; Anke, M. The biological and toxicological importance of molybdenum in the environment and in the nutrition of plants, animals and man: Part III. Molybdenum content of the food. Acta Aliment. **2009**, *38* (4), 471–481.

34. Anke, M.; Seifert, M. The biological and toxicological importance of molybdenum in the environment and in the nutrition of plants, animals and man: Part 1: Molybdenum in plants. Acta Biol. Hung. **2007**, *58* (3), 311–324.
35. Boswell, F.C. *Factors Affecting the Response of Soybeans to Molybdenum Application*; Corbin, F.T., Ed.; Westview Press: Boulder, Colorado. World Soybean Res. Conf. 1980, 417–432.
36. Karimian, N.; Cox, F.R. Adsorption and extractability of molybdenum in relation to some chemical properties. Soil Sci. Soc. Am. J. **1978**, *42* (5), 757–761.
37. Koval'skiy, V.V.; Yarovaya, G.A. Biogeochemical provinces rich in molybdenum provinces. Agrokhimia **1966**, *8*, 68–91.
38. Mandal, B.; Pal, S.; Mandal, L.N. Effect of molybdenum, phosphorus, and lime application to acid soils on dry matter yield and molybdenum nutrition on lentil. J. Plant Nutr. **1998**, *21* (1), 139–147.
39. Mortvedt, J.J. Sources and methods of molybdenum fertilization of crops. In *Molybdenum in Agriculture*; Gupta, U.C., Ed.; CRC Press: Boca Raton, FL, 1997; 171–181.
40. Adams, J.F. Yield response to molybdenum by field and horticultural crops. In *Boron and Its Role in Crop Production*; Gupta, U.C., Ed.; CRC Press: Boca Raton, FL, 1997; 182–201.
41. Gupta, U.C. Effect and interaction of molybdenum and limestone on growth and molybdenum content of cauliflower, alfalfa and brome grass on acid soils. Soil Sci. Soc. Am. Proc. **1969**, *33* (6), 929–932.
42. Fox, P.M.; Doner, H.E. Accumulation, release, and solubility of arsenic, molybdenum, and vanadium in wetland sediments. J. Environ. Qual. **2003**, *32* (6), 2428–2435.
43. Nautiyal, N.; Chatterjee, C. Molybdenum stress-induced changes in growth and yield of chickpea. J. Plant Nutr. **2004**, *27* (1), 173–181.
44. Chatterjee, C.; Nautiyal, N. Molybdenum stress affects viability and vigor of wheat seeds. J. Plant Nutr. **2001**, *24* (9), 1377–1386.
45. Majak, W.; Steinke, D.; McGillivray, J.; Lysyk, T. Clinical signs in cattle grazing high molybdenum forage. J. Range Manage. **2004**, *57* (3), 269–274.
46. Xiao-yun, S.; Guo-Zhen, D.; Hong, L. Studies of a naturally occurring molybdenum-induced copper deficiency in the yak. Vet. J. **2006**, *171* (2), 352–357.
47. McBride, M.B. Molybdenum and copper uptake by forage grasses and legumes grown on a metal-contaminated sludge site. Commun. Soil Sci. Plant Anal. **2005**, *36* (17/18), 2489–2501.
48. Neunhauserer, C.; Berreck, M.; Insam, H. Remediation of soils contaminated with molybdenum using soil amendments and phytoremediation. Water, Air, Soil Pollut. **2001**, *128* (1/2), 85–96.
49. Mullen, R.W.; Raun, W.R.; Basta, N.T.; Schroder, J.L.; Freeman, K.W. Effect of long-term application of biosolids on molybdenum content and quality of winter wheat forage. J. Plant Nutr. **2005**, *28* (3), 405–420.
50. Ladefoged, O.; Sturup, S. Copper deficiency in cattle, sheep and horses caused by excess molybdenum from fly ash: A case report. Vet. Hum. Toxicol. **1995**, *37* (1), 63–65.
51. Anke, M.; Seifert, M.; Holzinger, S.; Muller, R.; Schafer, U. The biological and toxicological importance of molybdenum in the environment and in the nutrition of plants, animals and man—Part 2: Molybdenum in animals and man. Acta Biol. Hung. **2007**, *58* (3), 325–333.
52. McBride, M.B.; Hale, B. Molybdenum extractability in soils and uptake by alfalfa 20 years after sewage sludge application. Soil Sci. **2004**, *169* (7), 505–514.
53. Telfer, S.B.; Kendall, N.R.; Illingworth, D.V.; Mackenzie, A.M. Molybdenum toxicity in cattle: An underestimated problem. Cattle Pract. **2004**, *12* (4), 259–263.
54. Smith, C.; Brown, K.W.; Deuel, L.E., Jr. Plant availability and uptake of molybdenum as influenced by soil type and competing ions. J. Environ. Qual. **1987**, *16* (4), 377–382.

Boron: Soil Contaminant

Rami Keren
Agricultural Research Organization of Israel, Bet-Dagan, Israel

Abstract
Boron is considered as a typical metalloid having properties intermediate between the metals and the electronegative non-metals. Boron has a tendency to form anionic rather than cationic complexes.

INTRODUCTION

Boric acid is moderately soluble in water. Its solubility increases markedly with temperature due to the large negative heat of dissolution. Boron is considered as a typical metalloid having properties intermediate between the metals and the electronegative non-metals. Boron has a tendency to form anionic rather than cationic complexes. Boron chemistry is of covalent B compounds and not of B^{3+} ions because of its very high ionization potentials. Boron has five electrons, two in the inner spherical shell ($1s^2$), two in the outer spherical shell ($2s^2$), and one in the dumbbell shaped shell ($2p_x^1$).[1] In the hybrid orbital state, the three electrons in the 2s and 2p orbitals form a hybrid orbital state ($2s^1\,2p_x^1\,2p_y^1$), where each electron is alone in an orbit whose shape has both spherical and dumbbell characteristics. Each of these three orbits can hold one electron from another element to form a covalent bond between the element and B (BX_3). This leaves one 2p electron orbit that can hold two electrons, which if filled would completely fill the eight electron positions (octet) associated with the second electron shell around B. BX_3 compounds behave as acceptor Lewis acids toward many Lewis bases such as amines and phosphines. The acceptance of two electrons from a Lewis base completes the octet of electrons around B. Boron also completes its octet by forming both anionic and cationic complexes.[1] Therefore, tri-coordinate B compounds have strong electron-acceptor properties and may form tetra-coordinate B structures. The charge in tetra-coordinate derivatives may range from negative to neutral and positive, depending upon the nature of the ligands.

For the unshared oxygen atoms bound to B, they are, probably, always OH groups. Thus, in accordance with the electron configuration of B, boric acid acts as a weak Lewis acid:

$$B(OH)_3 + 2H_2O = B(OH)_4^- + H_3O^+ \qquad (1)$$

The formation of borate ion is spontaneous. The first hydrolysis constant of $B(OH)_3$, K_{h1}, is 5.8×10^{-10} at 20°C,[2] and the other K_{h2} and K_{h3} values are 5.0×10^{-13} and 5.0×10^{-14}, respectively.[3] A dissociation beyond $B(OH)_4^-$ is not necessary to explain the experimental data, at least below pH 13.[4,5] Boron species other than $B(OH)_3$ and $B(OH)_4^-$, however, can be ignored in soils for most practical purposes. The first hydrolysis constant of $B(OH)_3$ varies with temperature from 3.646×10^{-10} at 178 K to 7.865×10^{-10} at 318 K.[6]

Both $B(OH)_3$ and $B(OH)_4^-$ ion species are essentially monomeric in aqueous media at low B concentration ($\leq 0.025\ mol\,L^{-1}$). However, at high B concentration, polyborate ions exist in appreciable amount.[7] The equilibria between boric acid, monoborate ions, and polyborate ions in aqueous solution are rapidly reversible. In aqueous solution, most of the polyanions are unstable relative to their monomeric forms $B(OH)_3$ and $B(OH)_4^-$.[8] Results of nuclear magnetic resonance[9] and Raman spectrometry[10] lead to the conclusion that $B(OH)_3$ has a trigonal-planar structure, whereas the $B(OH)_4^-$ ion in aqueous solution has a tetrahedral structure. This difference in structure can lead to differences in the affinity of clay for these two B species.

BORON–SOIL INTERACTION

The elemental form of boron (B) is unstable in nature and found combined with oxygen in a wide variety of hydrated alkali and alkaline earth-borate salts and borosilicates as tourmaline. The total B content in soils, however, has little bearing on the status of available B to plants.

Boron can be specifically adsorbed by different clay minerals, hydroxy oxides of Al, Fe, and Mg, and organic matter.[11] Boron is adsorbed mainly on the particle edges of the clay minerals rather than the planar surfaces. The most reactive surface functional group on the edge surface is the hydroxyl exposed on the outer periphery of the clay mineral. This functional group is associated with two types of sites that are available for adsorption: Al(III) and Si(IV), which are located on the octahedral and tetrahedral sheets, respectively. The hydroxyl group associated with this site can form an inner sphere surface complex with a proton at low pH values or with a hydroxyl at high pH values. The B adsorption process can be explained by the surface

Encyclopedia of Environmental Management DOI: 10.1081/E-EEM-120010211

complexation approach, in which the surface is considered as a ligand.[12] Such specific adsorption, which occurs irrespective of the sign of the net surface charge, can occur theoretically for any species capable of coordination with the surface metal ions. However, because oxygen is the ligand commonly coordinated to the metal ions in clay minerals, the B species $B(OH)_3$ and $B(OH)_4^-$ are particularly involved in such reactions. Possible surface complex configurations for B—broken edges of clay minerals—were suggested by Keren, Grossl, and Sparks.[12]

Keren and Bingham[11] reviewed the factors that affect the adsorption and desorption of B by soil constituents and the mechanisms of adsorption. Soil pH is one of the most important factors affecting B adsorption. Increasing pH enhances B adsorption on clay minerals, hydroxy-Al and soils, showing a maximum in the alkaline pH range (Fig. 1).

The response of B adsorption on clays to variations in pH can be explained as follows. Below pH 7, $B(OH)_3$ predominates and since the affinity of the clay for this species is relatively low, the amount of adsorption is small. Both $B(OH)_4^-$ and OH^- concentrations are low at this pH; thus, their contribution to total B adsorption is small despite their relatively strong affinity for the clay. As the pH is increased to about 9, the $B(OH)_4^-$ concentration increases rapidly. Since the OH^- concentration is still low relative to the B concentration, the amount of adsorbed B increases rapidly. Further increases in pH result in an enhanced OH^- concentration relative to $B(OH)_4^-$, and B adsorption decreases rapidly due to the competition by OH^- at the adsorption sites. Adsorption models for soils, clays, aluminum oxide, and iron oxide minerals have been derived by various workers.[13–17]

In assessing B concentration in irrigation water, however, the physicochemical characteristics of the soil must be taken into consideration because of the interaction between B and soil. Boron sorption and desorption from soil adsorption sites regulate the B concentration in soil solution depending on the changes in solution B concentration and the affinity of soil for B. Thus, adsorbed B may buffer fluctuations in solution B concentration, and B concentration in soil solution may change insignificantly by changing the soil-water content (Fig. 2). When irrigation with water high in B is planned, special attention should be paid to this interaction because of the narrow difference between levels causing deficiency and toxicity symptoms in plants.

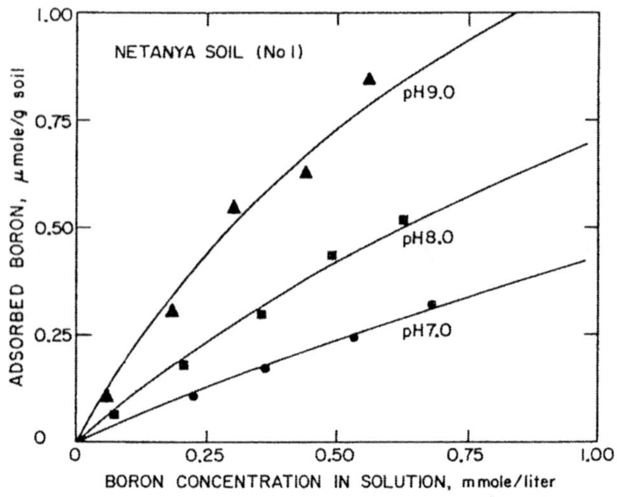
NETANYA SOIL (No I)
pH9.0
pH8.0
pH7.0
ADSORBED BORON, μmole/g soil
BORON CONCENTRATION IN SOLUTION, mmole/liter
0
0.25
0.50
0.75
1.00

Fig. 1 Boron adsorption isotherms for a soil as a function of solution B concentration and pH. Bold lines—calculated values. **Source:** Mezuman and Keren.[28]

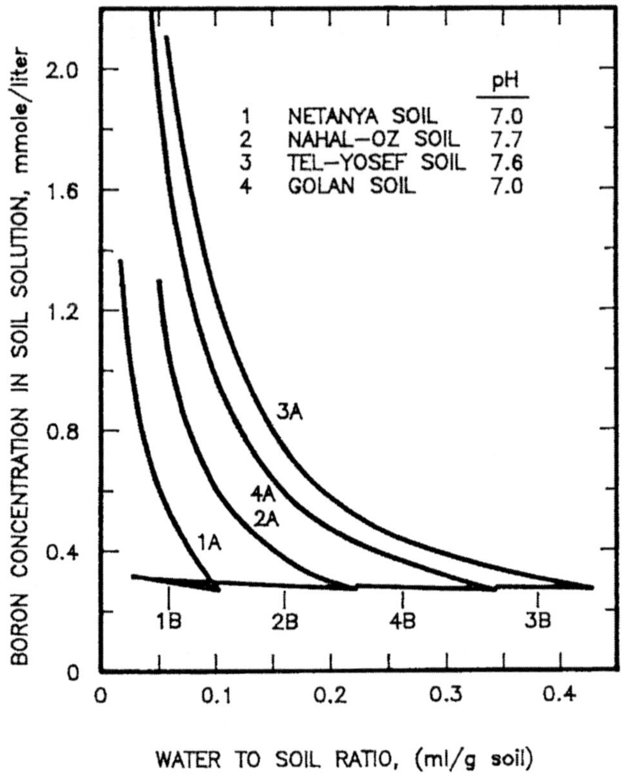
pH
1 NETANYA SOIL 7.0
2 NAHAL–OZ SOIL 7.7
3 TEL–YOSEF SOIL 7.6
4 GOLAN SOIL 7.0
3A
4A
2A
1A
1B
2B
4B
3B
BORON CONCENTRATION IN SOIL SOLUTION, mmole/liter
WATER TO SOIL RATIO, (ml/g soil)
0
0.4
0.8
1.2
1.6
2.0
0.1
0.2
0.3
0.4

Fig. 2 Boron concentration in soil solution as a function of solution-to-soil ratio for a given total amount of B. (A) No interaction between B and soil, (B) Boron adsorption account for. **Source:** Mezuman and Keren.[28]

BORON–PLANT INTERACTION

Boron is an essential micronutrient element required for growth and development of plants.

Many of the experimental data suggest that B uptake in plants is probably a passive process. There are clear evidences, however, that B uptake differs among species.[18] Several mechanisms have been postulated to explain this apparent paradox.[18–20] Boron deficiency in plants initially affects meristematic tissues, reducing or terminating growth of root and shoot apices, sugar transport, cell-wall synthesis and structure, carbohydrate metabolism and many biochemical reactions.[21,22] Tissue B concentrations associated with the appearance of vegetative deficiency symptoms have been identified in many crop species. It is essential to remember that for B, as for phosphorus and several other plant nutrient elements, deficiency may be present long before visual deficiency symptoms occur.

Excess and toxicity of boron in soils of semi-arid and arid areas are more of a problem than deficiency. Boron toxicity occurs in these areas either due to high levels of B in soils or due to additions of B in irrigation water. A summary of B tolerance data based upon plant response to soluble B is given by Maas.[23] Bingham et al.[24] showed that yield decrease of some crops (wheat, barley, and sorghum) due to B toxicity could be estimated by using a model for salinity response, suggested by Maas and Hoffman.[25]

There is a relatively small difference between the B concentration in soil solution causing deficiency and that resulting in toxicity symptoms in plants.[11] A consequence of this narrow difference is the difficulty posed in management of appropriate B levels in soil solution.

The suitability of irrigation water has been evaluated on the basis of criteria that determine the potential of the water to cause plant injury and yield reduction. In assessing the B in irrigation water, however, the physicochemical characteristics of the soil must be taken into consideration because the uptake by plants is dependent only on B activity in soil solution.[26,27] Boron uptake by plants grown in a soil of low-clay content is significantly greater than that of plants grown in a soil of high-clay content at the same given level of added B (Fig. 3). This knowledge may improve the efficacy of using water of different qualities, whereby water with relatively high B levels could be used to irrigate B-sensitive crops in soils that show a high affinity to B. Such water can be used for irrigation as long as the equilibrium B concentration in soil solution is below the toxic concentration threshold (the maximum permissible concentration for a given crop species that does not reduce yield or lead to injury symptoms) for the irrigated crop. The existing criteria for irrigation water, however, make no reference to differences in soil type.

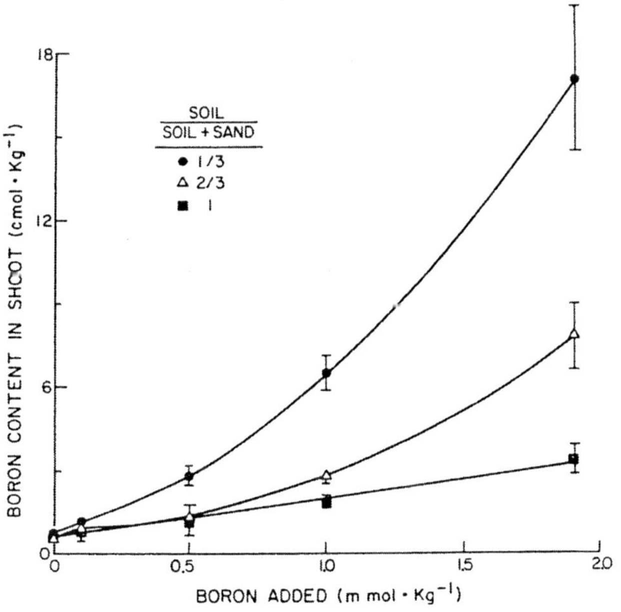
BORON CONTENT IN SHOOT (cmol · Kg^{-1})
18
12
6
0
SOIL / SOIL + SAND
● 1/3
△ 2/3
■ 1
0
0.5
1.0
1.5
2.0
BORON ADDED (m mol · Kg^{-1})

Fig. 3 Relationship between B content in wheat shoot and the amount of B added to soil, for three ratios of soil–sand mixtures. **Source:** Keren et al.[26]

REFERENCES

1. Cotton, F.A.; Wilkinson, G. *Advanced Inorganic Chemistry*, 5th Ed.; Wiley and Sons: New York, NY, 1988.
2. Owen, B.B. The dissociation constant of boric acid from 10 to 50°. J. Am. Chem. Soc. **1934**, *56*, 1695–1697.
3. Konopik, N.; Leberl, O. Colorimetric determination of PH in the range of 10 to 15. Monatsh **1949**, *80*, 420–429.
4. Ingri, N. Equilibrium studies of the polyanions containing B^{III}, Si^{IV}, Ge^{IV} and V^{V}. Svensk. Kem. Tidskr. **1963**, *75*, 199–230.
5. Mesmer, R.E.; Baes, C.F., Jr.; Sweeton, F.H. Acidity measurements at elevated temperature. VI. Boric acid equilibria. Inorg. Chem. **1972**, *11*, 537–543.
6. Owen, B.B.; King, E.J. The effect of sodium chloride upon the ionozation of boric acid at various temperatures. J. Am. Chem. Soc. **1943**, *65*, 1612–1620.
7. Adams, R.M. *Boron, Metallo-Boron Compounds and Boranes*; John Wiley and Sons: New York, 1964.
8. Onak, T.P.; Landesman, H.; Williams, R.E.; Shapiro, I. The B^{II} nuclear magnetic resonance chemical shifts and spin coupling values for various compounds. J. Phys. Chem. **1959**, *63*, 1533–1535.
9. Good, C.D.; Ritter, D.M. Alkenylboranes: II. Improved preperative methods and new observations on methylvinylboranes. J. Am. Chem. Soc. **1962**, *84*, 1162–1166.
10. Servoss, R.R.; Clark, H.M. Vibrational spectra of normal and isotopically labeled boric acid. J. Chem. Phys. **1957**, *26*, 1175–1178.
11. Keren, R.; Bingham, F.T. Boron in water, soil and plants. Adv. Soil Sci. **1985**, *1*, 229–276.
12. Keren, R.; Grossl, P.R.; Sparks, D.L. Equilibrium and kinetics of borate adsorption–desorption on pyrophyllite in aqueous suspensions. Soil Sci. Soc. Am. J. **1994**, *58*, 1116–1122.
13. Keren, R.; Gast, R.G.; Bar-Yosef, B. pH-dependent boron adsorption by na-montmorillonite. Soil Sci. Soc. Am. J. **1981**, *45*, 45–48.
14. Keren, R.; Gast, R.G. pH dependent boron adsorption by montmorillonite hydroxy-aluminum complexes. Soil Sci. Soc. Am. J. **1983**, *47*, 1116–1121.
15. Goldberg, S.; Glaubig, R.A. Boron adsorption on aluminum and iron oxide minerals. Soil Sci. Soc. Am. J. **1985**, *49*, 1374–1379.
16. Goldberg, S.; Glaubig, R.A. Boron adsorption on California soils. Soil Sci. Soc. Am. J. **1986**, *50*, 1173–1176.
17. Goldberg, S.; Forster, H.S.; Heick, E.L. Boron adsorption mechanisms on oxides, clay minerals and soils inferred from ionic strength effects. Soil Sci. Soc. Am. J. **1993**, *57*, 704–708.

18. Nable, R.O. Effects of B toxicity amongst several barley wheat cultivars: a preliminary examination of the resistance mechanism. Plant Soil **1988**, *112*, 45–52.
19. Nable, R.O.; Lance, R.C.M.; Cartwright, B. Uptake of boron and silicon by barley genotypes with differing susceptibilities to boron toxicity. Ann. Bot. **1990**, *66*, 83–90.
20. Brown, P.H.; Hu, H. Boron uptake by sunflower, squash and cultured tobacco cells. Physiol. Plant **1994**, *91*, 435–441.
21. Loomis, W.D.; Durst, R.W. Chemistry and biology of boron. BioFactors **1992**, *3*, 229–239.
22. Marschner, H. *Mineral Nutrition of Higher Plants*, 2nd Ed.; Academic Press: London, 1995.
23. Maas, E.V. Salt tolerance of plants. In *Handbook of Plant Science in Agriculture*; Christie, B.R., Ed.; CRC Press, Inc.: Cleveland, Ohio, 1984.
24. Bingham, F.T.; Strong, J.E.; Rhoades, J.D.; Keren, R. An application of the Maas–Hoffman salinity response model for boron toxicity. Soil Sci. Soc. Am. J. **1985**, *49*, 672–674.
25. Maas, E.V.; Hoffman, G.J. Crop salt tolerance—current assessment. ASCE J. Irrig. Drainage Div. **1977**, *103*, 115–134.
26. Keren, R.; Bingham, F.T.; Rhoades, J.D. Effect of clay content on soil boron uptake and yield of wheat. Soil Sci. Soc. Am. J. **1985**, *49*, 1466–1470.
27. Keren, R.; Bingham, F.T.; Rhoades, J.D. Plant uptake of boron as affected by boron distribution between liquid and solid phases in soil. Soil Sci. Soc. Am. J. **1985**, *49*, 297–302.
28. Mezuman, U.; Keren, R. Boron adsorption by soils using a phenomenological adsorption equation. Soil Sci. Soc. Am. J. **1981**, *45*, 722–726.

Buildings: Climate Change

Lisa Guan
School of Chemistry, Physics and Mechanical Engineering, Science and Engineering Faculty, Queensland University of Technology, Brisbane, Queensland, Australia

Guangnan Chen
Faculty of Engineering and Surveying, University of Southern Queensland, Toowoomba, Queensland, Australia

Abstract
The cycling interaction between climate change and buildings is of dynamic nature. On one hand, buildings have contributed significantly to the process of human-induced climate change. On the other hand, climate change is also expected to impact on many aspects of buildings, including building design, construction, and operation. In this entry, these two aspects of knowledge are reviewed. The potential strategies of building design and operation to reduce the greenhouse gas emissions from buildings and to prepare the buildings to withstand a range of possible climate change scenarios are also discussed.

INTRODUCTION

The greenhouse effect is a natural warming process of the earth. It is caused by the greenhouse gases, such as water vapor, carbon dioxide (CO_2), and methane (CH_4), which trap long-wave radiation and then radiate the energy in all directions, warming the earth's surface and atmosphere (Fig. 1).[1] Without the heat-trapping greenhouse gases, scientists estimate that the average earth surface temperature would likely to be some 30°C colder than it is today, or –18°C instead of the present mild average 15°C.[1]

The enhanced greenhouse effect is additional to the natural process of greenhouse effect and is mainly due to human activities (human induced), including burning fossil fuels, land clearing, and agriculture, which change the makeup of the atmosphere and lead to an increased concentration of greenhouse gases.[1] This has the potential to cause significant changes in the global climate system (referred to as climate change), including increased temperature, changed patterns of rainfall, tropical cyclone activities, and other extreme climatic events.

Climate change has now become one of the most important global environmental issues facing the world today. It is now widely recognized as having significant potential to seriously affect the integrity of our ecosystems and human welfare.[2] The effects, or impacts, of climate change may be in physical, ecological, social, and/or economic areas.

In this entry, the likely future climate change is first presented. The cycling interaction between climate change and buildings is then discussed, which includes both aspects of the implication of climate change on building performance and the contribution of buildings to the process of human-induced climate change. The potential strategies for building design and operation are then highlighted, in order to reduce the greenhouse gas emissions from buildings and to prepare the buildings to withstand a range of possible climate change scenarios.

LIKELY FUTURE CLIMATE CHANGE

Due to uncertainties in future emissions and concentrations of greenhouse gases, as well as the climate system's response to the changing conditions and the natural influences (e.g., changes in the sun and volcanic activity), it may be difficult to accurately predict the extent of climate changes.[2] However, the advancements in climate model simulations, combined with more and more observed data on climate changes, led the Intergovernmental Panel on Climate Change (IPCC) to predict the following likely scenarios of climate changes[2]:

- Average global surface temperature will likely rise a further 1.1–6.4°C (2.0–11.5°F) during the 21st century. It is expected that the average rate of warming is very likely to be at least twice as large as that experienced during the 20th century.
- Warming will not be evenly distributed around the globe and will vary with different seasons. For example, land areas will warm more than oceans, high latitudes will warm more than low latitudes, and winters will be warming more than summers in most areas.
- There will be significant changes to the amount and pattern of precipitation, including an increase in droughts,

Biological – Cadmium

Encyclopedia of Environmental Management DOI: 10.1081/E-EEM-120046134

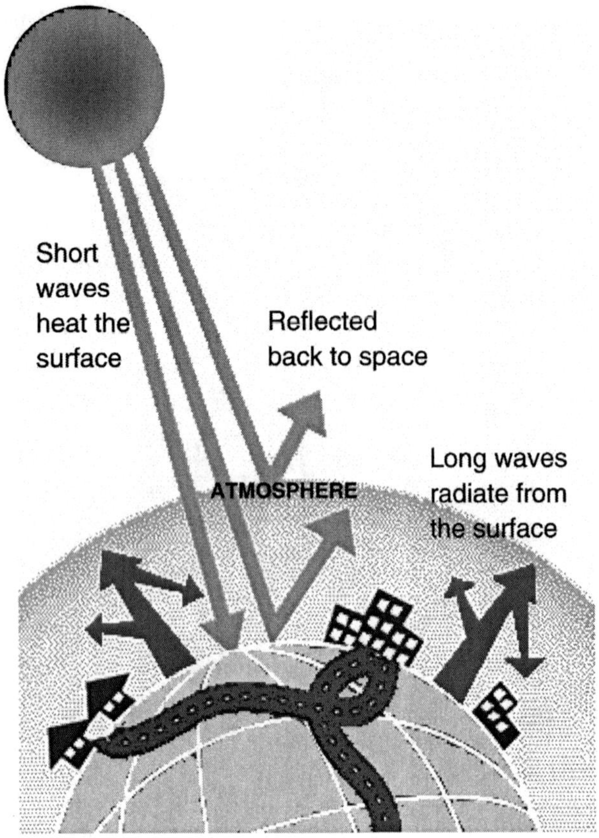
Short
waves
heat the
surface
Reflected
back to space
Long waves
radiate from
the surface
ATMOSPHERE

Fig. 1 Natural/enhanced greenhouse effect.

tropical cyclones, and extreme high tides. The changes in precipitation, either an increase or a decrease, will vary from region to region.

- The global average sea level is estimated to rise by 18–59 cm (7.2 to 23.6 inches) by 2100 relative to 1980–1999. Current model projections also indicate substantial variability in future sea level rise between different locations.
- Increases in the intensity of extreme weather events, such as storms, heat waves, and drought, are also predicted.

In regard to the projection of future climate change, it is noted that there are different levels of confidence (e.g., in terms of accuracies and reliability) for different climate parameters. For example, it is believed that there is higher confidence in the projection of increases in carbon dioxide concentrations and rises in sea level than in storminess or intense precipitation events. A list of climate and associated scenario variables, ranked subjectively in decreasing order of confidence, had been recommended by Hulme and Sheard[3] and is shown in Table 1.

IMPLICATION OF CLIMATE CHANGE ON BUILDINGS

Climate change is likely to affect both the performance of existing building stock and the design of new buildings.

Table 1 List of climate and associated scenario variables, ranked subjectively in decreasing order of confidence.

Climate variable	Confidence
Atmospheric CO_2 concentration	High
Global—mean sea level	↓
Global—mean temperature	
Regional seasonal temperature	
Regional temperature extremes	
Regional seasonal precipitation and cloud cover	
Regional potential evapotranspiration	
Changes in climatic variability (e.g., El Niño, daily precipitation regimes)	Low
Climate surprises (e.g., disintegration of the West Antarctic Ice Sheet)	Very low or unknown

Source: Hulmer and Sheard.[3]

For example, under climate change, buildings will have more overheating hours in summer and less underheating hours in winter, thus use more cooling energy and less heating energy. Where heating and cooling are provided by different fuels, this could have a significant influence on the design and operations of energy delivery systems.

Because climate change entails new climatic conditions for the building industries, it is expected that climate change will have a significant impact on the design, construction, and performance of buildings, as well as the health and productivity of people living and working inside them.[4] These impacts may include the following:

- *Higher building energy consumption*. Climate change may require higher capacity and more uses of air-conditioning equipment to provide comfort indoor environment. For example, it has been predicted that for air-conditioned office buildings in Australia, the cooling load may increase by 2% to 47%, depending on the assumed future climate scenarios, as well as different locations.[5] The increases of total building energy use would range from 0.4% to 15.1%. However, due to the potential decrease of heating energy in winter, skin-dominated buildings located in cold regions could receive some benefits from the climate change. Moreover, the expected increased stringencies in building energy codes around the world would also offset some increases of building energy use due to climate change.[6,7]
- *Deteriorating internal thermal environment*, such as more overheating in summer. It has been found that for air-conditioned office buildings in Australia, when the annual average temperature increase exceeds 2°C "threshold," the risk of current office buildings subjected to overheating will be significantly increased.[5] When the increase of external air temperature is more than 5°C, all the Australian office buildings would suffer from the

overheating problem regardless where they are located. This could have significant implications on people's health and capacity to work and productivity. It has been estimated that global warming is currently contributing to the death of about 160,000 people every year.[8]

- *Structural integrity*, such as more severe wind and snow loading, and foundation movement. For example, shrinkage or expansion in clay soils can lead to foundation movement and cracking of walls, which may cause damage in building structure. In Northwest Norway, severe damage to buildings was caused by the hurricane on January 1, 1992, and several buildings collapsed due to heavy snow loads on roofs during the 1999/2000 winter.[9]
- *External fabric*. Durability of external fabric becomes shorter due to increased storm, rain, flood, and other weather conditions. For example, the strength and durability properties of concrete may be influenced by the changes in its environment (e.g., temperature, humidity).
- *Construction process* may be disrupted due to adverse weather condition. This may have implication in the project planning and the associated challenges to complete the project on time and within the budget.
- *Service infrastructure* may become inadequate. For example, changing weather patterns and more frequent and intense storm may lead to the drainage problem. The existing drainage system in many parts of the world may not be able to cope with increased storm loads.

CONTRIBUTION OF BUILDINGS TO HUMAN-INDUCED CLIMATE CHANGE

The climate system is a dynamic system in transient balance.[10] A change of external and/or internal climate forcing imposed on the planetary energy balance would cause a corresponding change in global temperature.[10] Overall, scientists have now been able to reach a broad consensus and provide overwhelming evidence to suggest that human activities are having a discernible influence on the global climate.[2] In particular, it has been found that the recent rapid increase in global temperature is closely aligning with the strong growth in use of fossil fuel over the past 50 years.

The construction and operation of modern buildings consume a considerable amount of energy and materials and therefore contribute significantly to the process of human-induced climate change. Fig. 2 shows the world greenhouse gas emissions by sector, end use, activity, and gas types.[11] It can be seen that buildings, including both commercial buildings and residential buildings, account for 15.3% of world greenhouse gas emission, which is greater than the sectors of transportation, agriculture, and waste. Indeed, buildings are one of the most significant infrastructures in modern societies.

Worldwide, the Worldwatch Institute estimates that the construction and operation of buildings is responsible for 40% of the world's total energy use, 30% of raw materials consumption, 55% of timber harvests, 16% of freshwater withdrawal, 35% of global carbon dioxide (CO_2) emissions, and 40% of municipal solid waste sent to landfill.[12] In 2004, it was estimated that worldwide, the total emissions from the building sector, including the electricity consumed, were 8.6 Gt CO_2, 0.1 Gt CO_2–eq N_2O, 0.4 Gt CO_2–eq CH_4, and 1.5 Gt CO_2–eq halocarbons (including CFCs and HCFCs).[13]

Basically, the impact of buildings on the human-induced climate change is through three routes, including building operational energy, building embodied energy, and building-related refrigerants. Operational energy includes all energy used for mechanical services [e.g., heating, ventilation, and air conditioning (HVAC) systems], electrical services (e.g., lighting and other office appliances), and hydraulic services (e.g., pumping system). Embodied energy includes all energy used for the production of building materials, their transportation and handling, and building construction processes. Building-related refrigerants include the refrigerants used in air-conditioning systems (e.g., Freon used in compressors and chillers) and other building appliances (e.g., refrigerators and freezers), which may have ozone depletion potential (e.g., depleting the ozone in the upper atmosphere) and/or global warming potential that persists in the upper atmosphere, trapping the radiation emitted by the earth.

Previous studies of Life Cycle Assessment (LCA) have shown that the CO_2 emission from the sources of building embodied energy and building-related refrigerants is often fairly small for commercial office buildings. For instance, based on the assumption of a building having a 40 years life span, it has been shown that the embodied energy emissions contributed only approximately 8% to 10% of building total emission.[14] More than 90% of building energy consumption occurs during the use/operational phase, so the energy performance of the buildings is particularly important, especially for buildings with 24 hr occupancy.

RELATIONSHIP BETWEEN CLIMATE CHANGE AND BUILDINGS

As discussed above and also illustrated in Fig. 3, the cycling interaction between climate change and buildings is of dynamic nature. Climate change, for instance, would generally lead to more use of air conditioning, which leads to more greenhouse gas emission and then climate change. Therefore, both climate change and buildings are essentially the cause and the effect of each other.[15] On one hand, climate change is expected to impact on many aspects of building design, construction, and operation. On the other hand, buildings have also contributed significantly to the human-induced climate change. They have

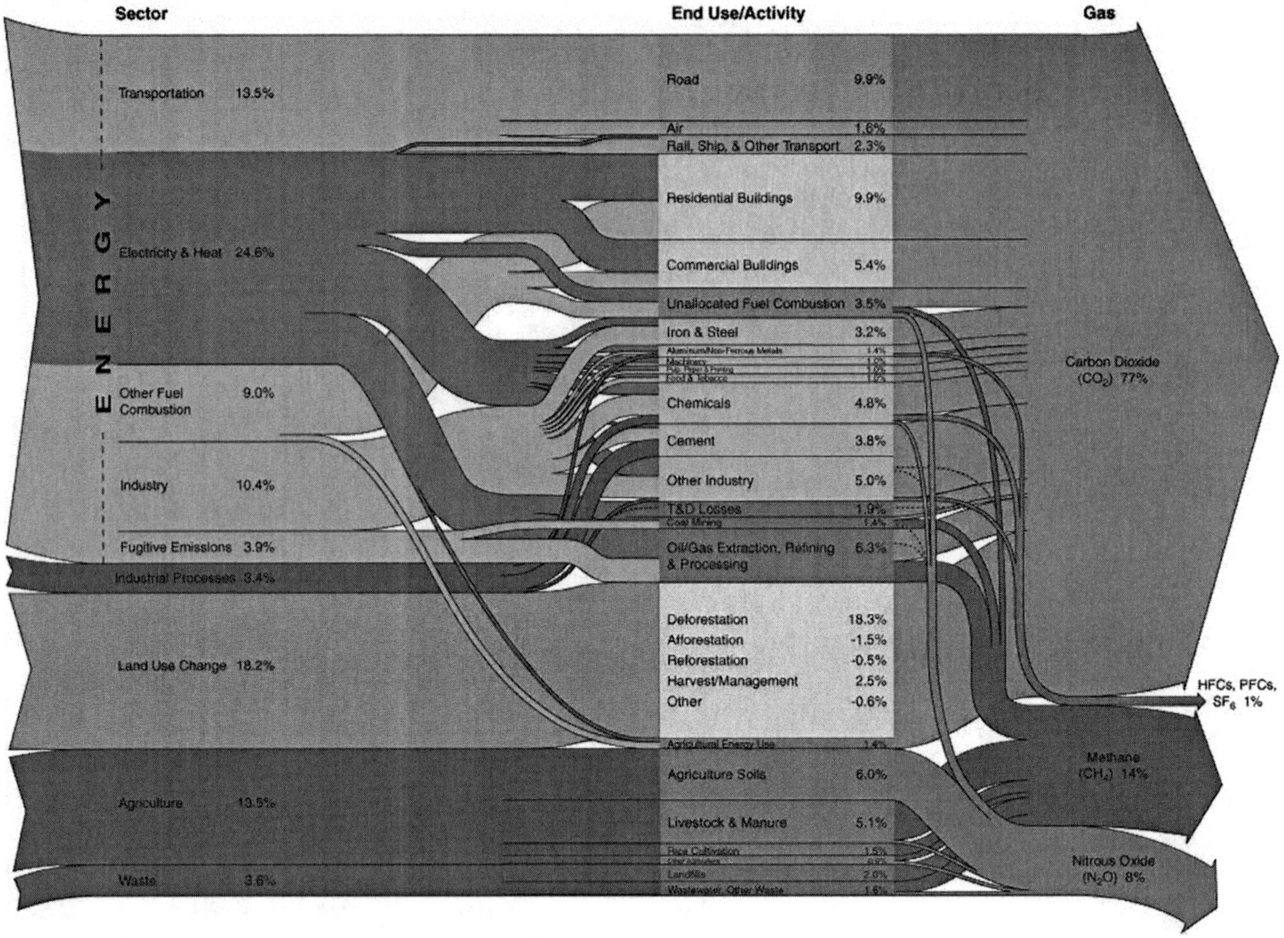

Sector
End Use/Activity
Gas
Transportation 13.5%
Road 9.9%
Air 1.6%
Rail, Ship, & Other Transport 2.3%
ENERGY
Residential Buildings 9.9%
Electricity & Heat 24.6%
Commercial Buildings 5.4%
Unallocated Fuel Combustion 3.5%
Iron & Steel 3.2%
Other Fuel Combustion 9.0%
Chemicals 4.8%
Cement 3.8%
Industry 10.4%
Other Industry 5.0%
T&D Losses 1.9%
Coal Mining 1.4%
Fugitive Emissions 3.9%
Oil/Gas Extraction, Refining & Processing 6.3%
Industrial Processes 3.4%
Carbon Dioxide (CO2) 77%
Deforestation 18.3%
Afforestation -1.5%
Reforestation -0.5%
Land Use Change 18.2%
Harvest/Management 2.5%
Other -0.6%
HFCs, PFCs, SF6 1%
Agricultural Energy Use 1.4%
Agriculture Soils 6.0%
Methane (CH4) 14%
Agriculture 13.5%
Livestock & Manure 5.1%
Rice Cultivation 1.5%
Landfills 2.0%
Waste 3.6%
Nitrous Oxide (N2O) 8%
Wastewater, Other Waste 1.6%

Fig. 2 World greenhouse gas emissions by sector, end use, activity, and gas types.
Notes: All data are for 2000. All calculations are based on CO_2 equivalents, using 100 years global warming potentials from the IPCC (1996), based on a total global estimate of 41,755 $MtCO_2$ equivalent. Land use change includes both emissions and absorptions. Dotted lines represent flows of less than 0.1% of total greenhouse gas emissions.
Source: World Resources Institute.[11]

produced more greenhouse gas emission than the sectors of transportation, agriculture, and waste.

Because greenhouse gas concentrations are still continuing to increase, the process of climate change will continue and may accelerate. This requires the building sector to develop

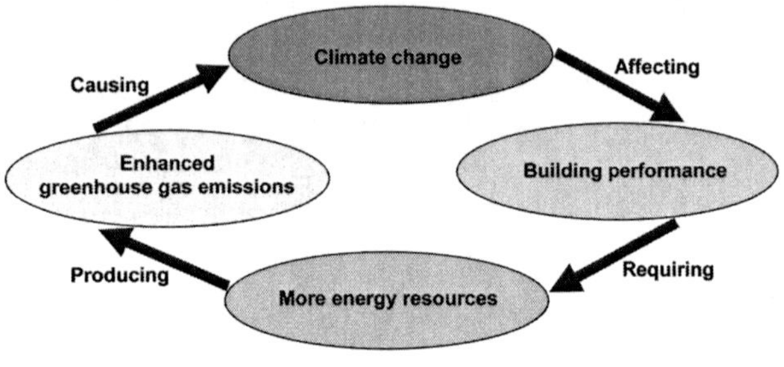

Climate change
Causing
Affecting
Enhanced greenhouse gas emissions
Building performance
Producing
Requiring
More energy resources

Fig. 3 The cycling interaction between global warming and buildings.

suitable strategies, including the enhancement in building energy codes, to mitigate the greenhouse gas emissions, which should be not only environmentally effective but also economically efficient. In parallel, appropriate adaptation strategies are also needed to prepare buildings to withstand the future inevitable climate change. Adaptation to climate change has now become one of the key requirements for buildings.

It is noted that there may be both potential synergy and conflict between adaptation and mitigation measures and strategies. This will require the development of integrated rather than separate responses.[16–18] Moreover, it seems that there is a tight intertwining between the issues of adaptation and mitigation.[19,20] The energy-efficient and renewable energy technologies, for instance, often can play dual roles in mitigating greenhouse gas emissions while increasing adaptive capacity by making buildings more disaster resilient. An effective response to climate change may not always be costly and could even be both profitable and socially beneficial.

POTENTIAL STRATEGIES FOR BUILDING DESIGN AND OPERATION

Buildings are one of the most significant infrastructures in modern societies, having significant impact on health and productivity of people living and working inside them. In particular, as buildings have a very long life span, typically 50–100 years, it is very important that all the current and future building stocks be designed and maintained to perform satisfactorily in future climates.

For both new and existing buildings, potential mitigation may include the utilization of renewable energy and low emission energy, energy efficiency, and energy conservation. Appropriate adaptation may also be achieved by focusing on major building design parameters for new buildings or retrofit of building envelope and/or internal heat sources such as lighting systems for existing buildings. In particular, those improved building energy codes currently adopted around the world, including the requirement of thermal conductance values for building envelope, lighting power densities for internal heat sources, and HVAC system efficiency for building mechanical services,[6,7] could play key roles in both the potential mitigation and adaptation of buildings. Overall, the aim is to ensure that all the existing and new buildings not only perform and operate satisfactorily in the new environment but also satisfy the environmental performance criteria of sustainability.

Building Envelope

The building envelope is a significant determinant of energy required to heat and cool a building. Building designers can use the building envelope as a "filter" to accept or reject solar radiation and outside air, depending on the need for heating, cooling, ventilation, and lighting.[13] The heat storage capacity of the building envelope can also be used to reduce peak thermal loads. As a result, a well-designed building shell can substantially reduce the building cooling and heating demands and lighting energy through the provision of daylighting, particularly at the perimeter zones. This can result in not only substantial capital cost savings by enabling smaller sizing of the air-conditioning plant to be used but also significant operating cost saving by reducing the quantity of air circulation and fan power, and partial load situation for the HVAC systems.

It is particularly important that building designers employ energy-efficient principles, as well as integrated building design approaches at design and construction of the building envelope. Overall, for the building envelope, the potential mitigation and adaptation strategies in the face of climate change may include the following:

- Installation of insulation in foundation, walls, and/or roofs to minimize conductive heat loss/gain.
- Optimum design of windows, including careful determination of window-to-wall ratio (WWR) and selection of types of glass and shading to optimize use of solar heat and light.
- Sealing gaps to prevent draughts and heat infiltration to minimize uncontrolled movement of air into the building.
- Utilizing thermal mass, reflective roofs, and trees for shade, etc., if it is appropriate.

Internal Heat Sources

The internal heat sources include the occupants, lights, electric appliances, and machines. Their influence may be not only directly on the total energy consumption of the building but also on the building thermal loads, which indirectly influence the energy consumption of HVAC systems. The selection of lights, for example, can have significant impact on the building energy and thermal performance. Advances in higher-efficiency light sources, lower-loss ballasts and control gear, and more efficient luminaries will therefore enable lighting energy usage to be significantly reduced. Developments in the technology of daylighting will further increase the potential lighting energy savings. The potential mitigation and adaptation strategies to manage internal heat sources may include the following:

- Optimum integration of lamp (lighting sources), ballasts (for electric-discharge lamps), and fixtures (e.g., reflectors, diffusers, and/or polarizing panels). It was found that to maintain the same lighting level of 320 lux, lighting power densities can vary from 2–5 W/m^2 (averaged over time) for having state-of-the-art lighting technology to 25–35 W/m^2 for having only mediocre level of efficiency.[21]
- Purchasing energy-efficient office equipment and running it wisely. Energy-efficient office equipment can use less than half the energy as the standard models—at no or little additional capital cost.[21] Recommendations may include the use of "energy star" office equipment and "energy labeling" appliance. It was found that every dollar invested in energy efficiency, at a 20%–30% saving rate, is equivalent to increasing net operating income by 3%–4% and net asset value by $2.50–$3.75.[22]

Using the building simulation technique, it has been shown that for Australian office buildings, if the building total internal load density, which includes all internal heat sources of lighting, plug load, and occupants, is reduced from 43 W/m^2 to 21.5 W/m^2, the building total energy use under the future 2070 high scenario (the worst case) can be reduced by up to 89 to 120 kWh/m^2 per annum and the overheating problem could be completely avoided and the office building will perform as good as at the current climate scenario.[23]

Building Mechanical Services

Building mechanical service or HVAC systems are used to control factors affecting thermal comfort, with the ultimate purpose being to provide a clean, noise-free, and efficient working environment.[24] Since air-conditioning systems account for around one-third of the total energy use in typical office buildings, the proper design and selection of HVAC systems are critical for the performance of these buildings.

Generally, according to the purpose of the building, as well as local site information and climate condition, a building may be designed as naturally ventilated, mechanically ventilated, air conditioned, or a combination of these three options (hybrid mixed-mode system). Potential mitigation and adaptation strategies for design and selection of building service systems include the following:

- Using passive design for heating, cooling, and ventilation. A building relying entirely on air conditioning should be avoided as much as possible.
- Utilization of renewable energy and/or low emission energy for the operation of the building. It was found that on clear days throughout the year, reductions of conventional power use of at least 60% can be achieved with optimum photovoltaic cladding densities targeted to lighting and small power load demands.[25]
- Selecting energy-efficient technologies. Such technologies may include evaporative cooling, desiccant cooling, chilled ceilings, displacement ventilation, cogeneration and district heating and cooling, active solar and heat pump systems, and underground earth pipes.[26]
- Good control of HVAC systems. It was found that the performance of the HVAC system is subject to the significant influence of its control strategy and maintenance. Building energy management systems (BEMSs) can also be a useful tool for energy saving in the range of 5%–40%.[26]
- Good commissioning, operation, and maintenance of HVAC systems. It was found that of the deficiencies found in the retro-commissioning of buildings, 85% were related to HVAC systems.[22] In the same study, it was also reported that energy savings between 7% and 29% may be achieved with paybacks ranging from 0.2 to 2.1 years.

Recommendations on Implementation of Strategies

Overall, it is recommended that the potential mitigation and adaptation strategies should be taken at the early planning and design stage. This is because the thermal loads gained/lost from the building envelope and internal heat sources are the "base load" for the selection of HVAC equipment. Appropriate selection and design of these systems are therefore vital for the energy efficiency of the building. Without a suitable choice, all later work will be built on an unsatisfactory foundation. It is also noted that the mitigation and adaptation strategies taken at the early design stage will future-proof the building and be more effective in the longer term. Although the initial capital cost of taking the measures may be higher, the overall cost in many cases would actually be much lower than having to do retrofit strengthening at a later stage.[16]

Building regulation and design codes can also play a crucial role in the anticipation and avoidance of risk.[27] Because climate change would have a major impact on the frequency of extreme weather events, building codes need to be reviewed regularly in order to maintain a proper level of reliability and be adhered to in practice.[9] For instance, changes in the frequency of storms will have building code implications.[28] However, it is argued that simply raising performance standards as a response to climate change, without dealing with the issue of noncompliance with existing standards, may not be desirable. It may run the risk of undermining the legitimacy of regulation generally.[27]

In the face of inevitable climate change, the shifting from a reliance on historical data to a reliance on predicted future data may be also necessary, as it is increasingly important to precisely predict the conditions under which buildings and other infrastructure will need to withstand in the future. It is understood that using different sets of design conditions will have significant implication on the cost and performance of buildings. For example, using the building simulation technique, it has been shown that for typical Australian office buildings, the required cooling capacity may increase from 28% to 59% if the new buildings were designed using the future project climate (e.g., 2070 high scenario) rather than the current climate condition.[5] In order to maintain comfort indoor condition, a further increase of 4%–10% may be required in addition to the possible increase of 27%–47% cooling load if the buildings were designed at current climate conditions.

CONCLUSION

The climate change induced by the emissions of greenhouse gases is one of the most important global environmental issues facing the world today. Buildings are one of the most significant infrastructures in modern societies, and they need to be prepared to withstand climate change. On one hand, climate change is going to impact on many aspects of buildings, including both building design and building operation. On the other hand, buildings have also contributed significantly to the process of human-induced climate change. In this entry, both aspects of knowledge have been discussed.

It has been suggested that the potential mitigation and adaptation strategies should focus on major energy-related factors, such as design and construction of building envelope, design and careful selection of the air-conditioning system, and selection of management of internal heat sources including both lighting systems and electrical equipment. In many cases, it has been demonstrated that the energy-efficient and renewable energy technologies can play dual roles in mitigating greenhouse gas emissions while increasing adaptive capacity by making buildings more disaster resilient.

It has also been recommended that energy efficiency principles, such as adopting passive design principles for heating, cooling, ventilation, and lighting, and integrated design approaches should be adopted at the early design stage and be maintained in the whole life of buildings. These strategies have the potential of not only being environmentally effective but also economically efficient and socially beneficial.

REFERENCES

1. Healey, J., Ed. *Climate Change—Issues in Society*; The Spinney Press: Sydney, Australia, 2003; Vol. 184, ISBN 1 876811 93 5.
2. IPCC, *Climate Change 2007: The Physical Science Basis*. Contribution of Working Group I to the Fourth Assessment Report of the Intergovernmental Panel on Climate Change; Solomon, S., Qin, D., Manning, M., Chen, Z., Marquis, M., Averyt, K.B., Tignor, M., Miller, H.L., Eds.; Cambridge University Press: Cambridge, United Kingdom and New York, NY, USA, 2007; 996.
3. Hulme, M.; Sheard, N. *Climate Change Scenarios for Australia Climatic Research Unit*; Norwich, U.K., 1999, 6 pp.
4. Guan, L. Global warming: Impact on building design and performance. Encycl. Energy Eng. Technol. **2009**, *1* (1), 1–6.
5. Guan, L. Implication of global warming on air-conditioned office buildings in Australia. Build. Res. Inf. **2009**, *37* (1), 1–12.
6. ANSI/ASHRAE/IESNA Standard 90.1-2007, Energy Standard for Buildings Except Low-Rise Residential Buildings.
7. Building Code of Australia (BCA)—Australian Building Codes Board (ABCB) 2010.
8. Bhattacharya, S. Global warming kills 160,000 a year. New-Scientist.com news service, 17:17 01 October 2003, available at http://www.newscientist.com/article.ns?id=dn4223.
9. Lisø, K.R.; Aandahl, G.; Eriksen, S.; Alfsen, K. Preparing for climate change impacts in Norway's built environment. Build. Res. Inf. **2003**, *31* (3), 200–209.
10. McGuffie K.; Henderson-Sellers, A. *A Climate Modeling Primer*, 3rd Ed.; John Wiley and Sons, Ltd., 2005.
11. World GHG Emissions Flow Chart, World Resources Institute, Retrieved on 9-2-2011 available at http://cait.wri.org/figures.php?page=/World-FlowChart.
12. Fenner, R.A.; Ryce, T. *A comparative analysis of two building rating systems. Part 1: Evaluation*, Proceedings of the Institution of Civil Engineers, Engineering Sustainability 161, March 2008, Issue ES1, 55–63, doi: 10.1680/ensu.2008.161.1.55.
13. Levermore, G.J. A review of the IPCC Assessment Report Four, Part 1: The IPCC process and greenhouse gas emission trends from buildings worldwide. Build. Services Eng. Res. Technol. **2008**, *29* (4), 349–361.
14. Australian Greenhouse Office, *Australian Commercial Building Sector Greenhouse Gas Emissions 1990–2010*; Canberra, 1999; ISBN 18 76536 195.
15. Degelman, L.O. Which came first—Building cooling loads or global warming? A cause and effect examination. Build. Services Eng. Res. Technol. **2002**, *23* (4), 259–267.
16. Lowe, R. Really rethinking construction. Build. Res. Inf. **2001**, *29* (5), 409–412.
17. Steemers, K. Towards a research agenda for adapting to climate change. Build Res. Inf. **2003**, *31*, 291–301.
18. Lowe, R. Lessons from climate change: A response to the commentaries. Build. Res. Inf. **2004**, *32* (1), 75–78.
19. Camilleri, M.; Jaques, R.; Isaacs, N. Impacts of climate change performance on building in New Zealand. Build. Res. Inf. **2001**, *29* (6), 440–450.
20. Mills, E. Climate change, insurance and the buildings sector: Technological synergisms between adaptation and mitigation. Build Res. Inf. **2003**, *31* (3–4), 257–277.
21. Sustainable Energy Development Authority (SEDA). *Tenant Energy Management Handbook—Your Guide to Saving Energy and Money in the Workplace,* 2000, ABN 80 526 465 581.
22. Cull, S.L. Commissioning: Retrocommissioning. Encycl. Energy Eng. Technol. **2007**, *1* (1), 200–206.
23. Guan, L. *Adaptation to global warming by changing internal loads of buildings.* Proceedings of International Conference on Building Energy and Environment (COBEE), Dalian, China, July 13–16, 2008.
24. ASHRAE. *ASHRAE Handbook: Fundamentals*; American Society of Heating, Ventilating and Air-Conditioning Engineers (ASHRAE): Atlanta, 2001.
25. Jones, A.D.; Underwood, C.P. Cladding strategies for building-integrated photovoltaics. Build. Services Eng. Res. Technol. **2002**, *23* (4), 243–250.
26. Levermore, G.J. A review of the IPCC Assessment Report Four, Part 2: Mitigation options for residential and commercial buildings. Build. Services Eng. Res. Technol. **2008**, *29* (4), 363–374.
27. Lowe, L. Preparing the built environment for climate change. Build. Res. Inf. **2003**, *31* (3), 195–199.
28. Larsson, N. Adapting to climate change in Canada. Build. Res. Inf. **2003**, *31* (3), 231–239.

Cabbage Diseases: Ecology and Control

Anthony P. Keinath
Coastal Research and Education Center, Clemson University, Charleston, South Carolina, U.S.A.

Marc A. Cubeta
Center for Integrated Fungal Research, Plant Pathology, North Carolina State University, Raleigh, North Carolina, U.S.A.

David B. Langston, Jr.
Rural Development Center, University of Georgia, Tifton, Georgia, U.S.A.

Abstract

Cabbage is an important component of the human diet and a source of chemoprotective phytochemicals. Six major diseases of cabbage found worldwide are black rot, clubroot, black spot (dark spot), downy mildew, watery soft rot (white mold), and wirestem. The pathogens causing black rot and black spot can be seedborne. The pathogens causing clubroot, watery soft rot, and wirestem are soilborne; the clubroot organism is remarkably difficult to eradicate from infested soils. Important control measures include seed treatment with hot water or fungicides, crop rotation to reduce survival of foliar pathogens, scouting to detect disease outbreaks, and judicious application of protectant fungicides. Further research is needed to clarify races of the downy mildew and club root pathogens and to find useful resistance to these two diseases.

INTRODUCTION

Cabbage (*Brassica oleracea* "Capitata Group") has long been cultivated as an important vegetable crop and a source of vitamins, minerals, and fiber, particularly during cold seasons in temperate climates. More recently, cabbage and other cruciferous vegetables (members of the Brassicaceae) have been recognized as important sources of chemoprotective phytochemicals in the diet. Cabbage is a productive vegetable based on biomass per area of cultivation. However, this crop is affected by many diseases, particularly those caused by fungi and bacteria. This entry focuses on six diseases of worldwide importance in cabbage production. These diseases also affect other cole crops, i.e., vegetables derived from *B. oleracea*, including broccoli, Brussels sprouts, cauliflower, collard, kale, and kohlrabi, and other genetically related cruciferous vegetables, such as turnip, rutabaga, Chinese cabbages, and mustards. Emphasis will be placed on stages in the life cycles of the pathogens that affect management. Control measures will be presented in an IPM context.

MAJOR DISEASES AND PATHOGEN ECOLOGY

Black Rot

Black rot is caused by the bacterium *Xanthomonas campestris* pathovar *campestris*. Because this bacterium can be seedborne, black rot is found in most areas of the world where cabbage and other crucifers are grown. The pathogen produces V-shaped chlorotic and necrotic lesions starting at the margins of leaves, but it also causes wilting of plants if it reaches the vascular system in the stem (systemic infection). Blackening of the leaf veins is a helpful diagnostic symptom. The pathogen survives in infested crop debris but can only live a few months in soil.

Clubroot

Clubroot is caused by the slime mold-like organism *Plasmodiophora brassicae*. This soilborne organism is an obligate parasite, completing its unique life cycle within the root cells of crucifers. Infected root cells enlarge and divide to produce the diagnostic swollen, club-like roots. The pathogen produces resting spores in the clubs that persist in soil for at least 10 years after the clubs decay. Isolates of *P. brassicae* differ in host range, and races have been found that are pathogenic on the few resistant cultivars of cabbage that have been bred.

Black Spot, Dark Leaf Spot

Two species of *Alternaria*, *A. brassicae* and *A. brassicicola*, infect cabbage and other crucifers. *A. brassicicola* has higher optimal temperatures for growth, sporulation, and spore germination (20–30°C.) than *A. brassicae* (18–24°C.). Both fungi can be seedborne and airborne, but do not survive apart from infested host debris in soil. Infested debris left on the soil surface can be a significant source of pathogen spores for up to 12 weeks after harvest.[1] Seedborne inoculum can lower seed germination and vigor but usually is not damaging to seedlings.

Encyclopedia of Environmental Management DOI: 10.1081/E-EEM-120041115

Downy Mildew

Crucifer downy mildew is caused by the Oomycete *Peronospora parasitica*. This fungus-like organism produces airborne sporangia on leaf undersides and oospores inside infected tissues. The pathogen is believed to survive as dormant oospores in roots and stems. Cabbage is affected by downy mildew particularly during the seedling and heading growth stages. High relative humidity, dew, and fog are favorable for infection. Separate host genes confer resistance in the cotyledon and adult plant stages of growth.[2] Interactions observed between resistant plant varieties and isolates of the pathogen suggest that races of the pathogen exist.

Watery Soft Rot, Sclerotinia Stem Rot, White Mold

Sclerotinia sclerotiorum has a wide host range, but is especially damaging to cabbage, because it not only infects the head in the field but also can cause decay in storage. The common names for this disease show that infection occurs primarily on heads or stems of cabbage, particularly at maturity when wrapper leaves shade the soil, providing a cool, moist environment that favors the pathogen. This fungus produces airborne spores that infect plants, but soilborne survival structures (sclerotia) also can cause infection when they germinate near a plant.

Wirestem

Wirestem, a postemergence disease, is caused by the soilborne fungus *Rhizoctonia solani* anastomosis groups (AG) 4 and 2-1. In soils cropped repeatedly to crucifers, AG 2-1 predominates. At low pathogen levels, wirestem is more prevalent or more severe than preemergence damping-off. Seedlings may be killed by wirestem when lesions girdle stems. Older plants may be killed later as a result of seedling infections or be stunted and fail to produce a marketable-sized head. Root rot also occurs when infection is severe but is absent when discrete stem lesions are the only symptoms.

CONTROL

General Control Principles

Exclusion

It is extremely important to prevent contamination of clubroot-free land by excluding the pathogen. Movement of transplants and equipment from clubroot-infested fields or farms should be avoided. Growers in clubroot-free areas should avoid purchasing field-grown transplants or equipment from infested areas.

Eradication

Outbreaks of black leg associated with seed have been reduced by testing seed for the pathogen *Phoma lingam*. Eradicate cruciferous weeds to eliminate sources of the pathogens causing black rot, downy mildew, and clubroot (Table 1). In addition, cruciferous ornamentals can be infected by the same species of *Alternaria*, *Peronospora*, *Plasmodiophora*, and *Xanthomonas* that infect cabbage.

Table 1 Management practices for common diseases of cabbage.

Disease	Plant resistant cultivars	Use healthy seed or transplants	Control weeds	Avoid wounding	Bury crop residue	Rotate with non-host	Apply protectant fungicide or bactericide
Black spot	+	+	+	–	+	+	+
Bacterial soft rot	–	–	–	+	–	–	–
Black leg	–	+	–	–	+	+	+
Black rot	+	+	+	+	+	+	–/+
Clubroot	–	+[a]	+	–	–	–	+/–
Downy mildew	–	–	+	–	+	+	+
Yellows	+	+[a]	–	–	–	–	–
Sclerotina stem rot	–	–	+	+	+	–	+/–
Damping-off	–	–	–	–	–	–	+
Wirestem	–	+[a]	–	–	+	–	+

Notes: +, practice can be used to manage the disease; –, practice is ineffective or inappropriate, based on the life cycle of the pathogen; +/–, practice may be useful under certain conditions.

[a]The pathogen is not seedborne, but can be spread on infected, field-grown transplants.

Avoidance

Do not plant susceptible cabbage in pathogen-infested fields. Wirestem is less severe when cabbage is planted into cool soils than into warm soils. In addition, using a shallow planting depth for transplants avoids contact of the susceptible hypocotyl with *Rhizoctonia*-infested soil. Avoid wounding plants to prevent black rot, bacterial soft rot, and watery soft rot.

Resistance

Host plant resistance is widely available in green (white) and red cabbage for yellows (caused by *Fusarium oxysporum* f. sp. *conglutinans*). Newer hybrid cultivars have partial resistance to black rot that restricts lesions to the wrapper leaves. A few cabbage cultivars (mostly red cabbage) have moderate resistance to *Alternaria*. Cabbage cultivars available in the U.S.A. are susceptible to *Sclerotinia*, downy mildew, wirestem, and clubroot.

Protection

Seed treatment is very effective to prevent damping-off caused by *Pythium* spp. and *R. solani*. Protectant fungicides are effective against foliar fungal pathogens and also are used against wirestem, clubroot, and black rot with varying degrees of success. Recently, the fungicide boscalid was registered in the U.S.A. to control *Sclerotinia* on cole crops.

Therapy

The only measure to control cabbage diseases postinfection is the application of systemic fungicides for downy mildew.

EXAMPLES OF INTEGRATED DISEASE MANAGEMENT

Controlling weeds, especially ragweed (*Ambrosia artemisiifolia*), can reduce incidence of watery soft rot. Ascospores of *Sclerotinia* infect ragweed flowers that then fall onto cabbage leaves and infect them, because flower parts provide nutrients for the pathogen.[3] Control flea beetles (*Phyllotreta cruciferae*), which carry conidia of *A. brassisicola* on their bodies and in their frass and transmit conidia while feeding.[4]

Private and public cabbage scouting programs have been developed and are useful for scouting production fields for diseases and insects. For example, the cabbage-scouting program in Suffolk County, New York, U.S.A., has operated for the past 20 years. In addition to insects, scouts record the presence, general severity, and field location of black rot, black spot, clubroot, downy mildew, viruses, watery soft rot, and yellows.

MANAGING SEEDBORNE PATHOGENS

Plant seed from seedlots that have tested negative for the presence of the pathogens that cause black rot and black leg. Hot water seed treatment is useful to control seedborne black rot bacteria, provided the water temperature is monitored carefully so it remains at 50°C for 25 minutes. Minimize leaf wetness periods when producing transplants in glasshouses, because of the ease of spreading pathogens. Apply protectant fungicides to seed crops to prevent infection of seed by *Alternaria*.

MANAGING SOILBORNE PATHOGENS

Soil fumigants generally are not used against soilborne pathogens in cabbage production because of the high cost, although they may be used to disinfest seedbeds and suppress clubroot. Field-grown transplants may be sources of the wirestem and clubroot pathogens and spread them to non-infested fields. Because of this risk, transplants should be produced in soilless mixes in glasshouses when possible. Do not plant any cruciferous vegetables in fields before or after cropping to cabbage. Use monocots as rotation crops, because *R. solani* AG 4 has a wide host range among dicotyledonous crops. The resting spores of the clubroot organism cannot be eradicated by rotation. Instead, liming soil to raise the pH above 7.2 with calcium oxide or hydrated lime prevents infection of roots in many soils.

MANAGING FOLIAR PATHOGENS

Diseases caused by foliar pathogens, such as *Xanthomonas* and *Alternaria*, can be managed with crop rotation during the period when infested host debris is decaying in affected fields, because these foliar pathogens of cabbage do not survive longer than one or two years in soil, respectively. Disk and bury or compost unmarketable cabbage heads. Apply protectant fungicides as needed based on environmental conditions and host susceptibility. Because *Alternaria* spp. require relatively long periods of leaf wetness for infection (a minimum of five to nine hours), disease can be reduced by increasing row width and plant spacing to promote air circulation that dries leaves.

CONCLUSIONS

The diseases black spot, downy mildew, watery soft rot, and wirestem often can be managed successfully using a combination of cultural, biological, and chemical control measures. The cultural and biological methods listed in Table 1 also are amenable to organic production systems. Management of black rot and clubroot remains more challenging. In the future, resistance to downy mildew and improved resistance to black rot may be available in cabbage cultivars. It may be possible to transfer downy mildew resistance from broccoli to cabbage using molecular genetics methods. Additional research is needed to clarify the identity of races of the downy mildew and clubroot organisms.

ACKNOWLEDGMENTS

Technical contribution No. 5100 of the Clemson University Experiment Station. We thank Richard Morrison, Sakata Seed Company, and J. Powell Smith, Clemson University, for reviewing this entry.

REFERENCES

1. Humpherson-Jones, F.M. Survival of *Alternaria brassicae* and *A. brassicicola* on crop debris of oilseed rape and cabbage. Ann. Appl. Biol. **1989**, *115* (1), 45–50.
2. Coelho, P.S.; Monteiro, A.A. Expression of resistance to downy mildew at cotyledon and adult plant stages in *Brassica oleracea* L. Euphytica **2003**, *133* (3), 279–284.
3. Dillard, H.R.; Hunter, J.E. Association of common ragweed with Sclerotinia rot of cabbage. Plant Dis. **1986**, *70* (1), 26–28.
4. Dillard, H.R.; Cobb, A.C.; Lamboy, J.S. Transmission of *Alternaria brassicicola* to cabbage by flea beetles (*Phyllotreta cruciferae*). Plant Dis. **1998**, *82* (2), 153–157.

Cadmium and Lead: Contamination

Gabriella Kakonyi
Kroto Research Institute, Sheffield University, Sheffield, U.K.

Imad A.M. Ahmed
Lancaster Environment Center, Lancaster University, Lancaster, U.K.

Abstract

The presence of cadmium and lead in the environment is greatly sourced from and affected by anthropogenic activities. These activities are severe and present global environmental and human health issues. The detection of these elements and the determination of their chemical state, thus toxicity, have challenged scientists and engineers for many years. On the basis of the knowledge gained, production is changing and regulations are being put in place to be able to reduce and control the effect of cadmium and lead in the environment.

Biological – Cadmium

INTRODUCTION

This entry summarizes the effects of cadmium (Cd) and lead (Pb) on the environment and on human health. The chemistry of Cd and Pb is first discussed to establish an understanding of their behavior, occurrence, and fate in the environment. This helps foresee toxicity and the possible natural attenuation or remediation strategies. Both Cd and Pb are widely produced and used. Therefore, awareness of the sources from where humans and the environment are exposed is described in some detail along with their epidemiology and regulatory measures taken to reduce or prevent their release into the environment. A brief section is also dedicated to the detection and analysis of Cd and Pb and their compounds within environmental and clinical samples.

CHEMICAL PROPERTIES

Cadmium (Cd) and lead (Pb) are transition metals that have no known vital or beneficial role in the human body or health. Cadmium (atomic number 48, atomic mass 112.4 g mol^{-1}) belongs to Group 12 of the periodic table, possessing somewhat similar chemical properties as zinc and mercury. Lead (atomic number 82, atomic mass 207.2g mol^{-1}) is a member of Group 14 in the periodic table, which also includes C, Si, Ge, and Sn.[1] The most predominant oxidation state of Cd and Pb under normal environmental conditions of temperature and pressure is +2. In organolead chemistry, the oxidation state (+4) of Pb is remarkably dominant.[2] Both Cd and Pb are considered to form stable oxidation states as divalent Cd^{2+} and Pb^{2+} ions in inorganic compounds. Their biological toxicity appears to be determined by their availability for ligand exchange and chelation properties.[3–6] The chemical similarity of Cd and Pb to certain alkaline earth metals such as calcium; their ability to form highly insoluble inorganic salts (e.g., phosphate, carbonate, sulfate), organometallic complexes, or free hydrated ions; and their increased affinity to biological donors (e.g., proteins) play an important role in their transport in the environment and toxicity in biological systems.[2,7,8]

OCCURRENCES

Cadmium and lead are trace elements in rocks and soils,[9] where the concentration of Cd is approximately 0.1 ppm and the concentration of Pb is about tenfold higher.[10,11] These concentrations are relatively low, and the presence of both Cd and Pb are normally associated with other more abundant elements. Therefore, both metals are inevitable by-products from the mining of zinc and copper ores. Lead and cadmium are chalcophilic elements, meaning that they have a tendency to form sulfide minerals.[12] Thus, the most abundant Cd and Pb minerals are greenockite (CdS) and galena (PbS). Most Cd is found associated with zinc ores such as sphalerite (ZnS) in which the Cd content is 0.5% of the Zn content.[11] Other important minerals of Pb are crocite ($PbCrO_4$), anglesite $PbSO_4$, massicot (PbO), and cerussite ($PbCO_3$).[13] Native (i.e., elemental) Cd and Pb are rare in the environment. Naturally occurring metallic Cd has been found in the Vilyuy River bedrock in Siberia.[14] Cadmium and lead are also found associated with clay and carbonate minerals. The Cd^{2+} ion has an ionic size of 95 pm close to that of Ca^{2+} (100 pm), whereas Pb^{2+} has an ionic size (119 pm) between K^+ (138 pm) and Ca^{2+}. Thus, during the formation of secondary minerals such as feldspars, mica, apatite, and calcite, major cations such as Ca^{2+} and K^+ can be substituted by Cd^{2+} and Pb^{2+} or other

Encyclopedia of Environmental Management DOI: 10.1081/E-EEM-120046485

trace metal ions of the same charge, sign, or of comparable ionic size.[15,16]

PRODUCTION AND USES

Cadmium

Cadmium is widely used in a number of industrial applications, the largest area being the production of nickel-Cd batteries.[17,18] Furthermore, Cd is increasingly used in solar panels, and still commonly used in pigments, coatings, corrosion-resistant plating, photography, as a fungicide and as a stabilizer and softener in plastics. It is also contained in coal and in rocks mined to produce phosphate fertilizers.[19] Cadmium atoms has the ability to absorb neutrons without fission or splitting, so it is used in nuclear reactor components such as control rods as a shield of neutrons and to control nuclear fission.[20]

Lead

The greatest use of Pb is lead acid batteries, which have been used extensively in automobiles since 1918 or so.[21] Lead is also commonly added to paint as Pb-chromate or Pb-carbonate to speed drying, increase durability, retain a fresh appearance, and resist moisture. Because elemental Pb has a low melting temperature (327°C), it enables easy casting and shaping and thus is commonly used in building constructions and joining metallic parts. Lead is the traditional base metal for domestic water pipes. One if its major uses is as a radiation shield in the glass of television and computer screens, and as a protecting shield from radioactive radiation such as x-ray and γ-ray in scientific and industrial instruments. Other uses are in infrared detectors, sheeting, cables, solders, Pb-crystal glassware, ammunitions, and as weight in sport equipment.

DETECTION

Because Cd is used as pigment and softener in plastics, the use of colored lids, tubes and certain plastic containers should be avoided during sampling and storage of environmental and toxicological samples.[22] It is well documented that significant amounts of trace metals can be lost on the walls of glass and some plastic containers and adsorption of Pb occurs on the walls of Pyrex, polypropylene and polyethylene containers. The loss of aqueous Cd and Pb by adsorption onto the wall of containers was not observed using Teflon. Therefore, particular attention must be paid in the collection, treatment and preservations of environmental samples, especially when Cd and Pb are present at submicromolar concentrations. Procedures required for total metal analysis are normally straight forward and state the acidification of samples to $pH < 1$ and preservation in inert containers such as Teflon. Furthermore, accurate determination of cadmium concentrations within sensitive samples may not be done by a smoking person in order to avoid cross-contamination.[23] Free cadmium in solution (Cd^{2+}) presents in water below pH values of 8. The highest concentration was measured to be 6 mM without hydroxide formation. Above pH 8, cadmium is expected to precipitate and hinder accurate determination of the concentration.[24] Removal of cadmium and lead from aerosol and fly ash samples is discussed in details by Lum[25] and Hlavay et al.[26] Total aqueous Pb and Cd concentrations can be measured using atomic absorption spectroscopy or inductively coupled plasma spectrometry, the former detecting ion concentrations at mg L^{-1} (ppm) levels and the latter at ng L^{-1} (ppt). Organometallic compounds of both Pb and Cd can be measured by inductively coupled plasma mass spectrometry, atomic absorption spectrometry, high-performance liquid chromatography, and gas chromatography analytical methods.[27–32]

ENVIRONMENTAL LEVELS

Cadmium and lead are released into the environment as a result of both industrial activities and natural processes. The main natural sources include the weathering of rocks releasing Cd and Pb into the hydrosphere, and volcanic activity increasing the atmospheric Cd and Pb concentration.[15] The eruption of Mount Pinatubo located on the Philippine island of Luzon is an example of the dramatic effects of volcanism on the distribution of elements in the lithosphere. During just 2 days in June 1991, Pinatubo ejected 10 billion metric tonnes of magma and 20 million tonnes of SO_2; the resulting aerosols influenced global climate for 3 years. This single event introduced an estimated 100,000 tonnes of Pb and 1000 tonnes of Cd to the surface environment.[15,33] From anthropogenic sources, humans are exposed to Cd through the atmosphere, hydrosphere, and the geosphere. Human uptake of Cd can happen through inhalation of air, soil, or dust containing fine Cd particles. Cadmium is introduced into the atmosphere through the combustion of fossil fuel (e.g., coal), municipal solid waste incineration, mineral smelting and as dust generated by recycling scrap iron and steel. The contamination of soils is apparent from zinc and phosphate ores and where Cd-containing phosphate fertilizers are used in agriculture.[34–37] Zinc and cadmium are mineralogically and geochemically linked to each other. Thus, large emissions of fumes containing both ZnO and CdO are produced from zinc smelters. However, these fumes are normally enriched in CdO because of its higher volatility in comparison to ZnO. It is well documented that phosphate fertilizers constitute a very diffuse source of Cd contamination. The quantity of Cd contained in a phosphate fertilizer depends on the source of the phosphate rock used in making it. In general, Cd content in phosphate fertilizers vary from 1-2 ppm for tertiery Ca-phosphate to 50-170 ppm for superphosphate. As a result, Cd is transported to aquatic environments, plants, animals, and finally to humans. Soil organic matter strongly adsorbs Cd, and acidic soils further enhance the Cd uptake by plants. Cadmium is also known to accumulate in aquatic organisms, especially in freshwaters. However, Chen et al.[38] reported that the addition of potassium fertilizers effectively reduce the phytoavailability of both Cd and Pb within soils.

Although Pb is a naturally occurring element, its major cycle in the environment is anthropogenically driven and a result of human activities. For example, Pb in the form of tetraethyl-lead has been in usage as additive in gasoline to avoid knocking effect of autoignition since the early 20th century.[39] Through this application, Pb has been released into the environment in the forms of Pb-chloride, -bromide, and -oxides from the car exhaust. Lead is also introduced into the environment from coal and solid waste combustion, a wide number of industrial and mining processes, and drinking water pipes containing Pb. When copper is present in (either from soldering or as contamination) a Pb-containing pipe, it accelerates corrosion of the Pb pipe by galvanic action. The water flowing in contact with these dissimilar metals serves as the electrolyte. Based on the electrochemical series, metallic Pb serves as the anode of this galvanic cell and is therefore oxidized (i.e., corroded) to form Pb^{2+} ions contaminating the drinking water. Lowering pH in water also promotes the dissolution and aqueous transport of Pb^{2+}. Evidences of galvanic corrosion of domestic Pb pipes had been given and discussed elsewhere.[40,41] Large Pb-containing particles released to the atmosphere settle quickly on the ground and then washed into soils or dissolve in aquatic phases, while the very fine particles remain in the atmosphere travelling long distances and fall back to the surface with rain. This cycle has caused an unnatural and extensive sequence exposing plants, aquatic environments, and humans to dangerous concentrations of Pb. Therefore, in 1978, it was forbidden for all European Union–member states to produce, import, or sell gasoline with more than 0.4 g Pb L^{-1}. Starting from the year 2000, the marketing of leaded petrol has been banned in Europe following Directive 98/70/EC and related acts of the European Parliament. In the United States, Pb was banned as fuel additive starting from 1996.[12,42–44] Unfortunately, the gasoline additives tri- and dialkyl Pb are stable compounds and are persistent in the environment; thus, the restriction in usage does not necessarily decrease contamination.[45–51] When Pb accumulates in living organisms, it becomes part of the food chain thereby creating further sources for human exposure. Beyond bioaccumulation, once released into surface waters, both Cd and Pb are deposited into the sediment, increasing the metal contamination by 10–100 times near the fallout areas. The type of sediment affects the severity of contamination, with carbonaceous, anoxic, and clay sediments being the most prone to high concentrations of metal uptake, while siltstones, shales, sandstones, limestones and marine evaporates are normally less affected.[52]

Biological – Cadmium

TOXICOLOGICAL EFFECTS

The International Agency for Research on Cancer (IARC) has classified Cd and Pb compounds as carcinogenic to humans.[87] This classification has been based mainly on epidemiological evidences of renal damage in rats and mice. The greatest concerns about the health effects of Cd and Pb arise from their tendency to form strong complexes, replacing essential elements (e.g., Ca^{2+}) and bioaccumulating in the human body. There is no known small enough amount of Pb uptake that would cause no harm to human beings. However, Reichlmayr-Lais et al.[53,54] found that the depletion of Pb resulted in hematological changes in rats. The concentration of Cd is approximately 0.4 μg kg^{-1} in a daily diet, which is said to be 10 times lower than the amount that can cause kidney damage.[55] The concentrations of Cd in human blood is usually between 0.1 and 2 μg L^{-1} and in urine <1 μg L^{-1}.[56] The concentration of Pb in human blood varies between 165 and 296 μg L^{-1}. The largest known catastrophe caused by Cd toxicity was identified in Japan in the 1940s, and it is referred to as the "itai-itai" disease, which is literary translated from Japanese to "ouch ouch" disease. Itai-itai refers to a syndrome that principally consists of a painful skeletal condition resulting from weak and deformed bones. The patients of itai-itai suffer from renal anemia, tubular nephropathy, and osteopenic osteomalacia, while 90% of the patients are postmenopausal women. The residents of the Jinzu River basin region were first exposed to Cd in the 1930s as a result of industrial contamination from nearby intensive mining activities, which caused serious pollution of the local river waters. This resulted in high Cd contaminations of rice fields. With rice being the principal dietary component, especially in rural Japan, and the bioaccumulating properties of Cd, residents were exposed to very high levels of Cd causing irreversible damage and poisoning.[57,58] Toxicity of Cd in living organisms occurs because of the substitution of essential elements such as Zn at the reactive centers of essential enzymes, which disrupts a wide range of metabolic functions. Further details on mechanisms of Cd toxicity are available elsewhere.[57,59–62]

Cadmium is taken into the body through food, drinking water, smoking, and particulate matter in air, especially near Cd-processing industrial fields and hazardous waste sites. Gastrointestinal absorption from contaminated water or food is the main source of internally deposited Cd in the general population. Only a small proportion of ingested Cd is transferred to the bloodstream while the unabsorbed Cd is excreted in the feces. The absorbed Cd binds to macromolecules, enzymes, and proteins, and the majority of it is being deposited in the liver and kidney. The Cd taken up by the kidney interferes with the filtering mechanism, causing the excretion of essential sugars and proteins, and damages the kidney's ability to remove acid from the blood. Another extremely painful effect of Cd is the softening of bones and decreasing their mineral density, resulting in fractures and paralysis in advanced cases. Furthermore, Cd damage of the central nervous system and psychological disorders were reported, as well as weakening of the immune system, reproductive failure, DNA damage, and the development of cancer.[63]

Lead can enter the human body through food, liquid, and air. Inorganic Pb compounds are known to pass through the skin.[64] It has been shown that the pollution of air with Pb particles from burning of fuel affects the cognitive behavior of children and adults living near busy roads or exposed to Pb contamination. Early experiments were carried out to study the dose-response to environmental levels of Pb. A comprehensive review of these studies can be found in U.S. Environmental Protection Agency (USEPA) 2005.[65] The absorbed Pb by the human body is distributed in blood, soft tissues, and in particular in bones and liver. Because Pb^{2+} can replace Ca^{2+}, ~9% of absorbed Pb ends up in the bones and teeth. The excess of Pb may cause several health effects, including damage to nervous system, chronic renal disease, anemia due to the inhibition of haem formation, damage of nervous system of unborn children, acute encephalopathy in young children, carcinogenicity and genotoxicity, and impaired reproductivity.[66–68] Even in low concentrations (1–10 μg dL^{-1} blood), Pb decrements neurocognitive abilities, intelligence measures, and perceptual–motor coordination. Lead toxicity of animals has been also studied extensively. For example, cattle showed poisoning symptoms within 6–8 wk when fed Pb-acetate at 7 mg Pb per kilogram body weight per day.[69–71] Accumulation of Pb in the liver and kidney was noted in calves fed 100 mg Pb per kilogram as Pb chromate for 100 days.[72] While the absorbed Cd and Pb could be excreted in urine, the daily excretion is <1% of the total body burden of Cd and Pb giving a biological half-life of Cd of more than 25 years[73,74] and 20–30 years for Pb in the skeleton.[75]

Cadmium and lead toxicity in animals is a major problem, especially in diary animals as Cd and Pb accumulate in the kidney, liver, and reproductive organs.[88] Many plants species are tolerant to certain amounts of heavy metals, which is likely to be achieved via metal–binding by specific proteins. Like all living organisms, at elevated concentrations of certain heavy metal plants start to show symptoms of toxicity. For example, Cd phytotoxicity can be identified in the form of stunting and chlorosis. Chlorosis is due to Cd interaction with foliar iron. A number of reports had shown evidences to Pb adverse affects on the growth and photosynthesis processes of plants. High concentration of Pb in soils is known to inhibit seed germination in a number of plant species and to induce abnormal morphologies.[89–91] It is interesting to note that a number of reports have shown that Cd accumulates in greater concentrations roots, tubers or leaves of plants. This means that Cd can present at higher concentrations in leafy and root vegetables than in fruits or grains.[91]

REGULATIONS AND CONTROL

In the European Union, Cd and Pb are on the list of the six hazardous substances that are banned in the manufacturing of various electrical and electronic components. This is enforced by the Restriction on Hazardous Substances Directive (RoHS, 2002/95/EC) requiring that the maximum concentration of Cd or Pb may not exceed 100 or 1000 ppm per weight of homogeneous material, respectively. The production, recycling, and disposal of batteries and accumulators are regulated by the 2006/66/EC directive. These metals are also regulated under the Registration, Evaluation, Authorisation, and Restriction of Chemicals (REACH, 1907/2006) by the European Commission. In the United States, the USEPA is the main regulatory body restricting and controlling the use of Cd and Pb. Generally, the legislations are separated by pollution of the atmosphere, geosphere, hydrosphere, or specific industrial activities. Furthermore, the World Health Organization (WHO) globally regulates the acceptable exposure levels and concentration limits in public areas (Table 1). The maximum permissible concentrations of Cd and Pb in drinking water determined by the USEPA, WHO, and Drinking Water Inspectorate-Department for Environment, Food and Rural Affairs (DWI-DEFRA, United Kingdom) is shown in Table 2. As a result of these regulations, the release of Cd and Pb into the environment and their human exposure has been lowered through recycling and safer, more conscious industrial processes. However, the need for such metals and the global consumption still persists. On an annual basis, 9.6 million tonnes of Pb and 19,000 tonnes of Cd are produced worldwide. To reduce and control exposure of the population through drinking water, regulations require the use of Pb-free (<0.2% Pb) pipe systems and continuous monitoring (USEPA). Cadmium and lead can be removed from water by reverse osmosis or ion exchange resins. For the treatment of groundwater and soil media, *in situ* precipitation techniques present a viable way to reduce the mobility of heavy metals. In the case of both Cd- and Pb phosphate–containing minerals (e.g., hydroxyapatite), calcium–carbonates and zeolites have been suggested as possible solutions.[76–82]

Table 1 General human exposure to cadmium and lead.

	Food	Water	Air
Cadmium	25 μg kg^{-1} body weight per month	3 μg L^{-1}	5 ng m^{-3} a^{-1}
Lead	0.8 μg kg^{-1} body weight per day	3.8–10 μg L^{-1} day^{-1}	0.5–4 μg m^{-3} day^{-1}

Source: Data from WHO.[83,84]

Table 2 Drinking water regulations for maximum permissible contaminant levels (USEPA, National Primary Drinking Water Regulations).

	USEPA[a] ($mg\ L^{-1}$)	WHO guideline value	DWI-DEFRA (UK)
Cadmium	5.0	3.0	5.0
Lead	zero action level: 0.015	10	25^b

[a] Maximum contaminant level goal.
[b] Maximum concentration is 25 mg L^{-1} until December 25, 2013, and 10 mg L^{-1} from December 25, 2012.
Source: Data from WHO[84,85] and DWI.[86]

CONCLUSION

Both Cd and Pb have played prominent roles in the industrial revolution and subsequent centuries, and are now included in a vast range of products. Both elements have also played important, -but highly contrasting- roles in terms of human, animal, and plant health. The increased environmental concentration of Cd and Pb poses irreversible effects on nature and the human body. The contamination problem of heavy metals -including Cd and Pb- remains a challenging issue to scientists and engineers. The question whether the dangers associated with Cd and Pb could be avoided or could be further lowered remains open. Production of Cd and Pb has grown to be a global need, which can only be compensated by the development and use of alternative technologies, advanced remediative solutions and preventive measures, sensitive monitoring systems, and the avoidance of human and animal exposures.

REFERENCES

1. Greenwood, N.N.; Earnshaw, A. *Chemistry of the Elements*; Elsevier: Oxford, 1997.
2. Grant, L.D. Lead and compounds. In *Environmental Toxicants*; Lippmann, M., Ed.; John Wiley and Sons, Inc.: Hoboken, NJ, 2009; 757–809.
3. Ahamed, M.; Siddiqui, M.K.J. Environmental lead toxicity and nutritional factors. Clin. Nutr. **2007**, *26* (4), 400–408.
4. Shukla, G.S.; Singhal, R.L. The present status of biological effects of toxic metals in the environment: Lead, cadmium, and manganese. Can. J. Physiol. Pharmacol. **1984**, *62* (8), 1015–1031.
5. Andersen, O. Chelation of cadmium. Environ. Health Perspect. **1984**, *54*, 249–266.
6. Deagen, J.T.; Oh, S.H.; Whanger, P.D. Biological function of metallothionein. VI. Metabolic interaction of cadmium and zinc in rats. Biol. Trace Elem. Res. **1980**, *2* (1), 65–80.
7. Huheey, J. *Inorganic Chemistry: Principles of Structure and Reactivity*, 4th Ed.; Harper Collins College Publishers: New York, 1993.
8. Lindsay, W.L. *Chemical Equilibria in Soils*; The Black Burn Press: Caldwell, NJ, 2001.
9. Turekian, K.K.; Wedepohl, K.H. Distribution of the elements in some major units of the earth's crust. Geol. Soc. Am. Bull. **1961**, *72* (2), 175–192.
10. Cox, P.A. *The Elements: Their Origin, Abundance, and Distribution*; Oxford University Press: Oxford, U.K., 1989.
11. Halka, M.; Nordstrom, B. *Periodic Table of the Elements: Transition Metals*; Facts on File: New York, 2011.
12. Kummer, U.; Pacyna, J.; Pacyna, E.; Friedrich, R. Assessment of heavy metal releases from the use phase of road transport in Europe. Atmos. Environ. **2009**, *43* (3), 640–647.
13. Halka, M.; Nordstrom, B. *Periodic Table of the Elements: Metals and Metalloids*; Facts On File: New York, 2011.
14. Fleischer, M.; Cabri, L.; Chao, G.; Pabst, A. New mineral names. Am. Mineral. **1980**, *65* (9–10), 1065–1070.
15. Garrett, R.G. Natural sources of metals to the environment. Hum. Ecol. Risk. Assess. **2000**, *6* (6), 945–963.
16. Sposito, G. *The Chemistry of Soils*; Oxford University Press, Inc.: Oxford, U.K., 1989.
17. Wiley-VCH. *Ullmann's Encyclopedia of Industrial Chemistry*, 7th Ed.; Wiley-VCH: Weinheim, Germany, 2011.
18. Tolcin, A. Cadmium. In *Minerals Yearbook Metals and Minerals 2009*; United States Government Printing Office, 2009.
19. Robert, U.A. Metals recycling: Economic and environmental implications. Resour. Conserv. Recycling **1997**, *21* (3), 145–173.
20. Harvey, T.; Thomas, B.; Mclellan, J.; Fremlin, J. Measurement of liver-cadmium concentrations in patients and industrial workers by neutron-activation analysis. Lancet **1975**, *305* (7919), 1269–1272.
21. Guberman, D.E. Lead. In *Minerals Yearbook Metals and Minerals 2009*; Guberman, D.E., Ed.; United States Government Printing Office: Reston, VA, 2009.
22. Sekaly, A.L.; Chakrabarti, C.; Back, M.; Gregoire, D.; Lu, J.Y.; Schroeder, W. Stability of dissolved metals in environmental aqueous samples: Rideau River surface water, rain and snow. Anal. Chim. Acta **1999**, *402* (1–2), 223–231.
23. Cornelis, R.; Caruso, J.; Crews, H.; Heumann K. *Handbook of Elemental Speciation I: Techniques and Methodology*; John Wiley and Sons, Ltd: Chichester, U.K., 2003.
24. Majidi, V.; Miller-Ihli, N.J. Potential sources of error in capillary electrophoresis-inductively coupled plasma mass spectrometry for chemical speciation. Analyst **1998**, *123* (5), 809–813.
25. Lum, K.; Betteridge, J.; Macdonald, R. The potential availability of P, Al, Cd, Co, Cr, Cu, Fe, Mn, Ni, Pb and Zn in urban particulate matter. Environ. Technol. Lett. **1982**, *3* (1–11), 57–62.
26. Hlavay, J.; Polyak, K.; Meszaros, E. Determination of the distribution of elements as a function of particle size in aerosol samples by sequential leaching. Analyst. **1998**, *123* (5), 859–863.
27. Bettmer, J.; Cammann, K. Transversely heated graphite atomizer–atomic absorption spectrometry (THGA AAS) in combination with flow injection analysis system–hydride generation (FIAS HG) as a reliable screening method for organolead compounds. Appl. Organomet. Chem. **1994**, *8* (7–8), 615–620.
28. Dunemann, L.; Hajimiragha, H.; Begerow, J. Simultaneous determination of Hg(II) and alkylated Hg, Pb, and Sn species in human body fluids using SPME-GC/MS-MS. Fresenius J. Anal. Chem. **1999**, *363* (5–6), 466–468.

29. Fragueiro, M.S.; Alava-Moreno, F.; Lavilla, I; Bendicho, C. Determination of tetraethyl lead by solid phase microextraction–thermal desorption–quartz furnaceatomic absorption spectrometry. J. Anal. At. Spectrom. **2000**, *15* (6), 705–709.
30. Infante, H.G.; Sanchez, M.L.F.; Sanz-Medel, A. Vesicle-mediated high performance liquid chromatography coupled to hydride generation inductively coupled plasma mass spectrometry for cadmium speciation in fish cytosols. J. Anal. At. Spectrom. **2000**, *15* (5), 519–524.
31. Moens, L.; De Smaele, T.; Dams, R.; Van Den Broeck, P.; Sandra, P. Sensitive, simultaneous determination of organomercury, -lead, and -tin compounds with head-space solid phase microextraction capillary gas chromatography combined withinductively coupled plasma mass spectrometry. Anal. Chem. **1997**, *69* (8), 1604–1611.
32. Yu, X.; Yuan, H.; Gorecki, T.; Pawliszyn, J. Determination of lead in blood and urine by SPME/GC. Anal. Chem. **1999**, *71* (15), 2998–3002.
33. Selinus, O.; Finkelman, R.B.; Centeno, J.A. Human health and ecosystems. In *Geology and Ecosystems*; Zektser, I.S., Marker, B., Ridgway, J., Rogachevskaya, L., Vartanyan, G., Eds.; Springer US: Boston, 2006; 197–218.
34. Grant, C.A.; Sheppard, S.C. Fertilizer impacts on cadmium availability in agricultural soils and crops. Hum. Ecol. Risk Assess. **2008**, *14* (2), 210–228.
35. Jiao, Y.; Grant, C.A.; Bailey, L.D. Effects of phosphorus and zinc fertilizer on cadmium uptake and distribution in flax and durum wheat. J. Sci. Food Agric. **2004**, *84* (8), 777–785.
36. Syers, J.K.; Mackay, A.D.; Brown, M.W.; Currie, L.D. Chemical and physical characteristics of phosphate rock materials of varying reactivity. J. Sci. Food Agric. **1986**, *37* (11), 1057–1064.
37. Taylor, M.D. Accumulation of cadmium derived from fertilisers in New Zealand soils. Sci. Total Environ. **1997**, *208* (1–2), 123–126.
38. Chen, S.; Sun, L.; Sun, T.; Chao, L.; Guo, G. Interaction between cadmium, lead and potassium fertilizer (K_2SO_4) in a soil–plant system. Environ. Geochem. Health **2007**, *29* (5), 435–446.
39. Hernberg, S. Lead poisoning in a historical perspective. Am. J. Ind. Med. **2000**, *38* (3), 244–254.
40. Nguyen, C.K.; Stone, K.R.; Dudi, A.; Edwards, M.A. Corrosive microenvironments atlead solder surfaces arising from galvanic corrosion with copper pipe. Environ. Sci. Technol. **2010**, *44* (18), 7076–7081.
41. Zhang, Y.; Triantafyllidou, S.; Edwards, M. Effect of nitrification and GAC filtration on copper and lead leaching in home plumbing systems. J. Environ. Eng. **2008**, *134* (7), 521–530.
42. Geivanidis, S.; Pistikopoulos, P.; Samaras, Z. Effect on exhaust emissions by the use of methylcyclopentadienyl manganese tricarbonyl (MMT) fuel additive and other lead replacement gasolines. Sci. Total Environ. **2003**, *305* (1–3), 129–141.
43. Hagner, C. European regulations to reduce lead emissions from automobiles? Did they have an economic impact on the German gasoline and automobile markets? Reg. Environ. Change **2000**, *1* (3–4), 135–151.
44. Kerr, S.; Newell, R.G. Policy-induced technology adoption: Evidence from the U.S. lead phase down. J. Ind. Econ. **2003**, *51* (3), 317–343.
45. Allen, A.G.; Radojevic, M.; Harrison, R.M. Atmospheric speciation and wet deposition of alkyllead compounds. Environ. Sci. Technol. **1988**, *22* (5), 517–522.
46. De Jonghe, W.R.A; Chakraborti, D.; Adams, F.C. Sampling of tetraalkyl lead compounds in air for determination by gas chromatography-atomic absorption spectrometry. Anal. Chem. **1980**, *52* (12), 1974–1977.
47. Hewitt, C.; Harrison, R.M. A sensitive, specific method for the determination of tetraalkyl lead compounds in air by gas chromatography/atomic absorption spectrometry. Anal. Chim. Acta **1985**, *167*, 277–287.
48. Hewitt, C.; Harrison, R.M.; Radojevic, M. The determination of individual gaseousionic alkyllead species in the atmosphere. Anal. Chim. Acta **1986**, *188*, 229–238.
49. Nerin, C.; Pons, B.; Martinez, M.; Cacho, J. Behaviour of several solid adsorbents for sampling tetraalkyllead compounds in air. Mikrochim. Acta **1994**, *112* (5–6), 179–188.
50. Nielsen, T.; Egsgaard, H.; Larsen, E.; Schroll, G. Determination of tetramethyllead and tetraethyllead in the atmosphere by a two-step enrichment method and gaschromatographic–mass spectrometric isotope dilution analysis. Anal. Chim. Acta **1981**, *124* (1), 1–13.
51. Radziuk, B.; Thomassen, Y.; Van Loon, J.; Chau, Y. Determination of alkyl lead compounds in air by gas chromatography and atomic absorption spectrometry. Anal. Chim Acta **1979**, *105*, 255–262.
52. Warren, L.J. Contamination of sediments by lead, zinc and cadmium: A review. Environ. Pollut. B. **1981**, *2* (6), 401–436.
53. Reichlmayr-Lais, A.M.; Eder, K.; Kirchgessner, M. The effect of lead supply on liver phospholipids of lactating rats. J. Anim. Physiol. Anim. Nutr. **1993**, *70* (1–5), 104–108.
54. Reichlmayr-Lais, A.M.; Eder, K.; Kirchgessner, M. Fatty acid composition of erythrocyte membranes, liver and milk of rats depending on different alimentary lead supply. J. Anim. Physiol. Anim. Nutr. **1993**, *70* (1–5), 109–116.
55. USEPA. *Cadmium Fact Sheet*; Agency of Toxic Substances and Disease Registry: Atlanta, GA, 2011.
56. Cornelis, R.; Heinzow, B.; Herbert, R.B.; Christensen, J.; Poulsen, O.; Sabbioni, E. Sample collection guidelines for trace elements in blood and urine (technical report). Pure Appl. Chem. **1995**, *67* (8/9), 1575–1608.
57. Ishihara, T.; Kobayashi, E.; Okubo, Y.; Suwazono, Y.; Kido, T.; Nishijyo, M. Association between cadmium concentration in rice and mortality in the Jinzu River basin, Japan. Toxicology **2001**, *163* (1), 23–28.
58. Kawano, S.; Nakagawa, H.; Okumura, Y.; Tsujikawa, K. A mortality study of patients with Itai-itai disease. Environ. Res. **1986**, *40* (1), 98–102.
59. Nogawa, K.; Yamada, Y.; Honda, R.; Ishizaki, M.; Tsuritani, I.; Kawano, S. The relationship between itai-itai disease among inhabitants of the Jinzu River basin and cadmium in rice. Toxicol. Lett. **1983**, *17* (3–4), 263–266.
60. Umemura, T.; Wako, Y. Pathogenesis of osteomalacia in itai-itai disease. J. Toxicol. Pathol. **2006**, *19* (2), 69–74.
61. Malcolm, E.; Morel, F. The biogeochemistry of cadmium. In *Metal Ions in Biological Systems*; Sigel, A., Sigel, H.,

Sigel, R., Eds.; Biogeochemical Cycles of Elements; CRC Press, Taylor and Francis Group: London, 2005; Vol. 43, 195–219.
62. Wilkinson, J.M.; Hill, J.; Phillips, C.J.C. The accumulation of potentially-toxic metals by grazing ruminants. Proc. Nutr. Soc. **2007**, *62* (02), 267–277.
63. Jarup, L.; Berglund, M.; Elinder, C.G.; Nordberg, G.; Vahter, M. Health effects of cadmium exposure—A review of the literature and a risk estimate. Scand. J. Work Environ. Health **1998**, *24* (Suppl 1), 1–51.
64. Sun, C.; Wong, T.; Hwang, Y.; Chao, K.; Jee, S.; Wang, J. Percutaneous absorption of inorganic lead compounds. AIHA J. **2002**, *63* (5), 641–646.
65. USEPA. *Ecological Soil Screening Levels for Lea*; U.S. Environmental Protection Agency, Office of Solid Waste and Emergency Response: Washington, D.C., 2005; OSWER Directive 9285.7-70.
66. Needleman, H.L.; Bellinger, D. The health effects of low level exposure to lead. Annu. Rev. Public Health **1991**, *12* (1), 111–140.
67. John, F.R. Adverse health effects of lead at low exposure levels: Trends in the management of childhood lead poisoning. Toxicology **1995**, *97* (1–3), 11–17.
68. Goyer, R.A. Lead toxicity: From overt to subclinical to subtle health effects. Environ. Health Perspect. **1990**, *86*, 177–181.
69. Aronson, A. Lead poisoning in cattle and horses following long-term exposure to lead. Am. J. Vet. Res. **1972**, *33* (3), 627–629.
70. Hammond, P.B.; Aronson, AL. Lead poisoning in cattle and horses in the vicinity of a smelter. Ann. N.Y. Acad. Sci. **1964**, *111*, 595–611.
71. Buck, W.B.; James, LF; Binns, W. Changes in serum transaminase activities associated with plant and mineral toxicity in sheep and cattle. Cornell Vet. **1961**, *51*, 568–585.
72. Dinius, D.A.; Brinsfield, T.H.; Williams, EE. Effect of subclinical lead intake on calves. J. Anim. Sci. **1973**, *37* (1), 169–173.
73. Kim, B.J.; Kim, M.; Kim, K.; Hong, Y.; Kim, I. Sensitizing effects of cadmium on TNF-a- and TRAIL-mediated apoptosis of NIH3T3 cells with distinct expression patterns of p53. Carcinogenesis **2002**, *23* (9), 1411–1417.
74. Golovine, K.; Makhov, P.; Uzzo, R.G.; Kutikov, A; Kaplan, D.J.; Fox, E. Cadmium down-regulates expression of XIAP at the post-transcriptional level in prostate cancer cells through an NF-κB-independent, proteasome-mediated mechanism. Mol. Cancer **2010**, *9* (1), 183.
75. Jarup, L. Hazards of heavy metal contamination. Br. Med. Bull. **2003**, *68* (1), 167–182.
76. Ahmed, I.A.M.; Young, S.; Crout, N. Time-dependent sorption of Cd^{2+} on CaX zeolite: Experimental observations and model predictions. Geochim. Cosmochim. Acta **2006**, *70* (19), 4850–4861.
77. Ahmed, I.A.M.; Young, S.; Crout, N. Ageing and structural effects on the sorption characteristics of Cd^{2+} by clinoptilolite and Y-type zeolite studied using isotope exchange technique. J. Hazard Mater. **2010**, *184* (1–3), 574–584.
78. Ahmed, I.A.M.; Crout, N.M.; Young, S.D. Kinetics of Cd sorption, desorption and fixation by calcite: A long-term radiotracer study. Geochim. Cosmochim. Acta **2008**, *72* (6), 1498–1512.
79. Xu, Y.; Schwartz, F.W.; Traina, S.J. Sorption of Zn^{2+} and Cd^{2+} on hydroxyapatite surfaces. Environ. Sci. Technol. **1994**, *28* (8), 1472–1480.
80. Ryan, J.A.; Zhang, P.; Hesterberg, D.; Chou, J.; Sayers, D.E. Formation of chloropyro-morphite in a lead-contaminated soil amended with hydroxyapatite. Environ. Sci. Technol. **2001**, *35* (18), 3798–3803.
81. Mavropoulos, E.; Rossi, A.M.; Costa, A.M.; Perez, C.A.C.; Moreira, J.C.; Saldanha, M. Studies on the mechanisms of lead immobilization by hydroxyapatite. Environ. Sci. Technol. **2002**, *36* (7), 1625–1629.
82. da Rocha, N.C.C.; de Campos, R.C.; Rossi, A.M.; Moreira, E.L.; Barbosa, A.dF.; Moure, G.T. Cadmium uptake by hydroxyapatite synthesized in different conditions and submitted to thermal treatment. Environ. Sci. Technol. **2002**, *36* (7), 1630–1635.
83. WHO. Exposure to cadmium: A major public health concern. Public Health and Environment, World Health Organization: Geneva, Switzerland, 2010.
84. WHO. *Lead in Drinking-Water: A Background Document for Development of WHO Guidelines for Drinking-Water Quality*; Public Health and Environment, World Health Organization: Geneva, Switzerland, 2011; WHO/SDE/WSH/03.04/09/Rev/1.
85. WHO. *Cadmium in Drinking-Water: A Background Document for Development of WHO Guidelines for Drinking-Water Quality*; Public Health and Environment, World Health Organization: Geneva, Switzerland, 2011; WHO/SDE/WSH/03.04/80/Rev/1.
86. DWI. Legislative background to the Private Water Supplies Regulations 2009: Section 9 (E&W). Drinking Water Inspectorate; 2010.
87. IARC. International Agency for Research on Cancer Monographs on the Evaluation of the Carcinogenic Risks to Humans—Beryllium, Cadmium, Mercury, and Exposures in the Glass Manufacturing Industry; WHO: Lyon, 1993; vol. 58, 119–237.
88. Neathery, M.W.; Miller, W.J. Metabolism and Toxicity of Cadmium, Mercury, and Lead in Animals: A Review. Journal of Dairy Science **1975**, *58* (12), 1767–1781.
89. Nagajyoti, P.; Lee, K.; Sreekanth, T. Heavy metals, occurrence and toxicity for plants: a review. Environmental Chemistry Letters **2010**, *8* (3), 199–216.
90. Das, P.; Samantaray, S.; Rout, G.R. Studies on cadmium toxicity in plants: A review. Environmental Pollution **1997**, *98* (1), 29–36.
91. Kabata-Pendias A. Trace Elements in Soils and Plants. CRC Press, Inc.: Boca Raton, FL, 2011.

Cadmium: Toxicology

Sven Erik Jørgensen
Institute A, Section of Environmental Chemistry, Copenhagen University, Copenhagen, Denmark

Abstract
Cadmium contamination of rice in Japan caused the dreadful itai-itai disease. Cadmium is mainly used for surface treatment, as a stabilizing agent in plastic and many alloys. The major sources of cadmium dispersion are cadmium content in phosphorus fertilizers and air pollutants from coal-fired power plants. Cadmium forms chloride complexes that are less toxic than the cadmium ions. Formation of the complexes increases the solubility of cadmium in water. It is therefore important to consider the cadmium–chloride complexes by examination of cadmium pollution cases. It is recommendable to set up regional mass balances for all major toxic substances and particularly for the most toxic heavy metals such as mercury, lead, cadmium, and copper. A mass balance for Danish agriculture land is shown to illustrate the use of regional mass balance in environmental management. Cadmium has a carcinogenic and teratogenic effect. It is also highly toxic, as indicated by the LD_{50} value for rats—70–90 mg/kg. The uptake efficiency of cadmium from food is about the same as for other heavy metals—7–10%. Cadmium is accumulated mainly in the kidneys. It is possible to express the cadmium accumulation in the human body as a function of time by the following differential equation: dCd/dt = daily uptake – excretion coefficient * Cd = $dCd/dt = 0.0025 - 0.0001 * Cd$ mg/24 hr. At steady state, the cadmium concentration becomes 0.0025/0.0001 = 25 mg, which (of course) is close to the average value at the age of 50 years.

INTRODUCTION: DISPERSION AND APPLICATION

From 1940 to 1960, Japanese in the Toyama Prefecture were poisoned by cadmium in their rice, because the river water used for irrigation was contaminated by cadmium from a cadmium mine.[1] Cadmium replaces calcium in the bones, which makes them soft. It causes extreme pain and the disease was named *itai-itai*, which means "ouch-ouch." When it was discovered that cadmium was causing the disease, the mine waste was controlled, but by then, several thousands (mainly farmers) were already suffering from the very painful disease.

Cadmium is used for metal surface treatment, as a stabilizing agent in plastic and many alloys. Phosphorus fertilizers have a relatively high cadmium concentration—in the order of 10–80 mg/kg. Together with the cadmium emitted from coal-fired power plants, it is the most important source of global cadmium dispersion. More than 1000 tons of cadmium is globally dispersed from coal-fired power plants. Cadmium was previously applied in ceramic, but this use of the very toxic metal is now banned in most industrialized countries.

Like other heavy metals, cadmium shows biomagnification and bioaccumulation and is accumulated in the sediment of aquatic ecosystems. It entails that the dispersion of cadmium in aquatic ecosystems can, in most cases, best be determined by analyses of the sediment where relatively higher concentrations are found.[2] Similar to lead (*Lead: Ecotoxicology*, p. 1651), it is possible by analyses of sediment cores to determine the history of cadmium emission. The phosphorus fertilizer industry has, for several decades, discharged cadmium-containing production waste to the Little Belt in Denmark. An environmental impact assessment was carried out for Little Belt by analyses of a large number of sediment cores, whereby the history of the contamination was determined. Fortunately, cadmium can form complexes with chloride, which are less toxic than the cadmium ions.[3,4] The complexes are at the same time more soluble, which means that less cadmium is transferred to the sediment and more to the open sea where it is diluted significantly. The discharge of cadmium waste was stopped as a result of the investigation, but it was also concluded that the contamination of the Little Belt was less than expected, probably due to the formation of cadmium–chloride complexes.

Due to the high cadmium concentration in phosphorus fertilizers, there is a risk for cadmium contamination of agricultural land, particularly by the application of intensive agriculture. Cadmium is taken up by plants (see also *Bioremediation*, p. 408). A cadmium model has been developed to relate the cadmium concentration in crops as a function of the cadmium contamination by the use of fertilizers.[5]

It is beneficial for all toxic substances and particularly for the most toxic heavy metals to make a regional mass balance to identify the sources and the dispersion to assess

Encyclopedia of Environmental Management DOI: 10.1081/E-EEM-120048406

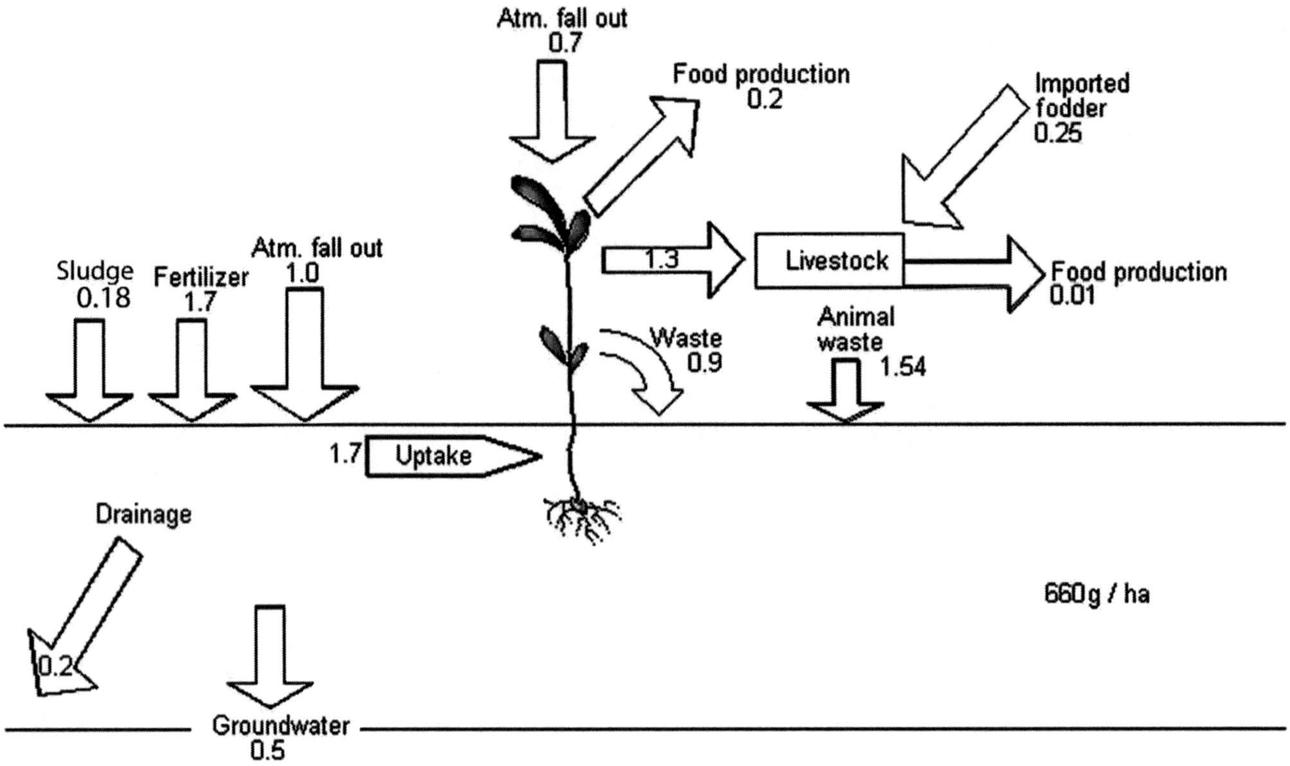
Atm. fall out
0.7
Food production
0.2
Imported
fodder
0.25
Atm. fall out
1.0
Sludge
0.18
Fertilizer
1.7
1.3
Livestock
Food production
0.01
Waste
0.9
Animal
waste
1.54
1.7
Uptake
Drainage
660g / ha
0.2
Groundwater
0.5

Fig. 1 Cadmium balance for Danish agricultural land. The accumulated amount is about 660 g/ha, while the annual input by fertilizers is 1.7 g/ha and by air pollution another 1.7 g/ha, of which 1 g is accumulated directly on the bare soil and the 0.7 g on the plants. The amount from fertilizers corresponds to the amount taken up by plants. 1.3 g/ha of the cadmium contamination of plants and 0.25 g/ha of cadmium in imported fodder give an accumulation of 1.55 g of cadmium per hectare per year in domestic animals, but only the 0.01 g/ha will end up as cadmium contamination of animal products.

whether a specified toxic substance would reach an unacceptably high concentration and thereby do the most harm. Jørgensen and Fath[5] have exemplified regional mass balances. Fig. 1 shows one example, namely, a cadmium balance for Danish agricultural land, but similar balances can also be found for lead, mercury, and copper, as well as in other regions. The cadmium pollution of the agricultural land comes from the use of fertilizers (1.7 g/ha/yr), from the use of sludge (0.18 g/ha/yr), and from air pollution, dry deposition, and rainwater (1.7 g/ha/yr). A total of 1.7 g/ha/yr is taken up from the soil by the plants and 0.2 g/ha/yr contaminates vegetable products, while only 0.01 g of cadmium per year contaminates the animal food, even when the cadmium in imported fodder is considered. By far, most of the cadmium-contaminating domestic animals go back to the agricultural land by animal waste. In total, about 3.8 g of cadmium is added to the agricultural land per year, but 0.7 g of cadmium per hectare per year is transported by drainage water and groundwater to other sites (ecosystems) and 0.2 g/ha/yr will be removed by the harvest of vegetables. The net accumulation in the agricultural soil is therefore 3.1 g/ha/yr. The mass balance can be used to calculate the effect of using fertilizers with less cadmium or the result of a reduction of cadmium discharged to the atmosphere. If more complex pollution abatement strategies are applied, it is necessary to apply an ecotoxicological model with the state variables, processes, and transfers as shown in Fig. 1 as the core of the model.

TOXICITY AND ECOTOXICITY

Cadmium has carcinogenic and teratogenic effects and is also highly toxic, as indicated by the LD_{50} value for rats—70–90 mg/kg. LD_{50} for cadmium in smoke (as cadmium oxide) by inhalation is 500 min/mg/m^3. The cadmium concentration in rice that caused the itai-itai disease in Japan was about 1 mg/kg.

Uptake from food of cadmium is about the same as for other heavy metals—7%–10%. Cadmium is accumulated mainly in the kidneys. The total cadmium concentration in the body increases with age, as the uptake is bigger than the excretion, mainly by the urine. The excretion follows a first-order reaction with a coefficient of 0.0001 $1/24h$. At an age of 50, the cadmium content in the body will be about 25 mg, of which the 8–9 mg is accumulated in the kidney and the 3 mg is accumulated in the liver.[6] The daily accumulation is about 2–3 μg.

It is possible to express cadmium accumulation in the body as a function of time by the following differential equation:

$$dCd/dt = \text{daily uptake} - \text{excretion coefficient} * Cd$$

Cd is the amount of cadmium in the body and the daily uptake is found as about 10% of the cadmium content in the food and 50% of the cadmium in the 20 m^3 of air used for respiration per 24 hr. The intake by food is about 25 μg/24 hr and the intake by respiration is very minor. Therefore, the equation can be written as:

$$dCd/dt = 0.025 * 10/100 - 0.0001 * Cd \text{ mg/24 hr}$$

At steady state, the cadmium concentration becomes 0.0025/0.0001 = 25 mg, which (of course) is close to the average value at the age of 50 years.

REFERENCES

1. Newman, M.C.; Unger, M.A. *Fundamentals of Ecotoxicology*, 2nd Ed.; CRC Press: Boca Raton, London, and New York, 2003.
2. Jørgensen, S.E. *Principles of Pollution Abatement*; Elsevier: Amsterdam, 2000; 520 pp.
3. Jørgensen, S.E.; Halling-Sørensen, B.; Mahler, H. *Handbook of Estimation Methods in Ecotoxicology and Environmental Chemistry*; Taylor and Francis Publ.: Boca Raton, FL; 230 pp.
4. Jørgensen, S.E.; Jørgensen, L.A.; Nors Nielsen, S. *Handbook of Ecological and Ecotoxicological Parameters*; Elsevier: Amsterdam, 1991; 1380 pp.
5. Jørgensen, S.E.; Fath, B. *Fundamentals of Ecological Modelling*, 4th Ed.; Elsevier: Amsterdam, Oxford, 2011; 396 pp.
6. *Om Metaller*, 2nd Ed.; Statens Naturvårdsverk: Sweden, 1980.

Carbon Sequestration

Nathan E. Hultman
University of Maryland, Maryland, U.S.A.

Abstract
Carbon dioxide (CO_2), a byproduct of hydrocarbon combustion and a natural emission from biomass burning, respiration, or decay, is a major greenhouse gas and contributor to anthropogenic climate change. Carbon sequestration describes the processes by which carbon can be either removed from the atmosphere (as CO_2) and stored, or separated from fuels or flue gases and stored. Carbon sequestration can thus be either technological (usually called carbon capture and storage) or biological (biological carbon sequestration). The viability of carbon sequestration depends on the cost of the process and the policy context that determines the value of sequestered carbon.

INTRODUCTION

The increasing likelihood of human-caused changes in climate could lead to undesirable impacts on ecosystems, economies, and human health and well-being. These potential impacts have prompted extensive assessment of options to reduce the magnitude and rate of future climate changes. Since climate changes are derived ultimately from increases in the concentrations of greenhouse gases (GHGs) in the atmosphere, such options must target either (a) reductions in the rate of inflow of GHGs to the atmosphere or (b) the removal of GHGs from the atmosphere once they have been emitted. Carbon sequestration refers to techniques from both categories that result in the storage of carbon that would otherwise be in the atmosphere as CO_2.

CO_2 is often targeted among the other GHGs because it constitutes the vast majority of GHG emissions by mass and accounts for three-fifths of the total anthropogenic contribution to climate change. Human emissions of CO_2 come primarily from fossil fuel combustion and cement production (80%), and land-use change (20%) that results in the loss of carbon from biomass or soil.

The rate of inflow of GHGs to the atmosphere can be reduced by a number of complementary options. For CO_2, mitigation options aim to displace carbon emissions by preventing the oxidation of biological or fossil carbon. These options include switching to lower-carbon fossil fuels, renewable energy, or nuclear power; using energy more efficiently; and reducing the rate of deforestation and land-use change. On the other hand, sequestration options that reduce emissions involve the capture and storage of carbon before it is released into the atmosphere.

CO_2 can also be removed directly from the atmosphere. While the idea of a large-scale, economically competitive method of technologically "scrubbing" CO_2 from the atmosphere is enticing, such technology currently does not exist. Policy has therefore focused on the biological process of carbon absorption through photosynthesis, either through expanding forested lands or, perhaps, enhancing photosynthesis in the oceans. This entry describes both the technological and biological approaches to carbon sequestration.

TECHNOLOGICAL SEQUESTRATION: CARBON CAPTURE AND STORAGE

The technological process of sequestering CO_2 requires two steps: first, the CO_2 must be separated from the industrial process that would otherwise emit it into the atmosphere; and second, the CO_2 must be stored in a reservoir that will contain it for a reasonable length of time. This process is therefore often referred to as carbon capture and storage (CCS) to distinguish it from the biological carbon sequestration that is described later.

Sources of Carbon

The best sites for CCS are defined by the efficiency of the capture technique, the cost of transport and sequestration, and the quantity of carbon available. The large capital requirements for CCS also dictate that large, fixed industrial sites provide the best opportunities. Therefore, although fossil-fueled transportation represents about 20% of current global CO_2 emissions, this sector presents no direct options for CCS at this time. The industrial sector, on the other hand, produces approximately 60% of current CO_2 emissions; most of these emissions come from large point sources which are ideal for CCS, such as power stations,

Encyclopedia of Environmental Management DOI: 10.1081/E-EEM-120041552

oil refineries, petrochemical and gas reprocessing plants, and steel and cement works.[1]

Separation and Capture

Carbon capture requires an industrial source of CO_2; different industrial processes create streams with different CO_2 concentrations. The technologies applied to capture the CO_2 will therefore vary according to the specific capture process.[2–4] Capture techniques can target one of three sources:

- Post-combustion flue gases
- Pre-combustion capture from gasification from power generation
- Streams of highly pure CO_2 from various industrial processes

Post-Combustion Capture

Conventional combustion of fossil fuels in air produces CO_2 streams with concentrations ranging from about 4 to 14% by volume. The low concentration of CO_2 in flue gas means that compressing and storing it would be uneconomical; therefore, the CO_2 needs to be concentrated before storage. Currently, the favored process for this task is chemical absorption, also known as chemical solvent scrubbing. Cooled and filtered flue gas is fed into an absorption vessel with a chemical solvent that absorbs the CO_2. The most common solvent for this process is monothanolamine (MEA). The CO_2-rich solvent is then passed to another reaction vessel called a stripper column. It is then heated with steam to reverse the process, thus regenerating the solvent and releasing a stream of CO_2 with a purity greater than 90%.

Scrubbing with MEA and other amine solvents imposes large costs in energy consumption in the regeneration process; it requires large amounts of solvents since they degrade rapidly; and it imposes high equipment costs since the solvents are corrosive in the presence of O_2. Thus, until solvents are improved in these areas, flue gas separation by this method will remain relatively costly: just the steam and electric load from a coal power plant can increase coal consumption by 40% per net kWh_e. Estimates of the financial and efficiency costs from current technology vary. Plant efficiency is estimated to drop from over 40% to a range between 24 and 37%.[2,5,6] For the least efficient systems, carbon would cost up to \$70/t CO_2 and result in an 80% increase in the cost of electricity.[5] Other studies estimate an increase in the cost of electricity of 25%–75% for natural gas combined cycle and Integrated Gasification Combined Cycle (IGCC), and of 60%–115% for pulverized coal.[4] A small number of facilities currently practice flue gas separation with chemical absorption, using the captured CO_2 for urea production, foam blowing, carbonated beverages, and dry ice production. In addition, several developments may improve the efficiency of chemical absorption.

Several other processes have been proposed for flue-gas separation. Adsorption techniques use solids with high surface areas, such as activated carbon and zeolites, to capture CO_2. When the materials become saturated, they can be regenerated (releasing CO_2) by lowering pressure, raising temperature, or applying a low-voltage electric current. A membrane can be used to concentrate CO_2, but since a single pass through a membrane cannot achieve a great change in concentration, this process requires multiple passes or multiple membranes. An alternative use for membranes is to use them to increase the efficiency of the chemical absorption. In this case, a membrane separating the flue gas from the absorption solvent allows a greater surface area for the reaction, thus reducing the size and energy requirements of the absorption and stripper columns. *Cryogenic* techniques separate CO_2 from other gases by condensing or freezing it. This process requires significant energy inputs and the removal of water vapor before freezing.

One of the main limitations to flue-gas separation is the low pressure and concentration of CO_2 in the exhaust. An entirely different approach to post-combustion capture is to dramatically increase the concentration of CO_2 in the stream by burning the fuel in highly enriched oxygen rather than air. This process, called oxyfuel combustion, produces streams of CO_2 with a purity greater than 90%. The resulting flue gas will also contain some H_2O that can be condensed and removed, and the remaining high-purity CO_2 can be compressed for storage. Though significantly simpler on the exhaust side, this approach requires a high concentration of oxygen for the intake air. While this process alone may consume 15% of a plant's electric output, the separated N_2, Ar, and other trace gases also can be sold to offset some of the cost. Oxyfuel systems can be retrofitted onto existing boilers and furnaces.

Pre-Combustion Capture

Another approach involves removing the carbon from fossil fuels before combustion. First, the fuel is decomposed in the absence of oxygen to form a hydrogen-rich fuel called synthesis gas. Currently, this process of gasification is already in use in ammonia production and several commercial power plants fed by coal and petroleum byproducts; these plants can use lower-purity fuels and the energy costs of generating synthesis gas are offset by the higher combustion efficiencies of gas turbines; such plants are called IGCC plants. Natural gas can be transformed directly by reacting it with steam, producing H_2 and CO_2. While the principle of gasification is the same for all carbonaceous fuels, oil and coal require intermediate steps to purify the synthesis fuel and convert the byproduct CO into CO_2.

Gasification results in synthesis gas that contains 35%–60% CO_2 (by volume) at high pressure (over 20 bar). While current installations feed this resulting mixture into the gas turbines, the CO_2 can also be separated from the gas before combustion. The higher pressure and concentration give a CO_2 partial pressure of up to 50 times greater than in the post-combustion capture of flue gases, which enables another type of separation technique of physical solvent scrubbing. This technique is well known from ammonia production and involves the binding of CO_2 to solvents that release CO_2 in the stripper under lower pressure. Solvents in this category include cold methanol, polyethelene glycol, propylene carbonate, and sulpholane. The resulting separated CO_2 is, however, near atmospheric pressure and requires compression before storage (some CO_2 can be recovered at elevated pressures, which reduces the compression requirement). With current technologies, the total cost of capture for IGCC is estimated to be greater than $25 per ton of CO_2; plant efficiency is reduced from 43 to 37%, which raises the cost of electricity by over 25%.[5]

Pre-combustion capture techniques are noteworthy not only for their ability to remove CO_2 from fossil fuels for combustion in turbines, but also because the resulting synthesis gas is primarily H_2. They therefore could be an important element of a hydrogen-mediated energy system that favors the higher efficiency reactions of fuel cells over traditional combustion.[7]

Industrial CO_2 Capture

Many industrial processes release streams of CO_2 that are currently vented into the atmosphere. These streams, currently viewed as simple waste in an economically viable process, could therefore provide capture opportunities. Depending on the purity of the waste stream, these could be among the most economical options for CCS. In particular, natural gas processing, ethanol and hydrogen production, and cement manufacturing produce highly concentrated streams of CO_2. Not surprisingly, the first large-scale carbon sequestration program was run from a previously vented stream of CO_2 from the Sleipner gas-processing platform off the Norwegian coast.

Storage of Captured CO_2

Relatively small amounts of captured CO_2 might be re-used in other industrial processes such as beverage carbonation, mineral carbonates, or commodity materials such as ethanol or paraffins. Yet most captured CO_2 will not be re-used and must be stored in a reservoir. The two main routes for storing captured CO_2 are to inject it into geologic formations or into the ocean. However, all reservoirs have some rate of leakage and this rate is often not well known in advance. While the expected length of storage time is important (with targets usually in the 100–1000 year range), we must therefore also be reasonably confident that the reservoir will not leak more quickly than expected, and have appropriate measures to monitor the reservoir over time. Moreover, transporting CO_2 between the point of capture and the point of storage adds to the overall cost of CCS, so the selection of a storage site must account for this distance as well.

Geologic Sequestration

Geologic reservoirs—in the form of depleted oil and gas reservoirs, unmineable coal seams, and saline formations—comprise one of the primary sinks for captured CO_2. Estimates of total storage capacity in geologic reservoirs could be up to 500% of total emissions to 2050 (Table 1).

Captured CO_2 can be injected into depleted oil and gas reservoirs, or can be used as a means to enhance oil recovery from reservoirs nearing depletion. Because they held their deposits for millions of years before extraction, these reservoirs are expected to provide reliable storage for CO_2. Storage in depleted reservoirs has been practiced for years for a mixture of petroleum mining waste gases called "acid gas."

A petroleum reservoir is never emptied of all its oil; rather, extracting additional oil just becomes too costly to justify at market rates. An economically attractive possibility is therefore using captured CO_2 to simultaneously increase the yield from a reservoir as it is pumped into the reservoir for storage. This process is called enhanced oil recovery. Standard oil recovery yields only about 30%–40% of the original petroleum stock. Drilling companies have years of experience with using compressed CO_2, a hydrocarbon solvent, to obtain an additional 10%–15% of the petroleum stock. Thus, captured CO_2 can be used to provide a direct economic benefit along with its placement in a reservoir. This benefit can be used to offset capture costs.

Coal deposits that are not economically viable because of their geologic characteristics provide another storage option. CO_2 pumped into these unmineable coal seams will adsorb onto the coal surface. Moreover, since the coal surface prefers to adsorb CO_2 to methane, injecting CO_2 into coal seams will liberate any coal bed methane (CBM) that can then be extracted and sold. This enhanced methane recovery is currently used in U.S. methane production, accounting

Table 1 CO_2 Reservoirs. Carbon dioxide storage capacity estimates. E is defined as the total global CO_2 emissions from the years 2000–2050 in IPCC's business-as-usual scenario IS92A. Capacity estimates such as these are rough guidelines only and actual utilization will depend on carbon economics.

	Storage capacity	
Reservoir type	**billion tonnes CO_2**	**% of E**
Coal basins	170	8%
Depleted oil reservoirs	120	6%
Gas basins	700	37%
Saline formations		
Terrestrial		276%
Off-shore		192%
Total	10.490	517%

Source: Dooley and Friedman.[8]

for about 8% in 2002. Such recovery can be used to offset capture costs. One potential problem with this method is that the coal, as it adsorbs CO_2, tends to swell slightly. This swelling closes pore spaces and thus decreases rock permeability, which restricts both the reservoir for incoming CO_2 and the ability to extract additional CBM.

Saline formations are layers of porous sedimentary rock (e.g., sandstone) saturated with saltwater, and exist both under land and under the ocean. These layers offer potentially large storage capacity representing several hundred years' worth of CO_2 storage. However, experience with such formations is much more limited and thus the uncertainty about their long-term viability remains high. Moreover, unlike EOR or CBM recovery with CO_2, injecting CO_2 into saline formations produces no other commodity or benefit that can offset the cost. On the other hand, their high capacity and relative ubiquity makes them attractive options in some cases. Statoil's Sleipner project, for example, uses a saline aquifer for storage.

Research and experimentation with saline formations is still in early stages. To achieve the largest storage capacities, CO_2 must be injected below 800 m depth, where it will remain in a liquid or supercritical dense phase (supercritical point at 31°C, 71 bar). At these conditions, CO_2 will be buoyant (a density of approximately 600–800 kg/m^3) and will tend to move upward. The saline formations must therefore either be capped by a less porous layer or geologic trap to prevent leakage of the CO_2 and eventual decompression.[9] Over time, the injected CO_2 will dissolve into the brine and this mixture will tend to sink within the aquifer. Also, some saline formations exist in rock that contains Ca-, Mg-, and Fe-containing silicates that can form solid carbonates with the injected CO_2. The resulting storage as rock is highly reliable, though it may also hinder further injection by closing pore spaces. Legal questions may arise when saline formations, which are often geographically extensive, cross national boundaries or onto marine commons.

Ocean Direct Injection

As an alternative to geologic storage, captured CO_2 could be injected directly into the ocean at either intermediate or deep levels. The oceans have a very large potential for storing CO_2, equivalent to that of saline aquifers (~10^3 Gt). While the ocean's surface is close to equilibrium with atmospheric carbon dioxide concentrations, the deep ocean is not because the turnover time of the oceans is much slower (~5000 years) than the observed increases in atmospheric CO_2. Since the ocean will eventually absorb much of the atmospheric perturbation, injecting captured CO_2 into the oceans can therefore be seen as simply bypassing the atmospheric step and avoiding the associated climate consequences. Yet little is known about the process or effects—either ecological or geophysical—of introducing large quantities of CO_2 into oceanic water.

At intermediate depths (between 500 and 3000 m), CO_2 exists as a slightly buoyant liquid. At these depths, a stream of CO_2 could be injected via a pipe affixed either to ship or shore. The CO_2 would form a droplet plume, and these droplets would slowly dissolve into the seawater, disappearing completely before reaching the surface. Depressed pH values are expected to exist for tens of km downcurrent of the injection site, though changing the rate of injection can moderate the degree of perturbation. In addition, pulverized limestone could be added to the injected CO_2 to buffer the acidity.

Below 3000 m, CO_2 becomes denser than seawater and would descend to the seafloor and pool there. Unlike intermediate injection, therefore, this method does not lead to immediate CO_2 dissolution in oceanic water; rather, the CO_2 is expected to dissolve into the ocean at a rate of about 0.1 m/y. Deep injection thus minimizes the rate of leakage to the surface, but could still have severe impacts on bottom-dwelling sea life.

The primary obstacles to oceanic sequestration are not technical but relate rather to this question of environmental impacts.[10] Oceanic carbon storage might affect marine ecosystems through the direct effects of a lower environmental pH; dissolution of carbonates on fauna with calcareous structures and microflora in calcareous sediments; impurities such as sulfur oxides, nitrogen oxides, and metals in the captured CO_2; smothering effects (deep injection only); and changes in speciation of metals and ammonia due to changes in pH. Few of these possibilities have been studied in sufficient detail to allow an informed risk assessment. In addition, the legality of dumping large quantities of CO_2 into the open ocean remains murky.

Overall Costs of CCS

The costs of CCS can be measured either as a cost per tonne of CO_2, or, for power generation, a change in the cost of electricity (Table 2). The total cost depends on the cost of capture, transport, and storage. Capture cost is mainly a function of parasitic energy losses and the capital cost of equipment. Transport cost depends on distance and terrain. Storage costs vary depending on the reservoir but are currently a few dollars per tonne of CO_2. The variety of approaches to CCS and the early stages of development make precise estimates of cost difficult, but current technology spans about \$25–\$85/t CO_2.

Table 2 Additional costs to power generation from CCS. Approximate capture and storage costs for different approaches to power plant sequestration.

Fossil type	Cost of CCS ¢ per kWh
Natural gas combined cycle	1–2
Pulverized coal	2–3
Coal IGCC	2–4

Source: Herzog and Golomb,[4] National Energy Technology Laboratory,[5] Dooley et al.,[11] and Freund and Davison.[12]

BIOLOGICAL SEQUESTRATION: ENHANCING NATURAL CARBON SINKS

The previous sections have described processes by which CO_2 could be technologically captured and then stored. Photosynthesis provides an alternate route to capture and store carbon. Enhancing this biological process is therefore an alternative method of achieving lower atmospheric CO_2 concentrations by absorbing it directly from the air.

Terrestrial Carbon Sinks

Carbon sequestration in terrestrial ecosystems involves enhancing the natural sinks for carbon fixed in photosynthesis. This occurs by expanding the extent of ecosystems with a higher steady-state density of carbon per unit of land area. For example, because mature forest ecosystems contain more carbon per hectare than grasslands, expanding forested areas will result in higher terrestrial carbon storage. Another approach is to encourage the additional storage of carbon in agricultural soils. The essential element in any successful sink enhancement program is to ensure that the fixed carbon remains in pools with long lives.

Afforestation involves planting trees on unforested or deforested land.[13,14] The most likely regions for forest carbon sequestration are Central and South America and Southeast Asia because of relatively high forest growth rates, available land, and inexpensive labor. However, the translation of forestry activities into a policy framework is complex. Monitoring the carbon changes in a forest is difficult over large areas, as it requires not only a survey of the canopy and understory, but also an estimate of the below-ground biomass and soil carbon. Some groups have voiced concern over the potential for disruption of social structures in targeted regions.

Soil carbon sequestration involves increasing soil carbon stocks through changes in agriculture, forestry, and other land use practices. These practices include mulch farming, conservation tillage, agroforestry and diverse cropping, cover crops, and nutrient management that integrates manure, compost, and improved grazing. Such practices, which offer the lowest-cost carbon sequestration, can have other positive effects such as soil and water conservation, improved soil structure, and enhanced soil fauna diversity. Rates of soil carbon sequestration depend on the soil type and local climate, and can be up to 1000 kg of carbon per hectare per year. Management practices can enhance sequestration for 20–50 years, and sequestration rates taper off toward maturity as the soil carbon pool becomes saturated. Widespread application of recommended management practices could offset 0.4 to 1.2 GtC/y, or 5%–15% of current global emissions.[15]

If sinks projects are to receive carbon credits under emissions trading schemes like that in the Kyoto Protocol, they must demonstrate that the project sequestered more carbon than a hypothetical baseline or business-as-usual case. They must also ensure that the carbon will remain in place for a reasonable length of time, and guard against simply displacing the baseline activity to a new location.

Ocean Fertilization

Vast regions of the open ocean have very little photosynthetic activity, though sunlight and major nutrients are abundant. In these regions, phytoplankton are often deprived of trace nutrients such as iron. Seeding the ocean surface with iron, therefore, might produce large phytoplankton blooms that absorb CO_2. As the plankton die, they will slowly sink to the bottom of the ocean, acting to transport the fixed carbon to a permanent burial in the seafloor. While some experimental evidence indicates this process may work on a limited scale, little is known about the ecosystem effects and potential size of the reservoir.[16]

PROSPECTS FOR CARBON SEQUESTRATION

Carbon sequestration techniques—both technological and biological—are elements of a portfolio of options for addressing climate change. Current approaches hold some promise for tapping into the geologic, biologic, and oceanic potential for storing carbon. The costs of some approaches, especially the improved management of agricultural and forest lands, are moderate (Table 3). Yet these opportunities are not infinite and additional options will be necessary to address rising global emissions. Thus, the higher costs of current technological approaches are likely to drop with increasing deployment and changing market rates for carbon.

Possible developments include advanced CO_2 capture techniques focusing on membranes, ionic (organic salt) liquids, and microporous metal organic frameworks. Several alternative, but still experimental, sequestration approaches have also been suggested. Mineralization could convert CO_2 to stable minerals. This approach seeks, therefore, to hasten what in nature is a slow but exothermic weathering process that operates on common minerals like olivine, forsterite, or serpentines (e.g., through selected sonic frequencies). It is possible that CO_2 could be injected in sub-seafloor carbonates. Chemical looping describes a

Table 3 Costs of carbon sequestration. Estimates for sequestration costs vary widely. Future costs will depend on rates of technological change.

Sequestration technique	Cost \$ per T CO_2
Carbon capture and storage	26–84
Tree planting and agroforestry	10–210
Soil carbon sequestration	6–24

Source: Herzog and Golomb,[4] National Energy Technology Laboratory,[5] Williams[7] Dooley et al.[11] Freund and Davison,[12] Van Kooten,[13] and Richards and Stokes.[14]

method for combusting fuels with oxygen delivered by a redox agent instead of by air or purified oxygen; it promises high efficiencies of energy conversion and a highly enriched CO_2 exhaust stream. Research also continues on microbial CO_2 conversion in which strains of microbes might be created to metabolize CO_2 to produce saleable commodities (succinic, malic, and fumeric acids). In addition, the nascent science of monitoring and verifying the storage of CO_2 will be an important element toward improving technical performance and public acceptance of sequestration techniques.

ACKNOWLEDGMENTS

The author thanks the anonymous reviewer for helpful comments on an earlier draft.

REFERENCES

1. Gale, J. Overview of CO_2 Emission Sources, Potential, Transport, and Geographical Distribution of Storage Possibilities. In *Proceedings of IPCC workshop on carbon dioxide capture and storage* Regina, Canada Nov 18–21, 2002 15–29. http://arch.rivm.nl/env/int/ipcc/pages_media/ccs2002.html (accessed April 2005). A revised and updated version of the papers presented at the Regina workshop is now available. See, Metz, B.; Davidson, O.; de Coninck, H.; Loos, M.; Meyer, L., Special Report on Carbon Dioxide Capture and Storage. Intergovernmental Panel on Climate Change: Geneva, http://www.ipcc.ch. (accessed May 2006), 2006.
2. Thambimuthu, K.; Davison, J.; Gupta, M. CO_2 Capture and Reuse, In *Proceedings of IPCC workshop on carbon dioxide capture and storage*, Regina, Canada, Nov 18–21, 2002 31–52 http://arch.rivm.nl/env/int/ipcc/pages_media/ccs2002.html (accessed April 2005).
3. Gottlicher, G. *The Energetics of Carbon Dioxide Capture in Power Plants*; U.S. Department of Energy Office of Fossil Energy Washington, DC, 2004; 193
4. Herzog, H.; Golomb, D. Carbon capture and storage from fossil fuel use *Encyclopedia of Energy* Cleveland, C.J. Ed.; Elsevier Science: New York, 2004; 277–287.
5. National Energy Technology Laboratory *Carbon Sequestration: Technology Roadmap and Program Plan* U.S. Department of Energy Washington, DC, 2004.
6. Gibbins, J.R.; Crane, R.I.; Lambropoulos, D.; Booth, C.; Roberts, C.A.; Lord, M. Maximizing the Effectiveness of Post Combustion CO_2 Capture Systems In *Proceedings of the 7th International Conference on Greenhouse Gas Abatement*, Vancouver, Canada, Sept 5–9, 2004 Document E2-2. http://www.ghgt7.ca (accessed June 2004).
7. Williams, R.H. Decarbonized Fossil Energy Carriers and Their Energy Technology Competitors In *Proceedings of IPCC workshop on carbon dioxide capture and storage* Regina, Canada, November, 2002; 119–135, http://arch.rivm.nl/env/int/ipcc/pages_media/ccs2002.html (accessed April 2005).
8. Dooley, J.J.; Friedman, S.J. *A Regionally Disaggregated Global Accounting of CO_2 Storage Capacity: Data and Assumptions* Battelle National Laboratory Washington DC, 2004; 15.
9. Kårsted, O. Geological Storage, Including Costs and Risks, in Saline Aquifers In *Proceedings of IPCC workshop on carbon dioxide capture and storage* Regina, Canada 2002 Nov 18–21 53–60 http://arch.rivm.nl/env/int/ipcc/pages_media/ccs2002.html (accessed April 2005).
10. Johnston, P.; Santillo, D. Carbon Capture and Sequestration: Potential Environmental Impacts, In *Proceedings of IPCC workshop on carbon dioxide capture and storage*, Regina, Canada, Nov 18–21, 2002; 95–110. http://arch.rivm.nl/env/int/ipcc/pages_media/ccs2002.html (accessed April 2005).
11. Dooley, J.J.; Edmonds, J.A.; Dahowski, R.T.; Wise, M.A. Modeling Carbon Capture and Storage Technologies in Energy and Economic Models. In *Proceedings of IPCC workshop on carbon dioxide capture and storage*, Regina, Canada, Nov 18–21, 2002; 161–172. http://arch.rivm.nl/env/int/ipcc/pages_media/ccs2002.html (accessed April 2005).
12. Freund, P.; Davison, J. General Overview of Costs. In *Proceedings of IPCC workshop on carbon dioxide capture and storage*, Regina, Canada, Nov 18–21, 2002 79–93. http://arch.rivm.nl/env/int/ipcc/pages_media/ccs2002.html (accessed April 2005).
13. Van Kooten, G.C.; Eagle, A.J.; Manley, J.; Smolak, T. How costly are carbon offsets? A meta-analysis of carbon forest sinks. Environ. Sci. Policy **2004**, *7*, 239–251.
14. Richards, K.R.; Stokes, C. A review of forest carbon sequestration cost studies: a dozen years of research. Climatic Change **2004**, *68*, 1–48.
15. Lal, R. Soil carbon sequestration impacts on global climate change and food security. Science **2004**, *304*, 1623–1627.
16. Buesseler, K.O.; Andrews, J.E.; Pike, S.M.; Charette, M.A. The effects of iron fertilization on carbon sequestration in the southern ocean. Science **2004**, *304*, 414–417.

Carbon: Soil Inorganic

Donald L. Suarez
U.S. Salinity Laboratory, Agricultural Research Service (USDA-ARS), U.S. Department of Agriculture, Riverside, California, U.S.A.

Abstract
Inorganic carbon (IC) in the form of calcite (calcium carbonate mineral) and dolomite (calcium, magnesium carbonate) constitutes the earth's major carbon source. Soil IC is also a major carbon pool in the earth's near-surface materials including soil. It is thus important to consider this C pool for quantification of the global C cycle as well as the impact of human activities on atmospheric carbon dioxide and potential mitigation actions. The link of this carbon pool to atmospheric carbon dioxide concentrations and climate change relates to the precipitation or dissolution of these minerals in the near-surface environment. The impact of changes in soil IC on atmospheric carbon dioxide concentration depends on the local environmental and hydrological conditions. Under most environmental conditions, dissolution of these minerals leads to net removal of carbon dioxide from the atmosphere. The environmental processes related to land use that impact IC storage are numerous, including land clearing, timber harvesting, cropping, tillage, irrigation, and fertilization practices.

INTRODUCTION

Concerns related to global climate change and greenhouse gas warming have led to extensive interest in changes in carbon storage and release of carbon dioxide into the atmosphere. While the burning of fossil fuels was a major contributor to the increase in atmospheric carbon dioxide over the last 200 years, there has been extensive research on the impact of other processes, such as changes in the organic carbon (OC) stocks in soils. Inorganic carbon (IC) storage and changes associated with land use have received relatively little attention, despite the fact that IC is a major carbon pool in the near-surface environment. At the global scale, total soil OC is estimated at 1550 Pg and soil IC at 950 Pg.[1] In addition, in arid regions, IC can comprise more than 90% of the total C in the soil. The IC in the soil is present in the minerals calcite and dolomite. Both minerals are relatively insoluble; however, dolomite dissolution is much slower than calcite dissolution at the intermediate pH values relevant to most soils. Also, dolomite does not generally precipitate under earth surface conditions. As a result, it can be considered that over the time frames of current interest related to climate change, the dolomite component in soils remains constant or decreases due to dissolution, while calcium carbonate content may increase or decrease.

This entry considers the soil processes and land-use changes that impact changes in soil IC and its implication for the atmospheric carbon dioxide budget. Among the land-use changes considered are land clearing, tillage, cropping, fertilization, irrigation in both humid and arid environments, irrigation management (leaching), and composition and source of the irrigation water. The impacts of these changes on atmospheric carbon dioxide concentrations are more difficult to analyze and differ with regard to long-term and short-term effects.

SOIL PROCESSES

In terms of OC at the land surface and in the soil, a decrease in OC is directly related to either a net removal of OC from the site such as timber or crop harvesting or input of carbon dioxide to the atmospheric in the case of oxidation of soil organic matter. In contrast, for dissolution of the carbonate minerals, a decrease in IC in the soil can result in either a release or an uptake of carbon dioxide from the atmosphere. The impact of these processes is entirely dependent on the local environmental and hydrologic conditions.

In the instance of dissolution under non-acidic conditions, the net reaction for dissolution of calcium carbonate is:

$$CaCO_3 + H_2O + CO_2 \rightarrow Ca^{2+} + 2HCO_3^-. \quad (1)$$

Thus, dissolution of calcium carbonate results in a sink for atmospheric carbon. Increased soil carbon dioxide concentrations also lead to a net increase in carbonic acid in solution (H_2CO_3). This carbonic acid is partially recharged to the groundwater and is thus a sink for atmospheric CO_2 that is greater than that indicated by Eq. 1 and generally not considered when evaluating organic matter decomposition.

Dissolution of calcium carbonate under acidic conditions results in the following net reaction:

$$CaCO_3 + 2H^+ \rightarrow Ca^{2+} + H_2O + CO_2, \quad (2)$$

Encyclopedia of Environmental Management DOI: 10.1081/E-EEM-120046426

thus leading to a net release of carbon dioxide to the atmosphere. In the case of dissolution of IC, the residence time in the groundwater will determine the net impact on atmospheric carbon dioxide in the time scale of most concern—tens to hundreds of years.

The majority of the alkalinity in the earth's near-surface environment is derived from dissolution of carbonate rocks. The net long-term impact is uptake of carbon dioxide from the atmosphere, dissolution of IC, movement of alkalinity in the surface and subsurface, and discharge to the oceans via rivers.

In terms of the overall C cycle, most of the carbon dioxide released from precipitation of carbonates in the soil (an important process in arid regions) results in a return of C to the atmosphere, as seen by the reverse reaction of Eq. 1. Precipitation of IC results in a net release of 1 mol of carbon dioxide for each mole of IC produced. Sequestration of C in the form of soil IC is carbon storage but the net result of the process is a net release of C to the atmosphere. The quantity released is less than indicated by Eq. 1 since a small amount of the carbon dioxide generated may be recharged with the groundwater, a process that occurs in the absence of precipitation of soil IC.

An important factor to consider when evaluating the implication of an increase in carbon storage is the origin of the calcium and bicarbonate ions. The above instance considered dissolution of IC from the soil. Weathering of silicate minerals is important over long time scales, resulting in release of calcium, sodium, magnesium, and potassium ions as well as bicarbonate. In this instance, the net reaction for dissolution is consumption of 2 mol of carbon dioxide to produce 2 mol of bicarbonate ions, in addition to the release of an equivalent quantity of cations into solution. Thus, silicate dissolution in the soil results in sequestering atmospheric carbon in the form of dissolved bicarbonate. It may be possible in arid regions that the resultant soil solution becomes concentrated, resulting in precipitation of IC. In this process, 1 mol of C is released during IC precipitation and 2 mol are sequestered from silicate dissolution. In this instance, the net result of the silicate weathering is still carbon sequestration from the atmosphere since not all of the alkalinity generated by silicate dissolution is reprecipitated. In terms of IC, the result is an increase in soil IC. Silicate dissolution has a relatively small role in terms of changes related to human activity.

DISSOLVED IC CYCLING

The net impact of soil IC changes on atmospheric carbon dioxide is critically related to the fate of alkalinity (primarily bicarbonate ions) transported into the subsurface below the root zone. If we consider that most of the recharge is ultimately discharged to surface systems, then the potential exists for reprecipitation in these systems, which are at lower carbon dioxide concentrations than the subsurface. It is estimated[2] that inland waters receive approximately 1.9 Pg C/yr, of which 0.8 is returned to the atmosphere as carbon dioxide while 0.9 Pg C/yr is discharged to the oceans, of which 0.45 Pg C/yr is IC. The direct degassing of groundwater is estimated at 0.01 Pg/yr with an uncertainty of 0.003–0.03 Pg/yr.[2] Combining soil liming as well as irrigation practices and other land-use changes, it is likely that the IC flux to groundwater has increased due to human activities. It is estimated that the dissolved IC (mostly bicarbonate) flux into groundwater is on the order of 0.2 Gg/yr.[3] The impact of changes in this subsurface flux, however, is likely relatively long term as the groundwater residence times are mostly on the order of hundreds to thousands of years.[3]

LAND USE

Land use, among other environmental factors, affects changes in soil IC with the full impact on atmospheric carbon dioxide expressed only over the long term. Both increases and decreases in IC storage are possible as a result of various management practices. In irrigated lands, there are various factors that cause either a decrease or an increase in soil IC. In most instances, multiple land-use changes have occurred as a result of human activity, and it is difficult to isolate the specific process that is dominant in explaining changes in soil IC. It is calculated that China has a total of 53 Pg of soil IC and that human activity has resulted in a net loss of 1.6 Pg of IC.[4]

Soil respiration processes include plant respiration and release of carbon dioxide via the roots, as well as microbial decomposition of soil organic matter. These processes result in an elevated carbon dioxide concentration in the soil relative to the atmospheric condition. The soil CO_2 concentration is typically in the range of 0.1%–5%, orders of magnitude higher than the atmospheric value. Increased plant biomass productivity due to crop selection irrigation and/or fertilization results in increased soil carbon dioxide production and concentration, thus contributing to increased weathering of both silicates and carbonates. Inadvertent or deliberate introduction of non-native species also impacts biomass productivity, soil CO_2 concentrations, water use, recharge, and erosion with subsequent impact on soil IC.

LAND CLEARING

Land clearing generally results in increased water runoff and soil erosion. This process or any other process such as tillage that increases erosion serves to remove the surface soil horizons. Since these horizons are generally depleted in IC relative to less weathered, deeper horizons, erosion causes an apparent increase in the IC content of the top 1–2 m of carbonate-containing soils. This may result in erroneous estimates of changes in soil IC when comparing the top 1 m of disturbed vs. native vegetation sites. In terms of carbonate dissolution, the impact of land clearing is not certain. After clearing, there is increased runoff, thus decreased surface infiltration, favoring less dissolution of carbonates from the surface horizons. This effect may be compensated by the decreased water consumption (lower evapotranspiration) after clearing, resulting in increased deep recharge and possibly greater carbonate dissolution or less reprecipitation at depth, depending on the volumes of water. Depending on how much biomass remains after clearing, there is likely a short-term increase in soil CO_2 followed by a longer-term reduction, favoring less carbonate dissolution in the soil.

CROPPING AND TILLAGE

The impact of cropping and tillage on IC storage is not certain and there are limited studies, but the sum of the studies suggests net IC accumulation. The impact of tillage and cropping on IC in the Northern Great Plains of the United States has been evaluated by comparing long-term cultivation (likely >80 years) with uncultivated grassland.[5] There was a statistically significant increase in IC with cropping for one soil type (at the 90% confidence level but not at the 95% confidence level) but no statistically significant differences for the other two soil types examined. The lack of a clear impact is likely related to the fact that these soils did not have significant quantities of IC in the upper portion of the profile where most of the water transport and root water uptake would occur.

In a study of Russian Chernozem soils, the soil IC was greater on a continuously cropped field and on a fallow field relative to a native grassland, with most of the approximately 100 Mg/ha increase in IC being in the upper portion of the profile.[6] A grassland field where hay was harvested showed no significant changes over the native vegetation (grassland) field. The authors attributed the increase in the cropped and fallow fields to cultivation, irrigation, and fertilization, with the calcium being supplied by fertilizer and manure additions. The manure also supplied alkalinity necessary for carbonate formation. The effects in this study may be more related to fertilizer application than changes due to cultivation per se.

When comparing IC in cropped soils to native grassland, it needs to be considered that grassland and native vegetative soils can also have net accumulation or loss of IC with net increase in pedogenic and recently formed IC. We cannot assume that increases in pedogenic carbonate necessarily correspond to increases in IC storage. For example, it is reported that for grasslands and forested soils in Saskatchewan, Canada, there is an increase in pedogenic IC of 1.2–1.8 kg/ha/yr,[7] and this has been cited as an increase in IC. However, the soils contain dolomite and recystallization of carbonates is listed by the authors as a major process producing the pedogenic carbonate.[7] Only one of the soils (Black Chernozem) had net IC accumulation based on comparison of the pedogenic carbonate accumulated and the depletion of the lithogenic carbonate.[7]

Land clearing, tillage, and overgrazing in arid lands serve to increase wind erosion and to redistribute soil in the landscape. In this manner, non-calcareous soils receive inputs of carbonates. This process likely increases net dissolution of carbonates, as it spreads carbonates across the landscape and into areas with non-calcareous soils.

In humid environments, IC is leached from the soil. The elevated CO_2 concentrations in the soil enhance IC solubility relative to earth surface conditions. In humid environments, carbonates are successively leached from the upper portions of the soil profile. Agricultural practices may serve to enhance or reduce the net removal of carbonates. Removal of vegetation from a site with cropping practices such as tree harvesting or crop or forage harvesting serves to remove base cations and causes net acidification of the upper portions of the soil profile. If carbonates are present deeper in the soil, this acidification increases IC dissolution. The impact of removal of vegetation in humid environments with carbonates in the subsoil can be calculated by assuming that the net harvested alkalinity is compensated by an equal increase in carbonate dissolution in the subsurface.

FERTILIZATION AND LIMING

Since optimum plant growth is generally at a pH lower than that observed in untreated calcareous soils, acid fertilizers are commonly applied in arid regions. Use of sulfur with subsequent oxidation to sulfate results in acid release to the soil (2 mol of protons per mole of sulfur). Application of ammonia salts with subsequent fixation into organic matter or oxidation to nitrous oxide or elemental nitrogen also releases protons (2 and 1 mol of protons per mole of ammonia ions, respectively). This acidification will increase carbonate dissolution proportionately and has a significant effect since ammonia salts are widely applied as fertilizers.

Application of fertilizer as urea or ammonia gas should have no net effect on carbonate dissolution (upon oxidation to nitrous oxide or elemental nitrogen) other than the indirect impact on soil CO_2. In contrast, use of nitrate fertilizers serves to increase pH and thus reduce carbonate dissolution. Generally, nitrate is not utilized on calcareous soils, so the impact on IC storage in soil is slight. The quantitative impact of fertilization on changes in IC is not easily calculated, as it depends on the extent of N incorporation into organic matter, mineralization, the extent to which the harvested biomass is removed from the site, and the occurrence of carbonates in the subsurface.

The practice of liming soils is very prevalent for crop production and urban landscaping in humid environments and represents a major anthropogenic input. The liming is done by application of calcium carbonate or dolomite to the soil. Addition of liming products, primarily calcite, are reported as 3.7 Tg C/yr in the United States for 1978.[8] This is a significant but temporary addition to the soil IC pool. Since liming is not generally needed in arid regions, it is reasonably assumed that the majority of the material is applied to acid soils and thus it is readily dissolved. The liming serves two functions, to increase the pH of acid soils and as an inexpensive calcium source in regions with leached, calcium-deficient soils. The net impact of liming is neutralization of most of the carbonate in the mineral and release of the generated carbon dioxide to the atmosphere, as represented in Eq. 2. The amount of carbon dioxide release to the atmosphere is less than indicted by Eq. 2 for all but the highly acidic soils, as some of the carbon is sequestered in the form of dissolved bicarbonate. In humid environments, most of the bicarbonate produced will likely not reprecipitate, even in the long term; instead, it will be discharged into the ocean via river drainage systems.

HUMID REGION IRRIGATION

Large quantities of water applied in excess of plant transpiration needs result in leaching and maintenance of elevated water content at or near the soil surface. The additional water applied as irrigation enhances dissolution of IC as does the increased soil carbon dioxide concentrations due to increased water content as in the soil. Irrigation in humid environments thus serves to increase the net recharge through the soils and thus removal of carbonates. These changes may be relatively difficult to detect in view of the limited amount of irrigation water added and the fact that irrigation in humid environments, although increasing rapidly, was very limited in the past. Field studies are needed to determine the impact of irrigation on changes in IC in humid environments.

ARID AND SEMIARID REGION IRRIGATION

Arid zone soils usually contain at least minor amounts of carbonates, even if they are not classified as calcareous. In the absence of irrigation, there may be redistribution of carbonates within the soil but likely little net precipitation. The majority of the pedogenic calcite is reprecipitated calcite with relatively small amounts added as a result of silicate mineral weathering. This process is of geological significance and central to explaining soil formation but of less importance to the time frames of interest with regard to recent changes in atmospheric CO_2. Significant amounts of carbonates are added to the surface of arid land soils as dust. IC is leached from the upper part of the soil profile by dissolution into the infiltrating rain and is mostly reprecipitated at depth after plant extraction of the available water.

Irrigation in arid and semiarid environments may result in a net increase or decrease in soil IC, depending on the water source and fraction of water applied that is leached (leaching fraction). There are various opposing effects. First, elevated CO_2 concentrations in the root zone relative to the atmospheric condition results in enhanced calcite and dolomite solubility and dissolution and, hence, depletion of soil IC. Second, plant water extraction and evaporation concentrate the soluble salts into a smaller volume of water and enhance calcite precipitation, thus increasing IC. Plant roots extract relatively dilute water from the soil, leaving most of the salts behind. At low leaching fractions (where the quantity of water applied plus rain is only 2%–10% greater than the amount of water transpired), the effect of concentration of salts due to plant water extraction and evaporation is greater than the enhanced CO_2 effect and there is net precipitation of calcite.

For a calcite-supersaturated surface water such as the Colorado River, it is estimated[9] that at a leaching fraction of 0.1, there is net precipitation of 125 kg/ha/yr of C, based on water consumption of 1.2 m/yr and an average soil CO_2 partial pressure of 3 kPa (approximately 3% of the soil atmosphere). Model simulations indicate that net precipitation of calcite occurs in the soil profile, with loss of IC in the upper portion of the root zone and precipitation of IC in the lower portion. At high leaching fractions, there is net dissolution of carbonates. Using a predictive calcite precipitation model that accounts for 3-fold calcite supersaturation in precipitating environments, it is predicted that at a leaching fraction of 0.4, there will be a net dissolution of IC of 70 kg/ha/yr of C. Again, this calculation is based on water consumption of 1.2 m/yr and soil CO_2 partial pressure of 3 kPa.[10] In all instances, there is a prediction of net dissolution in the upper portion of the soil root zone and net precipitation in the lower portions of the profile. Using average leaching fractions for the western United States, it is estimated that irrigation with surface waters on 12 million ha[11] results in an increase in soil IC of 1 Tg/yr,[10] or 80 kg/ha/yr. Considering the reverse reaction in Eq. 1, this increase in soil IC corresponds to a net release of almost 1 Tg/yr of C to the atmosphere.

Irrigation with groundwater saturated with respect to calcite will result in precipitation of carbonates at almost all leaching fractions, since the irrigation water is equilibrated at the groundwater CO_2 partial pressure and is highly supersaturated with respect to calcite upon degassing and application to the soil surface. Calcite-saturated groundwater is used for irrigation on an estimated 3.12 million ha in the United States.[11] It is estimated that irrigation on these soils results in a net IC precipitation of 1.3 Tg/yr or 420 kg/ha/yr.[10] This corresponds to slightly less than 1.3 Tg/yr of carbon dioxide release to the atmosphere.

The model calculations of IC precipitation are dependent on several assumptions regarding calcite precipitation and soil CO_2 concentrations and do not consider the effect of fertilizers. Generally, acidifying fertilizers are applied in arid regions, thus reducing the extent of expected increases in IC with irrigation. A partial validation of the modeling is available by comparing modeling results with data from Palo Verde, California. Using measured CO_2 partial pressure in the groundwater in Palo Verde Valley, there is prediction of no net change in IC at a leaching fraction of 0.5, obtained from the measured water diversions and drainage volumes. Consistent with these predictions, the groundwater composition draining back to the Colorado River from Palo Verde Valley shows no evidence for net precipitation or dissolution of carbonates in the soil.

There is some direct field evidence for the influence of irrigation on soil IC, but as expected from the above discussion, the results are site specific depending on water source, water composition, and leaching, as well as quantity and type of fertilizer used. It is also difficult to be certain that differences among sites are only related to changes in management. Researchers[12] observed a net decrease in the calcium carbonate content of three pairs of soil profiles taken from sites in Israel irrigated for approximately 40 years as compared to non-irrigated sites. The estimated input of 4.40 m of water per year at those sites is contrasted with the yearly potential evapotranspiration of 1.93 m. The observed trend is qualitatively consistent with model predictions if we account for the input of rain and the estimated high leaching fraction of 0.56. Isotopic evidence indicated that there was precipitation of pedogenic carbonate at depth despite a net decrease in IC content at depth.[12] This suggests solubilization and reprecipitation, but the measured net impact was still loss of IC in the soil profile, as expected from modeling this water budget data using surface water for irrigation.

In a study in the San Joaquin Valley in California, researchers compared samples of a soil taken from irrigated and native vegetation sites.[13] They also measured a net loss of carbonates attributed to 8 years of irrigation. Net carbonate loss was estimated as 7×10^3 kg/ha/yr (800 kg C/ha/yr). Leaching fractions at the site were not reported, but this value corresponds to approximately 10 times greater dissolution than expected based on model simulations. However, another study by the same author[14] found no change in total carbonate when comparing pedons with native vegetation and those irrigated for 5–25 years. In this instance, both gypsum and sulfur were applied as amendments for reclamation. Gypsum would tend to greatly increase precipitation of carbonates and thus increase IC, while sulfur would acidify the soil and cause net dissolution of carbonates.

In other studies, the results are also mixed as expected if we consider variations in irrigation water composition and leaching fractions. In a semiarid northwestern U.S. study, irrigated crops showed a net increase in IC while irrigated pasture showed a net decrease relative to native sagebrush vegetation.[15] In the San Joaquin Valley, California, there was a net gain in IC of 1.8 kg/m^2 after 30 years of irrigation and a net loss of IC after 55 years of irrigation at another site, while in Imperial Valley, there was a net gain of IC of 4 kg/m^2 or 40 Mg/ha after 85–90 years of irrigation.[16] Another, more extensive study in Imperial Valley[17] on paired soil cores (irrigated and adjacent non-irrigated sites) observed no changes in IC storage after 90 years of irrigation and no isotopic C shifts indicative of recrystallization. Accurately modeling the net global impact of irrigation on soil IC based on water types, water application quantities, and evapotranspiration will first require the ability to accurately simulate the impacts on specific locations.

SODIC SOIL RECLAMATION

Reclamation of sodic soils can result in either an increase or a decrease in soil IC. Gypsum application to a sodic and alkaline soil will increase the soil carbonate content, as the increased Ca will precipitate most of the soluble bicarbonate and carbonate. This will increase IC sequestration in the soil but have a net effect of also increasing the carbon dioxide flux to the atmosphere (and decreasing the net flux of C and alkalinity to the groundwater). Gypsum is also very commonly used in arid regions to improve water infiltration. At least some of the applied calcium from the gypsum application is reprecipitated as calcium carbonate, increasing the soil IC. Application of sulfuric acid, sulfur, or green manuring all serve to dissolve soil carbonates, thus depleting IC stocks in the soil. Addition of acid will result in an increase in atmospheric carbon dioxide equal to the IC removed.

Green manuring is a management practice of adding fresh plant organic matter to a calcareous soil, enhancing CO_2 concentrations in the soil and thus enhancing carbonate dissolution, providing a calcium source to replace the soil exchangeable sodium. This process will decrease soil IC and result in net release of CO_2 to the atmosphere. The dissolution process will deplete atmospheric carbon, but the effect is negated by the larger increase in carbon dioxide release related to the oxidation of the OC. It is estimated that this process can dissolve on the order of 400 to 800 kg/ha during a year of reclamation. Use of acid is currently a widespread and generally recommended practice to prevent emitter clogging in drip irrigation systems. This practice may result in total removal of carbonates within 10–20 years, for soils with less than 3% carbonate content.

IMPACT OF IC ON ATMOSPHERIC CARBON DIOXIDE

Dissolution of carbonates in neutral to alkaline environments results in consumption of CO_2 gas and formation of aqueous bicarbonate (HCO_3^-), while precipitation of carbonates results in release of CO_2. The net effect of dissolution or precipitation of soil IC on atmospheric CO_2 depends on the solution flow path. In regions irrigated with surface water, the dissolution of carbonates results in a net C sink. The high alkalinity drainage water usually flows back to the river, but even in shallow groundwater systems, this would take on the order of tens to several hundred years. The resultant degassing of carbonic acid in arid environments will result in reprecipitation of carbonate in the river or reservoir and releases CO_2 back into the atmosphere. If the water is recharged into deep aquifers, the net soil flux of dissolved IC is preserved as an IC sink. In acid environments, liming of soils results in CO_2 release to the atmosphere as there is little or no net alkalinity produced. Examination of records of volumes of water and concentrations of alkalinity (mostly bicarbonate) revealed that within the past 50 years, there has been a dramatic increase in net alkalinity export from the Mississippi River to the Gulf of Mexico.[18] The 17.7 Tg/yr increase in IC is attributed to increased rainfall in the basin and increased proton delivery to the land due to enhanced carbon dioxide production in the soil, resulting in increased weathering. It seems reasonable to consider that management changes for increased crop productivity may have increased soil liming and dissolution of IC. Such a drastic change does not likely reflect a response to a changing groundwater composition but is likely related to increased alkalinity in surface over land flow. The impact of a drier climate on the net export of alkalinity was not evaluated. However, it is reasonable to assume that decreased runoff would result in decreased mass of alkalinity transported to surface waters.

CONCLUSION

Land-use practices have a long-term impact on soil IC. Due to the large C pools in the soil, these impacts are not generally observed in short-term studies. Use of acidifying fertilizers such as ammonia and sulfur reduces soil IC. Practices that favor dissolution of carbonates include irrigation with surface waters and irrigation with water in large excess of plant transpiration. Practices that favor accumulation of IC in the soil include lower water applications relative to transpiration when irrigating in arid and semiarid regions (leaching less than 30% of the applied water), irrigation with groundwaters at elevated CO_2 concentrations, application of gypsum to alkaline soils, and use of nitrate fertilizer. Other factors that affect soil carbonate content include land clearing, cropping practices, and erosion.

In humid environments, the major anthropogenic impacts on inorganic C are liming of surface soils, use of fertilizers, removal of vegetation, and soil erosion. In semiarid and arid environments, increased IC is favored by use of groundwater vs. surface water for irrigation and application of gypsum amendments used to improve water infiltration. Decreased IC is favored by inefficient irrigation with surface water and application of NH_4 fertilizer. The net effect of irrigation on a global scale, neglecting the effects of fertilizer addition, is an estimated increase in soil inorganic C by 30 Tg/yr as well as a release of an almost equal amount of C to the atmosphere. Liming practices in humid regions throughout the world are estimated to have no net effect on inorganic soil C but release up to 85 Tg C/yr to the atmosphere. Of course this is the net effect from IC processes and does not include sequestration of OC by crop production.

REFERENCES

1. Batjes, N.H. Total carbon and nitrogen in soils of the world. Eur. J. Soil Sci. **1966**, *47*, 151–163.
2. Cole, J.J.; Prairie, Y.T.; Caraco, N.F.; McDowell, W.H.; Tranvik, L.J.; Striegl, R.G.; Duarte, C.M.; Kortelainen, P.; Downing, J.A.; Middelburg, J.J.; Melack, J. Plumbing the global carbon cycle: Integrating inland waters into the terrestrial carbon budget. Ecosystems **2007**, *10*, 171–184.
3. Kessler, T.J.; Harvey, C.F. The global flux of carbon dioxide into groundwater. Geophys. Res. Lett. **2009**, *28*, 279–282.
4. Wu, H.; Guo, Z.; Gao, Q.; Peng, C. Distribution of soil inorganic carbon storage and its changes due to agricultural land use activity in China. Agric. Ecosyst. Environ. **2009**, *129*, 413–421.
5. Cihacek, L.J.; Ulmer, M.G. Effects of tillage on inorganic carbon storage in soils of the Northern Great Plains of the U.S. In *Agricultural Practices and Policies for Carbon Sequestration in Soil*; Kimble, J.M., Lal, R., Follett, R.F., Eds.; CRC Press: Boca Raton, 2002; 63–69.
6. Mikhailov, E.; Post, C.A. Effects of land use on soil inorganic carbon stocks in the Russian Chernozem. J. Environ. Qual. **2006**, *35*, 1384–1388.
7. Landi, A.; Mermut, A.R.; Anderson, D.W. Origin and rate of pedogenic accumulation in Saskatchewan soils, Canada. Geoderma **2003**, *117*, 143–156.
8. Voss, R.D. What constitutes an effective liming material. In *National Conference on Agricultural Limestone*; National Fertilizer Development Center: Muscle Shoals, AL, 1980; 52–61.
9. Suarez, D.; Rhoades, J. Effect of leaching fraction on river salinity. J. Irrig. Drain. Div., Am. Soc. Civ. Eng. **1977**, *103* (2), 245–257.
10. Suarez, D. Impact of agriculture on CO_2 as affected by changes in inorganic carbon. In *Global Climate Change and Pedogenic Carbonates*; Lal, R., Kimble, J., Eswaran, H., Stewart, B., Eds.; Lewis: Boca Raton, 1999; 257–272.
11. Solley, W.B.; Pierce, R.R.; Perlman, H.A. *Estimated Use of Water in the United States in 1990*; U.S. Geol. Surv. Circular 1004; U.S. Gov. Printing Office: Washington, DC, 1993.
12. Magaritz, M.; Amiel, A. Influence of intensive cultivation and irrigation on soil properties in the Jordan Valley, Israel: Recrystallization of carbonate minerals. Soil Sci. Soc. Am. J. **1981**, *45*, 1201–1205.
13. Amundson, R.G.; Smith, V.S. Effects of irrigation on the chemical properties of a soil in the Western San Joaquin Valley, California. Arid Soil Res. Rehabil. **1988**, *2*, 1–17.

14. Amundson, R.D.; Lund, L. The stable isotope chemistry of a native and irrigated Typic Natrargid in the San Joaquin Valley of California. Soil Sci. Soc. Am. J. **1987**, *51*, 761–767.
15. Entry, J.A.; Sojka, R.E.; Shewmaker, G.E. Irrigation increases inorganic carbon in agricultural soils. Environ. Manage. **2004**, *33*, s309–s317.
16. Wu, L.; Wood, Y.; Jiang, P.; Li, L.; Pan, G.; Lu, J.; Chang, A.C.; Enloe, H.A. Carbon sequestration and dynamics of two irrigated agricultural soils in California. Soil Sci. Soc. Am. J. **2007**, *72*, 808–814.
17. Suarez, D.L. Impact of agriculture on soil inorganic carbon. Annual Meetings, Baltimore, MD, Oct. 18–23, 1998; Agronomy Abstracts; Soil Science Society of America: Madison, WI, 1998; 258–259.
18. Raymond, P.A.; Cole, J.J. Increase in the export of alkalinity from North America's largest river. Science **2003**, *301*, 88–91.

Chemigation

William L. Kranz
Northeast Research and Extension Center, University of Nebraska-Lincoln, Norfolk, Nebraska, U.S.A.

Abstract
Chemigation is the practice of distributing approved agricultural chemicals such as fertilizers, herbicides, insecticides, fungicides, nematicides, and growth regulators by injecting them into water flowing through a properly designed and managed irrigation system. The term chemigation was originally coined to describe the concept of applying commercial fertilizers that were needed for crop production. Field research, and advances in sprinkler and chemical injection technology have stimulated the use of chemigation as a major crop production tool. Today chemigation is one of the more efficient, economical, and environmentally safe methods of applying chemicals needed for successful crop, orchard, turf, greenhouse, and landscape operations.

INTRODUCTION

Chemigation began with the application of commercial fertilizers through irrigation systems in the late 1950s.[1] Later tests were initiated on sprinkler application of herbicides to selectively control weeds in field crops, fruit and nut orchards, rice, and potatoes.[2,3] These research efforts led the way for what has become a major research topic to identify management and equipment required for chemical application in agricultural and non-agricultural production settings.

The primary use of chemigation is to apply chemical directly to the soil using a range of irrigation water distribution systems. For example, drip/trickle, sprinklers, and some surface irrigation systems are commonly used to apply commercial fertilizers. However, federal regulations limit application of restricted use pesticides to systems that can safely and uniformly apply a chemical to a specific site at a rate specified on a chemical label. Though estimates vary greatly, chemigation is used to apply fertilizers on nearly four million hectares in the United States.[4] Specialists in Florida, Texas, and Wyoming report that more than 50% of their irrigated land received at least one chemigation application.[5]

ADVANTAGES OF CHEMIGATION

Chemigation offers producers of food and fiber many advantages that result from using existing equipment and timeliness of chemical applications. Advantages of chemigation include the following[6,7]:

- Uniformity of chemical application is equal to or greater than other means of application.
- Timeliness and flexibility of application are greater.
- Improved efficacy of some chemicals.
- Potential for reduced environmental risks.
- Lower application costs in some cases.
- Less mechanical damage to plants.
- Less soil compaction.
- Potential reduction in chemical applications.
- Reduced operator hazards.
- Application cost savings for multiple applications.

DISADVANTAGES OF CHEMIGATION

Chemigation also requires additional equipment and management to obtain successful results. Some of the disadvantages of chemigation include[6,7]:

- Chemical application accuracy depends on water application uniformity.
- Longer time of application than other methods.
- Some pesticide labels prohibit chemigation as a means of application.
- Potential for source water contamination.
- Additional capital costs for equipment.
- Potential for increased legal requirements in some states.
- Increased management requirements by the operator.

CHEMIGATION EQUIPMENT

Safe and efficient chemigation requires that the irrigation equipment, injection device, and safety equipment be properly installed and maintained. Fig. 1 provides an overview of equipment necessary for chemigation systems using groundwater. State and federal regulations specify the type of irrigation water distribution system that can be used and the required safety equipment. It is up to the irrigator to ensure the use of appropriate equipment and procedures.

Encyclopedia of Environmental Management DOI: 10.1081/E-EEM-120010031

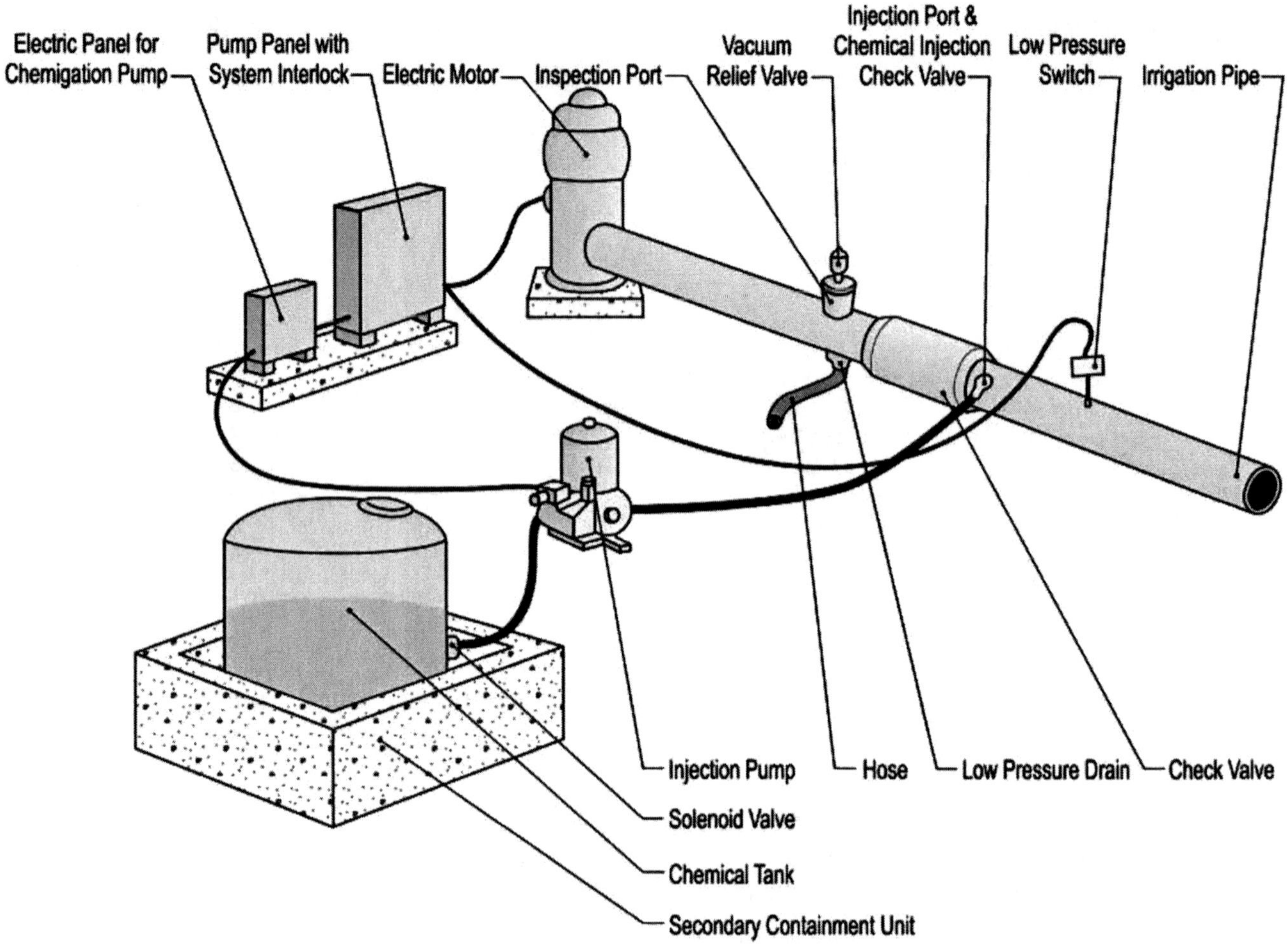
Electric Panel for Chemigation Pump
Pump Panel with System Interlock
Electric Motor
Inspection Port
Vacuum Relief Valve
Injection Port & Chemical Injection Check Valve
Low Pressure Switch
Irrigation Pipe
Injection Pump
Hose
Low Pressure Drain
Check Valve
Solenoid Valve
Chemical Tank
Secondary Containment Unit

Fig. 1 Chemigation injection and safety equipment commonly required when pumping groundwater. (Drawing courtesy of Midwest Plan Service, Ames, IA.)

Irrigation Equipment

Chemigation requires equipment capable of applying chemicals uniformly and with differing amounts of water, accurate and dependable injection equipment, and safety equipment for source water and worker protection. Appropriate sprinkler design and high application volumes can solve problems associated with canopy penetration and deposition that impact some aerial applications. Uniform water application can precisely place and incorporate chemicals in the soil and limit leaching of soluble chemicals from the zone of application.

Several different types of irrigation equipment can and are being used to distribute chemicals via chemigation. Most chemigation is conducted using either sprinkler or drip/trickle irrigation systems. Center pivot and linear-move systems are most commonly used for chemigation since prescription applications can be made with a high degree of uniformity. Drip/trickle systems are commonly used to place precise amounts of plant nutrients near the zone of plant uptake thus increasing chemical use efficiency.

In general, surface irrigation systems have limited potential for chemigation. Water distribution in furrow systems is typically non-uniform along the row and among rows. Thus, in-field variation in water infiltration results in chemical application uniformity that is below levels desired for chemigation. Development of surge-flow systems can improve distribution uniformity, however, the question remains whether consistent results are possible and whether producers have sufficient experience to make equipment adjustments when necessary. Level basin irrigation systems offer improved uniformity of water application, but water quality concerns have limited the use of chemigation.

Injection Equipment

Chemical injection can occur using either active or passive devices. Active devices use an external energy supply to create pressures at the injector outlet that exceed the irrigation pipeline pressure. Injection pumps are often powered by constant speed or variable speed electric motors. Typical examples include piston, diaphragm, rotary, and gear pumps. However, most new installations use either piston or diaphragm pumps (Fig. 2). These injection devices are relatively expensive. Component selection allows the

Fig. 2 Typical portable injection equipment for center pivot installations. (Photo courtesy of Agri-Inject, Inc., Yuma, CO.)

injection of commercial fertilizers, acids, or pesticides. Intermittent end guns and corner systems can lead to variable chemical application by constant rate injectors due to changes in the irrigation rate per hour.[8] Application errors of approximately 20% are possible when corner systems are used with a constant rate injection device. When the irrigation rate will change during a chemigation event, it is preferable to use a variable rate injection device.

Passive devices take advantage of pressure differentials that result from using a throttling valve or pitot tube unit to add chemical to water flowing through a pipeline. Chemicals are metered into the system using a venturi meter or orifice plate. These systems have low capital cost requirements. However, pumping cost may be greater since irrigation pump outlet pressure must be equal to the water distribution system pressure plus the friction loss associated with the throttling valve. In addition, changes in pumping pressure directly impact chemical injection rates which can lead to non-uniform chemical applications.

Selection criteria for injection devices include potential injection rates, available power supply, and the type of chemical to be injected. A single injection device is typically not capable of covering the range of injection rates and chemical types that could conceivably be applied via chemigation. Hence, if plant nutrients and pesticides are to be applied, two injection devices are desirable. Diaphragm injection devices offer greater chemical compatibility, ease of calibration, and precise injection rates which make them good choices for pesticide injection. Commercial fertilizers are less caustic and require relatively high injection rates which make high capacity piston and diaphragm injection devices good options. Research has noted that injection equipment calibration was necessary for each injection device and operating pressure.[9] Manufacturing tolerances and pipeline pressure impacted the rate of chemical injection. Further, performance tests conducted on new and used diaphragm pumps found that proper maintenance is required to ensure long-term accuracy of chemical injection rates.[10]

Safety Equipment

State and federal regulations differ regarding safety equipment that is required for chemigation. For example, the Nebraska Chemigation Act requires the safety equipment also found in many state regulations.[11] Most requirements are met through installation of a backflow protection device. Requirements typically include (Fig. 3):

1. A mainline check valve to prevent concentrated chemical and/or dilute chemical solution from flowing back into the water source.
2. A chemical injection line check valve to prevent flow of chemical from the chemical supply tank into the irrigation pipeline and to prevent flow of water through the injection system into the chemical supply tank.
3. Vacuum relief valve to prevent back siphoning of concentrated chemical and/or dilute chemical solution into the water source.
4. Low pressure drain to prevent back flow of chemical and/or dilute chemical solution into the water source should the mainline check valve fail.
5. An inspection port to ensure that the mainline check valve and low pressure drain are functioning properly.
6. An interlock between the injection system and the irrigation pumping plant to prevent injection of concentrated chemical into the irrigation pipeline should there be an unexpected shutdown of the irrigation pump.

American Society of Agricultural Engineers have published EP409.1 Safety Devices for Chemigation[12] which recommends the addition of a two-way interlock between the injection system and irrigation pumping plant and a normally-closed solenoid valve on the outlet of the chemical supply tank to prevent chemical spills attributed to chemical injection line or injection device failures. The engineering practice also encourages the positioning of a fresh water source near the chemical supply tank for washing chemicals that may contact skin, the use of a strainer on the chemical tank outlet to prevent fouling of injection equipment, the grading of the soil surface to direct flow away from the water supply, the location of mixing tanks and injection equipment safely away from potential sources of electrical sparks to prevent explosions, and the use of components that are well suited to a range of chemical formulations.

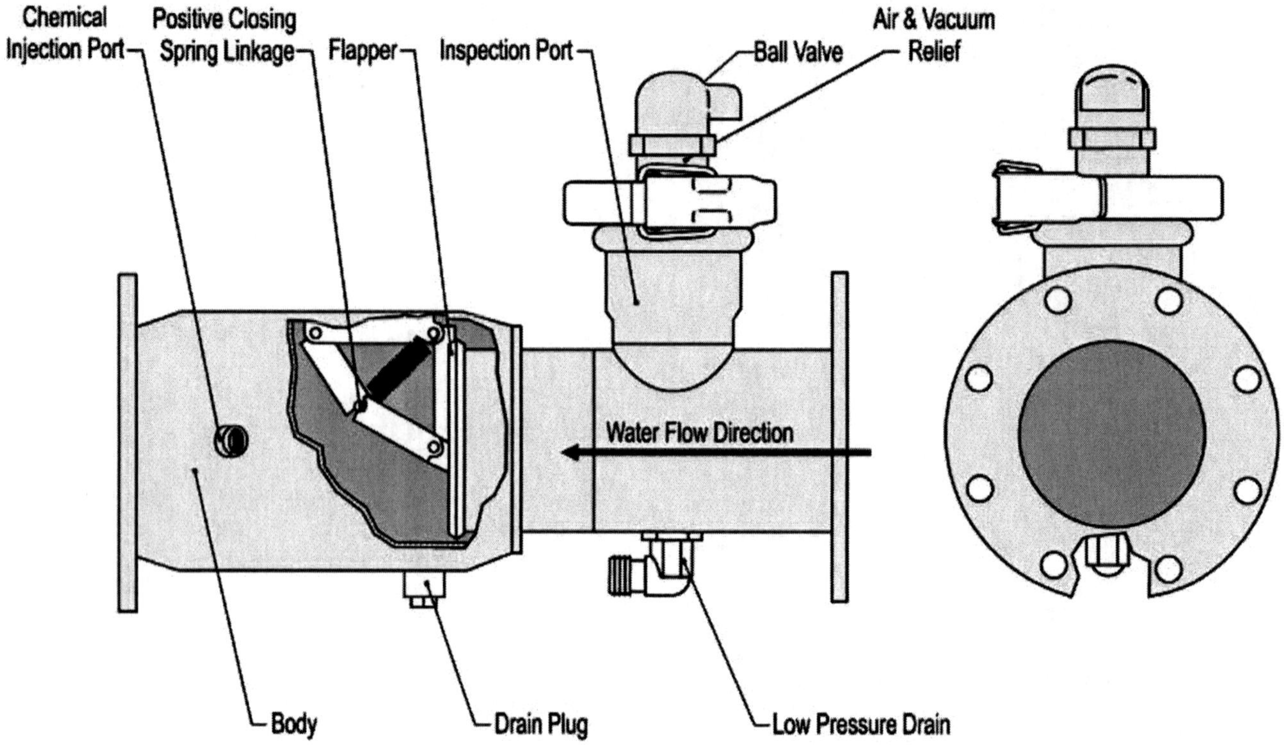
Chemical Injection Port
Positive Closing Spring Linkage
Flapper
Inspection Port
Ball Valve
Air & Vacuum Relief
Water Flow Direction
Body
Drain Plug
Low Pressure Drain

Fig. 3 Mainline check valve used to prevent flow of chemicals into a water source. (Drawing courtesy of Midwest Plan Service, Ames, IA.)

MANAGEMENT PRACTICES

Management flexibility based on chemical placement, application rate and mobility in the soil, water quality, application cost, and weather factors make chemigation a unique and effective production tool. Chemigation provides the opportunity to synchronize fertilizer applications to match plant needs and incorporate and, if needed, activate pesticides to increase efficacy. Equally important, chemigation provides the opportunity to reduce chemical applications by eliminating the need for insurance-type applications. Fields can be scouted for disease or pests and chemical applied only if damage or pest numbers exceed economic thresholds. Soil and plants can be monitored to determine fertilizer needs, making near real-time adjustments in the time of application and chemical formulation possible. Individual nozzle controls make site-specific applications well within reach.[13] However, a considerable amount of work remains to ascertain if site-specific applications are economical and to incorporate management tools into system controls.

CONCLUSION

Chemigation has gradually become one of the most effective means of chemical application available for crop production and landscape systems. Advantages of highly uniform prescription applications outweigh the potential disadvantages in most cases. Effective chemigation hinges on the selection of appropriate irrigation systems, chemical injection devices, and safety equipment. Through proper management, chemigation is poised to be a production practice that can help increase the quality and quantity of food produced worldwide.

REFERENCES

1. Bryan, B.B.; Thomas, E.L., Jr. *Distribution of Fertilizer Materials Applied with Sprinkler Irrigation Systems*, Agricultural Experiment Station Research Bulletin 598; University of Arkansas: Fayetteville, AK, 1958; 12 pp.
2. Ogg, A.G., Jr. Applying herbicides in irrigation water— a review. Crop Prod. **1986**, *5* (1), 53–65.
3. Cary, P.J. Applying herbicides and other chemicals through sprinkler systems. Proceedings of the 21st Annual Conference of the Washington State Weed Association, Yakima, WA, 1971.
4. USDA. Application of chemicals in irrigation and times irrigated by selected crop: 1998 and 1994. In *1998 Farm and Ranch Irrigation Survey: Table 24*; U.S. National Agricultural Statistics Service: Washington, DC, 1999; 102–124.
5. Adams business media; 2000 annual irrigation survey continues steady growth. Irrig. J. **2001**, *51* (1), 12–40.
6. Scherer, T.F.; Kranz, W.L.; Pfost, D.; Werner, H.D.; Wright, J.A.; Yonts, C.D. Chemigation. In *MWPS-30 Sprinkler Ir-*

rigation Systems; Midwest Plan Service: Ames, IA, 1999; 145–166.
7. Threadgill, E.D. Introduction to chemigation: history, development, and current status. In *Proceedings of the Chemigation Safety Conference, Lincoln, NE, April 17–18, 1985*; Vitzthum, E.F., Hay, D.R., Eds.; University of Nebraska Cooperative Extension: Lincoln, NE, 1985.
8. Eisenhauer, D.E. Irrigation system characteristics affecting chemigation. In *Proceedings of the Chemigation Safety Conference, Lincoln, NE, April 17–18, 1985*; Vitzthum, E.F., Hay, D.R., Eds.; University of Nebraska Cooperative Extension: Lincoln, NE, 1985.
9. Kranz, W.L.; Eisenhauer, D.E.; Parkhurst, A.M. Calibration accuracy of chemical injection devices. Appl. Engr. Agric. **1996**, *12* (2), 189–196.
10. Cochran, D.L.; Threadgill, E.D. Injection Devices for Chemigation: Characteristics and Comparisons, National Meeting of the American Society of Agricultural Engineers, Paper No. 86-2587; ASAE: St. Joseph, MI, 49805, 1986.
11. Vitzthum, E.F. *Using Chemigation Safely and Effectively*; University of Nebraska Cooperative Extension Division: Lincoln, NE, 2000; 1–59.
12. ASAE. *EP409.1: Safety Devices for Chemigation*; American Society of Agricultural Engineers: St. Joseph, MI, 2001.
13. Evans, R.G.; Buchleiter, G.W.; Sadler, E.J.; King, B.A.; Harting, G.H. Controls for Precision Irrigation with Self-propelled Systems. Proceedings of the 4th Decennial National Irrigation Symposium, Phoenix, AZ, 2000; 322–331.

Chesapeake Bay

Sean M. Smith
Ecosystem Restoration Center, Maryland Department of Natural Resources, Annapolis, Maryland, U.S.A.

Abstract
The Chesapeake Bay is a large estuary on the mid-Atlantic coast of the United States. The estuary receives fresh water inputs from a watershed encompassing six states and Washington, D.C. The many species of fish and wildlife that proliferate in the Bay have made it an important natural resource to human inhabitants for thousands of years. Changes in the estuary and its watershed resulting from the dramatic increases the human population surrounding it have had deleterious effects on aquatic habitat and related fisheries. In response, a large-scale restoration effort was formally initiated in 1983 to rehabilitate the Bay.

INTRODUCTION

The Chesapeake Bay is the largest estuary within the U.S.A.[1] Fresh water flows to the Chesapeake Bay from a watershed that covers an estimated 166,709 km^2, including portions of Delaware, Maryland, New York, Pennsylvania, Virginia, Washington, DC, and West Virginia (Fig. 1). The ecological productivity of the estuary has made it an important resource for Native Americans, European immigrants, and current residents in the region.[2] The population in the contributing watershed in recent years has swelled to over 15 million people, resulting in extensive direct and indirect impacts that are now a focus of a large-scale restoration effort led by the US Environmental Protection Agency.[1]

DISCUSSION

Located along the Mid-Atlantic coast of the U.S.A. within the limits of Maryland and Virginia, the Bay is approximately 304-km long, has an estimated surface area of 11,603 km^2, a width that ranges from 5.5 to 56 km, and an average depth of approximately 6.4 m.[1] Salinity in the tidal portions of the estuary transition from "fresh" conditions (i.e., 0–5 parts per thousand (ppt) salt concentration) at the northern-most end to "marine" conditions (30–35 ppt salt concentration) at the southern boundary with the Atlantic Ocean.

Evidence indicates that the modern Chesapeake Bay began forming approximately 35 million years ago with a meteorite impact in the proximity of what is now the confluence of the Bay with the Atlantic Ocean.[2,3] The impact created a topographic depression that influenced the location and alignments of several large river valleys, including those associated with the present day Susquehanna, Rappahannock, and James Rivers. Since then, the river valleys have been periodically exposed and flooded in response to cycles of global glaciation and associated fluctuations in sea level. The most recent, the Wisconsin glaciation, began retreating approximately 18,000 years ago. The retreat resulted in a rise in sea level by almost ninety meters, drowning the river valleys and forming the current Bay.

Eleven large rivers drain the Bay's watershed, the Susquehanna River from Pennsylvania and New York providing the largest contribution with an average of 98 million m^3/day flowing into the northern end of the estuary. The rivers drain one or more of five different physiographic provinces within the watershed, including the Appalachian Plateau, Ridge and Valley, Blue Ridge, Piedmont, and Coastal Plain (Fig. 1).[4] Each province's geologic composition and history creates dramatically different landscape settings from the western to eastern sides of the drainage basin. The Appalachian, Ridge and Valley, and Blue Ridge are characterized by mountainous terrain, a dominance of sandstone along ridge tops, and several carbonate valleys. The Piedmont has less relief, is dominated by metamorphic rocks, and is characterized by a surface that has been dissected by dendritic stream channel networks. Further to the east, the Coastal Plain is characterized by thick layers of unconsolidated geologic materials overlying bedrock deep beneath the surface. Waterways that flow from the Piedmont into the Coastal Plain traverse the "Fall Zone," a region that is easily distinguished by waterfalls coincident with an abrupt drop in the underlying bedrock elevations. Major ports and cities were developed along the Fall Zone, including Washington, DC, Baltimore, Maryland, and Richmond, Virginia, because of their locations at the upstream terminus of navigation from tidal waters and proximity to hydropower sources.

The Chesapeake Bay estuary is naturally dynamic and characterized by physical conditions that can be stressful to aquatic organisms. The salinity gradient broadly governs the spatial distribution of aquatic habitat types. Alterations in currents, wind, and freshwater inputs can cause salinity

Encyclopedia of Environmental Management DOI: 10.1081/E-EEM-120010130

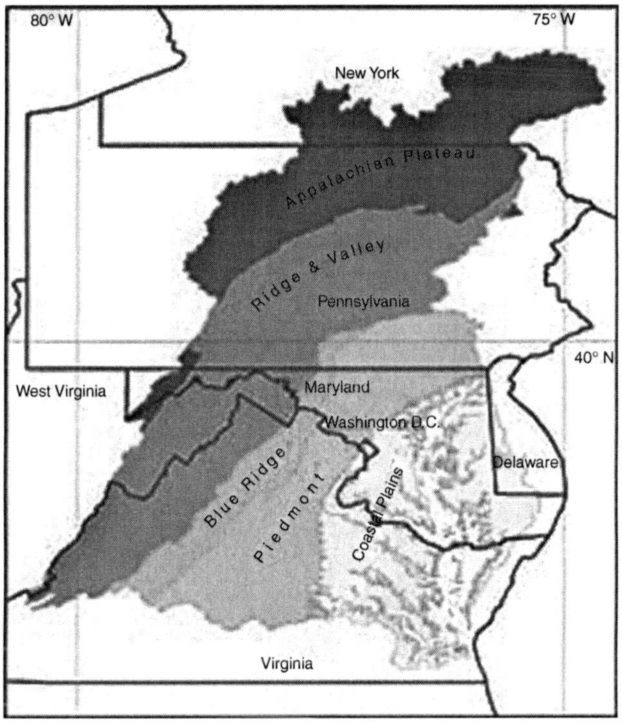
80° W
75° W
New York
Appalachian Plateau
Ridge & Valley
Pennsylvania
40° N
West Virginia
Maryland
Washington D.C.
Delaware
Blue Ridge
Piedmont
Coastal Plains
Virginia

Fig. 1 The Chesapeake Bay estuary and its watershed, including physiographic provinces and state boundaries. (Courtesy of M. Herrmann, Maryland Department of Natural Resources.)

conditions to vary over time. The shallow depths also cause colder winter and warmer summer water temperatures compared to the open ocean. These spatial and temporal fluctuations can create physiologically challenging conditions. However, many organisms have adapted and use the abundant nutrients and physical habitat in different portions of the estuary for specific periods of their life cycles or seasons of the year. As a result, the Bay supports an estimated 3600 species of plants, fish, and animals, including 348 finfish and 173 shellfish species. Some of the most notable of these include striped bass, American shad, blueback herring, blue crab, and the American oyster.[5] The name "Chesapeake" itself was coined from the Algonquin American–Indian word "Chesepiooc" meaning "great shellfish bay."[5]

Archaeologists estimate that Native American inhabitants first arrived in the Bay region from the south or west approximately 12,000 years ago as the ice sheets associated with the Wisconsin glaciation began to retreat and temperatures increased.[2] The first inhabitants are presumed to have been nomadic; however, archeological evidence suggests that selective food production started as early as 5000 years ago and settled towns began to be formed approximately 1300 years ago as the population density in the region increased. Recovered artifacts provide evidence of the extensive use of the Bay by the early inhabitants for travel, communication, tools, and food.

The first recorded European contact with the Chesapeake Bay region was by the Italian captain, Giovanni da Verrazano in 1524.[2] The English established one of the most well known early settlements at Jamestown, Virginia in 1607. English colonization expanded through expeditions to the north in the Bay, partly led by the famed Captain John Smith. Immigration to the region increased throughout the 1600s and much of the area was settled by the mid-1700s. The colonists made extensive use of the resources provided by the estuary, its wetlands, and tributaries. Shellfish, including oysters, blue crabs, and hard and soft clams, were harvested from shallow water areas.[2] The numerous piles of oyster shells that can be found near Coastal Plain tidal areas provide support for written claims of the extensive oyster beds that existed in the Bay when the European colonists arrived. Traps and nets were used to harvest finfish, including herring, striped bass, and shad. Migratory waterfowl, such as ducks and geese, were also plentiful food sources.

The rapid growth in the human population since European colonization of the Chesapeake Bay region dramatically increased the harvest of finfish, shellfish, waterfowl, and mammals naturally supported by the estuary. Extensive landscape alterations also caused direct and indirect physical changes to the Bay and its tributaries. The combination of overharvesting, pollution, and physical alterations has severely impacted the ecosystem and many of the species that historically flourished in the estuary.[1] Dramatic declines have been documented by the harvest records of popular commercial fisheries such as shad and striped bass. Records indicate a decline in the catch of blue crabs per unit of effort since the 1940s. The oyster harvest is currently at less than 1% of historic levels, although this reduction is partly attributed to disease. Many other species not harvested commercially have also been affected by the alterations in the Bay ecology that have accompanied European settlement and population growth.

One of the most important impacts to the Chesapeake Bay has been the increased erosion rates and downstream sedimentation caused by extensive deforestation of the watershed.[6] The influx of sediment into the tidal estuary has reduced the water depths in many embayments that once served as navigable ports.[7] Elevated suspended sediment inputs during storm events also increase turbidity in the tidal water column.[6] The resulting decrease in water clarity, which has been exacerbated by algal blooms associated with nutrient runoff pollution, reduces submerged aquatic vegetation (SAV) growth in shallow areas (i.e., depths less than 2 m). SAV coverage on the Bay bottom is estimated to have declined from approximately 80,900 hectares in 1937 to 15,400 hectares in 1984.[1] The loss has negative implications for a variety of species that use the vegetation for habitat, including blue crabs and juvenile finfish.

An extensive effort to restore the Bay has been undertaken by the US federal government in coordination with states in the watershed.[1,8] A large part of this effort has been focused on the recovery of the historic SAV distributions, as well as reversal of the abnormally low oxygen levels that now occur in the main stem of the Bay and its major tidal tributaries during summer months.[1,9] As with the water clarity problems, the low dissolved oxygen is related to excess nutrient inputs, mainly nitrogen and phosphorous, which stimulate algal production. The oxygen depletion occurs because of algal decomposition, resulting in, estuarine habitat degradation. Substantial reductions in nutrients from watershed runoff have been concluded to be necessary to achieve restoration goals related to both SAV and low dissolved oxygen.[1,9]

CONCLUSIONS

The Chesapeake Bay is a large and historically productive estuary on the Mid-Atlantic coast of the U.S.A., with extensive fisheries and wildlife resources. The Bay ecosystem has been impaired by watershed alterations and overharvesting accompanying human population growth in the region, thereby inspiring an extensive government-supported restoration effort. A large part of the restoration focuses on sediment and nutrient pollution associated with runoff from the contributing watershed.

REFERENCES

1. http://chesapeakebay.net (accessed July 11, 2005).
2. Grumet, R.S. *Bay, Plain, and Piedmont: A landscape history of the Chesapeake Heartland From 1.3 Billion Years Ago to 2000*; The Chesapeake Bay Heritage Context Project; U.S. Department of the Interior: Annapolis, MD, 2000; 183 pp.
3. Powars, D.S.; Bruce, T.S. *The Effects of the Chesapeake Bay Impact Crater on the Geological Framework and Correlation of Hydrogeologic Units of the Lower York-James Peninsula, Virginia*; U.S. Geological Survey Professional Paper 1612; United States Geological Survey: Reston, VA, 1999.
4. Smith, S.; Gutuierrez, L.; Gagnon, A. *Streams of Maryland-Take a Closer Look*; Maryland Department of Natural Resources: Annapolis, MD, 2003; 65 pp.
5. White, C.P. *Chesapeake Bay: Nature of the Estuary, A Field Guide*; Tidewater Publishers: Centreville, MD, 1989.
6. Langland, M.; Cronin, T. *A Summary Report of Sediment Processes in Chesapeake Bay and Watershed*; United States Geological Survey Water Resources Investigations Report 03-4123; United States Geological Survey: Reston, VA, 2003.
7. Gottschalk, L.C. Effects of soil erosion on navigation in upper Chesapeake Bay. Geogr. Rev. **1945**, *35*, 219–238.
8. Ernst, H.R. *Chesapeake Bay Blues*; Rowman and Littleford Publishers, Inc.: Lanham, MD, 2003.
9. Macalaster, E.G., Barker, D.A., Kasper, M., Eds.; *Chesapeake Bay Program Technical Studies: A Synthesis*; United States Environmental Protection Agency: Washington, DC, 1983.

Chromium

Bruce R. James
Dominic A. Brose
University of Maryland, College Park, Maryland, U.S.A.

Abstract

Chromium is a naturally occurring transition metal that is an essential nutrient in its trivalent oxidation state, but a toxicant in its hexavalent state. It has been shown to be a carcinogen by the U.S. National Toxicology Program (NTP), and its concentration in drinking water is regulated by the U.S. Environmental Protection Agency (EPA). High concentrations of Cr(VI) in natural waters are usually derived from industrial Cr-containing wastes, or possibly from the oxidation of certain forms of Cr(III) in soils or sediments. Chromite ore ($FeO{\cdot}Cr_2O_3$) is roasted under alkaline, high-temperature conditions to oxidize Cr_2O_3 to soluble Cr(VI), which is used as a starting material for production of stainless steel, pressure-treated lumber, chrome-tanned leather, pigments, chrome-plated metals, and other common products used in modern societies. Cr(VI) remaining in industrial by-products, such as chromite ore processing residue, chrome plating bath waste, paint aerosols, and other industrial wastes, may enrich soils and contaminate surface waters and groundwater that are water supplies for domestic uses, irrigation, and industrial processes.

INTRODUCTION

Chromium is a heavy metal that is essential for human health as a cofactor of insulin in its trivalent form [(Cr(III)] but may cause lung cancer if inhaled in the hexavalent oxidation state [Cr(VI)]. Trivalent Cr is only sparingly soluble in neutral to alkaline natural waters as $Cr(OH)_3$ or Cr_2O_3, but it can be oxidized to Cr(VI) by naturally occurring manganese (III,IV) (hydr)oxides and by hydrogen peroxide, ozone, chlorine gas, hypochlorite, and other electron acceptors used in the environmental remediation of water, sediments, and soils. Hexavalent Cr can be reduced to Cr(III) by elemental Fe and Fe(II), sulfides (H_2S and HS^-), easily oxidized organic C compounds, and other electron donors. The rates and extent of oxidation and reduction reactions of Cr are governed by the redox potential (Eh) and acidity (pH) of natural soil–water systems and of engineered environments, such as anthropogenic wetlands and wastewater treatment facilities.

HEALTH EFFECTS

Concerns surrounding the presence of chromium (Cr) in natural waters and drinking water supplies must address the contrasting solubilities and toxicities of its common oxidation states in natural environments: Cr(III) and Cr(VI). Currently, Cr regulation principally focuses on "total Cr," without distinguishing the oxidation states in a solid or water sample. The U.S. Environmental Protection Agency's (EPA) national standard for total Cr in drinking water is 100 μg/L (100 ppb), except that California has set its current drinking water standard at 50 μg/L.[1,2] In 1999, California set a Public Health Goal of 2.5 μg/L, based on a 1968 study in Germany that found stomach tumors in animals that drank chromium-enriched water. The U.S. EPA rejected that study as flawed and determined that there was no evidence it was carcinogenic in water, which resulted in California rescinding its goal in 2001 and reverting back to the 50 μg/L standard.[2] The point of contention regarding the 1968 study was whether Cr(VI) is reduced to Cr(III) in the stomach by gastric acids.

Chromium(VI) is genotoxic in a number of in vitro and in vivo toxicity tests.[3] Because the mechanisms of genotoxicity and carcinogenicity are not fully understood, the National Toxicology Program (NTP) conducted animal tests to assess the potential for cancer due to ingestion of Cr(VI).[4] Reduction of Cr(VI) to Cr(III) is hypothesized to occur primarily in the stomach, as a mechanism of detoxification. In this 2 years NTP study, no neoplasms or non-neoplastic lesions were observed in the forestomach or glandular stomach of rats or mice. However, observed increases in neoplasms, or abnormal growths of tissue, in the small intestine of mice, toxicity to red and white blood cells and bone marrow, and uptake of Cr(VI) into tissues of rats and mice suggest that at least a portion of the administered Cr(VI) was not reduced in the stomach.[4] The stepwise reduction of Cr(VI) to Cr(III) through Cr(V) and Cr(IV) creates short-lived oxidation states that may be the forms of Cr that are the actual carcinogens.

This finding, in addition to the absence of increases in neoplasms or non-neoplastic lesions of the small intestine in rats or mice exposed to chromium picolinate monohydrate (CPM), an organically bound form of Cr(III),[5]

Encyclopedia of Environmental Management DOI: 10.1081/E-EEM-120046622

provides evidence that Cr(VI) is not completely reduced in the stomach and is responsible for these carcinogenic effects. Additionally, it should be noted that Cr(III), like that found in CPM, is essential for human health in trace amounts as an activator of insulin,[6] but exists predominantly in nature in cationic forms that are typically only sparingly soluble in near-neutral pH soils, plants, cells, and natural waters.[7]

SOURCES OF CHROMIUM AND OCCURRENCE IN NATURAL WATERS AND WATER SUPPLIES

When soluble Cr is detected in natural waters, especially at high concentrations, it is usually Cr(VI) derived from industrial wastes containing Cr(VI) or possibly resulting from the oxidation of certain forms of Cr(III) in soils or sediments.[8,9] Chromium is the seventh most abundant metal on earth with an average content of 100 mg/kg in the earth's crust and 3700 mg/kg for the earth as a whole, principally as Cr(III) in unreactive, insoluble minerals, such as chromite ($FeO{\cdot}Cr_2O_3$).[10] Roasting chromite ore under alkaline, high-temperature conditions oxidizes Cr_2O_3 to soluble Cr(VI), a widely used starting material for production of stainless steel, pressure-treated lumber, chrome-tanned leather, pigments, chrome-plated metals, and other common products used in modern societies.[11] As a result, Cr(VI) remaining in industrial by-products, such as chromite ore processing residue, chrome plating bath waste, paint aerosols, and other industrial wastes, may enrich soils and contaminate surface waters and groundwater that are supplies for domestic uses, irrigation, and industrial processes.

In contrast to these concentrated, anthropogenic sources of Cr(VI); naturally occurring sources of Cr are predominantly Cr(III) and occur at low concentrations. Ultramafic and basaltic rocks (and soils developed from these parent materials), however, may contain up to 2400 mg Cr/kg, and can release small fractions of the Cr contained in them as Cr(VI), either through dissolution of Cr(VI) minerals or possibly via oxidation of Cr(III) by Mn(III,IV) (hydr)oxides. As a result, Cr(VI) has been detected in

Table 1 Oxidation states and forms of chromium in natural waters.

Oxidation state	Form	Name	Chemical conditions of water under which it is found and pertinent reaction in natural waters
Chromium (III) (trivalent chromium)	$Cr(H_2O)_6^{3+}$	Hexaquochromium(III)	pH < 3.5; strong affinity for negatively charged ions (e.g., phosphate) and colloid surfaces (e.g., living cells and phyllosilicate clays or fulvic and humic acids); green color
	$Cr(H_2O)_5OH^{2+}$	Monohydroxychromium(III)	First hydrolysis product formed at pH > 3.5 upon dilution of or addition of base to solution of Cr(III); green
	$Cr(H_2O)_4(OH)_2^+$	Dihydroxychromium(III)	Second hydrolysis product of Cr(III); may dimerize and polymerize to form high molecular weight cations in planes of octahedron; green
	$Cr(H_2O)_3(OH)_3^0$	Chromium hydroxide	Metastable, uncharged hydrolysis product that precipitates as the sparingly soluble $Cr(OH)_3$
	$Cr(H_2O)_2(OH)_4^-$	Hydroxochromate	Fourth hydrolysis product of Cr(III) that may form at pH > 11; may oxidize to Cr(VI) by O_2
	Cr(III)–organic acid complexes and chelates	For example: chromium citrate, chromium picolinate, chromium fulvate	Soluble complexes and chelates in which water molecules of hydration surrounding $Cr(H_2O)_6^{3+}$ are displaced by carboxylic acid and N-containing ligands; formation is pH and concentration dependent; blue–green–purple colors, depending on ligand binding Cr(III)
Chromium (VI) (hexavalent chromium)	H_2CrO_4	Chromic acid	Fully protonated form of Cr(VI) formed at pH < 1; see Fig. 2 for key Eh values for redox
	$HCrO_4^-$	Bichromate	Form of Cr(VI) that predominates at 1 < pH < 6.4; yellow; see Fig. 2 for key Eh values for redox
	CrO_4^{2-}	Chromate	Form of Cr(VI) that predominates at pH > 6.4; yellow; see Fig. 2 for key Eh values for redox
	$Cr_2O_7^{2-}$	Dichromate	Form of Cr(VI) that predominates at pH < 3 and in concentrated solutions (>1.0 mM); rapidly reverts to $HCrO_4^-$ or CrO_4^{2-} upon dilution or pH change

groundwater (0.05–0.5 mg/L) in arid regions dominated by these alkaline, Cr-rich rocks and soils. A concentration of Cr(VI) of 7.5 mg/L in pH 12.5 groundwater from Jordan is the highest known level that is not due to human influence. Naturally occurring Cr in alkaline, aerobic ocean water exists principally as Cr(VI) at concentrations in the range of 3–7.3 nM (0.16–0.38 μg/L).[12]

The balance of the different forms and the solubilities of Cr(III) and Cr(VI) in natural waters is governed by pH, aeration status, and other environmental conditions (Table 1). Understanding and predicting the oxidation state, solubility, mobility, and bioavailability of Cr in water are further complicated by the fact that Cr(III) can be oxidized (lose three electrons) to form Cr(VI), whereas Cr(VI) can gain three electrons and be reduced to Cr(III).[13,14] Natural variation and human-induced changes in pH and the oxidation–reduction status of soil and water can control the solubility of Cr. As a result, purification of drinking water supplies and treatment of wastewaters contaminated with Cr are possible through chemical and microbiological processes that modify the acidity and the relative abundance of oxidizing and reducing agents for Cr.[15,16]

SOLUBILITY CONTROLS OF CHROMIUM CONCENTRATIONS IN WATER

Most inorganic compounds of Cr(III) are less soluble in water than are those of Cr(VI) because Cr(III) cations have high ionic potentials (charge-to-size ratio) and hydrolyze to form covalent bonds with OH^- ions (Table 1). When three OH^- anions surround the Cr^{3+} cation, it is particularly stable in water as the sparingly soluble compound, $Cr(OH)_3$ (Table 2). Upon aging and dehydration, $Cr(OH)_3$ slowly converts to the more crystalline, less soluble Cr_2O_3.[12] Incorporation of Fe(III) or Fe(II) into solid phases and precipitates containing Cr(III) renders the Cr(III) less soluble, often by a factor of 10^3 in the solubility product (Ksp).[17,18] In the pH range of 5.5–8.0, Cr(III) reaches minimum solubility in water due to this hydrolysis and precipitation reaction, an important process that controls the movement of Cr(III) in soils enriched with industrial wastewaters and solid materials. Under strongly acidic conditions (pH < 4), unhydrolyzed $Cr(H_2O)_6^{3+}$ cations exist in solution, while $Cr(OH)_4^-$ forms under strongly alkaline conditions (pH > 11), particularly in response to adding base to solutions of soluble salts of Cr(III), e.g., $CrCl_3$, $Cr(NO_3)_3$, or $Cr_2(SO_4)_3$.

Other anions besides OH^- coordinate with $Cr(H_2O)_6^{3+}$ and displace water molecules of hydration to form sparingly soluble compounds and soluble chelates (Table 2). In water treatment facilities and in natural waters, phosphate ($H_2PO_4^-$, HPO_4^{2-}, PO_4^{3-}), arsenate ($H_2AsO_4^-$, $HAsO_4^{2-}$) and fluoride (F^-) may form low solubility compounds with Cr(III). Organic complexes of Cr(III) with carboxylic acids (e.g., citric, oxalic, tartaric, fulvic) remain soluble at pH values above which $Cr(OH)_3$ forms. By increasing the solubility of Cr(III) in neutral and alkaline waters, such organic complexes enhance the potential for absorption of Cr(III) by cells. Stable, insoluble complexes of Cr(III) also form with humic acids and other high molecular aggregate weight organic moieties in soils, sediments, wastes, and natural waters.[19]

With the exception of chromium(VI) jarosite (Table 2), Cr(VI) compounds are more soluble over the pH range of natural waters than are those of Cr(III), thereby leading to

Table 2 Solubility in water at pH 7 of selected chromium compounds.

Oxidation state	Compound name	Formula	Approximate solubility (mol Cr/L)
Chromium (III) (trivalent chromium)	Chromium(III) hydroxide	$Cr(OH)_{3\,(am)}$	10^{-12}
	Chromium(III) oxide	$Cr_2O_{3\,(cr)}$	10^{-17}
	Chromite	$FeO{\cdot}Cr_2O_{3\,(cr)}$	10^{-20}
	Chromium chloride	$CrCl_3$	Highly soluble
	Chromium sulfate	$Cr_2(SO_4)_3$	Highly soluble
	Chromium phosphate	$CrPO_4$	10^{-10}
	Chromium fluoride	CrF_3	1.2×10^{-3}
	Chromium arsenate	$CrAsO_4$	10^{-10}
Chromium (VI) (hexavalent chromium)	Potassium chromate	K_2CrO_4	3.2
	Sodium chromate	Na_2CrO_4	5.4
	Calcium chromate	$CaCrO_4$	0.14
	Barium chromate	$BaCrO_4$	1.7×10^{-3}
	"Zinc yellow" pigment	$3ZnCrO_4{\cdot}K_2CrO_4{\cdot}Zn(OH)_2{\cdot}2H_2O$	8.2×10^{-3}
	Strontium chromate	$SrCrO_4$	5.9×10^{-3}
	Lead chromate	$PbCrO_4$	1.8×10^{-6}
	Chromium(VI) jarosite	$KFe_3(CrO_4)_2(OH)_{6\,(cr)}$	10^{-30}

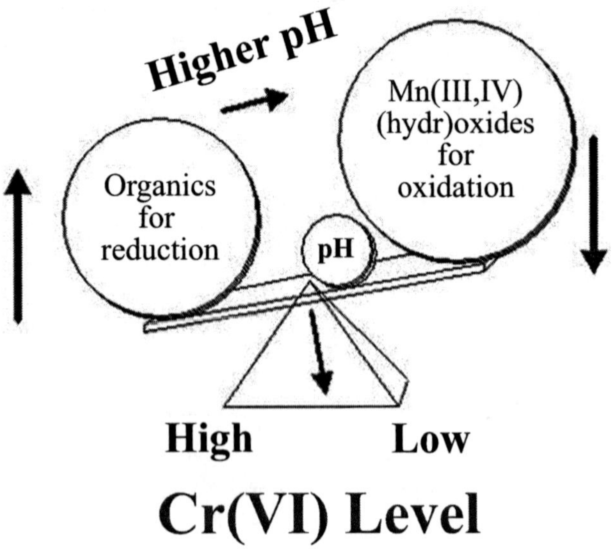
Higher pH
Mn(III,IV) (hydr)oxides for oxidation
Organics for reduction
pH
High
Low
Cr(VI) Level

Fig. 1 Seesaw model depicting a balance of the oxidation of Cr(III) by Mn(III,IV) (hydr)oxides and the reduction of Cr(VI) by organic compounds, with the pH acting as a sliding control (master variable) on the seesaw to set the redox balance for given quantities and reactivities of oxidants and reductants. The equilibrium quantity of Cr(VI) in the water is indicated by the pointing arrow from the fulcrum.

the greater concern about the potential mobility and bioavailability of Cr(VI) than Cr(III) in natural waters. The alkali salts of Cr(VI) are highly soluble, $CaCrO_4$ is moderately soluble, and $PbCrO_4$ and $BaCrO_4$ are only sparingly soluble. In colloidal environments containing aluminosilicate clays and (hydr)oxides of Al(III), Fe(II,III), and Mn(III,IV) (e.g., in soils and sediments), Cr(VI) anions may be adsorbed similarly to SO_4^{2-}. Low pH and high ionic strength promote retention of $HCrO_4^-$ and CrO_4^{2-} on positively charged sites, especially those associated with colloidal surfaces dominated by pH-dependent charge. Such electrostatic adsorption may be reversible, or the sorbed Cr(VI) species may gradually become incorporated into the structure of the mineral surface (chemisorption). Recently precipitated $Cr(OH)_3$ can adsorb Cr(VI) or incorporate Cr(VI) within its structure as it forms, thereby forming a Cr(III)–Cr(VI) compound.[20]

OXIDATION–REDUCTION CHEMISTRY OF CHROMIUM IN NATURAL WATERS

The paradox of the contrasting solubilities and toxicities of Cr(III) and Cr(VI) in natural waters and living systems is complicated by two reduction–oxidation (electron transfer) reactions: Cr(III) can oxidize to Cr(VI) in soils and natural waters, and Cr(VI) can reduce to Cr(III) in the same systems, and at the same time. Understanding the key electron transfer processes (redox) and predicting environmental conditions governing them are central to treatment of drinking water, wastewaters, and contaminated soils, and to predicting the hazard of Cr in natural systems.[21] The metaphor of a seesaw (Fig. 1) is useful in picturing the undulating nature of the changes in Cr speciation in water due to oxidation of Cr(III) and reduction of Cr(VI). A balance for the two redox reactions is achieved in accordance with the quantities and reactivities of reductants and oxidants in the system [e.g., organic matter and Mn(III,IV) (hydr)oxides, as modulated by pH and pe as a master variables].[15,22]

The thermodynamics (energetics predicting the relative stability of reactants and products of a chemical reaction) of interconversions of Cr(III) and Cr(VI) compared to other redox couples can be used to predict the predominance of Cr(III) or Cr(VI) in water supplies (Fig. 2). Certain electron-poor species may act as oxidants (electron acceptors) for Cr(III), especially soluble forms of Cr(III), in the treatment of water supplies or in soils enriched with Cr(III) (Nieboer and Jusys, Fig. 3).[19] Examples are those above the bold line for Cr(VI)–Cr(III) on the Eh–pH diagram: Cl_2, OCl^-, H_2O_2, O_3, and MnOOH. In contrast, electron-rich species may donate electrons to electron-poor Cr(VI) and reduce it to Cr(III): Fe^{2+} [or Fe(0)], H_2S, H_2, ascorbic acid (and organic compounds, generally), and SO_2. Sunlight may affect the kinetics of both oxidation and reduction reactions for Cr, a relevant fact for natural processes in lakes and streams and for treatment technologies for drinking water purification. Depending on pH, temperature, and the concentrations of oxidants and reductants, Cr(VI)-to-Cr(III) ratio in natural waters may be predicted.

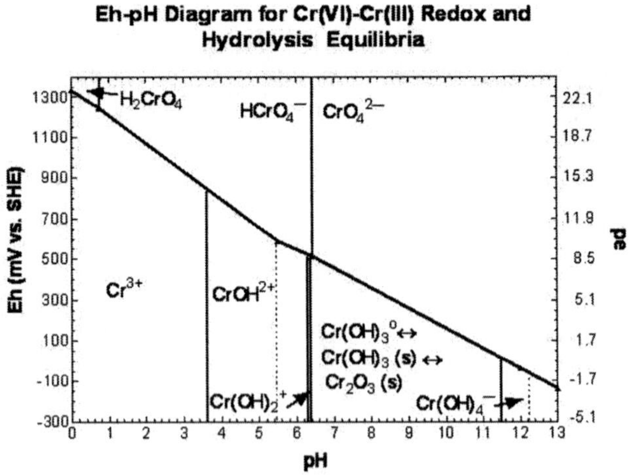
Eh-pH Diagram for Cr(VI)-Cr(III) Redox and Hydrolysis Equilibria
H2CrO4
HCrO4−
CrO4 2−
Cr3+
CrOH2+
Cr(OH)2+
Cr(OH)3° ↔
Cr(OH)3 (s) ↔
Cr2O3 (s)
Cr(OH)4−
Eh (mV vs. SHE)
pe
pH
1300
1100
900
700
500
300
100
-100
-300
22.1
18.7
15.3
11.9
8.5
5.1
1.7
-1.7
-5.1
0 1 2 3 4 5 6 7 8 9 10 11 12 13

Fig. 2 Eh–pH diagram illustrating the stability field defined by Eh (redox potential relative to the standard hydrogen electrode, SHE) and pH for Cr(VI) and Cr(III) at 10^{-4} M total Cr. The vertical dashed lines indicate semiquantitatively the pH range in which $Cr(OH)_3$ is expected to control Cr(III) cation activities in the absence of other ligands besides OH^-.

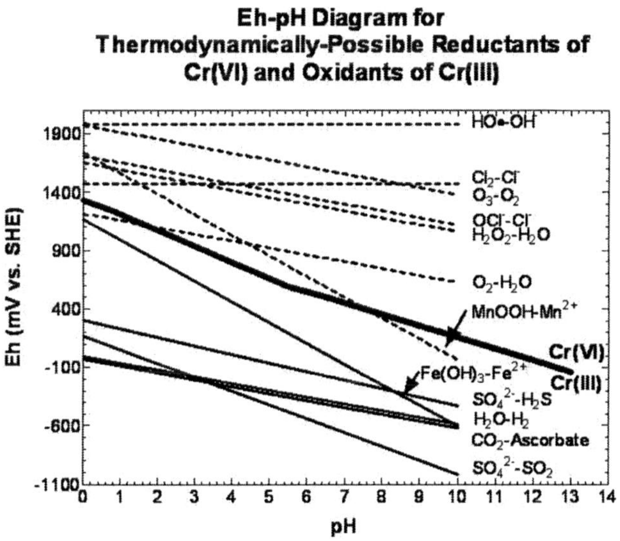
Eh-pH Diagram for
Thermodynamically-Possible Reductants of
Cr(VI) and Oxidants of Cr(III)
HO•-OH
Cl2-Cl⁻
O3-O2
OCl⁻-Cl⁻
H2O2-H2O
O2-H2O
MnOOH-Mn2+
Cr(VI)
Cr(III)
Fe(OH)3-Fe2+
SO4 2--H2S
H2O-H2
CO2-Ascorbate
SO4 2--SO2
Eh (mV vs. SHE)
1900
1400
900
400
-100
-600
-1100
0 1 2 3 4 5 6 7 8 9 10 11 12 13 14
pH

Fig. 3 Eh–pH diagram showing potential oxidants for Cr(III) in natural waters as dashed lines above the bold Cr(VI)–Cr(III) line and potential reductants for Cr(VI) below the line. Each line for an oxidant (first species of the pair) and reductant (second species) combination represents the reduction potential (in mV) at a given pH established by that oxidant–reductant pair (e.g., O_3–O_2). The oxidant member of a pair for a higher line is expected to oxidize the reductant member of the lower line, thereby establishing the area and species between the lines as thermodynamically favored to exist at chemical equilibrium.[23]

CONCLUSION

Hexavalent chromium remaining in industrial by-products may contaminate soils, surface waters, and groundwater that are supplies for domestic uses, irrigation, and industrial processes. Prediction of the likelihood of Cr(III) oxidation and Cr(VI) reduction occurring is important for water treatment and for establishing health-based regulations and allowable limits for Cr(VI) and Cr(III) in water supplies. In agricultural soil–plant–water systems, Cr(VI) added in irrigation water or formed via oxidation of Cr(III) will reduce to Cr(VI) if electron donors (e.g., Fe^{2+}, H_2S, and organic matter) and Eh–pH conditions are sufficiently reducing (Bartlett and James[8] and James and Bartlett[20]; Fig. 2). If not reduced, Cr(VI) may leach from surface soils to subsoils and groundwater. Therefore, prediction of Cr bioavailability and mobility in natural waters must consider redox reactions of this heavy metal.

REFERENCES

1. US Environmental Protection Agency (EPA). *Basic Information about Chromium in Drinking Water*, 2010. Available at http://www.epa.gov/safewater/contaminants/basicinformation/chromium.html#four. (Accessed on May 13, 2012).
2. California Department of Public Health (DPH). *Chromium-6: Timeline for Drinking Water Regulations*, 2009. Available at: http://www.cdph.ca.gov/certlic/drinkingwater/Pages/Chromium6timeline.aspx. (Accessed May 13, 2012).
3. International Agency for Research on Cancer (IARC). Chromium and chromium compounds. IARC Monogr. Eval. Carcinog. Risks Hum. **1990**, *49*, 49–256.
4. Stout, M.D.; Herbert, R.A.; Kissling, G.E.; Collins, B.J.; Travlos, G.S.; Witt, K.L.; Melnick, R.L.; Abdo, K.M.; Malarkey, D.E.; Hooth, M.J. Hexavalent chromium is carcinogenic to F344/N rats and B6C3F1 mice after chronic oral exposure. Environ. Health Perspect. **2009**, *117* (5), 716–722.
5. National Toxicology Program (NTP). Toxicology and carcinogenesis studies of chromium picolinate monohydrate (CAS No. 27882-76-4) in F344/N rats and B6C3F1 mice (feed studies). TR556. **2010**. Research Triangle Park, N.C.
6. Anderson, R.A. Essentiality of chromium in humans. Sci. Total Environ. **1989**, *86*, 75–81.
7. Kimbrough, D.E.; Cohen, Y.; Winer, A.M.; Creelman, L.; Mabuni, C.A. Critical Assessment of Chromium in the Environment. Crit. Rev. Environ. Sci. Technol. **1999**, *29* (1), 1–46.
8. Bartlett, R.; James, B. Behavior of chromium in soils: III. Oxidation. J. Environ. Qual. **1979**, *8*, 31–35.
9. James, B.R.; Petura, J.C.; Vitale, R.J.; Mussoline, G.R. Oxidation–reduction chemistry of chromium: Relevance to the regulation and remediation of chromate-contaminated soils. J. Soil Contam. **1997**, *6* (6), 569–580.
10. Nriagu, J.O. Production and uses of chromium. In *Chromium in the Natural and Human Environments*; Nriagu, J.O., Nieboer, E., Eds.; Wiley-Interscience: New York, 1988; 81–103.
11. Barnhart, J. Chromium chemistry and implications for environmental fate and toxicity. J. Soil Contam. **1997**, *6* (6), 561–568.
12. Ball, J.W.; Nordstrom, D.K. Critical evaluation and selection of standard state thermodynamic properties for chromium metal and its aqueous ions, hydrolysis species, oxides, and hydroxides. J. Chem. Eng. Data **1998**, *43*, 895–918.
13. Bartlett, R.J. Chromium redox mechanisms in soils: Should we worry about Cr(VI)? In *Chromium Environmental Issues*; FrancoAngeli: Milan, 1997; 1–20.
14. James, B.R. Remediation-by-reduction strategies for chromate-contaminated soils. Environ. Geochem. Health **2001**, *23* (3), 175–179.
15. James, B.R. The challenge of remediating chromium-contaminated soils. Environ. Sci. Technol. **1996**, *30*, 248A–251A.
16. Fendorf, S.; Wielenga, B.W.; Hansel, C.M. Chromium transformations in natural environments: The role of biological and abiological processes in chromium(VI) reduction. Int. Geol. Rev. **2000**, *42*, 691–701.
17. Rai, D.; Saas, B.M.; Moore, D.A. Chromium(III) hydrolysis constants and solubility of chromium(III) hydroxide. Inorg. Chem. **1987**, *26*, 345–369.
18. Sass, B.M.; Rai, D. Solubility of amorphous chromium(III)–Iron(III) hydroxide solid solutions. Inorg. Chem. **1987**, *26*, 2228–2232.

19. Nieboer, E.; Jusys, A.A. Biologic chemistry of chromium. In *Chromium in the Natural and Human Environments*; Nriagu, J.O., Nieboer, E., Eds.; Wiley-Interscience: New York, 1988; 21–80.
20. James, B.R.; Bartlett, R.J. Behavior of chromium in soils: VII. Adsorption and reduction of hexavalent forms. J. Environ. Qual. **1983**, *12*, 177–181.
21. James, B.R. Redox phenomena. In *Encyclopedia of Soil Science*, 1st Ed.; Lal, R., Ed.; Marcel Dekker, Inc.: New York, 2002; 1098–1100.
22. James, B.R.; Brose, D.A. Oxidation - Reduction Phenomena. In *Handbook of Soil Sciences: Properties and Processes*. 2nd edn.; Huang, P.M., Li, Y., Summer, M.E., Ed.; CRC Press: Boca Raton, 2012; 14-1–14-24.
23. James, B.R. Chromium. In *Encyclopedia of Water Science*; Stewart, B.A.; Howell, T., Ed., Marcel Dekker: New York, **2003**; 75–79.

Climate Policy: International

Nathan E. Hultman
University of Maryland, Maryland, U.S.A.

Abstract
Climate change is a long-term problem. Policy to address climate change faces the challenge of motivating collective action on global public goods in a world with no single international authority. International agreements can nevertheless aim to (1) reduce greenhouse gas emissions, possibly through an international emissions trading system (ETS) like that outlined in the Kyoto Protocol; (2) develop new low-emissions technology by providing incentives for cooperation on technology research and implementation; (3) provide adaptation assistance to countries and populations least able to cope with expected changes in climate.

INTRODUCTION

Anthropogenic climate change presents one of society's most vexing policy challenges. Because the problem stems largely from the fossil-fueled global economy, the costs of reducing emissions would accrue immediately and are easily quantified. On the other hand, the costs of potential damages are difficult to estimate and will likely be long term. In addition, governing the global atmospheric commons requires a large number of actors to agree on and comply with a mechanism of self-restraint; otherwise, even the relatively virtuous would tire of the rest of the world's free-riding. Moreover, understanding the complex risks and uncertainties of climate change is a challenge even for specialists. Communicating this information accurately and effectively to a marginally interested public is harder still.

Large, long-term changes would be required to reduce the risks of climate change. While the absolute amount of greenhouse gases (GHGs) in the atmosphere affects the degree of climate change, the only variable that society can easily control is the rate of GHG emissions. To simply stabilize the absolute amount of atmospheric GHGs, the emission rate must drop to about one-half to one-third of its present levels. Deciding on what concentrations constitute moderately safe levels is challenging, and requires deriving impact estimates (such as temperature change) from possible GHG stabilization levels,[1] as well as from the projected rate of emissions reduction.

An effective international regime to govern climate change policies must balance climate protection goals with the limited enforcement ability inherent in international law.[2–4] Because the atmosphere is a common resource, protecting the climate is in most people's best interest, but no country will want to burden itself unreasonably burden itself with the excess costs of a climate friendly policy unless it believes that other countries are making equivalent sacrifices. An effective climate regime must therefore minimize free-riding. It must also ensure that the participants are complying with their obligations, which requires systems of monitoring and enforcement. Finally, because of our evolving understanding of the science of climate change and its relationship to human societies, a sound policy must remain flexible enough to incorporate new information as it becomes available, with procedures for regular scientific re-evaluation and regular review of the adequacy of the policy.

CLIMATE POLICY OPTIONS

Addressing climate change requires coordination of domestic and international systems to reduce GHG emissions and assist countries in adapting to climate change. International treaties can set guidelines for action[5,6] that are then implemented via domestic legislation.

Reducing Emissions

Several policy mechanisms can address the free-ridership and overexploitation associated with public goods like the climate. Governments can regulate the common resource directly or stipulate specific technological approaches. This command-and-control approach is relatively simple in that the rules can be set by panels of experts, and it has the appearance of fairness, because everybody must attain the same goals. However, in a diverse economy with differing costs of pollution abatement to firms, it can lead to large imbalances in the cost of compliance.

Encyclopedia of Environmental Management DOI: 10.1081/E-EEM-120041553

Alternatively, a governmentally set price on the externality could allow producers more flexibility. This price can be set directly as a tax (for example, \$10 per ton of carbon dioxide (CO_2)), or indirectly by setting a total emissions limit and allowing entities to trade the rights to emit. These methods—taxes, emissions trading, or a hybrid of the two—can greatly reduce the total cost of compliance with the environmental target. Emissions trading systems come in two forms: in a baseline-and-credit (or permit) system, in which individual projects that result in a reduction of emissions below a pre-agreed baseline are granted credits that can be sold to firms that are not able to meet their reduction obligations. Alternatively, a cap-and-trade system sets an overall cap on emissions and then distributes, free or at auction, the entire amount of emissions allowances out to the producers.

Governments may also implement other policies to address market failures, for example by establishing minimum standards of efficiency or performance, supporting research and development of less-polluting technologies, or even guaranteeing a market for new technologies.

Equity Questions

One contentious question in developing a GHG trading system is how to allocate emissions quotas. Until now, all countries have had free access to the atmosphere; setting limits will inevitably lead to argument about who deserves a bigger slice. Three options for allocating these rights illustrate the policy challenge. The first method, usually called grandfathering, allocates permits according to what various countries or industries have emitted in the recent past. This method causes the minimum economic disruption, but it may also reward inefficient resource use and ignore the benefits that have already accrued to polluters. Alternatively, if one views the atmosphere as a universal resource or a life-support commons, then the quota may be allocated on a per-capita basis so that each person is assigned the same right to using the resource. If, on the other hand, one views the atmosphere as an economic input, the quota might be allocated to each country in proportion to its gross domestic product (GDP). The United States, for example, currently produces 24% of the world's GHG emissions,[7] has 5% of the world's population, and accounts for 21% of the world's GDP.

These simple formulae will likely not be used directly, but they do provide bases for negotiating commitments in the international community. One particularly large divide between developing and developed country positions is how to account for the cumulative GHGs emitted since the beginning of the Industrial Revolution by developed countries.[8] This atmospheric debt represents about 80% of the total anthropogenic GHG contribution, and less-developed countries' contribution will likely not equal that of developed countries until around 2100. These countries argue that richer countries should therefore move first to forestall further emissions. Developed countries often view the situation differently, pointing out that less-developed countries as a group, including China and India, will by 2010 emit GHGs at an annual rate equal to that of the developed world. Indeed China itself will soon be the world's largest GHG emitter, surpassing even the United States.

Whether generated through capture, biological sequestration, or mitigation, trading emissions requires measurement.[9] The most reliable statistics on GHG emissions relate to the burning of fossil fuels. Most countries, especially the industrialized countries that emit the most, keep detailed records of fossil fuel stocks and flows. Therefore, national emissions from fossil fuels are relatively precise and have a high amount of certainty.

Adaptation

Finally, given that some climate change is at this point inevitable, climate policy must encompass not only policies to reduce emissions of GHGs but also policies to enhance the resilience of countries to the expected changes. Often called adaptation measures, such activities include support for diversifying economies away from vulnerable crops, enhancing physical infrastructure, and bolstering institutional capabilities. Unlike policies focusing on GHGs, moreover, the benefits of adaptation accrue relatively quickly as they can immediately reduce suffering from hurricanes or floods regardless of the cause of these events.[10,11] Discussions about adaptation are often linked to questions about liability for climate damages.

EARLY INTERNATIONAL RESPONSE

Although Svante Arrhenius had postulated the existence of the greenhouse effect in 1896,[12] and significant scientific inquiry re-emerged in the 1950s, public concern about anthropogenic climate change was not significant until the late 1980s. Along with other simultaneously emerging global environmental problems like stratospheric ozone depletion and biodiversity loss, climate change moved quickly into the international arena.[13]

The international community's first concrete response was to refer the scientific questions to the World Meteorological Organization (WMO) and the United Nations Environment Programme (UNEP). In 1988, the WMO and UNEP established the Intergovernmental Panel on Climate Change (IPCC), which has since become the major international expert advisory body on climate change.[14] The IPCC divides thousands of experts into three working groups on climate science, impacts of climate change, and human dimensions. It produces comprehensive Assessment Reports every 5–6 years that describe the current state of expert understanding on climate, as well as smaller, targeted reports when they are requested by the international community.

U.N. FRAMEWORK CONVENTION ON CLIMATE CHANGE

The first international treaty to address climate change was the United Nations Framework Convention on Climate Change (UNFCCC), which entered into force in 1994 and has been ratified by 186 countries, including the United States.[15] Having emerged from the 1992 U.N. Conference on Environment and Development (the "Earth Summit"), the UNFCCC sets broad objectives and areas of cooperation for signatories. As the objective, it states that Parties to the Convention should cooperate to "prevent dangerous anthropogenic interference with the climate system." Here, dangerous is not defined explicitly but is required to include ecosystems, food supply, and sustainable economic development.

The UNFCCC identifies several important principles for guiding future treaty agreements. First, it endorses international equity. Second, it states that all signatories share a "common but differentiated responsibility" to address climate change. All countries must therefore participate, but they are allowed to do so in a way that depends on their domestic situation and historic GHG contributions. Third, the UNFCCC instructs the Parties to apply precaution in cases risking "serious or irreversible damage."

The UNFCCC also defined some emissions reduction goals for richer countries. Specifically, it grouped most developed countries into Annex I Parties (Annex I is a designation in the UNFCCC and reproduced in the Kyoto Protocol as Annex B. Annex I countries are: Australia, Austria, Belgium, Bulgaria, Croatia, Czech Republic, Denmark, Estonia, Finland, France, Germany, Hungary, Iceland, Ireland, Italy, Japan, Latvia, Liechtenstein, Lithuania, Luxembourg, Monaco, Netherlands, New Zealand, Norway, Poland, Portugal, Russian Federation, Slovakia, Slovenia, Spain, Sweden, Switzerland, U.K., Ukraine, U.S.A.) and urged them to stabilize their total emissions of GHGs at 1990 levels by the year 2000. These targets were non-binding, and in retrospect, most countries did not meet these initial goals. Finally, and most importantly, the Framework Convention established a system of national emissions reporting and regular meetings of Parties with the goal of creating subsequent, more significant commitments.[16,17]

Interim Negotiations

Subsequent debates focused, therefore, on negotiating a new treaty (called a Protocol) that could enhance international action. Yet, contentious debate arose over whether developing countries would be required to agree to any reductions in return for caps on Annex I emissions.

Several negotiating blocs were solidified during this period and remain active today. The broadest split between developed and developing countries was already evident in the UNFCCC. Within developing countries, the strongest advocates for action emerged in the Alliance of Small Island States (AOSIS), an association of low-lying coastal countries around the world that are extremely vulnerable to inundation due to sea-level rise. On the other hand, the Organization of Petroleum Exporting Countries (OPEC) has been reluctant to endorse any regulation of their primary export, fossil fuel, which when burned creates the GHG CO_2.

Developed countries also have several blocs: The European Union (EU), which functions as a single legal party to the convention, has tended to favor strong action on climate change, whereas the United States, Japan, Canada, New Zealand, and Australia have been more circumspect. Russia was never an enthusiastic advocate of action on climate, but the collapse of its economy in the 1990s means that its emissions decreased considerably, allowing it some flexibility in negotiating targets. A 1995 agreement (the Berlin Mandate) adopted by the Parties to the UNFCCC stated that developing countries should be exempt from any binding commitments, including caps on their emissions, in the first commitment period. The U.S. Senate disagreed and, in 1997, declared they would not ratify any Protocol to the UNFCCC that did not call for concrete targets from developing countries.

THE KYOTO PROTOCOL

After two years of preliminary negotiations, delegates to the UNFCCC met in Kyoto, Japan in 1997 to complete a more significant treaty calling for binding targets and timetables, eventually agreeing on the Kyoto Protocol to the UNFCCC. Maintaining the principle of the Berlin Mandate, delegates rejected language that required participation by developing countries, thus damping U.S. enthusiasm. Nevertheless, the Kyoto Protocol entered into force in 2005, having been ratified by EU countries, Canada, Japan, Russia, and most developing countries. The United States and Australia are currently not Parties to the Protocol. The Kyoto Protocol builds on the UNFCCC with specific and legally binding provisions.

Targets

First, it set legally binding emissions targets for richer countries. These targets oblige Annex I Parties as a group to reduce their emissions to a level 5.2% below 1990 levels by the target period (Fig. 1). This overall average reflects reductions of 8% for the EU, 7% for the United States, and 6% for Canada; as well as increases of 8% for Australia and 10% for Iceland. Russia, whose emissions had dropped significantly between 1990 and 2000 because of economic contraction, was nevertheless awarded a 0% change, effectively providing an effort-free bonus (often called hot air). The target period is defined as 2008–2012, and countries are allowed to take an average of their emissions over this period for demonstrating compliance.

Region	Greenhouse Gas Emissions billion tons CO_2e per year 1990	2000	KP Target
World	21.81	23.63	
Developing Countries	6.91	9.64	
Annex I	14.90	13.99	13.46
European Union	3.33	3.28	2.76
United States	4.98	5.76	4.55
Non-EU, Non-US OECD	1.84	2.20	2.05
Russia and Eastern Europe	4.75	2.74	4.19

Fig. 1 Historical emissions and Kyoto Protocol target emissions. Historical emissions from the United States Department of Energy[19] do not include land-use change emissions; Kyoto targets are based on net emissions reported to the UNFCCC and include land-use change emissions. Country-specific targets are available in the Kyoto Protocol text and from the UNFCCC Secretariat From UNFCC Secretariat Internet Resources.[15] Developing Countries are defined as countries not included in Annex I.
Source: United Nations.[15]

In addition, an individual country's emissions are defined as a weighted sum of emissions of seven major GHGs: CO_2, methane (CH_4), nitrous oxide (N_2O), hydrofluorocarbons (HFCs), perfluorocarbons (PFCs), chlorofluorocarbons (CFCs), and sulfur hexafluoride (SF_6). These gases are weighted according to a quantity called global warming potential (GWP) that accounts for different heat-trapping properties and atmospheric lifetimes of the seven gases. For a 100-year time horizon, example GWPs are: 1.00 for CO_2 (by definition), 21 for CH_4, 296 for N_2O, 100–1000 for a wide variety of halocarbons, and 22,200 for SF_6.[18] The resulting sum is reported in terms of "carbon-dioxide equivalent" or CO_2e. The Protocol allows countries to calculate a net emissions level, which means they can subtract any GHGs sequestered because of, for example, expansion of forested areas.

Implementation

The Protocol encourages countries to achieve their target primarily through domestic activities, usually called policies and measures. These include improved energy efficiency, increased use of renewable energy, and switching to lower-carbon forms of fossil fuels such as natural gas. In addition, the Protocol allows countries to offset emissions if certain domestic activities serve to absorb and sequester CO_2 from the atmosphere, thus reducing their net contribution to climate change. Allowable carbon sinks projects currently include afforestation, reforestation, forest management, cropland management, grazing land management, and revegetation. Conversely, deforestation is a process that must be counted as a cause of emissions.

The Kyoto Protocol also allows countries to obtain credits from other countries. In particular, it established three market-based mechanisms to provide states with flexibility in meeting their binding emissions reduction targets: emissions trading (ET), joint implementation (JI), and the clean development mechanism (CDM). Despite the different names, these three mechanisms are actually all forms of emissions trading—they create ways to reduce the overall cost of reaching the targets outlined above by allowing lower-cost reductions to be bought and sold on the market. All traded units are denoted in tons of CO_2e.

Emissions Trading (sometimes called Allowance Trading) is a cap-and-trade system under which the Annex I parties are assigned a maximum level of emissions (see Fig. 1), known as their assigned amount. They may trade these rights to emit through a UNFCCC registry. Only developed country Parties may participate in ET. Units of ET are termed *assigned amount units* (AAUs).

Joint Implementation is a baseline-credit system that allows trading of credits arising from projects coordinated between fully developed countries and countries in eastern Europe with economies in transition. This is a so-called project-based system, under which reductions below an independently certified baseline can be sold into the market. Joint implementation units are termed emissions reduction units (ERUs).

The CDM is another baseline-credit system that allows trading of credits arising from projects in developing countries. Another project-based system, the CDM will allow only projects that contribute both to sustainable development and to climate protection. Furthermore, they must provide benefits that would not have occurred in the absence of the CDM (so-called additionality). Post-Kyoto negotiations determined that acceptable projects include those that employ renewable energy, fossil-fuel repowering, small-scale hydroelectric power, and sinks; some projects (e.g., those under 15 MW_e) are also deemed to be "small-scale" and enjoy a streamlined approval process. Projects are subject to a process of public participation. Final acceptance of project proposals rests with the CDM Executive Board, which also approves methodologies and designates operational entities—NGOs, auditors, and other private developers—that implement and verify projects. A levy on each project will fund activities that help poor countries adapt to a changing climate. Clean development mechanism units are called certified emissions reductions (CERs).

Kyoto rules allow AAUs, ERUs, and CERs to be fungible or substitutable for each other. A final unit specific to sinks, called a removal unit (RMU), will be partially separate from this pool since it cannot be banked, or held from one commitment period to the next.

Other Provisions

The allowance assignments and flexibility mechanisms are the most significant elements of the Kyoto Protocol and its associated rules. Other noteworthy commitments include minimizing impacts on developing countries—primarily through funding and technology transfer—and establishing expert teams to develop monitoring, accounting, reporting, and review procedures.

The UNFCCC, Kyoto Protocol, and associated agreements establish three multilateral funds to assist poorer countries. The Adaptation Fund, financed through a levy on CDM transactions, is designed to help countries bolster their institutional and infrastructural capacity to manage changes in climate and damages from weather events. The Climate Change Fund focuses on technology transfer and economic diversification in the energy, transportation, industry, agricultural, forestry, and waste management sectors. Finally, the Least Developed Countries Fund exists to provide additional support in adaptation activities for the very poorest countries. The latter two funds are financed by voluntary contributions.

EUROPEAN UNION EMISSIONS TRADING SYSTEM

In 2005, the EU implemented what is to date the largest operational emissions trading system (ETS) for GHGs.[20,21] The EU–ETS is a cap-and-trade system for all 25 EU member countries, and it covers approximately 12,000 installations in six sectors (electric power; oil refining; coke ovens, metal ore, and steel; cement; glass and ceramics; paper and pulp). The plan regulates only CO_2 emissions until 2007, but thereafter other GHGs will be included. About one-half of the EU's CO_2 emissions will be regulated under the EU–ETS during this first 2005–2007 phase. The initial allocation of credits is based on individual countries' plans; countries are allowed to auction up to 5% of allowances in the first phase and 10% thereafter.

Notably, the EU–ETS contains more rigorous enforcement provisions than Kyoto. Compared to international law, the EU has far greater leverage to enforce legal provisions, and the EU–ETS imposes a steep fine (€40/tCO_2) for non-compliance. The EU–ETS replaces some national-level policies to control GHGs, notably the United Kingdom's pioneering ETS and other voluntary programs.[22] Accordingly, some facilities that had previously taken action to comply with pre-existing national laws are allowed limited exceptions to the EU–ETS.

Through a linking directive,[23] credits generated through Kyoto Protocol CDM or JI projects may be used to fulfill obligations under the EU–ETS, thereby providing an important market for these offsets. However, because of European concerns about the possible negative consequences of biological carbon sequestration (sinks) projects, CDM or JI credits generated within these categories are ineligible for the EU–ETS. The EU–ETS will thus form, by far, the largest trading program in the world and will likely set the standards for subsequent programs in other countries.

VOLUNTARY AND REGIONAL PROGRAMS

Trading programs outside national government legislation have also emerged. British Petroleum was the earliest major corporate adopter of an internal GHG trading system and, subsequently, has had a large consultative role in drafting both the Kyoto and U.K. emissions trading rules. Although the United States has declared its intention to ignore Kyoto,[24] many large American corporations have also adopted internal targets. The Chicago Climate Exchange has organized voluntary commitments from companies in the hope of establishing a position as the dominant American exchange. Yet voluntary programs, whether they derive from corporate or governmental initiatives, are ultimately constrained: often the most egregious polluters choose simply not to volunteer, and those firms that do participate may still not reduce emissions to socially desirable levels.

Many U.S. states, such as California and Oregon, have also passed or are considering legislation that would curb emissions directly or indirectly. While independent state initiatives are valuable in providing domestic innovation for the United States,[25] they are unlikely to add up to a significant reduction in global emissions. In addition, many of the companies most vulnerable to GHG regulation are asking for some guidance on what they can expect from regulators, and a state-by-state patchwork can never replace federal legislation for regulatory certainty.

CONCLUSION

International climate policy faces a period of uncertainty, innovation, and evolution over the next decade. The Kyoto Protocol remains the primary international agreement for addressing climate change globally, despite its imperfections and the continued absence of the United States and Australia. Yet, since Kyoto expires in 2012, attention has turned to negotiating a subsequent agreement that could re-engage the United States, involve China, India, and other developing countries more directly, and address concerns about compliance and enforcement.[26–28] Given the difficulties in forging an immediate, broad international consensus, the EU–ETS will likely foster the most institutional innovation and GHG market development in the near term.

The most likely interim solution, therefore, will consist of multiple, overlapping regimes that link domestic-level emissions reductions into one or more international markets.[29] In this model, the United States could, for example, institute a unilateral domestic program that addresses emissions and then allow Kyoto credits to be admissible for compliance as the EU has done. Additional agreements governing, for example, technology standards or adaptation policy may also emerge. From the perspective of energy, this evolving climate change policy will impose a carbon constraint on energy use,[30,31] most likely through a non-zero cost for GHG emissions.

ACKNOWLEDGMENTS

The author gratefully thanks the two anonymous reviewers for their thorough, detailed, and constructive comments on a previous draft.

REFERENCES

1. Mastrandrea, M.D.; Schneider, S.H. Probabilistic integrated assessment of "dangerous" climate change. Science **2004**, *304* (5670), 571–575.
2. Barrett, S. Environment and Statecraft: The Strategy of *Environmental Treaty-Making*; Oxford University Press: New York, 2003; 360.
3. Brown Weiss, E.; Jacobson, H.K. Getting countries to comply with international agreements. Environment **1999**, *41* (6), 16–20, see also 37–45.
4. Bodansky, D. International law and the design of a climate change regime. In *International Relations and Global Climate Change*; Luterbacher, U., Sprinz, D.F., Eds., MIT Press: Cambridge, Mass, 2001; 201–220.
5. Aldy, J.E.; Barrett, S.; Stavins, R.N. Thirteen plus one: a comparison of global climate policy architectures. Climate Policy **2003**, *3* (4), 373–397.
6. Sandalow, D.B.; Bowles, I.A. Fundamentals of treatymaking on climate change. Science **2001**, *292* (5523), 1839–1840.
7. United States Energy Information Administration. *Emissions of Greenhouse Gases in the United States 2003*, United States Department of Energy: Washington, DC, 2004; 108.
8. Smith, K.R. Allocating responsibility for global warming: the natural debt index. Ambio **1991**, *20* (2), 95–96.
9. Vine, E.; Kats, G.; Sathaye, J.; Joshi, H. International greenhouse gas trading programs: a discussion of measurement and accounting issues. Energy Policy **2003**, *31* (3), 211–224.
10. Pielke, R.A. Rethinking the role of adaptation in climate policy. Global Environ. Change **1998**, *8* (2), 159–170.
11. Wilbanks, T.J.; Kane, S.M.; Leiby, P.N.; Perlack, R.D. Possible responses to global climate change: integrating mitigation and adaptation. Environment **2003**, *45* (5), 28.
12. Rodhe, H.; Charlson, R.; Crawford, E. Svante Arrhenius and the greenhouse effect. Ambio **1997**, *26* (1), 2–5.
13. Bodansky, D. The history of the global climate change regime. In *International Relations and Global Climate Change*; Luterbacher, U., Sprinz, D.F., Eds., MIT Press: Cambridge, Mass, 2001; 23–40.
14. Intergovernmental Panel on Climate Change Secretariat. Sixteen Years of Scientific Assessment in Support of the Climate Convention: Geneva, 2004; 14.
15. United Nations Framework Convention on Climate Change Secretariat. UNFCCC Secretariat Internet Resources: Bonn, Germany, 2005.
16. Molitor, M.R. The United Nations climate change agreements. In *The Global Environment: Institutions, Law, and Policy*; Vig, N.J., Axelrod, R.S., Eds., CQ Press: Washington, DC, 1999; 135–210.
17. Schneider, S.H. Kyoto protocol: the unfinished agenda–An editorial essay. Climatic Change **1998**, *39* (1), 1–21.
18. Intergovernmental panel on climate change In *Climate Change 2001: The Scientific Basis*; Houghton, J.T., Ding, Y., Griggs, D.J., Noguer, M., van der Linden, P.J., Dai, X., Maskell, K., Johnson, C.A., Eds., Cambridge University Press: Cambridge, U.K., 2001; 881.
19. United States Energy Information Administration *World Carbon Dioxide Emissions from the Consumption and Flaring of Fossil Fuels, 1980–2000*; United States Department of Energy: Washington, DC, 2002.
20. European Parliament. Directive 2003/87/EC Establishing a Scheme for GHG Emission Allowance Trading within the EU: Brussels, 2003.
21. Pew Center on Global Climate Change *The European Union Emissions Trading Scheme (EU–ETS): Insights and Opportunities*; The Pew Center on Global Climate Change: Washington, DC, 2005; 20.
22. Lecocq, F. *State and Trends of the Carbon Market*; Development Economics Research Group, World Bank: Washington, DC, 2004; 31.
23. European Parliament. The Kyoto Protocol's Project Mechanisms: Amendment of Directive 2003/87/EC: Brussels, 2004.
24. Victor, D.G. *Climate Change: Debating America's Policy Options*; Council on Foreign Relations: New York, 2004; 165.
25. Rabe, B.G. *Statehouse and Greenhouse: The Emerging Politics of American Climate Change Policy*; Brookings Institution Press: Washington, DC, 2004; 212.
26. Stavins, R.N. Forging a more effective global climate treaty. Environment 2004, 23–30.
27. Beyond Kyoto, B.J. Foreign Affairs **2004**, *83* (4), 20–32.
28. Bodansky, D.; Chou, S.; Jorge-Tresolini, C. *International Climate Efforts Beyond 2012: A Survey of Approaches*; Pew Center on Global Climate Change: Washington, DC, 2004; 62.
29. Stewart, R.B.; Wiener, J.B. Practical climate change policy. Issues Sci. Technol. **2003**, *20* (2), 71–78.
30. Pacala, S.; Socolow, R. Stabilization wedges: solving the climate problem for the next 50 years with current technologies. Science **2004**, *305* (5686), 968.
31. Wirth, T.E.; Gray, C.B.; Podesta, J.D. The future of energy policy. Foreign Affairs **2003**, 132–155.

Coastal Water: Pollution

Piotr Szefer
Department of Food Sciences, Medical University of Gdansk, Gdansk, Poland

Abstract
As a result of human activities, different types of pollutants enter marine ecosystems, especially the coastal and estuarine areas. Persistent pollutants have become a global concern because of their accumulative and biomagnification abilities, along with successive levels of the aquatic trophic chain. Heavy metals, metalloids, and organic pollutants pose a huge health hazard to marine organisms and humans. In contrast to organic pollutants, the former are present at a natural background level, particularly in the abiotic and biotic components of marine ecosystems. Human industrial and agricultural activities result in the elevation of the natural content of chemical elements that are sometimes attained at significantly higher levels. As for nutrients, their elevated levels in coastal waters can cause eutrophication and proliferation of toxic algal blooms. The recently observed increase in tanker operations and oil use as well as marine tanker catastrophes has been responsible for the presence of excessively large amounts of oil spillage in coastal and marine ecosystems. Marine debris, especially plastics, is one of the most pervasive pollution problems. In spite of their knowledge on environmental pollutant toxicity, most developing countries still contribute to the significant amount of pollution in inshore waters.

INTRODUCTION

The anthropogenic activity of man in coastal regions and even in areas located far inland is responsible for generating a huge amount of pollutants that are transported to marine ecosystems directly or by means of coastal watersheds, rivers, and precipitation from air.

Therefore, water pollution is a key global problem that has threatened marine organisms, including edible ones, and marine life in general.

There are two types of water pollutants, i.e., point source and nonpoint source.

The point source type is attributable to harmful contaminants released directly to the aquatic environment while nonpoint source delivers pollutants indirectly to the site of their approach.

The former one is a single, well-localized source, e.g., directly discharging sewage or industrial waste to the sea, whereas in the latter, the source of pollution is not well defined. Examples of such nonpoint source are agricultural runoff and windblown debris. Nonpoint sources are considered to be much more difficult to control and regulate as compared to point source pollutants.

The following are the classifications of other sources of pollution in coastal waters:

- Discharge of sewage and industrial waste
- Exploration and exploitation of the seabed
- Accidental pollution by oil and other pollutants from the land via air and other routes

Among these sources of pollutants dominate those connected with the discharge of municipal sewage and industrial wastes into coastal or estuarine regions, especially in the case of their inadequate treatment to remove persistent and harmful compounds. However, natural (and not anthropogenic) phenomena (e.g., volcanoes, storms, algae blooms, earthquakes, and geysers) could also be responsible for polluting aquatic systems. Their influence causes crucial changes in the ecological status of aquatic ecosystems.

The following are factors that determine the severity of pollutants:[1]

- Chemical structure
- Concentration
- Persistence

Independent on their sources in the water, pollutants may be classified as those for which the environment has some or little/no absorptive capacity. They are named "stock pollutants" (e.g., persistent synthetic chemicals, non-biodegradable plastics, and heavy metals).[2]

Most marine pollutants have land origin. They are often transported via rivers from agricultural sources and also via atmospheric trajectory. A lot of pollutants may be taken up by various compartments (biotic and abiotic) of aquatic environments; some of them could be biomagnified along the successive members of the food chain. A good example of having such ability to biomagnify is mercury. Such biomagnification could have negative effect on the

Encyclopedia of Environmental Management DOI: 10.1081/E-EEM-120046080

quality of the water and hence on the health of the plants, animals, and humans whose lives depend on the quality of aquatic environments.

It should be emphasized that coastal areas are generally damaged from pollution, resulting in considerable impact on commercial coastal and marine fisheries. The pollution problem is very complex because of its interactions, interconnectedness, and uncertainty.[3–5] Pollutants, independent of their origin (e.g., air, water, land), enter the ocean, whether earlier or later.[3] Spatial distribution patterns of contamination concentrations exhibit a trend of their increase during transition from the south to the northern part of all oceans, i.e., to areas neighboring with both industrial centers and concentration of main pollution sources.[6]

The following are considered major pollutants:[6–11] fertilizers, pesticides, and agrochemicals; domestic and municipal wastes and sewage sludge; oil and ship pollution; trace elements; radionuclides; organic compounds; plastics; sediments; eutrophication and algal bloom; biological pollution; noise pollution; and light.

HEAVY METALS AND METALLOIDS

In contrast to organic pollutants, e.g., Polichlorinated Biphenyls (PCBs), heavy metals occur as natural elements of particular abiotic and biotic components of continental and aquatic ecosystems. They are present at a natural background level in rocks, soils, sediments, water, and biota. Human industrial and agricultural activities result in the elevation of this natural level to sometimes significantly higher values.

Typical metal concentrations are generally observed in open waters of marine ecosystems, although these remote regions can be affected by elevated levels of trace elements of anthropogenic or volcanic origin. For instance, the atmosphere affects the oceans and continental matter facilitating metal fluxes between these two compartments. Therefore, the atmosphere is a very important component as it makes it possible to transport metals that are natural in origin into distances far from their sources, e.g., from areas closest to forest fires as well as windblown dust, vegetation, and sea aerosols.[12,13] These sources are responsible for contributing metals to the lower troposphere, and therefore, their transport is associated to local and regional wind patterns, in contrast to specific sources such as volcano eruption, which can be responsible for injecting particulate metals not only into the troposphere but also into the stratosphere. In the latter case, particulate metals can be transported long distances under the appropriate circumstances.[13] Another example of long-distance transport of metals from their sources is dust carried from the Sahara Desert resulting in deposition of Fe, Mn, Al, and trace elements across the Mediterranean Sea, Atlantic Ocean, and Caribbean Sea.[13,14] Therefore, wind-driven dust transports particulate metals far offshore, in contrast to riverine flux carrying greater pole of mineral components from continental material to the coastal waters. These metals are promptly deposited to bottom sediments or taken up by biota, especially by phytoplankton in the surface waters, and transferred next to the food chain, recycles or settled to the bottom.[13]

The mass of metal of anthropogenic origin emitted to ecosystems is now equal to or greater than the mass introduced to the natural cycle on a global scale.[13] Some metals, e.g., Pb, Hg, and Cd, owing to their great toxicity, pose a high health risk; therefore, great attention has been paid to estimate their inputs to marine ecosystems, particularly to coastal waters. For instance, the largest masses of Pb are emitted to the atmosphere during processing of the metal (smelters) or from combustion-related sources like motor vehicles. Lead from motor vehicle exhaust was identified not only in the atmosphere but also in remote surface waters as well as in remote terrestrial areas. A ban on Pb usage in vehicle fuel has resulted in the effective reduction of metal inputs to the ecosystem since the 1970s. The decline in atmospheric Pb detected over a time scale of 10 years (from 1979 to 1989) because of the reduction of leaded gasoline in western European countries should be reflected in decreasing Pb levels in surficial water and biota. In fact, the temporal negative trend for cod in the Baltic Sea seems to support this argumentation.[15]

According to Mason et al.,[16] preindustrial fluxes and reservoirs of Hg pose one-third of its fluxes in the civilization era. It is suggested that modern emission of Hg to the atmosphere increased considerably, even 4 to 5 times, due to the human activities. The extremely elevated levels of Hg are frequently associated with Hg mining.

As has been reported, ca. 300 metric tons of dissolved Cd annually enter the oceans from rivers while ca. 400 to 700 metric tons of dissolved and particulate Cd are annually deposited to the oceans from the atmosphere.[17,18] It is estimated that human activities have contributed to increased Cd inputs to the ocean by 60% in the 1980s. It is also found that the higher proportion of land deposition of Cd is associated to the rapid removal of this metal from the atmosphere near inputs of air pollution. A substantial pole of Cd transported by rivers is deposited in estuaries and continental margins of oceans. It is found that increased concentrations of Cd occur locally, reflecting its mosaic contamination, especially near mining and industrial point sources—not managed.

A significant fraction of Zn entering the oceans is derived from atmospheric deposition.[18] Soils and sediments are main natural reservoirs of Zn. Zinc, like Cd, is not distributed evenly across the Earth's surface, since its increased concentrations occur locally in the vicinity of increased inputs, i.e., specific points of source inputs.

The concentrations of many trace elements, e.g., Pb, Cd, Hg, Cu, Zn, Se, and As in coastal and estuarine waters, especially in highly industrialized areas, are generally significantly greater than those in open oceanic waters.[13,18–20] The waters of harbors and marinas around the world contain variable concentrations of tributyltin (TBT),[13] but its extremely high levels may be characteristic for marinas in southwest England.[21]

Human industrial and agricultural activities affect inputs of several metals to reservoir/reservoirs and hence increase their concentrations, even sometimes very dramatically, above natural background levels. There are numerous examples of worldwide events leading to serious contamination of coastal waters by heavy metals and metalloids.[15] Therefore, relationships between man and ecosystem health have been explored, especially in relation to perturbed ecosystems. This includes the pollution status of coastal regions harmed by some catastrophes, large-scale pollution, environmental accidents and episodes, etc. High-risk groups consume extremely high quantities of trace metals present in specific assortment of seafood or offal and it concerns seaside populations. Marine fish and shellfish may be the dominant dietary sources of Hg for local populations.[22–25] A notable example of aquatic pollution by a toxic metal is the Minamata incident, commencing in 1953 and resulting in fish, shellfish, and bird mortalities in waters of the partially landlocked Minamata Bay.[15] Dogs, pigs, and especially cats were also victims of this incident. By the end of 1974, 107 of 798 officially verified patients had died. According to Tomiyasu et al.,[26] the sediments from the Minamata Bay contained levels of Hg that highly exceeded its background level. Among other incidents resulting in the release of Hg compounds to the environment, the most significant ones happened in the 1960s and early 1970s in Sweden, Canada and the United States, northern Iraq, Guatemala, Pakistan, and Ghana.[27–31] MeHg in aquatic ecosystems, especially those that bioaccumulated in fish, is a major public health problem all over the world.[32] Its levels in the hair of fishermen represent the critical group for dietary exposure. For instance, the concentrations of Hg (total and MeHg) in the hair of fishermen from Kuwait were 2 times higher than the "normal" level according to the World Health Organization.[33] Biomass burning in tropical forests also seems to have contributed significantly to the Hg input to the atmosphere. Approximately 31% of the Hg concentrations were associated with the vegetation fire component.[34] It is postulated (based on long-range air mass trajectory analyses) that Hg occurs in the Amazon basin over two main routes: to the South Atlantic and to the Tropical Pacific, over the Andes.[34]

Global emission flux estimates exhibited that biomass burning could be major contributor of heavy metals and black carbon to the atmosphere.[15] It is estimated that savannah and tropical forest biomass burning could emit huge amounts of Cu, Zn, and black carbon to the atmosphere, corresponding to 2%, 3%, and 12%, respectively, of the global level of these elements.[35]

The toxic effects of TBT were first indicated towards the end of the 1970s in Arcachon Bay, France, as the "TBT problem."[36,37] The release of TBT (from antifouling paints) to the area resulted in shell abnormalities and reduced growth and settlement in oysters, *Crassostrea gigas*, cultured in the vicinity of marinas. In much polluted water, oyster production was severely affected by the absence of reproduction, resulting in a strong decline in the marketable value of the remaining stock.[37] Imposex, i.e., the development of male sexual characteristics in female marine mesogastropods and neogastropods caused by TBT pollution, is a widespread phenomenon concerning several coastal species and, more recently, offshore species as well.[23,38,39] Subsequent regulations in 1990 that prohibit the use of TBT-based antifoulants on vessels less than 25 m in length have been highly effective in reducing TBT levels in coastal waters. However, larger vessels have continued the release of TBT, and major harbors remained pollution hot spots.[40] The Organotin Antifouling Paint Control Act restricted in the United States the use of TBT paints to vessels greater than 25 m in length.[41] The voluntary stoppage of TBT production in January 2001 by major U.S. and European manufacturers resulted in the decline of its presence in marine biota, but TBT paint is still being used in most Asian countries. The International Maritime Organization (IMO) imposed an international ban for the use of organotin compounds in antifouling treatments on ships longer than 25 m. The target is to prohibit their application starting 2003 and to require the removal of TBT from ships' hulls by January 1, 2008.[41,42]

The extensive flooding, especially occurred in river area of former or operating metalliferrous mining can be responsible for wide spreading of heavy metals and metalloids far distance from pollution source. An example of such environmental events is the flooding of the Severn catchment (United Kingdom) in January 1998.[43]

RADIONUCLIDES

Physicochemical aspects and applications of radioactivity in the environment were extensively presented in a book by Valcovic.[8] There are numerous papers reporting on problems resulting from radionuclide pollution and their sources in different ecosystems.[15,44] One of the first low-level emissions of radioactivity took place in the Hanford reactors (Columbia River, Washington, United States), which released radionuclides (mainly ^{60}Co, ^{51}Cr, and ^{65}Zn) to its environs from 1940 to 1971.[28] The nuclear reactors in Cumbria (northwest England) have also been responsible for discharging quantities of radioisotopes, i.e., ^{144}Ce, ^{137}Cs, ^{95}Nb, ^{106}Ru, and ^{95}Zr, to the marine environment. Although these emissions have been diminished recently, discharges from nuclear power stations such as Sellafield (formerly named Windscale) could still be identified, even in distances far away from their source.[45,46] Significant quantities of artificial radionuclides (^{137}Cs, ^{134}Cs, ^{90}Sr, ^{99}Tc) have been transported to the North Atlantic and Arctic from Sellafield, together with measurable amounts of Pu and Am.[46–48] The nuclear reprocessing plant at La Hague in France emitted ^{137}Cs and $^{239+240}Pu$ to the environment, although this plant mainly supplies ^{129}I and ^{125}Sb.[28,48,49] Besides the expected emission of radionuclides from nuclear and reprocessing facilities, significant quantities of radioisotopes contaminate aquatic and terrestrial environments from either nuclear weapons testing or nuclear reactor accidents.[15] For instance, the thermonuclear detonation that took place in 1954 at Bikini Atoll resulted in the contamination of a large area of the Marshall Islands. A number of atmospheric tests (520 in total) were mostly carried out in the Northern Hemisphere, including eight underwater tests, with a total yield of 542 Mt. Moreover, there have been a total of 1352 underground tests with a total yield of 90 Mt.[50]

A number of nuclear incidents were concernedly noted, including those affecting the crew of the Japanese fishing vessel "Fukuru Maru".[28] Plutonium released from the Kyshtym accident in the Urals has been much probably detected in deep basins of the Arctic Ocean.[51] In 1968, an aircraft from the U.S. Strategic Air Command crashed near the Thule Airbase in NW Greenland, releasing to the marine environment ca. 1 TBq $^{239+240}Pu$.[52] As a consequence, marine sediments as well as benthic organisms, i.e., bivalves, shrimps, and sea stars, have been contaminated by Pu, although their levels rapidly decreased.[53] A number of American and Russian nuclear submarines have been lost in the world's oceans. For instance, the Soviet Komsomolets submarine sank at a depth of 1700 m at Bear Island in the eastern part of the Norwegian Sea. The estimated radioactivity in the wreck was 2.8 PBq ^{90}Sr and 3 PBq ^{137}Cs.[54] Some nuclear powered satellites can incidentally be sources of radioactivity. They can burn up in the upper atmosphere, resulting in the contamination of the ocean. For instance, such an accident happened in 1964 when a SNAP-9A nuclear power generator containing 0.6 PBq ^{238}Pu aboard a U.S. satellite re-entered the atmosphere in the Southern Hemisphere. The estimated $^{238}Pu/^{239+240}Pu$ ratio in this region was higher than that in the ocean water from the Northern Hemisphere.[55,56]

Sea dumping was carried out since the late 1940s to mid-1960s mainly by the United States in the Atlantic Ocean and Pacific Ocean as well as by the United Kingdom in the Northeast Atlantic Ocean.[56] In 1967, an international operation was initiated by the former European Nuclear Energy Agency that contributed to the deposition of ca. 0.3 PBq solid waste at a depth of 5 km in the eastern Atlantic Ocean. Other international operations were continued until 1982 when ca. 0.7 PBq α activity, 42 PBq β activity, and 15 PBq tritium activity have been dumped in the North Atlantic.[57] It has been assessed that the radiological impact of the NEA (former European Nuclear Energy Agency) dumping activities resulted in some releases of Pu from the dumped waste.[15] This source would be responsible for only a part of the total body burden radioactivity in local benthic organisms, e.g., sea cucumbers; the remainder has been attributed to fallout.[58] According to Consortium for Risk Evaluation with Stakeholder Participation (CRESP) evaluation, the individual dose of a critical group consuming seafood such as molluscs from the Antarctic Ocean was estimated to be 0.1 $\mu Sv\ yr^{-1}$, in effect labeling ^{239}Pu and ^{241}Am as critical radionuclides. The indefinite collective dose to the world's population coming from sea dumping was estimated at 40,000 manSv with predominance of ^{14}C and ^{239}Pu.[56,58]

U.S. weapons production facilities account for a large fraction of radiocaesium discharges during the 1950s.[15] A striking incident occurred at Chernobyl in the former USSR where an explosion of a reactor core of the nuclear plant took place in April 1986. The Baltic countries and a large part of central and western Europe have been contaminated principally by ^{131}I, ^{134}Cs, and ^{137}Cs.[28,59] It is found that a significant part of the activity fell over the European marginal seas from which the Baltic Sea was the most affected by contamination.[56,60] It has been mainly responsible for additional inflow of the radioactive contaminants to the Northeast Atlantic Ocean.[56] Due to the Chernobyl accident, significant levels of ^{137}Cs were also found in the Black Sea. The outflow from this Sea has been the major source of additional ^{137}Cs in the Mediterranean Sea.[56] In the summer of 1987, the Chernobyl-derived ^{137}Cs was also detected in surficial waters of the Greenland Sea, Norwegian Sea, and Barents Sea as well as in the west coast of Norway and the Faroe Islands. According to Aarkrog,[56] the total Chernobyl ^{137}Cs input to the world's oceans was relatively significantly smaller than that estimated for nuclear weapons fallout because of the tropospheric nature of this accident that has contaminated the surrounding European continental areas.[15]

After the 2011 Tōhoku earthquake and tsunami, the radiation effects from the Fukushima Daiichi nuclear disaster resulted in the release of radioactive isotopes from the crippled Fukushima Daiichi Nuclear Power Plant. The total amount of ^{131}I and ^{137}Cs released into the atmosphere has been estimated to exceed 10% of the emissions from the Chernobyl disaster. Large amounts of radioactive isotopes have also been released into the Pacific Ocean.[61]

ORGANIC COMPOUNDS

The high lipophilicity of many persistent organic pollutants (POPs) enhances their bioconcentration/biomagnification, resulting in potential health hazards on predators at higher trophic levels, including humans. These xenobiotics occur widely in coastal waters and oceans from the Arctic to the Antarctic and from intertidal to abyssal. It should be emphasized that most of these compounds exist at a very low concentration level, and hence, their threat to marine biota is still not well recognized. However, it is well known that exposure to extremely low levels of halogenated hydrocarbons, e.g., PCBs, Dichlorodiphenyltrichloroethane (DDT), and TBT, may disrupt the normal metabolism of sex hormones in fish, birds, and marine mammals. Moreover, sublethal effects of these organic chemicals over long-term exposure may result in serious damage to marine populations since some of these POPs may impair reproduction functions of organisms while others may show carcinogenic, mutagenic, or teratogenic activity.[6] Some of the effects of these compounds have been reported by Goldberg.[62] For instance, very low levels of TBT (as endocrine disruptor) cause a significant disruption in sex hormone metabolism, resulting in the malformation of oviducts and suppression of oogenesis in female whelks, e.g., *Nucella lapillus*.[63] As a consequence, sex imbalance leads to species decline if not species extinction in some field populations.[64] Butyltins may be responsible for mass mortality events of bottlenose dolphins in Florida through suppression of the immune system.[65] Trace environmental levels of other compounds like chlorinated hydrocarbons, organophosphates, and diethylstilbestrol may be responsible for significant endocrine disruption and reproductive failure in different groups of animals, i.e., marine invertebrates, fish, birds, reptiles, and mammals.[6] For instance, high levels of DDT, PCBs, and organochlorines in the Baltic Sea significantly reduced the hatching rates of the fish-eating white-tailed eagle (*Haliaeetus albicilla*) in the 1960s and the 1970s.[66] Another example of the toxic impact of POPs is organochlorine contamination in different cetacean species dependent upon their diet, sex, age, and behavior. Many of these compounds, as endocrine disruptors, reduce reproduction and/or suppress immune function. DDT and PCBs are known as compounds affecting steroid reproductive hormones and can increase mammalian vulnerability to bacterial and viral diseases. Jepson et al.[67] reported a statistically significant relationship between elevated PCB level and infectious disease mortality of harbor porpoises (*Phocoena phocoena*).

The assessment and monitoring of existing and emerging chemicals in the European marine and coastal environment have been overviewed based on numerous, most recent worldwide references.[5] From this report, the extensive range of chemicals that are capable of disrupting the endocrine systems of animals can be categorized into the following: environmental estrogens (e.g., bisphenol A, methoxychlor, octylphenol, and nonylphenol), environmental anti-estrogens (e.g., dioxin, endosulfan, and tamoxifen), environmental anti-androgens [e.g., dichlorodiphenyldichloroethylene (DDE), procymidone, and vinclozolin], chemicals that reduce steroid hormone levels (e.g., fenarimol and ketoconazole), chemicals that affect reproduction primarily through effects on the central nervous system (e.g., dithiocarbamate pesticides, and methanol), and chemicals with multiple mechanisms of endocrine action (e.g., phthalates and TBT). There is a high level of international concern regarding developmental and reproductive impacts on marine organisms from exposure to endocrine-disrupting chemicals. This is the case for "new" substances such as alkylphenols; there is also renewed interest for some "old" organochlorines such as DDT and its metabolites. Brominated flame retardants (BFRs), particularly the brominated diphenyl ethers (BDEs) and hexabromocyclododecane (HBCD), have been detected in the European marine environment. It has been reported that the input of BDEs into the Baltic Sea through atmospheric deposition now exceeds that of PCBs by almost a factor of 40. BDEs are found in fish from various geographic regions. This resulted from the long-range atmospheric transport and deposition of these substances.[5] HBCD was detected in liver and blubber samples from harbor seals and harbor porpoises from the Wadden Sea and the North Sea. It is found that environmental concentrations of these BFRs in Japan and South China increased significantly during the last decades. PBDE levels in marine mammals and sediments from Japan, after showing peak concentrations in the 1990s, appear to have leveled off in recent years. Furthermore, in recent years, HBCD concentrations in marine mammals from Japanese waters appear to exceed those of PBDEs, presumably reflecting the increasing use of HBCDs over PBDEs. Pentabromotoluene (PBT) and Decarbomodiphenyl (DBDPE), for example, have been found in Arctic samples remote from sources of contamination. It is an indication of their potential for long-range atmospheric transport, showing a tendency for accumulation in top predators. Polymeric BFRs may be a source of emerging brominated organic compounds to the environment. Medium- and short-chain chlorinated paraffins (SCCPs) are ubiquitous in the environment and tend to behave in a similar way to POPs. They have been found in water as well as in fish and marine mammals.[5]

Perfluorinated compounds (PFCs), namely, perfluorooctane sulfonate (PFOS), have been detected in marine mammals.[5] They are globally distributed anthropogenic contaminants. PFCs, such as PFOS, have been industrially manufactured for more than 50 years and their production and use have increased considerably since the early 1980s. The main producer of PFOS voluntarily ceased its production in 2002. Furthermore, the large-scale use of PFOS has been restricted. PFOS has been used in many industrial applications such as fire-fighting foams and consumer applications such as surface coatings for carpets, furniture, and paper. PFCs are released into the environment during the production and use of products containing these compounds. About 350 polyfluorinated compounds of different chemical structures are known.[5] The most widely known are PFOS (C8F17SO3) and perfluorooctanoic acid (PFOA; C8F15O2), which are chemically stable and thus may be persistent (substance dependent). PFCs do not accumulate in lipid but instead accumulate in the liver, gallbladder, and blood, where they bind to proteins. PFCs have been detected worldwide, including the Arctic Ocean and Antarctic Ocean, in almost all matrices of the environment. High concentrations of PFCs have been found in marine mammals.[5] A screening project in Greenland and the Faroe Islands indicated high biomagnification of PFCs, with elevated concentrations in polar bear liver. A time trend study (1983–2003) showed increasing concentrations for all PFCs for ringed seals from East Greenland. In the United Kingdom, a study on stranded and by-catch harbor porpoise liver (1992 and 2003) found PFOS at up to 2420 $\mu g\ kg^{-1}$ wet weight. There is a decreasing trend going from south to north.[5]

Antifouling paint booster biocides were recently introduced as alternatives to organotin compounds in antifouling products.[5] These replacement products are generally based on copper metal oxides and organic biocides. Commonly used biocides in today's antifouling paints are as follows: Irgarol 1051, diuron, Sea-Nine 211, dichlofluanid, chlorothalonil, zinc pyrithione, TCMS (2,3,3,6-tetrachloro-4-methylsulfonyl) pyridine, TCMTB [2-(thiocyanomethylthio) benzothiazole], and zineb. It has been reported that the presence of these biocides in coastal environments around the world is a result of their increased use (notably in Australia, the Caribbean, Europe, Japan, Singapore, and the United States). For example, Irgarol 1051, the Irgarol 1051 degradation product GS26575, diuron, and three diuron degradation products [1-(3-chlorophenyl)-3,1-dimethylurea (CPDU), 1-(3,4-dichlorophenyl)-3-methylurea (DCPMU), and 1-(3,4-dichlorophenyl)urea (DCPU)] were all detected in marine surface waters and some sediments in the United Kingdom. Risk assessments indicate that the predicted levels of chlorothalonil, Sea-Nine 211, and dichlofluanid, in contrast to Irgarol 1051, in marinas represent a risk to marine invertebrates. Finally, non-eroding silicone-based coatings can effectively reduce fouling of ship hulls and are an alternative to biocidal and heavy-metal-based antifouling paints. Although polydimethylsiloxanes (PDMSs) are unable to bioaccumulate in marine organisms and their soluble fractions have low toxicity to marine biota, undissolved silicone oil films or droplets can cause physical–mechanical effects such as trapping and suffocation of organisms.[5]

Human and veterinary pharmaceuticals are designed to have a specific mode of action, affecting the activity of, e.g., an enzyme, ion channel, receptor, or transporter protein.[5] Clotrimazole, dextropropoxyphene, erythromycin, ibuprofen, propranolol, tamoxifen, and trimethoprim were detected in U.K. coastal waters and in U.K. estuaries. Concentrations of some pharmaceutical compounds are effectively reduced during their passage through a tertiary wastewater treatment works, while others are sufficiently persistent to end up in estuaries and coastal waters.[5] Compared with mammalian and freshwater organisms, there is a lack of experimental data on the impacts of pharmaceuticals in marine and estuarine species. However, there is experimental evidence that selected pharmaceuticals have the potential to cause sublethal effects in a variety of organisms. It has been concluded that antibiotic substances in marine ecosystems can pose a potential threat to bacterial diversity, nutrient recycling, and removal of other chemical pollutants. Although data on the occurrence of pharmaceuticals and antibiotics in the marine environment are becoming more available, the true extent of the potential risks posed by this group of contaminants cannot, at present, be assessed, mainly due the lack of effect data.[5]

Several studies showed that among personal care products (PCPs), synthetic musks (nitromusks, polycyclic musks, and macrocyclic musks) are widespread in marine and freshwater environments and bioaccumulate in fish and invertebrates.[5] There were identified products such as benzotriazole organic UV filters, namely, UV-320 [2-(3,5-di-*t*-butyl-2-hydroxyphenyl)benzotriazole], UV-326 [2-(3-*t*-butyl-2-hydroxy-5-ethylphenyl)-5-chlorobenzotriazole], UV-327 [2,4-di-*t*-butyl-6-(5-chloro-2*H*-benzotriazol-2-yl) phenol], and UV-328 [2-(2*H*-benzotriazol-2yl)-4,6-di-*t*-pentylphenol]. Their relatively high concentrations were found in marine organisms collected from waters of western Japan. There are indications that marine mammals and seabirds accumulate UV-326, UV-328, and UV-327. Benzotriazole UV filters were also detected in surface sediments from this area. The results suggest a significant bioaccumulation of UV filters through the marine food webs and a strong adsorption to sediments. Although a full risk assessment of some of these has been performed (e.g., musks), for most PCPs, there is little data on their occurrence and their effects in the marine environment.[5]

BIOLOGICAL POLLUTION

Eutrophication and Algal Bloom

Nutrient loadings in coastal waters cause direct responses such as changes in chlorophyll, primary production, macro- and microalgal biomass, sedimentation of organic matter, altered nutrient ratios, and harmful algal booms. The indirect responses of nutrient loadings are responsible for changes in benthos biomass, benthos community structure, benthic macrophytes, habitat quality, water transparency, sediment organic matter, sediment biogeochemistry, dissolved oxygen, mortality of aquatic organisms, food web structure, etc. Moreover, increase in phytoplankton biomass and attributing decrease in transparency and light intensity limit growth of submerged vascular plants.[6,68] Generally speaking, eutrophication leads to major changes in qualitative and quantitative species composition, structure, and function of marine communities over large areas. As for phytoplankton communities, such changes are connected with an increase in biomass and productivity.[69] For instance, a general shift from diatoms to dinoflagellates, as well as dominance of small-size nanoplankton (microflagellates, coccoids), has been reported. Similar trends were observed in the case of zooplankton communities, indicating replacement of herbivorous copepods by small-size zooplankton.[70,71] Some examples of consequences of eutrophication have been reported based on worldwide references.[15] The harmful deoxygenation of water giving rise to fish kills was producing nutrient-derived large mats of macroalgae in the Peel-Harvey Estuary, Western Australia.[72] Similar events took place in the northern Adriatic Sea where diatom blooming in summer resulted in the production of mucilage, affecting tourism in northeastern Italy and reducing fish catch.[28,73,74] Insufficient water exchange and increasing production of organic matter during this century caused depletion of O_2 in all deep waters of the Baltic Proper.[15] It resulted in devastating consequences for marine biota, leading to the replacement of O_2 by H_2S in these bottom waters.[75] Although eutrophication generally leads to an increase in fish productivity, it can also cause negative environmental changes in fish populations. Fish such as cod and plaice are threatened by O_2 depletion in Baltic deep basins, causing decreasing fish catch in Köge Bay in the Sound.[75]

The blooms of blue-green algae as well as *Nodularia* produce a toxic peptide hepatoxin under particular conditions, which can pass through the food web, affecting top consumers, e.g., man. The toxin is responsible for the degeneration of liver cells, promoting tumors and causing death from hepatic haemorrhage.[75] Paralytic shell poisoning (PSP) and/or ciguatera has/have been identified predominantly in the subtropical and tropical zones such as Australia[76–80] and especially in other Indo-Pacific regions, e.g., India, Thailand, Indonesia, Philippines, and Papua New Guinea.[81,82] Principal toxic dinoflagellate species, i.e., *Pyrodinium bahamense* var. *compressa*, killed many fish and shellfish from these regions.[15] The consumption of seafood in the Indo-Pacific area posed considerable public health problems.[28] The significant PSP incidences also took place in temperate zones. For instance, in May 1968, a poisoning episode affected 78 persons inhabiting Britain after consumption of soft tissue of the blue mussel *Mytilus edulis*.[83] Another dinoflagellate-poisoning event again happened in northeast England in the summer of 1990, possibly attributed to a specific combination of elevated nutrient inputs from rivers and exceptionally warm weather conditions, which could be favorable for algae growing.[28]

It has been reported that anthropogenically derived atmospheric N deposition to the North Atlantic Ocean was strictly responsible for harmful algal bloom expansion.[84] This event concerned especially the Eastern Gulf of Mexico, U.S. Atlantic coastal waters, the North Sea, and the Baltic Sea.[84–95] Expanding blooms of the noxious dinoflagellate *Alexandrium tamarense* have been observed along the Northeast U.S. Atlantic coastline.[84,92] There are numerous examples of specific harmful algal bloom expansions in coastal and off-shore waters in case of significant atmospheric deposition of N, e.g., in the North Sea, Adriatic Sea, Western Mediterranean Sea, and Baltic Sea.[84,96] Great attention has been paid to toxic hypoxia-inducing dinoflagellate blooms in the North Sea and the Western Baltic.[84] In the summer of 1991, a very extensive bloom of *Nodularia spumigena* in the open Baltic Sea and along the southern and southeastern Swedish coasts was observed. Dogs' mortalities caused by toxic *Nodularia* blooms have been observed in Denmark, Gotland, and the Swedish coastal waters.[15] In other Baltic areas, horses, cows, sheep, pigs, cats, birds, and fish also suffered from this event. *Nodularia* blooms have caused human health problems such as stomach complaints, headaches, eczema, and eye inflammation.[75] In the Skagerrak and Kattegat, harmful algal bloom expansion of toxic algae species such as *Prorocentrum*, *Dinophysis*, *Dichtyocha*, *Prymnesium*, and *Chrysochromulina* has taken place.[88] The recent blooms mostly killed pelagic organisms and the phyto- and zoobenthic organisms.

Invasive Species

The impacts of introduction and invasion of species throughout the world have recently been identified. There are an increasing number of reports that document this phenomenon taking place in coastal, estuarine, and marine waters.[6] For instance, the Chinese mitten crab (*Eriocheir sinensis*), as invasive species, now inhabits coastal regions in northwestern Europe, and it has caused damage to flood defense walls by burrowing, affecting local community structure.

Worldwide fish species introduction is connected with various consequences.[97] It has been pointed out that many aquaculture species are recently genetically modified. Such modified populations are frequently released and mixed with the natural populations and are breeding with them. It causes biological pollution from a molecular level to community and ecosystem levels. An example of such events is the flooding in Central Europe that caused the release of hybrid and modified fish like sturgeon (*Acipenser* spp.) from aquaculture installations.[98] The local populations of fish are generally not resistant to the pathogenic organisms carried by the introduced species and vice versa. Therefore, deliberate genetic selection and breeding for a long time may have numerous consequences in the aquaculture unit itself as well as the loss of the natural stock for numerous species in a global scale.[6,98,99]

FERTILIZERS AND PESTICIDES

Agricultural activity as an important pollution source has contributed to significant enrichment of nutrients (mainly ammonium ion and nitrates) in coastal marine waters. It is found[100] that wastes, manures, and sludges provide soils with significantly more hazardous substances as compared to fertilizers for achieving the equivalent plant nutrient content. The worldwide use of fertilizers, including organic fertilizers like manure, is huge. In the case of intensively monocultivated areas, a relatively small number of pesticides have been widely used in spite of their variety.[6]

The large mass of pesticide residues is accumulated in the environment since they are not rapidly degradable. The total global DDT production from the 1940s to 2004 was estimated as ca. 4.5 Mt.[101] Duursma and Marchand[102] estimated the world production of DDT to be ca. 2.8 Mt, of which 25% is assumed to be released to the ocean. According to Shahidul Islam and Tanaka,[6] the total emission of DDT through agricultural applications amounts to 1030 kt between 1947 and 2000. Organochlorine pesticides (OCPs) originating mostly in temperate and warmer areas of the world can be transported to coastal waters and even via atmospheric long-range transport and ocean currents to the Arctic. Owing to their bioaccumulative abilities (as lipophilic compounds) and biomagnification along the sequential trophic levels of the food chain, pesticides are classified as one of the most destructive agents for marine organisms. As a consequence, their very high levels can be observed among top predators, including man. Their toxic effects to marine organisms are often complex because they may be associated with the combination of exposure to pesticides and other POPs with environmental stresses such as eutrophication and pathogens.[6]

SEWAGE EFFLUENTS

Sewage effluents contain industrial, municipal, and domestic wastes; animal remains; etc. The huge amounts of these effluents generated in big cities are transported by drainage systems into rivers or other aquatic systems, e.g., coastal waters. It is estimated that the annual production of sewage amounts to ca. 1.8×10^8 m^3 for a population of 800,000. This load is equivalent to an annual release of 3.6×10^3 tons of organic matter.[6] Sewages pose significant effects on coastal marine ecosystems because they contain POPs (heavy metals/trace elements, organic pollutants) as well as viral, bacterial, and protozoan pathogens and organic substances subjected to bacterial decay. In case of such bacterial activity, the content of oxygen in water is reduced, resulting in the destruction of proteins and other nitrogenous compounds. Releasing hydrogen sulfide and ammonia exhibits toxic activity to marine biota, even at low levels. As for pathogens, domestic sewage released to coastal waters contains such harmful pathogens as *Salmonella* spp., *Escherichia coli*, *Streptococcus* sp., *Staphylococcus aureus*, *Pseudomonas aeruginosa*, the fungi *Candida*, and viruses such as enterovirus, hepatitis, poliomyelitis, influenza, and herpes.[6] Different bacteria and viruses can be transferred to some representatives of marine fauna, e.g., marine mammals.

OILS

The recently observed increase in tanker operations and oil use as well as marine tanker catastrophes has been responsible for the presence of excessively large amounts of oil spillage in coastal and marine ecosystems. It is estimated that ca. 2.7 million tons of oil pollution enter the ocean each year. The tanker accidents between 1967 and 2007 released ca. 4.5 million tons of oil to seawater. Notable examples of ecological catastrophes are the huge spill from a drilling platform in Gulf of Mexico (Mexico) in 1979 and the Deepwater Horizon drilling rig explosion in the Gulf of Mexico (United States) (April 20 to July 15, 2010), resulting in massive amounts of oil in the gulf. Another similar example took place during the Persian Gulf War in 1991, where ca. 2 million tons of oil was spilled, resulting to the death of many species of marine biota.[7,103,104] Therefore, oil pollution poses serious adverse effects on aquatic environment and marine organisms represented different trophic levels from primary producers to the top predators.[6] Although aerial and flying birds (e.g., gulls, gannets) are not seriously exposed to oil toxicity, birds that spend most of their time in contact with oil on the water surface (e.g., ducks, auks, divers, penguins) are at greater risk of oil toxicity. According to Smith,[105] the annual release of hydrocarbon can range from 0.6 to 1 million tons. Coastal refineries can be an important source of oil pollution since millions of gallons of crude oil and its fractions are processed and stored there. During their operation, pollutants are continuously released by way of leakages, spills, etc.

MARINE DEBRIS AND PLASTICS

Marine debris, especially plastics, is one of the most pervasive pollution problems. Nets, food wrappers, bottles, resin pellets, etc., have serious impacts on humans and marine biota. Medical and personal hygiene debris can enter coastal water through direct sewage outflows, posing a serious threat to human health and safety. Contact with water contaminated with these pollutants and pathogens (e.g., *E. coli*) can result in infectious hepatitis, diarrhea, bacillary dysentery, skin rashes, typhoid, and cholera.[106]

There are numerous reviews devoted to an important topic such as pollution by marine debris.[106–110] Entanglement in marine debris such as nets, fishing line, ropes, etc., can hamper an organism's mobility, prevent it from eating, inflict wounds, and cause suffocation or drowning. It was estimated that 136 marine species have been involved in entanglement incidents, including some species of seabirds, marine mammals, and sea turtles.[111] The decline in the population of the northern sea lion (*Eumetopias jubatus*), endangered Hawaiian monk seal (*Monachus schauinslandi*), and northern fur seal has been explained by entanglement of young specimens in lost or discarded nets and packing bands.[112] Abandoned fishing gear, e.g., fishing net, can contribute to catching and killing marine animals. This process called ghost fishing or ghost net can kill a huge number of commercial species.[108] An example of another serious pollution problem is ingestion of debris by marine animals. Plastic pellets and plastic shopping bags can be swallowed and lodged in animals' throats and digestive tracts, causing some animals to stop eating and slowly starve to death.[106] According to the U.S. Marine Mammal Commission,[111] ingestion incidents concerned 111 species of seabirds, 26 species of marine mammals, and 6 species of turtles. For instance, plastic cups were found in the gut of some species of fish from British coastal waters; the ingested cups were eventually responsible for their deaths.[112] Even Antarctic and sub-Antarctic seabirds, e.g., Wilson's storm-petrel (*Oceanites oceanicus*) and white-faced storm-petrel (*Pelagodroma marina*), are at risk for this ingestion hazard.[112–115] It is reported that the proportion of plastic debris among litter increases with distance from source because it is transported more easily as compared to a denser material like glass or metal and because it lasts longer than other low-density materials (paper). Floating plastic articles (material less dense than water, e.g., polyamide, polyterephthalate, polyvinyl chloride) pose a global problem because they can contaminate even the most remote islands.[107,116] Drift plastics can increase the range of some marine organisms or introduce unwanted and aggressive alien taxa species into an environment. It could be risky to littoral, intertidal zones, and the shoreline.[112,117] There is also potential danger to marine ecosystems from the accumulation of plastic debris (material more dense than water) on the seafloor. Such bottom accumulation of plastic can inhibit the gas exchange between overlying waters and the pore water. This process can result in hypoxia or anoxia in the benthic fauna, altering the makeup of life on the sea bottom.[6] Another threat is connected with potential entanglement and ingestion hazards for pelagic and benthic animals.[62,112,118] Plastic can adsorb and concentrate some pollutants in coastal waters, including PCBs, DDE, nonylphenyl, and phenanthrene. It has been reported that these sorbed POPs could subsequently be released if the plastics are ingested.[109,110] For instance, PCBs in tissues of great shearwaters (*Puffinus gravis*) were derived from ingested plastic debris.[119]

NOISE POLLUTION

In recent years, the marine biota has been affected by noise pollution. Natural sources of underwater noise may be physical and biological in character. Physical sources include wind, waves, rainfall, thunder and lighting, earthquake-generated seismic energy, and the movement of ice. Biological sources include marine mammal vocalizations and sounds produced by fish and invertebrates.[120,121]

Anthropogenic sound sources can be grouped into six categories, namely, shipping, seismic surveying, sonars, explosions, industrial activity, and miscellaneous.[122] Vessel traffic significantly contributes to underwater noise, mainly at low frequencies. Commercial shipping vessels generate noise mainly in areas confined to ports, harbors, and shipping lanes.[122] In contrast to wide geographic distribution of shipping industry, the oil and gas industry activities have taken place along continental margins in specific worldwide areas. Such resources exploration activities have been typically observed in shallow waters less than 200 m in depth. Other activities, in spite of their geographically widespread range, are also confined to near-shore coastal regions, namely, pile driving, dredging, operation of land- and ocean-based wind power turbines, power plant operations, and typical harbor and shipyard activities.[120] Offshore wind turbines may have significantly contributed noise to the underwater ecosystem bearing in mind that the relatively recent growth in offshore wind development has increased. It has been suggested that marine mammals may be indirectly affected by noise from offshore wind turbines, e.g., prey fish avoiding the sound source as well as the masking of marine mammals' mating and communication calls. On the other hand, a number of mass stranding of marine mammals, especially whales, found on worldwide beaches may be associated with the use of concurrent military sonar.[120] Another example of noise pollution affecting marine animals is continued exposure to anthropogenic noise pressure in vital sea turtle habitats, resulting in potential impact on its behavior and ecology. Brown shrimp exposed to higher pressure levels of noise in experimental area exhibited increased aggression, higher mortality rates, and significant reduction in their food uptake, growth, and reproduction. Sound exhibits measurable damage to sensory cells in the ears of fish.[123]

LIGHT POLLUTION

A remarkable recent interest concerns the introduction of light to the coastal zone and nearshore environment. It is estimated that at least 3351 cities in the coastal zones all over the world are illuminated. It is expected that artificial light will be continuously intensified not only by population growth but also by dramatically increasing the number of locations of high-intensity artificial light. According to the United Nations World Tourism Organization (UNWTO), there were ca. 900 million international tourist arrivals all over the world.[9] Tourist visits to beaches cause light pollution along the coastline since tracking the movement of population over time by research using satellite imagery showed that wherever human population density increases, the use of artificial light at night also increases. Living organisms are mostly sensitive to changes in the quality and intensity of natural light in the ecosystem. For instance, for algae and seaweeds, photosynthetic activity is highly dependent on available light, i.e., different cycles in natural light intensity and quality.[9] Light pollution takes place when biota is exposed to artificial light, especially in coastal areas, resulting in damaging effects on marine species in seas. The behavior, reproduction, and survival of marine invertebrates, amphibians, fish, and birds have been influenced by artificial lights. Light pollution disrupts the migration patterns of nocturnal birds and can result in hatchling sea turtles to head inland, away from the sea, which could be eaten by predators or run over by cars.[124] Ecological effects of light pollution concern disruption of predator–prey relationship. For instance, artificial light disturbs natural vertical migrations of zooplankton in the water column in accordance with the day–night cycle when natural light helps to reduce their predation by fish and other animals.[125]

CONCLUSION

The anthropogenic activity of man in coastal regions and even in offshore areas is responsible for emission of a huge amount of pollutants that are transported to marine ecosystems directly or by means of coastal watersheds, rivers, and precipitation from air. A lot of pollutants may be taken up by various compartments, i.e., biotic and abiotic, of aquatic environments and some of them could be biomagnified along the successive members of the food chain. Therefore, water pollution could have a negative effect on the quality of the water and hence on the health of the plants, animals, and humans whose lives depend on the quality of aquatic environments. Coastal areas are generally damaged from pollution, resulting in considerable impact on commercial coastal and marine fisheries.

There are numerous examples of worldwide events leading to serious contamination of coastal waters by persistent pollutants. Therefore, these areas have been extensively explored, especially in relation to perturbed ecosystems by heavy metals, radionuclides, POPs, oils, etc.

Elevated levels of nutrients in coastal waters resulted in eutrophication and proliferation of toxic algal blooms. The recently observed increase in tanker operations and oil use as well as marine tanker catastrophes has been responsible for the presence of excessively large amounts of oil spillage in coastal and marine ecosystems. Marine debris, especially plastics, is one of the most pervasive pollution problems. Marine pollutants are generally present in increased concentrations in the enclosed seas and coastal areas than in the open seawaters. Spatial distribution patterns of contamination concentrations exhibit a trend of their increase during transition from the south to the northern part of all oceans, i.e., in areas neighboring with industrial centers and concentration of main pollution sources.

REFERENCES

1. *Pollutant*; 2011, available at http://en.wikipedia.org/wiki/Pollution (accessed 2011).
2. *Pollutant*; 2011, available at http://en.wikipedia.org/wiki/Pollutant#a-Stock_pollutants (accessed 2011).
3. Williams, C. Combating marine pollution from land-based activities: Australian initiatives. Ocean Coastal Manage. **1996**, *33*, 87–112.
4. Falandysz, J.; Trzosińska, A.; Szefer, P.; Warzocha, J.; Draganik, B. The Baltic Sea, especially southern and eastern regions. In *Seas at the Millennium: An Environmental Evaluation, Vol. I: Europe, The Americas and West Africa*; Sheppard, C.R.C., Ed.; Pergamon, Elsevier: Amsterdam, 2000; 99–120.
5. Albaigés, J.; Bebianno, M.J.; Camphuysen, K.; Cronin, M.; de Leeuw, J.; Gabrielsen, G.; Hutchinson, T.; Hylland, K.; Janssen, C.; Jansson, B.; Jenssen, B.M.; Roose, P.; Schulz-Bull, D.; Szefer, P. *Monitoring Chemical Pollution in Europe's Seas-Programmes, Practices and Priorities for Research*; Position Paper 16, Marine Board—European Science Foundation: Ostend, 2011.
6. Shahidul Islam, Md.; Tanaka, M. Impacts of pollution on coastal and marine ecosystems including coastal and marine fisheries and approach for management: A review and synthesis. Mar. Pollut. Bull. **2004**, *48*, 624–649.
7. Laws, E.A. *Aquatic Pollution*; John Wiley and Sons: New York, 2000.
8. Valcovic, V. *Radioactivity in the Environment—Physicochemical Aspects and Applications*; Elsevier: Amsterdam, 2000.
9. Depledge, M.H.; Godard-Codding, C.A.J.; Bowen, R.E. Light pollution in the sea. Mar. Pollut. Bull. **2010**, *60*, 1383–1385.
10. Sheppard, C.R.C., Ed. *Seas in the Millennium: An Environmental Evaluation, Vol. I: Europe, The Americas and West Africa*; Pergamon, Elsevier: Amsterdam, 2000.
11. Tanabe, S., Ed. *Mussel Watch—Marine Pollution Monitoring in Asian Waters*; Center for Marine Environmental Studies, Ehime University: Japan, 2000.
12. Nriagu, J.O. Global inventory of natural and anthropogenic emissions of trace metals to the atmosphere. Nature **1979**, *279*, 409–411.
13. Luoma, S.N.; Rainbow, P.S. *Metal Contamination in Aquatic Environments: Science and Lateral Management*; Cambridge University Press: Cambridge, 2008.
14. Prospero, J.M. African dust in America. Geotimes. **2001**, 24–27, available at http://www.rsmas.miami.edu/assets/pdfs/mac/fac/Prospero/Publications/Prospero_Geotimes_African%20Dust_2001.pdf (accessed 2011).
15. Szefer, P. *Metals, Metalloids and Radionuclides in the Baltic Sea Ecosystem*; Elsevier Science B.V.: Amsterdam, 2002.
16. Mason, R.P.; Fitzegerald, W.F.; Morel, F.M.M. The biogeochemical cycling of elemental mercury: Anthropogenic influences. Geochim. Cosmochim. Acta **1994**, *58*, 3191–3198.
17. Jickells, T. Atmospheric inputs of metals and nutrients to the oceans: Their magnitude and effects. Mar. Chem. **1995**, *48*, 199–214.
18. Neff, J.M. *Bioaccumulation in Marine Organisms: Effects of Contaminants from Oil Well Produced Water*; Elsevier: Amsterdam, 2002.
19. Bryan, G.W.; Gibbs, P.E. Impact of low concentrations of tributyltin (TBT) on marine organisms: A review. In *Metal Ecotoxicology: Concepts and Applications*; Newman, M.C.; McIntosh, A.W., Eds.; Lewis Publishers: Ann Arbor, MI, 1991; 323–361.
20. Kabata-Pendias, A.; Mukherjee, A.B. *Trace Elements from Soil to Human*; Springer: Berlin, 2007.

21. Bryan, G.W.; Langston, W.J. Bioavailability, accumulation and effects of heavy metals in sediments with special reference to United Kingdom estuaries: A review. Environ. Pollut. **1992**, *76*, 89–131.
22. US EPA (US Environmental Protection Agency). *Health Effects Assessment of Mercury*; Environmental Criteria and Assessment Office: Cincinnati, Ohio, 1984.
23. dos Santos, M.M.; Vieira, N.; Santos, A.M. Imposex in the dogwhelk *Nucella lapillus* (L.) along the Portuguese coast. Mar. Pollut. Bull. **2000**, *40*, 643–646.
24. Gray, J.E.; Theodorakos, P.M.; Bailey, E.A., Turner, R.R. Distribution, speciation, and transport of mercury in stream-sediment, stream-water, and fish collected near abandoned mercury mines in southwestern Alaska, USA. Sci. Total Environ. **2000**, *260*, 21–33.
25. Maurice-Bourgoin, L.; Quiroga, I.; Chincheros, J.; Courau, P. Mercury distribution in waters and fishes of the upper Madeira rivers and mercury exposure in riparian Amazonian populations. Sci. Total Environ. **2000**, *260*, 73–86.
26. Tomiyasu, T.; Nagano, A.; Yonehara, N.; Sakamoto, H.; Rifardi; Oki, K; Akagi, H. Mercury contamination in the Yatsushiro Sea, south-western Japan: Spatial variations of mercury in sediment. Sci. Total Environ. **2000**, *257*, 121–132.
27. Förstner, U.; Wittmann, G.T.W. *Metal Pollution in the Aquatic Environment*, 2nd Ed.; Springer-Verlag: Berlin, 1983.
28. Phillips, D.J.H.; Rainbow, P.S. *Biomonitoring of Trace Aquatic Contaminants*; Elsevier Science Publishers Ltd.. London, 1993.
29. Akagi, H.; Malm, O.; Kinjo, Y.; Harada, M.; Branches, F.J.P.; Pfeiffer, W.C.; Kato, H. Methylmercury pollution in the Amazon, Brazil. Sci. Total Environ. **1995**, *175*, 85–95.
30. Harada, M. Characteristics of industrial poisoning and environmental contamination in developing countries. Environ. Sci. **1996**, *4*, 157–169.
31. Harada, M.; Nakachi, S.; Cheu, T.; Hamada, H.; Ohno, Y.; Tsuda, T.; Yanagida, K.; Kizaki, T.; Ohno, H. Monitoring of mercury pollution in Tanzania: Relation between head hair and health. Sci. Total Environ. **1999**, *227*, 249–256.
32. Wheatley, B.; Wheatley, M.A. Methylmercury and the health of indigenous peoples: A risk management challenge for physical and social sciences and for public health policy. Sci. Total Environ. **2000**, *259*, 23–29.
33. Al-Majed, N.B.; Preston, M.R. Factors influencing the total mercury and methyl mercury in the hair of the fishermen of Kuwait. Environ. Pollut. **2000**, *109*, 239–250.
34. Artaxo, P.; Calixto de Campos, R.; Fernandes, E.T.; Martins, J.V.; Xiao, Z., Lindquist, O.; Fernández-Jiménez, M.T.; Maenhaut, W. Large scale mercury and trace element measurements in the Amazon basin. Atmos. Environ. **2000**, *34*, 4085–4096.
35. Yamasoe, M.A.; Artaxo, P.; Miguel, A.H.; Allen, A.G. Chemical composition of aerosol particles from direct emissions of vegetation fires in the Amazon Basin: Water-soluble species and trace elements. Atmos. Environ. **2000**, *34*, 1641–1653.
36. Alzieu, C. TBT detrimental effects on oyster culture in France—Evolution since antifouling paint regulation. In Proceedings of Oceans 86 Conference Record. Organotin Symposium; 4 Institute of Electrical and Electronics Engineers: New York, 1986; 1130–1134.
37. Alzieu, C. Environmental impact of TBT: The French experience. Sci. Total Environ. **2000**, *258*, 99–102.
38. Shim, W.J.; Kahng, S.H.; Hong, S.H.; Kim, N.S.; Kim, S.K.; Shim, J.H. Imposex in the rock shell, *Thais clavigera*, as evidence of organotin contamination in the marine environment of Korea. Mar. Environ. Res. **2000**, *49*, 435–451.
39. Hung, T.-C.; Hsu, W.-K.; Mang, P.-J.; Chuang, A. Organotins and imposex in the rock shell, *Thais clavigera*, from oyster mariculture areas in Taiwan. Environ. Pollut. **2001**, *112*, 145–152.
40. Evans, S.M.; Nicholson, G.J. The use of imposex to assess tributyltin contamination in coastal waters and open seas. Sci. Total Environ. **2000**, *258*, 73–80.
41. Batt, J.M. *The world of organotin chemicals: Applications, substitutes, and the environment*; 2006, available at http://www.ortepa.org/WorldofOrganotinChemicals.pdf (accessed 2011).
42. Rumengan, I.F.; Ohji, M.; Arai, T.; Harino, H.; Arfin, Z.; Miyazaki, N. Contamination status of butyltin compounds in Indonesian coastal waters. Coastal Mar. Sci. **2008**, *32*, 116–126.
43. Zhao, Y.; Marriott, S.; Rogers, J.; Iwugo, K. A preliminary study of heavy metal distribution on the floodplain of the River Severn, U.K. by a single flood event. Sci. Total Environ. **1999**, *243/244*, 219–231.
44. Skwarzec, B. *Radiochemia Środowiska i Ochrona Radiologiczna* (in Polish), Environmental Radiochemistry and Radiological Protection; Wydawnictwo DJ sc.: Gdańsk, 2002.
45. ISSG. *The Irish Sea: An Environmental Review*; Report of the Irish Sea Study Group; Liverpool University Press: Liverpool, 1990.
46. Kershaw, P.J.; McCubbin, D.; Leonard, K.S. Continuing contamination of North Atlantic and Arctic waters by Sellafield radionuclides. Sci. Total Environ. **1999**, *237/238*, 119–132.
47. Aarkrog, A., Dahlgaard, H.; Hansen, H.; Holm, E.; Hallstadius, L.; Rioseco, J.; Christensen, G. Radioactive tracer studies in the surface waters of the northern North Atlantic including the Greenland, Norwegian and Barents Seas. Rit. Fiskideildar **1985**, *9*, 37–42.
48. Kershaw, P.J.; Baxter, A.J. The transfer of reprocessing wastes from north west Europe to the Arctic. Deep Sea Res. **1995**, *42*, 1413–1448.
49. Förstner, U. Inorganic pollutants, particularly heavy metals in estuaries In *Chemistry and Biogeochemistry of Estuaries*; Olausson, E., Cato, I., Eds.; John Wiley & Sons: New York, 1980; 307–348.
50. UNSCEAR. *Sources and effects of ionizing radiation*; Report to the General Assembly by the United Nations Scientific Committee on the Effects of Atomic Radiation; United Nations: New York, 1993.
51. Beasley, T.M.; Cooper, L.W.; Grebmeier, J.M.; Orlandini, K.; Kelley, J.M. Fuel reprocessing Pu in the Arctic Ocean Basin: evidence from mass spectrometry measurements. In Proc. Conf. on Environmental Radioactivity in the Arctic, Oslo, August 1995.
52. Aarkrog, A.; Dahlgaard, H.; Nilsson, K.; Holm, E. Studies of plutonium and americum at Thule, Greenland. Health Phys. **1984**, *46*, 29–44.
53. Smith, J.N.; Ellis, K.M.; Aarkrog, A.; Dahlgaard, H.; Holm, E. Sediment mixing and burial of the ^{239,240}Pu pulse from the 1968 Thule, Greenland nuclear weapons accident. J. Environ. Radioact. **1994**, *25*, 135–159.

54. Joint Russian–Norwegian Expert Group. *Radioactive Contamination of Dumping Sites for Nuclear Wastes in the Kara Sea. Results from the 1993 Expedition*; Norwegian Radiation Protection Authority: Østerås, 1994.
55. National Academy of Sciences. *Radioactivity in the Marine Environment*; National Academy of Sciences: Washington, DC, 1971.
56. Aarkrog, A. A retrospect of anthropogenic radioactivity in the global marine environment. Radiat. Prot. Dosim. **1998**, *75*, 23–31.
57. Commission of the European Communities. *The Radiological Exposure of the Population of the European Community from Radioactivity in North European Marine Waters—Project MARINA*, Report EUR 12483; EU, 1989.
58. NEA. *Co-ordinated Research and Environmental Surveillance Programme Related to Sea Disposal of Radioactive Waste*, CRESP Final Report, 1981–1995; OECD Paris, 1996.
59. INSAG. *Post Accident Review Meeting on the Chernobyl Accident*; Summary Report; International Atomic Energy Agency: Vienna, 1986.
60. WHO. Health hazards from radiocaesium following the Chernobyl nuclear accident. Report on a WHO Working Group. J. Environ. Radioact. **1989**, *10*, 257–259.
61. Radiation effects from Fukushima Daiichi nuclear disaster, available at http://en.wikipedia.org/wiki/Radiation_effects_from_Fukushima_Daiichi_nuclear_disaster (accessed 2011).
62. Goldberg, E.D. Emerging problems in the coastal zone for the 21th century. Mar. Pollut. Bull. **1995**, *31*, 152–158.
63. Gibbs, P.E. Oviduct malformation as a sterilising effect of tributyltin-induced imposex in *Ocenebra erinacea* (Gastropoda: Muricidae). J. Molluscan Stud. **1996**, *62*, 403–413.
64. Cadee, G.C.; Boon, J.P.; Fischer, C.V.; Mensink, B.P.; Tjabbes, C.C. Why the whelk *Buccinum undatum* has become extinct in the Dutch Wadden Sea. Neth. J. Sea Res. **1995**, *34*, 337–339.
65. Jones, P. TBT implicated in mass dolphin deaths. Mar. Pollut. Bull. **1997**, *34*, 146.
66. HELCOM. Batlic Sea Environment Proceedings No. 64B. Third Periodic Assessment on the State of the Marine Environment of the Baltic Sea, 1989–93 Background Document. Helsinki Commission, Baltic Marine Environment Protection Commission, 1996; 252.
67. Jepson, P.D.; Bennet, P.M.; Allchin, C.R.; Law, R.J.; Kuiken, T.; Baker, J.R.; Rogan, E.; Kirkwood, J.K. Investigating potential associations between chronic exposure to polychlorinated biphenyls and infectious disease mortality in harbour porpoises from England and Wales. Sci. Total Environ. **1999**, *243–244*, 339–348.
68. Cloern, J.E. Our evolving conceptual model of the coastal eutrophication problem. Mar. Ecol. Prog. Ser. **2001**, *210*, 223–253.
69. Riegman, R. Nutrient-related selection mechanisms in marine plankton communities and the impact of eutrophication on the plankton food web. Water Sci. Technol. **1995**, *32*, 63–75.
70. Kimor, B. Impact of eutrophication on phytoplankton composition. In *Marine Coastal Eutrophication*; Vollenweider, R.A., Marchetti, R., Vicviani, R., Eds.; Elsevier: Amsterdam, 1992; 871–878.
71. Zaitsev, Y.P. Recent changes in the trophic structure of the Black Sea. Oceanography **1992**, *1*, 180–189.
72. Birch, P.B.; Forbes, G.G.; Schofield, N.J. Monitoring effects of catchment management practices on phosphorus loads into the eutrophic Peel-Harvey Estuary, Western Australia. Water Sci. Technol. **1986**, *18*, 53–61.
73. Justic, D. Long-term eutrophication of the Northern Adriatic Sea. Mar. Pollut. Bull. **1987**, *18*, 281–284.
74. Degobbis, D. Increased eutrophication of the Northern Adriatic Sea. Second act. Mar. Pollut. Bull. **1989**, *20*, 452–457.
75. Forsberg, C. *Eutrophication of the Baltic Sea*; The Baltic Sea Environment: Uppsala, Sweden, 1993; 32 pp.
76. Gillespie, N. Ecological and epidemiological aspects of ciguatera fish poisoning. In Proc. of the Red Tide Workshop, Cronulla, June 18–20, 1984; Australian Department of Science: Canberra.
77. Hallegraeff, G.M.; Sumner, C. Toxic plankton blooms affect shellfish farms. Aust. Fish. **1986**, *45*, 15–18.
78. Holmes, P.R.; Lam, C.W.Y. Red tides in Hong Kong waters—Response to a growing problem. Asian Mar. Biol. **1985**, *2*, 1–10.
79. Phillips, D.J.H. Monitoring and control of coastal water quality. In *Pollution in the Urban Environment, POLMET 85*; Chan, M.W.H., Hoare, R.W.M., Holmes, P.R., Law, R.J.S., Reed, S.B., Eds.; Elsevier Applied Science Publishers: London, 1985; 559–565.
80. Morton, B.S. Pollution of the coastal waters of Hong Kong. Mar. Pollut. Bull. **1989**, *20*, 310–318.
81. Maclean, J.L. Indo-Pacific red tides, 1985–1988. Mar. Pollut. Bull. **1989**, *20*, 304–310.
82. Maclean, J.L.; White, A.W. Toxic dinoflagellate blooms in Asia: A growing concern. In *Toxic Dinoflagellates*; Anderson, D.M., White, A.W., Baden, D.G., Eds.; Elsevier: New York, 1985; 517–520.
83. Ayres, P.A. Mussel poisoning in Britain with special reference to paralytic shellfish poisoning. Environ. Health **1975**, *July*, 261–265.
84. Paerl, H.W., Whitall, D.R. Anthropogenically-derived atmospheric nitrogen deposition, marine eutrophication and harmful algal bloom expansion: Is there a link? Ambio **1999**, *28*, 307–311.
85. Paerl, H.W. Enhancement of marine primary productivity by nitrogen enriched rain. Nature **1985**, *315*, 747–749.
86. Paerl, H.W. Coastal eutrophication in relation to atmospheric nitrogen deposition: Current perspectives. Ophelia **1995**, *41*, 237–259.
87. Anderson, D.M. Toxic algal blooms and red tides. A global perspective. In *Red Tides: Biology, Environmental Science and Toxicology*; Okaichi, T., Anderson, D.M., Nemoto, T., Eds.; Elsevier Science Publishing Co., Inc.: New York, 1989; 11–21.
88. Aksnes, D.L.; Aure, J.; Furnes, G.K.; Skjoldal, H.R.; Saetre, R. Analysis of the *Chrysochromulina polylepis* bloom in the Skagerrak. Environmental conditions and possible causes. Bergen Scientific Centre Publication 1989, No. BSC 89/1.
89. Tester, P.A.; Stumpf, R.P.; Vukovich, F.M.; Fowler, P.K.; Turner, J.T. An expatriate red tide bloom: Transport, distribution, and persistence. Limnol. Oceanogr. **1991**, *36*, 1053–1061.

90. Buskey, E.J.; Stockwell, D.A. Effects of a persistent "brown tide" on zooplankton populations in the Laguna Madre of South Texas. In *Toxic Phytoplankton Blooms in the Sea*, Proc. 5th Intern. Conf. on Toxic Marine Phytoplankton; Elsevier, 1993; 659–665.
91. Hallegraeff, G.M. A review of harmful algal blooms and their apparent global increase. Phycologia **1993**, *32*, 79–99.
92. Anderson, D.M.; Kulis, D.M.; Doucette, G.J.; Gallagher, J.C.; Balech, E. Biogeography and toxic dinoflagellates in the genus *Alexandrium* from the northeastern United States and Canada. Mar. Biol. **1994**, *120*, 467–478.
93. ECOHAB. *The Ecology and Oceanography of Harmful Algal Blooms*; A National Research Agenda, US N.S.F./N.O.O.A. Publication; Woods Hole Oceanographic Inst.: Mass., USA, 1995.
94. Howarth, R.W.; Billen, G.; Swaney, D.; Townsend, A.; Jaworski, N.; Lajtha, K.; Downing, J.A.; Elmgren, R.; Caraco, N.; Jordan, T.; Berendse, F.; Freney, J.; Kudeyarov, V.; Murdoch, P.; Zhu, Z.-L. Regional nitrogen budgets and riverine N and P fluxes for the drainages to the North Atlantic Ocean: Natural and human influences. Biogeochemistry **1996**, *35*, 75–139.
95. Prospero, J.M.; Barret, K.; Church, T.; Detener, F.; Duce, R.A.; Galloway, J.N.; Levy, H.; Moody, J.; Quinn, P. Atmospheric deposition of nutrients to the North Atlantic basin. Biogeochemistry **1996**, *35*, 27–73.
96. Paerl, H.W. Coastal eutrophication and harmful algal blooms: Importance of atmospheric deposition and groundwater as "new" nitrogen and other nutrient sources. Limnol. Oceanogr. **1997**, *42*, 1154–1165.
97. Mills, E.L.; Holeck, K.T. Biological pollutants in the Great Lakes. Clearwaters **2001**, *31* (1), available at http://www.nywea.org/clearwaters/pre02fall/311010.html (accessed 2011).
98. Elliott, M. Biological pollutants and biological pollution—An increasing cause for concern. Mar. Pollut. Bull. **2003**, *46*, 275–280.
99. FAO. *Precautionary Approach to Fisheries. Part 1. Guidelines on the Precautionary Approach to Capture Fisheries and Species Introductions*; Food and Agriculture Organization of the United Nations: Rome, 1995.
100. Joly, C. Plant nutrient management and the environment. In *Prevention of Water Pollution by Agriculture and Related Activities*, Proceedings of the FAO Expert Consultation, Santiago, Chile, October 20–23, 1992; Water Report 1, FAO: Rome.
101. Li, Y.F.; Macdonald, R.W. Sources and pathways of selected organochlorine pesticides to the Arctic and the effect of pathway divergence on HCH trends in biota: A review. Sci. Total Environ. **2005**, *342*, 87–106.
102. Duursma, E.K.; Marchand, M. Aspects of organic marine pollution. Oceanogr. Mar. Biol. Ann. Rev. **1974**, *12*, 315–431.
103. *Oil spill*; 2011, available at http://en.wikipedia.org/wiki/Oil_spill (accessed 2011).
104. *Oil spills and disasters*; 2011, available at http://www.infoplease.com/ipa/A0001451.html (accessed 2011).
105. Smith, N. The problem of oil pollution of the sea. Adv. Mar. Biol. **1970**, *8*, 215–306.
106. Sheavly, S.B.; Register, K.M. Marine debris and plastics: Environmental concerns, sources, impacts and solutions. J. Polym. Environ. **2007**, *15*, 301–305.
107. Ryan, P.G.; Moore, C.J.; van Franeker, J.A.; Moloney, C.L. Monitoring the abundance of plastic debris in the marine environment. Philos. Trans. R. Soc. **2009**, *364*, 1999–2012.
108. Moore, C.J. Synthetic polymers in the marine environment: A rapidly increasing. Environ. Res. **2008**, *108*, 131–139.
109. Barnes, K.A.; Galgani, F.; Thompson, R.C.; Barlaz, M. Accumulation and fragmentation of plastic debris in global environments. Philos. Trans. R. Soc. **2009**, *364*, 1985–1998.
110. Teuten, E.L.; Saquing, J.M.; Knappe, D.R.U.; Barlaz, M.A.; Jonsson, S.; Björn, A.; Rowland, S.J.; Thompson, R.C.; Galloway, T.S.; Yamashita, R.; Ochi, D.; Watanuki, Y.; Moore, C.; Viet, P.H.; Tana, T.S.; Prudente, M.; Boonyatumanond, R.; Zakaria, M.P.; Akkhavong, K.; Ogata, Y.; Hirai, H.; Iwasa, S.; Mizukawa, K.; Hagino, Y.; Imamura, A.; Saha, M.; Takada, H. Transport and release of chemicals from plastics to the environment and to wildlife. Philos. Trans. R. Soc. B **2009**, *364*, 2027–2045.
111. US Marine Mammal Commission. *Marine Mammal Commission Annual Report to Congress*; Effects of Pollution on Marine Mammals: Bethesda, MD, 1996.
112. Derraik, J.G.B. The pollution of the marine environment by plastic debris: A review. Mar. Pollut. Bull. **2002**, *44*, 842–852.
113. Slip, D.J.; Green, K.; Woehler, E.J. Ingestion of anthropogenic articles by seabirds at Macquarie Island. Mar. Ornithol. **1990**, *18*, 74–77.
114. Van Frakener, J.A.; Bell, P.J. Plastic ingestion by petrels breeding in Antarctica. Mar. Pollut. Bull. **1988**, *19*, 672–674.
115. Bourne, W.R.P.; Imber, M.J. Plastic pellets collected by a prion on Gough Island, Central South Atlantic Ocean. Mar. Pollut. Bull. **2001**, *13*, 20–21.
116. Mato, Y.; Isobe, T.; Takada, H.; Kanehiro, H.; Ohtake, C.; Kaminuma, T. Plastic resin pellets as a transport medium for toxic chemicals in the marine environment. Environ. Sci. Technol. **2001**, *35*, 318–324.
117. Gregory, M.R. Plastics and South Pacific Island shores: Environmental implications. Ocean Coastal Manage. **1999**, *42*, 603–615.
118. Hess, N.A.; Ribic, C.A.; Vining, I. Benthic marine debris, with an emphasis in fishery-related items, surrounding Kodiak Island, Alaska 1994–1996. Mar. Pollut. Bull. **1999**, *38*, 885–890.
119. Ryan, P.G.; Connell, A.D.; Gardener, B.D. Plastic ingestion and PCBs in seabirds: Is there a relationship? Mar. Pollut. Bull. **1988**, *19*, 174–176.
120. Firestone, J.; Jarvis, C. Response and responsibility: Regulating noise pollution in the marine environment. J. Int. Wildl. Law Policy **2007**, *10*, 109–152.
121. Hatch, L.T.; Wright, A.J. A brief review of anthropogenic sound in the oceans. Int. J. Comp. Psychol. **2007**, *20*, 121–133.
122. National Research Council (NRC). *Ocean Noise and Marine Mammals*; National Academy Press: Washington, DC, 2003.
123. Samuel, Y.; Morreale, S.J.; Clark, C.W.; Greene, C.H.; Richmond, M.E. Underwater, low-frequency noise in coastal sea turtle habitat. J. Acoust. Soc. Am. **2005**, *117*, 1465–1472.
124. Gallaway, T.; Olsen, R.N.; Mitchell, D.M. The economics of global light pollution. Ecol. Econ. **2010**, *69*, 658–665.
125. Gliwicz, Z.M. A lunar cycle in zooplankton. Ecology **1986**, *67*, 883–897.

Cobalt and Iodine

Ronald G. McLaren
Soil, Plant, and Ecological Sciences Division, Lincoln University, Canterbury, New Zealand

Abstract
Cobalt (Co) and iodine (I) are two trace elements that are generally considered to be nonessential for the growth of higher plants.[1] However, both elements are essential nutrients for animals, particularly in the case of ruminants (sheep and cattle), and deficiencies of Co and I in grazing animals are not uncommon. Cobalt is also essential for nitrogen (N) fixation by micro-organisms such as rhizobium and blue-green algae, and can actually be toxic to plants if present at high concentrations in the soil. Most investigations of Co and I in the soil have concentrated on factors affecting the plant availability of these elements, with the aim of diagnosing potential deficiencies.

SOIL COBALT

The Co concentration in soils depends primarily on the parent materials (rocks) from which they were formed and on the degree of weathering undergone during soil development.[2] Cobalt tends to be most abundant as a substituent ion in ferromagnesian minerals, and therefore has relatively high concentrations in mafic and ultramafic rocks (rocks containing high or extremely high proportions of ferromagnesian minerals). Conversely, Co concentrations are relatively low in felsic rocks (rocks containing large amounts of silica-rich minerals) such as granite, and in coarse-textured quartz-rich sedimentary rocks (sandstones). Higher concentrations of Co may be associated with finer textured sediments (shales) in which Co has become surface adsorbed by, or incorporated into, secondary layer silicates by isomorphous substitution.[3] Typical concentrations of Co reported in different rock types are shown in Table 1. As a result of the large range in Co concentrations of soil parent materials, and variation in the degree of weathering, total soil Co concentrations also vary widely. However, the mean values reported for agricultural soils from many countries appear to have a somewhat restricted range of between approximately 2 and 20 mg/kg (Table 1).

Forms of Cobalt in Soils

Cobalt in soils, whether released from parent materials during soil development, or derived from anthropogenic contaminant sources, occurs in several different forms or associations. Cobalt may be present as 1) the simple Co^{2+} ion, or as complexes with various organic or inorganic ligands in the soil solution; 2) exchangeable Co^{2+} ions; 3) specifically adsorbed Co, bound to the surfaces of inorganic soil colloids (clays and oxides/hydrous oxides of Al, Fe, and Mn); 4) Co complexed by soil organic colloids; 5) Co occluded by soil oxide materials; and 6) Co present within the crystal structures of primary and secondary silicate minerals.[6] In some soils, there appears to be a particularly strong association between Co and manganese (Mn) oxides, especially in soils where Mn oxides occur as distinct nodules or coatings.[7,8]

Plant Availability of Soil Cobalt

The immediate source of Co for plant uptake is the soil solution. However, Co concentrations in the soil solution are extremely low, generally much less than 0.1 μg/mL.[1] Soil solution Co appears to be in equilibrium with Co adsorbed at the surfaces of soil colloids, and the distribution between solution and surface phases is strongly influenced by soil pH. As pH increases, soluble Co decreases.[6] Similarly, soils with high capacities to adsorb Co, particularly those soils with high Mn oxide contents, also have low solution Co concentrations.[9] Thus, in addition to soils with low total Co concentrations, the plant availability of Co may be low in soils with high pH or high Co adsorption capacities. Cobalt availability is also influenced by soil moisture status, availability increasing under waterlogged conditions.[10]

Determination of soil Co dissolved by various extractants is the most common way of assessing the plant availability of soil Co. Ideally, the forms of Co extracted should include any soluble and exchangeable Co, together with any solid-phase forms of Co that are able to move readily into the soil solution in response to changes in solution Co concentrations.[6] The two extractants used most commonly for this purpose are 2.5% acetic acid and solutions of ethylenediaminetetraacetic acid (EDTA). However, the ability of these extractants to accurately assess Co availability appears to be somewhat limited.[11,12]

Encyclopedia of Environmental Management DOI: 10.1081/E-EEM-120042647

COBALT DEFICIENCY

Cobalt deficiency in grazing sheep and cattle was first diagnosed in the 1930s, initially in New Zealand, Australia, and Scotland.[13–15] The condition causes animals to loose their appetite, become weak and emaciated, suffer severe anemia, and eventually die. Subsequently, it was shown that Co is a constituent of both vitamin B_{12} and a closely related coenzyme, and that Co deficiency is in effect a deficiency of vitamin B_{12}.[16] It has been concluded from field studies that pasture containing Co below 0.08 mg/kg for sheep or below 0.04 mg/kg for cattle is unlikely to meet nutritional requirements in terms of maintaining adequate serum and liver vitamin B_{12} concentrations, and healthy growth rates.[17,18]

Cobalt deficiency is commonly treated by applying Co-containing fertilizers to pastures, usually at very low rates, e.g., 350 g/ha/yr of $CoSO_4 \cdot 7H_2O$. However, on some soils such treatments appear to be ineffective.[11] Alternative treatments for Co deficiency involve injecting the animal with vitamin B_{12}, the use of Co drenches, or the insertion of slowly dissolving Co "bullets" in the animal's rumen.

SOIL IODINE

The range of I concentrations found in rocks and soils is shown in Table 1. Iodine occurs as a minor constituent in various minerals, where it can replace anions such as OH^-, SiO_4^{4-}, and CO_3^{2-}, and has relatively low concentrations in most types of rock.[19] Highest I concentrations are generally found in fine-grained sedimentary rocks (Table 1). However, soil I concentrations are generally higher than in the rocks from which they have been derived, a fact attributed predominantly to atmospheric accessions of I.[20] Iodine is known to be present in the atmosphere in vapor form and associated with particles of dust. In coastal areas, accession of I may also be related to sea spray.[1]

Table 1 Cobalt and iodine concentrations in rocks and soils.

	Co concentration (mg/kg)	I concentration (mg/kg)
Rock type		
Ultramafic (e.g., serpentinite)	100–300	0.01–0.5
Mafic (e.g., basalt, gabbro)	30–100	0.08–50
Intermediate (e.g., diorites)	1–30	0.3–0.5
Felsic (e.g., granites, gneiss)	<1–10	0.2–0.5
Sandstones	0.3–10	0.5–1.5
Shales/argillites	11–40	2–6
Limestones	0.1–3.0	0.5–3.0
Soils		
Complete range	0.1–300	<0.1–25.4
Range of mean values	2–21.5	1.1–13.1

Source: Kabata-Pendias and Pendias,[1] Aubert and Pinta,[4] and Swaine.[5]

Forms of Iodine in Soils

Information of the forms of I in soils is limited, most published analyses of soils have determined total I concentrations only. Of the three most common forms of I, elemental (I_2), iodide (I^-), and iodate (IO_3^-), it seems likely that most I in soils is present as iodide or possibly as elemental I. The presence of iodate in soils has also been postulated, but would require high oxidation conditions in neutral or alkaline soils.[20] Indeed, there is some evidence that when iodate is added to soils it is rapidly reduced to elemental I or iodide by soil organic matter.[21] There is also evidence that, under some conditions, elemental I can be volatilized from soils.[22]

Most I in soils appear to be associated with soil organic matter and iron (Fe) and aluminium (Al) oxides, materials by which both iodide and elemental I are known to be strongly adsorbed.[21,23] Indeed the distribution of I in soil profiles appears to follow the distribution of these soil components.[24] The atmospheric accessions of I and the affinity between I and organic matter often result in maximum concentrations of I in the surface horizons of soils.[20,24]

Plant Availability of Soil Iodine

Interest in soil analysis as a means of assessing plant availability of I has been minimal,[20] and relationships between the I-status of soils and the concentrations of I in plants appear to be poor.[1] Even soil extractants designed to determine the most soluble forms of I in soils do not provide a good indication of I availability to plants.[25] Plants are capable of absorbing I directly from the atmosphere,[1] and plant species or varietal differences appear to have a greater influence on plant I concentration than soil I status.[20,26] Dicotyledonous pasture species (e.g., clovers) generally have higher I contents than do grasses.[26] Concentrations of I in plants may be reduced by liming,[25] by the application of N fertilizer,[26] and by the application of farmyard manure.[25] Seasonal effects on pasture I concentrations have also been observed, with decreases in the summer, and slight increases in the autumn.[26]

IODINE DEFICIENCY

Low concentrations of I in food and water have been associated with the occurrence of endemic goitre (enlargement of the thyroid gland) in humans and farm livestock.[27] Early work suggested a close relationship between goitre incidence and low soil levels of I, however, it is now recognized that other factors are also involved. In particular, the presence of a group of substances known as goitrogens, which occur in various plant species, has been shown to reduce thyroid hormone synthesis and metabolism.[20] In the absence of goitrogens, it is considered that diets containing 0.5 mg I/kg DM will more than adequately meet the I requirements of all classes of animals, while levels as low as 0.15 mg/kg might be sufficient to meet the requirements for growing animals.[28] In the presence of goitrogens, I requirements may be as high as 2 mg I/kg DM.[28] Attempts to increase pasture I concentrations with I-containing fertilizers have been generally ineffective, with very low recoveries of the applied I.[25,26] Direct supplementation of livestock is normally the preferred way to increase I intakes.

CONCLUSIONS

Cobalt and Iodine deficiencies in livestock result from low soil concentrations and/or low availability of these trace elements for uptake by pasture plants. Plant availabilities of Co and I are determined by several factors including soil forms, soil sorption properties, soil pH, soil moisture status, season, plant species, and fertilizer applications. Deficiencies of Co and I can be prevented by application of fertilizers (Co) to the soil, or by direct treatment of livestock (Co and I).

REFERENCES

1. Kabata-Pendias, A.; Pendias, H. *Trace Elements in Soils and Plants*; CRC Press: Boca Raton, 1984; 238–246.
2. Mitchell, R.L. Cobalt in soil and its uptake by plants. In *Atti del IX Simposio Internazionale di Agrochimica su La Fitonutrizione Oligominerale*; Punta Ala: Italy, 1972; 521–532.
3. Hodgson, J.F. Chemistry of micronutrient elements in soils. Adv. Agron. **1963**, *15*, 119–159.
4. Aubert, H.; Pinta, M. *Trace Elements in Soils, Development in Soil Science 7*; Elsevier Scientific Publishing Co.: Amsterdam, 1977; 395 pp.
5. Swaine, D.J. *The Trace Element Content of Soils*; CAB: Harpenden, England, 1955; 157 pp.
6. McLaren, R.G.; Lawson, D.M.; Swift, R.S. The forms of cobalt in some Scottish soils as determined by extraction and isotopic exchange. J. Soil Sci. **1986**, *37*, 223–234.
7. Taylor, R.M.; McKenzie, R.M. The association of trace elements with manganese minerals in Australian soils. Aust. J. Soil Res. **1966**, *4*, 29–39.
8. Jarvis, S.C. The association of cobalt with easily reducible manganese in some acidic permanent grassland soils. J. Soil Sci. **1984**, *35*, 431–438.
9. Tiller, K.G.; Honeysett, J.L.; Hallsworth, E.G. The isotopically exchangeable form of native and applied cobalt in soils. Aust. J. Soil Res. **1969**, *7*, 43–56.
10. Adams, S.N.; Honeysett, J.L. Some effects of soil waterlogging on Co and Cu status of pasture plants grown in pots. Aust. J. Agric. Res. **1964**, *15*, 357–367.
11. McLaren, R.G.; Lawson, D.M.; Swift, R.S.; Purves, D. The effects of cobalt additions on soil and herbage cobalt concentrations in some S.E. Scotland pastures. J. Agric. Sci. Camb. **1985**, *105*, 347–363.
12. McLaren, R.G.; Lawson, D.M.; Swift, R.S. The availability to pasture plants of native and applied soil cobalt in relation to extractable soil cobalt and other soil properties. J. Sci. Food Agric. **1987**, *39*, 101–112.
13. Grange, L.I.; Taylor, N.H. Bush sickness. Part IIA. The distribution and field characteristics of bush-sickness soils. Bull. N. Z. Dept. Sci. Ind. Res. **1932**, *32*, 21.
14. Underwood, E.J.; Filmer, J.F. The determination of the biologically potent element (cobalt) in limonite. Aust. Vet. J. **1935**, *11*, 84–92.
15. Corner, H.H.; Smith, A.M. The influence of cobalt on pine disease in sheep. Biochem. J. **1938**, *32*, 1800–1805.
16. Smith, R.M.; Gawthorne, J.M. The biochemical basis of deficiencies of zinc, manganese, copper and cobalt in animals. In *Trace Elements in the Soil–Plant–Animal System*; Nicholas, D.J.D., Egan, A.R., Eds.; Academic Press: New York, 1975; 243–258.
17. Andrews, E.D. Observations on the thrift of young sheep on a marginally cobalt deficient area. N. Z. J. Agric. Res. **1965**, *8*, 788–817.
18. Gardner, M.R. *Cobalt in Ruminant Nutrition: A Review*; Department of Agriculture: Western Australia, 1977; 602–620.
19. Goldschmidt, V.M. *Geochemistry*; Oxford University Press: Oxford, 1954.
20. Fleming, G.A. Essential micronutrients. II. Iodine and selenium. In *Applied Soil Trace Elements*; Davies, B.E., Ed.; John Wiley and Sons, Ltd.: Chichester, 1980; 199–234.
21. Whitehead, D.C. The influence of organic matter, chalk, and sesquioxides on the solubility of iodide, elemental iodine, and iodate incubated with soil. J. Soil Sci. **1974**, *25*, 461–470.
22. Whitehead, D.C. The volatilisation, from soils and mixtures of soil components, of iodine added as potassium iodide. J. Soil Sci. **1981**, *32*, 97–102.
23. Whitehead, D.C. The sorption of iodide by soils as influenced by equilibrium conditions and soil properties. J. Sci. Food. Agric. **1973**, *24*, 547–556.
24. Whitehead, D.C. Iodine in soil profiles in relation to iron and aluminium oxides and organic matter. J. Soil Sci. **1978**, *29*, 88–94.
25. Whitehead, D.C. Uptake by perennial ryegrass of iodide, elemental iodine and iodate added to soil as influenced by various amendments. J. Sci. Food. Agric. **1975**, *26*, 361–367.
26. Hartmans, J. Factors affecting the herbage iodine content. Neth. J. Agric. Sci. **1974**, *22*, 195–206.
27. Underwood, E.J. *Trace Elements in Human and Animal Nutrition*, 3rd Ed.; Academic Press: New York, 1971; 543 pp.
28. Agricultural Research Council. In *The Nutrient Requirements of Ruminant Livestock*; Commonwealth Agricultural Bureau: Slough, U.K., 1980; 351 pp.

Community-Based Monitoring: Ngarenanyuki, Tanzania

Aiwerasia V.F. Ngowi
Tanzania Association of Public Occupational and Environmental Health Experts, and Department of Environmental and Occupational Health, Muhimbili University of Health and Allied Sciences (MUHAS), Dar-es-Salaam, Tanzania

Larama M.B. Rongo
Muhimbili University of Health and Allied Sciences, Dar-es-Salaam, Tanzania

Thomas J. Mbise
Tanzania Association of Public Occupational and Environmental Health Experts, Dar-es-Salaam, Tanzania

Abstract

Community-based monitoring was initiated in Ngarenanyuki, Tanzania, to study the impacts of pesticides on health and the environment. Twenty-five Ngarenanyuki community representatives were trained to monitor such impacts and to reduce the risks. Data collection tools were then developed and pretested at Mlangarini, training was conducted through seminars and public meetings, pesticide monitoring teams were established, and data were collected. Monitoring showed the use of Class 1b (chlorfenvinphos) and obsolete pesticides such as dichloro diphenyl trichloroethane (DDT), that pesticides were both available and affordable to farmers, and that more than three pesticides were often mixed per application. Two-thirds of the farmers showed pesticide poisoning, with cypermethrin–profenofos mixture and profenofos being the most prevalent. Observation revealed such symptoms as skin and eye irritation. This initiative helped farmers to assess their own pesticide health risks and to develop action plans.

INTRODUCTION

Community-based monitoring was initiated in Ngarenanyuki, Tanzania, to study the impacts of pesticides on health and the environment. This entry is organized and divided into the following main sections: "Introduction," "Methodology," "Results and Discussion," and "Conclusion." Illustrations are included in the "Introduction," "Methodology," and "Results and Discussion" sections. The main goal of the study was to reduce exposure to pesticides among the farmers in Ngarenanyuki by training the farmers on health impacts related to exposure to pesticides, how to monitor such impacts, and how to reduce the risks.

Incidences of poisoning from pesticides are estimated to be highest in developing countries, despite the higher use of pesticides in developed countries.[1,2] The monitoring of pesticides and their health impacts on farmers and the public in general, which is normally performed by qualified researchers, is not sufficiently practiced in many developing countries, owing to financial constraints and to competing research interests. It therefore makes more sense to empower communities themselves to monitor the impact of pesticides and to take decisions that might reduce the risks to themselves and to their environment. Community-based monitoring of the impacts from pesticides enable those communities to determine whether or not the chemicals they are already exposed to, or might be exposed to, present any sort of hazard to their health and a potential threat to their environment.

Community pesticide-surveillance methods have been successfully used in the Asia Pacific[3] and could therefore be considered appropriate in Tanzania and other Southern African countries, both for establishing better data on the extent of pesticide poisoning and to raise awareness among farmers themselves. Using their own system of observation and evaluation of risks, for example, Malaysian plantation workers have developed the Community Pesticides Action Monitoring (CPAM) approach, in which they succeeded in documenting the health effects of airborne pesticides, identifying paraquat in particular as a major problem.[4]

They then proceeded to take action to prevent further exposure of plantation workers to paraquat. Communities in Kasargod District, Kerala (India), after investigation, monitoring, and documentation, using a CPAM approach, identified endosulfan as the major pesticide causing health and environmental problems and subsequently called for the ban on endosulfan to prevent further exposure and damage to the communities.[5]

Incorrect pesticide handling and management is thus known to be unsafe to both human and environmental

Carbon – Cyanobacteria

Encyclopedia of Environmental Management DOI: 10.1081/E-EEM-120046906

health and jeopardizes biodiversity.[6] It is also further evident that most rural communities in Tanzania depend on farming and agribusinesses to earn their living. However, traditional agricultural production in the country has come under continuous pressure from globalization and other market forces, with the result that high-input agriculture has come to play an increasingly major role in the economies of rural communities. Although the use of pesticides in combating pests and diseases is widely encouraged among the farmers in these communities to promote production, less emphasis has been placed on safety practices and the proper handling and management of materials.

Ngarenanyuki Ward gives an example of a community in Tanzania where the majority of vegetable farmers believed that, without pesticide use, crop production would have been impossible. In a previous study carried out by the Work and Health in Southern Africa (WAHSA) team (unpublished), it was found that mixing three to five different types of pesticides in a single spray mix was a common practice in Ngarenanyuki, and that farmers did not understand what was written on the label or the meaning of the colors on the containers. They simply applied pesticides because a neighbor had applied them, and not because they had identified a particular pest problem. The farmers were also found to have mixed pesticides without following the doses recommended on the label, sometimes doubling or trebling the dosage regardless.

Retail outlets for pesticides in Ngarenanyuki were also found not to have been registered with the regulatory authority and therefore appeared to be selling pesticides illegally. It was further noted that, because the shop owners tended to repackage or dispense pesticides in other containers, they were observed on occasion to be left with empty pesticide containers, which they apparently destroyed by burning them at the marketplace. Farmers were also observed in some instances to have stocked substantial amounts of pesticides to cater for the whole year, owing to perceived shortages in the local village and the reported distances they had to travel in order to purchase pesticides in towns, frequently mentioned as Arusha or Moshi. Some of these stored pesticides were also observed to have become obsolete, which were then likely to create fresh problems of disposal as shown in Fig. 1.

The unintended outcomes of pesticide exposures are difficult to reverse once they have been established and are in themselves expensive. Although advances in acute pesticide-poisoning surveillance and treatment in developed countries have led to some achievements in control, pesticide poisoning remains a public health problem globally, particularly in developing countries.[7] Those applying pesticides need to understand the effects of these chemicals to the environment and to their own health and the resulting costs. Alternative pest management strategies that are cheaper and friendlier to end users and the environment need to be promoted.

A workshop was organized for Southern Africa Development Community (SADC) registrars of pesticides in Arusha on October 13–14, 2006. Participants suggested that WAHSA-TPRI (Table 1) should pilot the tool used in the Asia Pacific to establish a systematic mechanism for pesticide monitoring and data collection, with a view to determining the extent of pesticide exposures, injuries, and diseases at the community level. WAHSA-TPRI then selected Ngarenanyuki as the study area, based on their working experience in Northern Tanzania in health hazards posed by pesticides and on the knowledge that farmers in the area were especially at risk with regard to pesticide poisoning.

This entry reports on the process involved in the establishment of a community-monitoring team in Ngarenanyuki and on the preliminary results of the monitoring

Fig. 1 Hazardous practices observed in Ngarenanyuki, 2006–2007. (left) Haphazard disposal of empty containers. (right) Dispensing/repackaging of pesticides in retail shops.

Table 1 The WAHSA program was established in October 2004 as a regional initiative in Southern Africa to build capacity in the region in occupational health.

One of its key programs was its project *Action on Health Impacts of Pesticides*, which aimed to:

- Improve pesticide-safety materials for the SADC region
- Intervene to reduce pesticide usage
- Improve on agricultural policies and pesticide registration
- Enhance knowledge and improve surveillance about pesticide exposures and health impacts in the region
- Foster a strong regional network for information exchange and consultation

exercise. The authors hope that the findings will facilitate a process to identify those resources required to reduce pesticide use, the development of an action plan to access and mobilize these resources, and the further establishment of an effective system of communication among members of the community on pesticide use and access to any other information with regard to pesticide poisoning.

METHODOLOGY

This initiative was intended to pilot the CPAM approach that has already been used successfully in Asian countries. The WAHSA-TPRI Team was trained on the subject through their link with a non-governmental organization, AGENDA for Environment and Responsible Development, and was then employed to mobilize the community in Ngarenanyuki.

The initiative adopted a participatory research methodology by involving farmers in the collection and analysis of pesticide-related data. Data collection tools were developed by making use of Community Pesticide Action Kits (CPAKs), and training materials were developed in the regional language of Swahili. As a result of these activities, the community was sensitized and a subsequent rise in awareness was noted. CPAKs were produced by an ASEAN team of citizens' groups and farmer schools as a tool for action and advocacy, encouraging community education/empowerment. It contains modules that address various aspects of concern such as Warning! Pesticides are a Danger to Your Health; Breaking the Silence: Pesticides in Plantations; Profiting from Poisons: The Pesticides Industry; Drop Pesticides! Build a Sustainable World; Pesticides Destroy our World; Women and Pesticides; Keeping Watch: Pesticides Laws; How to say NO! to Pesticides: Community Organizing; and Seeking out the Poisons: A Guide to Community Monitoring. The modules are not complete in themselves but need additional materials in local languages.

After securing the community's consent and the involvement of farmers' representatives and communities in capacity building, the program of community-based data collection and analysis was started and the monitoring exercise was implemented. Selected farmers worked in collaboration with the WAHSA experts to monitor and record issues related to pesticide use and exposures in the Ngarenanyuki villages.

Through a series of village meetings, the ward government in collaboration with the ward extension officer invited the farmers to participate in the training. Thirty farmers were selected by the farmers themselves from two villages (Uwiro and Olkung'wado) to represent each subvillage. No farmers were selected from Ngabobo, Kisimiri Juu, and Kisimiri Chini as communication became difficult. Ngarenanyuki as a whole is situated between Mt. Kilimanjaro and Mt. Meru, the first and third tallest mountains in Africa, respectively. The terrain thus consists of rocks, hills, valleys, rivers, and streams, which make some areas impassable during the wet season. One of the villages left out did not cooperate well with the others as they are believed to grow cannabis, a plant that is illegal in Tanzania.

A 6-day training of 25 representative farmers from the Ngarenanyuki was conducted by WAHSA-TPRI scientists, who had expertise in agricultural extension, agronomy, toxicology, entomology, plant pathology, and environmental science, and covered the following topics:

1. Pesticide use and their impacts on human health and environment, where farmers learned about pesticide use around the world, including examples of the negative impacts of pesticides on human health and the environment.
2. Pesticide identification and classification according to their acute toxicity, where farmers learned to identify the types of pesticides used and how they are classified according to their acute toxicity by the World Health Organization (WHO). One such classification the farmers learned was according to the different chemical families, such as organochlorines, organophosphates, carbamates, pyrethroids, and so on.
3. Pesticide label identification and interpretation, where farmers learned to read and understand pesticide labels, the various pictograms, and colored warning signs on containers, including the interpretation of the various toxicity symbols.
4. Pesticide handling and management, where farmers learned how to handle pesticides properly to safeguard themselves, their families, their neighbors, and their surroundings. They learned to observe which pesticides are used in their area, how they are used, and to observe the protective measures that are taken during mixing and application. They also learned how they might get contaminated during the handling of pesticides.
5. Pesticide storage and disposal of empty pesticide containers, where farmers learned about proper storage and disposal of surplus pesticides and their empty containers. They also learned how the improper storage of pesticides and the careless disposal of empty containers could form a risk for children, foodstuff, freshwater supplies, farm animals, and so on.

6. Recognizing the signs and symptoms of pesticide poisoning, where farmers learned the different signs and symptoms of pesticide poisoning and how to recognize them. They also learned how to distinguish these from other signs and symptoms that are simply due to poor health.
7. Pest identification and management, where farmers learned how to identify different insects, distinguishing genuine pests from more beneficial insects and symptoms of common vegetable diseases. Farmers were introduced to the basic principles of pest control methods. They were thus equipped with a practical knowledge of insects as an important component in pest management and on how to protect their crops from insect attack with a view to reducing the insecticide load on the environment.
8. Introduction to Integrated Pest Management (IPM), where farmers learned the principles of IPM as a sustainable approach to managing pests by combining biological, cultural, physical, and chemical tools in a way that minimized financial, health, and environmental risks. Farmers were further informed that one of the primary missions of IPM was to assist them in producing profitable crops, using environmentally and economically sound approaches.
9. Spraying equipment and techniques, where farmers learned about spray equipment [such as the knapsack sprayer, the motorized ultra low volume (ULV) sprayer, and so on]. They also learned about the handling, maintenance, and spraying techniques with regard to this equipment.
10. Reducing pesticide costs, where farmers learned to assess actual costs of pesticide use to include direct and indirect costs.
11. Participatory data collection methodology, where farmers learned about methods of data collection and analysis. They were introduced to the kind of data needed and data collection procedures, using different techniques with different data collection tools. Demonstration and practical sessions on how to handle and record data were also held with the farmers.

Establishment of Community Pesticide Monitoring Team

The 25 trained representative farmers were divided into teams of at least three people each, who then became the focal point for monitoring and recording of all pesticide incidences in Ngarenanyuki, and who also worked closely with the WAHSA-TPRI Team, including those who had been working with communities in the Arumeru district in research and training in their respective fields.

Data Collection

Consent forms were developed to be completed by those individuals who agreed to participate in the Community Monitoring Project. Three data collection tools were developed: a questionnaire, a checklist, and a self-surveillance form. They were designed to cover all areas of interest in community pesticide monitoring through interviews, observation, and self-examination of pesticide exposure.

The tools were pretested for validity and consistence in Mlangarini ward with a sample of 30 farmers. Adjustments and other improvements were made to the tools prior to final data collection in Ngarenanyuki. A self-surveillance form without pictograms was preferred, owing to some confusion arising from the meaning of the pictograms.

Farmers were organized into teams to conduct crop surveys and recognize damages, assess losses, and collect insect pests for identification. The farmers went out into their respective villages to collect information on pesticides used, perceptions on pesticide hazards, poisoning, and symptoms using the questionnaire. They also used the checklist to observe and record pesticides available in the area, means of storage, and use of protective equipment. Each farmer contacted by the team members was asked to do self-surveillance and record pesticide use conditions and practices, as well as poisoning signs and symptoms experienced.

The Community Pesticide Monitoring Teams needed technical support and close follow-up to ensure consistency in data collection and in transferring the knowledge gained to the entire Ngarenanyuki community. However, the teams were fully prepared in getting the message across to the community and to involve them in providing relevant information regarding pesticide issues. Moreover, the village leaders were made responsible for making a close follow-up of the team and the villagers involved.

Data Analysis

Analysis of information collected was performed using two different approaches. Structured interviews were conducted by the farmers using questionnaires, and the information was tallied, before it was tidied up and analyzed with the aid of the SPSS (Statistical Package for the Social Sciences) computer software to obtain frequencies. The data that were collected through observation on the basis of checklists were manually analyzed by the farmers themselves using flip charts and colored pens.

RESULTS AND DISCUSSION

The training of the farmers was meant to prepare the community in taking responsibility themselves for monitoring the negative impacts of pesticides in their area. Subject matter specialists conducted the training with the aim of building the capacity of participants to make the right judgement and decisions when dealing with pest and pesticide issues. During the training sessions, farmers expressed keen interest in learning how to recognize insects (beneficial and harmful), to recognize the signs and symptoms of pesticide exposure, to practice safer pesticide handling and management, to understand proper spraying techniques and the maintenance of knapsacks, and to understand the benefits of participatory data collection techniques.

General Information

The data from the farmers in Ngarenanyuki Ward were collected between February and April 2007 by the farmers trained on pesticide monitoring and analyzed using the SPSS computer software. While the majority of the 120 farmers were males (90%), the average highest education level recorded was that of primary education (76%); hence, functional literacy was not a problem in this community. Agriculture (98.3%) was the major income-generating activity, although some farmers also kept livestock.

Pesticides Used in Ngarenanyuki

Thirty different types of pesticides commonly used in Ngarenanyuki were identified by the farmers. The major groups of pesticides used included insecticides, fungicides, and, to a lesser extent, herbicides. The most widely used insecticide and fungicide were chlorpyrifos (72.5%) and mancozeb (69.2%), respectively. Only 36.6% of the pesticides used in Ngarenanyuki had full registration, while some had provisional or experimental registration. It is mandatory for pesticides intended for use in Tanzania to go through a registration process, which involves efficacy and quality tests before they are approved for general, restricted, or experimental use. Pesticides under experimental use are not expected to be sold in retail shops. There was also the presence of one class 1b pesticide (chlorfenvinphos) and banned/restricted pesticides such as DDT.

Pesticide Availability, Affordability, and Application

The majority of pesticides (86.7%) were locally available in Ngarenanyuki, and a considerable proportion of farmers (65.8%) could afford to buy pesticides. Those unable to buy mostly obtained their pesticides on credit, and paid after harvesting. Most farmers (68.3%) claimed to have a pesticide application timetable, the most prominent approach being that of applying pesticides whenever insects or disease symptoms appeared.

Pesticide Mixing

The majority of farmers (90%) mixed more than one pesticide in a single application. The main reason given for mixing was to kill all pests and diseases at a go and to improve the quality of leaves and fruits (54%) in the field. A few (25%) said they preferred mixing to ease the workload and in order to cover larger areas with one treatment, while some (55%) said they had simply followed the pesticide retailer's advice. The mixing exercise was widely done in respective farms (89%). The common mixtures normally contained more than one fungicide and one insecticide, although some mixtures were found with around three fungicides and two insecticides.

Frequency of Pesticide Application, Number of Risk Days per Year, Spraying Equipment, and Pesticide Storage

The scale of environmental pollution was fairly evident as pesticides could be smelled all over the farms and in nearby residential areas, causing health problems (such as cough, sneezing, excessive difficulty breathing, and chest pains) to both sprayers and those who found themselves in the path of the sprays. The farmers worked out 52 risk days per year, as the majority (73%) of 120 farmers applied pesticides once a week and fewer (18%) applied the pesticides twice a week. The most common spraying equipment was the knapsack (76%), while in some cases (21%), buckets were also used.

Most respondents (57%) stored pesticides in a pesticide store, and in some cases, storage took place in sitting or living rooms (12%), in general stores (13%), and in bedrooms (7%). Pesticides were also found to be stored in toilets (1%). The choice of storage areas was often determined by their offering protection against thieves.

Adherence to Pesticide Label Instruction

The study by the farmers revealed that many of them did read the instructions on the pesticide containers, but only few actually followed the instructions as shown in Table 2. An example is the mixing of ULV formulations in water sprays while instructions given on the label are for direct application without dilution. The following were the arguments put forward for not following instructions: that some labels were only written in English, that the farmers were not familiar with conventional signs and symbols, and that some containers had no labels at all, having been dispensed from another container. The repackaging and dispensing of smaller quantities was found to be a common, albeit illegal, practice and it was felt that this needed greater attention, since this practice has negative implications for

Table 2 Adherence to pesticide label instruction by farmers in Ngarenanyuki during a previous farming season (December 2006 to March 2007).

Response toward pesticide label instructions	Number of farmers ($N = 120$)	% of farmers
Always read instructions	72	60
Follow instructions	45	38
Sometimes follow instructions	34	28
Sometimes read instructions	28	23
Trained on pesticide issues	16	13
Get information on pesticides	7	6

Table 3 Modes of disposal of empty pesticide containers in Ngarenanyuki (December 2006 to March 2007).

Mode of disposal	Number of farmers (*N* = 120)	% of farmers
Burn	41	34
Throw away on the farm	35	29
Bury in the farm surroundings	18	15
Sell back to pesticide vendors	8	7
Throw in the toilet	7	6
Use for other domestic uses	2	2

efforts to reduce the worst effects of pesticide poisoning by the implementation of proper labeling and instructions.

Disposal of Pesticide Containers

The major mode of disposal of empty pesticide containers by most of the farmers was by simply throwing the containers away in the farm surroundings and by burning. It was observed that some empty containers from the pesticide retail shops were also thrown or burnt at the marketplace. It was also revealed that some farmers did reuse empty pesticide containers for domestic purposes such as buying cooking oil and kerosene and for local brewing (Table 3).

Pesticide Poisoning

The self-surveillance form was used without pictograms, to record the signs and symptoms of pesticide poisoning, owing to the confusion arising from the use of pictograms. During the pretest, farmers did not understand what the pictograms meant, and since the majority were able to read and write, it was agreed that there would be no need to include pictograms in the surveillance form and that the list of signs and symptoms provided in the form was sufficient until proper research had been undertaken to determine what visual aid would be considered appropriate for the target audience.

The majority of the farmers (69.2%) had experienced pesticide poisoning in the previous farming season, owing to exposure, much of which had occurred more than 3 times to a single farmer. Pesticide poisoning was characterized by signs

Table 4 Pesticide poisoning, circumstances, and action taken in Ngarenanyuki (December 2006 to March 2007).

Event		Number of farmers (*N* = 120)	% of farmers
Pesticide poisoning	Affected by pesticides in the last farming season	83	69
	Not sure	18	15
Occurrence of effects in the last farming season	Once	11	9
	Twice	12	10
	Thrice	9	8
	More than three times	26	22
Pesticides used	Profenofos + cypermethrin	32	27
	Profenofos	25	21
	Mancozeb (Dithane)	14	12
	Endosulfan	14	12
	Triadimenol	12	10
	Chlorothalonil	12	10
	Lambda-cyhalothrin	8	7
	Mancozeb (Ivory)	8	7
	Deltamethrin	1	1
	Copper sulfate	1	1
Action taken after pesticide poisoning	Drank milk	52	43
	Went to the hospital	34	28
	No action taken	4	3
	Washed with water	2	2
Number of times admitted due to pesticide poisoning	Once	20	17
	Twice	23	19
	Thrice	8	7
	More than three times	69	58

and symptoms known from previous studies to be related to pesticide exposures. Cypermethrin–profenofos mixture and profenofos were mostly associated with poisoning, and the action taken by many of those exposed (43.3%) was to drink milk, while a few respondents had attended hospitals for a proper medical examination. A considerable high proportion (57.5%) had been admitted more than 3 times, owing to pesticide poisoning (Table 4). Validation of poisoning through biological monitoring was not possible during this pilot stage but has been planned in future surveillances.

The action taken by the 25 farmers who fully participated in the pilot study was to intensify the training by initiating capacity-building sessions in all villages in Ngarenanyuki. They held community pesticide monitoring training in every village meeting, gave feedback to the WAHSA-TPRI Team on the farmers' reaction, and suggested what further input they needed from the team.

CONCLUSION

This pilot enabled the building of Ngarenanyuki farmers' capacities to assess their own health and environment as far as pesticides were concerned, analyze the situation, develop a plan of action, and work toward improving their condition. It facilitated the farmers' capacity so that they could take control and work with pesticides more safely and so become healthier. This program therefore works to benefit not only the farming community in the long run but also those consumers who would otherwise be forced to eat contaminated crops, and it contributes to the health of the environment as a whole.

Relevant data relating to pesticides, such as their availability, their usage, the farmers' handling practices, risk perception, and behavior, all gathered during the pilot project, enabled farmers in Ngarenanyuki and the WAHSA-TPRI Team to properly document the incidents and adverse events resulting from pesticide use. The initial evaluation of the association of the observed adverse event and pesticide exposure revealed that different pesticide-related tasks gave rise to signs and symptoms of pesticide poisoning and that skin and eye problems, for example, needed more attention during interventions.

The impact of the sensitization and awareness-raising seminars has been dramatic. The disposal site at the marketplaces vanished and the mistake of mixing ULV formulations with water has also been abandoned by the trained farmers and their associates. The farmers realized that the formulation was suspended in water and they were spraying water in some areas instead of pesticide. The formation of the community monitoring teams enhanced the whole process of data collection and action being taken. This has also provided a base for the sustainability of the project as the team continues to be in close contact and collaboration with the WAHSA-TPRI Team through the training of other farmers and in responding to their queries on pests, pesticides, monitoring pesticide use, their application, and the disposal of obsolete pesticides and empty pesticide containers.

The project was well received by the Ngarenanyuki community and has shown that if it is applied elsewhere, it will help in changing risk behaviors and in reducing the negative impact of pesticide exposures in communities. It is therefore recommended that the program be implemented systematically in Ngarenanyuki and be extended to other communities in Tanzania such as Mang'ola, in Karatu District, where the current use of pesticides appears to be indiscriminate.

ACKNOWLEDGMENTS

We are indebted to the small-scale vegetable farmers in Ngarenanyuki for their cooperation. This project could not have been implemented without support from AGENDA for Environment and Responsible Development for which we are grateful. We appreciate the work done by the WAHSA-TPRI Team and the contribution of experts from Selian Agricultural Research Institute (Dr. Hussein Mansoor) and Mikocheni Research Institute (Dr. Ruth Minja) during the farmers' training.

REFERENCES

1. WHO. *Public Health Impact of Pesticide Used in Agriculture*; Geneva, 1990.
2. ILO Chemicals in the Working Environment. In *World Labour Report*; International Labour Office: Geneva, Switzerland, 1994.
3. Murphy, H.H.; Hoan, N.P.; Matteson, P.; Abubakar, A.L. Farmers' self-surveillance of pesticide poisoning: A 12-month pilot in northern Vietnam. Int. J. Occup. Environ. Health **2002**, *8* (3), 201–211.
4. Rengam, S. Breaking the silence: Women struggle for pesticide elimination. In *Silent Invaders*; Jacobs, M., Dinham, B., Eds.; Zed Books Ltd.: London, 2003.
5. Quijano, R.F. *Endosulfan Poisoning in Kasargod, Kerala, India: Report on a Fact Finding Mission*; Pesticide Action Network Asia and the Pacific: Penang, 2002.
6. Ngowi, A.V.F.; Mbise, T.J.; Ijani, A.S.M.; London, L.; Ajayi, O.C. Smallholder vegetable farmers in Northern Tanzania: Pesticides use practices, perceptions, cost and health effects. Crop Prot. **2007**, *26*, 1617–1624.
7. Ngowi, A.V.F. *Health Impact of Exposure to Pesticides in Agriculture in Tanzania*; PhD Thesis, Acta Universitatis Tamperensis 890; University of Tampere, 2002.

Composting

Nídia Sá Caetano
Chemical Engineering Department, School of Engineering (ISEP), Polytechnic Institute of Porto (IPP), and Laboratory for Process, Environmental and Energy Engineering, Porto, Portugal

Abstract

Biological aerobic treatment of organic waste (composting), as a way of recycling organic matter (OM) into soil, is of primary importance. Although a relatively simple and ancient process, it needs careful project and operational considerations. The process can be applied either at a small scale (home composting) or at a large scale (centralized composting), using a very simple and common technology (such as constructing a pile of OM and letting it degrade naturally) or a more sophisticated one (such as vertical or horizontal reactors, rotating drums, etc.), taking advantage of a consortia of microorganisms that develop and degrade the OM. Also, an alternative of biological biowaste degradation performed by worms (vermicomposting) is presented. In this entry, a brief history of composting will be presented, along with some details on the most important technologies that can be used and the most important operational parameters that can condition the production of good-quality compost.

INTRODUCTION: BIOLOGICAL WASTE (BIOWASTE)

Solid waste composition has varied since ancient times depending on the activity that originates it. One of the most important fractions is municipal solid waste (MSW), which represents a heterogeneous collection of wastes produced in urban areas, the nature of which varies from region to region.

Biowastes (biodegradable wastes) arise from living or once-living sources from several human, agricultural, horticultural, and industrial sources, in three groups—waste of directly animal origin (manures), plant materials (grass clippings and vegetable peelings), and processed material (food industry and slaughterhouse wastes and paper and paperboard wastes)—and include any waste that is capable of undergoing anaerobic or aerobic decomposition. Different terminologies have been applied to this kind of waste, such as *putrescible*, *green*, *food*, *yard*, *biosolids*, *garden*, or simply *organic wastes*, but chemically speaking, biowaste is characterized by high carbon content in the form of cellulose, hemicelluloses, and lignin, or even proteins and fat, that can be biologically degraded into carbon dioxide, methane, and water.[1]

DISPOSAL PROBLEMS ASSOCIATED WITH BIOWASTE

Disposal of biowaste either through uncontrolled landfill or if abandoned, presents some health and pollution issues that should be addressed. In fact, as biowaste is biologically degradable, it is a free food source for every kind of microorganism, including pathogens that could endanger an entire population.

Leachate

As water percolates through biowaste, it leaches out inorganic and organic compounds, with the risk of soil and groundwater contamination. Also, some persistent pathogens that can be found in long-term deposits of biowaste could endanger population. Landfill leachate is an organic-rich liquor that is an excellent food source for heterotrophic microorganisms, but being so concentrated, its treatment is hardly achieved. Also, the existence of heavy metal contamination is toxic to microorganisms that could otherwise be successful in performing the leachate treatment.

Methane

The second pollution issue from disposal of biowaste is that methane (CH_4) is produced under anaerobic conditions that naturally occur in landfills. The problems with methane are that it is a greenhouse gas, with more than 20 times the damaging effect of carbon dioxide (CO_2), and that it remains in the atmosphere for approximately 9–15 years.

REGULATORY ISSUES OF WASTE MANAGEMENT

Europe is committed to recycling as one of the main objectives of the waste management policy. Through recycling,

Encyclopedia of Environmental Management DOI: 10.1081/E-EEM-120047119

materials contained in solid waste are reintroduced in the production cycle, leading to raw materials and energy savings and reducing the cost of landfill disposal.[2] Current European Directive on solid waste management[3] demands that countries adopt appropriate waste treatment methods, aiming to reduce the amount of waste sent to its final destination—landfill. Taking this into consideration, waste valorization through reusing and/or recycling, or by using other processes (energetic, organic), is also intended. The organic fraction of solid waste can be valorized by composting or anaerobic digestion and, according to the established in the Council Directive on the landfill of waste,[4] should be diverted from the flux of wastes to landfill. Taking the year 1995 as baseline, member states should reduce landfilling of biodegradable MSW to 75% by 2010, to 50% by 2013, and to 35% by 2020.

The organic matter (OM) in solid waste (currently constitutes about 40% of the MSW in Portugal, 25% in the United States,[5] almost 50% in Abu Dhabi City,[6] 50% in France,[7] and almost 77% in Brazil[8]) can be recycled to useful products (compost, methane gas, etc.,) through biological treatment processes. MSW valorization is mainly achieved by recycling constituents such as glass, metals, plastics, paper and cardboard, and OM, which is only possible when these residues are mostly collected selectively, although construction of mechanical biological treatment (MBT) facilities, comprising screening and other physical separation units, can contribute to the achievement of the established targets.

Composting is seen as a valuable recycling process for the organic fraction of MSW (OFMSW) and, thus, is of particular importance given the already existing systems and the potential to grow. Nevertheless, in the European Union (EU), it was not yet possible to come to an agreement on biological treatment of biowaste, in spite of the long work that has been done and that resulted in the publication of a Working Document on Biological Treatment of Biowaste in 2001.[9] This is not only due to the enormous differences in the degree of development of waste management of the various member states but also, in part, due to the existing lobbies.

BIOLOGICAL AEROBIC WASTE TREATMENT: COMPOSTING

Composting is the biological process used most often for the controlled aerobic conversion of OFMSW and any kind of solid and semisolid organic waste to a humus-like material, known as compost. Overall, the composting process can be represented by the reaction in Fig. 1.

This exothermic process is realized in the presence of oxygen by a biological consortia of microorganisms and takes place in two distinct phases: a first phase, in which predominantly thermophilic biochemical degrading reactions occur (temperature rises as a result of the heat produced biologically), and a second phase, in which the humification/stabilization processes occur.[10,11] (Hogan and collaborators[12] suggested that the temperature rise results from the low thermal conductivity of waste.) Compost resulting from this process is a stable product, free of pathogens and plant seeds that can be applied, with benefits to the soil. This definition is intended to distinguish the composting process from the ordinary decomposition that occurs in nature.[13] Moreover, Bertoldi[14] clarified that the stabilization phase corresponds to a humification process that can be prevented under conditions of oxygen scarcity and substrate inadequacy.

Depending on the feedstock nature, the nitrogen, phosphorus, and potassium content of the compost may be insufficient for its classification as an organic fertilizer, allowing instead for its usage as a soil improver. This means that compost properties allow for soil pH amendment, acting as a source of OM that can enhance the water retention and cation exchange capacity of the soil and improve soil aeration.[15]

These are the reasons why composting is currently known as a process of recycling the OM in the solid waste and why using compost in soil represents the reintroduction of OM in soils, reducing erosion and thus desertification that is increased due to intensive land use.

Composting can be successfully applied to garden waste, separated MSW, mixed MSW, co-composting with sludge from urban and industrial wastewater treatment plants (WWTPs), and agricultural and livestock residues.

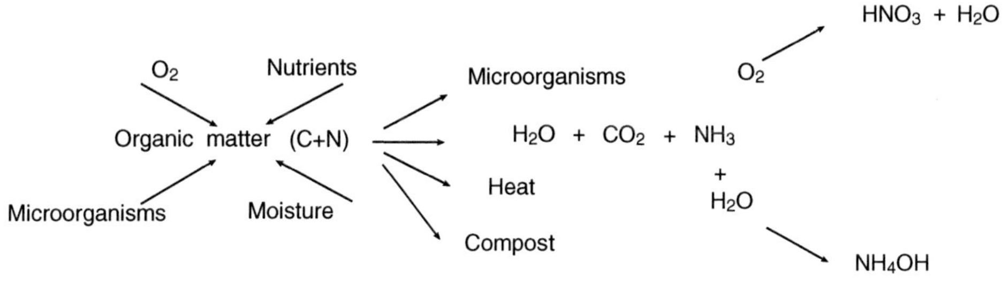
O2
Nutrients
Organic matter (C+N)
Microorganisms
Moisture
Microorganisms
H2O + CO2 + NH3
Heat
Compost
O2
HNO3 + H2O
+
H2O
NH4OH

Fig. 1 Schematics representation of the composting process.

Composting Backgrounds

Organic soil correction with agricultural and livestock waste dates back to the utilization of soil for crop production, having been the principal means of restoring the nutrient balance in soil.[16] Composting is known, for a long time, by farmers as a method that allows obtaining an organic fertilizer from domestic waste. There are records of composting in piles in China for more than 2000 years, and there are even biblical references on the practice of soil correction. About 1000 years ago, Abu Zacharia described these procedures that have been practiced 3000 years earlier in the manuscript of *El Doctor Excellente Abu Zacharia Iahia de Sevilla*, translated from Arabic into Spanish by the order of King Carlos V, and published in 1802 as *El Libro de Agricultura*. In this book, Abu Zacharia insisted that animal manure should not be applied too fresh or directly to the soil, but only after mixing with 5 to 10 times its weight of vegetable and animal bedding waste.[17]

In the growing cities in Europe, during the 18th and 19th centuries, farmers exchanged their products by MSW, using them as soil improver. Until the mid-20th century, MSWs were almost completely recycled through agriculture and did not pose an accumulation problem.[17] Composting of organic wastes and residues was envisaged as more of an art than a science until about the 1930s. By then, several developments of mechanical or intensive systems were achieved in Europe (Itano process in 1928, Beccari in 1931, and VAM in 1932). The Europeans continued to develop and install composting systems in Europe, South America, and Asia, and it was only in 1974 that the U.S. Department of Agriculture at Beltsville, Maryland, developed the "static pile" method that was currently used until the 1990s in the United States.[18]

By the end of 1960, composting was considered an attractive process for stabilization of the OFMSW with the final product being sold profitably as soil improver.[19] However, by the end of the same decade, composting was no longer that interesting for MSW management, not only due to the lower quality of solid waste but also due to the lack of market for compost. Recent stress on usage of less environmental impacting methods has redirected interest into the composting process, particularly concerning the recycling of MSW and urban and industrial WWTPs.

Despite the fact that implementing this process to agricultural wastes is ancient, scientific support was only established in the early 20th century, mainly due to the work done by an English agronomist (Sir A. Howard, 1924–1931) in Indore, India. This agronomist established the fundamental principles for the maintenance of a microbial population in optimal conditions of activity: the need to mix vegetable and animal wastes, the need for neutralization of the fermenting biomass, and the need for provision of adequate amounts of air and water. Thus, materials were stacked in piles (windrows) that have a dimension of 9 × 4.2 m and a height of 60 cm, which allows for maintaining adequate levels of heat and humidity.[20] Construction of these windrows was made by successive layers of waste (manure, soil, and straw) and moistening them conveniently. Windrows were revolved 16, 30, and 60 days after the start of the procedure that needed 90 days before completion and incorporation of the product in soil could be done.

Although windrow composting was the most common practice, during 1950–1960, there was a huge amount of publicity for projects involving composting in reactors.[21,22] However, these projects had almost universally poor results and had an abrupt end due to bad performance and high economic costs. Bad performance was generally the result of an inadequate project or operation and not of the process or technology itself. In the 1950s, there were already more than 20 patented composting systems.[23] In this period, mechanical separation for removal of any non-compostable from the waste stream was initiated.[14]

Co-composting of MSW and biosolids (composting simultaneously MSW and sludge) has attracted increasing attention.[24] MSWs are used as bulking agent and the biosolids act as readily available nitrogen and humidity source. This technique was investigated and applied fully and systematically in the 1950s,[25,26] but moisture content (96%) of the digested solids was a limiting factor as biosolids dehydration was not frequent by then.[27]

In the EU, production of agricultural and food wastes exceeds 1 billion tons/yr, which is 3 times larger than the production of sludge and 6 times higher than the production of MSW.[28] Intensive livestock farming incrementation worsened the problem through production of large amounts of animal waste, often in specific locations. Despite this, the relatively reduced application of composting in the management of this waste should be noticed, especially when compared with the application of this process in the management of MSW.[29] This is probably due to the fact that the agricultural wastes are often applied directly to soil, without any previous treatment, and that composting is generally considered a process for pollution control rather than a beneficial and efficient process for nutrient recycling.

Taking into account the requirements of EU legislation regarding waste management and environmental protection, and because there is nowadays a greater awareness of the importance of controlling the loss of nutrients in the waste treatment processes, recently, there has been a greater research effort to develop strategies that can control gaseous emissions, stabilize OM, and ensure nutrient retention and the absence of toxic products or pathogens in the compost. There are currently several groups of European researchers working on specific issues of composting of livestock and agricultural waste (such as kinetics of composting of MSW, effect of the composition of agricultural waste on compost quality and composting kinetics, effect of contaminants on composting process and on compost quality, etc.), aiming to produce good quality compost.[24,29–34] It is expected and highly desirable an increase in the world-wide application of composting for the treatment of these wastes and the use of its compost for agricultural purposes.

Composting has been successfully applied, including for the waste treatment of animal slaughter and carcasses.[35] Mesquita[36] enumerated some composting projects, where different raw materials were successfully used, and Williams[37] identified several large-scale separate composting schemes implemented in Europe.

Advantages and Drawbacks of Composting

If correctly used, the composting process for MSW treatment presents several advantages. When it is a part of a MSW integrated management system, these advantages are even more important; thus, valorization of the OFMSW by the integrated management can be achieved, and consequently, the final product quality can be improved.

A relevant issue is that besides being a process of effective organic waste treatment, composting is also an excellent recycling process, retaining in the final product (compost) macro- and micronutrients that may return to the soil and supplement other energy cycles in nature.[38]

The main advantages of the composting process are as follows:

- Economy and natural resource preservation and reuse of OM and of macronutrients (N, P, K, Ca, and Mg) and micronutrients (Fe, Mn, Cu, Zn, etc.,) owing to composting being a recycling process.
- Environmental benefits as a result of elimination of air, soil, and water pollution due to proper disposal of organic wastes or their appropriate confinement in controlled landfills. Also, there are additional benefits from using organic compost (instead of artificial fertilizers and soil improvers): in the recovery of degraded, contaminated sites and salty soil, in reforestation, in soil erosion control, etc.
- Public health benefits, due to the elimination of pathogens. In developing countries, this issue contributes to child mortality control.
- As an offshoot of health benefits, e.g., prevention and elimination of diseases, economic benefits arise, contributing to the reduction of costs of treatment and increased productivity. Another benefit is the elimination of costs for land remediation of uncontrolled dump sites.
- Social benefits arise from the fact that in addition to eliminating the practice of using organic waste as food for animals, composting promotes employment either in the treatment units (selective collection, sorting, processing) or in compost utilization and application.

The main disadvantages arising from composting are as follows:

- Being a labor-intensive process, composting gives rise to operational costs (compensated with positive social impact).
- Producing compost from selective organic collection entails high costs.
- The possibility of compost contamination (if the adopted process is not technologically appropriate and poorly conducted).

METHODS OF COMPOSTING

Composting is a simple technology that requires a fairly low intervention and has modest initial, operational, and support costs. It can be very attractive to authorities charged with biowaste management but with reduced budgets.

There are many factors that can influence the decision on the specific details of composting methods that should be adopted, but composting of OM in MSW or similar materials will most naturally be based in one of two options, either source-separated biowaste or biowaste recovered from MBT plants. Either of these options will ultimately be put in practice in home composting systems or in centralized composting facilities.

While home composting takes advantage of many distributed and local production units, centralized or large-scale facilities comprise two large groups of systems, both open and closed.[39] Among open systems of composting, three types can be distinguished: windrow (or revolved piles), the static aerated piles, and vermicomposting.[30] Closed systems of composting include reactors (either horizontal or vertical) of different sizes and shapes equipped with different feeding and aeration systems, aimed at accelerating the startup of the oxidation process, especially in cold climate countries, thus allowing for a more efficient control of odor emissions and minimizing the land area requirements.[37]

The choice of a specific type of system relies essentially on socioeconomic factors, on the amount of biowaste to compost, and on its final destination. Open systems are generally more suitable for developing countries[40,41] and for agro/livestock facilities[42] due to their easier operation and lower mechanization, building, maintenance, and operational costs.

Processing of agricultural and livestock waste using any of these technologies not only presents substantial benefits from the health, economic, and environmental points of view but also allows for a safe and potentially useful final product that can be used as a soil conditioner, valorized as a fertilizer, or for the floriculture and horticulture industries. Its use for agricultural purposes is important, both in countries where soil is extremely poor in OM (such as those of southern Europe) and in countries where extensive use of inorganic soil fertilizers has endangered its structure.[40]

The most common technologies, particularly for composting the OFMSW and similar biowaste, are presented below.

Home Composting

Home composting is one of the most interesting ways of managing biowaste. In fact, while people compost at home, they are guaranteeing the removal of OM from MSW flux, thus facilitating the achievement of the goals established in the EU Landfill and EU Waste Management Directives, as well as contributing to the reduction of waste management costs by reducing the total amount of waste that needs treatment. Also, by using home-produced compost, households do not need to use other fertilizers or soil conditioners; hence, they also have lower garden maintenance costs. Home composting demands the direct involvement of households, leading to higher separation rates of other recyclables as well.

One of the biggest issues in home composting is the choice of the compost bin. Although there are several configurations available in the market (Fig. 2), some of them are far more efficient than others and their cost is not always according to quality. In some cities, there are special programs that try to address this problem, reducing costs to the consumer. In Porto, Portugal, Lipor has been developing a project—Terra-à-Terra—that aims to promote organic waste reduction at the households of Lipor's municipalities. Householders with a garden or that work in companies with a garden in the project area can receive a free compost bin after attending a free composting training session. Through this project that started in 2007, more than 4500 composters have been delivered in 3 years, and it is estimated that there is a potential biowaste reduction of about 3000 tons/yr that will prevent the emission of 528 tons of CO_2 per year if this waste is treated at the energy recovery plant.[43]

Centralized Composting

Seasonal variation in the composition of the MSW requires a flexible and competent management of the composting system.[1,44] For this reason, careful choice of the system and a thorough control of the facilities are of primary importance.

Windrows

This technology is based on the Indore composting method. Rows of parallel piles of material with a height of 1.5–2 m and a width of 3–4 m are constructed until they reach 100 m or more. The shape of the piles is such that sufficient heat is generated to maintain temperatures while allowing for oxygen diffusion into the center of the pile, through

a)
b)

Fig. 2 Home composters: (a) Wood, homemade, with two composting chambers, 1 m^3 each. (b) Commercial plastic, with openings for air circulation (0.4 m^3), offered by the Terra-à-Terra project.

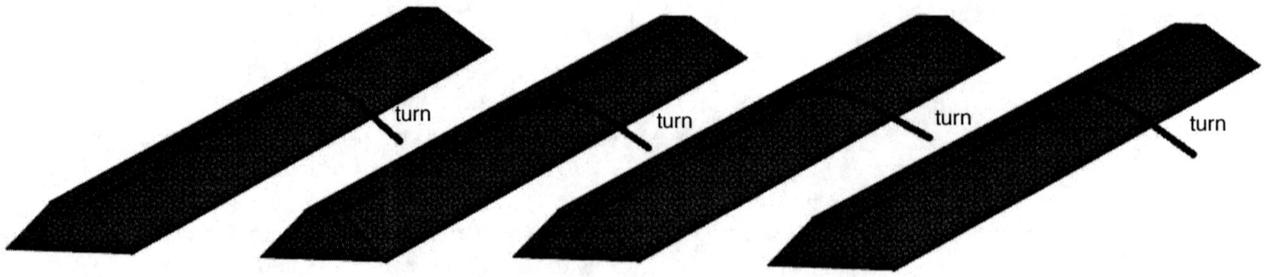
turn
turn
turn
turn

Fig. 3 Schema of windrow composting and how to turn composting piles; the upper material will stay in the bottom after turning.

natural convection. Batteries of piles must be constructed on an impermeable surface (e.g., cement) so as to facilitate the collection of leachate that eventually will form and so that piles can be easily turned. Windrows must be turned periodically (initially at the end of 5 days and then every 2 or 3 days), as shown in Fig. 3. The convex shape of the windrows allows rainwater to drip along the surface and not infiltrate the pile of material. This system has some drawbacks such as being prone to anaerobiosis due to layer compaction, reduction of free space for aeration, and production of leachate that fills the remaining voids between particles, which further contributes to reducing aeration.

Aerated Static Piles

Piles are similar to those in windrow composting. However, the material is not mechanically aerated. Temperature control is achieved by natural air convection or forced aeration, either by introduction of compressed air or by vacuum induction (Fig. 4).

The material for composting remains stacked during a very long period, which might result to collapsing and thus pore clogging. To prevent this, it is a common practice to mix biowaste with some bulking material (such as wood chips or sawdust) that should be more stable and have a higher particle size, thus providing a structure that will help prevent the material in the pile from collapsing. Aerated piles should be covered with a layer of compost or wood chips that confer some insulation, act as a biofilter, and reduce odor release.[42]

These composting technologies were traditionally implemented in open courtyards. Currently, and in regions where environmental conditions are particularly unsuited to this type of processing, these technologies are implemented in covered spaces or indoors (Fig. 5), in which case, it is possible to perform air treatment in biofilters, achieving a dramatic reduction in odorous emissions.

In-vessel

Due to being enclosed, in-vessel systems allow for tighter control of temperature, moisture, aeration, and biowaste mixing. In-vessel systems include different configurations such as tunnels, rotating drums, reactor tanks, silos or towers, agitated bays and beds, enclosed halls, or even containers.[37] Although the operating systems are essentially the same as for windrows or aerated piles, these reactors are much more efficient, thus needing much less area.[1]

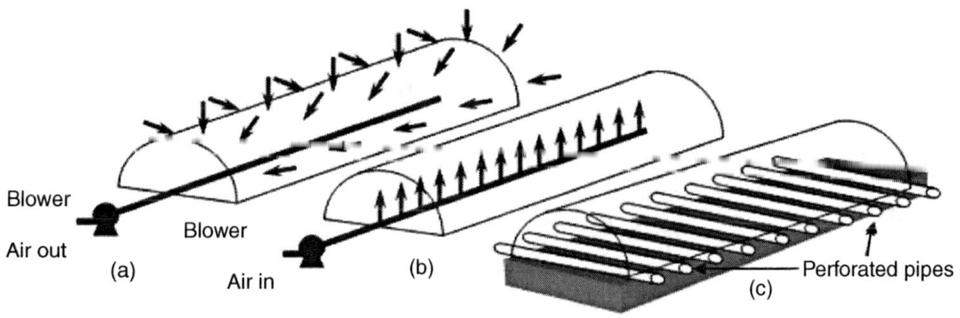
Blower
Air out
(a)
Blower
Air in
(b)
Perforated pipes
(c)

Fig. 4 Static pile composting: (a) with suction of air from the outer layers to the base of the piles; (b) with compressed air injection through the base of the pile; and (c) natural convection.
Sources: (a) from Risse and Faucette;[45] (b) and (c) adapted from Graves and Hattemer.[46]

Fig. 5 Indoor aerated static pile system with aeration through vacuum induction.

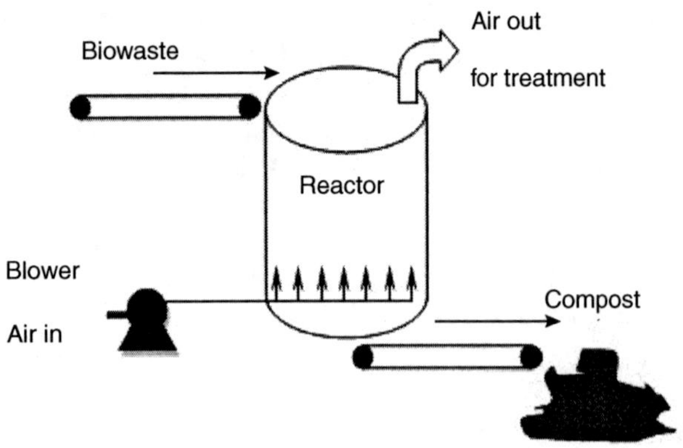
Air out
Biowaste
for treatment
Reactor
Blower
Air in
Compost

Fig. 6 Schematic drawing of vertical reactors system for composting.

Silos or Towers

This technology is based on the use of silos or other types of structures with a height exceeding 4 m. These reactors are fed at the top, via a distribution mechanism, and the material progresses towards the bottom by way of gravity, usually after 14 days of composting. After this period, a curing time of ca. 2 months is needed to stabilize the compost.[42] Process control is commonly performed by air injection at the base of the reactor, moving upwards countercurrent to the biowaste (Fig. 6).[47] Due to the great height of the reactor, it is necessary to use large airflow per unit surface area of distribution, which makes it difficult to control the process.[48] It is not easy to keep the temperature and oxygen at optimal levels, due to material compaction and to inexistence of mixing in the silo.[42] To minimize these problems, material should be thoroughly mixed prior to feeding the silo and air distribution and collecting systems must be improved, by changing the direction of airflow from vertical to horizontal between alternating sets of inlet and exhaust pipes.[44]

This type of reactor has been successfully used for composting of sludge from WWTPs (where the addition of bulking materials for porosity enhancement and a uniform feed facilitates process control) but is rarely used for processing heterogeneous materials, such as the OFMSW.[44]

Horizontal Reactors (Tunnels and Agitated Beds)

These reactors allow controlling the occurrence of high temperature, moisture, and oxygen gradients, which are frequent in vertical reactors, because air needs to travel lower distances through the material. There are a variety of configurations, comprising static or agitated reactors, with aeration by air injection or vacuum induction. Agitated systems usually take advantage of the revolving process to displace the material along the reactor continuously, while the static systems usually need loading and unloading mechanisms. Handling equipment can also grind the material, exposing new areas for decomposition, but excessive grinding can also reduce porosity. Typically, aeration systems are installed at the base of the reactor and can use temperature and/or oxygen concentration as control variables. The agitated systems with less than 2 to 3 m thick beds have been successfully used in heterogeneous MSW treatment (Fig. 7, left).

Tunnels are formed in long parallel channels (Fig. 7, left), made of concrete walls with a movable porous floor, like a grid to allow for greater airflow (usually forced aeration) under the compost, and covered by a roof. Biosolids are revolved by a turning machine (Fig. 7, right) running on tracks along the concrete walls between channels. The turner movement along the bed displaces the compost until it is ejected at the end of the bed.[49] The duration of the composting process is determined by the length of the bed and the turning frequency, ranging from 6 to 20 days, after which the compost must be further processed in windrows or aerated static piles for 1 to 2 months. Agitated bed systems operation ranges from 2 to 4 weeks; also, an extended curing period is needed to stabilize compost.[18]

Rotary Drums

In these reactors, biomaterial stays for only a few hours or days. Their effect is essentially homogenization and

Fig. 7 Horizontal composting systems: (left) tunnel composting system composed of horizontal parallel beds; (right) turning machine for a tunnel composting system.

Fig. 8 Composting rotary drum: (left) side view and (right) view from the waste admission side.

grinding of materials, only allowing start of composting by temperature control. As the reactors (diameter, 2 to 4 m; length, up to 45 m) have a slight inclination and have a rotating movement (0.5 to 2 rpm), the flow of material is continuous, countercurrent to the air supply; thus, the material leaving the reactor is cooled by fresh air and the material entering the drum makes contact with warm air, which favors bacterial growth (Fig. 8).[42] These reactors are particularly interesting in processing the OFMSW for composting[46] as they allow for a faster startup of the process while reducing the malodorous emissions to the atmosphere. The DANO drum is a commercial rotating cylinder that has retention times of about 3 days with eventual interest in modern MBT plants. Most of the biological process has to be realized after leaving the reactor.

All of the systems described above can be used for composting different types of materials. Obviously, capital and operational costs of reactor processes are significantly higher than those of a windrow or aerated static pile process; thus, the residence time in the reactors is rarely enough to obtain mature compost. For this reason, a reactor is used in the early stages of composting to allow an easier control of the process, which facilitates odor emission reduction. After this processing, the material that comes out of the reactor must be stabilized in windrows or aerated static piles.[44] As the OFMSW treatment is difficult due to high cellulosic carbon content and sometimes low moisture and low porous structure, it may take as long as 6 months for compost maturation and stabilization, unlike compost from WWTPs that usually only takes 2–3 mo to be stabilized.

Vermicomposting

Vermicomposting is an alternative technology to conventional composting in which selected species of earthworms (*Lumbricus terrestris* and *Dendrobaena veneta*) and red worms (*Eisenia foetida*, Fig. 9) eat organic material, absorb the nutrients they need, and excrete the rest, producing a humus-like material, known as vermicompost.[1,50] Vermicomposting should be applied preferentially to residues of fruit and vegetables, tea leaves, tea bags, coffee grounds, paper, and shredded green garden waste and has large application in agricultural wastes, including manure wastes.[51] Vermicompost produced from OFMSW is a "nutritive biofertilizer" 4–5 times more powerful than conventional composts and even better than chemical fertilizers for better crop growth and safer food production.[52]

Vermicompost systems can range from inexpensive wood or plastic boxes to sophisticated modular units (Fig. 9) that are self-contained and fully automated to keep controlled environmental conditions while processing waste

Fig. 9 *Eisenia foetida* in a home/commercial vermicomposter.

into earthworm castings. Open systems range from windrows or beds of variable scales to open field operations. Usually, windrows are built no more than 30 cm deep for easy aeration of the beds.[51] Vermicompost has the advantage of being a material that growers can produce "on-farm" and use as a biofertilizer, using own feedstock and implementing either a midscale vermicomposting unit technology or a modular unit.[1]

Worm beds can be as long as 50 m; feedstock can be the resulting product of a MBT plant, whereas OFMSW is processed in a rotary drum. Worms should be kept at temperatures of 10–35°C. Moisture is fundamental to worms; they breathe oxygen through their moist skin.[51] Resulting worm casts can be used as a biofertilizer as they contain 5 times more N, 7 times more P, 1.5 times more Ca, 11 times more K, and 3 times more exchangeable Mg than the soil. These casts are also rich in humic acids (which condition the soil), have a perfect pH balance, and have plant growth factors similar to those found in seaweed.[51]

Under the optimum temperature (20–30°C) and moisture (60%–70%) conditions for worm breeding, about 5 kg of worms (ca. 10,000) can process 1 ton of waste into vermicompost in 30 days.[53]

Worms can be easily recovered for further processing because they do not like light (a few hours of direct exposure to sunlight can cause paralysis or death), dryness, or even some specific odors (onions), migrating naturally for places where fresh wastes have been added. Worms can also be harvested by using a trommel screen.[54]

Vermicompost produced from organic wastes, such as food and yard wastes, have enormous economic potential for increasing crop yields, suppressing attacks from pests, and controlling the spread of diseases.[55–57]

Evaluation of a Composting System

Among the various criteria used for assessing the system efficiency, the most important for MSW composting are product quality, rejected ratio, and recycling rate. Compost quality can be evaluated by its appearance and by functional characteristics and the nature of the contaminants, characteristics that are critical for its commercialization. Rejected ratio and recycling rate affect the quality of the product, as they affect the contaminant's concentration and the amount of final waste that must be disposed off in landfill.

Compost quality should constitute the main objective of the composting facility. Some of the aspects of quality, such as the degree of maturation and size of particles, can be corrected at the end of the process by increasing the curing time and by using physical processes for particle size reduction. However, other issues may be far more difficult to remedy, as when there is chemical contamination. A composting factor should be not only thoroughly thought about when conceiving the biowaste management system but also observed during the initial phases of processing. Chemical contamination can be caused by heavy metals or other chemicals, often from domestic hazardous waste. Contamination can also be of physical type; that is, the product may be contaminated with inert materials (glass, plastics and metals, brick and concrete, etc.,) that arise from domestic or commercial waste either voluntarily or by mistake. Removal of physical contaminants may not be very difficult from a technological point of view but represents a high fraction of the economic costs of a composting facility, which cannot be completely eliminated because even source-separated waste can be contaminated. However, the use of a source-separated raw material is of primary importance for high-quality assurance of the compost.

PHASES OF COMPOSTING

Throughout the composting process, the microbial population will vary and different types of microorganisms will play distinct roles during the various stages of the process.[38] Thus, in the first step, there will be predominantly mesophilic bacteria that hydrolyze the easily fermentable organic material. The reactions are exothermic, leading to heat release and temperature rise. Thermophilic microorganisms (bacteria, fungi, and actinomycetes) begin to develop on a large scale—from 40°C.[58] The thermophilic microorganisms multiply, and as soon as the temperature reaches 55–60°C, the attack on complex molecules (carbohydrates, proteins, etc.,) starts, which will result in their transformation into simpler products (simple sugars, amino acids, etc.,) that are used by other microorganisms (Fig. 10). If there is no external control, the temperature of the composting material can reach 80°C, which represents the limit for the thermophilic population, resulting in microbiological activity inhibition and survival only of spores of bacteria. As degradation is reduced, and temperature lowers, mesophilic bacteria, actinomycetes, and fungi (mostly those that were in the outer areas of the pile) gain further activity, attacking the most resistant compounds (such as lignin and cellulose, the less readily biodegradable components of biowaste).[1] Complex enzymatic reactions are responsible for humus production, mainly by lignin and protein combination. At this stage, protozoa and some higher organisms (worms, nematodes, and millipedes) can be found in compost.

PROCESS PARAMETERS IN COMPOSTING

Preparation of a composting process is not a simple task, especially for achieving optimal results. For this reason, in commercial applications, mechanization is a key factor, allowing for efficiently controlling the most important project factors (temperature, pH, moisture, C/N ratio, aeration rate, particle size, and mixing/turning).[58]

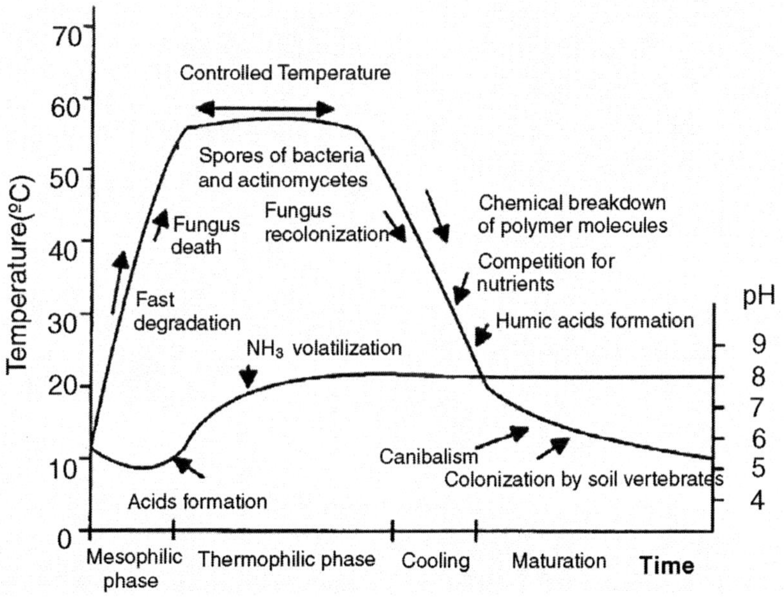
Temperature(ºC)
70
60
50
40
30
20
10
0
Controlled Temperature
Spores of bacteria and actinomycetes
Fungus death
Fungus recolonization
Fast degradation
NH_3 volatilization
Chemical breakdown of polymer molecules
Competition for nutrients
Humic acids formation
Canibalism
Colonization by soil vertebrates
Acids formation
pH
9
8
7
6
5
4
Mesophilic phase
Thermophilic phase
Cooling
Maturation
Time

Fig. 10 Microbiological transformations and temperature and pH profiles during a controlled composting process.
Source: Adapted from Neto and Mesquita.[38]

As any biological process, composting is conditioned by factors that affect the activity of the microorganisms involved in the process.

Temperature

Temperature is the process parameter that best describes the biological equilibrium and shows the efficiency of the process. Aerobic composting systems operate under mesophilic (20–40°C) and thermophilic (40–60°C) conditions that change along the process as a result of the respiration and metabolic activities of the organisms involved in the composting process.[1]

Temperature control of the composting material under static pile or in-vessel processes can be achieved through temperature monitoring and control of the air flow rate. In these processes, temperature can be controlled in the thermophilic range, around 60°C, which allows for the development of a more complex microbiological consortia, responsible for increasing the rate of decomposition of the OM (higher process efficiency), better compost sanitation or pathogen elimination, and elimination of weed seeds, insect larvae, and parasite eggs, among other advantages of the process.[38]

In windrow composting systems, temperature can only be controlled indirectly, by changing the turning frequency of the composting material. After turning, temperature lowers 5–10°C, rising again after only a few hours, as a result of the increase in oxygen availability. On the other hand, if temperature rises above 70°C, most of the microorganisms will die or enter the dormant phase, slowing the composting process and resulting to lower-quality compost. Also, high temperature can lead to humidity and nitrogen loss (through ammonia volatilization at a pH of 7.5).

In a system adequately controlled, temperature in the composting pile will rise up to 40–60°C from the second to the fourth day and will lower down to 35–38°C after 10 to 15 days because easily biodegradable organic material has already been converted. The increase in temperature also depends on factors such as nutrient availability, moisture content, particle size, aeration, turning of the pile, and the thermal insulation of the system (either exposed systems or confined systems).

Typical temperature profiles in non-controlled composting systems (with and without turning) are shown in Fig. 11.

It should also be noted that temperature is usually not uniform along the composting pile, the inner material having a higher temperature than the outer layers, as shown in Fig. 12.

Moisture

Decomposition of the OM mainly depends on moisture content to enable microbiological activity. Microorganisms have, in their structure, about 90% water that is needed not only for new cell production but also and specially for the dissolution of the nutrients needed for cellular metabolism.

Moisture content can be adjusted either through mixing different materials or through water addition to an optimum between 40% and 70%. Below 30% humidity, composting rate dramatically decreases, but more than 70% moisture content may lead to pore filling with water, thus prevent-

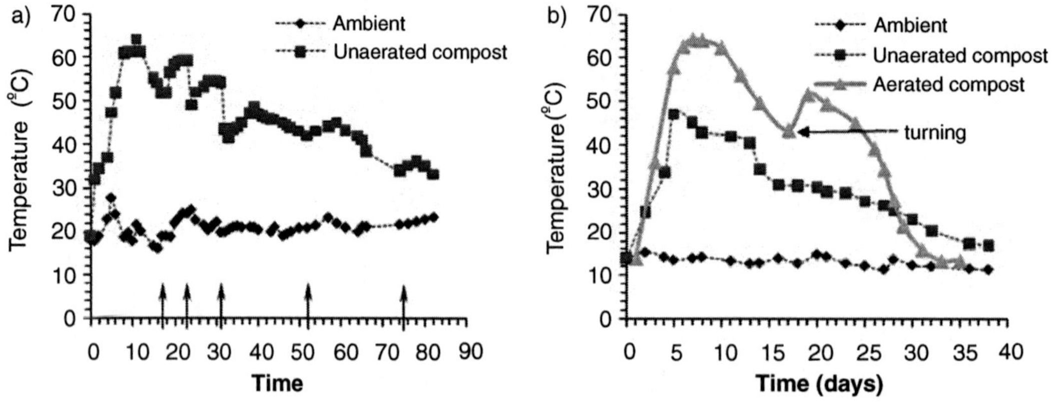
a)
Temperature (°C)
Ambient
Unaerated compost
Time
b)
Temperature (°C)
Ambient
Unaerated compost
Aerated compost
turning
Time (days)

Fig. 11 Typical temperature history observed in (a) windrow composting systems with turning[59] and (b) static piles with and without forced aeration.[59]

ing oxygen from reaching the OM and inducing anaerobic conditions. This is why the most recommended value is 60%.[1,42]

Also turning of the composting material can help control moisture content. If the organic material has 55%–60% humidity and the composting period is 15 days, the first turning should be done on the third day and then every other day.[58]

Aeration

Aeration is very important in composting and holds two main functions: the first, and most important one, is to supply aerobic microorganisms with the oxygen they need for their respiration and metabolic activity, and the second function is to control temperature and, consequently, the rate of oxidation of the OM as well as odor emission.

Theoretically, the optimal rate of aeration would be the one that would allow for biological oxygen demand (BOD) supply during all the different composting steps. Nevertheless, parameters such as the nature of the composting material, particle size, aeration technology, moisture content, and porosity of the material in the pile may prevent full satisfaction of this requirement.[23] On the other hand, excessive aeration may cause excessive loss of humidity and heat leading to the misidentification of the end of the composting process.

In static pile systems, initial oxygen concentration in the pores is similar to that in the surrounding air (ca. 17%), because this is the air that was trapped in these spaces when the pile was built, whereas CO_2 concentration is significantly lower (ca. 0.5%–5%). As aerobic degradation occurs, oxygen concentration will decrease and CO_2 concentration will increase. Anaerobic conditions will develop at O_2 concentration lower than 5%, which can be prevented through introduction of air, vacuum induction, or turning of the material in the composting pile (which will also allow for a more efficient distribution of nutrients and microorganisms).

In systems with forced aeration, air flowrate and total air needed in the process are fundamental project parameters that help maintain the level of available O_2 very close to the required level throughout the whole process. The total amount of oxygen required and the total amount of CO_2 produced in the process can be estimated through the mass balance in Eq. 1, where the composition of the OM

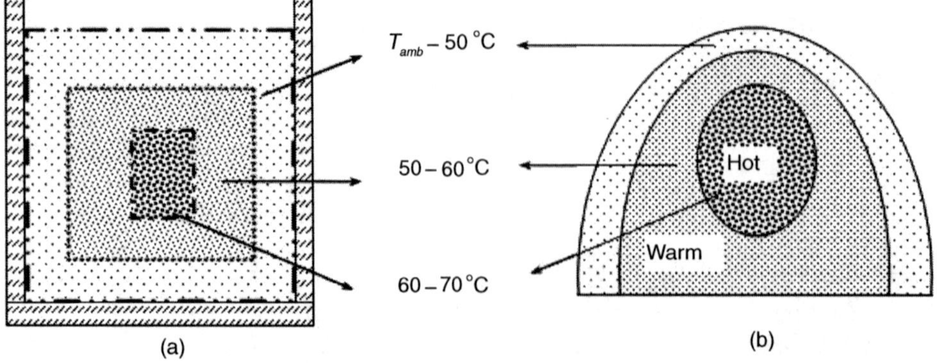
T_{amb} – 50 °C
50 – 60 °C
60 – 70 °C
Hot
Warm
(a)
(b)

Fig. 12 Temperature profile in a composting pile: (a) in-vessel/tunnel; (b) windrow.

for degradation and of the compost can be represented as $C_aH_bO_cN_d$ and $C_wH_xO_yN_z$, respectively.[58]

$$C_aH_bO_cN_d + 0.5\,(ny + 2s + r - c)\,O_2 \rightarrow n\,C_wH_xO_yN_z + sCO_2 + rH_2O + (d - nz)\,NH_3 \quad (1)$$

with $r = 0.5\,[b - nx - 3\,(d - nz)]$ and $s = a - nw$.

If ammonia produced in the process is converted into nitrate (nitrification), some more oxygen will be needed, as estimated using Eq. 2.

$$NH_3 + 2\,O_2 \rightarrow HNO_3 + H_2O \quad (2)$$

Particle Size

It could be expected that the lower the particle size of the organic material, the higher would be the rate of biochemical reactions, due to the increase in surface area exposed to microorganisms and to oxygen, which translates to the composting period being reduced. However, if particle size is too small, porosity will be reduced along with oxygen and CO_2 diffusion, which is fundamental during the thermophilic phase of the process, when oxygen consumption is higher.

An optimum particle size of 25–50 mm has been accepted for most of the materials, but this can vary between 13 and 75 mm depending on the nature of the composting material. A shredding operation may be needed prior to composting, which will entail an increased operational and financial cost.

Porosity of the composting material can be enhanced through introduction of bulking material that will allow for easier aeration while keeping structural characteristics that are essential for the construction of composting piles. Also, the height of the composting piles is based on particle size, in order to avoid excessive compaction during composting.

Feedstock Composition

The biomaterials used in composting have such a diversified composition that can usually supply all the elements that are fundamental for microbial metabolism (carbon, nitrogen, phosphorus, potassium, calcium, iron, copper, etc.).

Of these, the most important is undoubtedly the C/N ratio; an optimal ratio of 30/1 is recommended. Carbon is the source of energy and an essential factor for multiplication of cell material, and nitrogen is present (and available) in such an amount that it can be used for protein, amino acid, and nucleic acid formation.

Mixing of wastes with different compositions is usual since hardly ever a substrate has all the characteristics required for efficient and effective composting, leading to a compost with the recommended values of carbon/nitrogen ratio (C/N), moisture content, particle size, density, etc. Usually, these mixtures are prepared using carbon-rich vegetable waste and separate solid material (newsprint paper, manure, slurry or sludge, yard wastes) characterized by containing high levels of nitrogen, in order to obtain a C/N ratio of about 25/1 to 30/1[30] (Table 1). If the C/N ratio is lower than 30/1, there will be a loss of nitrogen as ammonia, causing malodorous emissions, but if C/N is higher than 80/1, nitrogen concentration will be so low that it will become a limiting factor. It must also be taken into consideration that in the OM, not all carbon will be in the readily available form, i.e., biodegradable, in contrast to nitrogen that is readily available. Thus, lignocellulosic material that is rich in carbon is very hardly biodegradable,

Table 1 Nitrogen content and C/N ratio of common materials used in composting (dry basis).

Material	*N* (%, dry weight)	C/N ratio[a]
Food processing wastes		
Fruit wastes	1.52	34.8
Mixed slaughterhouse wastes	7.0–10.0	2.0
Potato tops	1.5	25.0
Coffee grounds		20.0
Vegetable wastes	2.5–4.0	11–13
Manure		
Cow manure	1.7	18.0
Horse manure	2.3	25.0
Pig manure	3.75	20.0
Poultry manure	6.3	15.0
Sheep manure	3.75	22.0
Sludge		
Digested activated sludge	1.88	15.7
Raw activated sludge	5.6	6.3
Wood and straw		
Lumber mill wastes	0.13	170.0
Oat straw	1.05	48.0
Sawdust	0.10	200.0–500.0
Wheat straw	0.3	128.0
Wood (pine)	0.07	723.0
Paper		
Mixed paper	0.25	173
Newsprint	0.05	983
Brown paper	0.01	4490
Trade magazines	0.07	470
Junk mail	0.17	223
Yard wastes		
Grass clippings	2.15	20.1
Leaves (fleshly fallen)	0.5–1.0	40.0–80.0
Tree trimmings	3.1	16
Biomass		
Water hyacinth	1.96	20.0
Bermuda grass	1.96	24

[a] C/N ratio based on total dry weight.
Source: Rynk et al.[42] and Tchobanoglous et al.[58]

so when it is present, the most adequate value for the C/N ratio would be 35/1 to 40/1.[60]

During the composting period, C/N ratio slowly decreases because when OM is degraded, about 65% of the carbon is released as CO_2 and the remainder is used with nitrogen in cell, which is released only when there is cellular death. The final compost should have a C/N ratio of 15/1.

pH Control

Although authors agree that pH affects biological processes, composting studies using MSW and wastewater sludge where the initial pH of the composting material was varied have shown that a self-regulation of pH occurs;[10] thus, pH is not a critical factor in the composting process. Most of the times, there is no need for pH correction, although in few particular situations, lime can be added for pH correction.

Most of the bacteria involved in the process operate better in the pH range of 6 to 7.5, but fungi prefer a pH of 7.5 to 9. In the composting process, there is a pH variation across time. Thus, in the first phase of the process, simple organic acids are produced, lowering pH to 5 or even less. After only 3 days, pH starts rising up to 8–8.5, as organic acids are being further transformed. In the cooling phase, the pH of the compost will be slightly reduced until 7–8 (Fig. 13). Ideally, pH must be less than 8.5 so that nitrogen loss (as ammonia) is minimized.[58]

Odor Control

Odor problems arise from development of anaerobic conditions that allow for formation of organic acids, or ammonia release, most of which are extremely malodorous. Nevertheless, material in the pile can act as a biofilter that retains and degrades part of these compounds.

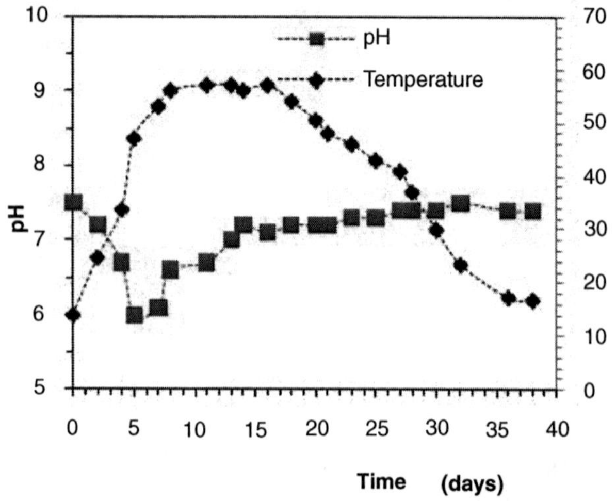
pH
Temperature
pH
Time (days)
10
9
8
7
6
5
70
60
50
40
30
20
10
0
0 5 10 15 20 25 30 35 40

Fig. 13 Typical temperature and pH range in windrow composting.
Source: Fonseca and Amorim.[59]

Pathogen Control

Pathogen destruction is of primary importance in composting systems as it is affected by temperature profile and aeration process. Pathogens can be eliminated under different conditions (*Salmonella* is destroyed in 15–20 min at 60°C, but at 55°C, it will take 1 hr). As a recommendation, if biowaste is kept at 70°C for 1–2 hr, all pathogens will be eliminated. It should be noted that in the second stage of the composting process (humification), the *Penicillium* fungus that destroys pathogens, acting as an antibiotic, appears.

Most European countries recommend a temperature of 55–65°C for 4–15 days to guarantee compost sanitization.[1,58]

EVALUATION OF MATURATION AND COMPOST MATURITY

As stated before, compost cannot be produced in record time; the length of the composting process depends on several factors, such as the feedstock material, particle size, nutrient balance, moisture content, and the composting technique. Although some authors and technologies claim to compost wastes in 7 or 20 days, this is not possible because compost produced in this time is not stabilized as total humification of the OM can only be achieved after about 110 days.[23] Thus, composting of garden or food waste materials can be successfully achieved in 3 mo under aerated, in-vessel, or even turned windrow systems but will take as long as 1 year in an unaerated static pile.

Compost maturation is fundamental in order to prevent ammonia liberation into the roots of cultures (at very low C/N ratios), biochemical reduction of soil nitrogen (when residual carbon in high C/N ratio compost is used by microorganisms), production of toxins that inhibit plant metabolism and seed germination, and microbiological activity that may lead to oxygen demand as well as other toxic effects to plants.[10]

The degree of maturation and stabilization of compost can be assessed using different methodologies: 1) final drop in temperature and self-heating capacity (the absence of temperature increase after turning and humidifying the biowaste up to 50% moisture content means that the process has ended); 2) amount of decomposable and resistant OM in the compost (lignin content higher than 30% means that compost is stable);[58] 3) chemical oxygen demand (COD) (a COD value below 350 mg O_2/g compost shows that the compost is stable); 4) oxygen uptake rate [it should be less than 40 mg of O_2 per kilogram of dry matter per hour for compost to be considered stable; a specific oxygen

Fig. 14 How compost looks like: (left) from OFMSW; (right) from co-composting of slaughterhouse with sawdust and agricultural waste

uptake rate (SOUR) test[61] can be performed]; 5) CO_2 production (it should be small, meaning that there is no significant microorganism activity); 6) C/N ratio (it usually decreases along the process, but it should be interpreted taking into account the initial characteristics of the OM and composting conditions. If OM has an initial C/N ratio of 35–40, the stabilized compost should have a C/N ratio of 18–20; however, if the initial C/N ratio is 10 or even lower as in manure wastes, an increase of the C/N ratio should be expected during the composting process);[17] 7) growth of the fungus *Chaetomium gracilis*;[58] and 8) the absence of ammonia in the compost and the presence of nitrates.[38]

COMMERCIALIZATION ISSUES

To be marketed, compost must have a uniform size; must be free of contaminants such as glass, plastics, and metals; and must be free from unpleasant odors. Typically, compost is a dark color product, has a wet soil smell, has a nice texture, and does not show the original raw material (Fig. 14). The most common treatments before compost packaging are crushing and screening, although sometimes compost is commercialized as pellets.

CONCLUSION

Methods that can be used to perform composting over biowastes of different nature have been presented, from the simpler and less expensive ones, to some more sophisticated and expensive technologies. Most of the composting processes are based on reactions involving microorganisms, but worms can also be used to perform the same task of degrading OM and producing a stabilized and mature product (that is sterilized and of good quality) that can be used as a soil conditioner or fertilizer. With CO_2 and H_2O being the only by-products, composting can be considered environmentally benign and can contribute to OM recycling. The influence of the main operational parameters has been discussed, and it was shown that under appropriate conditions, almost any OM can be composted.

REFERENCES

1. Evans, G. *Biowaste and Biological Waste Treatment*; James and James (Science Publishers) Ltd: London, 2001.
2. Gama, P.S. *Recolha Selectiva e Reciclagem de Resíduos Sólidos Urbanos*. Curso sobre Valorização e Tratamento de Resíduos. Prevenção, Recolha Selectiva, Compostagem e Confinamento em Aterro. LNEC/APESB, Lisbon, 10th–12th December, 1996.
3. Directive 2008/98/EC of the European Parliament and of the Council, of 19 November 2008 on waste and repealing certain Directives, Official Journal of the European Union, Brussels, 2008; L312/3-L312/30.
4. Directive 99/31/EC of the Council, of 26 April 1999 on the landfill of waste, Official Journal of the European Union, 1999; L182/1-L182/19.
5. Davis, M.L.; Masten, S.J. *Principles of Environmental Engineering and Science*, 1st Ed.; McGraw-Hill Companies, Inc.: New York, 2004.
6. Abu Qdais, H.A.; Hamoda, M.F.; Newham, J. Analysis of residential solid waste at generation sites. Waste Manage Res. **1997**, *15* (4), 395–405.
7. Francou, C.; Lineres, M.; Derenne, S.; Le Villio-Poitrenaud, M.; Houot, S. Influence of green waste, biowaste and paper–cardboard initial ratios on organic matter transformations during composting. Bioresour. Technol. **2008**, *99* (18), 8926–8934.
8. Lino, F.A.M.; Ismail, K.A.R. Energy and environmental potential of solid waste in Brazil. Energy Policy **2011**, *39* (6), 3496–3502.
9. Directorate-General Environment A.2, CEC. Working document on biological treatment of biowaste—2nd draft. European Commission, Brussels, 2001; 1–22, available at

http://www.ymparisto.fi/download.asp?contentid=5765 (accessed June 8, 2011).
10. Neto, J.T.P. On the Treatment of Municipal Refuse and Sewage Sludge Using Aerated Static Pile Composting—A Low Technology Approach. Ph.D. Thesis: University of Leeds, 1987.
11. Haug, R.T. *The Practical Handbook of Compost Engineering*; Lewis Publishers: Boca Raton, FL, 1993.
12. Hogan, J.A.; Miller, F.C.; Finstein, M.S. Physical modelling of composting ecosystem. Appl. Environ. Microbiol. **1989**, *55* (5), 1082–1092.
13. Golueke, C.G. *Biological Reclamation of Solid Wastes*; Rodale Press: Emmaus, PA, 1977; 249 pp.
14. Bertoldi, M. Compost quality and standard specifications: European perspective. In *Science and Engineering of Composting: Design, Environmental, Microbiological and Utilization Aspects*; Hoitink, H.A.J., Keener, H.M., Eds.; The Ohio State University, Renaissance Publications: Ohio, 1993; 523–535.
15. Epstein, E.; Taylor, J.M.; Chaney, R.L. Effects of sewage sludge and sludge compost applied to soil on some physical and chemical properties. J. Environ. Qual. **1976**, *5*, 422–426.
16. Avnimelech, Y. Organic residues in modern agriculture. In *The Role of Organic Matter in Modern Agriculture*; Chen, Y., Avnimelech, Y., Eds.; Martinus Nijhoff Publishers: Netherlands, 1986; 1–10.
17. Brito, L.M. Taxas de Mineralização da Matéria Orgânica nos Resíduos Sólidos Urbanos: Efeitos Agronómicos e Ambientais. In Seminário sobre Produção de Correctivos Orgânicos a Partir de Resíduos Sólidos Urbanos—Sua Importância para a Agricultura Nacional, Exponor (Matosinhos), April 8, 1997.
18. Corbitt, R.A. *Standard Handbook of Environmental Engineering*, 2nd Ed.; McGraw-Hill Book Company: New York, 1998; 1248 pp.
19. Mays, D.A.; Giordano, P.M. Landspreading municipal waste compost. BioCycle **1989**, *30* (3), 37–39.
20. Howard, A.; Wad, Y.D. *The Waste Products of Agriculture: Their Utilisation as Humus*; Oxford University Press: London, 1931.
21. Golueke, C.G. Composting refuse at Sacramento, California. Compost Sci. **1960**, *1* (3), 12–15.
22. McGauhey, P.H. Refuse composting plant at Norman, Oklahoma. Compost Sci. **1960**, *1* (3), 5–8.
23. Neto, J.T.P. Aspectos Tecnológicos da produção e Qualidade do Composto Orgânico a partir da Compostagem de Lixo Urbano. In Seminário sobre Produção de Correctivos Orgânicos a Partir de Resíduos Sólidos Urbanos—Sua Importância para a Agricultura Nacional, Exponor (Matosinhos), April 8, 1997.
24. Lu, Y.; Wu, X; Guo, J. Characteristics of municipal solid waste and sewage sludge co-composting. Waste Manage. **2009**, *29* (3), 1152–1157.
25. Black, R.J. The solid waste problem in metropolitan areas. California Vector Views **1964**, *11* (9), 51.
26. Diaz, L.F. Combining experience with common sense. BioCycle **1982**, *30* (10), 48–49.
27. Golueke, C.G.; Diaz, L.F. Historical review of composting and its role in municipal waste management. In Proceedings of the Science of Composting; Bertoldi, M., Sequi, P., Lemmes, B., Papi, T., Eds.; Chapman & Hall: Glasgow, 1996; 3–14.
28. Ferrero, G., L'Hermite, P. Composting—Progress to date in the European Economic Community and prospects for the future. In *Composting of Agriculture and other Wastes*; Gasser, J.K., Ed.; Elsevier Applied Science Pub.: Amsterdam, 1985; 3–10.
29. Lopez-Real, J.M. Composting of agricultural wastes. In Proceedings of the Science of Composting; Bertoldi, M., Sequi, P., Lemmes, B., Papi, T., Eds.; Chapman & Hall: Glasgow, 1996; 542–550.
30. Mesquita, M.M.F. Compostagem de material sólido—Principais vantagens e aspectos mais problemáticos. In Seminário para apresentação e discussão do Plano de adaptação à legislação ambiental pelo sector da suinicultura. LNEC e Federação Portuguesa das Associações de Suinicultores: LNEC, Lisbon, Jan. 22–23, 1996.
31. Hamoda, M.F.; Abu Qdais, H.A.; Newham, J. Evaluation of municipal solid waste composting kinetics. Resour., Conserv. Recycl. **1998**, *23* (4), 209–223.
32. John Paul, J.A.; Karmegam, N.; Daniel, T. Municipal solid waste (MSW) vermicomposting with an epigeic earthworm, *Perionyx ceylanensis* Mich. Bioresour. Technol. **2008**, *102* (12), 6769–6773.
33. Elango, D.; Thinakaran, N.; Panneerselvam, P.; Sivanesan, S. Thermophilic composting of municipal solid waste. Appl. Energy **2009**, *86* (5), 663–668.
34. Tintner, J.; Smidt, E.; Böhm, K.; Binner, E. Investigations of biological processes in Austrian MBT plants. Waste Manage. **2010**, *30* (10), 1903–1907.
35. Sims, J.T.; Murphy, D.W.; Handwerker, T.S. Composting of poultry wastes—Implications for dead poultry disposal and manure management. J. Sustainable Agric. **1992**, *2* (4), 67–82.
36. Mesquita, M.M.F. Aplicações Inovadoras da Compostagem e Especificações Técnicas para Avaliação da Qualidade do Composto. In Curso sobre Valorização e Tratamento de Resíduos. Prevenção, Recolha Selectiva, Compostagem e Confinamento em Aterro. LNEC/APESB, Lisbon, Dec 10–12, 1996.
37. Williams, P.T. *Waste Treatment and Disposal*, 2nd Ed.; John Wiley & Sons, Ltd.: Chichester, West Sussex, 2005.
38. Neto, J.T.P.; Mesquita, M.M.F. *Compostagem de Resíduos Sólidos Urbanos. Aspectos Teóricos, Operacionais e Epidemiológicos*; LNEC: Lisbon, 1992.
39. de Bertoldi, M.; Zucconi, F. Composting of organic residues. In *Bioenvironmental Systems*; Wise, D.L., Ed.; CRC Press: Boca Raton, Florida, 1987; Vol. III.
40. Loehr, R.C. *Pollution Control for Agriculture*, 2nd Ed.; Academic Press: Orlando, Florida, 1984.
41. Neto, J.P.; Stentiford, E.I. The main process constraints in composting. In Proceedings of the First Italian-Brazilian Symposium of Sanitary Engineering, Rio de Janeiro, Brazil, Mar 29–Apr 3, 1992.
42. Rynk, R.; van de Kamp, M.; Willson, G.B; Singley, M.E.; Richard, T.L.; Kolega, J.J.; Gouin, F.R.; Laliberty, Jr., L.; Kay, D.; Murphy, D.W.; Hoitink, H.A.J.; Brinton, W.F. *On Farm Composting*. Northeast Regional Agricultural Engineering Service. Cooperative extension. NRAES-54, Rynk, R., Ed., NRAES: Ithaca, New York, 1992, available at http://infohouse.p2ric.org/ref/24/23702.pdf (accessed May 20, 2012).

43. Lipor, Terra-à-Terra Project, 2007–2011, available at http://www.hortadaformiga.com/gb/conteudos.cfm?ss=8 (accessed August 2011).
44. Richard, T.L. MSW Composting: Biological Processing. Cornell Department of Agricultural and Biological Engineering, New York State College of Agriculture and Life Sciences: Cornell University, Ithaca, 1998, available at http://www.cals.cornell.edu/dept/compost/MSW.FactSheets/msw.fs2.html (accessed August 2011).
45. Risse, M.; Faucette, B. Food Waste Composting. Institutional and Industrial Applications. University of Georgia College of Agricultural and Environmental Sciences, 2000, available at http://www.caes.uga.edu/applications/publications/files/pdf/B%201189_2.PDF (accessed August 2011).
46. Graves, R.E.; Hattemer, G.M. Chapter 2—Composting, In *National Engineering Handbook*, Part 637—Environmental Engineering, United States Department of Agriculture, Natural Resources Conservation Service, 2010.
47. Misra, R.V.; Roy, R.N.; Hiraoka, H. *On-Farm Composting Methods*; Food and Agriculture Organization of the United Nations: Rome, 2003.
48. Diaz, L.F., Savage, G.M., Eggerth, L.L.; Golueke, C.G. *Composting and Recycling Municipal Solid Waste*; Lewis Publishers, CRC Press: Boca Raton, Florida, 1993.
49. Diaz, L.F.; Savage, G.M.; Eggerth, L.L. Solid Waste Management, United Nations Environment Programme, UNEP IETC and CalRecovery Inc., 2005, available at http://www.unep.or.jp/ietc/Publications/spc/Solid_Waste_Management/ (accessed August 2011).
50. Arancon, N.Q.; Edwards, C.A. The use of earthworms in the breakdown of organic wastes to produce vermicomposts and animal feed protein. In *Earthworm Ecology*; Edwards, C.A. Ed.; CRC Press: Boca Raton, Florida, 2004; 345–379.
51. Munroe, G. *Manual of On-Farm Vermicomposting and Vermiculture*; Organic Agriculture Centre of Canada, 2007.
52. Sinha, R.K.; Agarwal, S.; Chauhan, K.; Soni, B.K. Vermiculture technology: Reviving the dreams of Sir Charles Darwin for scientific use of earthworms in sustainable development programs. J. Technol. Investment **2010**, *1* (3), 155–172.
53. Sinha, R.K, Herat, S., Valani, D.; Chauhan, K. Earthworms—The environmental engineers: Review of vermiculture technologies for environmental management and resource development. Int. J. Global Environ. Issues **2010**, *10* (3/4), 265–292.
54. Bogdanov, P. *Commercial Vermiculture: How to Build a Thriving Business in Redworms*; VermiCo Press: Oregon, 1996.
55. Atiyeh, R.M.; Subler, S.; Edwards, C.A.; Bachman, G.; Metzger, J.D.; Shuster, W. Effects of vermicomposts and composts on plant growth in horticulture container media and soil. Pedobiologia **2000**, *44*, 579–590.
56. Arancon, N.Q.; Edwards, C.A. Effects of vermicomposts on plant growth. In Proceedings of the Vermi-Technologies Symposium for Developing Countries, Department of Science and Technology—Philippine Council for Aquatic and Marine Research and Development, Los Banos, Philippines, 2006, http://www.slocountyworms.com/wp-content/uploads/2010/12/EFFECTS-OF-VERMICOMPOSTS-ON-PLANT-GROWTH.pdf (accessed August 2011).
57. Arancon, N.A.; Edwards, C.A.; Babenko, A.; Cannon, J.; Galvis, P.; Metzger, J.D. Influences of vermicomposts, produced by earthworms and microorganisms from cattle manure, food waste and paper waste, on the germination, growth and flowering of petunias in the greenhouse. Appl. Soil Ecol. **2008**, *39* (1), 91–99.
58. Tchobanoglous, G.; Theisen, H.; Vigil, S. *Integrated Solid Waste Management. Engineering Principles and Management Issues*; McGraw-Hill Book Company: Singapore, 1993.
59. Fonseca, E.; Amorim, J. *Quantificação dos Resíduos Sólidos Gerados no ISEP: Projecto e Operação de Uma Unidade de Compostagem Para Tratamento da Fracção Biodegradável*, CESE em Engenharia Química—Tecnologias de Protecção Ambiental/Projecto, ISEP/IPP: Porto, 1999.
60. Cross, F.; O'Leary, P.; Walsh, O. Operation and maintenance considerations for waste-to-energy systems. Waste Age **1987**, *18* (8).
61. Lasaradi, K.E.; Stentiford, E.I. A simple respirometric technique for assessing compost stability. Water Res. **1998,** *32* (12), 3717–3723.

Composting: Organic Farming

Saburo Matsui
Graduate School of Global Environmental Studies, Kyoto University, Kyoto, Japan

Abstract
This entry examines four advantages of organic farming: it recycles nutrients by finding a use for organic waste; it produces compost that provides rich humus to enhance the soil; it reduces the need for synthetic chemical pesticides; and it provides us with tasty and authentic food. These four merits are made possible by the application of a probiotic principle to compost production. The probiotic approach in composting employs three major types of bacteria, separately or in combination, namely, *Bacillus* species, lactic acid bacteria species, and actinomycetous species. These bacteria are in general not harmful to human or plant health. The bacteria suppress the activities of organisms that cause continuous cropping hazards in the fields. Probiotic science is under development for human health, which utilizes beneficial/effective bacteria to support the human immune system. In the same manner, providing beneficial/effective bacteria to plant growth is called probiotics agriculture, in which pathogenic microorganisms are ecologically controlled in the fields. When grown via the endophytic or ectophytic manner, the beneficial/effective bacteria excrete auxin and/or cytokinin, which are major plant hormones for plant growth. Probiotics agriculture has great potential for further development, supporting the reduction in the application of chemical fertilizers and synthetic agrochemicals, thus providing safer food with authentic taste.

INTRODUCTION

Large-scale monocultures of major crops may offer one way to feed the world, but their production depends on the use of massive quantities of chemical fertilizers, pesticides, and herbicides. Modern agricultural methods practiced currently in major food supply countries are unsustainable, causing fundamental problems such as soil deterioration, land and water ecosystem destruction, groundwater contamination, and phosphorus depletion. Problems caused by chemical fertilizers have reached record levels worldwide, where the widespread use of nitrogen fertilizer is contaminating the global water supply. Lakes and reservoirs are eutrophying, ecosystem food webs are changing, and many precious species are becoming extinct. Efforts to reduce the level of eco-toxicity of agrochemicals or limit the amounts that are used are not enough to halt the ecological deterioration that is being reported. The future preservation of water resources—now an issue of global concern—depends increasingly on radical changes in agricultural practices.

The European Commission (EC) is one of a number of organizations that promotes transitioning from using agricultural chemicals toward applying organic farming methods in the European Union (EU). Defining organic farming as "an agricultural system that seeks to provide you, the consumer, with fresh, tasty and authentic food while respecting natural life-cycle systems",[1] EU organic farming policy guidelines call for strict limits on the use of synthetic chemical pesticides and fertilizers and advise farmers to take advantage of on-site resources, such as livestock manure for fertilizer or feed produced on the farm. These and other organic farming principles and objectives underpin a host of EU-recommended common farming practices designed to minimize the human impact on the environment, while ensuring that the agricultural system operates as naturally as possible. In Japan, the Ministry of Agriculture enacted the Law on Promotion of Organic Farming in 2006. This law introduced guidelines that accord closely with the standard practices for organic farming endorsed by the EU. Similar guidelines are endorsed or called for worldwide by other organizations and individuals rejecting synthetic chemical insecticides, herbicides, and fungicides in favor of natural methods to discourage insect predation and encourage plant growth.

PROBIOTIC AGRICULTURE

Plants need nitrogen, phosphate, and potassium for foliage growth, root development, and flowering to make fruit. Supplying them with these in the right quantities when they are needed most is perhaps the greatest challenge for organic farmers. Applying compost—the rich humus-like material made when microorganisms and fungi break

Encyclopedia of Environmental Management DOI: 10.1081/E-EEM-120046913

down organic matter—allows farmers to naturally supply plants with nitrogen, phosphorus, potassium, and the nutrients essential for plant growth while also enhancing soil structure. Plants grown on soils with compost are healthier and more resistant to diseases. The composting process also offers a valuable way to recycle nutrients by turning problematic organic wastes into a valuable resource for agriculture.[2]

Of the variety of composting processes available, one method currently gaining momentum across Japan uses a probiotic approach to introduce beneficial/effective bacteria during the composting process with the ultimate aim of increasing the supply of nutrients to the soil. The concept of "probiotics" is derived from human health science where lactic acid bacteria species and *Bifidobacterium* species are used to keep the intestinal gut system in a healthy condition and bolster the human natural immune system.[3,4] Lactic acid bacteria used to produce yogurt, cheese, pickles, sauerkraut, tsukemono, kimchi, and other fermented foods are known to aid the body's natural immune responses and digestion, as well as alleviate lactose intolerance. Bacteria that benefit the human body in these ways are also known to play a similar role in the digestive and immune systems of domesticated mammals, such as cattle, for which lactic-acid-fermented fodder proved to be an excellent probiotic food. In each case, "beneficial/effective" bacteria help the body—whether human or animal—protect itself.

Bacteria belonging to the *Bacillus*, lactic acid, and actinomycetous bacteria groups are proving particularly beneficial/effective to plant life when introduced during the composting process, both as composting agents and in the soil to which the compost is applied.[2] The practice of growing certain crops, vegetable, and fruits continually, without rotating them, has given rise to hazards from a variety of organisms including insects, nematodes, fungi, bacteria, and viruses that attack crops. Although plants, like humans, have a natural immune system to defend themselves when under attack from various organisms, they are not always strong enough to fend off the attack. They may therefore benefit from the provision of "friendly" bacteria that strengthen their immune systems through processes such as commensal or mycotrophic symbiosis. The concept of using "effective" bacteria to promote plant growth was first developed in Japan in the 1970s by eminent Japanese horticulturalist Dr. Teruo Higa, currently professor at the University of the Ryukyus, Okinawa. Higa's proposals for an "earth-saving revolution" based on the use of "effective microorganisms" (as he called them) in agricultural practice was distributed in Japanese and English by Sunmark Publishing in the 1990s.[5] Recent studies of endophytic microorganisms—organisms that reside in the living tissues of virtually all plants—show that their relationships with the host plants range from beneficial/effective and symbiotic to pathogenic (Subhash et al. 2010). Utilizing the endophytic function of *Lactobacillus* and *Bacillus* bacteria to stimulate plant growth while ecologically suppressing the activity of pathogenic endophytic microorganisms in the host plant helps the development of natural immune system of the host plant.

This is the essence of "probiotics agriculture." While the scientific understanding is new, using bacteria groups to compost organic waste is not: the use of human excrement or nightsoil on crops has a long history in Japan, revived most recently after the Second World War, when organic farmers added human excreta as well as animal dung—both rich sources of nitrogen, phosphate, and potassium—to the organic waste (leaves, roots, rice stalks, etc.,) generated in farming during the autumn harvest to make compost. (Starting after the autumn harvest, farmers would dig a large hole in a corner of their field and dump their farm waste into it, together with fermented human excreta. Natural composting would start with native *Bacillus* spp. and other aerobic bacteria active in the upper part of the hole and native anaerobic bacteria such as lactic acid and yeast active in the bottom part. The compost would mature in 5 mo—just in time for spring, when it would be added to new crops and paddy fields. The farmers using this method may not have recognized what types of aerobic and anaerobic bacteria and yeast were at work in their composting, but they were very sure of the quality compost that was its result!). Many prominent Japanese organic farmers practice farming: whether by adding rotten leaves—a prime source of *Bacillus* bacteria—into their organic waste when composting or by adding *Lactobacillus* spp. in a soluble jelly form directly on to their crops for use as a natural pesticide. Their efforts show that Gram-positive bacteria species in general can be used to protect plants and crops from the hazards of modern continuous cropping, which include the nematodes that can decimate potato crops and a number of insect types that attack soybeans and maize. They are also effective in controlling fungal diseases such as powdery mildew, downy mildew, and pythium blight that attack many plants, as well as club root—a common disease caused by *Plasmodiophora brassicae* that affects cabbages, turnips, and other plants belonging to the family Cruciferae (mustard family)—and *Fusarium* wilt, which can afflict solanaceous crop plants (tomato, potato, pepper, and eggplant) and banana plants.[2]

THREE MAJOR GROUPS OF BACTERIA

Typically, one bacteria group will be chosen over another in accordance to the type of waste that is being composted. Organic waste must therefore be carefully sorted to ensure that the bacteria chosen will work effectively. Efforts to control the oxygen supply, temperature, and level of moisture of the compost come next: such efforts are vital, if optimum conditions for bacterial activity are to be kept up throughout the composting process. Bacteria can be added in a wide variety of ways: introduced "naturally," in the case of brown rotten leaves carrying the *Bacillus* species in the simplest case, or applied more scientifically, in a powder or solution form administered through carefully measured inoculations, as pure cultures, or in a mixed culture containing all three bacteria groups.

Bacillus species such as *Bacillus subtilis*, *B. subtilis* var. *natto*, and *Bacillus thuringiensis* can rapidly decompose any type of organic waste including animal dung, raw garbage, and sewage sludge. As these aerobic bacteria require the presence of oxygen to live and grow, strict aerobic conditions must be observed throughout the composting process to ensure an optimum air supply. Good ventilation is also a priority: composting organic waste that contains protein produces high levels of ammonia, which occur when amino acids are broken down. Resulting odor problems can only be dealt with by keeping composting mounds or composting yards well ventilated. Rain protection above the mound is also necessary.

Fig. 1 shows *Bacillus* spp. aerobic composting carried out on an industrial scale in one of the several sewage works in Japan currently using this method to compost dewatered sewage sludge for use as natural fertilizer. Aerobic conditions are secured in concrete floored composting yards such as the one shown in the figure by channeling air into the compost through vinyl or cast iron pipes placed in shallow trenches underneath the mound. This ensures that an optimum air supply reaches the bottom layer of compost, as well as the top layer, which is exposed to the natural circulation of air. The bacteria give off heat when they begin work, decomposing organic matter. When the mound is big enough, temperatures rise within it, to between 85°C and 100°C. Vapor will start to rise off the mound when the temperature reaches 85°C or above, as can be seen in the third compartment from the left in the figure. When this happens, the life of the "friendly" bacteria within it ensures the continuance of hyperthermophilic composting. The preparation of seven compartments allows industrial-scale composters to lower the temperature of the compost mound by mixing and moving it from one compartment to the next—starting off new batches of compost as older ones reach maturity. In industrial-scale composting, the seven compartments correspond to the number of weeks required to stabilize compost in countries where temperatures drop sharply in winter—in Europe, for example, or in North America, China, or Japan. Small-scale composters using hyperthermophilic composting in aerobic conditions do not need to use compartments, but they must keep the compost well mixed, well aired, and within an acceptable temperature range at all times for composting to be successful.

It is important to note that hyperthermophilic composting at these high temperatures is enough to kill viruses, the eggs of worms and insects, and all bacteria in the Gram-negative group, which contains most of the bacterial pathogens to humans and animals. Anaerobic bacteria cannot grow because of the aerobic conditions observed in the composting process. Some Gram-positive spore-forming bacterium may survive, but these are in general not harmful to humans or to crops. Indeed, many of the spore-forming *Bacillus* species that remain in the compost where they germinate, long after the compost has been added to soil by farmers, function as microbial pesticide, herbicide, and fungicide and are as such beneficial/effective to the crops on which the compost is applied (Fig. 2). One notable exception is the dangerous bacteria *Bacillus anthracis*; a person who contracts this bacterium must be treated immediately by medical vaccination. Other than these extremely rare exceptions, compost produced using this method can be pronounced safe.

Lactic acid bacteria in the group including *Lactobacillus* spp., *Enterococcus* spp., *Lactococcus* spp., *Pediococcus* spp., and *Leuconostoc* spp. can work rapidly to

Fig. 1 Hyperthermophilic composting of dewatered sewage sludge in aerobic conditions using *Bacillus* species.

生産業者保証票
2.4%
2.2%
17%
6
問合せ先 鹿児島市南部処理場
099-268-3393

Fig. 2 Composted sewage sludge produced by the Kagoshima city south sewage works and certified for use as quality fertilizer.

Carbon – Cyanobacteria

decompose plant material and any type of organic waste in facultatively anaerobic conditions. Bacterial species in this group prefer more anaerobic condition to decompose organic waste that is rich in carbohydrate and sometimes work together with yeast species such as *Saccharomyces* spp. and *Schizosaccharomyces* spp. Rainwater protection over the mound of compost is a strict requirement, as is good drainage from the bottom of the mound. When the mound is big enough, temperatures within it rise, exceeding 60–70°C, and activating thermophilic composting. Although Gram-negative bacteria species and some of the spore-producing Gram-positive bacteria may survive at these temperatures, lactic acid produced by bacteria breaking down carbohydrate, fat, and oil brings the pH of the composting mound down below 5, creating acidic conditions in which most Gram-negative bacteria and pathogenic viruses cannot survive. In principle, only a number of eggs of worms and insects and some of the spore-producing Gram-positive bacteria may survive in these acidic conditions—none of which pose much of a problem. More problematic, perhaps, is the pathogenic yeast *Candida pneumocystis jiroveci*, which causes opportunistic respiratory tract infections in patients with compromised immune systems, and pathogenic lactic acid *Clostridium perfringens*, which can produce food-poisoning toxins that trigger abdominal cramps and diarrhea—both of which may also survive.

It takes 3–4 mo for compost to mature using this method, depending on the climate of the weather in the country where the compost is being made: composting in winter will take longer to mature in countries in the North Temperate Zone, for example. Lactic acid bacteria produce lactic acid when composting organic waste that contains carbohydrate, fat, and oil and amino acids when composting waste that contains protein: this drops the pH of compost and creates a low-pH, amino-rich compost that is particularly good for plant growth. Another benefit of this composting method is that odor problems due to ammonia are much less compared to *Bacillus* species composting (Fig. 3).

Fig. 3 Pig manure is composted by lactic acid bacteria. When some types of Lactobacillus spices are provided in pig feed for probiotic raising, they can survive in the pig intestine and be excreted in the manure in which the bacteria continue lactic acid and amino acid fermentation of substances in the manure. The manure does not smell ammonia and invite flies.

Actinomycetous species including *Actinoplanes* spp., *Ampullariella* spp., *Dactylosporangium* spp., and *Streptomyces* spp. play an important role in the decomposition of any type of organic waste, in particular oil, fat, and plant cell walls containing lignin, latex, and chitin. Aerobic conditions should be observed throughout the composting process, and both rainwater protection and good drainage must be provided to the compost mound. When the mound is big enough, relatively high temperatures of between 35°C and 65°C are reached within it. Mesophilic composting of this kind does not create high enough temperatures to kill off all Gram-negative bacteria, and some species of spore-producing Gram-positive bacteria may also survive, along with the eggs of worms and insects. The survived spore may wake up in soil where the condition is favorable to them, but most of them are not harmful to crops. Pathogenic viruses cannot survive a prolonged composting period, however, and they will die away—together with some Gram-negative bacteria—in the long period of starvation that results as the availability of organic food depletes over the course of composting.

It takes 3–5 mo depending on the composting temperature, air supply, and climate to get matured compost using this method. In Japan, where winters are cold, it will take 5 mo rather than 3 mo, for example, if composting outside, using this method. During composting organic waste that contains carbohydrate, fat, oil, and protein, actinomycetous bacteria produce low molecule organic acids and amino acids that drop the pH of the compost, and may help the growth of plants in the fields. The matured compost has a uniquely rich smell of earth or fungus that is created when the bacteria forms whitish lace-like filaments that cover the surface of the compost mound (Figs. 4 and 5). Bark compost produced by actinomycetous species composting is already commercially available. In addition, many types of antibiotics produced by the fermentation of the *Streptomyces* and other actinomycetous species are known to be effective in compost to control some Gram-negative pathogens to crops and vegetables. It is important to know that there are significant pathogenic actinomycetous species—including *Mycobacterium tuberculosis*, *Mycobacterium leprae*, *Corynebacterium diphtheriae*, and *Mycobacterium bovis*—that cause well-known and well-diagnosed human and domesticated mammal diseases.

ANOTHER ADVANTAGE OF PROBIOTIC AGRICULTURE

Selecting the correct species of bacteria as the catalytic function for composting organic waste is the key to successful

Fig. 4 Garbage compost showing a white zone development over the mound by actinomycetous species.

probiotic agriculture. This is because probiotic composting depends on choosing a catalytic bacterium or a mixture of catalytic bacteria that can foster the presence of friendly bacteria in the compost that is the end product. When this compost is applied to soil on which plants and crops are being grown, the bacteria can start to proliferate and work as microbial pesticides against bad microbes. For example, the application of bark compost made using actinomycetous bacteria suppresses problems for strawberry plants and watermelon caused by continuous cropping, while *Bacillus* species composting has proved effective in preventing the nematodes that feed on the roots and sieve tubes of potatoes.[2] In each case, the bacteria also actively improve the taste of the crops: bark compost greatly improves the sugar content of strawberries, for instance, and *Bacillus* species compost enhances the sweetness of potatoes.

The improvement in the taste of crops that have been grown with probiotically made compost is caused by the

Fig. 5 A magnified picture of the white zone showing a typical actinomycetous species growth over a compost.

plant hormones produced by the bacteria used in the composting process. Bochow et al.[6] has shown that *Bacillus* species produce auxin, a key plant hormone that regulates the direction of plant growth toward sunlight. Others have reported that actinomycetous species produce cytokinin, another key plant hormone for regulating plant growth and development.[7] Cytokinin promotes cell division in plant roots and shoots and regulates cell enlargement, chloroplast development, plant senescence or aging, and cell differentiation—all processes that relate to plants' ability to extend leaves and bear blossom and fruit.[7,8] My own studies have shown that *Lactobacillus* species produce both auxin and cytokinin.[9] When probiotically produced compost is introduced to the soil, the beneficial/effective bacteria in the compost can work in either endophytic or ectophytic ways. Ectophytic bacteria stimulate plants around the surface of roots and leaves, while endophytic bacteria live in sieve tubes where a symbiotic relationship develops. Endophytic bacteria also excrete auxin and/or cytokinin, which stimulate plant growth and enhance the plant's ability to bear blossom and fruit. The sweetness of fruit comes from its sugar content. *Lactobacillus* species stimulate the growth of any plant and hasten the bearing blossom and fruit. Many farmers in Japan and other countries who use friendly or beneficial/effective bacteria in their farming practices state that the taste of leaf vegetables is recognizably better and the sweetness of all fruits is enhanced. Their opinions would appear to be endorsed by the increasing numbers of consumers in Japan and elsewhere who appear willing to pay a higher price for organic farm products on account of their flavor as well as their perceived safety.

Fig. 6 Mr. N. Kaneko, a pioneer and leader of organic farming in Japan, who cultivates strawberries completely free from chemical fertilizer and synthetic pesticides by employing a species of beetles to kill louse and putting crab shells in the soil to increase the number of actinomycetous bacteria in addition of his original compost.

Fig. 7 Mr. Kaneko produces methane gas out of human excreta of his family and cow dung. The liquid production out of methane fermentation is rich in nitrogen, phosphate, and potassium, which is utilized for fertilizer.

Japanese Experience

Many Japanese farmers who apply synthetic chemical pesticide and herbicide to fields that they cultivate for commercial products say bluntly that they would not dream of applying those same chemicals to fields that are cultivated for their own consumption. I have met increasing numbers of small-scale growers eager to pioneer, practice, and promote organic farming in Japan such as Mr. N. Kaneko (Figs. 6–8). These changes in farming practices are welcomed by increasing numbers of citizens and consumers, predominantly Japanese women, who have been active agents of change in lobbying for agricultural practices that do not threaten the natural environment. At the national level, women in the organic farming movement have lobbied against genetically modified soya and Indian corn products and demanded safer food products to feed their families. The food industry is slowly but surely responding to calls such as these: in recent years, signs advertising "gen-nouyaku" (reduced chemical) and "mu-nouyaku" (no chemical, i.e., organically grown) products have appeared in high-street supermarkets and stores run by farmers' cooperatives all over Japan as increasing numbers of outlets sell organic products catering to consumers' changing tastes. Non-governmental organizations, such as the Japan Organic Agriculture Association (JOAA), Organic Farming Promotion Association, Organic Farming Research Council, and the Center of Japan Organic Farmers Group, are currently leading the way. In the future, city governments could tackle their organic waste problems by producing compost that not only is cheaper to use than chemical fertilizers but also brings significant environmental benefits.

Fig. 8 His original composting method is shown in his greenhouse, shaping a mound, utilizing all solid waste from agriculture. His method follows the principle of a traditional composting method in Japan, starting with the activity of *Bacillus* sp. bacteria naturally inoculated in the aerobic zone, followed by the activity of lactic acid bacteria naturally inoculated inside the mound of anaerobic zone. At the end of composting, all available organic substances are converted into the bacteria cells. However, plant cell walls that are rich in chitin are decomposed by actinomycetous species. During composting in winter, the temperature of the composting mound increases, providing heat for the germination of seeds that are placed on the top of the mound.

CONCLUSION

Composting organic waste for organic farming has a fourfold advantage: it recycles nutrients by finding a use for organic waste; it produces compost that provides rich humus to enhance the soil; it reduces the need for synthetic chemical pesticides; and it provides us with tasty and authentic food. These four merits, made possible by applying a probiotic principle to compost production, must be taken advantage of by greatly expanding the current scale of probiotic composting.

REFERENCES

1. European Commission, Organic farming: Good for nature, good for you, 2010, http://ec.europa.eu/agriculture/organic/organic-farming/what-organic_en (accessed December 3, 2010).
2. Matsui, S. Baiomas junkan katsuyo wo kanou ni suru purobaiotikusu nougyou genri [Probiotics principle that can help organic farming]. J. Environ. Sanit. Eng. Res. **2009** *23* (3), 81–87 (The Association of Environmental and Sanitary Engineering Research, Kyoto University).
3. Food and Agriculture Organization/World Health Organization Expert Consultation Report. Health and Nutritional Properties of Probiotics in Food including Powder Milk with Live Lactic Acid Bacteria, 2001, http://www.who.int/foodsafety/publications/fs_management/en/probiotics.pdf (accessed November 4, 2009).
4. Hamilton-Miller, J.M.T.; Gibson, G.R.; Bruck, W. Some insights into the derivation and early uses of the word 'probiotic'. Br. J. Nutr. **2003**, *90*, 845.

5. Higa, T. *An Earth-Saving Revolution: Solutions to Problems in Agriculture, the Environment and Medicine*; Sunmark Publishing Inc., 1993.
6. Bochow, H.; El-Sayed, S.F.; Junge, H.; Stavropoulu, A.; Schmiedeknecht, G. Use of *Bacillus subtilis* as biocontrol agent. IV. Salt-stress tolerance induction by *Bacillus subtilis* FZB24 seed treatment in tropical vegetable field crops, and its mode of action. Z. Pflanzenkrankh. Pflanzenschutz 2001, 108 (1), 21–30.
7. Nishimura, T. Housenkin wo riyo shita kofukakachi karumia sosikibaiyokin karano kogatahachimono sakushutsu gijutsu ni kansuru kenkyu [a study of a new technology for making high-price potted dwarf trees by a tissue culture using *Actinomycetes* species], PhD thesis, Graduate School of Bio-resources, Mie University, 2003.
8. Kirinuki, T.; Ofuji, H.; Suzuki, N. Chitin-lytic and -1, 3-glucanase activity of *Actinomycetes* and their effects for preventing fusarium-wilt of cucumber [in Japanese]. Sci. Rep. Fac. Agric., Kobe Univ. **1976**, *12* (1), 41–48.
9. Matsui, S.; Matsuda, T. An analysis of the production of auxin and cytokinin by *Lactobacillus* fermentum, 2011 (forthcoming publication, title provisional).
10. Higa, T. *An Earth-saving Revolution II: The Proven Effects of EM Technology*; Sunmark Publishing Inc., 1994.
11. Bhore, S.J.; Sathisha, G. Screening of endophytic colonizing bacteria for cytokinin-like compounds: Crude cell-free broth of endophytic colonizing bacteria is unsuitable in cucumber cotyledon bioassay. World J. Agric. Sci. **2010**, *6* (4), 345–352.

Copper

David R. Parker
Department of Soil and Environmental Sciences, University of California—Riverside, Riverside, California, U.S.A.

Judith F. Pedler
Department of Environmental Sciences, University of California—Riverside, Riverside, California, U.S.A.

Abstract

Copper (Cu) is a metallic trace element with two naturally occurring isotopes (^{63}Cu and ^{65}Cu), and is essential for the growth of all life. Naturally present in all soils, Cu is found in excessive and potentially toxic concentrations in mine spoils, in waste products from industrial and agricultural activities, and in some agricultural soils due to historic use of Cu sprays (e.g., Bordeaux mix) for disease control. Copper is deficient in many soils around the world, and addition of Cu fertilizer is required for productive crop growth.

COPPER IN SOILS

The average total concentration of Cu in the Earth's crust is estimated to be 70 mg/kg, although levels of 20–30 mg/kg are prevalent in average soils.[1] Common primary minerals include Cu sulfides, with Cu largely in the +I oxidation state, which dissolve by weathering processes. Secondary minerals of Cu(II) include oxides, carbonates (malachite), silicates, sulfates, and chlorides, most of which are relatively soluble. Copper(II) may substitute for Fe, Mg, and Mn in an assortment of minerals, especially silicates and carbonates.[1]

The Cu^{2+} ion can form strong inner-sphere complexes, and is thus immobilized by carboxylic, carbonyl, or phenolic functional groups, even at low pH. Exchangeable and weak acid-extractable Cu represent a small percentage of total Cu in most soils. The bulk of the Cu is complexed by organic matter, occluded in oxides, and substituted in primary and secondary minerals. Organic matter and Mn oxides are the most likely materials to retain Cu in a non-exchangeable form in soils. Alkali extraction techniques that remove organic matter from soils usually release large fractions of the total soil Cu.[2] Addition of organic matter to soils, and biological exudation of organic acids may both increase dissolved organic carbon, thus solubilizing Cu from mineral forms, increasing the total dissolved Cu in soil solution,[3] but predictive models of humic acid binding of Cu in soil solution are generally inadequate.[4] Overall, Cu is one of the least mobile of the trace elements, maintained in a form sufficiently available to plants but relatively resistant to movement by leaching.[1]

Free Cu^{2+} in soil solution decreases with increasing pH, reaching a minimum above pH 10. In the absence of organic ligands, Cu speciation is dominated by free Cu^{2+}, and increasingly by carbonate and hydroxy complexes as pH rises above 6.5.[1] The dissolved organic carbon found in most surface soils has a strong affinity for Cu, but estimates of the percentage of soluble Cu that is organically complexed can vary widely.

COPPER AND PLANTS

Plant uptake of Cu appears to be directly related to the concentration of the free ion, Cu^{2+}, but may also be influenced by the total concentration in soil solution, including organic complexes.[5] As with most trace metals, it is not known whether Cu is passively absorbed or actively taken up across the root-cell membrane.[6] Rates of absorption are generally low, on the order of 1 nmole h^{-1} (g root dry weight)$^{-1}$.[7] The activity of free Cu^{2+} required in nutrient solution for optimal plant growth is just 10^{-14}–10^{-16} M.[5] Copper absorption is generally halted by metabolic inhibitors and uncoupling agents which disrupt the normal trans-membrane potential.[7]

Uptake of Cu is strongly affected by pH: increasing concentration of hydrogen ions decreases the absorption of Cu ions by plant roots.[8] Uptake is also affected by the presence of Ca, and to a lesser extent by Mg, both of which compete with Cu for binding sites at the root plasmalemma.[9] The effects of other trace metals on Cu uptake have frequently been seen as inhibitory (Zn), or stimulatory (Mn), but not necessarily under well-defined conditions.[2]

COPPER AS AN ESSENTIAL ELEMENT

Copper is an essential element for plant growth and is a component of many enzymes, including plastocyanin, and thus is an indispensable prosthetic group in Photosystem 2. Cu-containing proteins are also important in respiration (cytochrome c oxidase is the terminal oxidase of the mitochondrial electron chain), in detoxification of superoxide radicals (super-

Encyclopedia of Environmental Management DOI: 10.1081/E-EEM-120001594

oxide dismutase), and in lignification (polyphenol oxidase). Ascorbate oxidase, which contains eight Cu^{2+} ions, has been proposed as an indicator of plant Cu status, although its relevant biological function has yet to be determined.[10]

The critical concentration of Cu in shoot tissue for optimal growth does not vary greatly between plant species, ranging from 1 to 6 µg/g dry weight of young leaf tissue.[11] Most crops are recorded as having a requirement of 3–5 µg/g.[11] The average concentration of Cu in plant parts varies with age and with the level of Cu and N supply. Translocation of Cu to plant shoots increases with an increasing supply of N. In the xylem and phloem saps, Cu is probably complexed by amino acids.[2] Copper is usually described as having "variable" phloem mobility in plants,[2] as the retranslocation of Cu from older tissues is regulated by both Cu-supply and by N-status. Lack of a sufficiently long-lived radioisotope makes study of Cu transport and translocation problematic.

PLANT GROWTH ON COPPER-DEFICIENT SOILS

Copper deficiency most often occurs on organic soils where excessive leaching has occurred, or on calcareous sands. In general, crops grown on mineral soils with Cu contents less than 4 to 6 µg/g, or on organic soils with less than 20 to 30 µg/g are the most likely to suffer Cu deficiency,[13] although this varies with specific soil type and the crop grown.

Symptoms of Cu-deficiency include chlorosis, necrosis, leaf distortion and terminal dieback, and are most evident in new leaf growth. Wilting can also occur, indicating structural weaknesses due to reduced lignification of the xylem elements. These symptoms are not entirely specific to Cu deficiency, and can be observed in plants under a variety of stresses. The most profound symptoms of Cu deficiency are those seen in the reproductive cycles of many sensitive species: delay of flowering, and/or reductions in seed and fruit yield as a consequence of sterile pollen or reduced floret numbers. Because these latter symptoms are not observed until maturity or harvest, Cu deficiency is often termed a "hidden hunger." Rice, citrus and cassava are sometimes referred to as indicator species that are more sensitive to Cu deficiency, but are still not reported to require more than 5–6 µg/g Cu to avoid Cu deficiency. Cereal rye and canola are crops more tolerant of Cu deficiency, requiring only 1–2 µg/g Cu.[11]

Copper deficiency in legumes depresses nodulation and N2 fixation, leading to N deficiencies. Unlike Mo and Co, there seems to be no specific Cu requirement for N2 fixation in nodules beyond that required for plant growth and the production of carbohydrates.[12]

Copper deficiency decreases polyphenol oxidase activity and thus lignification.[14] Susceptibility of Cu-deficient plants to pathogenic attack may be increased due to reduced lignification of xylem elements, or due to impaired lignification in response to pathogenic invasion (wounding response). Application of Cu to soil, at rates too low to directly affect the pathogen, has controlled powdery mildew in wheat.[15] It has also been suggested that, where Cu in soils is more than sufficient, the accumulation of additional Cu in roots provides a fungistatic defense against pathogens.[16] Conversely, where Cu is deficient, roots are more vulnerable to pathogenic invasion.

There are genetic differences in the absorption of Cu by plant roots. Rye is able to take up significantly more Cu from soil than wheat, and is thus viewed as being more Cu-efficient. Tritcale, the wheat–rye hybrid, inherits the efficiency factor.[7] Copper efficiency could be a useful trait in breeding crops for regions where soils are commonly Cu-deficient.[17] However, the mechanism of the efficiency factor is not clear.

PLANT GROWTH ON HIGH-COPPER SOILS

As some plant species have adapted to soils of low Cu status, others have evolved tolerance to Cu-toxic conditions. The 16th century author, Agricola, wrote of indicator plants that grow on naturally Cu-rich soils. Natural revegetation of mine spoils has been shown to reflect rapid genetic evolution of Cu tolerance by grasses and other plants.[14] There seem to be several possible mechanisms of Cu tolerance, although exclusion from shoots is a common feature. The exceptions are a few Cu-accumulator species which may contain in excess of 1000 µg/g Cu in shoot tissue.[12] In other Cu-tolerant species, root compartmentation or immobilization of Cu may be achieved by immobilization in cell walls, by complexation with intracellular proteins, or by removal of Cu to the vacuole.[14]

With nontolerant taxa, plant growth is likely to be depressed when Cu concentrations in the whole shoots exceed ~20 µg Cu g^{-1}. Symptoms of Cu toxicity include poorly developed and discolored root systems, reduced shoot vigor, and leaf chlorosis. Toxicity thresholds (e.g., for a 10% yield reduction) seem to vary widely, probably because of the low translocation of Cu from roots to shoots. Only when roots are overwhelmed by Cu rhizotoxicity does sufficient Cu reach shoots to affect growth and function. Other syndromes, such as Fe deficiency, can readily occur as secondary consequences of excess Cu.[14]

Exclusion of Cu from the shoot protects photosynthetic activity, which is highly sensitive to excess Cu. Photosynthetic electron transport is blocked by high levels of free Cu, at the oxidizing side of Photosystem 2, and at the reducing side of Photosystem 1. Excess Cu supply results in reduced lipid content and noticeable changes in the fatty-acid composition of tomato roots and primary leaves, indicating enhanced activity of enzymes which catalyze lipid peroxidation.[18]

Concentrated Cu sprays have historically been used to control foliar pathogens, especially in vineyard and orchard crops. These fungicidal sprays often included limestone to reduce their phytotoxicity, and to make them more rainfast. The accumulation of Cu in the soils under these crops has not regularly caused Cu toxicity, which indicates the remarkable ability of most plant roots to accumulate Cu, while regulating its flow to the shoots. Both Cu deficiency and toxicity may result in non-specific symptoms of plant stress. Assessment of the Cu status of a soil, or of crop plants, is most accurate when soil type, soil history, and soil and plant analyses for Cu are all considered.

REFERENCES

1. McBride, M.B. Forms and distribution of copper in solid and solution phases of soil. In *Copper in Soils and Plants*; Loneragan, J.F., Robson, A.D., Graham, R.D., Eds.; Academic Press: Sydney, Australia, 1981; 25–45.
2. Loneragan, J.F. Distribution and movement of copper in plants. In *Copper in Soils and Plants*; Loneragan, J.F., Robson, A.D., Graham, R.D., Eds.; Academic Press: Sydney, Australia, 1981; 165–188.
3. Sanders, J.R.; McGrath, S.P. Experimental measurements and computer predictions of copper complex formation by soluble soil organic matter. Environ. Pollut. **1988**, *49*, 63–79.
4. Robertson, A.P.; Leckie, J.O. Acid/base copper binding, and Cu^{2+}/H^+ exchange properties of a soil humic acid, an experimental and modelling study. Environ. Sci. Technol. **1999**, *33*, 786–795.
5. Bell, P.F.; Chaney, R.L.; Angle, J.S. Free metal activity and total metal concentrations as indices of micronutrient availability to barley (*Hordeum vulgare* L. Cv Klages). Plant Soil **1991**, *130*, 51–62.
6. Strange, J.; Macnair, M.R. Evidence for a role for the cell membrane in copper tolerance of Mimulus guttatus fischer Ex DC. New Phytol. **1991**, *119*, 383–388.
7. Graham, R.D. Absorption of copper by plant roots. In *Copper in Soils and Plants*; Loneragan, J.F., Robson, A.D., Graham, R.D., Eds.; Academic Press: Sydney, Australia, 1981; 141–163.
8. Lexmond, Th.M.; van der Vorm, P.D.J. The effect of pH on copper toxicity to hydroponically grown maize. Neth. J. Agric. Sci. **1981**, *29*, 217–238.
9. Parker, D.R.; Pedler, J.F.; Thomason, D.T.; Li, H. Alleviation of copper rhizotoxicity by calcium and magnesium at defined free metal-ion activities. Soil Sci. Soc. Am. J. **1998**, *62*, 965–972.
10. Maksymiec, W. Effect of copper on cellular processes in higher plants. Photosynthetica **1997**, *34*, 321–342.
11. Reuter, D.J.; Robinson, J.B. *Plant Analysis: an Interpretation Manual*; CSIRO Press: Melbourne, Australia, 1997; 83–566.
12. Römheld, V.; Marschner, H. Function of micronutrients in plants. In *Micronutrients in Agriculture*, 2nd Ed.; Mortvedt, J.J., Cox, F.R., Shuman, L.M., Welch, R.M., Eds.; SSSA: Madison, WI, 1991; 297–328.
13. Jarvis, S.C. Copper concentrations in plants and their relationship to soil properties. In *Copper in Soils and Plants*; Loneragan, J.F., Robson, A.D., Graham, R.D., Eds.; Academic Press: Sydney, Australia, 1981; 265–285.
14. Woolhouse, H.W.; Walker, S. The physiological basis of copper toxicity and copper tolerance in higher plants. In *Copper in Soils and Plants*; Loneragan, J.F., Robson, A.D., Graham, R.D., Eds.; Academic Press: Sydney, Australia, 1981; 235–262.
15. Graham, R.D. Susceptibility to powdery mildew (*Erisiphe graminis*) of wheat plants deficient in copper. Plant Soil **1980**, *56*, 181–185.
16. Graham, R.D.; Webb, M.J. Micronutrients and disease resistance and tolerance in plants. In *Micronutrients in Agriculture*; Mortvedt, J.J., Cox, F.R., Shuman, L.M., Welch, R.M., Eds.; Soil Science Society of America: Madison, WI, 1991; 329–370.
17. Owuoche, J.O.; Briggs, K.G.; Taylor, G.J. The efficiency of copper use by 5A/5RL wheat-rye translocation lines and wheat (*Triticum aestivum* L.) cultivars. Plant Soil **1996**, *180*, 113–120.
18. Ouariti, O.; Boussama, N.; Zarrouk, M.; Cherif, A.; Ghorbal, M.H. Cadmium-and copper-induced changes in tomato membrane lipids. Phytochemistry **1997**, *45*, 1343–1350.

Cyanobacteria: Eutrophic Freshwater Systems

Anja Gassner
Institute of Science and Technology, University of Malaysia-Sabah, Kota Kinabalu, Malaysia
Martin V. Frey
Department of Soil Science, University of Stellenbosch, Matieland, South Africa

Abstract
Aquatic systems in an urban environment are subjected to massive anthropogenic nutrient input in the form of either non-point (stormwater) sources or point sources (industry, sewage). Most of these water bodies progress from low-productivity or oligotrophic settings to productive mesotrophic conditions to overenriched eutrophic or hypertrophic conditions. The response to the so-called "cultural" eutrophication is excessive production of undesirable algae and aquatic weeds and oxygen shortages caused by their senescence and decomposition. Cyanobacterial (blue-green) algal blooms have become a serious water-quality problem around the world. From a human perspective it is desirable to prevent or minimize such processes for both aesthetic and health reasons. Algal blooms lower drinking water quality by production of often odorous and toxic compounds; the toxins produced in many blue-green algae have caused health problems for wildlife, livestock, pets, and humans in contact with contaminated water. Given the variety of uses of urban water bodies for recreation, housing development, fish farming, and nature reserves, management guidelines and increasing awareness are urgently needed. The objectives of this entry are to give a short introduction to freshwater blue-green algae, the key environmental factors that lead to their proliferation, and the subsequent environmental problems and to present management strategies.

FACTORS LEADING TO CYANOBACTERIAL DOMINANCE

The taxonomic composition of phytoplankton communities, the abundance and the relative dominance of the different species and groups present, undergo seasonal changes. This process of continuous community change is termed succession. Under undisturbed conditions, most phytoplankton populations are of relatively short duration. Typically, the growth and decline cycle of one specific population lasts, on average, 4 to 8 weeks. The "seasonal paradigm" of phytoplankton succession[1] describes the typical pattern of phytoplankton succession corresponding to the prevailing nutrient cycle in temperate, undisturbed lakes: a spring maximum of diatoms, sometimes followed by a second maximum in the autumn, an early summer maximum of Chlorophyceae (green algae) and a late summer maximum of Cyanophyta (blue-green algae).

It is generally accepted that with excess nutrients in the water column, in particular phosphorus, the phytoplankton flora deviates from the traditional seasonal community pattern with a shift toward cyanobacterial dominance. However, it must be stressed that nutrient limitation does not, in itself, provide cyanobacteria with the ability to become dominant; it is the combination of a multitude of abiotic and biotic factors. Enrichment experiments demonstrated that the maximum biomass of temperate lakes is ultimately limited by the phosphorus supply.[2] Increasing supplies of phosphorus lead to an increase of phytoplankton growth until other essential nutrients become limited. The first nutrient to become limited after phosphorus is usually nitrogen. Cyanobacteria are the only species that are able to fix atmospheric nitrogen. Whereas other algae become nitrogen limited, the ascendancy of nitrogen-fixing cyanobacteria is favored.

Apart from their ability of fixing atmospheric nitrogen, cyanobacteria feature some adaptations that enable them to outcompete other species. Eutrophic conditions result in large suspended stocks of phytoplankton, which reduce light penetration. Cyanobacteria possess gas vacuoles to control buoyancy. When subjected to suboptimal light conditions, they respond by increasing their buoyancy (regulated by the rate of photosynthesis) and move nearer to the surface and hence to the light.[2] Additionally, the possession of chlorophyll *a* together with phycobiliproteins allows them to harvest light efficiently and to grow in the shade of other species. Cyanobacteria are supposed to be more tolerant of high pH conditions and have an additional selective advantage at times of high photosynthesis because of their ability to use CO_2 as carbon source.[3] Some genera are able to offset the effects of photoinhibiting UV radiation encountered by near-surface populations. The resistance to photoinhibition is achieved by producing increased amounts of carotenoid pigments, which act as

Encyclopedia of Environmental Management DOI: 10.1081/E-EEM-120028862

"sunscreens."[4] Once established, cyanobacteria are able to inhibit the growth of other algae by producing secondary metabolites that are toxic to species of other genera.[5]

CONSEQUENCES OF CYANOBACTERIAL BLOOMS

Like any phytoplankton, bloom proliferation of blue-green algae reduces water quality in terms of human water use but also results in a reduction in diversity of the aquatic species assemblage at all trophic levels. The presence of "pea soup green" water, the accumulation of malodorous decaying algal cells, and the buildup of sediments rich in organic matter lead to user avoidance with the associated problems and implications for water quality management. The most obvious sign of an advanced blue-green algae bloom is the formation of green "scum," which leads to deoxygenation of underlying waters, subsequent fish kills, foul odors, and lowered aesthetic values of affected waters.[6] In addition, certain genera and species produce taste and odor compounds, typically geosmin and 2-methyl isoborneol, which cause non-hazardous but unpleasant problems for suppliers and users of potable water.[1]

The most serious public health concerns associated with cyanobacteria arise from their ability to produce toxins. Since the first published reported incidence of mammal deaths related to a toxic cyanobacterial bloom in 1978, more then 12 species belonging to nine genera of blue-green algae have been implicated in animal poisoning.[7] For human exposure, routes are the oral route via drinking water, the dermal route during recreational use of lakes and rivers, or consumption of algal health food tablets. Toxins produced in a random and unpredictable fashion by cyanobacteria are called cyanotoxins and classified functionally into hepatotoxins, neurotoxins, and cytotoxins. Additionally, some cyanobacteria produce the lesser toxic lipopolysaccharides (LPS) and other secondary metabolites that may be of potential pharamacological use.[8] One of the most tragic encounters of humans with cyanobacterial toxins led to the deaths of 60 dialysis patients due to contaminated water supply used in a hemodialysis unit.[9] Presently, a drinking water guideline of $1\,\mu g\,L^{-1}$ of toxin has been developed and implemented only for microcystin-LR.[4] Haider et al.[8] stress that the biggest challenge for water treatment procedures for the removal of cyanobacterial toxins is that one is faced with soluble and suspended substances. Thus, the most common treatment, chlorination, in general has been found not to be an effective process in destroying cyanotoxins.

MONITORING AND MANAGEMENT OF ALGAL BLOOMS

Drinking water treatment strategies are not always successful in removing algal toxins. Thus, detection of early-stage (emergent) blooms of cyanobacteria, especially if the bloom has not started to produce toxins, is important to allow municipalities and recreation facilities to implement a response plan. It has been shown that remote sensing technology can be used to estimate the concentration and distribution of cyanobacteria through measurement of the concentration of the pigment phycocyanin.[10]

Once detected, the growth of nuisance algae is prevented by the use of chemicals; the commonest is copper sulfate. Other algicides include phenolic compounds, amide derivatives, quaternary ammonium compounds, and quinone derivatives. Dichloronaphthoquinone is selectively toxic to blue-greens. The inherent problem of algicides is that on cell lysis, toxins contained in the algae cell are released into the surrounding water. In 1979, almost 150 people had to be hospitalized for treatment of liver damage after a reservoir contaminated with *Cylindrospermopsis* was treated with copper sulfate.[4] Biological control by zooplankton is, in principle, possible, although not always practical or effective because of the low nutrient adequacy, toxicity, and inconvenient size and shape of most blue-green algae. The only zooplankton reported to successfully graze on blue greens is *Daphnia* sp., but it tends to decrease with increasing nutrient content of the water[11] More effective is the use of microorganisms, as certain chytrids (fungal pathogens) and cyanophages (viral pathogens) specifically infest akinetes and other heterocysts, whereas Myxobacteriales (bacterial pathogens) can affect rapid lysis of a wide range of unicellular and filamentous blue-greens, although heterocysts and akinetes remain generally unaffected.[12]

The consensus regarding the management of blue-green algal blooms is the management of excess nutrient loads into receiving water bodies.[13,14] Management options can be divided into two broad categories: catchment management (decrease of nutrient export) or lake management (decrease of internal nutrient supply). Catchment options are, e.g., management of urban and agricultural runoff, biological and chemical treatment of wastewater, nutrient diversion, and implementation of legislation. Lake management options are dredging, chemical sediment treatment, and biomanipulation.[13]

CONCLUSIONS

Cyanobacteria pose a serious threat to ecosystem health and human livelihood. From a human perspective, the most serious threat associated with blue-greens are their toxins. Routes for human exposure are the oral route via drinking water, the dermal route during recreational use of lakes and rivers, or consumption of algal health food tablets. Removal of these algae and their toxins from water bodies poses a great logistical problem. However, it is important to understand that the proliferation of blue-greens and thus the presence of their toxins is a response to human-induced "cultural" eutrophication. Increasing awareness of the need of proper watershed management is urgently needed among municipalities and stakeholders, especially because chlorination has been shown not to be very effective in removing toxins from the water.

REFERENCES

1. Reynolds, C.S. *The Ecology of Freshwater Plankton*; Cambridge University Press: Cambridge, 1983.
2. Havens, K.E.; James, R.T.; East, T.L.; Smith, V.H. N:P ratios, light limitation, and cyanobacterial dominance in a subtropical lake impacted by non-point source nutrient pollution. Environ. Pollut. **2003**, *122* (3), 379–390.
3. Shapiro, J. Current beliefs regarding dominance by blue-greens: the case for the importance of CO_2 and pH. Proceedings of the International Association of Theoretical and Applied Limnology, Verh. Int. Verein. Theor. Angew. Limnol. **1990**, *24*, 38–54.
4. Chorus, I.; Bartram, J., Eds.; *WHO (World Health Organization) Toxic Cyanobacteria in Water: A Guide to Their Public Health Consequences, Monitoring and Management*, 1st Ed. E and F Spon: London, 1999.
5. Carr, N.G.; Whitton, B.A. The biology of blue-green algae. Bot. Monogr. Vol. 9. University of California Press, Berkeley.
6. Codd, G.A. Cyanobacterial toxins, the perception of water quality, and the prioritisation of eutrophication control. Ecol. Eng. **2000**, *16* (1), 51–60.
7. Carmichael, W.W. Health effects of toxin-producing cyanobacteria: "The cyanoHABs." Hum. Ecol. Risk Assess. **2001**, *7*, 1393–1407.
8. Haider, S.; Naithani, V.; Viswanathan, P.N.; Kakkar, P. Cyanobacterial toxins: a growing environmental concern. Chemosphere **2003**, *52* (1), 1–21.
9. Pouria, S.; de Andrade, A.; Barbosa, J.; Cavalcanti, R.; Barreto, V.; Ward, C. Fatal microcystin intoxication in haemodialysis unit in Caruaru, Brazil. Lancet **1998**, *352*, 21–26.
10. Richardson, L.L. Remote sensing of algal bloom dynamics. BioScience **1996**, *44*, 492–501.
11. Epp, G.T. Grazing on filamentous cyanobacteria by *Daphnia pulicaria*. Limnol. Oceanogr. **1996**, *41* (3), 560–567.
12. Philips, E.J.; Monegue, R.L.; Aldridge, F.J. Cyanophages which impact bloom-forming cyanobacteria. J. Aquat. Plant Manage. **1990**, *28*, 92–97.
13. Harding, W.R.; Quick, A.J.R. Management options for shallow hypertrophic lakes, with particular reference to Zeekoevlei, Cape Town. S. Afr. J. Aquat. Sci. **1992**, *18* (1/2), 3–19.
14. Herath, G. Freshwater algal blooms and their control: comparison of the European and Australian experience. J. Environ. Manage. **1997**, *51*, 217–227.

Desertification

David Tongway
Ecosystem Sciences, Commonwealth Scientific and Industrial Research Organization (CSIRO), Canberra, Australian Capital Territory, Australia

John Ludwig
Ecosystem Sciences, Commonwealth Scientific and Industrial Research Organization (CSIRO), Winnellie, Northern Territory, Australia

Abstract

Dealing with lands subject to desertification requires an analysis of the effect of climate or human disturbance on landscape functioning, to determine which biophysical processes have been affected. We propose a conceptual framework to enable the organization of data that explain landscape dysfunction. The rational selection of restoration procedures then follows, by selecting those which make good defective ecological processes. We also explain how to monitor the reversal of desertification, including the identification of a threshold of potential concern.

INTRODUCTION

The term *desertification* was coined in the 1970s to graphically represent the state of the Sahelian lands, on the southern margin of the Sahara Desert. This was a period when major drought accompanied by big increases in the human population served to cause the desert margins to apparently move into formerly more productive land.[1,2] The image of an encroaching desert is powerful and evocative and resulted in major international efforts to understand and deal with the problem. Since that time, the notion of desertification has been reworked in the light of additional information and improved conceptual frameworks to the extent that the desert is no longer seen as inexorably increasing in size, nor restricted to the Sahelian fringe of the Sahara.[3–5] Most rangeland areas in the world have suffered some sort of degradation due to the impact of disturbance regimes, and recent reviews[6] have shown the process to be not at all restricted to hot deserts or areas of high population density. This is not to deny, however, the major effects on the human populations using these lands, and no doubt, much human hardship has been endured. This entry describes a process whereby the degree of desertification can be assessed and then used to design restoration activities appropriate to local biophysical and socioeconomic constraints.

DESERTIFICATION REDEFINED

This entry focuses on the biophysical aspects of desertification. Traditionally, easy to measure structural and functional attributes of vegetation, such as species composition and productivity, were the means by which desertification was initially assessed. These methods served to show the effect of desertification but did not provide a predictive understanding of how to combat it. However, recent advances in landscape ecology and restoration technology[7,8,10] have led to generic and practical approaches to study the basic nature and reversibility of desertification. Principally, this involves treating the affected landscapes as biophysical systems, comprised of sequences of processes and feedback loops and summarized in a conceptual framework (Fig. 1).[8,9]

In this framework, vital resources such as water, nutrients, and topsoil are transported, utilized and cycled in space and time, and the processes are driven by processes such as runoff/runon, erosion/deposition, and plant litter decomposition.[8,11,12] Landscapes are said to be "functional," or non-desertified, if resources are substantially retained within the system and utilization and cycling processes are efficient. "Dysfunctional" or desertified landscapes are characterized by the depletion of the stock of some vital resources and the continued flow of these resources out of the system. This mind-set emphasizes the system attributes of processes acting in space, over time, in relation to applied stress and disturbance, rather than just changes in lists of species, or yields of marketable commodities. The role of vegetation as a significant regulator of energy and resources is integral to this approach.[13] Desertification should be viewed as a continuum ranging from slight to severe, rather than as a simple step function (Fig. 2).

ASSESSING THE DEGREE OF DESERTIFICATION

If field sites are characterized according to "resource regulation" capacity, not only can the degree of desertification be assessed, but also the critical pathway of resource loss may be identified. Tongway and Hindley[14] and Tongway and Ludwig[15] have designed and implemented monitoring programs to quickly provide information about biophysical processes and edaphic properties related to plant habitat

Encyclopedia of Environmental Management DOI: 10.1081/E-EEM-120046344

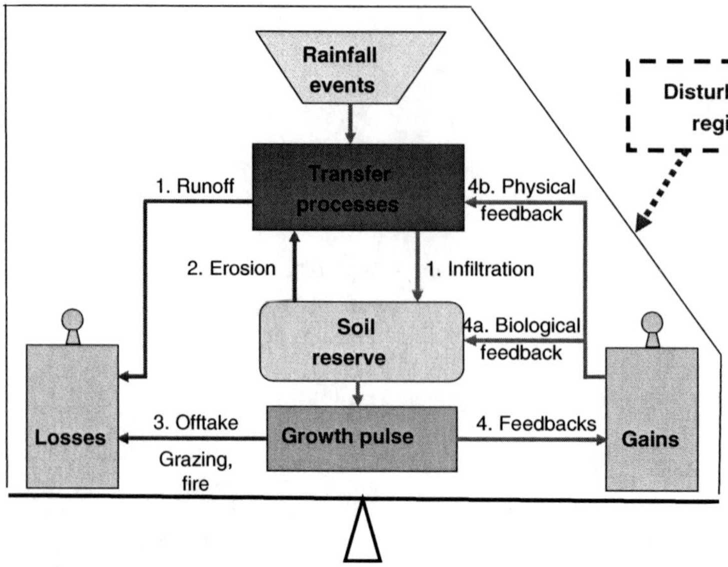
Rainfall events
Disturbance regime
Transfer processes
1. Runoff
4b. Physical feedback
2. Erosion
1. Infiltration
Soil reserve
4a. Biological feedback
Losses
3. Offtake
Grazing, fire
Growth pulse
4. Feedbacks
Gains

Fig. 1 A conceptual framework summarizing landscape functioning. Numbers refer to the recommended sequence of assessing practical function. The disturbance regime shown is generic, and it may impinge on a number of landscape processes at the same time.
Source: A modification of Fig. 2 in Tongway and Ludwig[9] (reproduced with permission).

favorability at both landscape and plot or patch scales. Typically, the initial analysis examines the fate of rainfall into infiltrated water and runoff water. The data gathered need an interpretational framework to facilitate generic application. Graetz and Ludwig[16] proposed that system response to desertification be represented by a four-parameter sigmoidal or logistic curve. The curve form acknowledges upper and lower asymptotic plateaux, at the non- and highly desertified ends of the spectrum, respectively, and a gradual transition between those plateaux. This approach permits questions about whether a system was "fragile," i.e., easily made dysfunctional, with low restoration potential or "robust," or rather able to withstand stress and disturbance with only low attenuation of biophysical processes (Fig. 3). Importantly, this curve type enables thresholds and milestones to be predicted and quantified using field indicators.[17]

PROCEDURES TO REVERSE DESERTIFICATION

Rehabilitation and restoration of desertified landscapes, under the functional biophysical system mind-set, require

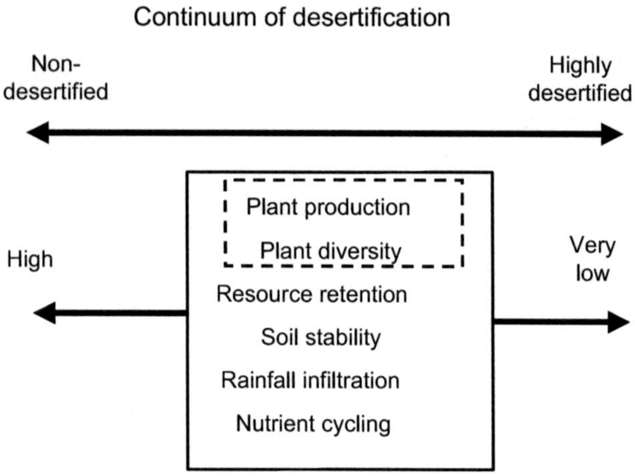
Continuum of desertification
Non-desertified
Highly desertified
High
Very low
Plant production
Plant diversity
Resource retention
Soil stability
Rainfall infiltration
Nutrient cycling

Fig. 2 Desertification as a continuum. The four new biophysical parameters (bold) are added to the two existing desertification descriptors (dotted box) to locate any given landscape on the continuum.

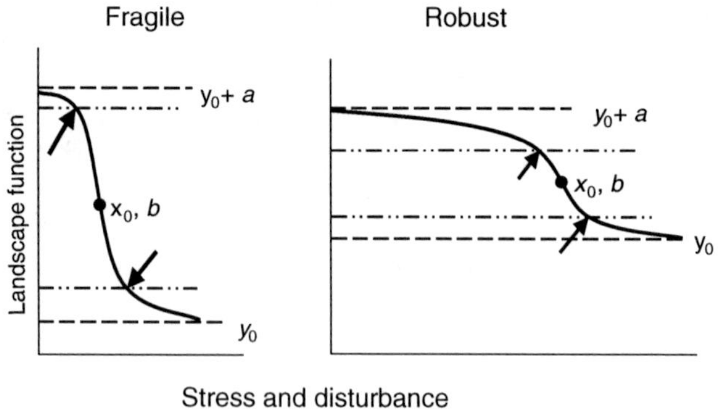
Fragile
Robust
Landscape function
y0+ a
x0, b
y0
Stress and disturbance

Fig. 3 Examples of response curves for fragile and robust landscapes. The initial response of landscape function to stress and/or disturbance is markedly different. A fragile landscape deteriorates with low applied stress and has a much lower base value (y_0) than the more robust landscape. Four-parameter sigmoid curves of the form $y = (y_0 + a)/1 + e^{-x-x_0)/b}$ provide four practical values reflecting the nature of the landscape. Critical thresholds (arrows) for each index of desertification can be determined by curve analysis.

that processes that accumulate resources be reintroduced or augmented, thus providing a rational procedure in the repair and functional recuperation of desertified landscapes. The approach explicitly seeks to retain vital resources by ecological processes.[18–20] Once the analysis of resource regulation system has been completed, and the most affected process identified, remedial efforts can begin. For example, Rhodes and Ringrose-Voase[20] deduced that ponding water for extended periods on clay soils with modest swell/shrink properties would eventually result in soils with high infiltration and water store through sequences of swelling and shrinking processes permitting infiltration into greater soil depths. Recolonization and establishment of plants then began spontaneously, eventually cycling organic matter, so that open friable soils developed, colonized by soil macrofauna that further improved soil properties. Tongway and Ludwig[18,19] used piles of branches arranged on the contour in gently sloping country to trap water, sediment, and organic matter to effect substantial improvements to a range of soil properties, permitting perennial grasses to self-establish. In each of these cases, an analysis of the different underlying causes facilitated the selection of the most appropriate techniques to reverse the observed desertification. Attempting to revegetate desertified areas by simply reseeding without understanding both the current and required edaphic properties needed for the desired vegetation mix frequently results in unexplained failure. In some instances, where the system function is close to the lower asymptote (Fig. 3), simple treatments such as exclusion from grazing will be too slow or ineffective and active intervention may be needed to improve one or more functional processes.

MONITORING REHABILITATION

It is important to monitor the progress of processes set in train by the rehabilitation activities. Essentially, the degradation curves such as those in Fig. 3 need to be driven "in reverse." The landscape assessment procedure proposed by Tongway and Hindley and Tongway and Ludwig[15,21] can also be used to follow the trajectory of improvement in ecosystem functioning. The procedures use simple, rapidly acquired indicators of processes of resource regulation. Data recording biota establishment and development are included in this procedure and interpreted in terms of the rising plane of delivery of goods and services to the whole system over time. It is important that monitoring should provide accurate information quickly and at low cost. Remotely sensed data, related to landscape function, is a cost-effective procedure,[22] and new products such as Google Earth may, in the future, provide synoptic assessment of restoration trends at coarse scales. The effect of rare stochastic events such as fires or storms may need to be assessed to see if the resultant stress and/or disturbance has set the system back beyond a critical threshold or not.

CONCLUSION

We have described an ecosystem-function-based set of data gathering processes by which the fundamental nature of desertification can be assessed and combated, using a framework that characterizes the biophysical status of the affected system. This can be simply expressed as "assessing the regulation of vital resources in space and time." In deploying the procedure, community groups can be easily instructed to "read the landscape." The procedure enables the user to design or adapt restoration or rehabilitation technologies appropriate to the problem at hand because of the predictive understanding acquired, rather than using a "recipe" from another type of landscape elsewhere. The information-gathering procedure can be adapted to a wide range of bioclimatic situations because it deals with the basic processes controlling the availability of vital resources to biota.

REFERENCES

1. UNEP. *United Nations Conference on Desertification: An Overview*; United Nations Environment Program: Nairobi, Kenya, 1977.
2. UNEP. *General Assessment of Progress in the Implementation of the Plan of Action to Combat Desertification 1978–84*; United Nations Environment Program: Nairobi, Kenya, 1984.
3. Arnalds, O. Desertification: An appeal for a broader perspective. In *Rangeland Desertification*, Arnalds, O., Archer, S., Eds.; Kluwer Academic: Dordrecht, 2000; 5–15.
4. Chen, Z.; Xiangzhien, L. In *People and Rangelands: Building the Future*, Proceedings of the VI International Rangeland Congress, Townsville, Australia, July 19–23, 1999; Eldridge, D., Freudenberger, D., Eds.; VI International Rangeland Congress: Townsville, 1999; 105–107.
5. Tongway, D.; Whitford, W. In *People and Rangelands: Building the Future*, Proceedings of the VI International Rangeland Congress, Townsville, Australia, July 19–23, 1999; Eldridge, D., Freudenberger, D., Eds.; VI International Rangeland Congress: Townsville, 1999; 89–142.
6. Archer, S.; Stokes, C. Stress, disturbance and change in rangeland ecosystems. In *Rangeland Desertification*; Arnalds, O., Archer, S., Eds.; Kluwer Academic: Dordrecht, 2000; 17–38.
7. Breedlow, O.A.; Voris, P.V.; Rogers, L.E. Theoretical perspective on ecosystem disturbance and recovery. In *Shrub-Steppe: Balance and Change in a Semi-Arid Terrestrial Ecosystem*; Rickard, W.H., Rogers, L.E., Vaughn, B.E, Liebetrau, S.F., Eds.; Elsevier: New York, 1988; 257–269.
8. Ludwig, J.A; Tongway, D.J. A landscape approach to rangeland ecology. In: *Landscape Ecology Function and Management: Principles from Australia's Rangelands*; Ludwig, J., Tongway, D., Freudenberger, D., Noble, J., Hodgkinson, K., Eds.; CSIRO: Melbourne, Australia, 1997; 1–12.
9. Tongway, D.; Ludwig, J. *Restoring Disturbed Landscapes: Putting Principles into Practice*; Island Press: Washington DC, 2010.

10. Whisenant, S.G. *Repairing Damaged Wildlands: A Process-Oriented, Landscape-Scale Approach*; Cambridge Univ. Press: Cambridge, England, 1999; 312 pp.
11. Jorgenson, S.E.; Mitsch, W.J. Ecological engineering principles. In *Ecological Engineering: An Introduction to Ecotechnology*; Mitsch, W.J., Jorgensen, S.E., Eds.; John Wiley and Sons: New York, 1989; 21–37.
12. Schlesinger, W.H.; Reynolds, J.F.; Cunningham, G.L.; Hueneke, L.F.; Jarrel, W.M.; Virginia, R.A.; Whitford, W.G. Biological feedbacks in global desertification. Science **1990**, *247*, 1043–1048.
13. Farrell, J. The influence of trees in selected agroecosystems in Mexico. In *Agroecology: Researching the Ecological Basis for Sustainable Agriculture*; Gliessman, S.R., Ed.; Springer-Verlag: New York, 1990; 167–183.
14. Tongway, D.J.; Hindley, N.L. *Landscape Function Analysis: Methods for Monitoring and Assessing Landscapes, with Special Reference to Minesites and Rangelands*; CSIRO Sustainable Ecosystems: Canberra, 2004, available at http://www.csiro.au/services/EcosystemFunctionAnalysis.htm.
15. Tongway, D.; Ludwig, J. *Restoring Disturbed Landscapes: Putting Principles into Practice*; Island Press: Washington DC, 2010.
16. Graetz, R.D.; Ludwig, J.A. A method for the analysis of piosphere data applicable to range assessment. Aust. Rangeland J. **1978**, *1*, 126–136.
17. Tongway, D.; Hindley, N. *Ecosystem Function Analysis of Rangeland Monitoring Data*; 2000, available at http://www.nlwra.gov.au/atlas.
18. Tongway, D.J.; Ludwig, J.A. Rehabilitation of semi-arid landscapes in Australia. I. Restoring productive soil patches. Restor. Ecol. **1996**, *4*, 388–397.
19. Ludwig, J.A.; Tongway, D.J. Rehabilitation of semi-arid landscapes in Australia. II. Restoring vegetation patches. Restor. Ecol. **1996**, *4*, 398–406.
20. Rhodes, D.W.; Ringrose-Voase, A.J. Changes in soil properties during scald reclamation. J. Soil Conserv. Serv. NSW **1987**, *43*, 84–90.
21. Tongway, D.; Hindley, N. Assessing and monitoring desertification with soil indicators. In *Rangeland Desertification*; Arnalds, O., Archer, S., Eds.; Kluwer Academic: Dordrecht, 2000, 89–98.
22. Kinloch, J.E.; Bastin, G.N; Tongway, D.J. (2000). Measuring landscape function in chenopod shrublands using aerial videography. In the Proceedings of the 10th Australasian Remote Sensing and Photogrammetry Conference, Adelaide, Australia 21–25 August 2000, 480–491; Causal Productions: Adelaide, South Australia.

Desertification: Extent of

Victor R. Squires
Dryland Management Consultant, South Australia, Australia

Abstract

There is now a general acceptance of the term desertification to encompass a process of land degradation in drylands induced by climatic factors and human use. Every inhabited continent has a problem with desertification, but in some it is quite acute. Land degradation induced by human activities (anthropogenic) is believed to be currently lowering the productivity of at least one fifth of the world's agricultural lands. The impacts are greatest in those regions with high population density and where the level of economic development is low.

INTRODUCTION

Desertification, as a *concept*, is one of the most complex to consider objectively. The word desertification unfortunately conjures up images of advancing sand dunes, which is only part of the problem. There is now a general acceptance of the term *desertification* to encompass a process of land degradation in drylands (those regions with a growing season of about 75–120 days per year) induced by climatic factors and human use. Fig. 1 shows the broad distribution of the world's drylands.

Whatever the definition, it is clear that large areas of the world's drylands are affected by land degradation and that this adversely affects the lives of about 1 billion people on almost every continent. To assess the global extent of desertification implies that there is universal acceptance of the term and that the monitoring and evaluation is somewhat uniform. Neither supposition is true. Until the decision by the United Nations Conference on Desertification[1] there was only broad agreement as to what was meant by it. It is only as signatories to the convention to combat desertification (CCD) have been preparing and implementing their respective action plans that are more accurate assessments have been made of the current status of desertification in each country. Prior to this, the maps and other documents produced by the UN through its various agencies were the product of expert panels who developed maps of the areas of potential desertification.[2] Every inhabited continent has a problem with desertification but in some it is quite acute. Land degradation induced by human activities (anthropogenic) is believed to be currently lowering the productivity of at least one fifth of the world's agricultural lands. Soil erosion from water and wind is a principal but by no means the only soil degrading process. Sub-humid and semi-arid areas in crops or pastures are particularly vulnerable to erosion. Land degradation in such areas leads to desertification by definition. Anthropogenic land degradation, however, also affects large areas of arable land and pasture in humid regions as well.

Depending on how it is measured exactly[3] up to 75% of some continents are affected. The impacts are greatest in those regions with high population density and where the level of economic development is low. For example, Australia is reported as having more than 75% of its land surface affected to a greater or lesser extent by desertification. Yet its low population density and the low level of dependence on these drier regions for its national prosperity means that desertification is perceived as a lesser threat in Australia than in say regions of comparable climate in the Indian sub-continent or in Africa.

ASSESSING THE EXTENT OF THE DESERTIFICATION PROBLEM

Desertification comprises two main types of degradation: vegetation degradation and soil degradation.

These can occur anywhere in dry areas and not just on desert fringes. Vegetation degradation involves a temporary or permanent reduction in the density, structure, species composition, or productivity of vegetation cover. It has been argued that degradation of vegetation should be given less weight in assessment of desertification because there is clear evidence that most change in vegetation is reversible over reasonable time scales whereas soil degradation is unlikely to occur within a human life span.

Desertification might be thought of as a situation where a landscape is stressed beyond its resilience. Most scientists would agree that, despite their apparent fragility, dryland ecological systems are quite resilient. The spatial and temporal variation in conditions means that coping mechanisms have developed to allow the system to continue in the face of adversity. Under the combined force of human-induced pressure and climatic forces, such as recurrent

Encyclopedia of Environmental Management DOI: 10.1081/E-EEM-120042660

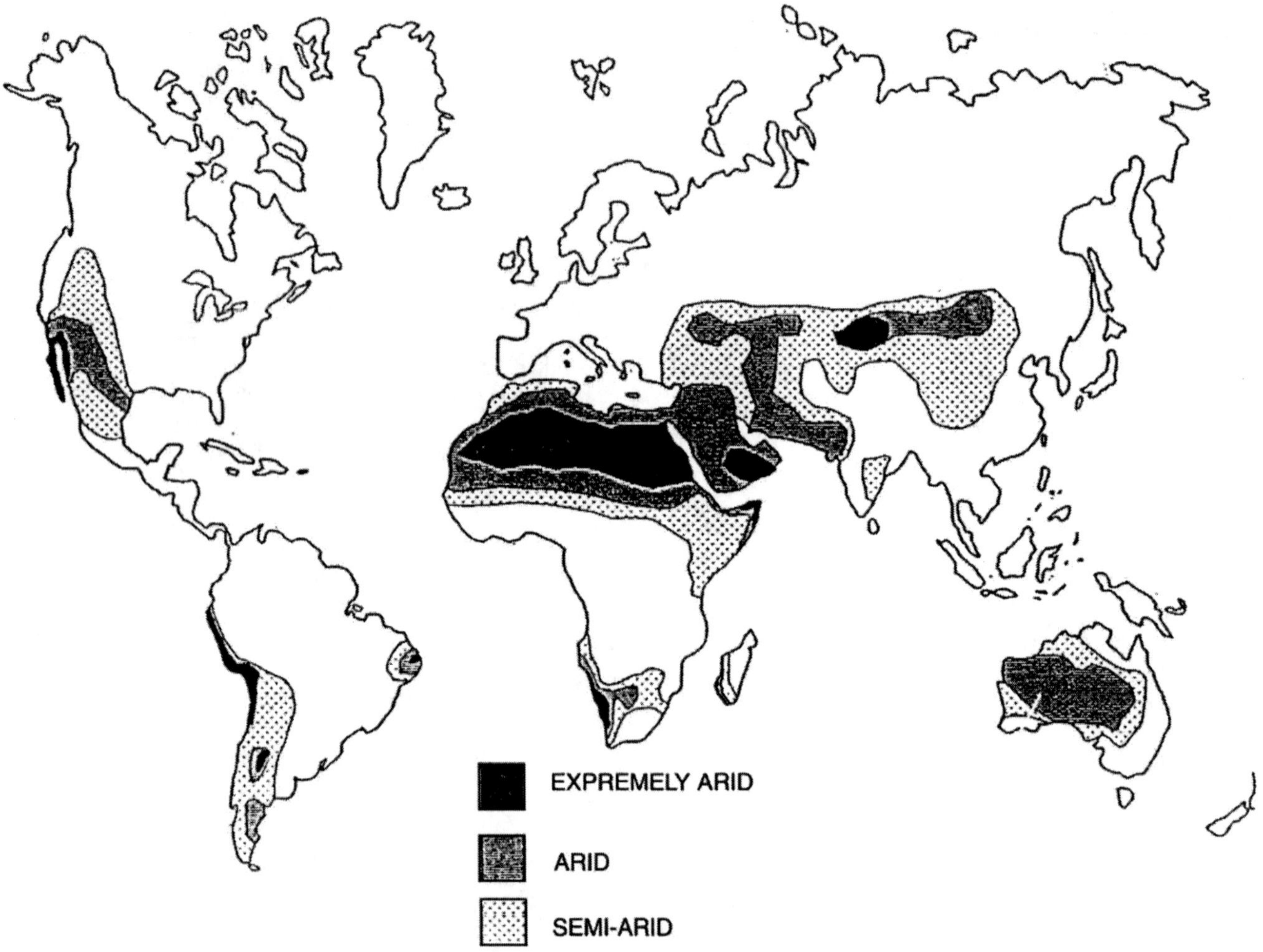
EXPREMELY ARID
ARID
SEMI-ARID

Fig. 1 Distribution of world's drylands.

drought, major changes in soils, and vegetation can occur. This is what has come to be recognized as desertification. This is what the CCD is designed to combat.

Desertification is a product of complex interactions between the social and economic systems (disease, poverty, hunger, and unreliable economy), and natural factors (drought, water erosion, soil salinization, degradation, and loss of vegetation cover). It is not surprising then that the criteria used to objectively assess the areal extent varies between regions. It is clear that a definition based purely on climatic indices does not give a true picture of the full extent of the problem.

No one has been able so far to monitor and document desertification and the resulting land cover change through reliable, verifiable, and repeated observation on a global, continental, regional, or even national scale. There is not full agreement on a single indicator of dryland degradation or an approach to assess and study desertification. Dryland degradation has many faces (driving forces as well as symptoms). It can only be assessed and understood through an interdisciplinary study of the changing characteristics and integrated trends of a variety of biological, agricultural, physical, and socioeconomic indicators over a long time period and at a variety of spatial scales.

Because of the scale of mapping and the fact that most desertification occurs as a mosaic of small, often-isolated, areas mapping defies the use of synoptic techniques such as remote sensing from Landsat and other satellites. Land cover/land use maps are being prepared by a number of countries and on a global scale land cover is being assessed by analysis of satellite data. These may help to clarify and quantify the extent of desertification, especially where local input helps in verification and ground truthing.

The use of the low-resolution NOAA/AVHRR satellite to generate a greenness index and to develop the normalized difference vegetation index might, on a broad scale, assist in monitoring change in plant cover and hence be an early warning of incipient change. But such change may only be a reflection of variability in space and time that characterizes the world's dryland regions. It is likely to be more useful in studies of drought and vegetation mapping. Its potential for use in desertification monitoring may be limited.[3] The most immediately detectable changes can be quite misleading.

Table 1 Some common manifestations of desertification.

Economic manifestations	Ecological manifestations	Social manifestations
Economic loss in cash	Loss of diversity in terms of wildlife, plants and ecosystems	Migration of population off affected areas
Decreased crop yield	Loss of inland lakes	Rural poverty
Loss of farmland due to desertification	Loss of topsoil in terms of organic matter, N, P, and K plant nutrients	Influx of ecological refugees into urban areas
Loss of rangeland due to desertification	Decreased ground water level, increasing salinity of water	
Decreased grazing capacity in terms of the number of livestock	Increased frequency of sand storms and associated losses of human lives and livestock	
Abandoned farmland		
Abandoned rangeland		
Drifting sand affects railway lines and highways		
Increase in suspended load raises river height and increases flood problems		

AREAL EXTENT OF DESERTIFICATION

The area affected by desertification in many developing countries is on the increase although the lack of sound baseline data make any attempt at quantification difficult. Because a lot of desertification is characterized by a more dispersed, patch-like process of degradation it is not amenable to remote sensing. It has long been recognized that operational monitoring of patchy soil and vegetation degradation in dry areas is inherently difficult, even with medium resolution satellite technology like Landsat MSS and that improved techniques are needed to make reliable monitoring and trend detection viable.

What is clear is that most of the more than 170 signatories to the CCD report growing concern about loss of productivity of their land in the face of rapid and continuing land degradation. The impacts are reflected in:

- a loss of land resource through erosion and transportation by wind and water
- a loss of productivity through land degradation
- eco-environmental change impacting on living conditions and health
- increased poverty and social instability in the rural hinterland
- flow-on impacts on cities and on industry
- effects on infrastructure and economic development

The principal manifestations of severe desertification are shown in Table 1.

UNEP's 1991[4] estimate of the area of desertified land is 3.6 billion ha, including 1 billion ha of dryland suffering from soil degradation and another 2.6 billion ha of rangeland with degraded vegetation.[5] But an external review claimed that its accuracy was limited by considerable subjectivity (so observations were not repeatable); lack of resolution (so comparisons through time were not possible); and the use of point assessments were unrepresentative of larger areas.[6]

As indicated above, there are few objective measures of the areas of land affected by desertification. The most reliable maps are those published by UNEP in the revised edition of the *World Atlas of Desertification*.[2]

Dr. Harold Dregne of Texas Tech University calculated the area of potential desertification (Table 2). Dregne determined the desertification status according to three factors taken as indicators, namely:

Table 2 Distribution of the area of land susceptible to desertification (by bioclimatic zones).

Susceptibility to desertification	Bioclimatic zone (000's km^2 %)						
	Arid	Semi-arid	Sub-humid				
Moderately	1144.5	6.6	12713.8	68.4	3345.6	23.4	
Highly	14585.8	82.5	2686.5	14.2		589.8	4.0
Very highly	1040.3	6.4	2157.5	12.4		173.5	1.3
Total	16770.6		17557.8		4108.9		

Source: Data from UNEP and other sources. Used with permission.

Table 3 Area of desertified lands by continent/region.

Degree of desertification	Africa	Asia	Australia	North America	South America	Europe	World total
Slight	12430	7980	2330	440	1340	—	24520
Moderate	1870	4480	3510	2720	1050	140	13770
Severe	3030	3210	520	1200	680	60	8700
Very severe	—	—	—	67	8	—	73
Total	17330	15670	6360	4427	3076	200	47063

Source: Data from UNEP, CCD and other sources. Used with permission.

- changes in the composition of the vegetation
- extent of erosion
- presence of soil salinity

He singled out four degrees of desertification: slight, moderate, strong, and very strong. Strong desertification is an irreversible process when it is impossible to restore the land. This classification comprises mobile sand dunes, heavily salt-affected land, and badlands. In Dregne's opinion such areas are not extensive and their area is about 50,000 km^2.[7]

The areas subject to severe desertification (Table 3) are in Africa, Asia, and South America and are mainly occupied by developing countries. There are a number of drivers of change. These include rapid population growth, the availability of modern technologies for land conversion, as well as loss of traditional land use controls. Political and economic policy decisions, market and trade arrangements, lack of environmental awareness, and lack of capacity to combat the problem also contribute.

CONCLUSIONS

The evidence from many sources is that desertification is a global problem. UNEP estimate that about 35% of the earth's land surface and about 20% of its population are affected by it. Clearly, according to these figures and the data presented here, desertification is a major environmental and social problem. The problem is worse for heavily populated countries on the desert margins. When the extent of desertification is broken down according to major types of land use we see that grazing land and rainfed cropping are most severely affected, each being desertified on over three-quarters of its area. Irrigated land suffers less so, with about one-fifth of the irrigated area in drylands affected by desertification (salinization and waterlogging).

The global overview of desertification provides a summary but recognition must be given to regional differences and impacts. A global summary of the kind presented here is a useful first step to take when looking at a large-scale issue. However, when working out a plan of action to combat desertification at the regional or local level, more detailed information needs to be gathered on the way desertification is happening and on its root causes. Such information can only be gathered by long term monitoring of a particular area.

REFERENCES

1. UNCOD, International convention to combat desertification in those countries experiencing serious drought and/or desertification, particularly in Africa, I.L.M. content summary. International Legal Materials **1995**, *33*, 1328–1382.
2. Middleton, N.; Thomas, D., Eds.; *World Atlas of Desertification*, 2nd Ed.; UNEP/Edward Arnold: London, 1997.
3. Hellden, U. Desertification monitoring—is the desert encroaching? Desertification Control Bulletin No. 17, **1998**, 8–12.
4. UNEP, *Status of Desertification and Implementation of the United Nations Plan of Action to Combat Desertification*; UNEP: Nairobi, 1991.
5. Dregne, H.M.; Kassas, M.; Rosanov, B. A new assessment of the world status of desertification. Desertification Control Bulletin **1991**, *20*, 6–18.
6. UNEP, *Draft Report of the Expert Panel Meeting on Development of Guidelines for Assessment and Mapping of Desertification/Land Degradation in Asia/Pacific*, UNEP: Nairobi, 1994.
7. Kassas, M. Desertification: a general review. Journal of Arid Environments **1995**, *30*, 15.

Desertification: Greenhouse Effect

Sherwood Idso
U.S. Department of Agriculture (USDA), Tempe, Arizona, U.S.A.

Craig Idso
Center for the Study of Carbon Dioxide and Global Change, Tempe, Arizona, U.S.A.

Abstract
What is the impact of the ongoing rise in the air's carbon dioxide (CO_2) concentration on the ecological stability of the world's deserts and the shrubs and grasslands that surround them? This question weighs heavily on the minds of many, as the nations of the Earth debate the pros and cons of the prodigious CO_2 emissions produced by the burning of fossil fuels. On the downside, there is concern that more CO_2 in the air will exacerbate the atmosphere's natural greenhouse effect, producing changes in climate that lead to desertification. On the upside, the aerial fertilization effect of additional atmospheric CO_2 may enhance plant prowess, increasing plant water use efficiency and enabling vegetation to reclaim great tracts of desert. The challenge, therefore, is to determine the relative merits of these competing phenomena.

THE WORLD IN TRANSITION

Climatical Changes

Questions surrounding the climatical effects of the ongoing rise in the air's CO_2 content are contentious. There is evidence the Earth has warmed significantly over the course of what is deemed the "Age of Fossil Fuels." Yet, most of this warming occurred well before the largest increases in the air's CO_2 content were recorded, peaking in the 1930s. Thereafter, the atmospheric CO_2 concentration rose at a much greater rate than it had previously, while temperatures stagnated before staging a hotly contested comeback in the 1980s and 1990s—a comeback more virtual than real. It is also possible that the atmospheric warming in the late 19th and early 20th centuries was nothing more that a natural recovery from the global chill of the Little Ice Age. Hence, although many scientists believe there has been a discernable human influence on the global climate of the past century, other scientists take serious issue with that contention.

Biological Changes

Climatical effects in the biological arena are also complicated, but not so much that certain facts cannot be used to draw some broad conclusions. Carbon dioxide, for example, is one of two main raw materials (the other being water) used by plants to produce the organic matter from which they construct their tissues. Thus, increasing the air's CO_2 content typically enables plants to grow better, as has been demonstrated in literally thousands of laboratory and field experiments.[1] In addition, higher concentrations of atmospheric CO_2 cause many plants to reduce the apertures of the small pores in their leaves, through which water vapor escapes to the air. Consequently, with more biomass production per unit of water lost, plant water use efficiency is significantly enhanced. In fact, it approximately doubles with a doubling of the air's CO_2 content,[2] allowing many species of plants to grow and reproduce where it had been too dry for them previously. Furthermore, the enhanced degree of groundcover resulting from this phenomenon reduces the magnitude of soil erosion caused by the ravages of wind and rain. And with greater plant growth, both above and below ground, there are significant increases in the amounts of organic matter that enter the soil. That matter enhances the soil's ability to sustain the more productive shrub and grassland ecosystems that come into being via this process of reverse desertification.

Although supported by a plethora of scientific studies, this scenario of vegetative transformation has been challenged on the assumption that resource limitations and environmental stresses encountered in nature might overpower the ability of atmospheric CO_2 enrichment to significantly enhance the vitality of plants. However, in a massive literature review designed to investigate this question, it was found that the percentage increase in plant growth produced by an increase in atmospheric CO_2 is generally not reduced by less-than-optimal levels of light, water, or nutrients, or by high temperatures, salinity, or gaseous air pollution.[3] In fact, the data demonstrate that the relative growth-enhancing effects of atmospheric CO_2 enrichment are typically greatest when resource limitations and environmental stresses are most severe.

Greening of the Earth Hypothesis

In light of these observations, there is reason to believe that the historical trend in the air's CO_2 concentration, which rose from a value of 280 parts per million (ppm) at the start of the 19th century to 370 ppm at the turn of

Encyclopedia of Environmental Management DOI: 10.1081/E-EEM-120001951

the millennium, may already be producing an ubiquitous "greening" of the Earth. This is especially true for bushes, shrubs, and trees, since woody plants are typically more positively affected by increases in the air's CO_2 content than herbaceous plants.[4] Consequently, the most readily documented aspect of this biological transformation should be seen in the spreading of woody vegetation onto grasslands.

WOODY PLANT RANGE EXPANSIONS

From Prehistory to Industrial Revolution

The savannas, grasslands, and deserts of the American southwest and the southern Great Plains got their start approximately 25 million years ago. About that time, the climate began to dry, and it continued to become more arid, particularly over the past 15 millennia. But when the engines of the Industrial Revolution began to pump CO_2 into the air at rates that exceeded natural geological processes, things began to change. As early as 1844, in fact, a trader from Santa Fe, New Mexico wrote in his memoirs[5] "there are parts of the southwest now thickly set with trees of good size, that, within the remembrance of the oldest inhabitants, were as naked as the prairie plains." He summarized the situation by saying "we are now witnessing the encroachment of timber upon the prairies." And in surveying the land a century later, Malin[6] would verify that trader's prescience by referring to the ecosystem it supported as a "tangled jungle."

Modern Studies

An especially good history of the vegetative transformation of the American Southwest was developed by Blackburn and Tueller[7] from a study of the growth rings of juniper and pinyon communities in east-central Nevada, where they determined that juniper began expanding its coverage of the land well before 1800, with more rapid increases in the densities and sizes of both species occurring subsequent to that time. But just as the rise in the air's CO_2 content developed most dramatically in the 20th century, the most remarkable increases in the presence and abundance of trees at these sites manifest themselves after 1920.

A parallel example involving other woody species has been documented on the Jornada Experimental Range in New Mexico.[8] When surveyed by the U.S. Land Office in 1858, no shrubs were evident on 60% of the land and only 5% of the range contained what could be described as dense stands of brush. However, 105 years later, none of the area was free of woody plants and 73% of it was dominated by thick stands of mesquite and creosote. Texas, too, had been thus transformed. By 1963, 88 million acres of former grasslands had been replaced by brush and shrubs. By 1982, the figure had risen to 105 million acres (Figs. 1 and 2).

In a comprehensive review of this subject, Idso[8] assembled many examples of woody-plant range expansions onto grasslands, citing studies conducted in California, Idaho, Kansas, Missouri, Montana, Nebraska, North Dakota, Oklahoma, Oregon, South Dakota, and a number of New England states. He also cited examples of the same phenomenon in South America, Europe, Asia, Africa, Australia, and New Zealand, augmenting this evidence with equally numerous and widespread reports of ever-accelerating woody-plant growth-rate increases.

Although a number of different hypotheses have been proposed to account for this ongoing vegetative transformation of the planet, many of which play significant roles in specific locations and circumstances, the ubiquitous nature of the phenomenon argues strongly for a single worldwide forcing factor that dominates the effects of most other influences. The ongoing rise in the air's CO_2 concentration is the only phenomenon that would appear to meet this global-scale criterion.

(a)

(b)

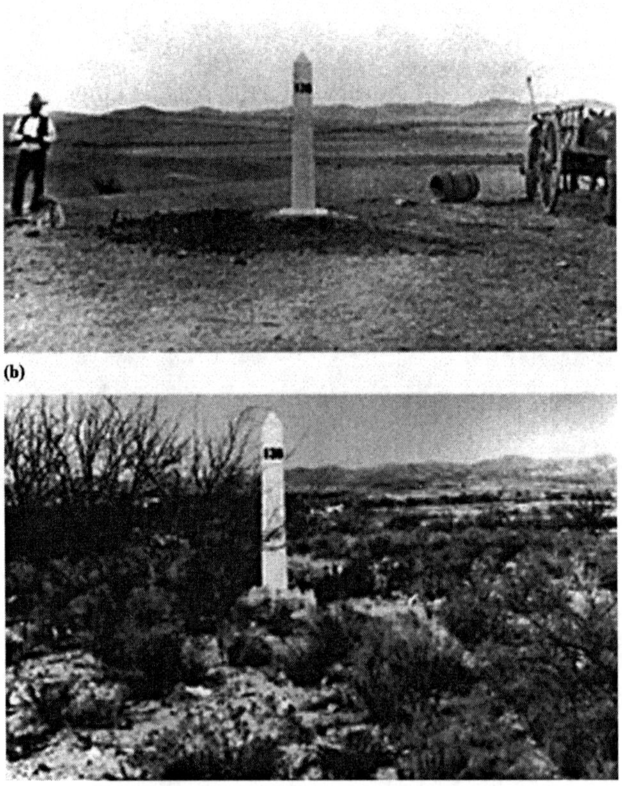

Fig. 1 Photographs of the U.S.–Mexico border just east of Sasabe, Arizona. (a) Taken in 1893 by D.R. Payne, the area was devoid of shrubs. (b) Same areas, taken in 1984 by R.R. Humphrey, was dominated by velvet mesquite, ocotillo, velvet-pod mimosa, snakeweed and burroweed.
Source: Humphrey, R.R. *90 Years and 535 Miles: Vegetation Changes Along the Mexican Border*; University of New Mexico Press: Albuquerque, 1987.

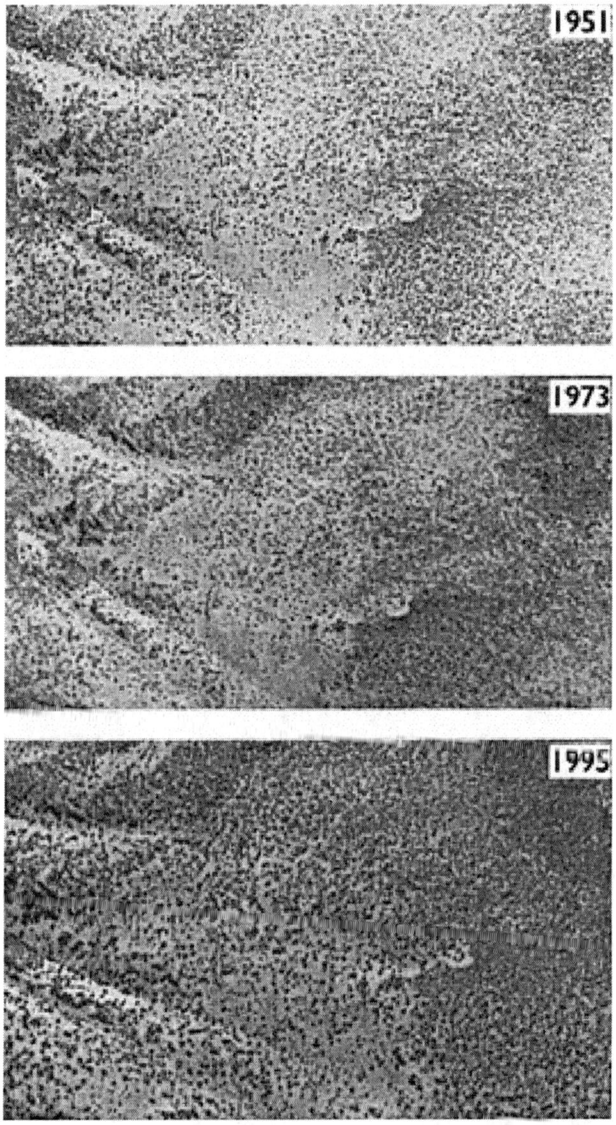
1951
1973
1995

Fig. 2 Aerial photographs of the Horse Ridge Research Natural Area in central Oregon depicting an increase in western juniper cover and density between 1951 and 1995.
Source: Knapp P.A.; Soule, P.T. Recent *Juniperus occidentalis* (Western Juniper) expansion on a protected site in central Oregon. Global Change Biology 1998, *4*, 347–357.)

CONCLUSIONS

What does the future hold? It will probably be more of the same. There are no signs that humanity's appetite for fossil-fuel energy will abate any time soon. Neither are there any indications the nations of the Earth will be able to reverse this trend by regulatory or legislative fiat. Therefore, if we do not completely cover the globe with concrete and asphalt before the Age of Fossil Fuels ends, we will probably see woody plants continue their invasion of grasslands, while grasslands intensify their assault upon the world's deserts.

With respect to global warming, it's anybody's guess. Even if it does occur, CO_2-induced increases in plant water use efficiency (a doubling for a doubling of atmospheric CO_2) and plant thermal tolerance (a 5°C increase in optimum growing temperature for a doubling of CO_2) should enable Earth's ecosystems to keep on responding as they have over the past two centuries. Thus, the greening of the Earth should continue, increasing vegetative productivity, the biomass of higher food-chain trophic levels, and ecosystem biodiversity concomitantly.[9]

REFERENCES

1. Idso, K.E. *Plant Responses to Rising Levels of Atmospheric Carbon Dioxide: A Compilation and Analysis of the Results of a Decade of International Research into the Direct Biological Effects of Atmospheric CO_2 Enrichment*, Climatological Publications Scientific Paper No. 23; Office of Climatology, Arizona State University: Tempe Arizona, 1992; 1–186.
2. Idso, S.B. *Carbon Dioxide and Global Change: Earth in Transition*; IBR Press: Tempe, Arizona, 1989; 1–292.
3. Idso, K.E.; Idso, S.B. Plant responses to atmospheric CO_2 enrichment in the face of environmental constraints: a review of the past 10 years' research. Agric. For. Meteorol. **1994**, *69*, 153–203.
4. Idso, S.B.; Kimball, B.A. Tree growth in carbon dioxide enriched air and its implications for global carbon cycling and maximum levels of atmospheric CO_2. Global Biogeochem. Cycles **1993**, *7*, 537–555.
5. Gregg, J. *Commerce of the Prairies: Or, the Journal of a Santa Fe Trader, During Eight Expeditions Across the Great Western Prairies, and a Residence of Nearly Nine Years in Northern Mexico*; Henry, C., Ed.; Langley: New York, 1844; Vol. 2, 202 pp.
6. Malin, J.C. Soil, animal, and plant relations of the grassland, historically reconsidered. Sci. Mon. **1953**, *76*, 207–220.
7. Blackburn, W.H.; Tueller, P.T. Pinyon and Juniper invasion in Black Sagebrush communities in East-central Nevada. Ecology **1970**, *51*, 841–848.
8. Idso, S.B. *CO_2 and the Biosphere: The Incredible Legacy of the Industrial Revolution; Department of Soil Water and Climate*, University of Minnesota: St. Paul, Minnesota, 1995; 1–60.
9. Idso, K.E.; Idso, S.B. Atmospheric CO_2 enrichment: implications for ecosystem biodiversity. Technology **2000**, *7S*, 57–69.

Desertification: Impact

David A. Mouat
Division of Earth and Ecosystem Sciences, Desert Research Institute, Reno, Nevada, U.S.A.
Judith Lancaster
Desert Research Institute, Reno, Nevada, U.S.A.

Abstract
Economic and political factors have changed the patterns and strategies of human occupation of dryland regions, where previous lifestyles of transhumance or nomadism and minimal involvement in a market economy permitted human occupation of arid and semiarid areas even in times of drought, without causing ecosystem stress. Areas of concern are those that are desertified, that are at risk for increases or acceleration of desertification processes, and that are on the threshold of change but where sustainable human life is still possible.

INTRODUCTION

The United Nations defines desertification as: "Land degradation in arid, semi-arid and dry sub-humid areas resulting from various factors, including climatic variations and human activities."[1] These dryland regions comprise 41% of the global land area, or 6150 million ha[2] and are present in every continent. Hyper-arid regions are almost rainless, natural deserts and not susceptible to desertification processes. Areas of concern are those that are desertified, that are at risk for increases or acceleration of desertification processes, and that are on the threshold of change but where sustainable human life is still possible.

Economic and political factors have changed the patterns and strategies of human occupation of dryland regions, where previous lifestyles of transhumance or nomadism and minimal involvement in a market economy permitted human occupation of arid and semiarid areas even in times of drought, without causing ecosystem stress.

MANIFESTATIONS OF THE DESERTIFICATION PROCESS

Dryland ecosystems are fragile, highly vulnerable to natural climatic fluctuations, and susceptible to desertification. Desertification impacts not only human populations and their livelihood, but also the landscape—vegetation, soils, hydrology—as well as insect, animal, and bird populations and biodiversity as a whole. Dryland soils are particularly at risk—their low levels of biological activity, organic matter, and aggregate stability can easily result in a breakdown or decline of soil structure, accelerated soil erosion, reduction in moisture retention, and increase in surface runoff.[3] Reduction in plant cover will exacerbate these processes, leading to a change of scale in the spatial distribution of soil resources,[4] an increasingly "patchy" landscape, and a decline in sustainability that is difficult or impossible to reverse.[5]

Dryland vegetation communities include grassland, shrubland, woodland, savanna, and steppe (Fig. 1) and vary according to climatic, edaphic, hydrologic, and anthropogenic factors.[3] Change in vegetation community structure and reduction in cover may be a catalyst for desertification processes.[6] A shift from grassland-dominated to shrub-dominated rangeland seems to be ubiquitous[7,8] and is generally associated with increased soil runoff. The main causes of this vegetation change seem to be grazing, fencing, and alterations to the natural fire regime, in association with climatic variability.[9]

The World Atlas of Desertification[10] suggests that overgrazing, agricultural practices, overexploitation of vegetation, and deforestation are the four most significant causes of soil degradation although bioindustry plays a locally important role in some countries. Soil degradation may result from displacement (by wind and water erosion) or internal deterioration by chemical or physical variables.[11] Erosion of soil by water results in loss of topsoil as well as changes to the landscape such as gullying and rilling. Loss of topsoil has an impact on both natural vegetation and cultivated crops,[12] and is the dominant erosion process in Africa.[10] Wind erosion is most severe in areas that have been disturbed by human activity—for example, by grazing pressure that reduces plant cover. Chemical deterioration can be divided into three classes: a loss of nutrients, salinization, and acidification.[10] A change in soil nutrients and a patchy distribution of nutrients at the landscape scale is a consequence of desertification[5] and in some cases may be partially due to erosion of fine particles by wind and water.[13] Physical changes in soil structure result from compaction, crusting and waterlogging, and—like salinization and acidification—are exacerbated by pressure of human land use and widespread irrigation.[10]

Desertification is linked to poverty[12,14] and tends to result in out-migration, initially of young people and men. This changes family structure and places burdens and constraints on the women, children, and old people remaining

Encyclopedia of Environmental Management DOI: 10.1081/E-EEM-120001732

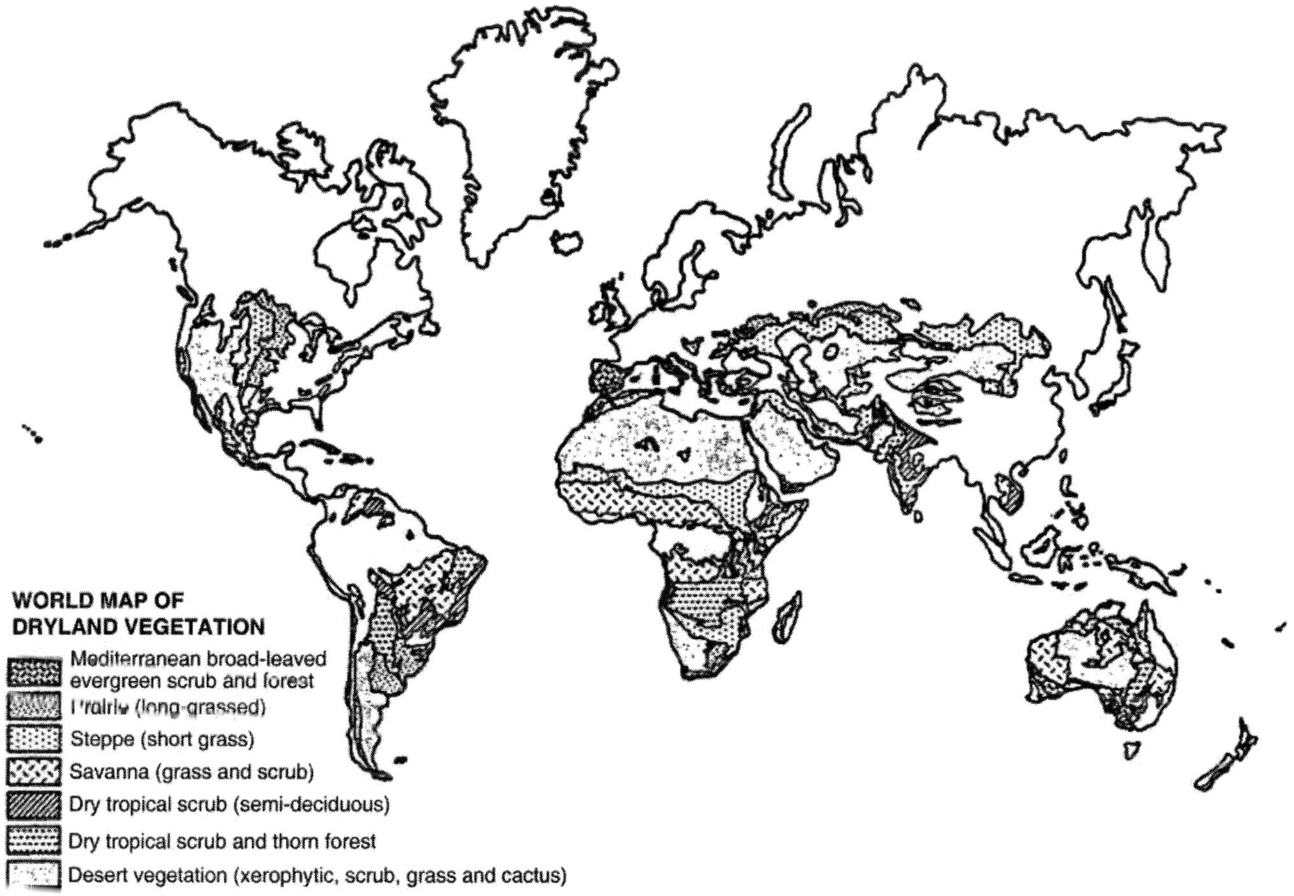
WORLD MAP OF
DRYLAND VEGETATION
Mediterranean broad-leaved evergreen scrub and forest
Steppe (short grass)
Savanna (grass and scrub)
Dry tropical scrub (semi-deciduous)
Dry tropical scrub and thorn forest
Desert vegetation (xerophytic, scrub, grass and cactus)

Fig. 1 Global distribution of vegetation communities.
Source: Williams and Balling[3]

in the settlements—which in turn changes patterns of land use and may exacerbate desertification.[12] Loss of human dignity is perhaps the most insidious consequence of the desertification process—it is difficult to measure and difficult to reverse.

Measurement of the extent and severity of desertification has been an ongoing process since the 1930s and is now well documented in many countries. However, in order to comply with UN treaties, management plans must be developed and implemented, education programs initiated, and local communities involved in decision making. To do this effectively, it is necessary to identify those areas that will most benefit from intervention, and where desertification processes may be stabilized or reversed. International conferences addressing these issues were held in Tucson, Arizona, U.S.A. in 1994 and 1997[15,16] at which indicators of desertification were much discussed.

INDICATORS OF DESERTIFICATION

Although some indicators may be relevant universally,[17] a standardized list of indicators is not possible as the climatic, geographic, and anthropomorphic conditions of desertification are variable, as are technology, human, and financial resources. Most indicators tend to be of a "snapshot" nature that provide measures of condition and trend, and a measure of what is "normal" only when compared in time and space. Process-based studies in semiarid and arid ecosystems that appear to be structurally and functionally intact[18] may provide baseline data on the expected and inherent variability of systems, to permit the establishment of thresholds indicative of irreversible change.[19]

Reporting on environmental indicators for the land, to meet Australia's UN obligations, Hamblin[20] selected indicators that would distinguish between anthropogenic interventions and natural causes, and recommended "key indicators" of condition, pressure, and response that would describe trend and impact upon the environment. The indicators are grouped by process in relation to change in erosion, habitat, hydrology, biota, nutrients, and pollution.

The 1994 International Desertification Conference in Arizona identified nutrient availability, water budget, energy balance, and biological diversity as key ecosystem processes affected by desertification which apply across international boundaries, and again recommended indicators that were grouped by process—in this case, nutrient availability, water budget, energy balance, and biological diversity. Out of a total of 28 recommended indicators, the following may be most significant.

- Infiltration, both under and between plants
- Cover type and distribution
- Erosion as indicated by rills, gullies, pedestalling, litter movement, flow patterns, and fetch length
- Depth to water table
- Normalized Difference Vegetation Index (NDVI) as a measure of greenness
- Surface temperature
- Ratio of native to exotic species for flora and fauna

There is a paucity of demographic or socioeconomic indicators in this list. Of the numerous indicators of desertification in general use, those concerning socio-economic factors are the least developed, and there is a disconnect in many areas between the communities affected by desertification processes, the scientists conducting research, and policy makers and managers.[21] The disconnect between communities, researchers, and policy makers has been a subject of concern in South Africa and Namibia for some time, and has been addressed by several recent projects.[22–24] The Karoo veld (rangeland) assessment handbook produced by Milton and Dean[25] is a successful example of orienting scientific data and their interpretation to a specific nonscience audience.

Soil is variable and complex, and the methods for assessing its condition are slow, tedious, and expensive; and to compound the issue—differences in soil type may be mistaken for differences in soil condition.[26] Also targeted to ranchers and managers, Tongway's rangeland soil condition assessment handbook[27] proposes an approach to assessing soil condition based on recognizing and classifying processes related to erosion, infiltration, and nutrient cycling.

The world's drylands provide critical habitats for migratory birds, and other wildlife,[12] and the maintenance of biodiversity in arid and semiarid areas is a major concern. Measures of floral, faunal, and avian diversity and reproductive success must be included in the list of indicators used to assess and monitor the status and extent of desertification. However, like demographic indicators, those involving biodiversity can be difficult to define, measure, and analyze and are frequently site specific. They may, however, provide the elusive "early warning" of a susceptibility for desertification that researchers and managers are seeking.

TECHNOLOGY, MEASUREMENT, AND ANALYSIS

The field data needed for desertification assessment and monitoring, such as soils and vegetation measurements, are typically collected at the plot scale, and many researchers, managers, and policy makers have questioned the use of remote sensing, because it provides a generalized measurement at scales of tens of meters to kilometers. However, there are numerous remote-sensing systems, operating at different spectral, spatial, temporal, and radiometric scales. The latest in the Earth Observing System (EOS) series of missions launched in December 1999 carried new sensors, two of which are the Advanced Spaceborne Thermal Emission and Reflection Radiometer (ASTER) and the Moderate Resolution Imaging Spectrometer (MODIS). The ASTER mission objectives focus on understanding the Earth as a system and the construction of models of Earth's global dynamics—including desertification.[28]

The long-term perspective critical to desertification assessment is also an essential component of the evaluation of future management options and development of remediation, land use, and management plans. A recently developed technique, alternative futures assessment, models potential changes to the landscape in a Geographic Information System (GIS) over several time scales, including projections for population growth. Models for basic environmental processes are then operated on the results of the change scenarios to show cumulative effects on the system as a whole.[29] Workshops and questionnaires are used to ascertain current and future attributes of the area stakeholders consider most desirable, to identify plausible change scenarios, and evaluate results.

Long-term ecological studies at plot, regional and local scales, remote sensing data, GIS, and alternative futures assessments are components of an integrated assessment, monitoring, and management strategy that is evolving in response to the need for greater interaction and communication between scientists, managers, and communities. This is a global concern, and the following case study for southeastern Arizona, discusses previous and ongoing research in the context of a strategy that involves local communities, land owners, and managers in the decision-making process.

CASE STUDY—THE SAN PEDRO RIVER BASIN

The San Pedro River rises in Mexico and flows north through the semiarid shrub steppe of southeastern Arizona. The Bureau of Land Management (U.S. Department of Interior) manages the San Pedro Riparian National Conservation Area immediately bordering the river, where there is one of the highest animal biodiversity totals in North America. Vegetation change, especially the deterioration of grasslands due in part to grazing and the suppression of fire has been a concern in the area for some years.[30] This change was quantified using aerial photography, remote sensing, and GIS to compare the extent and change over time between the eight vegetation classes in the region.[9] The most striking change between 1974 and 1987 was the considerable fragmentation of vegetation classes. This patchiness of the landscape is indicative of loss of biodiversity, and could be an early warning of desertification in the area.[9]

A multi-agency, multi-national, global change initiative to investigate the consequences of natural and human-induced environmental change in semiarid regions was started in 1995 in the San Pedro River Basin. Several multi-disciplinary projects have been initiated, and ASTER imagery is being acquired for the region. Among the studies currently being conducted in the San Pedro River Basin is an alternative futures assessment. This latter study has two objectives: to develop an array of plausible alternative patterns of land use and to assess their impacts on biodiversity, vegetation dynamics, fire regimes, hydrology, and aesthetics; and as a pilot study to develop a methodology applicable to other areas.[31]

CONCLUSIONS

Desertification impacts on landscapes, biodiversity, and human populations are moderately well documented, and there are numerous suites of indicators in use for assessment and monitoring.[20] Studies of the actual processes involved are valuable, particularly in areas that are marginal for desertification susceptibility, and are contributing to our understanding and therefore to our ability to meaningfully intervene to reduce or mitigate the effects of desertification on both the environment and human populations.[18]

In the long term, we still do not know what "desertification" really means, nor what the impacts to humans, flora, fauna, and the landscape will be. Desertification is a phenomenon that will not disappear as a result of human intervention, but may be reduced in severity or extent by a strategy including increasing understanding of the processes involved, monitoring, and investigating alternative management options. In particular, the societal impacts of desertification may be mitigated by involving local communities in the policy and decision making process.

REFERENCES

1. *Report of the United Nations Conference on Environment and Development; United Nations Conference on Environment and Development* (UNCED); United Nations: New York, 1992; Vol. 1, 486 pp.
2. Kassas, M. Desertification: a general review. J. Arid Environ. **1995**, *30*, 115–128.
3. Williams, M.A.J.; Balling, R.C., Jr. Interactions of Desertification and Climate; Arnold: London, 1996; 270 pp.
4. de Soyza, A.G.; Whitford, W.G.; Herrick, J.E.; Van Zee, J.W.; Havstad, K.M. Early warning indicators of desertification: examples of tests in the chihuahuan desert. J. Arid Environ. **1998**, *39* (2), 101–112.
5. Schlesinger, W.H.; Reynolds, J.F.; Cunningham, G.L.; Huenneke, L.F.; Jarrell, W.M.; Virginia, R.A.; Whitford, W.G. Biological feedbacks in global desertification. Science **1990**, *247*, 1043–1048.
6. Le Houe´rou, H.N. Climate change, drought and desertification. J. Arid Environ. **1996**, *34*, 133–185.
7. Ludwig, J.A.; Tongway, D.J. Desertification in Australia: an eye to grass roots and landscapes. In *Desertification in Developed Countries*; Mouat, D.A., Hutchinson, C.F., Eds.; Kluwer Academic Publishers: Dordrecht, 1995; 231–237.
8. Milton, S.J.; Dean, W.R.J. South Africa's arid and semi-arid rangelands: why are they changing and can they be restored? In *Desertification in Developed Countries*; Mouat, D.A., Hutchinson, C.F., Eds.; Kluwer Academic Publishers: Dordrecht, 1995; 245–264.
9. Mouat, D.A.; Lancaster, J. Use of remote sensing and GIS to identify vegetation change in the upper san pedro river watershed, Arizona. Geocarto Int. **1996**, *11* (2), 55–67.
10. Middleton, N.; Thomas, D. *World Atlas of Desertification*, 2nd Ed.; Middleton, N., Thomas, D., Eds.; Arnold: London, 1997; 182 pp.
11. Oldeman, L.R. *Guidelines for General Assessment of the Status of Human-Induced Soil Degradation*; International Soil Reference and Information Center: Wageningen, 1988.
12. http://www.unccd.int/main.php (accessed Dec. 2000).
13. Schlesinger, W.H.; Ward, T.J.; Anderson, J. Nutrient losses in runoff from grassland and shrubland habitats in Southern New Mexico: II. Field Plots. Biogeochemistry **2000**, *49*, 69–86.
14. Glantz, M.H. *Drought Follows the Plow: Cultivating Marginal Areas*; Glantz, M.H., Ed.; Cambridge University Press: Cambridge, 1994; 197 pp.
15. Mouat, D.A.; Hutchinson, C.F. *Desertification in Developed Countries*; Mouat, D.A., Hutchinson, C.F., Eds.; Kluwer Academic Publishers: Dordrecht, 1995; 363 pp.
16. Mouat, D.A.; McGinty, H.K. *Combating Desertification: Connecting Science with Community Action*; J. Arid Environ., Mouat, D.A., McGinty, H.K., Eds.; (Special Issue) Academic Press: London, 1998; 340 pp.
17. Mouat, D.A.; Lancaster, J.; Wade, T.; Wickham, J.; Fox, C.; Kepner, W.; Ball, T. Desertification evaluated using an integrated environmental assessment model. Environ. Monit. Assess. **1997**, *48*, 139–156.
18. Dean, W.R.J.; Milton, S.J. *The Karoo: Ecological Patterns and Processes*; Dean, W.R.J., Milton, S.J., Eds.; Cambridge University Press: Cambridge, 1999; 374 pp.
19. Tausch, R.J.; Wigand, P.E.; Burkhardt, J.W. Plant community thresholds, multiple steady states, and multiple successional pathways: legacy of the quaternary? J. Range Manag. **1993**, *46* (5), 439–447.
20. Hamblin, A. *Environmental Indicators for National State of the Environment Reporting: The Land*; Department of the Environment: Canberra, 1998; 124 pp.
21. Mouat, D.A.; McGinty, H.K.; McClure, B.C. Introduction. In *Combating Desertification: Connecting Science with Community Action*; Mouat, D.A., McGinty, H.K.,Eds.; Special Issue: J. Arid Environ. Academic Press: London, 1998; 97–99.
22. Seely, M.K. Can Science and community action connect to combat desertification? J. Arid Environ. **1998**, *39* (2), 267–278.

23. van Rooyen, A.F. Combating desertification in the Southern Kalahari: connecting science with community action in South Africa. J. Arid Environ. **1998**, *39* (2), 285–298.
24. Milton, S.J.; Dean, W.R.J.; Ellis, P.R. Rangeland health assessment: a practical guide for ranchers in arid karoo shrublands. J. Arid Environ. **1998**, *39* (2), 253–266.
25. Milton, S.J.; Dean, W.R.J. *Karoo Veld: Ecology and Management*; Agricultural Research Council Range and Forage Institute: Lynn East, 1996; 94 pp.
26. Tongway, D. Monitoring soil productive potential. J. Arid Environ. **1998**, *39* (2), 303–318.
27. Tongway, D.J. *Rangeland Soil Condition Assessment Manual*; Commonwealth Scientific and Industrial Research Organization: Melbourne, 1993; 69 pp.
28. http://asterweb.jpl.nasa.gov (accessed Dec. 2000).
29. Steinitz, C.; Binford, M.; Cote, P.; Edwards, T.; Ervin, S.; Forman, R.T.T.; Johnson, C.; Kiester, R.; Mouat, D.; Olson, D.; Shearer, A.; Toth, R.; Wills, R. *Biodiversity and Landscape Planning: Alternative Futures for the Region of Camp Pendleton, California*; Harvard University, Graduate School of Design: Cambridge, MA, 1996; 142 pp.
30. Hastings, J.R.; Turner, R.M. *The Changing Mile: an Ecological Study of Vegetation Change with Time in the Lower Mile of an Arid and Semi-Arid Region*; University of Arizona Press: Tucson, AZ, 1965.
31. http://www.gsd.harvard.edu/faculty/steinitz=sanpedroriver.html (accessed Dec. 2000).

Desertification: Prevention and Restoration

Claudio Zucca
Department of Agriculture, University of Sassari, Sassari, Italy
Susana Bautista
Department of Ecology, University of Alicante, Alicante, Spain
Barron Orr
Office of Arid Lands Studies, School of Natural Resources and the Environment, University of Arizona, Tucson, Arizona, U.S.A.
Franco Previtali
Department of Environmental Science, University of Milan—Bicocca, Milan, Italy

Abstract
Policy and management approaches and strategies to address desertification can be broadly grouped as prevention and reversal. Desertification is framed within multiscale, coupled human–environmental dynamics, and so must be the approaches for prevention and reversal. The development of integrated and participatory methods for evaluating and monitoring interventions is crucial in view of planning and implementing effective interventions. Recent approaches focus on indicators that relate to ecosystem integrity and services and to human well-being. The present entry reviews both the theoretical and operational issues related to the subject.

INTRODUCTION

This entry provides a brief overview of the most significant approaches to prevent and reverse land degradation in drylands, with a special emphasis on conceptual frameworks and on methods to monitor and evaluate their impacts.

The entry first discusses the main underlying concepts, in light of recent developments on the conceptual framework of desertification. The methodological developments related to integrated and participatory evaluation are then presented. Finally, the implementation of mitigation and restoration programs is addressed, with particular reference to the constraints and risk factors and to the practical lessons learned in the field.

DESERTIFICATION: THE UNDERLYING CONCEPTS

Desertification is defined here as "land degradation in drylands resulting from various factors, including climatic variations and human activities," in conformity with the UNCCD (United Nations Convention to Combat Desertification).[1] Before the entry into force of the UNCCD in 1996, the term "desertification" had been given a number of different definitions.[2–6] The UNCCD definition has been and still is the subject of scientific debate; the related evolving concepts have been reviewed by a number of papers.[7–12] The term "desertization" was also proposed as an alternative to indicate the irreversible arid land degradation (mostly man-made) resulting in desert-like land forms and landscapes in areas where they did not occur in the recent past.[10] The evolving and often changing use of the two terms, as well as the number of contrasting local translations they have been given, has undoubtedly created some confusion.

The Millennium Ecosystem Assessment (MA) defines desertification as a persistent reduction in the provision of ecosystem services over an extended period.[13,14] The scientific discussion promoted in 2009 by the UNCCD in view of the First Scientific Conference of its CST (Committee on Science and Technology), building on the MA definition, resulted in a proposal to redefine desertification as "an end state of the process of land degradation; this process is expressed by a persistent reduction or loss of biologic and economic productivity of lands that are under use by people."[15] The ongoing discussion also focuses on the integrated sets of indicators needed for monitoring and assessing desertification, and on the related conceptual frameworks that would help scientists, practitioners, and policy makers organize, use, and communicate the results of that monitoring.[16–18]

While the definition of desertification continues to evolve, ecosystem services are increasingly seen as a unifying supporting concept. The MA states that "desertification results from a long-term unbalance between demand for and supply of ecosystem services" and that measurement of persistent reduction in the capacity of ecosystems

Encyclopedia of Environmental Management DOI: 10.1081/E-EEM-120046343

to supply services provides a robust and operational way to quantify land degradation and desertification. The ecosystem services framework is increasingly thought to provide a basis to assess and value the impacts of land change and degradation, as well as the effects of the actions aimed at reversing it.[19–21]

Desertification manifests itself through different forms and processes in different ecosystems and socioeconomic contexts. Its direct causes are many and can be generally ascribed to different forms of land mismanagement, such as overgrazing, deforestation, overuse of irrigation, and non-resource-conservative agriculture and forest practices.[1,7,17] These are at the origin of the major land degradation processes that are globally affecting the provision of ecosystem services, including water and wind erosion, soil salinization, loss of vegetation cover and diversity, and degradation of the hydrological cycles.[13]

No satisfying estimates of the global extent and severity of desertification are available thus far; however, a new World Atlas of Desertification based on multiscale integrated sets of indicators is under development.[16,17] Assessment and monitoring of desertification, as well as of the impact of the prevention and restoration interventions, still constitute a major research challenge[15] and for this reason are a primary focus of this entry.

PREVENTION AND REVERSAL

Policy and management responses to desertification can be grouped under two major classes: prevention and reversal.[13] The boundaries between these ones are vague, as in practice they form a continuum of potential prevention, mitigation, and restoration actions, to be adapted to particular sites and dynamics through adaptive management approaches (Fig. 1).

Prevention actions can be considered as avoidance approaches, either through proactive management efforts or through changes in land use and management currently leading to desertification.

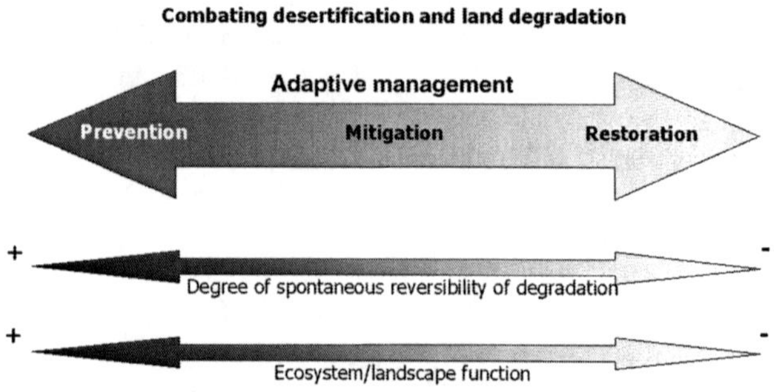

Combating desertification and land degradation
Adaptive management
Prevention
Mitigation
Restoration
+
-
Degree of spontaneous reversibility of degradation
+
-
Ecosystem/landscape function

Fig. 1 Continuum of actions to combat desertification and land degradation.

Prevention, however, is not enough to address the challenges posed by desertification. Vast areas of drylands are already severely degraded, with reduced plant cover and species diversity, falling productivity, depleted or eroded soils, and very low potential for spontaneous recovery of ecosystem functions, even when degradation forces are no longer stressing the system. Many of these systems have changed at a level at which restoration is the only viable option to recover ecosystem services that have been lost.[22–24]

Examples of prevention and mitigation actions include measures to improve water management and agricultural practices. These are often referred to as soil and water conservation (SWC) or sustainable land management (SLM) practices. The FAO–LADA (Food and Agriculture Organization of the United Nations/Land Degradation Assessment in Drylands)[25] and the WOCAT (World Overview of Conservation Approaches and Technologies)[26] projects provide a framework for classifying and evaluating SLM actions. In view of the needs of the UNCCD, SLM has been recently defined as "land managed in such a way as to maintain or improve ecosystem services for human wellbeing, as negotiated by all stakeholders."[21]

Examples include long-term crop rotations with cereals/legumes; more efficient use of fertilizers; improvements in water-use efficiency; conservation-minded tillage methods; traditional water-harvesting techniques; water storage; measures that protect soils from erosion, salinization, and other forms of soil degradation; improved crop–livestock integration, combining livestock rearing and cropping to allow a more efficient recycling of nutrients within the agricultural system; and in situ conservation of genetic resources and better resource use with efficient germplasm. Creating viable livelihood alternatives, including the creation of economic opportunities in urban centers, can also help reduce current pressures on drylands.

For extremely degraded lands, rehabilitation and restoration approaches often involve the improvement in the quantity and/or quality of vegetation cover through, for example, the establishment of seed banks, reintroduction of selected species, control of invasive species, and reforestation programs.

Desertification is driven by a combination of proximate causes and underlying forces, including their interactions and feedbacks; these vary from region to region and change over time.[27] Approaches and strategies to prevent and reverse desertification need therefore to address the dynamic causal patterns and multiplicity of actors, factors, and scales involved. In general, developing the appropriate engagement between scientific and local environmental knowledge is critically important for efforts to prevent and reverse land degradation.[28] Desertification is framed within multiscale, coupled human–environmental dynamics, and so must be the approaches for desertification prevention and reversal.[29,30] The relationships between land degradation and its causal agents are non-linear and complex. Degradative and aggradative trajectories commonly exhibit thresholds and rapid shifts, as well as hysteresis dynamics, where the trajectories of degradation and recovery differ.[31–33] Understanding and monitoring these relationships are critical in the design of strategies to combat desertification.

In addition, socioeconomic conditions impose limitations on the technology and inputs available. Therefore, the approaches to combat desertification should incorporate both current conditions and scenario projections of socioeconomic and environmental constraints and opportunities.[13]

Finally, there is growing evidence that land degradation in desertification-prone areas is likely to increase with climate change.[34] Desertification is linked to biodiversity loss and global climate change through the regulation of water and carbon fluxes. Therefore, interlinkages in policy formulations aimed at combating desertification, mitigating the effects of climate change, and conserving biodiversity must be beneficially exploited by developing multifunctional strategies that address the three global environmental goals.

IMPLEMENTING SOLUTIONS IN THE FIELD: LESSONS LEARNED

Despite the availability of technological, institutional, and even financial resources, efforts to combat desertification often fail because of poor implementation. A list of "lessons learnt" is presented below, to summarize a range of major constraints and risk factors that can hinder the successful implementation of the intervention projects, while highlighting the lessons learned and necessary improvements. Some of those issues stem from the concepts of adaptive management and multiscale human–environmental dynamics as introduced in the previous section. Others are more related to the issues of local participation and integrated assessment that will be discussed in the following section. Points 1 and 2 deal with the quality of the technical design and its degree of adaptation to local conditions and knowledge. Points 3 to 7 are related to the ability of the projects to cope with the long-term human–environmental dynamics. Points 8 to 11 are connected to the quality of the participatory processes implemented by the projects, while the last two points are linked to major, common organizational constraints.

1. Lack of awareness of actual land conditions. Sometimes, people (including decision makers) living in degraded areas do not realize that their land still maintains productive potential, or perhaps are unaware of how that potential can be exploited. Preliminary land surveys should be done to support project design. These baseline assessments are necessary to guide subsequent actions. In some cases, "no-action" options could be considered. In the case of degraded rangelands, simple and cost-effective "self-learning" tests based on grazing exclusion and rotation can be proposed to local communities as a means to demonstrate the effects of pressure mitigation and sustainable management. Such tests allow hands-on learning and practical experience and are much less risky than the direct introduction of often expensive "all or nothing" rehabilitation programs.
2. Schematic approach. Sometimes, big programs are extensively implemented by adopting schematic approaches, which are not able to adapt to specific local land characteristics and the needs of local people. In other cases, they address areas where they are not necessary or inadvertently lead to negative side effects. Implementation protocols with multiple technical solutions and local stakeholder input on the perceived benefits (and unintentional consequences) are needed.
3. Lack of long-time planning in restoration. Often the long-term dynamics of the "restored human–environmental systems" are not fully considered. This is particularly relevant when the interventions are based on introducing fast-growing, income-generating alien species that may require a "re-naturalization" strategy in the long term to balance ecological and social sustainability.
4. Inability to cope with natural crises. Especially in projects aimed at increasing plant cover, a poor or delayed wet season or recurrent droughts may cause the loss of part or all of the investment in a project. For example, this may happen in the case of fodder shrubs plantations, when farmers cannot avoid early grazing due to drought and unavailability of alternative feeding resources. Emergency/contingency funds, quick diagnostic and intervention mechanisms, and flexibility in project duration are necessary to mitigate against this risk.
5. Socioeconomic and demographic dynamics. Some projects have experienced labor shortfalls due to the out-migration of young people. In contrast, the return of people onto land previously "closed for restoration" may cause unsustainable pressure. Addressing economic constraints and associated demographic pressures such as migration prior to project implementation can reduce such kind of risks.
6. Market drivers. The dynamics of international and local market prices may completely and quite rapidly negate achievements produced by years of conservation programs.
7. Contrasting policies. A restoration initiative may be useful in practical terms and yet be overwhelmed by unrelated policies. A common example is farming incentives that come in conflict with the goals of mitigation programs, if not accompanied by adequate guidelines.

8. Passive community participation. Community participation should be strongly based on responsible awareness and sharing of project objectives, and participation in its planning, implementation, and evaluation. Cases where, for example, sectoral administration goals are implemented through prior agreement with land users (e.g., "I let you do on my land") may require or even force farmer action but, in the end, may not represent or address the key concerns or needs of local people.
9. Uncertain community commitment. Stakeholder engagement that fails to bring all parties to the table can have dire consequences. Comprehensive, balanced, and approved representation is not easy to obtain, but it is essential for success. The key question often raised is, "Who is committing on behalf of whom?" The commitment should include a community contribution or investment to cover the implementation costs, be linked to final results and be based on taking ownership and responsibility after a project ends, rather than simply for the completion of individual tasks. Community commitment should be clearly defined in a way that can cope with changing community priorities.
10. Institutional commitment. Complex projects, in which the implementation is based on the support of local administration staff, need a formal and clear institutional commitment. This may take time and very rarely can be established before project approval: projects should have an inception phase to set agreements.
11. Lack of transparency in the engagement process of the ultimate project beneficiaries. This is especially true (but not only) when the involvement of individual farmers or other land users in projects is not mediated by the community. Inappropriate or unrepresentative participation can lead to failure and loss of credibility. Transparent and objective selection mechanisms must be understood by all and be followed rigorously by project implementers.
12. Unrealistic project duration. Very often, the most common investment period imposed by donors (3–5 years) is too short to allow for adequate biophysical response and/or socioeconomic adaptation that would assure success. Of particular concern is enough time for effective "inception or learning phases" for community members to deal with the necessity to adapt to changing conditions and needs, to cope with unforeseen events, and most of all to monitor and assess impacts and sustainability.
13. Spending and reporting difficulties. Spending rules, procurement procedures, "exotic" rules imposed by the donors, or strict local rules, not well known by the local project management, often lead to delays and underperformance. More in general, the lack of flexibility can exacerbate the effects of most of the above-mentioned difficulties.

Finally, our capacity to design effective projects is often undermined by a lack of integrated assessment of the progress and success of the previous projects. Projects are often not adequately monitored and evaluated, and lessons learned come too late to be useful.

INTEGRATED EVALUATION OF PREVENTION AND RESTORATION ACTIONS

The development of integrated biophysical and socioeconomic analytical methods for evaluating progress and success, along with a framework for knowledge sharing and transfer, is crucial to combating desertification.[29] Furthermore, monitoring and evaluation are needed to demonstrate the benefits of sustainable dryland management, establish cost-effective thresholds for the various management alternatives, and identify priority areas where actions could be most effective.

In recent years, there have been a number of initiatives to develop common and comprehensive methodologies for assessing and evaluating the effectiveness of management and restoration programs.[35–39] These approaches focus on indicators that relate to ecosystem integrity and services, and human well-being (socioeconomic and cultural variables). Irrespective of the biophysical or socioeconomic attributes assessed, the selected indicators should be relevant, be sensitive to variations of environmental stress, have the capacity to respond to stress in a predictable manner, but also be simple and measurable with a reasonable level of effort and cost.[40,41]

Vegetation cover and composition are the most common metrics used for evaluating mitigation and restoration actions.[42] However, vegetation cover alone cannot always reflect how well an ecosystem is functioning. During the last decade, a variety of functional assessment approaches that relate to the spatial pattern of vegetation have been proposed.[43–45] Some of these functional assessment methods also incorporate properties relative to the soil surface condition.[46,47] The theoretical framework for these approaches considers that landscapes occur along a continuum of functionality from highly patchy systems that conserve all resources to those that have no vegetation patches and leak all resources.[48] For semiarid ecosystems, it has been hypothesized that vegetation patchiness could be used as a signature of imminent transitions[49] and that changes in patch-size distributions may be a warning signal for the onset of desertification.[50]

Evaluation frameworks must account for the cross-scale and social–ecological interactions affecting the response of degraded lands to mitigation/restoration actions.[13] A multiscale approach is always advisable. Farm- and project-scale assessments focus on local resources and productivity, and a private economic valuation perspective (market-priced goods and services), while landscape- and program-level indicators address environmental benefits and public/social welfare considerations (Fig. 2).

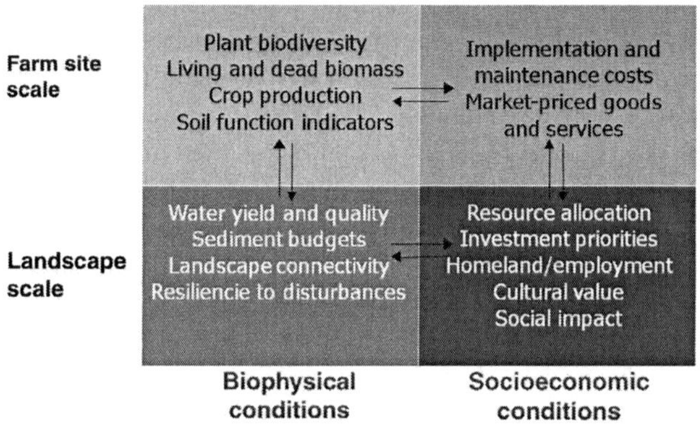
Farm site scale
Plant biodiversity
Living and dead biomass
Crop production
Soil function indicators
Implementation and maintenance costs
Market-priced goods and services
Landscape scale
Water yield and quality
Sediment budgets
Landscape connectivity
Resiliencie to disturbances
Resource allocation
Investment priorities
Homeland/employment
Cultural value
Social impact
Biophysical conditions
Socioeconomic conditions

Fig. 2 Example of a multiscale integrated framework and indicators for evaluating mitigation and restoration actions.

Because of the large spatial and temporal variability of ecosystems, particularly in drylands, it is critical to focus on "slow variables,"[51] so the assessment of long-term changes, and of the sustainability of land management, is not confused by short-term variations in land and socioeconomic conditions.[29] Assessment methods range from simple, qualitative assessments based on field observations to relatively complex protocols based on quantitative measurements of critical ecosystem and landscape attributes and socioeconomic surveys.[37] The development and accessibility of remote-sensing (RS) products have led some international bodies to recommend the integrated use of RS-based geospatial information with ground based observations to assess land degradation.[13,15,52]

PARTICIPATORY APPROACHES

While substantive progress has been made in developing more standardized and more relevant environmental assessment and monitoring approaches that reflect human–ecological interactions, the adoption of evaluation results at the local level remains challenging.[30,53] Though some suggest that local interests or even national policies outside the control of evaluators may partially explain this, environmental assessments tend to be independent, unilateral, and top–down, with results being delivered post-assessment. Stakeholder engagement is rare, or all-too-often limited to a period of "public comment" immediately before and/or after the assessment. In the same way that adaptive governance and ecosystem-based management approaches require local knowledge and continual stakeholder engagement, so too must the corresponding monitoring and evaluation if the goal is for results to be truly embraced and used.[54–56] Adaptive, ecosystems-based management is data intensive, and requires a commitment to a variety of data sources, including local knowledge, which in turn necessitates more attention to knowledge integration and methods of analysis at different scales. Adoption of best practices that in theory come from a well-executed evaluation of a given desertification mitigation or restoration action can fail due to poor or limited communication and knowledge exchange among the involved actors. Essentially ignoring local knowledge increases the risk of missing essential key local factors, opportunities, or constraints.

Engaging stakeholders from the outset of an environmental assessment and maintaining the interaction throughout the process can result in the integration of local people and their perceptions into management, planning, and evaluation, helping develop feelings of ownership and representation while giving voice to locals in the process.[28,57,58] The potential benefits go beyond the assessment itself. Analysis of environmental conflict resolution processes suggests that ensuring all parties are at the table and are effectively engaged is directly correlated with often-sought outcomes like reaching agreement, the quality of agreement, and improved working relationships among parties.[59,60]

An additional benefit to a participatory approach to evaluation and the exchange of ideas among stakeholders, including researchers, is the social learning that takes place. The collective self-reflection through interaction and dialogue among the diverse set of stakeholders involved with or affected by environmental challenge and the assessment of associated responses can result in the coproduction of knowledge.[61,62] Social groups that develop a shared understanding of a challenge can build up the experience necessary to improve linkages between knowledge and the environment, cope with change, and enhance adaptation, because social learning helps solidify knowledge systems made up of the relevant sets of actors, networks, or organizations.[63] While the majority of discussion about the benefits of encouraging social learning has been focused on improving adaptive management, it is clear that the wealth of information explored during an evaluation suggests that an assessment period is an ideal time to encourage stakeholder interaction and knowledge exchange. In this sense, the evaluation itself becomes a tool for outreach and *in*reach, where land users, natural resource managers, and scientists all stand to learn from and potentially benefit from each other's insights. The process has the potential to empower individuals, build relationships, expand networks, and thereby enhance the relevance and impact of decision making.

CONCLUSION

Desertification is one of the major global environment and sustainable development challenges. It affects the livelihoods of millions of people, threatening human well-being in drylands.[13] That risk, when considered relative to the foreseeable impacts of climate change, stands to grow significantly in the future, with estimates suggesting that as many as 50 million people will be in peril of physical displacement in the next 10 years.[64]

This entry underscores a range of conceptual and practical issues influencing the design and implementation of interventions, with a special emphasis on conceptual frameworks and associated methods to monitor and evaluate impacts.

Lessons learned in the field highlight several issues threatening the success of prevention/reversal actions. The capacity to design and implement effective restoration actions and other countermeasures to desertification is often undermined by a lack of assessment of the outcomes of previous projects.

The assessment of actions is complex and requires conceptual tools such as human–environmental frameworks for integrated assessment and participatory approaches to foster social learning. These should become the basis for the development of multifunctional mitigation strategies and the formulation of interlinked policies to address desertification, climate change, and biodiversity. The development of participatory approaches in the assessment of interventions is critical to capacity building and knowledge exchange. Encouraging social learning and a true sense of ownership is essential for successful adaptive management.

Other aspects, lying beyond the scope of the present entry, are crucial to combating desertification.

Primarily, the assessment of intervention programs should involve the policy makers to promote the adoption of sustainable rural development policies and to counteract the socioeconomic and policy-driven dynamics of desertification.[65] As a second priority, future interventions should be increasingly oriented to income-generating actions to strengthen social and economic sustainability in concert with environmental sustainability.[66] Finally, it is important to keep in mind that practices and interventions can be considered as "good" or "best" only with reference to their suitability in relation to specific human–environmental contexts. In this regard, to develop strategies tuned to the changing land features, land evaluation techniques could be updated based on the integrated assessment principles, as suggested by the new "anthroscape" concept.[67]

REFERENCES

1. UNEP. *United Nations Convention to Combat Desertification in Those Countries Experiencing Serious Drought and/or Desertification, Particularly in Africa*; UNEP: Geneva, 1994.
2. Aubreville, A. *Climats, forets et Désertification de l'Afrique tropicale*; Société d'éditions géographiques maritimes et coloniales: Paris, 1949.
3. UNEP. *Draft Plan of Action to Combat Desertification*. Document A/CONF.74/L.36. UNEP: Nairobi, 1977.
4. Dregne, H. *Desertification of Arid Lands*; Harwood Academic Publisher: London, 1983.
5. FAO/UNEP. *Provisional Methodology for Assessment and Mapping of Desertification*; FAO: Rome, 1983.
6. UNEP. *Status of Desertification and Implementation of the United Nations Plan of Action to Combat Desertification*; UNEP: Nairobi, 1991.
7. Thomas, D.S.G.; Middleton, N.J. *Desertification; Exploding the Myth*; John Wiley & Sons: Chichester, 1994.
8. Enne, G.; Zucca, C. *Desertification Indicators for the European Mediterranean Region. State of the Art and Possible Methodological Approaches*; ANPA: Rome, 2000.
9. Arnalds, O. Desertification: An appeal for a broader perspective. In *Rangeland Desertification*; Arnalds, O., Archer, S., Eds.; Advances in Vegetation Science, 19; Springer: Berlin, 2000.
10. Le Houérou, H.N. Man-made deserts: Desertization processes and threats. Arid Land Res. Manage. **2002**, *16* (1), 1–36.
11. Eswaran, H.; Lal, R.; Reich, P.F. Land degradation: An overview. In *Response to Land Degradation*; Bridges, E.M., Hannam, I.D., Oldeman, L.R., Pening de Vries, F.W.T., Scherr, S.J., Sompatpanit S., Eds.; Science Publishers, Inc.: Enfield (NH), 2001; 20–35.
12. Herrmann, S.M.; Hutchinson, C.F. The changing contexts of the desertification debate. J. Arid Environ. **2005**, *63* (3), 538–555.
13. MEA, Millennium Ecosystem Assessment. *Ecosystems and Human Well-Being: Desertification Synthesis*; Island Press: Washington, DC, USA, 2005.
14. Safriel, U.N. The assessment of global trends in land degradation. In *Climate and Land Degradation*; Sivakumar, M.V.N., Ndiang'ui, N., Eds.; Springer-Verlag: Berlin, 2007; 1–38.
15. DSD. Integrated Methods for Assessment and Monitoring Land Degradation Processes and Drivers (Land Quality). White paper of the DSD Working Group 1, version 2. DSD (Dryland Science for Development Consortium), 2009, available at dsd-consortium.jrc.ec.europa.eu/documents/WG1_White-Paper_Draft-2_20090818.pdf (accessed July 2010).
16. Sommer, S.; Zucca, C.; Grainger, A.; Cherlet, M.; Zougmore, R.; Sokona, Y.; Hill, J.; Della Peruta, R.; Roehrig, J.; Wang, G. Application of indicator systems for monitoring and assessment of desertification from national to global scales. Land Degrad. Dev. **2011**, *22* (2), 184–197.
17. Zucca, C.; Della Peruta, R.; Salvia, R.; Sommer, S.; Cherlet, M. Towards a World Desertification Atlas. Relating and selecting indicators and datasets to represent complex issues. Ecol. Indic. **2012**, *15*, 157–170.
18. Orr, B.J. Scientific review of the UNCCD provisionally accepted set of impact indicators to measure the implementation of strategic objectives 1, 2 and 3; White Paper—Version 1. UNCCD: Bonn, 2011, available at http://www.unccd.int/science/docs/Microsoft%20Word%20-%20White%20paper_Scientific%20review%20set%20of%20indicators_Ver1_31011%E2%80%A6.pdf (accessed February 2011).
19. Costanza, R.; d'Arge, R.; de Groot, R.S.; Farber, S.; Grasso, M.; Hannon, B.; Limburg, K.; Naeem, S.; O'Neill, R.V.; Paruelo, J.; Raskin, R.G.; Sutton, P.; van den Belt, M. The value of the world's ecosystem services and natural capital. Nature **1997**, *387* (6630), 253–260.
20. de Groot, R.S.; Wilson, M.A.; Boumans, R.M.J. A typology for the classification, description and valuation of ecosystem functions, goods and services. Ecol. Econ. **2002**, *41* (3), 393–408.

21. DSD. Monitoring and assessment of sustainable land management. White paper of the DSD Working Group 2, version 2. DSD (Dryland Science for Development Consortium), 2009, available at http://dsd-consortium.jrc.ec.europa.eu/documents/WG2_White-Paper_Draft-2_20090918.pdf (accessed July 2010).
22. Bell, R.W. Land restoration: Principles. In *Encyclopaedia of Soil Science*, 2nd Ed.; Taylor and Francis, 2006; 978–981.
23. Hobbs, R. Restoration ecology. In *Encyclopaedia of Soil Science*, 2nd Ed.; Taylor and Francis, 2006; 1513–1515.
24. Tongway, D.; Ludwig, J. Desertification: Reversal. In *Encyclopaedia of Soils Science*, 2nd Ed.; Taylor and Francis, 2006; 465–467.
25. FAO/LADA project website, available at http://www.fao.org/nr/lada/ (accessed July 2010).
26. WOCAT. *Where the Land Is Greener—Case Studies and Analysis of Soil and Water Conservation Initiatives Worldwide*; Centre for Development and Environment (CDE): Bern, 2007.
27. Geist, H.J.; Lambin, E.F. Dynamic causal patterns of desertification. BioScience **2004**, *54* (9), 817–829.
28. Reed, M.S.; Dougill, A.J.; Taylor, M.J. Integrating local and scientific knowledge for adaptation to land degradation: Kalahari rangeland management options. Land Degrad. Dev. **2007**, *18* (3), 249–268.
29. Reynolds, J.F.; Stafford Smith, D.M.; Lambin, E.F.; Turner, B.L., II; Mortimore, M.; Batterbury, S.P.J.; Downing, T.E.; Dowlatabadi, H.; Fernández, R.J.; Herrick, J.E.; Huber-Sannvald, E.; Leemans, R.; Lynam, T.; Maestre, F.T.; Ayarza, M.; Walker, B. Global desertification: Building a science for dryland development. Science **2007**, *316* (5826), 847–851.
30. Liu, J.; Dietz, T.; Carpenter, S.R.; Folke, C.; Alberti, M.; Redman, C.L.; Schneider, S.H.; Ostrom, E.; Pell, A.N.; Lubchenco, J.; Taylor, W.W.; Ouyang, Z.; Deadman, P.; Kratz, T.; Provencher, W. Coupled human and natural systems. Ambio **2007**, *36* (8), 639–649.
31. Scheffer, M.; Carpenter, S.R. Catastrophic regime shifts in ecosystems: Linking theory to observation. Trends Ecol. Evol. **2003**, *18* (12), 648–656.
32. Suding, K.N.; Gross, K.L.; Houseman, G.R. Alternative states and positive feedbacks in restoration ecology. Trends Ecol. Evol. **2004**, *19* (1), 46–53.
33. Suding, K.N.; Hobbs, R.J. Threshold models in restoration and conservation: A developing framework. Trends Ecol. Evol. **2009**, *24* (5), 271–279.
34. IPCC. *Climate Change 2007. Impacts, Adaptation and Vulnerability. 4th Assessment Report. Summary for Policymakers*; IPCC Secretariat: Geneva, 2007.
35. SER (Society for Ecological Restoration, Science & Policy Working Group). The SER Primer on Ecological Restoration, available at http://www.ser.org/ (accessed July 2010).
36. Adeel, Z.; King, C. *Development of an Assessment Methodology for Sustainable Development of Marginal Drylands*, Proceedings of the Third Project Workshop for Sustainable Management of Marginal Drylands (SUMAMAD), Djerba, Tunisia, Dec. 11–15, 2004; UNESCO-MAB: Paris, 2005; 13–22.
37. Bautista, S.; Alloza, J.A.; Vallejo, V.R. Conceptual framework, criteria, and methodology for the evaluation of restoration projects. The REACTION approach. CEAM Foundation: Valencia, 2004; 1–7, available at http://www.ceam.es/reaction/documents.htm (accessed September 2007).
38. Bautista, S.; Orr, B.J.; Alloza, J.A.; Vallejo, V.R. Evaluation of the restoration of dryland ecosystems in the northern Mediterranean: Implications for practice. In *Water in Arid and Semi-arid Zones. Advances in Global Change Research*; Courel, M.F.; Schneier-Madanes, G., Eds.; Springer: Dordrecht, the Netherlands, 2010; 295–310.
39. Ward, S.C. Restoration: Success and completion criteria. In *Encyclopaedia of Soil Science*, 2nd Ed.; Taylor and Francis, 2006; 1516–1520.
40. Dale, V.H.; Beyeler, S.C. Challenges in the development and use of ecological indicators. Ecol. Indic. **2001**, *1* (1), 3–10.
41. Jorgersen, S.E.; Xu, F.-L.; Salas, F.; Marques, J.C. Application of indicators for the assessment of ecosystem health. In *Handbook of Ecological Indicators for Assessment of Ecosystem Health*; Jorgensen, S.E., Costanza, R., Xu, F.-L., Eds.; Lewis Publishers, Inc.: USA, 2005; 5–66.
42. Ruiz-Jaen, M.C.; Aide, T.M. Restoration success: How is it being measured? Restor. Ecol. **2005**, *13* (3), 569–577.
43. Bastin, G.N.; Ludwig, J.A.; Eager, R.W.; Chewings, V.H.; Liedloff, A.C. Indicators of landscape function: Comparing patchiness metrics using remotely sensed data from rangelands. Ecol. Indic. **2002**, *1* (4), 247–260.
44. Ludwig, J.A.; Bastin, G.N.; Chewings, V.H.; Eager, R.W.; Liedloff, A.C. Leakiness: A new index for monitoring the health of arid and semiarid landscapes using remotely sensed vegetation cover and elevation data. Ecol. Indic. **2007**, *7* (2), 442–454.
45. Mayor, A.G.; Bautista, S.; Small, E.E.; Dixon, M.; Bellot, J. Measurement of the connectivity of runoff source areas as determined by vegetation pattern and topography: A tool for assessing potential water and soil losses in drylands. Water Resour. Res. **2008**, *44* W10423.
46. Tongway, D.J.; Hindley, N. *Landscape Function Analysis: Procedures for Monitoring and Assessing Landscapes*; CSIRO Publishing: Brisbane, 2004.
47. Herrick, J.E.; van Zee, J.W.; Havstad, K.M.; Whitford, W.G. *Monitoring Manual for Grassland, Shrubland and Savanna Ecosystems.* USDA-ARS Jornada Experimental Range, Las Cruces, New Mexico. University of Arizona Press: Tucson, 2005.
48. Ludwig, J.A.; Tongway, D.J. Viewing rangelands as landscape systems. In *Rangeland Desertification*; Arnalds, O., Archer, S., Eds.; Kluwer Academic Publishers: Dordrecht, the Netherlands, 2000; 39–52.
49. Rietkerk, M.; Dekker, S.C.; de Ruiter, P.C.; van de Koppel, J. Self-organized patchiness and catastrophic shifts in ecosystems. Science **2004**, *305* (5692), 1926–1929.
50. Kéfi, S.; Rietkerk, M.; Alados, C.L.; Pueyo, Y.; Papanastasis, V.P.; El Aich, A.; de Ruiter, P.C. Spatial vegetation patterns and imminent desertification in Mediterranean arid ecosystems. Nature **2007**, *449* (7159), 213–217.
51. Carpenter, S.R.; Turner, M.G. Hares and tortoises: Interactions of fast and slow variables in ecosystems. Ecosystems **2000**, *3* (6), 495–497.
52. ICCD/COP8/CST. Report of the Fifth Meeting of the Group of Experts of the Committee on Science and Technology. Addendum: Methodologies for the Assessment of

Desertification at Global, Regional and Local Levels. ICCD/COP(8)/CST/2/Add.6. 2008.
53. Turner, B.L., II; Kasperson, R.D.; Matson, P.A.; McCarthy, J.J.; Corell, R.W.; Christensen, L.; Eckley, N.; Kasperson, J.X.; Luers, A.; Martello, M.L.; Polsky, C.; Pulsipher, A.; Schiller, A. A framework for vulnerability analysis in sustainability science. Proc. Natl. Acad. Sci. U. S. A. **2003**, *100* (14), 8074–8079.
54. Slocombe, D.S. Defining goals and criteria for ecosystem-based management. Environ. Manage. **1998**, *22* (4), 483–493.
55. Folke, C.; Hahn, T.; Olsson, P.; Norberg, J. Adaptive governance of social–ecological systems. Annu. Rev. Environ. Resour. **2005**, *30*, 441–473.
56. Curtin, R.; Prellezo, R. Understanding marine ecosystem based management: A literature review. Mar. Policy **2010**, *34* (5), 821–830.
57. Johannes, R.E. Integrating traditional ecological knowledge and management with environmental assessment. In *Traditional Ecological Knowledge: Concepts and Cases. International Program on Traditional Ecological Knowledge*; Inglis, J., Ed.; Canadian Museum of Nature: Ottawa, Canada, 1993.
58. Brush, S.B.; Stabinsky, D. *Valuing Local Knowledge: Indigenous People and Intellectual Property Rights*; Island Press: USA, 1996.
59. Emerson, K.; Orr, P.J.; Keys, D.L.; McKnight, K.M. Environmental conflict resolution: Evaluating performance outcomes and contributing factors. Conflict Resolut. Q. **2009**, *27* (1), 27–64.
60. Orr, P.J.; Emerson, K.; Keys, D.L. Environmental conflict resolution practice and performance: An evaluation framework. Conflict Resolut. Q. **2008**, *25* (3), 283–301.
61. Pahl-Wostl, C.; Hare, M. Processes of social learning in integrated resources management. J. Community Appl. Soc. **2004**, *14* (3), 193–206.
62. Schusler, T.M.; Decker, D.J.; Pfeffer, M.J. Social learning for collaborative natural resource management. Soc. Nat. Resour. **2003**, *16* (4), 309–326.
63. Steins, N.A.; Edwards, V.M. Platforms for collective action in multiple-use common-pool resources. Agric. Hum. Values **1999**, *16* (3), 241–255.
64. Rechkemmer, A. Societal impacts of desertification: Migration and environmental refugees. In *Facing Global Environmental Change: Environmental, Human, Energy, Food, Health and Water Security Concepts*; Brauch, H., Berghof-Stiftung, K., Eds.; Springer: Berlin, 2009.
65. Zdruli, P.; Trisorio Liuzzi, G., Eds. *Managing Natural Resources through Implementation of Sustainable Policies*, Proceedings of the MEDCOASTLAND Euro-Mediterranean Conference, Beirut, Lebanon, Jun. 25–30, 2006; IAM: Bari, 2007.
66. Zdruli, P.; Trisorio Liuzzi, G., Eds. *Determining an Income-Product Generating Approach for Soil Conservation Management*, Proceedings of the MEDCOASTLAND International Seminar, Marrakech, Morocco, Feb. 12–16, 2004; IAM: Bari, 2005.
67. Eswaran, H.; Berberoğlu, S.; Cangir, C.; Boyraz, D.; Zucca, C.; Özevren, E.; Yazıcı, E.; Zdruli, P.; Dingil, M.; Dönmez, C.; Akça, E.; Çelik, I.; Watanabe, T.; Koca, Y.K.; Montanarella, L.; Cherlet, M.; Kapur, S. The anthroscape approach in sustainable land use. In *Sustainable Land Management—Learning from the Past for the Future*; Kapur, S., Eswaran, H., Blum, W.E.H., Eds.; Springer-Verlag: Berlin Heidelberg, 2011.

Desertization

Henry Noel LeHouerou
Center for Functional and Evolutionary Ecology (CEFE), National Center for Scientific Research (CNRS), France

Abstract
Desertization is the irreversible extension of desert-like landscapes and landforms to areas where they did not occur in the recent past. Desertization results, to a large extent, from land abuse by humans and livestock; it is not a consequence of climate fluctuation, albeit drought may worsen and accelerate a phenomenon that could have been triggered during high rainfall periods. The causes of desertization are both direct and indirect. Combating desertization first requires the discontinuation, or at least the serious mitigation of previous destructive activities that brought the situation about. Biological recovery may be achieved via centuries-old techniques. But such activities often necessitate a deliberate policy and adequate strategies of conservative land use; this, in turn, may require political and legal actions encouraging responsible land-use management to achieve sustainability.

INTRODUCTION

Desertization is the irreversible extension of desert-like landscapes and landforms to areas where they did not occur in the recent past.[1,2] Desertization results, to a large extent, from land abuse by humans and livestock; it is not a consequence of climate fluctuation, albeit drought may worsen and accelerate a phenomenon that could have been triggered during high rainfall periods (there are many examples of such seemingly paradoxical situations). It is estimated[3–6] that over 20% of the 42 million km^2 of dry lands are submitted to a greater or lesser extent to desertization processes.

The causes of desertization are both direct and indirect. Direct causes[3,7] are those that have a strong immediate impact on site soil characteristics and properties. The major one is the reduction of perennial plant cover and biomass that contributes to soil organic matter depletion. This reduction of soil organic matter content triggers a chain of actions and reactions affecting both soil and microclimate, leading to lowered biological activity, to compacted soils with high runoff, and fast erosion processes, both by water and wind; soil may thus ultimately be turned sterile or almost so in a few years of abuse.

The indirect causes are those that bring about the depletion of natural resources, such as long-standing overcultivation, overstocking, poor farming practices, deforestation, destructive collection of fuel wood, and other mismanagement practices;[1,2] all these result from excessive and permanent pressure on the land by exceedingly numerous human and livestock populations: continuous high densities and stocking rates.

Combatting desertization first requires the discontinuation, or at least the serious mitigation of previous destructive activities that brought the situation about. Biological recovery may not be feasible any more whenever degradation is extreme; artificial human-induced intervention may then be deemed necessary. On the other hand, in less extreme cases, artificial rehabilitation and restoration may be desirable to hasten the desired goal. Biological recovery may be achieved via centuries-old techniques of soil and water conservation, water harvesting, runoff farming, agroforestry, appropriate farming, stocking, range, and forest management. But such activities often necessitate a deliberate policy and adequate strategies of conservative land use; this, in turn, may require political and legal actions encouraging responsible landuse management to achieve sustainability, i.e., a profound land reform. This has actually been achieved by various centuries-old peasant civilizations in various parts of the world arid lands.

DEFINITION

Deserts are defined as regions where rainfed agriculture is not feasible because of excessive aridity.[8] Unlike the word "desertification," which has at least four distinct meanings, the term "desertization" is clearly defined. It implies irreversible extension of desert land-forms and conditions to regions where they did not occur in the recent past.[1,2] The desertization is primarily due to destructive human and livestock activities in arid environments. In spite of the UNEP Plan to Combat Desertification (as a follow-up to the UNCOD Conference of Nairobi in 1977, the Earth Summit of Rio de Janeiro in 1992, and the Convention of Paris in 1994), little has been done on a large scale, so desertization is likely to continue over the next 50 years as it has in the past 50 years.[9]

FACTS AND FALLACIES

Origin of Deserts

Deserts are regions where, because of excessive aridity, no agriculture is feasible without supplemental irrigation.[8,10]

Encyclopedia of Environmental Management DOI: 10.1081/E-EEM-120001691

On the basis of their origin, there are five types of deserts.[10,11] True climatic deserts result from general atmospheric circulation whereby the adiabatic subsidence of dried-up air masses occurs on the limits between the Ferrel and Hadley cells along the 30°N and 30°S parallels.[11] The origin of true deserts varies with local conditions (e.g., tropical and subtropical deserts). Rain-shadow deserts (a particular case of climatic deserts) are formed due to orographic obstacles cutting the trajectory of rain-bearing frontal depressions (e.g., the Death Valley and Colorado River Delta east of the Sierra Nevada of California, Patagonia). Coastal deserts are formed due to the atmospheric stability resulting from the upwellings of coastal cold currents, such as along the Western Continental shores of America and Africa (e.g., the Humboldt current in Chile and Peru, the Benguela current of Southwest Africa). Edaphic deserts result from geological and soil conditions that are inappropriate for agricultural activities and the development of natural vegetation, such as unweathered hard or toxic rocks, salinity, and alkalinity, etc. Man-made deserts result from anthropgenic activities and destructive land abuse. The entry is concerned with the formation of the human-induced desert.

Fluctuations and Trends of Climatic Variables in Historical Times

Rainfall

There is a consensus among scientists on the absence of trends in long-term rainfall since the beginning of history.[3,12–14] There is, however, evidence of considerable changes in the course of the Pleistocene and Holocene periods beginning 3000 and 10,000 years BP, respectively, with some 10 alternating periods of dry and humid climates in the Sahara and Arabia.[9,12,15,16] This evidence comes essentially from tree-ring studies, pollen analysis, and lake-level fluctuations. Minor fluctuations took place, however, in the course of history such as the recent 25 years period of drought in the African Sahel (1968–1983) and East Africa (1975–1984),[15,17] in Chile from ca. 1900 to date, and in Argentina between 1900 and 1950.

Temperature

Variations in temperature are more complex to analyze. There is a consensus on a global increase of 0.5°C over the past 100 years.[13,14] But this includes the effect of urbanization that may globally account for up to 0.2°C.[18] Fluctuations during the historical period are fairly well documented. The causes of these "random" fluctuations are not known with any certainty. The magnitude of increase over the past 100 years is thus of the same dimension or smaller than previous random fluctuations.

Air Pollution

There is general agreement on the fact that the atmospheric content CO_2 and other pollutants is increasing. The CO_2 content of the atmosphere was 280–300 ppmv (parts per million in volume) at the beginning of the industrial era some 150 years ago.[19] It is now 370 ppm, essentially due to the burning of fossil fuel.[14] Similarly, the content of methane has doubled from 800 to 1750 ppbv (parts per billion in volume) over the same period.[14] The main sources of CH_4 are swamps, rice fields, ruminants, forest fires, garbage disposal, oil industries, and termites.

But the present situation is different in as much as it is the increasing content of CO_2 and CH_4 in the atmosphere that is causing a temperature rise, while in the past 160,000 years, and presumably all the Quaternary, the situation was the opposite. It is the increase in temperature from orbital forcing that resulted in a rise of CO_2 and CH_4, by oxidation of soil organic matter content, melting of tundra permafrost soils, and release of CH_4 from the thaw of frozen swamps. The present undisputed 0.5°C/100 years of global temperature increase thus cannot, strictly speaking, be attributed to CO_2 or other atmospheric pollutants.

Land Surface Albedo

As vegetation is increasingly depleted and perennial plant cover recedes, the proportion of bare ground increases, enhancing its reflectivity, or the surface albedo.[20] Higher albedo reduces temperature, decreases evaporation, and creates a thermic depression, impeding cloud formation and possibly decreasing rainfall. This so-called surface albedo feedback theory is, however, highly debated—actual facts have not, so far, proved it; rather the opposite.[21,22]

DESERTIZATION

Definition

Desertization is the irreversible degradation of arid land resulting in desert-like land-forms in areas where they did not occur in a recent past.[1,2] This definition differs from the concept of desertification as accepted by UNEP in the 1977 UNCOD Conference of Nairobi.[4,23] It was then defined as "land degradation under arid, semiarid and dry subhumid climates that may ultimately lead to desert-like conditions." The term "desertification" is presently used for two distinct meanings. It implies the *irreversible* degradation of arid lands leading to formation of man-made deserts. In western Europe and EEC, desertification is synonymous with land abandonment, resulting in natural reafforestation and expansion of woodlands.[24,25] Land abandonment may be referred to as "land desertion" because the rural population migrates to towns.[25]

Causes of Desertization: Degradation Processes

The causes of desertization are both direct and indirect.

Direct Causes

The direct causes pertain essentially to the destruction of the perennial plant cover and biomass and its consequences on soil characteristics through erosion, sedimentation, water logging, and salinization.[1,2,16,26–28] The destruction of perennial biomass and cover results in the following degradative chain reaction:

1. Reduced litter production and return to the soil particles decrease soil organic matter content.
2. Reduced soil organic matter weakens or destroys soil structure and aggregate stability.
3. Unstable soil aggregates increase surface soil sealing from raindrop splash on silty or loamy soils, and biological crusting on coarse sandy soils.[29]
4. Sealed soil surfaces increase runoff, decrease soil water intake and storage, and create a drier environment.
5. Increased runoff causes flooding, water logging, and mass die-off of shrubs, trees, and crops downstream and in closed depressions.
6. High wind speed on the soil surface, a consequence of reduced perennial plant cover, leads to higher albedo, increased potential evapotranspiration, increased water and wind erosion, and therefore drier microclimatic conditions.
7. Higher soil surface daily maximum temperatures and therefore high evapotranspiration, due to the lack of shading, enhances oxidation of soil organic matter, a rapid depletion of water reserves, and a shorter growing season. Lower soil-surface night temperature causes a larger diurnal and seasonal temperature range (thermal amplitude), leading to less favorable conditions for germination, emergence, and establishment of seedlings, and to the restoration of the natural vegetation.
8. Decreases in the number, biomass, and activity of soil microflora (e.g., bacteria, actynomycetes, fungi, algae, and rhizospheric symbionts) as a result of decreasing organic matter content in the soil slow down nutrient turnover and reduce availability of major and trace geobiogene elements, leading to lower fertility and productivity.
9. Decreased productivity reduces the number, biomass, and activity of soil microfauna, mesofauna, macrofauna, and megafauna.
10. Wind erosion (creeping, saltation, suspension, corrosion) results in the development of desert pavements (regs, serirs, hamadas, gobis) rock mushrooms, and yardangs.
11. Wind deposition leads to the formation of sand clay dunes (lunettes), loess, nebkas, ramlas, shamo, ergs, edeyen, nefuds, sand sheets, and sand veils. Gravity erosion leads to dry colluvium deposits.
12. Water erosion (sheet, rill, gully, pipe, landslides, etc.), leads to flooding, sedimentation, and alluvium deposition.
13. Rising water tables and salinity result from higher runoff and lesser pumping by shrubs and trees.
14. Rising salinity and soil sterilization are caused by poor drainage and other mismanagements.

Indirect Causes

The indirect causes of desertization are those that exacerbate the development of the direct causes.

1. Rapid human population growth and quick increase of population density lead to high human and livestock pressure on the land. The human population in the arid lands of the developing countries has grown at an annual rate of 3.2% since 1950.[2,3,16,30]
2. Livestock populations in Africa and Southwest Asia have increased at an average annual rate of 2.2% since 1950.[3,12,16]
3. Soil degradation is accentuated by overcultivation, inappropriate farming methods, poor land management (e.g., cultivation of land which is too dry, too steep, too shallow), inappropriate tillage (up and downhill plowing), quasi-exclusive utilization of disc-plows, inappropriate crop rotation patterns, inadequate field size, careless harvesting, and postharvesting land management, etc.
4. Deforestation is exacerbated by excessive collection of firewood (1.5 kg per person per day). As arid land steppic vegetation of dwarf shrubs bears some 300–600 kg dry woody matter per hectare, each person depending on this type of fuel destroys about 1 ha of steppic vegetation per annum.[8]
5. Overcollection of medicinal and other useful plants contribute to the decline of plant cover and biomass.
6. Wildfires are particularly harmful in the tropical savannas. As they occur in the dry season, they leave a soil surface that remains bare and therefore vulnerable to erosion during the next rainy season.[17,31]
7. Vegetation arcs are particular patterns of distribution whereby plants are concentrated on contour strips alternating with bare and sealed interstrips. Seen from above, as on air photos or satellite images, the land looks like a tiger skin, hence the expression of "tiger bush."[17,24,27] This is a case of edaphic aridity when, owing to the thinning of vegetation from woodcutting and/or overbrowsing, and the consecutive runoff and erosion, the soil moisture regime is too low for the vegetation. Thus, vegetation is confined in strips that permit higher soil moisture due to runon.
8. Inadequate legislation or lack of enforcement of legislation on land use and erosion control measures.

Table 1 Desertization risk assessment from perennial plant canopy cover evaluation.

	Perennial plant canopy projection cover in percent of ground surface					
MAR nmm	**0–1**	**1–5**	**5–15**	**15–25**	**25–50**	**>50**
50–100	5	5	4	3	2	1
100–200	5	4	3	2	1	0
200–300	4	3	2	1	0	0
300–400	3	2	1	0	0	0
400–500	2	1	0	0	0	0
500–600	1	0	0	0	0	0

0 — Present hazard nonexistent or very low.
1 — Immediate risk low.
2 — Immediate hazard moderate.
3 — Immediate risk serious.
4 — Immediate hazard severe.
Source: Le Houérou.[33]

9. Inappropriate land-ownership systems under which land, grazing, and water are free communal resources while livestock are privately owned. Thus, it is in each stockman's interest to draw the maximum from the common resource without investing anything in it, the "tragedy of the commons."
10. Inappropriate marketing facilities and inadequate incentive to de-stock and cull animals.
11. Poverty and lack of education and inefficient extension services.

Diagnosis and Evaluation of Desertization

Desertization may be evaluated via various criteria, namely, measuring soil erosion and sedimentation, assessing perennial plant canopy cover and biomass or the basal cover of perennial species, and evaluating the rain-use efficiency (RUE = Kg DM/ha/yr/mm) and computing the production to rain variability ratio (PRVR = CV Ann. Prod./CV Ann. rainf.).[32] The evaluation of RUE is a particularly easy, quick, and clean method.

Low perennial plant cover is the principal factor of soil surface erosion by water and wind. In the Mediterranean steppelands there is virtually no soil-surface erosion when perennial plant canopy cover reaches 25% of soil-surface. In such cases, erosion is compensated by deposition and sedimentation.[26] In the North African steppes, for instance, each 1% of perennial canopy cover corresponds with a biomass of 30 to 60 kg DM per ha.[33] It is then easy to determine the stage of degradation and desertization from evaluation of perennial plant canopy cover (Table 1). The world regional distribution of arid lands is shown in Table 2.

Severity and Extent of Desertization

Can Desertization Occur without the Impact of Drought?

True climatic deserts cannot exist without permanent and extreme climatic aridity. But such is not the case with the occurrence and spread of man-made deserts. Contracted vegetation occurs in all world deserts with a mean annual rainfall isohyets of between 50 and 150 mm depending on the substrate, the anthropozonic impact, and other local factors (e.g., high atmospheric humidity and occult precipitation, or, conversely, continentality, low RH, and high PET). Desertization is often caused by droughts, but the primary causes have often been operating long before

Table 2 Regional distribution of world arid lands.

	Continental distribution (10^3 km^2)						
Geographic area	**Eremitic**	**Hyperarid**	**Arid**	**Semiarid**	**Total**	**%**[a]	
MAR mm[b]	<50	50–100	100–400	400–600			
100 P/PET mm[c]	<3	3–5	5–28	28–45			
Budyko's Aridity Index[d]	>50	10–50	3–10	2–3			
World	130,737	7,500	7,059	14,330	12,651	41,440	100
Africa	30,312	6,232	3,017	3,570	2,951	15,770	52
N. America	21,322	10	90	1,025	1,935	3,060	14
S. America	17,818	275	105	972	1,274	2,626	14
Asia	43,770	1,595	3,225	5,415	4,817	15,042	34
Europe (Spain)	500	NA[e]	NA	100	300	400	80

[a]Percent of geographic area;
[b]MAR = mean annual rainfall;
[c]100 P/PET = precipitation to potential evapotranspiration ratio, ×100 for easier handling. PET being evaluated via Penman Equation;
[d]Budyko's Index is the ratio between net energy budget and actual annual rainfall, in other words it evaluates how many times the energy budget could evaporate the MAR (which does not take into consideration of the aerodynamic parameter of Penman's equation);
[e]NA = not applicable.
Source: Le Houérou.[3]

Table 3 Extent and severity of arid and semiarid land degradation.

	Degradation intensity (10^3 km^2 and %)									
	Light		Moderate		Strong		Very strong		Total	
Bioclimatic zones	S	%	S	%	S	%	S	%	S	%[a]
Arid and semiarid zones	3,601	42	3,964	45	1,096	12	62	1	8,723	20
Water erosion	1,475	37	1,757	45	666	17	40	1	3,938	9
Wind erosion	1,662	46	1,815	50	152	4	15	0	3,644	8
Chemical Degradation	373	44	266	31	203	24	7	1	849	2
Physical Degradation	91	31	126	43	75	26	0	0	292	1

[a]Percent of dry lands (SA + A + HA.Er).
Source: Oldeman[5] and Middleton[6]

drought occurred. The effects may have been hidden by a period of better than average climatic conditions. This was the case in the West African Sahel.

The analyses of available information lead to the following conclusions:

1. Desertization may be triggered during nondrought periods.
2. Desertization cannot result from drought alone.
3. The main causative factor appears to be land mismanagement from excessive anthropogenic pressure on the land (land abuse), beyond the ecosystems' resilience.

Severity and Extent

The global extent of land degradation has been assessed by ISRIC/UNEP.[4–6,23] The conclusions are in good agreement with other independent studies.[3] Tables 3, 4, and 5 show that overstocking and overgrazing are the major causes of arid land degradation worldwide, followed by overcultivation and poor farming methods. The regional and local causes, however, may considerably alter this pattern. Overcultivation and shrubland clearing, for instance, are by far the prime causes in northern Africa, southwestern Asia, and northwestern China, while overgrazing is the foremost cause in the Sahel, East Africa, North and South America, and Australia (Table 5). Salinization affects 1.5 million ha annually out of a world irrigated land area of 2 million km^2. The total area lost to secondary salinity over the last century is about 25 million ha.[3,32]

Natural Amelioration

Natural amelioration of land management to restore or rehabilitate ecosystems and environment implies the elimination or substantial reduction of the causes that leads to degradation. This goal can be achieved by reducing stocking rates to carrying capacity; restricting farming to appropriate areas, such as depressions with deep soils benefiting from runoff or having a water table; improving grazing systems through control of livestock, fencing, rotational/controlled grazing, and installing temporary exclosures. These activities are usually cheap and efficient tools for restoring and rehabilitating arid lands.[34–36]

Managed Restoration

Human-assisted rehabilitation includes a large array of techniques from the simplest and cheapest to the most sophisticated and costly but not necessarily efficient.

Surface Roughness

Techniques to create surface roughness break the surficial crust (e.g., a silt seal or a biological crust), increase permeability to air and rainwater, and facilitate seed germination and seedling emergence. A simple technique for breaking crust involves spreading branches from spiny trees and shrubs on the soil surface (*Acacia* spp., *Balanites* spp., *Ziziphus* spp., *Prosopis* spp., *Rhus* spp., *Parkinsonia* spp., *Commiphora* spp., etc.). Soil-surface roughening may also be achieved with the plow harrow, hoe, cultivator, subsoiler, ripper, basin-lister, etc.[17]

Table 4 Causes of degradation in dry lands (10^3 km^2).

	Deforestation	Overgrazing	Overcultivation	Other	Total degradation	Nondegradation	Total dry lands
S	1,700	4490	2,450	83	8,723	34,832	43,554
%	20	51	28	1	100/20	80	100

Source: Le Houérou,[3] Le Houérou,[12] UNEP,[23] and Le Houérou.[38]

Table 5 Causes of desertization, % of desertized land.

Regions	Overcultivation	Overstocking	Fuelwood collection	Salinization	Urbanization	Other	Total
N.W. China	45	18	18	2	3	14	100
N. Africa and S.W. Asia	50	26	21	2	1	—	100
Sahel and E. Africa	25	65	10	—	—	—	100
Middle Asia	10	62	—	9	10	9	100
Australia	20	75	—	5	?	—	100
U.S.A.	22	73	—	5	1	—	100

Source: Le Houérou[7] and Le Houérou.[30]

Contour lining

This may be achieved by using stones or pieces of hardpan or duricrusts (e.g., gypsic pans, calcrete, iron pans, or lateritic pans). Light contour lining aims at reducing sheet erosion and improving water intake. It may also be achieved via mechanical equipment used for terracing and building contour banks, contour benches, or ditches.

Ripping and Subsoiling

These techniques use heavy equipment to break deep, hard lime, gypsum, silicium (silcrete), or iron pans or to loosen hard compacted soils to improve water intake and storage and deepen the root zone of protective plants, particularly shrubs and trees.

Termite-Assisted Hand-Pitting Across a Surficial Pan

The southern Sahelian technique of the "Zai" is a cheap and efficient practice. It involves digging pits (40 L × 40 W × 15 D) about 80 cm apart at a density of 2,500 pits/ha. These pits are dug a few months prior to the onset of the rainy season, and 1,000–1,500 kg/ha of organic matter (e.g., litter, straw, stovers, etc.,) is placed at the bottom of these pits. The organic matter attracts termites that dig deep galleries below the pan. The micro-environment thus created is then manured at a rate of 1,000 kg/ha of dry corral dung. Pearl millet or sorghum planted in these pits may reach maturity even in a dry year, owing to the better soil water regime. Each pit is a green island amid a devastated environment landscape, and the beneficial effect may last up to 30 years. The cost of making Zai is 150–300 human-days per hectare for a pit density of 2,500/ha, 1,000 kg of organic matter, and 1,000 kg of corral manure.[37]

Range Reseeding

Range reseeding and establishing improved pasture are also feasible in the semiarid zone, but not so in the arid zone.[7,28]

Agroforestry

Agroforestry is practiced on 3.2 million hectares in world arid lands. Agroforestry, using either native or exotic species, is a very potent and efficient tool for biological recovery, allowing productivity 3 to 10 times higher than pristine vegetation under the same ecological conditions. These lands are distributed as shown in Table 6.[35,36]

Wildlife Management, Wildlife Husbandry, Tourism

Wildlife management and husbandry are actively pursued in an increasing number of arid land countries: U.S.A., Australia, East Africa, Southern Africa, New Zealand, Argentina and Western Europe. Privately owned game ranches or game farms are quickly developing in Africa.

Adaptive Arid Land Strategy and Sustainable Development Policies

Mitigating desertization is a problem of sustainable arid land development. Sustainable development policies include

Table 6 World areas under agroforestry management.

Fodder cacti	1,200,000 ha	N. Africa, S. Africa, N. and S. America
Saltbushes	1,000,000 ha	N. Africa, S. Africa, S.W. Asia, N. and S. America, Australia
Acacias spp	600,000 ha	N. Africa, Sahel, S. Africa, Australia
Prosopis spp	50,000 ha	N. and S. America, India
Agave americana	100,000 ha	S. Africa
Miscellaneous (Saxaouls, Tamarix, Bohemia olive, poplars)	250,000 ha	Africa, Australia, N. and S. America, Europe, Central Asia

many facets (e.g., technical, social, economic, political, legislative, organizational, and educational). An important step toward sustainable development is reforming the land tenure system. This would require a fundamental land reform in order to shift from communal or tribal land ownership to secure private systems whereby long-term investment becomes possible and long-term planning can be implemented.

REFERENCES

1. Le Houérou, H.N. La Désertisation Du Sahara Septentrional Et Des Steppes Limitrophes. Ann. Algér. de Géogr. **1968**, *6*, 2–27.
2. Le Houérou, H.N. The nature and causes of desertization. Arid Lands Newsletter **1976**, *3*, 1–7.
3. Le Houérou, H.N. An overview of vegetation and land degradation in world arid lands. In *Degradation and Restoration of Arid Lands*; Dregne, H.E., Ed.; Texas Technical University: Lubbock, 1992; 127–163.
4. UNCOD, *United Nations Conference on Desertification*; UNEP, Nairobi and Pergamon Press: New York, 1977.
5. Oldeman, L.R.; Hakkeling, R.T.A.; Sombroek, W.G. World Map of Status of Human-Induced Soil Degradation An Explanatory Note, ISRIC, Wageningen and UNEP: Nairobi, 1990.
6. Middleton, N.; Thomas, D. *World Atlas of Desertification*, 2nd Ed.; Edward Arnold, Ed.; London and UNEP: Nairobi, 1997.
7. Le Houérou, H.N. Drought-tolerant and Water-Efficient trees and shrubs for the rehabilitation of tropical and subtropical arid lands. Journal of Land Husbandry **1996**, *1* (1–2), 43–64.
8. Meigs, P. *World Distribution of Arid and Semiarid Homoclimates*; 5 Maps. Reviews of Research on Arid Zone Hydrology; UNESCO: Paris, 1952.
9. Le Houérou, H.N. *Climate Change Drought and Desertification*; Report to IPCC; Washington, 1995.
10. McGinnies, W.G.; Goldman, B.G.; Paylore, P. *Deserts of the World: An Appraisal of Research into Physical and Biological Environments*; University of Arizona Press: Tuscon, 1968.
11. Monod, Th. *Les Déserts*; Horizons de France, Publ.: Paris, 1973.
12. Le Houérou, H.N. Vegetation and land-use in the mediterranean basin by the Year 2050: a prospective study. In *Climatic Change and the Mediterranean*; Jeftic, L., Milliman, J.D., Sestini, G., Eds.; Edward Arnold: London, 1992; 175–232.
13. Houghton, J.T.; Callander, B.A.; Varney, S.K. Climate Change, 1992: *The Supplementary Report to IPCC Scientific Assessment*; Cambridge University Press: London, 1992.
14. Houghton, J.T.; Jenkins, G.J.; Ephraums, J.J. Climate Change: *The IPCC Scientific Assessment*; Cambridge University Press: London, 1990.
15. Le Houérou, H.N. Changements Climatiques Et désertisation. Sècheresse **1993**, *4*, 95–111.
16. Le Houérou, H.N., La Méditerranée En 2050: Impacts Respectifs D'une éventuelle évolution Climatique et de la demographie Sur La Végétation, Les écosystèmes Et L'utilisation Des Terres. Etude Prospective. *La Météorologie*, VII Série, No. 36: 1991; 4–37.
17. Le Houérou, H.N. *The Grazing Land Ecosystems of the African Sahel*; Ecological Studies n. 75; Springer Verlag: Heidelberg, 1989.
18. Duplessy, J-C. Quand l'Ocean Se fâche. In *Histoire Naturelle du Climat*; Odile, Jacob, Ed.; Paris, 1996.
19. Keeling, C.D.; Bacastow, R.B.; Bainbridge, A.E.; Ekdahl, C.A.; Guenther, P.R.; Waterman, L.S. Carbon Dioxide variation at mauna loa observatory Hawaii. Tellus **1976**, *28*, 28.
20. Charney, J. Dynamics of deserts and drought in the sahel. Quat. J. Roy. Meteor. Soc. **1975**, *101*, 193–202.
21. Williams, M.A.J.; Balling, R.C., Jr. *Interactions of Desertification and Climate WMO*; Geneva and UNEP: Nairobi, 1994.
22. Courel, M.F. *Etude de L'évolution Recente Des Milieux Sahéliens à Partir Des Measures Fournies Par Les Satellites*; Thése Doct., Univ.: Paris I, 1985.
23. UNEP, World Atlas of Desertification. UNEP, Nairobi and Edward Arnold, London, 1992.
24. Le Houérou, H.N. Impact of Man and his Animals on Mediterranean Vegetaion. In *Mediterranean-Type Shrublands*; Di Castri, F., Goodall, D.W., Specht, R.L., Eds.; Ecosystems of the World, Elsevier: Amsterdam, 1981; Vol. 11, 479–521.
25. Le Houerous, H.N. Land Degradation in Mediterranean Europe: Can Agroforestry be a part of the Solution? A Prospective Review. Agroforestry Systems **1993**, *21*, 43–61.
26. Le Houérou, H.N. Recherches écologiques et Floristiques Sur La Végétation De La Tunisie Méridionale. Mém. H.S no. 6., Instit. Rech. Sahar. Univ. d'Alger, 1959.
27. Le Houérou, H.N. The Sahara from the bioclimatic view point. Definition and Limits. Ann. of Arid Zones **1995**, *34* (1), 1–16.
28. Le Houérou, H.N. Bioclimatologie Et Biogéographie Des Steppes Arides Du Nord De l'Afrique. In *Options-Mediterranéennes*; Sér. B (10): 1–396, CIHEAM: Paris, 1995.
29. Verrecchia, E.; Yaïr, A.; Kidron, G.J.; Verrecchia, K. Physical properties of the psammophile cryptogarnic crust and their consequences to the water regime of sandy soils, North-Western Negev Desert, Israel. J. Arid Environments **1995**, *29* (4), 427–438.
30. Le Houérou, H.N. Dégradation, Régénération Et Mise En Valeur Des Terres Séches D'Afrique. In *L'homme Peut-il Refaire ce Qu'il a Défait?*; Pontanier, R., M'Hiri, A., Aronson, J., Akrimi, N., Le Floc'h, E., Eds.; John Libbey Eurotext: Montrouge, 1995; 65–104.
31. Goldammer, J.G., Ed.; *Fire in Tropical Biota*; Ecological studies no. 84; Spring Verlag: Heidelberg, 1990.
32. Le Houérou, H.N. A Probabilistic approach to assessing arid rangelands productivity, carrying capacity and stocking rates (Ch. 12). In *Drylands: Sustainable Use of Rangelands into the 21st Century*; Squires, V.R., Sidahmed, A., Eds.; IFAD: Rome, 1998; 159–172.
33. Le Houérou, H.N., Aspects méteorologiques De la Croissance et du Dévelopment Végétal Dans Les Déserts et Les Zones Menacées De Désertisation. UNEP, Nairobi and WMO, Report no WMO/TD-No 194, Geneva, 1987.
34. Le Houérou, H.N. The Role of cacti (*Opuntia* Spp.) in land rehabilitation in the mediterranean basin. J. of Arid Environments **1996**, *32*, 1–25.

35. Le Houérou, H.N. Utilization of fodder trees and shrubs for the arid and semiarid zones of West Asia and North Africa. Arid Soil Research and Rehabilitation **2000**, *14*, 101–135.
36. Le Houérou, H.N. Restoration and rehabilitation of arid and semiarid Mediterranean ecosystems in North Africa and West Asia. A Review Arid Soil Research and Rehabilitation **2000**, *14*, 3–14.
37. Roose, E., Introduction à la Gestion Conservatoire de L'eau, de la Biomasse et de la fertilité Des Sols (GCES). Bulletin Pédologique no 70, FAO, Rome, 1994.
38. Le Houérou, H.N. Man-made deserts: desertization process and threats. Arid Land Research and Management **2000**, *16*, 1–36.

Developing Countries: Pesticide Health Impacts

Aiwerasia V.F. Ngowi
Tanzania Association of Public Occupational and Environmental Health Experts, and Department of Environmental and Occupational Health, Muhimbili University of Health and Allied Sciences (MUHAS), Dar-es-Salaam, Tanzania

Catharina Wesseling
Central American Institute for Studies on Toxic Substances (IRET), National University, Heredia, Costa Rica

Leslie London
Occupational and Environmental Health Research Unit, University of Cape Town, Observatory, South Africa

Abstract

The use of potentially hazardous chemicals is increasing in developing countries whose populations have the least capacity to protect themselves. The World Health Organization estimated three million cases of acute pesticide poisoning and 220,000 deaths annually worldwide a decade ago, the majority in developing countries with relatively few studies of long-term health effects of pesticide exposure among working populations in developing countries. Underestimations of acute and long-term effects of pesticide in developing countries occur due to under-diagnosis and/or underreporting. As a critical public health tool for the control of pesticide poisoning, surveillance in developing countries is bedeviled by multiple problems such as lack of access to health care for poisoning survivors, lack of human resources, diagnostic skills and equipment to identify cases, and weak information systems. The lack of human and technical resources in developing countries is aggravated by the global brain drain and weak economies. Vulnerable economies and weak infrastructure in developing nations hinder their ability to regulate the use of pesticides, particularly when macroeconomic pressures promote deregulation and restrict public spending required to implement regulatory controls. Pesticides banned or restricted in developed countries are often easily available in developing countries. Agricultural policies in many of these countries have emphasized short-term economic gains at the expense of environmental sustainability or human health. Farmers and farm workers rarely have access to adequate training in pesticide safety or advice on the complicated management of pesticides. This entry explores the impact of pesticides on health in developing countries.

INTRODUCTION

The use of potentially hazardous chemicals is increasing in developing countries whose populations have the least capacity to protect themselves. Hundreds of thousands of people die annually from the effects of use, misuse, or accidental exposures to pesticides.[1,2] Developing nations in Africa, Asia, and Latin America comprise more than 75% of the total world population, use 25% of the world's pesticides, yet account for 99% of deaths caused by these toxic agents.[3]

HEALTH IMPACTS

Acute Pesticide Poisoning

Two decades ago, the World Health Organization estimated that three million cases of acute pesticide poisoning resulting in 220,000 deaths occur worldwide each year, the majority in developing countries.[3] However, it is well recognized that these figures are an underestimate because of underdiagnosis and/or underreporting. Diagnostic difficulties are prominent in developing countries,[4–6] owing to insufficient medical training and high background levels of ill health.

Organophosphorus insecticides are the most common agents involved in acute pesticide poisonings, accounting for between 50% and 80% of all poisonings in Asia[7] and are a major public health concern in most African countries, where approximately 80% of the workforce is involved in agricultural work. In Central America, pesticides identified as causing most poisonings between 1992 and 2000 were paraquat; aluminum phosphide; the organophosphates methyl parathion, methamidophos, monocrotophos, chlorpyrifos, terbufos, and ethoprophos; the carbamates carbofuran, methomyl and aldicarb; and endosulfan.[8]

Part of the reason for this picture is the continued use in developing countries of pesticides no longer registered for use in the developed world, because of their high

Desertification – Ecosystems

Encyclopedia of Environmental Management DOI: 10.1081/E-EEM-120046235

toxicity, and the substitution of persistent organochlorines with organophosphate insecticides.

Fatality rates and lifelong disability resulting from pesticide poisoning in developing countries are exacerbated by poor diagnosis and delayed treatment, resulting in both human suffering and economic losses.

High rates of unintentional poisoning, mostly occupational, have been reported in rural agricultural working and urban populations worldwide.[9,10] Nearly 66,000 cases of acute pesticide poisonings occur annually in Nicaragua.[11] Mass poisonings by pesticides in developing countries have typically resulted in high numbers of fatalities.

In the remote Andean village of Tauccamarca in October 1999, 42 children were poisoned after eating a school breakfast contaminated with the organophosphate pesticide methyl parathion, resulting in 24 deaths before the children could reach medical treatment.[12]

However, it is only a limited number of the most extreme cases in developing countries, which appear to be documented. Less high-profile cases are common but unrecorded. For example, a methomyl-poisoning incident involving 11 female flower farm workers in Arusha, Tanzania, in March 2004 was reported in the press, but absence of adequate local investigation mechanisms prevented its documentation in the peer-reviewed literature.

Deliberate self-harm is a major problem in the developing world. Pesticides are commonly used as suicidal agents throughout developing nations and are associated with high mortality rates, causing an estimated 300,000 deaths annually in the Asia Pacific.[1,2,8,9,13] In India, suicide using aluminum phosphide was reported as so common that postmortem examinations on deceased bodies were said to be routinely conducted by staff wearing respirators for personal protection from released fumes.[14]

Underlying factors that make individuals at risk for self-harm are both social (including domestic problems, poverty, social isolation, and financial hardship) and medical.[1] Farmer indebtedness, widespread in many developing countries characterized by unequal economic systems, is an important factor driving high rates of suicide. More recent findings suggest that pesticides, particularly organophosphates, may be more than agents in suicidal attempts; they are also part of the causal pathway because of their neurotoxicity and the possible links between organophosphate exposure, depression, and impulsivity, mediated through effects on neurotransmitters such as serotonin.[9] In a context where the above social risk factors for depression are common in developing countries, further exposure to neurotoxic pesticides may substantially increase the risks of suicide.

Chronic Health Impacts Unknown

Although long-term consequences of pesticide poisoning are well recognized in the literature, relatively few studies of long-term health effects of pesticide exposure have been conducted among working populations in developing countries. Underdiagnosis is accentuated for long-term health consequences that require greater diagnostic capacity. Dermal exposure routes for developing country workers are also common but are an underdocumented yet critical pathway for systemic poisonings, both acute and chronic. Consequently, the extent of chronic health impacts of pesticides in developing country workers is poorly characterized.

However, there is little reason to believe that their impact would be any less than that in developed countries. Indeed, high levels of background morbidity and poor social conditions are likely to aggravate pesticide toxicity. For example, research among South African farm workers highlighted the link between chronic lifetime undernutrition, organophosphate exposure, and impaired neurological performance on tests of vibration threshold.[15]

Azoospermia (absent sperm), oligospermia (low sperm count in semen), and low fertility have been documented in more than 26,000 workers previously exposed to 1,2-dibromo-3-chloropropane (DBCP) on banana and pineapple plantations in more than 12 countries.[16]

WEAK SURVEILLANCE FOR HAZARDS AND IMPACT

Although a critical public health tool for the control of pesticide poisoning, surveillance in developing countries is bedeviled by multiple problems such as lack of access to health care for poisoning survivors, lack of human resources, diagnostic skills and equipment to identify cases, and weak information systems. Acute poisoning rates are consequently underestimated and may selectively undercount certain types of poisoning (occupational circumstances) and certain risk groups (women and migrant workers). Lack of professional competence and conflict of interest arising from compensation system levies may also lead occupational poisonings to be misreported as suicide.[10] As a result, inferences from review of flawed data may lead to mistaken policy decisions.[17]

To improve information on the extent of pesticide poisoning in developing countries, surveillance systems for acute health effects from pesticides are being established in developing nations. In 1998, almost 6000 pesticide poisonings were reported in five of the seven Central American countries generating an estimate (corrected for underreporting) of 30,000 pesticide poisonings annually in the region.[11] Poisoning rates reported in an intensified surveillance intervention in South Africa increased 10-fold in the study area compared to a control area.[17] Recently, WHO in collaboration with partners initiated a community intervention, the Global Public Health Initiative, to prevent self-harm by pesticide poisoning.[18]

WEAK REGULATION AND ENFORCEMENT

Vulnerable economies and weak infrastructure in developing nations hinder their ability to regulate the use of pesticides, particularly when macroeconomic pressures promote deregulation and restrict public spending required to implement regulatory controls. As a result, marketing and advertising of pesticides are often uncontrolled. Incorrectly labeled or unlabeled formulations, including ready-made solutions in soft drink bottles and other containers, are commonly sold at open stands. In South Africa, the repackaging of aldicarb granules into small-volume packets sold by street vendors[19] for domestic pest control has been linked to increasing numbers of poisonings in urban areas. Low retail prices, sometimes associated with subsidy policies, promote risky pesticide use. Weaknesses in sustainable international and national agricultural and chemical management policies manifest in a reliance on "safe-use" strategies. Yet, evidence has shown that the so-called "good agricultural practices" and "safe use" are ineffective in controlling risks in developing countries, principally because many measures assumed to enable safe use are not feasible in developing countries, particularly under tropical or adverse climatic conditions.[20]

LOW LEVELS OF WORKER AND COMMUNITY AWARENESS

Farmers and farm workers rarely have access to adequate training in pesticide safety or advice on the complicated management of pesticides. Hot climates are a disincentive to use of protective clothing, and many workers and farmers lack access to water for washing hands or exposed skin, increasing the risks of contamination. Recognition of pests and their predators is generally low, leading to over-reliance on routine pesticide applications to control pests; knowledge of product selection, application rates, and timing is poor; different products are often combined in the belief that the effect will be greater; re-entry periods after spraying are not known; and without knowledge of alternatives, farmers often assume that the only solution to pest problems is to spray more frequently.[21]

Pesticides are often stored improperly in or around farmers' homes, increasing family member's access.[21,22] In some instances, empty pesticide containers are reused to store water and food, resulting in serious poisonings.

IMPORT/EXPORT OF BANNED AND RESTRICTED COMPOUNDS

Pesticides banned or restricted in developed countries are often easily available in developing countries. These include pesticides causing significant acute and chronic morbidity (such as class I and II organophosphates and paraquat) and organochlorines earmarked for eradication under the Stockholm Convention on persistent organic pollutants (POPs) (particularly dieldrin, lindane, and chlordane). Endosulfan, a candidate pesticide for inclusion under the POPs treaty, has been responsible for a series of poisonings in Benin[21] and developmental impacts on children in Kerala, a state in India.[21,23]

The use of *p-p′*-dichlorodiphenyltrichloroethane (DDT) continues to be permitted for malaria control in developing countries, where malaria remains endemic, despite its known hazards for wildlife and controversial adverse effects on human health.[24,25] Ironically, DDT use in Africa has increased since the Stockholm Convention came into effect.[26] As a result, it is still produced for export in at least three countries. Because of its ongoing usage for public health vector control, unauthorized use for agricultural purposes remains a concern in developing countries, particularly where regulatory controls are weak. The presence and persistence of DDT and its metabolites worldwide are still problems of great global relevance to public health.

Although the Prior Informed Consent (PIC) procedure, upgraded in status from a voluntary agreement to an international convention known as the Rotterdam Convention, seeks to protect developing countries from harms arising from import of chemicals banned or restricted in exporting countries, the effectiveness of the Convention has been questioned. For example, the data requirement for a Severely Hazardous Pesticide Incident report, used under the Convention to add a pesticide onto the controlled list, lacks mechanisms suited to developing country conditions. This is usually because developing countries lack the infrastructure to collect the required data. Similarly, the process of adding a pesticide onto the POPs list (e.g., endosulfan) is often met in practice with strong resistance. Even when pesticides are restricted by the POPs and Rotterdam conventions, compliance with the obligations contained in the conventions may often be poor, despite developed countries ratifying the conventions. The PIC and POP secretariat could assist local contacts such as the Designated National Authorities (DNAs) in developing simple tools that will be used to collect relevant data and help in the establishment of suitable mechanisms for the flow of data and information in member countries.

Over the past few years, pressure from non-governmental organizations and discussions within the Food and Agriculture Organization and other intergovernmental bodies has recognized that greater effort must be put into restricting the availability of pesticides based not only on their acute toxicity, traditionally measured through the WHO classification system, but also on the capacity of particular pesticides to cause long-term toxic effects with chronic exposure. These initiatives, particularly the call for a progressive ban on highly hazardous pesticides, offer some hope for better protections for developing country populations, but are still in development.

LACK OF TECHNICAL AND LABORATORY CAPACITY

Many developing countries suffer from a lack of human and technical resources, aggravated by the global brain drain and weak economies. As a result, few developing countries are able to monitor pesticide residues. Most developing countries do not have laboratories capable of conducting analyses for pesticides and their residues, particularly at standards that meet good laboratory practice. Where laboratory capacity is available, it is usually to service residue testing of agricultural exports destined for consumers in developed countries. Produce grown for domestic consumption is rarely monitored.

Environmental media such as water and soil are rarely tested, and, even then, usually only on a research basis. Isolated studies of lactating women in Southern Africa have confirmed the presence of high levels of DDT metabolites in breast milk in populations living in malaria-endemic areas subject to DDT applications. Yet, despite provisions arising from the POPs treaty to undertake routine testing to monitor the impact of DDT use, there is no system for biological monitoring for DDT metabolites in place in Southern Africa. As a result, many infants in the region are substantially exposed through cross-placental transfer and breastfeeding, with potential adverse impacts on childhood neurodevelopment.

Research capacity to identify problems and develop prevention strategies is also constrained by limited investments in capacity building in relevant scientific fields in developing countries. As a result, there is neither proactive monitoring nor information systems usage to effect adequate responses to pesticide problems identified. It is critical to foster South–South learning to promote best practice because what applies in the north may be different to what happens in the south. There is a need to build southern capacity because reliance on the north at times perpetuates many of the problems. Indeed, there is much expertise in the south from which both north and south can learn.

PEST CONTROL POLICIES

Unlike many developed countries, agricultural policies in many developing countries have emphasized short-term economic gains at the expense of environmental sustainability or human health. Few developing countries have adopted integrated pest management or pest reduction strategies. The dominant "pesticide culture" assumes that the use of pesticides to control pest as the first option is the norm, is reinforced by advertising and marketing practices, and is often encouraged by agricultural credit policies and development aid. Much needs to be done to enhance research and development to support pesticide reduction for agriculture and public health, and to strengthen the capacity in developing countries to develop monitoring systems and research capacity to deal with the problems of pesticides in developing nations. Reducing deaths from pesticide poisoning through restrictions on the availability of pesticides can be accomplished based on a prior evaluation of national agricultural needs and the development of a plan to encourage substitution with less toxic pesticides without loss of agricultural output,[27] bearing in mind that policies aiming towards sustainable chemical-free agriculture would be the ideal long-term solution.

CONCLUSIONS

Underestimations of acute and long-term effects of pesticide in developing countries occur due to underdiagnosis and/or underreporting. The impact of pesticide poisoning is also unknown because of weak surveillance for hazards and impact, import/export of banned or restricted compounds, lack of technical and laboratory capacity, weak regulations and enforcement, low level of worker and community awareness, and inappropriate pest control policies. Enhancing research and development to support pesticide reduction for agriculture and public health and strengthening capacity to develop monitoring systems are critically important for developing countries to deal with problems concerning pesticides.

REFERENCES

1. Konradsen, F.; van der Hoek, W.; Cole, D.C.; Hutchinson, G.; Daisley, H.; Singh, S.; Eddleston, M. Reducing acute poisoning in developing countries—Options for restricting the availability of pesticides. Toxicology **2003**, *192* (2–3), 249–261.
2. Konradsen, F.; Dawson, A.H.; Eddleston, M.; Gunnell, D. Pesticide self-poisoning: Thinking outside the box. Lancet **2007**, *369* (9557), 169–170.
3. WHO. *Public Health Impact of Pesticide Used in Agriculture*; World Health Organization and United Nations Environment Programme: Geneva, 1990.
4. Corriols, M.; Marín, J.; Berroteran, J.; Lozano, L.M.; Lundberg, I.; Thörn, A. The Nicaraguan Pesticide Poisoning Register: Constant underreporting. Int. J. Health Serv. **2008**, *38* (4), 773–87.
5. Mbakaya, C.F.L.; Ohayo-Mitoko, G.J.A.; Ngowi, A.V.F.; Mbabazi, R.; Simwa, J.M.; Maeda, D.N.; Stephens, J.; Hakuza, H. The status of pesticide usage in East Africa. Afr. J. Health Sci. **1994**, *1*, 37–41.
6. London, L.; Myers, J.E. Critical issues for agrichemical safety in South Africa. Am. J. Ind. Med. **1995**, *27*, 1–14.
7. He, F.; Xu, H.; Quin, F. Intermediate myasthenia syndrome following acute organophosphates poisoning—An analysis of 21 cases. Hum. Exp. Toxicol. **1998**, *17*, 40–45.

8. Henao, S.; Arbelaez, M.P. Epidemiologic situation of acute pesticide poisoning in Central America, 1992–2000. Epidemiol. Bull. **2002**, *23* (3), 5–9.
9. London, L.; Flisher, A.J.; Wesseling, C.; Mergler, D.; Kromhout, H. Suicide and exposure to organophosphate insecticides: Cause or effect. Am. J. Ind. Med. **2005**, *47*, 308–321.
10. Wesseling, C. Multiple health problems in Latin America. In *Silent Invaders: Pesticides, Livelihoods and Women's Health*; Jacobs, M., Dinham, B., Eds.; ZED Books: London, 2003.
11. Corriols, M.; Marin, J.; Berroteran, J.; Lozano, L.M.; Lundberg, I. Incidence of acute pesticide poisonings in Nicaragua: A public health concern. Occup. Environ. Med. **2009**, *66*, 205–210.
12. Rosenthal, E. The tragedy of Tauccamarca: A human rights perspective on the pesticide poisoning deaths of 24 children in the Peruvian Andes. Int. J. Occup. Environ. Health **2003**, *9*, 53–58.
13. Chowdhury, A.N.; Banerjee, S.; Brahma, A.; Biswas, M.K. Pesticide poisoning in non-fatal deliberate self-harm: A public health issue. Study from Sundarban delta, India. J. Psychiatry **2007**, *49*, 262–266.
14. Levine, R.S.; Doull, J. Global estimates of acute pesticide morbidity and mortality. Rev. Environ. Contam. Toxicol. **1992**, *129*, 29–50.
15. London, L. Occupational epidemiology in agriculture: A case study in the Southern African context. Int. J. Environ. Occup. Health **1998**, *4*, 245–256.
16. Slutsky, M.; Levin, J.L.; Levy, B.S. Azoospermia and oligospermia among a large cohort of DBCP applicators in 12 countries. Int. J. Occup. Environ. Health **1999**, *5* (2), 116–122.
17. London, L.; Bailie, R. Challenges for improving surveillance for pesticide poisoning: Policy implications for developing countries. Int. J. Epidemiol. **2001**, *30* (3), 564–570.
18. WHO. *Safer Access to Pesticides: Community Interventions*; World Health Organization and International Association for Suicide Prevention: Geneva, 2006.
19. Rother, A. Falling through the regulatory cracks: Street selling of pesticides and urban youth in South Africa. Int. J. Occup. Environ. Health **2010**, *16*, 202–213.
20. Wesseling, C.; Ruepert, C.; Chavarri, F. Safe use of pesticides: A developing country's point of view. In *Encyclopedia of Pest Management*; Marcel Dekker, Inc.: New York, 2003.
21. Ngowi, A.V.F.; Maeda, D.N.; Wesseling, C.; Partanen, T.J.; Sanga, M.P.; Mbise, G. Pesticide handling practices in agriculture in Tanzania: Observational data on 27 coffee and cotton farms. Int. J. Occup. Environ. Health **2001**, *7*, 326–332.
22. Dinham, B.; Malik, S. Pesticides and human rights. Int. J. Occup. Environ. Health **2003**, *9*, 40–52.
23. Saiyed, H.; Dewan, A.; Bhatnagar, V.; Shenoy, U.; Shenoy, R.; Rajmohan, H.; Patel, K.; Kashyap, R.; Kulkarni, P.; Rajan, B.; Lakkad, B. Effect of endosulfan on male reproductive development. Environ. Health Perspect. **2003**, *111*, 1958–1962.
24. Bouwman, H.; Becker, P.J.; Schutte, C.H.J. Malaria control and longitudinal changes in levels of DDT and its metabolites in human serum from KwaZulu. Bull. W. H. O. **1994**, *72* (6), 921–930.
25. Eskenazi, B.; Chevrier, J.; Rosas, L.G.; Anderson, H.A.; Bornman, M.S.; Bouwman, H.; Chen, A.; Cohn, B.A.; de Jager, C.; Henshel, D.S.; Leipzig, F.; Leipzig, J.S.; Lorenz, E.C.; Snedeker, S.M.; Stapleton, D. The Pine River statement: Human health consequences of DDT use. Environ. Health Perspect. **2009**, *117*, 1359–1367.
26. van den Berg, H. Global status of DDT and its alternatives for use in vector control to prevent disease. Environ. Health Perspect. **2009**, *117*, 1656–1663.
27. Manuweera, G.; Eddleston, M.; Egodage, S.; Buckley, N.A. Do targeted bans of insecticides to prevent deaths from self-poisoning result in reduced agricultural output? Environ. Health Perspect. **2008**, *116* (4), 492–495.

Distributed Generation: Combined Heat and Power

Barney L. Capehart
Department of Industrial and Systems Engineering, University of Florida, Gainesville, Florida, U.S.A

D. Paul Mehta
Department of Mechanical Engineering, Bradley University, Peoria, Illinois, U.S.A.

Wayne C. Turner
Industrial Engineering and Management, Oklahoma State University, Stillwater, Oklahoma, U.S.A.

Abstract
Distributed generation (DG) is electric or shaft power generation at or near the site of use as opposed to central power station generation. Combined heat and power (CHP) takes advantage of this site location to recover the normally wasted thermal energy from power generation and utilizes it beneficially to increase the total system efficiency. This entry explores the rapidly developing world of DG and associated CHP. First the entry shows why DG is necessary in the U.S. power future and that DG is going to happen. Then, the entry briefly looks at the different technologies that might be employed and their relative advantages and disadvantages. The entry then explores who should be the major designers and implementers of DG and CHP technologies, and develops a strong argument that in many cases this should be an Energy Service Company (ESCO). Finally, the reasons for selecting either an independent ESCO or a local utility-affiliated ESCO are discussed, and in particular, opportunities for the local utility ESCO (the local grid) to be a major moving force in this effort are examined in depth.

This entry originally appeared as "Distributed Generation and Your Energy Future" in *Cogeneration and Distributed Generation Journal*, Vol. 18, No. 4, Fall 2003. Reprinted with permission from AEE/Fairmont Press.

INTRODUCTION

Distributed generation (DG) is electric or shaft power generation at or near the user's facility as opposed to the normal mode of centralized power generation and utilization of large transmission and distribution lines. Since DG is at or near the user's site, combined heat and power (CHP) becomes not only possible, but advantageous for many facilities. The CHP is the simultaneous production of electric or shaft power, and the utilization of the thermal energy that is "left over" and normally wasted at the central station generating site. Since DG means the power is generated at the user's site, CHP can be used to beneficially recover "waste" heat, and provide the facility with hot water, hot air, or steam, and also cooling through the use of absorption chillers.

Normal power generation using a steam Rankine cycle (steam turbine) is around 35% efficient for electric power production and delivery to the using site. The DG with its associated CHP potential means, the total system efficiency can be improved dramatically and sometimes even doubled. Thus, even though DG cannot usually beat the electrical generation efficiency of 35%–40% (at the central station), it can save the user substantial amounts of money through recovery of the thermal energy for beneficial use at the site. In addition, there are many other potential benefits from DG/CHP, discussed below.

Thus, the user can choose the objective he/she desires and will likely find a technology in this list that meets that objective. Objectives might include power production that is environmentally friendly, cost effective, more reliable, or yields better power quality. Each of these candidate technologies will be briefly explored in the sections that follow.

Why Distributed Generation?

To explore why DG should be (and is becoming) more popular, the question of why DG needs to be addressed from the perspective of the user, the utility, and society in general. Each of these perspectives is examined below.

The user might desire more reliable power. This can occur with DG, especially, when it is connected to and backed-up by the grid. The user might desire better quality power which can result, because there will be fewer momentary interruptions and possibly better voltage consistency. Often, the user desires better economics (cheaper power), which is quite possible with DG when CHP is employed. Finally, there could be a competitive advantage

Encyclopedia of Environmental Management DOI: 10.1081/E-EEM-120043089

during a utility power outage when the user has power and the competition does not.

The utility might desire less grid congestion and less future grid construction, both of which DG definitely yields. The utility may be able to hold on to a customer better, if the customer has DG and CHP at their site. Certainly, this is true if the utility constructs and runs the local power facility. Today's technology is capable of allowing the utility to remotely dispatch literally hundreds of DG units scattered in the grid, which could dramatically improve their ability to handle peak load and grid congestion problems.

Society in general likes DG/CHP, which can provide strong environmental advantages (when the appropriate technology is utilized). For example, CHP means less total fuel will be consumed placing less strain on the environment. Also, DG means less grid construction, and "green technology" (wind, photovoltaics, etc.,) can be used if desired. Since less total fuel is consumed, there will be reduced reliance on imported oil and gas, and an improved balance of payments for the United States.

Basic Philosophies

In the development to follow, some basic objectives are assumed. They are:

- Any expenditure of funds should be cost effective
- Any change should be good for all parties, or at least be of minimal harm to any party
- Existing partnerships that are working should be maintained if at all possible.

These assumptions should be clear as stated above; but more explanation will follow as the arguments are developed.

Existing and Future Markets for DG

Resource dynamics corporation has estimated that the installed base of DG units greater than 300 kW in size is at least 34 GW (34 × 10^9 W). The Gas Research Institute (GRI) estimates that the installed DG capacity in the United States of all sizes is 75 GW. Just over 90% of the installed units are under 2 MW (2 × 10^6 W) and well over 90% of the installed units are reciprocating engines. The use of DG power plants for back-up purposes is growing steadily at 7% per year while other DG applications for baseload and peaking requirements are growing at 11 and 17%, respectively. Resource Dynamics Corporation says there is the potential to double the installed DG capacity by adding as much as 72 GW by 2010. All these sources confirm that, there is already a large base of DG units nationally, and that the growth will be significant. Distributed generation is happening and will continue to happen.

Without trying to stratify it with exact numbers, the potential market could be broken into three components. They are (1) large and medium, (2) small, and (3) smaller. Each of these categories is examined briefly below.

The large and medium market is often 25 MW and larger (sometimes hundreds of megawatts) and is a mature market because there have been plants operating for many years. Typically these are in the larger process industries such as petroleum refining, pulp and paper, and chemical plants. Steam production may be in the range of hundreds of thousands of pounds per hour. While there are many such operating plants today, this mature market probably still offers the largest immediate growth potential. There is much more that could be done.

The small market will range somewhere between 50 (or 100) kW and 25 MW. These might be plants that need significant steam and could easily add a topping or bottoming steam turbine to become DG–CHP. Important to their success is the need for thermal energy and electricity (or shaft power) and the relative sizes of those needs might dictate which technology is appropriate. This market is virtually untouched today and the management/maintenance talents in these facilities might easily support this DG/CHP technological addition to their needs. Some facilities of this size will not have the management backing and maintenance talent that it takes to make these DG/CHP systems operate successfully. Those facilities would likely seek "outsourcing" for the power plant. The growth possibility here is extremely large, but will likely take a few years to realize its full potential.

The smaller market would include those small manufacturing plants or commercial facilities that need less than 50 (or 100) kW and do not have large thermal needs. These plants and facilities likely do not possess the management backing and desire or maintenance talent it takes to run them. The market potential here is tremendous in numbers of applications, but small in numbers of total megawatts. Finally, there is a significant drop in economies of scale somewhere around 200–500 kW, so the economics here would not be as exciting as the other two markets. Thus, the authors' opinion is that this market will not be as robust in the near to immediate future. Note that this could change overnight, if a local Energy service company (ESCO) such as the utility ESCO offered to design, install, run, and dispatch these units.

DG Technologies

There is a wide range of technologies possible for DG. They include:

- Reciprocating engines—diesel, natural gas, and dual fuel
- Gas turbines—microturbines, miniturbines, and large turbines
- Steam turbines
- Fuel cells
- Photovoltaic cells

- Wind turbines
- Storage devices (batteries or flywheels)

The following table briefly summarizes the pros and cons of these different DG/CHP technologies (Table 1). If more detailed information is needed, the authors recommend Capehart, et al;[1] Turner;[2] or Petchers.[3]

The table above demonstrates that there is a wide range of technologies available. Some are environmentally friendly, some are not. Some are more economically feasible, while others are extremely expensive. Some use mature technologies, while others are still somewhat of a gamble. What is badly needed for this market to mature are more ESCOs that are broadly experienced in DG/CHP applications and that are good at all of the above technologies, including the nontraditional approaches. Their tool sack contains all of these technologies and they know when and how to apply each of these technologies. To our knowledge, today only a few ESCOs can claim this broad a talent base.

Who?

Thus far, we have demonstrated that there is a significant market projected for DG/CHP systems and that this market needs to be satisfied. We have also shown there is a wide range of technologies that is available. What is needed is someone to "make this happen." Rather obviously, there are about three groups that could make this happen. They are

- The users themselves
- Energy service companies (ESCOs)
  - —Independent ESCOs (consultants)
  - —Utility affiliated ESCOs

This section examines each of these groups, and shows how they might contribute to the expanded need for DG/CHP. One consideration in evaluating the potential success of a DG/CHP project is the goal alignment of the participants, where the goals of the user or the facility are compared to the goals of the organization that is implementing the project. The closer these goals match up, or align, the more likely the DG/CHP project is to succeed.

The Group of Users Themselves

The users' goals are to have a DG/CHP project that provides an appropriate solution to their needs for electric or

Table 1 Overview of distributed generation (DG) technology.

Technology	Pros	Cons
Fuel cell	Very low emissions	High initial investment
	Exempt from air permitting in some areas	Only one manufacturer producing commercially available units today
	Comes in a complete "ready to connect" package	
Gas turbine	Excellent service contracts	Requires air permit
	Steam generation capabilities	The size and shape of the generator package is relatively large
	Mature technology	
Micro turbine	Low initial investment	Relatively new technology
	High redundancy with small units	Requires an air permit
	Low maintenance cost	Possible synchronization problems at large installations
	Relatively small size	
	Installation flexibility	
Engine	Low initial investment	High maintenance cost
	Mature technology	Low redundancy in large sizes
	Relatively small size	Needs air permit
Photovoltaics	Low operations and maintenance (O&M) costs	Very expensive initially
	Environmentally friendly	Very large footprint
		Sun must shine
		Battery storage usually needed
Wind	Low to medium O&M costs	Large footprint
	Environmentally friendly	Wind must blow

shaft power, and probably heat; works well for them both in the short term and the long term; and maximizes their economic benefit from this investment.

The user knows its process better than anyone else. This is a real advantage of doing DG/CHP projects in house and leads to the best economics if it works. Finally, the goal alignment for this group is the best of the three groups, as it is the user itself doing the job.

For this to work, the user must have a staff of technically qualified people who can analyze potential technologies, evaluate the options, select the best technology for their application, permit and install the equipment, and operate the DG/CHP system in a manner which produces the desired results. In addition, management and maintenance must both commit to the project. This often will not occur, if they wish to devote their time and efforts to building better products, delivering better services, or expanding into new products or services. Another disadvantage is that a very large capital investment is normally required and many plants and facilities simply do not have the necessary capital. Finally, these projects would involve grid interconnections and environmental permitting. Many plants and facilities are very unfamiliar with these requirements.

However, if the facility or plant does have a committed and skilled management and staff that can select, permit, install, finance and operate the DG/CHP system, this approach will most likely provide them with the highest rate of return for this kind of project.

The ESCO Group

For facilities that cannot or do not want to initiate and implement their own DG/CHP projects, the involvement of an ESCO, is probably their most appropriate alternative. Energy Service Companies bring a very interesting set of talents to DG/CHP projects. The right ESCO knows how to connect to the grid, what permits are required, and how to obtain those permits. The right ESCO knows all of the technologies available and how to choose the best type and size to utilize for this application. The right ESCO is a financial expert that knows all of the financing options available and which might be the best. Often, this means they have partnered with a financing source and have the money available with payback based on some mutually agreeable terms (interest bearing loan, shared savings, capital lease, true lease, etc.).

One of the disadvantages of using an ESCO is, they are sometimes "in a hurry." When this happens, the project design is not as well done as it should be, and they may leave before the equipment is running properly. Commissioning becomes extremely important here. Another disadvantage is that some ESCOs choose the same technology (cookie cutting) for all projects. A certain type turbine made by a particular company is always chosen, when this may not always be the best solution. If the ESCO approach is to work, all technologies must be considered and the best one chosen. For this group, goal alignment is not the best as the user no longer is in charge and the relationship is likely to be of limited duration (outsourcing being a possible exception).

However, if an "ideal" ESCO can be found and utilized, this approach offers a very satisfactory arrangement.

Independent ESCOs (Consultants)

The goals of the independent ESCO are typically to sell the customer a technology solution that the ESCO is familiar with; get the equipment installed and checked out quickly; maximize their profit on the project; and in the absence of a long term contract to provide maintenance or operating assistance to get out as quickly as they can. Sometimes the independent ESCOs goals do not line up that closely with the customer's. The independent ESCO may try to sell the customer a particular piece of equipment that they are most familiar with, and may be the one that gives them the largest profit. If there is not going to be any long-term contract for the ESCO, then they want to get the project completed as quickly as possible, and then get out. This may leave the user with a DG/CHP system that is not thoroughly checked out and tested, and leaves the user to figure out how to operate the system and how to maintain it. The project ESCO team may then depart the facility, and return to their distant office, which may be in a very different part of the country.

However, as long as the user is willing to pay the ESCO for continuing their support, the ESCO is almost always willing to do that. Unless the user is willing to pay for a part time or full time person to remain at their facility, they will have to deal with the ESCO by phone, FAX, FedEx, or Email.

One of the other potential problems with an independent ESCO is the question of its permanence. Will it be around for the long term? Historically, making the comparison of current DG/CHP ESCOs to the solar water heating companies of the 1970s and 1980s, leads to the concern that some of the DG/CHP ESCOs may not be around for the long term. Very few of the companies that manufactured, sold, or installed solar water heating systems at that time are still around today. Many of these solar water heating companies were gone within a few years of the customers purchasing the systems. Most of these companies were actually gone long before the useful lives of the solar systems had been reached. Repair services, parts, and operating advice were often no longer available, so many solar water heating system users simply stopped using them, or removed the systems. Based on this history, selection of an ESCO that is likely to be around for the long term is an important consideration.

Utility-Based ESCOs

Next, consider a utility-based ESCO. Utility-affiliated ESCOs have goals similar to the independent ESCOs' goals in many respects, but the big difference is that the utility is a permanent organization that is local is there for the long term, and is interested in seeing the user succeed, so that they will be an even better customer in the long run. Also, since the utility and the affiliated ESCO are local, they can send someone out periodically to check on the facility and the DG/CHP project to help answer questions and make sure the project is continuing to operate successfully. The utility is financially secure, stable, and, in most instances, is regarded by the community as an honest and trustworthy institution.

This ESCO now is an independent branch of the local utility. If they have the full set of tools (knowledge of all the technologies) then their advantages include all those listed above. In addition, there is much better goal alignment. They will be there as a partner as long as the wires are connected and that likely is almost forever. Thus, both parties want this to work. They are the grid, so the grid interconnection is not as much of a problem. The user and the utility have been partners for years. This would change the relationship; not destroy it. (The devil you know vs the one you do not.) Finally, this is what they do (almost).

One limitation of the utility-based ESCO is that they must change their mindset of "sell as much electricity as possible," and recognize that there is a lot of business and income to be captured from becoming an energy service organization. Someone is going to do these DG/CHP projects; the utility revenue base is most enhanced when they do it and when the project is successful. Their services could involve design, installation, start up, commissioning, and passing of the baton to the user or they might run it themselves (outsourcing).

Now, if the local utility company ESCO can take advantage of the opportunities they have, then they have a lot to offer to facilities and plants that are interested in working with them to put in DG and CHP systems. The old Pogo adage "We are surrounded by insurmountable opportunities" is always around in these situations. Another old saying would describe this DG/CHP opportunity for utility-affiliated ESCOs as "the business that is there for them to lose." The utility-affiliated ESCOs need to aggressively pursue these opportunities.

Some Local Utility ESCO Successes

A very good example of a local utility ESCO success story comes from the experiences shared by AmerenCILCO, a utility company in central Illinois, This company has experiences with both DG and CHP projects.

DG Only Projects

AmerenCILCO has extensive experience in using reciprocating engine generator sets as DG to meet peak load conditions on their system. The specifications of some of their DG projects are as follows:

- Hallock Substation, 18704 N. Krause Rd., Chillicothe, IL
- Eight reciprocating diesel engine generator sets
- Nominal capacity 1.6 MWe each, 12.8 MWe total
- Owned by AmerenCILCO
- Kickapoo Substation, 1321 Hickox Dr., West Lincoln, IL
- Eight reciprocating diesel engine generator sets
- Nominal capacity 1.6 MWe each, 12.8 MWe total
- Owned by Altorfer Inc.; power purchase agreement with AmerenCILCO, operating agreement provides for operations and maintenance (O&M)
- Tazewell Substation, 18704 N. Krause Rd., Chillicothe, IL
- Fourteen reciprocating diesel engine generator sets
- Nominal capacity 1.825 MWe each, 25.55 MWe total
- Owned by Altorfer Inc.; power purchase agreement with AmerenCILCO; operating agreement provides for O&M.

Although these DG power module facilities are primarily used as peaking facilities, they are also used to maintain system integrity in the event of an unanticipated outage at another AmerenCILCO generating station. They are unmanned and remotely operated from the company's Energy Control Center. They have proven to be a reliable and low cost option for the company to meet its peaking requirements. The power module sites were constructed at a cost of approximately $400/kW and have an operating cost of 75 dollars/MWh using diesel fuel at $0.85/gallon.

DG/CHP Projects

AmerenCILCO also has some successful CHP projects. A summary of two such DG/CHP projects are given below.

Indian Trails Cogeneration Plant

The Indian Trails Cogeneration Plant is owned and operated by AmerenCILCO. It is located on the property of MGP Ingredients of Illinois (MGP) in Pekin, Illinois and provides process steam to MGP and electricity to AmerenCILCO. The plant was constructed at a cost of $19,000,000 and went into full commercial operation in June 1995.

The plant consists of three ABB/Combustion Engineering natural gas-fired package steam boilers and one ABB STAL backpressure turbine-generator. Two of the boilers, boilers 1 and 2, are high-pressure superheat boilers rated at 185,000 lb/h of steam at 1250 psig and 900°F. Boiler 3 is a low-pressure boiler rated at 175,000 lb/h of steam at 175 psig and saturated temperature. Boilers 1 and 2 are normally in operation, with Boiler 3 on standby to insure maximum steam production reliability for MGP.

The high-pressure steam from boilers 1 and 2 passes through the ABB backpressure turbine-generator, which is rated at 21 MW. The steam leaving the turbine is at

175 psig and is desuperheated to 410°F to meet MGP's process steam requirements. The electricity produced goes to the AmerenCILCO grid to be used to meet utility system requirements.

The plant configuration provides significant operating efficiencies that benefit both MGP and AmerenCILCO. The Indian trails has an overall plant efficiency in excess of 80% and an electric heat rate of less than 5200 Btu/kWh. The construction of the Indian Trails by AmerenCILCO created an energy partnership with a valued customer. It allowed MGP to concentrate its financial and personnel resources on its core business. In turn, AmerenCILCO used its core business of producing energy to become an integral part of MGP's business, making AmerenCILCO more than just another vendor selling a product.

Medina Valley Cogen Plant

The Medina Valley Cogeneration Plant is owned and operated by AmerenCILCO. It is located on the property of Caterpillar and provides process steam and chilled water to Caterpillar, and electricity to AmerenCILCO. The plant was constructed at a cost of $64,000,000 and went into full commercial operation in September 2001.

The 40 MW electric generating plant consists of three natural gas-fired Solar Titan 130 model 18001S combustion turbines equipped with SoloNO$_x$ (low NO$_x$) combustion systems manufactured by Caterpillar driving electric generators rated at 12.2 MW (gross generating capacity) each. There are also two Dresser-Rand steam turbine-generators with a total rated capacity of 8.9 MW.

The 410,000 #/h steam plant consists of three Energy Recovery International (ERI) VC-5-4816SH heat recovery steam generators (HRSGs) equipped with Coen low NO$_x$ natural gas-fired duct burners and catalytic converters to reduce carbon monoxide (CO), rated at 109,000 lb/h at 600 psig each. There is also one Nebraska natural gas-fired steam generation boiler equipped with low NO$_x$ burners, rated at 100,000 lb/h at 250 psig.

The plant configuration provides significant operating efficiencies that benefit both Caterpillar and AmerenCILCO. Medina Valley has an overall plant efficiency in excess of 70% and an electric heat rate of less than 6400 Btu/kWh. The construction of Medina Valley by AmerenCILCO created an energy partnership with a valued customer, whereby competitive electricity and steam prices were provided as well as greater operational flexibility, improved quality control in manufacturing, and improved steam reliability. It also allowed Caterpillar to concentrate its financial and personnel resources on its core business. In turn, AmerenCILCO used its core business of producing energy to become an integral part of Caterpillars business, strengthening its ties with a major customer as well as adding additional efficient-generating capacity, and improving air quality (399 fewer tons pollutants/year).

CONCLUSIONS

Distributed generation and DG/CHP should, must, and will happen. The benefits to all parties when CHP is utilized are too much to ignore. Therefore, the question becomes who should do it, not should it be done.

If management is behind the project, the engineering and maintenance staff is capable, and financing is available, the project should be done in-house. Maximum economic benefits would result. However, the user must commit to this project.

If any of the above is not true, the best approach for the facility or plant is to seek the help of an ESCO. It is important that the ESCO chosen must be fully equipped with knowledge and experience in all of the technologies and be able to provide the financing package. Such ESCOs do exist today, but some of them need a better understanding of the different technologies required, as well as insuring that the DG/CHP project is successfully completed and turned over to a facility that can operate it and maintain it. Commissioning and baton passing must be part of the contract.

The authors believe there is a tremendous opportunity for local utility ESCOs to successfully participate in this movement to DG, and particularly to the use of CHP. A utility-affiliated ESCO, properly equipped as we have defined it, can do these projects, and can do them successfully. The local utility ESCO has entries with local facilities and plants that few other ESCOs have. If the utility can exploit this opportunity, they have the chance to help many facilities and to help themselves in the process. This is a true win–win opportunity for the utility affiliated ESCO. All utilities should be ready to fill this need or recognize that they will likely lose market share.

REFERENCES

1. Capehart, B.L., Turner, W.C., Kennedy, W.J., Eds.; *The Guide to Energy Management*, 5th Ed.; Fairmont Press: Lilburn, GA, 2006 (Chapter 14).
2. Turner, W.C. Ed.; *Energy Management Handbook*, 6th Ed.; Fairmont Press: Lilburn, GA, 2006 (Chapter 7).
3. Petchers, N. Ed.; *Combined Heating, Cooling and Power Handbook*, Fairmont Press: Lilburn, GA, 2003.

Drainage: Hydrological Impacts Downstream

Mark Robinson
Center for Ecology and Hydrology, Wallingford, U.K.

D.W. Rycroft
Department of Civil and Environmental Engineering, Southampton University, Southampton, U.K.

Abstract
Drainage systems may be broadly divided into surface drainage (comprising land grading and open ditches), shallow drainage (such as subsoiling to mechanically loosen the upper layer of soil), subsurface or groundwater drainage (buried perforated pipes or deep ditches), and the main drainage systems (commonly open channels) used to convey the drain water away. Drainage will inevitably affect the quantity and timing of water flows from the land and into the receiving watercourses. It is these downstream impacts of farmland drainage on the timing and magnitude of peak flows that are considered here, using the results of experimental studies and computer simulations, to present a coherent picture, and to answer most of the apparent anomalies and conflicts.

INTRODUCTION

Land drainage is the practice of removing excess water from the land, and it is one of the most important land management tools for improving crop production in many parts of the world. Drainage systems may be broadly divided into surface drainage (comprising land grading and open ditches), shallow drainage (such as subsoiling to mechanically loosen the upper layer of soil), subsurface or groundwater drainage (buried perforated pipes or deep ditches), and the main drainage systems (commonly open channels) used to convey the drain water away.[1] Drainage will inevitably affect the pattern of water flows from the land and into the receiving watercourses. It is these downstream impacts of farmland drainage on the timing and magnitude of peak flows that are considered here, using the results of experimental studies and computer simulations, to present a coherent picture, and to answer most of the apparent anomalies and conflicts.

HYDROLOGIC IMPACTS

Concern about the possible downstream effects of drainage is shown by many published papers worldwide, in North America,[2,3] Great Britain,[4,5] and continental Europe, including France,[6] Netherlands,[7] Ireland,[8,9] Finland,[10] and Germany.[11] The role of drainage has often been highlighted by serious flood events—for example, in the Midwest of the United States in 1993, and across Europe in 1997—which reawakened concerns that drainage could aggravate flooding downstream.

There has been a debate about the effects of drainage on streamflow for well over a century, but until recently, due to the lack of appropriate data, the debate has been largely speculation. Too often, the absence of evidence has erroneously been taken as evidence of an absence of effect. The earliest published account[12] was a report of a 4-day meeting held at the Institution of Civil Engineers in London in 1861. Many of the arguments and opinions expressed have resonance today, but due to the absence of objective measurements, the participants were unable to reach any conclusions and the meeting was inconclusive.

These conflicting opinions resulted from differences in the emphasis given to the two processes of water storage and routing. Considering the former, it may be argued that because drainage lowers the water table, the available storage capacity in the soil is enlarged and able to absorb more storm rainfall, thereby reducing peak flow rates. In contrast, according to the routing argument, the purpose of drainage is to "remove water from the land more quickly" than under natural conditions, so peak outflows must necessarily increase.

Probably more work has been carried out in Britain upon the effects of agricultural drainage upon streamflow than in any other country. Britain was the originator of modern field drainage[13] and so became the first country where concern arose about its downstream effects; it is also one of the most extensively drained countries in the world.

It is only in the last few years that it has been possible to obtain a coherent picture based on observations of field processes and supported and extended by computer modeling. This has shown that general statements that drainage "causes" or "reduces" flood risk downstream are

Encyclopedia of Environmental Management DOI: 10.1081/E-EEM-120046627

oversimplifications of the complex processes involved, and that any consideration of the impact of drainage on streamflow must identify the point of interest, whether at the outfall from the field, along the main channel, or a combination of both at the catchment scale.

Experimental studies indicate that the provision of surface drainage will result in higher peak flows downstream. This was shown by a long-term experiment at Sandusky in Northern Ohio[14] and is a result of the reduction/elimination of surface storage capacity, as well as the provision of more efficient faster flow routes. This has been demonstrated conclusively both by experimental studies and by computer simulations.

In contrast, there seems to be general agreement from experimental studies that subsurface drainage of waterlogged, poorly permeable clay soils reduces peak outflows.[15–17] Since this is one of the most common situations where artificial drainage is used, it might be considered to represent the most general result of field drainage.

There are, however, instances where even on heavy soils this result may not apply. Due to their low hydraulic conductivity, most water movement in clay soils is confined to flow through macropores, such as cracks. As a result of clay shrinkage and cracking in warm, dry summers, rapid macropore flow can result in larger peak flows from the drained land than from the undrained land. The role of macropores on the seasonality of peak flows from drained land was demonstrated in detail.[18]

More permeable, drier soils may also be drained where there is an economic justification—for example, drainage of land producing high-value crops. In contrast to clay soils, relatively few scientific field studies have investigated the impact of draining lighter, more permeable soils. This may be partly due to the emphasis on draining clay soils, but also, no doubt, results from the greater practical difficulty encountered in plot definition where the soils are more permeable. Nevertheless, data from several drainage experiments on permeable soils are available. At Withernwick,[19] flow peaks were increased in the first year after drainage and there was then a reduction in the following years due to the progressive deterioration of the secondary system of subsoiling designed to improve the soil structure. Supporting evidence of increased peak flows following the drainage of more permeable soils also comes from studies at Cockle Park in northern Britain[20] and Ellingen in Central Germany.[21]

To identify factors influencing drainage response, the results of field drainage experiments under temperate northern European climates were analyzed in terms of their site characteristics.[22,23] This included topography, precipitation, drainage depth and spacing, natural (i.e., predrainage) soil water regime, and the soil properties. The only characteristic distinguishing sites, where drainage increased peak flows from those where they were reduced, were those relating to the soil water regime before drainage. The experimental sites all had similar land practices on the drained and the undrained land.

Drainage reduced peak flows on sites that had wetter soils and with poor natural drainage, and significant amounts of storm runoff were generated as overland flow and near-surface flow in the thin upper layers of the soil. These sites had higher topsoil clay contents and shallower depths to a poorly permeable subsoil horizon. When artificially drained, the surface saturation was largely eliminated, greatly increasing the soil water storage capacity.

In contrast, at sites with more permeable, loamy soils that were not routinely saturated before drainage, natural storm flow occurred predominantly by slower subsurface flow, and the artificial drainage pipes provided more rapid flow routes leading to increases in peak outflows.

The findings are summarized in Fig. 1. This shows the topsoil texture, together with the effect of drainage on peak flows, and provides the engineer or conservationist with an initial guide to predict the effect on flows of the drainage of a site, based on knowledge of the predrainage site characteristics.

Further insights into the factors controlling the impact of drainage may be obtained by the application of modeling techniques to investigate the important interaction between soil properties and climate in determining soil water regimes. DRAINMOD[24] was applied to two of the field sites with similar climates: a heavy clay soil at Grendon and a more permeable loam at Withernwick. The model was applied to each site using actual field values of drain

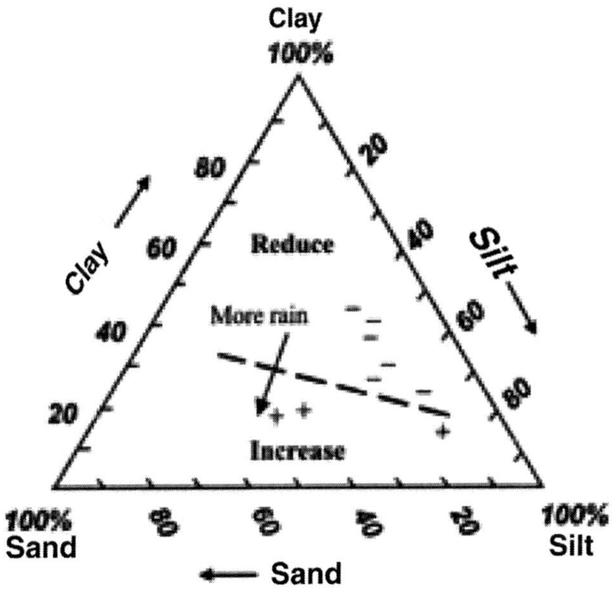
Clay
100%
80
60
40
20
Clay
20
40
60
80
Silt
Reduce
More rain
Increase
100%
Sand
80
60
40
20
100%
Silt
Sand

Fig. 1 Observed impact of pipe drainage on downstream peak flows (increase/reduce), showing the importance of soil texture. Model simulations of climate changes indicate that higher rainfall and wetter ground conditions will shift the balance towards drainage schemes reducing peak flows. See text for details.

and soil parameters, and the simulated peak flows from drained and undrained land were compared for similar rainfall inputs. The results showed a 70% lower median peak flow after drainage of clay soil and an increase of 40% in the median peak flow from the more permeable land.[23]

The modeled fluxes and water stores confirmed that the reduction in peaks from the clay soil after drainage was achieved by a change in storm runoff generation from overland flow (caused by soil saturation) to subsurface drainflow. For the loamy soil, the model indicates that the increase in peak subsurface flow rates was due to the steeper hydraulic gradients created by the closer-spaced artificial drains.

The model also demonstrated the effect of different climatic conditions. If the loam soil site at Withernwick had double the normal rainfall (1200 mm yr^{-1} instead of 600 mm yr^{-1}), the resulting increase in ground wetness would be sufficient to generate substantial amounts of overland flow on the undrained land. Artificial drainage in this case would then reduce peak flows—exactly as happens for a clay soil (where, in contrast, the ground wetness is caused by the low soil permeability). Using the model in this way enables these effects of site characteristics to be explored in an objective manner. The overall dominant criterion—the amount and frequency of surface runoff from undrained land—can be assessed in terms of both soil properties and climatic characteristics.

CONCLUSIONS

The effect of subsurface drainage on peak flows depends upon site wetness. If the water table is close to the surface (due to high rainfall or poor permeability), natural flows occur either over the surface or through the upper, more permeable layers of the soil. Drainage will increase soil water storage capacity and hence the amount of water that can infiltrate, thereby reducing surface runoff and peak storm flows. If the water table is deeper, due to a dry climate or due to more permeable soils, natural flows will occur through the body of the soil. In this case, artificial drainage will increase peak flows as a result of the shorter flow paths and steeper hydraulic gradients.

It must be noted that these conclusions depend upon the scale of the drainage considered. At the river catchment scale, main channel improvements will undoubtedly increase the speed of flow routing, and the timing of arrival of flows from different subcatchments will influence the peak discharge at the point of interest. The relative importance of field drainage and main drainage channels will vary with storm size: field drainage being dominant for small and medium storms, but main channel improvements becoming dominant for large events. In extreme situations where the rainfall intensity exceeds the infiltration capacity of the soil, the effects of the subsurface drains will be minimal, but the associated improved watercourses will rapidly carry away the surface runoff.

Overall, it seems likely that in large catchments, drainage schemes with substantial associated surface drainage and main channel improvements will lead to higher flow peaks downstream, even though locally, the effect of drainage may be to lower the peak flows.

REFERENCES

1. Skaggs, R.W.; van Schilfgaarde, J., Eds. *Agricultural Drainage*; Agron Monograph 38; ASA, CSSA, and SSSA: Madison, WI, 1999.
2. Whiteley, H.R. Hydrologic implications of land drainage. Can. Water Res. J. **1979**, *4*, 12–19.
3. Serrano, S.E.; Whiteley, H.R.; Irwin, R.W. Effects of agricultural drainage on streamflow in the Middle Thames River, Ontario, 1949–1980. Can. J. Civil Eng. **1985**, *12*, 875–885.
4. Bailey, A.D.; Bree, T., Eds. Effect of improved land drainage on river flows. *Flood Studies Report—5 Years on*; Thomas Telford: London, 1981; 131–142.
5. Rycroft, D.W., Ed. The hydrological impact of land drainage. 4th International Drainage Workshop, Cairo; ICID-CHD, CEMAGREF; 1990; 189–197.
6. Oberlin, G. Influence du drainage et de l'assainissement rural sur l'hydrologie. CEMAGREF Bull. **1981**, *285*, 45–56.
7. Warmerdam, P.M.M., Ed. The effect of drainage improvement on the hydrological regime of a small representative catchment area in the Netherlands. *Application of Results from Representative and Experimental Basins*; UNESCO Press: Paris, 1982; 318–338.
8. Burke, W. Aspects of the hydrology of blanket peat in Ireland. Int. Assoc. Hydrol. Sci. **1975**, *105*, 171–181.
9. Wilcock, D.N. The hydrology of a peatland catchment in N Ireland following channel clearance and land drainage. In *Man's Impact on the Hydrological Cycle in the U.K.*; Hollis, G.E., Ed.; Geo Abstracts: Norwich, 1979; 93–107.
10. Seuna, P.; Kauppi, L., Eds. *Influence of Subdrainage on Water Quantity and Quality in a Cultivated Area in Finland*; Water Research Institute Publ. No. 43; Nat. Board of Waters: Helsinki, Finland 1981.
11. Harms, R.W. The effects of artificial subsurface drainage on flood discharge. In *Hydraulic Design in Water Resources Engineering: Land Drainage*; Smith, K.V.H., Rycroft, D.W., Eds.; Computational Mechanics Publication: Southampton, 1986; 189–198.
12. Denton, J. Bailey On the discharge from underdrainage and its effects on the arterial channels and outfalls of the country. Proc. Inst. Civil Eng. **1862**, *21*, 48–130.
13. Van der beken. The development of the theory and practice of land drainage in the 19th century. In *Water for the Future*; Wunderlich, W.O., Prins, J.E., Eds.; A.A. Balkema: Rotterdam 1987; 91–99.
14. Schwab, G.O.; Thiel, T.J.; Taylor, G.S.; Fouss, J.L. Tile and surface drainage of clay soils 1. Hydrologic performance with grass cover USDA. Agric. Res. Serv. Bull. **1963**, 935.

15. Robinson, M.; Beven, K.J. The effect of mole drainage on the hydrological response of a swelling clay soil. J. Hydrol. **1983**, *63*, 205–223.
16. Harris, G.L.; Goss, M.J.; Dowdell, R.J.; Howse, K.P.; Morgan, P. A study of mole drainage with simplified cultivation for autumn sown crops on a clay soil. II. Soil water regimes, water balances and nutrient loss in drain water, 1978–80. J. Agric. Sci. **1984**, *102*, 561–581.
17. Armstrong, A.C.; Garwood, E.A. Hydrological consequences of artificial drainage of grassland. Hydrol. Process. **1991**, *5*, 157–174.
18. Robinson, M.; Mulqueen, J.; Burke, W. On flows from a clay soil—Seasonal changes and the effect of mole drainage. J. Hydrol. **1987**, *91*, 339–350.
19. Robinson, M.; Ryder, E.L.; Ward, R.C. Influence on streamflow of field drainage in a small agricultural catchment. J. Agric. Water Manage. **1985**, *10*, 145–148.
20. Armstrong, A.C., Ed. *The Hydrology and Water Quality of a Drained Clay Catchment—Cockle Park, Northumberland*; Report RD/FE/10; MAFF: London, 1983.
21. Schuch, M. Regulation of water regime of heavy soils by drainage, subsoiling and liming and water movement in this soil. In *International Institute for Land Reclamation and Improvement*, Proc. International Drainage Workshop; Wesseling, J., Ed.; Wageningen, Paper 1.14, 1978; 253–267.
22. Robinson, M., *Impact of Improved Land Drainage on River Flows*; Institute of Hydrology Report 113; Wallingford, U.K., 1990. ISBN 0-948540-24-9. http://nora.nerc.ac.uk/7349.
23. Robinson, M.; Rycroft, D.W. The impact of drainage on streamflow. *Agricultural Drainage*; Skaggs, R.W.; van Schilfgaarde, J., Eds.; Agron Monograph 38; ASA, CSSA, and SSSA: Madison, WI, 1999; Chap. 23, 767–800.
24. Skaggs, R.W. Drainage simulation models. *Agricultural Drainage*; Skaggs, R.W., van Schilfgaarde, J., Eds.; Agron Monograph 38; ASA, CSSA, and SSSA: Madison, WI, 1999; Chap. 13, 469–500.

Drainage: Soil Salinity Management

Glenn J. Hoffman
Biological Systems Engineering, University of Nebraska—Lincoln, Lincoln, Nebraska, U.S.A.

Abstract
Soil water must drain through the crop root zone when salinity is a hazard to prevent salts from increasing to levels detrimental to crop production. Drainage occurs whenever irrigation and rainfall provide soil water in excess of the soil's storage capacity. Regardless of the climate, if soluble salts are present, water in excess of that needed to satisfy the crop water requirement must be provided to leach excess salts. Leaching may be accomplished continuously or at intervals, depending on the degree of salinity control required. It may take decades or as little as one season, depending on the hydrogeology of the area, but without drainage, agricultural productivity cannot be sustained where salinity is a threat.

INTRODUCTION

Soil water must drain through the crop root zone when salinity is a hazard to prevent salts from increasing to levels detrimental to crop production. Drainage occurs whenever irrigation and rainfall provide soil water in excess of the soil's storage capacity. In humid regions, rainfall normally satisfies crop water requirements and precipitation infiltrating into the soil in excess of this requirement leaches (drains) salts present below the crop root zone. In subhumid areas, rainfall is often inadequate in amount or temporal distribution to satisfy crop needs and irrigation is implemented. For arid regions, rainfall is never abundant and the preponderance of the crop water requirement must be provided by irrigation. Regardless of the climate, if soluble salts are present, water in excess of that needed to satisfy the crop water requirement must be provided to leach excess salts. Leaching may be accomplished continuously or at intervals, depending on the degree of salinity control required. It may take decades or as little as one season, depending on the hydrogeology of the area, but without drainage, agricultural productivity cannot be sustained where salinity is a threat. For a more complete discussion on drainage design for salinity control, the reader is referred to Hoffman and Durnford.[1]

DRAINAGE CONDITIONS

All soils have an inherent ability to transmit soil water provided a hydraulic gradient exists. If the hydraulic gradient is positive downward, drainage occurs. Soils with compacted layers, fine texture, or layers of low hydraulic conductivity may be so restrictive to downward water movement that drainage is insufficient to remove excess salts. In some areas, the hydrogeology may be such that the hydraulic gradients are predominantly upward. This leads to water logging and salination.

Before designing a man-made drainage system, the natural drainage rate should be determined. If the natural hydraulic gradient causes soil water to drain out of the crop root zone, the capacity of the artificial system can be reduced, thereby decreasing the cost for drainage. In some situations, upward flow into the crop root zone from a shallow aquifer can significantly increase the drainage requirement. The upward movement of groundwater leads to salination as the water evaporates at the soil surface, leaving salts behind. If upward flow is ignored, the drainage system may be inadequate. Regardless of the source, an artificial drainage system will not function unless it is below the surface of the water table.

DRAINAGE REQUIREMENT

Saline Soils

The amount of drainage required to maintain a viable irrigated agriculture depends on the salt content of the irrigation water, soil, and groundwater; crop salt tolerance; climate; soil properties; and management. At present, the only economical means of controlling soil salinity is to ensure an adequate net downward flow of water through the crop root zone to a suitable disposal site. If drainage is inadequate, harmful amounts of salt can accumulate.

In irrigated agriculture, water is supplied to the crop from irrigation, rainfall, snow melt, and upward flow from groundwater. Water is lost through evaporation, transpiration, and drainage. The difference between water inflows and outflows is the change in soil water storage. A water balance, expressed in terms of equivalent depths (D) of water, can be written as

$$D_s = D_i + D_r + D_g - D_e - D_t - D_d \qquad (1)$$

where the subscripts s, i, r, g, e, t, and d designate storage, irrigation, rainfall and snow melt, groundwater, evapora-

Desertification – Ecosystems

Encyclopedia of Environmental Management DOI: 10.1081/E-EEM-120010112

tion, transpiration, and drainage, respectively. The corresponding salt balance, where S is the amount of salt and C is salt concentration, can be expressed as

$$S_s = D_iC_i + D_rC_r + D_gC_g + S_m + S_F - D_dC_d - S_p - S_c \quad (2)$$

with S_s being salt storage, S_m is the salt dissolved from minerals in the soil, S_f indicates salt added as fertilizer or amendment, S_p is precipitated salts, and S_c is the salt removed in the harvested crop.

Rarely do conditions prevail long enough for steady state to exist in the crop root zone. However, it is instructive to assume steady state to understand the relationship between drainage and salinity. If upward movement of salt, the term ($S_m + S_f - S_p - S_c$), and the change in salt storage are all essentially zero, then the salt balance Eq. (2) can be reduced to

$$D_dC_d = D_iC_i + D_rC_r \quad (3)$$

The leaching fraction, L, is the ratio of the amount of water draining below the crop root zone, D_d, and the amount applied, $D_i + D_r$. The ratio of the salt concentration entering and leaving the root zone can also be used to estimate L. Since C_r is essentially zero.

$$L = C_i/C_d = D_d/D_i + D_r \quad (4)$$

The concept in Eq. (4) is important because it illustrates the relationship between leaching fraction and salinity.

The minimum leaching fraction that a crop can endure without yield reduction is termed the leaching requirement, L_r. The leaching requirement is the minimum amount of drainage required to prevent excess accumulations of salt that result in loss of crop yield. Several models have been proposed to estimate the drainage (leaching) requirement. Of the four models tested,[2] the one presented in Fig. 1 agrees well with measured values of the drainage requirement through the range of agricultural interest. The drainage requirement given in Fig. 1 is the fraction of the volume of applied water that must pass through the crop root zone as a function of the salinity of the applied water and the salt tolerance of the crop.

Sodic Soils

A soil is said to be sodic if an excessive concentration of sodium causes a deterioration of soil structure. The impact of excess sodium is a reduction in hydraulic conductivity and crust formation. Sodic conditions decrease the rate of drainage. Before a sodic soil can be restored to full productivity the excess sodium in the soil must be replaced with calcium or magnesium. This process frequently requires copious amounts of leaching to reclaim the soil. The design of an artificial drainage system that may be required, however, is based upon the long-term

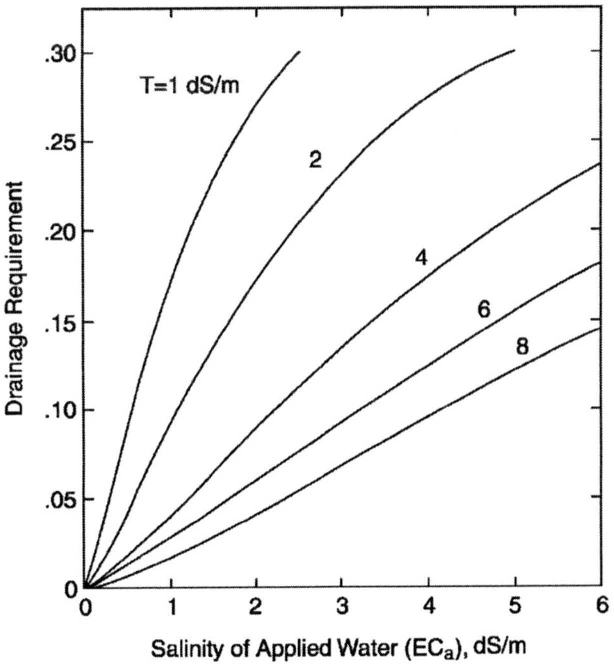

T=1 dS/m
2
4
6
8
Drainage Requirement
.30
.25
.20
.15
.10
.05
0
0
1
2
3
4
5
6
Salinity of Applied Water (EC_a), dS/m

Fig. 1 Drainage requirement as a function of the salinity of the applied water (reported as the volume weighted electrical conductivity) and the salt tolerance threshold value for the crop (T). **Source:** Adapted from Hoffman and Van Genuchten.[15]

requirement for drainage as estimated in Fig. 1 rather than the anticipated high drainage requirements for reclaiming a sodic soil.

DRAINAGE SYSTEM DESIGN

There are three types of subsurface systems used to control soil salinity: relief drains, shallow wells, and interceptor drains. Relief drains, usually consisting of perforated corrugated plastic tubes buried in a regularly spaced pattern, is the most common subsurface system. Laterals for relief drains are typically placed 2.0–3.5 m deep and are spaced horizontally ten to hundreds of meters apart where salinity is a hazard. Shallow wells, called tube wells in some regions, can also be used to lower the water table by allowing pumping from shallow, unconfined aquifers. Tube wells are spaced at distances of a few hundred meters to several kilometers and may be a few meters to a hundred meters deep. Interceptor drains are used to remove excess soil water from saline seeps. Frequently, one subsurface drain, properly located at the upslope side of the seep, is sufficient. Regardless of the type of drainage system, the depth of the water table must be maintained low enough that (1), salts in the soil profile move to the water table (2), the rate of water movement by capillary flow to the soil surface because of evaporation is minimal, and (3), upflow of saline groundwater into the root zone is prevented.

Relief Drains

A relief drainage system consists of a main drain, collector drains, and field drains (laterals). The main drain is frequently a surface stream or an open drainage canal. Collectors and laterals are usually buried in a regular parallel pattern. Either open ditches or perforated pipes can serve as collectors and laterals. Open ditches are not normally installed now because they occupy land, are difficult to maintain, and are only capable of shallow drainage. Laterals are up to 300 m long and terminate in a collector drain. Both single- and double-sided entries by laterals into a collector are common.

Drain Depth

Subsurface drains are installed much deeper for salinity control in arid regions than drains for water table control in humid regions. The goal for salinity control is to place the drains deep to limit salination of the root zone by capillary upflow. Drains are placed at depths of 2.0–3.5 m in arid regions.[3] The appropriate drain depth depends upon the depth capacity of the installation machinery, the location of a shallow soil layer that impedes water movement, and anticipated benefits compared to additional costs of deeper installation.

Drain Spacing

The spacing between laterals is often estimated using simple drainage design equations. Drain spacing determinations can be based on criteria of steady-state, falling-water-table, or fluctuating-water-table conditions.[4] For large drainage projects or where more accurate values are desired, computerized drainage design models are available. An early computer model developed by Skaggs[5] has been altered by several for irrigated conditions.[6,7] Other models present drainage designs for irrigated areas based on optimization,[8] decision support systems,[9] or reuse of drainage water.[10]

Drainage Wells

Shallow or tube wells offer a viable alternative to relief drains when the aquifer has sufficient transmissivity to provide a significant yield of drain water and the vertical permeability between the crop root zone and the aquifer is adequate. Under these conditions, tube wells have the advantages of being able to lower the water table to greater depths than relief drains and also provide supplemental water for irrigation if the quality is appropriate.

Because drainage wells can be installed at convenient locations within the area to be drained and can be operated either continuously or intermittently, the management of a system of drainage wells is more versatile than relief drains. Relief drains are typically a passive drainage system relying on gravity and designed to operate continuously.

Economic comparisons between the costs of drainage wells and relief drains vary. It is generally found that relief drains have lower construction and operation costs.[11] However, Mohtadullah[12] showed tube wells were a better economic choice than relief drains for the Indus Basin.

Saline Seeps

The occurrence of saline water at the soil surface downslope from a recharge area is referred to as a saline seep. Saline seeps can occur because of the reduction of evapotranspiration that occurs when grasses or forests are converted to cropland in the upland (recharge) areas of a watershed. Dryland farming practices that include fallow periods tend to aggravate the seepage problem. Salination occurs as water infiltrating in the upper elevations of the watershed moves through salt-laden substrate on its path to a discharge site at a lower elevation. In the discharge area of the seep, crop growth is reduced or the plants killed by an intolerable level of salinity. Saline seeps can be distinguished from other saline soil conditions by their recent origin, relatively local extent, saturated soil profile, and sensitivity to precipitation and cropping systems.[13] Saline seeps occur throughout the Great Plains of North American and in Australia, India, Iran, Turkey, and Latin America.[14]

Planting crops in the recharge area that consume soil water before it percolates below the crop root zone will prevent saline seeps. Failing this, improved drainage may provide a solution. Installing an interceptor subsurface drain immediately upslope from the saline seep is frequently a successful solution. Interceptor drains to control seepage should be installed as deep as practical. If the layer restricting soil water flow is not too deep, placing the interceptor drain just above this layer is the most effective location.

REFERENCES

1. Hoffman, G.J.; Durnford, D.S. Drainage design for salinity control. *Agricultural Drainage*; Agronomy Monograph No. 38; American Society of Agronomy: Madison, WI, 1999; 579–614; Chap. 17.
2. Hoffman, G.J. Drainage required to manage salinity. J. Irrig. Drain. Div., American Society of Civil Engineers, New York **1985**, *111*, 199–206.
3. Ochs, W.J. *Project Drainage Issues*, Proceedings of the 3rd International Workshop on Land Drainage, Columbus, OH, Dec 7–11, 1987; E-83–E-88. 202.
4. Bouwer, H. Developing drainage design criteria. *Drainage for Agriculture*; Agronomy Monograph No. 17; American Society of Agronomy: Madison, WI, 1974; 67–79.
5. Skaggs, R.W. *A Water Management Model for Artificially Drained Soils*, Report 267; Water Resources Research Institute, North Carolina State University: Raleigh, NC, 1980.
6. Chang, A.C.; Hermsmeir, L.F.; Johnston, W.R. *Application of DRAINMOD on Irrigated Cropland*, Paper No. 81-2543;

American Society of Agricultural Engineers: St. Joseph, MI, 1981.
7. Skaggs, R.W.; Chescheir, G.M. Application of drainage simulation models. *Agricultural Drainage*; Agronomy Monograph No. 38; American Society of Agronomy: Madison, WI, 1999; 529–556; Chap. 15.
8. Knapp, K.C.; Wichlens, D. Dynamic optimization models for salinity and drainage management. *Agricultural Salinity Assessment and Management*; American Society of Civil Engineers: New York, 1990; 530–548; Chap. 25.
9. Gates, T.K.; Wets, R.J.-B; Grisiner, M.E. Stochastic approximation applied to optimal irrigation and drainage planning. J. Irrig. Drain. Div., Am. Soc. Civil Eng. **1989**, *115*, 488–510.
10. El-Din El-Quosy, D.; Rijtema, P.E.; Boels, D.; Abdel-Khalik, M.; Roest, C.W.J.; Adbel-Gawad, S. Prediction of the quantity and quality of drainage water by the use of mathematical modeling. *Land Drainage in Egypt*; Drainage Research Institute: Cairo, Egypt, 1989; 207–241.
11. Zhang, W. Drainage inputs and analysis in water master plans. Proceedings of the 4th International Drainage Workshop, Cairo, Egypt, Feb 1990; 181–187.
12. Mohtadullah, K. Interdisciplinary planning, data needs and evaluation for drainage projects. Proceedings of the 4th International Drainage Workshop, Cairo, Egypt, Feb 1990; 127–140.
13. Brown, P.L.; Halvorson, A.D.; Siddoway, F.H.; Mayland, H.F.; Miller, M.R. *Saline-Seep Diagnosis, Control, and Reclamation*, USDA Conservation Research Report No. 30; 1983.
14. Halvorson, A.D. Management of dryland saline seeps. *Agricultural Salinity Assessment and Management*; American Society of Civil Engineers: New York, 1990; 372–392; Chap. 15.
15. Hoffman, G.J.; Van Genuchten, M.Th. Soil properties and efficient water use: water management for salinity control. *Limitations to Efficient Water Use in Crop Production*; American Society of Agronomy: Madison, WI, 1983; 73–85; Chap. 2C.

Ecological Indicators: Eco-Exergy to Emergy Flow

Simone Bastianoni
Luca Coscieme
Federico M. Pulselli
Ecodynamics Group, Department of Chemistry, University of Siena, Siena, Italy

Abstract

The holistic approach highlights the emergent properties that guide natural systems in their evolution and dynamics. Goal functions and orientors have been defined and successfully utilized to study and predict ecosystem behavior. Emergy can measure the inputs to an ecosystem from the environment in equivalent units, and it is an indicator of the energy and matter that the system needs to auto-maintain and develop. Eco-exergy represents the development of ecosystems by increase in work capacity. A combination of these two orientors, i.e., the ratio of eco-exergy to emergy flow, is proposed as an indicator of ecosystem efficiency in converting inputs of resources into ecosystem work capacity. This ratio can also be used in environmental management to assess the nutrient or pollutant effect of an input to a system and to indicate the health status of an ecosystem. Some new approaches are proposed, combining the ratio of eco-exergy to emergy flow together with the concept of ecosystem services.

INTRODUCTION

Living systems can be viewed as dissipative structures, self-organizing systems that dissipate energy for the maintenance of organization.[1] They exist by the transforming available energy potentials, building new structures as a consequence of the process.[2] To study the ecosystem in this view, we need proper tools able to take into account the biological time[3] that has been necessary for the "creative" planning and construction of dissipative structures in the system, through the concentration of energies and matter.

The study of the behavior of a single species in a system provides knowledge of the system's parts. Similar to a jigsaw puzzle, we need to know the single drawing on each piece, as well as the boundaries/shapes of the pieces, in order to properly choose where each piece must be placed, so that we can reconstruct the entire drawing. However, in a very complex context, we are not capable of reconstructing the whole drawing if we do not have a reference, a picture of the whole system that can guide us, e.g., the drawing on the box of the puzzle. Holistic indicators, orientors, or goal functions give us this kind of information, guiding the disciplinary research in a conjunction of efforts, in order to have the best description of the system. In fact, extending the study from a "simple" description of the system in a given time to a description of the system evolving through various states, the holistic indicators allow us to understand if the system under study is globally following a path that will take the system to a "better" or to a "worse" state.[4]

As maintained by Müller and Leupelt,[5] goal functions and orientors have been developed as holistic measures of the global performance of ecosystems, of what we could call, in a broad sense, complexity. Fath, Patten, and Choi[6] showed that 10 extremal principles involving orientors (power, storage, empower, emergy, ascendency, dissipation, cycling, residence time, specific dissipation, and empower/exergy ratio) can be unified by ecological network notation.

In its evolution, a system changes properties and behaviors, like human beings or other organisms change their "reactions" to the surrounding "signals" during their evolution. Every different holistic indicator is appropriate to describe the system in a particular state or phase of its evolution or a particular characteristic of the same system. When studying a dynamic system in evolution, holistic indicators (and the joint use of them) seem to be more powerful than nonsystemic indicators. Only this approach seems to be truly useful to describe the system and its dynamics, i.e., from a "young system" state, through intermediate states (i.e., of reorganization after possible perturbations), toward a climax mature stage.

This holistic approach is relatively new, but it is very efficacious and promising for ecosystem investigation and management.

In this entry, we describe the combined use of two holistic indicators, eco-exergy and emergy, composing a ratio

Encyclopedia of Environmental Management DOI: 10.1081/E-EEM-120047385

that is able to describe the system during its evolution. Moreover, it adds information to that obtained from the use of the two orientors separately. This kind of approach is in line with the book *A New Ecology: Systems Perspective* proposed by Sven Jørgensen and coauthors[4] and with what Fath, Patten, and Choi[6] maintained in their fundamental paper on the use of a plurality of "goal functions," highlighting their complementarity and interdependency.

A DESCRIPTION OF THE SYSTEM USING EMERGY

A class of indicators, based on the concept of emergy, are able to evaluate the convergence of matter and energy (several inputs) to a system. On the basis of a thermodynamic hierarchy of energy, and starting from solar energy input to the earth, emergy provides a measure of the environmental work necessary to generate an item or a flow (which could be, for example, an input to any system).

The concept of emergy derives from a reflection about the concept of energy quality.[7] The second principle of thermodynamics states that energy transformations imply an irreversible degradation of energy. In living systems and ecosystems, inputs of energy are transformed, and the energy quality changes. A portion of diluted sunlight is lost as heat, and a portion is concentrated into forms that are more able to do work, and/or more flexible to be used. According to Odum, many joules of low quality are needed for a few joules of high quality. In the case of a typical web of connections, like the food chain, "at each stage, energy is degraded as a necessary part of transforming a lower quality energy to a higher quality one in lesser quantity. The energy flows decrease as one goes up the food chain."[7] "A joule of sunlight, a joule of coal, a joule of human effort are of different quality and represent vastly different convergences of energy in their making."[8] Therefore, in many cases, it is not correct to use energy to describe the dynamics and behavior of systems. Solar energy can be considered the basis upon which energy transformations in the biosphere occur. Emergy, expressed in solar emergy joule (seJ or, according to a recent proposal, semj) is defined as the quantity of solar energy directly or indirectly necessary to produce a flow or a product. In the case of ecosystems, the emergy flow (empower) is considered, which is the quantity of solar energy directly or indirectly necessary to support the system and its level of organization.[9–11]

To compare all kinds of energy on the common basis of solar energy, solar transformity has been utilized, defined as the solar energy directly and indirectly required to generate 1 J of a product. It is a conversion factor that takes into account the position of one energy form in a sort of thermodynamic hierarchy in the biosphere. The solar transformity of the sunlight absorbed by the earth is 1 seJ/J by definition.[12]

To quantify the emergy of a product or system, all the inputs to the system or production process must be quantified and expressed in seJ by means of suitable transformities, which are used to convert different flows of energy into equivalent solar energy.[9] In case of matter inputs, the specific emergy (expressed in seJ/g or another unit) is used to convert mass into equivalent solar energy. Recently, the concept of unit emergy value has been used, independently of the unit in which the flows are expressed (energy or mass). The total emergy of a system (Em_S) is given, approximately, by the sum of the energy content (E_i) of the ith input to the system multiplied by the corresponding transformity (Tr_i), while avoiding double counting any inputs (see Bastianoni[13] for a thorough analysis of this calculation method):

$$Em_S = \sum E_i \cdot Tr_i \quad (1)$$

Every flow can be expressed by means of its solar equivalent, and a system of environmental accounting based on emergy can be implemented.[14]

Emergy represents the convergence of different kinds of energy to a system (E_i) times the quantity of solar energy that has been necessary to make available one unit (Tr_i). It is not a state function because it depends on the kinds of energy and the process that is used to obtain a certain item (that can be a product or a given state of the system). Therefore, emergy enables us to identify, quantify, and weigh the inputs that feed an ecosystem. If natural selection has been given time to operate, the higher the emergy flux necessary to sustain a system, component, or a process, the higher the hierarchical level and the usefulness that can be expected for that entity (maximum empower principle[7]).

In a pristine natural system, self-organizing from young states, this maximum empower principle (derived from the maximum power principle[15]) is realized, and the continuous increase of emergy is an indication of a proper evolution toward a mature system (climax stage).[16] Therefore, empower has been proposed as an ecosystem health indicator.[17]

In "real-world" cases, where natural systems interact with human systems and dynamics, the increase of emergy (as a consequence of the increase in energetic inputs that reach the ecosystem) is not always "good" in the sense that it will support the evolution of the system toward a climax stage. In fact, a portion of the inputs that the ecosystem receives is not used to build structures in order to maintain the nonequilibrium state (e.g., nutrient overflow, pollutants). Emergy flow alone cannot be used as a "reference direction" indicator[18] if the system is not a pure pristine system.

A DESCRIPTION OF THE SYSTEM USING ECO-EXERGY

The usable energy input to a system is converted into genetic information, biomass storages, and a relation network.[7,19,20] Structural complexity and biodiversity influence the possible evolutions of the system toward another (more or less) stable state.[21]

Eco-exergy expresses the development of ecosystems by the increase in the work capacity,[4] considering the biomass stored in a system and its genetic information.[22–25]

According to Jørgensen,[26,27] we can distinguish between technological exergy and eco-exergy: the former uses the environment as a reference state and is able to measure the useful energy provided by a production process; the latter uses as a reference state the same ecosystem at thermodynamic equilibrium. Eco-exergy thus estimates the distance of an ecosystem from thermodynamic equilibrium and is given by the formula

$$\text{Eco-Ex} = \sum \beta_i \cdot \boldsymbol{c}_i \qquad (2)$$

where $\boldsymbol{c}_i$ is the concentration of the ith component of the ecosystem and β_i is the weighting factor that accounts for the genetic information that the component carries (for a list of β values.[28]

When the available inputs to the system are used to build up new biomass and/or complexity, the system is tending to its climax stage (maximum exergy principle[4]). Eco-exergy is a state-based descriptor of a system's structure (and functions, networks, interactions) based on usable energy and information. It has also been used as an ecosystem health indicator.[24,25,29]

RATIO OF ECO-EXERGY TO EMERGY FLOW

Emergy and eco-exergy can be considered as complementary entities, with the former accounting for the amount of basic energy (solar) required to support a process or an ecosystem and the latter being the level of organization reached by a system.

The need to compare the emergy flow that supports an ecosystem with the consequent ecosystem reaction was already clear to Odum, who tried to assess the ecosystem response using the emergy-to-information ratio[7] as a measure of information hierarchy. Odum used the mass of DNA as information carrier and looked at the ratio of emergy to information measured in bits, considering that the emergy required per bit may give an indication on the usefulness of the information in an adapted system.[7,30] The emergy-to-information ratio, also used by Keitt,[31] has a problem that consists in the arbitrariness of the choice of the system's basic element: an atom or an individual of a species, a letter of an alphabet, or a gene as the basic "symbol." Bastianoni and Marchettini[30] have proposed joining emergy with eco-exergy in order to measure the ecosystem structure. The ratio between eco-exergy and emergy flow indicates the organization or structure of an ecosystem per unit of solar emergy flow required to produce or maintain it.[30] Bastianoni and Marchettini[30] first introduced thisrelation as the ratio of emergy (flow) to eco-exergy. This choice was made in order to maintain coherence with the definition of transformity (i.e., the emergy that contributes to a production system divided by the energy content of a product). In fact, the ratio of emergy flow to eco-exergy represents an empower converging to a certain system divided by the eco-exergy of the whole system. Actually, the inverse seems more comprehensible, where the effect (eco-exergy) is at the numerator and the requirement is at the denominator, as in any efficiency indicator.

The role of information and structure is fundamental when we approach the study of complex systems, such as an ecosystem. The use of eco-exergy adds something to the classical exergy approach (for an overview, see Jørgensen and Mejer,[22] Jørgensen,[27] and Sciubba and Wall[32]), which does not take into account information content. For instance, the difference between a living organism and a dead one is not related to the classical exergetic content that is in fact the same, but is related to the capability of the living system to use the information content in its DNA.[33]

The ratio of eco-exergy to emergy flow has also been considered as an indicator of the efficiency[30] in transforming available inputs, evaluated in emergy terms, into the structure, information, and ecosystem organization, evaluated in eco-exergy terms. In fact, it represents the state of the system (as eco-exergy) per unit input (as empower). Strictly speaking, its unit is $\text{J}\cdot\text{yr}\cdot\text{seJ}^{-1}$ (we maintain this representation even if β values may influence the pure thermodynamic nature of eco-exergy and its unit). Since the dimensions are those of time, it cannot be regarded as a real efficiency (which is dimensionless) but more as a proxy of efficiency. The higher its value, the higher the efficiency of the system; if the eco-exergy/empower ratio tends to increase (apart from oscillations due to normal biological cycles), it means that natural selection is making the system follow a thermodynamic path that will bring the system to a higher organizational level.[33,34] As an efficiency indicator, the ratio of eco-exergy to empower enlarges the viewpoint of a pure exergetic approach, where the exergy degraded and the eco-exergy stored for various ecosystems are compared: using emergy, there is a recognition of the fact that solar radiation is the driving force of all the energy (and exergy) flows on the biosphere, which is essential when important "indirect" inputs (of solar energy) are also present in a process, and must be identified, weighted, and finally, taken into account.

Fath, Patten, and Choi[6] have identified the ratio of eco-exergy to emergy flow as one of the possible orientors of an ecosystem: they link the emergy flow to the total system throughput and eco-exergy to the total system storage, therefore connecting the maximization of the ratio of eco-exergy to emergy flow with the maximization of residence time.

APPLICATIONS IN ENVIRONMENTAL INVESTIGATION AND MANAGEMENT

A Definition of Ecosystem Development

Ecological orientors and goal functions can indicate some aspects of the degree of naturalness of ecosystems; they provide a good basis for finding usable indicators for ecosystem health, ecological integrity, and sustainability;[5] they can also be used to evaluate the strength of human impact and an ecosystem's structural carrying capacity.

The ratio of eco-exergy to empower has often been applied in order to assess ecosystem health: in fact, ecosystems with different empower and eco-exergy can be compared with each other, also regarding their behavior and performance. In general, we can say that in natural systems, where selection has acted undisturbed for a long time, the ratio of eco-exergy to empower is higher and decreases with the progressive introduction of artificial inputs and stress factors that make the emergy flow higher and lower the eco-exergy content of the ecosystem. In the evolutionary process, close to the steady state (climax), the ratio of eco-exergy to empower tends to increase, which means that the system uses all the materials and energy available to reach a higher eco-exergy content. The same systems, once the climax has been reached, will remain in such a state for some time and can then grow/develop again only if further energy and/or materials are available. In the latter case, new sources of energy (or better emergy) can be used to build up new biomass and/or complexity of the ecosystem (stored eco-exergy). In terms of the ratio of eco-exergy to empower, when a system is relatively young and acquires new inputs, the ratio tends to be lower; when the system is developing toward the climax stage, the ratio tends to rise.[2,35]

This approach is helpful when different kinds of ecosystems are compared, natural and artificial. The former might have different quantities and qualities of available inputs, while the latter cannot be compared only on the basis of its "state" but also considering its requirements to develop and sustain the state itself. The results of a study on eight different aquatic ecosystems[30,35–37] demonstrated that the highest level of efficiency (in the exergy/empower sense) is obtained by a seminatural system within the lagoon of Venice, a farming basin developed over several centuries. Its efficiency is of the same order of magnitude as natural systems, but it is higher than systems with limited human input. Furthermore, the efficiency of the seminatural system is two orders of magnitude greater than artificial ecosystems by virtue of a higher level of organization and less need for external input.[36,38] Application to agroecosystems can be found in Bastianoni et al.[39] The ratios of eco-exergy to emergy and eco-exergy to empower have been also proposed to assess the self-organization efficiency of forest ecosystems.[40]

A Definition of Ecosystem Health and Pollution

A qualitative or quantitative change in the set of inputs can contribute to a change in a system's self-organization pattern and a system's different responses. Moreover, controlled human intervention can make a positive contribution to the system in terms of organization, information complexity, etc., (the eco-exergy of the system) that more than offset the environmental cost of the same intervention (the emergy flow corresponding to the human-induced inputs to the system).

If we consider the emergy flow to a system to vary between two equal and contiguous intervals, we will indicate the variation of emergy flow with ΔEM. Consequent changes in system organization can be measured by the variation of the exergy content of the system ΔEX. The quantity $\sigma = \Delta EX/\Delta EM$, with the dimensions of $J \cdot s \cdot seJ^{-1}$, represents the change of level of organization (exergy) of the system under study, when it is related to a change of the emergy flow. It is a quantity that is specific to the inputs that are subtracted or added. If σ is positive, the addition of emergy input gives rise to further organization (increasing eco-exergy), whereas a lowering of emergy has a negative effect on the system (decreasing eco-exergy). On the other hand, when σ is negative, a higher emergy flow implies a decrease in organization, whereas a lower quantity of one or more inputs causes increasing organization. In these two last cases, the inputs can generally be viewed as pollutants: in an evolutionary perspective, if they are removed, the system self-organizes; if they are added, the system is damaged. This provides a definition of pollution based on two holistic orientors representing system dynamics. A first-level observation of the behavior of σ (and of the system as a whole) gives information on the existence of pollutants; a deeper analysis can identify the intensity of pollution that is given by the sensibility of eco-exergy relative to a change in emergy flow.[41]

Holistic Interpretation of Ecosystem Services

Emergy and eco-exergy can be related to the concept of "ecosystem services." Ecosystem services are the benefits human populations derive, directly or indirectly, from ecosystem functions (e.g., food provision or waste assimilation).[42] This concept derives from a reconceptualization of ecosystem functioning from an anthropocentric viewpoint.[43]

In general, ecosystems utilize flows of energy and matter from the environment to maintain themselves as far as possible from thermodynamic equilibrium and to survive, grow, and/or develop. The degree of development and the efficacy with which these flows are used up and processed depend on the state/structure/organization of the ecosystem, which is a particular configuration of the abiotic–biotic system components, characterized by specific relationships between living organisms and nonliving surroundings. The outputs of an ecosystem are all flows of energy and matter moving from the system to the environment, as well as all goods and services useful for humans.

A relationship between the inflows of energy and matter supporting the ecosystem and the services it provides has been investigated by Pulselli, Coscieme, and Bastianoni,[21] who noted that in this input–output representation of ecosystems, the emergy flow (input) supporting an ecosystem and the value of the services (output) it provides are rather independent from each other, because the former depends on natural dynamics and the latter on the utility humans (decide to) draw from nature, which may vary from case to case. Despite this, at the global level, it has been calculated that nature contributes to humans not only more (as Costanza et al.[42] demonstrated) but also in a more efficient way than do all the world economic infrastructures. In fact, if we divide the world ecosystem service value by the emergy flow to the biosphere, we obtain the amount of money that is, on average, produced by 1 seJ of solar emergy feeding the global ecosystem. This ratio combines an amount of money that is not really circulating in the global economy and the flow of all renewable resources that feed the planet (sunlight, geothermal heat, rain, wind, etc.). It can be considered as a potential efficiency of the entire biosphere in providing a kind of economic wealth for humans (since at least a portion of it can be converted into real economic utility/benefit) based only on its natural functioning.[21]

The reciprocal of the above relation, i.e., the ratio of the global emergy flow to the total value of ecosystem services, is between 5.09×10^{11} and 1.51×10^{11} seJ/€ (depending on the minimum and maximum values calculated by Costanza et al.[42]). Both the maximum and minimum values are lower than values traditionally calculated for national economies, which in emergy theory are known as the emergy-to-money ratio (EMR). The EMR is given by the ratio of the emergy flow of a country (including both natural and commercial man-induced flows) to its GDP, and its order of magnitude is, in general, 10^{12} seJ/€ or more (for an overview of national values.[44] This means that the global ecosystem uses, on average, less emergy than a national economy per unit money provided to humans.

Jørgensen[26] proposed an approach to connect an ecosystem's structure and organization descriptor (eco-exergy) to a user-side measure (the value of ecosystem services), highlighting a relation between a biophysical and an economic evaluation of the environment. The calculation of ecosystem services through eco-exergy resulted in values higher than those proposed by Costanza et al.,[42] because eco-exergy represents the annual work capacity increase of an ecosystem that can be translated into the set of the possible services it can offer (not only the services that anthropic systems actually utilize) and can be compared with the actual flow of services utilized.

Ecosystem dynamics can be represented through a generic and complete input–state–output scheme. In this sense, we can imagine a kind of 3-D diagram, with the inflows of resources, measured in terms of solar emergy, on the x axis; the work capacity embodied in the system biota, expressed in terms of eco-exergy, on the y axis; and the useful services for humans, valuable in economic terms, on the z axis.[45] This multidimensional holistic approach makes it clear that inputs are used up, directly or indirectly, to produce services in output and/or to develop the system, and enables us to have an indication of changes in ecosystem dynamics, structure, and services. Within this framework, a thermodynamic/socio-ecological evolutionary time path can be acknowledged: from young systems, to climax-stage systems, to socio-ecological integrated systems.

CONCLUSION

The ratio of eco-exergy to emergy flow is the combination of two thermodynamics-based orientors: emergy flow, which quantifies the amount of resources necessary for the system to survive, and eco-exergy, which represents the actual state of the ecosystem in terms of work capacity and distance from thermodynamic equilibrium. The joint use of these two entities adds information that is useful for the investigation of the behavior of the system during its evolution. It is a measure of the ability of a system to reach and maintain a given structure (as eco-exergy) per unit input (as emergy); it is therefore a measure of the efficiency of the system in transforming available resources into organization. The use of the ratio of eco-exergy to emergy flow is important when investigating the evolution of an ecosystem or the effects on ecosystem dynamics of human intervention and infrastructures. Two further applications of this entity have been described too, which can be useful in the field of environmental management: according to the change in exergy due to a change in emergy flow, we can identify potential pollutants and define the intensity of pollution; the combined use of eco-exergy and emergy can also help in assessing the role of ecosystem services in human well-being.

REFERENCES

1. Prigogine, I. *From Being to Becoming: Time and Complexity in the Physical Sciences*; Freeman: New York, 1980.
2. Bastianoni, S.; Pulselli, F.M.; Rustici, M. Exergy versus emergy flow in ecosystems: Is there an order in maximization? Ecol. Indic. **2006**, *6*, 58–62.
3. Tiezzi, E. *The End of Time*; Wit Press: Southampton, U.K., 2003.
4. Jørgensen, S.E.; Fath, B.D.; Bastianoni, S.; Marques, J.C.; Müller, F.; Nielsen, S.N.; Patten, B.C.; Tiezzi, E.; Ulanowicz, R.E. *A New Ecology: Systems Perspective*; Elsevier: Amsterdam, 2007.
5. Müller, F.; Leupelt, M. *Eco Targets, Goal Functions, and Orientors*; Springer-Verlag: New York, 1998.
6. Fath, B.D.; Patten, B.C.; Choi, J.S. Complementarity of ecological goal functions. J. Theor. Biol. **2001**, *208*, 493–506.
7. Odum, H.T. Self organization, transformity and information. Science **1988**, *242*, 1132–1139.
8. Odum, H.T. Emergy and biogeochemical cycles. In *Ecological Physical Chemistry*; Rossi, C., Tiezzi, E., Eds.; Elsevier: Amsterdam, 1991; 25–26.
9. Odum, H.T. *Environmental Accounting. Emergy and Environmental Decision Making*; John Wiley and Sons: New York, 1996.
10. Odum, H.T. *Emergy of Global Processes, Folio #2. Handbook of Emergy Evaluation*; Center for Environmental Policy, University of Florida: Gainesville, FL, 2000.
11. Odum, H.T.; Brown, M.T.; Brandt-Williams, S. *Introduction and Global Budget, Folio #1. Handbook of Emergy Evaluation*; Center for Environmental Policy, University of Florida: Gainesville, FL, 2000.
12. Brown, M.T.; Ulgiati, S. Emergy analysis and environmental accounting. In *Encyclopedia of Energy*; Cleveland, C., Ed.; Academic Press, Elsevier: Oxford, 2004; 329–359.
13. Bastianoni, S.; Morandi, F.; Flaminio, T.; Pulselli, R.M.; Tiezzi, E.B.P. Emergy and emergy algebra explained by means of ingenuous set theory. Ecol. Modell. **2011**, *222*, 2903–2907.
14. Hau, J.L.; Bakshi, B.R. Promise and problems of emergy analysis. Ecol. Modell. **2004**, *178*, 215–225.
15. Lotka, A.J. Contribution to the energetics of evolution. Proc. National Acad. Sci. U. S. A. **1922**, *8*, 147–151.
16. Campbell, D.E. Proposal for including what is valuable to ecosystems in environmental assessments. Environ. Sci. Technol. **2001**, *35*, 2867–2873.
17. Campbell, D.E. Using energy systems theory to define, measure, and interpret ecological integrity and ecosystem health. Ecosyst. Health **2000**, *6* (3), 181–204.
18. Samhouri, J.F.; Levin, P.S.; James, C.A.; Kershner, J.; Williams, G. Using existing scientific capacity to set targets for ecosystem-based management: A Puget Sound case study. Mar. Policy **2011**, *35*, 508–518.
19. Jørgensen, S.E.; Patten, B.C.; Straškraba, M. Ecosystems emerging: 4. Growth. Ecol. Modell. **2000**, *126*, 249–284.
20. Ulanowicz, R.E. *Ecology, the Ascendent Perspective*; Columbia University Press: New York, 1997.
21. Pulselli, F.M.; Coscieme, L.; Bastianoni, S. Ecosystem services as a counterpart of emergy flows to ecosystems. Ecol. Modell. **2011**, *222*, 2924–2928.
22. Jørgensen, S.E.; Mejer, H.F. A holistic approach to ecological modelling. Ecol. Modell. **1979**, *7*, 169–189.
23. Jørgensen, S.E.; Mejer, H.F. Application of exergy in ecological models. In *Progress in Ecological Modelling*; Dubois, D., Ed.; Editions CEBEDOC: Liege, Belgium, 1981; 311–347.
24. Jørgensen, S.E. Application of holistic thermodynamic indicators. Ecol. Indic. **2006**, *6*, 24–29.
25. Jørgensen, S.E. Eco-exergy as an ecosystem health indicator. In *Encyclopedia of Ecology*; Jørgensen, S.E., Fath, B., Eds.; Elsevier: Amsterdam, 2008; 977–979.
26. Jørgensen, S.E. Ecosystem services, sustainability and thermodynamic indicators. Ecol. Complexity **2010**, *7*, 311–313.
27. Jørgensen, S.E. Exergy. In *Encyclopedia of Ecology*;, Jørgensen, S.E., Fath B.D., Eds.; Elsevier: Oxford, 2008; 1498–1509.
28. Jørgensen, S.E.; Ladegaard, N.; Debeljak, M.; Marques, J.C. Calculations of exergy for organisms. Ecol. Modell. **2005**, *185* (2–4), 165–175.
29. Zaldivar, J.M.; Austoni, M.; Plus, M.; De Leo, G.A.; Giordani, G.; Viaroli, P. Ecosystem health assessment and bioeconomic analysis in coastal lagoon. In *Handbook of Ecological Indicator for Assessment of Ecosystem Heath*; Jørgensen, S.E., Costanza, R., Xu, F.L., Eds.; CRC Press: Boca Raton, FL, 2005; 163–184.
30. Bastianoni, S., Marchettini, N. Emergy/exergy ratio as a measure of the level of organization of systems. Ecol. Modell. **1997**, *99*, 33–40.
31. Keitt, T.H. *Hierarchical Organization of Energy and Information in a Tropical Rain Forest Ecosystem*; M.S. Thesis, University of Florida: Gainesville, FL, 1991.
32. Sciubba, E.; Wall, G. A brief commented history of exergy from the beginnings to 2004. Int. J. Thermodyn. **2007**, *10* (1), 1–26.
33. Tiezzi, E. *Steps Towards an Evolutionary Physics*; Wit Press: Southampton, U.K., 2006.
34. Pulselli, F.M.; Gaggi, C.; Bastianoni, S. Eco-Exergy to emergy flow ratio for the assessment of ecosystem health. In *Handbook of Ecological Indicators for Assessment of Ecosystem Health*, 2nd Ed.; Jørgensen, S.E., Xu, F.L., Costanza, R., Eds.; CRC Press: Boca Raton, FL, 2010; 113–124.
35. Bastianoni, S. Eco-Exergy to emergy flow ratio. In *Encyclopedia of Ecology*; Jørgensen, S.E., Fath, B., Eds.; Elsevier: Oxford, 2008; 979–983.
36. Bastianoni, S. Use of thermodynamic orientors to assess the efficiency of ecosystems: A case study in the lagoon of Venice. Sci. World J. **2002**, *2*, 255–260.
37. Bastianoni, S. Emergy, empower and the eco-exergy to empower ratio: A reconciliation of H.T. Odum with Prigogine? Int. J. Ecodyn. **2006**, *1* (3), 226–235.
38. Bastianoni, S.; Marchettini, N.; Pulselli, F.M.; Rosini, M. The joint use of exergy and emergy as indicators of ecosystems performances. In *Handbook of Ecological Indicators for Assessment of Ecosystem Health*; Jørgensen, S.E., Costanza, R., Xu, F.L., Eds.; CRC Press: Boca Raton, FL, 2005; 239–248.
39. Bastianoni, S.; Nielsen, S.L.; Marchettini N.; Jørgensen, S.E. Use of thermodynamic functions for expressing some relevant aspects of sustainability. Int. J. Energy Res. **2005**, *29*, 53–64.

40. Lu, H.F.; Wang, Z.H.; Campbell, D.E.; Ren, H.; Wang, J. Emergy and eco-exergy evaluation of four forest restoration modes in southeast China. Ecol. Eng. **2011**, *37*, 277–285.
41. Bastianoni, S. A definition of 'pollution' based on thermodynamic goal functions. Ecol. Modell. **1998**, *113*, *2* (1–3), 163–166.
42. Costanza, R.; d'Arge, R.; De Groot, R.; Farber, S.; Grasso, M.; Limburg, K.; Naeem, S.; O'Neill, R.V.; Paruelo, J.; Raskin, R.G.; Sutton, P.; van den Belt, M. The value of the world's ecosystem services and natural capital. Nature **1997**, *387*, 253–260.
43. De Groot, R.S.; Wilson, M.A.; Boumans, R.M.J. A typology for the classification, description and valuation of ecosystem functions, goods and services. Ecol. Econ. **2002**, *41*, 393–408.
44. Sweeney, S.; Cohen, M.J.; King, D.; Brown, M.T. Creation of a global emergy database for standardized national emergy synthesis. In *Emergy Synthesis 4: Theory and Application of Emergy Methodology*; Brown, M., Ed.; University of Florida: Gainesville, FL, 2007; 23.1–23.15.
45. Pulselli, F.M.; Coscieme, L.; Jørgensen, S.E.; Bastianoni, S. Thermodynamics-based orientors for holistic interpretation of ecosystem services. In *Acts of the XXIV Congress of the Italian Chemical Society*; University of Salento: Lecce, Italy, 11–16 September 2011; 101.

Ecological Indicators: Ecosystem Health

Felix Müller
Ecology Center, University of Kiel, Kiel, Germany

Benjamin Burkhard
Marion Kandziora
Claus Schimming
Wilhelm Windhorst
Institute for the Conservation of Natural Resources, University of Kiel, Kiel, Germany

Abstract

This entry documents a literature review about approaches and concepts to indicate ecosystem health. After a short introduction, basic features and requirements of ecological indicators are sketched. The general concept of ecosystem health is introduced, and examples of ecosystem health indications are described. Ecosystem health is understood as the ability of an ecosystem to maintain its structure and function over time under external stress, safeguarding a sustainable provision of ecosystem goods and services contributing to human well-being. In the main part of this entry, several indicators that have been used to depict the health status of ecosystems are listed, distinguishing species and community-based indicators, holistic ecosystem theoretical indicators, indicators based on ecosystem analysis, and ending with ecosystem services that should be the target of the ecosystem health concept. Some applications of the ecosystem health indicator concept are summarized, referring to different ecosystem types, and finally, the state of the art is related to the initial requirements for ecosystem health indicators.

INTRODUCTION

Environmental management has to operate in an extremely complex framework, which can be characterized by a multitude of different components and an even higher number of interrelations between these parts.[1] Ecological as well as human and societal influences have to be taken into account, and the dominant role of indirect effects has to be realized.[2] Consequently, there is a very high demand for holistic management concepts, which approach the management object from an ecosystem-based starting point. The biggest challenge of such concepts arises from the enormous complexity of human–environmental systems.[3,4] To cope with this challenge, indicators can be suitable tools. They enable quantitative statements in spite of the complex environment, but this potential is attained due to simplification, aggregation, modeling, and abstraction. Therefore, indicators often are correlated with a high uncertainty, i.e., whether the indicated object is really well represented by the indicator.[5]

Ecosystem health is an environmental management concept that directly meets these problems.[6] It follows the important demand to manage the environment from a system-based viewpoint, acknowledging inherent complexities. Therefore, the selection of suitable indicators plays a major role for quantitative applications of the health concept.[7] In the following text, some focal approaches on how these challenging demands are met are documented. The leading questions are the following:

- What are the basic features and requirements for ecosystem-based indicators, and how are they related to the concept of ecosystem health?
- What are the focal ideas of ecosystem health, and how can they be translated into ecological indication concepts?
- What are the problems and experiences of health indication in different ecosystem types?
- How do recent indicator approaches cope with the challenging demands for ecosystem-based indication?

The entry starts with a short statement on the general requirements for ecological indicators. Thereafter, the actual indicandum—ecosystem health—is sketched, and a literature survey of established ecosystem health indicators is presented. In the final sections, indicator applications and aggregations are discussed, and demands for future development are formulated based on a comparison of identified requirements and the state of the art.

BASIC FEATURES OF ECOLOGICAL INDICATORS

Environmental management should be based upon qualitative or quantitative key variables that can be used to demonstrate the demand for management actions and the outcomes of the respective activities. Such indicators provide aggregated information on certain complex

Encyclopedia of Environmental Management DOI: 10.1081/E-EEM-120048415

Table 1 Scientific demands on good indicators.

Good indicator sets should provide **scientific correctness** basing upon the following:
• A clear representation of the indicandum by the indicator.
• Clear proof of relevant cause–effect relations.
• An optimal sensitivity of the representation.
• Information for adequate spatio-temporal scales.
• A very high transparency of the derivation strategy.
• A high degree of validity and representativeness of the available data sources.
• A high degree of comparability in and with indicator sets.
• An optimal degree of aggregation.
• Good fulfillment of statistical requirements concerning verification, reproduction, representativity, and validity.

Source: Wiggering an Müller.[9]

phenomena,[8–10] which often are not directly accessible.[11–13] Indicators are developed on the basis of specific management purposes; they often include an integrating, synoptical value, and they should be capable of showing the differences between an existing situation and an aspired-to target state.[14,15] Thus, indicators are signals for attracting attention on changes in complex human–environmental systems. Heink and Kowarik[16] propose the following indicator definition: "An indicator in ecology and environmental planning is a component or a measure of environmentally relevant phenomena used to depict or evaluate environmental conditions or changes or to set environmental goals."

Being applied for the management of human–environmental systems, indicators have to satisfy very different and challenging demands: On the one hand, scientific correctness is a major requirement, and on the other, transparency and public utility in the decision making processes are significant demands (see Tables 1 and 2). Therefore, indicator applications should be based on satisfying scientific hypotheses and relevant cause–effect relations, while they also have to translate the high complexity of ecosystems in a scientifically sound way to meet the needs of politicians and decision makers for common acceptance. Furthermore, the indicator should comprise an optimal sensitivity for the related disturbance, and it should be characterized by a clear representation of the indicandum by the indicator—in this case, it should represent the challenging properties of ecosystem health.

Table 2 Applied demands on good indicators.

Good indicator sets should provide **practical applicability** basing upon the following:
• Information and estimations of the normative loadings.
• High political relevance concerning the decision process.
• High comprehensibility and public transparency.
• Direct relations to management actions.
• An orientation toward environmental targets.
• A high utility for early warning purposes.
• A satisfying measurability.
• A high degree of data availability.
• Information on long-term trends of development.

Source: Wiggering an Müller.[9]

BASIC COMPONENTS OF ECOSYSTEM HEALTH

The pioneering ecologist Aldo Leopold's writings about land sickness[17] created the basic ideas for the ecosystem health concept. In the following decades, definitions of ecosystem health have been constantly evolving toward an increasing integration of human and societal contexts in order to understand what is considered to be a healthy ecosystem.[18,19] The concept has gained special popularity in the United States and in Canada, where ecosystem health has been integrated in legislation. Today, ecosystem health is part of various international political programs, like the Rio convention on sustainable development.[20] Here, it has been demanded in principle 7 that "states shall cooperate in a spirit of global partnership to conserve, protect and restore the health and integrity of the Earth's ecosystem." Ecosystem health does not only take into account ecological components but also requires a linkage with social, economic, and cultural dimensions (see Table 3).

De Kruijf and Van Vuuren[25] analyzed that "the present definitions of ecosystem health contain several of the following elements:

- Healthy ecosystems are free from ecosystem distress syndrome . . . ;
- Healthy ecosystems are resilient . . . they recover from natural perturbations and disturbances;
- Healthy ecosystems are self-sustaining and can be perpetuated without subsidies or drawing down natural capital;
- Healthy ecosystems do not impair adjacent ecosystems . . .;
- Healthy ecosystems are free from risk factors;
- Healthy ecosystems are economically viable;
- Healthy ecosystems sustain healthy human communities."

Summarizing the different approaches, a working definition of ecosystem health can be given as follows: Ecosystem health refers to the ability of an ecosystem to maintain its structure and function over time under external stress, safeguarding a sustainable provision of ecosystem goods and services contributing to human well-being.

The facts mentioned above have to be reflected while defining appropriate sets of indicators that can be applied for environmental management.[16] Suitable ecosystem health

Table 3 Different approaches defining ecosystem health.

Haskell et al.[21]: An ecological system is healthy and free from "distress syndrome" if it is stable and sustainable—that is, if it is active and maintains its organization and autonomy over time and is resilient to stress.'
Karr[22]: Ecosystem health is related to "the condition in which a system realizes its inherent potential, maintains a stable condition, preserves its capacity for self-repair when perturbed, and needs minimal external support for management."
Rapport et al.[23]: Ecosystem health refers to a "condition where the parts and functions of an ecosystem are sustained over time and where the system's capacity for self-repair is maintained, such that goals for uses, values, and services of the ecosystem are met."
Xu and Mage[24]: Ecosystem health refers to "the system's ability to realize its functions desired by society and to maintain its structure needed both by its functions and by society over a long time."

indicator sets have to consider ecological structures as well as ecological functions on different spatial and temporal scales. As shown by the Millennium Ecosystem Assessment,[26] most ecosystems on our planet have already been degraded under the pressure of increasing human demands. If ecosystem health shall be achieved on a long-term perspective, preventative and restorative environmental management strategies are needed. When looking at different typical ecosystem health indicators, like species diversity or water quality, many of the ecosystems on our planet can be considered unhealthy.[18] As a consequence, many ecosystem functions needed for the provision of ecosystem services have been altered. Ecosystem health refers to systems that are manipulated to satisfy human needs.[17] Therefore, ecosystem health provides a suitable conceptual framework describing the linkages between ecosystem functions, services, and human well-being.

The explicit integration of societal components makes it different from other ecosystem management concepts, for example, ecological integrity. Ecological integrity refers to the functioning of ecosystems based on self-organized processes, while ecosystem health also includes resilience and sustainability with regard to the provision of ecosystem services. Therefore, different ecological concepts like homeostasis, diversity, complexity, emergent properties, or hierarchy principles are closely related to the health concept.[19]

BASIC REQUIREMENTS FOR THE INDICATION OF ECOSYSTEM HEALTH

A focal problem of these concepts is the complexity of ecosystems that arises from the high number of components, relations, and interactions. Hence, for environmental practice and decision making, this complexity has to be reduced. Ecosystem theories provide an applicable basis for such a

Table 4 Axioms of ecosystem health.

Dynamism:	Nature is a set of processes, more than a composition of structures.
Relatedness:	Nature is a network of interactions.
Hierarchy:	Nature is built up by complex hierarchies of spatio-temporal scales.
Creativity:	Nature consists of self-organizing systems.
Different fragilities:	Nature includes various sets of different resiliences.

Source: Wiggering an Müller.[9]

reduction. Some of the respective theoretical fundamentals of the ecosystem health concept are listed in Table 4.

These requirements are summarized in the "V-O-R model,"[21] describing the ecosystem vigor, organization, and resilience. Vigor is indicated by activity, metabolism, or primary productivity, while organization represents the diversity and number of interactions between the system components. Resilience is understood as a system's capacity to maintain structure and function in the presence of (external) stress. When resilience is exceeded, the system can shift to an alternate state. By including this approach, ecosystem health is closely related to the concepts of stress ecology, where vigor, system organization, resilience, and the absence of ecosystem distress are the main factors for a system's condition.

This model is correlated with a very high demand for comprehensive data sets and long time series to determine resilience. Additionally, linkages of environmental and social–economic attributes and attributes representing structures as well as functions and organization have to be included. All these demands can hardly be fulfilled. Therefore, quantification deficits have to be expected as one of the main problems of the ecosystem health approach.

UTILIZED INDICATORS OF ECOSYSTEM HEALTH

In the following paragraphs, a short literature survey of health indicators is presented, whereby different approaches have been distinguished: In the beginning, community-based indicators are listed, followed by aggregated theoretical indicators and indicators based on ecosystem analysis. Finally, a link will be developed toward the indication of ecosystem services.

Species- and Community-Based Indicators and Indices

Biodiversity loss is one of the characteristic signs for ecosystems under stress[18] and thus is a major issue in environmental management.[26] Consequently, there are many initiatives and concepts to describe and assess biodiversity.[27] Biodiversity indicators and indices are based on the abundance, absence, or composition of selected species or communities. They vary from single-species indicators to complex composite indicators. Suitable indicator species have to be selected in order to be representative for certain phenomena or sensitive to particular environmental changes.[28] Therefore, the appearance and dominance of certain communities can be associated with states of ecosystem health.[7]

The parameters used to quantify respective indicators can be derived from direct measurements and observations of selected species' abundance. Species- and community-based indicators can be linked to numerous international and national policy instruments, for which biodiversity indicators need to be derived (e.g., Bern Convention 1979; Bonn Convention 1979; Convention on Biological Diversity CBD 1992; the Millennium Development Goals to be achieved by 2015). Most policy and decision makers rely on indices that aggregate biodiversity data across large numbers of species,[29] but also, a limited number of key taxa are frequently used to indicate ecosystem health.[30] Some of these biodiversity indicator concepts—mainly related to the indicator collections of Marquez et al.[31] and Joergensen et al.[6]—are presented in the following.

Species Richness

The most established way to indicate biodiversity is based on species counts and composite indices. One advantage is that the number of species in a certain area is a measure that is easily established and understood by a broad range of people. The Shannon–Wiener index and the Simpson index are the most commonly applied indicators.[29] The Shannon–Wiener index H' originates in information theory and integrates species' number and evenness:

$$H' = -\sum p_i \log 2p_i \qquad (1)$$

p_i is the proportion of individuals found in species i. The values of this index can vary between 0 and 5. H' has a maximum value if the individuals of all species occur with the same density. The Simpson index refers to the number of species present and the relative abundance of each species.[32] Species richness provides important information on ecosystem conditions. However, its application in ecosystem health assessments can be misleading, for example, concerning nonnative (exotic or "invasive") species. Their abundance will increase the value of standard biodiversity indicators, but their increased dominance in biotic communities can be a typical sign of ecosystem stress.[18]

Indicator Taxa

Indicator taxa (or bioindicators, indicator species) are species or higher taxonomic groups whose properties can be used as proxies for assessments of ecosystem health.[33,34] Respective species have to be selected in order to react on ecosystem alterations by changing their abundance, density, conditions, or activities. Therefore, indicator species have to be selected objectively and must represent clear indicator–indicandum relationships. Pollinators have been suggested as useful bioindicators for ecosystem health by Kevan[35] as they are crucial for the functioning of almost all terrestrial ecosystems. Moreover, pollinators are needed for the provision of manifold ecosystem services. Further indicators based on the abundance of selected species have been suggested by Jørgensen et al.[6,36] and Burkhard et al.[7]

- Saprobic Classification: The saprobe index gives information about the degree of water pollution.[37] The different saprogenic stages are related to certain indicator organisms like bacteria, fungi, algae, amoeba, mussels, worms, insect larvae, or fishes. The stages range from polysaprobic (very highly polluted), α-mesosaprobic (highly polluted), β-mesosaprobic (medium polluted), to oligosaprobic (rather clean and clear water).

- Bellan's Pollution Index: Aquatic species like *Platynereis dumerilii*, *Theosthema oerstedi*, *Cirratulus cirratus*, and *Dodecaria concharum* are used as indicators for water pollution, whereas species like *Syllis gracillis* and *Typosyllis prolifera* indicate clear water conditions.[38] Bellan's pollution index equation is

$$\text{IP} = \sum \text{dominance of pollution indicator species/clear water indicators} \qquad (2)$$

Index values higher than 1 indicate a pollution-based disturbance in the community.

- AZTI Marine Biotic Index: AZTI Marine Biotic Index (AMBI) distinguishes the soft bottom macrofauna into five groups in accordance with their sensitivity to increasing stres[39]:

I. Species very sensitive to organic enrichment and eutrophication, present only under unpolluted conditions.
II. Species indifferent to organic enrichment, occurring in low densities only and with no significant variations over time.
III. Species tolerant to excess organic matter enrichment, usually supported by organic enrichment, that can also be found under normal conditions.
IV. Second-order opportunist species.
V. First-order opportunist species (deposit feeders).

The coefficient is calculated as follows:

$$\text{AMBI} = \{(0 \times \%\,\text{I}) + (1.5 \times \%\,\text{II}) + (3 \times \%\,\text{III}) + (4.5 \times \%\,\text{IV}) + (6 \times \%\,\text{V})\} / 100 \qquad (3)$$

The AMBI values vary among the following: normal (0.0–1.2), slightly polluted (1.2–3.2), moderately polluted (3.2–5.0), highly polluted (5.0–6.0), or very highly polluted (6.0–7.0).

- BENTIX

BENTIX is based on the AMBI but uses three groups only[40]:

I. Species generally sensitive to disturbances.
II. Species tolerant to stress or disturbance. Populations may respond to organic enrichment or other pollution sources.
III. First-order opportunistic species (pioneer, colonizers, or species that are tolerant to hypoxia).

The indicator values are calculated as follows:

$$\text{Bentix} = \{(6 \times \%\ \text{I}) + 2\ (\%\ \text{II} + \%\ \text{III})\}\ /\ 100 \quad (4)$$

The results represent different states of aquatic ecosystems: normal (4.5–6.0), slightly polluted (3.5–4.5), moderately polluted (2.5–3.5), highly polluted (2.0–2.5), or very highly polluted (Bentix = 0).

- Macrofauna Monitoring Index: Twelve indicator species are included in the macrofauna monitoring index. Each indicator species is assigned a score, based on the ratio of its abundance. The actual index value is the average score of species that are present in the sample.[41]
- Umbrella, Flagship, and Keystone Species: Umbrella species have high demands for their habitat conditions with regard to habitat size and quality. When protecting these species, many other species will be supported automatically. Flagship (or charismatic) species are organisms whose necessity for protection can be easily communicated. Keystone species provide an extraordinary importance for the maintenance of ecosystem structures and functions as well as for other species in the same ecosystem. Therefore, the identification and protection of keystone species can be crucial for the management of ecosystem health.[42]

  Examples for the utilization of these indicator types are the Species Trend Index,[43] Red Lists,[44] or the Living Planet Index (LPI), which has been developed for land, freshwater, and marine vertebrate species. The average population trends over time are documented in the LPI. The actual calculations are based on a data set of more than 2500 species and 8000 population time series over the past 30 years. Three indices are calculated: 1) terrestrial species population index; 2) freshwater species population index, and 3) marine species population index. Each of these indices is set to a baseline of 100 in 1970, and all are given an equal weighting.[45,46]

Ratios Between Different Classes of Organisms or Elements

The increase or decrease of one species in relation to others provides information about changes in ecosystems, for example, Nygard's algal index[36] or the diatoms/non-diatoms ratio.[47]

Indicators Based on Ecological Strategies

Different ecological strategies are altered by human activities or during different stages of natural development. Hence, indicators for the distinct behavior of different taxonomic groups under environmental stress situations were developed, e.g., the nematodes/copepods index, the polychaetes/amphipods ratio, and the index of *r*/*k* strategists, which considers different taxa: Most communities in ecosystems in rather late developmental stages show dominance of *k*-selected or conservative species with large body sizes and long life spans. *R*-selected or opportunistic species have shorter life spans and are often numerically dominant. After a significant disturbance and during the following reorganization, the opportunistic species can become dominant in biomass as well as in number, whereas the conservative species are usually less favored.[36,48] Another strategy-related indicator is the trophic infaunal index, which refers to organisms' different feeding strategies (distinction of macrobenthos species into suspension feeders, interface feeders, surface deposit feeders, and subsurface deposit feeders).[49]

Additionally, there are several attempts aiming at harmonizing existing biodiversity indicator initiatives. Two examples are sketched here.

- Streamlining European 2010 Biodiversity Indicators: The Streamlining European Biodiversity Indicators (SEBI) were established in 2005 under the umbrella of the Convention on Biological Diversity (CBD). It is a process to select a set of biodiversity indicators to monitor progress toward the 2010 target of halting biodiversity loss and help achieve progress toward the target.[50,51] The SEBI is a regionally coordinated program that has been initiated in Europe as collaboration between the European Environment Agency and other European and United Nations institutions. The SEBI proposes a list of 26 indicators within the 7 CBD focal areas: status and trends of the components of biological diversity, threats to biodiversity, ecosystem integrity and ecosystem goods and services, sustainable use, status of access and benefit sharing, status of resource transfers, and public opinion.[50,51]
- Group on Earth Observations Biodiversity Observation Network: The Group on Earth Observations Biodiversity Observation Network (GEO BON) is a global partnership helping to collect, manage, analyze, and

report biodiversity data.[52] It is a voluntary partnership of 73 national governments and 46 participating organizations and was launched in 2002. The GEO BON aims at providing a framework to coordinate data and observations within the Global Earth Observation System of Systems (GEOSS). Biodiversity has been named as one of nine GEOSS priority societal benefit areas. The GEO BON will integrate key ecosystem functional parameters into a Terrestrial Ecosystem Function Index (TEFI). The TEFI will integrate data of measurements of the energy, carbon, and nutrient balance.[52]

Aggregated Theory-Based Indicators and Indices

In contrast to the biotic approaches, which mainly can be used as structural indicators, the health component "vigor" is included in ecosystem theory based-indicators and indicator sets. The following four aggregations stem from thermodynamics, network, and information theories. Their basic target is a holistic aggregation of ecosystem properties into one guiding variable.

- Exergy and Exergy Indices: Exergy is that energy fraction that can be transformed into useful mechanical work. In ecological terms, it can, for example, be measured by the total biomass of the system. Eco-exergy is a refinement of the exergy concept in which biomass is weighted by the genetic complexity of the species observed.[4,36] Further holistic indicators are the exergy index and the specific exergy.[4] In ecosystems, the captured exergy is used to build up biomass and structures during successions.[48] Therefore, more complex systems also have more built-in exergy than simpler ones. Both exergy and specific exergy have been used as indicators for ecosystem health.[6] Relations between the exergy values and other ecosystem health characteristics like diversity, structure, or resilience can be found. For example, a very eutrophic ecosystem has a very high exergy due to the high biomass, but the specific exergy is low as the biomass is dominated by algae with low β values. The combination of exergy index and specific exergy provides a satisfactory structural and holistic description of ecosystem health.

- Entropy: Entropy production is one result of any metabolic activity. It can be measured by the system's respiration or the total system's export. As life is a very effective producer of entropy, this indicator has been proposed as an ecological orientor to represent maturity as well as ecosystem stress.[53,54]

- Emergy: Emergy (embodied energy) accounts for the differences between distinct biomass fractions in ecosystems basing upon the energy that has been used to build up the respective structure.[55] Conversion values, called transformities, have been derived to allow the calculation of emergy values for many ecological entities as well as socio-ecological products.[56]

- Ascendency: Ascendency is a holistic indicator that is based upon the energy flows in ecological systems and the information associated with the particular network configuration.[57,58] It represents the total system throughput and the flow diversity as a result of the food web structure. The respective network configuration is indicated by the average mutual information. Ascendency is measured by the total system throughput times the average mutual information, providing helpful information on an ecosystem's energy flow schemes and efficiencies.

Ecosystem Analytical Indicators and Indices

While most of the approaches mentioned before are aiming at one focal dimension and one value to characterize the state of an ecosystem, the following indicators have been constructed as multidimensional approaches. They try to represent ecosystem structures (biotic and abiotic structures), functions (water, matter, energy flows), and (in some cases) their relevance for human systems.

- Integrity Indicators and Orientors: Several ecosystem assessments are based on the concept of ecosystem integrity, which is closely related to ecosystem health.[18,59] The focal difference can be found in the origins of the concepts: While integrity was related to wilderness as a target function, health has been referring to ecosystems under human pressures from the beginning.[60] Meanwhile, the core conceptions have become rather similar. Therefore, one approach of integrity indication will be included in the following paragraphs.

  Taking into account the focal ideas of the sustainable development concept, "meet the needs of future generations" means "keep available ecosystem services on a long-term, intergenerational and broad scale, intragenerational level." From a synoptic viewpoint, all ecosystem services are strongly dependent on the performance of the system's regulation capacity (see the section on "Ecosystem Service Indicators"). Taking into account that the integral of regulating ecosystem services represents self-organized processes,[61] it becomes clear that the respective benefits are dependent on the degrees and potentials of self-organization. To maintain these services, the ability for future self-organizing processes has to be preserved.[53,62] Under this viewpoint, Barkmann et al.[63] have defined ecological integrity as a "management target for the preservation against nonspecific ecological risks, that are general disturbances of the self-organising capacity of ecological systems. Thus, the goal should be a support and preservation of those processes and structures which are essential prerequisites of the ecological ability for self-organisation".

In ecosystem theory, many different approaches (see Joergensen[4]) are highly compatible with the theory of self-organization. The consequences have been condensed within the orientor approach,[64–66] a system-based theory about ecosystem development, which is founded on the ideas of nonequilibrium thermodynamics,[54,62] and network development on the one hand and succession theory on the other.[67,68] The basic idea is that throughout the undisturbed complexifying development of ecosystems, certain characteristics are increasing steadily, developing toward an attractor state, which is restricted by the specific site conditions. For instance, the food web will become more and more complex; heterogeneity, species richness, and connectedness will be rising; and many other attributes will follow a similar long-term trajectory.

Many of these orientors cannot be easily measured or modeled under usual circumstances. Therefore, the selected indicators have to be represented by variables that are accessible by "traditional" methods of ecosystem quantification. Furthermore, it has to be reflected that the number of indicators should be reduced as far as possible, providing a small set consisting of the most important items that can be calculated or measured in many local instances. The local subsystems that should be taken into account to represent ecosystem organization are ecosystem structures with the biotic and abiotic diversity and functions, represented by the energy, water, and matter balances (for a detailed justification, see Müller[61]).

On the basis of these features, a general indicator set to describe the ecosystem or landscape state in terrestrial environments has been derived. It is shown in Table 5. The basic hypothesis concerning this set is that a holistic representation of the degree of and the capacity for complexifying ecological processes on the basis of an accessible number of indicators can be fulfilled by these variables. They also represent the basic trends of ecosystem development; thus, they show the developmental stage of an ecosystem or a landscape. As a whole, this variable set represents the degree of self-organization in the investigated system. For quantifications, see Müller[69] or Müller and Burkhard.[70]

- The holistic ecosystem health indicator: An expansion toward an integration of human items is provided by the holistic ecosystem heath indicator (HEHI) system. It was developed in 1999 in Costa Rica as an integrative indicator that might be an appropriate tool for assessing and evaluating health of managed ecosystems.[73] The HEHI follows a hierarchical structure starting with three main branches: ecological, social, and interactive. The interactive branch includes measures relating to land use and management decisions that characterize the interactions between the human communities and the ecosystem. Furthermore, each branch is subdivided into categories or criteria (see Table 6). Each category is given a target or a benchmark, which is based on references available in scientific literature, policies, etc.

Table 5 Proposed indicators to represent the organizational state of ecosystems and landscapes.

Orientor group	Indicator	Potential key variable
Biotic structures	Biodiversity	Number of species
Abiotic structures	Biotope heterogeneity	Index of heterogeneity
Energy balance	Exergy capture	Gross or net primary production
	Entropy production	Entropy production[71]
		Entropy production[72]
		Output by evapotranspiration and respiration
	Metabolic efficiency	Respiration per biomass
Water balance	Biotic water flows	Transpiration per evapotranspiration
Matter balance	Nutrient loss	Nitrate leaching
	Storage capacity	Intrabiotic nitrogen
	Soil organic carbon	Soil organic matter

Source: Müller.[69]

Ecosystem Service Indicators

As the explicit target figure of ecosystem health indicators has more and more been moved toward human well-being, the respective indicanda—ecosystem goods and services—are mentioned here. Their implementation seems to be very significant as a criterion of success in ecosystem health management. Ecosystem services are the benefits people obtain from natural structures and functions. Since ecosystems are dynamic and complex units, the assessment of their services is strongly facilitated by the categorization into functional groups, which are exemplarily listed in Table 7, referring to the following:

- Provisioning services: products obtained from ecosystems, e.g., food, water.
- Regulating services: benefits from regulating ecosystem processes, e.g., flood regulation, disease regulation.
- Cultural services: nonmaterial benefits, e.g., recreation, spiritual benefits, information.

Ecosystem structures and functions can be indicated by ecological integrity, as described above. Regulating ecosystem services is strongly related to ecosystem functions, and some regulating services are even overlapping with ecological integrity processes (e.g., processes related to nutrient or water regulation.[74] Thus, clear definitions of ecological integrity variables and regulating ecosystem service indicators are mandatory. Most ecosystem functions are difficult to quantify under natural conditions, but

Table 6 Elements of the holistic ecosystem health indicator set.

Ecological elements	Social elements	Interactive elements
Soil quality	Income	Land use and distribution
Riparian zone	Access to	Watershed protection
Water quality	services	Land degradation
Biomass	Job stability	Citizen involvement
Land use	Gender roles	Implementation
Primary	Demographics	of legislation
production	Community	Environmental awareness
Regeneration	Strength	
Biodiversity		
Erosion		

Source: Aguilar.[73]

the application of ecological models can help. Perhaps the best data are available for provisioning ecosystem services. Normally, production and trade quantities and their market prices are used. Cultural ecosystem services again are rather difficult to quantify due to each individual's subjective and situation-dependent appreciation of related values.[26]

Ecosystem services are not a linear chain from means to ends because ecosystems as well as societal systems are complex, dynamic, and adaptive.[3] There exist multiscale relationships between services and benefits. When ecosystems are stressed and degraded, their service provision is affected too, which in turn has impacts on human activities and health.[75] The fact that there is a high correlation between decline in ecosystem health and service provision leads to the suggestion that ecosystem services are good integrative and aggregate measures, showing the consequences of the respective ecosystem health conditions.[76]

HEALTH INDICATORS IN DIFFERENT ECOSYSTEM TYPES

To illuminate the wide field of health indicators, in the following sections, some utilizations of the concept in different ecosystem types are presented.

Agroecosystem Health

An agroecosystem can be defined as "a socio-ecological system, managed primarily for the purpose of producing food, fiber and other agricultural products, comprising domesticated plants and animals, biotic and abiotic elements of the underlying soils, drainage networks, and natural vegetation and wildlife."[77–79] The health status of agro ecosystems has been described by Rapport,[80] and Rapport et al.[81] In this entry, a valuable overview about various approaches to indicate agroecosystem health on various spatial scales is provided. For example, Zhang et al.[82] merge geographical information systems (GIS)-based land use analysis data with the use of pesticides and their pathways through the environment. This approach allows the assessment of ecosystem health as a ratio between the amounts of pesticides applied in one grid to the maximum dose applied in the study area. Kaffka et al.[83] suggest that the capacity to retain nutrients like N and P might be useful to evaluate the ecosystem health of a catchment area, while Mitchell et al.[84] consider the content of soil organic matter (SOM) of agricultural areas as a focal indicator to assess ecosystem health in the foreground. Hopkins[85] indicates ecosystem health via the number and size of wildlife patches in an area dominated by intensive agriculture.

Table 7 List of ecosystem services.

Regulating ecosystem services	
Local climate regulation	Effects on temperature, wind, radiation, precipitation
Global climate regulation	For example, carbon sequestration, greenhouse gas emission
Flood protection	Extreme flood event dampening
Groundwater recharge	Runoff, flooding, aquifer recharge
Air quality regulation	Removal of toxic and other elements from the atmosphere
Erosion regulation	Soil retention and prevention of landslides
Nutrient regulation	(Re)cycling of, e.g., N, P, or other elements
Water purification	Removal of impurities from fresh water
Pollination	For example, by wind and bees
Provisioning ecosystem services	
Crops	Edible plants
Livestock	Edible animals
Fodder	Animal fodder
Capture fisheries	Fish accessible for fishermen
Aquaculture	Terrestrial or marine aquaculture
Wild foods	For example, berries, mushrooms, hunting, fishing
Timber	Trees or plants for construction
Wood fuel	Trees or plants for heating, cooking
Energy (biomass)	Trees or plants for energy generation
Biochemicals and medicine	Production of biochemicals, medicines
Freshwater	For example. for drinking, irrigation
Cultural ecosystem services (selection)	
Recreation and aesthetic values	Landscape and visual qualities
Intrinsic value of biodiversity	Value of nature and species themselves

Source: Belcher and Boehm,[87] Schönthaler et al.[111] and Reuter et al.[112]

In comparison to the aforementioned authors, Altieri and Nicholls[86] base their suggestions to achieve healthy agro-ecosystems on the avoidance capacity of diseases—indicated by optimal recycling of nutrients, closed energy flows, water and soil conservation, and biological pest regulation. Belcher and Boehm[87] use yield, soil N and P, soil water, SOM, soil erosion, and CO_2 emissions as major attributes in their sustainable agroecosystem model. However, the assessments of agroecosystem health regularly focus on resources that have to be classified as internal to the system.[88] Hence, the options to assess the health of an agroecosystem in relation to its ability to adapt to variations in its changing socio-economic and ecological context are rarely realized according to Waltner-Toews[79] or Ikerd.[89]

Forest Ecosystem Health

In the United States, 20 years ago, a sound definition of forest health had already been derived to sustain healthy conditions of ecosystem development and productivity in a long-term perspective.[90] The term "forest health" increasingly found evidence in mandates concerning environmental management and protection, mostly supported by the idea that the conventional measures for describing forest states (e.g., crown conditions, tree growth, loss of nutrients, soil potentials, biodiversity) can also be used to indicate forest ecosystem health. Regarding forests, only recently in Europe, Ecosystem Health has been advertised in the context of deposition of air pollutants[91] and the consequent change of chemical states, particularly the degree of eutrophication and acidification which relate to the holistic aspects of productivity and biogeochemical cycling. Information on nutrient recycling, imbalance with the inputs and outputs and on energy use are the essentials of the Ecosystem Assessment Health concept and the adequate indicators proposed by Jørgensen.[6] This kind of concept, basically productivity related and holistic, does greatly comply with ecosystem theories and opens on for indicating Forest Ecosystem Health combing both, utilitarian and ecosystem perspectives[92] respecting the conditions under which self-organization of forest ecosystems can take place.[50]

Whereas atmospheric deposition of acidifying air pollutants and eutrophic nitrogen is identified as one of the major environmental problems, an ecosystem process-orientated indicator has been demanded for quantifying marginal loads for damaging structures and for interference with ecosystem functions.[93–95] Since the respective concept of critical loads is a stoichiometric approach and a function of the load quantity on chemical effects on ecosystems, intensity criteria are needed to provide adequate threshold values.[96–98] The conduction of the concept depends on combined balances of mass and charge provided by mineral elements, nitrogen, and free acidity completing the ion composition of internal transfers and the matter–flux relationship with the abiotic environment. Imbalance between input and recycling respecting mass and ionic charge indicates the efficiency of nutrient use while the quantities can be related with the intensities of effective concentrations and free acidity. In this regard soil chemistry is a function of the extensity of production on the intensities of nutrient availability, free acidity and effectiveness of toxic concentrations under influence of ecosystem self-organization.

The performance of cycling and imbalance between input rates and the degree of recycling provide useful information on the Ecosystem Health aspect. Moreover, as imbalance of mineral element cycling and related losses from the ecosystem are irreversible processes for terrestrial ecosystems, decreasing alkalinity in combination with increasing free acidity are directing to maturity of ecosystems. Regarding abiotic structures, maturity emerges by development of soil structures, which is related to loss of potentials. Based on these principles, Ulrich[99] suggested the indication of stages of forest ecosystems, for instance, by the structural properties of soil constituents providing acid neutralization and buffer capacity.

Aquatic Ecosystem Health

Aquatic ecosystems (wetlands, marine and estuarine zones, lakes, groundwater, lagoons, and rivers) consist of complex structures and fulfill important functions for the provision of numerous ecosystem services.[26] Some of these aquatic ecosystems are heavily endangered, e.g., wetlands were turned into agricultural land with dramatic consequences due to the loss of the buffer capacity for pesticides, nutrients, and floods as well as the loss of habitat functions. Marine and estuarine ecosystems are heavily influenced by humans due to population growth and the associated consequences such as pollution, growing demands for resources, eutrophication, overfishing, and habitat modification (e.g., mangrove clearing). Estuarine and marine ecosystems are interdependent as estuarine areas are the nursery grounds for many species, which are then provided as successful functioning marine commercial stocks. Probably the most important function of lakes is the freshwater storage, which provides ecosystem services for society and economy, but these systems also are heavily endangered.

Aquatic ecosystems are focal areas of ecosystem health assessments. Therefore, many of the previously mentioned indicator concepts mainly provide information on aquatic ecosystem health (see Utilized indicators of ecosystem health or Joergensen et al.[6]). Besides these long indicator listings, Boesch and Paul[100] highlight the following traditional indicators: contaminant levels, material input (e.g., nutrients, sediments), water quality (e.g., dissolved oxygen), fish catch, extent of certain habitats (e.g., wetlands), community structure, toxicity biomarkers, and indicators of human pathogens).

Urban Ecosystem Health

A very special aspect is provided by the health concept in urban systems, i.e., because the human factor plays a dominant role in the relevant literature.[101–104,105] For instance, Su et al.[101] state that "an urban ecosystem consists of residents and their environment in certain time and space scales, in which, ecologically-speaking, consumers are the dominant component lacking producers and decomposers."

Therefore, urban ecosystem health must be assessed by very comprehensive, integrative indicator sets, which also include variables of human health. Consequently, Hancock[104] has determined six basic elements for healthy cities:

1. Population health and distribution.
2. Societal well-being.
3. Government, management, and social equity.
4. Human habitat quality and convenience.
5. Natural environment quality.
6. Impact of the urban ecosystem on the larger-scale natural ecosystem.

A similar approach can be found in the indicator sets of Su et al.,[101] who generally distinguish human and environmental subsystems (see Table 8).

Table 8 Some indicanda for urban health.

Human subsystem	Environmental subsystem
Public health	Provisioning services
Health expenditure	Environmental quality
Nutrition	Atmospheric quality
Budget and finance situation	Water quality
Urban infrastructure	Forest coverage
Human housing conditions	Farmland area
Education	Emergy density
Employment	Carrying capacity

Source: Su et al.[101]

INDICATOR APPLICATION AND AGGREGATION IN MANAGEMENT

Management toward ecosystem health is directed to improve human well-being. However, in order to environmentally manage, alternatives to achieve higher levels of human well-being and ecosystem health or to stabilize the present level in a changing world have to be identified. Hence, indicators have to facilitate the comparability of states in space and time, in order to allow informed decisions based on assessments of the present state and the state achievable by management options in respect to the power and competences in the hands of the manager. Concerning ecosystem health, at least, decision makers acting on the following levels have to be equipped with indicators:

1. Site management, e.g., field, forest, lake
2. Unit management, e.g., farm, forest district, catchment, nature sanctuary
3. Public management, e.g., community, county, nation
4. Public, e.g., citizens, interest groups (nongovernmental organizations)

All levels are interrelated. For example, 1) different kinds of pest management on the site level change not only the local ecosystem but also the food quality and availability. 2) The fodder quality available on the farm level depends on the productivity of the fields and impacts the economic efficiency of the farm. 3) The socio-economic state on the community level depends, on the one hand, on the economic viability of the hosted unit (item 2), but the provision of public services likewise constrains the range of activities of the units. 4) Adaptation to changing global constraints of public (item 3), unit (item 2), and site (item 1) management largely depends on the public awareness and level of satisfaction, e.g., human well-being realized at present and achievable in the future.

To account for all levels requires a nested approach in indication of ecosystem health plus mirroring the mutual synergistic and antagonistic interactions between the different levels. The need for such an approach has been articulated by Walter-Toewe and Wall[88] and Rapport and Singh.[106] However, a broadly accepted scheme meeting these requirements is not in place yet.[107] The present state of the art is largely influenced by the concept of ecosystem services and human well-being presented by the Millennium Assessment[26] and the TEEB Study.[108] With respect to "management," the further development of indicators should be constrained by the range of management options in space and time available to decision makers on the indicated levels (1–4) and be limited to parameters relatable to indicators of human well-being.

The need for indicator aggregation evolves out of the definition of ecosystem health. According to Waltner-Toews and Wall,[88] a nested approach is required in this context also. Another constraining aspect to be considered is societal interests, which are embedded in the definition as well. In addition, in case that the state of ecosystem health is considered to be poor, new or changed management activities have to be initiated and monitored by indicators. Thus, the goal of indicator aggregation is to transfer the information about the ecosystem state to those spatio-temporal scales on which management is possible and performed. The respective levels range from site management via communities, counties, states, nations up to the international institutions. Hence, a satisfying overlap between the spatial extent of the ecosystems at stake and the respective management unit is required. Systems theory and hierarchy theory provide a theoretical background to facilitate indicator aggregation for such nested systems.

To practically deal with this means to accept generalizations, including losses of information, as never can all interactions causing emergent properties relevant for human well-being be known. Hence, bottom-up aggregation is normally severely restricted. Furthermore, the focus on ecosystems enforces the integration of components, which are measured with parameters that cannot be aggregated on the base of concise units, e.g., the mortality rate of a specific species in forests of a watershed is already challenging to determine but analytically impossible to fuse with the number of pathogens endangering the fish population in a lake of the same catchment. A feasible approach to deal with this challenge is to work with relative indicator values, i.e., to indicate relative changes of the selected variables. A suitable option to facilitate spatio-temporal comparisons is to define a particular ecosystem state as reference state and to study the relative alterations from this reference state. Examples harnessing this approach have been presented by Windhorst et al.[109,110] A suitable method to facilitate aggregation and to deal with spatio-temporal interactions is to use simulation models (Belcher and.[87,111,112]

In any case, all ecosystem health indicators and respective aggregation approaches should be connectable to the different components of human well-being. The Millennium Assessment's[26] categories were "security, basic material for good life, health, good social relations and freedom of choice and action." They can be indicated for the management units at stake. Hence, multiple interactions to be considered take place and further are conceivable for future situations, creating a fuzzy environment, obstructing the development of generally applicable aggregation procedures. However, progress in identifying suitable aggregation procedures can be achieved by answering the following questions for each management unit at stake:

- Can the aggregated indicator of ecosystem health be addressed and modified by at least one management action?
- Are changed values of the aggregated indicator indicative for different ecosystem states of the management unit at stake?
- Do changed values of the aggregated indicator indicate betterment of at least one category of human well-being?

CONCLUSIONS

The previous paragraphs have shown that there is an enormous variety of indicators proposed to represent ecosystem health. Many of these indicators have been used to assess ecosystem health in different ecosystem types, and in many of those cases, the indicators have been useful tools for environmental management. On the other hand, these applications illuminate some general problems of health indicators that should be solved in the future:

- The health approach is very challenging, i.e., due to its comprehensive character. Therefore, health status can hardly be represented by one variable alone. Instead, comprehensive indicator sets are necessary to include the basic elements of vigor, organization, and resilience.
- Indicators or sets selected have often failed in fulfilling the comprehensive criteria of ecosystem health assessments because they are mostly specific for one particular environmental problem to be solved[6] and do not adequately represent ecosystem complexity. Thus, there are difficulties in satisfying the original ecosystem health idea advertised.
- Consequently, the meaningfulness of solely structural indicators is limited, as they do not reflect processes of ecosystem functions. Therefore, many of the listed indicators should be understood as elements of indicator sets, not as single indicators of ecosystem health.
- As the health concept has been outlined in strict contact with human systems, the linkage between man and environment should be included, at least in indicator selection. That linkage up to now can hardly be found in the literature.
- Due to the metaphoric character of the "health" approach, it has been very successful in some areas, while in other nations, it has not been applicable due to critical viewpoints on the concept's theoretical or even philosophical character.
- Furthermore, the indication of ecosystem health also covers the health status of the human population, as well as socio-economic and cultural dimensions in relation to the vigor, organization, and resilience of ecosystems.[18] Hence, parameters to indicate security, basic material for good life, health, and good social relations are indispensable to analyze overall ecosystem health, which is in line with the assessment of Rapport et al.[75] An integrative approach to bundle a suite of indicators and to attach meaningful values is the concept of ecosystem distress syndromes,[113] which can be seen as a forerunner of the environmental degradation syndromes elaborated by the German Advisory Council on Global Change (WBGU). The three major syndromes have been named by WBGU[105]: 1) utilization, which includes the overexploitation of marginal land; 2) development which includes the destruction of ecosystems as a result of large-scale projects; and 3) sink, comprising environmental degradation resulting from large-scale diffusion of long-lived substances.

Coming back to our initial questions, we can summarize the following:

- The basic features and requirements of the ecosystem health approach demand for comprehensive and integrative indicator sets, which is a big scientific and practical challenge.
- Therefore, several proposals exist, and the health concept is used as a reference in several cases, but very often, the interdisciplinary demands of the approach are not fulfilled.

- Applications can be found mainly in aquatic ecosystems, mostly being quite distant from the involvement of human factors.
- Good chances for future development can be seen by enhancing the integration with integrity and ecosystem services.

REFERENCES

1. Wilson, G.A.; Bryant, R.L. *Environmental Management: New Directions for the 21st Century*; Taylor and Francis Group: London, 2002.
2. Fath, B.D.; Patten, B.C. Review of the foundations of network environ analysis. Ecosystems **1999**, *2*, 167–179.
3. Allen, C.R.; Holling, C.S. *Discontinuities in Ecosystems and Other Complex Systems*; Columbia University Press, New York: 2008.
4. Joergensen, S.E.; Fath, B.; Bastianoni, S.; Marquez, J.; Müller, F.; Nielsen, S.N.;Patten, B.; Tiezzi, E.; Ulanowicz, R. *A New Ecology—The Systems Perspective*; Elsevier Publishers: Amsterdam, 2007.
5. Bossel, H. Indicators for sustainable development: Theory, method, applications. A report to the Balaton Group. International Institute for Sustainable Development: Winnipeg, 1999.
6. Jørgensen, S.E.; Xu, F.L.; Costanza, R., Eds. *Handbook of Ecological Indicators for Assessment of Ecosystem Health*, 2nd Ed. (Applied Ecology and Environmental Management); CRC Press: London, 2010.
7. Burkhard, B.; Müller, F.; Lill, A. Ecosystem health indicators: Overview. In *Ecological Indicators. Vol. [2] of Encyclopedia of Ecology*, 5 vols; Jørgensen, S.E., Fath, B.D., Eds.; Elsevier: Oxford, 2008; 1132–1138.
8. Dale, V.H.; Beyeler, S.C. Challenges in the development and use of ecological indicators. Ecol. Indic. **2001**, *1*, 3–10.
9. Wiggering, H.; Müller, F., Eds. *Umweltziele und Indikatoren*; Springer-Verlag: Berlin, Heidelberg, New York, 2003.
10. Müller, F.; Lenz, R. Ecological indicators: Theoretical fundamentals of consistent applications in environmental management. Ecol. Indic. **2006**, *6*, 1–5.
11. Walz, R. Development of environmental indicator systems: Experiences from Germany. Environ. Manage. **2000**, *26*, 613–623.
12. Turnhout, E.; Hisschemoller, M.; Eijsackers, H. Ecological indicators: Between the two fires of science and policy. Ecol. Indic. **2007**, *7*, 215–228.
13. Niemeijer, D.; De Groot, R.S. A conceptual framework for selecting environmental indicators sets. Ecol. Indic. **2008**, *8*, 14–25.
14. Girardin, P.; Bockstaller, C.; van der Werf, H.M.G. Indicators: Tools to evaluate the environmental impacts of farming systems. J. Sustainable Agric. **1999**, *13*, 5–21.
15. EEA. *Environmental Indicators: Typology and Use in Reporting*; EEA: Copenhagen, 2003; 20 pp.
16. Heink, U.; Kowarik, I. What are indicators? On the definition of indicators in ecology and environmental planning. Ecol. Indic. **2010**, *10*, 584–593.
17. Leopold, A. Wilderness as a land laboratory. Living Wilderness **1941**, *6*, 3.
18. Rapport D.J. Regaining healthy ecosystems: The supreme challenge of our age. In *Managing for Healthy Ecosystems*; Rapport, D.J., Ed.; Lewis Publisher: Boca Raton, 2003; 5–10.
19. Costanza, R.; Norton, B.G.; Haskell, B.D., Eds. *Ecosystem Health: New Goals for Environmental Management*; Island Press: Washington, DC, 1992; 279 pp.
20. UNCED (United Nations Conference for Environment and Development). *Agenda 21*; United Nations: New York, 1992.
21. Haskell B.D.; Norton, B.G.; Costanza, R. Introduction: What is ecosystem health and why should we worry about it? In *Ecosystem Health: New Goals for Environmental Management*; Costanza, R., Norton, B.G., Haskell B.D., Eds.; Island Press: Washington, DC, 1992; 3–20.
22. Karr, J.D. Measuring biological integrity: Lessons from streams. In *Ecological Integrity and the Management of Ecosystems*; Woodley, S., Kay, J., Francis, G., Eds.; CRC Press: Boca Raton, FL, 1993; 83–104.
23. Rapport, J.D.; Rolston, E.D.; Qualset, O.C.; Damania, B.A.; Lasley, L.W. *Managing for Healthy Ecosystems*; CRC Press: Boca Raton, London, New York, Washington, DC, 2002.
24. Xu, W.; Mage, J.A. A review of concepts and criteria for assessing agroecosystem health including a preliminary case study from southern Ontario. Agric., Ecosyst. Environ. **2001**, *83*, 215–233
25. De Kruijf, H.A.M.; Van Vuuren, D.P. Following sustainable development in relation to the North–South dialogue: Ecosystem health and sustainability indicators. Ecotoxicol. Environ. Safety **1989**, *40*, 414.
26. MA (Millennium Ecosystem Assessment). *Ecosystems and Human Wellbeing*: Synthesis. Island Press, World Resources Institute: Washington, DC, 2005.
27. Scholes, R.J.; Mace, G.M.; Turner, W.; Geller, G.N.; Jürgens, N.; Larigauderie, A.; Muchoney, D.; Walther, B.A.; Mooney, H.A. Toward a global biodiversity observing system. Science **2008**, *321*, 1044–1045.
28. van Strien, A.J.; van Duuren, L.; Foppen, R.P.B.; Soldaat, L.L. A typology of indicators of biodiversity change as a tool to make better indicators. Ecol. Indic. **2009**, *9*, 1041–1048.
29. Lamb, E.G.; Bayne, E.; Holloway, G.; Schieck, J.; Boutin, S.; Herbers, J.; Haughland, D.L. Indices for monitoring biodiversity change: Are some more effective than others? Ecol. Indic. **2009**, *9*, 432–444.
30. Rossi, J.P. Extrapolation and biodiversity indicators: Handle with caution! Ecol. Indic. **2010**, doi:10.1016/j.ecolind.2010.09.002.
31. Marquez, J.C.; Salas, F.; Patricio, J.; Teixera, H.; Neto, J.M. *Ecological Indicators for Coastal and Estuarine Environmental Assessment: A User Guide*; WIT Press: Southampton, 2009.
32. Simpson, E.H. Measurement of diversity. Nature **1949**, *163*, 688.
33. Hilty, J.; Merenlender, A. Faunal indicator taxa selection for monitoring ecosystem health. Biol. Conserv. **2000**, *92*, 185–197.
34. Pinto, R.; Patricio, J.; Baeta, A.; Fath, B.; Neto, J.M.; Marques, J.C. Review and evaluation of estuarine biotic indices to assess benthic condition. Ecol. Indic. **2009**, *9*, 1–25

35. Kevan, P.C. Pollinators as bioindicators of the state of the environment: Species, activity and diversity. Agric., Ecosyst. Environ. **1999**, *74*, 373–393.
36. Jørgensen, S.E. Introduction. In *Handbook of Ecological Indicators for Assessment of Ecosystem Health*, 2nd Ed. (Applied Ecology and Environmental Management); Jørgensen, S.E., Xu, F.L., Costanza, R., Eds.: CRC Press: London, 2010.Jørgensen, S.E.; Xu, F.L.; Salas, F.; Marques, J.C. *Application of Indicators for the Assessment of Ecosystem Health. Handbook of Ecological Indicators for the Assessment of Ecosystem Health*; CRC Press, Boca, Raton: 2005.
37. Kolkwitz, R.; Marsson, M. Grundsätze für die biologische Beurteilung des Wassers nach seiner Flora und Fauna. Mitt. aus d. kgl. Prüfungsanstalt für Wasserversorgung und Abwässerbeseitigung **1902**, *1*, 33–72.
38. Bellan, G. Pollution Indices. In *Encyclopedia of Ecology*, 5 vols; Jørgensen, S.E., Fath, B.D., Eds.; Elsevier: Oxford, 2008; 2861–2868.
39. Borja, A.; Franco, J.; Perez, V. A marine biotic index to establish the ecological quality of soft-bottom benthos within European estuarine and coastal environments. Mar. Pollut. Bull. **2000**, *40* (12), 1100–1114.
40. Simboura, N.; Zenetos, A. Benthic indicators to use ecological quality classification of Mediterranean soft bottom marine ecosystems, including a new biotic index. Mediterr. Mar. Sci. **2002**, *3* (2), 77–111.
41. Roberts, R.D.; Gregory, M.G.; Foster, B.A. Developing an efficient macrofauna monitoring index from an impact study—A dredge spoil example. Mar. Pollut. Bull. **1998**, *36* (3), 231–235.
42. Simberloff, D. Flagships, umbrellas, and keystones: Is single-species management passé in the landscape era? –Biol. Conserv. **1998**, *83* (3), 247–257.
43. Cocciufa, C.; Petriccione, B.; Framstad, E.; Bredemeier, M. Biodiversity Assessment in LTER sites. An EC Report (Deliverable 3.R2.D1) from ALTER Net (A Long Term Biodiversity, Ecosystem and Awareness Research Network); 2007.
44. Brito, D.; Ambal, R.G.; Brooks, T.; De Silva, N.; Foster, M.; Hao, W.; Hilton-Taylor, C.; Paglia, A.; Rodríguez, J.P.; Rodríguez, J.V. How similar are national red lists and the IUCN Red List? Biol. Conserv. **2010**, *143*, 1154–1158.
45. UNEP-WWF. *The Living Planet Report*; 2004.
46. Loh, J. *Living Planet Report 2000*. Gland, World Wide Fund for Nature; 2000.
47. Lenhart, H.J. Effects of river nutrient load reduction on the eutrophication of the North Sea, simulated with the ecosystem model ERSEM. Senckenbergiana maritima **2001**, *31* (2), 299–311.
48. Burkhard, B.; Fath, B.D.; Müller, F. Adapting the adaptive cycle: Hypotheses on the development of ecosystem properties and services. Ecol. Modell. **2011**, doi:10.1016/j.ecolmodel.2011.05.016.
49. Word, J.Q. Classification of benthic invertebrates into infaunal trophic index feeding groups. Coastal Water Research Project Report 1979–1980; Los Angeles, 1979; 103–121.
50. EEA. *Halting the Loss of Biodiversity by 2010: Proposal for a First Set of Indicators to Monitor Progress in Europe*, EEA Technical Report No 11/2007; 2007.
51. EEA. *Progress towards the European 2010 Biodiversity Target*, EEA Technical Report No 5/2009; 2009.
52. GEO BON (Group on Earth Observations Biodiversity Observation Network). Detailed Implementation Plan. GEO BON. Version 1.0, May 22, 2010.
53. Kay, J.J. On the nature of ecological integrity: Some closing comment. In *Ecological Integrity and the Management of Ecosystems*; Woodley, S., Kay, J., Francis, G., Eds.; St. Lucie Press: Delray, FL, 1993; 210–212.
54. Schneider, E.D.; Kay, J.J. Life as a manifestation of the second law of thermodynamics. Math. Comput. Modell. **1994**, *19*, 25–48.
55. Odum, H.T. *Environmental Accounting: Emergy and Environmental Policy Making*; John Wiley and Sons: New York, 1996; 370 pp.
56. Brown, M.T.; Ulgiati, S. Updated evaluation of exergy and emergy driving the geobiosphere: A review and refinement of the emergy baseline. Ecol. Modell. **2010**, *221*, 2501–2508.
57. Ulanowicz, R.E. *Growth and Development: Ecosystems Phenomenology*; Springer-Verlag: NY, 1986; 203 pp.
58. Ulanowicz, R.E. *Ecology the Ascendent Perspective*; Columbia University Press: NY, 1997; 201 pp.
59. Callicott, J.B. A review of some problems with the concept of ecosystem health. Ecosyst. Health **1995**, *1* (2), 101–112.
60. Westra, L. The ethics of ecological integrity and ecosystem health: The interface. In *Managing for Healthy Ecosystems*; Rapport, D.J., Ed.; Boca Raton, London, New York, Washington, DC, 2003; 31–40.
61. Müller, F. Ecosystem indicators for the integrated management of landscape health and integrity. In *Ecological Indicators for Assessment of Ecosystem Health*; Joergensen, S.E., Costanza, R., Xu, F.L., Eds.; Taylor and Francis: Boca Raton, FL, 2004; 277–304.
62. Kay, J.J. Ecosystems as self-organised holarchic open systems: Narratives and the second law of thermodynamics. In *Handbook of Ecosystem Theories and Management*; Joergensen, S.E., Müller, F., Eds.; Lewis Publishers: Boca Raton, 2000; 135–160.
63. Barkmann, J.; Baumann, R.; Meyer, U.; Müller, F.; Windhorst, W. Ökologische Integrität: Risikovorsorge im Nachhaltigen Landschaftsmanagement. Gaia **2001**, *10* (2), 97–108.
64. Bossel, H. Sustainability: Application of systems theoretical aspects to societal development. In *Handbook of Ecosystem Theories and Management*; Jørgensen, S.E., Müller, F., Eds.; CRC Press, Boca Raton: 2000; 519–536.
65. Müller, F.; Leupelt, M. *Eco Targets, Goal Functions, and Orientators*; Springer: Berlin, 1998; 619 pp.
66. Müller, F.; Jørgensen, S.E. Ecological orientors: A path to environmental applications of ecosystem theories. In *Handbook of Ecosystem Theories and Management*; Jørgensen, S.E., Müller, F., Eds.; CRC Publishers: New York, 2000; 561–576.
67. Odum, E.P. The strategy of ecosystem development. Science **1969**, *164*, 262–270.
68. Odum, E.P. *Fundamentals of Ecology*; WB Saunders WB: Philadelphia, 1971.
69. Müller, F. Indicating ecosystem and landscape organization. Ecol. Indic. **2005**, *5* (4), 280–294
70. Müller, F.; Burkhard. B. Ecosystem indicators for the integrated management of landscape health and integrity. In *Handbook of Ecological Indicators for Assessment of Ecosystem Health*, 2nd Ed.; Jorgensen, S.E., Xu, L., Costanza, R., Eds.; Taylor and Francis, 2010; 391–423.

71. Aoki, I. Entropy and exergy in the development of living systems: A case study of lake ecosystems. J. Phys. Soc. Jpn. **1998**, *67*, 2132–2139
72. Svirezhev, Y.M.; Steinborn, W. Exergy of solar radiation: Thermodynamic approach. Ecol. Modell. **2001**, *145*, 101–110.
73. Aguilar, B.J. Applications of ecosystem health for the sustainability of managed systems in Cost Rica. Ecosyst. Health **1999**, *5*, 1–13.
74. Burkhard, B.; Kroll, F.; Nedkov, S.; Müller, F. Mapping supply, demand and budgets of ecosystem services. Ecol. Indic. **2011**, doi:10.1016/j.ecolind.2011.06.019.
75. Rapport, D.J.; Costanza, R.; McMichael, A.J. Assessing ecosystem health. TREE **1998**, *13* (10), 397–402.
76. Rapport, D.J. Ecosystem services and management options as blanket indicators of ecosystem health. J. Aquat. Ecosyst. Health **1995**, *4*, 97–105.
77. Gallopin, G. The potential of agroecosystem health as a guiding concept for agricultural research. Ecosyst. Health **1995**, *1*, 129–140.
78. Waltner-Toews, D. Ecosystem health: A Framework for implementing sustainability in agriculture. BioScience **1996**, *46* (9), 686–689.
79. Waltner-Toews, D. Agro-ecosystem health: Concept and principles. In *Agro-Ecosystem Health*, Proceedings of a seminar held in Wageningen, Sep 26, 1996; van Bruchem, J., Ed.; NRLO-rapport nr. 97/31; 1997 9–22.
80. Rapport, D.J. Evaluating ecosystem health. J. Aquat. Ecosyst. Health **1992**, *1*, 15–24.
81. Rapport, D.J.; Lalsley, W.L.; Rolston, E.R.; Nielsen, N.O.; Qualset, C.Q.; Damania, A.B. *Managing for Healthy Ecosystems*, Proceedings of the 1st Ecosystem Management Congress, 1999; CRC Press LLC, 2003; 1510 pp.
82. Zhang, M.; Smallwood, K.S.; Anderson, E. Relating indicators of ecosystem health and ecological integrity to assess risks to sustainable agriculture and native biota—A case study of Yolo County, California. In *Managing for Healthy Ecosystems*, Proceedings of the 1st Ecosystem Management Congress, 1999; Rapport, D.J., et al., Eds.; CRC Press LLC, 2003; 757–768.
83. Kaffka, S.R.; Dhawan, A.; Kirby, D.W. Irrigation, agricultural drainage, and nutrient loading in the Upper Klamath Basin. In *Managing for Healthy Ecosystems*, Proceedings of the 1st Ecosystem Management Congress, 1999; Rapport, D.J., et al., Eds.: CRC Press LLC, 2003; 1011–1026.
84. Mitchell, J.P.; Lanini, W.T.; Temple, S.R.; Brostrom, P.N.; Herrrero, E.V.; Miyao, E.M.; Prather, S.; Hembree, K.J. Reduced-disturbance agroecosystems in California. In *Managing for Healthy Ecosystems*, Proceedings of the 1st Ecosystem Management Congress, 1999; Rapport, D.J., et al., Eds.; CRC Press LLC, 2003; 993–997.
85. Hopkins, D.H. Fallow LAND Patches and ecosystem health in California's Central Valley agroecosystem. In *Managing for Healthy Ecosystems*, Proceedings of the 1st Ecosystem Management Congress, 1999; CRC Press LLC, 2003; 981–991.
86. Altieri, M.A.; Nicholls, C.I. Ecologically based pest management: A key pathway to achieving agroecosystem health. In *Managing for Healthy Ecosystems*, Proceedings of the 1st Ecosystem Management Congress, 1999; Rapport, D.J., et al., Eds.; CRC Press LLC, 2003; 999–1010.
87. Belcher, K.W.; Boehm, M. Evaluating agroecosystem sustainability using an integrated model. In *Managing for Healthy Ecosystems*, Proceedings of the 1st Ecosystem Management Congress, 1999; Rapport, D.J., et al., Eds.; CRC Press LLC, 1209–1226.
88. Waltner-Toews, D.; Wall, E. Emergent perplexity of post-normal questions for community and agroecosystem health. Soc. Sci. Med. **1997**, *45* (11), 1741–1749.
89. Ikerd, J. Assessing the health of agroecosystems: A socioeconomic perspective. In Proceedings of the 1st International Ecosystem Health and Medicine Symposium, University of Missouri; 1994; 9 pp.
90. Kolb, T.E.; Wagner, M.R.; Covington, W.W. Utilitarian and ecosystem perspectives. Concepts of forest health. J. For. **1994**, *92* (2), 10–15.
91. Percy, K.E.; Ferretti, M. Air pollution and forest health: Towards new monitoring concepts. Environ. Pollut. **2004**, 130, 113–126.
92. USDA Forest Service. Healthy forests for America's future: A strategic plan, USDA Forest Service MP-1513; 1993; 58 pp.
93. Spranger, T.; Lorenz, U.; Gregor, H.D., Eds. *Manual on Methodologies and Criteria for Modelling and Mapping Critical Loads and Levels and Air Pollution effects, Risks and Trends*; Federal Environmental Agency (Umweltbundesamt): Berlin, 2004.
94. Gauger, T.; Haenel, H.D.; Rösemann, C.; Dämmgen, U.; Bleeker, A.; Erisman, J.H.; Vermeulen, A.T.; Schaap, M.; Timmermanns, R.M.A.; Builtjes, P.J.H.; Duyzer, J.H. *National Implementation of the UNECE Convention on Long-range Transboundary Air Pollution (Effects)—Part 1: Deposition Loads: Methods, Modelling and Mapping Results, Trends*, Environmental Research of the Federal Ministry of the Environment, Natural Conservation and Nuclear Safety Research Report 204 63 252, No. 38/2008, UBA-FB 001189E; Federal Environmental Agency (Umweltbundesamt): Dessau, 2010a.
95. Gauger, T.; Haenel, H.D.; Rösemann, C.; Nagel, H.D.; Becker, R.; Kraft, P.; Schlutow, A.; Schütze, G.; Weigelt-Kirchner, R.; Anshelm, F. Nationale Umsetzung UNECE-Luftreinhaltekonvention (Wirkungen) Teil 2: Wirkungen und Risiokoabschätzungen Critical Loads, Biodiversität, Dynamische Modellierung, Critical Levels Überschreitungen, Materialkorrosion, Environmental Research of the Federal Ministry of the Environment, Natural Conservation and Nuclear Safety Research Report 204 63 2521 No. 39/2008, UBA-FB 001189; Federal Environmental Agency (Umweltbundesamt): Dessau, 2010b.
96. UNECE Convention on Long-range Transboundary Air Pollution. Manual on methodologies for mapping critical loads/levels and geographical areas where they are exceeded, http://www.rivm.nl/thema/images/mapman-2004_tcm61-48383.pdf.
97. Warfinge, P.; Sverdrup, H. Calculating critical loads of acid deposition with profile—A steady soil chemistry model. Water, Air Soil Pollut. **1992**, *63*, 119–143.
98. Warfinge, P.; Sverdrup, H. Critical loads of acidity to Swedish forest soils, methods, data and results. Dep. Chem. Eng., Rep. Ecol. Environ. Eng. 5; Lund University: Lund, Sweden, 1995; 104 pp.

99. Ulrich, B. Stability, elasticity, and resilience of terrestrial ecosystems with respect to matter balance. In *Potentials and Limitations of Ecosystems Analysis*; Schulze, E.D., Zwölfer, H., Eds.; Springer: Berlin, 1987; 435 pp.
100. Boesch, D.F.; Paul, J.F. An overview of coastal environmental health indicators. Hum. Ecol. Risk Assess. **2001**, *7* (5), 1–9.
101. Su, M.; Fath, B.D.; Yang, Z. Urban ecosystem health assessment: A review. Sci. Total Environ. **2010**, *408*, 2425–2434.
102. Bell, M.L.; Cifuntes, L.A.; Davis, D.L.; Cushing, E.; Gusman Telles, A.; Gouveia, N. Environmental health indicators and a case study of air pollution in Latin American Cities. Environ. Res. **2011**, *111*, 57–66.
103. Alberti, M. Maintaining ecological integrity and sustaining ecosystem function in urban areas. Curr. Opin. Environ. Sustainability **2010**, *2*, 178–184.
104. Hancock, T. Health and sustainability in the urban environment. Environ. Impact Assess. Rev. **1996**, *16*, 259–277.
105. Tzoulas, K.; Korpela, K.; Venn, S.; Ylipelkonen, V.; Kazmierczak, A.; Niemela, J.; James, P. Promoting ecosystem and human health in urban areas using green infrastructure: A literature review. Landscape Urban Plann. **2007**, *81*, 167–178.
106. Rapport, D.J.; Singh, A. An EcoHealth-based framework for state of environment reporting. –Ecol. Indic. **2005**, *6*, 409–428.
107. Li, B.; Xie, H.L.; Guo, H.H.; Hou, Y. Study on the assessment method of agroecosystem health based on the pressure-state-response mode. In Proceedings of the IEEE International Conference on Industrial Engineering and Engineering Management, Hong Kong, Dec 8–11, 2009; 2458–2462.
108. Kumar, P., Ed. *The Economics of Ecosystems and Biodiversity: Ecological and Economic Foundations*; Earthscan, 2010; 456 pp.
109. Windhorst, W; Müller, F.; Wiggering, H. Umweltziele und Indikatoren für den Ökosystemschutz. In *Umweltziele und Indikatoren*; Müller, F., Wiggering, H., Eds.; Springer Verlag, 2003; 345–373.
110. Windhorst, W.; Colijn, F.; Kabuta, S.; Laane, R.P.; Lenhart, H. Defining a good ecological status of coastal waters—A case study for the Elbe plume. In *Managing European Coasts;* Vermaat, J.E., Bouwer, L., Turner, K., Salomons, W., Eds.; Springer Verlag, 2005; 59–74.
111. Schönthaler, K.; Meyer, U.; Pokorny, D.; Reichenbach, M.; Schuller, D.; Windhorst, W. *Ökosystemare Umweltbeobachtung—Vom Konzept zur Umsetzung*; Erich-Schmidt-Verlag: Berlin, 2004; 370 pp.
112. Reuter, H.; Middelhoff, U.; Schmidt, G.; Windhorst, W.; Schröder, W.; Breckling, B. Up-scaling the environmental effects of genetically modified plants—Assessing potential impact on nature conservation areas in Northern Germany. In *Multiple Scales in Ecology*; Schröder, B., Reuter, H., Reineking, B., Eds.; Peter Lang Internationaler Verlag der Wissenschaften: Frankfurt am Main. Band 13 edition. Theorie in der Ökologie, 2007; 95–109.
113. Rapport, D.J.; Regier, H.A.; Hutchinson, T.C. Ecosystem behavior under stress. Am. Nat. **1985**, *125* (5), 617–640.

Economic Growth: Slower by Design, Not Disaster

Peter Victor
Faculty of Environmental Studies, York University, Toronto, Ontario, Canada

Abstract
The prospects for comprehensive, global economic growth are increasingly in doubt as evidence mounts that the human economy is exceeding the biophysical capacity of the planet to support it. If poor countries are to benefit from economic growth, at least for a while, rich countries should replace the pursuit of economic growth with other more specific objectives to enhance human well-being and reduce competition with other species. What is entailed in an economy in which the rate of growth is lowered, even to zero? Some insights into this question are provided using the interactive, macro model LowGrow that simulates the Canadian economy to the year 2035. In this entry, scenarios are described in which there is full employment, no poverty, reductions in greenhouse gas emissions, and fiscal balance in Canada without economic growth. The entry concludes with a discussion of policy implications.

INTRODUCTION

Economic growth is a very recent phenomenon, dating back two or three centuries and limited in its extent to only some parts of the world. While economic growth has dramatically improved the lives of billions of people, billions more remain in extreme or serious poverty. The mainstream view is that all countries should strive for economic growth and, through a process of convergence where poorer countries grow the fastest, all the people of the world will eventually enjoy a high and secure material standard of living.

There are many problems with this vision of growth for all. Foremost is the mounting evidence that the world's economy is already bumping up against and even surpassing the biophysical limits of the planet. This is showing up in terms of global and local environmental degradation and resource scarcity. Consequently, a different vision is emerging, one in which rich countries no longer pursue economic growth as a primary objective, leaving room for the economies of poor countries to expand, at least temporally. An additional motivation for low and no growth is to reduce competition with other species with which humans share the planet.

This entry examines the possibility of rich countries managing without growth, by which is meant achieving improvements in well-being without economic growth. LowGrow, an interactive macro model of the Canadian economy, is described and several scenarios are discussed, which suggest that such an outcome is possible, in particular that objectives such as full employment, eradication of poverty, much reduced greenhouse gas emissions, and fiscal balance can be achieved in the absence of economic growth. The entry concludes with a discussion of policy directions implied by such a scenario with a special emphasis on employment policies and funding public services in a low/no-growth economy.

A BRIEF HISTORY OF ECONOMIC GROWTH

In terms of human history, economic growth is a very recent phenomenon. World gross domestic product (GDP) was only 14% higher at the end of the first millennium than when it began. GDP is the "total unduplicated value of the goods and services produced in the economic territory of a country or region during a given period" (Statistics Canada). All estimates of growth in world GDP and GDP per capita are from Maddison.[1] GDP per capita was no higher at all, slightly less even, owing to population growth. From year 1000 to 1400, world GDP increased 140%, but GDP per capita grew only 60% and at a barely perceptible rate, as population expanded. Over the next 400 years, world GDP increased a further 140% and although GDP per capita increased by 30% by 1700, it fell back again to 1800 to the same level as in 1400. Average world GDP per capita declined further in the early 1800s, after which economic growth accelerated. By 1900, world GDP was over 5 times greater than 100 years previously, growing at an average annual rate of 1.7%, and GDP per capita was more than 3 times greater. The rate of economic growth increased in the 20th century such that annual world GDP grew almost 9-fold, at an average annual rate of 2.2% and GDP per capita rose 2 1/2 times.

These figures, approximate as they are especially early on, show that economic growth has been exceptional in human history. The vast majority of people have lived in circumstances where they had no reason to believe that their children's lives would be materially any different from their own. To think otherwise is a very modern idea

Encyclopedia of Environmental Management DOI: 10.1081/E-EEM-120046918

dating back perhaps a dozen or so generations, and even then, only in some parts of the world.

Of course, global averages conceal much of the huge variation in experiences from country to country and region to region. In the 18th and 19th centuries, most of the world's economic growth was confined to Europe and ex-colonies of European powers. Some regions were impoverished in the process, 19th century India being a prime example through the deliberate destruction of cotton manufacturing at the behest of British industrialists. The rest of the world continued to experience lives in which economic growth had little or no impact and was not part of the lived experience of most people. Even in places where growth rates were highest, there were substantial areas of poverty. In the 20th century, economic growth spread to more parts of the globe but unevenly with 1.4 billion people still living in extreme poverty in the 21st century.[2] Contrary to earlier expectations that global economic growth would close the gap between rich and poor, the spread has increased, just as it has done within even the richest countries.[3]

All this economic growth has required a massive increase in the use of natural resources of all kinds. In the 20th century, the global use of construction materials, ores and industrial materials, fossil fuels, and biomass increased eightfold, almost as much as GDP.[4] The increasing extraction, use, and disposal of materials has resulted in large-scale, adverse environmental impacts. One prominent list includes the following: climate change, ocean acidification, stratospheric ozone depletion, overloading of the nitrogen and phosphorus cycles, global freshwater uses, land system change, biodiversity loss, atmospheric aerosol loading, and chemical pollution.[5] Then, there is the imminent threat of peak oil[6] and the more general concern that the age of cheap energy is coming to an end,[7] as well as an emerging scarcity of "critical" minerals for new technologies.[8] All this makes for a bleak outlook for real, long-term, comprehensive, global economic growth. With the world human population surpassing 7 billion, projected to rise to 9 billion by mid-century, the question arises as to whether economic growth for all is a viable option in the 21st century. If not, what are the alternatives?

One alternative is to remove the pursuit of economic growth in rich countries from its pre-eminent place among economic policy objectives. After all, it has only held this position since about 1960. "There is in fact hardly a trace of interest in economic growth as a policy objective in the official or professional literature of western countries before 1950."[9] And even when it was introduced, it was as a means to fulfill other policy objectives such as full employment, rather than as an end in itself. Now the pursuit of economic growth is deemed so important that it is customary for policy proposals across many domains including environment, education, and the arts to be judged in terms of their implications for growth, or one of its surrogates such as competitiveness or productivity.[10]

If the poor countries of the world are to benefit from economic growth, and the world economy is to function within the "safe operating space" of the planet,[5] then rich countries, those that have benefited the most from economic growth in the past, must be prepared to make room for them. Otherwise, disaster threatens, brought about by the excessive pressure on Earth systems.

One approach to this predicament based on principles of distributive justice would be to determine fair shares of access to global resources, accounting also for the interests of other species. This is the kind of dialogue that is underway in the ponderously slow international climate negotiations, illustrating most profoundly the difficulties in reaching agreement in a global world divided into national and regional power blocs. Nonetheless, a country or group of countries could adopt a view of its fair share independently of an international agreement, and set these as boundaries within which their economy must function.

Depending on the specification of the boundaries and the capability of economies to adjust, economic growth, measured in the conventional way as an increase in real (inflation) adjusted GDP, could conceivably continue, but incidentally rather than as a primary objective.

What would such growth entail? Total environmental (including resource) impact is the product of GDP and impact per unit of GDP (e.g., GDP × greenhouse gas emissions per unit of GDP or GDP × energy used per unit of GDP). Therefore, if environmental impact is to decline as GDP grows, environmental impact per unit GDP must decline faster than the rate of economic growth. This is one meaning of "green" growth and it underlies the downward sloping portion of the Environmental Kuznets Curve.[11] (The Environmental Kuznets Curve is an inverted "U" with a measure of environmental impact plotted on the y axis and GDP plotted on the x axis. The hypothesis is that in the early stages of economic growth, environmental impact rises to a maximum after which it declines as growth continues.) There are examples of obvious, local problems such as urban air quality whose history can be described by an Environmental Kuznets Curve, but it is a poor description of global materials and energy use or global environmental impacts over the past century.

Whether green growth defined in this way (i.e., impact/GDP declining faster than GDP increases) is possible at the global level remains an open question. If it is not possible to sufficiently decouple economic growth from its environmental and resource impacts, growth will have to cease and even turn negative for a time as proponents of degrowth argue.[12] Otherwise, it will not be possible to bring the global economy back within the planetary boundaries that it is currently exceeding and others which it will exceed if present trends are not reversed. Given what we know about the state of the environment, and concern over supplies of low-cost energy and critical materials, there is a very strong case on ethical and practical grounds for rich countries taking the lead in managing without growth.

ECONOMISTS QUESTION GROWTH

At the same time as economic growth was reaching the pinnacle of policy objectives, some dissenting voices were beginning to be heard. One of the most widely read was John Kenneth Galbraith. In *The Affluent Society*[13] published in 1958 and revised through multiple editions, Galbraith compared private affluence in the United States with public squalor. He also questioned the efficacy of dealing with poverty through a general rise in incomes. Many academic economists regarded Galbraith as more of a political commentator than a serious economist because of his disdain for theoretical economics, and on these tenuous grounds they resisted his arguments. The same could not be said of British economist Ezra Mishan who published *The Costs of Economic Growth* in 1967.[14] Mishan was a highly regarded and well-published expert in "welfare economics," the field within mainstream economics that is concerned with the relationship between economic activity and well-being. Thus, although Mishan's analysis of the costs of economic growth was aimed at a broad audience, no one could dismiss the author as not really understanding modern economic theory.

Perhaps this is one reason why Mishan's critique of economic growth, unlike Galbraith's, ignited a heated debate that went on for several years between him and Wilfred Beckerman, another well-established British economist. Beckerman wrote "Why We Need Economic Growth"[15] and *In Defence of Economic Growth*.[16] Later, Beckerman wrote *Small Is Stupid*[17] in response to Schumacher's widely read *Small Is Beautiful*,[18] Schumacher's critique of modern industrialized economies. Many of Schumacher's arguments about the optimal scale of an economy were anticipated, echoed, and augmented by other economists such as Kenneth Boulding in his seminal essay "The Economics of the Coming Spaceship Earth,"[19] N. Georgescu-Roegen,[20] who explored the implications of the second law of thermodynamics for economics and economic growth, and Herman Daly who has promoted a steady-state economy for more than three decades.[21] The publication in 1972 of *The Limits to Growth*[22] addressed similar themes using systems dynamics but was roundly and largely unfairly criticized especially by economists (Maddison,[1] pp. 90–94). It remains influential to this day. A useful summary of the state of the growth and no-growth debate, largely where it was left in the 1970s, can be found in Olson and Landsberg's collection of essays.[23]

After this flurry of publications in the 1960s and 1970s, the growth debate subsided. In the late 1990s, the criticisms of growth resurfaced stronger than ever with economists such as Douthwaite, *The Growth Illusion*,[24] Daly, *Beyond Growth: The Economics of Sustainable Development*,[25] and Booth, *The Environmental Consequences of Growth: Steady-State Economics*[26] leading the charge. By this time, the transdiscipline, ecological economics was almost 20 years old, with a dedicated peer-reviewed journal, *Ecological Economics*, publishing 12 times a year including many papers dealing with problematic aspects of economic growth. In the first part of the 21st century, it is impossible to keep up with the many papers, reports, blogs, conferences, media entries, even YouTube videos questioning economic growth. And there are many books such as Hamilton, *Growth Fetish*,[27] Booth, *Hooked on Growth*,[28] Victor, *Managing without Growth*,[29] Speth, *The Bridge at the End of the World*,[30] Brown, *Right Relationship*,[31] Jackson, *Prosperity without Growth*,[32] and Schor, *Plenitude*,[33] just to name a few in the English language alone.

The remainder of this entry describes an investigation into what might be possible in an economy in which economic growth ceases. In particular, the following question is addressed for the Canadian economy: is it possible to have full employment, reduced greenhouse gas emissions, eradication of poverty, and fiscal balance in the absence of economic growth and, if so, what policy frameworks or initiatives would be required? LowGrow, an interactive macroeconomic systems model, was developed for the Canadian economy specifically to help answer these questions. LowGrow is described in the next section followed by some simulation results for managing without growth. Brief comments on policy directions suggested by the simulations are followed by a more detailed consideration of employment in a no/low-growth economy and revenue generation for public services.

EXPLORING LOW AND NO GROWTH IN CANADA WITH LOWGROW

LowGrow is a quantitative model of the Canadian economy designed to make it easy to explore different assumptions, objectives, and policy measures.

Fig. 1 shows the simplified structure of LowGrow. Aggregate (macro) demand is determined in the normal way as the sum of consumption expenditure (C), investment expenditure (I), government expenditure (G), and the difference between exports (X) and imports (M). Their sum total is GDP measured as expenditure. There are separate equations for each of these components in the model, estimated with Canadian data from about 1981 to 2005 depending on the variable. Production in the economy is estimated by a Cobb–Douglas production function in which macro supply is a function of employed labor (L) and employed capital (K). The time variable (t) represents changes in productivity from improvements in technology, labor skills, and organization. The production function is shown as macro supply at the bottom of Fig. 1. It estimates the labor (L) and employed capital (K) required to produce GDP allowing for changes in productivity over time.

There is a second important link between aggregate demand and the production function. Investment expenditures (net of depreciation), which are part of aggregate

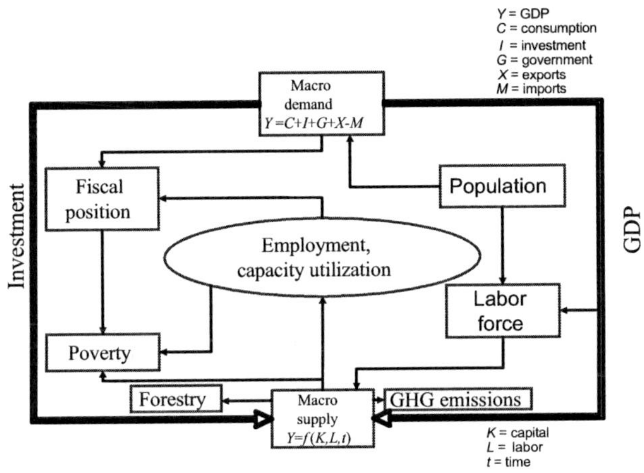
Y = GDP
C = consumption
I = investment
G = government
X = exports
M = imports
Macro demand
Y=C+I+G+X-M
Investment
Fiscal position
Population
GDP
Employment, capacity utilization
Labor force
Poverty
Forestry
Macro supply
Y=f(K,L,t)
GHG emissions
K = capital
L = labor
t = time

Fig. 1 High-level structure of LowGrow.
Source: Maddison.[1]

demand, add to the economy's stock of capital, increasing its productive capacity. Also, capital and labor become more productive over time.

It follows that, other things equal, without an increase in aggregate demand, these increases in capital and productivity reduce employment. Economic growth (i.e., increases in GDP) is needed to prevent unemployment rising as capacity and productivity increase.

Population is determined exogenously in LowGrow, which offers a choice of three projections from Statistics Canada. Population is also one of the variables that determines consumption expenditures in the economy. The labor force is estimated in LowGrow as a function of GDP and population.

There is no monetary sector in LowGrow. For simplicity, it is assumed that the Bank of Canada, Canada's central bank, regulates the money supply to keep inflation at or near the target level of 2% per year. LowGrow includes an exogenously set rate of interest that remains unchanged throughout each run of the model. A higher cost of borrowing discourages investment, which reduces aggregate demand. It also raises the cost to the government of servicing its debt. The price level is not included as a variable in LowGrow, although the model warns of inflationary pressures when the rate of unemployment falls below 4% (effectively full employment in Canada).

LowGrow includes features that are particularly relevant for exploring a low/no-growth economy. LowGrow includes emissions of carbon dioxide and other greenhouse gases, a carbon tax, a forestry submodel, and provision for redistributing incomes, and measures poverty using the UN's Human Poverty Index (i.e., HPI-2 for selected Organization for Economic Cooperation and Development [OECD] countries[34]). LowGrow allows additional funds to be spent on health care and on programs for reducing adult illiteracy (both included in HPI-2) and estimates their impacts on longevity and adult literacy with equations from the literature.

Implications of changes in the level of government expenditures can be simulated in LowGrow through a variety of fiscal policies including an annual percentage change in government expenditure that can vary over time and a balanced budget. LowGrow keeps track of the overall fiscal position of all three levels of government combined (federal, provincial, and municipal) by calculating total revenues and expenditures and estimating debt repayment based on the historical record. As the level of government indebtedness declines, the rates of taxes on personal incomes and profits in LowGrow are reduced endogenously, broadly consistent with government policy in Canada.

In LowGrow, as in the economy that it represents, economic growth is driven by the following: net investment, which adds to productive assets; growth in the labor force; increases in productivity; growth in the net trade balance; growth in government expenditures; and growth in population. Low- and no-growth scenarios can be examined by reducing the rates of increase in each of these factors singly or in combination.

Business as Usual

It is convenient to start analyzing low- and no-growth scenarios by establishing a base case with no new policy interventions. This is the "business as usual" case illustrated in Fig. 2.

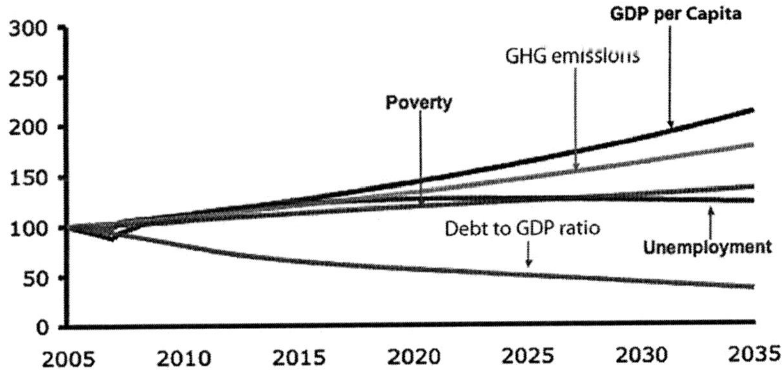
300
250
200
150
100
50
0
2005
2010
2015
2020
2025
2030
2035
GDP per Capita
GHG emissions
Poverty
Debt to GDP ratio
Unemployment

Fig. 2 Business as usual.

In the business-as-usual scenario, between the start of 2005 and 2035, real GDP per capita more than doubles, the unemployment rate rises then falls ending above its starting value, the ratio of government debt to GDP declines by nearly 40% as Canadian governments continue to run budget surpluses, the Human Poverty Index rises, largely due to the projected increase in the absolute number of unemployed people, and greenhouse gas emissions increase by nearly 80%.

A No-Growth Disaster

Economic growth is desired not only for what it offers in terms of increased living standards but also out of fear of what might happen if a modern economy deliberately tried to wean itself off growth. Such fears are well founded. Modern economies and their public, private and not-for-profit institutions, as well as individual citizens, have come to rely on growth. They expect it, they plan for it, they believe in it. Adjusting to life without economic growth could be a wrenching experience and a lot could go wrong as shown in Fig. 3. In this scenario, zero growth in GDP and GDP per capita is achieved around 2030 by eliminating growth in government expenditure, productivity, and population, and achieving zero net investment and net trade balance over a period of years starting in 2010. GDP per capita rises slightly until all the factors contributing to growth are extinguished and then drops back to the same level as at the start of 2005. Meanwhile, the unemployment rate literally goes off the chart, causing a dramatic rise in poverty. The debt-to-GDP ratio also rises to untenable heights largely because of the massive increase in income support paid to the rising number of unemployed. Certainly, the human misery entailed in such a scenario is to be avoided if at all possible.

A Better Low/No-Growth Scenario

A wide range of low- and no-growth scenarios can be examined with LowGrow. Some are not much better than the no-growth disaster just described but others offer more promise. One such promising scenario is shown in Fig. 4.

Compared with the business-as-usual scenario, GDP per capita grows more slowly, leveling off around 2028 at which time the rate of unemployment is 5.7%. The unemployment rate continues to decline to 4.0% by 2035. By 2020, the poverty index declines from 10.7 to an internationally unprecedented level of 4.9 where it remains, and the debt-to-GDP ratio declines to about 30% to be maintained at that level to 2035. Greenhouse gas emissions are 31% lower at the start of 2035 than 2005 and 41% lower than their high point in 2010. These results are obtained by slower growth in government expenditure, net investment and productivity, a positive net trade balance, cessation of growth in population, a reduced workweek, a revenue neutral carbon tax, and increased government expenditure on antipoverty programs, adult literacy programs, and health care.

POLICY DIRECTIONS FOR A LOW/NO-GROWTH SCENARIO

The contrast between the scenarios in Figs. 3 and 4 is striking and naturally raises questions about what makes the difference. The no-growth disaster scenario is based on a systematic elimination of all of the factors represented in LowGrow that contribute to growth without any compensating adjustments. The better no/low-growth scenario results from a wide range of policy measures, some more controversial than others, that would be required to transform the business-as-usual scenario in Fig. 2 into the kind of scenario illustrated in Fig. 4. In summary, these policy measures include the following:

- Investment: reduced net investment, a shift from investment in private to public goods through changes in taxation and expenditures.
- Labor force: stabilization through changing age structure of the population and population stabilization.
- Population: stabilization through changes to immigration policy.
- Poverty: trickle down replaced with focused antipoverty programs that address the social determinants of illness and provide more direct income support.

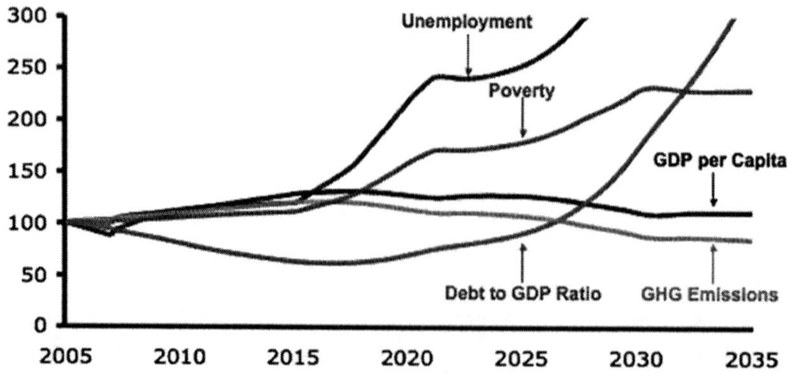
300
250
200
150
100
50
0
2005
2010
2015
2020
2025
2030
2035
Unemployment
Poverty
GDP per Capita
Debt to GDP Ratio
GHG Emissions

Fig. 3 A no-growth disaster.

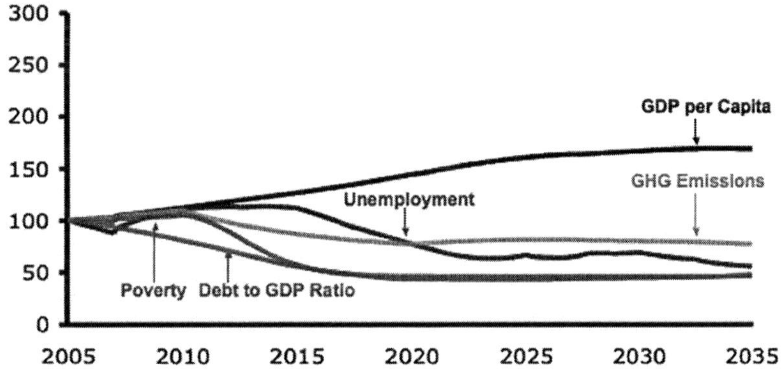
300
250
200
150
100
50
0
2005
2010
2015
2020
2025
2030
2035
GDP per Capita
GHG Emissions
Unemployment
Poverty
Debt to GDP Ratio

Fig. 4 A better low/no-growth scenario.

- Technological change: slower, more discriminating, preventative rather than end of pipe, through technology ssessment and changes in the education of scientists and engineers.
- Government expenditures: a declining rate of increase.
- Trade: a stable, positive net trade balance (and diversification of markets).
- Workweek: shorter, more leisure through changes in compensation, work organization and standard working hours, and active market labor policies.
- Greenhouse gases: a revenue neutral carbon tax.

To complement these policies:

- Consumption: more public goods, fewer positional (status) goods through changes in taxation and marketing.
- Environment and resources: limits on throughput and use of space through better land use planning and habitat protection and ecological fiscal reform.
- Localization: fiscal and trade policies to strengthen local economies.

The next two sections look more closely at two specific policy areas in relation to the low/no-growth scenario: strategies for full employment and funding government programs.

Economic Growth and Employment

In 1960, the UN World Economic Survey stated that "the reinterpretation of the objective of full employment under the United Nations Charter to embrace the goal of economic growth marks a second fundamental change in public policy thinking."[35] This statement from the UN is based on the insight derived from the early work on economic growth by Harrod, Domar, and others that if aggregate expenditure required for full employment in the short run expands the productive capacity of the economy, further increases in aggregate expenditure will be required in the future if full employment is to be maintained. This relationship between growth and employment is accentuated if the size of the labor force is increasing as well.

Eq. 1 expresses the relationship between GDP, productivity, the labor force, and unemployment:

$$GDP = P(1 - u)L \tag{1}$$

where GDP is real gross domestic product, P is productivity (real GDP per employed person), L is the labor force (employed plus unemployed persons), and u is the unemployment rate (unemployed/labor force).

Comparing Canada in 2009 and 1976, real GDP grew 134.5%, productivity increased 35.6%, the labor force rose by 75.1%, and the unemployment rate increased from 7.1% to 8.3%. While the increase in GDP had a positive impact on employment (and vice versa), it was more than offset by the increase in productivity (P). The net effect was an increase in the rate of unemployment (u).

This is a classic dilemma. A growing economy stimulates employment, an increase in productivity reduces it. How can the advantages of increased productivity be realized in an economy that is not growing without causing high unemployment? One way is to reap the benefits of increased productivity as more leisure rather than more goods and services. This can be accomplished by reducing the average number of hours worked by an employed person so that unemployment for a few becomes more leisure for the many. If more people worked fewer hours, it should be possible to have high levels of employment without relying on economic growth.

Some illustrations from the past show how this could work. From 1976 to 2009, the average hours worked per year by a Canadian employee decreased by 6.7%.[36] [2008 is the most recent year for which data are available in this publication (OECD Factbook 2010). Data for 2008 are used in this section for 2009.] If the decrease in average hours worked had been 10.9% rather than 6.7%, the rate of unemployment would have been 4% not 8.3% in 2009 given the same increases in GDP and the labor force. At an average of 1650 hours of work per employed Canadian, employees in Canada would still have been working about the same number of hours per year as the average employee in the U.K. and more than the average employee in 12 other OECD countries, in some cases substantially more. Had there been no decrease in the average hours worked between 1976 and 2009, the rate of unemployment would have been 14.5% not 8.3% for the same increases in GDP, productivity, and the labor force.

These calculations show that the average length of the work year, which includes vacation days, can have a marked impact on the rate of unemployment. By spreading the same amount of work among a larger number of employees, the unemployment rate can be lowered and the relationship, as shown by the above examples, is strong. It is for this reason that researchers have examined the potential for reductions in the average number of hours worked per employee to contribute to full employment. From the standpoint of managing without growth, the benefits of increased productivity would be experienced as increases in leisure and reduced impacts on the environment rather than as increases in output, consumption, and environmental impacts.

A shorter average work year is one of the factors included in the low/no-growth scenario illustrated in Fig. 3. Over the 30 years of the scenario, the work year declines by 14.1% from 2010 to 2035 so that the average hours spent in employment in Canada declines from 1737 in 2005 to 1492 in 2034. This compares with levels already surpassed in 2009 in Germany (1430), Norway (1422), and the Netherlands (1389).[36]

In general, European countries have been more proactive than Canada and the United States in reducing working time as an instrument of employment policy. During the 2008/2009 recession, some countries, Germany for example, mitigated the impact on employment by relying more heavily on reductions in work time.[37]

The arithmetic of reducing the rate of unemployment by reducing the average hours each employed person works is compelling. Achieving such gains in employment in the real world is another matter, but in a review of studies of the employment effects of working time reductions, Bosch finds that most show a gain of "25–70 percent of the arithmetically possible effect" (Bosch,[38] p. 180). Bosch has examined the European experience, and the six conditions he identified as particularly important for the success or failure of this policy are summarized in Table 1. He points out that the general political conditions must be suitable for a policy of reducing work time to reduce unemployment. There must be acceptance from employees, trade unions, and employers and support of the State.

Looking at working time policy in the future, Bosch concludes that "shorter working hours are an indicator of prosperity" (Bosch,[38] p. 192). They have been in the past, though more recently we have seen the emergence of a sector of the labor force that is "overemployed," working long hours and "failing to achieve a desired balance in their lives between paid work, family life, personal, and

Table 1 Policies for reducing the workweek.

1. Wage compensation—"If working time reductions and pay increases are negotiated as a total package, then the compensatory increase for the working time reduction can be offset by lower pay rises" (Beckerman,[17] p. 182). This could become more difficult with no or low growth.

2. Changes in work organization—"Larger reductions in working time generally have to be accompanied by changes in work organization" (Beckerman,[17] p. 183); otherwise, firms will rely on overtime and the employment effects will not materialize.

3. Shortages of skilled labor—"An active training policy is an indispensable supplement to working-time policy" (Beckerman,[17] p. 183) to ensure that there are people with the necessary skills to pick up the slack when skilled workers reduce their hours.

4. Fixed cost per employee—Such as benefits paid on a per-employee basis rather than an hourly basis are an obstacle to reducing working hours because it is costly to employers. Canada shares with most Western European countries the practice of financing statutory social programs through contributions that are usually a proportion of earnings or through taxation, minimizing this fixed cost problem.

5. The evolution of earnings—"The decreasing rate of real wage rises in most industrialized countries has reduced the scope for implementing cuts in working time and wage increases simultaneously" (Beckerman,[17] p. 184). This would be a serious obstacle unless there is widespread support for seeking prosperity without growth though it can be mitigated by a more equal distribution of income. "One fundamental precondition for the working time policy pursued in Germany and Denmark, for example, was a stable and relatively equal earning distribution" (Beckerman,[17] p. 185).

6. The standardization of working hours—Any reduction in standard working hours must strongly influence actual hours worked. If it merely generates more overtime for those already with jobs, it will fail to increase employment. Work reorganization will be required to allow more flexibility in hours worked.

Source: Maddison,[1] and Beckerman.[17]

civic time."[39] These are usually men with higher levels of education in management positions. Simultaneously, there are people who are underemployed and poorly paid, more often than not women. These circumstances contribute to and accentuate rising income inequality.

Layard in his work on economics and happiness concludes "that people over-estimate the extra happiness they will get from extra possessions" because of habituation. "The required correction is towards lower work effort and thus lower consumption."[40] This means that a shorter workweek not only would contribute to reducing unemployment but also may increase the general level of happiness for employees who find themselves better off working fewer hours, for less income and consuming at lower levels.

Funding Public Services in a Low/No-Growth Economy

Economic growth provides government with increasing resources without increasing tax rates. In times of rapid growth, receipts from corporation profits taxes, personal income taxes, and value-added taxes tend to increase faster than the economy as a whole, allowing governments to provide more services, invest more in infrastructure, redeem outstanding debt, reduce tax rates, or some combination of all these. These are circumstances that governments welcome and they have as much to gain from economic growth as anyone. How might this be different in an economy that eschewed economic growth as a policy objective?

We can gain some insight into this matter by examining the low/no-growth scenario shown in Fig. 4 and comparing it to the business-as-usual scenario.

Table 2 shows the values of some key variables in the base year (2004) and for 2028 and 2034 for each of the two scenarios.

In this illustrative low/no-growth scenario, GDP per capita rises by 48% from 2004 to 2028 after which it stabilizes. GDP follows a similar path, rising by 61% before stabilizing. The greater percentage increase in GDP is due to population growth, which ceases around 2025. Table 2 shows how the composition of GDP changes over time in each of the scenarios. In the business-as-usual scenario, the shares of each of the main components of GDP change very little, with a slight increase in the share of consumption expenditures matched by slight decreases in the shares of government and business investment expenditures. The net trade balance fluctuates between 0% and 2% of GDP.

In the low/no-growth scenario, the share of consumption expenditures remains about 58%, business investment expenditure declines to 12.2% as net investment declines towards replacement levels, government expenditure on goods and services including government investment rises to 24.5% of GDP, and the net trade balance rises to 5.3%. The increase in the share of government expenditures includes the increase in annual expenditures on adult literacy [$0 in 2028 and 2034, high of $831 million in 2012 (expenditures on adult literacy peak in 2012 and then decline on the assumption that adults enrolled in a 1 year literacy program become literate)] and health care (rising to $5 billion from 2019 onwards). In addition, poverty is essentially eliminated by increasing transfers to households by $10.3 billion in 2028, $9.6 billion in 2034, and a high of $15.0 billion in 2019 (i.e., people with incomes below the low-income cutoff as defined by Statistics Canada. The cost

Table 2 A comparison of business as usual and low/no-growth scenarios.

	Units	Base year 2004	BAU 2028	BAU 2034	Low/No growth 2028	Low/No growth 2034
GDP	$97m	1,121,318	2,425,258	2,951,727	1,801,544	1,800,0
GDP per capita	$97	35,053	63,201	74,474	51,894	51,87
Gov expenditure total	$97m	242,772	501,666	600,285	413,781	438,50
Gov expenditure capita	$97	7,589	13,708	15,152	11,919	12,63
Debt/GDP ratio	percent	62.1	27.2	22.1	28.5	30.0
Rate of income tax	percent	23.4	17.1	15.7	17.9	20.8
Rate of profits tax	percent	24.3	17.7	16.3	16.1	19.6
Carbon Tax	$97/tonne CO_2	0	0	0	200	200
GDP composition						
Consumption	percent	57.0	58.7	58.6	58.4	58.0
Business investment	percent	19.6	19.6	19.2	13.1	12.2
Government	percent	21.7	20.7	20.3	23.0	24.4
Net trade balance	percent	1.8	1.1	1.9	5.5	5.4

declines as unemployment is reduced). The sum total of these expenditures is at its highest in 2019 at $20.0 billion, declining to $15.3 billion in 2028 and $14.6 billion in 2034, representing 3.8%, 2.6%, and 2.5% of total government outlays, respectively. (Total government outlays include transfers to households, business and nonresidents, interest payments, and government expenditure on goods and services. Only the last item is included in GDP to avoid double counting.)

Table 2 also shows that the average rates of income and corporations tax required to generate revenues sufficient to maintain the debt-to-GDP ratio at about 30% are lower than they were in 2004 in the low/no-growth scenario. This is because of the inclusion of a revenue neutral carbon tax of $200/ton CO_2, phased in over 10 years starting in 2010.

These results suggest that a low/no-growth future does not present insurmountable fiscal obstacles though several caveats are in order: 1) In comparison with other G8 countries, Canada is starting from a comparatively strong fiscal position with annual budget surpluses for most of the past decade. (Canada emerged from the 2008/2009 recession with an overall fiscal deficit but maintained its relatively strong fiscal position compared to the other G8 countries.) 2) LowGrow analyzes Canada's overall fiscal situation in which the accounts of all three levels of government are combined. In doing so, it obscures some very real differences among the provinces and potential political obstacles pitting "have" and "have not" provinces against one another. 3) The implications of a stable population for government services are ambiguous. A stable population reduces the requirements for increases in publicly funded infrastructure and services. It also reduces the requirements for infrastructure and services aimed at the needs of young people such as new schools. Insofar as a stable population will also be an aging population, it increases the requirements for infrastructure and services needed by older people. 4) At the termination of the low/no-growth scenario in 2034, the economy is not in a steady state since some of the determinants of growth are still positive (i.e., net investment and net trade), and government expenditure is still rising albeit more slowly than in the business-as-usual scenario. Further adjustments will be needed to maintain GDP per capita at a constant level after 2034. The fiscal dimension of the low/no-growth scenario is only a partial representation of what might be accomplished through a more comprehensive program of ecological fiscal reform. 5) Implications for international capital flows and international competitiveness and monetary policy have not been considered and present additional challenges.

CONCLUSION

There are many reasons for considering how rich economies might manage without growth: biophysical constraints to continued growth are becoming more apparent, mounting evidence indicates that higher incomes do not make people happier beyond a level of per capita incomes far surpassed in rich countries, and despite decades of substantial economic growth many social and environmental problems remain. LowGrow provides a tool for examining possibilities for managing without growth specifically in Canada, but similar results would likely be obtained if the model was applied to other developed countries. Such work is underway or planned in New Zealand, Austria, U.K., and the United States. It will be interesting to compare the results.

AKNOWLEDGMENTS

This entry is based on Victor, P.A. *Managing without Growth: Slower by Design, Not Disaster*, 2008.

REFERENCES

1. Maddison, A. *The World Economy: A Millennial Perspective*; OECD: Paris, 2001.
2. World Bank, News and Broadcast, Press Release No. 2009/065/DEC, http://go.worldbank.org/CUQLLRX1Q0 (accessed August 29, 2010).
3. Jaumotte, F.; Lall, S.; Papageorgiou, C. Rising Income Inequality: Technology, or Trade and Financial Globalization? IMF Working Paper. WP/08/185, July 2008, http://www.imf.org/external/pubs/ft/wp/2008/wp08185.pdf (accessed August 29, 2010)
4. Krausmann, F.; Gingrich, S.; Eisenmenger, N.; K-H.; Haberl, H.; Fisher-Kowalski, M. Growth in global materials use, GDP and population during the 20th century. Ecol. Econ. **2009**, *68* (10), 2696–2705.
5. Rockstrom, J. et al. A safe operating space for humanity. Nature, **2009**, *24*, 461, 472–475.
6. Sorrell, S. *Global Oil Depletion*; Technology and Policy Research Centre: U.K., 2009.
7. Ayres, R.U.; Warr, B. *The Engine of Economic Growth*; Edward Elgar: Cheltenham, U.K., 2009.
8. *Critical Raw Materials for the EU* (June 2010), Report of the Ad-hoc Committee on Critical Raw Materials, http://ec.europa.eu/enterprise/policies/raw-materials/critical/index_en.htm (accessed July 1, 2010).
9. Arndt, H.W. *The Rise and Fall of Economic Growth: A Study in Contemporary Thought*; Longman Cheshire: Melbourne, 1978; p.13.
10. OECD. Economic Policy Reforms: Going for Growth; OECD Paris: France, 2010.
11. Dinda, S. Environmental Kuznets curve hypothesis: A survey. Ecol. Econ. **2004**, 49 (4), 431–455.
12. Schnieder, F.; Kallis, G.; Martinez-Alier, M. Crisis or opportunity? Economic degrowth for social equity and ecological sustainability. Introduction to this special issue. J. Cleaner Prod. **2010,** *18* (6), 511–518.
13. Galbraith, J.K. *The Affluent Society*; Houghton Mifflin: Boston, 1958.
14. Mishan, E.J. *The Costs of Economic Growth*; F.A. Praeger: New York, 1968.

15. Beckerman, W. *Why We Need Economic Growth.* Lloyds Bank Rev. 1971.
16. Beckerman, W. *In Defence of Economic Growth*; J. Cape: London, 1974.
17. Beckerman, W. *Small Is Stupid*; Duckworth: London, 1995.
18. Schumacher, E.F. *Small Is Beautiful: Economics as if People Mattered*; Harper and Row: London, 1973.
19. Boulding, K.E. The economics of the coming spaceship earth. In *Environmental Quality in a Growing Economy*; Jarrett, H., Ed.; Johns Hopkins Press: Baltimore, MD, 1966.
20. Georgescu-Roegen, N. *The Entropy Law and the Economic Process*; Harvard University Press: Cambridge, MA, 1971.
21. Daly, H. *Steady-State Economics: The Economics of Biophysical Equilibrium*; W.H. Freeman: San Francisco, 1977.
22. Meadows, D.H.; Meadows, D.L.; Randers, J.; Behrens III, W.W. *The Limits to Growth*; Earth Island Limited: London, 1972.
23. Olson, M.; Landsberg, H.H., Eds. *The No-Growth Society*; W.W. Norton and Company: New York, 1973.
24. Douthwaite, R. *The Growth Illusion*; New Society Publishers: Gabriola Island, Canada, 1994 (also, a second edition in 1999).
25. Daly, H. *Beyond Growth: The Economics of Sustainable Development*; Beacon Press: Boston, 1996.
26. Booth, D.E. *The Environmental Consequences of Growth: Steady-State Economics*; Routledge: London, 1998.
27. Hamilton, C. *Growth Fetish*; Allen and Unwin: Australia, 2003.
28. Booth, D.E. *Hooked on Growth: Economic Addictions and the Environment*; Rowman and Littlefield: Lanham, MD, 2004.
29. Victor, P.A. *Managing without Growth: Slower by Design, Not Disaster*; Edward Elgar Publishing: Camberley, U.K., 2008.
30. Speth, J.G. *The Bridge at the End of the World: Capitalism, the Environment, and Crossing from Crisis to Sustainability*; Yale University Press: New Haven, CT, 2008.
31. Brown, P.; Carver, G. *Right Relationship*; Berrett-Koehler Publishers Inc.: San Francisco, 2009.
32. Jackson, T. *Prosperity without Growth*; Earthscan: London, 2009.
33. Schor, J. *Plenitude: The New Economics of True Wealth*; The Penguin Press: New York, 2010.
34. United Nations Development Programme. *Human Development Report 2006*, New York, 2006.
35. United Nations. *World Economic Survey for 1959*; New York, 1960; p. 5 (quoted in Ref. 10, p. 62).
36. OECD (2010), *OECD Factbook 2010: Economic, Environmental and Social Statistics*, OECD Publishings. doi: 10.1787/factbook-2010-en.
37. OECD. Speech by Angel Gurria (April 28, 2010), http://www.oecd.org/document/6/0,3343,en_21571361_44315115_45085318_1_1_1_1,00.html (accessed August 25, 2010).
38. Bosch, G. Working time reductions, employment consequences and lessons from Europe. In *Working Time: International Trends, Theory and Policy Perspectives*; Golden, L., Figart, D.M., Eds.; Routledge: London, 2000; 177–211.
39. Figart, D.M.; Golden, L. Introduction and overview, understanding working time around the world. In *Working Time: International Trends, Theory and Policy Perspectives*; Routledge: London, 2000; 1–17.
40. Layard, R. Happiness and public policy: a challenge to the profession. Econ. J. **2006,** 116 (March), C24–C33.

Ecosystems: Large-Scale Restoration Governance

Shannon Estenoz
Office of Everglades Restoration Initiatives, U.S. Department of the Interior, Davie, Florida, U.S.A.

Denise Vienne
Alka Sapat
School of Public Administration, Florida Atlantic University, Boca Raton, Florida, U.S.A.

Abstract

Collaborative resource governance continues to expand as an approach to solving very complex environmental problems, including those arising in large-scale ecosystem restoration. Such problems transcend geographic and regulatory jurisdictions and involve cause-and-effect relationships that introduce layers of political and economic complexity. This entry focuses on the nature of collaborative resource governance as a means of managing complexity and cross-jurisdictional issues in large-scale ecosystem restoration. In doing so, we review the evolution, characteristics, challenges, and opportunities of various collaborative resource governance models for ecosystem management. We then discuss the task force as an institutional mechanism for collaborative resource governance and compare its application in the restoration of the Gulf Coast ecosystem following the 2010 Deepwater Horizon oil spill with its much longer-term use in the Florida Everglades ecosystem restoration program. We conclude by discussing how these programs can inform future research on governance structures and environmental management.

INTRODUCTION

Large-scale ecosystem restoration involves layers of complexity including, at the most fundamental level, ecological cause/effect relationships that may not be well understood, immediately observable, or measurable. Ecological complexity alone ensures that ecosystem restoration will be characterized by difficulty and uncertainty. However, in the context of large-scale ecosystem restoration, ecological complexity is compounded by other complexities including, for example, governance systems and regulatory frameworks not designed to approach restoration at the ecosystem scale and complex political interactions associated with resource consumption, private property rights, and economic development. Restoration leaders have created new governance structures, alternatively termed "collaborative resource governance" (CRG) or "collaborative environmental management," designed to work through and around such layered complexity.[1] While these structures vary in form and scope across restoration programs, there are common threads among them that include the scale and complexity of problems they are designed to address, emphasis on locally tailored solutions, the allowance for experimentation and dynamic adaption, and the importance of information sharing and collaborative problem solving at multiple overlapping levels.[2] The main focus of this entry is on the nature of CRG as a means of managing complexity and cross-jurisdictional issues in large-scale ecosystem restoration. In doing so, we review the evolution of these approaches in environmental and ecosystem management and the characteristics, challenges, and opportunities that such approaches present for governance and ecosystem restoration. More specifically, we discuss collaborative governance structures that have been adopted to manage two of the largest ecosystem restoration programs in the country, Gulf Coast ecosystem restoration and Everglades restoration.[3,4] Similar institutional mechanisms, i.e., ecosystem restoration taskforces, were put in place for both these ecosystems. These two programs present an opportunity for ascertaining whether synergies can be created in an ecosystem restoration program by cross-pollinating engagement at many levels with another restoration program.

We begin our discussion of this topic by examining how collaborative governance has emerged as one of the predominant mechanisms for addressing complex policy problems. Next, we note the evolution of environmental management in the United States toward large-scale ecosystem management through employment of collaborative governance. Although ecosystem restoration projects increasingly rely on this approach for governance and implementation, they are extremely diverse as a reflection of their political, administrative, and environmental contexts. Thus, the following section attempts to derive important lessons gleaned from these specific projects. Then, CRG will be defined as a distinct type of governance model that

Encyclopedia of Environmental Management DOI: 10.1081/E-EEM-120048165

exhibits several general characteristics. A specific example of CRG, the ecosystem restoration task force model is the focus of the final sections. The Florida Everglades and the Gulf of Mexico, two of the world's largest ecosystem restoration projects that employ the task force model, will be discussed and compared. The conclusion elaborates how experience with these models can serve as the basis for future comparative research and as a guide for governance structures and environmental management.

GOVERNANCE MODELS: COLLABORATIVE ENVIRONMENTAL MANAGEMENT

Since the 1980s, collaborative governance has emerged as a predominant paradigm to resolve various policy issues, and this shift has been part of a gradual and broader movement in both the theory and praxis of public administration. Public managers have increasingly been addressing complex problems through networks, strategic partnerships, alliances, coalitions, contractual relationships, committees, consortia, and councils; these mechanisms have been used to develop collaborative mechanisms across jurisdictions, governments, and sectors. Collaborative governance approaches are thus increasingly being used in response to public demands and to deal with boundary-spanning policy problems that cannot be addressed by traditional regulatory approaches relying on single agencies and jurisdictions.[5,6]

Governance models for natural resource management have mirrored and, in some instances, even led to the gradual shift in collaborative governance, due to the emergence of various forms of partnerships and civic environmentalism that emerged to deal with environmental problems.[7,8] Reasons for the emergence of collaborative environmental management are many, overlapping, and varied. Some types of collaborative environmental governance emerged as a result of federal cutbacks, leading to civic environmentalism and citizens pushing for environmental changes.[8] For the most part, though, forms of collaborative environmental management came about as a result of the problems stemming from command-and-control measures, which often resulted in protracted conflict or failure to achieve key objectives. This led to calls for change from bureaucratic, adversarial, technology-based regulatory approaches, which were the basis for many environmental policies in the 1970s and early 1980s, to "results-based" and voluntary approaches to regulation.[9] Similar to other policy problems, calls for more collaborative measures also stemmed from the complexity of environmental problems that were not effectively addressed by conventional rule-based, top-down, and hierarchical approaches. The latter were seen as problematic in that they did not allow for more democratic forms of participation, stymied potential innovation, were ineffective in addressing multimedia environmental hazards and those stemming from nonpoint sources of pollution, and relied on unrealistic models of administrative and individual rationality.[7,9–11] To deal with these cross-jurisdictional problems, there were calls for more holistic and integrated approaches to deal with ecosystems such as large watersheds and forests. The adoption of more holistic approaches through collaborative environmental management grew during the 1990s particularly the Clinton administration; during this period, the ecosystem approach was greatly expanded both to better administer large restoration projects (such as the Everglades and the Northwest Forest plans) and to address smaller ecosystem and habitat conservation planning.[12]

Collaborative environmental management takes many forms: some entail collaboration between multiple agencies only, while others are characterized by the presence of multiple stakeholders such as actors in the private and nonprofit sectors. In some cases of collaborative environmental management, government has led the efforts; in others, government has encouraged it through grants and other incentives; and, at times, the efforts at collaborative management have been pioneered by other stakeholders such as citizens or nonprofit groups.[13] What distinguishes these collaborative approaches from earlier forms of environmental management is the movement away from command-and-control policies toward those that are both more inclusive of and seek to incentivize public participation. The search for mutually beneficial policy solutions by encouraging broad participation from local stakeholders, underscoring the importance of consensus and voluntary approaches, and building trust-based policy networks are defining characteristics of collaborative environmental approaches. In addition, collaborative environmental governance also involves improvements in scientific understanding of how ecological processes affect resource outcomes across artificial jurisdictional and political boundaries[14] and careful scientific monitoring to allow for managerial adaptation as necessary.[12]

Since collaborative environmental governance is an institutional mechanism for natural resource management, its development and success may depend on the set of institutional rules applied in ecosystem management. As noted by Elinor Ostrom, successful institutions tend to have certain design principles, such as rules adapted to local circumstances, clearly defined resource boundaries, information about resource variability, monitoring and sanction mechanisms, and local conflict resolutions forums.[15] While these principles may not ensure success, they could be contributory factors, as noted in our discussion about CRG structures adopted for large-scale ecosystem restoration.

In the next section, we discuss how collaborative governance approaches have been used in ecosystem management, which has seen a shift from place-based to species-based conservation.

COLLABORATIVE GOVERNANCE APPROACHES AND ECOSYSTEM MANAGEMENT

Over the past 30 years, place-based and species-based approaches to environmental conservation have been giving way to more holistic and collaborative approaches to ecosystem management.[16] The earlier strategies involved prescriptive, or command-and-control, regulatory approaches that targeted discrete, identifiable sources of environmental damage.[17] Traditional regulatory approaches target the "low-hanging fruit" with respect to environmental problems, but they are insufficient for resource management and the restoration of ecosystems that are subject to multiple interdependent and interacting conditions, lacking clear or easily discernable sources of damage.[17] Collaborative ecosystem governance is a response to seemingly intractable environmental problems that do not fall neatly within the domain of any one governmental jurisdiction or agency.[1] Comparatively speaking, there is a great deal of variation in how such collaborative governing bodies are convened, their constellation of participants, the type and pattern of interactions, and the ecosystem problems they manage.

On the surface, these diverse institutional arrangements appear to be distinct problem-specific environmental responses because they conform to the particular resource systems they are designed to manage. However, collaborative ecosystem governance models share common characteristics that set them apart as a distinctive alternative to the traditional legal and institutional approaches relied upon for environmental management.[2] Furthermore, since they are characteristically more flexible, participatory, deliberative, and heterarchical forms of organization, they are considered to be more responsive, legitimate, and effective. Yet, little is known about the preconditions for their effectiveness or the ultimate consequences for democratic processes.[16]

The concern for democratic process is not inconsequential, and it is twofold. In the most general sense, the nature of these new governance structures precludes direct lines of political accountability. The decision-making process encompasses a diverse range of public and private actors along with nonelected administrators, and it supersedes jurisdictional authority. This not only threatens political accountability but also can be an impediment to collaboration to the extent that it undermines the political autonomy of participants. For example, governors are likely to be faced with choosing which to subordinate—the political will of their state or the will of the collaborative governance structure. In the case of interstate water compacts, this dilemma lies at the heart of legal conflicts. Despite commitment to interstate water agreements, legal recourse is often the only viable mechanism for conflict resolution when public officials are trapped between the terms of their compact agreements and evolving demands of their constituencies.[18] Given the long-term nature of these governance models, the challenge is to address both the issue of democratic process and strategies for maintaining broad-based public support and the political will necessary to sustain and inform them.

OPPORTUNITIES AND CHALLENGES OF LARGE-SCALE ECOSYSTEM RESTORATION

There are numerous challenges presented by collaborative ecosystem governance that stem from the complexity inherent both in the nature of large-scale ecosystem restoration and in the governance models themselves. Conceptualizing resource management as ecosystem management expands the range of considerations to include multiple competing uses and users, nonpoint sources of damage, nonlinear environmental reactions, and uncertain interactive effects and threshold points. The corresponding governance models are no less complex because they involve designing governing arrangements to accommodate specific environmental issues that are often framed by competing economic, political, and social contexts. As such, they necessarily overlap political jurisdictions and the functional boundaries of agencies, and they encompass multiple levels of governmental decisionmakers, regulatory bodies, and nongovernmental participants. Sustaining active cooperation, maintaining public and political support, and securing sufficient commitment resources necessary for uncertain and changing environmental problems over an indeterminate time frame are just some of the practical challenges. At the same time, embracing postsovereign governing arrangements, "rolling-rule" regulatory models, and adaptive management techniques represent paradigmatic shifts that entail philosophical challenges, which depend on more fundamental changes.[16,17,19,20] Reliance on intersovereign agreements and sovereign-dominated approaches has not been an adequate match for the scale of ecosystem problems particularly when they merely result in lowest-common-denominator agreements on uses and measures.[19] Nevertheless, expanding the range of issues addressed and the number of parties involved in governance raises questions about the role of government in facilitating effective decisions operating outside direct political control and in the "shadow of hierarchy."[16] In addition to overcoming philosophical and practical challenges, there is the additional challenge of designing the appropriate governance structure given the context and the particular ecosystem problem to be addressed.[20]

The propensity to focus on the challenges overlooks important opportunities presented by large-scale ecosystem governance. Collaborative environmental management offers a highly dynamic and participative approach to policy implementation that integrates localized knowledge and scientific learning into future decisionmaking.[2] Adaptive management approaches are more conducive to experimentation required in conditions of uncertainty, allowing for creativity, flexibility, and learning.[17] Broad-based participation improves the public and political buy-in necessary for credibility and legitimacy, the commitment and pooling of resources, and long-term collaboration.[20] Moreover, collaborative governance models offer an opportunity to address long-recognized challenges presented by the decentralization and devolution of government and related concerns about democratic control associated with the increasing discretion afforded networks of administrators and nongovernmental actors in performing functions of the state. The state has an important role to play in preventing dysfunction and facilitating these governance models by ensuring democratic processes and integrating technical and normative considerations toward effective implementation.[16] Distinct from networks of policy actors, public–private partnerships, or devolved federal responsibilities, CRG is a much narrower, issue-specific, model of democratic governance.

COLLABORATIVE RESOURCE GOVERNANCE

Collaborative resource governance can be defined as a diverse group of public and private stakeholders working together to address shared problems that extend beyond their individual capacities.[1,16] Unlike other policy configurations common in the literature, such as public–private partnerships, policy networks that seek to influence government, or networks of government actors that informally coalesce around "wicked problems," they are distinctive problem-solving, polyarchic governance models focused on complex ecological problems that overwhelm the capacity of the sovereign state.[17] Although examples of CRG vary in scale, focus, and structure, they are an increasingly important focus of research for the purpose of ascertaining factors that determine their effectiveness, the specific role of the state, and their level of success in managing large-scale ecological problems.[20,1]

The form a collaborative governance model assumes depends on environmental, political, and economic contexts. Comparative research is beginning to identify specific factors that contribute to how they function, as well as emergent concerns. For the most part, the diversity of these initiatives belies similarities among them. Some defining characteristics of CRG are its hybrid public–private structure, the scale and complexity problems they are designed to address, emphasis on locally tailored solutions, the allowance for experimentation and dynamic adaption, and the importance of information sharing and collaborative problem solving at multiple overlapping levels.[2] Success, on the other hand, is highly dependent on how they are organized, funded, and governed.[21] Integrating science into the decision-making process has also been seen as critically important, necessitating broad-based participatory processes that preclude science from being manipulated, or trumped, by the political process.[20] Process is just as important as structure for ensuring flexibility and adaptability in order to incorporate new scientific information and to make necessary course corrections. Despite the dependence on nongovernmental actors and decentralized and fragmented approaches, there is a critical role for government in providing definitional guidance, participatory incentives, and enforcement capabilities.[16] One common problem is that most initiatives severely underestimate the expense (financial, time, and personnel investment) of collaborative approaches to ecosystem management relative to traditional regulatory approaches.[20] Examples of CRG programs are indicative of specific lessons learned.

The Chesapeake Bay Program is generally acknowledged as one of the oldest and most organizationally successful of these programs despite questionable improvements in water quality.[17] It exemplifies the importance of having reliable mechanisms in place in order to facilitate participation, public outreach, and the integration of scientific information into decisionmaking.[20] Diffuse governance that is not contained by traditional political authority requires special effort not only to capitalize on the respective strengths of participants but also to cultivate legitimate and credible processes.

The California Bay-Delta Program (CALFED) has demonstrated the importance of generating broad-based scientific input, instituting internal and external peer review of scientific proposals, and devising conceptual models for communicating scientific information to decision makers and stakeholders. The CALFED has also shown the benefit of dedicating facilitators and planners to ensure vertical integration and secure long-term funding. However, problems did emerge with CALFED; for instance, several independent reports criticized its governance structure, stakeholders moved water management decisions outside the CALFED process using environmental litigation, the executive director and lead scientist resigned in 2004, and faced with mounting criticism, the CALFED Bay-Delta Authority voted to disband itself in 2005.[14]

The Comprehensive Everglades Restoration Plan (CERP) is an example of how clearly defined problems and agreement over the urgency of an environmental issue can coalesce political and financial support. However, CERP is also an example of how political tensions between state and federal levels can impede cooperation. These tensions stem in part from prematurely determining CERP's organizational structure and possible alternative solutions for Everglades restoration prior to scientific involvement, which constrained options by favoring methods preferred by the United States Army Corps of Engineers (USACE).[20] The CERP has demonstrated the benefit of adaptive monitoring in the early stages of new strategies and also, for the sake of long-term goals, the need for clarifying performance measures and indicators.

Since the 2010 oil spill, the Louisiana Coastal Area Ecosystem Restoration Program (LCA) has transitioned to the task-force model followed by CERP. However, problems with its previous model are instructive. The LCA, like CERP, experienced state–federal tensions for differing reasons but which were also exacerbated by restrictive control under the USACE. Options available to the LCA were severely constrained by political pressures by powerful stakeholders. Still, the LCA's structure and processes were not adequately designed to facilitate broad-based participation, generate public buy-in, or integrate science in ongoing decision-making.[20] The result was an excessive focus on local symptoms to the exclusion of root causes tied to resource practices in the Mississippi–Ohio–Missouri river basin.

Lastly, the Glen Canyon Adaptive Management Program exemplifies the potential for adaptive management in these new governance models.[20] Adaptive management emphasizes learning, adjustment, and the acceptance of rolling rules rather than grounding in static regulatory or managerial approaches. However, there is still much misunderstanding about these approaches that requires education in order to facilitate a transition from traditional, static approaches to adaptive methods of regulation and localized policy implementation.[17] The experiences of these CRG programs are indicative of jurisdictional issues relating to ecosystem management that do not easily correspond with traditional forms governance or regulatory approaches.

Cross-Jurisdictional Issues

Jurisdictional issues are of particular importance to CRG because the geographic dimensions of ecosystem restoration defy traditional political and functional boundaries, presenting new challenges for environmental regulation and management. The geographic scale of an ecosystem means that it often will encompass some combination of municipalities, counties, states, regions, and nations.[17] Similarly, resource users and sources of environmental degradation are not contained by any particular jurisdiction or subsumed under any one political authority. Thus, problems neither fall neatly within the control of any particular authority nor the functional realm of any particular agency. Nevertheless, the magnitude of environmental problems exceeds the capacity, resources, and expertise of any one governmental entity.

Traditional regulatory methods are not equipped to deal with uncertainty, complexity, and continuous change. They are appropriate for targeting point-source problems but cannot account for the diversity presented by numerous local circumstances.[16] Legal scholars are examining the implications of CRG for the future of environmental regulation and the use of adaptive rolling rules more suitable to uncertainty and change.[16,17,19] Similarly, the decentralized and fragmented governance structure necessitated by the nature and magnitude of resource problems presents an enormous challenge for management. There are numerous public and private parties involved, and addressing problems of this scope depends on long-term commitment and public support. This requires managing collaborative activities of numerous parties from multiple jurisdictions and diverse functional backgrounds that often have competing interests. The challenge is not only to coordinate participants but also to sustain cooperation. Although CALFED demonstrated the importance of broad participation and "bottom-up" policy approaches, the problems that emerged highlighted the importance of having a clear direction and identifiable goals. Authority may be diffuse, but leadership is imperative in order to maintain momentum, negotiate compromises among competing interests, and translate and communicate across functions (e.g., political decisionmakers and scientists, scientists and the public, and across agency missions and cultures). Adaptive management is appropriate for CRG because it allows for experimentation, learning, adaptation, and course corrections.[17,20] Adaptive management represents a paradigm shift, and possibly a hurdle, for politicians, administrators, and regulators used to operating within defined jurisdictions. As researchers consider the list of factors that determine the success or failure of CRG as a general approach to large-scale ecosystem restoration, it is helpful to highlight one specific CRG structure that has been both in place for a considerable period of time and recently replicated. As such, may provide important and unique opportunities to increase understanding of CRG.

FOCUS: ECOSYSTEM RESTORATION TASK FORCE MODEL

The restoration of the Gulf of Mexico and the restoration of the Florida Everglades are two of the largest and most complex ecosystem restoration programs in the world.[22–24] The highest level of intergovernmental coordination for both programs occurs under the auspices of intergovernmental task forces created by the federal government. This model warrants closer consideration because although the two task forces differ somewhat in constitution and scope, there are enough similarities between them to make the case that they are two examples of a single intergovernmental coordination and governance structure. This "duplication" of a specific CRG structure in two large-scale ecosystem restoration programs provides an opportunity to observe the successes and challenges for this institutional design and structure over time in a way that cannot occur when focused solely on a structure that is unique to a specific ecosystem. In addition, the model warrants closer consideration because of its longevity in the case of the Everglades and, in the case of the Gulf Coast, its relevance to the most current and high profile issues in large-scale ecosystem restoration.

South Florida Ecosystem Restoration Task Force

The origins of the South Florida Ecosystem Restoration Task Force (SFERTF) can be traced to the fractured state of intergovernmental relationships that existed in the late 1980s due to contentious litigation between the federal government and the state of Florida over degraded water quality in the Everglades.[23] The litigation, which began in 1988 and continues today, produced a settlement agreement in 1992 that, for the time being, opened a window of opportunity to improve intergovernmental relations in order to begin addressing a broader range of Everglades restoration issues.[23,25] At the beginning of the Clinton administration, then secretary of the U.S. Department of the Interior, Bruce Babbitt, exploited the litigation lull and ushered in a new era of collaboration on Everglades restoration issues, which included the first incarnation of the SFERTF, created by Bruce Babbitt.[23,25]

Membership of this original SFERTF was limited to six federal agencies, but 3 years later, Congress created a new task force (also called the SFERTF) and expanded its membership to include nonfederal representatives.[23] The new Task Force was to be chaired by the Secretary of the Department of the Interior and was to include the secretaries of the Commerce, Army, Agriculture, and Transportation departments, the Administrator of the Environmental Protection Agency (EPA), and the attorney general. Congress allowed these Presidential Cabinet members to designate appointees to represent them; however, the statute requires that designees be at the assistant secretary or equivalent level of authority.[4] The SFERTF also includes representation for the state of Florida, the Miccosukee Tribe of Florida, the Seminole Tribe of Florida, the South Florida Water Management District, and local governments. In the same legislation, Congress directed the Secretary of the Army to develop a "... proposed comprehensive plan for the purpose of restoring, preserving and protecting the South Florida ecosystem" [(4) §§ 528(b)(1)(A)(i)-528(f)(3)].

Initially, the SFERTF was charged with a number of coordination and oversight responsibilities during the restoration plan development phase; however the statute did not include a sunset provision to dissolve or transform the SFERTF at the completion of that phase. On the contrary, over the decade that followed, the Congress wove SFERTF oversight into the implementation phase of restoration, as did federal regulations governing the restoration program that were developed in 2002.[26,27] The responsibilities of the SFERTF include consultation with the Secretary of the Army; coordination of restoration policy, strategy, priorities, and programs; the exchange of information among task force members; facilitation of intergovernmental dispute resolution; coordination of restoration science; the support of implementing agencies; the coordination of financial reporting and budget requests; and reporting biennially to Congress on its own activities and on the progress of restoration efforts and results.[4]

Gulf Coast Ecosystem Restoration Task Force

In the spring of 2010, the Deepwater Horizon oil spill became one of the worst man-made environmental disasters in American history.[22] In total, an estimated 4.9 million gallons of oil was released into the Gulf of Mexico, resulting in short- and long-term environmental and economic impacts that may not be fully quantified or understood for years to come.[22] The emergency response to the Deepwater Horizon crisis required significant intergovernmental coordination to address rescue and recovery, well closure, and cleanup.[3,22] During the response period, U.S. President Barack Obama ordered the Secretary of the Navy, Ray Mabus, to develop a vision for moving from response to recovery and restoration in the Gulf of Mexico.

The so-called "Mabus Report" was released in September 2010 and took an expansive view of Gulf restoration that went beyond ecological damage caused by the oil spill and included consideration of the broad range of ecosystem challenges in the Gulf and Gulf Coast that preceded the Deepwater Horizon crisis.[22] Secretary Mabus issued a set of recommendations, which in part focused on short- and long-term intergovernmental coordination in the recovery and restoration effort. The secretary recommended that Congress establish the Gulf Coast Recovery Council to coordinate federal, state, and tribal restoration and recovery actions and to coordinate with and support activities conducted under the Natural Resources Damage Assessment (NRDA) process.[22] However, the Secretary also recommended the immediate designation of a lead federal restoration agency and the immediate creation of the Gulf Coast Restoration Task Force. The executive branch of government could carry out these recommendations without Congressional action. Mabus suggested that the task force initiate the development of a restoration and recovery strategy for the Gulf and pointed out that if Congress acted to establish the recommended council, it could subsume the task force.[22]

A month after the release of the Mabus Report, President Barack Obama created the Gulf Coast Ecosystem Restoration Task Force (GCERTF) by Executive Order.[3] The President's order states that "[t]o effectively address the damage caused by the BP Deepwater Horizon Oil Spill, address the [sic] longstanding ecological decline, and begin moving toward a more resilient Gulf Coast ecosystem, ecosystem restoration is needed" (Section 1). The Executive Order specifies federal membership of the GCERTF as including "senior officials" [Section 2 (a)(1)] of the Departments of Defense (Army Civil Works), Justice, Interior, Agriculture, Commerce, and Transportation; the EPA; the White House Offices of Management and Budget and Science and Technology Policy; and the White House Councils on Environmental Quality and Domestic

Policy. The task force includes representatives of the five Gulf coast states and has the authority to add representatives of affected tribal governments.[3] The GCERTF was created as an advisory body to coordinate intergovernmental restoration efforts, support the NRDA process, present to the President a strategy for Gulf restoration, coordinate scientific research, engage stakeholders and the public, and report to the President biennially on the progress of the restoration strategy.[3]

Common Threads Between the SFERTF and GCERTF

The similarities of structure and scope are evident between these two governance bodies. Every federal agency represented on the SFERTF is also represented on the GCERTF, a reflection of the significant jurisdictional and geographic overlap between the Everglades and the Gulf Coast; both include state representation, and both are responsible for strategic planning, the coordination of intergovernmental activities and science, stakeholder engagement, and biennial reporting on restoration progress.[3,4] Both task forces are administered by senior executives of the federal government who supervise full-time staff dedicated exclusively to task force administration.

More than 700 miles of Gulf coastline and 100% of the Everglades ecosystem are located in the state of Florida, and in fact, the Greater Everglades Ecosystem includes significant Gulf Coast resources such as the Caloosahatchee Estuary.[24,28] This shared geography not only ensures a great deal of intergovernmental cross-pollination at both the political and staff levels, but it also ensures high levels of cross-pollination among stakeholders, nongovernmental organizations, scientists, journalists, and elected officials involved in both programs. Operational or "day-to-day" collaboration, above and beyond the collaboration that flows from formal governance structures like the GCERTF and SFERTF, is an important factor in many large-scale ecosystem restoration programs.[1] Over time, the cross-pollination occurring between the Gulf Coast and Everglades restoration programs may have synergistic effects on operational collaboration for both. These two programs may help operationalize the premise that collaboration among ecosystem restoration efforts creates synergies, efficiencies, and benefits across programs, a premise central to the mission of efforts such as America's Great Waters Coalition.[29]

CONCLUSION

As discussed in this entry, large-scale ecosystem restoration, as an approach to solving complex, multidimensional environmental problems, presents challenges of scale, causality, and jurisdiction that can overwhelm the capacity of a single state, jurisdiction, or authority. Configurations of governmental, private sector, and nongovernmental actors have produced a variety of collaborative arrangements in response to addressing high levels of complexity. While models vary across restoration programs and tend to pursue strategies tailored to specific environmental problems, there are identifiable common features among models, such as the integration of adaptive management principles, stakeholder engagement processes, and the incorporation of science in decisionmaking. The government's comprehensive approach to ecosystem restoration in the aftermath of the Deepwater Horizon oil spill may be an indication that CRG is an approach to environmental management that is in its ascendancy. The Gulf ecosystem restoration program borrows its model of collaborative governance from a model that has been in place in the Everglades restoration program for 16 years. This is an opportunity for comparative research on collaborative resource management that can inform future applications of CRG about specific challenges and opportunities relating to large-scale ecosystem restoration. Importantly, future research must also account for several issues that are beyond the scope of this entry. For example, a nuanced understanding of the political, administrative, and logistical challenges requires an in-depth understanding of the histories and the legal challenges confronted in each of these cases, as well as assessments of the level of success in meeting respective environmental goals.

REFERENCES

1. Gerlack, A.; Heikkla, T. Comparing collaborative mechanisms in large-scale ecosystem governance. Nat. Resour. J. **2006**, *46* (Summer), 657–707.
2. Karkkainen, B. Collaborative Ecosystem Governance: Scale, Complexity, and Dynamism, available at http://www.law.virginia.edu/lawweb/lawweb2.nsf/0/2ba27078dc464a848 52569700060de96/$FILE/HDOCSscalecomplex.pdf.2002 (accessed September 2011).
3. Obama, B. Executive Order 13554 of October 5, 2010, Establishing the Gulf Coast Ecosystem Restoration Task Force. Federal Register. Vol. 75 (No. 195), Presidential Documents; 2010.
4. *Water Resources Development Act*, Pub. L. No. 104-303, 110 Stat. Section 528 3767 -3773; 1996.
5. Agranoff, R.; M. McGuire. *Collaborative Public Management: New Strategies for Local Governments*; Georgetown University Press: Washington, DC, 2003.
6. O'Leary, R.; Bingham, L., Eds. *The Collaborative Public Manager: New Ideas for the 21st Century*; Georgetown University Press: Washington, DC, 2009.
7. Koontz, T.M.; Thomas, C.W. What do we know and need to know about the environmental outcomes of collaborative management? Pub. Admin. Rev. **2006**, *66*, 111–121.
8. John, DeWitt. *Civic Environmentalism: Alternatives to Regulation in States and Communities*; CQ Press: Washington, DC, 1994.

9. Durant, R.F.; Fiorino, D.; O'Leary, R., Eds. *Environmental Governance Reconsidered: Challenges, Choices, and Opportunities*; The MIT Press: Cambridge, MA, 2004.
10. National Academy of Public Administration. *Resolving the Paradox: EPA and the States Focus on Results*; NAPA: Washington, DC, 1997.
11. O'Leary, R.; Durant, R.F.; Fiorino, D.; Weiland, P.S. *Managing for the Environment: Understanding the Legal, Organizational, and Policy Challenges*; Jossey-Bass: San Francisco, 1999.
12. Vig, N.J.; Kraft, M., Eds. *Environmental Policy: New Directions for the 21st Century*; CQ Press: Washington, DC, 2010.
13. Koontz, T.M.; Steelman, T.A.; Carmin, J.; Korfmacher, K.S.; Mosely, C.; Thomas, C.W. *Collaborative Environmental Management: What Roles for Government?* Resources for the Future: Washington, DC, 2004.
14. Lubell, M.; Segee, B. Conflict and cooperation in natural resource management. In *Environmental Policy: New Directions for the 21st Century*; Vig, N., Kraft, M., Eds.; CQ Press: Washington, DC, 2010; 171–196.
15. Ostrom, E. *Governing the Commons: The Evolution of Institutions for Collective Action*; Cambridge University Press: New York, 1990.
16. Gunningham, N. The new collaborative environmental governance: The localization of regulation. J. LawSoc. **2009**, *36* (1), 145–166.
17. Ruhl, J. Regulation by adaptive management—Is it possible? Minn. J. Law, Sci., Tech. **2006**, *7* (1), 21–57.
18. Heikkila, T.; Schlager, E.; Davis, M. The role of cross-scale institutional linkages. In common pool resource management: Assessing interstate river mangement. Policy Stud. J. **2011**, *39* (1), 121–145.
19. Karkkainen, B. Post-sovereign environmental governance. Global Environ. Polit. **2004**, *4* (1), 72–96.
20. Van Cleve, A.; Simenstad, C.; Geotz, F.; Mumford, T. Application of the "best available science" in ecosystem restoration: Lessons learned from large-scale restoration project efforts in the USA. Tech. Rep. **2004**, *1* (May), 1–29.
21. Wiley, H.; Canty, D. *Regional Environmental Initiatives in the United States: A Report to the Puget Sound Shared Strategy*; Evergreen Funding Consultants: Seattle, Washington, 2003, available at http://www.sharedsalmonstrategy.org/files/Final_regional%20initiatives.pdf (accessed October 2011).
22. Mabus, R. America's Gulf Coast: A Long Term Recovery Plan After the Deepwater Horizon Oil Spill 2010, available at http://www.restorethegulf.gov (accessed November 2011).
23. Salt, T.; Langton, S.; Doyle, M. The Challenges of Restoring the Everglades Ecosystem. In *Large-scale Ecosystem Restoration: Five case Studies from the United States*; Doyle, M., Drew, C., Eds.; Island Press: Washington, DC, 2008.
24. Gulf Coast Ecosystem Restoration Task Force. *Gulf of Mexico Regional Ecosystem Restoration Strategy* (Preliminary); 2011.
25. Grunwald, M. *The Swamp: The Everglades, Florida and the Politics of Paradise*; Simon and Schuster: New York, 2006.
26. *Water Resources Development Act*, Public Law 106–541, 114 Stat. Section 601; 2000; 2680–2693.
27. *Programmatic Regulations for the Comprehensive Everglades Restoration Plan*, Final Rule 33 CFR Part 385; 2002.
28. South Florida Ecosystem Restoration Task Force. *Strategy and Biennial Report*; July 2008–June 2010.
29. National Wildlife Federation. America's Great Waters Coalition, available at http://www.nwf.org/Wildlife/What-We-Do/Waters/Great-Waters-Restoration/Great-Waters-Coalition.aspx (accessed November 2011).

Ecosystems: Planning

Ioan Manuel Ciumasu
ECONOVING International Chair, University of Versailles Saint-Quentin-en-Yvelines, Guyancourt, France, and Center for Sustainable Exploitation of Ecosystems (CESEE), Alexandru Ioan Cuza University of Iasi, Iasi, Romania

Liana Buzdugan
Nicolae Stefan
Center for Sustainable Exploitation of Ecosystems (CESEE), Alexandru Ioan Cuza University of Iasi, Iasi, Romania

Keith Culver
Okanagan Sustainability Institute, University of British Columbia, Kelowna, British Columbia, Canada

Abstract
Ecosystems are complex natural systems, characterized by non-linear dynamics and thresholds. Contrarily, planning activities require predictability and clear cause–effect dynamics. Consequently, ecosystem planning emerges as a concept that captures the very essence of the current, tensioned relation between humans and ecosystems; a potentially valuable but highly difficult concept. Trying to negotiate various pulls and tensions between disciplines, this entry identifies ecosystem planning as an equilibrium-finding human activity of anticipating and inducing generation of a set of ecosystem goods and services, within the limits allowed by the intrinsic ecosystem dynamics. The entry scrutinizes the meanings of the definition and discusses the limits and potential uses of the concept with respect to the current priorities related to environmental changes and the transition to sustainable development. Additional focus is sought by following two priority threads of human impact upon nature: landscapes and cities.

INTRODUCTION

Humans have a long history of modifying their environment to suit their own perceived needs. Science and technology provide a certain, limited capacity to anticipate the behavior of ecosystems and plan a chosen set of desired ecosystem responses. However, because ecosystems are complex systems, i.e., governed by non-linear behavior via self-organization and thresholds, true planning activities are bound to have a very limited relevance and effect. In various ecosystem contexts, some planning may or may not be possible—depending on the type and scale of the intended ecosystem responses. Therefore, it is important to define the scope of ecosystem planning. The term "ecosystem planning" is very rarely used in the literature, and this formulation might well be regarded as arriving with negative connotations likely to hamper the future use of the concept in disciplinary contexts. However, its simplicity and straightforwardness have a certain appeal for environmental managers in their search for a synthetic vocabulary to help them in the necessary integration of knowledge from many fields and disciplines. The term "ecosystem planning" may, for this reason, be usefully adopted in practice, albeit in epistemologically controversial ways. An eventual adoption of the term by the growing community of problem-solving-oriented environmental managers will require some substantial definition clarification efforts and responses to epistemological difficulties. This entry proposes a definition with the prospect to conciliate disciplinary tensions and realities and discusses a set of potential applications.

CONTEXT

Any observer of the use of the expression "ecosystem planning" in the extant literature will see it frequently used as shorthand for "ecosystem management planning," most notably in situations of management of environmental resources. For example, silvicultural practices often employ the term "planning" in the context of forest management, sometimes referring to "forest planning" or even "ecological forest planning."[1] Such contexts are typically described in terms of managerial objectives such as "multi-objective managerial planning," to which descriptors such as "ecological" or "ecosystem" are often applied to account for the natural processes involved.[2] In this context, planning seems to refer to procedures applied to attain some desired, pre-established objectives in terms of exploitation of resources. One can observe that, despite occasional claims of

Encyclopedia of Environmental Management DOI: 10.1081/E-EEM-120047256

the contrary, the literature tends to recognize that planning refers to some empirical knowledge-related procedure to obtain a benefit *from* ecosystems, not a procedure to *make* ecosystems. Yet, this distinction is not always sufficiently clear, with certain confusions persisting in the authors' understanding of the fundamental differences between the aspiration to plan and the nature of ecosystems. Ecosystem planning is often mentioned in contexts that imply manipulating ecosystems to the purpose of obtaining desired outcomes. This fact, together with the lack of agreement on what "ecosystem planning" is supposed to designate, is a warning that work is needed to clarify the concept and the practice. This entry takes on this objective and proposes a definition that may constitute the basis for further developments and reviews.

To various extents, properties of ecosystems may be changed. At one end of the spectrum (gradient) of human intervention upon ecosystems is the domain of conservation biology, which mainly aims at protecting–preserving a given ecosystem (or components of it) as is.[3,4] At the other end of the gradient is "ecological engineering," which essentially pursues objectives related to "ecological reconstruction" or "ecosystem restoration" of heavily degraded areas or even aims at attaining the development of new ecosystems.[5] The term "ecosystem planning" incorporates an inbuilt contradiction between the self-organizing, non-linear character of ecosystems[6,7] and the linear, human-organizing character of planning. Despite this, the incontestable reality is that most terrestrial and water ecosystems are under some form of modification by humans and thus subject to management involving both natural ecosystem dynamics and planning. Even though nature conservation's original aim was the priority protection of pristine and near-pristine environments, and even though ecology seeks to understand natural processes, i.e., those undisturbed by human interventions, we have to deal every day, and in most situations, with ecosystems that are maintained under heavy human influences. Indeed, given the fact of global changes, we cannot just continue to talk about "untouched" or "wild" nature. The term "ecosystem planning" might therefore gain widespread use, in diverse circumstances, notwithstanding the epistemological difficulties it carries along.

DEFINITION

In order to reconcile the intrinsic nature of ecosystems with human planning reflexes and with the need for a practically useful, integrative understanding of the term "ecosystem planning," we propose a working definition: *ecosystem planning refers to the human activity of anticipating and inducing the generation of a set of ecosystem goods and services (EGS), within the limits allowed by the intrinsic dynamics of ecosystems.*

For our purposes, EGS are "the capacity of natural processes and components to provide goods and services that satisfy human needs, directly or indirectly." EGS involves the translation of ecological complexity (structures and processes) into a more restricted set of ecosystem functions that ground the provision of goods and services that are valued by humans. Admittedly, the term "ecosystem function" has been subject to conflicting interpretations, sometimes referring to internal functioning of an ecosystem (e.g., energy and matter fluxes, food chains, and food webs), and sometimes relating to the benefits derived by humans from the properties and processes of ecosystems (e.g., food production).[8,9] Such semantic hesitations illustrate or perhaps reproduce the epistemological difficulties that tend to be encountered at the human/nature interface. In offering our working definition, the concept of "ecosystem planning" is explicitly ascribed to, and constrained by, the limited human capacity to influence ecosystems. This attribute should be read as the crux of the concept and the take-home message of this entry.

Dynamics intrinsic to ecosystems, thresholds and non-linearity, are explained by the laws of physics and ecosystems can be understood as complex systems—with the support of complexity science. Ecosystem properties are subject to some limited influence, but they are not responsive to human planning imperatives. To use a somewhat less precise but more intuitive illustration, an ecosystem (as a multi-individual system, in terms of system biology) is analogous to a human body (an individual system) in that it can be trained and improved, and a certain number of its parts may be changed, but it cannot be changed as a system—any attempt to do so will cross organizational and functional thresholds that are steps towards its destruction. The system has its own character. It can be enlarged, squeezed, accelerated, or made more productive or more beautiful—but its essential characteristics as a system cannot be changed. Ecosystem planning is therefore like education of a human individual: it can aim at reaching the utilization of its maximum potential, but you cannot change its genetic endowment.

TERM USES

Viewed in isolation and out of context, the term "ecosystem planning" makes little sense. Its meaning and value become apparent only once it is considered in association with other types of planning, and with attention to the concept's carrying a specific understanding of intrinsic ecosystem dynamics.

Our working definition has several consequences and lessons. One is that the term "ecosystem planning" captures only partially the relation between humans and ecosystems. Employment of unduly broad or general meanings is harmful because they would allow (or inadvertently convey) the mistaken assumption that human action can compensate for ecosystem destruction by sheer planning. In reality, once destroyed (e.g., by overexploitation), the capacity of ecosystems to provide EGS (or "ecosystem carrying capacity" in the ecological economics vocabulary) cannot be simply restored by further planning. Carrying capacity is grounded in the ecosystem's intrinsic properties, and these develop during the long-term ecological history of the place. In economically equivalent terminology, the carrying capacity represents the "natural capital" (K_N), or, differently put, the ensemble of EGS extractable from a given ecosystem. Unlike the physical capital (K_P) and the human capital (K_H), natural capital is irreplaceable since it is dependent on ecological thresholds. At this point, we should recall that one fundamental characteristic of biological–ecological systems dynamics is their irreversibility, meaning that once an internal threshold is crossed, an essential set of characteristics of that ecosystem are irremediably lost because passage across biological–ecological thresholds cannot be reversed. The three forms of capital form together the total capital stock,[10] but the natural capital cannot be replaced by the other forms of capital. Even while total capital stock available may be seen increasing locally, its utility may nonetheless be decreasing. Thus, when the natural capital on which human activity depends is depleted irreversibly, the value of the other two major forms of capital also decreases—they are rendered unsustainable by the destruction of the underpinning natural capital.

As pointed out by Costanza et al.[11] in a widely cited entry, it is possible to imagine generating human welfare without natural capital/EGS in artificial "space colonies," but this possibility is too remote and impractical. One additional way to conceive the value of EGS is to determine what it would cost to replicate them in a technologically produced, artificial biosphere. Past experience with manned space missions and with Biosphere II in Arizona indicates that this would be an exceedingly complex and expensive project. Instead, Biosphere I (the Earth) is incomparably more efficient, least-cost provider of life support services for humans.

Discussions of costs open, of course, the issue of potential valuation of EGS in a market. The concept of "payment for ecosystem (goods and) services" (PES/PEGS) remains controversial,[12,13] but it has the virtue of pointing out that previously marginal concepts like "intangible values" (social and intellectual aspects—they are also connected with ecosystems) and "externalities" (ecosystem resources usually taken as "given" and not factored in by the current, neoclassical economics) must be taken into account. Moreover, they must be incorporated very quickly into our accounting, because they reflect the situation of ecosystems as life support systems, and humanity seems to be exceeding the natural limits of its support systems. In a review of the matter, Baggethun et al. (2010) have shown that the recent advances towards monetization of EGS indicate the slow progress of the concept. While monetization has helped draw policy and economic attention to ecosystems and EGS, it has also absorbed it into the logic of the old (but current) neoclassic theory of economics in ways that deprives the concept from many of its initial virtues related to ecological and social values. Reviewers of the literature may observe that the incomplete incorporation of EGS into mainstream policy decision-making is attributable to the prevalence of an approach that remains essentially disciplinary. EGS was repackaged or translated from ecology to current economics, along the way losing many of the virtues of EGC. Successful incorporation of EGS into policy decision-making will require a problem-driven rather than a tool- or discipline-driven transdisciplinary approach capable of synthesizing tools, skills, and methodologies. As their review suggests—reflecting wider opinions in the literature—valuation needs not always amount to an exclusively monetary valuation. This recommendation is naturally applicable to ecosystem planning as defined in this entry.

Planning efficiency is nonetheless dependent on financial costs; EGS and PES represent essential advancements towards a solid grounding and meaningful use of the concept of ecosystem planning. However, as the initial concept has become diluted into disciplinary habits, current developments are not real achievements. Instead, they may be regarded as important initial fathoms towards sustainability.[14] On the socioeconomically transformative journey that is the transition to sustainability, "ecosystem planning" needs to factor in how much exploitation of EGS ecosystems can bear. In other words, how much of the carrying capacity of ecosystems can be consumed by humans without risking the collapse of the ecosystem carrying capacity. In this sense, "ecosystem planning" can draw support from the concept of "ecological footprint," an accounting tool for assessing the natural capital and its degree of depletion at various scales—individuals, cities, nations, or the entire planet. This approach uses estimates of consumption of natural resources for food, shelter, transportation, personal care, pollution absorption capacity, and other uses of the natural capital and compares it with the carrying capacity of ecosystems. This is one highly relevant tool for overall assessment of ecosystem planning: a value of less than 1.0 indicates that the ecological footprint remains below the carrying capacity of the ecosystems (at the scale taken into consideration), which means that ecosystem exploitation is within the limits of sustainability. On the contrary, a value greater than 1.0 indicates that a person, city, nation, or the entire humanity lives beyond the natural support capacity limits of the ecosystems/biosphere.[15–17] Values that surpass 1.0 are of utmost importance because any excess represents "eating up" the regenerative capacity of the ecosystems/biosphere, with the consequence that the overall carrying capacity decreases. At the planetary scale, it appears that humanity is already close to this tipping point or already beyond it, and the possibility of collapse is very real and a reason for accelerating and improving our efforts to understand and respond to this possibility.

The solutions for an effective transition to sustainability are still to be developed. At epistemic and moral levels, one option is to aim for a steady-state economy. Such a steady state would amount to a kind of plateau of development—as permitted by the Earth's carrying capacity—and would maintain this constant value in the future.[18] Another approach is to commit to "degrowth," meaning a collective decision of humanity to consume less, especially in those affluent countries that consume beyond the ecosystem carrying capacity available to them.[19] This idea is likely to encounter major social acceptability obstacles, certainly in developed countries, and especially in developing countries. The degrowth proposal appears to express a chosen rejection of the excessive monetization of intangible assets (social and cultural values) and of ecosystem services. However, the grounds to be considered for sustainable development are first and foremost biophysical: natural limits cannot be overcome as they are thermodynamic realities already known for decades—since the origin of ecological economics.[20] The relevance of ecosystem planning relates to the immutable character of the physical laws: planning for a sustainable equilibrium of humans with ecosystems may be a matter of social decision-making and acceptance, but the consequences of those decisions are not. In other words, ecosystem planning will need to help humans accomplish sustainable development (and thus, survive). If this fails, collapse will follow. This is not science fiction; rather, it is a prediction grounded in history as civilizations have in fact overused their natural resources and faced collapse. At the planetary scale, this event is the natural outcome in the absence of a downward adjustment in the ecological footprint/carrying capacity balance.[21,22]

Taking this emergency seriously, Daily et al.[23] propose a conceptual frame for operationalizing the relation between humans and ecosystems, conceived as a never-ending circuit of five links (rendered in all caps here) and type of relations between them (rendered in italics here): ECOSYSTEMS → *biophysical models* → SERVICES → *economic and cultural models* → VALUES → *information* → INSTITUTIONS → *incentives* → DECISIONS → *actions and scenarios* → ECOSYSTEMS. Scenarios are developed for an applied case in Hawaii, called The Natural Capital Project, which tries to develop a scientific basis and connect it with policy and finance mechanisms, aiming to incorporate natural capital into resource- and land-use decisions on local and larger scales. This can be regarded as a useful example of current attempts to link ecosystem conservation with development and to render EGS concepts useful in policy and business. Yet at the same time, the entry shows that we are a long way from the desirable situation when this would become mainstream practice. In the light of these considerations, the relevance of the term "ecosystem planning" remains ambiguous in the literature dedicated to the theory and applications of environmental/ecosystem management.

Mac Nally et al.[24] use a more cautious expression. By "ecosystem-based planning," they mean planning activities for biodiversity conservation purposes. Often, the expression "conservation planning" is used to the same effect.[4] Margules and Pressey[25] identify different stages of biodiversity conservation planning in relation to ecosystem services. Conservation biology (a term designating the aim of biodiversity and ecosystem conservation) is progressively converging with the concept of EGS.[26–28] Sometimes, an equally cautious expression is used, "ecosystem-based management," where planning activities are discussed in terms of the constraints imposed by ecosystem properties and dynamics in general, and in specific contexts and case studies. In an extensive book on the matter,[29] Randolph uses terminology including "environmental protection," "land conservation," "environmental management," "environmental land use management," and "ecosystem management," as well as "land-use planning," "environmental planning," and "habitat conservation planning," to systematically describe the intricacies of human–nature interactions.

Among the rationales behind a potentially useful concept of "ecosystem planning," one can count earlier efforts to integrate environment planning and development via ecosystem approaches. Thus, the necessity to extend planning activities from human-created environment and modified environments to the natural environment arises from the expansion of human activities themselves. Therefore, the ideal situation of environmental planning would be one where it was not needed. This applies *mutatis mutandis* to "ecosystem planning" as well. The point we want to make here is that "ecosystem planning" may reveal itself to be not a goal per se, but a necessary practice toward an aspiration that is not yet concretized into a specific, readily measurable or quantifiable goal necessity: implemented well, it will embody a managerial compromise between human activity expansion and the need to control, limit, and minimize the impact of human activities on the carrying capacity of biophysical systems.

One common feature is evident throughout the literature surveyed above, namely, the avoidance of a direct association between "ecosystem" and "planning." Instead, the authors seem to be searching for new concepts to describe those actual situations where "ecosystem" and "planning" coincide. The authors surveyed are generally preoccupied with selection of the most apt definitions and ascription of the most appropriate meanings, hoping that the reality of human–nature interactions is being neither hastily misconceived nor ignored. For example, expressions such as "environmental planning" or "land-use planning" bear a lesser epistemological burden than "ecosystem planning"—such terms are general enough to avoid asserting an unnecessarily specific, contentious relation between planning and ecosystems, and yet they clearly do include reference to sheer planning of use/occupancy of natural resources, especially in the case of water or land resources. Instead of being merely a matter of semantics, the expression "ecosystem planning" cannot be a useful concept unless it is associated with a clear distinction between employing knowledge about ecosystems and employing knowledge about the effects of potential human action on ecosystems. The concept of "ecosystem planning" must help illuminate the border area between what humans know and what humans can do about environment and ecosystems. Various disciplines and approaches can contribute to such a new understanding of "ecosystem planning." However, the lack of a compact body of literature makes the epistemological and technical reviewing process a highly demanding endeavor—a potential subject for future examinations.

In practice, however, understanding of the relation between humans and ecosystems may not always precede norms and decisions, as social rules and behaviors may not wait for detailed scientific clarifications. Kagan[30] examines ecosystem planning from a legal perspective, using the contraction "ecosystem planning" when referring to human activities related to environmental management and decision process over land/resources use, seemingly unaware of the ecological implications of putting together the words "ecosystem" and "planning." The main point of his perspective, however, is that it reflects a situation where ecosystems are being viewed, consciously or not, as something "out there," exterior to social matters. This is not very different from the currently predominant (yet), neoclassic view in economics, according to which natural resources and social matters are external to the economic processes and associated accounts, in the sense that they are "externalities"—their values or influences need not be accounted for in setting and operating a core process toward a desired outcome. All those narrow perspectives, however, are now being challenged by the widening community of practitioners for sustainable development. Nicholas Stern, the author of the homonymous report on the cost of climate changes due to greenhouse gas emissions, has arguably provided the best synthesis of the man–nature relations in a conference at the Royal Economic Society in 2007: "Climate change is a result of the greatest market failure the world has seen. The evidence on the seriousness of the risks from inaction or delayed action is now overwhelming. We risk damage on a scale larger than the two world wars of the last century. The problem is global and the response must be collaboration on a global scale." The more urgent becomes, therefore, the need to reassemble disciplinary perspectives into a common understanding of the relations between natural, social, political, and economic realms.

At the borders between legislation, natural sciences, economy, and social sciences, one often-used terminology is "forest planning," meaning, in silvicultural practices, planning for forest ecosystem exploitation. The literature on forest management is among the earliest to acknowledge the tension between the deterministic approach of forest exploitation planning and the uncertainties intrinsic to ecosystem dynamics. Given the importance of forests as resource generators for human populations, uncertainty was "accommodated" in forest exploitation planning[31] in terms of managerial approaches that accounted for disturbances and other unpredictable, less deterministic phenomena. In this sense, forests represent a good case study for the relation between the management of EGS, ecosystems, and socioeconomic systems, where natural resource management planning is often seen as akin if not identical to "ecosystem planning." An extensive overview on the matter is beyond the scope of this entry. Some examples, however, may be highly illustrative. Exploitation of ecosystems is obviously related to the property bearing lands and waters. In Central and Eastern Europe, for example, Romania has served as a natural experiment in this regard. The recovering of the individual property rights to forests (lost at the time of military imposition of communism by the Soviet Union, when 23% of the country's forests were in private hands) in a context of a still-recovering economy and society has led to massive forest overexploitation. Large areas of forests have been clear-cut in the 90s for the immediate purpose of selling the wood (120,000 ha, almost half of the first wave of forest restitutions). Obviously, firm property rights are necessary for effective management of forests, because they provide marketability, i.e., options for future planning conservation of EGS and various uses—timber and other biomass, recreation, carbon sequestration, water retention and decontamination, and so forth. These property rights are therefore not in question, but their consequences are, in this perilous situation where the postcommunist socioeconomic transition has left a management and regulatory vacuum. Fortunately, the last decade has seen a gradual socioeconomic recovery of the country and improvements in the definition of the property rights, giving rise to remarkable improvements in forest management. Entrepreneurial activities are now related to uses of forests, based on 1) the right to access; 2) the (resource) withdrawal right; 3) the management right; 4) the exclusion right; and 5) the alienation right.[32] Along the way, a process of reforestation is taking place in the entire eastern half of the European Union. The region's forests are now recovering from some of the losses suffered during the previous years of harsh transition, with various countries being at various levels of reforestation—now matching the levels of socioeconomic convergence with the more developed western European countries.[33] In such situations, ecosystem planning bears the full weight of socioeconomic contexts.

A related example that is relevant to "ecosystem planning" is linked to the ecological reconstruction that may follow a cycle of ecosystem overexploitation and partial ecosystem recovery. Reforestation may be difficult on highly degraded and low-accessible lands, which may favor the installation of soil erosion processes. However, knowledge of the natural processes may provide the means to plan for revegetation of degraded slopes. To remain within our example from Central Europe, past experiences show how the natural vegetation succession can be used as a tool. Thus, field experiments with planting a certain shrub like the common sea-buckthorn (*Hippophae rhamnoides*) on degraded slopes in Romania have led to fast revegetation with the shrub as quasi-exclusive species. In later stages, this monospecific vegetation allows an evolution of the structure of the soil and the advent of other species (grasses and tree seedlings) to grow under its shade. In a next stage, oak (*Quercus*) and beech (*Fagus*) tree juveniles, as well as other shrubs, have grown taller and shaded the sea-buckthorn and have gradually replaced it as new forest, which, even if it is characteristic of the area, would not have grown there because the land was too degraded and the conditions were suboptimal prior to the establishment of the sea-buckthorn in the first place.[34] Application of such lessons from the field, however, is only possible under favorable socioeconomic conditions. The recent advances in the country and in the region may reopen the possibility of using older field ecological experiments. In such situations, "ecosystem planning" means planning for a desired favoring of certain natural processes, with the purpose of halting the loss of the natural capital and the recovery of certain ecosystem functions and services.

A fresh question now arises: to what extent can we take advantage of our knowledge to manipulate natural processes towards recovery, albeit partial, of some initial ecosystem structures and functions? The discipline of ecological engineering is one way to explore answers to this question. Its original principles seem to be updated to reflect the improving understanding of the limited planning powers of humans upon ecosystems[5,35]: 1) it is based on the ecosystems' self-organization capacity; 2) it can be a field test of ecological theory; 3) it relies on integrated system approaches; 4) it conserves non-renewable energy; and 5) it supports biological conservation. Such developments seem to indicate broad acceptance that planning, and for that matter "ecosystem planning," must incorporate the idea of adaptive management, both at the scales of landscapes and those of local settlements and despite epistemological uncertainties.[36,37]

LANDSCAPE PLANNING

The first and most obvious application of a future, broadly accepted concept of ecosystem planning is in the area of landscape planning. Others may understand ecosystem planning as "the process of land use decision-making that considers organisms and processes that characterize the ecosystem as a whole." In other words, "ecosystem planning" is used as shorthand for ecosystem-based planning of land uses across landscapes. The literature deals, it seems, with a concept that may have a certain utility in its own right, albeit narrow, yet it is often used in epistemically irregular and unstudied constructions. Musacchio et al.[38] use the expression "landscape ecological planning (LEP)." There are major challenges in integrating the concept of EGS in landscape planning and, for that matter, also in the concept of "ecosystem planning" as defined in this entry, simply because there is as yet no coherent and integrated approach for practical applications of EGS in planning, management, and decision-making in general.[39]

Research programs are currently under development, trying to tackle this obstacle. For example, De Groot et al.[40] have proposed and are currently developing a framework that involves a chain of five major links: 1) understanding and quantifying how ecosystems provide services; 2) valuing ecosystem services; 3) use of ecosystem services in trade-off analysis and decision making; 4) use of ecosystem services in planning and management; and 5) financing sustainable use of ecosystem services. Inevitably, such a succession of actions bears the intellectual virtue of a coherent synthesis of what we can imagine to be eventually done for successfully approaching the desired equilibrium between humans and nature. But it also carries a large extent of naivety with respect to what may be possible—for example, it sees ecosystems as "still poorly understood," and proposes more detailed quantification as the best way to achieve a better grasp of how ecosystems function and provide EGS, as if it is only a matter of time until we get it. This way of approaching the issue, however, seems to ignore 1) the serious epistemic hurdles involved in seeking total, comprehensive understanding complex systems, especially with a reductionist mindset; and 2) the practical feasibility of such an endeavor in terms of available time and resources. To give only one example, thousands of new chemicals are introduced to the market (and released into the broader environment) every year, and there is no chance that ecotoxicological test batteries could ever be developed for each of them to "better grasp" how ecosystems function, not to mention the inherently limited relevance of each bioassay. Some integrative, multitier studies can be done to account for changes in ecosystem parameters in space and time, for some specific indicators, e.g., water quality along a river basin.[41–43] But even those are very far from being sufficient for understanding the state of ecosystems and the quality status of EGS. What seems to be missing in the five-link type of framework proposed in the cited analysis is the acknowledgement of the facts that we only have a limited amount of time and material resources to be allotted to advancing towards the equilibrium between humanity and the ecosystems in which it is embedded, and that those resources are themselves dependent upon our mankind–ecosystems relation that we want to address. It still is a tool-based approach rather than a problem-based approach, being concerned more with understanding than understanding sufficient for remedial action. We cannot plan everything we would like to. Apart from this difficulty, and to be fair to the mentioned authors, the framework provided by De Groot et al.[40] does appear to us to have an essentially "color-test" value for the concept of ecosystem planning proposed hereby. Under the definition of "ecosystem planning" employed in their paper, the chain of actions proposed by the above-mentioned team at Wageningen University can actually be understood as applied to "ecosystem planning," as a useful organizer of what needs to be done by humans within the limits of the possible. Without the definition above, the same chain of actions proposed by De Groot et al. allows "everything and nothing" in a progressive demand of resources that we actually need to stop depleting. Thus, mapping landscape functions for planning the management of EGS is potentially a smart and justifiable approach, on the condition that it is integrated in a management process that leads to less destruction of EGS and thus interrupts the series of blind actions upon ecosystems.

A necessary feature of "ecosystem planning," as a synthetic concept, must be simplicity in the sense of operational readiness. Proposed methods should maintain sensitivity to the complexity of the phenomenon but avoid creating supplementary complications, since those would postpone and exacerbate the problem. Ecosystem planning should recognize limits of human knowledge and actions and seek genuinely integrative solutions, i.e., simple yet responding to a complexity of pressures. Humanity has not managed to integrate the fragmented knowledge we have produced so far, and the tension between humanity and ecosystem continues to increase. We cannot realistically hope to resolve this problem by producing more piles of disparate (and mutually untranslatable disciplinary) knowledge. Ecosystem planning, if effective, will need to consist in solutions that are relevant (adaptable or instructive) to situations as various as floodplains, polders, deserts, forests, marine and fluvial ecosystems, and so forth.

Although the literature explicitly using the concept of ecosystem planning is scarce, there is a large and eclectic body of knowledge on implicit ecosystem planning attempts. In a study upon the potential overlap between EGS and conservation policies in the coast eco-region of central California, Chan et al.[27] have extracted a set of key insights into the convergence of conservation and EGS planning: 1) both suitability of sites and demand from citizens are main drivers in what ecosystem planning can do for protecting biodiversity and EGS—near cities demand is higher but suitability is lower; 2) spatial scales may vary for optimum of biodiversity conservation or/and different types of EGS—pollination services result from a variety of small locations, while water quality from all river basins; 3) population centers yield tensions in planning between values and demand—for example, a low-value/high-demand situation can occur near cities; 4) data are usually lacking, and thorough research and analysis are necessary; 5) it must involve multi- and transdisciplinary teams, for the integration of theoretical and empirical expertise from diverse fields; 6) efficiency is conditioned by considering both trade-offs and side-effects of biodiversity conservation and EGS conservation—often trade-offs are the most efficient way, and often side effects reveal points of common relevance for various goals of planning.

While the concept of EGS appears to prioritize or privilege focus the human–ecosystem relations, challenges for future applications involving the concept of ecosystem planning can be related (without being equivalent) to situations as various as land-use planning; degraded landscapes management, saltmarshes and river regulation, flood prevention and management of river catchments, management of freshwater quality, silviculture (with related concepts like "ecosystem stewardship planning" and "forest stewardship plans," and "forest stewardship"), and planning for carbon sequestration in terrestrial and water ecosystems. In the context of mitigation and adaptation to climate change, forms of ecosystem planning have been imagined at planetary scale. Thus, geo-engineering approaches propose that the planetary ocean should undergo fertilization with limiting substances (like iron ions) to allow faster carbon sequestration in phytoplankton biomass. This, however, involves large-scale, unpredictable risks.[44]

URBAN PLANNING

Rees and Wackernagel[15] have proposed estimations of the ecological footprint of cities and show why cities cannot be sustainable per se, while they are actually central to sustainability. Cities can only be considered sustainable together with their hinterland, from where they draw their resources and on which they depend. In the current era of globalization, cities have become part of a planetary network of cities, with a "common hinterland," which is the Earth's biosphere. Cities play preeminent roles in the global exchanges of information and matter energy and have an immediate impact on, and are immediately affected by, the entire biosphere. Ecological footprint estimates allow cities to track the necessary resources and the plan for development according to this. Consequently, urban development planning for sustainability will require evaluations of the ecological demand of a given city (disaggregated by type of activities) and comparison with available ecosystem resources. In the same logic, ecological deficits may determine the way the city may develop, if it intends to remain sustainable.[17] Urban ecosystem services can be identified and described in urban areas just as well as in non-urban areas.[45] When proposing an integrated planning tool for sustainable cities, Rotmans et al.[46] implicitly refers to "ecosystem planning." In their paper, the concept of environmental capital seems to be equated with natural capital, as part of a wider planning for sustainable city.

Niemelä[47] describes urban ecology in the context of the relation between ecology and urban planning, where urban planning is a type of land-use planning. He makes the essential observation that urban nature has been regarded in prior studies as a true field experiment about human impact on ecosystems. This resonates with the wider preoccupations with the place of urban areas within the landscapes, e.g., with the degrees of vegetation cover[48] or habitat patch corridors to avoid excessive habitat fragmentation in highly urbanized areas and to allow landscape connectivity between local plant/animal populations as part of the wider metapopulations (network of local populations) within landscapes. In a recent communication concerned with the place of cities as a human social construct within local ecosystems, Ciumasu and Culver[49] describe the city in terms of an ecosystem disturbance. Applying current ecological theory—the intermediate disturbance hypothesis (IDH) and island biogeography theory (IB)—they identify an ecological taxonomy of cities and city areas as a function of disturbance intensity in a local biogeographical context: small city/city periphery (low disturbance intensity), medium-sized city/city near-center (intermediate disturbance intensity), and large city/city center (high disturbance intensity). Within this framework, any urban unit can be described on a gradient of ecological disturbance and in the local biogeographical context (for a review of urban areas in IB contexts,[50] with biodiversity theories and indices serving as proxies for the state and dynamics of ecosystems.

An abundant literature has emerged, treating urban biodiversity, providing the means for using urban ecology–biodiversity studies in urban planning and for planning for biodiversity conservation in urban areas.[51–54] Further literature points to the transformative impact of urbanization, particularly the reduction of native habitat and the creation of new habitats, and ecosystem homogenization.[55–56] These complex aspects must be accounted for by urban/ecosystem planning. In effect, city planning emerges as a form of planning the impact (types and intensity of disturbance) of humans upon ecosystems. The management of urban areas per se and the management of city hinterlands become a matter of planning for extraction of EGS for use in cities and human settlements in general. The place of the city within the surrounding ecosystems is therefore a matter of "ecosystem planning."

At the level of immediate impacts, urban planning may require intra-urban planning or peri-urban planning, which may involve a diversity of terrestrial/water ecosystems, coastal ecosystems, and inclusion of the effects of climate changes, grazing, and agricultural land. City planners must face the concomitant emergence of new knowledge domains and a plethora of new challenges. The central feature of urban ecology is that "cities are emergent phenomena of local-scale, dynamic interactions among socioeconomic and biophysical forces" that give rise "to a distinctive ecology and to distinctive forcing functions."[57]

The role of a city as disturbance of the local ecosystems is the result of a complex dynamic involving major concerns as disparate as urban energy systems[58] and global change,[59] solid waste and wastewater,[60,61] and infrastructure reliability engineering,[62] to cite only a few. All this requires effective community coordination, based on social multi-criteria conception and evaluation of urban sustainability policies.[63–65] In this context, the use of some multiple criteria decision aid (MCDA) methodologies seems to be unavoidable for any urban and associated ecosystem planning, and effective coordination is yet another step further. The best candidate to providing an organic integration of all these issues and perspective is the systemic view of sustainability, according to which any economic system is a subsystem of a social system, itself a subsystem of a biophysical system.[66,67] However, translation from such ecological–economic insight to management strategies, actually tried and revised, seems to be waiting.

Starting from the systemic conception of sustainability, Ciumasu et al.[68] have proposed an integrative approach for problem structuring and for identifying sustainable/unsustainable management scenarios for solid wastes. The method consists in translating the nested inclusion relationship between economy, society, and ecosystems into an ordered set of sustainability filters to be respected in managerial practices and proposed solutions *in illo ordine*: ecological sustainability filter (EcSF), social sustainability filter (SoSF), and economic sustainability filter (EnSF). Successfully passing all filters (i.e., meeting threshold values and indicators negotiated/listed in the technical definition of each filter) indicates that a given policy or solution proposed for the stated purpose (e.g., solving the problem of an old landfill) is consistent with sustainable development. This would be considered a sustainable scenario of action, under the conditions and knowledge available at the given place and time. Under this framework, the role of "ecosystem planning" can be recognized at the priority level of ecosystem carrying capacity (as reflected in EcSF) prior to action being taken towards further exploitation of EGS. Alternative methods can be identified to assess sustainability of solutions proposed for urban landfills, like the so-called sustainability potential analysis (SPA), and extension of the bio-ecological potential analysis (BEPA) that was originally developed for local and regional landscape analyses,[69] a reminder that any urban planning is part of planning for an equilibrium between humans and the support natural systems.

PERSPECTIVES

It is not clear yet whether the term "ecosystem planning" will acquire a widely recognized meaning in its own right. Nevertheless, the concept often appears to function as a convenient stylistic contraction employed in the description of complex, multilevel, multidiscipline-fed issues and topics related to environmental management. Should the term enter mainstream terminology, it will need to overcome some major epistemic difficulties. A variety of methodologies are expected to be developed in the future, and further redefinitions of the term are likely. However, all future developments and definitions of the concept will need to address the contradiction between the nature of planning and the characteristics of ecosystems. Ecosystems are an autonomous part of the environment, with their own, highly non-linear, dynamics. Planning regarding these dynamics cannot be made fully operational, if at all, by managerial task subdelegation let alone semantic reduction. The fact that planning may work effectively in certain sectors of environmental management does not translate into effective planning at the level of ecosystems. Rather, it typically means that 1) the concept of environmental planning has been limited to engineering- and management-tractable issues; 2) it has benefitted a certain capacity of environment-as-a-whole (i.e., ecosystem included) to absorb disturbances by human activity; and 3) environmental planning needs further refining to clarify the limitations of planning and account for the reality of non-linear behavior of ecosystems.

In the current context of vast and fast environmental changes, which are calling for coherent responses from science and management, certain knowledge domains may assimilate the concept of ecosystem planning. Notably, conservation biology aims at more effective nature protection, going well beyond the traditional focus on biodiversity conservation, with planning emerging as a necessary component of any comprehensive approach (for a review and discussion on the matter, see Reyers et al.[70]). The imperative of a global transition to sustainable development commands a profound transformation of the economy and society, which entails effective knowledge integration and management. A recent global survey that included experts in both social and natural sciences and that was carried out by the International Council for Science and the International Social Science Council has revealed a set of five grand challenges that need to be tackled within the next years[71] and which can be summarized in four points as follows: 1) higher management-relevant capacity to anticipate environmental changes; 2) coordinated observation of environmental changes; 3) comprehension and anticipation of disruptive environmental changes; 4) social–institutional transformation for the transition to sustainable development. As discussed in the entry hereby, ecosystem planning is a focal concept for current knowledge integration and management and, despite some challenging epistemic questions, has the potential to amalgamate into a core vocabulary of sustainability. Among the multiple approaches and contributions towards sustainability, landscape planning and urban planning emerge as particularly effective platforms for intellectual developments involving ecosystem planning.

REFERENCES

1. Zagas, T.D.; Raptis, D.I.; Zagas, D.T. Identifying and mapping the protective forests of southeast Mt. Olympus as a tool for sustainable ecological and silvicultural planning, in a multi-purpose forest management framework. Ecol. Eng. **2011**, *37*, 286–293.
2. Palahi, M.; Trasobares, A.; Pukkala, T.; Pascual, L. Examining alternative landscape metrics in ecological forest planning: A case for capercaillie in Catalonia. For. Syst. **2004**, *13* (3), 527–538.
3. Soulé, M.E. What is conservation biology? BioScience **1985**, *35* (11), 727–734.
4. Naidoo, R.; Balmford, A.; Ferraro, P.J.; Polasky, S.; Ricketts, T.H.; Rouget, M. Integrating economic costs into conservation planning. Trends Ecol. Evol. **2006**, *21* (12), 681–687.
5. Mitsch, W.J.; Jorgensen, S.E. *Ecological Engineering: An Introduction to Ecotechnology*; John Wiley & Sons: New York, 1989.
6. Koch, E.W.; Barbier, E.B.; Silliman, B.R.; Reed, D.J.; Perillo, G.M.E.; Hacker, S.D.; Granek, E.F.; Primavera, J.H.; Muthiga, N.; Polasky, S.; Halpern, B.S.; Kennedy, C.J.; Kappel, C.V.; Wolanski, E. Non-linearity in ecosystem services: Temporal and spatial variability in coastal protection. Front. Ecol. Environ. **2009**, *7* (1), 29–37.
7. Walther, G.R. Community and ecosystem responses to recent climate change. Philos. Trans. R. Soc. B **2010**, *365* (1549), 2019–2024.
8. De Groot, R.S. *Functions of Nature: Evaluation of Nature in Environmental Planning, Management and Decision Making*; Wolters-Noordhoff: Groningen, the Netherlands, 1992.
9. De Groot, R.S.; Wilson, M.A.; Boumans, R.M.J. A typology for the classification, description and valuation of ecosystem functions, goods and services. Ecol. Econ. **2002**, *41*, 393–408.
10. Barbier, E.B. *Natural Resources and Economic Development*; Cambridge University Press: Cambridge, U.K., 2005.
11. Costanza, R.; D'Arge, R.; De Groot, R.; Farber, S.; Grasso, M.; Hannon, B.; Limburg, K.; Naeem, S.; O'Neill, R.V.; Paruelo, J.; Raskin, R.G.; Sutton, P.; Van Den Belt, M. The value of the world's ecosystem services and natural capital. Nature **1997**, *387* (6630), 253–260.
12. Redford, K.H.; Adams, W.M. Payment for ecosystem services and the challenge of saving nature. Conserv. Biol. **2009**, *23* (4), 785–787.
13. Gómez-Baggethun, E.; De Groot, R.; Lomas, P.L.; Montes, C. The history of ecosystem services in economic theory and practice: From early notions to markets and payment schemes. Ecol. Econ. **2010**, *69*, 1209–1218.
14. Daily, G.C.; Matson, P.A. Ecosystem services: From theory to implementation. Proc. Natl. Acad. Sci. U. S. A. **2008**, *105* (28), 9455–9456.
15. Rees, W.; Wackernagel, M. Urban ecological footprints: Why cities cannot be sustainable—and why they are a key to sustainability. Environ. Impact Assess. Rev. **1996**, *16* (4–6), 223–248.
16. Wackernagel, M.; Onisto, L.; Bello, P.; Callejas Linares, A.; Lopez Falfan, I.S.; Mendez Garcıa, J.; Suarez Guerrero, A.I. National natural capital accounting with the ecological footprint concept. Ecol. Econ. **1999**, *29*, 375–390.

17. Wackernagel, M.; Kitzes, J.; Moran, D.; Goldfinger, S.; Thomas, M. The ecological footprint of cities and regions: Comparing resource availability with resource demand. Environ. Urbanization **2006**, *18* (1), 103–112.
18. Czech, B.; Daily, H.E. The steady state economy: What it is, entails, and connotes. Wildl. Soc. Bull. **2004**, *32* (2), 598–605.
19. Martínez-Alier, J.; Pascual, U.; Vivien, F.D.; Zaccai, E. Sustainable de-growth: Mapping the context, criticisms and future prospects of an emergent paradigm. Ecol. Econ. **2010**, *69*, 1741–1747.
20. Georgescu-Roegen, N. The steady state and ecological salvation: A thermodynamic analysis. BioScience **1977**, *27* (4), 266–270.
21. Liu, J.; Diamond, J. China's environment in a globalizing world. Nature **2005**, *435*, 1179–1186.
22. Costanza, R.; Graumlich, L.; Steffen, W.; Crumley, C.; Dearing, J.; Hibbard, K.; Leemans, R.; Redman, C.; Schimel, D. Sustainability or collapse: What can we learn from integrating the history of humans and the rest of nature? Ambio **2007**, *36* (7), 522–527.
23. Daily, G.C.; Polasky S.; Goldstein, J.; Kareiva, P.M.; Mooney, H A.; Pejchar, L.; Ricketts, T.H.; Salzman, J.; Shallenberger, R. Ecosystem services in decision making: time to deliver. Front. Ecol Environ **2009**, *7*, 21–28.
24. Mac Nally, R.; Bennett, A.F.; Brown, G.W.; Lumsden, L.F.; Yen, A.; Hinkley, S.; Lillywhite, P.; Ward, D. How well do ecosystem-based planning units represent different components of biodiversity? Ecol. Appl. **2002**, *12* (3), 900–912.
25. Margules, C.R.; Pressey, R.L. Systematic conservation planning. Nature **2000**, *405*, 243–253.
26. Kremen, C. Managing ecosystem services: What do we need to know about their ecology? Ecol. Lett. **2005**, *8*, 468–479.
27. Chan, K.M.A.; Shaw, M.R.; Cameron, D.R.; Underwood, E.C.; Daily, G.C. Conservation planning for ecosystem services. PLOS Biol. **2006**, *4* (6), 2138–2152.
28. Egoh, B.; Rouget, M.; Reyers, B.; Knight, A.T.; Cowling, R.M.; Van Jaarsveld, A.S.; Weltz, A. Integrating ecosystem services into conservation assessments: A review. Ecol. Econ. **2007**, *63*, 714–721.
29. Randolph, J. *Environmental Land Use Planning and Management*; Island Press: Washington, DC, 2004.
30. Kagan, R.A. Political and legal obstacles to collaborative ecosystem planning. Ecol. Law Q. **1997**, *24*, 871–876.
31. McCarthy, M.A.; Burgman, M.A. Coping with uncertainty in forest wildlife planning. For. Ecol. Manage. **1995**, *74*, 23–36.
32. Nichiforel, L.; Schenz, H. Property rights distribution and entrepreneurial rent-seeking in Romanian forestry: A perspective of private forest owners. Eur. J. For. Res. **2011**, *130* (3), 369–381.
33. Taff, G.N.; Müller, D.; Kuemmerle, T.; Ozdeneral, E.; Walsh, S.J. Reforestation in Central and Eastern Europe after the breakdown of socialism. Reforesting Landscapes **2010**, *10*, 121–147.
34. Stefan, N. Contributii la studiul sindinamicii asociatiei *Hyppophaetum rhamnoides* in Subcarpatii de Curbura. Academia Româna—Memoriile Sectiilor Stiintifice 1991, seria IV, tom XIV (1), 223–233.
35. Mitsch, W.J.; Jorgensen, S.E. Ecological engineering: A field whose time has come. Ecol. Eng. **2003**, *20*, 363–377.
36. Lessard, G. An adaptive approach to planning and decision-making. Landscape Urban Plann. **1998**, *40*, 81–87.
37. Williams, B.K. Adaptive management of natural resources—Framework and issues. J. Environ. Manage. **2011**, *92* (5), 1346–1353.
38. Musacchio, L.R.; Coulson, R.N.; Robert N. Landscape ecological planning process for wetland, waterfowl, and farmland conservation. Landscape Urban Plann. **2001**, *56*, 125–147.
39. ICSU, UNESCO, UNU. Ecosystem change and human wellbeing. Research and Monitoring Priorities Based on the Findings of the Millennium Ecosystem Assessment, 2008, Paris, International Council for Science, ISBN 978-0-930357-67-2, available at http://www.icsu.org/icsu-asia/news-centre/news/archive-2006-2010/ICSUUNESCO-UNU_Ecosystem_Report.pdf. (accessed December 12, 2011).
40. De Groot, R.S.; Alkemade, R.; Braat, L.; Hein, L.; Willemen, L. Challenges in integrating the concept of ecosystem services and values in landscape planning, management and decision making. Ecol. Complexity **2010**, *7*, 260–272.
41. De Pauw, N.; Heylen, S. Biotic index for sediment quality assessment of watercourses in Flanders, Belgium. Aquat. Ecol. **2001**, *35*, 121–133.
42. Neamtu, M.; Ciumasu, I.M.; Costica, N.; Costica, M.; Bobu, M.; Nicoara, M.N.; Catrinescu, C.; Becker van Slooten, K.; De Alencastro, L.F. Chemical, biological, and ecotoxicological assessment of pesticides and persistent organic pollutants in the Bahlui River, Romania. Environ. Sci. Pollut. Res. **2009**, *16*, S76–S85.
43. Von der Ohe, PC.; De Deckere, E.; Prüß, A.; Muñoz, I.; Wolfram, G.; Villagrasa, M.; Ginebreda, A.; Hein, M.; Brack, W. Toward an integrated assessment of the ecological and chemical status of European river basins. Integr. Environ. Assess. Manage. **2009**, *5*, 50–61.
44. Ciumasu, I.M.; Costica, M.; Secu, C.V.; Gurjar, B.R.; Ojha, C.S.P. Adapting to climate change. In *Greenhouse Gas Emissions and Climate Change*; Surampalli, R.Y., Zhang, T., Ojha, C.S.P., Gurjar, B.R., Eds.; ASCE Press - American Society of Civil Engineers, Reston, Virginia.
45. Bolund, P.; Hunhammar, S. Ecosystem services in urban areas. Ecol. Econ. **1999**, *29*, 293–301.
46. Rotmans, J.; Van Asselt, M.; Vellinga, P. An integrated planning tool for sustainable cities. Environ. Impact Assess. Rev. **2000**, *20*, 265–276.
47. Niemelä, J. Ecology and urban planning. Biodiversity Conserv. **1999**, *8*, 119–131.
48. Pauleit, S.; Duhme, F. Assessing the environmental performance of land cover types for urban planning. Landscape Urban Plann. **2000**, *52*, 1–20.
49. Ciumasu, I.M.; Culver, K. The city as disturbance of the local ecosystems. ISEE2011—Advancing Ecological Economics—Theory and Practice. The 9th International Conference of the European Society of Ecological Economics, June 14–17, 2011; Istanbul.
50. Marzluff, J.M. Island biogeography for an urbanizing world: How extinction and colonization may determine biological diversity in human-dominated landscapes. Urban Ecosyst. **2005**, *8*, 157–177.
51. Wittig, R. The origin and development of the urban flora of Central Europe. Urban Ecosyst. **2004**, *7*, 323–339.
52. Dearborn, D.C.; Kark, S. Motivations for conserving urban biodiversity. Conserv. Biol. **2010**, *24* (2), 432–440.

53. Tzoulas, K.; James, P. Making biodiversity measures accessible to non-specialists: An innovative method for rapid assessment of urban biodiversity. Urban Ecosyst. **2010**, *13*, 113–127.
54. MacGregor-Fors, I.; Morales-Perez, L.; Schondube, J.E. Does size really matter? Species–area relationships in human settlements. Diversity Distrib. **2011**, *17*, 112–121.
55. McKinney, M.L. Urbanization as a major cause of biotic homogenization. Biol. Conserv. **2006**, *126*, 247–260.
56. Kühn, I.; Klotz, S. Urbanization and homogenization—Comparing the floras of urban and rural areas in Germany. Biol. Conserv. **2006**, *127* (3), 292–300.
57. Alberti, M.; Marzluff, J.M.; Shulenberger, E.; Bradley, G.; Ryan, C.; Zumbrunnen, C. Integrating humans into ecology: Opportunities and challenges for studying urban ecosystems. BioScience **2003**, *53* (12), 1169–1179.
58. Manfred, M.; Caputo, P.; Costa, G. Paradigm shift in urban energy systems through distributed generation: Method and models. Appl. Energy **2011**, *88*, 1032–1048.
59. Grimm, N.B.; Faeth, S.H.; Golubiewski, N.E.; Redman, C.L.; Wu, J.; Bai, X.; Briggs, J.M. Global change and the ecology of cities. Science **2008**, *319* (5864), 756–760.
60. Dyson, B.; Chang, N.B. Forecasting municipal solid waste generation in a fast-growing urban region with system dynamics modeling. Waste Manage. **2005**, *25*, 669–679.
61. Muga, H.E.; Mihelcic, J.R. Sustainability of wastewater treatment technologies. J. Environ. Manage. **2008**, *88*, 437–447.
62. Zio, E. Reliability engineering: old problems and new challenges. Reliability Eng. Syst. Saf. **2009**, *94*, 125–141.
63. Lahdelma, R.; Saminen, P.; Hokkanen, J. Using multicriteria methods in environmental planning and management. Environ. Manage. **2000**, *26* (6), 595–605.
64. Munda, G. Social multi-criteria evaluation for urban sustainability policies. Land Use Policy **2006**, *23*, 86–94.
65. Ling, C.; Hanna, K.; Dale, A. A template for integrated community sustainability planning. Environ. Manage. **2009**, *44*, 228–242.
66. Giddings, B.; Hopwood, B.; O'Brien, G. Environment, economy and society: Fitting them together into sustainable development. Sustainable Dev. **2002**, *10* (4), 187–196.
67. Gowdy, J.; Erickson, J.D. The approach of ecological economics. Cambridge J. Econ. **2005**, *29*, 207–222.
68. Ciumasu, I.M.; Costica, M.; Costica, N.; Neamtu, M.; Dirtu, A.C.; De Alencastro, L.P.; Buzdugan, L.; Andriesa, R.; Iconomu, L.; Stratu, A.; Popovici, O.A.; Secu, C.V.; Paveliuc-Olariu, C.; Dunca, S.; Stefan, M.; Lupu, A.; Stingaciu-Basu, A.; Netedu, A.; Dimitriu, R.I.; Gavrilovici, O.; Talmaciu, M.; Borza, M. Complex risks from old urban waste landfills—A sustainability perspective from Iasi, Romania. J. Hazard., Toxic Radioact. Waste, **2012**, *16* (2), 158–168.
69. Lang, D.J.; Scholz, R.W.; Binder, C.R.; Wiek, A.; Stäubli, B. Sustainability Potential Analysis (SPA) of landfills—A systemic approach: theoretical considerations. J. Cleaner Prod. **2007**, *17*, 1628–1638.
70. Reyers, B.; Roux, D.J.; Cowling, R.M.; Ginsburg, A.E.; Nel, J.L.; O'Farrell, P. Conservation planning as a transdisciplinary process. Conserv. Biol. **2010**, *24* (4), 957–965.
71. Reid, W.V.; Chen, D.; Goldfarb, L.; Hackmann, H.; Lee, Y.T.; Mokhele, K.; Ostrom, E.; Raivio, K.; Rockström, J.; Schellnhuber, H.J.; Whyte, A. Environment and development. Earth system science for global sustainability: Grand challenges. Science **2010**, *330* (6006), 916–917.

Endocrine Disruptors

Vera Lucia S.S. de Castro
Ecotoxicology and Biosafety Laboratory, Environment, Brazilian Agricultural Research Corporation (Embrapa), São Paulo, Brazil

Abstract
Many pollutants are known to adversely affect development and physiology by interfering with normal endocrine functions. Among them, numerous pesticides are described as able to change the endocrine system, since the differentiation and development of the reproductive system are dependent of the action of hormones. Endocrine-disrupter chemicals (EDCs) are substances in the environment, food, and consumer products that interfere with hormone biosynthesis, metabolism, or action, resulting in a deviation from normal homeostatic control or reproduction. Exposure to an EDC may have different consequences for an adult compared with a developing fetus or infant. The developing organism interaction with these chemicals may lead to the development of a disease or dysfunction later in life. Results from animal models and epidemiological studies converge to implicate them as a significant concern to public health since their mechanisms involve different pathways that are present in wildlife and humans. Furthermore, effects of different classes of EDCs may be additive or even synergistic, but there are limited data on the interactions between them. An increased understanding of the potential human and environmental health risks of exposure to single and mixtures of EDCs as well as the efficient removal process of EDCs from water are important but remain understudied. This entry presents an overview of the environmental EDC contamination and its possible effects on different organisms. However, it is beyond the scope of this entry to describe all the possible disruption events of the endocrine system.

INTRODUCTION

Endocrine-disrupting chemicals (EDCs) refer to anthropogenic compounds that are able to mimic, antagonize, alter, or modify normal hormonal activity. Dichlorodiphenyltrichloroethane (DDT), an insecticide first produced on a wide scale in 1945, was used extensively during the 1960s and 1970s and was the first chemical found to be estrogenic. Subsequently, other organochlorine insecticides such as dieldrin, endosulfan, and methoxychlor were found to be estrogenic. Endocrine-disrupting chemicals include environmental estrogens such as *o,p*-DDT, endosulfan, non-planar polychlorinated biphenyl (PCB), octyl- and nonylphenols, the antiandrogens such as vinclozolin and DDE, and the thyroid hormone disrupters such as fenvalerate and benzene hexachloride.[1]

Endocrine-disrupting chemicals are a significant public health concern since these compounds interfere with normal function of pathways responsible for both reproduction and development and can affect the endocrine system, interfering in the production or action of hormones or compromising sexual identity, fertility, or behavior.[2,3] Besides, many of them are persistent in the environment, can be found in waters and sediments, and are easily transported long distances in the atmosphere.[4]

In recent years, numerous studies have suggested that many environmentally persistent chemicals have a potential to disrupt normal functions of the endocrine system. The field of endocrine disrupters, such as the special susceptibility of the developing organism and early induction of latent effects, has come a long way since its initial impetus in 1991.[5] Specially, the possible effects of EDCs on early events of proper gonadal development—which is dependent on intercellular signaling mechanisms—deserve attention since the early steps in mammalian sexual development are vulnerable to genetic and environmental perturbation.[6]

Exposure to EDCs is associated with dysfunctions of metabolism, energy balance, thyroid function, and reproduction, and an increased risk of endocrine cancers. These multifactorial disorders can occur through molecular epigenetic changes induced by exposure to EDCs early in life, the expression of which may not manifest until adulthood.[7] Effects attributed to the EDCs include developmental demasculinization and feminization in reptiles, mammals, amphibians, fish, and birds; reduced fecundity in reptiles, birds, and fish; and possibly increased breast cancer rates and reduced sperm counts in humans.[1]

Since hormones, in synergy with genes, are responsible for sex-related differences in anatomical, physiological, and behavioral traits, even if EDCs are present in minute amounts in environment, their effects in male and female physiology could be greater than before expected. They might also prejudice the sex steroid hormone–induced integrated physiological responses in women and men. In addition, differences in male and female susceptibility to

Encyclopedia of Environmental Management DOI: 10.1081/E-EEM-120047685

EDCs could be present even if there is still scarce information available on this aspect.[8]

Mechanisms of Action

Several EDCs may work by multiple mechanisms, including uncharacterized mechanisms of action. Because of cross talk between different components of the endocrine systems, effects may occur unpredictably in endocrine target tissues other than the system predicted to be affected. A few modes of action could contribute to the same outcome, including aromatase inhibition, antiestrogenicity, testosterone biosynthesis disruption, and antiandrogens that alter upregulation of aromatase in the target regions within the brain. More complex biological responses to EDCs will generally represent combinations of several physiological processes integrated through multiple biological pathways.[9,10]

Endocrine disrupters may interfere with the functioning of hormonal systems in at least three possible ways: 1) by mimicking the action of a naturally produced hormone, producing similar but exaggerated chemical reactions in the body; 2) by blocking hormone receptors, preventing or diminishing the action of normal hormones; and 3) by affecting the synthesis, transport, metabolism, and/or excretion of hormones, thus altering the concentrations of natural hormones.[11]

The first characterized mechanism of action of EDCs is to act directly as ligands to steroid hormone nuclear receptors (NRs), in particular, estrogen, androgen, and thyroid NRs. Nuclear receptors are a class of proteins found within cells. In response to the presence of hormones, these receptors work in concert with other proteins to regulate the expression of specific genes by a conformation change. Schematically, NRs may be classified into four classes according to their dimerization and DNA-binding proprieties.[10]

Cross talk between NR-mediated and other signal-transduction pathways is an important aspect of NR-based regulation. This so-called genomic or genotropic signaling is normally slow and sustained, taking hours before biological outcomes become manifest. For example, in the classic view of estrogen action, the effects of 17β-estradiol (E2) were thought of as mediated by the NRs estrogen receptor α and β, acting as ligand-dependent transcription factors, thereby regulating gene expression by binding estrogen response elements in the DNA.[12]

Another type of NR cross talk that has recently been recognized is the non-genomic action of several NRs. Some non-genomic actions of NR ligands are apparently mediated through membrane receptors that are not part of the NR superfamily.[10] For example, it has become clear that E2 can also rapidly and transiently trigger a variety of second messenger signaling events, including the induction of cyclic adenosine monophosphate (cAMP) and adenylate cyclase; the mobilization of intracellular calcium; and the stimulation of PI3K, PKB, and Src with consequent activation of the extracellular-regulated kinases Erk1 and Erk2 in the Src/Ras/Erk cascade. All these effects are believed to be mediated through a membrane-associated or cytosolic estrogen receptor (ER) and have, therefore, been termed non-genomic or extranuclear actions of E2.[12]

The cellular activities of estrogens and xenoestrogens are the result of a combination of extranuclear (non-genomic) and nuclear (genomic) events and highlight the need to take non-genomic effects and signaling cross talk into consideration when screening for environmental estrogen.[12]

Disruption of the endocrine system by xenobiotic compounds is consistently reported in humans and wildlife and is a matter of concern worldwide. A great variety of natural or synthetic chemicals such as EDCs are thought to exert an acute effect at different levels of the thyroid cascade. It is consensual that EDCs probably act by interfering with thyroid hormone (TH) synthesis, cellular uptake, and metabolism, at the level of TH receptors and also TH transport, by binding to thyroid hormone distributor proteins (THDPs). The TH transport system in particular may be quite susceptible to EDCs as many chemicals are structurally related to THs and may bind THDPs and disturb homeostasis of extracellular TH levels or even cellular uptake.[13]

Steroid hormone synthesis is controlled by the activity of several highly substrate-selective cytochrome P450 enzymes and a number of steroid dehydrogenases and reductases. Cytochrome P450 monooxygenases (CYPs) form a large group of enzymes found in most organisms from bacteria to mammals and can be grouped into 281 families. According to their function CYPs can be classified into enzymes metabolizing xenobiotics and enzymes that are part of key biosynthetic pathways, with narrow substrate specificity. Particularly, aromatase (CYP19), the enzyme that converts androgens to estrogens, has been the subject of studies into the mechanisms by which chemicals interfere with sex steroid hormone homeostasis and function, often related to (de)feminization and (de)masculinization processes.[14,15]

After all, several findings suggest that responses to EDCs cannot be assumed to be monotonic across a wide dose range and that change can occur in response to extremely low concentrations. In particular, low-dose effects may be mediated by endocrine-signaling pathways, evolved to act as powerful amplifiers, with the result that large changes can occur in response to extremely low concentrations. Dose–response relationship, however, is perhaps one of the most controversial issues in EDC studies. Reports on non-linearities in dose–response functions are highly controversial and the subject of intense research: non-monotonic, linear, and even threshold responses are all possible outcomes of low-dose exposure.[16] A non-monotonic response decreases testing efficiency and multiplies the time and other resources necessary to understand the potential hazard posed by a chemical. Because the issue of low-dose effects of EDCs was based on unknown and unexpected mechanisms, the actual features of these effects were not readily resolved.[17]

Low-dose effects of EDCs are based on unknown and unexpected mechanisms. Recent developments in the biological sciences, including homeostatic regulatory disturbance and epigenetic response, have aided in clarifying the mechanisms underlying the low-dose issue. Elucidating the xenobiotic effects of EDCs requires development of systems toxicology, i.e., deciphering the toxicity mechanisms underlying homeostatic regulatory disturbances.[17,18]

HORMONES, REPRODUCTIVE ASPECTS, AND EDCS

In vertebrates, the ability to attain reproductive competence in adulthood involves the organization of a complex, steroid-sensitive network in hypothalamic–preoptic–limbic brain regions during critical developmental windows. This process includes the establishment of the hypothalamic neural network of gonadotropin-releasing hormone (GnRH) cells, together with their regulatory inputs from other neuronal and glial cells in the brain, which enable feedback effects of steroid hormones on pulsatile GnRH release and the preovulatory GnRH/LH–luteinizing hormone surge in females. The anatomical development of this steroid-sensitive hypothalamic network occurs early in life, typically the late embryonic and early postnatal period in mammals, and its organization is important to the attainment and activation of appropriate reproductive functions in adulthood. Importantly, this same early developmental period is also a critical period for sexual differentiation of hypothalamic–limbic neural networks that must be organized perinatally to enable proper behavioral activation in adulthood. During mammalian development, the fetal organism is exposed to its own gonadal hormones, placental steroids, and maternal hormones that may cross the placental barrier. There are sex differences in exposures to androgens and estrogens that appear to underlie normal reproductive neuroendocrine development. Aberrations in these developmental patterns in females can cause masculinization (acquisition of a male-typical trait) or defeminization (loss of a female-typical trait) and in males may cause feminization or demasculinization (comparably defined). Perinatal hormones have permanent imprinting effects on the hypothalamus, manifested early on as morphological sex differences in the brain and manifested much later on as physiological and behavioral differences between the sexes.[19–21] Besides, androgens and estrogens can play a special role in the development of sexually dimorphic behaviors.[22]

Some populations exposed to chemicals from industrial accidents or chemical misadventures are of particular interest. Data from these select populations with higher levels of exposure than the common population seem to suggest that some of these chemicals have a role in genitourinary development as endocrine disrupters. Furthermore, animal studies of EDC effects on genitourinary development have confirmed that changes occur with exogenous manipulation of steroid levels or hormone receptors. These findings in animals have led to observational and epidemiological studies in humans to document a link between environmental exposure and human disease.[23]

Global declines in semen quality were suggested to be associated with enhanced exposure to environmental chemicals that act as endocrine disrupters as a result of increased use of pesticides, plastics, and other anthropogenic materials. A significant body of toxicology data based upon laboratory and wildlife animals studies suggests that exposure to certain endocrine disrupters is associated with reproductive toxicity, including the following: 1) abnormalities of the male reproductive tract (cryptorchidism, hypospadias); 2) reduced semen quality; and 3) impaired fertility in the adult.[24]

Recently, there has been increasing concern about the potential for environmental EDCs as fungicides to alter sexual differentiation in mammals. In this direction, observations demonstrate that vinclozolin (a systemic dicarboximide fungicide) can affect embryonic testicular cord formation in vitro. This transient in utero exposure to the fungicide increases apoptotic germ cell numbers in the testis of pubertal and adult animals. This effect is correlated with reduced sperm motility in the adult and putative effects on spermatogenic capacity later in adult life. In conclusion, transient exposure to this fungicide during the time of testis differentiation alters testis development and function.[25]

A higher prevalence of cryptorchidism and hypospadias was found in areas with extensive farming and pesticide use and in sons of women working as gardeners. Recently, a relation has been reported between cryptorchidism and persistent pesticide concentration in maternal breast milk.[26]

Other commonly used fungicides, such as the azoles, may also act as endocrine disrupters in vivo. They showed endocrine-disrupting potential when tested for endocrine-disruptive effects using a panel of in vitro assays. Overall, the imidazoles (econazole, ketoconazole, miconazole, prochloraz) were more potent than the triazoles (epoxiconazole, propiconazole, tebuconazole). The critical mechanism in vitro seems to be disturbance of steroid biosynthesis.[27]

Regarding in vivo effects, many of the commonly used azole fungicides act as endocrine disrupters, although the profile of action varies. Common features for azole fungicides are that they increase gestational length, virilize female pups, and affect steroid hormone levels in rat fetuses and/or dams.[28] For example, prochloraz causes reproductive malformations in androgen-dependent tissues of male offspring of exposed rats.[29] Also, tebuconazole has been found to demasculinize the male offspring and to possess some of the same endocrine effects as prochloraz. These effects strongly indicate that one major underlying mechanism for the endocrine-disrupting effects of azole fungicides is disturbance of key enzymes like CYP17 involved in the synthesis of steroid hormones.[28]

Also, triazole-induced male reproductive toxicity includes disruption of testosterone homeostasis. Elevated serum testosterone, increased testis, weights and anogenital distance, and hepatomegaly indicative of altered liver metabolism of steroids are the key events consistent with this mode of action.[30] Developmental exposure to triazole fungicides such as propiconazole, myclobutanil, and triadimefon can adversely impact reproduction in the female rat.[31] In this way, epoxiconazole and ketoconazole may be fetotoxic, increasing postimplantation loss and late resorptions.[32]

Aside from triazoles, the organic fungicide fenarimol possesses estrogenic properties[33] and acts both as an estrogen agonist and as an androgen antagonist.[34] In addition, fenarimol affects rat aromatase activity in vivo, inhibiting estrogen biosynthesis in rat microsomes[35] and in human tissues.[3] This compound also affects other enzymes of the cytochrome P450 gene family that are involved in the metabolism of steroids.[36]

Induction of reactive oxygen species (ROS) by environmental contaminants and associated oxidative stress also have a role in defective sperm function and male infertility, although there are some controversial data. This is evidence for the existence of a link between endocrine-mediated and ROS-mediated adverse effects of environmental contaminants on male reproduction. Another link is the antioxidant enzyme superoxide dismutase, which has been shown to have a superoxide scavenging effect as well as act as an alternate regulatory switch in testicular steroidogenesis.[37]

Endocrine-disrupting chemicals can also impact female fertility by altering ovarian development and function, purportedly through estrogenic, antiestrogenic, and/or antiandrogenic effects. These compounds may also cause transgenerational effects by targeting oocyte maturation and maternal sex chromosomes.[38]

In girls, earlier age at menarche was reported after exposure to PCBs, polybrominated biphenyls, persistent pesticides (DDT), and phthalate esters. However, several other studies found no effect of these compounds on age at menarche. In boys, exposure to PCBs, PCDFs, or the pesticide endosulfan was associated with delayed puberty or decreased penile length. Much of the results found in population studies are in accordance with experimental studies in animals. However, the mixture of different components with antagonistic effects (estrogenic, antiestrogenic, antiandrogenic) and the limited knowledge about the most critical window for exposure (prenatal, perinatal, and pubertal) may hamper the interpretation of results.[39]

In human and rodent models, EDCs also interfere with the development of cognition and behaviors. In this way, fenvalerate is a potential EDC and is a candidate environmental risk factor for cognitive and behavioral development, especially in the critical period of development.[40] Also, prenatal phthalate exposure was associated with childhood social impairment in a multiethnic urban population.[41]

Recently, the interference of EDCs with receptors regulating metabolism has been proposed especially in relation with the etiology of metabolic diseases such as obesity and diabetes. In particular, the harmful action of EDCs on normal adipocyte development, homeostatic control of adipogenesis, early energy balance, and, in turn, body weight has been demonstrated. Much remains to be studied about the endocrine pathways responding to EDC exposure, especially those controlling feeding behavior, as their impairment represents a real risk factor for metabolic and feeding disorders.[42]

SCREENING ASSAYS AND BIOMARKERS FOR EDCS

Environmental stresses as presence of EDCs due to human activities are increasingly likely to pose habitat disturbances that could have potential deleterious effects on physiological function in vertebrates. These effects could result in major impacts on the life cycles of organisms, affecting morphology, physiology, and behavior. However, because animals live in diverse habitats, there is variation in susceptibility to disruption of response systems to environmental cues. While some populations of vertebrates, from fish to mammals, temporarily resist environmental stresses and breed successfully, many others show varying degrees of failure, sometimes resulting in population decline.[43] The development of targeted bioassays in combination with adequate chemical analyses is important for EDC risk assessment.

It is acknowledged that EDCs can affect humans and animals at low exposure levels and that responses to EDCs are in many cases complex, activating a range of different molecular events, e.g., receptor agonism/antagonism and enzyme induction, in multiple hormone systems. As a result, regulatory testing for these effects and evaluating the results is complicated.[44,45] In the typical case of assessing human risk, a scientifically justified validation could only mean an experimentally validated mechanistic link between EDC assay results and human susceptibilities at environmental exposures, sustained by reliable sensitivity and specificity benchmarks. Analytical methods have long been used to determine concentrations of chemical residues that persist in environment and accumulated in biota. Although they are useful, the development of EDC screening and monitoring procedures may help in the establishment of EDC exposure and biological responses. In this way, several in vitro and in vivo procedures have been proposed to screen and monitor individual EDCs or their mixtures.[45]

As an in vitro model, the use of the bovine ovarian follicle has already been recommended as a valuable instrument to unravel reproductive events in women due to the similarities in ovarian follicular dynamics and endocrine control.[46]

As an in vivo test, some external biomarkers of prenatal androgen disruption may be used, including the anogenital distance and the juvenile nipple/areola number. The anogenital distance is defined as the distance between the genital papilla and the anus; male rodents have an anogenital distance that is approximately twice the length as that of females. Areolae are dark areas surrounding the nipple bud and. their presence as measured at postnatal day 2–3 is indicative of adult nipples. Adult female rats typically have 12 nipples, whereas males have none. Both of these biomarkers vary with prenatal exposure to androgens or antiandrogens in females and males, respectively. Reduction of anogenital distance and/or retention of nipples in male rats is indicative of prenatal exposure to antiandrogens.[47]

In this context, a wide spectrum of potential biomarkers also could be applied to the study of endocrine disruption in the aquatic environment. In fish, they include changes in hormone titres (steroid hormones, thyroid hormones), abnormal gonad development, low gamete viability, and alterations in some enzyme activities (i.e., aromatases) and protein levels (i.e., vitellogenin, zona radiata proteins, spiggin). Likewise, evidence is slowly growing that indicates that gamete development and vitellogenesis of marine bivalve mollusks are targets of EDCs.[48]

On the other hand, although it is known that aquatic invertebrates contain different classes of steroids,[49] a clear cause–effect relationship between exposure and specific responses for most EDCs is far from being established.[48]

In crustacean populations, the attribution of endocrine toxicity to observed disturbances requires the identification of definitive biomarkers of such toxicity. Mortality, reduced fecundity, lowered recruitment, and impaired growth all might serve as indicators of endocrine disruption in crustaceans; however, such end points are indicative of adversity involving a variety of mechanisms. An exception to this premise is excess males in parthenogenic branchiopod populations that normally exist predominantly as females.[50]

Other organisms, such as amphibians, may be used to study the endocrine system and can serve as sentinels for detection of the modes of action of EDCs. Recently, amphibians are being reviewed as suitable models to assess (anti)estrogenic and (anti)androgenic modes of action influencing reproductive biology as well as (anti)thyroidal modes of action interfering with the thyroid system.[51]

Biochemical end points can also be useful biomarkers since environmental toxicants can trigger biological effects at the organism level only after initiating biochemical and cellular events. The cellular response to stress is characterized by the activation of genes involved in cell survival to counteract the physiological disturbance induced by physical or chemical agents. As an example, Hsps are suitable as an early-warning bioindicator of environmental hazard by various pollutants such as EDCs, because of their sensitivity to even minor changes in cellular homeostasis and their conservation along the evolutionary scale.[52] In addition, a combined testing strategy, considering both markers of endocrine/hormonal maturation and behavioral end points under hormonal control, may evidence even subtle perturbations of the neuroendocrine homeostasis, which often go undetected.[53]

GUIDELINES FOR REGULATORY PURPOSES

In recent years, under the current European Union chemical regulation REACH (Registration, Evaluation, Authorization and Restriction of Chemicals), which revised plant protection product and biocide directives,[54] evaluation of endocrine-disrupting properties of chemicals has become a regulatory effort.

The initial framework for regulatory purposes has been revised by the Endocrine Disrupters Testing and Assessment (EDTA) Task Force at its meetings to reflect the Organization for Economic Cooperation and Development (OECD) member countries' views. The conceptual framework agreed upon by the EDTA6 in 2002 is not a testing scheme but rather a toolbox in which the various tests that can contribute information for the detection of the hazards of endocrine disruption are placed. The toolbox is organized into five compartments or levels, each corresponding to a different level of biological complexity (for both toxicological and ecotoxicological areas). Even though the conceptual framework may be full of testing tools, this does not imply that they all will be needed for assessment purposes. Tools will be added as they are validated in future. The conceptual framework is subject to further elaboration and discussion as the work on endocrine disrupters proceeds.[55] The OECD adopted in 2007 the uterotrophic bioassay as a standardized screening test with international regulatory acceptance. This assay may be used to screen for estrogenic properties of chemicals. However, generally, EDCs are handled as such only if their endocrine-disruption potential has been previously identified via, for example, academic research or is indicated by effects observed in required toxicity tests.[44]

The Endocrine Disruptor Screening Program (EDSP) of the United States Environmental Protection Agency (EPA) has been working to reach a consensus validation on a battery of screens and long-term tests for endocrine disrupters.[45] The Endocrine Disrupter Screening and Testing Advisory Committee (EDSTAC) was established by the EPA in 1996 as a federal advisory committee to provide advice in developing and implementing new screening and testing procedures for endocrine effects as mandated by the U.S. Congress (through the Food Quality Protection

Act of 1996) in response to public concern.[56] The EDSTAC assesses the current state of the science and assists the agency in developing an endocrine screening program. The EDSTAC consists of scientists and others representing various interests, including advocates of the endocrine-disruption theory and the regulated community.

The EDSTAC concluded that the assays necessary to determine the potential endocrine activity of chemical substances varied significantly in their degree of development and validation. Several screens had an extensive history, e.g., the uterotrophic and the Hershberger screens, but others were only partially developed or were only hypothetically useful as screens, e.g., the amphibian developmental screen and the fish gonadal recrudescence screen. The fundamental validation principles are to clearly state the purpose and biological basis for the assay and to verify the performance of the assay against validation criteria using a common set of test chemicals across multiple laboratories.[57]

At the same time, EDSTAC recommended that EPA develop an extensive program that would subject all chemicals to screening and testing for estrogenic, antiestrogenic, androgenic, antiandrogenic, and thyroid effects in both humans and wildlife. Specifically, EDSTAC recommended, among other things, that the EPA do the following: 1) adopt a two-tiered, hierarchical testing and evaluation framework; and 2) initiate a research program, composed of both basic and applied research, to develop, standardize, and validate the necessary endocrine test methods. The EPA's EDSP was implemented in 2009–2010 with the issuance of test orders requiring manufacturers and registrants of 58 pesticide active ingredients and 9 pesticide inert/high-production-volume chemicals to evaluate the potential of these chemicals to interact with the estrogen, androgen, and thyroid hormone systems. Despite this great effort, numerous questions and uncertainties remain as to the usefulness and limitations of the specific assays. Understanding the tests' strengths and limitations is critical for interpretation of the screening results and for decision making based on those results.[57,58]

During the time EDSTAC was meeting, OECD began collaborating with its member countries, including the U.S. EPA, to develop internationally harmonized test guidelines.[48] Although the EPA and OECD endocrine screening and testing methods have been substantially harmonized, the framework of OECD's endocrine screening and testing program differs significantly from EPA's two-tiered EDSP. The EPA screening will entail evaluation of responses in the Tier 1 Endocrine Screening Battery, consisting of 11 distinct in vitro and in vivo assays. The OECD framework provides the flexibility to enter and exit at any level depending on information needs and encourages the maximal use of all existing relevant information that may be equally predictive and reduce vertebrate testing. The screening results are collectively intended to identify chemicals for which subsequent Tier 2 testing is necessary. Tier 2 testing uses test methods that encompass reproduction and developmental life stages in several species to provide data on adverse effects and dose response for risk assessment.[57]

In the years that the EPA worked on developing, standardizing, and validating the EDSTAC-recommended assays and implementing the EDSP, significant advances have been made in both computational and molecular technologies for discerning potential endocrine activity.[57] Accordingly, there are efforts to model EDC effects using computational approaches by the development and validation of mechanistically based computational models of hypothalamic–pituitary–thyroid (HPT); hypothalamic–pituitary–gonadal (HPG); hypothalamic–pituitary adrenal (HPA) axes in ecologically relevant species to better predict accommodation and recovery of endocrine systems.[59]

OVERVIEW OF EDC EXPOSURE

Human Exposure Effects

Models for estimating human exposure to endocrine disruptor (ED) pesticides are an important risk management tool. Many of them are harmful at very low doses, especially if exposure occurs during sensitive stages of development, producing effects that may not manifest for many years or that affect descendants via epigenetic changes. The main requirement for the use of such models is more quantitative data on the sources and pathways of human ED pesticide exposure. Quantifying the risks posed by the different routes of exposure will play an important part in designing and implementing effective risk mitigation for ED pesticides. In fact, it is difficult to assess the relative importance of some routes of exposure because no data sets that would allow these to be calculated are available. Pesticide exposure from the use of pesticides for medicinal purposes and exposure from cut flowers and ornamental plants both need to be quantified, and better data sets are required for pesticide exposure from spray drift, home use, municipal use, and travel.[60]

Food and water are both chronic exposure routes affecting the entire population. Food residues are currently thought to be the most important exposure pathway, for although residue levels present in food tend to be below the maximum residue levels permitted by law, they do result in constant measurable low-level exposure.[51] Food as a major xenobiotic and heavy metal exposure route to humans is studied intensively. More than 100 chemicals have been identified as antiandrogens, including certain phthalates, widely used as plasticizers, pesticides, and various other chemicals found in food and consumer products.[61]

Indeed, typical food contaminants, like pesticides, dioxins, PCBs, methylmercury, lead, etc., are well characterized in food. In contrast, the role of food and beverage packaging as an additional source of contaminants has received much less attention, even though food packaging contributes significantly to human xenobiotic exposure. Especially, EDCs in food packaging are of concern since even at low concentrations, chronic exposure is toxicologically relevant. Thus, non-intentionally added substances migrating from food contact materials need toxicological characterization.[62]

Some chemicals used in food processing have an environmental endocrine-disrupting effect that affects reproduction in wildlife. For example, bisphenol-A is a monomer of polycarbonate plastics and a constituent of epoxy and polystyrene resins, which are used in the food cans and found as a contaminant not only in the liquid of the preserved foods but also in the water autoclaved in the cans. This chemical is also released from polycarbonate flasks during autoclaving. Moreover, it has been reported that significant amounts of bisphenol-A are detected in the saliva of dental patients treated with fissure sealants. The exposure to low doses of this chemical was reported to affect the rate of growth and sexual maturation, hormone levels in blood, reproductive organ function, fertility, immune function, enzyme activity, brain structure, brain chemistry, and behavior.[63]

Another important route of EDC exposure is occupational. Relatively high levels of exposure to environmental endocrine disrupters in the form of pesticides occur among people working in agriculture. Some pesticides are able to influence the synthesis, storage, release, recognition, or binding of hormones, which may lead to alterations in reproductive hormone levels. The issue of male infertility caused by occupational exposure is pertinent worldwide. A significant increase in the incidence of male infertility has been described in the international literature. Part of this effect may result from synthetic toxic substances acting on the endocrine system, many of which are routinely used in work processes. However, progress is needed in the knowledge of possible effects of exposure on male fertility since monitoring these effects requires sufficient time for the manifestations to occur. Such progress will allow the development of preventive measure within the field of workers' health.[64]

Apart from EDC effects on males, several studies on occupational exposure to pesticides and adverse effects on human reproduction have been performed, including end points such as prolonged time to pregnancy, spontaneous abortion or stillbirth, low birth weight, and developmental disorders.[65]

Complex EDC Mixtures

Concerns increase when humans are exposed to mixtures of similar-acting EDCs and/or during sensitive windows of development. It is difficult to predict biological effects directly from the composition of pollutant mixtures. In addition to simple additive effects, interactions between different chemicals in a mixture may result in either a weaker (antagonistic) or a stronger (synergistic) combined effect than would be expected from knowledge about the toxicity and mode of action of each individual compound. Such interactions may take place in the toxicokinetic phase (i.e., processes of uptake, distribution, metabolism, and excretion) or in the toxicodynamic phase (i.e., effects of chemicals on the receptor, cellular target, or organ). A chemical mixture may contain a number of xenoestrogens enhancing the response of endogenous estrogens, or it may contain xenoantiestrogens that inhibit the normal action of endogenous estrogens.[66] Substances of concern include certain phthalates, pesticides and chemicals used in cosmetics and personal care products. A lack of knowledge about relevant exposure scenarios presents serious obstacles for better human risk assessment. A disregard for combination-effect studies may lead to underestimations of risks. In this way, the study of EDC mixture effects by developing biomarkers that capture cumulative exposure to endocrine disrupters is needed.[67]

Doses of endocrine-disrupting pesticides that appear to induce no effects on gestation length, parturition, and pup mortality when alone induced marked adverse effects on these end points together with other pesticides. They can also affect the sexual differentiation of offspring.[68] Chemicals that act on different fetal tissues via diverse cellular mechanism of action may produce additive effects. This fact indicates that the current framework for conducting cumulative risk assessments should not only consider including chemicals from different classes with the same mechanism of toxicity but also include chemicals that disrupt differentiation of the same fetal tissue at different sites in the androgen signaling pathway.[5] Compounds that act by disparate mechanisms of toxicity to disrupt the dynamic interactions among the interconnected signaling pathways in differentiating tissues produce cumulative dose-additive effects, regardless of the mechanism or mode of action of the individual mixture component.[69] Predictive approaches are generally based on the mathematical concepts of concentration addition and independent action, both predicting the toxicity of a mixture based on the individual toxicities of the mixture components.[26]

In this sense, a combination of five pesticides with dissimilar mechanisms of action produced greater androgen-sensitive end-point responses than would be expected using response-addition modeling.[70] Deltamethrin, methiocarb, prochloraz, simazine, and tribenuron-methyl are all commonly used for agricultural and horticultural purposes. In vivo, the levator ani/bulbocavernosus muscle and adrenal gland weight changes indicated that the pesticides had an accumulating effect that was not observed for the individual pesticides. Several pesticide-induced gene expression changes were observed, indicating that these may be very sensitive antiandrogenic end points. In another study,[71] dexamethasone appeared to exacerbate the reproductive anomalies induced by in utero exposure of male rats to dibutylphthalate.

In a recent study, male Sprague Dawley rats were subchronically exposed to single doses of dibutyl phthalate, single doses of benzo(a)pyrene, and combined doses of both EDCs. Significant adverse effects were observed on the reproductive system, including decreased sperm count, increased production of abnormal sperm, changes in serum testosterone levels, and irregular arrangements of the seminiferous epithelium. It is also observed that biochemical analyses showed that the activities of superoxide dismutase and glutathione peroxidase decreased after exposure to these EDCs. Therefore, the data suggest that exposure to them, in either separate or combined doses, can affect the reproductive system of male rats adversely via oxidative stress-related mechanisms.[72]

Thus, assessment of risks posed by chemicals causing reproductive effects and protection of future generations are important public health tasks. To determine the levels of significant human exposure to a given chemical and associated health effects, the Agency for Toxic Substances and Disease Registry's (ATSDR's) toxicological profiles examine and interpret available toxicological and epidemiological data. The ATSDR categorizes the health effects according to their seriousness as serious (effects that prevent the organism from functioning normally or can cause death), less serious (changes that will prevent an organ or organ system from functioning in a normal manner but will not necessarily lead to the inability of the whole organism to function normally), or minimal (effects that reduce the capacity of an organ or organ system to absorb additional toxic stress but will not necessarily lead to the inability of the organ or organ system to function normally). The ATSDR uses the highest no-observed-adverse-effect level or the lowest-observed-adverse-effect level (LOAEL) in the available literature to derive a health-based guidance value called a minimal risk level (MRL). An MRL is defined as "an estimate of the daily human exposure to a substance that is likely to be without an appreciable risk of adverse, non-cancer effects over a specified duration of exposure." Minimal risk levels based on reproductive and endocrine effects were described in a review by Pohl et al.[73]

Some Examples of Animal Exposure Effects

There is widespread exposure to EDCs, which can disrupt the reproduction and development of various non-target organisms. Effects of EDCs have been shown by observed adverse reproductive and developmental effects. Indeed, most studies of potential EDC effects are based on indirect evidence of endocrine disruption rather than defined endocrine pathways.

Some domestic mammals may come into contact with EDCs by sewage exposure. As an example, sewage sludge is sometimes recycled to arable land or pasture and contains large amounts of a variety of pollutants, including EDCs and heavy metals, derived from industrial, agricultural, and domestic sources. A demasculinizing effect of exposure to higher pollutant concentrations with respect to exploratory sheep behavior was observed.[74] These observations demonstrate the need to take into account the effects of pollutant combinations, even at very low, environmental concentrations, and further highlight the usefulness of ethotoxicology for the study of biological effects of environmental pollutants.

Endocrine-disrupting chemicals have been found in sewage effluent in low concentrations (ng/L). Some of these estrogens bind with estrogen receptors in exposed organisms and have the potential to exert effects at extremely low concentrations. Data from laboratory experiments support the hypothesis that EDCs in the aquatic environment can impact the reproductive health of various fish species, but evidence in the aquatic environment is still weak and needs a dependable method or indicator to assess reproduction of fish in situ. The link between endocrine disruption and reproductive impairment that cause an ecologically relevant impact on the sustainability of fish populations remains to be better understood.[75]

Surface waters are the main sink of EDCs, which are mainly of anthropogenic origin. Thus, aquatic organisms, especially lower vertebrates such as fish and amphibians, are the main potential targets for EDCs, being at direct or indirect risk via ingestion and accumulation of EDCs via exposure or the food chain. These compounds may play an important role in the decline of the amphibian population.[51] Several incidents in the wildlife population strongly correlated decreased reproductive capacity with exposure to specific industrial chemicals, and the organisms may be viewed as sentinels of human health effects. Reported reproductive disorders in wildlife have included morphologic abnormalities, eggshell thinning, population declines, sex reversal, impaired viability of offspring, altered hormone concentration, and changes in sociosexual behavior.[76]

The ED are prevalent over a wide range of chemicals in the aquatic ecosystems, most of them being resistant to environmental degradation and considered ubiquitous contaminants.[48,77] Some imidazole (prochloraz, imazalil) and triazole (epoxiconazole) agricultural fungicides induced oocyte maturation in rainbow trout. Prochloraz, epoxiconazole, and imazalil strongly potentiated the induction of oocyte maturation by gonadotropin in a dose-dependent manner.[78] Above all, prochloraz caused responses consistent with aromatase inhibition, although there were indications that the fungicide may also be disturbing the balance between estrogens and androgens via effects elsewhere in the steroidogenic pathway.[77]

In U.K. rivers, a widespread feminization of wild fish was observed involving contributions from both steroidal estrogens and xenoestrogens and from other yet-unknown contaminants with antiandrogenic properties. The widespread occurrence of feminized male fish downstream of some wastewater treatment works has led to substantial interest from ecologists and public health professionals. This concern stems from the view that the effects observed have a parallel in humans and that both phenomena are caused by exposure to mixtures of contaminants that interfere with reproductive development.[79] Some authors reported the occurrence of fish feminization as well as reproduction and development interference with other aquatic organisms,[80] although there is no universally accepted bioassay or chemical technique to quantify EDCs in the aquatic environment.[81] Endocrine-disrupting chemicals can also promote disrupting effects in vitro on ovarian follicular cells exposed to environmentally relevant doses of mixtures of persistent organic pollutants extracted from marine and freshwater ecosystems.[66]

Population studies have revealed alterations in crustacean growth, molting, sexual development, and recruitment that are indicative of environmental endocrine disruption. However, environmental factors other than pollution (i.e., temperature, parasitism) also can elicit these effects and definitive causal relationships between endocrine disruption in crustacean field populations, and chemical pollution is generally lacking.[50] Also, temperature and photoperiod are the two most important environmental cues in the regulation of the annual cycles of circulating sex steroid hormones and reproduction in fish. Thus, these variables may alter the endocrine-disruption effects induced by EDCs.[82]

In contrast to mammals and birds, the mechanisms underlying sex determination and differentiation in fishes vary widely and are changeable or labile in response to environmental parameters. These environmental parameters include temperature, behavioral cues and demographic structure of the local population, and EDCs. Understanding the gender similarities and differences in how organisms respond following exposure to environmental chemicals is important to determine the relative risk of these agents to wildlife and human populations. Given the central role of sex steroid hormones in the sex determination and sexual differentiation of fishes, amphibians, and reptiles, future research that includes sex as a factor is recommended. Thus, the risk assessment can address the probable gender differences in effects from exposure to chemicals in the environment.[83]

DEALING WITH ENVIRONMENTAL EDC EMISSION

Municipal wastewater contains a complex mixture of EDCs originating from different sources. A number of organic pollutants, such as polycyclic aromatic hydrocarbons, PCBs, and pesticides, are resistant to degradation and represent an ongoing toxicological threat to both wildlife and human beings. Furthermore, recently, wastewater sludge has been subjected to reuse for production of value-added products. These facts have heightened the need for novel and advanced bioremediation techniques to effectively remove EDCs from a variety of contaminated environmental media including water, wastewater sludge, sediments, and soils. One possibility to solve this problem is the use of microbial potential to degrade or detoxify EDCs and other toxic intermediates.[80,84]

Also, there are physical methods such as absorption by activated carbon and rejection by membranes to remove EDCs. However, pollutant removal from wastewater is a process with high energy consumption, where cost and efficiency are the key considerations for their application. Biodegradation processes have proven to be the most cost-effective.[85]

Water companies became aware of the endocrine-disruption problem when a survey confirmed the observation by anglers of hermaphrodite fish in wastewater treatment plant lagoons after being exposed to significant levels of persistent man-made chemicals. The evolving regulatory context related to micropollutants in the environment may have a decisive impact on wastewater management and requires an increased knowledge of the fate of micropollutants during wastewater treatment. Advanced treatments such as oxidation (ozone) are known to be able to enhance the removal of micropollutants, but technical, economic, and environmental risk/benefit evaluations must be performed before implementing such additional processes. In any case, the reduction of the pollution at the source, i.e., upstream of the wastewater treatment plant, represents the most sensible option, which should be promoted.[86]

Although numerous studies have investigated degradation of individual EDCs in laboratory or natural waters, chemical-based analytical methods cannot represent the combined or synergistic activities between water quality parameters and/or the EDC mixtures at environmentally relevant concentrations since natural variations in water matrices and mixtures of EDC in the environment may confound analysis of the treatment efficiency. In conjunction with standard analytical approaches, bioanalytical assessments of residual estrogenic activity in treated water will enable estimates of the interactions and/or combined estrogenic activity among mixtures of EDCs and the water matrix in natural water.[87]

By contrast, the agricultural sector, a significant user of veterinary pharmaceuticals, has no such treatment—compounds are deposited straight to the ground in dung and urine or washed from hides in the case of topical applications. There has been little research as to whether any of these compounds leach into and persist in local soil and aquatic ecosystems. The extent to which the active ingredients of any of these chemicals (and their metabolites) leach into pastures, soil, runoff, and groundwater is a matter for field research. Also, much spraying of pesticides as herbicides and insecticides is done by ground crews. In such circumstances, it is not known whether they react with each other as well as pesticides and herbicides, forming further compounds which, either acting individually or in combination, could adversely affect bacteria, fungi, and higher organisms.[88]

Above all, a ranking system that could be customized for specific geographical locations will aid public policies in prioritizing EDCs that need monitoring and removal of aquatic sources as drinking water.[89] The establishment of simple but integrative screening assays for regulatory purposes is allowed by a strong correlation between xenoestrogen exposure and reproductive impairment. In fact, molecular screening assays could contain a battery of molecular targets allowing a more comprehensive approach in the identification of endocrine-disrupting compounds in fish and vertebrates in general.[90]

Different assays can be successfully employed as a battery of assays to screen environmental water samples for estrogenicity. The results obtained from this battery of assays should be interpreted as a first-tier screen for estrogenicity. Samples that test positive should be further investigated using second- and third-tier screens with routine sampling in order to monitor rivers for estrogenicity.[91] Complementarily, a fugacity-based model may be applied to simulate the distribution of EDCs in reservoirs of recycled wastewater,[92] or a fugacity-hydrodynamic model may be used for predicting the concentrations of the organic pollutants in surface water.[93]

CONCLUSION

Endocrine-disrupting chemicals can cause a wide range of reproductive damage and developmental, growth, immune, and behavior effects even in low doses and by different mechanisms of action. They encompass a variety of chemical classes, including hormones, plant constituents, pesticides, compounds used in the plastics industry and in consumer products, and other industrial by-products and pollutants. Some of them are widely dispersed in the environment. Exposure to EDCs can occur through direct contact with these chemicals or through ingestion of contaminated water, sediment, air, soil, and food and consumer products.

In humans, it is difficult to definitively link a particular EDC with a specific effect because the studies have inconclusive results. However, fetuses and embryos, whose growth and development are highly controlled by the endocrine system, are more vulnerable to exposure and may suffer reproductive abnormalities. The timing of exposure is also presumed to be critical, since different hormone pathways are active during different stages of development. Perinatal exposure, in some cases, can lead to permanent alterations that may be overt in adulthood.

Compared with humans, the evidence that wildlife has been affected adversely by exposures to EDCs is extensive. Available evidence seems to indicate that endocrine disruption caused by xenobiotics is primarily an ecotoxicologic problem. These chemicals may be extremely challenging for aquatic organisms and mammals that have a large habitat and that consume fish from many different areas throughout their lives. Low concentrations of endocrine disrupters can have synergistic effects in various organisms as amphibians. For removal of these compounds from aquatic sources, the most cost-effective process is biodegradation.

In spite of the need to manage the environmental, human health, and economic impacts of EDCs, most attention is focused on pharmaceutically active chemicals instead of those for agricultural use. The impact of these latter compounds is understudied.

The legal approach has been improved by new test protocols. Progress has been made in the identification and quantification of a wide array of chemicals with endocrine-active properties, especially those that persist and bioaccumulate in organisms and their environment. Studies with mammals have shown that exposure to endocrine-active compounds during early development may result in adverse health impacts that are not realized until adulthood. However, from a regulatory perspective, the ability of animals to recover from chemical insults is problematic because it complicates efforts to establish acceptable levels of exposure. Consequently, research to define the limits and biological cost recovery using standardized test designs is needed.[94]

However, exposure complexities, including transient and low-concentration exposure to EDCs, maternal metabolism of bioaccumulated EDCs, varying vulnerability and response by developmental stage, poorly understood exposure sources, mixtures and synergies, and cultural, social, and economic patterns, make it difficult for science to make solid exposure determinations. While there has been a great deal of research and effort in context with the hazard assessment and regulation of EDCs, there are also remaining uncertainties and issues. These include animal rights concerns due to significant increases in the use of animals to fulfill testing requirements; associated needs for alternative testing concepts such as *in vitro, in silico*, and modeling approaches; and lack of understanding of the relevance of exposure of humans and wildlife to EDCs.[95] Given the dynamic nature of the endocrine system, future efforts in the study of EDCs need more focus on the timing, frequency, and duration of exposure to these chemicals.

REFERENCES

1. Kretschmer, X.C.; Baldwin, W.S. CAR and PXR: Xenosensors of endocrine disrupters? Chem.-Biol. Interact. **2005**, *155*, 111–128.
2. Castro, V.; Mello, M.A.; Diniz, C.; Morita, L.; Zucchi, T.; Poli, P. Neurodevelopmental effects of perinatal fenarimol exposure on rats. Reprod. Toxicol. **2007**, *23*, 98–105.
3. Vinggaard, A.; Jacobsen, H.; Metzdorff, S.; Andersen, H.; Nellemann, C. Antiandrogenic effects in short-term in vivo

studies of the fungicide fenarimol. Toxicology **2005**, *207*, 21–34.
4. Waissmann, W. Health surveillance and endocrine disruptors. Cadernos de Saúde Pública **2002**, *18* (2), 511–517.
5. Hotchkiss, A.K.; Rider, C.V.; Blystone, C.R.; Wilson, V.S.; Hartig, P.C.; Ankley, G.T.; Foster, P.M.; Gray, C.L.; Gray, L.E. Fifteen years after ''wingspread''—Environmental endocrine disrupters and human and wildlife health: Where we are today and where we need to go. Toxicol. Sci. **2008**, *105* (2), 235–259.
6. Koopman, P. The delicate balance between male and female sex determining pathways: Potential for disruption of early steps in sexual development. Int. J. Androl. **2010**, *33*, 252–258.
7. Walker, D.M.; Gore, A.C. Transgenerational neuroendocrine disruption of reproduction. Nat. Rev. Endocrinol. **2011,** *7* (4), 197–207.
8. Bulzomi, P.; Marino, M. Environmental endocrine disruptors: Does a sex-related susceptibility exist? Front. Biosci. **2011,** *16* (7), 2478–2498.
9. Andersen, M.; Dennison, J. Mechanistic approaches for mixture risk assessments—Present capabilities with simple mixtures and future directions. Environ. Toxicol. Pharmacol. **2004**, *16*, 1–11.
10. Porcher, J.; Devillers, J.; Marchand-Geneste, N. Mechanism of endocrine disruptions—A tentative overview. In *Endocrine Disruption Modeling*; Devillers, J., Ed.; CRC Press: Boca Raton, 2009; 11–46.
11. Flynn, K. Dietary exposure to endocrine-active pesticides: Conflicting opinions in a European workshop. Environ. Int. **2011**, *37* (5), 980–990.
12. Silva, E.; Kabil, A.; Kortenkamp, A. Cross-talk between non-genomic and genomic signaling pathways—Distinct effect profiles of environmental estrogens. Toxicol. Appl. Pharmacol. **2010**, *245*, 160–170.
13. Morgado, I.; Campinho, M.A.; Costa, R.; Jacinto, R.; Power, D.M. Disruption of the thyroid system by diethylstilbestrol and ioxynil in the sea bream (*Sparus aurata*). Aquat. Toxicol. **2009**, *92*, 271–280.
14. Trosken, E.R.; Adamskab, M.; Arand, M.; Zarn, J.A.; Patten, C.; Volkel, W.; Lutz, W.K. Comparison of lanosterol-14α-demethylase (CYP51) of human and *Candida albicans* for inhibition by different antifungal azoles. Toxicology **2006**, *228*, 24–32.
15. Sanderson, J.T. The steroid hormone biosynthesis pathway as a target for endocrine-disrupting chemicals. Toxicol. Sci. **2006**, *94* (1), 3–21.
16. Belloni, V.; Dessì-Fulgheri, F.; Zaccaroni, M; . Di Consiglio, E.; De Angelis, G.; Testai, E.; Santochirico, M.; Alleva, E.; Santucci, D. Early exposure to low doses of atrazine affects behavior in juvenile and adult CD1 mice. Toxicology **2011,** *279*, 19–26.
17. Hirabayashi, Y.; Inoue, T. The low-dose issue and stochastic responses to endocrine disruptors. J. Appl. Toxicol. **2011**, *31*, 84–88.
18. Calabrese, E.J. Hormesis: A revolution in toxicology, risk assessment and medicine—Re-framing the dose–response relationship. EMBO Rep. **2004**, *5*, S37–S40.
19. Gore, A.C. Developmental programming and endocrine disruptor effects on reproductive neuroendocrine systems. Front. Neuroendocrinol. **2008,** *29*, 358–374.
20. Anway, M.D.; Skinner, M.K. Epigenetic transgenerational actions of endocrine disruptors. Endocrinology **2006**, *147* (6), S43–S49.
21. Crews, D.; McLachlan, J.A. Epigenetics, evolution, endocrine disruption, health, and disease. Endocrinology **2006**, *147* (6), S4–S10.
22. Li, A.A.; Baum, M.J.; McIntosh, L.J.; Day, M.; Liu, F.; Gray L.E., Jr. Building a scientific framework for studying hormonal effects on behavior and on the development of the sexually dimorphic nervous system. Neurotoxicology **2008**, *29*, 504–519.
23. Yiee, J.H.; Baskin, L.S. Environmental factors in genitourinary development. J. Urol. **2010**, *184*, 34–41.
24. Phillips, K.P.; Tanphaichitr, N. Human exposure to endocrine disrupters and semen quality. J. Toxicol. Environ. Health, Part B **2008**, *11* (3–4), 188–220.
25. Uzumcu, M.; Suzuki, H.; Skinner, M.K. Effect of the anti-androgenic endocrine disruptor vinclozolin on embryonic testis cord formation and postnatal testis development and function. Reprod. Toxicol. **2004**, *18* (6), 765–774.
26. Mnif, W.; Hassine, A.I.H.; Bouaziz, A.; Bartegi, A.; Thomas, O.; Roig, B. Effect of endocrine disruptor pesticides: A review. Int. J. Environ. Res. Pub. Health **2011,** *8* (6), 2265–2303.
27. Kjærstad, M.B.; Taxvig, C.; Nellemann, C.; Vinggaard, A.M.; Andersen, H.R. Endocrine disrupting effects in vitro of conazole antifungals used as pesticides and pharmaceuticals. Reprod. Toxicol. **2010,** *30*, 573–582.
28. Taxvig, C.; Hass, U.; Axelstad, M.; Dalgaard, M.; Boberg, J.; Andersen, H.R.; Vinggaard, A.M. Endocrine-disrupting activities in vivo of the fungicides tebuconazole and epoxiconazole. Toxicol. Sci. **2007**, *100* (2), 464–473.
29. Noriega, N.C.; Ostby, J.; Lambright, C.; Wilson, V.S.; Gray, L.E., Jr. Late gestational exposure to the fungicide prochloraz delays the onset of parturition and causes reproductive malformations in male but not female rat offspring. Biol. Reprod. **2005**, *72*, 1324–1335.
30. Goetz, A.K.; Ren, H.; Schmid, J.E.; Blystone, C.R.; Thillainadarajah, I.; Best, D.S.; Nichols, H.P.; Strader, L.F.; Wolf, D.C.; Narotsky, M.G.; Rockett, J.C.; Dix, D.J. Disruption of testosterone homeostasis as a mode of action for the reproductive toxicity of triazole fungicides in the male rat. Toxicol. Sci. **2007**, *95* (1), 227–239.
31. Rockett, J.C.; Narotsky, M.G.; Thompsona, K.E.; Thillainadarajah, I.; Blystone, C.R.; Goetz, A.K.; Rena, H.; Best, D.S.; Murrell, R.N.; Nichols, H.P.; Schmid, J.E.; Wolf, D.C.; Dix, D.J. Effect of conazole fungicides on reproductive development in the female rat. Reprod. Toxicol. **2006**, *22*, 647–658.
32. Taxvig, C.; Vinggaard, A.M.; Hass, U.; Axelstad, M.; Metzdorff S.; Nellemann, C. Endocrine-disrupting properties in vivo of widely used azole fungicides. Int. J. Androl. **2008**, *31*, 170–177.
33. Grünfeld, H.; Bonefeld-Jorgensen, E. Effect of in vitro estrogenic pesticides on human oestrogen receptor α and β mRNA levels. Toxicol. Lett. **2004**, *151* (3), 467–480.
34. Andersen, H.; Vinggaard, A.; Rasmussen, T.; Gjermandsen, I.; Bonefeld-Jørgensen, E. Effects of currently used pesticides in assays for estrogenicity, androgenicity, and aromatase activity in vitro. Toxicol. Appl. Pharmacol. **2002**, *179* (1), 1–12.

35. Hirsch, K.; Weaver, D.; Black, L.; Falcone, J.; Maclusky, N. Inhibition of central nervous system aromatase activity, a mechanism for fenarimol-induced infertility in the male rats. Toxicol. Appl. Pharmacol. **1987**, *91*, 235–245.
36. Paolini, M.; Mesirca, R.; Pozzetti, L.; Sapone, A.; Cantelli-Forti, G. Molecular non-genetic biomarkers related to fenarimol cocarcinogenesis, organ- and sex-specific CYP induction in rat. Cancer Lett. **1986**, *101*, 171–178.
37. Saradha, B.; Mathur P.P. Effect of environmental contaminants on male reproduction. Environ. Toxicol. Pharmacol. **2006**, *21*, 34–41.
38. Uzumcu, M.; Zachow, R. Developmental exposure to environmental endocrine disruptors: Consequences within the ovary and on female reproductive function. Reprod. Toxicol. **2007**, *23*, 337–352.
39. Hond, E.; Schoeters, G. Endocrine disrupters and human puberty. Int. J. Androl. **2006**, *29*, 264–271.
40. Meng, X.; Liu, P.; Wang, H.; Zhao, X.; Xu, Z.; Chen G.; Xu, D. Gender-specific impairments on cognitive and behavioral development in mice exposed to fenvalerate during puberty. Toxicol. Lett. **2011**, *203* (3), 245–251.
41. Miodovnik, A.; Engel, S.M.; Zhu, C.; Ye, X.; Soorya, L.V.; Silva, M.J.; Calafat, A.M.; Wolff, M.S. Endocrine disruptors and childhood social impairment. Neurotoxicology **2011**, *32* (2), 261–267.
42. Migliarini, B.; Piccinetti, C.C.; Martella, A.; Maradonna, F.; Gioacchini, G.; Carnevali, O. Perspectives on endocrine disruptor effects on metabolic sensors. Gen. Comp. Endocrinol. **2011**, *170*, 416–423.
43. Wingfield, J.C.; Mukai, M. Endocrine disruption in the context of life cycles: Perception and transduction of environmental cues. Gen. Comp. Endocrinol. **2009**, *163*, 92–96.
44. Beronius, A.; Rudén, C.; Hanberg, A.; Håkansson, H. Health risk assessment procedures for endocrine disrupting compounds within different regulatory frameworks in the European Union. Regul. Toxicol. Pharmacol. **2009**, *55*, 111–122.
45. Gori, G. Regulating endocrine disruptors. Regul. Toxicol. Pharmacol. **2007**, *48*, 1–3.
46. Campbell, B.K.; Souza, C., Gong, J.; Webb, R., Kendall, N.; Marsters, P., Robinson, G.; Mitchell, A.; Telfer, E.E.; Baird, D.T. Domestic ruminants as models for the elucidation of the mechanisms controlling ovarian follicle development in humans. Reprod. Suppl. **2003**, *61*, 429–43.
47. Wilson, V.S.; Blystone, C.R.; Hotchkiss, A.K.; Rider, C.V.; Gray, L.E., Jr. Diverse mechanisms of anti-androgen action: impact on male rat reproductive tract development. Int. J. Androl. **2008**, *31*, 178–187.
48. Porte, C.; Janer, G.; Lorusso, L.C.; Ortiz-Zarragoitia, M.; Cajaraville, M.P.; Fossi, M.C.; Canesi, L. Endocrine disruptors in marine organisms: Approaches and perspectives. Comp. Biochem. Physiol., Part C **2006**, *143*, 303–315.
49. Lafont, R.; Mathieu, M. Steroids in aquatic invertebrates. Ecotoxicology **2007**, *16*, 109–130.
50. LeBlanc, G.A. Crustacean endocrine toxicology: A review. Ecotoxicology **2007**, *16*, 61–81.
51. Kloas, W.; Lutz, I. Amphibians as model to study endocrine disrupters. J. Chromatogr. A **2006**, *1130*, 16–27.
52. Morales, M.; Planelló, R.; Martínez-Paz, P.; Herrero, O.; Cortés, E.; Martínez-Guitarte, J.L.; Morcillo, G. Characterization of Hsp70 gene in *Chironomus riparius*: Expression in response to endocrine disrupting pollutants as a marker of ecotoxicological stress. Comp. Biochem. Physiol. **2011**, *153*, 150–158.
53. Calamandrei , G.; Maranghi, F.; Venerosi, A.; Alleva, E.; Mantovani, A Efficient testing strategies for evaluation of xenobiotics with neuroendocrine activity. Reprod. Toxicol. **2006**, *22*, 164–174.
54. Danga, Z.C.; Ru, S.; Wang, W.; Rorije, E.; Hakkert, B.; Vermeire, T. Comparison of chemical-induced transcriptional activation of fish and human estrogen receptors: Regulatory implications. Toxicol. Lett. 2011, doi:10.1016/j.toxlet.2010.12.020.
55. OECO Conceptual Framework for the testing and Assessment of Endocrine Disrupting Chemicals, available at http://www.oecd.org/document/580.3343, en_2649_34377_2348794_1_1_1_1.00.html (accessed May 2011).
56. Juberg, D.R. An evaluation of endocrine modulators: Implications for human health. Ecotoxicol. Environ. Safety **2000**, *45*, 93–105.
57. Borgert, C.J.; Mihaich, E.M.; Quill, T.F.; Marty, M.S.; Levine, S.L.; Becker, R.A. Evaluation of EPA's Tier 1 Endocrine Screening Battery and recommendations for improving the interpretation of screening results. Regul. Toxicol. Pharmacol. **2011**, *59*, 397–411.
58. Vogel, J.M. Perils of paradigm: Complexity, policy design, and the Endocrine Disruptor Screening Program. *Environ. Health: Global Access Sci. Source* **2005**, *4*, 2, doi:10.1186/1476-069X-4-2.
59. Nichols, J.W.; Breen, M.; Denver, R.J.; Distefano, J.J.; Edwards, J.S.; Hoke, R.A.; Volz, D.C.; Zhang, X. Predicting chemical impacts on vertebrate endocrine systems. Environ. Toxicol. Chem. **2011**, *30* (1), 39–51.
60. McKinlay, R.; Plant, J.A.; Bell, J.N.B.; Voulvoulis, N. Calculating human exposure to endocrine disrupting pesticides via agricultural and non-agricultural exposure routes. Sci. Total Environ. **2008**, *398*, 1–12.
61. Kortenkamp A.; Faust M. Combined exposures to anti-androgenic chemicals: Steps towards cumulative risk assessment. Int. J. Androl. **2010**, *33*, 463–474.
62. Muncke, J. Exposure to endocrine disrupting compounds via the food chain: Is packaging a relevant source? Sci. Total Environ. **2009**, *407*, 4549–4559.
63. Narita, M.; Miyagawa, K.; Mizuo, K.; Yoshida, T.; Suzuki, T. Prenatal and neonatal exposure to low-dose of bisphenol-A enhance the morphine-induced hyperlocomotion and rewarding effect. Neurosci. Lett. **2006**, *402*, 249–252.
64. Queiroz, E.K.R.; Waissmann, W. Occupational exposure and effects on the male reproductive system. Cadernos de Saúde Pública **2006**, *22* (3), 485–493.
65. Bretveld, R.W.; Hooiveld, M.; Zielhuis, G.A.; Pellegrino, A.; van Rooij, I.A.L.M.; Roeleveld, N. Reproductive disorders among male and female greenhouse workers. Reprod. Toxicol. **2008**, *25*, 107–114.
66. Gregoraszczuk, E.L.; Milczarek, K.; Wojtowicz, A.K.; Berg, V.; Skaare, J.U.; Ropstad, E. Steroid secretion following exposure of ovarian follicular cells to three different natural mixtures of persistent organic pollutants (POPs). Reprod. Toxicol. **2008**, *25*, 58–66.
67. Kortenkamp A. Low dose mixture effects of endocrine disrupters: implications for risk assessment and epidemiology. Int. J. Androl. **2008**, *31*, 233–240.

68. Jacobsen, P.R.; Christiansen, S.; Boberg, J.; Nellemann C.; Hass, U. Combined exposure to endocrine disrupting pesticides impairs parturition, causes pup mortality and affects sexual differentiation in rats. Int. J. Androl. **2010**, *33*, 434–442.
69. Rider, C.V.; Furr, J.R.; Wilson, V.S.; Gray, L.E., Jr. Cumulative effects of in utero administration of mixtures of reproductive toxicants that disrupt common target tissues via diverse mechanisms of toxicity. Int. J. Androl. **2010**, *33*, 443–462.
70. Birkhoj, M.; Nellemann, C.; Jarfelt, K.; Jacobsen, H.; Andersen, H.R.; Dalgaard, M; Vinggaard, A.M. The combined antiandrogenic effects of five commonly used pesticides. Toxicol. Appl. Pharmacol. **2004**, *201*, 10–20.
71. Drake, A.J.; van den Driesche, S.; Scott, H.M., Hutchison, G.R.; Seckl, J.R.; Sharpe, R.M. Glucocorticoids amplify dibutyl phthalate-induced disruption of testosterone production and male reproductive development. Endocrinology **2009**, *150*, 5055–5064.
72. Chen, X.; An, H.; Ao, L.; Sun, L.; Liu, W.; Zhou, Z.; Wanga, Y.; Cao, J. The combined toxicity of dibutyl phthalate and benzo(a)pyrene on the reproductive system of male Sprague Dawley rats in vivo. J. *Hazard.* Mater. **2011**, *186* (1), 835–841.
73. Pohl, H.R.; Luukinen, B.; Holler, J.S. Health effects classification and its role in the derivation of minimal risk levels: Reproductive and endocrine effects. Regul. Toxicol. Pharmacol. **2005**, *42*, 209–217.
74. Erhard, H.W.; Rhind, S.M. Prenatal and postnatal exposure to environmental pollutants in sewage sludge alters emotional reactivity and exploratory behaviour in sheep. Sci. Total Environ. **2004**, *332*, 101–108.
75. Mills, L.J.; Chichester, C. Review of evidence: Are endocrine-disrupting chemicals in the aquatic environment impacting fish populations? Sci. Total Environ. **2005**, *343*, 1–34.
76. Fox, G.A. Wildlife as sentinels of human health effects in the Great Lakes–St. Lawrence Basin. Environ. Health Perspect. **2001**, *109* (Suppl. 6), 853–861.
77. Kinnberg, K.; Holbech, H.; Petersen, G.I.; Bjerregaard, P. Effects of the fungicide prochloraz on the sexual development of zebrafish (*Danio rerio*). Comp. Biochem. Physiol., Part C **2007**, *145*, 165–170.
78. Monod, G.; Rime, H.; Bobe, J.; Jalabert, B. Agonistic effect of imidazole and triazole fungicides on in vitro oocyte maturation in rainbow trout (*Oncorhynchus mykiss*). Mar. Environ. Res. **2004**, *58*, 143–146.
79. Jobling, S.; Burn, R.W.; Thorpe, K.; Williams, R.; Tyler, C. Statistical modeling suggests that antiandrogens in effluents from wastewater treatment works contribute to widespread sexual disruption in fish living in English rivers. Environ. Health Perspect. **2009**, *117* (5), 797–802.
80. Barnabé, S.; Brar, S.K.; Tyagi, R.D.; Beauchesne, I.; Surampalli, R.Y. Pre-treatment and bioconversion of wastewater sludge to value-added products—Fate of endocrine disrupting compounds. Sci. Total Environ. **2009**, *407*, 1471–1488.
81. Nelson, J.; Bishay, F.; van Roodselaar, A.; Ikonomou, M.; Law, F.C.P. The use of in vitro bioassays to quantify endocrine disrupting chemicals in municipal wastewater treatment plant effluents. Sci. Total Environ. **2007**, *374*, 80–90
82. Jin, Y.; Shu, L.; Huang, F.; Cao, L.; Sun, L.; Fu, Z. Environmental cues influence EDC-mediated endocrine disruption effects in different developmental stages of Japanese medaka (*Oryzias latipes*). Aquat. Toxicol. **2011**, *101*, 254–260.
83. Orlando, E.F.; Guillette, L.J., Jr. Sexual dimorphic responses in wildlife exposed to endocrine disrupting chemicals. Environ. Res. **2007**, *104,* 163–173.
84. Robinson, B.J.; Hellou, J. Biodegradation of endocrine disrupting compounds in harbour seawater and sediments. Sci. Total Environ. **2009**, *407*, 5713–5718.
85. Liu, Z.; Kanjo, Y.; Mizutani, S. Removal mechanisms for endocrine disrupting compounds (EDCs) in wastewater treatment—Physical means, biodegradation, and chemical advanced oxidation: A review. Sci. Total Environ. **2009**, *407*, 731–748.
86. Janex-Habibi, M.; Huyard, A.; Esperanza, M.; Bruchet, A. Reduction of endocrine disruptor emissions in the environment: The benefit of wastewater treatment. Water Res. **2009,** *43*, 1565–1576.
87. Chen, P.; Rosenfeldt, E.J.; Kullman, S.W.; Hinton, D.E.; Linden, K.G. Biological assessments of a mixture of endocrine disruptors at environmentally relevant concentrations in water following UV/H2O2 oxidation. Sci. Total Environ. **2007**, *376*, 18–26
88. Fisher, P.M.J.; Scott, R. Evaluating and controlling pharmaceutical emissions from dairy farms: A critical first step in developing a preventative management approach. J. Cleaner Prod. **2008**, *16*, 1437–1446.
89. Kumar, A.; Xagoraraki, I. Pharmaceuticals, personal care products and endocrine-disrupting chemicals in U.S. surface and finished drinking waters: A proposed ranking system. Sci. Total Environ. 2010, *in press*
90. Scholz, S.; Mayer, I. Molecular biomarkers of endocrine disruption in small model fish. Mol. Cell. Endocrinol. **2008**, *293*, 57–70.
91. Swart, J.C.; Pool, E.J.; van Wykb, J.H. The implementation of a battery of in vivo and in vitro bioassays to assess river water for estrogenic endocrine disrupting chemicals. Ecotoxicol. Environ. Saf. **2011**, *74*, 138–143.
92. Cao, Q.; Yu, Q.; Connell, D.W. Fate simulation and risk assessment of endocrine disrupting chemicals in a reservoir receiving recycled wastewater. Sci. Total Environ. **2010**, *408*, 6243–6250.
93. Zhang, Y.; Song, X.; Kondoh, A.; Xia, J.; Tang, C. Behavior, mass inventories and modeling evaluation of xenobiotic endocrine-disrupting chemicals along an urban receiving wastewater river in Henan Province, China. Water Res. **2011**, *45*, 292 – 302.
94. Nichols, J.W.; Breen, M.; Denver, R.J.; Distefano, J.J., III; Edwards, J.S.; Hoke, R.A.; Volz, D.C.; Zhang X. Predicting impacts on vertebrate endocrine systems. Environ. Toxicol. Chem. **2011**, *30* (1), 39–51.
95. Hecker, M.; Hollert, H. Endocrine disruptor screening: Regulatory perspectives and needs. Environ. Sci. Eur. **2011**, *23*: 15, doi:10.1186/2190-4715-23-15.

Energy Commissioning: Existing Buildings

David E. Claridge
Department of Mechanical Engineering, Energy Systems Laboratory, College Station, Texas, U.S.A.
Mingsheng Liu
Architectural Engineering Program, Peter Kiewit Institute, University of Nebraska—Lincoln, Omaha, Nebraska, U.S.A.
W.D. Turner
Department of Mechanical Engineering, Energy Systems Laboratory, College Station, Texas, U.S.A.

Abstract
Commissioning an existing building is referred to by various terms, including recommissioning, retro-commissioning, and continuous commissioning® (CC®). A comprehensive study of 182 existing buildings totaling over 22,000,000 ft^2 in floor area reported average energy savings of 18% at an average cost of 0.41 dollar/ft^2 after they were commissioned, producing an average simple payback of 2.1 years. The commissioning process for an existing building involves steps that should include building screening, a commissioning assessment to estimate savings potential and cost, plan development and team formation, development of performance baselines, detailed measurements and commissioning measure development, implementation, and follow-up to maintain persistence. Existing building commissioning has been successfully used in energy management programs as a standalone measure, as a follow-up to the retrofit process, as a rapid payback Energy Conservation Measure (ECM) in a retrofit program, and as a means to ensure that a building meets or exceeds its energy performance goals. Very often, it is the most cost-effective single energy management option available in a large building.

INTRODUCTION

Commissioning an existing building has been shown to be a key energy management activity over the last decade, often resulting in energy savings of 10, 20 or sometimes 30% without significant capital investment. It generally provides an energy payback of less than three years. In addition, building comfort is improved, systems operate better, and maintenance cost is reduced. Commissioning measures typically require no capital investment, though the process often identifies maintenance that is required before the commissioning can be completed. Potential capital upgrades or retrofits are often identified during the commissioning activities, and knowledge gained during the process permits more accurate quantification of benefits than is possible with a typical audit. Involvement of facilities personnel in the process can also lead to improved staff technical skills.

This entry is intended to provide the reader with an overview of the costs, benefits, and process of commissioning an existing building. There is no single definition of commissioning for an existing building, so several widely used commissioning definitions are given. A short case study illustrates the changes made when an existing building is commissioned, along with its impact. This is followed by a short summary of published information on the range of costs and benefits. The major portion of the this entry describes the commissioning process used by the authors in existing buildings so the reader can determine whether and how to implement a commissioning program. Monitoring and verification (M&V) may be very important to a successful commissioning program. Some commissioning-specific M&V issues are discussed, particularly the role of M&V in identifying the need for follow-up commissioning activities.

COMMISSIONING DEFINITIONS

The commissioning of a navy ship is the order or process that makes it completely ready for active duty. Over the last two decades, the term has come to refer to the process that makes a building or some of its systems completely ready for use. In the case of existing buildings, it generally refers to a restoration or improvement in the operation or function of the building systems. A widely used short definition of new building commissioning is the process of ensuring systems are designed, installed, functionally tested, and operated in conformance with the design intent. Commissioning begins with planning

Encyclopedia of Environmental Management DOI: 10.1081/E-EEM-120042061

and includes design, construction, start-up, acceptance, and training and can be applied throughout the life of the building. Furthermore, the commissioning process encompasses and coordinates the traditionally separate functions of systems documentation, equipment start-up, control system calibration, testing and balancing, and performance testing.[1]

Recommissioning

Recommissioning refers to commissioning a building that has already been commissioned at least once. After a building has been commissioned during the construction process, recommissioning ensures that the building continues to operate effectively and efficiently. Buildings, even if perfectly commissioned, will normally drift away from optimum performance over time, due to system degradation, usage changes, or failure to correctly diagnose the root cause of comfort complaints. Therefore, recommissioning normally reapplies the original commissioning procedures in order to keep the building operating according to design intent, or it may modify them for current operating needs.

Optimally, recommissioning becomes part of a facility's continuing operations and maintenance (O&M) program. There is not a consensus on recommissioning frequency, but some consider that it should occur every 3–5 years. If there are frequent build-outs or changes in building use, recommissioning may need to be repeated more often.[2]

Retrocommissioning

Retrocommissioning is the first-time commissioning of an existing building. Many of the steps in the retrocommissioning process are similar to those for commissioning. Retrocommissioning, however, occurs after construction, as an independent process, and its focus is usually on energy-using equipment such as mechanical equipment and related controls. Retrocommissioning may or may not bring the building back to its original design intent, since the usage may have changed or the original design documentation may no longer exist.[2]

Continuous Commissioning

Continuous Commissioning (CC®) (Continuous Commissioning® and CC® are registered trademarks of the Texas Engineering Expirement Station (TEES). Contact TEES for further information) is an ongoing process to resolve operating problems, improve comfort, optimize energy use, and identify retrofits for existing commercial and institutional buildings and central plant facilities. Continuous commissioning focuses on improving overall system control and operations for the building as it is currently utilized, and on meeting existing facility needs. Continuous commissioning is much more than an O&M program. It is not intended to ensure that a building's systems function as originally designed, but it ensures that the building and its systems operate optimally to meet the current uses of the building. As part of the CC process, a comprehensive engineering evaluation is conducted for both building functionality and system functions. Optimal operational parameters and schedules are developed based on actual building conditions and current occupancy requirements.

COMMISSIONING CASE STUDY—KLEBERG BUILDING

The Kleberg Building is a teaching/research facility on the Texas A&M campus consisting of classrooms, offices, and laboratories, with a total floor area of approximately 165,030 ft^2. A CC investigation was initiated in the summer of 1996 due to the extremely high level of simultaneous heating and cooling observed in the building.[4] Figs. 1 and 2 show daily heating and cooling consumption (expressed in average kBtu/h) as functions of daily average temperature. The pre-CC heating consumption data given in Fig. 1 show very little temperature dependence as indicated by the regression line derived from the data. Data values were typically between 5 and 6 MMBtu/h with occasional lower values. The cooling consumption is even higher (Fig. 2), though it shows more temperature dependence.

It was soon found that the preheat was operating continuously, heating the mixed air entering the cooling coil to approximately 105°F. The preheat was turned off, and heating and cooling consumption both dropped by about 2 MMBtu/h as shown by the middle clouds of data in Figs. 1 and 2. Subsequently, the building was thoroughly examined, and a comprehensive list of commissioning measures was developed and implemented. The principal measures implemented that led to reduced heating and cooling consumption were as follows:

- "Preheat to 105°F" was changed to "Preheat to 40°F."
- The cold deck schedule was changed from "55°F fixed" to "Vary from 62 to 57°F as ambient temperature varies from 40 to 60°F."
- The economizer was set to maintain mixed air at 57°F whenever the outside air was below 60°F.
- Static pressure control was reduced from 1.5 inH2O to 1.0 inH2O, and a nighttime set-back to 0.5 inH2O was implemented.
- A number of broken variable air volume terminal (VFD) boxes were replaced or repaired.
- Chilled water pump variable frequency drives (VFDs) were turned on.

These changes further reduced chilled water and heating hot water use as shown in Figs. 1 and 2 for a total annualized reduction of 63% in chilled-water use and 84% in hot-water use.

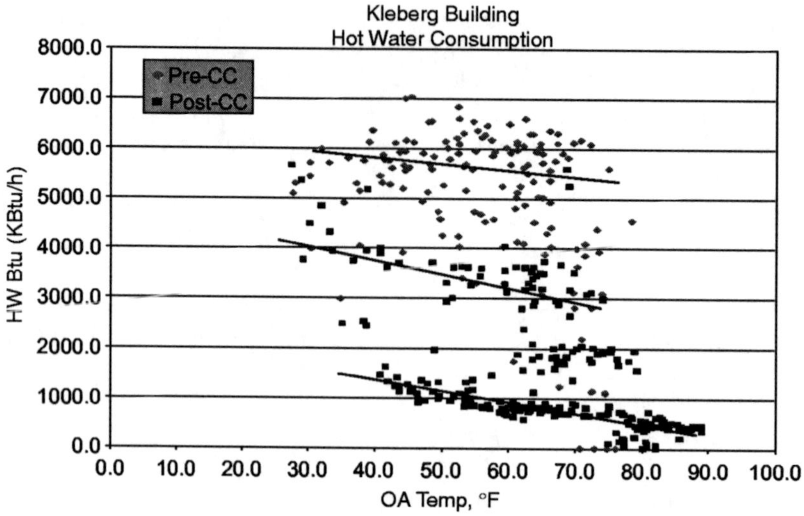
Kleberg Building
Hot Water Consumption
Pre-CC
Post-CC
HW Btu (KBtu/h)
8000.0
7000.0
6000.0
5000.0
4000.0
3000.0
2000.0
1000.0
0.0
0.0 10.0 20.0 30.0 40.0 50.0 60.0 70.0 80.0 90.0 100.0
OA Temp, °F

Fig. 1 Pre-CC and post-CC heating water consumption at the Kleberg building vs. daily average outdoor temperature. **Source:** Claridge et al.[3]

COSTS AND BENEFITS OF COMMISSIONING EXISTING BUILDINGS

The most comprehensive study of the costs and benefits of commissioning existing buildings was conducted by Mills et al.[4,5] This study examined the impact of commissioning 182 existing buildings with over 22,000,000 ft^2. The commissioning cost of these projects ranged from below $\$0.10/ft^2$–$\$3.86/ft^2$, but most were less than $\$0.50/ft^2$ with an average cost of $\$0.41/ft^2$. Savings ranged from essentially zero to 54% of total energy use, with an average of 18%. This range reflects not only differences among buildings in the potential for commissioning savings, but doubtless also includes differences in the level of commissioning applied and the skill of the commissioning providers. Simple payback times ranged from less than a month to over 20 years, with an average of 2.1 years. Fig. 3 illustrates the average payback as a function of building type and the precommissioning energy cost intensity. The sample sizes for office buildings and higher education are large enough that these averages for payback and energy savings may be representative, but the other sample sizes are so small that they may be significantly skewed by building specific and/or other factors.

Mills et al. concluded, "We find that commissioning is one of the most cost-effective means of improving energy efficiency in commercial buildings. While not a panacea, it can play a major and strategically important role

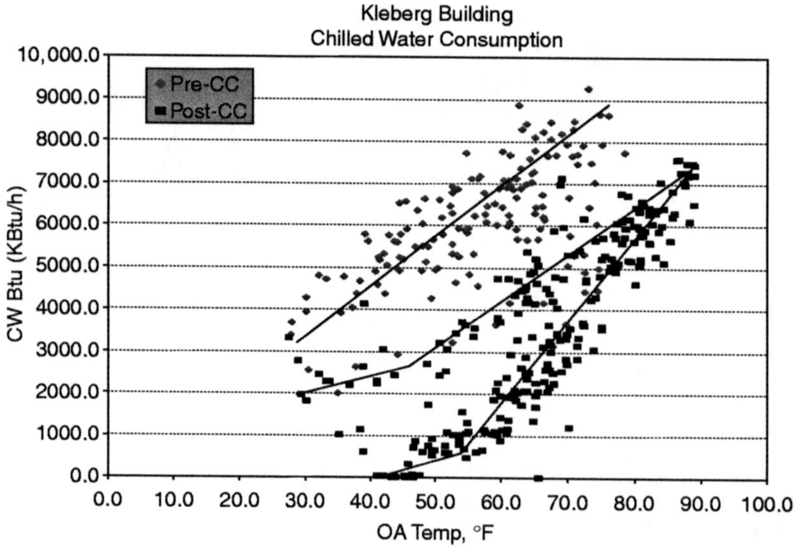
Kleberg Building
Chilled Water Consumption
Pre-CC
Post-CC
CW Btu (KBtu/h)
10,000.0
9000.0
8000.0
7000.0
6000.0
5000.0
4000.0
3000.0
2000.0
1000.0
0.0
0.0 10.0 20.0 30.0 40.0 50.0 60.0 70.0 80.0 90.0 100.0
OA Temp, °F

Fig. 2 Pre-CC and post-CC chilled water consumption at the Kleberg building vs daily average outdoor temperature. **Source:** Claridge et al.[3]

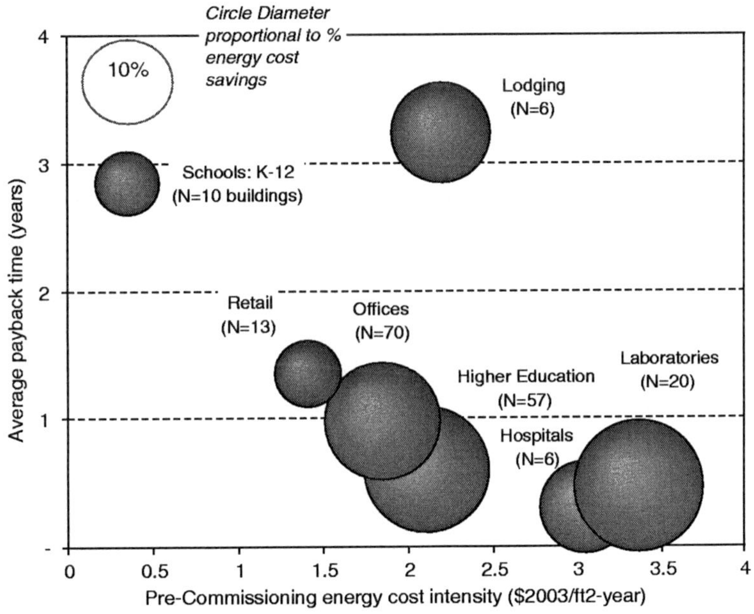
Circle Diameter proportional to % energy cost savings
10%
Lodging (N=6)
Schools: K-12 (N=10 buildings)
Retail (N=13)
Offices (N=70)
Higher Education (N=57)
Laboratories (N=20)
Hospitals (N=6)
Average payback time (years)
Pre-Commissioning energy cost intensity ($2003/ft2-year)
0 0.5 1 1.5 2 2.5 3 3.5 4
1 2 3 4

Fig. 3 Average simple payback time and percent energy savings from commissioning of existing buildings by building type. **Source:** Mills et al.[4]

in achieving national energy savings goals—with cost-effective savings potential of $18 billion per year or more in commercial buildings across the United States."

COMMISSIONING PROCESS IN EXISTING BUILDINGS

There are multiple terms that describe the commissioning process for existing buildings, as noted in the previous section. Likewise, there are many adaptations of the process itself. The same practitioner will implement the process differently in different buildings, based on the budget and the owner requirements. The process described here is the process used by the authors when the owner wants a thorough commissioning job. The terminology used will refer to the CC process, but many of the steps are the same for retrocommissioning or recommissioning. The model described assumes that a commissioning provider is involved, since that is normally the case. Some (or all) of the steps may be implemented by the facility staff if they have the expertise and adequate staffing levels to take on the work.

Continuous commissioning focuses on improving overall system control and operations for the building as it is currently utilized, and on meeting existing facility needs. It does not ensure that the systems function as originally designed, but ensures that the building and systems operate optimally to meet the current requirements. During the CC process, a comprehensive engineering evaluation is conducted for both building functionality and system functions. The optimal operational parameters and schedules are developed based on actual building conditions and current occupancy requirements. An integrated approach is used to implement these optimal schedules to ensure practical local and global system optimization and persistence of the improved operation schedules.

Commissioning Team

The CC team consists of a project manager, one or more CC engineers and CC technicians, and one or more designated members of the facility operating team. The primary responsibilities of the team members are shown in Table 1. The project manager can be an owner representative or a CC provider representative. It is essential that the engineers have the qualifications and experience to perform the work specified in the table. The designated facility team members generally include at least one lead heating, ventilating and air conditioning (HVAC) technician and an energy management control system (EMCS) operator or engineer. It is essential that the designated members of the facility operating team actively participate in the process and be convinced of the value of the measures proposed and implemented, or operation will rapidly revert to old practices.

Continuous Commissioning Process

The CC process consists of two phases. The first phase is the project development phase that identifies the buildings to be included in the project and develops the project scope. At the end of this phase, the CC scope is clearly defined and a CC contract is signed, as described in "Phase

Table 1 Commissioning team members and their primary responsibilities.

Team member(s)	Primary responsibilities
Project manager	1. Coordinate the activities of building personnel and the commissioning team
	2. Schedule project activities
Continuous commissioning (CC) engineer(s)	1. Develop metering and field measurement plans
	2. Develop improved operational and control schedules
	3. Work with building staff to develop mutually acceptable implementation plans
	4. Make necessary programming changes to the building automation system
	5. Supervise technicians implementing mechanical systems changes
	6. Project potential performance changes and energy savings
	7. Conduct an engineering analysis of the system changes
	8. Write the project report
Designated facility staff	1. Participate in the initial facility survey
	2. Provide information about problems with facility operation
	3. Suggest commissioning measures for evaluation
	4. Approve all CC measures before implementation
	5. Actively participate in the implementation process
CC Technicians	1. Conduct field measurements
	2. Implement mechanical, electrical, and control system program modifications and changes, under the direction of the project engineer

1: Project Development." The second phase implements CC and verifies project performance through the six steps outlined in Fig. 4 and described in "Phase 2: CC Implementation and Verification."

Phase 1: Project Development

Step 1: Identify Candidate Buildings. Buildings are screened to identify those that will receive a CC assessment. Buildings that provide poor thermal comfort, consume excessive energy, or have design features of the HVAC systems that are not fully used are typically good candidates for a CC assessment. Continuous commissioning can be effectively implemented in buildings that have received energy efficiency retrofits, in newer buildings, and in existing buildings that have not received energy efficiency upgrades. In other words, virtually any building can be a potential CC candidate. The CC provider should perform a preliminary analysis to check the feasibility of using the CC process on candidate facilities before performing a CC assessment.

The following information is needed for the preliminary assessment:

- Monthly utility bills for at least 12 months
- General building information—size, function, major equipment, and occupancy schedules
- O&M records, if available
- Description of any problems in the building, such as thermal comfort, indoor air quality, moisture, or mildew

An experienced engineer should review this information and determine the potential of the CC process to improve comfort and reduce energy cost. If the CC potential is good, a CC assessment should be performed.

Step 2: Perform CC Assessment and Develop Project Scope. The CC assessment involves a site visit by an experienced commissioning engineer who examines EMCS screens, conducts spot measurements throughout the building systems, and identifies major CC measures suitable for the building. The CC assessment report lists and describes the preliminary CC measures identified, the estimated energy savings from implementation, and the cost of carrying out the CC process on the building(s) evaluated in the assessment. Once a commissioning contract is signed, the process moves to Phase 2.

Phase 2: CC Implementation and Verification

Step 1: Develop CC Plan and Form the Project Team. The CC project manager and project engineer develop a detailed work plan for the project that includes major tasks, their sequence, time requirements, and technical requirements. The work plan is then presented to the building owner or representative(s) at a meeting attended by any additional CC engineers and technicians on the project team. Owner contact personnel and in-house technicians who will work on the project are identified.

Step 2: Develop Performance Baselines. This step should document all known comfort problems in individual rooms resulting from too much heating, cooling, noise,

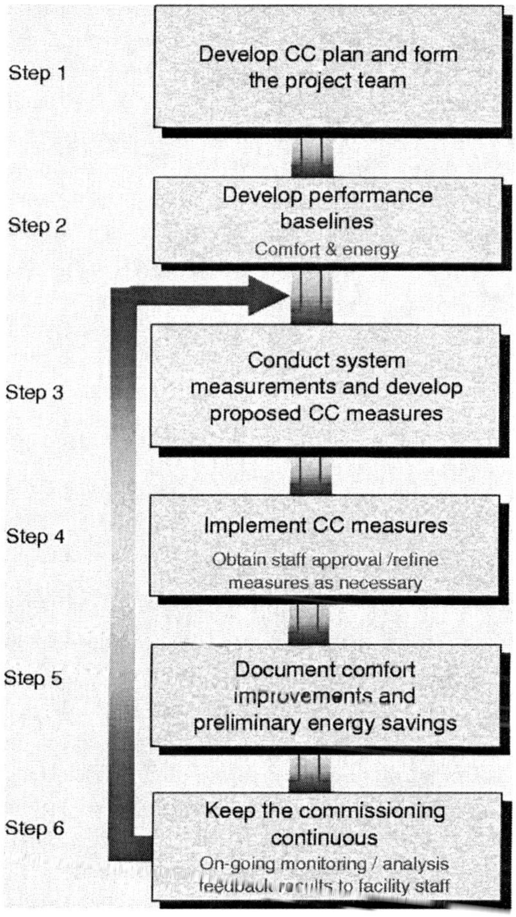
Step 1
Develop CC plan and form the project team
Step 2
Develop performance baselines
Comfort & energy
Step 3
Conduct system measurements and develop proposed CC measures
Step 4
Implement CC measures
Obtain staff approval /refine measures as necessary
Step 5
Document comfort improvements and preliminary energy savings
Step 6
Keep the commissioning continuous
On-going monitoring / analysis
feedback results to facility staff

Fig. 4 Outline of phase II of the CC process: implementation and verification.
Source: Energy Systems Laboratory.

humidity, or odors (especially from mold or mildew), or lack of outside air. Also, identify and document any HVAC system problems.

Baseline energy models of building performance are necessary to document the energy savings after commissioning. The baseline energy models can be developed using one or more of the following types of data:

- Short-term measured data obtained from data loggers or the EMCS system.
- Long-term hourly or 15-min whole building energy data, such as whole-building electricity, cooling, and heating consumption.
- Utility bills for electricity, gas, or chilled or hot water.

The baselines developed should be consistent with the International Performance Measurement and Verification Protocol,[6] with ASHRAE Guideline 14, or with both.[7]

Step 3: *Conduct System Measurements and Develop Proposed CC Measures*. The CC team uses EMCS trend data complemented by site measurements to identify current operational schedules and problems. The CC engineer conducts an engineering analysis to develop solutions for the existing problems; establishing improved operation and control schedules and set points for terminal boxes, air handling units (AHUs), exhaust systems, water and steam distribution systems, heat exchangers, chillers, boilers, and other components or systems as appropriate. Cost-effective energy retrofit measures can also be identified and documented during this step, if desired by the building owner.

Step 4: *Implement CC measures*. The CC project manager and/or project engineer presents the engineering solutions to existing problems and the improved operational and control schedules to the designated operating staff members and the building owner's representative to get "buy-in" and approval. Measures may be approved, modified, or rejected. A detailed implementation schedule is then developed by the CC engineer in consultation with the operating staff.

Continuous commissioning implementation normally starts by solving existing problems. Implementation of the improved operation and control schedules starts at the end of the comfort delivery system, such as at the terminal boxes, and ends with the central plant. The CC engineer closely supervises the implementation and refines the operational and control schedules as necessary. Following implementation, the new operation and control sequences are documented in a way that helps the building staff understand why they were implemented.

Step 5: *Document Comfort Improvements and Preliminary Energy Savings*. The comfort measurements taken in Step 2 (Phase 2) should be repeated at the same locations under comparable conditions and compared with the earlier measurements. The M&V procedures adopted in Step 2 should be used to determine the early post-CC energy performance and weather normalized to provide a preliminary evaluation of savings.

Step 6: *Keep the Commissioning Continuous*. The CC engineer should review the system operation after 6–12 months to identify any operating problems and make any adjustments needed. One year after CC implementation is complete, the CC engineer should write a project follow-up report that documents the first-year savings, recommendations or changes resulting from any consultation or site visits provided, and any recommendations to further improve building operations. Subsequently, the consumption should be tracked and compared with the first-year post-CC consumption during this period. Any significant and persistent increases in consumption should be investigated by the staff and/or CC engineer.

USES OF COMMISSIONING IN THE ENERGY MANAGEMENT PROCESS

Commissioning can be used as a part of the energy management program in several different ways:

- As a standalone measure. Commissioning is probably most often implemented in existing buildings because it is the most cost-effective step the owner can take to increase the energy efficiency of the building, generally offering a payback under three years, and often 1–2 years.
- As a follow-up to the retrofit process. Continuous commissioning has often been used to provide additional savings after a successful retrofit and has also been used numerous times to make an underperforming retrofit meet or exceed the original expectations.
- As an ECM in a retrofit program. The rapid payback that generally results from CC may be used to lower the payback of a package of measures to enable inclusion of a desired equipment replacement that has a longer payback in a retrofit package. This is illustrated by a case study in the next section. In this approach, the CC engineers conduct the CC audit in parallel with the retrofit audit conducted by the design engineering firm. Because the two approaches are different and look at different opportunities, it is very important to closely coordinate these two audits.
- To ensure that a new building meets or exceeds its energy performance goals. It may be used to significantly improve the efficiency of a new building by optimizing operation to meet its actual loads and uses instead of working to design assumptions.

CASE STUDY WITH CC AS AN ECM

Prairie View A&M University is a 1.7-million square foot campus, with most buildings served by a central thermal plant. Electricity is purchased from a local electric co-op.

University staff identified the need for major plant equipment replacements on campus. They wished to finance the upgrades through the Texas LoanSTAR program, which requires that the aggregate energy payback of all ECMs financed be ten years or less. Replacement of items such as chillers, cooling towers, and building automation systems typically have paybacks of considerably more than ten years. Hence, they can only be included in a loan if packaged with low payback measures that bring the aggregate payback below ten years.[8]

The university administration wanted to maximize the loan amount to get as much equipment replacement as possible. They also wanted to ensure that the retrofits worked properly after they are installed. To maximize their loan dollars, they chose to include CC as an ECM.

The LoanSTAR Program provides a brief walkthrough audit of the candidate buildings and plants. This audit is performed to determine whether there is sufficient retrofit potential to justify a more thorough investment grade audit.

The CC assessment is conducted in parallel with the retrofit audit conducted by the engineering design firm, when CC is to be included as an ECM. The two approaches look at different opportunities, but there can be some overlap, so it is very important to closely coordinate both audits. It is particularly important that the savings estimated by the audit team are not "double counted." The area of greatest overlap in this case was the building automation system. Considerable care was taken not to mix improved EMCS operation with operational improvements determined by the CC engineer, so both measures received proper credit.

The CC measures identified included the following:

- Hot and cold deck temperature resets
- Extensive EMCS programming to avoid simultaneous heating and cooling
- Air and water balancing
- Duct static pressure resets
- Sensor calibration and repair
- Improved start, stop, warm-up, and shutdown schedules

The CC engineers took the measurements required and collected adequate data on building operation during the CC assessment to perform a calibrated simulation on the major buildings. Available metered data and building EMCS data were also used. The CC energy savings were then written as an ECM and discussed with the design engineer. Any potential overlaps were removed. The combined ECMs were then listed and the total savings determined.

Table 2 summarizes the ECMs identified from the two audits:

The CC savings were calculated to be $204,563, as determined by conducting calibrated simulation of 16 campus buildings and by engineering calculations of savings from improved loop pumping. No CC savings were claimed for central plant optimization. Those savings were all applied to ECM #7, although it seems likely that additional CC savings will accrue from this measure. The simple payback from CC is slightly under three years, making it by far the most cost effective of the ECMs to be implemented. The CC savings represent nearly 30% of the total project savings.

Perhaps more importantly, CC accounted for two-thirds of the "surplus" savings dollars available to buy down the payback of the chillers and EMCS upgrade. Without CC as an ECM, the University would have had to delete one chiller and the EMCS upgrades, or some combination of chillers and a portion of the building EMCS upgrades from the project to meet the ten-year payback criteria—one chiller and the EMCS upgrades, or some combination of chillers and limited building EMCS upgrades. With CC, however, the university was able to include all these hardware items, and still meet the ten-year payback.

Table 2 Summary of energy cost measures (ECMs).

		Annual savings					
ECM #	**ECM**	**Electric kWh/yr**	**Electric demand kW/yr**	**Gas MCF/yr**	**Cost savings**	**Cost to implement**	**Simple payback**
#1	Lighting	1,565,342	5221	(820)	$94,669	$561,301	6.0
#2	Replace chiller #3	596,891	1250	-0-	$33,707	$668,549	19.8
#3	Repair steam system	-0-	-0-	13,251	$58,616	$422,693	7.2
#4	Install motion sensors	81,616	-0-	(44.6)	$3567	$26,087	7.3
#5	Add 2 bldgs. to CW loop	557,676	7050	-0-	$60,903	$508,565	8.4
#6	Add chiller #4	599,891	1250	-0-	$33,707	$668,549	19.8
#7	Primary/ secondary pumping	1,070,207	-0-	-0-	$49,230	$441,880	9.0
#8	Replace DX systems	38,237	233	-0-	$2923	$37,929	13.0
#9	Replace DDC/ EMCS	2,969,962	670	2736	$151,488	$2,071,932	13.7
#10	Continuous commissioning Assessment reports Metering M&V	2,129,855	-0-	25,318	$204,563	$ 605,000 $102,775 $157,700 $197,500	3.0
		9,606,677	15,674	40,440	$693,373	$6,470,460	9.3

SUMMARY

Commissioning of existing buildings is emerging as one of the most cost-effective ways for an energy manager to lower operating costs, and typically does so with no capital investment, or with a very minimal amount. It has been successfully implemented in several hundred buildings and provides typical paybacks of one to three years.

It is much more than the typical O&M program. It does not ensure that the systems function as originally designed, but focuses on improving overall system control and operations for the building as it is currently utilized and on meeting existing facility needs. During the CC process, a comprehensive engineering evaluation is conducted for both building functionality and system functions. The optimal operational parameters and schedules are developed based on actual building conditions. An integrated approach is used to implement these optimal schedules to ensure practical local and global system optimization and to ensure persistence of the improved operational schedules.

The approach presented in this entry begins by conducting a thorough examination of all problem areas or operating problems in the building, diagnoses these problems, and develops solutions that solve these problems while almost always reducing operating costs at the same time. Equipment upgrades or retrofits may be implemented as well, but have not been a factor in the case studies presented, except where the commissioning was used to finance equipment upgrades. This is in sharp contrast to the more usual approach to improving the efficiency of HVAC systems and cutting operating costs, which primarily emphasizes system upgrades or retrofits to improve efficiency.

Commissioning of new buildings is also an important option for the energy manager, offering an opportunity to help ensure that new buildings have the energy efficiency and operational features that are most needed.

ACKNOWLEDGMENTS

Two major sources of information on commissioning existing buildings are the *Continuous Commissioning^{SM} Guidebook: Maximizing Building Energy Efficiency and Comfort* (Liu, M., Claridge, D.E. and Turner, W.D., Federal Energy Management Program, U.S. Dept. of Energy, 144 pp., 2002) and *A Practical Guide for Commissioning Existing Buildings* (Haasl, T. and Sharp, T., Portland Energy Conservation, Inc. and Oak Ridge National Laboratory for U.S. DOE, ORNL/TM-1999/34, 69 pp.+App., 1999). Much of this entry has been abridged and adapted from the CC Guidebook.

The case studies in this entry have been largely abridged and adapted from Claridge et al.[3] and Turner.[8]

REFERENCES

1. ASHRAE. *ASHRAE Guideline 1–1996: The HVAC Commissioning Process*; American Society of Heating, Refrigerating and Air-Conditioning Engineers: Atlanta, GA, 1996.

2. U.S. Department of Energy. Building Commissioning: The Key to Quality Assurance, Washington, DC, 1999.
3. Claridge, D.E.; Turner, W.D.; Liu, M. et al. Is Commissioning Once Enough? Solutions for Energy Security and Facility Management Challenges: Proceedings of the 25th WEEC, Atlanta, GA, October 9–11, 2002, 29–36.
4. Mills, E.; Friedman, H.; Powell, T. et al. *The Cost-Effectiveness of Commercial-Buildings Commissioning: A Meta-Analysis of Energy and Non-Energy Impacts in Existing Buildings and New Construction in the United States*, Lawrence Berkeley National Laboratory Report No. 56637, 2004; 98. http://eetd.lbl.gov/EA/mills/emills.
5. Mills, E.; Bourassa, N.; Piette, M.A. et al. *The Cost-Effectiveness of Commissioning New and Existing Commercial Buildings: Lessons from 224 Buildings*. Proceedings of the 2006 National Conference on Building Commissioning, Portland Energy Conservation, Inc., New York, 2005. http://www.peci.org/ncbc/proceedings/2005/19_Piette_NCBC2005.pdf.
6. IPMVP Committee. International Performance Measurement and Verification Protocol: Concepts and Options for Determining Energy and Water Savings. U.S. Department of Energy, Vol. 1, DOE/GO-102001-1187, Washington, D.C., 2001.
7. ASHRAE Guideline 14-2002: Measurement of Energy and Demand Savings. American Society of Heating, Refrigerating and Air-Conditioning Engineers; Atlanta, GA, 2002.
8. Turner, W.D.; Claridge, D.E.; Deng, S.; Wei, G. *The Use of Continuous Commissioning*^{SM} *as an Energy Conservation Measure (ECM) for Energy Efficiency Retrofits*. Proceedings of 11th National Conference on Building Commisioning, Palm Springs, CA, CD, May 20–22, 2003.

Endocrine – Energy Conversion

Energy Commissioning: New Buildings

Janey Kaster
Yamas Controls West, San Francisco, California, U.S.A.

Abstract
Commissioning is the methodology for bringing to light design errors, equipment malfunctions, and improper control strategies at the most cost-effective time to implement corrective action. The primary goal of commissioning is to achieve optimal building systems performance. There are two types of commissioning: acceptance-based and process-based. Process-based commissioning is a comprehensive process that begins in the predesign phase and continues through postacceptance, while acceptance-based commissioning, which is perceived to be the cheaper method, basically examines whether an installation is compliant with the design and accordingly achieves more limited results. Commissioning originated in the early 1980s in response to a large increase in construction litigation. Commissioning was the result of owners seeking other means to gain assurance that they were receiving systems compliant with the design intent and with the performance characteristics and quality specified. Learn how commissioning has evolved and the major initiatives that are driving its growing acceptance. The general rule for including a system in the commissioning process is: the more complicated the system is, the more compelling is the need to include it in the commissioning process. Other criteria for determining which systems should be included are discussed. Discover the many benefits of commissioning, such as improved quality assurance, dispute avoidance, and contract compliance. Selection of the commissioning agent is key to the success of the commissioning plan. Learn what traits are necessary and what approaches to use for the selection process. The commissioning process occurs over a variety of clearly delineated phases. The phases of the commissioning process as defined below are discussed in detail: predesign, design, construction/installation, acceptance, and postacceptance. Extensive studies analyzing the cost/benefit of commissioning justify its application. One study defines the median commissioning cost for new construction as 1 dollar per square foot or 0.6% of the total construction cost. The median simple payback for new construction projects utilizing commissioning is 4.8 years. Understand how to achieve the benefits of commissioning, including optimization of building performance, reduction of facility life-cycle cost, and increased occupant satisfaction

INTRODUCTION

This entry provides an overview of commissioning—the processes one employs to optimize the performance characteristics of a new facility being constructed. Commissioning is important to achieve customer satisfaction, optimal performance of building systems, cost containment, and energy efficiency, and it should be understood by contractors and owners.

After providing an overview of commissioning and its history and prevalence, this entry discusses what systems should be part of the commissioning process, the benefits of commissioning, how commissioning is conducted, and the individuals and teams critical for successful commissioning. Then the entry provides a detailed discussion of each of the different phases of a successful commissioning process, followed by a discussion of the common mistakes to avoid and how one can measure the success of a commissioning effort, together with a cost–benefit analysis tool.

The purpose of this entry will be realized if its readers decide that successful commissioning is one of the most important aspects of construction projects and that commissioning should be managed carefully and deliberately throughout any project, from predesign to postacceptance. As an introduction to those unfamiliar with the process and as a refresher for those who are, the following section provides an overview of commissioning, how it developed, and its current prevalence today.

OVERVIEW OF COMMISSIONING

Commissioning Defined

Commissioning is the methodology for bringing to light design errors, equipment malfunctions, and improper control strategies at the most cost-effective time to implement corrective action. Commissioning facilitates a thorough understanding of a facility's intended use and ensures that the design meets the intent through coordination, communication, and cooperation of the design and installation team. Commissioning ensures that individual components function as a cohesive system. For these reasons, commissioning is best when it begins in the predesign phase of a construction project and can in one sense be viewed as the

Encyclopedia of Environmental Management DOI: 10.1081/E-EEM-120042217

most important form of quality assurance for construction projects.

Unfortunately, there are many misconceptions associated with commissioning, and perhaps for this reason, commissioning has been executed with varying degrees of success, depending on the level of understanding of what constitutes a "commissioned" project. American Society of Heating, Refrigerating and Air-Conditioning Engineers (AHSRAE) guidelines define commissioning as: the process of ensuring that systems are designed, installed, functionally tested, and capable of being operated and maintained to perform conformity with the design intent . . . [which] begins with planning and includes design, construction, startup, acceptance, and training, and is applied throughout the life of the building.[4] However, for many contractors and owners, this definition is simplified into the process of system startup and checkout or completing punch-list items.

Of course, a system startup and checkout process carried out by a qualified contractor is one important aspect of commissioning. Likewise, construction inspection and the generation and completion of punch-list items by a construction manager are other important aspects of commissioning. However, it takes much more than these standard installation activities to have a truly "commissioned" system. Commissioning is a comprehensive and methodical approach to the design and implementation of a cohesive system that culminates in the successful turnover of the facility to maintenance staff trained in the optimal operation of those systems.

Without commissioning, a contractor starts up the equipment but doesn't look beyond the startup to system operation. Assessing system operation requires the contractor to think about how the equipment will be used under different conditions. As one easily comprehended example, commissioning requires the contractor to think about how the equipment will operate as the seasons change. Analysis of the equipment and building systems under different load conditions due to seasonal conditions at the time of system startup will almost certainly result in some adjustments to the installed equipment for all but the most benign climates. However, addressing this common requirement of varying load due to seasonal changes most likely will not occur without commissioning. Instead, the maintenance staff is simply handed a building with minimal training and left to figure out how to achieve optimal operation on their own. In this seasonal example, one can just imagine how pleased the maintenance staff would be with the contractor when a varying load leads to equipment or system failure—often under very hot or very cold conditions!

Thus, the primary goal of commissioning is to achieve optimal building systems performance. For heating, ventilation, and air-conditioning (HVAC) systems, optimal performance can be measured by thermal comfort, indoor air quality, and energy savings. Energy savings, however, can result simply from successful commissioning targeted at achieving thermal comfort and excellent indoor air quality. Proper commissioning will prevent HVAC system malfunction—such as simultaneous heating and cooling, and overheating or overcooling—and successful malfunction prevention translates directly into energy savings. Accordingly, energy savings rise with increasing comprehensiveness of the commissioning plan. Commissioning enhances energy performance (savings) by ensuring and maximizing the performance of specific energy efficiency measures and correcting problems causing excessive energy use.[3] Commissioning, then, is the most cost-effective means of improving energy efficiency in commercial buildings. In the next section, the two main types of commissioning in use today—acceptance-based and processed-based—are compared and contrasted.

Acceptance-Based vs. Process-Based Commissioning

Given the varied nature of construction projects, contractors, owners, buildings, and the needs of the diverse participants in any building projection, commissioning can of course take a variety of forms. Generally, however, there are two types of commissioning: acceptance-based and process-based. Process-based commissioning is a comprehensive process that begins in the predesign phase and continues through postacceptance, while acceptance-based commissioning, which is perceived to be the cheaper method, basically examines whether an installation is compliant with the design and accordingly achieves more limited results.

Acceptance-based commissioning is the most prevalent type due to budget constraints and the lack of hard cost/benefit data to justify the more extensive process-based commissioning. Acceptance-based commissioning does not involve the contractor in the design process but simply constitutes a process to ensure that the installation matches the design. In acceptance-based commissioning, confrontational relationships are more likely to develop between the commissioning agent and the contractor because the commissioning agent and the contractor, having been excluded from the design phase, have not "bought in" to the design and thus may be more likely to disagree in their interpretation of the design intent.

Because the acceptance-based commissioning process simply validates that the installation matches the design, installation issues are identified later in the cycle. Construction inspection and regular commissioning meetings do not occur until late in the construction/installation phase with acceptance-based commissioning. As a result, there is no early opportunity to spot errors and omissions in the design, when remedial measures are less costly to undertake and less likely to cause embarrassment to the designer and additional costs to the contractor. As most contractors will readily agree, addressing issues spotted in the design or submittal stages of construction is typically much less costly than addressing them after installation, when correction often means tearing out work completed and typically delays the completion date.

Acceptance-based commissioning is cheaper, however, at least on its face, being approximately 80% of the cost of process-based commissioning.[2] If only the initial cost of commissioning services is considered, many owners will conclude that this is the most cost-effective commissioning approach. However, this 20% cost differential does not take into account the cost of correcting defects after the fact that process-based commissioning could have identified and corrected at earlier stages of the project. One need encounter only a single, expensive-to-correct project to become a devotee of process-based commissioning.

Process-based commissioning involves the commissioning agent in the predesign through the construction, functional testing, and owner training. The main purpose is quality assurance—assurance that the design intent is properly defined and followed through in all phases of the facility life cycle. It includes ensuring that the budget matches the standards that have been set forth for the project so that last-minute "value engineering" does not undermine the design intent, that the products furnished and installed meet the performance requirements and expectation compliant with the design intent, and that the training and documentation provided to the facility staff equip them to maintain facility systems true to the design intent.

As the reader will no doubt already appreciate, the author believes that process-based commissioning is far more valuable to contractors and owners than acceptance-based commissioning. Accordingly, the remainder of this entry will focus on process-based commissioning, after a brief review of the history of commissioning from inception to date, which demonstrates that our current, actively evolving construction market demands contractors and contracting professionals intimately familiar with and expert in conducting process-based commissioning.

History of Commissioning

Commissioning originated in the early 1980s in response to a large increase in construction litigation. Owners were dissatisfied with the results of their construction projects and had recourse only to the courts and litigation to resolve disputes that could not be resolved by meeting directly with their contractors. While litigation attorneys no doubt found this satisfactory approach to resolving construction project issues, owners did not, and they actively began looking for other means to gain assurance that they were receiving systems compliant with the design intent and with the performance characteristics and quality specified. Commissioning was the result.

While commissioning enjoyed early favor and wide acceptance, the recession of the mid-1980s placed increasing market pressure on costs, and by the mid-to late 1980s it forced building professionals to reduce fees and streamline services. As a result, acceptance-based commissioning became the norm, and process-based commissioning became very rare. This situation exists in most markets today; however, the increasing cost of energy, the growing awareness of the global threat of climate change and the need to reduce CO_2 emissions as a result, and the legal and regulatory changes resulting from both are creating a completely new market in which process-based commissioning will become ever more important, as discussed in the following section.

Prevalence of Commissioning Today

There are varying degrees of market acceptance of commissioning from state to state. Commissioning is in wide use in California and Texas, for example, but it is much less widely used in many other states. The factors that impact the level of market acceptance depend upon

- The availability of commissioning service providers
- State codes and regulations
- Tax credits
- Strength of the state's economy[1]

State and federal policies with regard to commissioning are changing rapidly to increase the demand for commissioning. Also, technical assistance and funding are increasingly available for projects that can serve as demonstration projects for energy advocacy groups. The owner should investigate how each of these factors could benefit the decision to adopt commissioning in future construction projects.

Some of the major initiatives driving the growing market acceptance of commissioning are:

- Federal government's U.S. Energy Policy Act of 1992 and Executive Order 12902, mandating that federal agencies develop commissioning plans
- Portland Energy Conservation, Inc.; National Strategy for Building Commissioning; and their annual conferences
- ASHRAE HVAC Commissioning Guidelines (1989)
- Utilities establishing commissioning incentive programs
- Energy Star building program
- Leadership in Energy Environmental Design (LEED) certification for new construction
- Building codes
- State energy commission research programs

Currently, the LEED is having the largest impact in broadening the acceptance of commissioning. The Green Building Council is the sponsor of LEED and is focused on sustainable design—design and construction practices that significantly reduce or eliminate the cradle-to-grave negative impacts of buildings on the environment and building occupants. Leadership in energy efficient design encourages sustainable site planning, conservation of water and water efficiency, energy efficiency and renewable energy, conservation of materials and resources, and indoor environmental quality.

With this background on commissioning, the various components of the commissioning process can be explored, beginning with an evaluation of what building systems should be subject to the commissioning process.[5]

COMMISSIONING PROCESS

Systems to Include in the Commissioning Process

The general rule for including a system in the commissioning process is: the more complicated the system is the more compelling is the need to include it in the commissioning process. Systems that are required to integrate or interact with other systems should be included. Systems that require specialized trades working independently to create a cohesive system should be included, as well as systems that are critical to the operation of the building. Without a commissioning plan on the design and construction of these systems, installation deficiencies are likely to create improper interaction and operation of system components.

For example, in designing a lab, the doors should be included in the commissioning process because determining the amount of leakage through the doorways could prove critical to the ability to maintain critical room pressures to ensure proper containment of hazardous material. Another common example is an energy retrofit project. Such projects generally incorporate commissioning as part of the measurement and verification plan to ensure that energy savings result from the retrofit process.

For any project, the owner must be able to answer the question of why commissioning is important.

Why Commissioning?

A strong commissioning plan provides quality assurance, prevents disputes, and ensures contract compliance to deliver the intended system performance. Commissioning is especially important for HVAC systems that are present in virtually all buildings because commissioned HVAC systems are more energy efficient.

The infusion of electronics into almost every aspect of modern building systems creates increasingly complex systems requiring many specialty contractors. Commissioning ensures that these complex subsystems will interact as a cohesive system.

Commissioning identifies design or construction issues and, if done correctly, identifies them at the earliest stage in which they can be addressed most cost effectively. The number of deficiencies in new construction exceeds existing building retrofit by a factor of 3.[3] Common issues that can be identified by commissioning that might otherwise be overlooked in the construction and acceptance phase are: air distribution problems (these occur frequently in new buildings due to design capacities, change of space utilization, or improper installation), energy problems, and moisture problems.

Despite the advantages of commissioning, the current marketplace still exhibits many barriers to adopting commissioning in its most comprehensive and valuable forms.

Barriers to Commissioning

The general misperception that creates a barrier to the adoption of commissioning is that it adds extra, unjustified costs to a construction project. Until recently, this has been a difficult perception to combat because there are no energy-use baselines for assessing the efficiency of a new building. As the cost of energy continues to rise, however, it becomes increasingly less difficult to convince owners that commissioning is cost effective. Likewise, many owners and contractors do not appreciate that commissioning can reduce the number and cost of change orders through early problem identification. However, once the contractor and owner have a basis on which to compare the benefit of resolving a construction issue earlier as opposed to later, in the construction process, commissioning becomes easier to sell as a win–win proposal.

Finding qualified commissioning service providers can also be a barrier, especially in states where commissioning is not prevalent today. The references cited in this entry provide a variety of sources for identifying associations promulgating commissioning that can provide referrals to qualified commissioning agents.

For any owner adopting commissioning, it is critical to ensure acceptance of commissioning by all of the design construction team members. Enthusiastic acceptance of commissioning by the design team will have a very positive influence on the cost and success of your project. An objective of this entry is to provide a source of information to help gain such acceptance by design construction team members and the participants in the construction market.

Selecting the Commissioning Agent

Contracting an independent agent to act on behalf of the owner to perform the commissioning process is the best way to ensure successful commissioning. Most equipment vendors are not qualified and are likely to be biased against discovering design and installation problems—a critical function of the commissioning agent—with potentially costly remedies. Likewise, systems integrators have the background in control systems and data exchange required for commissioning but may not be strong in mechanical design, which is an important skill for the commissioning agent. Fortunately, most large mechanical consulting firms offer comprehensive commissioning services, although the desire to be competitive in the selection processes sometimes forces these firms to streamline their scope on commissioning.

Owners need to look closely at the commissioning scope being offered. An owner may want to solicit commissioning services independently from the selection of the architect/mechanical/electrical/plumbing design team or, minimally, to request specific details on the design team's approach to commissioning. If an owner chooses the same mechanical, electrical, and plumbing (MEP) firm for design and commissioning, the owner should ensure that there is physical separation between the designer and commissioner to ensure that objectivity is maintained in the design review stages. An owner should consider taking on the role of the commissioning agent directly, especially if qualified personnel exist in-house. This approach can be very cost effective. The largest obstacles to success with an in-house commissioning agent are the required qualifications and the need to dedicate a valuable resource to the commissioning effort. Many times, other priorities may interfere with the execution of the commissioning process by an in-house owner's agent.

There are three basic approaches to selecting the commissioning agent:

Negotiated—best approach for ensuring a true partnership
Selective bid list—preapproved list of bidders
Competitive—open bid list

Regardless of the approach, the owner should clearly define the responsibilities of the commissioning agent at the start of the selection process. Fixed-cost budgets should be provided by the commissioning agent to the owner for the predesign and design phases of the project, with not-to-exceed budgets submitted for the construction and acceptance phases. Firm service fees should be agreed upon as the design is finalized.

Skills of a Qualified Commissioning Agent

A commissioning agent needs to be a good communicator, both in writing and verbally. Writing skills are important because documentation is critical to the success of the commissioning plan. Likewise, oral communication skills are important because communicating issues uncovered in a factual and nonaccusatory manner is most likely to resolve those issues efficiently and effectively. The commissioning agent should have practical field experience in MEP controls design and startup to be able to identify potential issues early. The commissioning agent likewise needs a thorough understanding of how building structural design impacts building systems. The commissioning agent must be an effective facilitator and must be able to decrease the stress in stressful situations. In sum, the commissioning agent is the cornerstone of the commissioning team and the primary determinant of success in the commissioning process.

At least ten organizations offer certifications for commissioning agents. However, there currently is no industry standard for certifying a commissioning agent. Regardless of certification, the owner should carefully evaluate the individuals to be performing the work from the commissioning firm selected. Individual experience and reputation should be investigated. References for the lead commissioning agent are far more valuable than references for the executive members of a commissioning firm in evaluating potential commissioning agents. The commissioning agent selected will, however, only be one member of a commissioning team, and the membership of the commissioning team is critical to successful commissioning.

Commissioning Team

The commissioning team is composed of representatives from all members of the project delivery team: the commissioning agent, representatives of the owner's maintenance team, the architect, the MEP designer, the construction manager, and systems contractors. Each team member is responsible for a particular area of expertise, and one important function of the commissioning agent is to act as a facilitator of intrateam communication.

The maintenance team representatives bring to the commissioning team the knowledge of current operations, and they should be involved in the commissioning process at the earliest stage, defining the design intent in the predesign phase, as described below. Early involvement of maintenance team representatives ensures a smooth transition from construction to a fully operational facility, and aids in the acceptance and full use of the technologies and strategies that have been developed during the commissioning process. Involvement of the maintenance team representatives also shortens the building turnover transition period.

The other members of the commissioning team have defined and important functions. The architect leads the development of the design intent document (DID). The MEP designer's responsibilities are to develop the mechanical systems that support the design intent of the facility and comply with the owner's current operating standards. The MEP schematic design is the basis for the systems installed and is discussed further below. The construction manager ensures that the project installation meets the criteria defined in the specifications, the budget requirements, and the predefined schedule. The systems contractors' responsibilities are to furnish and install a fully functional system that meets the design specifications. There are generally several contractors whose work must be coordinated to ensure that the end product is a cohesive system.

Once the commissioning team is in place, commissioning can take place, and it occurs in defined and delineated phases—the subject of the following section.

COMMISSIONING PHASES

The commissioning process occurs over a variety of clearly delineated phases. The commission plan is the set of documents and events that defines the commissioning process over all phases. The commissioning plan needs to reflect a systematic, proactive approach that facilitates communication and cooperation of the entire design and construction team.

The phases of the commissioning process are:

Predesign
Design
Construction/installation
Acceptance
Postacceptance

These phases and the commissioning activities associated with them are described in the following sections.

Predesign Phase

The predesign phase is the phase in which the design intent is established in the form of the DID. In this phase of a construction project, the role of commissioning in the project is established if process-based commissioning is followed. Initiation of the commissioning process in the predesign phase increases acceptance of the commissioning process by all design team members. Predesign discussions about commissioning allow all team members involved in the project to assess and accept the importance of commissioning to a successful project. In addition, these discussions give team members more time to assimilate the impact of commissioning on their individual roles and responsibilities in the project. A successful project is more likely to result when the predesign phase is built around the concept of commissioning instead of commissioning's being imposed on a project after it has been designed.

Once an owner has decided to adopt commissioning as an integral part of the design and construction of a project, the owner should be urged to follow the LEED certification process, as discussed above. The commissioning agent can assist in the documentation preparation required for the LEED certification, which occurs in the postacceptance phase.

The predesign phase is the ideal time for an owner to select and retain the commissioning agent. The design team member should, if possible, be involved in the selection of the commissioning agent because that member's involvement will typically ensure a more cohesive commissioning team. Once the commissioning agent is selected and retained, the commissioning-approach outline is developed. The commissioning-approach outline defines the scope and depth of the commissioning process to be employed for the project. Critical commissioning questions are addressed in this outline. The outline will include, for most projects, answers to the following questions:

What equipment is to be included?
What procedures arc to be followed?
What is the budget for the process?

As the above questions suggest, the commissioning budget is developed from the choices made in this phase. Also, if the owner has a commissioning policy, it needs to be applied to the specifics of the particular project in this phase.

The key event in the predesign phase is the creation of the DID, which defines the technical criteria for meeting the requirements of the intended use of the facilities. The DID document is often created based in part upon the information received from interviews with the intended building occupants and maintenance staff. Critical information—such as the hours of operation, occupancy levels, special environmental considerations (such as pressure and humidity), applicable codes, and budgetary considerations and limitations—is identified in this document. The owner's preference, if any, for certain equipment or contactors should also be identified at this time. Together, the answers to the critical questions above and the information in the DID are used to develop the commissioning approach outline. A thorough review of the DID by the commissioning agent ensures that the commissioning-approach outline will be aligned with the design intent.

With the commissioning agent selected, the DID document created, and the commissioning approach outline in place, the design phase is ready to commence.

Design Phase

The design phase is the phase in which the schematics and specifications for all components of a project are prepared. One key schematic and set of specifications relevant to the commissioning plan is the MEP schematic design, which specifies installation requirements for the MEP systems. As noted, the DID is the basis for creating the commissioning approach outline in the predesign phase. The DID also serves as the basis for creating the MEP schematic design in the design phase. The DID provides the MEP designer with the key concepts from which the MEP schematic design is developed.

The completed MEP schematic design is reviewed by the commissioning agent for completeness and conformance to the DID. At this stage, the commissioning agent and the other design team members should consider what current technologies, particularly those for energy efficiency, could be profitably included in the design. Many of the design enhancements currently incorporated into existing buildings during energy retrofitting for operational optimization are often not considered in new building construction. This can result in significant lost opportunity, so these design enhancements should be reviewed for incorporation into the base design during this phase of the commissioning process. This point illustrates the important principle that technologies important to retrocommissioning should be applied to new building construction—a point that is surprisingly often overlooked in the industry today.

For example, the following design improvements and technologies should always be considered for applicability to a particular project:

- Variable-speed fan and pumps installed
- Chilled water cooling (instead of DX cooling)
- Utility meters for gas, electric, hot water, chilled water, and steam at both the building and system level
- CO_2 implementation for minimum indoor air requirements

This list of design improvements is not exhaustive; the skilled commissioning agent will create and expand personalized lists as experience warrants and as the demands of particular projects suggest.

In addition to assisting in the evaluation of potential design improvements, the commissioning agent further inspects the MEP schematic design for:

- Proper sizing of equipment capacities
- Clearly defined and optimized operating sequences
- Equipment accessibility for ease of servicing

Once the commissioning agent's review is complete, the feedback is discussed with the design team to determine whether its incorporation into the MEP schematic design is warranted. The agreed-upon changes or enhancements are incorporated, thus completing the MEP schematic design.

The completed MEP schematic design serves as the basis on which the commissioning agent will transform the commissioning-approach outline into the commissioning specification.

The commissioning specification is the mechanism for binding contractually the contractors to the commissioning process. Expectations are clearly defined, including:

- Responsibilities of each contractor
- Site meeting requirements
- List of the equipment, systems, and interfaces
- Preliminary verification checklists
- Preliminary functional-performance testing checklists
- Training requirements and who is to participate
- Documentation requirements
- Postconstruction documentation requirements
- Commissioning schedule
- Definition for system acceptance
- Impact of failed results

Completion of the commissioning specification is required to select the systems contractor in a competitive solicitation. Alternatively, however, owners with strong, preexisting relationships with systems contractors may enter into a negotiated bid with those contractors, who can then be instrumental in finalizing the commissioning specification.

Owners frequently select systems contractors early in the design cycle to ensure that the contractors are involved in the design process. As noted above, if there are strong, preexisting relationships with systems contractors, early selection without a competitive selection process (described in the following paragraph) can be very beneficial. However, if there is no competitive selection process, steps should be taken to ensure that the owner gets the best value. For example, unit pricing should be negotiated in advance to ensure that the owner is getting fair and reasonable pricing. The commissioning agent and the MEP designer can be good sources for validating the unit pricing. The final contract price should be justified with the unit pricing information.

If the system selection process is competitive, technical proposals should be requested with the submission of the bid price. The systems contractors need to demonstrate a complete understanding of the project requirements to ensure that major components have not been overlooked. Information such as the project schedule and manpower loading for the project provide a good basis from which to measure the contractor's level of understanding. If the solicitation does not have a preselected list of contractors, the technical proposal should include the contractor's financial information, capabilities, and reference lists. As in the negotiated process described above, unit pricing should be requested to ensure the proper pricing of project additions and deletions. The review of the technical proposals should be included in the commissioning agent's scope of work.

A mandatory prebid conference should be held to walk the potential contractors through the requirements and to reinforce expectations. This conference should be held regardless of the approach—negotiated or competitive—used for contractor selection. The contractor who is to bear the financial burden for failed verification tests and subsequent functional-performance tests should be reminded of these responsibilities to reinforce their importance in the prebid meeting. The prebid conference sets the tone of the project and emphasizes the importance of the commissioning process to a successful project.

Once the MEP schematic design and commissioning specification are complete, and the systems contractors have been selected, the construction/installation phase begins.

Construction/Installation Phase

Coordination, communication, and cooperation are the keys to success in the construction and installation phase. The commissioning agent is the catalyst for ensuring that these critical activities occur throughout the construction and installation phase.

Frequently, value engineering options are proposed by the contractors prior to commencing the installation. The commissioning agent should be actively involved in the assessment of any options proposed. Many times, what appears to be a good idea in construction can have a disastrous effect on a facility's long-term operation. For example, automatic controls are often value engineered out of the design, yet the cost of their inclusion is incurred many times over in the labor required to perform their function manually over the life of the building. The commissioning agent can ensure that the design intent is preserved, the life-cycle costs are considered, and the impact on all systems of any value engineering modification proposed is thoroughly evaluated.

Once the design aspects are complete and value engineering ideas have been incorporated or rejected, the submittals, including verification checklists, need to be finalized. The submittals documentation is prepared by the systems contractors and reviewed by the commissioning agent. There are two types of submittals: technical submittals and commissioning submittals. Both types of submittals are discussed below.

Technical submittals are provided to document the systems contractors' interpretation of the design documents. The commissioning agent reviews the technical submittals for compliance and completeness. It is in this submittal review process that potential issues are identified prior to installation, reducing the need for rework and minimizing schedule delays. The technical submittals should include:

- Detailed schematics
- Equipment data sheets
- Sequence of operation
- Bill of material

A key technical submittal is the testing, adjusting, and balancing submittal (TAB). The TAB should include:

- TAB procedures
- Instrumentation
- Format for results
- Data sheets with equipment design parameters
- Operational readiness requirements
- Schedule

In addition to the TAB, other technical submittals, such as building automation control submittals, will be obtained from the systems contractors and reviewed by the commissioning agent.

The commissioning submittal generally follows the technical submittal in time and includes:

- Verification checklists
- Startup requirements
- Test and balance plan
- Training plan

The commissioning information in the commissioning submittal is customized for each element of the system.

These submittals, together with the commissioning specification, are incorporated into the commissioning plan, which becomes a living document codifying the results of the construction commissioning activities. This plan should be inspected in regular site meetings. Emphasis on the documentation aspect of the commissioning process early in the construction phase increases the contractors' awareness of the importance of commissioning to a successful project.

In addition to the submittals, the contractors are responsible for updating the design documents with submitted and approved equipment data and field changes on an ongoing basis. This update design document should be utilized during the testing and acceptance phase.

The commissioning agent also performs periodic site visits during the installation to observe the quality of workmanship and compliance with the specifications. Observed deficiencies should be discussed with the contractor and documented to ensure future compliance. Further inspections should be conducted to ensure that appropriate corrective action has been taken.

The best way to ensure that the items discussed above are addressed in a timely manner is to hold regularly scheduled commissioning meetings that require the participation of all systems contractors. This is the mechanism for ensuring that communication occurs. Meeting minutes prepared by the commissioning agent document the discussions and decisions reached. Commissioning meetings should be coordinated with the regular project meetings because many participants in a construction project need to attend both meetings.

Typical elements of a commissioning meeting include:

- Discussing field installation issues to facilitate rapid response to field questions
- Updating design documents with field changes
- Reviewing the commissioning agent's field observations
- Reviewing progress against schedule
- Coordinating multicontractor activities

Once familiar with the meeting process, an agenda will be helpful but not necessary. Meeting minutes should be kept and distributed to all participants.

With approved technical and commissioning plan submittals, as installation progresses, the contractor is ready to begin the system verification testing. The systems contractor generally executes the system verification independently of the commissioning agent. Contractor system verification includes:

- Point-to-point wiring checked out
- Sensor accuracy validated
- Control loops exercised

Each of the activities should be documented for each control or system element, and signed and dated by the verification technician.

The documentation expected from these activities should be clearly defined in the commissioning specification to ensure its availability to the commissioning agent for inspection of the verification process. The commissioning agent's role in the system verification testing is to ensure that the tests are completed and that the results reflect that the system is ready for the functional-performance tests. Because the commissioning agent is typically not present during the verification testing, the documentation controls how successfully the commissioning agent performs this aspect of commissioning.

In addition to system verification testing, equipment startup is an important activity during this phase. Equipment startup occurs at different time frames relative to the system verification testing, depending on the equipment and system involved. There may be instances when the system verification needs to occur prior to equipment startup to prevent a catastrophic event that could lead to equipment failure. The commissioning agent reviews the startup procedures prior to the startup to ensure that equipment startup is coordinated properly with the system verification. Unlike in verification testing, the commissioning agent should be present during HVAC equipment startup to document the results. These results are memorialized in the final commissioning report, so their documentation ultimately is the responsibility of the commissioning agent.

Once system verification testing and equipment startup have been completed, the acceptance phase begins.

Acceptance Phase

The acceptance phase of the project is the phase in which the owner accepts the project as complete and delivered in accordance with the specifications, and concludes with acceptance of the project in its entirety. An effective commissioning process during the installation phase should reduce the time and labor associated with the functional-performance tests of the acceptance phase.

Statistical sampling is often used instead of 100% functional-performance testing to make the process more efficient. A 20% random sample with a failure rate less than 1% indicates that the entire system was properly installed. If the failure rate exceeds 1%, a complete testing of every system may need to be completed to correct inadequacies in the initial checkout and verification testing. This random-sampling statistical approach holds the contractor accountable for the initial checkout and test, with the performance testing serving only to confirm the quality and thoroughness of the installation. This approach saves time and money for all involved. It is critical, however, that the ramifications of not meeting the desired results of the random tests are clearly defined in the commissioning specifications.

The commissioning agent witnesses and documents the results of the functional-performance tests, using specific forms and procedures developed for the system being tested. These forms are created with the input of the contractor in the installation phase. Involvement of the maintenance staff in the functional-performance testing is important. The maintenance team is often not included in the design process, so they may not fully understand the design intent. The functional-performance testing can provide the maintenance team an opportunity to learn and appreciate the design intent. If the design intent is to be preserved, the maintenance team must fully understand the design intent. This involvement of the maintenance team increases their knowledge of the system going into the training and will increase the effectiveness of the training.

Training of the maintenance team is critical to a successful operational handover once a facility is ready for occupancy. This training should include:

- Operations and maintenance (O&M) manual overview
- Hardware component review
- Software component review
- Operations review
- Interdependencies discussion
- Limitations discussion
- Maintenance review
- Troubleshooting procedures review
- Emergency shutdown procedures review

The support level purchased from the systems contractor determines the areas of most importance in the training and therefore should be determined prior to the training process. Training should be videotaped for later use by new maintenance team members and in refresher courses, and for general reference by the existing maintenance team. Using the O&M manuals as a training manual increases the maintenance team's awareness of the information contained in them, making the O&M manuals more likely to be referenced when appropriate in the future.

The O&M manuals should be prepared by the contractor in an organized and easy-to-use manner. The commissioning agent is sometimes engaged to organize them all into an easily referenced set of documents. The manuals should be provided in both hard-copy and electronic formats, and should include:

- System diagrams
- Input/output lists
- Sequence of operations
- Alarm points list
- Trend points list
- Testing documentation
- Emergency procedures

These services—including functional-performance testing, training, and preparing O&M manuals—should be included in the commissioning plan to ensure the project's successful acceptance. The long-term success of the project, however, is determined by the activities that occur in the postacceptance phase.

Postacceptance Phase

The postacceptance phase is the phase in which the owner takes beneficial occupancy and forms an opinion about future work with the design team, contractors, and the commissioning agent who completed the project. This is also the phase in which LEED certification, if adopted, is completed. Activities that usually occur in the acceptance phase should instead occur in the postacceptance phase. This is due to constraints that are not controllable by the contractor or owner. For example, seasonal changes may make functional-performance testing of some HVAC systems impractical during the acceptance phase for certain load conditions. This generally means that in locations that experience significant seasonal climate change, some of the functional-performance testing is deferred until suitable weather conditions exist. The commissioning agent determines which functional-performance tests need to be deferred and hence carried out in the postacceptance phase.

During the postacceptance phase, the commissioning agent prepares a final commissioning report that is provided to the owner and design team. The executive summary of this report provides an overall assessment of the design intent conformance. The report details whether the commissioned equipment and systems meet the commissioning requirements. Problems encountered and corrective actions taken are documented in this report. The report also includes the signed and dated startup and functional-performance testing checklists.

The final commissioning report can be used profitably as the basis of a "lessons learned" meeting involving the design team so that the commissioning process can be continuously improved and adaptations can be made to the owner's commissioning policy for future projects. The owner should use the experience of the first commissioned project to develop the protocols and standards for future projects. The documentation of this experience is the owner's commissioning policy. Providing this policy and the information it contains to the design and construction team for the next project can help the owner reduce budget overruns by eliminating any need to reinvent protocols and standards and by setting the right expectations earlier in the process.

Commissioning therefore should not be viewed as a one-time event but should instead be viewed as an operational philosophy. A recommissioning or continuous commissioning plan should be adopted for any building to sustain the benefits delivered from a commissioning plan. The commissioning agent can add great value to the creation of the recommissioning plan and can do so most effectively in the postacceptance phase of the project.

Fig. 1 depicts the information development that occurs in the evolution of a commissioning plan and summarizes the information presented in the preceding sections by outlining the various phases of the commissioning process.

With this background, the reader is better positioned for success in future commissioning projects and better prepared to learn the key success factors in commissioning

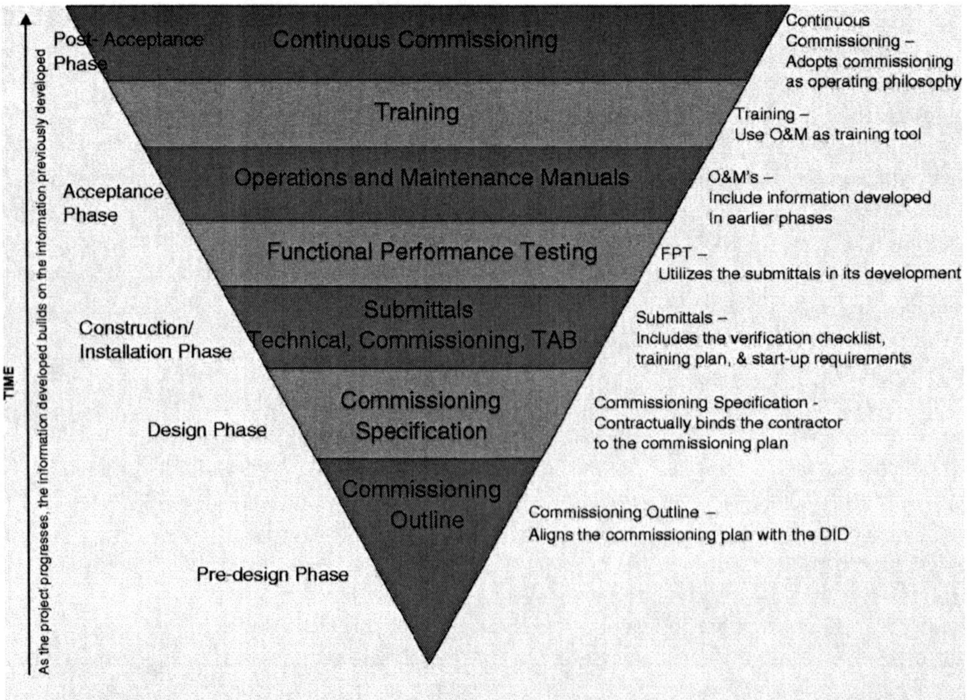
TIME
As the project progresses, the information developed builds on the information previously developed
Post-Acceptance Phase
Continuous Commissioning
Training
Operations and Maintenance Manuals
Acceptance Phase
Functional Performance Testing
Submittals
Technical, Commissioning, TAB
Construction/ Installation Phase
Commissioning Specification
Design Phase
Commissioning Outline
Pre-design Phase
Continuous Commissioning – Adopts commissioning as operating philosophy
Training – Use O&M as training tool
O&M's – Include information developed In earlier phases
FPT – Utilizes the submittals in its development
Submittals – Includes the verification checklist, training plan, & start-up requirements
Commissioning Specification - Contractually binds the contractor to the commissioning plan
Commissioning Outline – Aligns the commissioning plan with the DID

Fig. 1 The commissioning plan.

and how to avoid common mistakes in the commissioning process.

COMMISSIONING SUCCESS FACTORS

Ultimately, the owner will be the sole judge of whether a commissioning process has been successful. Thus, second only to the need for a competent, professional commissioning agent, keeping the owner or the owner's senior representative actively involved in and informed at all steps of the commissioning process is a key success factor. The commissioning agent should report directly to the owner or the owner's most senior representative on the project, not only to ensure that this involvement and information transfer occur, but also to ensure the objective implementation of the commissioning plan—a third key success factor.

Another key success factor is an owner appreciation—which can be enhanced by the commissioning agent—that commissioning must be an ongoing process to get full benefit. For example, major systems should undergo periodic modified functional testing to ensure that the original design intent is being maintained or to make system modification if the design intent has changed. If an owner appreciates that commissioning is a continuous process that lasts for the entire life of the facility, the commissioning process will be a success.

Most owners will agree that the commissioning process is successful if success can be measured in a cost/benefit analysis. Cost/benefit or return on equity is the most widely used approach to judge the success of any project. Unfortunately, the misapplication of cost/benefit analyses has been the single largest barrier to the widespread adoption of commissioning. For example, because new construction does not have an energy baseline from which to judge energy savings, improper application of a cost/benefit analysis can lead to failure to include energy savings technologies—technologies the commissioning agent can identify—in the construction process. Similarly, unless one can appreciate how commissioning can prevent schedule delays and rework by spotting issues and resolving them early in the construction process, one cannot properly offset the costs of commissioning with the benefits.

Fortunately, there are now extensive studies analyzing the cost/benefit of commissioning that justify its application. A study performed jointly by Lawrence Berkeley National Laboratory; Portland Energy Conservation, Inc.; and the Energy Systems Laboratory at Texas A&M University provides compelling analytical data on the cost/benefit of commissioning. The study defines the median commissioning cost for new construction as $1 per square foot or 0.6% of the total construction cost. The median simple payback for new construction projects utilizing commissioning is 4.8 years. This simple payback calculation does not take into account the quantified nonenergy impacts, such as the reduction in the cost and frequency of change orders or premature equipment failure due to improper installation practices. The study quantifies the median nonenergy benefits for new construction at $1.24 per square foot per year.[3]

While the primary cost component of assessing the cost/benefit of commissioning lies in whether there was a successful negotiation of the cost of services with the commissioning service provider, the more important aspect of the analysis relates to the outcomes of the process. For example, after a commissioning process is complete, what are the answers to these questions?

Are the systems functioning to the design intent?
Has the owner's staff been trained to operate the facility?
How many of the systems are operated manually a year after installation?

Positive answers to these and similar questions will ensure that any cost/benefit analysis will demonstrate the value of commissioning.

To ensure that a commissioning process is successful, one must avoid common mistakes. A commissioning plan is a customized approach to ensuring that all the systems operate in the most effective and efficient manner. A poor commissioning plan will deliver poor results. A common mistake is to use an existing commissioning plan and simply insert it into a specification to address commissioning. Each commissioning plan should be specifically tailored to the project to be commissioned.

Also, perhaps due to ill conceived budget constraints, commissioning is implemented only in the construction phase. Such constraints are ill conceived because the cost of early involvement of the commissioning agent in the design phases is insignificant compared with the cost of correcting design defects in the construction phase. Significant cost savings can arise from identifying design issues prior to construction. Studies have shown that 80% of the cost of commissioning occurs in the construction phase.[2] Also, the later the commissioning process starts, the more confrontational commissioning becomes, making it more expensive to implement later in the process.[2] Therefore, adopting commissioning early in the project is a key success factor.

Value engineering often results in ill-informed, last-minute design changes that have an adverse and unintended impact on the overall building performance and energy use.[3] By ensuring that the commissioning process includes careful evaluation of all value engineering proposals, the commissioning agent and owner can avoid such costly mistakes.

Finally, the commissioning agent's incentive structure should not be tied to the number of issues brought to light during the commissioning process, as this can create an antagonistic environment that may create more problems than it solves. Instead, the incentive structure should be outcome based and the questions outlined above regarding compliance with design intent, training results, and post-acceptance performance provide excellent bases for a positive incentive structure.

CONCLUSION

Commissioning should be performed on all but the most simplistic of new construction projects. The benefits of commissioning include:

- Optimization of building performance
  —Enhanced operation of building systems
  —Better-prepared maintenance staff
  —Comprehensive documentation of systems
  —Increased energy efficiency
  —Improved quality of construction
- Reduced facility life-cycle cost
  —Reduced impact of design changes
  —Fewer change orders
  —Fewer project delays
  —Less rework or postconstruction corrective work
  —Reduced energy costs
- Increased occupant satisfaction
  —Shortened turnover transition period
  —Improved system operation
  —Improved system reliability

With these benefits, owners and contractors alike should adopt the commissioning process as the best way to ensure cost-efficient construction and the surest way to a successful construction project.

REFERENCES

1. Quantum Consulting, Inc. *Commissioning in Public Buildings: Market Progress Evaluation Report*, June 2005. Available at http://www.nwalliance.org/resources/reports/141.pdf.
2. ACG AABC Commissioning Group. ACG Commissioning Guidelines 2005. Available at http://www.commissioning.org/commissioningguideline/ACGCommissioningGuideline.pdf.
3. Mills, E.; Friedman, H.; Powell, T. et al. *The Cost-Effectiveness of Commercial-Buildings Commissioning*, December 2004. Available at http://eetd.lbl.gov/Emills/PUBS/Cx-Costs-Benefits.html.
4. American Society of Heating. *Refrigerating and Air-Conditioning Engineers*, ASHRAE Guidelines 1–1996.
5. Green Building Council Web site. Available at http://www.usgbc.org/leed (accessed October 2005).

BIBLIOGRAPHY

1. SBW Consulting, Inc. *Cost-Benefit Analysis for the Commissioning in Public Buildings Project*, May 2004. Available at http://www.nwalliance.org/resources/documents/CPBReport.pdf (accessed).
2. Turner, W.C. *Energy Management Handbook*, 5th Ed.; Fairmont Press: Lilburn, GA, 2005.
3. Research News Berkeley Lab. *Berkeley Lab Will Develop Energy-Efficient Building Operation Curriculum for Community Colleges*, December 2004. Available at http://www.lbl.gov/Science-Articles/Archive/EETD-college-curriculum.html.
4. Interview with Richard Holman. *Director of Commissioning and Field Services for Affiliated Engineers, Inc.* Walnut Creek, CA.

Energy Conservation

Ibrahim Dincer
Faculty of Engineering and Applied Science, University of Ontario Institute of Technology (UOIT), Oshawa, Ontario, Canada

Adnan Midili
Department of Mechanical Engineering, Faculty of Engineering, Nigde University, Nigde, Turkey

Abstract

This study highlights these issues and potential solutions to the current environmental issues; identifies the main steps for implementing energy conservation programs and the main barriers to such implementations; and provides assessments for energy conservation potentials for countries, as well as various practical and environmental aspects of energy conservation.

INTRODUCTION

Civilization began when people found out how to use fire extensively. They burned wood and obtained sufficiently high temperatures for melting metals, extracting chemicals, and converting heat into mechanical power, as well as for cooking and heating. During burning, the carbon in wood combines with O_2 to form carbon dioxide (CO_2), which then is absorbed by plants and converted back to carbon for use as a fuel again. Because wood was unable to meet the fuel demand, the Industrial Revolution began with the use of fossil fuels (e.g., oil, coal, and gas). Using such fuels has increased the CO_2 concentration in the air, leading to the beginning of global warming. Despite several warnings in the past about the risks of greenhouse-gas emissions, significant actions to reduce environmental pollution were not taken, and now many researchers have concluded that global warming is occurring. During the past two decades, the public has became more aware, and researchers and policymakers have focused on this and related issues by considering energy, the environment, and sustainable development.

Energy is considered to be a key catalyst in the generation of wealth and also a significant component in social, industrial, technological, economic, and sustainable development. This makes energy resources and their use extremely significant for every country. In fact, abundant and affordable energy is one of the great boons of modern industrial civilization and the basis of our living standard. It makes people's lives brighter, safer, more comfortable, and more mobile, depending on their energy demand and consumption. In recent years, however, energy use and associated greenhouse-gas emissions and their potential effects on the global climate change have been of worldwide concern.

Problems with energy utilization are related not only to global warming, but also to such environmental concerns as air pollution, acid rain, and stratospheric ozone depletion. These issues must be taken into consideration simultaneously if humanity is to achieve a bright energy future with minimal environmental impact. Because all energy resources lead to some environmental impact, it is reasonable to suggest that some (not all) of these concerns can be overcome in part through energy conservation efforts.

Energy conservation is a key element of energy policy and appears to be one of the most effective ways to improve end-use energy efficiency, and to reduce energy consumption and greenhouse-gas emissions in various sectors (industrial, residential, transportation, etc.). This is why many countries have recently started developing aggressive energy conservation programs to reduce the energy intensity of the their infrastructures, make businesses more competitive, and allow consumers to save money and to live more comfortably. In general, energy conservation programs aim to reduce the need for new generation or transmission capacity, to save energy, and to improve the environment. Furthermore, energy conservation is vital for sustainable development and should be implemented by all possible means, despite the fact that it has its own limitations. This is required not only for us, but for the next generation as well.

Considering these important contributions, the energy conservation phenomenon should be discussed in a comprehensive perspective. Therefore, the main objective of this entry is to present and discuss the world's primary energy consumption and production; major environmental problems; potential solutions to these issues; practical energy conservation aspects; research and development (R&D) in energy conservation, energy

Encyclopedia of Environmental Management DOI: 10.1081/E-EEM-120042344

Table 1 World energy production and consumption models through statistical analysis.

Energy	Production (Mtoe)	Correlation coefficient	Consumption (Mtoe)	Correlation coefficient
World primary	= 148.70 × Year – 287,369	0.998	= 139.62 × Year – 269,953	0.998
World oil	= 44.47 × Year – 85,374	0.997	= 42.18 × Year – 80,840	0.998
World coal	= 20.05 × Year – 37,748	0.946	= 23.18 × Year – 43,973	0.968
World NG	= 45.73 × Year – 89,257	0.999	= 46.27 × Year – 90,347	0.999

Note: Mtoe, million tons of oil equivalent.

conservation, and sustainable development; energy conservation implementation plans; energy conservation measurements; and life-cycle costing (LCC) as an excellent tool in energy conservation. In this regard, this contribution aims to:

- Help explain main concepts and issues about energy conservation
- Develop relations between energy conservation and sustainability
- Encourage energy conservation strategies and policies
- Provide energy conservation methodologies
- Discuss relations between energy conservation and environmental impact
- Present some illustrative examples to state the importance of energy conservation and its practical benefits

In summary, this book contribution highlights the current environmental issues and potential solutions to these issues; identifies the main steps for implementing energy conservation programs and the main barriers to such implementations; and provides assessments for energy conservation potentials for countries, as well as various practical and environmental aspects of energy conservation.

WORLD ENERGY RESOURCES: PRODUCTION AND CONSUMPTION

World energy consumption and production are very important for energy conservation in the future. Economic activity and investment patterns in the global energy sector are still centered on fossil fuels, and fossil-fuel industries and energy-intensive industries generally have been skeptical about warnings of global warming and, in particular, about policies to combat it. The increase of energy consumption and energy demand indicates our dependence on fossil fuels. If the increase of fossil-fuel utilization continues in this manner, it is likely that the world will be affected by many problems due to fossil fuels. It follows from basic scientific laws that increasing amounts of CO_2 and other greenhouse gases will affect the global climate. The informed debate is not about the existence of such effects, but about their magnitudes and

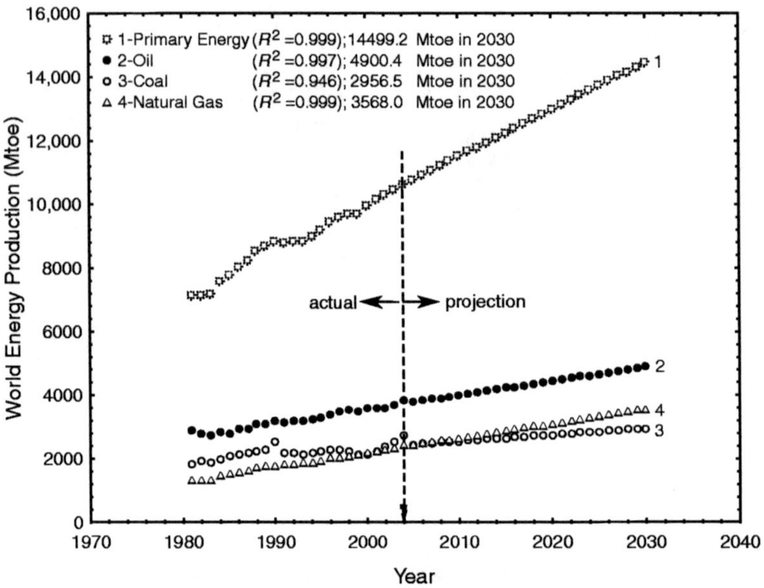

16,000
14,000
12,000
10,000
8000
6000
4000
2000
0
1-Primary Energy (R^2 =0.999); 14499.2 Mtoe in 2030
2-Oil (R^2 =0.997); 4900.4 Mtoe in 2030
3-Coal (R^2 =0.946); 2956.5 Mtoe in 2030
4-Natural Gas (R^2 =0.999); 3568.0 Mtoe in 2030
actual
projection
1
2
3
4
World Energy Production (Mtoe)
1970
1980
1990
2000
2010
2020
2030
2040
Year

Fig. 1 Variation of actual data taken from BP[3] and IEE,[4] and projections of annual world energy production. Mtoe, million tons of oil equivalent.
Source: BP[3] and IEE.[4]

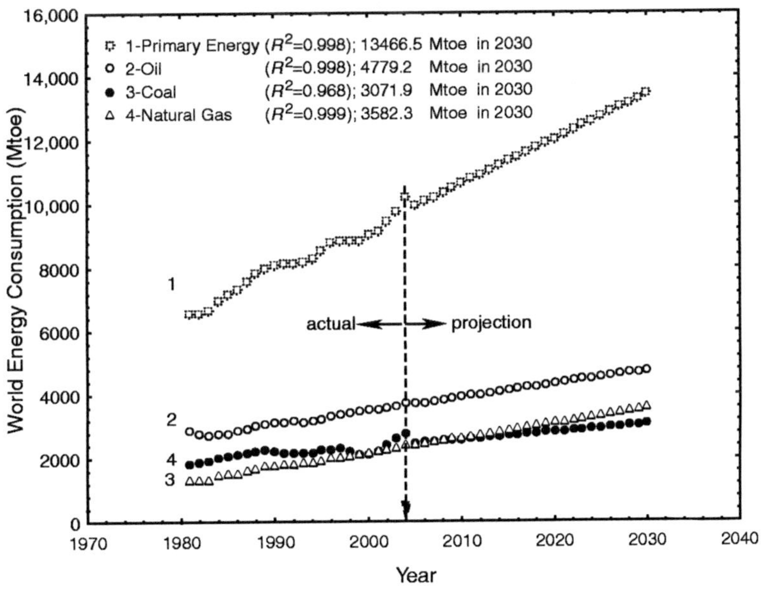
1-Primary Energy (R^2=0.998); 13466.5 Mtoe in 2030
2-Oil (R^2=0.998); 4779.2 Mtoe in 2030
3-Coal (R^2=0.968); 3071.9 Mtoe in 2030
4-Natural Gas (R^2=0.999); 3582.3 Mtoe in 2030
World Energy Consumption (Mtoe)
16,000
14,000
12,000
10,000
8000
6000
4000
2000
0
1
2
4
3
actual
projection
1970
1980
1990
2000
2010
2020
2030
2040
Year

Fig. 2 Variation of actual data taken from BP[3] and IEE,[4] and projections of annual world energy consumption. Mtoe, million tons of oil equivalent.
Source: BP[3] and IEE.[4]

seriousness. At present, the concentration of CO_2 is approximately 30% higher than its preindustrial level, and scientists have already been able to observe a discernible human influence on the global climate.[1]

In the past, fossil fuels were a major alternative for overcoming world energy problems. Fossil fuels cannot continue indefinitely as the principal energy sources, however, due to the rapid increase of world energy demand and energy consumption. The utilization distribution of fossil-fuel types has changed significantly over the past 80 years. In 1925, 80% of the required energy was supplied from coal, whereas in the past few decades, 45% came from petroleum, 25% from natural gas, and 30% from coal. Due to world population growth and the advance of technologies that depend on fossil fuels, reserves of those fuels eventually will not be able to meet energy demand. Energy experts point out that reserves are less than 40 years for petroleum, 60 years for natural gas, and 250 years for coal.[2] Thus, fossil-fuel costs are likely to increase in the near future. This will allow the use of renewable energy sources such as solar, wind, and hydrogen. As an example, the actual data[3,4] and projections of world energy production and consumption from 1980 to 2030 are displayed in the following figures, and the curve equations for world energy production and consumption are derived as shown Table 1.

As presented in Figs. 1 and 2, and in Table 2, the quantities of world primary energy production and consumption are expected to reach 14,499.2 and 13,466.5 Mtoe, respectively, by 2030. World population is now over six billion, double that of 40 years ago, and it is likely to double again by the middle of the 21st century. The world's population is expected to rise to about seven billion by 2010. Even if birth rates fall so that the world population becomes stable by 2030, the population still would be about ten billion. The data presented in Figs. 1 and 2 are expected to cover current energy needs provided that the population remains

Table 2 Some extracted values of world primary and fossil energy production and consumption.

Year	Primary energy production (Mtoe)	Primary energy consumption (Mtoe)	Oil prod. (Mtoe)	Oil cons. (Mtoe)	Coal prod. (Mtoe)	Coal cons. (Mtoe)	NG prod. (Mtoe)	NG cons. (Mtoe)
1994	8,996.9	8,310.1	3237.1	3204.4	2178.1	2185.5	1891.2	1876.7
2000	9,981.9	9,079.8	3614.0	3538.7	2112.4	2148.1	2189.9	2194.5
2006	10,930.3	10,115.7	3833.1	3767.0	2475.3	2515.7	2470.5	2471.8
2012	11,822.5	10,953.4	4099.9	4020.0	2595.6	2654.8	2744.9	2749.5
2018	12,714.7	11,791.1	4366.7	4273.1	2715.9	2793.8	3019.2	3027.1
2024	13,606.9	12,128.8	4633.5	4526.1	2836.2	2932.9	3293.6	3304.7
2030	14,499.2	13,466.5	4900.4	4779.2	2956.5	3071.9	3568.0	3582.3

Note: Mtoe, million tons of oil equivalent.

constant. Because the population is expected to increase dramatically, however, conventional energy resource shortages are likely to occur, due to insufficient fossil-fuel resources. Therefore, energy conservation will become increasingly important to compensate for shortages of conventional resources.

MAJOR ENVIRONMENTAL PROBLEMS

One of the most important targets of modern industrial civilizations is to supply sustainable energy sources and to develop the basis of living standards based on these energy sources, as well as implementing energy conservation measures. In fact, affordable and abundant sustainable energy makes our lives brighter, safer, more comfortable, and more mobile because most industrialized and developing societies use various types of energy. Billions of people in undeveloped countries, however, still have limited access to energy. India's per-capita consumption of electricity, for example, is one-twentieth that of the United States. Hundreds of millions of Indians live "off the grid"-that is, without electricity-and cow dung is still a major fuel for household cooking. This continuing reliance on such preindustrial energy sources is also one of the major causes of environmental degradation.[5]

After many decades of using fossil fuels as a main energy source, significant environmental effects of fossil fuels became apparent. The essential pollutants were from greenhouse gases (e.g., CO_2, SO_2, and NO_2). Fossil fuels are used for many applications, including industry, residential, and commercial sectors. Increasing fossil-fuel utilization in transportation vehicles such as automobiles, ships, aircrafts, and spacecrafts has led to increasing pollution. Gas, particulate matter, and dust clouds in the atmosphere absorb a significant portion of the solar radiation directed at Earth and cause a decrease in the oxygen available for the living things. The threat of global warming has been attributed to fossil fuels.[2] In addition, the risk and reality of environmental degradation have become more apparent. Growing evidence of environmental problems is due to a combination of factors.

During the past two decades, environmental degradation has grown dramatically because of the sheer increase of world population, energy consumption, and industrial activities. Throughout the 1970s, most environmental analysis and legal control instruments concentrated on conventional pollutants such as SO_2, NO_x, particulates, and CO. Recently, environmental concern has extended to the control of micro or hazardous air pollutants, which are usually toxic chemical substances and harmful in small doses, as well as to that of globally significant pollutants such as CO_2. Aside from advances in environmental engineering science, developments in industrial processes and structures have led to new environmental problems.[6,7] In the energy sector, for example, major shifts to the road transport of industrial goods and to individual travel by cars has led to an increase in road traffic and, hence, to a shift in attention paid to the effects and sources of NO_x and to the emissions of volatile organic compounds (VOC). In fact, problems with energy supply and use are related not only to global warming, but also to such environmental concerns as air pollution, ozone depletion, forest destruction, and emission of radioactive substances. These issues must be taken into consideration simultaneously if humanity is to achieve a bright energy future with minimal environmental impact. Much evidence exists to suggest that the future will be negatively impacted if humans keep degrading the environment. Therefore, there is an intimate connection among energy conservation, the environment, and sustainable development. A society seeking sustainable development ideally must utilize only energy resources that cause no environmental impact (e.g., that release no emissions to the environment). Because all energy resources lead to some environmental impact, however, it is reasonable to suggest that some (not all) of the concerns regarding the limitations imposed on sustainable development by environmental emissions and their negative impacts can be overcome in part through energy conservation. A strong relation clearly exists between energy conservation and environmental impact, because for the same services or products, less resource utilization and pollution normally are associated with higher-efficiency processes.[8]

Table 3 summarizes the major environmental problems-such as acid rain, stratospheric ozone depletion, and global climate change (greenhouse effect)-and their main sources and effects.

As shown in Fig. 3, the world total CO_2 production is estimated to be 18,313.13 million tons in 1980, 25,586.7 million tons in 2006, 27,356.43 million tons in 2012, and 29,716.1 million tons in 2020 whereas fossil-fuel consumption is found to be 6092.2 million tons in 1980, 8754.5 million tons in 2006, 9424.3 million tons in 2012, and 10,317.2 million tons in 2020. These values show that the CO_2 production will probably increase if we continue utilizing fossil fuel. Therefore, it is suggested that certain energy conversion strategies and technologies should be put into practice immediately to reduce future environmental problems.

The climate technology initiative (CTI) is a cooperative effort by 23 Organization for Economic Cooperation and Development (OECD)/International Energy Agency (IEA) member countries and the European Commission to support the objectives of the united nations framework convention on climate change (UNFCCC). The CTI was launched at the 1995 Berlin Conference of the Parties to the UNFCCC. The CTI seeks to ensure that technologies to address climate change are available and can be deployed efficiently. The CTI includes activities directed at the achievement of seven broad objectives:

Table 3 Major environmental issues and their consequences.

Issues	Description	Main sources	Main effects
Acid precipitation	Transportation and deposition of acids produced by fossil-fuel combustion (e.g., industrial boilers, transportation vehicles) over great distances through the atmosphere via precipitation on the earth on ecosystems	Emissions of SO_2, NO_x, and volatile organic compounds (VOCs) (e.g., residential heating and industrial energy use account for 80% of SO_2 emissions)	Acidification of lakes, streams and ground waters, resulting in damage to fish and aquatic life; damage to forests and agricultural crops; and deterioration of materials, e.g., buildings, structures
Stratospheric ozone depletion	Distortion and regional depletion of stratospheric ozone layer though energy activities (e.g., refrigeration, fertilizers)	Emissions of CFCs, halons (chlorinated and brominated organic compounds) and N_2O (e.g., fossil fuel and biomass combustion account for 65%–75% of N_2O emissions)	Increased levels of damaging ultraviolet radiation reaching the ground, causing increased rates of skin cancer, eye damage and other harm to many biological species
Greenhouse effect	A rise in the earth's temperature as a result of the greenhouse gases	Emissions of carbon dioxide (CO_2), CH_4, CFCs, halons, N_2O, ozone and peroxyacetylnitrate (e.g., CO_2 releases from fossil fuel combustion (w50% from CO_2), CH_4 emissions from increased human activity)	Increased the earth's surface temperature about 0.68C over the last century and as a consequence risen sea level about 20 cm (in the next century by another 28C–48C and a rise between 30 and 60 cm); resulting in flooding of coastal settlements, a displacement of fertile zones for agriculture and food production toward higher latitudes, and a decreasing availability of fresh water for irrigation and other essential uses

Source: Dincer,[9] Dincer,[10] and Dincer.[11]

- To facilitate cooperative and voluntary actions among governments, quasigovernments, and private entities to help cost-effective technology diffusion and reduce the barriers to an enhanced use of climate-friendly technologies
- To promote the development of technology aspects of national plans and programs prepared under the UNFCCC
- To establish and strengthen the networks among renewable and energy efficiency centers in different regions

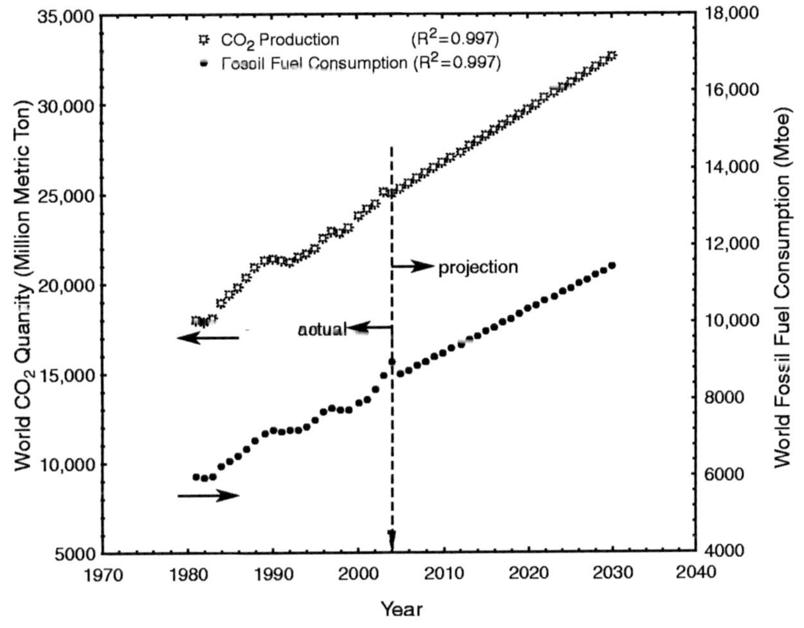
CO2 Production (R2=0.997)
Fossil Fuel Consumption (R2=0.997)
projection
actual
World CO2 Quantity (Million Metric Ton)
World Fossil Fuel Consumption (Mtoe)
Year
35,000
30,000
25,000
20,000
15,000
10,000
5000
18,000
16,000
14,000
12,000
10,000
8000
6000
4000
1970
1980
1990
2000
2010
2020
2030
2040

Fig. 3 Variation of world total fossil-fuel consumption and CO_2 production; actual data from BP[3] and projections. Mtoe, million tons of oil equivalent.
Source: BP.[3]

- To improve access to and enhance markets for emerging technologies
- To provide appropriate recognition of climate-friendly technologies through the creation of international technology awards
- To strengthen international collaboration on short-, medium-, and long-term research; development and demonstration; and systematic evaluation of technology options
- To assess the feasibility of developing longer-term technologies to capture, remove, or dispose of greenhouse gases; to produce hydrogen from fossil fuels; and to strengthen relevant basic and applied research

POTENTIAL SOLUTIONS TO ENVIRONMENTAL ISSUES

Although there are a large number of practical solutions to environmental problems, three potential solutions are given priority, as follows[11]:

- Energy conservation technologies (efficient energy utilization)
- Renewable energy technologies
- Cleaner technologies

In these technologies, we pay special attention to energy conservation technologies and their practical aspects and environmental impacts. Each of these technologies is of great importance, and requires careful treatment and program development. In this work, we deal with energy conservation technologies and strategies in depth. Considering the above priorities to environmental solutions, the important technologies shown in Fig. 4 should be put into practice.

PRACTICAL ENERGY CONSERVATION ASPECTS

The energy-saving result of efficiency improvements is often called energy conservation. The terms efficiency and conservation contrast with curtailment, which decreases output (e.g., turning down the thermostat) or services (e.g., driving less) to curb energy use. That is, energy curtailment occurs when saving energy causes a reduction in services or sacrifice of comfort. Curtailment is often employed as an emergency measure. Energy efficiency is increased when an energy conversion device-such as a household appliance, automobile engine, or steam turbine-undergoes a technical change that enables it to provide the same service (lighting, heating, motor drive, etc.) while using less energy. Energy efficiency is often viewed as a resource option like coal, oil, or natural gas. In contrast to supply options, however, the downward pressure on energy prices created by energy efficiency comes from demand reductions instead of increased supply. As a

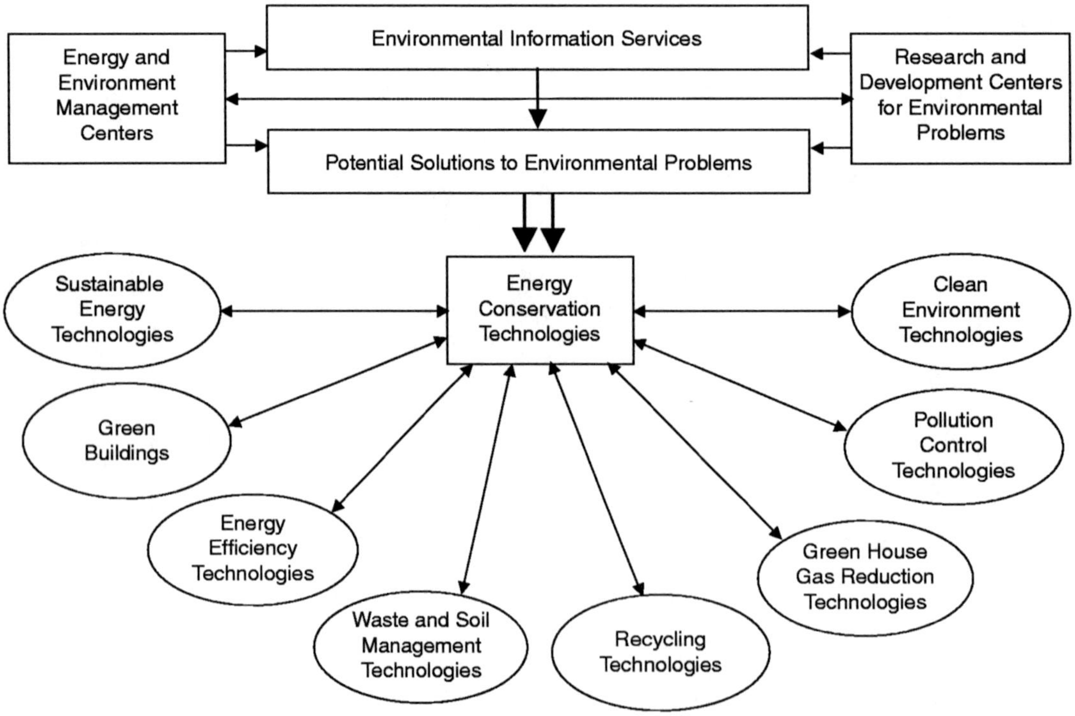
Environmental Information Services
Energy and Environment Management Centers
Research and Development Centers for Environmental Problems
Potential Solutions to Environmental Problems
Energy Conservation Technologies
Sustainable Energy Technologies
Clean Environment Technologies
Green Buildings
Pollution Control Technologies
Energy Efficiency Technologies
Green House Gas Reduction Technologies
Waste and Soil Management Technologies
Recycling Technologies

Fig. 4 Linkages between possible environmental and energy conservation technologies.

result, energy efficiency can reduce resource use and environmental impacts.[12]

The quality of a country's energy supply and demand systems is increasingly evaluated today in terms of its environmental sustainability. Fossil-fuel resources will not last indefinitely, and the most convenient, versatile, and inexpensive of them have substantially been used up. The future role of nuclear energy is uncertain, and global environmental concerns call for immediate action. OECD countries account for almost 50% of total world energy consumption: Current use of oil per person averages 4.5 bbl a year worldwide, ranging from 24 bbl in the United States and 12 bbl in western Europe to less than 1 bbl in sub-Saharan Africa. More than 80% of worldwide CO_2 emissions originate in the OECD area. It is clear, then, that OECD countries should play a crucial role in indicating a sustainable pattern and in implementing innovative strategies.[11]

From an economic as well as an environmental perspective, energy conservation holds even greater promise than renewable energy, at least in the near-term future. Energy conservation is indisputably beneficial to the environment, as a unit of energy not consumed equates to a unit of resources saved and a unit of pollution not generated.

Furthermore, some technical limitations on energy conservation are associated with the laws of physics and thermodynamics. Other technical limitations are imposed by practical technical constraints related to the real-world devices that are used. The minimum amount of fuel theoretically needed to produce a specified quantity of electricity, for example, could be determined by considering a Carnot (ideal) heat engine. However, more than this theoretical minimum fuel may be needed due to practical technical matters such as the maximum temperatures and pressures that structures and materials in the power plant can withstand.

As environmental concerns such as pollution, ozone depletion, and global climate change became major issues in the 1980s, interest developed in the link between energy utilization and the environment. Since then, there has been increasing attention to this linkage. Many scientists and engineers suggest that the impact of energy resource utilization on the environment is best addressed by considering exergy. The exergy of a quantity of energy or a substance is a measure of the usefulness or quality of the energy or substance, or a measure of its potential to cause change. Exergy appears to be an effective measure of the potential of a substance to impact the environment. In practice, the authors feel that a thorough understanding of exergy and of how exergy analysis can provide insights into the efficiency and performance of energy systems is required for the engineer or scientist working in the area of energy systems and the environment.[8] Considering the above explanations, the general aspects of energy conservation can be summarized as shown in Fig. 5.

RESEARCH AND DEVELOPMENT STATUS ON ENERGY CONSERVATION

Now we look at R&D expenditures in energy conservation to assess the importance attached to energy conservation in the long range. The share of energy R&D expenditures going into energy conservation, for example, has grown greatly since 1976, from 5.1% in 1976 to 40.1% in 1990 and 68.5% in 2002.[11] This indicates that within energy R&D, research on energy conservation is increasing in importance. When R&D expenditures on energy conservation are compared with expenditures for research leading to protection of the environment in the 2000s, the largest share was spent on environment research. In fact, it is not easy to interpret the current trends in R&D expenditures, because energy conservation is now part of every discipline from engineering to economics. A marked trend has been observed since the mid-1970s, in that expenditures for energy conservation research have grown significantly, both in absolute terms and as a share of total energy R&D. These expenditures also grew more rapidly than those for environmental protection research, surpassing it in the early 1980s. Therefore, if R&D expenditures reflect long-term concern, there seems to be relatively more importance attached to energy conservation as compared with environmental protection.

In addition to the general trends discussed above, consider the industrial sector and how it has tackled energy conservation.

The private sector clearly has an important role to play in providing finance that could be used for energy efficiency investments. In fact, governments can adjust their spending priorities in aid plans and through official support provided to their exporters, but they can influence the vast potential pool of private-sector finance only indirectly. Many of the most important measures to attract foreign investors include reforming macroeconomic policy frameworks, energy market structures and pricing, and banking; creating debt recovery programs; strengthening the commercial and legal framework for investment; and setting up judicial institutions and enforcement mechanisms. These are difficult tasks that often involve lengthy political processes.

Thus, the following important factors, which are adopted from a literature work[13] can contribute to improving energy conservation in real life. Fig. 6 presents the improvement factors of energy conservation.

ENERGY CONSERVATION AND SUSTAINABLE DEVELOPMENT

Energy conservation is vital for sustainable development and should be implemented by all possible means, despite the fact that it has its own limitations. This is required not only for us, but for the next generation as well.

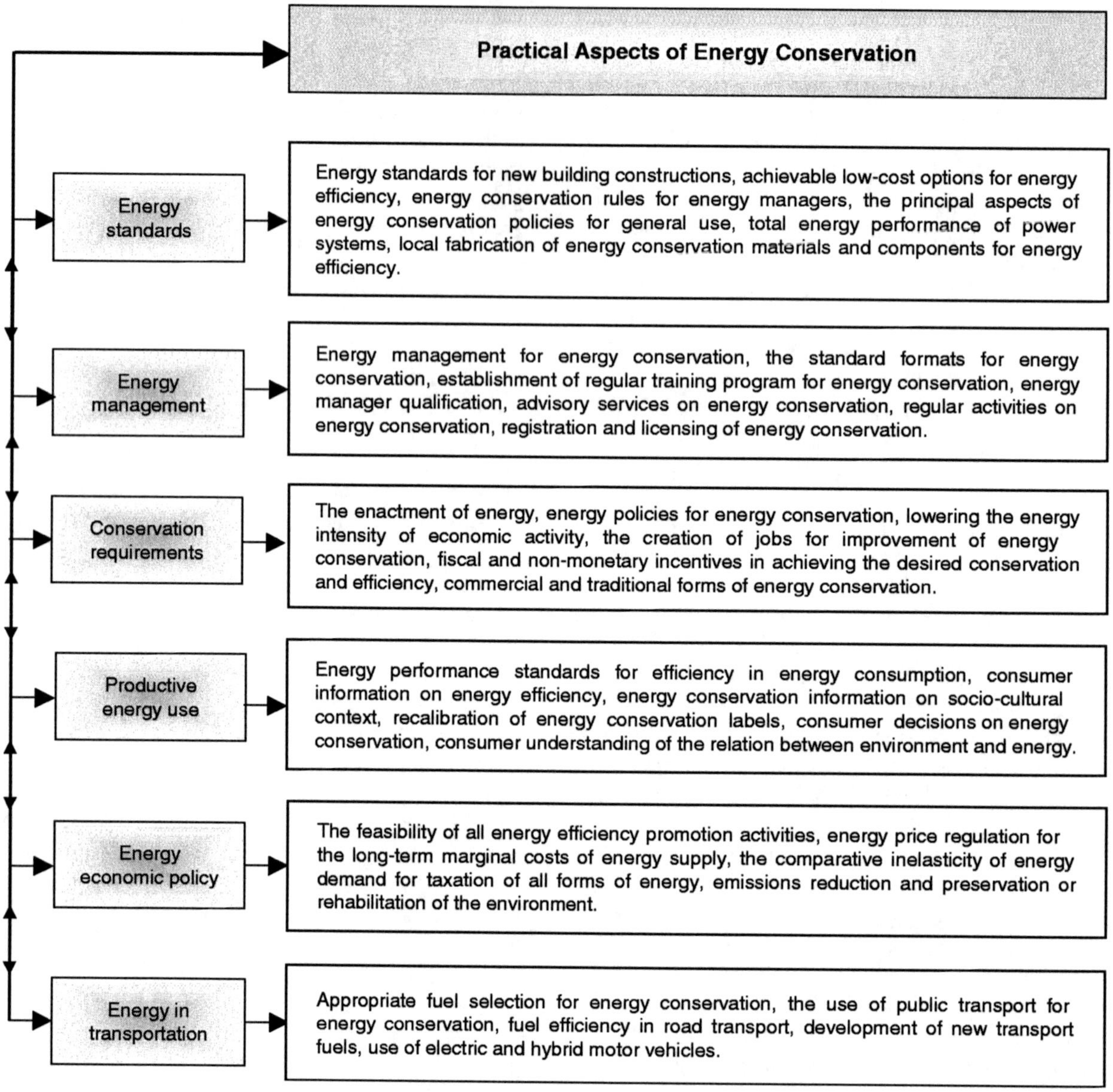
Practical Aspects of Energy Conservation
Energy standards
Energy standards for new building constructions, achievable low-cost options for energy efficiency, energy conservation rules for energy managers, the principal aspects of energy conservation policies for general use, total energy performance of power systems, local fabrication of energy conservation materials and components for energy efficiency.
Energy management
Energy management for energy conservation, the standard formats for energy conservation, establishment of regular training program for energy conservation, energy manager qualification, advisory services on energy conservation, regular activities on energy conservation, registration and licensing of energy conservation.
Conservation requirements
The enactment of energy, energy policies for energy conservation, lowering the energy intensity of economic activity, the creation of jobs for improvement of energy conservation, fiscal and non-monetary incentives in achieving the desired conservation and efficiency, commercial and traditional forms of energy conservation.
Productive energy use
Energy performance standards for efficiency in energy consumption, consumer information on energy efficiency, energy conservation information on socio-cultural context, recalibration of energy conservation labels, consumer decisions on energy conservation, consumer understanding of the relation between environment and energy.
Energy economic policy
The feasibility of all energy efficiency promotion activities, energy price regulation for the long-term marginal costs of energy supply, the comparative inelasticity of energy demand for taxation of all forms of energy, emissions reduction and preservation or rehabilitation of the environment.
Energy in transportation
Appropriate fuel selection for energy conservation, the use of public transport for energy conservation, fuel efficiency in road transport, development of new transport fuels, use of electric and hybrid motor vehicles.

Fig. 5 A flow chart of practical energy conservation aspects.

A secure supply of energy resources is generally considered a necessity but not a sufficient requirement for development within a society. Furthermore, sustainable development demands a sustainable supply of energy resources that, in the long term, is readily and sustainably available at reasonable cost and can be utilized for all required tasks without causing negative societal impact. Supplies of such energy resources as fossil fuels (coal, oil, and natural gas) and uranium are generally acknowledged to be finite. Other energy sources (such as sunlight, wind, and falling water) are generally considered to be renewable and, therefore, sustainable over the relatively long term. Wastes (convertible to useful energy forms through, for example, waste-to-energy incineration facilities) and biomass fuels usually also are viewed as being sustainable energy sources. In general, the implications of these statements are numerous and depend on how the term *sustainable* is defined.[14]

Energy resources and their utilization are intimately related to sustainable development. For societies to attain or try to attain sustainable development, much effort must be devoted not only to discovering sustainable energy resources, but also to increasing the energy efficiencies of processes utilizing these resources. Under these circumstances, increasing the efficiency of energy-utilizing devices is important. Due to increased awareness of the benefits of efficiency improvements, many institutes and agencies have started working along these lines. Many energy conservation and efficiency improvement programs have been developed and are being developed to reduce present levels of energy consumption. To implement these

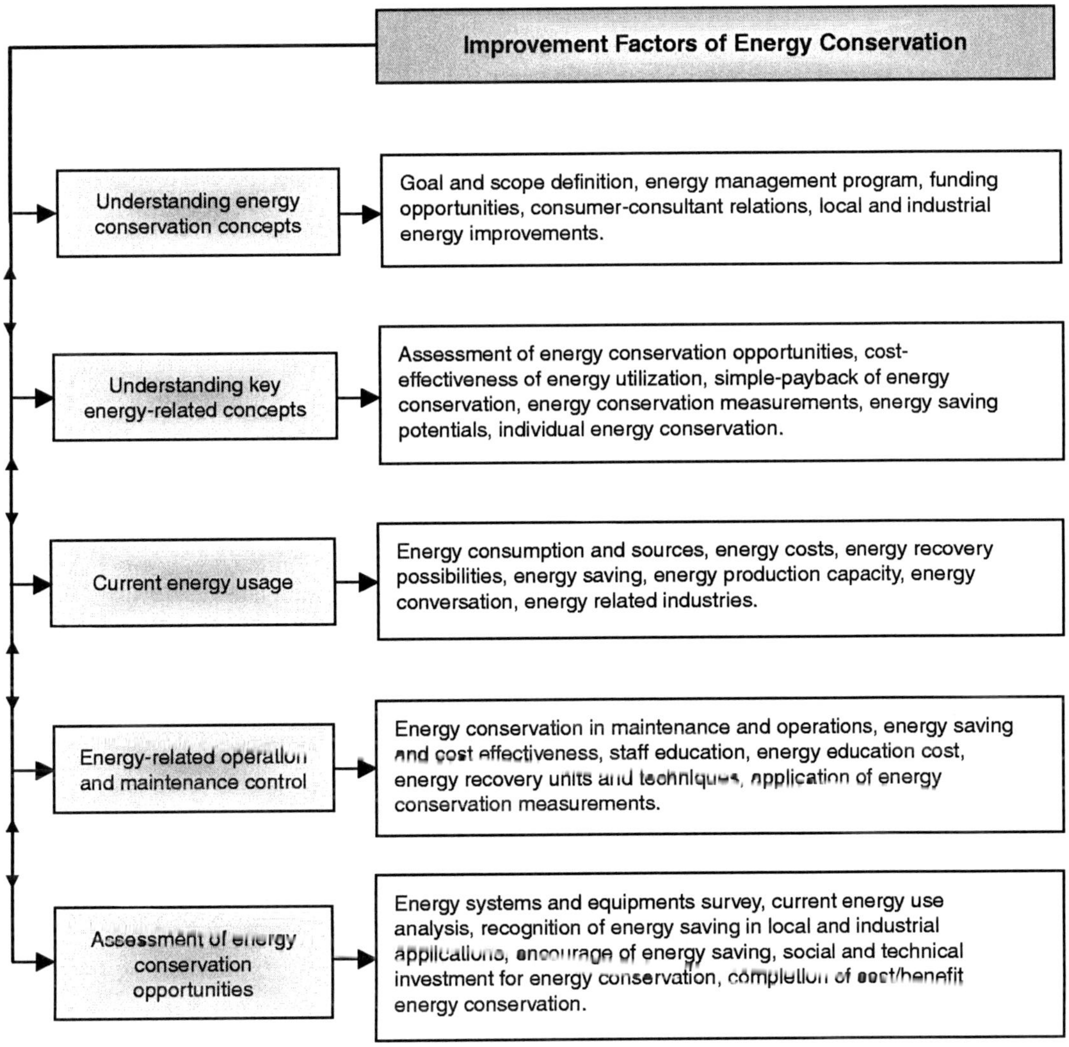

Improvement Factors of Energy Conservation
Understanding energy conservation concepts
Goal and scope definition, energy management program, funding opportunities, consumer-consultant relations, local and industrial energy improvements.
Understanding key energy-related concepts
Assessment of energy conservation opportunities, cost-effectiveness of energy utilization, simple-payback of energy conservation, energy conservation measurements, energy saving potentials, individual energy conservation.
Current energy usage
Energy consumption and sources, energy costs, energy recovery possibilities, energy saving, energy production capacity, energy conversation, energy related industries.
Energy-related operation and maintenance control
Energy conservation in maintenance and operations, energy saving and cost effectiveness, staff education, energy education cost, energy recovery units and techniques, application of energy conservation measurements.
Assessment of energy conservation opportunities
Energy systems and equipments survey, current energy use analysis, recognition of energy saving in local and industrial applications, encourage of energy saving, social and technical investment for energy conservation, completion of cost/benefit energy conservation.

Fig. 6 Improvement factors of energy conservation.

programs in a beneficial manner, an understanding is required of the patterns of "energy carrier" consumption-for example, the type of energy carrier used, factors that influence consumption, and types of end uses.[15]

Environmental concerns are an important factor in sustainable development. For a variety of reasons, activities that continually degrade the environment are not sustainable over time-that is, the cumulative impact on the environment of such activities often leads over time to a variety of health, ecological, and other problems. A large portion of the environmental impact in a society is associated with its utilization of energy resources. Ideally, a society seeking sustainable development utilizes only energy resources that cause no environmental impact (e.g., that release no emissions to the environment). Because all energy resources lead to some environmental impact, however, it is reasonable to suggest that some (not all) of the concerns regarding the limitations imposed on sustainable development by environmental emissions and their negative impacts can be overcome in part through increased energy efficiency. Clearly, a strong relationship exists between energy efficiency and environmental impact, because for the same services or products, less resource utilization and pollution normally are associated with increased energy efficiency.

Here, we look at renewable energy resources and compare them with energy conservation. Although not all renewable energy resources are inherently clean, there is such a diversity of choices that a shift to renewables carried out in the context of sustainable development could provide a far cleaner system than would be feasible by tightening controls on conventional energy. Furthermore, being by nature site-specific, they favor power system decentralization and locally applicable solutions more or less independently of the national network. It enables citizens to perceive positive and negative externalities of energy consumption. Consequently, the small scale of the equipment often makes the time required from initial design to operation short, providing greater adaptability in responding to unpredictable growth and/or changes in energy demand.

The exploitation of renewable energy resources and technologies is a key component of sustainable development.[11] There are three significant reasons for it:

- They have much less environmental impact compared with other sources of energy, because there are no energy sources with zero environmental impact. Such a variety of choices is available in practice that a shift to renewables could provide a far cleaner energy system than would be feasible by tightening controls on conventional energy.
- Renewable energy resources cannot be depleted, unlike fossil-fuel and uranium resources. If used wisely in appropriate and efficient applications, they can provide reliable and sustainable supply energy almost indefinitely. By contrast, fossil-fuel and uranium resources are finite and can be diminished by extraction and consumption.
- They favor power system decentralization and locally applicable solutions more or less independently of the national network, thus enhancing the flexibility of the system and the economic power supply to small, isolated settlements. That is why many different renewable energy technologies are potentially available for use in urban areas.

Taking into consideration these important reasons, the relationship between energy conservation and sustainability is finally presented as shown in Fig. 7.

ENERGY CONSERVATION IMPLEMENTATION PLAN

The following basic steps are the key points in implementing an energy conservation strategy plan[11]:

1. *Defining the main goals.* It is a systematic way to identify clear goals, leading to a simple goal-setting process. It is one of the crucial concerns and follows an organized framework to define goals, decide priorities, and identify the resources needed to meet those goals.
2. *Identifying the community goals.* It is a significant step to identify priorities and links among energy, energy conservation, the environment, and other primary local issues. Here, it is also important to identify the institutional and financial instruments.
3. *Performing an environmental scan.* The main objective in this step is to develop a clear picture of the community to identify the critical energy-use areas, the size and shape of the resource-related problems

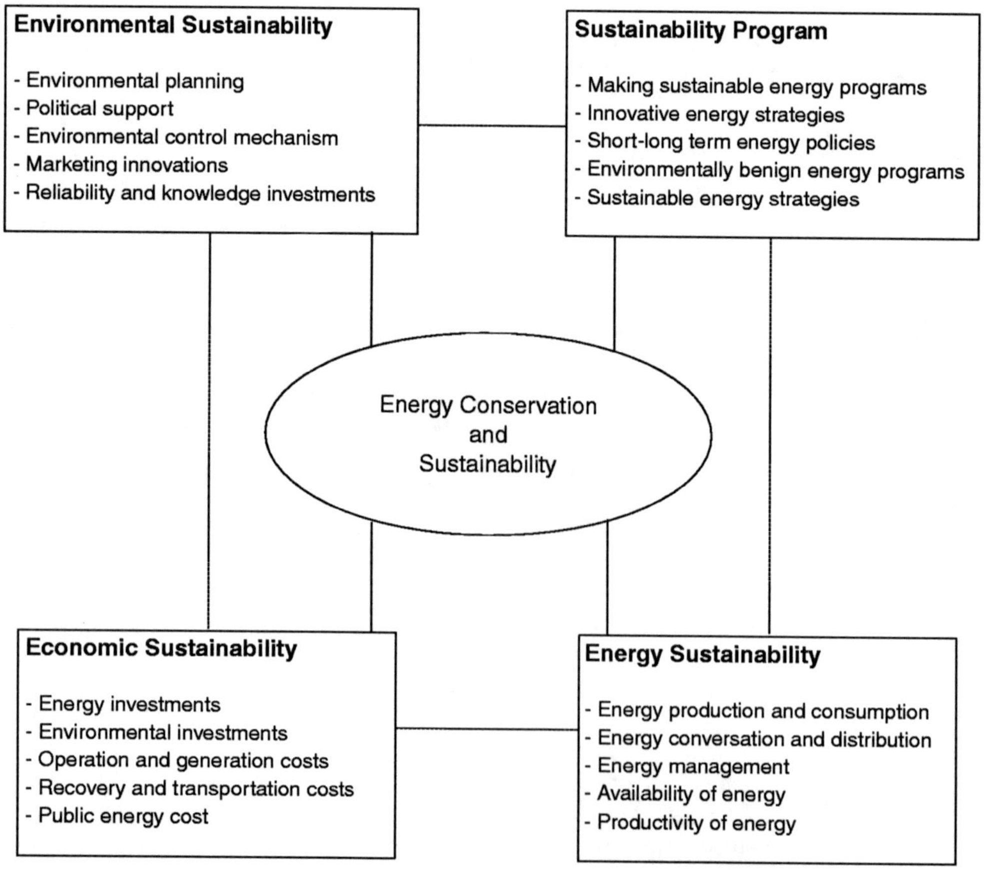
Environmental Sustainability
- Environmental planning
- Political support
- Environmental control mechanism
- Marketing innovations
- Reliability and knowledge investments
Sustainability Program
- Making sustainable energy programs
- Innovative energy strategies
- Short-long term energy policies
- Environmentally benign energy programs
- Sustainable energy strategies
Energy Conservation
and
Sustainability
Economic Sustainability
- Energy investments
- Environmental investments
- Operation and generation costs
- Recovery and transportation costs
- Public energy cost
Energy Sustainability
- Energy production and consumption
- Energy conversation and distribution
- Energy management
- Availability of energy
- Productivity of energy

Fig. 7 Linkages between energy conservation and sustainable development.

facing the city and its electrical and gas utilities, the organizational mechanisms, and the base data for evaluating the program's progress.

4. *Increasing public awareness.* Governments can increase other customers' awareness and acceptance of energy conservation programs by entering into performance contracts for government activities. They can also publicize the results of these programs and projects. In this regard, international workshops to share experiences on the operation would help overcome the initial barrier of unfamiliarity in countries.
5. *Performing information analysis.* This step carries out a wide range of telephone, fax, email, and Internet interviews with local and international financial institutions, project developers, and bilateral aid agencies to capture new initiatives, lessons learned, and viewpoints on problems and potential solutions.
6. *Building community support.* This step covers the participation and support of local industries and communities, and the understanding the nature of conflicts and barriers between given goals and local actors; improving information flows; activating education and advice surfaces; identifying institutional barriers; and involving a broad spectrum of citizen and government agencies, referring to the participation and support of local industrial and public communities.
7. *Analyzing information.* This step includes defining available options and comparing the possible options with various factors (e.g., program implementation costs, funding availability, utility capital deferral, potential for energy efficiency, compatibility with community goals, and environmental benefits).
8. *Adopting policies and strategies.* Priority projects need to be identified through a number of approaches that are best for the community. The decision-making process should evaluate the cost of the options in terms of savings in energy costs; generation of business and tax revenue; and the number of jobs created, as well as their contribution to energy sustainability and their benefit to other community and environmental goals.
9. *Developing the plan.* When a draft plan has been adopted, it is important for the community to review it and comment on it. The public consultation process may vary, but the aim should be a high level of agreement.
10. *Implementing new action programs.* This step involves deciding which programs to concentrate on, with long-term aims being preferred over short-term aims. The option that has the greatest impact should be focused on, and all details should be defined, no matter how difficult the task seems. Financial resources needed to implement the program should be identified.
11. *Evaluating the success.* The final stage is evaluating and assessing how well the plan performed, which helps identify its strengths and weaknesses and to determine who is benefiting from it.

ENERGY CONSERVATION MEASURES

For energy conservation measures, the information about the measure's applicability, cost range, maintenance issues, and additional points should be presented. Energy conservation involves efficiency improvements, formulation of pricing policies, good "housekeeping practices," and load management strategies, among other measures. A significant reduction in consumer energy costs can occur if conservation measures are adopted appropriately. The payback period for many conservation programs is less than 2 years.

In spite of the potentially significant benefits of such programs to the economy and their proven successes in several countries, conservation programs have not yet been undertaken on a significant scale in many developed and developing countries. Some reasons for this lack of energy conservation programs relate to the following factors:

- Technical (e.g., lack of availability, reliability, and knowledge of efficient technologies)
- Institutional (e.g., lack of appropriate technical input, financial support, and proper program design and monitoring expertise)
- Financial (e.g., lack of explicit financing mechanisms)
- Managerial (e.g., inappropriate program management practices and staff training)
- Pricing policy (e.g., inappropriate pricing of electricity and other energy commodities)
- Information diffusion (e.g., lack of appropriate information)

Reduced energy consumption through conservation programs can benefit not only consumers and utilities, but society as well. In particular, reduced energy consumption generally leads to reduced emissions of greenhouse gases and other pollutants into the environment.

Accelerated gains in energy efficiency in energy production and use, including those in the transportation sector, can help reduce emissions and promote energy security. Although there is a large technical potential for increased energy efficiency, there exist significant social and economic barriers to its achievement. Priority should be given to market forces in effecting efficiency gains. Reliance on market forces alone, however, is unlikely to overcome these barriers. For this reason, innovative and bold approaches are required by governments, in cooperation with industry, to realize the opportunities for energy efficiency improvements, and to accelerate the deployment of new and more efficient technologies.

Here, we look at energy conservation measures, which may be classified in six elements:

- Sectoral measures
- Energy conservation through systematic use of unused energy
- Energy conservation by changing social behavior

- International cooperation to promote energy conservation to counteract global warming
- Enhancing international and government-industry-university cooperation in developing technologies for energy conservation
- Promoting diffusion of information through publicity and education

The emphasis is on sectoral energy conservation. Table 4 presents some examples of such sectoral energy conservation measures. After determining which energy conservation measures are applicable, you should read the description of each of the applicable energy conservation measures. Information about the savings that can be expected from the measure, maintenance issues related to the measure, and other items to consider is provided for each energy conservation measure.

To evaluate the energy conservation measures, the following parameters should be taken into consideration[13]:

- *Cost estimation.* The first step is to estimate the cost of purchasing and installing the energy conservation measure. Cost estimates should be made for the entire development rather than for a single piece of equipment (e.g., obtain the cost for installing storm windows for an entire development or building, rather than the cost of one storm window). If you are planning to implement the energy conservation measure without the help of an outside contractor, you can obtain cost estimates by calling a vendor or distributor of the product. If, on the other hand, you will be using a contractor to install or implement the energy conservation measure, the contractor should provide estimates that include all labor costs and contract margins.
- *Data survey.* In this step, the questions on fuel consumption and cost should be listed for more than one possible fuel type (e.g., gas, oil, electric, or propane). The appropriate data for each fuel type should be selected and used accordingly for the cost estimation of each fuel.
- *Energy savings.* The amount of energy or fuel used should be estimated.
- *Cost savings.* This step determines the level of savings.
- *Payback period.* The last step in the cost/benefit analysis estimates the simple payback period. The payback period is found by dividing the cost of the measure by the annual cost savings.

LIFE-CYCLE COSTING

The term LCC for a project or product is quite broad and encompasses all those techniques that take into account

Table 4 Sectoral energy conservation measures.

Sector	Measures
Industrial	Strengthening of financial and tax measures to enhance adoption and improvement of energy saving technologies through energy conservation equipment investments
	Re-use of waste energy in factories and/or in surrounding areas
	Enhancing recycling that reduces primary energy inputs such are iron scraps and used papers, and devising measures to facilitate recycling of manufactured products
	Retraining of energy managers and diffusion of new energy saving technologies through them
	Creating database on energy conservation technologies to facilitate diffusion of information
Residential and Commercial	Revising insulation standards provided in the energy conservation law, and introducing financial measures to enhance adoption of better insulation
	Development of better insulation materials and techniques
	Developing' energy conservation' model homes and total energy use systems for homes
	Revising or adopting energy conservation standards for home and office appliances
	Developing more energy saving appliances
	Revising guidelines for managing energy use in buildings, and strengthening advisory services to improve energy management in buildings
Transportation	Because 80% of energy consumption of the sector is by automobiles, further improvement in reducing fuel consumption by automobiles is necessary together with improvement in transportation system to facilitate and reduce traffic flow
	Diffusion of information about energy efficient driving
	Adopting financial measures to enhance the use of energy saving transportation equipment such as wind powered boats

Source: Adapted from *Energy Conservation Policies and Technologies in Japan: A Survey.*[16]

Table 5 An example of life-cycle costing (LCC) analysis.

Cost of purchasing bulbs	Incandescent	Compact fluorescent
Lifetime of one bulb (hours)	1,000	10,000
Bulb price ($)	0.5	6.0
Number of bulbs for lighting 10,000 h	10	1
Cost for bulbs ($)	10 × 05 = 5.0	1 × 6 = 6
Energy cost		
Equivalent wattage (W)	75	12
Watt–hours (Wh) required for lighting for 10,000 h	75 × 10,000 = 750,000 Wh = 750 kWh	12 × 10,000= 120,000 Wh= 120 kWh
Cost at 0.05 per kWh	750 kWh × $0.05 = $37.5	120 kWh × $0.05 = $6
Total cost ($)	5 + 37.5 = 42.5	6 + 6 = 12

both initial costs and future costs and benefits (savings) of a system or product over some period of time. The techniques differ, however, in their applications, which depend on various purposes of systems or products. Life-cycle costing is sometimes called a cradle-to-grave analysis. A life-cycle cost analysis calculates the cost of a system or product over its entire life span. Life-cycle costing is a process to determine the sum of all the costs associated with an asset or part thereof, including acquisition, installation, operation, maintenance, refurbishment, and disposal costs. Therefore, it is pivotal to the asset management process.

From the energy conservation point of view, LCC appears to be a potential tool in deciding which system or product is more cost effective and more energy efficient. It can provide information about how to evaluate options concerning design, sites, materials, etc., how to select the best energy conservation feature among various options; how much investment should be made in a single energy conservation feature; and which is the most desirable combination of various energy conservation features.

A choice can be made among various options of the energy conservation measure that produces maximum savings in the form of reduction in the life-cycle costs. A choice can be made between double-glazed and triple-glazed windows, for example. Similarly, a life-cycle cost comparison can be made between a solar heating system and a conventional heating system. The one that maximizes the life-cycle costs of providing a given level of comfort should be chosen. The application of such techniques to energy conservation is related to determining the optimum level of the chosen energy conservation measure. Sometimes, energy conservation measures involve the combination of several features. The best combination can be determined by evaluating the net LCC effects associated with successively increasing amounts of other energy conservation measures. The best combination is found by substituting the choices until each is used to the level at which its additional contribution to energy cost reduction per additional dollar is equal to that for all the other options.

Illustrative Example

Here, we present an illustrative example on LCC to highlight its importance from the energy conservation point of view. This example is a simple LCC analysis of lighting for both incandescent bulbs and compact fluorescent bulbs, comparing their life-cycle costs as detailed in Table 5. We know that incandescents are less expensive (95% to heat and 5% to usable light) and that compact fluorescent bulbs are more expensive but much more energy efficient. So the question is which type of lighting comes out on top in an LCC analysis.

This example clearly shows that LCC analysis helps in energy conservation and that we should make it part of our daily lives.

CONCLUSION

Energy conservation is a key element in sectoral (e.g., residential, industrial, and commercial) energy utilization and is vital for sustainable development. It should be implemented by all possible means, despite the fact that it sometime has its own limitations. This is required not only for us, but for the next generation as well. A secure supply of energy resources is generally considered a necessary but not a sufficient requirement for development within a society. Furthermore, sustainable development demands a sustainable supply of energy resources that, in the long term, is readily and sustainably available at reasonable cost and can be utilized for all required tasks without causing negative societal impact.

An enhanced understanding of the environmental problems relating to energy conservation presents a high-priority need and an urgent challenge, both to allow the problems to be addressed and to ensure that the solutions are beneficial for the economy and the energy systems.

All policies should be sound and make sense in global terms-that is, become an integral part of the international process of energy system adaptation that will recognize the very strong linkage existing between energy requirements and emissions of pollutants (environmental impact).

In summary this study discusses the current environmental issues and potential solutions to these issues; identifies the main steps for implementing energy conservation programs and the main barriers to such implementations; and provides assessments for energy conservation potentials for countries, as well as various practical and environmental aspects of energy conservation.

REFERENCES

1. Azar, C.; Rodhe, H. Targets for stabilization of atmospheric CO2. Science **1997**, *276*, 1818–1819.
2. Midilli, A.; Ay, M.; Dincer, I.; Rosen, M.A. On hydrogen and hydrogen energy strategies: I: Current status and needs. Renew. Sust. Energy Rev. **2005**, *9* (3), 255–271.
3. BP. Workbook 2005. *Statistical Review—Full Report of World Energy*, British Petroleum, 2005, Available at http://www.bp.com/centres/energy (accessed July 25, 2005).
4. IEE. World Energy Outlook 2002. Head of Publication Services, International Energy, 2002, Available at http://www.worldenergyoutlook.org/weo/pubs/weo2002/WEO21.pdf.
5. Kazman, S. Global *Warming and Energy Policy*; CEI; 67–86, 2003. Available at http://www.cei.org.
6. Dincer, I. Energy and environmental impacts: Present and future perspectives. Energy Sources **1998**, *20* (4/5), 427–453.
7. Dincer, I. *Renewable Energy, Environment and Sustainable Development.* Proceedings of the World Renewable Energy Congress; September 20–25, 1998; Florence, Italy; 2559–2562.
8. Rosen, M.A.; Dincer, I. On exergy and environmental impact. Int. J. Energy Res. **1997**, *21* (7), 643–654.
9. Dincer, I. Renewable energy and sustainable development: A crucial review. Renew. Sust. Energy Rev. **2000**, *4* (2), 157–175.
10. Dincer, I. *Practical and Environmental Aspects of Energy Conservation Technologies.* Proceedings of Workshop on Energy Conservation in Industrial Applications; Dhahran, Saudi Arabia, February 12–14, 2000; 321–332.
11. Dincer, I. On energy conservation policies and implementation practices. Int. J. Energy Res. **2003**, *27* (7), 687–702.
12. Sissine, F. Energy efficiency: Budget, oil conservation, and electricity conservation issues. *CRS Issue Brief for Congress.* IB10020; 2005.
13. Nolden, S.; Morse, D.; Hebert, S. *Energy Conservation for Housing: A Workbook*, Contract-DU100C000018374; Abt Associates, Inc.: Cambridge, MA, 1998.
14. Dincer, I.; Rosen, M.A. A worldwide perspective on energy, environment and sustainable development. Int. J. Energy Res. **1998**, *22* (15), 1305–1321.
15. Painuly, J.P.; Reddy, B.S. Electricity conservation programs: Barriers to their implications. Energy Sources **1996**, *18*, 257–267.
16. Anon. *Energy Conservation Policies and Technologies in Japan: A Survey*, OECD/GD(94)32, Paris, 1994.

Energy Conservation: Benefits

Eric A. Woodroof
Profitable Green Solutions, Plano, Texas, U.S.A.

Wayne C. Turner
Industrial Engineering and Management, Oklahoma State University, Stillwater, Oklahoma, U.S.A.

Steven D. Heinz
Good Steward Software, State College, Pennsylvania, U.S.A.

Abstract

In addition to saving energy and reducing utility expenses, there are additional (often unreported) benefits from conserving energy. These financial and strategic *benefits extend beyond the utility budget*. These non-utility benefits contribute value worth an additional 18–50% of the energy savings, as demonstrated via a simple example. Calculations are shown on a spreadsheet, which can be downloaded for applications in your own facility. Download the spreadsheet (under the "Resources" tab) here: http://www.ProfitableGreenSolutions.com. Beyond illustrating these additional benefits, the goal of this entry is to motivate change towards saving more energy and money, while preserving more of our natural environment.

EXECUTIVE SUMMARY

"It's not the age . . . it's the mileage". . .

It is logical that a car driven 25% less each year will last longer. The same is true for most energy-consuming equipment, such as lights, motors, and even digital equipment. By turning "off" energy-consuming equipment when it is not needed, an organization can find a financial jackpot, which extends beyond the utility budget. *It doesn't matter how energy-efficient an organization is, there are savings from turning equipment "off" when it is not needed.* Listed here are some "secret" benefits of energy conservation and these are benefits that can be attained without a negative impact on productivity.

Budgetary Improvements

1. Efficient Net Income: When energy is conserved, utility budgets are reduced. This is no secret, but what is noteworthy is that conservation savings impact a bottom line far more efficiently than many other investment initiatives. *For example: an energy conservation program that saves $100,000 in operating costs is equivalent to generating $1,000,000 in new revenue (assuming the organization has a 10% profit margin). It is more difficult to generate $1,000,000 in new revenue, and would require more marketing, infrastructure, etc. Thus, the energy conservation/efficiency program is an investment with less risk and quickly improves cash flow.*
2. Extended Equipment Lives: If assets are lasting longer (owing to reduced operation per year), replacements are less frequent, thereby reducing capital budget requirements. *For example, if a lighting system is operating 30% fewer hours per year, it could last up to 30% longer. A 15-year replacement policy could be changed to 20 years.* Further savings could result from considering that if equipment lasts longer, then staff/engineering/project management time is reduced for reviewing new equipment proposals, evaluating competing bids, overseeing installation efforts, coordinating invoices and payables with accounting.
3. Reduced Maintenance Costs: When equipment runs fewer hours per year, maintenance material/labor requirements are reduced. *For example, if maintenance on a motor is done on a "run hour" basis and there are less "run hours" per year, there should be fewer maintenance visits. Further, if the motor is part of a ventilation system, air filter replacements would occur less often, reducing material and labor costs.* Predictive maintenance technologies can also assist in this strategy and reduce the cost per horsepower by 50%. (Ameritech)

Encyclopedia of Environmental Management DOI: 10.1081/E-EEM-120044952

4. Reduced Risk to Energy Supply Price Spikes: *For example, if less energy is consumed; the operational budget is less vulnerable when electric/gas/heating oil prices hit their seasonal spikes. The avoided costs can be worth millions to a large organization.*

Beyond the large financial benefits mentioned here, there are many strategic benefits of energy conservation, which can significantly add to your organization's "jackpot."

Long-Term Strategic Benefits

1. Ability to Sell "Carbon Credits": Organizations can claim emissions reductions from energy conservation. When energy is saved, power plants do not have to produce as much electricity, thereby reducing "smoke stack" emissions. *Emission benefits from energy conservation can be expressed in terms of "equivalent trees planted," or "equivalent barrels of oil not consumed."* There are environmental markets where "emissions credits" (from energy conservation) can be sold, generating revenue for an organization. These markets are already liquid in Europe (and are motivated by carbon-related legislation). California and other states already require emissions reporting and reductions, and federal regulations are in process that will open the door to a similar trading environment in the United States.[1]
2. Enhanced Public Image: Organizations that conserve/manage energy (thereby reducing emissions) can differentiate themselves as "environmentally friendly" and "good" members of a community. This can have tremendous political, strategic, competitive, and morale-building value for organizational leaders. Many benefits (such as attracting and retaining better employees, faculty, students, clients, suppliers, etc.,) result from being the "leader" in your field. A recent study showed that 92% of young professionals want to work for an organization that is environmentally friendly.[2] Even stock prices of corporations have been proven to improve dramatically (21.33% on average) when energy management programs are announced.[3]
3. Reduced Risk to Environmental/Legal Costs: If assets are replaced less frequently, an organization will generate less waste and be less vulnerable to environmental regulations governing disposal. (Disposal of batteries and fluorescent lamps is already regulated in most states.) Greater environmental regulations are inevitable and unforeseen legal costs can pose a significant expense and political risk.[4,5]

As will be shown in the example on the next page, benefits 2–7 represent a significant improvement (18%–50%) to the original savings estimates.

CLIMATE CHANGE AND ITS EFFECT ON ENERGY CONSERVATION APPROACHES

"The Writing is on the Wall"...

The glaciers are melting and climate change is here. Climate change is a result of changes in the Earth's atmosphere. The growth of "greenhouse gases" between 10,000 years ago and the 1800s was approximately 1% for that period. Since the 1800s, greenhouse gases have increased 33%. Thus, it is logical that this growth is due to human-caused activities with the dawn of the industrial age.

— *Time Magazine–Special Report, December 2007. Also quoted by the UN Intergovernmental Panel on Climate Change- February, 2007 as well as the US EPA website: http://www.epa.gov.*

The data is compelling and creating change in consumer choices.[6] Consumers are becoming more "green-minded" in their purchases, especially young people and college students. Studies show that more consumers are choosing to reduce their "carbon footprint," and thereby are choosing products, companies, and colleges that are more environmentally friendly. Carbon offset trading growth is greater than 200%: "in June 2007, the Chicago Climate Exchange reported that in the past 6 months, it had already traded 11.8 MtCO2e—more than had been traded in the entire year of 2006"[7] and 69% of consumers shop for brands aligned with a social cause.[8] Federal and state governments are introducing legislation that will mandate carbon emissions reporting and management.[9] In September 2006, Governor Schwarzenegger signed the California Global Warming Solutions Act, which mandates a 25% cut in emissions by 2020 and an 80% cut by 2050. In summary, the need for a "carbon diet" is driving activity in the energy-conservation industry.

The "Good" News . . .

Companies, colleges, and governments are responding to this growing "green" consumer market and competitors are innovating to be the "environmental leaders" of their fields. Hewlett-Packard says that in 2004, $6 billion of new business depended on answers to customer questions about the company's environmental record—a 660% growth from 2002.[10] "Sustainability efforts protect our license to grow" said Wal-Mart CEO Lee Scott in 2005.[11] Energy efficiency/conservation is ranked by corporate executives as the no. 1 way to reduce emissions in a cost-effective manner.[12] Because buildings contribute approximately 43% of the carbon emissions in the USA, an opportunity exists to reduce a large part of these emissions and become "environmental heros."[13] In addition, organizations perceived as more "environmentally friendly" can recruit better faculty, students, suppliers, and employees.[14] Finally, the "secret benefits" (discussed in the Executive Summary) are increasing in value and importance. *An energy conservation program is more valuable today because the material, waste, labor, emissions, and risk savings are more valuable in today's economy.*

A SIMPLE EXAMPLE TO DEMONSTRATE THE "SECRET BENEFITS"

A lighting conservation measure will serve as the example, although similar calculations could be applied towards motor systems. Motors and lights consume the majority of electricity in a typical building.[15] Computers and other digital equipment are also worth mentioning, because they can consume considerable amounts of energy, and it is noted that "plug loads" (computers, printers, and other digital equipment) have increased significantly during the past 20 years.

For this example, consider a large school with 10,000 light fixtures. Through a variety of energy conservation measures, it is common to reduce consumption by 25%.[16] First, we will calculate the dollar savings from electricity conservation. Then, we will show the "secret benefits," which have impacts beyond the utility budgets. A spreadsheet will illustrate the total savings/benefits.

Benefit 1: Reduced Utility Budget from Lighting Conservation

Assume the fluorescent lights are relatively new and consume 60 watts per 2-lamp fixture and operate 5000 hr/yr. (This example uses a standard T-8 lighting system, although the energy conservation savings would be even greater with a less efficient lighting system, such as a T-12.) Our baseline energy consumption is:

= (5000 hr/yr)(0.060 kW/fix)(10,000 fix)
= 3,000,000 kWh/yr

If the school pays approximately $.08/kwh, then the dollars spent on electricity for this lighting system:

= $240,000/yr

Thus, a 25% reduction from the baseline usage would equal: 750,000 kWh/yr, or $60,000/yr in savings, *which goes immediately to the bottom line* and improves cash flow. Note that we will not count demand (kW) savings as the electrical load reduction would most likely occur during non-occupied hours (off-peak electrical rates). *However, it is not unusual for conservation programs to reduce both kWh and kW.* In addition to "direct dollar savings," there are tax rebates and credits available that can further improve the financial results from energy conservation/efficiency programs/projects.

Benefit 2: The Value of Extended Equipment Lives (Reducing Capital Budgets)

If lights are used 25% less, the lighting system (ballasts) should last about 25% longer. Fluorescent fixture and wiring replacement costs are not included, as these components typically last longer than the ballast. We will address lamp life as a part of "maintenance costs" in Benefit 3. A lighting ballast is rated for 60,000 hr of operation. If the school operates the lights 5000 hr/yr, they would need to replace the ballasts at the 12th year and dispose of the old ballasts. If there are 5000 ballasts, each costing $25–$55 (material, installation, and disposal costs vary by geographic location), the replacement cost (minimum) at the 12th year would be:

= ($25/ballast)(5000 ballasts)
= $125,000

Annualized replacement cost would be:

= ($125,000)(1/12 yr)
= $10,417/yr

With a use rate of only 3750 hr/yr (a 25% reduction), the ballasts should last 16 years. If replacement occurs at failure or based on run time, these savings automatically occur. If replacements are planned in advance, planners should adjust their schedules to insure that savings are captured from extended equipment lives (not replacing assets pre-maturely). This would reduce the annualized replacement cost to:

= ($125,000)(1/16 yr)
= $7813/yr

Thus, the Annualized Savings (calculated as the difference between the original replacement cost minus the reduced replacement costs) are:

= $10,417/yr–$7813/yr
= $2604/yr(at $25 per ballast)

Using the same equations, at $55/ballast, the annualized savings, (from replacing at 16 years instead of 12 years) would be:

= $5729 per year

Thus, due to extended equipment life, we have reduced the annualized replacement cost by a minimum of $2604/yr to a maximum of $5729/yr.

Benefit 3: The Value of Reduced Maintenance Costs (Operating Expenses, Not Capital Replacements)

If the lights are used 25% less, the lamps should last about 25% longer. (Note that if lamps are turned "on" and "off" frequently at less than 3 hr intervals, the lamp's expected life will be reduced by approximately 25%, which would erode the savings in this category. Lamp life is rated at the factory by turning lamps on and off every 3 hr until they burn out. If the frequency of on/off cycling is less than 3 hr, lamp lives will decline by 25% on average. Therefore, turning a lamp off for longer periods is better than shorter periods. For example, it is better to find locations where you can turn off lamps for 5 hr out of 15 hr, instead of 1 min out of every 3 min, although the % time off is the same.) A typical fluorescent lamp life is 20,000 hr. With a use rate of 5000 hr/yr, the school would need to replace lamps at the 4th year. If there are 10,000 lamps, each costing \$3–\$5 (material, installation, and disposal costs vary by location), the replacement cost (minimum) at the 4th year would be:

= (\$3/lamp)(10,000 lamps)
= \$30,000

Annualized replacement cost would be \$30,000/4 = \$7500.

With a use rate of only 3750 hr, the lamps should last 5.3 years, thereby reducing the annualized replacement cost to:

= \$30,000/5.3 yr
= \$5660/yr

Thus, Annualized Savings are:

= \$7500–\$5660/yr
= \$1840/yr (at \$3/lamp)

Using the same equations, at \$5/lamp, the re-lamping cost would be \$50,000 and the annualized savings from replacing at 5.3 years instead of at 4 years would be = \$3066/yr.

Thus, due to extended lamp life, we have reduced the annualized maintenance cost by a minimum of \$1840/yr to a maximum of \$3066/yr.

Benefit 4: The Value of Reduced Risk to Energy Supply Price Spikes

Assume that on average, for one-quarter of the year, energy prices are 25%–50% higher (\$.02–\$.04 more per kWh) due to seasonal/supply spikes. Similar calculations could be used for systems that use natural gas, owing to its seasonal volatility.

If we are using less energy, we will pay less of a premium for the price spike. The avoided price spike premium is equal to:

= (price premium)(kWh saved)(premium period)
= (\$.02/kWh)(750,000 kWh/yr)(1/4)
= \$3750/yr

Using the same equations, a 50% price spike would represent an avoided premium worth:

= (price premium)(kWh saved)(premium period)
= (\$.04/kWh)(750,000 kWh/yr)(1/4)
= \$7500/yr

Thus, owing to reduced risk from price spikes, the avoided premiums are \$3750–\$7500/yr.

Benefit 5: The Value of Carbon Credits

According to the Environmental Protection Agency,[17] 1.37 lbs of CO_2 are created for every kWh burned. So, if we are saving 750,000 kWh/yr, the avoided power plant emissions would be equivalent to:

= (750,000 kWh saved)(1.37 lbs of CO_2/kWh)
= 1,027,500 lbs of CO_2 saved per year F

Translating lbs to Metric Tons:

= (1,027,500 lbs CO_2)(.000454 Metric Tons/lb)
= 466.5 Metric Tons of CO_2 saved per year

These avoided power plant emissions could be claimed as "carbon credits" and sold to another party who wants to buy "carbon credits." An "aggregator" may be required to trade carbon credits in small quantities. Note that as of this printing, European prices for carbon credits are well over five times the price of carbon credits in the USA. The US carbon market is expected to follow Europe's lead as US regulations begin to take effect. Therefore, it is logical that the US prices will approach the European prices, which are currently at \$34/metric ton.

Assuming for now a market price of \$6 per metric ton, the additional revenue generated by selling the carbon credits would be:

= (466 Metric Tons of CO_2/yr)(\$6/M-Ton)
= \$2799/yr

Using the same equations, at \$30 per metric ton, the additional revenue generated by selling the carbon credits would be:

= (466 Metric Tons of CO_2/yr)(\$30/M-Ton)
= \$13,980 per year

Thus, due to the new carbon market, there is a possible additional revenue stream worth a minimum of \$2799 to a maximum of \$13,980 per year from selling carbon credits. In addition, as carbon prices go higher . . . so does the value of this new revenue stream.

Benefit 6: The Value of Enhanced Public Image

Although calculation of this value is difficult and is not generalized here, it can be far greater than any of the benefits mentioned above. In today's "green-minded" economy, many organizations such as Patagonia, Google, GE, and Home Depot, have used "green" programs as a very effective marketing tool to differentiate themselves from the competition, achieve business objectives, secure and retain talent, improve productivity, and capture a greater market share.

The green-shaded area of Fig. 1 shows the "equivalent environmental benefits" from avoided power plant emissions. These reductions/benefits can be published in various places to improve the organization's green image with employees, clients, students, suppliers, distributors, shareholders, and other groups relevant to the success of an organization.

Thus, due to energy conservation program, the school can claim environmental benefits equivalent to removing 1008 cars off the road, thereby improving the school's public image. Although not calculated here, the benefits of attracting better faculty, students, employees, etc., could far out-weigh all the benefit estimates in this entry.

See Fig. 1 for additional expressions of environmental benefits.

Benefit 7: The Value of Reduced Risk of Environmental/Legal Costs

Although calculation of this value is also difficult and is not generalized here, it can be very significant. The risk is real, but unknown. This is demonstrated by the following environmental disasters that significantly crippled or destroyed the organizations deemed responsible:

- The Union Carbide accident in Bhopal
- Love Canal's hazardous waste
- Mercury poisoning at Alamogordo, NM

It is also interesting to note that Exxon's penalties and fees were four times the actual clean-up costs for the Valdez oil spill.

More relevant to this entry is that emissions regulations are quite likely to become a standard in the United States. Organizations that are implementing energy conservation programs will have a regulatory advantage over those that do not. Inaction could pose legal risks.

Thus, due to its energy conservation program, the school in this example can reduce its risk from unknown environmental and legal risks that may arise in the future.

Fig. 2 summarizes the dollar value from the benefits mentioned in this entry. The approach and calculations for these benefits could be used as a guide to identify the "secret benefits" of other energy consuming systems, such as HVAC and motors, etc.

CONCLUSION

This entry has presented additional benefits from energy conservation. The example described an energy conservation

PROFITABLEGREENSOLUTIONS

Complete Emissions Calculator

INSTRUCTIONS: Type in the kWh savings and see the emissions-environmental benefits in green-shaded areas. Insert your own $$ values for the Strategic Benefits in blue text.

Type the amount of electricity your program will save ⟹ 750,000 kWh/year

Emissions Reductions:	Annual Reductions	Reductions over 10 years
Conversion Factor: 1 kWh is worth 1.37 lbs of CO2 (Source: EPA 2006)		
GreenHouse Gas Reduction (in pounds of CO2)	1,027,500 lbs	10,275,000 lbs
or when converted to Metric Tons of CO2 >>>	466.5 Metric Tons	4,665 Metric Tons

Equivalent Environmental Benefits (mutually-exclusive):	Annual Reductions	Reductions over 10 years
Acid Rain Emission Reduction	5,625.0 lbs of SOx	56,250 lbs of SOx
Smog Emission Reductions	2,700.0 lbs of NOx	27,000 lbs of NOx
Barrels of Oil Not Consumed	1,085.0 Barrels	10,850 Barrels
Cars off the Road	100.8 Cars	1,008 Cars
Gallons of Gas not Consumed	53,130.3 Gallons	531,303 Gallons
Acres of pine trees reducing carbon	388.6 Acres	3,886 Acres

Strategic Benefits (quantifiable at site-specific level)	Annual Benefits	Benefits over 10 years
Annual Report to Shareholders,	?	?
Community Morale & "Green Image",	?	?
Productivity Improvements, Cost-Competitiveness	?	?
Avoided Future Capital Outlay	?	?
LEED Points, White Certificates, RECs	?	?
FREE Public Press (GREAT), Political/Strategic	?	?
Legal Risk Reduction, Avoided Penalties	?	?

Fig. 1 Complete emissions calculator.

Example: The "Secret Benefits" of Energy Conservation

	Additional Benefits Estimates Min	Additional Benefits Estimates Max
Assumptions: Baseline Electricity Expenses from the Lighting System = $240,000 per year. A 25% savings via basic energy conservation measures would yield $60,000 in savings/year	$/Year	$/Year
Value of "Secret Benefits" (most exist outside the utility budget)		
Benefit #2: Extended Equipment Lives (Avoided Annual Capital Costs)	$2,604	$5,729
Benefit #3: Reduced Maintenance Costs (Avoided Operational Expenses)	$1,840	$3,066
Benefit #4: Reduced Risk to Energy Price Spikes (Avoided Premium Costs)	$3,750	$7,500
Benefit #5: Selling Carbon Credits (emissions reductions via energy conservation)	$2,799	$13,980
Total Additional Value from Quantifiable "Secret Benefits">>>	**$10,993**	**$30,275**
% Savings of Baseline Electricity Expenses ($240,000/year) of the Lighting System	***4.6%***	***12.6%***
% Savings Improvement from Original Estimate of $60,000/year in Savings	***18.3%***	***50.5%***

Note: Estimates are Conservative because Dollar Values for Benefits #6 and #7 were not included here.

Fig. 2 Secret benefits of energy conservation.

project that was achieving a 25% reduction in electrical consumption from the lighting system. Beyond obvious energy savings, the "secret benefits" 2–5 yield additional value worth $10,993 to $30,275 per year. *In other words, if energy conservation saves 25% of a utility budget, the "secret benefits" are worth an additional 4.6%–12.6%.*

Looking at this in a different way, **the "secret benefits" contribute additional value worth a minimum 18% improvement from the original estimated savings of $60,000 per year.** *In other words, if we value the secret benefits as worth only an additional $10,993, this represents a minimum improvement of 18% to our energy savings of $60,000. In addition, there is a $4660 value improvement for each $10 rise in US carbon prices.*

Finally, all estimates in this entry **only included the quantifiable "secret benefits"** (benefits 2–5). Actual values could be much higher when accounting for enhanced public image and a reduction in legal and environmental risks (benefits 6 and 7).

We hope that this entry motivates additional action for energy conservation, dollar savings, and environmental benefits.

ACKNOWLEDGMENTS

This entry originally appeared as "Overcoming Barriers to Approval for Energy and Green Projects" in *Strategic Planning for Energy and Environment*, Vol. 27, No. 3, 2008. Reprinted with permission from AEE/Fairmont Press.

REFERENCES

1. Senate pases cap and trade legislation. New York Times, December 6, 2007.
2. How going green draws talent, cuts costs. Wall Street Journal, November 13, 2007.
3. Wingender, J.; Woodroof, E. When firms publicize energy management projects: Their stocks prices go up. Strategic Planning for Energy and the Environment **1997** (Summer Issue).
4. McCain-Lieberman Senate Proposal, 110th Congress, 8/2/2007.
5. Mayor Bloomberg calls for carbon tax. New York Times Articles, November 7, 2007.
6. If the entire world lived like North Americans, it would take three planet Earths to support the present world population (in 2006)—Source: Harvard Business Review on Business and the Environment.
7. State of the voluntary carbon market 2007, The Chicago Climate Exchange, July 18, 2007.
8. Survey data from United States Green Building Council.
9. Bingaman-Specter, Kerry-Snowe, Standers-Boxer Senate Proposals, 110th Congress, 8/2/2007.
10. Daniel, E. *Green to Gold*; Yale University Press, 2006.
11. Harvard Green Campus Initiative, UCSB Sustainability Program.
12. *Getting Ahead of the Curve: Corporate Strategies That Address Climate Change.* Pew Center on Global Climate Change, 2006.
13. Pew Center on Global Climate Change. *The U.S. Electric Power Sector and Climate Change Mitigation and Towards a Climate Friendly Built Environment.*
14. How going green draws talent, cuts costs. *Wall Street Journal*, November 13, 2007.
15. Association of Energy Engineers. Certified Energy Manager Program Workbook, 2008.
16. Gregerson, J. Cost effectiveness of commissioning 44 existing buildings, 5th National Conference on Building Commissioning.
17. EPA. Avoided power plant emissions calculations (updates October 2006. Available at http://www.ProfitableGreenSolutions.com ("Resources" tab).

Energy Conservation: Industrial Processes

Harvey E. Diamond
Energy Management International, Conroe, Texas, U.S.A.

Abstract
Energy Conservation in Industrial Processes will focus on energy conservation in industrial processes, will distinguish industrial processes and characteristics that differentiate them, will outline the analytical procedures needed to address them, will identify and discuss the main industrial energy intensive processes and some common ways to save energy for each of them, and will address managerial methods of conservation and control for industrial processes.

INTRODUCTION AND SCOPE

Energy conservation is a broad subject with many applications in governmental, institutional, commercial, and industrial facilities, especially because energy costs have risen so high in the last few years and continue to rise even higher. Energy conservation in industrial processes may well be the most important application—not only due to the magnitude of the amount of potential energy and associated costs that can be saved, but also due to the potential positive environmental effects such as the reduction of greenhouse gases associated with many industrial processes and also due to the potential of the continued economic success of all of the industries that provide jobs for many people.

This entry will focus on energy conservation in industrial processes—where energy is used to manufacture products by performing work to alter feedstocks into finished products. The feedstocks may be agricultural, forest products, minerals, chemicals, petroleum, metals, plastics, glass, or parts from other industries. The finished products may be food, beverages, paper, wood building products, refined minerals, refined metals, sophisticated chemicals, gasoline, oil, refined petroleum products, metal products, plastic products, glass products, and assembled products of any kind.

This entry will distinguish industrial processes and the characteristics that differentiate them in order to provide insight into how to most effectively apply energy conservation within industries. The level of applied technology, the large amount of energy required in many cases to accomplish production, the extreme conditions (e.g., temperature, pressure, etc.,) that are frequently required, and the level of controls that are utilized in most cases to maintain process control will be addressed in this entry.

This entry will outline the analytical procedures needed to address energy conservation within industrial processes and will comment on general analytical techniques that will be helpful in analyzing energy consumption in industrial facilities.

Many of the main energy intensive processes, systems, and equipment used in industries to manufacture products will be identified and discussed in this entry and some common ways to save energy will be provided.

This entry will cover main energy intensive processes, systems, and equipment in a general format. If more in-depth instruction is needed for explanation of a particular industrial process, system, or type of equipment or regarding the analytical procedures required for a specific process, then the reader should refer to the many other articles included in this Encyclopedia of Energy Engineering and Technology, to references at the Association of Energy Engineers, to the references contained in this entry, and if further detail is still needed, then the reader should contact an applicable source of engineering or an equipment vendor who can provide in-depth technical assistance with a specific process, system, or type of equipment.

In addition to the analytical methods of energy conservation, managerial methods of energy conservation will be briefly discussed. The aspects of capital projects versus managerial and procedural projects will be discussed. The justification of managerial efforts in industrial processes will be presented.

INDUSTRIAL PROCESSES—DIFFERENTIATION

Industrial processes require large amounts of energy, sometimes the highest level of technology, and often require very accurate process controls for process specifications, safety, and environmental considerations.

Industrial processes utilize an enormous amount of energy in order to produce the tons of production that are being produced within industrial facilities. Industrial processes utilize over one-third of the total energy consumed in America.[1] Consider the amount of energy that is

Encyclopedia of Environmental Management DOI: 10.1081/E-EEM-120042818

required to melt all of the metals being manufactured, to vaporize all of the oil and gasoline being refined, to dry all of the finished products that are made wet, to heat all of the chemicals that react at a certain temperature, to vaporize all of the chemicals that must be distilled for purity, to vaporize all of the steam that is used to heat industrial processes, to mechanically form all of the metal objects that we use, etc.,—this list is too long to be fully included in this entry. This is an enormous amount of energy that produces all of the things that humans need and use—food, clothes, homes, appliances, cars, municipal facilities, buildings, roads, etc.

The level of technology required by current industrial processes is the highest in many cases and it is always at a high level in most industrial processes. Most industrial processes are utilizing technology that has been developed in the last 100 years or so, and consequently it has been further improved in the most recent years. Industrial processes most often utilize aspects of chemistry and physics in a precise manner in order to produce the sophisticated products that benefit people in our culture today. Very often, industrial processes require a very high or low temperature or pressure. Often they require a very precise and sophisticated chemistry and commonly they require highly technical designed mechanical processes. The application of electrical equipment and facilities in industrial processes is the highest level of technology for electrical power systems.

Industrial processes often require the highest level and accuracy of controls in order to produce products that meet product specifications, keep processes operating in a safe manner, and maintain environmental constraints. Due to each of these requirements or due to a combination of these requirements, the process controls for the processes within industrial facilities are often real-time Distributed Control Systems (DCSs), that are of the most sophisticated nature. A typical DCS for industrial processes functions to control process variables instantaneously on a real-time basis, whereby each process variable is being measured constantly during every increment of time and a control signal is being sent to the control element constantly on a real-time basis. The accuracy of a DCS in an industrial facility today is comparable to that of the guidance systems that took the first men to the moon.

Most industrial facilities with DCS controls also utilize a process historian to store the value of most process variables within the facility for a certain increment or period of time. The stored values of these process variables are used for accounting purposes and technical studies to determine optimum operating conditions and maintenance activities.

ENERGY CONSERVATION ANALYSES FOR INDUSTRIAL PROCESSES

In any industrial facility, the first analysis that should be performed for the purpose of energy conservation should be that of determining a balance of the energy consumed for each form of energy. This balance is used to determine how much energy is consumed by each unit, area, or division of the plant or possibly by major items of equipment that are consuming major portions of the energy consumed by the plant. This balance should be determined for each form of energy, whether it is for natural gas, electricity, coal, fuel oil, steam, etc., (see Table 1 for an example of an energy balance). It might be best that this determination not be called a balance (in that the numbers might not exactly come to a precise balance) but that it sufficiently quantifies the amount of energy consumed by each unit, area, or division of the plant. A better term for this determination might be an "Energy Consumption Allocation." The term balance is more usually applied to chemical and thermodynamic processes where heat and material balances are worked together mathematically to determine a calculated variable and the numbers have to exactly balance in order to arrive at the correct mathematical solution.

Table 1 Energy balance.

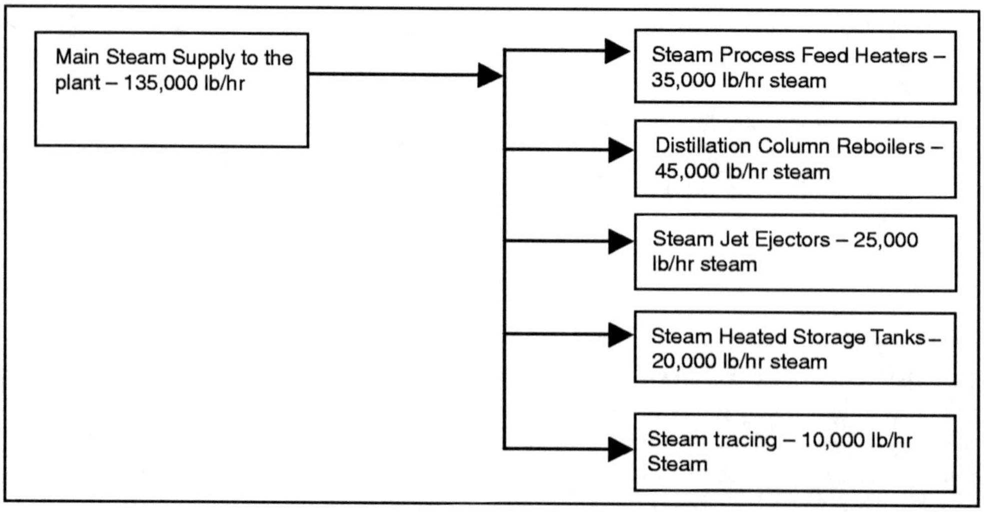
Main Steam Supply to the plant – 135,000 lb/hr
Steam Process Feed Heaters – 35,000 lb/hr steam
Distillation Column Reboilers – 45,000 lb/hr steam
Steam Jet Ejectors – 25,000 lb/hr steam
Steam Heated Storage Tanks – 20,000 lb/hr steam
Steam tracing – 10,000 lb/hr Steam

Once the amount of energy that is being actually consumed by each part of a plant has been determined, an energy consumption analysis should be performed for each item of energy consuming equipment and each major energy consuming facility in order to determine how much energy should realistically be consumed by each part of the plant. Notice that these calculations are called "realistic" as opposed to just theoretical because the object of these calculations is to determine as closely as possible how much actual energy each item of equipment or part of the plant should be consuming. By comparing these calculations with the actual energy consumption allocations mentioned above, it should be possible to obtain at least an initial indication of where energy is being wasted in a plant (see Table 2 for an example of an energy consumption analysis for a process feed heater). During the course of obtaining the values of process variables that are required to make these energy consumption analyses, it is possible that indications will be observed of energy wastage due to the presence of an inordinate value of some process variable, such as too high or low of a temperature or pressure. When this type of indication is discovered, it usually also provides insight into what is operating in the wrong way to waste energy. There are numerous instances of energy wastage that can be discovered during these analyses, such as the inordinate manual control of a process, loose operational control of a process variable, or simply not shutting down a piece of equipment when it is not needed.

The analyses discussed in the above paragraph encompass all technical engineering science subjects, such as chemistry, thermal heat transfer, thermodynamics, fluid mechanics, mechanical mechanisms, and electrical engineering.

The next set of energy conservation analyses that should be performed are used to calculate the efficiencies of each item of equipment or facility to which efficiency calculations would be applicable, such as boilers, fired heaters, furnaces, dryers, calciners, and all other thermodynamic processes (efficiency calculations for boilers and other combustion equipment is available in the *Energy Management Handbook* by Wayne C. Turner and Steve Doty[2] and in the *Guide to Energy Management* by Barney L. Capehart, Wayne C. Turner, and William J. Kennedy)[3]

Table 2 Example of energy consumption analysis: process feed HTR.

- A process feed heater heats 1,199,520 lb/day of liquid feed material with a specific heat of 0.92 Btu/lb-°F, from 67 to 190°F. A realistic heater efficiency for this type of heater is determined to be 88%. The amount of realistic heat required for this heater is calculated to be: Q = 1,199,520 lb/day × 0.92 Btu/lb-°F × (190°F − 67 °F) ÷ 0.88 = 154,247,367.3 Btu/day of realistic heat consumption.
- It is observed that this feed heater is consuming 186,143,720 Btu/day.
- This feed heater is being operated in a wasteful way and is wasting over 20% of its heat.

and for electrical and mechanical equipment such as motors, pumps, compressors, vacuum pumps, and mechanical machinery. Once the actual efficiencies of any of the above have been determined, these numbers can be compared to the realistic efficiency for the type of equipment or facility that is prevalent throughout industry. These calculations and comparisons will also reveal wastage of energy and will frequently identify the causes of energy wastage and the possible issues to be corrected.

The next level of energy conservation analysis that may be performed is process analysis that can be conducted on a particular chemical, thermodynamic, thermal, fluid flow, mechanical, or electrical process. These analyses are usually performed by experienced engineers to examine the process itself and the process variables to determine if the process is being operated in the most effective and efficient manner. Here again, an indication will be provided as to whether or not energy is being wasted in the actual operation of the process. Chemical, thermo-dynamic, thermal, fluid flow, and other processes, as well as combinations of any of these processes can often require process simulation software such as PROMAX by Bryan Research and Engineering, Inc.,[4] in order to properly analyze these processes. The analysis of distillation columns, evaporators, and dryers can fall into this category. A good example would be the process analysis of a distillation column to determine if an effective and efficient level of reflux to the column and reboiler duty is being used.

Another analysis that has been very useful in the past few years in identifying energy conservation projects is Pinch analysis. This analysis is performed on thermodynamic and thermal processes in order to identify sources of energy within existing processes that can be used to supply heat for these processes instead of having to add additional heat to the entire process. The net effect is to reduce the amount of energy required for the overall process. The performance of a Pinch analysis on a particular process or facility will usually identify capital projects where revisions to the facility can be made to decrease the total amount of energy required. These are very often waste heat recovery projects. See "Use Pinch Analysis to Knock Down Capital Costs and Emissions" by Bodo Linnhoff, Chemical Engineering, August 1994[5] and "Pinch Technology: Basics for the Beginners."[6]

MAIN INDUSTRIAL ENERGY PROCESSES, SYSTEMS, AND EQUIPMENT

This section provides an overview and a list of the more common energy intensive industrial processes that are used to manufacture products in industrial facilities. Most energy intensive industrial processes can be classified into about eight general process categories—process heating, melting, chemical reactions, distillation-fractionation, drying, cooling, mechanical processes, and electrical processes.

These processes are intended to be the main general energy intensive processes that are most commonly used and to which variations are made by different industries in order to make a specific product. In this regard, this is an overview—these processes are often not the specific process but a general category to which variations can be made to achieve the specific process.

In the following paragraphs, each process will be discussed by addressing its description, what systems it utilizes, what products are generally made, how it uses energy, and frequent ways that energy can be saved.

Common energy consuming systems and equipment that work to manufacture products in industrial facilities are also listed below and discussed in the same manner as the main industrial processes, as they are also common to industrial facilities and are related to these processes.

Process Heating

- *Description.* The addition of heat to a target in order to raise its temperature. Temperatures can range from the hundreds to the thousands in industrial process heating.
- *Energy form.* Heat must be generated and transferred to the intended object or medium.
- *Energy unit.* Btu, calorie, joule, therm or watt-hour.
- *Examples.* The application of heat in order to heat feed materials, to heat chemical processes, to heat metals for forming, to heat materials for drying, to heat materials in a kiln or calciner, to heat minerals and metals for melting.
- *Applied systems.* Combustion systems, steam systems, thermal systems, hot oil systems, heating medium systems such as Dowtherm[7] or Paracymene,[8] and electrical resistance or induction heating systems.
- *Common equipment.* Boilers, furnaces, fired heaters, kilns, calciners, heat exchangers, waste heat recovery exchangers, preheaters, electrical resistance heaters, and electrical induction heaters.
- *Common energy conservation issues.* Keeping the heat targeted at the objective—proper insulation, seals on enclosures, eliminating leakage, and eliminating unwanted air infiltration. Control issues—maintaining sufficient control of the heating process, temperatures, and other process variables to avoid waste of heat. Management issues—shutting down and starting up heating processes at the proper times in order to avoid waste of heat and management of important process variables to reduce the amount of heat required to accomplish the proper process. Application of Pinch Technology—identify process areas where heat can be recovered, transferred, and utilized to reduce the overall process heat requirement. Waste heat recovery.

Melting and Fusion

- *Description.* The addition of heat or electrical arc energy at a high temperature in order to melt metals, minerals, or glass. The melting process involves more than just process heating, it involves fluid motion, fluid density equilibrium, chemical equilibrium, cohesion, and sometimes electro-magnetic inductance. Reference: "The study showed that the fluid equations and the electromagnetic equations cannot be decoupled. This suggests that arc fluctuations are due to a combination of the interactions of the fluid and the electromagnetics, as well as the rapid change of the boundary conditions."[9]
- *Energy forms.* Heat at high temperatures or electrical arc energy in the form of high voltage and high current flowing in an arc.
- *Energy units.* Heat, Btu, calorie, joule, or therm.
- *Electrical arc.* kWhrs.
- *Examples.* Melting of ores in order to refine metals such as iron, aluminum, zinc, lead, copper, silver, etc. Melting of minerals in order to refine minerals such as silica compounds, glass, calcium compounds, potassium alum, etc.
- *Applied systems.* Combustion systems, chemical reactions, and electrical systems.
- *Common equipment.* Blast furnaces, arc furnaces, electrical resistance heaters, and electrical induction heaters.
- *Common energy conservation issues.* Pre-condition of feed material—moisture content, temperature, etc. Feed method—efficiency of melting process effected by the feed method, feed combinations, and feed timing. Control of electromagnetics during melting and use of magnetic fields during separation. Over-heating can waste energy without yielding positive process results. Heat losses are due to poor insulation, the failure of seals, or lack of shielding or enclosure.

Chemical Reactions

- *Description.* Chemicals react to form a desired chemical, to remove an undesired chemical, or to break out a desired chemical. The chemical reaction can involve heat, electrolysis, catalysts, and fluid flow energy.
- *Energy forms.* Heat, electrolysis, and fluid flow.
- *Energy units.* Heat, Btu, calorie, joule, or therm.
- *Electrolysis.* kWhrs.
- *Fluid flow.* ft-lbs or kg-m.
- *Examples.* Reaction of chemical feed stocks into complex chemicals, petrochemical monomers into polymers, the oxidation of chemicals for removal, dissolving of chemicals to remove them, reaction of chemicals with other chemicals to remove them, the reaction of lignin with reactants in order to remove it from cellulose, the electroplating of metals out of solution to refine them.
- *Applied systems.* Feed systems, catalysts systems, heating systems, cooling systems, vacuum systems, run-down systems, separation systems, filtering systems, and electroplating systems.
- *Common equipment.* Reactors, digesters, kilns, calciners, smelters, roasters, feed heaters, chillers, pressure vessels, tanks, agitators, mixers, filters, electrolytic cells.

- *Common energy conservation issues*. Close control of heating and cooling for chemical reactions. Close control of all reaction process variables—balance of all constituents, amount of catalyst, proper timing on addition of all components. Management of feedstocks, catalysts, and run-down systems for proper timing and correct balance for highest efficiency. Pinch analysis of feed heating, run-down products, cooling system, etc. Conservation of heating and cooling—proper insulation, sealing, and air infiltration. Waste heat recovery.

Distillation-Fractionation

- *Description*. A thermo-dynamic and fluid flow equilibrium process where components of a mixture can be separated from the mix due to the fact that each component possesses a different flash point.
- *Energy form*. Heat and fluid flow.
- *Energy units*. Heat, Btu, calorie, joule, or therm.
- *Fluid flow*. Ft-Lbs or KG-M.
- *Examples*. Distillation-fractionation of hydrocarbons in oil and gas refineries and chemical plants. Distillation of heavy hydrocarbons in gas processing plants where natural gas is processed to remove water and heavy hydrocarbons.
- *Applied systems*. Feed heating systems, over-head condensing systems, reflux systems, reboil systems, vacuum systems.
- *Common equipment*. Distillation columns or towers, over-head condenser heat exchangers and accumulators—vessels, reflux pumps, reboiler heat exchangers, feed pumps, feed—effluent heat exchangers, vacuum steam jet ejectors.
- *Common energy conservation issues*. Feed temperatures, reflux ratios, reboiler duty. Close control on pressures, temperatures, feed rates, reflux rates, and reboil duty. Management of overall operation timing—running only when producing properly. Concurrent use of vacuum systems—only when needed. Pinch analysis for feed and effluent streams and any process cooling systems. Proper insulation and elimination of lost heat for fired heater reboilers.

Drying

- *Description*. The use of heat and fluid flow to remove water or other chemical components in order to form a more solid product.
- *Energy forms*. Heat and fluid flow.
- *Energy units*. Heat, Btu, calorie, joule, or therm.
- *Fluid flow*. Ft-Lbs or KG-M.
- *Examples*. Spray dryers that dry foods, sugar, fertilizers, minerals, solid components, and chemical products. Rotary dryers that dry various loose materials. Line dryers that dry boards, tiles, paper products, fiberglass products, etc. Other dryers that dry all kinds of products by flowing heated air over finished products in an enclosure.
- *Applied systems*. Combustion systems, steam systems, thermal heating systems, cyclone systems, air filter systems, incinerator systems, Regenerative Thermal Oxidizer (RTO) systems.
- *Common equipment*. Spray dyers, spray nozzles, natural gas heaters, steam heaters, electrical heaters, blowers, fans, conveyors, belts, ducts, dampers.
- *Common energy conservation issues*. Efficient drying process for the components being eliminated. Proper amount of air flowing through dryer for drying. Proper insulation, seals, and elimination of lost heat due to infiltration. Waste heat recovery.

Process Cooling

- *Description*. The removal of heat by a cooling medium such as cooling water, chilled water, ambient air, or direct refrigerant expansion.
- *Energy form*. Heat.
- *Energy unit*. Btu, calorie, joule, or therm.
- *Examples*. Cooling water or chilled water circulated through cooling heat exchanger, an air cooled heat exchanger, or a direct expansion evaporator that cools air for process use.
- *Applied systems*. Cooling water systems, chilled water systems, refrigerant systems, thermal systems.
- *Common equipment*. Cooling towers, pumps, chillers, refrigeration compressors, condensers, evaporators, heat exchangers.
- *Common energy conservation issues*. Use evaporative cooling as much as possible. Keep chillers properly loaded. Restrict chilled water flow rates to where 10°F temperature difference is maintained for chilled water. Limit cooling water pumps to the proper level of flow and operation. Apply Pinch analysis to achieve most efficient overall cooling configuration. Proper insulation, seals, and elimination of air infiltration.

Mechanical Processes

- *Description*. Physical activities that involve force and motion that produce finished products. Physical activities can be discrete or can be by virtue of fluid motion.
- *Energy form*. Physical work.
- *Energy unit*. Ft-Lbs or KG-M.
- *Examples*. Machining of metals, plastics, wood, etc.; forming or rolling or pressing of metals, minerals, plastics, etc.; assembly of parts into products; pumping of slurries thru screens or filters for separation; cyclone separation of solids from fluids; pneumatic conveyance systems that remove and convey materials or products and separate out solids with screens or filters.
- *Applied systems*. Machinery, electrical motors, hydraulic systems, compressed air systems, forced draft or induced draft conveyance systems, steam systems, fluid flow systems.

- *Common equipment*. Motors, engines, turbines, belts, chains, mechanical shafts, bearings, conveyors, pumps, compressors, blowers, fans, dampers, agitators, mixers, presses, moulds, rolls, pistons, centrifuges, cyclones, screens, filters, filter presses, etc.
- *Common energy conservation issues*. Equipment efficiencies, lubrication, belt slippage, hydraulic system efficiency, compressed air system efficiency. Control of process variables. Application of variable speed drives and variable frequency drives. Management of system and equipment run times.

Electrical Processes

- *Description*. The application of voltage, current, and electromagnetic fields in order to produce products.
- *Energy form*. Voltage-current over time; electromagnetic fields under motion over time.
- *Energy units*. KWh.
- *Examples*. Arc welding, arc melting, electrolytic deposition, electrolytic fission, induction heating.
- *Applied systems*. Power generator systems, power transmission systems, amplifier systems, rectifier systems, inverter systems, battery systems, magnetic systems, electrolytic systems, electronic systems.
- *Common equipment*. Generators, transformers, relays, switches, breakers, fuses, plates, electrolytic cells, motors, capacitors, coils, rectifiers, inverters, batteries.
- *Common energy conservation issues*. Proper voltage and current levels, time intervals for processes, electromagnetic interference, hysteresis, power factor, phase balance, proper insulation, grounding. Infrared scanning of all switchgear and inter-connections.

Combustion Systems

Combustion systems are found in almost all industries in boilers, furnaces, fired heaters, kilns, calciners, roasters, etc. Combustion efficiency is most usually a prime source of energy savings.

Boilers and Steam Systems

Boilers and steam systems may well be the most widely applied system for supplying process heat to industrial processes. "Over 45% of all the fuel burned by U.S. manufacturers is consumed to raise steam."[10] Boiler efficiencies, boiler balances (when more than one boiler is used), and steam system issues are usually a prime source of energy savings in industrial facilities.

Flare Systems and Incinerator Systems

Flare and incinerator systems are used in many industrial facilities to dispose of organic chemicals and to oxidize volatile organic compounds. Proper control of flares and incinerators is an issue that should always be reviewed for energy savings.

Vacuum Systems

Vacuum systems are used to evaporate water or other solvents from products and for pulling water from products in a mechanical fashion. Vacuum systems are also used to evacuate hydrocarbon components in the petroleum refining process. Vacuum systems are frequently used in the chemical industry to evacuate chemical solvents or other components from a chemical process. Steam jet ejectors and liquid ring vacuum pumps are commonly used to pull vacuums within these systems. The efficiencies of the ejectors and the liquid ring vacuum pumps can be a source of energy savings as well as the management of vacuum system application to production processes. Pneumatic conveyance systems that utilize a fan or blower to create a low-level vacuum are sometimes used to withdraw materials or products from a process and separate the matter within a screen or filter. For large conveyance systems, the efficiencies of the equipment and the management of their operation can be a source of energy savings.

Furnaces, Fired Heaters, Kilns, Calciners

The above comments on combustion systems are applicable to these equipment items and additional energy savings issues can be found relative to them.

Centrifugal Pumps

Centrifugal pumps are used widely in industries. The flow rate being pumped is a primary determining factor for the amount of power being consumed and it is sometimes higher than required. Good control of the pumping rate is an important factor in saving energy in centrifugal pumps. The application of variable frequency drives to the motor drivers can be a good energy saving solution for this issue.

Fans and Blowers

The flow rate for fans and blowers is analogous to the pumping rate above for centrifugal pumps. Good control of the flow rate and the possible application of Variable Frequency Drive (VFD) apply here as well for fans and blowers.

Centrifugal Compressors

Compressors are used widely in industry. The above discussions of flow rates, control of flow rates, and application of VFDs apply here as well. Centrifugal compressors frequently will have a recycle flow that is controlled in order to prevent the compressor from surging. Close control of this recycle flow at its minimum level is very important for compressor efficiency.

Liquid Ring Vacuum Pumps

As mentioned above in several places, liquid ring vacuum pumps are used widely in industry. The amount of sealing liquid that is recycled to the pump and the temperature of the sealing liquid are important determinates of the efficiency of the Liquid Ring Vacuum Pump (LRVP).

The above overview and list of industrial processes, systems, and equipment has been general in nature due to the limitations of this entry. Greater and more specific familiarity with each of these industrial energy intensive processes, systems, and equipment will yield greater applicable and effective insight into ways to save energy related to each of these items.

CAPITAL PROJECTS VERSUS IMPROVED PROCEDURES

Energy conservation effort applied in industrial facilities can identify capital projects whereby the facilities can be changed in order to achieve greater overall energy efficiency or the efforts can identify changes to in day-to-day operating and maintenance procedures that can reduce waste of energy and also improve the overall efficiency of the facility. Frequently, energy saving procedural changes to day-to-day operations and maintenance activities within an industrial plant can be identified by taking and recording operating data once the processes, systems, and equipment have been studied and analyzed for energy consumption. Procedural changes to operations and maintenance within an industrial plant can often amount to low costs or possibly no costs to the facility. This aspect of energy conservation is often overlooked by highly technical personnel that have worked hard to design industrial facilities because they have technically designed the facility very well for energy consumption considerations and the more mundane activities related to day-to-day operation and maintenance tend to not register in their highly technical perspective. None-the-less, a considerable amount of energy can usually be saved within most industrial processes, systems, and equipment due to changes in the way they are operated and maintained. A general tendency within industrial plants is that operations will often operate the processes and systems at a point that provides a comfortable separation between an operating variable and its limitation in order to understandably ensure no upsets occur within the process or system. However, with the cost of energy being what it is today, it is frequently found that a significant amount of energy can be saved by operating processes and systems more tightly and efficiently, even though it may require more attention, increased control, and the monitoring of process variables.

EFFECTIVE ENERGY MANAGEMENT SYSTEMS

Another aspect of energy conservation that can be very productive in saving energy within industrial processes is that of an effective energy management system. An effective energy management system is comprised of operational and maintenance managers functioning in conjunction with an accurate and concurrent data collection system in order to eliminate waste and improve overall efficiency of industrial processes. It is not possible to manage any activity unless the activity is being properly monitored and measured with key performance metrics (KPMs). The data collection system part of an effective energy management system within any industrial facility provides the accurate and concurrent measurement data (KPMs) that is required in order to identify actions that are needed to eliminate waste of energy and improve overall efficiency of the facility. An effective energy management system is first built upon acquiring total knowledge of the facility down into every level of operation and maintenance of the facility. Such a level of thoroughness and complete analysis of energy consumption within a facility is sometimes referred to as *Total Quality Energy* management.[11] Once an effective energy management system has been established and is effectively controlling energy consumption of an industrial facility, it should be maintained, in effect, so that it will continue to monitor KPMs to maintain energy conservation for the facility. An effective energy management system within an industrial manufacturing facility can eliminate as much as two to three percent of the energy costs by eliminating waste of energy on a day-to-day operational and maintenance basis. In most industrial facilities, this level of cost reduction is significant and will justify an effective energy management system.

CURRENT NEED FOR GREATER ENERGY CONSERVATION IN INDUSTRY

With the present cost of all forms of energy today, it would certainly seem logical that all of industry would be seeking greater energy conservation efforts within their facilities. Unfortunately, many corporate industrial managers are not aware of the true potential of conserving energy within their processes and facilities. Greater awareness of the ability to conserve energy on the basis of increased efficiencies of processes, systems, and equipment is needed; and also due to the application of an effective energy management system. For the good of society and environment, corporate industrial managers should be more open to the possibility of the improvement of industry that will work to sustain their business and improve the world that we live in. This is in opposition to corporate political thinking, which does not want to consider making changes and wants no one to interfere with their present activities. Human beings should be willing to examine themselves and make changes that will make things better. The same outlook should be applied to businesses and industry in order to make things better. Greater management support is needed in industry today to accomplish greater and very much needed increased energy conservation.

CURRENT APPLICATION OF INCREASED ENERGY MANAGEMENT

With the recent technological advancements that have been made in digital computer and communications systems, data collection systems can be implemented in industrial facilities in a much more cost effective manner. Wireless communication systems for metering and data collection systems have advanced dramatically in the last few years and network-based computer communication has enabled whole new systems for measurement and control. With all of these new fields of configuration for data collection systems, with the increased technology, and with the lower costs to accomplish data collection systems, it is now possible to apply energy management systems to industry today with much greater applicability. Hopefully this will be recognized and result in greater applications of effective energy management systems.

From recent observations, it appears that most of industry today is a candidate for improved and more effective energy management systems. In conjunction with the increased technology and lower cost potentials, it seems that there is a definite match between supply and need for the application of increased energy management systems.

SUMMARY

Industrial processes have commonality in processes, systems, and equipment. There are logical and systematic analyses that can be performed in industrial processes that can identify ways to save energy. Effective energy management systems are needed in industry today and there are great possibilities to save energy in industrial processes. Energy can be conserved in industrial processes by analyses that will improve efficiencies, by implementation of procedures that eliminate waste, and by application of an effective energy management system.

REFERENCES

1. U.S. Department of Energy–Energy Efficiency and Renewable Energy, available at http://www.eere.energy.gov/EE/industry.html (accessed on 2006).
2. Turner, W.C.; Doty, S. *Energy Management Handbook*, 6th Ed.; Fairmont Press: Lilburn, GA, 2005.
3. Capehart, B.L.; Turner, W.C.; Kennedy, W.J. *Guide to Energy Management*, 5th Ed.; Fairmont Press: Lilburn, GA, 2004.
4. Bryan Research and Engineering, Inc. PROMAX; BRE, Bryan, TX; available at http://www.bre.com (accessed on 2006).
5. Linnhoff, B. *Use pinch analysis to knock down capital costs and emissions*. Chemical Engineering August 1994 http://www.che.com.
6. Solar Places Technology. *Pinch Technology: Basics for the Beginners,* available at http://www.solarplaces.org/pinchtech.pdf (accessed on 2006).
7. Dowtherm, Dow Chemical http://www.dow.com/heattrans/index.html (accessed on 2006).
8. Paracymene, Orcas International, Flanders, NJ 07836 available at http://www.orcas-intl.com (accessed on 2006).
9. King, P.E. *Magnetohydrodynamics in Electric Arc Furnace Steelmaking*. Report of Investigations 9320; United States Department of the Interior, Bureau of Mines, available at http://www.doi.gov/pfm/ar4bom.html (accessed on 2006).
10. U.S. Department of Energy–Energy Efficiency and Renewable Energy, available at http://www.eere.energy.gov/EE/industry.html the common ways (accessed on 2006).
11. Energy Management International, Inc. Total Quality Energy, available at http://www.wesaveenergy.com (accessed on 2006).

Energy Conservation: Lean Manufacturing

Bohdan W. Oppenheim
U.S. Department of Energy Industrial Assessment Center, Loyola Marymount University, Los Angeles, California, U.S.A.

Abstract
Productivity has a major impact on energy use and conservation in manufacturing plants—an impact often more significant than optimization of the equipment energy efficiency. This entry describes Industrial Processes, which represents the current state-of-art in plant productivity. A significant opportunity for energy savings by transforming production into single-piece Lean Flow is demonstrated. The impact of major individual productivity elements on energy is discussed. Simple metrics and models are presented as tools for relating productivity to energy. Simple models are preferred because productivity is strongly influenced by intangible human factors such as work organization and management, learning and training, communications, culture, and motivation, which are difficult to quantify in factories.

INTRODUCTION

At the time of this writing (2005), the world is experiencing strong contradictory global trends of diminishing conventional energy resources and rapidly increasing global demands for these resources, resulting in substantial upwards pressures in energy prices. Because the energy used by industry represents a significant fraction of the overall national energy use, equal to 33% in the United States in the year 2005, a major national effort is underway to conserve industrial energy.[1] The rising energy prices place escalating demands on industrial plants to reduce energy consumption without reducing production or sales, but by increasing energy density.

Optimization of industrial hardware and its uses, including motors and drives, lights, heating, ventilation and cooling equipment, fuel-burning equipment, and buildings, are well understood, have been practiced for years,[2] and are important in practice. However, they offer only limited energy conservation opportunities, rarely exceeding a few percent of the preoptimization levels. In contrast, the impact of productivity on energy use and energy density offers dramatically higher savings opportunities in energy and in other costs. In the extreme case, when transforming a factory from the traditional "process village" batch-and-queue system to the state-of-the-art, so-called Lean system, the savings in energy can reach 50% or more.

The best organization of production known at this time is called Lean, developed at Toyota in Japan.[3] It is the flow of value-added work through all processes required to convert raw materials to the finished products with minimum waste. Major elements of Lean organization include: steady single-piece flow with minimum inventories and no idle states or backflow; flexible production with flexible equipment and operators and flexible floor layouts ready to execute the order of any size profitably and just-in-time; reliable and robust supplies of raw materials; minimized downtime due to excellent preventive maintenance and quick setups; first-pass quality; clean, uncluttered, and well-organized work space; optimized work procedures; and, most importantly, an excellent workforce–well trained, motivated, team-based and unified for the common goals of having market success, communicating efficiently, and being well-managed. The Lean organization of production is now well understood among productivity professionals, but it is not yet popular among the lower tier suppliers in the United States. Its implementation would save energy and benefit the suppliers in becoming more competitive.

The engineering knowledge of energy conservation by equipment improvements is well understood and can be quantified with engineering accuracy for practically any type of industrial equipment.[2] In contrast, industrial productivity is strongly influenced by intangible and complex human factors such as management, work organization, learning and training, communications, culture, and motivation. These work aspects are difficult to quantify in factory environments. For this reason, the accuracy of productivity gains and the related energy savings are typically much less accurate than the energy savings computed from equipment optimization. Simple quantitative models with a conservative bias are therefore recommended as tools for energy management in plants. This entry includes some examples. They are presented in the form of energy savings or energy cost savings that would result from implementing a given productivity improvement, or eliminating a given productivity waste, or as simple metrics measuring energy density.

It is remarkable that in most cases, these types of energy savings occur as a natural byproduct of productivity improvements, without the need for a direct effort

Encyclopedia of Environmental Management DOI: 10.1081/E-EEM-120042824

centered on energy. Thus, the management should focus on productivity improvements. In a traditional non-Lean plant intending to transform to Lean production, the first step should be to acquire the knowledge of the Lean system. It is easily available from industrial courses and workshops, books,[3,4] and video training materials.[6] The next step should be the actual transformation of production to Lean. Most of the related energy savings will then occur automatically. Implementation of individual productivity elements such as machine setup time reduction will yield some energy savings, but the result will not be as comprehensive as those yielded by the comprehensive implementation of Lean production.

TRADITIONAL VS. LEAN PRODUCTION

The traditional organization of production still used frequently in most factories tends to suffer from the following characteristics:

- Supplier selection is based on minimum cost, resulting in a poor level of mutual trust and partnership, the need for receiving inspection, and often large inventories of raw materials (RM).
- Work-in-progress (WIP) is moving in large batches from process village to process village and staged in idle status in queues in front of each machine, while the machine moves one piece at a time. This work organization is given the nickname "batch-and-queue" (BAQ).[3]
- Finished goods (FG) are scheduled to complex forecasts rather than customer orders, resulting in large inventories.
- The floor is divided into "process villages" populated with large, complex, and fast similar machines selected for minimum unit cost.
- Minimum or no information is displayed at workstations, and the workers produce quotas.
- Work leveling is lacking, which results in a random mix of bottlenecks and idle processes.
- Unscheduled downtime of equipment occurs frequently.
- Quality problems with defects, rework, returns, and customer complaints are frequent.
- Quality assurance in the form of 100% final inspections attempts to compensate for poor production quality.
- The floor space is cluttered, which makes moving around and finding items difficult.
- The workforce has minimum or no training and single skills.
- The management tends to be authoritarian.
- A language barrier exists between the workers and management.
- There is a culture of high-stress troubleshooting rather than creative trouble prevention.

In such plants, the waste of materials, labor, time, space, and energy can be as much as 50%–90%.[3]

The Lean production method developed primarily at Toyota in Japan under the name Just-In-Time (JIT), and generalized in the seminal work[3] is the opposite of the traditional production in almost all respects, as follows:

- Raw materials are bought from reliable supplier–partners and delivered JIT in the amount needed, at the price agreed, and with the consistently perfect quality that obviates incoming inspection.
- Single-piece flow (SPF) of WIP is steadily moving at a common takt time (Takt time is the common rhythm time of the pieces moving from workstation to workstation on the production line. It is the amount of time spent on EACH operation. It precisely synchronizes the rate of all production operations to the rate of sales JIT.), from the first to the last process.
- The FG are produced to actual customer orders JIT resulting in minimum inventories.
- The floor is divided into flexible production lines with small simple machines on casters that can be pushed into position and setup in minutes.
- The labor is multiskilled, well motivated and well trained in optimized procedures.
- Quality and production status are displayed on large visible boards at each workstation, making the entire production transparent for all to see.
- Preventive maintenance assures no unscheduled downtime of equipment.
- All process operators are trained in in-line quality checks and variability reduction.
- No final inspection is needed, except for occasional sampled checks of FG.
- Defects, rework, returns, and customer complaints are practically eliminated.
- The floor space is clean and uncluttered.
- The workforce is trained in company culture and commonality of the plant mission, customer needs, workmanship, and quality.
- The culture promotes teamwork, multiple job skills, supportive mentoring management, and company loyalty.
- The management promotes trouble prevention and "stopping the line" at the first sign of imperfection so that no bad pieces flow downstream.

According to Womack et al. the transformation from traditional to Lean production can reduce overall cost, inventory, defects, lead times by 90%, and space by 50%, and vastly increase plant competitiveness, customer satisfaction, and workforce morale. The resultant energy savings can be equally dramatic. Liker,[4] contains interviews with industry leaders who have succeeded in this transformation.

IMPACT ON ENERGY

The impact of productivity on plant energy falls into the following two broad categories:

1. Productivity improvements that save infrastructure energy. These improvements reduce the energy consumed by all plant support systems, which tend to be energized regardless of the actual production activities, such as lights, space cooling and heating devices, cooling towers, combustion equipment (boilers, molten metal furnaces), air compressors, forklift battery chargers, conveyors, etc. To the first approximation, the infrastructure energy is reduced in proportion to the production time reductions, which can be huge in the Lean system. In order to perform more detailed estimates of the infrastructure energy savings, the management would have to conduct detailed energy accounting and understand how much energy is used by each support system under different production conditions. This knowledge is rarely available; therefore the former simplistic approach, combined with conservative estimates, offer useful tools.
2. Process energy savings. In this category, the energy savings of process equipment are obtained by improving the process productivity. Examples include the reduction of unscheduled machine downtime or setup time and the elimination of process variability, defects, rework, scrap, excessive labor time, etc.

Single Piece Flow (SPF)

Changing the traditional BAQ production to Lean production is by far the most effective productivity transformation a plant can undertake, creating dramatic savings in the overall throughput time, cost, quality, and energy. The example shown in Fig. 1 compares just one aspect of the transformation—a reduction of batch size from five to one, i.e., the SPF. In both cases, four processes of equal one-minute takt time are assumed. The benefits of the SPF alone are dramatic, as follows:

1. In BAQ, the batch is completed in 20 min and in SPF in only 8 min, a 60% reduction.
2. In BAQ, only one machine at a time produces value, while three others are idle. If the idle machines remain energized, as is the case, e.g., with injection molding, three of the four machines (75%) would be wasting energy, and doing it for 16 min each, adding up to 64 min of machine energy wasted. In the SPF system, no machine energy is wasted as no machine would be idle, except for the lead and tail of each process of 4 min, adding up to 16 min of machine energy wasted, a savings of 75% from BAQ.
3. Reducing the batch throughput time by 60% reduces the infrastructure energy by the same amount, assuming the production is completed faster and the plant is de-energized. Alternatively, the freed 60% time and energy could be used for additional production and profits.
4. An important additional benefit is that in SPF, a defect can be detected on the first specimen–as soon as it reaches the next process, while in the BAQ, the entire batch may be wasted before the defect is discovered and a corrective action undertaken, with the energy used for making the batch wasted.

This simple example clearly illustrates the dramatic impact of SPF on both overall productivity and energy consumption. Typically, as the factories transform to the Lean system, their sales, production, and profits increase simultaneously and the energy used decreases. A convenient metric to track the overall benefit is the gross energy density, ED_1 or ED_2:

$$ED_1 = \frac{EC_T}{P} \tag{1a}$$

$$ED_2 = \frac{EC_T}{AC} \tag{1b}$$

where EC_T is the overall annual cost of energy in the plant, P is the number of products produced per year and AC is the total annual costs (sales minus net profit). ED_1 should be used if similar products are made most of the time, and ED_2 should be used if the plant has a wide menu of dissimilar products. The ED ratios will decrease as progress is made from BAQ to SPF. If the volume of production remains constant during the transformation, energy savings and energy cost savings alone may be more convenient metrics to track plant energy efficiency.

Inventory Reduction

All inventories, whether in RM, WIP, or FG, beyond the immediate safety buffers, are detrimental. Inventory means that company capital is "frozen" on the floor; cutting into the cash flow; wasting labor for inventory control, storage, and security; wasting infrastructure energy for lights, forklift energy, and possible cooling or heating of the inventory spaces if the goods require temperature or humidity control; wasting space and the associated lease/mortgage fees and taxes; and becoming scrap if not sold (a frequent waste in large inventories). Inventory and inventory space reductions lead to infrastructure energy savings. Process energy can also be saved by not making the FG that end up in inventory, cannot be sold, and become scrap. Womack and Jones[3] and Liker[4] contain case studies for, among others, inventory reductions. A convenient nondimensional metric to track the overall impact of all inventories on energy savings is

$$EC_T \times \frac{I_T}{AC} \tag{2}$$

where I_T is the number of inventory turns per year.

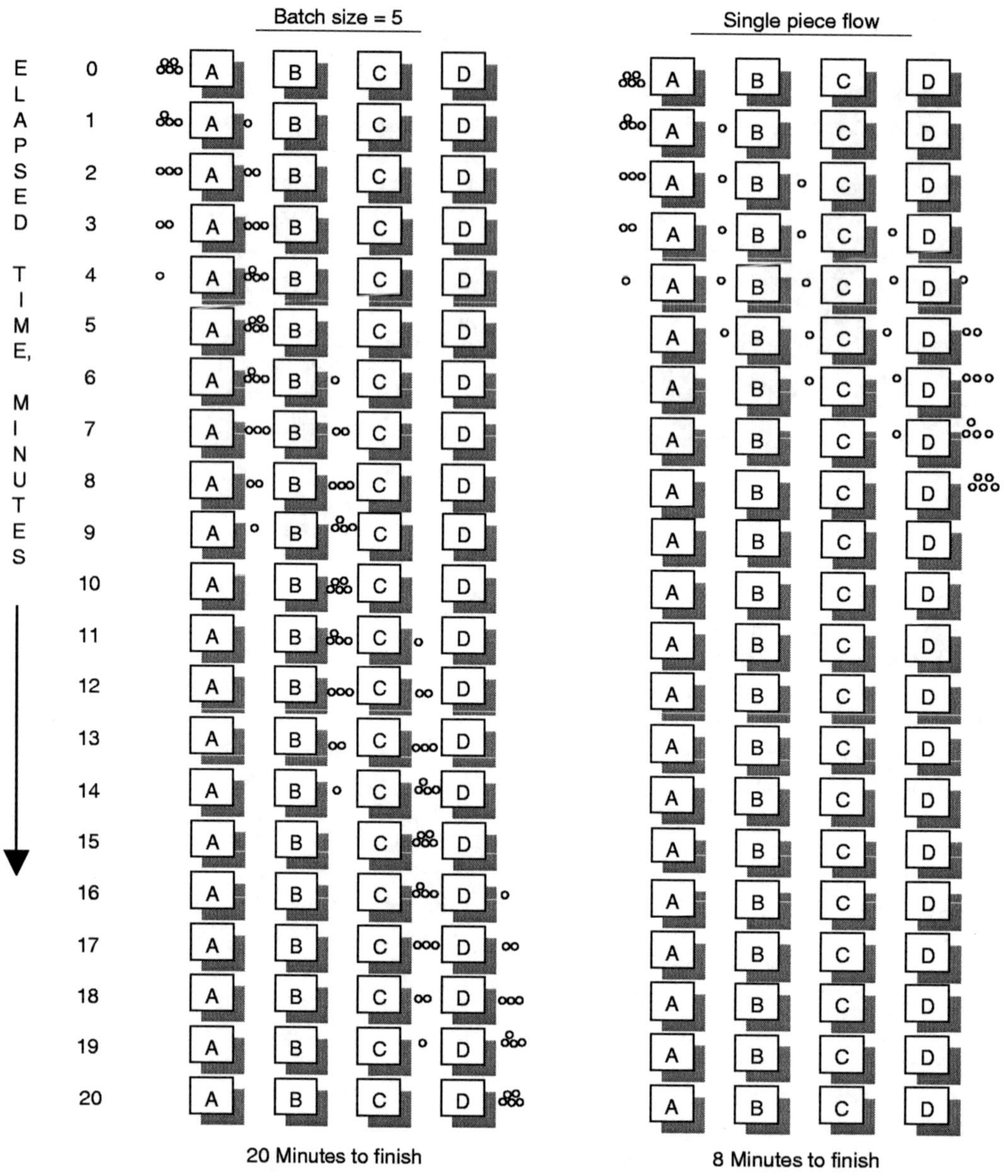
Batch size = 5
Single piece flow
ELAPSED TIME, MINUTES
A
B
C
D
20 Minutes to finish
8 Minutes to finish

Fig. 1 BAQ with batch size of five vs SPF.

Workmanship, Training, and Quality Assurance

In the ideal Lean system, the processes, equipment, procedures, and training are perfected to the degree that guarantees consistent and robust production with predictable effort, timing, quality, and cost; with no variability, defects, or rework, and with maximum ergonomics and safety. This is accomplished by a consistent long-term strategy of continuous improvement of all the above elements, including intensive initial training of the workforce and subsequent retraining in new procedures. A procedure must be developed for each process until it is robust and predictable and optimized for minimum overall cost, required quality, maximum ergonomics, and safety. Process operators must be trained in the procedures as well as in the process quality assurance, and they must be empowered to stop the process and take corrective action or call for help if unable to avoid a defect. Management culture must be supportive for such activities. Any departure from this ideal leads to costly penalties in quality, rework, delays, overtime or contract penalties, crew frustrations, and customer dissatisfaction. These, in turn, have negative impacts on energy as follows:

1. Defects require rework, which requires additional energy to remake or repair the part. The best metric to use here is the energy or energy cost per part used in the given defective process multiplied by the number of bad parts produced per year.
2. Variability in the process time or delays caused by defects mean that the production takes more time and

Table 1 Energy waste from poor workmanship.

A plant with \$20,000,000 in sales and \$2,000,000 in profits spends \$1,000,000 on energy per year. The typical order requires 10 processes of roughly equal enegy consumption. The production equipment consumes 60% and the supportive infrastructure consumes 40% of the plant energy. Sequential process #5 has the defect rate of 10%. In order to compensate for the defects, the first 5 processes must produce 10% extra pieces. The annual waste of energy cost (and the energy cost savings, if the defective process is fixed) is then:

(\$1,000,000/yr)(5/10 processes) (60% process energy) (10% defect rate) = \$30,000/yr (3)

The additional production time of 10% waste not only the cost of the process energy computed in (3) but also the infrastructure energy cost of:

(\$1,000,000/yr) (40% infrastructure energy) (10% defects) = \$40,000/yr (4)

Such delays also extend the promised delivery time and reduce customer satisfaction and factory competitiveness. Adding (1) and (2) together. (not counting the direct productivity losses), the wasted energy cost alone of \$70,000/yr represents 3.5% of the annual profits and 7% in annual energy costs. Based on the author's experience,[5] these numbers are not infrequent in industry. Fixing thr productivity of process #5 would eliminate these wastes.

more infrastructure and process energy for the same amount of value work and profits when compared with the ideal nonvariable process. Table 1 illustrates cases (1) and (2).

3. Defective processes usually require a massive final inspection to sort out the good products. Finding the finished goods defective is the most inefficient means of quality assurance because often the entire batch must then be remade, consuming the associated energy. The inspection space, labor, and energy represent a direct waste and should be replaced with in line quality assurance (ILQA) that detects the first bad piece (Governmental, medical, etc., orders usually require a 100% final inspection. In the Lean system, this is performed as a formality because everybody in the plant knows that all pieces will be perfect because all imperfections have been removed in real time before the inspection process.) and immediately undertakes a corrective action. Typically, the ILQA can be implemented in few days of operators' training and has the simple payback period measured in days or weeks.[5]

Overage Reduction

Many a plant compensates for its notorious defects by routinely scheduling production in excess of what the customer orders. Some minimum overage is usually justified for machine setups, adjustments, and QA samples. In a Lean plant this rarely exceeds a fraction of one percent. In a traditional plant, the value of 5%–15% is not infrequent. A 5% overage means that the plant spends 105% of the necessary costs. If the profit margin is 5%, the overage alone may consume the entire profit. The overall energy waste (and the opportunity to save energy) is simply proportional to the overage amount. Overage is one of the most wasteful ways of compensating for defective processes. The best remedy is to simply identify the defective process with ILQA, find the root cause (typically the lack of training, excessive work quotas, or bad process or material), and repair it.

Unintentional overage can also be destructive to profits and energy use. Example: A worker is asked to cut only a few small pieces from a large sheet of metal, but instead he cuts the entire sheet, thinking, "my machine is already setup and soon they will ask me to cut the rest of the sheet anyway, so I may as well do it now." The excessive pieces then move through all processes, unknowingly to the management, consuming energy, labor and fixed costs, to end up as excessive FG inventory and, in the worst case, find no buyer and end up as scrap. Uncontrolled and careless overage can easily consume all profits, and, of course, waste energy proportionately to the overage amount.

Downtime

Equipment downtime and idleness may occur due to scheduled maintenance, unscheduled breakdowns, machine setups, and poor process scheduling. The downtime may cause proportional loss of both profits and energy. The downtime may have fourfold impact on energy use, as follows:

1. When a process stops for whatever reason during an active production shift, the plant infrastructure continues to use energy and loosing money, as in Eq. 4. A good plant manager should understand what fraction of the infrastructure energy is wasted during the specific equipment downtime. With this knowledge, the energy waste can be estimated as being proportional to the downtime.
2. Some machines continue using energy during maintenance, repair, or setup in proportion to the downtime (e.g., the crucible holding molten metal for a die casting machine remains heated by natural gas while the machine is being setup or repaired). Reducing the

Table 2 Energy savings from setup time reduction.

A plant operates on two shifts, 260 days per year, performing on average of 20 two-hour setups per day on their electrically heated injection molding machines. Each machine consumes 20 kW when idle bur energized. By a focused continuous improvement system and training, the crew reduces the routine setup time 0.5 h, with few, if any expenses for additional hardware, thus saving:

(260 days/yr) (20 setups/day) (1.5 h saved/setup) = 7800 machine h/yr.

The resultant process energy saved will be:

$$(7800\ \text{h/yr})\ (20\ \text{kW}) = 156{,}000\text{kWh/yr} \tag{5}$$

In addition, infrastructure energy will be saved because of the reduced downtime. Using the data from Example 1, if the work is done in two shifts for 260 days per year (4160 h/yr.), the plant infrastructure uses 40% of the plant energy, and each machine consumes 2% of the plant infrastructure energy during the set up, the additional energy cost savings due to the setup time reduction will be:

$$(7{,}800\ \text{hr/yr})\ (0.02)\ (0.04)\ (\$1{,}000{,}000)/(4160\ \text{h/yr}) = \#15{,}000 \tag{6}$$

setup time or eliminating the repair time saves the gas energy in direct proportion to the downtime saved. In order to calculate energy savings in such situations, it is necessary to understand the energy consumption by the equipment per unit of time multiplied by the downtime reduction.

3. When a particular machine is down, additional equipment upstream or downstream of that machine may also be forced into an idle status but remain energized, thus wasting energy. In an ideal single-piece flow, the entire production line (As in the saying "In Lean either everything works or nothing works") will stop. In order to estimate the energy-saving opportunity from reducing this cumulative downtime, the energy manager must understand which equipment is idled by the downtime of a given machine and how much energy it uses per unit time while being idle.
4. Lastly, energized equipment should be well managed. A high-powered machine may be left energized for hours at a time when not scheduled for production. A good practice is to assign each of these machines to an operator who will have the duty of turning the machine off when not needed for a longer time, if practical, and to turn it back on just in time to be ready for production exactly when it is needed.

Preventive maintenance and setup time reduction have a particularly critical impact on both productivity and related energy use, as follows:

Preventive Maintenance

Practical and routine preventive maintenance should be done during the hours free of scheduled production (e.g., during night shifts, on weekends, or during layover periods). The maintenance should be preventive rather than reactive (The term "preventive" tends to be replaced with "productive" in modern industrial parlance). Well-managed "total" preventive[6] maintenance involves not only oiling and checking the machines per schedule but also ongoing training of the mechanics; developing a comprehensive database containing information on the particular use and needs of various machines; preparing a schedule of part replacement and keeping inventory of frequently used spare parts; and a well-managed ordering system for other parts, including vendor data so that when a part is needed it can be ordered immediately and shipped using the fastest possible means. Industry leaders have demonstrated that affordable preventive maintenance can reduce the unscheduled downtime and associated energy waste to zero. This should be the practical goal of well-run factories.

Setups

Modern market trends push industry towards shorter series and smaller orders, requiring, in turn, more and shorter setups. Industry leaders have perfected routine setups to take no more than a few minutes. In poorly managed plants, routine setups can take as long as several hours. In all competitive modern plants, serious efforts should be devoted to setup time reductions. The effort includes both training and hardware improvements. The training alone, with only minimal additional equipment (such as carts), can yield dramatic setup time reductions (i.e., from hours to minutes). Further gains may require a change of the mounting and adjustment hardware and instrumentation. Some companies organize competitions between teams for developing robust procedures for the setup time reductions. In a plant performing many setups, the opportunity for energy savings may be significant, both in the process and infrastructure energy, as shown in Table 2.

Flexibility

Production flexibility, also called agility, is an important characteristic of competitive plants. A flexible plant prefers small machines (if possible, on casters) that are easy to roll into position and plug into adjustable quick-connect electrical and air lines and that are easy to setup and maintain over the large fixed machines selected with large batches and small unit costs in mind (such machines are called "monuments" in Womack and Jones[3]). Such an ideal plant will also have trained a flexible workforce in multiple skills, including quality assurance skills. This flexibility allows for the setup of new production lines in hours or even minutes, optimizing the flow and floor layout in response to short orders, and delivers the orders JIT. The energy may be saved in two important ways, as follows:

- Small machines processing one piece at a time use only as much energy as needed. In contrast, when excessively large automated machines are used, the typical management choice is between using small batches JIT, thus wasting the large machine energy, or staging the batches for the large machine, which optimizes machine utilization at the expense of throughput time, production flow, production planning effort, and the related infrastructure energy.
- Small machines are conducive to flexible cellular work layout, where 2–4 machines involved in the sequential processing of WIP are arranged into a U-shaped cell with 1–3 workers serving all processes in the cell in sequence, and the last process being quality assurance. This layout can be made very compact—occupying a much smaller footprint in the plant compared to traditional "process village" plants, roughly a reduction of 50%[3,4]—and is strongly preferred by workers because it saves walking and integrates well the work steps. Such a layout also saves forklift effort and energy and infrastructure energy due to the reduction of the footprint.

Other Productivity Elements

The complete list of productivity elements is beyond the scope of this entry, and all elements have some leverage on energy use and conservation. In the remaining space, only the few most important remaining aspects are mentioned, with their leverage on energy. Descriptive details can be found in Ohno[7] and numerous other texts on Lean production.

- *Visual factory*: Modern factories place an increasing importance on making the entire production as transparent as possible in order to make any problem visible to all, which is motivational for immediate corrective actions and continuous improvements. Ideally, each process should have a white board displaying the short-term data, such as the current production status (quantity completed vs required); the rate of defects or rejects and their causes; control charts and information about the machine condition or maintenance needs; and a brief list and explanation of any issues, all frequently updated. The board should also display long-term information such as process capability history, quality trends, operator training, etc. Such information is most helpful in the optimization of, among other things, process time and quality, which leads to energy savings, as discussed above.
- *"Andon" signals*: The term refers to the visual signals (lights, flags, markers, etc.,) displaying the process condition, as follows: "green = all OK," "yellow = minor problem being corrected," and "red = high alarm, stopped production, and immediate assistance needed." The signals are very useful in identifying the trouble-free and troubled processes, which is conducive to focusing the aid resources to the right places in real time, fixing problems immediately and not allowing defects to flow downstream on the line. These features, in turn, reduce defects, rework, delays, and wasted costs, which improve overall productivity and save energy, as described above. It is also useful to display the estimated downtime (Toyota and other modern plants have large centrally located Andon boards that display the Andon signal, the workstation number, and the estimated downtime.). Knowing the forecasted downtime frees other workers to perform their pending tasks which have waited for such an opportunity rather than wait idle. This leads to better utilization of the plant resources, including infrastructure energy.
- *"5Ss"*: The term comes from five Japanese words that begin with the "s" sound and loosely translate into English as: sorting, simplification, sweeping, standardization, and self-discipline (many other translations of the words are popular in industry); and describes a simple but powerful workplace organization method.[8] The underlying principle of the method is that only the items needed for the immediate task (parts, containers, tools, instructions, materials) are kept at hand where they are needed at the moment, and everything else is kept in easily accessible and well-organized storage in perfect order, easy to locate without searching, and in just the right quantities. All items have their designated place, clearly labeled with signs, labels, part numbers, and possibly bar codes. The minimum and maximum levels of inventory of small parts are predefined and are based on actual consumption rather than the "just-in-case" philosophy. The parts, tools, and materials needed for the next shift of production are prepared by a person in charge of the storage during the previous shift and delivered to the workstation before the shift starts. The floor is uncluttered and marked with designated spaces for all equipment. The entire factory is spotlessly clean and uncluttered. Walls are empty except for the visual boards. In consequence of these changes, the searching for parts, tools, and instructions which can represent a significant waste of labor and time is reduced, and

Fig. 2 In this messy plant, the workers waste close to 20% of their time looking for items and scavenging for parts and tools, also wasting the plant energy.

this, in turn, saves energy. Secondary effects are also important. In a well-organized place, fewer mistakes are made; fewer wrong parts are used; less inspection is needed; quality, throughput time, and customer satisfaction are increased; and costs and energy are decreased. Fig. 2 illustrates a fragment of a messy factory, where the average worker was estimated to waste 20% of his shift time looking for and scavenging for parts and tools. This percentage multiplied by the number of workers yields a significant amount of wasted production time, also wasting plant energy in the same proportion. Sorting, cleaning, and organizing the workplace is one of the simplest and most powerful starting points on the way to improved productivity and energy savings.

CONCLUSION

Large savings in energy are possible as an inherent by-product of improving productivity. The state-of-the-art Lean productivity method can yield dramatic improvements in productivity. In the extreme case of converting from the traditional batch-and-queue and "process village" manufacturing system to Lean production, overall costs, lead times, and inventories can be reduced by as much as 50%–90%, floor space and energy by 50%, and energy density can be improved by 50%. The amount of energy that can be saved by productivity improvements often radically exceeds the savings from equipment optimization alone, thus providing a strong incentive to include productivity improvements in energy-reduction efforts.

Productivity strongly depends on human factors such as management, learning, and training, communications, culture, teamwork, etc., which are difficult to quantify, making accurate estimates of the cost, schedule, and quality benefits from various productivity improvements and the related energy savings difficult to estimate with engineering accuracy. For this reason, simple metrics and models are recommended, and some examples have been presented. If applied conservatively, they can become useful tools for energy management in a plant. The prerequisite knowledge includes an understanding of Lean Flow and its various productivity elements and a good accounting of energy use in the plant, including the knowledge of the energy used by individual machines and processes both when in productive use and in the idle but energized state, as well as the energy elements used by the infrastructure (various light combinations, air-compressors, cooling and heating devices, combusting systems, conveyers, forklifts, etc.). In the times of ferocious global competition and rising energy prices, every industrial plant should make every effort to improve both productivity and energy use.

ACKNOWLEDGMENTS

This work is a result of the studies of energy conservation using the Lean productivity method performed by the Industrial Assessment Center funded by the U.S. Department of Energy at Loyola Marymount University. The author is grateful to Mr. Rudolf Marloth, Assistant Director of the Center, for his help with various energy estimates included herein and his insightful comments, to the Center students for their enthusiastic work, and to his son Peter W. Oppenheim for his diligent editing.

REFERENCES

1. U.S. Department of Energy, Energy Efficiency and Renewable Energy, available at http://www.eere.energy.gov/industry/ (accessed on December 2005).
2. U.S. Department of Energy, available at http://eereweb.ee.doe.gov/industry/bestpractices/plant_assessments.html, (accessed on December 2005).
3. Womack, P.J.; Jones, D.T. *Lean Thinking*. 2nd Ed.; Lean Enterprise Institute: Boston, 2005; (http://www.lean.org), ISBN: 0-7432-4927-5.
4. Liker, J. *Becoming Lean, Inside Stories of U.S. Manufacturers*; Productivity Inc.: Portland, OR, 1998; service@productivityinc.com.
5. Oppenheim, B.W. *Selected Assessment Recommendations*, Industrial Assessment Center, Loyola Marymount University: Los Angeles, (boppenheim@lmu.edu), unpublished 2004.
6. *Setup Reduction for Just-in-Time*, Video/CD, Society of Manufacturing Engineers, Product ID: VT90PUB2, available at http://www.sme.org/cgi-bin/get-item.pl?VT392&2&SME&1990, (accessed on December 2005).
7. Ohno, T. *Toyota Production System: Beyond Large Scale Production*; Productivity Press: New York, 1988; info@productivityinc.com.
8. Hiroyuki, H. *Five Pillars of the Visual Workplace, the Sourcebook for 5S Implementation*; Productivity Press: New York, 1995; info@productivityinc.com.

Energy Conversion: Coal, Animal Waste, and Biomass Fuel

Kalyan Annamalai
Paul Pepper Professor of Mechanical Engineering, Texas A&M University, College Station, Texas, U.S.A.

Soyuz Priyadarsan
Texas A&M University, College Station, Texas, U.S.A.

Senthil Arumugam
Enerquip, Inc., Medford, Wisconsin, U.S.A.

John M. Sweeten
Texas A&M University, Amarillo, Texas, U.S.A.

Abstract

A brief overview is presented of various energy units; terminology; and basic concepts in energy conversion including pyrolysis, gasification, ignition, and combustion. Detailed sets of fuel properties of coal, agricultural biomass, and animal waste are presented so that their suitability as a fuel for the energy conversion process can be determined. It is also found that the dry ash free (DAF) heat values of various biomass fuels, including animal waste, remain approximately constant, which leads to a presentation of generalized results for maximum flame temperature as a function of ash and moisture contents. The cofiring technology is emerging as a cost-effective method of firing a smaller percentage of biomass fuels, with coal as the major fuel. Various techniques of cofiring are summarized. Gasification approaches, including FutureGen and reburn technologies for reduction of pollutants, are also briefly reviewed.

INTRODUCTION AND OBJECTIVES

The overall objective of this entry is to provide the basics of energy conversion processes and to present thermochemical data for coal and biomass fuels. Energy represents the capacity for doing work. It can be converted from one form to another as long as the total energy remains the same. Common fuels like natural gas, gasoline, and coal possess energy as chemical energy (or bond energy) between atoms in molecules. In a reaction of the carbon and hydrogen in the fuel with oxygen, called an oxidation reaction (or more commonly called combustion), carbon dioxide (CO_2) and water (H_2O) are produced, releasing energy as heat measured in units of kJ or Btu (see Table 1 for energy units). Combustion processes are used to deliver (i) work, using external combustion (EC) systems by generating hot gases and producing steam to drive electric generators as in coal fired power plans, or internal combustion (IC) engines by using the hot gases directly as in automobiles or gas turbines; and (ii) thermal energy, for applications to manufacturing processes in metallurgical and chemical industries or agricultural product processing.

Fuels can be naturally occurring (e.g., fossil fuels such as coal, oil, and gas, which are residues of ancient plant or animal deposits) or synthesized (e.g., synthetic fuels). Fuels are classified according to the phase or state in which they exist: as gaseous (e.g., natural gas), liquid (e.g., gasoline or ethanol), or solid (e.g., coal, wood, or plant residues). Gaseous fuels are used mainly in residential applications (such as water heaters, home heating, or kitchen ranges), in industrial furnaces, and in boilers. Liquid fuels are used in gas turbines, automotive engines, and oil burners. Solid fuels are used mainly in boilers and steelmaking furnaces.

During combustion of fossil fuels, nitrogen or sulfur in the fuel is released as NO, NO_2 (termed generally as NO_x) and SO_2 or SO_3 (termed as SO_x). They lead to acid rain (when SO_x or NO_x combine with H_2O and fall as rain) and ozone depletion. In addition, greenhouse gas emissions (CO_2, CH_4, N_2O, CFCs, SF_6, etc.,) are becoming a global concern due to warming of the atmosphere, as shown in Fig. 1 for CO_2 emissions. Global surface temperature has increased by 0.6°C over the past 100 years. About 30%–40% of the world's CO_2 is from fossil fuels. The Kyoto protocol, signed by countries that account for 54% of the world's fossil based CO_2 emissions, calls for reduction of greenhouse gases by 5% from 1990 levels over the period from 2008 to 2012.

The total worldwide energy consumption is 421.5 quads of energy in 2003 and is projected to be 600 quads in 2020, while U.S. consumption in 2004 is about 100 quads and is projected to be 126 quads in 2020. The split is as follows: 40 quads for petroleum, 23 for natural gas, 23 for coal, 8 for nuclear power, and 6 for renewables (where energy is renewed or replaced using natural processes) and others sources. Currently, the United States relies on fossil fuels

Encyclopedia of Environmental Management DOI: 10.1081/E-EEM-120041557

Table 1 Energy units and terminology.

The section on energy Units and Conversion factors in Energy is condensed from Chapter 01 of Combustion Engineering by Annamalai and Puri [2005] and Tables.

Energy Units

1 Btu (British thermal unit) = 778.14 ft lb_f = 1.0551 kJ, 1 kJ = 0.94782 Btu = 25,037 $lbmft/s^2$

1 mBtu = 1 k Btu = 1000 Btu, 1 mmBtu = 1000 k Btu = 10^6 Btu, 1 trillion Btu = 10^9 Btu or 1 giga Btu

1 quad = 10^{15} Btu or 1.05 × 10^{15} kJ or 2.93 × 10^n kW h

1 Peta J = 10^{15} J = 10^{12} kJ»0.00095 Quads

1 kilowatt-hour of electricity = 3,412 BTU = 3.6 Mj

1 cal: 4.1868 J, One (food) calorie = 1000 cal or 1 Cal

1 kJ/kg = 0.43 Btu/lb, 1 Btu/lb = 2.326 kJ/kg

1 kg/GJ = 1 g/MJ = 2.326 lb/mmBtu; 1 lb/mmBtu = 0.430 kg/GJ = 0.430 g/MJ

1 Btu/SCF = 37 kJ/m^3

1 Therm = 10^5 Btu = 1.055 × 10^5 kJ

1 m^3/GJ = 37.2596 ft^3/mmBTU

1 hp = 0.7064 Btu/s = 0.7457 kW = 745.7 W = 550 lbf ft/s = 42.41 Btu/min

1 boiler HP = 33475 Btu/h, 1 Btu/h = 1.0551 kJ/h

1 barrel (42 gal) of crude oil = 5,800,000 Btu = 6120 MJ

1 gal of gasoline = 124,000 Btu = 131 MJ

1 gal of heating oil = 139,000 Btu = 146.7 MJ, 1 gal of diesel fuel = 139,000 Btu = 146.7 MJ

1 barrel of residual fuel oil = 6,287,000 Btu = 6633 MJ

1 cubic foot of natural gas = 1,026 Btu = 1.082 MJ, 1 Ton of Trash = 150 kWh

1 gal of propane = 91,000 Btu = 96 MJ, 1 short ton of coal = 20,681,000 Btu = 21821 MJ

Emission reporting for pollutants: (i) parts per million (ppm), (ii) normalized ppm, (iii) emission Index (EI) in g/kg fuel, (iv) g/GJ, v) mg/ m^3 of flue gas:

Conversions in emissions reporting: (II) normalized ppm = ppm × (21-O_2% std)/(21-O_2% measured); (iii) EI of species k: C % by mass in fuel × mol Wt of *k* × ppm of species *k* × 10^{-3}/{12.01(CO_2% + CO%)}, (iv) g/GJ = EI/ {HHV in GJ/kg}; (v) mg/m^3 = ppm of species *k* × Mol Wt of *k*/24.5

Volume of 1 kmol (SI) and 1 lb mole (English) of an ideal gas at STP conditions defined below:
Pressure at 101.3 kPa (1 atm, 14.7 psia, 29.92 in.Hg, 760 Torr) fixed; *T* changes depending upon type of standard adopted

Scientific (or SATP, standard ambient *T* and *P*)	US standard (1976) or ISA (International standard atmosphere)	NTP (gas industry reference base)	Chemists-standard-atmosphere (CSA)
25°C (77°F)	15°C (60°F)	20°C(68°F), 101.3 kPa	0°C (32°F),
24.5 m^3/kmol (392 ft^3/lb mole); $\rho_{air,SATP}$ = 1.188 kg/ m^3 = 0.0698 lb_m/ft^3	23.7 m^3/kmol (375.6 ft^3/lb mole); $\rho_{air,ISA}$ = 1.229 kg/ m^3 = 0.0767 lb_m/ft^3	24.06 m^3/kmole or 385 ft^3/lb mole; $\rho_{air,NTP}$ = 1.208 kg/ m^3 = 0.0754 lb_m/ft^3	22.4 m^3/kmol (359.2 ft^3/lb mole), $\rho_{air,CSA}$ = 1.297 kg/ m^3 = 0.0810 lb_m/ft^3

for 85% of its energy needs. Soon, the U.S. energy consumption rate which distributed as electrical power (40%), transportation (30%), and heat (30%), will outpace the growth in the energy production rate, increasing reliance on imported oil. The Hubbert peak theory (named after Marion King Hubbert, a geophysicist with Shell Research Lab in Houston, Texas) is based upon the rate of extraction and depletion of conventional fossil fuels, and predicts that fossil-based oil would peak at about 12.5 billion barrels per year worldwide some time around 2000. The power cost and percentage use of coal in various U.S. states varies from 10 cents (price per kWh) at 1% coal use for power generation in California to 48 cents at 94% use of coal in Utah.

Biomass is defined as "any organic material from living organisms that contains stored sunlight (solar energy) in the form of chemical energy."[1] These include agro-based materials (vegetation, trees, and plants); industrial wastes (sawdust, wood chips, and crop residues); municipal solid wastes (MSWs), which contain more than 70% biomass (including landfill gases, containing almost 50% CH_4); and animal waste. Biomass is a solid fuel in which hydrogen is locked with carbon atoms. Biomass production worldwide is 145 billion metric tons. Biomass now supplies 3% of U.S. energy, and it could be increased to as high as 20%. Renewable energy sources (RES) include biomass, wind, hydro, solar, flowing water or hydropower, anaerobic

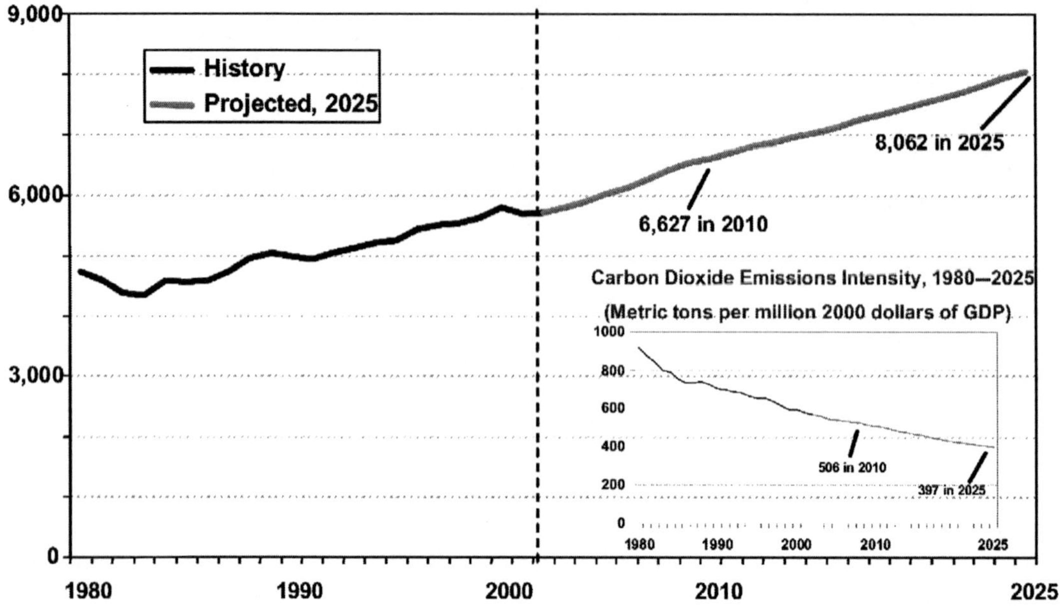
9,000
6,000
3,000
0
History
Projected, 2025
8,062 in 2025
6,627 in 2010
Carbon Dioxide Emissions Intensity, 1980—2025
(Metric tons per million 2000 dollars of GDP)
1000
800
600
400
200
0
506 in 2010
397 in 2025
1980
1990
2000
2010
2025

Fig. 1 Total CO_2 emission in million metric tons per year: History and Projected 1980–2025.
Source: DOE-EIA.[1]

digestion, ocean thermal (20°C temperature difference), tidal energies, and geothermal (a nonsolar source of energy), and these supply 14% of the world demand. The RES constitute only 6%, while coal, petroleum, and natural gas account for 23%, 40%, and 24%, respectively. About 9% of the world's electricity is from RES, and 65% of the electricity contributed by biomass. About 97% of energy conversion from biomass is by combustion. Many U.S. states have encouraged the use of renewables by offering REC (Renewable Energy Credits). One REC = 1 MW/h = 3.412 mmBtu; hence the use of 1 REC is equivalent to replacing approximately 1500 lb of coal, reducing emission of NO_x and SO_x by 1.5 lb for every 1 REC, assuming that emissions of NO_x and SO_x are 0.45 lb per mmBtu generated by coal. Several emission-reporting methods and conversions are summarized in Table 1. Recently, H_2 is being promoted as a clean-burning, non-global-warming, and pollution-free fuel for both power generation and transportation.

Fig. 2 shows a comparison between biomass and hydrogen energy cycles. In the biomass cycle, photosynthesis is used to split CO_2 into C and O_2, and H_2O into H_2 and O_2, producing Hydrocarbons (HC) fuel (e.g., leaves) and releasing O_2. The O_2 released is used to combust the HC and produce CO_2 and H_2O, which are returned to produce plant biomass (e.g., leaves) and O_2. On the other hand, in the hydrogen cycle, H_2O is disassociated using the photo-splitting process to produce H_2 and O_2, which are then used for the combustion process. The hydrogen fuel can be used

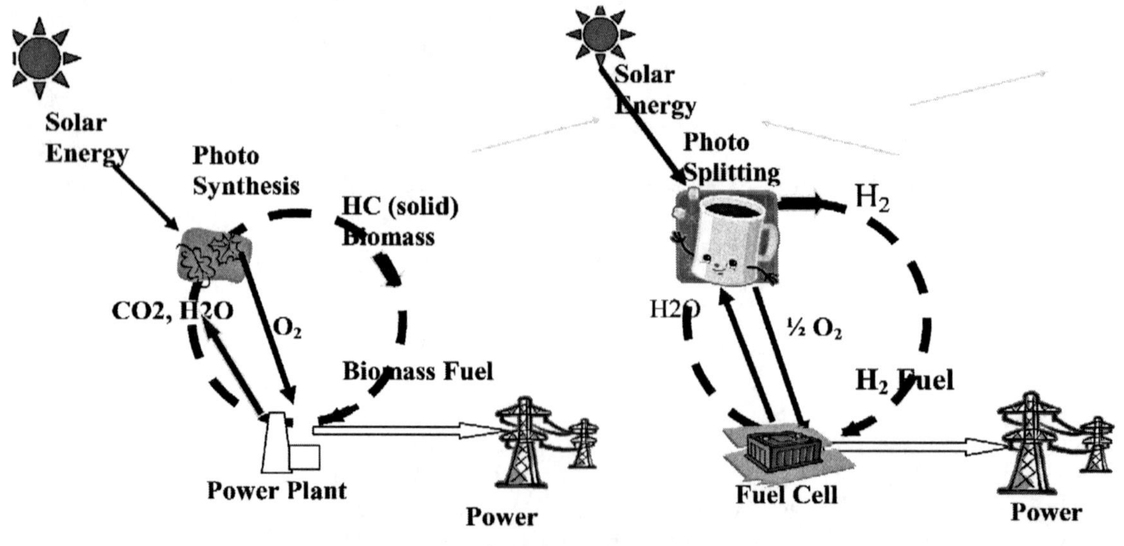
Solar Energy
Photo Synthesis
HC (solid) Biomass
CO2, H2O
O₂
Biomass Fuel
Power Plant
Power
Solar Energy
Photo Splitting
H₂
H2O
½ O₂
H₂ Fuel
Fuel Cell
Power

Fig. 2 Comparison of biomass energy and H_2 energy cycles.

in fuel cells to obtain an efficient conversion. Photosynthesis is water intensive; most of the water supplied to plants evaporates through leaves into the atmosphere, where it re-enters the hydrology cycle.

This entry is organized in the following format: (i) coal and biosolid properties; (ii) coal and biosolid pyrolysis (a process of thermal decomposition in the absence of oxygen), combustion, and gasification; (iii) combustion by cofiring coal with biosolids; (iv) gasification of coal and biosolids (a process that includes pyrolysis, partial oxidation due to the presence of oxygen, and hydrogenation); and (v) reburn for NO_x reduction.

FUEL PROPERTIES

Fuel properties play a major role in the selection, design, and operation of energy conversion systems.

Solid Fuels

The primary solid fuel widely used in power plants is coal containing combustibles, moisture, and intrinsic mineral matter originating from dissolved salts in water. During the "coalification" process, lignite, the lowest rank of coal (low C/O ratio), is produced first from peat, followed by subbituminous (black lignite, typically low sulfur, noncaking), bituminous (soft coal that tends to stick when heated and is typically high in S), and finally anthracite (dense coal; has the highest carbon content, >90%, low volatile <15%) with a gradual increase in the coal C/O ratio. The older the coal, the higher its rank. Anthracite (almost carbon) is the highest-ranked coal, with a high heating value. To classify coals and ascertain the quality of coal, it is essential to perform proximate and ultimate analyses according to American Society of Testing Materials (ASTM) standards.

Proximate Analysis (ASTM D3172)

A solid fuel consists of combustibles, ash, and moisture. Combustibles together with ash are called the solid content of fuel. A proximate analysis provides the following information: surface moisture (SM) or dry loss (DL), i.e., moisture in air-dried coal; the inherent moisture in the coal (M); volatile matter (VM; produced by pyrolysis, a thermal decomposition process resulting in release of water, gases, oil and tar); fixed carbon (FC; skeletal matter left after release of volatiles); mineral matter (MM; inert collected with solid fuel); and heating value (HV). On combustion, the MM may be partially oxidized or reduced, and the material left after combustion of C and H in the fuel is called ash (CaO, $CaCO_3$, Fe_2O_3, FeO, etc).

Table 2 shows comparative proximate analyses of coal, advanced feedlot biomass (FB, low-ash cattle manure; see "Coal and Bio-Solids Cofiring"), and litter biomass (LB, chicken manure).[2] Feedlot manure has higher moisture, nitrogen, chlorine, and ash content than coal. With aging or composting, the VM in manure decreases as a result of the gradual release of hydrocarbon gases or dehydrogenation, but fuel becomes more homogeneous.

Ultimate/Elemental Analysis (ASTM D3176)

Ultimate analysis is used to determine the chemical composition of fuels in terms of either the mass percent of their various elements or the number of atoms of each element. The elements of interest are C, H, N, O, S, Cl, P, and others. It can be expressed on an "as received" basis, on a dry basis (with the moisture in the solid fuel removed), or on a dry ash free (DAF) basis (also known as the moisture ash free basis MAF). Tables 3 and 4 show the ultimate analyses of various types of coal and biomass fuels.[3] While nitrogen is not normally present in natural gas, coal has 1%–1.5%; cattle manure and chicken waste contain high amounts of N (Table 2).

Heating Value (ASTM D3286)

The gross or higher heating value (HHV) of a fuel is the amount of heat released when a unit (mass or volume) of the fuel is burned. The HHV of solid fuel is determined using ASTM D3286 with an isothermal jacket bomb calorimeter. For rations fed to animals and animal waste fuels, the HHV for DAF roughly remains constant at about 19,500 kJ/kg (8400 Btu/lb),[4] irrespective of stage of decomposition of animal waste. The HHV can also be estimated using the ultimate analysis of the fuel and the following empirical relation from Boie[5]:

$$\begin{aligned} HHV_{fuel}(\text{kJ/kg fuel}) &= 35{,}160\, Y_C + 116{,}225\, Y_H \\ &\quad - 11{,}090\, Y_O + 6280\, Y_N + 10465\, Y_S \end{aligned} \quad (1)$$

$$\begin{aligned} HHV_{fuel}(\text{BTU/lb fuel}) &= 15{,}199\, Y_C + 49{,}965\, Y_H \\ &\quad - 4768\, Y_O + 2700\, Y_N + 4499\, Y_S, \end{aligned} \quad (2)$$

where Y denotes the mass fraction of an element C, H, O, N, or S in the fuel. The higher the oxygen content, the lower the HV, as seen in biomass fuels.

Annamalai et al. used the Boie equation for 62 kinds of biosolids with good agreement.[6] For most biomass fuels and alcohols, the HHV in kilojoules per unit mass of stoichiometric oxygen is constant at 14,360–14,730 kJ/kg of O_2 (6165–6320 Btu/lb of O_2).[7]

Estimate of CO_2 Emission

Using the Boie-based HVs for any fuel of known elemental composition, one can plot the CO_2 emission in g/MJ (Fig. 3) as a function of H/C and O/C ratios.[8] Comparisons for selected fuels with known experimental HVs are also shown in the same figure. Coal, with H/C ratio ≈ 0.5,

Table 2 Coal, advanced feedlot biomass (FB) and litter biomass (LB).

Parameter	Wyoming coal	Cattle manure (FB)	Chicken manure (LB)[a]	Advanced Feedlot biomass (AFB)[b]	High-ash Feedlot biomass (HFB)[b]
Dry loss (DL)	22.8	6.8	7.5	10.88	7.57
Ash	5.4	42.3	43.8	14.83	43.88
FC	37.25	40.4	8.4	17.33	10.28
VM	34.5	10.5	40.3	56.97	38.2
C	54.1	23.9	39.1	50.08	49.27
H	3.4	3.6	6.7	5.98	6.13
N	0.81	2.3	4.7	38.49	38.7
O	13.1	20.3	48.3	4.58	4.76
S	0.39	0.9	1.2	0.87	0.99
Cl	<0.01%	1.2			
HHV-as received (kJ/kg)	21385	9560	9250	14983	9353
$T_{adiab, Equil}$[c]	2200 K (3500°F)	2012 K (3161°F)			
DAF formula	$CH_{0.76}O_{0.18}N_{0.013}S_{0.0027}$	$CH_{1.78}O_{.64}N_{.083}S_{.014}$	$CH_{2.04}O_{0.93}N_{0.10}S_{0.012}$	$CH_{1.4184}O_{0.5764}N_{0.078}S_{0.0056}$	$CH_{1.4775}O_{0.5892}N_{0.083}S_{0.0076}$
HHV-DAF (kJ/kg)	29785	18785	18995	20168	19265
CO_2, g/GJ					
N, g/GJ					
S, g/GJ					

[a] Priyadarsan et al.[2]
[b] Priyadarsan et al.[37]
[c] Equilibrium temperature for stoichiometric mixture from THERMOLAB Spreadsheet software for any given fuel of known composition (Annamalai and Puri.[36] website http://www.crcpress.com/e_products/downloads/ download.asp?cat_no = 2553)

Table 3 Coal composition (DAF basis).

ASTM Rank	State (U.S.A.)	Ash, % (dry)	C	H	N	S*	O**	HHV_{Est} kj/kg	CO_2 kg/ GJ	N kg/GJ	S kg/GJ
Lignite	ND	11.6	63.3	4.7	0.43	0.98	30.5	24,469	94.8	0.196	0.401
Lignite	MT	7.7	70.7	4.9	0.8	4.9	22.3	28,643	90.4	0.279	1.711
Lignite	ND	8.2	71.2	5.3	0.56	0.46	22.5	28,782	90.7	0.195	0.160
Lignite	TX	9.4	71.7	5.2	1.3	0.72	21.1	29,070	90.4	0.447	0.248
Lignite	TX	10.3	74.3	5	0.37	0.51	19.8	29,816	91.3	0.124	0.171
Sbb. A	WY	8.4	74.3	5.8	1.2	1.1	17.7	31,092	87.6	0.386	0.354
Sbb. C	WY	6.1	74.8	5.1	0.39	0.3	18.9	30,218	90.7	0.295	0.099
HVB	IL	10.8	77.3	5.6	1.1	2.3	13.6	32,489	87.2	0.339	0.708
HVC	IL	10.1	78.8	5.8	1.6	1.8	12.1	33,394	86.5	0.479	0.539
HVB	IL	11.8	80.1	5.5	1.1	2.3	11.1	33,634	87.3	0.327	0.684
HVB	UT	4.8	80.4	6.1	1.3	0.38	11.9	34,160	86.2	0.381	0.111
HVA	WV	7.6	82.3	5.7	1.4	1.8	8.9	34,851	86.5	0.402	0.516
HVA	KY	2.1	83.8	5.8	1.6	0.66	8.2	35,465	86.6	0.451	0.186
MV	AL	7.1	87	4.8	1.5	0.81	5.9	35,693	89.3	0.420	0.227
LV	PA	9.8	88.2	4.8	1.2	0.62	5.2	36,153	89.4	0.332	0.171
Anthracite	PA	7.8	91.9	2.6	0.78	0.54	4.2	34,974	96.3	0.223	0.154
Anthracite	PA	4.3	93.5	2.7	0.24	0.64	2.9	35,773	95.8	0.067	0.179

Notes: HHV_{est}: Boie Equation. CO_2 in g/MJ or kg/GJ = C content in % ×36645/{HHV in kJ/kg}. CO_2 in lb per mmBtu = Multiply CO_2 in (g /MJ) or kg/GJ by 2.32. N in g/MJ or kg/GJ = N% × 10000/{HHV in kJ/kg}. For NO_x estimation, multiply N content in g/MJ by 1.15 to get NO_x in g/MJ which assumes 35% conversion of fuel N; For SO_2 estimation, multiply S content in g/MJ by 2 to get SO_2 in g/MJ assuming 100% conversion of fuel S (Multiply HHV in kJ/kg by 0.430 tc get Btu/lb); *Organic sulfur; **by difference.

Table 4 Ultimate analyses and heating values of biomass fuels.

Biomass	C	H	O	N	S	Residue	Measured HHV_M	[a]Estimated HHV	CO_2 g/MJ	N, g/MJ	S, g/MJ
Field crops											
Alfalfa seed straw	46.76	5.40	40.72	1.00	0.02	6.07	18.45	18.27	92.9	0.542	0.011
Bean straw	42.97	5.59	44.93	0.83	0.01	5.54	17.46	16.68	90.2	0.475	0.006
Corn cobs	46.58	5.87	45.46	0.47	0.01	1.40	18.77	18.19	90.9	0.250	0.005
Corn stover	43.65	5.56	43.31	0.61	0.01	6.26	17.65	17.05	90.6	0.346	0.006
Cotton stalks	39.47	5.07	39.14	1.20	0.02	15.10	15.83	15.51	91.4	0.758	0.013
Rice straw (fall)	41.78	4.63	36.57	0.70	0.08	15.90	16.28	16.07	94.0	0.430	0.049
Rice straw (weathered)	34.60	3.93	35.38	0.93	0.16	25.00	14.56	12.89	87.1	0.639	0.110
Wheat straw	43.20	5.00	39.40	0.61	0.11	11.40	17.51	16.68	90.4	0.348	0.063
Switchgrass[b]	42.02	6.30	46.10	0.77	0.18	4.61	15.99	15.97	96.3	0.482	0.113
Orchard prunings											
Almond prunings	51.30	5.29	40.90	0.66	0.01	1.80	20.01	19.69	93.9	0.330	0.005
Black Walnut	49.80	5.82	43.25	0.22	0.01	0.85	19.83	19.50	92.0	0.111	0.005
English Walnut	49.72	5.63	43.14	0.37	0.01	1.07	19.63	19.27	92.8	0.188	0.005
Vineyard prunings											
Cabernet Sauvignon	46.59	5.85	43.90	0.83	0.04	2.71	19.03	18.37	89.7	0.436	0.021
Chenin Blanc	48.02	5.89	41.93	0.86	0.07	3.13	19.13	19.14	92.0	0.450	0.037
Pinot Noir	47.14	5.82	43.03	0.86	0.01	3.01	19.05	18.62	90.7	0.451	0.005
Thompson seedless	47.35	5.77	43.32	0.77	0.01	2.71	19.35	18.60	89.7	0.398	0.005
Tokay	47.77	5.82	42.63	0.75	0.03	2.93	19.31	18.88	90.7	0.388	0.016
Energy Crops											
Eucalyptus											
Camaldulensis	49.00	5.87	43.97	0.30	0.01	0.72	19.42	19.19	92.5	0.154	0.005
Globulus	48.18	5.92	44.18	0.39	0.01	1.12	19.23	18.95	91.8	0.203	0.005
Grandis	48.33	5.89	45.13	0.15	0.01	0.41	19.35	18.84	91.5	0.078	0.005
Casuarina	48.61	5.83	43.36	0.59	0.02	1.43	19.44	19.10	91.6	0.303	0.010
Cattails	42.99	5.25	42.47	0.74	0.04	8.13	17.81	16.56	88.5	0.415	0.022
Popular	48.45	5.85	43.69	0.47	0.01	1.43	19.38	19.02	91.6	0.243	0.005
Sudan grass	44.58	5.35	39.18	1.21	0.08	9.47	17.39	17.63	93.9	0.696	0.046

Forest residue											
Black Locust	50.73	5.71	41.93	0.57	0.01	0.97	19.71	19.86	94.3	0.289	0.005
Chaparral	46.9	5.08	40.17	0.54	0.03	7.26	18.61	17.98	92.3	0.290	0.016
Madrone	48	5.96	44.95	0.06	0.02	1	19.41	18.32	90.6	0.031	0.010
Manzanita	48.18	5.94	44.68	0.17	0.02	1	19.3	18.9	91.5	0.088	0.010
Ponderosa Pine	49.25	5.99	44.36	0.06	0.03	0.3	20.02	19.37	90.1	0.030	0.015
Ten Oak	47.81	5.93	44.12	0.12	0.01	2	18.93	18.82	92.6	0.063	0.005
Redwood	50.64	5.98	42.88	0.05	0.03	0.4	20.72	20.01	89.6	0.024	0.014
White Fur	49	5.98	44.75	0.05	0.01	0.2	19.95	19.22	90.0	0.025	0.005
Food and fiber processing wastes											
Almond hulls	45.79	5.36	40.6	0.96	0.01	7.2	18.22	17.89	92.1	0.527	0.005
Almond shells	44.98	5.97	42.27	1.16	0.02	5.6	19.38	18.14	85.0	0.599	0.010
Babassu husks	50.31	5.37	42.29	0.26	0.04	1.73	19.92	19.26	92.5	0.131	0.020
Sugarcane bagasse	44.8	5.35	39.55	0.38	0.01	9.79	17.33	17.61	94.7	0.219	0.006
Coconut fiber dust	50.29	5.05	39.63	0.45	0.16	4.14	20.05	19.2	91.9	0.224	0.080
Cocoa hulls	48.23	5.23	33.09	2.98	0.12	10.25	19.04	19.56	92.8	1.565	0.063
Cotton gin trash	39.59	5.26	36.33	2.09		6.68	16.42	16.13	88.4	1.273	0.000
Macadamia shells	54.41	4.99	39.69	0.36	0.01	0.56	21.01	20.55	94.9	0.171	0.005
Olive pits	48.81	6.23	43.48	0.36	0.02	1.1	21.39	19.61	83.6	0.168	0.009
Peach pits	53	5.9	39.14	0.32	0.05	1.59	20.82	21.18	93.3	0.154	0.024
Peanut hulls	45.77	5.46	39.56	1.63	0.12	7.46	18.64	18.82	90.0	0.874	0.064
Pistachio shells	48.79	5.91	43.41	0.56	0.01	1.28	19.26	19.25	92.8	0.291	0.005
Rice hulls	40.96	4.3	35.86	0.4	0.02	18.34	16.14	15.45	93.0	0.248	0.012
Walnut shells	49.98	5.71	43.35	0.21	0.01	0.71	20.13	19.45	90.8	0.104	0.005
Wheat dust	41.38	5.1	35.19	3.04	0.19	15.1	16.2	16.78	93.6	1.877	0.117

[a] HHV based on Boie equation.
[b] Aerts et al.[20]; [Adapted from Ebeling and Jenkins[3] and Annamalai.[17] See foot note of Table 3.2 for conversions to English units and estimation of NO_x and SO_2 emissions.

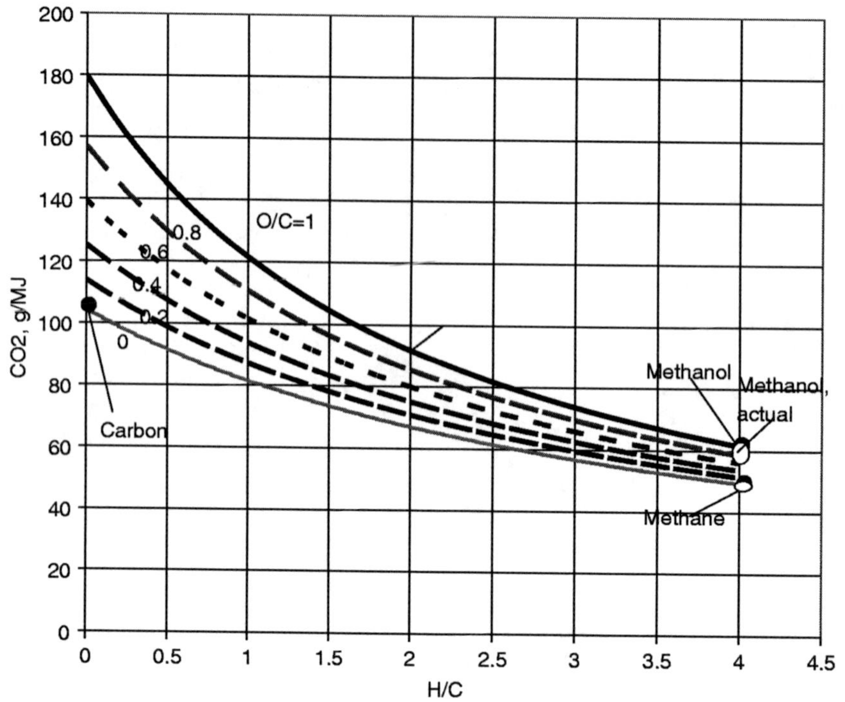
O/C=1
0.8
0.6
0.4
0.2
0
Carbon
Methanol
Methanol, actual
Methane
CO2, g/MJ
H/C

Fig. 3 Emission of CO_2 as a function of H/C and O/C atom ratios in hydrocarbon fuels.
Source: Adapted from Annamalai and Puri.[8]

releases the highest CO_2, while natural gas (mainly CH_4) emits the lowest CO_2. Because the United States uses fossil fuels for 86% of its energy needs (100 quads), the estimated CO_2 emission is 6350 million ton/year, assuming that the average CO_2 emission from fossil fuels is 70 kg/GJ (methane: 50 kg/GJ vs coal: 90 kg/GJ). Fig. 1 seems to confirm such estimation within a 10% error.

Flame Temperature

Fig. 4 shows a plot of maximum possible flame temperature vs moisture percentage with combustion for biomass fuels. The result can be correlated as follows[4]:

$$\begin{aligned} T(\mathrm{K}) = {} & 2290 - 1.89\ \mathrm{H_2O} + 5.06\ \mathrm{Ash} \\ & - 0.309\ \mathrm{H_2O\ Ash} \\ & - 0.180\ \mathrm{H_2O_2} - 0.108\ \mathrm{Ash}^2 \end{aligned} \tag{3}$$

$$\begin{aligned} T(^\circ\mathrm{F}) = {} & 3650 - 3.40\ \mathrm{H_2O} + 9.10\ \mathrm{Ash} \\ & - 0.556\ \mathrm{H_2O\ Ash} \\ & - 0.324\ \mathrm{H_2O_2} - 0.194\ \mathrm{Ash}^2 \end{aligned} \tag{4}$$

The adiabatic flame temperature decreases if the ash and moisture contents increase.

Flue Gas Volume

The flue gas volume for C–H–O is almost independent of O/C ratios. The fit at 6% O_2 in products gives the following empirical equation for flue gas volume (m^3/GJ) at SATP[8]:

$$\begin{aligned} & \text{Flue gas}_{\mathrm{vol}}\ (\mathrm{m^3/GJ}) \\ & = 4.96\left(\frac{\mathrm{H}}{\mathrm{C}}\right)^2 - 38.628\left(\frac{\mathrm{H}}{\mathrm{C}}\right) + 389.72 \end{aligned} \tag{5}$$

$$\begin{aligned} & \text{Flue gas}_{\mathrm{vol}}\ (\mathrm{ft^3/mmBtu}) \\ & = 184.68\left(\frac{\mathrm{H}}{\mathrm{C}}\right)^2 - 1439.28\left(\frac{\mathrm{H}}{\mathrm{C}}\right) + 14520.96 \end{aligned} \tag{6}$$

Liquid Fuels

Liquid fuels, used mainly in the transportation sector, are derived from crude oil, which occurs naturally as a free-flowing liquid with a density of $\rho \approx 780\ \mathrm{kg/m^3}$–$1000\ \mathrm{kg/m^3}$, containing 0.1% ash and 0.15%–0.5% nitrogen. Crude oil normally contains a mixture of hydrocarbons, and as such, the "boiling" temperature keeps increasing as the oil is distilled. Most fuel oils contain 83%–88% carbon and 6%–12% (by mass) hydrogen.

Gaseous Fuels

The gaseous fuels are cleaner-burning fuels than liquid and solid fuels. They are a mixture of HC but dominated by highly volatile CH_4 with very little S and N. Natural gas is transported as liquefied natural gas (LNG) and compressed

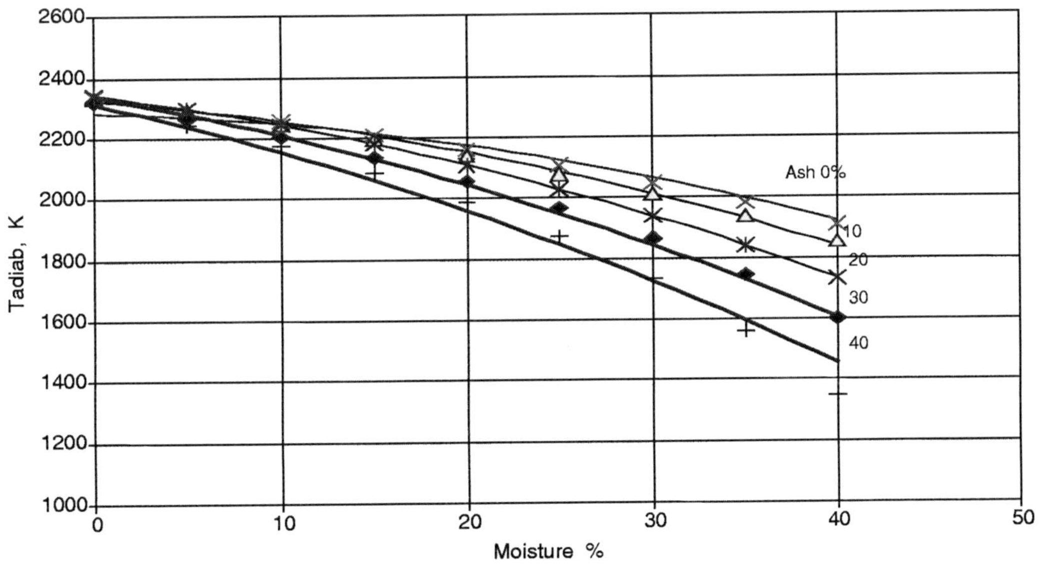
2600
2400
2200
2000
1800
1600
1400
1200
1000
Tadiab, K
0
10
20
30
40
50
Moisture %
Ash 0%
10
20
30
40

Fig. 4 Correlation of adiabatic flame temperature with moisture and ash contents.

natural gas (CNG), typically at 150–250 bars. Liquefied petroleum gas (LPG) is a byproduct of petroleum refining, and it consists mainly of 90% propane. A low-Btu gas contains 0–7400 kJ/SCM (Standard Cubic Meter, 0–200 Btu/SCF, standards defined in Table 1); a medium-Btu gas, 7400–14,800 kJ/SCM (200–400 Btu/SCF); and a high-Btu gas, above 14,800 kJ/SCM (more than 400 Btu/SCF). Hydrogen is another gaseous fuel, with a heat value of 11,525 kJ/SCM (310 Btu/SCF). Because the fuel quality (heat value) may change when fuel is switched, the thermal output rate at a fixed gas-line pressure changes when fuels are changed.

COAL AND BIOMASS PYROLYSIS, GASIFICATION, AND COMBUSTION

Typically, coal densities range from 1100 kg/m^3 for low-rank coals to 2330 kg/m^3 for high-density pyrolytic graphite, while for biomass, density ranges from 100 kg/m^3 for straw to 500 kg/m^3 for forest wood.[9] The bulk density of cattle FB as harvested is 737 kg/m^3 (CF) for high ash (HA-FB) and 32 lbs/CF for low ash (LA-FB).[10] The processes during heating and combustion of coal are illustrated in Fig. 5, and they are similar for biomass except for high VM. The process of release of gases from solid fuels in the absence of oxygen is called pyrolysis, while the combined process of pyrolysis and partial oxidation of fuel in the presence of oxygen is known as gasification. If all combustible gases and solid carbon are oxidized to CO_2 and H_2O, the process is known as combustion.

Pyrolysis

Solid fuels, like coal and biomass, can be pyrolyzed (thermally decomposed) in inert atmospheres to yield combustible gases or VM. While biomass typically releases about 70%–80% of its mass as VM (mainly from cellulose and hemicellulose) with the remainder being char, mainly from lignin content of biomass, coal releases 10%–50% of its mass as VM, depending upon its age or rank. Typically, a medium-rank coal consists of 40% VM and 60% FC, while a high-rank coal has about 10% VM. Bituminous coal pyrolyzes at about 700 K (with 1% mass loss for heating rates <100°C/s), as in the case of most plastics. Pyrolytic products range from lighter volatiles like CH_4, C_2H_4, C_2H_6, CO, CO_2, H_2, and H_2O to heavier molecular mass tars. Apart from volatiles, nitrogen is also evolved from the fuel during pyrolysis in the form of HCN, NH_3, and other compounds or, more generally, XN.

Sweeten et al. performed the thermogravimetric analysis (TGA) of feedlot manure.[4] The results are shown in Fig. 6. In the case of manure, drying occurred between 50 and 100°C, pyrolysis was initiated around 185°C–200°C for a heating rate of 80°C/min, and the minimum ignition temperature was approximately 528°C. The gases produced during biomass pyrolysis can also be converted into transportation fuels like biodiesel, methanol, and ethanol, which may be used either alone or blended with gasoline.

Volatile Oxidation

Once released, volatiles (HC, CO, H_2, etc.,) undergo oxidation within a thin gas film surrounding the solid fuel

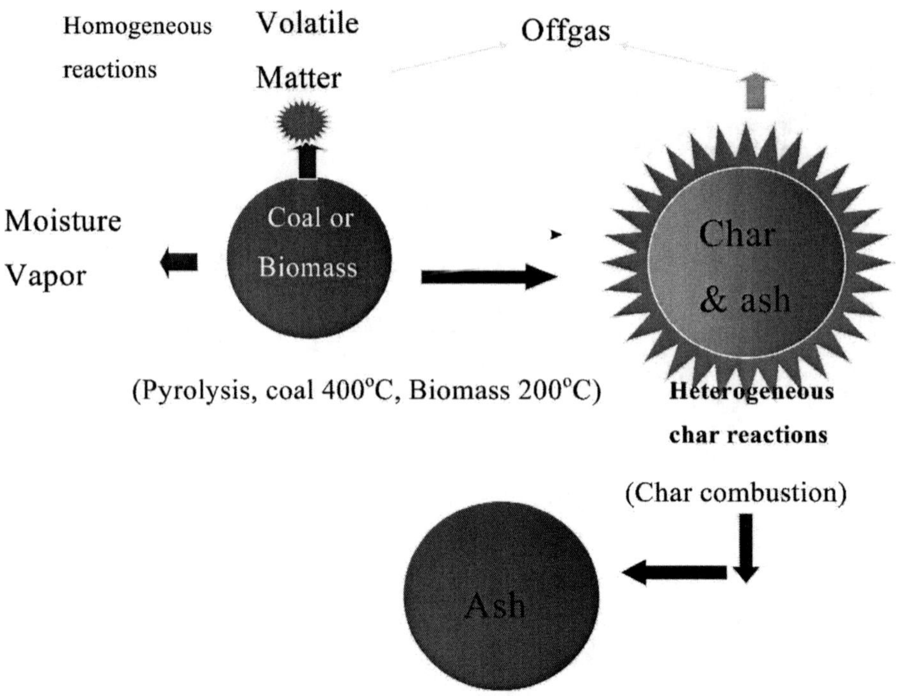
Homogeneous reactions
Volatile Matter
Offgas
Moisture Vapor
Coal or Biomass
Char & ash
(Pyrolysis, coal 400°C, Biomass 200°C)
Heterogeneous char reactions
(Char combustion)
Ash

Fig. 5 Processes during coal pyrolysis, gasification, and combustion.

particle. The oxidation for each HC involves several steps. The enveloping flame, due to volatile combustion, acts like a shroud by preventing oxygen from reaching the particle surface for heterogeneous oxidation of char. Following Dryer,[11] the one-step global oxidation of a given species can be written as

$$\text{Fuel} + \upsilon_{O_2} O_2 \xrightarrow{\text{oxidation}} \upsilon_{CO_2} CO_2 + \upsilon_{H_2O} H_2O \quad (7)$$

$$-\frac{d[\text{Fuel}]}{dt}, \frac{\text{kg}}{\text{m}^3\ \text{sec}} = \text{A}\ \exp\left(\frac{-E}{RT}\right)\left[Y_{\text{fuel}}\right]^a\left[Y_{O_2}\right]^b \quad (8)$$

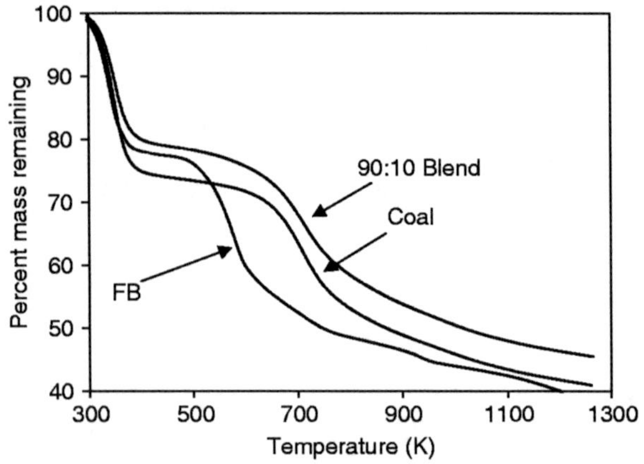
Percent mass remaining
100
90
80
70
60
50
40
300
500
700
900
1100
1300
Temperature (K)
90:10 Blend
Coal
FB

Fig. 6 Thermo-gravimetric analyses of Feedlot Biomass (FB or cattle manure), coal, and 90:10 coal: FB blends.
Source: Sweeten et al.[4]

where [ ] represents the concentration of species in kg/m^3, Y the mass fraction, A the pre-exponential factor, E the activation energy in kJ/kmole, and a and b the order of reaction; they are tabulated in Bartok and Sarofim for alkanes, ethanol, methanol, benzene and toluene.[11]

Char Reactions

The skeletal char, essentially FC, undergoes heterogeneous reactions with gaseous species. The heterogeneous combustion of carbon or char occurs primarily via one or more of the following reactions:

Reaction I. $C + \frac{1}{2}O_2 \rightarrow CO$

Reaction II. $C + O_2 \rightarrow CO_2$

Reaction III. $C + CO_2 \rightarrow 2\ CO$

Reaction IV. $C + H_2O \rightarrow CO + H_2$

Assuming a first-order reaction for scheme I, the oxygen consumption rate is given as

$$\dot{m}_{O_2} \approx \pi d_p^2 B_I T^n \exp\left(-\frac{E}{R_V T_p}\right) \rho_\infty Y_{O_2,w}. \quad (9)$$

The dominant oxygen transfer mechanism at high temperatures is via reaction I with an E/R (a ratio of activation energy to universal gas constant) of about 26,200 K, where $B_I = 2.3 \times 10^7$ m/s and $n = 0.5$ to 1. Reaction II has an E/R of 20,000 K, and $B_{II} = 1.6 \times 10^5$ m/s. Reaction III, the Boudouard reduction reaction, proceeds with an E/R of about

40,000 K. The reduction reactions, III and IV, may become significant, especially at high temperatures for combustion in boiler burners. Reaction with steam is found to be 50 times faster than CO_2 at temperatures up to 1800°C at 1 bar for 75–100 micron-sized Montana Rosebud char.[12] The combustible gases CO and H_2 undergo gas phase oxidation, producing CO_2 and H_2O.

Ignition and Combustion

Recently, Essenhigh et al. have reviewed the ignition of coal.[13] Volatiles from lignite are known to ignite at $T > 950$ K in fluidized beds. Coal may ignite homogeneously or heterogeneously depending upon size and volatile content.[14,15] A correlation for heterogeneous char ignition temperature is presented by Du and Annamalai, 1994.

Once ignited, the combustion of high volatile coal proceeds in two stages: combustion of VM and combustion of FC. Combustion of VM is similar to the combustion of vapors from an oil drop. The typical total combustion time of 100-micron solid coal particle is on the order of 1 s in boilers and is dominated by the time required for heterogeneous combustion of the residual char particle, while the pyrolysis time $\left(t_{pyr} = 10^6\,(\mathrm{s/m^2})d_p^2\right)$ is on the order of 1/10th–1/100th of the total burning time. Since bio-solid contains 70%–80% VM (coal contains 10% VM), most of the combustion of volatiles occurs within a short time (about 0.10 s)

For liquid drops and plastics of density ρ_c, simple relations exist for evaluating the combustion rates and times. If the transfer number B is defined as

$$B = \frac{cp\{T_\infty - T_w\}}{L} + \frac{Y_{O_2,\infty}}{\nu_{O_2}}\frac{h_c}{L}, \tag{10}$$

where $T_w \approx$ TBP for liquid fuels; $T_w = T_g$, the temperature of gasification for plastics; L is the latent heat for liquid fuel and $L = q_g$, heat of gasification for plastics; $Y_{O_2\infty}$ is the stoichiometric oxygen mass per unit mass of fuel (typically 3.5 for liquid fuels); and h_c is the lower heating value of fuel; then the burn rate ($\dot{m}$) and time (t_b) for spherical condensates (liquid drops and spherical particles of diameter d_p and density ρ_c) are given by the following expressions:

$$\dot{m} \approx 2\pi \frac{\lambda}{c_p} d_p \ln(1+B) \tag{11}$$

$$t_b = \frac{d_0^2}{\alpha_c}, \tag{12}$$

where

$$\alpha_c = 8\frac{\lambda}{c_p}\frac{\ln(1+B)}{\rho_c} \tag{13}$$

and c_p and λ are the specific heat and thermal conductivity of gas mixture evaluated at a mean temperature (approximately 50% of the adiabatic flame temperature).

The higher the B value, the higher the mass loss rate, and the burn time will be lower. The value of B is about 1–2 for plastics (polymers), 2–3 for alcohols, and 6–8 for pentane to octane. The burn time of plastic waste particles will be about 3–4 times longer than single liquid drops of pentane to octane (≈gasoline) of similar diameter.

COMBUSTION IN PRACTICAL SYSTEMS

The time scales for combustion are on the order of 1000, 10, and 1 ms for coal burnt in boilers and liquid fuels burnt in gas turbines and diesel engines. Coal is burnt on grates in lumped form (larger-sized particles, 2.5 cm or greater with a volumetric intensity on the order of 500 kW/m^3), medium-sized particles in fluidized beds (1 cm or less, 500 kW/m^3), or as suspensions or pulverized fuel (pf; 75 micron or less, 200 kW/m^3) in boilers.

Apart from pyrolysis, gasification, and combustion, another option for energy conversion (particularly if solid fuel is in slurry firm, such as flushed dairy manure), is the anaerobic digestion (in absence of oxygen) to CH_4 (60%) and CO_2 (40%) using psychrophylic (ambient temperature), mesophyllic (95°F) and thermophyllic (135°F) bacteria in digesters.[10] Typical options of energy conversion, indicated in Fig. 7, include anaerobic digestion (path 1, the biological gasification process), thermal gasification with air to produce CO, HC, CO_2 (path 2A) or with steam to produce CO_2 and H_2 (path 2B), cofiring (path 3), reburn (path 4; see "Reburn with Bio-Solids"), and direct combustion (path 5).

Suspension Firing

In suspension-fired boilers, solid fuel is pulverized into smaller particles (d_p = 75 μm or less) so that more surface area per unit mass is exposed to the oxidant, resulting in improved ignition and combustion characteristics. Typical boiler burners use swirl burners for atomized oil and pulverized coal firing, while a gas turbine uses a swirl atomizer in highly swirling turbulent flow fields. A swirl burner for pf firing is shown in Fig. 8. The air is divided into a primary air stream which transports the coal (10%–20% of the total air, heated to 70°C–100°C to prevent condensation of vapors and injected at about 20 m/s to prevent settling of the dust, loading dust and gas at a ratio of 1:2) and a secondary air stream (250°C at 60–80 m/s) which is sent through swirl vanes, supplying the remaining oxygen for combustion and imparting a tangential momentum to the air. In wall-fired boilers, burners are stacked above each other on the wall; while in tangential-fired boilers, the burners are mounted at the corners of rectangular furnaces.

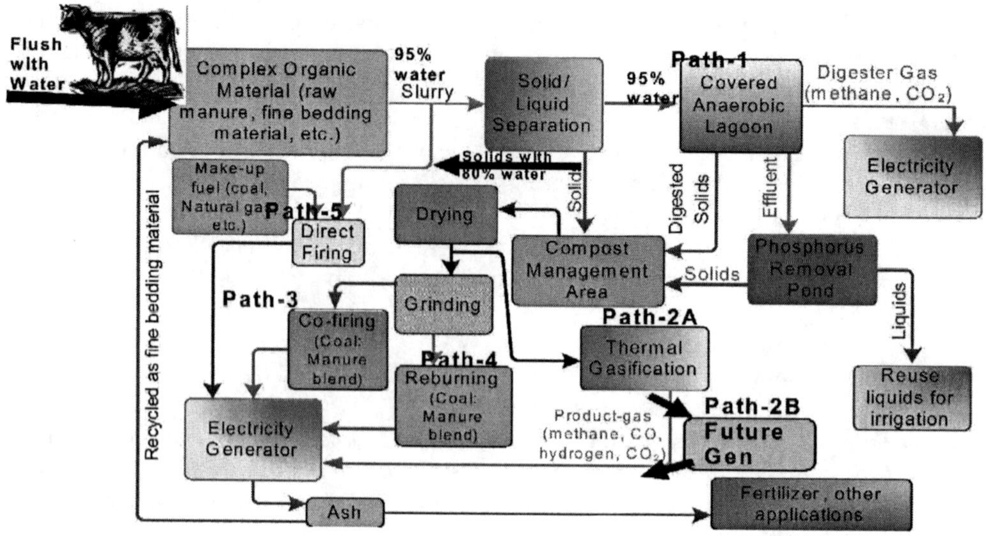
Flush with Water
Complex Organic Material (raw manure, fine bedding material, etc.)
95% water Slurry
Solid/ Liquid Separation
95% water
Path-1
Covered Anaerobic Lagoon
Digester Gas (methane, CO2)
Electricity Generator
Solids with 80% water
Make-up fuel (coal, Natural gas, etc.)
Path-5
Direct Firing
Drying
Solids
Digested Solids
Effluent
Compost Management Area
Solids
Phosphorus Removal Pond
Path-3
Grinding
Co-firing (Coal: Manure blend)
Path-4
Reburning (Coal: Manure blend)
Path-2A
Thermal Gasification
Liquids
Reuse liquids for irrigation
Electricity Generator
Product-gas (methane, CO, hydrogen, CO2)
Path-2B
Future Gen
Recycled as fine bedding material
Ash
Fertilizer, other applications

Fig. 7 Flow chart showing several energy conversion options for a typical dairy or cattle operation.

Stoker Firing

The uncrushed fuel [fusion temperature <1093°C (2000°F); volatile content >20%; sizes in equal proportions of 19 mm × 12.5 mm (3/4 in. × ½ in.), 6.3 mm × 3.2 mm (½ in. × ¼ in.), 3.2 mm × 3.2 mm (¼ in. × ¼ in.)][16] is fed onto a traveling chain grate below to which primary air is supplied (Fig. 9), which may be preheated to 177°C (350°F) if moisture exceeds 25%. The differential pressure is on the order of 5–8 mm (2–3 in.). The combustible gases are carried into an over-fire region into which secondary air (almost 35% of total air at three levels for low emissions) is fired to complete combustion. The over-fire region acts like a perfectly stirred reactor (PSR). It is apparent that solid fuels need not be ground to finer size.

Fixed Bed Combustor

The bed contains uncrushed solid fuels, inert materials (including ash), and processing materials (e.g., limestone to remove SO_2 from gases as sulfates). It is fed with air moving against gravity for complete combustion, but the velocity is low enough that materials are not entrained into the gas streams. Large solid particles can be used.

Fluidized Bed Combustor

When air velocity (V) in fixed bed combustor (FXBC) is increased gradually to a velocity called minimum fluidization velocity V_{mf}, the upward drag force is almost equal

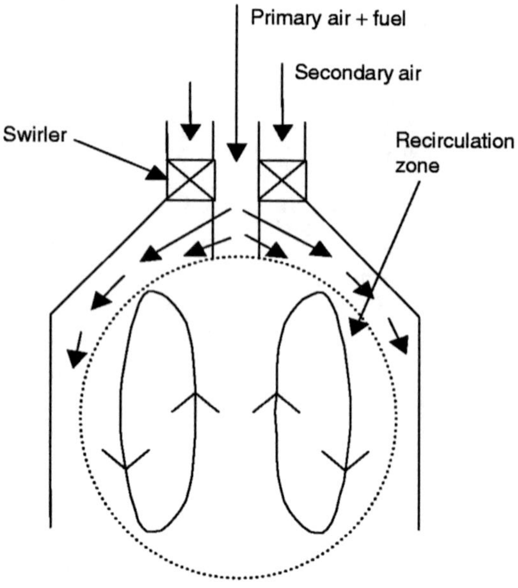
Primary air + fuel
Secondary air
Swirler
Recirculation zone

Fig. 8 Pulverized Fuel (pf) fired swirl burner.

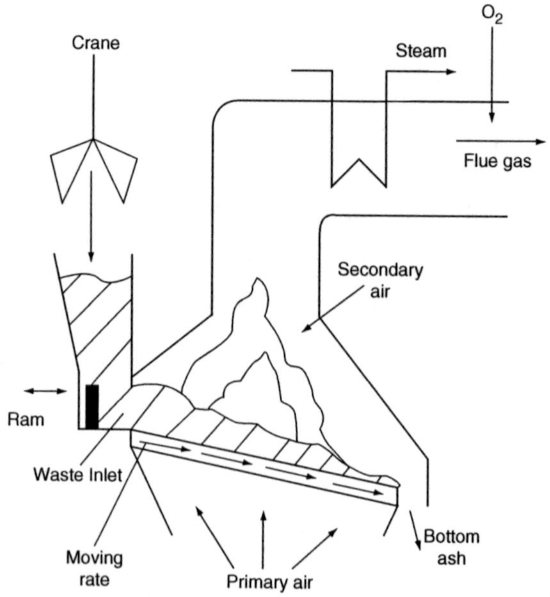
Crane
O2
Steam
Flue gas
Secondary air
Ram
Waste Inlet
Moving rate
Primary air
Bottom ash

Fig. 9 Schematic of stoker firing.
Source: Loo and Kessel.[39]

to the weight of the particle, so that solids float upward. The bed behaves like a fluid (like liquid water in a tank), i.e., it becomes fluidized. If $V > V_{mf}$, then air escapes as bubbles and is called a bubbling fluidized bed combustor (BFBC). The bed has two phases: the bubble phase, containing gases (mostly oxygen), and the emulsion phase (dense phase, oxygen deficient), containing particles and gas. Many times gas velocity is so high that gaseous combustibles produced within the bed burn above the bed (called free board region), while solids (e.g., char and carbon) burn within the bed. Fluidized Bed Combustor (FBC) is suitable for fuels which are difficult to combust in pf-fired boilers.

Circulating Fluidized Bed Combustor (CFBC)

When air velocity in FBC is increased at velocity $V >> V_{mf}$, particles are entrained into the gas stream. Since the residence time available to particles for combustion is shorter, unburned particles are captured using cyclones located downstream of the combustor and circulated back to the bed.

The residence time (t_{res}) varies from a low value for pf-fired burners to a long residence time for fixed bed combustors. The reaction time (t_{reac}) should be shorter than t_{res} so that combustion is complete. The reaction time includes time to heat up to ignition temperature and combustion. The previous section on fuel properties and the homogenous (e.g., CH_4, CO oxidation) and heterogeneous (e.g., carbon oxidation) reaction kinetics can be used to predict t_{reac} or burn time t_b.

COAL AND BIO-SOLIDS COFIRING

General Schemes of Conversion

Most of the previously reviewed combustion systems typically use pure coal, oil, or gas. The same systems require redesign for use with pure biomass fuels. A few of the technologies, which utilize bio-solids as an energy source, are summarized in Annamalai et al.[17] These technologies include direct combustion (fluidized beds), circulating fluidized beds, liquefaction (mostly pyrolysis), onsite gasification for producing low to medium Btu gases, anaerobic digestion (bacterial conversion), and hydrolysis for fermentation to liquid fuels like ethanol.[18,19]

Cofiring

Although some bio-solids have been fired directly in industrial burners as sole-source fuels, limitations arose due to variable moisture and ash contents in bio-solid fuels, causing ignition and combustion problems for direct combustion technologies. To circumvent such problems, these fuels have been fired along with the primary fuels (cofiring) either by directly mixing with coal and firing (2%–15% of heat input basis) or by firing them in between coal-fired burners.[20–24]

Cofiring has the following advantages: improvement of flame stability problems, greater potential for commercialization, low capital costs, flexibility of adaptation of biomass fuels and cost effective power generation, mitigated NO_x emissions from coal-fired boilers, and reduced CO_2 emissions. However, a lower melting point of biomass ash could cause fouling and slagging problems.

Some of the bio-solid fuels used in cofiring with coal are cattle manure,[25,26] sawdust and sewage sludge,[21] switch grass,[20] wood chips,[24,27] straw,[22,28] and refuse-derived fuel (RDF).[21] See Sami et al. for a review of literature on cofiring.[7]

Coal and Agricultural Residues

Sampson et al. reported test burns of three different types of wood chips (20%, HHV from 8320 to 8420 Btu/lb) mixed with coal (10,600 Btu/lb) at a stoker (traveling grate) fired steam plant.[24] The particulate emission in grams per SCF ranged from 0.05 to 0.09. An economic study, conducted

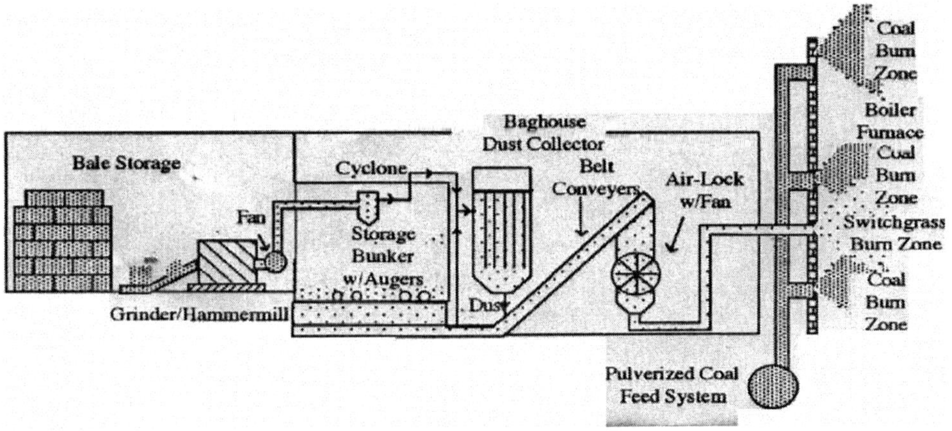
Coal
Burn
Zone
Boiler
Furnace
Coal
Burn
Zone
Switchgrass
Burn Zone
Coal
Burn
Zone
Baghouse
Dust Collector
Bale Storage
Cyclone
Belt
Conveyers
Air-Lock
w/Fan
Fan
Storage
Bunker
w/Augers
Grinder/Hammermill
Dust
Pulverized Coal
Feed System

Fig. 10 A cofiring Scheme for coal and biomass (Alternate Fuel Handling Facility at Blount St. Generating Station). **Source:** Aerts et al.[20]

Endocrine – Energy Conversion

for the 125,000 lb/h steam power plant, concluded that energy derived from wood would be competitive with that from coal if more than 30,000 tons of wood chips were produced per year with hauling distances less than 60 mi. Aerts et al. carried out their experiments on cofiring switch grass with coal in a 50-MW, radiant, wall-fired, pulverized coal boiler with a capacity of 180 tons of steam at 85 bar and 510°C (Fig. 10). The NO_x emissions decreased by 20%, since switchgrass contains lesser nitrogen (Table 4).[20] It is the author's hypothesis that a higher VM content of bio-solids results in a porous char, thus accelerating the char combustion process. This is validated by the data from Fahlstedt et al. on the cofiring of wood chips, olive pips and palm nut shells with coal at the ABB Carbon 1 MW Process Test Facility; they found that blend combustion has a slightly higher efficiency than coal-only combustion.[27]

Coal and RDF

Municipal solid waste includes residential, commercial, and industrial wastes which could be used as fuel for production of steam and electric power. MSW is inherently a blended fuel, and its major components are paper (43%); yard waste, including grass clippings (10%); food (10%); glass and ceramics (9%); ferrous materials (6%); and plastics and rubber (5%). Refer to Tables 5 and 6 for analyses. When raw waste is processed to remove non-combustibles like glass and metals, it is called RDF. MSW can decompose in two ways, aerobic and anaerobic. Aerobic decomposition (or composting) occurs when O_2 is present. The composting produces CO_2 and water, but no usable energy products. The anaerobic decomposition occurs in the absence of O_2. It produces landfill gas of 55% CH_4 and 45% CO_2.

Coal and Manure

Frazzitta et al. and Annamalai et al. evaluated the performance of a small-scale pf-fired boiler burner facility (100,000 Btu/h) while using coal and premixed coal-manure blends with 20% manure. Three types of feedlot manure were used: raw, partially composted, and fully composted. The burnt fraction was recorded to be 97% for both coal and coal-manure blends.[25,26]

NO_x Emissions

During combustion, the nitrogen evolved from fuel undergoes oxidation to NO_x; and this is called fuel NO_x to distinguish it from thermal NO_x, which is produced by oxidation of atmospheric nitrogen. Unlike coal, most of the agricultural biomass being burned is very low in nitrogen content (i.e., wood or crops), but manure has a higher N content than fossil fuels. A less precise correlation exists between cofiring levels on a Btu basis and percent NO reduction under cofiring. The following Eq. (valid between 3 and 22% mass basis cofiring) describes NO_x reduction as a function of cofiring level on a heat input basis:

$$\begin{aligned} &NO_x \text{ Reduction } (\%) \\ &= 0.0008\,(\text{COF}\%)^2 + 0.0006\,\text{COF}\% + 0.0752, \end{aligned} \tag{14}$$

Table 5 Chemical composition of solid waste.

	Percent	
Proximate analysis	**Range**	**Typical**
Volatile matter (VM)	30–60	50
Fixed carbon (FC)	5–10	8
Moisture	10–45	25
Ash	10–30	25

	Percent by mass (dry basis)					
Ultimate analysis	**C**	**H**	**O**	**N**	**S**	**Ash**
Yard wastes	48	6	38	3	0.3	4.7
Wood	50	6	43	0.2	0.1	0.7
Food wastes	50	6	38	3	0.4	2.6
Paper	44	6	44	0.3	0.2	5.5
Cardboard	44	6	44	0.3	0.2	5.5
Plastics	60	7	23			10
Textiles	56	7	30	5	0.2	1.8
Rubber	76	10		2		12
Leather	60	9	12	10	0.4	8.6
Misc. organics	49	6	38	2	0.3	4.7
Dirt, ashes, etc.	25	3	1	0.5	0.2	70.3

Table 6 Heat of combustions of municipal solid waste components.

	Inerts (%)		Heating values (kJ/kg)	
Component	**Range**	**Typical**	**Range**	**Typical**
Yard wastes	2–5	4	2,000–19,000	7,000
Wood	0.5–2	2	17,000–20,000	19,000
Food wastes	1–7	6	3,000–6,000	5,000
Paper	3–8	6	12,000–19,000	17,000
Cardboard	3–8	6	12,000–19,000	17,000
Plastics	5–20	10	30,000–37,000	33,000
Textiles	2–4	3	15,000–19,000	17,000
Rubber	5–20	10	20,000–28,000	23,000
Leather	8–20	10	15,000–20,000	17,000
Misc. organics	2–8	6	11,000–26,000	18,000
Glass	96–99	98	100–250	150
Tin cans	96–99	98	250–1,200	700
Nonferrous	90–99	96		
Ferrous metals	94–99	98	250–1,200	700
Dirt, ashes, etc.	60–80	70	2,000–11,600	7,000

where COF% is the percentage of co-firing on a heat input basis. The mechanisms used to reduce NO_x emissions by cofiring vary between cyclone firing and PC firing.

Fig. 11 shows the percentage reduction in NO with percentage cofiring of low-N agricultural biomass fuels. This relationship does not apply to high-N biofuels such as animal manure.

Fouling in Cofiring

Hansen et al. investigated the ash deposition problem in a multi-circulating fluidized bed combustor (MCFBC) fired with fuel blends of coal and wood straw.[25] The Na and K lower the melting point of ash. For ash fusion characteristics see Table 7. Rasmussen and Clausen evaluated the performance of an 80-MW co-generation power plant at Grenaa, Denmark, fired with hard coal and bio-solids (surplus straw from farming). Large amounts of Na and K in straw caused superheater corrosion and combustor fouling.[29] Annamalai et al. evaluated fouling potential when feedlot manure biomass (FB) was cofired with coal under suspension firing.[30] The 90:10 Coal:FB blend resulted in almost twice the ash output compared to coal and ash deposits on heat exchanger tubes that were more difficult

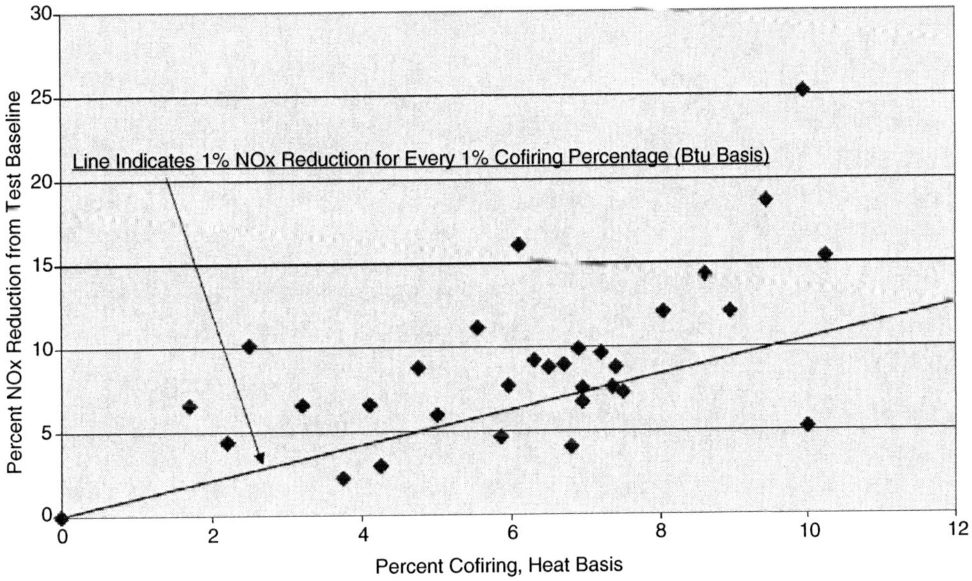
Line Indicates 1% NOx Reduction for Every 1% Cofiring Percentage (Btu Basis)
Percent NOx Reduction from Test Baseline
Percent Cofiring, Heat Basis
0
5
10
15
20
25
30
0
2
4
6
8
10
12

Fig. 11 NO_x reduction due to cofiring with low-N agricultural residues.
Source: Grabowski.[40]

Table 7 Ash fusion behavior and ash composition, fusion data: ASTM D-188.

	FB	PRB Coal	Blend
Ash Fusion, (reducing)			
Initial deformation, IT, °C (°F)	1140 (2090)	1130 (2060)	NA
Softening, °C (°F)	1190 (2170)	1150 (2110)	NA
Hemispherical, HT, °C (°F)	1210 (2210)	1170 (2130)	NA
Fluid, °C (°F)	1230 (2240)	1200 (2190)	NA
Ash fusion, (oxidizing)			
Initial deformation, IT, °C (°F)	1170 (2130)	1190 (2180)	NA
Softening, °C (°F)	1190 (2180)	1200 (2190)	NA
Hemispherical, HT, °C (°F)	1220 (2230)	1210 (2210)	NA
Fluid, °C (°F)	1240 (2270)	1280 (2330)	NA
Slagging Index, Rs, °C (°F)	1160 (2120)	1140 (2090)	
Slagging classification	High	Severe	
Ash composition (wt%)			
SiO_2	53.63	36.45	43.56
Al_2O_3	5.08	18.36	12.87
Fe_2O_3	1.86	6.43	4.54
TiO_2	0.29	1.29	0.88
CaO^+	14.60	19.37	17.40
MgO^+	3.05	3.63	3.39
Na_2O^+	3.84	1.37	2.39
K_2O^+	7.76	0.63	3.58
P_2O_5	4.94	0.98	2.62
SO_3	3.71	10.50	7.69
MnO_2	0.09	0.09	0.09
Sum	98.84	99.11	99.00
Volatile Oxides	30.77	28.25	
Basic oxides	32.73	35.51	
Silica ratio	0.73	0.53	
$Na_2O + K_2O$	11.60	2.00	5.97
Inherent Ca/S Ratio	6.71	1.86	2.48
kg alkali ($Na_2O + K_2O$)/GJ	5.37	0.06	0.29

to remove than baseline coal ash deposits. The increased fouling behavior with blend is probably due to the higher ash loading and ash composition of FB.

GASIFICATION OF COAL AND BIO-SOLIDS

Gasification is a thermo-chemical process in which a solid fuel is converted into a gaseous fuel (primarily consisting of HC, H_2 and CO_2) with air or pure oxygen used for partial oxidation of FCs. The main products during gasification are CO and H_2, with some CO_2, N_2, CH_4, H_2O, char particles, and tar (heavy hydrocarbons). The oxidizers used for the gasification processes are oxygen, steam, or air. However, for air, the gasification yields a low-Btu gas, primarily caused by nitrogen dilution present in the supply air. Syngas ($CO+H_2$) is produced by reaction of biomass with steam. The combustible product, gas, can be used as fuel burned directly or with a gas turbine to produce electricity; or used to make chemical feedstock (petroleum refineries). However, gas needs to be cleaned to remove tar, NH_3, and sulfur compounds. The integrated gasification combined cycle (IGCC) (Fig. 12), for combined heat and power (CHP), and traditional boilers use combustible gases from gasifiers for generation of electric power.

Typically in combined cycles, gaseous or liquid fuel is burnt in gas turbine combustors. High-temperature products are expanded in a gas turbine for producing electrical power; a low-temperature (but still hot) exhaust is then used as heat input in a boiler to produce low-temperature

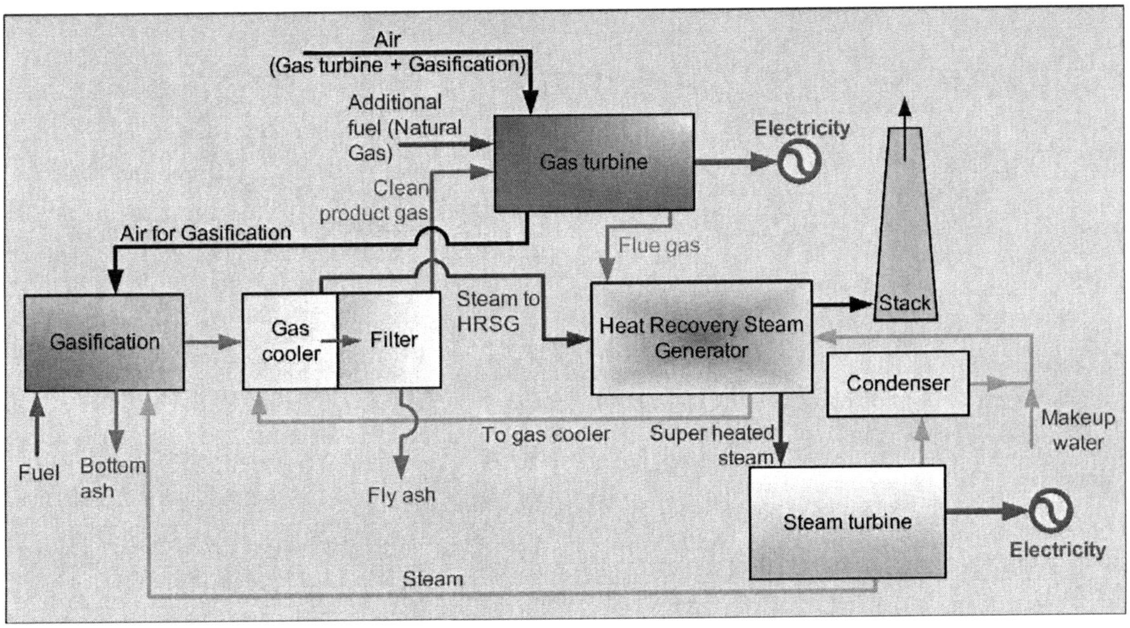
Air
(Gas turbine + Gasification)
Additional fuel (Natural Gas)
Clean product gas
Gas turbine
Electricity
Air for Gasification
Flue gas
Stack
Gasification
Gas cooler
Filter
Steam to HRSG
Heat Recovery Steam Generator
Condenser
To gas cooler
Super heated steam
Makeup water
Fuel
Bottom ash
Fly ash
Steam turbine
Electricity
Steam

Fig. 12 Fluidized-bed gasification for Integrated Gasification Combined Cycle (IGCC) Process.

steam, which then drives a steam turbine for electrical power. Therefore, one may use gas as a topping cycle medium, while steam is used as fluid for the bottoming cycle. The efficiency of a combined cycle is on the order of 60%, while a conventional gas turbine cycle has an efficiency of 42%.[31] Commercial operations include a 250-MW IGCC plant at Tampa, Florida, operating since 1996; a 340-MW plant at Negishi, Japan, since 2003; and a 1200-MW GE-Bechtel plant under construction in Ohio for American Electric Power, to start in 2010.[31]

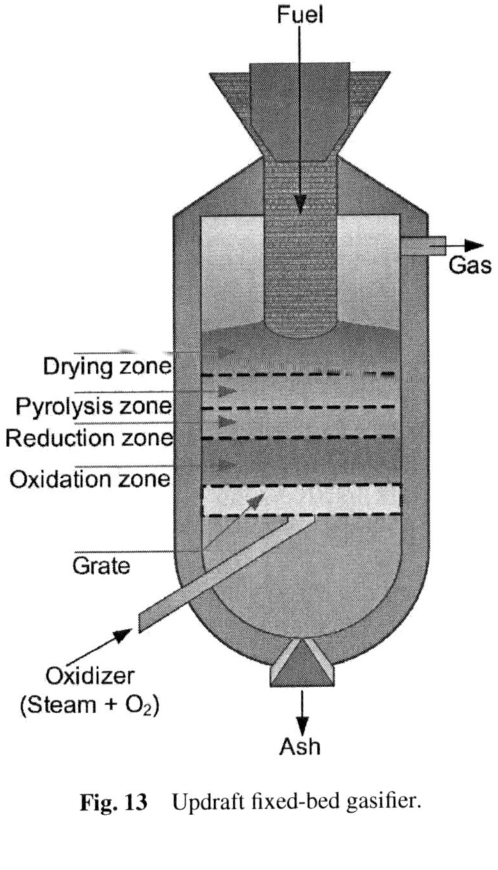
Fuel
Gas
Drying zone
Pyrolysis zone
Reduction zone
Oxidation zone
Grate
Oxidizer
(Steam + O2)
Ash

Fig. 13 Updraft fixed-bed gasifier.

There are three basic gasification reactor types: (i) fixed-bed gasifiers (Fig. 13); (ii) fluidized-bed gasifiers, including circulating-bed (CFB) or bubbling-bed; and (iii) entrained-flow gasifiers. The principles of operation are similar to those of combustors except that the air supplied is much below stoichiometric amounts, and instead of a combination of steam, air and CO_2, air can also be used. The oxidant source could also include gases other than air, such as air combined with steam in Blasiak et al.[32]

FUTUREGEN

FutureGen is a new U.S. initiative to build the world's first integrated CO_2 sequestration and H_2 production research power plant using coal as fuel. The technology shown in Fig. 14 employs modern coal gasification technology using pure oxygen, resulting in CO, C_nH_m (a hydrocarbon), H_2, HCN, NH_3, N_2, H_2S, SO_2, and other combinations which are further reacted with steam (reforming reactions) to produce CO_2 and H_2. The bed materials capture most of the harmful N and S compounds followed by gas-cleaning systems; the CO_2 is then sequestered and H_2 is used as fuel, using either combined cycle or fuel cells for electricity generation or sold as clean transportation fuel. With partial oxidation of gasification products and char supplying heat for pyrolysis and other endothermic reactions (i.e, net zero external heat supply in gasifier), the overall gasification reaction can be represented as follows for 100 kg of DAF Wyoming coal:

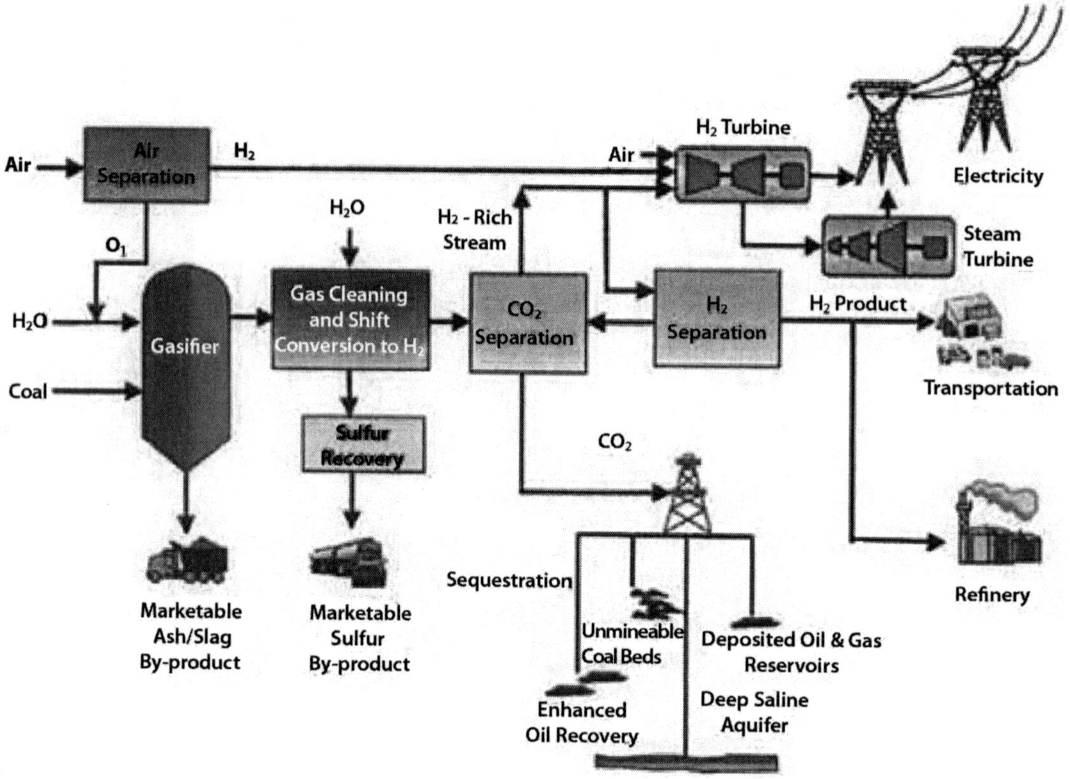
Air
Air Separation
H₂
O₁
H₂O
Coal
Gasifier
H₂O
Gas Cleaning and Shift Conversion to H₂
Sulfur Recovery
H₂ - Rich Stream
CO₂ Separation
Air
H₂ Turbine
H₂ Separation
H₂ Product
Steam Turbine
Electricity
Transportation
CO₂
Refinery
Sequestration
Unmineable Coal Beds
Deposited Oil & Gas Reservoirs
Enhanced Oil Recovery
Deep Saline Aquifer
Marketable Ash/Slag By-product
Marketable Sulfur By-product

Fig. 14 FutureGen layout.
Source: http://www.fe.doe.gov.

Reaction V $C_{6.3}H_{4.4}N_{0.076}O_{1.15}S_{0.014} + 8.14\ H_2O(\ell) + 1.67\ O_2 \rightarrow 6.3\ CO_2 + 10.34\ H_2 + 0.038\ N_2 + 0.014\ SO_2$

It is apparent that the FutureGen process results in enhanced production of H_2, using coal as an energy source to strip H_2 from water. For C–H–O fuels, it can be shown that theoretical H_2 production (N_{H_2}) in moles for an empirical fuel CH_hO_o is given as $\{0.4115\ h - 0.6204\ o + 1.4776\}$ under the above conditions. For example, if glucose $C_6H_{12}O_6$ is the fuel, then empirical formulae is CH_2O; thus, with h = 2, o = 1, $N_{H_2} = 1.68$ kmol, using atom balance, $CH_2O + 0.68\ H_2O(\ell) + 0.16\ O_2 \rightarrow CO_2 + 1.68\ H_2$.

REBURN WITH BIO-SOLIDS

NO_x is produced when fuel is burned with air. The N in NO_x can come both from the nitrogen-containing fuel compounds (e.g., coal, biomass, plant residue, animal waste) and from the N in the air. The NO_x generated from fuel N is called fuel NO_x, and NO_x formed from the air is called thermal NO_x. Typically, 75% of NO_x in boiler burners is from fuel N. It is mandated that NO_x, a precursor of smog, be reduced to 0.40–0.46 lb/mmBtu for wall and tangentially fired units under the Clean Air Act Amendments (CAAA). The current technologies developed for reducing NO_x include combustion controls (e.g., staged combustion or low NO_x burners (LNB), reburn) and post-combustion controls (e.g., Selective Non-Catalytic Reduction, SNCR using urea).

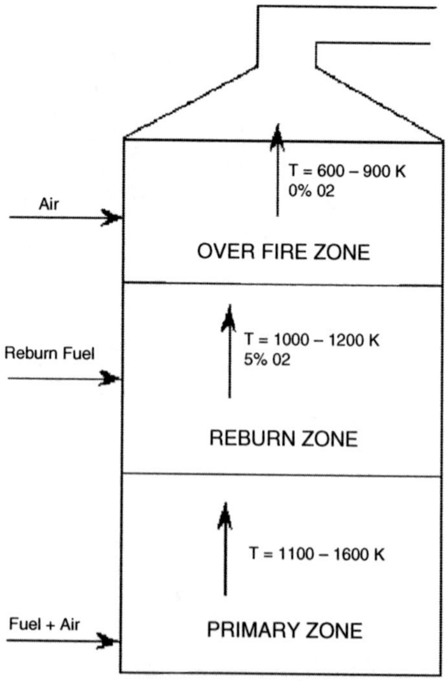
Air
T = 600 – 900 K
0% 02
OVER FIRE ZONE
Reburn Fuel
T = 1000 – 1200 K
5% 02
REBURN ZONE
Fuel + Air
T = 1100 – 1600 K
PRIMARY ZONE

Fig. 15 Schematic of reburn process.

Table 8 Percentage reduction in NO_x: demonstration and/or operating reburn installations on coal-fired boilers in the United States.

Type of burner	% Reburn heat in	% Reduction	NO_x with reburn lb/mmBtu[a]
Gas reburning			
Tangential	18	50–67	0.25
Cyclone	20–23	58–60	0.39–0.56
Wall without LNB	18	63	0.27
Coal reburn			
Cyclone (micronized)	30 (17)	52 (57)	0.39 (0.59)
Tangential (micron) with LNB	14	28	0.25

Note: LNB: Low NO_x Burners.
[a]1 lb per mmBtu = 0.430 kg/GJ.
Source: DOE.[38]

In reburning, additional fuel (typically natural gas) is injected downstream from the primary combustion zone to create a fuel rich zone (optimum reburn stoichiometric ratio (SR), usually between SR 0.7 and 0.9), where NO_x is reduced up to 60% through reactions with hydrocarbons when reburn heat input with CH_4 is about 10%–20%. Downstream of the reburn zone, additional air is injected in the burnout zone to complete the combustion process. A diagram of the entire process with the different combustion zones is shown in Fig. 15. There have been numerous studies on reburn technology found in literature, with experiments conducted, and the important results summarized elsewhere.[33] Table 8 shows the percentages of reduction and emission obtained with coal or gas reburn in coal-fired installations and demonstration units.

The low cost of biomass and its availability make it an ideal source of pyrolysis gas, which is a more effective reburn fuel than the main source fuel, which is typically coal. Recently, animal manure has been tested as a reburn fuel in laboratory scale experiments. A reduction of a maximum of 80% was achieved for pure biomass, while the coal experienced a reduction of between 10 and 40%, depending on the equivalence ratio.[34] It is believed that the greater effectiveness of the feedlot biomass is due to its greater volatile content on a DAF basis and its release of fuel nitrogen in the form of NH_3, instead of HCN.[35]

ACKNOWLEDGMENTS

Most of this work was supported in part by the U.S. Department of Energy of Pittsburgh, PA, the DOE of Golden, CO, the USDA, the Texas Advanced Technology Program, and the Texas Commission on Environmental Quality (TCEQ) of Austin, TX, through the Texas Engineering Experiment Station (TEES), Texas A&M University, and College Station, TX.

REFERENCES

1. U.S. DOE, Energy Information Administration, available at http://www.eia.doe.gov/ (accessed April 12, 2005).
2. Priyadarsan, S.; Annamalai, K.; Sweeten, J.M.; Holtzapple, M.T.; Mukhtar, S. *Co-gasification of blended coal with feedlot and chicken litter biomass*, Proceedings of the 30th Symposium (International) on Combustion; The Combustion Institute: Pittsburgh, PA, 2005; Vol. 30, 2973–2980.
3. Ebeling, J.M.; Jenkins, B.M. Physical and chemical properties of biomass fuels. Trans. ASAE **1985**, *28* (3), 898–902.
4. Sweeten, J.M.; Annamalai, K.; Thien, B.; McDonald, L. Co-firing of coal and cattle feedlot biomass (FB) fuels, part I: feedlot biomass (cattle manure) fuel quality and characteristics. Fuel **2003**, *82* (10), 1167 1182.
5. Bole, W. Wiss z Tech Hochsch. Dresden 1952/1953, 2, 687.
6. Annamalai, K.; Sweeten, J.M.; Ramalingam, S.C. Estimation of the gross HVs of biomass fuels. Trans. Soc. Agric. Eng. **1987**, *30*, 1205–1208.
7. Sami, M.; Annamalai, K.; Wooldridge, M. Co-firing of coal and biomass fuel blends. Prog. Energy Combust. Sci. **2001**, *27*, 171–214.
8. Annamalai, K.; Puri, I.K. 2006, *Combustion Science and Engineering*, Taylor and Francis: Orlando, FL, 2006.
9. Tillman, D.A.; Rossi, A.J.; Kitto, W.D. *Wood Combustion: Principles: Processes and Economics*; Academic Press: New York, 1981.
10. Sweeten, J.M.; Heflin, K.; Annamalai, K.; Auvermann, B.; Collum, Mc; Parker, D.B. *Combustion-Fuel Properties of Manure Compost from Paved vs Un-paved Cattle feedlots*; ASABE 06-4143: Portland, OR, July 9–12, 2006.
11. Bartok, W., Sarofim, A.F. 1991, *Fossil Fuel Combustion*, chapter 3: FL Dryer, John Wiley, and Hobokan, NJ. pp. 121–214.
12. Howard, J.B.; Sarofim, A.F. Gasification of coal char with CO_2 and steam at 1200–1800°C. *Energy Lab Report*; Chemical Engineering; MIT. MA, 1978.
13. Essenhigh, R.H.; Misra, M.K.; Shaw, D.W. Ignition of coal particles: a review. Combust. Flame **1989**, *77*, 3–30.
14. Annamalai, K.; Durbetaki, P. A theory on transition of ignition phase of coal particles. Combust. Flame **1977**, *29*, 193–208.
15. Du, X.; Annamalai, K. The transient ignition of isolated coal particles. Combust. Flame **1994**, *97*, 339–354.
16. Johnson, N. *Fundamentals of Stoker Fired Boiler Design and Operation*, CIBO Emission Controls Technology Conference, July 15–17, 2002.
17. Annamalai, K.; Ibrahim, Y.M.; Sweeten, J.M. Experimental studies on combustion of cattle manure in a fluidized bed

combustor. Trans. ASME, J. Energy Res. Technol. **1987**, *109*, 49–57.
18. Walawender, W.P.; Fan, L.T.; Engler, C.R.; Erickson, L.E. Feedlot manure and other agricultural wastes as future material and energy resources: II. Process descriptions. *Contrib.30*, Deptartment of Chemical Engineering, Kansas Agricultural Experiment Station: Manhattan, KS, 30, 1973.
19. Raman, K.P.; Walawander, W.P.; Fan, L.T. Gasification of feedlot manure in a fluidized bed: effect of temperature. Ind. Eng. Chem. Proc. Des. Dev. **1980**, *10*, 623–629.
20. Aerts, D.J.; Bryden, K.M.; Hoerning, J.M.; Ragland, K.W. *Co-firing Switchgrass in a 50 MW Pulverized Coal Boiler*, Proceedings of the 59th Annual American Power Conference, Chicago, IL, **1997**, *50* (2), 1180–1185.
21. Abbas, T.; Costen, P.; Kandamby, N.H.; Lockwood, F.C.; Ou, J.J. The influence of burner injection mode on pulverized coal and biosolid co-fired flames. Combust. Flame **1994**, *99*, 617–625.
22. Siegel, V.; Schweitzer, B.; Spliethoff, H.; Hein, K.R.G. Preparation and co-combustion of cereals with hard coal in a 500 kW pulverized-fuel test unit. Biomass for energy and the environment, *Proceedings of the 9th European Bioenergy Conference*, Copenhagen, DK, June 24–27, 1996; 2, 1027–1032.
23. Hansen, L.A.; Michelsen, H.P.; Dam-Johansen, K. Alkali metals in a coal and biosolid fired CFBC—measurements and thermodynamic modeling, *Proceedings of the 13th International Conference on Fluidized Bed Combustion*, Orlando, FL, May 7–10, 1995; 1, 39–48.
24. Sampson, G.R.; Richmond, A.P.; Brewster, G.A.; Gasbarro, A.F. Co-firing of wood chips with coal in interior Alaska. Forest Prod. J. **1991**, *41* (5), 53–56.
25. Frazzitta, S.; Annamalai, K.; Sweeten, J. Performance of a burner with coal and coal: feedlot manure blends. J. Propulsion Power **1999**, *15* (2), 181–186.
26. Annamalai, K.; Thien, B.; Sweeten, J.M. Co-firing of coal and cattle feedlot biomass (FB) fuels part II: performance results from 100,000 Btu/h laboratory scale boiler burner. Fuel **2003**, *82* (10), 1183–1193.
27. Fahlstdedt, I.; Lindman, E.; Lindberg, T.; Anderson, J. Cofiring of biomass and coal in a pressurized fluidized bed combined cycle. Results of pilot plant studies, *Proceedings of the 14th International conference on Fluidized Bed Combustion*, Vancouver, Canada, May 11–14, 1997; 1, 295–299.
28. Van Doorn, J., Bruyn, P., Vermeij, P. Combined combustion of biomass, fluidized sewage sludge and coal in an atmospheric Fluidized bed installation, Biomass for energy and the environment, *Proceedings of the 9th European Bioenergy Conference*, Copenhagen, DK, June 24–27, 1996; 1007–1012.
29. Rasmussen, I.; Clausen, J.C. ELSAM strategy of firing biosolid in CFB power plants, *Proceedings of the 13th International Conference on Fluidized Bed Combustion*, Orlando, FL, May 7–10, 1, 1995; 557–563.
30. Annamalai, K.; Sweeten, J.; Freeman, M.; Mathur, M.; O'Dowd, W.; Walbert, G.; Jones, S. Co-firing of coal and cattle feedlot biomass (FB) fuels, part III: fouling results from a 500,000 Btu/hr pilot plant scale boiler burner. Fuel **2003**, *82* (10), 1195–1200.
31. Langston, L.S. 2005 New Horizons. *Power and Energy* vol 2, June 2, 2005.
32. Blasiak, W.; Szewczyk, D.; Lucas, C.; Mochida, S. *Gasification of Biomass Wastes with High Temperature Air and Steam*, Twenty-First International Conference on Incineration and Thermal Treatment Technologies, New Orleans, LA, May 13–17, 2002.
33. Thien, B.; Annamalai, K. *National Combustion Conference*, Oakland, CA, March 25–27, 2001.
34. Arumugam, S.; Annamalai, K.; Thien, B.; Sweeten, J. Feedlot biomass co-firing: a renewable energy alternative for coal-fired utilities, Int. Natl J. Green Energy **2005**, *2* (4), 409–419.
35. Zhou, J.; Masutani, S.; Ishimura, D.; Turn, S.; Kinoshita, C. Release of fuel-bound nitrogen during biomass gasification. Ind. Eng. Chem. Res. **2000**, *39*, 626–634.
36. Annamalai, K.; Puri, I.K. *Advanced Thermodynamics*; CRC Press: Boca Raton, FL, 2001.
37. Priyadarsan, S.; Annamalai, K.; Sweeten, J.M.; Holtzapple, M.T.; Mukhtar, S. Waste to energy: fixed bed gasification of feedlot and chicken litter biomass. Trans. ASAE **2004**, *47* (5), 1689–1696.
38. DOE, *Reburning technologies for the control of nitrogen oxides emissions from coal-fired boilers* Topical Report, No 14, U.S. Department of Energy, May, 1999.
39. Loo, S.V.; Kessel, R., 2003, Optimization of biomass fired grate stroker systems, EPRI/IEA Bioenergy Workshop, Salt Lake City, Utah, February 19, 2003.
40. Grabowski, P. *Biomass Cofiring, Office of the Biomass Program*, Technical Advisory Committee, Washington, DC, March 11, 2004.

Index

2, 4-D. *See* 2, 4-Dichlorophenoxyacetic acid (2, 4-D)
2, 4-Dichlorophenoxyacetic acid (2, 4-D), 2005, 2556t, 2796t

A

Aarhus Protocol on Persistent Organic Pollutants, 1914
Abatement costs, 916, 917, 918, 918f, 919
Absorption
  herbicides, 1378–1379
Absorption chillers, 1356, 1357. *See also* Absorption heat pumps
  for building cooling applications, 1360
  for cogeneration, 1361
  for solar air-conditioning, 1361–1362
Absorption cycles, 1349, 1357
  bottoming and topping cycles, 1357
  type I, 1357, 1357f
  type II, 1357, 1358, 1358f
Absorption/desorption, of polychlorinated biphenyls, 2176
Absorption heat pumps, 1349, 1349f, 1357–1362. *See also* Absorption chillers; Heat pumps
  configurations, 1357
  heat source and heat sink configurations, 1357–1358
  performance, 1359–1360, 1360t
  refrigeration, operation of, 1358–1359, 1359f
  working fluid systems, 1358
Acanthoscelides obsoletus, 1116
Acaricides, 1, 2014
  amidines, 3
  antimetabolites, 3
  application technology, 4
  carbamates, 3
  organochlorine, 2–3
  organophosphates, 3
  organotins, 3
  ovicides, 3
  propargite, 3
  resistance to, combating of, 4
  synthetic pyrethroids, 3
  tetronic acids, 3–4
  use of, 2
Acatak. *See* Fluazuron
Accelerated solvent extraction, 2197
Acceptable daily intake (ADI), 1124, 1125f
Acceptance-based commissioning, 666–667
Accipiter nisus (Eurasian sparrowhawk), 417
Acetamide detection, 1319
Acetochlor, 1958
Acetogenesis, 2651
Acetogenin, 384
Acetylcholine, 1398
Acetylcholinesterase (AChE), 2005
AChE. *See* Acetylcholinesterase (AChE)
Acid–base accounting approach (ABA), 39, 40–41
Acid Deposition Control Program, 20
Acid deposition modeling, 1130–1131
Acidification, 1579
  and acid rains, 2359
Acidification of rivers and lakes, 2291, 2296, 2298. *See also* Lakes and reservoirs
  definition, 2291
  environmental problems caused by, 2297–2298
  historical perspectives, 2292
  measurement, 2291–2292
  solutions
    environmental legislation, 2298
    technology, 2298
  sources of freshwater
    anthropogenic, 2293f, 2296–2297
    natural, 2292–2294, 2293f
    redox reactions in water–sediment systems, 2294–2296
Acidithiobacillus ferrooxidans, 43, 1689, 1690, 2293
Acidithiobacillus thiooxidans, 1690
Acid mine drainage (AMD), 29, 269, 1694, 2293, 2296. *See also* Mine drainage
  and groundwater pollution, 1289
Acid-neutralizing capacity (ANC), 2291
Acidogenesis, 2651
Acid precipitation, 681
Acid Precipitation Act of 1980, 7, 20
Acid rain, 5, 916, 2304
  acidification of oceans, 13
  in Asian region, 8–9
  in Canada, 8
  control of, 15
  effects on aquatic ecosystem, 16
  effects on buildings and monuments, 16
  effects on vegetation and soil, 16
  in Europe, 8
  formation of, 5–6
  future projections through modeling, 14
  global sensitivity toward acidification, 13, 14f
  history of, 6
  natural acidity, 7
  precipitation scenario, regional comparison of, 14–15
  program, 2298
  regional acidity, 8–9
  sources of, 6
  spread and monitoring of, 7–8
  trends in acidity, 9–13
  in United States, 8
Acid rain, and N deposition, 20–24
  ecosystem impacts, 22–24
  human health effects, 22
  reducing effects of, 24
  sources and distribution, 20–22, 21f
  structural impacts, 22
Acid rock drainage (ARD), 29, 55, 2293
Acid soils, 2266
Acid sulfate soils (ASSs), 26, 55, 268, 2294
  active, 27
  classification
    Australian system, 28
    Entisols, 27
    FAO of United Nations, 28
    Histosols, 27
    Inceptisols, 27
    soil taxonomy, 27–28
    Sulfic Endoaquepts, 28
    Wassents suborder, 27
    WRB (World reference base), 28
  definition of, 26, 27
  diagnostic characteristics
    sulfidic materials, 27
    sulfuric horizon, 27
  drainage from, impacts of, 56
  extent of, in Australia, 56
  global extent of, 56
  global warming and, 56
  historical recognition, 26
  inland/upland *vs.* coastal, 28–29
  management of, 56–58
    acidity containment and neutralization in, 58–59
    Australia role in, 57–59
    avoidance in, 58
    education and assessment in, 58
    oxidation prevention in, 58
    Queensland's ASS Management Guidelines, 57–58
  postactive, 27
  potential, 26–27

Acid sulfate soils (ASSs) (*cont'd.*)
problems from, 55–56
Acid sulfate soil materials, 36
characteristic minerals and geochemical processes
oxidation processes, 42–43
reduction processes, 41–42
characteristics, 37–38
hazards arising from disturbance of acid sulfate soil materials, 43
acidification, 43
heavy metal and metalloid mobilization, 44
iron mobilization, 43–44
identification and assessment, 39–41
management for agriculture, forestry, and aquaculture, 49–50
scalded landscapes revegetation, 50
management for intensive developments
contaminants containment, 48
covering with fill, 47
dewatering, 48–49
general principles, 46–47
hydraulic separation, 47–48
neutralizing materials addition, 48
stockpiling, 48
strategic reburial, 48
occurrence, 38
physical behavior
consistence and strength, 45–46
permeability, 46
water bodies deoxygenation, 44–45
noxious gases production, 45
Acid Sulphate Soil Risk Maps, for New South Wales coastline, 58
A.C. modules. *See* Module inverters
A.C. switch disconnector, 223
Actinomycetous species, 531
Actinorrhizae (*Frankia*), 1515
Actino-uranium 235U decay series, 2238f
Activated sludge process (ASP), 2648, 2648f
operating parameters in, 2649
Activated sludge system, 2650
attached growth processes, 2650–2651
Activated sludge treatment, 2062
Active layer
permafrost zone, 1900, 1902f
Active solar heat storage, 2520–2521
Active solid matter, 149
Acute reference dose (acute RfD), 1124, 1401, 1402t
Acute RfD. *See* Acute reference dose (acute RfD)
Actual acidity, 40
Acyrthosiphon pisum, 1461
Adaptation Fund, 487
Additional energy, 770, 772, 773. *See also* Energy
categories of, 772
from chemical fuels, 773
life cycle considerations, 773
thermal energy, 772–773
types of, 771t
from wastes, 773
Additionality, 486
Adherence to pesticide label instruction, 509–510, 509*t*
ADI. *See* Acceptable daily intake (ADI)
Adiabatic flame temperature, 722, 723f, 725
Adipic acid, 1256
Admiral. *See* Pyriproxyfen
Adobe walls, 864
Adsorbents, 1908
Adsorption, 61–75, 1283, 1559
boron, 432, 432f
capacity, 62
chemical, 62
defined, 61
isotherms, 63–66, 432f
kinetics, 66–68
of REE, 2267
physical, 62–63
of strontium, 2427
techniques, 457
theory of, 61–68
in water purification, research progress on, 74–75
Adsorption–desorption, 2166
Advanced Spaceborne Thermal Emission and Reflection Radiometer (ASTER), 554, 2817t
Advanced thermal technologies (ATTs), 835–837, 843, 849
types, 836
Advanced very high resolution radiometer (AVHRR), 2817t
Advanced visible and near infrared radiometer (AVNIR), 2817t
Advanced wastewater treatment (AWT), 1583
Advection, 1282
Aerated concrete insulation, 1479
Aeration, 522–523
tanks, 2648
Aerial photography, 2819
Aerial ultra low volume (ULV) application, of mosquito insecticides
nontarget mortality by, minimization of, 1471
insecticide residue monitoring for, 1472–1473
novel application technologies for, 1472
right dose, 1472
right place, 1471
right time, 1471–1472
Aerobic biological waste treatment processes, 2648
Aerobic processes, 2647
Aerobic trickling filter, 2650f
Aerobic wetland conditions, 2864
Aerosols, 1196, 1196t
AerWay SSD system, 1674, 1674f
α-Fe_2O_3, 1712
Afforestation, 460
Aflatoxin, 1697
A-frequency-weighted sound pressure level, 2869
Africa
soil degradation in, 2379–2381, 2380f
distribution of, 2380f
African countries
pesticide poisonings incidence, 1408
Agelaius phoeniceus, 413
Agency capture theory, 926. *See also* Environmental policy innovations (EPI)
Agency for Toxic Substances and Diseases Registry (ATSDR), 650, 2049, 2052
Agency of Regional Air Monitoring of the Gdańsk Agglomeration (ARMAAG), 139
AGENDA for Environment and Responsible Development, 507
"Agent Orange," 1395
Agent success
weed biocontrol and, 2821–2822, 2822t
"Age of Fossil Fuels," 549, 551
Ageratum houstonianum, 1464
Aggregated theoretical indicators, 601, 604
Aggregation, 2752
Aging, 1398
Aglais urticae L., 1750
AGNPS model. *See* Agricultural nonpoint source pollution (AGNPS) model
Agricultural-based waste materials, 72–74, 74*t*
Agricultural and Environmental Information System for Windows (AEGIS/WIN), 2819
Agricultural ecosystems
acidic deposition and, 23
Agricultural experiment stations, 2009
Agricultural nonpoint source pollution (AGNPS) model, 966, 1164, 2762
Agricultural Policy/Environmental eXtender (APEX) model, 2761
Agricultural pollution, 2304. *See also* Pollution
Agricultural runoff, 81–84
BMPs and, 83
control of, 83
dissolved pollutants, 82–83
eroded sediments and, 82
IPM and, 83
quantity, 81–82
soil erosion and associated pollutants, 82
Agricultural soils, phosphorus in, 100–102
Agricultural sustainability
and agrobiodiversity, 353
farmers and consumers awareness for, 359–360
indicators for, 352. *See also* Bioindicators

Page numbers followed by f and t indicate figures and tables, respectively

Agricultural water quantity and quality management under uncertainty. *See* Inexact optimization models
Agriculture, 88
  benefits of herbicides to, 1383
  probiotic, 528–529, 531–533
  and public health, sustaining, 2003–2004
  sustainable. *See* Sustainable agriculture
  water quality for, 1566
  wastewater use in. *See* Wastewater use, in agriculture
Agrobacterium tumefaciens, 307
Agroecosystem biodiversity
  sustainable agriculture, 2457
Agroecosystems
  ecological infrastructure in, 372
  interaction with other ecosystems, 2434
Agroforestry, 129, 570, 1037, 1515
Agroforestry, WUE and, 129–130
  evaporation and, 130
    diffusion driven phase, 130
    energy driven phase, 130
  precipitation and, 130
Agro/horti ecosystem, pesticide pollution in, 2146
Agrostis spp., 1305
Aim. *See* Chlorfluazuron
Air, 916
  bioaccumulation in, 324–326
  carbon removal from, 1189–1190
    man role in, 1189–1190
    nature role in, 1190
  leakage, 866
  pesticides in, 1966, 1969–1972, 1971t, 1972
  sustainable agriculture, 2457
Air and Precipitation Network (APN), 7
Air-assisted nozzles, 1472
Airborne electromagnetic methods
  for saltwater detection, 1327
Air-conditioning, solar, 1361–1362
Air Drywall Approach, 861
Air gasification, 377
Air injection wells, 1335
Airlift bioreactor, 2626
Air pollutants, 1154–1155
  ambient and emission standards of, 1127–1129
  criteria, characteristics of, 136–138
  genotoxic effects of, 1155–1156
  sources of, 136, 137*t*
  transport and dispersion, 1129–1130
Air pollution, 1154–1159
  desertization and, 566
  genotoxic effects of, 1155–1156
  genotoxicity tests for, 1156–1157
  health risk associated with, 1157–1158
Air pollution index (API). *See* Air quality index
Air pollution legislation, history of, 133–136
Air pollution monitoring, 132–146
  biomonitoring using plants, 142–143
  geographical information system in, 143
  instruments
    classification of, 140–141
    requirements of, 139–140
  objectives of, 132–133, 132f
  program designing of, 134f
  quality of
    monitoring networks, designing of, 138–139
    obtained information, types of, 139, 139f
  remote monitoring techniques for, 143–144
  techniques of, 141–142, 141t
Air quality, 1019
  wind erosion effect on, 1019–1020
Air quality index (AQI), 134–136
Air quality modeling (AQM), 1130–1132
  acid deposition, 1130–1131
  basic ingredients of, 1130
  grid-type, 1130
  photo-oxidants, 1131–1132
  regional haze, 1131, 1132
  trajectory-type, 1130
Air quality standards, 133–134
Air sparging (AS), 1335
Air-stream bottom aerator (ASB), operation mode of, 1594, 1594f
Alachlor, 1318t, 1957, 1957t, 1958, 2142, 2806
Alauda arvensis (Eurasian skylark), 416
Albedos, defined, 1192
Albrecht effect, 1197t
Alcaligenes, 97
Alders, 1527t
Aldicarb, 3, 419, 1318t, 2806
Aldicarb sulfone, 1318t
Aldicarb sulfoxide, 1318t
Aldrin, 417
Alexandria Lake Maryut, 164–173
  current state of
    hydraulic functioning, 166–167
    hydrological functioning, 166–167
    social considerations and governance, 167, 169
    water quality and ecosystems, 167
  integrated action plan for
    defined, 169
    model development, 169–173
    scenario identification, 169
  location map of, 165f
Alfalfa (*Medicago sativa* L.), 371, 427, 1552
Alfisols, 427
Algal blooms, 475, 538, 2358
Algal control measures, 1584
Algal growth, 1559f
Alien plant species, 1182
Al(III)EDTA, 2245
Alkali earth metals, 1560
Alkaline bauxite refinery waste (red mud), 269
Alkaline fuel cells (AFCs), 1146–1148, 1152t
  applications of, 1146
  electrolytes in, 1147
  with liquid electrolyte, 1147–1148
  problems related to, 1147
  research groups working on, 1146–1147
  safety issues related to, 1147–1148
  with solid electrolyte membrane, 1146, 1148
  in space shuttle missions, 1146
  technical applications and demonstration projects on, 1147
  types, 1146
Alkaline rains, 9f
Allelochemics, 175–178
  natural occurrence, 175–177, 176t, 177f
  overview, 175
  practical applications, 177–178
  push–pull application, 178
  terminology, 175
Allelopathy, 2031
Allievi-Michaud formula, 211–212
Alliance of Small Island States (AOSIS), 485
Allolobophora caliginosa, 409
Allomones, 175
Allowance trading, 486
α-Al_2O_3, 1712
Alopex lagopus, 2596
Alpha-chloralose, 413–414
Alpha (α) decay, 2234–2235
Alphitobius diaperinus, 1971
Alsystin. *See* Triflumuron
Alternanthera philoxeroides (Malancha), 426
Alternaria, 442, 444
Alternative energy
  photovoltaic solar cells, 226–240
Alternative prey/hosts
  natural enemies and, 370
Altica carduorum, 2822
Altosid. *See* Methoprene
Alum, 267
Aluminosilicate clay minerals, 268
Aluminum (Al), 267
  alleviation of
    in soil, 269
    in water, 270
  in atmosphere, sources of, 270, 270t
  chemistry of, 267
  in cosmetics, 267
  in drinking water, 270
  in foodstuff, 270, 270t
  in pharmaceuticals, 270
  production of, 268, 268t
  in soil, sources of
    Al minerals, 268
    forms of soluble Al, 268
    mobilization of Al in soil, 268–269

Aluminum (Al) (*cont'd.*)
soluble forms of, 267
toxicity of, 43
in aquatic organisms, 271
in humans, 271–272
in plants, 270–271
uses of, 267, 268
in water, sources of
acid deposition, 269
acid mine drainage, 269
acid sulfate soils, 269
Al from water purification, 269–270
Aluminum oxyhydroxide particles, 1714, 1714f
Alum sludges, 270
Alzheimer's disease, 271
Amaranthus, 1936
Amazonian rainforest, 333
Ambient air quality and SO_2, 2437–2439
Ambient air quality standards (AAQS), 2438
Ambient energies, 1346
Ambient energy fraction (AEF), heat pump system, 1352–1353, 1352f, 1353t
Ambient geothermal energy, 772
Ameliorants, 1684
chemical, 2370
reclamation of sodic soils and
optimal supply, 2373
rate of supply, 2373–2374
without addition of, 2373
Amelioration
erosion and, 1059
Amendments, soil. *See* Organic soil amendments
American Carbon Registry, 1774
American elm. *See Ulmus americana*
American Petroleum Institute (API) gravity, 2042, 2045
American Society of Agricultural Engineers, 471
American Society of Heating, Refrigerating, and Air-Conditioning Engineers, Inc. (ASHRAE), 1365, 1366
Ames test, 1156
Amide herbicides, 1319
4-Aminopyridine (Avitrol®), 413, 415
Aminosulfonyl acids, 1318
Amitraz, 3
Ammonia (NH_3)
as refrigerant, 1358
soil quality and, 2390
volatilization. *See* Volatilization, ammonia
Ammonia-rich wastewater, 2653
Ammonia volatilization, 93, 94, 1517
Ammonia wet scrubber, 2441
Ammonium ($NH4^+$), 97, 1768, 2736
Ammonium nitrate, 125
Amorphous silicon, 232
Ampelomyces quisqualis, 2107
Amplitude-modulated sound, 2872–2873
Amylase, 89
Anaerobic ammonium oxidation, 2653
Anaerobic digester systems, 1674
Anaerobic digestion, 377
Anaerobic earthen lagoon system, for manure storage, 1672, 1672f
Anaerobic methane-oxidizing bacteria, 391
Anaerobic ponds, 2636
Anaerobic processes, 2647
Anaerobic wastewater treatment, 2651
advantages, 2651
digestion process, 2651
disadvantages, 2651
operating parameters in anaerobic reactors, 2652
pH, 2652
UASB reactor, 2651–2652
Anaerobic wetland conditions, 2864
Anagrus, 371
Anagrus epos, 1936
Analytical cost estimation (ACE) method, 840–843
Analytical semi-empirical model (ASEM), 837–840
Analytical models, for, pest management, 1950
Anammox, 2653
ANC. *See* Acid-neutralizing capacity (ANC)
Andalin. *See* Flucycloxuron
Andasol solar power station, 2521, 2521f
flow schematic, 2522f
Andisols, 1854
Anilide herbicides, 1319
Animal exposure effects, of endocrine-disrupting chemicals, 650–651
Animal husbandry
and organic farming, 1107
Animal manure, 1667
Animal pests
regulation of, 1115–1117
anthropogenic measures, 1116–1117
biotic (natural) resistance of environment and, 1115–1116
strategy of controlling in organic farming, 1115
Animals
biomonitoring, 2128
toxocity of herbicides to, 1385–1386
Anisopteromalus calandrae, 2424
Annoyance, 2871–2873, 2873t, 2878f
Anoxic processes, 2647
ANS-118. *See* Chromafenozid
Antagonistic microorganisms
stimulation of, 1879
Antagonistic plants, 288–290
benefits and risks, 289
in cultural pest control, 288–289, 289t
defined, 288
nature of, 288–289
overview, 288
Antarctica, pollution in, 2138
Antarctic ozone hole, 1887, 1888f
Anthemis arvensis, 1928
Anthonomous grandis grandis, 2009
Anthracite, 717
Anthraquinone, 415
Anthropogenic acid deposition, 268
Anthropogenic climate change, 483
Anthropogenic factors for acidification, 2293f, 2296–2297, 2297t
acid mine drainage, 2296
air pollution with acidifying compounds, 2296
Anthropogenic forcings, 1199f
Anthropogenic greenhouse gases, 1192
Anthropogenic measures
animal pests, 1116–1117
Anthropogenic modification of global climate, 955
Anthropogenic pollutants, 2303
Anthropogenic sources, removal from SO_2, 2440
ammonia wet scrubber, 2441
magnesium wet scrubber, 2441
wet techniques, 2440–2441
Antibiotics, in fish farms, 2061
Anti-erosion drop replacement, 2310
Antiperspirants, AL in, 270
Antireflective coatings (ARCs), in solar cells, 234
Apera spica venti, 1928
Apex, Dianex. *See* Methoprene
Apex solar cells, 232
Aphis fabae, 1878
Apollo missions, 1716
Applaud. *See* Buprofezin
Application efficiency, 1546, 1547t, 1564
AQM. *See* Air quality modeling
Aquaprene. *See* Methoprene
Aqua Regia extractable phosphorus, 102f
Aquatic biochemical equilibrium, 2731f
Aquatic ecosystems, 1576
acidic deposition and, 24
effects of acid rain on, 16
health, indication of, 607
Aquatic environment, pesticides in, 1965–1966, 2146–2147
Aquatic organisms, Al toxicity in, 271
Aquatic plants, 2753
Aquatic systems, 538
Aquatic weed management by fish, in irrigation systems, 1922
Aquifers, 1282, 1290
coastal, 1321
remediation, surfactant-enhanced, 1844
storage, 1285, 1286, 1286f
Arachis hypogaea L. (Peanuts), 424
Aral Sea, degradation of, 1438–1440
Aral Sea disaster, 300–302, 300t, 301t, 302f

Aralsk and, 301–302
average water supply and demand in, 300, 300t
environmental problems, 302
human tragedy, 302
irrigation and cotton, 301
Muynak and, 301–302
overview, 300
Aralsk
Aral Sea disaster and, 300–302
ARD. *See* Acid rock drainage (ARD)
Areal loading approach, 2865
Areal Nonpoint Source Watershed Environment Response Simulation (ANSWERS) model, 966, 2761
Argillan, 1627
Arid region irrigation
and soil IC, 465–466
Aridity index map, 2376–2377
Arion ater, 383
Arion distinctus, 383
Arion intermedius, 383
Arion silvaticus, 383
Armillaria mellea Vahl ex.fr, 1879
Armillaria spp., 1879
Array inverters, 222, 222f
Arrhenius' equation, 161
Arsenic, 1263
as contaminant in groundwater, 1281
content in soil, 1263
in drinking water, 1262, 1264. *See also* Arsenic-contaminated groundwater
drinking water standard for, 1281
in food chain, 1274, 1275f
groundwater contamination by, 1292
health impacts of, 2118–2119
inorganic forms, 1263
occurrence of, 1271
source of, 1262
toxic level of, in groundwater, 1626
use of, 1263
Arsenic-contaminated groundwater
in Assam, 1268
in Bangladesh, 1268–1271, 1270f
in Bihar, 1267
in Chhattisgarh, 1268
cohort study on, 1274
effect on children, 1275, 1275f
in GMB Plain, 1262, 1264, 1264f, 1265t
impacts of, on human health, 1271
cancer effects, 1274, 1274f
cardiovascular effects, 1272
dermal effects, 1271–1272, 1272f
endocrinological effects, 1273
gastrointestinal effects, 1272–1273
neurological effects, 1273
reproduction and developmental effects, 1273–1274
respiratory effects, 1272, 1272f
incidents of, 1263–1264, 1263f
in India and Bangladesh, 1269t, 1271t
for irrigation, 1274, 1275f
in Jharkhand, 1268
in Manipur, 1268
safe water options
deep tube wells, 1276
dug wells, 1276
rainwater harvesting, 1276
watershed management and purification, 1276
socioeconomic effects, 1276
source of, 1276
in Uttar Pradesh, 1267–1268
in West Bengal, 1265–1267, 1266f, 1267f
Arsenic trioxide, 2016t
Arsenopyrite, 1281
Arthropod parasitoids/predators
artificial diets for, 1746
Artificial destratification, 1592t
Artificial diets
for predators and parasitoids, 1746–1747
successes and failures with, 1746
types of, 1746
Artificial impoundments, 1576
Artificial rearing of natural enemies, 1747
Artificial stream ecosystem (ASE), 2154
Arundo donax (Giant Reed), 1182
AS. *See* Air sparging (AS)
Ascendency, 604
ASEAN, 875t
Ash composition, 730t
Asia
acid rain in, 8–9
soil degradation in, 2378–2379, 2379f
distribution of, 2379f
trends in acidity, 13
distribution of, 2379f
trends in acidity, 13
Asia, environmental legislation in, 874–876
biodiversity laws
enforcement, 881–882
regional cooperation, 882
biodiversity laws at national level, 877
Cambodia, 881
China, 877
Hong Kong, 879–880
Indonesia, 878–879
Japan, 878
Korea, 877
Lao People's Democratic Republic, 880
Malaysia, 878
Philippines, 879
Singapore, 879
Thailand, 880
Vietnam, 881
development in international law and its consequences on national legislations, 877
environmental assessment law
Cambodia, 887–888
China, 885
Hong Kong, 886
Indonesia, 886
Japan, 886
Lao PDR, 887
Malaysia, 886
Philippines, 887
Republic of Korea, 886
Singapore, 887
Thailand, 887
Vietnam, 888
e-waste management law, 882
Cambodia, 884
China, 882–883
Hong Kong, 883
implementation and enforcement, 885
Japan, 883
Lao People's Democratic Republic, 884
Malaysia, 883
Philippines, 883–884
Republic of Korea, 883
Singapore, 884
Thailand, 884
Vietnam, 884
e-wastes and environmental assessment, 882
law on biodiversity conservation, 876
transboundary movements of hazardous waste, 884
Asia Arsenic Network (AAN), 1269
Asnæs Power Station (APS), 1434
ASP. *See* Activated sludge process (ASP)
Asparagus officinalis, 288
Aspen, 1527t
Aspergillus carbonarius, 1697
Aspergillus flavus, aflatoxin by, 1697
Aspergillus niger, 1697
Aspergillus parasiticus, aflatoxin by, 1697
Assam, groundwater arsenic contamination in, 1268
Assigned amount units (AAU), 486
Association of Energy Engineers (AEE), 1366
Association of Higher Education Facilities Officers (APPA), 1493
ASTER. *See* Advanced Spaceborne Thermal Emission and Reflection Radiometer (ASTER)
Atabron. *See* Chlorfluazuron
Atlas of Australian Acid Sulfate Soils, 58
Atmosphere
aluminum in, 270, 270t
sustainable agriculture, 2458
Atmospheric aerosol forcing, 1196
Atmospheric deposition
of nitrogen, 1769–1770
Atmospheric nanoparticles, 1715, 1715t
Atmospheric ozone, 1883

Atmospheric pollution, 2359
Atmospheric trace species, field measurements of, 1891
Atomic numbers of REE, 2267t
Atrazine, 1282, 1318t, 1957, 1958, 2114, 2142, 2806
Attached growth process, 2647
Attenuation, 1843f
  chemical, 1846
  physical, 1846
Attic floor insulation techniques, 1487–1488
  batt attic, 1488
  loose-fill, 1488
Attic ventilation, 1486–1487
  powered attic ventilator, 1487
  vent selection, 1487
Augmentation, 1474. *See also* Biological control
Augmentative biological control, 1747
Australia
  ASSs management in, 56–59. *See also* Acid sulfate soils (ASSs)
  soil degradation in, 2383–2384, 2384f
    distribution of, 2384f
Australian Biocontrol Act of, 2823
Australian bush fly management, in micronesia, 1920
  bush fly origins and habits, 1920–1921
  management efforts worldwide, 1921
Australian Pesticide Act (1999), 1961
Australian Soil Classification, 41
Australian Soil Resource Information System (ASRIS), 58
Autoclaved aerated concrete (AAC), 1483
Automated ribosomal RNA intergenic spacer analysis (ARISA), 2617
Automated sampling
  of GHG, 1202
  suspended sediment concentration, 965, 966f
Auto-oxidation, 1907
Auxiliary, size of, 254
Available soil water (ASW), 2341
Avena fatua L.
  landscape pattern of, 2038, 2039f
Avena sativa (Oat), 1307
Avermectins, 3
Avicides, 2014
Avigrease®, 413
Avis Scare®, 413
Axor. *See* Lufenuron
Azadirachta indica, 1460, 1769
Azadirachtin, 1460, 2029, 2029f
Azadirachtin indica, 1878
Azafenidin (Milestone), 2016
Azardirachta indica, 1930
Azimuth and tilt angles, 251f
Azocosterol (Ornitrol®), 415
Azolla, 1515, 1516
Azolla–Anabaena system, 350
Azospirillum (AZS), 350, 351, 1515, 1516
Azotobacter (AZT), 350, 426, 1515, 1516
AZTI Marine Biotic Index (AMBI), 602–603

B

Bacillus, 321, 350, 382
Bacillus alvei, 323
Bacillus brevis, 323
Bacillus megaterium, 350
Bacillus sphaericus, 323, 389, 1991
Bacillus subtilis (strain QST71383), 382, 392, 530
Bacillus thuringiensis (Bt), 127, 295, 321, 322–323, 322t, 382, 407, 1112, 1474, 1476, 1749–1750, 1930, 1953, 2029, 2112, 2141
  culture and control, 323
  for insect control, 411–412, 411t
Bacillus thuringiensis (Bt) crops, 307, 315
  benefits, 311
    conventional insecticides, reduced risk compared with, 311
    economic savings, 311
    global food security, 311
  BT toxins, 308
    consumption of prey containing, 310–311
    indirect consumption of, 310
      soil contamination via root exudates, 310
    and transgene escape, Reduction in, 314
  commercialized, 309t
  compatibility with biological control, 314
  consumption of, 308
    live plant tissue, 308–309
  defined, 307
  detritus, consumption of, 310
  environment, impact on, 311
  environmental management of, 313
  global prevalence, 308
  large-scale integrated pest and resistance management, 314
  pests, targeted, 308
  pollen feeding, 310
  resistance management techniques, 314
  risks, 312
    gene escape, 313
    non-target organisms, impact on, 312
    pleiotropic effects of genetic transformation, 312
    presence in human food supply, 313
  sources and fate of, 308
  within-plant modifications, 313
    low-risk promoters, 313
    tissue- and time-specific expression, 313
BACT. *See* Best available control technologies (BACT)
Bacteria, 521–526
  Actinomycetous species, 531
  Bacillus species, 530, 532
  lactic acid bacteria, 530–531, 531f
  in waste and polluted water, 2698
  metal resistance mechanisms in, 393
  in mine drainage systems, 1689–1690
Bacterial pest controls, 321–323
  Bacillus thuringiensis, 322–323
  overview, 321
  Paenibacillus (Bacillus) popilliae (Dutky), 321–322
  potential agents, 323
  Serratia entomophila, 322
Bactericidal effect, of noble metal nanoparticles, 1740
Bactericides
  derived from nature, 2029–2030, 2030f
Baffle scrubbers, 154
Bahiagrass (*Paspalum notatum* Fluegge), 1305
Baker Lake, CW in, 2670
Baltic Sea, oil pollution in, 1826–1840
  observations of, 1828
  oil spill drift, numerical modeling of, 1833–1835, 1835–1838*f*
  operational satellite monitoring of, 1830–1833
  problems and solutions of, 1838–1839
  sources of, 1827–1828
  statistics of, 1828–1830
    tendencies of, 1830
Bamboo-reinforced tanks, 2264
Bamboo roofs, 2263
Banding, of pesticides, 1983–1985
  advantages, 1983–1984
  disadvantages, 1984
  equipment requirements, 1984
  future perspectives, 1984–1985
  usage, 1983–1984, 1984t
Bangladesh, groundwater arsenic contamination in, 1268–1271, 1270f
Bangladesh Arsenic Mitigation and Water Supply Project (BAMSWSP), 1269
Banned and restricted compounds import/export, 575–576
Banned pesticides
  banning exports of, 1616
Barium (Ba), 1560
Barium chromate, 479t
Barley (*Hordeum vulgare* L.), 425
Barnyard grass (*Echinochloa crus-galli*), 426, 1526t
Basel Convention on the Control of Transboundary Movements of Hazardous Wastes and Their Disposal, 1914
Baseline-credit system, 486
Baseload, 1790
Basin irrigation efficiency, 1365–1366, 1565f

Basta®, 2030
Bates, 1943–1944
Batteries, solar, 218
Batts, insulating walls with, 1485f
Bauxite, 267, 271. *See also* Aluminum (Al)
Baycidal. *See* Triflumuron
BCAs. *See* Biocontrol agents (BCAs)
BCF. *See* Bioconcentration factor (BCF)
Beans (*Phaseolus* spp.), 426
Beauvaria bassiana, 382, 1930, 1988
Beauveria brongniartii, 382
Becquerel (Bq), 2235
Bedload flux, 965, 965f
Bee dancing
  pesticides effects on, 2005
Bee extracts, 1746
Bellan's pollution index, 602
Bell's vireo (*Vireo bellii pusillus*), 1186
Bench terraces, 968–969, 969f, 2536, 2538, 2764, 2764f
Bendiocarb, 419
Beneficial water use, 1547
Benomyl, 2806
Bensulide, 2015, 2024t
Bentgrass species (*Agrostis* spp.), 1305
Benthic eco-system technology (BEST), 2154
BENTIX, 603
Benzene
  drinking water standard for, 1282
Benzene EN 14662, online gas chromatography for, 142
Benzene, toluene, ethylbenzene, and xylene (BTEX), 1336, 1337, 1339
Benzoic acid, aromatic hydroxyacid derivatives of, 2079f, 2080t
Best available control technologies (BACT), 1128
Best management practices (BMPs), 2808, 2813
  agricultural runoff and, 83
  evaluation of, 2750
17β-estradiol, 1405
Beta minus (β^-) radiation, 2234–2235
Beta-Poisson model, 2700
Beta vulgaris (Sugar beet), 1307, 1552, 1983
BET measurements. *See* Brunauer–Emmett–Teller (BET) measurements
BFBG. *See* Bubbling fluidizedbed gasifier (BFBG)
Bialaphos, 2030–2031, 2031f
Bifenazate, 3
Bifenthrin, 3, 11
Big hydro plants, 202, 212
Big Spring Number Eight (BSNE) samplers, 1022, 1023f
Bihar, groundwater arsenic contamination in, 1267
Bioaccumulation, 324–326
  applications and future perspectives, 326
  defined, 324
  environment and, 324–326, 325f
    air, 325–326
    soil, 325
    water, 324–325
  mechanism in biota, 326
  molecular properties, 324
  of polychlorinated biphenyls, 2178
Bioassay procedure, at water treatment plants, 2017
Bioaugmentation, 395f
Bioavailability, of polychlorinated biphenyls, 2176
Bioavailability of toxic organic compounds, 401–403
  contaminant state of, 402
  factors influenced, 402
  low water solubility, 402
  microbial adaptations, 403
  pore size distribution, 403
  sorption on solid phase of soil, 402
Bioavailable nutrient
  defined, 1817
Biobarrier systems, 1336–1337
  contaminated groundwater treatment, 1339–1342
    aerobic and anaerobic biodegradation, 1339–1340
    for chlorinated-hydrocarbon, 1340–3141
    for petroleum-hydrocarbon, 1339–1340
    treatment train system, 1342, 1342f
  design factors, 1338
    electron acceptor/donor, 1338
    groundwater flow, 1338
    location, 1338
    natural environment, 1338
    time scales, 1338
  groundwater contamination, 1333
  groundwater remediation technologies, 1334–1336
  limitations of, 1338–1339
  microorganisms, effects of, 1339
  nonaqueous phase liquid, 1333–1334
Biobeds, 2020, 2023, 2023f
Biocapacity, 2468, 2469–2470
  deficit, 2470
Biocapsules, 397
Biocatalysts, 1258
Biochar, 843–846
  sorbent properties of, 844–845
Biochemical oxygen demand (BOD), 2632, 2864, 2865t
Biocides, 2005, 2014
  effect on animals and human beings, 2005
Biocolloid formation methods, 396
Bioconcentration factor (BCF), 324, 403
Bioconcentration, of polychlorinated biphenyls, 2178
Biocontrol agents (BCAs), 2107, 2108t. *See also* Fungi, biological control by
  application, 2109
  mechanisms, 2107–2109
Biodegradable municipal waste (BMW), 908–909, 2401
  reduction, 93
Biodegradation, 328, 330–332, 387. *See also* Bioremediation
  estimation of, 331–332
  of polychlorinated biphenyls, 2176–2178, 2177f, 2177t, 2178f
Biodiesel, 377
Biodimethylether (DME), 378
Biodiversity, 333–334
  care of, 335–336
  definition of, 333
  destruction of, 335
  ecosystem services by, 334
    cultural, 334
    production, 334
    regulation, 334
  of food web, 334
  importance of, 334–335
  indicators, 601–602
  maximum level, 334
  nature use of, 334
  number of species on Earth, 333
  quality approach, 356
  species diversity, occurrence of, 333
  species richness in tropics, 333–334
  support service by, 334
  and wildlife, 1186
Bioenergy, 373, 374f, 378–379, 843–846. *See also* Biomass
  benefits from, 378
  resources, 373
  share in world primary energy mix, 378, 379f
Bioenergy crops
  benefits of, 345, 346t
  carbon (C) sequestration, potential of, 346–347, 347f
  influence of (at temporal scale), 345, 346f
  overview of, 345
  soil organic carbon (SOC) pool, 345, 346
Bio-ETBE (ethyl-tertio-butyl-ether), 378
Bioethanol, 377–378
Biofertilizers, 349–351, 350t, 1514–1516
  background, 349
  biologically fixed nitrogen, 1514
  crop residues, 1515
  leguminous green manures, 1515
  manures and composts, 1515
  market potential for, 349, 350t
  mechanisms of growth promotion, 349–351
  nitrogen-fixing, 1516
  nutrient-mobilizing, 1516
  overview, 349

Biofertilizers (*cont'd.*)
phosphate-solubilizing, 1516
phytohormones, 349–350
plant nutrient acquisition, 350–351
biological nitrogen fixation, 350
microbial siderophore uptake, 351
other nutrients, 351
phosphorus solubilization, 350
sewage sludge, 1515
Biofilm formation, 393
Biofilters (BF), 2612f, 2621t, 2622, 2623t, 2726. *See also* Biobeds
Biofuel-based power production, 1251
Biofuels, 377–378
decentralized electricity from, 1246–1250
importing, 1249–1250
Biofungicides, 382–383
Biogas, 377
from anaerobic digestion of biomass, 377
composition of, 377
from landfills, 377
use of, 377
Biogenic model, of carbonate formation, 1446
Bioherbicide/inundative method, of biological weed control, 1525, 1525f
Bioherbicides, 383
Bioindicators, 352
concept of, 353–354
definition of, 353
demand for, 355
development of, 352
in EU agroecosystems, 356, 357t–358t
indicators of biodiversity change as tool for, 356, 359f
in shaded coffee in Latin America, 356, 359
directions for future study on, 352
features of, 354t
history of use of, 353
importance of, 352
steps in developing of, 354
usefulness of, at different levels of organization, 355
vs. abiotic indicators, 354–355
Bioinformatics, 397
Bioinsecticides, 382
Bio-liquid fuels, 846–847
markets and energy prices, 831t
properties of, 846t
Biological aerated filters, 2724. *See also* Filter(s)
Biological carrying capacity, 339
Biological control
advantage of, 2595
augmentation, 1474
for cane toads, 2604
of cats, 2602–2603. *See also* Feline panleukopenia virus (FePV)
classical, 1474, 2595
conservation, 1474
conservation of, 370–372. *See also* Natural enemies
agroecosystems, ecological infrastructure in, 372
favorable conditions enhancement for natural enemies, 371–372
harmful conditions reduction, 370–371
defined, 1474
definition of, 2595
general principles, 366
improved efficacy, evidence for, 366–367, 367t
natural enemies. *See* Natural enemies
of nematodes, 1752–1753
future perspectives, 1753
nonspecific biocontrol agents, 1753
overview, 1752
specific biocontrol agents, 1752–1753
new associations, 366
citrus leafminer (*Phyllocnistis citrella*), 368
Eurasian watermilfoil (*Myriophyllum spicatum*), 367
Lantana camara, 367
southern green stink bug (*Nezara viridula*), 367–368
tarnished plant bug, 368
tarnished plant bug (*Peristenus digoneutis*), 368
triffid weed (*Chromolaena odorata*), 367
of plant pathogens
by fungi, 2107–2110
of plant pathogens (viruses), 2111–2113
in practice, 1476–1477
of rabbits, 2597–2602. *See also* Myxomatosis; Rabbit hemorrhagic disease virus (RHDV)
strategies for using, 1474
successes and failures, 366–369, 367t
of vertebrate pests, 2595. *See also specific examples*
biotechnology for, 2604–2605
natural history of, 2597–2603
parasites for, 2596–2597
potential agents for, 2603
predators for, 2595
of weeds. *See* Weeds, biological control of
Biological control, 1584
Biological degradation of organics, 2646
Biological effects, of ozone depletion, 1893
Biological fixation
of nitrogen, 1770
Biological integrity sampling, 1582
Biological invasions, of species, 1531–1533, 1532t
characteristics of recipient community and, 1532
controlling, 1533
impacts, 1532–1533
lags and surprises, 1533
need for, 1531
overview, 1531
propagule pressure, 1531
species traits and, 1531
steps, 1532t
Biological material
elements in, 1455
Biological methods, for contaminant remediation, 1293
Biological nitrogen fixation (BNF), 350, 1515
Biological oxygen demand (BOD), 331, 522
reduction in municipal wastewater, 2709–2712, 2710–2713f
Biological pesticides (BP), 1986–1987. *See also* Chemical pesticides (CPs); Integrated pest management (IPM)
Biological phosphorus removal, 2654
Biological oxygen demand (BOD), 331, 522
reduction in municipal wastewater, 2709–2712, 2710–2713*f*
Biological pollution, in coastal waters. *See also* Pollution
algal bloom, 495
eutrophication, 494–495
invasive species, 495
Biological removal of nitrogen, 2653
anaerobic ammonium oxidation, 2653
Canon process, 2654
combined nitrogen removal, 2654
Sharon process, 2653
Biological sequestration, 460
Biological treatment processes, 2654
Biological waste, 512
aerobic treatment for, 513–515
disposal problems associated with, 512
gas treatment techniques, 2611
management, regulatory issues of, 512–513
Biomagnification, of polychlorinated biphenyls, 2178–2179
Biomal®, 1111
Biomanipulation, 1584, 1593t
Biomarkers, 1068–1069
for EDC risk assessment, 647
Biomass, 373–379
benefits of, 378
biochemical conversion of, 377
biofuel, 377–378
biogas, 377
as carbon-neutral energy resource, 373, 378–379
combustion, 375
co-combustion/direct cofiring, 375–376
indirect cofiring, 376
parallel combustion, 376
problems in, 376

definition of, 373
energy production from, 373–374, 374f. *See also* Bioenergy
plant, 373
thermochemical conversion of, 376
comparison of technologies of, 377t
gasification, 376–377
pyrolysis, 376
types of, for energy use, 374t
use of, as fuel, 373–374
mechanical processes related to, 374–375
problems in, 374
Biomass, 192, 824
defined, 715
gasification and direct firing, 1248–1249
and hydrogen energy cycles (comparison), 716, 716f
low cost of, 733
sources, 824, 825, 826
Biomass and wastes, 2482
gasification of, 2488–2495
advances in, 2491–2494
gasifiers, 2488–2491
plants for, 2494t
reaction involved in, 2490
syngas, composition and heating value of, 2493t
syngas for downstream application, 2492t
technologies for, 2491
world scenario, 2495
pyrolysis of, 2483–2488, 2486t
features for fast, 2487
pathways, 2486f
products, 2484
reactions involved in, 2485f
world scenario of, 2489t–2490t
scopes for, 2483
Biomass burning
ammonia volatilization, 87
N_2O emissions, 97
Biomass energy, 772
Biomass fuels, 345
Biomethanol, 378
Biomolluscicides, 383
bacteria-based, 383–384
combination, 384
from nematodes, 383
of plant origin, 384
use of, 383
Biomonitoring
animals, 2129
human, 2129
plants, 142–143, 2128
Bio-MTBE (methyl-tertio-butyl-ether), 378
Bio-oils, 376, 2485, 2488
Bio-oxidation, 2152
Biopesticides, 381
classification of
biochemical pesticides, 381
microbial biopesticides, 381
plant-incorporated protectants, 381
definition of, 381
formulations of, 381
organisms used in, 381
range of
bioacaricides, 382
biofungicides, 382–383
bioherbicides, 383
bioinsecticides, 382
biomolluscicides, 383–384
use of, 381
Bioprocesses, 2152t, 2553
Bio-reduction, 2152t
Bioreactors, for waste gas treatment, 2611
airlift bioreactor, 2626
biofilters, 2612f, 2620t, 2621, 2626
bioscrubbers, 2611f, 2621, 2623
biotrickling filters, 2614t, 2621, 2622–2623
critical factors and development requirements, 2625–2626
mass transfer aspects, 2611–2615
membrane biofilm reactors, 2612f, 2624–2625, 2624t
microbial kinetics, 2611–2621
monolith bioreactor, 2625, 2626
process engineering and performance parameters, 2618–2619
rotating biological contactors, 2625, 2626f
two-phase bioreactors, 2612f, 2625
waste gas stream, characteristics of, 2620–2621
Bioregulators
in organic farming, 1115–1116
Bioremediation, 387, 392, 1335, 1340
of bioaugmentation and biostimulation, 395f
for environmental cleanup, 388f
factors affecting, 390
of inorganic contaminants, 392–393, 2120
microbial degradation of organic pollutants, 388–392
molecular probes in, 397–398
mycoremediation, 392
of organic contaminants, 2116–2117
phytoremediation, 394–395
principle of, 387–388
redox reaction
leading to immobilization, 393
leading to solubilization, 393–394
restoration of contaminated soils
bioavailability of toxic organic compounds, 401–403
biodegradation, 403
uptake of heavy metals by plants, 403–405
technology, 395–396
Bioscrubbers (BSs), 2611, 2614t, 2621, 2623
Biosolids, 1807–1808
reburn with, 732–733, 732f, 733t
Bio-sorption, 2152
Biostimulation, 388f, 395f
Biotechnology, 407–412
advantages, 412
disease resistance in crops, 407, 408t
genetic engineering in pest control, benefits of, 407
impact of, on pest management, 1949–1953
transgenic virus-resistant potatoes in Mexico, 409–412
for vertebrate pest control, 2604–2605. *See also* Virally vectored immunocontraception (VVIC)
Biotic community, 291
Biotrickling filters (BTFs), 2611, 2612f, 2614t, 2621, 2623t, 2625
BioVECTOR, 382
BIPV (building-integrated photovoltaics), 223–224
Bird control chemicals, 413–415, 414t
future developments, 415
immobilizing agents, 414
lethal stressing agents, 414
repellents, 414–415
reproductive inhibitors, 415
toxicants, 413–414
Bird impacts, from pesticide use, 416
farmland bird species, decline in, 416
measuring of
field testing, 421
incident monitoring, 420–421
modeling, 422
surveys, 421
mechanisms of, 417
effects on reproduction, 418
lethal effects, 417–418
persistent organochlorine pesticides, 417
secondary poisoning, 418
sublethal effects and delayed mortality, 418
routes of, pesticide exposure, 418
abuse and misuse, 418–419
forestry insecticides, 420
granular formulations, 419
liquid formulations on insect prey, 420
liquid formulations on vegetation, 419–420
treated seed, 419
vertebrate control agents, 420
Bisacylhydrazines, 1465
Bisphenol A, 1405
Bistrifluron, 1461
Black blizzards, 2381
Black box approach, 1722
Black carbon. *See* Pyro-char
Black rot, 442, 443t
Black spot, 442, 443t
Blackwater, 44, 45
Bladder cancer, by NO_3-N exposure, 1303

Blattella germanica (German cockroach), 1462, 1988
Blown foam insulation, 1484, 1485f
Blown sidewall insulation, 1484, 1485
Blue baby syndrome, 1282, 1302, 2736
Bluegrass (*Poa pratensis* L.), 1305
Blue green algae (BGA), 1511, 1516
Blue-green algal blooms, 538, 539
"BLUE" scenario, 201
BMPs. *See* Best management practices (BMPs)
BNF. *See* Biological nitrogen fixation (BNF)
Bog, 2825t
Bonneville Power Administration
  load and generation, 1420f
Borehole, 1367–1368, 1368f
Borehole thermal energy storage (BTES), 2518
Boric acid, 431
Boron (B), 424
  adsorption isotherms, 432f
  availability of, 424
  deficiencies, 424
  deficiency in plants, 432
  elemental form of, 431
  Lewis bases, 431
  properties of, 431
  as soil contaminant, 431–433
  soil interaction with, 431–432
  for soil–sand mixtures, 433f
  solution-to-soil ratio, 432f
  sources of, for humans, 424
  toxicity. *See* Boron toxicity
  uptake in plants, 432–433
Boron toxicity, 424, 425
  causes of, 425
  levels in crops, 425
  management/reduction of
    boron-tolerant cultivars, 425
    leachates recirculation, 426
    phosphorus, zinc and silica, use of, 426
    phytoremediation, 426
    waste materials use in, 426
  sources of, 425
  symptoms of, 425
BotaniGard®, 382
Botrytis cinerea, 383, 1991, 2109
Botswana, 2690
Bottomland, 2825t, 2826f
Bottoming cycles, 1357
Bottom-up estimate approach, 1792
Boudouard reaction, 2490
Boundary conditions
  and nutrient entry, 1821–1822
Boundary element method (BEM), 1314, 1314f
Box equation, 965
BPH. *See* Brown-plant-hopper pest (BPH)
Brain
  development of, 1756
  maturation of, 1756
Brassica napus, 1380
Brassica oleracea (Broccoli), 1307
Brassica oleracea (Cabbage), 442
Brassica species, 2435
Brayton cycle, 819
Brevicoryne brassicae (L.), 1928
Brine, 1569t
Briquettes, 375t
Britain, work on field drainage in, 584
British Co-op Group, 1401
British Food Standards Agency, 1403
British thermal units (BTUs), 1363, 1493, 1652
  loss of, 1493
Broccoli (*Brassica oleracea* L.), 1307
Brodifacoum, 420
Broiler litter, subsurface band application of, 1674
Bromadiolone, 420
Bromoxynil, 409
Brown pelican (*Pelecanus occidentalis*), 417
Brown-plant-hopper pest (BPH), 2034, 2141
Brüel and Kjær multi-gas monitor, 1203
Brunauer–Emmett–Teller (BET) isotherm, 65. *See also* Isotherm(s)
Brunauer–Emmett–Teller (BET) measurements, 1715
Brundtland report, 338
BT α-endotoxin, 411–412
BTEX, 1281–1283. *See* Benzene, toluene, ethylbenzene, and xylene (BTEX)
Bubbling fluidized bed combustor (BFBC), 727
Bubbling fluidizedbed gasifier (BFBG), 2492–2493
Bubo virginianus (Great horned owl), 420
Buffer capacity of soils, 1375
Buffers, vegetative, 995–1000, 995f, 996–997t, 998–999f
  hydraulic resistance, 998–1000
  types, 995, 998
Building. *See also* Climate change
  climate change on, impact of, 436–437
  component storage, 2519–2520
  contribution of, to human-induced climate change, 437
    building-related refrigerants, 437
    embodied energy, 437
    operational energy, 437
  cooling applications, 1360
  design and operation, strategies for, 439–440
    building envelope, 439
    building mechanical service, 440
    internal heat sources, 439
    recommendations on, 440
  life-cycle cost of, 1648, 1648f
Building energy management systems (BEMSs), 440
Building envelope, 439
Building envelope insulation, 1479
Building Performance Initiative (BPI), 1650
Building research establishment environmental assessment method (BREEAM), 1655
Buildings
  life-cycle cost of, 1648, 1648f
  rating systems for, 1655
  simulation technique, 439, 440
  types of, 860
Buprofezin, 1460, 1463
Business, role in ecological footprint, 2477–2478
Business-as-usual scenario, 617–618, 621t
Butachlor, 2142
Butane process, 1256

C

Ca/(HCO_3 + SO_4) ratios
  for saltwater detection, 1327
Ca/Mg ratios
  for saltwater detection, 1327
Cables
  in photovoltaic systems, 222–223
Cabbage disease (ecology and control)
  cabbage as human diet, 442
  control principles
    avoidance, 444
    eradication, 443, 443t
    exclusion, 443
    protection, 444
    resistance, 444
    therapy, 444
  diseases and pathogen ecology
    black rot, 442
    black spot, 442
    clubroot, 442
    dark leaf spot, 442
    downy mildew, 443
    sclerotinia stem rot, 443
    watery soft rot, 443
    white mold, 443
    wirestem, 443
  foliar pathogen management, 444
  integrated disease management (examples), 444
  seedborne pathogen management, 444
  soilborne pathogen management, 444
Cacao swollen shoot virus, 2111
Cactoblastis cactorum, 2821
Cadmium (Cd), 453, 1370
  contamination, in coastal waters, 490
  dispersion and application, 453–454
  health impacts of, 2118
  toxicity and ecotoxicity, 454–455

Cadmium and lead, 446
chemical properties, 446
detection, 447
environmental hazards, 447–448
human exposure to, 449t
human health effects, 448–449
occurrences, 446–447
production and uses, 447
regulations and control, 449
Cadmium telluride (CdTe) solar cells, 232
Calcareous sodic soils. *See also* Sodic soils
reclamation of, 2370
Calcic soils
isotopes in, 1452–1453
Calcite, 88, 462
Calcium (Ca)
montmorillonite clay, swelling of, 2338, 2339f
Calcium chromate, 479t
Calcium-lactate extractable phosphorus, 102f
Calibration, groundwater modeling, 1298, 1298f, 1299
Calibration methods, 1663
California Bay-Delta Program (CALFED), 627
California, wastewater reuse in, 1571
Calispermon spp., 2031
Caloric theory of heat, 815
Calorific value, of fuel, 374
Camelina sativa, for bio-liquid fuels, 847
Canada
acid rain in, 8
soil degradation in, 2383
Canadian Air and Precipitation Monitoring Network (CAPMoN), 7
Canadian Organic Products Regulation (2006), 1109
Cancer
arsenic toxicity and, 1274, 1274f
exposure to pesticides and, 1415
Cancers, from pesticides, 1394–1396
animal studies, 1394–1395
epidemiological studies, 1395–1396
overview, 1394
pesticide exposure, 1394
risk of, 1394, 1395t
Candida spp., 2109
Cane toads (*Bufo marinus*), 2596, 2605
Canine Herpesvirus-1 (CHV), 2604
Canola (rapeseed), for bio-liquid fuels, 846
Canon process, 2654
Capillaria hepatica, 2596
Capital costs for nuclear and coal-fired powerplants, 1790
Capping, 2315t
Capsella, 1936
Capsella bursa pastoris, 1928
Capsicum annuum L. (Peppers), 425
Captan, 1318
Capture cost, 459
Capturing plate, 242
Carbamate insecticides, 2804
impact on birds, 417
Carbamate pesticides, 1317
Carbamates, 3, 277, 2005, 2141
Carbaryl (Sevin), 3, 2016t, 2805, 2142, 2806
Carbazates, 3
Carbendazim, 2142
Carbofuran (Furadan), 417, 419, 1318t, 2805, 2806
Carbohydrates, plant, 373
Carbon (C), 88
cycle, 88–90, 89f
decomposition and mineralization, 90
organic substrate quality, effect of, 92–93
soil microorganisms, effect of, 90–92
fixation, 88–89
inorganic. *See* Inorganic carbon
keeping in soil, 1190–1191
occurence of, 88
removal from air, 1189–1190
man role in, 1189–1190
nature role in, 1190
sources of, 456–457
Carbon adsorbents, 68–69
Carbonate, 1451
models of formation, 1446
biogenic model, 1446
per ascensum model, 1446
per descensum model, 1446
in situ model, 1446
pedogenic *vs.* geogenic, 1445–1446
simulations of pedogenic accumulation, 1451–1452
Carbonate bonded REE, 2268t
Carbon balance assessment of bioenergy crops. *See* Bioenergy crops
Carbon capture and storage (CCS). *See* Carbon sequestration
Carbon dioxide
atmospheric, IC impact on, 466–467
pollution, control methods, 157
Carbon cycle, 2389f
Carbon dioxide (CO_2), 486, 1202
emission estimation, 717, 722, 722f
emissions, 774
emissions per year, 716f
erosion and, 934–937
conceptual mass balance model, 935–936, 935f
emissions, estimates of, 936, 936t
SOC, 935–936
greenhouse effect. *See* Greenhouse effect
soil quality and, 2388
storage of, 458
Carbon footprint, 2468
Carbon gases. *See also specific* entries
soil quality and, 2388–2389
Carbon hydrides pollution, problems of, 156–157
control methods, 157
Carbonization, 2485–2486
Carbon losses from soil
mineralization processes and, 1857
soil erosion processes and, 1857–1858
Carbon monoxide (CO)
characteristics of, 136
EN 14626, non-dispersive infrared for, 142
pollution, problems of, 156–157
control methods, 157
Carbon sequestration, 1858–1861, 2833–2834
biological sequestration
ocean fertilization, 460
terrestrial carbon sinks, 460
carbon capture and storage (CCS)
carbon, sources of, 456–457
geologic sequestration, 458–459, 458t
industrial carbon dioxide capture, 458
ocean direct injection, 459
overall costs of, 459, 459t
post-combustion capture, 457
pre-combustion capture, 457–458
separation and capture, 457
storage of carbon-di-oxide, 458
costs of, 460
creation of wetland, 2835
global importance of, 1858
greenhouse gases (GHG), 456
human activities, impact of, 2835
long-term field experiments, role of, 1858
mechanisms, 1858, 1860
potential of, 346–347, 347f
with primary production, 2833
prospects for, 460–461, 460t
in soils, 2833–2834
SOM, energy, and full C accounting, 1861
Carbon tetrachloride, 1317, 1318, 1318t
Carboxylic herbicides, 1318
Carcinogens, 1570
Carcinops troglodytes, 1921
Cardiovascular effects, of arsenic toxicity, 1272
Carduus nutans, 2822
Caribbean Environment Programme, 2383
Caribbean Islands
soil degradation in, 2382f, 2383
Carnot cycle, 819f, 820
efficiency, 820, 2530
Carnot engine, 2529, 2530f
Carnot law, 1356
Carnot ratio equality, 820
Carnot HCOP, 1347
Carnot heat engine, 683

Carpophilus pilosellus, 1921
Carry-over soil moisture (SMco), 1563
Carson, Rachel, 2010
Carter, Jimmy, 1521
Cartridge filter, 2723–2724. *See also* Filter(s)
Carum carvi, 384
Carya aquatica, 2830
Cascade. *See* Flufenoxuron
Cassava mosaic virus, 2111
Casten, Thomas, 1505
Catalytic WO reactions, 1909, 1909t
  using heterogeneous catalysts, 1910
    NS-LC, 1910
    Osaka Gas, 1910
  using homogeneous catalysts, 1909–1910
Catchment surface, 2263–2264
Cat-clays. *See* Potential acid sulfate soils
Catechol, 391
Cathedral ceilings
  building R-30, 1489
  exposed rafters, insulating, 1490
  insulation techniques, 1489
  options, 1489t
  scissor trusses, 1489–1490
Cation exchange capacity, 1374
Catolaccus grandis, 1747
CATOx2 process, 1910
Cats, biological control of, 2602–2603. *See also* Feline panleukopenia virus (FePV)
Cattle manure, 718t
Cauliflower (*Brassica oleracea* var. Botrytis L.), 425
Cauliflower mosaic virus (CaMV 35S), 308
Causal chain frameworks, 354f, 356
CBEP. *See* Community-based environmental partnership (CBEP)
CC. *See* Continuous Commissioning
CCPR. *See* Codex Committee on Pesticide Residues (CCPR)
Cd. *See* Cadmium (Cd)
Cd model, 404f, 405f
CEC. *See* Commission of the European Communities (CEC)
Ceilings and roofs, 1486–1490
Celestite (SrSO4), 2426
Cellulase, 89
Cellulose insulation, 1478
  in attics, 1488t
  health impacts of, 1479
Celtis laevigate, 2830
Center pivot irrigation, 1547t, 1554
Center pivots, 1551–1553
Central America
  soil degradation in, 2383, 2382f
    distribution of, 2383f
Central inverters, 221–222, 222f
Centralized composting. *See also* Composting
  horizontal reactors, 518, 518f
  in-vessel, 517
  rotary drums, 518–519, 519f
  silos/towers, 518
  static pile, 517, 517f
  windrows, 516–517, 516f
Central nervous system development, 281
Centre for Alternative Wastewater Treatment of Fleming College, 2671
Centrocercus urophasianus (Sage grouse), 420
CENTURY SOM model, 1875
Ceratophyllum demersum, 2830
Cerium (Ce) content in grains/seeds, 2269t
Cernuella virgata, 383
Certified emissions reductions (CER), 486
Certified geothermal designers (CGD), 1366
"Certified organic" crops, 125
C factor. *See* Climatic (C) factor
CFB. *See* Circulating fluidized bed (CFB)
CFBG. *See* Circulating fluidized-bed gasifier (CFBG)
CH_4. *See* Methane (CH_4)
Chamber scrubbers, 154
Chamber techniques
  GHG measuring fluxes, 1202, 1203, 1204f
Channel erosion, 2756
Charcoal, 376, 844
Charged-droplet scrubbers, 156
Char reactions, 724–725
Chelatobacter heintzii, 2246
Chemical deterioration map, 2377
Chemical exergy, 1084, 2531
Chemical industry sludge, 2267t
Chemical methods, for contaminant remediation, 1293
Chemical mixtures, 2572–2579
  concentration–response surface analyses for, 2576–2578, 2576–2578f
  future challenges to, 2579
  multicomponent mixtures, reduced mixture toxicity designs for, 2579
  strength and weakness of, 2579
  toxicities of. *See* Concentration addition; Independent action
Chemical nitrification inhibitors, 94
Chemical oxidation, 1335
Chemical oxygen demand (BOD), 524
Chemical pesticides (CPs), 1986
  and biological pesticides, integration of, 1987–1988
    examples of, 1989t
    impact on ecosystems, 1991–1994
    negative, 1988
    positive, 1988, 1989t
  reducing use of, 1986
Chemical pollutants, 2357
Chemical Review Committee, 1616
Chemical risk assessment, 1721
Chemicals, dissolved, 2753
Chemicals, runoff, and erosion from agricultural systems (CREAMS) model, 991
Chemical sensitivities. *See also* Pesticide sensitivities
  causes, 1411–1412
  impact of, 1412
  mechanism, 1412
Chemical speciation of phosphorus, 100
Chemigation, 469–472
  advantages, 469
  background, 489
  disadvantages, 469
  equipment, 469–471
    injection, 470–471, 470f, 471f
    irrigation, 470
    safety, 470f, 471
  management practices, 472
Chemiluminescence, for NO EN 14211, 141–142
Chemisorption, 63
Chemotaxis, 392
Chenopodium, 1936
Chernobyl accident, 2304
Chesapeake Bay, 474–476, 475f
  human population, growth in, 475
  meaning of name, 475
  physical conditions, 474–476
  sedimentation and, 475
Chesapeake Bay Program, 627
Chesterfield Inlet's tundra treatment wetland, 2670
Cheyletus eruditus, 2424, 2424t
Chhattisgarh, groundwater arsenic contamination in, 1268
Chicago Climate Exchange, 487
Chicken cholera bacteria (*Pasturella multocida*), 2596
Chicken manure, 718t
Chickpeas (*Cicer arietinum* L.), 424, 427
Children
  arsenic toxicity effect on, 1275, 1275f
  health effects of pesticides on, 1121–1122
  pesticides effects on, 1415–1416
Chili pepper powder, 2021
Chilled water storage, 2510
Chilo suppressalis (Walker), 311
China
  trends in acidity, 13
Chips, 375t
Chitin, 72, 1461
Chitin synthase, 1461
Chitin synthesis inhibitors (CSIs), 1460, 1461
  benzoylphenyl ureas, 1461–1463
    bistrifluron, 1461
    chemical structure of, 1462f
    chlorbenzuron, 1461
    chlorfluazuron, 1461
    diflubenzuron, 1461

fluazuron, 1461
flucycloxuron, 1461
flufenoxuron, 1461
hexaflumuron, 1461, 1462
lufenuron, 1462
novaluron, 1462
noviflumuron, 1462
teflubenzuron, 1462–1463
triflumuron, 1463
non-benzoylphenyl ureas, 1462f, 1461
buprofezin, 1463
chemical structure of, 1462f
cyromazine, 1463
dicyclanil, 1463
etoxazole, 1463
Chitosan, 72, 73t
Chlorbenzuron, 1461
Chlordane, 1318t, 1394
Chlorfenapyr, 3
Chlorfluazuron, 1461
Chlorinated hydrocarbons, 1317
Chlorinated solvents
degradation pathways of, 1843, 1843f
metabolic pathways of, 1841, 1842f
oxidation of, 1844
Chlorination of storage tanks, 2264
Chlorofluorocarbons (CFCs), 486, 1893–1894
destruction of, 1886
properties of, 1885–1886
-ozone depletion hypothesis, 1887f
Chlorophyll meter readings, 1306
Chlorpyrifos, 2015, 2015f, 2020, 2142
Chlorpyrifos, 1318
Cholera, 2696
Chondrostereum purpureum, 383
Chontrol, 383
Chorioptes bovis, 2
Christiansen uniformity coefficient (UCC), 1542, 1548
Chromafenozid, 1466
Chromite, 478, 479t
Chromium (Cr), 477, 1370, 2585–2586, 2588
carcinogenic effects, 478
as cofactor of insulin, 477, 478
forms and sources of, in natural waters, 478–479, 478t
health effects, 477–478
health impacts of, 2118–2119
hexavalent, 477
occurrence of, 478
oxidation and reduction reactions of, 480, 480f, 481f
solubility controls of, in water, 479–480, 479t
trivalent, 477
uses of, 478
Chromium arsenate, 479, 479t
Chromium chloride, 479t
Chromium fluoride, 479t
Chromium(III). *See* Chromium (Cr)
Chromium(III) hydroxide, 479t
Chromium(III) oxide, 479t
Chromium jarosite, 479, 479t
Chromium phosphate, 479, 479t
Chromium picolinate monohydrate (CPM), 477
Chromium sulfate, 479t
Chromium(VI). *See* Chromium (Cr)
Chromolaena odorata, 367
Chromosomal aberrations (CAs) test, 2127–2128
Chrysanthemum, 2028
Chrysolina quadrigemina, 2821
Chrysomya megacephala, 1920
Chrysopa perla. See Golden-eyed fly
Chrysoperla carnea, 1750
Cicer arietinum L. (Chickpeas), 424
Cingular Wireless, 2226–2227
Cinnamic acid, aromatic hydroxyacid derivatives of, 2080f, 2081t
Circuit breaker, 223
Circulating fluid bed dry scrubber, 2442
Circulating fluidized bed (CFB), 2492–2493
boilers, 375
Circulating fluidized bed combustor (CFBC), 727
Circulating fluidized-bed gasifier (CFBG), 2492–2493
Cirsium spp., 2822
CITEAIR (Common Information to European Air), 135, 136
Citizen monitoring, 1582–1583
Citrus leafminer. *See Phyllocnistis citrella*
Citrus limon (Lemon), 426
Citrus tristeza virus, 2111
Ciudad Juarez Valley (Mexico), 2345
Civic environmentalism, 924
Cl/Br ratios
for saltwater detection, 1327
Clausius law, 818
Clausius statement, 818
Clay flocculation, 960, 960f
Classical biological control, 1474. *See also* Biological control
Classical weed biocontrol
history and impact of, 2821–2822
legislation, 2823
Clays, 1627
coatings, 1627
illuviation, 1627
minerals, 69–70, 71t
soils, water movement in, 585
1970 Clean Air Act, 20
Clean Air Act, 133, 157–158
Clean Air Act Amendments (CAAA), 732
Clean Air for Europe (CAFE), 136
Clean development mechanism (CDM), 486
CleanSeaNet, 1828, 1830
Clean-up techniques, 2198–2199
Clean water, 2752
Clean Water Act (CWA), 1064, 1289, 1942, 1943, 2805, 2808, 2809, 2750, 2753. *See also* Total maximum daily load (TMDL)
overview, 2809f
Clik. *See* Dicyclanil
Climate
defined, 1192
feedbacks, 1194
model, 1199
organic matter decomposition and, 1853
Climate change, 333, 435, 636, 804
and buildings, relationship between, 437–438, 438f
global, sulfur and, 2434–2435
greenhouse gases and, 1894–1895
hydrocarbons and, 1895–1896, 1896t
impacts of, 435
on ozone layer, 1896–1897
on wind erosion, 1019
implication of, on building, 436–437
building energy use, increase in, 436
construction process, 437
external fabric durability, 437
internal thermal environment, deterioration in, 436–437
service infrastructure, 437
structural integrity, 437
likely future, 435–436, 436t
Climate Change Fund, 487
Climate forcings and feedbacks, 1194–1197, 1194f, 1195f
atmospheric aerosol forcing, 1196, 1197t
greenhouse gas forcing, 1195, 1196t
land-use change forcing, 1196–1197
natural climate forcings (solar and volcanic variability), 1194–1195, 1196t
Climate models, 1867
Climate policy (International)
adaptation, 484
early international response, 484
emmission reduction, 483–484
equity questions, 484
European Union emissions trading system (EU-ETS), 487
greenhouse gases (GHG), 483
Kyoto protocol, 485–487
implementation, 486
targets, 485–486
UN framework convention on climate change (UNFCCC), 485
interim negotiations, 485
voluntary and regional programs, 487
Climate protection, 182–183
Climate system, 1193f

Climate technology initiative (CTI), 680
  activities, 680
Climatic conditions
  leaching and, 1622
Climatic (C) factor, 1012
Clofentazine, 3
Clomazone, 2014, 2024f
Clonal propagation, 1182
Closed-cell, spray polyurethane insulation, 1479
Closed circuit systems, 248
Closed ecosystems, 1935
Closed-loop pumped storage systems, 1423–1424
Clostridium, 321
Cloud condensation nuclei (CCN), 1020, 1196
Cloud point extraction, 2194
Cloud-reflectivity feedback, 1195f
Clouds earth's radiation energy system (CERES), 2818
Clovers (*Trifolium* spp.), 427
Clubroot, 443, 444t
Co. *See* Cobalt (Co)
CO_2. *See* Carbon dioxide (CO_2)
^{60}Co, 2243, 2244
Coal, 718t
  and bio-solids
    gasification of, 730–731, 731f
  deposits, 458
  and manure, 728
  reburn, 733t
  and refuse-derived fuel (RDF), 728, 728t
Coal agricultural residues, 727–728, 727f
Coal bed methane (CBM), 458
Coal-fired electrical generating station, 1086–1089
  condensation, 1087
  energy
    efficiency, 1087
    values, 1087
  exergy
    consumption values, 1089
    efficiency, 1087
    flows, 1090
    values, 1088
  flow diagram, 1087
  net energy flow rates, 1090t
  power production, 1087, 1089
  preheating, 1087, 1089
  steam generation, 1086–1087, 1088
  thermodynamic characteristics of, 1089
Coalification process, 717
Coal mining, 2240
Coastal aquifers, 1321
Coastal deserts, 566
Coastal waters, pollution in, 489–498
  biological pollution
    algal bloom, 495
    eutrophication, 494–495
    invasive species, 495
  fertilizers, 495–496
  heavy metals, 490–491
  light pollution, 497–498
  marine debris, 496–497
  metalloids, 490–491
  noise pollution, 497
  oils, 496
  organic compounds, 492–494
  pesticides, 495–496
  plastics, 496–497
  radionuclides, 491–492
  sewage effluents, 496
Cobalt (Co), 1370
Cochlicella acuta, 383
CODESA-3D
  for saltwater intrusion, 1329
Codex Alimentarius, 1108, 1612, 2704
Codex Alimentarius Commission, 1126, 2036
Codex Committee on Pesticide Residues (CCPR), 1126
Coefficient of performance (COP), 1364
Coefficient of variation (CV), 1542
CO_2 emissions, 799t
"CO_2-fertilization effect," 1858
Cofiring (coal and bio-solids). *See* Energy conversion
Cofiring, in fossil-fired power stations, 375, 378
^{60}Co(III)EDTA, 2244, 2245
Cogeneration, 1361
  and cool storage, 2515, 2515f
Cohort models, for organic matter modeling, 1864
Cold air distribution, 2517
Coleomegilla maculata, 1936
Coliform counts in wastewater, 1571
Colinus virginianus, 2005
Coliphage, 1626
Collaborative environmental management, 625
Collaborative governance approaches
  and ecosystem management, 626
Collaborative resource governance (CRG), 624, 627–628
Collego®, 1111
Columba livia, 413
Combined heat and power (CHP), 730. *See also* Cogeneration
  plants, 375, 375t
Combined nitrogen removal, 2653
Combined pumped storage systems, 1423
Combustion, 161, 774. *See* Energy conversion
  biomass, 375–376
  energy and exergy in, 1093, 1093f
Combustion turbine inlet air cooling (CTIAC), 2516–2517
  ambient temperature effect on, 2516f
  waste heat recovery in, 1360–1361
Cometabolism, 2176
Comet assay, 1156, 2126–2127
Command-and-control approach, 483
Command and control strategy
  quantity-based, 919
  technology-based, 919
Commercial Buildings Energy Consumption Survey (CBECS), 793
Commercial use of energy, 793
Commercial WO, 1908
Commissioning agent
  selection of, 668–669
  skills of qualified, 669
Commissioning of existing buildings. *See also* Continuous Commissioning (CC®); Recommissioning
  case study, 657
  costs and benefits of, 658–659
  definitions, 656–657
  measures, 656, 657
  monitoring and verification (M&V), 656
  process, 659–661
  team, 659
  uses in energy management process, 661–662
    as ECM in retrofit program, 662
    to ensure building meets or exceeds energy performance goals, 662
    as follow-up to retrofit process, 662
    as standalone measure, 662
Commissioning of new buildings
  acceptance-based vs process-based, 666–667
  barriers to, 668
  benefits of, 676
  cost/benefit of, 675
  defined, 665–666
  goal of, 666
  history of, 667
  importance of, 668
  market acceptance of, 667
  meetings, 672
  O&M manuals, 673
  phases
    acceptance, 673–674
    construction/installation, 671–673
    design, 670–671
    postacceptance phase, 674–675
    predesign, 670
  prevalence of, 667–668
  process, 668–670
  success factors, 675–676
  systems to include, 668
  team, 669–670
Commissioning plan, 668, 674, 675
Commission of the European Communities (CEC), 1409
Commission on Environmental Law of IUCN (World Conservation Union, The), 1620
Common air quality index (CAQI), 135–136

Common Operational Webpage (COW), 136
Common property resources, destruction of, 914
Community-based environmental partnership (CBEP), 925
Community balance. *See* Pesticide impacts, on aquatic communities balance
Community commitment, 560
Community level, 312
Community participation, 560
Community Pesticide Action Kits (CPAKs), 507
Community Pesticides Action Monitoring (CPAM), 505
Compartment model
  inorganic carbon, 1451–1452
Compensation point, 1952
Competitive crops, 1111
Competitive water markets, 2780–2781
  scarcity rents, 2782
  shadow prices, 2782–2783
Comply. *See* Fenoxycarb
Compost, 961
Composting, 512–525, 513f
  advantages/disadvantages of, 515
  background of, 514–515
  centralized
    horizontal reactors, 518, 518f
    in-vessel, 517
    rotary drums, 518–519, 519t
    silos/towers, 518
    static pile, 517, 517f
    windrows, 516–517, 516f
  commercialization issues of, 525
  home, 516
  maturation of, 524–525
  phases of, 520
  process parameters in, 520–521
    aeration, 522–523
    feedstock composition, 523–524, 523t
    moisture, 521–522
    odor control, 524
    particle size, 523
    pathogen control, 524
    pH control, 524
    temperature, 521, 522f
  system, evaluation of, 520, 521f
  vermicomposting, 519–520, 519f
Compost wetlands, 1690
Comprehensive Environmental Response, Compensation, and Liability Act (CERCLA), 2114–2115, 2116t, 2118, 2119t
Comprehensive Everglades Restoration Plan (CERP), 627
Comprehensive Everglades Restoration Plan (CERP), 1080, 1081
Compressed natural gas (CNG), 723
Compressive strength, 864
Computer-derived packages, of IPM, 1522
Computer models
  for saltwater intrusion, 1329–1330
Concentrated cookers, 2503
Concentration addition (CA), 2572–2574, 2573t
  applications of, 2575–2576
  characteristics of, 2575
  predictive power of, 2576
Concentrating solar power (CSP), 2520–2521, 2521–2522f
Concentration–response surface analyses, for chemical mixtures, 2576–2578, 2576–2578f
Conceptualization, groundwater modeling, 1297–1299, 1298f
Concrete
  block and poured concrete, 863–864
  block cores, insulating, 1482
  poured, 863–864
  precast, 864
Conditioned attic assemblies, 1487
Confined animal feedlot operation (CAFO), 2752
Confirm. *See* Tebufenozide
Coniothyrium minitans, 2107
Conjugative gene transfer, 387
Conotrachelus nenuphar, 2009
Conservation, 1474. *See also* Biological control
  agricultural systems, 91, 119–120, 335
  biology, 633, 635
  biological controls. *See* Biological controls, conservation of
  of natural enemies, 1749–1751
    habitat and environmental manipulation, 1750
    harmful pesticides practices avoidance, 1749–1750
    overview, 1749
    overwintering and shelter sites, 1750
  prioritization, 972
Conservation farming
  in United States, 2382
Conservation program performance, 1774
Conservation Reserve Program (CRP), 346
Conservation structures, to control soil erosion, 970, 970f
Conservation tillage, 2762–2763, 2763f
Conservation tillage, 119
  to control wind erosion, 1028, 1027f
Constitutive equations
  for saltwater intrusion, 1328–1329
Constrained-equilibrium models, for exergy analysis, 1094
Constructed treatment wetlands (CTW), 2155
Constructed wetlands (CW), 2020–2021, 2663, 2730, 2862, 2863t
  definition, 2670
  and ecofiltration, 2154–2155
  potential for, 2670–2671
Construction methods, 860
Construction standards, 1655
Consult. *See* Hexaflumuron
Consumer concerns, 1401–1404
  'cocktail' effects from multiple residues, 1401, 1402t
  food residues, 1403
  government action plans, 1403
  NGO initiatives, 1404
  overview, 1401
  pests/pesticide safety in homes and gardens, 1401–1403
  retailer initiatives, 1403–1404
  towards residue-reduced food, 1403–1404
Contaminants
  environmental, 1626
  interactions in soil and water, 2166
    inorganic chemicals, 2166, 2169
    organic chemicals, 2169
  levels in wastewater, 2865t
  sources and nature of, 2166, 2167t–2169t
Contamination
  arsenic. *See* Arsenic; Arsenic-contaminated groundwater
  definition of, 1281
  groundwater. *See* Groundwater contamination
  with heavy metals, 1374
  of waterways, 2754
Continuous Commissioning (CC®), 657
  case study, 662
  process, 659–661
    assessment, 660
    building screening, 660
    CC measures development, 661
    implementation and follow-up, 661
    performance baselines, development of, 660–661
    plan development and team formation, 660
Continuous liquid–liquid extraction, 2193
Continuous simulation erosion models, 991–992
Continuous simulation models
  water quality, 2749
Contour farming, 2763
Contouring, 968, 968f
Contour lining, 570
Contour ridges, 969
Contour stone terraces, 969, 969f
Contour stripcropping, 2763, 2763f
Contour tillage, 2539
Control-A-Bird®, 413
Controlled-release fertilizers (CRFs), 1306
Controlled-release nitrogen fertilizers, 1809
Controlled traffic, 120

Control volume (CV), 815, 815f
Conventional activated sludge (CAS) systems, 2062
"Convention on the Prior Informed Consent (PIC) Procedure for Certain Hazardous Chemicals and Pesticides in International Trade," 2036
Conversion factors and the values of universal gas constant, 2527t, 2534
Conveyance efficiency, 1545–1546
Conveyance systems, 2264
Cooling system in dairy, 1664
Cool storage, 2511–2517
  application and design features, 2514–2517
  building certification, 2514, 2514f
  power generation and, 2511–2514
    efficiency and emissions, 2512–2514, 2513f
    electrical demand, 2511–2512
  technologies, 2509–2510
Coontail. *See Ceratophyllum demersum*
Co-oxidation/co-reduction, 2152t
Copidosomopsis plethorica, 1919
Copper (Cu), 535–537, 1370
  concentration, 1375
  deficiency
    plant growth and, 536
    in yaks, 427
  as essential element, 535–536
  high-copper soils, plant growth on, 536–537
  plants and, 535
  in soils, 535
Copper-indium selenide (CIS) solar cells, 232
Copper sulfate (for algae removal), 539
Coptotermes curvignathus, 1461
Coquis (*Eleutherodactylus coqui*), 2596
Corn (*Zea mays* L.), 1303
Corn oil, 415
Correlation coefficients, 2427
Corrosion, 190
Corvus spp., 415
Cosmetics, aluminum in, 270
Cosmetic standards, 1120
  environmental and public health impacts of pesticides and, 1121–1122
  health effects of eating insects in food and, 1122
  history of, 1120–1121
  increase in, of fruits and vegetables, 1120, 1121
  pesticide use and, 1120
Cosmic ray, 2235
Cosmogenic radionuclides, 2235, 2239
Costa Rica
  pesticide poisonings incidence, 1408
Cost-benefit analysis (CBA), 917–918, 917f, 918f
  for crop water use, 2220
Cost-effective analysis, 918–919
Costelytra zealandica, 322
Costs
  pesticide poisonings, 1416
  weed biocontrol and, 2822–2823
Cotton (*Gossypium hirsutum* L.), 425
Cottonwood. *See Populus deltoides*
Courant number, 1315
Courier. *See* Buprofezin
Cover crops, 1110
Coversoil resources
  open pit mining, 1692
Cow washing, 1663–1664
Cr. *See* Chromium (Cr)
Cradle to grave approach, 2418
Crailsheim borehole system, 2518
CREAMS model. *See* Chemicals, Runoff, and Erosion from Agricultural Systems (CREAMS) model
Creep samplers, 1023f
Critical flocculation concentration (CFC), 960
Critical loads, 2292
Crops
  depletion, 1663–1664
  detritus, 310
  disease resistance in, 407, 408t
  response to REE, 2269, 2269t
  water use, 1563
  yield and water requirement, 1664t
  yield increase by REE, 2269, 2269t
Crop-border diversity
  effects on pests, 1928
Crop diversity, for pest management, 1925–1929
  border diversity, 1928
  crop rotation
    effects on diseases and pests, 1925, 1926t–1927t
    weed abundance and, 1928
  decoy and trap crops, 1925, 1928
  overview, 1925
  weed diversity, 1928
CropLife America, 2014
Crop model (CM) techniques, 2818, 2818f
Cropping systems
  ammonia volatilization, 86–87, 86t
Crop rotation
  effects on diseases and pests, 1925, 1926t–1927t
  multifunctional, 1508
  sulfur and, 2432–2433
  weed abundance and, 1928
Crop rotations, 1018–1019
  organic farming systems and, 126
Crops/cropping
  impact on soil IC, 464
  pesticides in, 1966
  selection and diversification, 2675–2676
  water harvesting for, 2739
  yield
    potentials of, 2676t
    and water use, 2218–2219
Crop tolerance
  classification of, 2347t
  defined, 2346
  salt-affected soils and, 2346–2347
Crop water use, 2218–2219
  landscape, 2219
  spatial analysis of, 2219–2221, 2219–2220f
"Cross protection," 2111
Crotalaria juncea (Sunnhemp), 1511
Crotalaria spp., 289
Crucifer downy mildew, 443
Crude oil, chemical composition of, 2042–2046
Cry9C contamination, 313
Cry9C proteins, 312
Cryogenic techniques, 457
Crysoperla carnea, 412
Crystal iron oxides, 2266, 2268t
Crystalline (Cry) proteins, 308
Crystalline silicon solar cells, 231–232
  amorphous, 232
  monocrystalline, 231–232
  polycrystalline, 231
  structure and manufacture of, 232–234
    ARC, deposition of, 234
    diffusion formation of a p–n junction, 233
    formation of p–n junction by ion implantation, 233
    metal contacts, 233–234
    passivation of silicon surface, 233
    surface preparation, 232
Ctenopharyngodon idella, 1922
CTI. *See* Climate technology initiative
Cu. *See* Copper (Cu)
Cucumber (*Cucumis sativus* L.), 426
Cucurbita pepo, 407
Culex annulirostris, 2598
Culex tarsalis, 1922
Cultivar selection, 1112
Cultural eutrophication, 538
Cultural pest control
  antagonistic plants in, 288–289, 289t
Cultural practices
  effects on natural enemies, 370
Curie (Ci), 2235
Curve Number method, 1958
Customer demand, of electricity, 1418, 1419f
Cuticle, 1459
CV. *See* Coefficient of variation (CV)
CWA. *See* Clean Water Act (CWA)
Cyanide
  groundwater contamination by, 1291–1292
Cyanobacteria, 394
Cyanobacteria in eutrophic freshwater systems
  algal blooms, monitoring/management of, 539
  consequences of, 539

dominance of cyanobacteria, 538–539
extent of problem, 538
Cyanobacterial (bluegreen) algal blooms, 538
Cyanobacterial toxins, 2137
Cyanophyta, 1515
Cyclonic scrubbers, 154
Cydia pomonella, 2009
Cyhexatin, 3
Cylindrospermopsis, 539
Cyperine, 2031, 2031f
Cypermethrin, 2142
Cyprus, 2689
Cyperus rotundus, 2031
Cyprinid Herpes virus 3 (CyHV-3), 2603
Cyromazine, 1460, 1463
Cytochromes P450 (CYP), 281
Czochralski's method, 231

D

2,4-D, metribuzin, 2806
Dacthal, 2014
Dactylopus ceylonicus (mealy bug), 2821
Daily water intake of dairy cattle, 1664t
Dairy, water use in
cow washing, 1663–1664
drinking water requirements, 1663, 1664t
estimation of water use, 1663
flushing manure, 1664
milking equipment/parlor, washing of, 1664
rainwater from roofs, 1665
recycling dairy wastewater, 1664, 1665t
sprinkling and cooling, 1664
water budget development, 1665t, 1666
Dairy wastewater treatment, 2863f
Dalapon, 1318t
Dams and weirs, 208–209
Danaus plexippus (L.), 313
Dandelion (*Taraxacum*), 1527t
Danish International Development Agency (DANIDA), 1269
Daphne, 271
Daphnia magna, 409
Darcy's law, 1312
for saltwater intrusion, 1329
Dark leaf spot, 442, 443t
Dart. *See* Teflubenzuron
Databases
GIS in land use planning and, 1163
Data loggers, 736
Datura metel L., 289
Datura stramonium L., 288, 289
Dayoutong, 1464
DBCP. *See* Dibromochloropropane (DBCP)
2,4-Dbe. *See* 2,4-D butyl ester (2,4-Dbe)
2,4-D butyl ester (2,4-Dbe), 2006
D.c.–a.c. converter. *See* Inverters, in photovoltaic systems
D.c. main switch, 222–223
DCE. *See* Dichloroethene (DCE)
DDD, 1399
DDE (dichlorodiphenyldichloroethylene), 417
DDT (dichlorodiphenyltrichloroethane) 417, 1397, 1399, 1403, 1414, 2015, 2015f
discovery, development, and impact, 2009–2010
effects of, 2005
Dead-band concept, 738
Dead Sea, degradation of, 1438
Dead state, 1084
Dead zones, 1302
Decision guidance document (DGD), 1616
Decision support system engine (DSSE), 2819
Decision support system for agro-technology transfer (DSSAT), 2819
Declarations (on soil conservation), 1619
Decomposition
of organic matter, 90, 1625
Decoy crops
effects on pests, 1925, 1928
Decreasing block tariff (DBT) pricing strategies, 2786
Deep sea mining, 2359
Defect action levels (DALs), 1120–1122
Defoliators, effect on photosynthetic rate, 1951
Deforestation
erosion, historical review, 1050, 1051f
Degradation process, 2136, 2863
soil. *See* Soil degradation
Delia antiqua, 2021
Deltamethrin, 2142
DEM. *See* Digital elevation models (DEM)
Denaturing gradient gel electrophoresis (DGGE), 2618
Denitrification, 94, 1770–1771
nitrous oxide from, 96–97
Denitrification, 2864
Denmark
wind output and net electricity flows in, 1427f
Dense NAPL (DNAPL), 1334
Densification, 374
Denver Arsenal, 1333
Deoxynivalenol, 1696
Department of Energy (DOE), 1794
Department of Health and Human Services, 2049
Depletion, defined, 1563
Deposition
erosion and, 1046–1047, 1047f
soil erosion, 958
Deposit–refund system, 903
Depth-and-width integrated sampling, 965
Depth stratification , of soil organic matter, 91–92, 91f
Dermacentor variabilis (dog ticks), 2
Dermal effects, of arsenic toxicity, 1271–1272, 1272f
Deroceras caruanae, 383
Deroceras laeve, 310
Deroceras reticulatum, 383
Derxia, 350
DES. *See* Diethylstilbestrol (DES)
Desertification, 541, 557–558
assessment of degree of, 541–542. *See also* Landscapes
biophysical aspects of, 541
causes of, 558
as continuum, 542f
defined, 552, 557
in dry lands, 335
and greenhouse effect, 549–551
biological changes, 549
climatical changes, 549
greening of the Earth hypothesis, 549–550
modern studies, 550–551, 550f–551f
from prehistory to industrial revolution, 550
woody plant range expansions, 550–551
indicators, 553–554
landscape ecology and restoration approach to, 541, 542f
manifestations of process, 552–553
monitoring, 543
prevention and reversal of, 558–559, 558f
agricultural practices in, 558
climate change and, 559
evaluation of, 560–561, 561f
intervention projects for, implementation of, 559
issues affecting, 558–559
multiscale human–environmental dynamics and, 558
participatory approach to, 561
rehabilitation and restoration approach in, 558
socioeconomic conditions and, 559
water management in, 558
response curve to, 542, 542f
and revegetation, 543
reversal of, procedures for, 542–543
San Pedro River (case study), 554–555
technology, measurement, and analysis, 554
Desertization, 557, 565

Deserts/desertization
air pollution and, 566
causes of, 565, 569t, 570t
direct, 565, 567
indirect, 565, 567–568
defined, 565, 566
degradation processes, 567–568
diagnosis/evaluation, 568
fluctuations and trends of climatic variables in historical times, 566
land surface albedo, 566
managed restoration, 569–570
natural amelioration, 569
origin of, 565–566
overview, 565
rainfall and, 566
severity and extent of, 568–569, 569t
temperature and, 566
Desiccation, 1625
Designated National Authorities (DNAs), 575, 1616
Designated uses, of navigable waters, 2810
Design for the Environment (DfE), 1253
Design intent document (DID), 669
Desorption, 432
of contaminant, 1626
Detachment, soil erosion, 958, 984, 985
Deterministic uncertainty analysis, 1722
Deterministic water quality models, 2749
Detoxification, 2153
Detritus, 310
Developing countries, pesticide health impacts in, 573
banned and restricted compounds import/export, 575–576
health impacts
acute pesticide poisoning, 573–574
unknown chronic health impacts, 574
lack of technical and laboratory capacity, 576
low levels of worker and community awareness, 575
pest control policies, 4
weak regulation and enforcement, 575
weak surveillance for hazards and impact, 574
Developmental neurotoxicity (DNT), behavioral aspects of, 1756
Developmental toxicants, 280
Developmental toxicology bioassays, 283
DeVine®, 383, 1111
DGD. *See* Decision guidance document (DGD)
Diabetes mellitus
arsenic ingestion and, 1272
Diabrotica virgifera virgifera LeConte, 310
Diamond. *See* Novaluron; Teflubenzuron
Diatraea saccharalis, 367
Diazinon, 419, 1318, 1998t
Dibromochloropropane (DBCP), 2015, 2015f
reproductive problems from exposure, 277–278, 278t
Dibrotica spp., 1983
Dichloroethene (DCE), 1334
Dichlorodiphenyltrichloroethane (DDT), 575, 1916, 1917, 2114
global alliance for alternatives to, for disease vector control, 1917
Dichloronaphthoquinone, 539
Dicyclanil, 1463
DID. *See* Design intent document (DID)
Dieldrin, 417, 1283
Diesel cycle, 819
Diethylstilbestrol (DES), 1405, 1406
Difenacoum, 420
Difethialone, 420
Differential absorption lidars, 144. *See also* Lidars
Diffuse double layer (DDL), in soil, 2338
Diffuse melanosis, 1271
Diffusion, 1817–1820
driven phase, evaporation, 130
Fick's first law of, 1818
nutrient movement by, 1819–1820, 1820f
Diflubenzuron, 1461
Digestion process, 2651
Digital elevation models (DEM), 1164
Digital orthophoto quadrangles (DOQ), 2272
Dikrella cruentata, 1936
Dilution-extractive system, for monitoring stack gases, 144–145, 145*f*
Dimethoate, 2016t, 2142
Dimethyl sulfide (DMS), 2439
Dimilin. *See* Diflubenzuron
Dinghy, 209
Dinitroaniline herbicides, 1318
Dinitrophenols (DNOC), 2009
Dinoseb, 1318t
Diofenolan, 1464
Dipel, 382
Diplococcus, 321
Diquat, 1318t
Direct aqueous injection–electron capture detection (DAI-ECD), 2795–2796
Direct aqueous injection–gas chromatography–electron capture detection (DAI–GC–ECD), 1847
Direct contamination of surface water, 1576
Direct current (DC) resistivity
for saltwater detection, 1327
Disc filters, 2731. *See also* Filter(s)
Direct measurement, of soil redistribution by wind, 1021, 1021f, 1022f
Direct methanol fuel cells (DMFCs), 1146, 1151–1152, 1151f, 1152t
applications of, 1152
carbon monoxide and, 1151
contamination levels, acceptable, 1151
hydrogen sulfide and, 1151
and methanol crossover problem, 1152
operating principle of, 1151, 1151f
technological status of, 1152
Direct radiative forcings, 1194, 1196t
Direct toxicity, 312
Dirty water, 2304
Discount rates, 1795–1796, 1800f, 1801f
Discretization, groundwater modeling, 1298f, 1299
Disease resistance
in crops, 407, 408t
Disinfection methods, for cooking, 2705t
Dispersible clay content (DC), 2336, 2337f
Dispersion, 1627
Dispersion ratio of clay (DRC), 2336, 2337f
Dispersion, soil, 2334, 2339, 2340f
effect of clay–cation interaction on swelling and dispersion, 2338–2340
management of, 2341, 2342t
Dispersive liquid–liquid microextraction (DLLME), 1970t, 2193
Displacement, 809f
Displacement law, 2234
Disposal
of pesticide containers, 510, 510t
problems, associated with biowaste, 512
Dissolved chemicals, 2753
Dissolved organic N (DON), 1304
leaching from agricultural soil, 1304
Dissolved oxygen, 2753
concentration, 2649
Dissolved pollutants, agricultural runoff and, 82–83
fertilizers, 83
nitrogen, 82–83
pesticides, 83
Distribution cycle exergetic ratio, 1175
Distribution reservoir, 1577
Distribution uniformity (DU), 1548, 1564
District cooling, 2515–2516, 2516f
District energy systems, 773
Disulfoton, 419
Diuron, 1318, 2806
Diversification
and organic farming, 1106
Diversifying cropping systems, for reducing NO3-N leaching, 1307
Diversion/graded terraces, 968, 969f
Diversity
functional, in intercropping, 1937
"Diversity–stability hypothesis," 1934–1935
DNAPL, 1282. *See* Dense NAPL (DNAPL)
DNAs. *See* Designated national authorities (DNAs)
DNOC. *See* Dinitrophenols (DNOC)
Dobson unit (DU), 1885

2-Dodecanone, 2021
DOE. *See* U.S. Department of Energy (DOE)
Dolomite, 88, 465
Domestic hot water (DHW), 242
Doppler lidars, 144. *See also* Lidars
DOQ. *See* Digital orthophoto quadrangles (DOQ)
Dose-response models, 916
Dose–response relationship, 1705
Double catalyst system, 157
Double pumping, 1330
Downy mildew, 443, 443t
Drain
  depth, 590
  spacing, 590
Drainable system (wall), 864
Drainage, 584, 1559
  artificial, 585
  climatic conditions effect on, 586
  definition of, 585
  downstream effects of, 584–586
  of drier soils, 585
  experimental studies on, 585
  factors influencing response to, 585
  and flood events, 584
  main drainage systems, 584
  of permeable soils, 585
  purpose of, 584
  role in crop production, 584
  shallow, 584
  simulation model, 585
  subsurface/groundwater, 584
    reduced peak flows from, 585
  surface, 584
    higher peak flows from, 585
  topsoil texture and, 585
Drainage, soil salinity management and conditions, 588
  requirement
    saline soils, 588–589, 589f
    sodic soils, 589
  system design, 589–590
    drainage wells, 590
    drain depth, 590
    drain spacing, 590
    relief drains, 590
    saline seeps, 590
Drainage lakes, 1577
Drainage water, 1547
Drainage wells, 590
DRC-1339 (3-chloro-*p*-toluidine hydrochloride) (Starlicide®), 413
Dredging, 1592t, 1692, 2315t
Drinking water, 2790–2800
  analysis, microbiological research methods for, 2799–2800
  bacteria and pathogens in, characteristics of, 2797–2799, 2799t
  inorganic components and pollutants of, 2791–2793, 2791t, 2792t
  and inorganic pollutants, 2120
  intake in dairy, 1663, 1664t
  organic pollutants of, 2793–2797, 2795t, 2796t, 2797f
  requirements of, 2791
  water treatment systems, 539, 2062–2063
    removal efficiencies, 2066–2067
Drip irrigation, 2704
  systems, 2677
Drip irrigation system, 1542–1544
  hydraulic design of, 1542–1543
  for optimal return, water conservation, and environmental protection, 1543–1544
  uniformity of water application and design considerations, 1542
Driving force–pressure–state–impact–response (DPSIR) matrix, 2385, 2385f
Dry adiabatic lapse rate, 1129
Dry ash free (DAF) basis, 714, 717, 719t
Dry ash gasifier, 2491
Dry deposition, 149
Dry loss (DL), 717, 718t
Dry matter intake (DMI), 1663
Dry reservoirs, 1588. *See also* Reservoir(s)
Dry techniques and SO_2, 2442
  circulating fluid bed dry scrubber, 2442
  magnesium oxide process, 2443
  sodium sulfite bisulfite process, 2442–2443
Dual fluidized-bed gasifier (DFBG), 2493
Dual-path-type in situ system, for monitoring stack gases, 145–146, 145f
Dubinin–Radushkevich isotherm, 65. *See also* Isotherm(s)
Duckweed (*Lemna gibba*), 426
Duct sorbent injection, 2442
Dust, 149
Dug wells, 1276
Durum wheat (*Triticum aestivum* L.), 425
Dusicyon culpaeus, 2596
Dust, windblown, 1019, 1019f
Dust Bowl of the Great Plains, 1010
Dust Production Model, 1025
Dust storms, 1010
  in Asia, 1020
  in Bodele Depression, 1020
  erosion by, 1035
DustTrak, 1024, 1024f
Dyer's woad (*Isatis tinctoria*), 1527t
Dye-sensitized cells (DSCs), 234–239
  advantages, 238–239
  electron process, 237f
  mechanism of operation, 236–238
  photogeneration of charge carriers in, 236f
  principles, 235
  structure, 235, 235f
  titania solar cells, 238
  titanium dioxide, 235–236

E

Earias insulana, 2021
Early site permit (ESP), 1789
Earth
  coupled systems, 1363
  energy budget, 242f
  heat exchangers, types of, 1365–1366, 1366f
  nanoparticles beyond, 1716
Earth Observing System (EOS), 554, 2818
Earth–sun energy balance, 771, 772f
  and global warming, 772
Earth system response (global climate change)
  attribution of, to human influence, 1199, 1199f
  climate forcings and feedbacks, 1194–1197, 1194f, 1195f
    atmospheric aerosol forcing, 1196, 1197t
    greenhouse gas forcing, 1195, 1196t
    land-use change forcing, 1196–1197
    natural climate forcings (solar and volcanic variability), 1194–1195, 1196t
  future projections, 1199–1200, 1200f
  human-induced, evidence of, 1197–1198, 1198f
  impacts on natural systems, 1192
  and natural greenhouse effect, 1192–1193, 1193f
ECD. *See* Electron capture detector (ECD)
Ecdysone, 1465
Ecdysone agonists (EAs), 1460, 1465–1466
  chromafenozid, 1466
  furan tebufenozide, 1466
  halofenozide, 1466
  methoxyfenozide, 1466
  tebufenozide, 1466
Ecdysteroids, 1465
ECETOC, 1770
Echinochloa crus-galli (Barnyard grass), 426
ECMs. *See* Energy cost measures (ECM)
Eco-exergy, 594, 604, 778, 780–787, 787f, 1096
  to emergy flow, ratio of, 594–595
  illustrative examples of, 787–788
Ecofiltration, constructed wetlands and, 2154–2155
Eco-filtration systems, 2154–2155, 2730
Ecological approach, of rehabilitation, 1684–1685, 1685t
Ecological economics
  Ems-axis, lower saxony, growth and booming region, 196–199
  renewable energy, 186–195
  renewable energy in Germany and planned nuclear exit, 195

Ecological footprint, 634, 2467–2478
applications of
business, role of, 2477–2478
products and services, 2477
terrestrial systems, 2476–2477
fundamentals of, 2468–2469
limitations of, 2478
of nation, 2469–2471
of product, 2471–2473
three-dimensional geography of, 2475–2476
weakness of, 2478
Ecological infrastructure management, 1509
Ecological pest management, 1930–1932
diversity for croping systems, 1931
environmental and economic benefits, 1932
overview, 1940
risks in adopting, 1932
spreading practices of, 1930
strategies, 1930
Ecological risk assessment (ERA), 1992–1993, 1993f
Ecological sustainability filter (EcSF), 639
Economically optimum N level (EON), 1302
Economic capital, 2465
Economic costs
pesticide poisonings, 1416
Economic growth, 614–622
economists debate against, 616
employment and, 619–621
history of, 614–615
low/no-growth scenario, 616–622
business-as-usual scenario, 617–618, 621t
funding public services in, 621–622
high-level structure of, 617f
policy directions for, 618–622
as policy objectives, 615
Economic incentives, 923–924
Economic loss
virus and viroid infections, 2111
Economic pressure
integrated farming, 1507
Economics considerations
pest management and, 2011
Economic sustainability filter (EnSF), 639
Ecosystem(s), 333
acidic deposition impacts on, 22–24
closed, 1935
development of, 595
estuarine, environmental assessment of, 1064–1069
functioning of, 1068
global distribution of organic matter, 1851–1854
health and population, 595
and humans, relation between, 635, 636
open, 1935
planning, 632–640
challenges to, 640
context of, 632–633
defined, 633
landscape planning, 637–638
perspectives of, 639–640
term uses of, 633–637
urban planning, 638–639
services, holistic interpretation of, 595–596
structure of, 1068
Ecosystem functioning, of Alexandria Lake Maryut, 167, 170–173
Ecosystem health, 599
axioms of, 601t
basic components of, 600–601
definition of, 600–601
ecological indicators, basic features of, 599–600
health indicators in
agroecosystem health, 606–607
aquatic ecosystem health, 607
forest ecosystem health, 607
urban ecosystem health, 607–608
indication of, basic requirements for, 601
management, indicator application and aggregation in, 608–609
utilized indicators of, 601
aggregated theory-based indicators and indices, 604
ecosystem analytical indicators and indices, 604–605
ecosystem service indicators, 605–606
species- and community-based indicators and indices, 601–604
Ecosystem models, 1867
Ecosystem restoration task force model, 628
Gulf Coast Ecosystem Restoration Task Force (GCERTF), 629–630
South Florida Ecosystem Restoration Task Force (SFERTF), 629, 630
Ecosystem services, 334, 557–558. *See also* Biodiversity
degradation of, 335
restoration, 335–337
Ecosystem services, 604–605, 605t
Ecosystem services valuation (ESV), 634
Ecotechnology, 2150, 2152–2153
and climate change, 2156
and developed countries, 2157–2158
and developing countries, 2158–2160
history of, 2153–2154
and India, 2160–2162
potential and scope of, 2730–2731
scope of, 2153
Ecotones, 2824
Ecotoxicity, 1704
elements with, 1455, 1457t
by inorganic compounds, 1455–1458
EC Regulation 2078/92, 1108
Ectoparasitoids, 1747, 2424
Edaphic deserts, 566
EDC. *See* Endocrine disruptor chemical (EDC)
Edge-defined film-fed growth (EFG) method
solar cell production from, 232
Edwards aquifer (of Texas), 1284, 1285, 1285f, 1287f
Effect summation, 2574–2575
Egg parasitoids, 1746, 2424
Egypt, 2682, 2689
EI_{30}, rainfall erosivity index, 943
Ejector scrubber, 155
Electric air conditioning, 793
Electrical conductivity, 1547, 1559, 1568, 1569t
Electrical demand
and thermal storage, 2511–2514
Electrical resistance space heater, 1086
exergy efficiency of, 1086
Electric heat pump, energy flow of, 1352, 1352f
Electric utility system, 1418
operation, dynamics of, 1418–1419
Electricity, 799
generation, 791–792, 792t
methods, estimated costs of, 122t
technologies for, 773t, 774t
Electricity sector
customers in, 1493–1494
electric utilities in, 1493–1494
integrated systems methodology in, 1494, 1504f. *See also* Integrated energy systems, case study from ISU
lowest cost to consumer in, 1506
questions related to, 1504
savings potential in, 1504–1505
investments in, 1493, 1494
problems of, 1493
Electric power industry, 792
Electric Power Research Institute (EPRI), 1652
Electric utilities, 1493
"Electrolyte effect," 2370, 2371
Electromagnetic radiation, 1192
Electromagnetic radiation, spectrum of, 1883f
Electron capture detector (ECD), 1203, 1975
Electron–hole pair, generation of, 227f
Electronic waste, 1431
Electrostatic precipitator (ESP), 153–154, 1128
Elk (*Cervus elephus*), 2596
Elodea nuttallii, 294
Elovich equation, 67
Eluviation/illuviation, 1627
environmental implications, 1627–1628

Embodied energy, of farm chemicals, 121, 121t
Emergy, 593, 604, 1097
to eco-exergy, ratio of, 594–595
Emergy–exergy ratio for flow, 1097
Emergy-to-money ratio (EMR), 596
Emissions, 916
nitrogen in fuels, 728–729, 729f
reduction mandates, 919
reduction of, 483–484, 485
tax on, 919
Emissions reduction units (ERU), 486
Emissions trading (ET), 486
Emissions trading system (ETS), 483, 484, 487
Emission uniformity, 1548–1549
EMP. *See* Energy master planning
Empirical model
Empirical models, 974
advantages of, 974
development of, 974–975, 976t
in pest management strategies, 1949
for water erosion. *See* Water erosion models, empirical
for wind erosion. *See* Wind erosion equation (WEQ)
Empirical water quality models, 2749
Employment, 619–621
EN 14791:2006. Stationary source emissions, 2440
Encapsulation, 2510–2511
Encarsia formosa, 1475
Endangered Species Act (ESA), 1942–1943
Endocrine Disrupter Screening and Testing Advisory Committee (EDSTAC), 647–648
Endocrine Disrupters Testing and Assessment (EDTA) Task Force, 647
Endocrine-disrupting chemicals (EDC), 643–652, 1405, 1942
animal exposure effects of, 650–651
emission of, 651–652
hormones effect on, 645–646
human exposure effects of, 648–650
mechanisms of action, 644–645
regulatory purposes, guidelines for, 647–648
reproductive toxicity of, 646
risk assessment of
using biomarkers, 647
using screening assays, 646–647
Endocrinological effects, of arsenic toxicity, 1273
Endocrine Disruptor Screening Program (EDSP), 647, 648
End-of-life tires (ELTs), 896, 909–910
management of, 896
End-of-life vehicles (ELVs), 896, 897, 902–903, 911
reuse/recovery rates of, 903, 903f, 903f
Endoparasitoids, 2424
Endosulfan, 575, 2016t, 2019, 2142
Endothall, 1318t
Endothermic reaction, 2527
δ-endotoxin, 2029
Endrin, 1318t
ENERCON, 195
Energetic reinjection ratio, 1173
Energetic renewability ratio, 1173
work, 809, 809f, 811
Energy, 770, 808, 809f, 810, 811, 821
additional, 770. *See also* Additional energy
carriers of, 770
conservation, first law of, 813–816
degradation, second law of, 816
efficiency, 775. *See also* Energy efficiency
entropy and second law of thermodynamics, 817–819
forms of, 770
life cycle of, 773–774, 773t, 774t
forms and classifications, 811–813, 812t
heat, 809–810, 811
heat engines, 819
exergy and the second-law efficiency, 820
impact on environment, 1095
natural, 770. *See also* Natural energy
renewable, 774
reversibility and irreversibility, 816–817
resources, 770
selection, 774
environmental considerations in, 774–775
solar, 771–772
sustainability, 776
and sustainable development, 798–799
use of, 774–775
Energy audits, 118
assessment process, 118–119
level 1, 118
level 2, 118
level 3, 118
Energy balance of collector, 245–246
collector efficiency curves, 245–246
instantaneous efficiency of a collector, 246
Energy carriers, 685, 770
vs. energy sources, 770
Energy conservation
and efficiency improvement programs, 684
environmental problems, 680–682
potential solutions, 682
example, 689
first law of, 813–816
implementation plan, 686–687
importance of, 689
improvement factors of, 685t
life-cycle costing in, 688–689
measures, 687–688
elements, 687
evaluation of, 642
practical aspects of, 682–683
flow chart, 684t
programs, 677
renewable energy resources and, 685–686
research and development (R&D) in, 683
sectoral, 688
and sustainable development, 683–686
technical limitations on, 683
technologies, 682
world energy resources, 678–680
Energy consumptions circuit exergetic ratio, 1175
Energy conversion
biomass and hydrogen energy cycles (comparison), 716, 716f
carbon-di-oxide emissions per year, 716f
char reactions, 724–725
cofiring (coal and bio-solids)
coal agricultural residues, 727–728, 727f
coal and manure, 728
coal and RDF, 728, 728t
conversion schemes, 727
fouling in, 729–730, 730t
NOx emissions, 728–729, 729t
combustion
circulating fluidized bed combustor (CFBC), 727
defined, 723, 724f
fixed bed combustor (FXBC), 726
fluidized bed combustor, 726–727
ignition and, 725
in practical systems, 725–727, 726f
stoker firing, 726, 726f
suspension firing, 725, 726f
energy units and terminology, 715t
fuel properties
gaseous fuels, 722–723
liquid fuels, 722
solid fuels, 717–722, 718t, 719t, 720t, 722f, 723f
FutureGen layout, 731–732, 732f
gasification, defined, 723, 724f
gasification of coal and bio-solids, 730–731, 731f
ignition, 725
objectives of, 714–717
pyrolysis, 723, 724f
reburn with biosolids, 732–733, 732f, 733t
technologies, 774
volatile oxidation, 723–724
Energy cost measures (ECMs), 662, 663t
continuous commissioning as, 662

Energy curtailment, 682
Energy degradation, second law of, 816
  reversibility and irreversibility, 816–817
  entropy and second law of thermodynamics, 817–819
Energy driven phase, evaporation, 130
Energy efficiency, 682, 775
  methods for improving of
    advanced energy systems, use of, 775
    audits, on irrigation systems, 120
    building envelopes, improving of, 775
    energy leak and loss prevention, 775
    energy storage, 775
    energy supplies and demands, matching of, 775
    examples of, 776, 776t
    exergy analysis, use of, 776
    improved monitoring, control, and maintenance, 775
    passive strategies, use of, 775
    use of high-efficiency devices, 775
Energy efficiency ratio (EER), 1364
Energy engineering for developing countries, 2465
Energy flows, 862
Energy Information Administration (EIA), 790
Energy intensity, 774–775
Energy management, 764
Energy Management System (EMS), 735
Energy master planning (EMP)
  American approach, origins of, 765
  business as usual, improving, 765
  optimization, 764
  steps to, 765–766
    business approach development, 766
    casting wide net, 767
    communicate results, 768
    energy team, creation of, 767
    ignite spark, 766
    obtain and sustain top commitment, 766
    opportunity recognition, 766
    organization's energy use, 767
    set goals, 767
    upgrades, implementation of, 768
    verify savings, 768
  tips for success, 768–769
  unexpected benefits of, 763–764
  vs. energy management, 764–765
Energy Master Planning Institute (EMPI), 765, 766t
Energy production and consumption, impacts of, 915
Energy report cards, 768
Energy return factor (ERF), of photovoltaic systems, 215
Energy security, 799–803
  risk assessment, 800–801
    depleting oil reserves, 801–802
    supplies from Middle East, 803
Energy Standard for Buildings Except Low-Rise Residential Buildings, 1656
Energy Star program, 1655
  for windows, 867, 867f, 867t
Energy storage
  benefits, 1420–1421
  technologies, 1419–1421
    development status, 1421t
    types, 1421
Energy transfer, 808, 809
  and disorganization, 816–817
  versus energy property, 811–813
  reversible, 819, 821
Energy units and terminology, 715t
Energy use, 118. *See also* Energy audits
  commercial, 793
  for different crops in different countries, 120–121, 121t
  distribution of, by industry groups, 795f
  electricity generation, 791–792, 792t
  farming systems, effect of, 119–121
    conservation farming practices, 119–120
    irrigation methods, 120
    machinery operation, 119
  and greenhouse gas emissions, 118, 120
  industrial, 794–795, 794f
  residential, 793
  saving of, 122
  by sectors, 790–791, 791t
  transportation, 722, 723, 725
  by type of energy, 791, 794f
  US overview, 790
Energy utilization and environment, 683
Energy Valuation Organization (EVO), 1506
Engineered nanoparticles, 1721
Engineered terraces, 2538–2539
Engineering approach, of rehabilitation, 1683–1684
Enhanced Thematic Mappers (ETM), 2817t, 2818
Enhydra fluctuans (Water cress), 426
Entamoeba histolytica, 2696
Enterobacter, 321
Enthalpy, 2526
  of formation, 2526–2527
  of reaction, 2526–2527
Enthalpy change of combustion, 2527
Entomobyroides dissimilis, 1879
Entomopathogenic nematodes, 382
Entomopathogenic viruses, as bioinsecticides, 382
Entrained bed type gasifiers, 2490–2491, 2491f
Entropy, 604, 821, 1097, 2529
  and exergy, 816–819
  generation, 816–817
  and second law of thermodynamics, 817–819
Environmental costs and regulations, 1793–1795
Environment 2010: Our Future, Our Choice, 136
Environmental compatibility value (ECV), 1986
Environmental contaminants, 2136
Environmental contamination, 2584
Environmental Defense Fund, 924, 2010
Environmental degradation, 2307
Environmental degradation and national security, 2461–2462
Environmental genotoxicity, 2124–2125
Environmental health, 336
  environmental conditions and, 336
  importance of, 336–337, 336f, 337f
  indicators of improvement of, 337
  steps to be implemented for, 337
  tools for, 337–338, 338f
Environmental interest
  elements of minor, 1455
Environmental issues
  tailings and, 1683
Environmental Kuznets curve (EKC), 1221
Environmental legislation in Asia, 874, 876
  biodiversity laws
    enforcement, 881–882
    regional cooperation, 882
  biodiversity laws at national level, 877
    Cambodia, 881
    China, 877
    Hong Kong, 879–880
    Indonesia, 878–879
    Japan, 878
    Korea, 877–878
    Lao People's Democratic Republic, 880
    Malaysia, 878
    Philippines, 879
    Singapore, 879
    Thailand, 880
    Vietnam, 881
  development in international law and its consequences on national legislations, 877
  environmental assessment law
    Cambodia, 887–888
    China, 885
    Hong Kong, 886
    Indonesia, 886
    Japan, 886
    Lao PDR, 887
    Malaysia, 886
    Philippines, 887
    Republic of Korea, 886
    Singapore, 887
    Thailand, 887
    Vietnam, 888
  e-waste management law, 882
    Cambodia, 884
    China, 882–883
    Hong Kong, 883

implementation and enforcement, 885
Japan, 883
Lao People's Democratic Republic, 884
Malaysia, 883
Philippines, 883–884
Republic of Korea, 883
Singapore, 884
Thailand, 884
Vietnam, 884
e-wastes and environmental assessment, 882
law on biodiversity conservation, 876
transboundary movements of hazardous waste, 884
Environmental legislation, to control soil erosion, 970–971
Environmentally relevant phenomena, 353, 354f
Environmentally sound technologies (EST), 2671
Environmental management, 1242
systematic approach, 1242–1243
quality characterization, 1245
standard protocols, 1243
uncertainty assessment, 1243
Environmental matrices, chemical analyses in, 1065
Environmental policy
burden of proof, 916–917
common property resource destruction, 914
economic framework
cost benefit analysis (CBA), 917–918, 917f, 918f
cost-effective analysis, 918–919
impacts of energy production and consumption
air, 916
land use, 916
water, 916
multidisciplinary approach, 914–915
policy instruments
emission reduction mandates, 919
intrinsic/nonuse benefits, 920
liability rules, tightening, 919
redistributive effects, 920
renewable energy resources, 920
specific control technology, 919
sustainable development, 920
tax on emissions, 919
tax on polluting good, 919
tradable emissions permits, 919
pollution, 914
Environmental policy innovations (EPI), 922–924, 925t
economic incentives, 923–924
nature of, 923–925
civic environmentalism, 924
policy entrepreneurs, 923
"policy reinvention," 923
resources, neds, politics, and determinants
institutional factors, 925–926
interest group support, 926–927
need/problem severity, 925
regional diffusion, 927
state policy innovations, 924
Environmental pollution, 1654, 2123–2130. *See also* Pollution
biomonitoring
animals, 2129
human, 2129
plants, 2128
genotoxicity
environmental, 2124–2125
human, 2124–2125
pesticide, 2125–2128
by nanomaterials, 2129
reproductive toxicity, 2124
Environmental pollution, by pesticides, 2013–2014. *See also* Pesticides
mitigation of, 2018
CW microcosms, use of, 2020–2021
natural products for pest control, use of, 2021–2022, 2021f
slot-mulch biobed systems, use of, 2020
soil microorganisms activity, enhancing of, 2020
SOM content, increasing of, 2018–2019
pesticide residues detection and measurement of, 2015–2017
cleanup, 2018
confirmation, 2018
extraction, 2018
quantification, 2018
sampling, 2018
Research Farm, agricultural practices at, 2022–2024
Environmental problems
nitrogen leaching and, 1761–1762
Environmental Protection Agency (EPA), 133, 135–136, 629, 1118, 1363, 1394, 1405, 1570, 1571, 1612, 1793, 1940–1946, 1961, 1991, 2010, 2805, 2809–2810, 2811, 2812, 2813
DNT guidelines of, 1757
drinking water standards, 1730
Endocrine Disrupter Screening and Testing Advisory Committee, 647–648
Endocrine Disruptor Screening Program, 647, 648
Guidelines for Drinking Water Quality, 2793
mutagenicity testing battery, 2124–2125
on total Cr in drinking water, 477
Environmental Protection Agency (US), 919
Environmental Quality Indicators Program (EQIP), 127
Environmental risk assessment, 1723
Environmental safety
herbicides and, 1380–1381
Environmental services programs, payments for, 1773
Environmental simulation models
for surface water quality, 2761–2762
Environmental state, 1084
Environmental stewardship, 57
Environmental Working Group, 1404
Environment Protection Agency (EPA), 1609, 1611
Environomics, 1097
Enzymes, function of, 89
Enzyme variation
and pesticides, 1400
EOS. *See* Earth Observing System (EOS)
EPA. *See* Environmental Protection Agency (EPA)
EPA Review of Chlorpyrifos Poisoning Data (1997), 1411
EPIC (Erosion Productivity Impact Calculator) model, 976
Epidinicarsis lopezi, 1475
Epofenoname, 1464
EP409.1 Safety Devices for Chemigation, 471
Epworth Sleepiness Scale (ESS) scores, 2875f
EQIP. *See* Environmental Quality Indicators Program (EQIP)
Equilibrium models, for exergy analysis, 1093
Equine intoxication, 428
Equine leukoencephalomalacia (ELEM), 1697
Equity and fairness, 2785–2786
Eroded sediments
agricultural runoff and, 82
Erodibility, soil, 980–989
effect of particle travel rates on, 986–987, 986–987f
Erosion, 2754
and carbon dioxide, 934–937
conceptual mass balance model, 935–936, 935f
emissions, estimates of, 936, 936t, 937f
SOC, 934–935
carbon losses from soil and, 1857–1858
continuous simulation models, 991–992
definition of, 943
effects of rain on, 945
precipitation and, 945. *See also* Precipitation
primary agents for, 943
process-based models, 991–992
and rainfall erosivity, 943
snowmelt. *See* Snowmelt erosion
by water, 943, 991–992
wind, 943, 945. *See also* Wind erosion
Erosion, accelerated, 1044–1048
deposition, 1046–1047

Erosion, accelerated (*cont'd.*)
gullies, 1046, 1046f
modeling, 1048
overview, 1044
processes, 1047–1048, 1047f, 1048f
rainfall, 1044–1045, 1045f
rills, 1045–1046, 1046f
transport, 1045, 1045f
Erosion by water, 951
future of soil, 955
direct effects of future climate change, 955
indirect effects of future climate change, 955
global problem of accelerated, 954
impacts of, 954–955
processes, 951–953
spatial and temporal scale, 953
Erosion control, 1040–1043
mulch tillage, 1042, 1042f
no-till, 1040–1041, 1041f, 1042f
overview, 1040
ridge tillage, 1041–1042, 1042f
strip tillage, 1042–1043
tillage, 1040, 1041f
on soil properties, 1040
EROSION-2D/3D, 1958, 2761
Erosion hazard maps, 1037
Erosion pins, 1021, 1021f
Erosion plots, 974–975, 978
Erosion problems
historical review, 1050–1052
deforestation, 1050, 1051f
sedimentation rates, 1050–1052
Erosion-Productivity Impact Calculator (EPIC) model, 2761
Erosivity, rainfall, 980–990
Error analysis, groundwater modeling, 1299
E-sampler, 1024
Escherichia coli, 1626, 2696
ESP. *See* Exchangeable sodium percentage (ESP)
EST. *See* Environmentally sound technologies (EST)
Esteem. *See* Pyriproxyfen
Estimation of Ecotoxicological Properties (EEP), 332
Estrogen-like substances, 281
Estuaries, environmental problems of, 1063–1070
assessment of, 1064–1069
bioassay methods, 1067
biological responses at community structure level, 1066–1069
chemical analyses, in environmental matrices, 1065
tools for, 1068–1069
origin of, 1063–1064
solving, 1069–1070
Estuarine and marine ecosystems, 607
Estuarine intertidal emergent wetlands, 2827
Estuarine intertidal unconsolidated shore, 2827
Estuarine quality paradox, 1065–1066
Estuarine turbidity maximum (ETM), 1063, 1064
Ethanol, from biomass, 377–378
Ethinylestradiol, 1405
Ethiopia, 2683
Ethylene dibromide (EDB), 1317, 1318, 1318t, 2015
Ethylene thiourea (ETU), 1394
Etoxazole, 1463
ETU. *See* Ethylene thiourea (ETU)
Eucalyptus spp., 177
Euhrychiopsis lecontei, 367
Eulerian–Lagrangian Localized Adjoint Method, 1316
Eulerian methods, 1315
Euphorbia esula, 2822
Eurasian skylark *(Alauda arvensis)*, 416
Eurasian sparrowhawk *(Accipiter nisus)*, 417
Eurasian watermilfoil. *See Myriophyllum spicatum*
Europe
acid rain in, 8
control policy, 15
natural erosion across, 2378
soil degradation in, 2377–2378, 2377f
distribution of, 2378f
trends in acidity, 9, 11–13
European agroecosystems, bioindicator development in, 356, 357t–358t
European carp (*Cyprinus carpio*), 2603
European Commission (EC), 215, 528, 1406
Directive 96/62/EC, 138
Directive 99/30/EC, 138
Directive 2000/69/EC, 138
Directive 2002/134/EC, 138
Waste Framework Directive, 2399
European Community (EC)
Directive 76/403/EEC, 2181
Directive 96/59/EC, 2181
Guidelines for Drinking Water Quality, 2793
Water Framework Directive, 1064
European rabbits (*Oryctolagus cuniculus*), 2596
European red foxes (*Vulpes vulpes*), 2596, 2604
European Soil Erosion Model (EUROSEM), 966, 2761
European Union (EU), 355, 1612, 1613
Landfill, 516
organic forming policy guidelines of, 528
Soil Thematic Strategy, 2375
Waste Management Directives, 516
Working Document on Biological Treatment of Biowaste, 513
European Union (EU) Drinking Water Directive, 424
European Union emissions trading system (EU-ETS), 487
European Union's REACH legislation, 284–285
European Water Framework Directive, 1961
EUROSEM model, 985–986, 992
Eutrophication, 494–495, 538, 1074, 1456, 1559, 1578–1579, 2093–2094, 2102–2103, 2137, 2360, 2753
model framework, 1585
phytoplankton, growth of, 1075–1078
problem, 1074–1075
solutions to, 1078–1079
Evacuated tube collectors, 244, 244f
Evacuated tube panels, 242, 244f
Evapoconcentration effect, 1559
Evaporation, 1564t
diffusion driven phase, 130
energy driven phase, 130
Evapotranspiration (ET), 1184, 1547, 1549, 1559, 1566
Event erosivity index, 982
Event sediment yield, 982
Everglades (of Florida)
Comprehensive Everglades Restoration Plan (CERP), 1080, 1081
features of, 1080
hydrologic modifications in, 1080
water management in
environmental issues, 1080–1081
history of, 1080
restoration, 1081
Everglades Agricultural Area (EAA), 1080
Evergreen field, defined, 2762
Excavation
of inorganic contaminants, 2120
Exchangeable Na ions, displacement of, 2370, 2371f
Exchangeable sodium content (ES), 2335, 2336, 2337f
Exchangeable sodium percentage (ESP), 1693, 2335, 2335f, 2336, 2337, 2337f, 2345
Exchangeable sodium ratio (ESR), 2335
Exergetic reinjection ratio, 1175
Exergetic renewability ratio, 1173
Exergy, 683, 778, 780, 780f, 785f, 821, 1083, 1092–1093, 1176, 2531–2532
advantages over energy analysis, 1083
analysis, 1092, 1093. *See* Exergy analysis
in environmental policy development, 1097
reference environment in, 1093–1095
for sustainable development, 1097
applications of, 1099–1100
macroinvertebrate communities, 1100
water bodies, 1100

balances, 1084
in combustion process, 1093, 1093f
consumption, 1084
defined, 1083, 1084
definition of, 1084
and economics, 1089
and ecosystems, 1095–1097, 1100
efficiencies *vs.* energy efficiencies, 1091
of emission, 1092
entropy and, 816–819
and environment, 1089
and environmental impact, relations between, 1097–1099
order destruction and chaos creation, 1097–1098
resource degradation, 1098
waste exergy emissions, 1098, 1098f
equation for, 1092
evaluating, 1083
impact on environment, 1095
reducing of, 1097
indices, 604
losses, 1095
vs. energy losses, 1090–1091
and lost work, 2532
method, 1083
of natural environment, 1093
to reduce waste exergy emissions, 1099
and reference environment, 1084
and reference-environment models, 1093
constrained-equilibrium models, 1094
equilibrium models, 1094
natural-environment-subsystem models, 1094
process-dependent models, 1095
reference-substance models, 1094
and second-law efficiency, 820
of stream, 2531
Exergy analysis, 1083, 1084–1085, 1176–1177, 2531
applications of, 1085–1086
and efficiency, 1085
practical limitations, 1085
theoretical limitations, 1085
Exergy-based ecological indicators
biodiversity, 1096
buffering capacity, 1096
dissipation, 1096
ecological process efficiency, 1096
health and quality, 1096
maturity, 1096
optimization, 1096
structure, 1096
Exothermic reaction, 2527
Exposure assessment, 1705
Ex situ techniques, 2170
Extended producer responsibility (EPR), 893
Exterior foam insulation, 1482, 1483f
Exterior insulation finish systems (EIFS), 864
External casing, 242
External combustion (EC), 714
External contamination, sources of, 2264
External costs, 917, 917f
reduction of, 917f
External feeders, 2423, 2424t
Externalities, 634, 636
Externally feeding pests, biocontrol of, 2424–2425
Extraction wells, 1293
Extruded polystyrene (XPS), 1478

F

F. oxysporum f. sp. *udum*, 1925
Fabric Filter (FF), 1128
FACE experiment. *See* Free-Air CO_2 Enrichment (FACE) experiment
"Facilitated transport," 1622
Factory-built modular building, 860
Facultative ponds, 2633–2634, 2636
BOD removal, 2633f
kinetics, 2634
Falcon. *See* Methoxyfenozide
Falco peregrinus (Peregrine falcon), 417
Fallout, radioisotopic, 1024–1025
Fall nitrogen applications, 1808
False negatives, 1721
False positives, 1721
Famphur, 420
FAO. *See* Food and Agriculture Organization (FAO)
FAO International Code of Conduct, 1612
FAO–LADA (Food and Agriculture Organization of the United Nations/ Land Degradation Assessment in Drylands), 558
FAO/WHO Food Standards Program, 1126
Farmed wetland, 2825t
Farm energy calculators, 118–119
Farmer Field Schools (FFS), 1930
Farming practices, sustainable, 353, 353t
potential bioindicators for
arthropods, 357t
birds, 358t
plants, 358t
soil fauna, 357t
soil microbiota, 357t–358t
Farming systems
integrated. *See* Integrated farming systems
organic. *See* Organic farming
Farm machinery, 119
Farm nutrient management plans, 1813–1814
compliance with standards, rules, and regulations, 1814
consistency with, 1814
manure inventory, 1814
manure spreading plan, 1814
nutrient crediting, 1813–1814
on-farm nutrient resources, assessment of, 1813
soil test reports, 1813
Farm subsystem, 1561t
FDA. *See* Food and Drug Administration (FDA)
Fe. *See* Iron (Fe)
Fecal bacteria, 2635
Fecal coliform bacteria, 2753, 2865t
Fecal pollution, 2699
Feces by livestock, 2753
Federal Communications Commission (FCC), 2228
Federal Environmental Pesticide Control Act (FEPCA), 1940
Federal Food, Drug, and Cosmetic Act (FFDCA), 1118, 1940
Federal Insecticide, Fungicide, and Rodenticide Act (FIFRA), 1118, 1612, 1940, 1942, 2125
Federal Pesticide Act, 1612
Federal Plant Pest Act of 1957, 2823
Federal Water Pollution Control Act (FWPCA) (1948), 2808
Feedback effects, 1134
Feedlot biomass (FB), 717, 718t, 729
Feedstock(s), 794
gasification of, 2483t
gasifier, compositions and heating values of, 2484t
Feedstock composition, of composting materials, 523–524, 523t
Fe(III)EDTA, 2245
Feldspar, weathering of, 1625
Feline immunodeficiency virus, 2603
Feline leukemia virus, 2603
Feline panleukopenia virus (FePV)
for biological control of cats, 2603–2604
on Marion Island, 2603
Felis catus, 2596
FEMWATER
for saltwater intrusion, 1330
Fen, 2825t
Fenbutatin-oxide, 3
Fenitrothion, 420
Fenoxycarb, 1461
Fenpropathrin, 3
Fensulfothion, 419
Fenthion (Queletox®), 413, 420
Fermentation, 2152t
Ferrets (*Mustela furo*), 2596
Ferrihydrite, 1689
Ferrocement tanks, 2264
Fertigation, 1809
Fertility, potassium
assessing and managing, 2211
Fertilization
excessive, 1510

Fertilization (*cont'd.*)
  mineral and organic nitrogen, 1770
  and soil IC, 464–465
Fertilizers
  agricultural runoff and, 83
  biofertilizers, 1514–1516
  in coastal waters, 495–496
  consumption and nitrous oxide production, 98, 98t
  environmental problems with, 1516t
  micronutrient, 1514
  mineral, 1512–1513
    nitrogen, 1513
    phosphorus, 1513–1514
    potassium, 1514
  role of, 1512
  sulfur, 2434
Ferula asafoetida, 384
FFDCA. *See* Federal Food, Drug, and Cosmetic Act (FFDCA)
FFS. *See* Farmer Field Schools (FFS)
Fiber-bed scrubber, 156
Fiberglass insulation, 1478
  health impacts of, 1479
Fick's first law of diffusion, 1818
Fick's Second Law, 1452
Field erosion plots, 964–965
FIFRA. *See* Federal Insecticide, Fungicide, and Rodenticide Act (FIFRA)
Fill factor (FF), of solar cells, 230
Filter-based samplers, 1024
Filter(s)
  biofilters, 2724
  biological aerated, 2724
  cartridge, 2723–2724
  disc, 2731
  granular, 2722–2723
  rapid gravity filter with coagulant aid, 2722
  sand
    slow, 2722
    pressure, 2722
  soil escape, 2731
  trickling, 2724–2725
  types of, 2723f
Filter strips, 1553
Filth fly abundance management, in dairies and poultry houses, 1922–1923
Filtration, 1282, 2721–2722
  eco-filtration, 2730
  membrane, 2725–2727, 2725f
    applicability of, 2727
    future developments of, 2726–2727
  ultrafiltration
    performance of, 2727
    system design, 2727
    in water treatment, 2727–2728
Fimbristylis miliacea (Joina), 426
Finite differences (FD), 1314, 1314f
Finite element method (FEM), 1314, 1314f
Finite volumes, 1314, 1314f
Fin-tube heat exchanger, 873
First elements, 1455
First-flush (or foul-flush) device, 2264
First law of thermodynamics, 2525–2528
  formulation, 2526
Fish catch, 1593t
Fish depletion, 1579–1580
Fishery management
  eutrophication and, 2102
Fission, 2235
  radionuclides, 2237
Fitch, Asa, 2009
Fixed bed combustor (FXBC), 726
Fixed-bed gasifiers, 2490–2491, 2491f
Fixed carbon (FC), 717
Flame temperature, 722
Flash pyrolysis, 2487
Flat plate collectors, 242
Flat-plate solar collector, 242–243
Flatwood wetland, 2825t, 2826f
Flavonoids
  chemical composition and structure of, 2082t
  general structure of, 2081f
Flaxseed, for bio-liquid fuels, 846
Flea beetles, 444
Floating fern (*Salvinia molesta),* 1527t
Float zone method, 231–232
Flocoumafen, 420
Flood control reservoir, 1577
Flooded soils
  N_2O emissions, 97
Floodgates, 209
Flooding, 1184
Flood irrigation, 2676
Floor insulation, 1482
  raised floor, 1482
  slab-on-grade, 1482
Flow-driven saltation (FDS), 987, 989
Flow-through reservoirs, 1588. *See also* Reservoir(s)
Fluazuron, 1461
Flucycloxuron, 1461
Flue gas cleaning, of sulfur dioxide, 158
Flue gas desulfurization (FGD), 2440
Flue gases, concentration in SO_2, 2440
Flue gas volume, 722
Flufenoxuron, 1461
Flufenzine, 1463
Fluid bed dry scrubber, 2442
Fluidized bed combustor, 726–727
Fluidized-bed extraction (FBE), 2197–2198
Fluidized-bed gasification, 731f
Fluidized bed gasifiers, 2491–2492, 2491f
Fluorescent pseudomonads, 382
Fluorine, 1887
Flushing, 1592t
Flushing manure, 1664
Flush systems, 1665, 1665t
Flux chambers
  GHG measurement, 1202, 1203, 1204f
Fly ash, 426, 428, 1432
Foam insulation strategies, 1482
Foam products and chlorofluorocarbons (CFCs)
  health impacts of, 1479
FOB. *See* Functional observational battery (FOB)
Foliar pathogen management, 444
Fonofos, 419, 1318
Food
  tolerance limits
    for natural or unavoidable defects in, 1610t
    for pesticide residues in, 1610t
Food and Agricultural Organization (FAO), 28, 1615, 1661, 1930, 1932, 1986, 2005, 2034
Food and Drug Administration (FDA), 1120, 1401, 1609, 1611, 2051
  on insect fragments in processed foods, 1609
  on pesticide residues in processed foods, 1609
Food chain
  arsenic in, 1274, 1275f
Food, Conservation, and Energy Act (2008), U.S., 1961
Food contamination
  with pesticide residues. *See* Pesticide residues, food contamination with
Food industry sludge, 2267t
Food laws and regulations, 1609–1611. *See also* Food and Drug Administration (FDA)
  food manufacturers responsibility, 1610–1611
  potential consumer benefits, 1611
  tolerance limits
    for insect fragments, 1609–1610
    for pesticide residues, 1609
Food manufacturers
  responsibility of, 1610–1611
Food processing
  pesticide residue and, 1125
Food Quality Protection Act (FQPA), 284, 1118–1119, 1613, 1941, 2010, 2028, 2750
  consumer right-to-know, 1119
  endocrine disruption, 1118–1119
  overview, 1118
  potential impacts, 1119
  "risk cup," 1118
  tolerances, 1118
Food Quality Protection Act of 1996, 1416, 1609
Food residues, 1403
Food Safety and Inspection Service (FSIS), 1609
Food sources
  alternative, natural enemies and, 371

Page numbers followed by f and t indicate figures and tables, respectively

Foodstuff, aluminum in, 270, 270t
Food-to-microorganism ratio, 2649–2650
Food waste pyrolysis, 847
Food-web models, 1864
Forage crops, irrigation of, 1664, 1665t
Force, 809f
Forced circulation systems, 247, 248f
Forebay tank, 210
Forest ecosystems
  acidic deposition and, 23–24
  health, indication of, 607
Forest management
  benefits of herbicides to, 1383
Forest planning, 636
Forestry insecticides, 420
Forest Stewardship Council, 1657
Formula of Manning, 205
Fossil fuels, 226, 345, 772
  combustion, 1127–1136, 1195
  depletion of, 180–181
    data and predictions, 180f
    oil, delivery and detection of, 182f
    reserves/resources, regional distribution of, 180, 181f
  as energy source, 679
  environmental effects, 2461–2462
  environmental effects of, 677, 680
  production and consumption, 679, 681
Fouling in cofiring, 729–730, 729f
Fourier Transform Infrared (FTIR), 1205
Four-way exchange valve, 1350–1351
Foxes (*Dusicyon griseus*), 2596
FQPA. *See* Food Quality Protection Act (FQPA)
Fraction of water, 1547
Framboids, 38f, 42
Fraxinus pennsylvanica, 2830
Free-Air CO_2 Enrichment (FACE) experiment, 1190
Free-water surface (FWS) wetlands, 2863f, 2864f, 2865
Frequency-domain electromagnetic methods (FDEMs)
  for saltwater detection, 1327
Freshwater pollution, 2303
Freshwater/saline lakes, 1577–1578
Freundlich isotherms, 64–65, 2427. *See also* Isotherm(s)
Friction period, 1638
Friedman, Thomas, 1503
Frit fly, 1116
Fruits, pesticides in, 1972–1974, 1973f, 1973t
FSIS. *See* Food Safety and Inspection Service (FSIS)
FTIR. *See* Fourier transform infrared (FTIR)
Fuel cell, 1145
  advantages of, 1145
  intermediate- and high-temperature, 1138
    applications of, 1138
    high-temperature proton exchange membrane fuel cells, 1138–1139
    molten carbonate fuel cells, 1140
    operational characteristics and technological status of, 1143t
    phosphoric acid fuel cells, 1139
    solid oxide fuel cells, 1141–1143
  low-temperature, 1145–1146, 1152t
    alkaline fuel cells, 1146–1148
    direct methanol fuel cells, 1151–1152
    proton exchange membrane fuel cells, 1148–1151
  stack, 1145
  types of, 1145–1146
  use of, 1145
Fuel properties
  gaseous fuels, 722–723
  liquid fuels, 722
  solid fuels, 717–722, 720t, 722, 723
Fuels
  bio-liquid, 846–847
  classification of, 714
  in groundwater, 1281
  properties, 832t
  solid, 830–834
  sources, 831t
The Fukushima (Japan), 183
Fukushima nuclear disaster impact, on nuclear power cost, 1803–1804
Full cost prices, 2784, 2784f
Full economic cost, 2784
Full supply cost, 2784
Fully extractive system, for monitoring stack gases, 144, 144f
Fulvic acids, 1570, 1625
Fumazon, 288
Fumigants, 1318, 2016t
Functional Observational Battery (FOB), 2006
Funding public services, in low/no-growth economy, 621–622
Fungi, as mycoherbicides, 383
Fungi, biological control by, 2107–2110
  application, 2109
  mechanisms, 2107–2109
Fungicides, 2014, 2029, 2115t
  derived from nature, 2029–2030, 2030f
  effects on natural enemies, 370
  used in homes and gardens, 1401
Furan tebufenozide, 1466
Furnace sorbent injection, 2442
Furrow irrigation, 2676–2677
Furrow-irrigation erosion, 1552
Furrows, 50
Fusarium crookwellense, 1696
Fusarium culmorum, 1696
Fusarium graminearum, 1696
Fusarium head blight, 1696
Fusarium kernel rot, 1697
Fusarium oxysporum, 2108t, 2109
Fusarium oxysporum f. sp. *conglutinans*, 444
Fusarium proliferatum, fumonisins by, 1697
Fusarium solani f.sp. *phaseoli*, 1878
Fusarium verticillioides, fumonisins by, 1697
Fusegates, 209
FutureGen process, 731–732, 732f
Fuzzy mathematical programming (FMP), 107
FWS wetlands. *See* Free-water surface (FWS) wetlands

G

γ-Al_2O_3, 1712
Gallium arsenide (GaAs) solar cells, 232
Gamma (γ) radiation, 2235
Ganga–Meghna–Brahmaputra (GMB) Plain
  arsenic contamination in, 1262
GAP Analysis Program, 2272
Garden insecticide, 1318
Gas absorption, 159–161
Gas chromatograph (GC), 1473, 2200
  capillary, 1974
  gas sampling, 1203
Gas chromatographic/mass selective detection (GC/MSD) analysis, 2017, 2018, 2019f
Gas flux, 2618
Gas compression heat pump, 1349–1350, 1350f. *See also* Heat pumps
Gaseous pollutants, industrial, 1430
Gasification, 376–377, 457
  definition of, 376
  stages in, 376
  temperature requirement, 376
Gasification, biomass and wastes, 2488–2495
  advances in, 2491–2494
  gasifiers, 2488–2491
  plants for, 2494t
  reaction involved in, 2490
  syngas, composition and heating value of, 2493t
  syngas for downstream application, 2492t
  technologies for, 2491
  world scenario, 2495
Gasification of coal and bio-solids, 730–731, 731f
Gasifiers, 2488–2491
  biomass and coal gasification, parameters for, 2488, 2493t
  conventional type, comparison of, 2488–2492, 2492t
  selection for, 2494–2495

Gas–liquid chromatography (GLC), 1126
Gas-phase bioreactor models, 2613, 2619
Gas reburning, 733t
Gas reserves, 803
Gastrointestinal effects, of arsenic toxicity, 1272–1273
GC. *See* Gas chromatograph (GC)
GCPF. *See* Global Crop Protection Federation (GCPF)
General circulation models (GCMs), 1867
Generalized analytic cost estimation (GACE) approach, 843
Generator junction box
  in photovoltaic systems, 219
Generic groundwater model, 1312–1313, 1313f
Genetically engineered organisms, 1110
Genetically modified crops, benefits of, 1951, 1953
Genetically modified organisms (GMOs), 125, 126, 1106
Genetic effects
  of pesticide, 1399
Genetic engineering
  environmental problems associated, 412
  in pest control, benefits of, 407
Genetic transformation, 307
Genistein, 1405
Genotoxicity
  environmental, 2124–2125
  human, 2124–2125
  pesticide, 2125–2128
  tests, for air pollution, 1156–1157
Geochemical investigation
  of saltwater intrusion, 1327–1328
Geoexchange systems, 1363
Geogenic carbonate, 1445–1446
Geogenic contaminants, 2166
Geographic information system (GIS), 554, 2215
  in air quality monitoring, 143
  inventorying and monitoring, 2816, 2817
  in land use planning, 1163–1165
    GIS/model interface, 1164–1165, 1165f
    LRIS, development of, 1164
    overview, 1163
    reliability of results, 1164–1165
    site suitability analyses, 1164
    spatial databases development, 1163–1164
  for regional pest management, 1952f, 1953
  remote sensing and. *See* RS/GIS integration
  research and development, 2818–2819, 2818f, 2818t
  software, 2039
  types of, 2818t
  for watershed management, 2816–2819, 2818t
Geographic information systems (GIS)-based indicators, 356
Geological erosion, 951
Geological investigation
  of saltwater intrusion, 1321–1330
Geologic materials, 1560
Geologic sequestration, 458–459
Geophysical investigation
  of saltwater intrusion, 1326–1327
    advantages and disadvantages, 1326
Geotextiles, 970–971
Geothermal brine specific exergy utilization index, 1175
Geothermal electricity, 825
Geothermal energy, 772, 1167
  activities for adopting technology of, 1169
  direct use of, 1167
  environmental benefits of, 1169
  history of use of, 1168
  hydrogen production from, 1168
  source of, 1167
  for sustainable development, 1169–1170
Geothermal energy resources, 1168–1169
  classification of, 1170
    by energy, 1170–1171, 1170t, 1171t, 1172t
    by exergy, 1171, 1173
    Lindal diagram, 1173–1176, 1174f, 1176f
Geothermal energy systems, performance assessment procedure for, 1176–1178
  case study on, 1178–1179, 1178f
Geothermal fluid, 1169
Geothermal gradient, 1168
Geothermal Heat Pump Consortium (GHPC), 1366
Geothermal heat pump (GHP) systems, 1363
  and conventional HVAC system, 1365, 1365t
  in cool weather, 1364
  development of
    borehole, 1367–1368, 1368f
    building, 1366–1367
    pond/lake heat exchanger, 1368, 1368f
    thermal mass, 1367
  earth heat exchangers, 1365–1366
  efficiency of, 1364
  environmental benefit of, 1364–1365
  operation of, 1364
    cooling mode, 1364, 1365f
    heating mode, 1364, 1364f
  organizations related to, 1366
  piping system in, 1368–1369
  principle of, 1364
  terms related to, 1366
  in warm weather, 1364
Geothermal resources, 825
Geothermal system, 1169
GeoWEPP model, 2761
German cockroach (*Blattella germanica*), 1988
German Democratic Republic, 1939
German Federal Immission Control Act (2002), 1961
German Soil Protection Act (1999), 1961
γ-Fe_2O_3, 1712
Ghana, 2683
GHG. *See* Greenhouse gases (GHG)
GHG pollution, 1364
Ghyben–Herzberg relation, 1323, 1326f, 1328
  saltwater detection using, 1325–1326, 1326f
Giant reed (*Arundo donax*)
  biology and ecology, 1182–1183, 1183f
    effects on streams/water resources, 1183
    biodiversity and wildlife, 1186
    control methods, restoration, revegetation, 1186
    flooding, 1184, 1184f
    water use, 1184
    wildfire, 1184–1186, 1185f
  hand clearing methods, 1186
  mechanical clearing methods, 1186
Gibberella ear rot, 1696
Gibbs free energy or Gibbs function, 62, 2532
Gilbert, J.H., 1858
GIS. *See* Geographic Information System (GIS)
GIS/model interface
  in land use planning, 1163–1165, 1165f
GIS software. *See* Geographic information systems (GIS) software
Glaciation indirect effect, 1197t
GLASOD (The global assessment of soil degradation), 1011
GLC. *See* Gas–liquid chromatography (GLC)
Gliocladium spp., 2107, 2109
Global Agrochemical Market, 2140
Global Atmospheric Watch (GAW), 8
Global Change and Terrestrial Ecosystem/ Soil Erosion Network (GCTE/SEN) exercise, 1026, 1027
  Soil Organic Matter Network (SOMNET) database, 1863, 1865–1866t
Global commerce
  pest management and, 2010
Global Crop Protection Federation (GCPF), 1409
Global earth observation system of systems (GEOSS), 604
Global emission, of ammonia, 87
Global Footprint Network, 2469, 2476–2477
  Ecological Footprint Standards, 2478

Globalization, 1218
global environment, future of, 1222–1223
global governance, 1221–1222
and sustainable economic growth and environmental impacts, 1220–1221
understanding, 1218–1220
Global-mean temperature, 1192, 1193
Global NPP map, 2376
Global ozone layer, depletion of, 1891–1893, 1892f
Global positioning systems (GPS), 2039, 2213–2214, 2272, 2818f
Global primary energy supplies, 834
Global Programme of Action for the Protection of the Marine Environment from Land-Based Activities (GPA), 1914
Global stability, problems affecting, 1227, 1228f
Global unrest
effects of, 1228
and peace, 1236
Global warming, 333, 772, 798, 804–805, 1132–1134, 1189–1191, 1197
and acid sulfate soils, 50, 56
carbon removal from air, 1189–1190
man role in, 1189–1190
nature role in, 1190
cause of, 772
CO_2 emission reductions by
demand-side conservation and efficiency improvements, 1136
shift to non-fossil energy sources, 1136
supply-side efficiency measures, 1136
effects of
climate changes, 1134–1135
sea level rise, 1134
global average surface temperature, 1134
health-related implications of, 805
keeping carbon in soil, 1190–1191
overview, 1189
projected earth surface temperature increase, 1134
threats for developing countries, 804–805
Global warming potential (GWP), 486, 1195, 1196t
Globodera spp., 1752
Glycine max, 1983
Glycine max L. Merr. (Soybeans), 425
Glyphosate, 1318t, 1387, 1399, 2142
GMOs. *See* Genetically modified organisms (GMOs)
GMP. *See* Good manufacturing practice (GMP)
Goddard Institute for Space Sciences (GISS), 1198f
Goethite, 1689
Golden-eyed fly, 1116
Gold mining, and groundwater contamination, 1291–1292
Gold nanoparticles
bactericidal effect of, 1740
pollutant removal, mechanism in, 1739–1740
Gompertz model, 2634
Goniozus emigratus, 1919
Goniozus legneri, 1919, 1920
Good manufacturing practice (GMP), 1610
Google Earth, 543
GOSSYM/COMAX model, 2221
Gossypium hirsutum L. (Cotton), 425, 1983
Goulden large-sample extraction, 2193
Government action plans, 1403
GPS. *See* Global positioning systems (GPS)
Graded terraces, 2764
Gradient terraces, 2538–2539, 2538–2539f
Gradual agents, 1456
Grain borer, 2423
Gram (*C. arietinum* L.), 425
Granular filters, 2722–2723. *See also* Filter(s)
Granular insecticides, 419
Granulated activated carbon adsorption systems, 2063
Granule deterioration, 2653
Granulosis viruses (GVs), 382
Graphical user interface (GUI), 1164
Grass filter strips, 2014
Grasshopper effect, 2017
Grass waterways, 969–970, 969f, 2763, 2764f
Grate furnaces, 375
Grätzel cells, 234–235. *See also* Dye-sensitized cells (DSCs)
Gravel sludge, 1431
Gray wolves (*Canis lupus*), 2596
Grazing
birds, 419
management, 2752, 2754
and pathogen contamination, 2753
Great horned owl (*Bubo virginianus*), 420
Great Lakes
degradation of, 1438
pesticide pollution in, 2146
Greece, 2684
Green ash. *See Fraxinus pennsylvanica*
Green bridge–horizontal eco-filtration system, 2155–2156
Green bridge technology (GBT), 2154, 2731
Green building certification, 1647
benefits of, 1648
Green Building Certification Institute (GBCI), 1651
Green Building Council (GBC), 1654
Green buildings, 1655
benefits of, 1651–1652
energy efficiency potential of, 1652
Green chemistry, 1253
green products, to produce, 1255
catalysts, 1255–1257
nanomaterials, 1257
solvents, 1257–1258
manufacturing process, application, 1256f, 1258
objective, 1254f
principles of, 1254–1255
sustainability, 1255
Green energy, 1227
analysis, 1235–1236
applications, 1234–1235
based sustainability ratio, 1235
benefits of, 1230
case study, 1236–1241
challenges, 1230
defined, 1227
and environmental consequences, 1227–1229
environment and sustainability, 1229–1231
essential factors for, 1233–1234
exergetic aspects of, 1234
resources, 1232–1233
and sustainable development, 1227, 1231
technologies, 1231–1232
commercial potential, 1232–1233
progress on, 1232
utilization ratio, 1235
Green engineering, 1255
Greenhouse effect, 435, 436f, 681, 1132, 1133
climate change due to, 435
desertification and, 549–550
biological changes, 549
climatical changes, 549
greening of the earth hypothesis, 549–550
modern studies, 550–551, 550f–551
from prehistory to industrial revolution, 550
woody plant range expansions, 550–551
enhanced, 435
human activities and, 435
Greenhouse gas concentrations trends, 1135–1136
CO_2 concentration, 1135
Greenhouse gas (GHG) emissions, 118, 1132
from building sector, 1353
reduction of, 378
by heat pumps, 1353–1354, 1354t
by sector, end use, activity, and gas types, 438f
Greenhouse gases (GHG), 435, 456, 483, 772, 798, 803, 1192, 1193, 1364
and climate change, 1894–1895
measuring fluxes, 1202–1205
chamber techniques, 1202, 1203t
closed and open chambers, 1202, 1204f

Greenhouse gases (GHG) (*cont'd.*)
gas chromatography, 1203
manual/automated sampling and analysis, 1202–1203
micrometeorological methods, 1203–1204, 1203t, 1205f
nonisotopic tracer methods, 1204
photo-acoustic-infrared detector, 1203
ultra-large chambers with long-path infrared spectrometers, 1204–1205
offset programs, 1773–1774
Greenhouse gas forcing, 1195, 1196t
"Greening" of the Earth, 549–550
Green manure
and organic farming, 1107
Green manuring, 466
Green manures
of leguminous crops, 1515
Green marketing incentive
integrated farming and, 1507
Greenockite, 446
Greenpeace, 1612
Green power, 1657
Green products, 1255
catalysts, 1255–1257
nanomaterials, 1257
solvents, 1257–1258
"Green revolution" technologies, 126
Green technology, 1255
Green Water Credits, 1038
Greigite, 42
Grey partridge (*Perdix perdix*), 416
Grid-connected photovoltaic systems, 218, 219, 219f, 220
configuration, 223
Griffith University Erosion System Template (GUEST), 966, 1036
Grimm particle sampler, 1024
Gross domestic product (GDP), 484, 614, 615, 618, 2468, 2475
Gross head, measurements of, 206, 207f
Ground catchment techniques, 2263, 2263f
Ground coupled heat pumps, 1363, 2519
Ground source heat pumps, 1363
Groundwater, 1302, 1714
contaminants in, 1302
contamination. *See also* Groundwater contamination
by arsenic, 1292
cyanide and, 1291–1292
by mining activities. *See* Mining
from nitrogen fertilizers, 1302–1307
non-point sources of, management of, 1319
point sources of, management of, 1319
uranium and, 1292
impact analysis, 1571
movement, 1290
nitrate-nitrogen in, 1302. *See also* Nitrate-nitrogen (NO_3-N), in groundwater
outflow, 1284
pesticide contamination in. *See* Pesticides in groundwater
pollution, 1570
remediation by nZVI, 1737–1739
as renewable resource, 1284
resources, 1289–1290
saltwater intrusion in, 1321–1330
Groundwater contamination, 1281
by arsenic, 1281
biological contaminants, 1281
chemical contaminants
inorganic, 1281
organic, 1281
climate, effect of, 1282
fate of contaminants, 1282–1283
plume of, 1282
sources of, 1281–1282
chloride, 1282
degreasers, 1282
fuels, 1282
human and animal wastes, 1281
nitrate, 1282
non-point sources, 1281
pesticides, 1282
point sources, 1281
toxic salts, 1282
wood-treating chemicals, 1282
transport of contaminants, 1282
treatment, biological techniques for, 2014
Groundwater drained lakes, 1577
Groundwater irrigation
and soil IC, 465
Groundwater Loading Effects of Agricultural Management Systems (GLEAMS) model, 2761
Groundwater mining
defined, 1284
Edwards aquifer, 1284, 1285, 1285f, 1287f
mass curve, 1287, 1287f
pumping, 1284, 1285, 1285f
and sustainable aquifer use, 1285–1287, 1287f
and water balance, 1284–1285, 1285f, 1286f
Groundwater modeling, 1295–1301
generic models, 1296
overview, 1295
phenomena, 1295–1296, 1296f
process, 1297–1301, 1298f, 1299f
calibration, 1298f, 1299
conceptualization, 1297–1299, 1298f
discretization, 1298f, 1299
error analysis, 1299
model selection, 1299–1300
predictions and uncertainty, 1300–1301
site-specific models, 1296–1297
usage of, 1296–1297, 1297f
Groundwater problems, numerical methods for
boundary element method (BEM), 1314, 1314f
finite differences (FD), 1314, 1314f
finite element method (FEM), 1314, 1314f
generic numerical method for solving groundwater flow, 1312–1313, 1313f
integrated finite differences (IFD), 1314
simulating solute transport, 1314–1315, 1315f
Groundwater remediation technologies, 1334–1336
bioremediation, 1335
passive PRB system configurations, 1336f
pump-and-treat technology, 1335
SVE and AS, 1334–1335
Groundwater Ubiquity Score (GUS), 2765
Group on Earth Observations Biodiversity Observation Network (GEO BON), 603–604
Grout, 1367
GUEST model, 992
GUI. *See* Graphical user interface (GUI)
Guidance on Pest and Pesticide Policy Development, 1986
Gulf Coast Ecosystem Restoration Task Force (GCERTF), 629–630
Gulf Coast Restoration Task Force, 629
Gulf War Veterans
pesticide sensitivities and, 1412
Gullies
erosion and, 1046, 1046f
Gully erosion, 958, 964
Gully stabilization structures, 970, 970–971f
Gutters for rainwater storage, 2264
Gypsum, 466, 959–960, 2370, 2372t
for management of soil dispersion, 2342

H

Habrobracon hebetor, 2424, 2424t
Haematobia irritans, 1460
Halocarbons, 1195
Halofenozide, 1466
Halogenated compounds (volatile), 1318
Halogenated organic compounds, 1841–1848
analytical methods of, 1847
applications of, 1842
degradation pathways of, 1843–1844, 1843f
in environmental compartments, 1842–1844, 1845–1846t
occurrence of, 1844

- metabolic pathways of, 1841, 1842f
- remediation strategies for, 1844, 1846–1847
- sources of, 1842

Halons, 1886
Hand clearing methods for *A. donax*, 1186
Hardtack Quince test, 187
Hardwood swamp, 2826f
Harris, T.W., 2008
Harrowing, 1111
Harvest of perennial energy crop, delaying, 1249
Hastened oxidation, 49
Hatch Act, 2009
Hazardous air pollutants (HAP), 1129
Hazardous waste, 893, 902
Hazard prevention, definition of, 1961
HCOP (heating performance of heat pump), 1347
Headspace (HS) analysis, 1847
Health
- definition of, 2871
- effects, of polychlorinated biphenyls, 2179, 2181
- guidelines for irrigation, 1571
- indicators in ecosystem
  - agroecosystem health, 606–607
  - aquatic ecosystem health, 607
  - forest ecosystem health, 607
  - urban ecosystem health, 607–608

Health advisory levels (HALs), 1317
Heat capacity at constant pressure, 2532
Heat, 809–810, 811
Heat engine, 819, 1346, 1347f
- exergy and the second-law efficiency, 820

Heating, ventilation, and air conditioning (HVAC) systems, 437, 439–440, 666
Heating energy efficiency ratio (HEER), heat pump, 1351
Heating value (HV), 717, 2527–2528
- of biomass fuels, 720t–721t
- of common fuels, 2528t
- higher, 2527
- lower, 2527

Heat of fusion, 2509
Heat pipes, 873
Heat pumps, 1346
- absorption, 1349, 1349f
- advantages of, 1348
- applications of
  - commercial and industrial, 1353
  - domestic, 1353
- energy flow of, 1346–1347, 1347f
- environmental benefits of, 1348
- fundamentals of, 1346–1347
- gas compression, 1349–1350, 1350f
- GHG saving potential of, 1353–1354, 1354t
- heating performance of, 1347
- heat sources for, 1347, 1347t
- heat supply and value of, 1348
- history of, 1346
- magnetic, 1348, 1348f
- performance parameters of, 1351–1353
  - ambient energy fraction, 1352–1353, 1352f, 1353t
  - COP and EER, 1351
  - primary energy ratio, 1351–1352, 1352f
- and refrigerators, 1347
- short-term/long-term potential use of, 1353–1354, 1354t
- thermoelectric, 1348–1349, 1348f
- types of, 1348–1351
- for underground thermal storage, 2519
- vapor compression, 1350–1351, 1350f, 1351f

Heat rates, 2527–2528
Heat recovery, 2517
Heat recovery steam generator (HRSG), 873
Heat source/heat sink configurations, 1357–1358, 1357f
Heat storage, 2510–2511
- encapsulation, 2510–2511
- system, 252–254
  - dimensioning, 253–254

Heat transfer, resistance to, 860
Heat wheels, 873
Heavy metal balance of agroecosystems and organic fertilizers. *See* Organic fertilization on heavy metal uptake
Heavy metal concentrations in organic fertilizers, 1375t
Heavy metals, 1370–1373, 1456–1457, 1560. *See also* specific entries
- as air pollutants, 161–163
- biological effects, 1372–1373, 1372f, 1373t
- in coastal waters, 490–491
- contamination with, 1374
- general chemistry, 1370, 1371t
- overview, 1370
- plant uptake of, 1375–1376
- in rocks and soils, 1370–1372, 1372f
- sources, 1372t

Height above-average-terrain (HAAT), 2229
Helicobacter pylori, 2696, 2698
Helicoverpa (Heliothis) zea, 382
Helis aspersa, 383
Helix. *See* Chlorfluazuron
Helminth egg removal, 2636
Helminthioses, 2696, 2699
Helminths, in waste and polluted water, 2698, 2699
Helminthosporium solani, 394
Helophytes, 2829
Helsinki Convention, 1828
Henry's Law, 328
Heptachlor, 1318t, 2015, 2015t
Heptachlor epoxide, 1318t
Herbicide-resistant crops (HRCs), 407, 409, 410t, 1379–1380
- economic impacts of, 411
- toxicity of herbicides and, 409, 411
- use of, 1953

Herbicides, 1186, 1318–1319, 1378–1381, 1382–1390, 2014, 2142, 2761t
- background, 1397
- benefits to agriculture and forest management, 1383
- chemical structure, 1378
- classification, 1378
- components of action, 1378–1379
  - absorption, 1378–1379
  - interaction at the target site, 1379
  - metabolic degradation, 1379
  - translocation, 1379
- derived from nature, 2030–2031
  - allelopathy, 2031
  - bialaphos, 2030–2031, 2031f
  - cyperine, 2031, 2031f
  - monoterpene cineoles, 2031, 2031f
  - phosphinothricin, 2030–2031, 2031f
  - triketones, 2031, 2031f
- discovery, 1378
- effects
  - at ecosystem/trophic level, 1388
  - factors interacting with, 1388–1389
  - on natural enemies, 370
  - at population/community level, 1387–1388
  - at species level, 1386–1387
- exposure to primary producers, 1383–1385
- history of, 1382
- glyphosate, 1399
- mechanisms of action, 1382–1383
- method/timing of application, 1378
- mode of action, 1378, 1379, 1380t
- overview, 1378
- paraquat, 1399, 1627–1628
- phenoxy acids, 1399
- phytotoxicity of, 1389–1390, 1389f
- resistance in weeds, 1379
- resistant crops, 1379–1380
- safety and environmental fate of, 1380–1381
- selectivity, 1379
- in surface waters, 2805
- toxicity of
  - to animals, 1385–1386
  - HRCs and, 409, 411
  - to humans, 1385–1386
  - to plants, 1386
- types of, 1383
- used in homes and gardens, 1401

Herbivorous insects, health effects of, 1122
Heterobasidium annosum, 2109
Heterodera spp., 1752

Heterogeneous TiO_2 photocatalysis, 2556–2565
  applications of, 2561, 2564
  mechanism of, 2557–2558
  operational parameters of, 2558, 2561, 2562–2563t
    oxygen concentration, 2561
    pH value, 2561
    temperature, 2561
    TiO_2 loading, 2561
  optimum conditions and rate of, 2559–2560t
  for water treatment, research trends in, 2564–2565
Heterotrophic bacteria, 2633
Hexachlorocyclohexane (HCH), 390
Hexaflumuron, 1461
Hexythiazox, 3
Hg. *See* Mercury (Hg)
"Hidden hunger," 536
High-copper soils
  plant growth on, 536–537
High-density polyethylene (HDPE) pipe, 1367
Higher heating value (HHV), 374
  of fuel, 717, 720t–721t, 831–834
High-intensity rainfall, 952
High performance liquid chromatography (HPLC), 1126, 1473, 1974
High-pressure nozzles, 1472
High-pressure reactor, 1905
High resolution visible and middle infrared (HRVIR), 2817t
High-rise buildings, 860
High-rise building's dewatering system, 1365–1366
High-temperature proton exchange membrane fuel cells (HT-PEMFCs), 1138–1139
  operating temperature range of, 1138
  for stationary applications, 1138
  technical challenges related to, 1138
High-temperature Winkler (HTW), 2491
"High throughput," 2028
Hilgardia, 1521
Hirsutella thompsonii, 382
H isotopes
  for saltwater detection, 1327–1328
Histosols, 1854
Hodgkin's lymphoma, 1399
Holistic ecosystem heath indicator (HEHI), 605, 606t
Home composting, 516. *See also* Composting
Homes and gardens
  pests and pesticide safety in, 1401–1403
Homogeneous photo-Fenton reaction, 2548–2556
  applications in water treatment, 2553–2556, 2554–2556t
  classification of, 2548f
  contaminant concentrations and characteristics, 2552
  iron concentration, impact of, 2549, 2552
  mechanism of, 2548–2549
  optimum conditions and rate of, 2550–2551t
  oxident concentration, impact of, 2552
Homo sapiens, 1717
Hordeum vulgare L. (Barley), 425
Horizontal flow systems, for constructed wetlands, 2841–2843
Horizontal mass flux (HMF), 1021–1022
  samplers, 1022–1024, 1023f, 1024f
Horizontal reactors, 518, 518f
Hormonal disruption, in humans, 1405–1406
  mechanisms, 1405–1406
  in men, 1406
  pesticides test management, 1406
  in women, 1406
Hormones, effect on endocrine-disrupting chemicals, 645–646
Host plant effects
  natural enemies and, 371
Host plant resistance, 444
Host-specificity tests and
  weed biocontrol and, 2822
Hot water tanks, 2520
Household vehicle fuel economy, 792f
House mice (*Mus musculus*), 2596
HPLC. *See* High performance liquid chromatography (HPLC)
HPPD. *See* Hydroxyphenylpyruvate dioxygenase (HPPD)
HRCs. *See* Herbicide-resistant crops (HRCs)
HRSG. *See* Heat recovery steam generator
HRT. *See* Hydraulic residency time (HRT)
HTW. *See* High-temperature Winkler (HTW)
H_2S, 42, 45
Hubbert peak theory, 715
HUD. *See* U.S. Department of Housing and Urban Development (HUD)
Hughes Plant 44, 1333
Human(s)
  activities impact on carbon storage, 2835
  Al toxicity in, 271–272
  biomonitoring, 2129
  and ecosystem, relation between, 635, 636
  exposure effects, of endocrine-disrupting chemicals, 648–650
  genotoxicity, 2124–2125
  pesticide effect on, 2147–2148
  role in carbon removal from air, 1189–1190
  toxocity of herbicides to, 1385–1386
Human development index (HDI), 798, 2468, 2473, 2474f
Human-induced global warming, 1192
Human pesticide poisonings, 1408–1410
  global incidence, 1408–1409
    African countries, 1408
    comparison, 1409
    Costa Rica, 1408
    developing and developed countries, 1409
    regional reports, 1408–1409
    Sri Lanka, 1408
    Taiwan, 1408
  international support and, 1409–1410
  overview, 1408
Humic acid, 269
Humid region irrigation
  and soil IC, 465
Humification, 1625
Hybrid sorbent injection, 2442
Hydraulic conductivity, 46
  water retention and, relationship between, 1621
Hydraulic energy, 772
Hydraulic functioning, of Alexandria Lake Maryut, 166–167, 169–170
Hydraulic gradient, 46
Hydraulic residency time (HRT), 2671
Hydraulic short circuit, 1424. *See also* Pumped storage hydro
Hydrilla verticillata, 1922
Hydrocarbons, 716, 717, 836
  characteristics of, 136–137
  chlorinated, 1317
  and climate change, 1895–1896, 1896t
Hydroelectric dams, 916
Hydroelectric resources, 825
Hydrochlorofluorocarbons (HCFCs), 1894
Hydroelectric power generation, 1421–1422
Hydrofluorocarbons (HFC), 486
Hydrogen, 723
Hydrogeomorphic approach (HGM), 2827
Hydrograph modification. *See* Irrigation and river flows
Hydrolases, 89
Hydrological conditions
  leaching and, 1621, 1622f
Hydrological functioning, of Alexandria Lake Maryut, 166–167
Hydrology. *See* Timber harvesting
Hydrolysis, 2651
  constant, 431
Hydropower, 201
  classification, 202–203
  gross head, measurements of, 206
  instream flow and environmental impact, 206–208
  mini hydropower, 203
    civil works in MHP plants, 208–212
  water flow, measure of, 204–206
  water resource, 203–204
  water turbines, 212
Hydroprene, 1464

Hydropyrolysis, 2487
20-hydroxyecdysone, 1465, 1465f
Hydroxy herbicides, 1318
Hydroxyphenylpyruvate dioxygenase (HPPD), 2031
Hyperecdysonism, 1465
Hypericum perforatum. See St. John's wort
Hypolimnion water removal, 1592t
Hypoxia, 1303, 1559

I

IAEA. *See* International Atomic Energy Agency (IAEA)
IARC. *See* International Agency for Research on Cancer (IARC)
Ice melting, 1199
Ice nuclei, 1196, 1197t
Ice storage, 2509–2510, 2509f, 2510f
ICM. *See* Integrated crop management (ICM)
ICRC. *See* Interim Chemical Review Committee (ICRC)
Ideal gas law, 2525, 2526
Ideal gas state, 2526
Ideal-solution approximation, 2533
IDGCC. *See* Integrated drying and gasification combined cycle (IDGCC)
IDW. *See* Inverse distance weighting (IDW)
IEA Heat Pump Centre, 1348, 1353–1354
IFOAM. *See* International Federation of Organic Agriculture Movements (IFOAM)
Ignition and combustion, 725
IGRs. *See* Insect growth regulators (IGRs)
Illinois State University (ISU), 1494
Illuviation. *See* Eluviation/illuviation
ILO. *See* International Labour Organisation (ILO)
Imazamox, 2020
Imidazolinones, 409
Immobile nutrient, 100
Immobilization (of heavy metals), 1375
Immobilized biomass reactor (IBR), 2553
Immobilizing agents
  for bird control, 414
INC. *See* Intergovernmental Negotiating Committee (INC)
Incident monitoring, 420–421
Increasing block tariff (IBT) pricing strategies, 2786–2788
Incubation methods, 41
Independent action (IA), 2573t, 2574
  applications of, 2575–2576
  characteristics of, 2575
  predictive power of, 2576
Index of sustainable economic welfare (ISEW), 2468, 2475
India, 2683
  acid rain in, 9, 10f
  trends in acidity, 13
Indian Council of Medical Research, 2142
Indian ecotechnological pollution treatment systems, 5
Indicator in ecology, definition of, 600
Indicator taxa, 602–603
Indicators
  desertification, 553–554
Indigenous technical knowledge (ITK), 2385
Indirect aerosol effect, 1197t
Indirect consumption of soil contamination via root exudates, 310
Indirect radiative forcings, 1194
Indium tin oxide (ITO), 235
Indoor air quality (IAQ), 1655
Indoor environmental quality, 1657
Indoor residual spraying (IRS), 1915
Indoor water storage system, 2263f
Induced erosion, 1010–1012
Industrial air pollution, 159
  absorption, 159–160
  adsorption, 160–161
  combustion, 161
Industrial carbon-di-oxide capture, 458
Industrial desertification, 2377–2378
Industrial ecology, 1253
Industrialization and pollution, 2170
Industrial network
  example of, 1434
  benefits of, 1435
Industrial pollution, 2304
Industrial Revolution, 1189
  from prehistory to, 550
Industrial use of energy, 794–795, 794f
Industrial waste, 2137
  soil pollution by, 1430–1432
    quantities, 1431t
    reuse possibilities, 1432f
    soil amelioration, 1431
    soil remediation, 1431–1432
    types, 1430–1431
Industrial wastewaters, 1570
Industrial wind turbines, 2867–2868. *See also* Wind turbine noise
Inexact double-sided fuzzy chance-constrained programming (IDFCCP) model, 113–116
Inexact optimization models, 106
  case study, 108–116
  fuzzy mathematical programming, 107
  interval linear programming, 107–108
  stochastic mathematical programming, 106–107
Inexact stochastic water management model, 111–112
Infestations of *A. donax,* 1182
Infiltrated water, 1559
Infiltration, 2314, 2752
  by soil, vegetation and, 995
Information technology, 2213. *See also* Precision agriculture techniques
Infrared absorption spectrometers, 1204–1205
Ingot, silicon, 231
Injection equipment
  chemigation, 470–471, 470f, 471f
Injection wells, 1293
Inhalation, of inorganic pollutants, 2120–2121
Inland marsh, 2826f
Inoculation/classical approach, of biological weed control, 1525, 1525f
Inorganic carbon, 462–467, 1451–1453
  arid and semiarid region irrigation, 465–466
  compartment model, 1451–1452
  composition, 1444, 1445f
  cropping and tillage, 464
  cycling, dissolved, 463
  fertilization and liming, 464–465
  formation, 1444–1446
    biogenic model, 1446
    pedogenic *vs.* geogenic carbonate, 1445–1446
    per ascensum model, 1446
    per descensum model, 1446
    in situ model, 1446
  humid region irrigation, 465
  impact on atmospheric carbon dioxide, 466–467
  isotopes in calcic soils, 1452–1453
  land clearing, 463–464
  land use, 463
  pedogenic carbonate accumulation, simulations of, 1451–1452
  sodic soil reclamation, 466
  soil processes, 462–463
Inorganic chemicals, in groundwater, 1281
Inorganic contaminant interactions, 2166, 2169
Inorganic contaminants
  bioremediation of, 392–393
Inorganic mulches, 961–962
Inorganic nitrogen, 1559
Inorganic pollutants, 2117–2121, 2118t, 2119t
  classes and concentration ranges of, 2118
  of drinking water, 2791–2793, 2791t, 2792t
  pathways of exposure, 2120–2121
  potential health impacts of, 2118–2120
  remediation of, 2120
Inorganic soil amendment, 958–959
Insect control
  Bacillus thuringiensis for, 411–412, 411t
Insect fragments
  tolerance limit for, 1609–1610
    in food (examples), 1610t

Insect growth regulators (IGRs), 1459
administration of, by feed-through method, 1460
advantages of, 1466
chitin synthesis inhibitors, 1460, 1461–1463
classification of, 1460
contact/oral application, 1460
disadvantages of, 1466–1467
ecdysone agonists, 1460, 1465–1466
and integrated pest management, 1467
juvenile hormone analogs, 1460, 1464
naming of, 1460–1461
overview, 1459–1460
resistance to, 1467
selectivity of, 1460
for social insects, 1460
use of, in insect control, 1459
Insecticide Act of 1910, 1940
Insecticide residue monitoring, 1472–1473
Insecticide-resistant malaria mosquitoes, combating of, 1990
Insecticides, 1317–1318, 1609, 1697, 2014
derived from nature, 2028–2029
azadirachtin, 2029, 2029f
Bacillus thuringiensis, 2029
juvenile-hormone mimics and pheromones, 2029
milbemycins/avermectins, 2029, 2029f
nereistoxin derivatives, 2029, 2029f
nicotine, 2029, 2029f
pyrethrins, 2028–2029, 2029f
spinosyns, 2029, 2029f
effects of repeated exposure to OPs, 1399
effects on natural enemies, 370
intermediate syndromes, 1398
mechanisms of action, 1398
OPIDIN, 1399
organophosphorus compounds, 1398
in surface waters, 2805
used in homes and gardens, 1401
Insects, beneficial, 2423
Insegar. *See* Fenoxycarb
In situ acid minesoil remediation, 1693
In situ ground water biodegradation, 396t
In situ model, of carbonate formation, 1446
In situ sodic minesoil remediation, 1693
In situ soil reclamation
at open cut mine, 1693
In situ techniques, 2170
Instantaneous field of view (IFOV), 2817t, 2818
Institutional commitment, 560
Instituto Nacional de Investigaciones Forestales Agricolas y Pecuaris (INIFAP), 1763, 1765
In-stream flow requirements, 1566
Insulated concrete forms (ICFs), 1483, 1484f
walls, 864, 864f
Insulated material, 242
Insulation
building envelope, 1479f
ceilings and roofs, 1486–1490
and environment, 1479
floor. *See* Floor insulation
foam, 1482
guidelines, critical, 1482
materials, 1478–1479
comparison of, 1479t, 1480–1481t
strategies, 1479–1482
wall. *See* Wall insulation
Intangible values, 634
Integrated collector storage, 242, 243–244, 244f
"Integrated control," concept of, 1521
Integrated crop management (ICM), 1507, 1509
Integrated disease management (examples), 444
Integrated drying and gasification combined cycle (IDGCC), 2491
Integrated energy systems, case study from ISU, 1494, 1503–1504
asset management, 1501–1502
planning methodology, 1495, 1497–1498, 1500–1501, 1501f
profits, 1503
project assessment/analysis, 1495
chilled water operating calculations, 1500t
cogeneration operating calculations, 1498t
ISU executive summary, 1496t
ISU pro forma, 1497f
sensitivity analysis, 1501t
steam operating calculations, 1499t
related history, 1494
risk management, 1502–1503
stakeholders in, 1494–1495
Web-based plan, assumptions in, 1495, 1497–1498, 1500–1501
Web-based tools, for communication and reports, 1495
Integrated environmental management, 164–173
of soil water erosion, 971–972
Integrated farming systems, 1507–1509
agronomic, 1508
defined, 1507
development, 1508, 1508t
economic pressure, 1507
environmental, 1507–1508
green marketing incentive, 1507
legislation, 1507
overview, 1507
principles, 1508–1509
ecological infrastructure management, 1509
implementation, 1509
integrated crop management, 1509
integrated nutrient management, 1508–1509
minimum soil cultivation, 1509
multifunctional crop rotation, 1508
Integrated finite differences (IFD), 1314
Integrated gasification combined cycle (IGCC) process, 457, 730, 731f
Integrated membrane system (IMS), 2729
Integrated nutrient management (INM), 1508–1509, 1510
adoption of, at farm, 1516–1517
aim of, 1511–1512
approach of, 1510–1511
components of, 1511–1516, 1511f. *See also* Biofertilizers; Fertilizer
biofertilizers, 1514–1516
mineral and synthetic fertilizers, 1512–1514
concept of, 1510
environmental concerns of, 1517–1518
negative effects, 1517
nutrient losses, 1517
problems with fertilizer use, 1516t
soil carbon sequestration, 1517–1518
toxic accumulation, 1518
fertility degradation and, 1518–1519
nutrient cycling in soil–plant–air–water systems, 1512f
steps in, 1511
use of, need for, 1510
Integrated pest management, 1521
Integrated pest management (IPM), 409, 508, 1467, 1507, 1521–1523, 1524, 1878, 1930, 1931t, 1986, 2010–2011, 2033, 2143, 2148. *See also* Chemical pesticides (CPs)
agricultural runoff and, 83
BP and CP compatibility in, 1987–1988
examples of, 1990
increasing of, approach for, 1990–1991
practical indications for promotion of, 1992t
selectivity in, 1991
building blocks, 1522
computer-derived packages, 1522
defined, 1521
derivation of, 1521–1522
development of, 1522–1523
menu systems, 1522
need for, 1522
overview, 1521
practices, 314
protocols, 1522–1523
strategies, 295
Integrated Pollution Prevention and Control (IPPC) Directive of 1996, 1255
Integrated solid waste management (ISWM), 2419
Integrated weed management, 1526t–1527t. *See also* Weed control

Integrity indicators and orientors, 604–605
Intelligent Decision Support System (IDSS), 2819
Intensity factor, 100
Interaction at the target site
  herbicides, 1379
Intercropping, 1937–1939
  functional diversity, 1937
  future perspectives, 1939
  monoculture and, 1937, 1938t
  in practice, 1939
  protection mechanisms, 1937–1938, 1938t
Intergovernmental Negotiating Committee (INC), 1616
Intergovernmental Panel on Climate Change (IPCC), 96, 98, 435, 484, 934, 1197
  Third Assessment Report (TAR), 1200f
Interill
  defined, 958
  erosion, 2756
Interim Chemical Review Committee (ICRC), 1616
Interior foam insulation, 1482, 1483t
Interior framed wall insulation, 1482, 1483f
Intermediate disturbance hypothesis (IDH), 639
Intermittent energy resources, 1419
Internal energy, 2523
Internal feeders, 2424t
Internally feeding pests, biocontrol of, 2423–2424
International Agency for Cancer Research (IARC), 1264, 1394, 1395t
International ASS conference, Australia, 55
International Atomic Energy Agency (IAEA), 183
International Biochar Initiative, 844
International Biosphere Reserve, 1082
International climate policy. *See* Climate policy (International)
International Code of Conduct on the Distribution and Use of Pesticides, 1615
1994 International Desertification Conference (Arizona), 553
International Energy Agency (IEA), 201, 215, 1348
  projections, 378
International environmental law, 1618–1619
International Erosion Control Association, 1035
International Federation of Organic Agriculture Movements (IFOAM), 125, 1103, 1108, 1109, 1116
International Ground Source Heat Pump Association (IGSHPA), 1366
International Group of National Associations of Agrochemical Manufacturers (GIFAP), 1409
International initiative for a sustainable built environment, 1655
International Institute for Land Reclamation and Improvement (ILRI), 55
International Labour Organisation (ILO), 1409, 2036
International Land Reclamation Institute (ILRI), 26
International Measurement and Verification Protocol (IPMVP), 1657
International Organization of Biological Control (IOBC), 1991
International Performance Measurement and Verification Protocol (IPMVP or MVP), 768
International Programme on Chemical Safety (IPCS), 1409
International Trade, 1613
Interval linear programming (ILP), 107–108
Interval parameter water quality management model (IPWM), 109–111
Interval-stochastic chanceconstrained programming (ISCCP) model, 111–113
Intra-organizational listings/scorecards, 768
Intraparticle diffusion model, 67–68
Intrepid. *See* Methoxyfenozide
Invasion biology. *See* Biological invasions
Inversion, 1129
Inverters
  in photovoltaic systems, 219, 220–221
    functions, 220
    without transformers, 220
    types, 221–222
In-vessel composting, 517. *See also* Composting
Ion exchange, 1282–1283
Ion exchange application, 2734, 2735f
  in water and wastewater treatment, 2736–2737
Ionic liquids, 1257
Ion implantation, formation of p–n junction by, 233
Iowa P index, 2764–2765, 2765t
IPCC. *See* Intergovernmental Panel on Climate Change (IPCC)
IPCS. *See* International Programme on Chemical Safety (IPCS)
IPM. *See* Integrated pest management (IPM)
Iran, 2682
IRF. *See* Irrigation return flows (IRF)
Iron (Fe), 351, 1370
Iron concentration, effect on homogeneous photo-Fenton reaction, 2549, 2552
Iron-oxidizing bacteria, 1690
Irreversibility, 816–817
Irrigation
  definition of, 120
  frequency, 2677
  importance of, 1551
  management, 1319
  method, 1547t, 1564
  methods, 120
    energy consumption for, 120, 120f
  overview, 1572
  sewage effluent for. *See* Sewage effluent for irrigation
  and soil IC, 465–466
  soil salinity and, 1572–1574
    causes, 1573
    deleterious effects, 1572–1573
    management strategies, 1574
  sprinkler, 1551
  surface, 1551, 1552
  wastewater
    management, 2676–2677, 2679t
    methods, parameters for evaluation of, 2678t. *See also* Wastewater irrigation
  water, 433
  water management, 1813
Irrigation and river flows
  basin irrigation efficiency, 1565–1566, 1565f
  depletion, 1563
  environmental concerns
    in-stream flow requirements, 1566
    salt loading pick-up, 1566
    water quality implications for agriculture, 1566
  hydrograph modification
    irrigation efficiencies, 1564–1565
    irrigation methods, 1564
    irrigation return flows, 1564
    reservoir storage, 1563–1564
  hydrologic studies, 1563
Irrigation efficiency, 1564–1565
  basin, 1565–1566, 1565f
  definition of, 1545
  performance efficiency
    application efficiency, 1546, 1547t
    seasonal irrigation efficiency, 1546–1547
    storage efficiency, 1546
    water conveyance efficiency, 1545–1546
  uniformity in irrigation
    Christiansen's uniformity coefficient, 1548
    emission uniformity, 1548–1549
    low-quarter distribution uniformity, 1548
  water transport components, 1545, 1546f
  water use efficiency (WUE), 1549

Irrigation equipment
chemigation, 470
Irrigation erosion, 1551–1552
sprinkler-irrigation systems and, 1553–1555
surface irrigation and, 1552–1553
Irrigation furrows, 1552, 1552f
Irrigation pumping, energy savings on, 120
Irrigation return flows (IRF), 1564, 1565f
components of, 1557–1558, 1558f, 1558t
consequences of, 1557
defined, 1557
off-site water quality impact reduction, 1560–1561, 1561t
water quality constituents in, 1558–1560
nitrogen, 1559
pesticide contamination in, 1560
phosphorus, 1559–1560
salts, 1559
trace elements, 1560
Irrigation system
drip. *See* Drip irrigation system
subsurface drip irrigation (SDI), 1535–1538
Irrigation water use efficiency (IWUE), 1549
Irrigation with saline water. *See* Saline waters
Isaria fumosorosea, 382
Ishipron. *See* Chlorfluazuron
Island biogeography (IB) theory, 639
Islanding, in photovoltaic systems, 221
detection
active methods, 221
passive methods, 221
Isokinetic sampling, 965, 965f
Isoproturon, 2142
ISO 7934:1989. Stationary source emissions, 2440
ISO 7935:1992. Stationary source emissions, 2440
ISO 10396:2007. Stationary source emissions, 2440
ISO 11632:1998. Stationary source emissions, 2440
Isotherm(s)
adsorption, 63–64
Brunauer–Emmett–Teller, 65
Dubinin–Radushkevich, 65
Freundlich, 64–65
Langmuir, 64
Redlich–Peterson, 65–66
Temkin, 65
Isotopes
in calcic soils, 1452–1453
Israel, 2682, 2684
saltwater intrusion in, 1322
wastewater reuse in, 1571
"Itai-itai" disease, 448, 453
Italy, 2690
saltwater intrusion in, 1322
Ivermectins, 3
Ixodes scapularis (deer ticks), 2

J

Japan
acid rain in, 9
participatory biodiversity inventory in, 359–360, 360f
Jarosite, 40, 42, 43, 1689
Jharkhand, groundwater arsenic contamination in, 1268
Joina (*Fimbristylis miliacea*), 426
Joint Expert Committee on Food Additives and Contaminants (JECFA), 1696
Joint implementation (JI), 486
Jordan, 2682
Jornada Experimental Range in New Mexico, 550, 550f
J.R. Geigy Co., 2009
Jupiter. *See* Chlorfluazuron
Juvenile hormone (JH), 1463, 2029
anti-JH agents, 1464–1465, 1465f
in insects, 1463–1464, 1463f
Juvenile hormone analogs (JHAs), 1460, 1464
dayoutong, 1464
diofenolan, 1464
fenoxycarb, 1464
hydroprene, 1464
kinoprene, 1464
methoprene, 1464
pyriproxyfen, 1464

K

Kairomones, 175
Kara–Bogaz–Gol Bay, degradation of, 1440–1441
Kasugamycin, 2030, 2030f
Kellogg–Rust–Westinghouse (KRW), 2491
Kelvin–Planck law, 818
Kelvin–Planck statement, 818
Kinetic energy
rainfall, 983–984
Kentucky State University (KSU)/ Water Quality and Environmental Toxicology Research Program, 2022–2024
Keys to Soil Taxonomy, 28
"K-fabric," 1444
Killo. *See* Chromafenozid
Kilowatt (kW), 1364
KINEROS model, 992
Kinoprene, 1464
Klebsiella oxytoca, 394
Knack. *See* Pyriproxyfen
Koi Herpes virus. *See* Cyprinid Herpes virus 3 (CyHV-3)
Koinobiontic Hymenoptera, 1747
Kolleru Lake, pesticide pollution in, 2147
KRW. *See* Kellogg–Rust–Westinghouse (KRW)
Kullback's measure of information, 783
Kuwait, 2682
Kyoto Protocol, 183, 485–487, 1896
target emissions, 486f
Kyoto targets, 486f

L

Labeling, pesticide, 1941
Labidura riparia, 1921
Lactic acid bacteria, 530–531, 531f
Lagoons, 2663. *See also* Wastewater treatment in arctic regions, wetlands usage
Lagrangian method, 1315
Lagrange multipliers. *See* Shadow prices
Lake(s), 1577. *See also* Lakes and reservoirs
aeration, 1584
degradation of, 1436–1438
sampling, 1582
Lake Chad, degradation of, 1437
Lake Eyre Basin, 1020
Lake management options, 539
Lake Mead, degradation of, 1438
Lake Ohlin, degradation of, 1437
Lake recultivation, 1588–1598
methods of, 1590–1597, 1592–1593t
phases of, 1590–1591, 1591f
Lakes and reservoirs, 1576–1577
classification of, 1577
freshwater/saline lakes, 1577–1578
trophic status, 1578
conventions for protection, 1585
The Protocol on Water and Health, 1585
Ramsar Convention, 1585
pollution, sources of, 1581
pollution issues of, 1581t
problems associated with, 1578
acidification, 1579
eutrophication, 1578–1579
fish depletion, 1579–1580
sedimentation, 1579
stratification, 1580–1581
toxic materials, 1579
protective and restorative measures, 1583–1585
eutrophication model framework, 1585
restoration measures, 1584t
water quality monitoring, 1581–1582
biological integrity sampling, 1582
citizen monitoring, 1582–1583
lake sampling, 1582

tributary mass load sampling, 1582
tributary water quality sampling, 1582
Lake Victoria, degradation of, 1437
Land
as finite resource, 1600
and river runoffs, 2358
surveys, 559
Landcare program, in Australia, 1037
Landcare Trust, 2385
Land clearing
and soil IC, 463–464
Land cover map, 2376
Land evaluation and site assessment (LESA), 1164
Landfill Allowance Trading Scheme (LATS), 900
Landfill gas, 377
Landfill leachates, 2482
Land Grant University System, 2009
Land Information System (LANDIS), 2819
Land resource information systems (LRIS), 1163
development of, 1164
Landscape(s)
assessment procedure, 541–542, 542f
dysfunctional, 541
fragile, 542, 542f
functional, 541
functioning. conceptual framework of, 542f
robust, 542, 542f
Landscape patterns, pest and, 2038–2039, 2039f
larger scale, 2038
small patches, 2038, 2039f
temporal stability of, 2038–2039
Landscape planning, 637–638
Land surface catchments, 2263, 2263t
Land use, 916
and soil IC, 463
Land Use Analysis System (LUCAS), 2819
Land-use change forcing, 1196–1197
Land use planning
GIS in, 1163–1165
GIS/model interface, 1164, 1165f
LRIS, development of, 1164
overview, 1163
reliability of results, 1164–1165
site suitability analyses, 1164
spatial databases development, 1163–1164
Langmuir isotherm, 64, 2427. *See also* Isotherm(s)
Lantana camara, 367, 2821
Large-scale ecosystem restoration governance, 624
collaborative environmental management, 625
collaborative resource governance, 627–628
ecosystem restoration task force model, 628
Gulf Coast Ecosystem Restoration Task Force (GCERTF), 629–630
South Florida Ecosystem Restoration Task Force (SFERTF), 629, 630
opportunities and challenges, 626–627
Larvadex. *See* Cyromazine
Latent heat storage, 2509
Lateral dispersion, 1282
Lateritic soils, 427
Latin America, bioindicator development in, 356, 359
Latium perenne (Ryegrass), 1305
Lawes, J. B., 1858
Law on Promotion of Organic Farming (Japan), 528
Laws, on pesticides use. *See* Pesticide regulation
Leachate
problems associated with biowaste, 512
Leaching, 1621–1623, 1625–1626, 1771, 2137, 2138, 2433
of acidity and salt, 49
climatic conditions and, 1622
environmental implications of, 1626
experiments (REE), 2268
"facilitated transport" and, 1622
factors influencing, 1621–1623
management practices, 1622–1623
overview, 1621
pesticide properties and, 1622, 1622t
pollution prevention strategies, 1622–1623
preferential flow and, 1621
soil and hydrological conditions, 1621, 1622f
Leaching and chemistry estimation (LEACHM) model, 1164
Leaching fraction (LF), 1547, 1559
LEACHM. *See* Leaching and chemistry estimation (LEACHM) model
Lead (Pb), 1370
abatement methods for reduction of, pollution, 1633–1634
and cadmium, 446
chemical properties, 446
detection, 447
environmental hazards, 447–448
human exposure to, 449t
human health effects, 448–449
occurrences, 446–447
production and uses, 447
regulations and control, 449
dispersion and application, 1630
in food, 1631t
heavy metal pollution in River Rhine, 1631t
and earnings, 1640f
ecotoxicity and environmental problems of, 1631–1633
biomagnifications in aquatic ecosystem, 1633f
concentration in glacial ice, 1631f
contaminated aquatic ecosystems, 1631
ingestion in birds, 1631
toxicological and ecotoxicological effects, 1632–1633
uptake from food, 1631, 1632f
integrated environmental management, 1634–1635
Lead chromate, 479t, 480
Leadership in Energy and Environmental Design (LEED), 667, 1363, 1647, 1648
and existing buildings, 1648–1649
Online, 1651
rating systems, 1648
Version 1.0, 1648
Version 2.0, 1648, 1649
Version 2.1, 1648
Version 2.2, 1648
Leadership in Energy and Environmental Design for Existing Buildings: Operations and Maintenance (LEEDEB O&M), 1647, 1649, 1650, 1650t. *See also* Green buildings
building evaluation categories, 1649
certification levels, 1650, 1651t
credits and points, 1650, 1650t
Green Building Rating System, 1648
implementation process, 1651
issues addressed by, 1649
minimum program requirements, 1649–1650
and other LEED products, 1649
overview of, 1649–1651
prerequisites, 1650, 1650t
registration process, 1650–1651
US Green Building Council (USGBC), 1647–1648
Leadership in Energy and Environmental Design for New Construction (LEED-NC)
assessment of, 1657–1659
construction standards, 1655
ecosystem disruption, 1654
environmental pollution, 1654
green buildings, 1655
LEED-NC rating system, 1655–1656
LEED prerequisites categories and criteria, 1656–1657
rating systems for buildings, 1655
sustainability, 1654
Lead regulations, 1636
benefits assessed, 1638–1639
cardiovascular risk reductions, 1642–1643

Lead regulations (*cont'd.*)
challenges and oportunities, 1643–1644
general framework, 1636–1638
IQ-related benefits, 1639–1642
Lead toxicity of animals, 449
LEED. *See* Leadership in Energy and Environmental Design (LEED)
Legislation
integrated farming, 1507
pesticide control, 2035–2036
elements, 2035
implementation, 2035–2036
international conventions, 2036–2037
overview, 2035
problems related to, 2036
weed biocontrol, 2823
Legionella exposure risk, 255f
prevention and control of, 254–255
Legume green manuring, 1515
Legumes, 1110, 1807
Lemna gibba (Duckweed), 426
Lemon (*Citrus limon*), 426
Lens culinaris Medic (Lentil), 425
Lentic oxygenation technology system (LOTS), 2154
Lentil (*Lens culinaris* Medic), 425
Leptinotarsa decemlineata, 1461, 1928, 2009
Leptospermone, 2031, 2031f
Leptospirillum ferrooxidans, 43, 1689
LESA. *See* Land evaluation and site assessment (LESA)
Lethal stressing agents
for bird control, 414
Leucomelanosis, 1272
Levuana iridescens, 367
Lewis acid, 431
Liberty®, 2030
Lichens, biomonitoring using, 143
LIDAR. *See* Light detection and ranging (LIDAR)
Life
nanoparticles and, 1716–1718, 1717t
Life cycle assessment (LCA), for solid waste management, 2399–2412
analysis and interpretation, 2405–2406
goal and scope of, 2404–2405, 2405t
importance of, 2411–2412
ranking of, 2406–2411
study area and system description of, 2401–2403
weighting factors, determination of, 2406
Life cycle assessment (LCA) study, 437
Life cycle costing (LCC) analysis, 688–689
Life cycle inventory (LCI), 2405
Ligand-to-metal charge transfer (LMCT) reaction, 2549
Light detection and ranging (LIDAR), 2215
differential absorption, 144
Doppler, 144
range finder, 144
Lighting efficiency, 775, 776, 776t
Light NAPL (LNAPL), 1334
Lightning, 1185
Light non-aqueous phase liquid (LNAPL), 1281, 1282
Light pollution. *See also* Pollution
in coastal waters, 497–498
Lightweight concrete products, 1483, 1484
Lignin, 373
Lignocellulases, 89
Limburg soil erosion model (LISEM), 966, 2761
Lime, 959, 2370
applications on soil, 2428f
for management of soil dispersion, 2342
Liming
of soil, 269
and soil IC, 464–465
Limit of quantitation (LOQ), 1126
Lindane, 2–3, 1318t, 1957, 2015, 2015f
Linear aggression analysis
for crop water use, 2220
Linear imaging self scanner system (LISS), 2817t, 2818
Linuron, 1318
Lipopolysaccharides (LPS), 539
Liquefied natural gas (LNG), 722
Liquefied petroleum gas (LPG), 723, 793
Liquid chromatography (LC), 2200–2201
Liquid film diffusion model, 68
Liquid fuels, 714
Liquid–liquid extraction (LLE), 1970t, 2192–2193
Liquid–liquid/mass spectrometric (LC/MS) analysis, 2018
Liquid manure injection system, 1674
AerWay SSD system, 1674, 1674f
drag-hose system, 1674f
Listeria, 1696
Lithium (Li), 1560
Litter biomass (LB), 718t
Litter deposition, 2752, 2753
Livestock
impacts on soil, 2752
impacts on vegetation, 2753
impacts to streams/riparian areas, 2754
from range and pasture lands, 2752
Livestock drinking water
water harvesting for, 2738–2739
Livestock feed
manure management, 1812–1813, 1812f
Living Planet Index (LPI), 603
Living systems
in action for pollution treatment, 2152
in treatment of pollution, 2151–2152
Lixophaga diatraege, 1930
Lixophaga sphenophori, 1928
LNAPL. *See* Light NAPL (LNAPL)
LOAEL. *See* Lowest-observed-adversed-effect level (LOAEL)
Local Environment Plans (LEPs), 58
Logic. *See* Fenoxycarb
Lolium spec., 1925
Longitudinal dispersion, 1282
Long-range transboundary air pollution (LRTAP), 2298
Long-wave radiation, 865
LOQ. *See* Limit of quantitation (LOQ)
Los Angeles smog, 1130f
Louisiana Coastal Area Ecosystem Restoration Program (LCA), 628
Love Canal, 1333
Low-cost/no-cost energy-saving projects
ambient air temperature, reset on, 739–740
comfort and safety, 735
corporate payback analyses, 735
cost-accounting system, 740
dedicated AC Units to cool server closets, 739
domestic hot water heater, installing, 737
fan speeds, reducing, 739
gross system overcapacity, reducing, 737–738
lag/lead, reducing, 738–739
longer-term consequences of, 735
lugs on time clocks, replace, 736
manual light switches, optimize strategy for, 740
systems fighting, resolve, 738
thermostats, relocating, 738
tighten leaky outside air dampers, 739
tighten schedules, 736
update schedules, 736–737
Low-cost treatment systems, 2157
"Low-dose effects," 1405
Lower heating value (LHV), 374
Lowest-observed-adversed-effect level (LOAEL), 2006
Low/no-growth economy, 616–622
business-as-usual scenario, 617–618, 621t
funding public services in, 621–622
high-level structure of, 617f
policy directions for, 618–622, 620t
Low-quarter distribution uniformity, 1548
Low-rise multifamily buildings, 860
Low-temperature solar thermal technology, 242
Loxodonta africana, 2595
LRIS. *See* Land resource information systems (LRIS)
LRTAP. *See* long-range transboundary air pollution (LRTAP)
Lufenuron, 1462
Lungworms (*Rhabdias* spp.), 2596
Lyases, 89

Lycopersicon esculentum (Tomato), 425, 1983
Lygus, 1936
Lygus lineolaris, 368
Lymantria dispar, 2009
Lymantria monacha, 382
Lymnaea acuminata, 384
Lymnaea stagnalis, 383
Lynchets
  ancient, 2536
  contemporary, 2536–2537

M

Maas–Hoffman equation, 2676
Mach II. *See* Halofenozide
Mackinawite, 42
Macrofauna monitoring index, 603
Macroinvertebrate communities, exergy of, 1100
Macrophyte(s), 2863f, 2864f
Macrophyte reintroduction, 1592t
Macropores, 1318
Magnesium oxide process, 2443
Magnesium wet scrubber, 2441
Magnetic heat pump, 1348, 1348f. *See also* Heat pumps
Magnetic nanoparticles, 1715
Maize (*Zea mays* L.), 426
"Maize-ley" system, 1939
Malancha (*Alternanthera philoxeroides*), 426
Malathion, 2805, 2806
Management system for energy (MSE) 2000, 765
Mancozeb, 2142, 2806
Maneb, 1318, 2806
Manganese (Mn), 1370
Mange mites, 2
Mangrove swamp (Mangal), 2825t
Manipur, groundwater arsenic contamination in, 1268
Manual sampling
  of greenhouse gases (GHG), 1202–1203
Manufacturing Energy Consumption Survey (MECS), 794
Manure
  application, 1668
    and NO_3-N concentrations, 1304
    rates, 1810
    site considerations for, 1811
    timing, 1811
  livestock feed management, 1812–1813
  management, 1810–1813. *See also* Dairy, water use in
  poultry. *See* Poultry manure
  storage, 1811–1812
Manure inventory, 1814
Manure phosphorus concentration reduction, 1667
Marginal benefits, 918f
Marginal cost of abatement, 918f
Marine and estuarine ecosystems, 607
Marine debris, 496–497, 2360
Marine pollution, 2357, 2358
  acidification and acid rains, 2359
  atmospheric pollution, 2359
  deep sea mining, 2359
  eutrophication, 2360
  land and river runoffs, 2358
  noise pollution, 2361
  oil and ship pollution, 2358–2359
  plastic debris, 2360–2361
  radioactive waste, 2359–2360
Marine Strategy Directive, 2187
Market potential
  for biofertilizers, 349, 350t
MARPOL Convention, 1828
Mars, 1716
Marsh, 2825t, 2826f
Masonry walls
  adobe walls, 864
  concrete block and poured concrete, 863
  insulated concrete forms (ICF), 864, 864f
  precast concrete, 863–864
Mass and energy, 808
Mass curve, 1287, 1287f
Mass flow
  nutrient transfer by, 1819–1820, 1820f
Mass rearing entomophagous insects, 1716
Mass spectrometry (MS), 1975
Massachusetts Department of Environmental Protection (MA DEP), 2049, 2050, 2050t, 2052
Match. *See* Lufenuron
Mathematical modeling
  for saltwater intrusion, 1328–1329
Mathematical water quality models, 2749
Matric. *See* Chromafenozid
Matricaria perforate, 1928
Matricaria spp., 1928
Matrix solid-phase dispersion (MSPD), 2198
Maturation ponds, 2634–2635
  for fecal coliform removal, 2636
Maximum achievable control technologies (MACT), 1129
Maximum contaminant levels (MCL), 916, 1289, 1317, 1318t
  of contaminants from mining, 1290t
  for nitrate, 1303
Maximum permissible concentrations (MPCs), 1518
Maximum power point (MPP)
  in photovoltaic systems, 219, 220
Maximum power point tracking (MPPT), 220–221
Maximum residue limit (MRL), 1124, 1125–1126, 1125t, 1403
MCS. *See* Multiple chemical sensitivities (MCS)
Meals ready to eat (MRE) waste, 847
Mean annual rainfall (MAR), 944
Meandering, 2313
Mean residence time (MRT). *See also* Soil organic matter (SOM) turnover
  defined, 1872, 1873f
  of total soil organic C, 1874, 1874t
    effect of tillage practices, 1875t
Mechanical, electrical, and plumbing (MEP), 669
Mechanical clearing methods for *A. donax*, 1186
Mechanical scrapers, for manure removal, 1672, 1672f
Mechanical scrubbers, 156
Medicago sativa L. (Alfalfa), 427, 1552
Mediterranean-type climates, 1183
Melaleuca spp., 49
Meloidogyne spp., 1752
Member states (MSs), 903, 904
Membrane biofilm reactors (MBfRs), 2611, 2612f, 2622, 2624–2625
Membrane extraction, 2195–2196
Membrane filtration, 2725–2727, 2725f. *See also* Filtration
  applicability of, 2727
  future developments of, 2726–2727
Membrane inlet mass spectrometry (MIMS), 1847
Men, hormonal disruption in, 1406
Mental component score (MCS), 2875f
Menu systems, of IPM, 1522
MEP. *See* Mechanical, electrical, and plumbing
MEPS. *See* Molded expanded polystyrene
Mercuric chloride, 2016t
Mercury (Hg), 1370, 1676
  contamination, in coastal waters, 490, 491
  health impacts of, 2119
  important processes, 1676–1677
  pollution
    abatement of, 1677–1678
    effects, 1676
    sources of, 1676
Metabolic degradation
  herbicides, 1379
Metabolic engineering, 397
Metalaxyl, 2020
Metal contacts, in solar cells, 233–234
  rear contacts, 234
Metal framing, 1486
Metallothionein, 394
Metal retention processes, 1374
Metal surface treatment
  cadmium for, 453
Metalloids, in coastal waters, 490–491
Metamorphosis, 1459
Metaphycus helvolus, 1750

Metarhizium anisopliae, 382, 1988
Methane (CH_4), 486, 1203
emissions from rice fields, mitigating options for, 1679–1682, 1681t
organic matter management, 1680
problems and feasibility of the options, 1681–1682
processes controlling, 1679, 1680f, 1680t
soil amendments and mineral fertilizers, 1680
water management, 1679, 1680f
problems associated with biowaste, 512
soil quality and, 2388–2389
Methanogens, 1679, 2651
Methanol, 188
from biomass, 378
Methanopyrolysis, 2487
Methemoglobinemia, 1302, 1303, 1559
Methiocarb (Mesurol®), 415
Method of characteristics (MOC), 1315–1316
Methoprene, 1460, 1464
Methoxychlor, 1318t, 1405, 2015, 2015f
Methoxyfenozide, 1466
Methyl anthranilate, 415
Methyl bromide, 1886–1887, 2015, 2015f, 2016t, 2114
Methyl esters, 2021
Methylparathion, 2005, 2015, 2015f, 2805
Methyl tert-butyl ether (MTBE), 916
Metschnikowia pulcherrima, 1991
Mexican bean beetle, 1935
Mexico Nitrogen Index, 1763
Mexico
saltwater intrusion in, 1322
wastewater reuse in, 1571
Micelles, 2194
Microarrays, 2130
Microbial biopesticides, 381
Microbial community, 2616
Microbial degradation
of organic pollutants, 388–392
Microbial inoculants, 1515
Microbial isomerization reactions, 390
Microbial siderophore uptake, 351
Microbiological research methods, for drinking water analysis, 2799–2800
Microclimate for natural enemies, 371–372
Microflora in soil, 2169
Micro hydro plant, 202
Microirrigation, 1547t, 1548, 1549
Microlife DCB series bioremediation products, 397f
Micro–meso–macro method, 1906
Micrometeorological methods
for GHG, 1203–1204, 1203t, 1205f
Micromite. *See* Diflubenzuron
Micronucleus assay, 1156–1157, 2127
Micronutrients, 1374
Microorganisms, 2170
nanoparticles and, interactions of, 1716–1718, 1717t
Microrills, 953
Microscale solvent extraction (MSE), 2193–2194
Microwave-assisted extraction (MAE), 2197
Milardet, Pierre, 2009
Milbemycins/avermectins, 2029, 2029f
Milking equipment, washing of, 1664
"Milky disease," 321
Millennium Assessment, 608
Millennium Ecosystem Assessment (MA), 557
Miller Amendments, 1940
Mill tailings, 1683
Mimic. *See* Tebufenozide
Mined aquifer, 1285f
Mine drainage, 1688, 1689f
causes of, 1688
chemistry of, 1688–1689
control of, 1690
prevention, 1690
treatment, 1690–1691
environmental impacts of, 1690
microbiology related to, 1689–1690
mineralogy of, 1689
problems by, 1688
total acidity of, 1690
Mineral fertilizers
CH_4 emissions from rice fields and, 1680, 1680t
Mineralization, 90, 460, 2176
carbon losses from soil and, 1857
influence of heavy metals on, 162
sulfur, 2432
Mineral matter (MM), 717
Mineral weatherability, 1625
Mineral wool insulation, 1478
health impacts of, 1479
Minimum efficiency reporting value (MERV), 1657
Mini hydro plant, 202
Mini hydropower (MHP), 203
civil works in MHP plants, 208–212
Mining
classification, 1692
for coal, 1289, 1291
groundwater pollution by, 1289
abandoned mine sites and, 1291
acid mine drainage and, 1291
biological methods for remediation, 1290, 1293
chemical methods for remediation, 1293
containment of contaminants in, 1293
gold mining and, 1291–1292
groundwater analysis, interpretation of, 1292
groundwater resources, 1289–1290
hydrogeological characteristics of site and, 1290, 1290t
metal contaminants, 1291
organic contamination, 1291
physical methods for remediation, 1293
and pump-and-treat method, 1293
and remediation strategies, 1292–1293, 1293f
and in situ remediation, 1293
tests for constituents, 1292, 1292t
transport of contaminants, 1290
water contaminants, 1290–1292, 1291t
water movement, 1290
impact on groundwater supplies, 1289
surface, 1289
Missouri River ("the Big Muddy"), 1714
Mites, 1
damage from mite feeding, 2
general description of, 1
population outbreaks, control of. *See* Acaricides
spider, 1–2
Varroa, 2
Mitigating desertization, 570–571
Mitigation, 2877
ammonia volatilization and, 87
regulating permissible noise level, 2877–2879
Mitsui Babcock ABGC (air-blown gasification cycle), 2491
Mixed liquor, 2648
Mixed-liquor suspended solids (MLSS), 2649
Mixed-liquor volatile suspended solids (MLVSS), 2649
Mixing height, 1130
MLSS. *See* Mixed-liquor suspended solids (MLSS)
MLVSS. *See* Mixed-liquor volatile suspended solids (MLVSS)
Mn. *See* Manganese (Mn)
Mo. *See* Molybdenum (Mo)
MOCDENSE
for saltwater intrusion, 1329
Mode of governing, 2421
Moderate resolution imaging spectrometer (MODIS), 554, 2817t, 2818
Modified Fournier index, 943, 944
Modified universal soil loss equation (MUSLE), 975–976, 982, 2759
Modified Wilson and Cooke (MWAC) samplers, 1022, 1023f
MODIS. *See* Moderate resolution imaging spectrometer (MODIS)
Module cables, 222
Module inverters, 222, 222f
Modulus of rupture (MOR), 2341f
Moisture, carry-over, 1563
Moisture ash free (MAF) basis, 717
Moisture content
biomass feedstocks, 374
in composting materials, 521–522

Page numbers followed by f and t indicate figures and tables, respectively

Molded expanded polystyrene (MEPS), 1478
Molecular nitrogen
  emission of, 1770–1771
Molinate, 2806
Molten carbonate fuel cells (MCFCs), 1139–1141
  acceptable contamination levels, 1140
  applications of, 1141
  fuel for, 1140
  operating principle, 1140, 1140f
  operating temperature of, 1139
  problems related to, 1140–1141
    cathode dissolution, 1141
    corrosion of separator plate, 1141
    electrode creepage and sintering, 1141
    electrolyte loss, 1141
    reforming catalyst poisoning, 1141
  technological status of, 1141
Molting, 1459, 1465
Molybdenosis, 428
Molybdenum (Mo), 424, 1370, 2586, 2588–2589
  deficiencies, 424
  monitoring of, in soils and waste materials, 428
  in soils, 426–427
  sources of, 426
  toxicity, 424
    cattle grazing and, 428
    cause of, 427
    clinical signs of, 427
    Cu:Mo:S ratio and, 428
    irrigation and, 428
    prevention of, 428
    sources of, 427
    symptoms and levels, 427–428
  use of, 426
Monacha cantiana, 383
Mongooses (*Herpestes auropunctatus*), 2596
Monitoring wells
  for saltwater intrusion, 1325
Monocrotophos, 417, 1993
Monocrystalline silicon, 231–232
Monoculture, 1935
  defined, 1937
  intercropping and, 1937, 1938t
  resistance gene, 1937
Monod equation, 2634
Monolith bioreactor, 2625
Monosulfides, 42
Monosulfidic black ooze (MBO), 44, 45
Monoterpene cineoles, 2031, 2031f
Monothanolamine (MEA), 457
Monte Carlo uncertainty analyses, 1721, 1722
Montevideo Programme, 1619
Montmorillonite clay, swelling of, 2338, 2339f
Montreal Protocol, 1882, 1893–1894, 1897
Moon, 1716
Moore's law, 788
Moraxella osloensis, 383
Morocco, 2682
Morrell Act (1862), 2009
Mosquito control, 1471
  nontarget mortality during, 1471
    minimization of. *See* Aerial ultra low volume (ULV) application, of mosquito insecticides
  ULV application for, 1471
Motile algae, 2633
Mousepox virus (Ectromelia virus), 2604
MRL. *See* Maximum residue limit (MRL)
MRT. *See* Mean residence time (MRT)
MSW. *See* municipal solid waste (MSW)
MTBE (methyl tertiary butyl ether), 157
Mucus melanosis, 1272
Mud for roof catchment, 2263
Mueller, Paul, 2009
Mulches, 961–962, 1110
Mulch tillage, 1042, 1042f
Multi-circulating fluidized bed combustor (MCFBC), 729
Multicomponent mixtures, reduced mixture toxicity designs for, 2579
Multicriteria decision making (MCDM), 2399–2400
Multidisciplinary approach for developing environmental policy, 914–915
Multifunctional crop rotation, 1508
Multiple-attribute decision making (MADM), 2400, 2403–2404
Multiple chemical sensitivities (MCS), 1411. *See also* Pesticide sensitivities
  causes, 1411–1412
  symptoms, 1412, 1412t
Multiple-objective decision making (MODM), 2400
Multipurpose reservoir, 1577
Multiresidue methods, 1126
Municipal sewage, 1570
Municipal solid waste (MSW), 512, 715, 728, 895, 902, 902f, 2415, 2482
  characteristics and waste quantities, 2416–2418
  co-composting of, 514
  composition, 2483
  heat of combustions of, 729t
  historical overview, 2415–2416
  management practices, 2418
  modes of governing, 2420t
  organic fraction of, 513, 514
  organic matter in, 513, 519, 520
  valorization, 513
  Municipal wastewater, 2709–2718. *See also* Wastewater
  composition of, 2709, 2709*t*
  recycling of, 2717–2718
  treatment methods for
    biological oxygen demand, reduction of, 2709–2712, 2710–2713f
    nitrogen concentration, reduction of, 2715–2717, 2716f, 2717f
    phosphorous concentration, reduction of, 2712–2715, 2713–2714f
Municipal WWTP systems, removal efficiencies, 2063–2066
Murine cytomegalovirus (MCMV), 2604
Murray–Darling basin (MDB), 56
Musca domestica, 1460, 1920, 1921
Musca sorbens, 1920, 1921
Muscidifurax, 1922
Muynak, Aral Sea disaster and, 301–302
Mycalesis gotama Moore, 311
Mycoherbicide, 383
Mycoremediation, 392
Mycorrhizae, 349, 1110, 1516
Mycotoxins, 1696
  aflatoxin, 1697
  deoxynivalenol, 1696
  fumonisin, 1697
  ochratoxin, 1697
  and pesticides, 1698
  workplace exposure to, 1698
  zearalenone, 1696
Mycotrol®, 382
Myriophyllum spicatum, 367, 1922
Myriophyllum spp., 2830
Myxomatosis, 2597. *See also* Myxoma virus (MyxV)
  in Australia, 2597–2600
  in Europe and other regions, 2600
Myxoma virus (MyxV), 2597
  as biological control for rabbits, 2597–2599, 2598f
  genetic resistance to, 2599, 2599f
  Glenfield strain of, 2600
  Lausanne strain of, 2600
  standard laboratory strain (SLS), introduction of, 2600
  vector of, introduction of, 2599–2600

N

Na/Cl ratios
  for saltwater detection, 1327
Nafion, 1148
Nanocrystalline films
  model of photocurrent generation in, 236, 237f
Nanofood regulation, 1703–1704
Nanomaterials, 1700
  dose–response relationship, 1705
  environmental pollution by, 2129
  exposure assessment, 1705
  nanofood regulation, 1703–1704
  pharmaceutical regulation, 1702–1703

Nanomaterials (*cont'd.*)
REACH, 1700–1701
risk assessment of, 1704
technical guidance, 1706
risk characterization, 1705–1706
WFD, 1701–1702
Nanoparticles, 1711–1718, 1720
aluminum oxyhydroxide particles, 1714, 1714f
in atmosphere, 1715, 1715t
beyond the Earth, 1716
"critical zone," 1711
defined, 1711
environmental risks, 1723–1725
health effects, 1717
and life, 1716–1718, 1717t
magnetic, 1715
microorganisms and, interactions of, 1716–1718, 1717t
overview, 1711
physical chemistry of, 1711–1712, 1711f, 1712f, 1712t, 1713f
in sediments, rocks, and the deep Earth, 1715–1716
in soil and water, 1712–1715, 1713t, 1714f
"X-ray amorphous," 1712, 1713f
Nano-semiconductor catalysts
decontamination of toxic pollutants by, 1733–1736, 1734–1735t
pollutant removal, mechanism in, 1731–1733
Nanotechnology, 1720–1721, 1730–1741
significance of, 1730–1731
NAPAP. *See* U.S. EPA National Acid Precipitation Assessment Program (NAPAP)
Naphthalene, 388
NAPL. *See* nonaqueous phase liquids (NAPL)
Napropamide, 2024f
National Acid Precipitation Assessment Program (NAPAP), 7
National Ambient Air Quality Standards (NAAQS), 133, 1639
National Arsenic Mitigation Information Centre (NAMIC), 1269
National Atmospheric Deposition Program (NADP), 7
National Environmental Policy Plans (NEPP), 2446–2454
National Fenestration Rating Council (NFRC), 866, 867f
National Organic Standards Board (NOSB), 125
National pesticide assessment, 1317
National Pesticide Field Program, 1945
National Pollutant Discharge Elimination System (NPDES), 1289, 1943, 2808
permit program, 2809–2810, 2810f
National Toxicology Program (NTP), 477
National waste regulation in EU countries, 894
economic instruments, 897–900
regulatory instruments, 897
National Water Quality Assessment (NAWQA), 963, 1317, 2804
Natural acidity, 7
Natural amelioration, 569
Natural capital, 2465
Natural circulation systems, 247, 247f
Natural climate forcings (solar and volcanic variability), 1194–1195, 1196t
Natural cycling of materials, 2863
Natural enemies. *See also* Biological control
alternative food sources and, 371
alternative prey/hosts, 371
for augmentation, 1474
conservation of, 1749–1751
habitat and environmental manipulation, 1750
harmful pesticides practices avoidance, 1749–1750
overview, 1750
overwintering and shelter sites, 1750
cultural practices and, 370
host plant effects and, 371
pesticides effects on, 370
secondary enemies and, 371
shelter and microclimate for, 371–372
strategies for using, 1474
types of, 1475–1476
parasitoids, 1475, 1475f
pathogens, 1475–1476, 1476f
predators, 1475, 1475f
Natural enemies, rearing of
arthropod parasitoids/predators, 1746
artificial diets
for predators and parasitoids, 1746–1747
successes and failures with, 1747
quality control of, 1747
Natural energy, 770, 771. *See also* Energy
non-solar-related energy, 772
solar energy, 771–772
air-based, 772
land-based, 772
types of, 771t
water-based, 772
Natural-environment-subsystem models, for exergy analysis, 1094
Natural erosion, 1010
Natural gas, 793, 794
Natural gas combined cycle (NGCC), 840–843
Natural greenhouse effect, 1192–1193, 1193f
Natural mortality, 1115–1116
Natural occurrence
allelochemics, 175–177, 176t, 177f
Natural pesticides, 2028–2031
fungicides and bactericides, 2029–2030
kasugamycin, 2030, 2030f
polyoxins, 2030, 2030f
strobilurins, 2030, 2030f
validamycin, 2030, 2030f
herbicides, 2030–2031
allelopathy, 2031
bialaphos, 2030–2031, 2031f
cyperine, 2031, 2031f
monoterpene cineoles, 2031, 2031f
phosphinothricin, 2030–2031, 2031f
triketones, 2031, 2031f
historical use for pest management, 2028
insecticides, 2028–2029
azadirachtin, 2029, 2029f
Bacillus Thuringiensis, 2029
juvenile-hormone mimics and pheromones, 2029
milbemycins/avermectins, 2029, 2029f
nereistoxin derivatives, 2029, 2029f
nicotine, 2029, 2029f
pyrethrins, 2028–2029, 2029f
spinosyns, 2029, 2029f
overview, 2028
prospect, 2031
structural diversity, 2028
Natural products, for pest control, 2021–2022, 2021f
Natural Resources Conservation Services (NRCS), 127
Natural Resources Damage Assessment (NRDA) process, 629
Natural resources degradation, 2816
Natural riverbed protection, 2313
Natural sources of freshwater acidification, 2292–2294, 2293f
geochemical factors, 2293–2294
geological factors, 2292–2293
Natural wetlands, 2862, 2863t
Nature, role in carbon removal from air, 1190
Natuur and Milieu, 1404
Navel orangeworm management in almond orchards, 1919–1920
NAWQA. *See* National Water-Quality Assessment (NAWQA)
NDVI. *See* Normalized difference vegetation index (NDVI)
Necrotrophic/hemibiotrophic fungi, 383
Negative lynchet, 2537
Negligence, pesticide use and, 1944
Nemagon, 288
Nemaslug, 383–384
Nematicides, 2014
Nematodes, biological control of, 1752–1753
future perspectives, 1753
nonspecific biocontrol agents, 1753
overview, 1752
specific biocontrol agents, 1752–1753

Neporex. *See* Cyromazine
Neptunium-237 decay series, 2236, 2239f
Nereistoxin derivatives, 2029, 2029f
Netherlands, The
- saltwater intrusion in, 1322

Net primary productivity (NPP), 2376, 2833
Network of industries, 1434
Neural tube birth (NTD), by fumonisin, 1697
Neuroesterase (NTE), 1399
Neurological effects, of arsenic toxicity, 1273
Neurotoxicants evaluation, 1755–1758
- protocols for, 1757–1758
- vulnerability, periods of, 1756–1757

Neutralization, of acidity, 49
New Orleans Regional Medical Center (NORMC)
- district cooling, 2515–2516, 2516f

New Waste Framework (NWF) Directive, 892, 911
New York Nitrate Leaching Index, 1763
New Zealand
- soil degradation in, 2384–2385, 2384f
  - distribution of, 2384f

Nezara viridula, 367–368
N_2-fixing cyanobacteria, 349
Ngarenanyuki, Tanzania
- pesticide impact in, community-based monitoring of, 505–511
  - adherence to pesticide label instruction, 509–510, 509t
    - community pesticide monitoring team, establishment of, 508
    - data analysis, 508
    - data collection, 508
    - disposal of pesticide containers, 510, 510t
    - methodology of, 507–508
    - pesticide affordability, 509
    - pesticide availability, 509
    - pesticide mixing, 509
    - pesticide poisoning, 510–511
    - pesticides, application of, 509

NGO initiatives, 1404
NH_3. *See* Ammonia (NH_3)
Ni. *See* Nickel
Nickel (Ni), 1370
Nickel-cadmium batteries, 447
Nicotine, 2029, 2029f
Night soil sludge, 2267t
1985 Helsinki Protocol, 15
1999 Gothenburg Protocol, 15
Niobium, 2585, 2587–2588
Nitrapyrin, use of, 1306
Nitrate, 1805, 2753
- in groundwater, 1282, 1283
- leaching, 1559
- pollution, 2304
- reductase, 426
- runoff, 345

Nitrate fertilizer
- and soil IC, 464

Nitrate Leaching Hazard Index, 1763
Nitrate leaching index, 1761–1765, 1764f
- assessment, 1762–1763
- components, 1763–1765
- environmental problems, 1761–1762
- quick tools and indicators, 1762

Nitrate-nitrogen (NO_3-N), in groundwater, 1302
- agricultural practices contributing to
  - containerized horticultural crops, 1306
  - grasslands/turf, 1305–1306
  - manure application, 1304
  - nitrogen fertilizer, 1304
  - row crops, 1303–1304
- elevated levels of, 1302
- environmental impacts, 1303
- health problems by, 1302, 1303
- by leaching of N fertilizer, 1302–1303
- nitrogen sources of, 1303
- occurrence of, 1302–1303
- strategies, for reducing NO_3-N leaching, 1306–1307
  - cover crops, 1306–1307
  - diversified crop rotation, 1307
  - grassland/turf management, 1307
  - nitrification inhibitors, 1306
  - reduced tillage, 1307
  - soil testing and plant monitoring, 1306

Nitric oxide (NO)
- soil quality and, 2389–2390

Nitrification, 93, 1770–1771
- defined, 97
- inhibitors, 1306, 1809
- nitrous oxide from, 97
- process, 1768–1769, 1769f

Nitrifiers, 2862
Nitrite, 1559
Nitrobacter, 97
Nitrogen (N), 88, 350, 1074–1075, 1304, 1513, 1559, 1761, 2434, 2753, 2865t
- acid rain and, 20–24
  - ecosystem impacts, 22–24
  - human health effects, 22
  - reducing acidic deposition effects, 24
  - sources and distribution, 20–22, 21f
  - structural impacts, 22
- agricultural runoff and, 82–83
- ammonia volatilization, 1770
- atmospheric deposition, 1769–1770
- biological fixation, 1770
- concentration, reduction in municipal wastewater, 2715–2718, 2716f, 2717f
- cycle, 88–90, 89f, 1513, 1768, 2389f
- decomposition and mineralization, 90. *See also* Soil organisms activity, environmental influences on
  - organic substrate quality, effect of, 92–93, 93f
  - soil microorganisms, effect of, 90–92
- effect on surface water quality, 2756–2757, 2758t
- fixation, 88–89, 426
- gases. *See also specific* entries
  - air pollution problems of, 158–159
  - soil quality and, 2389–2390
- input processes, 1769–1770
- leaching, 94, 1771
- losses of, 89–90, 93
  - measurements, need for, 1772
  - mitigation of, 93–94
  - pathways for, 1777–1778
- loss processes, 1770–1771
- manure
  - application of, 1810, 1810f
- mineral and organic fertilization, 1770
- and NO_3-N contamination, 1303
- overview, 1768
- runoff losses of, 94
- transformations in soil, 1768–1769, 1769f
- uptake by plants, 1770

Nitrogenase, 88, 89, 426
Nitrogen cycle, 1777
- conceptual diagram of
  - in an aquatic ecosystem, 1076f

Nitrogen dioxide (NO_2)
- soil quality and, 2389–2390

Nitrogen fertilizer, NO_3-N contamination from, 1303
Nitrogen-fixing bacteria, 88
Nitrogen-fixing biofertilizers, 1516
Nitrogen (N) nutrient
- applications, 1806, 1806f
  - additional tests for fine-tuning, 1806
- controlled-release nitrogen fertilizers, 1809
- fall, 1808
- nitrification inhibitors, 1809
- preplant, 1808
  - side-dress, 1808–1809
  - split, 1809
  - timing, 1808–1809, 1808t
- placement, 1809–1810

Nitrogen oxides (N_xO_y), 1154–1155, 2304
- characteristics of, 137
- EN 14211, chemiluminescence for, 141–142

Nitrogen phosphorus detector (NPD), 1975
Nitrogen trading tools (NTTs), 1772–1782
- description of, 1779–1780
- examples of, 1775–1776, 1781–1782
- functionality of, 1780
- limitations of, 1778
- outputs of, 1780
- role of, 1774–1775

Nitrosomonas europaea, 97

Nitrous oxide (N_2O), 486, 1202
emissions, 96–98, 1770–1771
from agriculture, 96–98
biomass burning, 97
denitrification, 96–97
fertilizer consumption and, 98
flooded soils, 97
management practices to reduce, 98
nitrification, 97
overview, 96
soil quality and, 2389–2390
NO. *See* Nitric oxide (NO)
NO_2. *See* Nitrogen dioxide (NO_2)
N_2O. *See* Nitrous oxide (N_2O)
NOAEL. *See* Nonobservable-adverse-effect level (NOAEL)
Noble metal nanoparticles
bactericidal effect of, 1740
pollutant removal, mechanism in, 1739–1740
Nodding thistle *(Carduus nutans),* 1526t
No-growth disaster, 618
Noise pollution, 2361
Noise pollution *See also* Pollution
in coastal waters, 497
industrial wind turbines, 2867–2868
mitigation, 2877
regulating permissible noise level, 2877–2879
regulating setback distances, 2879
wind turbine noise, acoustic profile of, 2868–2870
wind turbine noise, human impacts of, 2870
and annoyance, 2871–2873
health impacts, quantifying, 2870–2871
and low-frequency/infrasound components, 2876–2877
psychological description, 2870
and sleep, 2873–2876
wind turbine syndrome, 2876
Nomolt. *See* Teflubenzuron
Nonaqueous phase liquids (NAPL), 1334
Non-catalytic WO, 1908–1909
commercial, 1909t
Nonchemical/pesticide-free farming, 1103–1108
basic principles, 1103
definitions, 1103
global growth, 1103–1106, 1104t–1106t
objectives, 1103
organic farming. *See* Organic farming
producer-consumer driven movement, 1107–1108
Non-condensable gases, 2485
Non-dispersive infrared, for CO EN 14626, 142
Non-energy input, 791
Nongovernmental organizations (NGOs), 1221
Non-Hodgkin's lymphoma, 1399
Nonisotopic tracer methods, 1204
Non-metals, 1560
Non-motile algae, 2633
Nonobservable-adverse-effect level (NOAEL), 2006
Non-point source of water pollution, 1580
Non-point source pollution (NPSP), 2167t–2169t
chemicals, associated, 2137t
contaminant interactions, 2136
contributors to, 2137t
example of, 2136
global effect, 2138
management of, 2137–2138
rainwater as source of, 2136
remediation of, 2137–2138
soil/environmental quality, 2136–2137
urban sewage and, 2136
water pollution prevention, 2138
Non-potable uses of water, 2262
Non-saline water, 1569t
Non-steroidal anti-inflammatory drug (NSAID), 2062
Non-threshold agents, 1456
Nonylphenol, 1405
Normalized difference vegetation index (NDVI), 2398, 2397f
North America
soil degradation in, 2381–2383, 2382f
distribution of, 2383f
Northern Agricultural Catchments Council of Australia, 1028
Northern jointvetch *(Aeschynomene virginica),* 1526t
NOSB. *See* National Organic Standards Board (NOSB)
NO_3 test, late-spring, 1306
No-till practices
adoption of, 1036–1037
conservation farming system, 119–120
erosion process and, 1040–1041, 1041f, 1042f
Novaluron, 1460, 1462
Noviflumuron, 1462
Nozzles, 1472
NPP. *See* Net primary production (NPP)
NPSP. *See* Non-point source pollution (NPSP)
NRCS. *See* Natural Resources Conservation Services (NRCS)
NTE. *See* Neuroesterase (NTE)
Nuclear and coal-fired power plant overnight costs, 1790–1792
Nuclear atmospheric weapons tests, 2304
Nuclear power and wind power, 183–185
Nuclear power economics, 1789
capital costs for nuclear and coal-fired powerplants, 1790
cost of alternative, 1796–1797
discount rates, 1795–1796, 1800f, 1801f
environmental costs and regulations, 1793–1795
Fukushima nuclear disaster impact on, 1803–1804
nuclear and coal-fired power plant overnight costs, 1790–1792
results, 1797–1803
total project costs derivation, 1792–1793
Nuclear reaction, 2235
Nuclear Regulatory Commission (NRC), U.S., 1789, 1794
Nuclear Waste Policy Act (NWPA) (1982), 1794
Nucleopolyhedroviruses (NPVs), 382
Nuisance algae growth prevention, 539
Nuisance claim, 1945
Nunavut, 2671
Nutrient(s), 1805–1814
application
rates, 1805–1808
timing, 1808–1809
credits, 1807–1808
effect on surface water quality, 2756–2757, 2758t
dissolved load from different continents to world's oceans, 2756t
farm nutrient management plans, 1813–1814
irrigation water management, 1813
management in soil, 2217–2222
manure management, 1810–1813
placement, 1809–1810
soil conservation practices, 1813
variable-rate fertilizer technologies, 1810
Nutrient crediting, 1813–1814
Nutrient loss from grazing lands, 2753
Nutrient management
diet modification on, 1668–1669, 1668f
integrated, 1508–1509
progress in practices, 1667
Nutrient mining, 1510
Nutrient-mobilizing biofertilizers, 1516
Nutrient reduction, 1584
Nutrient removal in WSPS, 2635
Nutrient runoff, 1080
Nutrient spiralling, 2101
Nutrient supply
and organic farming, 1107
Nutrient tracking tool (NTrT), 1776, 1782
Nutrient transfer
boundary conditions and, 1821–1822, 1821f
diffusion, 1817–1820
by mass flow and diffusion from soil to plant roots, 1819–1820, 1820f
overview, 1817
root hairs, importance of, 1820–1821, 1821f
soil plant system, 1817, 1818f
Nutsedges (*Cyperus* spp.), 1526t
Nylar. *See* Pyriproxyfen

O

Oak Ridge National Laboratory, 2243
Oat (*Avena sativa* L.), 1307
Oat-frit fly (*Oscinella frit* L.) system, 1939
OC. *See* Organochlorines
Occupational Safety and Health Administration, 2051
Ocean
  acidification of, 13
  currents, 824, 826
  fertilization, 460
  sources, 825–826
Ocean color and temperature scanner (OCTS), 2817t
Ocean direct injection, 459
Ocean thermal energy conversion (OTEC) devices, 772
Ochratoxin, 1697
Ochratoxin A (OTA), 1697
Octanol–water partition coefficient, 1283
Octylphenol, 1405
Odor control, of composting materials, 524
Oedaleus asiaticus, 1990
Office of Pesticide Programs (OPP), EPA, 1941–1942
Office of Prevention, Pesticides, and Toxic Substances, 1406
Off-site movement of pesticides, 2804–2805
Off-site pollution, 1559
Off-site water quality impact reduction, 1557, 1560–1561, 1561t
Off-the-farm discharge, 1557
OFPA. *See* Organic Food Production Act (OFPA)
Oil and ship pollution, 2358–2359
Oil pollution, in Baltic Sea, 1826–1840
  observations of, 1828
  oil spill drift, numerical modeling of, 1833–1835, 1835–1838f
  operational satellite monitoring of, 1830–1833
  problems and solutions of, 1838–1839
  sources of, 1827–1828
  statistics of, 1828–1830
  tendencies of, 1830
Oils, in coastal waters, 496
Oil spills, 2042, 2043t
  drift, numerical modeling of, 1833–1835, 1835–1838f
O isotopes
  for saltwater detection, 1327–1328
Oka, I.N., Dr., 2034
Oman, 2682, 2690
Omnivorous feedstock converters (OFCs), 830, 832f
Oncomelania hupensis, 384
On-farm nutrient resources, assessment of, 1813
Online gas chromatography, for benzene EN 14662, 142
On-site pollution, 1559
On-site residential wastewater, 2862, 2865f
Onthophagus gazella, 1461
Opassess.com, 1493
Open-cell, low-density polyurethane insulation, 1479
Open circuit systems, 248
Open collectors, 243, 243f
Open ecosystems, 1935
Open pit mining, 1692–1694
  acid mine drainage, 1694
  coversoil resources, 1692
  coversoil thickness requirements, 1693
  landscape regrading, 1692, 1693f
  in situ soil reclamation, 1693
  steep slope reclamation, 1694
Open-tubular trapping (OTT), 2195
Operational satellite monitoring, of oil pollution, 1830–1833
Operations and maintenance (O&M) program, 657
Operations research methods in pest management, 1950
OPIDIN, 1399
OPIDP. *See* Organophosphate induced delayed poly-neuropathy (OPIDP)
Opportunity cost, 2784
OPs. *See* Organophosphates (OPs); Organophosphorus compounds (OPs)
Optical sensors, 1024
Optimizing model, 2781–2783
  scarcity rents, 2782
  shadow prices, 2782–2783
Opuntia vulgaris, 2821
Oreochromis (*Sarotherodon*) *hornorum*, 1922
Oreochromis (*Sarotherodon*) *mossambica*, 1922
Organic agriculture, 125–127
  crop and pest performance in, 126–127
  crop rotations and, 126
  defined, 125
  global statistics, 126
  history, 125–126
  key issues requiring additional research, 127
  overview, 125
Organic carbon partition coefficient (K_{OC}), 2015
Organic chemicals, 2169
  in groundwater, 1281
Organic compounds, in coastal waters, 492–494
Organic farming
  animal husbandry, 1107
  animal pests
    regulation of, 1115–1117
    strategy of controlling, 1115
  basic principles, 1103
  bioregulators in, 1115–1116
  composting organic waste for, 528–533
    bacteria, 529–531
    probiotic agriculture, 528–529, 531–533
  defined, 528, 1115
  diversity, 1106
  genetically engineered organisms, 1110
  global growth, 1103–1106, 1104t–1106t
  green manure, 1107
  as holistic and systematic approach, 1106–1107
  as model for sustainable development, 1108
  nutrient supply, 1107
  objectives, 1103
  pest management, 1107, 1110–1111, 1115–1117
    diseases, 1112–1113
    insects and invertebrates, 1112
    weeds, 1111–1112
  producer-consumer driven movement and, 1107–1108
  soil building, 1110
  soil management, 1107
  standards, 1109–1110
  weed management, 1107
Organic fertilization on heavy metal uptake
  about heavy metals, 1374
  contamination with heavy metals, 1374
  heavy metal concentrations in organic fertilizers, 1375t
  plant uptake of heavy metals, 1375–1376
  sources of organic fertilizers, 1375, 1375t
Organic Food Production Act (OFPA), 125, 1108, 1109
Organic loading rate, 2650, 2652
Organic matter (OM), 1374, 1375
  bonded REE, 2267, 2268t
  sludge-derived, 1375
  in solid waste, 513
  turnover. *See* Soil organic matter (SOM) turnover
Organic matter (OM), global distribution of, 1851–1855
  andisols, 1854
  decomposition
    climate, 1853
    quality of matter, 1853
  histosols, 1854
  inputs
    placement, 1852–1853
    quantity, 1851
    species composition, 1852
  overview, 1851
  physical and chemical influences, 1853–1854
Organic matter (OM) management, 1857–1861

Organic matter (OM) management (*cont'd.*)
carbon losses from soil
mineralization processes and, 1857
soil erosion processes and, 1857–1858
CH_4 emissions from rice fields and, 1680
soil carbon sequestration, 1858–1861
global importance of, 1858, 1859t–1860t
long-term field experiments, role of, 1858
mechanisms, 1858, 1860
SOM, energy, and full C accounting, 1861
SOM role in 21st century, 1861
Organic matter modeling, 1863–1868, 1865–1866t
application, examples of, 1864–1867
approaches, 1863–1864
challenges in, 1867–1868
performance, 1864
recent advances in, 1867
turnover, factors affecting, 1864, 1865–1866t
Organic mulches, 961–962
Organic nitrogen, 1559
Organic pollutants, 388, 2114–2117, 2115t, 2116t
bioremediation of, 2116–2117
degradation of, 1906
of drinking water, 2066, 2793–2797, 2795t, 2796t
determination of, 2795–2797, 2797f
potential impacts of, 2115–2116
soil contamination from, 2114–2115
Organic production and processing, 1103
Organic sludge, 961
Organic soil amendments, 959, 1878–1880
antagonistic microorganisms stimulation, 1879
future perspectives, 1880
impacts on plant health and weeds, 1878
mechanisms of action, 1878–1879
overview, 1878
plant resistance and, 1878, 1879t
release of compounds toxic to insects and plant pathogens, 1878
in 21st century, 1879–1880
Organization for Economic Co-operation and Development (OECD), 355, 647, 1037, 1219, 2124
endocrine screening and testing program, 648
DNT guidelines of, 1757
Organization of Petroleum Exporting Countries (OPEC), 485
Organochlorine acaricides, 2–3
Organochlorine insecticides, 1399
Organochlorine pesticides, 390, 1966–1967. *See also* Pesticides
chemical structure of, 1966f
Organochlorines (OC), 277, 2804, 2805, 2806
Organometallic complexes, precipitation of, 1625–1626
Organonitrogen pesticides, 1967. *See also* Pesticides
chemical structure of, 1967f
Organophosphate hydrolase (OPH), 390
Organophosphate induced delayed polyneuropathy (OPIDP), 1415
Organophosphates (OPs), 3, 277, 2005
Organophosphorus, impact on birds, 417
Organophosphorus compounds (OPs), 1398
background, 1397
effects of repeated exposure to, 1399
intermediate syndromes, 1398
mechanisms of action, 1398
Organophosphorus (OP) insecticide, 2804, 2806
Organotin acaricides, 3
Orientation of the intake, 210
Ornithonyssus sylviarum (northern fowl mite), 2
Orthophosphate, 100
Ortstein, 1625
Oryza sativa (Rice), 425
Osaka gas process, 1910
Oscinella frit. See Frit fly
Ostrinia nubilalis, 412, 1116
Otto cycle, 819
"Ouch ouch" disease, 448
Ovarian cancer, by NO_3-N exposure, 1303
Overflow, 1557, 1558t
Overgrazing and soil erosion, 1034
Overhead irrigation, 2677
Overnight cost, 1790
Overwintering, natural enemies conservation and, 1750
Ovicides, 2014
Owens Lake, degradation of, 1437–1438
Owner-operated rainwater harvesting, 2262
Owners' costs, 1791
Oxamyl (Vydate), 1318t
Oxidation reaction, 714
Oxidative biodegradation, 1337
Oxidative stress, 281
Oxident concentration, effect on homogeneous photo-Fenton reaction, 2552
Oxydemeton methyl, 2142
Oxydia trychiata, 367
Oxyfluorfen, 2142
Oxyfuel combustion, 457
Oxygen, 2862, 2864
consumption rate, 724
dissolved, 2753
starvation, 1559
transfer mechanism, 724
Oxygenation, 1592t, 1593–1594
Oxyreductases, 89
Ozone (O_3)
characteristics of, 137
EN 14625, ultraviolet photometry for, 142
Ozone-depleting substances (ODSs), 1886–1887
Ozone layer, 1882–1898
atmospheric ozone, 1883
CFC-ozone depletion hypothesis, 1885–1886, 1887f
climate change impact on, 1896–1897
depletion, biologinal effects of, 1893
origin of, 1883–1885
stratospheric ozone depletion
antarctic ozone hole, discovery of, 1887
atmospheric trace species, field measurements of, 1891
global ozone layer, depletion of, 1891–1893, 1892f
measurements and distribution of, 1885
polar ozone chemistry, 1887–1891
Ozone problems, 916

P

PA-14 (Tergitol®), 414
Packaging waste, 903–906
Packaging Waste (PW) Directive, 894
recovery rate of, 905f
recycling rate of, 905f
Packed-bed scrubbers, 156
Paecilomyces lilacinus, 1752, 1753, 1879
Paecilomyces lilacinus Thom (Sam), 288
Paenibacillus lentimorbus, 321
Paenibacillus (Bacillus) popilliae (Dutky), 321–322
culture and control, 321–322
saltwater intrusion in, 1322
PAH. *See* polycyclic aromatic hydrocarbons (PAH)
PAID. *See* Photo-acoustic-infrared detector (PAID)
Paleoenvironmental data and fly-ash particle analysis, 2292
Palestine
saltwater intrusion in, 1322
Palustrine emergent wetlands, 2827
Palustrine forested wetlands, 2827
Palustrine scrub-shrub wetlands, 2827
PAN. *See* Pesticides Action Network (PAN)
Panonychus ulmi, 2141
Papaver rhoeas, 1925
Papaya ringspot virus (PRV), 2111
Papilio polyxenes F., 313
PAR. *See* Photosynthetically active radiation (PAR)
Paraben, general formula of, 2083f
Paradichlorobenzene (PDB), 2009

Paraffin oil, 415
Paraquat, 1397, 1399
Parasarcophaga misera, 1920
Parasites, as biological control agents, 2596–2597
Parasitoids, 1460, 1475, 1475f, 2423, 2424. *See also* Natural enemies
  intercropping and, 1938
Parathion, 3, 419, 2806
Participatory approach, 561
Particle size, of organic materials, 523
Particulate matter (PM), 133, 142
  characteristics of, 137–138
Particulate pollution
  control methods, 149–156
  cyclones, 152
    distribution patterns, modifying, 150–151
    electrostatic precipitator, 153–154
    equipment, 151
    filters, 152–153
    settling chambers, 151
    wet scrubbers, 154–156
  problem, 149
  sources of, 149
Paspalum notatum (Bahiagrass), 1305
Passer domesticus, 413
Passivation, of silicon surfaces, 233
Passive heat storage
  solar, 2519, 2519f
Passive PRB system, 1336
Pasteuria penetrans, 1752–1753
Pasture lands. *See* Range and pasture lands
Pathogen, 1475–1476, 1476f. *See also* Natural enemies
  with livestock, types of, 2753
  pathogen control, of composting materials, 524
  protection mechanisms acting in intercropped systems, 1937, 1938t
  reductions, on-farm options for, 2678t
  in sewage effluent, 1570
Pathogenic bacteria, in drinking water, 2797–2799, 2799t
Paulatuk treatment wetland, 2664–2670
  aerial view, 2665f
  background information, 2664–2665
  discussion, 2669–2670
    natural ultraviolet radiation, 2670
  facultative lake to Arctic Ocean, 2666f, 2667f, 2668f
  influent and effluent concentrations, 2665t
  methods, 2665–2666
  results, 2666–2669
    cBOD5 effluent, expected, 2668, 2669f
    E. coli, concentration gradients of, 2669f
    effluent concentrations, 2666
    NH3 $^{+}$-N, concentration gradients of, 2668, 2669f
Pay-as-you-throw (PAYT), 895, 900, 911
Payment for ecosystem (goods and) services (PES/PEGS), 634
Pb. *See* Lead (Pb)
Pb contamination, in coastal waters, 490
PCB. *See* Polychlorinated biphenyls (PCB)
PCE. *See* Perchloroethylene (PCE)
PDB. *See* Paradichlorobenzene (PDB)
Peanuts (*Arachis hypogaea* L.), 424
Peas (*Pisum sativum* L.), 427
Peatlands, 1868, 2826f, 2827
Peclet number, 1315
Pedogenic carbonate, 1445–1446
Pelecanus occidentalis (Brown pelican), 417
Pellet(s), 375t
Pellet furnaces, 375
Peltier effect, 1348
Penicillium aurantiogriseum, 1697
Penicillium belaji, 1516
Penicillium expansum, 1991
Penicillium nordicum, 1697
Penicillium spp., 2109
Penicillium verrucosum, 1697
Pennisetum glaucum, 1012
Penstock, 210–211
Pentachlorophenol (PCP), 388
Pepper fruit extracts, 2021
Peppers (*Capsicum annuum* L.), 425
Per ascensum model, of carbonate formation, 1446
Percent tree cover map, 2376
Perchloroethylene (PCE), 1283, 1334, 1340
Percolation, 1564, 1566, 1570
Per descensum model, of carbonate formation, 1446
Perdix perdix (Grey partridge), 416
Peregrine falcon (*Falco peregrinus*), 417
Perennial pastures, 94
Perfectly stirred reactor (PSR), 726
Perfluorocarbons (PFC), 486
Performance efficiency of irrigation. *See* Irrigation efficiency
Peristenus digoneutis, 368
Permafrost, 1900–1902
  active layer, 1900, 1902f
  characteristics, 1900–1901, 1901f
  living with, 1902
  surface energy balance, 1901–1902
Permeable reactive barrier (PRB), 1335–1336
Permethrin, 3, 1318, 2005
Peronospora parasitica, 443
Persistent organic pesticides (POPs), 575, 1913–1914, 2186
  alternative approaches, 1916–1917
  global alliance for alternatives to DDT for disease vector control, 1917
  issues concerning, 1915
  under Stockholm Convention, 1914–1915
Persistent organic pollutants (POP), 324–325, 1904, 2036–2037
Pest(s)
  border diversity effects on, 1928
  crop rotation effects on, 1925, 1926t–1927t
  decoy crops effects on, 1925, 1928
  plant resistance effects on, 304–305
  trap crops effects on, 1925–1928
  weed diversity effects on, 1928
Pest control
  Bacterial. *See* Bacterial pest controls
  cultural, antagonistic plants in, 288–289, 289t
  genetic engineering in, 407
Pesticide, 2013–2014
  abuse, 418–419
  active ingredients in, 2013
  acute effects, 1414–1415, 1415f
  agricultural runoff and, 83
  background, 1397
  band applications, 1983–1985
    advantages, 1983–1984
    disadvantages, 1984
    equipment requirements, 1984
    future perspectives, 1984–1985
    usage, 1983–1984, 1984t
  cancers from. *See* Cancers, from pesticides
  carbamates, 2016t
  categories, 277
  characteristics of, 1963–1965
  chlorinated hydrocarbons, 2016t
  chronic effects, 1415
  chronic intoxication/poisonings statistics, 1397
  circulation in environment, 1965–1966, 1965f
    air, 1966
    aquatic environment, 1965–1966
    crops, 1966
    soil, 1966
  classification of
    based on applications, 1964t
    based on chemical class, 1964t
    based on chemical structure, 1964t
    based on toxicity, 1965t
  in coastal waters, 495–496
  concentration in water, 2017
  consumption trend, 2140
  contamination in water, 1560
  defined, 1397
  derivatization of, 1974
  degradation, 389t–390t
  development of, 1998
  dispensing/repackaging of, 506f, 509
  effects, 2005–2006
    of biopesticides, 381. *See also* Biopesticides
    of synthetic pesticides, 381
    on children, 1415–1416

Pesticide (*cont'd.*)
on natural enemies, 370
on surface water quality, 2757, 2759, 2760t, 2761t, 2765t
environmental effects of, 1121
enzyme variation, 1400
epidemiological studies, 1400
exposure to, 1414
fumigants, 2016t
genetic effects of, 1399
genotoxicity assessment, 2125–2128
global usage, 1414
in groundwater, 1281
grouped by use, 1398t
groups of, 2015f
group I, 2015
group II, 2015
group III, 2015
group IV, 2015
hazards, avoiding, 1999–2002
judicious use, encouraging, 2001
nonchemical strategies, promoting, 2001–2002
personal protection, provision of, 2002
health effects of, 1121–1122
herbicides, 1399
history of, 2008–2012, 2145–2146
ascendancy, 2008–2009
DDT, 2009–2010
inevitable conflict, 2008
IPM, 2010–2011
rebuff and reassessment, 2010–2011
impact in Ngarenanyuki, Tanzania community-based monitoring of, 505–511
inert ingredients in, 2014
inorganic pesticides, 2016t
insecticides, 1398–1399
laws and regulations, 1612–1614
available international guidelines, 1612–1613
future global policy, 1613
historical perspectives, 1612
implementation problems, 1613
need for, 1612
present scenario and probable remedies, 1613
steps undertaken, 1613
toxicological and other data requirements for registration, 1614
mechanisms of action, 277
methodologies for determining, 1968–1976, 1968f, 1969t, 1970t
in air, 1969–1972, 1971t, 1972t
in fruits, 1972–1973, 1972t, 1973f, 1973t
in soil, 1974
in vegetables, 1972–1973, 1972t, 1973f, 1973t
in water, 1968–1969, 1970t
misuse, 419
mobility of, after release, 2016–2017, 2017f
natural. *See* Natural pesticides
natural products, 2016t
non-polar, 1967
observed effects
in experimental systems, 277
in humans, 277–278
in native animals, 277
oil/fat soluble, 2015
organochlorine, 1966–1967
organochlorine insecticides (DDT, DDD), 1399
organonitrogen, 1967
organophosphates, 2016t
poisonings, 1414–1415, 1415f. *See also* Human pesticide poisonings
economic costs, 1416
symptoms of, 1968
poisoning surveillance, 2002–2004
agriculture and public health, sustaining, 2003–2004
polar, 1967
pollution, in natural ecosystems, 2145–2148
agro/horti ecosystem, 2146
aquatic environment, 2146–2147
humans, 2147–2148
soil environment, 2146
terrestrial environment, 2147
possible long-term effects, 1397
prevention, 1400
problems associated with, 2146
alternatives for, 2148
properties, leaching and, 1622, 1622t
regulation, 1940, 1998–1999
common law on, 1944
negligence, 1944
nuisance, 1945
strict liability, 1945
trespass, 1945
current law on, 1943–1944
federal laws on, 1940–1943
enforcement of pesticide use, 1941–1942
interaction with other federal regulations, 1942–1943
overview of, 1940–1941
registration of pesticide, 1941
tolerance level for pesticide residue, 1942
preemption, 1943
regulatory trends, 1945–1946
remediation techniques, objectives of, 2014
reproductive effects of, 1399
residues, in agro-horticultural ecosystems, 2142–2143
risk assessment, 2750
risk to human health, 1964, 1967–1968
role in farming and food production, 2014
Rotterdam Convention and, 1615–1617
sensitivities, 1411–1413
case example, 1412
causes, 1411–1412
impact of, 1412–1413
mechanism, 1412
overview, 1411
prevalence, 1411
symptoms, 1412
soil adsorption of, 2015–2016, 2016f
solutions to reduce release of, 2014
sterility caused by, 277–278
stewardship, 2001t
synthetic pyrethroids, 1399
transport and paterns of occurence in surface waters, 2804–2805
use, and BT transgenics, 2141–2142
use impact on natural enemies, 2141
use impact on pollinators, 2140–2141
use in weeds, 2142
use reduction approaches, 2143
water and soil contamination by, 2014
water-soluble, 2015
Pesticide control legislation, 2035–2037
elements, 2035
implementation, 2035–2036
international conventions, 2036–2037
overview, 2035
problems related to, 2036
Pesticide Data Program, 1609
Pesticide-free farming. *See* Nonchemical/pesticide-free farming
Pesticide groundwater database (PGWDB), 1317
Pesticide impacts, on aquatic communities
balance, 291
examples, 294
measuring impacts, 291–292
recent advances and outstanding issues, 295–297
risk assessment and, 292–294
risk reduction, 295
The Pesticide Index, 1461
The Pesticide Manual, 1461
Pesticide persistence, 1957
Pesticide–plant combinations, 1319
Pesticide residues
food contamination with, 1124–1126
after registration, 1124–1125
analytical methods for, 1126
food processing, 1125
MRL, 1124, 1125–1126, 1125t
overview, 1124
before registration (risk assessment), 1124, 1125f
trade issues, 1125–1126, 1125t
in foods from organic, integrated and conventional production, 1403

tolerance limit for, 1609
in food (examples), 1610t
regulatory inspection and enforcement, 1609
Pesticides Action Network (PAN), 1612
Pesticides in groundwater
groundwater contamination, 1319
management of, 1319
irrigation management, 1319
maximum contaminant levels (MCL), 1317, 1318t
in soils and water
fumigants, 1318
fungicides, 1318
herbicides, 1318–1319
insecticides, 1317–1318
use of, 1317
Pesticide test management
hormonal disruption and, 1406
Pesticide use, reduction in
overview, 2033–2034
successes in, 2033–2034
Pest management, 1319, 1919, 1949
in agro-ecosystem, 1949
aquatic weed management by fish in irrigation systems, 1922
Australian bush fly management in micronesia, 1920
bush fly origins and habits, 1920–1921
management efforts worldwide, 1921
biotechnology impact on, 1952–1953
crop diversity for, 1925–1929
definition of, 1949
ecological. *See* Ecological pest management
ecological aspects, 1934–1936
examples, 1935–1936
monoculture and polyculture, 1935
open and closed ecosystems, 1935
overview, 1934
pest population dynamics and species diversity, 1934–1935
economics considerations, 2011
filth fly abundance management, in dairies and poultry houses, 1922–1923
global commerce and, 2011
integrated. *See* Integrated pest management (IPM)
intercropping for. *See* Intercropping
modeling approaches in, 1949–1950
analytical, 1950
empirical, 1949
operations research, 1950
physiological, 1950–1951, 1951f
simulation, 1950–1951, 1951f
statistical, 1949
navel orangeworm management in almond orchards, 1919–1920
and organic farming, 1107
in organic farming, 1115–1117
in organic farming systems, 126–127
population pressure and, 2011–2012
public attitude and, 2011
regional, 1953
research, components of, 1950f
Pest regulation
objectives, 1115
PET. *See* Potential evapotranspiration (PET)
Petroleum hydrocarbons
abatement of, in natural wetland, 2855f, 2855–2856
chemical composition of, 2042–2046
constituents of, 2044t
contamination, 2040–2057
definitions, 2854
environmental fate of, 2046–2048
in environmental media, determination of, 2051–2056
environmental relevance of, 2040–2042
ranges of, 2051t
toxicity of, 2048–2051, 2050t
treatment of, in engineered/constructed wetland, 2856–2857
Petroleum reservoir, 458
Petroleum. *See* Petroleum hydrocarbons
PGPR. *See* Plant growth-promoting rhizobacteria (PGPR)
Phacelia tanacetifolia, 1750
Phacelia tanacetifolia Bentham, 371
Phaenerochaete chrysosporium, 392
Pharmaceuticals, 2062
aluminum in, 270
antibiotic resistance, 2778
in aquatic environment, 2776–2778
in aquatic systems, treatment, 2062
alternative systems for water treatment, 2062–2063
drinking water treatment systems, removal efficiencies, 2066–2067
municipal WWTP systems, removal efficiencies, 2063–2066
concentration ranges of, 2777
disposal of, 2776
ecotoxicity for, 2064
effects of, 2777–2778
in environment, 2060–2062, 2061f
formulation facilities, 2060
in human medicine, 2776
metabolism, 2776
occurrence and fate, 2776–2777
regulation, 1702–1703
risk and risk management, 2778
in sewage and sewage treatment plants (STPs), 2776
sources, distribution, and sinks of, 2777
in veterinary medicine and animal husbandry, 2776
Phascolarctos cinereus, 2595
Phase change materials (PCMs), 2509, 2520
Phase equilibrium, 2533
Phaseolus vulgaris L.cv. Monel, 1878
Phasmarhabditis hermaphrodita, 383
PH control, of composting materials, 524
Phenolic compounds
chemical composition and structure of, 2072–2075t
as indicators of original plant matter, 2077t
natural derivatives of, 2076t
threshold limit values, 1905t
Phenols, 2071–2088
anthropogenic origin of, 2078–2079
chemical composition and structure of, 2079–2080
conversion of, 2085–2086
distribution of, 2085–2086
environmental impact of, 2086
identification of, 2086–2087
natural origin of, 2071, 2075–2078
nomenclature of, 2079
physicochemical properties of, 2080–2084, 2083t
quantification of, 2087
transport of, 2085–2086
uses of, 2084–2085
Phenoxy acids, 1399
Phenylurea herbicides, 1318
Pheromones, 175, 2029
Pheromone traps, 1112
Philosophical Transactions of the Royal Society of London, 1010
Phlebotomus papatasi, 1460
Phloem-feeding insects, 308
Phoma, 1111
Phorate, 419
Phosphamidon, 420
Phosphate fertilizers, 1513–1514, 1518, 2266
strontium in, 2429
Phosphate rocks, 2240
Phosphate-solubilizing bacteria (PSB), 1511
Phosphate-solubilizing biofertilizers, 1516
Phosphinothricin, 2030–2031, 2031f
Phosphogypsum, 2266
Phosphoric-acid-doped PBI membrane, 1139
Phosphoric acid fuel cells (PAFCs), 1139
advantage of, 1139
carbon monoxide tolerance of, 1139
components of, 1139
fuel for, 1139
operating temperature range of, 1139
for terrestrial applications, 1139
uses of, 1139
Phosphorus (P), 1559–1560, 2137, 2753
autocorrelation parameters, 101t
availability
critical concentration and fertilization, 2095
residual phosphorus, 2095–2096
behavior in soils, 2094–2096

Phosphorus (P) (*cont'd.*)
calibrated soil tests for, 1806–1807
chemical speciation of, 100
concentrations, 1081
dynamic nature in soils, 2096
effect on surface water quality, 2757, 2758t
fertilizer use, 2096–2097
forms and amounts, 2094–2095
loss
implications for, 2097
and mitigation strategies, 2097
in manure
application of, 1810, 1811f
diet modification, effects of, 1668–1669, 1668f
livestock feed management, 1812–1813, 1812f
manure phosphorus concentration reduction, 1667
nutrient management practices, progress in, 1667
mobility in agricultural soils, 100–101
nutrient
applications, timing, 1809
placement, 1810
phosphorus mirabilis, 100
potentially harmful effects, 2093–2094
-related impairment in flowing and lake waters for targeted remediation, 2101–2102, 2104f, 2104t
solubilization, 350
sources, 2096–2097
spatial speciation of, 101–102, 101t, 102f, 103f
transport, in riverine systems, 2100–2105
land use impacts, integration of, 2101, 2102–2103f
use, and B toxicity, 426
Phosphorous concentration, reduction in municipal wastewater, 2712–2715, 2713–2715f
Phosphorus cycle, 1075, 1076f
Phosphorus-loading models, 1585
Photo-acoustic-infrared detector (PAID), 1203
Photochemical smog, 156
Photo-electronic pins, 1021
Photoinhibition, 538
Photorhabdus, 321
Photosensitive windows, 775
Photosynthesis, 88, 345, 716
and biomass production, 373
Photosynthetically active radiation (PAR), 2833
Photovoltaic cell, 215
Photovoltaic devices, 771, 772f
Photovoltaic effect, 226–229
Photovoltaic electricity, 826
Photovoltaic generators, 215
Photovoltaic modules, 219
price of, 224f
series connection, 216, 216f
structure of, 216–217, 217f
Photovoltaic solar cells, 226–240
beginnings of, 226t
characteristics, 229–230
current–voltage characteristics of, 229f
different types of, comparison, 239t
electrical model of, 228–229, 228–229f
materials, 230–239
power–voltage characteristics of, 229f
production procedure, 231
structure and functioning of, 232f
temperature dependence of, 230f
Photovoltaic systems, 217–219
advantages, 217–218
applications, 218
BIPV, 223–224
components of, 219–223
costs of investment in, 224–225, 225t
large scale power plants, 223, 223f, 224f
types of, 218
Phragmites australis, 2830
Phyllocnistis citrella, 368
Phyllosilicates, 1627
Phyllotreta cruciferae, 444
Phylloxera vittifolae, 304
Physical exergy, 1084, 2531
Physical methods, for contaminant remediation, 1293
Physical solvent scrubbing, 458
The Physics of Blown Sand and Desert Dunes, 1013
Physisorption, 62–63
Phytase enzymes, 1813
Phytase supplementation, 1668, 1668f
Phytohormones, 349–350
Phytolacca americana, 2031
Phytophtora infestans, 409
Phytoplankton, growth of, 1075–1078
Phytoremediation, 394, 426, 1376, 2021
of inorganic contaminants, 2120
of organic contaminants, 2117
Phytosarcophaga gressitti, 1920
Phytoseiulus persimilis, 1750
Phytotoxicity, of herbicides, 1389–1390, 1389f
PIC. *See* Prior informed consent (PIC)
Picloram, 1318
PIC system. *See* Prior informed consent (PIC) system
"Pilot Safe Use Projects," 1409
Pimentel, David, 1401, 1403
Pinatubo, 447
Pipe drainage, impact of, on downstream peak flows, 585, 585f
Piperonyl butoxide (PBO), 2019
Pirata subpiraticus, 311
Pisum sativum L. (Peas), 427
Pit storage, energy
underground, 2518
Pittsburgh Sleep Quality Index (PSQI) scores, 2874f
PJM Interconnection Grid, 2511, 2511f
load duration curve, 2512f
Placic horizon, 1625
Planning for sustainability, 2447–2448
actions and responsibility, 2449–2450
adaptive planning and transition management, 2452–2453
appraisal and monitoring, 2450
effects of, 2450–2452
environmental policy integration, 2452
futurity and uncertainty, 2452
participation and expertise, 2452
illustrative forms of, 2448t
objectives and interpretation, 2448–2449
politics, 2453–2454
Plant(s)
Al toxicity in, 270–271
biomass, 373
biomonitoring, 142–143, 2128
Cu and, 535
demand-side pests, 1951–1952
growth on copper-deficient soils, 536
growth on high-copper soils, 536–537
potassium in, 2208
sulfur supply and, 2435
toxocity of herbicides to, 1386
Plant growth, organic soil amendments and, 1878
Plant growth-promoting rhizobacteria (PGPR), 351
Plant pathogens (viruses). *See* Virus and viroid infections
Plant Protection Act of 1990, 2823
Plant resistance, 304–305
advantages, 304
economic benefits, 305
effects on pest populations, 304–305
organic soil amendments and, 1878, 1879t
overview, 304
percentage of crops having, 304, 305t
social benefits, 305
Plant-soil interactions
potassium in, 2210–2211
Plant toxicity and heavy metals, 162
Plasma-enhanced chemical vapor deposition (PECVD), 234
Plasmodiophora brassicae, 442
Plasmopora viticola, 2009
Plastic(s)
in coastal waters, 496–497
debris, 2360–2361
mulch, 2014
recycling of, 847–848
Platanus occidentalis, 2830
Plate and frame heat exchanger, 873
Pleated filtration panel innovation, 2732
Pluralist theorists, 926

Plutella xylostella (L.), 314, 1928
P–n junction
equilibrium in, 228f
formation of, 227f, 233
by ion implantation, 233
Poa pratensis (Bluegrass), 1305
Podisus maculiventris, 310
Podzolization, 1625–1626
Podzols, 426–427, 1625–1626
Point source (PS) pollution
contaminants
interactions in soil and water, 2166, 2169
sources and nature of, 2166, 2167t–2169t
defined, 2166
and environmental quality, 2169
impact assessment, 2170
remediation process, 2170
sampling for, 2169–2170
on soil microorganisms, 2169
through industrial emissions, 2170
Point-type in situ system, for monitoring stack gases, 145, 145f
Poisonings, pesticide, 510–511. *See* Pesticide poisonings
Polar ozone, 1887–1891
characteristics of, 1887–1889
destruction of, 1891
Polar stratospheric clouds (PSCs), 1887, 1889–1891, 1890f
Political economy, 2419
Pollen feeding, 310
Pollinators, 602
Pollinators, effects of IGRs on, 1466
Pollutants, agricultural runoff and associated, 82
dissolved, 82–83
of drinking water
inorganic, 2791–2793, 2791t, 2792t
organic, 2793–2797
soil, 2114–2122
defined, 2114
inorganic, 2117–2121, 2116t
organic, 2114–2117, 2115t
Pollution, 914
in Antarctica, 2138
biological, 494–495
in coastal waters, 489–498
defined, 2166
environmental. *See* Environmental pollution
light, 497–498
localized, 2166. *See also* Point source (PS) pollution
noise, 497
pesticides, in natural ecosystems, 2145–2148
transport, 915
Polyacrylamide (PAM), 960, 961, 1553, 1555
Polyaromatic hydrocarbons (PAHs), 2116
Polybenzimidazole (PBI) membranes, 1138–1139
Polychlorinated biphenyls (PCB), 388, 1334, 2172–2182, 2357
applications of, 2173, 2174t
in aquatic environment, 2188–2189
chemical identity of, 2172
determination of, 2201–2202
disposal of, from environment, 2182
distribution of, 2173–2175
extraction techniques, 2192
health effects of, 2179, 2181
impact on environment
absorption/desorption on organic matter, 2176
bioaccumulation, 2178
bioavailability, 2176
bioconcentration, 2178
biodegradation, 2176–2178, 2177f, 2177t, 2178f
biomagnification, 2178–2179
transformation, 2178
volatilization, 2175–2176
isolation techniques
from sediment samples, 2196–2199
from water samples, 2192–2196
physicochemical properties of, 2172–2173, 2173f
properties and fate of, in aquatic environment, 2189–2190
regulations for, 2181–2182
sampling, transport and storage of the samples, 2191–2192
sources and transport of, 2175t
speciation of, in sediments, 2190–2191
structure of, 2173f
wastes, destruction of, 2180–2181t
Polycrystalline silicon, 231
Polyculture, 1935
Polycyclic aromatic hydrocarbons (PAH), 388, 1334, 2186
biological and health effects of, 2188
carcinogenic classifications, 2189t
determination of, 2199–2201
extraction techniques, 2192
Isolation techniques
from sediment samples, 2196–2199
from water samples, 2192–2196
in marine sediments, 2188t
sampling, transport and storage of the samples, 2191–2192
sources and fate of, in aquatic environment, 2187–2188
speciation of, in sediments, 2190–2191
Polyethylene, 838
recycling of, 847–848
tanks, 2264
Polygonum spp., 2038
Polyisocyanurate insulation, 1479
Polymerase chain reaction (PCR), 2616
Polymers, 960–961
Polyoxins, 2030, 2030f
Polyphosphate-accumulating organisms (PAOs), 2654
Polypogon monspeliensis (Rabbit foot grass), 426
Polysaccharide, 960
Pomatomus saltatrix, 293
Pond/lake heat exchanger, 1368, 1368f
POP. *See* Persistent organic pollutants (POP)
Popillia japonica, 321
POPs. *See* Persistent organic pollutants (POP)
Population pressure, pest management and, 2011–2012
Populus deltoides, 2830
Porcine pulmonary edema (PPE), 1697
Pore space, in soils, 1627
Pork freezing, 2704
Porosity of soil, 2752
Positive lynchet, 2537
Post-combustion carbon capture, 458
Postirrigation measurement, 1548
Potamogeton pectinatus, 1922
Potassium (K), 2208–2211
calibrated soil tests for, 1806–1807
fertility, assessing and managing, 2211
overview, 2208
in plants, 2208
in plant–soil interactions, 2210–2211
in soil, 1514, 2208–2210, 2209f
Potassium (K) nutrient
applications
timing, 1809
Potassium chromate, 479t
Potato (*Solanum tuberosum* L.), 1304, 1307
Potato tubers (*Solanum tuberosum* L.), 425
Potential acid sulfate soils, 2294
sulfide mineral formation (sulfidization), 31–33
and accumulation, 32–33
oxidizable organic carbon, 31
reactive iron, 32
reducing/saturated conditions, 31–32
sulfate, 31
sulfate-reducing bacteria, 32
sulfuricization, 33–34, 34f
Potential evapotranspiration (PET), 1853
Pothole marsh, 2826f
Poultry manure, 1670
cleaning/removing frequency, 1671
liquid, 1670, 1672
application system, 1674, 1674f
storage of, 1672, 1672f
litter use and cleanout, 1671, 1671f
management of, 1670–1674
production of, 1670, 1670t
solid, 1670
high-rise layer facilities for, 1671, 1671f

Poultry manure (*cont'd.*)
subsurface application of, 1673–1674, 1674f
surface application of, 1672–1673, 1673f
storage of, 1671–1672, 1671f
indoor storage, 1671–1672, 1671f
nutrient losses during, 1672–1673
outdoor storage, 1672
utilization of
compost pits, 1672, 1673f
energy uses, 1674
as fertilizer, 1672
liquid manure application, 1674
subsurface application, 1673–1674, 1674f
surface application, 1673, 1673f
Poured concrete, 863–864
Power cycle, 2527, 2528
efficiency, 2527–2528
Power high-temperature Winkler (PHTW) gasifier, 2492–2493
Practical application impact ratio, 1235
Praseodymium, 2267t, 2268
PRB. *See* permeable reactive barrier (PRB)
Precast autoclaved aerated concrete (PAAC), 1483
Precast concrete, 864
Precaution, definition of, 1961
Precipitation, 1593t
agroforestry and, 130
erosion and, 1057
method, 918
or chelation, 2152
and water erosion, 945
and wind erosion, 945
Precision Agricultural-Landscape Modeling System (PALMS), 2221
Precision agriculture (PA), 2213–2215, 2217–2223
crop yield, 2218–2219
crop water use, 2218–2219
landscape, 2219
spatial analysis of, 2219–2221
future perspectives, 2215, 2221–2222
geographic information systems (GIS), 2215
global positioning systems (GPS), 2213–2214
overview, 2213
paradigm for, 2213–2214, 2214f
sensing for, 2214–2215
Precision conservation, 1037
Precocenes, as anti-JH agents, 1464–1465, 1464f
Pre-combustion carbon capture, 457–458
Precor. *See* Methoprene
Predators, 1475, 1475f, 2423, 2424. *See also* Natural enemies
as biological control agents, 2596
intercropping and, 1938
Preemption, 1943
Preferential flow
leaching and, 1621
Pregnancy, and arsenic exposure, 1273f, 1274
Prenatal developmental toxicity, 283
Preplant nitrogen applications, 1808
Preplant N tests, 1306
Pre-side-dress N tests, 1306
Pressure Drop, 210–211
Pressured sprinkler systems, 120
Pressure sand filter, 2722–2723. *See also* Filter(s); Sand filter
Pressurized liquid extraction (PLE). *See* Accelerated solvent extraction
Primary energy ratio (PER), heat pump system, 1351–1352, 1352f
Primary pests, 2424t
Primary productivity
carbon sequestration and, 2833
Primary radionuclides, 2239
Primer effect, 175
Primordial radionuclides, 2235–2236
Principal response curves (PRCs), 292
Prior informed consent (PIC), 1612, 1616t, 1617
history of, 1615–1616
procedure, 575
Private costs, 917, 917f
Private nuisance, 1945
Probabilistic hazard assessment (PHA), 296
Probabilistic risk assessment (PRA), 291, 296
Probabilistic uncertainty analysis, 1722
Probiotic agriculture, 528–529
advantage of, 531–532
Process-based commissioning, 666–667
Process-based erosion models, 991
continuous simulation, 991–992
Process-based models
multicompartment organic models, 1863–1864
for soil erosion prediction, 984–986, 986f
Process-dependent models, for exergy analysis, 1095
Process engineering of biological waste gas purification, 2619
Prodigy. *See* Methoxyfenozide
Producer-consumer driven movement and organic farming, 1107–1108
Producer gas, 376
Program for Water Quality and Quantity Improvement in Rural Catchments, 971
Project-based system, 486
Propagule pressure
biological invasions and, 1531
Propargite, 3
Propineb, 2142
Prostephanus truncatus, 2423
Protective and restorative measures, 1583–1584
Proteinases, 89
Proteins, 786
Proteomes, 786
Proteus, 321
Prothoracic glands, 1465
The Protocol on Water and Health, 1585
Protocols
IPM, 1522–1523
Proto-imogolite complexes, 1626
Proton exchange membrane fuel cells (PEMFCs), 1145, 1148, 1152t
applications of, 1150–1151
cold start of, 1149
contamination levels, acceptable, 1149
CO poisoning effect on, 1149
description of, 1148, 1149f
operating principle of, 1148–1149
operating temperature of, 1148
research on problems related to, 1149–1150
stacks and components, 1150
technological status of, 1150
Protozoa, in waste and polluted water, 2698
Provisional maximum tolerable daily intake (PMTDI), 1696
Proximate analysis (ASTM D3172), 717, 718t
PRV. *See Papaya ringspot virus* (PRV)
Pseudo–first-order model, 66
Pseudo–second-order model, 66–67
Pseudomonas, 96, 321
Pseudomonas fluorescens, 323
Pseudomonas paucimobilis UT26, 390
Pseudomonas putida, 351
Pseudomonas spp., 392, 2030
Psoroptes ovis, 2
P-spiralling, 2101
PS pollution. *See* Point source (PS) pollution
Public attitude, pest management and, 2011
Public health dangers and concerns, 1414–1415
acute effects (pesticide poisonings), 1414–1415, 1415f
chronic effects (cancer and other health concerns), 1415
economic costs, 1416
effects on children, 1415–1416
exposure to pesticides, 1414
global usage of pesticides and, 1414
Public support, for wind erosion control, 1028
Public utility commissions (PUCs), 1791, 1792
Pulverized fuel (pf) fired swirl burner, 726f
Pump-and-treat method, 1293, 1844
Pump-and-treat technology, 1334–1335
Pumped storage hydro, 1418, 1421–1422, 1423

benefit for renewable energy resources, 1425–1427
countries with, 1424t
electric utility usage of, 1424–1425
environmental issues, 1424
facility description, 1423–1424
ramping characteristics of, 1427
schematic diagram of, 1422f
summer operation of, 1425f
technology description, 1422–1423
winter operation of, 1426f
Public utility commissions (PUCs), 1791, 1792
Pumping
groundwater, 1284, 1285, 1285f
to manage saltwater intrusion, 1330
Pure pumped storage systems, 1423–1424
Pushbutton setup, 736, 737f
Pyramid Lake, degradation of, 1437
Pyrethrin, 2016t, 2019, 2028–2029, 2029f
Pyrethroid, 3, 2141, 2805
insecticides, 1318
Pyricularia oryzae, 2030
Pyridaben, 3
Pyridazinones, 3
Pyriproxyfen, 1460, 1464
Pyrite, 42-43, 55
Pyrite, oxidation of, 1688–1689
Pyro-char, 844. *See also* Biochar
Pyro-gas, 376
Pyrolysis, 376, 723, 724f, 836, 844, 845
biomass and wastes, 2488
features for fast, 2487
pathways, 2486f
products, 2484
reactions involved in, 2485f
technologies, conditions, and major products, 2486t
world scenario of, 2489t–2490t
commercial applications of, 376
definition of, 376
oils, 846
process of, 376
products, organization of, 837–840
Pyrroles, 3
Pythium spp., 2107, 2109

Q

Qinghai Hu Lake, degradation of, 1437
QMRA methodology, 2707
Q-SOIL, 1864
Qualitative uncertainty analysis, 1722
Quantitative indicators, of soil water erosion, 967
Quantity factor, 100
Quasi-equilibrium processes, 816
Quasi-kinetic energy, 811
Quasi-potential energy, 811
Quaternary N herbicides, 1318
QuEChERS method, 1973, 1973f
Queensland Acid Sulfate Soils Manual, 58
Quelea quelea, 413
“Quesungual” agro-forestry method, 2382
Quiet zone, 129
Quinalphos, 2142

R

Rabbit Calicivirus Australia-1 (RCV-A1), 2602
Rabbit foot grass (*Polypogon monspeliensis*), 426
Rabbit hemorrhagic disease (RHD), 2596
as biological control for European rabbits, 2600
Rabbit hemorrhagic disease virus (RHDV), 2596, 2600–2601
apathogenic rabbit caliciviruses and, 2601–2602
as biological control agent
in Australia, 2601
future for, 2602
impact of, 2601
in New Zealand, 2601
Rabbits, biological control of, 2597–2602. *See also* Myxomatosis; Rabbit hemorrhagic disease virus (RHDV)
Radiant heat barriers (RHBs), 1490–1491
configuration, 1491f
mechanism, 1491
Radiation of energy transport, 865
Radiative forcings, estimated, 1194f
Radioactive cesium, 1628
Radioactive decay, 2234–2235
alpha (α) decay, 2234
beta minus (β^-) radiation, 2234–2235
gamma (γ) radiation, 2235
Radioactive pollution, solution to, 2240
Radioactive strontium, 1628
Radioactive waste, 2136, 2359–2360
Radioactivity, 2234
radioactive decay, 2234–2235
radioactive pollution, solution to, 2240
radionuclides
anthropogenic, 2236–2240
in environment and pollution problem, 2240
natural and artificial, 2235–2236
Radio frequency towers, 2224, 2225f
with elaborate camouflage, 2227f
ethics and public administration, 2229–2230
impact and study of RF hazards, 2224
background research, 2225
recent trends, 2224–2225
RF research on animals and humans, 2225–2226
voluntary initiative, 2226
RF tower placement process, 2228–2229
Students Against Cell Towers (SACT), 2226–2228
Radioisotopic techniques, 1024–1025
Radiometers, 1194
Radionuclides, 2243–2246, 2304, 2359
anthropogenic, 2236–2240
bioremediation, 390t
in coastal waters, 491–492
in environment and pollution problem, 2240
fate and transport of, 2243–2246
geochemical processes, 2244, 2245f
hydrologic processes, 2243–2244, 2244f
microbial processes, 2244–2246
natural and artificial, 2235–2236
overview, 2243
Raindrop impact on soil, 2752, 2753
Raindrop-induced saltation (RIS), 987, 989
Rainfall
desertization and, 566
erosion and, 1044–1045, 1045t
erosivity, 943, 980–989
and rain amount, 944, 944t
erosivity index, 943
kinetic energy, 983–984
Rain measurements and hydrology, 203–204
Rainwater
pipes, 2264
as source of NPSP, 2136
Rainwater harvesting, 1276
advantages of, 2262
design/maintenance of
catchment surface, 2263–2264
conveyance systems, 2264
storage tanks, 2264
importance of, 2262
types of
land surface catchments, 2263, 2263f
large systems, 2262, 2263f
rooftop collection systems for high-rise buildings, 2263
simple rooftop collection systems, 2262, 2263f
stormwater collection in urbanized catchment, 2263
Ramsar Convention, 1585
Ramsar Convention Bureau, 2825
Ranavirus, 2605
Random walk method, 1315
Range and pasture lands
biological characteristics, 2753–2754
chemical characteristics
dissolved chemicals, 2753
dissolved oxygen, 2753

Range and pasture lands (*cont'd.*)
clean water, 2752
livestock, 2752
physical characteristics
livestock impacts on soil, 2752
livestock impacts on vegetation, 2753
suspended sediment, 2752
riparian site protection, 2754
runoff/erosion, limit on, 2754
Range finder lidars, 144. *See also* Lidars
Range reseeding technique, 570
Rankine cycle, 819
Rapid gravity filter with coagulant aid, 2722. *See also* Filter(s)
Rare earth elements (REE)
atomic numbers/symbol of, 2266, 2267t
binding forms in soils, 2266
chemical speciation of
adsorption of, 2267
species of, 2266, 2268t
total content, 2266, 2268t
translocation of, 2268
crop response to, 2269, 2269t
crop yield increase by, 2269t
heavy, 2267t, 2268t
light, 2266, 2267t, 2268t
mean content of, in soils, 2268t
physico-chemical soil properties, 2266
in sludges (concentrations/coefficients), 2267t
soil REE, origins of, 2266, 2267t
uptake of, by plants, 2268, 2268t, 2269t
Rating systems for buildings, 1655
RDF. *See* Refuse-derived fuel
REACH (Registration, Evaluation, Authorization and Restriction of Chemicals), 647
Reactor system, 1906
Reactive oxygen species (ROS), 646, 1155
Real fluids, 2532–2533
Real-time kinematic (RTK-GPS) systems, 2213
Real-time polymerase chain reaction (RTPCR), 2130
Real-time sensor, 2214–2215
Reburn with biosolids, 732–733, 732f, 733t
Recessed lighting insulation, 1490
Recharge flux, 1284–1285, 1285f
Reclamation
open cut mine, 1693
of sodic soils. *See* Sodic soils, reclamation of
Recommended maximum concentrations (RMCs)
of selected metals and metalloids in irrigation water, 2677t
Recommissioning, 657
Recruit II. *See* Hexaflumuron
Recruit III. *See* Noviflumuron
Recruit IV. *See* Noviflumuron
Recultivation, 2312
Recuperator, 873
Recycling
of dairy wastewater, 1664, 1665t
and SWEATT, 847–848
Redevance d'Enlèvement des Ordures Ménagères (REOM), 895
Red Lists, 603
Red mud, 1431–1432
Redlich–Peterson isotherm, 65–66. *See also* Isotherm(s)
Redox reactions
leading to immobilization, 393
leading to solubilization, 393–394
in water-sediment systems, 2294–2296
biological factors, 2295, 2295t
climatic factors, 2295–2296, 2296t
surface inland waters, 2295t
Reduced inorganic sulfur (RIS), 36, 37, 39, 40, 42
Reduced tillage, 1307
Reductive biodegradation, 1338
REE. *See* Rare earth elements (REE)
Reference dead state, 816
Reference environment, 1084, 1093
Reference state, 1084
Reference-substance models, for exergy analysis, 1094–1095
Reflective insulation, 1479
Refrigerants, 1358
Refuse-derived fuel (RDF), 727, 836
Regional Air Pollution in Developing Countries (RAPIDC) program, 7
Registration, Evaluation, and Authorization of Chemicals (REACH), 1700–1701, 1704
Registration, Evaluation, Authorisation, and Restriction of Chemicals (REACH, 1907/2006), 449
Regulations
of animal pests, 1115–1117
anthropogenic measures, 1116–1117
biotic (natural) resistance of environment and, 1115–1116
pesticides. *See* Pesticide control legislation
Regulations/laws
pesticides, 1612–1614
available international guidelines, 1612–1613
future global policy, 1613
historical perspectives, 1612
implementation problems, 1613
need for, 1612
present scenario and probable remedies, 1613
steps undertaken, 1613
toxicological and other data requirements for registration, 1614
Rehabilitation, 2308, 2316t
after open cut mines, 1692–1694
acid mine drainage, 1694
coversoil resources, 1692
coversoil thickness requirements, 1693
landscape regrading, 1692, 1693f
in situ soil reclamation, 1693
steep slope reclamation, 1694
of minerals processing residue, 1683–1687. *See also* Tailings
soil
case study, 2397–2398, 2397f
overview, 2396
soil biological indicators, 2396–2397, 2397t
Reinfestation of *A. donax*, 1186
Relative agronomic effectiveness (RAE), 1514
Relay logic, 736
Releaser effect, 175
Reliability
of GIS-based analysis results, 1164–1165
Relief drains, 590
Remediation, 2316t
Remote monitoring techniques, for air quality monitoring, 143–144
Remote sensing (RS)
advances in, 2818
GIS integration and. *See* RS/GIS integration
inventorying and monitoring, 2816–2817
in precision agriculture, 2215
research and development, 2817t, 2818–2819, 2818f
satellites with sensors, 2817t
as source of spatial data, 2271–2272, 2272f
uses of, 2273t
in watershed management, 2816–2819, 2817t
Remote sensing/GIS integration
applications, 2272
compatibility issues, 2271
linking, 2271–2273
for site-specific farming, 2272–2273
Remote sensing imagery, 1164, 2272
Remote-sensing products, use of, 561
Removal unit (RMU), 486
Renaissance of natural theology, 2008
Renaturization, 2309–2310
Renewable energy, 774, 806
adequacy of, 828–829
conversion efficiencies, 825
costs, 825–826
energy efficiency gains, 828
forms, 824
intermittent sources, 826–827
portfolios, 2462
present energy use, 827, 828t
resources, 825
thermal storage for, 2517–2521
active solar heat storage, 2520–2521, 2522f

  building component storage, 2519–2520
  solar passive heat storage, 2519
  underground thermal energy storage, 2518–2519
 use of, 122
Renewable energy sources (RES), 715–716, 920, 1419. *See also* Pumped storage hydro
 pumped storage hydro benefit for, 1425–1427
Repellents
 for bird control, 414–415
 used in homes and gardens, 1401
Report on the Insects of Massachusetts Injurious to Vegetation, 2008
Reproductive effects
 of pesticide, 1399
Reproductive inhibitors
 for bird control, 415
Reproductive toxicity, 2124
 of endocrine-disrupting chemicals, 646
Research
 precision agriculture-related, 2221–2222
Reseda odorata, 1925
Reservoirs, 1577–1578. *See also* Lakes and reservoirs
 dry, 1588
 flow-through, 1588
 retention, 1588
 tillage, 1554
 storage, 1563–1564
Reservoir specific exergy utilization index, 1175
Residential Energy Consumption Survey (RECS), 792–793
Residual current devices (RCDs), 223
Residue-reduced food, 1403–1404
Resistance ratio (RR), 1467
Resource allocation, 915
Resource concentration hypothesis, 1935
Respiratory effects, of arsenic toxicity, 1272, 1272f
Response curves, for fragile and robust landscapes, 542, 542f
Responsibility
 of food manufacturers, 1610–1611
Restoration, 2316t
 of native terrestrial plants, 2316t
Restricted use product (RUP), 1941
Restriction on Hazardous Substances directive (RoHS, 2002/95/EC), 449
Retailer initiatives, 1403–1404
Retained acidity, 40
Retention, 2314
Retention/absorption terraces, 968, 968f
Retention reservoirs, 1588. *See also* Reservoir(s)
Retrocommissioning, 657
Retrospective Analysis of the Clean Air Act, 1638
Revegetation, 1186
Revenue stability, 2786
Reverse cycle air conditioner, 1347
Reverse osmosis, 2728–2729
 pretreatment for, 2729, 2729f
Reversibility, 2528, 2531
 and irreversibility, 816–817
Reversible circuit heat pumps, 1350
Reversible energy transfer, 819, 821
Reversible heat transfer, 817, 817f
Reversible thermodynamic process, 2531
Revised universal soil loss equation (RUSLE), 941, 943, 975, 977, 981, 1035, 1048, 1059, 2759
Revised Wind Erosion Equation (RWEQ), 1026, 1026t, 1959
Revitalization, 2316t
Rewashing, 1592t
R-factor, in Universal Soil Loss Equation, 943
RHBs. *See* Radiant heat barriers
Rhinocyllus conicus, 2822
Rhizobia, 1110
Rhizobium (RHZ), 88, 349, 1515, 1516, 1770
Rhizoctonia-infested soil, 444
Rhizoctonia solani, 2030, 2107, 2109
Rhizoctonia solani anastomosis groups, 443
Rhizodegradation, 1847
Rhizome or culm, 1182
Rhodnius prolixus, 1462
Ribulose bisphosphate carboxylase (rubisco), 89
Rice (*Oryza sativa*), 425
Rice fields
 CH_4 emissions from, mitigating options for, 1679–1682, 1681t
  organic matter management, 1680
  processes controlling, 1679, 1680f, 1680t
  soil amendments and mineral fertilizers, 1680
  water management, 1679, 1680f
 problems and feasibility of the options, 1681–1682
Richness factor, 100
Ricinus communis L., 288, 289
Ricinus spp., 1878
Rid-A-Bird®, 413
Ridge tillage, 1027
 erosion process and, 1041–1042, 1042f
Ridges, 50
Right-to-farm laws, 1945
Riley, C. V., 2009
Rills, 964
 defined, 958
 erosion and, 1045–1046, 1046f
Rimon. *See* Novaluron
Rio Agenda 21, 1654
Riparian ecosystems, 1182, 1183, 2830
Riparian site protection, 2754
Riplox method, 1595
Ripping, 570
Risk assessment, 1722
 new/existing chemicals, 1070
Risk shifting and risk reduction, distinction between, 1796
River
 ecosystems, 2314
 functionality analysis, 2310
 recultivation, 2309–2316
  multitasking for, 2309, 2311f
  planning and development, 2310f
 restoration, 2307–2308
River flows, irrigation and. *See* Irrigation and river flows
Riverine systems
 abiotic process, 2101
 biotic process, 2101
 and land use impacts on P transport, integration of, 2101, 2102–2103f
 nutrient spiralling, 2101
 physical process, 2100–2101
 P-related impairment in flowing and lake waters for targeted remediation, 2101–2102, 2104f, 2104t
 transport in, phosphorus, 2100–2105
 water quality response, implications for, 2103, 2105
 watershed management, implications to, 2102–2103
River pollution, 2303–2304
 removal of pollutants, 2304–2305
 sources, 2304
  acid rain, 2304
  agricultural pollution, 2304
  industrial pollution, 2304
  radionuclides, 2304
Road runoff, 2321, 2771
Rock-plant filters, 2863, 2864f, 2865f
Rocks
 heavy metals in, 1370–1373, 1372f
 nanoparticles in, 1715–1716
Rock-soil-water interactions, 2293–2294
Rock wool, 1478
Rocky Mountain hydroelectric plant, 1422f
Rodenticides, 420, 2014
 used in homes and gardens, 1401
Romdan. *See* Tebufenozide
Roof(s)
 preventing air flow restrictions, 1488–1489
 soffit air ventilation, 1489
 stick-built, 1489
Roof catchment, materials used for, 2263
Rooftop collection systems
 for high-rise buildings in urban area, 2263
 simple, 2262, 2263f
Root-absorbing power, 1821, 1821f
Root biomass, 346
Root hairs
 importance of, 1820–1821, 1821f
Rootshield, 383
Root zone storage capacity, 1546

Ro-Pel®, 415
Rotary drums, 518–519, 519f
Rotating biological contactors (RBCs), 2625, 2626f
Rotating rainfall simulator, 964f
Rotation
crop, 1111, 1112
livestock, 1111
Rotterdam Convention, 575, 1615–1617, 2036
banning exports of banned pesticides, 1616
building capacity/improving regulations, 1617
Chemical Review Committee, 1616
Designated National Authorities, 1616
import decisions, information, and website, 1617
information exchange, 1617
notifying regulatory actions, 1616
overview, 1615
PIC, 1616t, 1617
history of, 1615–1616
from voluntary to legally binding, 1616
Rotterdam Convention on the Prior Informed Consent Procedure for Certain Hazardous Chemicals and Pesticides in International Trade, 1914
Row crops, 1303–1304
RS. *See* Remote sensing (RS)
RTK-GPS systems. *See* Real-time kinematic (RTK-GPS) systems
Ruminants, copper deficiency in, 428
Run-around coils, 873
Runner. *See* Methoxyfenozide
Runoff
Agricultural. *See* Agricultural runoff
erosion rates and, 1058–1059
limited, 2754
water, 1546
Run-of-river, 1423
plants, 202
Rural and urban water resources management, 2157
RUSLE. *See* Revised universal soil loss equation (RUSLE)
Ruth's type storage system, 2521, 2522f
R-value, 1478
Rye, 1110
Ryegrass (*Latium perenne* L.), 1305

S

Safe Drinking Water Act (SDWA), 1289
Safety equipment
chemigation, 470f, 471, 472f
Sage grouse (*Centrocercus urophasianus*), 420
Saissetia oleae, 1750
Salihli Geothermal District Heating System (SGDHS), case study analysis of, 1178–1179, 1178f
Saline formations, 459
Saline seeps, 590
Saline soils
drainage requirement, 588–589, 589f
Saline waters
classification of, 1569t
effect on soil, 1568
irrigation with, 1568–1569
low-salt and salty waters, mixing, 1569
Salinity, 2137, 2137t
primary source of, 1557–1558, 1558t
Salinity Control Act of 1974, 1566
Salinity control projects, 1566
Salt(s), 1559
concentration, 1569t
in soil, 1568
Salt-affected soils, 2345–2347
classification, 2346t
crop tolerance, 2346–2347, 2347t
defined, 2345
extent and distribution, 2345–2346, 2346f
overview, 2345
Saltation, 1019
defined, 1014
Salt hydrates, 2509
Saltiphone sensor, 1023, 1024f
Salt loading pick-up, 1566
Salt loading values, 1559
Salt marsh, 2825t
Saltwater intrusion
in groundwater, 1321–1332
combating, 1330
computer models, 1329–1330
Israel and Palestinian territories, 1322
Italy, 1322
mathematical modeling, 1328–1329
mechanisms, 1322–1325
Mexico, 1322
monitoring and exploration of, 1325–1328
Netherlands, The, 1322
planning and management, 1330
transition zones, 1324–1325, 1325f, 1328
into unconfined aquifer, 1322f
United States, 1321–1322
vertical cross sections of, 1323f
Sand filter. *See also* Filter(s)
pressure, 2723–2724
slow, 2723
San Pedro River (case study), 554–555
Saprobe index, 602
SAR. *See* Sodium adsorption ratio (SAR)
Sarcoptes scabiei, 2
Sarritor, 383
Saudi Arabia, 2682, 2689
SAV. *See* Submerged aquatic vegetation (SAV)
Save the Planet groups, 1612
Sawgrass wetlands, 1081
Scarcity rents, 2782, 2783
Scarites subterraneus (F.), 310
Scatterbird®, 413
Scattering, 1196
Scheduling and Network Analysis Program (SNAP), 2819
School of Environmental Studies (SOES), survey of, 1265
Schwertmannite, 42–43, 1689
Sclerotinia, 443, 444
Sclerotinia minor, 383
Sclerotinia sclerotiorum, 443
Sclerotinia spp., 2107
Sclerotinia stem rot, 443, 443t
Scotia segetum, 1116
Screen printing, 233
Scrubbing, 456, 457
SDI. *See* Subsurface drip irrigation (SDI)
SDSS. *See* Spatial decision support systems (SDSS)
Seasonal energy efficient ratio (SEER), 1364
Seasonal heating energy efficiency ratio (SHEER), 1351
Seasonal irrigation efficiency, 1546–1547
Seasonal performance factor (SPF), 1351
SEAWAT
for saltwater intrusion, 1329
Seawater expansion, 1199
Seawater scrubbing, 2441
Secale cereale (Winter rye), 1306–1307
Secondary enemies
effects on natural enemies, 371
Secondary pests, 2424t
Secondary salinity, defined, 2335
Second-law efficiency, exergy and, 820
Second law of thermodynamics, 819, 2528–2531
entropy and, 817–819
formulation, 2529–2530
ideal work, 2530–2531
lost work, 2530
Sectoral energy conservation, 688
Sectoral impact ratio, 1235
Sedimentation, 953, 964, 1579
tank, 210
Sedimentation rates
erosion, historical review, 1050–1052
Sediment composition
effect of particle travel rates on, 986–987, 987f
Sediment control, 958–962
Sediment delivery ratio (SDR), 966
Sediment flux, 965
Sediment ponds, 1553
Sediments, 2863
effect on surface water quality, 2755–2756
isolation, 1593t
nanoparticles in, 1715–1716
nutrient deposition/deactivation in, 1592t

Page numbers followed by f and t indicate figures and tables, respectively

Sediment transport
  in watersheds, 965, 965f
Sediment trap, 994, 2264
  efficiency, 1000–1001, 999f
Sediment yield, 965–966
Seed and Fertilizer Approach, 2816
Seedborne inoculum, 442
Seedborne pathogen management, 444
Seed dressings, 419
Seeding rate, 1111
Seedlings killing, 443
Seed treatment, 444
Seepage lakes, 1577
Seize. *See* Pyriproxyfen
Selectivity, herbicides, 1379
Selenium, toxic level of, in wetlands, 1626
Selenium toxicosis, 1560
Self-balancing concept
  erosion by wind and, 1015, 1015f
Self-organization, 604–605
Semiarid region irrigation
  and soil IC, 465–466
Semidirect effect, 1197t
Semidry techniques, 2441
  duct sorbent injection, 2442
  furnace sorbent injection, 2442
  hybrid sorbent injection, 2442
  and SO_2, 2441
    duct sorbent injection, 2442
    furnace sorbent injection, 2442
    hybrid sorbent injection, 2442
    spray dry scrubbers, 2441–2442
  spray dry scrubbers, 2441–2442
Semiochemicals, 175
Semivariance, 103f
Sensible heat storage, 2509
Sensible thermal energy, 811
Sensing for precision agriculture, 2214–2215
Sensitivities, pesticide. *See* Pesticide sensitivities
Sensit, 1023, 1024f
Sentricon. *See* Hexaflumuron
Sequential extraction methods, 41
Serenade®, 382
Serratia, 321
Serratia entomophila, 322
Service stability, 2786
Sesbania (*Sesbania aculeata*), 1511
Seston removal/catch, 1593t
Settling/terminal velocity, 151
Severely hazardous pesticide incident report, 575
SEVIN, 2033
Sewage effluents for irrigation
  in California, 1571
  coliform counts in, 1571
  damages, 1570
  in Israel, 1571
  long-term effects of, 1570
  in Mexico, 1571
  in Middle East, 1570
  monitoring guidelines for, 1571
  objectives of, 1570
  reuse standards, 1571
  untreated, 1571
Sewage effluents, in coastal waters, 496
Sewage sludge, 269, 1515
  for bio-liquid fuels, 847
  and INM approach, 1515
  as sources of REE in soils, 2266, 2267t
  treatment, 1908
SF_6. *See* Sulphur hexafluoride (SF_6)
Shadow prices, 2782–2783
Shadow-Voltaic systems, 224
Shakeback disease, of yaks, 427–428
Shaking-assisted extraction, 2196
Shannon–Wiener index, 602
SHARP
  for saltwater intrusion, 1329
Sharon process, 2653
Sheathing, 862, 863
Shell and tube heat exchanger, 873
Shell Chemicals, 2031
Shelter for natural enemies, 371–372
Shelter sites
  natural enemies conservation and, 1750
Shewanella alga, 2245
Short-rotation woody crops (SRWC), 345
Sick building syndrome, 1655
Side-dress nitrogen applications, 1808–1809
Signal-to-noise ratio, 2817t, 2018
Silent Spring, 2010
Silicon, 230–231. *See also* Crystalline silicon solar cells
  crystalline, solar cells, 231–232
Silos/towers composting, 518. *See also* Composting
Silt, 1716
Silver nanoparticles
  bactericidal effect of, 1740
  pollutant removal, mechanism in, 1739–1740
Silvicultural practices variation and water quality effects, 2772–2773
Simazine, 1318t, 1957t
Simpson index, 602
Simulation models
  for crop management, 2221
  of crop systems, 1950–1951, 1951f
  for surface water quality, 2759–2762, 2762t
    environmental models, 2761–2762
    soil erosion models, 2759–2761
Simulation of solute transport, 1315–1316, 1315f
Simulator for Water Resources in Rural Basins (SWRRB), 992
Single-drop microextraction (SDME), 2193
Single-path-type in situ system, for monitoring stack gases, 145–146, 145f
Single photon emission computed tomography (SPECT), 1412
SIPs. *See* structural insulated panels
Sister chromatid exchange (SCE) analysis, 2128
Site-specific agriculture. *See* Precision agriculture techniques
Site-specific farming (SSF)
  remote sensing/GIS integration for, 2272–2273
Site-specific management. *See* Precision agriculture (PA)
Site-specific models
  groundwater modeling, 1296–1297
Site suitability analysis
  GIS in land use planning and, 1164
Slagging gasifier, 2491
Slag wool, 1478
Slaking, 1627, 2334, 2335f, 2339, 2340f
  management of, 2341, 2342t
Slow sand filter, 2722. *See also* Filter(s); Sand filter
Sludge volume index (SVI), 2649
Sluicing, 47
Small hydro plants, 202
Small hydropower (SHP) plants, 201–202
Smectitic soils, 1628
Smith, John, 475
Smith-Lever act (1914), 2009
Smith, Robert 6
Smoothing natural riverbed, 2313
Snowmelt erosion, 1057–1059
  amelioration, 1059
  modeling, 1059
  precipitation, 1057
  runoff events, 1058–1059
  soil, 1057–1058, 1058f
SOC. *See* Soil organic carbon (SOC)
Social capital, 2464
Social cost, 917
Social sustainability filter (SoSF), 639
Soddy-Fajans periodic method, 2234
Sodicity, 2137, 2137t
Sodic soils
  chemistry of clay–cation interaction and its effect on swelling and dispersion, 2338–2339, 2339f
  defined, 2335–2337
  drainage requirement, 589
  global distribution of, 2335t
  management of, 2341, 2342t
  physical property and behavior, effect of sodicity on, 2339–2342, 2340f
    hydrologic properties, 2340–2341
    mechanical properties, 2341
  sodicity, measurement of, 2335–2338

Sodic soils, reclamation of, 2370–2374
ameliorants and
optimal supply of, 2373
rate of supply of, 2373–2374
without addition, 2373
biology and organic matter, role of, 2371–2373
chemical ameliorants, 2370
exchangeable Na ions displacement, 2370, 2371f
plant growth, effects of, 2371
and soil IC, 466
strategies, 2373–2374
water flow, improvement in, 2370
Sodium (Na)
montmorillonite clay, swelling of, 2336, 2336f, 2338, 2339f
Sodium adsorption ratio (SAR), 1559, 1693
Sodium chromate, 479t
Sodium salts, 1568
Sodium sulfite bisulfite process, 2442–2443
Sodium tolerance of crops
classification, 2347t
SOES's recommendations, for arsenic mitigation, 1277
SOFCs. *See* Solid oxide fuel cells (SOFCs)
Soil
acidification, 22
adsorption, 1957
aluminum in, 268–269
amelioration, 1431
amendments, 2137
CH_4 emissions from rice fields and, 1679
gypsum, 959–960
lime, 959
manure, compost, and organic sludge, 961
synthetic polymer, 960–961
bioaccumulation in, 325
biodiversity, 334–335, 353
biotechnology, 2730
carbon losses from
mineralization processes and, 1857
soil erosion processes and, 1857–1858
carbon sequestration in, 2833–2834
compaction, sulfur and, 2433–2434
conservation
approaches to, 1036–1037, 1036t
services, 1037
widespread adoption of, 1037–1038
contamination
boron in, 431–433
treatment, biological techniques for, 2014
via root exudates, 310
copper in, 535
cultivation, minimum, 1509
degradation, 1618
ecological services of, 1624
enzyme, 2020
erosion. *See* Erosion
fertility, defined, 2431
formation, 1034
fumigants, 444
health-based interventions, for wastewater irrigation, 2678–2679
heavy metals in, 1370–1372, 1372f
leaching and, 1621, 1622f
livestock impacts on, 2752
management, 1619, 1620
management, and organic farming, 1106–1107
matrix, 1570
microbial biomass, 2396
microbial diversity, 388
microorganisms, 2014, 2020
PS pollution on, 2169, 2170
nanoparticles in, 1712–1715, 1713t
nitrogen transformations in, 1768–1769, 1769f
permafrost, 1900–1902
pesticide effect on, 2147–2148
pesticides in, 1966, 1974
pH
and boron availability, 424
and Mo availability, 427
potassium in, 2208–2211, 2209f
primary function of, 1570
quality
assessment of, 2458–2459, 2458f
sustainable agriculture. *See* Sustainable agriculture
rehabilitation. *See* Rehabilitation
remediation by nZVI, 1737–1739
retention, 2314
salt-affected. *See* Salt-affected soils
scape filter, 2731, 2731f
sodification, 1561
strontium and interaction with soil matrix, 2426–2427, 2428f
structure, 2752
sulfur. *See* Sulfur
sustainable agriculture, 2457
sustainable use of, 1618
tillage methods, 119t
translocation processes in, 1624. *See also* Eluviation/illuviation; Leaching
winter erosion process, 1057–1058, 1058f
Soil and water assessment tool (SWAT) model, 966, 2762
Soil and water conservation (SWC), 558
Soil-aquifer recharge systems with dewatering, 1571
Soilborne pathogen management, 444
Soil building, 1110
Soil carbon sequestration, 460
Soil conditioner, 959
Soil conservation, 2762
practices, 967–968, 967f, 1813
Soil degradation, 2375–2385
in Africa, 2379–2381, 2380f
in Asia, 2378–2379, 2379f
in Australia, 2383–2385, 2384f
background, 2375–2377
in Europe, 2377–2378, 2377f, 2378f
modeling, data sets used in, 2376t
in North America, Central America, and the Caribbean Islands, 2381–2383, 2382f, 2383f
in South America, 2381, 2381f
Soil-dwelling fauna, 310
Soil erodibility, 995
Soil erosion, 345, 1034. *See also* Soil degradation
assessment of, 974
causes of, 1034
control, 1036–1037, 1036t
government involvement in, 1037–1038
widespread approach to, 1037–1038
extent of, 1034–1035
impact of, 1034–1035
from irrigation. *See* Irrigation erosion
and losses of C and N, 93
models for measurements of, 974. *See also* Empirical models
monitoring and modeling, 1958
water erosion models, 1958–1959
wind erosion models, 1959
pesticide translocation assessment, 1957–1958
and pesticide translocation control, 1959–1961
productivity impacts of, 1035–1036
and runoff control, 1961
scope of, 1034–1035
variability in, 974, 975f
water erosion and, 1955–1956
and water quality, 2755–2766
indicators, 2764–2766
mitigation strategies, 2762–2764, 2763f
nutrients, 2756–2757, 2758t
pesticides, 2757, 2759, 2760t, 2761t
sediments, 2755–2756
simulation models, 2759–2762
wind erosion and, 1956,1956f
Soil law
basis of, 1618
Commission on Environmental Law of IUCN (World Conservation Union, The), 1620
effectiveness of, 1619–1620
international and national, 1618
International environmental law, 1618–1619

declarations, 1619
international conventions, covenants, treaties, and agreements, 1619
soil
degradation, 1618
sustainable use of, 1618
Soil Loss Estimation Model for Southern Africa (SLEMSA), 976–977
Soil loss tolerance, 967, 1035
Soil map, 2376
Soil organic carbon (SOC), 345, 346, 934–935, 1857
conceptual mass balance model, 935–936, 935f
emisions, estimates of, 936, 936t, 937f
worldwide land use and management impacts, 1859t–1860t
Soil organic matter (SOM), 93, 345, 1518, 1857, 1861, 2013, 2018–2019
role in 21st Century, 1861
turnover. *See* SOM turnover
Soil organic matter (SOM) turnover
defined, 1872, 1873f
of different pools, 1875–1876
factors controlling, 1874–1875, 1875t
first-order model, 1872–1873
measuring, 1872–1874
MRT. *See* Mean residence time (MRT)
overview, 1872
range and variation in estimates, 1874, 1874t
Soil organisms activity, environmental influences on, 90
organic substrates distribution, 91–92, 91f, 92f
soil texture, 91
temperature, 90, 90f
water content, 90–91, 91f
SOILOSS, computer program, 977
Soil plant system, 1817, 1818f
Soil–plant transfer of contaminants, 2137
Soil pollutants. *See* Pollutants
Soil pollution migration, 2136, 2137
Soil P testing, 2097
Soil remediation, 1431–1432
Soil ripening, 45
Soil salinity
irrigation and, 1572–1574
causes, 1573
deleterious effects, 1572–1573
management strategies, 1574
management, drainage and. *See* Drainage, soil salinity management and
Soil-surface roughening, 569
Soil test phosphorus (STP), 1667
Soil test reports, 1813
Soil vapor extraction (SVE), 1335
Soil water erosion, 958–972
assessment, 964–967
indicators, 966–967
models, 966
monitoring, 964–966
categorization of, 958
control, 967–971
defined, 958
erosivity and erodibility, 980–9
integrated environmental management, 971–972
problems, 963–964
sedimentation, 964
soil amendments, 958–962
sources, 964
vegetation for controlling, 994–1001
Solanum tuberosum (Potato), 1304, 1983
Solanum tuberosum L. (Potato tubers), 425
Solar air conditioning, 1361–1362, 2503–2504
closed cycle systems, 2504
designs of, 2504
open cycle systems, 2504
Solar and volcanic variability, 1194–1195, 1196t
Solar-assisted heat pumps, 1351
Solar batteries, 218
Solar-boosted heat pumps, 1351
Solar box cookers, 2503
Solar cells, 2498. *See also* Photovoltaic solar cells
Solar central receiver or solar tower, 2504t, 2505
Solar chimney, 2504
Solar collection stations, 916
Solar collectors, 2504t
operational characteristics of, 2505
Solar cookers
advantages of, 2502
types of, 2503
Solar cooking, 2502–2503
Solar cooling, 249
Solar crop drying, 2502
Solar distillation, 2502
Solar dryers
advantages of, 2502
classification of, 2502
Solar energy, 192, 771–772, 2498
Solar forcings, 1199f
Solar fraction, 1352
of solar system, 1352
Solar heat gain coefficient (SHGC), 867, 867t
Solar house, 2499
Solar industrial process heat, 826
Solar integrated collector storage system innovations, 256
Solar inverters. *See* Inverters, in photovoltaic systems
Solar panel cookers, 2503
Solar parabolic cookers, 2503
Solar parabolic dish, 2505
Solar parabolic trough, 2504–2505, 2506f
Solar passive heat storage, 2519, 2519f
Solar photovoltaic (SPV) technology, 2498
Solar photo-Fenton, 2553
Solar ponds, 2500–2502
artificial, 2501
convective, 2501
nonconvective, 2501
power generation system, 2501f
zones
lower convective, 2501
nonconvective, 2501
upper convective, 2501
Solar radiation, 824, 826, 1192
resources, 825
Solar resources, of electricity, 1419
Solar space heating, 2498–2499
active, 2499
hybrid, 2499
passive, 2499, 2500f
direct solar gain, 2499
indirect solar gain, 2499
isolated solar gain, 2499
Solar stills, 2502, 2503f
Solar thermal energy, 241
auxiliary, size of, 254
closed circuit systems, 248
comparison between different types of collectors, 246–247
energy balance of collector, 245–246
collector efficiency curves, 245–246
instantaneous efficiency of a collector, 246
energy balance of solar thermal collector, 245f
forced circulation systems, 247, 248f
heat storage system, 252–254
dimensioning, 253–254
large systems, 255
Legionella exposure risk, 255f
prevention and control of, 254–255
natural circulation systems, 247, 247f
open circuit systems, 248
preliminary analysis and solar thermal plant project, 249–252
design phase, 249–252
matching energy availability and thermal energy need, 249
solar collector technology, 242
evacuated tube collectors, 244
flat-plate solar collector, 242–243
integrated collector storage, 243–244
unglazed/open collectors, 243
solar cooling, 249
solar integrated collector storage system innovations, 256
Solar thermal plant project, preliminary analysis and, 249
design phase, 249–252

collector field surface, sizing of, 251–252
DHW need, estimation of, 251
logistic aspects, 250
on-site investigation, 250
saving energy interventions, 250
solar plant type, choice of, 250–251
users' consumptions, analysis of, 250
matching energy availability and thermal energy need, 249
Solar thermal power generation, 2504–2505
Solar thermal technologies
broad economic bandwidth, 2498
high-temperature, 2498
low-temperature, 2498
market growth and trends, 2506
medium-temperature, 2498
Solar water heaters, 2499
built-in-storage, 2499
direct system, 2499
efficiency of, 2499–2500
elements of, 2499
forced circulation, 2499, 2501f
indirect active, 2499, 2501f
indirect system, 2499, 2500f
in industrial applications, 2500
operating principles, 2499
thermosyphon, 2499
Solar water heating, 826, 2499–2500
Solenopsis invicta, 1464, 1983
Solid catalysts, 1907
Solid contaminants
in water
classification of, 2720f
sizes of, 2720f
Solid fuels, 715–722, 722f, 723f, 830–834
CO_2 emission estimation, 717, 722, 722f
flame temperature, 722
flue gas volume, 722
heating value (ASTM D3286), 717
proximate analysis (ASTM D3172), 717, 718t
ultimate analysis (ASTM D3176), 717, 719t, 720t–721t
Solid oxide fuel cells (SOFCs), 1141–1143
acceptable contamination levels of, 1142
applications of, 1143
fuel for, 1141
operating principle, 1142, 1142f
operating temperature of, 1141
problems related to, 1142–1143
technological status of, 1143
Solid phase extraction (SPE), 1970t, 2194–2195
Solid-phase extraction (SPE) cartridges, 2018
Solid phase microextraction (SPME), 1847, 1970t, 1971, 2194–2195
Solid retention time (SRT), 2649f
Solid waste, 830–834, 895, 900–902
chemical composition of, 728t
industrial, 1431
management
characteristics, 2417t
normative principles in, 2418–2419
policy drivers, policy regimes, and modes of governance in, 2419–2421
management, life cycle assessment, 2399–2412
goal and scope of, 2404–2405, 2405t
importance of, 2411–2412
multiple-attribute decision making in, 2403–2404
ranking of, 2406–2411
study area and system description of, 2401–2403
weighting factors, determination of, 2406
Solid waste-integrated gasification-combined cycle (SW-IGCC) system, 840–843
cost of electricity (COE) vs. the cost of fuel (COF) for, 841–842, 841f
Solid waste to energy by advanced thermal technologies (SWEATT), 830, 831
bioenergy and biochar, 843–846
bio-liquid fuels, 846–847
Biomass Alliance with Natural Gas, 848
recycling and, 847–848
Solubilization, 393–394
Solutes
precipitation of, 1625–1626
production of
by decomposition of organic matter, 1625
by weathering of minerals, 1625
translocations of. *See* Leaching
Solute transport, simulation of, 1314–1316, 1315f
Sonication. *See* Ultrasound-assisted extraction (UAE)
Sorption-desorption process, 2136, 2169
"Sound level," 2869
South America
soil degradation in, 2381, 2381f
distribution of, 2381f
Southern green stink bug. *See Nezara viridula*
South Florida Ecosystem Restoration Task Force (SFERTF), 629, 630
South Florida Water Management District, 1321
Soxhlet extraction, 2197
Soybeans (*Glycine max* L. Merr.), 425, 1303, 1935
Space shuttle missions, AFCs in, 1146
"Space weathering," 1716
Spain, 2682, 2690
Spalangia, 1922
Sparging, 1592t, 1593
Spartina alterniflora salt marshes, 2835
Spatial decision support systems (SDSS), 1163, 1164
Spatial speciation of phosphorus, 101–102, 101t, 102f, 103f
Special Report on Emissions Scenarios (SRES), 1200f
Species- and community-based indicators and indices, 601–604
indicator taxa, 602–603
indicators based on ecological strategies, 603–604
ratios between different classes of organisms or elements, 603
species richness, 602
Species diversity, 1934–1935
Species invasion, 495
Species Trend Index, 603
Specific eco-exergy, 778
Specific exergy, 782, 785f, 786–787
Specific exergy index (SExI), 1171
Specific heat, 725
SPECT. *See* Single photon emission computed tomography (SPECT)
Spider mites, effect on photosynthetic rate, 1951
Spillway, 209
Spilopsyllus cuniculi, 2597, 2599–2600
Spin-on technique, for ARCs, 234
Spinosyns, 2029, 2029f
Spiromesifen, 3–4
Spirotetramat, 3, 4
Splash erosion, 958, 985
Split nitrogen applications, 1809
Spodic horizon, 1625
Spodoptera exigua, 310
Spodoptera frugiperda (Smith), 314
Spodosols, 1625
Sporodesmium sclerotivorum, 2107
Spotted knapweed (*Centaurea maculosa*), 1526t
Spray cloud, 1472
Spray dry scrubbers, 2441–2442
Spray nozzles, 1472
Spray polyurethane foam (SPF), 1479
Spray window, 1471
Spring flow, 1285, 1285f
Sprinkler irrigation system, 1564, 1565f, 2677, 2678t
Sprinkler-irrigation erosion, 1553–1555
Sprinklers, 1319, 1547t, 1568, 1664
Sri Lanka, pesticide poisonings incidence, 1408
SSF. *See* Site-specific farming (SSF)
St. John's wort (*Hypericum perforatum*), 1527t, 2821
Stable isotope fractionation analysis, 1846
Stack gases, continuous emission monitoring systems for, 144–146
dilution-extractive system, 144–145, 145f
dual-path-type in situ system, 145–146, 145f

  fully extractive system, 144, 144f
  point-type in situ system, 145, 145f
  single-path-type in situ system, 145–146, 145f
Stafilinid, 409
Stand-alone cost. *See* Levelized cost
Stand-alone photovoltaic systems, 218–219, 218f, 220
Standardization, 374–375
Standard oil recovery, 458
Standard test conditions (STC), of photovoltaic systems, 217
State policy innovations, 923
State-space analysis, for crop water use, 2218, 2219
Static pile composting, 517, 517f. *See also* Composting
Stationary/bubbling fluidized bed (SFB) boilers, 375
Statistical models, in pest management strategies, 1949
Statistics
  for landscape crop water use, 2219
Steam gasification, 2490
Steel-framed walls, 863
Steel shots, 1432
Steep slope reclamation, 1694
Steinerema feltiae, 409
Stellaria media, 1928
Stem borers, effect on photosynthetic rate, 1951
Sterility
  caused by pesticides, 277–278
Sterling motor, 157
Sticky traps, 1112
Stir bar sorptive extraction (SBSE), 1970t, 2195
Stoats (*Mustela erminea*), 2596
Stochastic chance-constrained programming (SCCP), 111
Stochastic mathematical programming (SMP), 106–107
Stochastic (or random) models
  water quality, 2749
Stockholm Convention, 575
  on Persistent Organic Pollutants, 1914
Stoichiometric combustion, 774
Stoker firing, 726, 726f
Stokes' law, 151
Stone backfill, 2313
Storage, 2314
  and conservation reservoirs, 1577
  efficiency, 1546
  energy. *See* Energy storage
  of groundwater, 1284
  matrix, 1312
  tanks for rainwater, 2264
  variation (groundwater), 1312
  of water, 1563
Storage plants, 202–203
Stored-product pests, biological control of
  classical biological control, 2423
  description of, 2423
  examples of pests and natural enemies, 2424t
  externally feeding pests, biocontrol of, 2424–2425
  factors for, 2423
  internally feeding pests, biocontrol of, 2423–2424
Stories from a Heat Earth—Our Geothermal Heritage, 1167
Storm energy, 943
  and rain amount, 943–944, 944t
Stormwater
  collection in urbanized catchment, 2263
  treatment, 1081
Storm windows, 866
Stranglervine (*Morrenia odorata*), 1526t
Strategic environmental assessment (SEA), 2450
Strategus, 321
Stratification, 1580–1581
Stratification ratio of soil organic C, 92, 92f
Stratified chilled water storage tank, 2510, 2510f
Stratospheric ozone depletion, 681
  antarctic ozone hole, discovery of, 1887
  atmospheric trace species, field measurements of, 1891
  global ozone layer, depletion of, 1891–1893, 1892f
  measurements and distribution of, 1885
  polar ozone chemistry, 1887–1891
Straw bale walls, 865
Stream erosion, 958, 964
Streamflow depletion, 1563
Streamlining European Biodiversity Indicators (SEBI), 603
Streptomyces, 2029
Streptomyces anulatus, 1879
Streptomyces hygroscopis, 1378, 2030
Streptomyces kasugaensis, 2030
Streptomyces violaceoruber, 384
Strict liability, 1945
String cables, 222
String diodes
  in photovoltaic systems, 220
String fuses
  in photovoltaic systems, 219–220
String inverters, 222, 222f
Strip mining, 1692
Stripper column, 457
Strip tillage
  erosion process and, 1042–1043
Strobilurins, 2030, 2030f
Strontianite (SrCO3), 2426
Strontium
  adsorption of, 2427
  chemical speciation, 2427
  geogenic origin, 2427
  isotopes of, 2426
  soil parameters, correlation between, 2428f
  in soils, 2426–2427, 2428t
    interaction with soil matrix, 2426–2427, 2428f
  sorption, 2427
  uptake by soils, 2428–2429
    impact of fertilizer use on, 2428f, 2429
Strontium chromate, 479t
Structural insulated panels (SIPs), 865, 1484–1486
Strychnine, 413
Students Against Cell Towers (SACT), 2226–2228
Sturnus vulgaris, 413
Submerged aquatic vegetation (SAV), 475
Submerged-macrophyte systems, 2863
Submerged orifice scrubbers, 154–155
Subsoiling, 570
Subsurface band applicator implement, for poultry litter, 1673–1674, 1674f
Subsurface drainage, 1557, 1558, 1558f, 1558t
Subsurface drip irrigation (SDI), 1535–1538
  air entry and flushing, 1537
  chemical injection, 1537
  defined, 1535
  development of, 1535
  laterals and emitters, 1537
  lateral type, spacing, and depth, 1536
  maintenance, 1538
  operation, 1538
  pumps, filtration, and pressure regulation, 1537
  site, water supply, and crop, 1535–1536
  special requirements, 1536–1537
  system components, 1537
  system design, 1535–1537
Subsurface-flow constructed wetlands, 2841–2846
  hybrid systems, 2845
  horizontal flow systems, 2841–2843
  vertical flow systems, 2843–2845
  use for wastewater, 2845–2846, 2845t
Subsurface flow wetlands. *See* Vegetated submerged bed (VSB) wetlands
SubWet, 2671
  2.0, 2672
Sugar beet (*Beta vulgaris* L.), 1307, 1552
Sugarberry. *See Celtis laevigate*
Sugarcane industry, New South Wales, 58, 59
Sulfate, 31
  in atmosphere, 14–15
Sulfate-reducing bacteria, 32
Sulfic Endoaquepts, 28
Sulfide mineral formation (sulfidization), 31–32

and accumulation, 32–33
oxidizable organic carbon, 31
reactive iron, 32
reducing/saturated conditions, 31–32
sulfate, 31
sulfate-reducing bacteria, 32
Sulfidic materials, 27
identification and assessment, 39
Sulfonamide antibiotics, 2061
Sulfonylureas, 409
Sulfur, 2431–2435
agro-ecological aspects, 2434–2435
global change, 2434–2435
interaction with other ecosystems, 2434
plant health and, 2435
sustainability, 2434
emisions, 2434
fertilizers, 2434
overview, 2431
in soils, 2431, 2432f
biological aspects, 2431–2433
capillary rise/leaching, 2433
crops and crop rotation, 2432–2433
mineralization, 2432
physico-chemical aspects, 2433
soil compaction, 2433–2434
soil water regime, 2433
Sulfur dioxide (SO_2), 1154, 2304, 2437
air pollution of, 157–158
ambient air quality, 2437–2438
anthropogenic sources, removal from, 2440
ammonia wet scrubber, 2441
characteristics of, 137
concentration of, from natural sources, 2439
anthropogenic sources, 2439–2440
dry techniques, 2442
circulating fluid bed dry scrubber, 2442
magnesium oxide process, 2443
sodium sulfite bisulfite process, 2442–2443
EN 14212, ultraviolet fluorescence for, 142
flue gases, concentration in, 2440
semidry techniques, 2441
duct sorbent injection, 2442
furnace sorbent injection, 2442
hybrid sorbent injection, 2442
spray dry scrubbers, 2441–2442
sources, 2438
biogenic, 2439
Sulfur hexafluoride, 486
Sulfuric horizon, 27
Sulfuricization, 33–34, 34f
Sulfuric material identification and assessment, 39
Sulphur hexafluoride (SF_6), 1204
Sumilarv. *See* Pyriproxyfen
Sunday soil, 2341
Sunnhemp (*Crotalaria juncea*), 1511
"Super bugs," 391
Supercritical fluid (SCF), 1257–1258
Supercritical fluid extraction, 2198
Superfund. *See* Comprehensive Environmental Response, Compensation, and Liability Act (CERCLA)
Supply-side pests, 1951
Supply side thermal storage, 2520–2521, 2521–2522f
Surface air temperature, global annual-mean, 1198f
Surface albedo feedback theory, 566
Surface area (SA), of biochar, 844
Surface coalmines, 916
Surface drainage, 1560
Surface energy budget effect, 1197t
Surface-flow wetlands. *See* Free-water surface (FWS) wetlands
Surface irrigation, 120, 1547t, 1551–1553, 1564, 1565f
Surface Mining Control and Reclamation Act (SMCRA), 1289
Surface mining methods
classification, 1692
Surface moisture (SM), 717
Surface roughness
desertization and, 569
Surface runoff, 2262
limitation, 2314
Surface seals, soil, 960
Surface waters, 1576, 2804–2806
direct contamination of, 1576
effects of pesticides, 2805–2806
knowledge gaps, 2806
pesticide transport and paterns of occurence, 2804–2805
pesticide use and, 2804
runoff, 2358
Suspended growth processes, 2647
Suspended particle display (SPD), 867
Suspended particulate matter, 149
Suspended sediment, 2752
Suspended sediment concentration (SSC), 965, 965f
SUSpended Sediment TRAp (SUSTRA), 1022
Suspension firing, 725, 726f
Sustainability, 333, 2446
Brundtland Report on, 338
components of, 1950f
concept of, 1654
definition of, 338
filters, 639
importance of, 338–339
indicators for, 339
meaning of, 338
potential analysis, 639
reports, 2461, 2463
benefits of, 2463
steps for, 339
sulfur and, 2434
tools for, 340
Sustainability Assessment of Farming and the Environment (SAFE), 356
Sustainable agriculture, 2457–2459. *See also* Ecological agriculture
agroecosystem biodiversity, 2458
air and atmosphere, 2458
assessment of soil quality, 2458–2459, 2458f
management strategies, 2459t
soil, 2457
water, 2457–2458
Sustainable aquifer use, 1285–1287, 1287f
Sustainable communities, aspects of, 2464–2465
Sustainable development, 920, 1169–1170, 2446
aspects of, 2461
for community, 2464
creative, cooperative, design and planning teamwork, 2462–2463
defined, 2461
in developing countries, 2465
education for, 2463
geothermal energy for, 1170
renting vs. buying, 2463–2464
social interactions, 2462
Sustainable energy future, 2462
Sustainable forest management, 339
Sustainable land management (SLM), 558
SUTRA
for saltwater intrusion, 1329
SVE. *See* soil vapor extraction (SVE)
Swamp, 2825t, 2826f
Swedish Forest Agency, 2298
Swedish Poisons Information Centre, 1401, 1403
Swelling, clay
effect of clay–cation interaction on swelling and dispersion, 2338–2339
Swine, effects of deoxynivalenol on, 1696
Swirl burners, 725, 726
Swishing noise, 2871–2872
Switchgrass (SWG), 345
environmental benefits of, 346
SWRRB. *See* Simulator for Water Resources in Rural Basins (SWRRB)
Sycamore. *see Platanus occidentalis*
Synanthedon exitiosa, 1919
Synaptogenesis, 281
Synergism, in agroecosystems, 1990
Synergists, 1992
Synergy, 1988
Syngas, 376–377, 730
Synomone, 175
Synthesis gas, 457, 458
Synthetic aperture radar (SAR), 1827, 1830–1831, 1833, 2817t
advanced, 1827

Page numbers followed by f and t indicate figures and tables, respectively

Synthetic pesticide use, in United States, 1120, 1121f
Synthetic polymers, 960–961
Synthetic pyrethroids, 277, 1399
Syria, 2686
Syrphidae, 1116
System Analyzer software, 1365
System coefficient of performance (SCOP), heat pump, 1351
System energy, 811
System management approach, of biological weed control, 1525, 1525f
System networks, importance of, 1434
Syzygium aromaticum, 384

T

Tachinaephagus zealandicus, 1922
Tachinid species, 1747
Tagetes erecta L., 288, 289
Tagetes patula L., 288, 289
Tagetes spp., 288, 289, 1878
Tailings, 1683–1687
characteristics, 1685–1687
chemical properties, 1685–1686
defined, 1683
physical properties, 1686–1687
potential environmental problems, 1683
rehabilitation practices
ecological approach, 1684–1685, 1685t
engineering approach, 1683–1684
Tailwater, 1557, 1558t, 1564
Taiwan, pesticide poisonings incidence, 1408
Taiwan National Poison Center, 1408
Tall fescue (*Festuca arundinacea* cv. Kentucky 31), 425
Tandonia budapestensis, 383
Tandonia sowerbyi, 383
Tank storage, energy
underground, 2518
Tannins
condensed, 2082f
hydrolyzable, 2082f
Tantalum, 2585, 2587–2588
Tapered element oscillating microbalance (TEOM) dust sampler, 1024, 1025f
Tapered element oscillating microbalance (TEOM), of particulate matter, 142
Tarnished plant bug, 368
Taum Sauk upper reservoir, 1422f
Tax
on emissions, 919
on polluting good, 919
Taxe d'Enlèvement des Ordures Ménagères (TEOM), 895
TCD. *See* Thermal conductivity detector (TCD)
TCDD. *See* 2,3,7,8-Tetrachlorodibenzo-paradioxin (TCDD)
TCE. *See* Trichloroethylene (TCE)
TDL. *See* Tuneable diode laser (TDL) technique
Tebufenozide, 1466
Technological exergy, 778–780, 779f
Technological impact ratio, 1235
Technologically enhanced naturally occurring radioactive materials (TENORM), 2240
Technological sequestration. *See* Carbon capture and storage (CCS)
Technology-forcing strategy, 919
Teflubenzuron, 1462
TEM. *See* Terrestrial Ecology Model (TEM)
Temkin isotherm, 65. *See also* Isotherm(s)
Temperature
Antarctic, 1889, 1889f
Arctic, 1889, 1889f
control, of composting materials, 521, 522f
desertization and, 566
Temporal stability
of landscape patterns, 2038–2039
TEPP (tetraethyl pyrophosphate), 3
Terbufos, 419, 1318
Teretrius nigrescens, 2423, 2424t
Terminal insecticide concentration (TIC), 1472
Terminal restriction fragment length polymorphism (TRFLP), 2616–2617
Terminator gene, 314
Termite-assisted hand-pitting, 570
Terra preta, 844
Terraces, 968–969, 968–969f, 2764, 2764f
benefits and problems, 2539–2540
engineered, 2537–2539
formation, by tillage erosion, 2536–2540
ancient lynchets, 2536
contemporary lynchets, 2536–2538
Terrestrial carbon sinks, 460
Terrestrial ecology model (TEM), 2272
Terrestrial Ecosystem Function Index (TEFI), 604
Terrestrial environment, pesticide effect on, 2147
Tersilochus heterocerus, 1116
Tetanops yopaeformis, 1983
2,3,7,8-Tetrachlorodibenzo-paradioxin (TCDD), 1394, 1395
Tetrahedral oxyanion, 100
Tetrahedral structure, 431
Tetranychus urticae, 310, 1750, 2021
Tetronic acids, 3–4
Theba pisana, 383
The International Journal of Exergy, 1084
Thematic Mapper (TM), 2818
Therioaphis maculata, 1521
Thermal conductivity detector (TCD), 1203
Thermal design
bottoming and topping cycles, 1357
fundamentals, 1356–1357
Thermal energy, 772–773
storage system, 1086
energy efficiency of, 1086
evaluation of, 1086
Thermal energy storage (TES), 2508–2522
cool storage, 2511–2517
technologies, 2509–2510
heat storage, 2510–2511
and renewable energy, 2517–2521
types of, 2509
Thermal environmental conditions for human occupancy, 1657
Thermal exergy, 1084
Thermal mass, 861
Thermal power cycle, 1357
Thermal technologies
advanced, 835–837
Thermally impaired waters, 2813
Thermionic specific detector (TSD), 1975
Thermochemical decomposition, 2483–2484. *See also* Pyrolysis
Thermo-chemical process, 730
Thermocline, 2510
Thermodynamic concepts as superholistic indicators, 778
Thermodynamic effect, 1197t
Thermodynamic information, 782
Thermodynamic process, 2526
Thermodynamic properties, calculation of
ideal gas, 2532–2533
phase equilibrium, 2533–2534
real fluids, 2532–2533
rigorous equations for, 2531–2532
solutions, 2533
Thermodynamics, 2525–2534
Thermoelectric heat pump, 1348–1349, 1348f. *See also* Heat pumps
Thermogravimetric analysis (TGA), 723
Thermostatic expansion (T-X) valves, 1350
Thermotolerant coliforms, 2699
The World is Flat, 1503
Thin-film solar cells, 232
Thiobacillus ferrooxidans, 33, 43, 1694
Thiobencarb, 2806
Thiocarbamate herbicides, 1318
Thiocarbamate pesticides, 1317
"Third–party certification," 125
Thorium-232 decay series, 2237f
Three-dimensional ecological footprint geography, 2475–2476
Threshold agents, 1456
Ticks, 1, 2
control of. *See* Acaricides
Tidal energy, 772
Tidal marsh, 2825t, 2826f

Tidal power, 824, 825
TIE. *See* Toxicity identification evaluation
Tiger. *See* Pyriproxyfen
Tilapia zillii, 1922
Tile drainage, nitrate-nitrogen concentrations in, 1303–1304
Tillage, 1111
  erosion process and, 1040, 1041f
    patterns of erosion and deposition, 2537f
      terrace formation by, 2536–2540
  impact on soil IC, 464
Tillage berms, 2539–2540, 2540f
Timber harvesting, 2770
  annual yield and storm flow response to planting and, 2773–2774
  hydrological behavior of forests and, 2770–2771
  modern forestry hydrology history, 2772
  variation in silvicultural practices and water quality effects, 2772–2773
  water quality effects mitigation through BMPs, 2773
Time-domain electromagnetic methods (TDEMs)
  for saltwater detection, 1327
Titania. *See* Titanium dioxide
Titanium dioxide (TiO_2), 235–236
  environmental pollution by, 2129
  semi conductor, 236
  solar cells, 238
    advantages, 238–239
Titanium dioxide (TiO_2) catalyst, 1731–1732
  decontamination of toxic pollutants by, 1733, 1735
  forms of, 1733f
Titanium white, 235
Tithonia diversifolia, 1878
T-max. *See* Noviflumuron
TMDL. *See* Total maximum daily load
Tolerance(s), 1124, 1125–1126, 1125t. *See also* Maximum residue limit (MRL)
  FQPA and, 1118
Tolerance limit
  defined, 1609
  for insect fragments, 1609–1610
    in food (examples), 1610t
  for pesticide residues, 1609
    in food (examples), 1610t
    regulatory inspection and enforcement, 1609
Tomato (*Lycopersicon esculentum*), 425
Ton, 1363–1364
Topping cycles, 1357
Topsoil, 1019
  loss of, 1019, 1019f
Total dissolved solids (TDS), 918, 1559
  for saltwater intrusion, 1325, 1329
Total maximum daily load (TMDL), 2750, 2752, 2808–2809
  in achieving water quality standards, 2812
    section 303 impaired waters list, 2812
    state water planning and, 2812–2814
  application of, 2814
  nonpoint source state management plans, 2811
    water quality standards before TMDL process, 2811–2812
  point source regulation and
    NPDES permit program, 2809–2810
    pollutant discharge, 2809
    section 404 dredge and fill permit program, 2810
  water quality standards in point source permitting before TMDL process, 2810–2811
Total petroleum hydrocarbons (TPHs), 2051–2056, 2054t
Total Petroleum Hydrocarbons Criteria Working Group (TPHCWG), 2049, 2052
Total primary energy supply (TPES), 834, 835f
Total project costs derivation, 1792–1793
Toxic Substances Control Act of 1976 (TSCA), 2125, 2181
Toxic substances removal, photochemistry of, 2547–2565
  heterogeneous TiO_2 photocatalysis, 2556–2565
    applications of, 2561, 2564
    mechanism of, 2557–2558
    operational parameters of, 2558, 2561, 2562–2563t
    optimum conditions and rate of, 2559–2560t
    for water treatment, research trends in, 2564–2565
  homogeneous photo-Fenton reaction, 2548–2556
    applications in water treatment, 2553–2556, 2554–2556t
    classification of, 2548f
    contaminant concentrations and characteristics, 2552
    iron concentration, impact of, 2549, 2552
    mechanism of, 2548–2549
    optimum conditions and rate of, 2550–2551t
    oxidant concentration, impact of, 2552
  technologies of, 2547–2548
Total suspended solids (TSS), 2865t
Toxaphene, 1318t
Toxicants
  for bird control, 413–414
Toxic chemicals, 280–281, 1570
  data quality control of experimental studies, 283
  developmental animal experimental studies, 282–283
  developmental impairment in children, 281–282
  international protocols, 283–284
    European Union's REACH legislation, 284–285
    Food Quality Protection Act, 284
  nervous system development, effects on, 281
  on non-target organisms in environment, 285–286
  oxidative stress, 281
  perspectives, 285
  reproductive and developmental protocols, 283
Toxicity, 1704, 2653
  of boron, 432, 433
  expressions for, 1457t, 1458
Toxicity identification evaluation (TIE), 2806
Toxic materials, 1579
Toxicological assays, 2170
Toxicological requirements
  for pesticides registration, 1614
Toxicological synergy, 1991
Toxins, 2357
Trace element contamination
  in groundwater, 1560
  sources of, 1560
  in subsurface drainage water, 1560
  in surface runoff, 1560
Trace elements, 1455. *See also* Heavy metals
  defined, 1370
Tradable emissions permit, 919
Trade issues
  pesticide residue, 1125–1126, 1125t
Trail formation, 2752
Trampling, 2752
Transformation, of polychlorinated biphenyls, 2178
Transgenic crops, 307. *See also Bacillus thuringiensis* (BT) Crops
Transgenic virus-resistant potatoes (Mexico), 409–412
Transition metals, 1560
Transition zone
  between saltwater and freshwater regions, 1324–1325, 1325f, 1328
Translocation
  herbicides, 1379
Transmission/distribution (TD) system, 1796
Transmutation, 187
Transparent conducting oxide (TCO), 236
Transparent coverage, 242
Transpiration ratio, 1549
Transpiration stream concentration factor (TSCF), 2117, 2117f
Transport, biofuels for, 377–378
Transport, soil erosion, 958
Transportation, energy use in, 716, 716f

Transport capacity, 1021
Trans stilbenes, general formula of, 2081f
Transuranic elements, 2237–2239
Trap crops
effects on pests, 1925, 1928
Treated seeds, 419
Treatment systems use of wetlands. *See* Wetlands as treatment systems
Tree cover, permanent, 335
Trespass, 1945
Triazines, 1319
Tributary mass load sampling, 1582
Tributary water quality sampling, 1582
Tributyltin (TBT), 491
1281,1,1-trichloroethane (TCA), 1282
Trichloroethylene (TCE), 388, 1282, 1334, 1340, 2114, 2117
federal drinking water standard for, 1282
Trichoderma, 2107, 2109
Trichoderma harzianum, 383
Trichoderma viride, 383, 1879
Trichogramma, 1930, 2424
Trichogramma semblidis, 1750
Trichogrammatidae, 2424
Trichoplusia ni (Hubner), 2021
Trickle bed reactor system, 1906
Trickle irrigation, 1564
Trickling filters, 2724–2725. *See also* Filter(s)
2-Tridecanone, 2021
Trifid weed. *See Chromolaena odorata*
Triflumuron, 1462
Trifluralin, 1957, 2024f, 2806
Trigard. *See* Cyromazine
Trihalomethanes (THMs), formation of, 2793, 2794f
Triketones, 2031, 2031f
Tripene, 1464
Triple bottom line, 2461
Triple cropping forage systems, 1664, 1665t
Trissolcus vasilievi, 1750
Triticum aestivum (Wheat), 1307, 1552
Triticum aestivum L. (Durum wheat), 425
Tritrophic interactions, 312
Trombe wall, 2499, 2519, 2520, 2520f
Trophic infaunal index, 603
Trophic status, lakes and reservoirs, 1578
Trueno. *See* Hexaflumuron
Tube wells, 1276
Tuneable diode laser (TDL) technique, 1204
Tungsten, 2586–2587, 2589–2590
Tunisia, 2682, 2689
Tunnel erosion, 964
Turf, NO_3-N contamination from, 1305–1306
Turning Off the Heat: Why America Must Double Energy Efficiency to Save Money and Reduce Global Warming, 1505
Tweed River fish kill, 56
Twomey effect, 1197t
Two-phase bioreactors (TPBRs), 2612f, 2625
Two-phase reactors, 1908
Tylenchulus semipenetrans, 288, 1878
Typhlodromus occidentalis, 2141
Typhlodromus pyri, 1750

U

$^{235/238}$U, 2243, 2244
U(VI), 2244, 2245
UASB reactor, 2651–2652, 2652f
UCC. *See* Christiansen uniformity coefficient
UGas, 2491
Ulmus americana, 2830
Ultimate analysis (ASTM D3176), 717, 719t, 720t–721t
Ultrafiltration. *See also* Filtration
performance of, 2727
system design, 2727
in water treatment, 2727–2728
Ultramafic rocks, 2426
Ultrapyrolysis process, 2187
Ultrasound-assisted extraction (UAE), 2196
Ultraviolet–diode array detector (UV–DAD), 1975
Ultraviolet fluorescence, for SO_2 EN 14212, 142
Ultraviolet photometry, for O_3 EN 14625, 142
Ulva, 788
Umbrella, flagship, and keystone species, 603
UNCCD (United Nations Convention to Combat Desertification), 557, 558
UNCED. *See* United Nations Conference on Environment and Development
Uncertainty, 1720–1721
in systems analysis, 1245
three dimensions of, 1723f
Uncertainty analysis, 1720–1721
alternative approaches, 1722–1723
nanoparticles, within environmental characterizing, 1723–1725
2-Undecanone, 2021
Underground thermal energy storage (UTES), 2518–2519
aquifer storage, 2518–2519
borehole storage, 2518
pit storage, 2518
tank storage, 2518
Understoker furnaces, 375
UN Economic Commission for Europe (UNECE)
Convention on Long-Range Transboundary Air Pollution (LRTAP), 1914
UN Environment Programme (UNEP), 1615
UNEP. *See* United Nations Environment Programme
Unglazed collectors, 242, 243, 243f
Uniformity in irrigation. *See* Irrigation efficiency
UNITAR. *See* United Nations Institute for Training and Research
United Arab Emirates, the, 2682
United Nations (UN)
Human Poverty Index, 617
United Nations Children's Fund (UNICEF), 1265
United Nations Commission on Sustainable Development, 1654
United Nations Conference on Environment and Development (UNCED), 182, 1524, 1612, 1616
United Nations Convention on Biological Diversity, 333
United Nations Convention on Climate Change (UNCC), 1223
United Nations Convention to Combat Desertification (UNCCD), 1035
United Nations Environmental Programme (UNEP), 484, 1893, 2036, 2181, 2416–2417, 2671
United Nations Environment Programme (UNEP) to Combat Desertification, 565
United Nations Food and Agriculture Organization (FAO), 1108, 1397, 2035, 2036
United Nations framework convention on climate change (UNFCCC), 485, 486f, 845–846
United Nations Institute for Training and Research (UNITAR), 1409
United States, 2682, 2684
acid rain in, 8
control policy, 15
saltwater intrusion in, 1321–1322
trends in acidity, 9, 11–13
United States Army Corps of Engineers (USACE), 627
United States Department of Agriculture (USDA), 1940
United States Endangered Species Act, 1532
United States Environmental Protection Agency (USEPA), 924, 963, 1264, 1636–1643, 2013
United States Geological Survey (USGS) study, on quality of groundwater, 1283
United States Green Building Council (USGBC), 1647–1648
LEED green building certification system, 2514
Unit energy costs, 1495

Unit energy equation, 943
Universal Index of Onchev, 943
Universal soil loss equation (USLE), 934, 943, 964, 966, 967, 975–977, 980–981, 991, 1048, 1059
  variants of, 981–983, 981–983f
University of Arizona
  district cooling, 2515, 2515f
Up-coning, 1323–1324, 1324f
Upland erosion, defined, 958
Uranium
  groundwater contamination by, 1292
Uranium-238 decay series, 2236f
Urban agriculture, 2695
Urban ecosystem health, indication of, 607–608, 608t
Urban planning, 638–639
Urban sewage management, 2150–2151
Urban wastes treatment, 1908
Urinary biomarkers, for deoxynivalenol, 1696
Urtica spp., 1750
Uruguay, 2774
U.S. Congress, 2010
USDA-NASS (U.S. Department of Agriculture–National Agricultural Statistics Service) report, 2142
U.S. Department of Agriculture, 1609, 1611
  Food Safety and Inspection Service (FSIS), 1609
  on pesticide residues in processed foods, 1609
U.S. Department of Agriculture (USDA), 1120
U.S. Department of Agriculture–Agricultural Research Service (USDA-ARS), 1673
U.S. Department of Energy (DOE), 215, 2243
U.S. Department of Housing and Urban Development (HUD), 1413
U.S. Endangered Species Act of 1973, 2823
U.S. Energy Information Administration (EIA), 1652
U.S. EPA National Acid Precipitation Assessment Program (NAPAP), 20
U.S. Geological Survey (USGS), 1983, 2272, 2804
U.S. National Cancer Institute, 1394
U.S. National Toxicology Program study, 1697
U.S. primary energy supplies, 834
U.S. Social Security Administration, 1413
USDA National Organic Standards Board, 125
USDA Natural Resources Conservation Services, 127
USEPA. *See* United States Environmental Protection Agency (USEPA)
U.S. Green Building Council (USGBC), 1654, 1655, 1658
USGS. *See* U.S. Geological Survey (USGS)
USLE. *See* Universal soil loss equation (USLE)
Utility-operated rainwater harvesting, 2262
Uttar Pradesh, groundwater arsenic contamination in, 1267–1268

V

Vaccinia virus, 2604
Vacuum pyrolysis, 2487
Validamycin, 2030, 2030f
Vanadium, 2585, 2587
Vanadium and chromium groups, 2582
  environmental levels, 2584
    chromium, 2585–2586
    molybdenum, 2586
    niobium, 2585
    tantalum, 2585
    tungsten, 2586–2587
    vanadium, 2585
  geochemical occurrences, 2582–2583
  metabolism and health effects
    chromium, 2588
    molybdenum, 2588–2589
    niobium, 2587–2588
    tantalum, 2587–2588
    tungsten, 2589–2590
    vanadium, 2587
  uses, 2583–2584
Vapor compression heat pump, 1350–1351, 1350f. *See also* Heat pumps
  classification of, 1351
  components in, 1350, 1350f
  compressors in, 1350
  condensing temperature effect on ideal COP, 1351f
  evaporating temperature effect on ideal COP, 1351f
  phase-changing processes in, 1350
Vapor compression refrigeration, 1358–1359, 1359f, 1360
Vapor pressure, 2533–2534
Variable-rate fertilizer technologies, 1810
Variogram maps, 103f
Varroa mites, 2
VC. *See* Vinyl chloride (VC)
Vector control technique, combining CP and BP for, 1990
Vegetable oil, 377
Vegetables
  pesticides in, 1972–1973, 1972t, 1973f, 1973t
  oils, for liquid fuels, 846
Vegetated submerged bed (VSB) wetlands, 2862, 2864f, 2865, 2865f
Vegetated wetlands, 2824
Vegetation, 2754
  for controlling soil water erosion, 994–1001
    buffers, 995–1000, 995f, 996–997t, 998–999f
    increased infiltration of water into soil, 995
    reduced soil erodibility, 995
    sediment trapping efficiency, 1000–1001, 999f
    slower runoff, 994–995
  cover, 560
  livestock impacts on, 2753
Vegetative deficiency, 433
Vegetative insecticidal (VIP) proteins, 308
Velvetleaf (*Abutilon theophrasti*), 1526t
Venturia canescens, 2424
Venturi scrubber, 155–156
Verdale-Simi Fire, 1185f
Vermicomposting, 519–520, 519f. *See also* Composting
Vermiculitic soils, 1628
Vertebrate pests, 2595
  biological control of, 2595–2605. *See also* Biological control, of vertebrate pests
  conventional control measures for, 2595
Vertical flow systems, for constructed wetlands, 2843–2845
Vertical mixing, 49
Verticillium chlamydosporium, 1752, 1753, 1879
Vesicular–arbuscular mycorrhizae (VAM), 1516
Veterinary pharmaceuticals, 2061
Vibrio cholerae, 2696, 2698
Vibroacoustic disease, 2877
Vicia faba, 411
Vienna Convention for the Protection of the Ozone Layer, 1893
Vinclozolin, 1405
Vinyl chloride (VC), 1334
Viola tricolor arvensis, 1928
Virally vectored immunocontraception (VVIC), 2604
  of foxes, 2604
  of mice, 2604
  prospects for, 2604–2605
  of rabbits, 2604
Vireo bellii pusillus, 1186
Viron/H, 382
Virus and viroid infections, 2111–2112
  biological control of, 2111–2112
  control measures, 2111–2112
  economic loss, 2111
  global impact, 2111
  prospects, 2112

Viruses, in waste and polluted water, 2698
Virus-resistant crops, 407, 408t
Visible infrared scanner (VIRS), 2817t
Visual deficiency, 433
Visual indicators, of soil water erosion, 966–967
Volatile matter (VM), 717, 723, 725
Volatile organic compounds (VOCs), 136, 2616
  in biofilters, 2616t
Volatile oxidation, 723–724
Volatilization, 1560, 2753
  of inorganic pollutants, 2120
  of polychlorinated biphenyls, 2175–2176
Volatilization, ammonia, 1770
  from agricultural soils, 85–87
  emissions
    animals and their wastes, 86
    biomass burning, 87
    cropping systems, 86–87, 86t
    global significance, 87
    mitigation, 87
  measurement, 86
  mechanism, 85–86
Volcanic forcing, 1195, 1199f
Volcanic sulfur-bearing gases, 2439
Volcanic variability, 1194–1195, 1196t
Volcanoes on weather and climate, 1196t
Volicitin, 177
Voluntary/regional programs for climate change, 487
V-O-R model, 601
VSB wetlands. *See* Vegetated submerged bed (VSB) wetlands

W

Wagon Wheel Gap Study, 2774
Wake zone, 129
Wall insulation, 1482–1486
  2 × 4, 1483–1484
    materials required, 1483–1484
    problems and solutions, 1484
  2 × 6, 1486
  blown foam insulation, 1484, 1485
  blown loose-fill insulation, 1484, 1485
  concrete, 1482
  concrete block cores, 1482
  exterior foam insulation, 1482, 1483
  insulated concrete forms, 1483
  interior foam, 1482, 1483
  interior framed, 1482, 1483
  lightweight concrete products, 1483, 1484
  side stapling, avoiding, 1484, 1485
Walls, 860–865, 861f. *See also* Windows
  building types, 860
  exterior insulation finish systems (EIFS), 864–865
  masonry, 863–864, 864f
  steel-framed, 863
  straw bale, 865
  structural insulated panels (SIPS), 865
  utility of, 860
  wood-framed, 861–862, 862f, 863f
Wsalsh, B. D., 2009
Washing
  of cows, 1663–1664
  of milking equipment, 1664
Waste, 2415
  batteries and accumulators, 902
  composition, 2652
  hierarchy, 2418f
  life cycle assessment (LCA), 2418
  management, institutionalization of, 2416
  oils, 908
Waste from electrical and electronic equipment (WEEE), 900, 906–907
  recovery rates of, 908t
  recycling rates of, 909t
Waste emissions, exergy contents of, 1095
Waste gas stream, characteristics of, 2620–2621
Waste gas treatment, bioreactors for. *See* Bioreactors, for waste gas treatment
Waste heat, 1356
  high-grade, 869
  low-grade, 869
  medium-grade, 869
  quality of, 869
  recovery, applications, 1356–1362
Waste heat recovery
  engineering concerns in, 870–871
  equipment
    selection of, 871
    types of, 872–873
  quality vs quantity, 869–870
  sample calculations, 871–872
Waste heat stream
  cleanliness and quality of, 871
  determining value of, 870
  dilution of, 870
  mass flow rate for, 870, 872
  quality of, 869
  quantifying, 870
  recovery, 873
Waste load allocation, 2813
Waste management, impacts on
  future perspectives, 911–912
  hazardous waste, 902
  solid waste, 900–902
  waste streams, 902
    biodegradable municipal waste, 908–909
    end-of-life tires, 909–910
    end-of-life vehicles, 902–903
    packaging waste, 903–906
    waste batteries and accumulators, 902
    waste electrical and electronic equipment, 906–907
    waste oils, 908
Waste stabilization pond system (WSPS)
  design of, 2635
    anaerobic ponds, 2636
    design parameters, 2635
    facultative ponds, 2636
    Helminth egg removal, 2636
    maturation ponds for fecal coliform removal, 2636
    water flows and BOD concentrations, 2636
  nutrient removal in, 2635
  oxygen tension in, 2634
  processes in, 2633
    anaerobic ponds, 2633
    facultative ponds, 2633–2634
    facultative ponds, kinetics, 2634
    maturation ponds, 2634–2635
  types of, 2632
  water quality and, 2636–2637
Waste streams, 902
  biodegradable municipal waste, 908–909
  end-of-life tires, 909–910
  end-of-life vehicles, 902–903
  packaging waste, 903–906
  waste batteries and accumulators, 902
  waste electrical and electronic equipment, 906–907
  waste oils, 908
Wastewater, 2681–2682
  exposure, motivating safe practices along, 2688–2689
  industrial, 1430–1431
  municipal. *See* Municipal wastewater
  policy interventions and risk reduction, 2686
    agricultural communities, 2686–2687
    farmers and families, 2686
    farm product consumers, 2687–2688
  policy issues
    in developed countries, 2684
    in developing countries, 2684–2685
  policy requirement, 2684
  public policy examples, 2689–2690
  purification of, 2729
  as resource in water-scarce settings, 2682
    in developed countries, 2682
    in developing countries, 2683
  treatment of, 2720–2721
    and non-treatment alternatives, 2685–2686
Wastewater irrigation, 2675–2679
  crop selection and diversification, 2675–2676
  irrigation management, 2676–2677
  soil-health-based interventions, 2678–2679

Wastewater recycling in dairy, 1664, 1665t
Wastewater treatment, 2645–2646, 2862
activated sludge process, 2648, 2648f
operating parameters in, 2649
aeration tanks, 2648
aerobic biological waste treatment processes, 2648–2650
agricultural practices, 2701, 2704
attached growth process, 2647–2648
biological phosphorus removal, 2654
biological removal of nitrogen, 2653
anaerobic ammonium oxidation, 2653
Canon process, 2654
combined nitrogen removal, 2653–2654
Sharon process, 2653
biological treatment options, 2646
aerobic processes, 2647
anaerobic processes, 2647
anoxic processes, 2647
characteristics of, 2702t–2703t
chemotherapy and immunization, 2705
cleaner production and pretreatment discharge programs, 2700–2701
crop selection restrictions, 2701
educational and awareness campaigns, 2705
food preparation, 2704–2705
granule deterioration, 2653
irrigation methods, 2701
loading rate, 2652
local technologies, 2704
marketing, 2704
pretreatment, 2646
primary treatment, 2646
retention time, 2652–2653
secondary clarification, 2646
secondary clarifiers, 2648–2649
secondary treatment, 2646
selection criteria for intervention measures, 2705
solid retention time, 2648
subsurface-flow constructed wetlands for, use of, 2845–2846, 2845t
suspended growth processes, 2647
temperature, 2652
tertiary treatment, 2646
toxicity, 2653
transportation, 2704
washing, packing, and on-site storage, 2704
waste composition, 2652
Wastewater treatment/disposal. *See* Sewage effluent for irrigation
Wastewater treatment in Arctic regions, wetlands usage, 2662–2663
Arctic Canada and its regions, map of, 2664f
knowledge and practice, state of, 2663
performance, 2663
constructed wetlands, potential for, 2670–2671
modeling treatment wetlands, 2671–2672
Paulatuk treatment wetland, 2664–2670
Wastewater treatment plants (WWTP), 2060
Wastewater use, in agriculture, 2694
assessment
indicators, 2699
monitoring, 2699
risk assessment, 2699–2700
future perspectives, 2707
negative health impacts, 2695–2697
diseases related to chemical exposure, 2696–2697
exposed populations, 2695–2696
exposure routes, 2695, 2696t
infectious diseases, 2696
secondary health problems, 2697
pollutant sources, 2698–2699
positive impacts, 2697–2698
present situation, 2694–2695
solutions, 2700–2707
multiple-barrier concept. *See* Wastewater treatment
policy framework, 2706–2707
standards, setting, 2705–2706
Water, 916
agricultural uses of, 1281
aluminum in, 269–270
budget in dairy farms, 1665t, 1666
delivery subsystem, 1561t
desalination, 2502
efficiency, 1656
environment, bioaccumulation and, 324–325
erosion. *See* Erosion by water
of soils, 1034
erosion models, empirical
ABAG (German USLE), 977
erosion predictions, accuracy of, 977–978, 978f, 978t
evolution of, 974–975, 976t
Modified universal soil loss equation, 975–976
Revised universal soil loss equation, 977
Soil Loss Estimation Model for Southern Africa (SLEMSA), 976–977
SOILOSS, 977
Universal Soil Loss Equation, 975
flow meters, 1663
loss, 1184
management, 1560–1561
management in soil, 2217–2222
nanoparticles in, 1712–1715, 1713t, 1714f
nutrient deposition/deactivation in, 1592t
pesticides in, 1968–1969, 1970t
purification of, 2729
as refrigerant, 1358
soil erosion by. *See* Soil water erosion
surface, soil erosion effect on. *See* Soil erosion, effect on surface water quality
sustainable agriculture, 2457–2458
treatment of, 2720–2721
treatment, alternative systems for, 2062–2063
usage minimization, 1665t, 1666
use, 1184
use in dairy farms. *See* Dairy, water use in
Water balance and groundwater mining, 1284–1285, 1285f, 1286f
Water cress (*Enhydra fluctuans*), 426
Water erosion, 1955–1956
Water Erosion Prediction Project (WEPP), 984–985, 966, 992, 1048 1959, 2761
Water erosion vulnerability map, 2376
Water-filled pore space (WFPS), 1770
Water filtering, 2704
Water flows and BOD concentrations, 2635–2636
Water Framework Directive (WFD), 1701–1702
Water harvesting
classifications, 2738
definition, 2738
for domestic use, 2739
failure, 2739
for growing of crops, 2739
for livestock drinking water, 2738–2739
Water hickory. *See Carya aquatica*
Water hyacinth (*Eichhornia crassipes*), 1527t
Water intake, 209–210
Water leaching index, 1763
Waterlogging, of sodic soils, 2341
Water management
CH_4 emissions from rice fields and, 1679, 1680f
Water milfoil. *See Myriophyllum* spp.
Water pollution, 333, 2013, 2020. *See also* Environmental pollution, by pesticides; *See also* Pesticides
prevention, 2138
Water Pollution Control Act in US, 1557
Water prices, 2779
competitive markets and, 2780–2781
multiple-criteria framework, 2785f
optimizing model, 2781–2783
scarcity rents, 2782
shadow prices, 2782–2783
structuring, 2783
economic efficiency, 2785
equity and fairness, 2785–2786
full cost, 2784
full economic cost, 2784
full supply cost, 2784
opportunity cost, 2784

simplicity, 2786
stability and quality, 2786
substitutability, 2784–2785
sustainability, 2786
water tariffs and pricing strategies, 2786–2788
Water Productor Program, 971
Water quality. *See also* Range and pasture lands
for agriculture, 1566
constituents in IRF, 1558–1560
nitrogen, 1559
pesticide contamination in, 1560
phosphorus, 1559–1560
salts, 1559
trace elements in, 1560
criteria, 2811
effects mitigation, through BMPs, 2773
monitoring, 1581–1582
biological integrity sampling, 1582
citizen monitoring, 1582–1583
lake sampling, 1582
trading programs, 1772–1773
tributary mass load sampling, 1582
tributary water quality sampling, 1582
wind erosion, effect of, 1020
and WSPS, 2636–2637
Water Quality Act (1965), 2808
Water quality modeling, 2749–2750
BMPS, 2750
classification, 2749–2750
large-scale systems behavior, 2750
overview, 2749
risk assessment of pesticides, 2750
roles, 2749
sources/impacts of pollutants, evaluation of, 2750
uses, 2750
Water removal subsystem, 1557, 1561t
Water retention
hydraulic conductivity and, relationship between, 1621
Watersheds
management
Geographic information system (GIS) for, 2816–2819, 2818t
implications to, 2102–2103
remote sensing (RS), 2816–2819, 2817t
water quality response, implications for, 2103, 2105
sediment transport in, 965, 965f
Water solubility, 1957
Water-stable aggregates, 91
Water supply buffer, 2262
Water table control, 1547t
Water table management, 49–50
Water transport components, 1545, 1546f
Water turbines, 212
Water use efficiency (WUE), 1549
agroforestry and, 129–130. *See also* Agroforestry, WUE and
Water vapor, greenhouse properties of, 1194, 1195f
Waterways, to control soil erosion, 969–970, 969f
Watery soft rot, 443, 443t
Wave energy, 772, 826
Weasels (*Mustela nivalis*), 2596
Weathering effect, 1559
Weather-resistive barrier on wall, 864
Web-based tools, for integrated energy investments. *See* Integrated energy systems, case study from ISU
Wedge Dust Flux Gauge (WDFG) samplers, 1022
Weed
diversity
effects on pests, 1928
herbicides resistance in, 1379
intercropping and, 1938
management and organic farming, 1107
organic soil amendments and, 1878
science, 1524
Weed(s), biological control of, 2821–2823
costs and agent success, 2822–2823, 2822t
history and impact of classical approach, 2821–2822
host-specificity tests and, 2822
legislation, 2823
Weed abundance, crop rotation and, 1928
Weed control, 119, 1524
biological, methods of, 1525f
inoculative/classical approach, 1525, 1525f
inundative/bioherbicide method, 1525, 1525f
system management approach, 1525, 1525f
integrating biological control with other methods, 1525, 1526t–1527t
ecological integration, 1525, 1528
horizontal integration, 1525
physiological integration, 1528
purpose-specific approaches, 1525
vertical integration, 1525
Well-posed initial and boundary value problem
for saltwater intrusion, 1329
WEPP. *See* Water Erosion Prediction Project (WEPP) equation
Werneckiella equi, 1462
West Bengal, groundwater arsenic contamination in, 1265–1267, 1266f, 1267f
Wet deposition, 149
Wetland microcosms, 426
Wetland(s), 607, 2854
and carbon sequestration, 2833–2835
case studies
effectiveness, evaluation of, 2854–2855
engineered/constructed wetland, treatment of petroleum hydrocarbons in, 2856–2857
petroleum hydrocarbons, abatement of, 2855f, 2855–2856
definitions, 2824, 2854
extent of, 2827, 2827t
fauna, 2831
flora, 2829–2831, 2830f, 2831f
hydrological patterns, 2829, 2830f
landscape perspective, 2829
as natural resources, 2824
types of, 2824–2825, 2825t, 2826f, 2827
Wetlands as treatment systems
defined, 2862
design considerations, 2864–2865, 2866t
FWS wetlands, 2865
VSB wetlands, 2865
influent concentrations, 2865t
in North America (1994), 2863t
operation and maintenance, 2865–2866
treatment processes, 2862–2863, 2865f
treatment wetland types
constructed *vs.* natural wetlands, 2862, 2863t
free-water *vs.* submerged-bed wetlands, 2863, 2863f
use of wetlands, 2862
wastewater, types of, 2865t, 2866t
Wet meadow, 2825t, 2826f
Wet oxidation (WO), 1905–1906, 1906t
advantages, 1907
applications, 1908
catalytic, 1909
using heterogeneous catalysts, 1910
using homogeneous catalysts, 1909–1910
challenges of, 1910–1911
commercial, 1908
limitations, 1907–1908
mechanism, 1906–1907, 1907f
non-catalytic, 1908–1909
POP removal using, 1904
Wet oxidation reactors, 1908
Wet scrubbers, 154–156
Wet techniques, 2440–2441
WFD. *See* Water Framework Directive
WFPS. *See* Water-filled pore space
Wheal Jane metal mine, Cornwall, 1291
Wheat (*T. aestivum* L.), 427, 1307, 1552
WHEELS, 1959
White mold, 443, 443t
WHO. *See* World Health Organization (WHO)
Whole systems thinking, 2461
Wildfire, 1184–1186, 1185f
Wildlife, biodiversity and, 1186
Wildlife management and husbandry, 570
Wild oat. *See Avena fatua* L.

Willingness to accept compensation (WTA), 1637
Willingness to pay (WTP), 1637–1638
Wind energy, 772, 824
  resources, 824
Wind erosion, 1013–1015, 1956
  affects of, 1017
  causes of, 1017–1019
  control strategies, 1027–1028
  definition of, 1017
  global hot spots, 1010–1012, 1011f
  implications of, 1019–1020
  induced, 1010–1012
  monitoring of, tools for, 1020–1021
    direct measurement, 1021, 1021f, 1022f
    horizontal mass flux, 1021–1024
    radioisotopic techniques, 1024–1025
    wind erosion models, 1025–1027
  natural, 1010
  overview, 1013
  particle entrainment, 1014, 1014f
  particles movement by wind and, 1017–1018
  processes, 1013
  self-balancing concept, 1015, 1015f
  source regions of, 1020, 1021f
  wind dynamics, 1013–1014, 1014f
Wind Erosion Assessment Model (WEAM), 1026t
Wind Erosion Equation (WEQ), 1025–1026, 1026t
Wind Erosion on European Light Soils model, 1025
Wind Erosion Prediction System (WEPS), 1025, 1026–1027, 1026t, 1959
Wind erosion vulnerability map, 2376
Windfarm noise, 2867, 2868
Wind generators, 916
Windows. *See also* Walls
  building types, 860
  energy transport, 865–866, 865f, 866f
  future improvements, 867
  solar heat gain through, 866
  utility of, 860
  window rating system, 866–867, 866f, 867t
Wind park, 2867
Wind power, 179, 192
  capacity distribution, 259
  capacity factor, 259–260
  capacity growth, 259
  climate protection, 182–183
  costs, 258–259, 260
  depletion of fossil fuels, 180–182
    data and predictions, 180f
    oil, delivery and detection of, 181f
    reserves/resources, regional distribution of, 180, 181f
  ecological economics
    Ems-axis, lower saxony, growth and booming region, 196–199
    renewable energy, 188–195
    renewable energy in Germany and planned nuclear exit, 195–196
  electrical production, 258
  future
    hydrogen economy, 261
    maximum production, reaching, 261
    other issues, 261
    projected cost, 261
    projected growth, 261
    projected production, 261
  geographical distribution, 258
  governments and regulation, role of, 265
    environmental regulation, 265
    grid interconnection issues, 265
    improving wind information, 265
    subsidies, tax incentives, 265
  history, 258
  location
    favored geography, 260
    maximum production limits, 260–261
    sizing, 260
  nuclear power, role of, 183–187
  site, 260
  strengths
    costs, 262
    environment, 261–262
    local and diverse, 262
    quick to build, easy to expand, 262
    renewable, 262
  technology, 263–264
  weaknesses
    bird impact, 263
    connection to grid, 262
    local resource shortage, 263
    natural variability, 262
    noise, 263
    visual impact, 263
Wind protection
  plant water status and, 130
Wind resources, of electricity, 1419
Windrows composting, 516–517, 516f. *See also* Composting
Wind strips, 1027–1028
WINTOX, 332
Wind turbine
  blades, 264
  components
    blade diameter, 263–264
    controls and generating equipment, 264
    tower height, 263
  control mechanisms, 264
  generators, 264
  nacelles, 264
  wind sensors, 264
Wind turbine noise
  acoustic profile of, 2868–2870
  horizontal-axis wind turbine, 2868f
  human impacts of, 2870
    and annoyance, 2871–2873
    and low-frequency/infrasound components, 2876–2877
    psychological description, 2870
    quantifying the health impacts, 2870–2871
    and sleep, 2873–2876
    wind turbine syndrome, 2876
Winter cover crops, for reducing NO_3-N leaching, 1306–1307
Winter rye (*Secale cereale* L.), 1306–1307
Wirestem, 443, 443t
WO. *See* Wet oxidation
WOCAT (World Overview of Conservation Approaches and Technologies), 558
Women
  hormonal disruption in, 1406
Wood
  fuels, 375t
  preservatives
    used in homes and gardens, 1401
Wood-framed walls, 861–862, 862f, 863f
Wood pellets, 375, 375t
Work, 809, 811
Work and Health in Southern Africa (WAHSA), 506
  Action on Health Impacts of Pesticides, 507t
Work capacity, 778
Work–energy principle, 813
Work–heat–energy principle, 813–816
Working fluid, 243
World Atlas of Desertification, 552
World Bank, 2034
World Business Council for Sustainable Development project (WBCSD), 2477
World Commission on Environment and Development, 338
World energy production and consumption, 678–680
World Federation of Associations of Clinical Toxicology Centers and Poison Control Centers, 1409
World fossil fuel consumption (WFFC), 1229
World Geothermal Congress 2010, 1167
World green energy consumption (WGEC), 1229
World Health Organization (WHO), 132, 424, 507, 1263, 1264, 1397, 1400, 1408, 1409, 1917, 2001, 2033, 2686, 2687
  air quality guidelines, 134
  Guidelines for Drinking Water Quality, 2793
  pesticides chronic intoxication (statistics), 1397, 1398f

Page numbers followed by f and t indicate figures and tables, respectively

World Meteorological Organization (WMO), 484
World primary energy consumption (WPEC), 1229
World Reference Base (WRB), 28
World total fossil-fuel consumption and CO_2 production, 680, 681
World War II
  use of pesticides after, 1397
Worldwatch Institute, 437
World Wide Web (WWW), 1164
WUE. *See* Water use efficiency (WUE)
WWF International, 2477
WWW. *See* World Wide Web
Wyoming coal, 718t

X

Xanthobacter autotrophicus, 384
Xanthomonas campestris pathovar *campestris*, 442
Xenobiotics, 281
  effect of, 2005
Xenohormones, 1405
Xenopsylla cunicularis, 2597, 2600
Xenorhabdus, 321, 382
XPS. *See* Extruded polystyrene
"X-ray amorphous," 1712, 1713f
Xylocoris flavipes, 2424

Y

Year Average Common Air Quality Index (YACAQI), 136
Yellow boy, 1689
Yellow River, 2884–2886
  distributions of runoff, 2885
  harnessing of, 2885
  irrigation area, 2885
  overview, 2884
  terrain of, 2884
Yield goals, realistic, 1807
Yield sensors, 2215

Z

Zabrus gibus, 1116
Zea mays, 426, 1303, 1983
Zearalenone, 1696–1697
Zeolites, 70–72
"Zero tolerance," 1126
Zero-valent iron nanoparticles (nZVI)
  pollutant removal, mechanism in, 1736–1737
  soil and groundwater remediation by, 1737–1739, 1738t, 1739t
"Zero waste" initiatives, 2418
Zetzellia mali, 2141
Zinc (Zn), 1370
  contamination, in coastal waters, 490
  phytotoxicity, 2138
  use, and B toxicity, 426
Zinc yellow pigment, 479t
Zineb, 1318, 2806
Ziram, 415, 2806
Zn. *See* Zinc (Zn)
Zooplankton, 539